STANDARD HANDBOOK OF ENVIRONMENTAL SCIENCE, HEALTH, AND TECHNOLOGY

Jay H. Lehr Editor
Bennett & Williams, Inc.

Janet K. Lehr Associate Editor
Environmental Education Enterprises, Inc.

McGRAW-HILL

New York   San Francisco   Washington, D.C.   Auckland   Bogotá
Caracas   Lisbon   London   Madrid   Mexico City   Milan
Montreal   New Delhi   San Juan   Singapore
Sydney   Tokyo   Toronto

Library of Congress Cataloging-in-Publication Data

Standard handbook of environmental science, health, and technology / Jay H. Lehr, editor.
     p.   cm.
   Includes index.
   ISBN 0-07-038309-X
   1. Environmental sciences—Handbooks, manuals, etc. 2. Environmental
protection—Handbooks, manuals, etc. 3. Environmental health—Handbooks, manuals,
etc. I. Lehr, Jay H., date

GE123.S73   2000
363.7—dc21   00-036172

McGraw-Hill

A Division of The McGraw·Hill Companies

1 2 3 4 5 6 7 8 9 0 DOC/DOC 0 6 5 4 3 2 1 0

ISBN 0-07-038309-X

*The production supervisor for this book was Pamela A. Pelton. It was set in
Times Roman by TechBooks.*

Printed and bound by R. R. Donnelley & Sons Company.

McGraw-Hill books are available at special quantity discounts to use as premiums and sales promotions, or for use in corporate training programs. For more
information, please write to the Director of Special Sales, Professional Publishing,
McGraw-Hill, Two Penn Plaza, New York, NY 10121-2298. Or contact your local
bookstore.

This book is printed on acid-free paper.

To: Kevin Wolka (1949–1999)

A brilliant environmental scientist whose dedication to his profession is evidenced through the hundreds of students he affected in his tireless drive to leave the earth a better place and its inhabitants capable of preserving their heritage.

STANDARD HANDBOOK OF ENVIRONMENTAL SCIENCE, HEALTH, AND TECHNOLOGY

CONTENTS

Chapter 11. Assessment Sampling and Monitoring 11.1

Chapter 12. Toxicology and Risk 12.1

Chapter 13. Control Technologies 13.1

Chapter 14. Remediation Technologies 14.1

Chapter 15. Ubiquitous Environmental Contaminants   15.1

Part 3   Place Based Environmental Science, Health and Technology

Chapter 16. Managing Place Scale Problems   16.3

Chapter 17. System Level Risk Assessment   17.1

Chapter 18. Environmental Science in the Legal System   18.1

Chapter 19. Sensitive Environmental Systems   19.1

CONTRIBUTORS

Tom Aley

Tom Aley was awarded B.S. and M.S. degrees in forestry from the University of California at Berkeley in 1960 and 1962. Since 1973, he has been president of the Ozark Underground Laboratory, Inc., a contract studies firm that specializes in groundwater and land use issues in karst and fractured rock regions, with special emphasis on groundwater tracing with fluorescent dyes. Tom holds certification as a professional hydrogeologist from the American Institute of Hydrology and as a certified forester from the Society of American Foresters. He is a registered professional geologist in Kentucky and Arkansas and is a registered geologist in Missouri. He has taught numerous professional short courses in karst hydrology and karst resource management and has authored about 70 professional publications.

Ozark Underground Laboratory, 1572 Aley Lane, Protem, Missouri 65733
oul@tri-lakes.net

Alaa H. Aly

Alaa Aly, P.E., Ph.D., worked on the projects described in this *Handbook* while a graduate student and research engineer with USU/DBIE. Dr. Aly has been involved with many groundwater investigation projects, including; evaluation, modeling, calibration and remediation management projects for the U.S. Air Force, the Department of Justice and several private sector clients.

Alaa Aly is currently Senior Engineer at Waterstone Environmental Hydrology and Engineering in Boulder, Colorado, and his current work involves the development of global optimization algorithms using adaptive neural networks for highly nonlinear systems, remedial design optimization, stochastic optimization, and groundwater flow and transport modeling. He received a B.S. degree in Civil Engineering from Cairo University and an M.S. and Ph.D. in Irrigation Engineering from Utah State University. He also completed an MS degree in Mathematical Statistics from Utah State Universtiy.

Waterstone Environmental Hydrology and Engineering, 1650 38th Street, Boulder, Colorado 80301
AlaaHAly@crosswinds.net

Bruce Ames

Dr. Ames is Professor of Biochemistry and Molecular Biology and Director of the National Institute of Environmental Health Sciences Center, University of California at Berkeley. He has been a member of the National Academy of Sciences since1972 and was a member of the Board of Directors of the National Cancer Institute, the National Cancer Advisory Board, from 1976 to 1982. His numerous awards include the General Motors Cancer Research Foundation Prize (1983); the Tyler Prize (1985); the Gold Medal Award of the American Institute of Chemists (1991); the Glenn Foundation Award of the Gerontological Society of America (1992); the Lovelace Institutes Award for Excellence in Environmental Health Research (1995); the Achievement in Excellence Award of the Center for Excellence in Education (1996); the Honda Prize of the Honda Foundation, Japan (1996); the Japan Prize (1997), the Kehoe Award, American College of Occupational and Environmental Medicine (1997); the Medal of the City of Paris (1998); the Joseph Priestley Award (1998); and the U.S. National Medal of Science (1998). Dr. Ames has authored over 400 publications.

Environmental Health Science Center, University of California Berkeley, One Cycleotron Rd. MS-90K Berkeley California 94720
BNAMES@uclink4.berkeley.edu

Dennis Avery

Dennis Avery grew up on a Michigan dairy farm and studied agricultural economics at Michigan State and the University of Wisconsin. He is an internationally known expert on the dynamic changes taking place in the world's ability to feed itself and still protect environmental resources. He was senior agricultural analyst in the U.S. Department of State for nearly a decade; he has also done policy analysis for the U.S. Department of Agriculture and served on the staff of President Johnson's National Advisory Commission on Food and Fiber. He travels about 200,000 miles a year to tell audiences worldwide that farmers are the true hope for the world's environment. His weekly column is distributed by Bridge Information Services to newspapers across America and financial news subscribers worldwide. He is author of *Saving the Planet with Pesticides and Plastics: The Environmental Triumph of High-Yield Agriculture.*

Center for Global Food Issues, PO Box 202, Churchville, Virginia 24421
cgfi@rica.net

George Ball

George Ball is a geographic information systems (GIS) analyst/programmer at Condor Earth Technology. He has over 9 years of experience in GIS development, focusing on local government. While employed by Tuolumne County, Mr. Ball supervised and trained technicians, interns, and temporary employees in developing and maintaining GIS data layers. He develops custom applications and programs for data automation using ARC/INFOO and ArcViewO in support of Condor's wide range of consulting activities. He is also responsible for implementing a GIS Internet server and creating three-dimensional models and animations with GIS. At Condor he has provided on-site services for a telecommunications firm, performing complex marketing analyses.

Condor Earth Technology, 21663 Brian Lane, Sonora, California 95370
gball@condorearth.com

Kirk R. Barrett

Dr. Barrett is an assistant professor and research director of the Meadowland Environmental Research Institute at Rutgers University, New Jersey. He has expertise in surface and wetland hydrology, hydraulics, and water quality processes. He holds a doctoral degree in environmental engineering from Northwestern University, a M.S. in computer science, and a B.S. in chemical engineering. He is a professional wetland scientist, certified by the Society of Wetland Scientists, and a registered professional engineer. He has published numerous journal articles.

Meadowlands Environmental Research Institute (MERI), Rutgers University, 180 University Avenue, Newark, New Jersey 07102
kbarrett@compuserve.com

Douglas Beal

Mr. Beal has over 19 years of experience in the field of geology, over 12 years of which involves soil and groundwater remedial investigations and remediation. He is Senior Project Manager at the Chatham, New Jersey, office of BEM Systems, Inc., where he is responsible for client services within BEM's Industrial Client Division. Mr. Beal has extensive experience in the design, construction, operation, and maintenance of various types of remediation systems, including groundwater extraction, bioremediation, and in situ chemical oxidation. He has been involved with the investigation and remediation at many large industrial DNAPL sites and has provided litigation support for a number of environmental projects.

BEM Systems Inc., 100 Passaic Avenue, Chatham NJ 07928
dbeal@bemsys.com

Milovan S. Beljin

Milovan S. Beljin has a Dipl. Ing. degree in hydrogeology and engineering geology from Belgrade University and a Ph.D. in groundwater hydrology from Ohio University. He has over 20 years of experience in consulting and teaching. He is a consulting hydrologist and president of M. S. Beljin & Associates; he also serves as adjunct faculty at the University of Cincinnati and at Wright State University in Dayton, Ohio. His interests include development and application of mathematical models to groundwater flow and mass transport, horizontal well technology, and application of geographic information systems (GIS) in groundwater hydrology.

mbeljin@cinci.rr.com

Richard C. Benson

Richard Benson is a certified geologist and a registered geologist in Florida and Georgia and has a B.S. degree in geology and geophysics and in coastal engineering. Mr. Benson has extensive experience in engineering geology, hydrogeology, and environmental work. He specializes in site characterization of complex geologic conditions including karst and fracture flow. His work with karst includes settlement of structures, subsidence due to underground mines, the impact of sinkholes on highways and airport runways, water resources development, and environmental issues. He has pioneered the applications of many of the contemporary geophysical methods in use today. In 1971 he formed Technos, Inc., which has become a world-class leader in the field of site characterization and monitoring. Mr. Benson has conducted several short courses and lectures and has authored numerous books and professional papers.

Technos, Inc., 3333 NW 21st Street, Miami, Florida 33142
info@technos-inc.com

Patrick V. Brady

Patrick V. Brady graduated from U. C. Berkeley in Geology in 1984 and received his M.S. (1987) and Ph.D. (1989) at Northwestern University. He did post doctoral work at the Swiss Federal Institute for Water Resources and Water Pollution Center (EAWAG-ETH) and was an Assistant Professor at Southern Methodist University (Departments of Geological Sciences and Civil Engineering) for 2.5 years before joining the Geochemistry Research Group at Sandia National Laboratories. Brady's expertise is in applying mineral surface chemistry to remediate hazardous waste and understand global geochemical change. He has edited one book on mineral surface chemistry and written two others on natural attenuation, and acts as a referee for manuscripts submitted to Science, Environmenal Science and Technology, Geochimica Cosmochimica Acta, Chemical Geology, and the Journal of Soil Contamination as well as a number of American and international funding agencies.

Geochemistry Department, MS0750, Sandia National Labs, 1515 Eubank SE, Albuquerque, New Mexico 87123-0750
pvbrady@sandia.gov

Warren Brady

Warren Brady is a project geochemist in the Technology Applications Group of IT Corp. He has a M.S. degree in soil chemistry from Louisiana State University and a B.S. degree in hydrogeology from Louisiana Tech University. Mr. Brady has worked on various environmental remediation and site investigation projects with an emphasis on monitored natural attenuation evaluation/implementation, contaminant transport processes, and risk-based corrective action programs. He has provided technical leadership for natural attenuation investigations and implementation of natural attenuation performance monitoring programs. Mr. Brady has also assisted the American Society of Civil Engineers and the Department of Energy in developing natural attenuation guidance documents. Mr. Brady is active in conducting seminars and workshops on natural attenuation and has presented papers on this topic at numerous national and state environmental meetings.

IT Corporation, 8281 Goodwood Boulevard, Baton Rouge, Louisiana 70806-7742
Wbrady@theitgroup.com

Peter Bruce

Peter Bruce is managing director for Resampling Stats, Inc., and marketing director for Cytel Software Corp., the makers of StatXact, LogXact, and other software for exact inference. He has taught resampling statistics at the University of Maryland as well as short courses for the American Statistical Association and the National Technological University. He is lead investigator for a program sponsored by the National Science Foundation to teach computer-intensive resampling methods to undergraduate statistics instructors. He is co-developer of the Windows and Macintosh versions of Resampling Stats and principal developer of the Excel version of Resampling Stats. Peter has authored numerous articles on resampling statistics and for several years has been the primary instructor for Resampling: The "New Statistics" at the Institute for Professional Education.

Resampling Stats, 612 N. Jackson Street, Arlington, Virginia 22201
pbruce@resample.com

C. R. Brunner

C. R. "Charles" Brunner is a consulting engineer and president of Incinerator Consultants, Inc. He has over 30 years of experience in the evaluation, selection, design, operation, and formatting of thermal systems for destruction of wastes. He has written 12 widely used textbooks and dozens of technical articles on waste disposal. He has lectured on this subject for technical and nontechnical audiences and gives seminars on waste disposal technology throughout the United States and overseas. His major clients include chemical and pharmaceutical firms, hospitals, waste disposal firms, equipment manufacturers, regulatory agencies, and other governmental and nongovernmental organizations. He has helped develop operating standards and regulations for medical waste and hazardous waste disposal systems as well as regulations for the disposal of municipal solid and sludge wastes.

Incinerator Consultants Inc., 11204 Longwood Grove Drive, Reston, Virginia 20194
Cbrunner@clark.net

Janusz Z. Byczkowski

Dr. Byczkowski began his scientific career in 1967 as a research assistant in toxicology at the Academy of Medicine in Gdansk, Poland. He has been a cancer research scientist at Roswell Park Cancer Institute in Buffalo, New York; a research scientist and assistant professor at the College of Public Health at the University of South Florida in Tampa; a project scientist and study director at ManTech Environmental Technology, Inc., at Wright-Patterson Air Force Base, Ohio; and a consultant and senior toxicologist at TN & Associates, Inc., in Cincinnati, Ohio. He has authored over 100 peer-reviewed articles, technical reports, and book chapters and has presented over 100 short communications. Since August 1999 he has worked as an independent consultant in Fairborn, Ohio, providing expertise to governmental agencies and the private sector.

Human Health Risk Assessments, 212 N. Central Avenue, Fairborn, Ohio 45324
januszb@aol.com

Mark P. Cal

Dr. Cal has a B.S. in chemical engineering and a Ph.D. in environmental engineering from the University of Illinois at Urbana-Champaign. He has 13 years of experience in air quality engineering and science; his research interests include air pollution control, indoor air quality, gas-phase adsorption, plasma processing of gas streams, and gas separation. Dr. Cal has held appointments at the University of Nevada-Desert Research Institute, the Illinois State Geological Survey, the University of Illinois at Urbana-Champaign, and New Mexico Tech, where he is currently an assistant professor of environmental engineering. He is a member of the American Institute of Chemical Engineers, Air and Waste Management Association, Association of Environmental Engineering Professors, and Sigma Xi. Dr. Cal has authored more 40 journal publications, book chapters, conference proceedings, and project reports.

New Mexico Tech, Department of Environmental Engineering, 801 Leroy Place, Socorro, New Mexico 87801
mcal@nmt.edu

Bejurin J. Cassady

Bejurin Cassady has a B.A. in social science from Eckerd College in St. Petersburg. He is a writer and editor with more than 25 years of experience preparing technical documents for a various disciplines. Since 1992, he has specialized in writing and editing documents on topics in the environmental sciences. He has written, edited, and coordinated the production of environmental publications of all types and scales, for both technical and general audiences. His extensive editing and document coordination experience enables him to work cooperatively and proactively with technical authors, manage publications production staff effectively, and organize multiple projects to meet deadlines. His editorial specialty and great love is adapting and presenting technical information for general audiences.

Windward Enironmental LLC, 200 West Mercer Street, Seattle, Washington 98119
bejurinc@windwardenv.com

Leonard W. Casson

Leonard W. Casson, Ph.D., P.E., is associate professor of environmental engineering in the Department of Civil and Environmental Engineering at the University of Pittsburgh. Dr. Casson's research focuses on the adsorption, fate, and transport of particles, chemicals, and environmental pathogens in unit processes and the natural environment. He is a professional engineer in Florida and Pennsylvania. He has authored over 30 papers in the field of water and wastewater treatment.

University of Pittsburgh, Department of Civil and Environmental Engineering, Pittsburgh, 944 Bendum Hall, Pittsburgh Pennsylvania 15261-2294
casson@engrng.pitt.edu

Bernard L. Cohen

Dr. Cohen received a Ph.D. in physics from Carnegie-Mellon University in 1950, was a group leader in charge of cyclotron research at Oak Ridge National Laboratory for 6 years, and, except for short leaves of absence, has been at the University of Pittsburgh since 1958, where he is professor emeritus of physics and astronomy and of environmental and occupational health. He has authored 6 books, over 300 papers in scientific journals, and about 75 articles in nontechnical journals. He has presented invited lectures in the United States, Canada, Japan, Australia, and 24 other countries. He has received the American Physical Society Bonner Prize, the Health Physics Society Distinguished Scientific Achievement Award, and the American Nuclear Society Walter Zinn Award, Special Award, and Public Information Award.

Physics Department, University of Pittsburgh, Pittsburgh, Pennsylvania 15260
blc+@pitt.edu

Peter Collopy

Mr. Collopy has a master's degree in environmental engineering from Rensselaer Polytechnic Institute and is certified in comprehensive practices by the American Board of Health Physics and the American Academy of Industrial Hygienists. Mr. Collopy has served as the radiation protection manager for several facilities licensed by the Nuclear Regulatory Commission, including power plants and universities. Mr. Collopy's major endeavors are participation in the remediation and environmental assessment of radioactively contaminated CERCLA and industrial facilities. Mr. Collopy currently works as Director of Environmental Health and Safety Services for MJW Corp. in Buffalo, New York.

MJW Corporation, 338 Harris Hill Road, Williamsville, New York 14221
mjwcorp.com

C. Richard Cothern

C. Richard (Rick) Cothern received his B.A. from Miami University (Ohio), his M.S. from Yale University, and his Ph.D. from the University of Manitoba. He has spent the past 19 years with the U.S. Environmental Protection Agency at the Center for Environmental Statistics, Science Advisory Board, and the Offices of Drinking Water and Pollution Prevention and Toxic Substances. He is professor of chemistry at Hood College. Dr. Cothern has taught physics at the University of Manitoba and the University of Dayton. He has been involved in research in nuclear spectroscopy, environmental physics, and surface analysis. He has authored over 100 scientific articles and has written and edited 14 books.

Rcothern@nova.umuc.edu

Michael Crowe

Dr. Crowe, a retired corporate executive and engineer, has had extensive experience in project management of social and environmental systems development. He has designed and developed extensive land use change projects. He has conducted confidential environmental reviews and represented corporations and individuals during regulatory agency contact. His environmental and ecological natural systems design includes wetland restoration and creation, riparian or buffer zone creation and restoration, and ecological systems for stormwater collection, treatment, and flood control; air and noise buffer zones in high traffic areas; and water quality enhancement through passive and/or mechanical systems. Dr. Crowe has organized and conducted public information meetings and acted as a corporate and private agent to state and federal regulatory and planning agencies. His expertise lies in combining science and engineering to meet human socioeconomic needs while enhancing the natural environment.

MRB Environmental Services, Inc., 12261 Springfield Road, Springfield, Ohio 44443
mrbcorp.com

George F. Czapar

George Czapar received a B.S. in agronomy and a M.S. in weed science from the University of Illinois and a Ph.D. in weed science from Iowa State University. He is an extension educator in integrated pest management (IPM) for the University of Illinois Extension at the Springfield Center. He is also state coordinator for the Illinois Council on Best Management Practices. From 1985 to 1991 he was on the staff of the Iowa State University working on weed science, water quality, and IPM. His current research includes evaluating best management practices to protect surface water, using global positioning systems for weed management, IPM adoption, using subsurface drainage tiles to estimate chemical movement to groundwater, and decision-making tools for weed management.

University of Illinois Extension Center, PO Box 8199, Springfield, Illinois 62791
g-czaper@uiuc.edu

James M. Davidson

James Davidson is a hydrogeologist and president of Alpine Environmental, Inc. He has extensive experience investigating and remediating petroleum releases and since 1984 has been involved with more than 400 contamination projects across the United States and internationally. Mr. Davidson is a nationally recognized expert on the subsurface occurrence, movement, and remediation of the gasoline additive MTBE. Since 1995 he has authored 12 publications on a variety of MTBE environmental issues and has taught MTBE training sessions on more than 45 occasions for a variety of regulatory agencies and professional organizations.

Alpine Environmental Inc., 203 W. Myrtle Street, Suite C, Fort Collins, Colorado 80521
JDavidsonalpine@cs.com

Marlowe Dawag

Marlowe Dawag has a Bachelor's degree in chemical engineering from Washington State University. She is an environmental engineer with the U.S. Army Corps of Engineers, Alaska District. Her environmental engineering experience includes various remediation technologies, remedial action contracting, petroleum remediation, hazardous waste, environmental regulations, and site closeouts. Her current projects include soil vapor extraction applications, risk evaluation, environmental compliance, and closeout of 14 formerly used defense sites throughout the Seward Peninsula.

U.S. Army Corps of Engineers, Alaska District, PO Box 898, Anchorage, Alaska 99506

William Deutsch

Bill Deutsch holds B.S. and M.S. degrees in geological sciences from the University of Washington, Seattle. He has been a hydrogeochemist with research and consulting firms for over 20 years, and his project experience includes environmental assessments and investigations of landfills, refineries, pesticide plants, distributorships, military bases, mines and mills, federal weapons facilities, and various other industrial sites. He has also participated in remedial designs of sites contaminated with metals, radionuclides, pesticides, solvents, petroleum hydrocarbons, and ordnance compounds. Since 1985 Mr. Deutsch has taught courses in groundwater geochemistry and geochemical modeling for a number of professional organizations. He is the author of *Groundwater Geochemistry* and is an independent consultant.

Geochemistry Consulting
Billdeutsch@acadia.net

Frank Dillon

Frank Dillon has a B.A. in zoology from Southern Illinois University and a M.S. in biology from Western Illinois University. He is a partner at Windward Environmental LLC. He is an aquatic ecologist and toxicologist with over 16 years of professional experience investigating anthropogenic effects on aquatic ecosystems. Mr. Dillon has served as project and technical manager of ecological risk assessments at a variety of sites, including several Superfund sites. His interests are primarily in the problem formulation and study design phases of the risk management process. Mr. Dillon is currently serving as technical manager of a major sediment-driven NRDA site in the Great Lakes. He has served as technical manager of the Commencement Bay Phase I NRDA as well as coprincipal investigator of two Phase II injury assessment studies.

Windward Environmental LLC, 200 West Mercer Street, Suite 401, Seattle, Washington 98119
frankd@windwardenv.com

Jack W. Dini

Jack W. Dini is a metallurgical engineering graduate of Cleveland State University. He spent his entire career in the electroplating industry; for the last 15 years he concentrated on waste minimization. He was employed at Battelle Memorial Institute, Sandia Laboratories, and Lawrence Livermore National Laboratory (LLNL). He has authored over 200 technical publications and a book titled *Electrodeposition: The Materials Science of Coatings and Substrates*. He is a past president and a fellow of the American Electroplaters and Surface Finishers society (AESF). He retired from LLNL in 1997 and currently serves as a consultant. He also writes a monthly column on environmental issues for *Plating & Surface Finishing* (the technical journal of the AESF).

jdini@earthlink.net

Martha L. Doemland

Martha Doemland received her doctorate in epidemiology from the School of Medicine, State University of New York at Buffalo in 1993. Her work is primarily in the field of environmental and occupational epidemiology. She focuses her efforts on the health effects of PCBs. Dr. Doemland is currently an independent consultant.

Doemland@erols.com

Daryl J. Doyle

Dr. Doyle earned his B.S. (1971) and Ph.D. (1981) degrees in chemistry from North Dakota State University in Fargo. He taught for 4 years at Valley City State College and for 1 year at North Dakota State University before he joined the faculty at GMI Engineering & Management Institute (now Kettering University) in Flint, Michigan, in 1982. Dr. Doyle has been involved in a variety of research while at Kettering University including analyzing chemical by-products formed from the laser cutting of polymers, using lasers to surface prepare polymers for adhesive bonding, and assisting corporations in the preparation of ISO 14000 certification. Dr. Doyle has taught several courses, including a course on handling hazardous materials. He is a member of the American Chemical Society, the Society of Automotive Engineers, the Society of Manufacturing Engineers, and the Adhesive Society.

Science and Math Department, Kettering University, 1700 W.3rd Avenue, Flint, Michigan 48504
ddoyle@kettering.edu

Varadarajan Dwarakanath

Varadarajan Dwarakanath holds a Ph.D. in petroleum and geosystems engineering from the University of Texas at Austin. Dr. Dwarakanath has been a senior geosystems engineer with Duke Engineering and Services since 1997 and manages the DE&S Geosystems Laboratory in Austin, Texas. His primary duties include design and selection of surfactants and partitioning tracers for remediation and characterization of aquifers.

Duke Engineering and Services, 9111 Research Boulevard, Austin, Texas 78758
vxdwarak@dukeengineering.com

David A. Dzombak

David A. Dzombak teaches and conducts research in water and soil quality engineering at Carnegie Mellon University, with focus on physical-chemical processes, groundwater contamination, hazardous waste site remediation, and industrial waste treatment. He has extensive experience with experimental investigations and modeling of reactions of metals with mineral surfaces, organic matter, and whole soils. Dr. Dzombak holds a Ph.D. degree in Civil-Environmental Engineering from the Massachusetts Institute of Technology, and B.S. and M.S. degrees in Civil and in Civil-Environmental Engineering from Carnegie Mellon. He is the recipient of the W.L. Huber Civil Engineering Research Prize from ASCE (1997), the Harrison Prescott Eddy Medal from WEF (1993), and a Presidential Young Investigator Award from NSF (1991).

Carnegie-Mellon Department of Civil & Environmental Engineering, Pittsburgh, Pennsylvania 15213-7813
dzombak@cmu.edu

J. Gordon Edwards

J. Gordon Edwards received his B.S. degree in botany from Butler University in 1942. He then spent nearly 4 years in the army, including time as a combat medic in Europe. He earned a Ph.D. in entomology from Ohio State University in 1949. He taught entomology at San Jose State University for 49 years and is now curator of the University's large entomology collection. His publications include *Coleoptera, or Beetles, East of the Great Plains* and *The Climber's Guide to Glacier National Park, Montana.* He has been active in many biological organizations and is a lifetime Fellow of the California Academy of Sciences.

San Jose State University, Department of Biology, One Washington Square, San Jose, California 95192-0100

Hugh W. Ellsaesser

Hugh Ellsaesser earned a S.B. in meteorology in 1943 and a Ph.D. in theoretical meteorology in 1964 from the University of Chicago and an M.A. in dynamic meteorology in 1947 from the University of California at Los Angeles. He was an air weather officer in the U.S. Air Force for 21 years, with the permanent rank of Lieutenant Colonel when he retired in 1963. He was employed as a physicist by Lawrence Livermore National Laboratory from 1963 to 1986 and continued his work there from 1986 to 1997 as a participating guest scientist. He has authored numerous meteorological papers for presentation as invited lectures and/or in technical publications, including invited chapters in several books. His research convinced him that establishment views on air pollution, threats to the ozone layer, and the climatic effect of carbon dioxide cannot be supported scientifically; the hazards have been greatly exaggerated.

Lawrence Livermore National Laboratory, 4293 Stanford Way, Livermore CA 94550
Hughel@home.com

Cynthia R. Evanko

Cynthia R. Evanko holds M.S. and Ph.D. degrees in Civil-Environmental Engineering from Carnegie Mellon University, and a B.S. degree in Civil-Environmental Engineering from the Massachusetts Institute of Technology. Her expertise lies in aquatic and environmental chemistry, with thesis research that focused on better understanding organic-solid interactions in order to improve models for contaminant fate and transport in surface water and groundwater systems. Dr. Evanko is currently employed with GeoSyntec Consultants, where she provides expertise in metals fate, transport, and remediation for project teams throughout the company. She has performed extensive modeling of metals behavior in natural systems in order to support conceptual site models and remedial designs for several project sites. Dr. Evanko has also designed and conducted experimental investigations in order to demonstrate the feasibility of metals remediation technologies.

GeoSyntec Consultants, 1100 Lake Hearn Dr., Ste. 200, Atlanta, Georgia 30342

Donald R. Fawn Jr.

Don Fawn is a senior member of the Texas Natural Resource Conservation Commission Emergency Response Unit. His experience spans two decades and his training and expertise include nuclear, chemical, and biological response operations; managing the consequences of terrorism; hazardous materials control; hazardous waste site personal protection and safety; confined space entry/rescue; hydrocarbon firefighting; alternative treatment methods for Superfund sites; risk assessment; incident command system; bioremediation; railroad tankcars/intermodal tanks; dispersant use/effect; and chemical safety audits. His firefighting work pioneered the use of environmentally benign microbially active firefighting foams on hydrocarbon fires. He has medical training in industrial toxicology, autonomic pharmacology, ballistics and kinetics of trauma, pediatric trauma, pathology and treatment of acute diseases of the nervous system, penetrating and blunt trauma injuries, and primary and secondary spinal cord injuries.

TNRCC, 8110 Little Deer Crossing, Austin, Texas 78736
DFWN@TNRCC.state.tx.us

Geoff Freeze

Geoff Freeze, P.E., has a M.S. in agricultural engineering from Texas A&M University and a B.A.Sc. in civil engineering from the University of British Columbia. He is a civil/environmental engineer who specializes in probabilistic applications of groundwater flow and transport models. He has applied model results to radioactive waste disposal performance assessments, remediation and reclamation studies, and decision and risk analyses. He has experience in field hydrogeology and project and personnel management and has authored over 20 research papers and technical reports. Mr. Freeze has taught several short courses in computer solutions to groundwater problems. He has supported commercial and government clients in many countries including the United States, Canada, Germany, Switzerland, and Japan.

Duke Engineering & Services, 1650 University Boulevard, NE., Suite 300, Albuquerque, New Mexico 87102
gafreeze@dukeengineering.com

Ralph R. Fullwood

Ralph Fullwood has a B.S. in physics from Texas Tech University (1952) and an A.M. in physics from Harvard University (1954). He did graduate study at the University of Pennsylvania (1956–1960), and earned a Ph.D. in nuclear science and engineering from the Rensselaer Polytechnic Institute. He joined Los Alamos Scientific Laboratory in 1966 and in 1972 he joined the Science Applications International Corp. in Washington, DC, where he participated in the Rasmussen study of nuclear power safety using probabilistic safety analysis (PSA). In 1985 he joined Brookhaven National Laboratory to apply PSA to optimize nuclear safety. He wrote interactive computer codes for PSA studies of aging and reliability as well as teaching physics and chemistry for the Department of Energy (DOE). He was editor of the DOE *Risk Management Quarterly* and has written over 200 reports, papers, book articles, and books.

rrfullwood@yahoo.com

Richard C. Gaskins Jr.

Richard Gaskins has a law degree from Harvard Law School (cum laude) and a B.S. in mechanical engineering from Duke University (summa cum laude). He also worked as an engineer and received a Fels Research Fellowship to study public policy issues at the University of Pennsylvania. He is past chair of the Environmental and Natural Resources Law section of the North Carolina Bar Association, a member of the Executive Committee of the ASTM Environmental Assessment Committee, chair of the ASTM Global Sustainability and Pollution Prevention Subcommittee, and former chair of the ASTM Phase II Training Task Group. He is a member of the Grievance Committee for the 26th judicial district and the North Carolina Bar Foundation CLE Committee and was an appointee to the Legislative and Legal Issues Subcommittee of the Governor's Waste Management Board for the state of North Carolina. He is listed in *The Best Lawyers in America* for environmental law. He has published papers and has spoken at seminars and conferences around the country on environmental and litigation topics.

Kirkpatrick Stockton, LLP, 3500 One First Union Center, Charlotte, North Carolina 28202
rgaskin@kilstockton.com

Sharron C. Gaudet

Sharron Gaudet received a B.S. in Engineering and Applied Science from Caltech in 1982 and a M.S. in Civil Engineering from MIT in 1984. From 1985 to 1987 she worked for ERT as a Water Resources Engineer doing computer modeling of surface and ground water systems. In 1988 she initiated a professional exchange program between ENSR and the Delft Hydraulic Laboratory (DHL) in the Netherlands and spent nine months working at DHL. In 1991, she returned to Massaschusetts and began working at HydroAnalysis. She has a P.E. in Civil Engineering in Massachusetts

Lois Swirsky Gold

Dr. Gold is Director of the Carcinogenic Potency Project at the Environmental Health Sciences Center, University of California, Berkeley, and a senior scientist at the E. O. Lawrence Berkeley National Laboratory. She has published 110 papers on the methodology of risk assessment, analyses of animal cancer tests, and implications for cancer prevention and regulatory policy. Her carcinogenic potency database, published as a CRC Handbook, analyzes the results of 5,500 chronic long-term cancer tests on 1,400 chemicals. Dr. Gold has served on the panel of expert reviewers for the National Toxicology Program, the Board of the Harvard Center for Risk Analysis, and the Harvard Risk Management Group. She is on the editorial board of *Regulatory Toxicology and Pharmacology*. Dr. Gold is recipient of the 1999 Annapolis Center Award for Risk Communication.

Carcinogenic Potency Project, Environmental Health Sciences Center, One Cyclotron Rd. MS-90-K, Berkeley, California 94720
lois@potency.berkeley.edu

Leah S. Goldberg

Leah Goldberg received her B.A. in 1985 from Colorado State University in political science with a minor in economics. She received her J.D. from the University of California, Hastings College of the Law in 1991, where she was awarded the Tony Patião Fellowship for academic achievement, demonstrated leadership ability, and substantial contributions to the community. Ms. Goldberg practices law in San Francisco, where she focuses on environmental and land use law. She specializes in reuse and redevelopment of contaminated properties, including purchase and sale of environmentally impacted properties; strategies for redevelopment; and working with cities, regulators, developers, lenders, and concerned citizens to develop creative solutions to complex reuse problems. She is vice chair of the California Redevelopment Association Brownfields Committee and serves on the San Joaquin County PHS/EHD technical review committee.

Hanson, Bridgett, Marcus, Vlahos & Rudy, LLP, 333 Market Street, 23rd Floor, San Francisco, California 94105
lgoldberg@hansonbridgett.com

Henk Haitjema

Dr. Haitjema received his Master's degree in civil engineering from the Technical University of Delft, The Netherlands (1976) and his Ph.D. in civil engineering from the University of Minnesota (1982). In 1984 he joined Indiana University, where he is a professor at the School of Public and Environmental Affairs. He teaches groundwater flow modeling, soil mechanics and science, and applied mathematics, for which he received awards in 1992 and 1998. Dr. Haitjema's research emphasizes applying analytic functions in modeling groundwater flow, the analytic element method, as opposed to more numerical approaches such as finite differences or finite elements. In 1994 he published an analytic element model GFLOW. In his 1995 book *Analytic Element Modeling of Groundwater Flow*, he provides a practical introduction to the theory and application of the analytic element method, using as educational version of GFLOW as a training vehicle.

Indiana University, School of Public & Environmental Affairs, MS-418, Bloomington, Indiana, 47405
haitjema@indiana.edu

David Hanneman

David Hanneman received his B.S. in water resources from the University of Wisconsin at Stevens Point, with a minor in chemistry. He received his M.S. in environmental sciences from the University of Alaska at Anchorage. He is an environmental scientist with the U.S. Army Corps of Engineers, Alaska District. He has been involved with environmental chemistry for over 18 years, dealing with hazardous waste cleanups, monitoring, and analysis of environmental samples through employment in private consulting, manufacturing, regulatory agency, and the federal government. His past projects include development of a groundwater monitoring system, PCB removals in soil and concrete, and various RCRA and fuel remediation projects in remote areas.

U.S. Army Corps of Engineers, Alaska District, PO Box 898, Anchorage, Alaska 99506

Blayne Hartman

Blayne Hartman received his Ph.D. in geochemistry from the University of Southern California, where he studied a variety of onshore and offshore geochemical projects, including petroleum characterization and tracing, utilization of stable and radiogenic isotopes as tracers, and gas transfer between water and air. Since he entered private industry, he has worked extensively with surface geochemical methods for environmental and exploration applications. He is the founder and director of Transglobal Environmental Geochemistry, a private company that offers on-site laboratory analysis, direct push environmental sampling, and soil vapor surveys. Dr. Hartman has been an instructor in several environmental-related continuing education programs, including the University of California and University of Wisconsin. He has provided geochemical training to state regulatory agencies in over 20 states and to all the EPA regions.

TEG, 432 N. Cedros Avenue, Solana Beach, California 92075
bh@tegenv.com

Robert Alan Haviland

Robert A. Haviland graduated from the U.S. Military Academy at West Point, New York, in 1975. In 1983 he received a Master's degree in civil engineering from the University of Alaska at Anchorage. He became a registered professional engineer in Alaska in 1984 and in California in 1991. Mr. Haviland is a senior engineer with the Alaska District, U.S. Army Corps of Engineers. He has served as the District's innovative technology advocate and the District's risk assessor. Mr. Haviland was instrumental in formulating the District's risk-based approach to environmental remediation, specializing in petroleum contamination and in the application of indefinite delivery type remedial action contracts for District environmental restoration.

U.S. Army Corps of Engineers, Alaska District, PO Box 898, Anchorage, Alaska 99506
robert.a.havilland@USACC.army.mil

Jenifer S. Heath

Jenifer Heath received her Ph.D. in environmental toxicology from Cornell University. She has a M.S. with a double major in toxicology and nutrition from North Carolina State University, an M.A. in public policy from Duke University, and a B.S. in public health from the University of Massachusetts at Amherst. Before she started her own business, Dr. Heath was a pesticide registration chemical safety analyst for Bayer, a public health toxicologist for the state of North Carolina, and an environmental consultant with Geraghty & Miller and Woodward-Clyde. Her work focuses on contaminated sites, environmental permitting, and toxic torts. She has testified at jury and bench trials as an expert witness and has authored over five dozen articles and book sections. She has extensive training in mediation and arbitration and is on the commercial panel of the American Arbitration Association.

Woven Egg Consulting, PO Box 9134, Denver, Colorado 80209-0134
jheath@mediate.com

David A. Hill

David Hill graduated from Franklin College in Franklin, Indiana, in 1971. He has a B.A. in English and a B.S. in organic chemistry. During the 1970s Mr. Hill worked in the field of waste disposal (construction and solid wastes) and gained experience in sales, operations, and management. In the 1980s these activities expanded to include composting organic wastes and financing waste operations. By the 1990s Mr. Hill was working exclusively in composting; he developed and managed a large composting operation. Odor was always an issue in both operations and management and in 1994 Mr. Hill left the composting industry and began working full time in odor identification and odor management. He has worked with odor problems on five continents. In 1996 he joined Global Odor Control Technologies with the intent of developing a network of reliable, trained professionals to work with odor problems throughout the world.

Global Odor Control Technologies, Inc., 4901 N. Mount Gilead Road, Bloomington, Indiana 47408
goctindy@kiva.net

William A. Hollerman

William Hollerman is an experimental physicist who works in accelerator-based physics, environmental technology, and materials science. He has a Ph.D. in applied physics from Alabama A&M University (1996) and Master's degrees in physics from Western Michigan University and Purdue University. Dr. Hollerman completed a group major Bachelor's degree in mathematics and physics from St. Joseph's College in Indiana. He earned the CHMM designation from the Institute of Hazardous Materials Management in 1992. He is currently an assistant professor of physics at the Acadiana Research Laboratory on the campus of the University of Louisiana at Lafayette. He has been a postdoctoral research associate with the H. J. Foster Center for Irradiation of Materials at Alabama A&M University, training director at the Center for Environmental technology, and staff scientist at Nichols Research Corp. in Huntsville, Alabama. He has published more than 18 journal articles and has given lectures, courses, and talks on various topics.

University of Louisiana at Lafayette, PO Box 44210, Lafayette, Louisiana 70504
hollerman@louisiana.edu

Robert M. Hordon

Robert Hordon is a physical geographer and hydrologist in the Geography Department at Rutgers University. His B.A. from Brooklyn College was followed by service in the U.S. Navy. He has an M.A. and a Ph.D. from Columbia

University. He is a professional hydrologist, certified by the American Institute of Hydrology. Dr. Hordon has expertise in physical geography, surface and groundwater hydrology, and application of quantitative techniques, such as multivariate analysis, to issues in water resources management. He has published several articles in professional journals and has presented papers at numerous professional meetings. He has been an invited participant in several geography and earth science review panels for the U.S. Environmental Protection Agency in Washington, DC, and has served as one of the Rutgers University representatives to the Advisory Committee for the New Jersey Water Supply Master Plan.

Department of Geography, Rutgers University, 54 Joyce Kilmer Avenue, Piscataway, New Jersey 08854-8045
hordon@rci.rutgers.edu

William Huber

Dr. Huber maintains a private practice serving clients by interpreting and presenting environmental data. He specializes in environmental statistics and geostatistics, decision analysis, and geographic information systems (GIS). He holds a Ph.D. in mathematics from Columbia University in New York and a B.A. in philosophy and mathematics with high honors from Haverford College, Pennsylvania.

Quantitative Decisions, 539 Valley View Road, Merion Station, Pennsylvania 19066
whuber@quantdec.com

Herbert Inhaber

Dr. Inhaber holds several degrees in physics and mathematics: a B.Sc. from McGill University, an M.S. from the University of Illinois, and a Ph.D. from the University of Oklahoma in Norman. He has been a principal scientist at Westinghouse Savannah River Co. and a senior risk assessment expert at the University of Nevada in Las Vegas. He has authored eight books and has published about 150 scientific articles and reviews. His work has been cited over 790 times in the scientific literature, including 160 books and monographs. He has lectured on risk analysis in many states and in 14 countries. While he was in Canada his work on risk was the most-requested study of the agency for which he worked. Dr. Inhaber is a Fellow of the American Nuclear Society and he was elected to its International Board of Directors. He has been on editorial board of *Risk Analysis, Risk Abstracts*, and *Scientometrics*.

Risk Concepts, 3920 Mohigan Way, Las Vegas, Nevada 89119
hinhaber@hotmail.com

Richard Jackson

Richard Jackson is manager of the geosystems section of Duke Engineering and Services in Austin, Texas. He is responsible for characterizing and remediating sites contaminated by fuels, solvents, and other nonaqueous phase liquids. He has degrees in hydrology, civil engineering, and hydrogeology; the latter is a Ph.D. from the University of Waterloo. Before he joined Duke Engineering and Services, Dr. Jackson was chief of the groundwater contamination project at the National Water Research Institute of Canada in Burlington, Ontario.

Duke Engineering & Services, 9111 Research Boulevard, Austin, Texas 78758
rejacks1@dukeengineering.com

James A. Jacobs

James Jacobs has an M.A. in geology from the University of Texas at Austin and a B.A. in geology and English from Franklin and Marshall College. He is a registered geologist in several states and a certified hydrogeologist in California. Mr. Jacobs is founder and president of FAST-TEK Engineering Support Services, a firm that offers a variety of direct push environmental sampling, in situ remediation technologies, excavation, and GPS/GIS services to leading consulting firms. He is past president of the Groundwater Resources Association of California, San Francisco Chapter, and is on their board of directors. He is past president of the American Institute of Professional Geologists, California Section, and a member of their national advisory board. He is an incorporator and director of the California Council of Geoscience Organizations and cofounder and president of the Independent Environmental Technical Evaluation Institute.

FAST-TEK Engineering Support Services, 247 B. Tewksbury Avenue, Pt. Richmond, California 94801
augerpro@jps.net

David Jeffrey

David Jeffrey has over 10 years of professional experience as a risk assessment consultant and 10 years of experience in the chemical sciences. He is an environmental organic chemist with expertise in fate and transport modeling, risk assessment, life cycle assessment, and litigation support. Dr. Jeffrey has managed and performed numerous human health risk assessments, ranging from screening level evaluations to complex, multipathway/receptor/scenario sites.

He has managed and conducted site risk assessments under various RBCA protocols for sites containing petroleum-derived contaminants, including ASTM and TNRCC. He has published and presented papers on numerous technical topics. Dr. Jeffrey's professional activities include extensive research laboratory experience and experience in predictive computer modeling and information retrieval software.

SECOR International, 1390 Willow Pass Road, #360, Concord, California 94520

Zhen-Gang Ji

Zhen-Gang Ji received his B.S. in atmospheric physics from the University of Science and Technology of China (1982); his M.S. (1985) and Ph.D. (1987) in atmospheric sciences from the Institute of Atmospheric Physics, Chinese Academy of Science; and his Doctor in Engineering Science in civil engineering from Columbia University (1993). He is a licensed professional engineer in civil engineering. Dr. Ji specializes in the study of hydrodynamics, water quality, sediment transport, and toxics in surface water systems for water resources management. He has authored more than 30 papers and technical reports and has provided consulting services to private and government clients in the United States, Europe, and Asia. Dr. Ji is a diplomate of the American/Academy of Environmental Engineers and he is a senior environmental engineer with Tetra Tech, Inc.

Tetra Tech, Inc., 10306 Eaton Place, Fairfax, Virginia 22030
jijetetratech-ffx.com

Michael Johns

Michael Johns received a B.S. in biology from the Citadel, a M.S. in zoology, and a Ph.D. in oceanography from the University of South Carolina. He also has a M.B.A. in management from the University of Rhode Island. Dr. Johns is an aquatic scientist specializing in aquatic ecological risk assessments, particularly those associated with contaminated sediment. The emphasis of his 21 years of professional experience has been in the effects of toxic pollutants on aquatic organisms. He is a recognized expert on the use of bioassessment techniques to evaluate sediment contamination. He has been responsible for the development of several bioassays including long-term toxicity tests designed to determine the effects of contaminated sediment on the growth and reproductive success of marine benthic species. He has published more than 30 technical papers in peer-reviewed journals and symposia.

Windward Environmental LLC, 200 West Mercer Street, Suite 401, Seattle, Washington 98119
mikej@windwardenv.com

Paul Johnson

Dr. Johnson is an Associate Professor in the Department of Civil and Environmental Engineering at Arizona State University. His research, teaching, and consulting activities focus on developing a better understanding contaminant fate and transport mechanisms, so that this knowledge can be applied to risk assessment, mitigation and residuals management. His current research focuses specifically on the in situ bio-treatment of MTBE-impacted aquifers, predicting the fate and transport of soil gas vapors, in situ air sparging, and assessing the longevity and impacts of subsurface contaminant residuals.

Department of Civil and Environmental Engineering Arizona State University, Tempe, Arizona 85287-5306
paul.c.johnson@asu.edu

Richard R. Jurin

Richard Jurin has an associate degree in science, engineering processes, and chemistry; a degree in biology; and the equivalent of a Master's degree in biochemistry from England. He has a Master's degree in environmental communications and a Doctorate in environmental education, communication, and interpretation, with a minor in adult/community education from The Ohio State University. Dr. Jurin has over 25 years of experience in biochemistry research and has published numerous papers and chapters in books. He has taught in universities and medical schools in England and the United States. His current research is on barriers to user understanding of information, factors affecting how information is used, and how information is interpreted from visual media. He has had his own environmental communications consulting company since 1995.

Jurin@eceic.com

William B. Katz

William Katz has a Bachelor's degree from the University of Illinois and a Master's degree from the Massachusetts Institute of Technology, both in chemical engineering. He is a registered professional engineer in Illinois, Indiana, Michigan, and Wisconsin and is recently retired from an environmental business he founded over 25 years ago. Mr. Katz has authored several publications on environmental subjects. He is an Emeritus member and a Fellow of the American Institute of Chemical Engineers, a member of the American Chemical Society, a Fellow of the American Association for the Advancement of Science, and a former member of the National Environmental

Training Association. He is also a member of the National Society of Environmental Consultants, an organization of real estate environmental auditing professionals. He was a founding member and has been elected an honorary director for life of the Spill Control Association of America. Mr. Katz also teaches environmental science (for non-science majors) at Roosevelt University and Oakton Community College.

billk@oakton.edu

W. Scott Keys

W. Scott Keys has Bachelor's and Master's degrees in geology. He spent most of his career with the U.S. Geological Survey and has 40 years of experience in applying borehole geophysics to solving problems related to waste migration from radioactive, industrial, and municipal storage and disposal sites. He started a consulting company in 1983, specializing in the computer interpretation of geophysical logs in the environmental field. He has worked as a consultant at the world's largest landfill and at municipal landfills, industrial sites, metal mines, and military bases. He has published numerous papers on borehole geophysics in scientific journals and has written several books on the subject. Mr. Keys has taught courses on borehole geophysics for private industry, professional organizations, and government agencies for many years. He has also received several awards from professional organizations for his investigations.

GEOKEYS, Inc., PO Box 1447, Kremmling, Colorado 80459
ScottKeys@compuserve.com

Renate D. Kimbrough

Renate Kimbrough received her M.D. from the Ernst August University in Goettingen, Germany. She has been a senior medical associate at the Institute for Evaluating Health Risks for the past 8 years. She served for 4 years in the administrator's office of the U.S. Environmental Protection Agency as an advisor for medical toxicology and risk evaluation. Before that she was a medical toxicologist for the Centers for Disease Control and Prevention. Dr. Kimbrough is a fellow of the American Academy of Clinical Toxicology and an honorary fellow of the American Academy of Pediatrics. She received the Clinton H. Thienes award of the American Academy of Clinical Toxicology, the Stockinger Award of the American Conference of Government Industrial Hygienists, and the Award for Meritorious Civilian Service of the Department of the Air Force.

P.O. Box 15452 Washington DC 20003-0452
Rkimbrough@erols.com

Mark L. Kram

Mark Kram is a Ph.D. candidate at the University of California at Santa Barbara. He earned a B.A. degree in chemistry from the University of California at Santa Barbara and a M.S. degree in geology from San Diego State University. Mr. Kram has over 15 years of experience with environmental assessment techniques and has authored papers, national standards, articles, and book chapters on the subject. He has invented several tools for measuring hydrogeologic properties and determining well design parameters and has served as the chief scientist for innovative research and development projects. Mr. Kram has been focusing on real-time assessment of halogenated organic contaminant source zones using direct push techniques, the use of direct push installed wells for long-term monitoring, and evaluation and development of innovative sampling devices.

School of Environmental Science Management, 4670 Physical Sciences North, University of California, Santa Barbara, California 93106-5131
mkram@bren.ucsb.edu

John H. Kramer

John Kramer, Ph.D., is a California certified hydrogeologist who directs project work on mine closure activities, subsurface investigations, environmental monitoring, and groundwater remediation for Condor Earth Technologies, Inc. In conjunction with consulting tasks, he integrates computerized field mapping methods with geographic information systems. He chairs technical sessions on digital field mapping for the Geological Society of America and provides short courses and training on digital data collection tools through a number of organizations. He implements the exciting new real-time Hydra 3-D motion monitoring systems. John holds a U.S. patent on wick-enhanced vadose zone monitoring systems that greatly extend the range of leak detection monitoring instruments such as the neutron probe. He manages a neutron-monitoring program using two 600-meter-long horizontal access pipes to successfully demonstrate containment of hazardous wastes. A similar system developed by Dr. Kramer resulted in a significant reduction of the performance bond at a municipal landfill.

Condor Earth Technology, 21663 Brian Lane, Sonora, California 95370
jkramer@condorearth.com

Robert T. Lackey

Dr. Lackey has a doctoral degree from Colorado State University, a Master's degree from the University of Maine, and a Bachelor's degree from Humboldt State University. He is Associate Director for Science of EPA's Western Ecology Division. His research focuses on ecosystem management, ecological risk assessment, and the interface between science and policy. Dr. Lackey is courtesy professor of fisheries science and adjunct professor of political science at Oregon State University. He has published 80 scientific papers and a book, edited 3 books, and been associate editor of three scientific journals. He frequently lectures at professional and scientific societies and universities. Before he joined the EPA, Dr. Lackey was a tenured associate professor at Virginia Polytechnic Institute and State University.

Environmental Protection Agency, 200 SW 35th Street, Corvallis, Oregon 97333
lackey.robert@epamail.epa.gov

David C. Lager

Mr. Lager has a Master's degree in urban and regional planning from the University of Illinois at Champaign-Urbana. He is president of NETCo., a professional services organization specializing in providing environmental compliance services to business and industry. His experience includes over 15 years helping industry meet RCRA, CERCLA, and CWA requirements, including expert witness testimony. He has provided project management assistance to corporations and developers in developing site remediation engineering and construction programs, provided on-site supervision of hazardous waste remediation activities, supervised the development of compliance documentation, and negotiated project changes with federal and state environmental authorities. Mr. Lager has helped several companies negotiate compliance agreement with the U.S. EPA and has provided training programs to meet compliance requirements, including authoring several hazardous waste interactive training programs.

NETCo, 6 Philbrook Terrace, Lexington, Massachusetts 02421
Lexlager@aol.com

Jay H. Lehr

Dr. Lehr received the nation's first Ph.D. in ground water hydrology from the University of Arizona in 1962, following a degree in geological engineering from Princeton University and a few years in the U.S. Navy's Civil Engineering Corps in the Western Pacific. After graduate school he taught both at the University of Arizona and The Ohio State University before serving 25 years as head of the Association of Ground Water Scientists and Engineers, where he was editor of the Journals of Ground Water and Ground Water Monitoring Review. During that period he continued to perform academic sponsored research in many areas of environmental science.

From 1968 through 1982 he assisted the federal government in establishing a safety net of environmental regulations involving surface water, ground water, air pollution and waste disposal, testifying before congressional committees on more than three dozen occasions.

Dr. Lehr has published 12 books and over 400 journal articles relating to ground water science. He is an outspoken proponent of sane environmental regulation that does not overly distort problems to the economic detriment of society. His textbook *Rational Readings on Environmental Concerns*, published in 1992 by Van Nostrand Reinhold, established a milestone for the publication of objective science by the world's leading environmental experts.

Lehr is currently a consultant with Bennett & Williams, Inc. and senior scientist with Environmental Education Enterprises, which teaches advance technology short courses for environmental professionals. He spends a considerable amount of time on the road, consulting and addressing meetings and conventions on environmental and motivational issues.

Environmental Education Enterprises, Inc., 6011 Houseman Road, Ostrander, Ohio 43061
Jaye3power@aol.com

Maurice E. LeVois

Maurice LeVois received his B.A. in math education from the University of Iowa in 1968 and his Ph.D. in health psychology from the University of California, San Francisco, in 1984. He was formerly director of the Veterans Administration's Office of Agent Orange Research and Education; a research scientist in the Agent Orange Study Unit, Centers for Disease Control and Prevention; and senior scientist at the Institute for Evaluating Health Risks in Washington, DC. He has designed and conducted large cohort studies and occupational mortality studies, research on problems of artifact in epidemiologic methods, research on cancer incidence and reproductive health effects in populations exposed to agricultural and industrial chemicals, epidemiologic risk modeling, and failure analysis of toxic waste management facilities. He is principal scientist at LeVois and Associates and provides epidemiological research and analysis services in the San Francisco Bay area.

Melevois@pacbell.net

Stephen H. Lieberman

Stephen Lieberman received a Ph.D. in chemical oceanography from the University of Washington, a M.S. in chemistry from San Diego State University, and a B.A. in chemistry from the University of California at San Diego. He was a postdoc at the California Institute of Technology. Dr. Lieberman is a senior research scientist in the Environmental Sciences Division at the Space and Naval Warfare Systems Center San Diego (SSC San Diego). He leads the optical chemical sensor group at SSC San Diego, whose efforts are directed at developing real-time chemical sensor systems for measuring environmental contaminants in soils, groundwater, and marine systems. The group has developed and fielded a laser-induced fluorometer sensor for petroleum hydrocarbons and a laser-induced breakdown spectroscopy sensor for metals.

Environmental Sciences Division, Space & Naval Warfare Systems Center, San Diego (D361), San Diego, California 92152
lieberman@spawar.navy.mil

Walter Wei-To Loo

Walter Loo is a certified hydrogeologist, a certified environmental manager, and a certified engineering geologist. He is an associate member of the American Society of Civil Engineers and a member of the Long Beach Petroleum Club. He has over 25 years of experience as a principal geohydrologist and extensive management and consulting experience in property audit and divestiture, in situ mining, groundwater modeling, and groundwater and soil investigation and remediation. He has prepared and presented over 100 technical papers and reports nationwide and internationally. He is a pioneer in soil and groundwater cleanup using electrokinetic treatment, bioventing, soil vapor extraction, and ultraviolet light treatment in combination and a leader in bioremediation of chlorinated solvents and petroleum hydrocarbons using co-metabolic processes.

Environment & Technology Services, 4690 Tompkins Avenue, Oakland, California 94619
ETSLOO@aol.com

T. D. Luckey

T. D. (Don) Luckey is Professor Emeritus at the University of Missouri, Columbia, and Honorary Professor of Olde Herborn University. He has a B.S. degree in chemistry from Colorado State University and M.S. and Ph.D. degrees from Wisconsin University. He completed 8 years of research in gnotobiology at Notre Dame University followed by 30 years as a professor of biochemistry, 14 years as chair, at the University of Missouri. Dr. Luckey has lectured extensively on nutrition, intestinal microecology, germ-free life, thymic hormones, and radiation hormesis. In 1979 he received a Humboldt Senior Scientist Award for research in Germany. In 1984 he was knighted, Ritter von Greifenstein. In 1996, he became a member of the Board of Directors of Radiation, Science and Health, an international group of independent individuals knowledgeable in radiation science and public policy.

Steve Maloney

Steve Maloney has a B.S. (1971) and a M.S. (1974) in biology from Middle Tennessee State University. He has worked in environmental consulting for 25 years. In 1988 he and William Griggs founded Griggs & Maloney, Inc., an engineering and environmental consulting firm in Murfreesboro, Tennessee. He is a certified hazardous materials manager and a certified environmental inspector, maintains OSHA's hazardous waste operations and emergency response certification, and is a member of the Tennessee Association of Business's Environmental Committee. Mr. Maloney's experience includes hazardous waste site investigations, RCRA and Superfund site remediation projects, environmental site assessments, groundwater remediation projects, bioremediation, underground storage tank investigations, and biological/ecological studies.

Griggs & Maloney, 745 S. Church Street, Belmont Park, Suite 205, Murfreesboro, Tennessee 37133
griggs.maloney@nashville.com

Stanley E. Manahan

Stanley Manahan received his A.B. in chemistry from Emporia State University in 1960 and his Ph.D. in analytical chemistry from the University of Kansas in 1965. He is Professor of Chemistry at the University of Missouri, Columbia, where he has been on the faculty since 1965; he is also president of ChemChar Research, Inc., a firm that develops nonincinerative thermochemical waste treatment processes. Since 1968 his primary research and professional activities have been environmental chemistry, toxicological chemistry, and waste treatment. He teaches courses on environmental chemistry, hazardous wastes, toxicological chemistry, and analytical chemistry; he has lectured on these topics throughout the United States as an American Chemical Society Local Section tour speaker, and he has written a number of books on these topics.

Department of Chemistry, 125 Chemistry Building, University of Missouri, Columbia, Missouri 65211
manahans@missouri.edu

Margaret McBrien

Margaret (Peg) McBrien received a M.S. in environmental engineering from Northwestern University and a B.A. in geology from Mount Holyoke College. She is a certified professional wetlands scientist and professional engineer with experience in wetlands and water resources management. She is a former regulator at the U.S. Army Corps of Engineers, U.S. Environmental Protection Agency, and Wisconsin Department of Natural Resources. Ms. McBrien has extensive federal and state regulatory experience including delineating wetland boundaries, assessing wetland functions and values, restoring and creating wetlands, obtaining state and federal permits, writing comprehensive environmental impact assessments, analyzing municipal and industrial wastewater treatment, and managing stormwater. Ms. McBrien specializes in designing wetlands for treating wastewater and stormwater, wetlands permitting and mitigation, environmental assessments, habitat evaluations, and watershed management.

Louis Berger & Associates, Inc., 100 Halsted Street, East Orange, New Jersey 07019
mmcbrien@louisberger.com

John J. McKetta Jr.

John J. McKetta spent much of his career at the University Texas at Austin, where he began as an assistant professor of chemical engineering and has since served as chairman of the Chemical Engineering Department, Dean of the College of Engineering, and Executive Vice Chancellor of the entire University of Texas system. He was also a holder of the Joe C. Walter Chair in Chemical Engineering. In 1984 the John J. McKetta Centennial Energy Chair was established at The University of Texas. He serves on the board of directors of six companies, is a member of the National Academy of Engineering, serves on numerous national advisory boards, and has held 50 local and national offices in eight professional, educational, and technical societies. He has published over 500 technical articles and was co-editor of *Advances in Petrochemicals and Refining*, which has been translated into nine languages. He has served several U.S. presidents in energy and environmental matters and has won numerous awards.

mcketta@che.utexas.edu

Takehiko Murayama

Takehiko Murayama received her Doctor of Engineering in environmental policy and planning in 1989 from the Tokyo Institute of Technology. Her major field is decision-making processes in risk management and communication. She is on the editorial advisory board of the journal *Risk, Health, Safety and Environment*. Dr. Murayama is an associate professor in the School of Science and Engineering, Waseda University.

School of Science & Engineering, Waseda University, 51-04-02B, 3-4-1, Ohkubo, Shinjuko, Tokyo 169-8555 Japan

Erin Mutch

Erin Mutch is a GIS administrator for Condor Earth Technology, where she oversees GIS-related activities and associated software operations using ARC/INFO and ArcView software. She develops and analyzes databases, customizes ESRI products, interfaces GIS items with PenMapO, and does other GIS custom programming in support of Condor's wide range of consulting activities. She also provides training, technical support, and GIS demonstrations. She directs Condor's ongoing GIS integration, analysis, and support of the Stanislaus Watershed FERC Relicensing Project for the Tri-Dam Project and the Pacific Gas And Electric Co. and the GIS for the largest open pit mining operation in California.

Condor Earth Technology, 21663 Brian Lane, Sonora, California 95370
emutch@condorearth.com

Kent S. Novakowski

Kent Novakowski is an associate professor in the Department of Earth Sciences at Brock University, St. Catharines, Ontario. He teaches undergraduate and graduate courses in introductory hydrogeology, groundwater modeling, and the hydrogeology of fractured rock. Dr. Novakowski is currently conducting research in the area of aqueous-phase gasoline and solvent migration in fractured rock and in the fundamental processes governing groundwater flow and solute transport in discrete fractures and fracture networks.

Department of Earth Sciences, Brock University, 500 Glenridge Avenue, St. Catharines, Ontario L2S 3A1, Canada
kent@craton.geol.brocku.ca

Raúl O'Ryan

Raúl O'Ryan received a Ph.D. in economics from the University of California, Berkeley, and is an Engineer of the Universidad de Chile. He is an associate professor at the Industrial Engineering Department of the Universidad de Chile. His research focuses on environmental economics, including choice of instruments for environmental regulation, international trade and the environment in developing contexts, computable general equilibrium analysis of environmental policies, and evaluation of policies to improve air quality in Santiago. He is counselor for Chile's Environmental Commission and other ministries in issues such as arsenic regulation, tradeable permits for Santiago, Santiago's decontamination plan, strategy for ozone depleting substances, and global warming. Dr. O'Ryan is also a consultant for The World Bank and has published over 20 articles and book chapters.

Dept de Ingenieria Industrial, Universidad de Chile, Republica 701, Santiago, Chile
roryan@dii.uchile.cl

Jane M. Orient

Jane Orient obtained her undergraduate degrees in chemistry and mathematics from the University of Arizona in Tucson and her M.D. from Columbia University College of Physicians and Surgeons in 1974. She completed an internal medicine residency at Parkland Memorial Hospital and University of Arizona Affiliated Hospitals and then became an instructor at the University of Arizona College of Medicine. She has been in solo private practice since 1981 and has served as Executive Director of the Association of American Physicians and Surgeons since 1989. She has authored several papers and books in the scientific and popular literature and is editor of *AAPS News*, *Doctors for Disaster Preparedness Newsletter*, and *Civil Defense Perspectives*. She is also a member of the editorial board of *The Medical Sentinel*.

Association of American Physicians & Surgeons, 1601 N. Tucson Boulevard, Suite 9, Tucson, Arizona 85716
71161.1263@compuserve.com

M. Alice Ottoboni

M. Alice Ottoboni received her B.A. degree (summa cum laude) from the University of Texas, Austin, in 1954. She received a Ph.D. in comparative biochemistry from the University of California, Davis, in 1959 and went to work in the pharmacology laboratory of the USDA Western Regional Research Laboratory, where her research interest turned to toxicology. In 1962 she became the first toxicologist employed by the state of California when she assumed the position of public health toxicologist in the Department of Public Health. She retired in 1985 to lecture and write. She is a member of Phi Beta Kappa, Iota Sigma Pi, and the Society of Toxicology. She has authored numerous scientific papers and a book on toxicology. She and her husband are currently working on a book on nutrition.

ottoboni@reno.quik.com

Ioan C. Paltineanu

Ioan C. Paltineanu has Engineer of Agronomy and Ph.D. diplomas from the Agronomic Institute in Bucharest, Romania. He is founder of an independent research and technology transfer company involved in real-time natural resources monitoring studies over large areas. As research associate in the NCSU Soil Science Department he was the final report coordinator of the North Carolina Nitrogen Budget project, using GIS databases maintained by academic institutions and state and federal agencies. He has authored more than 100 research and technical papers, 3 books, and 7 patents. He has been a research leader of multidisciplinary teams contracting with the FAO/IAEA and director of experimental stations and research institutes for irrigation and drainage in Romania. He has been an invited lecturer and visiting scientist to the United States, United Kingdom, Sweden, and Germany. He is a member of the ASA, SSSA, IUSS, IA, and ASAS-Romania.

PALTIN International, Inc., 6309 Sandy Street, Laurel, Maryland 20707
icpaltin@bellatlantic.net

Richard Peralta

Richard Peralta, P.E., is a full professor at Utah State University in Logan. He develops and applies simulation/ optimization groundwater models. His current research emphasis is on optimizing pump and treat (maximum efficiency or effectiveness or minimum cost or time). These models help optimize management of groundwater, surface water and soil moisture addressing contaminants from point to non point sources. Dr. Peralta has received five awards for outstanding research. He has chaired the Utah Extension Water Quality Task Force coordinating interagency actions and education. As a U.S. Air Force Reserve Bio-environmental Engineer attached to the Pentagon, he negotiates with environmental regulators and provides environmental remediation training and advice.

Biological and Irrigation Engineering Department, Utah State University, Building EC-216, Logan, Utah 04322-4105
peralta@cc.usu.edu

Jefferson Phillips

Mr. Phillips has worked with Fast-Tek for several years specializing in GIS and GPS. He is currently posted to the Buenos Aires office of Fast-Tek. He has a B.A. degree in Environmental Studies and Environmental Biology from the University of Colorado, Boulder.

FAST-TEK Engineering Support Services, 247 B. Tewksbury Avenue, Pt. Richmond, California 94801

Elizabeth Power

Elizabeth (Beth) Power is an environmental scientist who specializes in aquatic bioassessment, ecological risk assessment, and contaminated sediments. Ms. Power received her M.Sc. in zoology in 1987 from the University of British Columbia, Vancouver, Canada. She has over 13 years of experience in environmental consulting. She has managed projects involving watershed-level investigations, sediment and water quality assessment at contaminated sites, liaison with engineers and regulatory agencies in remedial planning, and environmental training. Her clients include mines, port facilities, petrochemical industries, waste management facilities, land developers, consulting engineers, and local and national governments. She has worked in Canada, the United States, Aruba, Tanzania, and Southeast Asia. She revels in public speaking and training and has a strong interest in regulatory affairs and alternative dispute resolution.

EVS Envionmental Consultants, 195 Pemberton Avenue, North Vancouver, British Columbia V7P 2R4
bpower@evsenvironment.com

Rumana Riffat

Rumana Riffat obtained her Master's and Ph.D. degrees in civil and environmental engineering from Iowa State University, Ames, in 1991 and 1994. She obtained her Bachelor's degree in civil engineering from Bangladesh University of Engineering and Technology in 1988. She is an assistant professor in the Department of Civil and Environmental Engineering at George Washington University, Washington, DC. She is investigating removal of petroleum hydrocarbons from industrial wastewater by anaerobic biological treatment. She recently completed a project for the Water Environment Research Foundation to identify physical and chemical hazards at wastewater treatment plants. She has received numerous awards including nomination for the Harrison Prescott Eddy Medal.

Civil & Environmental Engineering Department, George Washington University, 801 22nd Street, NW, Room 663, Washington, DC 20052
rriffat@seas.gwu.edu

Ana Maria Sancha

Ana Maria Sancha is a chemist of the Pontificia Universidad Católica de Chile. Mrs. Sancha is an associate professor at the Civil Engineering Department of the Universidad de Chile. Her interests include research, teaching, and consulting work in topics related to water quality and wastewater treatment. She has taught in Chile and Latin America. She is also counselor for Chile's Environmental Commission and the National Environmental Center on issues such as drinking water standards. Mrs. Sancha is also a consultant for the World Health Organisation and Interamerican Development Bank. She has published over 70 articles in international and national congresses and seminars.

Water Resources & Environment Division, Department of Civil Engineering, Facultad de Ciencias Fsicas y Matematicas, Universidad de Chile Blanco Encalada 2002, Santiago, Chile
amsancha@cc.uchile.cl

John M. Shafer

John Shafer received his B.S. degree in earth science from Penn State University, MS. in resource development from Michigan State University, and Ph.D. in civil engineering from Colorado State University. Dr. Shafer is director of the Earth Sciences and Resources Institute at the University of South Carolina and sole proprietor of the environmental/groundwater consulting concern GWPATH. Before he joined the faculty at the University of South Carolina he was head of the hydrology section of the Illinois State Water Survey in Champaign and before that he was a senior researcher at Battelle's Pacific Northwest National Laboratory. He is a professional hydrologist certified by the American Institute of Hydrology. He is co-recipient of the 1991 John C. Frye Memorial Award of the Geological Society of America and was awarded the 1993 Illinois Groundwater Science Award for outstanding achievement in groundwater science. He has published many peer-reviewed papers and reports on a wide range of hydrologic and geohydrologic topics.

Earth Sciences & Resources Institute (ESRI), University of South Carolina, 901 Sumter Street, Columbia, South Carolina 29208
jshafer@esri.esri.sc.edu

Peter Shanahan

Peter Shanahan holds a Ph.D. in environmental engineering from the Massachusetts Institute of Technology and a M.S. in environmental earth sciences from Stanford University. He is a consulting environmental engineer and hydrologist with HydroAnalysis, Inc., of Acton, Massachusetts. He is also a lecturer in the Department of Civil and Environmental Engineering at the Massachusetts Institute of Technology. Dr. Shanahan specializes in computer modeling, environmental fluid dynamics, and transport of pollutants in surface water and groundwater.

HydroAnalysis, Inc., 481 Great Neck Road, Suite 3, Acton, Massachusetts 01720
shanahan@ma.ultranet.com

Thomas T. Shen

Thomas Shen received his Ph.D. in environmental engineering from RPI, New York. He is an international advisor and has more than 40 years of environmental research and teaching experience. He was a senior research scientist with New York State and taught graduate courses at Columbia University. He has lectured at over 30 universities and research institutes and conducted environmental seminars and workshops in the United States and in Asian and European countries. He was a consultant to the United Nations Development Program and World Health Organization on environmental projects and a member of the U.S. EPA's Science Advisory Board. He was an advisor of the Taiwan EPA's Environmental Planning Program and the Taiwan Industrial Technology Research Institute's Environmental Research Program. In 1993 he received the Frank Chamber Scientific Award from the Air and Waste Management Association in Denver, Colorado.

cs.tt.shen@worldnet.att.net

Donald I. Siegel

Donald Siegel, received a B.S. from the University of Rhode Island, a M.S. from Penn State University, and a Ph.D. from the University of Minnesota. He has been a professor of earth sciences at Syracuse University since 1982. He was a hydrologist with the U.S. Geological Survey from 1976 to 1982. He has been chairman (1994–1995), first vice president (1993–1994), and second vice president (1992–1993) of the hydrogeological division of the GSA; a GSA Fellow (1995); and a Birdsall Distinguished Lecturer of the GSA (1992–1993). He was a member of the National Research Council Committee on Techniques for Assessing Ground Water Contamination (1991–1993), Associate Editor of *Water Resources Research* (1992–1995), and Associate Editor of *Wetlands* (1994–1997). He has written about 60 peer-reviewed papers on paleohydrogeology, wetland hydrology and geochemistry, and contaminant hydrology and geochemistry.

Department of Earth Sciences, Syracuse University, 307 Heroy Geology Laboratory, Syracuse, New York 13244-1070
disiegel@mailbox.syr.edu

S. Fred Singer

Dr. Singer, Professor (Emeritus) of environmental sciences at the University of Virginia, is founding president of the Fairfax-based Science & Environmental Policy Project, a nonprofit educational association of scientists concerned with providing a sound scientific base for environmental policies. Singer has held several academic and government positions, including the first director of the U.S. Weather Satellite Service (now part of NOAA), deputy assistant administrator for policy of the Environmental Protection Agency, and chief scientist of the U.S. Department of Transportation. He also devised the instrument used to measure stratospheric ozone from satellites. He is author of many scientific articles and books.

The Science & Environmental Policy Project, 4084 University Drive, Suite 101, Fairfax, Virginia 22030
singer1@gmu.edu

James L. Starr

James Starr has 26 years of experience in interdisciplinary research linking the physical and hydrologic properties of soils with transport of reactive and nonreactive chemicals through soils and with growth and metabolic processes of soil microorganisms and growing crops. His current research focuses on identifying and characterizing the interactive effects of soils, cropping systems, and tillage on real-time soil water dynamics and the associated reaction and transport of nutrients. He is a member of the American Geophysical Union, the Soil Science Society of America, and the International Union of Soil Sciences.

U.S. Department of Agriculture, Agricultural Research Service, Beltsville Agricultural Research Center, Environmental Chemistry Laboratory, NRI Building 007, BARC-W, Beltsville, Maryland 20705
Jstarr@asrr.arsuda.gov

Jan Swider

Dr. Swider is a researcher in the Mechanical and Aerospace Engineering Department at UCLA. He has a Ph.D. in engineering and his main research interests are in human health, ecological, and engineering systems risk assessments. He is particularly interested in the instabilities in interspecific interactions caused by environmental agents and has been investigating the effects of air pollution on interspecific interactions in the scope of ecological risk assessment. Another area of interest is application of probabilistic risk assessment methods to accident prevention and management. He has also served as a consultant to industry. Dr. Swider was an International Atomic Energy Agency Fellow in 1990 and is a member of the Society for Rick Assessment.

swider@hej.cnchost.com

Stephen M. Testa

Steve Testa is currently president of Testa Environmental Corp., Mokelumne Hill, California, and editor-in-chief of the American Association of Pétroleum Geologists, Division of Environmental Geosciences journal *Environmental Geosciences*. He is past president of the American Institute of Professional Geologists. Mr. Testa has over 20 years of consulting experience in the fields of engineering and environmental geology and has been an instructor at the University of Southern California and California State University at Fullerton. Mr. Testa has published over 100 papers and abstracts and 7 books.

Testa Environmental Corporation, 19814 Jesus Maria Road, Mokelumme Hill, California 95245
stesta@goldrush.com

Rolf R. von Oppenfeld

Rolf R. von Oppenfeld received his B.A. (honors) in 1978 from Lawrence University and a B.S. in chemistry and biology from the American University in 1979. He received his J.D. (summa cum laude) in 1982 from the George Washington University, where he was Order of the Coif and received the U.S. Law Week Award as the outstanding graduate of the 1982 class. He is managing partner of the Team for Environmental, Science and Technology Law (TESTLaw), operating under the law firm von Oppenfeld, Hiser & Freeze, P.C. He has worked as a chemist for the U.S. EPA and was associated with EPA's Superfund program and with the Office of Hazardous Waste Enforcement. He regularly lectures on various environmental, natural resource, and water issues.

The TESTLAW Practice Group, VonOppenfeld, Hiser & Freeze, P.C., 4201 N. 24th Street, Suite 300, Phoenix, Arizona 85016
VHFPC@TESTLAW.com

Lisa Wadge

Lisa Wadge is a civil engineer with 17 years of experience in environmental engineering. Her career focuses on identification of risks through extensive experience in site audits, assessments, and information technology. She started and ran an environmental consulting firm for 10 years, specializing in audits and assessments. She sold the company and started a GIS and database service for New England. This information tool provided detailed, technical, online data to scientists, lawyers, and bankers. She sold the company to Vistainfo and her current position there is in the area of new product development, which focuses on creating data and GIS layers for use by customers from various disciplines. This includes natural GIS data for NEPA compliance, automated scoring systems, and international data.

VistaInfo, 67 N. Main Street, Essex, Connecticut 06426
Lisaw@vistainfo.com

Douglas Walker

Douglas Walker earned a B.S. in watershed science (1983), a M.S. in civil engineering (1986), and a Ph.D. in agricultural engineering, all from Colorado State University. His professional experience includes field hydrogeology, geostatistical analysis, groundwater modeling, and model development. He specializes in development and application of geostatistical approaches to groundwater contaminant transport. He has authored over 30 papers and technical reports and has provided consulting services to private and governmental clients throughout the United States, Europe, and Asia. He is currently a hydrogeologist with Duke Engineering and Services, Inc.

Duke Engineering & Services, 1650 University Boulevard, NE, Suite 300, Albuquerque, New Mexico 87102
ddwalker@dukeengineering.com

Elizabeth Whelan

Elizabeth Whelan holds Masters' degrees in public health from the Yale School of Medicine and the Harvard School of Public Health and a doctorate in public health from Harvard. She is president and founder of the American

Council of Science and Health (ACSH), a consumer education and advocacy group based in New York City. ACSH is directed and advised by over 350 scientists and physicians from the United States, Canada, Great Britain, and South Africa. ACSH publishes a quarterly magazine, *Priorities*, and maintains a Web site (www.acsh.org). ACSH's mission is to provide the media and consumers with science-based information about the relationship between nutrition, the environment, and lifestyle factors and human health. Dr. Whelan is the author of over 20 books.

Whelan@acsh.org

Martin Whittaker

Martin Whittaker has a Ph.D. in environmental science from the University of Edinburgh and Bachelor and Master of Science degrees in chemistry from the University of St. Andrews and McGill University, respectively. He is an environmental consultant, researcher, and senior analyst with Innovest Strategic Value Advisors, Inc., specializing in environmental risk, strategy, and finance. Before he joined Innovest he spent 2 years with Golder Associates, a leading international environmental consultancy, where he focused on industrial environmental risk assessment, liability analysis, environmental strategy, and contaminated site management. Dr. Whittaker has also worked as a technical and environmental assistant with Elf Oil. He is a member of the Society for Risk Analysis and the International Association of Water Quality and a past Committee Member of the Royal Society of Chemistry's Environmental Chemistry Group.

Innovest 225 E. Beaver Creek Rd. Ste 300 Richmond Hill, (Toronto) Ontario L4B 3P4 Canada
mwhittaker@home.com

Kevin Kent Wolka

Kevin Wolka (deceased) received a B.S. (1972) and a M.S. (1974) in civil engineering from Purdue University. He earned a Ph.D. in water resources in 1986 from Iowa State University. He was a professional engineer registered in Illinois and Michigan and a certified underground storage tank professional. He had over 20 years of experience in a variety of environmental engineering topics, including stormwater management, wastewater treatment, water supply, and solid waste management. He was a member of the American Society of Civil Engineers, the American Water Works Association, the Water Environment Federation, and the Engineering Society of Detroit. He published several technical papers. At the time of his death he was an associate professor of civil engineering at Carroll College. He previously taught at Iowa State University and at Lawrence Technological University and worked for several engineering firms.

Lynn Yuhr

Lynn Yuhr has a B.S. in environmental sciences with a minor in geology from Florida International University, with graduate studies in geology/hydrogeology at Florida International University. She has 21 years of experience including work throughout the United States and the Caribbean. Her work involves geotechnical site investigations for foundations, mine stability, dam assessments, and sinkhole collapse as well as environmental site characterization work. She specializes in interpreting complex geologic, hydrologic, and geophysical data. Her effort has been facilitated by the use of unified databases with graphical capabilities. She established Technos computer software and hardware for comparing and contrasting large amounts of diverse data and integrating this information for interpretation and presentation. She is author of over 20 technical publications.

Technos, Inc., 3333 NW 21st Street, Miami, Florida 33142
info@technos-inc.com

FOREWORD

The term environmentalist was first introduced in 1916. Yet the modern environmental movement was born in the 1960s in response to public demands for cleaner air and water, and protection of natural resources such as forests, mountains, and streams.

The Merriam-Webster Dictionary defines environmentalist as "an advocate of environmentalism . . . ." one who advocates "the preservation or improvement of the natural environment, especially the movement to control pollution." The American Heritage Dictionary concurs, noting an environmentalist is "one concerned about environmental quality, especially of the human environment with respect to the control of pollution."

All Americans can embrace the original goals of environmentalism, especially as they are expressed by current dictionary definitions. According to these parameters of environmentalism, who among us is not an environmentalist? Who is in favor of polluted water, foul air, or the destruction of beaches and other public recreation sites? When the environmental movement originated, it seemed that there was only one possible vote for environmental protection—and that was yes!

Now, after environmentalism has evolved over the past four decades, Americans have come to understand that protecting the environment is a complex, multifaceted and unique endeavor. Environmental policy must take into account not only the intrinsic value of nature, but real-life human factors, such as the need for food, water, housing, energy, commerce and the ongoing pursuit of a higher standard of living.

Modern day environmental issues do not fall neatly into dichotomies of polluted vs. pristine, contaminated vs. pure, hazardous vs. harmless, or even natural vs. synthetic. Classic environmentalism may have started as a universally embraced, loosely defined concept. But it soon developed myriad facets and alleys—alleys which led off in many divergent directions.

For some environmental activists, environmental preservation and cleanup efforts became a vehicle for achieving unrelated and unidentified ideological or economic goals. The American who would once proudly claim to be an environmentalist and support efforts to curb dumping of raw sewage into rivers and lakes soon became entangled by association in deviant efforts to remove minuscule levels of alleged toxins from the same waters—even though there was no evidence that the chemicals in question posed either ecological or public health threats. Under the original environmental banner, more radical environmentalists redefined pollution and contamination using increasingly lower measurements of parts per billion or less. Often the mere ability to detect the presence of a chemical sufficed to fit this new-fangled definition of pollution.

Regretfully, during the last three decades of the 20th century, the image of an environmentalist has evolved to include some new, undesirable characteristics: hyperbole about risk; misguided characterizations of nature as benign and manmade as malignant; a determination to advance protection of the environment, even in response to hypothetical risks, no matter what the economic or even human health costs.

Under this new and dubious profile of environmentalism, American consumers have witnessed calls for widespread bans of agricultural chemicals (like Alar, a growth regulator used on apples, and EDB, a fumigant used on grains and citrus); demands for stringent controls of air and water emissions; strident predictions of ecological crises (i.e., global warming), and alarming claims of human disaster (i.e., cancer epidemics) allegedly caused by energy production. The public debates between environmentalists and those they termed polluters or polluter-apologists became deafening and rhetoric-laden. Understandably, the public was frequently bewildered by such conflicting information.

As we enter a new century, we need to reconstitute the definition of environmentalism and redirect its efforts by emphasizing an essential element that in recent years has been overlooked: sound, peer-reviewed science.

A scientific approach is essential to rationally evaluating the frequently dire predictions of today's environmentalists. Science is critical to avoiding environmental policies founded on irrationality, undocumented hypotheses, and emotional mantras. A full spectrum of scientific expertise is critical if we are to avoid what some have called junk science but others more accurately, have termed unscientific junk.

Toward this end, the *Standard Handbook of Environmental Science, Health, and Technology* should become the focal point for sensible solutions to environmental concerns based on rational, scientific assessments of environmental risks and benefits. Prepared by the most distinguished scientists in the world, this handbook successfully bridges the gaps among a variety of disciplines, including math, chemistry, and engineering in its careful analysis of essential philosophical questions on how to manage the environment in the best interest of mankind. In short, the handbook represents a comprehensive state-of-the art description of environmental science for the 21st century.

Dr. Elizabeth M. Whelan
President, American Council on Science and Health

PREFACE

Over the past three decades the United States and many of its global partners have wrought the most successful grass roots effort to improve and preserve our environment, which had fallen into disrepair after the second World War and the subsequent industrial revolution. We began as a nation divided among those who were part of the problem and those who were part of the solution. However, by the end of the first decade of effort, nearly all Americans as well as citizens of most of the developed world could count themselves as environmentalists—all of them working hard to protect and improve the quality of our air, water, and soil. Although the job is not now and never will be complete, the gains we have made as a population pulling together are truly amazing.

As we begin a new century it is an appropriate time to take stock of the scientific advances that have fostered our many environmental successes. In this Handbook, we have collected the wisdom of the very best scientists and engineers who have been toiling in this multidisciplinary field since the beginning of our national effort. We have asked them to recount the methodologies that have been most effective in carrying out our tasks as well as the knowledge that has become evident about the state of our nation's environment.

The Handbook begins with the basic knowledge needed to operate within environmental science (Part 1) and then moves to the solutions of problems in the field at the local level (Part 2) and concludes by dealing with broad environmental issues (Part 3). Although some of the knowledge was available previously, much more of it is only now on the cutting edge. Some sections have broad general appeal and others are intended only for experienced practitioners. Among its 1650 pages, virtually everyone will find treasures of environmental knowledge.

The concept of the Handbook and its outline was initially conceived for McGraw-Hill by Charles Menzie, Richard Brown, George Hoag, and myself. It was left to me to carry out their grand plan with the assistance of 100 of the world's leading environmental professionals. They worked without compensation to assist in the gathering of what we hope will be seen as the most up-to-date source of scientific and engineering knowledge on the environment available anywhere between two covers. It should serve as an outstanding reference for a host of readers for the next decade and beyond.

The authors of the individual pieces that make up this treatise are all active in their profession. They have enthusiastically participated on the team that has created this Handbook. Most important of all, they remain ready and willing to advance your knowledge further if you have the desire and need to contact them.

As stated above, the job of protecting our environment is never complete, but we should occasionally take the time to celebrate our significant achievements. The environmental principles that are a part of conducting our lives and doing business have become more and more clear to corporations, municipalities, and individual citizens, all of whom are adopting a new sense of environmental responsibility. This is evident in new products and manufacturing processes, local government affairs, and the attitudes of most American families.

Industry today tries to find ways to cut down on waste, save energy, and make their products more environmentally attractive. They look for cooperation with traditional adversaries, such as the government and environmental advocacy groups, in order to achieve a win-win alliance.

What motivates us all today and moves us to act in a responsible manner is societal pressure to make positive changes in behavior. A free-market system brings the solution itself, by focusing on people's desire and self-interest for a better quality of life without sacrificing the freedom of choice.

This book is our collective effort to further enhance the environment in this new century in a manner that will preserve both the earth's grandeur and our personal freedom to enjoy it.

Jay H. Lehr
Editor

ABOUT THE EDITORS

Jay H. Lehr, Ph.D., is among the country's most frequently quoted authorities on groundwater hydrology and hydrogeology. He is a prolific author on a broad range of environmental science subjects having authored a dozen books and more than 300 journal articles. Lehr currently divides his time between working with Bennett & Williams, Inc., environmental consultants in Columbus, Ohio, and the Heartland Institute in Chicago, as well as serving as senior scientist with Environmental Education Enterprises, Inc. (E^3), one of the nation's largest environmental short course providers. Lehr served on the faculty of the University of Arizona and the Ohio State University and was editor of the prestigious journal *Ground Water* for 25 years. He played a significant role in the development of many of the United States' primary environmental laws.

Janet K. Lehr has been director of Education Operations for Environmental Education Enterprises, Inc. (E^3), for the past decade. She has facilitated the development of more than 85 short courses that have been presented across the United States. Many of the prestigious faculty members she assembled to teach these courses contributed significantly to this handbook.

THE INTERACTION OF BASIC SCIENTIFIC DISCIPLINES ON CONTAMINANT FATE AND TRANSPORT IN THE ENVIRONMENT

CHAPTER 1
CHEMISTRY

SECTION 1.1

AQUATIC CHEMISTRY

Stanley Manahan

Dr. Manahan is a professor of chemistry at the University of Missouri, Columbia, where he has taught since 1965. He is the author of the leading college textbook on environmental chemistry.

Aquatic chemistry is a function of the properties of water molecules, which in turn determine its physical and chemical behavior.[1] The water molecule, H_2O, is polar; an oxygen atom provides one end with a partial negative charge and two hydrogen atoms provide a partial positive charge on the other end. Therefore, in acting as a solvent the water molecule has a strong tendency to interact with polar solute molecules as well as with charged ions. The water molecule also has the ability to form hydrogen bonds in which the two hydrogen atoms bond simultaneously to two atoms, usually oxygen, that have a partial negative charge. This phenomenon tends to link the molecules of liquid water together and enhances the solvating ability of water for other species that are capable of undergoing hydrogen bonding.

Because of its molecular structure water has a number of physical properties that influence aquatic chemistry. Among the most important of these physical properties are very high heat capacity, very high heat of vaporization, and very high heat of fusion. Water also has a very high surface tension and it is transparent to light, which means that photosynthesis can occur in water. As a result of the temperature density behavior of water, in bodies of water during the summertime, a layer of warmer water floats on top of a layer of cooler water. Water in the top layer is exposed to atmospheric oxygen and becomes oxidizing, whereas water in the bottom layer remains out of contact with air and becomes chemically reducing.

Aquatic chemistry is the result of a number of different types of chemical and biochemical processes. These include acid-base reactions, solubility and precipitation phenomena, oxidation-reduction processes, and complexation processes. The influences of microorganisms on aquatic chemistry are very strong, especially with respect to oxidation-reduction processes.

A number of acid-base reactions occur in water. The water molecule itself undergoes dissociation to produce the H^+ ion, which is characteristic of acids, and the OH^- ion, which is characteristic of bases:

$$H_2O \rightleftharpoons H^+ + OH^-$$

In nature water commonly contains dissolved carbon dioxide (CO_2), which acts as a weak acid. The capacity of solutes in water to neutralize strong acid is an important characteristic known as alkalinity. Alkalinity is normally due to the presence of bicarbonate ion, HCO_3^-; at higher pH values carbonate, CO_3^{2-}, and hydroxide, OH^-, ions also contribute to the alkalinity of water. Normally alkalinity is established in water by the reaction of dissolved CO_2 with solid calcium carbonate ($CaCO_3$) mineral limestone as follows:

$$CaCO_3(s) + CO_2(aq) + H_2O \rightleftharpoons Ca^{2+} + 2HCO_3^- \qquad (1.1.1)$$

where (s) indicates solid and (aq) indicates aqueous.

Dissolved metal ions play an important role in the chemistry and biology of water. Although the formula of dissolved metal ions is commonly written in the form M^{n+} (such as Na^+ for dissolved sodium), metal ions dissolved in water are always bound with water molecules, usually 6 per metal ion, as indicated by the formula $Fe(H_2O)_6^{2+}$. Hydrated dissolved metal ions, especially those with a higher charge, tend to lose a hydrogen ion and act as acids as shown below for the iron(III) ion:

$$Fe(H_2O)_6^{3+} \rightleftharpoons Fe(H_2O)_5OH^{2+} + H^+ \tag{1.1.2}$$

Metal ions are hydrated because of the tendency of the charged metal species to bind with electron pairs on water molecules. Stronger electron pair donors, known as ligands, may displace water molecules in a process known as complexation to form metal complexes:

$$Fe(H_2O)_6^{2+} + CN^- \rightleftharpoons FeCN(H_2O)_5^+ + H_2O \tag{1.1.3}$$

When a ligand has two or more electron donor functional groups located in the ligand structure, so that they may simultaneously bind with the same metal ion, a chelate is formed and the ligand is known as a chelating agent. Some synthetic chemical species, e.g., ethylenediaminetetraacetic acid (EDTA), which can bind to a metal ion in up to six places, are strong chelating agents. These substances are used in metal plating baths, industrial cleaning solutions, and other applications and may occur as water pollutants. Such chelating agents are of concern because of their ability to dissolve and mobilize toxic heavy metal ions, such as lead. Humic and fulvic acids, which are formed by partial biodegradation of plant material, are naturally occurring chelating agents. Fulvic acid chelates of iron can discolor water.

Oxidation-reduction reactions are responsible for changing the oxidation states of species in water.[2] The species most often responsible for making water oxidizing is molecular oxygen (O_2). Oxidized species such as iron(III), nitrate, and CO_2 predominate at the surface of water in contact with atmospheric O_2, whereas the reduced forms of the elements—iron(II), ammonium ion, and methane—predominate in the bottom regions of bodies of water and in sentiments away from contact with molecular oxygen. A key aspect of the oxidation-reduction processes in water is that they are usually mediated by microorganisms.

The degree to which water is oxidizing or reducing can be expressed by a parameter called pE, where a high pE value denotes oxidizing conditions with low electron activity and a low pE value denotes reducing conditions with high electron activity. Electron activities are closely related to the activities of the hydrogen ion H^+. It is useful to construct pE/pH diagrams, various regions of which denote areas of high or low electron activity (low or high pE) and high or low pH (low or high H^+ activity). Such a diagram would show, for example, that in a region of low pE and low pH, a reducing acidic medium, NH_4^+ would be the predominant inorganic nitrogen species. At a higher pE value, denoting oxidizing conditions, nitrate ion, NO_3^-, would predominate as an inorganic nitrogen species.

1.1.1   PHASE INTERACTIONS

Processes that occur between water and other phases are an important aspect of aquatic chemistry.[3] Phases other than water can be divided into sediments and substances that are suspended in water. The atmosphere is a source of crucial gases, particularly oxygen and CO_2. A very thin oily film consisting of hydrocarbon materials, such as those from spilled boat motor fuel, is often found floating on the surface of water. The sediments serve as repositories of nutrients, pollutants, and exchangeable ions. Several processes are responsible for sediment formation, the simplest of which is mechanical movement of solids into the body of water and deposition there. Biological substances including detritus from dead organisms can contribute to sediments. Sedimentary material can also result from chemical reactions of species dissolved in water, as indicated below for the formation of solid calcium carbonate from dissolved calcium bicarbonate:

$$CaCO_3(s) + CO_2(aq) + H_2O \rightleftharpoons Ca^{2+} + 2HCO_3^- \tag{1.1.4}$$

A particularly important phase found in suspended matter in water consists of colloidal particles, which are of the order of a micrometer or less in size. Because of their very high surface area/volume ratio, colloidal particles are very active and can strongly affect chemical and biological processes in water. There are three major types of colloidal particles in water. Hydrophobic colloids have no affinity for water and remain suspended because they have the same electrical charge, which causes them to repel each other. Colloidal minerals commonly form hydrophobic colloids. Clay minerals consisting of hydrated silicon and aluminum oxides are the most common mineral colloids. Hydrophilic colloidal particles, which are often of biological origin, remain suspended because of their affinity for water. Association colloids are formed by species such as soap anions that have a dual organic and ionic nature. In a group of these ions, the hydrocarbon ion tails cluster inside the association colloid particle with the charged ionic heads on the outside of the particle in contact with water.

An important aspect of colloids in water is their aggregation to form precipitates. This process is known as coagulation or flocculation, depending on the mechanism involved. It is important in water treatment because many of the impurities in water are present in a colloidal form. It is also responsible for formation of sediments from colloids suspended in water.

1.1.2 *MICROORGANISMS IN WATER*

Microorganisms play a very strong role in aquatic chemistry, particularly as catalysts of oxidation-reduction reactions.[4] In general, aquatic microorganisms are single-celled bacteria, fungi, protozoa, and viruses. Of these, viruses grow only within the cells of other organisms and have no effect on aquatic chemistry. They are of concern, however, because of their potential to cause disease, and they are an important consideration in drinking water supplies. Fungi grow mostly on land, but their products, such as humic substances from partial biodegradation of plant material, can be important in water. Protozoa are generally capable of movement and in a sense can be regarded as small single-celled animals, although some are capable of photosynthesis. They participate in some significant aquatic chemical processes and may be responsible for formation of major mineral deposits containing calcium and silicon. In addition, some protozoa, such as those responsible for amebic dysentery, can cause disease in humans and other animals. Algae produce biomass by photosynthesis and thus provide the food upon which the aquatic food chain is based, as shown by the following reaction:

$$CO_2 + H_2O \xrightarrow{h\nu} \{CH_2O\} + O_2(g) \tag{1.1.5}$$

where $\{CH_2O\}$ represents a unit of carbohydrate, $h\nu$ is the energy of a quantum of light (h is Planck's constant and ν is frequency of radiation), and (g) indicates gas.

The most important microorganisms insofar as aquatic chemistry is concerned are the bacteria. Bacteria are single celled organisms with an amazing variety of shapes and functions. They can be divided into two major classifications: autotrophic bacteria and heterotrophic bacteria. The autotrophic bacteria derive all their energy and carbon for biomass from simple inorganic materials. A special class of autotrophic bacteria consists of those that are capable of photosynthesis. Bacteria can also be classified on the basis of their need for molecular oxygen. Aerobic bacteria require molecular oxygen, anaerobic bacteria do not require it and are usually poisoned by it, and facultative bacteria can use molecular oxygen if it is available and other oxidants if it is not.

A major function of bacteria in water is degradation of biomass as shown by the following reaction:

$$\{CH_2O\} + O_2(g) \rightarrow CO_2 + H_2O \tag{1.1.6}$$

This process of aerobic respiration is the one responsible for biodegradation of waste biomass and even some synthetic chemicals in the environment. In the absence of oxygen, specialized bacteria growing under oxygen-free highly reducing conditions can generate methane as shown by the following reaction for anaerobic respiration:

$$2\{CH_2O\} \rightarrow CH_4 + CO_2 \tag{1.1.7}$$

In addition to the important processes discussed above, bacteria mediate a number of significant oxidation-reduction reactions that occur in water. Important examples of such processes are illustrated by the following reactions:

β–oxidation of hydrocarbons:

$$CH_3CH_2CH_2CH_2CO_2H + 3O_2 \rightarrow CH_3CH_2CO_2H + 2CO_2 + 2H_2O \qquad (1.1.8)$$

$$\text{Nitrogen fixation:} \quad 3\{CH_2O\} + 2N_2 + 3H_2O + 4H^+ \rightarrow 3CO_2 + 4NH_4^+ \qquad (1.1.9)$$

$$\text{Nitrification:} \quad 2O_2 + NH_4^+ \rightarrow NO_3^- + 2H^+ + H_2O \qquad (1.1.10)$$

$$\text{Denitrification:} \quad 4NO_3^- + 5\{CH_2O\} + 4H^+ \rightarrow 2N_2 + 5CO_2 + 7H_2O \qquad (1.1.11)$$

$$\text{Sulfate reduction:} \quad SO_4^{2-} + 2\{CH_2O\} + 2H^+ \rightarrow H_2S + 2CO_2 + 2H_2O \qquad (1.1.12)$$

$$\text{Sulfide oxidation:} \quad H_2S + 2O_2(g) \rightleftharpoons SO_4^{2-} + 2H^+ \qquad (1.1.13)$$

In mediating reactions such as those shown above, microorganisms play a key role in the biology of chemical cycles of various elements. The nitrogen cycle, in which nitrogen is exchanged among elemental nitrogen in the atmosphere, biological nitrogen in the biosphere, nitrate ion under oxidizing conditions, and ammonium ion under reducing conditions, is a prime example of a biogeochemical cycle in which microorganisms are strongly involved.

Bacteria are involved in a number of processes involving metals and metalloids that result in formation of water pollutants. An important example of such a process is the formation of soluble and mobile methylated mercury species from insoluble inorganic mercury by the action of anaerobic bacteria:

$$HgCl_2 \xrightarrow{\text{Methylcobalamin}} CH_3HgCl + Cl^- \qquad (1.1.14)$$

Another another important bacterially mediated process involving a metal that results in formation of a water pollutant is the formation of acid mine water from mineral pyrite. In the presence of molecular oxygen, bacteria oxidize mineral pyrite to produce sulfuric acid and soluble iron(II):

$$2FeS_2(s) + 2H_2O + 7O_2 \rightarrow 4H^+ + 4SO_4^{2-} + 2Fe^{2+} \qquad (1.1.15)$$

Other bacteria oxidize the iron(II) to iron(III),

$$4Fe^{2+} + O_2 + 4H^+ \rightarrow 4Fe^{3+} + 2H_2O \qquad (1.1.16)$$

which can chemically oxidize additional pyrite as follows:

$$FeS_2(s) + 14Fe^{3+} + 8H_2O \rightarrow 15Fe^{2+} + 2SO_4^{2-} + 16H^+ \qquad (1.1.17)$$

Among the bacteria that are involved in acid mine water formation are *Thiobacillus ferrooxidans*, *Metallogenium*, *Thiobacillus thiooxidans*, and *Ferrobacillus ferrooxidans*.

Bacteria are also involved in the corrosion of metals, particularly iron. In so doing the bacteria may form small structures called tubercles on the surface of the metal in which they mediate the reactions that lead to oxidation of the metal.

1.1.3   WATER POLLUTION

A wide variety of substances are involved in water pollution.[5] These substances may be directly toxic or they may produce conditions that lead to deterioration of water quality.

Toxic elements are an important class of water pollutants. The most prominent of these are heavy metals, such as cadmium, lead, and mercury. Among the nonmetals, fluoride may be present at toxic levels as the fluoride ion, and chloride is associated with saline pollution.

Several inorganic species may be significant water pollutants. Among these are cyanide, which is very toxic as the cyanide ion CN^-. Other potentially harmful inorganic pollutants include ammonia (NH_4^+), hydrogen sulfide (H_2S), free CO_2, nitrite ion (NO_2^-), and oxygen-consuming sulfite (SO_3^{2-}). Algal nutrients are an important class of inorganic pollutants, primarily phosphate, nitrate, and potassium ion. These substances tend to increase the growth of algae in water to an excessive degree, a condition known as eutrophication. Of these, phosphate is generally the limiting nutrient and the one that is commonly reduced in efforts to control eutrophication.

Excess acidity, alkalinity, and salinity may be significant water pollutants. Harmful pollutant acid is normally due to the presence of strong acids, usually sulfuric acid such as that from acid mine water described above. Although alkalinity is a normal constituent of water, it may be present at excessive levels, usually from leaching of mineral formations. Salinity is usually due to the presence of NaCl dissolved in water, but it can be due to other salts is well. Water for domestic use and irrigation water may pick up excess salinity.

The presence of substances in water that consume dissolved molecular oxygen, usually by the action of bacteria, constitute pollutant biochemical oxygen demand (BOD). When biological material degrades microbially in water, dissolved oxygen is consumed as shown by the following reaction:

$$\{CH_2O\} + O_2(g) \rightarrow CO_2 + H_2O \tag{1.1.18}$$

There are several common sources of BOD including sewage, livestock feedlot runoff, and biomass produced by eutrophication (see above).

A variety of organic compounds are encountered as water pollutants. Biodegradable organic compounds that consume oxygen are mentioned above. Because of their widespread application to crops and soil, pesticides are common water pollutants. The use of biodegradable pesticides has significantly reduced this problem, however. Arguably the most troublesome pesticides are the herbicides, which are applied rather heavily to control weeds.

In some cases, radioactive materials can be significant water pollutants. Of these, radium, which can enter groundwater from mineral formations, is the most likely to cause a problem.

1.1.4 WATER TREATMENT

Water treatment can be divided into the three major categories: (1) purification for domestic use, (2) treatment for industrial applications, and (3) wastewater treatment.[6] A number of standard water treatment techniques are used in these three applications.

The most common water resources are groundwater, rivers, and reservoirs or lakes. Water from these sources may require a variety of kinds and degree of treatment processes depending on the water quality and the intended use. Aeration of water is commonly used to drive out volatile contaminants that cause taste or odor problems. Aeration may be used to oxidize dissolved iron(II), which precipitates as gelatinous iron(III) hydroxide, removing impurities as it settles. Water that contains suspended matter is commonly treated with coagulants, allowed to settle, and filtered. Excess water hardness in the form of dissolved calcium bicarbonate can be treated with lime:

$$Ca^{2+} + 2HCO_3^- + Ca(OH)_2 \rightarrow 2CaCO_3(s) + 2H_2O \tag{1.1.19}$$

For domestic use water must be disinfected, usually by treatment with Cl_2 or chlorine dioxide (ClO_2). Additional treatment measures may be required for water to be used in specialized applications. For example, boiler feedwater requires removal of scale-forming calcium or dissolved silica.

The major objective of treating municipal wastewater (sewage) is removal of BOD, mentioned above. The most common means for doing this is biological treatment in which microorganisms degrade oxygen-demanding organic matter and convert it to biomass and CO_2. In the activated sludge process a large quantity of suspended microorganisms is exposed to the waste in the presence of air in an aeration chamber, thus removing BOD. The wastewater is then taken to a settling chamber where the microorganisms settle out as sewage sludge and are pumped back into the aeration chamber. A second major type of biological treatment process uses fixed films of microorganisms held so that they contact both air and wastewater. Such films may be placed on rocks in a trickling filter or they may be coated on rotating disks in a rotating biological contactor.

Advanced wastewater treatment may be used for water that has been treated to remove BOD. Activated carbon in a granular or powdered form is commonly used to remove residual organic matter. Reverse osmosis can be used to remove ionic contaminants. Special measures may be taken to remove algal nutrients that can cause eutrophication. Nutrient phosphate can be precipitated with lime and nutrient nitrate can be reduced biologically to nitrogen gas with degradable organic matter in the absence of oxygen.

REFERENCES

1. Manahan, S., "Fundamentals of Aquatic Chemistry," in *Environmental Chemistry*, 7th ed., CRC Press, Boca Raton, FL, 1999, Chapt. 3, pp. 55–98.

2. Manahan, S., "Oxidation-Reduction," in *Environmental Chemistry*, 7th ed., CRC Press, Boca Raton, FL, 1999, Chapt. 4, pp. 99–119.

3. Manahan, S., "Phase Interactions," in *Environmental Chemistry*, 7th ed., CRC Press, Boca Raton, FL, 1999, Chapt. 5, pp. 121–145.

4. Manahan, S., "Aquatic Microbial Biochemistry," in *Environmental Chemistry*, 7th ed., CRC Press, Boca Raton, FL, 1999, Chapt. 6, pp. 147–185.

5. Manahan, S., "Water Pollution," in *Environmental Chemistry*, 7th ed., CRC Press, Boca Raton, FL, 1999, Chapt. 7, pp. 187–227.

6. Manahan, S., "Water Treatment," in *Environmental Chemistry*, 7th ed., CRC Press, Boca Raton, FL, 1999, Chapt. 8, pp. 229–263.

CHAPTER 1
CHEMISTRY

SECTION 1.2

ENVIRONMENTAL CHEMISTRY OF THE ATMOSPHERE

Stanley Manahan

Dr. Manahan is a professor of chemistry at the University of Missouri, Columbia, where he has taught since 1965. He is the author of the leading college textbook on environmental chemistry.

1.2.1 THE ATMOSPHERE AND ATMOSPHERIC CHEMISTRY

The atmosphere is divided into layers based on temperature. From Earth's surface to an altitude of 11–15 km is the troposphere, which has a relatively uniform gas composition, except for a variable water vapor content, and is characterized by decreasing temperature with increasing altitude. Above the troposphere and up to an altitude of about 50 km is the stratosphere, where the temperature increases with increasing altitude because of the absorption of solar energy by molecules in the stratosphere, especially ozone (O_3). The atmosphere serves as a source of oxygen, which is required by organisms for their metabolism, and carbon dioxide (CO_2), which is required for photosynthesis. The atmosphere is exposed to a number of waste products and pollutants.

A major aspect of atmospheric chemistry is the predominance of photochemical reactions that occur as a result of molecules absorbing photons of light and breaking down to produce radicals with unpaired electrons.[1] The formation of stratospheric O_3 is a photochemical process initiated by the photodissociation of molecular oxygen by the action of ultraviolet radiation ($h\nu$, where h is Planck's constant and ν is the frequency of the radiation),

$$O_2 + h\nu \rightarrow O + O \tag{1.2.1}$$

followed by a combination of molecular and atomic oxygen in the presence of a third body, M (usually a molecule of N_2), to carry away excess energy from the reaction:

$$O + O_2 + M \rightarrow O_3 + M \tag{1.2.2}$$

Although the atmosphere is very thin and has a negligible mass compared with the rest of Earth, it serves two protective functions that are essential to life on Earth. The first of these is the action of O_3 in the stratosphere, which acts as a filter to remove harmful ultraviolet radiation from the sun. The O_3 layer has been threatened by chlorofluorocarbons (Freons), such as CCl_2F_2, in the stratosphere, until recently widely used as refrigerant fluids, for foam blowing, and other applications. The chlorofluorocarbons, which are completely stable in the troposphere, are photodissociated by short-wavelength ultraviolet radiation in the stratosphere to produce Cl atoms, which take part in the catalytic destruction of O_3:

$$Cl + O_3 \rightarrow ClO + O_2 \tag{1.2.3}$$

$$ClO + O \rightarrow O_2 + Cl \tag{1.2.4}$$

$$O_3 + O \rightarrow 2O_2 \tag{1.2.5}$$

By participating in the chain-reaction process above, a single Cl atom may destroy as many as 10,000 molecules of O_3.

The second major protective function of the atmosphere is its stabilization of Earth's surface temperature by reabsorbing infrared radiation by which Earth loses energy that it has absorbed from the sun. CO_2 released in the combustion of fossil fuels is a very effective "greenhouse gas," and increasing CO_2 levels in the atmosphere are believed to contribute to global warming. Whereas the problem with destruction of the O_3 layer can be largely alleviated by replacing chlorofluorocarbons with substitutes, the problem with atmospheric CO_2 may prove to be much more difficult to solve.

1.2.2 PARTICLES IN THE ATMOSPHERE

Small particles in the atmosphere, called aerosols, play an important role in atmospheric chemistry and are important in atmospheric quality. The particles of greatest concern are the very small ones—less than a few micrometers in size. These particles are normally condensation aerosols produced by the reaction of gases. For example, the oxidation of sulfur dioxide (SO_2) in the atmosphere

$$SO_2 + \frac{1}{2}O_2 + H_2O \rightarrow H_2SO_4 \tag{1.2.6}$$

produces aerosol droplets of sulfuric acid (H_2SO_4).

Aerosols are involved in a number of processes in the atmosphere. Chemical reactions commonly occur on particle surfaces and some chemical processes, most notably the oxidation of SO_2 to H_2SO_4, occur within liquid water droplets. Small particles serve as condensation nuclei for removal of water vapor from the atmosphere as liquid water droplets. Particles move around in the atmosphere, they coagulate together to form larger particles, and they settle from the atmosphere. Particles suspended in the atmosphere scatter light, reducing visibility and detracting from atmospheric aesthetics. Such light-scattering particles are the end result of photochemical smog formation and are one of the detrimental effects of smog.

1.2.3 INORGANIC OXIDES

Gaseous inorganic oxides of carbon, nitrogen, and sulfur, all produced in the combustion of fossil fuels, are common atmospheric pollutants.[2] The potential of CO_2 to cause greenhouse warming was mentioned above. Carbon monoxide (CO), is a toxic gas produced by the incomplete combustion of carbonaceous fuels. Nitric oxide (NO) and nitrogen dioxide (NO_2), collectively known as NO_x, can be produced by the direct reaction of elemental nitrogen and oxygen under the conditions that obtain in an internal combustion engine,

$$N_2 + O_2 \rightarrow 2NO \tag{1.2.7}$$

and can also be produced by the combustion of organic nitrogen in fossil fuels. NO released by combustion is converted to NO_2 by photochemical processes in the atmosphere.

SO_2 is produced by the combustion of sulfur in fossil fuels, particularly coal:

$$S(\text{fuel}) + O_2 \rightarrow SO_2 \tag{1.2.8}$$

An air pollutant in its own right, SO_2 can undergo a series of atmospheric chemical reactions, the net result of which is the production of droplets of H_2SO_4 as shown above. This process is the main route to production of acidic precipitation or acid rain.

1.2.4 PHOTOCHEMICAL SMOG

One of the most common air pollution problems is photochemical smog, which is formed by the action of sunlight on nitrogen oxides and organic air pollutants under low-humidity conditions.[3] The initial event in the formation of photochemical smog is photodissociation of NO_2

$$NO_2 + h\nu \rightarrow NO + O \qquad (1.2.9)$$

to produce reactive oxygen atoms. This initiates a series of chain reactions that result in production of atmospheric oxidants, including O_3, noxious organic substances including aldehydes and organic oxidants, and visibility-obscuring atmospheric particulate matter. Atomic oxygen produced by photodissociation of NO_2 reacts with a hydrocarbon to abstract hydrogen,

$$
\begin{array}{ccc}
\underset{\underset{H}{|}}{\overset{\overset{H}{|}}{H-C}} - \underset{\underset{H}{|}}{\overset{\overset{H}{|}}{C}} - \underset{\underset{H}{|}}{\overset{\overset{H}{|}}{C}} - H + O & \rightarrow & \underset{\underset{H}{|}}{\overset{\overset{H}{|}}{H-C}} - \underset{\underset{H}{|}}{\overset{\overset{H}{|}}{C}} - \overset{\overset{H}{|}}{C^{\bullet}} + HO^{\bullet}
\end{array} \qquad (1.2.10)
$$

producing reactive hydroxyl radical and a reactive hydrocarbon radical species. In turn these undergo a series of chain reactions leading to formation of photochemical smog. A common reaction of the alkyl radical product, such as the propyl radical above, is with molecular oxygen,

$$
\begin{array}{ccc}
\underset{\underset{H}{|}}{\overset{\overset{H}{|}}{H-C}} - \underset{\underset{H}{|}}{\overset{\overset{H}{|}}{C}} - \overset{\overset{H}{|}}{C^{\bullet}} + O_2 & \rightarrow & \underset{\underset{H}{|}}{\overset{\overset{H}{|}}{H-C}} - \underset{\underset{H}{|}}{\overset{\overset{H}{|}}{C}} - \overset{\overset{H}{|}}{C} - OO^{\bullet}
\end{array} \qquad (1.2.11)
$$

to produce an alkylperoxyl radical. This strongly oxidizing species plays a key role in formation of smog by oxidizing NO back to NO_2,

$$
\begin{array}{ccc}
\underset{\underset{H}{|}}{\overset{\overset{H}{|}}{H-C}} - \underset{\underset{H}{|}}{\overset{\overset{H}{|}}{C}} - \overset{\overset{H}{|}}{C} - OO^{\bullet} + NO & \rightarrow & \underset{\underset{H}{|}}{\overset{\overset{H}{|}}{H-C}} - \underset{\underset{H}{|}}{\overset{\overset{H}{|}}{C}} - \overset{\overset{H}{|}}{C} - O^{\bullet} + NO_2
\end{array} \qquad (1.2.12)
$$

which can undergo photochemical dissociation again, thus reinitiating the process by which smog is formed.

Hydroxyl radical, $HO^{\bullet}$, is a key reactive species in maintaining the chain reactions by which photochemical smog is formed. The reactivity of organic species in smog formation is expressed by their rates of reaction with the hydroxyl radical. Hydroxyl radical can undergo two major kinds of reactions with hydrocarbons, an abstraction reaction, such as the reaction with propane,

$$
\begin{array}{ccc}
\underset{\underset{H}{|}}{\overset{\overset{H}{|}}{H-C}} - \underset{\underset{H}{|}}{\overset{\overset{H}{|}}{C}} - \underset{\underset{H}{|}}{\overset{\overset{H}{|}}{C}} - H + HO^{\bullet} & \rightarrow & \underset{\underset{H}{|}}{\overset{\overset{H}{|}}{H-C}} - \underset{\underset{H}{|}}{\overset{\overset{H}{|}}{C}} - \overset{\overset{H}{|}}{C^{\bullet}} + H_2O
\end{array} \qquad (1.2.13)
$$

and an addition reaction in which the radical adds across an unsaturated bond:

$$
\begin{array}{ccc}
\underset{\underset{H}{|}}{\overset{\overset{H}{|}}{H-C}} - \underset{\underset{H}{|}}{\overset{\overset{H}{|}}{C}} = C\overset{\diagup H}{\diagdown H} + HO^{\bullet} & \rightarrow & \underset{\underset{H}{|}}{\overset{\overset{H}{|}}{H-C}} - \underset{\underset{H}{|}}{\overset{\overset{O}{|}}{C}} = C\overset{\diagup H}{\diagdown H}
\end{array} \qquad (1.2.14)
$$

The ability of unsaturated organic compounds to undergo relatively facile addition reactions with hydroxyl radical contributes to their higher reactivity in the atmosphere compared with saturated alkanes.

REFERENCES

1. Manahan, S., "The Atmosphere and Atmospheric Chemistry," in *Environmental Chemistry*, 7th ed., CRC Press, Boca Raton, FL, 1999, Chapt. 9, pp. 265–305.

2. Manahan, S., "Gaseous Inorganic Air Pollutant," in *Environmental Chemistry*, 7th ed., CRC Press, Boca Raton, FL, 1999, pp. 329–352.

3. Manahan, S., "Photochemical Smog," in *Environmental Chemistry*, 7th ed., CRC Press, Boca Raton, FL, 1999, Chapt. 13, pp. 379–404.

CHAPTER 1
CHEMISTRY

SECTION 1.3

SOIL ENVIRONMENTAL CHEMISTRY

Stanley Manahan

*Dr. Manahan is a professor of chemistry at the University of
Missouri, Columbia, where he has taught since 1965. He is the
author of the leading college textbook on environmental chemistry.*

Soil is a complex medium in which a number of important chemical and environmental chemical processes occur.[1] There many different types of soil. It is normally formed by weathering processes that act on rocks, which are broken down physically and altered chemically. Typically, soil contains 5 percent organic matter, although organic contents may be as high as 95 percent in some soils. In addition to organic and inorganic solids, soil contains air spaces and water, which may be rather tightly bound to the soil. Healthy soils support a variety of life. In addition to plant root structures, which make up a significant fraction of the viable biomass in most soils, soil supports bacterial and fungal growth.

Clay minerals are generally an important constituent of the mineral fraction of soil. Clays are secondary mineral aluminosilicates, consisting of generally hydrated aluminum and silicon oxides existing in layered structures. Clays have anionic groups that enable them to act as cation exchangers and to hold cationic plant nutrients, such as calcium ion or ammonium ion. They also bind water and hold water in the soil.

Except for carbon dioxide from the atmosphere and some atmospheric nitrogen fixed by bacteria living synergistically with some kinds of plants, water and all the other nutrients that plants require must come from the soil. These nutrients enter through the roots and are carried throughout the plant by water, which is lost to the atmosphere by a process called transpiration. A crucial aspect of soil in this process is the soil solution, which enables transfer of chemical species between soil and plant roots. Of utmost importance for plant nutrition are the chemical species commonly regarded as fertilizers. These include nitrogen as nitrate, potassium as ionic K^+, and phosphorus as orthophosphate ions. In addition, plants require a variety of micronutrients such as iron.

Soil is a cation exchanger, holding and releasing cations such as potassium (K^+) and calcium (Ca^{2+}). In taking up cations through their roots, plants exchange hydrogen ion (H^+) for the cations. This results in a net acidification of soil to the extent that it may become unproductive as a medium for supporting plant life. Such soil is treated with agricultural lime, usually as limestone ($CaCO_3$), which neutralizes the acid and replaces H^+ with Ca^{2+}.

A key chemical component of most healthy soils is humus. Humus is the residue that remains after fungal and bacterial decay of plant matter in soil. It contains a water-soluble fulvic acid fraction and a humic acid fraction that can be extracted by a strong base. It also contains a humin fraction that does not dissolve. Humus is responsible for a significant fraction of the cation exchange capacity of soil. It also strongly retains water and lends desirable qualities to the texture of the soil.

An important aspect of soil environmental science is loss and deterioration of the soil. Productive soil can be lost by water and wind erosion, which carry productive topsoil way. Fortunately, in recent years much progress has been made in reducing erosion. One of the most successful approaches has been to use cultivation techniques in which the soil is not tilled and crops are planted in the residue of previous years' crops, with light applications of herbicides to prevent competing weeds from crowding out the crops during the early stages of growth.

Two related conditions involving soil degradation are desertification and deforestation. Desertification is associated with drought, loss of fertility, and deterioration of topsoil quality so that the soil no longer supports plant life and turns to desert, a process that has occurred in the Middle East, the southwestern United States, and other parts the world. Deforestation is the loss of tree cover characteristic of productive forests. In recent decades it has become a particularly acute problem in tropical regions, which have lost rain forests at an alarming rate. In some parts of the world, however, such as regions of the northeastern United States, marginal farmland has been allowed to return to forests, thus reversing some forest loss.

REFERENCE

1. Manahan, S., "Soil Environmental Chemistry," in *Environmental Chemistry*, 7th ed., CRC Press, Boca Raton, FL, 1999, Chapt. 16, pp. 473–507.

CHAPTER 1
CHEMISTRY

SECTION 1.4

ENVIRONMENTAL BIOCHEMISTRY

Stanley Manahan

*Dr. Manahan is a professor of chemistry at the University of
Missouri, Columbia, where he has taught since 1965. He is the
author of the leading college textbook on environmental chemistry.*

As the chemistry of life processes, biochemistry plays a very important role in environmental chemistry and toxicological chemistry.[1] The specific application of biochemistry to the environment is called environmental biochemistry. The role of biochemistry in acting on toxic substances is covered in the section "Toxicological Chemistry." Biochemistry and its environmental subdivision are complex areas, and only the briefest introduction is given here.

Biochemistry is very much involved with environmental chemical processes. Production of biomass, degradation of organic matter, and transitions between inorganic species that occur in water and soil are generally biochemical processes that are carried out by microorganisms. Furthermore, as discussed in the section on toxicological chemistry, the toxic effects of environmental pollutants occur by biochemical mechanisms.

Basic to biochemistry is the cell. Although the human cell is very small, of the order of a few micrometers in size, it has a complex structure. The cell is enclosed by a cell membrane that regulates passage of substances into and out of the cell. The interior of a human cell is filled with water-soluble protein called cytoplasm. Suspended in the cytoplasm are various organelles such as mitochondria, ribosomes, and golgi bodies. Central to the cell is deoxyribonucleic acid (DNA), which directs reproduction and protein synthesis. Damage to the cell membrane and to cellular DNA are common modes of toxicant action.

Several important kinds of biomolecules are synthesized in living systems. Carbohydrates, with the general formula CH_2O, consist of simple sugars, starch, cellulose, and related compounds. Lipids are defined in terms of their solubilities in organic solvents, which can be used to extract the lipids from tissue. There are several chemical types of lipids, including common fats, oils, waxes, and steroids, such as cholesterol. A number of environmental pollutants, such as polychlorinated biphenyls, are lipophilic, which means they dissolve and become concentrated in the fat tissue of organisms. This is one of the main mechanisms by which organic pollutants undergo biomagnification in the food chain. The characteristic of lipophilicity enables organic compounds to traverse cell membranes readily, which adds to their toxicity hazard.

Proteins are polymers of amino acids that occur in a large variety of structures. Nutrient proteins, such as casein in milk, are a food source. Ferritin, a storage protein, enables iron to be stored in animal tissue. Structural proteins include collagen in tendons and keratin in hair. Contractile proteins are found in muscles, enabling movement to occur. Hemoglobin is a transport protein that moves molecular oxygen in the blood. Defense proteins are found in antibodies in the immune system. Insulin and human growth hormone are regulatory proteins that control biochemical processes.

A particularly important class of proteins consists of enzymes that act as biochemical catalysts to carry out the vast number of complex chemical transitions performed by organisms. Enzymes act on molecules called substrates, which they recognize on the basis of molecular shape. One of the common modes of action of toxic substances is to bind to enzymes, altering the shapes of enzymes so that they no longer interact properly with substrates. One of the most significant enzyme systems is the cytochrome P-450 system, which is responsible for the initial attack of an organism on foreign substances (see phase 1 reactions in the section "Toxicological Chemistry").

Enzymes are named for where they occur and what they do. Thus, *gastric proteinase* is a protein that occurs in the stomach (*gastric*) and hydrolyzes proteins (*proteinase*). In addition to hydrolyzing enzymes that break down large biomolecules with the addition of water, other classes of enzymes include oxidoreductase enzymes that carry out oxidation and reduction reactions, transferase enzymes that move chemical groups from one molecule to another, ligase enzymes that link molecules together [working with adenosine triphosphate (ATP), a high-energy molecule that is essential in energy-yielding, glucose-oxidizing metabolic processes], and lyase enzymes that remove chemical groups without hydrolysis.

The shapes of enzymes and other proteins are crucial for proper functioning. Alteration of protein structures is called denaturation. It is one of the most common modes of action of toxic substances.

Nucleic acids are huge polymers of nitrogenous bases, phosphate, and sugars. As noted above, DNA, plays a crucial role in cell reproduction and protein synthesis. DNA has the famed double-helical structure consisting of a double α-helix of oppositely wound polymeric strands held together by hydrogen bonds between opposing pyrimidine and purine groups. The concept of this structure was first advanced by James D. Watson and Francis Crick in 1953,[2] a discovery for which they were awarded a Nobel Prize in 1962. The two strands of DNA wound around each other in a spiral in the double helix are complementary, so that if they are pulled apart, each generates its complementary strand. This process occurs in cell reproduction. DNA is modified deliberately to produce transgenic organisms, such as pest-resistant crops, a subject of some controversy that has some significant environmental implications.

Hormones are substances that act as messengers from one part of the body to another, starting and stopping essential body functions. Lipid steroids constitute some of the essential hormones, including male and female sex hormones (estrogens). Hormones are generated by endocrine glands in the body. These glands include the pituitary, thyroid, parathyroid, thymus, and adrenal glands as well as the ovaries in females and testes in males.

A key aspect of biochemistry is metabolism in which enzymes act on various substances. Catabolism is a metabolic process in which molecules are broken down to provide energy and building materials for new biomolecules. Anabolism occurs when biochemical metabolism builds small molecules into large biomolecules, such as proteins. Organisms can derive energy by several types of metabolic processes. Aerobic respiration uses molecular oxygen for catabolism of organic compounds to yield energy, whereas anaerobic respiration occurs in the absence of molecular oxygen. The Krebs cycle functions in aerobic respiration to provide energy by the following process for the oxidation of a simple sugar:

$$C_6H_{12}O_6 + 6O_2 \rightarrow 6CO_2 + 6H_2O + \text{energy} \tag{1.4.1}$$

This process releases short-term stored chemical energy in the form of high-energy ATP molecules. Energy can be stored longer term in the form of glycogen or starch polysaccharides, and still longer term as storage lipids (fats) generated and retained by the organism. Fermentation is an energy-yielding process that does not have an electron transport chain as does respiration. A common example of fermentation is production of ethanol from simple sugars by yeasts:

$$C_6H_{12}O_6 \rightarrow 2CO_2 + 2C_2H_5OH \tag{1.4.2}$$

Photosynthesis is a metabolic process represented as,

$$6CO_2 + 6H_2O + h\nu \rightarrow C_6H_{12}O_6 + 6O_2 \tag{1.4.3}$$

in which plant and algal chloroplasts capture light energy, synthesize sugars from carbon dioxide and water, and release oxygen.

Metabolic processes are of the utmost importance in the environmental chemistry and toxicological chemistry of pollutants and xenobiotics. Metabolic processes in microorganisms degrade environmental pollutants in water and soil. Organisms metabolize xenobiotic compounds to render them nontoxic and in some cases produce toxic species (toxicants) from nontoxic ones (protoxicants), as discussed in the section "Toxicological Chemistry."

REFERENCES

1. Manahan, S., "Environmental Biochemistry," in *Environmental Chemistry*, 7th ed., CRC Press, Boca Raton, FL, 1999, Chapt. 21, pp. 669–694.
2. Watson, J. D. and Crick, F. H. C., "Molecular Structure of Nucleic Acids: A Structure for Deoxyribose Nucleic Acids," *Nature*, 171:737–738, 1953.

CHAPTER 1
CHEMISTRY

SECTION 1.5

TOXICOLOGICAL CHEMISTRY

Stanley Manahan

*Dr. Manahan is a professor of chemistry at the University of
Missouri, Columbia, where he has taught since 1965. He is the
author of the leading college textbook on environmental chemistry.*

Toxicology is the science of poisons, in which a poison, or toxicant, is a substance that has some sort of harmful effect on biological processes, tissues, or organs that results in harm to an organism. Toxicological chemistry is the science that deals with the chemical nature and reactions of toxic substances including their origins, uses, and chemical aspects of exposure, fates, and disposal. It relates the toxic effects of chemicals to their chemical nature.[1]

There are several ways a human can be exposed to toxic chemicals: through inhaled air, ingestion with food or water, and absorption through the skin. Exposure can be local, such as skin harmed by direct contact with concentrated nitric acid. Of much greater importance, however, is systemic exposure in which a chemical species enters the body, is transported (usually in the blood), undergoes biochemical transformations, and eventually is eliminated (usually in urine). In this process, foreign substances, called xenobiotic substances, are normally metabolized, in some cases to a more toxic form.

Various types of tissue in the body have enzyme systems capable of metabolizing xenobiotic substances. This usually occurs by cometabolism in which an enzyme system that normally acts on naturally occurring substances also metabolizes xenobiotic materials. The liver is the most active organ for xenobiotic metabolism.

Two classes of metabolic processes may occur during xenobiotic metabolism. A phase 1 reaction attaches a functional group, such as a hydroxyl group, to a xenobiotic compound. Phase 1 reactions normally involve oxidation and are carried out by the cytochrome P-450 enzyme system. A phase 2 reaction acts on a phase 1 reaction product or on a xenobiotic compound with a suitable functional group to attach an endogenous conjugating agent, such as glucuronide, glutathione, or sulfate. Normally these phase 2 reaction products are more soluble and more readily eliminated from the body than is the parent compound. They are usually, though not invariably, less toxic than the parent species.

Two major phases of xenobiotic metabolism are the kinetic phase and the dynamic phase. In the kinetic phase, phase 1 and phase 2 reactions may occur that detoxify a toxic xenobiotic compound or that convert a nontoxic protoxicant to a toxic form. The dynamic phase of toxic action may be divided into three parts. First is the primary reaction in which a toxicant or a toxic metabolite reacts with a receptor, such as an enzyme system, to produce a modified receptor. Second, as a result of this modification, there is a detrimental biochemical effect, such as enzyme inhibition. The third part of the dynamic phase is a behavioral or physiological response, such as alteration of vital signs or carcinogenesis.

Many biochemical effects can result in a toxic response. The action of enzymes may be impaired by the toxicant binding to the enzyme, coenzymes, metal activators of enzymes, or enzyme substrates. Cell membranes or carriers in the membranes may be altered. There may be interference with the metabolism of proteins, carbohydrates, and lipids; in the case of lipids the result may be excess lipid accumulation, a condition known as "fatty liver." Respiration, the overall process by which electrons are transferred to molecular oxygen in the biological oxidation of energy-yielding substrates, may be impaired. Toxicants may act on DNA to stop or interfere with protein biosynthesis. There may be interference with regulatory processes mediated by hormones or enzymes.

A key aspect of toxicity is the dose–response curve in which the percentage of test organisms that give a specific response (such as death) is plotted against log of the dose, usually in units of milligrams per kilogram of body mass. Ideally, the result is an S-shaped curve, the inflection point of which represents the dose at which 50 percent of the test subjects give the specified response. When death is the response being measured, this inflection point is referred to as an LD_{50} value, the dose required to kill 50 percent of test organisms.

There is an enormous range of kinds of toxic compounds, modes of action and effects of toxic compounds, and degrees of toxicity, commonly expressed as the mass of the toxicant per unit mass of the organism causing a specific effect. Toxicities are often expressed as LD_{50} values, which vary over many orders of magnitude for different compounds. For example, as expressed by LD_{50}, tetrodotoxin from the puffer fish is about a million times as toxic as diethylhexylphthalate, which is used as a plasticizer or to make plastics flexible.

Toxicities may be strongly affected by chemical interactions between substances, one or both of which is a poison. Such interactions may occur, for example, because of competition between substances for the same binding site on the receptor they affect. Two substances having the same toxic effect may act in an additive fashion, or they may act synergistically, which means the total effect exceeds the sum of the individual effects. Nontoxic substances may influence the toxicities of toxic ones. Enhancement of the toxic effect of a substance by one that is not toxic by itself is called potentiation, whereas when a toxic effect is reduced by a nontoxic substance, the effect is called antagonism.

Toxic effects may be reversible or irreversible. For example, exposure to carbon monoxide, which binds with blood hemoglobin thus depriving body tissues of oxygen, is a reversible effect so long as the brain is not damaged by oxygen deprivation. After the subject is removed from exposure to carbon monoxide, the hemoglobin reverts back to the oxygenated form, leaving no permanent effects. On the other hand, exposure to a corrosive acid may irreversibly damage exposed tissue.

One of the most harmful and most difficult to study modes of toxic action is that of carcinogenesis resulting in cancer from long-term systemic exposure to compounds that can be metabolized to a cancer-causing species. Most commonly, carcinogenesis results from the attachment of methyl (CH_3) and related groups to DNA, causing it to malfunction and allow unrestricted cell growth.

There are a variety of possible toxic effects from different substances. Some examples[2] are the following:

- Elemental white phosphorus is a systemic poison that can cause anemia, gastrointestinal system dysfunction, bone brittleness, and eye damage. Exposure to white phosphorus can cause the jawbone to deteriorate, a condition known as phossy jaw.

- Elemental halogens, especially fluorine (F_2) and chlorine (Cl_2), are corrosive irritants that attack skin, eye tissue, and the mucous membranes of the respiratory tract.

- Heavy metals, such as cadmium, mercury, and lead, are sulfur-seeking compounds that may bind with enzyme active sites and cause a variety of toxic effects.

- Cyanide as HCN or CN^-, bonds to iron(III) in iron-containing ferricytochrome oxidase, which is involved in respiration, the chemical process by which molecular oxygen is used by the body. This stops the respiration process and can cause sudden death.

- Nitrogen dioxide (NO_2) causes severe irritation of the innermost parts of the lungs, resulting in pulmonary edema (fluid accumulation), which can be fatal.

- Inhalation of asbestos fibers, a group of fibrous silicate minerals, typically those of the serpentine group, for which the approximate chemical formula is $Mg_3(Si_2O_5)(OH)_4$, can cause asbestosis (a pneumonia condition), mesothelioma (tumor of the mesothelial tissue lining the chest cavity adjacent to the lungs), and bronchogenic carcinoma (cancer originating in the air passages in the lungs).

- Inhalation of hydrogen sulfide (H_2S) in levels more than about 1000 parts per million in air can be rapidly fatal because of asphyxiation from respiratory system paralysis. Lower doses damage the central nervous system and cause symptoms that include headache, dizziness, and excitement.

- Gaseous alkanes, such as methane and propane, are simple asphyxiants; they form mixtures with air that contain insufficient oxygen to support respiration.

- Benzene is oxidized in the body with the addition of an oxygen atom to form benzene oxide as a reactive intermediate, which causes damage to bone marrow and may be involved in causing leukemia.

- Because it has an oxidizable methyl side group that benzene does not have, toluene is metabolized to hippuric acid, which is eliminated with the urine; it is much less toxic than benzene.

- Benzo(*a*)pyrene, a five-ring polycyclic aryl compound, is oxidized metabolically to the 7,8-diol-9,10-epoxide, which binds with DNA and may cause cancer.

- Methanol is metabolized to formic acid and damages the optic nerve; blindness may result.

- Ethylene glycol is metabolically oxidized to glycolic acid ($HOCH_2CO_2H$), a cause of acidemia, and then to oxalic acid, which may precipitate in the kidneys as solid calcium oxalate (CaC_2O_4) and cause clogging of the kidney tubules.

- Phenol is a protoplasmic poison, which, because of its former widespread use as a disinfectant, has caused a large number of poisonings.

- *N*-nitroso compounds (nitrosamines) containing the $N–N = O$ functional group can be metabolized to alkylating agents that attach groups such as the methyl group to DNA and cause cancer.

- Aniline and nitrobenzene cause the iron(II) in hemoglobin to be oxidized to iron(III), forming methemoglobin, which does not carry oxygen in the blood.

- Carbon tetrachloride (CCl_4) is metabolized to radical species such as $CCl_3OO^\bullet$, which cause lipid peroxidation, consisting of the attack of free radicals on unsaturated lipid molecules, followed by oxidation of the lipids through a free radical mechanism.

- Vinylchloride (C_2H_3Cl) is one of the few known human carcinogens; cases of a rare angiosarcoma of the liver have been observed in exposed workers in the polyvinylchloride plastics industry.

- Organophosphate esters, such as Sarin developed as a military poison "nerve gas," inhibit the acetylcholinesterase enzyme so that nerve impulses do not stop, which leads to respiratory system paralysis and death.

Biological monitoring—measurement of xenobiotic substances and their metabolites in blood, urine, breath, and other samples of biological origin to determine exposure to toxic substances—is an important aspect of toxicological chemistry.[3] Two main considerations in biological monitoring are the type of sample and the type of analyte.

For biological monitoring, breath samples are confined to volatile xenobiotics and volatile metabolites, so they are limited in their application. Miscellaneous samples include nails and hair (for trace elements, such as selenium), adipose tissue (largely limited to cadavers), milk (limited to lactating females), and perspiration. This leaves blood and urine as the two most common sample sources for biological monitoring. Blood is a useful source because it is the means by which xenobiotics and their metabolites are distributed around the body; thus, it is likely to contain evidence of exposure to all systemic poisons. The disadvantage of using blood is that it is a rather complex sample matrix and a relatively difficult sample to procure. Urine is the main pathway of elimination of xenobiotic metabolites, so it is likely to contain evidence of exposure, and it is a sample that is relatively easy to obtain. However, corrections must be made for widely variable concentrations of metabolites in the urine because of differences in water intake among individuals and over time.

A variety of analytes are actually measured as evidence of exposure. The most straightforward of these is parent species. This is obviously true of exposures to elements, such as toxic arsenic or lead. Parent species are also commonly determined for nonmetabolized volatile organic compounds, such as nonmetabolized fractions of cyclopropane, methyl chloride, toluene, trichloroethane, and ethylene oxide. In some cases, these compounds may be measured by direct injection of blood or urine into a gas chromatograph, although this procedure causes rapid column deterioration, or chromatographic analysis of headspace air collected from above the liquid sample. Solvent extraction and purge-and-trap techniques can also be used to determine parent compounds in blood or urine.

Phase 1, and in some cases phase 2, metabolites may be determined as biological monitors of exposure. A common example of a phase 1 metabolite so measured is *trans,trans*-muconic acid

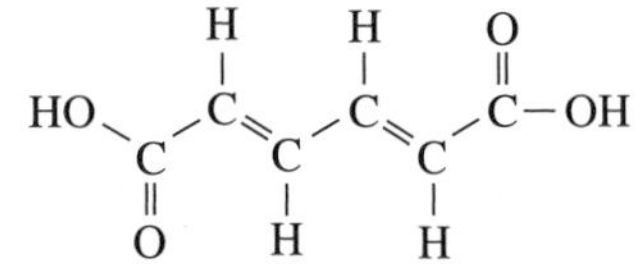

trans,trans-muconic acid

used to monitor exposure to benzene. A classic example of a phase 2 metabolite used for biological monitoring is hippuric acid, a metabolite of toluene. It has the disadvantage of being produced naturally by metabolic processes, which can cause significant interference. In recent years mercapturic acid phase 2 conjugates have gained favor as analytes to use as evidence of exposure. These substances consist of the xenobiotic species attached to an *N*-acetylcysteine amino acid as shown below, where the original xenobiotic material can be represented in general as H-X-R:

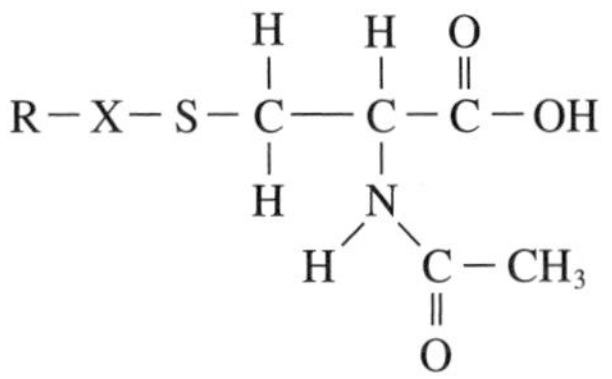

mercapturic acid conjugate of a xenobiotic, HXR

One reason for the popularity of these substances as analytes is their relatively facile and sensitive determination by high-performance liquid chromatographic separation and fluorescence detection of their *o*-phthaldialdehyde derivatives.

Adducts of xenobiotics or their metabolites with biomolecules endogenous to the body are gaining popularity as biological monitors of exposure. The most common endogenous substances for which such measurements are made are blood hemoglobin, blood plasma albumin protein, and DNA. A simple example of a hemoglobin adduct is carbon monoxide, which can be determined by direct spectrophotometric measurement. Other xenobiotics that have been measured by their hemoglobin adducts are acrylamide, acrylonitrile, 1,3-butadiene, 3,3′-dichlorobenzidine, ethylene oxide, and hexahydrophthalic anhydride. Toluene diisocyanate, benzo(*a*)pyrene, styrene, styrene oxide, and aflatoxin B1 have been determined as their blood plasma albumin adducts. The DNA adduct of styrene oxide has been measured to test for exposure to carcinogenic styrene oxide. Measurement of DNA adducts is likely to gain in popularity because DNA adduct formation is a common mechanism of carcinogenicity.

In some cases exposure to a xenobiotic substance is inferred by measurement of an endogenous substance that does not contain the original xenobiotic but that is formed in the body as a result of exposure. A good example of this is measurement of methemoglobin, the product that results from iron(II) in hemoglobin oxidizing to iron(III) after exposure to nitrobenzene, aniline, or related compounds.

Exposure to some xenobiotic substances causes alterations in enzyme activity, which can be measured. Inhibition of acetylcholinesterase is used to measure exposure to organophosphates or carbamate insecticides.

Immunoassay techniques are attractive for biological monitoring and have been used in simple test kits for blood glucose and pregnancy testing. Immunoassay techniques use reagents formed by antibodies to synthetic substances produced in rabbits or other mammals exposed to the synthetic substances bound with a protein. When applicable, immunoassay methods offer simplicity and specificity. Their applications to biological monitoring have been limited in part because of interferences in complex biological systems. However, polychlorinated biphenyls have been measured in blood plasma by immunoassay, and additional applications can be anticipated.

REFERENCES

1. Manahan, S., "Toxicological Chemistry," in *Environmental Chemistry*, 7th ed., CRC Press, Boca Raton, FL, 1999, Chapt. 22, pp. 695–719.

2. Manahan, S., "Toxicological Chemistry of Chemical Substances," in *Environmental Chemistry*, 7th ed., CRC Press, Boca Raton, FL, 1999, Chapt. 23, pp. 721–751.

3. Manahan, S., "Analysis of Biological Materials and Xenobiotics," in *Environmental Chemistry*, 7th ed., CRC Press, Boca Raton, FL, 1999, Chapt. 27, pp. 809–823.

CHAPTER 1
CHEMISTRY

SECTION 1.6

GEOCHEMISTRY

William J. Deutsch

Mr. Deutsch is president of Geochemistry Consulting, Brooksville, Maine. He is a hydrogeochemist with 20 years of experience in research and consulting on environmental problems.

1.6.1 INTRODUCTION

As water flows through the vadose and saturated zones of the subsurface, its composition changes and evolves in response to the environmental conditions encountered along the flow path. An important component of the subsurface environment is its geochemistry. Geochemistry includes those aspects of fate and transport that involve reactions between the dissolved constituents in groundwater and the geologic media comprising the subsurface material. The subsurface solid material consists of minerals (solids with a crystalline structure and fixed composition), amorphous solids, and organic matter. Movement of dissolved inorganic constituents in the subsurface can be affected by adsorption onto the surfaces of the solid material or by precipitation and incorporation into the three-dimensional structure of the inorganic crystalline or amorphous solids. Figure 1.6.1 shows the distribution of a compound between the various phases. The relative distribution of a compound between the solution and solid phases in a particular environment is a function of the solubility of minerals containing the compound and the degree to which the solid surfaces adsorb the compound. Evaluating this distribution is the focus of understanding geochemical fate and transport mechanisms.

Transfer of a compound from one phase to another (such as mineral formation by precipitation from a solution) is a chemical reaction that involves a change in chemical bonding. For the case of mineral formation by precipitation of barite from solution, the reaction can be written as follows:

$$Ba^{2+} + SO_4^{2-} \rightarrow BaSO_4(s)$$

This reaction says that dissolved barium (Ba^{2+}) and sulfate (SO_4^{2-}) can bind together and form the solid-phase mineral barite [$BaSO_4(s)$]. If barite is precipitating from solution and the solid itself is not mobile in the subsurface, then the movement of barium and sulfate will be affected by the formation of this solid phase containing these compounds. As a corollary, the presence of precipitated barite provides a source of barium and sulfate to infiltrating water percolating by the mineral in the future. If the water is undersaturated with respect to the mineral, some of the barite will dissolve, releasing barium and sulfate to the water.

The subsurface also contains solid phases that attract dissolved constituents to the surface. The dissolved constituents are attached to the solid surface by adsorption reactions. If the surface of the solid is represented by $\equiv$SOH, the adsorption reactions of Ba^{2+} and SO_4^{2-} onto the surface are:

$$\equiv SOH + Ba^{2+} \rightarrow \equiv SOHBa^+ + H^+$$
$$\equiv SOH + SO_4^{2-} \rightarrow \equiv SOHSO_4^{2-}$$

In the case of adsorption, dissolved Ba^{2+} and SO_4^{2-} react with the surfaces of minerals in contact with the solution and some of the dissolved species attach themselves to the surface. As a result of this

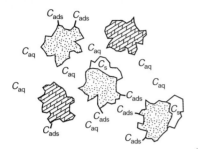

FIGURE 1.6.1 Distribution of a compound (C) between the aqueous phase (C_{aq}), solid phase (C_s), and adsorbed on solid surfaces (C_{ads}).

adsorption to the surface of an immobile solid, movement of the dissolved species is inhibited. As with mineral precipitation, the presence of adsorbed species provides a potential future source of the species to water that comes into contact with the solid and its reservoir of adsorbed species.

Figure 1.6.2 shows the concentration ranges or zones where adsorption and mineral equilibrium can control the dissolved concentration of a constituent, thereby influencing its mobility in the subsurface. At relatively low solution concentration, adsorption is the controlling reaction because the species is not present in a high enough concentration for a mineral containing the species to precipitate. In this zone, the line that relates dissolved to solid-phase concentrations has a positive slope, reflecting the fact that the adsorbed concentration increases as the dissolved concentration increases. As the dissolved concentration increases, the solution becomes saturated with the components of a mineral. At that concentration the mineral begins to precipitate and limit additional increases in solution concentration. This is represented in Figure 1.6.2 by the horizontal line in the zone of mineral control on solution concentration. As long as the environmental conditions remain constant, mineral solubility will provide an upper limit on dissolved concentration. This chapter focuses on the effect of adsorption and mineral equilibrium on solution composition and the effect of site-specific conditions on these geochemical processes.

1.6.2 ADSORPTION PROCESSES

Attachment of a dissolved species to the surface of a solid is a complex reaction that involves both electrostatic processes (attraction of oppositely charged species) and chemical reactions (bonding). The properties of the solid surface that cause adsorption change with the condition of the system [temperature, pH, redox potential (Eh), ionic strength, solution composition, etc.] and there are a wide variety of types of solid surfaces (minerals and other solid phases) in the environment. Because of this complexity, our ability to define the importance of adsorption on fate and transport for a particular site is a difficult task. For instance, we may sample an aquifer and determine from laboratory tests the amount of a species that will adsorb onto the solid material. Will the same amount adsorb onto the solid phase if a low pH front moves through the aquifer? Also, the solid-phase material in an aquifer is generally very heterogeneous and adsorption affinity may not be uniform. Nevertheless, adsorption will affect the transport of dissolved species and we need a way to estimate this effect. We start with the most simple representation of adsorption and work up to more complicated models.

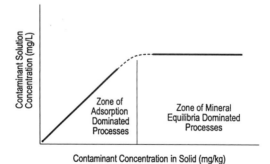

FIGURE 1.6.2 Adsorption and mineral equilibrium effects on contaminant solution concentration/mobility.

$$C_{aq} + X \longleftrightarrow C_{ads}$$

$$K_d = \frac{C_{ads}}{C_{aq}} = \frac{\text{(contaminant concentration on solid, mg/Kg)}}{\text{(contaminant concentration on solution, mg/L)}}$$

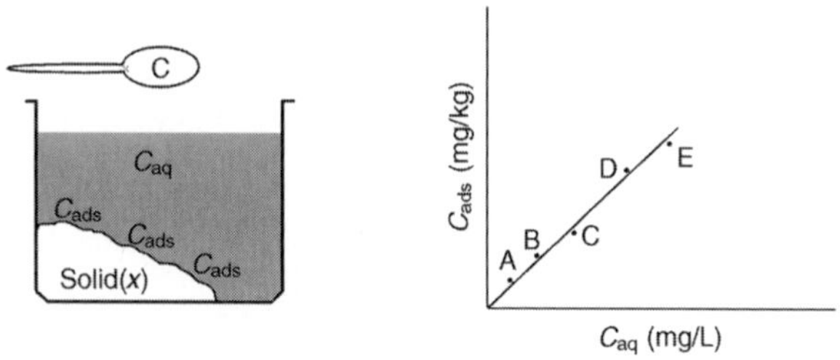

FIGURE 1.6.3   Linear adsorption isotherm.

1.6.3 ADSORPTION ISOTHERMS

The simplest representation of adsorption is shown by the beaker system in Figure 1.6.3. This water/rock system represents the interaction between dissolved species in water and the solid phase in contact with the water. When the water with its dissolved species initially comes into contact with the rock, the surfaces of the minerals and other solids comprising the rock attract and adsorb the dissolved species. The process can be represented for barium or any other dissolved species by the adsorption reaction:

$$\equiv SOH + Ba^{2+} \rightarrow \equiv SOHBa^+ + H^+$$

After equilibrium between the dissolved and adsorbed concentrations of barium is reached, the solution and solid concentration will not change. This concentration is shown in Figure 1.6.3 by location A. If additional dissolved barium is added to the beaker, a disequilibrium condition is generated because there will be too much barium in solution compared with the amount adsorb. This disequilibrium will drive the barium adsorption reaction to the right, causing some of the added dissolved barium to adsorb onto the solid surface. Equilibrium will be reestablished at a higher dissolved and adsorbed barium concentration, reflecting the barium added to the system. This new equilibrium state is shown by location B in Figure 1.6.3. The addition of barium can be repeated many times, adding a point to the plot each time. If the adsorption response is linear, we will be able to connect all the points by a straight line that intersects the axes at the origin.

This is a simplified representation of adsorption because it assumes that adsorption affinity is constant at all solution/solid concentrations and that there is an infinite amount of adsorbent. The linear isotherm is represented by the distribution coefficient (K_d), which is simply the ratio of the amount adsorbed onto a solid phase and the dissolved concentration.

$$K_d = C_{ads}/C_{aq}$$

where K_d = distribution coefficient (liters of solution/kg of rock)
$\quad C_{ads}$ = concentration on the solid (mg/kg of rock)
$\quad C_{aq}$ = concentration in solution (mg/liter of solution)

The K_d can be calculated from any point on the linear isotherm (Figure 1.6.3); it is also the slope of the linear isotherm because the line intersects the origin.

As a species moves through an aquifer it is constantly adsorbing and desorbing from the solid surfaces as the groundwater concentration changes. As a consequence, the velocity of a compound through an aquifer is retarded relative to the velocity of the water. This is represented by the retardation factor R.

$$R = v_{gw}/v_c = 1 + (\rho_b/\theta)K_d$$

where v_{gw} = average linear groundwater velocity
v_c = average linear contaminant velocity
ρ_b = bulk density
θ = porosity
K_d = distribution coefficient (L/kg)

The ratio of bulk density to porosity (ρ_b/θ) is the mass of rock in contact with a liter of groundwater. It is a conversion factor for K_d, which has units of liters of solution (L) per kilogram of rock (kg). If the bulk density is 2 kg/L, the porosity is 0.25, and the K_d for a contaminant is 5, the retardation factor is 41. This contaminant in this aquifer would move at a rate of only 2 percent of the groundwater velocity. The greater the K_d the higher the tendency of a species to be adsorbed than to be in solution. Species with a high K_d tend to spend more time adsorbed than dissolved in groundwater; however, even a small K_d has a retardation effect on transport of a species through an aquifer.

The linear isotherm may be an accurate representation of adsorption at low concentrations, but as the adsorbing surfaces of the solids become filled with the adsorbing species of interest or another species in the system, adsorption response may become nonlinear. As concentrations increase, adsorption affinity decreases. The result is that solution concentrations increase faster than they would if the adsorption response remained linear. Figure 1.6.4 shows nonlinear response curves. In the plot on the left, the adsorption response is fairly linear at low concentrations of the species in solution and on the solid, but as concentrations increase the line curves toward the dissolved concentration axis. This response curve is called a Freundlich isotherm and the equation representing the line is:

$$C_{ads} = K_F C_{aq}^{1/n}$$

where C_{ads} = concentration on the solid (mg/kg of rock)
C_{aq} = concentration in solution (mg/L)
K_F = Freundlich adsorption constant (L/Kg)
n = curve-fitting parameter (unitless)

To determine whether adsorption of a species for a particular system can be represented by the Freundlich isotherm, the logarithm of aqueous and solid-phase concentration data are plotted [log C_{aq} (X axis) versus log C_{ads} (Y axis)]. If the transformed data plot along a straight line, then the curve fitting parameter (n) can be calculated from the slope of the line and K_F can be calculated

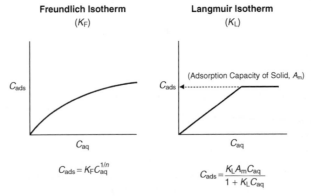

FIGURE 1.6.4   Nonlinear adsorption response curves.

from the Y intercept. The retardation factor for species that follow the Freundlich response curve is as follows[1]:

$$R = 1 + (\rho_b/\theta)(K_F n)\left(C_{aq}^{n-1}\right)$$

The Freundlich isotherm provides a way to more accurately estimate the effect of adsorption on contaminant migration than the linear isotherm for a broader range of concentrations; however, it still assumes an unlimited number of adsorption sites on the solid. To account for sites where the contaminant completely fills the adsorption sites, the Langmuir isotherm is used. The right hand plot in Figure 1.6.4 shows the Langmuir isotherm response curve that includes an adsorption capacity term (A_m) beyond which adsorption becomes zero. The equation for this curve is as follows:

$$C_{ads} = [K_L A_m C_{aq}]/[1 + K_L C_{aq}]$$

To determine whether adsorption of a species for a particular system can be represented by the Langmuir isotherm, C_{aq} (X axis) can be plotted versus C_{aq}/C_{ads} (Y axis)]. If the data plot along a straight line, then A_m can be calculated from the slope of the line and K_L can be calculated from the Y intercept. The retardation factor for species that follow the Langmuir response curve is as follows[1]:

$$R = 1 + (\rho_b/\theta)[K_L A_m/(1 + K_L C_{aq})^2]$$

As we progress from the linear (K_d) isotherm to the nonlinear (K_F and K_L) isotherms, we are adding complexity to our adsorption process, which more closely approximates reality. In the simple K_d case we say very little about the solid surface other than it has a certain attraction for a species. For K_F and K_L, we acknowledge some property (adsorption capacity, change in affinity with loading) about the surface of the solid, but we still are not addressing the effect of major changes in system variables (pH, Eh, ionic strength, etc.) on adsorption. To evaluate the more complicated situation we need a more detailed model of the adsorbing surface and of the process of adsorption.

1.6.4 *ADSORPTION SURFACE COMPLEXATION MODEL*

Removal of a species from solution and its attachment to a solid surface can be considered a complexation reaction analogous to aqueous complexation reactions. In aqueous complexation, a molecule such as carbonate (CO_3^{2-}) can complex with the hydrogen ion to form the aqueous species bicarbonate (HCO_3^-) and carbonic acid (H_2CO_3). In surface complexation, the functional groups on the surface of the solid attract solutes that complex with the functional group. The degree of surface complexation (adsorption) for a particular species depends on the chemical affinity of the surface for the species and the electrostatic (physical) attraction or repulsion of the species by the charged solid surface. This adsorption model accommodates changes in adsorption affinity as the species fills more and more of the surface sites, because the electrostatic characteristics of the surface changes as solutes adsorb to it.

The surface complexation model also includes changes in adsorption due to changes in the system master variables, such as pH. Figure 1.6.5 is a model of a solid with available surface adsorption sites. The surface sites carry an electrical charge because the surface atoms M (metal) and O (oxygen)] are not fully coordinated with (surrounded by) balancing atoms. In water, H^+ and OH^- are attracted to the surface and form surface complexes with the solid. As shown in Figure 1.6.5, if the pH of the solution changes, surface charge is developed by (1) complexation of H^+ with surface OH^- groups, (2) stripping of H^+ from surface OH^- groups, and (3) loss of OH^- groups from the surface.

At low pH values, the surface becomes more positively charged, whereas at higher pH values the surface is predominantly negatively charged. Because of the predominance of positively charged surface sites at lower pH values, the solid preferentially adsorbs anions under these conditions. At

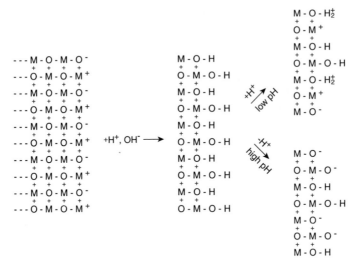

FIGURE 1.6.5   Model of solid adsorbent and surface complexation.

higher pH values, the predominantly negatively charged surface becomes more attractive to cations. This pH effect on adsorption may be observed in experimental data when the amount of adsorption is plotted versus pH. As shown in Figure 1.6.6, as the pH increases the amount of cation adsorption increases in response to the increasing number of available negatively charged surface sites. For anions, adsorption increases as the pH decreases because more of the surface sites are positively charged at lower pH values.

The effect of pH on adsorption is not part of the simple isotherm adsorption models; therefore, the surface complexation approach must be used to adequately estimate adsorption for a range of environmental site conditions. This method is valid over a wide range of pH conditions (as well as Eh and competing species conditions) because it includes the adjustment of surface conditions on the adsorbent solids for changes in the system. Also, if associated aqueous redox and complexation calculations are made, the appropriate species will compete for the available surface sites.

Currently, a set of consistent adsorption constants has been developed only for hydrous ferric oxide.[2] This solid is similar to the naturally occurring mineral ferrihydrite, which is a common weathering product of iron-containing minerals. The adsorption constants for hydrous ferric oxide and a surface complexation model have been incorporated into the computer codes MINTEQ[3] and PHREEQE.[4] If hydrous ferric oxide is the dominant adsorbent at a site, these codes can be used to develop a computer model of adsorption for a variety of species including zinc, cadmium, copper, nickel, lead, calcium, barium, sulfate, phosphate, and arsenic. It is anticipated that similar data sets will be developed in the future for other important adsorbents.

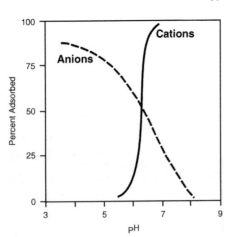

FIGURE 1.6.6   Effect of pH on anion and cation adsorption.

1.6.5   MINERAL PRECIPITATION/DISSOLUTION

Minerals and amorphous solids that comprise the solid phases of earth materials may form by precipitation reactions from the dissolved constituents in water. The solids can also provide their

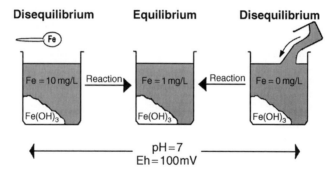

FIGURE 1.6.7   Mineral equilibrium control on dissolved iron concentration.

constituents to water by dissolution reactions. We can represent the precipitation and dissolution of a solid like ferric hydroxide [$Fe(OH)_3$] by the reactions:

$$Fe^{3+} + 3H_2O \rightarrow Fe(OH)_3 + 3H^+ \qquad \text{(precipitation reaction)}$$
$$Fe(OH)_3 + 3H^+ \rightarrow Fe^{3+} + 3H_2O \qquad \text{(dissolution reaction)}$$

The combination of precipitation and dissolution of ferric hydroxide can be represented by the composite reaction:

$$Fe^{3+} + 3H_2O \longleftrightarrow Fe(OH)_3 + 3H^+$$

The double-barbed arrow in the reaction represents a condition in which ferric hydroxide dissolves and precipitates from solution to maintain a fixed amount of dissolved iron in solution as long as other system parameters, such as pH, do not change. The middle beaker in Figure 1.6.7 shows a simplified system in which ferric hydroxide has dissolved in pure water to produce a dissolved concentration of total iron (ferric and ferrous iron) of 1 mg/L. At equilibrium, the dissolved iron concentration does not change, because for each amount of ferric hydroxide that dissolves into the water a corresponding amount of iron precipitates from solution.

The presence of ferric hydroxide and its equilibrium with water has a major effect on the fate of iron in this system. If the dissolved concentration of iron is increased by adding iron to the solution, the system reacts by precipitating more ferric hydroxide and lowering the dissolved iron level to the equilibrium concentration (Figure 1.6.7). If the solution in contact with ferric hydroxide is replaced by fresh water without any dissolved iron, the system reacts by dissolving some of the ferric hydroxide until the equilibrium dissolved iron concentration is once again attained (Figure 1.6.7).

The effect of mineral dissolution and precipitation on solution concentrations is different from the effect of adsorption and desorption reactions in which dissolved concentrations increase and decrease in response to the concentration of the species on the surface of the solid. In the case of mineral equilibrium, as long as some of the mineral is present in contact with the solution, the dissolved concentration will not vary. It does not matter whether the mineral makes up 100 percent or 0.01 percent of the solid; the resulting equilibrium dissolved concentration remains the same. For example, calcite ($CaCO_3$) in a limestone aquifer may make up most of the solid, whereas in a sandstone aquifer, calcite concentrations may be low. As long as the major system variables (pH, ionic strength, etc.) are similar and the carbonate concentrations are the same in each groundwater, the dissolved calcium concentration of water in contact with the two rock types will be the same.

Because of the potential for solid phases to have such an important effect on solution concentrations, it is useful to determine the minerals that are in equilibrium with the water. This can be done fairly simply by collecting a water sample, analyzing the dissolved concentrations, and measuring other system parameters that might affect mineral solubility. These parameters may include temperature, pH, and Eh. It is necessary to measure not only the dissolved constituents of interest but also all the major dissolved ions because mineral solubility is affected by solution complexation reactions and the ionic strength of the solution as discussed below. The solution data can be entered into equilibrium speciation computer codes (such as MINTEQ and PHREEQC) that calculate whether the solution composition corresponds to equilibrium with minerals in the database of the computer code. Without actually identifying the minerals in the rock, it is possible to predict which ones influence water composition because of the equilibrium relationship between the water and some of the minerals in the rock.

1.6.6 MINERAL SOLUBILITY

The amount of a mineral that will dissolve in a particular solution is the mineral's solubility. Conversely, the dissolved concentrations of the components of a mineral are limited by mineral solubility; thus, it is necessary to quantify this feature of a mineral if we are to understand the effect of mineral equilibrium on fate and transport. As mentioned above, mineral solubility is affected by temperature, pH, Eh (for redox-sensitive species), and the concentrations of major ions in solution. For most natural systems, to adequately consider all these variables it is necessary to calculate mineral equilibrium and solubility by using a competent equilibrium speciation computer code. The MINTEQ code (available from EPA http://www.epa.gov/CEAM/) was used to calculate the mineral solubility discussed in this section.

The solubility of many minerals increases with temperature; however, carbonate minerals become less soluble as temperature increases. Table 1.6.1 shows the effects of increasing temperature on simple systems of gypsum/water and calcite/water. Increasing the temperature from 10 to 75 °C raised the solubility of gypsum by a factor of 1.3 but lowered the solubility of calcite by a factor of 3.4. Because many groundwaters are in equilibrium with calcite, this effect may influence sampling results if the temperature of the sample is raised. This effect should be considered when well water is used for industrial purposes that might include large increases in temperature.

The pH of water can affect mineral solubility because it governs aqueous complexation. For example, the carbonate in carbonate minerals forms several ion species (H_2CO_3, HCO_3^- and CO_3^{2-}) in water. Table 1.6.2 shows the solubility of the calcite/water system closed to the exchange of CO_2 gas, which is representative of groundwater below the water table. Because the speciation of inorganic carbon is pH dependent and favors the carbonate ion at high pH, the solubility of calcite is depressed as pH increases. The solubility decreases by a factor of almost 50 as pH increases from 6.5 to 9.5. Other minerals are not as strongly affected by pH changes. Table 1.6.2 shows that the solubility of gypsum ($CaSO_4 \cdot 2H_2O$) is not strongly affected by a pH change in the same range. The reason for this is that the calcium and sulfate complexes ($CaOH^+$ and HSO_4^-) that form in water as gypsum dissolves are not present in high concentration over the selected pH range.

Many elements that form minerals are naturally present in more than one redox state. For example, iron occurs as ferric iron [Fe(III)] and as ferrous iron [Fe(II)]. If the system is in redox

TABLE 1.6.1 Effect of Temperature on Gypsum and Calcite Solubility (pH 7.5)

Temperature (°C)	Gypsum solubility (mg/L)	Calcite solubility (mg/L)
10	1460	250
25	1570	190
50	1720	120
75	1920	76

TABLE 1.6.2 Effect of pH on Calcite and Gypsum Solubility (25 °C)

pH	Calcite solubility (mg/L)	Gypsum solubility (mg/L)
6.5	890	1570
7.5	190	1570
8.5	54	1570
9.5	18	1689

TABLE 1.6.3   Effect of Eh on Ferric Hydroxide Solubility (pH 7.5)

Eh(mV)	Ferrihydrite solubility (mg/L)
300	0.0009
200	0.002
100	0.06
0	3.0

TABLE 1.6.4   Effect of Ionic Strength on Barite Solubility

Ionic strength (NaCl solution)	Barite solubility (mg/L)
0.00004	2.5
0.01	3.7
0.1	7.4
1	13

equilibrium, the distribution of iron between its two redox states is fixed by the redox potential (Eh) of the system. Because minerals typically contain only one redox state, changes in Eh can affect mineral solubility. Table 1.6.3 lists the solubility of ferric hydroxide at pH 7.5 and several Eh values. Because the dominant iron redox state changes from $Fe(III)$ to $Fe(II)$ over the selected redox range, the solubility of ferric hydroxide [an $Fe(III)$ mineral] increases by a factor of over 3000 as the redox potential decreases from $+300$ to 0 mV. Other elements that are redox sensitive include, chromium, arsenic, sulfur, selenium, mercury, copper, manganese, molybdenum, and uranium.

The major ions in solution can shield each other from reactions thereby increasing the solubility of minerals containing these ions. The ionic strength of a solution is a measure of the total dissolved ions and it is used to estimate the impact of ion shielding. Ion shielding lowers the activity (effective concentration) of dissolved species; therefore, the higher the ionic strength the greater the shielding and the greater the solubility of minerals in contact with the solution. Table 1.6.4 shows the effect of increasing ionic strength on the solubility of barite $(BaSO_4)$. The solubility increases by a factor of about 5 if the ionic strength increases from 0.00004 to 1. For reference, the ionic strength of most dilute groundwater is in the range 0.01–0.001, whereas seawater has a value of 0.7.

1.6.7   MINERAL REACTIVITY

For mineral dissolution and precipitation to be an important fate and transport process, the mineral must equilibrate with the solution composition during the solution's residence time in the system. Some minerals equilibrate with water relatively quickly—on the order of minutes to days. For example, water infiltrating a limestone aquifer will equilibrate quickly with calcite, and groundwater from a monitoring well with a high content of dissolved iron and a low redox potential will precipitate ferric hydroxide and equilibrate with this mineral on exposure to oxidizing, atmospheric conditions. On the other hand, many minerals commonly found in the aquifer solid phase do not equilibrate with groundwater because they are either slow to equilibrate or the pressure and temperature conditions are not appropriate for their formation. Hematite (Fe_2O_3) is found in aquifers but forms so slowly that it does not limit the iron concentration of the groundwater. Most of the mafic minerals (biotite, hornblende, pyroxene, olivine) form only at high temperatures $(>400°C)$ and, although present in the aquifer solid matrix, do not equilibrate with groundwater.

To develop accurate conceptual geochemical models of water and rock systems it is important to include only those phases that react rapidly enough to control the concentrations of their components. Table 1.6.5 provides a list of potential solid phases that might limit solution concentrations for the various elements. Because of the variability of solubility, the solution concentration of the element limited by the solid must be determined for the environmental conditions specific to a site. As discussed above, equilibrium speciation computer codes are commonly used to make the necessary calculations and include all the important solution and solid-phase reactions.

TABLE 1.6.5   Solid Phases That Might Limit Solution Concentrations

Element	Potential reactive minerals
Ca^{2+}	Calcite ($CaCO_3$), gypsum ($CaSO_4^-2H_2O$)
Mg^{2+}	pH $<$ 7.5, none
	pH $>$ 7.5, possibly dolomite and clays (montmorillinite, chlorite)
Na^+	pH $<$ 9, none. pH $>$ 9, possibly albite ($NaAlSi_3O_8$)
	Evaporative conditions: mirabilite ($Na_2SO_4 \cdot 10H_2O$), bloedite [$MgNa_2(SO_4)_2 \cdot 4H_2O$]
K^+	Illite
HCO_3^-/CO_3^{2-}	pH $>$ 6, calcite
	pH $>$ 7.5, rhodocrosite ($MnCO_3$), otavite ($CdCO_3$), cerrusite ($PbCO_3$), witherite ($BaCO_3$)
	Reducing: siderite ($FeCO_3$)
Cl^-	Halite (NaCl) at very high concentration ($>$200,000 parts per million)
SO_4^{2-}	Gypsum ($CaSO_4 \cdot 2H_2O$), sulfate reduction to sulfide
NO_3^-	Nitrate reduction to nitrogen, plant uptake
Si	Amorphous silica (SiO_2am), clays
Al	$Al(OH)_3$am, gibbsite [$Al(OH)_3$]; clays, kaolinite ($Al_2Si_2O_5(OH)_4$)
	Low pH, high sulfate, basalunite, alunite $KAl_3(SO_4)_2(OH)_6$
Fe	Oxidizing, pH $>$ 5.5, ferrihydrite [$Fe(OH)_3$]
	Oxidizing, acidic, jarosite [$KFe_3(SO_4)_2(OH)_6$], alunite
	Reducing, alkaline, siderite [$Fe(CO_3)$]
	Reducing, sulfide present, amorphous ferrous sulfide (FeS), mackinawite (FeS), pyrite (FeS_2)
Mn	Oxidizing: Nsutite ($MnO_{1.9}$)
	Reducing: Manganite (MnOOH), Mn_2O_3, Mn_3O_4
	Reducing: alkaline, rhodocrosite ($MnCO_3$)
As	Oxidizing, $FeAsO_4$, $Pb_3(AsO_4)_2$, $Mn_3(AsO_4)_2$
	Reducing, AsS_2
B	None
Ba	pH $<$ 9, barite ($BaSO_4$)
	pH $>$ 9, witherite ($BaCO_3$)
Be	$Be(OH)_2$
Cd	High pH, otavite ($CdCO_3$)
	Near neural pH, $Cd_3(PO_4)_2$
	Reducing, greennokite: CdS
Cr	Oxidizing, Cr_2O_3, $PbCrO_4$
	Reducing, $(Fe, Cr)(OH)_3$, $FeCr_2O_4$, Cr_2O_3
Cu	Oxidizing, $CuFe_2O_4$ (cupric ferrite)
	Reducing, $Cu_2Fe_2O_4$ (cuprous ferrite)
F	Fluorite (CaF_2)
Hg	Reducing, HgS
Mo	Oxidizing, $Fe_2(MoO_4)_3$, wulfenite ($PbMoO_4$), $CaMoO_4$
	Reducing, molybdenite (MoS_2)
Ni	Oxidizing, $NiFe_2O_4$
	Reducing, NiS
Pb	Lead phosphates, cerrusite ($PbCO_3$), $Pb(OH)_2$
PO_4^{3-}	Apatite {$Ca_3[(PO)_4]_2$}, variscite ($AlPO_4 \cdot 2H_2O$),
	Strengite ($FePO_4 \cdot 2H_2O$), $MnHPO_4$, plants
Se	Oxidizing, $Fe(OH)_4(SeO_3)$
	Reducing, Se, $FeSe_2$
U	Reducing, uraninite (UO_2)
Zn	Franklinite ($ZnFe_2O_4$)

Source: Lindsay, W. L., *Chemical Equilibria in Soils*, Wiley, New York, 1979. Appelo, C. A. J., and D. Postma, *Geochemistry, Groundwater and Pollution*, Balkema, Rotterdam, 1994. Rai, D., and J. Zachara, *Chemical Attenuation Rates, Coefficients, and Constants in Leachate Migration, Vol. 1: A Critical Review*, EPRI EA-3356, Electric Power Research Institute, Palo Alto, CA, 1984.

REFERENCES

1. Fetter, C. W., *Contaminant Hydrogeology*, Macmillan, New York, 1993.

2. Dzombak, D. A., and Morel, F. M. M., *Surface Complexation Modeling. Hydrous Ferric Oxide*, Wiley, New York, 1990.

3. Allison, J. D., Brown, D. S., and Noro-Gradac, K. J., *MINTEQA2/PRODEFA2, A Geochemical Assessment Model for Environmental Systems*, U.S. Environmental Protection Agency, 1991.

4. Parkhurst, D. L., *User's Guide to PHREEQC*, U.S. Geological Survey, 1995.

CHAPTER 2
CLASSES OF CHEMICALS

SECTION 2.1

HYDROCARBONS

David Jeffrey

Dr. Jeffrey is senior risk assessment scientist with SECOR International, Inc. He performs and manages health risk assessments, as well as fate and transport evaluations in support of risk assessments.

2.1.1 INTRODUCTION

Hydrocarbons are chemicals that are comprised of only carbon and hydrogen atoms. Hydrocarbons are a frequently encountered chemical class in many environmental situations, because hydrocarbons are the principal components of petroleum and petroleum-derived products, which are ubiquitous throughout the world. The names of some of these chemicals are familiar: benzene, octane, butane, propane, methane. Benzene and octane occur in gasoline; butane and propane are used for heating and lighting as liquefied gases maintained under pressure in cylinders. Methane is a natural, biological breakdown product of living matter, including plants (e.g., "swamp gas"). Figure 2.1.1 shows the structures and names of some hydrocarbons.

As Figure 2.1.1 shows, there are several groups within the hydrocarbon class. The largest distinction to make is between aliphatic and aromatic hydrocarbons. The simplest definition of these two groups may be expressed as follows: aliphatic hydrocarbons lack a benzene-type structure, aromatic hydrocarbons contain a benzene-like structure. This is why cyclooctane, although having a cyclic structure like benzene, is classified as an aliphatic hydrocarbon but *n*-butylbenzene is aromatic (Figure 2.1.1). Polyaromatic hydrocarbons (PAHs), also referred to as polynuclear aromatic hydrocarbons (PNAs), are a special class of aromatic hydrocarbons. The distinction between aliphatic and aromatic hydrocarbons is very important in terms of the properties and therefore fate of these chemicals. In general, aromatic hydrocarbons tend to be more stable, less reactive, and more long-lived than aliphatic hydrocarbons. Any good textbook in introductory organic chemistry will provide a discussion of the theoretical basis for these differences in properties. For a much more in-depth and rigorous explanation in terms of classical organic molecular orbital theory, the reader is referred to the classic treatise by Woodward and Hoffmann (1970).

The next major subgroups for hydrocarbon classification are cyclic and acyclic hydrocarbons, meaning those that contain a cyclic structure and those that do not. In general, no special properties are associated with cyclic structures except perhaps that, for two hydrocarbons with the same carbon chain length, the cyclic hydrocarbon tends to be more volatile, more water soluble, and to sorb less strongly to natural soils than the acyclic hydrocarbon.

Saturated and unsaturated hydrocarbons are another group. A saturated hydrocarbon contains no double or triple carbon-carbon bonds. An unsaturated hydrocarbon contains one or more double or triple bonds. Although some chemists consider cyclic structures to constitute units of unsaturation, for the purposes of environmental chemistry it is more useful to not follow this classification scheme. One can observe significant differences in chemical properties between saturated and unsaturated hydrocarbons. In general, unsaturated hydrocarbons are more reactive than saturated chemicals. Saturated hydrocarbons always follow the molecular formula C_nH_{2n+2}.

AROMATIC HYDROCARBONS

Volatile Hydrocarbons (e.g., BTEX)

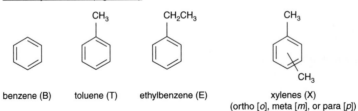

benzene (B)   toluene (T)   ethylbenzene (E)   xylenes (X)
                                                          (ortho [*o*], meta [*m*], or para [*p*])

Less Volatile ("Semi-volatile") Hydrocarbons (e.g., polynuclear aromatic hydrocarbons [PAHs])

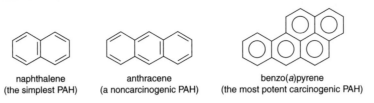

naphthalene   anthracene   benzo(*a*)pyrene
(the simplest PAH)   (a noncarcinogenic PAH)   (the most potent carcinogenic PAH)

ALIPHATIC HYDROCARBONS

Volatile Hydrocarbons
Saturated Hydrocarbons
Acyclic

$CH_3(CH_2)_nCH_3$

$n = 0$ (ethane) to $n = 6$ (*n*-octane)

Cyclic

CH₃

methyl cyclopentane

Unsaturated Hydrocarbons
Acyclic

1-hexene

Cyclic

cyclohexene

FIGURE 2.1.1   Examples of hydrocarbons.

Finally, within the subcategory unsaturated hydrocarbons one finds alkenes and alkynes. Alkenes are hydrocarbons with one or more carbon-carbon double bond. Alkynes have one or more triple bond.

2.1.2 FATE AND TRANSPORT MECHANISMS

Once hydrocarbons have been released into the environment, via oil spills, gasoline leaks, vehicular exhaust, and other sources, six main fate and transport mechanisms may be operative, depending on the nature of the hydrocarbons and the type of release. These six mechanisms are listed below:

1. volatilization, for low molecular weight (MW) hydrocarbons;

2. sorption, for higher MW hydrocarbons;

3. oxidation;

4. photodegradation;

5. biodegradation; and

6. leaching.

Intermolecular reactivity under typical environmental conditions is not included in this list. Although benzene and alkylbenzenes may self- or cross-react to ultimately give rise to PAHs, this occurs only at elevated temperatures. Additionally, the bimolecular kinetic nature of these types of reactions cause their ranking as significant fate and transport processes to be low. The six fate and transport mechanisms listed above are discussed in the following sections.

Volatilization

General Considerations.   Volatilization is the process by which chemicals release vapors to the environment. These vapors consist of molecules of the chemical in the gas phase. The characteristic odor of gasoline is due to volatilization processes. Volatilization of hydrocarbons in the environment may occur from soil, sediment, surface water, groundwater, or even biota media. As the MW of a hydrocarbon increases, its tendency to volatilize decreases. A good rule of thumb is that hydrocarbons with a MW > 200 g/mol are not likely to significantly volatilize [U.S. Environmental Protection Agency (USEPA), 1998]. Another important parameter that determines a chemical's volatilization potential is its Henry's law constant (H), which measures a chemical's tendency to volatilize from an aqueous environment, as defined in Equations 2.1.1 and 2.1.2.

$$H_d = C_a/C_w \tag{2.1.1}$$

where H_d = the dimensionless Henry's law constant, which is 41.7 times H for 20°C;
  C_a = the air (or gas phase) concentration of a chemical in contact with water; and
  C_w = the water concentration of a chemical in contact with air.

$$H = P_v/S \tag{2.1.2}$$

where P_v = the chemical's vapor pressure [atmospheres (atm)]; and
  S = the chemical's water solubility (mol/m^3).

Note that H_d, the so-called dimensionless H, is not really dimensionless but has units of volume-water/volume-air. Ignoring this fact can lead to significant frustration when checking units for various environmental transport equations. USEPA (1998) has suggested rule-of-thumb values of 1×10^{-5} atm-m^3/mol for H and 200 g/mol for the MW, where any chemical with an H value greater than this and with a MW less than this cutoff value is considered to have potentially significant volatilization-from-water [or soil water (i.e., moisture)] potential. Remember that for many environmental contamination problems involving soil impacts, because of the presence of soil water, and the generally relatively low concentrations of a contaminant, the H parameter is an important determinant of soil volatilization. Other sources have also recommended rule-of-thumb type values to identify chemicals that may have significant volatilization potential, as follows:

- For a MW equal to 200 g/mol, an H value $>3 \times 10^{-8}$ to 3×10^{-6} atm-m^3/mol (derived from medium ranges for vapor pressure and solubility from Ney Jr., 1990);
- H value $>5 \times 10^{-6}$ atm-m^3/mol (Dragun, 1988); and
- H value $>1 \times 10^{-5}$ atm-m^3/mol (Lyman et al., 1990).

As the information listed above shows, rule-of-thumb values can differ substantially from one source to another. For example, there is more than a 300-fold difference between the lowest value for H (3×10^{-8} atm-m^3/mol) and the highest (1×10^{-5} atm-m^3/mol). Those hydrocarbons that have MW or H values below or above these rule-of-thumb values not only may be categorized as having potentially significant volatilization potential but also belong to a large and very important group of environmental chemicals: volatile organic chemicals (VOCs).

Quantifying the Volatilization Rate.   Because much of the hydrocarbon contamination in the environment is due to gasoline releases or leaks of one kind or another [e.g., leaking underground fuel storage tanks (USTs)], and the main toxic chemicals of concern that comprise gasoline are VOCs [e.g., benzene, toluene, ethylbenzene, xylenes (known as BTEX)], volatilization is often a very important fate and transport mechanism for hydrocarbons. Because of its importance, and because of the frequent need to know the volatilization rate of a chemical from a source medium (e.g., soil) to the atmosphere, various mathematical models have been developed for estimation of a vapor flux from a source containing volatile hydrocarbons. It should also be mentioned that, with today's advanced field measurement techniques and sensitive and selective air analytical methods, it is often possible to measure a chemical's volatilization rate. These two main methods for estimating a volatilization rates for hydrocarbons are discussed in the following sections.

Various models have been developed for quantification of the volatilization rates of hydrocarbons. Models have been developed for volatile hydrocarbons releasing vapors from subsurface soil, groundwater, and surface water. All these models use the source concentration of the hydrocarbon (e.g., mg of chemical per kg of soil), chemical-specific properties of the hydrocarbon, and properties of the source (and occasionally of the air) medium to derive a flux. These fluxes are computed in units of mass of chemical released per unit time per unit emissions area [e.g., mg of chemical (vapor) per s per m^2].

Many different models of the general type described in the preceding paragraph have appeared in the scholarly environmental science literature over the past 15–20 years. The reader may be familiar with the names of some of these: Shen, Jury, Karimi, Farmer, and Mackay. In comparing one model to another for the purposes of selecting the most appropriate model for a given application, one may first differentiate on the basis of conservation of mass. It may seem surprising, but many volatilization models in common use today do not conserve mass. What does this mean?

For the great majority of environmental sites—that is, physical locations where some type and magnitude of chemical release has occurred—the original source of the release has been removed. For example, for a gasoline service station, there may still be residual impacts to local soil and/or groundwater after the discovery of a leaking UST, but at some point in the site management process the tank is removed. This means a finite supply of chemical contaminant remains in site media. For volatile contaminants, this means the vapor flux must follow a pseudo-exponential curve, as shown in Figure 2.1.2; that is, the flux decreases with time because there is less contaminant remaining in the soil, groundwater, or surface water. This is a consequence of Fick's law of diffusion, as presented below:

$$F = -D \times dc/dx \qquad (2.1.3)$$

where F = surface vapor flux;
$\quad D$ = air diffusion coefficient; and
$\quad dc/dx$ = concentration gradient along the axis of diffusive transport (e.g., X axis)

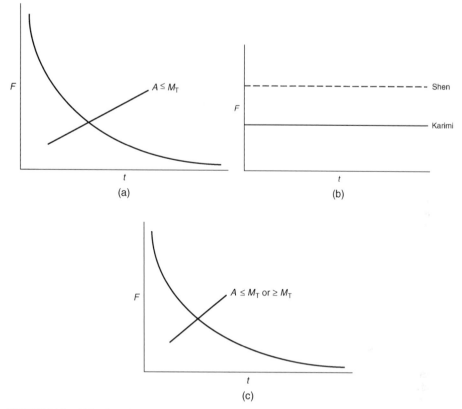

FIGURE 2.1.2 (a) Jury's model. $F = f(t, C_T, D_a, D_w, P_a, P_w, J_w, \mu, z, L)$, where $C_T = \rho_b C_s + P_w C_w + P_a C_g$, $F =$ chemical vapor flux at soil surface (in mg/sec-m^2), $t =$ time, $C_T =$ total chemical concentration, $\rho_b =$ soil bulk density, $C_s =$ sorbed soil phase concentration, $P_w =$ water-filled soil porosity, $C_w =$ soil pore water phase concentration, $P_a =$ air-filled soil porosity, $C_g =$ soil gas phase concentration, $D_a =$ air diffusion coefficient, $D_w =$ water diffusion coefficient, $J_w =$ soil water evaporation rate, $\mu =$ first order degradation rate, $z =$ depth of chemical incorporation, $L =$ height of clean soil layer, $A =$ integrated area under curve, $M_T =$ total available chemical mass for volatilization (b) Shen's/Karimi's Model. $F = [C_g D_a (P_a^{10/3}/P_t^2)]/L, K_{sa} C_s]/(\pi \alpha t)^{1/2}$, where $K_{sa} =$ soil-to-air partition coefficient, $C_i =$ soil chemical concentration, $\alpha = f$ (soil bulk density, soil porosity, K_{sa}) other terms as defined previously.

Because the concentration gradient dc/dx is continually decreasing as mass is transported from the source medium to the air, the flux also must decrease, following the behavior shown in plot A in Figure 2.1.2. However, many volatilization models do not conserve mass and instead conservatively assume an infinite supply of volatile contaminant. For these models, the flux is independent of time and follows the behavior of plot B in Figure 2.1.2. Figure 2.1.2 plot C shows the flux behavior of models that may recognize the depleting mass principle but that mathematically can still lead to overestimates of the actual mass available if run out on a long enough timescale. Although mass conservation models are clearly more realistic of most environmental site situations, the attraction of using models that do not conserve mass lies in the simplicity of the computations and the added degree of health protectiveness a regulatory agency may require, especially when it is unclear that a limited amount of contaminant remains at a site. For the experienced fate and transport modeler, selection of an infinite source model for a site with relatively low volatile hydrocarbon contamination, and perhaps other mitigating site features, may represent the best choice for a given environmental assessment.

TABLE 2.1.1 Volatilization from Soil and Groundwater, a Comparison of Some Available Models

Model	Source depletion?	P_a?	D_w?	Soil sorption?	Water evaporation?	Degradation?	Cap fringe?	Agency acceptability	Applications	Inputs
Shen/soil	No	No	No	No	No	No	NA	USEPA SEAM (1988)	$\geqslant$1 cm clean soil (CLFs w/o LFG)	C_s or C_{sg}, $H(C_s)$, SM, P_t, L
1-2 OM<Shen Karimi/soil	No	Yes	No	No	No	No	NA	Generally recognized as simple refinement of Shen's model	Same as Shen's model	Shen + P_a
3-4 OM<Karimi Hwang and Falco/soil	No; but $\frac{1}{\sqrt{t}}$	Yes	No	Yes	No	No	NA	Current USEPA, Cal/EPA	no clean cover	Karimi + K_{oc}, DSBD, f_{oc}, p_D
1-4 OM<H&F Jury's BAM/soil	Yes	Yes	Yes	Yes	Yes	Yes	NA	May require pre-buy off	with or w/o clean cover over subsurface	H&F + D_w, v_p, S, u, J_w, RH, z (+L)
USEPA SEAM/LFG	No	NA	NA	No	Yes	No	NA	USEPA	co-disposal LFs	C_{sg} (assumes avg. lit. LFG velocity)
Shen/GW	No	No	No	No	No	No	No	Modification of soil model	NA	Same as soil except C_w
Karimi/GW	No	Yes	No	No	No	No	No	Modification of soil model	NA	Soil except C_w
1-2 OM<Karimi Army/GW1-2	No	Yes	Yes	No	No	No	Yes	Similar to Jury's model	NA	Karimi + D_w, WSBD, h_c (+ n_{tz}, n_{uz})

Volitization from soil — Increasing LOE (rows Shen/soil through USEPA SEAM/LFG)

Volitization from groundwater — Increasing LOE (rows Shen/GW through Army/GW1-2)

OM = Order of magnitude (of model results); BAM = Behavior Assessment Model; SEAM = Superfund Exposure Assessment Manual; 1-2, 3-4, = vapor flux; LOE = Level of Effort; H&F = Hwnag and Falco; LFG = Landfill Gas; GW = Groundwater; LF = Land Fill; Cal/EPA = California Environmental Protection Agency; DSBD = Dry soil bulk density; lit. = Literature; WSBD = Wet soil bulk density; NA = Not applicable.

It is beyond the scope of this *Handbook* to describe in detail the various volatilization models that may be applied at volatile hydrocarbon sites. Instead, the reader is referred to USEPA (1988), Lyman et al. (1990), Dragun (1988), The American Petroleum Institute's (API) Decision Support System (DSS), and PC GEMS (USEPA's Graphical Exposure Modeling System for the PC).

Table 2.1.1 compares some of the models contained in the sources listed above. As mentioned earlier in this section, which model is selected to quantify the volatilization of a hydrocarbon depends on certain criteria. The following seven criteria should always be considered when selecting an appropriate volatilization (or any other) model:

1. Goals of the overall assessment,

2. Site concentration of the hydrocarbon,

3. Physicochemical properties of the hydrocarbon,

4. Toxicity of the hydrocarbon,

5. Characteristics of the site,

6. Particular regulatory environment, and

7. Project schedule and budget.

One may make field measurements of the volatilization rate of a hydrocarbon from soil or ground-water sources; there are currently no approved methods for measuring the volatilization rate from a surface water source. USEPA (1986) has developed the surface flux isolation chamber (flux chamber) for making these types of measurements. The flux chamber consists of a metal bowl fitted with several inlet and outlet ports. The bowl is sealed to the site surface (generally by pushing the lip of the chamber down into the soil) and a known flow rate of ultrapure air is passed over the enclosed site surface. Summa canisters, small metal cylinders used for the collection of chemical vapors, fitted to the bowl part of the chamber collect any chemical vapors that may be emitting from the site. By knowing the surface emission area (i.e., the footprint of the chamber), the flow rate, and the concentration of chemical vapor in the canister (determined off site by an analytical laboratory using USEPA Method TO-14), it is possible to compute the surface vapor flux of each chemical analyzed. Figure 2.1.3 shows the flux chamber.

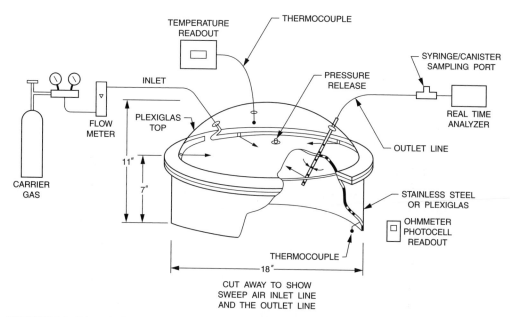

FIGURE 2.1.3   Schematic diagram of the flux chamber (reproduced from USEPA, 1986).

TABLE 2.1.2   Flux Chamber Computations

Surface vapor flux for unpaved emission areas and continuous pavement	
Summa gas cylinder concentration measured/reported by laboratory	y (ppbv)
Laboratory result converted to mass units	$y \times (MW/24.45) = n\ (\mu g/m^3)$
Flow rate of sweep air through flux chamber	p (L/s)
Area of flux chamber	$a\ (m^2)$
Associated flux	$F_{co}\ (\mu g/m^2\text{-s}) = (n \times p \times 0.001\ m^3/L)/a$
Surface vapor flux for cracked pavement	
Lineal feet of cracks	z (ft)
Area of cracked pavement	$x\ (ft^2)$
Associated flux	$F_{cr}\ (\mu g/m^2\text{-s}) = F_{cr}\ (\mu g/ft\text{-min}) \times z/x \times 10.764\ ft^2/m^2 \times 0.0167\ min/s$

ppbv = parts per billion by volume.

Table 2.1.2 shows equations used to compute surface vapor fluxes from data obtained from a flux chamber sampling and analysis program. The flux chamber may be used over paved as well as soil surfaces because of a recently developed technique involving the use of Teflon tape (Schmidt and Zdeb, 1999). With method detection limits for USEPA Method TO-14 that are usually below levels associated with significant inhalation exposures, the flux chamber protocol is a useful tool for the environmental site assessor. Results obtained from a flux chamber sampling program are also usually of sufficiently high precision and accuracy (i.e., overall data quality) for direct use in a risk assessment.

Deciding whether to model or measure volatilization rates for volatile hydrocarbons involves considering various factors—the pros and cons of modeling versus measuring in general with some additional factors unique to the flux chamber. A list of pros and cons follows:

- Modeling pros: predictive, generally less costly, and no detection limts;
- Modeling cons: at best an approximation of actual site phenomena, and more complex models difficult to explain;
- Flux chamber pros: measure actual site fluxes; and
- Flux chamber cons: not predictive, prone to field errors, and generally more expensive.

Perhaps the most significant drawback of using a model is that, even with the best of models, one is still approximating actual site phenomena. This leaves a site assessment that relies at least partially on modeling open to potential criticism by those who may not like the model used in the assessment or by those who may not understand it. This point can be particularly problematic in environmental litigation situations, when real measurements may be more well received by a lay jury than the black box results of a model. That is, actual field measurements get around the criticism of approximation in a very direct and obvious way.

However, field measurements, using the flux chamber or other techniques, have the inherent limitation of no future predictability; that is, a model can tell you in principle not only what the flux is at a site today but can also predict the flux 1 week, 1 year, or 100 years from now. Field measurements are, in contrast, site snapshots of phenomena that are likely continuous, ongoing processes. For human health risk assessments that are typically concerned with chemical doses over 25- to 30-year periods of exposure, it can be very important to know how the flux will change over a given time frame. Because of this limitation, it usually is necessary to conservatively assume that the flux measured over 1 day of sampling at a site is constant over the long exposure periods typically used in risk assessment. One can sometimes partially compensate for the conservativeness of this approach by collecting a sufficiently large number of flux chamber samples from different site locations (especially for a larger site) and applying a statistical analysis to the flux chamber result dataset [i.e., by computing the 95 percent upper confidence limit of the arithmetic mean detected flux value (95UCL)]. One then uses the 95UCL value for each detected chemical for all further risk-assessment computations. If such a statistical analysis is not done, the maximum detected flux value must be used, which may be a full order of magnitude or more above the 95UCL. In computing the 95UCL values, one generally assumes a value of one-half the method reporting limit (MRL) for those analytes that are not detected in certain samples. This type of statistical analysis is recommended by USEPA for conducting reasonable maximum exposure (RME) assessments and therefore should be acceptable to most state and local agencies.

Sorption

Sorption is the process involving the interaction of a chemical with solid environmental media, usually soil or sediment. This interaction is not as strong as an actual chemical bond, but it may be significant enough to retard the mobility of a chemical in the environment. For example, a chemical that sorbs strongly to the organic matter in subsurface soils within a saturated groundwater zone may not pose a risk to downgradient users of the groundwater because of retardation of mobility (in this case, flowing groundwater) by sorption. However, chemicals that sorb strongly to soils or sediments may be immobile with respect to environmental processes such as groundwater migration (or leaching) but may be associated with increased site exposures for certain receptors because of dust-related exposure pathways. That is, a chemical that sorbs strongly to soil is also a chemical that binds tightly to dust particles derived from site soil, for sites where surface spills or releases may have occurred, and where portions of the site soil surface are exposed. In practice, however, exposure pathways associated with dusts usually are not the main contributors to a particular receptor's overall site exposure. This means that, in most cases, stronger sorption means lower environmental mobility and lower potential chemical exposures.

For hydrocarbons, there is a strong relationship between sorption ability and MW, although useful rule-of-thumb values are not as accessible as for volatilization potential. In general, one may say that hydrocarbons of fewer than about nine carbons do not strongly sorb to most soils or sediments. Using the $2n + 2$ rule gives us a MW of about 128 g/mol. This means that hydrocarbons of lower MW such as BTEX (Figure 2.1.1, MW of 78–106) sorb weakly to soils and sediments and consequently are relatively mobile in the environment, whereas PAHs such as benzo(a)pyrene (Figure 2.1.1), with a MW of 252 g/mol, sorb strongly and are relatively immobile.

The most common measure of a chemical's tendency to sorb to soils or sediments is its "soil-water partition coefficient," or K_d, which is defined as follows:

$$K_d = C_s / C_w \qquad (2.1.4)$$

where C_s = equilibrium soil concentration of a chemical in contact with water, and
$\quad C_w$ = equilibrium water concentration of a chemical in contact with soil.

K_d, in turn, is a function of a chemical's intrinsic ability to sorb to organic matter contained in soil or sediment and the organic content of a soil or sediment. This is represented as:

$$K_d = K_{oc} \times f_{oc} \qquad (2.1.5)$$

where K_{oc} = organic soil-water partition coefficient, and
$\quad f_{oc}$ = fraction of organic matter in a soil or sediment.

The mechanism implied by Equations 2.1.4 and 2.1.5 is one in which a chemical (hydrocarbon) interacts only with the soil or sediment organic matter. Dragun (1988) gives a value of 0.001, or 0.1 percent, for the organic content of native soils (i.e., f_{oc}). USEPA (1996) gives a default value of 0.006 (0.6 percent) for use in exposure modeling. Although the USEPA values are meant to be conservative default values, it is unlikely that many actual sites will have f_{oc} values much higher (i.e., more than one order of magnitude) than this. This means that a very large fraction of site soils are likely composed of inorganic clays and sands. These clays and sands, in turn, are composed largely of aluminates and silicates [i.e., (M_xO_y), where M = aluminum (Al), silicon (Si)]. A potential problem with the sorption mechanism involving only organic matter lies in the much larger fraction of inorganic aluminates and silicates and the fact that aluminates and silicates are common solid-phase media used for chromatographic separations; that is, these aluminates and silicates are known to interact with and retard organic compounds, including hydrocarbons. Otherwise they would have no utility in chromatography.

For whatever reason, however, for most environmental practitioners, including most regulatory agencies, organic content is still the only criterion used to predict the sorption of an organic chemical, including hydrocarbons. However, one may measure the K_d of a chemical as an alternative to computing it based on Equations 2.1.4 and 2.1.5. When this is done, one should find that the sorption potential of a soil or sediment is significantly greater than that predicted from these equations and therefore the mobility of chemicals detected in site soil or sediment is significantly less than what one would normally predict. Because most site assessments, including risk assessments, use calculated, not measured, K_d values, this means there is some degree of conservatism or health protectiveness already built in to most assessments.

Rule-of-thumb values for the K_{oc} parameter are available in the literature. Dragun (1988) gives a K_{oc} cutoff of 2000 as corresponding to immobile compounds and anything below 50 as very mobile. Ney Jr. (1990) uses cutoff values of 10,000 and 1000 to represent the same categories. Benzene, a low-MW volatile hydrocarbon with a K_{oc} of 59 L/kg (USEPA, 1996), is classified as mobile, well within Ney Jr.'s classification scheme and close to the very mobile classification of Dragun. Anthracene, a higher-MW PAH with a K_{oc} of 29,500 L/kg (USEPA, 1996), is classified as immobile. Both classifications are consistent with general site experience. That is, benzene generally leaches from soil to groundwater and migrates within a groundwater zone much more rapidly and extensively than anthracene. Of course, rule-of-thumb cutoff values are generally derived from general site experience, so this consistency is not surprising or significant. It also should be emphasized that these rule-of-thumb values work well only with chemicals and sites clearly within one end of a recommended range or extreme. For chemicals or sites having fate and transport parameter values only close to these recommended rule-of-thumb values, predictions are more risky.

Oxidation

Hydrocarbons are, by definition, compounds that contain carbon atoms in a highly reduced form. This means that hydrocarbons should be prone to oxidation. For many hydrocarbons, especially those with π-electrons (aromatic and unsaturated hydrocarbons), oxidation is a thermodynamically favored reaction. For anthracene, for example, the half-cell one-electron oxidation (to give a radical cation) has a potential of +0.73 V relative to the standard calomel electrode (CRC, 1992). When this potential is adjusted to refer to a standard hydrogen electrode (by adding +0.241 V) and then added to one-quarter the potential for the reduction of oxygen [+0.401 V (neutral conditions) to +1.23 V (acidic conditions)], a total reaction potential change of from +1.07 (neutral) to +1.27 (acidic) V is obtained, indicating a highly favorable reaction. The *CRC Handbook of Chemistry and Physics* [CRC, 1992 (and more recent editions)] is a source for redox potentials for some organic chemicals, but it does not list many common organic contaminants.

The products of the oxidation of hydrocarbons are alcohols, ketones, aldehydes, and carboxylic acids. These types of compounds are all significantly more water soluble than the parent hydrocarbons they are derived from—that is, oxidation increases the environmental mobility of hydrocarbons. In addition, the products of oxidation tend to be more chemically reactive than the parent hydrocarbons. These increases in mobility and reactivity may reduce the level of concern over this type of contamination, as the oxidation products may be excreted and metabolized more efficiently than the hydrocarbon parents. However, increases in chemical reactivity can always lead to increases in toxicity.

For various petroleum derivatives such as gasoline, diesel, and motor oil, exposure to oxidative conditions over extended periods leads to a weathering of these materials. However, one must be careful with the terminology. Weathering generally refers to a crude or refined petroleum product that has been chemically changed by one or more of the following four processes:

1. ambient oxidation,
2. photodegradation,

3. volatilization, and

4. biodegradation.

Of these four processes, ambient oxidation and volatilization have been discussed. Photodegradation and biodegradation are discussed in the following sections. The weathering process is an important basis behind the concept of natural attenuation (ASTM, 1996), which in turn supports the risk-based corrective action (RBCA, pronounced Rebecca) protocol. The RBCA protocol has become very important and widespread in the U.S. for assessing and managing petroleum impacted sites (ASTM, 1995).

Photodegradation

Photodegradation is a process in which a chemical absorbs energy in the form of light, either visible or ultraviolet, and, as a result of this light absorption, is transformed to another chemical(s). Photodegradation is an important loss mechanism for many hydrocarbons, especially for airborne hydrocarbons (chemical vapors or dusts). Three general comments may be made about the photodegradation of hydrocarbons.

- PAHs are most prone to photodegradation, followed by alkylbenzene and polyunsaturated aliphatic compounds, then monounsaturated hydrocarbons, and finally saturated hydrocarbons.
- Photodegradation processes are potentially important for surface soil, sediment, and surface water releases or impacts from higher-MW, less volatile hydrocarbons and for any type of release of volatile hydrocarbons.
- Most photodegradation transformations of hydrocarbons lead to oxidation.

The first point can be explained in terms of basic chemical molecular orbital and bonding theory. Molecular systems that have extended π-electron orbital configurations, such as PAHs and benzene derivatives, tend to have relatively small energy gaps between the highest occupied molecular orbital (HOMO) and the lowest unoccupied molecular orbital (LUMO). Therefore, these systems are well set up to absorb light, as many available environmental electromagnetic wavelengths will be sufficiently energetic for absorption to occur. This arrangement can be contrasted with saturated hydrocarbons, which have no π-electrons and have lower-lying HOMOs (from σ bonding) and consequently larger HOMO-LUMO gaps. This means that these types of chemicals are less capable of absorbing light that occurs at typical environmental wavelengths.

One can gain a semiquantitative, semipredictive understanding of the environmental photochemistry potential of common hydrocarbons by referring to Table 2.1.3, which lists the molar extinction coefficient (ϵ_{max}) and the wavelength (λ_{max}) for the maximum light absorption of a hydrocarbon. The higher the values of ϵ_{max} and λ_{max} the more prone a chemical is to photodegradation; the lower the values, the less prone.

The third point listed above means that oxidation of hydrocarbons can be an important overall fate mechanism for surface impacts, because ambient, chemical oxidation will occur in parallel with, and therefore add to, photooxidation.

Biodegradation

The fifth and final important fate and transport mechanism for hydrocarbons is biodegradation. As mentioned previously, biodegradation is a component of natural weathering/attenuation, which is one of the main supporting assumptions behind the RBCA process and remediation by natural attenuation (RNA; ASTM, 1996).

In general, ambient, natural biodegradation occurs more rapidly under aerobic than under anaerobic conditions. This means that surface and shallow subsurface soil impacts by hydrocarbons are biodegraded more efficiently than deeper impacts, as more oxygen is available at or near the surface than at deeper depths. Although many naturally occurring soil microbes will utilize electron recipients

TABLE 2.1.3   Typical Chromophoric Values for Selected Organic Groupings

Chromophore	λ_{max} (nm)	ϵ (approx)	Transition
C–H or C–C	<180	1000	$\sigma \rightarrow \sigma^*$
C=C	180	10,000	$\pi \rightarrow \pi^*$
C=C–C=C	220	20,000	$\pi \rightarrow \pi^*$
Benzene	260	200	$\pi \rightarrow \pi^*$
Naphthalene	310	200	$\pi \rightarrow \pi^*$
Anthracene	350	10,000	$\pi \rightarrow \pi^*$
Phenol	275	1500	$\pi \rightarrow \pi^*$
Aniline	290	1500	$\pi \rightarrow \pi^*$
RS-SR	300	300	$n \rightarrow \sigma^*$
C=O	280	20	$n \rightarrow \pi^*$
Benzoquinone	370	500	$n \rightarrow \pi^*$
C=C–C=O	320	50	$n \rightarrow \pi^*$
C=N	<220	20	$n \rightarrow \pi^*$
N=N	350	50	$n \rightarrow \pi^*$
N=O	300	100	$n \rightarrow \pi^*$
Ar–NO$_2$	280	7000	$\pi \rightarrow \pi^*$
Indole	290	5000	$\pi \rightarrow \pi^*$

Reproduced from Larson and Weber (1994); Original Source: Scott (1964) and Turner (1948) as cited in Larson and Weber (1964).

other than oxygen (e.g., sulfate, nitrate) for their normal metabolic functions, and the availability of these alternative oxidants may be relatively high at some sites, oxygen is still the oxidant of choice and significantly accelerates biodegradation relative to other oxidants.

It is difficult to generalize about the effect of chemical structure and bonding on relative rates of biodegradation. However, for aromatic hydrocarbons, benzene and PAHs lacking alkyl substitution degrade at a slower rate than alkyl-substituted derivatives. This is shown in Table 2.1.4, which lists degradation half-life data for BTEX, naphthalene, and benzo(*a*)pyrene. The half-lives for benzene, naphthalene, and benzo(*a*)pyrene are in general longer than for TEX, which are alkyl-substituted benzene derivatives. This trend may be due to the high energetic price paid in disrupting the stable aromatic ring system of the unsubstituted parents relative to "chewing" down the alkyl chains attached to the aromatic rings.

A chemical half-life is defined as the time required to reduce the concentration of a chemical in some medium to one-half its original value. Half-lives are extremely useful data, because if the half-life of a chemical under certain conditions is known, and if the chemical process associated with the half-life follows simple, first-order kinetics, then one may easily predict the concentration of that chemical at any future point in time according to the following equations.

$$k = 0.693/t_{1/2} \qquad (2.1.6)$$

$$C_t = C_{t=0} \times e^{-kt} \qquad (2.1.7)$$

where k = first-order rate constant,

$t_{1/2}$ = reaction (degradation) half-life,

C_t = chemical concentration at time t, and

$C_{t=0}$ = chemical concentration at time zero.

First-order kinetics means that the reaction rate depends on only one molecular participant in the reaction. Because most environmental transformations involve a contaminant chemical, which is usually present at much lower concentrations than water or oxygen, the time-dependent concentrations of the contaminant change much more significantly than the concentrations of water or oxygen and the reaction is called pseudo-first-order and follows Equations 2.1.6 and 2.1.7. One must exercise some caution when considering transformations that involve biological systems that use enzymes to accelerate all their metabolic reactions.

TABLE 2.1.4 Reported Degradation Rates for Petroleum Hydrocarbons

Source of data	Chemical decay rates [day^{-1}, (half-life, days)]							
	Benzene	Toluene	Ethylbenzene	Xylenes	o-Xylene	MTBE	Naphthalene	Benzo (a)Pyrene
Borden aquifer, Canada	0.007 [99]	0.011 [63]	...	...	0.014 [50]	...	...	...
Eastern Florida aquifer	0.0085 [82]	...	...	...	...	...	...	...
Northern Michigan aquifer	0.095 [7]	...	...	...	...	...	...	...
Traverse City, MI, aquifer	0.007 to 0.024 [99] to [29]	0.067 [10]	...	0.004 to 0.014 [173] to [50]	...	...	...	...
Literature	0.0009 [730] to 0.069 [10]	0.025 [28] to 0.099 [7]	0.003 [228] to 0.116 [6]	0.0019 [365] to 0.0495 [14]	...	0.0019 [365] to 0.0866 [8]	0.0027 [258]	0.0007 [1058] to 0.0061 [114]

MTBE = Methyl *t*-butyl ether; ... = Data not available (reproduced from ASTM 1995).

Some excellent sources of half-life data are available. In particular, Howard et al. (1991) and, to a lesser extent, the Michigan Department of Environmental Quality (1997) have compiled useful collections of these data.

One final word of warning about half-life data typically encountered in the environmental scientific literature. Many half-lives reported in the literature are really measures of more than one major environmental loss mechanism. For example, for benzene a reported half-life may actually include loss from volatilization, oxidation, and photodegradation in addition to biodegradation. It is important to know the conditions under which the half-life determination was made in order to properly use this data for site assessment.

Leaching

Leaching is the process in which chemicals present in the unsaturated soil zone (the vadose zone) migrate downward under the influence of infiltrating rainwater and reach the water table [the saturated zone (the groundwater)]. In general, because of the potential to affect groundwater, which may serve as a drinking water supply, or hydraulically connected surface water bodies (i.e., ecological impacts), leaching must be carefully considered as a potentially significant fate and transport mechanism for all soil chemical impacts.

For hydrocarbons, the tendency to leach to groundwater closely follows the tendency to volatilize and sorb to soils. That is, those hydrocarbons that have relatively low MWs and potentially significant volatilization potential but only weakly sorb to soil tend to leach to groundwater under certain conditions. BTEX is a good example of these types of hydrocarbons. As important components of gasoline, BTEX have been found in groundwater at thousands of sites across the United States because of leaking USTs. In contrast, hydrocarbons that have higher MWs, are relatively nonvolatile, and strongly sorb to soils do not tend to leach to groundwater at environmentally significant rates. PAHs are a good example of these types of hydrocarbons.

Various models have been developed to quantify the rate of leaching for hydrocarbons (and other chemicals). The two most commonly used models are SESOIL and VLEACH. SESOIL is a complex, sophisticated model developed for the USEPA. The model allows for the introduction of multiple soil lithologies as distinct layers, conserves mass, and even subtracts the upward diffusion of chemical vapors due to volatilization from the overall leaching rate. VLEACH is less sophisticated than SESOIL but it is an appropriate choice for projects in which leaching is less of a concern.

Both SESOIL and VLEACH are run with computer programs. A Windows version of SESOIL has recently become available (see the USEPA internet web site: http://www.epa.gov/), which should reduce the effort required to set up and run the model. However, both models still require a significant investment of time for setting up, running, and interpreting results. A much simpler alternative to either SESOIL or VLEACH is provided by ASTM (1995). A volatilization factor (VF) equation is provided that has far fewer input requirements than either SESOIL or VLEACH and it can easily be set up and used in spreadsheet form. However, in simplifying these types of computations significant conservatism has been introduced, which is a general principle of fate and transport modeling. For instance, the ASTM VF equation assumes that a soil contaminant is already in contact with the water table and ignores any distance between the contaminant in the soil column and the water table. Nevertheless, for site assessments involving low soil concentrations of relatively immobile hydrocarbons and requiring some sort of quantitative leaching evaluation, this approach may be most appropriate.

Summary

Six important fate and transport mechanisms have been discussed for hydrocarbons released into the environment. Can some generalizations be made about these mechanisms, particularly the overall effects of more than one mechanism operating simultaneously? The answer is yes.

In general, there are two major types of hydrocarbon releases to the environment: surface releases and subsurface releases. Surface releases mostly occur at gasoline and/or diesel service stations during the fueling of various vehicles. Subsurface releases mostly occur from leaking USTs.

Of these two types of releases, subsurface releases are more problematic than surface releases because hydrocarbons in surface or shallow soils or in surface water will volatilize, oxidize, photodegrade, and biodegrade more readily than the same chemicals located at depth. In addition, hydrocarbons present at depth, especially the lower MW, more mobile chemicals (e.g., benzene), have a greater potential to affect groundwater by leaching processes than the same chemicals present at or near the site surface.

Another general aspect to fate and transport evaluations involving hydrocarbons, or any chemicals for that matter, is the choice of chemical-specific property values—that is, values for Henry's law constants, organic soil-water partition coefficients, and other related parameters. These parameters occur throughout every fate and transport model regardless of its level of complexity and whether it conserves mass. Although the sensitivity of these parameters on the model output results may vary from model to model, or even within the same model, the selection of values for these parameters is a critical part of the overall modeling exercise.

The major concern about selection of chemical-specific property values is best illustrated in Table 2.1.5, which represents a page from Mackay et al.'s important series (Mackay et al., 1993). This series is by far the most definitive work on property values for chemicals of environmental significance. The main feature that sets this work apart from all other similar efforts with one possible exception (Montgomery, 1996) is the compilation of multiple values for a given property and chemical. That multiple values generally exist for most parameters and most chemicals makes sense because these values are often determined by separate research groups working in parallel and unaware of each other's efforts. In addition, different groups use different methods to estimate chemical property values. Therefore, if one exhaustively searches the literature for these properties, as Mackay et al. have done, what one gets is a distribution of values for individual properties and chemicals.

As Table 2.1.5 shows, if one assumes equal weighting for each chemical property determination, regardless of the method used to obtain a particular value, one observes a wide variation of values. Depending on the sensitivity of a particular parameter(s) in a particular model, one may obtain

TABLE 2.1.5 Multiple Chemical Property Values for 1,1,1-TCA

16529	(quoted, Dilling, 1977; Suntio et al., 1988)
16663	(selected, Nathan, 1978)
16615	(quoted, Cowen and Baynes, 1980)
12797	(selected, Mills et al., 1982; Neely, 1982)
13373	(estimated as per Perry and Chilton, 1973; Arbuckle, 1983)
13330	(20°C, quoted, Verschueren, 1983)
12797	(20°C, quoted, McNally and Grob, 1984)
17768	(quoted, Kamlet et al., 1986)
16490	(quoted, Riddick et al., 1986)
16491	(calculated-Antoine eqn., Stephenson and Malanowski, 1987)
17023	(interpolated between two data points, Warner et al., 1987)
13300	(selected, Suntio et al., 1988)
16529, 960	(quoted, calculated-solvatochromic p. & UNIFAC, Banerjee et al., 1990)
13330	(20°C, selected, Gillham and Rao, 1990)
13149	(20°C, quoted from DIPPR, Tse et al., 1992)
20519	(30°C, quoted from DIPPR, Tse et al., 1992)

Henry's law constant (Pa m^3/mol)

1638	(calculated as $1/K_{AW}$, C_w/C_A, reported as exptl., Hine and Mookerjee, 1975; quoted, Nirmalakhandan and Speece, 1988)
3433	(20°C, McConnell et al., 1975; quoted, Jones et al., 1977/1978)
3495	(calculated as per McConnell et al., 1975 reported as exptl., Neely, 1976)
3473	(calculated-P/C, Neely, 1976)
2800	(exptl., Dilling, 1977; quoted, Suntio et al., 1988)
2975	(calculated, Dilling, 1977; quoted, Love Jr. and Eilers, 1982)
2025	(20°C, batch stripping, Mackay et al., 1979; quoted, Yurteri et al., 1987)

(Continued)

TABLE 2.1.5   Multiple Chemical Property Value (Continued)

1996	(concentration ratio, Leighton and Calo, 1981)
3039	(calculated-P/C, Mabey et al., 1982)
1520	(20°C, batch stripping, Munz and Roberts, 1982; quoted, Roberts and Dändliker, 1983; Roberts et al., 1985; Yurteri et al., 1987)
1824	(calculated-P/C, Thomas, 1982)
728.9	(quoted actual value from Kavanaugh and Trussell, 1980, Arbuckle, 1983)
606.9	(20°C, predicted-UNIFAC, Arbuckle, 1983; quoted, Yurteri et al., 1987)
1743	(20°C, EPICS, Lincoff and Gossett, 1983; Gossett, 1985; quoted, Yurteri et al., 1987)
1337	(20°C, EPICS, Lincoff and Gossett, 1984)
1358	(20°C, batch stripping, Lincoff and Gossett, 1984)
1652	(estimated-P/C, Lyman, 1985; Mackay, 1985)
1621	(estimated as per Hine and Mookerjee, 1975, Lyman, 1985)
1013	(estimated as per Cramer, 1980; Lyman, 1985)
498.5	(adsorption isotherm, Urano and Murata, 1985)
1735	(20°C, EPICS, Ashworth et al., 1986; quoted, Yurteri et al., 1987)
1360	(20°C, multiple equilibration, Munz and Roberts, 1986; quoted, Yurteri et al., 1987)
399.9	(20°C, calculated-P/C, McKone, 1987)
498.5, 413.4	(quoted, calculated-P/C, Warner et al., 1987)
3223	(quoted, Ryan et al., 1988)
1345	(20°C, EPICS, Gossett, 1987; quoted, Yurteri et al., 1987; Tse et al., 1992)
1572	(20°C, EPICS, Yurteri et al., 1987)
1735	(EPICS, Ashworth et al., 1988)
1743	(EPICS, quoted from Gossett, 1987; Suntio et al., 1988; Thoms and Lion, 1992)
2464	(calculated-P/C, Suntio et al., 1988)
1413	(EPICS, Gossett, 1987; quoted, Grathwohl, 1990)
1638, 430.8	(quoted, calculated-Quantitative Structure Activity Relationship (OSAR), Nirmalakhandan and Speece, 1988)
1317	(20–25°C and low ionic strength, quoted, Pankow and Rosen, 1988; Pankow, 1990)
3619	(selected, Jury et al., 1990)
1276	(20°C, Tse et al., 1992)
2026	(30°C, Tse et al., 1992)

Octanol/water partition coefficient, log K_{ow}

2.17	(Tute, 1971; quoted, Callahan et al., 1979; Schwarzenbach et al., 1983)
2.49	(Hansch and Leo, 1979; quoted, Iwase et al., 1985)
2.47	(shake flask-LSC, Banerjee et al., 1980; quoted, Karickhoff, 1981, 1985; Davies and Dobbs, 1984; Neely and Blau, 1985; Banerjee and Howard, 1988; Suntio et al., 1988; Olsen and Davis, 1990)
2.47	(shake flask-LSC, Veith et al., 1980)
2.51	(calculated-f const., Mabey et al., 1982)
2.18	(selected, Mills et al., 1982)
2.6	(selected, Neely, 1982)
2.47	(Veith and Kosian, 1982; quoted, Saito et al., 1992)
2.47	(quoted actual exptl. value from Banerjee et al., 1980; Arbuckle, 1983)
2.35	(calculated-UNIFAC with octanol and water mutual solubility not considered, Arbuckle, 1983)
2.29	(calculated-UNIFAC with octanol and water mutual solubility considered, Arbuckle, 1983)
2.49	(calculated, Hansch and Leo, 1985; quoted, Howard, 1990; Thoms and Lion, 1992)
2.49, 2.39	(quoted, calculated-hydrophobicity const., Iwase et al., 1985)
2.49, 2.39	(quoted, calculated-molar volume and solvatochromic p., Leahy, 1986)
2.49	(Abernethy et al., 1988)
2.49, 1.96	(quoted, calculated-UNIFAC, Banerjee and Howard, 1988)
2.47	(quoted, Isnard and Lambert, 1988, 1989)
2.49, 2.36	(quoted, calculated-molar volume and solvatochromic p., Kamlet et al., 1988)
2.48	(selected, Suntio et al., 1988)
2.49	(quoted from Hansch and Leo, 1979; Grathwohl, 1990)
2.48	(calculated, Müller and Klein, 1992)

Reproduced from Mackay et al. (1993).

widely varying results from the model by using the actual distribution of chemical property values, instead of the single-point values generally listed by regulatory agencies (Jeffrey, 1999). These factors become particularly important when one is conducting a probabilistic risk assessment, which are becoming increasingly popular and important in site assessment efforts. A probabilistic risk assessment mathematically mixes the statistical distributions of input parameters to give a distribution of risk results. This allows for the uncertainty inherent in any risk assessment to be quantitatively and, if desired, pictorially expressed. The risk manager (generally the regulatory agency) then decides at what percentile (distribution of the risk results) to manage chemical exposures due to site contaminants. These probabilistic methods allow for a more realistic, less conservative assessment of the actual risks posed by the presence of site contaminants.

REFERENCES

American Society for Testing and Materials (ASTM), *Standard Guide For Risk-Based Corrective Action Applied at Petroleum Release Sites*, Designation E 1739–95, American Society for Testing and Materials, West Conshohocken, PA, November, 1995.

ASTM, *ASTM Draft Guide for Remediation by Natural Attenuation at Petroleum Release Sites*, March 8, 1996.

CRC, *Handbook of Chemistry and Physics*, 71st ed., Lide, D. R., ed., CRC Press, Boca Raton, FL, 1992.

Dragun, *The Soil Chemistry Hazardous Materials*, J. Dragun, pub. Hazardous Materials Control Resources Institute, Silver Spring Maryland, 1988.

Howard et al., *Handbook of Environmental Degradation Rates*, Howard, P. H., Boethling, R. S., Jarvis, W. F., Meylan, W. M., and Michanlenko, E. M., eds., Lewis Publishers, Chelsea, MI, 1991.

Jeffrey, *Chemical Property Variability, Quantitative Uncertainty Analysis, and Consequences in Risk Assessment*, presented by D., Jeffrey (SECOR International Incorporated) at American Society of Testing and Materials (ASTM) Committee E-47 (Biological Effects and Environmental Fate) Conference, Seattle, Washington, April 21, 1999.

Kemblowski.

Lyman et al., *Handbook of Chemical Property Estimation Methods*, Lyman, W. J., Reehl, W. F., and Rosenblatt, D. H., eds., McGraw-Hill, Inc., New York, 1990.

Mackay et al., *Illustrated Handbook of Physical-Chemical Properties and Environmental Fate for Organic Chemicals; Volume III: Volatile Organic Chemicals*, eds., Mackay, D., Shiu, W. Y., and Ma, K. C., Lewis Publishers, Chelsea, MI, 1993.

MDEQ, *Operational Memorandum No. 10: Presentation of Tier 2 and Ier 3 Groundwater Modeling Evaluations*, Michigan Department of Environmental Quality, Underground Storage Tank Division, November 4, 1997.

Montgomery, *Groundwater Chemicals Desk Reference*, 2nd ed., Montgomery, J. H., eds., CRC Press, Boca Raton, FL, 1996.

Ney, Jr., *Where Did That Chemical Go? A Practical Guide to Chemical Fate and Transport in the Environment*, Ney, R. E. Jr., eds., Jr. Van Nostrand Reinhold, New York, 1990.

Schmidt and Zdeb., *Direct Measurement of Indoor Infiltration Through a Concrete Slab Using the USEPA Flux Chamber*, Schmidt, C. E., and Zdeb, T. F., eds., to be submitted for publication, 1999.

USEPA, "Measurement of Gaseous Emission Rates from Land Surfaces Using an Emission Isolation Flux Chamber, Users Guide," EPA Environmental Monitoring Systems Laboratory, Las Vegas, Nevada, EPA Contract No. 68-02-3889, Work Assignment No. 18, February 1986.

USEPA, *Superfund Exposure Assessment Manual*, U.S. Environmental Protection Agency, Office of Remedial Response, Washington, DC., EPA/540/1-88/001, April, 1988.

USEPA, *Soil Screening Guidence: User's Guide*, United States Environmental Protection Agency, Office of Solid Waste and Emergency Response, Washington DC., Publication 9355.4-23, July 1996.

USEPA, *Region IX Preliminary Remediation Goals (PRGs) 1998*, Memorandum from Smucker, S. J., USEPA Region IX, San Francisco, California, March, 1998.

Woodward and Hoffmann, *The Conservation of Orbital Symmetry*, Woodward, R. B., and Hoffmann, R., eds., Academic Press, New York, 1970.

CHAPTER 2
CLASSES OF CHEMICALS

SECTION 2.2

PESTICIDES AND POLYCHLORINATED BIPHENYLS

David Jeffrey

Dr. Jeffrey is senior risk assessment scientist with SECOR International, Inc. He performs and manages health risk assessments, as well as fate and transport evaluations in support of risk assessments.

2.2.1 INTRODUCTION

Pesticides and polychlorinated biphenyls (PCBs) comprise an important class of environmental chemicals/contaminants. The persistence (long half-lives), relatively high toxicity (especially carcinogenic potential), and common occurrence of many of these chemicals underscore their significance.

Pesticides, which include insecticides, fungicides, herbicides, and fumigants, comprise a structurally diverse group of compounds. As many as 18 separate subclasses of pesticides may be identified, as illustrated in Figure 2.2.1. For the purposes of this fate and transport discussion, these eighteen subclasses are collected into four groups, groups I, II, III, and IV.

Group I consists of four types of pesticides derived from natural products (i.e., from natural plant defense mechanisms): nicotine and derivatives, pyrethroids (e.g., Allethrin and Pyrethrin I), rotenone and derivatives, and derivatives of valeric acid (e.g., Fenvalerate).

Group I also includes bipyridilium compounds (e.g., Diquat and Paraquat), substituted amides (e.g., Propanil and Alachlor), substituted ureas [e.g., 3-(4-chlorophenyl)-1,1-dimethyl urea], pentachlorophenol (PCP), nitrophenols and nitroanalines [e.g., DNOC and Trifluralin (Treflan)], N-heterocycles [e.g., Atrazine and Metrabuzin (Sencor)], organomercury compounds [e.g., ethoxyethylmercuric chloride], alkylphosphate and alkylthiophosphate compounds [e.g., disulfotone, paraoxon, parathion, chlorpyrifos (Dursban), and malathion], alkylphosphonates [e.g., glyphosphate (Roundup)], carbamates and thiocarbamates [e.g., 4-(N'-methyl-2'-propynolamino)-3,5-dimethylphenyl-N-methyl carbamate and sodium N-methyldithiocarbamate], organoarsenic compounds [e.g., hydroxydimethylarsine oxide (cacodylic acid)], and chlorophenoxy compounds [e.g., 2,4,5-trichlorophenoxyacetic acid and esters (2,4,5-T) and 2,4-dichlorophenoxyacetic acid and esters (2,4-D)].

Group II consists of aromatic organochlorine pesticides (OCPs) that are substituted ethane derivatives [e.g., methoxychlor and dichlorodiphenyltrichloroethane (DDT)]. Group III consists of cyclic, aliphatic OCPs some of which are terpene-derived (e.g., endosulfone, aldrin, dieldrin, toxaphene, chlordane, and lindane). Group IV is comprises of acyclic, aliphatic halogenated compounds (e.g., methyl bromide, ethylene dibromide, ethylene dichloride, 1,2-dibromo-3-chloropropane (DBCP), 1,3-dichloro-propene, and 1,2,3-tribromopropane).

Group I contains those pesticides that have reactive functional groups that tend to decrease the persistence of these compounds in the environment, lower molecular weights, and higher water solubilities. Group II contains those pesticides that are less mobile (including volatilization potential), are more persistent, and tend to bioconcentrate or bioaccumulate. This is a useful classification scheme because it allows pesticides and PCBs to be discussed together; group II pesticides have a

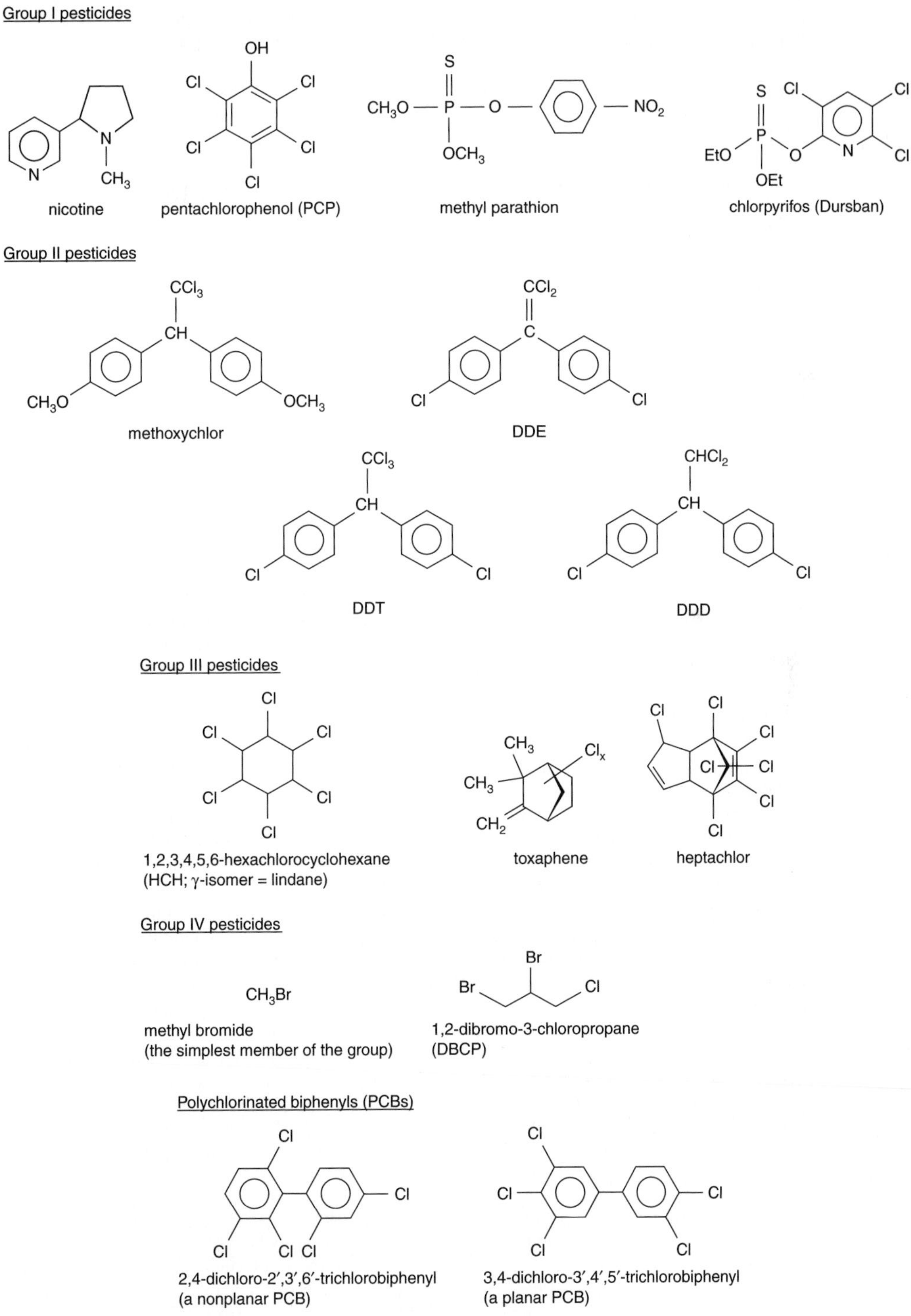

Group I pesticides

nicotine   pentachlorophenol (PCP)   methyl parathion   chlorpyrifos (Dursban)

Group II pesticides

methoxychlor   DDE

DDT   DDD

Group III pesticides

1,2,3,4,5,6-hexachlorocyclohexane   toxaphene   heptachlor
(HCH; γ-isomer = lindane)

Group IV pesticides

CH₃Br

methyl bromide   1,2-dibromo-3-chloropropane
(the simplest member of the group)   (DBCP)

Polychlorinated biphenyls (PCBs)

2,4-dichloro-2′,3′,6′-trichlorobiphenyl   3,4-dichloro-3′,4′,5′-trichlorobiphenyl
(a nonplanar PCB)   (a planar PCB)

FIGURE 2.2.1   Examples of pesticides and PCBs.

fate and transport profile very similar to PCBs. Group III pesticides have properties similar to group II with one important exception: group III pesticides tend to have significant volatilization potential. Group IV pesticides are halogenated or polyhalogenated compounds like groups II and III, but they are significantly more mobile, water soluble, and volatile, and they have low bioconcentration and bioaccumulation potential.

There are two subclasses of PCBs: planar or coplanar, and nonplanar PCBs, also illustrated in Figure 2.2.1.

There is too much structural diversity between the four pesticide groups, and even within certain large groups (e.g., group I), to be able to make general statements about differences in chemical properties versus differences in toxicities. Planar and nonplanar PCBs, on the other hand, have similar chemical but very different toxicological properties, as has recently been observed. In fact, the toxicities of some planar PCBs have been shown to approach that of dioxin (2,3,7,8-tetrachloro dibenzo-*p*-dioxin), which has by far the highest cancer slope factor of any chemical listed by the U.S. Environmental Protection Agency (USEPA) (2000). Although this section focuses on fate and transport behavior, which excludes toxicological differences, the distinction between planar and nonplanar PCBs is still important, especially given the overall implications of these toxicity differences for site assessment and the recent great attention paid to planar PCBs.

The following sections discuss the important fate and transport mechanisms for the various group I, II, and III pesticides and PCBs.

2.2.2 FATE AND TRANSPORT MECHANISMS

Once pesticides or PCBs have been released into the environment, via direct pesticide use around home gardens and commercial agricultural areas and from various electrical components (e.g., transformers and capacitors) that may contain PCBs as components of heat sink fluids (askarel), as well as other potential sources, nine main fate and transport mechanisms may be operative, depending on the specific nature of the pesticides or PCBs and the type of release. The nine mechanisms are sorption, for some pesticides and all PCBs; uptake and bioconcentration/bioaccumulation in biota (i.e., food chain mechanisms, for some pesticides and all PCBs); reduction, for some pesticides and all PCBs; hydrolysis, for some pesticides; surface runoff, for some pesticides; photodegradation, for PCBs and group II pesticides; volatilization, for group III and IV pesticides; leaching, for some pesticides; and intermolecular and intramolecular reactivity, for PCP and some PCBs.

Note that the difference between this list of fate and transport mechanisms and the list for hydrocarbons presented in the section "Hydrocarbons" lies in the replacement of oxidation by reduction, the lack of biodegradation, the introduction of surface runoff and biota uptake, the introduction of hydrolysis, and the introduction of intermolecular self-reactivity. This is mostly consistent with the common understanding of these chemicals as highly persistent (low degradation potential of some pesticides and all PCBs), relatively mobile (leaching, runoff, and volatilization potential of some pesticides), and posing a potential threat to the food chain (biota uptake of some pesticides and all PCBs).

Sorption

Group I and IV Pesticides. Because environmental sorption to soils or sediments usually involves competing solubilization by a nearby aqueous phase, and most of the group I and IV pesticides have high affinities for water, they do not tend to sorb very strongly. Such weak sorption potential (and high water solubility) means that these chemicals are prone to leaching and surface runoff.

Group II and III Pesticides and PCBs. In contrast to the group I pesticides, the group II pesticides and PCBs represent some of the most highly sorbing chemicals known to occur in the environment. This means that these chemicals are rarely a threat to groundwater via leaching processes, especially because most environmental releases of these types of chemicals are surface releases. Surface runoff is also rarely significant. However, because of the relatively high toxicity and persistence of these chemicals, they are often included in a groundwater impact (i.e., leaching) evaluation, especially if there is a potential drinking water opportunity at or near a site containing these chemicals.

Bioconcentration/Bioaccumulation

Although members of the group I and IV pesticides have not in general been implicated in significant biota uptake/bioconcentration/food chain exposures for humans or plant and animal life, this cannot be claimed for members of the group II and III pesticides and PCBs. Pesticides such as DDT, aldrin, chlordane, toxaphene, and others have been shown to be easily taken up and bioconcentrated or bioaccumulated by a variety of biota. In fact, the exposure of raptors to DDT via food chain mechanisms involving soil-to-plant-to-small mammal (prey)-to-raptor (predator) pathways, resulting in both severe mutations in raptor hatchlings and in egg shell thinning, was a classic study that turned much of the current public attention toward the environment.

Because there is so much confusion about the three terms bioconcentration, bioaccumulation, and biomagnification, with some terms used interchangeably (and incorrectly), these terms need to be defined. Bioconcentration refers to partitioning of a chemical in water, soil, sediment, or plant media between that medium and the tissue of a plant or animal exposed to the chemical. Fish bioconcentration studies are probably most common. Bioaccumulation is a conceptual term pertaining to bioconcentration and related processes. Biomagnification refers to a food chain or food web of successive food source-ingester (or prey-predator) paths that relates the chemical concentration at the lowest food source level (e.g., soil chemical concentration) to the concentration in a higher predator (e.g., a hawk ingesting ground squirrels). Biomagnification may refer to an aquatic or terrestrial food chain or some combination of the two.

Bioconcentration is defined by a chemical's bioconcentration factor (BCF), which for aquatic organisms may be represented as:

$$BCF = C_b/C_{sw} \tag{2.2.1}$$

where C_b = chemical concentration in aquatic biota (usually fish) exposed to surface water chemicals, and
C_{sw} = chemical concentration in surface water.

Those compounds that tend to bioconcentrate or bioaccumulate are usually highly lipophilic (with very high K_{oc} and K_{ow} values), and the fatty tissues of a predator ingestor are often capable of storing very high concentrations of these chemicals. In this way, these ingestors can accumulate very high concentrations of these chemicals in much the same way a relatively small sponge draws up liquid spilled over a large area.

The environmental literature contains compilations of BCFs. A good source of these data is Mackay et al. (e.g., 1992, 1993). Mackay et al. (1992) lists log BCFs for PCBs, and Mackay et al. (1993) for some group IV pesticides. These BCFs are not only chemical specific but also species specific. Table 2.2.1 provides some BCF values for pesticides and PCBs as well as for chemicals known to have low BCF potentials (e.g., volatile organic chemicals [VOCs]). A BCF value of 100 is considered a rule-of-thumb threshold value by some. However, other sources (ATSDR, 1997; Great Lakes, 1999) suggest a cutoff value closer to 1000. As Table 2.2.1 shows, pesticides belonging to class II and III have BCFs exceeding 1000, but those in class I do not, consistent with the known tendencies of these chemicals.

Reduction

OCPs (i.e., class II and III pesticides), class IV pesticides, and PCBs may undergo reduction under certain conditions; the carbon-halogen bond(s) in these compounds is broken and a carbon-hydrogen bond is made in its place (i.e., a hydrogenolysis reaction). However, class II and III pesticides and PCBs, being relatively immobile with strong soil sorption tendencies, tend to be found close to the surface of affected sites, near their original release point. Class IV pesticides, being significantly more mobile, may

TABLE 2.2.1   BCFs of Pesticides

Pesticide	Group	BCF (L/kg)
Aldrin	III	28
(Benzene)	VOC	5.2
Chlordane	III	**14000**
DDT	II	**54000**
Dieldrin	III	**4760**
Heptachlor	III	**15700**
Heptachlor epoxide	III	**14400**
Lindane	III	130
Malathion	I	0
Methyl parathion	I	45
PCP	I	770
Toxaphene	III	**13100**

Bold values exceed 1000, the recommended (ATSDR, 1997; Great Lakes, 1999) cutoff value for chemicals that may significantly bioaccumulate. VOC, volatile organic chemical. All BCFs from USEPA (1986).

infiltrate the soil column to greater depths. One finds exceptions to this general scenario with, for example, buried transformers containing PCBs.

Because of these factors, class II and III pesticides and PCBs tend to be present in oxygen-rich environments where reduction is less likely. However, class IV pesticides may be present in oxygen-deficient environments where reduction may occur. Reduction may occur abiotically or biotically. Abiotic reduction occurs for many group IV pesticides because of their polyhalogenated composition, which tends to increase the demand for electrons (i.e., increase the reduction potential). Although the deeper subsurface environments are generally oxygen deficient, other reductants (e.g., sulfide, ferrous species, nitrite) are usually present. And biotic reduction is a common environmental fate of such structures because of the general availability of suitable microorganisms. Such reduction chemistry may either detoxify or increase the toxic potential of these chemicals depending on their exact structure and reactivity. Complete reductive dehalogenation to give hydrocarbon products is detoxifying, but partial reduction may lead to more mobile, more volatile organochlorine or oganobromine products that pose greater inhalation risks.

Hydrolysis

Hydrolysis is a major mode of the environmental fate and reactivity of some of the class I pesticides. Organophosphates, thiophosphates, and phosphonates react with water to give alcohols (or thiols) and an acidic organophosphorous residue. In most cases this is a detoxifying mechanism, as the results of hydrolysis, free phosphoric acid esters, are no longer capable of phosphorylating key metabolic biomolecules (e.g., nucleotide bases and certain amino acid side chains of proteins and enzymes). However, there are exceptions to this general rule. Figure 2.2.2 shows an example of such an exception as reported by Lydy et al. (1990), as parathion, with an oral rat median lethal dose (LD_{50}) of 4–13 mg/kg, is hydrolyzed to paraoxon, having a lower LD_{50} (and therefore a higher toxicity). Mabey and Mill (1978) give numerous examples of the effect of structure on hydrolysis rate for phosphorous-derived pesticides.

Surface Runoff

Surface runoff is an important transport mechanism for group I and, to a lesser extent, group IV pesticides. This behavior tracks directly with the low soil sorption and high water solubility of chemicals belonging to these classes. Stormwater runoff will contact and solubilize these chemicals, transporting

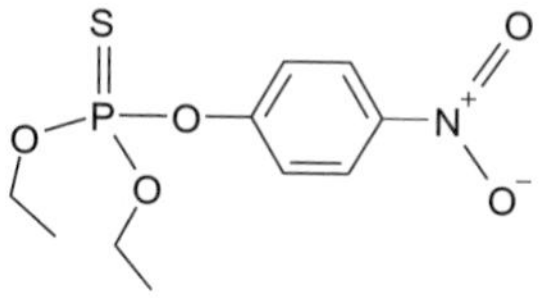

Parathion
a thiophosphonate ester (of phosphoric acid)

$LD_{50}^{\text{oral rat}} = 4-13$ mg/kg

hydrolysis

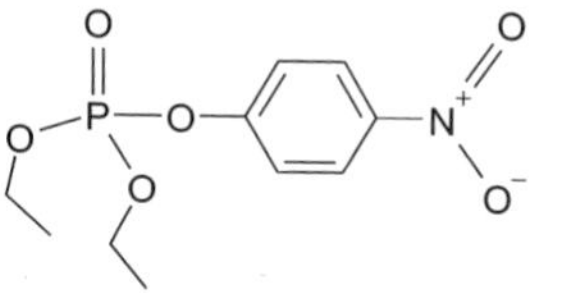

Paraoxon

$LD_{50}^{\text{oral rat}} = 2$ mg/kg

Lydy *et al.*, 1990:

$t_{1/2,\text{hyd}} = 24$ hrs
hydrolysis rate largely independent of pH but optimum pH = 6
consistent with hydrolysis mechanism involving either acid or base catalysis

However, strong temperature dependence:

pH	T(°C)	EC$_{50}$ (μg/L)
4	10	77
	20	5.8
	30	0.6
6	10	60
	20	6.9
	30	3.2

Arrenhius: $k = Ae^{-E^*/RT}$

k = rate constant
A = Arrenhius preexponential factor
E = activation energy
R = gas constant
T = temperature

FIGURE 2.2.2 Hydrolysis of a pesticide and increase in toxicity.

them to stormwater discharge locations. These discharge locations can be drinking water reservoirs, lakes, rivers, streams, or marine environments. Such discharges represent a potentially significant source of exposure to chemicals of these types for both human and ecological receptors.

Although the type of runoff described above, solubility driven, is probably most significant in terms of the total loadings to the environment, one must recognize that a completely separate mechanism exists for surficial runoff of chemicals in general, especially those that have properties very different from the group I and IV pesticides. Under certain severe storm conditions, and depending on the type of soil and terrain involved, soil particles can be moved significant distances by storm waters. Under this mechanism, chemicals that tend to sorb strongly to soils and have low water solubilities, that is the group II and III pesticides and PCBs, tend to migrate.However, the distances traveled under this sorption-driven mechanism are generally less than for the solubility-driven mechanism. On the other hand, the toxicities and persistence of some of these group II and III pesticides and PCBs (e.g., DDT) is such that even low concentrations reaching a surface water body or ingestable plants (i.e., food chain exposures via plant uptake) can be cause for concern.

Photodegradation

Photodegradation involves the initial absorption of energy in the form of either ultraviolet or visible light, followed by various intra- or intermolecular modes of reactivity. The tendency of a chemical to undergo photodegradation depends on three main factors: availability of light of the appropriate wavelengths, intensity of light of appropriate wavelengths, and molecular structure of the chemical.

If all three factors are met, photodegradation can be the most significant environmental fate loss mechanism or sink acting on an environmental contaminant, with half-lives on the order of days or even, in some cases, hours.

In terms of the first two criteria listed above, the chemical must be present either in the atmosphere, on the soil or sediment surface, or in surface water; that is, chemicals that are present in the subsurface (soil or groundwater) will not tend to undergo photodegradation because of the absence of light of appropriate wavelengths or because of the low intensity of light of appropriate wavelengths. Therefore, pesticides in group I, and to a lesser extent group IV, may initially be prone to photodegradation from surface releases but with time may leach to subsurface locations where they will be protected. On the other hand, photodegradation is almost always important for group II and III pesticides and PCBs,as these chemicals tend to be released at or close to the surface and do not tend to leach to the subsurface at significant rates.

Regarding the third criterion listed above, the structure of the chemical, most chemicals with benzene ring substructures tend to easily and efficiently absorb light energy, whereas those that do not have such substructures do not tend to absorb much light. Table 2.1.3 lists various chemicals and substructures along with their λ_{max} and ϵ values. Figure 2.2.3 shows the distribution of electromagnetic wavelengths that strike the earth's surface. λ_{max} is the wavelength at which the compound absorbs the maximum amount of light. ϵ is the molar extinction coefficient at λ_{max} and is a quantitative measure of the maximum efficiency of light absorption for a chemical. In general, the higher the values of λ_{max} and ϵ, the more prone that chemical (or substructure) is to light absorption and therefore photodegradation. Without a benzene ring substructure—that is without some aromatic character—a molecule will not tend to absorb much available light.

Some of the group I and all of the group II pesticides (e.g., DDT), and all PCBs, have benzene ring substructures. As a result, these pesticides and PCBs tend to be rapidly photodegraded under many typical environmental conditions, especially given their tendency to be close to the surface, as discussed previously. Table 2.2.2 lists air and surface water half-lives for some group II pesticides and some PCBs. Based on molecular structure, group III and IV pesticides tend to be less readily photodegraded than members of groups I and II.

To summarize the effects of both environmental location (i.e., the tendency to be present at or near surface environments) and structure, group II pesticides and PCBs tend to be most prone to photodegradation, with group IV pesticides least prone, and groups I and III somewhere in between.

TABLE 2.2.2   Photodegradation Half-Lives for Pesticides and PCBs

Chemical	Half-life (days)	Medium	Source
Pentachlorobiphenyls	0.6–1.4	Air	Mackay et al. (1992)
DDT	56–110	Surface water	USEPA (1986)
PCBs	58	Air	USEPA (1986)
	2–12.9	Surface water	USEPA (1986)

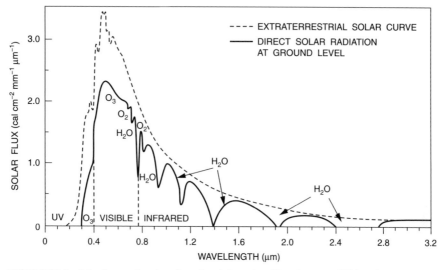

FIGURE 2.2.3   Solar flux as a function of wavelength (reproduced from Larson and Weber 1994).

Predicting the nature of the products of photodegradation can be very difficult and inexact. When a molecule efficiently absorbs light, it is raised to an electronically excited state, where electrons previously occupying ground state molecular orbitals are promoted to higher lying orbitals. In such a state, any organic molecule with a reasonably complex structure can rearrange, lose certain molecular fragments, or undergo other transformations. This is why predictions of the photoproducts of such processes is so difficult.

However, one main mode of photoreactivity and transformation is photooxidation. Many molecules in an excited state react with molecular oxygen to ultimately give photoproducts corresponding to oxidation of the original molecule. Such processes tend to convert some of the less mobile pesticides and PCBs to more soluble, mobile compounds. This, in general, has a detoxifying effect with respect to potential exposures to various pesticides and PCBs. In these more soluble forms, these compounds tend to bioconcentrate less, so that food chain exposures, which can be important for these types of chemicals, become less significant.

Volatilization

Although not generally known for their volatilization potential, some pesticides do volatilize from soil, surface water, or groundwater at environmentally significant rates. Some group III pesticides and all group IV pesticides have moderate volatilization potentials. In fact, heptachlor, a group III pesticide, has a Henry's constant of 8×10^{-4} atm-m^3/mol (USEPA, 1996), 80 times above USEPA Region 9's rule-of-thumb threshold of 1×10^{-5} atm-m^3/mol (USEPA, 1998). Table 2.2.3 lists some common pesticides and PCBs, along with their Henry constants and an indication of whether USEPA (1996) considers these chemicals to have significant volatilization potentials.

As Table 2.2.3 shows, there are exceptions to the 1×10^{-5} atm-m^3/mol rule-of-thumb for Henry constant values. dichlorodiphenyldichloroethene (DDE) and methoxychlor have H values greater than this rule-of-thumb but are not listed by USEPA as having significant volatilization potential, although their H values are only two times above the cutoff value. Chemicals that have H values at or below the cutoff but still listed as volatile include dieldrin, heptachlor epoxide, α-BHC, and toxaphene. But again, these four chemicals have H values at or just below the cutoff value, so it is not too surprising that they still may be considered to have significant volatilization potential. Pesticides in group IV

TABLE 2.2.3   Henry Constant Values for Pesticides and PCBs

	Henry's law constant (atm-m^3/mol)	Listed as volatile?
Pesticide		
Group II		
DDD	4.E-06	No
DDE	**2.E-05**	No
DDT	8.E-06	No
Methoxychlor	**2.E-05**	No
Group III		
Aldrin	**2.E-04**	Yes
Chlordane	**5.E-05**	Yes
Dieldrin	1.E-05	Yes
Endosulfan	1.E-05	No
Endrin	7.E-06	No
Heptachlor	**8.E-04**	Yes
Heptachlor epoxide	9.E-06	Yes
α-HCH (α-BHC)	1.E-05	Yes
β-HCH (β-BHC)	7.E-07	No
γ-HCH (lindane)	1.E-05	No
Toxaphene	6.E-06	Yes
Group IV		
Chlorodibromomethane	**8.E-04**	Yes
1,2-Dichloropropane	**3.E-03**	Yes
1,2-Dichloropropene	**2.E-02**	Yes
Methyl bromide	**6.E-03**	Yes
PCBs		
General PCBs	—	No

Values in bold type exceed 1×10^{-5}, the recommended (USEPA, 1998) cutoff value for chemicals considered volatile. Data are from USEPA (1996).

have the greatest volatilization potentials based on their H values. This is not surprising, because these chemicals tend to be lower molecular weight, simpler halogenated structures.

Leaching

Pesticides in groups I and IV have potentially significant leaching potential because of their generally high water solubility and their low soil sorption tendencies. These chemicals may leach from surface soils or sediments and reach surface waters during stormwater runoff events. Alternatively, group I or IV pesticides in surface or subsurface soil may leach to groundwater. Some of these chemicals—for example, the organophosphates and phosphonates—may undergo hydrolysis more rapidly than leaching to give the corresponding phosphorous acids. The soil sorption potential of these acid hydrolysis products then depends strongly on the soil pH. In general, most native soils are slightly alkaline. Because many of the phosphorous acid hydrolysis products are strong acids, these chemicals exist predominantly in the anionic charged form, which should increase their leaching potential from alkaline (i.e., anionic) soils.

Important Environmental Transformations

There are two significant modes of reactivity for group II pesticides and for PCBs: DDT →DDE and PCBs → polychlorodibenzofurans (PCDFs). These two types of environmental transformations are discussed in the following sections.

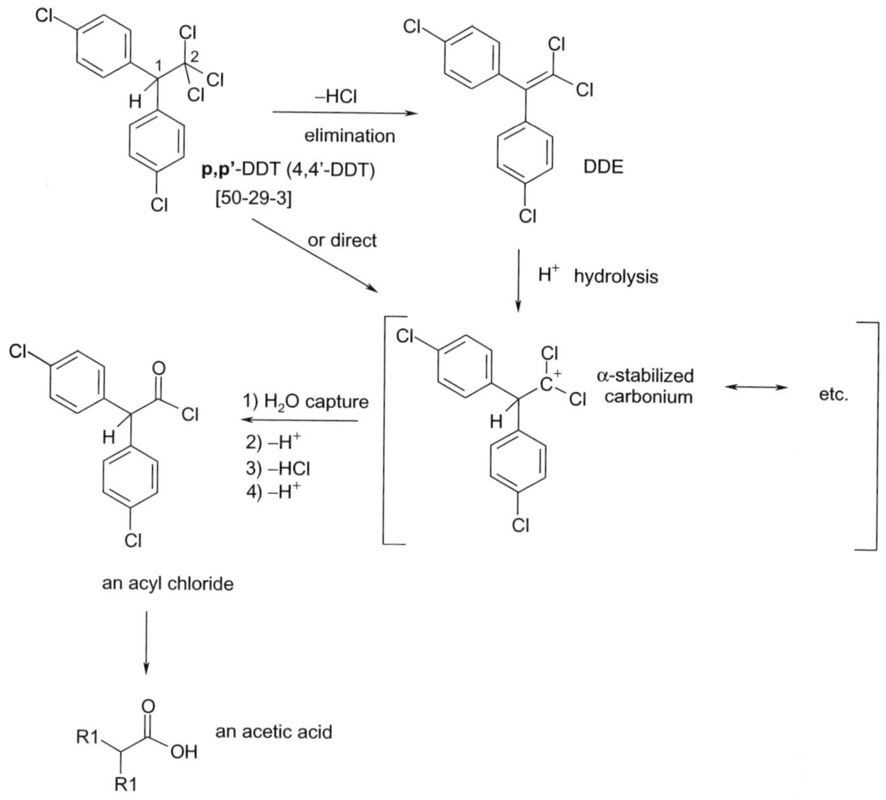

p,p'-DDT (4,4'-DDT)
[50-29-3]

−HCl
elimination

DDE

or direct

H⁺ hydrolysis

α-stabilized
carbonium

etc.

1) H₂O capture
2) −H⁺
3) −HCl
4) −H⁺

an acyl chloride

an acetic acid

FIGURE 2.2.4   Transformation of DDT to DDE.

Formation of DDE from DDT.   Figure 2.2.4 shows the transformation of DDT to DDE. This type of transformation (a dehydrochlorination) probably occurs via a bimolecular elimination reaction mechanism or an E_2 process. This means that the rate of the reaction is a function of both the concentration of DDT and the concentration of hydroxide ion. As the concentration of hydroxide ion is directly related to the pH of the environment, this transformation occurs more rapidly at higher pH values. For instance, at pH 7 (neutral conditions), the half-life for hydrolytic degradation in water at 27°C is approximately 5600 days, whereas the half-life at pH 9 is only 100 days (Larson and Weber, 1994). Another possible mechanism for this transformation involves redox steps (Quirke et al., 1979 as cited in Schwarzenbach et al., 1993), but the strong pH dependence of the transformation (Larson and Weber, 1994) suggests that the E_2 mechanism is dominant.

USEPA (1999) lists the same value for the oral cancer slope factor (0.34 reciprocal mg.kg^{-1}.day^{-1}) for both DDT and DDE. Therefore, the transformation of DDT to DDE can not be considered a detoxifying mechanism.

Formation of Polychlorinated Dibenzofurans from PCBs.   Figure 2.2.5 shows the transformation of PCBs to PCDFs. The chemistry involved is nucleophilic aromatic substitution by water (to replace a benzene ring chlorine atom with a hydroxyl group) followed by intramolecular oxidation (to close the third ring, the new furan ring). Alternatively, the sequence of these steps may be reversed. Or, for PCBs that lack chlorine atoms at the ortho ring positions, oxidation may be responsible for both the initial introduction of oxygen and the subsequent ring closure process.

It has been known for some time that fires involving electrical components (e.g., transformers, capacitors) can result in formation of PCDFs, which are formed from the PCBs contained in the

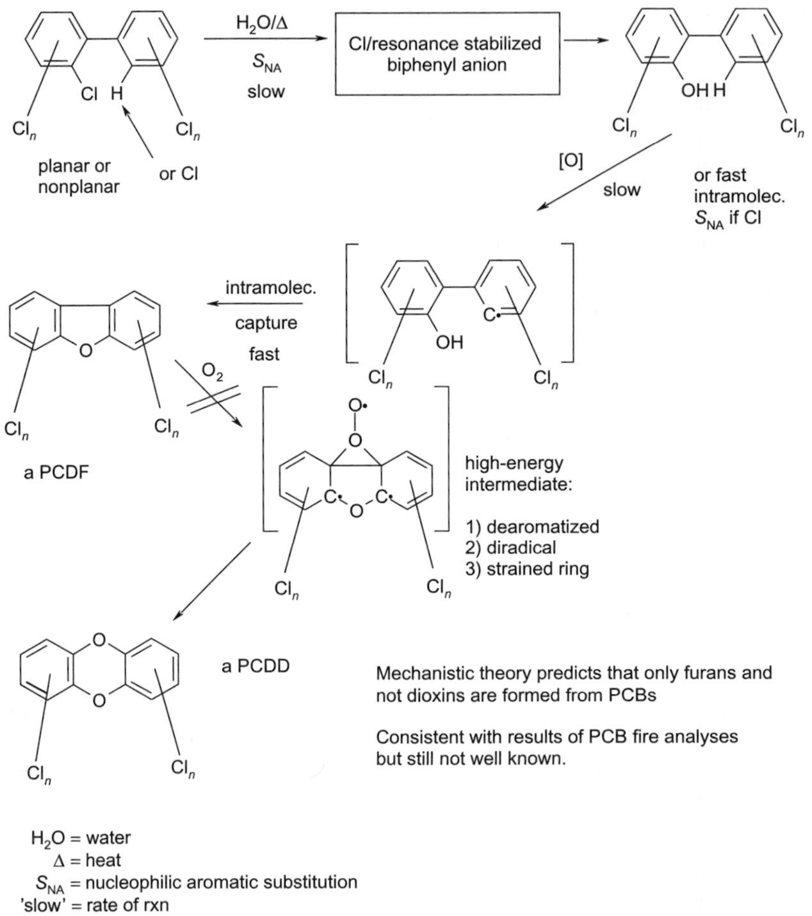

FIGURE 2.2.5   Transformation of PCBs to PCDFs.

aroclor mixtures used as liquid heat sinks to dissipate high temperatures resulting from high electrical currents (Milby et al., 1985). These aroclor liquids (e.g., aroclor 1260) contain mixtures of PCBs in a high-boiling mineral oil often referred to as askarel. Interestingly, however, and contrary to popular assumptions about this chemistry, polychlorinated dibenzodioxins (PCDDs) generally are not formed under these conditions (des Rosiers and Lee, 1986). As Figure 2.2.5 shows, formation of PCDDs from PCBs requires formation of a very high energy intermediate. Various papers have been published identifying safe worker reentry levels [i.e., surface (wipe sample) concentrations of PCDFs] after some of the larger PCB fires that have occurred over the years (Kim and Hawley, 1985; Wade and Woodyard, 1987; NIOSH, 1986).

In general, formation of PCDFs from PCBs increases the toxicity of an environmental system (Milby et al. 1985). For instance, the maximum oral cancer slope factor for nonplanar PCBs is 2 reciprocal $mg.kg^{-1}.day^{-1}$ (USEPA, 2000), whereas some PCDF congeners have cancer potencies as high as one-tenth that of 2,3,7,8-tetrachlorodibenzodioxin (2,3,7,8-TCDD, or Dioxin) or 15,000 reciprocal $mg.kg^{-1}.day^{-1}$. So-called coplanar PCBs already have much of the potency of PCDFs and tend to be less readily transformed to PCDFs. In general, lack of chlorine atom substitution at the ortho ring positions of PCB stends to result in coplanar PCBs, whereas the presence of chlorine atoms at these positions gives nonplanar PCBs.

REFERENCES

ATSDR, *U.S. Public Health Service Agency for Toxic Substances and Disease Registry's Toxicological Profiles on CD-ROM*, CRC Press, Boca Raton, FL, 1997.

des Rosiers, and Lee, "PCB Fires: Correlation of Chlorobenzene Isomer and PCB Homolog Contents of PCB Fluids with PCDD and PCDF Contents of Soot," *Chemosphere*, 15(9–12): p. 1313, 1986.

Great Lakes, U.S. Environmental Protection Agency 40 CFR Parts, 9, 122, 123, 131, and 132, Final Water Quality Guidance for Great Lakes System, 1999.

Kim and Hawley, *Re-entry Guidelines, Binghampton State Office Building*, Bureau of Toxic Substances Assessment, Division of Environmental Health Assessment, New York State Department of Health, Albany, New York, July 1985.

Larson and Weber, *Reaction Mechanisms in Environmental Organic Chemistry*, Larson, R. A., and Weber, E. J., eds., Lewis Publishers, Boca Raton, FL, 1994.

Lydy et al., *Influence of pH, Temperature, and Sediment Type on the Toxicity, Accumulation, and Degradation of Parathion in Aquatic Systems*, Lydy, M. J., Lohner, T. W., and Fisher, S. W., eds., Aquatic Tox, 17: 27–44, 1990.

Mabey and Mill, "Critical Review of Hydrolysis of Organic Compounds in Water Under Environmental Conditions," Mabey, W., and Mill, T., eds., *J. Phys. Chem. Ref. Data*, 7(2): 383–415, 1978.

Mackay et al., *Illustrated Handbook of Physical-Chemical Properties and Environmental Fate for Organic Chemicals; Volume I: Monoaromatic Hydrocarbons, Chlorobenzenes, and PCBs*, eds., Mackay, D., Shiu, W. Y., and Ma, K. C., Lewis Publishers, Chelsea, MI, 1992.

Mackay et al., *Illustrated Handbook of Physical-Chemical Properties and Environmental Fate for Organic Chemicals; Volume III: Volatile Organic Chemicals*, Mackay, D., Shiu, W. Y., and Ma, K. C., eds., Lewis Publishers, Chelsea, MI, 1993.

Milby et al., "PCB-Containing Transformer Fires: Decontamination Guidelines Based on Health Considerations," *J. Occup. Med.*, 27(5): 351, 1985.

NIOSH, *Polychlorinated Biphenyls (PCBs): Potential Health Hazards from Electrical Fires or Failures*, National Institute of Occupational Safety and Health Current Intelligence Bulletin 45, February 24, 1986.

Quirke et al., "The Degradation of DDT and its Degradative Products by Reduced Iron (II) Porphyrins and Ammonia," Quirke, J. M. E., Marci, A. S. M., and Eglington, G., eds., *Chemosphere*, 3: 151–155, 1979.

Schwarzenbach et al., *Environmental Organic Chemistry*, Schwarzenbach, R. P., Gschwend, P. M., and Imboden, D. M., eds., Wiley, New York, 1993.

USEPA, *Soil Screening Guidance: User's Guide*, United States Environmental Protection Agency, Office of Solid Waste and Emergency Response, Washington DC., Publication 9355.4-23, July 1996.

USEPA, *Region IX Preliminary Remediation Goals (PRGs) 1998*, Memorandum from Smucker, S. J., USEPA Region IX, San Francisco, California, March, 1998.

USEPA, *Integrated Risk Information System*, U.S. Environmental Protection Agency on-line database: http://www.epa.gov/iris/subst/index.html, 2000.

Wade and Woodyard, Sampling and Decontamination Methods for Buildings Contaminated with Polychlorinated Dibenzodioxins. Chapter 30 in Solving Hazardous Waste Problems: Learning from Dioxins. American Chemical Society (ACS) Symposium Series 338, Exner, J. H., eds., 1987.

CHAPTER 2
CLASSES OF CHEMICALS

SECTION 2.3

CHLORINATED SOLVENTS

Daryl Doyle

Dr. Doyle is professor of chemistry, Kettering University, Flint, Michigan. He has been involved in research, including the analysis of chemical by-products formed from the laser cutting of polymers and the use of lasers to surface prepare polymers for adhesive bonding. Recently he has assisted corporations in the preparation of ISO 14000 certification.

Chlorinated solvents are used extensively as solvents in a variety of materials such as paints, adhesives, solvents for washing and cleaning parts, etc. Because methylchloride (CH_3Cl) is a gas at room temperature (boiling point, $-24°C$), there are three single-carbon chlorinated solvents: dichloromethane (CH_2Cl_2), chloroform ($CHCl_3$), and carbon tetrachloride (CCl_4).

Dichloromethane or methylene chloride is a suspected human carcinogen. It is nonflammable but produces phogene gas when heated excessively. It has a median lethal dose (LD_{50}) of 0.5–5.0 g/kg and a TLV (Threshold Limit Value)-TWA (Time Weighted Average) of 50 ppm. The boiling point of dichloromethane is 104°F (40°C). It is more dense than water (specific gravity, 1.33) and is insoluble (1.3 g/100 mL of water). It is considered rather volatile with a vapor pressure of 350 mm Hg and a vapor density of 2.9 (vapor density of air, 1.0). Dichloromethane has a National Fire Protection Association (NFPA) rating of health = 2, flammability = 1, and reactivity = 0 (Pohanish and Greene, 1996, pp. 515–516; Lewis, 1997, p. 753; International Chemical Safety Card, ICSC: 0058, dichloromethane, Commission of the European Communities; IPCS, CEC, 1993).

Chloroform or trichloromethane is a potential human carcinogen. It is a noncombustible liquid but produces phogene gas when heated excessively. It has an LD_{50} of 0.5–5.0 g/kg and a TLV-TWA of 10 ppm. The boiling point is 142°F (61°C). It is more dense than water (specific gravity, 1.48) and is insoluble (0.8 g/100 mL of water). It has a vapor pressure of 160 mm Hg and a vapor density of 4.1. Chloroform has a NFPA rating of health = 2, flammability = 0, and reactivity = 0 (Pohanish and Greene, 1996, pp. 357–358; Lewis, 1997, p. 267; International Chemical Safety Card; ICSC: 0027 chloroform, Commission of the European Communities; IPCS, CES, 1993).

Carbon tetrachloride is considered to be a potential human carcinogen. It is a noncombustible liquid. It has an LD_{50} of 0.5–5.0 g/kg with a TLV-TWA of 5 ppm. The boiling point is 170°F (77°C). It is more dense than water (specific gravity, 1.59) and is insoluble (0.1 g/100 mL of water). It has a vapor pressure of 91 mm Hg and a vapor density of 5.3. Carbon tetrachloride has a NFPA rating of health = 3, flammability = 0, and reactivity = 0 (Pohanish and Greene, 1996, pp. 337–338; Lewis, 1997, p. 235; International Chemical Safety Card, ICSC: 0024 carbon tetrachloride, Commission of European Communities; IPCS, CES, 1993).

In general all three single-carbon chlorinated solvents pose a significant health risk. All three are more dense than water and are insoluble; thus, they are expected to sink to the bottom of a waterway should a spill occur. In addition, the vapor density is considerably greater than that of air; thus, ventilation systems should be used that extract fumes from below instead of from above.

The two-carbon chlorinated solvents include 1,1-dichlorethane, 1,2-dichloroethane, 1,1,1-tri-chloroethane, 1,1,2-trichloroethane, 1,1,2,2-tetrachloroethane, 1,1-dichloroethylene, 1,2-dichloro-ethylene, and trichloroethylene.

1,1-Dichloroethane is considered to be a flammable liquid. Its explosion limits are 5.4–11.4 per-cent with a flash point of 2°F (-17°C). It has an LD_{50} of 0.5–5 g/kg and a TLV-TWA of 100 ppm (405 mg/m^3). The boiling point of 1,1-dichloroethane is 135°F (57°C). It is more dense than water (specific gravity, 1.17) and is insoluble (0.6 g/100 mL of water). It has a vapor pressure of 182 mm Hg and a vapor density of 3.42. 1,1-Dichloroethane has a NFPA rating of health = 2, flammabil-ity = 3, and reactivity = 0 (Pohanish and Greene, 1996, pp. 505–507; International Chemical Safety Card, ICSC: 0249, 1,1-dichloroethane, Commission of the European Communities: IPCS, CEC, 1993).

1,2-Dichloroethane is a flammable liquid. Its explosion limits are 6.2–16 percent with a flash point of 55°F (13°C). It is more dense than water (specific gravity, 1.23) and is insoluble (0.87 g/100 mL of water). It has a vapor density of 3.42. 1,2-Dichloroethane has a NFPA rating of health = 2, flammabil-ity = 3, and reactivity = 0 (J. T. Baker MSDS (Material Safety Data Sheet) Number D2440, effective date 09/08/97; International Chemical Safety Card; ICSC: 0250 1,2-dichloroethane, Commission of the European Communities; IPCS, CES, 1993).

1,1,1-Trichloroethane is considered to be a chemical mutagen. It is a combustible liquid with explosive limits of 8–16 percent. It has an LD_{50} of 5–15 mg/kg and a TLV-TWA of 350 ppm or 1910 mg/m^3. The boiling point is 165°F (74°C). It is more dense than water (specific gravity, 1.34) and is insoluble in water. It has a vapor pressure of 100 mm Hg and a vapor density of 4.6. 1,1,1-Trichloroethane has a NFPA rating of health = 2, flammability = 1, and reactivity = 0 (Pohanish and Greene, 1996, pp. 1565–1567; International Chemical Safety Card, ICSC: 0079, 1,1,1-trichloroethane, Commission of the European Communities; IPSC, CEC, 1993).

1,1,2-Trichloroethane is thought to be a potential human carcinogen. It has an LD_{50} of 580 mg/kg and a TLV-TWA of 10 ppm or 55 mg/m^3. Its explosion limits are 6–15.5 percent. The boiling point of this liquid is 237°F (114°C). It is more dense than water (specific gravity, 1.44) and is insoluble in water. It has a vapor pressure of 19 mm Hg and a vapor density of 4.6. 1,1,2-Trichloroethane has a NFPA rating of health = 2, flammability = 1, reactivity = 1 (Pohanish and Greene, 1996, pp. 1567–1569; International Chemical Safety Card, ICSC: 0080, 1,1,2-trichloroethane, Commission of the European Communities; IPCS, CEC, 1993).

1,1,2,2-Tetrachloroethane is not considered a fire hazard but is generally considered the most toxic of the common chlorinated hydrocarbons. It has a TLV-TWA of 1 ppm or 6.9 mg/m^3. The boiling point is 295°F (146°C). It is more dense than water (specific gravity, 1.6) and is insoluble (0.29 g/100 mL of water). It has a vapor pressure of 5 mm Hg and a vapor density of 5.8. 1,1,2,2-Tetrachloroethane has a NFPA rating of health = 3, flammability = 0, and reactivity = 1 (J. T. Baker MSDS Number T 0760 effective date 09/08/97; International Chemical Safety Card, ICSC: 0332, 1,1,2,2-tetrachloroethane, Commission of the European Communities; IPSC, CEC, 1993).

1,1-Dichloroethylene or vinylidene chloride is an extremely flammable liquid with a flash point of 42°F (5.6°C). Its explosion limits are 5.6–16 percent in air. It has an LD_{50} of 200 mg/kg and a TLV-TWA of 5 ppm or 20 mg/m^3. The boiling point of vinylidene chloride is 90°F (32°C). It is more dense than water (specific gravity, 1.2) and is considered insoluble (0.25 g/100 mL of water). It has a vapor pressure of 500 mm Hg and a vapor density of 3.3. This substance can easily form explosive peroxides and polymerizes when in contact with copper or aluminum (Supelco MSDS, 48526, 1,1-dichloroethylene 5G, reviewed Aug. 24, 1995; International Chemical Safety Card, ICSC: 0083, vinylidene chloride, Commission of the European Communities, IPCS, CEC, 1993).

1,2-Dichloroethylene is a flammable liquid with a flash point of 36°F (2.2°C) and explosion limits of 9.7–12.8 percent. It has an LD_{50} of 770 mg/kg and a TLV-TWA of 220 ppm or 793 mg/m^3. The boiling point is 140°F (60°C). It is more dense than water (specific gravity, 1.27) and is considered slightly soluble in water. It has a vapor pressure of 400 mm Hg and a vapor density of 3.34. 1,2-Dichloroethylene has a NFPA rating of health = 2, flammability = 3, and reactivity = 2 (Pohanish and Greene, 1996, pp. 507–508; PPG Industries, MSDS for *trans*-1,2-dichloroethylene, cleaning compound, solvent; MSDS Serial Number BYHCZ, prepared Nov. 14, 1994 and reviewed March 19, 1996).

The last chlorinated solvent to be considered is trichloroethylene, which is thought to be a potential human carcinogen. It has a flash point of 90°F (32°C) and explosion limits of 8–10.5 percent. It has an LD_{50} of 50–500 mg/kg and a TLV-TWA of 50 ppm or 269 mg/m^3. The TLV–STEL for trichloroethylene is 200 ppm or 1070 mg/m^3. The boiling point is 189°F (87°C). It is more dense than water (specific gravity, 1.46) and is slightly water soluble (0.1 g/100 mL of water). Its vapor pressure is 58 mm Hg and it has a vapor density of 4.5. Trichloroethylene has a NFPA rating of health = 2, flammability = 1, and reactivity = 0 (Pohanish and Greene, 1996, pp. 1569–1571; International Chemical Safety Card, ICSC: 0081, trichloroethylene, Commission of the European Communities; IPCS, CEC, 1993).

REFERENCES

Lewis, *Hazardous Chemicals Desk Reference*, 4th ed., Wiley, New York, 1997.

Pohanish and Greene, *Hazardous Materials Handbook*, Van Nostrand Reinhold, 1996.

CHAPTER 2
CLASSES OF CHEMICALS

SECTION 2.4

RADIONUCLIDES

Peter Collopy

Mr. Collopy currently works as the director of Environmental Health & Safety Services for MJW Corporation in Buffalo, New York. His major endeavors are participation in the remediation and environmental assessment of radioactively contaminated CERCLA and industrial facilities.

2.4.1 RADIOACTIVITY

Radioactive Decay

With the exception of normal hydrogen, all atoms, whether they exist naturally or are produced artificially, consist of protons, electrons, and neutrons. As denoted by the symbol Z, the atomic number for each element is defined by the number of positively charged protons within the nucleus. The sum of the protons and neutrons within the nucleus constitutes nearly all the mass of an element (for the most part the electrons are ignored because an electron weighs only a little more than 1/2000 of a proton or neutron). The symbol A is used to designate the mass of an element as approximated by the total number of neutrons and protons within the nucleus. Conventionally, the atomic symbol for an element is expressed as $^A_Z X$ where X is the atomic symbol for that element. For example $^{23}_{12}$Na is sodium and has a nucleus with 12 protons and 11 neutrons. Atoms or elements with the same atomic number but a different number of neutrons are known as isotopes.

The ratio of the neutrons to protons within the nucleus determines the mass-energy state of the nucleus; for some elements, this state is unbalanced or unstable. To achieve a more balanced mass-energy state, the nucleus will undergo a transformation that changes the number of neutrons and protons and results in the release of energy from the nucleus in the form of α- or β-particles or as high energy photons (γ-rays). Those nuclei capable of undergoing this nuclear transformation are said to be radioactive, and the atoms are referred to as radionuclides, radioactive isotopes, or radioisotopes. Isotopes of the same atomic number have essentially the same transport and fate characteristics within the environment. More than 1500 such nuclides have been identified; some elements having as many as 30 isotopes.

Mass effects on chemical reaction kinetics can occur with the smaller Z numbered elements such as tritium (^{3}H) and carbon-14 (^{14}C) and should be considered when evaluating the different ratios of isotopes within a particular medium. However, in most environmental assessment situations these mass effects have a negligible influence on transport and fate characteristics because of their low concentrations.

Whenever a radionuclide undergoes a transformation or decay that changes the number of protons within the nucleus, the resulting product or progeny becomes a different element with a different atomic number and chemical characteristics. For most of the lower-atomic-number radionuclides ($Z < 82$), the progeny are stable and will not undergo further transformation. For the higher-atomic-numbered radionuclides, the decay product is radioactive and will undergo transformations to yet another radioactive progeny.

TABLE 2.4.1 Typical Radionuclides and Half-Lives

Radionuclide	Half-life	Radionuclide	Half-life
^{37}K	1.2 s	^{90}Sr	28.5 years
^{18}F	1.8 h	^{137}Cs	30 years
^{212}Pb	10.6 h	^{14}C	5730 years
^{131}I	8.1 days	^{36}CI	3×10^5 years
^{55}Fe	2.6 years	^{129}I	1.7×10^7 years
^{60}Co	5.3 years	^{40}K	1.3×10^9 years
^{3}H	12.4 years	^{238}U	4.5×10^9 years

Groups of radioactive elements that transform to different elements that are also radioactive are considered to form a decay chain. With sufficient passage of time, the elements within this chain reach a state of equilibrium in which, for each decay of the parent, each progeny within the chain also undergoes an almost simultaneous decay. Consequently, when evaluating the environment in the presence of radionuclides certain inferences can be made for all radionuclides within the chain by measuring only one or two of the chain radionuclides.

Disequilibrium can occur when the precursor decay progeny are chemically removed from the chain or the medium is disturbed so that the radon progeny diffuse from the medium at a greater rate than in the natural, undisturbed state. For example, ^{234}U can have a slightly higher mobility than either ^{235}U or ^{238}U in near-surface soils. This is because ^{234}U derives from the decay of ^{238}U and hence tends to reside in mineral sites that have been damaged by the decay process. Solutions that pass through soils therefore will leach a disproportionately larger amount of ^{234}U than the other uranium isotopes. In addition, changes to the physical state of the sampled medium, such as soil, can cause a greater degree of radon off-gassing than in the in situ state and must be accounted for when sampling and analyzing soil or water for uranium and its progeny.

Although it is known that an unstable nucleus will undergo a transformation or decay, the exact time when a particular radioactive atom will do this is unknown. However, if a statistically significant number of these unstable atoms are present together, transformations will occur at a definite and predictable rate that is characteristic of that radioisotope. This rate of decay is often expressed by the radionuclide's half-life, or the amount of time for 50 percent of the population of radioactive atoms to undergo a transformation.

For example, $^{90}_{38}Sr$ (strontium-90) has a half-life of 28.5 years, so that over 285 years, the original population of $^{90}_{38}Sr$ atoms will have been reduced by a factor of >1000 or 10 half-lives (2^{10}). Table 2.4.1 shows some typical radionuclides and their respective half-lives.

The terms used to describe the amount of a particular radionuclide(s) within the environment is not always expressed in the traditional volumetric concentration metrics such as milligrams per kilogram or parts per million (ppm). Radionuclide concentration is often expressed as an activity per mass or volume. In this case, activity is expressed as the total number of transformations per unit of time and the SI unit for this is the becquerel (Bq). A becquerel is defined as one disintegration or transformation per second.

For example, activity at 1 MBq means there are 10^6 transformations per second occurring for that population of radioactive atoms. Before SI units were used the historical unit of activity was the curie (Ci), which is equivalent to 3.7×10^{10} Bq. The curie is still used extensively in the United States. Table 2.4.2 provides a comparison chart for becquerels and curies.

Conversion of activity to mass can be achieved by the following equation:

$$\text{Bq/g or specific activity (SpA)} \cong \frac{1.323 \times 10^{16}}{(t_{1/2})(\text{atomic mass})}$$

where $t_{1/2}$ is the half-life (years).

For example, the half-life of ^{90}Sr is ~28.5 years so the SpA for ^{90}Sr is

$$\text{SpA (Bq/g)} = \frac{1.323 \times 10^{16}}{28.5 \times 90} \approx 5.2 \times 10^{12}$$

TABLE 2.4.2   SI Unit Conversion

1 becquerel (Bq) $= 2.7 \times 10^{-11}$ curie (Ci)
1 Ci $= 3.7 \times 10^{10}$ Bq $= 37$ GBq
1 mCi $= 37$ MBq
1 μCi $= 37$ kBq
1 Bq $= 27$ pCi

Common Prefixes for SI Units

10^{-3}	milli	m
10^{-6}	micro	μ
10^{-9}	nano	n
10^{-12}	pico	p
10^{3}	kilo	k
10^{6}	mega	M
10^{9}	giga	G
10^{12}	tera	T

TABLE 2.4.3   Specific Activity for Some Radionuclides

Radionuclide	Half-life	Bq/g	Radionuclide	Half-life	Bq/g
^{24}Na	15 h	3.22×10^{17}	^{137}Cs	30 years	3.22×10^{12}
^{3}H	12.3 years	3.59×10^{14}	^{226}Ra	3×10^{5} years	1.95×10^{8}
^{55}Fe	2.6 years	9.25×10^{13}	^{36}Cl	3.1×10^{6} years	1.19×10^{8}
^{90}Sr	28.5 years	5.16×10^{12}	^{232}Th	1.41×10^{10} years	4.04×10^{3}

This equation represents the specific activity for the pure form of the radionuclide and should not be confused with concentration in a medium such as soil or water, where the activity is divided by the mass of measured medium. Table 2.4.3 provides some examples of fairly common radionuclides found in the environment and their activity to mass conversion factors. Note that specific activity is inversely proportional to the half-life of the radionuclide; i.e., the longer the half-life, the lower the specific activity.

The level of activity is not a perfect descriptor of the potential hazard involved with a radionuclide. To adequately assess the potential hazard, one must know not only the activity level or amount of radioactivity but also the exposure pathway, the radiation emission types, and the energy of those emissions. All these factors can provide information to calculate the dose or energy imparted to an individual or organism in the affected environment. The SI units used for dose are the gray (Gy) and the sievert (Sv); their historical counterparts are the rad and rem, respectively. One hundred rads equals 1 Gy and 100 rem equals 1 Sv. The difference between the gray and the sievert is that the gray represents absorbed energy in a unit of mass, whereas the sievert is the dose equivalent in tissue. The sievert can actually be thought of as the potential for biological damage. It is the unit used for risk expression in defining the hazards when an individual is exposed to a radiological environment. Unless highly concentrated, environmental media containing radioactive materials have dose levels on the order of nanosieverts per hour ($\sim$10–100 microrem/h).

Radionuclide Material Classes

Radionuclides can be divided into three major classes of materials related to their origin or means of production:

1. NORM, naturally occurring radioactive materials;

2. Anthropogenic, manmade by fission or high-energy accelerator processes; and

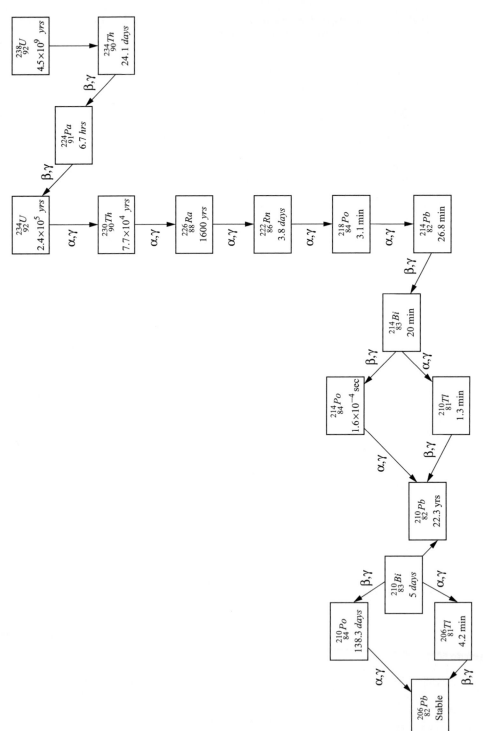

FIGURE 2.4.1 Uranium decay series.

3. TENORM, technologically enhanced naturally occurring radioactive materials—natural in origin but their concentrations are increased by manmade activities.

Regardless of origin, the fate and transport of these radionuclides depends on their physical and chemical forms within the environment. For materials with shorter half-lives, the decay rate must be included in determining the rate of removal from and eventual transport through environmental media.

Naturally Occurring. Naturally occurring radionuclides are either primordial or cosmogenic in origin. Primordial radionuclides were produced at the creation of the universe. Cosmogenic radionuclides are created by interaction of the atmosphere and the earth with high-energy nucleons originating from outside the earth's atmosphere.

The primordial radionuclide category is dominated by heavy elements that principally belong to one of three radioactive chain series headed by uranium-238 (uranium series), uranium-235 (actinium series), and thorium-232 (thorium series). The relationships among the members of the uranium and thorium series are shown in Figures 2.4.1 and 2.4.2, respectively. Note that all the series eventually decay to stable lead as the last member of the series or chain.

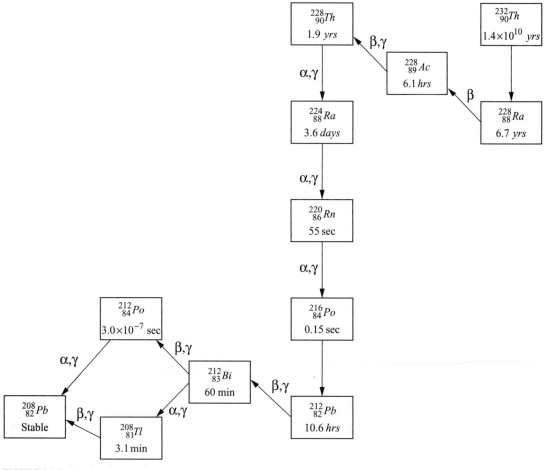

FIGURE 2.4.2   Thorium decay series.

TABLE 2.4.4 Primordial Radionuclides and Relative Abundance

Nuclide	Half-life (years)	Isotopic abundance (%)	Concentration Bq/kg (pCi/g)	
			Rock	Soil
Uranium 238 ($^{238}_{92}$U)	4.5×10^9	99.2745	36 (0.97)	66 (1.78)
Thorium 232 ($^{232}_{90}$Th)	1.4×10^{10}	~100	44 (1.19)	37 (1.0)
Rubidium 87 ($^{87}_{37}$Rb)	4.8×10^{10}	27.9	100 (2.7)	50 (1.45)
Potassium 40 ($^{40}_{19}$K)	1.3×10^9	0.0118	850 (23)	400 (10.8)

The primordial chain radionuclides uranium and thorium and their associated progeny are ubiquitous in the environment. Uranium has a higher abundance (4 ppm) in the earth's crust than boron, tantalum, bromine, mercury, silver, and gold. Uranium is commonly found in either a +VI or +IV valence state. The +VI state is generally more soluble, so uranium tends to deposit at the boundary between an oxidizing and a reducing environment. Consequently, elevated concentrations of uranium can be found in coal, oil, granite, and other similar mineral deposits. In the higher-grade ores such as autunite and uraninite, U_3O_8 can be found at concentrations up to 4 percent.

Thorium-232 is nearly as abundant as lead (16 ppm) and is more abundant than cesium, germanium, beryllium, and arsenic. Thorium is usually found in a +IV valence state and is generally insoluble in water. Hot spots of highly concentrated thorium are found in the monazite sands of Brazil and India. Other primordial radionuclides of significance within the environment are potassium-40 and rubidium-87. The isotopic abundance of ^{40}K in natural potassium is only 0.0118 percent, but potassium is so widespread that ^{40}K is at significant concentrations within the geosphere. The isotopic abundance of ^{87}RB is somewhat higher, but the concentration of elemental rubidium in the earth's crust, is 2 orders of magnitude less than that of potassium. Some other primordial radionuclides are ^{50}V, ^{87}Rb, ^{113}Cd, ^{115}In, ^{123}Te, ^{138}La, ^{142}Ce, ^{144}Nd, ^{147}Sm, ^{152}Gd, ^{174}Hf, ^{176}Lu, ^{187}Re, ^{190}Pt, ^{192}Pt, and ^{209}Bi. Table 2.4.4 shows typical concentrations of the more abundant primordial radionuclides within various media.

Cosmic radiation permeates all of space, the source is primarily outside of our solar system. The radiation is in many forms, from high-speed heavy particles to high-energy photons and muons. As shown in Figure 2.4.3, high-energy nucleon interactions with the atmosphere, the earth's crust, and the oceans produce numerous cosmogenic radionuclides. Cosmogenic radionuclides arise from the collision of highly energetic cosmic ray particles with stable elements in the atmosphere and in the ground. The entire geosphere, the atmosphere, and all parts of the earth that directly exchange material with the atmosphere contain cosmogenic radionuclides. The major production of cosmogenic radionuclides results from the interaction of cosmic rays with atmospheric gases.

The outermost layer of the earth's crust is another area where reactions with cosmic rays occur. However, the rate at which they occur is several times smaller than the atmospheric component because most of the cosmic rays are attenuated in the atmosphere. The result is that the contribution to background dose is minimal.

The most important radionuclide produced is ^{14}C. However, many others, such as ^{3}H, ^{22}Na, and ^{7}Be, occur. Carbon-14 produced in the atmosphere is quickly oxidized to CO_2. The equilibrium concentrations of ^{14}C in the atmosphere are controlled primarily by the exchange of CO_2 between the atmosphere and the ocean. The oceans are the major sink for removal of ^{14}C from the atmosphere.

Most of the other cosmogenically produced radionuclides in the atmosphere are oxidized and become attached to aerosol particles. These particles act as condensation nuclei for formation of cloud droplets and eventually coagulate to form precipitation. About 10–20 percent of cosmogenically produced radionuclides are removed from the atmosphere by dry deposition on the earth's surface.

Concentrations of cosmogenic radionuclides vary in the atmosphere with time and location. Variations are day-to-day, seasonal, longitudinal, and sunspot-cycle related. The concentrations of some cosmogenic radionuclides, such as ^{3}H, ^{14}C, ^{22}Na, and ^{37}Ar, have increased during nuclear tests.

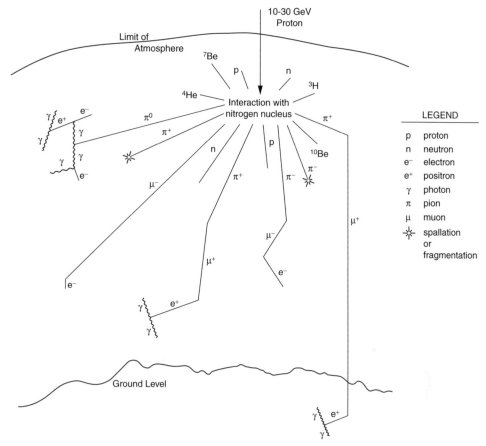

FIGURE 2.4.3 Cosmic interactions.

Reactors also generate ^{14}C that eventually will be distributed in the atmosphere, but it is estimated to be 2 orders of magnitude lower than the natural concentration. Some other cosmogenic radionuclides of lesser environmental significance are $^{10}Be, ^{18}F, ^{22}Na, ^{24}Na, ^{38}Mg, ^{26}Al, ^{31}Si, ^{32}Si, ^{32}P, ^{33}P, ^{35}S, ^{38}S, ^{37}Ar, ^{39}Ar, ^{34m}Cl, ^{36}Cl, ^{38}Cl, ^{39}Cl,$ and $^{86}Kr.$ Table 2.4.5 provides a listing of the more important cosmogenic radionuclides and their typical concentrations detected within environmental media.

Manmade. Manmade radionuclides are primarily produced by either high-energy accelerators or fissioning of uranium and plutonium. Accelerators tend to produce neutron-deficient nuclei by accelerating a charged particle at an energy high enough to react with and alter the target element's nucleus. The fission process produces radionuclides, which are typically neutron rich and with a much longer half-life than accelerator-produced radionuclides. Table 2.4.6 lists several important radionuclides resulting from each process.

The nuclear fuel cycle is the most significant manmade or anthropogenic source relative to the production and potential deposition of radionuclides in the environment. The fuel cycle starts with mining of uranium and ends with disposal or reuse of the fuel. The various stages of the cycle are depicted in Figure 2.4.4. In the United States, uranium and plutonium in fuel are not recycled and therefore the United States system is considered open ended, with all materials destined for long-term storage and disposal in geologic repositories. In countries such as Japan and France, the useful

TABLE 2.4.5 Important Cosmogenic Radionuclides

Nuclide	Half-life	Source	Typical background concentration		
			Soil	Water	Air
Tritium ($^3_1$H)	12.3 years	Cosmic-ray interactions with N and O; spallation from cosmic rays, ^{6}Li(n, α)^{3}H	0.099–2.7 Bq/g (3.7–100 pCi/g)	0.1–0.6 Bq/L (3–16 pCi/L)	1.6×10^{-3} Bq/m^3 (0.06 pCi/m^3)
Carbon 14 ($^{14}_6$C)	5730 years	Cosmic-ray interactions, ^{14}N(n, p)^{14}C;	~0.04 Bq (4–6 pCi) of ^{14}C/gC	Bq/L 1.3–1.8 (34–49 pCi/L)	51 mBq/m^3 (1.9 pCi/m^3)
Beryllium 7 ($^7_4$Be)	53.3 days	Cosmic-ray interactions with N and O;	<1 mBq/g	<1 Bq/L	0.014 Bq/m^3 (0.5^1 pCi/m^3)

TABLE 2.4.6 Manmade Radionuclides

Radionuclide	Half-life	Production method
^{137}Cs	30 years	Fission of ^{238}U
^{131}I	8.1 days	Fission of ^{238}U
^{198}Au	6.2 days	^{197}Au(n, γ)^{198}Au (neutrons produces from fission or accelerator process)
^{32}P	14.3 days	^{32}S(n, p)^{32}P or ^{31}P(n, γ)^{32}P (neutrons produced from fission or accelerator process)
^{22}Na	2.6 years	^{25}Mg(p, α)^{22}Na accelerator
^{7}Be	53 days	^{7}Li(p, n)^{7}Be accelerator
^{57}Co	267 days	^{60}Ni(p, α)^{57}Co accelerator
^{51}Cr	27.8 days	^{51}V(p, n)^{51}Cr accelerator
^{18}F	1.8 h	^{19}F(p, pn)^{18}F accelerator

plutonium is extracted from the spent fuel and is recycled by incorporation into new fuel for use in power reactors. Even with the recycling of the fissionable materials, a considerable quantity and variety of radionuclides remain as waste products and need to be treated and disposed of safely.

Uranium-bearing ores are either deep or open-pit mined. The ore is then milled by grinding into fine sand and separated by alkaline or acid leaching to produce U_3O_8, more commonly known as yellow cake. The mill tailings are considered a waste product, which contains the unseparated uranium and radioactive progeny. These mill tailings are then stabilized in earthen mounds to prevent rapid transport of the remaining uranium and their progeny through the environment. During the 1950s, mail tailings were used as backfill in housing and other industrial developments. This practice led to increased public exposure to radon emanations and direct γ the emissions from the mill tailings. The federally funded program known as the UMTRA (Uranium Mill Tailings Remedial Action) project was implemented to remediate affected areas throughout the United States.

Radium concentrations in the dry mill tailings vary between 2 and 40 Bq/g (50–1000 pCi/g). The remaining progeny are in relative equilibrium with ^{226}Ra—i.e., their concentrations are approximately equal. Erosion by wind or surface water as well as natural and manmade disturbances of the mill tailing impoundments can cause transport of these materials to the environment. Currently, the major concern is off-gassing of the radon progeny from the tailing pile and subsequent elevations in local area radon air concentrations.

The yellow cake extracted during the mining and milling process is then transported to facilities where the material is converted to a metal or an intermediate uranium compound such as orange oxide (UO_3) or green salt (UF_4). These compounds are converted to uranium hexafluoride (UF_6) and shipped to gaseous diffusion plants. The gaseous uranium is then isotopically enriched in ^{235}U by passing the gas through cascades of porous barriers. This enrichment is obtained by using the

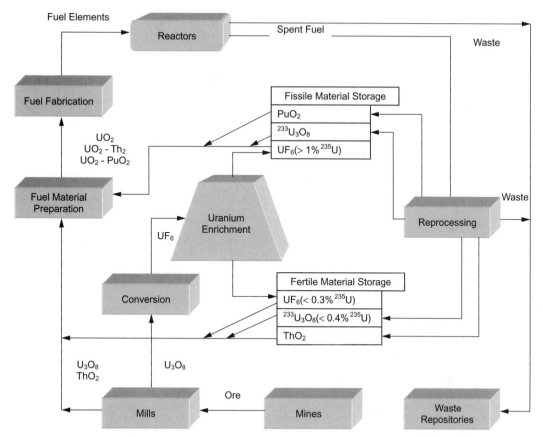

FIGURE 2.4.4 Nuclear fuel cycle.

different diffusion rates of the uranium isotopes due to their mass differences. Normal power reactor fuel uses a ^{235}U enrichment of 3–5 percent. Nuclear submarine's typical enrichment is on the order of 90 percent. Because there is a significant use of hazardous chemicals during these processes, the resulting contaminated byproducts are typically mixed wastes with both radiological and chemical hazardous characteristics.

In most cases, environmental contamination is contained within the facility, although during the early days of the process, uranium dusts were exhausted in large amounts. Because uranium is so widespread in the environment, it is difficult to detect elevated concentration away from the facility. In rare cases, discharges to the water can result in reconcentration of the uranium in water-treatment facilities. Further incineration of the sludge may increase uranium concentrations further.

The enriched uranium fuel is then used for one of three types of reactors: power reactors in which steam is produced to create electrical power or to directly power vessels such as submarines, research reactors such as those located at university facilities, and industrial reactors used for testing the effect of radiation on materials. In the United States most power reactors are light-water reactors, where the water used for production of steam is not enriched in deuterium ($^2_1$H), a heavier isotope of hydrogen. Canadian reactors use deuterium or heavy water as their fuel uses natural uranium and is not enriched in ^{235}U. Other reactor types include gas-cooled graphite reactors and liquid metal breeder reactors. In each case, the basic types of radionuclides produced are the same, either fission products or activation products. Table 2.4.7 shows typical fission and activation products and their production rates for 1000 MWe light-water reactor.

TABLE 2.4.7   Predicted Reactor Coolant Inventory of Fission Products and Corrosion Products [1000–MW(e) (Megawatts electric) PWR (Pressurized Water Reactor) at 578°F][a] $-32(\times \frac{5}{9}) = 303.3°C$

Noble gas fission products		Fission products	
Isotope	μCi/mL	Isotope	μCi/mL
^{85}Kr	1.11	^{84}Br	3.0×10^{-2}
^{85m}Kr	1.46	^{88}Rb	2.56
^{87}Kr	0.87	^{89}Rb	6.7×10^{-2}
^{88}Kr	2.58	^{89}Sr	2.52×10^{-3}
^{133}Xe	1.74×10^2	^{90}Sr	4.42×10^{-5}
^{133m}Xe	1.97	^{90}Y	5.37×10^{-5}
^{135m}Xe	0.14	^{91}Y	4.77×10^{-4}
^{138}Xe	0.36	^{92}Sr	5.63×10^{-4}
Total nobale gases	187.3[b]	^{92}Y	5.54×10^{-4}
		^{95}Zr	5.04×10^{-4}
		^{95}Nb	4.70×10^{-4}
		^{99}Mo	2.11
		^{131}I	1.55
Corrosion products		^{132}Te	0.17
		^{132}I	0.62
Isotope	μCi/mL	^{133}I	2.55
		^{134}Te	2.2×10^{-2}
^{54}Mn	4.2×10^{-3}	^{134}I	0.39
^{56}Mn	2.2×10^{-2}	^{134}Cs	7.0×10^{-2}
^{58}Co	8.1×10^{-3}	^{135}I	1.4
^{59}Fe	1.8×10^{-3}	^{136}Cs	0.33
^{60}Co	1.4×10^{-3}	^{137}Cs	0.43
Total corrosion products	3.7×10^{-2}	^{138}Cs	0.48
		^{144}Ce	2.3×10^{-4}
		^{144}Pr	2.3×10^{-4}
		Total fission products	12.8[b]

[a] Contamination concentration corresponding to 1 percent failed fuel near end of fuel life.

[b] 187.3 μCi/mL = 6.9 GB$_q$/L, 12.8 μCi/mL = 0.47 Bq/L.

Source: Eichholz, G., *Environment Aspects of Nuclear Power*, Ann Arbor Science Publishers, Ann Arbor MI, 1997.

Fission products are the by-product of the separation or cleaving of the uranium nucleus during the fission process. These fission products typically have mass numbers that vary between 85 and 144, with half-lives as long as 30 years.

Fission products are retained within the fuel cladding but leaks through the cladding or fissioning of tramp uranium on the exterior of the fuel element result in small amounts of fission products being introduced into the reactor cooling system. Under catastrophic conditions, such as those experienced at the Three Mile Island and Chernobyl reactors, destruction of the fuel cladding and the interior fuel pellets can result in release of all the fission product gasses and a significant proportion of the nongaseous elements. Depending on the safeguard systems and secondary containment systems, these fission products may or may not be released to the environment in significant quantities.

Activation products result from the direct interaction of neutrons with the nucleus of an atom. This interaction causes a change in the atomic state of the nucleus and a resulting unstable or radioactive element. In light-water reactors neutron interaction with the water, coolant system component corrosion products and reactor structural components produce most of the activation production.

Provided the fuel has minimal leakage, the corrosion activation products such as ^{58}Co, ^{60}Co, and ^{54}Mn are the principal radionuclides along with fission product gasses (^{85}Kr, ^{133}Xe) that are of environmental concern.

Because there is some release of activation and fission products into the light-water reactor cooling system, all reactors use filters and ion-exchange systems along with systems designed to hold up primary system coolants and gasses for radioactive decay to prevent or minimize release of radionuclides to the environment. Those radionuclides captured in the filter and ion-exchange systems of the plant are then immobilized as waste products and disposed of in approved shallow land burial facilities. Deactivation or decommissioning of plants no longer considered useful for electrical production or research represents the largest source of radioactive contaminants available for disposal and eventual transport into the environment. During the decommissioning process the fuel is normally removed and the major source of fission product inventory is relocated to a storage or fuel reprocessing facility. Consequently, only activated structural material remains. Delays in the decommissioning process result in reduction of activity for the shorter-lived radionuclides. However, activation radionuclides such as ^{59}Ni and ^{94}Nb have half-lives on the order of 10,000 years and, as such, their disposal is a longer-term environmental concern.

The spent or used fuel is the principal high-level source of radioactive material and is considered high-level waste. This waste must be properly treated, immobilized, and then stored in geologic repositories. These repositories, along with the immobilization techniques, are designed to minimize leakage of radionuclides up to and beyond 10,000 years. Currently, there are no active commercial waste repositories, and utility companies are seeking to store the fuel in above-ground, naturally air-cooled containers until a repository becomes available.

In fuel reprocessing the cladding from spent fuel is sheared into small segments and then the fuel within is chemically dissolved (usually using nitric acid). The useful uranium and plutonium is then extracted by a solvent extraction process known as PUREX. Destruction of the fuel cladding and chemical treatment result in the release of large quantities of fissionable materials with the noble gas radionuclides presenting the largest potential release to the environment. All these fission waste products are considered high-level waste because of the intense radioactivity associated with the materials. Because of the large quantities and types of radioactive materials present in the fuel, special effluent control systems are needed during fuel reprocessing operations. Barring any catastrophic events, the environmental releases associated with these reprocessing operations are primarily due to fission product noble gasses such as ^{85}Kr. Clean-up and eventual decommissioning of these facilities will obviously cause the environmental concerns to focus on fission products and fissionable materials such as uranium and plutonium. Currently, no commercial reprocessing is done in the United States and it does not appear that any commercial reprocessing is planned for the near term (the West Valley, New York, facility operated briefly during the early 1970s but is now being decommissioned).

Radionuclides are used extensively in industrial operations and medical applications and are typically in either a sealed form designed to prevent leakage of contaminants or in an unsealed form in which the radionuclide's chemical properties are essential to the particular application. Typical uses of radioactive materials in industrial and medical applications are shown in Table 2.4.8

Sealed source applications include industrial radiography, borehole (geological) logging, radiation gauge measurements, smoke detectors, sterilization of foods and medical equipment, and therapeutic medical treatment (typically cancer therapy). Use of sealed sources typically does not represent an environmental contaminant problem unless the source encapsulation fails and the chemical compound containing the radionuclide is released to uncontrolled areas.

All sealed sources eventually lose their activity through radioactive decay and the intensity or emission strength needed for the process being conducted diminishes. For sealed sources with a shorter half-life, storage for decay or even burial will result in minimal source activity available should the source containment fail. For longer-lived sources, the source encapsulation or containment will eventually fail and the surrounding environment will then be the means for preventing transport through the geosphere. Sealed sources are usually encased in concrete before final land disposal, thus providing an additional barrier to prevent contaminant dispersal.

Occasionally, sources are lost in the process and deliberately sealed into their last known position (e.g., borehole logging) or source containment failures occur because of poor design or from accidents.

TABLE 2.4.8 Industrial and Medical Radionuclide Applications

Application	Radionuclides	Typical source strengths
Sealed sources		
Industrial radiography	^{192}Ir, ^{137}Cs, ^{170}Tm, ^{60}Co	0.4–4 TBq (10–100 Ci)
Borehole logging	^{137}Cs, ^{60}Co, Pu-Be, Am-Be, ^{252}Cf	0.4–70 GBq (10 mCi–2 Ci) 1.9–700 GBq (50 mCi–20 Ci) ~4 MBq (100 μCi)
Satellite isotopic generators	^{90}Sr, $^{238/239}$Pu	37–3700 TBq (10^3–10^5 Ci)
Radiation gauges, automatic weighing equipment	^{90}Sr, ^{147}Pm, ^{144}Ce, ^{137}Cs, ^{60}Co	1.9–700 GBq (50 mCi–20 Ci)
Smoke detectors	^{241}Am	200 kBq (5 μCi)
Luminous signs	^{3}H	20 GBq (0.5 Ci)
Mossbauer analysis	^{57}Fe, ^{57}Co	0.4 MBq (2–50 μCi)
Unsealed applications		
Hydrological tracers	^{3}H, ^{82}Br	4 TBq (1–100 Ci), 4 GBq (10–100 mCi)
Reservoir engineering	^{85}Kr	7 GBq (200 mCi)
Biological research tracers	^{3}H, ^{14}C, ^{32}P, ^{35}S, ^{125}I,	3.7 kBq–3.7 GBq (0.1–100 mCi)
Nuclear medicine Angiogram, bone scans, blood volume, thyroid therapy	^{99m}Tc, ^{18}F, ^{51}Cr, ^{131}I,	370 kBq–740 MBq (10 μCi–20 mCi)
Metal alloy, ceramics, lens manufacture production processes	Th, U	3.7 MBq–37 GBq (0.1mCi–1 Ci)

Adapted from U.S. Environmental Protection Agency, *Environmental Radiation Dose Commitment: An Application to the Nuclear Power Industry*, EPA-520/4-73-002, Washington, DC, 1974. Leventhal, L., et al. "Assessment of Radiopharmaceutical usage Release Practices by Eleven Western Hospitals," in *Effluent and Environmental Radiation Surveillance*, ed. J. J. Kelly, STP 698, American Society for Testing and Materials, Philadelphia, 1980.

TABLE 2.4.9 Sealed Source Containment Failures

Radionuclide	Activity	Incident description
^{60}Co	930 GBq (25 Ci)	Auburn, NY (1983): Gauge source in scrap metal feed melted in steel plant electric arc furnace. Extensive contamination of furnace through baghouse components. ~$4,000,000 to remediate facility.
^{60}Co	16.7 TBq (420 Ci)	Juarez, Mexico (1983): Teletherapy source inadvertently dismantled and device sold for scrap steel. Melted into rebar (6600 tons) and table legs (3000) sold throughout United States; material eventually recovered.
^{137}Cs	51.8 TBq (1400 Ci)	Goiania, Brazil (1987): Teletherapy source removed from device and encapsulation broken. CsCI salt dispersed through houses and parts of town; 244 persons contaminated, several fatalities resulted.
^{137}Cs	150 TBq (4500 Ci)	Decatur, GA (1988): Irradiation facility source encapsulation failed and ^{137}Cs leakage spread throughout the facility. Significant property contamination (6000 ft^2) with some contaminants transferred to automobiles and property of employees before discovery made. Facility decommissioned and no longer used; $45,000,000 remediation costs.

Some of the more significant events that result in sealed source leakage to the environment are shown in Table 2.4.9.

The use of unsealed or dispersible forms of radioactive materials is primarily done in laboratory operation, field tracer studies, or in medical diagnostic and therapeutic applications. For laboratory and tracer operations, radionuclides are used as tags to trace chemical compounds in biological, chemical, and environmental research. The primary radionuclides used as tracers are ^{3}H, ^{14}C, ^{32}P, ^{33}P, ^{35}S, ^{51}Cr, ^{55}Fe, and ^{125}I.

Because of the short half-lives involved, most research radionuclides do not result in a significant build-up of radioactive inventory. However, ^{3}H and ^{14}C, with half-lives of 12.3 and 5730 years, respectively, can and do cause environmental contamination problems when improperly disposed. Before recent disposal restrictions by the Environmental Protection Agency and the Nuclear Regulatory Commission, it was not uncommon for large quantities of research-related radionuclides to be disposed of in sanitary drain systems or by shallow land trench disposal on the research group's property.

Additional disposal and environmental problems can occur with these research radionuclides, because they are often combined with chemicals or result in a by-product that contains a hazardous chemical constituent. Because there are restrictions on disposal of materials considered to be both chemically and radiologically hazardous, these mixed wastes often end up as an orphan waste stream in storage.

Other non-laboratory-related uses of unsealed radioactive materials include production of high-strength, heat-resistant metals or ceramics by incorporation of uranium or thorium into the ceramic or alloy matrix, the use of thorium-containing refractory powders for repair of refractory brick in kilns and furnaces, and the use of ^{3}H or ^{147}Pm in luminescent products such as exit signs and watch dials. Environmental contaminants can result from these operations but the contamination is usually restricted to the process area, the immediate property, and the drain disposal system. The largest use of unsealed or dispersible forms of radioactive material outside the fuel cycle facilities is for medical purposes. The major effect from use of radionuclides in medicine is from their use as radiotracers in nuclear medicine. It is important to note that the major pathway to the environment from these radionuclides is through patient excreta. The environmental impact from the excreta is from transport through the sewer system and it can occur from both in-patient and outpatient excreta. Most of these materials are produced by accelerators and are short-lived and thus do not represent a long-term environmental hazard. These compounds can often affect an environmental analysis that is not directly associated with these facilities, particularly for air or water collected downwind or downstream of a hospital.

The introduction of non-naturally occurring radionuclides into the global environment has been primarily through nuclear weapons testing and, to a lesser extent, by large-scale nuclear accidents such as at Chernobyl. Atmospheric and above-ground testing peaked in the early 1960s and, because of the 1963 Nuclear Test Ban Treaty, weapons testing is done mostly underground. Consequently, the global inventory of radionuclides from this testing has been steadily decreasing since 1963 with the exception of small increases from Chinese, Indian, and French above-ground tests. Fallout radionuclides are virtually indistinguishable from cosmogenically produced radionuclides and fuel cycle fission product radionuclides of the same isotope. In some cases, inferences can be made by the ratios of existing radionuclides to determine whether the fission products are fresh and from a recent fission event. In environmental work, fallout radionuclides such as ^{90}Sr, ^{137}Cs, and ^{238,239}Pu will always be found in the background analysis and must be accounted for when determining environmental transport or impacts from operations producing similar radionuclides. Table 2.4.10 lists important fallout radionuclides and typical concentrations to be found in environmental media.

TABLE 2.4.10 Fallout Radionuclides

Radionuclide	Typical background concentrations	
	Soil, Bq/g (pCi/G)	Surface water, Bq/L (pCi/L)
^{90}Sr	0.73–50.2 (0.027–1.86)	0.078–0.36 (2.1–9.6)
^{137}Cs	0.35–41.9 (0.013–1.55)	0.27–0.33 (7.4–8.8)
^{239,240}Pu	0.13–2.6 (0.00486–0.098)	1.9×10^{-4}–8.1×10^{-4} (0.005–0.022)

Note: These concentrations are from representative site background data and vary based on geographic location, depth and type of soil, and seasonal sampling time.

TABLE 2.4.11 TENORM Producing Processes

Process/Materials	Radionuclides	Concentrations, Bq/g (pCi/g)
Coal burning: ash and slag byproducts	^{226}Ra	0.11–0.14 (3.1–3.9)
	^{232}Th	0.077 (2.1)
	^{238}U	0.12 (3.3)
Geothermal energy production: waste sludges	^{226}Ra	4.9 (132)
Metal mining and processing; slag, leachate, and tailings	^{226}Ra	0.18–33.3 (5–900)
	^{232}Th	0.7–144 (19–3900)
	^{238}U	1–666 (43)
Drinking water treatment sludges	^{226}Ra	0.59 (16)
	^{238}U	0.15 (4.0)
Oil and gas scale and sludge	^{226}Ra	3.33 (90)
Phosphogypsum tailings and phosphate fertilizers	^{226}Ra	0.31–0.29 (8.3–35)
	^{232}Th	0.03 (0.77)
	^{238}U	0.92 (25)
	^{40}K (fertilizer)	26 (696)

Technologically Enhanced Naturally Occurring Radioactive Materials (TENORM). Manmade processes that use minerals, metals, and ores with high uranium and thorium contents tend to result in inadvertent concentration of these materials. Table 2.4.11 provides a summary of the various processes that produce TENORM and the typical radionuclides and concentrations found in the process byproducts and components.

The radioactivity of coal can vary over 2 orders of magnitude depending on the type of coal and the region from which it is mined. The concentrations of ^{238}U and ^{232}Th in coal average about 0.022 and 0.018 Bq/g (0.6 and 0.5 pCi/g), respectively. Electric utility boilers generate ash at a rate of 10 percent of the original volumes of coal with over 95 percent of the ash retained.

About 70–80 percent of the coal ash is disposed of in landfills and ponds. Coal ash is also used in concrete, cement, and roofing materials and as structural fill for road construction. The radon emanation for ash is low because the ash is vitrified. However, compared with coal the concentrations of ^{226}Ra and ^{238}U tend to be enriched in ash.

Radionuclides (principally ^{226}Ra and its progeny) are concentrated with minerals that precipitate out of solution and form scale or sludge on the inside surfaces of the drilling and production equipment (e.g., steam turbines, heat exchangers, process lines, valves, turbines, and fluid-handling equipment) used to extract geothermal heat. Concentrations of NORM in geothermal wastes vary with the geology and mineralogy of a geothermal resource area along with the physical and chemical changes that occur during energy extraction.

The metals extraction industry typically generates 1.5 billion MT of waste per year including about 1.0 MT of waste rock and overburden, 0.4 billion MT of ore tailings, and fewer than 0.1 billion MT of smelter slag. Depending on the original ores and processing methods, some of these wastes contain elevated TENORM (see Table 2.4.12 for metals associated with TENORM).

It is generally believed that the level of NORM found in ores depends more on the geologic formation or region than on the particular type of mineral being mined. These ores often contain many different minerals, and the radionuclide content of one type of ore or mining operation or its wastes is not representative of other mines or waste types.

One study by the U.S. Environmental Protection Agency (1993) describes naturally occurring radioactive byproducts from the mining and processing of three categories of metals: rare earth metals, special application metals, and metals produced in bulk quantities (i.e., large volumes) by industrial extraction processes. Rare earth (or lanthanide) metals comprise 16 chemical elements, including those with atomic numbers 57 (lanthanum) through 71 (lutetium) as well as yttrium (atomic number 39) with similar chemical properties. Special application metals are regarded as metals that

TABLE 2.4.12 TENORM Metal and Mining Industries

Bauxite	Lead	Thorium
Beryllium	Molybdenum	Tin
Columbium	Nickel	Uranium
Copper	Rare Earths	Titanium
Gold	Silver	Zinc
Iron	Tantalum	Zirconium

Source: U.S. Environmental Protection Agency, Office of Radiation and Indoor Air, *Diffuse NORM Wastes–Waste Characterization and Preliminary Risk Assessment*, Draft, RAE-9232/1-2, SC & A, Inc., and Rogers & Associates Engineering Corporation, Salt Lake City, UT, May 1993.

have unique commercial and industrial uses and include hafnium, tin, titanium, and zirconium. Metals mined and processed in bulk for industrial applications include aluminum, copper, iron, lead, zinc, and precious metals such as gold and silver.

A small portion of the public water supply systems in the United States treat water that contains elevated naturally occurring radionuclide concentrations—most significantly, uranium and radium. Many water-treatment technologies typically used for removing solids from water for softening and purification can significantly increase the concentration of uranium and its decay progeny in the removal system components. If the water-treatment system processes water effluents from facilities discharging manmade radionuclides, then these also can be reconcentrated within the treatment medium and sludge.

Increased concentration of naturally occurring radioactivity from municipal water treatment can be found in sludge and solids that include filter sludge, spent ion-exchange resins, spent granular activated carbon, and water from filters. It is estimated that approximately 260,000 MT of water treatment sludge containing elevated levels of TENORM, including spent resins and charcoal, are generated annually.

Some oil and natural gas production and processing activities generate TENORM-contaminated components and materials. Radium is a major contributor to the radioactivity found in pipe scale and sludge from production and processing operations. Uranium and thorium compounds are mostly insoluble and, as oil and natural gas are brought to the surface, these compounds tend to remain embedded in underground geologic formations. However, some radium and radium daughter products are slightly soluble in water and can become mobilized when groundwater (containing dissolved mineral salts) is brought to the surface from production and processing. When this occurs, some radium and its daughters may precipitate out of solution because of geologic chemical changes and reduced temperature and pressure. Radium concentrations from geologic formations can precipitate out in sludge and on the internal surfaces of oil and natural gas piping and production and processing equipment.

TENORM is generated from the mining and processing of phosphate rock (phosphorite) needed to produce phosphate fertilizers, detergents, animal feed, food products, pesticides, and other phosphorous chemicals. Phosphogypsum and scale are the principal waste by-products generated during production of phosphoric acid and fertilizers. Scale is deposited in small quantities in process piping and in filtration receiving tanks. Even though these phosphate scale ^{226}Ra concentrations are associated with small volumes of materials, the radium concentrations have been found to be as high as 137 Bg/gram.

Ferrophosphorus and phosphate slag are the principal waste by-products from production of elemental phosphorous, which is produced by reduction of phosphate rock in large electric furnaces that use carbon and silica as catalysts. Both ferrophosphorus and slag are found in the residual solids that remain from furnace processing.

Monitoring, Analysis and Modeling. The nature of radioactivity and associated emissions provides unique opportunities for field monitoring and identification of both radionuclides and concentrations

TABLE 2.4.13 Radiological Environmental Modeling Codes

Model	Description	Location
GENII	Estimates potential radiation doses to individuals or populations from both routine and accidental releases of radionuclides to air or water and residual contamination from spills or decontamination operations. Exposure pathways include direct exposure via water (swimming, boating, and fishing), soil (surface and buried sources), air (semi-infinite and finite cloud geometry), inhalation, and ingestion.	http://www.pnl.gov/health/health_prot/genii.html
CAP88	Atmospheric transport model for assessing dose and risk from radioactive air emissions. Applicable to Department of Energy compliance with Clean Air Act.	http://www.er.doe.gov/production/er-80-cap88/
RESRAD	Designed to calculate site-specific residual radioactive material guidelines and radiation dose and excess cancer risk to an on-site resident (maximally exposed individual). Nine environmental pathways are considered: direct exposure; inhalation of dust and radon; and ingestion of plant foods, meat, milk, aquatic foods, soil, and water	http://www.ead.anl.gov/~web/resrad/index.html
PRESTO	Multimedia model for assessing low-level and low activity wastes, NARM (Natural Occurring on Acceleration Produces Radioactive Material), and uranium mill tailing waste.	http://www.epa.gov/radiation/assessment/presto.html
COMPLY	Atmospheric screening model for assessing dose from radioactive air emissions. Applicable to Department of Energy compliance with Clean Air Act, and Nuclear Regulatory Commission 10 CFR 20.	http://www.epa.gov/radiation/assessment/comply.html
MILDOS-AREA	Calculates radiation doses received by individuals and the general population within an 80-km radius of an operating uranium recovery facility.	http://web1.ead.anl.gov/mildos/miltitle.html
AIRDOSPC	Designed to calculate the effective dose equivalent to maximally exposed individuals at Department of Energy facilities, as required by 40 CFR 61.93(a), and to prepare a two-page compliance report suitable for submission to EPA.	http://www-rsicc.ornl.gov/ALPH.html
DandD	Defines radiation exposure scenarios to address residual radioactive contamination inside buildings, in soils, and in groundwater.	http://techconf.llnl.gov/radcri/java.html

within environmental media. In some cases an incorrect approach to monitoring can cause inappropriate expenditures and misleading data collection.

Because radioactivity from natural and manmade sources may be found in all media, it is important to establish background with sufficient accuracy and statistical precision to ensure that nonbackground contributions can be detected. Radionuclides, whose decay scheme includes a γ photon emission, provide an energy signature that can be used to determine the activity and concentration in an environmental medium. Some commercially available equipment allows for rapid and accurate analysis of soil without having to extract samples from the ground. For radionuclides at depths greater than 18 cm, borehole logging with special detectors can also provide information on the nature, depth, and activity of the environmental sources. Other media such as water and air can be field analyzed with reasonable accuracy and precision. Depending on data quality objectives, these types of surveys may be sufficient for determining future actions relative to the radioactive constituents found.

Radionuclides such as ^{3}H, ^{14}C, and ^{90}Sr emit only β radiation and accurate field identification is not possible. With sufficient activity and β emission energy one may detect gross levels of surface contaminants or activity on filter media, but radionuclide identification must be done by chemical separation processes in a laboratory environment. Normally direct identification of α-emitting radionuclides such as ^{238}U, ^{232}Th, ^{239}Pu, and ^{241}Am, cannot be obtained by field measurements. In the case of the chain radionuclides such as U and Th, their γ-emitting progeny can be identified and then inferences can be made about the level of activity of the parent. The disadvantage to this type of identification is that assumptions about the level of equilibrium may cause erroneous conclusions. For ^{239}Pu and ^{241}Am, the α emission is also accompanied by a weak γ- or x-ray photon emission, and that photon energy can be used, if concentrations are great enough, to identify the radionuclide in situ.

In most cases, laboratory analysis of collected media will be needed. The Environmental Protection Agency through environmental monitoring laboratory programs has defined procedures (HASL 300) for field collection and analysis of various radionuclides. It is important that analytical data be validated to ensure that laboratory or collection errors do not result in misapplication of the results.

Numerous screening and modeling codes for radionuclides within the environment are available. Some of the more widely used codes, their application, and where they can be found are listed in Table 2.4.13

BIBLIOGRAPHY

API Bulletin E2 (BUL E2), "Bulletin on Management of Naturally Occurring Radioactive Materials (NORM) in Oil and Gas Production," American Petroleum Institute, April 1, 1992.

Castellano, S. D., and Dick, R. P., "Measurement of Tritium Activity in Soils," *Health Physics*, 65(5): 539–540, 1993.

"Carbon-14 in the Environment," *NCRP Report No. 81*, National Council on Radiation Protection and Measurements, May 15, 1985.

Dehmel, J. C., *Naturally Occurring Radioactivity and Refractory Materials, Technical and Regulatory Issues*, February 1994.

Egidi, P., and Hull, C., *NORM and TENORM—Producers, Users and Proposed Regulations*, presented at Health Physics Society 1998 Annual Meeting, Minneapolis, MN, July 1998.

Eichholz, G., *Environmental Aspects of Nuclear Power*, Ann Arbor Science Publishers, Ann Arbor, MI, 1977.

Eisenbud, M., *Environmental Radioactivity*, 4th edition, Academic Press, New York, NY, 1997.

Environmental Radiation Dose Commitment: An Application to the Nuclear Power Industry, EPA-520/4-73-002, Washington, DC, 1973.

"Environmental Radiation Measurements," *NCRP Report No. 50*, National Council on Radiation Protection and Measurements, Dec. 27, 1976.

"Exposures from the Uranium Series with Emphasis on Radon and its Daughters," *NCRP Report No. 77*, National Council on Radiation Protection and Measurements, March 15, pp. 89, 1984.

"Exposures from the Uranium Series with Emphasis on Radon and its Daughters," *NCRP Report No. 77*, National Council on Radiation Protection and Measurements, March 15, 1984 (reprinted January 15, 1987).

"Exposure of the Population in the United States and Canada from Natural Background Radiation," *NCRP Report No. 94*, National Council on Radiation Protection and Measurements, Dec. 30, pp. 61–70, 1987.

Frazier, J. R., Prichard, H. M., and Hull, C., *Health Physics Aspects of Environmental Norm*, presented at Health Physics Society 1995 Annual Meeting, Boston, MA, July 1995.

Kereiakes, J. G., Shalek, R. J., and Waggener, R. G., *Handbook of Medical Physics*, Vol. I, CRC Press, Boca Raton, FL, 1982.

Kereiakes, J. G., Shalek, R. J., and Waggener, R. G., *Handbook of Medical Physics*, Vol. II, CRC Press, Boca Raton, FL, 1982.

Kereiakes, J. G., Shalek, R. J., and Waggener, R. G., *Handbook of Medical Physics*, Vol. III, CRC Press, Boca Raton, FL, 1982.

Klement, A. W., *Handbook of Environmental Radiation*, CRC Press, Boca Raton, FL, 1982.

Leventhal, L., et al., "Assessment of Radiopharmaceutical Usage Release Practices by Eleven Western Hospitals," in *Effluent and Environmental Radiation Surveillance*, Kelly, J. J., ed., STP 698, American Society for Testing and Materials, Philadelphia, 1980.

Levin, I., Kromer, B., Barabas, M., Munnich, and K. O., "Environmental Distribution and Long-Term Dispersion," *Health Physics*, 54(2): 149–156, 1988.

Meyer, R. H., and Till, J. E., *Radiological Assessment, A Textbook on Environmental Dose Analysis*, U.S. Nuclear Regulatory Commission, Washington, DC, 1983.

Miller, W. H., et al., "Radioisotopes in Municipal Sewage Sludge," *Health Physics*, 71(3): 286–289, 1996.

"Natural Background Radiation in the United States," *NCRP Report No. 45*, National Council on Radiation Protection and Measurements, Nov. 15, 1975.

"Public Radiation Exposure from Nuclear Power Generation in the United States," *NCRP Report No. 92*, National Council on Radiation Protection and Measurements, Dec. 30, 1987.

"Radiation Exposure from Consumer Products and Miscellaneous Sources," *NCRP Report No. 56*, National Council on Radiation Protection and Measurements, Nov. 1, 1977.

"Radiological Assessment: Predicting the Transport, Bioaccumulation, and Uptake by Man of Radionuclides Released to the Environment," *NCRP Report No. 76*, National Council on Radiation Protection and Measurements, March 15, pp. 96–135, 148, 204, 206, 230–231, 1984.

"Radiation Exposure of the U.S. Population from Consumer Products and Miscellaneous Sources," *NCRP Report No. 95*, National Council on Radiation Protection and Measurements, Dec. 30, 1987.

"Radiation Protection in the Mineral Extraction Industry," *NCRP Report No. 118*, National Council on Radiation Protection and Measurements, Dec. 30, p. 16, 1993.

Schultz, V., and Whicker, F. W., *Radioecology: Nuclear Energy and the Environment*, Vol. I, CRC Press, Boca Raton, FL, 1982.

Schultz, V., and Whicker, F. W., *Radioecology: Nuclear Energy and the Environment*, Vol. III, CRC Press, Boca Raton, FL, 1982.

Shleine, B., Slaback, L. A. Jr., and Birkey, B. K., *Handbook of Health Physics and Radiological Health*, Williams & Wilkins, Baltimore, MD, 1998.

Tries, M. A., et al., "Environmental Monitoring: Incinerator Operations," *Health Physics*, 71(3): 384–394, 1996.

"Tritium in the Environment," *NCRP Report No. 62*, National Council on Radiation Protection and Measurements, March 9, 1979.

U.S. Environmental Protection Agency, Office of Radiation and Indoor Air, *Diffuse NORM Wastes—Waste Characterization and Preliminary Risk Assessment*, Draft, RAE-9232/1-2, SC & A, Inc., and Rogers & Associates Engineering Corporation, Salt Lake City, UT, May 1993.

U.S. Environmental Protection Agency, Office of Radiation and Indoor Air Pollution, *Diffuse NORM Wastes—Waste Characterization and Preliminary Risk Assessment*, Draft, RAE-9232/1-2, SC & A, Inc., and Rogers & Associates Engineering Corporation, Salt Lake City, UT, May 1993.

Wang, Y., *Handbook of Radioactive Nuclides*, The Chemical Rubber Co., Cleveland, OH, 1969.

CHAPTER 3
GEOLOGY

Stephen M. Testa
President of Testa Environmental Corporation, past president of the American Institute of Professional Geologists, and current editor of the peer review journal Environmental Geosciences.

James A. Jacobs
Hydrogeologist and president of FAST-TEK Engineering Support Services. He has over 20 years of experience and is registered in several states, including California. He specializes in assessment methods and in situ remediation technologies.

SECTION 3.1

GEOLOGIC PRINCIPLES[*]

3.1.1 INTRODUCTION

Successful subsurface characterization, detection monitoring, and ultimate remediation are predicated on a solid conceptual understanding of geology and hydrogeology. The factors affecting fate and transportation of contaminants, determination of possible adverse risks to public health, safety, and welfare or the degradation of groundwater resources is largely controlled by regional and local subsurface conditions. Successful resolution of significant subsurface contamination requires adequate geologic and hydrogeologic characterization leading to insights of the preferential migration pathways, followed by a development of an appropriate remediation strategy.

Geologic factors control the vertical and lateral movement, distribution, and quality of groundwater as well as contaminants through physical, chemical, and biological processes (Back et al., 1988). Fate and transport of contaminants, both in the liquid and vapor form in soils and aquifers are, to a large part, controlled by the lithology, stratigraphy, and structure of the geologic deposits and formations. The rate of migration of liquid and gas or vapor-phase contaminants in the subsurface depends on numerous factors. Nonetheless, the concepts of porosity, permeability, and preferred migration pathways applies to both. Most of the examples in this article focus on migration of liquid-phase contaminants in aquifers.

Modeling of contaminant migration through preferred pathways relies on both regional and local scale evaluations and understanding. Regional scale might describe a fault-bounded structural basin containing dozens of vertically and laterally adjacent depositional facies. Each depositional facies might be composed of rocks containing some of the same building blocks: gravels, sands, silts and clays, in different configurations and juxtaposition. Facies distribution and changes in distribution are dependent on a number of interrelated controls: sedimentary process, sediment supply, climate, tectonics, sea level changes, biological activity, water chemistry, and volcanism. The relative importance of these regional factors ranges between different depositional environments (Reading, 1978). However, on the local and sublocal scale, the porosity and permeability of a particular sandstone

---

[*] Modified and updated from Testa, Stephen M., Chapter 3, Geologic Principles from *Geologic Aspects of Hazardous Waste Management*, pp. 55–99, Lewis Publishers, Boca Raton, FL, 1994.

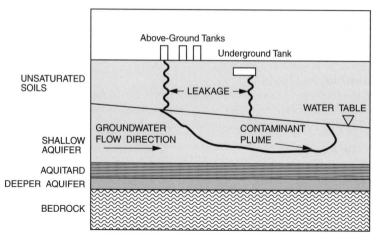

FIGURE 3.1.1   Interaction of geologic control and contaminant plume migration.

aquifer in a specific site in that basin might depend on factors such as individual grain size, grain size sorting, and primary and secondary cementation.

Lithology and stratigraphy are the most important factors affecting contaminant movement in soils and unconsolidated sediments. Stratigraphic features, including geometry and age relations between lenses, beds, and formations, as well as lithologic characteristics of sedimentary rocks, such as physical composition, grain size, grain packing, and cementation, are among the most important factors affecting groundwater and contaminant flow in sedimentary rocks. Igneous rocks, which include volcanic rocks such as basalts (or lava flows) and tuffs, and plutonic rocks such as granites, are produced by cooling of magmas during emplacement of the melt into the country or host rock. Metamorphic rocks are produced by deformation after deposition or crystallization. Groundwater and contaminant flow in igneous and metamorphic rocks is most affected by structural features such as cleavages, fractures, folds, and faults (Freeze and Cherry, 1979).

The largest percentage of environmentally contaminated sites in the world lies on alluvial and coastal plains consisting of complex interstratified sediments. The majority of these contaminated sites have some component of impacted shallow soils or unconsolidated sediments. Shallow groundwater contamination can result from surface or near-surface activities, including unauthorized releases or leaching from landfills, repositories, underground and above-ground storage tanks and pipelines, wells, septic systems, and accidental spills (Figure 3.1.1). Contamination can also occur from the application of agricultural chemicals to the land surface, or from nonpoint sources. Deeper soils or rocks may also become impacted because of preferred flow pathways along fault zones, or a lack of a competent aquitard to stop migrating contaminants or unintended man-made conduits, such as abandoned mines or improperly designed or abandoned wells.

Data collected from subsurface investigations can be compiled to obtain a three-dimensional framework used in developing a corrective action plan. Contamination potential maps can be developed based on several parameters including:

- Depth to shallow aquifers (i.e., 50 ft or less)
- Hydrogeologic properties of materials between the aquifers and ground surface
- Relative potential for geologic material to transmit water
- Description of surface materials and sediments
- Soil infiltration data
- Presence of deeper aquifers
- Potential for hydraulic intercommunication between aquifers

Development of other types of geologic and hydrogeologic maps can be used for the preliminary screening of sites for the storage, treatment, or disposal of hazardous and toxic materials. Such maps focus on those parameters that are evaluated as part of the screening process. Maps exhibit outcrop distribution of rock types that may be suitable as host rocks, distribution of unconsolidated, water-bearing deposits, distribution and hydrologic character of bedrock aquifers, and regional recharge/discharge areas. These maps can thus be used to show areas where special attention needs to be given for overall waste management including permitting of new facilities, screening of potential new disposal sites or waste management practices, and increased monitoring of existing sites and activities in environmentally sensitive areas.

Overall understanding of the regional geologic and hydrogeologic framework, characterization of regional geologic structures, and proper delineation of the relationship between various aquifers is essential to implementing both short- and long-term aquifer restoration and rehabilitation programs, and assessing aquifer vulnerability. Often, maximum contaminant levels (action levels) are used to dictate the level of effort required for remediation with no or little conception of risk, thus, misutilizing limited manpower and financial resources. Understanding of the regional hydrogeologic setting can servein designation of aquifers for beneficial use, determination of the level of remediation warranted, implementation of regional and local remediation strategies, prioritization of limited manpower and financial resources, and development of future management strategies.

3.1.2 POROSITY, PERMEABILITY, AND DIAGENESIS

Contaminants in the subsurface follow the preferred fluid and vapor migration pathways are influenced by porosity and permeability, sedimentary sequences, facies architecture, and fractures. *Porosity* is a measure of pore space per unit volume of rock or sediment and can be divided into two types: absolute porosity and effective porosity (Scholle, 1979a, 1979b). *Absolute porosity* (n) is the total void space per unit volume and is defined as the percentage of the bulk volume that is not solid material. The equation for basic porosity is as follows:

$$n = \frac{\text{bulk volume} - \text{solid volume}}{\text{bulk volume}} \times 100 \qquad (3.1.1)$$

Porosity can be individual open spaces between sand grains in a sediment or fracture spaces in a dense rock (Table 3.1.1). A fracture in a rock or solid material is an opening or a crack within the material. Matrix refers to the dominant constituent of the soil, sediment, or rock, and is usually a finer-sized material surrounding or filling the interstices between larger-sized material or features. Gravel may be composed of large cobbles in a matrix of sand. Likewise, a volcanic rock may have large crystals floating in a matrix of glass. The matrix will usually have different properties than the other features in the material. Often, either the matrix or the other features will dominate the behavior of the material, leading to the terms matrix-controlled transport, or fracture-controlled flow.

Effective porosity (N_e) is defined as the percentage of the interconnected bulk volume (i.e., void space through which flow can occur) that is not solid material. The equation for effective porosity is

TABLE 3.1.1   Porosities for Common Consolidated and Unconsolidated Materials

Unconsolidated sediments	η (%)	Consolidated rocks	η (%)
Clay	45–55	Sandstone	5–30
Silt	35–50	Limestone/dolomite (original &	
Sand	25–40	secondary porosity	1–20
Gravel	25–40	Shale	0–10
Sand & gravel mixes	10–35	Fractured crystalline rock	0–10
Glacial till	10–25	Vesicular basalt	10–50
		Dense, solid rock	<1

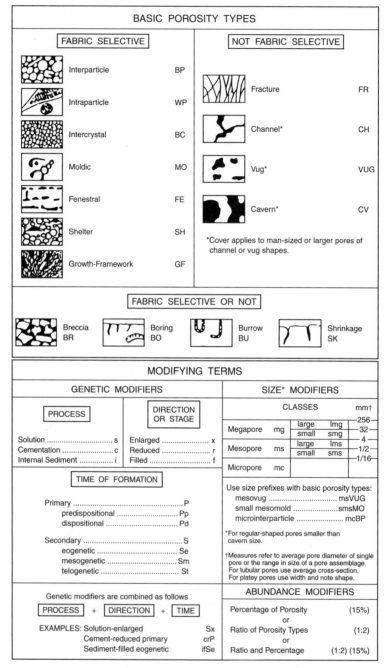

FIGURE 3.1.2 Classification of porosity types (after Choquette and Pray, 1970).

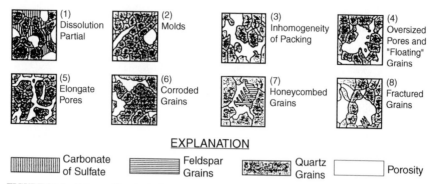

FIGURE 3.1.3 Petrographic criteria for secondary porosity (Schmidt et al., 1977).

as follows:

$$N_e = \frac{\text{interconnected pore volume}}{\text{bulk volume}} \times 100 \qquad (3.1.2)$$

Effective porosity (N_e) is of more importance and, along with permeability (the ability of a material to transmit fluids), determines the overall ability of the material to readily store and transmit fluids or vapors. Where porosity is a basic feature of sediments, permeability is dependent upon the effective porosity, the shape and size of the pores, pore interconnectiveness (throats), and properties of the fluid or vapor. Fluid properties include capillary force, viscosity, and pressure gradient.

Porosity can be primary or secondary. Primary porosity develops as the sediment is deposited and includes inter- and intraparticle porosity (Figure 3.1.2). Secondary porosity develops after deposition or rock formation and is referred to as diagenesis (Choquette and Pray, 1970).

Permeability is a measure of the connectedness of the pores (Figure 3.1.3). Thus, a basalt containing many unconnected air bubbles may have high porosity but no permeability, whereas a sandstone with many connected pores will have both high porosity and high permeability. Likewise, a fractured, dense basaltic rock may have low porosity but high permeability because of the fracture flow. The nature of the porosity and permeability in any material can change dramatically through time. Porosity and permeability can increase, for example, with the dissolution of cements or matrix, faulting, or fracturing. Likewise, porosity and permeability can decrease with primary or secondary cementation and compaction (Tables 3.1.2 and 3.1.3).

Once a sediment is deposited, diagenetic processes begin immediately and can significantly affect the overall porosity and permeability of the unconsolidated materials. These processes include compaction, recrystallization, dissolution, replacement, fracturing, authigenesis, and cementation (Schmidt et al., 1977). Compaction occurs by the accumulating mass of overlying sediments called

TABLE 3.1.2 Parameters Affecting Permeability of Sediment Following Deposition

Depositional processes	Diagenetic processes
Texture	Compaction
Grain size	Recrystallization
Sorting	Dissolution
Grain slope	Replacement
Grain packing	Fracturing
Grain roundness	Authigenesis
Mineral composition	Cementation

(After Testa, 1994).

TABLE 3.1.3   Summary of Diagenesis and Secondary Porosity

Process	Effects
Leaching	Increase n and K
Dolomitization	Increase K; can also decrease n and K
Fracturing joints, Breccia	Increase K; can also increase channeling
Recrystallization	May increase pore size and K; can also
decrease n and K	
Cementation by calcite,	Decrease n and K
dolomite, anhydrite,	
pyrobitumen, silica	

(After Testa, 1994).

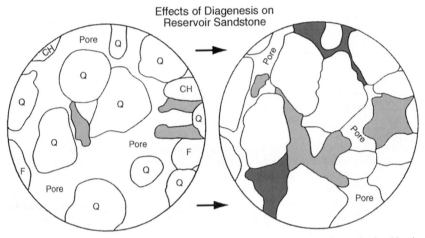

FIGURE 3.1.4   Reduction in porosity in sandstone as a result of cementation and growth of authigenic minerals in the pores affecting the amount, size, and arrangement of pores (modified after Ebanks, 1987).

overburden. Unstable minerals may recrystallize, changing the crystal fabric but not the mineralogy, or they may undergo dissolution and/or replacement by other minerals. Dissolution and replacement processes are common with limestones, sandstones, and evaporites. Authigenesis refers to the precipitation of new mineral within the pore spaces of a sediment. Lithification occurs when cementation is sufficient in quantity, such that the sediment is changed into a rock (Figure 3.1.4). Examples of lithification include sands and clays changing into sandstones and shales, respectively.

The most important parameters influencing porosity in sandstone are age (time of burial), mineralogy (i.e., detrital quartz content), sorting, and the maximum depth of burial, and to a lesser degree, temperature. Compaction and cementation will reduce porosity, although porosity reduction by cement is usually only a small fraction of the total reduction. The role of temperature probably increases above a geothermal gradient of 4°C/100 m. Uplift and erosional unloading may also be important in the development of fracture porosity and permeability. Each sedimentary and structural basin has its own unique burial history, and the sediments and rocks will reflect unique temperature and pressure curves versus depth (Figure 3.1.5).

3.1.3   SEDIMENTARY SEQUENCES AND FACIES ARCHITECTURE

Analysis of sedimentary depositional environments is important because groundwater resource usage occurs primarily in unconsolidated deposits formed in these environments (Selley, 1978). Aquifers

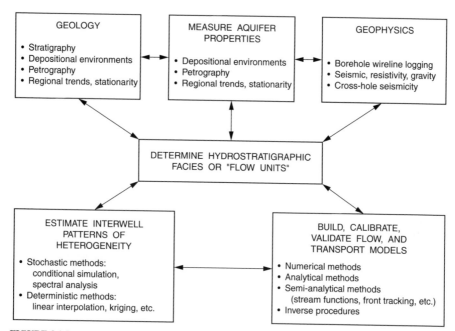

FIGURE 3.1.5 Schematic depicting the various components of an integrated aquifer description (after Testa, 1994).

are water-bearing zones often considered of beneficial use and warrant protection. Most subsurface environmental investigations conducted are also performed in these types of environments.

The literature is full of examples where erroneous hydraulic or contaminant distribution information has been relied upon. In actuality, however, (1) the wells were screened across several high-permeability zones or across different zones creating the potential for cross-contamination of a clean zone by migrating contaminants from impacted zones, (2) inadequate understanding of soil-gas surveys and vapor-phase transport in heterogeneous environments allowing for an ineffective vapor extraction remedial strategy prevailed, (3) wells screened upward-fining sequences with the technically unsound expectation of remediating hydrocarbon-affected soil and groundwater within the upper fine-grained section of the sequence, or (4) the depositional environment was erroneously interpreted. Heterogeneities within sedimentary sequences can range from large-scale features associated with different depositional environments that further yield significant large- and small-scale heterogeneities via development of preferential grain orientation. This results in preferred areas of higher permeability and, thus, preferred migration pathways of certain constituents considered hazardous.

To adequately characterize these heterogeneities (Fogg, 1989), it becomes essential that subsurface hydrogeologic assessment include determination of the following:

- Depositional environment and facies of all major stratigraphic units present

- Propensity for heterogeneity within the entire vertical and lateral sequence and within different facies of all major stratigraphic units present

- Potential for preferential permeability (i.e., within sand bodies)

The specific objectives to understanding depositional environments as part of subsurface environmental studies are to: (1) identify depositional processes and resultant stratification types that cause heterogeneous permeability patterns, (2) measure the resultant permeabilities of these stratification types, and (3) recognize general permeability patterns that allow simple flow models to be generated. Flow characteristics in turn are a function of the types, distributions, and orientations of the internal stratification. Because depositional processes control the zones of higher permeability within unconsolidated deposits, a predictive three-dimensional depositional model to assess potential connections or intercommunication between major zones of high permeability should also be developed. A schematic depicting the various components of an integrated aquifer description has been developed.

Understanding of facies architecture is extremely important to successful characterization and remediation of contaminated soil and groundwater. Defining a hydrogeologic facies can be complex. Within a particular sedimentary sequence, a hydrogeologic facies can range over several orders of magnitude. Other parameters, such as storativity and porosity, vary over a range of only one order of magnitude. A *hydrogeologic facies* is defined as a homogeneous, but anisotropic, unit that is hydrogeologically meaningful for purposes of conducting field studies or developing conceptual models. Facies can be gradational in relation to other facies, with a horizontal length that is finite but usually greater than its corresponding vertical length. A hydrogeologic facies can also be viewed as a sum of all the primary characteristics of a sedimentary unit. A facies can thus be referred to with reference to one or several factors, such as lithofacies, biofacies, geochemical facies, among others. The importance of facies cannot be understated. For example, three-dimensional sedimentary bodies of similar textural character are termed lithofacies. It is inferred that areas of more rapid plume migration and greater longitudinal dispersion correlate broadly with distribution and trends of coarse-grained lithofacies and are controlled by the coexistence of lithologic and hydraulic continuity. Therefore, lithofacies distribution can be used for preliminary predictions of contaminant migration pathways and selection of a subsurface assessment and remediation strategy. However, caution should be taken in proximal and distal assemblages where certain layered sequences may be absent because of erosion, and the recognition of cyclicity is solely dependent on identifying facies based simply on texture. Regardless, the facies reflects deposition in a given environment and possesses certain characteristics of that environment. Sedimentary structures also play a very important role in deriving permeability-distribution models and developing fluid-flow models.

Nearly all depositional environments are heterogeneous, which for all practical purposes restricts the sole use of homogeneous-based models in developing useful hydraulic conductivity distributions data for assessing preferred contaminant migration pathways, and for developing containment and remediation strategies. There has been much discussion in the literature regarding the influences of large-scale features such as faults, fractures, significantly contrasting lithologies, diagenesis, and sedimentalogical complexities. Little attention, however, has been given to internal heterogeneity within genetically defined sand bodies caused by sedimentary structures and associated depositional environment and intercalations. In fact, for sand bodies greater variability exists within bedding and lamination pockets than between them. An idealized model of the vertical sequence of sediment types by a meandering stream shows the highest horizontal permeability (to groundwater or contaminants) to be the cross-bedded structure in a fine- to medium-grained, well-sorted unconsolidated sand or sandstone (Figure 3.1.6).

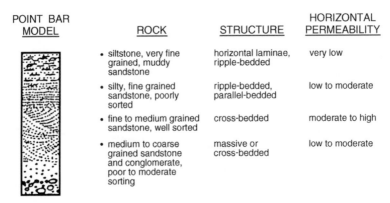

POINT BAR MODEL	ROCK	STRUCTURE	HORIZONTAL PERMEABILITY
	• siltstone, very fine grained, muddy sandstone	horizontal laminae, ripple-bedded	very low
	• silty, fine grained sandstone, poorly sorted	ripple-bedded, parallel-bedded	low to moderate
	• fine to medium grained sandstone, well sorted	cross-bedded	moderate to high
	• medium to coarse grained sandstone and conglomerate, poor to moderate sorting	massive or cross-bedded	low to moderate

FIGURE 3.1.6   Point bar geologic model showing the influences of a sequence of rock textures and structures in an aquifer consisting of a single point bar deposit on horizontal permeability excluding effects of diagenesis (modified after Ebanks, 1987).

The various layers illustrated affect the flow of fluids according to their relative characteristics. For example, in a point-bar sequence, the combination of a ripple-bedded, coarser-grained sandstone will result in retardation of flow higher in the bed, and deflection of flow in the direction of dip of the lower trough crossbeds (Ebanks, 1987).

Hydrogeologic analysis is conducted in part by the use of conceptual models. These models are used to characterize spatial trends in hydraulic conductivity and permit prediction of the geometry of hydrogeologic facies from limited field data. Conceptual models can be either site specific or generic. Site- specific models are descriptions of site-specific facies that contribute to understanding the genesis of a particular suite of sediments or sedimentary rocks. The generic model, however, provides the ideal case of a particular depositional environment or system (Scholle and Spearing, 1982; Scholle et al., 1983; Reading, 1978; Potter and Pettijohn, 1977; Nilsen, 1982). Generic models can be used in assessing and predicting the spatial trends of hydraulic conductivity and, thus, dissolved contaminants in groundwater. Conventional generic models include either a vertical profile, which illustrates a typical vertical succession of facies, or a block diagram of the interpreted three-dimensional facies relationships in a given depositional system.

Several of the more common depositional environments routinely encountered in subsurface environmental studies are discussed below. Included is discussion of fluvial, alluvial fan, glacial, deltaic, eolian, carbonate, and volcanic-sedimentary sequences. Hydrogeologic parameters per depositional environment are available from the literature and summarized in Table 3.1.4.

Fluvial Sequences

Fluvial sequences are difficult to interpret because of their sinuous nature and complexities of their varied sediment architecture reflecting complex depositional environments (Larkin and Sharp, 1992; Miall, 1988). Fluvial sequences can be divided into high-sinuosity meandering channels and low-sinuosity braided channel complexes. Meandering stream environments (i.e., Mississippi River) consist of an asymmetric main channel, abandoned channels, point bars, levees, and floodplains (Figure 3.1.7).

Usually developed where gradients and discharge are relatively low, five major lithofacies have been recognized:

- Muddy fine-grained streams
- Sand-bed streams with accessory mud
- Sand-bed stream without mud

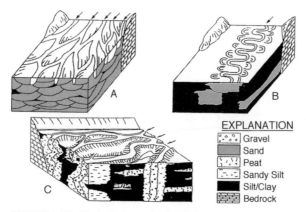

FIGURE 3.1.7 Fluvial facies model illustrating contrasting patterns of heterogeneity in (A) braided rivers, (B) meandering rivers, and (C) anastomosing rivers (Anderson, 1989).

TABLE 3.1.4 Summary of Hydraulic Properties for Certain Depositional Environments

Depositional environment[a]	Hydrogeologic facies	Hydraulic conductivity[b,c]		Porosity in percent[c]	Reference
		Horizontal	Vertical		
Eolian	Dune sand	5–140 (54)		42–55 (49)	Pryor (1973)
	Interdune/extra-erg	0.67–1800			Chandler et al. (1989)
	Wind-ripple	900–5200			Chandler et al. (1989)
	Grain flow	3700–12,000			Chandler et al. (1989)
Fluvial	River point bar	4–500 (93)		17–52 (41)	
	Beach sand	3.6–166 (68)		39–56 (49)	
Glacial	Meltwater streams	10^{-1}–10^{-5} cm/s			Anderson (1989)
	Outwash drift	10^{-3}–10^{-4} [d]			Sharp (1984)
	Basal till	10^{-4}–10^{-9} cm/s	10^{-11} [e]		Anderson (1989)
	Esker sediment	10^{-1}–10^{-3} cm/s			Casewell (1988a; 1988b); De Gear (1986); Patson (1970)
	Supraglacial sediments	10^{-3}–10^{-7} cm/s			Stephanson et al. (1989)
Deltaic	Distributary channel sandstone	(436)		(28)	Tillman and Jordan (1987)
	Splay channel sandstone	(567)		(27)	Tillman and Jordan (1987)
	Wave-dominated sandstone within prodelta and shelf mudstone	(21)		(21)	Tillman and Jordan (1987)
Volcanic-Sedimentary	Basalt (CRG)[f]	0.002–1600 (0.65)	10^{-8}–10		Lindholm and Vaccaro (1988)
	Basalt (SRG)[f]	150–3000			Lindholm and Vaccaro (1988)
	Basalt (CRG)	1×10^{-8}–10^{-5} cm/s	1×10^{-7}–2×10^{-7} cm/s		Testa (1988); Wang and Testa (1989)
	Sedimentary interbed (SRG)	3×10^{-6}–3×10^{-2} [e]			Lindholm and Vaccaro (1988)
	Tuffaceous siltstone (interbeds; CRG)	1×10^{-6}–2×10^{-4} cm/s	1×10^{-8}–1×10^{-3} cm/s [e]	27–68 (42)	Testa (1988); Wang and Testa (1989)
	Interflow zone (CRG)	2×10^{-4} cm/s			Testa (1988); Wang and Testa (1989)

[a] Carbonate not represented but can have permeabilities ranging over 5 orders of magnitude.
[b] Values are in millidarcy (mD) per day unless otherwise noted; cm/s = centimeters per second; 1 mD = .001 darcy; 1 cm/s = 1.16×10^{-3} darcy.
[c] Values shown in parentheses are averages.
[d] Field.
[e] Laboratory.
[f] CRG = Columbia River Group; SRG = Snake River Group.

- Gravelly sand-bed stream
- Gravelly stream without sand

Meandering streams can also be subdivided into three subenvironments: floodplain subfacies, channel subfacies, and abandoned channel subfacies. Floodplain subfacies comprises very fine sand, silt, and clay deposited on the overbank portion of the floodplain, out of suspension during flooding events. Usually laminated, these deposits are characterized by sand-filled shrinkage cracks (subaerial exposure), carbonate caliches, laterites, and root holes. The channel, subfacies is formed as a result of the lateral migration of the meandering channel, which erodes the outer concave bank, scours the riverbed, and deposits sediment on the inner bank, referred to as the point bar. Very characteristic sequences of grain size and sedimentary structures are developed. The basal portion of this subfacies is lithologically characterized by an erosional surface overlain by extraformation pebbles and intraformational mud pellets. Sand sequences with upward fining and massive, horizontally stratified and trough cross-bedded sands overlie these basal deposits. Overlying the sand sequences are tabular, planar, cross-bedded sands that grade into microcross-laminated and flat-bedded fine sands, grading into silts of the floodplain subfacies. The abandoned channel subfacies are curved fine-grained deposits of infilled abandoned channels referred to as oxbow lakes. Oxbow lakes form when the river meanders back, short-circuiting the flow. Although lithologically similar to floodplain deposits, geometry and absence of intervening point-bar sequences distinguishes it from the abandoned channel subfacies.

Braided river systems consist of an interlaced network of low-sinuosity channels and are characterized by relatively steeper gradients and higher discharges than meandering rivers. Typical of regions where erosion is rapid, discharge is sporadic and high, and little vegetation hinders runoff, braided rivers are often overloaded with sediment. Because of this sediment overload, bars are formed in the central portion of the channel around which two new channels are diverted. This process of repeated bar formation and channel branching generates a network of braided channels throughout the area of deposition.

Lithologically, alluvium derived from braided streams are typically composed of sand and gravel-channel deposits. Repeated channel development and fluctuating discharge results in the absence of laterally extensive cyclic sequences as produced with meandering channels. Fine-grained silts are usually deposited in abandoned channels formed by both channel choking and branching, or trapping of fines from active downstream channels during eddy reversals.

The degree of interconnectedness is important in addressing preferred migration pathways in fluvial sequences. Based on theoretical models of sandbody connectedness, the degree of connectedness increases very rapidly as the proportion of sand bodies increases above 50 percent. Where alluvial soils contain 50 percent or more of overbank fines, sand bodies are virtually unconnected.

Alluvial Fan Sequences

Alluvial fan sequences accumulate at the base of an upland area or mountainous area as a result of an emerging stream (Cehrs, 1979). These resulting accumulations form segments of a cone with a sloping surface ranging from less than 1° up to 25° averaging 5°, and rarely exceeding 10°. Alluvial fans can range in size from less than 100 m to more than 150 km in radius, although typically averaging less than 10 km. As the channels shift laterally through time, the deposit develops a characteristic fan shape in plane view, a convex-upward cross-fan profile, and concave-upward radial profile as illustrated in Figure 3.1.8 (Spearing, 1974).

Facies analysis of alluvial fans require data on morphology and sediment distribution, and can be divided into four facies: proximal, distal, and of lesser importance, outer fan, and fan fringe facies. Proximal facies are deposited in the upper or inner parts of the fan near the area of stream emergence and are composed of relatively coarser-grained sediments (Nilsen, 1982). The proximal facies comprising the innermost portion of the fan (i.e., apex or fan-head area) contains an entrenched straight valley that extends outward onto the fan from the point of stream emergence. This inner fan region is characterized by two subfacies: a very coarse-grained, broad, deep, deposit, of one or several major channels, and a finer-grained channel-margin level and interchannel deposits, which

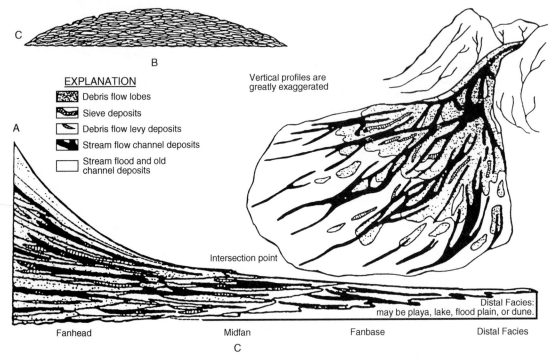

FIGURE 3.1.8 Generalized model of alluvial fan sedimentation showing (A) fan surface, (B) cross-fan profile, and (C) radial profile (after Spearing, 1974).

may include coarse-grained landslides and debris flows type material. Distal facies are deposited in the lower and outer portion of the fan, and are composed of relatively finer-grained sediments. The distal facies typically comprises the largest area of most fans, and consists of smaller distributary channels radiating outward and downfan from the inner fan valley. Hundreds of less-developed channels may be present on the fan. Depending on fan gradient, sediment time and supply, and climatic effects among other factors, commonly braided but straight, meandering, and anastomosing channel systems may also be present. Outer fan facies are composed of finer-grained, laterally extensive, sheet-like deposits of nonchannelized or less-channelized deposits. These deposits maintain a very low longitudinal gradient. The fanfringe facies is composed of very fine-grained sediments that intertongue with deposits of other environments (i.e., eolian, fluvial, lacustrine, etc.). Most deposits comprising alluvial fan sequences consist of fluvial (streamflow) or debris flow types.

Alluvial fans are typically characterized by high permeability and porosity. Groundwater flow is commonly guided by paleochannels, which serve as conduits, and relatively less permeable and porous debris and mud-flow deposits. The preponderance of debris-flow and mud-flow deposits in the medial portion of fans may result in decreased and less predictable porosity and permeability in these areas. Aquifer characteristics vary significantly with the type of deposit and relative location within the fan. Pore space also develops as intergranular voids, interlaminar voids, bubble cavities, and desiccation cracks.

Deltaic Sequences

Deltas are abundant throughout the geologic record with 32 large deltas forming at this time and countless others in various stages of growth. A delta deposit is partly subaerial built by a river into

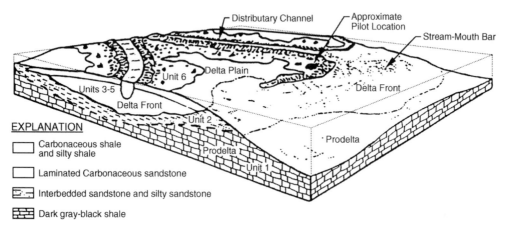

FIGURE 3.1.9   Block diagram showing vertical and aerial distribution of units in a typical modern delta (modified after Harris, 1975).

or against a body of permanent water. Deltaic sedimentation requires a drainage basin for a source of sediment, a river for transport of the material, and a receiving basin to store and rework it. During formation, the outer and lower parts are constructed below water, and the upper and inner surfaces become land reclaimed from the sea. Deltas form by progradation or a building outward of sediments onto themselves.

As the delta system progrades further and further, the slope and discharge velocity lessens and the carrying capacity of the delta is reduced by its own sediment load. Once a particular pathway in a delta system is no longer available because of the sediment buildup and upward vertical migration of the channel bed and the adjacent levees, another delta system forms in a different location, usually by a break in a levee wall in an upgradient position. This break in the wall and shift in the locus of deposition increases the slope, sediment carrying capacity, and discharge velocity of the new delta system. Any given delta is thus a composite of conditions reflecting initiation of delta development to its ultimate abandonment of a particular deposition center (Figure 3.1.9).

Delta sequences reflect condition of source (volume and type of available sediment) and distribution and dispersal processes. Two general classes or end members have been defined: high- and low-energy deltas. High-energy deltas or sand deltas are characterized by few active meandering distributary channels, with shoreline composed of continuous sand (i.e., Nile, Rhone, or Brazos-Colorado). Low-energy or mud deltas are characterized by numerous bifurcating or branching, straight to sinuous, distributary channels, with shorelines composed of discontinuous sands and muds.

No two deltas are exactly alike in their distribution and continuity of permeable and impermeable sediments. The most important parameters controlling the distribution is size and sorting of grains. All delta systems form two parts during development: a regressive sequence of sediment produced as the shoreline advances seaward and a system of distributary channels. These two parts result in two main zones of relatively high permeability: channel sands and bar sands.

Typical deltaic sequences from top to bottom include marsh, inner bar, outer bar, prodelta, and marine (Table 3.1.5). Depositional features include distributary channels, river-mouth bars, interdistributary bays, tidal flats and ridges, beaches, eolian dunes, swamps, marshes, and evaporite flats (Figure 3.1.10).

The geometry of channel deposits reflects delta size, position of the delta in the channel, type of material being cut into, and forces at the mouth of the channel distributing the sediments. For high-energy sand deltas, channels can be filled with up to 90 percent sand, or clay and silt. For example, those sequences of high permeability in bar deposits is at or near the top with decreasing permeability with depth and laterally away from the main sand buildup. Within bar deposits, permeability generally increases upward and is highest at or near the top, and is similar parallel to the trend of the bar (i.e.,

TABLE 3.1.5   Summary of Deltaic Sequences and Characteristics

Sequence	Sediment type	Permeability
Marsh	Clay, silt, coal with a few aquifer sands of limited extent	Relatively low
Inner bar	Sand with some clay/silt intercalations	Moderate to high
Outer bar	Sand with many clay/silt intercalations	Moderate
Prodelta	Clay and silt	Low
Marine	Clay	Low

(After Testa, 1994).

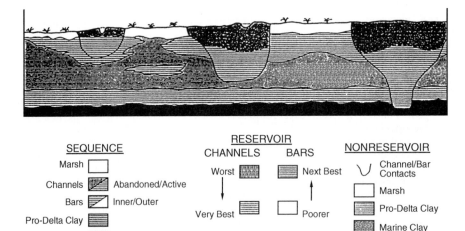

FIGURE 3.1.10   Deltaic facies model illustrating heterogeneity and continuity (after Testa, 1994).

highest at the top near the shore, decreasing progressively seaward). Porosity is anticipated to be well connected throughout the bar with the exception of the lower part. In channels, however, high-permeability sequences occur in the lower part with decreasing permeability vertically upward. Within channel deposits, porosity and permeability is high at the lower part of the channel, decreasing verti-cally upward, with an increase in the number, thickness, continuity, and aerial extent of clay interbeds.

Low-energy mud deltas differ from sand deltas in that sediments are carried into a basin via numerous channels during flooding events, and the fine-grained silts and clays are not winnowed from the sand before new sediment circulates from the next flooding event. The coarse-grained sediment is thus more discontinuous with more numerous particles and continuous clay and silt interbeds (less than 1 in thick). Along the perimeter or within bar deposits, grain size increases with sorting; improves vertically upward; and clay and silt interbeds decrease in number, thickness, and aerial extent. Within individual bars, overall permeability decreases laterally away from the coarse-grained sand depositional pathways. The highest permeability is at or near the top of the bar decreasing vertically downward and in a seaward direction. Sand continuity, thus permeability, is poor because of the numerous shifting distributary channels forming widespread clay and silt interbeds ranging from less than 1 ft to more than 12 ft in thickness. Coarse-grained sands predominate within the lower portion of the distributary channels with clay and silt interbeds typically less than 1 ft in thickness, and range from a few feet to a few tens of feet in maximum aerial extent, thus, not providing a barrier to vertical flow. The number, thickness, and aerial extent of these fine-grained interbeds generally increases vertically upward depending on how fast and where the channel was abandoned. Overall, permeability and porosity continuity is high only in the upward portion of the bar. Within the channels, however, permeability and porosity continuity is high at the basal portion of the channels, but the amount and quality of coarse-grained sand (high permeability zones) is dependent upon the location and rate of channel abandonment.

Average porosity and permeability based on a broadly lenticular wave-dominated deltaic sandstone (e.g., Upper Cretaceous Big Wells aquifer, which is one of the largest oil fields located in south Texas) increased in prodelta and shelf mudstones, averaging 21 percent and 6 mD, respectively (Tyler et al., 1987). Studies on the El Dorado field located in southeastern Kansas, a deltaic sequence containing the 650-ft-thick Admire sandstone, has reported porosity and permeability averaging 28 percent and 436 mD, respectively, within the distributary channel sandstones (Sneider et al., 1978). Thinner and discontinuous splay channels sandstones average 27 percent porosity and 567 mD in permeability. The variation in porosity and permeability reflect diagenetic processes (i.e., deformation, secondary leaching of feldspar, and formation of calcite cement and clay laminae).

Glacial Sequences

Sequences derived from glacial processes include four major types of materials: tills, ice-contact, glacial fluvial or outwash, and delta and glaciolacustrine deposits. Glacial tills make up a major portion of a group of deposits referred to as diamictons, which are defined as poorly sorted, unstratified deposits of nonspecific origin. Tills and associated glaciomarine drift deposits are both deposited more or less directly from ice without the winnowing effects of water. Till is deposited in direct contact with glacial ice and, although substantial thickness accumulations are not common, tills make up a discontinuous cover totaling up to 30 percent of the earth's continental surface.

Glaciomarine drift, however, accumulates as glacial debris melts out of ice floating in marine waters. These deposits are similar to other till deposits but also includes facies that do not resemble till or ice-contact deposits. A lesser degree of compaction is evident because of a lack of appreciable glacial loading.

Tills can be divided into two groups based on deposition: basal till or supra glacial till, or three groups based upon physical properties and varying depositional processes: lodgement, ablation, and flow. Lodgement tills are deposited subglacially from basal, debris-laden ice. High shear stress results in a preferred fabric (i.e., elongated stones oriented parallel to the direction of a flow) and high degree of compaction, high bulk densities, and low void ratios of uncemented deposits. Ablation tills are deposited from englacial and superglacial debris dumped on the land surface or the ice melts away. These deposits lack significant shear stresses and thus are loosely consolidated with a random fabric. Flow tills are deposited by water-saturated debris flowing off glacial ice as mudflow. Flow tills exhibit a high degree of compaction, although less than that of lodgement tills, with a preferred orientation of elongated stones resulting from flowage.

Till is characterized by a heterogeneous mixture of sediment sizes (boulders to clay) and a lack of stratification. Particle size distribution is often bimodal with predominant fractions in the pebble-cobble range and silt-clay range, both types being massive with only minor stratified intercalations. Other physical characteristics of till include glaciofluvial deposits or outwash deposits having strong similarities to sediments formed in fluvial environments because of similar transportation and deposition mechanisms. These types of deposits are characterized by abrupt particle-size changes and sedimentary structures reflecting fluctuating discharge and proximity to glaciers. Characteristics include a downgradient fining in grain size and down-gradient increase in sorting, therefore a decrease in hydraulic conductivity. Outwash deposits can be divided into three facies: proximal, intermediate or medial, and distal. Outwash deposits are typically deposited by braided rivers, although the distal portions are deposited by meandering and anastomosing rivers. Proximal facies are deposited by gravel-bed rivers while medial and distal facies are deposited by sand-bed rivers. Thus, considerable small-scale variability within each facies assemblage exists. Vertical trends include fining-upward sequences as with meandering fluvial sequences. Within the medial portion, a series of upward fining or coarsening cycles is evident depending on whether the ice front was retreating or advancing,

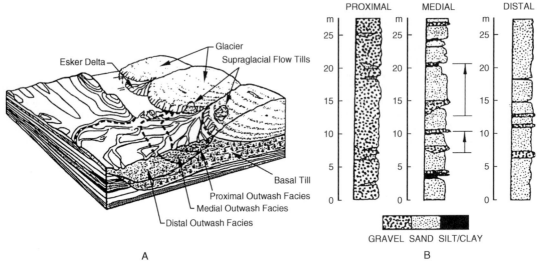

FIGURE 3.1.11   Hydrogeologic facies model for glacial depositional environment (after Testa, 1994).

respectively. Layered sequences within the gravel-dominant proximal facies and sand-dominant distal facies are either absent or hindered by the relatively large-grain size component of the proximal facies. A hydrogeologic facies model and their respective vertical profile has been developed (Figure 3.1.11).

Delta and glaciolacustrine deposits are formed when meltwater streams discharge into lakes or seas. Ice-contact delta sequences produced in close proximity to the glacier margins typically exhibit various slump-deformation structures. Delta sequences produced a considerable distance from the glacier margins exhibit no ice-collapse structures, variable sediment discharge, and particle-size distribution and structures (i.e., graded bedding, flow rolls, varies, etc.) similar to that of meltwater streams.

Also associated with till deposits are ice-contact deposits, which form from meltwater on, under, within, or marginal to the glacier. Detritus deposits formed on, against, or beneath the ice exhibit better sorting and stratification, a lack of bimodal particle-size distribution, and deformational features such as collapse features (i.e., tilting, faulting, and folding).

Hydrogeologically, hydraulic conductivity of basal tills facies are on the order of 10^{-4} cm/s with horizontal hydraulic conductivities on the order of 10^{-3} to 10^{-7} cm/s reflecting locations and degree of interconnected sand and gravel channel deposits contained within the till. Drift deposits can vary from about 10^{-11} m/s (laboratory tests) to 10^{-6} to 10^{-7} (field) when permeable sand lenses or joints are intersected (Sharp, 1984).

Eolian Sequences

Eolian or wind-deposited sediments are complex, highly variable accumulations. They are characterized as well-sorted, matrix-free, well-rounded sediments with a dominance of sand-sized fractions, and are perceived as essentially lithologically homogeneous with irregular plan and cross-sectional geometries, with exception to the linear trends of coastal dunes. Unlike many other sedimentary facies, eolian deposits have no predictable geometry and/or cyclic motif of subfacies. Recent studies have provided a better understanding of the stratigraphic complexity and thus flow regime within these deposits.

Small-scale forms of eolian deposits include wind sand ripples and wind granule ripples. Wind sand ripples are wavy surface forms on sandy surfaces whose wavelength depends on wind strength and remains constant with time. Wind granule ripples are similar to wind sand ripples but are usually produced in areas of erosion. Excessive deflation produces a large concentration of grains 1 to 3 mm in diameter, which are too big to be transported via saltation under the existing wind conditions.

Larger-scale eolian sand forms include sand drifts, sand shadows, gozes, sand sheets, and sand dunes. Sand drifts develop by some fixed obstruction that lies in the path of a sand-laden wind. When sand accumulates in the lee of the gap between two obstacles, a tongue-shaped sand drift develops. As the wind velocity is reduced by an obstruction, a sand shadow develops. Gozes are gentle large-scale undulatory sand surfaces associated with sparse desert vegetation. Sand sheets are more or less flat, with slight undulations or small dune tile features, and encompassing large areas.

Sand dunes are the most impressive features and develop whenever a sandladen wind deposits in a random patch. This patch slowly grows in height as a mound, until finally a slip face is formed. The sand mound migrates forward as a result of the advance of the slip face, but maintains its overall shape providing wind conditions do not change. Sand dunes are characterized by wind conditions, sand type, and sand supply (Figure 3.1.12).

Dune Type	Definition and Occurrence	
Barchan Dune	A crescent-shaped dune with horns pointing downward. Occurs on hard, flat floors of desserts. Constant wind and limited sand supply. Height 1 m to more than 30 m.	
Transverse Dune	A dune forming a asymmetrical ridge transverse to wind direction. Occurs in area with abundant sand and little vegetation. In places grades into barchans.	
Parabolic Dune	A dune of U-shape with the open end of the U facing upwind. Some form by piling of sand along leeward and lateral margins of areas of deflation in older dunes.	
Linear Dune	A long, straight, ridge-shaped dune parallel with wind direction. As much as 100 m high and 100 km long. Occurs in deserts with scanty sand supply and strong winds varying within one general direction. Slip faces vary as wind shifts direction.	
Star Dune	An isolated hill of sand having a base that resembles a star in plan view. Ridges converge from basal points to a central peak as high as 100 m. Tends to remain fixed in place in an area where wind blows from all directions.	

FIGURE 3.1.12   Dune types based on form (modified after McKee, 1979).

FIGURE 3.1.13   Basic eolian bed forms as related to the number of slip facies (after Reineck and Singh, 1975).

Eolian deposits are stratigraphically complex because of (1) differing spatial relationships of large-scale forms such as dunes, interdunes, and sand-sheet deposits relative to one another and to ectradune (noneolian) sediments and (2) varying dune types, each with its own cross-bedding patterns and different degrees of mobility; thus, there are different fluid-flow properties when consolidated or lithified. Sedimentary structures within eolian deposits include ripples, contorted sand bedding, cross-bedding with great set heights, normally graded beds, inversely graded beds, evenly laminated beds, discontinuously laminated beds, nongraded beds, and lag deposits along boundary surfaces and sets. Basic eolian bed forms as related to a number of slip facies is well documented (Figure 3.1.13).

Relatively recent eolian deposits are presumed to have high porosity and permeability, and are typically well-rounded, well-sorted, and generally only slightly cemented. Regional permeability is usually good due to a lack of fine-grained soils, shales, interbeds, and others, and thus constitutes important aquifers. Studies conducted on several large eolian deposits (i.e., Page sandstone of northern Arizona) have shed some light on preferred fluid migrations in such deposits. For example, fluid flow is directional dependent because of inverse grading (Chandler et al., 1989) within laminae. Permeability measured parallel to wind-ripple laminae has been shown to be from two to five times greater than that measured perpendicular to the laminae. Four common cross-set styles based on bulk permeability and directional controls of each stratum type for the Page sandstone in northern Arizona (Figure 3.1.14).

Page sandstone is poorly cemented, has high porosity, and has a permeability outcrop, which has never been buried. In Case A, the cross-set is composed exclusively of grain-flow strata; thus, the permeability of each grain-flow set is high in all directions with significant permeability contrasts,

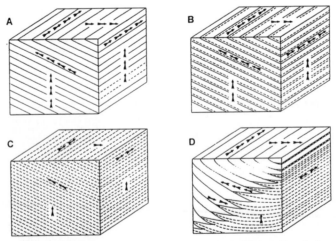

FIGURE 3.1.14   Common styles of cross-strata in Page sandstone: (A) grain-flow set, (B) wind-ripple set, (C) interlaminated grain-flow and wind-ripple set, and (D) grain-flow foresets toeing into wind-ripple bottom sets. Directional permeability indicated by arrows; more arrows denote higher potential flow (after Reineck and Singh, 1975).

EXPLANATION

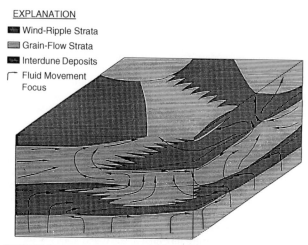

FIGURE 3.1.15   Fluid flow through idealized eolian sequences based on relative permeability values of stratification types and bounding surfaces, assuming a vertical pressure field (after Chandler et al., 1989).

flow barriers, or severely directional flow. Because of inverse grading, fluid flow is greater parallel to the grain-flow strata than across it. In Case B, laminae occurring in wind-ripple cross-sets have low permeability throughout, thus inhibiting flow and imparting preferred flow paths. Case C illustrates bulk permeability based on the ratio of grain-flow strata to wind-ripple strata in the cross set. Higher ratios are indicative of a greater capacity to transmit fluids, while low permeability logs created by wind-ripple sets act to orient fluid migration parallel to the stratification. Case D shows that a cross-set that exhibits grain-flow deposits grading into windripple laminae is more permeable at the top of the set because of the dominance of grain-flow strata. Fluid flow is thus reduced downward throughout the set because of the transition from high-permeability grain-flow cross strata to low permeability wind-ripple laminae, with greater ease of flow occurring in the cross-set from the wind-ripple laminae into the grain-flow strata.

Overall, interdune or extra-erg deposits are least permeable (0.67 to 1800 mD), wind-ripple strata moderately permeable (900 to 5200 mD), and grain-flow strata the most permeable (3700 to 12,000 mD) as illustrated in Figure 3.1.15.

Compartmentalization develops because of bounding surfaces between the cross-sets, with flow largely channeled along the sets. Fluid flow is especially great where low permeability interdune or extra-erg deposits overlie bounding surfaces (i.e., along sets). Flow windows, however, occur where low permeability strata were eroded or pinched out. Because of high-permeability grain-flow deposits relative to wind-ripple strata, fluid migration between adjacent grain-flow sets would be more rapid than across bounding surfaces separating sets of wind-ripple deposits. Flow through the sets themselves would have been dictated by internal stratification types.

Carbonate Sequences

Carbonate sequences are important in that aquifers within such sequences are often heavily depended upon for drinking water, irrigation, and other uses. Carbonate rocks are exposed on over 10 percent of the earth's land area. About 25 percent of the world's population depends on fresh water retrieved from Karst aquifers. The Floridan aquifer of Cenozoic age, for example, is the principal aquifer in the southeastern United States, and is encountered in Florida, Georgia, Alabama, and South Carolina. However, although its major resource is as a potable water supply, the nonpotable part of the aquifer in southern Florida serves as a disposal zone for municipal and industrial wastewater via injection wells.

Karst terrains are the foremost examples of groundwater erosion. Karst terrains can be divided into two types: well-developed and incipient. Well-developed karst terrains are marked by surface features such as dolines or sinkholes, which can range from 1 to 1000 m in maximum dimension. Other features are closed depressions, dry valleys, gorges, and sinking streams and caves, with local groundwater recharge via both infiltration and point sinks. The subsurface systems of connected conduits and fissure openings, enlarged by solution (i.e., cave systems), serve a role similar to that of stream channels in fluvial systems, and reflect the initial phases of subsequent surface karst landscape features (i.e., sinkholes and depressions). Cave systems actually integrate drainage from many points for discharge at single, clustered, or aligned springs. Incipient karst terrains differ from well-developed terrains in that few obvious surficial features exist, and recharge is limited primarily to infiltration. Several other approaches to classification of karst terrains exist: holokarst, merokarst or fluviokarst, and parakarst. In holokarst terrains, all waters are drained underground, including allogenic streams (i.e., those derived from adjacent nonkarst rocks), with little or no surface channel flow. In fluviokarst terrains, major rivers remain at the surface, reflecting either large flow volumes that exceed the aquifer's ability to adsorb the water or immature subsurface development of underground channels. Parakarst terrains are a mixture of the two reflecting mixtures of karst and nonkarst rocks. Covered karst reflects the active removal of carbonate rocks beneath a cover of other unconsolidated rocks (i.e., sandstone and shales); whereas, mantled karst refers to deep covers of unconsolidated rocks or materials. Paleokarst terrains are karst terrains or cave systems that are buried beneath later strata and can be exhumed or rejuvenated. Pseudokarst refers to karst-like landforms that are created by processes other than rock dissolution (i.e., thermokarst, vulcanokarst, and mechanical piping) as illustrated in Figure 3.1.16.

Carbonate sediments can accumulate in both marine and nonmarine environments (Beck, 1989; Cayeux 1970; Scholle et al., 1983). The bulk of carbonate sediments are deposited in marine environments and in tropical and subtropical seas, with minimal or no influx of terrigenous or land-derived

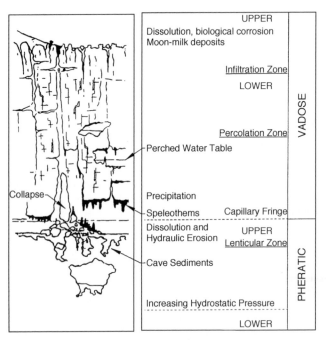

FIGURE 3.1.16 Idealized authigenic karst profile (after Scholle et al., 1983).

detritus. Marine depositional environments include tidal flat, beach and coastal dune, continental shelf, bank, reef, basin margin and slope, and deeper ocean or basin. Lakes provide the most extensive carbonate deposits on land, regardless of climate, although carbonate can occur or caliche (i.e., soil-zone deposits) and travertine (i.e., caves, karst, and hot-spring deposits).

Carbonate rocks are defined as containing more than 50 percent carbonate minerals. The most common and predominant carbonate minerals are calcite ($CaCO_3$) and dolomite [$CaMg(CO_3)_2$]. Other carbonate minerals include aragonite ($CaCO_3$), siderite ($Fe_{12}CO_3$), and magnesite ($MgCO_3$). The term *limestone* is used for those rocks in which the carbonate fraction is composed primarily of calcite, whereas the term *dolomite* is used for those rocks composed primarily of dolomite.

Overall, carbonate rocks serve as significant aquifers worldwide and are not limited by location or age of the formation. Carbonate rocks show a total range of hydraulic conductivities over a range of ten orders of magnitude. The broad diversity in hydrogeologic aspects of carbonate rocks reflects the variable combination of more than 60 processes and controls. Hydrogeologic response is related to rock permeability, which is affected most by interrelated processes associated with dynamic freshwater circulation and solution of the rock. Dynamic freshwater circulation is controlled and maintained primarily by the hydraulic circuit: maintenance of the recharge, flowthrough, and discharge regime. Without these regimes, the overall system is essentially stagnant and does not act as a conduit. The primary controls on solutions include rock solubility and chemical character of the groundwater; secondary controls include diagenetic, geochemical, and chronologic aspects.

Carbonate aquifers are characterized by extremely heterogeneous porosity and permeability, reflecting the wide spectrum of depositional environments for carbonate rocks and subsequent diagenetic alteration of the original rock fabric. Pore systems can range from thick, vuggy aquifers in the coarse-grained skeletal-rich facies of reef or platform margin, to highly stratified, often discontinuous aquifers in reef and platform interiors, and nearshore facies. Because of the brittle nature of limestones and dolomites, most exhibit extensive joint or fault systems because of uneven isostatic adjustment and local stresses produced by solution effects and erosion.

Rainwater commonly absorbs carbon dioxide from the air and forms carbonic acid, a weak acid. Once exposed at or near the surface, limestone and dolomite can be easily dissolved by acidic rainwater (Driscoll, 1986).

Karst terrains are highly susceptible to groundwater contamination. When used for waste disposal, these areas are susceptible to potential failure because of subsidence and collapse, which in turn can result in aquifer compartmentalization. To assess secondary porosity and potential contamination susceptibility, characterization of carbonate or karst aquifers include generation of data regarding percent rock core recovery, mechanical response during drilling, drilling fluid loss, and drilling resistance.

Volcanic-Sedimentary Sequences

Volcanic-sedimentary sequences are prevalent in the northwestern United States. The Columbia Lava Plateau, for example, encompasses an area of 366,000 km^2 and extends into northern California, eastern Oregon and Washington, southern Idaho, and northern Nevada; the Snake River Group encompasses 40,400 km^2 in southern Idaho. The geology of this region consists of a thick, accordantly layered sequence of basalt flows and sedimentary interbeds.

The basalt flows can range from a few tens of centimeters to more than 100 m in thickness, averaging 30 to 40 m. From bottom to top, individual flows generally consist of a flow base, colonnade, and entablature (Figure 3.1.17).

The flow base makes up about 5 percent of the total flow thickness and is typically characterized by a vesicular base and pillow-palagonite complex of varying thickness if the flow entered water. The colonnade makes up about 30 percent of the flow thickness and is characterized by nearly vertical 3- to 8-sided columns of basalt, with individual columns about 1 m in diameter and 7.5 m in length. The colonnade is usually less vesicular than the base. The entablature makes up about 70 percent of the flow thickness. This upper zone is characterized by small diameter (averaging less than 0.5 m) basalt columns, which may develop into a fan-shape arrangement; hackly joints, with cross joints less consistently oriented and interconnected, may be rubbly and clinkery, and the upper part is vesicular.

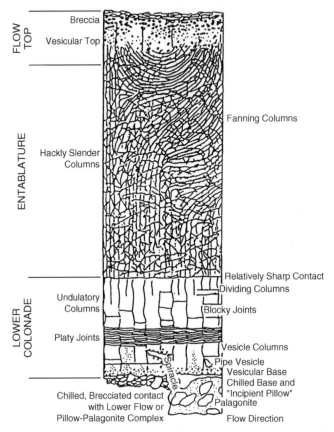

FIGURE 3.1.17   Generalized schematic diagram showing intraflow structure of a basalt flow (modified after Swanson and Wright, 1978).

Following extrusion, flows cool rapidly, expelling gases and forming vesicles and cooling joints. These upper surfaces are typically broken by subsequent internal lava movement resulting in brecciated flow tops. The combination of the superposed flow base and vesicular upper part of the entablature is referred to as the interflow zone. Interflow zones generally make up 5 to 10 percent of the total flow thickness.

Groundwater occurrence and flow within layered sequences of basalt flows and intercalated sedimentary interbeds is complex. Such sequences typically consist of multiple zones of saturation with varying degrees of interconnection. Principle aquifers or water-bearing zones are associated with interflow zones between basalt flows. These interflow zones commonly have high to very high permeability and low storativity because of the open nature, but limited volume, of joints and fractures.

Furthermore, because of the generally impervious nature of the intervening tuffaceous sediments and dense basalt, stratigraphically adjacent interflow zones may be hydraulically isolated over large geographic areas (Figure 3.1.18). This physical and hydraulic separation is commonly reflected by differences in both piezometric levels and water quality between adjacent interflow aquifers.

Recharge occurs mainly along outcrops and through fractures, which provide hydraulic communication to the surface. Interflow zones generally have the highest hydraulic conductivities and can form a series of superposed water-bearing zones. The colonnade and entablature are connected better vertically rather than horizontally, which allows for the movement of groundwater between interflow zones, although overall flow is three-dimensional. Multiple interflow zones can result in high total

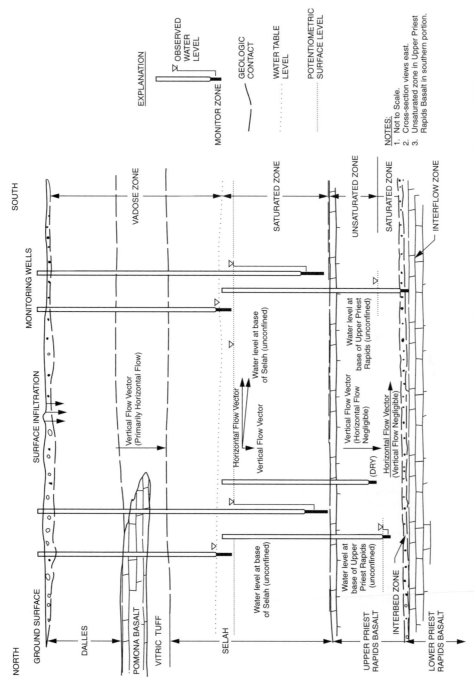

FIGURE 3.1.18 Conceptual flow model showing zones of saturation flow vectors in relationship to observed water levels (after Testa et al., 1988).

horizontal transmissivity. Position of the basalt flow within the regional flow system and varying hydraulic conductivities create further head differences with depth that can be very large in comparison to other sedimentary sequences. Horizontal hydraulic conductivities range from 0.65 up to 1600 or even 3000 m/day, whereas vertical hydraulic conductivities range from 10^{-8} to 10 m/day depending on the structural elements present (i.e., degree of fracturing joints, presence of sedimentary interbeds, etc.).

Sedimentary interbeds are typically composed of tuffaceous sediments of varying thickness, lateral extent, lithology, and degree of weathering. These interbeds usually impede groundwater movement in many areas. Groundwater flow within the more prominent interbeds are affected by the thickness and anisotropy of each hydrostratigraphic unit, and the position and continuity of each layer within the units. As with any layered media, the hydrostratigraphic unit with the lowest vertical hydraulic continuity is the controlling factor for groundwater flow in the vertical direction (normal to bedding). In the horizontal direction (parallel to bedding), groundwater flow is controlled by the hydrostratigraphic unit with the highest horizontal hydraulic continuity (Wang and Testa, 1989). Horizontal hydraulic conductivities based on pumping tests range from about 1×10^{-6} to 1×10^{-4} cm/s. Vertical hydraulic conductivities based on laboratory tests range from about 1×10^{-6} to 1×10^{-4} cm/s. Both methods showed a variance of two orders of magnitude.

3.1.4 STRUCTURAL STYLE AND FRAMEWORK

Structural geologic elements that can play a significant role in the subsurface environment related issues include faults, fractures, joints, and shear zones. Faults can be important from a regional perspective in understanding their impact on the regional groundwater flow regime, and delineation and designation of major water-bearing strata. Faults are usually less important in most site specific situations. Fractures, joints, and shear zones, however, can have a significant role both regionally and locally in fluid flow and assessment of preferred migration pathways of dissolved contaminants in groundwater in consolidated and unconsolidated materials. Regional geologic processes that produce certain structural elements, notably fracture porosity, include faulting (seismicity), folding, uplift, erosional unloading of strata, and overpressing of strata (Table 3.1.6).

Tectonic and possibly regional fractures result from surface forces (i.e., external to the body as in tectonic fractures); contractional and surface-related fractures result from body forces (i.e., internal to the forces). Contractional fractures are of varied origin resulting from desiccation, syneresis, thermal gradients, and mineral phase changes. Desiccation fractures develop in clay and silt-rich sediments upon a loss of water during subaerial drying. Such fractures are typically steeply dipping, wedge-shaped openings that form cuspate polygons of several nested sizes. Syneresis fractures result from a chemical process involving dewatering and volume reduction of clay, gel, or suspended colloidal material via tension or extension fractures. Associated fracture permeability tends to be isotropically distributed because developed fractures tend to be closely and regularly spaced. Thermal contractional

TABLE 3.1.6 Classification of Fractures

Fracture type	Classification	Remarks
Experimental	Shear	
	Extension	
	Tensile	
Natural	Tectonic	Due to surface forces
	Regional	Due to surface forces (?)
	Contractional	Due to body force
	Surface-related	Due to body force

(After Testa, 1994).

fractures are caused by the cooling of hot rock, as with thermally induced columnar jointing in fine-grained igneous rocks (i.e., basalts). Mineral phase-change fracture systems are composed of extension and tension fractures related to a volume reduction that results from a mineral phase change. Mineral phase changes are characterized by irregular geometry. Phase changes, such as calcite to dolomite or montmorillonite to illite, can result in about a 13 percent reduction in molar volume. Surface-related fractures develop during unloading, release of stored stress and strain, creation of free surfaces or unsupported boundaries, and general weathering. Unloading fractures or relief joints occur commonly during quarrying or excavation operations. Upon a one-directional release of load, the rock relaxes and spalls or fractures. Such fractures are irregular in shape and may follow topography. Free or unsupported surfaces (i.e., cliff faces and banks) can develop both extension and tensional fractures. These types of fractures are similar in morphology and orientation to unloading fractures. Weathering fractures are related to mechanical and chemical weathering processes such as freeze-thaw cycles, mineral alluation, diagenesis, small-scale collapse and subsidence, and mass-wasting processes.

Faults

Faults are regional structures that can serve as barriers, partial barriers, or conduits to groundwater flow and contaminant transport. The influence and effect of faults on fluid flow entrapment depend on the rock properties of strata that are juxtaposed and the attitude or orientation of the strata within their respective fault blocks. The influence of regional structural elements, notably faults, can have a profound effect on groundwater occurrence, regime, quality and usage, and delineation and designation of water-bearing zones of beneficial use.

The Newport-Inglewood Structural Zone in southern California exemplifies this important role (Testa et al., 1988; Testa, 1989 (Figure 3.1.19)). The structural zone is characterized by a northwesterly trending line of gentle topographic prominences extending about 40 miles.

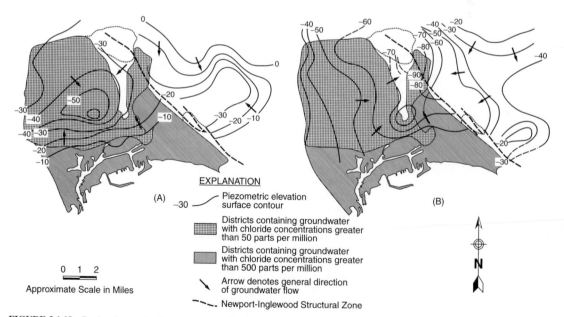

FIGURE 3.1.19 Regional groundwater contour maps showing the Newport-Inglewood Structural Zone in relation to major water-bearing units (after Testa, 1994).

This belt of domal hills and mesas, formed by the folding and faulting of a thick sequence of sedimentary rocks, is the surface expression of an active zone of deformation. An important aspect of this zone is the presence of certain fault planes that serve as effective barriers to the infiltration of seawater into the severely downdrawn groundwater aquifers of the coastal plain. These barriers also act as localized hydrogeologic barriers for freshwater on the inland side of the zone, reflected in the relatively higher water level elevations and enlarged effective groundwater aquifers.

The structural zone separates the Central groundwater basin to the northeast from the West Coast groundwater basin to the southwest. In the West Coast Basin area, at least four distinct water-bearing zones exist. In descending stratigraphic position, these zones are the shallow, unconfined Gaspur aquifer, the unconfined Gage aquifer of the upper Pleistocene Lakewood Formation, the semiconfined Lynwood Aquifer, and the confined Silverado aquifer of the lower Pleistocene San Pedro Formation.

Groundwater conditions are strikingly different on opposing sides of the structural zone and are characterized by significant stratigraphic displacements and offsets, disparate flow directions, as much as 30 ft of differential head across the zone, and differences in overall water quality and usage. Shallow water-bearing zones situated in the area south of the structural zone have historically (since 1905) been recognized as being degraded beyond the point of being considered of beneficial use because of elevated sodium chlorides. Groundwater contamination is also evident by the localized but extensive presence of light nonaqueous phase liquid (LNAPL) hydrocarbon pools and dissolved hydrocarbons resulting from the presence of 70 years of industrial development including numerous refineries, terminals, bulk liquid-storage tank farms, pipelines, and other industrial facilities on opposing sides of the structural zone.

The underlying Silverado aquifer has a long history of use, but has not been significantly impacted thus far by the poor groundwater quality conditions that have existed for decades in the shallower water-bearing zones where the Lynwood aquifer serves as a "guardian" aquifer. This suggests a minimal potential for future adverse impact of the prolific domestic-supply groundwater encountered at depths of 800 to 2600 ft below the crest of the structural zone. South of the structural zone no direct communication exists between the historically degraded shallow and deeper water-bearing zones. The exception is in areas where intercommunication or leakage between water-bearing zones or heavy utilization of groundwater resources may exist (i.e., further to the northwest within the West Coast Basin). In contrast, north of the structural zone, shallow groundwater would be considered beneficial as a guardian aquifer due to the inferred potential for leakage into the deeper water-supply aquifers.

The beneficial use and clean-up standards thus are different north and south of the structural zone, with lower standards to the south. The overall environmental impact on groundwater resources, regardless of the ubiquitous presence of LNAPL pools and dissolved hydrocarbon plumes in certain areas relative to the structural zone, is minimal to nil. Within the structural zone, structures such as folds and faults are critical with respect to the effectiveness of the zone to act as a barrier to the inland movement of saltwater. An early continuous set of faults is aligned along the general crest of the structural zone, notably within the central reach from the Dominguez Gap to the Santa Ana Gap. The position, character, and continuity of these faults are fundamental to the discussion of groundwater occurrence, regime, quality, and usage. In addition, delineation and definition of aquifer interrelationships with a high degree of confidence is essential. The multifaceted impact of the structural zone is just one aspect of the level of understanding required prior to addressing certain regional groundwater issues.

Another important issue is the assessment of which aquifers are potentially capable of beneficial use versus those that have undergone historic degradation. Those faults that do act as barriers with respect to groundwater flow may, in fact, be one of several factors used in assigning a part of one aquifer to beneficial-use status as opposed to another. A second issue, based on the beneficial-use status, is the level of aquifer rehabilitation and restoration deemed necessary as part of the numerous aquifer remediation programs being conducted in the Los Angeles Coastal Plain. This example illustrates that, relative to aquifer remediation and rehabilitation efforts, clean-up strategies should not be stringent, nor should they be applied uniformly on a regional basis. Clean-up strategies should, however, take into account the complex nature of the hydrogeologic setting, and clean-up standards should be applied appropriately.

Fractured Media

Fractured media in general can incorporate several structured elements including faults, joints, fractures, and shear zones. These structural elements, as with faults, can serve as a barrier, partial barrier, or conduit to the migration of subsurface fluids. Most fractured systems consist of rock or sediment blocks bounded by discrete discontinuities. The aperture can be open, deformed, closed, or a combination thereof. The primary factors to consider in the migration of subsurface fluids within fractured media are fracture density, orientation, effective aperture width, and nature of the rock matrix. Fracture networks are complex three-dimensional systems. The analysis of fluid flow through a fractured media is difficult because the only means of evaluating hydraulic parameters is by means of hydraulic tests. The conduct of such tests requires that the geometric pattern or degree of fracturing formed by the structural elements (i.e., fractures) be known. Fracture density (or the number of fractures per unit volume of rock) and orientation are most important in assessing the degree of interconnection of fracture sets. Fracture spacing is influenced by mechanical behavior (i.e., interactions of intrinsic properties). Intrinsic and environmental properties include load-bearing framework, grain size, porosity, permeability, thickness, and previously existing mechanical discontinuities. Environmental properties of importance include net overburden, temperature, time (strain rate), differential stress, and pore fluid composition (Hitchon and Bachu, 1988). Fracturing can also develop under conditions of excessive fluid pressures. Clay-rich soils and rocks, for example, are commonly used as an effective hydraulic seal. However, the integrity of this seal can be jeopardized if excessive fluid pressures are induced, resulting in hydraulic fracturing. Hydraulic fracturing in clays is a common feature in nature at hydrostatic pressures ranging from 10 KPa up to several MPa. Although hydraulic fracturing can significantly decrease the overall permeability of the clay, the fractures are likely to heal in later phases because of the swelling pressure of the clay.

Several techniques have been used to attempt to characterize fracture networks. These techniques have included field mapping (i.e., outcrop mapping, and lineation analysis), coring, aquifer testing, tracer tests, borehole flowmeters, statistical methods, geophysical approaches, and geochemical techniques to evaluate potential mixing. Vertical parallel fractures are by far the most difficult to characterize for fluid flow analysis because of the likelihood of their being missed during any drilling program (Aquilera, 1988; Schmelling and Ross, 1989). This becomes increasingly important because certain constituents, such as solvents and chlorinated hydrocarbons, which are denser than water, are likely to migrate vertically downward through the preferred pathways, and may even increase the permeability within these zones.

Within a single set of measured units of the same lithologic characteristics, a linear relationship is assumed between bed thickness and fracture spacing (Narr and Lerche, 1984). A typical core will intercept only some of the fractures.

In viewing the schematic block diagram of a well bore through fractured strata of varying thicknesses, the core drilled in the upper and lower beds would intersect fractures, but the cores drilled within the two central beds do not encounter any fractures (Figure 3.1.20). Closer fracture spacing is, however, evident in the two upper thinner beds.

The probability of intercepting a vertical fracture in a given bed is given by

$$P = \frac{D}{S} = \frac{DI}{T(\text{average})} \tag{3.1.3}$$

where P = probability, D = core diameter, S = distance between fractures, T(average) is average thickness, and I is a fracture index given by

$$I = \frac{T_i}{S_i} \tag{3.1.4}$$

where the subscript i refers to the properties of the bed. In other words, T_i is thickness of the ith bed and S_i is fracture spacing in the ith bed. A fracture index must also be determined independently for each set of fractures in the core, and must be normal to the bedding.

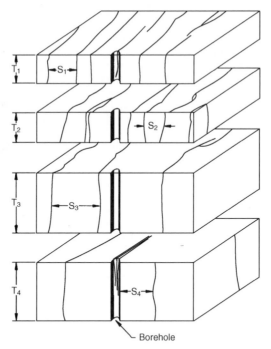

FIGURE 3.1.20   Schematic block diagram of well bore through fractured strata of two different thicknesses. Cores cut in upper and lower beds would intersect fractures, but core from two central beds is unfractured. Note closer fracture spacing in two upper, thinner beds (after Narr and Lerche, 1984).

Based on probability, a core has a finite chance of intersecting a vertical fracture in a bed of a given thickness depending on core diameter, bed thickness, and the value of the fracture index (I). Thus, a sparsely fractured region has a small value of I (i.e., large spacing between parallel fractures), and thicker beds have a larger spacing between fractures for a given I.

Seismicity

Earthquakes can cause significant changes in water quality and water levels, ultimately enhancing permeability. During the Loma Prieta, California earthquake of October 17, 1989 (magnitude 7.1), ionic concentrations and the calcite salination index of streamwater increased, streamflow and solute concentrations decreased significantly from within 15 minutes to several months following the earthquake, and groundwater levels in the highland parts of the basins were locally lowered by as much as 21 m within weeks to months after the earthquake (Rojstaczer and Wolf, 1992). The spatial and temporal character of the hydrologic response sequence increased rock permeability and temporarily enhanced groundwater flow rates in the region as a result of the earthquake.

REFERENCES

Anderson, M. P., "Hydrogeologic Facies Models to Delineate Large-Scale Spatial Trends in Glacial and Glaciofluvial Sediments," *Geological Society of America Bulletin*, 101: 129–147, and 501–511, 1989.

Aquilera, R., "Determination of Subsurface Distance between Vertical Parallel Natural Fractures based on Core Data," *American Association of Petroleum Geologists Bulletin*, 72(7): 845–851, 1988.

Back, W., Rosenshein, J. S., and Seaber, P. R., *Hydrogeology, The Geology of North America*, Vol. 0–2, Geological Society of America, Boulder, CO, 1988.

Barker, J. F., Barbash, J. E., and Labonte, M., "Groundwater Contamination at a Landfill Sited on Fractured Carbonate and Shale," *Journal of Contaminant Hydrology*, 3: 1–25, 1988.

Beck, B. F., *Engineering and Environmental Impacts of Sinkholes and Karst*. A. A. Balkema, Rotterdam, Netherlands, 1989.

Berg, R. C., and Kempton, J. P., "Potential for Contamination of Shallow Aquifers from Land Burial of Municipal Wastes," Illinois State Geological Survey Map, Scale I: 500,000, 1984.

Caswell, B., Time-of-Travel in Glacial Aquifers: *Water Well Journal*, March: 48–51, 1988a.

Caswell, B., Esker Aquifers: *Water Well Journal*, July: 36–37, 1988b.

Cayeux, L., *Carbonate Rocks*: (translated by A. T. Carozzi), Hafner Publishing, Darien, CT, 1970.

Cehrs, D., "Depositional Control of Aquifer Characteristics in Alluvial Fans, Fresno County, California," *Geological Society of America Bulletin*, 90(8): Part I, 709–711, Part II, 1282–1309, 1979.

Chandler, M. A., Kocurek, G., Goggin, D. J., and Lake, L. W., "Effects of Stratigraphic Heterogeneity on Permeability in Eolian Sandstone Sequence, Page Sandstone, Northern Arizona," *American Association of Petroleum Geologists Bulletin*, 73(5): 658–668, 1989.

Chocquette, P. W., and Pray, L. C., "Geologic Nomenclature and Classification of Porosity in Sedimentary Carbonates," *American Association of Petroleum Geologists Bulletin*, 54: 207–250, 1970.

Driscoll, Fletcher G., *Groundwater and Wells*, Second Edition, Johnson Filtration Company, Inc., St. Paul, Minnesota, 1986.

Ebanks, W. J., Jr., "Geology in Enhanced Oil Recovery," in *Reservoir Sedimentology*, Tillman, R. W. and Weber, K. J. eds., Society of Economic Paleontologists and Mineralogists, Special Publication: No. 40, pp. 1–14, 1987.

Esteban, M., and Klappa, C. F., "Subaerial Exposure Environment," in *Carbonate Depositional Environments*, Scholle, P. A. et al., eds., American Association of Petroleum Geologists, Memoir 33, 1983.

Fogg, G. E., "Emergence of Geologic and Stochastic Approaches for Characterization of Heterogeneous Aquifers," in *Proceedings of the U.S. Environmental Protection Agency*, Robert S. Kerr Environmental Research Laboratory Conference on New Field Techniques for Quantifying the Physical and Chemical Properties of Heterogeneous Aquifers, Dallas, pp. 1–17, March 20–23, 1989.

Freeze, R. A., and Cherry, J. A., *Groundwater*, Prentice Hall, New York, 1979.

Fryberger, S. G., "Stratigraphic Traps for Petroleum in Wind-Laid Rocks," *American Association of Petroleum Geologists Bulletin*, 70(12): 1765–1776, 1986.

Harris, D. G., "The Role of Geology in Reservoir Simulation Studies," *Journal of Petroleum Technology*, May, 625–632, 1975.

Hitchon, B., Bachu, S., and Sauveplane, C. M., eds., "Hydrogeology of Sedimentary Basins – Application to Exploration and Exploitation," in *Proceedings of the National Water Well Association Third Canadian/American Conference on Hydrogeology*, Banff, Alberta, Canada, June 22–26, 1986.

Hitchon, B., and Bachu, S., eds., "Fluid Flow, Heat Transfer and Mass Transport in Fractured Rocks," in *Proceedings of the National Water Well Association Fourth Canadian/American Conference on Hydrogeology*, Banff, Alberta, June 22–26, 1988.

Larkin, R. G., and Sharp, J. M., Jr., "On the Relationship Between River-Basin Geomorphology, Aquifer Hydraulics, and Ground-Water Flow Direction in Alluvial Aquifers," *Geological Society of America Bulletin*, 104: 1608–1620, 1992.

Lindholm, G. F., and Vaccaro, J. J., "Region 2, Columbia Lava Plateau," in *The Geology of North America, Vol. 0-2 Hydrogeology*, Geological Society of America, Boulder, CO, 1988.

McKee, E. D., "A Study of Global Sand Seas," *United States Geological Survey Professional Paper*, No. 1052, 1979.

Miall, A. D., "Reservoir Heterogeneities in Fluvial Sandstones: Lessons from Outcrop Studies," *American Association of Petroleum Geologists Bulletin*, 72(6): 682–697, 1988.

Narr, W., and Lerche, I., "A Method for Estimating Subsurface Fracture Density in Core," *American Association of Petroleum Geologists Bulletin*, 68(5): 637–648, 1984.

Nilsen, T. H., "Alluvial Fan Deposits," in *Sandstone Depositional Environments*, Scholle, P. A., and Spearing, D., eds., American Association of Petroleum Geologists, Memoir No. 31, Tulsa, OK, 1982.

Pettijohn, F. J., Potter, P. E., and Siever, R., *Sand and Sandstone*, Springer-Verlag, New York, 1973.

Potter, P. E., and Pettijohn, F. J., *Paleocurrents and Basin Analysis*, Springer-Verlag, New York, 1977.

Pryor, W. A., "Permeability-Porosity Patterns and Variations in Some Holocene Sand Bodies," *American Association of Petroleum Geologists Bulletin*, 57(1): 162–189, 1973.

Reading, H. G., ed., *Sedimentary Environments and Facies*, Blackwell Scientific Publications, Oxford, 1978.

Reineck, H. E., and Singh, I. B., *Depositional Sedimentary Environments With Reference to Terrigenous Clastics*, Springer-Verlag, New York, 1975.

Rojstaczer, S., and Wolf, S., "Permeability Changes Associated with Large Earthquakes: An Example from Loma Prieta, California," *Geology*, 20: 211–214, 1992.

Scholle, P. A., *A Color Illustrated Guide to Carbonate Rock Constituents, Textures, Cements and Porosities of Sandstones and Associated Rocks*, American Association of Petroleum Geologists, Memoir No. 27, Tulsa, OK, 1979a.

Scholle, P. A., *A Color Illustrated Guide to Constituents, Textures, Cements and Porosities of Sandstones and Associated Rocks*, American Association of Petroleum Geologists, Memoir No. 28, Tulsa, OK, 1979b.

Scholle, P. A., and Spearing, D., *Sandstone Depositional Environments*, American Association of Petroleum Geologists, Memoir No. 31, Tulsa, OK, 1982.

Scholle, P. A., Bebout, D. G., and Moore, C. H., *Carbonate Depositional Environments*, American Association of Petroleum Geologists, Memoir No. 33, Tulsa, OK, 1983.

Schmelling, S. G., and Ross, R. R., *Contaminant Transport in Fractured Media: Models for Decision Makers*, U.S. Environmental Protection Agency EPA/5401489/004, 1989.

Schmidt, V., McDonald, D. A., and Platt, R. L., "Pore Geometry and Reservoir Aspects of Secondary Proposity in Sandstones," *Bulletin of the Canadian Petroleum Geology*, 25: 271–290, 1977.

Selley, R. C., *An Introduction to Sedimentology*, Academic Press, London, 1976.

Selley, R. C., *Ancient Sedimentary Environments*, Chapman and Hall, London, 1978.

Sharp, J. M., "Hydrogeologic Characteristics of Shallow Glacial Drift Aquifers in Dissected Till Plains (North-Central Missouri)," *Ground Water*, 22(6): 683–689, 1984.

Sneider, R. M., Tinker, C. N., and Meckel, L. D., "Deltaic Environment Reservoir Types and Their Characteristics," *Journal of Petroleum Technology*, 1583–1546.

Spearing, D. R., "Alluvial Fan Deposits – Summary Sheet of Sedimentary Deposits," *Sheet 1*, Geological Society of America, Boulder, CO, 1974.

Stephenson, D. A., Fleming, A. H., and Mickelson, D. M., 1974. "The Hydrogeology of Glacial Deposits," in ~fydrogeology, Back, W., Rosenshein, J. S., and Seaber, P. R., eds., Geological Society of America Decade of North American Geology, Boulder, CO, Vol. 0-2, pp. 301–304, 1989.

Swanson, D. A., and Wright, T. L., "Bedrock Geology of the Southern Columbia Plateau and Adjacent Areas," in *The Channeled Scabland*, Baker, V. R., and Numendal, D. eds., National Aeronautical and Space Administration Planetary Geology Program, Washington D.C., p. 37–57, 1978.

Testa, S. M., *Geological Aspects of Hazardous Waste Management*, CRC Press/Lewis Publishers, Boca Raton, FL, 537 p., 1994.

Testa, S. M., Henry, E. C., and Hayes, D., "Impact of the Newport-Inglewood Structural Zone on Hydrogeologic Mitigation Efforts – Los Angeles Basin, California," in *Proceedings of the Association of Groundwater Scientists and Engineers FOCUS Conference on Southwestern Groundwater Issues*, Albuquerque, NM, pp. 181–203, 1988.

Testa, S. M., "Regional Hydrogeologic Setting and its Role in Developing Aquifer Remediation Strategies," in *Proceedings of the Geological Society of America Annual Meeting*, Abstracts with Programs, St. Louis, pp. A96, November 6–9, 1989.

Testa, S. M., "Site Characterization and Monitoring Well Network Design, Columbia Plateau Physiographic Province, Arlington, North-Central Oregon," Geological Society of America, Abstract, 1991.

Tyler, N., Gholston, J. C., and Ambrose, W. A., "Oil Recovery in a Low Permeability, Wave-Dominated, Cretaceous, Deltaic Reservoir, Big Wells (San Miguel) Field, South Texas," *American Association of Petroleum Geologists Bulletin*, 71(10): 1171–1195, 1987.

Wang, C. P., and Testa, S. M., "Groundwater Flow Regime Characterization, Columbia Plateau Physiographic Province, Arlington, North-Central Oregon," in *Proceedings of the Association of Groundwater Scientists and Engineers Conference on New Field Techniques for Quantifying the Physical and Chemical Properties of Heterogeneous Aquifers*, pp. 265–291, 1989.

CHAPTER 4
HYDROLOGY

SECTION 4.1

THE HYDROLOGIC CYCLE

Robert M. Hordon
Dr. Hordon is a physical geographer and hydrologist in the
Geography Department at Rutgers University in New Jersey.

4.1.1 GENERAL CONCEPTS

The *hydrologic cycle* (or water cycle) is defined as the continuous circulation of water and water vapor over the entire globe. This never-ending circulation penetrates the three parts of the total earth system: the *atmosphere* (the gaseous envelope above the hydrosphere), the *hydrosphere* (the water covering the surface of the earth), and the *lithosphere* (the solid rock beneath the hydrosphere). The energy for the circulation is provided by solar energy and gravity.

Speculation about the hydrologic cycle goes back to ancient times. Various philosophers, such as Homer, Plato, and Aristotle in Greece, and Lucretius, Seneca, and Pliny in Rome considered the concept of the cycle (Chow, 1964, 1-1–1-7). Perhaps the most poetic concept of the hydrologic cycle was expressed by Ecclesiastes in the bible over 2200 years ago in the famous, often-quoted verse:

> All the rivers run into the sea,
> Yet the sea is not full;
> Unto the place whither the rivers go,
> Thither they go again.

Another 1000 years passed before the concept was accepted and discussed in Western Europe during the Middle Ages. Indeed, it was only during the Renaissance of the fifteenth and sixteenth centuries that the concepts of the hydrologic cycle slowly changed from a purely philosophical practice to a more observational science. Examples of this change can be found in the works of Leonardo da Vinci and Bernard Palissy.

Hydrological measurements increased during the seventeenth century. Precipitation, evaporation, and capillarity were measured by Perrault in the Seine watershed in France; channel cross-section and stream velocity measurements in the Seine River in Paris enabled Mariotte to calculate river discharge; and Halley measured stream discharge and evaporation in his Mediterranean Sea study (Chow, 1964, 1-1–1-7).

There is no beginning or end to the hydrologic cycle. Water is evaporated from the oceans and land, with the oceans providing the largest amounts, covering 71 percent of the earth's surface and accounting for 86 percent of atmospheric moisture (Jones, 1997, 23–25). The evaporated water, carried into the atmosphere, usually drifts tens to hundreds of kilometers before being returned to the earth as rain, snow, hail, or sleet (McDonald, 1962, 168–170; 172–177). This precipitated water may be intercepted and transpired by plants, may run over the ground surface and into streams, or may infiltrate into the ground. Much of the intercepted and transpired water and the surface runoff returns to the air by evaporation. The infiltrated water may seep down to deeper water-bearing zones of the earth, forming groundwater storage. This storage may later flow out to streams as the base flow component of runoff and finally evaporate into the atmosphere to complete the hydrologic cycle.

Thus, the hydrologic cycle is described generally in terms of precipitation (P), infiltration (I), evaporation (E), transpiration (T), surface runoff (R), and groundwater flow (G). The cycle behaves as a closed system, but rates and storage times vary regionally, seasonally, and annually. The variation in turnover rates ranges over many orders of magnitude. For example, the turnover rates or residence times range from days for atmospheric water and streams to thousands of years for ice caps and glaciers, as follows (Hornberger, Raffensperger, Wiberg, and Eshleman, 1998, 7–10; Jones, 1997, 21–24; Lvovich, 1977, 13–21; Maidment, 1993, 1-1–1-15; Spiedel and Agnew, 1988, 27–36):

Atmosphere	8 days
Streams	17 days
Lakes	10 years
Shallow groundwater	330 years
Oceans	3000 years
Deep groundwater	5000+ years
Ice caps and glaciers	8000–15,000 years

The quantity of water going through the hydrologic cycle during a given period for an area can be evaluated by the hydrologic equation (or continuity equation):

$$I - O = delta\ S$$

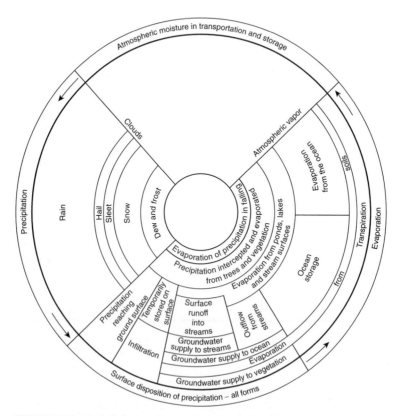

FIGURE 4.1.1 The hydrologic cycle—a qualitative representation (Horton, 1931).

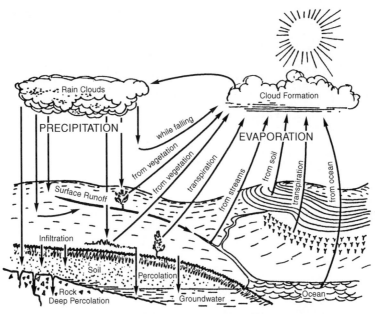

FIGURE 4.1.2   The hydrologic cycle—a descriptive representation (Ackermann, Colman, and Ogrosky, 1955).

where I is the total inflow of surface runoff, groundwater, and total precipitation; O is the total outflow, which includes evapotranspiration, as well as subsurface and surface runoff from the area; and *delta S* is the change in storage in the various forms of retention and interception.

The hydrologic cycle may be illustrated qualitatively (Figure 4.1.1), descriptively (Figures 4.1.2 and 4.1.3), and quantitatively (Figure 4.1.4).

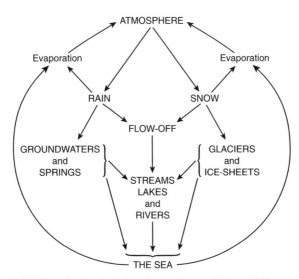

FIGURE 4.1.3   The circulation of meteoric water (Holmes, 1965).

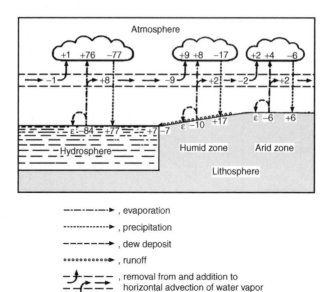

----·--·--▶ , evaporation

·············▶ , precipitation

------▶ , dew deposit

oooooooooo▶ , runoff

, removal from and addition to
horizontal advection of water vapor

ε, values less than 0.5 relative unit.

FIGURE 4.1.4   The hydrologic cycle—a quantitative representation (Lettau, 1954).

4.1.2   THE MAGNITUDE OF THE HYDROLOGIC CYCLE

The quantities of precipitation, evaporation, and runoff involved in the hydrologic cycle are shown in Table 4.1.1. An estimated 500,000 km^3 (120,000 mi^3) of water evaporates from the earth's surface annually. Of this amount, 430,000 km^3 (103,000 mi^3), or 86 percent, is accounted for by the oceans,

TABLE 4.1.1   Hydrologic Cycle: Quantities of Precipitation, Evaporation, and Runoff

	Volume (1000 cubic km)	Percentage of total water
Annual evaporation		
From oceans	430	0.025
From land	70	0.005
Total	500	0.031
Annual precipitation		
On oceans	390	0.024
On land	110	0.007
Total	500	0.031
Annual runoff to oceans		
From rivers and ice caps	38	0.003
Groundwater outflow to oceans	2	0.0001
Total	40	0.0031

Notes: 1. Evaporation is a measure of the total amount of water that is annually involved in the hydrologic cycle.
2. The groundwater outflow to oceans is estimated to be about 5 percent of surface runoff.
Source: U.S. Geological Survey.

TABLE 4.1.2   Global Distribution of Water

Location	Surface area (1000 sq km)	Water volume (1000 cubic km)	Percentage of total water
World oceans	361,000	1,340,000	97
Ice caps and glaciers	16,000	24,000	2.1
Subsurface water			
Soil moisture and vadose water	130,000	67	0.005
Groundwater less than 1 km deep	130,000	4200	0.3
Groundwater greater than 1 km deep	130,000	4200	0.3
Subtotal subsurface water (rounded)	130,000	8500	0.61
Surface water			
Fresh water lakes	1236	91	0.009
Saline lakes and inland seas	822	85	0.008
Average (instantaneous) in stream channels		1	0.0001
Subtotal surface water		177	0.02
Total water on land areas (rounded)	130,000	8700	0.62
Atmosphere (at sea level)	510,000	13	0.001
Fresh water	149,000	35,000	2.5
Total (rounded)	510,000	1,400,000	100

Sources: Shiklomanov and Sokolov (1983); UNESCO (1978); and others.

with inland water bodies and wet soils making up the remaining 70,000 km^3 (17,000 mi^3), or 14 percent.

The bulk of the inland evaporation occurs into relatively dry air masses. Much of the water evaporated from the oceans is transported by maritime air masses (which can hold considerably more water vapor than continental air masses) to the continents, where total precipitation amounts to 110,000 km^3/yr (26,000 mi^3). This is enough water to cover the entire state of Texas (692,408 km^2, 267,339 mi^2) to a depth of 150 m (492 ft). Of the 110,000 km^3/yr (26,000 mi^3) precipitated, 40,000 km^3 (9600 mi^3), or 36 percent, returns to the sea as runoff to balance the excess of precipitation over inland evaporation (see Table 4.1.1).

The global distribution of water is shown in Table 4.1.2. The oceans contain an estimated 97 percent of all the water in the world, or 1340×10^6 km^3 (320×10^6 mi^3). If the earth were a uniform sphere, this volume of water would cover the entire earth to a depth of 244 m (800 ft). Ice caps and glaciers account for another 2.1 percent of the world's water. One humbling conclusion drawn from these facts is that 99.1 percent of all of the world's water is either too salty or frozen to be of immediate use for human consumption.

Subsurface water constitutes another 0.61 percent of the total water, 50 percent (or 0.3 percent of the total) is located in aquifers deeper than one km (0.62 mi). This further reduces the amount of water available for human use because of the substantial pumping costs for deep aquifers.

Surface water, which includes lakes (both fresh and saline) and inland seas, amounts to only 0.02 percent of the total water. More impressively, the average water volume in all of the world's streams amounts to only 0.0001 percent of the total water (see Table 4.1.2). Finally, water in the atmosphere accounts for the remaining 0.001 percent.

Reichel (1952, 155) calculated that the mean annual precipitation for the world is 864 mm (34 in), which is balanced by a comparable amount of evaporation. The average annual precipitation over the continental United States would amount to 76.2 cm (30 in) if it were spread evenly. However, topographic configurations and patterns of atmospheric circulation result in uneven distribution of precipitation, ranging from a few tens of mm (inches) in the arid Southwest to over 254 cm (100 in) in parts of the Pacific Northwest. The 17 western states receive only 25 percent of the total precipitation but contain 60 percent of the land area.

The 76.2 cm (30 in) of water for the United States represents 1.63×10^{13} l/day (4.8×10^8 acre-ft/yr; 4.3×10^{11} gal/day). Of this amount, 54.6 cm (21.5 in), or 71.7 percent, is returned to the atmosphere by the processes of evapotranspiration. The remaining 21.6 cm (8.5 in), or 28.3 percent, becomes surface and groundwater runoff into the oceans (Robinove, 1963, 375–389).

TABLE 4.1.3   Global Distribution of Fresh Water

Location	Percentage of total
Ice caps and glaciers	75
Groundwater deeper than 762 m	14
Groundwater less than 762 m	11
Lakes	0.3
Soil moisture	0.06
Atmosphere	0.035
Streams	0.03
Total (rounded)	100

Source: UNESCO (1978).

The total volume of fresh water on the earth is estimated to be 35×10^6 km^3 (8.4×10^6 mi^3) or about 2.5 percent of the total amount of water in the world (see Table 4.1.2). The global distribution of fresh water is shown in Table 4.1.3. Note that the figures are stationary estimates of distribution. Thus, although huge amounts of water pass through the atmosphere, the water content is relatively small at any given moment.

Ice caps and glaciers account for 75 percent of the fresh water total. Another 14 percent is located in deep aquifers between 762 and 3810 m deep (2500 and 12,500 feet). Taken together, 89 percent of the total amount of fresh water in the world is either frozen or is located at depths in the ground that are too great for pumping.

Groundwater located at less than 762 m (2500 ft) in depth, another 11 percent of the fresh water total, is at least partially available for human use. Lakes account for only 0.3 percent of fresh water (see Table 4.1.3), even though individual lakes are renowned for their depth and volume. For example, Lake Baikal in Siberia is the deepest lake in the world at 1741 m (5712 ft) and contains an astonishing 25 percent of all of the fresh water in all of the world's lakes.

The average volume of water in streams is lower than in fresh water lakes, accounting for only 0.03 percent of the total amount of water (see Table 4.1.3). Thus, the two most immediately available sources of fresh water for human use, fresh water lakes and streams, represent not only small fractions of the total global supply of fresh water but are often located at great distances from major population centers (Lake Baikal, Amazon River, and others).

4.1.3   THE ROLE OF THE HYDROLOGIC CYCLE

If the earth and the atmosphere are viewed as separate entities, radiation and conduction fail to provide balanced heat budgets, because the earth's surface has a net gain and the free atmosphere a net loss. The hydrologic cycle provides the link between the gain and loss.

Some of the heat absorbed by the earth's surface is used in evaporation, and therefore transferred to latent heat (the quantity of heat absorbed or emitted without change in temperature during a change of state of a unit mass of a material, "hidden heat"). This latent heat is later realized as sensible heat (the heat added to a body when its temperature is changed) and released to the atmosphere when the vapor condenses to clouds. Evaporation is high where relatively cool air moves over warmer oceans. The highest evaporation values found in the northern hemisphere occur in the Atlantic and Pacific trade wind belts south of 30°N. High evaporation values also occur over the northwestern Pacific and North Atlantic Oceans during the northern hemisphere winter when cold, dry continental air masses (cP and cA) move over warmer waters.

The average life of water vapor molecules in air varies from an hour to several days. Latent heat is usually liberated far from the regions where evaporation occurred. This is particularly true of evaporation in the trade wind belts, which supply much of the vapor that eventually precipitates in

TABLE 4.1.4   Runoff to the Oceans

Ocean	Water area (million sq km)	Runoff area (million sq km)	Runoff (1000 cubic km)	Percentage of total runoff
Pacific	178.7	24.8	14.1	31.6
Atlantic	91.7	50.7	19.8	44.4
Indian	76.2	20.9	5.6	12.6
Arctic	14.7	22.4	5.1	11.4
Totals	361.3	118.8	44.6	100

Notes: 1. The world ocean in this table consists of only four units. Marginal seas, such as the Mediterranean, are considered as being attached to the nearest ocean.

2. The boundaries of the oceans as determined by the International Hydrographic Bureau are summarized in Fairbridge (1966).

3. The areas of the continents that have internal drainage (endorheic) are estimated to be 30,200,000 square km, receive 8700 cubic km of precipitation, and evaporate 8100 cubic km.

4. The total land area (including islands) is 149,000,000 square km, 20 percent of which has internal drainage.

Source: UNESCO (1978).

middle and high latitudes. Thus, the circulation of water is a key part of heat transfer from low to high latitudes and from oceans to continents (Petterssen, 1964, 3-1–3-39).

The ecological importance of the hydrologic cycle was recognized by the U.S. National Research Council, which considers the cycle to be a fundamental physical template for biological processes. In this context, ecological hydrology, which is defined as the study of relationships among hydrologic, climatologic, and ecologic processes in a human framework, will greatly assist in global water resources management (Post, Grant, and Jones, 1998, 517, 526).

4.1.4   THE RETURN OF WATER TO THE OCEANS

In spite of the relatively uniform pattern of evaporation in the various latitudinal belts of the ocean, there is a marked regional imbalance in the return flow of water to the oceans. The explanation for this discrepancy lies in the concentration of major rivers (e.g., Amazon, Mississippi, Congo, Niger, St. Lawrence, Danube, Po, Nile, and Rhine), which drain into the Atlantic Ocean and its marginal seas (Black Sea, Gulf of Mexico, and the Mediterranean Sea). In contrast, the Pacific has only a limited number of major discharge outlets (Yangtze, Hwang-Ho or Yellow, Yukon, Columbia, and Colorado). Table 4.1.4 provides further evidence that the Atlantic Ocean not only drains the largest portion of the earth's land surface, but also has the highest proportion of land area draining into the oceans. Specifically, out of the total runoff area of $118.8 \times 10^6 \, \text{km}^2$ ($45.9 \times 10^6 \, \text{mi}^2$) draining into the world's oceans, $50.7 \times 10^6 \, \text{km}^2$ ($19.6 \times 10^6 \, \text{mi}^2$), or 42.7 percent, drain into the Atlantic. Thus, the Atlantic Ocean receives 44.4 percent of the total runoff going into the oceans (see Table 4.1.4).

As water is evaporated from the ocean, it becomes desalinated, but storm-generated wave bubbles carry salt into clouds, thereby contaminating the rainfall ("cyclic salts"). Both natural and anthropogenic contaminants change the quality of the water at every stage of the cycle. Anthropogenic CO_2 has increased dramatically in the twentieth century. Model projections reported by the National Research Council (1983) indicate that this should result in a small but noticeable increase in the intensity of the hydrologic cycle.

REFERENCES

Ackermann, W. C., Colman, E. A., and Ogrosky, H. O., "From Ocean to Sky to Land to Ocean," in *Water, Yearbook Agr.*, (U.S. Dept. Agr.), pp. 41–51, 1955.

Allaby, M., *Water: Its Global Nature*, Facts on File, New York, 1992.

Barry, R. G., and Chorley, R. J., *Atmosphere, Weather, and Climate*, Routledge, New York, 6th edition, 1992.

Berner, E. K., and Berner, R. A., *Global Environment: Water, Air, and Geochemical Cycles*, Prentice Hall, Upper Saddle River, New Jersey, 1996.

Camp, T. R., and Meserve, R. L., *Water and Its Impurities*, Dowden, Hutchinson & Ross, Stroudsburg, PA, 2nd edition, 1974.

Chow, V. T., *Handbook of Applied Hydrology*, McGraw-Hill, New York, 1864.

Dingman, S. L., *Physical Hydrology*, Prentice Hall, Upper Saddle River, New Jersey, 1994.

Emiliani, C., *Planet Earth*, University Press, Cambridge, 2nd edition, 1995.

Fairbridge, R. W., *The Encyclopedia of Oceanography*, "Encyclopedia of Earth Science Series," Vol. 1, New York: Reinhold Publishing Corp., 1966.

Falkenmark, M., *Hydrological Phenomena in Geosphere-Biosphere Interactions: Outlooks to Past, Present and Future*, Wallingford, Oxfordshire, UK: International Association of Hydrological Sciences (IASH) in cooperation with the International Institute for Hydraulic and Environmental Engineering, 81 pp, IASH Monographs and Reports No. 1, 1989.

Holmes, A., *Principles of Physical Geology*, rev. edition, Ronald Press, New York, 1965.

Hornberger, G. M., Raffensperger, J. M., Wiberg, P. L., and Eshleman, K. N., *Elements of Physical Hydrology*, The Johns Hopkins University Press, Baltimore, 1998.

Horton, R. E., "The Field, Scope and Status of the Science of Hydrology," *Trans. Am. Geophys. Union*, 12: 189–202, 1931.

Jones, J. A. A., *Global Hydrology*, Longman, Essex, England, 1997.

Lettau, H., "A Study of the Mass, Momentum and Energy Budget of the Atmosphere," *Arch. Meteorol. Geophys. Bioklimatol.*, Ser. A, 7: 131–153, 1954.

Livingston, D. A., "Chemical Composition of Rivers and Lakes," *U.S. Geol. Surv. Profess. Paper 440-G*, 1963.

Lovich, M. I., "World Water Resources Present and Future," *Ambio*, 1977.

Maidment, D. R., "Hydrology," Chapter 1 in *Handbook of Hydrology*, Maidment, D. R., ed., McGraw-Hill, New York, 1993.

McDonald, J. E., "The Evaporation-Precipitation Fallacy," *Weather*, London, 1962.

Moore, J. W., *Balancing the Needs of Water Use*, Springer-Verlag, New York, 1989.

National Research Council, *Changing Climates: Report of a Carbon Dioxide Assessment Committee*, National Academy Press, Washington, D.C., 1983.

National Research Council, *Opportunities in the Hydrological Sciences*, National Academy Press, Washington, D.C., 1991.

Peixoto, J. P., and Oort, A. H., *Physics of Climate*, Amer. Inst. of Physics, New York, 1992.

Petterssen, S., "Meteorology," Chapter 3 in *Handbook of Applied Hydrology*, Chow, V. T., ed., McGraw-Hill, New York, 1964.

Post, D. A., Grant, G. E., and Jones, J. A., "New Developments in Ecological Hydrology Expand Research Opportunities," *EOS Transactions*, American Geophysical Union, 79(43): 517, 526, October 27, 1998.

Reichel, E., "Der Stand des Verdunstungs problems," (The Status of the Evaporation Problem), *Ber. Deut. Wetterdienst*, Bad Kissingen, 35: 155, 1952.

Robinove, C. J., "What's Happening to Water," *Smithsonian Inst. Ann. Rept.*, 375–389, 1962, 1963.

Spiedel, D. H., and Agnew, A. F., "The World Water Budget," Chapter 3 in *Perspectives on Water*, Spiedel, D. H., Ruedisili, L. C., and Agnew, A. F., eds., Oxford University Press, New York, 1988.

UNESCO, *World Water Balance and Water Resources of the Earth*, UNESCO, Paris, (Prepared by the USSR Committee for the International Hydrological Decade), 1978.

Viessman, W., Jr, Lewis, G. L., and Knapp, J. W., *Introduction to Hydrology*, Harper & Row, New York, 3rd edition, 1989.

Wetzel, R. G., *Limnology*, Saunders, Philadelphia, 2nd edition, 1983.

Wolman, A., "Water Resources," A Report to the Committee on Natural Resources of the National Academy of Sciences—National Research Council, *Natl. Acad. Sci. Natl. Res. Council, Publ. 1000-B*, 1962.

CHAPTER 4
HYDROLOGY

SECTION 4.2

PRECIPITATION AND RUNOFF

Robert M. Hordon

Dr. Hordon is a physical geographer and hydrologist in the Geography Department at Rutgers University in New Jersey.

4.2.1 THE HYDROLOGIC CYCLE

Precipitation and runoff are two of the major components of the hydrologic cycle—the continuous circulation of water between the atmosphere and the lands and oceans of the planet. Precipitation represents water in either solid or liquid form that falls to the earth's surface from the atmosphere. Runoff pertains to that part of precipitation that appears in surface streams and groundwater, which, in most instances, flows down gradient to the ultimate sink on earth—the oceans.

Because oceans contain the overwhelming amount of the water in the world (97.2 percent), discussion about the hydrologic cycle will begin with evaporation from the oceans, which totals about 419,000 cu km (100,000 cu mi) per year. Annual evaporation from soil, plants, and lakes on land areas of the world totals about 69,000 cu km (17,000 cu mi) per year. The total evaporation from the earth's surface is then 488,000 cu km (117,000 cu mi) per year, with the oceans accounting for 86 percent of the total.

The source of precipitation is water vapor, which although present at all times in the atmosphere, albeit in varying amounts, constitutes less than one percent of the volume of the atmosphere. Because the oceans cover 71 percent of the earth's surface, they receive most of the precipitation: 382,000 cu km (92,000 cu mi) or 78 percent of the total. Land surfaces receive 106,000 cu km (25,000 cu mi) per year of precipitation, which is 37,000 cu km (9000 cu mi) more than is evaporated. This excess of precipitation over evaporation provides the runoff component of the hydrologic cycle, whereby the extra water on the lands is returned to the oceans by surface (streams) and subsurface (groundwater) flow paths (Strahler and Strahler, 1992).

Thus, the global water balance can be expressed as:

$$P = E + R$$

where P = precipitation, E = evaporation, and R = runoff.

4.2.2 PRECIPITATION PROCESSES

Almost all clouds that produce precipitation form as a consequence of the upward motion of air. Rising air expands and cools as it moves to lower pressure levels. When the temperature of the ascending air reaches the dew point (the temperature at which air is saturated), the water vapor in the air condenses onto minute hygroscopic particles in the air, which are called condensation nuclei, thereby forming cloud droplets. Sulfate and sea salt particles make up most of the nuclei.

The moisture in a cloud is made up of tiny water droplets or ice crystals, which have an average diameter of about 0.01 mm (0.0004 in). Air currents, even if slight, keep the cloud droplets suspended.

Coalescence or ice crystal growth allows the droplets to become heavy enough to fall to the earth as some form of precipitation. It takes about a million cloud droplets to make one raindrop, which averages about 1 mm (0.04 in) in diameter. If we hypothetically assumed that all moisture in the atmosphere were to be released suddenly as precipitation, it is estimated that only about 25 mm (1 in) would fall to the earth. Because it is known that much more than 25 mm (1 in) can fall in an hour during a storm, moist air must be continuously recharging the clouds.

4.2.3 FORMS OF PRECIPITATION

Water that falls to the earth is called precipitation. Accordingly, forms of condensed atmospheric moisture, such as dew, fog, and frost, are not considered precipitation even though they are critical to the sustenance of many organisms.

Drizzle is a fine mist that usually falls from stratus clouds. The drops are slightly larger than heavy fog, with a diameter of about 0.1 to 0.5 mm (0.004–0.02 in). Precipitation intensity from drizzle is low, generally less than 1 mm (0.04 in) per hour.

All liquid precipitation heavier than drizzle is classified as *rain*. The average diameter of raindrops is about 1 mm (0.04 in), but they can range from 0.5 to 5 mm (0.02–0.2 in).

Ice pellets, also known as *sleet*, occur when raindrops freeze when they fall through air that is below 0°C (32°F). These transparent or translucent grains of ice usually have a diameter smaller than 5 mm (0.2 in).

Snow forms when ice crystals in the atmosphere move together and stick to one another. Ice crystal coalescence produces snowflakes, which can become fairly large when the ground temperature is below 0°C (32°F).

Hail is a special type of ice pellet. It forms when strong vertical updrafts within high, large cumulonimbus clouds carry the ice pellets to elevations where they contact supercooled drops. Hailstones range in diameter from 5 mm to over 10 cm (0.2 to over 4 in). Hail will not occur when the ground temperature is below freezing.

4.2.4 CONDITIONS CAUSING PRECIPITATION

Moist air must be cooled below its dew point for condensation and precipitation to occur. The amount of moisture in the air and the rapidity of its cooling will determine the onset of precipitation. Air can be cooled quickly if it is forced to ascend rapidly where the atmospheric pressure will be lower and the air can expand. There are three ways in which this can be done: (1) warm air can be lifted over cold air (cyclonic), (2) air columns can rise within a thunderstorm or cumulonimbus cloud (convective), and (3) air can be forced to rise over a mountain (orographic).

Cyclonic precipitation occurs along the fronts separating warm and cold air masses. Warm air is less dense than cold air and can hold much more moisture. Conversely, cold air is denser than warm air and does not hold as much moisture. Because these air masses have different temperatures, barometric pressures, moisture levels, and densities, the boundaries between them are called *fronts*. Historically, the concept of fronts between differing air masses was developed by Norwegian meteorologists at the time of World War I as an analogy to the fronts between the opposing armies in Europe.

Warm fronts form when warm, moist air is forced to rise over a cold air mass. Because the air masses do not mix, the warm air is cooled so that condensation and precipitation can occur. The slope of the warm front is gentle (1 in 200 to 1 in 80; or 1 km vertical rise for every 200 km of horizontal distance), so the precipitation intensity is light to moderate, although the area of precipitation is extensive. For example, precipitation can extend out for 300 to 500 km (200–300 mi) from the surface location of the front. Typical cloud forms associated with a warm front and ranked in terms of proximity to the front are nimbostratus (when rain or snow are falling), stratus, cirrostratus, and cirrus at very high altitudes (6–12 km; 20,000–40,000 ft).

An advancing cold air mass pushes the warm air it is displacing to ascend fairly rapidly. The slope of the cold front is much steeper than that of the warm front (1 in 40), resulting in a smaller zone of precipitation but greater intensity. Cumulonimbus clouds usually form near the front, but if the cold air is moving rapidly, thunderstorms can occur 160 km (100 mi) in advance of the front. Clear conditions, and in the winter, very cold weather generally follows the passage of the front.

Occluded fronts occur when the faster moving cold front catches up with a warm front and forces the less dense warm air to rise. This type of front is quite common in the midlatitudes and is often associated with some form of precipitation.

Convective precipitation pertains to vertical movement of air. Uneven heating of the ground by the sun allows a convection cell in the atmosphere to develop and rise. The ascending air expands and cools until it attains the temperature of the surrounding air. Condensation occurs when the dew point is reached. Because condensation is a heating process, the convection cell continues to ascend. Large cumulonimbus (thunderstorm) clouds can then develop very quickly (within an hour or so) if there is enough moisture in the air and heat continues to be available. Thunderstorm intensity can be very substantial and result in hail.

Orographic precipitation is associated with mountains. Moist air moving over a mountain is forced to rise. The ascending air cools, condenses to form clouds, and if conditions are favorable, precipitation occurs. The higher the mountain, the greater the amount and intensity of precipitation. The Sierras in California provide a good example of orographic precipitation. Moist air from the Pacific Ocean moves inland as part of the prevailing westerlies circulation. Substantial precipitation occurs as the air is forced to rise over the windward (western) side of the Sierras. Upon reaching the summits of the Sierras, which have elevations approaching 4000 m (14,000 ft), the air descends and heats at the dry adiabatic rate of 10°C/1000 m (5.5°/1000 ft). The result is a rainshadow desert that covers much of Nevada and eastern California on the leeward side of the mountain range. Annual precipitation on the windward slopes of the Sierras ranges from 100 to 180 cm (40–70 in) as compared to the leeward slopes, which receive only about 25 cm (10 in). The pronounced difference in vegetation along Interstate 80 between the heavily wooded slopes of the Sierras in California and the northern desert plants of the Reno, Nevada area on the leeward side provide an excellent indication of the importance of orographically induced precipitation on the environment.

An even more exaggerated form of orographic precipitation occurs in Hawaii. If these islands did not exist, the annual average rainfall on the Pacific at this location would be about 635 mm (25 in). Instead, the presence of mountains in the path of the northeast trade winds raises the annual average to about 1780 mm (70 in), an increase of 1145 mm (45 in). Indeed, rainfall totals in the mountainous parts of the state exceed 6100 mm (240 in). For extremes, the annual average reaches an astonishing 12,345 mm (486 in) at Mt. Waialeale on the island of Kauai (Blumenstock and Price, 1974, pp. 614–639).

4.2.5 DISTRIBUTION OF PRECIPITATION

The amount, rate, and duration of precipitation varies enormously over the earth's surface. On a global scale, great spatial and temporal differences exist in atmospheric water vapor, which is the critical element in precipitation. For example, the precipitable water content of the atmosphere averages 25 mm (1 in) with an equatorial maximum of 44 mm (1.7 in) and a polar minimum of 2 to 8 mm (0.08–0.3 in) depending on the season. Although water vapor in the atmosphere is essential for precipitation, it is the lifting forces that are necessary for it to occur.

The average annual precipitation for the earth is about 864 mm (34 in), but the distribution is highly variable (Reichel, 1952, p. 155). Differences are related to air mass source regions and the movements of these air masses. The equatorial zone (10°N–10°S) is characterized by large amounts of precipitation. The average annual totals are over 2000 mm (80 in) and are caused by warm moist air masses that meet and rise along the intertropical convergence zone, resulting in numerous thunderstorms throughout the year. The driest areas in the world occur under the subtropical high pressure cells where the descending air is heated and dried, resulting in very unfavorable conditions for precipitation. The great tropical deserts (i.e., Sahara, Arabian, Kalahari, Australian) are located approximately along the Tropic of Cancer and Tropic of Capricorn where the warm, dry continental tropical (cT) air mass is dominant. Average annual precipitation totals are less than 250 mm (10 in) and in many places less than 50 mm (2 in). In terms of extremes, Arica in the Atacama Desert in northern Chile on the border with Peru (18.5°S) had an annual average rainfall of only 0.75 mm (0.03 in), based on a 59-year record. Indeed, Arica had no measurable rainfall for 14 consecutive years from 1903 to 1918 (Riordan and Bourget, 1985). Arctic and polar deserts are located poleward of 60° and are characterized by cold dry continental air masses that hold little moisture (cA and cP). Average annual precipitation is under 300 mm (12 in) but the evaporation rates are low.

The midlatitudes have moderate to heavy amounts of precipitation depending on the movement of the jet stream, which guides storms and the prevailing westerlies. The humid subtropical region from 25 to 45°N and S is located in the southeastern sections of the continents. Warm moist maritime tropical (mT) air is dominant during the summer or high-sun period. Annual precipitation averages 1000 to 1500 mm (40–60 in). Midlatitude deserts and steppes are located in the continental interiors of Asia and North America between 35° and 50°N. Annual average precipitation totals range from less than 100 mm (4 in) in the deserts to 500 mm (20 in) in the relatively moister steppes and grasslands. The paucity of precipitation is attributed to sheer distance from sources of moisture in the ocean and to being located in the rainshadow of mountains. Midlatitude coasts and islands on the western sides of continents in the 35 to 65°N and S latitudinal belt, such as the British Isles, the states of Washington and Oregon, and southern Chile are dominated by the cool moist maritime polar (mP) air mass, which provides ample precipitation of about 1000 mm/yr (40 in/yr).

Annual totals of precipitation are of course useful values to consider in categorizing global climates, but they can be misleading if there are marked seasonal patterns. Three major patterns cover the seasonal distribution of precipitation: (1) precipitation evenly distributed throughout the year, (2) a maximum of precipitation during the summer or high sun period, and (3) a maximum of precipitation during the winter or low sun period. The first pattern can vary from ample to minimal precipitation in every month. For example, Singapore in the rainy tropics at 1.5°N has an annual total of 2410 mm (95 in), with all monthly totals above 150 mm (6 in). On the drier side, Cairo at 30°N has only an average of 28 mm (1.1 in) of rain for the entire year, with monthly totals ranging from 0.0 to 5 mm (0.0 to 0.2 in). The second pattern is illustrated by Wilmington, North Carolina at 34°N in the humid subtropics. The annual total is 1200 mm (47 in), but the monthly totals range from 50 mm (2 in) in November to 180 mm (7 in) in July, clearly showing the summer maximum when the warm, moist maritime tropical (mT) air mass is dominant. Winter (low-sun) precipitation maximums are characteristic of midlatitude locations on the west coasts of continents when maritime polar (mP) air masses are prevalent. A typical example of this category of precipitation regime is Sacramento at 39°N where the monthly totals vary from practically zero in July and August to 100 mm (4 in) in December and January. Because this type of precipitation pattern is very common in the countries that border the Mediterranean, the climate is often referred to as Mediterranean.

4.2.6 RUNOFF

In order to maintain a global water balance, the excess of precipitation over evaporation on land surfaces results in water returning to the oceans by surface and subsurface means. This return flow to the oceans is called *runoff*.

The surplus water on the land may flow over the land surface and is simply referred to as overland flow. It is distinguished from streamflow (or channel flow) where water is moving in a defined channel with lateral banks. There are several forms of overland flow, such as sheet flow, which forms a thin, continuous film of water over smooth land surfaces. Other forms of overland flow may not even be seen, such as the movement of water below the layers of organic matter and leaves on wooded slopes.

Overland flow is measured in units (inches of water per hour), which is similar to precipitation and infiltration. The rate of production of overland flow (OV) is equal to the precipitation rate (P) minus the infiltration rate (I). This relationship is expressed in a simple formula as:

$$OV = P - I$$

Interflow represents another form of runoff, and occurs where a less permeable soil or rock layer impedes infiltration. Many soils have a layer of clay accumulating beneath a permeable upper soil horizon. Because the infiltrating water has difficulty penetrating the restrictive clay layer, the water is forced to move downslope in a direction parallel with the ground surface. Because of friction with the soil particles, interflow is slower in movement than overland flow.

Water that eventually reaches the water table becomes groundwater, which is the third form of runoff. This water moves very slowly down the hydraulic gradient and eventually seeps into a stream. Depending on the depth and the type of soil and bedrock, the groundwater portion of runoff can return to the stream (or the ocean directly in coastal areas) in days, weeks, months, years, or centuries. The movement of groundwater and interflow into a stream channel is called base flow or fair-weather flow because it sustains streamflow during dry periods.

All of the unevaporated surplus water on land is returned to the oceans by a network of streams that flows into larger and larger streams that eventually empties into the oceans. Thus, ranked in terms of rapidity of movement, runoff consists of overland flow, interflow, and groundwater flow.

4.2.7 *FACTORS AFFECTING RUNOFF*

The magnitude of flow of the surface and subsurface components of runoff is affected by many factors. These factors include such natural variables as local climatological mechanisms, soil characteristics, geology, vegetation, and the increasingly important anthropogenic influence of river control and urbanization.

Precipitation is an obvious critical component of the runoff process, having enormous spatial and temporal variation in location, amount, intensity, duration, and areal extent. Precipitation for a storm event is often plotted on a hyetograph, a histogram or vertical bar chart with the amount of precipitation on the y-axis and time on the x-axis.

Infiltration is another significant hydrologic mechanism in runoff. The more water that can be absorbed into the soil, the less there is for overland flow, which can get into the stream channel much more quickly and lead to possible flooding. Infiltration rates range over two orders of magnitude for different types of soil. The National Resource Conservation Service (formerly the Soil Conservation Service) classifies soils into four hydrologic soil groups (A–D) based on the average infiltration rate when thoroughly wetted. The infiltration rates range from 1.5 mm/hr (0.06 in/hr) for soils with a high clay content (D group) to 160 mm/hr (6.3 in/hr) for soils with a high sand or gravel content (A group). Hydrologic soil groups B and C have moderate 5 to 160 mm/hr (0.2–6.3 in/hr) to slow 1.5 to 5 mm/hr (0.06–0.2 in/hr) infiltration rates, respectively. The runoff potential from different parts of a watershed therefore varies substantially depending upon the different types of soils and infiltration rates.

Runoff is also affected by the amount of antecedent precipitation. If the soils are already saturated from previous precipitation, additional rain would only add to the runoff and flooding potential for an area. For example, the unusual occurrence of Hurricanes Connie and Diane in 1955 coming so close together in time caused extensive flooding along the Delaware River in New Jersey and Pennsylvania.

Land cover plays an important role in the runoff process. Heavily vegetated areas can intercept precipitation and evapotranspire large amounts of water. Indeed, evapotranspiration can account for 50 percent or more of the average annual precipitation in many watersheds. The relationship among the variables precipitation (P), evapotranspiration (ET), and runoff (RO) can be expressed in a simple formula as:

$$P = ET + RO$$

As evapotranspiration changes, runoff must also change in order to balance the equation. For example, when dairy farms were abandoned in New York State during the Depression of the 1930s, the cleared land started to revert to the climax vegetation of trees. Because evapotranspiration is much greater for trees than for grass, the regional runoff started to decrease. The opposite effect occurs dramatically when the land cover is changed from a vegetated condition to a paved landscape where there is minimal opportunity for infiltration.

The volume of runoff depends primarily on the amount of precipitation and infiltration characteristics, which include soil type, land cover, and land use. Runoff timing considerations include slope, distance, and surface roughness. Watershed slopes can be changed by constructing terraces for building lots, parking areas, and roads. In urban areas, the slopes of storm sewers, gutters, and roads are very important in determining travel times for runoff. The distance water has to travel may be decreased if the natural meanders of streams are channelized to form straight paths. On the other hand, the travel distance for water increases if overland flow is diverted through swales, storm sewers, or street gutters. The velocity of runoff increases when the flow over rough surfaces of forests, grassy areas, and natural channels is changed to move over the smooth surfaces of parking lots, storm sewers, gutters, and channels that are lined.

Peak rates of runoff are based on the aforementioned timing considerations, antecedent precipitation, drainage area size, the location of buildings and roads within the watershed, the existence and storage capacity of lakes, reservoirs, and wetlands in the basin, and the intensity and spatial distribution of precipitation.

The hydrologic effects of urbanization on runoff are substantial. The increase in impervious cover associated with urbanization (and suburbanization) is directly related to development intensity. For example, the estimated extent of impervious cover for single family homes on lot sizes of 15,000 (1/3-acre) and 6000 (1/8-acre) sq ft is 25 and 80 percent, respectively (Antoine, 1964). Industrial facilities range in impervious cover from 60 to 80 percent and suburban shopping center have coverages of about 95 percent (Metropolitan Washington Council of Governments, 1979).

The changes in watershed hydrology resulting from urbanization can be summarized as follows (Anderson, 1970; Leopold, 1968; Schueler, 1987):

1. Postdevelopment peak discharges are about 2 to 5 times higher than predevelopment values.

2. A moderately developed area may experience a 50 percent increase in runoff volume from the same storm as compared to a wooded area.

3. The time of concentration, which is the travel time for runoff to reach the stream, may be reduced by up to 50 percent if major drainage changes are made.

4. Flooding severity and frequency commonly increase.

5. Streamflow becomes lower during dry weather periods to the extent that discharge in perennial headwater streams may cease.

6. Runoff velocities increase during storm events because of increased peak discharges, faster time of concentration, and smoother surfaces for the water to flow over.

4.2.8 GEOGRAPHICAL DISTRIBUTION OF RUNOFF

The runoff from land areas varies from continent to continent as a function of precipitation and evaporation. Asia accounts for the largest proportion of land area in the world (29.6 percent) and the largest absolute amount of runoff (30.7 percent of the total). At the other extreme, Australia and Antarctica are the driest continents, with runoff accounting for 6.0 and 5.0 percent of the world total, respectively. In ranked order, the other continents have a percentage of total runoff as follows: South America (27.8), North America (14.7), Africa (8.6), and Europe (7.1). However, in terms of runoff intensity as expressed as so many inches of runoff per square mile, South America ranks first with 9.4 in/sq mi. Africa, even though it is the second largest continent with 20.0 percent of the total land area, has the lowest runoff intensity (1.7 in/sq mi) of any of the continents. The presence of the vast Sahara and Egyptian Deserts in Africa accounts for the low intensity. The runoff intensity in inches/square mile for the other continents, in ranked order, is as follows: Europe (4.3), Asia (4.2), Australia (4.1), North America (3.7), and Antarctica (2.1) (Spiedel and Agnew, 1988).

There is also wide variation in the discharge from the major rivers of the world. By virtue of its location in the rainy tropics and sheer size (5,778,000 sq km; 2,231,000 sq mi), the Amazon ranks first in the world in discharge (212,250 cu m/sec; 7,500,000 cu ft/sec). The Amazon alone accounts for 15 percent of the total discharge to the oceans of the world. The second largest discharger in the world is the Congo with an average of 39,600 cu m/sec (1,400,000 cu ft/sec). The third and fourth largest dischargers are the Yangtze in China and the Brahmaputra in India and Bangladesh with 21,800 and 20,000 cu m/sec (770,000 and 700,000 cu ft/sec). The Mississippi River ranks seventh in the world with an average discharge of 17,300 cu m/sec (611,000 cu ft/sec) (Leeden, Troise, and Todd, eds., 1990, 181).

Given the wide diversity of climatic regimes in the United States, it is reasonable to expect wide variation in runoff rates. The average annual runoff ranges from 0 to 25 mm (0–1 in) in large parts of the southwestern states (southeastern California, almost all of Nevada, western Utah, Arizona, and New Mexico) and the Great Plains to well over 1000 mm (40 in) in the coast ranges of Oregon and Washington and the Sierras in California. The runoff in the central and eastern parts of the nation ranges from 130 to 1000 mm/yr (5–40 in/yr).

4.2.9 INTERIOR DRAINAGE

Interior (or internal) drainage occurs when surface drainage does not reach the ocean. It is common in arid and semiarid regions, such as the Great Basin in the western United States. The runoff that does occur from precipitation or snowmelt from surrounding mountains collects in the lower parts of the basin and either evaporates directly into the atmosphere or infiltrates the soil to become part of the groundwater system.

Land with interior drainage is unevenly distributed among the continents. The estimated amounts in ranked order is Australia (50 percent), Africa (40 percent), Asia (30 percent), Europe (20 percent), South America (10 percent), and North America (5 percent) (Spiedel and Agnew, 1988).

REFERENCES

Allaby, M., *Water: Its Global Nature*, Facts on File, New York, 1992.

Anderson, D. C., "Effects of Urban Development on Floods in Northern Virginia," *U.S. Geological Survey Water-Supply Paper 2001-C*, 1970.

Antoine, L. H., "Drainage and Best Use of Urban Land," *Public Works*, 95: 88–90, 1964.

Barry, R. G., and Chorley, R. J., *Atmosphere, Weather, and Climate*, Routledge, New York, 6th edition, 1992.

Berner, E. K., and Berner, R. A., *Global Environment: Water, Air, and Geochemical Cycles*, Prentice Hall, Upper Saddle River, New Jersey, 1996.

Blumenstock, D. I., and Price, S., "The Climate of Hawaii," in *Climates of the States*, Water Information Center, Port Washington, New York, Vol. 2, 1974.

Camp, T. R., and Meserve, R. L., *Water and Its Impurities*, Dowden, Hutchinson & Ross, Stroudsburg, PA, 2nd edition, 1974.

Dingman, S. L., *Physical Hydrology*, Prentice Hall, Upper Saddle River, New Jersey, 1994.

Emiliani, C., *Planet Earth*, University Press, Cambridge, 2nd edition, 1995.

Falkenmark, M., *Hydrological Phenomena in Geosphere-Biosphere Interactions: Outlooks to Past, Present and Future*, Wallingford, Oxfordshire, UK: International Association of Hydrological Sciences (IASH) in cooperation with the International Institute for Hydraulic and Environmental Engineering, 81 pp, IASH Monographs and Reports No. 1, 1989.

Hornberger, G. M., Raffensperger, J. M., Wiberg, P. L., and Eshleman, K. N., *Elements of Physical Hydrology*, The Johns Hopkins University Press, Baltimore, 1998.

Leeden, F. v. d., Troise, F. L., and Todd, D. K., eds., *The Water Encyclopedia*, 2nd edition, Lewis Publishers, Chelsea, MI, 1990.

Leopold, L. B., "Hydrology for Urban Land Planning—A Guidebook on the Hydrologic Effects of Urban Land Use," *U.S. Geological Survey Circular 554*, 1968.

Maidment, D. R., "Hydrology," Chapter 1 in *Handbook of Hydrology*, Maidment, D. R., ed., McGraw-Hill, New York, 1993.

Metropolitan Washington Council of Governments, *Guidebook for Screening Urban Nonpoint Pollution Management Strategies*, Washington, D.C., 1979.

Moore, J. W., *Balancing the Needs of Water Use*, Springer-Verlag, New York, 1989.

National Research Council, *Changing Climates: Report of a Carbon Dioxide Assessment Committee*, National Academy Press, Washington, D.C., 1983.

National Research Council, *Opportunities in the Hydrological Sciences*, National Academy Press, Washington, D.C., 1991.

Peixoto, J. P., and Oort, A. H., *Physics of Climate*, Amer. Inst. of Physics, New York, 1992.

Post, D. A., Grant, G. E., and Grant, J. A., "New Developments in Ecological Hydrology Expand Research Opportunities," *EOS Transactions*, American Geophysical Union, 79(43): 517, 526, October 27, 1998.

Reichel, E., "Der Stand des Verdunstungs problems," (The Status of the Evaporation Problem), *Ber. Deut. Wetterdienst*, Bad Kissingen, 35: 155, 1952.

Riordan, and Bourget, *World Weather Extremes*, U.S. Army Corps. of Engineers, 1985.

Robinove, C. J., "What's Happening to Water," *Smithsonian Inst. Ann. Rept.*, 375–389, 1962, 1963.

Schueler, T. R., *Controlling Urban Runoff: A Practical Manual for Planning and Designing Urban BMPs*, Washington Metropolitan Water Resources Planning Board, Washington, D.C., 1987.

Spiedel, D. H., and Agnew, A. F., "The World Water Budget," Chapter 3 in *Perspectives on Water*, Spiedel, D. H., Ruedisili, L. C., and Agnew, A. F., eds., Oxford University Press, New York, 1988.

Strahler, A. H., and Strahler, A. N., *Modern Physical Geography*, Wiley, New York, 4th edition, 1992.

UNESCO, *World Water Balance and Water Resources of the Earth*, UNESCO, Paris, (Prepared by the USSR Committee for the International Hydrological Decade), 1978.

Viessman, W., Jr, Lewis, G. L., and Knapp, J. W., *Introduction to Hydrology*, Harper & Row, New York, 3rd edition, 1989.

Wetzel, R. G., *Limnology*, Saunders, Philadelphia, 2nd edition, 1983.

Wolman, A., "Water Resources," A Report to the Committee on Natural Resources of the National Academy of Sciences—National Research Council, *Natl. Acad. Sci. Natl. Res. Council, Publ. 1000-B*, 1962.

CHAPTER 4
HYDROLOGY

SECTION 4.3

STREAMFLOW AND FLOOD CONTROL

Robert M. Hordon

*Dr. Hordon is a physical geographer and hydrologist in the
Geography Department at Rutgers University in New Jersey.*

4.3.1 INTRODUCTION

The oceans account for 97.2 percent of all of the water in the world. Ice sheets and glaciers account for another 2.15 percent. This leaves only 0.65 percent for the other storage repositories on earth, with groundwater accounting for most of it (0.62 percent of the remaining total). Stream channels hold an astonishingly small fraction of the world's water—only 0.0001 percent. Yet, this small amount has been so critical to the development of human societies.

From the beginning of human history, rivers have played a vital role in the development of civilization. For centuries, they have been actively used for irrigation, domestic, commercial, and industrial water supply, navigation, hydropower, powerplant cooling, fire fighting, and wastewater disposal. Consider the major cities of the world that developed on or near rivers—London on the Thames, Paris on the Seine, New York on the Hudson, Shanghai on the Yangtze, Philadelphia on the Delaware, Montreal on the St. Lawrence, New Orleans and St. Louis on the Mississippi, Vienna on the Danube, Cairo on the Nile. The list could go on and on, but the important thing is that most of the world's settlements are associated with proximity to a waterbody, and most of these waterbodies are streams.

4.3.2 STREAM DEVELOPMENT

Several things happen to precipitation as it approaches the earth's surface. In drier climates, some of the precipitation actually evaporates in the atmosphere and never reaches the surface. Another portion of precipitation is intercepted by vegetation, forms a small pool on leaf surfaces, and may evaporate. The portion that does reach the earth's surface can either infiltrate the soil or flow downslope over the land as overland flow. Over time, this downslope movement of water coalesces and erodes the land surface to a point where permanent channels start forming. Streams become perennial when they have cut a channel deep enough to be supplied by groundwater. Indeed, this movement of groundwater to the stream, called *base flow*, can easily account for half of the total streamflow. For example, it is estimated that 50 to 70 percent of the water in the Delaware River was originally groundwater. Thus, surface runoff consists of both overland flow and groundwater, even though they move at very different rates (groundwater being much slower).

Streams move from higher to lower elevations, with the oceans being the ultimate sink. Headwater streams are generally characterized by steep gradients, rapids, and waterfalls. With time, streams remove the irregularities in the channel and the profile becomes smooth with a progressively gentler slope. Downward cutting of the channel is replaced by lateral cutting where the stream widens its bed and valley. Meanders develop and stream sinuosity increases. The floodplain, a depositional feature of a stream, widens with time and distance downstream.

4.3.3 STREAM HYDRAULICS

Water is a fluid and, consequently, cannot resist stress. Any stress, even if it is very small, will cause the water to move in either laminar or turbulent flow. If the velocity is very low (fractions of a millimeter or inch per second), the water moves as laminar flow. In most streams, this is only possible at the boundary layer next to the banks and bed where friction reduces the velocity to extremely low values. Velocities are higher in the rest of the stream channel. Thus, turbulent flow, which is characterized by the chaotic movement of individual water particles, is the dominant type of flow in almost all streams. As a consequence of this natural turbulence, the bulk (95 to 97 percent) of the energy in a stream is lost as it is converted to heat by inner turbulence and by friction as it flows along the channel banks and bed. This leaves only a small fraction of the initial energy in the stream available to do work, and yet this small amount is sufficient to carve up the landscape and slowly but surely move the continental landmass to the sea, particle by particle. Streams also transport the excess of precipitation over evaporation on the land surfaces of the world to the oceans as runoff, a major component of the hydrologic cycle.

4.3.4 MEASUREMENT OF STREAMFLOW

Discharge is the volume of water flowing past a cross-section of the stream during a specified period of time. It is calculated by multiplying the cross-sectional area of the stream channel by the mean velocity. This relationship is expressed as:

$$Q = A \times V$$

where Q is stream discharge (cu m/sec; cu ft/sec); A is cross-sectional area (sq m; sq ft); and V is mean velocity (m/sec; ft/sec).

It is readily apparent that in order to balance the equation, if the cross-sectional area of the stream is reduced, the velocity must increase. Conversely, expanding the area results in a decrease in velocity.

The earliest known records of streamflow measurement go back about 5000 years to the ancient Egyptians. The early staff gages, called nilometers, were permanently attached to the banks of the Nile and allowed the Egyptians to see the rise and fall of the river level. Their primary purpose was to alert downstream farmers when to expect the crucial flood wave that was expected each summer from the headwaters of the Nile in the Ethiopian highlands. Staff gages are still used for river, lake, and canal levels. The earliest stream gaging station in the nation was established in 1889 by the U.S. Geological Survey (USGS) on the upper Rio Grande at Embudo, New Mexico (Manning, 1997).

One of the standard methods of measuring streamflow is to take velocity measurements with a current meter at a number of depths and at different locations across the width of the channel. The mean velocity is simply the sum of all the individual velocity readings at each location in the cross-sectional grid, divided by the total number of observations. Discharge is then readily obtained by using the $Q = AV$ formula.

The USGS measures discharge at over 7300 gaging stations on the major rivers and their tributaries. Regrettably, this number is gradually being reduced by either budget cuts or urbanization. In some highly urbanized states, such as New Jersey with the highest population density in the nation, some streams are so affected by surface and groundwater diversions, by wastewater flows out of the basin, by regulation by dams, and by imports of water from other basins that the "natural flow" can no longer be obtained and the gaging station is discontinued.

4.3.5 STREAM LENGTH AND DISCHARGE

The length and discharge of streams vary enormously. The three longest rivers in the world are the Nile (6650 km; 4130 mi), the Amazon (6400 km; 4000 mi), and the Yangtze (6300 km; 3915 mi).

The Mississippi ranks fourth in the world in length (5971 km; 3710 mi). Other major rivers include the Yenisei in Siberia (5540 km; 3440 mi), the Yellow or Huang Ho in China (5464 km; 3395 mi), the Ob in Siberia (5410 km; 3360 mi), and the Parana in Argentina (4880 km; 3303 mi) (Leeden, Troise, and Todd, 1990, 179–180).

The Amazon in Brazil ranks first in the world in terms of watershed area (6,000,000 sq km; 2,300,000 sq mi, or an area 74 percent that of the conterminous 48 states) and average discharge at the mouth (175,000 cu m/sec; 6,100,000 cu ft/sec) (Oltman, 1968, 15). This astounding discharge is approximately 10 times that of the Mississippi, which ranks seventh in the world in discharge (17,300 cu m/sec; 611,000 cu ft/sec). The Congo River in central Africa ranks second in the world with an average discharge at the mouth of 39,600 cu m/sec (1,400,000 cu ft/sec). Note the profound influence of the heavy precipitation regime in the rainy tropics with over 2000 mm/yr (80 in/yr) on the flow in the Amazon and Congo. Even though the Nile is the longest river in the world, its discharge ranks thirty-third in the world (2830 cu m/sec; 100,000 cu ft/sec) because it flows for much of its length through the desert portions of northern Sudan and Egypt (Leeden, Troise, and Todd, 1990, 181).

Streamflow fluctuates with seasonal variation of either precipitation, evapotranspiration, or both. Typically, streams in the eastern United States have their highest and lowest average flows in late winter–early spring and late summer–early fall, respectively. For example, the Ohio River at Metropolis, Illinois with a drainage area of 236,100 sq km (91,160 sq mi) has an average monthly flow of 14,900 and 2180 cu m/sec (526,000 and 77,000 cu ft/sec) in March and October, respectively. The nearly seven-fold difference in average flow is attributed to seasonal variations in evapotranspiration rather than variations in precipitation. The seven-fold difference also shows up for the average flows in the Susquehanna River near Harrisburg, Pennsylvania with a drainage area of 62,400 sq km (24,100 sq mi): the flow is 2210 and 306 cu m/sec (78,000 and 10,800 cu ft/sec) in March and September, respectively.

In some cases, streams have large natural lakes that regulate the flow and even out the monthly average discharge. An excellent example of this is the St. Lawrence River, which drains the Great Lakes. The highest and lowest average monthly flows at Ogdensburg, New York (drainage area of 764,600 sq km; 295,200 sq mi) are 7340 and 6150 cu m/sec (259,000 and 217,000 cu ft/sec) in June and February, respectively. The high to low ratio of only 1.2 is attributed to the enormous storage volume of the Great Lakes, which account for an astonishing 18 percent of the water in all of the freshwater lakes in the world.

4.3.6 THE GEOLOGIC WORK OF STREAMS

The three closely interrelated activities of streams are erosion, transportation, and deposition. Erosion is the removal of material from the sides and bed of the stream channel. This occurs in both alluvial and bedrock channels, although the removal rates are much faster in alluvium because it is much less resistant to particle entrainment and capture. As the particles erode, they are carried by the stream in solution, suspension, or bed load. Dissolved matter cannot be seen because it is mostly in the form of chemical ions. A very common form of material in solution in all streams are salts from mineral alteration. Suspended load consists of clay and silt particles that are kept afloat by stream turbulence. The term "the Big Muddy" as applied to the Mississippi-Missouri River is based on the highly visible soil particles that have eroded from the cultivated fields in the Midwest and transported to the stream. Larger particles include sand, gravel, and cobbles, which move as bed load close to the channel floor. The third activity of streams is deposition, which occurs on the streambed, floodplain, or on the bottom of a body of water, such as an ocean, that the stream empties into.

Suspended load increases dramatically as discharge increases. For example, if discharge increased from 1 to 10 cu m/sec (a 10-fold increase), suspended sediment load would increase approximately from 100 to 10,000 metric tons/day (a 100-fold increase). There is an enormous range in sediment load among the major rivers in the world. For example, the average annual sediment yield varies from a high of 2600 to a low of 4 metric tons/sq km for the Yellow River in China and the Yenisei River in Siberia, respectively. This difference of over three orders of magnitude is attributed to the large soil-erosion rate in the highly cultivated silts of the Yellow River basin as compared to the mostly forested and uncultivated Yenisei watershed. Most of the annual sediment yield of the Mississippi River of 97 metric tons/sq km is attributed to the Missouri River, which flows through subhumid and semiarid grasslands, large portions of which have been cultivated.

The sediment load carried by a stream is extremely important in the design of reservoirs and irrigation canals. Sediment that is trapped behind a reservoir dam will eventually fill the reservoir and eliminate the storage capacity. For example, several small reservoirs that were used to supply Santa Barbara, California filled up with sediment in only a few decades and had to be abandoned. Even major reservoirs that are located in semiarid to arid regions where the slopes in the watershed are not completely covered with vegetation have limited lifetimes due to sediment input.

4.3.7   *EXOTIC STREAMS*

Streams that have their source in humid, well-watered headwater areas and that flow through deserts where there is little precipitation and maximum evaporation are called exotic streams. The best examples are the (1) Colorado River, which begins in the Colorado Rockies and flows 2330 km (1450 mi) into the Gulf of California and (2) the Nile River, which starts in the Ethiopian highlands and the lake region of east-central Africa and flows for a long distance through the deserts of northern Sudan and Egypt before emptying into the Mediterranean.

4.3.8   *FLOOD CONTROL—INTRODUCTION*

Floods are one of the most common and damaging of natural hazards, ranking first in the number of fatalities (40 percent of the total). They occur anywhere in the world, but are most prevalent in valleys in humid regions when streams overflow their banks. The water that cannot be accommodated within the channel flows out over the floodplain, a low, flat area on one or both sides of the channel. Floodplains have attracted human settlement for thousands of years as witnessed by the ancient civilizations that developed along the Nile, the Yangtze, and Yellow Rivers in China, and the Tigris and Euphrates Rivers in Mesopotamia (now Iraq). More people are living in river valleys than ever before and the number is increasing. About 7 percent of the United States consists of floodplains that are subject to inundation by a 100-year flood. Most of the largest cities of the nation are located within this relatively small area. Even as flood damages increase, development in the floodplain has been growing by about 2 percent a year.

Large areas of the floodplain are not only inundated but are also subjected to rapidly moving water, which has enormous capacity to move objects such as cars, buildings, and bridges. If the flood is large enough, the water will inundate even stream terraces—older floodplains that are higher than the stream. For example, Hurricane Agnes in 1972 caused the Susquehanna River to rise nearly 5 m (16 ft) above flood stage, inundating the downtown portion of Harrisburg, Pennsylvania, which is built on a terrace.

Floods can also be devastating in semi-arid areas where sparsely vegetated slopes offer little resistance to large volumes of overland flow generated by storms. For example, the winter storms and flooding in January, 1969, in southern California resulted in 100 deaths.

Floods also occur in coastal and estuarine environments. A winter storm in 1953 resulted in severe coastal flooding in eastern England and northwestern Europe (particularly The Netherlands), causing the deaths of over 2000 people, the destruction of 40,000 homes, and the loss of thousands of cattle. Bangladesh is particularly vulnerable to coastal flooding. Most of the country (115 million people in an area about the size of New York State) live in the downstream floodplain of the Brahmaputra and Ganges Rivers. In addition, its location at the head of the Bay of Bengal only accentuates the tropical storm surges that frequently occur in this area. For example, coastal storm surges killed 225,000 people in 1970 in Bangladesh.

4.3.9 FLOOD DYNAMICS

Two aspects that are instrumental in flood occurrence are: (1) the amount of surface runoff, and (2) the uniformity of runoff from different parts of the watershed. If the response and travel times are very uniform, then the flow would more likely result in a flood. Conversely, watersheds that have soils with high infiltration rates are less prone to flooding. Flood magnitude depends on the intensity, duration, and areal extent of precipitation in conjunction with the condition of the land. If the soils in the watershed have been saturated because of antecedent precipitation, the flooding potential would be much greater. For example, the unusual occurrence of Hurricanes Connie and Diane in the same season of 1955 resulted in severe flooding after Diane along the Delaware River in New Jersey and Pennsylvania because of previous ground saturation from Connie.

Floods are caused by the following factors:

1. *Climatological:* Heavy rain from tropical storms and hurricanes, severe thunderstorms, midlatitude cyclones and frontal passages; rapid snow and ice melt
2. *Part-Climatological:* Tides and storm surges in coastal areas
3. *Other:* Ice and log jam breakups, earthquakes, landslides, and dam failures

Flood-intensifying conditions include:

1. *Fixed basin characteristics:* Area, shape, slope, aspect (north or south facing), and elevation
2. *Variable basin characteristics:* Water storage capacity and transmissibility in the soil and bedrock, soil infiltration rates, extent of wetlands and lakes
3. *Channel characteristics:* Length, slope, roughness, and shape
4. *Human effects:* River regulation, conjunctive use of groundwater, interbasin transfers, wastewater release, water diversion and irrigation, urbanization and increases in impervious cover, deforestation and reforestation, levees, and land drainage

Because floods are capable of such extensive damage, knowledge of their magnitude and frequency is very useful. Accordingly, hydrologists use statistical methodology to estimate the probability that a flood of a certain size will occur in a given year. The estimates are based on the use of historical streamflow records and special graph paper. The peak discharge for each year of record is plotted on the y-axis, which is scaled arithmetically. The bottom x-axis is scaled in probability terms, which provides the percent probability that a given discharge will be equaled or exceeded. The plotting position for each annual peak flow is calculated by using a special equation. A straight line is drawn through the plotted points, forming a flood frequency graph for the particular gaging station. The upper x-axis at the top of the graph provides the return period in years (or recurrence interval), which is the inverse value of the probability percentage on the bottom x-axis. Thus, a discharge associated with a value of 5 percent means that this discharge is expected to be equaled or exceeded in 5 out of 100 years. In this example of a 5 percent probability value, the return period is 20 years, implying that on the average, this discharge is estimated to be equaled or exceeded once every 20 years. This frequency estimate is also called the 20-year flood.

The flood estimate is stated in probability terms. This has confused some people who believe that if a 20-year flood event occurred, then the next flood of that magnitude will not occur again for another 20 years. This is incorrect, as two 20-year floods can occur in the same year even though the probabilities are low. A 50-year and 100-year flood have a probability of being equaled or exceeded of only 2 and 1 percent, respectively. The longer the record, the greater the confidence level. However, the historical period of record for many streams of 50 or even 75 years is considerably shorter than the total time the stream has been around. In addition, many watersheds have been so extensively changed by urbanization, farming activities, logging, and mining that previous discharges may not be in accord with current conditions. Also, climatic change, particularly near large metropolitan areas in the form of urban heat islands, increased particulate matter, and greater thunderstorm frequency, may have been great enough to produce quantifiable changes in discharge. Thus, forecasts that are based on past flows may or may not be suitable for estimating future flows.

The hydrologic effects of urbanization on flooding potential are substantial. Roofs, driveways, sidewalks, roads, and parking lots greatly increase the amount of impervious cover in the watershed. For example, it is estimated that residential subdivisions with lot sizes of 0.06 ha (6000 sq ft; 1/8-acre) and 0.14 ha (15,000 sq ft, 1/3-acre) have impervious areas of 80 and 25 percent, respectively (Antoine, 1964, 88–90). Industrial sites have impervious areas in the range of 60 to 80 percent and suburban shopping centers with their large expanse of parking area have about 95 percent impervious cover (Metropolitan Washington Council of Governments, 1979).

As the impervious cover increases, infiltration is reduced and overland flow is increased. Consequently, the frequency and flood peak heights are increased during large storms. Another change related to urbanization is the installation of storm sewers, which route storm runoff from paved areas directly to streams. This short-circuiting of the hydrologic cycle reduces the travel time to the stream channel, reduces the lag time between the precipitation event and the ensuing runoff, and increases the height of the flood peak. In essence, an increasing volume of water is sent to the stream channel in a shorter period of time.

4.3.10   FLOOD CONTROL MEASURES

Societies have made many attempts over the years to prevent floods, most of which involve some form of structural control. For example, the Flood Control Acts of 1928 and 1936 assigned the U.S. Army Corps of Engineers the responsibility of building reservoirs, levees, channels, and stream diversions along the Mississippi and its major tributaries. As a result of this legislation, 76 reservoirs and 3500 km (2200 mi) of levees in the upper Mississippi basin alone were built by the Corps since the late 1930s. State and local governments constructed an additional 9300 km (5800 mi) of levees in the same watershed. The Natural Resources Conservation Service (formerly the Soil Conservation Service) built over 3000 reservoirs on the smaller tributaries in the basin. These very expensive efforts on the Mississippi and similar efforts on other watersheds still did not prevent the disastrous floods in 1993 on the Mississippi and in 1997 on the Red River in North Dakota and Minnesota.

Flood control measures can be divided into structural and nonstructural approaches. The structural approach consists of technical and engineering techniques that attempt to either hold back runoff in the watershed or change the lower reaches of the river where inundation of the floodplain is most probable. The nonstructural approach is best illustrated by zoning regulations.

One type of structural measure is to treat watershed slopes by planting trees or shrubs to increase infiltration opportunity, which thereby decreases overland flow. This measure, when combined with the building of storage dams in the valley bottoms, can substantially reduce flood peaks and increase the lag time between the storm event and the runoff. Another very common type of structural measure is to build artificial levees along the channel. They are usually earthen and high enough to contain the design flood. During high water on the Mississippi at New Orleans, it is possible to see ships sailing above you if you are standing at the foot of the levee. Starting in 1879 with the Mississippi River Commission, a large system of levees was constructed in the hope of containing all floods. This levee system has been expanded and improved so that it now totals many thousands of km (mi) in length and is as high as 9 m (30 ft) in some places. One problem with the levees is that the river aggrades or builds up its channel over time so that it is higher than the bordering floodplain. If the levees fail, the water in the channel will rush into the lower-lying floodplain and cause a disastrous flood. For example, the Yellow River in China flooded an area nearly the size of England (130,000 sq km; 50,000 sq mi) in 1887, which resulted in the direct death of nearly one million people and the indirect death of millions more by the famine that came later.

Another structural measure that has been tried by the Corps on the Mississippi is to cut channels (or cutoffs) across the wide meander loops, which shortens the river length. This reduction in length increases the river slope, which correspondingly increases the average velocity. As velocity increases, more water can move through the channel and flood peaks can be reduced. Although the technique was initially successful, the river responded by developing new meanders, which only increased the length.

Where feasible, selected portions of the floodplain are established as temporary basins and are deliberately flooded in order to reduce the flood peak in the main channel. A related structural measure, found in the delta region of the lower Mississippi, is to use spillways that divert water from the channel directly to the ocean. For example, the Bonnet Carre spillway 32 km (20 mi) upstream of New Orleans has the capacity to divert 7100 cu m/sec (250,000 cu ft/sec) of floodwaters into Lake Pontchartrain which is connected to the Gulf of Mexico (Matthai, 1990, 110).

As an alternative to structural flood control measures, which involve engineering solutions, the nonstructural approach is to view floods and the damages they cause as natural events that have occurred in the past and will continue in the future even after expensive and elaborate engineering structures have been built. Levees that were designed for a 100-year flood will fail if a flood of greater magnitude, say a 200-year flood, occurs. The real problem is the ongoing urban, commercial, and industrial development in the floodplain. Consequently, the nonstructural measure that has received increasing attention in recent years is floodplain zoning, which restricts development in flood-prone areas. This type of planning would allow some agricultural activity and recreation in the floodplain as these types of land use could handle occasional flooding. However, permanent structures such as houses, schools, shops, and industries would not be permitted in flood-prone areas.

Another nonstructural technique that has developed over the years in the United States to reduce flood losses is flood insurance. The notion of an insurance program that provides money for flood losses appeared to be reasonable because the pooling of risks, collection of annual premiums, and payments of claims to those property owners who suffered losses was similar to other types of programs such as fire insurance. The first attempts to begin a federal flood-insurance program began, as expected, after flooding in 1951. Legislation was introduced over the years and resulted in the National Flood Insurance Act of 1968 (Public Law 90-448). This Act created the National Flood Insurance Program, administered through the Federal Emergency Management Agency (FEMA). The major objectives of the program include: (1) nationwide flood insurance for all communities that have a flooding potential and (2) prevention of future development in flood-prone areas. A useful outcome of the program has been the preparation of maps showing the approximate delineation of the 100-year flood area. The flood insurance program is meant to be self-supporting.

However, there is some controversy about flood insurance. For example, although one of the intents of the program was to assist people who could not buy insurance at private market rates, the effect has been to encourage building on the floodplain because people feel that the government will help them no matter what happens. Some people would even like to move their homes and commercial facilities to higher land in another location, but property owners can only collect for damages if they rebuild in the same flood-prone location. If these criticisms are correct, the policy of rebuilding on the floodplain will only perpetuate the problem.

As distinct from an earthquake that occurs without warning, major storms and potential flooding can be predicted in advance. Satellites and airplanes can track storms and provide early warning of potentially heavy rainfall and storm surges. The River and Flood Forecasting Service of the U.S. National Weather Service maintains 85 offices along major rivers. These offices issue flood forecasts to local communities who can then close roads and bridges and recommend evacuation of flood-prone areas.

REFERENCES

Allaby, M., *Water: Its Global Nature*, Facts on File, New York, 1992.

Antoine, L. H., "Drainage and Best Use of Urban Land," *Public Works,* 95: 88–90, 1964.

Dingman, S. L., *Physical Hydrology*, Prentice Hall, Upper Saddle River, NJ, 1994.

Dunne, T., and Leopold, L. B., *Water in Environmental Planning*, Freeman, New York, 1978.

Dzurik, A. A., and Theriaque, D. A., *Water Resources Planning*, Rowman and Littlefield, Lanham, MD, 2nd edition, 1996.

Gleick, Peter H., *The World's Water: 1998–1999*, Island Press, Washington, DC, 1998.

Hornberger, G. M., Raffensberger, J. M., Wiberg, P. L., and Eshleman, K. N., *Elements of Physical Hydrology*, The Johns Hopkins University Press, Baltimore, 1998.

Jones, J. A. A., *Global Hydrology: Processes, Resources and Environmental Management*, Longman, Essex, England, 1997.

Leeden, F. v. d., Troise, F. L., and Todd, D. K., eds., *The Water Encyclopedia*, 2nd edition, Lewis Publishers, Chelsea, MI, 1990.

Manning, J. C., *Applied Principles of Hydrology*, Prentice Hall, Upper Saddle River, NJ, 3rd edition, 1997.

McDonald, A., and Kay, D., *Water Resources: Issues and Strategies*, Longman, New York, 1988.

Marsh, W. M., *Landscape Planning: Environmental Applications*, Wiley, New York, 3rd edition, 1998.

Mather, J. R., *Water Resources: Distribution, Use, and Management*, Wiley, New York, 1984.

Matthai, H. F., "Floods," Chapter 5 in *Surface Water Hydrology*, Wolman, M. G., and Riggs, H. C., eds., Vol. 0-1, The Geological Society of America, Boulder, CO, 1990.

Metropolitan Washington Council of Governments, *Guidebook for Screening Urban Nonpoint Pollution Management Strategies*, Washington, D.C., 1979.

Newsome, M., *Hydrology and the River Environment*, Oxford University Press, New York, 1994.

Oltman, R. E., "Reconnaissance Investigations of the Discharge and Water Quality of the Amazon River," *U.S., Geological Survey Circular 552*, 1968.

Owen, L. A., and Unwin, T., eds., *Environmental Management: Readings and Case Studies*, Blackwell, Malden, MA, 1997.

Speidel, D. H., Ruedisili, L. C., and Agnew, A. F., eds., *Perspectives on Water: Uses and Abuses*, Oxford University Press, New York, 1988.

Strahler, A. H., and Strahler, A. N., *Modern Physical Geography*, Wiley, New York, 4th edition, 1992.

Viessman, W., Jr., Lewis, G. L., and Knapp, J. W., *Introduction to Hydrology*, Harper and Row, New York, 3rd edition, 1989.

White, G. F., *Choice of Adjustment to Floods*, Department of Geography Research Paper No. 93, University of Chicago, Chicago, 1964.

Wilby, R. L., *Contemporary Hydrology: Towards Holistic Environmental Science*, Wiley, New York, 1997.

Wolman, M. G., and Riggs, H. C., eds., *Surface Water Hydrology*, Vol. 0-1, The Geological Society of America, Boulder, CO, 1990.

CHAPTER 4
HYDROLOGY

SECTION 4.4

GROUNDWATER FLOW

Milovan S. Beljin

*Dr. Beljin is a consulting hydrologist and president of M.S. Beljin
& Associates. He serves as adjunct faculty at the University of
Cincinnati and Wright State University, Dayton, Ohio, where he
has taught groundwater modeling.*

Without water life is not possible. Earth is unique among the planets of the solar system because it has a full hydrological cycle. Though water exists on the planet in all three forms, the majority of the total mass of water is in liquid form. Two thirds of the earth's surface area is covered by oceans, and at any one time over half of the planet is covered with clouds. The atmosphere contains only about 0.001 percent of the total water in the system, mainly in vapor form. The salty oceanic water account for 96.5 percent of total water. Considering that about 70 percent of fresh water is locked in polar ice and snow, a relatively small percentage of the earth's water is available for drinking. Lakes and rivers contain only 0.26 and 0.006 percent, respectively, of total fresh water. So most of the world's fresh water supply is stored in subsurface as groundwater. Groundwater is a significant component of the hydrological cycle. Understanding how groundwater moves through the subsurface is important in predicting fate of contaminants as they are introduced at the various points of the hydrological cycle.

This section will review the basic definitions and concepts as used in groundwater hydrology. The list of major texbooks on the topic is provided at the end of the section.

4.4.1 BASIC DEFINITIONS

Groundwater is water in the zone of saturation, i.e., the water that completely fills the pore space. The term *subsurface water* is used for all water below the ground surface (within the saturated and unsaturated zones).

Aquifer is a geologic formation that contains water and permits significant amount of water to move through it under natural field conditions. The second part of the definition is ambiguous because "significant amount" is a relative term. One should immediately ask, "Significant compared to what?" An "insignificant" aquifer in Ohio could be a "significant" aquifer in the arid parts of the country.

Aquitard or *leaky confining layer* is a semipervious geologic formation that can store and transmit water at a rate an order or several orders of magnitude slower compared to the aquifer.

Aquiclude is a geologic formation (e.g., a clay layer) that may contain water in appreciable quantities, but is incapable of transmitting significant quantities under ordinary field conditions. It is important to realize that no geologic formation is completely impervious; it is only a matter of degree how much water can flow through it.

Aquifuge is an impervious formation (e.g., nonfractured granite), which neither contains nor transmits water.

Confined aquifer is an aquifer bounded above and below by confining units of lower permeability than the aquifer material. When a well is installed in a confined aquifer, the water level in the well rises above the top of the confined aquifer. If the water rises above the ground surface it is called *artesian water*.

Leaky aquifer is a confined or water table aquifer that loses or gains water through adjacent semipermeable confining units. A leaky aquifer may behave as a confined aquifer for a relatively short period of time at the begining of the pumping.

Perched aquifer is a phreatic aquifer that occurs above an impervious or semipervious layer, often of limited areal extent, between the water table of an unconfined aquifer and the ground surface.

Potentiometric surface is a surface that represents the level to which water will rise in wells penetrating a confined aquifer.

Porous medium is a portion of space occupied by heterogeneous or multiphase matter. The solid phase or *solid matrix* is distributed throughout the porous medium. The space within the porous medium that is not part of solid matrix is called *pore space* or *void space*. The openings of the void space are relatively narrow. The void space is occupied by gas or liquid phases.

Water table is the upper surface of a zone of saturation, where the water pressure in the porous medium is equal to the atmospheric pressure. A water table or unconfined aquifer is bounded above with the water table surface and below by an impermeable unit.

4.4.2 PHYSICAL PROPERTIES OF GROUNDWATER

Density

The *fluid density* (ρ) is defined as the mass of the fluid m per unit volume V of the fluid

$$\rho = \frac{m}{V}.$$ (4.4.1)

The fluid density is function of pressure and temperature. The dimensions of the fluid density are expressed as ML^{-3} (e.g., g/cm^3, kg/m^3, lb/ft^3).

Specific Weight

The *specific weight* or *weight density* (γ) is defined as the weigth of the fluid per unit volume of the fluid. It is related to the fluid density ρ by the following relationship

$$\gamma = \rho g$$ (4.4.2)

where g is the *gravitational accelaration*. The dimensions of γ are $ML^{-2}T^{-2}$ and often used units are dynes/cm^3, N/m^3, b$_f$/ft^3.

Viscosity

The *fluid viscosity* (μ) is a measure of fluid resistance to relative motion and shear deformation during flow. The more viscous the fluid, the greater the shear stress τ at any given velocity gradient dv/dy

$$\tau = \mu \frac{dv}{dy}$$ (4.4.3)

where the constant of proportionality μ is the *dynamic viscosity*. This equation is known as the Newton's law of viscosity. The dimension of dynamic viscosity is $ML^{-1}T^{-1}$. The unit of dynamic viscosity is g cm^{-1} sec^{-1} called the *poise*. Water has a dynamic viscosity of 1.0 *centipoise* ($= 0.01$ poise) at 20°C.

The *kinematic viscosity* (v) is given as

$$v = \frac{\mu}{\rho}$$ (4.4.4)

where ρ is the fluid density. The dimension of kinematic viscosity is L^2T^{-1}. The basic unit is cm^2/sec called the *stoke* and a more common used unit called *centistoke* ($= 0.01$ stoke).

Fluid Compressibility

The *fluid compressibility* (β) is the measure of volume V and density ρ changes of the fluid resulting from changes in pressure p under isothermal conditions

$$\beta = -\frac{1}{V}\frac{dV}{dp} = \frac{1}{\rho}\frac{d\rho}{dp}$$ (4.4.5)

The negative sign indicates a decrease in volume as pressure increases. The inverse of compressibility is the *module of elasticity*. The dimensions for β are LT^2M^{-1}, the inverse of the pressure or stress dimensions. For water, the value of β is 4.4×10^{-10} m^2/N.

Fluid Pressure

The *fluid pressure* (p) is defined as the weight of the column of a standing body of fluid over a unit area. The fluid pressure p divided by the fluid specific weight γ is called the *pressure head* (ψ)

$$\psi = \frac{p}{\rho g} = \frac{p}{\gamma}.$$ (4.4.6)

The dimension of the pressure head is L.

4.4.3 *PHYSICAL PROPERTIES OF SOLID MATRIX*

Grain-size Distribution

The distribution of soil particle sizes affects hydrogeologic and geomechanical behavior of the soil. There are three parameters that can be used to define the property: the effective grain-size, the uniformity coefficient, and the coefficient of curvature.

The *effective grain size* (d_{10}) is the grain diameter greater than that of 10 percent of the particles by weight. In other words, it is the size corresponding to the 10 percent line on the grain-size curve.

The *uniformity coefficient* (C_u) is defined as the ratio of the grain-size that is 60 percent finer by weight, d_{60}, and the effective grain-size d_{10}

$$C_u = \frac{d_{60}}{d_{10}}$$ (4.4.7)

A gravel sample with a C_u less than 4 is considered well sorted (poor gradation); if the value of C_u is greater than 4, it is poorly sorted (good gradation). For sand samples, the cut-off point is approximately 6.

The *coefficient of curvature* or *coefficient of concavity* is defined as

$$C_c = \frac{d_{30}^2}{d_{10}d_{60}}$$ (4.4.8)

For well-graded gravels and sands the coefficient of curvature should be between 1 and 3.

Porosity

The *porosity* (n) is defined as the ratio of volume of voids V_v to the bulk volume V_b of a porous medium (Freeze and Cherry, 1979):

$$n = \frac{V_v}{V_b}. \tag{4.4.9}$$

If V_v is the void space of all pores, the porosity is also called the *total porosity*. Because only interconnected pores are important for the flow, the *effective porosity*, n_e, is defined as the ratio of the interconnected void space and the bulk volume of the porous medium.

The *void ratio* (e) is defined as the ratio of volume of voids to the volume of solids V_s

$$e = \frac{V_v}{V_s}. \tag{4.4.10}$$

The porosity and the void ratio are related as:

$$n = \frac{e}{1+e} \qquad e = \frac{n}{1-n} \tag{4.4.11}$$

The porosity of unconsolidated sediments depends on packing of the grains, shape of grains, and size distribution.

Based on the origin of pores, porosity can be *primary* and *secondary*. The primary porosity was formed at the same time when the rock was formed. The secondary porosity is due to changes in rock after the rock is formed (e.g., fractures, openings in limestone).

Bulk Density

The *bulk density* (ρ_b) is defined as the mass of dry soil (a soil sample after 24 hours of drying in an oven at 105°C) per bulk volume of soil:

$$\rho_b = \frac{m_s}{V_b}. \tag{4.4.12}$$

The *density of solids* or the *mineral density* (ρ_s) is the density of the solid material comprising the soil matrix and it is defined as the mass of dry soil per volume of solids:

$$\rho_s = \frac{m_s}{V_s}. \tag{4.4.13}$$

The relationship between ρ_b and ρ_s can be derived using the above definitions:

$$\rho_b = \frac{m_s}{V_b} = \frac{m_s}{V_s}\frac{V_s}{V_b} = \rho_s(1-n). \tag{4.4.14}$$

where n is the porosity.

Specific Surface

The *specific surface* (M) is defined as the total surface area of the pores per unit bulk volume of the porous medium. In literature, the specific surface is also defined with respect to "unit volume of solid material" or "unit mass of porous medium."

Effective Stress

The *total stress*, (σ) acting on a horizontal plane placed in the saturated porous medium is due to the weight of overlying rocks and water. The stress is borne by the grains of the porous medium (the *effective stress* σ_e), and by the fluid pressure p (Todd, 1980):

$$\sigma = \sigma_e + p \tag{4.4.15}$$

If the total stress is constant, $d\sigma = 0$, then:

$$d\sigma_e = -dp \tag{4.4.16}$$

or simply stated, if the fluid pressure decreases, the effective stress must increase.

Porous Medium Compressibility

The *coefficient of bulk compressibility of a porous medium* (α) is defined for a saturated porous medium as the change in the bulk volume of the porous medium V_b with a unit change in effective stress, σ_e (Verruijt, 1982):

$$\alpha = -\frac{1}{V_b}\frac{dV_b}{d\sigma_e} \tag{4.4.17}$$

An increase in effective stress results in reduction of the bulk volume of porous medium. Although the grains can be compressible, the reduction in volume is usually due to the grain rearrangement and the compressibility α can be expressed in terms of porosity as:

$$\alpha = \frac{1}{1-n}\frac{dn}{d\sigma_e}. \tag{4.4.18}$$

The compressibility is not a constant; it is a function of the applied stress, and it is dependent on the previous loading history.

4.4.4 FLUID POTENTIAL

Fluid potential can be expressed as the mechanical energy per unit mass (*Hubbert's potential*) or per unit weight of fluid (*hydraulic head*).

Hubbert's Potential

Hubbert (1940) defined fluid potential as "a physical quantity, capable of measurement at every point in a flow system, whose properties are such that flow always occurs from regions in which the quantity has higher values to those in which it has lower, regardless of the direction in space." The fluid potential for flow through porous media is therefore the mechanical energy per unit mass of fluid. The Hubbert's potential ϕ^* is defined as the work required to bring a unit mass of fluid from some arbitrary reference state (z_0, p_0, ρ_0) to a new state (z, p, ρ). For compressible fluids, the Hubbert's potential is given as (McWhorter and Sunada, 1977):

$$\phi^* = gz + \int_{p_0}^{p}\frac{dp}{\rho(p)} \tag{4.4.19}$$

For incompressible fluid, the density of fluid ρ is not a function of pressure and the Hubbert's potential is given as:

$$\phi^* = gz + \frac{p - p_0}{\rho} \tag{4.4.20}$$

In practice, it is much easier to measure water levels than water pressure, and for that reason, instead of the Hubbert's potential, we use the concept of the hydraulic head.

Hydraulic Head

The *hydraulic head* (h) is a potential expressed in terms of energy per unit weight of fluid (Freeze and Cherry, 1979):

$$h = z + \frac{p}{\gamma} = z + \frac{p}{\rho_w g} = z + \psi \tag{4.4.21}$$

where z is the elevation of the point, p is the pressure, and γ ($= \rho g$) is the specific weight of fluid. Thus, the hydraulic head is the sum of the *elevation head*, z, and the *pressure head*, ψ. The dimensions of h, z, and ψ are length [L].

Hydraulic Gradient

The hydraulic gradient (i) is defined as the change in the hydraulic head per unit distance (Heath, 1989):

$$i = -\frac{h_2 - h_1}{\Delta L} = -\frac{\Delta h}{\Delta L} \tag{4.4.22}$$

where Δh is the *head loss* measured in the direction of the maximum decrease in hydraulic head, and ΔL is the distance between the measuring points.

4.4.5 DARCY'S LAW

Henry Darcy, a French sanitary engineer, conducted a series of experiments with a column filled with different sand filters (Darcy, 1856). His experiments lead to a conclusion that the rate of flow Q is proportional to the cross-section area of the column A (i.e., the area perpendicular to the flow), and to the hydraulic gradient i:

$$Q = -KA\frac{h_2 - h_1}{\Delta L} = KAi \tag{4.4.23}$$

where K is the coefficient of proportionality called the *hydraulic conductivity*.

Hydraulic Conductivity

The *hydraulic conductivity* (K) of a porous medium is a measure how easy fluid can flow through the medium. It is defined as (Davis and De Wiest, 1966):

$$K = k\frac{\rho g}{\mu} \tag{4.4.24}$$

where ρ is the fluid density, g is the gravitational constant, μ the dynamic viscosity of the fluid, and k is the *intrinsic permeability*. While the hydraulic conductivity includes both properties of the porous matrix and the fluid, the permeability depends only on the properties of the porous medium:

$$k = Cd_{10}^2 \qquad (4.4.25)$$

where C is a dimensionless *"packing"* or *"shape"* factor, and d_{10} is the effective grain diameter. The dimension of the hydraulic conductivity is $[L/T]$, commonly expressed as ft/day, m/d, cm/s or in English units $[gal/ft^2]$. The dimension of the permeability is $[L^2]$ and the units are cm^2 or darcy.

Specific Discharge

The *specific discharge* (q) is the volume of water per unit time through a unit cross-sectional area:

$$q = \frac{Q}{A} = Ki \qquad (4.4.26)$$

The dimension of q is $[LT^{-1}]$, so it is sometimes called *Darcy's velocity*. However, this term should not be used to avoid confusion with the groundwater velocity.

Groundwater Velocity

The *average groundwater velocity* or *seepage velocity* (V) is given as:

$$V = \frac{q}{n_e} = \frac{K}{n_e}i \qquad (4.4.27)$$

where n_e is the effective porosity. The dimension of groundwater velocity V is $[LT^{-1}]$, usually expressed as ft/day, ft/year or m/day, cm/s.

Range of Validity of Darcy's Law

The range of validity of Darcy's law can be determined using the *Reynolds number* (R_e) defined as (Bear, 1979):

$$R_e = \frac{qd}{\nu} \qquad (4.4.28)$$

where q is the specific discharge, d is the characteristic length of the porous medium, and ν is the *kinematic viscosity*. The effective grain diameter d_{10} is often taken as the characteristic length d.

Darcy's law is valid if the flow is laminar, which is the case if the Reynolds number is smaller than some value between 1 and 10. This is the upper limit of Darcy's law. Experiments in the low permeability soils indicate that there is a lower limit of Darcy's law. Fortunately, most groundwater flows obey Darcy's law, except in the close vicinity of pumping wells or in limestone and other aquifers with large caverns.

Extended Darcy's Law

Darcy's law given by Equation (4.4.23) is limited to one-dimensional flow of a homogeneous incompressible fluid. In a three-dimensional flow field, the Darcy's flow equation is given as:

$$\mathbf{q} = K\mathbf{i} = -K\,\text{grad}\,h = -K\nabla h \qquad (4.4.29)$$

where $\mathbf{i}$ is the hydraulic gradient vector with the components i_x, i_y, i_z in the Cartesian coordinate system

$$i_x = -\frac{\partial h}{\partial x} \qquad i_y = -\frac{\partial h}{\partial y} \qquad i_z = -\frac{\partial h}{\partial z} \tag{4.4.30}$$

If the components of the hydraulic gradient vector are known, the resultant vector $\mathbf{i}$ is:

$$\mathbf{i} = \sqrt{i_x^2 + i_y^2 + i_z^2} \tag{4.4.31}$$

The specific discharge vector $\mathbf{q}$ has three components in a three-dimensional flow field:

$$q_x = K i_x = -K \frac{\partial h}{\partial x}$$

$$q_y = K i_y = -K \frac{\partial h}{\partial y} \tag{4.4.32}$$

$$q_z = K i_z = -K \frac{\partial h}{\partial z}$$

If the components of the specific discharge vector are known, the resultant vector $\mathbf{q}$ can be computed as:

$$\mathbf{q} = \sqrt{q_x^2 + q_y^2 + q_z^2} \tag{4.4.33}$$

The specific discharge values are then used to determine the groundwater velocity components V_x, V_y, V_z:

$$V_x = -\frac{q_x}{n_e} \qquad V_y = -\frac{q_y}{n_e} \qquad V_z = -\frac{q_z}{n_e} \tag{4.4.34}$$

where n_e is the effective porosity. Finally, the resultant vector $\mathbf{V}$ is computed as:

$$\mathbf{V} = \sqrt{V_x^2 + V_y^2 + V_z^2} \tag{4.4.35}$$

If a contaminant particle is being transported by the bulk volume of groundwater only (the advection process) and the groundwater velocity components are known in the flow field, then the pathline as well as the location of the contaminant particle can be predicted for a given time of travel.

4.4.6   THE MASS BALANCE EQUATION

The basic mass balance equation for flow in porous media is given as follows (Bear, 1972):

$$-\nabla \cdot (\rho \mathbf{q}) = \frac{\partial}{\partial t}(n\rho) \tag{4.4.36}$$

or assuming that density of water ρ is constant in space:

$$-\nabla \cdot (K \nabla h) = \frac{1}{\rho}\frac{\partial}{\partial t}(n\rho) \tag{4.4.37}$$

The rate of change of the mass of the fluid per unit volume of porous medium can be expressed in terms of the *specific storage* S_s defined as (Domenico and Schwartz, 1990):

$$S_s = \alpha \rho g + n \beta \rho g \tag{4.4.38}$$

The three-dimensional transient groundwater flow through heterogeneous, anisotropic confined aquifer can be mathematically described with the following partial differential equation:

$$\frac{\partial}{\partial x}\left(K_x \frac{\partial h}{\partial x}\right) + \frac{\partial}{\partial y}\left(K_y \frac{\partial h}{\partial y}\right) + \frac{\partial}{\partial z}\left(K_z \frac{\partial h}{\partial z}\right) = S_s \frac{\partial h}{\partial t} \tag{4.4.39}$$

In steady-state flow there is no change in hydraulic head with time and the flow is described by the equation known as the *Laplace equation* (Fetter, 1994):

$$\frac{\partial}{\partial x}\left(K_x \frac{\partial h}{\partial x}\right) + \frac{\partial}{\partial y}\left(K_y \frac{\partial h}{\partial y}\right) + \frac{\partial}{\partial z}\left(K_z \frac{\partial h}{\partial z}\right) = 0 \tag{4.4.40}$$

If the groundwater flow is two-dimensional and the aquifer is homogeneous and isotropic, Equation (4.4.40) becomes:

$$\frac{\partial^2 h}{\partial x^2} + \frac{\partial^2 h}{\partial y^2} = \frac{S}{T}\frac{\partial h}{\partial t} \tag{4.4.41}$$

where $S = S_s b$ is the *storage coefficient* [dimensionless], $T = Kb$ is the transmissivity $[L^2 T^{-1}]$, and b is the aquifer thickness $[L]$.

REFERENCES

Bear, J., *Dynamics of Fluids in Porous Media*, Elsevier Publishing Co., 1972.

Bear, J., *Hydraulics of Groundwater*, McGraw-Hill Inc., 1979.

Bouwer, H., *Groundwater Hydrology*, McGraw-Hill Inc., 1978.

Darcy, H., Les Fontaines Publiques de la ville de Dijo, V. Dalmount, Paris, 1856.

Davis, S. N., and De Wiest, R. J. M., *Hydrogeology*, John Wiley & Sons Inc., 1966.

de Marsily, G., *Quantitative Hydrogeology*, Academic Press Inc., 1986.

De Wiest, R. J. M., *Geohydrology*, John Wiley & Sons Inc., 1965.

Domenico, P. A., and Schwartz, F. W., *Physical and Chemical Hydrogeology*, John Wiley & Sons Inc., 1990.

Driscoll, F. G., *Groundwater and Wells*, Johnson Filtration Systems Inc., St. Paul, Minnesota, 1989.

Fetter, C. W., *Applied Hydrogeology*, Third Edition, Macmillan College Publishing Company, New York, 1994.

Freeze, R. A., and Cherry, J. A., *Groundwater*, Prentice-Hall Inc., 1979.

Harr, M. E., *Groundwater and Seepage*, McGraw-Hill Inc., 1962.

Heath, R. C., *Basic Ground-Water Hydrology*, USGS Water-Supply Paper 2220, Denver, Colorado, 1989.

Hubbert, M. K., "The Theory of Ground Water Motion," *J. Geolo.*, 48: 785–944, 1940.

Hunt, B., *Mathematical Analysis of Groundwater Resources*, Butterworth & Co., 1983.

McWhorter, D. B., and Sunada, D. K., *Ground-Water Hydrology and Hydraulics*, Water Resources Publications, Fort Collins, Colorado, 1977.

Strack, O. D. L., *Groundwater Mechanics*, Prentice-Hall Inc., 1989.

Todd, D. K., *Groundwater Hydrology*, John Wiley & Sons Inc., 1980.

Verruijt, A., *Theory of Groundwater Flow*, MacMillan, London, 1982.

Walton, W., *Principles of Groundwater Engineering*, Lewis Publishers Inc., Chelsea, Michigan, 1991.

CHAPTER 4
HYDROLOGY

SECTION 4.5

WELL HYDRAULICS

Milovan S. Beljin

Dr. Beljin is a consulting hydrologist and president of M.S. Beljin & Associates. He serves as adjunct faculty at the University of Cincinnati and Wright State University, Dayton, Ohio, where he has taught groundwater modeling.

Well hydraulics is one of the most important topics in groundwater hydrology. Every hydrology textbook dedicates at least one chapter to the topic, and well hydraulics continues to be a popular subject of many scientific papers and books (Kruseman and de Ridder, 1970; Batu, 1998). The reason is simple; a well is the only available device for extracting groundwater, and almost all hydrogeologic field work involves pumping wells. To understand or predict the effects of pumping a well or a wellfield on the water levels in an aquifer, a quantitave analysis must be employed. The basis for such an analysis are mathematical models.

This section focuses on the analytical models of groundwater flow toward a vertical well in an idealized aquifer. The mathematical formulation of the flow problem is significantly simplified by assuming that the flow toward the well is two-dimensional horizontal and the aquifer is homogeneous, isotropic, and of uniform thickness. The additional assumptions regarding the well itself are: (1) The well is a line sink (a discharge well) or source (an injection well), i.e., the wellbore storage is negligible, (2) the well screen fully penetrates the aquifer, and (3) the pumping rate is constant.

When a well is pumped, the water level in the aquifer will be lowered and a *cone of depression* will develop around the well. At a given point, within the cone of depression, the distance between the original water level and the new water level is called *drawdown*. The distance from the discharge well to the outer limit of the cone of depression is called the *radius of influence* of the well.

The groundwater flow regime can be transient or steady-state, depending on whether or not the hydraulic head is changing with time. In theory, the flow will be transient until a recharge boundary (e.g., a lake or a river) is intercepted by the cone of depression. However, from a practical point of view, the flow around the pumping well can be assumed to be steady when no significant additional drawdown occurs with time.

The section is organized based on the type of flow regime: The first part focuses on the steady-state flow equations, and in the second part, basic transient flow models are described. A list of references and other suggested readings is provided at the end of the section.

4.5.1 STEADY FLOW TO A WELL

The two-dimensional steady-state flow equation in the Cartesian coordinates is given as (Bear, 1979):

$$\frac{\partial^2 h}{\partial x^2} + \frac{\partial^2 h}{\partial y^2} = 0 \tag{4.5.1}$$

where h is the hydraulic head. This is the *Laplace equation*, a well-known partial differential equation in other engineering disciplines.

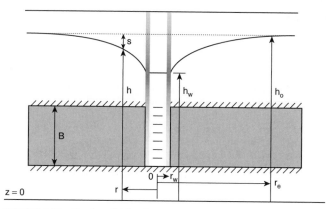

FIGURE 4.5.1   Steady radial flow in a confined aquifer.

The flow toward a well in a homogeneous and isotropic aquifer is radial and symmetric, and the radial coordinate system is often used to study well hydraulics. For the simplest radial flow, Equation (4.5.1) becomes:

$$\frac{1}{r}\frac{\partial}{\partial r}\left(r\frac{\partial h}{\partial r}\right) = 0 \tag{4.5.2}$$

or

$$\frac{\partial^2 h}{\partial r^2} + \frac{1}{r}\frac{\partial h}{\partial r} = 0 \tag{4.5.3}$$

where r is the distance from the well to the observation point.

Steady Flow in a Confined Aquifer

Figure 4.5.1 shows a circular island with a discharge well located in the center of the island. The well is discharging water from a confined aquifer at a constant rate. Because the well is located in the center of an island, the radius of the influence of the well is equal to the the radius of the island. The boundary conditions can be expressed as:

$$\begin{aligned} h &= h_w \qquad r = r_w \\ h &= h_e \qquad r = r_e \end{aligned} \tag{4.5.4}$$

where r_w is the radius of the well, h_w is the hydraulic head in the well, r_e is the *radius of influence* of the well, and h_e is the hydraulic head at the radius r_e. The radius of the well is often assumed to be equal to the radius of the borehole. However, for a gravel-packed well or a well with an increased permeability zone in the vicinity of the well, the *effective well radius* should be applied instead. The effective radius is equal to the distance from the center of the well at which the theoretical drawdown equals the actual drawdown.

The general solution to Equation (4.5.2) or (4.5.3) can be obtained by integrating the equation twice and applying the boundary conditions to determine the constants of integration. The resulting solution has the following form:

$$h = h_w + \frac{h_e - h_w}{\ln(r_e/r_w)}\ln(r/r_w). \tag{4.5.5}$$

The specific discharge q_r is given as:

$$q_r = -K\frac{\partial h}{\partial r} = -K\frac{h_e - h_w}{\ln(r_e/r_w)}\frac{1}{r} \tag{4.5.6}$$

where K is the hydraulic conductivity and $\partial h/\partial r$ is the hydraulic gradient along the radial coordinate. The negative sign indicates that the flow is toward the well.

The hydraulic head, h, and the specific discharge, q_r, in Equations (4.5.5) and (4.5.6), respectively, are expressed in terms of known hydraulic heads. However, it is more convenient to express those equations in terms of the pumping rate Q:

$$Q = q_r A = 2\pi Kb\frac{h_e - h_w}{\ln(r_e/r_w)} \tag{4.5.7}$$

where b is the thickness of the aquifer, and A ($= 2\pi rb$) is the cross-sectional area perpendicular to the radial flow. The drawdown in the pumping well, s_w, is then:

$$s_w = h_e - h_w = \frac{Q}{2\pi Kb}\ln(r_e/r_w) \tag{4.5.8}$$

The hydraulic head h at any distance r from the production well is given as (Thiem, 1906):

$$h = h_w + \frac{Q}{2\pi T}\ln(r/r_w) \tag{4.5.9}$$

where T ($= Kb$) is the aquifer transmissivity. The equation indicates that the hydraulic head is not bounded because the hydraulic head will increase indefinitely as r increases. However, considering the physical reality, it is obvious that the equation can be valid only within the radius of influence of the well.

Let us assume now that two piezometers are located within the radius of influence at distances r_1 and r_2 from the pumping well. If the drawdowns s_1 and s_2 are measured in the observation wells, respectively, then Equation (4.5.9) takes the following form:

$$s_1 - s_2 = h_2 - h_1 = \frac{Q}{2\pi T}\ln(r_2/r_1) \tag{4.5.10}$$

This equation is known in the groundwater literature as the *Equilibrium* or *Thiem equation*. If the drawdowns around a pumping well are measured, the Thiem equation enables hydrologists to determine the transmissivity of a confined aquifer.

The specific discharge q_r is then given as:

$$q_r = -\frac{Q}{2\pi rb} \tag{4.5.11}$$

and the average groundwater velocity V_r is:

$$V_r = \frac{q_r}{n_e} = -\frac{Q}{2\pi n_e rb} \tag{4.5.12}$$

where n_e is the effective porosity of the aquifer. The equation indicates that the velocity and the drawdown are inversely proportional (i.e., the velocity of a water molecule is increasing more rapidly as the molecule approaches the pumping well).

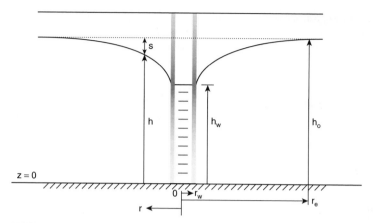

FIGURE 4.5.2   Steady radial flow in a unconfined aquifer.

Steady Flow in an Unconfined Aquifer

The two-dimensional, steady-state *unconfined* flow equation in Cartesian coordinates is given as (Bear, 1979):

$$\frac{\partial^2 h^2}{\partial x^2} + \frac{\partial^2 h^2}{\partial y^2} = 0 \tag{4.5.13}$$

or in radial coordinate system:

$$\frac{\partial^2 h^2}{\partial r^2} + \frac{1}{r}\frac{\partial h^2}{\partial r} = 0 \tag{4.5.14}$$

which is linear in h^2.

Figure 4.5.2 illustrates a well pumping groundwater from an unconfined aquifer (note that the datum for the hydraulic head is the aquifer bottom elevation). If the Dupuit assumptions are valid, the well discharge rate Q is given as:

$$Q = 2\pi r K h \frac{\partial h}{\partial r} \tag{4.5.15}$$

which, after separation of variables and integration between $h = h_w$ at $r = r_w$ and $h = h_e$ at $r = r_e$, yields:

$$h_e^2 - h_w^2 = \frac{Q}{\pi K}\ln\left(r_e/r_w\right) \tag{4.5.16}$$

or integrating from any distance r to r_e:

$$h_e^2 - h^2 = \frac{Q}{\pi K}\ln\left(r_e/r\right) \tag{4.5.17}$$

Combining Equations (4.5.16) and (4.5.17) yields an expression for computing the hydraulic head within the radius of influence of the pumping well:

$$h^2 = h_e^2 - \frac{h_e^2 - h_w^2}{\ln\left(r_e/r_w\right)}\ln\left(r_e/r\right) \tag{4.5.18}$$

If the drawdowns are small relative the total saturated thickness, h_e, an unconfined aquifer may be treated as a confined aquifer and the steady-state equation for a confined flow can be applied (Equation 4.5.10). In case the drawdowns are appreciable, the following equation can be used to determine the transmissivity of an unconfined aquifer:

$$T = Kh_e = \frac{Q}{2\pi(s_1' - s_2')}\ln(r_2/r_1) \tag{4.5.19}$$

where s_1' and s_2' are the corrected drawdowns or drawdowns that would occur in an equivalent confined aquifer. The measured drawdown, s, can be corrected as (Jacob, 1950):

$$s' = s - s^2/2h_e \tag{4.5.20}$$

Steady Flow in a Leaky Aquifer

The two-dimensional steady-state flow equation in the Cartesian coordinates for a leaky confined aquifer is given as (Hantush and Jacob, 1955):

$$\frac{\partial^2 h}{\partial x^2} + \frac{\partial^2 h}{\partial y^2} + \frac{h_0 - h}{B^2} = 0 \tag{4.5.21}$$

or in the radial coordinate system as:

$$\frac{1}{r}\frac{\partial}{\partial r}\left(r\frac{\partial h}{\partial r}\right) + \frac{h_0 - h}{B^2} = 0 \tag{4.5.22}$$

where B is the *leakage factor*:

$$B = \sqrt{T/(K'/b')}, \tag{4.5.23}$$

h_0 is the hydraulic head in the adjacent aquifer that acts as a source, K'/b' is the *leakance*, and K' and b' are the hydraulic conductivity and the thickness of the semipervious layer, respectively.

When pumping occurs in a leaky aquifer (Figure 4.5.3), the head difference between the two aquifers creates leakage through the semipervious layer that separates the aquifers. After certain period of pumping at a constant rate Q, a steady-state flow is reached. The flow is almost entirely sustained by the leakage. The assumption is made that the adjacent aquifer maintains a relatively constant hydraulic head, h_0. It is also assumed that the leaky confined aquifer is homogeneous, isotropic, and of infinite extent. The drawdown (s) in an observation well, placed in the leaky aquifer

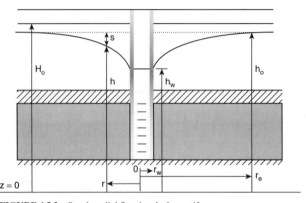

FIGURE 4.5.3   Steady radial flow in a leaky aquifer.

at some distance r from the discharge well, can be computed as:

$$s = \frac{Q}{2\pi T} K_0 \left(\frac{r}{B} \right)$$

(4.5.24)

where K_0 is zero-order Bessel function.

When $x \ll 1$ then $K_0(x) \approx \ln(1.123/x)$, with an error less than 5 percent and thus Equation (4.5.24) can be written as:

$$s = \frac{Q}{2\pi T} \ln \left(\frac{1.123B}{r} \right)$$

(4.5.25)

A comparison of Equations (4.5.25) and (4.5.8) reveals that the radius of influence of a well in a leaky aquifer is approximately $1.123B$. It can be shown that 95 percent of water enters the aquifer cylinder of radius $4B$ through the semipervious layer.

4.5.2 TRANSIENT FLOW TO A WELL

The two-dimensional transient-state flow equation in Cartesian coordinates is given as (Bear, 1979):

$$\frac{\partial^2 h}{\partial x^2} + \frac{\partial^2 h}{\partial y^2} = \frac{S}{T} \frac{\partial h}{\partial t}$$

(4.5.26)

where S is the storage coefficient and T is the aquifer transmisivity. For radial groundwater flow, it is more convenient to write Equation (4.5.26) in the radial coordinate system:

$$\frac{1}{r} \frac{\partial}{\partial r} \left(r \frac{\partial h}{\partial r} \right) = \frac{S}{T} \frac{\partial h}{\partial t}.$$

(4.5.27)

Transient Flow in a Confined Aquifer

Let a single well continuously pump groundwater from an infinite, homogeneous and isotropic confined aquifer at a constant rate. The well radius is assumed to be small so the storage effect (the volume of water in the well) is negligible. The initial potentiometric surface is horizontal. As the water is being discharged, a cone of depression will develop around the well and it will continue to expand as long as pumping continues.

The flow problem can be described mathematically, in terms of drawdown s, with the following partial differential equation (Bear, 1979):

$$\frac{\partial^2 s}{\partial r^2} + \frac{1}{r} \frac{\partial s}{\partial r} = \frac{S}{T} \frac{\partial s}{\partial t}$$

(4.5.28)

and the initial and boundary conditions:

$$s(r, 0) = 0$$

$$s(\infty, t) = 0$$

$$\lim_{r \to 0} r \frac{\partial s}{\partial r} = -\frac{Q}{2\pi T}$$

(4.5.29)

Theis (1935) recognized that the analogy between the heat and groundwater flow and the solution to a similar problem of heat transfer yields:

$$s = \frac{Q}{4\pi T} \int_u^\infty \frac{e^{-u}}{u} du$$

(4.5.30)

where

$$u = \frac{r^2 S}{4Tt} \tag{4.5.31}$$

Equation (4.5.30) is known as the *nonequilibrium* or *Theis equation* and it is considered by hydrologists the most important well hydraulics equation.

The integral in the Theis equation is known as the *exponential integral function*, $-E_i(-u)$

$$-E_i(-u) = \int_u^\infty \frac{e^{-u}}{u} du = \gamma - \ln u + u - \frac{u^2}{2.2!} + \frac{u^3}{3.3!} - \frac{u^4}{4.4!} + \cdots \tag{4.5.32}$$

where γ is *Euler's constant*:

$$\gamma = \int_0^\infty \frac{e^{-u}}{u} du = -0.57721566\ldots \tag{4.5.33}$$

The Theis equation is often written as:

$$s = \frac{Q}{4\pi T} W(u) \tag{4.5.34}$$

where $W(u)$ is the exponential integral function, better known as the *Theis well function* for a confined aquifer. The integral function can be computed using polynomial and rational approximations given by Abramowitz and Stegun (1972, 231) or estimated from the tabulated values (Freeze and Cherry, 1979, 318).

Cooper and Jacob (1946) observed that for $u \leq 0.01$, the series terms in Equation (4.5.32) become negligible after the first two terms, and thus can be approximated with the following expression:

$$s \approx \frac{Q}{4\pi T} \ln\left(\frac{2.25Tt}{r^2 S}\right) \tag{4.5.35}$$

Equation (4.5.35) is valid within 6 percent for $u \leq 0.1$ and within 0.5 percent for $u \leq 0.01$. It should be noted that u will be small for small distance r and/or large time.

The Effective Radius of Influence

The Theis equation (4.5.30) indicates that at any observation point at a given time drawdown is proportional to the discharge rate and inversely proportional to transmissivity and storage coefficient. The cone of depression increases in depth and extent with increasing time. As time approaches infinity the Theis equation predicts that the cone of depression around the well approaches infinity. However, for practical purposes, the drawdown becomes negligible at a finite radius known as the *effective radius of influence*, r_e. At early times, the radius of influence expands rapidly and more slowly as the time becomes large. The *effective radius of influence*, r_e, can be regarded as the distance at which the drawdown, s_e, is some arbitrary small drawdown.

Equation (4.5.34) can be written in terms of the the well function as:

$$W(u) = \frac{4\pi T s_e}{Q} \tag{4.5.36}$$

where

$$u = \frac{r_e^2 S}{4Tt} \tag{4.5.37}$$

After computing $W(u)$, the value of u can be looked up in the well function table. The radius of influence is then found from Equation (4.5.37) as:

$$r_e = \left(\frac{4uTt}{S} \right)^{1/2} \tag{4.5.38}$$

If the Cooper-Jacob logarithmic approximation is used to express drawdown s_e at the distance r_e:

$$s_e = \frac{Q}{4\pi T} \ln \left(\frac{2.25Tt}{r_e^2 S} \right) \tag{4.5.39}$$

then for s_e set to zero the radius of influence, r_e, is computed with the following equation:

$$r_e = \left(\frac{2.25Tt}{S} \right)^{1/2} \tag{4.5.40}$$

In an infinite aquifer that is not recharged, r_e varies as $\sqrt{t}$. If t is large, r_e varies slowly as if the steady state was achieved.

Transient Flow in a Leaky Aquifer

The governing groundwater flow equation in a leaky aquifer under transient condition is given as (Hantush, 1964):

$$\frac{\partial^2 s}{\partial r^2} + \frac{1}{r} \frac{\partial s}{\partial r} + \frac{s}{B^2} = \frac{S}{T} \frac{\partial s}{\partial t}. \tag{4.5.41}$$

In addition to the assumptions made in developing the Theis equation, Equation (4.5.41) is based on the assumptions that the storage of the semipervious layer is negligible and that the hydraulic head in the adjacent (source) aquifer is constant.

Drawdown in a leaky confined aquifer can be computed with the following equation (Hantush and Jacob, 1955):

$$s = \frac{Q}{4\pi T} \int_u^\infty \frac{\exp\left[-u - (r^2/4B^2 u)\right]}{u} du = \frac{Q}{4\pi T} W(u, r/B) \tag{4.5.42}$$

where $W(u, r/B)$ is the *well function for a leaky aquifer*. For the given arguments, this function can be estimated from the tabulated values of the function (Fetter, 1994, 620).

Radial Flow in an Unconfined Aquifer

A well pumping from an unconfined aquifer creates a cone of depression in the water table and thus, in addition to horizontal components, there will be vertical components of groundwater. The water is released initially from the storage because of expansion of the water in the aquifer and the compaction of the aquifer material. The second mechanism for releasing water from the aquifer is the actual dewatering of the unconfined aquifer.

Analysis of the flow in an unconfined aquifer that involves a saturated–unsaturated flow system is rarely used. The research has shown that the location of the water table is not significantly affected by the flow in the unsaturated zone.

A simpler approach is to use the same governing equation as for the confining flow, with the storativity defined in terms of the specific yield, S_y, and the transmissivity T defined as the product of the hydraulic conductivity and the initial saturated thickness:

$$\frac{\partial^2 s}{\partial r^2} + \frac{1}{r} \frac{\partial s}{\partial r} = \frac{S_y}{T} \frac{\partial s}{\partial t} \tag{4.5.43}$$

Yet another approach used in the practice is based on the concept of *delayed yield*. Boulton (1963) and Neuman (1972, 1973, 1975) developed a theory in which the response of the water table to a single well pumping can be described with a three-segment time-drawdown curve.

Well Interference

A group of pumping wells (a wellfield) will cause a drawdown in an observation well that will be, from the principle of superposition, equal to the sum of the drawdowns caused by each well individually. Thus,

$$s = \sum_{i=1}^{n} s_i = \sum_{i=1}^{n} \frac{Q_i}{4\pi T} W\left(\frac{r_i^2 S}{4T t_i}\right) \tag{4.5.44}$$

where s_i is the drawdown resulting from pumping of the i-th well at a constant pumping rate Q_i for a duration t_i, located at the distance r_i from the observation well.

Recovery in a Well

At the end of a pumping, once the pump is turned off, the water levels will start to rise. During the recovery of the water levels, the drawdown is called the residual drawdown. The drawdown can be computed by using the principle of superposition (i.e., by placing two wells at the same location). The assumptions are that the first well has been continuously discharged at a constant rate, and the second well has been recharged at the same rate since the pump was shut off.

The drawdown because of pumping of the discharge well can be computed as:

$$s_1 = \frac{+Q}{4\pi T} W\left(\frac{r^2 S}{4T t}\right) \tag{4.5.45}$$

and the rise of water level because of the injection well is:

$$s_2 = \frac{-Q}{4\pi T} W\left(\frac{r^2 S}{4T t'}\right) \tag{4.5.46}$$

where t and t' represent elapsed time since the start of pumping and the recovery, respectively.

Theis (1935) showed that the residual drawdown s' can be computed as:

$$s' = s_1 + s_2 = \frac{Q}{4\pi T}\left[W\left(\frac{r^2 S}{4T t}\right) - W\left(\frac{r^2 S}{4T t'}\right)\right] \tag{4.5.47}$$

For small distance r and/or large time t', the well functions can be approximated by the Cooper-Jacob approximation and Equation (4.5.47) can be written as:

$$s' = \frac{Q}{4\pi T} \ln\left(\frac{t}{t'}\right) \tag{4.5.48}$$

Pumping Near Aquifer Boundaries

A frequent practical flow problem is that of a well close to a recharge boundary (e.g., a river) or to a no-flow boundary (e.g., the valley wall of a buried valley aquifer). However, one of the assumptions applied to all previous solutions was that the aquifer is of infinite areal extent. If a pumping well is located near such a boundary, analysis is based on the principle of superposition, which states that the drawdown caused by two or more pumping wells is the sum of the drawdowns because of each discharge well and the buildups because of each recharge well. By introducing imaginary wells that pump at the *same rate* as the corresponding "real" wells, an aquifer of finite extent is transformed from a semi-infinite into an infinite aquifer.

Surface water bodies are often considered as infinite sources of water and thus with constant water level. An image well is introduced on the other side of the boundary and is located on a line perpendicular to the boundary at the same distance from the boundary as the real well. The image well is injecting water at the same rate as the discharge well is withdrawing water. The drawdown s at a point located the distance r from the real well and the distance r_i from the image well, can be computed with the following equation:

$$s = \frac{Q}{4\pi T}\left[W\left(\frac{r^2 S}{4Tt}\right) - W\left(\frac{r_i^2 S}{4Tt}\right)\right] \tag{4.5.49}$$

The first term represents the drawdown because of the discharge well, and the second term represents the buildup because of the image well.

If Cooper-Jacob approximation can be used (i.e., the logarithmic approximation is valid for both well functions), then the drawdown is given as:

$$s = \frac{Q}{4\pi T}\left[\ln\left(\frac{2.25Tt}{Sr^2}\right) - \ln\left(\frac{2.25Tt}{Sr_i^2}\right)\right] \tag{4.5.50}$$

or

$$s = \frac{Q}{2\pi T}\ln\left(\frac{r_i}{r}\right) \tag{4.5.51}$$

The drawdown along the line representing the stream ($r = r_i$) will be obviously zero. The assumption is that the stream level is not (significantly) affected by the pumping (i.e., the stream represents a constant head boundary).

The drawdown in an observation well at a distance r from a discharge well in a semi-infinite aquifer bounded by an impervious boundary is given as:

$$s = \frac{Q}{4\pi T}\left[W\left(\frac{r^2 S}{4Tt}\right) + W\left(\frac{r_i^2 S}{4Tt}\right)\right] \tag{4.5.52}$$

The second term represents the drawdown due to an image *discharge* well located at the same distance from the boundary as the real well. The radius r_i is the distance from the observation point to the image well.

When logarithmic approximations of well functions are possible, then:

$$s = 2\left\{\frac{Q}{4\pi T}\left[\ln\left(\frac{2.25Tt}{Sr^2}\right) + \ln\left(\frac{r}{r_i}\right)\right]\right\} \tag{4.5.53}$$

The drawdown along the line representing the impervious boundary, where $r = r_i$, will be twice the drawdown caused by the real well.

Drawdown Resulting from Variable Pumping Rates

Let a discharge well pump groundwater from a confined aquifer at a constant rate Q_0. At $t = t_1$ there was a step change in discharge $\Delta Q_1 = Q_1 - Q_0$. The Theis solution to the step change in discharge ΔQ_1 at $t = t_1$ is given as:

$$s = \frac{Q_0}{4\pi T}W\left(\frac{r^2 S}{4Tt}\right) + \frac{\Delta Q_1}{4\pi T}W\left(\frac{r^2 S}{4T(t - t_1)}\right) \tag{4.5.54}$$

The equation indicates that the drawdown in an observation well some distance r from the pumping well can be computed as the sum of the drawdown caused by the well discharged at the *constant* rate Q_0 over the time duration t, and the additional drawdown caused by the well discharging at the rate ΔQ_1 for the duration $t - t_1$.

The drawdown at any time $t \geq t_n$ in response to n step changes in pumping rate is:

$$s = \frac{Q_0}{4\pi T} W\left(\frac{r^2 S}{4Tt}\right) + \frac{1}{4\pi T} \sum_{i=1}^{n} \Delta Q_i W\left(\frac{r^2 S}{4T(t - t_i)}\right) \tag{4.5.55}$$

The Cooper-Jacob form of this equation is:

$$s = \frac{Q_0}{4\pi T} \ln\left(\frac{2.25Tt}{r^2 S}\right) + \frac{1}{4\pi T} \sum_{i=1}^{n} \Delta Q_i \ln\left(\frac{2.25T(t - t_i)}{r^2 S}\right) \tag{4.5.56}$$

REFERENCES

Abramowitz, M., and Stegun, I. A., *Handbook of Mathematical Functions*, Dover, New York, 1972.

Batu, V., *Aquifer Hydraulics*, John Wiley & Sons, New York, 1998.

Bear, J., *Hydraulics of Groundwater*, McGraw-Hill, Inc., New York, 1979.

Boulton, N. S., "Analysis of Data from Nonequilibrium Pumping Tests Allowing for Delayed Yield from Storage," *Proc. Inst. Civil Engrs.*, 26: 469–482, 1963.

Carslaw, H. S., and Jaeger, J. C., *Conduction of Heat in Solids*, Oxford University Press, Oxford, England, 1946.

Cooper, H. H., Jr., and Jacob, C. E., "A Generalized Graphical Method for Evaluating Formation Constants and Summarizing Well Field History," *Trans. Am. Geophys. Union*, 27: 526–534, 1946.

Davis, S. N., and De Wiest, R. J. M., *Hydrogeology*, John Wiley & Sons Inc., 1966.

de Marsily, G., *Quantitative Hydrogeology*, Academic Press Inc., 1986.

Driscoll, F. G., *Groundwater and Wells*, Johnson Filtration Systems Inc., St. Paul, Minnesota, 1989.

Fetter, C. W., *Applied Hydrogeology*, Third Edition, Macmillan College Publishing Company, New York, 1994.

Freeze, R. A., and Cherry, J. A., *Groundwater*, Prentice-Hall Inc., 1979.

Hantush, M. S., and Jacob, C. E., "Non-steady Radial Flow in an Infinite Leaky Aquifer," *Trans. Am. Geophys. Union*, 36, 1: 95–100, 1955.

Hantush, M. S., "Hydraulics of Wells," *Adv. Hydrosci.*, 1964.

Jacob, C. E., "Flow of Groundwater" in *Engineering Hydraulics*, Rouse, H., ed., John Wiley & Sons, New York, pp. 321–386, 1950.

McWhorter, D. B., and Sunada, D. K., *Ground-Water Hydrology and Hydraulics*, Water Resources Publications, Fort Collins, Colorado, 1977.

Neuman, S. P., "Theory of Flow in Unconfined Aquifers Considering Delayed Response of the Watertable," *Water Resources Research*, 8: 1031–1045, 1972.

Neuman, S. P., "Supplementary Comments on Theory of Flow in Unconfined Aquifers Considering Delayed Response of the Watertable," *Water Resources Research*, 10: 303–312, 1973.

Neuman, S. P., "Analysis of Pumping Test Data from Anisotropic Unconfined Aquifers Considering Delayed Gravity Response," *Water Resources Research*, 11: 329–342, 1975.

Theis, C. V., "The Relation Between the Lowering of the Piezometric Surface and the Rate and Duration of Discharge of a Well using Groundwater Storage," *Trans. Am. Geophys. Union*, 16: 519–524, 1935.

Thiem, G., *Hydrologische Methoden.*, Gebhardt, Leipzig, 1906.

Todd, D. K., *Groundwater Hydrology*, John Wiley & Sons Inc., 1980.

Walton, W., *Principles of Groundwater Engineering*, Lewis Publishers Inc., Chelsea, Michigan, 1991.

CHAPTER 4
HYDROLOGY

SECTION 4.6

REAL-TIME SOIL WATER DYNAMICS

Ioan C. Paltineanu
Based on his international education and experience in soil physics irrigation, Dr. Paltineanu founded an independent research and technology transfer company, PALTIN, International, Inc., involved in real-time natural resources monitoring studies over large areas.

James L. Starr
Dr. Starr is a soil scientist with the USDA, Agricultural Research Service, Beltsville, MD. His research is largely focused on identifying and characterizing the interactive effects of soil, cropping systems, and management practices on real-time soil water dynamics and associated reaction and transport of nutrients.

4.6.1 INTRODUCTION

Water is the major vehicle for moving chemicals through and over soils. Reliable predictions of the environmental impact of different management practices require accurate measures of vadose zone soil water content in both space and real-time. The vadose zone is the aerated region of soil above the permanent water table [Glossary of Soil Science Terms, *Soil Sci. Soc. Am.*, 121 (1996)]. Soil properties, climatic factors, and management factors influence water storage and movement. Better understanding of the interactions between factors controlling the fate of water in soil is needed to minimize losses of water and dissolved constituents by leaching and surface runoff [Starr and Paltineanu, *Soil Sci. Soc. Am. J.*, 62: 114–122 (1998a)]. A continued need exists for better methods and techniques to perform accurate, real-time nearly continuous soil water measurement at specific depths, with minimal soil disturbance, and covering field scale areas with few cables and data loggers. Monitoring soil water content in real-time, and covering micro- to large-scale hydrobasins can provide critical "ground truth" for accurately predicting and managing the fate of organic and inorganic chemicals applied to the land by private and public sectors.

4.6.2 THEORY AND INSTRUMENTATION

A survey of soil water measurement methods in current use showed gravimetric analysis as the most common method, followed by neutron thermalization, and, far behind, time domain reflectometry (TDR) and gamma attenuation [Green and Topp, *SSSA Spec. Publ.*, 30: 281–288 (1992)]. High labor costs and radioactive risk hazards and regulations, along with the neutron's probe individual discrete data collection, make the neutron thermalization method unsuitable for real-time soil water dynamics across large areas.

Two methods (TDR and Capacitance) have been developed, based on the correlation between the apparent dielectric constant (K_a) of the soil-air-water mixture surrounding the sensor and the soil water content at different electromagnetic field frequencies. These methods are increasingly used as part of permanent or semipermanent soil water monitoring systems for assessing real-time soil water dynamics in the vadose zone. Capacitance is especially suited for monitoring soil water content over large areas [Paltineanu and Starr, *Soil Sci. Soc. Am. J.*, 61: 1576–1585 (1997), and Starr and Paltineanu, *Soil Sci. Soc. Am. J.*, 62: 114–122 (1998a)].

Relationship of Apparent Dielectric Constant and Soil Water Content

Both TDR and Capacitance utilize the K_a of the soil surrounding the sensors in order to measure θ_v, which is an intrinsic characteristic of the soil-water-air mixture. At radio frequencies, the dielectric constant of pure water (K_w) at 20°C and atmospheric pressure is 80.4, that of soil solids is 3 to 7, and that of air is 1 [Weast, R. C., *Handbook of Chemistry and Physics*, 61st edition, CRC, Boca Raton (1980)]. Electromagnetically, a soil can be represented as a dielectric mixture of air, bulk soil, bound water, and free water [Hallikainen et al., *IEEE Trans. Geosci. Remote Sens.*, 23: 25–34 (1985)].

The Time Domain Reflectometry Method (TDR).   The TDR method measures the velocity of propagation of a high-frequency signal reflected back from the end of a transmission line or wave guides in the soil. Wave guides (with two, three, or more rods) may be installed vertically or horizontally in the soil profile. Although vertical placement of probes may conduct heat into the soil via the metallic probes and possibly results in preferential flow of water along the probes [Zegelin et al., *SSSA Spec. Publ.*, 30: 187–208 (1992)], the magnitude of the problem is not well documented. The same authors show that horizontal placement of probes requires excavation of a pit, which must then be back-filled to a condition as close as possible to the soil's initial condition. From laboratory experiments at frequencies of 1MHz to 1GHz [Topp et al., *Water Res.*, 16: 574–582 (1980)], a third-degree polynomial relationship was determined between K_a and θ_v, with a standard error of estimate of about 1.3 percent for most mineral soils. Soils with either a high organic matter content or high 2:1 clay content require site-specific calibrations [Herkelrath et al., *Water Resour. Res.*, 27: 857–864 (1991); Dirksen and Dasberg, *Soil Sci. Soc. Am. J.*, 57: 660–667 (1993); Bridge et al., *Aust. J. Soil Res.*, 34: 825–835 (1996); Evett, *Trans. ASAE*, 41(2): 361–369 (1998)]. Soil water measurement by TDR is largely confined to a quasi-rectangular area of approximately 1000 mm^2 surrounding the waveguides, with sensitivity decreasing logarithmically with distance from the wave guides [Baker and Lascano, *Soil Sci.*, 147(5): 378–384 (1989)]. No significant variation in sensitivity was observed along the length of the waveguides. Special attention has been given to the spatial and temporal distribution of soil water by designing different automated TDR systems of data collection and monitoring techniques [van Wesenbeek and Kachanoski, *Soil Sci. Soc. Am. J.*, 52: 363–368 (1988); Baker and Allmaras, *Soil Sci. Soc. Am. J.*, 54: 1–6 (1990); Heimovaara and Bouten, *Water Resour. Res.*, 26: 2311–2316 (1990); Herkelrath et al., *Water Resour. Res.*, 27: 857–864 (1991)]. These systems often work quite well in small areas, but are limited by cable length for measuring real-time soil water dynamics across large areas. Water measuring accuracy degrades quickly with cables longer then about 25 m [Heimovaara, *Soil Sci. Soc. Am. J.*, 57: 1410–1417 (1993)]. A new TDR technique is reported to extend the cable distance up to 100 m [Hook et al., *Soil Sci. Soc. Am. J.*, 56: 1384–1391 (1992)].

TDR Instrumentation.   A comprehensive study on the TDR instrumentation is presented in [O'Connor, K. M., and Dowding, C. H., *Geomeasurements by Pulsing TDR Cables And Probes*, CRC Press, Boca Raton (1999)].

The Capacitance Method.   The dielectric constant of soils can be measured by capacitance. This method includes the soil as part of a capacitor, in which the permanent dipoles of water in the dielectric medium are aligned by an electric field and become polarized. To contribute to the dielectric constant, the electric dipoles, of any origin, must respond to the frequency of the electric field. The freedom of the dipoles to respond is determined by the local molecular binding forces so that the overall response

TABLE 4.6.1 Textural fractions and dry bulk densities of soils used for calibration of multisensor capacitance probe under laboratory conditions

Location	Soil	Dry bulk density (g cm^{-3})	Sand (%)	Silt (%)	Clay (%)
Beltsville, MD	Silt loam	1.24 to 1.58	35	56	9
Fresno, CA	Sand	1.3	100	—	—
	Sandy loam	1.3 to 1.64	59	22	19
	Clay	1.0 to 1.16	6	35	49
Adelaide,	Sandy loam	1.1 to 1.3	66.3	16.4	17.2
So. Australia	Loamy sand	1.46 to 1.54	82.3	7.7	10.0
	Sandy loam	0.95 to 1.97	66.5	18.3	16.2
	Loamy sand	1.0 to 1.39	82.7	9.4	75

Source: Paltineanu and Starr, 1997.

is a function of molecular inertia, the binding forces, and the frequency of the electric field [Dean et al., *J. Hydrol.*, 93: 67–78 (1987)]. Measurement of the capacitance gives the dielectric constant, hence the volumetric water content of the soil.

Capacitance Probes.   Information on portable, semi-permanent, or permanent capacitance probes and monitoring systems can be found in: [Kuraz, V., *Geotech. Test. J.*, 4: 111–116 (1982); Dean et al., *J. Hydrol.*, 93: 67–78 (1987); Campbell, J. E., *Soil Sci. Soc. Am. J.*, 54: 332–341 (1990); Paltineanu and Starr, *Soil Sci. Soc. Am. J.*, 61: 1576–1585 (1997); Hilhorst, M. A., *Dielectric Characterization of Soil*, DLO, Wageningen (1998)].

Calibration of Multisensor Capacitance Probes.   Calibration experiments with the multisensor capacitance probes show a highly significant, nonlinear relationship between the volumetric soil water content (θ_v) and the scaled frequency (SF) for soils of Beltsville, MD, Fresno, CA, and Adelaide, South Australia, as presented in Table 4.6.1 and Figure 4.6.1 [Paltineanu and Starr, *Soil Sci. Soc. Am. J.*, 61: 1576–1585 (1997)]:

$$\theta_v = aSF^b$$

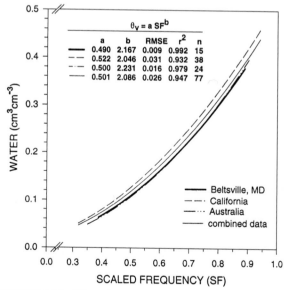

FIGURE 4.6.1 Volumetric water content versus scaled frequency at three sites (Paltineanu and Starr, 1997).

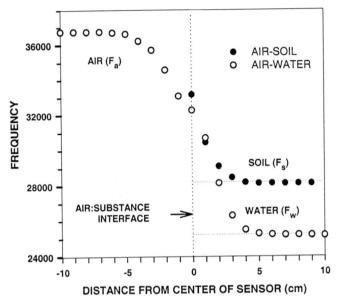

FIGURE 4.6.2   Axial sensitivity of sensors to air:water and air:soil interfaces (Paltineanu and Starr, 1997).

The *SF* represents the ratio of frequencies measured by each sensor (inside the PVC pipe) in the soil (F_s) compared with sensor responses in the air (F_a) and in nonsaline water (F_w) at room temperature (~22°C),

$$SF = (F_a - F_s)/(F_a - F_w).$$

Axial and radial sensitivity of sensors near multisensor capacitance probes is presented in Figures 4.6.2 and 4.6.3. These figures show: (1) that the center of measurement (i.e., at the plastic ring between

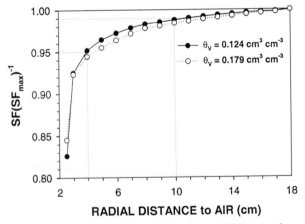

FIGURE 4.6.3   Relative radial sensitivity of sensors (SF/SF$_{max}$) as a function of radial thickness of soil around the access pipe containing the capacitance sensors (Paltineanu and Starr, 1997).

the two metal rings) has a 5-cm axially symmetric zone of accurate influence; and (2) that 99 percent of the sensor's response was obtained within a radial distance of 10 cm from the wall of the PVC pipe wall. These observations suggest the importance of the PVC access pipe installation techniques that ensures a tight contact with the surrounding soil. As with TDR, more calibration research is needed with these capacitance sensors for special soils (e.g., swelling 2:1 clays or high organic matter content) and for extremes of soil temperature that can occur diurnally with bare surface soils and seasonally in many climates.

Sensor response to cable length. Unlike the TDR method, the multisensor capacitance probes readings are not affected by cable lengths up to 500 m [Starr and Paltineanu, *Soil Sci. Soc. Am. J.*, 62: 114–122 (1998a)].

Multisensor Capacitance Probes and Real-time Soil Water Content Monitoring System for Large Areas.   A detailed description of a monitoring system (EnviroSCAN®, Sentek Pty. Ltd., Kent Town, South Australia) that was used for real-time soil water dynamics studies at the USDA-ARS Beltsville Agricultural Research Center is given in [Starr and Paltineanu, *Soil Sci. Soc. Am. J.*, 62: 114–122 (1998a)].

4.6.3 NEAR REAL-TIME NETWORKS FOR IN-SITU SOIL WATER DYNAMICS

The USDA-NRCS national "near" real-time monitoring soil moisture/soil temperature (SM/ST) pilot project (formally called Global Climate Change Project) is described in [Shaefer et al., CGU-AGU Annual Meeting, SMST.doc., Banff, Canada (1995)]. This project has been operating since 1992, and now has 33 sites in 25 states, as schematically shown in Figure 4.6.4. Soil water content, soil

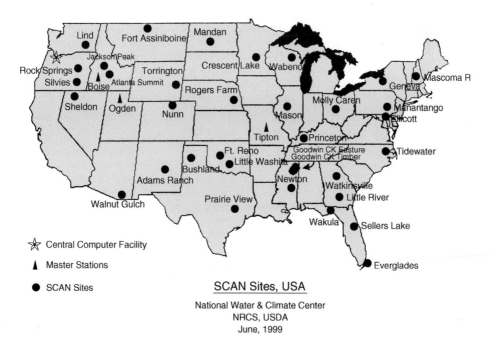

FIGURE 4.6.4 Locations of the National Water and Climate Centers (NRCS, USDA) "near" real-time soil moisture/soil temperature pilot project sites, three master stations, and the central computer facility as of June 1999 (Schaeffer et al., 1996).

temperature, and salinity are measured at each site with four-rod capacitance sensors (Hydra—Vitel Inc., Chantilly, VA, USA) at soil depths of 2, 4, 8, 20, and 40 inches. The SM/ST Pilot Project uses meteor burst communication techniques for obtaining remote site information, which bounces radio signals off cosmic dust. Remote sites, master stations, and the central computer facility are shown in Figure 4.6.4. The data is available 24 hours a day via the internet (http://www.wcc.nrcs.usda.gov/scan/index2.html). User access requires only a signed agreement between the individual/agency and the NRCS.

4.6.4 REAL-TIME SOIL WATER DYNAMICS MONITORING SYSTEMS OVER LARGE AREAS—CASE STUDIES

Automatic real-time monitoring systems using TDR sensors for small-size plots have been reported by several researchers [Baker and Allmaras, *Soil Sci. Soc. Am. J.*, 54: 1–6 (1990); Heimovaara and Bouten, *Water Resour. Res.*, 26 (10): 2311–2316 (1990); Herkelrath et al., *Water Resour. Res.*, 27(5): 857–864 (1991)]. Applications in agriculture, environmental protection, hydrology (global climate change), waste disposal sites (nuclear and nonnuclear), and remote sensing of natural resources all need continued development of better methods and techniques to perform accurate, real-time and nearly continuous soil water content monitoring at specific depths, with minimal soil disturbance, and covering field-scale areas.

Applications in Agriculture and Environmental Protection

Research with multisensor capacitance probes and monitoring systems have been conducted at the USDA-ARS Beltsville Agricultural Research Center, Beltsville, Maryland, since June 1995. Real-time soil water dynamics was studied for several large field experiments with different cropping systems (comparative corn tillage methods, field crops sustainable agricultural methods, herbicide leaching to water table under corn, comparative soil cover technologies [plastic versus vetch mulch cover tomatoes] for minimizing ground and surface water pollution, and others). Results of short- and long-term studies at a field site that was incrementally changed from moldboard plow till (PT) to no-till (NT) corn during a 4-year period are presented by several researchers [Starr and Paltineanu, *Soil Sci. Soc. Am. J.*, 62: 114–122 (1998a); Starr and Paltineanu, *Soil and Till. Res.*, 47: 43–49 (1998b); Paltineanu and Starr, *Soil Sci. Soc. Am. J.*, 64: 44–54 (2000)].

Multisensor Capacitance Probes and Monitoring System Installation.   A detailed description of proper field installation of the EnviroSCAN® soil water monitoring system, used in the comparative corn tillage study (0.5 ha), along with the weather station and real-time evapotranspiration (ET) instrumentation is shown in Figure 4.6.5. The three main parts of this installation are: (1) eight semipermanently installed probes with four capacitance sensors each, centered at 10-, 20-, 30-, and 50-cm depths; (2) a data logging station, and two five-wire cables (<500 m) connected to probes; (3) an internal power supply charged by a solar panel (or alternatively by external battery or AC power) [Starr and Paltineanu, *Soil Sci. Soc. Am. J.*, 62: 114–122 (1998a)]. In these studies, soil water contents were measured every 10 minutes.

Field Determination of the in situ Apparent Water Holding Capacity (aWHC).   Repeating patterns of nearly constant maximal water contents following early-season rainfall events (when corn plants were small—6–7 leaves—with low to moderate evaporative demand), caused temporary rises in the soil water content as the macro pores filled and quickly drained, as shown in Figure 4.6.6. These observations indicate an *in situ* apparent water-holding capacity (aWHC) at each probe, and sensor depth [Starr and Paltineanu, *Soil and Till. Res.*, 47: 43–49 (1998b)]. The aWHC is very important for establishing irrigation and drainage strategies for minimizing leaching of agrochemicals to deeper soil layers.

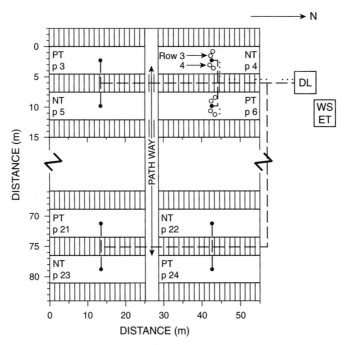

FIGURE 4.6.5 Schematic of plot layout, with position of capacitance sensors (●), cables (---), data logging station (DL), weather station (WS), ET gauge (ET), and irrigation-water collectors (C) between no-till (NT) and plow-till (PT) plots near to the probes (Starr and Paltineanu, 1998a).

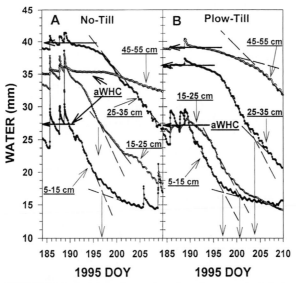

FIGURE 4.6.6 Real-time soil water dynamics at four sensor depths in the corn-row, and showing the apparent water holding capacity and breaking points (Starr and Paltineanu, 1998b).

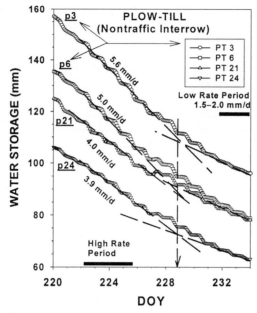

FIGURE 4.6.7   Soil water dynamics (cumulative for 5 to 55 cm) during a period of high ET and full canopy for all eight plots. The vertical arrows approximate the breaking points between high and low rates of water loss, and the dashed lines represent the mean rates of water losses (Starr and Paltineanu, 1998b).

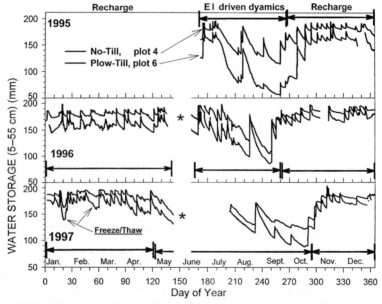

FIGURE 4.6.8   Year around cumulative soil water dynamics in one pair of NT-PT plots, with multisensor capacitance probes placed at the nontraffic interrow position; ★, Probes removed for spring tillage operations.

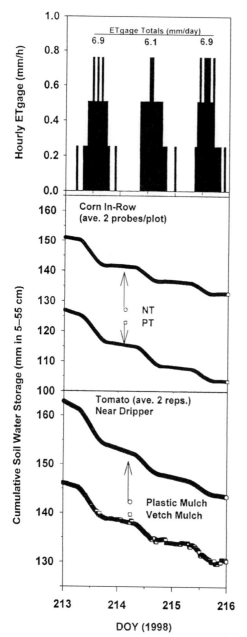

FIGURE 4.6.9 Real-time ETgage and soil water dynamics in NT and PT corn and plastc- and vetch-covered tomatoes.

Establishing "Breaking Points" Between Periods of High and Low Rates of Water Loss. Real-time soil water dynamics in the soil profile (5–55 cm) of four PT corn plots (across the 0.5-ha field site), under high evaporative demand showed a diurnal response (high soil water consumption during the day and low to none during the night); see Figure 4.6.7. Initial high rates of daily water loss suggest that the soil water supply was adequate. However, with continued high atmospheric water demand, the soil water reserves became depleted, resulting in slower, multidaily rates of water loss in the vadose zone. "Breaking points" developed at the intersection of the slopes of quasi-steady rates of water loss in all four replicates [Starr and Paltineanu, *Soil and Till. Res.*, 47: 43–49 (1998b)]. Viewing the soil water data in real-time allows the user to anticipate when the breaking points will occur and to provide irrigation at the right time. The amount of water to apply can then be easily determined as the difference between the aWHC and the "breaking point" for each probe location and capacitance sensor depth.

Multiannual Real-Time Soil Water Dynamics. Multiannual, year around changes in real-time soil water dynamics in the vadose zone of corn under NT versus PT tillage methods, shown in Figure 4.6.8, demonstrate that soil water content responds quickly to rainfall events throughout the year. Seasonal effects were largely controlled by the high evaporative demand during the growing season, and soil and groundwater recharge events from late fall to early spring [Starr et al., *Agron. Abs.*, 237 (1998)].

Influence of Soil Cover on Real-time Soil Water Content Dynamics. Soil cover (biological mulch or plastic) influences real-time water dynamics under different crop's agricultural technologies versus the real-time evaporative demands data as shown in Figure 4.6.9. Real-time soil water content data, taken by the multisensor capacitance probes and monitoring systems, under corn (no-till versus plow till) and tomato (plastic versus vetch mulch cover), show a decreasing step-like water uptake pattern. Real-time internal drainage after rainfall and irrigation events should be considered when calculating crop coefficients for irrigation scheduling [Paltineanu et al., *Agron. Abs.*, 198 (1998)].

Influence of Drip Irrigation Application and Soil Cover on Real-time Soil Water Dynamics. Placing drip irrigation lines at 10 cm depth in dry soil under plastic cover has a large effect on surface water content as shown in Figure 4.6.10. Dominant direction of irrigation water was downward under plastic cover, resulting in excessively dry soil in the 5–15 cm depth interval. By contrast, placing drip irrigation lines on the surface of vetch cover resulted in better horizontal water movement (at 10- and 30-cm locations of multisensor capacitance probes from the drip lines) [Paltineanu et al., *Agron. Abs.*, 198 (1998)].

Real-time studies of capillary fringe variation in the vadose zone. Capillary fringe variation in the vadose zone was studied in real-time by using covered and uncovered multisensor capacitance probes, showing good agreement with the real-time data obtained from pressure transducers installed in piezometers. In Figure 4.6.11, the time lag between the initiation of rainfall and the rise of water table and the capillary fringe (comparing covered and uncovered plots) was about eight hours [Paltineanu et al., *Agron. Abs.*, 198 (1998)].

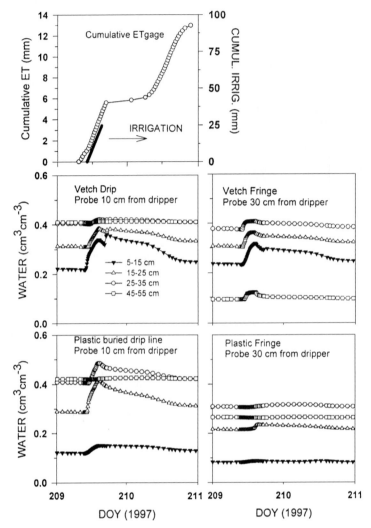

FIGURE 4.6.10 Real-time soil water distribution under plastic- and vetch-covered tomatoes in relation to cumulative irrigation and Etgage values.

Applications in Climate Change Studies—Forestry

A key issue in assessing climate change scenarios is the response of forests to decreased water availability or increased occurrence of drought. Recent review articles [Hanson et al., Oak Ridge National Laboratory, Oak Ridge, Tennessee, ORNL/TM-13586, pp. 36 (1998)] have called for large-scale experiments to adequately address the impacts of changing climates on ecosystems. These authors describe a field-scale experiment (1.92 ha) in an upland oak forest in eastern Tennessee. This research seeks to identify important ecosystem responses that might result from future precipitation changes by monitoring soil water content with TDR sensors (0–35 and 37–70 cm) placed at 310 locations across the site. Statistically significant differences in soil water content were observed in years having both dry and wet conditions.

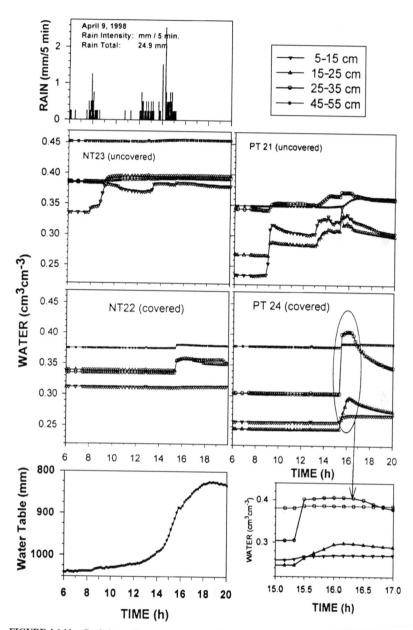

FIGURE 4.6.11.   Real-time soil water dynamics and direction of water movement under covered and uncovered plots in relation to rainfall distribution.

Applications in Monitoring Soil Contamination by Radioactive Materials

Actinide contamination of soils around Rocky Flats, CO, USA, resulted from leaking drums of Pu-contaminated oil at an outdoor site. Contaminate transport through the soil to groundwater was studied using an advanced monitoring system gathering real-time data on soil water content, groundwater level, and timing of gravitationally flowing water [Litaor et al., *J. Environ. Qual.*, 25: 671–683 (1996); Litaor et al., *J. Environ. Radioactivity*, 38: 17–46 (1998); Ryan et al., *Environ. Sci. Technol.*, 32: 476–482 (1998)]. Soil water content was monitored in five pits (5 m long × 1 m × 1 m) with TDR probes manufactured by Campbell Scientific Inc., UT, USA, using two wave guides connected to parallel cable lines, linked to a cable tester (Tektronix 1502B) through a hierarchy of multiplexers.

Field experiments to evaluate subsurface monitoring systems at low-level radioactive waste disposal sites and for monitoring decommissioned facilities has been performed by the University of Arizona in cooperation with the Bureau of Economic Geology at the University of Texas at Austin, and Stephens and Associates in Albuquerque, New Mexico [Young et al., NUREG/CR-6462, University of Arizona and University of Texas at Austin for the U.S. Nuclear Regulatory Commission, pp. 49 (1996)]. Irrigation was applied to a 50-m by 50-m plot, and real-time (hourly) soil water content dynamics studied by using TDR probes. The probes were placed along a 70-m trench transect (1.5 m depth) and at islands going down to 3 m. This report describes in detail the design of the experiment and the methodology proposed for evaluating the data.

Applications in Reclamated Dumps and the Acceptability of Soil for Earthworking

Increased interest in reclamation of dumps from mining industry and the use of biologic origin covers prompted the use of dielectric constant probes to study the dynamics of soil water content. A dielectric soil moisture meter was used to monitor the time and space distribution of soil moisture under a surface dump containing cellulose sludge [Kuraz and Matusek, International Symposium on Soil System Behaviour in Time and Space, Comm. of Austrian Soil Science Society, p. 67–70 (1997)]. The "dielectric soil moisture meter" was shown to be highly suitable for monitoring the water regime in the condition of very heterogeneous artificial soil profiles in reclamation of surface dumps.

The acceptability of soil for earthworking was studied in the field for 56 weeks with the aid of soil moisture capacitance probes in 10 access tubes (at 2-cm intervals down to 1.7 m) to validate the FORE-SALT model [Smith et al., *Electronic Journal of Geotech.* Eng., http://geotech.civen.okstate.edu/ejge/ppr9701/index.htm, issue 2 (1997)]. Information on recent and ongoing research studies involving multisensor TDR probes and real-time monitoring systems (Moisture-Point®, ESI, Victoria, BC, Canada) in landfill test cover studies can be obtained at (http://www.esica.com).

Applications in Real-time Geotechnical Site Characterization

Soil water measurement is a key parameter in any environmental or geotechnical site investigation. Development of a cone penetration testing probe that measures both electrical resistivity, soil temperature, and the volumetric soil water content (capacitance) for geotechnical site characterization is presented by several researchers [Shin II et al., *Field Analytical Chemistry and Technology*, 2(2): 103–109 (1998a); Rose et al., *Geotechnical Site Characterization*, Robertson & Mayne, eds., Balkema, Rotterdam, p. 589–595 (1998); Shinn II et al., *Geotechnical Site Characterization*, Robertson & Mayne, eds., Balkema, Rotterdam, p. 595–599 (1998b)].

4.6.5 FURTHER RESEARCH AND DEVELOPMENT IN REAL-TIME SOIL WATER DYNAMICS MONITORING

Aqteplatics—An Integrated Concept of Real-time Water Dynamics Over Large Areas

Recent developments in instrumentation for real-time water dynamics studies can lead to great improvement in our understanding of the soil-plant-atmosphere continuum not only in situ, but over large areas. Whole systems studies, **aqua-terra-planta-at**mosphere dynamics, over large areas—

including the lower atmosphere, plant communities, and vadose zone hydrology, should be integrated in a unifying concept of Aqteplatics [Paltineanu et al., Agron. Abstr., p. 286 (1998)].

A Global Network for Monitoring Real-time Soil Water Content Dynamics

Accurate "ground truth" data provided by independent real-time soil water content monitoring systems, covering micro- to large-scale hydrobasins, can have a great impact on the remote sensing estimates of water resources at local, regional, and global scales. New developments in precision agriculture, remote sensing, soil surface, and subsurface preferential water flow patterns, simulation models for soil-water-plant-atmosphere interrelationships over large areas and permanent watch of leakage from waste material depositing sites can all benefit from real-time soil water dynamics data. The relatively new multisensor (Capacitance or TDR) probes and monitoring systems can provide basic real-time data for GPS (Global Positioning System) and GIS (Geographic Information System) designed network. Reliable "ground truth" is critical for validation and real-time calibration of the actual and future remote sensing sensors installed on orbital platforms.

GENERAL REFERENCES

Hilhorst, A. M., *Dielectric Characterisation of Soil*, Doctoral Thesis. Wageningen Agricultural University, IMAG-DLO, Wageningen, The Netherlands, 1998.

Hillel, D., *Environmental Soil Physics*, Academic Press, San Diego, CA, 1998.

Hook, H. R., Livingston, N. J., Sun, Z. J., and Hook, P. B., "Remote Diode Shorting Improves Measurement of Soil Water by Time Domain Reflectometry," *Soil Sci. Soc. Am. J.*, 56: 1384–1391, 1992.

Kuráz, V., "Testing of a Field Dielectric Soil Moisture Meter," ASTM, *Geotech. Test. J.*, 4: 111–116, 1982.

O'Connor, K. M., and Dowding, C. H., *GeoMeasurements by Pulsing TDR Cables and Probes*, CRC Press LLC, Boca Raton, Florida, 1999.

Paltineanu, I. C., and Starr, J. L., "Real-time Soil Water Dynamics Using Multisensor Capacitance Probes: Laboratory Calibration," *Soil Sci. Soc. Am. J.*, 61: 1576–1585, 1997.

Starr, L. J., and Paltineanu, I. C., "Soil Water Dynamics Using Multisensor Capacitance Probes in Nontraffic Interrows of Corn," *Soil Sci. Soc. Am. J.*, 62: 114–122, 1998a.

Starr, L. J., and Paltineanu, I. C., "Real-time Soil Water Dynamics Over Large Areas Using Multisensor Capacitance Probes and Monitoring System," *Soil and Till. Res.*, 47: 43–49, 1998b.

Topp, G. C., Davis, J. L., and Annan, A. P., "Electromagnetic Determination of Soil Water Content: Measurement in Coaxial Transmission Lines," *Water Resour. Res.*, 16(3): 574–582, 1980.

CHAPTER 4
HYDROLOGY

SECTION 4.7

FATE AND TRANSPORT MODELING—EFFECTIVELY USING ISOTOPES OF WATER TO SOLVE PRACTICAL HYDROGEOLOGICAL PROBLEMS

Donald I. Siegel
Dr. Siegel is professor of earth sciences at Syracuse University and the winner of the Birdsall Distinguished Lecturer Award from the Geological Society of America.

4.7.1   INTRODUCTION

Determining the fate and transport of contaminants in groundwater systems can be difficult as part of environmental assessments that are guided by regulatory concerns. In the regulatory environment, financial and protocol limitations often result in gathering insufficient hydraulic and geochemical data to clearly resolve questions related to groundwater flow and water quality. In this chapter, we present how many typical uncertainties related to flow system hydraulics and groundwater geochemistry in "applied" hydrogeologic studies can be effectively and inexpensively addressed by analysis of water isotopes.

Academic and governmental scientists have used isotopic geochemistry for decades as a powerful tool, *independent of standard chemical analyses*, to characterize solute sources and chemical reactions; (Hoefs, 1987; Fritz and Fontes, 1980; Faure, 1986; Mazor, 1991). Most recently, Clark and Fritz (1997) published the first comprehensive text that addresses the diversity and richness of isotopic techniques used specifically in hydrogeologic studies.

The information gained from isotopic analysis of even a small number of water samples often provides critical information on: (1) whether groundwater flow systems are dominated by fracture or nonfracture flow, (2) whether there are multiple sources of contamination in groundwater systems, (3) the extent to which surface waters and groundwaters mix and disperse, (4) the general oxidation-reduction environment of the subsurface geochemical environment, and (5) the age of groundwater.

Furthermore, the costs of isotopic analyses of water and most dissolved elements of environmental interest are a fraction of the cost of routinely analyzing a standard suite of priority organic contaminants. Contract laboratories have provided inexpensive isotopic analyses of water and solutes for many years, notably the University of Waterloo Environmental Isotopic Laboratory (http://www.science.uwaterloo.ca/~rkhmskrk/), Geochron Laboratories (http://www.geochronlabs. com/), and Global Geochemistry (http://www.ggclab.com). The reader should refer to http://geology. uvm. edu/geowww/isogeochem.html#anchor559545) for additional information on contract and university laboratory facilities.

4.7.2   WHAT ARE THE ISOTOPES OF WATER?

What are isotopes? Elements are defined by the number of protons whereas isotopes of elements are defined by their number of neutrons. The two types of isotopes are radiogenic and stable. The nuclei of radiogenic isotopes spontaneously decay to daughter elements, atomic particles, and energy. In contrast, the nuclei of stable isotopes do not decay spontaneously. For example, the most common isotopic form of oxygen, ^{16}O, has an atomic mass of 16 atomic-mass units, consisting of eight protons and eight neutrons. The two next most common isotopes of oxygen are ^{17}O, which has one additional neutron, and ^{18}O, which has two additional neutrons. These isotopes of oxygen are *stable isotopes*, which do not radioactively decay to atomic byproducts and daughter elements. Hydrogen also has three isotopes: ^1H, ^2H (deuterium), and ^3H (tritium), the most common form being ^1H. Tritium, in contrast to the other two isotopes of hydrogen, is a *radiogenic* isotope, spontaneously decaying to ^3He. Tritium is naturally formed in the high atmosphere when ^{14}N is bombarded by cosmic radiation.

The stable isotopes of water naturally are fractionated or "sorted" by their atomic masses by physical, chemical, and biochemical processes. The sources of different waters and contaminants can be often determined by the extent of isotopic fractionation observed. The stable isotopes of carbon, nitrogen, sulfur, chlorine, lead, and other elements are also used to trace their cycling in hydrogeologic systems and their sources, but these applications are used less in regulatory-driven hydrogeologic studies compared to the isotopes of water.

Stable isotopes are measured by isotopic ratio mass spectroscopy. In this analytical method, the isotopes of interest are placed into an ionized beam and separated by a magnetic field according to their atomic masses. Isotopes are measured by the mass spectrometer as ratios rather than directly counting the numbers off atoms of different mass. These ratios are then compared to internationally accepted standards to derive an isotopic composition or value. In the case of radiogenic isotopes that decay rapidly, such as tritium (^3H), the isotope is determined by counting the rates of radiogenic particle release.

The isotopic ratios of stable isotopes are reported in "DELTA" notation:

$$\delta R_{sample} = [(R_{sample}/R_{standard}) - 1] \cdot 1000\%o \qquad (4.7.1)$$

where:

δR_{sample} is the "DELTA" isotopic ratio of the sample
R_{sample} is the isotopic ratio of the heavier to lighter, most common isotope, in the sample
$R_{standard}$ is the isotopic ratio of the heavier to lighter, most common, isotope in the standard

Because the stable isotopic fractionation does not cause large variations in the isotopic ratios, the δ–values are conventionally multiplied by 1000 and expressed in parts-per-thousand or "permil" differences from the standard. For example,

$$\delta^{18}O_{sample} = \left[\left(^{18}O/^{16}O_{sample}\right)/\left(^{18}O/^{16}O_{standard}\right) - 1\right] \cdot 1000\%o_{VSMOW}$$
$$= +5\%o_{VSMOW} \qquad (4.7.2)$$

In this case, the ratio of the heavier isotope of oxygen, ^{18}O, to the lighter and more common isotope, ^{16}O, in the sample is 5 parts-per-thousand heavier, or enriched, compared to the standard VSMOW. VSMOW refers to Vienna Standard Mean Ocean Water, the internationally accepted standard. The $\delta^{18}O$ value of VSMOW (and δ values for other standards) is defined arbitrarily as zero, such that the ratios of samples will be either greater in the heavier isotope (enriched) or smaller in the heavier isotope (depleted). Details on the internationally accepted standards used in stable isotopic analysis can be found at websites for the United Nations International Atomic Energy Agency (www.iaea.org) and the United States National Institute of Standards and Technology (www.nist.gov).

In contrast to the stable isotopes, radiogenic isotopic compositions of elements are reported differently, depending upon the origin of the radiogenic element and its rate of decay. The decay of radiogenic isotopes can be conveniently expressed by its "half-life," the time it takes for half of the isotope to decay to its daughter element. The half-life of tritium is $\sim$12.4 years. The atmospheric testing of atomic weapons in the 1950s and 1960s on islands in the Pacific Ocean injected huge amounts of tritiated water into the atmosphere. Once the tritium in precipitation recharges groundwater flow systems, its radioactive decay to ^{3}He makes it a very useful tool to at least semiquantitatively date water in the hydrologic cycle up to about 50 years before present (BP). The concentrations of tritium are expressed in tritium units (TU), equal to one tritium atom per 10^{18} atoms of hydrogen. If tritium is measured in water, one TU per liter of water is equal to a radioactivity of 3.2 pico-curries per liter (pCi/L).

The other radiogenic isotope commonly used to directly determine groundwater age is carbon-14 (in dissolved inorganic carbon). Because the half-life of this isotope is $\sim$5730 years, groundwater up to about 40,000 years old can be age-dated. Using ^{14}C to age-date groundwater can be very complicated because of multiple natural carbon sources and in-situ chemical reactions that fractionate carbon, and the reader is referred to Clark and Fritz (1997) for additional information.

4.7.3   WHY DO STABLE ISOTOPES FRACTIONATE?

Stable isotopes fractionate because certain thermodynamic properties of molecules depend on the masses of the atoms that constitute the molecule. As a result, one isotope is concentrated relative to the other on one side of the isotopic equilibrium reaction (Clark and Fritz, 1997).

$$\alpha = R_A / R_B \tag{4.7.3}$$

where α is the fractionation factor, R_A is the ratio of the heavy to the light isotope in molecule (or phase) A, and R_B is the ratio of the heavy to the light isotope in molecule (or phase) B. One consequence of isotopic fractionation is that the molecular bonds formed by isotopically lighter atoms are thermodynamically weaker and more easily broken, making isotopically light molecules more reactive than isotopically heavier ones. Isotopic fractionation is also temperature dependent. The lower the temperature, the greater the isotopic fractionation.

The phase transformation between water vapor and liquid water can be used as a convenient example to illustrate isotopic fractionation:

$$H_2O_{liquid} = H_2O_{vapor} \tag{4.7.4}$$

$$\alpha^{18}O_{liquid-vapor} = \left(^{18}O/^{16}O_{liquid}\right)/\left(^{18}O/^{16}O_{vapor}\right) \tag{4.7.5}$$

At low temperatures, the evaporation of water should result in the concentration of ^{18}O in the liquid phase because the isotopically lighter molecules with more ^{16}O enter the vapor phase.

The isotopes of water are fractionated during changes in state (e.g., when water vapor changes to liquid), chemical reactions occur (e.g., when methane is formed in landfills), and when concentration gradients occur (e.g., when tritiated water diffuses into dead pore spaces in clay). *Isotopic equilibrium* occurs when chemical reactions equilibrate in well-mixed geochemical systems. *Isotopic nonequilibrium* occurs when forward reactions are accelerated because products are removed or physiochemical conditions, such as temperature change.

Water in the atmosphere is effectively "distilled" under nonequilibrium conditions when it moves in the atmospheric circulation system from the equator to the poles and from the oceans to the continental regions. Simply stated, water vapor that evaporates from equatorial oceans contains more ^{16}O than ^{18}O (and ^{1}H than ^{2}H) because it takes less solar energy to evaporate water molecules that have smaller total mass than those having greater mass. This water vapor is isotopically "lighter" or "depleted" relative to the original pool of water from which it evaporated. Conversely, as water vapor moves northward into colder regions, the heavier fractions of the vapor pool are successively incorporated into precipitation because is takes less energy to condense water molecules that have greater mass. The precipitation is isotopically "heavier" or "enriched" compared to the vapor water pool from which it was derived.

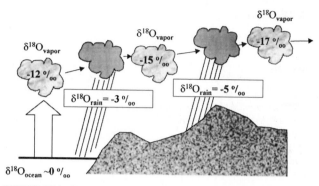

FIGURE 4.7.1   How Rayleigh distillation works in the hydrologic cycle. Note how water vapor is isotopically depleted (more negative number) relative to the ocean and how condensed precipitation over continents is enriched (less negative number) compared to the water vapor pool from which it precipitated.

This evaporation-condensation process is modeled by the equation for Rayleigh distillation as the isotopic water vapor reservoir becomes increasingly depleted:

$$R = R_o f^{(\alpha - 1)} \tag{4.7.6}$$

where R_o is the initial isotopic ratio in the water reservoir, R is the isotopic ratio in the depleting isotopic reservoir, f is the remaining volume fraction of the initial water reservoir, and α is the fractionation factor (R_{liquid}/R_{vapor}). Figure 4.7.1 schematically illustrates how the isotopic composition of oxygen in water changes because of Rayleigh distillation as air masses from oceans move over continental regions.

Because of the nonequilibrium Rayleigh distillation of water vapor, there is a distinctive latitudinal gradient in the isotopic composition of volume-weighted precipitation from the equator to the poles. As an example, Figure 4.7.2 shows the distribution of $\delta^{18}O$ in volume weighted precipitation across North America. Note on this figure how the $\delta^{18}O$ of average precipitation near the equatorial regions is only a few parts-per-mil smaller than the standard VSMOW (assigned as zero). In contrast, the isotopic ratio of $\delta^{18}O$ in volume-weighted annual precipitation near the poles is less than $-26‰$, reflecting the removal of ^{18}O during multiple rainouts. The International Atomic Energy Agency has collected and tabulated stable and radiogenic isotopic data of monthly precipitation at hundreds of sites around the world for decades. These data can be digitally accessed at (www.iaea.org).

The fractionation of the hydrogen and oxygen isotopes in precipitation co-vary in precipitation to define the most famous linear relationship in isotopic geochemistry, the Global Meteoric Water Line (*GMWL*; Craig, 1961; Dansgaard, 1964) which nominally is:

$$\delta^{18}O = 8\,\delta D + 10 \qquad \text{(standard-VSMOW)} \tag{4.7.7}$$

This equation defines and incorporates the isotopic composition of all global precipitation. Locally, the isotopic composition of precipitation defines local meteoric water lines (LMWL), the intercepts (called the deuterium excess), and the slopes of which are partly controlled by elevation (which modifies air temperature), humidity, and distance from the ocean. For example, in Antarctica and near the equator, where average annual air temperatures are $<-35°$ and $>+20°$ respectively, the $\delta^{18}O$ of annual precipitation varies from less than $-40‰$ to about $-2‰$. Globally, the average annual air temperature (degrees celsius) and $\delta^{18}O$ are very well correlated, nominally as (Dansgaard, 1964):

$$\delta^{18}O = 0.69\,T - 13.6 \tag{4.7.8}$$

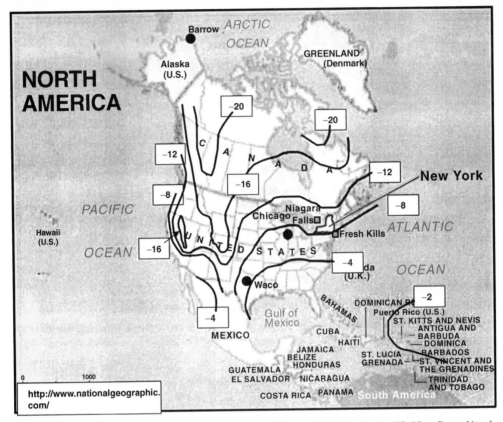

FIGURE 4.7.2   The $\delta^{18}O$ composition of volume weighted precipitation over North America (modified from Rozanski et al., 1993). Also shown on the figure are location of major case studies presented.

Seasonally, the isotopic composition of precipitation changes depending on the sources of air masses providing the moisture. Figure 4.7.3 shows examples of the spread of isotopic values for monthly precipitation at Waco (Texas), Chicago (Illinois), and Barrow (Alaska), separated by about 45 degrees of latitude (see Figure 4.7.2 for locations).

Note how the values for the $\delta^{18}O$ for precipitation in most months become increasing smaller (more depleted) from Waco to Chicago to Barrow because of Raleigh distillation from the equator towards the poles and with lower monthly temperatures. Furthermore, the range in $\delta^{18}O$ values during the year is greatest at Chicago, which seasonally receives precipitation from different air masses. In the summer, isotopically heavy Gulf Coast air provides most of the rain, whereas during the winter, isotopically lighter air masses from the Pacific Northwest contribute most of the snow. In contrast, at Waco the precipitation mostly comes from tropical air masses in the Gulf of Mexico and consequently, has less variation in its isotopic composition. The volume weighted average $\delta^{18}O$ values for precipitation at Waco, Chicago, and Barrow are about -6.3, -7.2, and $-19.5\permil$ respectively (Rozanski et al., 1993).

In contrast to many dissolved solutes, the stable isotopes of water are chemically "conservative," or nonreactive when two waters with different isotopic compositions mix. The nonreactive nature of the isotopes makes them very useful to determine the proportions of waters in mixtures. Equation 4.7.9 is used to determine the percentages (out of one liter) of two mixing waters, water-1 and water-2, that have different isotopic (or also, chemical) compositions:

$$n_{\text{water-1}} = (\delta_{\text{mixture}} - \delta_{\text{water-2}})/(\delta_{\text{water-1}} - \delta_{\text{water-2}}) \tag{4.7.9}$$

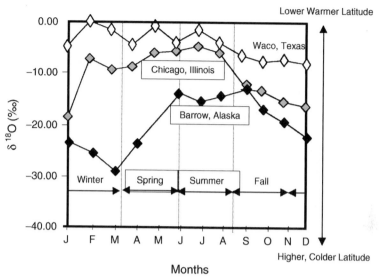

FIGURE 4.7.3 Distribution of $\delta^{18}O$ in monthly precipitation at Waco, Texas (1976), Chicago, Illinois (1976), and Barrow, Alaska (1966). Data from www.iaea.org.

where n is the fraction of water-1 in the mixture. By difference, $(1 - n)$ is the fraction of water-2. As an example, assume that water-1 and water-2 in Equation 4.7.10 have $\delta^{18}O$ compositions of $-10\%o$ and $-6\%o$, respectively, and that their mixture has a value of $-7\%o$. The percentage of water-1 in the mixture is:

$$n_{water-1} = [(-7\%o - -6\%o)/(-10\%o - -6\%o)] = 0.25 \text{ or } 25\% \qquad (4.7.10)$$

4.7.4 APPLICATIONS

Are Aquifers Fractured or is Groundwater in Aquifers Well-mixed?

The seasonal differences in the isotopic composition of precipitation can be effectively used to determine when precipitation recharges groundwater systems and whether fractures control the hydrodynamics of groundwater flow in aquifers. In temperate regions, most recharge to consolidated and unconsolidated, nonfractured aquifers occurs at spring snowmelt and during spring and fall rains when plant transpiration is at a minimum. Consequently, the isotopic value of groundwater in these types of aquifers usually is a few parts-per-mil depleted relative to the average annual value for precipitation in the aquifer's recharge area.

Furthermore, dispersion in granular, unfractured, porous media mixes recharge waters such that their isotopic values *cluster at about the same place* on the GMWL (Fritz et al., 1987b). In contrast, the isotopic composition of groundwater in fractured aquifers in temperate regions will often *spread* over a range of several $\%o$ $\delta^{18}O$ (and commensurate δD) on the LMWL, providing the means to geochemically identify the presence of an active fracture system recharged in spring and fall after plants become dormant.

Figure 4.7.4 shows the isotopic compositions of groundwater in the Lockport Dolomite near industrial Niagara Falls, New York (Noll, 1989; Siegel, unpublished data), noted for multiple sites of groundwater contamination (e.g., "Love Canal"), and in unconsolidated and sandstone aquifers near a coal fly-ash leachate lagoon near Lake Ontario, New York and located about 60 km northeast of Niagara Falls (Stearns and Wheler, 1991). The isotopic data for the Lockport formation were obtained specifically

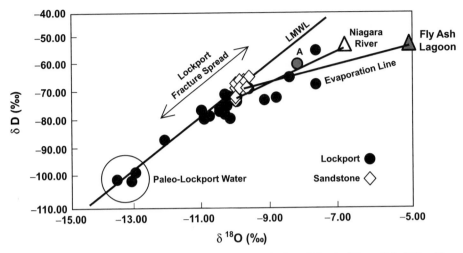

FIGURE 4.7.4   The isotopic composition of groundwater from the Lockport Dolomite (Niagara Falls, NY) and from sandstone and unconsolidated aquifers near southern Lake Ontario. The dolomite is fractured (Yager and Kappel, 1997), whereas the unconsolidated deposits and sandstone are not (Stearns and Wheler, 1991).

to determine if multiple horizontal bedding plane fractures in the formation (e.g., Yager and Kapel, 1997) were hydraulically connected by vertical fractures, whereas that near the fly-ash leachate lagoon was obtained as part of a remedial investigation to determine the extent of possible contamination.

The isotopic composition of the Lockport groundwater on or near the Local Meteoric Water Line spreads over a range of ~5‰ $\delta^{18}O$, indicating that the Lockport is a fractured porous media that was recharged at different times. The average volume weighted value for $\delta^{18}O$ in precipitation at Niagara Falls is about −10‰ (Figure 4.7.2), where much of the Lockport data on the LMWL plot. However, many Lockport samples plot at much smaller (more negative) isotopic values on the LMWL. In contrast, the isotopic composition of groundwater near the fly-ash leachate lagoon plot ~+/−1‰ $\delta^{18}O$ at the volume weighted average value of precipitation, indicating less fracture control over hydrodynamics.

Is Groundwater Thousands of Years Old?

Groundwater that has unusually depleted $\delta^{18}O$ values compared to the modern value for recharge, may indicate that the water is thousands of years old and recharged under colder conditions during the last period of continental glaciation. Figure 4.7.4 also shows a cluster of Lockport groundwater at a $\delta^{18}O$ composition of about −14‰, ~4‰ more depleted ("lighter") than the rest of the Lockport waters. This anomalous depletion suggests that part of this water may consist of either glacial meltwater originally precipitated at much high latitudes or perhaps local recharge water at much colder times in the late Pleistocene era (e.g., Noll, 1989; Perry et al., 1982; Siegel, 1989, 1991). Paleo-recharge water has been widely identified throughout the world by isotopic compositions of groundwater depleted in $\delta^{18}O$ by more than ~3‰ compared to average modern recharge (e.g., Clark and Fritz, 1997; Rozanski et al., 1997; Figure 4.7.2).

Are Aquifers Recharged from Surface Waters?

When standing water bodies evaporate, the rate of diffusion of $H_2^{16}O$ is faster than the diffusion of $^2H_2^{16}O$, $^2H\,H^{16}O$, $^2H_2^{18}O$, or $^2H\,H^{18}O$ from the water surface to vapor phase. Because of the relative differences in diffusion rates among the different isotopic water types, local evaporation causes the

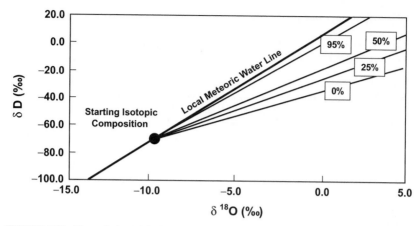

FIGURE 4.7.5   Theoretical trends in the isotopic composition of water evaporating under different conditions of humidity (modified from Gonfiantini, 1986).

residual water reservoir to become relatively more enriched in the heavier isotope of oxygen, ^{18}O, than the heavier isotope of hydrogen, deuterium. Consequently, the isotopic composition of evaporating water will plot at an acute angle *below* the LMWL, intersecting it at the starting isotopic composition.

The lower the relative humidity of the atmosphere, the greater the acute angle the data will describe from the LMWL. Figure 4.7.5 shows selected theoretical trends for evaporating water under different conditions of relative humidity in still air.

When relative humidity is low, evaporation is maximized and the slope of the isotopic evaporative trend of the residual water pool is much smaller than that of the LMWL for precipitation. When relative humidity is high, evaporation is minimized and the slope of the evaporative trend is close to that of the LMWL. Note how all the evaporation lines intersect the LMWL at the value of the surface water before evaporation began.

Referring again to Figure 4.7.4, note how some of the Lockport Groundwater trend on a line below the LMWL that intercepts the isotopic value for the Niagara River. The Niagara River is affected by evaporation and, consequently, its isotopic composition plots below the LMWL. Those Lockport waters plotting between the average value of recharge and the river water are mixtures of river water and native groundwater. In this case, the river water was used in drilling the wells and the isotopic data showed that the wells were incompletely developed before sampling (Yager and Kappel, 1997). How much drilling water is in each sample on the trend can be determined from the mixing Equation 4.7.9, modifed for the apppropriate end members:

$$n_{river\ water} = (\delta_{mixture} - \delta_{groundwater})/(\delta_{river\ water} - \delta_{groundwater}) \qquad (4.7.11)$$

Selecting the grey-colored sample identified by "A" on the mixing line shown on (Figure 4.7.4):

$$n_{river\ water} = (\delta_{mixture} - \delta_{groundwater})/(\delta_{river\ water} - \delta_{groundwater})$$
$$n_{river\ water} = [(-8.7 - (-10.5))]/[-6.8 - (-10.5)] = 1.8/3.7 = 0.49 = 49\%$$

Figure 4.7.4 also shows a white triangle data point marking the isotopic composition of water in a lagoon of leachate from a fly-ash landfill (Stearns and Wheler, 1991). Because of elevated dissolved iron concentrations, a regulatory agency hypothesized that this leachate was contaminating the underlying groundwater system of sandstone bedrock (Stearns and Wheler, 1991). The isotopic composition of the lagoon water shows substantial evaporation because it plots at an acute angle below the LMWL. However, the isotopic composition of virtually none of the groundwater samples (the white diamonds) sampled near the lagoon fall on the evaporative mixing line, unequivocally showing no significant contamination from the lagoon. The elevated iron concentrations in the groundwater were probably caused, rather, by the natural dissolution of iron hydroxides in the aquifer matrix (Stearns and Wheler, 1991).

The proportions of different water sources contributing to storm discharge, precipitation, and groundwater discharge from shallow and deep sources can also be determined from isotopic mixing models (e.g., Bazemore, Eshleman, and Hollenbeck, 1994; Buttle, 1994; Hinton, Schiff, and English, 1994; McDonnell, Stewart, and Owens, 1991), as can be the residence of water in lakes (e.g., Krabbenhoft et al., 1990; Michell and Kraemer, 1995) and wetland-groundwater interaction (e.g., Huddart, Longstaffe, and Crowe, 1999; Hunt, Krabbenhoft, and Anderson, 1996; Kehew et al., 1998).

Is Groundwater Anoxic or Has It Passed Through an Anoxic Environment?

Bacterially mediated biochemical reactions that form: (1) methane (CH_4) from organic matter or dissolved carbon dioxide, and (2) hydrogen sulfide (H_2S) from dissolved sulfate will fractionate only hydrogen, rather than oxygen, from pore water or other hydrogen sources. The gas that leaves the system consequently is highly enriched in ^1H. Conversely, the residual water reservoir becomes enriched in deuterium (e.g., Baedecker and Back, 1979; Clark and Fritz, 1997; Fritz et al., 1987a; Hackley et al., 1996; Siegel et al., 1990; Sugimoto and Wada, 1995). The water does not similarly become enriched in ^{18}O because oxygen is not involved in the isotopic exchange.

Groundwater that has been in a methanogenic or sulfide-reducing environment will thus isotopically plot on a vertical trend *upward* from the LMWL, intersecting it at the isotopic value of the groundwater before methanogenesis or sulfate reduction occurred. So, the isotopic value for groundwater *itself* can be used to determine whether groundwater has passed or come into contact with highly reducing conditions and to identify groundwater that has mixed with water, such as landfill leachate, associated with methanogenesis or sulfate reduction (Siegel et al., 1990).

Figure 4.7.6 shows the stable isotopic data of groundwater from a set of monitoring wells installed proximate to the world's largest active methanogenic landfill, the Fresh Kills Landfill (Staten Island, New York, International Technology Corporation, 1993). This landfill was originally built on a estuarine mud flat. Three sets of data are shown; shallow (~5 m) groundwater in quaternary and glacial

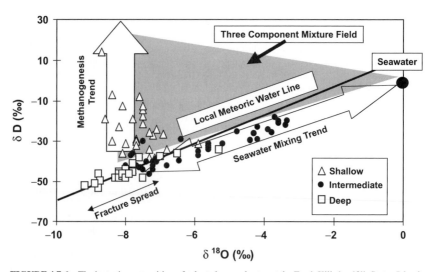

FIGURE 4.7.6   The isotopic composition of selected groundwater at the Fresh Kills landfill, Staten Island, New York (from International Technology Corporation, 1993).

deposits, groundwater in partially lithified sands (~20 m) under the Quaternary marine and glacial deposits, and finally, deep groundwater (~40 m) in underlying sedimentary and crystalline bedrock.

The isotopic composition of the shallow groundwater is clearly enriched in deuterium, trending upward from the LMWL and reflecting the combination of sulfate reduction and methanogenesis notable at the landfill. Were there no dissolved solute data or if dissolved solute concentrations became diluted below regulatory action limits, the stable isotopes would still document that the shallow groundwater had passed through a contaminated setting.

The proportions of leachate in leachate-groundwater mixtures can be determined from Equation 4.7.9 if a fixed value for the isotopic composition of the leachate is assumed. This assumption, however, needs to be used with great care because leachates can vary substantively in their chemical and isotopic compositions depending on the types of refuse, the rates of biochemical reactions and the age of the refuse (Rees, 1980). Note also that the isotopic spread of $\delta^{18}O$ in water from deep bedrock wells spans ~3‰, suggesting, as in the Lockport Dolomite case, that the bedrock is probably fractured. None of the waters from deep wells plot above the LMWL, indicating deep groundwater is not contaminated by leachate.

Many of the samples from intermediate depths and some from shallow depths plot on a line connecting the isotopic compositions of uncontaminated groundwater on the LMWL and sea water. These samples identify where seawater has intruded into the natural groundwater system. Finally, a triangle is drawn connecting the three isotopic end members in the Fresh Kills geochemical system: leachate, native groundwater, and sea water. Waters with isotopic content that falls within this triangle consist of mixtures of all three water sources, and the relative proportions can be determined by three component end-member mixing approaches (e.g., Christophersen and Hooper, 1992).

Can Isotopes of Water Document Water-rock Interactions?

Isotopic exchange between groundwater and minerals in the aquifer matrix is a minor or negligible process associated with most applied groundwater problems concerned with potable aquifers at near surface or moderate temperatures. The isotopic exchange between water and minerals under high temperatures (geothermal conditions) enriches the $\delta^{18}O$ in the water, while depleting the $\delta^{18}O$ of minerals in the aquifer. Therefore, mixtures of meteoric waters and geothermal waters plot *parallel* to the $\delta^{18}O$ axis on a δD versus $\delta^{18}O$ diagram and intersect the LMWL at the average value for meteoric water recharge (Figure 4.7.7).

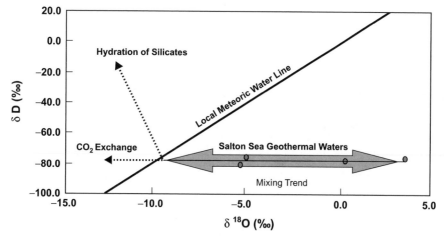

FIGURE 4.7.7   Isotopic mixing of geothermal waters and shallow groundwaters and theoretical trends for hydration of silicate minerals and exchange of oxygen isotopes with carbon dioxide. Example data for hydrothermal mixing is from the Salton Sea geothermal system, California (Craig, 1966).

The opposite trend, depletion in the $\delta^{18}O$ of groundwater (Figure 4.7.7), can occur at low temperatures when oxygen in the water isotopically exchanges with carbon dioxide or with hydrated secondary minerals derived from silicate minerals. Figure 4.7.7 shows the isotopic directions of such trends as dotted lines beginning at an assumed value for average recharge water on the LMWL. However, these isotopic fractionations proceed very slowly and measureable isotopic differences take millions of years (Clark and Fritz, 1997).

Can Isotopes of Water Be Used to Determine the Sources of Salinity in Groundwaters?

Saline groundwater is common in most continents at depth, and in many places within 30 meters of the land surface (Feth, 1965). There are three major sources of salinity in groundwater: (1) seawater, (2) brines (water with dissolved solids composition >100,000 mg/L) in crystalline rocks or sedimentary basins, and (3) dissolving natural halite deposits or rock salt used as road deicer. The isotopic composition of water associated with these three sources of salinity is markedly different and can be used to fingerprint the salinity source with ionic solute ratios as necessary (e.g., Mazor, 1992; Siegel et al., 1990).

Figure 4.7.8 shows additional isotopic data of saline groundwater from intermediate-depth wells located proximate to estuarine distributary channels at the Fresh Kills landfill. These samples were collected where seawater has intruded into the intermediate-depth formations under the landfill, and they have an isotopic composition that plots on the linear mixing trend connecting the isotopic value for seawater with the isotopic value of groundwater unaffected by seawater or landfill leachate. Once again, the proportions of seawater and native groundwater in any sample can be determined from Equation 4.7.9 using seawater and native groundwater isotopic end members.

The isotopic composition of a brine from the Salina Formation near Syracuse, New York (Siegel, unpublished data) is also plotted on Figure 4.7.8. The Salina formation contains extensive bedded salt deposits that were once hydraulically mined as a chlorine source. The mining induced fractures and

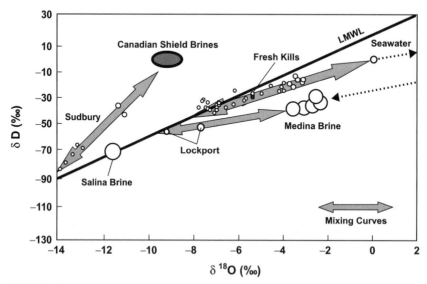

FIGURE 4.7.8 Isotopic content of some New York state saline waters (smallest circles, TDS ~1000 to 10,000 mg/L; small circles, TDS ~10,000–70,000 mg/L and large circles, TDS >100,000 mg/L. Dashed line shows the beginning and the end of a curved evaporative trajectory of the isotopic composition of a brine generated by the evaporation of sea water).

subsidence in the overlying bedrock and allows modern recharge to reach the formation and dissolve salt (Rubin et al., 1991). The isotopic value of the Salina brine sample plots directly on the GMWL at the isotopic values for modern recharge. The isotopic composition of meteoric recharge water *does not change when halite is dissolved*, either naturally or when deposited as road salt because there is no oxygen or hydrogen in halite to be involved in isotopic exchange. Similarly, the isotopic composition of landfill leachate or other contaminated water will not change by the addition of chloride.

In contrast to the Salina Brine, sedimentary basin brines produced from oil and gas fields that have little bedded halite have isotopic compositions enriched in both ^{18}O and D and plot below the LMWL. The location of these brines below the LMWL largely reflects the evaporative concentration of dissolved salts when the brines initially formed. When sea water evaporates, the stable isotopic composition of the water first moves along a theoretical evaporation trajectory below the LMWL, but then reverses in direction back toward the line. This reversal is caused when water molecules form sheaths around the concentrated solutes, hydrated salts precipitate, and diagenetic water-mineral reactions occur after the brine is buried (Clark and Fritz, 1998; Knauth and Beeunas 1986). The dotted line on Figure 4.7.8 shows the general trend in the isotopic evolution of such brines from evaporating sea water, resulting in Medina brines formed in the Appalachian Basin of eastern United States (Siegel and Szustokowski, 1990). These brines now mix with modern recharge water in the deepest part of the Lockport Dolomite. Two examples of these mixtures are also shown on Figure 4.7.8 (Noll, 1989; Siegel, unpublished data). The mixtures plot on a linear trend connecting the recharge water on the LMWL to the position of the brines.

Finally, brines found in crystalline rocks in the continental shields commonly have an isotopic composition that plots above the meteoric water line. The origin of these brines is unknown, but may be related to extensive water-rock interactions over geologic time. Figure 4.7.8 shows the isotopic field where Canadian Shield brines in crystalline rocks typically plot, along with the mixing linear trend between the brines and meteoric water defined by groundwaters and minewaters near Sudbury, Ontario (Frape and Fritz, 1982).

How Can Tritium Be Used to Date Groundwater in Applied Hydrogeological Studies?

Two major questions often raised in contaminated groundwater problems are: (1) When did the contamination occur? and (2) How old is the water? These questions are not mutually exclusive because both relate to the residence time of water along flow paths. The age of groundwater is usually calculated from the seepage velocity, obtained from assuming a value for interconnected porosity and measured values of the hydraulic gradient and hydraulic conductivity along chosen flow paths:

$$v = KI/n_e \qquad (4.7.12)$$

where v is the seepage velocity (L/T), K is the hydraulic conductivity (L/T), I is the hydraulic gradient (L/L) and n_e is the effective porosity (L^3/L^3).

Being able to directly determine the age of groundwater from tritium activity is an important way to independently check calculations of seepage velocities made from hydraulic measurements alone, which can have substantive uncertainty (e.g., Rovey and Cherkauer, 1995). Often, many groundwater contamination problems can be solved by simply knowing if the contamination could have historically reached the receptor (well or stream) within the time frame that the contaminant source existed.

When thermonuclear weapons were tested atmospherically from 1952 to 1963, mostly in the northern hemispheric, the tritium activity of precipitation increased to as high as ~8000 TU (maximum at about 45°N, decreasing to ~100 TU near the equator) compared to background levels of <5 TU. Figure 4.7.9 shows the measured tritium activity in precipitation at Ottawa, Canada. Note how tritium activity in 1963 was in the thousands of TU.

Along groundwater flow paths, "peaks" of tritium activity should occur marking the maximum tritiated recharge centered about 1963. Water older than ~1952 should have tritium activity less than ~5 TU, and younger water should have values less than the maximum value found, the activity of which depends on the activity of the recharge waters during from 1962–1972. The approximate age of

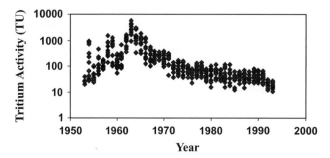

FIGURE 4.7.9   Tritium activity near Ottawa, Canada (data from www.iaea.org).

the water with the maximum tritium activity is the present year minus 1963. Given this approximate age, the groundwater seepage velocity can be directly determined by dividing the distance along the flow path from the recharge area by the number of years since the 1963 recharge +/− ~5 years. Once the seepage velocity is determined from the tritium values and flowpath length, Darcy's Law also can be used to calculate the average hydraulic conductivity along the flow path considered, if the hydraulic gradient is known and assuming an effective porosity (Equation 4.7.13).

$$K = vn_e/I \tag{4.7.13}$$

There are many approaches that have been attempted to more precisely determine groundwater age from tritium, including directly using the radioactive decay equation coupled with multiple tritium input functions (reflecting the different thermonuclear test signals) and time-series analysis (Clark and Fritz, 1997). However, in most applied groundwater problems, there is insufficient three-dimensional subsurface hydraulic and isotopic data to merit such a sophisticated analysis and a more qualitative approach is more appropriate. Futhermore, bomb tritium activity is decreasing by radiogenic decay to helium-3 since the mid-1960s to the point that tritium in samples usually has to be chemically concentrated in the laboratory before precise measurements can be made. Nonetheless, Clark and Fritz (1997) present useful guidelines to qualitatively use tritium activity (see Figure 4.7.10).

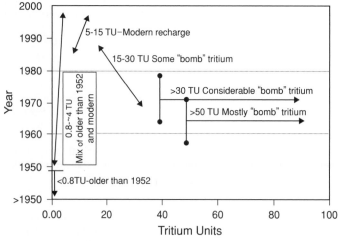

FIGURE 4.7.10   A qualitative use of tritium activity to radiogenically date groundwater (modified from Clark and Fritz, 1998).

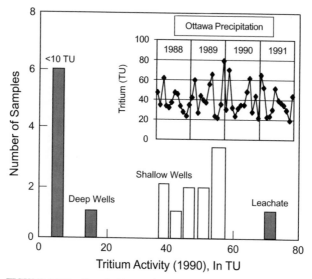

FIGURE 4.7.11   Use of tritium to determine the age of groundwater near a fly ash landfill and leachate lagoon, western New York (from Stearns and Wheler, 1991).

An example showing how qualitative measurements of tritium activity can be effectively used in applied groundwater problems is presented in Figure 4.7.11, a bar graph showing the tritium activity in: fly ash landfill leachate, water from unconsolidated surficial materials, and water from deep bedrock at the fly ash facility previously discussed (see figure and discussion related to Figure 4.7.4). Inserted in Figure 4.7.11 is the monthly tritium activity in precipitation from 1988–1991 in Ottawa (Canada), bracketing 1990, the time period when the wells were sampled. (The IAEA Ottawa Station was the closest one to the landfill site in New York.)

The 1990 tritium activity in the leachate exceeded the tritium activity in most precipitation, suggesting that some "bomb" tritium may have mixed with modern water. However, the tritium activity in shallow wells was about half that in the leachate and similar in range to that in precipitation at the time. This similarity between the shallow groundwater and precipitation tritium activity suggested the water in the shallow ground aquifers consisted mostly of modern recharge (Stearns and Wheler, 1991). Except for one sample, the tritium activity in water from the deep wells was below the detection method (10 TU), indicating that water at this depth is mostly over 40 years old. Because the landfill was only about 20 years old, leachate from it could not have plausibly contaminated the deeper aquifers.

Tritium activity also can be very useful to test whether presumed aquitards are, in fact, confining (e.g., Desaulniers, Cherry, and Fritz, 1981; Ruland, Cherry, and Feenstra, 1991). Unfortunately, using tritium activity to age-date groundwater will be become increasingly difficult as tritium activity continues to halve every 12.4 years. However, it is possible in at least confined aquifers, to measure the daughter product of the radiogenic decay, ^{3}He, so long as it can be assumed that the helium (as a gas) is not lost by volatilization (Clark and Fritz, 1997).

REFERENCES

Balabane, M., Galimov, E., Hermann, M., and Letolle, R., "Hydrogen and Carbon Isotope Fractionation During Experimental Production of Bacterial Methane," *Org. Geochemic*, 2: 115–119, 1987.

Baedecker, M. J., and Back, W., "Hydrogeological Processes and Chemical Reactions at a Landfill," *Ground Water*, 17: 429–437, 1979.

Bazemore, D. E., Eshleman, K. N., and Hollenbeck, "The Role of Soil Water in Stormflow Generation in a Forested Headwater Catchment: Synthesis of Natural Traver and Hydrometric Evidence," *Journal of Hydrology*, 162: 47–75, 1994.

Buttle, J. M., "Isotope Hydrograph Separations and Rapid Delivery of Pre-event Water from Drainage Basins," *Progress in Physical Geography*, 18: 16–41, 1994.

Burke, R. A. Jr., "Possible Influence of Hydrogen Concentration on Microbial Methane Stable Hydrogen Isotopic Composition," *Chemosphere*, 26: 55–67, 1993.

Cherkauer, D. S., Rovey, C. W., "Assessing Flow Systems in Carbonate Aquifers Using Scale Effects in Hydraulic Conductivity," *Ground Water*, 24: 224–253, 1995.

Clark, I., and Fritz, P., *Environmental Isotopes in Hydrogeology*, Lewis, CRC Press, 1997.

Craig, H., "Isotopic Variations in Meteoric Waters," *Science*, 133: 1702–1703, 1961.

Craig, H., "Isotopic Composition and Origin of the Red Sea and Salton Sea Brines," *Science*, 154: 1544–1547, 1966.

Christophersen, N., Hooper, R. P., "Multivariate Analysis of Stream Water Chemical Data: The Use of Principle Components Analysis for the End-member Mixing Problem," *Water Resources Research*, 28: 99–107, 1992.

Daniels, L., Fulton, G., Spencer, R. W., and Orme-Johnson, "Origin of Hydrogen in Methane Produced by Methanovacterium Thermoautotrophicum," *J. of Bacteriology*, 145: 694–698, 1980.

Dansgaard, W., "Stable Isotopes in Precipitation," *Tellus*, 16: 436–468, 1964.

Desaulniers, D. E., Cherry, J. A., and Fritz, P., "Origin, Age and Movement of Pore Water in Argillaceous Quaternary Deposits at Four Sites in Southwestern Ontario," *Journal of Hydrology*, 50: 231–257, 1981.

Feth, J. H., "Preliminary Map of the Conterminous United States Showing Depth of and Quality of Shallowest Ground Water Containing More that 1,000 Parts of Million Dissolved Solids," *U.S. Geological Survey Investigations Atlas HA-199*, 1965.

Frape, S. K., and Fritz, P., "The Chemistry and Isotopic Composition of Saline Groundwaters from the Sudbury Basin," *Canadian Journal of Earth Sciences*, 19: 645–661, 1982.

Fritz, P. G., Matthes, G., and Brown, R. M., "Deuterium and Oxygen-18 as Indicators of Leachwater Movement from a Sanitary Landfill," in *Interpretation of Environmental Isotope and Hydrochemical Data in Groundwater Hydrology*, International Atomic Energy Agency, Vienna, 131–142, 1987a.

Fritz, P., Drimmie, R. J., Frape, S. K., and O'Shea, O., "The Isotopic Composition of Precipitation and Groundwater in Canada," in *Isotope Techniques in Water Resources Development*, IAEA Symposium 299, Vienna, 539–550, 1987b.

Gonfiantini, R., "Environmental Isotopes in Lake Studies," in Fritz, P. and Fontes, J-Ch., eds., *Handbook of Environmental Isotope Geochemistry*, Vol. 2., The Terrestrial Environmental, Elsevier, Amsterdam, The Netherlands, 113–168, 1986.

Hackley, K. C., Liu, C. L., and Coleman, D. D., "Environmental Isotope Characteristics of Landfill Leachates and Gases," *Ground Water*, 34: 827–836, 1996.

Hinton, M. J., Schiff, S. L., and English, M. C., "Examining the Contributions of Glacial Till Water to Storm Runoff Using Two- and Three-Component Hydrograph Separations," *Water Resources Research*, 30: 983–993, 1994.

Huddart, P. A., Longstaffe, F. J., and Crowe, A. S., "Delta D and Delta ^{18}O Evidence for Inputs to Groundwater at a Wetland Coastal Boundary in the Southern Great Lakes Region of Canada," *Journal of Hydrology*, 214: 18–31, 1999.

Hunt, R. J., Krabbenhoft, D. P., and Anderson, M. P., "Groundwater Inflow Measurements in Wetland Systems," *Water Resources Research*, 32: 495–507, 1996.

International Technology Corporation, *Final Hydrogeological Report*, Fresh Kills Leachate Mitigation System Project, Somerset, NJ, 1993.

Kehew, A. E., Passero, R. N., Krishnamurthy, R. V., Lovett, C. K., Betts, M. A., and Dayharsh, B. A., "Hydrogeochemical Interaction Between a Wetland and an Unconfined Glacial Drift Aquifer, Southwestern Michigan," *Ground Water*, 36: 849–856, 1998.

Knauth, L. P., and Beeunas, M. A., "Isotope Geochemistry of Fluid Inclusions in Permian Halite with Implications For the Isotopic History of Ocean Water and the Origin of Saline Formation Waters," *Geochemica Cosmochimica Acta*, 50: 419–433, 1986.

Krabbenhoft, D. P., Bowser, C. J., Anderson, M. P., and Valley, J. W., "Estimating Groundwater Exchange with Lakes I. The Stable Isotope Mass Balance Method," *Water Resources Research*, 26: 2445–2453, 1990.

Mazor, E., *Applied and Chemical Isotopic Groundwater Hydrology*, Halsted Press, New York, 1991.

McDonnell, J. J., Bonell, M., Stewart, M. K., and Pearce, A. J., "Deuterium Variations in Storm Rainfall: Implications for Hydrograph Separation," *Water Resources Research*, 26: 455–458, 1990.

McDonnell, J. J., Stewart, M. K., and Owens, I. F., "Effect of Catchment-scale Subsurface Mixing on Stream Isotopic Response," *Water Resources Research*, 27: 3065–3073, 1991.

Michell, R. L., and Kraemer, T. F., "Use of Isotopic Data to Estimate Water Residence Times of the Finger Lakes, New York," *Journal of Hydrology*, 164: 1–18, 1995.

Noll, R. S., *Geochemistry and Hydrology of Groundwater in the Lockport Dolomite, Near Niagara Falls, New York*, MS Thesis, Syracuse University, 1989.

Perry, E. G. Jr., Gilkenson, R. J., and Grundle, T. J., "H, O, and S. Isotopic Study of the Ground Water in the Cambiran-Ordovician Aquifer System of Northern Illinois," in *Isotope Studies of Hydrologic Processes*, Perry, E. G. Jr., and Montgomery, C. W., eds., Northern Illinois University Press, De Kalb, 35–45, 1982.

Rees, J., "The Fate of Carbon Compounds in the Landfill Disposal of Organic Matter," *Journal Chemical Technology and Biotechnology*, 30: 161–175, 1980.

Rovey, C. W., and Cherkauer, D. S., "Scale Dependency of Hydraulic Conductivity Measurements," *Ground Water*, 33: 769–780, 1995.

Rozanski, K., Araguas-Araguas, L., Gonfiantini, R., "Isotopic Patterns in Modern Global Precipitation," in *Continental Isotope Indicators of Climate*, American Geophysical Union Monograph, 78: 1–36, 1997.

Rubin, P. A., Ayers, J. C., and Grady, K. A., "Solution Mining and Resultant Karst Development in Tully Valley, New York," in *Conference on Hydrology, Ecology, Monitoring, and Management of Groundwater in Karst Terraines*, Quinlan, J. F., and Stanley, A., eds., Nashville, TN, 1991.

Ruland W. W., Cherry, J. A., and Feenstra, S., "The Depth of Fractures and Active Ground-water Flow in a Clayey Till Plain in Southwestern Ontario," *Ground Water*, 29: 405–417, 1991.

Siegel, D. I., "Evidence for Dilution of Deep, Confined, Ground Water by Vertical Recharge of Isotopically Heavy Pleistocene Water," *Geology*, 19: 433–436, 1991.

Siegel, D. I., "The Hydrogeochemistry of the Cambrian-Ordovician Aquifer System, North-Central United States," *U.S. Geol. Survey Prof. Paper*, 1405-D, 1989.

Siegel, D. I., and Szustokowski, R. J., "Regional Appraisal of Brine Chemistry in the Albion Group Sandstones (Silurian) of New York, Pennsylvania and Ohio," *Association of Petroleum Geochemical Explorationists Bulletin*, 6: 66–77, 1990.

Siegel, D. I., Stoner, D., Bynes, T., and Bennett, P., "A Geochemical Process Approach to Identify Inorganic and Organic Ground-water Contamination," *Groundwater Management, No. 2*, National Groundwater Association, Columbus, Ohio, 12891-1301.

Stearns and Wheler Environmental Engineers and Scientists, Hydrogeochemical investigation of the NYSEG Kintigh Solid Waste Disposal Area, September 1991, Cazenova, New York, 1991.

Sugimoto, A., and Wada, E., "Hydrogen Isotopic Composition of Bacterial Methane: CO_2/H_2 Reduction and Acetate Fermentation," *Geochimica Cosmochimica Acta*, 59: 1329–1337, 1995.

Yager, R. M., and Kappel, W. W., Infiltration and Hydraulic Connections from the Niagara River to a Fractured-dolomite Aquifer in Niagara Falls, New York, *Journal of Hydrology*, 206, 84–97, 1997.

CHAPTER 4
HYDROLOGY

SECTION 4.8

FATE AND TRANSPORT IN FRACTURED ROCK

Kent Novakowski

Dr. Novakowski is associate professor in the Department of Earth Sciences at Brock University, St. Catharines, Ontario. He teaches undergraduate and graduate courses in the areas of introductory hydrogeology, groundwater modeling, and the hydrogeology of fractured rock. Dr. Novakowski is presently conducting research on aqueous-phase gasoline and solvent migration in fractured rock and on the fundamental processes governing groundwater flow and solute transport in discrete fractures and fracture networks.

4.8.1 INTRODUCTION

There are many locations throughout the world where the only source of potable water is a bedrock aquifer. These aquifers may take many forms depending on rock type, although in almost all, the migration of groundwater is controlled entirely by the presence of interconnected discontinuities. The term fracture or joint is usually used to refer to these discontinuities. In many cases, these aquifers are overlain by only a thin veneer of overburden, offering little, if any protection, from potential sources of groundwater contamination. For example the sedimentary Karoo aquifer of central South Africa, and the crystalline rock of northern Brazil are particularly susceptible to contamination from human waste in rural areas.

In industrialized countries, fractured rock aquifers may become contaminated through a myriad of possible mechanisms. Recently, the presence of contamination from dense nonaqueous phase liquids (DNAPLs) has been discovered in the bedrock at many industrial and waste sites. Leaking underground storage tanks are also a common source of groundwater contamination in bedrock aquifers. Examples of gasoline-contaminated sites abound along the northeastern seaboard of the United States and in eastern Canada. Because groundwater velocities can be relatively rapid in fractured rock, the migration of nitrates and other agricultural chemicals may be very significant where the soil zones are relatively thin and permeable. The transport of bacteria and viruses can be particularly rapid in heavily populated resort areas where significant topography results in strong hydraulic gradients within the fracture network.

Despite the fact that many bedrock aquifers provide water for communities around the world, it is also true that many of these provide only modest amounts of sometimes very poor quality water. Thus, bedrock has been viewed historically as a relatively poorly-permeable environment, through which only a limited amount of groundwater might pass. In fact, intact blocks of rock such as might be found in a granitic pluton can indeed possess only minimal permeability, and have been proposed as ideal host environments for nuclear and other kinds of waste. The fear remains, however, that should the integrity of the intact rock fail, the transport of nuclear or other waste constituents through open fractures in the surrounding bedrock would lead to rapid and widespread contamination of the subsurface.

In the past 15 years, considerable effort has been spent on gaining a greater understanding of the processes of flow and contaminant migration in fractured rock environments. The objective of this chapter section is to review the findings of some of this work and provide the reader with a basic understanding of the principal mechanisms governing flow and transport. The issues that arise as a result of the inappropriate use of models and interpretive methods intended for porous media will also be discussed.

4.8.2 *THE OCCURRENCE OF FRACTURES*

The occurrence and formation of fractures in various rock environments have been studied by geologists and engineers for many decades (Atkinson, 1987). The process of fracture formation is now well understood and can be classified into three broad categories. The categories are defined on the basis of mode of formation. Figure 4.8.1 illustrates a commonly used schematic diagram that depicts each of the fracture types, namely, Mode I or extension fractures, Mode II or shear fractures, and Mode III or tear fractures. Mode I fractures are the most common form and occur in almost all rock types. Examples include cooling fractures that occur in basalt, sheeting fractures that are common to granitic rock, and bedding-plane fractures that occur in layered sedimentary rock. Mode II and Mode III fractures are more common to bedrock that has undergone deformation through tectonic activity. Thus, shear fractures and shear zones occur frequently in the metamorphic terrains associated with contemporary and paleo-orogenic belts. Examples of these fractures can also be found in deformed stratigraphic rock, such as the Fold and Thrust belt of eastern North America. It is important to note that although Mode I fractures open as a result of an extension mechanism, many of these features also undergo a small degree of shearing, sometimes repeated on many occasions. This offsets the fracture walls, allowing for open space through which groundwater may pass. This opening is referred to as fracture aperture and often given the nomenclature, $2b$.

Figure 4.8.2 shows a cross section of an idealized model of fracture distribution in a monzonitic gneiss. The cross section was constructed on the basis of a very detailed study of the subsurface occurrence of fractures open to groundwater flow and the interconnection between these as determined by hydraulic means (Raven, 1986). The results of this study suggested that horizontally oriented extension fractures of significant areal extent dominate the flow system. The total groundwater flux is limited by the smaller aperture fractures that interconnect the sheeting fractures. Similarly, in the idealized cross section of a horizontally stratified carbonate rock depicted in Figure 4.8.3 the groundwater flow is dominated by extension fractures following bedding planes. In this case, these fractures are of wider areal extent and the vertical exchange of fluid is governed by the occurrence

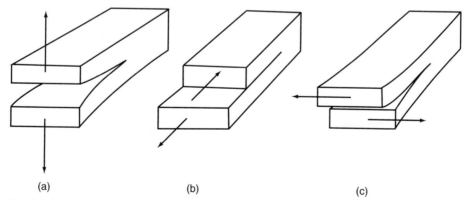

(a)   (b)   (c)

FIGURE 4.8.1 Modes of fracture formation: (a) extension type or Mode I, (b) shearing type or Mode II, and (c) mixed type or Mode III (from Atkinson, 1987, with permission).

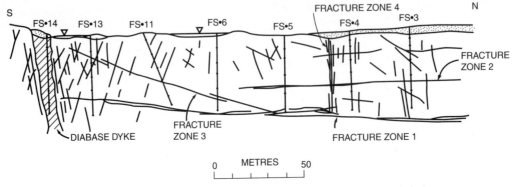

FIGURE 4.8.2   Example of sheeting fractures in a monzonitic gneiss (from Raven, 1986, with permission).

of through-going vertical fractures and vertical hydraulic gradients. The former environment (i.e., crystalline rock) provides for more difficulties in deciphering the groundwater flow and contaminant transport system.

4.8.3   *SITE CHARACTERIZATION*

A considerable amount of information on fracture distributions and orientations can be obtained from indirect sources such as local rock outcrop, quarries, surface geomorphology, and surface and borehole geophysics. This will not, however, circumvent the need for direct subsurface investigation conducted using core drilling. For example, one of the most difficult parameters to obtain is fracture trace length. This is particularly difficult where only subsurface information (i.e., from boreholes) is available. In this case, inter-borehole studies conducted using pumping tests and tracer experiments, combined with numerical simulation of the hypothesized fracture zone are virtually the only means by which to infer fracture length and interconnection.

Site characterization remains an evolving practice in the hydrogeology of fractured rock. Often, inappropriate methods intended for use in porous media are applied to fractured rock environments. Where the investigations are focused on the assessment of groundwater usage or abstraction in these

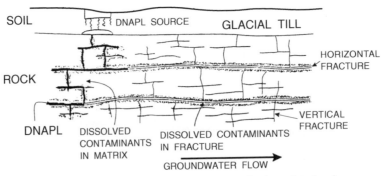

FIGURE 4.8.3   A conceptual model for fracture arrangement in a layered stratigraphy.

environments, the error incurred by using equivalent porous medium (epm) techniques, is minimal. Where solute or contaminant transport is the focus, however, the use of epm methods at the site scale may lead to substantial error. In the following sections, the type and magnitude of the error will be introduced.

4.8.4 FATE AND TRANSPORT MECHANISMS

Groundwater Flux and Velocity

The most fundamental groundwater parameters required by hydrogeologists in making a prediction of contaminant migration, are groundwater flux and velocity. In fractured rock, the flux and velocity of the fluid carried in open fractures is governed by the Navier-Stokes (N-S) equation. When formulated in one dimension, and assuming a relatively incompressible fluid and smooth-walled fractures, an analytical expression for volumetric flux, Q, can be derived from the N-S equation (Witherspoon et al., 1980):

$$\frac{Q}{\Delta H} = \frac{\rho g}{12\mu} \frac{W}{L} (2b)^3 \tag{4.8.1}$$

where ρg is the specific weight of water (9.8×10^3 N/m^3), μ is dynamic viscosity (1.386×10^{-3} N·s/m^2), W is the width of the flow domain, L is the length of the flow domain, and ΔH is the difference in hydraulic head along the length of the domain. Thus, we see immediately that the flow rate of groundwater passing through a fracture of given aperture, $2b$, is proportional to the cube of that aperture.

Equation 4.8.1 can be directly related to the epm term, transmissivity, T, through the following expression (Novakowski, 1988):

$$T = \frac{\rho g (2b)^3}{12\mu} \tag{4.8.2}$$

Thus, the results of slug or pumping tests conducted in fractured aquifers can be expressed in terms of an equivalent single fracture aperture. For example, supposing a slug test was conducted in an open well completed into 6 m of bedrock and intersected by one fracture. The results of the slug test were interpreted using the Cooper, Bredehoeft, and Papadopolus (1967) method, and a T of 4.0×10^{-5} m^2/s was calculated. Using Equation 4.8.2 above, we can convert this estimate of T into an estimate of $2b$, which, for our example, is 410 μm or 0.410 mm. This is generally referred to as an "hydraulic aperture" in that it was determined from a hydraulic test of the domain.

There are two observations to make about this calculation. First, the estimate of fracture aperture when calculated from T in this way, is independent of the length of the well over which the hydraulic test was conducted (i.e., the length of the testing interval is irrelevant because the value of T is dominated by the fracture aperture), and second, fracture apertures don't have to be large to provide considerable permeability.

It is also possible to relate hydraulic conductivity to fracture aperture, through the following expression:

$$K_f = \frac{\rho g (2b)^2}{12\mu} \tag{4.8.3}$$

Hydraulic conductivity, however, must be related to the length of the testing interval for the interpretation of slug or pumping tests, and thus is more restrictive in the expression of results. For example, in the case for a single fracture discussed above, the K_f would be determined from T by dividing 4.0×10^{-5} m^2/s by 6 m. If the testing interval were diminished by 3 m (assuming that the fracture remained in the center of the testing interval), and another slug test was conducted, the same value of T would result, but the value of K_f would increase by a factor of 2. The same relation prevails when the hydraulic test is interpreted directly for hydraulic conductivity, such as with the Hvorslev method. In summary, it is entirely defensible to use an equivalent porous medium parameter to describe flow through a fractured medium without knowing the exact location and property of each fracture, provided the meaning and use of that parameter is clearly understood.

The average velocity of groundwater in a discrete fracture is found by dividing Equation 4.8.1 by the fracture aperture and setting a unit width that yields:

$$v_{avg} = -\frac{\rho g}{12\mu}(2b)^2 \frac{dh}{dx} \qquad (4.8.4)$$

for flow in the x-direction. Thus, groundwater velocity is proportional to the square of the fracture aperture. Using the example fracture discussed above (410 μm) and a modest hydraulic gradient of 0.001, a v_{avg} of 8.5 m/day is calculated using Equation 4.8.4. For a sand aquifer of 10 m thickness, a porosity of 0.30, and the same T as the fracture, the velocity as calculated using Darcy's Law is 0.0012 m/day. Thus, the propensity for contaminants to migrate very rapidly over great distances in fractured rock is very significant.

Many hydrogeologists underestimate groundwater velocity in fractured rock by attempting to use Darcy's Law with what seem like reasonable estimates of porosity. Although intuitively attractive, this method will yield estimates of velocity significantly in error, in some cases by many orders of magnitude. To illustrate this, we will use Darcy's Law to estimate velocity for the example described above. To do so, we would first calculate an hydraulic conductivity from T (recall $T = 4.0 \times 10^{-5}$ m^2/s), resulting in a K_f of 6.6×10^{-6} m^2/s, and assume a very small value of porosity, a value of 0.005, for example. The resulting calculation for velocity yields a v_{avg} of 0.1 m/day, an estimate almost two orders of magnitude less than the actual velocity as determined from the cubic law. Were we to use the actual porosity (fracture aperture over interval length) of 6.8×10^{-5}, we would calculate the correct velocity as given by the cubic law. Unfortunately, it is impractical to measure effective porosity on a regular basis, thus another means must be sought by which to correctly estimate groundwater velocity in a network of fractures. For the case of a single fracture, this is elementary; simply convert to $2b$ from the estimate of T and use Equation 4.8.4.

For the case where more than one fracture intersects the isolated interval, the issue becomes more complicated. It is very rare that the specific number of fractures (and their hydraulic properties) intersecting a given interval are known. To illustrate the problem, consider the example above, but now assume we know that 5 fractures intersect the 6-m interval and contribute evenly to the measured transmissivity of 4.0×10^{-5} m^2/s. Because transmissivity is additive, we can calculate the T of each fracture, which is 8.0×10^{-6} m^2/s, and using Equation 4.8.2, we determine the aperture of each fracture to be 240 μm. Using Equation 4.8.4 to calculate the velocity in each fracture, assuming the same hydraulic gradient as used previously, we obtain v_{avg} equal to 2.9 m/day, or about 1/3 the velocity for the single fracture. This estimate is closer to the v_{avg} calculated using Darcy's Law (the conditions for calculation using Darcy's Law did not change), and in physical terms, the actual porosity increased by almost 300 percent, closer to our arbitrary estimate of 0.005. Clearly, we can continue these calculations until we have enough fractures intersecting the interval to yield a porosity of 0.005 and at that point the estimates of velocity will converge. Fortunately, in nature, it is most common for only a few fractures to dominate the hydraulic conditions over short vertical distances such as the 6-m distance used in our example. Therefore, likely the best approach for estimating groundwater velocity in fractured media is to calculate an equivalent single fracture aperture from the T obtained hydraulically, and use the cubic law (Equation 4.8.4) to estimate v_{avg}, recognizing that some small degree of error will accrue with this approach.

Finally, it should noted that, as anyone who has observed fractures in outcrop can attest, natural fractures are physically complex features, having undulating walls, points of contact between the walls, and areas of larger pore space. Figure 4.8.4 depicts a hypothetical cross section of a natural fracture. Clearly, the use of a conceptual model based on perfectly parallel walls must lead to some error, and indeed, considerable research has been conducted on the use of the cubic law to

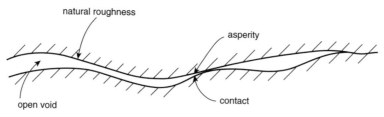

FIGURE 4.8.4   A schematic diagram depicting a natural fracture.

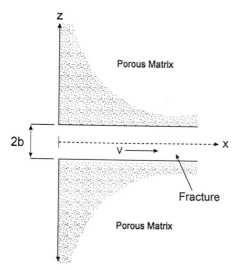

FIGURE 4.8.5   A schematic diagram illustrating the process of matrix diffusion in a parallel-plate fracture (from Lapcevic et al., 1999, with permission).

approximate flow and velocity in natural fractures (Brown, 1995; Oron and Berkowitz, 1998; Witherspoon et al., 1980). Although the results of this research has provided us with corrections and caveats for the use of the cubic law, considering the error inherent in field measurement, use of the cubic law in the form presented above is probably defensible for most practitioners.

Matrix Diffusion

Although groundwater velocity in discrete fractures may be quite rapid in comparison to porous media of equivalent permeability, this is mitigated by processes that act to slow the migration of solutes and contaminants carried in the moving fluid. For many environments, the single most significant process that influences the rate of contaminant migration is matrix diffusion. Except for some porous rocks, such as Cretaceous and younger-aged sandstones, there is very little advective movement of the fluid contained in the porosity of the unfractured rock. Matrix diffusion is the process whereby contaminants are transferred from the fluid migrating in the fracture to the stagnant fluid held in this porosity. The driving force governing matrix diffusion is the thermal-kinetic energy, which repel solutes away from one another. Thus, the concentration gradient between the fluid in the fracture and that in the matrix is the solitary means of solute transfer.

Figure 4.8.5 depicts the process of matrix diffusion in an idealized fracture having parallel walls with fluid traveling at a given velocity. By inspecting this diagram, the factors that control the rate of matrix diffusion are apparent. As the solute front travels forward in the fracture, solute is transferred to the matrix from the front at a rate governed by the porosity of the rock, the velocity of the fluid, and the aperture of the fracture. The governing equation for the solute concentration in the fracture is written as:

$$R_f \frac{\partial c}{\partial t} + v_{avg} \frac{\partial c}{\partial x} - D_x \frac{\partial^2 c}{\partial x^2} - \frac{2\theta_m D'}{2b} \left. \frac{\partial c'}{\partial z} \right|_{z=\frac{2b}{2}} = 0 \qquad 0 \leq x \leq \infty \qquad (4.8.5)$$

where x is the coordinate in the direction of the fracture and z is the coordinate perpendicular to the fracture as depicted in Figure 4.8.5. The first term on the left-hand side (LHS) of the equation describes the change in the storage of concentration with respect to time. The coefficient R_f is a retardation factor that accounts for the exchange of solute between the fluid and the solid phase on the fracture walls. This exchange could be governed by electrochemical sorption or organic-to-organic affinity. The second term on the LHS accounts for advective velocity in the fracture. The third term accounts for hydrodynamic dispersion, and the fourth accounts for the diffusive exchange between the fracture and the unfractured matrix. The coefficient D' is the effective molecular diffusion coefficient and θ_m is the effective porosity of the matrix. The concentration in the matrix is given by c'. In addition to Equation 4.8.5, an equation is required to describe the diffusion process in the matrix. This equation is given by:

$$\frac{\partial c'}{\partial t} - \frac{D'}{R_m}\frac{\partial^2 c'}{\partial z} = 0 \qquad \frac{2b}{2} \le z \le \infty \qquad (4.8.6)$$

where R_m is the retardation factor for the matrix. The relation between R_f and R_m will be discussed more thoroughly in the next section. Equations 4.8.5 and 4.8.6 are coupled by a continuity condition between c and c' expressed at the fracture wall. An analytical solution to this boundary value problem with a step function input in concentration was first obtained by Tang et al. (1981). A numerical solution has recently been developed (Therrien and Sudicky, 1996), with which solute transport in a fully three-dimensional fracture network can be simulated.

To illustrate the influence of matrix diffusion, the results of several simulations conducted using the Tang solution are shown in Figures 4.8.6. and 4.8.7. In Figure 4.8.6, the concentration of a solute in a fracture located 15 m from a source of constant concentration is plotted for two groundwater velocities and three values of matrix porosity. The fracture has an aperture of 500 μm, and a dispersivity of 5.0 m was used. Of immediate note is the profound influence of matrix porosity at slower groundwater velocity. This occurs because of the increase in the exposure time of the moving solute front to the fracture walls. The matrix porosity is the limiting parameter controlling the rate of solute transport across the fracture walls, particularly at slower velocity. As the velocity increases, there is less time available for the solute to be transferred.

Figure 4.8.7 illustrates the position of a solute front traveling in two fractures of different aperture under approximately the same hydraulic gradient (0.002). The position of the front is defined by a relative concentration of 0.05. Clearly, groundwater velocity and fracture aperture also have a significant influence on the rate of solute migration in a discrete fracture. In the case of the smaller-aperture fracture, the front moves rapidly over the first approximately 100 days and then slows to a constant advance, which is approximately 0.02 m/day, a mere fraction of the groundwater velocity in the fracture. The migration of the front in the larger-aperture fracture also occurs most quickly in the initial days, and settles to a constant rate at a later time. The constant rate in this case, however, is considerably

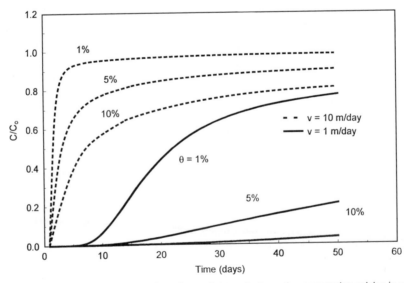

FIGURE 4.8.6   The effect of porosity and groundwater velocity on the concentration arriving in a single 500-μm fracture at a location 15 m from the source (from Lapcevic et al., 1999, with permission).

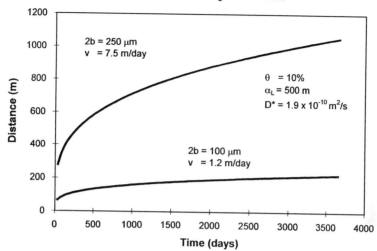

FIGURE 4.8.7   The location of a solute front (as defined by c/c_0 of 0.05) for two fracture examples.

larger than for the smaller fracture (0.1 m/day approximately), although, again, much less than the groundwater velocity in the fracture. From this comparison, it is evident that there is a combination of fracture aperture and groundwater velocity at which the front will achieve a steady position. The aperture and velocity are likely to be only marginally less than the smaller example used here.

Although the danger of using a porous medium approach to estimate groundwater velocity has been shown, there are also temptations to use solute transport models developed for porous media, to simulate transport in fractured rock. To illustrate the kind of error that can accrue by doing this, Figure 4.8.8 shows a comparison between the arrival of solute in a 500 μm fracture at a distance 500 m from a constant source as predicted by the Tang et al. (1981) solution and the traditional Ogata-Banks solution for porous media. The Tang solution was executed first and then the Ogata- Banks solution fit to the resulting curve. For this example, a very conservative condition was used, that of a large-aperture fracture, small dispersivity (5.0 m), and a small matrix porosity (1 percent). The groundwater velocity in the fracture was 10 m/day. To fit to the simulated data with the Ogata-Banks solution, a dispersivity of almost 500 m and a groundwater velocity of 27 m/day, were required. The large dispersivity is required to account for the spreading of the front, which results from the matrix diffusion process. The high groundwater velocity was required to account for the retardation of the velocity of the front. For larger values of matrix porosity and smaller fracture aperture, this disparity is much greater.

Therefore, because we cannot use models intended for porous media to simulate transport in fractured rock and the use of numerical discrete fracture models is not yet widespread, we have a dilemma. Probably the best approach is to simulate transport using a simple analytical model, such as Tang et al. (1981), for a discrete pathway defined through site characterization. Alternatively, dual continuum models (Huyakorn et al., 1983; Rowe and Booker, 1990; and Sudicky, 1990), which approximate the fracture system and matrix system as two separate domains, can be employed. These models have limitations with respect to fracture geometry, although the correct transport processes are incorporated.

Sorption and Exchange

For some ionic compounds and organic contaminants, the solutes transported in the fluid may exchange with solutes sorbed on the fracture walls and in the matrix. This process can be specifically simulated

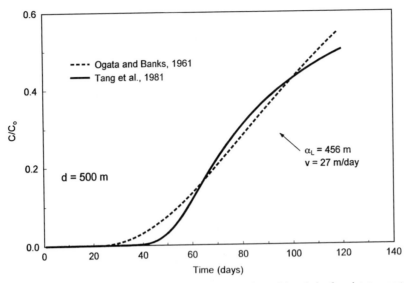

FIGURE 4.8.8   A comparison between the Ogata-Banks solution and the solution for solute transport in a discrete fracture at a location 500 m from a constant source. The fracture is 500 μm in aperture width and groundwater velocity in the fracture was 10 m/day. The porosity of the matrix was 1 percent.

using the boundary value problem expressed in Equation 4.8.5 and 4.8.6, where the effect of the sorption and exchange is accounted for using the retardation factors, R_f and R_m. In the following, we will look more closely at the sorption processes for organic compounds, although the principles are approximately the same for ionic species.

For sorption of organic contaminants to occur, natural organic carbon must be present in the rock (although some electrochemical exchange will occur with charged minerals in the rock). Obviously, igneous and high-grade metamorphic rock do not contain organic carbon, and the primary mode of sorption for these rocks is through ionic interaction. Some sedimentary rocks, however, have abundant organic carbon. For example, some shales may have as much as 40 percent organic carbon (Gehman, 1962). Carbonate rocks, which are more commonly used as aquifers in North America, may have organic carbon content ranging on average from 0.2 to 0.3 percent (Gehman, 1962; Hunt, 1962).

To explore the potential retardation imparted by organic carbon in the rock, a hypothetical value of the retardation factor can be calculated using the percent organic carbon and published estimates of the octanol-water partitioning coefficient, K_{ow}, for the specific organic compound of interest. Using the following equation (Schwarzenbach and Westall, 1981), an estimate of the distribution coefficient, K_d, is obtained:

$$\log K_d = 0.72 \log K_{ow} + \log f_{oc} + 0.49 \qquad (4.8.7)$$

where f_{oc} is the weight fraction of organic carbon. To calculate the retardation factor, R_m, for the rock matrix, the standard expression for the retardation equation is used:

$$R_m = 1 + \frac{\rho_b}{\theta_m} K_d \qquad (4.8.8)$$

where ρ_b is the bulk density of the rock. The distribution coefficient, K_d, is usually expressed in mL/g or cm^3/g. To calculate the equivalent retardation factor for the fracture walls, R_f, the internal specific surface area must be defined and a new retardation equation used. The retardation equation is given by:

$$R_f = 1 + \frac{2K_f}{2b} \qquad (4.8.9)$$

where $2b$ is the aperture of the fracture and the K_f is the distribution coefficient for the fracture, which is related to K_d by the internal specific surface area of the porous medium, γ:

$$K_d = \gamma K_f \qquad (4.8.10)$$

The coefficient, γ, can be defined using a simple conceptual model (Bickerton and Novakowski, 1993), as:

$$\gamma = \frac{2\theta}{\alpha \rho_b} \qquad (4.8.11)$$

where α is the geometric factor that defines the arrangement of the pore space. The geometric factor may range from 0.01 to 0.5 (Freeze and Cherry, 1979), and must be determined experimentally.

It should be noted that biotransformation and radioactive decay may also play a role in retarding the migration of a solute front in a fractured rock environment. Although radioactive decay is well-understood, very little is known about the processes of the degradation of organic contaminants in fractured media.

4.8.5 CASE STUDY

During the late 1970s and early 1980s, a small transfer site for poly-chlorinated biphenyls (PCBs) was operated in the town of Smithville, located centrally in the Niagara Peninsula of southern Ontario. In the period from 1985 to 1988, PCB and solvent oils were found to have penetrated 6-m of clay till overburden and invaded the upper horizons of the underlying Lockport dolomite. Unfortunately, the Lockport formation is a Silurian-aged dolomite, approximately 40 m in thickness, which was used to supply the town with drinking water. The nearest municipal water supply well was located less than 500 m down-gradient from the site and had become contaminated with trichloroethylene (TCE). Thus, the municipal well was shut down and a pump-and-treat system installed in the upper 4 m of bedrock to control the migration of the aqueous phase leaving the site boundaries. By the time the pump-and-treat system had begun operation, separate aqueous plumes of TCE, trichlorobenzene (TCB), and PCB had developed in the fracture system pervading the Lockport. The pump-and-treat system has been in continual operation since that time.

In 1995, a new study was initiated to develop a detailed conceptual model for flow and transport in the fracture framework of the Lockport formation. The objective was to obtain a thorough understanding of the pathways and processes governing the transport of the aqueous phase contaminants at this site. To conduct the study, 18 new boreholes were drilled in the site vicinity. The boreholes were hydraulically tested at several different scales, and several were completed with multilevel piezometer systems. In addition, a laboratory study was undertaken to investigate the properties of the rock matrix, including effective porosity, the geometric factor, and the organic carbon content. Novakowski et al. (1999) provides a complete description of this study.

Figure 4.8.9 illustrates the results of the hydraulic testing conducted on a series of boreholes forming a cross section of about 225 m length, located immediately to the south of the contaminated site. The hydraulic testing was conducted using the constant-head injection method (Novakowski, 1988, 68–80), and the results are expressed in terms of transmissivity. The testing was conducted by isolating a test interval using a pair of packers, injecting water into the test interval, and measuring the flow rate into the interval at steady conditions. For the results shown here, a testing interval of 0.5 m was used. The lithology log shown on the left-hand side of Figure 4.8.9, illustrates the contacts between the four members of the Lockport Formation. It is immediately apparent that the degree of permeability

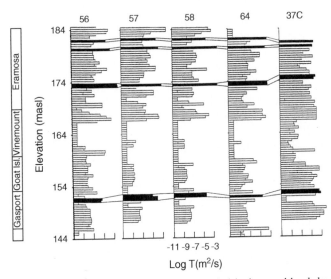

FIGURE 4.8.9   The vertical distribution of transmissivity in several boreholes forming a cross-section down-gradient of the Smithville PCB site. The testing interval for this distribution was 0.5 m. The black shading indicates potential correlation of discrete features between boreholes.

is extremely variable over the thickness of the formation. At least four fractures (or fracture zones) however, can be identified and correlated over the 225-m distance. These discrete features are of significant aperture ranging from 300 to 2000 μm. Groundwater velocity was measured directly in several of the discrete fractures using the point dilution method. Typical velocities, depending on the aperture of the fracture, range from 2 to 40 m/day. These results agree to within an order of magnitude with estimates of velocity obtained using Equation 4.8.4 and estimates of the local hydraulic gradient.

It is important to note that although a considerable amount of effort was required to obtain the information necessary to produce Figure 4.8.9, if the hydraulic testing had been conducted using a larger (and more economical) testing interval, the specific features would not have been identified. However, further to the discussion above, provided some information is obtained on approximately how many fractures might contribute to a given interval, then use of a larger test interval can be justified. Thus, some hydraulic testing at the scale of discrete fractures is necessary as part of any site characterization program.

Figure 4.8.10 shows the results of several simulations conducted using the Tang et al. (1981) solution to estimate what the concentration of TCE might be at Twenty Mile Creek, a potential area of discharge located about 1.5 km down-gradient from the contaminated site. The simulations were conducted using an average aperture size (500 μm) for the upper horizons of the Lockport formation. The simulations were conducted for an input pulse of 500 ppb TCE over five years. Three different velocities with two values of porosity were simulated, to explore the potential uncertainty in any loading estimates conducted for the Creek. Based on the results of the conceptual model development, the results for the lower velocity and higher porosity, are most defensible. The results illustrate, again, the significant sensitivity of the solute transport process to both velocity and matrix diffusion.

Note that the simulations shown in Figure 4.8.10 were conducted in the absence of retardation, which occurs because of solute exchange. Measurements of the organic carbon content of the Lockport Formation were obtained using a well-verified method based on carbon combustion. The organic carbon content for the upper horizons ranged from 0.00 to 0.20 percent with a mean of 0.07 percent. Although the organic carbon content is minimal, we can calculate theoretical values of R_f and R_m for TCE and TCB using the equations given above. A fracture aperture of 500 μm was assumed and a measured geometric factor of 0.1 (Novakowski et al., 1999) was used. Thus, the calculated

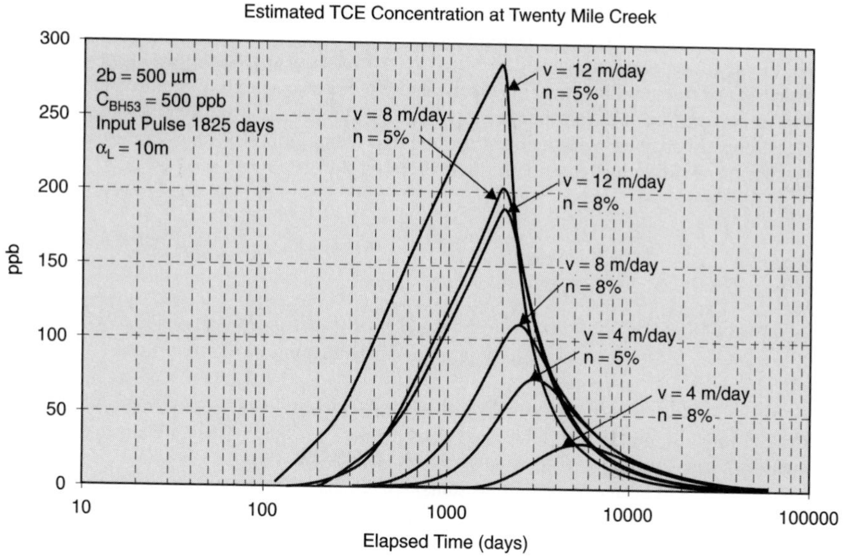

FIGURE 4.8.10 Simulations conducted using the Tang et al. (1981) solution to investigate the potential concentration of TCE in groundwater that might discharge to a creek located down-gradient from the Smithville PCB site.

retardation factor for TCE in the bulk rock is approximately 8 and for TCB, approximately 86. For both the TCE and the TCB, the R_f is about twice the value of the R_m. For fractures of smaller aperture (i.e. less than the 500 μm example used), the value of R_f will increase proportionally. It is recognized (Schwarzenbach and Westall, 1981), that theoretical calculations conducted using Equation 4.8.7 tend to underestimate retardation factors measured directly from plume migration, at least in porous media. Thus, the actual field migration of chlorinated compounds at Smithville may undergo even greater degrees of retardation than suggested by these calculations. Unfortunately, sufficient historical plume data are unavailable to verify this for this site.

4.8.6 CONCLUSIONS

To assess the migration of contaminants in fractured rock, it is necessary to recognize that the primary conduits for groundwater flow and solute transport are the discontinuities prevalent in the rock mass. These discontinuities may be large-scale structural features such as faults, or more local-scale features such as fractures. In any case, it is critical to identify at least some of the primary pathways through the framework of structural elements and fractures. For studies where only flow is important, this need not be conducted on a discrete basis, but rather averaged over the entire rock mass. For contaminant transport and NAPL migration, however, it is necessary to characterize the discrete pathways directly, and to incorporate this into any modeling approach or calculation used to predict groundwater velocity and transport.

Although it is recognized that fracture networks can be very permeable and that groundwater velocity in the discrete features may be very rapid, there are several processes that act to mitigate against the widespread distribution of contamination in fractured rock. Principal among these is the process of matrix diffusion, whereby contaminants migrate from the fluid traveling in the fracture to the immobile fluid in the matrix under a concentration gradient, alone. Retardation in the form of solute exchange between the liquid and solid phases may also play a significant role. When all of the contributing processes combine during contaminant transport, considerable diminishment of the rate of migration and peak concentration results.

Finally, it is essential to recognize that the processes of solute migration are specific to the physical construct of fractured media. This means that the use of conceptual or numerical models intended for use in porous media to interpret transport in fractured media will lead to significant error. This can be demonstrated even with the most rudimentary calculations, such as the use of Darcy's Law, in the prediction of groundwater velocity in discrete fractures.

REFERENCES

Atkinson, B. K., (ed.), *The Fracture Mechanics of Rock*, Academic Press, London, 1987.

Bickerton, G., and Novakowski, K., "Measuring Adsorption in Low Porosity Rock," Unpublished Manuscript, 1993.

Brown, S. R., "Simple Mathematical Model of a Rough Fracture," *J. of Geophys. Res.*, 100(B4): 5941–5952, 1995.

Cooper, H. H., Jr., Bredehoeft, J. D., and Papadopulos, S. S., "Response of a Finite-diameter Well to an Instantaneous Charge of Water," *Water Resour. Res.*, 3(1): 263–269, 1967.

Gehman, H. M., "Organic Matter in Limestone," *Geochimica et Cosmoschimica Acta*, 26: 885–897, 1962.

Hunt, J. M., "Some Observations on Organic Matter in Sediments, Paper Presented at the Oil Scientific Session Twenty-five Years Hungarian Oil," October 8–13, Budapest, 1962.

Huyakorn, P. S., Lester, B. H., and Mercer, J. W., "An Efficient Finite Element Technique for Modeling Transport in Fractured Porous Media, 1. Single Species Transport," *Water Resour. Res.*, 19(3): 841–854, 1983.

Novakowski, K. S., "Comparison of Fracture Aperture Widths Determined from Hydraulic Measurements and Tracer Experiments," *Proceedings of 4th Canadian/American Conference on Hydrogeology*, Hitchon, B., and Bachu, S., eds., Nat. Water Well Assoc., Dublin, Ohio, 1988.

Novakowski, K., Lapcevic, P., Bickerton, G., Voralek, J., Zanini, L., and Talbot, C., "The Development of a Conceptual Model for Contaminant Transport in the Dolostone Underlying Smithville, Ontario." Final Report submitted to the Smithville Phase IV Bedrock Remediation Program, 1999.

Oron, A., and Berkowitz, B., "Flow in Rock Fractures: The Local Cubic Law Assumption Reexamined," *Water Resour. Res.*, 34(11): 2811–2825, 1998.

Raven, K. G., "Hydraulic Characterization of a Small Ground-water Flow system in Fractured Monzononitic Gneiss," *Nat. Hyd. Res. Inst. Scientific Series No. 149*, No. 30, 1986.

Rowe, R. K., and Booker, J. R., "Contaminant Migration Through Fractured Till into an Underlying Aquifer," *Can. Geotech. J.*, 27(4): 484–495, 1990.

Schwarzenbach, R. P., and Westall, J., "Transport of Non-polar Organic Compounds from Surface Water to Groundwater: Laboratory Sorption Studies," *Environ. Sci. Tech.*, 15(11): 1360–1367, 1981.

Sudicky, E. A., "The Laplace Transform Galerkin Technique for Efficient Time-continuous Solution of Solute Transport in Double-porosity Media," *Geoderma*, 46: 209–232, 1990.

Tang, D. H., Frind, E. O., and Sudicky, E. A., "Contaminant Transport in Fractured Porous Media: Analytical Solution for a Single fracture," *Water Resour. Res.*, 17(3): 555–564, 1981.

Therrien, R., and Sudicky, E. A, "Three-dimensional Analysis of Variably-saturated Flow and Solute Transport in Discretely-fractured Porous Media," *J. of Cont. Hydrol.*, 23(1–2): 1–44, 1996.

Witherspoon, P. A., Wang, J. S. Y., Iwai, K., and Gale, J. E., "Validity of Cubic Law for Fluid Flow in a Deformable Rock Fracture," *Water Resour. Res.*, 16(6): 1016–1024, 1980.

CHAPTER 5
PHYSICAL TRANSPORT

SECTION 5.1

AIR POLLUTANTS AND FUGITIVE DUST

Zhen-Gang Ji

*Dr. Ji is a senior environmental engineer with Tetra Tech, Inc.
He specializes in the study of hydrodynamics, water quality,
sediment transport, and toxics in surface water systems.*

5.1.1 AIR

The air quality in a given area is determined by pollutant sources and weather conditions. When pollutant sources are relatively constant, weather conditions control the air quality in the area. The determining factors of weather conditions include wind speed, wind direction, solar radiation, and vertical temperature profile. Gaseous and particulate air contaminants are primarily dispersed into ambient air through physical transport of wind action and atmospheric turbulence. The physical transport in the atmosphere depends on wind adection and atmospheric dispersion in horizontal and vertical directions. Wind is one of the most important vehicles in the distribution, transport, and dispersion of air pollutants. Wind velocity determines the travel time of a particulate to a receptor and, to a larger degree, the dispersion rate of the pollutants. Atmospheric dispersion is the result of atmospheric turbulence and molecular diffusion.

Atmospheric Stability

The stability of the atmosphere largely depends on the vertical temperature profile of ambient air. The atmosphere is unstable (stable) when the temperature profile leads to acceleration (deceleration) of air parcels in the direction of the displacement (i.e., an unstable atmosphere enhances vertical movement, while a stable atmosphere retards vertical movement). Stable atmospheric conditions confine pollutants near the ground level and, therefore, significantly reduce atmospheric dilution. Atmospheric stability greatly influences the vertical motion of air parcels, and is essential in determining the dispersion process in the atmosphere.

Point Source Gaussian Plume Model

The tool most commonly used to describe time-averaged pollutant concentrations in the atmosphere is the Gaussian plume model. The Gaussian plume model can be applied to single-point sources (such as smokestacks), line sources (such as automobile emissions on a highway), and area sources (such as a heavily polluted industrial area, or a large number of point sources).

Figure 5.1.1 is a schematic representation of a Gaussian plume resulting from a smokestack emission. The location of the smokestack is set to be the origin of the coordinates, the downwind

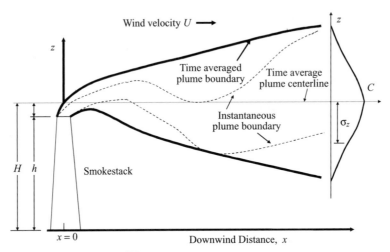

FIGURE 5.1.1 Typical Gaussian plume distribution.

direction is the x direction, and the vertical direction is the z direction. Other symbols in Figure 5.1.1 will be discussed later. Figure 5.1.1 illustrates that even though at any given moment, the instantaneous plume boundary (dash lines) has an irregular shape, the time-averaged concentration envelope (solid lines) has the Gaussian distribution shown in Figure 5.1.1. The plume centerline indicates the peak pollutant concentration averaged over time, and the Gaussian curve on the right indicates the spatial distribution of the pollutant in the vertical direction.

The point source Gaussian equation relates atmospheric pollutant concentrations to source emission rate, wind speed, effective stack height, and atmospheric conditions. Its form is derived from the principle of the conservation of mass. The derivation of the Gaussian equation assumes that: (1) wind speed and wind direction are constant with respect to time and space, (2) the point source has a constant emission rate and the pollutant is conservative, and (3) the ground level is flat and there are no significant mountains or large buildings in the area. Under these assumptions, the Gaussian concentration equation for pollutant concentration on the ground level can be stated as (Seinfeld, 1986):

$$C(x, y) = \frac{Q}{\pi U \sigma_y \sigma_z} \exp\left(\frac{-H^2}{2\sigma_z^2}\right) \exp\left(\frac{-y^2}{2\sigma_y^2}\right) \qquad (5.1.1)$$

where $C(x, y)$ = pollutant concentration at ground level in $\mu g/m^3$ at location (x, y), x = distance in m in the mean wind direction, y = crosswind distance in m from the plume centerline, Q = pollutant emission rate in $\mu g/s$, H = effective stack height in m, U = average wind speed in m/s, σ_y = horizontal dispersion coefficient in m, and σ_z = vertical dispersion coefficient in m. The effective stack height (H) is the actual stack height (h) plus the plume rise (Δh) because of the initial emission velocity of the pollutant, as shown in Figure 5.1.1.

The two dispersion coefficients, σ_y and σ_z, represent the dispersion characteristics of the atmosphere. Small values of σ_y and σ_z mean less dispersion, a narrower Gaussian curve, and higher peak concentration. Large values of σ_y and σ_z represent more dispersion, a wider Gaussian curve, and lower peak concentration. In general, σ_y and σ_z are functions of downstream distance and atmospheric stability. Greater downstream distance and less stable atmospheric stratification lead to larger values of σ_y and σ_z. More information on σ_y and σ_z can be found in Turner (1970).

According to the Gaussian plume equation, the ground-level pollutant concentration has the following features:

1. Ground level concentration $C(x, y)$ is proportional to the emission rate Q. It means that to cut ground concentration to a desired level, the emission rate needs to be reduced proportionally.

2. $C(x, y)$ is inversely proportional to wind speed. As expected, strong wind leads to low pollutant concentration.

3. The actual stack height influences $C(x, y)$ greatly. For a given emission rate and weather conditions, $C(x, y)$ decreases exponentially with H^2.

4. $C(x, y)$ reduces exponentially with the square of crosswind distance y.

In summary, the Gaussian plume model is a simple and useful tool for estimating air pollutant concentration. Most of the current air quality models are based on Gaussian plume models similar to the one presented here.

Atmospheric Deposition

Atmospheric deposition is one of the primary processes of pollutant transport and removal, and has significant impact on surface waters and ecosystems. For example, high lead concentration in the atmosphere caused great concern for human health in the past. The introduction of unleaded gasoline in the United States has dramatically reduced levels of lead in the atmosphere to well below the National Ambient Air Quality Standards (NAAQS).

Atmospheric pollutants are deposited in either dry or wet form. Dry deposition is the settling of particulate matters because of gravity. Wet deposition occurs when particulate matters are removed from the atmosphere by precipitation. Wet deposition accounts for the majority of fine particulate matters removed from the atmosphere.

Once emitted into the atmosphere, pollutants may be deposited locally or may travel long distances before deposition. Many industrial and urban centers in the central United States emit pollutants that are not only deposited locally downwind, but also as far away as the East Coast of the United States.

5.1.2   *FUGITIVE DUST**

Sources of air contaminants may be classified as stationary, mobile, or fugitive. Fugitive dust is largely generated by traffic on unpaved and paved roads, construction activities, agricultural sources, wind erosion, and mining operations. Fugitive dust is the dominant source of PM-10 and PM-2.5 emissions in the United States. PM-10 includes particulate matters with aerodynamic diameters smaller than 10 μm, which is approximately one-seventh the diameter of a human hair. PM-2.5 consists of those particles that are less than 2.5 μm in diameter. PM-10 and PM-2.5 comprise one of the six criteria pollutants for which the Clean Air Act sets standards.

The original standard for particulate matter was a Total Suspended Particulate (TSP) standard, established in 1971. In 1987, EPA replaced the TSP standard with a PM-10 standard to focus on these smaller particles, which cause the greatest human health concern. In 1997, EPA revised the particulate matter standards by adding new standards for PM-2.5 and by adjusting the form of the PM-10 24-hour standard.

Fugitive dust may contain hundreds of different chemical elements from a variety of sources. Fine particles (PM-10 and PM-2.5) may contain substantial quantities of sulfate, ammonium, nitrate, elemental carbon, and condensed organic compounds. Carcinogenic compounds and heavy metals such as arsenic, selenium, cadmium, and zinc are also concentrated in these particles. Larger particles, such as soil particles, fly ash, road aggregate, wood ash, soot, and pollen are primarily composed of minerals, including silicon, aluminum, potassium, iron, calcium, and other alkaline elements.

Generally, fugitive dust is generated by two physical processes: (1) pulverization and abrasion of surface materials by application of mechanical force, such as trucks on unpaved roads, and (2) entrainment of dust particles by atmospheric turbulence, such as erosion of an exposed surface by strong wind.

---

* Some discussions on fugitive dust are based EPA (1995).

EPA (1996) estimated that fugitive dust sources contributed to 92 percent of the PM-10 emissions in the United States in 1995. These fugitive dust sources and their primary source contributions are unpaved roads (28 percent), construction (23 percent), agricultural sources (19 percent), paved roads (15 percent), wind erosion (5 percent), and mining (1 percent).

Unpaved roads are a major source of airborne particulate matter. Vehicle traffic generates dust from resuspension of the roadway surface dust, by pavement and tire wear, by wake disturbance of road shoulders, and by resuspension of dust on vehicle undercarriages.

Construction activities often constitute a major source of fugitive dust in urban areas. The three most common activities are debris removal, site preparation, and construction. Fugitive dust can originate from trucks tracking on construction sites, and from construction equipments that transfer soil and bulk aggregate materials.

Agricultural sources include agricultural fields, agricultural tilling, and agricultural processing. Among these sources, agricultural crops, grain transport, grain elevators, and feed mills contribute significantly to the agricultural fugitive dust source.

Paved roadways are a major source of fugitive dust in urban areas, through the emission of antiskid materials, abraded pavement particles, rubber tire fragments, and dirt tracked from unpaved areas. High levels of PM-10 in late winter and early spring in snow areas are related to residual road sand loadings from wintertime antiskid controls.

The impact of a fugitive dust source on air pollution depends on the quantity and transport potential of the dust particles injected into the atmosphere. In addition to large dust particles that settle out near the source, considerable amounts of fine particles, such as PM-10 and PM-2.5, are emitted and dispersed over much greater distances from the source. These fine particles do not easily settle to the ground as a result of atmospheric turbulence.

The transport of fugitive dust is determined by the initial emission height of the particle, the particle settling velocity, and the atmospheric turbulence. Because of their settling velocities, fine particles from fugitive dust may be transported 1000 kilometers or more from their source. Under the influence of gravity, larger particles of fugitive dust do not remain suspended and tend to settle out, creating localized areas of high particle deposition. According to EPA (1995), for a typical mean wind speed of 16 km/hr, particles larger than about 100 μm are likely to settle out within 6 to 9 meters from the edge of the road or other point of emission. Particles that are 30 to 100 μm in diameter are likely to undergo impeded settling. These particles, depending upon the extent of atmospheric turbulence, are likely to settle within a few hundred feet from the road. Smaller particles, such as PM-10, have much slower gravitational settling velocities and are much more likely to have their settling rate retarded by atmospheric turbulence.

5.1.3 POLLUTANT SOURCES AND EFFECTS

The Clean Air Act sets National Ambient Air Quality Standards (NAAQS) for six criteria pollutants that are not to be exceeded in outdoor air. These six criteria pollutants are: SO_2 (sulfur dioxide), NO_2 (nitrogen dioxide), Pb (Lead), CO (carbon monoxide), O_3 (ozone), and PM (particulate matter).

Sulfur dioxide (SO_2) is a colorless, nonflammable gas with a pungent odor, which is formed primarily by the combustion of fossil fuels and is detectable by the human nose. SO_2 is considered one of the major causes of pollution problems worldwide. Sulfur dioxide can affect the respiratory tract as well as vegetation and other materials. Coal, with amounts of sulfur, must usually be cleaned before or during burning in order to meet clean air standards.

Nitrogen dioxide (NO_2) is a reddish-brown, highly reactive gas present in all urban air. NO_2 is a strong oxidizing agent that reacts in the air to form corrosive nitric acid, as well as toxic organic nitrates. It also plays a major role in the atmospheric reactions that produce ground-level O_3. Individuals with asthma, respiratory disorders, and lung diseases are more sensitive to the effects of NO_2.

Lead (Pb) is a heavy, soft, bluish metal, and occurs in nature in the form of ores. Once Pb is mined, processed, and introduced into the environment, it is a potential problem forever. No technology will destroy it nor render it permanently harmless. Children and fetuses are especially susceptible to low doses of Pb. Low-level Pb poisoning may have nonspecific symptoms such as headaches, abdominal pain, and irritability. High blood-lead levels in children may cause permanent deficiencies in growth and intelligence. In adults, high blood-lead concentrations may cause kidney disorders, infertility, and cancer.

Carbon monoxide (CO) is a colorless, odorless, tasteless, poisonous gas. CO is a by-product formed when carbon in fuels is not completely burned. It occurs naturally in the air as the result of processes such as agricultural fires, oxidation of methane, plant growth and decay, and other natural processes. Man-made sources of CO are responsible for high concentrations often found in urban areas. CO affects the central nervous system by depriving the body of oxygen. In small amounts it can impair alertness, and cause fatigue and headaches. In large amounts it can be fatal to humans. People with heart conditions and respiratory ailments are especially susceptible.

Ozone (O_3) is formed when nitrogen oxides and certain hydrocarbons combine in the presence of sunlight. Produced mainly from automobiles, ozone is one component of photochemical smog. Even though O_3 production is a year-round occurrence, peak O_3 levels typically occur from May to August when the reactions are stimulated by sunlight and temperature. O_3 can affect the tissues in plants and is irritating to the eye, nose, throat, and respiratory systems.

Particulate matters (PM-10 and PM-2.5) are very small or liquid particles in the air. They can aggravate respiratory diseases, reduce sunlight, and damage plants. PM-10 and PM-2.5 can be found in dust, smoke, fumes, mist, spray, and fog. The major sources in the United States are traffic on unpaved and paved roads, construction activities, agricultural fields, wind erosion, and mining operations.

REFERENCES

EPA, "Section 13.2: Fugitive Dust Sources," in *Compilation of Air Pollutant Emission Factors, Vol. 1: Stationary Point and Area Sources*, Fifth Edition, AP-42. United States Environmental Protection Agency, Office of Air Quality Planning and Standards, Research Triangle Park, North Carolina, 1995.

EPA, *National Air Pollutant Emission Trends, 1900–1995*. EPA-454/R-96-007. United States Environmental Protection Agency, Office of Air Quality Planning and Standards, Research Triangle Park, North Carolina, 1996.

Seinfeld, J. H., Atmospheric Chemistry and Physics of Air Pollution, John Wiley & Sons, New York, 1986.

Turner, D. B., *Workbook of Atmospheric Dispersion Estimates*, U.S. Environmental Protection Agency, Washington, D.C., 1970.

CHAPTER 5
PHYSICAL TRANSPORT

SECTION 5.2

MIXING AND TRANSPORT OF POLLUTANTS IN SURFACE WATER

Peter Shanahan
Dr. Shanahan is president of HydroAnalysis, Inc., a hydrology and water quality consulting practice in Acton, Massachusetts, and a lecturer in the Department of Civil and Environmental Engineering at the Massachusetts Institute of Technology, Cambridge. He holds a Ph.D. in environmental engineering from MIT and an M.S. in environmental earth sciences from Stanford University.

Sharron C. Gaudet
Ms. Gaudet is a registered professional engineer in Massachusetts specializing in water resources. She received a B.S. in engineering and applied science from the California Institute of Technology and an M.S. in civil engineering from MIT. She has worked as a specialist in computer modeling of surface and ground water with ENSR Corporation, the Delft Hydraulics Laboratory in the Netherlands, and HydroAnalysis, Inc.

TABLE OF NOMENCLATURE AND UNITS

a = coefficient in Equations 5.2.11 and 5.2.12 $[L^{2/3}/T]$

a = empirical coefficient in Equation 5.2.15b

A = cross-sectional area $[L^2]$

A_{jk} = cross-section area between boxes j and k in Equation 5.2.8 $[L^2]$

b = empirical coefficient in Equation 5.2.15b [dimensionless]

c = speed of sound in water $[L/T]$

C = concentration of mass in the water $[M/L^3]$

C_k = concentration in box k in Equation 5.2.8 $[M/L^3]$

d = depth of channel $[L]$

D = coefficient of proportionality known as the molecular diffusion coefficient $[L^2/T]$

E_L = longitudinal dispersion coefficient $[L^2/T]$

E_t = transverse dispersion coefficient $[L^2/T]$

E_T = Taylor dispersion coefficient $[L^2/T]$

E_{jk} = dispersion coefficient between boxes j and k in Equation 5.2.8 $[L^2/T]$

f = friction factor [dimensionless]

F = rate of mass flux $[M/L^2/T]$

$F(l)$ = neighbor diffusivity in Equation 5.2.10 $[L^2/T]$

g = gravitational acceleration $[L/T^2]$

k = constant in Equation 5.2.13 [dimensionless]

K = parameter in Equation 5.2.26 $[L^2/T]$

l_{jk} = distance separating the centroids of boxes j and k in Equation 5.2.8 $[L]$

l = neighbor separation in Equation 5.2.10 $[L]$

l = cross-sectional mixing length in Equation 5.2.20 [L]

L = distance downstream of discharge point in Equation 5.2.20 [L]

L = length of estuary in Equation 5.2.26 [L]

m = constant in Equation 5.2.26 [dimensionless]

N = Brunt-Väisälä frequency [$1/T$]

Q_{jk} = flow between box k and neighboring box j in Equation 5.2.8 [L^3/T]

$q(l)$ = neighbor concentration function in Equation 5.2.10 [dimensionless]

r = distance from the center of mass [L]

R = radius of curvature [L]

R_H = hydraulic radius [L]

s = slope of water surface [dimensionless]

S = cross-sectionally averaged salinity [M/L^3]

S_{crit} = critical water column stability [$1/L$]

S_{max} = maximum salinity at the ocean entrance to an estuary [M/L^3]

t = time [T]

u, v, w = velocity components in the x, y, z coordinate directions [L/T]

U = cross-sectional average velocity [L/T]

$\overline{U}, \overline{V}$ = depth-averaged velocities [L/T]

U_{max} = maximum tidal velocity [L/T]

u^* = shear velocity [L/T]

V_k = volume of box k in Equation 5.2.8 [L^3]

w = length scale in Equation 5.2.12 [L] in meters

W = wind speed [L/T]

$\overline{W}$ = mean channel width [L]

z_t = depth to the thermocline [L]

x, y, z = spatial coordinates [L]

α = coefficient in Equation 5.2.16 [L]

α_0, α_2 = empirical coefficients in Equation 5.2.25

β = coefficient in Equation 5.2.16 and Equation 5.2.22 [dimensionless]

$\varepsilon_x, \varepsilon_y, \varepsilon_z$ = turbulent diffusion coefficients in the x, y, z coordinate directions [L^2/T]

ε_H = horizontal turbulent diffusion coefficient [L^2/T]

ε_v = vertical diffusion coefficient [L^2/T]

ε_{min} = minimum value of ε_v [L^2/T]

ε_{max} = maximum value of ε_v [L^2/T]

ρ = water density [M/L^3]

ρ_0 = reference water density [M/L^3]

$\overline{\rho}$ = average water density [M/L^3]

τ_0 = bed shear stress [ML/T^2]

5.2.1 INTRODUCTION

Most environmental engineers are familiar with this description of pollutant mixing: "Dilution is the solution to pollution." It is an irreverent but apt description: Mixing indeed plays a very important role in the environment's ability to assimilate pollutants and in protecting humankind from its own wastewater. This section describes the basic principles of mixing and transport in surface water, with an emphasis on developing quantitative estimates of dispersion or diffusion for use in predictive modeling of environmental water quality.

Transport Processes

For the purposes of this section, *transport* may be broadly defined as the ability of moving water to convey mass or other properties (such as heat or momentum) from place to place. *Mixing* is one aspect of transport, and describes the dilution of mass when different parcels of water commingle. The following sections describe the transport processes significant in surface water.

Advection.   The key to transport is movement of the water, and different types of movement create different types of transport. Conceptually simplest is advective motion, the organized motion of water over a large scale. The flow of a river from upstream to downstream is a good example of advection. Advective flow dominates the movement of pollutants in most situations in surface water—pollutants are carried by the flowing water in roughly the same direction and speed as the water itself.

Nonetheless, even a cursory look at a river reveals far more complicated motion than advection alone. The water is often highly turbulent, swirling behind rocks and along river banks, splashing through riffles, and rolling from bank to bank and top to bottom downstream of curves in the river. These deviations from simple advective flow create mixing processes known as diffusion and dispersion.

Molecular Diffusion.   Diffusion in surface water bodies is most appropriately called turbulent diffusion, to differentiate it from its namesake, molecular diffusion. Molecular diffusion is a well-defined physical process in which matter is transported by random molecular motions. It is described by Fick's Law of Diffusion, which states that the rate of mass movement resulting from molecular diffusion is inversely proportional to the gradient of mass concentration (Crank, *The Mathematics of Diffusion*, 2nd Edition, Oxford Univ. Press, 1975):

$$F = -D \frac{\partial C}{\partial x} \qquad (5.2.1)$$

where F is the rate of mass flux (the mass of concentrate crossing a unit area per unit time) $[M/L^2/T]$; C is the concentration of mass in the water $[M/L^3]$; D is a coefficient of proportionality known as the molecular diffusion coefficient $[L^2/T]$; and x is length $[L]$. (Within brackets are shown the units of terms in the equation using the convention of M for mass units, L for length units, and T for time units.) Equation 5.2.1 states, in simple terms, that mass will naturally move from areas of high concentration to areas of low concentration, and that the rate of that movement is greatest when the greatest change in concentration occurs over the shortest distance.

Turbulent Diffusion.   The process of molecular diffusion is generally unimportant to pollutant transport in water bodies. Nonetheless, the movement of mass by random molecular motion in molecular diffusion is an attractive analog for the movement of mass by random water movement in turbulent flow. This analogy is the basis for the concept of turbulent diffusion, the transport of mass by turbulent water motion in accordance with Fick's Law. The extension of Fick's Law to turbulent transport holds that mass is transported by turbulence in the same way as molecular diffusion: from areas of high concentration to areas of low concentration at a rate proportional to the concentration gradient. Though described using the same law as molecular diffusion, turbulent diffusion typically results in vastly greater rates of transport.

The representation of turbulent mass transport using Fick's Law is an imperfect—but highly successful—approximation. That it is an approximation can be understood by considering a microscale view of turbulent water movement. If the turbulent flow field was considered in sufficient detail in time and space, it would be entirely described as advection: There would be no random (turbulent) component and thus no diffusion. However, in fact, it is impossible to consider a real flow field in such detail, and there always remains a random component. It can be shown mathematically (see for example: Daily and Harleman, *Fluid Dynamics*, Addison-Wesley, p. 432, 1966) that turbulent transport arises from the interaction of this random component in the flow velocity with the similar random component in the concentration distribution. Based on the mathematical derivation, turbulent transport is formally defined as the residual transport that remains after averaging the transient field of velocity and concentration in turbulent flow over a short, but finite, period of time. However, therein lies the inherent approximation in applying Fick's Law to turbulent transport: The averaging period is not intrinsically defined, and the coefficient of turbulent diffusion can vary depending on the time period of averaging. We return to this limitation and its implications in the sections below.

To summarize, turbulent diffusion is the mixing that results from random turbulent motion in flowing water; turbulent diffusion formally derives from averaging velocities and concentrations over time; and turbulent diffusion behaves similarly to molecular diffusion and is well described by Fick's Law.

Dispersion.   A subtle distinction differentiates dispersive transport from diffusive transport. Consider once again a river and suppose a quantity of colored dye is dropped as a line across the river. In the central core of the river, where the flow is fastest, dye will move ahead, outpacing the dye placed in slower areas near the streambank. Thus, the dye spreads out—disperses—along the river much as if it had been mixed longitudinally. This apparent mixing is known as dispersion.

Just as turbulent diffusion can be shown mathematically to arise from averaging over time in a turbulent flow field, dispersion can be shown to arise from averaging over space in a spatially nonuniform flow field. Further, just as with turbulent diffusion, dispersion is usually approximated using Fick's Law. The analogy with Fick's Law becomes still more imperfect when applied to dispersive transport, but a Fickian model remains a practical and widely used approximation for all manner of dispersive transport phenomena in a wide range of water bodies.

In summary, dispersion is the apparent mixing that results from spatial variations in advective velocity; dispersion formally derives from spatial averaging of the flow field and concentration distribution; and dispersion is usually represented using Fick's Law of Diffusion.

Mathematical Representation

Diffusion Relations.   The most fundamental mathematical relation of mixing is Fick's Law of Diffusion, which is stated as Equation 5.2.1. Fick's Law is extended through mass balance considerations to the one-dimensional diffusion equation:

$$\frac{\partial C}{\partial t} = D \frac{\partial^2 C}{\partial x^2} \tag{5.2.2}$$

where t is time $[T]$. The equation is further extended to the general case of three spatial dimensions in Equation 5.2.3:

$$\frac{\partial C}{\partial t} = D \left(\frac{\partial^2 C}{\partial x^2} + \frac{\partial^2 C}{\partial y^2} + \frac{\partial^2 C}{\partial z^2} \right) \tag{5.2.3}$$

where x, y, z are the three spatial coordinates $[L]$.

Advection-Diffusion Equation.   Equation 5.2.3 gives the rate of change in mass concentration because of diffusion alone. A more general representation of mass transport considers the contribution of advection as well. (A still more general equation would consider biological, physical, and chemical reactions also—topics not covered in this section.) The equation for mass transport resulting from advection and molecular diffusion is:

$$\frac{\partial C}{\partial t} + u \frac{\partial C}{\partial x} + v \frac{\partial C}{\partial y} + w \frac{\partial C}{\partial z} = D \left(\frac{\partial^2 C}{\partial x^2} + \frac{\partial^2 C}{\partial y^2} + \frac{\partial^2 C}{\partial z^2} \right) \tag{5.2.4}$$

where u, v, w are velocity components in the x, y, z coordinate directions $[L/T]$.

When applied to turbulent diffusion, the molecular diffusion coefficient D is replaced with the turbulent diffusion coefficient ε. Turbulence may vary from one coordinate direction to another, and therefore, the turbulent diffusion coefficient is most generally represented as dimensionally dependent, as indicated by the subscripts in Equation 5.2.5:

$$\frac{\partial C}{\partial t} + u \frac{\partial C}{\partial x} + v \frac{\partial C}{\partial y} + w \frac{\partial C}{\partial z} = \varepsilon_x \frac{\partial^2 C}{\partial x^2} + \varepsilon_y \frac{\partial^2 C}{\partial y^2} + \varepsilon_z \frac{\partial^2 C}{\partial z^2} \tag{5.2.5}$$

where $\varepsilon_x, \varepsilon_y, \varepsilon_z$ are turbulent diffusion coefficients in the x, y, z coordinate directions $[L^2/T]$.

Equation 5.2.5 is the general expression for the transport of mass resulting from advection and turbulent diffusion in three dimensions.

Averaging Equation 5.2.5 over one or more spatial dimensions gives rise to dispersion as described above. The most commonly used form is the one-dimensional equation applied to rivers and estuaries:

$$\frac{\partial C}{\partial t} + U \frac{\partial C}{\partial x} = \frac{1}{A} \frac{\partial}{\partial x} \left(A E_L \frac{\partial C}{\partial x} \right) \tag{5.2.6}$$

where U is the cross-sectional average velocity $[L/T]$; A is the cross-sectional area $[L^2]$; and E_L is the longitudinal dispersion coefficient $[L^2/T]$.

In wide rivers and estuaries, and in lakes, the spread of pollutants across, as well as along the direction of flow is important. For these problems, the two-dimensional horizontal dispersion equation is used:

$$\frac{\partial C}{\partial t} + \overline{U} \frac{\partial C}{\partial x} + \overline{V} \frac{\partial C}{\partial y} = \frac{1}{A} \left[\frac{\partial}{\partial x} \left(A E_L \frac{\partial C}{\partial x} \right) + \frac{\partial}{\partial y} \left(A E_t \frac{\partial C}{\partial y} \right) \right] \tag{5.2.7}$$

where $\overline{U}, \overline{V}$ are depth-averaged velocities $[L/T]$; and E_t is the transverse dispersion coefficient $[L^2/T]$. In rivers and estuaries, $\overline{V}$ is usually taken as zero, but the transverse dispersion term is retained.

Finally, we consider an averaged form of Equation 5.2.5 commonly used in modeling large lakes. This is the multiple-box model (or simply box model) in which the lake is subdivided into a number of fully-mixed volume elements or "boxes." Each box is represented by a three-dimensionally averaged form of the advection-diffusion equation:

$$V_k \frac{dC_k}{dt} = \sum_j \left(Q_{jk} C_j - Q_{jk} C_k + \frac{E_{jk} A_{jk}}{l_{jk}} (C_j - C_k) \right) \tag{5.2.8}$$

where C_k is the concentration in box k $[M/L^3]$; V_k is the volume of box k $[L^3]$; Q_{jk} is the flow from box j to downstream box k $[L^3/T]$; E_{jk} is a dispersion coefficient between boxes j and k $[L^2/T]$; A_{jk} is the cross-section area between boxes j and k $[L^2]$; and l_{jk} is the distance separating the centroids of boxes j and k $[L]$.

5.2.2   HORIZONTAL MIXING IN OCEANS AND LARGE LAKES

The Nature of Turbulent Diffusion

Oceanic mixing is typically considered to be a diffusive rather than a dispersive process. There are conditions in which velocity varies with depth or horizontal distance, known as shear currents, and which cause dispersion. In most situations however, the turbulent diffusion of the ocean environment is presumed to dominate mixing.

Carl Eckart (*J. Mar. Res.*, 7, p. 265, 1948) has given a very intuitive explanation of the turbulent diffusion processes that cause oceanic mixing. He uses the analogy of the mixing of coffee and cream. He defines three stages after the cream has been poured into a cup of coffee. At first, the two fluids are distinctly separated: sharp interfaces separate large areas of the coffee and cream. Although the concentration gradient of the cream is large at the interface, averaged over the entire cup surface the gradient is fairly small. This situation could persist for a long time if stirring did not occur. In the second stage, this pattern is disturbed when the coffee is stirred. The coffee and cream bodies are quickly distorted and the interface area consequently increases. Over the whole cup, the average concentration gradient is increased. In the final stage, molecular diffusion acts along the extensive interface areas to complete the mixing process quite abruptly, producing the liquid you wanted to drink.

The same basic process as mixing in a coffee cup is at work in the oceans. Turbulent motions produce a mixing roughly equivalent to stage two in the coffee and cream mixture. The motion is not nearly as simple as that in a coffee cup, however. The one simple whirlpool motion in the coffee cup is replaced by a complicated series of eddy motions of all sizes in the ocean. Although an eddy is usually thought of as a small swirling motion, a more comprehensive definition is useful in the analysis of diffusion. Stommel (*J. Mar. Res.*, 8, p. 199, 1949) states that the essential concept of an eddy is the "area over which turbulent velocities are similar or correlated." This is quantified through a Fourier analysis of the motion that identifies Fourier components with various wave numbers. These wave numbers are then taken to indicate eddy sizes (Okubo, *Impingement of Man on the Oceans*, (Hood, ed.), Wiley-Interscience, p. 89, 1971).

Analysis of the characteristics of the series of eddies leads to the concept of the eddy spectrum and energy transfer between eddies of different sizes. The largest eddies, the ocean-sized circulation patterns, constantly draw their energy from the wind. Because the kinetic energy of the ocean remains relatively constant, some form of energy dissipation must take place. What happens is that some of the energy of the general circulation is lost to form eddies of about 100 kilometer in size. These, in turn, pass energy down to smaller eddies, and so on. Eventually, the smallest eddies dissipate into heat generated by viscous friction. Viscosity also acts on all of the larger eddies in the spectrum (Stommel, 1949). The whole process has been very neatly summarized by L.F. Richardson (quoted in Batchelor, *Proc. Cambridge Phil. Soc.*, 43, p. 33, 1947):

> Big whorls have little whorls,
> which feed on their velocity;
> Little whorls have smaller whorls,
> and so on into viscosity.

The eddy spectrum is quite significant in diffusion. If a bucket of dye is dumped in the open ocean, diffusion processes will enlarge it to produce a constantly spreading cloud. Figure 5.2.1 gives a

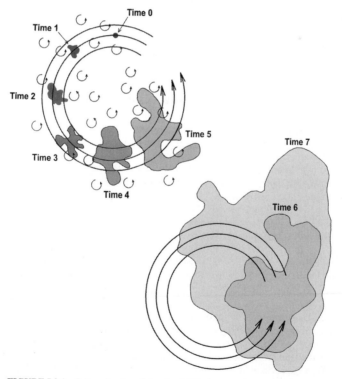

FIGURE 5.2.1   Schematic view of dye cloud diffusion over time under the influence to two different sized eddies.

schematic view of what would happen to such a cloud in an ocean with only two scales of turbulence. At first, only the smallest eddies would act to spread the cloud; the larger eddy appears only as advection, moving the entire cloud. With spreading, however, the cloud eventually reaches a size comparable to the large eddy, and then it too acts to spread the dye. In the ocean, with a continuous spectrum of motions, the dividing line between advection and turbulence constantly changes as the cloud enlarges. Eddies that happen to be about the same size as the cloud have a short period during which they produce a large change in the cloud. This process continues with larger and larger eddies until the dye is widely dispersed.

Descriptions of Horizontal Diffusion

Fickian Diffusion.   The process of dispersion can be cast in mathematical terms using the advection-diffusion equation based on Fick's Law, Equation 5.2.4. For now, we will simplify by considering the diffusion terms alone, separate from advection. This is done by transforming to a Lagrangian coordinate system in which the coordinate origin moves with the center of mass of the advecting pollutant. This mathematical construct also has practical application. For example, Adams et al. (*R. M. Parsons Lab. Report No. 205*, MIT, 1975) developed an ocean diffusion model based on such a Lagrangian model.

In the transformed coordinate system, horizontal diffusion about the center of pollutant mass is given by:

$$\frac{\partial C}{\partial t} = \varepsilon_H \left(\frac{\partial^2 C}{\partial x^2} + \frac{\partial^2 C}{\partial y^2} \right) \tag{5.2.9a}$$

or in radial coordinates:

$$\frac{\partial C}{\partial t} = \frac{\varepsilon_H}{r} \frac{\partial^2}{\partial r^2} \left(r \frac{\partial C}{\partial r} \right) \tag{5.2.9b}$$

where r is distance from the center of mass [L]; and ε_H is the horizontal turbulent diffusion coefficient [L^2/T].

The expression ε_H is also called the horizontal turbulent diffusivity or eddy diffusivity. It is used in the presumption that horizontal diffusion is the same in the x and y directions (that is, $\varepsilon_x = \varepsilon_y = \varepsilon_H$). Equation 5.2.9b is the model used in the earliest studies of oceanic diffusion, studies that pointed out some major deviations from Fick's Law.

When field experiments with dye clouds were carried out in the ocean, the diffusivity was found to increase with time. The reasons for this increase are related to the distribution of eddy sizes, as discussed above. Those eddies approximately equal in size to the cloud contribute the most to its mixing. As the cloud expands the dominant eddies shift to the larger portion of the eddy spectrum, and the mixing power consequently increases. Thus, ε_H is an increasing function of time and cloud size (Deacon, *Inter. J. Air Water Poll.*, 2, p. 92, 1959). This is in direct conflict with the concepts underlying the Fickian approach. Fick's Law assumes that the velocity of a particle is independent of the velocities of nearby particles. With turbulence, however, the closer two particles are, the more similar their velocities will be. Thus, the basic predicate of the Fick equation is violated.

Statistical Description of Diffusion.   Further insight to the limitations in applying Fick's Law to turbulent diffusion is gained by considering the statistics of the random movement of a single particle of pollutant. The mathematical derivation, which is complicated, is given by Fischer et al. (1979). The derivation considers the statistical correlation between the particle's motion at one instant in time with motion at previous instants in time—the particle's "Lagrangian autocorrelation." An important related concept resulting from the statistical derivation is the Lagrangian time scale. The Lagrangian time scale can be thought of as the period of time for the autocorrelation to go to zero, in essence for the particle to forget its past motion. The statistical theory shows that for time periods shorter than the Lagrangian time scale the diffusion coefficient changes with time and a Fickian description is therefore invalid. For time periods that are long relative to the Lagrangian time scale, the diffusion coefficient becomes constant and Fick's Law can be used. Unfortunately, there is no *a priori* method to determine the Lagrangian time scale for a particular situation. Related derivations show, however, that if the scale of the diffusing plume is greater than the scale of turbulent motion, the Fickian relation holds. This criterion has proven to be most useful in bounded water bodies, where the depth or width can be assumed to be an upper limit on the scale of turbulence. In the ocean, experiments with length scales as great as 40 kilometers fail to show a limit on the turbulent diffusion coefficient.

Richardson Neighbor Diffusion

Richardson (*Proc. Royal Soc. London*, 110A, p. 709, 1926) addressed the apparent relation between the diffusion coefficient and length scale for time periods shorter than the Lagrangian time scale. Noting that the diffusivity of the atmosphere varied by a billion orders of magnitude, he sought an expression for the diffusion coefficient that would account for this variation, but still not be a function of position or time. The relevant parameter was, he said, the distance l by which two particles are separated. He then went on to define a theory based on this "neighbor separation." The equation proposed was modeled after the Fick equation:

$$\frac{\partial C}{\partial t} = \frac{\partial}{\partial l}\left(F(l)\frac{\partial q}{\partial l}\right) \tag{5.2.10}$$

where l is the neighbor separation $[L]$, equal to the instantaneous distance between two particles; $q(l)$ is the neighbor concentration function [dimensionless], equal to the number of particle pairs separated by the distance l; and $F(l)$ is the neighbor diffusivity $[L^2/T]$.

$F(l)$ is of the same nature as ε_H and has the same units. For diffusion of the Fickian type, when $F(l)$ is a constant, it can be shown that $F(l) = 2\,\varepsilon_H$. For oceanic turbulent diffusion, $F(l)$ is a function of l. From experimental data, Richardson proposed:

$$F(l) = al^{4/3} \tag{5.2.11}$$

Taylor (1959) attributes the choice of 4/3 rather than 1.3 or 1.4 to a remarkable intuitive insight on Richardson's part. He states that Richardson felt that $F(l)$ was related to the energy transfer from larger to smaller eddies, and that a simple universal rule must therefore govern the process. Fifteen years later the eddy diffusivity was proven analytically to be proportional to the 4/3 power of the eddy length by Kolmogoroff through a dimensional analysis of Equation 5.2.10 (Batchelor, *Proc. Cambridge Philosophical Soc.*, 43, p. 533, 1947).

Richardson's 4/3 Law may be used in practical problems to define a diffusivity based on the width of a diffusing plume or cloud. The following empirical formula is based on field data:

$$\varepsilon_H = aw^{4/3} \tag{5.2.12}$$

where a is a coefficient $[L^{2/3}/T]$ with a value typically between 4×10^{-6} and 2×10^{-5} m$^{2/3}$/sec; and w is a pertinent length scale such as the plume width $[L]$.

The formula is based on field data collected and analyzed by Okubo (1971) as reported by Fischer et al. (1979). In the ocean, the relation has been found to hold for length scales from 10 meters to over 10 kilometers. It should be applied with caution in bounded water bodies where free turbulence is impeded by shorelines and the bottom.

Murthy and Okubo (*Symp. Modeling Transport Mechanisms in Oceans and Lakes*, Manuscript Report Series No. 43, Mar. Science Directorate, Dept. Fisheries and Environ. Ottawa, Canada, p. 129,

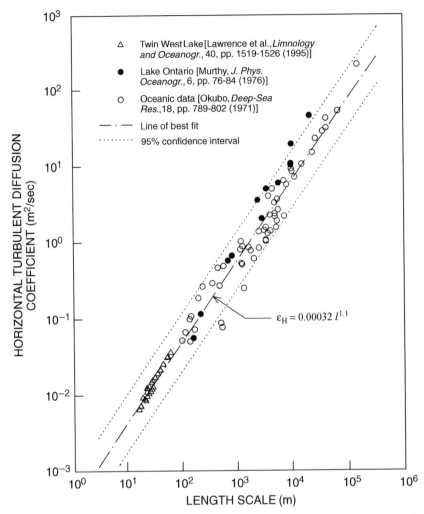

FIGURE 5.2.2   Horizontal turbulent diffusivity in oceans and large lakes as a function of length scale (based on Lawrence et al., *Limnology and Oceanogr.*, 40, pp. 1519–1526, 1995, reproduced with permission).

1977) compiled field data from ocean and lake diffusion experiments and arrived at a slightly different dependency on the scale of diffusion. Their results (Figure 5.2.2), show ε_H to be proportional to $w^{1.1}$. The minor difference between the 4/3 power in Equation 5.2.12 and the 1.1 power in Figure 5.2.2 may arise from spatial intermittency in the turbulence field (Lawrence et al., *Limnology and Oceanogr.*, 40, p. 1519, 1995).

Application.   A classic paper by Brooks (*Proc. Inter. Conference on Waste Disposal Marine Environ.*, p. 246, 1960), which is also described by Fischer et al. (1979), considers the mixing of sewage effluent discharged to the ocean. Brooks' paper gives the mathematical solution for a sewage plume subject to turbulent diffusion that conforms to the 4/3 Law. The original paper or Fischer et al. (1979) should be consulted for a description of the solution and its application.

Horizontal Dispersion in Lake Models

Transport and mixing in large lakes is often modeled using box models, previously defined using Equation 5.2.8. A dispersion coefficient is defined in these models to capture the effects of nonadvective mixing between adjacent boxes. This is an essentially empirical formulation, however, and there are no direct means to compute the dispersion coefficient based on the geometric or physical characteristics of the water body. Shanahan and Harleman (*J. Environ. Eng. ASCE*, 110, p. 42, 1984) describe a procedure to define the dispersion coefficient for box models.

5.2.3   *VERTICAL MIXING IN THE OCEAN AND LARGE LAKES*

The effects of vertical mixing are much less pronounced in the oceans and large lakes than are the effects of horizontal mixing. The reason is the existence of the thermocline, a depth interval over which temperature decreases rapidly with depth. (Formally, the thermocline is the interval of depth in which the change in temperature with depth exceeds 1°C per meter.) The thermocline is found below a warmer, mixed layer at the water's surface. The cool waters below the thermocline are denser than those above; thus, the thermocline acts as a physical barrier to mixing. In the oceans particularly, the thermocline is relatively shallow and horizontal mixing dominates vertical mixing as a dilution mechanism. Nonetheless, vertical mixing can be a very important water quality process, particularly in lakes and reservoirs where water quality varies greatly in the vertical dimension.

Vertical Mixing in the Ocean

Vertical mixing in the ocean is reviewed by Okubo (1971) and Murthy and Okubo (1977). They report that most experimental values of the vertical turbulent diffusion coefficient found in ocean experiments range between 1 and 100 cm²/sec, with the smallest values representative of conditions below the thermocline. Kullenberg (*Tellus*, 23, p. 129, 1971) relates the value of the vertical diffusion coefficient to the wind speed and water-column stability through this relation:

$$\varepsilon_v = k \frac{W^2}{N^2} \left| \frac{du}{dz} \right| \tag{5.2.13}$$

where k is a constant [dimensionless] with a value of between 2×10^{-8} and 8×10^{-8}; W is the mean wind speed [L/T]; N is the Brunt-Väisälä frequency [$1/T$]; and u is the horizontal water velocity [L/T].

The variable N is related to the density stability of the water column. Physically, it represents the natural frequency of oscillation that occurs in the water column following a disturbance. It is defined as (Phillips, *The Dynamics of the Upper Ocean*, 2nd Edition, Cambridge Univ. Press, 1977):

$$N = \left(-\frac{g}{\rho_0} \frac{\partial \rho}{\partial z} - \frac{g^2}{c^2} \right)^{\frac{1}{2}} \tag{5.2.14}$$

where g is gravitational acceleration [L/T^2]; ρ is the water density as a function of z [M/L^3]; ρ_0 is the reference density at $z = 0$ [M/L^3]; and c is the speed of sound in water [L/T], equal to approximately 1500 m/sec.

Figure 5.2.3 is based on Murthy and Okubo's (1977) plot of Equation 5.2.13 against data from the ocean and Lake Ontario.

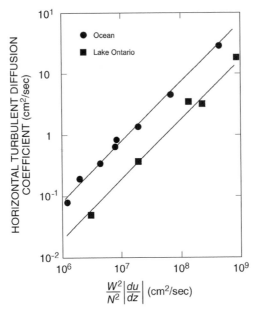

FIGURE 5.2.3 Empirical relation for the vertical turbulent diffusivity (based on Murthy and Okubo, *Symp. on Modeling of Transport Mechanisms in Oceans and Lakes*, Manuscript Report Series No. 43, Mar. Sciences Directorate, Dept. of Fisheries and Environ., Ottawa, Canada, p. 129, 1977).

Vertical Mixing in Lakes and Reservoirs

Vertical mixing in temperate lakes is strongly influenced by seasonal cycles in the lake's thermal structure. At the end of winter, a lake is typically mixed throughout its depth and shows a vertically isothermal temperature profile. As the lake surface is warmed by the sun and the atmosphere through the spring, the shallowest water warms relative to the deeper water. Soon, a pronounced thermocline separates the warm surface water, the epilimnion, from the cold deeper water, the hypolimnion. Through the summer, the thermocline becomes stronger (that is, the difference in temperature over the vertical distance of the thermocline increases). Finally, with surface cooling in the fall, the temperature stratification weakens until the fall overturn, when the lake mixes throughout its depth.

Vertical mixing is of greatest interest during the summer stratification. The strong temperature and density gradient across the thermocline separates zones of distinctly different water quality. The epilimnion receives sunlight and atmospheric oxygen, and thus supports phytoplankton growth. Typically, dissolved oxygen is high and nutrient concentrations low in the epilimnion. Algal growth is limited in the cooler, darker hypolimnion, and dissolved oxygen is reduced or absent. Organic matter that settles into the hypolimnion and chemical constituents diffused from the lake bottom increase nutrient and other constituent concentrations, creating a dramatically different water quality than in the epilimnion. The distinct water qualities in these two layers make vertical mixing a very important water-quality process in lakes.

The vertical diffusion coefficient in lakes is typically on the order 0.1 to 0.01 cm^2/sec, but may range from as high as 1 cm^2/sec to as low as 0.001 cm^2/sec. Table 5.2.1, which is drawn from Schnoor (*Environmental Modeling*, John Wiley & Sons, 1996), lists observed vertical diffusion coefficients from a number of published studies.

Deep Mixing. Mixing across a well-established thermocline or within the hypolimnion is very limited. Lake modeling studies by Wang and Harleman (*R. M. Parsons Lab. Report No. 270*, MIT,

TABLE 5.2.1   Summary of Vertical Diffusion Measurement in Lakes

Lake	Vertical diffusion coefficient (cm^2/sec)	Remarks
Whole lake average vertical diffusion		
Lake Erie[a]	0.58	
Lake Erie[b]	102	Unstratified
	0.05–0.25	Stratified
Lake Erie[c]	15	
Lake Huron[a]	1.16	
Lake Huron[d]	1.16	Unstratified
Lake Ontario[a]	3.47	
Wellington Reservoir, Australia[e]	1.00	
White Lake, Michigan[f]	0.4	
Lake LBJ, Texas[g]	0.18	Feb.–April
	0.12	May–June
	0.01	July–Jan.
Cayuga Lake, New York[h]	2.31	
Lake Greifensee, Switzerland[i]	0.2	April
	0.15	May–Aug.
	0.05	Sept.–Nov.
Vertical diffusion across the thermocline		
Lake Erie[a]	0.21	
Lake Ontario[j]	0.125, 0.063	
Cayuga Lake, New York[j]	0.178, 0.25	
Cayuga Lake, New York[k]	0.253	
Lake Greifensee, Switzerland[i]	0.25	
Lake Zurich, Switzerland[l]	0.71	April
	0.14	May
	0.064	June
	0.039	July
	0.026	Aug.
	0.020	Sept.
	0.074	Oct.
Lake Zurich, Switzerland[j]	0.03	
Lake Mendota, Wisconsin[j]	0.025	
Onondaga Lake, New York[m]	0.04	May
	0.09	June
	0.03	July
	0.005	Aug.
	0.008	Sept.
	0.015	Oct.
Lake Tahoe, Nevada[k]	0.178	
Lake Luzern, Switzerland[j]	0.10	
Lake Washington, Washington[j]	0.03	
Lake Sammamish, Washington[j]	0.03	
Linsley Pond, Connecticut[j]	0.003	
Lake Baikal, USSR[j]	2.5–7.4	

[a] (Torgersen et al., *Limnology and Oceanogr.*, 22, p. 181, 1977).
[b] (DiToro and Connolly, Report EPA-600/3-80-065, U.S. EPA, 1980).
[c] (Heinrich et al., *J. Great Lakes Res.*, 7, p. 264, 1981).
[d] (DiToro and Matystik, Jr., Report EPA-600/3-80-056, U.S. EPA, 1980).
[e] (Imberger et al., *J. Hydraulics Div.*, ASCE, 104, p. 725, 1978).
[f] (Lung and Canale, *J. Environ. Eng. Div.*, ASCE, 104, p. 663, 1978).
[g] (Park and Schmidt, *Water Resourc. Bulletin*, 9, p. 932, 1973).
[h] (Bedford and Babajimopoulus, *J. Environ. Eng. Div.*, ASCE, 103, p. 133, 1977).
[i] (Imboden and Emerson, *Limnology and Oceanogr.*, 23, p. 77, 1978).
[j] (Snodgrass and O'Melia, *Environ. Sci. Technol.*, 9, p. 937, 1975).
[k] (Powell and Jassby, *Water Resourc. Res.*, 10, p. 191, 1974).
[l] (Li, *Sch. Zeit. Hydr.*, 35, p. 1, 1973).
[m] (Wodka et al., *J. Environ. Eng. Div.*, ASCE, 109, p. 143, 1983).

1982) show that diffusion across and below the thermocline of stratified lakes is at or near the rate of molecular diffusion (1.6×10^{-5} cm²/sec for salt [NaCl] in water at 25°C). Hypolimnetic mixing may be higher if there is significant flow or other motion within the hypolimnion. For example, there may be flow in a reservoir from stream inflows to a deep dam outlet. Another source of motion is an internal seiche, the back-and-forth oscillation of the thermocline in a type of motion similar to sloshing in a bathtub.

Surface Mixing.   The epilimnion of a lake or reservoir typically is well-mixed owing to a nearly constant input of mixing energy from the wind. A more subtle and interesting question than mixing within the epilimnion is the rate at which mixing from the wind causes the surface layer to mix into the thermocline and become deeper. This is a far more important mechanism for transport from the hypolimnion to the epilimnion than diffusion across the thermocline.

The deepening of the surface layer, and related questions about lake thermal structure through an annual cycle, have been the subject of water-quality modeling studies over the last twenty years (Fischer et al., 1979; Harleman, *J. Hydraulics Div.*, ASCE, 108, p. 302, 1982). The earliest models (Dake and Harleman, *Water Resourc. Res.*, 5, p. 2, 1969; Water Resources Engineers, Inc., "Mathematical Models for the Prediction of Thermal Energy Changes in Impoundments," report to U.S. EPA, 1969) relied on specification of a single vertical diffusion coefficient to compute vertical mixing. The empiricism and lack of generality of this approach led to definitions of the diffusion coefficient in terms of the lake's thermal structure and/or input mixing energy.

Through the 1970s, descriptive formulae for the vertical diffusion coefficient in lakes were improved, but with what was still an essentially empirical approach. The widely-used WQRRS Model (Smith, "Water Quality for River-Reservoir Systems," HEC, US Army Corps of Eng., 1978) includes two diffusion coefficient formulae typical of this approach. For deep, well-stratified lakes, the diffusion coefficient is determined from the water column stability:

$$\varepsilon_v = \varepsilon_{max} \quad \text{when } \frac{1}{\rho}\frac{\partial \rho}{\partial z} \le S_{crit} \tag{5.2.15a}$$

$$\varepsilon_v = a\left(\frac{1}{\rho}\frac{\partial \rho}{\partial z}\right)^b \quad \text{when } \frac{1}{\rho}\frac{\partial \rho}{\partial z} > S_{crit} \tag{5.2.15b}$$

where ε_v is the vertical diffusion coefficient [L^2/T]; ε_{max} is an empirically determined maximum value of ε_v [L^2/T] with a recommended value ranging from 0.2 to 10.0 cm²/sec and an average value of about 2.5 cm²/sec; ρ is the density of the water as a function of z [M/L^3]; $\overline{\rho}$ is the average density of the water [M/L^3]; S_{crit} is a critical water column stability [$1/L$], typically between 1×10^{-6} and 1×10^{-5}m^{-1}; a is an empirical coefficient with units that depend on the value of b; and b an empirical coefficient [dimensionless] with a typical value of -0.7.

This equation predicts a diffusion coefficient that is constant and equal through the epilimnion and hypolimnion, but which decreases through the thermocline.

For lakes in which wind mixing is a dominant mechanism, WQRRS provides this relation:

$$\varepsilon_v = \varepsilon_{min} + \alpha W \exp\left(-\frac{\beta z}{z_t}\right) \tag{5.2.16}$$

where ε_{min} is the minimum value of ε_v [L^2/T] with a recommended value of 0.1 to 0.5 cm²/sec in well-mixed reservoirs and 0.01 to 0.001 cm²/sec in stratified reservoirs; W is the wind speed [L/T]; z_t is the depth to the thermocline [L]; α is an empirical coefficient [L] with a recommended value of 0.01 to 0.02 cm in well-mixed reservoirs and 0.001 to 0.005 cm in stratified reservoirs; and β is an empirical coefficient (dimensionless) with a recommended value of 4.6.

Equation 5.2.16 predicts a diffusion coefficient that has its maximum value at the surface, where there is the greatest wind energy, decreasing exponentially with depth to a minimum value. The values recommended in the WQRRS manual for ε_{max} in Equation 5.2.15 and ε_{min} in Equation 5.2.16 are probably high based on more recent research that shows vertical diffusivities between 1 and 50 times molecular diffusivity at depth in lakes.

Equations 5.2.15 and 5.2.16 are just two widely used examples of a large number of empirical equations for vertical diffusivity found in the literature. Good reviews of the many equations proposed are given by Henderson-Sellers (*J. Environ. Eng. Div.*, ASCE, 102, p. 517, 1976) and McCutcheon (*Proc. Conference on Frontiers in Hydraulic Eng.*, ASCE, p. 15, 1983). Table 5.2.2 provides a summary of the literature addressing vertical dispersion in lakes.

Wind-Mixing Models. The most sophisticated and successful models of vertical mixing in lakes are those that simulate the effects of surface wind mixing, surface heat transfer, vertical density structure, and hydraulic forces through time. Harleman (1982) reviews these models with an emphasis on the first three mechanisms; the review by Fischer et al. (1979) places greater emphasis on the last mechanism.

5.2.4 DISPERSION IN RIVERS

River Dispersion Processes

If a mass of material is introduced into a river, it will be mixed largely by two mechanisms. Firstly, and most importantly, the mass is dispersed because of variations in flow velocity across the river. Some portions of the river travel faster than others, and thus will carry some of the material further downstream. The net effect of this advective dispersion is to spread the material along the river. If no other mechanisms were important, this would produce quite a high rate of dispersion. However, this potentially high rate is lowered somewhat by the second process: turbulent diffusion. The effect of this process is to shift quantities of material between zones of different velocity so that none remains in one zone indefinitely. This decreases the spread of pollutant by shifting the material in low velocity zones to flowing zones, and that in the maximum velocity zones into slower moving sectors (Glover, *U.S. Geo. Survey Prof. Paper 433-B*, 1964; Fischer, *U.S. Geo. Survey Prof. Paper 582-A*, 1968).

In practice, analysis of mass transport in rivers is frequently limited to a one-dimensional representation of longitudinal dispersion, using Equation 5.2.6. When necessary, the two-dimensional advective-dispersion equation (Equation 5.2.7) can be used to explicitly consider transverse dispersion. Vertical transport is rarely modeled explicitly because rivers are usually well-mixed vertically in the time periods of interest.

Descriptions of Longitudinal Dispersion in Rivers

Classical Descriptions. In order to understand the mathematical descriptions of longitudinal dispersion in rivers, we begin by reviewing Taylor's (*Proc. Royal Soc. London*, 219A, p. 186, 1953; *Proc. Royal Soc. London*, 220A, p. 446, 1953) work concerning dispersion in pipes. Neglecting reactions and molecular diffusion, and assuming homogeneous turbulent diffusion in a uniform velocity field, longitudinal dispersion can be represented by the one-dimensional diffusion equation:

$$\frac{\partial C}{\partial t} + u \frac{\partial C}{\partial x} = \varepsilon_x \frac{\partial^2 C}{\partial x^2} \qquad (5.2.17)$$

The solution of this equation for a mass of pollutant injected at $t = 0$ is an ever-spreading Gaussian distribution moving downstream with velocity u, with u a function of y and z (Figure 5.2.4a).

Taylor (1953) realized that a formulation limited to turbulent diffusion alone would not describe the spread of a solute with much accuracy. His studies on flow in a pipe recognized that the variation of velocity from zero at the pipe walls to a maximum at the center would also cause a spreading of the pollutant. Figure 5.2.4b shows the difference between this mechanism and diffusion as seen in Figure 5.2.4a. The small amounts of concentrate that remain at the injection point and the greater spread of the solute "cloud" are noteworthy. Taylor concluded from his studies that dispersion could be represented by an expression identical to Equation 5.2.17. The only difference is that u would be replaced by U, the cross-sectionally averaged velocity, and ε_x would be replaced by a simple sum of the turbulent

TABLE 5.2.2 Summary of References on Vertical Diffusion in Lakes

Reference	Factors considered							Comments
	Epilimnion	Hypolimnion	Thermocline	Wind mixing	Through flow	Field data	Formula for E_v	
DiGiano et al.[a]	×			×		×	×	Dispersion in shallow lakes
Henderson-Sellers[b]	×		×	×			×	Evaluation of literature formulae
Powell and Jassby[c]		×					×	Evaluation of formulae for hypolimnetic diffusion
Kullenberg[d]	×		×	×		×	×	E_v from stability, wind
Imberger[e]	×			×		×		River underflow mixing
Adams et al.[f]			×		×	×	×	Diffusion in stratified cooling pond
Bloss and Harleman[g]	×		×	×		×	×	Wind-mixing model of thermocline

[a] (DiGiano, Liklema and Van Straten, *Ecol. Modelling*, 4, p. 237, 1978).
[b] (Henderson-Sellers, *J. Environ. Eng. Div.*, ASCE, 102, p. 517, 1976).
[c] (Powell and Jassby, *Water Resourc. Res.*, 10, p. 191, 1974).
[d] (Kullenberg, *Tellus*, 23, p. 129, 1971).
[e] (Imberger, *J. Hydraulic Eng.*, ASCE, 113, p. 697, 1987).
[f] (Adams et al., *J. Hydraulic Eng.*, ASCE, 113, p. 293, 1987).
[g] (Bloss and Harleman, *R. M. Parsons Lab. Report No. 249*, MIT 1979).

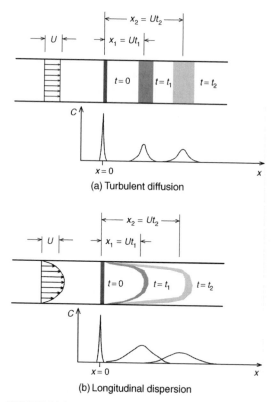

FIGURE 5.2.4 Mixing processes in circular pipe flow (a) turbulent diffusion, (b) longitudinal dispersion (based on Daily and Harleman, *Fluid Dynamics*, Addison-Wesley, 1966, reproduced with permission).

diffusion coefficient and an additional dispersive term: $E_L = E_{\text{DISPERSION}} + E_{\text{TURBULENT DIFFUSION}}$. E_L is called the longitudinal dispersion coefficient and is approximately two orders of magnitude larger than the corresponding turbulent diffusivity. It includes the dispersive effects attributable to the nonuniform velocity field.

Taylor's work was later extended by Elder (*J. Fluid Mech.*, 5, p. 544, 1959) to an infinitely wide open channel. Here, the velocity variation was with depth, rather than with radius from the pipe centerline as in Taylor's experiments. Nevertheless, a similar behavior was predicted: a Gaussian concentration cloud advecting downstream at the average flow velocity. Elder was able to deduce an expression for E_L:

$$E_L = 0.23 \, du^* \tag{5.2.18}$$

where d is the depth of the channel [L]; and u^* is the shear velocity [L/T].

The shear velocity, u^*, is related empirically to the bed friction factor and the mean stream velocity, U, as:

$$u^* = \sqrt{\frac{\tau_0}{\rho}} = \sqrt{\frac{f}{8}U^2} = \sqrt{gsR_H} \tag{5.2.19}$$

where U is the cross-sectionally averaged stream velocity [L/T]; τ_0 is the bed shear stress [M/LT^2];

ρ is the density of the fluid $[M/L^3]$; f is a friction factor (dimensionless) equal to approximately 0.02 for natural, fully turbulent flow; s is the slope of the stream water surface [dimensionless]; and R_H is the stream's hydraulic radius $[L]$.

Dispersion in Natural Rivers.   As would be expected, the theoretical developments of Taylor (1953) and Elder (1959), based on straight pipes and channels with regular geometries and near ideal conditions can be translated to streams only imperfectly. The differences between the theoretical developments based on ideal systems and the behavior of natural streams are quite important.

There is a fundamental difference in the cross-sectional velocity distribution between the infinitely wide open channel and a natural stream. In the infinitely wide channel, the velocity is uniform across the channel and only varies with depth. In a natural stream, there is usually a central zone of high velocity and slower velocity zones along one or both banks. These lateral variations are more important for dispersion than the vertical variations that control in Elder's experiments. In streams vertical mixing occurs fairly quickly. However, velocity zones are further apart horizontally than vertically, and lateral mixing therefore takes a good deal longer (Fischer, *J. Hydraulics Div.*, ASCE, 93, p. 187, 1967).

Godfrey and Frederick (*U.S. Geo. Survey Prof. Paper 433-K*, 1970) studied dispersion in natural streams and reported E/du^* between 140 and 500 (versus 0.23 predicted by Elder). As shown in Table 5.2.3, subsequent experiments have shown that E/du^* varies widely for natural streams, but is always much greater than Elder's result. The larger values are caused by the variations in velocity across the stream, which are not accounted for in Elder's analysis of vertical velocity variations. The longitudinal dispersion coefficient is proportional to the square of the distance over which the shear profile extends. Because the width to depth ratio for natural streams is usually greater than 10, it is not surprising that the transverse velocity profile is much more important than the vertical profile in producing longitudinal dispersion.

Fischer (1967) extended Taylor's and Elder's classical studies and showed the limits of their applicability to natural streams. Fischer pointed out that pure dispersion (that is, the effect of differential advection alone) does not conform with Fick's Law. Fick's Law is met through the additional influence of vertical and lateral diffusion. These cross-sectional mixing processes create a more-or-less uniform cross-sectional concentration that behaves much like Fickian diffusion. However, cross-sectional mixing processes are not instantaneous—the pollutant must travel a certain distance (or an equivalent travel time) downstream of the discharge point before a uniform cross-sectional distribution is achieved. Based on this concept, Fischer derives the following criterion for use of Equation 5.2.17, the one-dimensional advective-dispersion equation:

$$L > 1.8 \frac{l^2}{R_H} \frac{U}{u^*} \qquad (5.2.20)$$

where L is distance downstream of the discharge point $[L]$; and l is the cross-sectional mixing length $[L]$, taken as the distance between the thread of maximum velocity and the furthest bank.

According to Equation 5.2.20, a diffusion-type relation is an invalid description of stream dispersion for a distance less than the value of the equation's right-hand side.

A second major discrepancy between the idealized and natural systems arises with respect to geometry. Taylor and Elder dealt with straight channels of regular and constant cross-sectional geometry. This, of course, is not the case with natural streams. Among the nonuniformities that can be found in streams are bends, islands, dead zones, structures, reservoirs, water falls, expanding sections, navigation locks, density currents, diversions to industrial plants for process water, braided sections, and overbank flow during floods. These nonuniformities can all contribute to deviations from predicted behavior. For example, irregularities along the banks increase the length of time before the concentration distribution becomes Gaussian. A long "tail" of slightly elevated concentration is produced because of the initial detention and subsequent slow release of small amounts of the concentrate in dead zones. Bends, on the other hand, accentuate velocity differences across the stream, and therefore greatly increase the longitudinal dispersion coefficient. Sooky (*J. Hydraulics Div.*, ASCE, 95, p. 1327, 1969) gives a mathematical formulation for this effect.

TABLE 5.2.3 Summary of Longitudinal Dispersion Measurements in Rivers

River	Longitudinal dispersion coefficient (m²/sec)	$\dfrac{E_L}{du^*}$
Missouri River, Indiana–Nebraska[a]	56,000	
Missouri River, Indiana–Nebraska[b]	610–2500	2300–9500
Missouri River, Indiana–Nebraska[c]	1500	7500
White River, Indiana[d]	30	1200
Chattahoochee River, Georgia[d]	33	380
Susquehanna River, Pennsylvania[d]	90	1000
Elkhorn River, Nebraska[d]	9–21	670–1100
John Day River, Oregon[d]	14–65	150–180
Comite River, Louisiana[d]	7–14	600–620
Amite River, Louisiana[d]	23–30	410–550
Yadkin River, North Carolina[d]	110–260	480–530
Muddy Creek, North Carolina[d]	14–33	213–275
Noosak River, Washington[d]	35–153	98–170
Monocracy River, Maryland[d]	5–40	300–600
Antietam Creek, Maryland[d]	9–26	390–440
Bayou Anacoco, Louisiana[d]	14–40	510–740
South Platte River, Nebraska[e]	16.2	510
Green-Duwamish River, Washington[f]	6.5–8.4	120–160
Clinch River, Virginia-Tennessee[g]	14	235
	46	210
	54	245
	8	280
Sabine River, Texas[d]	320	2800
	670	1700
Coachella Canal, California[g]	9.6	140
Powell River, Tennessee[g]	9.5	200
Nemadji River, Wisconsin[h]	16.9	492
Coraville Reservoir, Iowa River, Iowa[i]	0.5–9	50–900
Colorado River, Arizona[j]	55–243	19–127

[a] (Sayre, in *Environ. Impact on Rivers*, Shen, ed., Water Resourc. Pubs., Ft. Collins, Colorado, 1973).
[b] (Yotsukura et al., *U.S. Geo. Survey Water-Supply Paper 1899-G*, 1970).
[c] (Fischer, *J. Environ. Eng. Div.*, ASCE, 101, p. 453, 1975).
[d] (McQuivey and Keefer, *J. Environ. Eng. Div.*, ASCE, 100, p. 997, 1974).
[e] (Glover, *U.S. Geo. Survey Prof. Paper 433-B*, 1964).
[f] (Fischer, *U.S. Geo. Survey Prof. Paper 582-A*, 1968).
[g] (Fischer, *J. San. Eng. Div.*, ASCE, 94, p. 927, 1968).
[h] (Hibbs et al., *J. Environ. Eng.*, ASCE, 124, p. 752, 1998).
[i] (Mossman et al., *Water Res.*, 25, p. 1405, 1991).
[j] (Graf, *Water Resourc. Bull.*, 31, p. 265, 1995).

Dispersion Coefficient Formulae.   Many investigators have developed equations for estimating longitudinal dispersion in rivers. Table 5.2.4 lists available technical references related to longitudinal dispersion. Two of the most frequently used formulae are by Fischer (*J. Environ. Eng. Div.*, ASCE, 101, p. 453, 1975) and Liu (*J. Environ. Eng. Div.*, ASCE, 103, p. 59, 1977). Fischer developed the equation:

$$E_L = \frac{0.011 U^2 \overline{W}^2}{du^*}$$

(5.2.21)

where U is the mean stream velocity [L/T]; $\overline{W}$ is the mean channel width [L]; d is the mean depth [L]; and u^* is the shear velocity [L/T] defined in Equation 5.2.19.

TABLE 5.2.4 Summary of References on Longitudinal Dispersion in Rivers

Reference	Velocity	Channel irregularity	Dead zones	Mountain streams	Other	Field data	Formula for E_v	Comments
Fischer[a]	×					×	×	Equation for E_L from streamflow measurements
Wu[b]	×				×		×	Effect of wind on E_L in wide channels
McQuivey and Keefer[c]					×	×	×	Empirical function of streamflow, slope, width
Fischer[d]	×					×	×	Discussion of McQuivey and Keefer[c]
Liu[e], Liu and Cheng[f]	×						×	Widely-used formula
Bencala and Walters[g]			×	×		×		Model of dispersion in pool-and-riffle streams
Yu and Wenzhi[h]			×					Dead zone model
Magazine et al.[i]		×					×	Bed and side roughness
Beltaos[j]				×			×	Steep streams
Sobol and Nordin[k]			×			×		Dead zone model
Valentine and Wood[l, m]			×				×	Dead zone model, lab data
Beer and Young[n]			×			×	×	E_L by time series analysis
Bajraktarevic-Dobran[o]				×		×		Mountain stream storage model
Jobson[p]								E_L by numerical routing

[a] (Fischer, *J. San. Eng. Div.*, ASCE, 94, p. 927, 1968).
[b] (Wu, *Water Resourc. Res.*, 5, p. 1097, 1969).
[c] (McQuivey and Keefer, *J. Environ. Eng. Div.*, ASCE, 100, p. 997, 1974).
[d] (Fischer, *J. Environ. Eng. Div.*, ASCE, 101, p. 453, 1975).
[e] (Liu, *J. Environ. Eng. Div.*, ASCE, 103, p. 59, 1977).
[f] (Liu and Cheng, *J. Hydraulics Div.*, ASCE, 106, p. 1021, 1980).
[g] (Bencala and Walters, *Water Resourc. Res.*, 19, p. 718, 1983).
[h] (Yu and Wenzhi, *Proc. 1988 Conf. Hydraulic Eng.*, ASCE, p. 1026, 1988).
[i] (Magazine et al., *J. Hydraulic Eng.*, ASCE, 114, p. 766, 1988).
[j] (Beltaos, *J. Hydraulics Div.*, ASCE, 108, p. 591, 1982).
[k] (Sobol and Nordin, *J. Hydraulics Div.*, ASCE, 104, p. 695, 1978).
[l] (Valentine and Wood, *J. Hydraulics Div.*, ASCE, 103, p. 975, 1977).
[m] (Valentine and Wood, *J. Hydraulics Div.*, ASCE, 105, p. 999, 1979).
[n] (Beer and Young, *J. Environ. Eng.*, ASCE, 109, p. 1049, 1983).
[o] (Bajraktarevic-Dobran, *J. Environ. Eng. Div.*, ASCE, 108, p. 502, 1982).
[p] (Jobson, *Water Resourc. Res.*, 23, p. 169, 1987).

Liu's equation is similar:

$$E_L = \frac{\beta U^2 W^2}{d u^*} \tag{5.2.22}$$

where $\beta = 0.5 \, u^*/U$ depends on the dimensionless bottom roughness.

These equations typically predict E_L within a factor of 4 to 6 (Schnoor, 1996; Fischer et al., 1979).

Descriptions of Transverse Dispersion in Rivers

Transverse Dispersion in Straight Channels. In the absence of secondary currents, transverse dispersion in open channels is due to turbulent diffusion. Thus, the magnitude of the transverse dispersion coefficient is generally much lower than that of the longitudinal coefficient. Fischer et al. (1979)

report over 75 separate experiments conducted by various investigators using straight rectangular laboratory channels, all yielding E_t/du^* between 0.1 and 0.2. They suggest the following estimate of E_t:

$$E_t = 0.15 \, du^* \qquad (5.2.23)$$

Detailed analyses by Okoye (*Report KH-R-23*, California Inst. Technol., 1970) and Lau and Krishnappen (*J. Hydraulics Div.*, ASCE, 103, p. 1173, 1977) indicate that the width of the channel affects the transverse dispersion no matter how wide the channel. However, it is not yet clear exactly what role the width plays. Equation 5.2.23 is usually correct within 50 percent for uniform straight rectangular channels (Fischer et al., 1979).

Transverse Dispersion in Natural Rivers. Of course, natural rivers differ from uniform rectangular channels in several ways: The depth varies irregularly, the channel is likely to curve, and there may be irregularities along the banks such as man-made structures or points of land. Holley et al. (*J. Hydraulic Res.*, 10, p. 27, 1972) report the impact of regular cross-sectional depth variations on transverse dispersion, but generally this is not a major factor.

Sidewall irregularities of various types are very common in natural channels and have a major effect on transverse dispersion because of the presence of secondary lateral flows around the irregularities. Generally, the larger the irregularity, the faster transverse mixing occurs. Fischer et al. (1979) report that sidewall irregularities increase E_t/du^* to between 0.3 and 0.7.

Of the three characteristics affecting transverse dispersion in natural channels, channel curvature is the best understood. Bends increase E_t/du^* with reported values ranging from 0.4 to 0.8 for slowly meandering rivers and even higher values in sharply curving channels (Fischer et al., 1979). This increase is due to lateral flows (often called secondary currents) that occur along the bend. Centrifugal forces induce flow toward the outside bank at the water surface with a compensating reverse flow near the channel bottom. Fischer (*Water Resourc. Res.*, 5, p. 496, 1969) used the velocity distribution around a curve to determine that:

$$E_t \propto \left(\frac{U}{u^*}\right)^2 \left(\frac{d}{R}\right)^2 \qquad (5.2.24)$$

where R is the radius of curvature [L]. He reported that the constant of proportionality was approximately 25. However, Sayre and Yeh (*Report No. 145*, Iowa Inst. of Hydraulic Res., Univ. of Iowa, 1973) subsequently found that E_t varies along the bend, from a low of one-half its average value in the upstream portion of the bend, to twice the average in the downstream portion. Finally, Holley and Jirka (1986) point out that stream bends must be sufficiently long and curved to produce a secondary current that affects dispersion. They give general guidelines on classifying meandering streams according to their potential influence on transverse mixing.

In summary, curves and sidewall irregularities increase transverse dispersion in natural streams, so that E_t/du^* is rarely less than 0.4. Table 5.2.5 presents a summary of field and experimental values of E_t/du^* based on compilations by Bowie et al. (1985) and Fischer et al. (1979). Fischer et al. (1979) suggest using $E_t/du^* = 0.6 \pm 50\%$ for slowly meandering streams with moderate sidewall irregularities, and higher values for sharply curving channels or rapid changes in channel geometry. A summary of references addressing transverse dispersion is provided in Table 5.2.6.

Recommended Practice

The complications of longitudinal and transverse dispersion in rivers are mitigated by an important fact: Dispersion is not a practical concern in many problems. For example, the most common problem in riverine water quality is the development of a wasteload allocation, the determination of allowable rates of discharge for flow and oxygen-demanding wastes from a continuous wastewater discharge. Typically, wasteload allocations are developed using a one-dimensional steady-state model of the river. Fortunately, the predicted longitudinal distribution of pollutant is highly insensitive to the dispersion

TABLE 5.2.5 Summary of Transverse Dispersion Measurements in Rivers

River	Transverse dispersion coefficient (m²/sec)	$\dfrac{E_t}{du^*}$
Missouri River, Nebraska[a]	0.07–0.13	0.3–0.65
Missouri River, Nebraska[b]	1.1	3.3
South River, Virginia[a]	0.005	0.3
Aristo Feeder Canal, New Mexico[a]	0.009	0.22
Bernado Conveyance Channel, New Mexico[a]	0.013	0.3
Athabasca River, Alberta, Canada[c]	0.041	1.16
	0.093	0.75
	0.067	0.41
	0.010	0.56
Beaver River, Alberta, Canada[c]	0.043	1.01
	0.020	2.54
Grand River, Ontario, Canada[d]	0.009	0.26
Danube River, Hungary[e]	0.038	0.25
Mississippi River, Minnesota[f]	0.171	2.03
Coraville Reservoir, Iowa River, Iowa[g]	0.5	50

[a] (Yotsukura and Cobb, *U.S. Geo. Survey Prof. Paper 582-C*, 1972).
[b] (Sayre and Yeh, *Report No. 145*, Iowa Inst. Hydraulic Res., Univ. Iowa, 1973).
[c] (Beltaos, *J. Hydraulics Div.*, ASCE, 106, p. 1607, 1980).
[d] (Lau and Krishnappan, *J. Hydraulics Div.*, ASCE, 107, p. 209, 1981).
[e] (Somlyódy, *J. Hydraulic Res.*, 20, p. 203, 1982).
[f] (Demetracopoulos and Stefan, *J. Environ. Eng.*, ASCE, 109, p. 685, 1982).
[g] (Mossman et al., *Water Res.*, 25, p. 1405, 1991).

TABLE 5.2.6 Summary of References on Transverse Diffusion in Rivers

Reference	Factors considered				Field data	Formula for E_t	Comments
	Turbulence	Channel irregularity	Bends	Other effects			
Yotsukura and Sayre[a]		×	×		×	×	Streamtube model
Holley and Abraham[b]				×	×		Computation from dye dispersion data
Holley and Jirka[c]			×				Analysis of bends
Beltaos[d]				×	×		Effects of ice cover
Weber and Schatzmann[e]	×	×		×			Laboratory experiments for several parameters
Lau and Krishnappan[f]	×					×	Laboratory experiments for several parameters

[a] (Yotsukura and Sayre, *Water Resourc. Res.*, 12, p. 695, 1976).
[b] (Holley and Abraham, *J. Hydraulics Div.*, ASCE, 99, p. 2313, 1973).
[c] (Holley and Jirka, *Tech. Report E-86-11*, Environ. and Water Quality Operational Studies, U.S. Army Corps of Eng., 1986).
[d] (Beltaos, *J. Hydraulics Div.*, ASCE, 106, p. 1607, 1980).
[e] (Weber and Schatzmann, *J. Hydraulic Eng.*, ASCE, 110, p. 423, 1984).
[f] (Lau and Krishnappan, *J. Hydraulics Div.*, ASCE, 103, p. 1173, 1977).

coefficient. In most cases, to predict the *steady-state* pollutant distribution from a *continuous* discharge, dispersion can be neglected altogether!

The reason that this general recommendation can be made is that pollutant concentration does not change rapidly along the river if the discharge is continuous. If the water at one location in the river is not drastically different from the water at a neighboring location, it makes little difference if the waters mix. The exceptions to this recommendation are several. The exceptions are those cases when concentration changes rapidly with position in the river as in sudden discharges, such as spills. Here, the concentration distribution is a small area of high concentration in an otherwise unpolluted river. Dispersion may be the single most important mechanism in reducing the concentration in the river as the spill moves downstream. Another exception occurs for a pollutant that degrades rapidly as it flows downstream. Even with a continuous discharge, the rapid loss of pollutant leads to significant concentration differences along the river and longitudinal mixing may be important.

A final important exception is the case of a shoreline discharge into a wide river. Here, the mixing of the pollutant across the river needs to be addressed and thus transverse mixing must be considered as well as longitudinal dispersion. A one-dimensional model could greatly underpredict concentrations in the river with the model's implicit assumption that the discharge has mixed throughout the river cross section. In fact, the shoreline plume is diluted by a smaller portion of the river's flow and is therefore more concentrated.

In those cases where dispersion is an important process, and a dispersion coefficient is required, we recommend the following approach. First, the validity of the Fickian diffusion model should be checked using Equation 5.2.20. Passing that criterion, a dispersion coefficient may be developed and applied. If there exist field measurements of mixing from the river in question, those data are generally preferred to any of the empirical formulae for predicting dispersion coefficients. Such data rarely exist, however, in which case the modeler should select and apply at least several pertinent relations from the references in Tables 5.2.4 and 5.2.6. These will yield a range of results, presenting the modeler with a judgment. If the results generally converge on a narrow range of results, an average of those would be a good choice. If there is scatter, a conservative approach would be to use the result giving the lowest dispersion coefficient, and thus the highest predicted concentrations.

5.2.5 DISPERSION IN ESTUARIES

Estuarine Mixing

The dispersive characteristics of a river become further complicated where a river nears the sea and becomes an estuary. In an estuary, the significant influence of tidal flows is added to the list of mechanisms that affect dispersion in rivers. Still another mechanism is added in stratified estuaries. In a stratified estuary the fresh river water flows atop the denser salt water that intrudes into the estuary from the sea. Not all estuaries are stratified, however: In estuaries in which the back-and-forth tidal flow is significant compared to the freshwater river flow, the salt and fresh water mix more or less completely through the depth of the estuary. Such estuaries are known as well-mixed or partially mixed depending on their degree of vertical stratification.

Fischer (*Inter. Symposium on Discharge of Sewage from Sea Outfalls*, London, 1974, Pergamon Press, 1975) and Fischer et al. (1979) describe the many physical phenomena that cause mixing in estuaries. The major mechanisms are described in the sections to follow.

Tidal Pumping.   *Tidal pumping* is the residual circulation pattern created by the different patterns of flow that exist between the flood tide and the ebb tide. Different flow patterns are set up by the geometry and bathymetry of the estuary. For example, in a bay with a narrow inlet, flow on the flood tide creates a jet-like flow into the bay when it enters the inlet. On the ebb tide, however, flow in the bay tends to flow radially into the inlet. Averaged over a tidal cycle, the incoming jet flow and outgoing radial flow create a net circulation pattern (Figure 5.2.5).

Gravitational Circulation.   *Gravitational circulation* refers to the net transport created by the flow of different density waters in a partially stratified estuary. In the surface layers of the estuary, there is a net seaward transport resulting from the river's freshwater flow. In the deeper layers, there is a net landward flow of saline water from the sea. At mid-depth, there is a vertically upwards flow associated with underlying salt water being entrained and mixed into the freshwater flow. The gravitational circulation

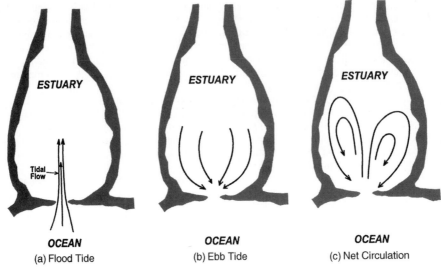

FIGURE 5.2.5   Illustration of tidal pumping in a bay.

also creates a net horizontal circulation with a landward flow in the deeper parts of the estuary and a seaward flow in the shallow sections.

Shear Flow Dispersion.   Longitudinal dispersion caused by cross-sectional velocity variations occurs in an estuary in much the same fashion as it occurs in a river. It is modified, however, by the effects of the oscillating tidal flow. If the tidal period (the time period of one tidal oscillation) is shorter than the time period for cross-sectional mixing, then the dispersion coefficient is effectively reduced. Holley et al. (*J. Hydraulics Div.*, ASCE, 96, p. 1691, 1970) discuss this effect in detail and give a relation for determining the amount of decrease in the dispersion coefficient.

Dead Zones.   Just as in rivers, *dead zones* increase longitudinal dispersion in estuaries. The effect may be magnified in estuaries by what Fischer et al. (1979) call "tidal trapping." Because of tidal oscillations, pollutants trapped in dead zones and tributary branches during the flood tide may rejoin the mainstem flow during the ebb tide in a different part of the flow than that from which they originated. The effect is an important dispersion mechanism in some estuaries.

Wind Mixing.   Mixing induced by the wind may be important in some large estuaries and bays. It is most important in situations where a steady wind sets up current patterns that create large-scale mixing.

Descriptions of Estuarine Dispersion

Capturing all of the complicated mechanisms described above in a single dispersion coefficient is an ambitious undertaking. It is not surprising that the results are less than fully successful. Reviews of the state of the art in modeling dispersion in estuaries by Fischer (1976), Chatwin and Allen (1985) both end with pessimistic assessments of the ability to predict appropriate dispersion coefficients for estuaries. A ready-to-use formula for the prediction of dispersion coefficients in estuaries simply does not exist.

Tidally Averaged Dispersion.   A confusing aspect of determining the dispersion coefficient in estuaries is the choice of the operating time period. The most common approach is the tidally averaged model. In these models, concentration predictions are intended to represent the average over the tidal cycle. Flow in the model is the freshwater flow only; the oscillatory flow of the tide is not explicitly considered. The dispersion coefficient for these models must capture all dispersive effects of the oscillatory tidal flow and is typically much larger than the dispersion coefficients seen in rivers.

Real-Time Dispersion. A second approach to predicting the dispersion equation is to use a time-varying, or "real-time" model. In this approach, the tidal flow is included in the advective term and the dispersion coefficient need capture only the effects of velocity shear and gravitational circulation. This approach is more deterministic than the tidally averaged approach, and some empirical formulae have been developed to predict the time-varying dispersion coefficient, E_L. This dispersion coefficient is appropriate for use in an equation such as Equation 5.2.6, the one dimensional advection-dispersion equation. However, when Equation 5.2.6 is used with the time-varying dispersion coefficient, the velocity term must include both the freshwater flow and a tidal flow that varies sinusoidally through the tidal cycle.

Prediction of the Dispersion Coefficient. There are few formulae in the literature to use in determining the longitudinal dispersion coefficient in an estuary. Chatwin and Allen (1985) give the following general equation for the longitudinal dispersion coefficient that accounts for both shear dispersion and gravitational circulation:

$$E_L = \alpha_0 + \alpha_2 \left(\frac{\partial S}{\partial x} \right)^2 \tag{5.2.25}$$

where S is the cross-sectionally averaged salinity $[M/L^3]$, used as an indicator of density; and α_0, α_2 are empirical coefficients.

Chatwin and Allen (1985) do not give guidelines for determining α_0 and α_2, which are best found on a case-specific basis.

Thatcher and Harleman (*J. Environ. Eng. Div.*, ASCE, 107, p. 11, 1981) use a formula with a different dependence on the salinity (density) gradient. It is shown here in a modified form that is more directly comparable to Equation 5.2.25:

$$E_L = m E_T + K \frac{L}{S_{\max}} \left(\frac{\partial S}{\partial x} \right) \tag{5.2.26}$$

where m is a constant [dimensionless] that accounts for the effects of modified shear flow dispersion and channel irregularities; E_T is the dispersion coefficient $[L^2/T]$ based on a Taylor-like dispersion coefficient, for example as computed by Equation 5.2.21 or 5.2.22; K is a parameter $[L^2/T]$ that depends upon the stratification in the estuary; $S_{\max}$ is the maximum salinity at the ocean entrance $[M/L^3]$; and L is the length of the estuary $[L]$ from the ocean entrance to the head of tide. $S_{\max}$ and L are simply normalizing constants to preserve the equation's dimensionality. The stratification parameter varies between about $4 \times 10^{-4} U_{\max} L$ for an unstratified estuary to $1.1 \times 10^{-3} U_{\max} L$ for a strongly stratified estuary, where $U_{\max}$ is the maximum tidal velocity.

Thatcher and Harleman (1981) used Equation 5.2.26 to produce a very satisfactory simulation of long-term salinity in the Delaware River Estuary, a demanding test of the model. Nonetheless, Fischer (*J. Environ. Eng. Div.*, ASCE, 107, p. 874, 1981) severely criticized the formulation for its departure from the form of Equation 5.2.25. Thatcher and Harleman (*J. Environ. Eng.*, ASCE, 109, p. 253, 1983) responded with a physical basis for the different formulation. This exchange of views is important as an illustration of the considerable uncertainty in predicting longitudinal dispersion coefficients in estuaries.

In the face of this uncertainty, a recommended procedure for developing a coefficient value for a particular study is to either: (1) conduct field experiments to determine the coefficient by direct experimentation, or (2) calibrate the dispersion coefficient in a transport model of the salt distribution in the estuary. Salt is useful in this regard because it is a conservative substance and not subject to the complicating factors of degradation or decay. For general background, Table 5.2.7 summarizes field observations of tidally averaged dispersion coefficients.

TABLE 5.2.7   Summary of Tidally Averaged Longitudinal
Dispersion Measurements in Estuaries

Estuary	Longitudinal dispersion coefficient (m^2/sec)
Hudson River, New York[a]	600
Hudson River, New York[b]	450–1500
Hudson River, New York[c]	160
Delaware River[a]	150
Delaware River[c]	500–1500
East River, New York[a]	300
Cooper River, South Carolina[a]	900
Savannah River, Georgia, S.C.[a]	300–600
Lower Raritan River, New Jersey[a]	150
South River, New Jersey[a]	150
Houston Ship Channel, Texas[a]	810
Cape Fear River, North Carolina[a]	60–300
Potomac River, Virginia[a]	30–300
Potomac River, Virginia[b]	6–60
Potomac River, Virginia[c]	55
Potomac River, Virginia[d]	20–100
Compton Creek, New Jersey[a]	30
Fishkill Creek, New Jersey[a]	15–30
South San Francisco Bay[b]	20–200
North San Francisco Bay[b]	50–1900
Hudson River, New York[e]	47 ± 6
	162 ± 22
Conwy Estuary, Wales[f]	0–35
Bremer River, Australia[g]	27.2
Oxley Creek, Australia[g]	7–12

[a] (Hydroscience, Inc., "Simplified Mathematical Modeling of Water Quality," U.S. EPA, 1971).
[b] (Officer, *Physical Oceanography of Estuaries (and Associated Coastal Waters)*, Wiley and Sons, 1976).
[c] (Thatcher and Harleman, *R. M. Parsons Lab. Report No. 144*, MIT, 1972).
[d] (Hetling and O'Connell, *Water Resourc. Res.*, 2, p. 825, 1966).
[e] (Clark et al., *Environ. Sci. Technol.*, 30, p. 1527, 1996).
[f] (Guymer and West, *J. Hydraulic Eng.*, ASCE, 118, p. 718, 1992).
[g] (Steele, *Water Resourc. Res.*, 27, p. 839, 1991).

GENERAL REFERENCES

Bowie, G. L., Mills, W. B., Porcella, D. B., Campbell, C. L., Pagenkopf, J. R., Rupp, G. L., Johnson, K. M., Chan, P. W. H., Gherini, S. A., and Chamberlin, C. E., "Rates, Constants, and Kinetics Formulations in Surface Water Quality Modeling," 2nd Edition, Report EPA/600/3-85/040, U.S. Environmental Protection Agency, Athens, Georgia, 1985, http://www.epa.gov/ORD/WebPubs/surfaceH2O/surface.html.

Chatwin, P. C., and Allen, C. M., "Mathematical Models of Dispersion in Rivers and Estuaries," *Annual Review of Fluid Mechanics* 17: 119, 1985.

Crank, J., *The Mathematics of Diffusion*, 2nd Edition, Oxford University Press, Oxford, 1975.

Fischer, H. B., "Mixing and Dispersion in Estuaries," *Annual Review of Fluid Mechanics* 8: 107, 1976.

Fischer, H. B., List, E. J., Koh, R. C. Y., Imberger, J., and Brooks, N. H., *Mixing in Inland and Coastal Waters*, Academic Press, New York, 1979.

Holley, E. R., and Jirka, G. H., "Mixing in Rivers," Technical Report E-86-11, Environmental and Water Quality Operational Studies, U.S. Army Corps of Engineers, Vicksburg, Mississippi, 1986.

Taylor, G. I., "The Present Position of Turbulent Diffusion," *Advances in Geophysics* 6: 101, 1959.

CHAPTER 6

BIOLOGY

J. Gordon Edwards

Dr. Edwards is emeritus professor of biology at San Jose State University. He taught entomology there for 49 years and is currently the curator of the University's entomology collection.

SECTION 6.1

BIOLOGY

Modern science in sixteenth century Europe produced a phenomenal growth of human understanding of chemistry and physics. Copernicus, Bruno, and Galileo were magnificent, and they were followed by Isaac Newton and his scientific laws. Biological concepts took longer to become established because of the great multiplicity and complexity of principles.

As naturalists studied the forms of life around them, and travelers noticed distinct local differences between very similar populations, it became more difficult to describe all the different types of plants and animals well enough to distinguish them from all other kinds. Natural species were believed to have been divinely created, and therefore immutable, until investigators such as Buffon, Lamarck, Wallace, and Darwin dared to suggest that species had developed from ancestors unlike themselves and were still evolving by means of natural selection. (It was already known that domesticated dogs, sheep, horses, and cattle could produce offspring unlike themselves as a result of selective breeding.) Eventually, enough evidence accumulated to prove that natural evolution occurred rather frequently, causing the formation of "new" species. Instead of clinging to the concept of immutable "typological species," naturalists began to theorize that new forms had resulted from natural selection. Studies and observations continued, and educated scientists arrived at the species concept of the modern era. Species were observed to be composed of individuals that share a common gene pool and are "reproductively isolated" from other such species. (That is, they are not capable of normally and freely interbreeding with members of other species.)

In 1758 Swedish naturalist Carolus Linnaeus produced his *Systema Naturae*, and almost all scientists since that time have followed the prescribed procedure for assigning a genus and species name to each truly different kind of animal or plant. It was realized that, to be of value to other scientists worldwide, the name applied to a species must be permanently preserved and never used for any other species. Biologists devoted years and decades to defining terms, explaining difficulties, and establishing international procedures that must be followed in order to provide animals and plants with appropriate scientific names. They must be described so accurately that scientists around the world can recognize them by the correct genus and species names. Without such restrictive criteria, no scientist can be sure what is meant by the names other scientists use, and it would be impossible for anyone to identify animals or plants with any certainty. The names must have the same meaning to everyone. To illustrate the necessity and the value of the system, consider just one group of homopterous insects (leaf hoppers and other "hoppers"). Linnaeus listed only 42 species of them, but by 1930 more than 30,000 species had been described and named. Three of the greatest authorities in the world, Drs. Ernst Mayr, Gorton Linsley, and Robert Usinger, in their classic book *Methods and Principles of Systematic Zoology* reminded us that many kinds of individuals have striking differences in structure "owing to sexual dimorphism, polymorphism, age differences, and other forms of individual variation . . . Such forms have often been described as different species, but as soon as they were found to be members of a single interbreeding population they were deprived of their species status, regardless of the degree of morphological difference."

They also note that "some species have several hundred populations that differ from each other significantly (in the statistical sense). Naming all of these was not realistic, because completely uniform population groups do not exist in sexually reproducing species."

Leading biologists everywhere cooperated, and every nation in the world joined in the International Commission on Zoological Nomenclature. They agreed on an international "Code," for all to follow when describing or discussing species and other aspects of taxonomy, systematics, and zoological nomenclature. The Code specifies that "To be regarded as published, the description must be reproduced in ink, on paper, by some method that assures numerous identical copies will be issued for scientific public permanent record, and be obtainable by purchase or free distribution." The Code was repeatedly improved during the next century by international conferences of the leading systematists and taxonomists in the world. In 1961 J. Chester Bradley, President of the International Commission, presided over an extensive updating of the Code, and the results were published by the International Trust for Zoological Nomenclature in London. That Code is still considered to be the only legitimate guide to taxonomic and nomenclatural procedures by scientists worldwide. Without such universal criteria, no scientist could be sure what is meant by the names used by other scientists, and it would be impossible for anyone to identify animals or plants to species with any certainty.

Incredible as it seems, all those successes by generations of brilliant scientists during a century of progress were recently attacked by a few activists who decided to refer to what they call "species" of animals (and plants) without complying with the International Commission's Code and without even publishing descriptions or providing scientific names. Those activists wanted indistinct subspecies and local populations to also be called "species," for political purposes. They avoided acceptance of the scientific definitions and descriptions of species and referred to many animals and plants by vague, imprecise, unscientific names. They even refer to some populations as "endangered species" if they are only uncommon locally. With such procedures in mind, they wrote the Endangered Species Conservation Act, and their scheme was published in the *Federal Register* 35: 8491–8497, 2 June 1970. They wanted the U.S. Fish and Wildlife Service to make decisions about their vague proposals. A Bill in the U.S. House of Representatives proposed limiting such protection to real species and would require scientific standards to be used when enforcing the Act (*Federal Register* 41: 47197, 27 Oct 76). In 1979 the General Accounting Office urged Congress to amend the Endangered Species Act and redefine species to exclude local populations (*Washington Balance* 27 July 1979). The U.S. Fish and Wildlife Service rejected that, and eventually more than 30 percent of their listed "endangered species" were indistinct, nonscientific, local populations. Tom Foley, Democrat Speaker of the House, warned that the Act reinforces an environmentalist utopian view that we might by federal law guarantee that no species would ever become extinct. A *Wall Street Journal* editorial (15 January 1992) stated: "The plain intent was that extinction be stopped in all cases, regardless of cost."

The Act also devised a definition of "taking" endangered species. U.S. Federal Code, Title 16, Section 1532(19) specified: "The term '*take*' means to harrass, harm, pursue, hunt, shoot, wound, kill, trap, capture, or collect, or to attempt to engage in any such conduct." It also prohibits destruction or alteration of the "critical habitat" of any endangered or threatened species, subspecies, or local population. Anyone convicted of such a "taking," on either public or private land, faces severe consequences. The alleged goal was to save the natural environments where plants and animals live, by preventing humans from altering them.

Title 16 also specified that "the Secretary of Interior shall publish in the *Federal Register* a list of all species determined to be threatened, and specify for each any critical habitat within such range." So, the information is all available to everyone who has access to the *Federal Register*. (The 1996 edition was 64,592 pages long, and contained 4937 final rules.)

The general public is unaware of the situation described above, and when an "endangered species" is mentioned, most people assume the population really is a legitimate biological species. Even some biologists may believe the reports refer to genetically unique entities that have been described by scientists. That belief is very wrong! The word "species" has been misused to gain public and political support and to generate more funding.

The strong relationships between members of distinct biological species are maintained because they do not interbreed with members of other species. Each local population may be referred to as a "deme," and within most species or subspecies there are numerous demes, but they are very difficult to distinguish. Mating occurs more freely within infraspecific demes than between members of isolated populations belonging to the same species. Demes are not likely to persist indefinitely as distinct units. Some species, as well as distinct subspecies, have been listed as "endangered" because they contain demes that are very scarce. There is a tremendous difference between the activists' "endangered species" and legitimate species that are endangered. Discussions and publications by conscientious scientists must differentiate between "true biological species" and "Endangered Species Act species," to avoid being misunderstood. Biologists around the world (except for those activists) still define species as being reproductively isolated and genetically distinctive natural populations. American scientists have not interfered with the requirements for publishing legitimate species descriptions in accordance with the established rules of the International Commission on Zoological Nomenclature.

A famous conflict involving the Endangered Species Act dealt with the snail darter and the Tellico Dam. A suit was filed to prevent the dam from being closed, because "it would flood the only known home" of the tiny fish, but they were then found living in four other streams! The Tennessee Valley Authority objected, pointing out that the fish was not a described species. The U.S. Fish and Wildlife Service acknowledged the lack of a legitimate published species description but said "To delay listing as endangered until a species description is published would thwart the purpose of the endangered species act." (Which was to halt developments.) The General Accounting Office pointed out that the government's "Endangered Species" funds were being used improperly, "to acquire habitats for species that are not highly threatened." A U.S. Supreme Court decision ruled that Congress could permit a species to become extinct if the benefits clearly outweighed the importance of conserving that species. A Congressional committee, locally called "The God Squad," said that environmentalists "applied the Act solely to delay major energy and other projects" and quickly removed the snail darter from endangered species protection. The dam was soon completed.

6.1.1 NATIONAL BIOLOGICAL SURVEY

During the 1992 Earth Summit in Brazil, Agenda 21 proposed a "National Biological Survey" to be created under our U.S. Department of the Interior as a permanent agency. It was to be in place by October 1993, with $180 million provided by Interior Secretary Bruce Babbitt. The survey would map the whole country's biology and regulate it's development "because that is our obligation as set forth in the Endangered Species Act," said Babbitt [1]. The director of the Sierra Club's Legal Defense Fund said that "environmental atrocities are leading to humanity's reckless slide toward environmental suicide." He joined Vice President Al Gore in singling out automobiles and single-family homes as "the American way of life that cannot be sustained."

6.1.2 SIMILARITY OF APPEARANCE CASES

A very important modification of the Endangered Species Act is included in U.S. Code, Title 16, Section 1533 (8)(2)(e). The issue is referred to there as SIMILARITY OF APPEARANCE CASES (capitalized in the original document). It specifies "The Secretary may treat any species as an endangered species or threatened species, even though it is *not* listed, if he finds that such species so closely resembles in appearance a listed species that *enforcement personnel* would have *substantial difficulty in attempting to differentiate between* the "listed" and "unlisted" species (emphasis added).

There are large numbers of scientifically legitimate "sibling species" of birds and mammals, which are really different species but look so similar that they often cannot be distinguished, even by specialists. However, they are "reproductively isolated," which means they are different species and cannot interbreed. If one of these sibling species is officially listed as endangered or threatened, the other is equally protected, by the Act, even if it is a common species.

The U.S. Department of the Interior proposed adding eight common species of crocodiles to the endangered species list simply because they look like eight other species that are believed to be endangered (*Federal Register* 42: 18287, 22 April 1977).

In the field of entomology, great numbers of specimens are indistinguishable except by experts. Many species of junebugs (beetles in the scarab genus *Phyllophaga*) are so similar that specialists must dissect the male genitalia to determine the species, and the females can only be identified according to the males with which they consort. Similar difficulties occur in hundreds of other genera of insects. Very few "enforcement personnel" in the U.S. Department of the Interior have the ability to distinguish between species of similar ground beetles, click beetles, or scarab beetles, or even to know in which of the 120 different families of beetles individual specimens belong. The same is true of flies, bees, wasps, ants, moths, spiders, mites, and thousands of other invertebrates. It is even more impossible for them to identify the immature stages of arthropods, and those comprise five-sixths of the animal kingdom! It is unfortunate that the framers of the Endangered Species Act were apparently not aware of this. They might then have limited their proposals to "Endangered Vertebrates," which would have avoided problems with unidentifiable animals. The U.S. Fish and Wildlife Service should have avoided "listing" species of insects, spiders, mussels, and other animals and plants that cannot be identified by enforcement personnel. They could still delist those that have already been listed as endangered or threatened. Otherwise, how far will Endangered Species Act enforcers go to protect the thousands of nonendangered arthropods and plants that "resemble in appearance" so-called "endangered species" but cannot be identified by "enforcement personnel?"

6.1.3 ECOSYSTEMS

After tremendous failures of the Endangered Species Act, Bruce Babbitt must have realized that he could not achieve further success with that Act. He then proposed a shift from protecting endangered species to seeking long-term protection of whole ecosystems and all of their inhabitants, explaining "That would protect *all* populations in the ecosystem, even if *none* are rare or endangered" (*Santa Rosa Press Democrat*, 17 February 1993). The World Wildlife Fund, the Nature Conservancy, and the World Bank immediately set their sights on saving whole "endangered ecosystems," rather than individual species [2]. U.S. Interior Department spokesmen agreed with sincere scientists who were concerned that "ecosystem management" was simply another move toward massive federal land-use regulation.

But what **is** an ecosystem? Interior Secretary Babbitt was asked that, in a joint hearing before the 103rd Congress, on 15 July 1993.

A question, by Representative Jay Dickey (Ark) was: "In your mind, Secretary Babbitt, what is an ecosystem, how will one be defined, and how will you differentiate one from another?"

Answer: "To some degree, an ecosystem is in the eye of the beholder."

Q. "Would you like to elaborate?"

A. "I can put it in specific context. The timber problem and the salmon problem drives you to an ecosystem which essentially runs from the crest of the Cascades to the Pacific Ocean from approximately Puget Sound to the beginning of the Sierra Nevada in California. It is characterized by a lot of the commonalities. Stream drainage is certainly a big one. So that is an ecosystem."

Later, Babbitt stated that: "An ecosystem is any unit of landscape. It might be a watershed, a river basin, an estuary, a forest—it could be a unit of desert. It could be the entire biosphere. On the other hand, it might be the organisms on a Petri plate.... We usually think of life as extending from the depths of the ocean to perhaps the top of the stratosphere, 25,000 feet up, so an ecosystem could be that inclusive." Obviously, it would not be difficult for Babbitt to revamp the Endangered Species Act, and change it to the "Endangered Ecosystem Act," with greatly increased funding.

The General Accounting Office observed that "ecosystems may be any size or shape, from a drop of water to the entire planet. They change constantly in time and space. Regardless of human activities, the planet displays an endless series of overlapping, intermingling, constantly-changing ecosystems."

The National Research Council pointed out that: (a) there is no agreed-upon classification system for ecosystems; (b) no accepted list of core ecosystem attributes exists; (c) protocols for sampling, measuring, and recording data are not defined; and (d) scientists cannot predict which species and which interactions are key to determining the makeup and location of them.

6.1.4 BIODIVERSITY

In the 1980s the term "biodiversity," was introduced at a conference sponsored by the Smithsonian Institution and the National Research Council. E. O. Wilson, an outstanding ant specialist, referred to biodiversity as "the variation in the entirety of life on the planet." He expressed concern about the loss of natural habitats as a result of human population increases, which he believed would cause the extinction of an infinite number of species.

The Great Smoky Mountains National Park in Tennessee is participating in an "All Taxa Biodiversity Inventory," sponsored by the National Park Service and the nonprofit group Discover Life in America. They estimate that there are about 100,000 species in the park, but fewer than 10,000 have been discovered there. The program is similar to one operated by Professor Daniel Janzen in Costa Rica, which was very successful for several years. Permitting, and actually encouraging, such valuable biological research is a giant step toward analyzing biodiversity in the United States. Perhaps other national parks (and state parks) might similarly encourage, rather than continuing to discourage, the collection and identification of plant and animal species. Unfortunately, most parks, both here and abroad, oppose the collecting of insect and plant specimens. More than 40 percent of the animal species in each area are insects, and most of them cannot be accurately identified to species unless they are examined under a microscope by specialists with extensive entomological libraries.

There are great differences of opinion about biodiversity and about the rates of species evolution and extinction. There appear to be abundant examples of past evolution, based on present population levels and apparent phylogenetic relationships. However, details of what the future may bring are impossible to accurately hypothesize. There are too many environmental variables and too little stability of living populations to permit much confidence in predictions. Current biodiversity is exciting to study and attempt to interpret, but all conclusions about future changes are exceedingly tenuous.

A biodiversity meeting was held at Stanford University in 1998, sponsored by the Global Invasive Species Program. At the meeting, it was concluded that the major threats to biodiversity are (a) habitat changes, (b) invasion by exotic species, (intentional or accidental), and (c) climate shifts. Other threats include global warming, greenhouse effects, global cooling, the ozone hole, and extreme ultraviolet exposure [3].

At the 1992 United Nations Earth Summit in Rio de Janeiro, Brazil, the United States, under President Bush, was the only major developed nation to balk at signing the BioDiversity Treaty (also called the Convention on Biological Diversity). Agenda 21 of the Treaty sought to transfer wealth from developed nations to third world nations under the guise of preserving the environment. The cost was estimated at hundreds of billions of dollars. There was no agreed-upon definition of "biodiversity," but it generally dealt with variability between and within species, infraspecific groups, and their ecosystems.

Dr. Terry Erwin, of the Smithsonian Institution, collected insects from the tropical American rainforest canopy for years, by fogging the trees with natural insecticides and collecting the specimens that fell onto tarps on the ground. More than 1200 species of beetles fell from 19 trees in one small area, and beetles were fewer than half of the total number of insects obtained there. Erwin repeated that process in a dozen other areas in Central and South America. As a result he estimates that the number of real species of insects in the world approaches 30 million. That estimate is accepted by most scientists doing research in tropical biology (*Smithsonian* magazine, September 1986, 80–90).

On 4 June 1993, U.S. Secretary of State Madeleine Albright signed the International Biodiversity treaty, but without congressional approval. The U.S. Senate refused to ratify it, but the Clinton administration's "ecosystem management policy" agreed with the Convention's determination to control the use of both public and private land in order to achieve the objectives of the United Nations. The worldwide environmental plan is under the control of the United Nations, as outlined in 1977 by G. O. Barney, in *The Unfinished Agenda* [4] and in 1988 by T. A. Comp, in *Blueprint for the Environment* (Salt Lake City, Howe Brothers) [5]. Those books were prepared by members of more than a dozen nongovernment environmental organizations, referred to by the United Nations as NGOs.

At the Earth Summit Reed Noss, a member of the board of directors of the Wildlands Project, stated in the Project's book, *Wild Earth*, that "The collective needs of non-human species must take precedence over the needs and desires of humans" [6] (*Ecologic* Jan/Feb 1998, page 19). John Davis, editor of *Wild Earth* and a director of the Wildlands Project board, stated that "people would not be required to relocate, if they would refrain from any use of motors, guns, or cows" (In Special Issue of *Wild Earth* 1992, page 3). Davis later wrote that paradise means "the end of industrialization" and "everything civilized must go" [7]. The Wildlands Project was prepared under contract with the National Audubon Society and the Nature Conservancy. It is discussed in detail in *The Global Biodiversity Assessment* (1400 pages), published in 1995 by the Cambridge University Press, funded by the United Nations Environment Program [8]. Details about Agenda 21 and the U.N. Conference on Environment and Development fill 800 pages [9].

Maurice Strong, the head of the 1992 United Nations Earth Summit in Rio de Janeiro, asked "Isn't it the only hope for the planet that the industrial civilizations collapse? Isn't it our responsibility to bring that about?" (repeated in *British Columbia Report*, 18 May 1992).

Les Knight, of Population Crusade, stated "Total human extinction is the only answer to the planet's mounting environmental woes."

Many others are frustrated by human fecundity. Jacques Cousteau, in an interview that appeared in the November 1991 *UNESCO Courier*, said that "in order to stabilize world population, we need to eliminate 350,000 people per day. It is a terrible thing to say, but it's just as bad not to say it."

6.1.5 BIOTRANSFORMATION

What happens to the body tissues and chemicals in each of the plants and animals in food chains or food webs? Some chemicals are excreted or otherwise eliminated, but others may be converted into other chemical compounds or simply stored temporarily in other forms of life.

Food chains almost always begin with photosynthesis and end with degradation, but there may be dozens of "links" in the food chains. So many ramifications occur that the designation of "food webs" is much more realistic than "food chains" Some environmentalists alleged that fish-eating birds were endangered because of pesticides (especially DDT) in their food chains, but actual conditions showed that to be false. On Funk Island (in the North Atlantic) the number of gannets increased from 200 pairs in 1945 (when use of DDT began) to 2000 pairs in 1958 and 3000 pairs by 1970 (before DDT was banned). Murres increased there also, from 15,000 pairs in 1945 to 150,000 pairs in 1956 and a million and a half pairs by 1971 [10]. The presence of DDT in the environment obviously did not adversely affect the productivity of those fish-eating birds, or of any others.

6.1.6 BIOACCUMULATION

Green plants store energy from inorganic sources and then die and degrade because of the activities of insects, bacteria, and fungi. This is an example of a very simple food chain. In more complex food chains, the plants are eaten by herbivores, which then die and degrade, or the herbivores may be consumed by carnivores, which then die and decay. Other modifications involve herbivorous insects that are eaten by predaceous insects, birds, mammals, or amphibians, all of which ultimately die and degrade.

The green parts of plants contain chlorophyll, which causes water molecules to break down and release oxygen (O_2). Through the process of photosynthesis, leaves are able to produce sugars from carbon dioxide (CO_2) and water, in the presence of energy provided by light.

Energy from the light is transferred to chemical energy in the green plant, which is then utilized by animals that eat the plants (and by other animals that eat those animals).

$$6CO_2 + 6H_2O + light\ energy \longrightarrow C_6H_{12}O_6 + 6O_2\ (sugar\ and\ carbon\ dioxide)$$

In the process of respiration, both plants and animals alter the sugar by adding oxygen, which produces CO_2 and H_2O, along with chemical energy which can be either used or stored. Some of the chemical energy may remain available for centuries, in the form of coal.

Much of a plant's chemical energy is stored in an unstable compound called ATP (adenosine triphosphate). It is then capable of causing the rapid synthesis of sugar. Sugar is the primary source of energy used by plants to continue the transformation of many structures. Sugar may be changed into cellulose, which provides strong supportive tissues in the plants. Cellulose cannot be altered further by the plants. When it is eaten by animals, however, it may be digested by symbiotic bacteria or protozoans, or it may simply be passed through the digestive system unaltered.

In larger plants there must be mechanisms to move water and sugar from one part to other parts. Water is taken in by the roots and moved to every living part of the plant. Slender root hairs take water particles from the soil by the process of osmosis, which is the passage of water through semipermeable membranes from purer water into areas that contain dissolved substances such as sugars and salts. The water is moved upward by "root pressure," as a result of effective osmosis. In trees the water moves up through long slender tubes that form the xylem of the cambium layer (just under the bark). It has been found that a 6 percent saline solution can exert an osmotic pressure of 50 atmospheres, and one atmosphere equals 15 pounds of pressure per square inch. Capillary action inside the narrow xylem tubes is caused by the physical adherence of molecules to each other. Inside the tubes, water molecules cohere, because of the strong attraction between hydrogen and oxygen atoms. The "pull" of water up the narrow tubes is actually hundreds of pounds per square inch. High in trees water is released into the air through tiny pores (stomata) in the leaves, by the process of "transpiration." In this manner, a single corn plant has been found to transpire more than 2 gallons (7.5 liters) of water daily! The amount of water vapor released into the air around a tree top may amount to many gallons of water daily.

In addition to water, foods and other chemicals must also be moved about within the plant. Chains of living phloem cells, with porous "sieve tubes" through which liquid foods are passed, transport such essentials from the leaves where they abound, into the lower cells, including the roots.

Cool weather causes great changes in the leaves of some plants, most notably in deciduous trees. Chlorophyll is no longer produced, so the green colors fade. Orange, yellow, and red pigments in the leaves were previously obscured by the dense green pigments, but they now become visible. As a result, entire forests of deciduous trees take on resplendent colors, before the leaves fall to the ground and the tree becomes dormant until the following spring. The old leaves, as well as all other dead plant remains, begin to decay and produce nutrients for bacteria, fungi, and for new generations of plants.

6.1.7   *DEGRADATION OF LIFE FORMS*

When plants die, their cells and tissues are degraded in the environment, and the resultant chemicals are available for reuse. Trees and large animals would remain relatively unchanged after death, were it not for the activities of other forms of life. Wood-boring beetles gnaw networks of tunnels in dead wood, which affords access for fungi and bacteria; thus, in a few years even large tree trunks are pulverized and their stored nutrients become available to many other forms of life. Dead vertebrate animals are quickly invaded by carrion beetles and fly maggots, followed in a few days by fungi, bacteria, and other detritivores. Soon only the bones remain, and even those are soon devoured by creatures that need more calcium.

Much of the chemical energy stored by living plants and animals becomes available as nutrients in the soil, which helps sustain microorganisms and nourish new plant and animal growth. From the detritus, animals also obtain essential vitamins and minerals that they cannot synthesize themselves.

Fungi abound in dead organic matter, gleaning energy, carbon, and nitrogen from it (by enzyme actions). Fungi are eaten, in turn, by insects and their larvae. Those insects are the preferred food of predaceous invertebrates as well as many kinds of vertebrates. Most animals cannot immediately use the complex molecules in their food. They must digest it, breaking it down into amino acids, fatty acids, and sugars, which they can then synthesize into the nutrient molecules from which they extract their energy sources.

6.1.8 BIOMAGNIFICATION

A concern was developed that levels of chemicals, including pesticides, were magnified at each step up the food chain. Scientists who were aware of the physiological processes at work in living creatures realized that the concept of "biological magnification," or "biomagnification," was untenable. It was refuted by experimental results as well as by field analyses. If there were no metabolism of ingested chemical compounds, no urinary excretion of metabolites, and no fecal elimination of waste products, then all animals would forever retain all indigestible chemical residues they swallowed, inhaled, or absorbed through their skin. That would indeed result in biological magnification, but, except for heavy metals, it does NOT happen! Each animal loses those residues constantly by defecation, metabolism, and excretion.

As long as animals take in more chemicals than they lose, the amounts in their body temporarily increase. Concentrations of most insecticidal compounds in the environment are incredibly low, exposures to them are brief, and the amounts are seldom sufficient to adversely affect any vertebrates. Pesticide concentrations retained in blood and muscle tissues will be much lower than they were in the food that animal ate; however, in fatty tissues the concentrations may become temporarily higher than that in the animal's food. It is misleading to warn that "biomagnification up the food chain" has occurred, when only the fat tissue of the eater temporarily contains amounts that exceed the concentration in their food.

J. L. Hamelink wrote that "a hypothesis that biological magnification of pesticides was dependent upon passage of the residues through a food chain was rejected and a hypothesis that accumulation depends upon adsorption and solubility differences was proposed" [11].

A committee of the National Academy of Sciences (NAS) chaired by J. W. Kanwisher (30 January 1973) concluded that "the absence of concentration increase going up the food chain can be due either to varying solubilities or to biodegradation in a pool of pesticides relatively sequestered from their surroundings." Another NAS committee that year, chaired by G. R. Harvey, pointed out that "the measured concentrations of DDT in various organisms of the open Atlantic and the Gulf of Mexico give no support for food web concentration as a general phenomenon."

DDT concentrations in marine algae in Monterey Bay were found to be 0.8 part per million. Other researchers reported that the anchovies from that bay also contained 0.8 ppm of DDT. Because the major food of those anchovies was the marine algae, some degree of biomagnification might have been expected, but there was none. Sea lions in the bay also contained much lower concentrations of DDT than did the fish they ate [12].

D. J. Jefferies and B. N. Davis also considered the possibility of biomagnification. They reared earthworms for 20 days in soil containing 25 ppm of the organochlorine pesticide dieldrin. The worms then contained 18 to 25 ppm of the chemical. Those worms were used to provide captive thrushes with daily diets containing up to 5.6 ppm of dieldrin. Thrushes ate nothing but those earthworms for 6 weeks, after which the birds were found to have body residues of only 0.09 to 4.03 ppm of dieldrin. Instead of "biomagnification" there was less dieldrin in their body than was present in their daily diet [13]. This was also the case during DDT feeding experiments by many other researchers.

Diagrams in some controversial articles showed only 0.04 ppm DDT in wet-weight zooplankton, and increasing levels up a food chain through fish and ducks, to a hawk with 25 ppm of DDT in its brain. The increase from 0.04 to 25 ppm is only 600 times, but various authors have stated that DDT levels in the food chain increased 300,000 times or more. How did they get those figures?

They *started* the "food chain" with 3 parts per trillion of DDT in the water where the zooplankton lived, rather than with the 0.04 ppm in the zooplankton at the bottom of the food chain. Obviously "biological" magnification does not start until the chemical enters into some biological part of the food chain. When wet weight analyses of the zooplankton or phytoplankton are compared with the dry weight muscle sample of the fish that ate them, the difference in dilution causes an increase in the concentration of DDT in the dry sample. If both samples are equally diluted, there is very little increase. When the DDT in lean fish muscle is compared with that in fatty tissue of ducks that ate the fish, an increase appears because DDT always concentrates in lipids. (If lean muscles of fish and ducks are compared, there is little difference in DDT concentration.) By comparing different kinds of samples in each step of the food chain, some researchers have suspected biomagnification, but the differences were in different tissues, rather than in animals at different steps in the food chain, and did not really indicate significant biomagnification [14].

6.1.9 ECOCATASTROPHE

C. F. Wurster put some marine algae in large tanks of sea water and then added large amounts of DDT to see if it would harm the algae. Because DDT goes into solution in water at only 1.2 parts per billion, he said he "added ethanol in order to get a concentration of 500 parts per billion DDT" [15]. Ukeles Ravenna had earlier reported that 600 ppb DDT (without solvents) did not affect the algae [16], and D. W. Menzel found a *beneficial* effect on the algae after 700 ppb of DDT was added [17]. Wurster, however, alleged that DDT reduced photosynthesis and warned that "Since a substantial part of the world's photosynthesis is performed by phytoplankton, interference with this process could be important to the biosphere."

Paul Ehrlich expanded on Wurster's article and wrote what he called a "scenario" based upon it [18]. Titled "Ecocatastrophe," his scenario was published in *Ramparts* magazine in August 1969 and began with: "The end of the ocean came late in the summer of 1979, and it came even more rapidly than the biologists had expected. It was announced in a short paper in *Science*, but to ecologists it smacked of doomsday." Thousands of school children were required to read Ehrlich's article, with warnings that "humans are endangering the earth with pesticides." Ehrlich later published that identical article in a British journal called *The Year's Best Science Fiction* [19].

6.1.10 EXTINCTION

The Sierra Club listed 15 animals that they said had become extinct in the United States since 1980 [20]. None of the 15 were actually biological "species." It was pointed out by The National Wilderness Institute that 32 kinds of birds have become extinct in the United States, but only 6 of them were native to North America. In Hawaii 40 percent of the bird species have become extinct, and another 40 percent are now endangered. Much of this extinction was caused by bird diseases transmitted by nonnative mosquitoes (*Culex fatigans*). Other extinctions of birds and other native Hawaiian life were hastened by the introduction of pigs, sheep, cattle, and rats.

Of the six large mammals that have become extinct in North America, four were merely subspecies. They are:

- Eastern elk (*Cervus canadensis canadensis*) (last seen in 1880),
- Badlands bighorn sheep (*Ovis canadensis auduboni*) (last seen in 1910),
- Plains wolf (*Canis lupus nubilus*) (last seen in 1926),
- Florida black wolf (*Canis niger niger*) (a variant of the red wolves, which are still in a zoo. Both variants are now believed to be "coyotes"),

- The species, Steller sea cow (*Hydrodamalis gigas*) was last seen in Bering Sea, about 1768 (killed by Russian seal hunters),
- The species, Merriam's elk (*Cervus merriami*) was last seen in the Arizona Mountains about 1906.

The most famous extinct species in the United States was the Passenger pigeon (*Ectopistes migratorius*), which was destroyed for food, to protect crops, and to reduce the nuisance they caused. The last one died in the Cincinnati zoo in 1914. The Carolina parakeet (subspecies *Conuropsis carolinensis carolinensis*) was hunted out of existence for its brilliant feathers. The heath hen (subspecies *Tympanuchus cupido cupido*) was protected beginning in the 1880s and had increased to nearly 2000 birds by the late 1920s. A wildfire then swept the range, killing most of them, and the last one died in 1932.

Paul Ehrlich wrote "Only three forest birds went extinct during the destruction of the Eastern United States, but the deforestation was only transitory, and regrowth had occurred by the middle of this century" (*San Francisco Chronicle*, 21 May 1993).

6.1.11 PEREGRINE FALCONS

The eastern peregrine falcon (subspecies *Falco peregrinus anatum*) was driven to extinction by hunters, falconers, egg collectors, and people seeking to protect other birds from being slaughtered by them. William Hornaday, the president of the New York Zoological Society, wrote (in *Our Vanishing Wildlife*, 1913) that "they deserve death, but they are so rare that we need not take them into account." He also urged that "persons finding a nest should first shoot the male and female birds, then destroy the eggs or young" [21]. Tom Cade, the founder of the Peregrine Fund, said, during an interview with Gannett News Service, that "peregrine falcons completely disappeared from east of the Rocky Mountains in the 1960s, and the Fund has not been able to reestablish the subspecies that lived in the east. It is probably extinct" [22]. Egg collectors liked peregrine eggs. One collector in Philadelphia had several drawers full of those eggs and a Boston collector had over 700, according to Roger Tory Peterson, in *Birds Over America*, 1948, Dodd, Mead & Co., New York [23]. D. D. Berger cited "the prevalence, for 70 years, of fanatic egg collecting" [24], and J. N. Rice said "53 sets of eggs were taken in Vermont before 1934, and most nest eyries were deserted by 1940" [25] (reported in J. J. Hickey's *Peregrine Falcon Populations*, 1969). That was 5 years before any DDT was used in North America! DDT was never shown to harm peregrines, either in the wild or in caged birds, and was never responsible for eggshell thinning in the natural environment.

After the eastern subspecies became extinct (long before DDT was present), thousands of nonnative peregrines were produced in captivity in the United States by birds imported from Mexico, Canada, Scotland, and Europe and released into the eastern environment. Hundreds have also been produced in captivity in western states and released. Brian Walton, the director of the west coast Peregrine Fund, reported in *Chevron World* (Spring 1985) that each peregrine produced in captivity costs $1500.00 to $3000.00 and millions of dollars have been spent rearing and releasing more than 4000 of those foreign peregrines in the United States [26]. Now that so many of them have been released in this country, many biologists are urging that the species be taken off of the endangered species list. The National Peregrine Fund's director complained in 1985 that they were having trouble raising money "because 50 million people were starving in Ethiopia." The Fund, he said, "needs $1 million a year for the peregrine recovery program."

It seems untruthful to have the offspring of those foreign peregrines touted as proof of "the recovery of the eastern peregrines," because **none** of them were members of the extinct eastern subspecies *Falco peregrinus anatum*. Nevertheless, many environmentalists (and most news media) continue to say that the eastern peregrine population recovered simply because DDT was banned. Environmentalists who know the truth have avoided correcting those false statements (and their funding continues). Cade was disappointed in 1977 when a regional director of the U.S. Fish and Wildlife Service ordered that no more European peregrines be released in the eastern United States, and complained that "we are left with a large number of Spanish and Scottish peregrines on our hands" [27].

Frank Beebe, Canada's leading authority on raptors, wrote in his 1971 book, *The Myth of the Vanishing Peregrines*, that "It appears that the Canadian Arctic peregrines, not knowing how gravely ill they are, go right on reproducing in blissful unconcern of their desperate plight" [28].

6.1.12 BALD EAGLES

In 1921 W. G. VanName published an article in *Ecology* titled: "Threatened Extinction of the Bald Eagle" [29]. Alaska paid bounties on 128,000 bald eagles, up to 1952, 8 years after the use of DDT began. In 1930 (15 years before DDT) ornithologists reported that there were only 10 nesting bald eagles in Pennsylvania, 15 in the Washington, DC, area, and none in most of New England. *Bird Lore* (32: 165, 1930) reported "this will give you some idea of the rarity of the eagle in eastern U.S" [30]. So, bald eagles were nearly extinct long before DDT or other man-made pesticides were produced.

The Hawk Mountain Sanctuary Association reported that the number of bald eagles migrating through Pennsylvania more than doubled during the first 6 years of heavy DDT applications in eastern North America (J. W. Taylor gave details in *Summaries of Hawk Mountain Migrations*, 1934 to 1970) [31]. In 1941, before DDT was used, the Audubon Christmas Bird Counts recorded only 197 bald eagles [32], but after years of heavy DDT use they recorded 891 bald eagles in 1961 [33].

In 1960–1964 the U.S. Fish and Wildlife Service Center in Patuxent, Maryland, autopsied 76 bald eagles that were found dead in the United States and reported that 71 percent had died violently (shot, electrocuted, or flew into towers and buildings, and four died of diseases, but none were poisoned by pesticides. It was their conclusion that "the role of pesticides has been greatly exaggerated" [34]. From 1964 to 1980 the Fish and Wildlife Service autopsied 652 more dead bald eangles. Most had been shot, and the majority of the others also died violently. There were a few cases of suspected dieldrin poisoning, but no deaths attributed to DDT [35]. They confessed that no analyses were made for mercury, lead, or PCBs, all of which are much more destructive to birds and their eggshells than DDT, DDD, or DDE. The Fish and Wildlife Service fed large amounts of DDT to caged bald eagles for 112 days (up to 4000 mg/kg), with no adverse effects [36].

In 1973 Everglades National Park biologist William Robertson studied 50 to 55 breeding pairs in the Park and said: "I know of no evidence that the region ever supported a larger number of nesting bald eagles" [37].

From 1973 to 1988 millions of dollars was spent in the United States for eagle breeding and rearing programs. As a result more eagles were seen by people in almost every part of the United States. In 1983 New York state had only three active bald eagle nests, and then they imported 150 eagles from Alaska. Peter Nye reported in *Natural History* magazine (May 1992) that in 1940 there were only a few pairs of bald eagles, "yet the oftmentioned culprit DDT wasn't there until the 1950s, when the last few nesting eagles were already struggling for survival." It is remarkable that many environmentalists, and most media sources, continue to mistakenly report that the noble birds "have recovered from the effects of DDT, which was banned in 1972." The truth is that the recovery was not correlated with either the presence or the later absence of DDT. Truth about environmental matters has become exceedingly rare!

6.1.13 FAILURE OF THE ENDANGERED SPECIES ACT

In the past 20 years, 7 of the more than 900 listed endangered species of animals and plants in the continental United States have become extinct. Others were removed from the endangered species list after it was found that they had been listed erroneously and additional populations were found. Such errors have been documented and publicized by the National Wildlife Institute [38]. More than 5000 additional animal and plant "species" await listing as "endangered" or "threatened." Representative Gerry Studds introduced a Bill in the House that would hasten the listing of the first three thousand of those. The cost of just listing one as endangered or threatened is about $60,000.00 and if it is later determined that the listing was an error, it costs another $30,000.00 to "delist" it. The National Wilderness Institute has carefully researched the claims made by the Environmental Protection Agency, the U.S. Fish and Wildlife Service, and others, regarding "endangered species." They point out that the Endangered Species Act does not require any valid scientific or reliable proof that a species is actually endangered before listing it. Instead, the Fish and Wildlife Service may simply rely on "the best available data" of endangerment. The peregrines that were cage-reared by the thousands and released were finally delisted in mid-August 1999 "to prove that the endangered species act works." Rather than explain that peregrines became extinct in eastern United States *before* DDT was used, the news media have alleged that the peregrines became endangered *because* of DDT!

Six of the first 18 "species" to be delisted were, according to the Fish and Wildlife Service, listed by error. Each case is fully documented by the National Wilderness Institute (25766 Georgetown Station, Washington, DC 20007). Five others were delisted because they were said to have recovered. Three of them live only on a small island 400 miles from the Philippines, and local experts say they were never uncommon in the first place (but remember the dependence of the U.S. Fish and wildlife service on "the best available data"). Twenty other "species" of plants and animals were also erroneously stated to have become extinct recently. No individuals of one of them (the Simpson's pearly mussel) has been found alive during the last 90 years. The dusky seaside sparrow was determined by experts to be genetically identical to the common seaside sparrow living near it on the Atlantic coast. The "last one" died when its cage blew over. The Tecopa Pupfish was morphologically identical to the common Amargosa River population and therefore should not have been listed at all.

The listing goes on, with fewer than half of the "listed" populations being true biological species. Instead, the majority have just been local populations or subspecies.

After more than 20 years, the Endangered Species Act floundered. Fewer than 10 percent of the "species" were improving, 40 percent were declining, and the Fish and Wildlife Service did not know the status of another 30 percent. The Fish and Wildlife Service said, however, "The track record is encouraging, considering the limited funding of this agency." (Their budget exceeded $100,000,000.00 a year.)

The National Wildlife Institute also referred to 554 "endangered or threatened species" that are being provided federal support. They cited the expenses involved in preserving several subspecies needing support. Some examples of the annual expenditures on nonspecies are as follows. (Notice that they are only subspecies.)

- Spotted owl (*Strix occidentalis caurina*) $9,687,200;
- Least bell's vireo (*Vireo belli pusillus*) $9,168,800;
- Florida panther (*Felis concolor coryi*) $4,387,400.

The National Wildlife Institute also listed some of the organizations that bore the costs of supporting the hundreds of so-called "endangered and threatened species," in fiscal year 1992, including the following:

- U.S. Fish and Wildlife Service, $23,591,358;
- Federal Highway Administration, $13,757,050;
- U.S. Corps of Engineers, $83,368,400;
- U.S. Department of Defense, $10,470,000;
- U.S. Bureau of Reclamation, $23,248,000;
- U.S. National Oceanographic and Atmospheric Administration, $4,061,000;
- U.S. National Park Service, $3,830,200;
- U.S. Forest Service, $21,808,300;
- Plus Many others, $15,532,149;
- TOTAL FOR THE YEAR, $68,483,412.00.

6.1.14   *ESTIMATES OF EXTINCTION RATES*

In the Sierra Club's *Avocet* it was estimated that "species are being irrevocably lost at the dizzying rate of one per hour." (That means about 60,000 per year, but no names were mentioned.)

The World Watch Institute estimated in 1978 that one species a year was becoming extinct because of humans and that by the year 2000 the number would be "hundreds of thousands." The year 2000 arrived, and fortunately very few species are known to have become extinct.

Thomas Lovejoy of the Smithsonian Institution and the World Wildlife Fund, who was Bruce Babbitt's assistant, said that "government inaction is likely to lead to the extinction of 13 to 20 percent of all species (nearly 10 million) before the year 2000. That deadline has passed quietly, with no names of species that became extinct recently.

In *The Sinking Ark*, Norman Myers reported the extinction of one species of animal every 4 years during the years 1600 to 1900 and about one species a year between 1900 and 1979. He then escalated figures and stated that "as many as 40,000 species will likely die out annually in the last decade of this century." (We have no record of such extinctions, as of the year 2000.)

Extending hypotheses far beyond rationality, E. O. Wilson guessed that there are "thousands of species becoming extinct each year, before they have even been discovered by scientists." He estimated the current extinction rate is "1,000 to 10,000 times higher than the background rate." That appears to be a rather loose estimate and has not been accompanied by any cited examples.

Julian Simon and Aaron Wildavsky made realistic predictions in *The Resourceful Earth, A Response to the Global 2000 Report* (pages 171–183, 1993). They stated that Myers' 1979 conjecture about the rate of extinction was being "expanded wildly and used as the basis for the projections quoted everywhere" [39]. They concluded: "A fair reading of the available data suggests a rate of extinction not even one-thousandth as great as the doomsayers claim. If the rate were any lower, evolution itself would need to be questioned."

All of these estimates of extinction rates of hypothetical species lack confirmation! They are pure guesswork, but those guesses have been published in newspapers and read by millions of people who might think they are legitimate. This certainly provides some frightening propaganda; however, Americans appear to have become adapted to unsupported propaganda. Many are obviously aware that we are already beyond the end of the period of time (2000 AD) specified by those propagandists, but no lists of thousands, or even hundreds or dozens of names of extinct species have appeared.

6.1.15   *BIOLOGICAL CONTROL OF PESTS*

Biological control programs, including integrated pest management (IPM), have provided employment for thousands of entomologists over the past 30 years. Simply stated, those programs do not depend primarily on insecticides to protect crops.

Some chemicals, called "insect growth regulators," alter the physiology of the pests so much that they cannot mature. Some prevent formation of a healthy cuticle, so the immature stages die while molting or pupating, and muscles cannot become firmly attached. One problem is that insect growth regulators are not specific to pest species, so some harmless or beneficial insects can be killed as quickly as pests.

For the past 30 years, the U.S. Department of Agriculture has spent more than half of its budget on biological control methods. Of those funds, 67 percent was devoted to the control of pest insects and of that amount 76 percent was spent on biological control programs. P. S. Messenger of the University of California found that there was a total of 38 cases of substantially successful biocontrol in United States, with 32 of those being in Hawaii and California.

Entomologists go to the homelands of destructive insect pests and determine why they are not very destructive there. They collect samples of parasites and predators that appear to be keeping the pests under control there and bring them back to United States to be reared in insectaries. Great numbers of parasites and predators are then reared. They must then be intensively studied to determine if they might cause harm to non-pests when released in this country.

In the 1860s the destructive cottony cushion scale was accidentally imported on ornamental plants from Australia. By 1880 they had killed hundreds of thousands of California citrus trees. In 1888 a California entomologist went to Australia to see why the scales were not that destructive there. He found ladybugs feeding on them, so he brought back specimens of *Rodolia cardinalis* and others. The *Rodolia* spread rapidly through the citrus groves, eating the scales, and fewer trees were killed. Similar control campaigns were later effective against red scales, black scales, oyster shell scales, and a dozen other serious pests.

Biocontrol programs directed against aphids, mealybugs, and scale insects have been tremendously successful. Those pests are not very mobile, so it is easy for the enemies to concentrate on them. In addition to the parasitic wasps, good control of such pests has also been provided by ladybugs and green lacewings.

Between 1880 and 1990 more than 400 species of predators and parasites were imported into the United States for biological control programs. Most of them failed to become established here, but about 90 showed promise, and many have been extremely valuable in the battle against agricultural pests. About 12 percent of our native insects also parasitize other insects, and 16 percent are predaceous on other insects.

Thousands of species of tiny wasps in the families of braconids, ichneumonids, and chalcids are totally parasitic on other insects. Insect pests that have been attacked by such imported parasitic insects include cotton boll weevils, pink bollworms, Mediterranean fruit flies, and many types of caterpillars. Few of those major pests have actually been controlled by the parasites alone. Eventually insecticides may have to be used, because each kind of parasite is specific to only one or two kinds of pests, and most crops are attacked by any of a dozen kinds of pests.

Extensive testing is necessary to attempt to determine what, if any, nontarget species the imported predator or parasite might attack. If no harm appears likely, masses of imported insects will be released with the goal of controlling the pests here as they were doing in their native lands. It should have also been determined that they will not also destroy beneficial insects or native nontarget species.

Many gardeners and agriculturists express concerns about biological control measures. Natural-born killers, mainly insects, are emerging as "green alternatives" to pesticides. When released, these parasitic wasps have the potential to destroy the pests as well as the capability to perpetuate themselves. Therein lies the paradox of biological control. Nature Conservancy president John Sawhill's July 1999 editorial in *Nature Conservancy* magazine stated: "Sometimes their activities are destructive to innocent species, and the attacks, once begun, can seldom be called off." In that same issue the Nature Conservancy included an article titled *The Peril and Potential of Biological Control*" [40]. In 1999 *Ecology* magazine asked: "How risky is biological control?" and concluded that "It should be viewed as a method of last resort, rather than as a first resort." If application of an insecticide causes unexpected harm, it does not continue to be destructive for very long. However, if imported parasites "go bad" and switch to nontarget native hosts, they may continue to increase, forever. They might be considered to be self-replicating insecticides.

Other biocontrol programs rely on protozoans, bacteria, or viruses to control pests.

Sterile male techniques have been very effective in a few cases, the most notorious being the eradication of screwworm flies (*Cochliomyia hominivorax*) from Florida, Texas, and much of Mexico. Those flies deposit eggs in wounds, even scratches, on livestock. The maggots eat live flesh, and enlarged wounds attract even more flies to lay eggs. The infected animals are doomed. Screwworms killed thousands of horses and cattle every year, and in Florida alone the financial loss was over $20 million annually. In the southwest there was an additional $100 million annual loss. Scientists had already determined that the adult flies are harmless, females mate only once, and eggs fertilized by sterile males do not hatch. If enough sterile males could be produced and liberated in the fields, and they mated with most of the wild females, there would soon be a large decrease in the numbers of screwworm flies, and most livestock and wildlife would be saved. The U.S. Department of Agriculture developed huge facilities in an empty airplane hangar in Sebring, Florida, to house three million fertile flies as breeding stock, raise 50 million sterile flies a week, sterilize the pupae with radioactive cobalt, and release the sterile adults from a fleet of 20 small planes. After 18 months, there were no more screwworm cases east of the Mississippi. The total cost was about $11 million. A similar program in Mission, Texas, produced another 150 million sterile flies per week. They were fed 40 tons of horse and whale meat each week, and 4500 gallons of beef blood. The operation covered five states and was completed in 2 years, with a saving of over $100 million annually to the cattle industry there.

Another use of the sterile male technique involved the Mediterranean fruit fly (*Ceratitis capitata*), which invaded central California in 1975. Twenty million sterile male flies were released every week (at a cost of $170,000 per million). The infestation spread to four more counties anyway. In 1981 the sterile fly proponents gave up, and the state sprayed malathion-laced bait from fleets of helicopters every night. During the daytime state entomologists sprayed SEVIN (carbaryl) in almost every yard and attached poison baits to trees and poles. The pesticide treatment cost over $60 million, but it eradicated the pests and temporarily saved California agriculture.

6.1.16 INVASIONS BY EXOTIC PLANTS AND ANIMALS

A great many destructive animals and plants have already invaded this hemisphere. More than 4500 foreign species are now established in the United States. "America's natural heritage is under siege from an invading army of non-native plants, birds, and fishes. But when you ask the general public to rank environmental threats, problems posed by alien species appear nowhere on the list." In *BioScience* (October 1998) it was stated that "Invasive exotics constitute a greater threat to endangered species than do pollution, over-harvesting, and diseases, combined." Imported exotics hurt the economy, too, costing Americans $122 billion annually, according to a 1998 Cornell University study.

Invading destructive animals that quickly come to mind are English sparrows, starlings, mongoose, lampreys, zebra mussels, European earwigs, European bark beetles (with Dutch elm disease), gypsy moths, Mediterranean fruit flies, screwworm flies, Japanese beetles, Colorado potato beetles, whiteflies, mealybugs, and dozens of species of destructive aphids and scale insects.

Many introduced species of plants have also caused great environmental problems, including Russian thistle, kudzu vines, tumbleweeds, Klamath weed (Saint John's-wort), tansy ragwort, spurge, broom, water hyacinth, alligator weed, knapweed, yellow star thistle, opuntia (prickly pear cactus), and malaleuca trees.

Two species of attractive European leaf beetles in the genus *Chrysolina* are restricted to eating the leaves of Klamath weeds, which otherwise would crowd out all native grasses in the Pacific Northwest.

Tansy ragwort has become a major pest in northern California and western Oregon. These European weeds spread rapidly and cause $10 million in annual losses in Oregon by poisoning cattle and crowding out good forage. In 1960, 5000 larvae and pupae of Cinnabar moths were shipped from France. Their spectacular orange and black larvae devour tansy ragwort but do not eat anything else. They now are eradicating tansy ragwort in North America.

Opuntia, or prickly pear cactus, is an attractive plant with edible fruit that was introduced into Australia in the 1800s. They escaped cultivation and by 1925 had formed an impenetrable thicket covering more than 20 million acres of former grazing land. Tractors and flame-throwers had little effect on those thickets. An entomologist reported that the larvae of a small moth in Argentina, *Cactoblastis cactorum*, feed on opuntia. Therefore many of them were transported to Australia, and in two decades they killed 99 percent of the cactus and opened the country to successful grazing.

In the 1950s the *Cactoblastis* moths were introduced into Central America for the same purpose, but soon they were also destroying at least 15 other species of *Opuntia*. The rare semaphore opuntia cactus (*Opuntia spinosissima*) was thriving on Florida lands purchased by the Nature Conservancy in 1989, but they were also being destroyed by the moth larvae. The only surviving semaphore cacti are now in greenhouses.

Another plant disaster involved small weevils, (*Rhinocyllus conicus*), which have already destroyed 12 native species of thistles, including three that were candidates for protection under the Endangered Species Act.

Euglandina rosea, a cannibal snail that combats the giant African snail, was deliberately released on islands in the Pacific and Indian Oceans. Unfortunately, they preferred native mollusks, and they have already eliminated more than 50 species of land snails.

Bufo marinus, a huge toad from Central America, was deliberately imported to Australia in the 1930s to combat sugar cane scarab beetles. The beetles still abound, and the toads have become a major plague, poisoning people who touch them and eating a wide range of native wildlife.

The Zebra mussel was accidentally introduced into our Great Lakes in 1988. These small freshwater pests attach themselves to underwater surfaces and quickly plug up every sort of opening. They also form a heavy crust on other forms of life, such as clams, which are quickly killed. There appears to be no hope of controlling them without eradicating all other forms of life in the lakes.

6.1.17   *AMERICAN AGRICULTURE IS STILL SURVIVING*

Food production in the United States is still thriving The American Farm Bureau Federation recently compared U.S. agricultural productivity with agriculture in Cuba, where pesticide and fertilizer availability is limited. Cuba uses 58 percent of its land area for agriculture but cannot feed its own population of 11 million people. In the United States 42 percent of the land area is used for agriculture, producing enough food for the population of 270 million, as well as millions more, overseas, and provides an agricultural trade surplus of over $21 billion a year.

Despite such success, some Americans oppose the use of chemicals that make such productivity possible! They support "organic" farming and prefer "organic" foods, seldom realizing that organic farming only avoids synthetic (man-made) pesticides. Based on California's Health and Safety Code, Sections 26569, 11–17, organic farmers can still use natural chemicals such as arsenic, cyanide, borax, lime, and sulfur, and they usually use chemicals produced by plants, such as nicotine, rotenone, and pyrethrum [41].

Reportedly, people who eat organic food are more likely to become infected with food-borne bacteria, however the CDC has not published the results of their study [42].

6.1.18   *THE FOOD QUALITY PROTECTION ACT*

The Food Quality Protection Act (FQPA) enacted by the EPA in 1996 poses great threats [43]. This remarkable mandate states that the EPA may ban any chemical "unless they believe there is a reasonable certainty of no harm from the total amount of that chemical in the aggregate of food, water, or residential use." (Isn't it still difficult to reasonably "prove a negative?") Why not simply require that there be "no significant danger of serious harm to non-target organisms"? There is no indication of what the EPA may mean by the word "reasonable." They provide even less assurance of what they may mean by the word "harm," or what they mean by *no* harm. If one in a million or one in a billion test rodents is harmed, the chemical would fail the "no harm" requirement of the FQPA and the EPA could therefore ban the chemical.

This Act replaced the Delaney Clause of the Food and Drug Administration, which specified that a chemical might be banned if it is found, "after tests *which are appropriate* for the evaluation of the safety of food additives, to induce cancer in man or animals" (emphasis added). Anti-pesticide activists seldom, if ever, relied on tests that were even remotely "appropriate" for such purposes [44]!

6.1.19   *MEDICAL CATASTROPHES*

During the middle ages great epidemics of plague, transmitted by flea bites, killed one-quarter of the entire population of Europe and nearly two-thirds of the population of Britain. Chlorinated insecticides have prevented such plagues in recent decades. During World War I (1917–1921) Russia lost about 3 million people to louse-borne typhus, and millions more died of that illness in the Balkans, Poland, and Germany. In World War II, 80,000 people died of typhus in North Africa and a major epidemic was developing in Italy before DDT, a simple chlorinated hydrocarbon, controlled body lice and halted the epidemic.

Science magazine (9 June 1972) reported that at least 80 percent of all human infectious disease is still transmitted by insects, mites, and ticks [45]. Chlorinated insecticides still provide the greatest protection against such ailments, despite the activities of Rachel Carson and of Greenpeace. Malaria still kills 3 million people a year, and without chlorinated insecticides the figure would be at least three times that high! In 1970 the National Academy of Sciences stated in their book, *The Life Sciences*: "To only a few chemicals does man owe as great a debt as to DDT. In little more than two decades DDT has prevented 500 million human deaths, due to malaria, that would otherwise have been inevitable" [46].

6.1.20   MERCURY

Mercury is widespread in the lakes and streams of North America. Scientific studies around the world have found no adverse health effects in people who eat fish that contain twice as much mercury as the World Health Organization's "safe" guidelines.

Michael Gough, a director at the CATO Institute, reported in *Environment News*, January 1998, that "fish are high in proteins and low in saturated fats," and that "eating more fish may reduce risks of heart disease." Other health officials share his view that the EPA's proposed restrictions, based on hypothetical risks, will be especially harmful to poor people who rely on fish as a regular part of their diet.

In the Seychelles, people consume about 12 fish meals per week and they have an average 10 to 20 times higher mercury concentration in their hair than in U.S. fish eaters. Obviously any adverse effects due to mercury will show up there long before in America. In the Seychelles, researchers tested six different measures of child performance for 66 months, including language, drawing, letter reading, and general behavior. No negative correlations with mercury levels were found [47].

The U.S. Environmental Protection Agency, however, seeks to frighten people about eating fish. They insist that levels considered harmless by the U.S. Food and Drug Administration and the World Health Organization are five times too high for safety. Health officials in Alaska have pointed out (*Science* 12 December 1997) that the EPA's restrictions have harmed the health and welfare of many poor people who rely on catching fish for their sustenance.

The EPA compared data from 370 U.S. sites and found only 2 percent of the sites exceeded the Food and Drug Administration's action level of 1 ppm of mercury in the hair. The long-standing World Health Organization guideline for a safe concentration of methyl mercury intake is equivalent to a concentration in hair of 5 ppm. In the Seychelles study, hair contained concentrations of 7 ppm mercury and higher [48]. Although mercury is difficult to remove from the environment, scientists at the University of Georgia have produced genetically altered plants that absorb mercury through their roots. They believe that an acre of those plants could clean up to 10 metric tons of mercury in 9 weeks, converting it to elemental mercury, which is then released into the air as a vapor.

6.1.21   LEAD

Lead is a neurotoxin and is especially hazardous to young humans with developing nervous systems. It is linked to impairments of fetal development, birth defects, miscarriages, learning impairment, and behavioral problems.

Increased industrialization has resulted in large-scale consumption of coal, which releases lead when it is burned. Some scientists express fear that even low-level exposures may have serious irreversible effects on brain function, such as lowering IQ levels and diminishing academic performance.

The U.S. Communicable Disease Center called lead poisoning a common pediatric health problem in the United States but said it was entirely preventable. The source of most lead in children is lead-based paint in houses that were painted between 1884 and 1978, after which it was banned for house-painting purposes. Many children eat flakes and chips of that old paint (some say it has a sweet taste). A more common source of lead in children is contaminated soil and dust. According to the H. W. Mielke, in "Lead in the Inner city," 50 to 70 percent of the children in some U.S. inner cities have lead levels in their blood above the current guideline of 10 μg/dl (micrograms/deciliter) [49].

Many cosmetics and hair dyes also contain substantial quantities of lead, some of which will be absorbed through the skin. The Food and Drug Administration formerly allowed up to 6000 ppm of lead acetate in such products.

An environmental group (The Natural Resources Defense Council) has charged that antacids and calcium supplements contain amounts of lead that exceed 0.5 μg of lead exposure per day. Many exceeded 4 μg/day, and one brand contained more than 210 μg. (*San Francisco Chronicle*, 4 February 1997). The National Osteoporosis Foundation accused environmentalists of "scare tactics" that might discourage women from taking needed amounts of calcium (*San Francisco Chronicle*, 12 August 1999). Makers of those calcium supplements said that calcium absorbs lead and keeps it from entering the bloodstream. Because of that a judge refused to order warnings to pregnant women that lead in calcium tablets can harm their fetuses (*San Francisco Chronicle*, 7 February 1997).

In 1992, child psychologists sought to repudiate reports by Herbert Needleman, a paid expert witness who said that adverse mental effects were caused by lead in the diet. One group of 18 students who Needleman excluded from his report had high concentrations of lead but absolutely normal IQ levels. He implied that the 58 other students with high lead readings appeared to be mentally impaired but refused to divulge his data to other scientists, even after he was ordered to do so by a judge (*Accuracy in Media*, March 1992). The EPA's Expert Panel on Pediatric Neurobehavior Evaluations concluded that Needleman's studies "neither support nor refute the hypothesis that low or moderate levels of lead exposure lead to cognitive or other behavioral impairments in children." Ignoring those experts, the Health and Human Services secretary then relied on Needleman's work and lowered the level of lead he considered hazardous. He issued his ruling in a press release without including any scientific references to support it.

Before 1970, lead poisoning was indicated by blood lead levels greater than 60 μg/dl. The Communicable Disease Center later lowered the threshhold level to just 10 μg/dl, after use of lead had been banned in gasoline and paint and for lead-soldering cans containing food.

Lead exposure is cumulative and its harmful effects usually are not reversible. Another serious threat is the presence of contaminated soil and dust in cities and on floors inside houses in poor neighborhoods, resulting in blood concentrations of 10 to 25 μg/dl.

In 1986, when lead was banned from automobile fuel, it was the most common source of lead in the United States. About 75 percent of gasoline lead was from exhaust pipes and 25 percent was from the oil adhering to the bottom of cars.

An EPA news release in June 1998 defined their level of statutory concern as between 1 and 5 percent probability of a child having a blood lead concentration in excess of 10 μg/dl and proposed a floor-lead standard of 50 μg/ft^2. Scientists have estimated that 5 percent of children will have a blood lead level concentration of more than 10 μg/dl when the lead level on house floors is 0.46 μg/m^2 (about one-tenth the proposed EPA floor standard.)

6.1.22 SELENIUM

Selenium is a naturally occurring element in the soil. It is found in many plants that live in selenium-rich soil and in animals that eat those plants. An average concentration of about 0.07 ppm occurs in soil, but many areas are selenium-deficient and other areas contain several parts per million of the element.

In mammals, including humans, selenium may be either life-threatening or life-saving, depending on the concentration in the body. In 1943 the Food and Drug Administration reported that 10 ppm of selenium in the diets of laboratory rats resulted in liver cirrhosis and liver tumors. National Institutes of Health researchers surprised most scientists when they proved that low levels of selenium in mammals prevent liver degeneration and other health problems. Selenium in the human diet also has been found to have some protective effect against both pancreatic and colon cancer. In some selenium-deficient countries, sodium selenite is added to table salt or is otherwise made available for human diets.

Analyses of soil and vegetation throughout the United States reveal that mideastern states, New England, Florida, and the Pacific Northwest have concentrations of selenium that are lower than recommended for farm animals, but the states between the Mississippi River and the Rocky Mountains provide adequate concentrations for healthy farm animals. There are local areas in the north central states where selenium is so abundant that animals that eat the vegetation or drink the water may be poisoned.

In California a notorious example of problems caused by high levels of selenium in birds has been carefully studied. In the western San Joaquin Valley, natural selenium deposits are high. Rainfall washes it into the valley, where it enters irrigation systems. The runoff from 42,000 acres of agricultural land was not permitted to be drained into the rivers or the San Francisco Bay, so it flows into Kesterson Reservoir for storage. As the water evaporates, the concentration of selenium naturally increases, and it is believed to have caused deformities, such as "crossed bills," in about 20 percent of the ducks living there.

6.1.23 CHLORINE

Chlorine is the 11th most abundant element in the earth's crust. More than 2400 naturally occurring organohalogens are discussed in *Today's Chemist at Work* (March 1995). Over 1700 of those are organochlorine compounds.

Nearly 2000 chlorinated compounds are produced in natural environments, including 5 million tons from decaying wood and 400,000 tons of chlorinated phenols from Swedish peat bogs. Such "pollution" dwarfs the 26,000 tons emitted by all human activities!

Seawater evaporation puts 600 million tons of chlorine into the atmosphere annually. Thirty million more tons were added to the stratosphere by Mt. Pinatubo's eruption in June 1991.

Chlorine-based disinfectants are responsible for the prevention of disease transmission in homes, restaurants, hospitals, and many industries. They purify 98 percent of our drinking water and probably have saved more human lives from disease than any other chemical. In 1991 Peru stopped chlorinating their water and more than a million people soon developed cholera. Over 19,000 people died as a result. A *Science* editorial (26 August 1994) cited the World Health Organization's estimate that 25,000 children die each day from drinking water that has not been chlorinated [50].

6.1.24 DIOXIN

A major chlorine compound is the dioxin (2,3,7,8-tetrachlorodibenzo-*p*-dioxin) also known as TCDD. This is one of hundreds of organohalogens formed by living plants. It is produced even more abundantly by the burning of forests, by firewood, by motor vehicle exhaust, by steel mills, and by smelters. It also resulted when chlorine was used to bleach paper pulp, but that was probably less than one pound annually from the entire paper pulp industry in United States [51]. Chlorine dioxide is now used instead by most companies and produces even fewer emissions.

A few nonchemists claim that dioxins are among the most deadly poisons known. Fortunately, no human deaths have ever been recorded, not even in the notorious exposures caused by the explosion in Seveso, Italy, and the Times Beach fiasco in Missouri. The soil at Times Beach had levels up to 100 ppb, which was 100 times higher than what EPA considered a safe level. That level might cause one case of cancer per million people during 70 years of exposure [52]. G. W. Gribble, in "*Confronting Chemophobia*," recalled that "In 1976 the worst dioxin accident in history occurred at Seveso, Italy, but it has not resulted in serious harm to the health of the exposed population of 37,000 people. A skin disorder, called chloracne, is the only medically-documented human health effect of exposure to dioxin." At Seveso, some people had dioxin levels as high as 56,000 parts per trillion, but the only adverse health effect was 193 cases of chloracne, and the skin lesions later disappeared in all but one of those people. Pregnancies, fetal development, immune responses, chromosomal aberrations, nervous system functions, blood and liver conditions, and rates of sickness and death all remained within the range normal for nonexposed populations nearby [53].

In 1971 waste oil was spread on a street of Times Beach, Missouri, for dust control. The oil was contaminated with dioxin as a trace by-product of the manufacture of trichlorophenol. After the road was paved, traces of dioxin were found beside it. Misled by television propaganda, Dr. Vernon Houk, assistant Surgeon General, ordered evacuation of the town, and the government bought all of the property there for $37 million. A $200 million demolition of the buildings and removal of the topsoil began.

By 1991 Dr. Houk had second thoughts (*Chemical and Engineering News*, 12 August 1991: 7–14). He announced that "in retrospect, it looks as though the evacuation was unnecessary; if dioxin is a carcinogen it's a very weak one, and federal policy needs to reflect that." He commented that "federal government standards for dioxin are based on chemophobia, and it turns out we were in error in evacuating Times Beach." Asked if this meant the costly cleanup would now be halted, Dr. Houk said there was little choice but to go ahead with it, "because we've got the public so riled up." Reed Irvine, *Accuracy In Media* editor, reminded Lawrence Tisch (CEO of CBS) that his network had been largely responsible for the faulty Times Beach propaganda and suggested that CBS might halt the waste by publicizing the facts, but Mr. Tisch only responded "That's not our job" (*Washington Inquirer*, 23 August 1991). (The soil at Times Beach had up to 100 parts per billion of dioxin, which was 100 times higher than the EPA considered a level that might cause cancer in one person per million if exposed to that concentration for 70 years.)

During the Vietnam war a group of servicemen called the Ranch Hands handled the herbicide known as Agent Orange. They loaded it onto planes daily and sprayed it over the jungle to thin the foliage so that enemy troop movements would not be hidden from view. Because they had the most frequent and heaviest exposure to dioxins of any men, they were the subject of thorough medical examinations. A 20-year study was made of the health of the 995 Ranch Hand veterans, who had been hundreds of times more heavily exposed than anyone in the jungle or elsewhere on the ground. There were no immune system effects, no increases in melanoma or any other cancers, and not even chloracne in any of those exposed veterans.

Other studies were made, including analyses of thousands of workers in dioxin plants. Again, there were no cancer increases, no excess miscarriages, and no increase of fetal deaths or birth defects in the children of those workers. Gribble reports that "over 40,000 scientific articles have discussed dioxins, and the evidence does not support claims that dioxins are a major health threat" [54]. The news media generally ignore all of those scientific reports and persist in printing frightening articles on the subject. Such behavior may have led H. L. Menken to write: "The whole aim of practical politics is to keep the populace alarmed (hence clamoring to be led to safety) by menacing it with an endless series of hobgoblins, all of them imaginary."

6.1.25 CHLOROFLUOROCARBONS (CFCs)

In 1956 Cambridge University meteorologist Gordon Dobson and his colleagues discovered the ozone hole near the Antarctic and measured it at 150 Dobson units. It was measured almost every year thereafter. In 1958, French researchers measured it at 120 Dobson units, which was thinner than at any time since then. Dr. Dixy Lee Ray reminded us that "that was 30 years ago, before the widespread use of chlorofluorocarbons, which are now being blamed for ozone depletion." For nearly 40 years other scientists have claimed to be the discoverer of the "hole." They seldom mention Dobson's 1956 research; however, the temporarily thinned layer of ozone, which is erroneously referred to as a "hole," is still measured in Dobson units. In 1980 the National Academy of Sciences predicted a future 18 percent decrease in the ozone layer there. Four years later they reduced that number to just 7 percent and 2 years later to only a 2 percent anticipated decrease. Eventually they settled on a prediction of "a 5 percent decline over the next 100 years."

Dr. Ray also observed that "The assumption that stratospheric chloride comes from CFCs is based upon hypotheses only. No breakdown products from freon have been identified in the stratosphere. World production has been only 1.1 million tons of CFCs per year. The 1813 eruption of Mt. Tambora in Indonesia ejected 211 million tons of chloride, and it would take our society 282 years to produce as much chloride-yielding CFCs as that one eruption. Mt. Erebus, just 15 kilometers upwind from McMurdo Sound (where the ozone measurements are made) has been constantly erupting for the last 100 years, ejecting 50 times more chlorine daily than an entire year's production of CFCs" [55].

It is frequently pointed out that there are too many people living in tropical countries and that they are multiplying too rapidly. If they can be deprived of the refrigeration needed to preserve their food and keep it safe until needed, thousands of humans will die. A massive campaign aimed at depriving people of inexpensive refrigeration is under way, with the stated objective of reducing the amount of chlorine in the atmosphere. Banning CFC refrigerants and air conditioning units helps to accomplish that. In North America expensive substitutes are available to replace the CFCs, but such replacements are unavailable or unaffordable in most poverty-stricken tropical countries. In the United States not much mention is now made of the anticipated human illnesses and deaths in the tropics.

6.1.26 *ENDOCRINE AND HORMONE MODIFICATION*

A Penguin Book by T. Colborn, titled *Our Stolen Future*, was published in 1996 by Dutton, New York. Jessica Mathews, in the *Washington Post*, 11 March 1996 stated. "We have been too obsessed with the obvious risks of toxic chemicals, cancer and birth defects. Immune suppression and hormone disruption, if proved, could be more dangerous... Hormone disrupters can do their damage in infinitesimal concentrations of one part per trillion... There are thousands of organic pollutants on the market, of which 50 are known to be hormone disrupters." She contends that the charges "will make earlier struggles over nitrates, saccharin, formaldehyde, Times Beach, Love Canal, cholesterol, alar, and even tobacco look like kids' stuff." The book alleges that "synthetic compounds found in pesticides and industrial chemicals may be wreaking havoc with endocrine systems, decreasing fertility and compromising immune systems in humans, as well as in wildlife" [56].

Book coauthor Carol Dumanoski admitted in 1994 that she "manipulated facts about the hole in the ozone layer" while working for the *Los Angeles Times* in order to get top billing for her story, which therefore ran on page one. She also stated that "There is no such thing as objective reporting, and I've become even more crafty about finding voices to say the things I think are true. That's my subversive mission" [57].

In a 1996 review of the book, the *American Council on Science and Health* reported in a position paper "The scientific evidence is extremely tentative but the potential for arousing fear in non-scientists is great. It is an alarmist tract, crafted for political impact." *Science* magazine observed that "there was no discrimination between anectodal reports and scientific studies," and the book "raises questions about the scientific judgement of the authors." However, the Environmental Protection Agency welcomed the unconfirmed allegations and unsupported hypotheses and is spending millions of dollars in support of its antiscience stance. The frightening allegations have been refuted by experts, scientific data, and common sense. Ms. Colborn admitted in the book that "Often the needed information simply does not exist or it is unavailable."

The book suggests that human sperm counts have plummeted, and there is an epidemic of undescended testicles and shrinking penises. Their claim that human sperm counts declined by almost 50 percent between 1938 and 1990 was dashed when it was revealed that the data were based on erroneous statistical data [58]. When correctly reanalyzed, those data indicated a possible increase in sperm counts. Other experts documented that there was no evidence of a decline in sperm and semen quality between 1938 and 1996 [59]. Dr. Robert Golden (of Environmental Risk Sciences in Washington) pointed out that estrogens must function by interactions with specific receptors, and only a few organochlorines have the chemical ability to bind with estrogen receptors [60]. Toxicologists and physiologists proved that DDT and similar organochlorines have no significant estrogenic activity [61]. A National Academy of Sciences report in 1996 listed 36 categories of natural foods that contain endocrine disrupters. More than 300 kinds of plants, in 16 common families, contain chemicals that may have significant estrogenic activity in humans or wildlife.

A spectacular charge in the book was that alligator penises in Lake Apopka, Florida, were only one-third to one-half normal size and a local biologist said he "knew" this had to be the result of a 1980 spill of the miticide dicofol on the lake shore. *Wilderness* magazine (May 1986) reported that Apopka was a "cesspool" in the 1950s. *National Observer* (21 June 1971) wrote that "Apopka is a fetid, shallow body of water with human waste dumped into it from Winter Garden's sewage treatment plant," and "effluent from a citrus processing plant on the shore still goes into the lake." A common birth control chemical, ethynylestradiol, was in the Winter Garden sewage and is hormonally active at concentrations as low as 0.1 nanogram (a nanogram is a billionth of a gram) [62].

Audubon magazine (September 1971) reported the first known die-off of alligators in the badly polluted lake, and said "thousands of turtles and fish also died." The cause was "a bacterium, *Aeromonas liquifaciens*, that dissolves the internal organs of aquatic animals." These difficulties occurred *before* the dicofol spill.

6.1.27 HORMESIS

Long ago it was noticed that there appears to be a linear relationship between the size of a chemical dose and the severity of its effect. It was generally assumed that if people were made very ill from a large dose of a chemical, they would become less ill from smaller doses. Sometimes, though, this linear relationship does not hold true. J. B. Gresham discussed the health benefits of low-level radiation, comparing it to the effects of vaccines. He mentions vitamin tablets containing selenium, boron, chromium, and manganese, which (in low doses) stimulate our defense systems. Background radiation comes from the sun, from minerals in the earth, and from other sources in nature. He says "a typical adult body may receive 30 million radioactive disintegrations each hour, and in a year nearly every cell in your body has been hit, with no evidence of harm."

Hormesis refers to any physiological effect that occurs at low doses and that cannot be anticipated by extrapolation from the toxic effects that occur at high doses. If adverse effects are elicited only at or above a specific dose level, that level is the "threshold dose." Doses below the threshold level may have no effect or may even have beneficial effects. For example, vitamins are usually beneficial, but very large doses of some vitamins may be harmful. In the sixteenth century, a physician named Paracelsus was so impressed by such situations that he tested the principle with many poisons. "What is not poison?" he wondered, and concluded that "All things are poison, and no things are not poison," and it is the size of the dose that determines the effect.

Even pure water can harm people who drink too much of it at one time. In 1979 a man drank 17 liters of water within a short time and died because of cerebral edema and a severe electrolytic imbalance. On the other hand, small amounts of arsenic may cause no adverse effects in people. Residents of certain parts of the world live well despite exposure to large amounts of local poisons that would make an outsider very ill if he or she ingested or breathed it. The human body is to some extent physiologically adaptable to certain poisons. (That obviously applies to certain alcoholic beverages!)

In a 1943 experiment, caged rats were exposed to high levels of uranium dust. The dosed rats outlived the controls and also produced more young. Some things that are very poisonous to humans (or other animals) at high doses become nonpoisonous at lower concentrations and may even become beneficial at dosages below their threshold level. The most impressive examples of this phenomenon have been documented where radiation is involved, but it also occurs when certain chemicals are involved.

In 1980 Professor Thomas Luckey published the results of 1239 separate studies that dealt with ionizing radiation. In 1989 the CRC Press published an updated book titled *Radiation Hormesis*. It was evident that there is a threshold dose below which the radiation is harmless and often appears to be beneficial. The effects were often found to be stimulating and resulted in increased activity. Other effects were resistance to disease, greater reproductive success, and greater longevity In 1989 the CRC Press published an updated book by Luckey, with the same title [63].

Traces of pesticides are often too weak to kill certain insects, but almost certainly are not directly beneficial to them. However, if insects that are beneficial to humans are sensitive to a pesticide spray while destructive pests are not, the results might be difficult to interpret. It may *appear* that the chemical is beneficial to the pests, but further study may reveal that its primary effect was actually detrimental to the predaceous insects (therefore beneficial to the pests). For example, if a pesticide has an adverse effect on ladybugs or lacewings, instead of eradicating aphids in the field, the aphids might appear to have been stimulated by the pesticide. Complicated environmental relationships are not always easy to interpret.

6.1.28   BIOTECHNOLOGY

Every year plant breeders run field tests on as many as 50,000 new genotypes, in their search for favorable genetically altered crops. Biotechnology (genetic engineering) greatly speeds this research, helping to develop DNA molecules with new arrangements of their genes. The process of "gene splicing" (with genes from one organism placed into another organism) can quickly transfer desirable qualities into the recipient. The process has resulted in plants that are less susceptible to physical disabilities, more capable of survival, better protected from natural enemies, resistant to herbicides, and capable of producing better yields.

Germ-line gene therapy involves the removal, alteration, and insertion of individual genes into the chromosomes of other organisms. If a piece of DNA had significant properties in the original donor, the "new" form can pass those beneficial traits to its offspring.

Other biotechnological advances have engineered soybeans to produce plants that have built-in chemical ability to resist commonly used herbicides. A field of those herbicide-resistent soybeans can be sprayed with an herbicide that kills the weeds in the field but does not bother the soybeans. The crops need not be sprayed with herbicides, and the cost of producing the food is therefore reduced. The less expensive productivity results in a more affordable source of food for humanity.

Many newer techniques have been developed. Genetically engineered plants may now contain built-in insecticides, such as the toxins from *Bacillus thuringiensis*. Those toxins have been experimentally incorporated into the genetic makeup of corn plants, potato plants, and cotton plants. It kills European corn borer larvae, corn rootworms, Colorado potato beetle adults and larvae, and several cotton pests. As a result, cornfields can produce thousands of tons of grain that otherwise would have been destroyed, healthy potatoes will be available because of fewer Colorado potato beetles, and better cotton yields will aid all local economies.

Greenpeace called a press conference to warn that pollen from genetically engineered corn might fall on milkweed leaves and adversely affect monarch butterfly larvae that eat those leaves. There are very few milkweed plants in good corn fields, and perhaps some concern should be felt for peasants who may starve if their corn crops are destroyed by corn earworms and other pests. Corn pollen is too heavy to spread more than 15 or 20 feet; therefore monarch butterfly larvae on milkweed plants not adjacent to a corn field will not be harmed.

Because frost often destroys fruits it was wonderful when a bacterial strain of *Pseudomonas syringae* that reduces frost damage was discovered. Using genetic engineering (biotechnology), scientists were able to implant the ingredients of that bacterium into the genes of strawberry plants. As a result the fruit became more resistent to frost damage. To avoid losing control of agricultural productivity the EPA quickly declared frost to be an agricultural pest, and therefore pest-resistant strawberries are "pesticides," and must of course be regulated by the EPA. Because these genetically altered plants develop effective natural chemical insecticides within themselves, the EPA is expected to classify each individual plant as a pesticide! New regulations drafted by the EPA in 1999 can establish their right to regulate all genetically altered plants as if the plants themselves are potentially hazardous "chemicals."

Dr. Bruce Ames and his colleagues have demonstrated that 5 to 10 percent of the dry weight of every green plant (whether wild or cultivated) contains natural pesticides, and about half of those natural pesticides have been found to be carcinogenic [64].

How can the EPA deal with the proven fact that every plant contains natural pesticides and about half of them are carcinogens? Must *every* such plant be called a "pesticide"? Must they all therefore be registered by the EPA, as other pesticides are? Will they be regulated, and perhaps banned by the EPA, under the Food Quality Protection Act? If the United States is forced, as a result, to purchase agricultural products from other countries (where there is no EPA) will restrictions be placed on food imports that have been genetically altered? Many problems loom ahead as a result of biotechnology.

6.1.29   THE ENVIRONMENTAL PROTECTION AGENCY

The first Earth Day, in 1970, helped bring about the establishment of the EPA. Most scientists assumed that this would be an agency composed of qualified scientists who would establish legitimate procedures having a sound scientific basis. As it turned out, *none* of the administrators in the following 29 years have had any such background. Almost every one of them has been an attorney!

In September 1971, the combined forces of the EPA and the nongovernmental Environmental Defense Fund went to federal court in an effort to ban DDT. After 7 months of hearings, which generated more than 9000 pages of transcript and hundreds of exhibits, the expensive spectacle came to an end. After considering the testimony for those seven months EPA Hearings Judge Edmund Sweeney concluded that DDT should *not* be banned. In his final official decision, issued on 26 April 1972, he pointed out that: "DDT is not a carcinogenic, mutagenic, or teratogenic hazard to man. The uses of DDT under the regulations involved here do not have a deleterious effect on freshwater fish, estuarine organisms, wild birds, or other wildlife. The evidence in this proceeding supports the conclusion that there is a present need for the essential uses of DDT."

EPA Administrator William Ruckelshaus had never attended a single day of those 7 months of hearings, and his Special Assistant (Marshall Miller) told reporters that Ruckelshaus did not even read the transcript (*Santa Ana Register* 23 April 1972). He disregarded the hearings and overruled the judge, then proceeded to ban DDT, all by himself. Attorney Ruckelshaus wrote to Allan Grant (American Farm Bureau Federation president) on 26 April 1979, stating: "Decisions by the government involving the use of toxic substances are political with a small '*p*.' The ultimate judgment remains political" (emphasis added).

Mr. Ruckelshaus refused to release EPA data to the U.S. Department of Agriculture, under the Federal Freedom of Information Act. He also refused to file Environmental Impact Statements regarding the anticipated environmental effects of the DDT ban. He recommended replacing DDT with methyl parathion (a much more toxic organophosphate, which farm workers and Rachel Carson strongly opposed) [65].

A suit by environmental groups later urged that dieldrin also be banned, but on March 28, 1972, the EPA Science Advisory Committee unanimously recommended that it not be banned. That echoed the similar recommendations of the following authorities: the U.S. Food and Drug Administration; the National Academy of Sciences ; the U.S. Department of Agriculture; the Mrak Commission of the Department of Health, Education and Welfare; and the World Health Organization Food and Agriculture Committee. None of those science-oriented organizations influenced the EPA administrator. He quickly banned dieldrin. Because that was the only chemical that could halt the huge locust invasions that were destroying African grain crops, the EPA ban had drastic effects on millions of humans, causing widespread malnutrition, starvation, and death. Such procedures by the government of the United States are destructive to humanity worldwide, as well as being disastrous to our own populace.

CITED REFERENCES

1. *The Free Market*, June 1993.
2. *Science*, 274, 816–918, 8 November 1996.

3. *Federal Ecosystem Management*, CATO Institute Policy Analysis, 26 October 1994.

4. *The Unfinished Agenda*, New York, Crowell, 1977.

5. *Blueprint for the Environment*, Salt Lake City, Howe Brothers.

6. *Wild Earth*, special edition, 1992.

7. *New American*, 25 October 1999.

8. *Global Biodiversity Assessment*, Cambridge University Press, 1995.

9. *Agenda 21* and the *UNCED Proceedings*, Conference on Environment and Development, Oceana Publications, New York.

10. *Animals Magazine*, 554–555, April 1971.

11. *Trans Amer. Fisheries Society*, 109, 207–214, 1971.

12. *Science*, 170, 71–72, 1970.

13. *J. Wildlife Mgmt.*, 32, 441–456, 1968.

14. J. Lehr, *Rational Readings on Environmental Concerns*, 1992, pp. 125–131, Van Nostrand Reinhold, New York.

15. *Science*, 159, 1474, 29 March 1968.

16. *Applied Biology*, 10, 532–537, 1962.

17. *Science*, 167, 1724–1726, 27 March 1970.

18. *Ramparts Magazine*, 8, 24–30, August 1969.

19. *The Year's Best Science Fiction*, 1970, pp. 66–79, Sphere Books, Aylesbury, England.

20. *Avocet*, Sierra Club, January 1991.

21. *Our Vanishing Wildlife*, N.Y. Zoological Society, 1933, p. 26.

22. *Audubon Magazine*, November 1973.

23. *Birds Over America*, 1948, p. 27, Dodd, Mead & Co., New York.

24. *Peregrine Falcon Populations*, University of Wisconsin Press, 1969.

25. *Peregrine Falcon Populations*, University of Wisconsin Press, 1969, pp. 155–163.

26. *Chevron World*, Spring 1985.

27. *Audubon Magazine*, November 1977.

28. *The Myth of the Vanishing Peregrines*, Canadian Raptor Society, 1971.

29. *Ecology*, 2, 76, 1921.

30. *Bird Lore*, 32, 165, 1930. (Audubon Society journal)

31. *Summaries of Hawk Mountain Sanctuary Association Counts*, 43, 2, 1971.

32. *Audubon Magazine*, January/February 1944, pp. 1–75.

33. *Audubon Field Notes*, 15(2): 84–300, 1961.

34. *Pesticide Monitoring Journal*, 9, 12–13, 1975.

35. *Bulletin Environmental Contamination & Toxicology*, 28, 1982.

36. *Trans. 31st N. A. Wildlife Conference*, 190–200, 1966.

37. *Autumn of the Eagle*, G. Laycock, 1973, pp. 156–157.

38. *National Wilderness Institute*, 25766 Georgetown Station, Washington, DC.

39. *On Species Loss, The Absence of Data, and Risks to Humanity*, 1993.

40. *The Peril and Potential of Biological Control, Nature Conservancy Magazine*, July 1999.

41. *California Farmer* July 1991, pp. 40–41.

42. Telephone communication with Communicable Disease Center in Atlanta, 22 March 2000.

43. *Speak Up America*, 13 March 1995.

44. *21st Century Science & Technology*, 5–7, Winter 1995–1996.

45. *Science*, 176, 1153–1155, 9 June 1972.

46. *The Life Sciences*, 432, 1970.

47. *Neurotoxicology*, 16, 677, 1995.
48. *Neurotoxicology*, 16, 597, 1995.
49. *American Scientist*, 62–73, January 1999.
50. *Science*, 265, 1155, 26 August 1994.
51. *Heartland Institute Journal*, May 1996.
52. *Science News*, 367–368, 28 January 1983.
53. *Dartmouth Alumni Magazine*, 12–13, winter 1991.
54. *Chemical & Engineering News*, 7–14, 12 August 1991.
55. Ray, D. L., *New American*, 7, fall 1993.
56. *Our Stolen Future*, Dutton, New York, 1996.
57. *American Spectator*, July 1991.
58. *Fertility and Sterility*, 63, 887–893, 1995.
59. *Proc. International Environmental Conference*, 1995.
60. *Environmental Health Perspectives*, 103, 346–351, 1995.
61. *Endocrine Disrupters in Natural Environment*, 1996.
62. *21st Century Science & Technology*, fall 1996.
63. *Radiation Hormesis*, London, 1980; and CRC Press in Florida, in 1989.
64. *Science*, 236, 271–280.
65. *Congressional Record*, Senate, S11545–11547, 24 July 1972.

CHAPTER 7
CONTAMINANT EFFECTS

SECTION 7.1

TOXICITY TESTING

Peter Bruce

*Mr. Bruce is president and managing director of Resampling Stats,
Inc., and marketing director for Cytel Software Corp. (makers of
StatXact, LogXact and other software for exact inference). He has
taught statistics at the University of Maryland.*

7.1.1 STATISTICS

The subject of statistics could consume many volumes on its own. The purpose of this section is quite limited. It is to convey a basic understanding of the hardest-to-understand concept in statistical analysis—hypothesis testing, also called significance testing.

The art and science of statistics can be divided into several areas:

- Examining data,
- Collecting scientific data (including design of experiments and sampling procedures), and
- Coping with variability in data and the uncertainty it produces.

Hypothesis testing addresses the last item—could something have occurred by chance—and is the subject of this section. The computer-intensive simulation method used below has recently come into favor as a flexible and easy-to-understand alternative to the traditional formula approach.

7.1.2 IS EFFLUENT TOXIC?

An environmental monitoring agency wants to establish that the effluent discharged from a sewage plant is not toxic. The procedure is to sample small vials of effluent in which 20 small water fleas are placed. If water fleas survive in sufficient number after a given period, the sample is deemed nontoxic. The expected survival rate of fleas in clean (nontoxic benchmark) water is 70 percent. Our sample had only 55 percent survival (11 fleas).

The difficulty is this: Even in nontoxic water where water fleas that are known to thrive, the number of fleas that survive can differ greatly from one sample to the next. In the same way, if 10 people in a room toss a coin 10 times each, some will get 5 heads and some will get as many as 7 or 8 or as few as 2 or 3 heads.

If our sample shows a survival rate at or above the expected survival rate in clean water, the decision is clear—the sample passes. However, if we have a sample with a low water flea count, how do we know whether it's low because of toxicity or because we just happened to get a sample with a small number of surviving fleas?

We answer this question by analogy to the coin tossing. We explicitly recognize that there is a range of outcomes that might be consistent with nontoxic samples and reason as follows:

1. We start with the presumption that the effluent is nontoxic unless our sample demonstrates otherwise.

2. Therefore, if the sample outcome is consistent with nontoxicity, the sample passes. In other words, if the number of surviving water fleas is within the range of normal chance variation, the sample is okay.

3. If the sample has too few water fleas to be explained by chance variation, we conclude that it is toxic.

Clearly, the next step is to determine the normal range of chance variation. How much might one sample of 20 fleas differ from another? Once we know that, we can determine whether our observed sample falls within that range of normal chance variation.

The following procedure models the 70 percent survival rate; it uses random digits (0–6 = survive, 7–9 = die) (see Figure 7.1.1):

1. Generate 20 random digits (0–9).

2. Count how many fall between 0 and 6, inclusive, record.

3. Repeat steps 1 and 2 many times—say 1000 times.

4. How often did we get as few "survives" as the observed sample?

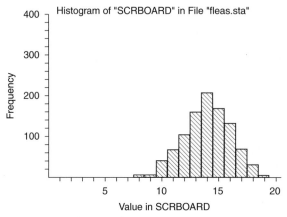

# "survive"	Freq	Cum Pct	Pct
8	6	0.6	0.6
9	6	0.6	1.2
10	40	4.0	5.2
11	68	6.8	12.0
12	104	10.4	22.4
13	160	16.0	38.4
14	207	20.7	59.1
15	169	16.9	76.0
16	132	13.2	89.2
17	70	7.0	96.2
18	31	3.1	99.3
19	6	0.6	99.9
20	1	0.1	100.0

FIGURE 7.1.1 Simulating samples from a survival rate of 70 percent.

7.1.3 INTERPRETING RESULTS

What do we make of our observed sample with only 11 survivals? We see that **simulated nontoxic** samples (samples in which each flea has a 70 percent chance of surviving) produce samples as poor as our observed sample about 12 percent of the time. (In statistical jargon, the probability or P value is 0.12.) We cannot rule out the possibility that the effluent is okay and that random chance produced this poor sample—our sample is consistent with nontoxic effluent. In statistical jargon, the result is not **statistically significant**.

The above paragraph summarizes perhaps the most difficult and subtle concept in statistics, so it is worth restating. Our benchmark universe—a 70 percent survival rate—was found to be capable of producing samples with as few as 11 survivors (some had even fewer). Therefore, we cannot reject this benchmark universe.

Suppose we had found that simulated nontoxic samples produce samples as poor as our observed sample about 10 percent of the time? 7 percent of the time? 3 percent of the time? When do we cross the line and say our sample is too rare to be consistent with nontoxic effluent?

This is an arbitrary decision—typically, statisticians use a 5 percent cutoff. If the simulated nontoxic samples produce samples as poor as our observed sample 5 percent of the time (or less), we say that the difference is "statistically significant."

Note that our analysis proceeded from the judgment at the beginning that the effluent was considered to be nontoxic unless the sample proved otherwise. This was not a statistical judgment but a legal, scientific, and managerial judgment.

We consider the facility to be innocent until proven guilty—in many cases that is the way the law about toxicity testing is written. Other factors—the costs of false alarms, the costs of toxicity, and the toxicity levels that are required to trigger these costs of toxicity—should go into this decision.

7.1.4 SUPPOSE WE PRESUME TOXICITY AND REQUIRE PROOF OF NONTOXICITY?

Statistically speaking, there is nothing that says we could not have set up the problem in the reverse fashion—presume the effluent is toxic unless it is proven otherwise by the sample. Here's how we would deal with this situation:

1. Determine the minimal survival rate decline we mean when we say "toxic." Let's say we mean that the survival rate declines to 60 percent.

2. Simulate this 60 percent survival rate to learn how robust a sample must be before we can say, with 95 percent certainty, that the sample almost surely does not come from a 60 percent survival rate.

Figure 7.1.2 shows the distribution of samples produced by a 60 percent survival rate (same simulation procedure as above, except we count 0–5 as survive, 6–9 as die).

From this, we see that a sample with as many as 16 survivors still could be **consistent** with a decline in the survival rate to 60 percent. (This comes from applying the 5 percent cutoff to the top of the above frequency distribution. Seventeen or more survivors is judged **inconsistent** with a survival rate of 60 percent because it happens less than 5 percent of the time. Sixteen or more survivors is judged **consistent** with a survival rate of 60 percent because it happens more than 5 percent of the time.)

7.1.5 WHICH WAY TO PROCEED?

Usually the analyst does not have the choice of deciding whether to presume toxicity or nontoxicity in conducting statistical analysis. In testing sewage effluent, the law is written with an assumption

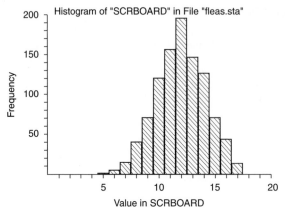

# "survive"	Freq	Cum Pct	Pct
5	1	0.1	0.1
6	4	0.4	0.5
7	14	1.4	1.9
8	40	4.0	5.9
9	70	7.0	12.9
10	120	12.0	24.9
11	156	15.6	40.5
12	195	19.5	60.0
13	147	14.7	74.7
14	126	12.6	87.3
15	70	7.0	94.3
16	44	4.4	98.7
17	13	1.3	100.0

FIGURE 7.1.2   Simulating samples from a survival rate of 60 percent.

of innocence, and a requirement that samples be presumed nontoxic unless it is so far out of line that it demonstrates toxicity. When one goes into testing, it must be clear up front what the operative assumption is in a particular area.

7.1.6   THE GRAY AREA

You will have noted that there is a considerable gray area in sample results—samples that could be consistent either with toxic effluent or nontoxic effluent.

If you presume toxicity, you will flunk a lot of innocent facilities. If you presume nontoxicity, and there is toxic effluent, you stand a greater chance of missing it. However, note that even if you presume nontoxicity, you will flunk up to 5 percent of all innocent facilities (look again at the results of the 70 percent simulation).

The only way to reduce this gray area is to gather more information—take larger samples. This will reduce sampling variability and, hence, reduce the proportion of samples that could be consistent with either toxicity or nontoxicity. For more on sample size, see Section 7.1.1 on statistics.

7.1.7 DOSE-RESPONSE RELATIONSHIP

Often an investigator is interested in the concentration or dose level at which a substance is toxic. Something that is harmless at very low concentrations may be toxic at higher concentrations. To learn more about the relationship between concentration and toxicity, several samples are taken, each at different concentrations. Extending the above problem, let us consider six samples, taken at various concentrations (with a maximum of 100 percent effluent).

Percent effluent	0	5	10	20	50	100
Number surviving	15	12	16	11	12	7

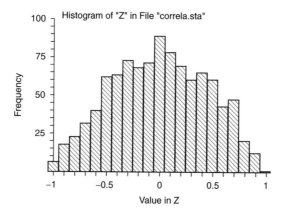

Correlation Coefficient	Freq	Pct	Cum Pct
−1	6	0.6	0.6
−0.9	18	1.8	2.4
−0.8	23	2.3	4.7
−0.7	32	3.2	7.9
−0.6	40	4.0	11.9
−0.5	62	6.2	18.1
−0.4	63	6.3	24.4
−0.3	73	7.3	31.7
−0.2	68	6.8	38.5
−0.1	71	7.1	45.6
0	89	8.9	54.5
0.1	78	7.8	62.3
0.2	69	6.9	69.2
0.3	60	6.0	75.2
0.4	65	6.5	81.7
0.5	60	6.0	87.7
0.6	43	4.3	92.0
0.7	47	4.7	96.7
0.8	20	2.0	98.7
0.9	12	1.2	99.9
1	1	0.1	100.0

FIGURE 7.1.3 Interpretation: A correlation coefficient as low as −0.84 is very rare when all six samples are drawn from the same urn. Specifically, it occurred in only 2.8 percent of all resamples. We conclude that there is indeed a statistically significant dose-response relationship.

The data appear to suggest that, as concentration increases, the fewer fleas survive. There are various ways to measure how much survival declines as concentration increases. The Pearson correlation coefficient measures the extent to which two variables move in tandem in a linear relationship. Linear regression fits the straight line to the above points in two-dimensional space that minimizes vertical deviations between the above points and the line. Other measures can describe a nonlinear relationship.

Typically, the measure required is specified in the testing regime for a given situation. However, the tricky problem of random variation remains. Do these survival numbers differ from what we might get by chance if there were no dose-response relationship? Let's answer that question in the context of the correlation coefficient, the formula for which is as follows:

$$\text{Correlation of } x \text{ and } y: \quad \rho = \frac{\Sigma xy}{\sqrt{\Sigma x^2}\sqrt{\Sigma y^2}}$$

The correlation coefficient can take on values ranging from -1 (perfect inverse correlation) through 0 (no relationship) to $+1$ (perfect positive correlation). For the above data, the correlation is -0.84 (it is negative because the relationship is an inverse one—as concentration increases, survival decreases.

Might this simply be a chance result? Could all concentrations share the same survival rate and the apparent negative correlation be due merely to a random grouping of the low survival rates in the high concentration samples?

We can test by simulating samples from a common survival rate and observing how often we obtain a correlation coefficient as low as 0.84. What survival rate should we use? If we are now asking about only the dose-response relationship (not whether the survival rate has remained at 70 percent), we should use the average survival rate among the six samples. Here is a procedure:

1. Constitite an urn that contains 73 ones (survive) and 47 zeroes (die).

2. Draw 20 numbers and replace each number in the urn after you draw it.

3. Count the number of survivals among the 20 and record.

4. Repeat steps 2 and 3 five more times.

5. Calculate the correlation coefficient between the number of survivors in each of the six resamples and the concentration levels.

6. Repeat steps 2 to 5 many times—say, 1000—and record how often you get a correlation coefficient of -0.84 or less.

The results of this procedure are shown in Figure 7.1.3.

CHAPTER 7
CONTAMINANT EFFECTS

SECTION 7.2

ECOLOGICAL RISK ASSESSMENT

Jan Swider

Dr. Swider is a researcher in the Mechanical and Aerospace Engineering Department, UCLA, School of Engineering and Applied Science. His main interests are in human health and ecological and engineering systems risk assessments.

7.2.1 BACKGROUND

Risk assessment is a systematic process of providing information about the outcome of undesirable events. Usually such information is presented in terms of the probability of loss or injury to the public, property, or environment. Its origin is associated with gambling and insurance practices. Later, the development of technological systems led to its application, along with systems analysis, to resolve concerns about safety and profitability (Gratt, 1996). Finally, the regulatory decision-making process led to development of human health and environmental risk assessments. However, there is a disparity in the stage of development of human risk assessment methods compared with ecological risk assessment practices. Methodologies for assessing human health risks are quite well developed. This is a result of more than three decades of involvement of the nuclear industry in development of methods to evaluate catastrophic emission, transport and fate of radioactive contaminants in the environment, and risks posed by such emissions to an individual human being and the entire population. Ecological risk assessment is a relatively new field that started gaining momentum in the 1980s. The major forces behind development of ecological risk theory were regulatory requirements associated with the cleanup of hazardous waste sites. Regulatory agencies needed some methods similar to human health risk assessments for use in making more and more complex environmental decisions. Methods were needed to assess the threats posed to the environment from the polluted site, to assign cleanup priorities, and to compare the cleanup costs and effectiveness.

The ecological risk assessment (ERA) methodology is based on the human health risk assessment approach but it is significantly expanded to incorporate the much greater complexity of ecological systems. The objective of ecological risk assessment is to estimate, with available information, the risks posed to ecological receptors exposed, usually, to chemical contaminants. In other words, the ecological risk assessment process uses available toxicological and ecological information to determine the likelihood that exposure to, for example, a toxic chemical, will have adverse effects on the environment and to estimate quantitatively the anticipated magnitude of the effects. Chemical stress is only one class of disturbances to ecological systems; however, analyses of this type of disturbance prevail in ecological risk assessments. Because of the complexity of ecological systems, ecological risk assessment is a multidisciplinary science that combines a broad range of disciplines including ecology, toxicology, chemistry, physics, and systems analysis among others. Table 7.2.1 presents some definitions for risk assessment-related terms based on various publications of the U.S. Environmental Protection Agency (EPA). It is meant to help in understanding basic terms commonly used in risk assessment activities.

TABLE 7.2.1   Basic ERA Definitions

Adverse ecological effects—Undesirable changes of the structural or functional characteristics of ecosystems or their components considered in terms of the type, intensity, and scale of the effect.

Assessment endpoint—An explicit expression of the environmental value that is to be protected, operationally defined by an ecological entity and its attributes. For example, species and communities identified for a site are ecological entities; reproduction is one of their important attributes.

Characterization of exposure—A portion of the analysis phase of ecological risk assessment that evaluates the interaction of the stressor with one or more ecological entities. Exposure can be expressed as contact and interaction between stressor and ecological receptor.

Ecological risk assessment—The process that evaluates the likelihood that adverse ecological effects may occur or are occurring as a result of exposure to one or more stressors.

Exposure—The contact or co-occurrence of a stressor with a receptor.

Receptor—The ecological entity exposed to the stressor. This term may refer to tissues, organisms, populations, communities, and ecosystems.

Risk—The probability of harm (injury, disease or death, loss to the public or the environment) under specific circumstances. Quantitatively, risk is expressed in probabilistic terms ranging from 0.0 (no harm will occur) to 1.0 (harm definitely will occur).

Risk assessment—The determination of the kind and degree of hazard posed by a stressor, the extent to which a particular endpoint has been or may be exposed to the stressor, and the present or potential risks that exist because of the stressor.

Risk characterization—A phase of ecological risk assessment that integrates the exposure and stressor response profiles to evaluate the likelihood of adverse ecological effects associated with exposure to a stressor.

Risk management—A decision-making process that combines political, social, economic, and engineering information with risk-related information to develop, analyze, and compare available alternatives and to select the appropriate response to potential hazards.

Source—An entity or action that releases to the environment or imposes on the environment a chemical, physical, or biological stressor or stressors.

Source term—As applied to chemical stressors, the type, magnitude, and patterns of chemical(s) released.

Stressor—Any physical, chemical, or biological entity that can induce an adverse response.

7.2.2   APPLICATIONS

Recent awareness about human health and environmental risks has dramatically increased the interest in risk assessment as a tool to evaluate and solve a broad range of ecological problems. This interest has stemmed, in part, from the need to make critical environmental decisions in the face of limited financial resources. As a result, various risk assessment procedures are being used in regulatory decision-making processes. Ecological risk assessments require one to clearly define the ecological problem, collect and analyze data to describe the problem, and combine the results of the risk assessment so that ecological consequences of an action can be incorporated into the decision-making process. The ecological risk assessment process allows for evaluation, interpretation, and constructive communication of potential options and consequences to the public and decision makers. It can be used to evaluate the likelihood of future adverse effects or the effects of past exposures.

Ecological risk assessments are used in a variety of different applications. Typical applications range from helping identify effects related to remedial actions to ecological evaluations performed for the Comprehensive Environmental Response, Compensation, and Liability Act, Superfund (CERCLA) and the Resource Conservation and Recovery Act (RCRA) hazardous waste sites. Ecological risk assessments performed for CERCLA and RCRA hazardous waste sites usually require a comprehensive, multichemical, multipathway approach. The assessments include characterizations of potentially affected habitats and identification of ecological receptors as well as exposure and toxicity assessments. Chemical fate characteristics are considered along with feeding habits, migration patterns, and population dynamics to obtain credible estimates of species-specific exposures. Exposure assessment data are then combined with receptor-specific toxicity data to characterize potential ecological risks. In addition, the ecological risk assessment process provides data useful for performing Natural Resource Damage Assessments (NRDAs) as required by CERCLA NRDA regulations (61 FR 20559, May 7, 1996).

Another application is the ecological risk-based evaluation of remedial alternative actions. This process routinely evaluates potential impacts associated with waste site remediation, treatment, and disposal activities. The evaluations consider the particular habitat characteristics of potentially affected ecological receptors as well as the characteristics of the waste to provide the most accurate evaluation of potential ecological threats or benefits associated with a specific action.

Estimation of ecological impacts of habitat modification and restoration is yet another application of ecological risk assessment. Evaluation of the impacts of habitat modification resulting from activities such as wetlands development, timbering operations, bridge and highway construction, dredging activities, and coastal development is often conducted on the basis of ecological risks. The results of such studies are helpful in developing options to predict and avoid potential ecological impacts or to mitigate the impacts.

Ecotoxicity assessments are often conducted in support of ecological risk assessments. Such assessments focus on comprehensive toxicity evaluations of the ecological effects of chemical pollutants. Depending on the needs, the evaluations range in complexity from literature review and data appraisal to quantitative structure activity analysis or development of species-specific toxicity criteria.

7.2.3 GUIDELINES

Originally, ecological risk assessments used the same four-step approach that was used in human health risk assessments—i.e., hazard identification, dose-response modeling, exposure assessment, and risk characterization. That was in part due to the absence of specific guidance. The developments in this field led to a distinctive set of paradigms outlined in the *Framework for Ecological Risk Assessment* (EPA, 1992). Recently, EPA issued a new document, the *Guidelines for Ecological Risk Assessment* (EPA, 1998) specifically for risk assessors and risk managers. The *Guidelines*, developed by the EPA's Risk Assessment Forum, appear to be currently the most comprehensive document describing principles of the ecological risk assessment process. As the EPA states, the *Guidelines* are provided to improve the quality and consistency of ecological risk assessments among various EPA programs; however, this is also a road map for other agencies and ecological risk assessors. The *Guidelines* are based on a wide range of source documents including case studies previously conducted by EPA's Risk Assessment Forum. The *Guidelines* expand on and replace the 1992 *Framework for Ecological Risk Assessment* (EPA, 1992). Because of the rapidly evolving field of ERA, the *Guidelines* provide a generic approach. EPA intends to follow these *Guidelines* with a set of detailed documents that address specific ecological risk assessment issues.

Another EPA document, *Ecological Risk Assessment Guidance for Superfund: Process for Designing and Conducting Ecological Risk Assessments* (EPA, 1997) describes the fundamentals of ecological risk assessment as applied to the Superfund cleanup process. This document provides guidance to site managers and remedial project managers about how to prepare and conduct technically justifiable ecological risk assessments for the Superfund Program. It is also based on the previously published *Framework* (EPA, 1992) and is specifically tailored toward the Superfund Program. This procedure supersedes EPA's 1989 *Risk Assessment Guidance for Superfund* as guidance on how to design and conduct an ecological risk assessment under Superfund.

Another guideline worth mentioning is an ecological risk assessment methodology developed by the Ecosystem Conservation Directorate of Environment Canada under the National Contaminated

Sites Remediation Program (NCSRP). The NCSRP guidance recommends a three-tier methodology of gradually more detailed and complex evaluations for use in site assessment and remediation in Canada (Gaudet, 1994). This approach is conceptually similar to the human health risk assessment but it includes receptors and ecological effects of a stressor on the ecosystems and is highly site specific. A tiered approach allows for flexibility in planning and conducting ERAs depending on the level of concern. Level 1 ERAs, or a screening level, are simple, usually qualitative or semi-quantitative evaluations (e.g., quotient methods), which are adequate for an initial assessment of risk. Higher tiered ERAs usually are more sophisticated and may involve exposure-response methods and models or probabilistic approaches to data propagation and can be used for more refined risk assessments.

7.2.4   RISK ASSESSMENT PROCESS

Ecological risk assessment is a part of a larger context. As described in the EPA *Guidelines* (EPA, 1998), the process begins with thorough planning to determine the value of risk assessment in a given decision-making process. Also, the type and scope of the risk assessment is chosen to address identified problems. Next, the risk assessment is conducted to systematically evaluate all the risks. The results of the risk assessment need to be communicated to risk managers, who conduct the risk management and communicate the results to other parties.

The evaluation of ecological risks is based on characterization of its two key components: exposure and effects. The whole ecological risk assessment process is built around these elements. EPA recommends a three-phase process: problem formulation, analysis, and risk characterization. The risk assessment framework is shown in Figure 7.2.1.

In the problem formulation phase, the objectives, scope of the effort, and the problem itself are defined and preliminary hypotheses about the ecological effects are generated and evaluated. All available information on stressor sources, ecosystem characteristics, receptors, and effects are gathered and integrated into the problem formulation. The outcome of the formulation process is (1) a set of assessment endpoints reflecting the ecosystem and the management goals, (2) conceptual models describing interactions between stressors and assessment endpoints, and (3) an analysis plan, which is the final product of the problem formulation.

Analysis is an essential part of a risk assessment, linking problem formulation with risk characterization. It uses the information from the problem formulation to characterize exposure to stressors and to determine what type of ecological effects might occur as a result of such exposure. Analysis produces an exposure profile and a description of relationships between a stressor and ecological response to the stressor under the defined exposure profile. In the analysis phase, the data and models developed in problem formulation are thoroughly examined. Field survey data are desirable but laboratory studies can also provide useful data. Verification of data applicability and uncertainty evaluation must be carried on throughout the analysis. Exposure characterization describes how a stressor and a receptor get in contact. The questions of how, when, and where exposure occurs are addressed and the source of the stressor, its distribution in the environment, and the transport pathways are described. In addition, secondary stressors may be considered. The final exposure profile should identify the receptor, exposure pathways, and intensity and extent of contact. Variability and uncertainty analysis of the exposure estimates should be included and a conclusion about the likelihood of exposure occurrence should be drawn. During characterization of ecological effects, the stressor-induced effects are linked to the assessment endpoints and evaluated for different stressor levels to define the effects of interest. An ecological response analysis is conducted to evaluate the magnitude of the effects and to verify that the stressor indeed causes the effect. The stressor-response profile summarizes the results and describes, among others, the affected ecological units, the nature and intensity of the effects, and the associated uncertainty.

In the risk characterization phase, the results of the analysis phase are used to obtain an estimate of the risks to the assessment endpoints identified in problem formulation. The exposure and stressor-response profiles are combined in the risk estimation. Finally, risk description presents the

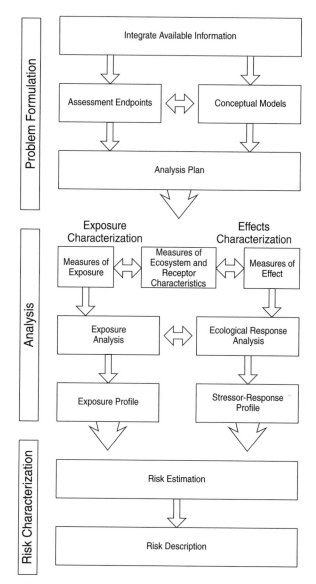

FIGURE 7.2.1   Ecological risk assessment framework (adapted from EPA, 1998).

risk estimates, evaluates the evidence and interprets the results—e.g., the importance of the harmful effects on the assessment endpoints. Presentation of the risk assessment results, their interpretation, and discussion are no less important than the risk evaluation process.

The risk assessment process may differ slightly from the one described above, especially when it is tailored toward specific requirements (see, e.g., a framework for Superfunds, Figure 7.2.2.), but the major steps of the process are valid. Some operational procedures and practical applications of the ERA, mostly based on the EPA's *Framework* (EPA, 1992) can be found in the literature (e.g., Calabrese and Baldwin, 1993; Suter, 1993).

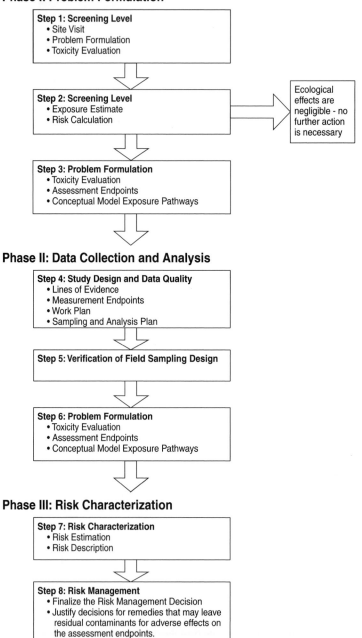

FIGURE 7.2.2   Ecological risk assessment framework for Superfunds (adapted from EPA, 1997).

REFERENCES

Calabrese, E. J., and Baldwin, L. A., *Performing Ecological Risk Assessments*, Lewis Publishers, Boca Raton, FL, 1993.

Gaudet, C., *A Framework for Ecological Risk Assessment at Contaminated Sites in Canada: Review and Recommendations*, Scientific Series No. 199, Ecosystems Conservation Directorate, Ottawa, Ontario, 1994.

Gratt, L. B., *Air Toxic Risk Assessment and Management: Public Health Risk from Normal Operations*, Van Nostrand Reinhold, New York, 1996.

Suter, G. W., *Ecological Risk Assessment*, Lewis Publishers, Boca Raton, FL, 1993.

U.S. EPA, *Framework for Ecological Risk Assessment*, U.S. Environmental Protection Agency, EPA/630/R-92/001, 1992.

U.S. EPA, *Ecological Risk Assessment Guidance for Superfund: Process for Designing and Conducting Ecological Risk Assessments*, U.S. Environmental Protection Agency, EPA/540/R-97/006, 1997.

U.S. EPA, *Guidelines for Ecological Risk Assessment*, U.S. Environmental Protection Agency, EPA/630/R-95/002F, 1998.

CHAPTER 7
CONTAMINANT EFFECTS

SECTION 7.3

ECOLOGICAL EFFECTS
OF TOXICANTS

Jan Swider

Dr. Swider is a researcher in the Mechanical and Aerospace Engineering Department, UCLA, School of Engineering and Applied Science. His main interests are in human health and ecological and engineering systems risk assessments.

All ecological systems are subject to various kinds of disturbances as a result of either natural processes or human activities, such as fires, storms, toxic spills, or air pollution. In particular, human-induced impacts can be a cause of rapid, irreversible changes to the environment. The full consequences of such changes are rarely predictable, are mostly not perceived at the time of action, and often come to light when there is no possibility of reversing the action. Toxic contaminants, habitat destruction, and overharvesting are considered to be the major problems of the recent years (EPA, 1990). Spatial and temporal scales of the stresses range from the habitat to the global level. Ongoing research focuses on developing methods to quantify magnitudes of such stress impacts and on new tools for comparing risks due to those stresses.

A full-scale assessment of ecological effects of toxic pollution requires specific toxicological data, characteristics of habitats, and many hours of in-field observations, among others. Changing concentrations of pollutants in the environment affect physical, chemical, and biological processes in ecosystems. Biological changes, when observed, are often apparent only in the vicinity of the pollution sources. The alterations in visible appearance and biological performance can be measured and quantified. More difficult to quantify are subtle chronic changes that occur over large regions and many years of development.

7.3.1 ENVIRONMENTAL STRESS

Stress occurs everywhere in nature, often causing evolutionary changes. Therefore, it is necessary to differentiate between natural and anthropogenic factors involved in ecological mechanisms. Any stress reduces performance but the effect of the stress depends on its severity. Stress can put forth a selection process in the population. It can reduce performance—e.g., growth rate—and increase a coefficient of selection. Although animals can react to stress by relocating to a more suitable area, plants have to stay and are subject to a strong selection process. Evolution is the way plants adapt to man-made stress; resistant plants are able to grow in areas that are lethal to plants from habitats that do not experience such stress. The evolution of metal resistance and the growth of herbicide resistance in agricultural weeds prove that some species can adapt to the stress whereas others follow the pathway of extinction.

Ecotoxicology is the science that focuses on the fate and effects of toxic substances in ecosystems. Effects of toxic chemicals on ecosystems are often analyzed at various levels from the molecular to the whole ecosystem level. Ecological effects of toxicants can be classified based on the stress concept. A few classes of toxicant effects can be defined based on stress or nonstress effects as proposed by Newman (1995). Following is a brief description of such effects.

Stress Effects

- Specific response or Selyean stress is a characteristic response to an external toxic stressor and consists of three phases: the alarm reaction, adaptation, and exhaustion.
- Preadaptive stress is similar to a specific response but it includes a previous adaptation to the stressor; therefore, the system responds to events that can be expected or anticipated.
- Damage or distress is the adverse effect of a stressor presence that is not a result of a system response. It is a process of system's modification by a toxicant without a system's active response.

Nonstress Effects

- A beneficial effect is a stimulatory effect of low doses of a toxicant on a system, also called a hormetic effect.
- A neutral effect is a measurable change that has no adverse or beneficial effect on a system.
- An ambiguous effect is an effect of undefined qualities relative to the degree of damage or benefit.

Other classifications may be based on mortality (lethal or sublethal), carcinogenicity, time or temporal context (acute or chronic), and type of exposure (direct or indirect). The range of ecological effects depends on the complexity of a receptor. Some features of ecological effects of toxicants related to the receptor's level of organization are presented in Figure 7.3.1.

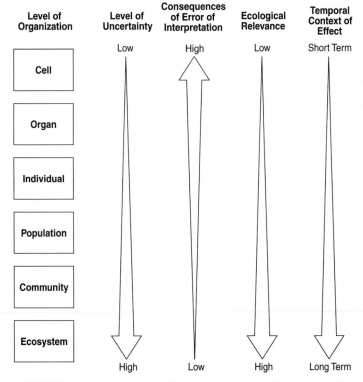

FIGURE 7.3.1   Features of toxicants effects based on level of ecological organization; adapted from (Newman, 1995).

7.3.2   *ECOLOGICAL ASSESSMENT*

In human health risk assessment, toxicants are of particular interest because of their carcinogenic nature. In ecological risk assessment (ERA), their effect is at least twofold: they may cause illness and death or weakening of some species and may change the ecological balance of an ecosystem. To estimate risks associated with toxic pollution, a detailed analysis of impacted ecosystems must be conducted. It is obvious that such an analysis involves the use of parameters and models with inherent uncertainty. These uncertainties are associated with the collection and analysis of data as well as with imperfections of mathematical models used in such assessments. The uncertainties come from emission quantification, from modeling of environmental transport and fate, from determining exposure routes, and, of course, from our understanding of ecological processes and systems. One has to realize that these uncertainties may induce overestimation or underestimation of the risks, and they must be a factor in the decision-making process.

Current methods used to test the effects of a toxicant on wildlife are based on the use of individual organisms of a single species. They are used for laboratory tests and are chosen either based on their similarity to the human biological system or on ease of rearing in laboratory conditions. Usually, in such studies, there is no estimation of the relevance of the laboratory species used in an assessment of a pollutant effect on a particular ecosystem. An ecosystem is not just a group of species at a specific site—it is a product of complex interactions among species and physical environment. The interactions are functional elements of the ecosystems—e.g., the structure of the ecosystem reflects the response of the species to physical environments as well as intra- and interspecific interactions. Transfer of energy through an ecosystem is determined by the feeding relationships of species within the ecosystem. Vivacity or productivity of the ecosystem depends on nutrient flow, predation, and reproduction. An example of the air toxics stress propagation through a terrestrial ecosystem is shown in Figure 7.3.2. These interactions may have a tremendous impact on the response of an ecosystem to toxicant exposure, which cannot be predicted from the outcome of the single-species laboratory tests.

By combining the complexity of the air pollution emissions with the complexity of ecological systems—it is estimated that in California alone there are about 400 types of habitats (Bakker, 1984)—one may end up with an unresolved puzzle. Therefore, there is a need for simplification, or breaking down the problem, and a systematic approach. By investigating the impacts of pollution-caused disturbances on those subsystems, an understanding of the ecosystem's response to the locally or globally emitted pollution can be achieved. One important element of such an analysis is an effect of air pollution on a delicate interspecific interaction taking place in most habitats (ecosystems). It has been shown in many experiments and simulations (Birch, 1953; Safe, 1980; Hoffman et al., 1995) that a slight

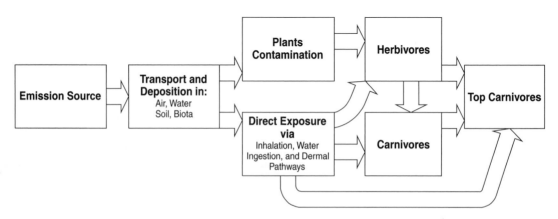

FIGURE 7.3.2   Air toxics stresses in a terrestrial ecosystem.

change in environment may have a tremendous impact on coexisting populations and the interaction processes such as competition and predation.

In the ERA it is essential to gain knowledge and comprehension of how ecosystems that are exposed to pollution function and develop. The question of how to quantify ecological effects due to toxic emissions is one of the key issues in a toxics risk analysis. The development of measures of specific ecosystem disturbances and their impact for use in risk management is sought. There is also a need for methods to quantify and value ecological impacts of toxic emissions, possibly by estimating the reduction in population size of indicator species.

An ecological assessment is an estimation of contaminant exposure experienced by an ecosystem. This type of study includes evaluation of contaminant pathways, the biological fate, and correlations between the contaminant concentration and biological effects. Various well-established ecological techniques are used to investigate ecological effects. One must evaluate the structure of the ecological community, species composition, habitat, density, and diversity to detect anomalies. Some contaminant effects can be identified by a field investigation. The field measurements and observations made to estimate the risks due to toxic emissions are exposure to the contaminant and its effects on wildlife. Edwards (1990), discussing the assessment of wildlife survival methods applied to pesticide analysis, writes "Exposure may be described as the product of pesticide availability, numbers of wildlife and their behaviour; while effects on wildlife can be measured by assessment of mortality, survival and behaviour."

Other methods involve numerical analyses of contaminant transport and its concentrations in various media to find the pathways. These methods are similar to human health risk assessment. The cause and effect characteristics are crucial in ERA. The contaminant concentration in the biological target along with its concentration in environmental media can be used to evaluate cause and effect, as well as pathways. A correlation between a concentration of contaminant in environmental media and in the biological resource is strong evidence of contamination effects. The most common analysis of cause and effect is analysis of tissue concentration in the given organism. However, other methods and measurements such as biomass, reproduction, and presence or absence of given species can be even more useful because they relate directly to assessment endpoints (Maughan, 1993).

However the prevailing methods, most often used on an after-the-fact basis, in estimating an effect of a toxicant on wildlife are mortality methods. They are simple and include collecting corpses and examining them to identify the cause of death. The other approach used in pesticide hazard assessment is a survival method. This method accounts for a loss of live animals by comparing the number of living animals in treated areas with the number in similar nearby control areas. These methods deal only with the acute consequences of toxic emissions. However, there are many questions about long-term consequences—for instance, what is the impact of a toxicant on the natural balance of the ecosystem? Is a species one wants to protect really endangered or is there some other species whose extinction can lead to dramatic changes of the ecosystem in the long run? Furthermore, if development of a population is heavily affected by a toxicant for a certain period of time, say one season, is it able to return to its original level? How does this fluctuation affect other members of that ecosystem? The hidden impacts of long-term exposure may have a more significant effect on the population than the acute effects.

7.3.3 POLLUTION INDICATORS

Ecological endpoints is a term in ERA for defining cleanup criteria or acceptable risk for some undesirable events. The endpoints are not limited to the numeric values. An endpoint can also be establishing a community with ecological functions as in similar systems in the region or an ultimate goal such as breeding a species of concern or egg production by a bird species. Typical ecological endpoints can reflect any ecological hierarchy from populations to communities to ecosystems (Suter, 1990).

The health of a wildlife population depends on longevity, reproduction rate, time to maturity, and interspecific interactions. However, in toxicological studies, investigations of effects of toxic pollution on populations are based on acute mortality estimates. Recently, it has been recognized that acute mortality methods are a poor estimator of ecological endpoints such as the population growth rate (Walthall and Stark, 1997). The measure of changes in the population growth rate due to sublethal or long-term effects is considered to be more relevant than lethal exposure methods in determining the impacts of a toxicant on a population. In other words, using acute mortality methods alone does not allow prediction of how a population's growth rate will respond to a toxic stress. Moreover, the experimental results suggest that the assessments based on acute mortality estimates lead to wrong conclusions about a population response.

Animal biomarkers are also heavily used as pollution indicators to assess the health of organisms in the environment (Peakall, 1992). Biomarkers are typically changes in cellular or biochemical processes and allow characteristic biochemical response to a chemical exposure to be measured. They use the field samples obtained by nondestructive sampling methods to provide a link between exposure and effect. Major difficulties with this approach arise from a large number of potential toxicants and the even more complex problem of chemical mixtures.

Quantification of contaminant transfer to biota or biological concentration factors through measurements is, so far, the best method for ecological assessments. By comparing concentrations of contaminant in the given organism and in media such as water, soil, or food it is possible to identify the critical areas where the remediation is necessary. Quantification of the transfer from the media to the organisms may result in the risk-based remediation criteria. However, the population level exposure response can be estimated only through population dynamics.

7.3.4   *TERRESTRIAL ECOSYSTEMS*

The impact of air pollution on the ecosystems can be divided into three categories:

- lethal effects (direct or short term)—for example, an acute dose of a pollutant;
- sublethal effects (long term), changing the competitiveness of the species within the habitat; and
- subtle effects, causing a perturbation in the wildlife interactions and/or changing the patterns of intra- and interspecific interactions.

The effect of a toxic pollutant on the population can also be categorized according to its severity as acute, chronically accumulated, impairment, behavioral effects, and biochemical changes (Sheehan et al., 1984). The first two categories ultimately cause death, whereas the third category causes damage to an individual. This damage can lead to weakening of population survivability and in turn to changes at the ecosystem level.

It is difficult to investigate subtle effects of pollution on ecosystems. Ecological systems do not behave in constant or repeatable ways. They are marked by fluctuations in the species population levels and their distribution. For a researcher trying to investigate the effect of pollution on the ecosystems, such features pose a question of how to distinguish whether the change in the system's behavior is due to a pollutant or whether it is part of a natural fluctuation. Often, one must estimate whether the change indicates that the pollution affects the ecosystem level without even knowing how the system behaves without pollution. For ecosystems, it is essential not only to investigate the effect of a toxic pollutant on an individual or even one species but also to understand its effect on the overall structure and functioning of the ecosystem.

Chronic pollution effects often depend on time since exposure and it is difficult to observe the effects right after the exposure took place. Because of biological complexity, the time before the effects become observable is longer for each increasing trophic level (Sheehan et al., 1984). Under long-term anthropogenic pollution stress, ecosystem structure is usually simplified; in addition, severely stressed ecosystems may become energetically unbalanced. There may be changes in the species composition and/or physiological characteristics, such as decreased life span, reproduction, feeding patterns, etc. The system becomes less stable.

Assessment of exposure to contaminants requires translating the contaminant concentrations at the receptor point into quantitative estimates of the amount of toxicant that comes in contact with an individual organism from the population at risk. Exposure of terrestrial wildlife to contaminants differs considerably from human exposure. This pattern leads to differences in exposure time, mobility, and exposure pathways. Also, there is a scarcity of available data if one takes into account the biological richness of the environment. The extrapolation from laboratory studies to field assessments, and even from a normalized ecosystem to the investigated ecosystem or from one ecosystem to another, is loaded with high uncertainty. Laboratory experiments do not predict every potential effect on an ecosystem. They are usually designed either to investigate the maximum adverse effects or to screen for known effects. At this stage of the ERA's development only a risk assessor's experience and application of good science can result in some sort of quantified risk assessment.

REFERENCES

Bakker, E., *An Island Called California*, University of California Press, Berkeley, 1984.

Birch, L. C., "Experimental Background to the Study of the Distribution and Abundance of Insects. III. The Relations Between Innate Capacity for Increase and Survival of Different Species of Beetles Living Together on the Same Food," *Evolution*, 7: 136–44, 1953.

Edwards, P. J., "Assessment of Survival Methods Used in Wildlife Trials," in *Pesticide Effects on Terrestrial Wildlife*, Somerville, L., and Walker, C. H., eds., Taylor & Francis, Washington, DC, 1990.

Hoffman, D. J., Rattner, B. A., Burton, G. L. Jr., and Cairns, J. Jr., *Handbook of Ecotoxicology*, Lewis Publishers, Boca Raton, FL, 1995.

Maughan, J. T., *Ecological Assessment of Hazardous Waste Sites*, Van Nostrand Reinhold, New York, 1993.

Newman, M. C., *Quantitative Methods in Aquatic Ecotoxicology*, Lewis Publishers, CRC Press, Boca Raton, FL, 1995.

Peakall, D., *Animal Biomarkers as Pollution Indicators*, Chapman & Hall, 1992.

Safe, S., "Metabolism, Uptake, Storage and Bioaccumulation," in *Halogenated Biphenyls, Terphenyls, Naphthalenes, Dibenzodioxins and Related Products*, Kimbrough, R. D., ed., Elsevier/North-Holland, New York, 1980.

Suter, G. W. II, "Endpoints for Regional Ecological Risk Assessments," *Environ. Man.*, 14(1): 9–23, 1990.

Sheehan, P. J., Miller, D. R., Butler, G. C., and Bourdeau, P., eds., *Effects of Pollutants at the Ecosystem Level*, John Wiley & Sons, New York, 1984.

U.S. Environmental Protection Agency, *Performance Evaluation of Full-Scale Hazardous Waste Incinerator*, NTIS, PB 85-129500, EPA, 1990.

Walthall, W. K., and Stark, J. D., "A Comparison of Acute Mortality and Population Growth Rate as Endpoints of Toxicological Effect," *Ecotox. Env. Safety*, 37(1): 45–52, 1997.

CHAPTER 7
CONTAMINANT EFFECTS

SECTION 7.4

AIR POLLUTANT SOURCES AND EFFECTS

Jan Swider

Dr. Swider is a researcher in the Mechanical and Aerospace Engineering Department, UCLA, School of Engineering and Applied Science. His main interests are in human health and ecological and engineering systems risk assessments.

7.4.1 INTRODUCTION

Air pollution is one of the most important environmental risks. Pollutants are emitted into the atmosphere mostly as the result of a variety of human activities and they cause a number of undesirable effects. In humans these emissions are responsible for some lung diseases and overall they can be mutagenic, carcinogenic, teratogenic, or otherwise harmful. The impact of air pollution on ecological systems is less understood, although it may be disastrous. It causes not only the forest death syndrome but also dramatic changes in the whole ecosystem, leading to extinction of many species and, in a roundabout way, threatens human beings. Although there are state and federal regulations, such as the Clean Air Act (CAA), which set emission standards, these standards have been developed to determine acceptable levels of exposure and a range of risk to humans and usually do not address risks posed to the other members of an ecological system. Moreover, there are no exposure limits for all members of an ecosystem imperiled by such pollution and the impact of air pollution on ecosystems has not been adequately investigated.

Air Pollution

Air pollution is often defined as a concentration of certain elements or chemicals above their normal ambient levels. The amount of air pollution is determined by emission sources. The source term is, in general, a function of the amount, rate, and mode of release. The major sources of outdoor air pollution are transportation, electric power plants, industrial and domestic fuel burning, and industrial processes. These pollutants are emitted to the atmosphere; then transport processes, and often chemical and physical transformations, take place. The released chemical ability to produce toxic effects varies widely and depends on the following:

- Chemical toxicity,
- Volume of released chemicals,
- Degradation and persistence of the chemical in the environment,
- Bioconcentration, and
- Terrain configuration and meteorological factors

Classification of Air Pollutants

Air pollutants may be classified according to emission sources, pollutant origin, and state of matter. Air pollution emission sources describe

- Natural sources—e.g., erosion, volcanoes, pollen, forest, and brush fires; and
- Anthropogenic sources—the result of human activities such as transportation, agriculture, and industry.

Considering pollutant origin, the air pollutants may be classified as

- Primary—those that meet a receptor in their original form as they were emitted from the source, or
- Secondary—those formed in the atmosphere after undergoing various transformations on the way.

Based on the state of matter, air pollutants can be

- Gases—they tend to diffuse and not to settle out, such as ozone; and
- Particulate matter—fine pieces of solids or droplets of liquids (the heaviest settle out around the source but small particles can remain suspended in the air and be transported with the winds—e.g., sulfates).

The 1970 CAA, with amendments passed by Congress in 1990, distinguishes two general classes of air pollutants. The first group is called criteria pollutants and includes pollutants emitted in millions of tons per year, spread across broad areas of the country, and affecting human health and the environment over those areas. This group includes six pollutants: carbon monoxide, lead, nitrogen oxides, ozone, PM_{10} (particulate matter of 10 μm or less in diameter), and sulfur dioxide. The effect of such pollutant emissions on humans is relatively well known and the U.S. Environmental Protection Agency (EPA) has set national standards for each of them.

The second group of pollutants is described as hazardous air pollutants (HAPs) or air toxics. These pollutants usually are more localized with regard to the emission source, and they can pose a hazard to human health and the environment. Title III of the CAA, as amended in 1990 under Section 112(b), identifies a group of 189 hazardous air pollutants for regulatory purposes. In addition, the CAA divides these pollutants into several subclasses according to their chemical properties.

The air toxics risk assessment differs considerably from the criteria pollutants risk estimations. First, there are a large number of chemicals with diverse chemical and physical properties. Second, there is limited knowledge about health effects due to such exposures, and there are no thresholds and no regulations for many HAPs. Additionally, many air toxics undergo chemical transformation in the atmosphere, resulting in different airborne products often with different health effects than the originally emitted HAPs. Therefore, the source data alone make toxics exposure estimates and risk assessment difficult.

Air Pollution Modeling

Many studies have attempted to understand the phenomena associated with air pollutant emissions and their effect on human health as well as the fate and effects of toxicants in various ecosystems. There is an extensive literature describing such effects and methods of their modeling:

- Seinfeld (1986) studied the chemistry and physics of air pollutants. In his air pollution monograph, he presented a comprehensive description of the formation, growth, and dynamics of air pollutants as well as their global behavior and concentration statistics.
- Zannetti (1990) reviewed mathematical models, which are important in estimation of the transport and diffusion processes, and presented some computer packages for meteorological and air-quality simulations.

- Jørgensen et al. (1996) presented a comprehensive overview of environmental models covering, among others, atmospheric pollution, ecotoxicological, and population dynamics models.
- The Electric Power Research Institute report (EPRI, 1985) on assessing the risks of airborne carcinogens presents a method for analyzing the risks of airborne pollutants from fossil fuel plants. A step-by-step procedure based on emission, air quality, exposure, and risk was applied for benzo(*a*)pyrene and arsenic.
- The PRA Procedures Guide issued by the U.S. Nuclear Regulatory Commission (NRC, 1982) was the first to describe extensively the inclusion of uncertainties in the process of radionuclide release and transport resulting from core-melt accidents in a nuclear reactor. In this guide, uncertainty analysis in probabilistic risk assessment (PRA) for nuclear power plants was introduced with emphasis on a quantitative estimation of uncertainty.

7.4.2   *EMISSION CHARACTERIZATION*

Quantification of risks associated with air pollution emissions starts with analysis of displacement of pollutants to the atmosphere. Different wind and terrain condition affect the way pollutants are mixed and transported in the atmosphere and therefore pollutant concentrations. A proper description of local emissions is important for understanding the chemicals' fate. There are various physical and chemical processes that affect transport and distribution of pollutants, such as advection, dilution, mixing, washout, deposition, and chemical transformation. Often the most harmful air pollutants are not those emitted directly by the source but those that are formed in the atmosphere as a result of chemical reactions. There are a wide variety of airborne substances. They include compounds of carbon, nitrogen, sulfur, and halogens.

The way pollutants are emitted into the atmosphere has considerable effects on the final receptor exposure calculation. Usually, three types of emission sources are considered:

- Point sources—emission is assumed to come from a single point such as a stack or chimney.
- Line sources—emission is assumed to be evenly distributed along a line and is a good approximation for describing roads or railways.
- Area sources—emission is assumed to be uniformly distributed over an area such as a gas station where the concentration of hydrocarbons can be approximated as being constant over the whole area.

For computational purposes grid layers are used and background emission levels are combined into a grid. Often the time variations in the local emission sources are essential and dynamic air-quality simulation models are frequently used.

In modeling combustion emissions, especially from coal power plants, two classes of pollutants have to be distinguished: organic and inorganic. They behave differently during the combustion process. Because of difficulties in predicting generation and destruction rates of organic pollutants, one has to rely on empirical data. Unfortunately, data on such emissions are limited; therefore, theoretically derived estimates are often used for organics. For inorganics, a mass balance can be applied because there is no generation or destruction.

Given a steady-state condition, the mass balance of the input streams is equal to the sum of output streams. In the case of a coal-fired plant, the input is the amount of coal used by the plant. There are a number of the output pathways—e.g., fly ash and combustion gases released through the stack, bottom ash via the disposal system, etc. Usually, fly ash is the primary source of airborne emissions from the coal power plants (EPRI, 1985).

For the purpose of risk assessment, coal consumption must be determined directly from the power plant records or must be estimated based on average load, plant heating rate, and coal heating values. From mass conservation, fly ash emission rates can be estimated. One must take into account the influence of various control devices on emission rates. It is also important to know the flow rate and velocity of the gases exiting the stack.

Air Toxics Inventories

Toxic emissions data are sometimes available from the other studies. For instance, in California, the South Coast Air Quality Management District prepared the air toxics emissions inventory for stationary and mobile sources, such as automobiles (SCAQMD, 1987), and California's Air Toxics Hot Spots Bill (AB2588) has been a driving force for air pollutant release inventories in California.

During the SCAQMD study, 1244 point sources were identified and emission data were collected. Most of those sources were associated with emissions of arsenic, beryllium, mercury, methyl bromide, nickel, and vinyl chloride. Metal emissions were predominately due to combustion, plating, and other industrial processes. Methyl bromide emission was due to its use as a fumigant and in organic synthesis, whereas vinyl chloride was emitted from production of polyvinyl chloride and from municipal and hazardous waste landfills. Area sources were responsible for most methylene chloride, perchloroethylene, and trichloroethylene emissions because of their use in metal degreasing, solvent extraction, and dry cleaning. Mobile sources, in this case only automobiles were considered, contributed the major sources of cadmium, ethylene dibromide, ethylene dichloride, lead, toluene, and xylenes. Benzene emissions were about even between automobile emissions and area sources such as gasoline service stations, stationary gasoline engines, crude oil production, and agricultural burning.

AB2588 requires emissions inventory to be generated for major point sources. Toxics inventory reports were developed for all emission sources at individual facilities emitting more than 10 tons per year of criteria pollutants. It should be noted, that although California's AB2588 forced development of emissions inventories, the inventories were done only for a selected set of facilities. It does not necessarily mean that it is an accurate representation of all such emission sources. Also, mobile sources and other off-road vehicles were not included and available data for emissions from area sources were limited. Such inventories, therefore, do not represent the best known or even the worst-case concentrations of air toxics substances; there were significant uncertainties in the inventory process for both point and area sources. Not every state has prepared the air toxics emission inventories. In such a case, the EPA's Toxics Release Inventory (TRI) is a major source of data.

TRI has been published by the EPA Office of Pollution Prevention and Toxics. TRI contains annual state reports about the release and transfers of more than 300 toxic chemicals and compounds. The top states for air pollutant releases for 1996 were Texas with 127.2 million lb (57.7 million kg) (fugitive plus stack emissions), Tennessee with 84.4 million lb (38.28 million kg), and Louisiana with 83.9 million lb (38.06 million kg) (see Figures 7.4.1 and 7.4.2). The TRI Public Data Release Report is

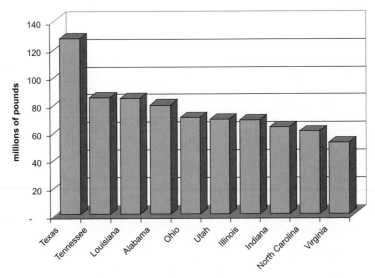

FIGURE 7.4.1   Top 10 states for air toxic pollutant emission in 1996. (*Source: U.S. Environmental Protection Agency: 1996 Toxics Release Inventory Public Data Release Report*)

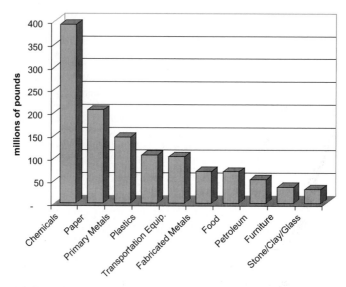

FIGURE 7.4.2   Top industries with largest air toxic pollutant emissions in 1996. (*Source: U.S. Environmental Protection Agency: 1996 Toxics Release Inventory Public Data Release Report*)

publicly available and can be accessed through the Wide World Web at *http://www.epa.gov/opptintr/tri/* or ordered directly from the National Service Center for Environmental Publications. Free access to TRI data, and other related databases from several federal agencies, is provided because of the Emergency Planning and Community Right-to-Know Act provisions enacted to promote emergency planning and provide the public with information on toxic chemicals released in their communities (EPA, 1998).

7.4.3   SIGNIFICANCE OF ATMOSPHERIC POLLUTION TO TERRESTRIAL ECOSYSTEMS

It is difficult to build accurate cause and effect characteristics, especially in terrestrial ecosystems, because (1) the wildlife is mobile and consumes various diets, and (2) contamination usually is not spatially homogeneous. The difficulties also arise from a complex mechanism of contaminant transport through an ecosystem and problems with measuring the effects. Many of the hazardous air pollutants have an impact on terrestrial ecosystems through effects on plant growth, biomass quality, and changes in ecosystem processes. Vegetation damage is an indicator of pollutants such as fluoride, sulfur dioxide, and nitrogen dioxide. Animal health has been affected in particular by fluoride from chemical industry emissions and by arsenic emitted from smelters. These factors supplement a set of natural stresses that occur in terrestrial ecosystems.

Evaluation of air toxic pollution impacts on terrestrial ecosystems is a challenge and needs to be thoroughly investigated. Most of the research has been concentrated on atmospheric pollution and terrestrial vegetation, whereas its effect on animal health, behavior, and population changes is less known. Large spatial scales, seasonal changes, and number and variability of exposure pathways are areas that must be considered while conducting, for instance, a terrestrial ecological risk assessment for local vertebrates. There are also changes in multiple pollutants or complex mixtures from diverse anthropogenic sources interacting with each other, often forming secondary pollutants. Potential effects can be similar to natural environmental stresses or can amplify them. Some pollutants may serve as nutrients in some cases and may have poisonous effects in other cases. The impacts of air toxics emission on ecosystems are poorly understood and described (McLaughlin and Norby, 1991). There is a body of evidence on the adverse effects of pollution on terrestrial vegetation. Pollution stress, sensitivity of vegetation species to particular pollutants, genetic changes, and other effects of pollutants on evolutionary processes in plants have been observed and investigated. A new subdivision of science, chemical ecology, has evolved to study chemical interactions between organisms and their environment. Pitelka (1988) showed that air pollution stress causes alterations in plant species, populations, and communities. It turns out that plant competition and repair mechanisms are the most important factors in plants' battle for survivorship under the pressure of air pollution.

Unfortunately, even less research has been done with wild animals. Most studies have concentrated on aquatic organisms. This is partly due to the EPA's program of pesticide regulations that require conducting an ecological risk assessment for pesticides. Consequently, the amount of research on ecological risks from agrochemicals is astounding. New models for aquatic organisms, integrating chemical and environmental information into the ecological risk assessment, have been developed. Bioconcentration, bioaccumulation, and even pharmacokinetic models relating the uptake of chemicals from aquatic reservoirs to fish exposures have been investigated and modeled. Terrestrial risk assessment, however, has been not pushed forward by a program similar to the pesticide regulations; therefore, the scope of the research is limited.

From the terrestrial risk assessment standpoint, we are interested in two effects of air pollutants on wildlife: direct and indirect. Assessment of direct exposure consists of steps similar to those applied to humans: inhalation and dermal exposure and absorption of air pollutants by every member of an ecosystem. Indirect exposure is estimated based on particular food webs. A combination of these exposure pathways allows doses for each member of the ecosystem to be estimated. Because this is a dynamic system, the impacts of these toxicants cannot be analyzed separately. However, the mechanisms behind these processes ought to be investigated and understood in order to limit the uncertainties in modeling such systems.

Although there are many uncertainties about linkages between atmospheric emissions and environmental damage, there is also a considerable amount of evidence indicating that pollution-related chronic stress has an adverse effect on terrestrial ecosystems.

REFERENCES

Electric Power Research Institute, *Assessing the Health Risks of Airborne Carcinogens*, EA-4021, EPRI, 1985.

Jørgensen, S. E., Halling-Sorensen, B., and Nielsen, S. N., eds., *Handbook of Environmental and Ecological Modeling*, Lewis Publishers, Boca Raton, FL, 1996.

McLaughlin, S. B., and Norby, R. J., "Atmospheric Pollution and Terrestrial Vegetation: Evidence of Changes, Linkages, and Significance to Selection Processes," in *Ecological Genetics and Air Pollution*, Taylor, G. E. Jr., Pitelka, L. F., and Clegg, M. T., eds., Springer-Verlag, 1991.

Pitelka, L. F., "Evolutionary Responses of Plants to Anthropogenic Pollutants," *Trends Ecol. Evol.*, 3(9): 233–236, 1988.

South Coast Air Quality Management District, *Toxic Emissions Data for the South Coast Air Basin*, SCAQMD, 1987.

Seinfeld, J. H., *Atmospheric Chemistry and Physics of Air Pollution*, John Wiley & Sons, New York, 1986.

U.S. Environmental Protection Agency, *Toxic Release Inventory*, EPA/745/R-98/005, EPA, 1996.

U.S. Nuclear Regulatory Commission, *PRA Procedure Guide*, NUREG/CR-2300, NRC, Washington, DC, 1982.

Zannetti, P., *Air Pollution Modeling, Theories, Computational Methods and Available Software*, Van Norstrand Reinhold, New York, 1990.

CHAPTER 8
ANALYSIS AND MODELING

SECTION 8.1

GEOCHEMICAL MODELING

William J. Deutsch

Mr. Deutsch is president of Geochemistry Consulting, Brooksville, Maine. As a hydrogeochemist with research and consulting firms for more than 20 years, his project experience includes environmental assessments and investigations of landfills, refineries, pesticide plants/distributorships, military bases, mines and mills, federal weapons facilities, and a wide variety of additional industrial sites.

8.1.1 INTRODUCTION

Modeling the natural environment is an attempt to simulate the processes that are active in the environment in order to mimic and explain our observations. These observations may include the composition, pH, or redox potential of water; the gas concentrations of soil vapor; or the presence of mineral weathering products. To model the natural environment, we define a system of phases to represent that portion of the environment we want to simulate. Each phase has uniform, distinct properties. In geochemical modeling, the environmental system usually consists of solution, solid, and vapor phases. Figure 8.1.1 shows a system consisting of a vapor phase in contact with a solution phase, which is also in contact with various solid phases composed of minerals and amorphous compounds. Our geochemical model of this system simulates the processes that occur between the various phases and predicts the outcome of these processes on the composition and other properties of the phases.

There are two primary reasons for developing a geochemical model of a site. The first reason is to better understand the processes that are active at the site and be able to explain observed compositions of the various phases and trends in composition across the site. If the model developed for a site accurately simulates existing conditions, you will be able to make a strong case that you understand the processes that affect the movement of chemicals in the environment. At a hazardous waste site, this is useful for evaluating human health and ecological risk. The second reason for developing a geochemical model is to aid in the remedial design for cleanup of a contaminated site. An accurate geochemical model of a contaminated site can be used to estimate the outcome of various remedial actions. The model can be used to select the best remediation technologies and to optimize application of the technology.

Development of a geochemical model of a system can be represented by the flowchart shown in Figure 8.1.2. Available information on water composition, presence of solid phases, hydrogeology, etc. is compiled. The existing site conditions are evaluated in terms of the known types of geochemical and physical processes that occur in the environment. These processes include mineral dissolution/precipitation, oxidation/reduction, adsorption/desorption, groundwater recharge, and mixing of groundwaters from two or more aquifers. The combination of site-specific conditions and known, common geochemical/physical processes leads to development of a conceptual model of how the site operates. The conceptual model can be combined with a computer code to calculate the effects of the various processes and provide quantitative estimates of these impacts on the composition of

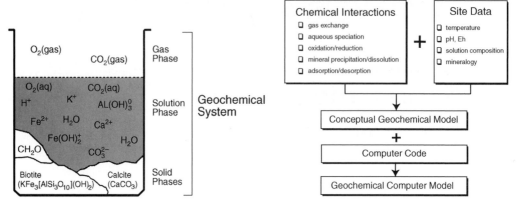

FIGURE 8.1.1   Phases of a geochemical system.

FIGURE 8.1.2   Geochemical modeling process.

the various phases. The result is a geochemical computer model consisting of the conceptual model and the computer code that together can simulate existing conditions and predict the impact on the system of changes in environmental conditions. The remainder of this section provides details about understanding and developing the various components of a geochemical computer model.

8.1.2   GEOCHEMICAL MODELING DATA COLLECTION REQUIREMENTS

Geochemical modeling of water/rock/vapor interactions requires site-specific data on the composition of the various phases. Table 8.1.1 provides a list of solution data that might be useful in developing a geochemical model for a site. Dissolved concentrations of major ions (cations and anions) are always necessary to evaluate mineral equilibrium for solids that might form from these ions. Data on major ions are also needed because they form solution complexes with minor/trace dissolved species that affect the solubility of minerals containing the species present at relatively low concentrations. The pH of the water is a necessary parameter because of its role in solution complexation, its effect on mineral solubility, and the impact of hydrogen ions on adsorption/desorption. The redox potential (Eh) is a master variable like pH and it should be measured if redox-sensitive elements such as Fe, Mn, Cr, As, Se, Hg, Cu, and U are being modeled. The dissolved concentration of oxygen gas can be useful for evaluating the oxidizing capacity of the groundwater and it may be a qualitative indicator of redox potential under high Eh conditions. Dissolved carbon dioxide (CO_2) gas may be measured or calculated from the alkalinity and pH. Its concentration is useful for predicting whether groundwater will degas CO_2 or dissolve CO_2 when it is exposed to atmospheric conditions.

To calculate clay mineral equilibrium the dissolved concentrations of silicon and aluminum must be known, and the concentrations of iron and manganese are necessary for calculating equilibrium with the commonly important oxyhydroxide minerals of these metals. Data on the other dissolved trace elements in groundwater should be collected if reactive minerals containing the element may be present (e.g., lead in lead carbonate), the element forms strong complexes (e.g., fluoride and phosphate), or the element competes for adsorption sites (e.g., As, Se, Cu, Pb).

Naturally occurring organic compounds in water, such as fulvic and humic acids, can complex with inorganic elements thus affecting mineral solubility and the mobility of the element. Organic compounds in water, both dissolved and particulate or natural and anthropogenic, also provide a strong reductant in the system that can consume dissolved oxygen and lower the redox state of other dissolved species.

Stable and unstable isotopes of the elements can be used to date water as well as to discern some of the physical and chemical processes that might affect water composition. For example,

TABLE 8.1.1   Site Groundwater Data for Geochemical Modeling

Data	Use
Major ions	Calculation of solution complexes
(Ca, Mg, Na, K, $HCO_3/CO_3/Cl/NO_3$)	Calculation of ionic strength and solute activity Saturation indices for minerals with these components
pH	Ion speciation/complexation and mineral solubility
Eh	Ion speciation/complexation and mineral solubility of redox-sensitive elements
Dissolved gases	
(O_2, CO_2)	O_2: qualitative measure of redox potential CO_2: stability of groundwater pH in contact with atmospheric air
Minor/trace elements	
(Si, Fe, Mn, Al)	Clay and oxyhydroxide mineral equilibria
Trace metals	
(Ba, V, Cr, Mo, Pb, Cu, Zn, Hg, Cd, B, etc.)	Mineral equilibria, competitive adsorption
Trace semimetals	
(As, Se)	Mineral equilibria, competitive adsorption
Trace nonmetals	
(F, Br, I, P)	Complexation, mineral equilibria, competitive adsorption
Organic compounds	
(Humic/fulvic acids, etc.)	Complexation, oxygen consumption, sorption reactions
Stable isotopes	
($^{18}O/^{16}O$, D/H $^{34}S/^{32}S$)	Water signature, mineral reactions
Unstable isotopes	
(^3H, ^{14}C, ^{36}Cl)	Age dating

evapotranspiration changes the $\delta^{18}O$ and δD values from that consistent with the meteoric water line (Thomas, J. M., Welch, A. H., and Preissler, A. M., "Geochemical Evolution of Ground Water in Smith Creek Valley—A Hydrologically Closed Basin in Central Nevada, U.S.A.," *Appl. Geochem.*, 4:493–510, 1989), and $\delta^{13}C$ can be used to validate mass transfer predictions between groundwater, carbonate minerals, and CO_2 gas (Plummer, L. M., and Back, W., "The Mass Balance Approach: Application to Interpreting the Chemical Evolution of Hydrologic Systems," *Am. J. Sci.*, 280: 130–142, 1980).

Because the solid material in contact with the solution is the source of most of the dissolved constituents it is very important to characterize the minerals and other solids in the system. Table 8.1.2 provides a list of the more common reactive minerals and characteristics of the solid phase that are potentially useful in developing a geochemical model of the system. Other reactive minerals that may be important in particular systems are listed in Table 1.6.5 of Chapter 1 Section 6 (Geochemistry). In addition to general composition, other features of the solid material that may be useful in developing a geochemical model include (1) variations in composition and concentration along the flow path, (2) secondary minerals forming in the system, (3) minerals that are being dissolved or leached by the water, (4) exchangeable cations on the clay minerals and trace elements adsorbed onto metal oxyhydroxides and organic compounds, and (5) isotopic composition of the minerals and variations along the flow path.

8.1.3   GEOCHEMICAL PROCESSES TO BE MODELED

The presence of dissolved constituents in water is a result of chemical interactions between water (a solvent) and the vapor and solid phases in contact with the water. A geochemical model of a site simulates these processes.

TABLE 8.1.2   Site Solid Phase Data for Geochemical Modeling

Constituent of solid phase	Potential impact on system
Calcite	Mineral solubility control on solution concentration, partial measure of neutralization capacity
Gypsum	Mineral solubility control on solution concentration
Dolomite	Source of constituents to solution, partial measure of neutralization capacity
Clay mineral identification, concentration, and exchange capacity	Exchange sites for major cations, mineral solubility control on solution concentration
Ferric and manganese oxyhydroxide	Mineral solubility control on solution concentration, adsorption substrates for minor/trace elements
Pyrite	Mineral solubility control on solution concentration, source of acidity under oxidizing conditions
Silicate minerals	Sources of many dissolved constituents
Organic carbon	Adsorbent medium for organic and inorganic compounds, reducing agent, source of dissolved carbon

Solution/Vapor Processes

Surface water and groundwater in the unsaturated zone are in contact with atmospheric air and soil vapor, respectively. The concentrations of the constituents in these two phases tend to equilibrate by either dissolving gases from the air/vapor into the solution or exsolving gases from the solution into the vapor phase. The two most common gases of geochemical interest are O_2 and CO_2. Equilibration between a solution and these gases can be written as:

$$O_2(g) \leftrightarrow O_2(aq)$$

$$2CO_2(g) + H_2O \leftrightarrow CO_2(aq) + H_2CO_3$$

O_2 gas [$O_2(g)$] dissolving into water forms aqueous O_2 [$O_2(aq)$], which creates a high redox potential and provides oxidizing capacity to the system. This is an important process to model if redox-sensitive metals or organic compounds are being modeled. CO_2 gas [$CO_2(g)$] dissolves into water, producing aqueous CO_2 and carbonic acid (H_2CO_3). This can be a very important process to model because of its effect on acid/base conditions and the solubility of most minerals. Henry's law constants are used to proportion constituents between the vapor and solution phases Schwarzenbach, R. P., Gschwend, P. M., and Imboden, D. M., *Environmental Organic Chemistry*, J. Wiley & Sons, Chapter 6, 1993).

Solution Phase Processes

Within the solution phase, the dissolved constituents interact with each other to form solution complexes, which affects mineral solubility and other properties of the system. Solution complexation can be represented by the following reactions:

$$Fe^{3+} + H_2O \leftrightarrow Fe(OH)^{2+} + H^+$$

$$Mg^{2+} + SO_4^{2-} \leftrightarrow MgSO_4$$

The first reaction shows that some of the dissolved ferric iron (Fe^{3+}) reacts with water to form the iron hydroxyl species $Fe(OH)^{2+}$ and hydrogen ion. This reaction lowers the effective concentration of dissolved ferric iron and makes minerals containing ferric iron more soluble. The reaction is pH dependent because the hydrogen ion is part of the reaction. In the second reaction, between dissolved sulfate and magnesium, both ions are bound, thus increasing the solubility of minerals that contain these species. The reaction also lowers the ionic strength of the solution, which also has an effect on mineral solubility.

Mineral Dissolution/Precipitation

Most of the dissolved constituents in water result from dissolution of minerals in contact with the water. Mineral dissolution can be represented for calcite as:

$$CaCO_3 \leftrightarrow Ca^{2+} + CO_3^{2-}$$

The double barbed arrow in the reaction implies that calcite can dissolve to produce its components in solution and that it may precipitate from solution when the concentrations of its components reach saturation with the mineral. Consequently, minerals provide both a source of dissolved constituents and a potential sink for those constituents. Details on the impact of mineral dissolution/precipitation on solution concentration are provided in Chapter 1 Section 6 (Geochemistry).

Oxidation/Reduction

Redox processes may be important in modeling a geochemical system because of the large number of elements that occur naturally in more than one valence state. Oxidation is an electrochemical reaction in which the valence state of an element is increased; reduction is a decrease in valence. The following half reactions show the reduction of ferric iron to ferrous iron and the oxidation of carbon (−I) in benzene to carbon (+IV) in CO_2.

$$Fe^{3+} + e^- \rightarrow Fe^{2+} \quad \text{(reduction reaction)}$$

$$C_6H_6 + 12H_2O \rightarrow 6CO_2 + 30H^+ + 30e^- \quad \text{(oxidation reaction)}$$

If benzene is the reductant for ferric iron and ferric iron is the oxidant for benzene, then these two reactions can be added together to eliminate the electron and produce the complete reaction:

$$30Fe^{3+} + C_6H_6 + 12H_2O \rightarrow 30Fe^{2+} + 6CO_2 + 30H^+$$

Oxidation/reduction reactions can have a major impact on the dominant species in solution, the pH, solubility of materials, stability of organic compounds, and concentrations of dissolved gases.

Adsorption/Desorption

Adsorption is the attachment of a dissolved species to the surface of a solid phase, and desorption is the release of the adsorbed species back into the solution. Adsorption can be represented by the following reaction:

$$\equiv FeOH_2^+ + Ag^+ \leftrightarrow \equiv FeOHAg^+ + H^+$$

The reaction shows dissolved Ag^+ being adsorbed onto the solid surface represented by $\equiv FeOH_2^+$. Each adsorbed silver ion releases a hydrogen ion into solution. Surface complexation is a representation of adsorption/desorption onto metal oxyhydroxides that are commonly present in the environment. These solids often have an important effect on the mobility of trace metals. Cation exchange is a form of adsorption/desorption that involves mainly clay minerals and the major cations in typical water. Details on the adsorption/desorption process are provided in Chapter 1 Section 6 (Geochemistry).

8.1.4   CONCEPTUAL GEOCHEMICAL MODELS

A conceptual model of a geochemical system is the investigator's mental construct of what phases interact and how they interact to produce the observed system. For example, the pH of pure water in contact with atmospheric air is about 5.7. To explain why the pH is acidic and not neutral (pH 7), we might consider the gases in contact with the water. Gases in air in high concentrations are nitrogen and oxygen. These gases dissolve into the water to some degree on the basis of their solubility at the temperature of the system; however, they do not produce hydrogen ions in the water and thus do not directly affect the pH. Minor gases in atmospheric air include CO_2 and nitrogen and sulfur oxides. When these gases dissolve in water they react with the water to form other carbon, nitrogen, and sulfur species; these reactions form additional hydrogen in the solution, thereby lowering its pH. Nitric and sulfuric acids are extremely strong acids and produce a very low pH if they are present even in low quantities. The pH of our water/air system is not very low at 5.7; therefore, it is unlikely that the nitrogen or sulfur gases are present at important concentrations in the atmospheric air of our system. CO_2 forms a weak acid (carbonic acid) in water and it is the likely candidate for explaining the pH of our water. Our conceptual model for the water/air system is that (1) CO_2 in the air dissolves into the water, (2) the dissolved CO_2 forms carbonic acid, and (3) the acid dissociates sufficiently to lower the pH of the water from 7 to 5.7. Note that we have ignored some components of the system (oxygen and nitrogen) that are present in high concentrations but do not affect pH. We can check the reasonableness of our conceptual model by calculating the pH of water in contact with CO_2 at atmospheric partial pressure conditions assuming equilibrium between the two phases. The calculation results in a pH value of 5.7 under typical CO_2 concentrations in atmospheric air.

As shown in the flow diagram (Figure 8.1.2), our conceptual model of a particular geochemical system is based on measured water and gas compositions if our solution is in contact with a vapor phase. It is also based on the solid phases that may dissolve into the solution or precipitate from the solution if they become oversaturated. Although all solid phases in contact with the water are potential sources of constituents to the solution, they may not precipitate from the solution because the near-surface environment is not conducive to their formation. For example, the pyroxene minerals typically form at temperatures above several hundred degrees Celsius from a rock melt or strong metamorphic conditions. They may be present near the earth's surface now because of tectonic activity, but they will never form in the near-surface environment. For this reason, our conceptual model of interaction with solid phases should not allow the formation of this type of mineral from surface water or groundwater. A list of minerals that have been identified as forming in the near surface environment may be found in Table 1.6.6 of Chapter 1 Section 6 (Geochemistry).

In developing conceptual geochemical models, it is often useful to calculate the minerals in equilibrium with the water in the system. This is commonly done by analyzing the composition of the water and calculating the saturation indices of the minerals by using an aqueous speciation computer code. Minerals that calculate to be close to equilibrium with the solution may be forming in the system and limiting the solution concentration of at least one of their components. Minerals that calculate to be undersaturated cannot be forming in the system and should not be considered as a sink for dissolved constituents. Minerals that are oversaturated are not forming fast enough in the environment to limit the concentrations of their constituents in solution. They should not be considered reactive minerals in the conceptual model.

There may be a very large number of conceptual models that can be developed for a system that will explain some or most of our observations of that system. For example, acidic conditions in water can be produced by naturally occurring CO_2, acid rain, oxidation of sulfide minerals, organic acids from decaying vegetation, and other possible reactions. When developing a conceptual model, it is important to initially consider all the possible alternatives and develop a number of models that might explain the observations. Each model produces its own set of secondary reactions and side effects. The most plausible model can be selected by evaluating the results of each model against reality.

8.1.5 GEOCHEMICAL MODELING CODES

It is possible to calculate by hand the equilibrium pH of the simple water/CO_2 gas system described in the previous section on conceptual models. The only requirements are the partial pressure of CO_2 gas in the vapor in contact with the water, the temperature of the system and the temperature-dependent equilibrium constants for CO_2 dissolving in water, and the various carbonate species in water. However, even this relatively simple calculation may require several hours to complete if the thermodynamic data are not readily available. Furthermore, the calculations necessary to evaluate a complex natural system in which many of the typical geochemical processes are occurring would require an exceedingly long period of time to do by hand and would be fraught with errors. For these reasons, computer codes have been developed to make the necessary aqueous speciation calculations and allow for simulating the important geochemical processes that impact natural and contaminated systems. There are three basic types of geochemical modeling codes: mass balance, equilibrium aqueous speciation/saturation, and equilibrium mass transfer.

Mass balance modeling codes are useful for evaluating the geochemistry of a system when solution data are available from at least two points along the flow path. The data for these two locations (or more locations if mixing of water types is to be considered) are used as input to the computer codes along with the composition of vapor and solid phases that might impact water composition as the water moves between the two points along the flow path. The output of the code is the amount of each gas or solid phase that must dissolve into or exsolve/precipitate from the solution to cause the composition to change along the flow path. The most commonly used mass balance computer codes are BALANCE (Parkhurst, D. L., Plummer, L. N., and Thorstenson, D. C., *BALANCE—A Computer Program for Calculation of Chemical Mass Balance*, U.S. Geological Survey, 1982) and its successor NETPATH (Parkhurst, D. L., Plummer, L. N., and Prestemon, E. C., *An Interactive Code (NETPATH) for Modeling Net Geochemical Reactions Along a Flow Path*, U.S. Geological Survey, 1991). These codes can be downloaded from the U.S. Geological Survey website at http://h2o.usgs.gov/software/geochemical.html.

To evaluate the equilibrium state of a solution with respect to mineral and vapor phases, a method of calculating aqueous speciation is required. This is necessary because equilibrium is related to the activity (effective concentration) of the dissolved species, and the activity of a species in a given solution is a function of speciation reactions involving that species. If the activities of the dissolved species have been calculated, then equilibrium can be evaluated with respect to mineralogy, oxidation/reduction, gas exchange, and other interactions among the phases present in the system. Equilibrium aqueous speciation computer codes take the composition of a water sample and determine the activities of all the dissolved species in solution. Codes with a geochemical application then calculate the saturation indices for all possible minerals that might be present in the system. The saturation index of a mineral is used to evaluate whether the mineral is undersaturated, at equilibrium, or oversaturated with respect to the concentration of its composition in a solution. One of the most commonly used speciation codes is WATEQF, which is available as part of the NETPATH computer package at the U.S. Geological Survey website.

To model changing water composition along a flow path or predict the impact of changing a system variable (such as pH or Eh) on solution composition, the calculation method must be able to transfer mass between the various phases in the system. The mass balance codes transfer mass but do not consider thermodynamic equilibrium between the phases. To consider the equilibrium condition, mass transfer routines are added to the aqueous speciation computer codes.

Figure 8.1.3 shows a flow path for a typical equilibrium mass transfer calculation. The composition of the water is entered into the aqueous speciation subroutine that calculates the concentrations and activities of all possible species in the solution. The activities are then used to calculate saturation indices for all possible minerals in the system. At this point the modeler can stipulate the type of mass transfer allowed. Mass transfer between phases may consist of mineral dissolution/precipitation, adsorption/desorption, and/or gas dissolution/exsolution. When enough mass has been transferred to reach equilib-

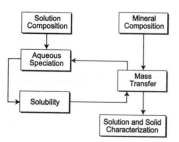

FIGURE 8.1.3 Equilibrium mass transfer calculations.

rium, or some other preset condition, calculations are stopped and the output shows the amount of mass transfer that occurred and the resulting solution composition. Two of the most commonly used equilibrium mass transfer codes are MINTEQ (Allison, J. K., Brown, D. S., and Novo-Gradac, K. J., *MINTEQA2/PRODEFA2, a Geochemical Assessment Model for Environmental Systems*, U.S. Environmental Protection Agency, 1991) and PHREEQE (Parkhurst, D. L., *User's Guide to PHREEQC, a Computer Model for Speciation, Reaction Path, Advective Transport and Inverse Geochemical Calculations*, U.S. Geological Survey, 1995). MINTEQ is available at the U.S. Environmental Protection Agency website at ftp://ftp.epa.gov/epa_ceam/wwwhtml/softwdos.htm and PHREEQ is available at the U.S. Geological Survey website http://h2o.usgs.gov/software/geochemical.html.

8.1.6 APPLICATIONS OF GEOCHEMICAL COMPUTER MODELS

Development of a computer model of the geochemistry of a site allows investigators to better understand the processes that are active at the site and also provides a tool for stimulating the impact of changing conditions on the system. A common application of modeling for understanding an existing system is to evaluate the mobility of contaminants in the environment. For example, the fate and transport of iron and other metals from an acidic source area through a neutralizing environment and then into a zone of changing redox conditions can be simulated with a computer model (Deutsch, W. J., *Groundwater Geochemistry*, CRC Press, 1997, p. 103). Additional examples of the use of geochemical models for site characterization and contaminant migration are listed in Table 8.1.3.

The ability to simulate changing environmental conditions opens the entire realm of remediation design to geochemical modeling. Once an accurate model of an existing contaminated site has been developed, it is relatively simple to vary site parameters and evaluate the effect of the new condition on contaminant concentrations in the solution and solid phases. The parameters that might be varied include pH, redox potential, ionic strength, and concentrations of complexing compounds. For remediation purposes, it is possible to evaluate both *in situ* fixation and immobilization of contaminants as well as enhanced mobility and removal of contaminants from a system. For example, arsenic is

TABLE 8.1.3   Applications of Geochemical Models

Site characterization
- Identifying reactive solid phases
- Quantifying acid neutralizing and acid producing capacities
- Quantifying oxidation/reduction capacities
- Determining degree of solution complexation
- Evaluating seawater intrusion
- Simulating aquifer thermal energy storage
- Estimating mine pit lake chemistry

Contaminant migration
- Landfill environment (metals)
- Wood-treating compounds (Cu, Cr, As)
- Pesticide plants (As)
- Acid mine drainage (metals)
- Mill tailings (metals and radionuclides)
- High-level nuclear waste repository (radionuclides)

Remediation design
- Air stripping towers (encrustation, clogging)
- Reactive barriers (secondary products)
- Pump and treat (effectiveness)
- Soil washing (effectiveness)
- *In situ* immobilization (mineral precipitation, adsorption)

strongly adsorbed onto iron oxyhydroxide minerals in shallow, oxidizing aquifers. If the dissolved arsenic concentration must be reduced in order to attain a cleanup goal, then adsorption of arsenic and removal from the groundwater might be enhanced by adding additional iron minerals to the aquifer, or arsenic might be desorbed and flushed from the aquifer by circulating a sulfate solution through the aquifer. Both of these remediation alternatives can be simulated with a geochemical model.

One of the additional benefits of modeling remediation is that secondary reactions that might not be immediately obvious to the modeler may become clear in reviewing the modeling results. For example, adding iron to an aquifer to precipitate as iron oxyhydroxide and provide additional adsorption sites may change the pH of the solution because the iron precipitation reaction produces hydrogen ions. If the system is not well buffered, the pH will decrease, which in turn may mobilize other metals. The aqueous speciation computer codes used to develop the geochemical models will accurately take into account the formation of hydrogen ions during iron precipitation and will allow for buffering by carbonate/bicarbonate ions in solution. If carbonate minerals are also present in the system, the modeler may want to allow them to interact with the solution as iron precipitates. Additional examples of uses of geochemical models for remediation design are listed in Table 8.1.3.

8.1.7 LIMITATIONS OF GEOCHEMICAL MODELING

The utility of a geochemical model for a particular site is limited by errors introduced during development of the model and our ability to model certain processes. Following the general steps in developing a geochemical model (Figure 8.1.1), the first opportunity for introducing error into the process is data collection. The model developed is highly dependent on accurate physical and chemical data from the site. We need to know flow direction to determine whether concentrations are increasing or decreasing along the flow path and we need accurate measurements of temperature, pH, Eh, and concentrations of dissolved constituents. For example, an incorrect pH measurement will affect the calculated saturation indices of most minerals and may lead to an inappropriate conceptual model for the site. Identification of reactive minerals along the flow path is also very useful in developing the geochemical model—for example, not knowing that pyrite is present in an aquifer could lead to a totally incorrect conclusion about the source of acidity in the system.

Most of the geochemical processes that affect natural systems are well known and incorporated into the modeling codes. However, certain systems may be relatively unique and some of the processes that occur may not be standardized in the available computer codes. For example, some of the elements that may be of interest for evaluating radionuclide migration (such as plutonium and americium) are not in the standard databases of MINTEQ or PHREEQE. Also, a rare mineral or amorphous solid may limit solution concentrations of a contaminant that does not normally reach high concentrations in the natural environment. If this solid phase is not in the code database, its formation cannot be simulated. Fortunately, most of the available computer codes allow the user to input new thermodynamic data for elements, solution complexes, and solid phases. This allows the user to develop an appropriate model if concentration and thermodynamic data are available.

In the discussion of the development of conceptual models, the necessity for generating multiple hypotheses for evaluating the data was explained. Developing multiple hypotheses minimizes the chances that a preconceived notion will become the de facto model for a site. There are other ways of introducing error during the conceptual modeling stage. A mineral that is unlikely to form under surface or aquifer conditions might be selected to equilibrate with water. Minerals of this type include the olivines, pyroxenes, hornblendes, and biotites. If water is forced to equilibrate with these minerals, the dissolved concentrations of the components of the mineral will not be realistic. A conceptual model may also be in error if the flow direction is not well known. A process might be selected to remove constituents from the water when, in fact, concentrations increase across a site.

The computer code used to make the calculations required by the model has its own limitations. Most of the equilibrium thermodynamic codes do not have a method of calculating reaction rates. If a mineral forms or dissolves slowly in a system, the model developed from these codes will not account for these kinetic effects. This is not a major limitation for most aquifer systems where residence times are measured in years; however, kinetic effects can become more important in modeling surface water systems.

The standard versions of MINTEQ and PHREEQE use the ion association theory to calculate solution complexation. This theory is appropriate to an ionic strength of about 0.5, which is well above most groundwater systems. However, seawater and groundwater impacted by soluble salt deposits have a higher ionic strength. A modified version of PHREEQE called PHROPITZ (Plummer, L. N., Parkhurst, D. L., Fleming, G. W., and Dunkle, S. A., *A Computer Program Incorporating Pitzer's Equation for Calculation of Geochemical Reactions in Brines*, U.S. Geological Survey, 1988) has been developed to model high ionic strength systems.

These limitations should be considered in developing a geochemical model; however, they are rarely a sufficient reason for not developing a model for a site. Considering the number of useful applications for a geochemical model in site characterization, contaminant transport, and remediation design, it often is beneficial to develop this type of model for a site.

CHAPTER 8
ANALYSIS AND MODELING

SECTION 8.2

GROUNDWATER FLOW MODELING

Henk Haitjema

*Dr. Haitjema teaches groundwater flow modeling, soil mechanics
and science, and applied mathematics at Indiana University where
he is a professor at the School of Public and Environmental Affairs.*

8.2.1 GROUNDWATER FLOW MODELING

Groundwater flow modeling is almost as old as the science of groundwater hydrology. Shortly after Darcy (1856) published his now famous equation of flow through soils, solutions to practical groundwater flow problems were offered (Dupuit, 1863). These mathematical solutions offered quantitative answers to questions of groundwater flow: the purpose of groundwater flow modeling. Dupuit also proposed to ignore the presence of vertical components of flow, which led to what is now known as the Dupuit-Forchheimer approximation (Dupuit, 1863; Forchheimer, 1886). All horizontal two-dimensional groundwater flow models rest on Dupuit's assumption of 1863, although we no longer have to ignore vertical flow in these models.

Groundwater flow modeling is an established science, in which output consists of the potentiometric head distribution (e.g., groundwater table) and groundwater fluxes in the aquifer, which may be used to draw pathlines of (hypothetical) groundwater particles (see Figure 8.2.1). Heads and fluxes can be predicted with reasonable accuracy, at least in sand and gravel aquifers. Groundwater flow in fractured rock is much more difficult to describe. In many cases, detailed knowledge of the position and properties of all (thousands) fractures is required for a reliable solution. In some cases, when fractures are very wide (karst aquifers), Darcy's law does not even apply. In practice, when the fracture network is relatively dense compared with the distance between streams and lakes, the aquifer is often thought of as a continuum and traditional groundwater flow models are used. Modeling flow in sparse discrete fracture systems is outside the scope of this section. Groundwater flow modeling often serves as the first step in modeling the fate of contaminants in groundwater, a subject that is discussed in the subsection Contaminant Transport Modeling. Anderson and Woessner (1992) provide detailed discussions on procedures for groundwater modeling, which they call "Simulation of Flow and Advective Transport." Haitjema (1995) discusses a stepwise modeling approach, gradually increasing model complexity until the modeling objective has been satisfied.

Below I summarize some groundwater flow modeling objectives, methods, and procedures.

Objectives

There are many different applications of groundwater flow modeling, and each requires different approaches and often different modeling tools. Some possible modeling objectives, in order of modeling complexity, are as follows:

1. *Illustration of known groundwater flow principles.* This may be used in the classroom and perhaps in court. It involves the use of simple conceptual models of flow often described by closed form analytic solutions.

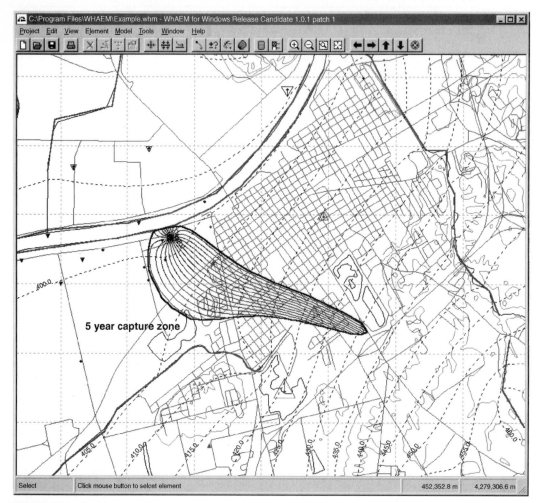

FIGURE 8.2.1   Graphics screen of the U.S. Environmental Protection Agency program WhAEM used for capture zone delineation. Dotted lines are equipotentials (lines of constant head). Lines inside the 5-year capture zone are particle traces, which are traced back in time from the well.

2. *Determination of aquifer parameters.* Evaluation of pumping tests, regional modeling to obtain global values for the aquifer transmissivity, etc. Groundwater solutions are obtained from graphs or relatively simple computer models.

3. *Understanding the groundwater flow regime.* Matching calculated heads for different conceptual models to measured heads in the field. The modeling result is often an optimal conceptual model, which explains the basic hydrogeology of the area.

4. *Predicting future groundwater flow patterns.* For instance, assessing the long-term change in groundwater levels due to a projected well field.

5. *Predicting future contaminant movement.* Timing the arrival of a contaminant plume at some target (e.g., drinking water well).

6. *Basic research on flow or transport phenomena.* Numerical experiments using research codes.

It is important to consider the purpose of a modeling project before deciding on the approach, software, and data acquisition. There is a growing realization that "appropriate technology" should be used for groundwater flow modeling—i.e., appropriate for the purpose of the modeling. Simple hand calculations, or the use of highly interactive (but basic) groundwater modeling software, often provide sufficient insight into a groundwater flow problem at low cost. On the other hand, if detailed questions about groundwater travel times in a complex hydrogeologic setting must be resolved, a very sophisticated groundwater flow model may be needed. Such complex models, however, are difficult to use and require major efforts in field data acquisition.

Methods

There are many groundwater flow modeling methods, including closed-form exact analytic solutions, electrical analogs, viscous flow models, and numerical computer models. Today almost all groundwater models are mathematical models implemented in a computer program. These computer models have evolved from simple codes run in batch mode with numerical input and output to user-friendly interactive programs with a modern graphical user interface (GUI). The modern groundwater models present modeling results in the form of potentiometric contour plots and groundwater pathlines in two or three dimensions. All groundwater flow models construct solutions to the *governing differential equation* for groundwater flow. This equation may differ depending on the type of flow system or, more accurately, the conceptual model of that flow system. Two of the most common governing differential equations are presented below.

Transient Dupuit-Forchheimer flow (two-dimensional horizontal flow) in an isotropic aquifer is described by Poisson's equation:

$$\frac{\partial^2 \phi}{\partial x^2} + \frac{\partial^2 \phi}{\partial y^2} = \frac{S}{T}\frac{\partial \phi}{\partial t} - \frac{R}{T} \tag{8.2.1}$$

where S is the aquifer storage coefficient, R (m/day) is the areal recharge rate due to precipitation, and T (m^2/day) is the aquifer transmissivity. All three parameters may vary with position. The parameter ϕ (m) is the potentiometric head (also called the piezometric head) or briefly *head* at a point (x, y) in the aquifer.

Transient three-dimensional flow in an isotropic aquifer is described by:

$$\frac{\partial^2 \phi}{\partial x^2} + \frac{\partial^2 \phi}{\partial y^2} + \frac{\partial^2 \phi}{\partial z^2} = \frac{S}{k}\frac{\partial \phi}{\partial t} \tag{8.2.2}$$

where k (m/day) is the hydraulic conductivity of the aquifer, which may vary with location. The recharge rate R is missing from the equation because in a model of three-dimensional flow the recharge due to precipitation is a boundary condition to the flow regime: a given value for the vertical groundwater flux q_z at the aquifer top. In fact, for an average recharge rate R the boundary condition at the aquifer top is $q_z = -R$.

Steady-state flow is described by the same equations, except the storage term (containing S) is missing. The term with recharge rate R in Equation 8.2.1 is a production term; without it (and without transient flow) the Poisson Equations 8.2.1 and 8.2.2 reduce to the equation of Laplace:

$$\nabla^2 \phi = 0 \tag{8.2.3}$$

where ∇^2 is the operator of Laplace, which stands for the second derivatives with respect to the two- or three-coordinate directions in Equations 8.2.1 and 8.2.2, respectively.

The differential equation, subject to its boundary conditions, may be solved in different ways. The simplest and most well-known exact analytic solutions are those for one or more wells near a stream or no-flow boundary, which are obtained with the method of images and are discussed in almost all textbooks on groundwater hydrology. More sophisticated analytic solutions have been obtained by use of complex variables and conformal mapping (e.g., Harr, 1962; Strack, 1989). Approximate analytic solutions to the groundwater flow problem are offered by the *analytic element method* (AEM), which uses superposition of integrated singularities (Green's functions) along interior boundaries of the flow domain (Strack, 1989 and Haitjema, 1995). Most analytic element solutions exactly satisfy the differential equation but approximate the boundary conditions of the flow problem. The classic *boundary integral equation method* (BIEM) is mathematically equivalent to the AEM but it uses numerical integration of the singularities along exterior boundaries of the flow domain (Liggett and Liu, 1983). The *finite difference method* (FDM) replaces the differential equation by a set of linear algebraic equations, which are solved approximately (McDonald and Harbaugh, 1988). The *finite element method* (FEM) replaces the differential equation by a variational formulation of the groundwater flow problem, which again leads to (approximately) solving a system of linear algebraic equations (Zienkiewicz, 1977). The FDM and the FEM, therefore, approximate both the differential equation and the boundary conditions.

The fact that an *approximate* solution technique is being used does not necessarily imply an inferior modeling result. Often a simpler conceptual model is necessary to facilitate an exact solution. For instance, a well near a stream may be modeled with a well and an image well. The solution is exact, but the conceptual model may be a severe simplification. A conceptual model that takes the actual stream geometry into account may lead to an approximate mathematical solution but result in a more realistic description of the groundwater flow regime. On the other hand, it is not true that the more complicated (realistic) the conceptual model the better the groundwater flow solution. The more realistic the conceptual model, the more must be known about the (hydrogeologic) conditions in the field. Yet field data are usually scarce and hard to come by. A detailed conceptual model that lacks the field data to back it up may lead to a false sense of accuracy when it is solved on a computer.

Although there is much debate about the pros and cons of various modeling techniques (mathematical solution methods), particularly among model developers, the crux of the matter is not which modeling technique to select but which conceptual model.

Conceptual Models

The design of an appropriate conceptual model requires good understanding of both the modeling objectives and the hydraulic principles of groundwater flow. This is why groundwater modeling requires a skilled groundwater hydrologist and not just skilled computer specialists. To arrive at a manageable conceptual model, from the perspective of both field data collection and model complexity, reality often must be reduced to what seems a caricature. Simplifying the world into a conceptual model for groundwater flow is often the most controversial part of any modeling project. Most reviewers of a modeling exercise are quick to point out where the modeling is unrealistic or where it over-simplifies reality. In almost all cases the critique is based on *intuition*. A Dupuit-Forchheimer model (horizontal flow model) is quickly condemned because "everybody knows that the world is three-dimensional." A steady-state model cannot be used "in the presence of transient flow." The aquifer is clearly "not homogeneous," and so on and so forth. Each of these critical statements seems self-evident and most groundwater flow modelers present their models with great fear—how to defend the many simplifications?

In truth, however, intuition often fails us when we are judging the adequacy of a conceptual model. It is counterintuitive that most Dupuit-Forchheimer models (two-dimensional models) provide nearly the same results as their three-dimensional counterparts, yet they do. It is also counterintuitive that steady-state flow may provide insight in what is inherently a transient flow system, but it is often true. That the groundwater flow rates and potentiometric heads in a stratified aquifer may be nearly the same as in a properly selected single aquifer is also not obvious to every reviewer. Instead of relying on intuition, we should *test* the adequacy of a conceptual model in view of the modeling objectives. By comparing many different conceptual models and their groundwater flow solutions some *rules of thumb* have been developed, which are presented below.

Two-Dimensional Versus Three-Dimensional Flow. Three-dimensional effects are local effects. But how local? As a general rule, groundwater flow becomes nearly horizontal at distance d from a three-dimensional feature (e.g., a partially penetrating well), which may be estimated from:

$$d = 2\sqrt{k_h/k_v}\,H \tag{8.2.4}$$

where H (m) is the saturated aquifer thickness near the flow feature and where k_h and k_v (m/day) are the horizontal and vertical hydraulic conductivity in the aquifer, respectively. However, the vertical resistance to groundwater flow near a partially penetrating well or a shallow creek can significantly add to the overall resistance to flow (horizontal plus vertical) unless the horizontal distance between boundary conditions is an order of magnitude larger than the aquifer thickness. In other words, a horizontal flow model (Dupuit-Forchheimer model), which ignores vertical resistance to flow, can be used if the distance L between boundary conditions is larger than:

$$L > 10\sqrt{k_h/k_v}\,H \tag{8.2.5}$$

It is noted that (approximate) vertical components of flow can be computed in a "horizontal flow" model (see Kirkham, 1967; Strack, 1984). These vertical fluxes are derived from conservation of mass considerations rather than from Darcy's law and appear accurate for most practical cases of regional flow (Haitjema, 1987). Consequently, even when three-dimensional streamlines are required it still is possible to use a Dupuit-Forchheimer model, provided Equation 8.2.5 is satisfied.

Steady-State Versus Transient Flow.   There are two different transient groundwater flow problems: (1) initial boundary value problems and (2) problems involving periodic (e.g., seasonal) variations in flow. An example of the first type is a well that starts pumping at a particular time. To describe its effect on the groundwater flow regime a fully transient groundwater flow solution is called for. The second type of transient flow problems (e.g., seasonal effects) are often ignored when so-called *average conditions* are presented with a steady-state groundwater flow model. The question, of course, is do these average conditions actually occur at any one time? Also, how different are (dry) summer conditions from (wet) winter conditions? Can these summer and winter conditions be modeled with a steady-state model? The answers to these questions depend on the response time of the groundwater flow system to a (periodic) forcing function. If the flow system adjusts quickly to changes in water levels in streams or changes in recharge due to precipitation, steady-state solutions can be constructed that represent reasonable average conditions or summer/winter conditions. To assess whether a system responds quickly enough to use "successive steady state solutions" the following condition must be met (Townley, 1995; Haitjema, 1995):

$$\frac{SL^2}{4TP} < 1 \tag{8.2.6}$$

where S is the storage coefficient, L (m) is the average distance between head-specified boundaries (streams and lakes), T (m^2/day) is the average aquifer transmissivity, and P (days) is the period of the forcing function (e.g., 365 days for seasonal variations). The smaller the factor in Equation 8.2.6 the faster the groundwater regime adjusts to changing boundary conditions. The value of 1 in Equation 8.2.6 may be marginal, depending on the specific circumstances and the modeling objectives. A value near 0.1 or less guarantees that steady-state solutions with summer or winter boundary conditions will reasonably reflect summer and winter flow patterns and potentiometric head distributions.

Single-Aquifer Versus Multiaquifer Flow.   Aquifer stratification can cause differences in the head with depth. If these vertical head gradients are substantial the stratified aquifer behaves as a multi-aquifer system and may have to be modeled as a multiaquifer system. I write "may," because whether the multiaquifer system can be lumped into a single *comprehensive* aquifer or whether it is necessary to model the flow and heads in each aquifer individually depends on the model objectives. This issue is illustrated in Figure 8.2.2. Part of the aquifer in Figure 8.2.2 is separated by a leaky layer into an upper and a lower aquifer. The well in the upper aquifer generates different flow patterns in the upper and lower aquifer. However, the effect of the well on the single aquifer zone is not influenced by the presence of the clay layer. In other words, the effect of the well on the flow in the single aquifer zone

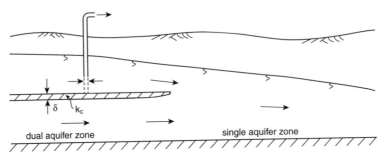

FIGURE 8.2.2   Flow in a single and dual aquifer zone.

is the same whether the well is screened in the lower aquifer, the upper aquifer, or both aquifers, or whether the clay layer is present at all. To fully understand the groundwater flow mechanics under these circumstances the reader is referred to the *comprehensive potential* concept discussed elsewhere (Strack, 1989; Haitjema, 1995).

Leaky layers can separate the flow in adjacent aquifers only over a limited distance. This means that the flow patterns in the upper aquifer and lower aquifer in Figure 8.2.2 will become the same at some distance from the well even if the clay layer is still present. For the case of the well in Figure 8.2.2 the flow patterns and heads in the upper and lower aquifer become indistinguishable at a distance of 4λ from the well, where λ (m) is the "characteristic leakage length" and is defined as:

$$\lambda = \sqrt{Tc} \tag{8.2.7}$$

where T (m^2/day) is the largest transmissivity of the two aquifers and where c (days) is the resistance of the leaky layer, defined as:

$$c = \frac{\delta}{k_c} \tag{8.2.8}$$

where δ (m) is the thickness of the clay layer and k_c (m/day) is the hydraulic conductivity of the separating clay layer.

The same is true for the effects of linear features, such as streams that penetrate only the upper aquifer. The effects in the lower aquifer of such streams become indistinguishable from the effects in the upper aquifer at about 3λ from the stream, a little earlier than for the case of a well.

The implication of the above observation is that if boundary conditions in multiaquifer systems are far apart, the system may well behave as a single aquifer. For each resistance layer a characteristic leakage length λ may be computed. In case λ is always orders of magnitude smaller than the average distance between streams and lakes, multiaquifer modeling may prove futile (Bakker et al., 1999).

Selection of a simpler conceptual model over a more complex one is not just a matter of efficiency. Complex conceptual models require computational resources that may be better invested elsewhere. This issue is further addressed below.

Model Resolution

The accuracy with which groundwater flow models represent the conceptual model depends in large part on the model resolution. For the FDM or FEM, model resolution is defined by the cell size or element size, respectively. For the AEM, model resolution is defined by the number of degrees of freedom (collocation points) along rivers, lake boundaries, and domains with different aquifer properties. The analytic element method is discussed in more detail in the section Analytic Element Modeling. Here I focus on resolution issues for the popular finite difference method.

Haitjema and Kelson (1998) investigated the performance of the most widely used finite difference program, MODFLOW. The study focused on the capability of MODFLOW to accurately represent the groundwater flow field, which forms the bases for contaminant transport simulations (see also the section Contaminant Transport Modeling). MODFLOW (a modular three-dimensional finite difference code) was developed by the U.S. Geological Survey (McDonald and Harbaugh, 1988) and is in the public domain. In recent years many GUIs have been developed that use MODFLOW as their computational engine. Also, MODFLOW has been expanded with new *modules* for added functionality, such as the river package, which adds surface water and groundwater interactions.

Haitjema and Kelson compared a number of MODFLOW solutions to exact or nearly exact analytic solutions. The study resulted in some *rules of thumb* for minimum grid resolutions (maximum cell sizes). These are as follows:

1. Narrow surface waters (creeks, small rivers, canals, etc.) that end inside the model domain must be represented by at least four or five cells along their length.

2. Wide surface water features (large rivers, lakes, etc.) with resistance to flow into or out of the aquifer (bottom resistance) must be represented by cells that are at least as small as the characteristic leakage length λ. For this case λ is defined as $\lambda = (Tc)^{1/2}$, where T (m^2/day) is the aquifer transmissivity and c (days) is the resistance of the stream or lake bottom (see Equation 8.2.8).

3. When modeling flow in aquifers bounded by one or two aquitards the cell size must be at least as small as the characteristic leakage length λ. For this case λ is defined as $\lambda = (Tc)^{1/2}$, where T (m^2/day) is the aquifer transmissivity and c (days) is the lowest resistance of the bounding aquitards (see Equation 8.2.8).

4. Domains with different aquifer transmissivity must be represented by 50 or more cells in case the abrupt change in transmissivity across the domain boundary is to be modeled accurately. Fewer cells lead to local errors in the velocity field near the domain boundaries, particularly near corners.

The proper cell sizes may be obtained by locally refining the grid. MODFLOW facilitates local grid refinement in a limited fashion. Depending on circumstances, however, the proper grid resolution may require the use of very many cells. Many finite difference models of regional groundwater flow problems therefore do not comply with the minimum cell sizes defined above. Consequently, they may produce significant errors (in excess of 5 to 10 percent), particularly in the groundwater flow field (Haitjema and Kelson, 1998).

It is obvious that multiple grid layers, necessary for modeling three-dimensional or multiaquifer flow, will quickly strain the computer resources when applying these minimum cell size criteria. In view of this, a simple conceptual model may be preferable over a more complex one. For instance, a seven-layer three-dimensional regional groundwater flow model requires seven times as many cells as its two-dimensional counterpart. This leaves a lot of room for increased grid resolution (smaller cells) in the two-dimensional model. The two-dimensional model may well be the only one that can satisfy the cell size criteria listed above. Consequently, the two-dimensional model may actually provide a superior solution to the three-dimensional model, particularly when answering questions about regional flow.

Calibration

Groundwater flow models are *calibrated* by comparing calculated (predicted) potentiometric heads in the aquifer with observed heads (measured in the field). The idea is to make adjustments in the conceptual model to better match model predictions with known conditions in the field. In some cases, when the model solves surface and groundwater flow conjunctively—e.g., the river package in MODFLOW or stream features in the analytic element model GFLOW—calibration may also involve comparing calculated stream flows with measured stream flows. There are a number of pitfalls associated with this calibration process. Anderson and Woessner (1992) describe the process in detail for finite difference or finite element models. Additional calibration issues are addressed by Haitjema

(1995). A few key issue are summarized below:

1. Calibration to fine tune the conceptual model is best done with a *steady-state* model of groundwater flow. Transient models are sensitive to the initial conditions, which usually are obtained from observed potentiometric head distributions. As a result, transient models are less sensitive to the conceptual model (transmissivities, resistance to flow at boundary conditions, etc.) than steady-state models. A steady-state model may be calibrated by comparing it with averages of time series of measured potentiometric heads. The calibrated steady-state model then may be used as the initial condition for a transient simulation (Franke et al., 1987).

2. Calibration is effective only if sufficient *far-field features* are included in the model. Far-field features are true hydrogeologic boundaries (streams, lakes, etc.) outside the area of interest in the model. The purpose of far-field features is to nullify the effect of assumed specified head or no-flow conditions on the grid boundaries. Consequently, the model area must extend well beyond the immediate area of interest to facilitate the inclusion of far-field features. The modeler may test the inclusion of sufficient far-field features by making arbitrary changes in the conditions on the grid boundary and checking that they do not alter the groundwater flow solution in the *near field* (area of interest). Often it is difficult to maintain the desired grid resolution when the model domain must be enlarged to incorporate far-field features. Consequently, many models lack sufficient far-field features, which makes the modeling results sensitive to the artificial model (grid) boundaries.

3. *Groundwater divides*, observed or estimated in the field, should not be used as specified no-flow boundaries in a groundwater flow model. Groundwater divides are sensitive to the conceptual model and instead should be used as a calibration target. Also, if the model is used to make predictions—for instance, adding a new well field to assess its impact on the groundwater flow regime—the groundwater divide must be allowed to adjust to these new conditions.

4. Groundwater mounding, which is the increase in head $\Delta\phi$ (m) between surface waters due to areal recharge R (m/day), is roughly proportional to the following factor:

$$\Delta\phi \doteq \frac{RL^2}{T} \tag{8.2.9}$$

where L (m) is the distance between head specified boundaries (streams) and T (m²/day) is the average aquifer transmissivity. Assuming that the location of surface waters is known (L is known), Equation 8.2.9 indicates that calibrating a model to measured heads provides insight only in the ratio of recharge to transmissivity (R/T). Calibration, therefore, does not provide any one of these parameters directly, unless one of them is already known.

5. Equation 8.2.9 also indicates that a model is relatively *insensitive* to variations in R/T if there are many surface water features in contact with the aquifer, so distance L between surface water features is small. Under these circumstances, different choices for R/T in the model may show only small changes in the heads ($\Delta\phi$ is small to begin with).

6. Equation 8.2.9 also indicates that a model is relatively *insensitive* to variations in recharge R if the aquifer transmissivity T is large, particularly if there also are many surface water features (L is small).

7. Models that support (steady state) *conjunctive* surface water and groundwater flow solutions can be calibrated to both the heads and the stream flows to resolve both regional transmissivity T and average regional recharge rate R.

8. Comparing modeled head *contours* with contour plots of field data may be misleading. It is preferable to compare individual heads measured in wells with heads calculated by the model in the same locations. Most contour plots of field data are subjective interpretations. Even when these contour plots are generated with computer software (e.g., SURFER) or geostatistical methods (e.g., kriging) (see subsection Geostatistics), interpretations are added and data are omitted or partially discounted.

9. When comparing modeled heads with observed (measured) heads, the modeler should look for *trends* rather than an exact match. Available field data often consist of heads that are measured at different times and different depths with uncertain accuracy. An optimal calibration may consist of modeled heads that are both higher and lower than those measured in the field, provided these differences are somewhat randomly distributed over the model domain. Clusters of modeled heads that are too high or too low may indicate (local) errors in the conceptual model.

10. Heads measured in *deep wells* tend to be lower than modeled heads, except in the immediate vicinity of surface waters, where this trend is reversed.

11. Heads measured in *shallow wells* tend to be higher than modeled heads, except near surface waters where the differences should vanish.

Traditionally, calibration has been a trial-and-error procedure. The modeler tests a range of values for various parameters, such as the hydraulic conductivity and areal recharge rate. Using his or her insight into the possible hydrogeologic conditions the modeler also may try various zones of differing aquifer properties and various assumptions on surface and groundwater interactions (e.g., stream and lake bottom resistances). Modern software packages have been developed to automate this calibration procedure and to provide quantitative (statistical) error bands on the calibration results. An example of a universally applicable calibration program is UCODE (Poeter and Hill, 1998).

8.2.2   ANALYTIC ELEMENT MODELING

In the groundwater flow modeling world, the AEM is a relatively new kid on the block. Although it is increasingly applied to a great variety of groundwater flow problems, many groundwater hydrologists remain uncertain about what this new method has to offer. In this subsection I try to provide some answers to frequently asked questions.

What It Is

The AEM solves groundwater flow problems by superposition of (many) analytic solutions to elementary groundwater flow problems: the analytic elements. Examples of analytic elements are line sinks to model flow into stream sections, surface sinks (areal elements) to model spatially varying recharge or flow into lake sections, and line doublets (double layers) to model discontinuities in aquifer properties. Analytic element models do not use a grid or an element network in the flow domain; instead they directly represent streams and lake boundaries with strings of line elements. Consequently, analytic element models are intuitive to use and are relatively scale independent. The method is uniquely suited for regional groundwater flow. Perhaps still small in numbers, analytic element modelers are true believers. Thus, the method is rapidly gaining popularity.

Where It Comes From

Although superposition of analytic solutions to groundwater flow is an old idea, it was not until the end of the 1970s before it was being used in computer simulations of complex regional groundwater flow problems. In 1977 Strack proposed superimposing hundreds of analytic solutions (line sinks, line doublets, and wells) to model the groundwater flow regime in the Divide Cut Section of the Tennessee–Tombigbee Waterway, then under construction. Line doublets (double layers) were used to model the transition between areas with constant but different hydraulic conductivity. The first analytic element computer program SYLENS (SteadYflow with LENSes) used the analytic element method to solve groundwater flow in a system of two interconnected aquifers (Strack and Haitjema, 1981a,b). The code is no longer operational, but it demonstrated the power of the AEM to produce realistic solutions to groundwater flow at a large regional scale and maintained local detail where it was needed.

During the 1980s the method was further developed and applied to a variety of practical problems, in both the United States and Europe. In 1989 Strack published a textbook (*Groundwater Mechanics*) in which he offers a detailed mathematical expose of the method and its (analytic) elements. Shortly thereafter Geraghty and Miller published a basic analytic element program QUICKFLOW, which was based on the public domain code SLWL enclosed in the textbook. QUICKFLOW became a success because it was easy to use and facilitated quick modeling analyses of many groundwater flow problems. In the early 1990s several more commercial analytic element programs were available: SLAEM (Strack, 1989), GFLOW (Haitjema, 1995), and TWODAN (Fitts, 1994). During the late 1980s and the 1990s Strack developed a sophisticated multiaquifer model, MLAEM, which was funded by the Dutch government and used to create a national groundwater flow model of the Netherlands (see applications discussed below). In 1994 the U.S. Environmental Protection Agency published the public domain analytic element model WhAEM (Wellhead analytic element model) developed in cooperation with the University of Minnesota (Strack) and Indiana University (Haitjema). The original DOS program was replaced by a Windows version in early 2000.

Over the years the method became more sophisticated. New analytic elements were developed, such as higher order curvilinear line sinks and line doublets and higher order area sinks (Strack, 1989, 1991, 1992b, 1992c). Higher order in this context means that the sink densities or doublet densities vary as higher-order polynomials (higher than order zero). Three-dimensional elements (partially penetrating well, sink disc) were developed and embedded in the two-dimensional flow model GFLOW (Haitjema, 1985; Haitjema and Kraemer, 1988), Additional three-dimensional analytic elements were developed by Fitts (1989, 1991), Luther and Haitjema (1999) and Steward (1996). A first attempt to model transient flow with the AEM resulted in the research code UCATES (Haitjema and Strack, 1986). Zaadnoordijk proposed an alternative method based on integrals of Theis' solution (Zaadnoordijk, 1988). In 1992 Strack implemented density flow in MLAEM (Strack, 1992a). Numerous other mathematical and software engineering developments continue to improve the power and efficiency of the analytic element codes (see *Journal of Hydrology* special issue volume 226, number 3–4, 1999).

Recent research deals with implementation of new ways to model transient flow as well as flow in aquifers with continuously varying properties. The latter development, including varying aquifer properties, is a fundamental one. To date, analytic element codes can handle aquifer heterogeneity only when the properties are piecewise constant. Analytic element codes by their nature are very suitable for object-oriented software design. At the time of this writing the analytic element developers are forming a consortium to create public domain software based on standardized classes for the various analytic elements. Public domain AEM codes may be downloaded from the web site: http://www.indiana.edu/~aem.

How It Works

For a detailed discussion of the method reference is made to the literature. Strack (1989) offers detailed mathematical discussions on the various analytic functions and how they are used in the context of the analytic element method. Haitjema (1995) offers a more introductory description of the method and explains how to apply it to practical problems. A brief expose of the AEM follows.

Mathematically speaking the method is not new at all. It may be viewed as a variant of BIEM (Liggett and Liu, 1983), which uses Green's functions based on Green's second identity. In the classic BIEM, Green's functions are distributed over the boundaries of a closed domain and the associated integrals are evaluated numerically. In contrast, Green's functions in the AEM are distributed along interior boundaries inside an infinite domain and the associated integrals are evaluated analytically. In aerospace engineering the AEM is known as the panel method (e.g., Hess and Smith, 1967; Hess, 1973). In essence, an analytic element model uses analytic solutions to basic features of the groundwater flow regime and adds them together to represent the entire (regional) flow problem. For instance, a narrow stream is subdivided into a number of straight sections, each of which is represented by a line sink of constant sink density. The sink density distribution along the line sink string approximates the groundwater inflow distribution along the stream, which is a priori not known. What are considered known, however, are the (average) water levels in the stream sections.

For each unknown sink density one known head can be assigned—for instance, to the center of each line sink (Dirichlet boundary condition). The simplest approach is to assume that the head underneath the stream equals the water level in the stream. It also is possible to express the head in terms of the water level in the stream, a resistance to flow into the stream, and the groundwater inflow rate or sink density (Cauchy boundary condition). Either way, there are as many unknown sink densities as there are known boundary conditions, which allows the sink densities to be computed by solving a system of equations. Once all sink densities are known, the head at any point in the aquifer can be calculated by superimposing the influence on the head of all line sinks. Use of more sophisticated analytic elements and introduction of more complicated boundary conditions leads to a more complicated solution procedure, which may include nonlinear sets of equations and non-square coefficient matrices. However, once unknown sink densities, doublet densities, etc. have been found, the field quantities (head, discharge vector, or groundwater velocity) can be calculated anywhere through a straight forward superposition procedure.

In fact, analytic element models use functions that generate a contribution to the *discharge potential* instead of the head. Although input and output occur in terms of heads, the internal discharge potential formulation simplifies the mathematics. There are several reasons for this. First, solutions to groundwater flow in terms of the discharge potential are exactly the same under confined and unconfined flow conditions, which is not true for solutions in terms of the head. Although the governing differential equation for steady unconfined flow is nonlinear in terms of the head, it is linear in terms of the discharge potential, which is a prerequisite for superposition of solutions. Another advantage of using the discharge potential is the existence of an associated *stream function*. Most analytic elements are represented by harmonic functions: solutions to Laplace's equation. These solutions may be formulated in terms of a complex potential function. The real part of that function is the discharge potential and the imaginary part is the stream function, which is constant along streamlines. However, most practical groundwater flow problems involve areal recharge or leakage to an adjacent aquifer, or both. These features are represented by solutions to Poisson's equation for which no stream function exists. However, by distinguishing between the solutions to Poisson's equation and to Laplace's equation it remains possible to use the stream function for tracing streamlines or defining leaky or no-flow boundaries in analytic element models (Haitjema and Kelson, 1994).

Three-dimensional models use solutions to Laplace's equation only in three dimensions, which are always real functions. Three-dimensional analytic elements may be embedded in regional two-dimensional models (Haitjema, 1985; Luther and Haitjema, 1999), which is much more efficient than creating a three-dimensional model of the entire (regional) groundwater flow system.

The groundwater discharge vector is the gradient of the discharge potential. Analytic element codes, therefore, also contain the spatial derivatives of the discharge potential for each feature. With the same sink densities, these derivative functions are superimposed to obtain the groundwater discharge vector at any point in the aquifer. Division by the saturated aquifer thickness and (effective) porosity yields the average groundwater flow velocity.

Most commercial analytic element codes use relatively simple functions, which are exact solutions to the governing differential equation. Because the analytic element solution to the groundwater flow problem is merely the sum of these exact solutions, analytic element models do not suffer from local or global water balance errors, whereas the discharge vector or velocity vector is everywhere continuous and fully consistent with the potentiometric head field.

In the absence of a grid or element network, analytic element models can include large regional domains without undue computational or organizational effort. In fact, in the absence of a boundary around the model domain (grid or element mesh boundary) analytic element models must include remote hydrological features to act as a natural boundary around the area of interest. This circumstance encourages proper model calibration practices (see the far-field and near-field concepts discussed in the subsection calibration).

Screening Models.   An analytic element model of a groundwater flow system usually is easy to construct and, more importantly, easy to change. This is due to the absence of a grid or element network. The introduction or modification of line sink strings to represent streams and lake boundaries is an intuitive task. With a modern GUI, model construction is reduced to a series of mouse clicks to put line sink strings on top of streams on a background map. In general, only few numerical data need to be entered to define the (global) aquifer properties and water levels in the streams. Rapid deployment is a recognized feature of analytic element models, as illustrated by the name of one of the commercial codes: QUICKFLOW.

The relative ease with which analytic element models can be set up makes them very suitable screening models. A screening model provides a first cut at a modeling project and yields much insight into the groundwater flow regime at little cost. The screening model also can easily be refined, modified, or expanded. Line sinks may be added to improve the resolution or to include additional features, and polygons of line doublets may be entered to define areas of different aquifer properties. If a numerical model is desired, a large-scale analytic element model may be used to define the boundary conditions on the grid or element network boundary (Hunt et al., 1998). Kelson (1999) developed a procedure for extracting local MODFLOW models from regional analytic element models, which is implemented in the commercial code GFLOW. By their nature, analytic element models are very suitable for a stepwise modeling approach. Combined with clearly defined study objectives (specific questions to be answered), stepwise modeling leads to robust solutions at a minimum cost (Haitjema, 1995).

What It Can Do

Analytic element models can be and are used for nearly all groundwater flow modeling tasks. Below three example applications are presented for which analytic element modeling appeared uniquely suitable.

Wellhead Protection.   In 1995 the Marion County Health Department instituted a moratorium on business development near all public water supply well fields in Marion County (Indiana) until appropriate wellhead protection zones had been delineated for each of the well fields. Wittman of the Center for Urban Policy and the Environment in Indianapolis proposed to complete all five capture zone delineations in a 6-month time frame by the AEM. He used a stepwise approach, first representing all five (interacting) well fields in a single model and then zooming in on each one for added detail (see Figure 8.2.3) (Wittman and Haitjema, 1995). Proposals based on traditional (numerical) modeling

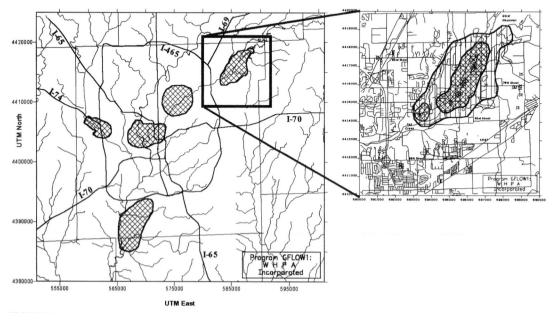

FIGURE 8.2.3   Capture zones for five well fields in Marion County, Indiana. Inset shows detailed particle traces for the Geist well field.

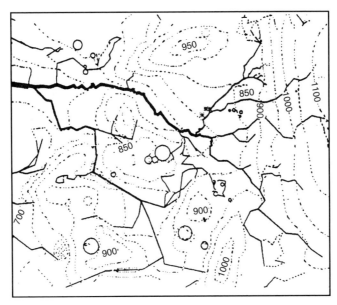

FIGURE 8.2.4 Conjunctive surface water and groundwater flow solution of the Pere Marquette River basin in Michigan. Line thickness is proportional to base flow rate in the river and its tributaries.

procedures could meet these time constraints only at great cost. Wittman used different modeling approaches at different well fields, depending on local hydrogeological circumstances. At one of the well fields, he used a local MODFLOW model extracted from the regional analytic element model. A more detailed discussion on the technical and political aspects of the work are found in Lindsey et al., 1997.

Modeling Surface Water and Groundwater Conjunctively. Sherry Mitchell-Bruker (1993) pioneered conjunctive surface water and groundwater flow solutions in analytic element models. The surface water and groundwater interactions are steady state and are based on water levels in the streams that are constant in time and independent of the modeled stream flow rate. Although representing only average conditions, the conjunctive solutions appear very helpful in model calibration. The dashed headwaters of some of the tributaries in Figure 8.2.4 indicate that these stream sections do not receive groundwater, at least not on average during the year. For the case of Figure 8.2.4 modeled base flow was compared with measured base flow at various gauging stations in the Pere Marquette watershed. Combined with comparisons of modeled and measured heads, this led to significantly improved model calibration compared with heads calibration only (Mitchell-Bruker and Haitjema, 1996).

Modeling Groundwater Flow Underneath the Netherlands. Toward the end of the 1980s the Dutch Government Institute for Inland Water Management and Waste Water Treatment (RIZA), in cooperation with the Delft research institute TNO, started development of a National Groundwater Model (NAGROM) (de Lange, 1991). The NAGROM model is implemented with the multiaquifer analytic element model MLAEM developed by Strack (1989). The model domain is about 250×400 km and is underlain by three interconnected aquifers. Grid-based or element-based numerical models were deemed impractical at this supraregional scale. The AEM made this national model possible, which serves as a framework for many local groundwater flow studies (de Lange and van der Meij, 1994). Whenever a local groundwater study is called for, one of the supraregional models is refined in the area of interest by simply replacing some large elements with many smaller elements (see Figure 8.2.5). The NAGROM model has been coupled to several other models, including an unsaturated flow

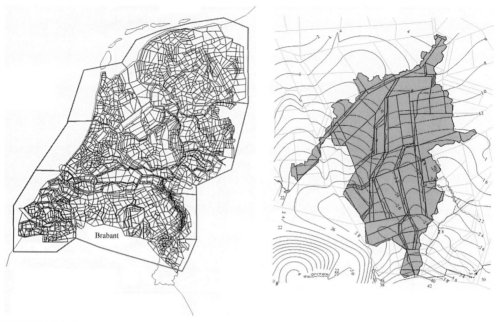

FIGURE 8.2.5   Set of supraregional analytic element models covering the Netherlands (left) and a refined model of part of the supraregion Brabant (courtesy of RIZA).

model and an ecology model, to form a multimedia modeling system. A similar multimedia modeling system, with the AEM simulating saturated groundwater flow, is currently under development by the U.S. Environmental Protection Agency.

REFERENCES

Anderson, M. P., and Woessner, W. W., *Applied Groundwater Modeling*, Academic Press, 1992.

Approximate Analytic Solutions to 3D Unconfined Groundwater Flow Within Regional 2D Models, Luther, K., and Haitjema, H., *J. of Hydrol.*, in press 2000.

Bakker, M., Anderson, E. I., Olsthoorn, T. N., and Strack, O. D. L., "Regional Groundwater Modeling of the Yucca Mountain Site Using Analytic Elements," *J. Hydrol.*, in press 1999.

Darcy, H., *Les Fountaines Publiques de la Ville de Dijon*, Victor Dalmont, 1856.

de Lange, W. J., *A Groundwater Model of the Netherlands*, Technical Report Note 90.066, The National Institute for Inland Water Management and Waste Water Treatment, 1991.

de Lange, W. J., and van der Meij, J. L., *Reports on NAGROM*, Technical Report, TNO-GG Delft, RIZA Lelystad, 1994.

Dupuit, J., *Études Théoriques et Practiques sur le Mouvement des Eaux dans les Canaux Decouverts et à Trauvers les Terrains Perméables*, 2nd ed., Dunod, 1863.

Fitts, C., "Simple Analytic Functions for Modeling Three-Dimensional Flow in Layered Aquifers," *Water Resour. Res.*, 25(5): 943–948, 1989.

Fitts, C. R., "Modeling Three-Dimensional Flow About Ellipsoidal Inhomogeneities with Application to a Gravel-Packed Well and Flow Through Lens-Shaped Inhomogeneities," *Water Resour. Res.*, 27(5): 815–824, 1991.

Fitts, C. R., "Well Discharge Optimalization Using Analytic Elements," *Ground Water*, 32(4): 547–550, 1994.

Forchheimer, P., "Ueber die Ergiebigkeit von Brunnen-Anlagen und Sickerschlitzen," *Architekt. Ing. Ver. Hannover*, 32: 539–563, 1886.

Franke, O. L., Reilly, T. E., and Bennet, G. D., *Definition of Boundary and Initial Conditions in the Analysis of Saturated Groundwater-Flow Systems: an Introduction*, Technical Report 03-B5, U.S. Geological Survey, 1987.

Haitjema, H. M., "Comparing a Three-Dimensional and a Dupuit-Forchheimer Solution for a Circular Recharge Area in a Confined Aquifer," *J. Hydrol.* 91: 83–101, 1987.

Haitjema, H. M., *Analytic Element Modeling of Groundwater Flow*, Academic Press, 1995.

Haitjema, H. M., and Kelson, V. A., *Assessing the Suitability of MODFLOW for Modeling Advective Transport*, Technical Report, for the Ministry of Transport, Public Works and Water Management, Institute for Inland Water Management and Waste Water Treatment (RIZA), 1998.

Haitjema, H. M., and Kraemer, S. R., "A New Analytic Function for Modeling Partially Penetrating Wells." *Water Resour. Res.*, 24(5): 683–690, 1988.

Haitjema, H. M., and Strack, O. D. L., *An Initial Study of Thermal Energy Storage in Unconfined Aquifers*, Technical Report PNL-5818 UC-94e, Pacific Northwest Laboratories (operated by Battelle Memorial Institute), Contract DE-AC06-76RLO 1830 for the Department of Energy, Subcontract B-E1234-A-0, 1986.

Harr, M. E., *Groundwater and Seepage.* McGraw-Hill, 1962.

Hess, J. L., "Higher Order Numerical Solution of the Integral Equation for the Two-Dimensional Neumann Problem," *Computer Meth. Appl. Mech. Eng.*, 2: 1–15, 1973.

Hess, J. L., and Smith, A. M. O., "Calculation of Potential Flow About Arbitrary Bodies," *Prog. Aeronaut. Sci.*, 8: 1–138, 1967.

Hunt, R. J., Anderson, M. P., and Kelson, V. A., "Improving a Complex Finite-Difference Groundwater-Flow Model Through the Use of an Analytic Element Model," *Ground Water*, 36(6): 1011, 1998.

Kelson, V. A., *Practical Advances in Groundwater Modeling with Analytic Elements*, PhD thesis, Indiana University, 1999.

Kirkham, D., "Explanation of Paradoxes in Dupuit-Forchheimer Seepage Theory," *Water Resour. Res.*, 3: 609–622, 1967.

Liggett, J., and Liu, P.-F., *The Boundary Integral Equation Method for Porous Media Flow*, George Allen and Unwin, 1983.

Lindsey, G., Wittman, J., and Rummel, M., "Using Indices in Environmental Planning: Evaluating Policies for Wellhead Protection," *J. Environ. Plan. Manage.*, 40(6): 685–703, 1997.

Luther, K. H., and Haitjema, H. M., "An Analytic Element Solution to Unconfined Flow Near Partially Penetrating Wells," *J. Hydrol.*, 1999.

McDonald, M. G., and Harbaugh, A. W., *A Modular Three-Dimensional Finite-Difference Ground-Water Flow Model*, Techniques of Water Resources Inverstigations of the United States Geological Survey, U.S. Geological survey, 1988.

Mitchell-Bruker, S., *Modeling Steady State Groundwater and Surface Water Interactions*, PhD thesis, School of Public and Environmental Affairs, Indiana University, 1993.

Mitchell-Bruker, S., and Haitjema, H. M., "Modeling Steady State Conjunctive Groundwater and Surface Water Flow with Analytic Elements," *Water Resour. Res.*, 32(9): 2725–2732, 1996.

Modeling Three-Dimensional Flow in Confined Aquifers by Superposition of Both Two- and Three-Dimensional Analytic Functions, Haitjema, H. M., *Water Resources Res.*, Vol. 21, No. 10, Oct. 1995.

Poeter, E. P., and Hill, M. C., *Ucode: A Computer Code for Universal Inverse Modeling*, Technical Report 98-4080, U.S. Geological Survey, 1998.

Special Issue of the "Analytic Based Modeling of Groundwater flow," *Journal of Hydrology*, Vol. 226, No. 3–4, 1999.

Steward, D. R., *Vector Potential Functions and Stream Surfaces in Three-Dimensional Groundwater Flow*, PhD Thesis, University of Minnesota, 1996.

Strack, O. D. L., "Three-Dimensional Streamlines in Dupuit-Forchheimer Models," *Water Resour, Res.*, 20(7): 812–822, 1984.

Strack, O. D. L., *Groundwater Mechanics*, Prentice Hall, 1989.

Strack, O. D. L., *An Area-sink for Modeling Leakage Near Wells*, Report Pursuant to Contract DB-612, for RIZA, 1991.

Strack, O. D. L., *An Area Element for Leakage Near Streams*, Report pursuant to Contract DB-612, for RIZA, 1992a.

Strack, O. D. L., *A Multi-Quadratic Area Sink*, Report pursuant to Contract DB-612, for RIZA, 1992b.

Strack, O. D. L., and Haitjema, H. M., "Modeling Double Aquifer Flow Using a Comprehensive Potential and Distributed Singularities 1. Solution for Homogeneous Permeabilities," *Water Resour. Res.*, 17(5): 1535–1549, 1981a.

Strack, O. D. L., and Haitjema, H. M., "Modeling Double Aquifer Flow Using a Comprehensive Potential and Distributed Singularities 2. Solution for Inhomogeneous Permeabilities," *Water Resour. Res.*, 17(5): 1551–1560, 1981b.

Townley, L. R., "The Response of Auifers to Periodic Forcing," *Adv. Water Res.*, 18(3): 125–146, 1995.

Townley, L. R., "The Response of Quifers to Periodic Forcing," *Adv. Water Res.*, 18(3): 125–146, 1995.

Using the Stream Function for Flow Governed by Posson's Equation, Haitjema, H. M., and Kelson, V. A., *J. of Hydrol.*, 187: 367–386, 1996.

WhAEM: Program Documentation for the Wellhead Analytic Element Model, Haitjema, H. M., Wittman, J., Kelson, V., Bauch, N., EPA report, EPA/600/R-94/210, Project Manager Stephen R. Kraemer, December 1994.

Wittman, J., and Haitjema, H. M., *Delineation of Wellhead Protection Areas in Marion Country, Indiana*. Technical Report 95-E14, Center for Urban Policy and the Environment (CUPE), Indianapolis, 1995.

Zaadnoordijk, W. J., *Analytic Elements for Transient Groundwater Flow*, PhD thesis, University of Minnesota, 1988.

Zienkiewicz, O. C., *The Finite Difference Method*, 3rd ed., McGraw-Hill, 1977.

CHAPTER 8
ANALYSIS AND MODELING

SECTION 8.3

SOLUTE TRANSPORT MODELING

Geoff Freeze
Mr. Freeze is Computational Sciences Section Manager, Duke Engineering & Services, Albuquerque, New Mexico. He is a civil/environmental engineer who specializes in probabilistic applications of groundwater flow and transport models.

Douglas D. Walker
Dr. Walker is a hydrogeologist with Duke Engineering & Services, Inc. He specializes in the development and application of geostatistical approaches to groundwater contaminant transport and has authored more than 40 papers and technical reports.

Milovan S. Beljin
Dr. Beljin is a consulting hydrologist and president of M.S. Beljin & Associates. He also serves as adjunct faculty at University of Cincinnati and Wright State University, Dayton, Ohio.

8.3.1 INTRODUCTION

Solute transport through porous media may involve several different processes: advection, mechanical dispersion, diffusion, biological and radioactive decay, sorption, and geochemical and biochemical reactions. A conceptual understanding of these processes is critical to the development of representative solute transport models. Several excellent textbooks are available that provide this conceptual basis (Bear, 1972; de Marsily, 1986; Domenico and Schwartz, 1990; Fetter, 1993; Freeze and Cherry, 1979). Solute transport models combine mathematical representations of selected transport processes to predict contaminant movement through the subsurface. Generalized components of a solute transport model are shown in Figure 8.3.1.

Mathematical Models

Solute transport (in one dimension) can be expressed mathematically by the following partial differential equation, typically referred to as the advection-dispersion equation (Freeze and Cherry, 1979):

$$\frac{\partial}{\partial x}\left(D\frac{\partial C}{\partial x}\right) - \frac{\partial}{\partial x}(Cv_x) + \frac{r}{n} = \frac{\partial C}{\partial t} \tag{8.3.1}$$

where x = principal direction of groundwater flow,
    C = concentration of the solute (M/L^3),
    v_x = average linear groundwater velocity in the x direction (L/T),

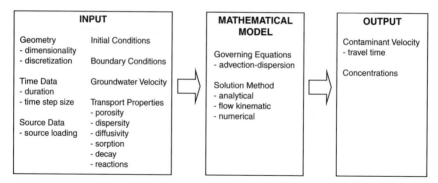

FIGURE 8.3.1   Components of a solute transport model.

$$D = \text{coefficient of hydrodynamic dispersion (L}^2/\text{T}),$$
$$= \alpha \, v_x + D^*,$$
$$\alpha = \text{dispersivity (L)},$$
$$D^* = \text{coefficient of molecular diffusion (L}^2/\text{T}),$$
$$n = \text{porosity ( )},$$
$$r = \text{mass rate of solute production (M/L}^3\text{T), and}$$
$$t = \text{time (T)}.$$

Equation 8.3.1 represents the effects of advection and hydrodynamic dispersion in an isotropic medium. Advection is the transport of solutes (e.g., dissolved contaminant species) by the bulk motion of groundwater, typically at the average velocity of and in the principal direction of flow (see subsection *Heat and Density Dependence* for exceptions). Hydrodynamic dispersion describes the combined effects of mechanical dispersion (the spreading of solutes because of microscale variations in flow direction and velocity resulting from small-scale heterogeneities) and molecular diffusion. The dispersivity is typically described by a longitudinal component (α_L, for spreading in the direction of flow) and a transverse component (α_T, for spreading perpendicular to flow), although more complex conceptualizations are possible (see Scheidegger, 1961). The coefficient of molecular diffusion is commonly much smaller than the coefficient of mechanical dispersion (αv_x); thus diffusion frequently is neglected in solute transport modeling. The formulation of the coefficient of hydrodynamic dispersion in Equation 8.3.1 approximates mechanical dispersion based on an analogy to Fick's law for diffusion. This approximation is often referred to as Fickian dispersion. Figure 8.3.2 schematically illustrates of the effects of advection, longitudinal, and transverse dispersion.

Additional processes, such as decay and sorption, typically are included in the advection-dispersion equation (ADE) through the mass rate term (r/n):

$$\frac{\partial}{\partial x}\left(D\frac{\partial C}{\partial x}\right) - \frac{\partial}{\partial x}(Cv_x) - R_d\lambda C = R_d\frac{\partial C}{\partial t} \qquad (8.3.2)$$

where $\lambda = $ decay constant ($1/T$), and
$R_d = $ retardation factor.

Equation 8.3.2 includes the effects of first-order decay (through the decay constant) and an equilibrium linear sorption isotherm (through the retardation factor). The decay constant controls the decrease in solute mass due to radioactive or biological decay. The retardation factor quantifies the effect of sorption in retarding or delaying the spread of contaminants. The relationship between retardation factor and sorption (also referred to as adsorption) and the various mathematical representations of sorption are summarized in Chapter 1, Section 6 (Geochemistry).

Equations 8.3.1 and 8.3.2 are the one-dimensional (1-D) form of the ADE. Multidimensional forms of the ADE include additional left-hand-side derivative terms with respect to the principal y and z

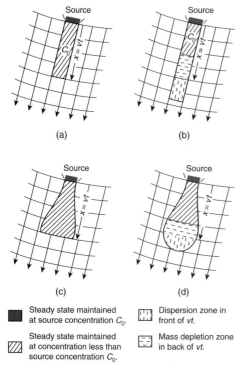

(a)   (b)

(c)   (d)

▦ Steady state maintained at source concentration C_0.

▨ Steady state maintained at concentration less than source concentration C_0.

▥ Dispersion zone in front of vt.

▦ Mass depletion zone in back of vt.

FIGURE 8.3.2 Schematic diagram showing effects of (a) advection; (b) advection and longitudinal dispersion; (c) advection and transverse dispersion; and (d) advection, longitudinal dispersion, and transverse dispersion (from Domenico and Schwartz, 1990).

directions. Mathematical solutions to the ADE provide solute concentration as a function of space and time [$C(x, y, z, t)$]. Depending on assumptions about initial and boundary conditions, the mathematical solutions may be analytical (subsection 8.3.2), flow kinematic (subsection 8.3.3), or numerical (subsection 8.3.4). Typically, analytical solutions involve significant simplifying assumptions but they may provide a low-cost, approximate answer. Conversely, numerical solutions may address more complex problems in multiple dimensions, but they may be more costly to obtain. The appropriate choice of mathematical solution depends on the answers to two key questions: (1) Given the project budget and technical needs, how complex a solution is necessary? and (2) Given the data available, how complex a solution is possible?

Equations 8.3.1 and 8.3.2 (or their multidimensional equivalents) are applicable to most solute transport problems. However, they assume that (1) concentration and fluid density gradients do not affect the groundwater flow field, (2) groundwater chemistry follows linear local equilibrium, and (3) the system is isothermal. Modeling approaches to account for deviations from these assumptions are presented in subsection 8.3.6.

Model Input

Transport model input comes from a combination of a site-specific conceptual model, groundwater- and soil-dependent transport properties, and possibly flow model results.

The conceptual model and the groundwater flow field (typically calculated by using a flow model) provide the basis for groundwater movement. Selection of an appropriate conceptual model, properly representing the site-specific geometry and processes, is perhaps the most important step in solute transport modeling. Guidance in selecting values for the key transport properties, which govern how contaminant movement differs from groundwater movement, is presented in subsection 8.3.5. A particular solution to the ADE also requires additional information about the system in the form of initial and boundary conditions. The initial and boundary conditions can have a significant effect of transport model results and must be defined appropriately.

Initial conditions describe the distribution of solute concentration $C(x, y, z, t)$ in the modeled domain at some initial time ($t = 0$) that represents the beginning of the time period to be modeled. The initial concentration, denoted $C(x, y, z, 0)$ or C_0, may be zero or some other defined distribution (i.e., nonzero in the area defining the source of contamination, referred to as the source term).

Boundary conditions define the mathematical relationship between the modeled domain and the surrounding region (which is not otherwise included in the model). Where possible, model domains are selected so that anomalous effects from the boundaries are small. Table 8.3.1 summarizes three

TABLE 8.3.1 Types of Boundary Conditions for Solute Transport Models

Name	Type	Condition
Dirichlet	First	Specified concentration
Neumann	Second	Specified concentration gradient
Cauchy	Third	Specified mass flux

types of boundary conditions commonly used to represent different physical conditions. Note that each of these results in a different mathematical solution.

Model Output

A solute transport model solves one or more governing equations (such as the ADE) using a specified mathematical solution to calculate specific output. Most commonly, transport models use a forward solution, where the input parameters for the flow and transport properties are known and the specified outputs are calculated. The output of the ADE forward solution is the solute concentration (for one or more solutes) as a function of space and time $C(x, y, z, t)$. This information is most usefully viewed with a graphical postprocessing package. In addition to the concentration output, graphic displays of velocity vector output (both groundwater and contaminant) and flow paths (with associated travel times) are informative.

An alternative approach to solute transport modeling is to calculate an inverse solution, where the solute concentrations and velocities are known and the flow and transport model is used to back-calculate the corresponding flow and transport properties. The inverse approach is most useful for parameter estimation and model calibration (see subsection *Model Calibration*).

8.3.2   ANALYTICAL SOLUTIONS

Analytical solutions of the ADE are developed for specific flow regimes, boundary conditions, media characteristics, and source terms, resulting in closed-form mathematical expressions. Mathematical development of these solutions commonly employs either the Green's function approach or Laplace and Fourier transformation methods (Wexler, 1989). Many of these solutions were originally developed to address heat transport problems and later were adapted to contaminant transport in porous medium (Carslaw and Jaeger, 1959). These analytical solutions are exact solutions to the ADE under specific, relatively simple flow, initial, and boundary conditions and commonly are used to represent the idealized conditions of laboratory and numerical experiments. Analytical solutions are useful when approximate answers are required for designing more complex numerical models, in scoping calculations, and in decision support software. Many analytical solutions may be superimposed (combined additively) in space and time to create relatively complex models of multiple sources and complex source histories.

Advection and Dispersion in One Dimension

The simplest case is 1-D transport in a homogeneous porous medium with a steady groundwater flow velocity and zero initial concentration. For a spill that occurs at a point in an infinite domain, such that a mass of contaminant instantaneously appears at $x = 0$, the boundary and initial conditions are (in mathematical terms):

Zero initial concentration:	$C(x, 0) = 0$
Boundary condition:	$C(\infty, t) = 0$
Instantaneous source term:	$C(0, 0) = C_0$

Bear (1972) adapted a diffusion equation solution to describe advection and dispersion of the mass pulse from the point source and found:

$$\frac{C(x, t)}{C_0} = \frac{V_0/n}{(4\pi D't)^{1/2}} \exp\left[-\frac{(x - v't)^2}{4D't}\right] \tag{8.3.3}$$

where $V_0 =$ volume of spill with concentration C_0 (L^3),

$D' = D/R_d$, retarded coefficient of hydrodynamic dispersion (L^2/T),

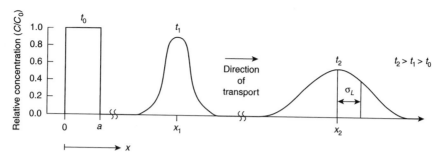

FIGURE 8.3.3 Advection and dispersion in one dimension from an instantaneous source (from Fetter, 1993).

$v' = v/R_d$, retarded average linear groundwater velocity (L/T), and
R_d = retardation factor.

The relative concentration $C(x, t)/C_0$ is an exponential function that decays with time. If we use the moving coordinate for the center of the plume $\bar{x} = v't$ and define $\sigma_L^2 = 2D't$, then this solution can be seen as a Gaussian (normal) bell-shaped distribution function, with mean and variance $\bar{x}$ and σ_L^2, respectively (Figure 8.3.3).

The solution given by Equation 8.3.3 assumes an instantaneous point source. Ogata and Banks (1961) developed the following solution for a continuous source at $x = 0$:

$$\frac{C(x, t)}{C_0} = 1/2 \left\{ \text{erfc} \left[\frac{x - v't}{2(D't)^{1/2}} \right] + \exp \left[\frac{xv'}{D'} \right] \text{erfc} \left[\frac{x + v't}{2(D't)^{1/2}} \right] \right\} \tag{8.3.4}$$

The complementary error function, erfc, is related to the error function via erfc $= 1 - $ erf. Both the error function and its complement are tabulated in many mathematical references (e.g., Freeze and Cherry, 1979) and are library functions in many software packages and compilers. Additional one-dimensional solutions are presented by van Genuchten and Alves (1982).

Advection and Dispersion in Multiple Dimensions

One useful property of the Green's function approach to analytical solutions is that the source term can be separated into x, y, and z components and solved independently. The resulting 1-D solutions may then be recombined to create a solution in multiple dimensions. This method can be used to extend the 1-D instantaneous point source solution from subsection *Advection and Dispersion in One Dimension* to a three-dimensional (3-D) infinitely thick aquifer (Figure 8.3.4a). If a point source is centered at (x_0, y_0) and groundwater flow is in the x direction, then the 1-D components of the analytical solution are:

$$\frac{C(x, t)}{C_0} = \frac{V_0/n}{2(\pi D'_x t)^{1/2}} \exp \left[-\frac{(x - v't)^2}{4D'_x t} \right] \tag{8.3.5a}$$

$$\frac{C(y, t)}{C_0} = \frac{1}{2(\pi D'_y t)^{1/2}} \exp \left[-\frac{y^2}{4D'_y t} \right] \tag{8.3.5b}$$

$$\frac{C(z, t)}{C_0} = \frac{1}{(\pi D'_z t)^{1/2}} \exp \left[-\frac{z^2}{4D'_z t} \right] \tag{8.3.5c}$$

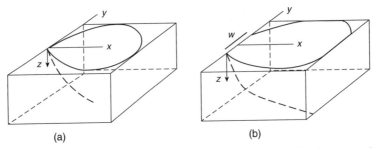

FIGURE 8.3.4   Schematic conceptualization of (a) a point source, and (b) a line source of length W in a 3-D flow and transport system (from Domenico and Schwartz, 1990).

The full 3-D solution is the product of Equations 8.3.5a, -b, and -c:

$$\frac{C(x, y, z, t)}{C_0} = \frac{V_0/n}{4(\pi t)^{3/2}(D'_x D'_y D'_z)^{1/2}} \exp\left[-\frac{(x - v't)^2}{4D'_x t} - \frac{y^2}{4D'_y t} - \frac{z^2}{4D'_z t}\right] \tag{8.3.6}$$

If the source is a line in y of width W (Figure 8.3.4b), then Equation 8.3.5b is replaced by:

$$\frac{C(y, t)}{C_0} = \frac{1}{2W}\left\{\text{erf}\left[\frac{\left(\frac{W}{2} + y\right)}{(4D'_y t)^{1/2}}\right] + \text{erf}\left[\frac{\left(\frac{W}{2} - y\right)}{(4D'_y t)^{1/2}}\right]\right\} \tag{8.3.7}$$

Various authors have compiled useful solutions of this type, allowing a wide variety of source term geometries (Codell et al., 1982; Domenico and Schwartz, 1990; Yeh and Tsai, 1976). Typically, source terms are defined as uniform distributions of mass over simple geometries (e.g., points, lines, rectangular areas, or parallelepiped volumes). However, analytical solutions to the ADE are not limited to uniform mass distributions. Cleary and Ungs (1978), for example, present a solution for a Gaussian line source in the y axis, perpendicular to the direction of flow.

Solutions such as Equation 8.3.6 are for an instantaneous source term—that is, a one time introduction of a slug of contaminant. Although this is useful for modeling brief contaminant spills, such solutions may be adapted to long duration, time-variable sources by integrating the instantaneous source solutions over time:

$$C(x, y, z, t) = \int_0^t C_i(x, y, z, t - \omega)\partial\omega \tag{8.3.8}$$

That is, the solution for a time-continuous source term is the superposition (convolution) in time of the solutions for series of instantaneous sources. This integral (Equation 8.3.8) is usually solved numerically, requiring either computer codes or symbolic math software.

Comments

Decision support models and screening models commonly represent the transport processes of interest using a chain of relatively simple analytical solutions. The models are coupled through a simple mass flux balance (e.g., equating the mass flux through the vadose zone with the mass flux in the source term of the saturated zone). Although this coupling is crude relative to a strict continuum approach used by multiphase numerical models (see subsection *Multiphase Flow and Transport*), it is a widely used approach for scoping calculations used in risk assessment (Huyakorn et al., 1987).

There are a large number of analytical solutions to the ADE, some developed for very specific problems. Wilson and Miller (1978), for example, provide a useful solution for the case of a two-dimensional (2-D) system with a steady source, which can be used to simulate an injection well. Leij et al. (1991) present a series of solutions for third-type boundary conditions. Harada et al. (1980) include the daughter products of radioactive chain decay. Aral and Liao (1996) present solutions that include time-variant dispersion. Several solutions have been developed for nonhomogeneous media, including solutions for flow in parallel fractures (Sudicky and Frind, 1982) and lognormal hydraulic conductivity distributions (Serrano, 1995).

Analytical solutions have certain advantages. They are probably the most efficient approach when data are sparse or uncertain, and only order-of-magnitude accuracy is required. Where the assumptions regarding boundary conditions and media properties are well-met, analytical solutions may be the correct, exact solution for solute transport. Analytical solutions generally do not require complex computer software, and widely available software packages can be adapted easily to evaluate the required numerical integrations or special functions. The disadvantages of analytical solutions are that they are limited to cases of simple boundary conditions and simple approximations of flow and transport media.

8.3.3 FLOW KINEMATIC SOLUTIONS

A large number of transport models are neither purely analytical solutions (using exact mathematical solutions) nor purely numerical, relying on approximate solutions for the differential equations. Although various authors use different classification schemes (Domenico and Schwartz, 1990; Fetter, 1988), all these hybrid solutions will be discussed together in this subsection as flow kinematic solutions. Flow kinematic transport solutions first determine groundwater flow patterns in the field of interest (from an analytical or a numerical flow model) and then use approximate solutions for contaminant transport along streamlines of the flow field. This class includes:

- Semianalytical solutions for advective transport in streamlines created by potential theory flow solutions,
- Particle tracking to numerically determine the advective flow paths of hypothetical particles in velocity fields generated by flow models, and
- Extensions of particle tracking solutions to include hydrodynamic dispersion.

Unlike purely analytical solutions of the ADE, the flow fields can be quite complex, depending on the level of detail and method used to solve the flow equations.

Semianalytical

This approach is referred to as semianalytical because it uses exact analytical solutions for streamlines in the flow field, followed by numerical approximation for advective transport along the streamlines. The streamlines are determined by superimposing potential theory solutions for 2-D, steady-state flow in homogeneous aquifers (Javandel et al., 1984; Strack, 1989). The streamline solutions describing specific sources or sinks are developed from complex velocity potential equations (a.k.a analytic elements) that may be superimposed to represent regional flow, wells, streams, etc., in the field of interest. The combination of analytic elements leads to a set of equations for:

1. The stream functions ψ that describe the streamlines of the steady-state groundwater flow field, and
2. The velocity potentials ϕ that describe the product of hydraulic head and hydraulic conductivity in the steady-state groundwater flow field.

The stream functions give the exact flow paths from contaminant sources. The velocity potentials can be manipulated to determine the Darcy velocity q_i via the relationship $-\partial\phi/\partial x_i = q_i = -K\,\partial h/\partial x_i$. Together with the aquifer thickness and porosity, this yields the advective velocity at all points in the flow field. The travel time is then approximated by numerically integrating the travel time along each streamline, following the contaminant from the source to the sink. The contaminant mass arriving at each sink is simply the flow per unit width times the concentration for each streamline arriving at the sink (Javandel et al., 1984).

The advantage of the semianalytical approach is that if one has solutions for the analytic elements that describe sources and sinks of interest then the computer programs to solve these equations are simple to write and very efficient. The disadvantage is that the method is applicable only to advective transport in 2-D, homogeneous, steady-state flow. If the transport in the field of interest has significant heterogeneity, transient behavior, or dispersive mixing, then semianalytical solutions will be accurate only within an order of magnitude.

Chapter 8, Section 2 further discusses the use of analytic elements in groundwater modeling.

Particle Tracking

Unlike semianalytical methods that directly determine the streamlines, particle tracking determines the groundwater velocities between calculation nodes in a hydraulic head field created by an analytical or numerical model. Particle tracking algorithms then interpolate between these velocities to numerically solve for the flow path followed by a hypothetical particle. The head fields may be generated by any convenient flow model; thus, particle tracking may be used to calculate advective transport in 3-D, transient, and heterogeneous fields.

Several numerical methods have been developed to solve for particle paths. In steady-state, homogeneous fields, the particle tracking algorithm is nothing more than simple linear interpolation between known velocities and analytical solutions for the resulting particle locations. In the case of transient flow and highly heterogeneous fields, more sophisticated numerical solutions may be required to determine the particle position at the end of each time step (Anderson and Woessner, 1992).

Extensions to Particle Tracking

There are extensions of particle tracking that add the mixing effects of hydrodynamic dispersion to the advective transport of particle tracking. The simplest is the random walk approach (Prickett et al., 1981), which computes the particle tracks as described in subsection *Particle Tracking* and then perturbs the particle position after each time step to simulate hydrodynamic dispersion. The size of the perturbation is randomly chosen from a normal distribution whose variance is a function of dispersivity.

Another extension of particle tracking is based on the method of characteristics (MOC), a mathematical technique that breaks the ADE into a characteristic equation representing the advective transport:

$$\frac{v_i}{n} = \frac{dx_i}{dt} \tag{8.3.9a}$$

And a differential equation representing the dispersive transport:

$$\frac{1}{n}\frac{\partial}{\partial x_i}\left(D_{ij}\frac{\partial C}{\partial x_j}\right) + F = \frac{dC}{dt} \tag{8.3.9b}$$

where D_{ij} is a second-order tensor representing the coefficient of hydrodynamic dispersion and F is a term to account for sources and sinks (Huyakorn and Pinder, 1983). If the model axes are aligned with the principal direction of groundwater flow, the dispersivity tensor simplifies to longitudinal and transverse components as described in subsection *Mathematical Models*. Particle tracking provides the change in position versus the change in time dx_i/dt, thus solving Equation 8.3.9a. A numerical solution for Equation 8.3.9b provides the dispersive transport contribution, which is added to each particle track (Konikow and Bredehoeft, 1978). In practice, the MOC algorithm places several particles in each cell of a finite difference grid, solves a flow model to determine the velocity field, and moves the particles to new cells to represent the advective transport. The concentration in each cell is averaged from the mass associated with each particle in a cell, and the concentrations in each cell are changed according to the numerical solution for dispersion.

These extensions to particle tracking are fast relative to a full numerical solution for transport, but they have several drawbacks. Transport near impermeable boundaries is problematic, because the particles may be projected erroneously outside the model; a similar phenomenon arises with source and sink terms, where particles need to be created and destroyed. Nevertheless, such models may yield results in good agreement with analytical solutions and models that are more complex.

Comments

Several variants of simple advective particle tracking have been developed. One useful variant is reverse particle tracking, which attempts to find the starting location of particles arriving at a certain location. This is particularly useful in defining the capture zone of a well and in analyzing complex flow fields. Even when transport is not of interest, particle tracking also is useful to help visualize the flow field (Anderson and Woessner, 1992).

Particle tracking has many advantages: it is generally a much more computationally efficient solution than a full numerical solution of the ADE, and it allows for a much more complex flow field than the analytical and semianalytical solutions. The disadvantage of particle tracking is that it is primarily limited to modeling advection-dominated problems even when extended to include hydrodynamic dispersion.

8.3.4 NUMERICAL SOLUTIONS

Numerical methods are used for the most complex solutions of the ADE. Whereas analytical and flow kinematic solutions may be possible for simplified situations, numerical solutions are required for problems with heterogeneous and/or isotropic property distributions, complex geometry, complex boundary conditions, and variable historical flow gradients and/or mass loading. In other words, the complexities of many real-world problems often require a numerical solution.

Numerical solutions to the ADE must approximate both spatial and time derivatives (i.e., the terms containing $\partial^2 C/\partial x^2$ and $\partial C/\partial t$ in Equation 8.3.1). Common solution methods for these derivatives are described in the following subsections.

Solution of Spatial Derivatives

The most commonly used spatial derivative approximation methods are finite difference (FD), integrated finite difference (IFD; also known as integrated finite volume), and finite element (FE). Each of these methods is summarized below and in Table 8.3.2.

TABLE 8.3.2 Summary of Methods for Spatial Solution of Solute Transport

Method	Grid	Advantages	Disadvantages
FD	Regular	• Simple gridding and input • Easy to understand conceptually	• Square blocks coarsely approximate curved boundaries • Numerical dispersion
IFD	Irregular	• Grid flexibility can represent most geometries	• Complex input/grid specification, • Numerical dispersion
FE	Irregular	• Grid flexibility can represent most geometries, • More numerically versatile	• Non-intuitive solution, • Numerical instability

The FD method requires regularly spaced nodes, resulting in square- or rectangular-sided grid blocks (Figure 8.3.5b). With the FD method, spatial gradients and derivatives are calculated based on the difference between adjacent nodal values. Various conventions for differencing and weighting are possible [e.g., the difference may be calculated as (Node 1) − (Node 2) or as (Node 2) − (Node 1)]. The most commonly used conventions are central weighting, which is simple but may produce numerical oscillations in the solution, and upstream weighting, which is more robust but produces numerical errors in the solution called numerical dispersion. Solute transport modelers must be aware of the magnitude of numerical dispersion in their model because it has the same effect on results as hydrodynamic dispersion. In poorly designed models or inappropriate applications, numerical dispersion can lead to an inaccurate approximation of the spatial derivatives and an incorrect solution. Subsection *Other Transport Properties* provides some rules of thumb to minimize numerical dispersion and related numerical problems.

The IFD method allows for polyhedron-shaped grid blocks, which gives the flexibility of constructing an irregular mesh. With the IFD method, spatial gradients and derivatives are approximated from an integral form of the ADE. Solution requires input of grid block volumes, surface areas, and distances to interfaces for each grid block connection. If square or rectangular grid blocks are used, the IFD solution simplifies to the FD solution. As with the FD method, numerical dispersion can be problematic in poorly posed IFD problems.

The FE method also allows for irregular mesh construction, with each grid block or element delineated by corner nodes (Figure 8.3.5c). A FE approximation is derived from a set of piecewise continuous functions (called basis functions) that approximate an integral form of the ADE between corner nodes of an element. The integral is typically formulated with one of two approaches: weighted residual Galerkin or variational. Numerical instability (oscillations, overshoot, or undershoot) can be a problem but can be minimized with careful mesh design or upstream basis functions (Huyakorn and Pinder, 1983).

Solution of Time Derivatives

Mathematically, approximation of the time derivative is accomplished by stepping the model through specified time increments Δt. Regardless of the spatial derivative solution method, the time derivative $\partial C / \partial t$ is calculated as the difference between values at the "new" time $t + \Delta t$ and at the "previous" time t. The time derivative typically is calculated by using either an explicit or an implicit scheme. The solution scheme identifies the time when the spatial derivative terms on the left-hand-side of the ADE (e.g., Equation 8.3.1) must be evaluated.

With an explicit scheme the spatial derivative terms are evaluated at time t and therefore are known from solution of the previous time step. Explicit solutions (also referred to as forward-in-time) are computationally fast but require small Δt and large grid block sizes; otherwise, the solution will magnify errors and become unstable.

With an implicit scheme, the left-hand-side spatial derivatives are evaluated at time $t + \Delta t$. Because the spatial derivatives at time $t + \Delta t$ are unknown (and must be calculated), an implicit solution is more computationally intensive than an explicit solution. However, an implicit solution (also referred to as backward-in-time) is usually unconditionally stable and therefore, within reason, does not impose restrictions on time step and grid block size.

Explicit and implicit schemes represent the theoretical endpoints for solution of the time derivative. Other options are available, such as Crank-Nicolson, which evaluates the left-hand-side spatial derivative at the average of times t and $t + \Delta t$.

Another option for evaluating the ADE over a specified time is the Laplace transform approach (Sudicky, 1989). This method begins by taking the Laplace transform of the ADE with respect to time thus temporarily removing the time derivative term. The transformed equation of spatial derivatives is solved by the FE method and then inverse-transformed at the time of interest to get the solution for concentration versus location. This method completely avoids the time-stepping associated with the direct solution for the time derivative and thus avoids numerical dispersion. However, the inverse Laplace transform must be computed numerically, requiring the same computational effort as several

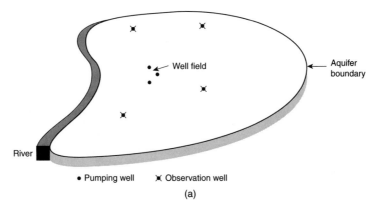

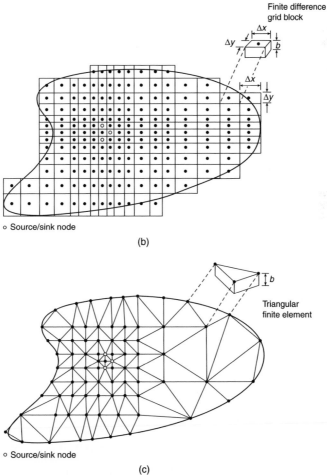

FIGURE 8.3.5   Aquifer with sources, sinks, and physical boundaries (a), and its representation by (b) FD and (c) FE numerical models (from Wang and Anderson, 1982).

time derivatives (depending on the number of nodes in the mesh). Thus, the Laplace transform method is most commonly used when transport times are very long (i.e., many time steps would be required) or for problems in which the results otherwise would be significantly affected by numerical dispersion.

Matrix Solvers

Regardless of which numerical methods are used, the result of the approximations of the spatial and time derivatives is a set of simultaneous equations, where the unknown values (concentrations) are defined in terms of the known values (dispersion coefficients, velocities, etc.). The resulting solution of the set of simultaneous equations typically is accomplished through one of two matrix solution techniques: direct elimination or iteration.

Direct elimination methods (e.g., Gaussian elimination, lower-upper decomposition) are used to reduce the solution matrix to a triangular or banded form, which can be solved more efficiently. The efficiency of direct elimination methods can be enhanced in some cases by optimal grid block or element numbering schemes, which produce banded or tridiagonal matrices. Direct elimination methods yield the exact solution to the system of equations and are computationally advantageous for small to moderate sized matrices.

Iterative methods provide approximate matrix solutions, successively improving from a first approximation until specified convergence criteria are met. Iterative methods include Gauss-Seidel, successive overrelaxation, and others. Iterative methods are advantageous for solving large matrices.

8.3.5   DETERMINATION OF MODEL PROPERTIES

The most important thing to remember for numerical modeling is "garbage in, garbage out." A complex model will yield a meaningless solution unless all input data are appropriate, site specific, and well understood by the modeler. The most common inputs required for solute transport models include dispersivity, retardation factor, porosity, decay constant, time stepping, and grid block size. Guidance for these inputs is provided in the following subsections. Other important input for solute transport models include initial (source) concentrations, groundwater velocities, and boundary conditions. These inputs are discussed in subsection *Model Input*.

Dispersivity

Dispersivities (both longitudinal and transverse) are the most difficult transport parameters to quantify and are an area of great research activity. Researchers have noted that dispersivities determined from laboratory tests are not representative of field (and model) scale dispersivities (Lallemand-Barres and Peaudecerf, 1978; Pickens and Grisak, 1981; Gelhar, 1986). This scale dependence is generally attributed to differences between the effects of small-scale variations in the pore structure compared with the effects of large-scale heterogeneity. Conceptually, dispersivity may be thought of as consisting of two components, $\alpha = \alpha^* + A$, where α^* represents the small-scale microdispersivity and A represents the larger macrodispersivity arising from aquifer-scale heterogeneity (Frind et al., 1987). Various theories also suggest that the scale dependence of dispersivities may be quite complex (Gelhar and Axness, 1983; Dagan, 1988; Jaekel and Vereecken, 1997).

Reasonable estimates of site-specific dispersivities may be obtained in two ways. Field-scale tracer tests, where a nonreactive tracer (e.g., iodine or bromide) is injected and then downgradient samples are taken to monitor transport, can be analyzed to obtain dispersivity values. Alternatively, inverse modeling of existing contaminant plumes can be used to estimate the dispersivities. Where site-specific values are not available, two general guidelines are helpful in selecting model dispersivities. First, Lallemand-Barres and Peaudecerf (1978) suggest that the longitudinal dispersivity can be approximated as $\alpha_L = 0.1x$, where x is the distance from the contaminant source to the center of the

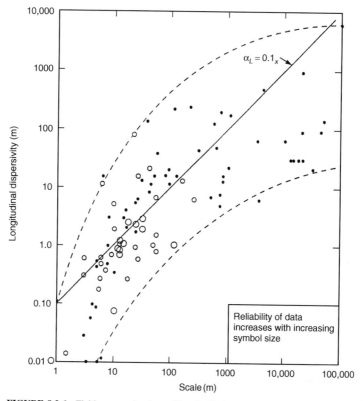

FIGURE 8.3.6   Field-measured values of longitudinal dispersivity as a function of the scale of measurement (from Fetter, 1993).

plume (Figure 8.3.6). Second, Gelhar (1986) suggests that longitudinal dispersivities are on the order of 10 times greater than horizontal transverse dispersivities.

A further consideration in selecting a model dispersivity is the model conceptualization. An analytical or numerical solution for a homogeneous medium applied to aquifer-scale transport should use a relatively large dispersivity value $\alpha = \alpha^* + A$ to simulate the hydrodynamic dispersion arising from the unmodeled heterogeneity (Figure 8.3.7a). In contrast, a numerical solution that includes heterogeneity should use a much lower value of dispersivity $\alpha = \alpha^*$, because the macro dispersivity A is addressed by modeled heterogeneity (Figure 8.3.7c). In practice, dispersivity values used in aquifer-scale modeling range widely and usually are confirmed by model calibration (see subsection *Model Calibration*). The dispersivity of a heterogeneous numerical model ($\alpha = \alpha^*$) typically is less than the grid block dimension and must be checked versus model-specific criteria for numerical dispersion and stability.

Other Transport Properties

The retardation factor attempts to quantify the effect of sorption to retard or delay the spread of contaminants. It is defined as (groundwater velocity)/(solute velocity). Therefore, a retardation factor of 1 implies that the solute does not sorb to the soil (i.e., conservative transport). As the degree of sorption (and retardation) increases, retardation factors increase above one. Retardation factors depend on several variables related to the rock, fluid, and contaminants present. They are site and

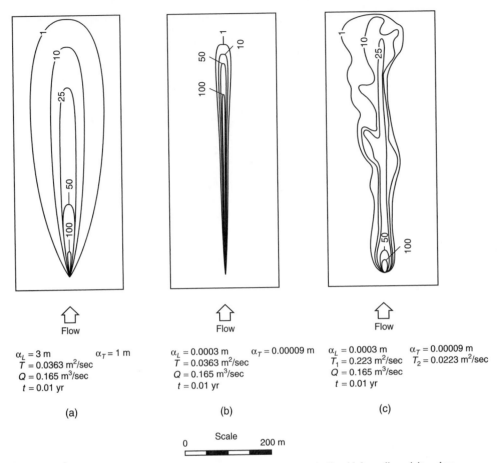

FIGURE 8.3.7 Comparison of solute transport model results for (a) homogeneous media with large dispersivity values, (b) homogeneous media with small dispersivity values, and (c) heterogeneous media with small dispersivity values (from Fetter, 1993).

contaminant specific and should be determined by laboratory testing with the appropriate soils and contaminants. Chapter 1, Section 6 provides a more detailed discussion of retardation factors, sorption, and associated properties.

For advective-dispersive transport of solutes, effective porosity is the dominant component of total porosity, because it determines the groundwater velocity. The effective porosity represents the volume of interconnected pore space available for transport. Dead-end pores or fractures do not contribute to advective-dispersive transport but can create large volumes that participate in diffusion. Table 8.3.3, reproduced from Freeze and Cherry (1979), lists ranges of total porosity for various rock types. Remember that the effective porosity may be much less than the total porosity for fine-grained materials and some fractured rocks.

Radioactive or biological decay processes can lead to a decrease in solute mass. Decay typically is expressed as a first-order equation (Domenico and Schwartz, 1990):

$$\frac{\partial C}{\partial t} = \lambda C \tag{8.3.10}$$

The half-life $t_{1/2}$ of a contaminant is defined as the time needed for the initial concentration C_0 to

TABLE 8.3.3 Ranges of Porosity for Common Consolidated and Unconsolidated Materials (from Freeze and Cherry, 1979)

Unconsolidated sediments	Porosity (%)	Consolidated rocks	Porosity (%)
Clay	40–70	Sandstone	5–50
Silt	35–50	Limestone/dolomite	0–20
Sand	25–50	Shale	0–10
Gravel	25–40	Fractured crystalline rock	0–10
Sand/gravel mixture	10–35	Fractured basalt	5–50
Glacial till	10–25	Dense crystalline rock	0–5

TABLE 8.3.4 Numerical Criteria for Solute Transport (after Reeves et al., 1986)

Spatial derivative differencing scheme	Time derivative differencing scheme	Criteria to minimize numerical dispersion*	Criteria to minimize overshoot-undershoot*
Central weighting (centered-in-space)	Crank-Nicolson (centered-in-time)	None	$\Delta x + 2\alpha \leq 2\,\Delta x^2/v\Delta t$ $\Delta x \leq 2\alpha$
Upstream weighting (backward-in-space)	Crank-Nicolson (centered-in-time)	$\Delta x \ll 2\alpha$	$\Delta x + 2\alpha \leq 2\,\Delta x^2/v\Delta t$
Central weighting (centered-in-space)	Implicit (backward-in-time)	$v\Delta t \ll 2\alpha$	$\Delta x \leq 2\alpha$
Upstream weighting (backward-in-space)	Implicit (backward-in-time)	$\Delta x + v\Delta t \ll 2\alpha$	None

*Criteria assume that molecular diffusion is negligible relative to mechanical dispersion. Δx represents any grid block dimension (x, y, or z) but direction of evaluation should be consistent with direction of α and v.

decay to $C_0/2$. The decay constant λ can be calculated from the half-life as:

$$\lambda C = \frac{\ln(2)}{t_{1/2}} = \frac{0.693}{t_{1/2}} \tag{8.3.11}$$

Depending on the methods selected to solve the spatial derivatives (subsection *Solution of Spatial Derivatives*) and time derivatives (subsection *Solution of Time Derivatives*), limitations on the time step size (Δt) and grid block sizes (Δx, Δy, Δz) may be required to reduce numerical dispersion and instability (i.e., overshoot-undershoot). Table 8.3.4, based on Reeves et al. (1986), summarizes numerical criteria for FD and IFD formulations.

8.3.6 *ADVANCED TOPICS*

Model Calibration

As mentioned in subsection *Model Output*, transport models generally use a forward solution, where the input parameters for the flow and transport properties are known and the flow velocities and solute concentrations are calculated. As a practical matter, however, it can be difficult to measure the key parameters of the groundwater system (e.g., boundary flows, hydraulic conductivity, porosity, and dispersivity). Thus it is relatively common for modelers to want to adjust these key model parameters to reproduce measured flows, observed heads, and measured concentrations. This adjustment, or calibration, of the model parameters to match an observed output is also known as the inverse problem.

In groundwater flow and transport, the inverse problem is difficult to solve because it is frequently ill-posed. Ill-posedness is the result of three possible characteristics of the inverse solution (Carrera and Neuman, 1986):

- Unidentifiability, when no solution can be found—this can occur when there are too many parameters being estimated for too few observations;
- Nonuniqueness, when many combinations of hydraulic conductivity, porosity, recharge rate, etc., can be found that reproduce the same set of observations; and
- Instability, where minor changes (such as small observation errors) can result in wildly different inverse solutions for parameters.

Although many groundwater models may be ill-posed, there are several options that can make the problem tractable. These include reducing the number of parameters to be estimated, increasing the number of observations to be matched, and providing restrictions on the distribution of parameters (e.g., prior assumptions about parameter values). Nonunique solutions also may be presented stochastically as a distribution of parameter values (subsection *Stochastic Simulation*).

There are two methods for calibrating a model: (1) manually adjusting the model parameters until the model simulated values agree with the observed values, and (2) automated calibration with a rigorous mathematical framework to minimize the disagreement between the simulated and the observed values. Comparison of the model simulated values and the observed values (typically hydraulic heads and/or concentrations) is commonly expressed in terms of an objective function, J, to be minimized, such as with the sum of squared errors:

$$J = \sum^i (\text{observed value } i - \text{ simulated value } i)^2 \tag{8.3.12}$$

Although automated calibration is generally more efficient, uncritical use of automated calibration can lead to physically unrealistic parameter sets. Anderson and Woessner (1992) provide a good overview of model calibration; McLaughlin and Townley (1996) review various approaches; and Chu et al. (1987) provide examples of transport model calibration.

Heat and Density Dependence

All the solution techniques presented previously in this subsection have assumed that the solute is at relatively low concentrations, and thus the groundwater velocity field (whether steady state or transient) is not affected by changes in groundwater chemistry. Although this is usually a reasonable assumption, it is possible for concentration gradients or chemical reactions to dramatically change the fluid density and thus influence the groundwater flow field (i.e., the fluid pressure or hydraulic head distribution). The influence of the chemical composition on the flow field therefore may require a coupled solution that solves the pressure (flow) equation and the transport (ADE) equation at each time step and provides feedback between the two equations. Common uses of density-dependent models are simulation of seawater intrusion in coastal areas and disposal of industrial effluents with high salt content.

The above solutions also have assumed that the modeled system is isothermal—i.e., that there are minimal temperature changes. Although it is frequently acceptable to assume that a shallow groundwater system is isothermal, temperatures may vary widely in some flow and transport problems (e.g., geothermal systems, high-level radioactive waste disposal, petroleum reservoirs). Temperature changes can affect the density and viscosity of groundwater, changing the groundwater flow field, chemical reaction rates, and biological degradation rates. Simulating heat dependence requires solving a third equation for the transport of heat, along with the flow and chemical transport equations.

Reactive Transport

The previously mentioned solutions typically assume that all reactions are in instantaneous equilibrium and simple linear functions of the concentration of a single solute. Reactive transport models explicitly simulate the geochemical reactions in relation to the groundwater flow and contaminant transport rates. Such models allow simulation of more complex sorption processes (e.g., nonlinear, competitive sorption), biological decay (e.g., rate-limited hydrolysis), radioactive and chemical decay with ingrowth of daughter products, and detailed geochemistry (e.g., kinetic reactions, mineral precipitation/dissolution, oxidation/reduction).

Such reactive transport models generally are developed on an as-needed basis to address specific processes of interest and may be exceedingly complex.

Although these more complex solutions have the potential to produce a more accurate solution, they require more detailed input and greater computational resources. As with any modeling exercise, a simple model with good input invariably will yield a better solution than a complex model with insufficient or erroneous input.

Stochastic Simulation

In many cases, site knowledge is insufficient to specify a model input parameter as a single (i.e., deterministic) value. Instead, the uncertainty can be accounted for by specifying a range or distribution within which the value is expected to fall (e.g., porosity has a uniform distribution with mean and variance taken from core samples). A common method for obtaining useful results with uncertain input is Monte Carlo analysis, in which the model is run repeatedly with input values selected (sampled) from the range or distribution of parameter values. Thus, instead of a single deterministic result, the model produces a stochastic set of results, reflecting the uncertainty of the input parameters. Monte Carlo analyses can be quite complex, with many stochastic input parameters, correlation between parameters, and model calibration for each simulation.

Multiphase Flow and Transport

All the solute transport solutions described previously in this subsection have dealt with transport of a dissolved contaminant species in a flow system fully saturated with groundwater (i.e., a single liquid phase containing two components: water and contaminant). However, groundwater contamination investigations must also sometimes examine the vadose zone (where the second phase is gaseous, composed of air and possibly water vapor components) and non-aqueous-phase liquids (NAPLs), which exist as additional liquid phases (Figure 8.3.8).

Mathematical models of multiphase, multicomponent flow and transport were first developed in the petroleum industry. A good reference textbook is Aziz and Settari (1979). The key factors that differentiate multiphase flow from single-phase flow (with associated model parameters listed in parentheses) are:

- The flow of each fluid phase i is governed by a separate set of flow equations (phase fluid properties and pressures, P_i).

- The pore space is occupied by more than one immiscible fluid (phase saturations, S_i).

- The flows of the multiple phases are coupled by interfacial tension between phases (capillary pressure, P_{ci}), and impedance in the flowpaths (relative permeability, k_{ri}).

- There is mass transfer of components between phases (solubilization and volatilization).

For single-phase flow and transport, the fluid saturation is inherently 1. For multiphase flow and transport, the sum of the saturations of all phases is 1. The hydraulic conductivity of each phase K_i is calculated from:

$$K_i = k_{ri} \frac{k \rho_i g}{\mu_i} \tag{8.3.13}$$

where ρ_i = fluid density of phase i,
   μ_i = fluid viscosity of phase i,
   g = acceleration of gravity, and
   k = intrinsic permeability (medium's resistance to flow, independent of fluid).

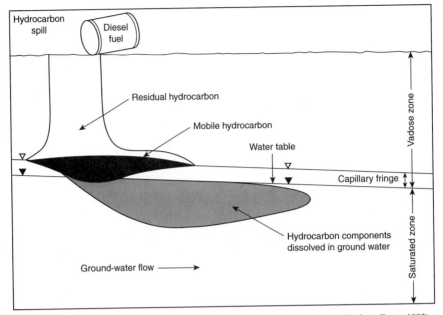

FIGURE 8.3.8 Anatomy of a multiphase transport problem arising from a NAPL spill (from Fetter, 1993).

The relative permeability k_{ri} ranges from 0 to 1 and changes with saturation. As the saturation of phase i decreases (and the saturations of the other phases increase), the cross section of the pore space available for phase i flow decreases. This phenomenon is quantified by a reduction in the relative permeability and a corresponding reduction in the hydraulic conductivity of phase i.

Because each of the immiscible fluid phases may move in response to pressure and gravitational gradients, the governing equations for multiphase flow are nonlinear—that is, the fluid pressure is a function of the flow, the flow is a function of the permeability, the permeability is a function of the saturation, and the saturation is a function of the fluid pressure. The governing flow equation for each phase is also coupled to the pressure and flow of the other phases through capillary pressure relationships. In the most general case, each component (e.g., water, air, contaminant) can volatilize into the gaseous phase or dissolve into one of the other liquid phases. For example, at a single point in time, trichloroethane (which is a NAPL component) may exist as a liquid in the NAPL phase, as a dissolved species in the water phase, and as a gas in the gaseous phase. Simulating the transport of this type of contaminant conceivably may require solving the coupled nonlinear equations for fluid flow, coupled with the (nonlinear) equations for transport.

Given the complexity and nonlinearity of multiphase transport, it is not surprising that such problems are computationally difficult to solve. Currently, numerical models of multiphase transport are just beginning to see use outside the research environment; the interested reader is referred to Pinder and Abriola (1986) for more details.

REFERENCES

Anderson, M., and Woessner, W., *Applied Groundwater Modeling—Simulation of Flow and Advective Transport*, Academic Press, San Diego, CA, 1992.

Aral, M. M., and Liao, B., "Analytical Solutions for Two-Dimensional Transport Equation with Time-Dependent Dispersion Coefficients," *J. Hydrol. Eng.*, 1(1): 20–32, 1996.

Aziz, K., and Settari, A., *Petroleum Reservoir Simulation*, Elsevier Applied Science, New York, 1979.

Bear, J., *Dynamics of Fluids in Porous Media*, American Elsevier, New York, 1972.

Carrera, J., and Neuman, S. P., "Estimation of Aquifer Parameters under Transient and Steady State Conditions: 1. Maximum Likelihood Method Incorporating prior Information," *Water Resources Res.*, 22(2): 199–210, 1986.

Carslaw, H. S., and Jaeger, J. C., *Conduction of Heat in Solids*, Oxford University Press, New York, 1959.

Cleary, R. W., and Ungs, M. J., *Analytical Methods for Groundwater Pollution and Hydrology*, Water Resour. Prog. Rep. 78-WR-15, Department of Civil Engineering, Princeton University, Princeton, NJ, 1978.

Codell, R. B., Key, K. T., and Whelan, G., *A Collection of Mathematical Models for Dispersion in Surface Water and Groundwater*, U.S. Nuclear Regulatory Commission, NUREG-0868, 1982.

Chu, W., Strecker, E., and Lettenmaier, D., "An Evaluation of Data Requirements for Groundwater Contaminant Transport Modeling," *Water Resources Res.*, 23(3): 408–424, 1987.

Dagan, G., "Time-Dependent Macro-Dispersion for Solute Transport in Anisotropic Heterogeneous Aquifers," *Water Resources Res.*, 24(9): 1491–1500, 1988.

de Marsily, G., *Quantitative Hydrogeology*, Academic Press, Orlando, FL, 1986.

Domenico, P. A., and Schwartz, F. W., *Physical and Chemical Hydrogeology*, John Wiley & Sons, New York, 1990.

Fetter, C. W., *Applied Hydrogeology*, MacMillan, New York, 1988.

Fetter, C. W., *Contaminant Hydrogeology*, Macmillan, New York, 1993.

Freeze, R. A., and Cherry, J. A., *Groundwater*, Prentice-Hall, Englewood Cliffs, NJ, 1979.

Frind, E., Sudicky, E., and Schellenberg, S., "Micro-Scale Modelling in the Study of Plume Evolution in Heterogeneous Media," *Proceedings, NATO Advanced Research Workshop on Advances in Analytical and Numerical Groundwater Flow and Quality Modelling*, Lisbon, Portugal, 1987.

Gelhar, L., "Stochastic Subsurface Hydrology from Theory to Applications," *Water Resources Res.*, 22(9): 1355–1455, 1986.

Gelhar, L. W., and Axness, C. L., "Three-Dimensional Stochastic Analysis of Macrodispersion in Aquifers," *Water Resources Res.*, 19(1): 161–180, 1983.

Harada, M., Chambré, P., Foglia, M., Higashi, K., Iwamoto, F., Leung, D., Pigford, T., and Ting, D., *Migration of Radionuclides through Sorbing Media Analytical Solutions - I*, Lawrence Berkeley Laboratory, University of California, Berkeley, 1980.

Huyakorn, P. S., and Pinder, G. F., *Computational Methods in Subsurface Flow*, Academic Press, Orlando, FL, 1983.

Huyakorn, P. S., Ungs, M. J., Mulkey, L. A., and Sudicky, E. A., "A Three-Dimensional Analytical Method for Predicting Leachate Migration," *Ground Water*, 25(5): 588–598, 1987.

Jaekel, U., and Vereecken, H., "Renormalization Group Analysis of Macrodispersion in a Directed Random Flow," *Water Resources Res.*, 33(10): 2287–2299, 1997.

Javandel, I., Doughty, C., and Tsang, C. F., eds., *Groundwater Transport: Handbook of Mathematical Models*, Water Resour. Monogr. Ser., Vol. 10, American Geophysical Union, Washingtion, DC, 1984.

Konikow, L. F., and Bredehoeft, J. D., *Computer Model of Two-Dimensional Solute Transport and Dispersion in Ground Water.* Techniques of Water-Resources Investigations, Book 7, Chapter C2, U.S. Geological Survey, Reston, VA, 1978.

Lallemand-Barres, P., and Peaudecerf, P., "Recherche des Relations Entre la Valeur de la Dispersivité Macroscopique d'un Milieu Aquifere, ses Autres Caracteristiques et les Conditions de Mesure, Etude Bibliographique," *Bull. Bureau de Recherches Géologiques et Miniéres*, 3/4: 277–287, 1978.

Leij, F., Skaggs, T., and van Genuchten, M., "Analytical Solutions for Solute Transport in Three-Dimensional Semi-infinite Porous Media," *Water Resources Res.*, 27(10): 2719–2733, 1991.

McLaughlin, D., and Townley, L. R., "A Reassessment of the Groundwater Inverse Problem," *Water Resources Res.*, 32(5): 1131–1161, 1996.

Ogata, A., and Banks, R. B., "A Solution of the Differential Equation of Longitudinal Dispersion in Porous Media," *U.S. Geological Survey Professional Paper 411-A*, 1961.

Pickens, J. F., and Grisak, G. E., "Scale-Dependent Dispersion in a Stratified Granular Aquifer," *Water Resources Res.*, 17(4): 1191–1211, 1981.

Pinder, G., and Abriola, L., "On the Simulation of Nonaqueous Phase Organic Compounds in the Subsurface," *Water Resources Res.*, 22(9): 1095–1195, 1986.

Prickett, T., Naymik, T., and Lonnquist, C., *A "Random-Walk" Solute Transport Model for Selected Groundwater Quality Evaluations*, Illinois State Water Survey, Bulletin 65, 1981.

Reeves, M., Ward, D. S., Johns, N. D., and Cranwell, R. M., *Theory and Implementation for SWIFT II*, U.S. Nuclear Regulatory Commission, NUREG/CR-3328, SAND83-1159, 1986.

Scheidegger, A. E., "General Theory of Dispersion in Porous Media," *J. Geophys. Res.*, 66(4): 3273–3278, 1961.

Serrano, S., "Analytical Solutions of the Nonlinear Groundwater Flow Equation in Unconfined Aquifers and the Effect of Heterogeneity," *Water Resources Res.*, 31(11): 2733–2742, 1995.

Strack, O., *Groundwater Mechanics*, Prentice-Hall, Englewood Cliffs, NJ, 1989.

Sudicky, E. A., and Frind, E. O., "Contaminate Transport in Fractured Porous Media: Analytical Solution for a System of Parallel Fractures," *Water Resources Res.*, 18(6): 1634–1642, 1982.

Sudicky, E. A., "The Laplace Transform Galerkin Technique: A Time-Continuous Finite Element Theory and Application to Mass Transport in Groundwater," *Water Resources Res.*, 25(8): 1833–1846, 1989.

van Genuchten, M. Th., and Alves, W. J., *Analytical Solutions to the One-Dimensional Convective-Dispersive Solute Transport Equation*, Tech. Bull. 1661, US. Salinity Lab., Riverside, CA, 1982.

Wang, H. F., and Anderson, M. P., *Introduction to Groundwater Modeling*. W. H. Freeman and Company, San Francisco, 1982.

Wexler, E. J., *Analytical Solutions for One-, Two-, and Three-Dimensional Solute Transport in Ground-Water Systems with Uniform Flow*, U.S. Geological Survey Report, 89–56, 1989.

Wilson, J. L., and Miller, P. L., "Two-Dimensional Plume in Uniform Ground-Water Flow," *J.Hydraul. Dov. Proc. Am. Soc. Cov. Eng.*, 104(HY4), 503–514, 1978.

Yeh, G.-T., and Tsai, Y.-J., "Analytical Three-Dimensional Transient Modeling of Effluent Discharges," *Water Resources Res.*, 12: 533–540, 1976.

CHAPTER 8
ANALYSIS AND MODELING

SECTION 8.4

FATE AND TRANSPORT IN RIVERS, LAKES, AND ESTUARIES

Zhen-Gang Ji

Dr. Ji is a senior environmental engineer with Tetra Tech, Inc. He specializes in the study of hydrodynamics, water quality, sediment transport, and toxics in surface water systems for water resources management, and air quality modeling.

8.4.1 INTRODUCTION

The fate and transport processes of contaminants in rivers, lakes, and estuaries are controlled by two factors: their reactivity and their physical transport. Reactivity includes chemical processes, biological processes, and biouptakes. Physical transport has three mass transport processes: transport by advection of water current, transport by diffusion and turbulent mixing within the water column, and transport of particles due to settling and resuspension on the water-sediment bed interface.

Based on the principle of conservation of mass, the change in concentration of a reactant can be calculated with mass balance equations. A mass balance is simply an accounting of mass inputs, outputs, reactions, and net change as follows:

$$\frac{\partial C}{\partial t} = -U\frac{\partial C}{\partial x} + \frac{\partial}{\partial x}\left(A\frac{\partial C}{\partial x}\right) + S + R \qquad (8.4.1)$$

Netchange of concentration   Advection   Mixing   Settling   Reactivity

where C = reactant concentration,

t = time,

x = distance,

U = advection velocity in x direction,

A = mixing and dispersion coefficient,

S = sources and sinks due to settling and resuspension, and

R = reactivity of chemical and biological processes.

External loadings to the aquatic system from point and nonpoint sources are not included in Equation 8.4.1.

In Equation 8.4.1, the advection term accounts for the mass inputs and mass outputs by water current. The mixing term describes net transport due to molecular diffusion and turbulent mixing. The settling term represents particle settling to and resuspension from the bed. The reactivity term refers to chemical and/or biological processes that take place within the water column. All these processes directly or indirectly affect the environmental behavior of contaminants.

Most contaminants are associated, to a greater or lesser degree, with suspended and deposited particles in natural systems. Adsorption of metal and organic toxicants to particulate substrates is one of the most important processes affecting their fate, transport, and bioavailability. Therefore, accurate treatment of sediment processes is essential to the fate and transport of sediment-associated toxics, such as heavy metals and hydrophobic organics. Processes to be considered include (1) accurate prediction of hydrodynamic transport (advection mixing) and settling of suspended solid in the water column; (2) particulate exchange between water column and bed sediment compartments, with consideration of important hydrodynamic and physicochemical controls on deposition and remobilization; and (3) diffusive exchange between the water column and bed sediment compartments.

Settling velocities in the water column are influenced by the size, shape, relative density, and concentration of the particulates. The slow settling velocity of dispersed fine sediment, with which many toxicants are associated, suggests the critical importance of accurate hydrodynamic description as part of any toxicant transport and fate study. Bed sediment is scoured and brought into the water column when critical shear stresses at the bed-water interface are exceeded. Deposition occurs when gravitational forces exceed turbulent forces and particles reach the sediment-water interface.

Kinetic processes of reactants may be grouped into two broad categories: interphase exchange and chemical reactions. Interphase exchange covers adsorption/desorption to mineral, colloidal, and organic particulates; bioaccumulation and bioconcentration; and air-water exchange (e.g., volatilization and diffusion). Chemical reactions, which may be reversible or irreversible, involve fundamental change in the molecular structure of the compound and usually in its environmental behavior as well.

In this section, the discussion on fate and transport in rivers, lakes, and estuaries is based on the principle of conservation of mass and its representations in Equations 8.4.1 and 8.4.2. Subsection 8.4.2 discusses kinetics in aquatic systems (the fourth term on the right side of Equations 8.4.1 and 8.4.2). Subsections 8.4.3, 8.4.4, and 8.4.5 discuss physical transport processes (the first three terms on the right side of Equations 8.4.1 and 8.4.2) in rivers, lakes, and estuaries, respectively.

8.4.2   KINETICS IN AQUATIC SYSTEMS

Contaminants in aquatic systems include nutrients, organic toxicants, heavy metals, and communicable disease organisms. If no degradation reactions occurred in nature, every contaminant discharged in the past still would be polluting the environment. Fortunately, natural purification processes dilute, transport, remove, and degrade contaminants. It is essential to understand the kinetics of reactants and to describe them mathematically.

Major nutrients in aquatic systems include phosphorus, nitrogen, carbon, and silica. Excessive amounts of nutrients from municipal and industrial discharges and agricultural and urban runoffs lead to eutrophication problems in rivers, lakes, and estuaries.

Organic toxicants can be categorized based on their usage and chemical classes (Council on Environmental Quality, 1978). The common organic toxicants in aquatic systems include pesticides, polychlorinated biphenyls, and dioxins. They persist in the environment and bioaccumulate and magnify in the food web.

Heavy metals usually refer to those metals between atomic numbers 21 and 84. In contrast to organic pollutants, heavy metal pollutants are pervasive and, perhaps, more persistent. They frequently have natural background sources from dissolution of rocks and minerals. Heavy metals can be in the particulate-adsorbed or dissolved phases in the sediment or overlying water column. Interchange between sorbed metal ions and aqueous metal ions occurs via adsorption-desorption mechanisms. Depending on the concentration gradient, metal ions in pore water of the sediment can diffuse to the overlying water column and vice versa. Heavy metals are a serious pollution problem when their concentration exceeds water-quality standards. In addition, volatile metals (such as mercury) are emitted from industrial stack gases and can be directly deposited to surface water.

Bacteria and other organisms that may cause communicable disease are controlled by the growth rate of the contaminant, solar radiation, temperature, salinity, predation, nutrient supply, toxicity in the water column, and settling of the contaminant.

Although reaction kinetics in aquatic systems can be described in numerous ways, as represented in the fourth term on the right side of Equation 8.4.1, the form for single reactants is generally expressed as follows:

$$\frac{dC}{dt} = -kC^m \tag{8.4.2}$$

where m = order of reaction and k = rate constant of the m-order reaction. In natural aquatic systems, the commonly used forms of the above equation are with $m = 0$, 1, and 2.

Zero-order Reactions

In aquatic systems, a zero-order reaction ($m = 0$) represents irreversible degradation of a reactant that does not depend on the reactant concentration, and the solution to Equation 8.4.2 is

$$C = C_0 - kt \tag{8.4.3}$$

where C_0 = initial concentration at $t = 0$. In this case, a plot of concentration versus time should yield a straight line with a slope of k, as shown in Figure 8.4.1 (left). Methane production and release of hydrolysis products from anaerobic sediment are examples of zero-order reactions (Schooner, 1996).

First-order Reactions

First-order reactions ($m = 1$) have reaction rates proportional to the concentration of the reactant and most commonly are used to describe chemical and biological reactions. For first-order reactions, the solution to Equation 8.4.2 is

$$C = C_0 e^{-kt} \tag{8.4.4}$$

Equation 8.4.4 indicates that, for first-order reactions, reactant concentration decreases exponentially with time. In this case, a plot of logarithm concentration versus time should yield a straight line with slope k, as shown in Figure 8.4.1 (middle). Examples of first-order reactions include biochemical oxygen demand in surface water systems, death and respiration rates for bacteria, and production reaction of algae (Thomann and Mueller, 1987). It should be noted that, although most kinetic formulations are parameterized by first-order reaction, derivation of reaction rate constant k might require a significant amount of data.

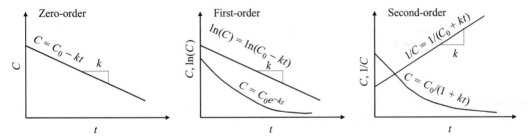

FIGURE 8.4.1   (Left): Concentration versus time for zero-order reaction. (Middle) Concentration and logarithm concentration versus time for first-order reaction. (Right) Concentration and inverse concentration versus time for second-order reaction.

Second-order Reactions

For second-order reactions ($m = 2$), the solution to Equation (8.4.2) is

$$\frac{1}{C} = \frac{1}{C_0} + kt \tag{8.4.5}$$

Therefore, if a reaction is indeed second-order, a plot of inverse concentration of C ($1/C$) with time should yield a straight line with slope k [Figure 8.4.1 (right)]. Equation 8.4.5 can also be expressed as

$$C = \frac{C_0}{1 + kC_0 t} \tag{8.4.6}$$

which reveals that, like the first-order reaction, the resulting concentration of a second-order reaction also decreases and approaches zero as time increases. Processes that can be described by second-order reactions include atmospheric gas reactions and zooplankton death rates.

8.4.3 *PHYSICAL TRANSPORT PROCESSES IN RIVERS*

Kinetics plays an important role in a chemical's fate in the environment, but an equally important aspect is physical transport of the chemical in the aquatic environment. Physical transport in aquatic systems includes three processes, presented by the first three terms on the right side of Equation 8.4.1. They are (1) advection of water current, (2) dispersion due to diffusion and turbulent mixing within the water body, and (3) particle settling and resuspension within the water column and on the water-sediment bed interface. This subsection and the next two subsections discuss physical transport in rivers, lakes, and estuaries, respectively.

Rivers and streams vary widely by morphological, hydraulic, and ecological characteristics, which include river width, depth, and slope; river velocity and flow rate; water temperature; and sediment transport and river bed configurations. A river often acts as a sink for contaminants distributed along the length of the river. An example of an external contaminant source is an effluent from wastewater treatment plants that empties nutrients, heavy metals, and/or organic toxicants into the river.

The dominant transport process in rivers and streams is advection due to river water flow, represented by the first term on the right side of Equations 8.4.1 and 8.4.2. The water flow rate at a given point in a river depends on watershed and hydrological characteristics, such as the drainage area and precipitation; slope of the river and bed roughness; man-made structures such as dams, reservoirs, and ship locks; and water discharges and withdrawals along the river.

The mixing process in rivers, represented by the second term on the right side of Equation 8.4.1, is often less important in transport of reactants. Advection results in the reactant's moving downstream, whereas longitudinal mixing leads to spreading or smearing in the longitudinal dimension. Lateral and vertical mixing processes determine how long it takes for a reactant to be completely mixed across a river. Compared with advective transport, transport by dispersion in rivers and streams is relatively small and often is not included in simplified hydraulic calculations.

The principal processes that control spatial distribution of dissolved oxygen (DO) in a river include oxidation of the biochemical oxygen demand (BOD), reaeration of DO from the atmosphere, and transport due to the river flow. A first-order reaction equation can be used to describe these processes as

$$U\frac{dC}{dx} = -k_d B + k_a(C_s - C) \tag{8.4.7}$$

where C = DO concentration,
$\quad B$ = BOD concentration,
$\quad C_s$ = saturated dissolved oxygen concentration,

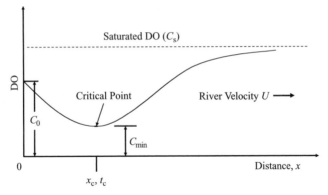

FIGURE 8.4.2   DO sag curve in a river.

k_d = deoxygenation rate constant of BOD, and
k_a = first-order reaeration rate constant of DO.

By assuming that BOD has a first-order degradation reaction with decay rate constant k_r, the solution to Equation 8.4.7 is the famous Streeter-Phelps (1925) equation:

$$C = C_s - \frac{k_d L_0}{k_a - k_r} \left(e^{-k_r x/U} - e^{-k_a x/U}\right) - (C_s - C_0)e^{-k_a x/U} \qquad (8.4.8)$$

The DO sag curve in Figure 8.4.2 shows distribution of DO concentration with distance downstream as given by the Streeter-Phelps equation. DO decreases to minimum C_{min} at critical distance x_c (or critical time $t_c = x_c/U$) and then increases downstream. Between the discharge point ($x = 0$) and the critical distance ($x = x_c$) deoxygenation exceeds the rate of reaeration and DO concentration decreases in the river. Beyond the critical distance ($x > x_c$), the rate of reaeration exceeds deoxygenation and DO concentration increases and approaches its saturated concentration.

8.4.4 PHYSICAL TRANSPORT PROCESSES IN LAKES AND RESERVOIRS

Even though the terms *lakes* and *reservoirs* are sometimes used interchangeably, a lake is commonly referred to as a natural surface water system, and a reservoir is often referred to as an artificial water body formed by engineering structures to impound water for flood control, navigation, recreation, power generation, and/or water supply.

The physical features of a lake include length, depth, volume, surface water area, drainage area, and inflow and outflow rates. The distinguishing characteristics of lakes include relatively low inflows and outflows and development of vertical stratification. Lakes also often become sinks of nutrients, sediments, toxins, and other substances originating from point and nonpoint discharges.

The major difference between rivers and lakes or reservoirs is in the speed of the water current. Water speeds are much smaller in lakes and reservoirs than in rivers. The flowing nature of rivers often results in well-mixed profiles in the vertical and lateral directions and rapid downstream transport, whereas the deeper, slower moving water in lakes tends to have stratified profiles in the vertical direction and horizontal gradient in the lateral direction. Thus, on the right side of Equation 8.4.1, the first term (advection term) is generally much larger than the second term (mixing term) in rivers, and the first term may be comparable to or even smaller than the second term in lakes.

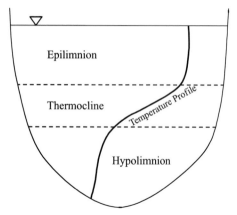

FIGURE 8.4.3 Typical temperature profile in a lake in summer.

Because of its relatively large velocity, a river usually can be well represented one-dimensionally. By contrast, a lake often has much more complicated circulation patterns, and its flow and mixing processes are largely affected by lake geometry, vertical stratification, hydrological conditions, and meteorological conditions. The ratio of lake volume (V) to lake outflow rate (Q_{out}) is the hydraulic detention time (V/Q_{out}), the time required to empty the lake water with outflow rate Q_{out}. Lakes and reservoirs tend to store water over seasons and years. Such a long detention time often makes internal chemical and biological processes in the lake water column and on the sediment bed significant, whereas these processes may be negligible in rapid-flowing rivers and streams.

Generally, lakes are not well mixed. A typical vertical temperature profile of a lake in summer is shown in Figure 8.4.3. The lake has strong stratification and forms three distinct layers. According to the Hutchinson (1957, 1967), the epilimnion is the uppermost layer where the temperature is generally constant with depth and is usually well-mixed by wind action. Beneath the epilimnion is the thermocline. The thermocline exhibits the maximum rate of decrease of temperature with respect to the depth and the minimum vertical mixing, as shown in Figure 8.4.3. The hypolimnion is the layer that extends to the bottom of the lake where the temperature steadily decreases. Water in the hypolimnion is much colder than epilimnion water. In simplified applications, lakes may be considered to have two layers: epilimnion and hypolimnion. Thermal behavior in lakes and reservoirs significantly affects engineering applications, such as water-quality management, power plant siting considerations, and thermal effects of power plant discharges on ecosystems.

The vertical temperature profiles of lakes vary according to season. At the end of winter, a lake is often well mixed from top to bottom as the result of winter meteorological conditions (e.g., cold air temperature, strong wind, and weak solar radiation). Lake stratification begins in spring and peaks in midsummer, as shown in Figure 8.4.3. Surface water temperature decreases gradually from the end of the summer and through the winter; eventually the lake temperature becomes vertically homogenous in winter.

8.4.5 PHYSICAL TRANSPORT PROCESSES IN ESTUARIES

An estuary is defined as a semiclosed coastal body of water that is freely connected to the open sea and within which seawater is measurably diluted with freshwater derived from land drainage (Pritchard, 1967). This classic definition has been extended to include certain areas of inland lakes that receive riverine water. For example, the backwater reaches draining into the Great Lakes are considered to be estuaries where the lake water intrudes upstream into the mouth of the rivers.

The primary factors controlling transport processes in estuaries are tides and freshwater inflows. Rivers are the primary source of freshwater to an estuary, which mixes with saline water as tidal elevation rises and falls. Tides are the fluctuation of the water surface above and below a datum plane. Tidal currents are the associated horizontal movement of the water into and out of an estuary. At high tide at the estuary mouth, the slope of the water surface forces water to rush into the estuary. At low tide, the reversal slope flushes water out of the estuary. In the estuary system depicted in Figure 8.4.4, a tidal river is a reach where, although some current reversal occurs, seawater has not penetrated to the region and the tidal water remains fresh.

As an example of tidal transport processes in estuaries, Figure 8.4.5 depicts the tidal elevation, tidal velocity, and salinity in Morro Bay, California, for 31 days (March 12 to April 11, 1998). It is clear that the tidal elevation [Figure 8.4.5 (top)] shows strong diurnal variability and has spring-neap variability of 15 days or so. Tidal elevation variabilities control tidal velocity [Figure 8.4.5 (middle)] and therefore control the transport processes of substances in the estuary, such as the salinity time series shown in Figure 8.4.5 (bottom).

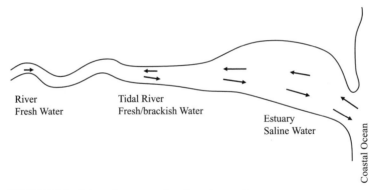

FIGURE 8.4.4   Schematic representation of an estuary system.

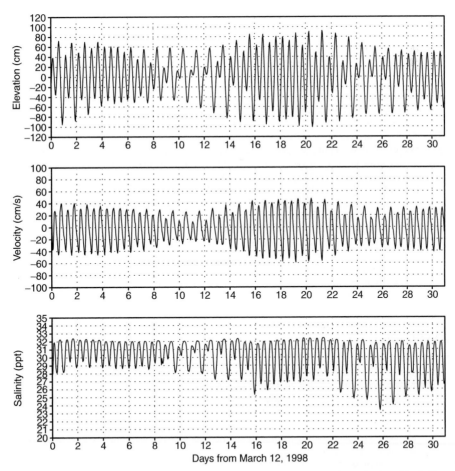

FIGURE 8.4.5   Tidal elevation, tidal velocity, and salinity in Morro Bay, California, for 31 days (March 12 to April 11, 1998). ppt = parts per trillion.

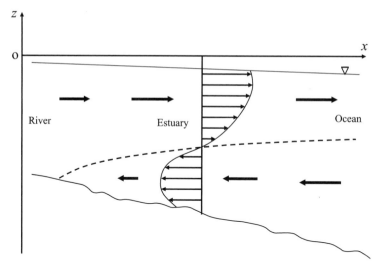

FIGURE 8.4.6   Vertical velocity profiles along the length of an estuary.

A principal feature of estuaries is the tidal-averaged circulation of two-layer flow in the vertical profile. Figure 8.4.6 indicates that the mean circulation consists of a net seaward transport in the surface layer and a net landward transport in the bottom layer. The surface flow is fresh to brackish, and the bottom flow is more saline and also may be colder. Differences in salinity and temperature result in differences in density and complex circulation patterns and significantly affect dissolved oxygen, toxic transport, and nutrient cycling.

In Figure 8.4.6, the seaward transport of freshwater in the surface layer and the landward transport of seawater in the bottom layer result in a convergence zone around the end of the saline water intrusion area, which creates vertical velocities in the estuary. This is a region in which the suspended solid concentrations are increased because of the sediment resuspension.

The net (tidal-averaged) velocity in an estuary can be less than 10 or 20 percent of the tidal velocity. The net flow in the estuary over a tidal cycle or a given period of several days can flush material out of the estuary and is a significant parameter in estimation of the fate and transport processes in the estuary.

The characteristics of estuaries are often determined by the differences in the chemical and biological characteristics of riverine water and seawater. In addition to the tides and river freshwater inflows, transport and mixing processes in an estuary are also affected significantly by wind action and estuarine geometry.

8.4.6   CONCLUDING REMARKS

The fate and transport in rivers, lakes, and estuaries are very complicated processes that include physical transport and chemical and biological kinetics. In this section, only the primary features and basic concepts have been discussed. To better describe and understand these processes, a mathematical model is indispensable. In past decades, significant progress has been made in numerical model development, in environmental data collection, and in computer software and hardware. These developments have helped environmental models become reliable tools for environmental management and engineering applications.

Environmental models of rivers, lakes, and estuaries can help us

1. Gain a better understanding of the fate and transport of chemicals by quantifying their reactions and transport: to describe the behaviors of pollutants after they are discharged into surface waters, we must use mathematical models to describe their fate and transport.

2. Determine the effect of exposure to chemical concentrations on aquatic organisms and/or humans in the past, present, and future.

3. Predict future conditions under various loading scenarios or management action alternatives: regardless of the quantity of monitoring data available, estimates of chemical concentration under different conditions are always desirable where field data do not exist and/or for future wasteloading scenarios.

Martin and McCutcheon (1999) summarized the mathematical models available for aquatic system modeling. Among the comprehensive three-dimensional mathematical models, the Environmental Fluid Dynamics Code is the public-domain model supported and sponsored by the U.S. Environmental Protect Agency (Hamrick and Wu, 1997).

REFERENCES

Council on Environmental Quality, *Environmental Quality, the Ninth Annual Report of the Council on Environmental Quality*, U.S. Government Printing Office, Washington, DC, 1978.

Hamrick, J. M., and Wu, T. S., "Computational Design and Optimization of the EFDC/HEM3D Surface Water Hydrodynamic and Eutrophication Models," in *Next Generation Environmental Models and Computational Methods*, Delich, G., and Wheeler, M. F., eds., Society of Industrial and Applied Mathematics, Philadelphia, pp. 143–156, 1997.

Hutchinson, G. E., *A Treatise on Limnology, Vol. I. Geography, Physics and Chemistry*, Wiley, New York, 1957.

Hutchinson, G. E., *A Treatise on Limnology, Vol. II. Introduction to Lake Biology and the Limnoplankton*, Wiley, New York, 1967.

Martin, J. L., and McCutcheon, S. C., *Hydrodynamics and Transport for Water Quality Modeling*, Lewis Publishers, Boca Raton, FL, 1999.

Pritchard, D. W., in *What Is an Estuary: Physical Viewpoint. Estuaries*, Lauff, G. H., ed., Publication 83, American Association for the Advancement of Science, Washington, DC, 1967.

Schooner, J. L., *Environmental Modeling: Fate and Transport of Pollutants in Water, Air, and Soil*, Wiley, New York, 1996.

Streeter, H. W., and Phelps, E. B., *Bulletin No. 146*, U.S. Public Health Service, 1925.

Thomann, R. V., and Mueller, J. A., *Principles of Surface Water Quality Modeling and Control*, Harper and Row, New York, 1987.

CHAPTER 8
ANALYSIS AND MODELING

SECTION 8.5

GEOSTATISTICS

Douglas D. Walker

*Dr. Walker is a hydrogeologist with Duke Engineering & Services,
Inc. He specializes in the development and application of
geostatistical approaches to groundwater contaminant transport
and has authored more than 40 papers and technical reports.*

8.5.1 INTRODUCTION

Geostatistics, as the name implies, is the application of statistical methods to earth science data (Matheron, 1960, as cited by Journel and Huijbregts, 1978). It borrows heavily from classic statistical analysis (i.e., that which is usually taught in undergraduate statistical courses), including descriptive statistics, hypothesis testing, confidence intervals, analysis of variance (ANOVA), multiple regression, etc. Geostatistics differs from classic statistics in that geostatistics includes specialized methods for spatial and temporal data in the earth sciences. In addition, classic statistics commonly assume that the data are independent, or uncorrelated, whereas geostatistics specifically addresses data that are correlated in space or time or both. Kriging, a class of geostatistical estimation methods for continuous spatial data, is so widely used that kriging is virtually synonymous with geostatistics. This artificially narrow view is unfortunate, because geostatistics includes a variety of methods for analysis, estimation, and simulation of various data types, such as continuous and discrete processes, lattice data, series analysis, multivariate ANOVA, and principal component analysis.

Geostatistical methods have found use in a variety of fields, including geosciences, engineering, geographical information systems, environmental pollution, biology, etc. Summarizing all geostatistical methods would be nearly impossible; thus this section skips over classic (univariate) statistical methods, for which many texts are available (e.g., Gilbert, 1987). Similarly, although ANOVA and multiple ANOVA, multiple linear and nonlinear regression, and principal component analysis are very useful in analyzing earth science data, they are not unique to geostatistics. This section is limited somewhat arbitrarily to methods for spatial data as continuous, discrete, or lattice processes. Readers interested in a wider survey of geostatistical methods should consult the very approachable text by Davis (1986).

Geostatistics has a rich toolbox of methods to analyze spatial data, including methods that address the following tasks:

- Data analysis (for populations, clusters or groups, distributions, outliers, moments and trends)
- Estimation (for the expected value of points/blocks and their uncertainty)
- Simulation (for alternative images, can be used in subsequent Monte Carlo studies)

The emphasis of this section is on what might be considered classic or mining geostatistics, which is the analysis of regionalized (continuous spatial) variables. This is followed by a summary of geostatistical methods for discrete processes.

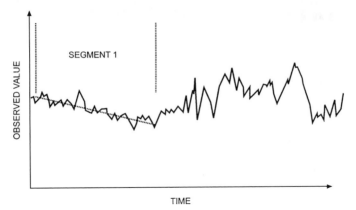

FIGURE 8.5.1   Conceptual relationship between scale and correlation. Samples from within segment 1 might tend to give a biased view of the entire sequence.

What, then, is so different about spatial data that geostatistical methods are required? Consider a one-dimensional sequence of measurements v_j, $j = 1, 2, 3, \ldots N$, as shown in Figure 8.5.1. If each measurement is independent, then each measurement gives important information and reduces the uncertainties about the entire sequence V. However, dependence in the form of spatial correlation is common in the earth sciences—that is, samples v_j tend to vary in relation to each other (i.e., they are correlated). In this case, the probability distribution of V is conditional on where one is looking in the sequence v_j, $j = 1, 2, 3, \ldots N$, called the multivariate or conditional probability case. Thus, spatial correlation means that the measurements clustered together in a finite interval such as segment 1 are somewhat redundant. If additional measurements are taken within segment 1, each new measurement adds little information about the characteristics of the entire sequence. In short, spatial correlation changes the statistical properties of the variables, affecting estimators and uncertainties (see Cressie, 1991, Chapter 1). The chief concerns of geostatistics are analysis of these spatial relationships and their incorporation into estimation and simulation for characterization and modeling of earth science data.

8.5.2   CONTINUOUS SPATIAL DATA

Data Analysis

The first task of the geostatistician is exploratory data analysis (EDA), whose objectives are to identify the data characteristics and to determine suitable models for use in subsequent prediction and simulation. EDA is "like numerical detective work, trying to confirm what we know of about the field being measured and looking for clues as to the internal structure of the measured variables" (Tukey, 1983). EDA is essentially organizing, plotting, and summarizing the data in various ways to let the data speak for itself. Classic statistics offers many EDA tools, including histograms, box plots, scatter plots of one variable versus another, etc., to help the analyst understand the measurements and gain insight into the underlying process. Geostatistics has several simple EDA tools to investigate the special characteristics of spatial data:

Post Plots.   These are simple maps of the measured value v plotted at s, the location of the measurement (Figure 8.5.2). The objectives of a post plot are to see if neighboring measurements have similar values (suggesting spatial correlation), to see if any one value is anomalously different than

81 +	77 +	103 +	112 +	123 +	19 +	40 +	111 +	114 +	120 +
82 +	61 +	110 +	121 +	119 +	77 +	52 +	111 +	117 +	124 +
82 +	74 +	97 +	105 +	112 +	91 +	73 +	115 +	118 +	129 +
88 +	70 +	103 +	111 +	122 +	64 +	84 +	105 +	113 +	123 +
89 +	88 +	94 +	110 +	116 +	108 +	73 +	107 +	118 +	127 +
77 +	82 +	86 +	101 +	109 +	113 +	79 +	102 +	120 +	121 +
74 +	80 +	85 +	90 +	97 +	101 +	96 +	72 +	128 +	130 +
75 +	80 +	83 +	87 +	94 +	99 +	95 +	48 +	139 +	145 +
77 +	84 +	74 +	108 +	121 +	143 +	91 +	52 +	136 +	144 +
87 +	100 +	47 +	111 +	124 +	109 +	0 +	98 +	134 +	144 +

FIGURE 8.5.2 Example of a post plot (from Issaks and Srivastava, 1989, used with permission).

the surrounding values (suggesting a possible error and further investigation), and to see if there are any field-wide trends. Post plots can be refined by marking or coloring the measurements that fall within certain ranges (e.g., the 10 percent highest or lowest values). A variation of the post plot is to transform the $j = 1, \ldots, N$ measurements to a zero or a one depending on whether it is greater than a certain cutoff value v_c:

$$i(s_j; v_c) = \begin{cases} 1 & \text{if } v(s_j) \le v_c \\ 0 & \text{if } v(s_j) > v_c \end{cases} \tag{8.5.1}$$

known as the indicator transform (Journel, 1989). A post plot of the indicator tranforms can quickly reveal data patterns and, if cutoff v_c is chosen to be a critical value such as a regulatory compliance limit, then the resulting map can also help guide additional sampling and management decisions. The indicator transform also has an equivalent probablistic definition of $i[s_j, v_c] = $ Prob$[V(s_j) \le v_c]$—that is, the indicator transform is the probability that the variable at location s_j is less than the cutoff value. This probabilistic statement of the indicator transform can be quite powerful. If, for example, the measurements are on a regular grid (lattice), then indicator transforms can be shaded for each block in the grid, resulting in a pixelated map (Figure 8.5.3). The indicator transform for several cutoff values can be used, creating a sequence of maps summarizing the probabilities of exceeding critical cutoff values at various locations in the field. This sort of probability mapping is related to estimation via indicator kriging and indicator simulation, which are discussed later in this section.

Clustering and Global Estimation. Suppose that a sampling program detects contaminated areas on an initial pass, and follow-up sampling is performed to confirm the contamination levels and refine the areas. This sampling scheme can result in clustering samples in areas of high contamination that can give the arithmetic mean of the samples an upward bias. This bias can be reduced by appropriately downweighting the clustered observations. One approach to declustering weights is cell declustering (Journel, 1983), which begins by overlaying the field with a grid of square cells. The declustering weight for each sample is 1/(number of samples in the cell) and is calculated for several

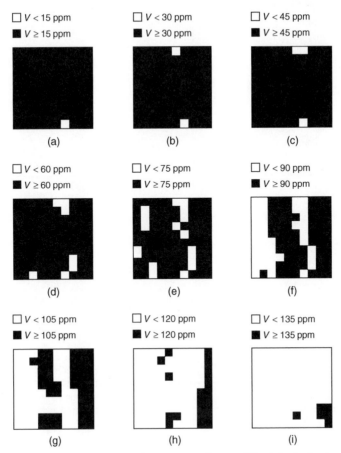

FIGURE 8.5.3 Indicator transform maps at various cutoff levels for the data in Figure 8.5.2 (from Issaks and Srivastava, 1989, used with permission). ppm = parts per million.

alternative grid placements and cell sizes. The optimal cell declustering weights are those whose cell size minimizes the field-wide weighted mean, correcting the initial bias introduced by preferentially clustering samples in areas of high contamination. Declustering weights can also be used to correct estimates of the variance, median, and the histogram. Other declustering algorithms include Thiessen polygons (Issaks and Srivastava, 1989) and radial declustering (Henley, 1981).

h-scattergram. This graphic EDA tool is a method to examine spatial correlation. The measurements are sorted into pairs separated by the distance $h = |s_j - s_k|$, sometimes referred to as the lag spacing. The set of lag pairs $[v(s_j), v(s_j + h)], j = 1, 2, \ldots, N - 1$, are plotted as cartesian coordinates, and, if the plotted values fall close to a line with a 1:1 slope, this suggests that the measurements are spatially corrrelated. The plot can be repeated for various lag distances $2h, 3h, \ldots$ to see how spatial correlation decays with increasing distance (Figure 8.5.4). Note that measurements commonly are not on an even grid spacing, and so the data pairs are classed by using tolerance Δ for lag window $H = h + \Delta$. For two- and three-dimensional data, the data pairs can be sorted by direction to see if the correlation is anisotropic. The h-scattergram can also be used to check for outliers in the set of measurements (Issaks and Srivastava, 1989, p. 58).

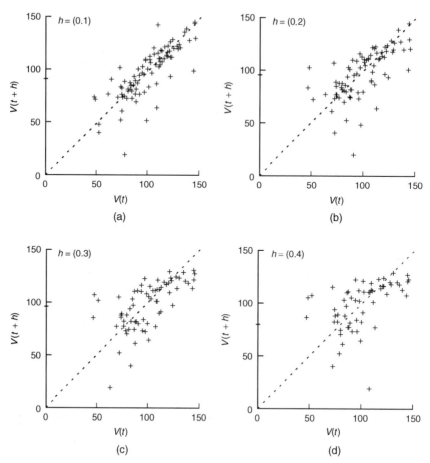

FIGURE 8.5.4   *h*-scattergrams for the data in Figure 8.5.2 (from Issaks and Srivastava, 1989, used with permission).

Note about Outliers.   Some authors advocate removing anomalous high or low measurements from a data set on the basis they that are outliers—i.e., that they do not conform to the pattern established by the remainder of the data. Outliers can strongly affect estimators of the mean, variance, and spatial correlation, and they might reasonably be edited from the set of measurements in order to estimate the general behavior of the variable. Unfortunately, removing outliers may remove the most important feature in the field (e.g., the maximum concentration). Outlier management and the related problem of censoring via measurement limits are beyond the scope of this section. Generally speaking, nonparametric methods are more resistant (handle outliers better) than parametric methods (Gilbert, 1987, Section 15). The indicator approach is currently the most viable nonparametric method for continuous spatial variables.

Variography.   Although the *h*-scattergram is a useful EDA tool, subsequent estimation and simulation require determining the spatial correlation function. In mining geostatistics, many geostatistical estimation methods are based on minimizing the squared error of the estimate. One consequence of this minimization is that the resulting estimators require knowing the squared difference of data pairs as a function of lag spacing, commonly expressed as the variogram:

$$2\gamma(h) = \text{Var}[V(s) - V(s + h)] = E[\{V(s) - V(s + h)\}^2] \tag{8.5.2}$$

i.e., the variance of pairs of measurements separated by distance h. In practice, the variogram is never known and must be estimated from the sample $v(s)$ for each lag using:

$$\hat{\gamma}(h) = \frac{1}{2N_h} \sum_{j=1}^{N_h} \{v(s_j) - v(s_j + h)\}^2 \tag{8.5.3}$$

Equation 8.5.3 is the traditional estimator of the experimental semivariogram and is computed for number of lag distances h to examine the decay of correlation with distance. The constant $1/2$ arises from the mathematical development of the variogram and is the cause of much confusion [$2\gamma(h)$ is the variogram, and $\gamma(h)$ is the semivariogram]. Note that because $s = (x, y, z)$, the experimental semivariogram can be calculated in any direction for any pair separated by h, or it can be computed for pairs oriented in specific directions. The latter option allows the analyst to examine the possibility of anisotropy in the correlation. Several alternative estimators have been proposed to assist in estimating the semivariogram for specific data types (Journel, 1988).

Variography does not end with computing the experimental semivariogram for each lag. The next step is to fit a continuous function to the estimates, called the model semivariogram. There are several traditional functional forms for the model semivariogram, such as the spherical and exponential models (Figure 8.5.5). The model parameters (range, a, and sill, C_1) are chosen so that the model closely follows the experimental semivariogram. Nonzero behavior of a semivariogram near the origin is called the nugget, which represents small-scale variability (e.g., sampling error). The models also may be combined to create a nested model semivariogram (e.g., nugget plus spherical model) and may be anisotropic as well. Representing the semivariogram as continuous function is not just a convenience; it is necessary to ensure that the resulting kriging system of equations can be solved (i.e., that the resulting covariance matrix will be positive definite, and therefore a solution for the kriging system of linear equations will exist). It is also possible to analyze the correlation of two or more variables with each other by using the cross-variogram (the simultaneous variogram of two variables as a measure of spatial cross-correlation).

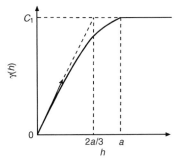

(a) spherical model $\gamma(h) = C_1 \left[\dfrac{3}{2}\left(\dfrac{h}{a}\right) - \dfrac{1}{2}\left(\dfrac{h}{a}\right)^3 \right]$ for $h < a$

$\gamma(h) = C_1$ for $h > a$

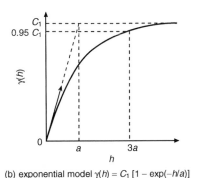

(b) exponential model $\gamma(h) = C_1 [1 - \exp(-h/a)]$

FIGURE 8.5.5 Model semivariograms and their functional form. (a) Spherical model with practical range $= a$. (b) Exponential model with practical range $= 3a$.

The variogram is a measure of the spatial correlation versus separation distance, and it is related to the important concept of stationarity. If the process and all possible functions of the process are invariant over the field, the process is said to be strictly or first-order stationary (i.e., statistically homogeneous or without trends). When only the mean and covariance are homogeneous, the process is second-order stationary. Second-order stationarity is important because it limits the theoretical maximum of the semivariogram to the field-wide variance—i.e. $\lim_{h \to \infty} \gamma(h) = \sigma^2$. Thus, if the analyst believes the process is second-order stationary (has no trend), then the stable maximum (sill) of the model semivariogram should equal the field-wide variance. The correlation length, also known as the effective or practical range, is the lag distance where the semivariogram reaches this maximum level. The correlation length is a function of the range parameter, a, of the chosen functional form of the model semivariogram (Figure 8.5.5).

In mining geostatistics, the process of fitting a model semivariogram to the experimental estimates is also known as structural analysis, which implies that the model semivariogram should be compatible with the underlying geologic structure. For example, the geologic process may create deposits that are elongated in a particular direction; in this case, correlation may persist over a longer distance in that same direction, and consequently the model semivariogram would have a longer range in that direction (so-called geometric anisotropy). Similar sorts of underlying structure can be found in rainfall data, hydraulic conductivities, and contaminant deposition, which may not be indicated by the experimental semi-variogram when the data sets are small or when outliers are present. Thus, although it is tempting to rely on automatic algorithms to fit the model semivariogram (e.g., nonlinear regression or iterative cross-validation), there is no substitute for an intimate understanding of the measured process. The final test of variography is not necessarily the best possible model fit to the experimental semivari-ogram but that the model semivariogram results in estimates and simulations that are compatible with the process being measured.

Estimation

Regardless of how densely a field is sampled, it is always necessary to interpolate between measured locations or determine the average value within specific areas. As a statistical question, we want an optimal estimate of a point or a block value and some idea of its uncertainty. At first glance, it is tempting to fall back on the answer provided by classic statistics: use the arithmetic mean (or median) of the sample and the variance of the mean as its uncertainty. As discussed above, the arithmetic mean ignores spatial correlation in the data, which can lead to a biased estimate. Another possible estimator is the arithmetic mean of the measurements within a finite window around the location to be estimated. This moving average allows some local bias to affect the estimate but still ignores the spatial correlation among data in the window. One alternative is to weight the arithmetic mean:

$$\hat{V}(s_o) = \sum_{j=1}^{N} w_j v(s_j) \tag{8.5.4}$$

that is, a weighted average of the measurments, whose weights w_j would be chosen to emphasize the measurements $v(s_j)$ that are closest to the point of interest s_o. This simple idea of a weighted average is the basis of the most widely used estimators in mining geostatistics, as discussed in the following paragraphs.

Ordinary Kriging (OK). Kriging uses a weighted average of the form given in Equation 8.5.4 and attempts to determine the optimal weights by minimizing the squared estimation error:

$$\sigma_{OK}^2 = E[\{V(s_o) - \hat{V}_{OK}(s_o)\}^2] \tag{8.5.5}$$

subject to the unbiasedness condition:

$$\sum_{j=1}^{N} w_j = 1 \tag{8.5.6}$$

Optimizing the weights by minimizing the squared error (Equation 8.5.5) leads to the system of linear equations:

$$\sum_{j=1}^{N} w_j \gamma(s_j - s_k) + \lambda = \gamma(s_o - s_k) \qquad \text{for } k = 1, N \tag{8.5.7}$$

(Journel and Huijbregts, 1978, p. 304). λ is a Lagrange multiplier, a constant that allows incorporating the unbiasedness constraint of Equation 8.5.6. Equations 8.5.6 and 8.5.7 are known as the ordinary kriging (OK) system. If we solve this system of equations, e.g., by gauss elimination, we get the least-squares optimal weights for the estimator $\hat{V}_{OK}(s_o)$. The semivariograms on the right hand side of Equation 8.5.7 account for the spatial correlation between the estimate and the measured values, whereas the left-hand-side of Equation 8.5.7 accounts for spatial correlation among measurements. The weights w_j provide the intuitive results: the estimator is more heavily weighted to the closest measurements, the weights reflect the anisotropy and range of correlation represented by the model semivariogram, and clustered observations are downweighted relative to uniformly spaced observations. Unlike most alternative estimators, the OK system also yields the estimation error, expressed as a variance:

$$\sigma_{OK}^2(s_o) = \sum_{j=1}^{N} w_j \gamma(s_o - s_j) + \lambda \tag{8.5.8}$$

Through the semivariogram term $\gamma(s_o - s_j)$, the estimation variance is a function of the configuration of samples around location s_o rather than being constant for the entire field. The constraint 8.5.6 allows for simultaneous estimation of the mean of the process from the set of measurements, which is the distinguishing characteristic of OK (as opposed to simple kriging, universal kriging, etc.). When OK is applied to a subset of the data within a finite neighborhood around the estimation location, the OK estimates include local trends in the mean in the estimates. This attribute of OK comes at an often-unadvertised price: the OK estimator depends on the choice of neighborhood and the spatial arrangement of the measurements. There are various methods of constructing neighborhoods by different search strategies, but ultimately the choice of the neighborhood is somewhat subjective and not always well-documented. The size of the neighborhood and search strategy are usually justified by cross-validation and the appearance of the resulting estimated surface.

OK is an exact interpolator (Delhomme, 1978), meaning that if an OK estimate coincides with the location of a measurement, the OK estimate is equal to the measured value. That is, $\hat{V}_{OK}(s_o) = v(s_j)$ when $s_o = s_j$, even when the model semivariogram indicates sampling error (has a nonzero nugget variance) (Journel, 1989, p. 16). This exactitude property also results in zero OK variance at all measurements. Although it is possible to define the variogram and to allow the OK estimates to be inexact (i.e., to explicitly include measurement error), this should be used cautiously because it can lead to unstable solutions of the OK system of equations (Cressie, 1991).

Kriging is sometimes referred to by the acronym BLUE (best linear unbiased estimator). Kriging yields the best estimates in the sense that weights are optimal, minimizing the estimation variance (Equation 8.5.8). The form of the estimator is simple addition (linear combination) of weighted values, and the unbiasedness condition (8.5.6) ensures that the mean error is zero. Acronyms aside, it is reasonable to ask if minimizing the squared error is sufficient criterion to say that we are using the best of all possible estimators. In general, if the variable is normally distributed, the minimum squared-error criterion is appropriate and ensures that the kriging weights are uniformly optimal. Suppose, however, that variable U has a lognormal distribution; in this case, we would work with the transformed variable $Y = \ln(U)$, computing the estimates as the weighted linear combination of the transformed measurements y_j. In this rather common case, OK using the untransformed variable is not the minimum-varianced best estimator, and OK using the transformed variable Y is not a linear estimator. Thus, although the acronym BLUE is a correct description of kriging in many circumstances, it is not a guarantee that the OK estimator is always the best choice (Weber and Englund, 1994).

Indicator Kriging (IK). It is not unusual for the underlying distribution of the variable to be either simply unknown or not normal. In such cases, OK may not be an appropriate estimator for $V(s_o)$. Rather than trying to directly estimate $V(s_o)$, IK first estimates the conditional distribution of $V(s_o)$ and then takes the expected value of that estimated conditional distribution as a nonparametric estimator of $V(s_o)$. In mathematical terms, the IK estimator of $V(s_o)$ is

$$\hat{V}_{IK}(s_o) = \hat{E}[V(s_o) \mid v(s_j)] = \int_{-\infty}^{+\infty} v \cdot d\hat{F}[s_o; v \mid v(s_j)] \tag{8.5.9}$$

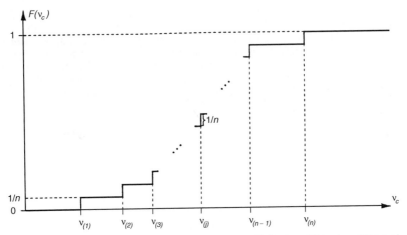

FIGURE 8.5.6 ccdf discretely approximated by indicator cutoffs (Issaks and Srivastava, 1989, used with permission).

That is, the IK estimator of $V(s_o)$ is $\hat{E}[\cdot]$, the estimate of the expected value of $V(s_o)$ conditional to the surrounding measurements $v(s_j)$. This conditional expected value, or conditional mean, is just the expected value of V computed with $\hat{F}[s_o; v \mid v(s_j)]$, the estimated conditional cumulative distribution function (ccdf) of V at s_o. The trick of IK lies in recognizing that the ccdf of V at s_o is related to the probabilistic interpretation of the indicator transform:

$$i[s_o, v_c] = \text{Prob}[V(s_o) \leq v_c] = F[s_o; v_c \mid v(s_j)] \tag{8.5.10}$$

That is, the indicator transform of $V(s_o)$ with respect to cutoff level v_c is one point on the ccdf curve (Figure 8.5.6). We don't know $i[s_o, v_c]$, but we can estimate it as a weighted average of the indicator transforms of the surrounding measured values $v(s_j)$:

$$\hat{i}[s_o, v_c] = \sum_{j=1}^{N} w_j i[s_j, v_c] \tag{8.5.11}$$

Weights w_j can be determined by solving the kriging system of equations developed from $2\gamma(v_c; h)$, the model variogram of indicator transforms of the data using cutoff v_c. IK is repeated for several cutoff values v_c so that the integral of Equation 8.5.9 can be evaluated numerically. In short, the IK approach is to transform the measurements, fit model semivariograms, and krige several indicator transforms to discretely approximate the conditional distribution; then post-process the discrete ccdf to estimate the expected value at unmeasured locations (Deutsch and Journel, 1998, section IV.1.9).

The IK approach has many advantages: indicator variograms are robust and resistant, and IK gives both an estimate and the distribution of estimation errors without distributional assumptions. However, the indicator transform results in a loss of information about $V(s)$ that can be recovered only by laboriously repeating the transformation, variography, and kriging process for many cutoff values. Further, the range and density of observations must be such that the chosen cutoff values create an increasing estimated ccdf whose maximum value is 1.0—that is, order relations must be satisfied. Even in the most ideal of data sets, many ccdf's will require order relations corrections, typically an average correction of 1 percent for 1/2 to 2/3 of the estimated field (Deutsch and Journel, 1998, p. 81).

Other Estimators. The OK and IK estimators are widely used in the earth sciences but are by no means the only approaches to estimation. Various types of kriging estimators have been developed to address cross-correlation to a second variable (cokriging), data with large-scale trends (universal kriging and intrinsic functions of order K), etc. There are also a variety of methods that initially appear to be completely unrelated to kriging but can be shown to be mathematically equivalent under certain conditions (e.g., splines, spectral methods, autoregressive methods) (see Cressie, 1989).

Analysts also use several methods that predate the development of kriging. One such estimator is inverse distance weighting (ID), which uses the same weighted average as Equation 8.5.4 but assumes that the weights are of the form $1/h^a$. Typically, $0 \leq a \leq 2$, so the weights decrease with increasing distance from location of the estimate. The ID estimator is quite robust under a variety of test cases but the power a is arbitrary and the weights do not account for clustering among the measurements (Weber and Englund, 1994). Trend-surface fitting is another early estimation method; it fits a simple polynomial function (a trend surface) to the data by least-squares regression (either globally or within neighborhoods). Yet another approach is to divide the region into triangles whose vertices correspond to measurement locations (a Delauney triangulation). The estimate for the variable within each triangle is determined by the plane equation passing through the measured values at the vertices (Watson and Philip, 1984). And, if all these methods are not sufficient, each application of an estimator requires subjective judgments on the part of the analyst; thus it can be said that there are as many estimation methods are there are analysts (Englund, 1990).

Each estimation method can be shown to be better in some sense than all the rest under some circumstances. However, kriging methods have the advantage of being based in mathematical probability instead of being developed on an ad hoc basis (Journel, 1986; Weber and Englund, 1992 and 1994). This theoretical basis gives kriging both optimal estimates and the uncertainty of those estimates (expressed as as the kriging variance). Kriging also benefits from well-established methods for structural analysis and simulation.

Special Issues.   The preceding paragraphs have presented estimators in terms of the interpolated value at a point between measured values (also called a punctual estimate). Managers may be more interested in knowing $V_B(s_o)$, the average over a portion of a field centered at location s_o, (e.g., the average contaminant concentration for a block of soil to be treated). In the case of the OK and IK approaches, the point estimators can be extended to block estimates using $\bar{\gamma}_B(s_o - s_k)$, the semivariogram averaged over the block. A related volume-averaging problem is that of regularization, which is a method for reducing a variogram based on point samples to a variogram averaged over larger blocks. The reader is referred to Journel and Huijbregts (1978, p. 77) for details about volume or area averaging and the associated regularization.

Block estimates, block variances, and the underlying distributional assumptions give managers a powerful tool to estimate the expected concentration of a volume considered for remediation [a so-called volume of selective remediation (VSR)] (Desbarats, 1996). Decision and risk analysis for sampling and remediation of VSRs are active research areas in geostatistics, and they attempt to use estimation and simulation to evaluate regulatory compliance and limit risk. Although a number of geostatistical sampling and risk assessment approaches have been proposed, there still is no widely accepted approach (Rouhani et al., 1996; McKenna 1996). Note also that the simulation of block-scale hydraulic conductivities for numerical groundwater models is a rather complex problem, because a simple average of small-scale conductivities may not be adequate in representing flow at a larger block scale. This is problem is referred to as upscaling and is currently a topic of intense research interest (Renard and de Marsily, 1997).

Given the bewildering array of available estimation methods, the analyst needs a quantitative method to compare alternative estimators and parameter choices. Unlike the regression analysis of classical statistics, very few examples of hypothesis testing for model parameters can be found in mining geostatistics. This is because the spatially correlated probability models do not easily lend themselves to simple analytical expressions for test statistics. Geostatisticians usually compare estimators and parameter choices by cross-validation to the set of measurements. Cross-validation is a sample reestimation exercise, where each measured value is estimated by using the remaining measurements. The cross-validation errors of measured versus estimated values are compiled for each candidate estimation method, and the method with least error is assumed to be the best with respect to the measured data. Cross-validation is somewhat subjective, however, because the decision is sensitive to choice of error statistic. (Weber and Englund, 1992, 1994; Delhomme, 1978). Additionally, cross-validation can reveal only which estimator is most appropriate for the measured data and not which estimator is appropriate for the underlying process. Consequently, cross-validation alone is insufficient to discriminate between candidate estimators (Issaks and Srivastava, 1989).

Simulation

The preceding paragraphs examined the estimators of a variable at particular locations in the field of interest. Such estimators may be used to map the expected value of the variable for use in planning additional sampling or remediation efforts. But intuitively we know that the best estimate is only a partial image of the field. For example, it is easy to anticipate some variability around these estimated concentrations and, knowing that some areas are just below the compliance limit, plan on remediating slightly more than the areas expected to exceed that limit. But effective decision making and project planning require that we accurately estimate how much more is necessary to remediate to reduce risks while minimizing costs. One solution to this problem is stochastic simulation, which creates multiple images with a realistic level of variability that can be used in "what if" analyses. For example, a stochastic simulation of soil contamination may be used to create alternative images of possible distributions of contamination, each of which is compatible with (conditioned on) the observed data. Managers or regulators can then look at a set of such simulations and decide to remediate (or resample) those areas that exceed a compliance limit in a certain proportion of the simulations.

Stochastic simulation creates alternative images of a process that sacrifices the optimality of estimates to retain a realistic level of variability and spatial patterns. In geostatistics, the problem is complicated somewhat by both spatial correlation and the need to have the simulated fields incorporate the existing measurements. This is called conditional geostatistical simulation and is solved by combining kriging and classic statistics to create fields that both honor (are conditioned on) the existing measurements and have a realistic noise level. The result is a set of equally likely images (realizations) of the field of interest (Figure 8.5.7) that can be used to get an understanding of the variability of the field. One important use of simulation is to generate alternative images of input parameters for Monte Carlo simulation of complex process models. The effects of spatial variability and spatial correlation in the input parameters can then be propagated through the process model to assess the probability distribution of possible outcomes. A typical example is stochastic simulation of hydraulic conductivity in numerical groundwater models, where multiple simulations of conductivity fields are used to estimate uncertainties about travel times, travel paths, and concentrations. In the specific case of groundwater contaminant transport modeling, the advective and dispersive transport are functions of the variability of hydraulic conductivity, so that transport simulations based on smooth kriging estimates may yield biased results. Geostatistical simulation is appropriate for simulating the hydraulic conductivity fields for model input because it creates a realistic level of variability in the simulated fields.

The process of geostatistical simulation is deceptively simple: generate the expected field of best estimates via kriging, and then add noise to the field based on the kriging variance (Equation 8.5.8). This is complicated by the fact that the error distributions are not independent—that the random error at any simulated point is conditional on the surrounding measurements and simulated data. Thus, the methods for simulation are frequently rather complex and, to the uninitiated, rather intimidating. There are a wide variety of simulation methods that may be more or less appropriate for the field size, data density, and underlying process. Methods can be combined to create precisely the image of interest; for example, a categorical simulation can divide the field into two media, each of which can then be filled in with different stochastic processes. An initial simulation may also be post-processed with a second method to superimpose additional features, smooth the initial simulation, or train the initial simulation to resemble a reference image. Various approaches to simulation have been developed, and this is currently an active area of research. The details of particular algorithms are beyond the scope of this section; readers interested in learning about specific methods should refer to Deutsch and Journel (1998).

a - Reference Image

b - Location Map of Sample Data

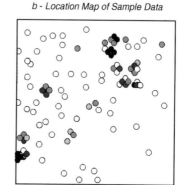

c - Kriging

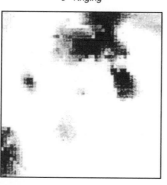

d - Simulation

d - Simulation

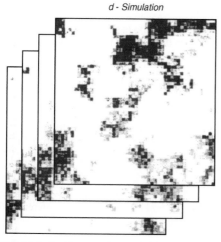

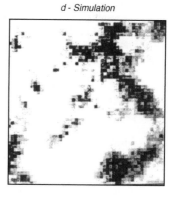

FIGURE 8.5.7   Reality, sampling, estimation, and simulation. Reference image (a) has been sampled (b) and then estimated by kriging (c) and simultated (d). Comparing these images, note that kriging yields a smoothed image (a versus c), where stochastic simulation has a realistic level of variability or roughness (a versus c versus d). Note also that the sampled locations (b) remain the same between realizations (d), a result of conditioning the simulation to the observed data. (Deutsch and Journel, 1998, used with permission).

In general, simulation algorithms may have one or more distinguishing characteristics. Simulations may be conditional (honor the existing measurements) or unconditional (governed only by the input process model). Many simulation methods produce fields with normal probability distributions (e.g., matrix decomposition, turning bands, sequential gaussian, fast Fourier transformation), and others may produce nonparametric fields (sequential indicator simulation, simulated annealing). Each of these simulation methods can create fields with remarkably dissimilar characteristics; the only thing they have in common is that, if misused, the resulting simulated fields may be unlike the intended process—and sometimes spectacularly so. Similar to fitting a variogram model, the choice of the simulation algorithm is ultimately judged by the similarity of the resulting fields to the process of interest.

Note about Lattice Processes.   The preceding comments on numerical models bring up a specific type of spatial data known as lattice processes, which share the characteristics of both continuous and discrete processes. A lattice process consists of discrete values assigned to a fixed patchwork or lattice of discrete areas. The lattice may be regular and uniform, such as the colors of pixels or voxels in remote sensing and Geographic Information System images, or they may be irregular, such as the populations of regions on a political map. Unlike continuous spatial processes where an infinite number of locations can be considered, the locations of a lattice process are finite, equal to the number of cells in the lattice. Although lattice processes might deserve a separate section of their own, many of the exploratory data analysis and modeling approaches used for continuous spatial processes are also used for lattice processes (e.g., post-plots, trend analysis, variography, block kriging, etc.).

Lattice processes occur frequently as uniform regular grids, so the locations and spatial relationships between data are entirely defined by the row and column indices of the lattice. This organization is particularly well-adapted to a class of models borrowed from time series analysis, known as autoregressive (AR) models. For example, suppose we are interested in analyzing a city district map to see if air quality in one district affects the air quality of neighboring districts. A simple AR model for a lattice process of the districts can be represented as:

$$V_j = \mu_v + \sum_{k=1}^{N} \delta_{jk} \, \rho_{jk} \, v_k + \varepsilon_j \tag{8.5.12}$$

That is, the air quality V in district j is the combination of the mean air quality μ_v, the measured air quality in district k, the correlation ρ_{jk} between districts j and k, plus random error ε_j. The identity matrix δ_{jk} indicates which of the districts are neighbors:

$$\delta_{jk} = \begin{cases} 1 & \text{if cells } j \text{ and } k \text{ are neighbors} \\ 0 & \text{if otherwise} \end{cases} \tag{8.5.13}$$

Neighbor relationships can be more complex than adjacent lattice cells (e.g., the artificial links between inner-city districts and bedroom communities created by expressways). Such relationships can be tested for significance by carefully defining the elements of δ_{jk} and then testing for the significance of the resulting model (Cressie, 1991, pp. 427, 440). One special lattice process property is connectivity, or the tendency for high (or low) values to be adjacent to each other. For example, consider the block hydraulic conductivities of numerical groundwater models. The occurrence of high hydraulic conductivities in adjacent numerical grid blocks can create preferential flow transport simulations paths that may dominate. Individual lattices may be categorized by the number and extent of neighboring cells with similar values for further understanding how this connectivity affects the flow and transport (Deutsch, 1998). A special case of lattice processes are data sequences classed into mutually exclusive states, such as rock types along boreholes (sandstone, shale, or limestone) or transects of soil samples (over or under compliance limits). This type of lattice process might be modeled as a Markov chain, a method that examines the probability of changing from one state to another. Similar to AR models, we can propose various models for a Markov chain and test them versus the observed process (Davis, 1986, p. 150).

8.5.3 DISCRETE PROCESSES

Although geostatistics emphasizes continuous processes, earth science data occur frequently as discrete processes. Discrete processes can be categorized as point processes, such as the positions of trees in a forest, and marked processes, such as the occurrence and magnitude of earthquakes. The chief tasks in working with discrete processes are exploratory data analysis (i.e., determining the underlying process or pattern) and simulation. Estimation of unsampled objects is unusual but is simply the expected frequency of the inferred underlying process (i.e., the expected number of objects in an unsampled portion of a field is the same as the frequency observed in the sampled portion of the field).

Point Processes

Point processes are sets of objects that occur in space without an associated magnitude; most of the analyst's effort is spent evaluating the underlying process that determines the objects' locations. Although the underlying physical process may never be fully understood, we may represent the resulting spatial distribution by a simple stochastic function. For example, the cause of a cancer death may be heredity, occupation, or bad luck. However, a simple point process model of the spatial and temporal pattern of cancer deaths might help identify important risk factors. Note that physical objects may not be points in the mathematical sense; point process models are simply models of the reference location of the object, typically the center of mass.

For the sake of convenience, point processes are categorized as purely random, aggregated (clustered), or uniform. Spatially random point processes can be modeled by using the expected frequency of occurrence, $\lambda = E$ [(number of objects)/(unit area)]. Purely random spatial patterns are also known as Poisson processes because the probability of r objects within given area A is given by the Poisson distribution:

$$\text{Prob}[r \mid A] = \frac{e^{-\lambda A}(\lambda A)^r}{r!} \tag{8.5.14}$$

Poisson processes have no fixed pattern and the relative location of an object is independent of the location of all other objects; thus clusters are as likely as empty areas. Complete spatial randomness can be deceiving at first glance, because the occasional clusters may lead the analyst to believe that there is an underlying pattern (Figure 8.5.8).

At the opposite extreme from Poisson processes are uniform point patterns. As the name implies, objects are distributed uniformly over the field on a grid (triangular, square, hexagonal, etc.). Although such extreme organization is rare in nature, uniform point patterns are found in city housing, manufactured materials, agricultural fields, cooling joints in rock formations, etc. The location of each object in a pure, uniform point process is completely determined by the underlying grid that defines the position of each object; there is no randomness.

Between the two extremes of Poisson and uniform point processes lie aggregated point processes, where the objects occur in clusters of various densities and sizes. Aggregated point processes are very common in biological and earth sciences. Examples include animal and plant populations (e.g., parents and offspring) and contagious disease (carriers and infected individuals). Depending on the degree of randomness and organization, aggregated point processes can be represented by combining models of randomness and uniformity. As an example, consider a set of randomly located clusters in which the number of objects in a cluster are independent and logarithmically distributed. Under these conditions, the probability of r objects in area A is given by the negative binomial distribution:

$$\text{Prob}[r \mid A] = \frac{(k+r-1)!}{r!(k-1)!} p^k (1-p)^r \tag{8.5.15}$$

If the clustering parameter k is positive and small, the process is highly clustered; if k is 0, then there is no clustering (i.e., purely random locations) and the negative binomial reduces to the Poisson

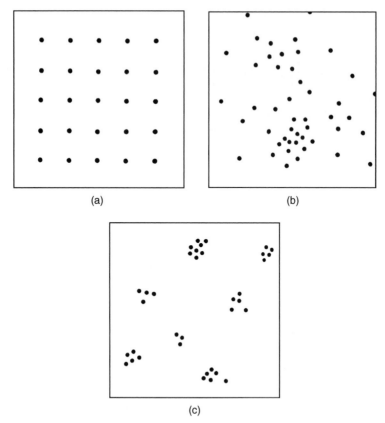

FIGURE 8.5.8   Typical point processes: (a) uniform, (b) random (Poisson), and (c) clustered (aggregate) (Davis, 1986, used with permission).

distribution (Davis, 1986, p. 305). There are many compound models that may be used to represent aggregated processes (e.g., inhibition processes, Neyman processes, etc.). Likewise, the process need not be stationary (i.e., the frequency and clustering parameters may vary in space).

Exploratory data analysis of point processes focuses on comparing the observed process to various candidate models. Thus the analyst estimates the parameters of frequency and clustering and uses a χ^2 goodness-of-fit test to try to reject the hypothesis that "the observed process is an X process." For practical purposes, one would like to have the minimum necessary number of aggregated models and parameters. Consequently, the analysis begins by testing the simplest of models (the Poisson process) and then progresses to more complex models until the hypothesis cannot be rejected (i.e., the observed process is not significantly different from the candidate process). The final model selected to represent the process depends on the imagination and patience of the analyst in selecting candidate models.

There are several methods for estimating the parameters of point processes. One method is to randomly sample small areas (quadrats) and then estimate the frequency λ as the arithmetic mean number of objects in the quadrats. In practice, random quadrat estimators are sensitive to the quadrat size; it is possible to choose a quadrat size that will detect a desired distribution. If the process is Poisson, then the optimal quadrat area should be approximately $1.6/\lambda$. If the process is clustered, the quadrat should be approximately the same size as the clusters. However, both of these rules of thumb imply that we know something about the process a priori. An alternative to random quadrats is to use a telescoping mesh of contiguous quadrats covering the entire domain, but this requires exhaustively sampling the entire field (Cressie, 1991, p. 592). The preferred alternative to quadrat methods is to estimate the frequency by one of several distance methods. For example, the frequency can be estimated from the distance between objects and their nearest neighbors or the distances from an object to a randomly drawn transect. A large variety of test statistics and indices have been developed for the distance methods, but their discussion is beyond the scope of this section (see Cressie, 1991, p. 602). Distance methods generally require correcting the estimates for the effect of encountering the edges of the field (Davis, 1986, p. 309).

Marked Processes

Marked processes are discrete processes in which the objects have one or more associated variables of interest, or marks, which can be measured only at the location of interest (e.g., the spatial relationship between earthquake magnitudes). Marked processes can be seen as extensions of point processes and thus commonly are represented as aggregates of point process models. For example, suppose that we are interested in discrete regions of contamination in a field; a point process could represent the center of contamination and the contaminant concentration could be an independent random function. The similarity of marked processes to point processes also implies that many of the point process estimation and testing techniques can be used to estimate and test marked process models. For example, suppose we are interested in the occurrence of linear features in a field (e.g., faults, ore bodies, etc.). The frequency of the features can be evaluated by distance methods, examining the centers of the features as a point process. Boolean processes are a subset of marked processes that examine only the existence (the location and size) of an object. Generally, Boolean processes are simple geometric objects such as disks or ellipsoids, whose size and occurrence are independent random functions (Deutsch and Journel, 1998, p. 156). Although Boolean models appear to be quite simple, their mathematical basis is quite complex and is beyond the scope of this section (Cressie, 1991, Chapter 9).

Marked processes can be very complex, consisting of aggregate point process models for the occurrence of the object and multivariate process models for the mark. For example, rock fractures can be represented with a Poisson process for the fracture centers, the lengths of the fractures can be an independent power law distribution, and the fracture orientation can be randomly distributed. Although the simplest representation is that the fracture location, length, and orientation are independent of any other fracture, rock stresses may cause the occurrence of small fractures to be more likely where large fractures are present (e.g., a fracture swarm). This implies that a spatially correlated marked process model may be more appropriate (i.e., that the fracture length law is a multivariate function of the surrounding fractures).

Parameter inference for complex marked process models may require indirect estimation by Monte Carlo simulation of a candidate process model followed by comparison with the observed process, iteratively updating the model parameters until the observed process model is adequately represented. In such Monte Carlo estimation, simulations of the candidate process model and parameters are compared with the observed process with a goodness-of-fit test of the hypothesis that "the observed process is indistinguishable from the candidate model with the current parameters." The parameters are refined until the hypothesis fails to be rejected. Again using the fracture example, suppose a geologist has observed the frequency and orientation of fractures in an access tunnel. A candidate model of fracture frequency, lengths, and orientation could be used to simulate the fractures intersecting the tunnel, refining the model parameters until there is no statistical difference between the candidate model and the observed fracture intersections (Doe and Geier, 1990; Deutsch and Journel, 1998, p. 156). Although the Monte Carlo procedure is robust and flexible, it can result in many candidate models that are statistically indistinguishable from the observed process. In this case, additional constraints must be used to determine an acceptable model (e.g., the rock fracture distribution may require groundwater tracer tests to eliminate some models and parameter values).

Special Issues. A frequent concern of site characterization studies is that the sampling program will detect discrete objects—e.g., waste drums, alluvial channels, etc. It is relatively straightforward

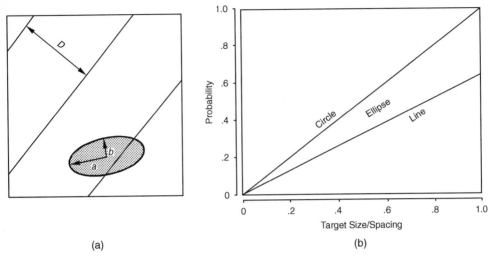

FIGURE 8.5.9 Search for an elliptical target. (a) Search schematic and (b) probability versus relative size (Davis, 1986, used with permission).

to determine the probability of detecting simple geometric objects of known size and shape with a particular sampling grid (Davis, 1986, p. 289). For example, suppose we want to detect an elliptical target with major and minor axes of lengths a and b, respectively, using a series of parallel transects of spacing D. The probability of finding the target is $P = Q/D$, where

$$Q = 2\sqrt{\frac{a^2 + b^2}{2}} \tag{8.5.16}$$

Note that the probability of detection changes with both the line spacing and the shape and size of the target (Figure 8.5.9). In the case of complex objects, (e.g., fractures of various orientations and lengths), the probabilities can be estimated by Monte Carlo simulation. In such a case, the proposed sampling network is used to detect the object of interest in multiple realizations of the hypothesized process, with the probability of detection estimated from the number of detections over the Monte Carlo set.

Analysts often must evaluate clustering of discrete objects—either the spatial clustering of points or the similarity of mark values. This may be addressed by a nonparametric technique that uses multiresponse permutation procedures (MRPP) (Mielke, 1991). For example, consider lead concentrations in soils taken in and around a major city, where we suspect that high lead concentrations may occur in the inner city because of increased air pollution. This may be examined by using the spatial separation among members of the group of soil samples with abnormally high lead levels. Using MRPP, we examine whether the group of high lead levels is meaningful (significant) by testing the hypothesis that "the separation distance among group members is no different than any random group." The MRPP statistic δ_0 is computed as the arithmetic mean of spatial distances among pairs of data in the high-lead group. The significance of the test is the rank of δ_0 versus the separation distance of any random pair of data δ divided by the total number of possible pairings:

$$P = \frac{(\text{no. of } \delta \leq \delta_0)}{N!} \tag{8.5.17}$$

Thus if the mean separation among sample locations within the high lead group is small (i.e., high lead samples tend to be close to each other) then the P value would tend to be very small and we reject the hypothesis that the high lead samples are spatially unclustered. MRPP is quite versatile and has been applied to a variety of questions in the earth sciences (Mielke, 1991; Orlowski et al., 1993) and environmental pollution (Mielke et al., 1983).

The complexity of marked processes sometimes requires special EDA plots that usually are developed on a case-by-case basis for the type of mark being examined. Examples include post plots of stars as dots whose sizes vary with the magnitude of the star or projections of three-dimensional vectors into spheres (e.g., Schmidt net projections of fracture orientations). Such plots generally are highly specialized and are beyond the scope of this section (Davis, 1986, p. 339).

REFERENCES

Cressie, N. A. C., *Statistics for Spatial Data*, Wiley-Interscience, New York, 1991.

Cressie, N., *The Many Faces of Spatial Prediction, Geostatistics*, Vol. 1, 1989, pp. 163–176.

Davis, J. C., *Statistics and Data Analysis in Geology*, Wiley, New York, 1986.

Delhomme, J. P., "Kriging in the Hydrosciences," *Adv. Water Res.*, 1(5): 251–266, 1978.

Desbarats, A. J., "Modeling Spatial Variability Using Geostatistical Simulation," in *Geostatistics for Environmental and Geotechnical Applications*, ASTM STP 1283, Srivastava, R. M., Rouhani, S., Cromer, M. V., Johnson, A. I., and Desbarats, J. A., eds., American Society for Testing and Materials, 1996.

Deutsch, C. V., *Fortran Programs for Calculating Connectivity of Three-Dimensional Numerical Models and for Ranking Multiple Realizations, Computers & Geosci.*, 24(1): 26–76, 1998.

Deutsch, C. V., and Journel, A. G., *GSLIB, Geostatistical software library and user's guide*, 2nd ed., Oxford University Press, New York, 1998.

Doe, T. W., and Geier, J. E., *Interpretation of Fracture System Geometry Using Well Test Data*, Swedish Nuclear Fuel and Waste Management Company, SKB Stripa Project Report 91-03, 1990.

Englund, E. J., "A Variance of Geostatisticians," *Math. Geol.*, 22(4): 417–455, 1990.

Gilbert, R. O., *Statistical Methods for Environmental Pollution Monitoring*, Van Nostrand Reinhold, New York, 1987.

Henley, S., *Nonparametric Geostatistics*, Wiley, New York, 1981.

Issaks, E. H., and Srivastava, R. M., *An Introduction to Applied Geostatistics*, Oxford University Press, New York, 1989.

Journel, A. G., and Huijbregts, C. J., *Mining Geostatistcs*, Academic Press, London, 1978.

Journel, A. G., "Nonparametric Estimation of Spatial Distributions," *Math. Geol.*, 15(3): 445–468, 1983.

Journel, A. G., "Geostatistics: Models and Tools for the Earth Sciences," *Math. Geol.*, 18(1): 119–140, 1986.

Journel, A. G., "New Distance Measures: the Route Toward Truly Non-Gaussian Geostatistics," *Math. Geol.*, 20(4): 459–475, 1988.

Journel, A. G., *Fundamentals of Geostatistics in Five Lessons: Short Course in geology*, Vol. 8, American Geophysical Union, Washington, DC, 1989.

McKenna, S. A., *Geostatistical Analysis of PU-238 Contamination in Release Block D, Mound Plant, Miamisburg, Ohio*, Sandia Report SAND97-0270, Sandia National Laboratories, 1996.

Mielke, H. W., Anderson, J. C., Berry, K. J., Mielke, P. W., Chaney, R. L., and Leech, M., "Lead Concentrations in Inner-City Soils As a Factor in the Child Lead Problem," *Am. J. Publ. Health*, 73: 1366–1369, 1983.

Mielke, P. W. Jr., "The Application of Multivariate Permutation Methods Based on Distance Functions in the Earth Sciences," *Earth Sci. Rev.*, 31: 55–71, 1991.

Orlowski, L. A., Grundy, W. D., Mielke, P. W. Jr., and Schumm, S. A., "Geological Applications of Multi-Response Permutation Procedures," *Math. Geol.*, 25(4): 483–500, 1993.

Renard, Ph., and de Marsily, G., "Calculating Equivalent Permeability: a Review," *Adv. Water Resources*, 20(5-6): 253–278, 1997.

Rouhani, S., Srivastava, R. M., Desbarats, A. J., Cromer, M. V., and Johnson, A. I., *Geostatistics for Environmental and Geotechnical Applications*, ASTM, Pennsylvania, 1996.

Tukey, P., *Proceedings of Symposia in Applied Mathematics*, Vol. 28, American Mathematical Society, pp. 8–46, 1983.

Watson, D. F., and Philip, G. M., "Triangle-Based Interpolation," *Math. Geol.*, 16(8): 779–795, 1984.

Weber, D., and Englund, E., "Evaluation and Comparison of Spatial Interpolators," *Math. Geol.*, 24(4): 381–391, 1992.

Weber, D., and Englund, E., "Evaluation and Comparison of Spatial Interpolators II," *Math. Geol.*, 26(5): 589–603, 1994.

BIBLIOGRAPHY

Davis, J. C., *Statistics and Data Analysis in Geology*, Wiley, New York, 1986.

Deutsch, C. V., and Journel, A. G., *GSLIB, Geostatistical Software Library and User's Guide*, 2nd ed., Oxford University Press, New York, 1998.

Englund, E., and Sparks, A., *GEO-EAS 1.2.1 (Geostatistical Environmental Assessment Software) User's Guide*, U.S. Environmental Protection Agency, Las Vegas, 1991.

Issaks, E. H., and Srivastava, R. M., *An Introduction to Applied Geostatistics*, Oxford University Press, New York, 1989.

Rouhani, S., Srivastava, R. M., Desbarats, A. J., Cromer, M. V., and Johnson, A. I., *Geostatistics for Environmental and Geotechnical Applications*, ASTM, Pennsylvania, 1996.

CHAPTER 8
ANALYSIS AND MODELING

SECTION 8.6

PHARMACOKINETIC/DYNAMIC MODELING

Janusz Z. Byczkowski

Dr. Byczkowski is an independent consultant with more than 30 years of experience teaching and performing research in toxicology, biochemistry, and pharmacology.

8.6.1 INTRODUCTION

Pharmacokinetics and Toxicodynamics

Pharmacokinetics (PK) may be defined as "the study of the time course of drug and metabolite levels in different fluids, tissues, and excreta of the body, and of the mathematical relationships required to develop models to interpret such data" (Gibaldi and Perrier, *Pharmacokinetics*, 2nd ed. Marcel Dekker, New York, 1982). In toxicology, which is not limited to the study of drugs but deals with all variety of chemical compounds, sometimes in harmful or toxic concentrations, an equivalent term *toxicokinetics* is in use (Kantrowitz and Yacobi, *Pharmacokinetics of Drugs*, Springer-Verlag, 1994, chapt. 14, pp. 383–403). However, the toxicity of a given chemical compound (drug or environmental pollutant alike) is a function of its dose, or more precisely, its internal dose or local concentration in the target organ. Moreover, most known potentially toxic chemicals have thresholds, below which no adverse or toxic effect is observed [no-observable adverse-effect level (NOAEL)]. Thus, instead of toxicokinetics, term *pharmacokinetics* is used throughout this chapter to depict time courses of concentrations and disposition of all chemicals, drugs and environmental pollutants alike, because the same quantitative description and the same algorithms are being used to follow and to model disposition of chemicals below and above their toxic thresholds. On the other hand, the term *pharmacodynamics* (PD) should be restricted to depict effects and the handling of beneficial drugs by the body, as it does not apply to the toxic effects of environmental pollutants. Therefore, in this chapter the quantitative descriptions linking exposure or internal dose with toxic responses of the body (above the threshold) are referred to as *toxicodynamics* (TD). However, in the literature both terms, pharmacodynamics and toxicodynamics, are often used as synonyms.

PK/TD Modeling

The goal of PK/TD modeling is to describe quantitatively the relationship that links the external dose or concentration (exposure level) of a chemical compound (input) with the effective internal dose or local concentration of the chemical (or its active metabolite) in the target organ (internal dose surrogate, IDs) and/or with toxic response of the body (output). A general concept of PK/TD modeling is depicted schematically in Figure 8.6.1. The PK/TD models are useful for exposure and dose-response assessment, route-to-route extrapolation, interspecies extrapolation, derivation of toxicity values (acceptable daily intake, benchmark dose, minimal risk level, reference concentration,

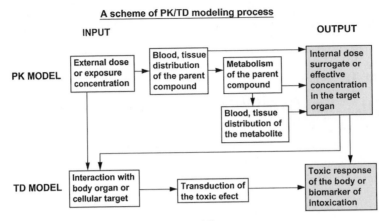

FIGURE 8.6.1 General concept of PK/TD modeling.

reference dose, threshold limit value), etc. Because biological responses of tissues and organs are mechanistically linked to the local concentration of the active form of a chemical compound, the internal dose of chemical that reached a particular physiological compartment must be used for any meaningful risk characterization.

Several modeling and computing methods are used for analysis and presentation of PK/TD data, estimation of PK parameters, extrapolations between the experimental data points and simulations of *disposition* of chemicals (absorption, distribution, metabolism, and excretion). Mathematical methods used in PK/TD modeling can be subdivided into *descriptive* and *predictive* methods.

Descriptive PK/TD Models

A nonlinear regression analysis is an example of the descriptive mathematical method, in which parameters of the equation for a continuous curve are fit into the experimental data points. The pitfall of the curve-fitting in pharmacokinetics is that, even if one succeeds in finding equations that duplicate the behavior of the PK/TD system, their internal workings may be very different from the internal mechanism of the real biological system. The simulations only mimic kinetic behavior of the chemical and do not provide insight into the quantitative or mechanistic relations between the internal components of the real system. Parameters of the fitted equations are often artificial and arbitrary, without physicochemical and/or physiological meaning. Descriptive models may be valuable for extrapolating the system variables between the experimental data points, but they often fail in extrapolations beyond the range of experimental calibrations. Calculations of PK/TD parameters based on phenomenological data analysis, such as curve fitting, curve stripping or feathering, area under the curve (AUC) calculation, etc., are sometimes called *model independent* because they are apparently free of any assumption about underlying compartmental model that the chemical obeys. Examples of some computerized software useful for curve fitting, descriptive modeling, and noncompartmental phenomenological PK/TD data analysis are listed in Appendix A.

Predictive PK/TD Models

A physiologically based pharmacokinetic (PBPK) model is an example of the predictive method, in which parameters of the PBPK model quantitatively describe relations between the internal workings and correspond to the physicochemical and physiological properties of the PK system. The PBPK model provides insight into the mechanism of disposition of the chemical within the biological system and reflects as well as describes the real physiological phenomena of interaction between the chemical and the organism, usually expressed by nonlinear relationships. In contrast, classic compartmental PK models describe linear systems in which the *rate of transfer* of the chemical from one compartment to another is directly proportional to the total mass of the chemical in this compartment. In reality, biological systems are nonlinear and cannot be accurately described by a single elimination rate constant for a wide range of concentrations. These nonlinearities are well handled by PBPK models but not by classic compartmental PK models.

In PBPK modeling there is no need for assumptions of a steady state or first-order kinetics. These predictive models, when properly calibrated and validated, may be used beyond the range of experimental data points, and thus they can still reliably predict the behavior of PK/TD systems in regions where no information is available. This is often impossible with classic PK models, which are data based and easily fail in extrapolations. Examples of the computerized software useful for both compartmental modeling and predictive PK/TD simulations are listed in Appendix B.

8.6.2 CLASSIC PHARMACOKINETIC MODELING

In classic linear pharmacokinetics, the concentration of a chemical compound in the one-compartment system (Figure 8.6.2) is calculated in the following way: the change in the amount of a chemical (dA in mg) over time interval (dt in h) is proportional to the amount (A in mg) present in compartment i. Therefore, if the apparent volume of distribution (V in L) stays constant, the concentration of a chemical (C in mg/L) may be described by a simple linear relationship, analogous to the amount (Figure 8.6.2). By integrating this simple proportionality equation, one may calculate the concentration of a chemical within the compartment at a given time (t in h). The constant of proportionality (k in 1/h), usually referred to as the *elimination rate constant*, is inversely related to the half-life ($t_{1/2}$ in h; Figure 8.6.2). The logarithmic form of the integrated equation:

$$\log C_{i(t)} = \log C_{i0} - k \times t/2.303 \qquad (8.6.1)$$

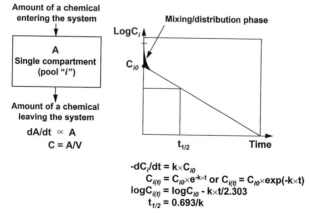

FIGURE 8.6.2   Basics of classic PK modeling (according to Byczkowski and Fisher, *Computer Meth. Progr. Biomed.*, 46: 155–163, 1995). *A*, amount of a chemical present in compartment i (mg); t, time (h); C, concentration of a chemical (mg/L); V, apparent volume of distribution (L); $C_{i(t)}$, concentration of a chemical in compartment i at time t (mg/L); C_{i0}, theoretical concentration at $t = 0$ (mg/L); k, rate constant of elimination of a chemical from compartment i (1/h); $t_{1/2}$, half-life (h).

may be rewritten in the form of the straight line equation:

$$y = b + a \times t \tag{8.6.2}$$

where b is the intercept (equivalent to $\log C_{i0}$ in mg/L), a is the slope (equivalent to $-k/2.303$ in 1/h), and t is time (h). This equation may be easily fitted to the data with any linear regression program.

This basic model becomes more complicated when additional compartments are considered simultaneously. For the multicompartment linear system, the concentration in the i-th compartment ($C_{i(t)}$ in mg/L) is a sum of exponential terms of the form:

$$C_{i(t)} = A_i \times e^{-\alpha t} + B_i \times e^{-\beta t} + \cdots + N_i \times e^{-\Gamma t} \tag{8.6.3}$$

where each term represents a partial contribution to the total concentration in the i-th compartment; A, B, and N are proportionality constants (macroconstants in mg/L); α, β, and Γ are rate constants (microconstants in 1/h) and t is time (h). Equations of this type may be fitted to the data with nonlinear regression programs or resolved into two or three or more linear components by using a method of residuals or curve stripping or feathering.

One disadvantage of the classic approach is that experimental resolution of these biexponential, triexponential, etc., equations becomes progressively difficult with increasing numbers of compartments. Another disadvantage is that it requires an interpretation of several abstract parameters, which

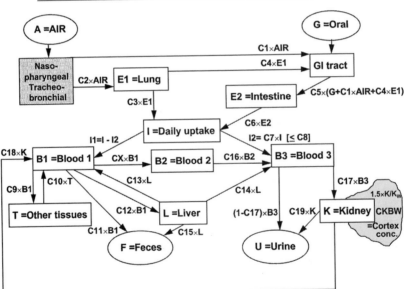

Schematic representation of Cd mass flow in the linear PK model

FIGURE 8.6.3 Flow scheme of the linear classic PK model for cadmium (Cd) disposition (according to Kjellström and Nordberg, in: *Cadmium and Health: A Toxicologic and Epidemiological Appraisal, Vol. I. Exposure, Dose and Metabolism*, CRC Press, Boca Raton, FL, 1985 pp. 179–197). C1–C19, coefficients (rate constants and ratios); A and G, inputs (μg, of Cd/d); $E1$, $E2$, $B1$, $B2$, $B3$, T, L, K, linear rates of change of Cd in body compartments (μg of Cd/d); I, estimate of internal daily uptake (μg of Cd/d); F and U, outputs (μg of Cd/d); $CKBW$, kidney cortex concentration (μg/g of tissue); arrows show direction of Cd mass flow. C17 decreases linearly from age 30 to age 80 by 33 percent; C19 increases each year from age 30 with a rate constant 1.1×10^{-6} per d. Model parameters are from Kjellström and Nordberg (*Environ. Res.* 16: 248–269, 1978). Model was encoded in ACSL (for codes and numerical values of all parameters, see Appendix C).

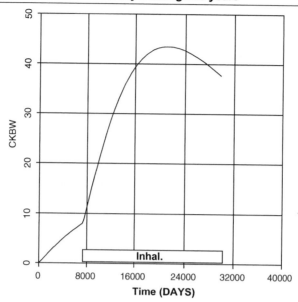

PK model: renal cortical Cd concentration (CKBW in μg/g), oral exposure to 20 μg Cd/day for 82 years plus inhalation of 20 μg Cd/day from age 20 years

FIGURE 8.6.4   Example of PK model simulation of kidney cortex cadmium (Cd) concentration (CKBW in μg/g wet weight) under the following exposure scenario: lifetime oral intake of 20 μg of Cd/d (0.3 μg of Cd kg^{-1} d^{-1} with the introduction of exposure to airborne soluble cd particulate (e.g., cadmium oxide, MMAD = 5 μm) at a concentration of 1 μg/m^3 (yielding 20 μg of Cd/d) starting at age 20 years. The combined inhalation and oral intake in this scenario after age 20 years was 40 μg of Cd/d (0.6 μg of Cd kg^{-1} d^{-1}. A complete set of equations for Cd mass transfer (shown in Figure 8.6.3) was solved simultaneously with the ACSL Tox software from $t = 0$ to $t = 29,930$ d (82 years).

usually are not directly available for experimental physical measurements and often are counterintuitive (e.g., intercompartmental transport rate constants, compartmental volumes, volume of distribution, clearance, half-life, etc.).

These disadvantages may be partially overcome by dividing a complex PK system into smaller manageable blocks, experimentally determining or estimating mass transfer rates between these blocks (compartments), setting differential equations describing in a stepwise manner the mass transfer between these blocks, and, finally, simultaneously solving the whole set of differential equations with the aid of a computer.

An example of a complex classic linear compartmental model, developed to simulate cadmium PK, is shown in Figure 8.6.3. This PK model (Kjellström and Nordberg, "A Kinetic Model of Cadmium Metabolism in the Human Being," *Environ. Res.*, 16: 248–269, 1978) is a multicompartment expansion of the linear one-compartment model developed by Friberg et al. (Friberg, Piscator, Nordberg, and Kjellström, *Cadmium in the Environment*, CRC Press, Cleveland, OH, 1974, pp. 79–89). The model simulates uptake, accumulation, and elimination of cadmium in eight compartments that represent the lung, gastrointestinal tract, blood (three compartments), liver, kidney, and other tissues. The model accepts input of daily cadmium intakes (μg of Cd/per d) to the respiratory tract and gastrointestinal tract. The PK model and its key parameters have been adequately described in the updated toxicological profile for cadmium [Agency for Toxic Substances and Disease Registry, *Toxicological Profile for Cadmium (Draft for Public Comment, Update)*, U.S. Department of Health and Human Services, Public Health Service, Atlanta, GA, 1998]. This PK model was based on data collected from humans and it was intended for human risk assessment applications. Most of the parameters used in this PK model do not have a direct physiological meaning and were fitted to the available data. This descriptive model does not address any potential for interspecies extrapolations and is based on parameters derived from and/or fitted to human data. The result of a PK simulation of the kidney cortex Cd concentration (CKBW in μg/g wet weight) in humans under the occupationally relevant scenario is shown in Figure 8.6.4. The computer codes and numerical values of all parameters used in this simulation are listed in Appendix C.

8.6.3   PHYSIOLOGICALLY BASED PHARMACOKINETIC MODELING

In contrast to the classic PK models, PBPK models use experimentally verifiable parameters. In the PBPK modeling, all compartments of the system are physiologically defined (Yang and Andersen, "Pharmacokinetics," in *Introduction to Biochemical Toxicology*, Hodgson, E., and Levi, P., eds., 2nd ed., Appleton & Lange, Norwalk, CT, 1994, chapt. 3, pp. 49–73). The basic assumption in PBPK modeling is that blood flow to the tissue or diffusion is limiting the chemical compound delivery (Medinsky and Klaassen, "Toxicokinetics," in *Casarett and Doull's Toxicology, the Basic Science of Poisons*, Klaassen, C. D., Amdur, M. O., and Doull, J., eds., 5th ed., McGraw-Hill, New York, 1996, chapt. 7, pp. 187–198). A schematic representation of a single tissue compartment in the blood flow limited PBPK model is shown in Figure 8.6.5. Because the chemical is retained by the tissue according to its tissue/blood partition coefficient (P_i in L/kg), which may be measured *in vitro*, the concentration of the chemical compound in venous blood, leaving the tissue (CV_i in mg/L), during the equilibration phase is lower than the concentration in arterial blood (CA in mg/L). Therefore, the rate of change in the amount of a chemical in tissue (RA_i in mg/h) is given by a simple difference between concentration in blood entering and exiting the tissue ($CA - CV_i$) multiplied by the blood flow (Q_i in L/hr) (Figure 8.6.5). Integrating this equation over a given time, one can calculate the amount of chemical (A_i in mg) present in tissue; therefore, if the actual volume of tissue (V_i in kg) is known, one can calculate the concentration of a chemical in the tissue (C_i in mg/kg) at any time (Figure 8.6.5).

Multicompartmental PBPK models are built by interlinking several tissue compartments with arterial blood flow as an input and venous blood as an output (Figure 8.6.6). The difference between classic and PBPK modeling approaches becomes apparent in the multicompartmental systems. PBPK models use experimentally verifiable physiological values for ventilation rate and blood flows to the organs and the physicochemical affinity (solubility) of the toxicant to the particular bodily fluid or tissue. The solubility of a chemical in a tissue may be experimentally measured at steady state by its distribution (partitioning), for instance, between blood and air, liver and blood, milk and blood, and so on, and is called a *partition coefficient*. Metabolism of the substance can be measured and expressed in a way analogous to the Michaelis-Menten description of enzymatic activity. In addition, PBPK models use physically real volumes of organs, expressed as fractions of the body weight, that may be allometrically scaled up or scaled down, according to the size of the individual (Krishnan and Andersen, "Interspecies Scaling in Pharmacokinetics," in *New Trends in Pharmacokinetics*, Rescigno, A., and Thakur, A. K., eds., Plenum Press, New York, 1991, pp. 203–226).

Thus, PBPK models require experimental determination of basic physiological parameters: blood flows, ventilation rates, organ volumes (Brown, Delp, Lindstedt, Rhomberg, and Beliles, "Physiological Parameter Values for Physiologically Based Pharmacokinetic Models," *Toxicol. Ind. Health*, 13: 407–484, 1997); physicochemical parameters: blood/air, and tissue/blood partition

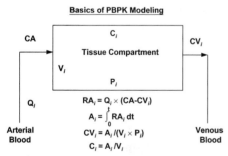

Basics of PBPK Modeling

$RA_i = Q_i \times (CA-CV_i)$

$A_i = \int_0^t RA_i \, dt$

$CV_i = A_i /(V_i \times P_i)$

$C_i = A_i /V_i$

FIGURE 8.6.5   Basic concepts of PBPK modeling (according to Byczkowski and Fisher, *Computer Meth. Progr. Biomed*, 46: 155–163, 1995). Symbols used: Q_i, blood flow to tissue i (L/h); CA, concentration of a chemical in arterial blood (mg/L); C_i, concentration of chemical in tissue i (mg/kg); V_i, volume of tissue i (kg); P_i, partition coefficient of chemical between tissue i and blood (L/kg); CV_i, concentration of chemical in blood leaving tissue i (mg/L); RA_i, rate of change in amount of chemical in tissue i (mg/h); A_i, amount of chemical present in tissue i (mg); t, time (h).

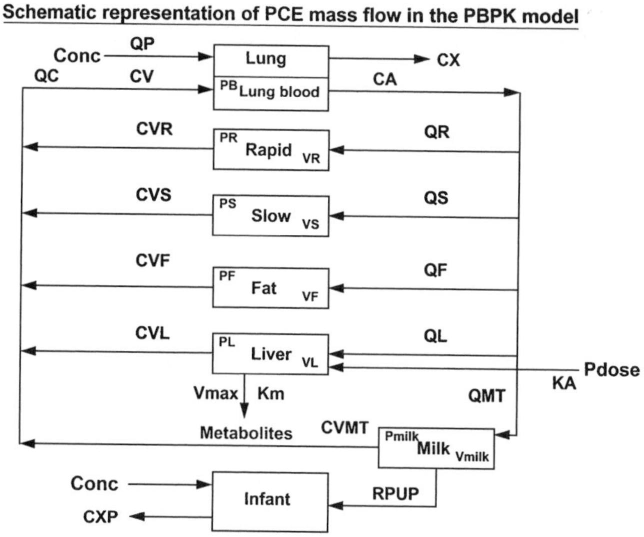

Schematic representation of PCE mass flow in the PBPK model

FIGURE 8.6.6 Flow scheme of nonlinear PBPK model for PCE disposition and lactational transfer to infant (according to Byczkowski, *Drug Inf. J.*, 30: 401–412, 1996). Model was encoded in ACSL (for codes and numerical values of all parameters, see Appendix C).

coefficients (Poulin and Krishnan, "Tissue-Based Algorithm for Predicting Tissue: Air Partition Coefficients of Organic Chemicals," *Toxicol. Appl. Pharmacol.*, 136: 126–137, 1996, and Poulin and Krishnan, "Molecular Structure-Based Prediction of the Partition Coefficients of Organic Chemicals for Physiological Pharmacokinetic Models," *Toxicol. Methods* 6: 117–137, 1996); and biochemical parameters: apparent Michaelis-Menten constant of chemical metabolism, pseudomaximal velocity, and/or first-order metabolism rate for its biotransformation (Gargas, Andersen, and Clewell, III, "A Physiologically Based Simulation Approach for Determining Metabolic Constants from Gas Uptake Data," *Toxicol. Appl. Pharmacol.* 86: 341–352, 1986). Even appropriately parameterized PBPK models should be further calibrated with a coherent, experimentally measured data set.

PBPK models are useful for mathematical description of pharmacokinetics and prediction of the biologically active doses in tissues, which, in turn, may be used in risk assessment. It is also possible with PBPK models to extrapolate between high and low concentrations, different routes of administration, high and low body weight, and between different species. PBPK models are indispensable for constructing meaningful physiologically based TD models and for mechanistic dose-response modeling.

An example of the complex PBPK model for perchloroethylene (PCE) is shown in Figure 8.6.6. This PBPK model includes human physiological parameters and it is a modification of the earlier PBPK model that simulated lactational transfer of PCE, experimentally developed and originally calibrated in rats (Byczkowski, Kinkead, Leahy, Randall, and Fisher, "Computer Simulation of Lactational Transfer of Tetrachloroethylene in Rats Using a Physiologically Based Model," *Toxicol. Appl. Pharmacol.*, 125: 228–236, 1994). The PCE mass flow between compartments, depicted schematically in Figure 8.6.6, was described by a set of differential equations in the general form:

$$V_i \times dC_i/dt = Q_i \times (CA - CV_i) \tag{8.6.4}$$

where V_i is volume of the tissue compartment (in kg), C_i is concentration of PCE in the tissue (in mg/kg), t is time (in h), Q_i is blood flow to the tissue (in L/hr), CA is PCE concentration in arterial blood (in mg/L), CV_i is PCE concentration in venous blood leaving the tissue (in mg/L), and i depicts tissue compartments: rapidly perfused, slowly perfused, fat, liver, and mammary glands.

In this blood flow-limited model, the PCE concentration in venous blood leaving the tissue (CV_i in mg/L) was calculated from experimentally measured tissue/blood partition coefficients (P_i in L/kg):

$$CV_i = C_i / P_i \qquad (8.6.5)$$

where C_i is the PCE concentration in the tissue (in mg/kg).

For lung compartments with two mass inputs (mixed venous blood and inhaled air) and two outputs (arterial blood and exhaled air), at steady state the amount in alveolar air is in equilibrium with the amount in lung blood. Therefore,

$$QP \times (CI - CX) = QC \times (CA - CV) \qquad (8.6.6)$$

$$CX = CA/PB \qquad (8.6.7)$$

where QP is air flow through the lungs (alveolar ventilation rate in L/hr), CI is PCE concentration in inhaled air, calculated as $CI = \text{concentration} \times 166./24450$. [to convert from parts per million (ppm) of PCE to mg/L], concentration is the PCE concentration in inhaled air (in ppm), CX is the PCE concentration in alveolar air (in mg/L), CA is the PCE arterial blood concentration (leaving the lungs, in mg/L), PB is blood/air partition coefficient (ratio), QC is blood flow through the lungs (rate of cardiac output in L/hr), CV is PCE venous blood concentration (entering the lungs, in mg/L). This equation was solved for CA.

For the liver compartment with mass input from blood and two outputs (venous blood and metabolism, without biliary excretion, which, in the case of PCE, was negligible), the toxicant mass transfer is given by the equations:

$$dAL/dt = QL \times (CA - CVL) - dAM/dt \qquad (8.6.8)$$

$$dAM/dt = (V_{\max} \times CVL)/(K_m + CVL) \qquad (8.6.9)$$

$$CVL = AL/(VL \times PL) \qquad (8.6.10)$$

where AL is amount of PCE in the liver (in mg), t is time (in h), QL is blood flow through the liver (in L/h), CA is PCE arterial blood concentration (in mg/L), CVL is PCE venous blood concentration (in mg/L), AM is amount of PCE metabolized (in mg), $V_{\max}$ is pseudomaximal velocity of PCE metabolism (in mg/h), K_m is apparent Michaelis-Menten metabolism constant (in mg/L), VL is liver volume (in kg), and PL is liver/blood partition coefficient (in L/kg). The PBPK model was written in Advanced Continuous Simulation Language (ACSL; Mitchell and Gauthier Associates, Inc., Concord, MA). The PBPK simulations were done with ACSL Tox software on a Pentium personal computer. The computer codes and numerical values of all parameters used in this simulation are listed in Appendix C.

Experimental Calibration, Verification, Confirmation, and Validation

Once the PBPK model has been constructed and parameterized, the key parameters [identified by the sensitivity analysis (Evans and Andersen, "Sensitivity Analysis and the Design of Gas Uptake Inhalation Studies," *Inhal. Toxicol.*, 7: 1075–1094, 1995)] should be calibrated with a coherent experimentally determined data set. Parameters of the PBPK model for PCE have been calibrated with an extensive set of experimental data from lactating rats and pups exposed to PCE via mother's milk. The time- and dose-dependent model predictions were compared with the measured concentrations of PCE in blood of dams and pups, milk, tissue of whole pups, and their gastrointestinal tract. At this stage, the key modeling parameters were adjusted and fine tuned (optimized) to fit the whole set of the experimental data.

Next, the PBPK model was allometrically scaled to humans and verified with the available data from controlled experiments in human subjects. Although it is not feasible to measure xenobiotic concentration in each physiological compartment, at least venous blood and/or exhaled air and one or two target tissue compartments should be sampled at several exposure doses of xenobiotic and at different times. Figure 8.6.7(A) shows an example of the verification of the PBPK model with

PBPK model: PCE concentration in exhaled air (CX in ppm), inhalation exposure to 194 ppm of PCE for +1.5 hr or ▲3 hr

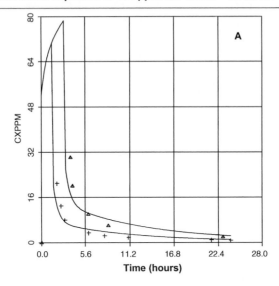

PBPK model: PCE concentration in ▲venous blood (CV in mg/L) and + breast milk (CMAT in mg/L), inhalation exposure to 600 ppm of PCE for 1 hr

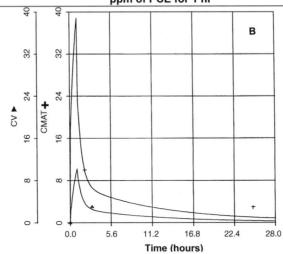

FIGURE 8.6.7 Examples of (A) PBPK model verification with experimentally measured data for PCE concentration in exhaled air (CX in ppm), and (B) confirmation of PBPK model by simulation of time-dependent concentrations of PCE in blood and milk of lactating mother (according to Byczkowski and Fisher, *Risk Anal.*, 14: 339–349). Curves represent PBPK model simulations, symbols represent experimentally measured data points. (A) Human subjects were exposed to 194 ppm of PCE for 1.5 h (+) and 3 h (Δ) (according to Stewart, Baretta, Dodd, and Torkelson, *Arch. Environ. Health*, 20: 224–229, 1970). (B) Lactating mother was allegedly exposed to 600 ppm of PCE for 1 h; Δ, concentration in blood (CV in mg/L); +, concentration in milk (CMAT in mg/L) (according to Bagnell and Ellenberger, *Can. Med. Assoc. J.*, 117: 1047–1048, 1977). A complete set of equations for PCE mass transfer (shown in Figures 8.6.5 and 8.6.6) was solved simultaneously with the ACSL Tox software from $t = 0$ to $t = 28$ h.

measured time-dependent concentrations of PCE in exhaled air in human subjects. Some limited adjustment of one or two modeling parameters still may be required at this stage.

Finally, the PBPK model was confirmed with another set of results from the real-life exposure scenario [Figure 8.6.7(B)]. Even limited data, measured in blood and milk of a mother incidentally exposed to PCE, were enough to confirm that the PBPK model was capable of simulating the environmentally relevant scenario. Obviously, no parameter adjustments or any changes to the model can be performed at this stage.

Idealistically, if enough available human data exist, the PBPK model that is intended for use in risk assessment still should be validated with values measured in the general population to account for interindividual variability of modeling parameters and population heterogeneity of outputs of the modeled variables. If only the distributions of the key modeling parameters in the human population are known, the PBPK model can be interfaced with a Monte Carlo simulation module (Thomas, Lytle, Keefe, Constan, and Yang, "Incorporating Monte Carlo simulation into Physiologically Based Pharmacokinetic Models Using Advanced Continuous Simulation Language (ACSL): A Computational Method," *Fundam. Appl. Toxicol.*, 31:19–28, 1996), which would randomly sample the input values (within the range predetermined from the population) and run multiple PBPK simulations that would return a predicted distribution of output variables expected for the population.

8.6.4   BIOLOGICALLY BASED TOXICODYNAMIC MODELING

Whereas PK modeling essentially simulates what an organism does to the chemical, TD modeling simulates what the chemical can do to the organism. From the theoretical point of view, both PK and TD models describe dynamic systems in which internal concentration and the effect of the chemical depend on the state of the system. However, PK models predict the response of the system describing the fate of a chemical, whereas TD models predict the response of the system describing the effect of a chemical. The relationship between the chemical concentration at the site of action (delivered concentration in the target organ) and toxic effects is usually nonlinear, often involves discontinuities (threshold) and delays, and may vary depending on the mechanism of action.

Whereas PK models are always time dependent, the traditional pharmacodynamic models based on classical receptor theory are time independent. Usually, they describe the equilibrium relationship between concentration at the site of action ($C_{i(t)}$ in mg/L) and effect at steady state ($E_{(C)}$ either in units of the effect measurement, in percent of maximum response, or in probability units of quantal effect). Sometimes, tissue concentration is not an appropriate input into the TD model; instead the receptor occupancy, AUC over time (in h × mg/L), or time-weighted-average tissue concentration (TWA/d) have to be used. Moreover, the effect at the same chemical concentration may not be the same at different times (hysteresis or proteresis). In any case, the TD models, analogous to traditional pharmacodynamic models (Holford and Ludden, "Time Course of Drug Effect," in *Handbook of Experimental Pharmacology*, Welling, P. G., and Balant, L. P., eds., Springer-Verlag, Berlin, 1994, chapt. 11, pp. 333–352), have to be considered as empirical descriptions of the observed effect (Lauffenburger, and Linderman, *Receptors—Models for Binding, Trafficking, and Signaling*, Oxford University Press, New York, 1993), and thus they require a lot of experimental data for calibration and toxicological expertise for correct interpretation.

Immediate Tissue Response Models

Direct response of the tissue due to chemical binding to the receptor site can be described by an E_{max} model, a linear model, or a sigmoid model. The equilibrium relationships between the internal dose surrogate ID_S ($C_{i(t)}$, AUC, TWA, etc.) and effect E may be described by some variation of the classic Hill equation (Hill, "The Possible Effects of the Aggregation of the Molecules of Haemoglobin on Its Dissociation Curves," Proceedings of the Physiological Society, January 22, *J. Physiol. London*,

40: iv–vii, 1910):

$$E_{(C)} = E_0 + E_{\max} \times \mathrm{ID}_S^h / \left(\mathrm{EC}_{50}^h + \mathrm{ID}_S^h \right) \tag{8.6.11}$$

where $E_{(C)}$ is effect at steady state, E_0 is the baseline effect without a chemical (for most toxic end points $E_0 = 0$), $E_{\max}$ is the maximum effect achievable by the chemical, EC_{50} is the chemical concentration producing 50 percent of the maximum effect, and h is a Hill coefficient. In this sigmoid model, the steepness of the curve is modified by the Hill coefficient h (the larger the value of h, the steeper the curve in a semilogarithmic plot).

When the Hill coefficient h is 1 (for example, the Michaelis-Menten Equation 8.6.9 is a specific case of the Hill equation where $h = 1$), the equation in the form:

$$E_{(C)} = E_0 + E_{\max} \times \mathrm{ID}_S / (\mathrm{EC}_{50} + \mathrm{ID}_S) \tag{8.6.12}$$

is known as the $E_{\max}$ model. When concentration of the chemical is much less than EC_{50} (which is the case in many environmentally relevant exposure scenarios), this equation may be collapsed to a simple linear model:

$$E_{(C)} = E_0 + S \times \mathrm{ID}_S \tag{8.6.13}$$

where S is the slope of the dose-effect relationship (the change in the effect per unit of internal dose surrogate).

Kinetic Delay Models

If the tissue response to the chemical has a kinetic delay, it may be because some time may be required for a chemical to move from the site where it was measured to the site where receptors reside (non-steady-state TD models). When the effect increases with time at the same measured concentration, the TD response exhibits hysteresis. However, when the effect diminishes with time at the same measured concentration, the TD response may exhibit proteresis due to chemical leaking out from the receptor site. These phenomena may be modeled by subdividing the tissue compartment i into two functional subcompartments—pk (predicted by PK model) and e, (effect compartment)—where receptors reside. Fitting the rate constants for chemical movement into the effect compartment (K_{ein} in 1/h) and out of the effect compartment (K_{eo} in 1/h) may help fit the TD model to the data that exhibit hysteresis or proteresis:

$$dA_e/dt = K_{ein} \times A_{pk} - K_{eo} \times A_e \tag{8.6.14}$$

where A_e is the actual amount of chemical in the effect compartment (in mg), A_{pk} is the amount of chemical predicted as a time course by the PK model (in mg), and t is time (in h).

Linearized Multistage Model

Several quantitative models have been devised to describe the dynamics of chemically initiated tumor formation (Sielken, "Useful Tools for Evaluating and Presenting More Science in Quantitative Cancer Risk Assessments," *Toxic Subst. J.*, 9: 353–404, 1989). Models used for cancer risk assessment assume that there is no threshold in the carcinogenic action of chemicals, that chemical carcinogens act irreversibly, that effects are additive, and that all individuals have the same risk of tumor development (these health-protective generalizations originally imposed by the U.S. Environmental Protection Agency (EPA) are invalid for many particular chemicals. The conservative assumptions were based on our early understanding of effects of radiation on DNA, and the proposed new revised U.S. EPA guidelines call for inclusion of the mode of action in assessing carcinogenic risk of chemicals, U.S. EPA, *Federal Register*, 61: 17960–18011, 1996). The linearized multistage model, still utilized by the U.S. EPA for cancer risk assessment, is based on the general algorithm, developed by Armitage and Doll (Armitage and Doll, "Stochastic Model for Carcinogenesis," *Proceedings of the Fourth Berkeley Symposium on Mathematical Statistics and Probability*, In Lecam and Heyman, (eds.) Vol. 4, University of California Press, Berkeley, CA, pp. 19–38, 1961), in the form (for the two-stage mechanism):

$$Pd = 1 - \exp(-q_0 - q_1 \times D - q_2 \times D^2) \tag{8.6.15}$$

where Pd is the probability that chemical at dose D (in mg kg^{-1} d^{-1}) will cause a tumor over the animal's life span, q_0 is the probability of cancer at dose $D = 0$, q_1 is the slope of the dose-dependent probability function (in d $\times$ kg/mg), and q_2 is the second-order dose-dependent parameter. Equation 8.6.15 may be fitted into the experimental results of carcinogenicity bioassays in sensitive animals (usually rodents) and values for q_1 and q_2 may be estimated.

When q_1 is replaced with slope of the upper 95 percent confidence interval ($q_1^* = \Delta Pd/\Delta D_{95LCL}$), the whole expression collapses into the linear estimate of 95 percent bound on cancer risk (*CR*):

$$CR = q_1^* \times D \tag{8.6.16}$$

where D_{95LCL} is the lower 95 percent confidence limit on a dose associated with 10 percent extra risk (according to U.S. EPA guidelines). This equation was used in the PBPK/TD model for PCE to estimate the extra cancer risk for a nursing infant due to PCE received with mother's milk (Appendix C).

Mechanistic Dose-Response Models

In cases in which the mechanism of action of a chemical is exactly known and may be explicitly described mathematically, it is possible to construct a mechanistic, biologically based dose-response model (BBDR). The BBDR models link internal concentrations of active metabolites with their biological effects at a given time. An example of such a model, linking the internal dose of trichloroethylene (a weak prooxidant chemical) to liver concentration of the metabolically generated reactive free radicals and to resulting lipid peroxidation (Byczkowski, Channel, and Miller, "A Biologically Based Pharmacodynamic Model for Lipid Peroxidation Stimulated by Trichloroethylene in Vitro," *J. Biochem. Mol. Toxicol.*, 13: 205–214, 1999), is available on the Internet (http://members.spree.com/tnatox/PBPK/PBPK.htm). This biologically based toxicodynamic model essentially describes a buildup of the lipid peroxidation products over time, whereas at a fixed time it may be applied as a BBDR to predict a dose-response of the liver tissue at a given time.

8.6.5 CONCLUSION

PK/TD modeling is a rapidly growing field, in which current developments in computing hardware and software are opening new possibilities and applications. The importance of the appropriate PK/TD modeling in risk assessment has been stressed by the National Research Council and the use of PBPK models in environmental health and risk assessments is widely recommended by U.S. governmental agencies, such as ATSDR and the U.S. EPA. Several courses and advanced workshops on PK/TD modeling are currently offered (e.g., Workshop on PBPK/PD Modeling and Risk Assessment, Colorado State University at Fort Collins, CO), and it has been a subject of numerous publications in the peer-reviewed literature. More information on the PK/TD modeling and some critical reviews of its application in the environmental and health sciences may be found in the Suggested Readings.

SUGGESTED READINGS

Gargas, M. L., Burgess, R. J., Voisard, D. E., Cason, G. H., and Andersen, M. E., "Partition Coefficients of Low-Molecular Weight Volatile Chemicals in Various Liquids and Tissues," *Toxicol. Appl. Pharmacol.*, 98: 87–99, 1989.

Gibaldi, M., and Perrier, D., *Pharmacokinetics*, 2nd ed. Marcel Dekker, New York, 1982.

Goodman and Gilman, *The Phamacological Basis of Therapeutics*, 9th ed. McGraw-Hill, New York, 1996.

Lauffenburg, D. A., and Linderman, J. J., *Receptors—Models for Binding, Trafficking, and Signaling*, Oxford University Press, New York, 1993.

Medinsky, M. A., and Klaassen, C. D., "Toxicokinetics," in *Casarett and Doull's Toxicology, the Basic Science of Poisons*, Klaassen, C. D., Amdur, M. O., and Doull, J., eds., 5th ed., McGraw-Hill, New York, 1996, chapt. 7, pp. 187–198.

National Research Council, *Pharmacokinetics in Risk Assessment, Drinking Water and Health*, Vol. 8, National Academy Press, Washington, DC, 1987.

Ramsey, J. C., and Andersen, M. E., "A Physiologically Based Description of the Inhalation Pharmacokinetics of Styrene in Rats and Humans," *Toxicol. Appl. Pharmacol.*, 73: 159–175, 1984.

Welling, P. G., and Balant, L. P., *Pharmacokinetics of Drugs*, Springer-Verlag, Berlin, 1994.

Yang, R. H., and Andersen, M. E., "Pharmacokinetics," in *Introduction to Biochemical Toxicology*, Hodgson, E., and Levi, P. E., eds., 2nd ed., Appleton & Lange, Norwalk, CT, 1994, chapt. 3, pp. 49–73.

SELECTED COMPUTER CODES, SOFTWARE DESCRIPTION AND AVAILABILITY

Software for phenomenological PK/TD data analysis (model independent):

1. Curve-fitting programs and statistical methods (e.g., Excel, Crystal Ball, TopFit, SAS). These are general use commercial programs, developed and distributed by business and scientific software manufacturers (Microsoft, SAS Institute, etc.). Although very useful, they are not specifically designed for PK/TD modeling.

2. Noncompartmental methods of analysis (e.g., PK Solutions, NCOMP, WinNonlin, NONMEM).

   • PK Solutions is a Microsoft Excel version 5.0 or higher workbook, which provides a non-compartmental pharmacokinetics data analysis. It was developed by David S. Farrier, Summit Research Services, Ashland, OH <http://www.SummitPK.com>.

   • NCOMP in conjunction with Microsoft Excel version 5.0 or higher, provides an interactive and graphical environment for noncompartmental analysis of pharmacokinetic data (Laub and Gallo, "NCOMP–A Windows-Based Computer Program for Noncompartmental Analysis of Pharmacokinetic Data," *J. Pharm. Sci.*, 85: 393–395, 1996). Developed by Paul B. Laub, Fox Chase Cancer Center, Philadelphia, PA <plavb@incyte.com>.

   • WinNonlin—Input and Output Data Management. WinNonlin's input and output data are managed with Excel-compatible spreadsheet files. The data interface allows you use formulas and functions to create and/or modify WinNonlin data files. Import and export of ASCII and Excel files, missing value codes, column names, column formats, cut and paste, sort, merge, and powerful editing capabilities are just a few of the features included in this module. WinNonlin's Descriptive Statistics module allows you to generate summary statistics of variables in input or output datasets. In addition to standard descriptive statistics, Geometric Mean, Harmonic Mean, Mean and Standard Deviation of the logs, Percentiles, and Confidence Intervals are included. WinNonlin also supplies weighted summary statistics. These weighted statistics include mean, standard deviation, standard error, coefficient of variation, confidence interval, and variance. WinNonlin is a powerful program for solution of nonlinear regression problems, constrained estimation problems, systems of differential equations, and mixtures of differential equations and functions. Using WinNonlin's simulation capabilities, you can see the effect of different dosing regimens and changes in the pharmacokinetic parameters. WinNonlin's library of models includes PK, PD, NCA, PK/PD Link models, Indirect Response Models, and Simultaneous PK/PD Link models. In addition to selecting library models, WinNonlin allows you to write your own models. WinNonlin supports both ASCII user models and compiled user models. It was developed by Scientific Consulting/Pharsight, Inc., Palo Alto, CA <http://www.sciconsulting.com/wnlstd.htm>.

   • The Noncompartmental Analysis module (NCA) computes derived PK parameters (AUC, AUCINF, C_{max}, Cumulative excretion, etc.) from blood or urine data. Parameters for steady-state data also can be calculated. You may use the mouse to select the terminal elimination phase while viewing a semilogarithmic plot, or let WinNonlin select it for you. Exclusions of individual data points for estimation of the lambda z calculation can also be specified. AUC may be calculated by using linear or linear/log trapezoidal rules. WinNonlin also computes partial AUCs, even when the endpoints of the desired range are not included in the data set.

- NONMEM is an evolving program, reflecting tested methodological and programming improvements. The software consists of three parts: (i) NONMEM itself, the basic and very general nonlinear regression program; (ii) PREDPP, a package of subroutines handling population PK data; and (iii) NM-TRAN, a preprocessor allowing control and other needed inputs to be specified. Both NONMEM and NM-TRAN are batch-type programs. It was developed by NONMEM Project Group, University of California, San Francisco, CA <http://c255.ucsf.edu/nonmem1.html>.

SOFTWARE FOR COMPARTMENTAL MODELS

1. Classic linear pharmacokinetics (e.g., SAAM/CONSAAM, XLMEM)

   - WinSAAM and SAAM II—Research tools to aid in design of experiments and analysis of data. SAAM II is a new compartmental and numerical modeling program developed to help researchers create models, design and simulate experiments, and analyze data. Compartmental models are constructed graphically by creating a visual representation of the model. Numerical models are built by entering algebraic equations directly. SAAM II creates systems of ordinary differential equations from the compartmental model structure, permits simulation of complex experimental designs on the model, solves the model and fits it to data by using mathematical and statistical techniques. SAAM/CONSAAM was developed by Loren A Zech. WinSAAM is maintained at NCI NIH < http://www-saam.nci.nih.gov/WinSAAM/ >. SAAM II is maintained at SAAM Institute, Inc., Seattle, WA <http://weber.u.washington.edu/~rfka/htm/s2sftwr.htm>.
   - XLMEM—A nonlinear mixed effect modeling tools for Microsoft Excel spreadsheet under Windows for 1, 2, and 3-compartment systems. It was developed by C. Minto and T. Schnider at Stanford University, CA <http://pkpd.icon.palo-alto.med.va.gov/pkpd.htm>.

2. Physiologically based pharmacokinetics and pharmacodynamics (e.g., SCoP, ACSL, ACSL Tox).

   - SCoP—A simulation control program is a general-purpose simulation package that builds a stand-alone simulation program from a mathematical description of a model provided by the user, solves whole sets of differential equations, displays results with graphic interface, provides comparison with and fitting to experimental data, and allows model parameters to be estimated and optimized. It was developed originally at the Department of Physiology, Duke University Medical Center, Durham, NC, and it is copyrighted and supported by Simulation Resources, Inc., Berrien Springs, MI <http://www.simresinc.com/>.
   - ACSL and ACSL Tox—Developed by MGA Software, ACSL Model includes a dedicated graphical user interface. Predefined Powerblocks (icons) coupled with a set of drawing and annotation tools, enable you to visualize a program, execute, and analyze simulation models with greater ease and speed. ACSL Optimize is a product suite available from MGA Software, which provides support for mathematical simulation, parameter estimation, and sensitivity analysis. It is designed for researchers and engineers in toxicology, pharmacology, chemical engineering, nutrition, and related fields.<http://www.mga.com/>.

COMPUTER CODES

COMPUTER CODES OF LINEAR CLASSIC PK MODEL FOR CADMIUM

```
PROGRAM: CdKN

'Kjellström and Nordberg (1978) MODEL FOR Cd IN HUMANS     '
'program allows you to estimate pharmacokinetics of        '
'Cd IN LINEAR MULTICOMPARTMENT SYSTEM IN HUMAN,            '
'TIME IS IN [DAYS], Cd IS IN [micrograms] PER AVERAGE      '
'INDIVIDUAL 70 kg BW                                       '
'ACSL codes by Janusz Z. Byczkowski, TN & A, 09/5/1998     '
'----------------------------------------------------------'
INITIAL

CONSTANT TSTOP =25550. $'time for run termination [day]       '
CONSTANT POINTS =100.  $'points per run                       '
 CINT = TSTOP/POINTS   $'communication interval [day]          '
'Kjellstrom and Nordberg (1978) PARAMETERS                     '
CONSTANT AIR = 3.  $'Cd intake from smoking 20 cig/day (ug/day) '
CONSTANT SMOK= 1.  $'Defaulted inhalation exposure (switch)     '
CONSTANT ENDSM=9855.  $'Stop inhalation exposure (day)          '
CONSTANT G = 16.        $'Cd intake from diet and water (ug/day) '
CONSTANT C1  = 0.1      $'cigarette smoke, or factory dust =0.7  '
CONSTANT C2  = 0.4      $'cigarette smoke, or factory dust =0.13 '
CONSTANT C3  = 0.05     $'rate constant (1/day)                  '
CONSTANT C4  = 0.005    $'rate constant (1/day)                  '
CONSTANT C5  = 0.048    $'ratio                                  '
CONSTANT C6  = 0.05     $'rate constant (1/day)                  '
CONSTANT C7  = 0.25     $'ratio                                  '
CONSTANT C8  = 1.       $'microgram                              '
CONSTANT C9  = 0.44     $'ratio                                  '
CONSTANT C10 = 0.00014  $'rate constant (1/day)                  '
CONSTANT C11 = 0.27     $'ratio                                  '
CONSTANT C12 = 0.25     $'ratio                                  '
CONSTANT C13 = 0.00003  $'rate constant (1/day)                  '
CONSTANT C14 = 0.00016  $'rate constant (1/day)                  '
CONSTANT C15 = 0.00005  $'rate constant (1/day)                  '
CONSTANT C16 = 0.012    $'rate constant (1/day)                  '
CONSTANT C17 = 0.95     $'decreases from age 30 to 80 by 33%     '
CONSTANT C18 = 0.00001  $'rate constant (1/day)                  '
CONSTANT C19 = 0.00014  $'rate constant (1/day) increases from   '
                        'age 30 with C21 each year               '
CONSTANT CX  = 0.04     $'ratio                                  '
CONSTANT C20 = 0.1      $'ratio                                  '
CONSTANT C21 = 0.0000011 $'rate constant (1/day)                 '
```

```
CONSTANT CORTX=1.51    $'Ratio of kidney/cortex concentration  '
CONSTANT ALW = 1500.   $'Average liver weight (g)               '
CONSTANT AKW = 281.    $'Average kidney weight (g)              '
CONSTANT ABV = 70.     $'Average blood volume (ml/kg)           '
CONSTANT ABSG= 1.06    $'Average blood specific gravity (g/ml)  '
CONSTANT ADUA= 1.0     $'Average daily urine excretion adult(L)'
CONSTANT ADUO= 0.9     $'Average daily urine excretion old (L) '
CONSTANT ADUC= 0.5     $'Average daily urine excretion child(L)'
CONSTANT BW  = 70.     $'Body weight (kg)                       '
CONSTANT ATW = 67467.  $'Weight of other tissues (g)            '
         Asm = AIR * SMOK  $'Smoking/nonsmoking switch            '

END                        $'end of initial                     '
DYNAMIC

ALGORITHM IALG = 2  $'Gear method for stiff systems            '
DERIVATIVE

PROCEDURAL
IF (t.LE.10950.) C17t = C17
IF (t.GT.10950.) C17t = C17*(1.198 - t*(1.8E-05))
IF (T.LT.7300.) A = 0.          $'no smoking at age < 20 y    '
IF (T.GE.7300.) A = Asm         $'start smoking at age 20 y   '
IF (T.GT.ENDSM) A=0.        $'stop smoking at age ENDSM/365 >=y'
IF (T.LE.3650.) ADU = ADUC
IF (T.GT.3650.) ADU = ADUA
IF (T.GE.21900.)ADU = ADUO
END                        $'END OF PROCEDURAL                  '
'*****************MODEL FOR CADMIUM **************************************************'
'LUNG                                                                '
 E1 = C2*A                   $'rate exposing the lung (ug/day)    '
E1t = E1 -C3*AE1t -C4*AE1t  $'rate of loading (ug/day)            '
AE1t= INTEG(E1t,0.)         $'amount remaining in the lung (ug)  '
AAt = INTEG(E1,0.)          $'amount introduced (ug)             '

 Rgo= AE1t*(C3+C4)          $'rate of absorption (ug/day)        '
 Gone=Integ(Rgo,0.)         $'amount gone (ug)                   '
E1bal = AAt - AE1t - Gone $'check of mass conservation: E1bal=0'

'INTESTINE                                                          '
E2 = C5*(G + C1*A +C4*AE1t)  $'rate loading GI tract (ug/day)    '
E2t= E2 - C6*AE2t            $'rate of change GI tract (ug/day)  '
AE2t = INTEG(E2t,0.)        $'amount remaining in the GI (ug)    '
  Gt = INTEG(G,0.)          $'amount introduced with diet (ug)  '

'DAILY ABSORBED AMOUNTS OF Cd                                       '
  I = C3*AE1t +C6*AE2t      $'rate of internal uptake (ug/day)   '
AIt = INTEG(I,0.)           $'internal I amount received (ug)    '

'Daily absorbed amounts at steady state in a long term exposure'
'                                               (ug/day)'
IssAG = A*C2*C3/(C3+C4)+G*C5+(A*C1+A*C2*C4/(C3+C4))*C5  $'A & G  '
```

```
IssA = A*(C2*0.91+C5*C1+C5*0.09) $'by inhalation only (ug/day)     '

PROCEDURAL
IF (I2t.GE.C8) I2t = C8
END                              $'END OF PROCEDURAL             '

I2t = C7*I                 $'rate loading MTN for values =<C8     '
I1t = I - I2t              $'rate unloading to plasma (ug/day)    '

'BLOOD COMPARTMENT 1 PLASMA                                       '
B1=I1t +C10*ATt +C13*ALt +C18*AKt $'rate loading plasma (ug/day)'

B1t= B1 - B1*CX                   $'rate of change in 1 (ug/day)'
AB1t = INTEG(B1t,0.)              $'amount in plasma (ug)         '

'Blood Compartment 2 including ERYTHROCYTES                       '
  B2 = B1t*(1-C9-C11-C12)  $'rate loading blood 2 (ug/day)        '
 B2t = B2 - C16*AB2t    $'rate of change in blood 2 (ug/day)      '
AB2t = INTEG(B2t,0.)    $'amount in blood 2 (ug)                  '
BLDCd= AB2t/5000        $'estimate of Cd blood conc. (ug/ml)      '

'BLOOD COMPARTMENT 3 METALLOTHIONEIN                              '
B3t = I2t +C16*AB2t +C14*ALt      $'rate loading MTN (ug/day)     '
AB3t =INTEG(B3t,0.)  $'amount bound to metallothionein etc.(ug)'

'AMOUNT IN WHOLE BLOOD (ug)                                       '
B4t = B2t +C20*(B1t + B3t) $'rate of change in blood ug/day       '
AB4t= INTEG(B4t,0.)        $'amount in whole blood (ug)            '
CBt = AB4t/(ABV*BW*ABSG)   $'Conc in whole blood (ug/g)            '

'OTHER TISSUES                                                    '
 OT = C9*B1t          $'rate loading other tissues (ug/day)       '
 Tt = OT- ATt*C10     $'rate of change in other tissues (ug/day)  '
ATt = INTEG(Tt,0.)    $'amount in other tissue (ug)               '
ATBW= ATt/ATW         $'concentration in other tissue (ug/g)      '

'LIVER                                                            '
  L = C12*B1t                  $'rate loading liver (ug/day)       '
  Lt = L -ALt*(C13+C14+C15) $'rate of change in liver (ug/day)     '
ALt = INTEG(Lt,0.)            $'amount in liver (ug)               '
ALBW= ALt/ALW                 $'concentration in liver (ug/g)      '
PROCEDURAL
IF (t.LT.10950.) n = 0.
IF (t.GE.10950.) n = (t - 10950.)/365
END                   $'END OF PROCEDURAL                         '

'KIDNEY                                                           '
C19t= C19 + C21*n
 K = C17t*B3t              $'rate loading kidney (ug/day)          '
Kt = K - C18*AKt -C19t*AKt $'rate of change in kidney (ug/day)     '
AKt = INTEG(Kt,0.)         $'amount in kidney (ug)                 '
AKBW= AKt/AKW              $'concentration in kidney (ug/g)        '
CKBW= AKBW*CORTX           $'concentration in kidney cortex (ug/g) '
```

```
'WHOLE BODY                                                      '
Wt = B4t +Lt +Kt +Tt +E1t +E2t   $'amount in whole body (ug/day)'
WBW= Wt/BW                        $'concentration (ug/kg)         '

'FECES EXCRETION                                                 '
Ft = C11*B1t + C15*ALt       $'rate of excretion (ug/day)         '
AFt = INTEG(Ft, 0.)          $'amount excreted (ug)               '

'URINARY EXCRETION                                               '
Ut = B3t* (1-C17t) +AKt*C19    $'rate of urinary excret.(ug/day)'
AUt = INTEG(Ut,0.)             $'cumulated amount in urine (ug) '
CUt = Ut/ADUA                  $'conc. in daily urine (ug/L)    '

PROCEDURAL
IF (CKBW.GT.0.) URCO=Ut/CKBW   $'ratio urinary Cd/cortex Cd     '
END                            $'end procedural                 '

TERMT(T.GE.TSTOP)
END                  $'end of derivative                        '
END                  $'end of dynamic                           '
END                  $'end of program                           '
```

COMPUTER CODES OF NONLINEAR PBPK/TD MODEL FOR PCE

```
PROGRAM: LACTMAN

'Byczkowski (1996) PBPK MODEL FOR PCE IN HUMANS          '
'program allows you to simulate pharmacokinetics of      '
'PCE IN MULTICOMPARTMENT SYSTEM IN HUMAN, INCLUDING      '
'BREAST MILK AND LACTATIONAL TRANSFER TO INFANT          '
'TOXICODYNAMIC MODULE IS FOR CANCER RISK ASSESSMENT      '
'TIME IS IN [hours], PCE IS IN [miligrams] PER AVERAGE   '
'ADULT INDIVIDUAL 60 kg BW AND INFANT 7.2 kg BW          '
'ACSL codes by Janusz Z. Byczkowski, TN & A, 11/5/1998   '
'---------------------------------------------------------------'
INITIAL

'*Physiological parameters                                       '
CONSTANT  BWA = 60.     $'INITIAL BODY WT [KG]                    '
CONSTANT  BWC = 1.      $'MATERNAL BODY WT GAIN [RATIO]           '
CONSTANT BWPI = 7.2     $'INITIAL BODY WT INFANT [KG]             '
CONSTANT  QLC = 0.25    $'FRACTIONAL BLOOD FLOW TO LIVER          '
CONSTANT  QFC = 0.05    $'FRACTIONAL BLOOD FLOW TO FAT            '
CONSTANT  QPC = 19.7    $'ALVEOLAR VENTILATION RATE [L/HR/KG]     '
CONSTANT QPCP = 25.2    $'ALVEOLAR VENTILATION RATE INFANT [L/HR] '
CONSTANT  QCC = 18.0    $'CARDIAC OUTPUT [L/HR]                   '
CONSTANT QCCP = 22.0    $'CARDIAC OUTPUT INFANT [L/HR]            '
CONSTANT QMTC = 0.1     $'BLOOD FLOW TO MAMMARY TISSUE [L/HR]     '
CONSTANT  VFC = 0.2     $'VOLUME OF FAT [L/L BW]                  '
CONSTANT VMATC= 0.05    $'VOLUME MAMMARY TISSUE                   '
CONSTANT  VLC = 0.04    $'FRACTION LIVER TISSUE                   '
CONSTANT   PB = 19.8    $'BLOOD/AIR PARTITION COEFFICIENT, PCE    '
CONSTANT  PPB = 8.0     $'PART. COEFF.INF BLOOD/AIR, PCE          '
```

```
CONSTANT   PL= 6.83    $'LIVER/BLOOD PARTITION COEFFICIENT, PCE        '
CONSTANT   PF= 159.03 $'FAT/BLOOD PARTITION COEFFICIENT, PCE           '
CONSTANT   PS= 7.77    $'SLOWLY PERFUSED TISSUE/BLOOD PART.COEFF.,PCE'
CONSTANT   PR= 6.83    $'RICHLY PERFUSED TISSUE/BLOOD PART.COEFF.,PCE'
CONSTANT   PPt= 6.596 $'TISSUE INF/BLOOD PC, PCE                        '
CONSTANT VtCP= 0.9     $'FRACTION INF TISSUE [based on 75ml blood/kg]'
CONSTANT   PNO = 1.    $'NO. OF INFANTS                                  '
CONSTANT   PMILK = 2.8      $'MILK BLOOD PC FOR PCE                     '
CONSTANT   VMILK = 0.02917  $'VOLUME OF MILK [L]                        '
CONSTANT    OUTI = 0.02917  $'MILK YIELD [L/HR]                         '
CONSTANT    VGI = 0.0836    $'FRACTION GI TRACT, INFANT                 '

'* Toxicant                                                              '
CONSTANT   MW = 166. $'MOLECULAR WEIGHT [G/MOL]                          '
CONSTANT   KM = 0.315$'MICHAELIS-MENTEN CONSTANT [MG/L]                  '
CONSTANT   KFC = 0.    $'FIRST ORDER METABOLISM RATE CONSTANT[/HR-1KG]'

'* Exposure                                                              '
CONSTANT   PDOSE= 0.0    $'ORAL DOSE [MG/KG]                             '
CONSTANT   VMAXC= 0.151 $'MAXIMAL VELOCITY METABOLISM, PCE[MG/KG/HR]'
CONSTANT    KA= 0.00    $'ORAL UPTAKE RATE [/HR]                        '
CONSTANT    KAP= 0.50   $'ORAL UPTAKE, INF [/HR]                        '
CONSTANT IVDOSE= 0.00   $'IV DOSE [MG/KG]                               '
CONSTANT   CONC= 50.    $'INHALED CONCENTRATION [PPM]                   '
CONSTANT    BCK= 0.0041$'BACKGROUND INHALATION EXPOSURE [PPM]           '
CONSTANT   INFEX= 0.    $'INFANT NOT EXPOSED TO BCK (if so INFEX=1) '
CONSTANT       Y= 1.    $'YEARS OF EXPOSURE TO PCE [YEARS]              '

IF (PDOSE.EQ.0.) KA = 0.

'* Timing commands                                                       '
CONSTANT   WDAYS = 5.,WEDAYS = 2.,DAYS = 27.
CONSTANT   PDAYS = 0.
INTEGER DAY
           TSTOP = (DAYS+PDAYS)*24.  $'LENGTH OF SIMULATION [HRS]  '
CONSTANT   TCHNG = 8.0   $'LENGTH OF INHALATION EXPOSURE [HRS]    '
CONSTANT   TINF = 0.01  $'LENGTH OF IV INFUSION [HRS]              '
CONSTANT   CINT  = 0.1   $'COMMUNICATION INTERVAL                  '
           DAY=-1  $'TO START ON MONDAY -1, TUES 0, WEDN 1, ETC.  '

END    $'END OF INITIAL                                             '
'-----------------------------------------------------------------'
DYNAMIC

'*Discrete schedule for CI and OUTX                                '

'    CI = CONCENTRATION IN INHALED AIR (MG/L)                      '
'   OUTX = SUCKLING MILK                                           '
DISCRETE CAT1
   INTERVAL CAT = 24.                    $'EXECUTE CAT1 EVERY 24 HR  '
           DAY=DAY+1
         IF(MOD(DAY,7).GE.5) GOTO OUT

         CI = CONC*MW/24450.  $'START INHAL. EXPOSURE [MG/L]  '
         OUTX = 0             $'NOT FEEDING INFANT             '
```

```
        SCHEDULE CAT2 .AT. T + TCHNG    $'SCHEDULE END OF EXPOSURE      '
                 OUT.. CONTINUE

END              $'END OF CAT1

DISCRETE CAT2
             CI = BCK*MW/24450       $'BACKGROUND INHALATION EXPOSURE'
             OUTX = OUTI             $ 'KEEP FEEDING INFANT           '

END              $'END OF CAT2                                        '
'-----------------------------------------------------------------'

ALGORITHM IALG = 2        $ 'GEAR METHOD FOR STIFF SYSTEMS          '
DERIVATIVE

'* Scaled parameters                                                '
      VL = VLC*BW             $ 'VOLUME LIVER [L]                    '
      VS = 0.79*BW-VF-VMAT    $ 'VOLUME SLOWLY PERF. TISSUE [L]      '
      VR = 0.12*BW-VL         $ 'VOLUME RAPIDLY PERFUSED TISSUE [L]  '
      QL = QLC*QC             $ 'BLOOD FLOW TO LIVER [L/HR]          '
      QS = 0.24*QC-QF         $ 'BLOOD FLOW TO SLOWLY PERF.TISS.[L/HR]'
      QR = 0.76*QC-QL-QMT     $ 'BLOOD FLOW TO RAPIDLY PERF.TISS[L/HR]'
      QF = QFC*QC             $ 'BlOOD FLOW TO FAT [L/HR]            '
    VMAX = VMAXC*BW**.74      $ 'Vmax FOR PCE [MG/HR]                '
      KF = KFC/BW**0.3        $ 'FIRST ORDER METABOL. RATE PCE [/HR] '
    DOSE = PDOSE*BW           $ 'ORAL DOSE, MOTHER [MG]              '
     IVR = IVDOSE*BW/TINF     $ 'RATE INTRAVENOUS DOSING [MG/HR]     '

   BW=BWC*BWA        $ 'BODY WEIGHT [KG]                             '
   QMT=QMTC*QC       $ 'MAMMARY BLOOD FLOW [L/HR]                    '
   VF=VFC*BW         $ 'FAT VOLUME [L]                               '
 VMAT=VMATC*BW       $ 'MAMMARY VOLUME [L]                           '
   QC=QCC*BW**0.74   $ 'CARDIAC OUTPUT [L/HR]                        '
   QP=QPC*BW**0.74   $ 'ALVEOLAR VENTILATION RATE [L/HR]             '

'* Scaled parameters for infant                                     '
      BWP = BWPI*PNO         $ 'TOTAL BODY WEIGHT OF ALL INFs [KG]   '
      QCP = QCCP*BWP**0.74   $ 'CARDIAC OUTPUT INFANT [L/HR]         '
      QPP = QPCP*BWP**0.74   $ 'ALVEOLAR VENTILATION RATE INF. [L/HR]'
      VtP = VTCP*BWP         $ 'VOLUME SOLID TISSUE INFANT [L]       '
      GIW = VGI*BWP          $ 'GASTROINTESTINAL TRACT WEIGHT INF[KG]'
      INF = BCK*MW/24450.    $ 'CONCENTR.INFANT BREATHING ZONE [MG/L]'
'-----------------------------------------------------------------'
'* CA = CONCENTRATION IN ARTERIAL BLOOD [MG/L]                      '
     CA=(QC*CV+QP*CI)/(QC+(QP/PB))
     AUCB=INTEG(CA,0.)

'* AX = AMOUNT EXHALED [MG]                                         '
     CX=CA/PB
   CXPPM=(0.79*CX+0.21*CI)*24450./MW
     RAX=QP*CX
      AX=INTEG(RAX, 0.)
    CEXP=(0.79*CX+0.21*CI)
```

```
'* AS = AMOUNT IN SLOWLY PERFUSED TISSUES [MG]

        RAS=QS*(CA-CVS)
         AS=INTEG(RAS,0.)
        CVS=AS/(VS*PS)
         CS=AS/VS

'* AR = AMOUNT IN RAPIDLY PERFUSED TISSUES [MG]
        RAR=QR*(CA-CVR)
         AR=INTEG(RAR,0.)
        CVR=AR/(VR*PR)
         CR=AR/VR

'* AF = AMOUNT IN FAT TISSUE [MG]
        RAF=QF*(CA-CVF)
         AF=INTEG(RAF,0.)
        CVF=AF/(VF*PF)
         CF=AF/VF

'* AL = AMOUNT IN LIVER TISSUE [MG]
        RAL=QL*(CA-CVL)-RAM
         AL=INTEG(RAL,0.)
        CVL=AL/(VL*PL)
         CL=AL/VL

'* AM = AMOUNT METABOLIZED [MG]
        RAM=(VMAX*CVL)/(KM+CVL) + KF*CVL*VL
         AM=INTEG(RAM,0.)

'* IV = INTRAVENOUS INFUSION RATE [MG/HR]
         IV=IVR*(1.-STEP(TINF))

'* CV = MIXED VENOUS BLOOD CONCENTRATION [MG/L]
         CV=(QF*CVF+ QL*CVL+ QS*CVS+ QR*CVR+QMT*CVMT)/QC
      AUCCV=INTEG(CV,0.)
       AVEN=INTEG(CV*QC,0.)

'* PCEMAS = MASS BALANCE IN MOTHER [MG]
     PCEMAS=AF+AL+AS+AR+AM+AX+AMAT+APUP
       AINH=INTEG(CI*QP,0.)
     INHBAL=PCEMAS-AINH                      $'MASS BALANCE CHECK

'* AMAT = AMT OF PCE IN MILK [MG]
       RMAT=QMT*(CA-CVMT)-RPUP
       AMAT=INTEG(RMAT,0.)
       CVMT=AMAT/(VMILK*PMILK)
       CMAT=AMAT/VMILK
     AUCMAT=INTEG(CMAT,0.)

'* ELIM. RATE FOR TCE FROM MILK TO INF [MG/HR]
  RPUP=OUTX*CMAT
  APUP=INTEG(RPUP,0.)
DOSEP=APUP/BWP     $'DOSE RECEIVED BY INF [MG/KG]
'-----------------------------------------------------------------'
```

```
'   MODULE CALCULATING INTERNAL DOSE SURROGATE FOR CARCINOGENICITY'
PROCEDURAL

  IF (T.GE.24.) MOMID=(AINH*24)/(BW*T)
  IF (T.GE.24.) IDM=(DOSEP*24)/T
  IF (T.GE.24.) IDI=(AINF*24)/(BWP*T)
END                $'END OF PROCEDURAL, MOMID,IDM,IDI [MG/KG/DAY] '
'-----------------------------------------------------------------'
'* AMOUNT REMAINING IN GI. TRACT, INFANT [MG]                     '
  MR=INTEG(RMR,0.)
 RMR=RPUP-RAP        $'RATE OF GI. LOADING, INFANT [MG/HR]         '
 RAP=MR*KAP          $'RATE OF GI. ABSORPTION, INFANT [MG/HR]      '
 AAP=INTEG(RAP,0.)   $'AMOUNT ABSORBED, INFANT [MG]                '
CGIT=MR/GIW          $'CONCENTRATION IN GI. TRACT, INFANT [MG/KG]  '
GIBAL=APUP-(MR+AAP)                      $'MASS BALANCE CHECK      '

'* CAP = CONCENTRATION IN ARTERIAL BLOOD INFANT [MG/L]            '
    CAP=(QCP*CVP+QPP*INF*INFEX)/(QCP+(QPP/PPB))
   AUCBP=INTEG(CAP,0.)
    AINF=INTEG(INF*INFEX*QPP,0.) $'AMOUNT INHALED FROM BCK,INF.[MG]'

'* AXP = AMOUNT EXHALED INFANT [MG]                               '
    CXP=CAP/PPB
  CXPPMP=(0.7*CXP)*24450./MW
   RAXP=QPP*CXP
    AXP=INTEG(RAXP,0.)

'* ATP = AMOUNT IN SOLID TISSUES EXCEPT GI. TRACT, INFANT [MG]    '
   RATP=QCP*(CAP-CVP)+RAP
    ATP=INTEG(RATP,0.)
    CTP=ATP/VTP

'* CPUPT = CONCENTRATION IN A WHOLE INFANT [MG/KG]                '
   CPUPT=(MR+ATP)/VTP

'* CVP = CONCENTRATION IN VENOUS BLOOD INFANT [MG/L]              '
    CVP=ATP/(VTP*PPT)

'* TMASSP = MASS BALANCE INFANT [MG]                              '
   TOTINF=IDM+IDI  $'TOTAL INFANT DOSE [MG/KG/DAY]                 '
   TMASSP=AtP+AXP
   PUPBAL=TMASSP-AAP    $'BALANCE IN TISSUES EXCEPT GI. TRACT, INF.'
    TOTP=TMASSP+MR
   TTBALP=TOTP-(APUP+AINF)     $'BALANCE IN A WHOLE INFANT         '
   '-----------------------------------------------------------------'
'*****      TOXICODYNAMIC MODULE - CARCINOGENESIS IN INFANT     *****'
'* Risk assessment: ECRI - FROM PCE IN MILK, INFANT              '
'                   TCRI - FROM BOTH INGESTED AND INHALED, INFANT '
    ECRI = 0.051*IDM*Y/70    $'1 YEAR NURSING OVER LIFESPAN OF 70 '
    TCRI = 0.051*TOTINF*Y/70 $'1 YEAR NURSING OVER LIFESPAN OF 70 '
    '-----------------------------------------------------------------'
'* CONDITION FOR TERMINATION OF RUN TO PREVENT JUMPING OVER TSTOP '
    TERMT(T.GE.TSTOP)
END        $'END OF DERIVATIVE                                    '
END        $'END OF DYNAMIC                                       '
  '-----------------------------------------------------------------'
END        $'END OF PROGRAM LACTMAN.CSL                           '
```

CHAPTER 8
ANALYSIS AND MODELING

SECTION 8.7

GEOGRAPHIC INFORMATION SYSTEMS

John Kramer
Dr. Kramer is a California certified hydrogeologist who directs project work on mine closure activities, subsurface investigations, environmental monitoring, and groundwater remediation for Condor Earth Technologies, Inc., a leading provider of innovative methods with Geographic Information Systems (GIS).

Erin Mutch
Ms. Mutch is GIS administrator for Condor. She oversees GIS-related activities and associated software operations, develops and analyzes databases and customizes ESRI products. She also provides training, technical support, and GIS demonstrations.

George Ball
Mr. Ball is a GIS analyst/programmer at Condor. He has more than nine years of GIS development experience. His primary responsibilities are in developing custom applications and programs for data automation and he has also implemented a GIS Internet server and created 3-D models and animations using GIS.

8.7.1 INTRODUCTION

A Geographic Information System (GIS) is a computer-based tool for mapping and analyzing things that exist and events that happen on earth (ESRI, 1999). Effective GIS depends on people to operate the computer system according to policies and methods established by the managing organization. GIS includes a computer system with hardware and software that is used to house, maintain, and process files containing data that have a geographic address (latitude and longitude or other coordinates). A key feature of GIS is the ability to view and process data in map formats. A GIS can process resident data and/or acquire data from outside sources to answer questions about relationships between points, lines, or areas. For example, a GIS could be used to search previously unrelated databases of business permits, water well records, geology maps, property parcels maps, and roadways for which the only common data link is geography. The GIS could create a map of all known wells located in limestone terrain within 1000 ft (304.8 m) of any service station in a county. Further, the map could indicate which wells are located on parcels smaller than 1 acre (4047 m^2) (most likely to be domestic) and could show the shortest route to visit each well. GIS is used increasingly by industry and government to manage complex sets of environmental data that, until the advent of GIS, could not be related.

All conceivable phases of research and project management can use GIS, from initial resource inventory, to project siting, planning, design, construction, management, and closure. GIS has become a commonplace tool for industries with large land holdings such as railroads, forest products, and farms. Municipalities are quickly becoming dependent on GIS technology to deliver faster, better, more dependable services, such as emergency response, planning, recording, and facilities management. A fundamental understanding of the issues and power of GIS are essential for environmental professionals. This chapter presents information on GIS fundamentals, followed by sources of data, and a description of how GIS is used as a core planning and management tool for a new city. Additional information about GIS, including links to information can be located on the Internet at Geo Information Systems Web Resources, (1999) and Geographic Information Systems (GIS) www Resource List (1999).

8.7.2 GIS FUNDAMENTAL CONCEPTS

A GIS is a relational database keyed to geography. In a GIS, information is stored in thematic layers, also referred to as coverages, that can be displayed as maps. An example theme might be "environmental samples." Within each thematic layer data are stored in records organized by reference address. Thus, the data in different layers are linked by address.

Attributes are the field headers that describe what type of data are contained in the records. Following our example, the environmental-sample coverage might contain attribute fields like sample type, date, collector, and measurement value. GIS software has powerful preprogrammed utilities for managing the geographic data. Utilities include sophisticated queries, logical searches, data processing, map algebra (mathematical operations applied to whole maps), and map-guided reference to any form of digital data. Photos, charts, graphs, text, videos, and essentially anything that can be viewed on a computer screen is accessible at the click of the mouse—so long as it is properly entered and referenced by the people running the system. Coverages can be combined and displayed as maps in infinite ways, limited only by the imagination and need of the user.

8.7.3 RASTER AND VECTOR DATA

Two fundamentally different types of data are used in GIS: raster data and vector data. Both types have unique advantages and disadvantages. Historically GISs developed as either raster based or vector based. Most common GISs today have evolved to handle both types of data.

Raster data are organized like a photograph or a pointillist painting. Data are stored in an array of pixels, like the dots of shade or color in a picture. Each pixel has a value assigned to the discrete area within the pixel boundaries. That value could represent a color code, a hydraulic head, a chemical concentration, a shade of light, or anything. When the raster image is viewed the color you see is from the values that are stored in each pixel. When magnified to near the dimensions of the pixel, a raster image displays as a checkerboard pattern of squares. The smaller the pixel, the sharper the detail, but the larger the memory required to store and process the information. Raster data are useful for describing fuzzy boundaries or continuously varying features like potential fields (pressure, chemical, electromagnetic, thermal). Raster data are perfect for representing finite-difference model results, where each pixel value represents a piece of the discretized domain.

Vector data are organized to depict point, line, or area data by specifying x, y, and z coordinates that make up the feature itself. In general, a point feature such as an environmental sample is specified by a single (x, y, z) address. A line feature is defined by two points with x, y, and z values and the line itself is defined as connecting from point α to point β. Curved lines, areas, and volumes are expressed mathematically by the appropriate formulas for defining them. Vector data use far less computer memory than raster information and produce sharp lines no matter how much the image is magnified. Most GISs in common use today employ two-dimensional vector addresses. The third dimension is carried as an attribute and software utilities handle calculations of slope distances, areas, and volumes.

8.7.4 POINTS, LINES, AREAS

All map data can be related to points, lines, or areas. Points indicate a single location on earth, like a power pole. Lines are features like roads. Lines are made up of points that are linked together in some defined fashion, usually specifying a direction start to finish. By preserving the direction, a GIS can uniquely define different attributes to the right and left of a line, creating a boundary. Line data are also used in conjunction with a concept called measure, which is described below. Areas such as a lake are defined by polygons on a map. Polygons are made up of lines and points in a closed loop, with an interior and an exterior. In a GIS three different types of coverages (point, line, and polygon) can exist at the same location.

Instead of using x, y measurements, data can also be located by specifying a measure along linear features. The measure can represent distance, time, address, or any other event at a given point along a linear feature. A measure, therefore, can be defined as "a value that defines a discrete location along a linear feature" (ESRI, 1999). Measures allow mapping geographic data that are not in spatial data format. The locations derived from measures are known as events. Events can include attributes along sections of linear features, like paved canal lining versus unlined canals, but can also include instantaneous features, like sampling events. Measures have many applications for routing applications and managing events.

8.7.5 SPATIAL AND ATTRIBUTE QUERIES

A GIS user queries the database to find, process, and extract information. The GIS queries are designed to answer geographic questions or find data embedded in the GIS. Two types of queries are spatial queries and attribute queries. Spatial queries ask questions that pertain to measurements or objects and their relationship to each other: what houses are within 300 ft (91.4 m) of a project site? Attribute queries ask questions that pertain to the attributes, such as what color house or who owns it. Attribute queries can also be used to create graphs, charts, and other representations of data from tables.

8.7.6 MAP OPERATIONS

There are many kinds of map (spatial) operations that can be done in GIS. An example is feature extraction, which cuts out a piece of data to keep and display. One can also split information into a number of coverages and logically select features to delete, keep, or merge. Merging features or adjacent maps is used to combine two or more adjacent maps into one data set. Probably the most common spatial operation that is done is the buffer operation (Figure 8.7.1). In environmental work, the buffer operation might be used in a risk management plan to identify property owners within a defined distance of a bulk chemical storage area. All the previously mentioned operations could be combined to create data for making maps, producing tables, or creating new data sets.

8.7.7 METADATA

When a data set is made, characteristics about the source, methods, and accuracy of the data are retained. This is called metadata—data about the data. Metadata normally list projection information

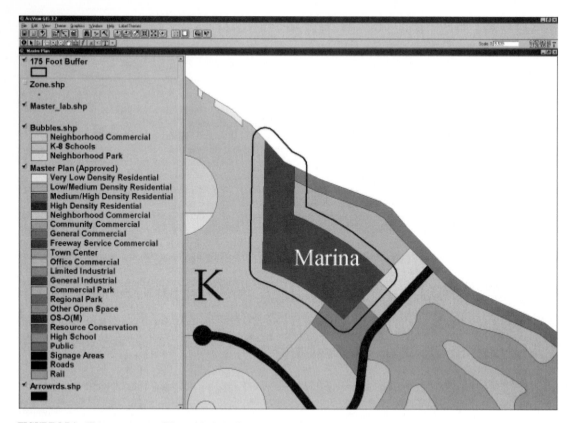

FIGURE 8.7.1   The most common GIS query is the buffer operation that retrieves information from an area around a polygon.

(if locational information is used to display spatial data), intended use of data, technical contact for the data set, digital processing environment (what kind of hardware and software were used to make the data set), positional accuracy, data sources, feature listing (points, lines, or polygons), and attribute descriptions for each different attribute that is assigned to the data set. Metadata help the GIS user determine things such as where the data originated, how current they are, how accurate they are, and what they represent.

Government agencies promote the useful and accurate interchange of data by promoting standard metadata formats. An example of guidance for preparing metadata files was prepared by the U.S. Geological Survey in conjunction with development of the draft digital geologic map data model (U.S. Geological Survey, 1997).

8.7.8   CONVERTING EXISTING DATA INTO GIS FORMAT

Data sources for GIS include paper and digital maps and images as well as tabulated data that are converted to GIS formats. Conversion methods should be listed in the metadata file for the data. The most typical methods for converting existing data are described there.

Digitizing consists of placing a paper map on a digitizer tablet and tracing lines or points by clicking the puck when the target points on the paper map are in the puck cross hairs. Accuracy varies greatly based on the original paper material, the equipment used, and the operator's skill. This method is most useful for capturing lines and points of selected data.

Paper media can be scanned electronically to produce a digital image. This process can generate very large raster files.

Digital images are a graphic representation of a place or thing produced by an optical or electronic device. Examples include satellite data, photographs, and scanned images or documents. These are raster files that can be located in the GIS and used for direct display or georeferenced to facilitate analysis, navigation, and GIS query.

Georeferencing occurs when data are taken from a known or unknown coordinate system and converted into the current standard format of the GIS database with four or more reference points.

Rubber sheeting is a process that adjusts a data set in a nonuniform way with links that define the to and from locations for adjustment.

The typical GIS user should know many possible conversion techniques that would be specific to their GIS environment. Many GIS data sources include options for converting diverse data formats into the current standard format of the GIS database. One example is to import computer-aided design drawing files into a GIS by using the drawing exchange format file option. Special programs for accessing, analyzing, and organizing multiple types of data sets for use by GIS have been developed (e.g., GeoMedia, Intergraph, 1999).

The Internet is a great resource for GIS data sets. There are many Internet sites where data can be found and downloaded. Some of the data are free and some must be purchased. A good starting point for map information is the U.S. Geological Survey (U.S. Geological Survey National Mapping Information, 1999).

8.7.9 FIELD DATA ACQUISITION

For many GIS projects existing data are supplemented by collection of original field data. Examples are resource inventories, sample locations, and area mapping. Traditional methods for field data collection include mapping on paper base and transferring field notes to digital formats compatible with GIS. Data loggers also have been used to blindly create addressed information to upload into a GIS via mutually compatible digital formats. Such data loggers are frequently attached to global positioning system receivers and used to collect coded data. Field data can be acquired directly in GIS formats with field computers. Field mapping and data collection software allow the user to visually check location while data are collected (e.g., PenMap, 1999). The data can be viewed, checked for completeness, processed, and used for navigation while in the field (Figure 8.7.2).

8.7.10 GIS FROM THE GROUND UP: AN EXAMPLE PLANNED CITY

GIS technology, being relatively new, typically has been applied to problems related to existing assets. Of major concern to those familiar with metadata issues has been the accuracy and reliability of GISs that are based on the rubber sheeting of various preexisting data sets collected at different scales and resolutions. Professional planners and environmental scientists should be aware that a crisp line on a well-printed map may not speak with the authority of its appearance. In the late 1990s, however, new urban development opportunities and the continued availability of GIS have overlapped, resulting in the first of a new suite of GIS projects georeferenced to survey-grade accuracy. A complete GIS is being developed for the first new town chartered in California in the past 40 years, a planned city at a 5000-acre (20,235-km^2) "greenfields" site in California. Formerly agricultural land, the townsite has no significant preexisting data layers to incorporate into the GIS. GIS coverages based on survey-grade

FIGURE 8.7.2   Computer field mapping directly into the GIS format with PenMap.

data, will be used for general planning, geologic characterization, watershed analysis, transportation cost estimates, and marketing. Future uses include the GIS database for monitoring and maintenance of activities associated with public services (Mutch and Skaggs, 1999) (Figure 8.7.3).

8.7.11 SUMMARY

GIS is a computer-based technology for storing and processing geographic data. GIS can be implemented effectively only if people maintain and use it according to procedures and policies set up by the organization implementing the GIS. GIS uses a relational database that links diverse data by geography. The future of GIS is bright not only because it is a powerful enabling technology for solving a host of problems but also because complementing information technologies such as remote sensing, the Internet, and computerized data collection tools have evolved to feed it information. As a result, GIS is experiencing a stellar rise in activity and application.

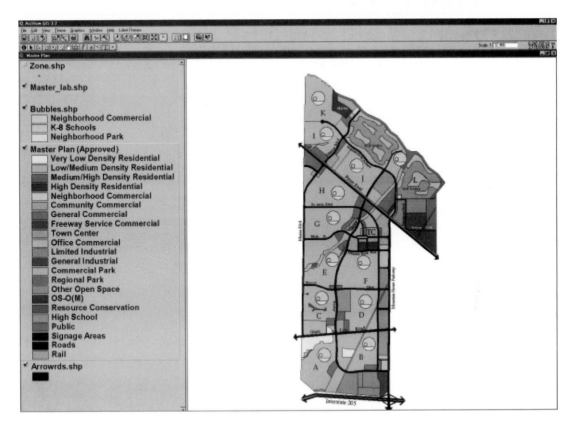

FIGURE 8.7.3   Survey-based GIS used for land planning and marketing (after Mutch and Skaggs, 1999).

REFERENCES

ESRI, www.esri.com/library/gis/abtgis/what gis.html, 1999.

Geo Information Systems Web Resources, www.geoinfosystems.com/resource.htm, 1999.

Geographic Information Systems (GIS) www Resource List, www.geo.ed.ac.uk/home/GISwww.html, 1999.

Intergraph, http://www.intergraph.com/geomedia/default.asp, 1999.

Mutch, E., and Skaggs, R., "Mountain House New Town GIS—From the Ground Up," in *Proc.*, ESRI Users' Conference, San Diego, CA, 1999.

PenMap, http://www.condorearth.com/Products/main.htm, 1999.

U.S. Geological Survey National Mapping Information, *Guidelines for Writing Metadata*, http://ncgmp.usgs.gov/ngmdbproject/standards/metadata/metaWG.html, 1997.

U.S. Geological Survey National Mapping Information, http://mapping.usgs.gov/, 1999.

CHAPTER 8
ANALYSIS AND MODELING

SECTION 8.8

COMPUTING OPTIMAL PUMPING STRATEGIES FOR GROUNDWATER CONTAMINANT PLUME REMEDIATION

Richard C. Peralta
Dr. Peralta specializes in optimization modeling as a professor in the Department of Biological & Irrigation Engineering at Utah State University in Logan.

Alaa H. Aly
Dr. Aly specializes in groundwater modeling at Waterstone Environmental Hydrology and Engineering in Boulder, Colorado.

8.8.1 INTRODUCTION

Groundwater remediation by pumping has been widely applied during the last decade. Extracting contaminated water and injecting clean (treated) water to the groundwater aquifer (termed pump and treat) has been the selected remediation technology for many groundwater contaminant plumes. Generally, pump-and-treat systems are designed to achieve plume containment, plume cleanup, or a combination of the two. Plume containment is achieved by preventing further spreading of the plume into clean areas of the aquifer. Plume cleanup is achieved by reducing contaminant concentrations to below acceptable levels, such as the maximum contamination limit (MCL). Figures 8.8.1 and 8.8.2 clarify typical use of containment and cleanup. Figure 8.8.1 shows a finite difference grid for a groundwater flow model superimposed over an idealized plume. One sees the outer nondetect contour and the inner MCL contour. Figure 8.8.2 uses the same plume to represent how plume management might be mandated. To achieve plume containment, one does not want contaminant concentrations to reach any new cells at any time during the management period. To achieve plume cleanup, one wants concentrations at all cells to not exceed MCL at the end of the management period.

A pumping strategy refers to a spatially and possibly temporally distributed set of pumping rates. It can include both extraction and injection rates. Moving a well location, albeit slightly, or even changing a pumping rate at one well constitutes creating a different pumping strategy. Simulation/optimization (S/O) models greatly facilitate the process of finding pumping strategies that satisfy a number of management constraints while predicting the best (optimal) performance for prescribed management goals.

S/O models can be used to greatly speed the process of computing desirable groundwater pumping strategies for plume management. They make the process of computing optimal strategies fairly straightforward and can help minimize the labor and cost of groundwater contaminant cleanup and/or containment.

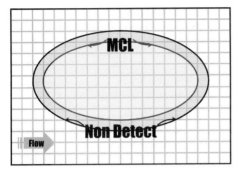

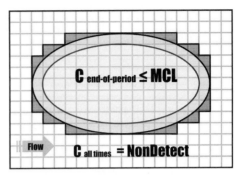

FIGURE 8.8.1 Typical nite-dif ference grid and idealized plume (Peralta, 1999).

FIGURE 8.8.2 Contaminant plume management: containment and cleanup (Peralta, 1999).

The remainder of this chapter is organized as follows. Subsection 8.8.2 introduces S/O models. Subsection 8.8.3 describes how state and decision variables can be related inside S/O models and S/O model formulations for groundwater remediation design. Subsection 8.8.4 discusses different optimization algorithms used in S/O models. Subsection 8.8.5 describes and solves a simple optimization problem and Subsection 8.8.6 describes S/O model application to two field problems. Subsection 8.8.7 discusses S/O model application under uncertainty.

8.8.2 S/O MODELS

Suppose a manager wants to minimize the cost of containing a plume by using existing wells. In that case, minimizing pumping can be a practical surrogate for minimizing cost. The objective function is the sum of pumping rates from all existing wells. Assuming constant pumping, the objective function includes as many pumping rates as there are wells.

The pumping strategy that achieves the best value for the objective function from among all feasible strategies is termed the optimal strategy. By definition, an optimal strategy satisfies all constraints imposed by management.

How does one go about creating an optimal strategy? To develop a pumping strategy for a specific goal by using a typical simulation (S) model such as MODFLOW (McDonald and Harbaugh, 1988), one would probably use the following process (Peralta and Aly, 1997).

1. Specify what you want the pumping strategy to achieve (i.e., what system responses—heads, gradients, etc.) are acceptable.
2. Assume a reasonable pumping strategy that you think might achieve those goals.
3. Simulate system response to the pumping strategy with the simulation model.
4. Evaluate acceptability of the strategy and its consequences.
5. Based on the evaluation of step 4, repeat steps 2–4 until you think you should stop.

When an S model is used, the process of assuming, predicting, and checking might have to be repeated many times. As the numbers of possible pumping sites and system response requirements increase, the likelihood that you have assumed the optimal strategy decreases, and it might become difficult to assume even a feasible strategy.

On the other hand, a groundwater S/O model directly computes the pumping strategy that best satisfies your goals. It contains both simulation equations and an operations research optimization algorithm. The simulation equations permit the model to appropriately represent aquifer response to

TABLE 8.8.1   Partial Comparison of Inputs and Outputs of S/O Models

Model type	Input values	Computed values
S	Physical system parameters	
	Initial conditions	
	Some boundary flows	Some boundary flows
	Some boundary heads	Heads at variable head cells
	Pumping rates	
S/O	Physical system parameters	
	Initial conditions	
	Some boundary flows	Optimal boundary flows
	Some boundary heads	Optimal heads at variable head cells
	Bounds on pumping, heads, and flows	Optimal pumping, heads, and flows
	Objective function (equation)	Objective function value

Both types of models also require as input descriptors and parameters defining the physical system.

hydraulic stimuli and boundary conditions. The optimization algorithm permits the specified management objective to serve as the function driving the search for an optimal strategy.

Table 8.8.1 summarizes differences in inputs and outputs of groundwater flow S and S/O models. Both model types require data describing the physical system. However, model capabilities differ and other inputs and outputs differ.

The familiar S models compute aquifer responses to assumed boundary conditions and pumping values. The boundary conditions and pumping values are all used as data inputs. System response is the output.

On the other hand, S/O models directly calculate the best pumping strategies for the specified management goals. The goals and restrictions are specified via the objective function, constraint equations, and bounds. Data needed to formulate these goals represent additional input required by S/O models (Table 8.8.1). Outputs include optimal pumping rates and the resulting system responses.

Although S/O models require additional data, they are the data needed to ensure that the computed strategy indeed satisfies all your management goals. For example, upper or lower bounds of pumping rates, heads or gradients reflect the ranges of values that management considers acceptable. The model automatically considers those bounds while calculating optimal pumping strategies. One might impose lower bounds on head at a specific distance below current water levels or above the base of the aquifer. Upper bounds on head might be the ground surface or a specified distance below the ground surface.

Contaminant transport S and S/O models can both consider contaminant mass or concentration. Again, the most important difference between S and S/O models is that one must input a pumping strategy to an S model, whereas S/O model finds an optimal pumping strategy.

Many researchers have developed S/O models for specific problems. Generally applicable S/O models for hydraulic (flow) optimization include MODMAN, REMAX, and MODOFC. No generally applicable S/O model is commercially available at this time for transport optimization, although a WINDOWS version of REMAX has that capability. Appendix B lists some unique features of REMAX.

8.8.3   STATE AND DECISION VARIABLES

An S/O model is considered to contain two types of variables: decision and state variables. Decision variables represent variables that can be controlled by the decision maker in the field. State variables represent physical system responses to the decision variables. State variable values are controlled by adjusting the values of decision variables. Contaminant transport S/O models can include as decision variables water injection rate, concentration and rate of injected nutrients, and oxygen or carbon sources. State variables include groundwater gradients, contaminant concentrations, and plume mass.

S/O models must have a means of quantifying the relationship between decision and state variables. For example, an S/O model must be able to predict the concentration that will result at a monitoring (compliance) point from a particular set of pumping rates. These quantitative relationships are sometimes termed simulation constraints.

Several exact and approximation approaches are used to form simulation constraints. One exact approach (the embedding method) uses separate constraint equations to represent each flow and transport equation for each stress period (Aguado and Remson, 1980; Gharbi and Peralta, 1994). Although preferred for some situations, the embedding method can sometimes result in an impracticably huge optimization problem.

Another exact approach calls for a full simulation model (such as MODFLOW or MT3D) during each iteration of an optimization algorithm (Gorelick, 1983; McKinney et al., 1994, Wang and Zheng, 1998). If calling for a full simulation model thousands of times within an optimization algorithm, this approach can require more computer time than is practicable.

A third method uses linear systems theory and superposition via a discretized convolution expression. This response matrix method is exactly applicable for linear (confined) aquifers and has been implemented in several software packages—MODMAN (Greenwald, 1998), MODOFC (Ahlfeld and Riefler, 1998), and REMAX (Peralta and Aly, 1997)—and applied widely (Hegazy and Peralta, 1997; and many others). Sometimes this method can be reasonably applied to slightly nonlinear aquifers (those having saturated thickness that is large by comparison with the change in head with time). REMAX includes an adaptation of the superposition approach to accurately address nonlinear (unconfined) aquifers and piecewise linear (e.g., river-aquifer seepage) flows.

Except for some cases of injecting contaminated water, transport optimization problems usually are nonlinear. S/O models addressing transport optimization problems sometimes apply surrogate expressions to describe physical system response to stimuli. Lefkoff and Gorelick (1990), Ejaz and Peralta (1995), Cooper et al. (1997), and Aly and Peralta (1999a) used polynomial functions to simulate contaminant concentrations. A promising approximation function is an artificial neural network applied to simulate contaminant peak concentration as a function of pumping rates (Aly and Peralta, 1999b).

The previous discussion describes how linear and nonlinear equations can be used as simulation constraints in an S/O model. Depending on whether linear or nonlinear equations are used, different optimization algorithms can be used to solve the formulated optimization problem. The following subsection describes different kinds of optimization problems and the corresponding optimization algorithms.

8.8.4 OPTIMIZATION PROBLEMS: TYPES AND ALGORITHMS

Figure 8.8.3 lists common terminology descriptive of optimization problem types. Linear optimization problems usually are solved by simplex or other techniques that check vertices of the feasible solution

Obj. Fnctn	Constraints	Optimiz Pblm	Optimality
Linear	Linear	Linear (LP)	Global
Some quadratic	Linear	Quadratic (QP)	Global
Linear & integer	Linear & integer	Mixed integer (MIP)	Generally local
Linear or nonlinear	Nonlinear	Nonlinear (NLP)	Local
Linear, integer & nonlinear	Linear, integer or nonlinear	Mixed integer nonlinear (MINLP)	Local

FIGURE 8.8.3 Optimization problem types (modified from Peralta, 1999).

space. Historically, nonlinear problems were most commonly solved by gradient search methods. Branch-and-bound techniques were most commonly used for MIP problems, and outer approximation methods were applied to MINLP techniques. Currently, evolutionary optimization techniques are used increasingly for NLP, MIP, and MINLP problems.

Several researchers applied nonlinear optimization to aquifer contamination problems (Ahlfeld, 1990; Gharbi and Peralta, 1994; Gorelick et al., 1984; Peralta et al., 1995, Peralta and Aly, 1996). Nonlinear programming techniques cannot guarantee global optimality when applied to large non-convex problems. For real problems, where the time required to simulate the groundwater system is significant, nonlinear programming methods may need prohibitive amounts of central processing unit time.

The limitations of mathematical programming have motivated researchers to use alternative optimization techniques such as simulated annealing (Rizzo and Dougherty, 1996; Shieh and Peralta, 1997b) and genetic algorithms (GAs) (McKinney and Lin, 1994; Ritzel et al., 1994; Rogers and Dowla, 1994; Shieh and Peralta, 1997a). Aly and Peralta (1999a) found that a GA performed better than mathematical programming for nonlinear and mixed-integer nonlinear problems. McKinney et al. (1994) found that using a GA to compute the starting point for a nonlinear gradient-based optimization algorithm provided significant advantages and allowed them to locate solutions that are approximately globally optimal. Aly and Peralta (1999b) used neural networks and a GA to design an aquifer cleanup system to reduce the concentrations of two contaminants simultaneously.

8.8.5 *A SIMPLE OPTIMIZATION PROBLEM*

For relatively simple problems, one can manually determine the optimal pumping strategy. Figure 8.8.4 and the following example (after Peralta and Aly, 1997) illustrate a graphical approach for a simple containment problem in a confined aquifer patterned after a real problem from the northeastern United States. Assume the potentiometric surface is initially horizontal and at equilibrium. The box in the upper right corner depicts a plan (map) view of the study area and management problem. There are two existing extraction wells and four observation wells. The observation wells are paired to allow head difference between members of a pair to be described.

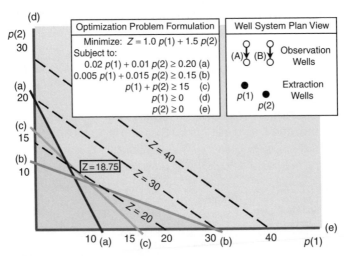

FIGURE 8.8.4 Graphical solution to linear two-well containment optimization problem (modified from Peralta and Aly, 1997).

The goal is to minimize the cost of pumping from the two wells necessary to cause the groundwater level at the head of arrow A to be at least 0.2 m lower than at the tail (a), and the groundwater level at the head of arrow B to be at least 0.15 m lower than at the tail (b). The strategy is also constrained in that the sum of pumping rates from wells 1 and 2 must be at least 15 units (c), and pumping at wells 1(d) and 2(e) must be extraction.

The upper left inset box in Figure 8.8.4 describes the optimization problem mathematically. It includes the objective function (top unnumbered equation) and five constraint equations. The goal is to minimize the value of the objective function Z, which is a sum of pumping rates times unit pumping costs. The 1.5 indicates that a unit pumping at well 2 costs 150 percent of that at well 1.

Pumping rates are considered decision variables. These are variables that management can control directly. Groundwater heads or head differences are state variables—variables defining the state of the physical system.

Equations (d) and (e) are sometimes termed bounds in that they are simply limits on acceptable variable values. They also can be termed constraints. Equation (c) ensures that total pumping is at least 15 units. Equations (a) and (b) constrain final head differences between observation well pairs. The linear superposition Equations (a) and (b) are applicable because confined aquifers are linear. Those expressions employ the additive and multiplicative properties of linear systems theory [Appendix A discusses formation of Equations (a) and (b)].[*]

As described below, restrictions (a)–(e) represent constraints that define the set of feasible solutions (the feasible solution space). The feasible solution space is two dimensional because there are two pumping rates being optimized (i.e., two degrees of freedom).

Figure 8.8.4 illustrates how the constraints restrict the two-dimensional solution space. Because Equation (a) is a ≥ constraint, all points in the graph to the right of line (a) satisfy that equation. All points to the right of lines (b) and (c) satisfy Equations (b) and (c), respectively. Equation (b) is similar to Equation (a)—it describes the effect of pumping on head difference across control pair B.

Bound Equations (d) and (e) prevent decision variables $p(1)$ and $p(2)$ from being negative (i.e., representing injection). Thus, only positive values of $p(1)$ and $p(2)$ are acceptable.

Only points inside or on the boundaries of the region formed by all five constraint or bound lines satisfy all five equations. These points constitute the feasible solution space. The optimization problem goal is to find the smallest combination of $p(1) + 1.5p(2)$ in the solution space. Because all involved equations are linear, that optimal combination will lie on the boundary between the feasible solution region and the infeasible region. In fact, it will be at a point where two or more lines intersect (a vertex of the solution space). For this simple problem of only two decision variables, a graphical or manual solution (evaluating Z at the intersections of the lines) is simple—the minimum value of Z is 18.75. $p(1)$ and $p(2)$ both equal 7.5. Note how the Z isocontour lines decrease as one moves toward the optimal solution.

Optimization problems can become complex. For example, if we want to optimize three pumping rates in the above problem, we must solve the problem within three space (i.e., three dimensions, one for each pumping rate). Problems can rapidly become difficult or impossible to solve manually.

8.8.6 S/O MODELS FOR CONTAMINANT PLUME CONTAINMENT AND CLEANUP

Plume Containment

Norton Air Force Base (NAFB) lies in the San Bernardino Valley, California, a graben filled with deep unconsolidated alluvial material (Figure 8.8.5). Peralta and Aly (1995a) provide a more-detailed

---

[*] Equations (a) and (b) are applications of Equation A8.8.4. In Equation (a), both $p^{ut}(1)$ and $p^{ut}(2)$ equal 1.0. Also, $\delta^{\Delta h}(1, 1)$ and $\delta^{\Delta h}(1, 2)$ are 0.02 and 0.01, respectively. The 0.02 coefficient describes the effect of pumping $p(1)$ on the difference in head between the two observation wells at control location A. Each unit of $p(1)$ causes a 0.02 increase in head difference between the two observation points of control pair A (i.e., an increase in gradient toward pumping well 1). Each unit of $p(2)$ causes a 0.01 increase in head difference toward well 1 at the same location.

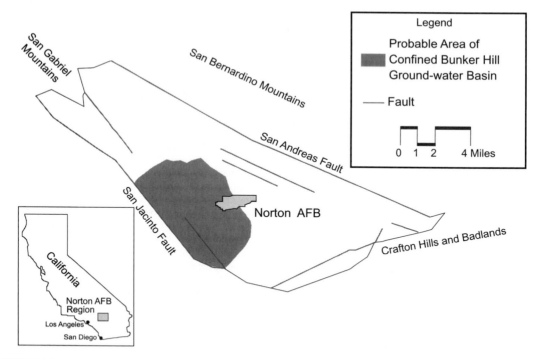

FIGURE 8.8.5   Regional aquifer Norton Air Force Base (from Peralta and Aly, 1995a) (1 mi = 1.6 km).

description for the site. In the NAFB vicinity are three groundwater-bearing zones—the upper two are semiconfining. The top layer is contaminated by dissolved trichloroethylene (TCE), which is migrating from NAFB toward water supply wells. A record of decision mandates that NAFB is to "maintain hydraulic control to the extent possible of the plume while extracting contaminated groundwater, and reinjecting treated groundwater into the contaminant plume or the clean portion of the aquifer." NAFB addressed this goal by installing two pump-and-treat systems—one in the central base area near the TCE plume source (for cleanup) and the other near the southwestern base boundary (for containment). Development of a cleanup pumping strategy (after Peralta and Aly, 1995b) is described in the next subsection.

REMAX was used to compute optimal pumping strategies that would achieve plume containment by preventing any contaminated groundwater from migrating outside base boundaries. Figure 8.8.6 shows the candidate wells. Figure 8.8.7 shows the final pumping strategy and rates.

The well locations shown in Figure 8.8.7 are a subset of those in Figure 8.8.6. REMAX indicated that there should be no pumping at the other wells shown in Figure 8.8.6. The locations of subset well locations were checked to ensure physical feasibility of installation wells at the selected locations.

Figure 8.8.7 shows that total extraction equals total injection. All extracted water is treated and reinjected. Figure 8.8.7 shows how extraction and injection can be used together to prevent contaminated groundwater from reaching off-base supply wells. Injection is used to split the plume and direct contaminated water toward extraction wells. Without this coordinated application of injection and extraction, much more extraction would be required. The presented optimal pumping strategy required 2250 gpm of extraction while satisfying all management criteria. This is 10 percent below the 2500 gpm upper limit of the originally envisioned treatment equipment. The 10 percent reduction

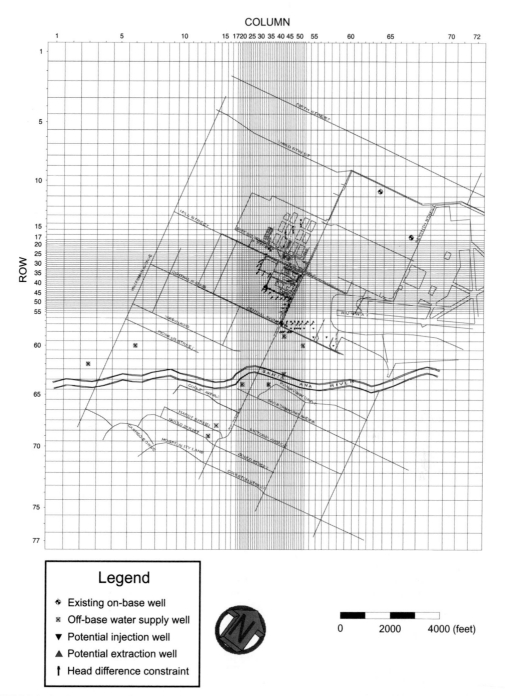

FIGURE 8.8.6 Candidate wells, gradient control locations, and finite difference grid: Norton Air Force Base, southwest boundary area (Peralta and Aly, 1995a) (1 ft = 0.3 m).

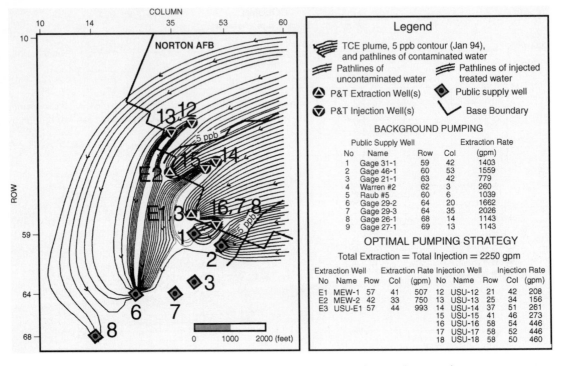

FIGURE 8.8.7 Groundwater pathlines computed to result from implementing the optimal pump and treat pumping strategy.

provided some capacity for future pumping strategy modification, should that be necessary, without requiring additional treatment capacity. Table 8.8.2 shows the savings.

Plume Cleanup

As mentioned in the previous subsection, a pump-and-treat system was needed to maximize contaminant removal from the central base area near the plume source. Peralta and Aly (1995b) provide a more-detailed description for the site. A consultant specified fixed injection well locations to be placed along existing pipelines. The consultant also proposed locations for extraction wells. Because of time

TABLE 8.8.2 Example Cost Savings Resulting from Pump-and-Treat Optimization

	Original	Optimized	Reduction in cost after optimization
Injection Wells	8	7	$100K
Extraction Wells	4	3	$150K
Auxillary, Construction (Pipelines, etc.)	$8M	$6M	$2M
Extraction Rates (gpm)	3500	2250	
O & M Costs (per year)	$1.6M	$1.25M	$350K
O & M Costs (project life)	$24M	18.75M	$5.25M
Operation Time	15 years	15 years	

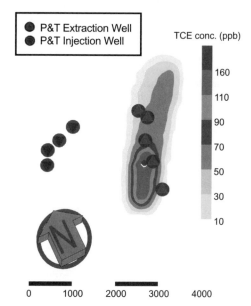

- ● P&T Extraction Well
- ● P&T Injection Well

TCE conc. (ppb)

160
110
90
70
50
30
10

0   1000   2000   3000   4000

FIGURE 8.8.8 Potential pumping locations and initial TCE concentration: Norton Air Force Base, central base area (Peralta and Aly, 1995b) (ppb = parts per billion).

restrictions on accomplishing this optimization effort, REMAX was used to utilize the proposed well locations, assume 100 gpm injection rate at each of four proposed injection locations, and determine optimal extraction rates for five proposed extraction locations. A REMAX precursor determined the optimal (maximum mass of contaminant extraction) strategies needed to achieve cleanup. Figure 8.8.8 shows the potential pumping locations.

Results showed that about 31 percent of the original plume mass can be removed with a treatment facility size of 400 gpm. At 2000 gpm, about 50 percent removal could be achieved.

Other sites at which optimization has been formally applied to plume remediation include March Air Force Base (containment), Mather Air Force Base (cleanup), Travis Air Force Base (containment), and Wurtsmith Air Force Base (cleanup and containment).

Optimization methods rely on the prediction accuracy of flow and transport simulators. Because accurate modeling of any aquifer can be very difficult, developed optimal strategies may not be optimal for the real aquifer system. There is a growing attention to considering the stochastic nature of aquifer parameters while designing remediation strategies. Gorelick (1990) discusses some techniques used to account for uncertainty in designing groundwater management systems. In the following subsection, we describe the most significant proposed approaches and discuss their applicability.

8.8.7 S/O MODELS FOR PLANNING UNDER UNCERTAINTY

Groundwater remediation system design is often complicated by the random nature of aquifer parameters. Three general techniques have been used for solving groundwater management problems under uncertainty. In the first, the sources of uncertainty are not defined but it is assumed that optimal pumping rates can be modified after a period of implementation and monitoring (Jones et al., 1987; Whiffen and Shoemaker, 1993). In this technique, the differences between variable values predicted via optimization and the measured variable values (obtained from the field after the optimal strategy is implemented) are used to guide subsequent modification of the optimal strategy. The relation used to modify the computed optimal strategies is termed a feedback law. The process is continued as the modified optimal strategy is implemented.

In the second technique, a probability distribution is either derived or assumed for the variables of interest. Then analytical relations are developed to relate the quantiles of this distribution to the decision variables. These analytical relations are used as constraints in the optimization problem. These constraints are termed chance constraints and the resulting optimization model is known as the chance-constrained model (Cantiller and Peralta, 1989; Peralta and Ward, 1991).

In the third stochastic groundwater management technique, a group of constraints is formulated—each for a different realization of the uncertain aquifer parameters (Wagner and Gorelick, 1987). A realization is a set of the uncertain parameter values. Typically, each realization is generated from the probabilistic model of the uncertain parameters. The resulting optimal strategy must satisfy all (or some) of the realizations simultaneously. The idea is to find optimal strategies that are robust (satisfy all management constraints) for a range of the uncertain parameters. Several studies tried to estimate the reliability of optimal strategies computed by the multiple-realization technique (Chan, 1993, 1994; Morgan et al., 1993).

All cited studies concluded that to ensure a design that has a high level of reliability at least 50 to 100 realizations are needed (Chan, 1993, 1994; Morgan et al., 1993). For large problems, where the time required to simulate the system is significant, the time required to generate all the constraint equations can be prohibitive. However, because the response surfaces for different realizations can be evaluated simultaneously, one can greatly speed the process by computing them in parallel. Another possible remedy is to determine whether some realizations can be dropped without having to carry out the optimization (Gomez-Hernandez and Carrera, 1994; Ranjithan et al., 1993). Aly and Peralta (1999b) present and apply an approximation method that develops the tradeoff curve between the treatment facility size (total groundwater extraction) and estimated reliability.

REFERENCES

Aguado, E., and Remson, I., "Groundwater Management with Fixed Nonlinear Optimization Problems Charges," *J. Water Resour. Plan. Manage.*, 106(WR2): 375–382, 1980.

Ahlfeld, D. P., "Two-Stage Groundwater Remediation Design," *ASCE J. Water Resour. Plan. and Manage.*, 116(4): 517–529,1990.

Ahlfeld, D. P., and Riefler, R. G., *Documentation for MODOFC: A Program for Solving Optimal Flow Control Problems Based on MODFLOW Simulation*, Department of Civil and Environmental Engineering, University of Massachusetts, Amherst, MA, 1998.

Aly, A. H., and Peralta, R. C., "Comparison of a Genetic Algorithm and Mathematical Programming to the Design of Groundwater Cleanup Systems," *Water Resour. Res.*, 35(8): 1999a.

Aly, A. H., and Peralta, R. C., "Optimal Design of Aquifer Cleanup Systems Under Uncertainty Using a Neural Network and a Genetic Algorithm," *Water Resour. Res.*, 35(8): 1999b.

Belaineh, G., Peralta, R. C., and Haghes, T. C., "Simulation/Optimization Modeling for Water Resources Management," *J. of Water Resour. Plan. and Manage.*, 125(3): 154–161, 1999.

Cantiller, R. R. A., and Peralta, R. C., "Computational Aspects of Chance-Constrained Sustained Groundwater Yield Management," *Trans. ASAE*, 32(3): 939–944, 1989.

Chan, N., "Robustness of the Multiple Realization Method for Stochastic Hydraulic Aquifer Management," *Water Resour. Res.*, 29(9): 3159–3167, 1993.

Chan, N., "Partial Infeasibility Method for Chance-Constrained Aquifer Management," *J. Water Resour. Plan. Manage.*, 120(1): 70–89, 1994.

Cooper, G., Peralta, R. C., and Kaluarachchi, J. J., "Optimizing Separate Phase Light Hydrocarbon Recovery from Contaminated Unconfined Aquifers," *Adv. Water Resour.*, 21: 339–350, 1997.

Ejaz, M. S., and Peralta, R. C., "Modeling for Optimal Management of Agricultural and Domestic Wastewater Loading to Streams," *Water Resour. Res.*, 31(4): 1087–1096, 1995.

Gharbi, A., and Peralta, R. C., "Integrated Embedding Optimization Applied to Salt Lake Valley Aquifers," *Water Resour. Res.*, 30(3): 817–832, 1994.

Gomez-Hernandez, J. J., and Carrera, J., "Using Linear Approximations to Rank Realizations in Groundwater Modeling: Application to Worst Case Selection," *Water Resour. Res.*, 30(7): 2065–2072, 1994.

Gorelick, S. M., "A Review of Distributed Parameter Groundwater Management Modeling Methods," *Water Resour. Res.*, 19(10): 305–319, 1983.

Gorelick, S. M., "Large Scale Nonlinear Deterministic and Stochastic Optimization: Formulations Involving Simulation of Subsurface Contamination," *Math. Prog.*, 48: 19–39, 1990.

Gorelick, S. M., Voss, C. I., Gill, P. E., Murray, W., Saunders, M. A., and Wright, M. H., "Aquifer Reclamation Design: the Use of Contaminant Transport Simulation Combined with Nonlinear Programming," *Water Resour. Res.*, 20(4): 415–427, 1984.

Greenwald, R. M., *Documentation and User's Guide: MODMAN, an Optimization Module for MODFLOW, Version 4.0*, HSI-Geotrans, Freehold, NJ, 1998.

Hegazy, M. A., Peralta, R. C., *Feasibility Considerations of an Optimal Pumping Strategy to Capture TCE/PCE Plume at March AFB, CA*, Report SL/OL 97-1, Systems Simulation/Optimization Laboratory, Department of Biological and Irrigation Engineering, Utah State University, Logan, UT, 1997.

Jones, L., Willis, R., and Cottle, R. W., "Optimal Control of Nonlinear Groundwater Hydraulics Using Differential Dynamic Programming," *Water Resour. Res.*, 30(7): 2065–2072, 1987.

Lefkoff, L. J., and Gorelick, S. M., "Simulating Physical Processes and Economic Behavior in Saline, Irrigated Agriculture: Model Development," *Water Resour. Res.*, 26(8): 1359–1369, 1990.

McDonald, M. G., and Harbaugh, A. W., *A Three-Dimensional Finite-Difference Groundwater Model*, U.S. Geological Survey, Open File Report 83-875, 1988.

McKinney, D. C., Gates, G. B., and Lin, M.-D., "Groundwater Resource Management Models: a Comparison of Genetic Algorithms and Nonlinear Programming," in *Computational Methods in Water Resources*, Vol. 2, Peters, A., et al., eds., Kluwer Academic Publishers, the Netherlands, 1994.

McKinney, D. C., and Lin, M.-D., "Genetic Algorithm Solution of Groundwater Management Models," *Water Resour. Res.*, 30(6): 3775–3789, 1994.

McKinney, D. C., and Lin, M.-D., "Approximate Mixed-Integer Nonlinear Programming Methods for Optimal Aquifer Remediation Design," *Water Resour. Res.*, 31(2): 847–860, 1995.

Morgan, D. R., Eheart, J. W., and Valocchi, A. J., "Aquifer Remediation Design Under Uncertainty Using a New Chance Constrained Programming Technique," *Water Resour. Res.*, 29(3): 551–561, 1993.

Peralta, R. C., "Transport Optimization for Contaminant Plume Management, Workshop Materials," *DOD Conference on Reducing Long Term Remediation Operation and Maintenance Costs*, St. Louis, MO, June, 1999.

Peralta, R. C., and Aly, A. H., *Optimal Pumping Strategy to Capture TCE Plume at Base Boundary*, ERC Report No. 1, U.S. Air Force, Air Force Center for Environmental Excellence, Environmental Restoration Directorate, Norton AFB, CA, 1995a.

Peralta, R. C., Aly, A. H., *Optimal Pumping Strategies to Maximize Contaminant Extraction of TCE Plume at Central Base Area*, ERC Report No. 2, U.S. Air Force, Air Force Center for Environmental Excellence, Environmental Restoration Directorate, Norton AFB, CA, 1995b.

Peralta, R. C., Aly, A. H., *Optimal Pumping Strategies to Maximize Contaminant Extraction of TCE Plume at Mather AFB, California*, ERC Report No. 3, U.S. Air Force, Air Force Center for Environmental Excellence, Environmental Restoration Directorate, 1996.

Peralta, R. C., and Aly, A. H., *REMAX: Simulation/Optimization Model for Groundwater Management Using the Response Matrix and Related Methods, Version 2.70*, Software Engineering Division, Biological and Irrigation Engineering Department, Utah State University, Logan, UT, May 1997.

Peralta, R. C., Solaimanian, J., and Musharrafiah, G. R., "Optimal Dispersed Groundwater Contaminant Management: MODCON Method," *ASCE J. Water Resour. Plan. Manage.*, 121(6): 490–498, 1995.

Peralta, R. C., and Ward, R., "Short-Term Plume Containment: Multiobjective Comparison," *Ground Water*, 29(4): 526–535, 1991.

Ranjithan, S., Eheart, J. W., and Garrett, J. H. Jr., "Neural Network-Based Screening for Groundwater Reclamation Under Uncertainty," *Water Resour. Res.*, 29(3): 563–574, 1993.

Ritzel, B. J., Eheart, J. W., and Rajithan, S., "Using Genetic Algorithms to Solve a Multiple Objective Groundwater Pollution Containment Problem," *Water Resour. Res.*, 30(5): 1589–1603, 1994.

Rizzo, D. M., and Dougherty, D. E., "Design Optimization for Multiple Management Period Groundwater Remediation," *Water Resour. Res.*, 32(8): 2549–2561, 1996.

Rogers, L. L., and Dowla, F. U., "Optimization of Groundwater Remediation Using Artificial Neural Networks with Parallel Solute Transport Modeling," *Water Resour. Res.*, 30(2): 457–481, 1994.

Shieh, H. J., and Peralta, R. C., *Optimal in Situ Bioremediation System Design Using Parallel Recombinative Simulated Annealing*, Report 97-6, Systems Simulation/Optimization Laboratory, Department of Biological and Irrigation Engineering, Utah State University, Logan, UT, 1997a.

Shieh, H. J., and Peralta, R. C., *Optimal in Situ Bioremediation System Design Using Simulated Annealing*, Report 97-7, Systems Simulation/Optimization Laboratory, Department of Biological and Irrigation Engineering, Utah State University, Logan, UT, 1997b.

Wagner, B. J., and Gorelick, S. M., "Optimal Groundwater Quality Management Under Parameter Uncertainty," *Water Resour. Res.*, 23(7): 1162–1174, 1987.

Wang, P. P., and Zheng, C., "An Efficient Approach for Successively Perturbed Groundwater Models," *Adv. Water Resour.*, (21)6: 499–508, 1998.

Whiffen, G. J., and Shoemaker, C. A., "Nonlinear Weighted Feedback Control of Groundwater Remediation Under Uncertainty," *Water Resour. Res.*, 29(9): 3277–3289, 1993.

FORMATION OF CONSTRAINT EQUATIONS FOR RESPONSE MATRIX MODELS (DERIVED FROM PERALTA AND ALY, 1997)

A particular type of S/O model, termed a response matrix model, utilizes the multiplicative and additive properties of linear systems. The following equation illustrates use of the multiplicative property in groundwater head computation. Here we assume that the initial water table is horizontal and at equilibrium. Groundwater is extracted at a single well, index number $\hat{e}$.

$$\Delta h(\hat{o}) = \delta^h(\hat{o}, \hat{e}) \frac{p(\hat{e})}{p^{ut}(\hat{e})} \tag{A8.8.1}$$

where $\Delta h(\hat{o})$ = change in steady-state aquifer potentiometric surface elevation at observation location $\hat{o}$ (L),

$\delta^h(\hat{o}, \hat{e})$ = influence coefficient describing effect of steady groundwater pumping at location $\hat{e}$ on steady-state potentiometric surface elevation at location $\hat{o}$ (L),

$p(\hat{e})$ = pumping rate at location $\hat{e}$ (L^3/T),* and

$p^{ut}(\hat{e})$ = magnitude of steady unit pumping stimulus in location $\hat{e}$ used to generate the influence coefficient (L^3/T). This does not necessarily equal 1.

Assume that a unit steady pumping extraction rate of 1 m^3/min at well $\hat{e}$ causes a drawdown of 1 m at observation point $\hat{o}$. In that case, $p^{ut}(\hat{e}) = 1$ and $\delta^h(\hat{o}, \hat{e})$ equals (-1). Equation A8.8.1 shows that if $\delta^h(\hat{o}, \hat{e})$ and $p^{ut}(\hat{e})$ are known, the change in head caused by any pumping rate can be easily computed. If pumping $p(\hat{e})$ equals 2 m^3/min, head change will equal $(-1)(2)/(1)$ or -2. This linear response is typical of confined aquifers (or approximates behavior of unconfined aquifers where the change in transmissivity due to pumping is small by comparison with the original transmissivity).

Similarly, the effect caused by a unit pumping at location $\hat{e}$ on the final difference in potentiometric surface elevation between locations 1 and 2 of a pair of locations, $\hat{u}$, can be expressed as:

$$\delta^{\Delta h}(\hat{u}, \hat{e}) = \delta^h(\hat{o}_{\hat{u},1}, \hat{e}) - \delta^h(\hat{o}_{\hat{u},2}, \hat{e}) \tag{A8.8.2}$$

where $\hat{o}_{\hat{u},1}$ = index referring to point 1 of pair of locations $\hat{u}$, and

$\hat{o}_{\hat{u},2}$ = index referring to point 2 of pair of locations $\hat{u}$.

For example, if $\delta^h(\hat{o}_{1,x}, \hat{e})$ for locations $x = 1$ and $x = 2$ of pair A are (-1) and (-1.02), respectively, $\delta^{\Delta h}(\hat{o}, \hat{e})$ equals 0.02.

Assume that pumping at M^p locations affect head at location $\hat{o}$. The cumulative effect at $\hat{o}$ is simply the result of adding the effect of M^p pumping rates. The following summation expression illustrates this application of the additive property, with the same assumptions as above.

$$\Delta h(\hat{o}) = \sum_{\hat{e}=1}^{M^p} \delta^h(\hat{o}, \hat{e}) \frac{p(\hat{e})}{p^{ut}(\hat{e})} \tag{A8.8.3}$$

where M^p = total number of locations at which water is being pumped from the aquifer.

---

* For clarity and ease of explaining this example, pumping to extract groundwater is treated as positive in sign, and the δ^h influence coefficients are negative. In REMAX those signs are reversed to be consistent with MODFLOW.

Similarly, the additive property can be used to describe the effect on head difference due to pumping at M^p locations.

$$\Delta\Omega(\hat{u}) = \sum_{\hat{e}=1}^{M^p} \delta^{\Delta h}(\hat{u},\hat{e}) \frac{p(\hat{e})}{p^{ut}(\hat{e})} \qquad (A8.8.4)$$

where $\Omega(\hat{u})$ = difference in potentiometric surface elevation between locations 1 and 2 of pair $\hat{u}$ (L). Here, because the initial steady-state potentiometric surface is horizontal, $\Omega(\hat{u})$ also equals the change in the difference $\Delta\Omega(\hat{u})$ due to pumping.

UNIQUE FEATURES OF REMAX (FROM PERALTA AND ALY, 1997)

1. Well-proven, diverse simulation modules to address porous and fractured media. REMAX is appropriate for optimizing flow and transport management in heterogeneous multilayer porous or fractured aquifers. To develop influence coefficients describing hydraulic head or flow response to stimuli, REMAX uses MODFLOW and MT3D for porous media simulation or SWIFT for fractured media. Other simulation models are easily added as necessary.

2. Robust and proven optimization solvers. REMAX contains all software needed to solve the described optimization problems.

3. Easily maintained data sets. For any particular problem, REMAX reads all data files from a user-specified subdirectory (or folder in WIN95/NT). This allows REMAX users to save all problem-specific input and output in a distinct location.

4. User-friendly data files, error checking, and diagnostics. Innovative REMAX input file organization allows users to write comments, use blank lines, or use blank spaces as desired. This permits thorough data set documentation. REMAX also checks every input file entry and generates error messages with diagnostic explanations.

5. Compatibility with other software. REMAX can read standard MODFLOW, MT3D, and SWIFT data sets. Users can prepare these files with their preferred preprocessor and use the generated input files within REMAX.

6. Ability to compute head at well casing or at cell center. This feature is useful for managing unconfined aquifers of small saturated thickness and for computing hydraulic lift costs.

7. Ability to address systems in which pumping cells or head control cells might initially be or might become fully dewatered. This nonlinear or piecewise linear problem is not addressed by normal response matrix models.

8. Automatic cycling and postoptimization simulation. This enables users to accurately address nonlinear systems (unconfined aquifers and stream/aquifer systems). Cycling proceeds until user-specified maximum number of cycles or convergence criteria for decision variables are achieved. Postoptimization simulation verifies that the results in the nonlinear physical system should be like those in the optimization model.

9. Almost infinite flexibility in addressable problem types. Any of the different types of objective functions can be combined into composite objective functions. Any type of the mentioned constraints can be used with any of the objective function types.

10. Optimization under uncertainty or for risk management. Optimization can satisfy constraints for an unrestricted number of sets of assumed boundary conditions and aquifer parameters (realizations) simultaneously. Reliability of computed strategies is determined by Monte Carlo postoptimization simulation. This feature can be used with any combination of objective function(s) and constraints.

11. Ability to develop cost-reliability tradeoff curves. This ability is provided by using the following REMAX features:

    - optional use of head at well casing instead of average cell heads;
    - use of quadratic objective function including pumping rate, volume, and cost;
    - use of binary and mixed integer variables to include cost of well installation or water treatment plant sizing within the optimization; and
    - coupled use of cost optimization with the multiple realization option.

12. Adaptability for special situations (available within a special REMAX version). Additional constraints can be added as needed, such as those for: (1) managing reservoir releases and conjunctive water delivery to a system of irrigation unit command areas (Belaineh et al., 1998); or (2) ensuring that legal water right priorities are satisfied (assume two adjacent surface water users: user 1 has a higher legal water right than user 2; special constraints can ensure that user 2 does not receive any water unless all of user 1 water right is satisfied).

CHAPTER 8
ANALYSIS AND MODELING

SECTION 8.9
ECONOMICS

Kevin Wolka

*Dr. Wolka was assisstant professor with the Department of
Mathematics, Engineering, Computer Sciences, and Physics at
Carroll College in Helena, Montana. Dr. Wolka passed away while
working on this text. This handbook is dedicated to him.*

8.9.1   INTRODUCTION

This chapter discusses the topic of economic cosiderations in decision making for environmental issues. Economics plays an increasingly important role in environmental decision making as the problem-solving techniques in this scientific discipline mature. The resolution of virtually every environmental issue contains an economic component, either explicitly or implicitly.

There are four principal objectives for this discussion. The first objective is to present the reader with a description and summary of various environmental decision-making processes—e.g., standards, site-specific risk assessment, economic analysis techniques. Providing the reader with a summary of the fundamentals of economic analyses, including the time value of money, the concept of equivalence, and the definition of compound interest is the second objective. The third objective is to give the reader an appreciation for typical economic analysis techniques, such as present worth analysis and benefit-cost analysis, which are used in environmental decision making. The final objective is to show the reader how the viewpoint (perspective) of the economic analysis can affect the results.

8.9.2   ENVIRONMENTAL ISSUES

Environmental issues may be divided into two general categories: (1) compliance with existing environmental regulations and (2) concerns about human health, safety, welfare, and ecological issues that are not currently or are only partly regulated. Many environmental issues do not fit mutually exclusively into one category. A particular issue may encompass both these categories—e.g., effects of rain forest depletion or thinning of the ozone layer. To make this discussion practical and use historical case data, this discussion concentrates on economic aspects associated with the first category—compliance with existing environmental regulations.

8.9.3   ENVIRONMENTAL REGULATIONS

Generally, environmental regulations concentrate on the compliance of operating facilities, especially those in the industrial, commercial, and public sectors. There are many examples of legislation authorizing this type of environmental regulation, including the Resource Conservation and Recovery Act, which addresses hazardous waste management; the Occupational Safety and Health Act, which tackles on-the-job safety and health topics; and the Safe Drinking Water Act, which attempts to ensure that all permitted drinking water supplies are safe for consumption. Some legislative efforts, most notably the Comprehensive Environmental Response Compensation and Liability Act, address environmental issues resulting from abandoned facilities. A few environmental issues, such as radon and lead-based paint, occur primarily in residential scenarios.

Environmental regulations address the protection of human health and safety and ecological concerns. The human health and safety aspects primarily include those for on-site workers at a facility and those working or living off-site of the facility. Ecological concerns may deal with the health and safety of terrestrial flora and fauna and/or aquatic species on or off the facility site.

8.9.4 ENVIRONMENTAL DECISION MAKING

Facilities that have environmental permits must comply with regulatory agency guidelines that specify criteria needed to obtain and to maintain operating permits. Economic analysis most often is used by these facilities to compare specific pollution control strategy alternatives (including equipment) that might be used by the facilities to meet regulatory criteria and to monitor environmental compliance.

At sites of environmental contamination, various methods have been used to determine the degree to which cleanup must be accomplished, an important factor in subsequent economic analyses. In the past, the most common method has been to use a standard maximum concentration value for certain elements, chemical compounds, or chemical groups allowed to remain in the soil, groundwater, or air after cleanup. These standard values originally may have been selected based on previous laboratory detection limits, generic risk assessments, or other experience-based or practical considerations.

An increasingly popular method for determining cleanup levels, especially for complex sites with multiple sources of contamination and large projected remediation costs, is a site-specific risk assessment. The advantage of using site-specific risk assessment remediation goals for the party responsible for paying for the remediation is that the concentration values typically are greater than those values determined by a generic risk assessment, which has several very conservative assumptions associated with it. The outcome of a site-specific risk assessment typically results in a less costly remediation at the site. Chapter 12 of this Handbook contains a detailed discussion of site-specific risk assessment procedures and the differences between them and generic risk assessment procedures.

After cleanup levels have been determined, there are two economic components in the process of determining a preferred remediation approach. First, there is an implicit economic benefit to the public associated with control or removal of hazardous substances from the contaminated site. Second, there is a cost to the party responsible for execution of the remediation. Various methods of comparing alternative remediation technologies and selecting a recommended technology to achieve the remediation goals have been used. Quite often these methods contain an economic analysis procedure component. The economic analysis results are compared along with other criteria, such as implementabilty and effectiveness of the remediation technologies and public acceptance of the remediation technologies, in developing a selected remediation technology.

8.9.5 BASICS OF ECONOMIC ANALYSIS*

Time Value of Money

That there is a "time value of money" is demonstrated by the willingness of individuals, businesses, and financial institutions to pay interest for the use of money at the present time instead of waiting for it at a later date. Interest can be thought of as rent or as a charge for the use of money over time.

---

* This subsection is a summary of selected subparts of Chapters 3 and 4 of *Engineering Economic Analysis* (Newnan and Lavelle, 1998). The definition of terms and mathematical symbols for the terms used in the equations are consistent with those of the textbook.

In the case of a loan, the borrower pays interest to the lender. In the case of savings, the bank pays interest to the savings account holder for the use of the money in the account.

The time value of money is the reason someone would rather have $1000 now than $1000 a year from now. Whether the same person would rather have $1000 now or $1100 a year from now (representing an annual interest rate of 10 percent) depends on what interest rate that person believes he/she could achieve from alternative uses of the $1000 over the 1-year time period.

Simple Interest

Interest computed from a present sum of money is called "simple" interest. Mathematically, simple interest is expressed as follows:

$$I = P \times i \times n$$

where I = interest earned,
  P = present sum of money,
  i = simple interest rate per unit time, and
  n = number of time units.

The mathematical formula that determines the total amount of money (F), the present sum of money plus interest earned, is expressed as follows:

$$F = P(1 + i \times n)$$

Example 1: simple interest—An environmental consultant buys a portable gas chromatograph for $5000 by obtaining a loan at a simple interest rate of 8 percent per year. The consultant repays the loan plus interest 2 years later.

The interest earned is

$$I = (\$5000) \times (0.08/\text{year}) \times (2\,\text{years}) = \$800$$

The total amount is

$$F = (\$5000) \times [1 + (0.08/\text{year}) \times (2\,\text{years})] = \$5800$$

Equivalence

When one is indifferent toward a present sum of money now or some future sum of money, the two sums of money are said to be equivalent. Equivalence depends on the interest rate, which can be shown by referring back to example 1. The loan transaction occurred because both parties agreed on an 8 percent interest rate ($5000 present sum; $5800 future sum). If the borrower was willing to pay only a 6 percent interest rate ($5000 present sum; $5600 future sum), the loan would not have occurred. If the lender had insisted on a 10 percent return on the money ($5000 present sum; $6000 future sum), the loan would not have occurred. The loan occurred only because both parties considered the present sum and the future sum to be equivalent at an 8 percent interest rate.

Equivalence is an important factor in economic analysis. The concept of equivalence is used in economic analysis techniques to compare the economic values of alternatives.

Compound Interest

In actual practice, the compound interest method is almost always used to compute interest. In this method, interest is charged (obtained) at the end of an interest period on both unpaid debt (savings) and unpaid interest (interest on savings).

Compound interest formulae generally are more mathematically complex than simple interest formulae. A series of compound interest formulae that are used in economic equivalence computations are described below.

Single Payment Compound Amount.   A single payment compound amount is a future sum of money (F) equivalent to a present sum of money (P) compounded over n interest periods at an interest rate per interest period of i. The compound interest effect occurs when interest is accrued on the present sum plus previous interest. Mathematically, the single payment compound amount is expressed as follows:

$$F = P(1 + i)^n$$

This equation can also be written in a functional notation form as:

$$F = P(F/P, i, n)$$

where $F/P = (1 + i)^n$, the single payment compound amount factor.

The value of the single payment compound amount (F) therefore can also be determined by multiplying the present sum (P) by the single payment compound amount factor (F/P). The functional notation method has been preferred by practitioners in the past because it allowed compound interest factors to be condensed in reference tables, eliminating the duplication of repetitive calculations and saving time before computers and electronic calculators were extensively used.

Example 2: single payment compound amount—Referring back to example 1, the loan amount was \$5000, and the annual interest rate was 8 percent. Using the single payment compound amount formula, the single payment compound amount would be as follows:

$$F = (\$5000)\ (F/P, 8 \text{ percent/year}, 2 \text{ years})$$

where

$$F/P = (1 + 0.08)^2 = 1.1664$$

then

$$F = (\$5000) \times (1.1664) = \$5832$$

The single payment compound amount is \$32 greater than the total amount of example 1 because interest in the second year is charged on the loan amount plus the interest in the first year when the compound interest formula is used. The simple interest formula charges interest only on the loan amount for both years.

Single Payment Present Worth.   Sometimes the value of a future sum of money—i.e., the single payment compound amount (F)—is known, and the equivalent present sum of money (P) is desired. Rearranging the single payment compound amount formula to have only the present sum on the left side of the equation, the single payment present worth formula is obtained, as shown below.

$$P = F(1 + i)^{-n}$$

The present sum of money is also called the single payment present worth amount. In functional notation the single payment present worth formula is

$$P = F(P/F, i, n)$$

where

$$F/P = (1 + i)^{-n}, \text{ the single payment present worth factor}$$

Example 3: single payment present worth—Referring back example to 2, the future sum (single payment compound amount) after 2 years was $5832 with an annual interest rate of 8 percent. Using the single payment present worth formula, the present sum would be as follows:

$$P = (\$5832) \times (P/F, 8 \text{ percent/year}, 2 \text{ years})$$

where

$$F/P = (1 + 0.08)^{-2} = 0.8573$$

then

$$P = (\$5832) \times (0.08573) = \$5000$$

Uniform Series Compound Interest

In the preceding section, compound interest formulae were used to compute equivalent sums of money at different points in time. Many situations are more complex—e.g., mortgage payments or retirement annuities—and they require a uniform series of receipts or disbursements. The uniform series amount (A) is defined as an end-of-period cash receipt or disbursement in a uniform series, continuing for n periods, with the entire series equivalent to P or F at interest rate i.

Uniform Series Compound Amount.   Referring back to the discussion of the single payment compound amount, the present sum (P) at one time increases to a future sum (F) in n time periods. If the present sum were replaced by a uniform series amount (A) at the end of each of n time periods and all of the uniform series amounts were compounded at interest rate i to one future time, the following equation would result after mathematical rearrangement:

$$F = A\{[(1 + i)^n - 1]/i\}$$

or (in functional notation)

$$F = A(F/A, i, n)$$

where

$$F/A = [(1 + i)^n - 1]/i, \text{ the uniform series compound amount factor}$$

Example 4: uniform series compound amount—Using again a future sum of $5832, an annual interest rate of 8 percent, and a total time of two years, the annual series amount (representing two equal amounts at the end of the first and second years) would be

$$\$5832 = A(F/A, 8 \text{ percent/year}, 2 \text{ years})$$

where

$$F/A = [(1 + 0.08)^2 - 1]/0.08 = 2.080$$

then

$$A = \$5832/2.080 = \$2804$$

Rather than an equipment loan, this example might better represent the future sum one would obtain at the end of 2 years in a savings account to purchase a gas chromatograph after depositing a uniform series amount at the end of the first and second years.

Uniform Series Sinking Fund.   A sinking fund is a separate fund into which one makes a uniform series of deposits (A) with the goal of accumulating some desired future sum (F) at some future time. If one wants to multiply the future sum in example 4 by a factor instead of dividing by the uniform series compound amount factor, its inverse could be used. This inverse value is known as the uniform series sinking fund factor (A/F) and can be represented in example 4 as follows:

$$A/F = i/[(1+i)^n - 1] = 1/2.080 = 0.4808$$

$$A = (\$5832) \times (0.4808) = \$2804$$

Uniform Series Capital Recovery.   If one uses the uniform series compound interest formula including the sinking fund factor and substitutes for the future sum (F) the single payment compound amount formula [$P(1+i)^n$], one obtains an equation for determining the value of a uniform series of end-of-period receipts or disbursements (A) when the present sum (P) is known. The equation is as follows:

$$A = P\{[i(1+i)^n]/[(1+i)^n - 1]\}$$

or (in functional notation)

$$A = P(A/P, i, n)$$

where

$A/P = [i(1+i)^n]/[(1+i)^n - 1]$, the uniform series capital recovery factor

Example 5: uniform series capital recovery—If the present sum is \$5000, the interest rate is 8 percent per year, and the time period is 2 years (similar to the values in the previous example problems), the uniform series capital recovery amount is determined as follows:

$$A = \$5000\,(A/P, 8\text{ percent/year, 2 years})$$

where

$$A/P = [0.08(1+0.08)^2]/[(1+0.08)^2 - 1] = 0.5608$$

$$A = (\$5000) \times (0.5608) = \$2804$$

Uniform Series Present Worth.   The inverse of the uniform series capital recovery factor is the uniform series present worth factor. If the uniform series capital recovery formula is solved for the present sum (P), the uniform series present worth formula is obtained as follows:

$$P = A[(1+i)^n - 1]/[i(1+i)^n]$$

or (in functional notation)

$$P = A(P/A, i, n)$$

where

$P/A = [(1+i)^n - 1]/[i(1+i)^n]$, the uniform series present worth factor

Example 6: uniform series present worth—If a uniform series of end-of-period payments of \$2804 are made for 2 years with an annual interest rate of 8 percent, the present sum can be computed as follows:

$$P = \$2804\,(P/A, 8\text{ percent/year, 2 years})$$

where

$$P/A = [(1 + 0.08)^2 - 1]/[0.08(1 + 0.08)^2] = 1.783$$

then

$$P = (\$2804) \times (1.783) = \$5000$$

The uniform series present worth amount is often used to determine the value at the present time of a uniform series of receipts or disbursements in the economic comparison of alternatives.

Nominal and Effective Interest

Previously, the interest i was defined as the interest rate per interest period. The interest i is now more precisely defined as the effective interest rate per interest period. Using this definition, the effective interest rate per year i_a is the annual interest rate taking into account the effect of any compounding during the year. The nominal interest rate per year r is the annual interest rate without considering the effect of any compounding. The effective interest rate per year and the nominal interest rate per year are related as follows:

$$i_a = (1 + r/m)^m - 1$$

where m = number of compounding subperiods per year.

In the typical annual effective interest computation, there are multiple compounding subperiods per year ($m > 1$). Problems that use compound interest factors to compare alternatives typically make evaluations using effective interest rates for the computations.

Example 7: nominal and effective interest—In the previous examples, the effective interest rate was 8 percent per year, because the interest period was assumed to be 1 year. If instead the interest rate is 8 percent per year, compounded quarterly ($m = 4$ subperiods per year), 8 percent is the nominal interest rate. The effective interest rate is

$$i_a = (1 + 0.08/4)^4 - 1 = 0.0824 = 8.24 \text{ percent per year}$$

8.9.6 ECONOMIC ANALYSIS TECHNIQUES

Three economic analysis techniques that often have been used previously to evaluate environmental project alternatives are summarized below. There are many other economic analysis techniques available, but these three techniques frequently have been used in the past to evaluate environmental project alternatives. Those individuals who want to investigate these or other economic analysis techniques in more detail should seek a basic textbook on engineering economic analysis techniques.

Present Worth Analysis

Present worth analysis is an economic analysis technique for comparing alternatives. This technique takes future receipts and disbursements of alternatives and converts them to present values for comparison. Present worth analysis is the most widely used economic analysis technique for environmental projects.

There are three general scenarios for the use of present worth analysis: fixed input, fixed output, and neither input nor output fixed. Fixed input represents the scenario of fixed input resources and different output for each alternative. Fixed output represents the scenario of a fixed benefit or task accomplished and different input for each alternative. Neither input nor output fixed represents the scenario of alternatives with different input resources and different benefits.

The economic efficiency criterion is different for each scenario. The fixed input scenario has the criterion of maximizing the present worth of outputs. The fixed output scenario uses the criterion of minimizing the present worth of inputs. When neither inputs nor outputs are fixed, the economic efficiency criterion is maximizing the present worth of the difference between outputs and inputs.

Environmental projects typically are considered to be fixed output scenarios. Meeting the environmental regulatory directives is the fixed output. The chosen alternative is that one that meets the directives at the least present value cost (input). For example, a groundwater remediation project may have a fixed output, such as remediating groundwater to specified concentrations of pollutants. The selected remediation approach is that which obtains these concentrations at minimum present value costs.

The present worth analysis procedure varies depending on the time period for analysis. There are three general situations involving the time period for analysis. The first situation occurs when the useful life of each alternative equals the analysis time period. The second situation occurs when the useful life of each alternative is different than the analysis time period. The third situation occurs when the analysis time period is infinite. Environmental projects typically assume that the useful life of each alternative equals the analysis time period, which is the most simple situation to analyze.

Example 8: present worth analysis—Remediation of petroleum constituents in groundwater at a specific site can be accomplished by either air sparging technology or in situ bioremediation. An air sparging system has a construction cost of $250,000, has an operational cost of $150,000 per year, and can obtain target cleanup levels in 2 years with no costs in years 3–5. The in situ bioremediation system has a construction cost of $150,000, has an operational cost of $100,000 per year, and can obtain target cleanup levels in 5 years. Using an interest rate of 6 percent per year and an analysis time period of 5 years, compute the present value costs of both alternatives and recommend a remediation technology for the site.

Solution

Air sparging

$$P_{as} = \$250,000 + \$150,000\,(P/A, 6\text{ percent}, 2) = \$250,000 + \$150,000\,(1.833) = \$524,950$$

In situ bioremediation

$$P_{ib} = \$150,000 + \$100,000\,(P/A, 6\text{ percent}, 5) = \$150,000 + \$100,000\,(4.212) = \$571,200$$

Since both technologies can obtain target cleanup levels, air sparging should be chosen over in situ bioremediation because it has the lesser present worth cost.

Annual Cash Flow Analysis

Annual cash flow analysis is a second economic analysis technique for comparing alternatives. It is similar to present worth analysis in that it has the same three general scenarios and economic efficiency criteria that present worth analysis has. It differs from present worth analysis in that results are given as equivalent uniform annual costs (EUAC) or equivalent uniform annual benefits (EUAB) instead of the present values computed in present worth analysis. Annual cash flow analysis also yields the same conclusions as present worth analysis.

Example 9: annual cash flow analysis—Given the same conditions as presented in example 8, compare air sparging and in situ bioremediation by annual cash flow analysis.

Solution

Air sparging

$$EUAC_{as} = \$524,950\,(A/P, 6\text{ percent}, 5) = \$524,950\,(0.2374) = \$124,623$$

In situ bioremediation

$$EUAC_{ib} = \$150{,}000\,(A/P,\,6\ \text{percent},\,5) + \$100{,}000 = \$150{,}000\,(0.2374) + \$100{,}000 = \$135{,}610$$

Annual cash flow analysis also determined that air sparging is the lower cost alternative over a 5-year analysis period.

Benefit-Cost Ratio Analysis

A third economic analysis technique that has been used with evaluation of environmental project alternatives is benefit-cost analysis. In this method, one calculates the present worth of benefits and costs (or EUAB and EUAC) and then states them as a ratio of benefits to costs, a benefit-cost ratio (B/C). Only those alternatives with B/C > 1.00 are selected for further evaluation. For fixed input or fixed output scenarios, the economic efficiency criterion is to select the alternative that has the maximum B/C. When nether benefits nor costs are fixed, an incremental analysis technique must be used. Benefit-cost analysis yields the same conclusions as present worth and annual cash flow analysis.

Example 10: benefit-cost analysis—Given the same conditions as presented in example 8, compare air sparging and in situ bioremediation by benefit-cost analysis. Assume that the environmental regulatory agency has determined that the benefits of remediation of this site have a present worth of $750,000 ($EUAB = \$178{,}050$).

Solution

Air sparging

$$B/C_{as} = \$750{,}000/\$524{,}950 = 1.43 \quad \text{or} \quad B/C_{as} = \$178{,}050/\$124{,}623 = 1.43$$

In situ bioremediation

$$B/C_{ib} = \$750{,}000/\$571{,}200 = 1.31 \quad \text{or} \quad B/C_{ib} = \$178{,}050/\$135{,}610 = 1.31$$

Benefit-cost analysis has also chosen air sparging, which has B/C > 1.00 with B/C ratio greater than that for in situ bioremediation.

8.9.7 ECONOMIC ANALYSIS FOR PUBLIC SECTOR PROJECTS

A branch of economics called welfare economics was developed to deal with the complexities associated with public investment decisions that *"promote the general welfare,"* as described in the U.S. Constitution. Environmental projects may be considered as public investment decisions because benefits and/or costs for the project may be borne by the public.

The public investment characteristics of environmental projects generally complicate the alternative evaluation process. For example, a municipal groundwater supply remediation project resulting from an industrial spill will be viewed very differently by the private responsible party who pays most of the costs of remediation, by the direct beneficiaries of the cleanup in the public, and by members of the general public who do not directly benefit from the remediation. The responsible party typically has minimization of remediation costs as a primary objective. The direct beneficiaries typically have maximizing benefits as their primary objective. The general public typically desires that the remediation adequately protect the public and the environment and while minimize utilization of public resources. The conflicting objectives from different perspectives make these evaluations significantly more complex than most evaluations in the private sector. The proper approach is to take a viewpoint at least as broad as those who pay the costs and as those who receive the benefits.

Selection of an interest rate for economic analyses in environmental projects also can be a complex process. Government investment decisions may use various measures such as cost of capital, a government opportunity cost, or a taxpayer opportunity cost to determine an interest rate for analysis. Environmental projects usually have the interest rate set by the responsible environmental regulatory agency.

Finally, the public nature of environmental projects invites political considerations to enter into the decision-making process, which significantly complicate the evaluation process. Most feasibilty study procedures contain a public comment component that is included in the decision-making process. Selection of an environmental project alternative is rarely made without considering the attitude of the public toward the decision.

8.9.8   CASE STUDY—A MICHIGAN MANUFACTURING COMPANY

A case study in which engineering economics techniques were actually used is presented to put the methods previously discussed in a real-world perspective. How other economic techniques could be used is also discussed.

Site Description

The case study involves a small company in Michigan that fabricated aluminum and copper tubing for refrigerators and air conditioners at a single facility on a 5-acre (20.235-km^2) site. Low concentrations (fewer than 10 parts per billion) of trichloroethylene (TCE) were detected in samples collected from an onsite drinking water well during routine testing to meet public health requirements. TCE was used at the facility as a degreasing compound for the fabricated tubing. Environmental enforcement actions subsequently were initiated against the company.

Remedial Investigation

After several phases of remedial investigation activities, the environmental consultant for the company determined that a spill of 200–250 gal (757–946.25 liters) of TCE had occurred at the facility several years before the TCE had been detected in the well water. The TCE had contaminated groundwater in a shallow, permeable, complex geologic formation at the facility extending less than 20 ft (6.1m) below the ground surface. The shallow permeable formation was underlain by a relatively impermeable layer of silt, which had prevented further downward migration of the TCE.

Risk Assessment

A risk assessment of the spill was also conducted to determine the appropriate cleanup levels at the facility. The carcinogenic risk to a child in a future residential land use scenario due to the existing TCE contamination was determined to be approximately 10^{-3}, significantly higher than the regulatory maximum carcinogenic risk of 10^{-6}. An industrial land use scenario yielded a carcinogenic risk value slightly below 10^{-6}. The lead regulatory agency concluded that evaluation of alternative remedial actions at the facility was necessary.

Feasibility Study

A feasibility study of alternative remedial actions was performed by the environmental consultant for the company. Two general remedial action approaches were evaluated. The first remedial approach was institutional controls (deed restrictions) that prevented future residential land uses of the facility site and long-term groundwater monitoring at the facility site. This approach would meet minimum

TABLE 8.9.1   Michigan Manufacturing Company Case Study: Present Worth Analysis.

Alternative	Capital cost ($)	Operation & maintenance costs ($/year)	Present worth ($)
Monitor/controls	36,000		
First 3 years		88,000	
Next 27 years		23,000	
			565,000
Active remediation	452,000		
First 3 years		134,000	
Next 27 years		73,000	
			1,738,000

regulatory requirements. The second approach was active remediation of the soil and groundwater contaminated by TCE by dual-phase (soil vapor and ground water) extraction techniques. This approach would also meet minimum regulatory requirements.

A present worth analysis of both approaches was calculated with an interest rate of 6 percent. Table 8.9.1 summarizes the results of the present worth analysis.

The lead regulatory agency selected the groundwater monitoring and institutional controls approach. In general, environmental regulatory agencies prefer active remediation over other remedial approaches because the contaminants are permanently removed from the site. The lead regulatory agency made its decision based on four site-specific factors. First, the shallow permeable formation that had been contaminated was not used as a drinking water supply. Second, the next permeable geologic formation below the shallow unit showed no signs of contamination. Third, the future residential land use scenario was not likely to be realized because the site had been in industrial use for the previous 50 years. Last, the effectiveness of the active treatment alternative was questionable because of the complex geology of the shallow unit. Although the groundwater monitoring and institutional controls alternative had a smaller present worth cost than the active remediation alternative, the economic efficiency criterion was not explicitly identified by the lead regulatory agency as a primary factor in choosing the selected remedial approach.

Benefit-Cost Analysis

Some further insight might be gained if one looks at this situation with a hypothetical benefit-cost analysis approach. Reduction of carcinogenic risk at the facility site can be thought of as benefits. Costs consist of the capital, operation, and maintenance costs of the remediation alternatives.

Table 8.9.2 shows estimates of the remaining carcinogenic risk levels and the potential exposed population size for risk scenarios associated with each remedial alternative. The no action scenario is used as a reference condition.

If the remaining carcinogenic risk is multiplied by the exposed population, the product is the number of exposed individuals with carcinogenic effects, which will be called the carcinogenic exposure at the site. Table 8.9.3 shows the carcinogenic exposure, present worth cost, and B/C of each remedial alternative. Because neither the benefits nor the costs are fixed, we must use an incremental benefit-cost analysis. Therefore, the benefit-cost analysis comparing alternatives uses a $\Delta B / \Delta C$ ratio.

TABLE 8.9.2   Michigan Manufacturing Company Case Study: Remedial Alternatives.

Alternative	Risk scenario	Cancer risk	Exposed population
No action	Residential, child	10^{-3}	6
Monitor/controls	Industrial worker	10^{-6}	150
Active remediation	Commercial	10^{-8}	2500

TABLE 8.9.3   Michigan Manufacturing Company Case Study: Carcinogenic Exposure, Present Worth Cost, and B/C of Each Remedial Alternative

Alternative	Exposure (10^{-3})	Present worth cost ($\$10^6$)	$\Delta B / \Delta C (10^{-9}/\$)$
No action	6.0	0	—
Monitor/controls	0.150	0.565	10.4
Active remediation	0.025	1.738	3.44

Table 8.9.3 indicates that the groundwater monitoring and institutional controls alternative has a better B/C than active remediation. However, only alternatives with B/C >1 should be considered. The units for the B/C in Table 8.9.3 is exposures per dollar. When the value of reducing one cancer exposure exceeds $96.2 million, B/C >1.00.

In conclusion, this benefit-cost analysis exercise also might choose monitoring and controls over active remediation. Naturally there would be some disagreement between interested parties about whether the B/C of monitoring and controls was >1.00. The greatest difficulty with using benefit-cost analysis to evaluate environmental project alternatives is that agreement by interested parties about the value of the benefits is rarely achieved in practice.

BIBLIOGRAPHY

Newnan, D. G., and Lavelle, J. P., *Engineering Economic Analysis*, 7th ed., Engineering Press, 1998.

Wolka, K. K., "Emerging Ideas: Site-Specific Benefit-Cost Analysis for Environmental Remediation Projects," in *Practice Periodical of Hazardous, Toxic, and Radioactive Waste Management*, American Society of Civil Engineers, 1997, pp. 47–49.

SITE BASED ENVIRONMENTAL SCIENCE, HEALTH AND TECHNOLOGY

CHAPTER 9
POLLUTION PREVENTION CONCEPTS AND POLICIES

SECTION 9.1

POLLUTION PREVENTION SCIENCE AND TECHNOLOGY

Thomas T. Shen

Dr. Shen is an independent international advisor in environmental research and engineering.

9.1.1 INTRODUCTION

In the rush toward industrialization, it became expedient for people to produce more products and to consume more. Now the extent of environmental destruction has forced us to recognize that the scale of industrial production worldwide seems set for inexorable growth. Ecosystems in all parts of the world have declined severely. Developing countries clearly aim to achieve the levels of material prosperity enjoyed in the developed countries, and they intend to do it by industrializing. We are witnessing the evolution of a fully industrialized world, with global industrial production, global markets, global telecommunication, global transportation, and global prosperity. This prospect brings the realization that past technology development has led to deterioration of the natural environment and the quality of human life. We need to develop new technologies to minimize stress on the environment. Current environmental management technologies and patterns of industrial production will not be adequate to sustain environmentally safe growth unless a pollution prevention concept is faithfully implemented in all fields: government, industry, agriculture, mining, land use, transportation, and energy use and consumption.

This section calls for a deeper understanding of pollution problems and presents the concept and practices of pollution prevention. It explains the definitions of pollution, pollution prevention (P2), and P2 technology. Emphasis is given in preventive technologies in industrial production design and manufacturing, energy efficiency and alternative sources, and cleaner transportation. Discussions include P2 assessment, benefits, barriers, challenges, and opportunities. The purpose of this section is to provide readers with knowledge and information about P2 technologies with various related issues to protect our environmental quality and natural resources.

Pollutants and Wastes

Pollutants are unused raw materials or by-products resulting from the production process and from various human activities. Once pollutants enter the environment, they cycle throughout the air, water, and soil. Unless they are destroyed, pollutants will continue to transfer from one medium to another as shown in Figure 9.1.1 (Shen and Sewell, 1986). Many pollutants and wastes have multiple risks and adverse effects. There also may be synergistic interactions and cumulative exposures from different pollutants and wastes in the environment. For many harmful pollutants and chemicals, there are no safe levels of exposure; this means that even very small exposures eventually can result in serious and perhaps fatal health effects. In our environment, there are hundreds of toxic pollutants that are not regulated at all.

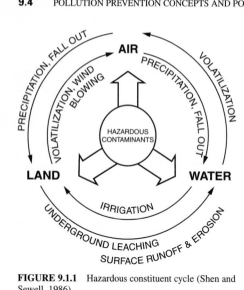

FIGURE 9.1.1   Hazardous constituent cycle (Shen and Sewell, 1986).

Wastes (liquid, gaseous, and solid) contain pollutants, which are discarded process materials or chemicals. These pollutants, released into the environment beyond the assimilation capacity of the environment, cause pollution. Industry is the primary source of all waste generators because of the quantity and toxicity of its source, although not all industrial waste contains toxic pollutants. Industrial production, distribution, transport, storage, consumption of goods, and environmentally unfriendly products and services have known to be the most critical sources of environmental pollution problems. Only in recent years have we begun to realize that environmentally unfriendly products and services can cause even greater pollution in terms of risks to human health and the environment. Some examples of environmentally unfriendly products are explosives, toxic chemicals, pesticides, herbicides, food and drugs, plastics, and short-life disposal items. Unfriendly environmental services such as engineering design and construction, management practices, and education and training can cause long-term poor administration and wasteful human behavior. Environmental laws, regulations, and management strategies in developed countries are being modified or expanded to cover not only pollutants and wastes but also environmentally unfriendly products and services.

All sectors of our society generate waste. Today's rapidly changing technologies, industrial products, and practices may generate wastes that, if improperly managed, can threaten public health and the environment. Many wastes, when mixed, can produce a hazard through heat generation, fire, explosion, or release of toxic substances. Environmental problems, or risks to health and safety, often are difficult to foresee. From a technical viewpoint, the increase in toxic chemicals and hazardous wastes in our environment has become the most urgent environmental pollution problem both now and in the future. To minimize this urgent problem, we must take immediate pollution prevention action on a worldwide scale of a magnitude never before undertaken by mankind. Waste generators must describe and characterize their wastes accurately by including information about the type and the nature of the waste, chemical composition, hazardous properties, and special handling instructions (Shen, 1997a).

Understanding Pollution Problems

Traditionally, pollution focused primarily on pollutants and wastes, as evidenced by our environmental laws, regulations, and management strategies. Throughout the years, environmental management strategies have emphasized controlling pollutants and waste released from manufacturing facilities. There has been little concern about managing industrial product transport, storage, distribution, and consumption, all of which have proved to be as important as waste management. Certain useful products, while leaving the manufacturing plant for distribution through transport and storage (explosives and toxic liquid chemicals) as well as for consumption (drugs and unhealthy food) and after-use disposal (plastics and hazardous materials), also may cause pollution (Shen, 1999).

Although there have been considerable advancements in the development and application of environmental control technologies, our environmental problems are becoming more severe and far-reaching. Considerable progress has been made during the past three decades in developing and applying technologies for controlling wastes and pollutants from major sources by reducing, for example, air emissions from fossil fuel combustion and motor vehicle sources as well as wastewater releases from industrial plants. At the same time, there are large numbers of waste and pollutant sources from small business and commercial operations that are very difficult or impossible to address by the traditional application of end-of-pipe treatment and disposal technologies and the regulatory programs through which they are implemented.

There are three stages of industrial pollution problems, which must be solved if our environmental quality is to be protected. The first stage of industrial pollution problems is release of waste within the plant facility; the second stage deals with industrial product use including product transport, storage, and distribution; and the third stage concerns disposal of used and unused products. The increased releasing of environmentally unfriendly toxic chemicals and hazardous wastes from various sectors of society into the environment has created a serious pollution problem. The commonsense, precautionary response to growing pollution problems is to seek to prevent pollution before it happens. Where it is already occurring, the aim should be to eliminate the source of the problem rather than to attack symptoms by often-expensive "end-of-pipe" methods such as filters, scrubbers, precipitators, cyclones, treatment plants, and incineration (Ling, 1998).

Current pollution problems are also the result of unintentional mismanagement by previous decision makers who did not consider effects on the natural environment while they were pursuing socioeconomic development. These decisions "borrowed" from the environment, and the debt now has come due in the form of expensive cleanups and other remedial action. Formation of toxic or hazardous waste constitutes a problem of its own. We are dealing with toxic chemicals and hazardous wastes that present a real danger to human health or the environment. We must deal with the underlying causes of the problem and not just the symptoms. Because we must have natural assets to support our economy and a healthy human population, decision makers need to follow a sustainable development concept and a pollution prevention strategy. Such strategies will integrate environmental and economic considerations based on a total cost system with sound pollution cost information, including costs of pollution control and pollution damage.

Concept and Practice of P2

P2 is the act of taking advance measures against possible or probable pollution. P2 seeks not only to reduce pollutants and wastes but also to optimize the total materials cycle from virgin materials, to finished materials, to component, to product, to obsolete product, to eliminate disposal, and to various other services. P2 is generally contrasted with pollution control. Generally speaking, P2 requires less effort, time, and money than pollution control. P2 involves a broad environmental management strategy, which includes practicing of waste minimization, source reduction, cleaner production, and design for the environment. Waste minimization and source reduction deal with pollutants and wastes, whereas cleaner production and design for the environment deal not only with pollutants and wastes but also with products and services. In practice, P2 approaches can be applied not only to industrial sectors but to all pollution-generating activities, including energy production and consumption, transportation, agriculture, construction, land use, city planning, government activities, and consumer behavior.

In 1976, Dr. Joseph Ling of 3M Company talked about a program of "Pollution Prevention Pays" or 3P program during the first United Nations Economic Commission for Europe Seminar on "Principles and Creation of Non-waste Technology" held in Paris. The 3P program has been based on technological and management advances that reduced environmental releases of wastes and resulted in production costs lower than those associated with the previous pollution control method. Dr. Ling believes that pollution controls solve no problem. Pollution controls only alter the problem, shifting it from one form to another, contrary to one of the immutable law of nature—the form of matter may be changed but matter does not disappear. Conventional controls, at some point, create more pollution than they remove and consume resources out of proportion to the benefits derived. It takes resources to remove pollution; pollution removal generates residue; it makes more resources to dispose of this residue and disposal of residue also produces pollution. Although the United Nations meeting set out principles of P2 in 1976, it was not until 1990 that the U.S. Congress enacted the Pollution Prevention Act. It has taken the world a long time to work into the P2 mode. The P2 approach has been gaining ground in policy and corporate circles as the most cost-effective way to achieve environmental and economic efficiency. More companies are realizing the pollution they produce is a sign of inefficiency and that waste reflects raw materials not used in final products.

Environmental management strategies are gradually transforming as more and more professionals accept the P2 concept. The difference between pollution control technology and P2 technology is that the former emphasizes controlling pollutants in wastes and minimizing waste generation at the sources; the latter is not just to control pollutants and wastes but also to search for better ways to manufacture and consume environmentally friendly products and services in a broader sense. Sometimes the pursuit of P2 produces subtle changes in institutional trends, changes that in small ways inspire considerable improvements. Therefore, environmental management strategy needs to upgrade beyond waste management to a broader approach of pollution management.

The U.S. Pollution Prevention Act of 1990 identified P2 as a national objective and established the following waste management hierarchy: prevention, recycling, treatment, and disposal. All these options are valid, and the current state of economic and environmental knowledge supports using a mix of them. Nonetheless, P2 is the preferred approach to protecting environmental quality for a variety of reasons:

- Individual pollutant controls do not always address cross-media impacts; for example, the sludge created by treatment of air or water pollution becomes a waste disposal problem.

- Both nonpoint pollution sources, such as urban storm water runoff, and small, dispersed point sources, such as commercial establishments, contribute a significant portion of society's harmful pollutants. However, such sources are difficult to regulate with traditional large-source emission standards.

- P2 embodies efficiency in resource and energy use and therefore, in most cases, offers a more cost-effective solution in the long run than direct regulation. This is particularly relevant at a time when economic competitiveness is a national priority and total pollution control costs to businesses, industries, and public agencies have grown annually [U.S. Environmental Protection Agency (U.S. EPA), 1998].

Many organizations in the United States define pollution prevention in terms of the Pollution Prevention Act hierarchy or Resource Conservation and Recovery Act waste management hierarchy, although there is much debate about whether to include recycling in the definition. Some define P2 more narrowly as source reduction or toxics use reduction; others view P2 more conceptually as any process that involves continuous improvement and movement up the environmental management hierarchy. Many terms similar to P2 are in use today such as waste minimization, waste reduction, clean technology, and design for the environment. There is no universal consensus about what these terms mean. As far as their objectives are concerned, this section uses a broad term of P2 to cover the following four terms:

- Waste minimization refers to reduction or elimination of generation of waste (the total volume or toxicity) at the source, usually within a process.

- Source reduction is defined as any practice that reduces the amount of any hazardous substance, pollutant, or contaminant entering any waste stream or otherwise released into the environment before recycling, treatment, or disposal. It includes practice, which reduces the hazards to public health, and the environment associated with the release of such substances, pollutants, or contaminants. Source reduction includes equipment or technology modifications; process or procedure modifications; reformulation or redesign of products; substitution of raw materials; and improvements in housekeeping, maintenance, training, and inventory control.

- Design for the environment is a process that aims at minimizing or preventing environmental damage as a design objective in the first place.

- Clean technology aims at product design and manufacturing processes that use less raw materials, energy, and water; generate less or no wastes (gas, liquid, and solid); and recycle waste as useful materials in a closed system. It may alter existing manufacturing processes to reduce generation of wastes. Clean technology includes the use of materials, processes, or practices that reduce or eliminate the creation of pollutants or wastes (Shen, 1999).

In the United States, P2 legally means source reduction—that is, any practice that (1) reduces the amount of any hazardous substance, pollutant, or contaminant entering any waste stream or otherwise released into the environment before recycling, treatment, or disposal; (2) reduces the hazards to public health and the environment associated with the release of such substances, pollutants, or contaminants; (3) increases efficiency in the use of raw materials, energy, water, or other resources; or (4) protects natural resources by conservation. P2 also includes equipment or technology modifications; process or procedure modifications; reformulation or redesign of products; substitution of raw materials for cleaner production; and improvements of managerial services in housekeeping, maintenance, training, or inventory control. As the P2 concept has evolved, successful case studies reveal key characteristics of P2, which contrast sharply with the concept of pollution control, the dominant environmental policy for the past three decades worldwide. The Canadian Ministry of Environment defines P2 as any action that reduces or eliminates the creation of pollutants or wastes at the source, achieved through activities that promote, encourage, or require changes in the basic behavioral patterns of industrial, commercial, and institutional generators or individuals.

9.1.2 P2 TECHNOLOGY

This section defines P2 technology and selects three critical sectors (industry, energy, and transportation) as illustrative examples to explain why and how P2 technology applies. Technology makes useful things that enrich our lives. Computers, television sets, cars, buildings, aircraft, tractors, and various machines are products of technology. The application of technology has been the central means of greater human productivity and consequent increases in standards of living. As technology has increased in sophistication and complexity, so have its adverse and widespread impacts on the environment. For example, the increased production of electricity from fossil fuels has resulted in increased emissions of air pollutants that damage human health and the ecology; plastic products have complicated the disposal of solid wastes.

What Is P2 Technology?

P2 technology is a tool for minimizing undesirable effluents, emissions, and wastes from manufacturing processes as well as reducing environmentally unfriendly products and services in production and consumption. Closed technological systems produce the least waste and use the fewest resources. P2 technology requires a total system approach that prevents pollutants from being created in the first place or that minimizes undesirable wastes and obviates the need for treatment and control. A preventive approach involves using fewer or nonpolluting materials, designing processes that minimize waste products and pollutants or that direct them to other useful purposes, and creating recyclable products. A total environmental management approach requires P2 technology to examine the full life cycle of products and practices. P2 technologies mean profound change in thinking and action in environmental strategy priority from pollution control technologies to P2 technologies.

Because P2 is a very broad term and aims for all sectors of society, many organizations and institutions can apply P2 concept and technology, which not only reduces the generation of waste materials but also produces environmentally friendly products and services. These are ways the technical design of the P2 process may change to accommodate the new criterion of environmental performance. Clearly, each design discipline will evolve differently, but in general terms we should see a broader interpretation of the design process, encompassing the cradle-to-grave aspects of the design and therefore necessarily covering many technical and managerial aspects. Information gathering and interpretation will become more important. The critical need of P2 technology must focus on (1) industrial toxic chemicals (increasing knowledge about chemicals, improving information, and substituting toxic chemicals with less toxic ones), (2) energy efficiency, and (3) a system approach of preventive technology. P2 technology involves process waste minimization, recycling and reuse, modifying input materials and energy, and rethinking output products. However, both industry and government must assess the effectiveness of new technologies in meeting the goals of pollution prevention and economic efficiency to ensure that any new P2 technology is an environmentally sound technology.

Industrial Product Design and Manufacturing

Industry involves thousands of products and production processes, resulting in a decentralized enterprise system. Current P2 technology generally employs conventional engineering approaches. P2 techniques for industrial manufacturing facilities such as waste minimization and source reduction can be understood by observing the path of material as it passes through an industrial site. Even before materials arrive at the site, we could avoid toxic materials, when less toxic substitutes exist. P2 technologies for industries can be categorized into five groups:

- Improved plant operation,
- In-process recycling,
- Process modification,
- Materials and product substitutions, and
- Materials separations (U.S. EPA, 1992).

P2 technology embraces waste minimization and source reduction. Figure 9.1.2 shows a diagram of source reduction methods, which can be accomplished through product changes or process changes to meet new environmental requirements.

In designing products, manufacturing processes, and consuming products, new preventive technologies will give us a competitive edge as well as a healthier environment. The worldwide market for such technologies will continue to grow as the connections between environmental and economic well-being become more apparent. The separation of concept and execution will become increasingly difficult when environmental considerations have to be at the forefront. Many environmental issues

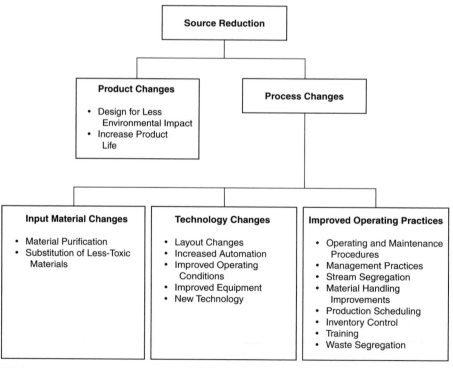

FIGURE 9.1.2   Source reduction methods (U.S. EPA, 1992).

will relate to aspects of execution—for example, the materials used, the production process used, how to dispose of a product. These questions have to be addressed during the development and design of a product and not introduced as an afterthought once the product has been manufactured.

According to the findings of the Industrial Pollution Prevention Project, the four most important general motivators for P2 in industry are economics, technical and financial assistance, open communication, and flexibility (especially regulatory flexibility). The project found that the key trigger for P2 is a stringent regulation or enforcement action (U.S. EPA, 1995). The desire to avoid being subject to environmental pressures from regulations and from consumers and professional advocacy provided the most critical impetus for P2, not only motivating source reduction initiatives but also ensuring their success in the marketplace.

Product Design.   P2 in product design includes practices that reduce or eliminate the creation of pollutants through increased efficiency in the use of raw materials, energy, water or other resources, or protection of natural resources by conservation. It calls for the judicious use of resources by source reduction, energy efficiency, reuse of input materials during production, and reduced water consumption. P2 design technologies that can be used in any industrial facility are material separation, process modifications, and product changes, which essentially aim for cleaner production. They reduce the volume and toxicity of waste production and of end products during their product life cycle and at disposal.

Product life-cycle assessment is an objective tool or method to identify and evaluate all aspects of environmental effects associated with a specific product of any given activity from the initial gathering of raw material from the earth until the point at which all residuals are returned to the earth. The method can also be used to evaluate the types and quantities of product inputs, such as raw materials, energy, and water, and of product outs, by-products, and waste that affects various resources management options to create sustainable systems (Vigon et al., 1993). Life-cycle analysis (LCA) as shown in Figure 9.1.3, is composed of four components: (1) goal definition and scope, (2) inventory analysis, (3) impact analysis, and (4) improvement analysis.

Corporate and government attention to the environmental impacts of products has existed since the beginning of environmental concern. Some products—petrochemicals, for example—have had to be closely regulated to minimize inherent risks to health and the environment. Others, such as cars, have faced classic end-of-pipe treatment with the mandating of catalytic converters to reduce nitrogen oxides and hydrocarbon exhaust emissions. The state of California has begun the process of requiring the sale of zero emission vehicles that prevent pollution by using alternative fuels. Managing a product

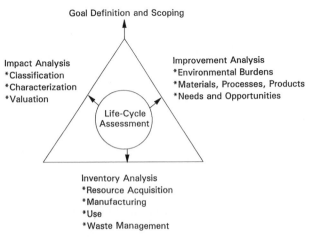

FIGURE 9.1.3   Product (LCA) framework.

life cycle for minimal environmental impacts poses tough conceptual and operational challenges for businesses. Each step in the life of a product has implications for the environment, often giving rise to a number of issues. Business, research institutes, and governments are working to develop LCAs. But LCAs go far beyond studies of energy balance. They also should be used to evaluate resource requirements and environmental impacts. Each LCA has three parts: (1) an inventory of energy, resource use, and emissions during each step of the product's life; (2) assessment of the impact of these components; and (3) an action plan for improving the product's environmental performance. As of today, LCA still is not widely applied.

It is important to understand that LCAs are seen as a technical tool and not as a panacea for resolving the complex environmental issues involved with every product. The purpose is to stimulate action and improvement. Procter & Gamble is one of the leaders in LCA and has used it to illuminate a number of controversial issues, such as the relative merits of disposable diapers (which it produces) and cloth diapers. The company has also used LCAs to identify ways to reduce packaging waste. This has led to packaging innovations, such as maximizing the use of recycled materials, selling detergents in a concentrated form, and introducing refillable containers. LCA implies life-cycle responsibility. A combination of increasing external pressures and growing internal commitment has made some leading companies ensure that their products are made, used, and disposed of in the most environmentally compatible ways.

Asbestos, for example, was once seen as an ideal construction material—cheap, versatile, resistant, and usable in all types of buildings. During the 1970s, however, evidence was growing that asbestos fibers, when inhaled, present a serious health risk. As scientists were unable to indicate a safe threshold of exposure to asbestos dust, in 1980 the Swiss Eternit Group decided to stop using asbestos altogether. In a far-reaching technology cooperation program between the Swiss parent and its affiliates, the problem was turned into an advantage through innovation. In some countries, fiber cement could be produced with locally available renewable materials at lower cost and was able to gain considerable market shares, because the new products were better adapted to local needs. At Ciba-Geigy, finished products represented only 30 percent of all outputs in 1979, the rest being waste. By 1988, the company's efficiency had increased to 62 percent; a goal of 75 percent efficiency has been set for the end of this decade. In one particular process at Ciba-Geigy, producing 1 metric ton (1000 kg) of a chemical called amide traditionally required 3 tons of highly corrosive phosphorous trichloride and 12 tons of water; 14 tons of resulting effluent had to be treated. This has been replaced by a system that uses only 1.9 tons of raw materials and no water; the by-products are 0.6 ton of pure acetic acid, which can be recycled in other processes, and 0.3 ton of solid organic waste, which is incinerated (Schmidheiny, 1992).

Product reformulating has been seen largely as a way to reduce waste and pollution. But companies are realizing that to capture and retain customers in an increasingly environmentally conscious marketplace, the products themselves have to be cleaner. According to 3M's Joe Ling, the challenging question is "What waste are we creating for our customers?" Thus 3M has reexamined some of its products. Nevertheless, Dr. Ling believes that "environmental constraints have helped 3M focusing its efforts and developing better and more-efficient products." For example, the need to eliminate the use of polyvinyl chloride in products destined for the computer industry has resulted in a product reformulation that is "less expensive, better quality, and better for the environment." The reformulation of single-use plastic bags, containers, and packaging products by many industries is another example of successful waste reduction.

All outputs from a manufacturing facility can be put into two classifications: product and waste. Anything that the customer pays for is a product; all else that leaves the facility is waste. Industry must strive to prevent or reduce the waste from manufacturing because waste represents an inefficient use of scarce resources. Industry must understand that pollution represents a loss of profits in manufacturing. The future factory must consider the environmental implications of a new product and/or service. It must design pollution management strategies into its research experiments. It must insist on pollution prevention in the acquisition of new components or systems from others so that waste generation in the life cycle of a product can be minimized. The future factory must constantly use preventive techniques to reduce waste and seek solutions. It is critical for people to know what to do. If behavior changes are required, the responsibility for change rests on thoughtful management.

Planning and Implementation. Implementation of P2 technology in any industrial facility requires careful planning. An industrial facility is a complex entity, which needs many different skills to operate properly (i.e., technical, management, financing, marketing, purchasing, scheduling, and labor). Planning a P2 project is just as complex, because the P2 concept and ethic must permeate every activity conducted at the facility. An industrial P2 plan consists of organizing a P2 team, establishing goals, collecting and analyzing data, identifying P2 opportunities, heightening employee awareness and involvement, promoting education and training, and presenting P2 project proposals.

According to the U.S. EPA's toxic release inventory, more than 20 industries release toxic wastes, such as food, tobacco, textiles, apparel, lumber, furniture, paper, printing, chemicals, petroleum, plastics, leather, stone/clay, primary metals, fabricating metals, machinery, electrical, transportation, photography, and others. Those industries can choose from a broad range of options to apply P2 technologies and improve their products' environmental performance. The options are:

- Eliminate or replace product
- Eliminate or reduce harmful ingredients
- Substitute environmentally preferred materials or process
- Decrease weight or reduce volume
- Produce concentrated product
- Produce in bulk
- Combine the functions of more than one product
- Produce fewer models or styles
- Redesign for more efficient use
- Increase product life span
- Reduce wasteful packaging
- Improve reparability
- Redesign for consumer reuse
- Remanufacture the product

Product engineers who accept new challenges must consider P2 issues such as recyclability, durability, and repairability in choosing technologies and materials. Clearly, the most radical solution is to remove a product from the market. In fact, some chemical companies have voluntarily withdrawn products they believed posed too great a risk to the environment. Ciba-Geigy, for example, has taken about 40 dyes off the market because it was unable to change the production process sufficiently to make the product both economically and environmentally efficient. More companies have been and are using more environmentally compatible materials in the end product.

Technology has historically moved in the direction of dematerialization, which can benefit the environment. Optical fibers have replaced copper wires for communications, silicon chips have replaced vacuum tubes for electronic uses, and composite materials have replaced metals in many applications because of weight and strength advantages. These and other technological substitutions have increasingly reduced the volume of materials needed to perform industrial and societal functions, thereby reducing the amount of natural material used. Engineers have to consider a wide range of criteria as part of the design process: marketing, production, and financial and technical considerations. Compared with these, environmental considerations could be even more complex and hard to handle. Often there are no clear answers, information is hard to find, guidelines may not be available, and much original research and thinking may be necessary. Increasingly, businesses and institutions are assessing their environmental performance comprehensively with tools such as environmental auditing and recognizing that good environmental performance does not come simply from attempting to improve in one or two areas. Development of a product that is environmentally safer in use or disposal may well be desirable, but if its production consumes four times as much energy the net result may be less beneficial.

In summary, the conceptual and technological foundations for achieving P2 have already been laid. There is considerable and growing experience of P2, and companies are beginning to forge ahead with product life-cycle responsibility. But P2 will be challenged to achieve zero pollution emissions from production plans and to redirect product development to meet basic needs, including those of the poor. P2 also will be challenged to prevent pollution from product designing and manufacturing as well as from environmentally unfriendly products and services.

Energy Efficiency and Alternatives Sources

Of all the technological areas that need to be put on a sustainable basis, energy is the most critical. Energy is essential for economic production and a high standard of living. Efficiency is the most cost-effective and environmentally beneficial approach for resolving the major energy problems at this time. Although a totally nonpolluting, nonfossil energy system may be feasible in the future, at present we need technologies to increase energy efficiency, especially in developing countries. Generation and use of energy are responsible for a large portion of almost all forms of pollution. For this reason alone, sustainable development will be impossible without new energy technologies. Moreover, if cheap and nonpolluting energy technologies were in place, a host of new, less-polluting, and economically attractive industrial, agricultural, and transportation technologies could be used. In the energy sector, P2 technologies can reduce environmental damages from extraction, processing, transport, and combustion of fuels.

Fossil fuel, the dominant source of energy supply sustaining modern civilization, creates some of the most threatening environmental problems, ranging from global-scale changes in climate to regional effects of acid deposition to local urban air pollution problems and health and safety risks in mining and production. Environmental concerns and sources of fuel require the development of technologies that increase the efficiency of energy production and end use and the further development and use of alternative energy technologies based on nonfossil fuels.

Our current systems of producing and consuming energy must be changed. Energy activity is closely interwoven with the environment. A progressive shift away from fossil fuels as quickly as possible in both the energy and transportation sectors is crucial. Economic incentives dramatically different from those now in force are required. Coal, oil, and gas prices, for instance, must reflect the environmental costs associated with their combustion. Over the long term, we must shift to alternative nonpolluting sources of energy, principally renewable sources. In the interim, we must develop technologies that use energy more efficiently and thereby consume less fuel.

The technological path of reducing dependence on fossil fuel while increasing energy productivity has three steps. The first step is to improve overall energy efficiency. Existing energy-efficient technologies, from light bulbs to recycling, can dramatically increase a nation's production of goods and services and reduce costs without increasing energy consumption. The second step is to make the transition to existing clean-fuel technologies. Fossil fuels are not all alike; energy produced from natural gas is much cleaner than energy produced from oil or coal. The transition away from fossil fuels should begin with a shift toward greater use of natural gas and increased reliance on available nonfossil fuel technologies. The third step is to develop renewable sources and other advanced energy technologies that can harness renewable resources cost-effectively, use fossil fuels cleanly and efficiently, and transform energy into products and services with minimal waste (WB, 1998).

Energy Efficiency.   Energy efficiency is the generic term that, to some degree, embodies energy intensity, energy savings, and energy rationalization. Energy intensity is consumption of energy per unit of product output; energy savings represent a reduction in this intensity; and energy rationalization is a combination of energy savings and switching from higher cost to lower cost fuels. The efficiency concept also applies to production, transmission, distribution, delivery, and consumption of energy.

Energy-efficient technologies available today can perform the same tasks, produce the same products, and provide comfort and convenience comparable to traditional technologies while using a fraction of the energy. For example, compact fluorescent lamps are "off-the-shelf" devices so efficient that an 18-watt fluorescent light provides as much light as a 75-watt incandescent bulb and the fluorescent lamp lasts 10 times longer and uses only one quarter as much electricity. Expanded use of recycling, another energy-efficient practice, could save enormous quantities of energy. The manufacture of paper, glass, and metal products from recycled materials generally uses less energy than is needed to produce them from raw materials.

Energy efficiency can be improved by closing the gaps between technically and economically possible energy-efficient technologies and their practical achievement. Support may come from

- Sectorwide promotion through industry associations,
- Tax incentives or new accounting practices encouraging replacement of old equipment with more efficient technology,
- More open market economies with effective price signals, and
- Competition among and privatization of utilities.

In the energy sector, for example, preventive technology and management can reduce environmental damage from extraction, processing, transport, and combustion of fuels. Preventive technology and management approaches include:

- Increasing efficiency in energy production, distribution, and use;
- Substituting environmentally benign fuel sources; and
- Making design changes that reduce the demand for energy.

Increasingly more developing countries realize that the traditional approaches used by industrialized countries to spur economic growth will no longer work, because such approaches cost too much economically and environmentally. Because social and economic development of these countries still depends on providing more and more energy services, the global community must identify, explore, and implement alternative paths that are better for the environment and the economy.

Transition to Cleaner Fuel.   To ease the transition from today's high-energy consumption path to a sustainable path based on renewable resources and other advanced energy technologies, we must make effective use of the cleanest fuels available right now. Of all the fossil fuels available today, natural gas is the cleanest. Sulfur dioxide and particulate emissions from natural gas are negligible. In addition, natural gas emits two-thirds of the carbon dioxide emitted from oil and slightly more than half the carbon dioxide from coal. Current forecasts about natural gas fuel prices and availability indicate that supplies will be abundant and economical for years to come. As advanced energy technologies push power-plant efficiencies from the present 30 percent range to a projected 50 to 60 percent, the use of natural gas as a transition fuel to generate power to supply efficient, new electric end-use technologies makes sense both economically and environmentally.

Renewable Energy.   Renewable energy resources will be an increasingly important part of the power generation system over the next several decades. Not only do these technologies help reduce sulfur oxides, nitrogen oxides, carbon oxides, and particulate emissions that cause air pollution, they also add some much needed flexibility to the energy resource by decreasing our dependence on limited reserves of fossil fuels. Experts agree with that solar energy, wind power, photovoltaics (PV), hydropower, and biomass are on track to become strong players in the energy market of the next century.

Solar energy makes possible a number of promising renewable energy technologies, including solar thermal, solar PV, wind, ocean thermal, biomass, and others. Geothermal energy is another renewable source that draws on the earth's interior heat. Some of these sources are now nearly competitive in cost with fossil-fuel energy. Renewable technologies could be competitive in meeting needs within 15 to 20 years if given adequate support through economic incentives and funding of applied research, development, and demonstration projects.

Wind power is the fastest-growing electricity technology currently available. Wind-generated electricity is already competitive with fossil fuel-based electricity. Wind power is available for small industrial units and for homes in some remote locations. Meanwhile, PV electricity is seeing impressive growth worldwide. Although PV currently costs three to four times more than conventional delivery electricity, it is particularly attractive for applications not served by the power grid. Many developing countries are attracted to the PV technologies' modular nature, located close to the user, and the units are far cheaper and quicker to install than central station power plants and their extensive lengths of transmission line (EPRI, 1998).

Other Advanced Energy Technologies.   Nuclear power is still the biggest question mark in the energy picture over the next few decades. To date, the choice between fossil fuels and nuclear power is a choice between pollution and the adverse health consequences of fossil fuel-generated power on the one hand and fear of a nuclear accident and problems with nuclear waste disposal on the other. Two technological developments could result in making nuclear power safer. First, within the next decade, a new family of nuclear generators described as "inherently safe" is likely to be developed. Second, fusion power may become commercially feasible. These developments, combined with concern about the inevitable emissions of air pollutants in urban areas and carbon dioxide from fossil fuel combustion, could make nuclear power a desirable energy option. Therefore, research should continue on improved models of fusion reactors and on nuclear fusion to keep nuclear options open (U.S. EPA, 1997).

Fuel cells offer compact, modular packaging, high efficiency, fuel and siting flexibility, and pollution-free operation. They could become widely used as distributed premium power sources at industrial sites and in manufacturing plants, office buildings, institutional settings, and perhaps eventually homes. In many areas, fuel cells are expected to provide strong competition to commercial and industrial electricity rates at the point of end use. Advanced fuel cell technologies also are expected to find many off-grid applications as lightweight, compact, remote portable power generators. Because of their high efficiency in converting natural gas, methanol, hydrogen, and even gasoline into electricity, fuel cells offer the lowest carbon dioxide emissions of any fossil fuel power system.

Power electronics based on silicon semiconductor switching and converter devices are transforming the ability to manage the power delivery system in real time. They are analogous to the low-power transistors and integrated circuits that brought about the computer age, but they operate at multi-megawatt power levels. They can switch electricity to a wide range of voltages, frequencies, and phases with minimal electrical loss and component wear. Advanced high-power electronics based on new types of semiconductor materials are expected to enable precise control and turning of all circuits—even gigawatt-scale power systems. And packaged devices made with these new materials could be as much as 100 times smaller and lighter than today's silicon devices. As a result, advanced power electronics promise unprecedented increases in the efficiency and cost-effectiveness of devices for a wide range of electricity production, delivery, and end-use applications. For electricity providers, power electronics represent the critical enabling technology for improving power system performance, offering value-added services to customers, and succeeding in a competitive marketplace (EPRI, 1998). For semiconductor manufacturers, Silicon Valley in California is highly polluted by toxic metals and organic vapors. The new pollution-free technology of advanced power electronics could trigger a second electronics revolution and unlock an entirely new multi-billion-dollar market.

Cleaner Transportation

Transportation accounts for a large amount of fossil fuel consumption today and produces a large percentage of air pollution. The most environmentally beneficial future applications of electric technologies are likely to be in the field of transportation. Although vehicle fleet fuel efficiency has improved dramatically in recent years, the larger number of vehicles on the road traveling more miles each year offsets efficiency gains per vehicle-mile. In transportation, the long-range need is a shift in automobile technology to alternative sources of energy that will pollute our cities less. This requires incentives for more fuel-efficient automobiles and for fewer miles driven as well as for new technologies such as electricity and, in the longer term, hydrogen. Electric vehicles will use electricity generated by cleaner fuels and by renewable energy. A number of experts believe that the world eventually will shift to a hydrogen-based economy. Hydrogen is an entirely nonpolluting fuel with an inexhaustible supply; it can readily be used as either a liquid or a gas. The catch is that hydrogen, like electricity, must be produced by another energy source. Most experts who favor this fuel envision nuclear fusion as the method used to manufacture hydrogen, but clearly a great deal remains unknown about both its feasibility and its environmental effects.

Clean, energy-efficient electric vehicles and electrified mass transit offer promising solutions to growing transportation and urban environmental problems. Electric vehicles can be up to 97 percent cleaner than their gasoline-powered counterparts and up to 88 percent cleaner than vehicles that burn methanol, natural gas, or propane. An electric vehicle in California produces less than half as much carbon dioxide as a gasoline-powered vehicle. Electric transportation technologies offer the promise of large-scale, cost-effective, environmentally sustainable commercial applications.

Product design changes can have far-reaching implications for preventing pollution. Consider, for example, substituting electric motors for gasoline engines in automobiles. This would result in significant environmental impacts beyond the automobile itself. It likewise would influence the way fuels are produced and used and how energy is generated and distributed. Research should be carried out to evaluate the design and use of vehicles, their parts, and other related products from a P2 practice. There is a need to evaluate product design options that will reduce waste by raw material substitutions, improved product durability, and reduce packaging in the field of transportation (U.S. EPA, 1997).

9.1.3 P2 ASSESSMENT AND P2 BENEFITS

This subsection describes how to assess life-cycle effects of a product to the environment based on P2 concepts and practices. It also highlights various benefits of P2 in environmental management.

Pollution Prevention Assessment

P2 assessment analyzes product life-cycle impacts to the environment. It considers all aspects of product design, manufacturing, and consumption as well as recycling, reuse, and disposal of discarded products. The level of required assessment depends on the complexity of the considered P2 project. A simple, low-capital cost improvement, such as preventive maintenance, would not need much assessment to determine whether it is technically, environmentally, and economically feasible. On the other hand, input material substitution could affect a product specification, and a major modification in equipment could require large capital expenditures. Such changes also could alter the quantity and composition of waste, thus requiring more systematic evaluation.

Detailed P2 assessment includes evaluation of technical, environmental, economical, and institutional feasibilities. It is important to note that many of the issues and concerns during P2 feasibility assessments are interrelated (Shen, 1999).

Technical Feasibility Assessment.   Technical feasibility assessment requires comprehensive knowledge of P2 techniques, vendors, relevant manufacturing processes, and resources and limitations of the facility. The assessment can involve inspection of similar installations, obtaining information from vendors and industry contacts, and using rented test units for bench-scale experiments when necessary. Some vendors will install equipment on a trial basis and accept payment after a prescribed time, if the user is satisfied.

Technical assessment should determine which technical alternative is the most appropriate for the specific P2 project in question. It requires considerations of a number of factors and asks very detailed questions to ensure that the P2 technique will work as intended. Examples of facility-related questions to be considered, include:

- Will it reduce waste?
- Is space available?

- Are utilities available or must these be installed?
- Is the new equipment or technique compatible with current operating procedures, work flow, and production rates?
- Will product quality be maintained?
- How soon can the system be installed?
- How long will production be stopped in order to install the system?
- Is special expertise required to operate or maintain the new system?
- Will the vendor provide acceptable service?
- Will the system create other environmental problems?
- Is the system safe?
- Are there any regulatory barriers?

Options that can affect production or quality need careful study. Although an inability to meet the above constraints may not present insurmountable problems, it will likely add to the capital or operating costs.

Environmental Feasibility Assessment.   The environmental feasibility assessment weighs the advantages and disadvantages of each option with regard to the environment. Most housekeeping and direct efficiency improvements have obvious advantages. Some options require a thorough environmental evaluation, especially if they involve product or process changes or substitution of raw materials. The environmental option of pollution prevention is rated relative to the technical and economical options with respect the criteria that are most important to the specific facility. The criteria may include:

- Reduction in waste quantity and toxicity
- Risk of transfer to other media
- Reduction in waste treatment or disposal requirements
- Reduction in raw material and energy consumption
- Impact of alternative input materials and processes
- Previous successful use within the company or in other industry
- Low operating and maintenance costs
- Short implementation period and ease of implementation
- Regulatory requirements

Economic Feasibility Assessment.   Economic feasibility assessment is a relatively complex topic, which deals with allocation of scarce, limited resources to various P2 modifications and compares various investments to help determine which investments will contribute most to the company. Usually a benefit is defined as anything that contributes to the objectives of the P2 project; costs are defined as anything that detracts from the achievement of a project's objectives. Normally, benefits and costs are evaluated from the perspective of whether they contribute to (or detract from) maximization of a company's income.

When measuring savings, it is important to look at not only the direct savings but also the indirect savings of P2, such as workers' health and future liability costs. In addition, there are noncost (or intangible) benefits of P2 such as plant reputation and better public relations. In many cases, the indirect savings and noncost benefits of P2 are difficult to quantify in financial terms; nevertheless, they are an important aspect of any P2 project and should be factored into the decision-making process. The economic feasibility assessment of P2 alternatives examines the incremental costs and savings that will result from each P2 option. Typically, P2 measures require some investment on the part of the operator, whether in capital or operating costs. The purpose of economic feasibility analysis is to compare those additional costs with the savings (or benefits) of P2.

Institutional Feasibility Assessment.   Institutional assessment is concerned with evaluating the strengths and weaknesses of a company's involvement in implementation and operation of investment in P2 projects. For example, it includes:

- Staffing profiles and skill levels,
- Task analysis and definitions of responsibility,
- Processes and procedures,
- Information systems and flows for decision making, and
- Policy positions on pollution prevention priorities.

The analysis should cover managerial practices, financial processes and procedures, personnel practices, staffing patterns, and training requirements. Issues of accountability need to be addressed. Proper incentives, in terms of money and career advancements, will encourage employees to achieve P2 goals.

P2 Benefits

Realization of P2 benefits must be based on recognition that regulatory programs and activities could include environmental incentives to encourage voluntary action while providing for improvements in environmental protection, including:

- Long-term commitment
- Comprehensive environmental accounting
- Variable environmental capability of regulated parties
- P2 limited by government constraints
- Need of regulatory stimuli to promote innovation and diffusion

Benefits of P2 are many, but high-level government and industrial decision makers who really understand the benefits are few. P2 helps national environmental goals and coincides with the interests of industry. Reducing wastes provides upstream benefits because it reduces ecological damage due to raw material extraction and pollutants released during the production process and during waste recycling, treatment, and disposal operations. A company with an effective, ongoing P2 plan may well be the lowest-cost producer and, as a result, may have a significant competitive edge. Furthermore, costs per unit produced will drop as P2 measures reduce liability risks and operating costs; P2 measures enhance a company's public image, public health, and overall environmental benefits.

Reduced Risk of Liability.   Industry can decrease the risk of both civil and criminal liabilities by reducing the volume and potential toxicity of gaseous, liquid, and solid wastes released into the environment. Because toxicity definitions and regulations change, reducing the total volume of wastes, including nonhazardous wastes, P2 is a sound long-term management policy for the following reasons (U.S. EPA, 1992):

- Environmental regulations require that industrial plants document waste release that must comply with emission/effluent standards. Companies that produce excessive waste risk heavy fines, and their managers may be subject to fines and imprisonment for mismanaging potential pollutants.
- Civil liability increases for generating hazardous waste and other potential pollutants, because handling of waste affects public health and property values in the communities that surround production and disposal sites. Even if current waste regulations do not cover certain materials, they may present a future risk for civil litigation.
- Workers' compensation costs and risks directly relate to the volume of waste materials produced—not only hazardous waste but also nonhazardous waste.

Cost Savings. An effective P2 program can yield cost savings that will more than offset program development and implementation costs. Cost reductions may involve immediate savings that appear directly on the balance sheet, or they may involve anticipated savings in terms of avoiding potential future costs. Cost savings are particularly noticeable when the costs result from treatment, storage, or services that produce the waste, such as:

- Adopting production and packaging procedures that consume fewer resources, thereby creating less waste, can reduce materials costs. As wastes are reduced, the percentage of raw materials converted to the finished product increases, with a proportional decrease in material costs.

- Waste management and disposal costs are obvious and readily measured potential savings to be realized from P2. Environmental regulations mandate special in-plant handling procedures and specific treatment and disposal methods for toxic wastes. The costs of complying with regulatory requirements and reporting on waste disposition are direct costs to businesses. Higher taxes for public services such as landfill management represent indirect costs to businesses. These costs will continue to increase at higher rates, but P2 will reduce waste management costs.

- A P2 assessment can find out where to reduce production costs. When a multidisciplinary group examines production processes, it may uncover unrealized opportunities to increase efficiency. Production scheduling, material handling, inventory control, and equipment maintenance can be optimized to reduce all types of waste and production costs.

- P2 will lower energy costs in various production lines. A thorough assessment of how various operations interact also will reduce the energy used to operate the overall factory.

- Complying with future regulations may result in costly plant cleanups, selling production facilities, selling off-site waste storage, or selling disposal sites. P2 will minimize these future costs because less waste will be generated.

Improved Company Image. As society increasingly emphasizes environmental quality, a company's policies and practices for controlling waste will increasingly influence employees' and the community's attitudes.

- Employees will have a positive feeling toward their company when management provides a safe working environment and acts as a responsible member of the community. By participating in P2 activities, employees can interact positively with each other and with management. Helping to implement and maintain a P2 program will increase their sense of identity with company goals. This positive atmosphere will retain a competitive workforce and attract new highly qualified employees.

- The community will endorse companies that operate and publicize a thorough P2 program. Most communities actively resist having new waste disposal facilities located in their areas. In addition, they are growing more conscious of the monetary costs of treatment and disposal. When a company creates environmentally compatible products and avoids excessive consumption and discharge of material and energy resources, instead of concentrating solely on treatment and disposal, it will greatly enhance its image within the community and toward potential customers.

Public Health and Environmental Benefits. It should be noted that, in specifying acceptable limits for specific pollutants, many of our regulatory standards rely on hypothetical extrapolations of laboratory data of health risks. When we translate laboratory results to public health with average values and specific conditions or factors, we are making hypotheses and general assumptions. The use of hypothetical arguments as the basis for restrictive regulatory action is acceptable only when there is a real threat to public health. However, these hypotheses and limits must be readily modified if new evidence or data so demand. Toxic release inventory and material accounting surveys can evaluate public health and environmental benefits of P2. Such inventory and surveys provide data to indicate primarily the progress of P2 by calculating the reduction of material use for production processes and reduction of pollutants entering the environment.

Among all the benefits, the economic benefits of P2 have proven to be the most compelling argument for industry and business to undertake prevention projects. Cost savings from P2 come not only from avoiding environmental costs such as hazardous waste disposal fees but also from avoiding costs that are often more challenging to count, such as those resulting from injuries to workers and ensuing losses in productivity. In that sense, prevention is not only an environmental activity but also a tool to promote worker health and safety (Shen, 1999).

9.1.4 POLLUTION PREVENTION BARRIERS

Despite the need for P2 technologies, many potential barriers currently exist that prevent their widespread use. A P2 technology may not be used because it is not commercially viable (i.e., because it is more expensive than competing technologies); because its existence is not widely known among potential users; or because there are regulatory, cultural, or institutional barriers to its adoption. Although various government incentives for P2 are available, sometimes barriers can inhibit implementation. Exact circumstances differ from facility to facility; some of the most common barriers identified include reluctance to change, lack of technical information, lack of funds, and lack of management support. This subsection analyzes the current concern of potential P2 barriers and provides suggestions about how to avoid the barriers. Barriers can be overcome if they are clearly understood and analyzed. Ignoring obstacles to P2 only reduces the likelihood of success (Shen, 1997b).

Mental Barrier

The first and most important of all potential barriers to P2 is mental reluctance or resistance to change. Many people do not like to learn new ideas or do not care for P2 technology, which can help waste generators to reduce the toxicity and volume of the waste they generate. Many people do not know that there are strong economic incentives for using P2 technology. Furthermore, they do not realize that P2 measures enhance a company's public image, public heath, liability, and overall environmental benefits. Business leaders as well as politicians, and indeed most people, focus on immediate concerns. They consider short-term success to be important. The hardships of developing countries give short-term profits a high priority. There are always those for whom temporary success is enough. Many managers believe that environmental protection inevitably costs money and that it is a peripheral issue and a diversion from basic corporate goals.

For mental change to come about, knowledge, information, and education with creative and probing personalities is necessary. It is no longer acceptable to rest on environmental experiences gained decades ago in wastewater treatment or air pollution control systems. The factory of the future will consider the environmental implications of a new product or service. It will insist on P2 in the acquisition of new components or systems from others so that waste generation in the life cycle of its product will be minimized. The factory of the future constantly use preventive techniques to reduce waste and seek solutions.

Technical Barrier

P2 technologies need to be taught in higher education and training workshops. Information is needed on preventive technologies and procedures such as how to integrate them in the production process and what side effects are possible. Product quality or customer acceptance concerns might cause resistance to change. Potential product quality degradation can be avoided by contacting customers and verifying customer needs, testing the new process or product, and increasing quality control during manufacture.

Limited flexibility in the manufacturing process may be another technical barrier. A proposed P2 option may involve modifying the workflow or the product or installing new equipment; implementation could require a production shutdown, with loss of production time. The new operation might not work as expected or might create a bottleneck that could slow production. In addition, the production facility might not have space for P2 equipment. These technical barriers can be overcome by having design and production personnel take part in the planning process and by using tested technology or setting up pilot operations.

For the past several years, new preventive technologies are being developed to cope with changing waste characteristics, such as focusing on industrial toxics, energy efficiency, and a system approach. Industry needs to establish or strengthen information networks and to encourage employees to watch for information in technical journals, in newsletters, and on the Internet. However, information resources could be a problem for small- or medium-sized industries and businesses. They may not have ready access to central sources of information about P2 techniques. Government must help provide technical assistance and specific P2 technology information for small- and medium-sized industries and businesses.

Financial Barrier

Financial barriers include inaccurate market signals, lack of a full accounting system to illustrate P2 benefits, incomplete cost/benefit analyses, inappropriately short time horizons, fear of market share loss and/or consumer pressure, limited access to necessary resources, and workers' fear of job loss. Changing existing ways of doing business can be seen by industry as being more risky. Although substantial gains can be achieved with improved efficiency and better technology, there comes a time when significant technological change and investment are required. Capital is scarce in all industries, and investments in P2 must compete with other seemingly more profitable projects for funding.

The lack of immediately obvious economic benefits is a common barrier even though such benefits exist. Saving and making money is an immutable motivation for all people. But because the true cost of waste has been at the end of the pipe, instead of upstream where wastes are produced, and because many costs have been externalized to society as a whole, workers in industry may not see the economic benefits of P2. Providing incentives such as rewards for successful efforts is a powerful tool for workers who have reduced generation of wastes and cut costs for the company. Incentives may be classified as economic benefits, enhanced public image and relations, regulatory compliance, and reduction in liability. New patterns of value systems, accounting systems, and effective communication with appropriate feedback and interaction that can clearly illustrate economic benefits of P2 may reduce those barriers. Information needs to be shared and employees need to know that sharing information is crucial in implementing P2.

Economic investment in existing technologies is a principal roadblock to the induction of technological improvements or P2 technologies that have environmental benefits. Manufacturers are reluctant to abandon the long years of experience and considerable expertise invested in and gained from existing technology. Another roadblock is failure of the economy to incorporate environmental damage and other externalities into the cost of products. In a free market, if the price of a technology reflects societal as well as private costs, society will quite literally buy the amount of environmental protection it wants.

Regulatory Barrier

The major regulatory functions of environmental programs are still administered under media-specific, commend-and-control regulatory structure. The exiting end-of-pipe pollution control regulations have caused delay in implementation of P2 programs. These regulations have consumed a great portion of environmental management resources and have reduced short-term pollution but have not really eliminated pollutants. These regulations might transfer pollutants from one medium to another. Regulatory inflexibility and uncertainty also can be barriers to some P2 opportunities. For example, changing to another feed material in a manufacturing plant may require changing the existing permits. In addition, it may be necessary to learn what regulations apply to proposed alternative input materials.

The use of price regulations (such as raw materials, goods and services) and green tax to achieve environmental objectives can provide strong incentives for technological innovation and behavioral change. They offer good prospects for achieving environmental objectives in a cost-effective manner, but the price should better reflect their full environmental and social costs. Industry should work with government agencies and other concerned people to identify and reduce legislative and regulatory barriers. Industry also should make positive constructive recommendations for change, which can improve environmental quality and enhance industrial competitiveness. Working with the appropriate regulatory bodies early in the planning process will help overcome many regulatory barriers. It is advisable to amend the existing laws, regulations, and implementation processes that reflect sustainable development policy, multimedia P2 strategy, incentives, and a partnership approach, as well as product life-cycle concept, value system, and full cost accounting.

Institutional Barrier

General resistance to change and friction among elements within an organization can result from many factors such as lack of top management support, lack of awareness of corporate goals and objectives, lack of clear communication of priorities, organizational structures separating environmental decisions, and individual or organizational resistance to change. The development and implementation of a P2 program can involve a significant change in a company's traditional business practices. P2 must involve employees from all areas of the plant and levels of management, including hourly workers.

Lack of clarity about how to share leadership in various aspects of P2 technology in a multimedia approach creates some confusion. For example, it is widely acknowledged that it is difficult to share data across medium-specific offices, but it is not clear whose role it is to take the lead in resolving this issue. Lack of clarity about one's leadership role in prevention can result in little risk taking and a reluctance to elevate P2 policy issues. This, in turn, can limit the organization's creativity in finding ways to define the functional relationships among offices as a matrix that promotes cross-media and P2 outcomes.

Summary

Five potential barriers to implementing P2 programs are identified as mental, technical, financial, regulatory and institutional barriers. They are closely interrelated and can be avoided or reduced by:

- Strengthening education, training, and outreach programs;
- Providing information and a communication network;
- Amending regulations and implementation processes;
- Applying environmental value and full costing systems;
- Emphasizing incentives for pollution prevention;
- Improving organizational structure and management system to ensure interaction and cooperative spirit; and
- Gaining the support of decision makers and staff at all levels in very early stages of the P2 effort.

There is consensus about many of the changes needed to reduce potential barriers and increase incentives for preventing pollution. Some decision makers believe that much remains to be learned such as the regulatory amendments, institutional changes, preventive technologies, and integration of P2 into the current regulatory and environmental management systems. We must educate students and train working professionals to gain knowledge and information that are critical to understand clearly the causes of potential barriers and to learn how to avoid such barriers (Shen, 1997b).

9.1.5 P2 CHALLENGES AND OPPORTUNITIES

P2 must flow from action by all people, and not just government action. The government plays a crucial role in research, public information, standard setting, and enforcing the law. But government action alone cannot substitute for a universal symbiosis with economic and aesthetic, public and private actions. Market incentives and voluntary partnerships can spur actions that regulation alone cannot.

We need socioeconomic development, which is necessary to improve the standard of living and quality of life, but development must carefully consider the quality of the environment. We are challenged to help develop and recommend new developmental and environmental strategies not only in the present but also in the future. P2 is the current environmental strategy to minimize pollutants and wastes and also to provide environmentally friendly products and services that prevent ever-increasing pollution problems. P2 is a management tool to establish a society-oriented approach toward sustainable development.

Role of Government

Environmental policy should encourage replacing preventive technologies with those that are environmentally friendly and increasing investment in innovation through performance-based incentives and other mechanisms. Current pollution laws, regulations, and programs are aimed at the effects and control of waste instead of at preventing pollution. Laws are seldom a positive, driving force for incorporating environmental considerations into the design and use of technologies in industry, agriculture, mining, transportation, and other sectors. To become a positive force, laws and programs must be based on P2 instead of on taking care of pollution after the problem has been created. Until P2 occurs, our environment will not be cost-effectively protected.

Regulatory requirements have produced innovative solutions to environmental problems. Labeling major appliances for energy efficiency, for example, has increased public understanding and helped direct market forces toward more environmentally benign technologies. Regulations that affect technology should be based on extensive and unbiased assessments of the technological possibilities and associated costs as well as their effectiveness in promoting technological development and use. Regulatory obstacles to development and introduction of P2 technology should be removed. There is virtually no area of the economy that cannot be motivated to renovate, update, or replace environmentally harmful technologies.

The role of the government in technological development is crucial. There is an opportunity to implement policies to stimulate investment in new technologies and to encourage phasing out outmoded capital equipment, and replacing it with new, environmentally beneficial technologies and equipment. Government may establish a variety of measures, such as new tax and other policies, enactment of carefully crafted investment tax credits, and green research and development tax incentives to stimulate savings and investment in environmentally advantageous technologies.

Changes must be developed to move the environment from the periphery to the center of decision making. Doing so will require new organizational entities and responsibilities. At a minimum, each government agency must have a top-level manager for environmental issues, develop environmental goals, and establish a strategy for reaching those goals—reporting on progress, and setting up a means for regular interaction with various governmental environmental agencies. Cooperation and collaboration between government and the private sector can inspire positive technological change. This is particularly true in areas where individual companies cannot directly recoup the benefits of investing in research and development. The government should substantially increase funding for research and development and participate in partnerships with private industry to explore new and innovative technologies that are environmentally sound.

The government can use P2 technology more effectively in four priority areas of economic and environmental concern by:

- Encouraging the application of preventive technologies. To reduce wastes and promote environmentally friendly products and services, the government must facilitate flexible, voluntary commitments and performance-based incentive approaches.

- Making industry more competitive with world standards. With the new economic strength of the global market, the government must support technological policies that will enhance economic performance.

- Safeguarding the environment. Choices and trade-offs must be made that will affect economic growth, energy use, and the quality of the human habitat. There will also be opportunities to develop new P2 technologies for manufacturing processes and products that safeguard the environment efficiently. Scientific and technological information is critical.

- Restructuring education and research. Ensuring the technical capability to address P2 and sustainable development goals will require substantial reforms in college education and research into preventive science and technology. Education and research programs should ensure that future P2 and pollution reductions can be met at minimal cost and in ways that contribute to sustainable development. This undertaking has a long lead time, and early action is important. Information about the availability and benefits of preventive technologies should be available to everyone.

These areas of concern are interdependent and all require attention at the highest policy levels of government. Strengthening economic performance and the environmental technology base, for example, will require simultaneous consideration of economic, energy, environmental, regulatory, trade, and technological policies. Future progress in all these areas depends on a national commitment to maintain and replenish the nation's storehouse of basic knowledge in science.

Looking at Industry

Many essential human needs can be met only with goods and services provided by industry. Industry has the power to enhance or degrade the environment; it invariably does both. Industry can no longer afford to ignore environmental needs. Profit becomes pointless without quality of life. Financial accounts tell many stories but not all, and measuring performance by profit alone does not suffice. However, a greener future remains an idealistic dream unless industrialists and environmentalists meet to transform it into a reality by talking and sharing problems (Schmidheiny, 1992).

If industry continues to react to environmental injuries and try to repair them, the quality of our environment will continue to deteriorate, and eventually our economy will decline as well. However, if industry applies preventive technologies, shifts its policies, makes bold economic changes, and embraces a new ethic of environmentally responsible behavior, it is far more likely that the coming years will bring a higher quality of life, a healthier environment, and a more vibrant economy for all. Corporate environmental responsibility no longer ends at the factory gate; it extends from cradle to grave. Ultimately, this means manufacturing only products that can be used with an environmental management system that minimizes environmental impacts and maximizes environmental efficiency. This requires construction of new commercial infrastructures and new relationships between producers, consumers, and governments.

By adopting a P2 approach, companies can start to take control of the process of environmental change in ways that make economic and operating sense instead of seeing their own processes controlled by tightening regulations and expectations. Environmental considerations must be fully integrated into the heart of the production process, affecting the choice of raw materials, operating procedures, technology, and human resources. P2 means that environmental efficiency becomes, like profitability, a cross-functional issue that everyone is involved in promoting.

The myriad of industrial P2 possibilities can be divided into four categories: good housekeeping, materials substitution, manufacturing modifications, and resource recovery. Often companies employ a number of these management and technical approaches simultaneously to resolve a particular problem.

1. Good housekeeping requires attention to detail and constant monitoring of raw material flows and impacts. Many companies still have no idea how much or what type of wastes and pollution they produce. P2 starts from the basis of accurate measurement, identifying and then separating and recycling wastes. Improvement in information technology has also made environmental monitoring more affordable. Waste can also be prevented with more efficient inventory control. Surplus raw materials that are no longer needed can be sold to third parties.

2. Identifying and eliminating sources of pollution often implies restructuring for both producers and consumers. Full or partial phaseouts of lead, mercury, dichloro diphenyl trichloro ethane (DDT), and chlorofluorocarbons (CFCs) have been implemented in various parts of the world as the only effective ways to solve the problems they cause. Substituting one material for another offers the prospect of completely eliminating a given pollution problem. One of today's largest issues for companies is how to deal with emissions of volatile organic compounds (VOCs), which are linked to two of the industrial world's major air pollution problems—photochemical smog and global warming. A major industrial source of VOCs is solvents, particularly in the paint and coatings industry. Industry, not the government, will be the primary developer of new technologies in response by signals given by prices. Typically, several technologies are available to perform a particular function, and the least expensive one is most likely to be chosen. The existence of a potentially lucrative market for new technologies such as CFC substitutes will attract researchers and investors. Economic measures will encourage development and adoption of new technologies.

3. Often companies can considerably reduce emissions by simplifying production technology through lowering the number of process stages. Switching to closed-loop processing can also conserve resources and cut noxious emissions; by installing a closed-loop decanter system that separates solvent from water, the solvents can then be distilled and reused. Sometimes more fundamental changes, such as moving from a chemical to a mechanical process, can help prevent pollution. Vulcan Automotive Equipment, a small Canadian automobile engine remanufacturer, cut raw material, labor, and waste management costs while improving product quality by replacing its traditional inorganic caustic cleaner with a high-velocity aluminum shot system. Caustic soda had been used to clean caked oil and grime from old engines, which created health hazards for workers and a large amount of waste sludge. In its place, Vulcan introduced a two-step system in which the metal parts are initially baked to remove volatile oils and grease and then sprayed with a high-velocity stream of aluminum shot to remove the remaining dirt and rust. The dividing line between process and product is artificial; many products are inputs for other processes (such as plastics for automobile components and steel for auto chassis). Furthermore, the industrial ecosystem of the manufacturing process is only one part of a much wider ecosystem, which contains all flows of materials and wastes produced through the production, consumption, and disposal of goods and services. But the present design of the industrial ecosystem is flawed; instead of acting according to the circular principles of natural ecosystems, the flow of goods and services is essentially linear. Products are produced, purchased, used, and dumped, with little regard for environmental efficiency or impact.

4. Car companies now regularly recycle production waste. General Motors separates scrap polyvinylchloride plastic by color. Then it grinds, melts, and reuses it, along with new polyvinylchloride. At Volkswagen, waste thermosetting plastics used in the production line are collected and returned to the supplier. They are then reconditioned and mixed with 20 percent new material and reused as soundproofing material in Volkswagens.

Waste exchange systems have been established in many developed and developing countries to overcome the information gap between waste producers and potential customers. Organic wastes now are often being considered for use as agricultural fertilizers. In China, organic wastes from thousands of small straw pulp mills are used for this purpose; in Denmark, the Novo-Nordisk biotechnology company has decided to convert nitrogen wastes from its Kalundborg plant into fertilizer, which it distributes free to farmers. In India, more than 30,000 metric tons of solid waste known as willow dust is produced by the textile industry each year. This is now being used to produce bio-gas, cutting energy consumption within the industry. In Germany, BASF reports that it already produces 71.5 percent of the steam it needs with reclaimed heat from chemical processes or by burning residuals (Schmidheiny, 1992).

Every industrial activity is linked to hundreds of other transactions and activities that can have environmental impacts. A large manufacturing industry, for example, may have hundreds of suppliers. The industry actually may manufacture hundreds of different products for a wide variety of customers with different needs and cultural characteristics. Accordingly, each customer may use the product very differently, which is a big consideration if use and maintenance of the product may be a source of potential environmental impact. When the product is finally disposed of, it may end up in a landfill, an incinerator, on the roadside, or in a river that supplies drinking water to a small community. The success of market-driven P2 initiatives depends in large part on consumer awareness and knowledge of P2 issues. To use the product and service market effectively as a P2 tool, there must be some assurance that P2 benefit claims made about processes and products are truthful and result in real improvements in environmental quality. The recent rush to market green products has achieved some success because of environmentally conscious consumers who translate their environmental values and fears into purchasing decisions.

One of the most fundamental changes that has occurred over the past 10 years is the response of businesses and industries to their environmental responsibilities. Many corporations today are interested not simply in their legal responsibilities to control pollution before it escapes to the environment but also in the broad corporate benefits that result from efforts to prevent and reduce pollution at its source. Yet experience of the past decade also suggests that people will respond to environmental problems with energy, creativity, and a deep-seated sense of responsibility for future generations. Most people believe environmental quality is an essential component of their long-term health and economic prosperity.

Educational and Research Institutions

Educators and researchers must anticipate and prevent pollution problems and devise inspired solutions. They must educate students and assist working professionals to understand and accept P2 concepts, equip them with knowledge and information, and stimulate their thinking and use of P2 technologies for sustainable development. Understanding environmental pollution subjects stretches far beyond the boundaries of traditional pollution control technology to P2 technology and also beyond waste management to pollution management. Educating people about P2 technology and pollution management and many other social and economical areas may come to be seen as a necessary part of the higher education curriculum along with computer programming and marketing.

Unfortunately, many design courses in higher education institutes continue to neglect P2 technology issues, thus equipping students poorly even for current demands. There can be little doubt that it is essential to include information about preventive technology issues and their relationship to design process in the core curriculum. Equally important is development of an understanding that the environment is not a separate subject that engineers can choose to be interested in or not, but rather it is one basic criterion against which all design work should be assessed. Breadth of experience and vision will be valued as we attempt to cope with large-scale, apparently impenetrable and highly complex pollution problems. Redesigning our society to minimize environmental problems demands an integrated approach.

The relative newness of P2 technologies means that good P2 technical curriculum material and training manuals are only just emerging. Educators and researchers are responsible for assisting students, government and industry, and community leaders to strengthen their capability and capacity for formulating and implementing P2 strategies, plans, and programs. They are challenged to create professional and institutional linkages between various governments and nongovernmental organizations and their counterparts in the international community through P2 technology workshops, seminars, and conferences. Educators and researchers are also challenged to develop an electronic information network and provide expertise and educational materials in support of P2 projects or programs.

Sources of Information

Worldwide P2 information has been increasing rapidly in recent years. This P2 information is being stored in and manipulated by computers in academic and research institutions as well as public and private libraries. Many of the P2 data are being gathered by automated systems and fed electronically into various networks. The ability to use these P2 data depends on the ability to locate, access, combine, compare, and collate the data and on collaboration and communication between data gathers and data users. The Internet provides information from all over the world and it is a powerful tool for searching and evaluating P2 data and technology. Accessing the Internet is becoming easier and more efficient every year. People who are searching for P2 ideas and information on P2 technology can use an e-mail listserve, which is a system that allows users to send e-mail to a large group of subscribers.

Internet information provided by various organizations worldwide is frequently updated. Some of the organizations that provide important pollution prevention information are listed below:

- United Nations Environmental Programme (UNEP)
- U.S. Environmental Protection Agency's Office of Pollution Prevention & Toxics (OPPT)
- National Pollution Prevention Center for Higher Education (NPPC)
- National Technical Information Services (NTIS)
- World Resources Institute (WRI)
- Center for Technology Transfer and Pollution Prevention (CT2P2)
- The Tellus Institute

9.1.6 CONCLUSIONS

P2 technology seeks lasting and complete solutions to environmental pollution problems. It is a tool to benefit present and future generations without transferring an environmental problem from one medium (air, water, or land) to another and without producing environmentally unfriendly products and services. Although the concept of P2 is well accepted, it is not vigorously implemented. The fundamental reason for slow implementation is lack of knowledge and information, which are required for action. Adequate education and training programs are essential.

Environmental regulatory agencies must establish understandable, reasonable, implementable, and affordable P2 laws and regulations. There must be a preventive approach that will improve economic growth, energy use, and the quality of the human habitat. Government policy must encourage applying preventive technologies to reduce pollutants and wastes and also must promote environmentally friendly products and services through flexible, voluntary commitments and performance-based incentive approaches. New P2 technologies are needed for manufacturing processes and products that safeguard the environment efficiently. P2 technology must take its place alongside cost, safety, and health as a guiding criterion for technology development.

Some P2 technologies, skills, and tools are already in existence and information is available in government documents, professional journals, and on the Internet. New scientific and technological information for P2 technologies and programs should be developed and available to all. Government agencies should substantially increase funding for P2 education, research, and development as well as participate in partnerships with private industry to explore new and innovative P2 technologies. It is essential that economic signals reflect environmental values. Breaking through the barriers to developing and using P2 technologies requires government leadership, private-sector ingenuity, and public support. This necessitates new pricing mechanisms, investment incentives, regulatory changes, and collaborative undertakings.

REFERENCES

EPRI, "Technologies for Tomorrow," *Electric Power Res. Inst. J.*, Jan./Feb.: 37–40, 1998.

Ling, J. T., "Industrial Waste Management," *Vital Speeches of the Day*, 64(9): 284–288, Feb. 18, 1998.

PPIC, Pollution Prevention Information Clearinghouse, U.S. EPA Office of Pollution Prevention and Toxics, Washington, DC, EPA/742/F-98/004, summer 1998.

Schmidheiny, S., *Changing Course*, MIT Press, Cambridge, MA, 1992, pp. 90–98.

Shen, T. T., "Educational Aspects of Multimedia Pollution Prevention," in *Proceedings of International Conference on Pollution Prevention held in Geneva, Switzerland and also in the U.S. EPA' publication of Clean Technologies and Clean Products Environmental Challenge of 1990s*, EPA/600/9-90/039, Sept. 1990.

Shen, T. T., "Environmental Management in China," presented at *AWMA 90th Annual Meeting and Exhibition*, Toronto, Canada, Air and Waste Management Association, Pittsburgh, PA, June 8–13, 1997, 1997a.

Shen, T. T., "Five Categories to Avoid Industrial Waste Minimization and Pollution Prevention Barriers," presented at *Asian-Pacific Conference on Industrial Waste Minimization and Sustainable Development*, Taipei International Convention Center, Foundation of Taiwan Industry Service, Taipei, Taiwan, Dec. 14–18, 1997, 1997b.

Shen, T. T., *Industrial Pollution Prevention*, 2nd ed., Springer-Verlag, Heidelberg, Germany, 1999, Chaps. 1, 4, 6.

Shen, T. T., and Sewell, G. H., "Control of Toxic Pollutants Cycling," *Proceedings of the International Symposium on the Environmental Pollution and Toxicology*, HongKong Baptist College, HongKong, 1986.

U.S. EPA, *Facility Pollution Prevention Guide*. Office of Research and Development, U.S. EPA, Washington, DC, EPA/600/R-92/088, 1992.

U.S. EPA, *Industrial Pollution Prevention Project (IP3): Summary Report.* Document No. EPA-820-R-95-007, U.S. EPA, Washington, DC, 1995.

U.S. EPA, *Pollution Prevention 1997—A National Progress Report*, EPA-742-R-97-000, U.S. EPA, Washington, DC, 1997, pp. 105–119.

U.S. EPA, *Office of Pollution Prevention and Toxics: 1997 Annual Report*, U.S. EPA, Washington, DC, EPA 745-R-98-003, 1998.

Vigon, B. W. et al., "Life-Cycle Assessment: Inventory Guidelines and Principles," *Contract Report of Battelle Research Institute*, Columbus, OH. Document No. EPA/600/R-92–245, 1993

World Bank, *Energy Issues in the Developing World. Industry and Energy Department Paper, Energy Series Paper No. 1*, Washington, DC, Feb. 1998, pp. 1–19.

CHAPTER 9
POLLUTION PREVENTION CONCEPTS AND POLICIES

SECTION 9.2

PLANT OPERATIONS

William B. Katz
Mr. Katz is a registered professional engineer in Illinois, Indiana,
Michigan and Wisconsin. Retired from an environmental business
he founded in 1966, he is actively engaged as a consultant in
performing environmental site assessments and in teaching at
Oakton Community College and Roosevelt University.

9.2.1 INTRODUCTION

In a plant facility context a spill of any solid, liquid, or gaseous material, hazardous or not, is a potential pollution problem. Essentially such spills are point-source problems in that the source of the spill or emission can be identified. Such spill events are accidents, and spill prevention should be approached in the same fashion as accident prevention: identify as many probable causes of spills as possible, and eliminate them. When similar events have occurred over a period of time, common factors can often be identified. Changing or correcting those factors can reduce the chance of a future spill, but one can never be certain that all possible causes of spills have been identified; zero spill potential is probably unattainable.

Definition

Spill prevention can be defined as determining the probable factors and events that can lead to a spill and, by changing those factors and events, effecting a reduction in the odds that such a specific type of spill will occur again.

Assumptions

One can make some assumptions about the conditions that lead to spills:

1. Large manufacturing plants usually have a good idea of the problems involved in handling hazardous materials and do a fairly good job of seeing that obvious preventative measures are taken. Small plants where hazardous materials are used occasionally may not have a good idea of the problems involved. Personnel in such plants may need to be educated in spill prevention.

2. When there has been some recognition of potential spill problems, the obvious causes usually have been recognized and some action has been taken. Hence, one usually is concerned with finding and dealing with low-probability, not-so-obvious causes of spills.

3. Some spill probabilities can only be estimated (guesstimated may be a better word).

9.2.2 BENEFITS OF A PREVENTION PROGRAM

There are a number of benefits to be derived from an active spill prevention program. One immediate benefit is reduced loss of product, which is a direct economic gain. If contamination is the result of a spill, there are immediate costs for direct cleanup, and there may be a loss from downtime if the spill site is close enough to an active production area to require evacuation of personnel or curtailment of production activities.

More indirect benefits include a reduction in capital investment for cleanup equipment and materials, lowered out-of-pocket charges for cleanup contractors' services, lowered insurance rates, and a reduction in fines and penalties associated with a spill. Frequently, increased consciousness of spill potential, as with accidents of any kind, makes workers more careful in their operations, resulting in a reduction in injuries and lost-time accidents. The fact that none of these benefits can be calculated accurately is one of the difficulties in trying to justify a particular level of expense for prevention.

9.2.3 REQUIREMENTS FOR A PREVENTION PROGRAM

When people operate in a daily set routine, they develop an attitude of acceptance rather than an attitude of questioning. What is needed to identify the potential causes of spills and to suggest solutions that will reduce the chance of spill events occurring is a "why" attitude or a "what if" type of approach to surroundings, jobs, procedures, and policies.

Records of past spill events should always be made available to those responsible for spill prevention. An accurate record-keeping and reporting procedure should be established as one of the first steps in spill prevention. There are two reasons for this. The most obvious is the need to have available a record to study the causes, with a view to making corrections that will eliminate those causes. A less obvious reason is protection in the event of lawsuits, hearings, or fine appeals. In the excitement and turmoil of a spill problem, chronology often is forgotten or warped. Memories become dim with the passage of time, and, when a lawsuit finally is brought to court, defense becomes difficult without some written, legally acceptable reminder of what occurred.

It is not always possible, but it certainly is highly desirable, to have one person involved with a spill event as a technically qualified observer whose function is to record on film or tape and in writing what is occurring. Such records should be made in the form determined by legal counsel to be valid for legal purposes.

Personnel

Personnel for a prevention program may come from the plant staff or from another plant whose personnel are not familiar with the plant being studied. In either event, the persons selected should have the type of mind that sets them apart as "different" in routine situations. Such people are not easy to find, because in routine production jobs they tend to upset applecarts and often are not around very long. Frequently, such people are found in research and development activities. Once a prevention program has been established, these people usually are not the best people to run it, but a regular reapplication of their special approach may prove useful and profitable.

Outside Help

Consulting firms that specialize in spill and risk management have been increasing in number and experience over the years. Such firms, both large and small, local and with nationwide branches, advertise extensively in the many environmental journals now being published.

Cleanup contractors often can be hired to do this type of survey work. If they are employed on a retainer to provide the cleanup service needed in the event of a spill, they may do survey work and help to set up a prevention program as part of their retainer fee. Many firms that specialize in this field are members of the Spill Control Association of America.*

Insurance agencies and suppliers may provide engineering service, free or for a fee, as a means of reducing their risk exposure.

Federal and state environmental agency personnel may be helpful in making suggestions, although normally they are not available to do this type of work while employed by the government.

Avoidance of Upsets

One of the normal results of an active prevention program is change—in procedures, in equipment, in policies, and occasionally in jobs. Such change usually is resisted. Management, especially middle management, tends to resist because a prevention program requires spending money and interrupting procedures that they hope are efficient and because the net result is a reduction in profit with no visible signs of gain. Plant workers whose jobs or accustomed ways of doing things are threatened may find ways to be uncooperative or even try to stop installation of a spill prevention program. Hence, it is necessary first to have the active support of top management, who must understand the aims and goals of such a program and agree to the cost and the timetable. Unions, if involved, must be informed and sold on the positive gains for their members: safer jobs, better working conditions, and anything else that is applicable. Any changes that result from a prevention program must be planned carefully and, if possible, implemented with the help of personnel involved in the planning stages.

9.2.4 FACTORS TO BE CONSIDERED IN PLANNING A PREVENTION PROGRAM

Personnel

The human element is almost always a factor in spills. This involvement can be direct (as a result of carelessness, indifference, or sabotage) or indirect (as a result of inadequate knowledge of job or procedures, poor training, physical inability, or another cause). Hence any good spill prevention program must start with a careful evaluation of personnel and correction of deficiencies in attitude, motivation, training, and knowledge.

Equipment

Accidental spills are frequently the result of inherent flaws in equipment design. Much equipment now in use was designed, built, and installed with ends in mind very different from minimizing spills. The changed climate of regard for spill control may well necessitate modifications in equipment.

As one example from the past, the filling of tank trucks with flammable materials was often done with little regard for vapor emission. Tight vapor emission control has dictated the design of new loading equipment and procedures for using that equipment to eliminate or at least to decrease drastically vapor emission during tank loading.

As another example, hose connections to the bottom of tank cars containing toxic or dangerous products often were made with loose control of spillage from hose drainage. Frequently the hoses were just drained onto the ground. Modern improvements call for installing drain pans beneath rail unloading facilities, in addition to using quick shutoff fittings on hoses and piping to prevent such spills and to collect those that do occur.

---

* Spill Control Association of America, 8631 W. Jefferson, Detroit, MI 48209.

Environment

Proper security for both equipment and personnel is a major need in spillage control. This includes preventing nontrained personnel (such as outside delivery drivers) from operating equipment, controlling personnel movements into unauthorized areas, and protecting equipment from vehicular accidents that could cause permanent damage (because of driveways located too close to tanks, for example). Protection against deliberate acts of vandalism, sabotage, and terrorism also must be considered in planning for security protection.

Prevailing weather must be considered as a contributing cause of spills and must be studied to determine possible effects on spill occurrence. Examples are plant location in regular tornado or hurricane areas or in areas with heavy winter snowfalls or regularly occurring ice and sleet storms. Some attempt must be made to reduce the possible impact of weather on facilities that could result in spills, usually under conditions that make effective control and cleanup of a spill difficult or impossible. National Oceanic and Atmospheric Administration records on rainfall, snowfall, winds, and floods all can be useful in this regard.

Plant geography must be studied (or at least considered) in planning to prevent or control spills. Ground elevations affect water drainage patterns. Water-table depth and soil porosity may be such that a spill will be essentially nonrecoverable (even at tremendous expense) unless seepage into the water table can be prevented. Proximity of storm sewers and watercourses, either flowing rivers or creeks or ditches that may contain water during thaw or rainfall conditions, may determine in part the preventative measures required. Close proximity of watercourses obviously increases exposure to rapid spreading of a spill, with a vastly increased potential recovery cost. Hence more effort and money may be justified to anticipate spill sites and to install control measures of a possible spill flow path. The location of all storm drains should be entered on the facility plot plan.

9.2.5 FACILITY INSPECTION

The first step in starting a program of spill control is a detailed facility inspection to survey potential problems. The questions below should be answered as the facility is systematically covered completely. These questions may not be complete (such a list never can be complete), but they should give the user a point of view that will stimulate additional questions as the inspection proceeds.

Records

Written records should be kept of facility inspections, especially because reinspections (which should be made at perhaps 6-month intervals) will show by comparison with past records what progress is being made in reducing spill exposure. Such inspections can profitably be made by a small team (consisting of no more than three persons), which includes both operating personnel (with some authority) and the specialist (in-plant or outside consultant) working together. Experience has shown that despite the demands on his or her time and attention a plant manager or an assistant plant manager makes a better team member than an operating supervisor, because on-scene inspection impresses the need for action better than secondhand reports.

It is difficult to design an inspection report form that will suffice for all plants. However, a specific form prepared for a specific facility will ensure that a series of inspections over a period of time will cover the same ground. Any such form should be reviewed for completeness once a year or so as plant conditions change. Photographs attached to the report will help those who have not participated in the actual survey to understand the records.

Product Knowledge of Personnel

It is fundamental that every person who has anything to do with handling products has full and complete product knowledge. This must be true for every product handled in the facility. Current "right-to-know" legislation requires the maintenance and availability to plant personnel of material safety data sheets (MSDSs). Training sessions for all personnel to familiarize them with these MSDSs is essential for plant safety.

MSDSs answer the following questions and many others: What are the products' physical properties? Are they heavier than water? Lighter? Soluble or nonsoluble? Are the products toxic? Flammable? Explosive? Are the products reactive; if so, with what do they react? If they come into contact with commonly used substances (such as oil, gasoline, or water), will there be an adverse effect? Will special safety precautions be required in the event of a spill? What about on-scene personnel not involved in the spill? Is evacuation necessary? Is special protective clothing necessary? Is it available?

9.2.6 OUTSIDE SURVEY

Roadways Inside Facility, Access Roads, Parking Areas, and Storm Drains

Drainage. Are all roads paved and/or guttered? Would a spill be retained or drained off onto the ground? Are there storm drains; if so, where do they lead? Would a spill on the roadway reach a watercourse via these storm drains? Can provision be made to shut the drains in an emergency or to bypass the flow through a retention basin that might hold a spill? Are all the drains really necessary? Are they in the right locations to prevent spread of a spill because rainfall does not properly flow away from a high-potential spill area? Are access ways to public roads near public storm drains?

Traffic Control. Are roads protected against traffic accidents, or should controls be installed to protect the movement of vehicles carrying hazardous materials both within the plant and at the point where they enter and leave public roads? Is there sufficient clearance that all vehicles using the roads cannot possibly damage buildings or equipment?

Vehicle Inspection

What kind of inspection is made of employees', suppliers', and visitors' vehicles to ensure that there is no leakage from transmissions or fuel tanks? If the facility has a visitors' center, are visitors' actions controlled while their cars are parked there? (Visitors have been known to drain motor oil onto a parking area while they are visiting a public information center and then to refill and drive away from the mess created.)

Are tank trucks and other supply or delivery vehicles inspected as they enter or leave the facility to be sure that there is no leakage from the tanks or from containers on the vehicles?

Loading and Unloading Facilities

Included in this category are rail switch tracks and rail tank car loading racks; tank truck loading racks, both overhead and bottom-loading; building loading platforms (docks); river docks; pipeline connections; valve manifolds; and perhaps other areas where products are loaded or unloaded in bulk quantities or in drum containers. These areas are high-probability sources of spills, mostly small but occasionally large.

An immediate consideration at any location is whether a spill will require cessation of operations, evacuation of the area, and emergency procedures to control or contain the spill. Each location should be inspected with that basic thought in mind. Loading areas especially should have some means of such containment and control, because outside personnel without the high degree of training of in-plant personnel may be involved.

Are there materials at hand for spill control and containment? Is there safety equipment for personnel who may be involved? Where will a spill go if it occurs? Is drainage provided that leads to a controlled collection area? What kind of communication is provided to stop transfer action in the event of a spill or to summon help? Are there automatic shutoff devices? Can a driver pull away from a loading area while the vehicle is still connected to the loading equipment?

How much material could be spilled if the worst-case accident occurred, and is the containment provision adequate to handle such a spill? [A mile of 254-mm (10-in.) diameter pipeline, for example, holds approximately 83.3 m^3 (22,000 gal) of product, which could all be lost in the event of a valve failure and gravity drainage of the line.]

Are the materials used for construction of the facilities proper for the materials being handled? (This situation may have changed between the time when the facility was built and the time of the survey.) Could products handled or stored in the area react in a dangerous fashion in the event of a spill? If so, where else can they be stored, or what procedures can be instituted to minimize such contact? What maintenance procedures are used for loading and unloading equipment? How often are safety inspections made? What criteria are used to test equipment? Are proper test records kept? What rules and regulations must be complied with?

Tank Storage

Are tank areas properly diked in accordance with existing codes covering size, height, and distance? Are tank dikes and dike areas impervious to a spill? If not, can they be made so by use of an impervious plastic lining, clay, or other means? Is the dike itself safe from penetration by burrowing animals that might impair its integrity? Are small horizontal tanks far enough from dike walls that a leak high on the tank will not spray over the dike?

How often are tanks inspected for integrity? How often are they cleaned and inspected internally? How are below-ground tanks inspected, if at all? Are underground tanks and piping pressure tested (local codes vary widely in this regard)? Do any tanks have cathodic protection? Is it necessary? Do dike areas have provisions for water drainage? Is there a separator or other protective device to ensure that there is no discharge of product from the dike area while water is being discharged? New regulations covering underground storage tanks became effective in December 1998.

What provision is made for overfill control such as high-level alarms or cutoffs? Are valve connections such that operator error is minimized or eliminated, reducing the chance that a product will be put into a full tank or that products that can react dangerously cannot accidentally be mixed? Are there valve interlocks, padlocked valves (with proper control of keys), color-coded valving and piping, noninterconnectable fittings, and proper supervision and connection inspections before transfers are started?

Pipe Alleys

Pipelines into a facility and pipe alleys within are low-potential sources of spills. However, they should be protected if possible by diking when above ground. This diking serves both as a containment for a spill if it occurs and as protection against damage from contact. Are pipelines protected cathodically against corrosion? Are the lines pressure tested? If so, how often? If the pipelines are no longer in use, are they empty of product (blown with an inert gas and blind flanged)?

Are there shutoff valves at both ends of the lines for use if either end is inaccessible because of a spill or a fire?

Drum Storage

If the plant product is packaged in drums, is the outdoor storage area for filled drums roofed or open? Is there protection against leakage and/or damage (as from a fork truck) in the actual storage area (such as a concrete pad drained through a containment device of some sort) and also along the path by which drums must be transported from filling to storage to shipping?

Are drums for filling, if new, properly empty and clean? If reconditioned, are they inspected before filling to ensure that they are clean and empty?

Are drums accepted as returns from any source? If so, are they empty, properly closed, and nonleaking? Is there a policy in this regard that has been properly communicated to customers, to those making pickups, and to those accepting the drums at the facility? What happens to rejected returns in handling, storage, and disposal? Is there inspection to ensure that the returned drums are not mislabeled and do not contain hazardous materials either in themselves or from the standpoint of being mixed accidentally with products handled by the facility?

Where are returns stored? If on a drained pad, is it roofed to minimize rainwater contamination of a product containment device and overload of its capacity? For both inbound and outbound drums, are the storage pads properly curbed and drained? Are water-soluble products stored with water-insoluble ones so that an accident might upset separator operation?

Retention Basins, Containment Ponds, and Product Separators

Are there ponds, basins, and/or separators that might accumulate products by design or accident? What provision has been made for removal of accumulated products manually, automatically, continually, or at irregular intervals? If outside contractors are used for this purpose, are they fully aware of product properties? If there is a separator, is it of adequate capacity? Is it functioning properly, and what provision has been made for performance testing? Is there any problem with suspended solids that might affect separator performance? Do retention ponds have baffles or weirs designed to prevent accidental escape of product in the event of flooding or massive product discharge?

Temporary Construction and Other Facilities

Is there contractor activity that might affect or be affected by routine plant operations? Are the contractor's personnel properly trained and informed about spill potential affected by their operations? Are the contractor's storage facilities (usually fuel, stored high for gravity feed close to roadways) properly protected against accident?

9.2.7   INSIDE SURVEY

Small Leaks from Fittings, Valve Packing Glands, Bearings, and Seals

Where do such leaks go? If they go into a floor drain, does the drain lead to a sump from which product can be recovered or to a storm sewer or drain? Can such leaks be controlled by better maintenance or with drip pans, sorbents, or other means? If the product flows to a sump, where does the sump effluent go? If to an in-plant separator, is that working properly, and how is proper working determined? Can a surge of water or product overload the separator or sump? Are pumps used to transfer sump contents of a kind that will form emulsions or dispersions of product?

Vapor Exhausts: Safety Blowoff Areas

Where do vents from compressors, exhaust blowers, and emergency vents (rupture disks) lead? If they lead to outside the building, do the discharges condense onto the ground or onto roofs, where rainfall might wash them into sewers? If they are discharged into the in-plant atmosphere, will they condense into accumulations that can come in contact with water or steam and be washed into a plant sewer or drain?

Floor Drains

Most plants have floor drains. In the event of an accident, where will these drains take spilled product? Are all the drains really necessary? Can they be closed or relocated so that only the really essential drains are open for use? Can areas where floor drains are essential be subdivided with curbing to segregate water drains from possible contact with spilled product?

Tanks

Are inside storage tanks, including reservoirs in equipment, protected against accidental leaks of product in large amounts? If a rupture occurs, where will the product flow? Can such tanks be curbed, diked, or segregated behind walls to contain spills? If the diked or curbed areas have drains, are they necessary? Could they be closed and spills cleaned up by hand? Are the tanks vented; if so, to where? If they are vented to outside the buildings, do the tanks have high-level alarms or shutoffs to prevent spillage out of the vent caused by an overfill? Are the vents high enough to provide a head greater than the head capacity of the pump feeding the tank? If a discharge through a vent to a roof cannot assuredly be prevented, can the roof itself be safely used as a containment basin? Can it be equipped for such a purpose?

Are process and storage tanks inspected on a regular basis and are repairs made promptly when required?

Piping

Are walls and floors free from holes caused by removal of old equipment and piping? Are there cracks alongside pipes extending through walls or floors? Are such openings for pipes properly caulked to prevent leaks if they represent potential exit points of a spill? Is plant piping well marked, and are operators trained so that an operating error cannot result in an accidental product discharge? Is piping regularly tested for leaks and inspected for corrosion?

Drum Product and Raw-Material Storage

Is internal drum storage of finished product or raw materials so segregated that an accident in the storage area will not spread through the operating area? Is it possible to install curbing around storage areas to prevent this?

Valves and Fittings

Are valves properly identified, by color coding or naming, so that operating errors are minimized? Are valves not in daily use locked against accidental change in setting? In the event of an accident, are there alternative locations where valving or controls can be operated to minimize the spill by shutting off the flow?

Along the normal work path within the facility (where drums are moved on hand trucks, fork trucks travel, or equipment and machinery may be moved) is all piping and valving safe from accidental contact that could cause a spill by damage to that piping and valving? Are there drains, floor drainage slopes, or other conditions that could cause an accident along the normal work paths to be uncontrollable?

Maintenance Operations

Is handling of product collected in the maintenance of building and machinery (flushing of equipment, washing of walls or machines, or cleaning of storage areas) properly controlled to prevent discharge of product? (In other words, does the cleanup porter dump a bucket of dirty water containing toxic chemicals down the most convenient plant floor drain?) How is waste handled and disposed of to prevent discharges? Are the solutions normally used in cleaning and routine building maintenance such that, if improperly disposed of, they might upset the operation of a separating system? Can specific disposal areas be provided for such solutions and segregation of such solutions be ensured?

Available Control Materials

It is assumed that, for specific chemicals that require neutralization or counteraction of some sort if spilled, a plant will have proper materials and instructions available to carry out such counteraction. The plant survey should ensure that this is so.

There are some general types of materials for use in many product spill situations, which should be considered during a survey. If these are applicable, the survey should indicate which kinds arc usable and where they should be located.

Sorbents

Despite some problems with definitions (whether products are absorbents, adsorbents, or just plain sorbents), when there is a spill, sorbents are useful tools to mop up and help dispose of spilled material. There are several classes of sorbents, as described below.

Natural Products. This class includes materials such as hay, straw, rice bran, peanut hulls, cotton linters, pine boughs, and other naturally occurring products (in the sense that they are not processed products designed for sorbent use). These materials can be used in spill cleanup but generally have rather poor efficiency as well as handling and storage problems; in an emergency, when one is (literally) grasping at straws, one uses what is available.

Products made for specific sorbent use, however, are generally more efficient, are less costly per unit of recovered product, are easier to store, and are easier to use and dispose of.

Selective Sorbents: Removing Product from Water. These sorbents are designed to remove product from on top of water and leave the water behind. Several different types are available.

Cellulosic sorbents are made from cellulose fiber treated to render it water repellent. They come in pads, sheets, particulates, and rolls. They generally possess good sorption for oil, but they present difficulties in long-term storage (generally covered storage is required). Although they will sorb hazardous materials, care in handling and disposal is required.

Synthetic fibers generally are made from polypropylene fiber that is felted into rolls and sheets or shredded into particles used to form small bags or longer boom sections. Sorption is good for oil and generally is good for a range of hazardous materials. Storage characteristics are good, and no special storage protection is required.

Synthetic foams generally are polyurethane foams but occasionally are polyethylene foams. They offer good pickup of oil but generally pick up more water than the sorbents described above. Storage is fairly good.

Imbibitive polymers are copolymers that react with certain classes of organic product, "drinking" them into the molecular structure; first they form sticky masses and then, as saturation nears, they form a jellylike material that can be cut without losing any product. These materials can be used to reduce the vaporization rate and to solidify, and hence make easier to handle, a wide variety of hazardous and toxic chemicals. In proper form, valves can be constructed that will pass water but will swell and stop the flow when product is encountered. They are very useful for draining water from tanks and out of dike areas. These products cannot be used as in-line filters, because swelling will stop the flow.

Selective Sorbents: Product Dissolved in Water.   Many hazardous materials are water-soluble. Activated-charcoal filters have been used successfully in removing such dissolved materials, usually at high cost.

Nonselective Sorbents.   A highly porous nonreactive silicate has come onto the market. This product, with a density of about 32 kg/m^3 (2 lb/ft^3), will pick up whatever it touches. Therefore, it cannot be used to remove a product floating on water, but it can be used on land to sorb water, aqueous solutions of any pH, and pure product including oil and hazardous materials.

Sorbent Availability.   Sorbents are produced by several manufacturers and are generally stocked and sold through local distributors both in the United States and in a number of other countries. Most commercial sorbents are fairly light weight, are easy to handle, and require only moderately careful storage. Some, as noted, require enclosed space for long-term storage.

Sorbent Disposal.   It is not always easy to dispose of sorbent saturated with product. Although most manufacturers make a claim of reusability, the cost of squeezing out recovered product, combined with the inherent messiness of such an operation and, depending on the product, the possible handling hazard, makes this characteristic more of a paper benefit than a real one.

Generally, saturated sorbent is put into drums, plastic bags, or, occasionally, open trucks for disposal either by burning (in a properly designed incinerator) or by burial in an approved landfill. Occasionally, especially with oil, a product may be burned for fuel if it is not too heavily contaminated with nonburnable solids. Even more infrequently, synthetics can be dissolved in hot oil and used for fuel as part of the oil.

Special Sorbent.   Depending on circumstances, sand, snow (in cold climates), dirt, Fuller's earth, and a variety of other materials occasionally have been used as sorbents, mostly because of quick availability in a situation that demands immediate action. Proper spill control planning dictates storage of some sort(s) of sorbents and prior arrangement for quick supplies of additional sorbents if needed.

Chemicals

Chemicals can be a general term covering dispersants, neutralizers, gelling agents, firefighting foams, sinking agents, and a host of other products. Committee F-20 of the American Society for Testing and Materials (ASTM)[1] has a division that has been grappling with the problems of chemical control of chemical spills and has issued a number of standards in this regard.

Equipment

In many instances specialized equipment is available to contain and recover spills of hazardous materials. Most of this equipment was developed for use with spills of petroleum products; some is usable on other types of product spills. New equipment is reaching the market all the time. Many members of ASTM F-20 and of the Spill Control Association of America represent companies engaged in development and manufacture of such equipment in the United States. Contact with these two organizations should provide specific names of such companies.

A detailed discussion of the many types of equipment available to manage product spills is beyond the scope of this chapter. However, it is proper to discuss the use of some types of equipment to be used for prevention, and that is done here.

9.2.8 PREVENTION

Containment Materials

Preventing the spread of spilled product greatly simplifies cleanup provided it can be done quickly and safely. In particular, prevention of the spread of spills on water greatly reduces the contaminated area and also reduces the time, cost, and hazard of recovery.

Materials at Hand. If a spill occurs on nonporous ground (paved areas, frozen ground, clay soil), it frequently can be surrounded by a dike of earth dug from the nearest available area with shovels, a backhoe, or some other mechanical device depending on need. Often advance provision can be made in an area of high risk by prefilling sandbags and storing them outside, properly protected by tarpaulins, near the potential use site. The same sandbags can be used to shut off storm drains ahead of a spill, provided there is sufficient time after the spill occurs.

Existing ditches, normally used for draining water, can be used to confine a spill by piling earth across the ditch and compacting it to form a dam. Floating product can be retained if there is water in the ditch and the product is insoluble by supporting a pipe on stones with the upstream end lower than the downstream end and covering this with dirt to form an underflow dam or weir.

Manufactured Materials: Quick-Setting Foams. Several manufacturers of insulation sell aerosol containers of polyurethane foam. Allthough rather expensive and with a relatively short shelf life (perhaps a year), these foams rapidly swell and solidify (usually within 1 min of application) to form an impervious barrier that resists many chemical products and that is nominally fireproof (although most foams will burn if they are ignited at a thin cut edge or come into contact with a hot fire).

Manufactured Materials: Containment Booms. There are a number of manufacturers of containment booms, which generally consist of a flotation section supporting a skirt, at the bottom of which is ballast, usually link chain. The flotation section often is made of polyethylene and hence is relatively impervious to a wide range of hazardous materials. Skirt material is usually a webbing of woven synthetic fiber, such as nylon or polyester, coated with a material such as polyvinyl chloride. These booms are quite impervious to petroleum products and to a large range of hazardous chemicals. Other materials are available for special uses, as are special metals for boom fittings, ballast, and connectors.

Water-immiscible products can be confined very well on still water such as a pond, a lake, or a quiet bay or, with a difference in technique and boom configuration, on flowing water such as a river (within limits). Thus, booms can be used to reduce the spread of a spill reaching water and to aid in cleanup.

Safety Gear

Handling a hazardous material spill requires proper safety equipment. Anyone engaging in containment, control, or recovery activities should be provided with the proper equipment and should be well trained in its use.

Storage of Control Material

Material to be used for spill control should be stored as near as practicable to the site where it may be used. Proper protection from deterioration during storage, from theft and vandalism, from pilferage, and from use for other purposes must be provided. Some estimate of possible quantities must be made so that emergency materials are available in sufficient quantity to function efficiently if the need arises. Generally, it is wise to provide separate storage for such materials either in a permanent building or in a small emergency trailer if several areas need protection and this method proves more economical than duplication of materials and storage facilities. Care must be taken to ensure that security in storage is not so tight that entry to the stored materials is difficult when the need arises. Several persons should be authorized to have keys and access to the stored material.

9.2.9   EMERGENCY PROCEDURES

Contingency Planning

Contingency planning is what spill prevention is all about. The "what if" attitude should disclose most foreseeable (but never all) problems and should set forth in detail some method of dealing with them. Things never occur as planned, but thinking about them helps one to prepare for the unexpected.

A contingency plan should be in written form. Copies should be available at potential spill sites and should be used as training aids for all personnel.

Of particular importance in responding to any spill emergency is response time. If a spill can be found while it is small and response is rapid, its effects usually can be minimized. Thus an essential part of any contingency plan is provision for regular security checks of both high- and low-risk areas so that, if spills do occur, they are found promptly.

Training for Action

Regular training drills are necessary for all personnel who are to be involved in cleanup and control. A training manual should be developed, updated as often as necessary, and combined with regular class-room and field hands-on training, including carefully controlled actual spills of hazardous materials. One must learn to handle problems with advance practice and not by on-the-job training. Response time to a spill can be greatly reduced by such regular training.

Notification

A schedule of notification procedures to be followed in the event of a spill should be established. It should include all legally required notifications (federal, state, and local authorities) with an indication of the person who is to make the call, plant personnel, company personnel not at the plant, and outside cleanup contractors. Notification lists should be updated and reviewed on a regular schedule and reissued whenever any changes are required. Normally this is a task to be done on a monthly basis.

Outside Help

The equipment and expertise to handle large hazardous material spills are rarely available when needed within a plant or company. Unless there is a continual number of such large spills, it simply is not economically feasible to be prepared for eventualities of low potential. There are a number of highly competent cleanup contractors who specialize in this work, and their ranks are growing with increasing awareness of the need for such services. Competent contractors are expensive; the investment in equipment and training maintained by good contractors is very high; hence, their charges also are high. It is a prudent decision to make advance contact with such contractors and to have advance understanding of their charges, capabilities, and availability. In many instances, a yearly retainer will ensure preferred service plus the help of the contractor in assessing in-plant exposures, corrective measures, and possibly assistance in training personnel.

9.2.10 DECISION MAKING

Spill prevention, like accident prevention, is a matter of identifying potential causes and taking corrective action. Rarely is it possible to base a decision on a mathematical procedure, balancing the cost of taking some action against the savings that accrue because of reduced cleanup costs, product savings, reduced downtime, or perhaps no spill at all.

What types of decisions are needed? They are those that will reduce the probability (not determinable) of having a spill by an amount in keeping with the money expended to accomplish that result.

Legal Decisions

There is not much choice in legal matters. Despite the best of goodwill on the part of someone who might have a spill of hazardous material and the deepest commitment (moral and financial) to comply with the law, there still will be legitimate areas of disagreement about what must be done. If there is a court decision, there is no legal problem, although perhaps a financial one. What the court determines you must do, you must do (after you have exhausted all legal appeals). Obviously, a public relations problem is also involved. Even though you may be legally right in a position you take, a battle through the court system, with attendant publicity, may cost more in the long run in public ill will and reduced sales than spending the money to do what is not legally required. The type of decision then moves from a legal one to a logical one.

Logical Decisions

Logical decisions are based on experience, intuition, and perhaps even engineering appraisal. They tend to be the types of decisions that all persons make, one way or another, during most of their lives.

For example, suppose that a plant has an undiked tank situated in a large open area near a road, with a storm drain (leading directly to a river) located a short distance downhill on the road. The tank contains a highly toxic chemical. Logic dictates that the gain from installing a relatively simple earthen dike to protect the river from the result of a low-probability tank leak far outweighs the relatively small cost of installing the dike.

Move that tank inside or to a location where space is at a premium and there is no room for a simple dike properly distanced from the tank, and the cost of protection rises. The decision requires other input. How often does a tank of this sort leak? When was it last inspected? Can it be safely inspected and some sort of regular program established to reduce the chance of an accidental leak? What will a high concrete dike close to the tank cost, and how will it affect the ability to operate? If there is a leak in these close quarters, will the only loss be product and the only damage environmental, or will personnel be injured or perhaps killed?

Under these conditions the decision, although still based on logic, may be different. The cost may be so high and the risk (based on past experience) may be low enough that no dike is installed. The tank may be relocated. Or nothing may be done at all.

What is absolutely necessary in such instances is that whatever the decision, it be established as a part of the company records. In today's climate of environmental responsibility, it is essential that decisions be deliberate and recallable. One simply cannot afford to be in the position of having to say "We didn't realize this was a problem." In balancing the many needs for money, one must be able (or at least try) to justify a decision not to do something because of the inability to do everything and because of a greater need or greater danger to be allayed elsewhere.

Economic Decisions

Economic considerations are a part of most business decisions. In any endeavor with some potential for spills of hazardous materials, a budget must contain some amount for spill control training, planning, and prevention. It may not be possible to reach numerical answers on specific questions. It is possible, as a matter of business judgment, to allocate a portion of an operating budget to spill control matters, just as is done for research and development, sales effort, public relations, and advertising. Spill control is a necessary part of today's world. A good way to approach a decision about how much money to allocate to spill prevention is to make the surveys previously suggested.

It is rare that such a program will not disclose many ways to spend money on fixing up, repairing, or replacing equipment. A program can be established within the normal maintenance schedule of any facility, over a reasonable period of time (up to 5 years, with modification as experience and resurvey dictate), and an annual amount allocated in the budget for spill prevention maintenance. The personnel to be involved and their time for training on a regular basis are relatively easy to determine, and the cost can be added to the maintenance amount. If an outside contractor is to be retained, that fee also can be determined and added to the total.

Inclusion of these amounts in a budget may be a great help, if a spill occurs, in reducing fines and obtaining the cooperative assistance of governmental authorities, because one will have demonstrated a commitment to spill prevention and a commitment that is actively in progress.

Technical Decisions

Technical decisions are often involved in choosing between alternative methods of reducing spill probabilities. For example, should one install a high-level alarm to prevent a tank overfill or raise the overflow pipe height and reduce the capacity of the feed pump head? Both cost and effectiveness of method enter into such decisions, and, provided the proper information is available on which to base such decisions, technical decisions are reached more easily than legal, economic, and logical ones.

Much of the equipment now available for cleanup, control, and prevention of spills and the techniques of properly using that equipment have been developed over the past couple of decades by ASTM Committee F-20. That committee has issued many standards for testing such equipment and standard practices for using it.

REFERENCE

ASTM Committee F-20, *Hazardous Substances and Oil Spill Response*, American Society for Testing and Materials, West Conshohocken, PA.

CHAPTER 9
POLLUTION PREVENTION CONCEPTS AND POLICIES

SECTION 9.3

URBAN STORMWATER MANAGEMENT AND NONPOINT SOURCE POLLUTION CONTROL

Kirk R. Barrett

Dr. Barrett is a research assistant professor and the research director of the Meadowlands Environmental Research Institute (MERI) at Rutgers University, Newark, New Jersey. He has expertise in surface and wetland hydrology, hydraulics and water quality processes.

The author would like to acknowledge the helpful reviews of Margaret McBrien, Randell Greer, Gordon England, Kimberly Davis, and Thomas O'Connor.

TABLE OF NOMENCLATURE

AMC = antecedent (before a storm) moisture conditions (in soil)

BOD = Biochemical Oxygen Demand

BMPs = best management practices (techniques for managing stormwater quantity and quality)

CAFO = concentrated animal feeding operations (feedlots for livestock)

CN = curve number, an index related to infiltration capacity in TR-55

COD = chemical oxygen demand

CSOs = Combined Sewer Overflows

DO = dissolved oxygen

FHWA = Federal Highway Administration

NO_3-N = nitrate nitrogen

NPDES = National Pollutant Discharge Elimination System (a U.S. federal discharge permitting program)

NPSP = nonpoint source pollution

NRCS = U.S. Natural Resource Conservation Service (formerly Soil Conservation Service)

SCS = U.S. Soil Conservation Service, renamed the Natural Resource Conservation Service

SSOs = sanitary sewer overflows

TKN = total Kjeldahl nitrogen (organic and ammonia N)

TMDL = total (allowable) maximum daily load (of contaminants to a water body)

TP = total phosphorus

TR-55 = a method for estimating runoff, developed by the SCS and described in "Urban Hydrology for Small Watersheds," Technical Release 55

TSS = total suspended solids

U.S. EPA = United States Environmental Protection Agency

USEFUL WEBSITES

U.S. Federal government sites
  Federal Highway Administration (FHWA) (hydraulics publications and software)
    www.fhwa.dot.gov/bridge/hyd.htm
  U.S. Army Corps of Engineers Waterways Experiment Station (WES)
    wes.army.mil/
  U.S. EPA sites
    Storm water permitting program
      www.epa.gov/owm/stormw.htm
    Office of Wetlands, Oceans, & Watersheds, Total Maximum Daily Loads
      www.epa.gov/OWOW/tmdl
    Center for Exposure Assessment Modeling (CEAM; computer models)
      ftp://ftp.epa.gov/epa_ceam/wwwhtml/ceamhome.htm
  U.S. Geological Survey's Water Resources Information Server
    h2o.er.usgs.gov/
  National Oceanic and Atmospheric Administration (NOAA) National Climatic Data Center (precipitation data)
    www.ncdc.noaa.gov
  Natural Resources Conservation Service (Soil Conservation Service)
    TR-55, Urban Hydrology for Small Watersheds
      www.wcc.nrcs.usda.gov/water/quality/text/hydrolog.html
    National Water and Climate Center
      www.wcc.nrcs.usda.gov/

U.S. state government stormwater and nonpoint source pollution sites
  California Department of Water Resources
    wwwdwr.water.ca.gov/
  Florida DEP Nonpoint Source Page
    www2.dep.state.fl.us/water/Slerp/Nonpoint_Stormwater/stormh2o.htm
  Maryland stormwater manual
    www.mde.state.md.us/environment/wma/stormwatermanual/mdswmanual.html
  Massachusetts stormwater manual
    www.magnet.state.ma.us/dep/brp/ww/wwpubs.htm
  Michigan nonpoint source program
    www.deq.state.mi.us/swq/nps/npshome.htm
  Texas nonpoint source book
    www.txnpsbook.org

Private hydrologic software
  Alan A. Smith (MIDUSS98 program)
    www.alanasmith.com
  Applied Microcomputer Systems (HydroCAD)
    www.hydrocad.net
  Boss International (DAMBRK, XP-SWMM, others)
    www.bossintl.com
  Computational Hydraulics International (SWMM, PCSWMM)
    www.chi.on.ca
  Dodson and Associates (HydroCalc and others)
    www.dodson-hydro.com
  Haested methods (StormCAD, Pond Pak, others)
    www.haested.com
  Water Resources Consulting Services (Hydraflow, DAMBRK others)
    www.waterengr.com

Professional societies
  American Public Works Association
    www.apwa.net
  American Society of Agricultural Engineers
    asae.org
  American Society of Civil Engineers
    www.asce.org
  American Storm Water Institute
    www.stormwater.org
  American Water Resources Association
    www.awra.org
  International Association on Water Quality
    www.iawq.org.uk
  International Erosion Control Association
    www.ieca.org
  Water Environment Federation
    www.wef.org/

Universities and research institutes
  Center for Watershed Protection (Publisher of Watershed Protection Techniques)
    www.cwp.org
  Colorado Water Resources Research Institute
    cwrri.colostate.edu/
  Illinois State Water Survey
    www.sws.uiuc.edu
  National Institutes for Water Resources
    wrri.nmsu.edu/niwr/
  North Carolina State University Water Quality Group (includes bibliography and links)
    www.bae.ncsu.edu/bae/programs/extension/wqg/
  Old Dominion University, Civil/Environmental Model Library, Civil & Environmental Engineering
    Department,
    www.cee.odu.edu/cee/model/
  University of Texas Center for Research in Water Resources (includes links)
    www.ce.utexas.edu/centers/crwr/
  Western Regional Climate Center (precipitation data for western US states)
    www.wrcc.dri.edu/pcpnfreq.html

Miscellaneous sites
  Stormwater News (includes bibliography)
    www.stormwater-resources.com
  Hydrology Web, Hydrology Group, Pacific Northwest National Laboratory
    terrassa.pnl.gov:2080/EESC/resourcelist/hydrology.html
  Natural Resources Defense Council glossary of stormwater terms
    www.nrdc.org/nrdcpro/storm/gloss.html
  Rouge River National Wet Weather Demonstration Project
    www.waynecounty.com/rougeriver/
  US Geological Survey glossaries of stormwater terms
    wwwrvares.er.usgs.gov/nawqa/glos.html
    ga.usgs.gov/edu/dictionary.html
  William James' stormwater bibliography (includes bibliography)
    www.eos.uoguelph.ca/webfiles/james/homepage/Research/Bibliographies.html
  Metropolitan Washington Council of Governments (stormwater publications)
    www.mwcog.org
  Local Government Environmental Assistance Network (Links to state environmental agencies)
    www.lgean.org

9.3.1 INTRODUCTION

Flow of stormwater occurs when rainfall or snowmelt exceeds the land's storage and infiltration capacity. Stormwater (generally equivalent to the term runoff) flows over or just under the ground surface, soon reaching small channels or storm drain inlets and eventually discharging into receiving waters (i.e., streams, rivers, lakes, and oceans).

Both rainfall and snowmelt contain contaminants, sometimes in significant concentrations. As stormwater water flows over the land surface, it picks up additional contaminants. These contaminants are commonly known as nonpoint source pollution (NPSP), which refers to pollution other than end-of-pipe discharges from sanitary sewers, industries, or wastewater treatment plants. However, the federal statutory definition of point sources such as in the Water Quality Act of 1987 includes many traditional nonpoint sources. The terms runoff pollution, diffuse pollution, and wet weather flow pollution are frequently synonymous with NPSP, without the regulatory confusion. Atmospheric deposition and hydrologic modification are also generally considered nonpoint pollution sources, as are overflows from combined sanitary/storm (or separate sanitary) sewers during wet weather.

Stormwater quantity (volume and flow rate) and quality are both important. Nearly half of the 325 public agencies surveyed in 1980 reported serious stormwater flooding problems (American Public Works Association, 1980). The U.S. Environmental Protection Agency reported in the 1994 National Water Quality Inventory that nonpoint sources were the primary causes of water-quality impairment (agricultural runoff for rivers and lakes, urban runoff for estuaries). They also reported that hydrologic/habitat modification was one of the top five causes of impairment in rivers and lakes.

This discussion of stormwater intends to provide an inventory and a brief overview of relevant subtopics and to point the reader to useful, comprehensive, authoritative discussions of subtopics and to computerized tools and data, especially those that are immediately available (and often free) on the World Wide Web.

Much of the concern about stormwater involves the effects of human modification of the land. Specifically, a frequent concern is evaluating and controlling the effects of development or agriculture by best management practices (BMPs) to reduce peak flow rates and reduce pollution. Land modifications and subsequent activities can have major and harmful effects on the magnitude, frequency, and duration of both high and low flows (affecting flooding, ecosystems, channel stability, recreation, and water availability) and water quality (affecting water supply, recreation, aesthetics, and human and ecological health).

Modern stormwater management seeks both to reduce the risk posed by flooding to life and property and to protect and enhance the aesthetic, recreational, and ecological value of water bodies, upland watersheds, wetlands, shore areas, and stream corridors—at reasonable cost. On a scale of tens of hectares or larger, stormwater management is tightly linked with the emerging field of watershed management.

9.3.2 REGULATION/PERMITTING

In the United States, stormwater is regulated at the federal, state, and local levels. Regulations commonly apply to discharge (volumetric flow rate) and water quality and less often to runoff volume.

U.S. Federal Regulations

The National Pollutant Discharge Elimination System (NPDES), established by the Clean Water Act of 1972, regulated traditional point sources. The Water Quality Act of 1987 (Sec. 502-14) initiated the NPDES stormwater Phase I program, which expanded regulation to include sources such as storm drain outfalls in service areas with a population of more than 100,000, combined sewer overflows, runoff from construction sites larger than 2 hectares (5 acres), and runoff from animal feed lots (CAFOs).

NPDES stormwater program, Phase II, scheduled for implementation in 2000, addresses additional stormwater discharges, including smaller urban areas, smaller construction sites, and certain retail, commercial, and residential activities. Updates on the U.S. Environmental Protection Agency's (U.S. EPA) stormwater permitting program are available at www.epa.gov/owm/sw.

A NPDES permit is required for regulated activities; the permit program is administered by state environmental agencies in many states. In regulated cities, many industries also must obtain a permit for stormwater discharge. Refer to the web site associated with the state for specific information. Many states, counties, and municipalities also have stormwater regulations. A directory of agencies and regulations is provided by the Local Government Environmental Assistance Network (www.lgean.org). NPDES permitting for stormwater is fully covered by Dodson (1998).

U.S. EPA's total maximum daily load (TMDL) program under Section 303(d) of the Clean Water Act is also relevant to stormwater. States are required to identify waters that are not "fishable and swimmable" and to develop a TMDL for each. The TMDL process involves quantitative assessment of water-quality problems, contributing sources, and load reductions or control actions needed to restore and protect bodies of water. Often, much of a water body's load is NPSP. The TMDL program is discussed fully at www.epa.gov/OWOW/tmdl.

General Regulartory Concerns

Initially, concern about stormwater was focused on controlling peak runoff discharge to prevent and reduce flooding resulting from increased imperviousness (pavement and rooftops) during urbanization. Concern about increases in runoff volume (also resulting from increased imperviousness) is more recent and is less common. Increasing runoff volume lowers the amount of infiltration and groundwater recharge, sometimes reducing the baseflow of streams during droughts, which affects water supply, ecosystems, and recreation (Simmons and Reynolds, 1982).

A frequent regulatory requirement for development is to analyze a storm (or the stormwater flow conditions) of a given magnitude or frequency (return period) and duration—i.e., the design storm.

It should be understood that frequency of a storm and of the runoff it produces are not necessarily equivalent (Marino and Bradley, 1986). Although this is a theoretically acomplicating problem, design conditions set by state or local regulations usually overlook this discrepancy.

Regulations commonly require that the peak flow rate from a developed site be no larger than that in the predevelopment condition; a set of runoff controls are required to achieve this goal. Typical design conditions are from the 2 to 100-year storm for peak discharge control. A 24-h storm duration is common. Regulations on runoff volume may specify, for example, retention (keeping on site) and infiltration into the ground of all the runoff for up to a 2-year storm. Massachusetts requires infiltration of between 0 and 1 cm (0 and 0.4 in.) of runoff from impervious surfaces, depending on soil conditions (Massachusetts Department of Environmental Protection, 1997).

Water quality control is also a regulatory concern. For water quality control, a much smaller volume (and therefore a more frequent storm) than for flooding control forms the design conditions. Typical requirements include treating the first 1 to 2.5 cm (0.4 to 1 in.) of runoff (from the entire site or from impervious areas only), treating 90 percent of all the runoff, and treating the runoff from the 2-year storm. Alternatively, regulations may specify a degree of treatment (e.g., 80 percent removal of total suspended solids TSS), or a limit on off-site mass discharge (e.g., in kg of TSS per hectare per year).

9.3.3   QUANTITY OF STORMWATER RUNOFF—FACTORS AND PROCESS

Runoff Quantity: General

Full analyses of procedures for determining runoff volume and flow rates are found in standard hydrology texts (e.g., Linsley et al., 1982; Ponce, 1989). Also refer to the hydrology section in this publication. Web-available introductions to basic stormwater hydrology and common methods are supplied at the sites of Applied Microcomputer Systems and Dodson and Associates.

The stormwater story begins with precipitation, either liquid or frozen. As rain falls from the sky, some of it will be intercepted and retained by foliage—this is called interception storage. Interception storage is generally on the order of 0.25 to 1.25 mm (0.01 to 0.05 in.) (Ponce, 1989), which is significant in an annual hydrologic budget but insignificant in large-storm analysis.

The rain that does reach the surface will infiltrate into the soil in pervious areas (impervious areas, comprising pavement and most rooftops, have zero infiltration). Infiltration capacities vary tremendously based on soil type, ranging from over 50 cm/h on sand soils to only 0.05 cm/h on clay soils (Horn, 1971).

If rainfall rate exceeds infiltration capacity, then water begins to accumulate on the surface, first filling low areas in the landscape (depression storage). Depression storage ranges from about 1.5 mm (0.06 in.) in impervious areas to around 7 mm (0.3 in.) for grasslands, and up to 80 mm (3 in.) for flat, plowed fields (Novotny and Olem, 1994, p. 107).

The total of interception, evaporation, infiltration and depression storage before runoff begins is called the initial abstraction or initial loss. In calculations, after the initial loss has been satisfied, the infiltration rate often is assumed to be constant throughout a storm.

Once depression storage is filled, water starts to flow downgradient over the surface (overland flow) until it reaches a channel—a distance typically of up to a few hundred meters. Forested watersheds generally produce the least runoff, and impervious areas produce the most runoff. Runoff flow rate and velocity also depend on the slope of the land.

Effects of Urbanization

Urbanization (development) has a profound effect on runoff. Development replaces pervious surfaces with impervious ones, removes vegetation that intercepts rain, and, often most importantly, accelerates the movement and concentration of runoff by replacing hydraulically rough pervious surfaces with smoother impervious surfaces and by increasing the number and length of channels. Thus, a developed basin generates more runoff and responds more rapidly to intense, brief periods of rainfall, thereby increasing both volume and peak flow rate. Runoff from paved surfaces generally is considered to carry more contaminants than unpaved.

In hydrographs (flow versus time curve) for developed and undeveloped areas for identical storms, the developed area will show more "flashy" hydrology, with a rapid climb to a high peak, followed by a rapid decline (Figure 9.3.1).

Because uncontrolled development would lead to flooding, site development regulations often specify that postdevelopment peak off site discharge cannot exceed predevelopment peak discharge for a storm of specified design. The peak flow rate is most commonly controlled through detention basins that temporarily store high rates of incoming runoff and discharge it at lower rates (Figure 9.3.1).

9.3.4 COMPUTING STORMWATER FLOW RATES AND VOLUMES

Precipitation

Stormwater flow analysis begins with precipitation, which occurs in both liquid and frozen form. Obviously, snow causes no runoff until it melts. Precipitation data can be obtained in a raw form from a specific location (depth per unit time) or can be compiled into maps of rainfall depths of specific frequencies or into intensity-duration-frequency (IDF) curves (see below). The most comprehensive source for raw data is the National Weather Service's network of about 13,000 rain gauges throughout the United States. These data are available from the National Oceanic and Atmospheric Administration's (NOAA) National Climatic Data Center (NCDC) (www.ncdc.noaa.gov) in a variety of frequencies of measurements, ranging from hourly to monthly. If a station is not near enough

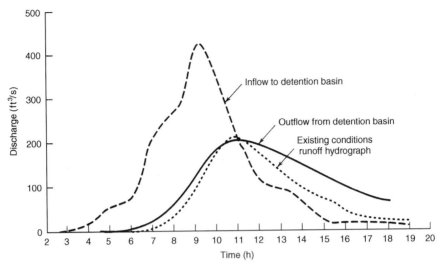

FIGURE 9.3.1 Typical runoff hydrographs for undeveloped and urbanized (with and without detention) watersheds (Akan, 1993).

to a particular site of interest, additional gauges are often operated by water and wastewater treatment utilities, flood control and irrigation districts, airports, environmental research stations, and universities.

Compiled frequency data are often presented as maps with lines of equal rainfall (isopluves or isohyets) for a storm of specific frequency (return period) and duration. The U.S. National Weather Service (NWS) publishes *NOAA Atlas 2* for the 11 coterminous states west of the 103 meridian (i.e., west of the Dakotas). Scanned images of maps are available at the web site of the Western Regional Climate Center (http://www.wrcc.dri.edu/pcpnfreq.html).

The standard references for the eastern 37 coterminous states are NWS Hydro-35 and the older USWB TP-40. Other publications cover Alaska, Hawaii, and the Caribbean. These publications are available from the NWS Hydrometeorological Design Studies Center (www.nws.noaa.gov/oh/hdsc). The maps are published in TR-55 (see below) and are available in electronic format from several sources, including NCDC. Additional data are often available from state agencies. *Rainfall Frequency Atlas of the Midwest* (covering nine states) is available from the Illinois State Water Survey.

Compiled data can also be presented in IDF curves, which plot, on log-log scales, the rainfall intensity (typically in millimeters or inches per hour) versus rainfall duration for storms of various return periods (Figure 9.3.2). IDF curves are available from many state and local agencies. For easy use in computerized analysis, these curves can be described with the following equation (Ponce, 1989, p. 25):

$$i = a/(t_r + b)^c \tag{9.3.1}$$

where i = rainfall intensity (depth per time); t_r = rainfall duration; and a, b, and c are curve-fitting parameters. Free computer programs are available on the web that accept depth or intensity for different time increments and compute the parameters (Alan A. Smith, Inc. web site).

More sophisticated methods for computing runoff, including the common curve number method (used in TR-55) require a hyetograph (rainfall depth versus time) in addition to the total amount of rainfall. The method's developer, the Natural Resources Conservation Service (NRCS), established four standard storm rainfall distributions that cover the entire United States. These distributions are discussed below in the section on the curve number method. Some state and local agencies have established other distributions that are more relevant to specific areas.

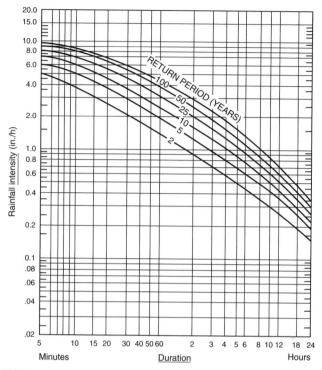

FIGURE 9.3.2   Rainfall intensity-duration-frequency (IDF) curve (Browne, 1998).

Rational Method

The rational method is a simple, common method for computing peak flow rates and volumes for small watersheds. The assumptions behind the rational method are that the rainfall intensity is constant over the entire basin and throughout the storm and that the basin's storage and infiltration characteristics are constant at all times. The assumptions are more valid in small, highly impervious watersheds (because infiltration characteristics become more constant) and for storms of short duration. The equation is

$$Q = c\,i\,Af \qquad\qquad (9.3.2)$$

where Q = peak flow rate at the drainage basin outlet (m^3/s or ft^3/s),
$\quad c$ = runoff coefficient (unitless),
$\quad i$ = rainfall intensity (cm/h or in./h),
$\quad A$ = drainage area (hectares or acres), and
$\quad f$ = unit conversion factor (0.028 for SI units, 1.01 for U.S. units).

Estimating Runoff Coefficient.   The runoff coefficient c is essentially the fraction of rainfall that is converted into runoff. Quantifying the runoff coefficient is difficult because the assumption on which it is based (that a basin's storage and infiltration characteristics are constant) is not very accurate in many cases. Studies have shown that c can vary considerably with intensity and duration of a storm (Rantz, 1971).

Nevertheless, the rational method is commonly used for planning purposes and for small watersheds. Tables of c are widely available (Table 9.3.1). Runoff coefficients can also be estimated by correlation with percent imperviousness in the watershed (Figure 9.3.3).

TABLE 9.3.1 Runoff Coefficients for the Rational Method

Land use	Coefficient range
Business	
Downtown	0.70–0.95
Neighborhood	0.50–0.70
Residential	
Single family	0.30–0.50
Multiunit detached	0.40–0.60
Multiunit attached	0.60–0.75
Suburban resident	0.25–0.40
Apartment	0.50–0.70
Industrial	
Light	0.50–0.80
Heavy	0.60–0.90
Parks and Cemeteries	0.10–0.25
Playgrounds	0.20–0.35
Pavement	
Asphalt and concrete	0.70–0.95
Brick	0.70–0.85
Roofs	0.70–0.95
Grass	
Flat, 0–2 percent	0.05–0.15
Moderate, 2–7 percent	0.10–0.20
Steep, 7 percent	0.15–0.35

Adapted from ASCE/WEF (1992).

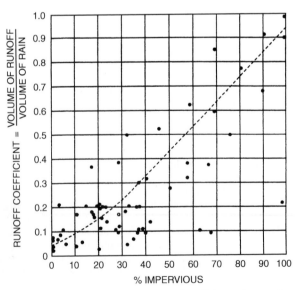

FIGURE 9.3.3 Relationship between runoff coefficient and percent imperviousness in the watershed (U.S. EPA, 1983).

If a basin is not homogenous, it should be split into more-or-less homogenous subbasins, and a weighted c value can be obtained based on the areal proportion of each subbasin.

Time of Concentration. When computing peak flow via the rational method, the rainfall intensity is taken from an IDF curve using a duration equal to the time of concentration for the drainage area. The time of concentration t_c is defined as the time for runoff from the hydraulically most remote point (in terms of travel time, not distance) in the drainage basin to reach the basin's outlet. Using this time as the duration of the storm ensures that the entire watershed is contributing flow to the outlet.

The NRCS (see TR-55 below) recommends methods for analyzing sheet flow and channel flow. Travel times for each type of flow can be computed and summed to give t_c. One may need to compute travel time for several points in the basin to ensure that the largest travel time (i.e., t_c) has truly been determined. Generally, the rational method should be used only when the time of concentration is 30 min or less (Novotny and Olem, 1994, p. 145).

Urbanization decreases t_c by replacing hydraulically rough pervious surfaces with smoother impervious surfaces and by creating shorter and faster flow paths.

NRCS Curve Number Method (TR-55)

Methods more sophisticated than the rational method are required for predicting a stormwater hydrograph, which is used to size detention basins. A very common method for computing hydrographs for development projects is the Soil Conservation Service (SCS; since renamed NRCS) curve number method. The method is fully described in "Urban Hydrology for Small Watersheds," Technical Release 55 (hence the shorthand name TR-55), available at www.wcc.nrcs.usda.gov/water/quality/text/hydrolog.html. TR-55 is a simplified version of the method described in TR-20. McCuen (1982) published an excellent resource for the use of these methods.

When using TR-55 to generate a runoff hydrograph, the required inputs are drainage area, curve number for the area, time distribution of rainfall, total amount of rainfall over 24 h, time of concentration, and travel time. The curve number method is based on the NRCS runoff equation:

$$Q = (P - 0.2S)^2/(P + 0.8S) \qquad (9.3.3)$$

where Q is runoff depth (in.), P is precipitation depth (in.), and S is potential maximum retention (in.). The variable S is then related to a runoff curve number (CN), which is an index of the infiltration capacity of the basin:

$$S = (1000/CN) - 10 \qquad (9.3.4)$$

A CN value of 100 indicates no infiltration; low CN values indicate high infiltration. CN depends mostly on antecedent moisture conditions (AMC) or antecedent runoff condition (ARC), soil type, and land cover.

The NRCS defined four hydrologic soil groups (HSG) for use in TR-55, ranging from group A (low runoff potential, high infiltration) to group D (high runoff potential, low infiltration). Most county soil surveys in the United States list the hydrologic soil groups. A Hydrologic Soils Group List is available on the web at the same site as TR-55. Tables of curve numbers for a wide variety of different hydrologic soil groups and land covers are available in TR-55; these tables also appear in many other publications.

The CN values listed in these tables are for the average AMC, termed AMC II, which are usually used for design calculations. Chapter 4 of the *National Engineering Handbook* (USDA, 1985) and Rallison and Miller (1981) provide guidance with adjusting the CN to reflect other AMCs.

For basins with slightly nonhomogenous land covers, an area-weighted average curve number can be calculated. For dramatic variation and/or large areas, the basin should be split into more homogenous subbasins to reduce errors resulting from the nonlinearity of the fundamental equation. The fundamental equation produces runoff in units of depth. One can obtain runoff volume by multiplying runoff depth by basin area. Runoff amounts are transformed into a hydrograph using specific routing procedures of TR-55.

Rainfall amounts for use in the methods can be found in the data sources discussed previously. As for hyetographs, TR-55 contains four standard, normalized, cumulative distributions for 24-h rainfall that cover the entire United States. More geographically specific distributions are also sometimes available. For example, *Illinois State Water Survey Circular 173* (1990) presents a distribution based on Illinois data.

Computerized versions of the TR-55 method are available from the NRCS web site and are implemented in numerous models from commercial vendors (see below); these programs largely make obsolete the graphic and manual tabular methods described in TR-55.

Numerous caveats are in order for using TR-55. They are discussed in the TR-55 document and by McCuen (1982). The *CN* method is best applied to small basins with concentration times less than 3 h. The *CN* method is less accurate when runoff is less than 0.5 in. If *CN* is small (less than 40), results are also less accurate. This limitation may be important in predicting runoff in a predevelopment condition. (TR-55 typically is used to predict hydrographs and control flooding from large storms in urban or semiurban areas.) Results are also sensitive to antecedent moisture conditions. Runoff from snowmelt or frozen ground cannot be predicted with TR-55.

According to Ponce (1989) "experience with the TR-55 method has shown that results are sensitive to curve number. The method should be used judiciously, with particular attention paid to its capabilities and limitations."

9.3.5   *METHODS OF STORMWATER QUANTITY MANAGEMENT*

The quantity part of stormwater management concerns controlling peak flow rates and, less commonly, runoff volume. Stormwater quantity management is typically mandated for new developments and improvements are often necessary for existing developed areas to prevent or reduce flooding. The terminology for BMPs for such management is not standard, and the same term sometimes is applied to markedly different techniques. The treatment train management approach refers to an integrated, multilevel approach, including source reduction, on-site control, and detention.

Source Reduction of Runoff: Nonstructural Approaches

The old adage "an ounce of prevention is worth pound of cure" applies here; stormwater quantity management begins with efforts to reduce runoff and control it close to the source. Source reduction is best accomplished by careful site planning using nonstructural measures as much as possible. Such measures are discussed by the Center for Watershed Protection (1995) and by Ferguson and Debo (1987). Quantity reduction measures also are generally good for improving runoff quality.

Site-planning techniques that promote infiltration and localized detention include the following:

- minimization of impervious areas;
- open-celled pavers, porous pavement, and grass or gravel automobile parking areas instead of conventional asphalt or concrete;
- Placing new impervious areas atop low-infiltration soils to minimize the decrease in infiltration;
- decreasing directly connected impervious areas by using grassed swales instead of storm sewers or concrete channels and not connecting roof drains directly to storm sewers;
- avoiding steep slopes;
- maintaining and protecting existing vegetation;
- using rooftop detention and/or vegetation; and
- using on-site retention (cisterns) to reuse water for landscaping.

One important advantage of source reduction over structural controls is that they usually require less maintenance–often a major problem with structural controls. They work well when planned into new developments; it is often difficult to retrofit source reduction measures into existing developments.

In agricultural areas, source control methods include no-till planting, contour plowing, terraces and maintaining cover crops; among others.

Stormwater Conveyance Systems

Typical components of a stormwater conveyance system are curbs and gutters, open channels (ditches, swales, and small channels), and underground conduits (called culverts, storm drains, or storm sewers). Systems typically are designed to carry the flow from a storm of specified rainfall depth or return period. Conveyance systems are fully described by the ASCE/WEF (1992) and by Metcalf and Eddy (1972).

Traditionally, the approach has been to collect and transport stormwater as quickly as possible to minimize flooding. Although this approach reduces flooding upstream, it can increase downstream where high flows from rapidly draining tributary areas combine.

Also, lining streams with concrete and placing them in culverts has been common practice. Although this increases safety and reduces flooding in the immediate vicinity, it reduces recreational, aesthetic, and ecological value. Modern stormwater conveyance practices seek to protect or enhance these values while still transporting flows.

Conventional Urban/Suburban Systems: Inlets, Curbs, Gutters, and Storm Drains.   In urban and suburban areas, a conventional system consists of a short distance for overland flow leading to an inlet into a storm drain. Drainage in streets collects in a gutter confined by a curb and is directed to an inlet. Rooftops often are directly connected to a storm drain, or they may discharge onto a lawn or gutter. Lawns are graded to discharge into a street or into a swale, which then may enter a storm-drain inlet.

Advantages of the conventional urban system are its efficiency in removing runoff and reducing land area for stormwater transport. Disadvantages include accumulation of water on roads during large storms, limited opportunity for infiltration into and treatment by soil/vegetation, and rapid concentration of flow, which may increase flooding downstream.

Flow in gutters generally is limited to a velocity of 3.0 m/s (10 ft/s) and a depth of 15 cm (6 in.) for safety reasons. Also, flow or standing water should be less than 0.6 m (2 ft) wide to allow pedestrians to cross (Browne, 1998, p. 7.90). Spacing of inlets depends on design flows and specific street conditions.

Stormwater drains generally are laid with a constant slope. A manhole is inserted at a change in slope or horizontal alignment. Storm drains can be constructed of a variety of materials, including concrete, iron, polyvinyl chloride, reinforced concrete (RCP), vitrified clay, and corrugated metal.

Where flow enters a pipe, the flow rate inside the pipe may be controlled by the geometry of the pipe entrance—this is called inlet control. If a pipe does not discharge freely (e.g., because it is submerged), then the flow is under outlet control. In these circumstances, the flow conditions inside the pipe cannot be fully described by simple formulae such as Manning's equation.

The Federal Highway Administration (FHWA) has published several comprehensive works, including software, on culvert design (e.g., Normann et al., 1985). These publications and/or the implementing software are collected in a downloadable package at the FHWA web site (search for HYDRAIN and HY8 Culvert Analysis). Numerous commercially developed computer programs also aid in storm drain design (see list of web sites at start of section).

Open Channels: Grassed Swales and Small Channels.   Grassed swales (shallow channels with gently sloping sides) can be used to effectively transport stormwater while providing some detention, infiltration, and water quality improvement. They are best suited for situations with adequate available land. Design is fully covered by Temple et al. (1987) and by Temple (1983).

TABLE 9.3.2 Approximate Values of Manning's
Friction Coefficients for Various Channel Materials

Material	Manning (n)
Excavated earth channels	
Clean and straight	0.016–0.020
Gravelly	0.022–0.030
Weedy	0.025–0.040
Stony, Cobbles	0.025–0.040
Grassed swales	
Turf grasses	0.03–0.12

Source: Chow (1959).

Velocities up to 2.4 m/s (8 ft/s) with flow depths of about 0.3 m (1 ft) are permissible in swales with robust vegetation and erosion resistant soils. If the swale is going to be mowed (mostly practiced merely for landscaping appearance), side slopes should be 1V:3H or less.

The discharge capacity of grassed swales and other open channels is described by Manning's equation:

$$Q = (x/n)R^{2/3}S^{1/2}A \qquad (9.3.5)$$

where Q = flow rate (ft^3/s or m^3/s),
  x = constant (1 for SI units, 1.49 for U.S. units),
  n = Manning's friction coefficient,
  R = hydraulic radius (cross sectional area of flow/wetted perimeter, ft or m),
  S = water surface slope (unitless), and
  A = cross-sectional area of flow (ft^2 or m^2).

The primary difficulty in applying this equation for grassed swales is estimating the friction coefficient (n), which theoretically depends on the species, density, and height of vegetation. Approximate values of n for different channel linings are listed in Table 9.3.2.

Numerous free programs or spreadsheets that solve Manning's equation for regular and irregular channels are available on the web (e.g., http://www.waterengr.com; FHWA's HY22 Urban Drainage Design Programs).

Design and analysis of large channels are conventionally outside the realm of stormwater management. Standard references on open channel flow include Montes (1998) and the classic treatment by Chow (1959).

Infiltration and Retention Structures

In infiltration structures, water is retained until it soaks into the ground or evaporates. Infiltration can be practiced on a small scale as a form of source control. At a larger scale, stormwater can be conveyed to an infiltration (also called exfiltration) BMP, such as a basin or gravel-filled trench. A comprehensive treatment of the subject is provided by Ferguson (1994), who reported more than 20,000 infiltration basins extant in the United States.

Where practical, infiltration is an ideal method for stormwater management. It eliminates surface discharge, so there are no flooding or surface water quality concerns. By recharging groundwater, infiltration helps sustain baseflow in streams during droughts.

There is potential for contamination of groundwater from infiltration (Pitt et al., 1995). This should be a concern especially over shallow drinking water aquifers and when managing highly contaminated stormwater.

Infiltration is often difficult in practice because of the restrictive soil or groundwater conditions under which infiltration is suitable and potential high maintenance requirements. Browne (1998) recommends a minimum soil permeability of 1.5 cm/h (0.6 in./h) through a depth of 1.5 m (5 ft). Shallow impermeable layers or high groundwater elevation may limit infiltration rate regardless of surface soil types. Urbonas and Stahre (1993) present a point system for evaluating the suitability of a site for infiltration.

Also, the size of a basin may need to be very large. The area is based on the depth of water to be infiltrated and the infiltration rate of the basin's soils. A bypass typically is needed if the infiltration plus storage capacity might be exceeded by high flows.

A forebay to settle out a portion of the suspended sediment is typically recommended. Even so, clogging, formation of impermeable crusts, and resulting failure of infiltration BMPs remain a concern. Studies have found that over 50 percent of infiltration trenches and basins were not performing properly within 5 years of installation (Galli, 1992). On the other hand, Ferguson (1994) reports that hundreds of infiltration basins on Long Island, New York, (sandy soils) have been functioning since the 1960s.

Detention for Peak Flow Reduction

Detention has been the standard practice in the United States for peak flow reduction since the 1970s and it remains a common BMP. The goal of detention basins (or ponds) is to temporarily store high rates of incoming runoff and discharge at lower rates, thereby preventing or reducing downstream flooding. For new development, the design discharge rate is typically equal to that of the site under predevelopment conditions for a specified (by regulation) design storm.

Even with detention, development of subwatersheds may increase downstream flow rates—developed areas discharge at high rates for longer durations, which can lead to higher flows where subwatersheds combine (McCuen, 1974).

Detention basins can either be wet (that is, a permanent pool formed by locating the outlet above the bottom of the pond; these are sometimes called retention ponds because a portion of the inflow is retained, at least until the next storm) or dry (outlet flush with pond bottom, so the pond empties between storms). Wet basins are effective at improving water quality by, most importantly, settling solids, whereas dry basin are not, unless designed with slow-release outlets, which makes them "extended detention" dry ponds (Figure 9.3.4).

A detailed example of the design of a wet pond was published by the Center for Watershed Protection (Claytor, 1995), which provides step-by-step instructions with sample computations. Basin sizing and outlet design are also well explained by Akan (1993). A comprehensive treatment of detention is given by Urbonas and Stahre (1993).

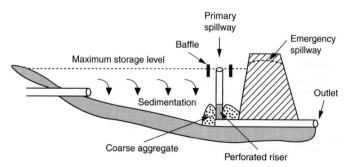

FIGURE 9.3.4 Slow-release outlet design of an extended detention dry basin (Maidment, 1993).

Sizing and Designing Detention Basins for Peak Flow Control. Sizing begins with calculation of the inflow hydrograph, typically by a method such as TR-55. TR-55 also contains design specifications relating the ratio of the design maximum outflow rate to the peak inflow rate (Q_o/Q_i) to the ratio of pond storage volume to the runoff volume (V_s/V_r). With this relationship and with Q_i and V_r known from hydrologic analysis and Q_o established by design, one can calculate V_s. One can then design the pond's shape and side slopes, giving a stage-storage (depth versus volume) relationship.

The elevation, type, and size of outlet structure(s) must then be chosen to produce the design peak discharge when the pond contains the computed runoff volume (plus any permanent pool volume).

Most detention basins discharge through a submerged pipe or a perforated riser pipe during normal conditions and have a spillway over which water flows during extreme conditions (Figure 9.3.4). Details on outlet design are given by ASCE/WEF (1992).

A submerged pipe generally behaves like an orifice, and the discharge is proportional to the square root of the head. The spillway acts as a weir and the discharge is proportional to the head raised to the 1.5 power. If both structures are present, the equations can be combined to give an overall stage-discharge curve (rating curve).

A dam breach analysis also may be warranted based on pond size and downstream conditions or required by regulation. The National Weather Service's Dam Break model DAMBRK (enhanced versions from commercial software vendors such as Water Resources Consulting Services and Boss International) is commonly used.

Routing Flow Through Detention Basins. Routing is the procedure used to track an inflow hydrograph through channel reaches, detention basins, and reservoirs to determine the outflow hydrograph, as in Figure 9.3.1. One can use routing to check the basin peak outflow calculated by a simpler method and fine tune outlet structures and basin morphology. One can also examine the behavior of a basin under different inflow conditions or with different source control measures.

Hydraulic routing combines the continuity equation (conservation of mass or volume of water) with the equation of conservation of momentum, and it is used in complex situations such as backwater effects. Hydraulic routing is covered in comprehensive open-channel flow texts, such as those by Montes (1998) and Chow (1959).

Hydrologic routing methods are simpler and are based on the equation of continuity coupled with an empirical relationship between storage, inflow, and/or outflow. Details can be found in the texts referenced above and in Akan (1993). The continuity equation can be written as

$$\text{rate of change in storage in a basin} = \text{inflow rate} - \text{outflow rate} \tag{9.3.6}$$

Over time interval Δt, the equation can be written in finite difference form as

$$\frac{S_{t+\Delta t} - S_t}{\Delta t} = \frac{I_t + I_{t+\Delta t}}{2} + \frac{O_t + O_{t+\Delta t}}{2} \tag{9.3.7}$$

where
S = storage volume (m^3 or ft^3),
I = inflow rate (m^3/s or ft^3/s),
O = outflow rate (m^3/s or ft^3/s),
Δt = time increment (s), and
t and $t + \Delta t$ = values at the beginning and end, respectively, of a time increment.

Several methods are used to solve this equation including the Muskingum Method for channel routing and the Storage Indication Method for basin or reservoir routing. The Storage Indication Method requires stage-storage and stage-discharge relationships. With these defined, the inflow hydrograph can be routed through a basin, starting at time 0 with I_0 and S_0 known, computing a new storage and thus a new stage and discharge at the end of each time step. Numerous computer programs automate the Storage Indication Method (see below). It can also be easily programmed on a computer spreadsheet.

9.3.6 COMPUTERIZED MODELS OF STORMWATER HYDROLOGY

The are numerous computerized models available. Several are based on NRCS method, such as NRCS's TR-55 software. Numerous private companies have developed their own implementations. These programs extend TR-55's capabilities by adding, for example, the rational and/or other hydrologic calculation methods, additional hyetographs and design and analysis of culverts, and channels and detention ponds, including hydrologic routing. They also improve the user interface by, for example, providing online help, input file generators, and graphic output of hydrographs.

Software in this category includes the following:

- MIDUSS, Alan A. Smith, Co, www.alanasmith.com
- HydroCalc, Dodson and Associates, www.dodson-hydro.com
- HydroCAD, Applied Microcomputer Systems, www.hydrocad.net
- Hydraflow, Water Resources Consulting Services, www.waterengr.com
- PondPack and StormCAD, Haested Methods, www.haested.com.

There are also numerous programs that are designed for large-scale simulations and that simulate hydrologic processes in detail, such as HSPF and SWMM. These programs also simulate water quality processes, so their discussion is deferred to the end of the subsection on water quality.

9.3.7 NONPOINT SOURCE POLLUTION CONTROL: RUNOFF QUALITY

General

Controlling stormwater peak discharge (and even volume) is only half the battle. Control of runoff quality is also important. Excess contaminants in runoff can lead to negative effects on human health, ecological and economic systems, and quality of life. Rainfall and snowmelt contain contaminants. As runoff flows over the land surface, it picks up additional contaminants. Human activity, such as urbanization, septic systems, mining, solid-waste landfills, and agriculture (including row crops, forestry, animal grazing, and feed lots), can increase the amount of the contaminants above their natural level and lead to undesirable effects.

Runoff entrains these contaminants in a dissolved or a particulate form and transports them to a water body. Such NPSP is by its nature diffused through a watershed and therefore difficult to control with traditional treatment plants. Atmospheric deposition (which can be the primary source of some pollutants), hydrologic modification, and streambank erosion are also classified as nonpoint pollution. A list of contaminants, their sources, and their effects is presented in Table 9.3.3.

Agricultural Runoff Pollution

Diffuse pollution from agricultural areas is different than that from urban areas (Novotny and Olem, 1994, p. 44). In agricultural areas, runoff can erode soils yielding high-sediment loads on overgrazed land, clearcut forests, bare plowed fields, and between rows of crops. Other primary concerns are fertilizers and pesticides from crop lands, fecal matter from livestock (especially dairy barns and feedlots or concentrated animal feeding, CAFOs), and erosion caused by livestock trampling streambanks.

Urban Runoff Pollution

Urban runoff quality problems are different from agricultural. Impervious surfaces do not erode and pervious areas (except for construction sites) generally are covered with vegetation, which minimizes

TABLE 9.3.3 Major Nonpoint Source Contaminants and Their Causes and Effects

Contaminant(s)	Cause(s)/mechanism(s)	Effect(s)
Biodegradable matter/ biochemical oxygen demand (BOD)/chemical oxygen demand (COD)	Decaying plants and algae. Fecal matter from humans, livestock, pets or wildlife.	Biodegradation consumes dissolved oxygen (DO), reducing DO to levels that can stress or kill biota, especially sensitive, highly valued fish (e.g., trout).
Increased temperature	Runoff from hot paved surfaces; heating in detention ponds; removal of shade along streams.	Some organisms require/prefer lower temperature. Oxygen solubility decreases as temperature increases, leading to low DO.
Turbidity/suspended solids	Construction sites, row-crop agriculture, eroding streambanks, fallout washed from impervious surfaces, road sand.	Ugliness, decreased light penetration depressing aquatic plant growth, poor visibility affects sight-feeding fish, settled solids (sediment) cover spawning/feeding areas and fill in reservoirs, ponds, and catch basins.
Floatables	Trash, paper, wood, plastic, leaves.	Ugliness, certain types are harmful to humans, can be ingested by and harm animals. Can clog grates and cause flooding.
Acidity (pH)	Mine drainage, waste sites, illegal discharges.	Depressed function, injury, illness, death, and/or reproductive problems in plants, animals, and humans
Nutrients (nitrogen and phosphorus)	Fertilizers from agricultural fields, lawns, and golf courses. Fecal matter from livestock, pets, urban wildlife (e.g., geese), failing septic systems, and incompletely treated sewage.	Eutrophication (overgrowth of aquatic plants, especially algae) causing ugly green water, bad odors, and tastes; interference with boat travel; increased turbidity; and low DO from respiration and decomposition.
Pathogenic (disease causing) microorganisms (bacteria, viruses, etc.)	Fecal matter from livestock, pets, urban wildlife, failing septic systems, combined sewer overflows, sanitary sewer overflows, and poorly treated sewage.	Spread of diseases (e.g., cholera) in humans and wildlife. Contamination of fin/shell-fisheries. Closed beaches. Closed water recreation sites.
Hydrocarbons (oil and grease), toxic metals (Cu, Pb, Zn etc.), and toxic complex organic compounds (pesticides, polychlorineted biphenyls, polyaromatic hydrocarbons, etc.), salts	Pesticides, industrial and automotive fluid spills and leaks, automobile body and part wear, brake wear, tire wear, waste sites, deposition of air pollution, metal roofs, fallout of combustion products, road salt, uncovered salt piles, and illegal discharges.	Depressed functions, injury, illness, death, and/ or reproductive problems in plants, animals, and humans

erosion. On the other hand, the myriad of human activities deposit a host of contaminants onto the land. Paved surfaces are often used in activities that generate contaminants, especially automobile travel and parking. Pollutants that accumulate on impervious surfaces are not degraded by biological processes at the same rate that occurs on vegetated surfaces.

Although runoff from paved surfaces generally is considered to carry more contaminants than that from unpaved surfaces, there is a large amount of variability in concentration, both from location to location and from storm to storm (see below). Even if concentrations from paved surfaces are not significantly larger than those from unpaved surfaces, the fact that paved surfaces generate larger volumes of runoff means that they also generally yield larger contaminant loads (concentration times water volume per unit time, giving mass per unit time). Also, high runoff rates can lead to streambank erosion and hydroperiod modification.

Rooftops are the other main type of impervious area and they cover a significant area in urban areas. Rooftop runoff is usually considered clean, but some studies have found high concentrations of metals from some surfaces—for example, zinc from galvanized roofs (*Is Rooftop Runoff Really Clean?, Watershed Protection Techniques*, Center for Watershed Protection, Vol. 1, No. 2, 1994; available at http://www.epa.gov/OWOW/NPS).

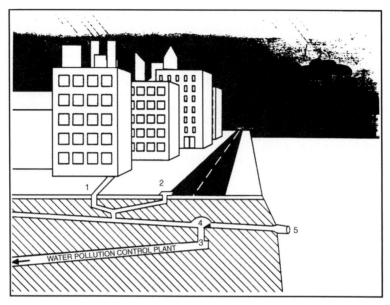

FIGURE 9.3.5 Configuration of combined sewers, which leads to overflows during heavy rain (*Final Comprehensive Conservation and Management Plant*, New York–New Jersey Harbor Estuary Program, 1996). 1, wastewater flow; 2, stormwater flow; 3, during dry weather wastewater is conveyed to treatment plant; 4, during wet weather capacity of treatment plant is sometimes exceeded and some flow overflows toward waterways; 5, outfall for combined sewer overflow.

Combined Sewer Overflows (CSOs) and Sanitary Sewer Overflows (SSOs)

Many older communities have combined sewers that collect and convey both stormwater runoff and sewage in the same pipes. During dry weather, a combined sewer system transports sewage to a sewage treatment plant. During heavy rain or snowmelt, however, the capacity of the treatment plant (or the sewers themselves) may be exceeded, and a mixture of stormwater and sewage overflows through bypasses into waterways (Figure 9.3.5). The major problem caused by CSOs is contamination by pathogenic microorganisms, which makes water unfit for swimming, fishing, and sometimes even boating.

Novotny and Olem (1994, p. 37) noted an interesting fact about combined sewer systems. The concentration of pollutants is higher in CSOs than in discharges from separate storm discharges. However, combined sewers convey all their flow to a treatment plant in the frequent, small storms (whereas separate storm sewer systems typically discharge to waterways), so pollutant loads from combined sewers are, for many constituents, about the same as those from separate storm sewers.

In the United States in the late 1990s, combined sewer systems served roughly 950 communities with about 40 million people, mostly in the Northeast and Midwest.

A SSO occurs when stormwater inadvertently flows into sanitary sewers (e.g., through cracks) and exceeds the capacity of sewers or treatment plants. The flow is bypassed (as in a CSO) or overflows into basements or out of manholes and eventually into waterways.

Control and prevention of CSOs and SSOs are discussed below.

Pollutant Forms and Effects

Accurately predicting or even assessing the effects of NPSP on the biological and chemical health of receiving waters is not simple. Often, multiple, competing factors are in play, and neither the reletive size of the causes of the problems nor the solutions are obvious.

Lakes, ponds, and reservoirs have long residence times. Contaminants remain in the system for a long time even if the input load is reduced. Rivers and streams have short residence times, so they are quicker to recover after a pollution event and they respond more quickly to input changes.

Many pollutants, including nitrogen, phosphorus, metals, and organics, are attached to solid particles and are transported in particulate form. In this form, many contaminants are nonreactive and nonbioavailable, and they have little effect on biota. Therefore some have argued that the importance of controlling these pollutants does not justify the expense (Lee and Jones-Lee, 1995). However, under certain conditions (often related to pH or reduction-oxidation condition), these contaminants can be transformed to more bioavailable forms (Reddy and D'Angelo, 1994).

Many contaminants end up in lake or river sediments. Often, concentrations in the sediment are much larger than those in the water column. Contaminants can become more concentrated (bioaccumulation) in benthic (bottom dwelling) organisms that live in and/or ingest the sediment or that filter large amounts of water, such as mussels and clams. As these animals are consumed by other animals, the contaminants can be concentrated further (biomagnification) (Levine and Miller, 1994).

In addition to carrying a load of attached contaminants (albeit sometimes nonreactive), suspended solids are an important pollutant in their own right (Table 9.3.3). Therefore, solids are one of the pollutants of most concern, along with nutrients.

9.3.8 MONITORING OF RUNOFF

Monitoring Programs

A monitoring program can be conducted for a number of purposes, such as to:

- Determine the potential effects of runoff on receiving waters,
- Determine loads and concentrations from various sources of nonpoint pollution, and
- Calibrate or verify computer models.

Monitoring can be performed in a receiving water or at storm drain outfalls or inlets and/or outlets of water control structures (e.g., detention basins).

Monitoring is more fully discussed by Singh et al. (1998) and by Bartram and Balance (1998). References specifically related to permit-required sampling include U.S. EPA (1993) and Dodson (1998).

Monitoring runoff is difficult and expensive—runoff events are unpredictable and the locations are many and dispersed. To monitor on a large time or geographic scale, it may be appropriate to enlist volunteer groups (e.g., "Friends of the Murky River") to reduce costs. The data from volunteer efforts, however, may be viewed suspiciously by reviewers or regulators unless quality control is very rigorous.

To calculate pollutant loads, both runoff flow rates and concentrations must be quantified. Monitoring flow rate only is comparatively simple. An outflow control device, such as a weir, may have a known stage-discharge relationship, in which case only the stage needs to be monitored. This can be done manually by installing and observing a staff gauge or with an automatic water level sensor.

For a channel, one typically measures water depth and flow velocity at one or more locations along a cross section and then uses a stage versus cross-sectional area relationship and the basic continuity equation (flow rate = average velocity times cross-sectional area) to obtain flow rate. Doing so at multiple stages allows a site-specific stage-discharge relationship to be constructed, allowing subsequent monitoring of stage alone.

Monitoring quality is more difficult because multiple parameters are involved, many of which are not easily measured. Furthermore, runoff quality can be highly variable during a single runoff event, from one event to another, and from one site to another.

To capture samples throughout a storm, an automatic sampler is very helpful [some manufacturers are ISCO (www.isco.com) and American Sigma (www.americansigma.com)]. These manufacturers and others [e.g., Hydrolab (www.hydrolab.com), YSI (ysi.com), and In-situ (www.in-situ.com)] also make sensors that measure water quality parameters (temperature, pH, turbidity, dissolved oxygen, and conductance are the most common) in real time, with the data storable in an electronic logger.

Nationwide Urban Runoff Program (NURP)

NURP, funded by the U.S. EPA, in the early 1980s is the most comprehensive monitoring study to date (U.S. EPA, 1983). At least 2300 storms at 81 sites in 23 cities were monitored for parameters including total suspended solids, chemical oxygen demand, total Kjeldahl nitrogen (organic and ammonia nitrogen), nitrate nitrogen, total phosphorus, soluble phosphorus, copper, lead, and zinc. Much of the data are currently downloadable at http://www.eng.ua.edu/~awra/. Another good source of NURP data is Mustard et al. (1987).

To characterize a storm with one concentration statistic, NURP employed the "event mean concentration" (EMC), defined as the total mass of a contaminant contained in runoff for an event divided by the total volume of runoff for the event. EMC has been called "the most appropriate variable for evaluation of the impact of urban runoff " (Novotny and Olem, 1994, p. 484).

Important conclusions from NURP are that

- The distribution of EMCs for a series of storms at a specific site follow a lognormal distribution,
- Heavy metals (especially copper, lead, and zinc) are the most prevalent "priority pollutants" in urban runoff (since leaded gasoline was phased out, lead concentrations have declined) (Novotny and Olem, 1994, p. 219), and
- Bacteria are present at high levels and will exceed criteria levels during and just after storms in many surface waters.

Wide variations in concentrations were found for every contaminant, both from storm to storm and from site to site. Despite the strong intuitiveness of a correlation between land use type and runoff quality, NURP did not find a statistically significant difference between EMCs among sites of residential, commercial, and mixed land use (undeveloped land was significantly less than the other three). The NURP report concluded that "if land use category effects are present, they are eclipsed by the storm to storm variabilities and that, therefore, land use category is of little general use to aid in predicting urban runoff quality at unmonitored sites or in explaining site-to-site differences where monitoring data exists (U.S. EPA, 1983)."

For simple predictive purposes, then, it is usually best to use the NURP concentration data aggregated from all developed sites (Novotny and Olem, 1994, p. 492), given in Table 9.3.4. Note that contaminant loads (mass per unit area per year) can be computed for different land use types because of the different volumes of runoff from the different land use types (load = concentration × volume).

Combined Sewer Overflows

CSOs generally have higher concentrations of biochemical oxygen demand (BOD), phosphate, and coliforms than "regular" stormwater, as shown in Table 9.3.5. Because the overflows occur at varying frequencies, the significance of this contamination is variable. Fecal matter is usually the most significant contaminant.

First Flush Effect

The first flush effect refers to higher pollutant concentrations at the beginning of a runoff event. It is based on the theory that pollutants build up and then wash off surfaces, so the first flush of runoff is more contaminated. Some studies have found a pronounced first flush effect (Mercer et al. 1996), and while others have not. Barrett et al. (1996) recommend "in considering the potential

TABLE 9.3.4 Summary of Event Mean Concentrations (EMCs) Found in the Nationwide Urban Runoff Program (All Storms and Monitoring Sites Aggregated)

Contaminant (units)	Median EMC for median urban site	Mean EMC for median urban site (use for computing pollutant loads)	Typical coefficients of variation (standard deviation divided by mean) for individual sites
TSS (mg/l)	100	180	1–2
BOD (mg/l)	9	12	0.5–1
Chemical oxygen demand (mg/l)	65	82	0.5–1
Total P (mg/l)	0.33	0.42	0.5–1
Soluble P (mg/l)	0.12	0.15	0.5–1
Kjeldahl N (organic + ammonia) (mg/l)	1.5	1.90	0.5–1
Nitrate/nitrite N (mg/l)	0.68	0.86	0.5–1
Copper (ug/l)	34	43	0.5–1
Lead (ug/l)	144	182	0.5–1
Zinc (ug/l)	160	202	0.5–1

Source: U.S. EPA (1983).

TABLE 9.3.5 Nationwide Average Characteristics of Combined Sewer Overflows

Contaminant (concentration units)	Average concentration
Total suspended solids (mg/l)	115
BOD (mg/l)	370
Phosphate (mg/l)	10
Total N (mg/l)	2
Lead (ug/l)	0.4
Total coliforms (MPN CFUs per 100 ml)	100–10,000

Source: U.S. EPA, *Report to Congress of Control of Combined Sewer Overflow in the United States*, EPA 460/9-78-006, 1978.

effectiveness of stormwater treatment systems, constant concentrations for individual storm events should be assumed." Urbonas and Stahre (1993, p. 315) echo: "lacking local data, it is safest to assume that there is no first flush."

9.3.9 TECHNIQUES FOR URBAN RUNOFF QUALITY CONTROL AND IMPROVEMENT

Numerous techniques are available and generally are divided into structural and nonstructural.

Nonstructural Techniques: Source Control/Pollution Prevention

The nonstructural techniques for control of runoff quantity apply to runoff quality as well—reducing runoff quantity is an excellent way to reduce pollutant loads. Source control is more fully discussed by

Ferguson and Debo (1987), by Center for Watershed Protection (1995) and by Delaware Department of Natural Resources (1997).

In addition to the quantity control techniques are the following best management practices (BMPs):

- Street sweeping and good solid waste management (Sutherland and Jelen, 1996),
- Street flooding (for combined sewer systems only)
- Transportation demand management (TDM), also called transportation control measures (TCMs), which reduces the number and length of automobile trips that are the source of much NPSP,
- Waste oil redemption programs to prevent dumping,
- Education/regulation to reduce over/misapplication of fertilizer and pesticide,
- Septic system inspection programs, and maintenance
- Air pollution control to reduce atmospheric deposition,
- Erosion control practices on construction sites and agricultural fields, and
- Animal waste control, both pets and urban wildlife such as geese.

Structural Techniques: Overview

Schueler et al. (1992) reviewed 11 classes of structural, urban BMPs, including dry basins, wet ponds, wetlands, infiltration structures, sand filters, grassed swales, filter strips, and oil/grit separating chambers. Characteristics of each type of BMP are reported in Table 9.3.6. Schueler et al. reported the following:

- Long-term performance of infiltration structures is not reliable because of clogging.
- Performance of oil/grit separating chambers (so-called water quality inlets) are not reliable because of insufficient design basis.
- Pond BMPs are the most widely applicable type of BMP.
- Infiltration practices are limited to applicable soils and water table conditions.
- Limited data existed on the cost-effectiveness of various BMPs. Specific BMPs are usually selected on the basis of longevity and feasibility. Maintenance costs may rival construction costs for many BMPs.
- Many jurisdictions did not have very active BMP maintenance programs.

Typical percent removal characteristics of structural BMPs for numerous constituents are presented in Table 9.3.7.

The American Society of Civil Engineers and U.S. EPA are developing a National Stormwater BMP Database, which includes a large bibliography (www.asce.org/peta/tech/nsbd01.html). The Center for Watershed Protection has also published *National Pollutant Removal Performance Database for Stormwater BMPs–1997*, which contains summaries of 123 urban BMP pollutant removal monitoring studies.

An excellent design reference, which applies to 10 types of BMPs, is *Design of Stormwater Filtering Systems* (Claytor and Schueler, 1996), published by the Center for Watershed Protection. They also published *BMP Design Supplements for Cold Climates* in 1997.

Structural BMPs are most effective at removing solids by settling. Because sediment is an important pollutant in its own right and many other pollutants are attached to sediment particles, removal of sediment is an important water quality goal. For these reasons, sediment removal is often the focus of stormwater quality regulations and the primary factor on which the effectiveness of BMPs is judged.

A good reference for cost estimating is Cost Estimating Guidelines: Best Management Practices and Engineered Controls for the Rouge River National Wet Weather Demonstration Project (see web site list).

TABLE 9.3.6 Comparison of Characteristics of Urban, Structural BMPs

BMP type	Performance and reliability of pollutant removal	Longevity (based on current design and typical maintenance)	Site applicability	Inappropriate climate(s)	Environmental concerns	Relative cost	Special considerations
Modified (extended detention) dry basins	Moderate; not always reliable	20+ years, but frequent clogging and short detention common	Widely applicable	Very few	Stream warming and habitat destruction	Lowest cost for large sites	Recommended with design improvements, such as micropools and wetlands
Wet ponds	Moderate to high	20+ years	Widely applicable	Arid areas	Stream warming, habitat destruction	Moderate to high	Recommended with careful site evaluation; possible safety hazard
Wetlands	Moderate to high	20+ years	Space may be limiting	Arid and areas with short growing seasons	Stream warming, habitat destruction	Slightly higher than wet ponds	Recommended
Combination pond/wetlands	Moderate to high; more reliable because of redundancy	20+ years	Many options	Arid regions	Stream warming, habitat destruction	More expensive than single-component systems	Recommended
Infiltration trenches	Presumed moderate	50 percent failure rate within 5 years	Highly restricted, based on soils, depth to groundwater, slope, sediment load	Arid and cold regions	Groundwater contamination	Reasonable for smaller sites; rehab costs can be high	Recommended with pretreatment and verification of appropriate site conditions
Infiltration basins	Presumed moderate, if working	60 to 100 percent failure within 5 years	Highly restricted, like infiltration trenches	Arid and cold regions	Groundwater contamination	Construction cost moderate; rehab costs high	Not recommended until longevity improves
Porous pavement	High, if working	75 percent failure within 5 years	Extremely restricted	Cold regions	Groundwater contamination	Comparable to convention asphalt	Recommended in special situations
Sand filters	Moderate to high	20+ years	Applicable for smaller areas	Few	Minor	High construction costs and frequent maintenance	Recommended in appropriate situations
Grassed swales	Low to moderate, but unreliable	20+ years	Low density development and roads	Arid and cold regions	Minor	Lower than curbs and gutters	Recommended, with check dams, as part of a system
Vegetated filter strips	Unreliable in urban setting	Unknown, but may be limited	Low density areas	Arid and cold regions	Minor	Low	Recommended as part of a system
Grit and oil separators (water quality inlets)	Presumed low	20+ years	Small, highly impervious catchments	Few	Disposal of residuals	Higher than infiltration trenches and sand filters	Not recommended as primary BMP

Source: Adapted from Schueler et al. (1992).

TABLE 9.3.7 Typical Percent-Removal Characteristics of Structural Stormwater BMPs (Each Cell Lists Probable Minimum, Average, and Probable Maximum; "?" Means Insufficient Data)

Structure type	Total suspended solids	Total phosphorus	Total nitrogen	Chemical oxygen demand	Lead	Zinc
Infiltration basin	50–75–100	50–65–100	50–60–100	50–65–100	50–65–100	50–65–100
Infiltration trench	50–75–100	50–60–100	50–55–100	50–65–100	50–65–100	50–65–100
Grass swale	20–60–?	20–20–40	10–10–30	?–25–?	10–70–?	10–60–?
Oil and grit separator (water quality inlet)	10–35–?	5–5–10	5–20–?	5–5–10	10–15–25	5–5–10
Modified (extended detention) dry pond	?–45–90	10–25–60	20–30–60	?–30–40	20–50–60	?–20–60
Wet pond	50–60–90	20–45–90	10–35–90	10–40–90	10–75–95	20–60–95
Wetland	50–65–90	0–25–80	0–20–40	?–50–?	30–65–90	?–35–?

Source: Adapted from U.S. EPA (1993).

Catch Basins/Sumps

Catch basins are inlets, usually located in curbs and parking lots, that admit runoff into a storm drain system. Catch basins are equipped with a sump or basket to retain grit, detritus (decaying plant matter), and sediment. Historically, the purpose of the sump was to prevent debris from clogging the culvert. Catch basins are generally considered only slightly effective at removing suspended solids, but studies have found various degrees of effectiveness, ranging up to 97 percent removal (Novotny and Olem, 1994, p. 613). To maintain effectiveness, the sump must be cleaned periodically (twice per year is a common recommendation), usually by a vacuum truck. A hood or trap hung over the outlet pipe of a catch basin helps prevent introduction of floatable material (paper, plastic bottles, wood) in to the storm drains. For an example, see the "SNOUT" by Best Management Products, www.bestmp.com.

Floatable Controls

Floatables are the most visible and obvious type of NPS. Although they usually do not affect the chemical quality of water, they are an important aesthetic concern, especially in urban areas, and certain types can be a health threat. Floatables can also clog grates and cause flooding.

Furthermore, animals (e.g., turtles) may ingest or get tangled in plastic products, which may cause death. Because plastic is essentially nondegradable, it is very persistent and can accumulate over time. Plastics in the ocean—the ultimate repository for floatables—has become a major problem for certain marine life (Carr, 1987).

Floatables are controlled by vortex treatment units (see below), trash racks on inlets to storm drains and/or outlets of detention basins, and floating booms around storm drain outfalls and hooded catch basins. All floatable control systems must be cleaned periodically to maintain performance and prevent clogging.

Grey and Oliveri (1999) (http://www.hydroqual.com/presents.htm) report that catch basin hoods (see preceding subsection) were the most cost-effective floatable control system and that a hood installation program in New York reduced floatable pollution by 70 percent.

In-Line Vortex Treatment Units (Swirl Concentrators)

These commercially produced units are made of metal, concrete, and/or plastic and are installed in-line with storm drains (Figure 9.3.6). Some common manufacturers/products include StormTreat, Stormceptor, Vortechnics, HIL Downstream Defender, Baffle Boxes, and Stormwater Filter. Most of these companies have web sites that describe their units in detail; most are accessible from the

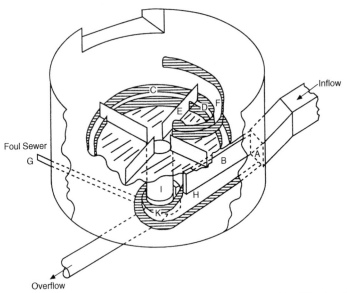

Inflow

Foul Sewer
G

Overflow

FIGURE 9.3.6   Typical swirl/vortex treatment unit (*Swirl and Helical Bend Regulator/ Concentrator for Stormwater and CSO Control,* U.S. EPA 600/2-84/15, 1982). A, inlet ramp; B, flow detector; C, scum ring; D, overflow weir and weir plate; E, spoiler; F, floatables trap; G, foul sewer outlet; H, floor gutters; I, downshaft; J, secondary overflow weir; K, secondary gutter.

"stormwater management" category at www.yahoo.com. This industry has developed rapidly in the past few years, with more manufacturers entering the market every year.

The units generally have no moving parts; they have a chamber with baffles that direct inflow in a helical pattern. Suspended solids are drawn by currents created by the vortex to the center of the chamber. There they settle to the bottom and are trapped by baffles. Floatables are also drawn to the center and are similarly trapped. The units must be periodically cleaned to maintain effectiveness. The units are sized to treat a given flow rate (e.g., flow from the 2-year storm), with a bypass for higher flows.

Manufacturers report very high removal efficiencies (over 90 percent) depending on the size of the units and the flow rate. The Massachusetts Executive Office of Environmental Affairs Strategic Envirotechnology Partnership reviewed reports of the StormTreat and Stormceptor systems and concluded that performance data suggested that the systems can remove about 80 percent of incoming total suspended solids when sized accordingly (www.magnet.state.ma.us/step/techtoc.htm).

Modified Dry Ponds, Wet Ponds and Wetlands

General.   Ponds are probably the most common stormwater quality control structure. As mentioned above, conventional dry ponds (those with outlet flush with invert, so the pond drains completely between events) are not effective at improving water quality because the runoff from the small, frequent events passes directly through them. However, if the outlet of the pond is restricted to discharge slowly, a dry pond can provide extended detention (at least several hours, typically 24), which can remove settleable solids (Figure 9.3.4). For this reason, these ponds are called modified or extended detention dry ponds.

Wet ponds, with their permanent pool, provide buffering and detention for inflows, allowing solids to settle before discharge. Wet ponds are sometimes called retention ponds, because a portion of the inflow is retained indefinitely. Urbonas and Stahre (1993) produced a comprehensive work on pond design.

TABLE 9.3.8 Distribution of Particle Settling Velocities in Urban Stormwater

Percentage of particles with this settling velocity or less	Settling velocity, m/h	Settling velocity, ft/h
10	0.01	0.03
30	0.1	0.3
50	0.5	1.5
70	2.1	7.0
90	21.0	70.0

Source: Adapted from Driscoll, E. D. (*Performance of Detention Basins for Control of Urban Runoff Quality, International Symposium on Urban Hydrology, Hydraulics and Sediment Control*, University of Kentucky, 1983) and U.S. EPA (*Methodology for Analysis of Detention Basins for Control of Urban Runoff*, EPA 440/5-87-001, U.S. EPA, 1986).

Stormwater wetlands are essentially ponds with the addition of vegetation. The two facilities can also be combined into one, with a wetland fringe (vegetated shallow area) along the edge and deeper water (a pond) in the middle. They can also be linked in series (Barrett and Goldsmith, 1998).

The most popular wetland design manual is by Schueler (1993). Parts of the manual have been incorporated into state stormwater manuals (e.g., Massachusetts and Delaware), including its nomenclature (e.g., pocket wetlands). Kadlec and Knight (1996) produced a comprehensive work on water quality processes in wetlands.

Sizing. Sizing can be theoretically based on detention time and pond depth or the overflow rate (outflow rate per unit bottom area, also referred to as hydraulic loading rate) and settling velocity of particles, because removal of suspended solids usually is the design criterion. For discrete particles, settling velocity can theoretically be determined from Stokes law. Then, if the particle size distribution is known, it is theoretically possible to compute the distribution of settling velocities. However, the particle size distribution is seldom known with any certainty. Moreover, some studies have determined that particle-size distributions are difficult to manipulate into settling velocities because of the complexity of the settling environment (Whipple and Hunter, 1981).

Generalized estimates of settling velocity are probably more useful, such as Table 9.3.8, which resulted from studies of approximately 50 urban runoff samples. Note that settling velocities can vary over several orders of magnitude from more than 1000 cm/h for coarse sand, down to 0.01 cm/h or less for clay particles.

With a distribution of settling velocities and a given pond depth, one can compute the detention time need to remove a certain percentage of particles. For example, if 70 percent removal is desired, then the fastest settling 70 percent of particles will be removed, corresponding to a minimum settling velocity of 0.1 m/h. If the pond were 1 m deep, then 10 h of detention time would be required to settle out 70 percent of the particles.

The calculation can be expressed as

$$t = d/v \tag{9.3.8}$$

where t is the required detention time (h), d is pond depth (m), and v is the settling velocity of the slowest particle to be removed (m/h).

The detention time can be used compute the required volume of the pond

$$\Psi = Qt \tag{9.3.9}$$

where Ψ is pond volume (m³), Q is flow rate (m³/h), and t is detention time (h).

Similar reasoning can be applied based on the overflow rate, as is done for settling tanks in wastewater treatment plants (Novotny and Olem, 1994, p. 624):

$$OR = Q/A \tag{9.3.10}$$

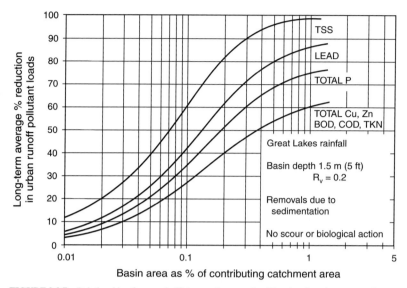

FIGURE 9.3.7   Relationship of removal efficiency of wet ponds with ratio of pond area to catchment area, by analysis of NURP data (Novotny and Olem, 1994; after Driscoll, 1988). BOD, biochemical oxygen demand; COD, chemical oxygen demand; TKN, total Kjeldahl nitrogen (organic and ammonia nitrogen); TSS, total suspended solids.

where OR is overflow rate (m/h), Q is flow rate (m^3/h), and A is pond bottom area (m^2). In this approach, a particle is removed if its settling velocity is greater than or equal to the overflow rate. In practice, the use of such a theoretical approaches is frustrated by the fact that flow rate is highly variable and that pond hydraulics may not be ideal.

Alternatively, sizing of ponds/wetlands is based on empirical correlations or regulatory mandate, where the primary sizing variable is one of the following:

- storage volume (based on a certain depth of runoff from the entire watershed or only the impervious area),
- mean detention time,
- overflow rate, or
- ratio of pond area to watershed (or impervious) area.

Driscoll (1988) presents an empirical curve for wet ponds studied in NURP, relating removal efficiency of total suspended solids and other constituents to the ratio of pond area to catchment area (Figure 9.3.7).

Regulations may specify a certain treatment volume—volume equal to the first 1.3 to 2.5 cm (0.5 to 1 in.) of runoff from the entire site or from impervious areas only, or the runoff from to a storm of a given recurrence interval (typically up to 2 years), or a percentage of all runoff.

Often, once the sizing criteria are met, a typical percentage removal of solids is merely assumed. For example, the Massachusetts Department of Environmental Protection (1997) allows applicants to claim 80 percent of total suspended solids removal for an adequately sized wet pond.

Other Design Considerations.   If not designed or maintained properly, ponds and wetlands can pose a drowning hazard to children, generate odor problems and become eyesores (covered with thick algae mats), and become mosquito breeding grounds. On the positive side, a U.S. EPA study (1995) reported that well-maintained wet ponds enhanced the value of surrounding property. Design considerations for peak flow rate control are also relevant to water quality control.

To address concerns about safety (i.e., children drowning), pond sides should be benched or gently sloping, at least near the pond edge. That way, if a child does fall in, he or she can easily get up again. Also, outlets should be designed for low velocities and configured to ensure that no one can be trapped at the outlet. The design of the inlet should dissipate flow energy and avoid erosion, spread out inflow to involve the entire cross section of pond/wetland (i.e., avoid dead zones and short-circuiting), and allow access for maintenance. A pond should include a forebay for coarse sediment removal that can be easily reached by dredging equipment.

Maintenance.   Inadequate maintenance is a common problem affecting long-term pond and other BMP performance (Pazwash, 1991). Proper operation and maintenance of structural treatment facilities are critical for effectiveness. Responsibility for maintenance should be assigned before construction is complete.

Because the structures are designed to accumulate solids, sediment will eventually need to be removed (design life between removal is typically between 20 and 30 years, but unanticipated loading, often resulting from extensive upstream construction, can greatly decrease this time). The Northern Virginia Planning District Commission's publication *Investigation of Potential Sediment Toxicity From BMP Ponds* (1997) discusses appropriate disposal options, including onsite use as a soil amendment (metals and organics were below "hazardous" levels).

A study of 51 urban runoff treatment facilities in Ocean County, New Jersey, determined that the major cause of inadequate performance was lack of adequate maintenance. The causes of these failures varied, including factors such as lack of funding, lack of manpower, and lack of equipment; uncertain or irresponsible ownership; unassigned maintenance responsibility; and ignorance or disregard of potential consequences of maintenance neglect (Ocean County Planning & Engineering Department, 1989).

The study also found the following:

- Bottoms, side slopes, trash racks, and low-flow structures were the primary sources of maintenance problems.

- Infiltration facilities seemed to be more prone to maintenance neglect and were generally in the poorest condition overall.

- Retention facilities appeared to receive the greatest amount of maintenance and generally were in the best condition overall.

- Publicly owned facilities were usually better maintained than those that were privately maintained.

- Facilities located at office developments were better maintained than those at commercial or institutional sites; facilities in residential areas received intermediate maintenance.

- Highly visible urban runoff facilities were generally better maintained that those in more remote, less visible locations.

The study also recommended program elements to ensure the proper design, implementation, and operation and maintenance of stormwater controls.

Cost.   The U.S. EPA Guidance Manual U.S. EPA (1993, p. 4–31) reports probable construction cost (not including land purchase) for wet ponds between $0.5 and $1 per ft^3 ($0.014 to $0.028 per m^3) for ponds less than 10^6 ft^3 (8990 m^3) and between $0.1 and $0.5 per ft^3 ($0.00028 to $0.014 per m^3) for ponds over 10^6 ft^3 (8990 m^3). Probable annual operation and maintenance costs range from 1 to 5 percent of capital cost.

The NURP final report (U.S. EPA, 1983) reports a best-fit correlation for dry pond construction cost of $C = 77.4\ V^{0.51}$ where C is the cost in 1980 dollars and V is pond volume in ft^3.

Combined Sewer Overflows

Among the techniques for managing CSOs is reducing runoff peaks and volumes by the methods covered previously to reduce the number and magnitude of overflows. The inline, swirl/vortex treatments units mentioned above can treat combined sewer flow if they are sized accordingly. Other methods include sewer separation, storage of combined flows for subsequent treatment, and wetland treatment.

Installing a new sewer system to separate sewage from stormwater eliminates CSO, but it is very costly and disruptive (most CSOs occur in older, intensely developed areas). Furthermore, as noted by Novotny and Olem (1994, p. 37) the water quality benefits of sewer separation are variable (in frequent, small storms, combined sewers convey all their flow to a treatment plant and separate storm sewer systems discharge to waterways).

The storage and treatment approach is also very expensive because of the large volumes that must be conveyed (some times over large distances) and stored. An example of this approach is the Metropolitan Water Reclamation District of Greater Chicago's Tunnel and Reservoir Project, which cost more than a billion dollars.

Wetlands can also be effective at treating combined sewer overflows, although there are problems and expenses of land availability in developed areas and conveyance to the wetland(s). Wetlands have the advantage over storage reservoirs and tunnels of providing additional benefits, such as wildlife habitat and recreation.

9.3.10   MINING AND AGRICULTURE RUNOFF

Novotny and Olem (1994) and the U.S. EPA NPS guidance document (1993) both devote chapters to agricultural issues. Source control techniques are particularly applicable to agriculture, because the land owner usually has full control over a large area. To reduce erosion and sediment loads, techniques include no-till planting (which may require more herbicide), planting cover crops, contour plowing and terraces to slow and infiltrate runoff, filter strips, fencing to keep livestock from eroding streambanks, and good grazing practices. For nutrient and pesticide control from cropland, techniques include those listed above (because these contaminants are often attached to soil particles) and integrated pest management to minimize pesticide application, and minimal/appropriate fertilization.

Ponds and such structural BMPs are less applicable in row-crop agriculture because of the large volumes of water that must be treated and lack of many owners to share the cost of construction. To control manure pollution from livestock (e.g., feedlots, chicken/swine farms, and dairy barns), pond-like systems are effective, as is fencing to keep livestock out of waterways.

Conserving natural wetlands and restoring wetlands by restoring natural hydrology (i.e., removing/plugging drainage tiles) are low-cost agricultural NPSP control measures. Sediment pollution from strip mines and tailings can best be remedied with aggressive revegetation. Wetlands have proven very effective in remediating acid mine drainage (Fennessy and Mitsch, 1989; Smith, 1997).

9.3.11   COMPUTER MODELS OF STORMWATER QUANTITY AND QUALITY

Overview

The issues and methods in the preceding subsections are integrated into many computerized models of stormwater quantity and quality. These models are explicitly discussed in this subsection. Flow-only models have already been discussed.

An excellent reference on this subject by Donigian and Huber (1991) can be downloaded from the web site of the U.S. EPA's Center for Exposure Assessment Modeling. The report contains detailed reviews of numerous complex models as well as simple procedures (Table 9.3.9). It also gives an overview of modeling fundamentals and a guide to model selection. Another excellent reference is U.S. EPA (1997).

Models can be classified based on various characteristics. First, a model can address only hydrology, or it can address both hydrology and water quality; it can calculate contaminant loadings, effects in receiving waters, or both. A model can be oriented toward urban processes or agricultural processes.

TABLE 9.3.9   Models Reviewed by Donigian and Huber (1991)

Urban	Non-urban	
SWMM	EPA screening procedures	HSPF
STORM	AGNPS	PRZM
DR3M-QUAL	ANSWERS	SWRRB
	CREAMS/GLEAMS	UTM-TOX

Source: Donigian, and Huber, *Modeling of Nonpoint Source Water Quality in Urban and Non-urban Areas*, 1991.

A deterministic model produces one set of outputs for a given set of inputs, whereas a stochastic model produces a distribution. In spatial terms, a model can be lumped or distributed. In temporal terms, a model can be steady state or dynamic. A dynamic model can be event oriented or continuous.

Models come in a wide spectrum of complexity. Complex models, such as U.S. EPA's Hydrologic Simulation Program-FORTRAN (HSPF), can require hundreds of parameters and extensive data on watershed conditions and meteorology. In theory, sophisticated, complex models more accurately depict watershed processes. However, in many cases, such parameters and data cannot be accurately supplied, and the added sophistication is of little or no value. In fact, using a sophisticated model with suspect data or default parameters values can give less accurate results than a simpler model.

Choosing the most appropriate model is not simple. The simplest model adequate for the task is generally the best. Features to look for in any model include good documentation (user's manual, explanation of theory and numerical procedures, data needs, data input format, etc.), a track record of use and acceptance, and available support from the commercial vendor, a user's group, or a government agency.

Numerous common, public-domain models (both documentation and executable programs) are available from the U.S. EPA's Center for Exposure Assessment Modeling. However, one should remember that a program purchased from a commercial vendor usually comes with more support than one downloaded free from a government web site.

Simple Models

The simplest models can be run on a computerized spreadsheet. They do not simulate geographic- or time-dependant variations. Table 9.3.10 lists some well-known simple models.

Typically, simple models compute contaminant loading (usually on an annual basis) based on a unit loading rate (mass of contaminant per unit area per unit time, a.k.a. export coefficient) or based on a runoff volume (e.g., from the rational method) multiplied by an average contaminant concentration. Monitoring has shown that export rates and concentrations are highly variable and therefore very difficult to predict for modeling purposes. Mean concentrations for urban areas are listed in Table 9.3.4. Export coefficients for agricultural areas have been published (Reckhow et al., 1980). Urban export coefficients are published in the NURP final report (U.S. EPA, 1983). An online version of such a method (Pollutant Budget Estimation Form) is available from the North Carolina State University water quality group (http://h2osparc.wq.ncsu.edu/lake/bass/spread1.html).

The effect of a control measure can easily be simulated by applying a constituent removal percentage, such as those listed in Table 9.3.7.

Intermediate Models

Intermediate models add simulation of geographic- and time-dependant variations. They have greater data requirements than simple models (e.g., precipitation time series). They lack detailed simulation of transport and transformation. Examples of such models are P8, AGNPS, and SLAMM (U.S. EPA, 1997).

TABLE 9.3.10 Well-Known Simple Models of Stormwater Runoff

Model name	Method of computing pollutant loads	Reference
Simple method	Based on runoff volume (found by a modified rational method) and EMC (from NURP data)	Schueler, T. R., *Controlling Urban Runoff: A Practical Manual for Planning and Designing Urban BMPs*, Metropolitan, Washington Council of Governments, 1987.
U.S. EPA's screening procedure	For agricultural areas, uses the universal soil loss equation to predict sediment load and an "enrichment factor" specifying contaminant concentrations in sediment; for urban areas uses buildup/washoff approach. Also include methods for analyzing impacts to receiving waters.	Mills, W. B. et al., *Water Quality Assessment: A Screening Procedure for Toxic and Conventional Pollutants in Surface and Groundwater*, EPA/600/685/002a,b, U.S. EPA, 1985.
U.S. Geological survey regression method	Uses regressions of watershed characteristics and climatic parameters versus NURP monitoring data.	Tasker, G. D. et al., "Estimation of Mean Urban Stormwater Loads at Unmonitored Sites by Regression," in *Symposium Proceedings on Urban Hydrology*, American Water Resources Association, 1990, pp. 127–138.
FHWA method	Inputs include statistical properties of rainfall (mean and coefficient of variation of storm event depth, duration, intensity and interevent time), area, and runoff coefficient for the hydrologic component, plus EMC median and coefficient of variation for a constituent of interest.	Driscoll, E. D., Shelley, P. E., and Strecker, E. W., *Pollutant Loadings and Impacts from Highway Stormwater Runoff, Vols. I to IV*, FHWA-RD-88-006 through -009, 1990.

P8 simulates generation and transport of stormwater runoff pollutants in small urban catchments (Palmstrom and Walker, 1990). It can perform continuous simulation of water and suspended solids (five size fractions), total phosphorus, total Kjeldahl nitrogen, copper, lead, zinc, and total hydrocarbons, driven by continuous hourly rainfall, and daily air temperature time series. It is useful for evaluating, selecting, and siting BMPs and calculating their pollutant removal efficiencies. Default input values come from NURP data as a function of land use, land cover, and soil properties. U.S. EPA (1997) recommends that, without calibration, model results be used for relative comparisons only. The model, data files and documentation are available at www2.shore.net/~wwwalker/p8/.

The Source Loading and Management Model (Pitt, 1998) performs continuous simulation of hydrology and particulate and dissolved contaminants, addressing numerous BMPs (infiltration practices, wet detention ponds, porous pavement, street cleaning, catch-basin cleaning, and grass swales). SLAMM is used mostly as a planning tool, to better understand sources of urban runoff pollutants and their control. Details and software available at wi.water.usgs.gov/slamm/.

The Agricultural Non-Point Source Pollution Model (AGNPS) simulates surface runoff, sediment, and nutrient transport, primarily from agricultural watersheds. It uses the SCS curve number method to simulate hydrology; upland erosion and sediment transport is estimated by using a modified form of the Universal Soil Loss Equation. AGNPS is often linked with Geographic Information Systems (GIS) (Tim and Jolly, 1994). Program and documentation are available at www.cee.odu.edu/cee/model/ and from the NRCS at www.wcc.nrcs.usda.gov.

Complex Models

Complex models most closely simulate watershed processes according to current scientific understanding. In theory, these models are the most accurate. However, large effort is necessary to obtain the data and parameters to effectively use these models to their fullest extent.

One of the most visible complex models is the U.S. EPA's BASINS system (www.epa.gov/OST/ BASINS/), which integrates the models NPSM, QUAL2E, and TOXIROUTE in a GIS environment. The Nonpoint Source Model (NPSM) estimates land-use-specific nonpoint source loadings for selected pollutants, putting a Windows-based interface onto the HSPF model. HSPF is one of the most comprehensive watershed models available, simulating impervious and pervious runoff hydrology and contaminant transport and in-stream hydraulic and sediment-chemical interactions (Donigian et al., 1994). The simulation can produce a time history of runoff flow rate, sediment load, and nutrient and pesticide concentrations at desired points in a watershed. It requires extensive data for calibration or verification and great expertise to use properly (U.S. EPA, 1997).

BASINS also supplies assessment tools for evaluating water quality and point source loadings, utilities for data import, land-use classification, watershed delineation, management of water quality observation data, and postprocessing tools for interpreting model results.

Another well-known and often used model in USEPA's Stormwater Management Model, SWMM (Huber and Dickinson, 1988). SWMM consists of several modules or blocks designed to simulate most quantity and quality processes on a continuous or single event basis. Channels, culverts, reservoirs and water quality processes are simulated with the Runoff Block. The Extran Block simulates more complex hydraulics. The Storage/Treatment Block may be used for storage-indication flow routing. Data requirements are, at a minimum, precipitation, area, imperviousness, slope, roughness, depression storage, and infiltration characteristics.

SWMM has the advantage of extensive support. Computational Hydraulics International (CHI, www.chi.on.ca) runs an electronic mailing list (listserv) for SWMM users and hosts an annual conference on stormwater modeling. The program is available through USEPA, CEAM and CHI.

There are several extensions to SWMM which provide enhanced interfaces, such as MIKESWMM and XPSWMM. CHI has developed PCSWMM, a graphical decision support system, which provides file management, model development, model calibration, output interpretation/presentation, and reference tools. A graphical user interface and postprocessor for SWMM is available from CAiCE (http://www.drainage.com).

REFERENCES

Akan, A. O., *Urban Stormwater Hydrology: A Guide to Engineering Calculations*, Technomic Publishers, 1993.

American Public Works Association, *Stormwater Update No. 3*, APWA Institute for Water Resources, Chicago, 1980.

ASCE/WEF Task Committee, "Design and Construction of Urban Stormwater Management Systems," *ASCE Manuals and Reports of Engineering Practice, No. 77*, American Society of Civil Engineers and the Water Environment Federation, 1992.

Barrett, K. R., and Goldsmith, W., "Bioengineered System in Existing Stream Channel," *Proceedings of the ASCE Conference on Wetlands Engineering and River Restoration*, ASCE, 1998.

Barrett, M. E. et al., *Characterization of Highway Runoff in the Austin, Texas Area*, FHWA/TX-DOT report 96/1943-4, FHWA, 1996.

Bartram, J., and Balance, R., *Water Quality Monitoring: Practical Guide to the Design and Implementation of Freshwater Quality Studies and Monitoring Program*, Routledge, 1998.

Browne, F. X., "Stormwater Management," in *Standard Handbook of Environmental Engineering*, 2nd ed., edited by Corbitt, R. A., McGraw-Hill, 1998, pp. 7.1–7.130.

Carr, A., "Impact of Nondegradable Marine Debris on the Ecology and Survival Outlook of Sea Turtles," *Mar. Pollut. Bull.*, 18(6B): 352–356, 1987.

Center for Watershed Protection, *Site Planning for Urban Stream Protection*, CWP, Silver Springs, MD, 1995.

Chow, V. T., *Open-Channel Hydraulics*, McGraw-Hill, 1959.

Claytor, R. A., *Stormwater Management Pond Design Example*, Center for Watershed Protection, 1995.

Claytor, R., and Schueler, T., *Design of Stormwater Filtering Systems*, Center for Watershed Protection, 1996.

Claytor, R., and Schueler, T., *BMP Design Supplements for Cold Climates*, Center for Watershed Protection, CWP, Silver Springs, MD, 1997.

Delaware Department of Natural Resources, *Conservation Design for Stormwater Management*, Delaware Department of Natural Resources, 1997.

Dodson, R. D., *Storm Water Pollution Control: Municipal, Industrial and Construction Compliance*, 2nd ed., McGraw-Hill, 1998.

Donigian, A. Jr., Bicknell, B., and Imhoff, J., "Hydrologic Simulation Program-FORTRAN," in *Computer Models of Watershed Hydrology*, edited by Singh, V. P., Water Resources Publications, 1994.

Donigian, A. S. Jr., and Huber, W. C., *Modeling of Nonpoint Source Water Quality in Urban and Non-Urban Areas*, USEPA EPA/600/3-91/039, (ftp://ftp.epa.gov/epa_ceam/wwwhtml/ceamhome.htm), 1991.

Driscoll, E. D., "Long Term Performance of Water Quality Ponds," in *Design of Urban Runoff Quality Controls, Proceedings of an Engineering Foundation Conference*, edited by Roesner, L. A. et al., ASCE, 1988, pp. 145–162.

Driscoll, E. D., Shelley, P. E., and Strecker, E. W., *Pollutant Loadings and Impacts from Highway Stormwater Runoff, Vols. I to IV*, Federal Highway Authority, FHWA-RD-88-006 through 009, 1990.

Fennessy, S., and Mitsch, W. J., "Design and Use of Wetlands for Renovation of Drainage from Coal Mines," in *Ecological Engineering: An Introduction to Ecotechnology*, edited by Mitsch, W. J., and Jorgensen, S. E., Wiley, New York, 1989, pp. 232–252.

Ferguson, B. K., *Stormwater Infiltration*, Lewis Publishers, Boca Raton, FL, 1994.

Ferguson, B. K., and Debo, T. N., *On-Site Stormwater Management: Applications for Landscape and Engineering*, PDA Publishers Corp., 1987.

FHWA, "Urban Drainage Design Manual," *Hydraulic Engineering Circular 22*, FHWA, U.S. Department of Transportation, FHWA-SA-96-078, 1996.

Galli, F. J., *Preliminary Analysis of the Performance and Longevity of Urban BMPs Installed in Prince George County, Maryland*, Department of Environmental Resources, Prince George County, 1992.

Grey, G. M., and Oliveri, F., "The Role of Catch Basins in a CSO Floatables Control," *Proceedings of the WEFTEC '99 Conference*, Water Environment Federation, 1999.

Horn, M. E., "Estimating Soil Permeability Rates," *J. Irrig. Drainage Div. ASCE*, 97: 263–274, 1971.

Huber, W. C., and Dickinson, R. E., *Storm Water Management Model User's Manual, Version 4*, EPA/600/3-88/001a, 1988.

Illinois State Water Survey Circular 173, Time Distributions of Heavy Rain Storms in Illinois, ISWS, Champaign, IL, 1990.

Kadlec, R. H., and Knight, R., *Treatment Wetlands*, Lewis Publishers, 1996.

Lee, G., and Jones-Lee, A., *Stormwater Runoff Management: The Need for a Different Approach*, Water Engineering and Management, 1995, pp. 36–39.

Levine, J. S., and Miller, K. R., *Biology: Discovering Life*, D.C. Heath, 1994, Chapt. 7.

Linsley, R., Kohler, M., and Paulhus, J., *Hydrology for Engineers*, McGraw-Hill, 1982.

Maidment, D. R., *Handbook of Hydrology*, McGraw-Hill, 1993

Marino, L., and Bradley, A., *Precipitation Frequency-Runoff Frequency Relationship in Hydrologic Design*, AGU Fall Meeting, 1986.

Massachusetts Department of Environmental Protection, *Stormwater Management*, Vols. 1 and 2, Boston,1997.

Massachusetts Department of Environmental Protection, *Stormwater Management: Volume 2: Stormwater Technical Manual*, 1997.

McCuen, R., *A Guide to Hydrologic Analysis Using SCS Methods*, Prentice-Hall, Englewood Cliffs, NJ, 1982.

McCuen, R. H., "A Regional Approach to Urban Stormwater Detention," *Geophys. Res. Lett.*, 1(7): 321–322, 1974.

Mercer, G. et al., "Comprehensive Watershed Analysis Tools: The Rouge Project—A Case Study," *Proceedings, Watershed '96*, U.S. EPA, 1996.

Metcalf, and Eddy, *Wastewater Engineering: Collection, Treatment, Disposal*, McGraw-Hill, NYC, 1972.

Montes, S., *Hydraulics of Open Channel Flow*, ASCE Press, 1998.

Mustard, M. H., Driver, N. E., Chry, J., and Hansen, B. G., *Urban-Stormwater Data Base of Constituent Storm Loads; Characteristics of Rainfall, Runoff, and Antecedent Conditions; and Basin Characteristics*, U.S. Geological Survey Water-Resources Investigation 87-4036, U.S. Geological Survey, 1987.

National Pollutant Removal Performance Database for Stormwater BMPs—1997, Center for Watershed Protection.

Normann, Houghtalen, and Johnston, *HDS 5—Hydraulic Design of Highway Culverts*, FHWA, Washington, DC, 1985.

Novotny, N., and Olem, H., *Water Quality: Prevention, Identification, and Management of Diffuse Pollution*, Van Nostrand Reinhold, 1994.

Ocean County Planning & Engineering Department, *Ocean County Demonstration Study: Stormwater Management Facilities Maintenance Manual*, New Jersey Department of Environmental Protection, Division of Water Resources, 1989.

Palmstrom, N., and Walker, W., "The P8 Urban Catchment Model for Evaluating Nonpoint Source Controls at the Local Level," *Enhancing States' Lake Management Programs*, U.S. EPA, 1990.

Pazwash, H., "Maintenance of Stormwater Management Facilities: Neglects in Practice," *Hydraulic Engineering, Proc., 1991 National Conference*, ASCE, 1991, pp. 1072–1077.

Pitt, R., "Unique Features of the Source Loading and Management Model (SLAMM)," in *Advances in Modeling the Management of Stormwater Impacts—Vol. 6*, edited by James, W., CHI, Guelph, ONT., CN, 1998.

Pitt, R. et al., *Groundwater Contamination from Stormwater Infiltration*, Ann Arbor Press, Ann Arbor, NI, 1995.

Ponce, V. M., *Engineering Hydrology: Principles and Practices*, Prentice-Hall, Englewood Cliffs, NJ, 1989.

Rallison, and Miller, "Past, Present and Future SCS Runoff Procedure. Rainfall-Runoff Relationships," *Proc. Int. Symp. Rainfall-Runoff Modeling*, edited by Singh, P., Mississippi State University, Oxford, 1981, pp. 353–364.

Rantz, S. E., "Suggested Criteria for Hydrologic Design of Storm-Drainage Facilities in the San Francisco Bay Region, California," *U.S. Geological Survey Professional Paper 422-M*, U.S. Geological Survey, Reston, VA, 1971.

Reckhow, K. H., Beaulac, M. N., and Simpson, J. T., *Modeling Phosphorous Loading and Lake Response Under Uncertainty: a Manual and Compilation of Export Coefficients*, U.S. EPA 440/5-80-011, U.S. EPA, 1980.

Reddy, K. R., and D'Angelo, E. M., "Soil Processes Regulating Water Quality in Wetlands," in *Global Wetlands: Old World and New*, edited by Mitsch, W., Elsevier, 1994, pp. 309–324.

Schueler, T., *Design of Stormwater Wetland Systems: Guidelines for Creating Diverse and Effective Stormwater Wetlands in the Mid-Atlantic Region*, Metropolitan Washington Council of Governments, 1993.

Schueler, T. R., Kumble, P. R., and Heraty, M. A., *A Current Assessment of Urban Best Management Practices: Techniques for Reducing Non-Point Source Pollution in the Coastal Zone*, The Metropolitan Washington Council of Governments, Washington, DC, 1992.

Simmons, D., and Reynolds, R., "Effects of Urbanization on Baseflow of Selected South-Shore Streams, Long Island, NY," *Water Res. Bull.*, 18(5): 797–805, 1982.

Singh, V. P. et al., *Water Quality Monitoring Network Design*, Kluwer Academic Publishers, 1998.

Smith, K., *Constructed Wetlands for Treating Acid Mine Drainage*, Restoration and Reclamation Review, www.hort.agri.umn.edu/h5015/rrr.htm.html, Vol. 2, Spring, 1997.

Sutherland, R., and Jelen, S., "Studies Show Sweeping Has Beneficial Impact on Stormwater Quality," *APWA Reporter*, Nov. 1996.

Temple, D. M., "Design of Grass-Lined Open Channels," *Trans. ASAE*, 26(4): 1064–1069, 1983.

Temple, D. M., Robinson, K. M., Ahring, R. M., and Davis, A. G., *USDA Agricultural Handbook No. 667*, U.S. Department of Agriculture, Washington, DC, 1987.

The Northern Virginia Planning District Commission's Publication, *Investigation of Potential Sediment Toxicity from BMP Ponds*, 1997.

Thomann, R. V., and Mueller, J. A., *Principles of Surface Water Quality Modeling and Control*, Harper and Row, 1987.

Tim, U., and Jolly, R., "Evaluation of Agricultural Nonpoint-Source Pollution Using Integrated Geographic Information Systems and Hydrologic/Water Quality Models," *J. Environ. Quality*, 23: 25–25, 1994.

Urbonas, B., and Stahre, P., *Stormwater: Best Management Practices and Detention for Water Quality, Drainage, and CSO Management*, PTR Prentice Hall, 1993.

Urbonas, B. R., and Rooster, L. A., "Hydrologic Design for Urban Drainage and Flood Control," in *Handbook of Hydrology*, edited by Maidment, D. R., McGraw-Hill, New York, 1993.

USDA, *National Engineering Handbook*, Soil Conservation Service, U.S. Department of Agriculture, Washington, DC, 1985.

U.S. EPA, *Results of the Nationwide Urban Runoff Program: Final Report*, U.S. EPA Water Planning Division, 1983.

U.S. EPA, *Guidance Specifying Management Measures for Sources of Nonpoint Pollution in Coastal Waters*, U.S. EPA, EPA-840-B-93-001, www.epa.gov/owowwtr1/NPS/MMGI/index.html, 1993.

U.S. EPA, *Economic Benefits of Runoff Controls*, U.S. EPA, EPA 841-S-95-002, U.S. EPA, 1995.

U.S. EPA, *Compendium of Tools for Watershed Assessment & TMDL Development*, EPA 841-B-97-006, 1997.

U.S. EPA, NPDES, *Storm Water Sampling Guidance Document*, U.S. EPA, 1993.

Wanielista, M. P., and Yousef, A.Y., *Stormwater Management*, Wiley, New York, 1993.

Water Environment Federation, *Combined Sewer Overflow Pollution Abatement MOP FD-17*, Water Environment Federation, 1989.

WEF/ASCE, *Urban Runoff Quality Management, ASCE Manuals & Reports of Engineerings Practice No. 87*, Water Environment Federation and American Society of Civil Engineers, 1998.

Whipple, and Hunter, "Settleability of Urban Runoff Pollution," *J. Water Pollution Control Fed.*, 53(12): 313–325, 1981.

CHAPTER 10

ENVIRONMENTAL REGULATIONS

Richard Gaskins

*Mr. Gaskins is a partner with the law firm of Kilpatrick Stockton
in Charlotte, North Carolina, where he handles environmental
litigation, compliance advice and environmental aspects of
business transactions. He is active in the Environmental and
Natural Resources Law Section of the North Carolina Bar
Association and the Toxic and Environmental Torts Committee
of the American Bar Association.*

SECTION 10.1

BASICS OF ENVIRONMENTAL LAW

10.1.1 RELATIONSHIP OF FEDERAL AND STATE LAWS

Almost all statutory environmental programs have some basis in federal law, although there are
some notable exceptions from state to state. In general, Congress enacts statutes, such as those for
clean air and water, that establish national minimum standards of environmental protection, including
specifying acceptable ambient levels of contaminants or effluent concentrations. However, Congress
determined that the states would want to administer these programs themselves. Therefore, these
statutes generally contain provisions that also specify that the states can be designated to administer
the programs within their own boundaries in lieu of administration by Environmental Protection
Agency (EPA). For example, Section 3006 of Subchapter III of the federal Solid Waste Disposal
Act describes the bases upon which a state becomes authorized to administer the hazardous waste
program. This designation as an approved state program is often referred to as "primacy." The statutes
typically specify the minimum requirements that a state program must meet in order to be eligible
to achieve primacy, including such things as adequate provisions on permitting and enforcement and
adequate staffing. Often, these minimum standards and the procedures for meeting them are set out
in greater detail in rules promulgated by the EPA.

Once a state demonstrates that it meets the minimum standards for primacy, it is given the responsibility for permitting, advice, monitoring, and enforcement within its boundaries. It is expected that
the state agency will maintain these functions unless there is a serious breakdown in its powers or
abilities. If the state agency suffers a sufficiently serious lapse in its environmental protection efforts
in the particular program, EPA may initiate proceedings to "take the program back" (i.e., to assume
responsibility for that program within the state). A recent example in the Carolinas was the effort
by EPA to take back North Carolina's hazardous waste management program, which so far has not
occurred.

To the extent that state law is in conflict with federal law, under the Supremacy Clause of the United
States Constitution, federal law controls. Federal law generally controls interstate pollution. Disputes
between states, including environmental disputes, can go directly to the United States Supreme Court.
However, much of the environmental protection activity and initiative is occurring at the state level.

10.1.2 *NATURE OF LAWS, RULES, PERMITS, GUIDANCE DOCUMENTS, AND MEMOS*

In judging one's legal duties, it is often said, "ignorance of the law is no excuse." In the area of the legal protection of the environment, the universe of statutes, laws, rules, regulations, interpretative memoranda, guidance documents, permits, rulings, and other documents and declarations presents a complicated and ever-changing array of what the regulator may claim is the "law" that the regulated entity should have known about. Even given the fact that an awareness of all of these types of pronouncements may be advisable—if not necessary—in order to conduct one's business safely and legally, it is nevertheless advisable to be aware of the difference between and among the various types of standards, which are discussed briefly below.

The phrase *environmental law* is usually applied to all requirements imposed by various governmental bodies to protect the environment. They can take the form of statutes passed by Congress or state legislatures. Technically, in order to be a "law," the requirement must be passed as such a statute. However, the phrase *environmental law* is usually also applied to all of the rules and regulations adopted and enforced by federal and state environmental agencies, pursuant to those statutes as well as unwritten, common law theories of liability, such as nuisance and trespass. There are numerous interpretations, guidance documents, advisories, and similar documents and policies that such agencies and their personnel follow, and which the agency will often treat as "law," even though their applicability and enforceability should often be examined (and perhaps questioned) before blind acquiescence. The phrase *environmental law* also includes local ordinances adopted by cities, counties, or other municipal governmental bodies. Finally, the phrase should also include any permits, registrations, or other types of permissions received from governmental agencies to satisfy the requirements of those laws, rules, or ordinances.

As in any other area of the law, Congress or a state legislature identifies a problem requiring protection of the environment, holds hearings and debates on bills that address the particular environmental issue. Eventually both houses of Congress, or the state legislature, enact a bill, and the result is what is known as an environmental statute. Statutes create regulatory agencies and empower them. They also create permitting programs, provide for enforcement of those programs, mandate the establishment of standards of environmental quality, and provide guidance to agencies for implementation of the statutory scheme. These environmental statutes generally provide the penalties for violation and sometimes provide for novel enforcement schemes, such as creating specific causes of action for the violation of the statute or its implementing regulations, which private parties may enforce as private attorneys general.

Administrative agencies or administrative officers of the federal or state governments adopt rules. Rules are the province of the executive branch operating in its quasi-legislative role. Rules are designed to interpret and implement environmental statutes and, theoretically, cannot expand the authority of the administrative agency beyond the authority granted by the legislature or Congress. It has been a common experience in environmental regulation that legislatures have to periodically catch up with regulators, and have amended environmental laws to conform to the scope of authority assumed by rules adopted by the agencies. Rules can be adopted by legislatively created boards or commissions or by specified officials within the administrative department.

Many environmental laws establish programs for permitting various activities. In order to undertake certain activities, persons must apply for a permit from the administrative agency in which the authority is vested under the Act. Generally, that agency will have adopted rules to govern procedures for applying for and issuing the permits provided by statute. The requirements of the permit cannot go beyond the authority granted by the statute, as interpreted by the rule. Again, many environmental statutes provide sanctions for violations of the terms of permits issued under the legislatively created permit program.

In discussing rules and permits, it is important to note that *permits* generally are specifically directed at one party and one situation. They are issued, generally, to one particular discharger or operator. *Rules*, on the other hand, are statements of general applicability. In a sense, the permit is a document by which the general application of the rule is made specific to the particular permittee. As rules are confined by the bounds of the statute, permits, theoretically, are confined by the bounds of the rules. Another way of explaining the concept is to say that a permit cannot require of a permittee something that is not provided by rule, and a rule cannot require of the regulated public something with which the agency is not empowered to require under the environmental statute.

The enactment of statutes, the promulgation of rules, and the issuance of permits are all reasonably formal procedures. Obviously, administrative agencies deal with a regulated public in informal ways as well. These informal means of "regulating" can be as informal as a phone call or a discussion during a site inspection, or a meeting requested by a regulated party.

A step above that degree of informality, but less formal than the permitting process, is the issuance of guidance documents and memoranda. A guidance document is the opinion of the agency and an expression of its policy with regard to how it will approach certain issues within its authority. There is no presumption of validity to a guidance document such as there is to rules and permits. Permits, for instance, may be challenged only within a specified time after issuance. If the permit is not challenged within that period of time, its terms become binding on the permittee to the extent that the agency had the authority to impose them.

Rules, which are properly promulgated, are also presumed to be valid. Under some federal acts, promulgated rules may be challenged as to their validity and as to whether they are within the authority of the agency granted to it by the statute within a specified time following promulgation. Failure to challenge the rule within that specified time results in a presumption of its validity. Under other administrative schemes, there may be no method by which a rule can be directly challenged, but the rule would be subject to challenge in the context of an enforcement proceeding, or through the vehicle, for instance, of a declaratory ruling or declaratory judgment. A properly promulgated rule need not have its validity proven by the agency that promulgated it. Guidance documents, however, being simply expressions of opinion or policy, do not shift the burden of proof to the regulated party and are not presumed to be binding on the regulated community.

Memoranda are similar to guidance documents but, as a general rule, are attended by even less procedural formality involved with their preparation. Frequently, guidance memoranda are prepared and circulated for comment, and consequently carry greater persuasive force as an expression of the expert opinion of the agency on the meaning of certain provisions of the environmental statutes or rules. Memoranda tend to be more specific and more limited to particular factual situations than do guidance documents. These again have no presumption attaching as to their validity and do not shift the burden of proving them invalid to regulated parties.

Frequently, guidance documents and memoranda are among the most helpful agency documents in terms of explaining exactly how an agency intends to approach a certain piece of legislation or how the agency elects to interpret its authority, and in some cases how the agency will elect to limit its otherwise potentially broader authority to act. Interestingly, some EPA regional offices do not consider themselves bound by guidance documents or memoranda issuing from EPA headquarters. This is not a phenomenon that is customarily seen in the regional offices of the state environmental agencies, but EPA regional offices have been known to issue regional guidance documents or memoranda that depart from or in some cases differ from national guidance documents and memoranda.

10.1.3   *MAJOR ENVIRONMENTAL AGENCIES*

The preeminent environmental regulatory agency in the federal government is, of course, the United States Environmental Protection Agency (EPA). EPA administers the Clean Air Act; the Clean Water Act; the Resource Conservation and Recovery Act (RCRA), as amended by the Hazardous and Solid Waste Amendments of 1984 (HSWA); the Comprehensive Environmental Response Compensation and Liability Act (CERCLA), as amended by Superfund Amendment and Reauthorization Act (SARA); the Toxic Substances Control Act (TSCA); and the Federal Insecticide, Fungicide and Rodenticide Act (FIFRA). EPA is headquartered in Washington D.C. and has eleven regional offices.

Section 404 of the Clean Water Act provides for permitting of activities involving dredging or filling in waters of the United States. This includes the deposition of fill material in what are known as wetlands. The implementation of Section 404 involves agencies other than the EPA. The Corps of Engineers of the Department of the Army (Corps) is the primary permitting and enforcement agency for dredge or fill activities. EPA retains concurrent jurisdiction over all so-called 404 activities. EPA and the Corps of Engineers have separate but parallel regulations for the administration of the Section 404 wetlands program. These agencies operate under a Memorandum of Agreement to divide primary responsibility for the regulatory provisions. For instance, the Corps of Engineers refer interpretations of 404(f) exemptions directly to EPA. However, the permitting program, including the issuance of nationwide and after-the-fact permits, is initially within the province of the Corps of Engineers. The Corps of Engineers operates through district offices, which in the recent past were redrawn to conform to state boundaries.

Particularly in wetlands permitting issues, questions under the Endangered Species Act sometimes arise. The Corps of Engineers is required to consider the effect of its permitting activities on threatened or endangered species. The federal agency that promulgates rules designating threatened or endangered species and describing critical habitats for those species is the United States Fish and Wildlife Service (USF&W). The USF&W is not a regulatory agency in the strictest sense, and its dealings with individual permit applicants under the 404 program is indirect. The USF&W advises the Corps of Engineers with respect to endangered species impacts, and in some situations, the Corps of Engineers may be required to engage in what is known as a Section 7 consultation with the USF&W. Permit applicants in areas with populations of, for instance, the red cockaded woodpecker, have come to understand the role of the USF&W in environmental regulation.

Finally, the Council on Environmental Quality (CEQ) merits some mention. This body, appointed by and answering directly to the president, is the rule-making agency under the National Environmental Policy Act. This agency writes the rules for environmental impact statements and reviews statements prepared by federal agencies. The CEQ was extremely prominent in environmental issues in the early days of environmental regulation. Environmental impact statement questions arising under National Environmental Protection Act (NEPA) do not seem to be quite so newsworthy today, partly because the program has matured and federal agencies have a more complete understanding of what is required of them in terms of examining the environmental impacts of their activities that significantly affect the human environment.

10.1.4   FEDERAL LAWS

Statutes, regulations, or the common law can create liabilities and obligations. Federal law is the most publicized source of environmental liability. Federal law is important both because it is directly applicable to many situations and because many state laws are patterned after a federal law. However, state laws may be more important than federal laws to the day-to-day operation of most facilities. This section of the paper discusses federal laws, and subsequent sections address state laws and the common law.

Comprehensive Environmental Response, Compensation and Liability Act of 1980 (CERCLA or Superfund)

Congress enacted CERCLA in 1980, and subsequently amended it in 1986, in an effort to remedy the thousands of abandoned hazardous waste sites where the owners could not be located. Although the Resource Conservation and Recovery Act (RCRA) existed prior to 1980, RCRA was intended to provide for the remediation of hazardous waste on active, as opposed to abandoned, sites where responsible parties could be more easily located. Pursuant to the provisions of CERCLA, EPA must identify those sites requiring priority cleanup. The sites that pose the greatest danger to human health and the environment are placed on the National Priorities List (NPL), published as Appendix B to the National Contingency Plan.

CERCLA allows the federal government to force private parties to clean up or pay for the federal government to clean up waste sites that threaten human health or the environment. In 1991, the average Superfund cleanup cost approximately $25 million, excluding litigation costs. Currently, the average Superfund cleanup requires thirteen to fifteen years from listing on the NPL until completion of cleanup, although remediation of some sites may take as long as 41 years.

To be eligible for a full cleanup under the Superfund program, a site must be on the NPL. However, many aspects of the Superfund program are applicable to sites that have not been placed on the NPL. There are three ways that a site may be listed on the NPL: (1) a site assessment, (2) priority designation by a state, or (3) the Agency for Toxic Substances and Disease Registry issues a public health advisory recommending that individuals be isolated from the release, the EPA determines that the release poses a significant threat to human health or the environment, and remedial action would be more effective than a removal.

Under CERCLA, the government or private parties may perform cleanups and recover the cost from liable parties. A party is liable if: (1) there has been a release or threatened release (2) of hazardous substances (3) from a facility (4) that caused the incurring of response costs and (5) the defendant is a potentially responsible party (PRP) as defined by statute. With certain exceptions, a release is any "spilling, leaking, pumping, pouring, emitting, discharging, injecting, escaping, leaching, dumping, or disposing into the environment (including the abandonment or discarding of barrels, containers, and other closed receptacles containing any hazardous substance or pollutant or contaminant)."

The term *hazardous substance* includes those substances designated under Sections 307(a) or 311 of the Clean Water Act, any waste listed as a RCRA hazardous waste or that possesses a RCRA hazardous waste characteristic, any waste identified under Section 12 of the Clean Air Act, any substance on which EPA has taken action under the Toxic Substances Control Act, and any other substance that EPA has designated under Section 102 of CERCLA.

A *facility* is defines in CERCLA as any "building, structure, installation, equipment, pipe or pipeline (including any pipe into a sewer or publicly owned treatment works), well, pit, pond, lagoon, impoundment, ditch, landfill, storage container, motor vehicle, rolling stock or aircraft or any site or area where hazardous substances have been deposited, stored, disposed of, or placed or otherwise come to be located. 'The term does not include "any consumer product in consumer use or any vessel."'

Response costs are costs incurred because of removal or remedial action. *Removal action* is defined as actions that are "necessary to prevent, minimize, or mitigate damages to the public health or welfare as to the environment." *Remedial action* includes actions necessary to "prevent or minimize the release of hazardous substances so that they do not migrate to cause substantial danger to present or future public health or welfare or the environment." In order to recover response costs under CERCLA, removal or remedial actions must be consistent with the National Contingency plan.

CERCLA imposes strict liability, or liability regardless of fault, on four classes of PRPs:

- Present owners or operators of the facility where the release or threatened release occurs
- Past owners or operators of the facility at the time of the release
- Persons who arranged for disposal of the waste
- Persons who accepted hazardous substances for transportation to a waste disposal site

There is no exception for owners holding only legal title, such as trustees or executors. In addition, CERCLA allows joint and several liability. Courts have allowed PRPs to show, however, that apportionment of liability should be allowed based on relative responsibility. Specific CERCLA defenses are discussed below.

CERCLA Defenses

There are very few useful exceptions to CERCLA liability. CERCLA contains defenses for acts of God, acts of war, and acts of third parties, but these defenses are rarely asserted successfully. The following defenses are more likely to be of use to fiduciaries:

- Third party defense
- Innocent landowner defense
- Innocent inheritor defense
- Secured creditor defense

Resource Conservation and Recovery Act (RCRA)

The Solid Waste Disposal Act, as amended by the Resource Conservation and Recovery Act (RCRA), governs both "solid wastes" (under subtitle D) and "hazardous wastes" (under subtitle C). In addition, RCRA governs underground storage tanks (under subtitle I). *Hazardous wastes* are a subset of solid wastes. Both categories of wastes are defined at great length in the statute and regulation. In short, a substance is a hazardous waste if it satisfies each of the following four conditions:

- The substance must be a "waste."
- The substance must be a "solid waste" under the applicable statutes and rules.
- The substance must be defined as "hazardous" under the applicable statutes and rules.
- The substance must be "generated" or otherwise handled in a manner that causes it to become subject to the hazardous waste requirements.

These conditions are discussed in greater detail below. It should be noted that these conditions are definitely not self-explanatory. For example, a *solid waste* is defined to include gaseous and liquid materials.

RCRA and its implementing regulations detail a comprehensive "cradle to grave" system governing the generation, transport, storage, treatment, and final disposal of hazardous waste. The RCRA system is structured to allow individual states to assume principal enforcement responsibility under authorization and supervision from the federal EPA. EPA delegates enforcement authority to individual states on a case-by-case basis, and currently not all states are authorized to enforce RCRA. In these states, EPA operates the RCRA program directly. The federal statutes and regulations present baseline national enforcement standards for hazardous wastes, but these rules do not "preempt" the field. Each state (whether authorized to enforce RCRA or not) may adopt more stringent rules for hazardous wastes if it chooses to do so.

In addition to authorizing the regulation of hazardous wastes, RCRA also authorizes EPA to regulate underground storage tanks (USTs). UST regulations required owners of existing tanks to implement a method of corrosion protection and spill/overfill protection by December 1998. The deadlines for implementing a method of leak detection ranged from December 1989 to December 1993, depending on the age of the tank. Tank owners are required to report and investigate any suspected releases. If a release is confirmed, the tank owner must submit a corrective action plan and implement the corrective action plan. Finally, the UST regulations establish closure requirements that apply to any tank that has been out of service for more than 12 months.

10.1.5   DEFINITION OF WASTE

The bulk of RCRA regulates only *wastes*. The statute does not, as a general rule, regulate stored or transported raw (i.e., unused) materials or chemicals. The underground storage tank provisions are an exception to the general scope of RCRA.

Generally speaking, a *waste* is defined as any material that is discarded, abandoned, disposed of, spilled, or leaked, whether or not that activity occurs with the owner's knowledge or intent. The act of leaving raw industrial materials at an inactive manufacturing facility for more than 90 days constitutes "abandonment" sufficient to consider those materials as "waste" for purposes of RCRA. Similarly, any raw materials that have leaked or seeped from storage sites or containers also are wastes. Recycled materials that are used in a manner constituting disposal, burned for energy recovery, reclaimed, or accumulated speculatively also are wastes. Finally, some material has been designated by the EPA as "inherently waste-like" when certain criteria are met. Among the materials that have been declared inherently waste-like are chlorinated dioxins, dibenzofurans, and phenols.

10.1.6 DEFINITION OF SOLID WASTES

The basic operative phrase for determining the applicability of the hazardous waste statutes and rules is the phrase *solid waste*. For purposes of RCRA, *hazardous waste* is a subcategory of solid waste. Accordingly, hazardous wastes in liquid, gaseous, or solid form are all regulated by RCRA. If a certain material is not a solid waste for purposes of RCRA, then that material cannot be a hazardous waste under RCRA.

The phrase *solid waste* is defined in RCRA to mean: "any garbage, refuse, sludge from a waste treatment plant, water supply treatment plant, or air pollution control facility and other discarded material, including solid, liquid, semisolid, or contained gaseous material resulting from industrial, commercial, mining, and agricultural operations, and from community activities...."

Thus, "solid waste" is a misnomer, because it covers not only solid wastes, but liquid, semisolid, and gaseous matter in contained form.

The definition of *solid waste* in EPA's RCRA rules is much more structured than is the statutory definition. Under the rules, the phrase *solid waste* is defined as "any discarded material" that is neither excluded by rule or variance. This phrase *discarded material* is, in turn, explicitly defined in the rules, as discussed below.

10.1.7 DISCARDED MATERIALS

Under the RCRA rules, materials are *discarded* if they are:

- Abandoned
- Recycled
- "Inherently waste-like"

Materials are *abandoned* if they are:

- Disposed of
- Burned or incinerated
- Accumulated, stored, or treated

Materials are *recycled* if they are:

- Used in a manner constituting disposal
- Burned for energy recovery
- Reclaimed, or accumulated speculatively (i.e., with the hope of selling it as a product or co-product, rather than managing it as a waste)

Finally, materials are hazardous wastes if they are recycled and are "inherently waste-like." A material is "inherently waste-like" if it is normally disposed of, or if it contains toxic constituents, and EPA determines that the material poses a threat to human health or the environment when recycled.

10.1.8   *EXCLUSIONS FROM SOLID WASTE DEFINITION*

Some wastes are explicitly excluded from RCRA's definitions of solid wastes. These include:

- Domestic sewage and other wastes handled by publicly owned treatment works
- Effluent discharges regulated under the Clean Water Act
- Irrigation return flows
- Source special nuclear or by-product material as defined by the Atomic Energy Act of 1954
- Materials subjected to in-situ mining techniques that are not removed from the ground as part of the extraction process
- Pulping liquors reclaimed in a recovery furnace and reused in the pulping process (unless accumulated speculatively)
- Spent sulfuric acid used to produce virgin sulfuric acid (unless accumulated speculatively)
- Secondary materials reclaimed and returned to the original process in which they were generated (with certain restrictions)

RCRA does not regulate stored or transported raw (i.e., unused) materials or chemicals. Rather, RCRA regulates only wastes that are generated, stored, transported, treated, or disposed of. Variances from the "solid waste" classification can be obtained for some recyclable materials on a case-by-case basis. For example, if some recyclable materials are accumulated speculatively or are reused in the original process in which they were generated, an application for a variance from the "solid waste" classification may be submitted.

Residues from *empty containers* are not regulated. For these purposes, a container is deemed "empty" if it either: (1) has been rinsed out appropriately three times ("triple rinse"), or (2) contains less than one inch of hazardous waste material in the bottom.

10.1.9   *HAZARDOUS WASTES*

Under RCRA, a solid waste is considered *hazardous* one of two ways. A solid waste is hazardous by virtue of being specifically listed (a "listed hazardous waste"), or by virtue of its displaying one of four specific characteristics (a "characteristic hazardous waste"). Part 261 of the RCRA Rules are divided into four subparts to identify these listings and characteristics. As noted in 261.1(a), the listed wastes are contained in Subpart D and the four groups of characteristic hazardous wastes are contained in Subpart C. In keeping with EPA's practice of defining regulatory terms by listing exclusions, some solid wastes are excluded from the definition of *hazardous waste*. Those materials include:

- Household waste
- Solid waste generated by agricultural activity
- Mining overburden returned to the mine site
- Fly ash waste
- Bottom ash waste
- Slag waste
- Flue gas emission control waste from the burning of coal or other fossil fuels
- Drilling fluids
- Some wood or wood product wastes
- Certain wastes which fail the TCLP test because of chromium content alone
- Some wastes from mining ores and minerals
- Cement kiln dust waste

10.1.10 CHARACTERISTIC HAZARDOUS WASTES

The Four Characteristics

Presently there are four groups of characteristic hazardous wastes:

- Ignitable wastes
- Corrosive wastes
- Reactive wastes (i.e., that it is volatile or produces a violent chemical reaction when heated or mixed with water)
- Toxic wastes, as measured by the TCLP test

These characteristics were established based on criteria contained in 40 CFR 261.10, which indicate a possibility of increasing death or illness, or causing a substantial present or potential hazard to human health or the environment when improperly handled. Under 40 CFR 261.10(a)(2), these characteristics can be detected either through analytical testing, or "reasonably detected by generators of solid waste through their knowledge of their wastes." This latter provision is particularly important, because it allows industrial facilities to characterize unchanging waste streams based on initial, representative sampling.

10.1.11 THE TCLP TESTING RULES

On March 29, 1990, EPA replaced the "extraction procedure" (EP) test for waste analysis with the TCLP. The TCLP test is designed to more closely reproduce the leaching activity in landfills and provide an accurate basis for establishing regulatory levels for toxic constituents of hazardous wastes. In addition to the new testing procedure, EPA added 25 organic constituents, as well as keeping the 14 metals and pesticides that the EP test had tested for. The implementation of these new TCLP requirements has significantly increased the amount of materials that are considered hazardous. This TCLP is now being challenged in the courts. TCLP did not displace or supersede EP completely, however, if a material passes TCLP, but still fails EP, it is still considered hazardous.

10.1.12 FINANCIAL CONSIDERATIONS

On October 26, 1988, the EPA published the final financial responsibility requirements applicable to owners/operators of UST's containing petroleum. These requirements are not applicable to hazardous substance tanks. Under these requirements, owners/operators are required to maintain a certain amount of financial coverage to handle potential damages caused by a release from their UST systems. The amount required ranges from $500,000 to $2,000,000. These requirements are not applicable to tanks that have been closed in accordance with the Technical Standards. However, the regulations are silent as to whether tanks closed or abandoned improperly prior to December 22, 1988 require financial assurance.

The amount of coverage required is based on the number and types of tanks owned or operated by an entity or person. The EPA allows the owner/operator to choose from a number of acceptable mechanisms to provide this assurance, including surety bonds, letters of credit, guaranty, self-insurance, state funds or other state assurance, trust funds, liability insurance, or risk retention group coverage. A combination of these mechanisms may also be used as long as they provide the appropriate amount of assurance.

Documentation of financial assurance must be submitted after a suspected release occurs, when a provider becomes incapable of providing insurance, when a provider revokes a mechanism and the owner/operator is unable to obtain alternative coverage, or when requested by the implementing agency. Records of financial assurance must be maintained at the place of business.

10.1.13   CLEAN AIR ACT

The Clean Air Act provides the framework for federal and state regulation of air emission sources. The Act is divided into the following sections:

- National Ambient Air Quality Standards (NAAQS)—Section 109, 40, C.F.R. Part 50. (Title I, 1990 Amendments)
- State Implementation Plans (SIPs)—Section 110, 40 C.F.R. Part 51
- New Source Performance Standards (NSPS)—Section 111, 40 C.F.R. Part 60
- National Emission Standards for Hazardous Air Pollutants (NESHAPs)—Section 112, Air Toxics provisions—Title III of the 1990 Amendments, 40 C.F.R. Part 61
- Enforcement—Sections 113 and 120, Title VII of the 1990 Amendments, 40 C.F.R. Parts 65, 66 and 67
- Prevention of Significant Deterioration (PSD)—Sections 160–169, C.F.R. 9~ 5 1.166 and 52.21
- New Source Review in Non-Attainment Areas (NSR)—Sections 171–178, Title I of the 1990 Amendments, 40 C.F.R. ~~ 51.160, 52.24, and Part 51, Appendix S
- Emission Standards for Mobile Sources, Sections 202–234, Title II of the 1990 Amendments, 40 C.F.R. Parts 85, 86 and 87
- Permits, Title V of the 1990 Amendments
- Acid Rain, Title IV of the 1990 Amendments
- Stratospheric Ozone, Title VI of the 1990 Amendments

10.1.14   HISTORY OF THE ACT

The Air Quality Act of 1967

Prior to 1967, the U.S. Department of Health, Education and Welfare (HEW) had the authority to abate air pollution, but the process was cumbersome, consisting only of requiring conferences with company and state and federal officials. The 1967 Act established no clear definition of air pollution, but directed HEW to promulgate a list of air pollutants, to identify control technologies, and to publish air quality criteria that states would use to create air quality control regions. Enforcement was generally left to the states, or through conferences between federal authorities and industry.

Clean Air Amendments of 1970

The 1970 Amendments directed EPA to build on existing air quality criteria and to create national health-based ambient air quality standards. States were directed to achieve compliance with national ambient air quality standards (NAAQS) through SIPs, and EPA was authorized to promulgate NSPS and NESHAPs.

Clean Air Act Amendments of 1977

The 1977 Amendments basically retained the structure of 1970 amendments, but added the PSD permitting program for major new and modified sources in areas meeting NAAQS (attainment areas). The 1977 law also created the new source review-permitting program for new sources locating in areas not meeting NAAQS (nonattainment areas).

The Amendments also extended attainment deadlines to 1982 or to 1987 for nonattainment areas that could demonstrate serious smog problems. A number of the nonattainment provisions of the law were also strengthened. Nonattainment area SIP revisions were required to impose "reasonably available control technology" (RACT) on existing sources to demonstrate "reasonable further progress" toward attainment, and mobile source inspection and maintenance programs were added for states seeking to extend nonattainment deadlines to 1987. For areas that remained in nonattainment, major new construction projects were banned, and EPA was authorized to cut off federal highway funding to severely noncompliant nonattainment areas.

Clean Air Act Amendments of 1990

The 1990 Amendments include considerably more stringent control measures for nonattainment areas, mobile sources, and enforcement. New programs were included for air toxic emissions control, acid deposition control, permitting, and stratospheric ozone protection.

10.1.15 REGULATION OF EMISSION SOURCES: FRAMEWORK OF THE ACT—NATIONAL AMBIENT AIR QUALITY STANDARDS (NAAQS)

Introduction

Section 108(a)(l) requires that EPA list and revise the list of air pollutants that "may reasonably be anticipated to endanger public health or welfare." Once listed, EPA must prepare and revise periodically "criteria documents" for each pollutant. These are voluminous documents covering primarily health effects. Under Section 109, EPA must then establish NAAQS for each "criteria" pollutant.

Substance of Standards

Courts have interpreted NAAQS to be health based with no consideration of cost or technology. The standards are based on a concentration of pollutants in the ambient air averaged over several time periods.

Primary and Secondary Standards

Primary NAAQS standards protect public health; secondary NAAQS standards must, in addition to protecting public health, "protect the public welfare," which includes "effects on soils, water, crops, vegetation, manmade materials, animals, wildlife, weather, visibility, and climate, damage to and deterioration of property, and hazards to transportation, as well as effects on economic values and on personal comfort and well being." NAAQS have been established for the following:

- Sulfur dioxide
- Particulate matter
- Carbon monoxide
- Ozone
- Nitrogen dioxide
- Lead

State Implementation Plans

States must meet the standards through control measures contained in State Implementation Plans (SIPs). A SIP is a state plan for attainment of NAAQS. The State submits its SIP to EPA for approval. If EPA disapproves, the State must revise the SIP, or EPA must prepare a SIP for the State—known as a "federal implementation plan."

Each state must submit a SIP within nine months of promulgation or revision of NAAQS. EPA must approve the original SIP within four months, however, EPA is subject to no deadline for approval of SIP revisions. A SIP must provide for attainment of primary and secondary standards, and must contain enforceable emission limitations. Under the 1990 amendments, each state's SIP must also contain other control measures, means, or techniques (including economic incentives such as fees, marketable permits, and auctions of emissions rights), schedules, timetables for compliance and other measures, including transportation controls, air quality maintenance plans, and preconstruction review for new sources.

A SIP must contain enforcement provisions, and assurances that the State has the authority to implement the SIP, PSD, and nonattainment permitting programs. The SIP must also establish permitting fees to cover the cost of reviewing permits. Furthermore, each state's SIP must prohibit emissions from stationary sources that will prevent other states from meeting the NAAQS, and must provide for revision as necessary to meet revised NAAQS.

States may apply to EPA for authority to implement state programs to enforce the new source performance standards (NSPS) and national emissions standards for hazardous air pollutants (NESHAPs) regulations. EPA, however, retains concurrent authority to enforce both regulatory programs. Within nine months of enactment of the 1990 amendments, EPA must promulgate "completeness criteria." The agency must also act on all complete SIP submittals within one year. SIP revisions are not effective until approved by the agency, and EPA may conditionally approve a SIP revision, based on a state commitment to adopt specific enforceable measures by a date certain.

10.1.16 NEW SOURCE PERFORMANCE STANDARDS (NSPS)

NSPS are technology-based nationally uniform standards applicable to new, modified, or reconstructed sources, regardless of where the sources are located.

Substance of NSPS

NSPS have been established for nearly 60 categories of industries including fossil fuel fired steam generators, electric utility steam generating units, industrial/commercial and institutional steam generating units, municipal solid waste incinerators, petroleum refineries, sewage treatment plants, smelters, and many other facilities. The content of NSPS must be based on the "best technological system of continuous emission reduction which (taking into consideration the cost of achieving such emission reduction, any non-air quality health and environmental impact, and energy requirements) the Administrator has determined has been adequately demonstrated."

EPA must generally promulgate an emission limitation, and the source may choose whatever technology it desires to comply. The agency may, however, promulgate a design, equipment, work practice, or operational standard if it is not feasible to prescribe a numerical emission limitation. NSPS represents the minimum for the PSD "BACT" determination.

Applicability of the NSPS

The term *new source* means any source, the construction or modification of which is commenced after the publication of *proposed* regulations. Commencement of construction means that the owner or operator has either undertaken a continuous program of construction or modification or has entered into a contractual obligation to undertake and complete such a program. A *modification* is any physical change in, or change in the method of operation of, a stationary source that results in new or increased emissions of any air pollutant.

Permitting and Enforcement

States may apply to EPA to implement the NSPS for sources located in the state, and most states have done so through their SIP. EPA, however, retains authority to enforce the NSPS even in those states that have submitted a satisfactory enforcement plan. EPA may grant a waiver from NSPS for up to seven years to encourage innovative technology. EPA may also issue guidelines under Section 111(d) of the Act, directing the states to apply the NSPS for new facilities to existing facilities, for pollutants not regulated elsewhere under the Act, using a process similar to the SIP process. The Act allows a state to take into account, in addition to the factors listed above, the remaining useful life of the facility. States have nine months to implement the guidelines.

10.1.17 NATIONAL EMISSION STANDARDS FOR HAZARDOUS AIR POLLUTANTS OR "AIR TOXICS" (NESHAPs)

Introduction

Section 112 of the Act directs EPA to list hazardous air pollutants and set emission limitations for those pollutants. A hazardous air pollutant is one that may reasonably be anticipated to result in an increase in mortality or an increase in serious irreversible, or incapacitating irreversible, illness. EPA has established NESHAPs for eight pollutants: arsenic, asbestos, benzene, beryllium, mercury, radionuclides, radon-222, and vinyl chloride. EPA has announced its intent to regulate the following additional pollutants: 1,3-butadiene, cadmium, carbon tetrachloride, chloroform, chromium, ethylene dichloride, ethylene oxide, methylene chloride, perchloroethylene, and trichloroethylene.

Vinyl Chloride Standard

After the D.C. Circuit's opinion regarding the vinyl chloride NESHAP, EPA began to engage in a two-step process to set emission limitations: (1) determine what level is safe based solely on risks to health; (2) set the emission limitation at a level that guarantees an ample margin of safety taking into account health impacts, costs and feasibility. EPA now uses a "multifactor" risk assessment test.

Under the first step, EPA considers the estimated risk to the most exposed individual. If the risk is no higher than approximately 1 in 10,000, it is acceptable. The agency weighs this factor with other health measures, including the overall incidence of cancer within the exposed population, the policy assumptions and uncertainties associated with the risk measures, and the evidence that a pollutant is harmful to health.

Under the second step, EPA sets an emission limitation at a level that provides an ample margin of safety, considering all health, technological, and economic information. The benzene and radionuclides standards are currently being challenged in court. Although this standard-setting process will be given less emphasis under the 1990 Amendments, residual risk assessment under the 1990 Amendments will still require this two-step analysis.

Applicability of NESHAPs

NESHAPs apply to both existing and new sources. Standards apply to existing sources 90 days after promulgation, and to new sources upon promulgation. "New sources" are defined as stationary sources, the construction or modification of which is commenced after EPA proposes regulations. EPA may also promulgate design, equipment, work practice, and operational standards under Section 112 if it is not feasible to prescribe a numerical emission limitation.

Air Toxics Provisions of 1990 Amendments

Under the 1990 Amendments, EPA must promulgate technology-based emission standards for sources of approximately 200 hazardous air pollutants. Most of these pollutants are currently regulated under Section 313 of the Superfund Amendments and Reauthorization Act (SARA). Major stationary sources will be regulated: A major stationary source is defined as a source that emits IO tons per year (TPY) of any listed hazardous air pollutant, or 25 TPY for a combination of listed hazardous air pollutants. *Area sources* will also be regulated under the Act. The amendments include new provisions for the evaluation of residual risk, the control of accidental releases, and the regulation of emissions from municipal waste incinerators.

10.1.18   *PREVENTION OF SIGNIFICANT DETERIORATION (PSD)*

Introduction

EPA's PSD program is designed to preserve air quality in regions that meet the NAAQS (attainment areas), through preconstruction review and the application of Best Available Control Technology (BACT) requirements to the construction or modification of major sources.

Application of PSD Requirements

Preconstruction PSD permitting requirements apply to major stationary sources and major modifications in attainment areas. Major sources are those listed by EPA that emit or have the potential to emit 100 tons per year (TPY) of any air pollutant regulated under the Act, and any source with the potential to emit 250 TPY or more of any regulated pollutant. A *major modification* is a physical change in, or change in the method of operation of, a major stationary source that would result in a "significant net emissions increase" in any regulated pollutant.

Structure of PSD Requirements

PSD regulations apply to attainment areas; an area may be attainment for one criteria pollutant, but not for another. Attainment areas are designated under the Act as Class I, II, or III for purposes of determining the amount of "increment" that may be used up by a particular source. Class I (national parks and wilderness areas) only permit minor air quality deterioration. Class II designation allows for moderate deterioration and more growth and deterioration is allowed in Class III areas. An *increment* is an amount of deterioration allowable for each pollutant, measured as the maximum allowable increase from all sources above a baseline concentration before there is "significant" deterioration. The baseline is generally the air quality concentration existing on date of first major source permit application. Once increments are consumed, no further deterioration is allowed unless offsets are obtained.

No construction or major modification is allowed unless the source obtains a permit that establishes emission limitations that protect the PSD increment, and demonstrate compliance with NAAQS, NESHAPs, and NSPS. An application for a PSD permit must contain provisions for the development of monitoring data for each pollutant emitted in a "significant" amount.

Best Available Control Technology (BACT)

New sources, or modifications to major stationary sources, must apply BACT. BACT is defined as an emission limitation which EPA, taking into account energy, environmental, and economic impacts and other costs, determines is achievable. BACT must be at least as stringent as NSPS and NESHAPs. EPA establishes BACT through a procedure know as "top-down" analysis. First, the Agency determines the most stringent emission limitation currently in effect for a particular source category. If the PSD permit applicant does not desire to implement that emission limitation, it is the applicant's burden to demonstrate why extraordinary economic, technical, or environmental factors justify a less stringent emission limitation.

10.1.19 NONATTAINMENT AREAS

Introduction

The 1970 Amendments to the Clean Air Act set July 1975 as the deadline for attainment of NAAQS. However, by 1977, approximately 160 AQCRs were not in compliance with NAAQS for at least one pollutant. Under the 1977 Amendments, by July 1, 1979, states had to revise their SIPS to provide for attainment by December 31, 1982, or no new construction or modification of major sources would be permitted in nonattainment areas. Extensions were granted until December 31, 1987 for attainment for ozone and CO if certain conditions were met.

Nonattainment Area Requirements

Nonattainment area SIPS must include the following:

- Attainment of the NAAQS "as expeditiously as practicable" and, for primary standards, no later than December 31, 1982. Attainment of the ozone and CO NAAQS by December 31, 1987
- Implementation of all "reasonably available control measures" as expeditiously as practicable
- Reasonable further progress toward attainment of the NAAQS, including reductions in emissions from existing sources through the adoption, at a minimum, of reasonably available control technology (RACT)
- A comprehensive inventory of actual emissions from all sources in the area
- Preconstruction permitting requirements for new and modified sources. The new source or modification must utilize control technology in order to attain the lowest achievable emission rate (LAER). LAER is the more stringent of: (1) The most stringent limitation contained in a SIP for the source category, or (2) The most stringent limitation achieved in practice by such source category

Permitting requirements also include a demonstration that all major stationary sources owned or operated by the applicant in the state are in compliance with the SIP and other Clean Air Act requirements, and a demonstration of a reduction in emissions (through offsets) from existing sources in the area so that "reasonable further progress" will be made toward attainment of the NAAQS.

10.1.20 NONATTAINMENT PROVISIONS OF THE 1990 AMENDMENTS

The 1990 Amendments significantly change the regulation of nonattainment areas under the Act. The following new categories are established for ozone nonattainment areas: extreme, severe, serious, moderate, and marginal. Additional requirements are included for ozone, CO, and PM-10 nonattainment areas, including progress requirements toward achievement of NAAQS for ozone, ranging from 3 to 20 years.

The definition of a major source changes under the Amendments, and in some ozone nonattainment areas, sources that emit as little as 10 TPY are regulated. The 1990 Amendments also include new provisions regarding:

- Automotive tailpipe emissions and alternate fuels
- Federal Implementation Plans (FIPs) (mandatory if state fails to submit adequate SIP)
- Sanctions for failure to attain limited to a cutoff of highway funds, or a 2 to 1 offset requirement
- RACT is not restricted to existing CTGs
- Inclusion of "collar counties" in nonattainment areas

10.1.21 TOXIC SUBSTANCES CONTROL ACT (TSCA)

Introduction

The Toxic Substances Control Act (TSCA or the Act), enacted on January 1, 1977, gives the Administrator of the U.S. Environmental Protection Agency (EPA or the agency) broad regulatory authority to regulate commerce and protect human health and the environment by requiring the testing of, and imposing use restrictions on, certain chemical substances and mixtures, including new chemicals and PCBs. Under TSCA, EPA has the authority to identify potentially harmful chemical substances, to regulate the manufacture, import, and export of such substances, and to require industry to keep thorough records, submit reports to EPA, and conduct tests on chemicals of concern. The Act also provides EPA with authority to take appropriate control action to regulate a chemical that is determined to pose an "unreasonable risk" to human health and the environment. Such control actions include prohibitions on disposal, processing, manufacture, or distribution; limits on use; or requirements for specific labeling of certain chemical substances and mixtures. TSCA applies to all chemical substances except pesticides, tobacco, nuclear material, firearms and ammunition, food, food additives, drugs, and cosmetics.

Once having determined through testing required under TSCA that a chemical poses an "unreasonable risk" to human health and the environment, EPA has a number of regulatory options under Section 6(a) of the Act including:

- Prohibitions or limitations of the manufacture, processing, distribution or disposal of the chemical
- Prohibitions on the use of the chemical in excess of specified concentration limits
- Requirements for adequate warnings and instructions with respect to chemical use, distribution in commerce or disposal
- Requirements for record keeping
- Prohibitions or other forms of regulation on the disposal of a chemical
- Requirements to replace or repurchase a chemical substance or mixture or a requirement to notify the purchasers or the general public about the risks involved

EPA is statutorily required to choose the "least burdensome" of the regulations that is adequate to protect human health and the environment against the risk.

In addition, Section 6(b) provides EPA with the authority to regulate the quality control of chemical manufacturing. Section 6(b) provides the agency with the authority to review a manufacturer's quality control procedures. If these procedures are determined to be inadequate, EPA can order a manufacturer to modify its procedures, notify the public of the risk, and to replace or repurchase any substance produced under these procedures. While Section 6 provides the authority for EPA to actually regulate existing hazardous chemicals, the "existing chemicals program" under TSCA includes all TSCA activities dealing with existing chemicals, including premanufacture notification under Section 5, the testing program under Section 4, and information gathering activities under Section 8.

Although TSCA might appear to be of interest only to chemical manufacturers, the broad language of TSCA applies to virtually every type of business activity that uses chemicals. A manufacturer of office equipment, a large business printer, and an international communications and computer giant each recently paid over $1 million in fines for violating the Toxic Substance Control Act (TSCA). Many companies choose to ignore TSCA, believing that it does not apply to them or that EPA does not enforce TSCA vigorously. For a large and diverse segment of American industry, neither assumption holds true. More and more companies are discovering, in a most painful way, that many of the provisions of TSCA do indeed apply to their operations. In general, TSCA may regulate the manufacture, importation, processing, distribution in commerce, and use and disposal of chemical substances. Consequently, companies that engage in any of these activities, whether they make widgets or window cleaner, are subject to TSCA.

Because TSCA and EPA regulations define most key terms very broadly, TSCA requirements apply to many activities occurring as a normal part of producing hard goods and distributing products. For example, under current TSCA regulations, a company that buys a trade name solvent mixture directly from a foreign vendor is considered to be a manufacturer of every constituent in the product. Similarly, a company that transfers chemicals from its Canadian facility to its U.S. facility is considered to be a U.S. manufacturer of these products. Under the same logic, a company that brings waste back to the United States from its factory in Mexico is also considered to be a manufacturer of every chemical in the waste. Why? Because TSCA defines "import" to be synonymous with "manufacture." As a result, companies that import chemical products, whether as commodity feed stock chemicals or under conditions described above, must meet the same requirements as a traditional chemical manufacturer.

The story is similar for another major activity regulated by TSCA—chemical processing. Once again, TSCA not only regulates traditional chemical processing activities, such as formulating paint products, but also includes many activities where chemicals play a role in the manufacture of other kinds of products. For example, a company that assembles machinery receives drums from lubricants on its central warehouse. The company "processes" the lubricant when it repackages it into smaller containers for distribution along with its machinery. Similarly, a manufacturer of plastic articles becomes a "processor" when it melts plastic pellets and forms its products. An office products manufacturer is also a "processor" when it uses resins to coat the metal part of its products, or simply paints a desk. Companies engaging in these types of activities are chemical "processors" and subject to a number of regulatory requirements under TSCA.

The traditional term "manufacturer" is defined in a similarly broad manner, pulling into the fine net of regulatory control companies that simply use chemicals or manufacture or distribute chemical-containing products. As described above, manufacturer includes importing activities, as well as extraction, purification, and almost any reaction between chemicals, whether or not the chemicals formed are in products.

Once a company is caught in the myriad regulatory requirements of TSCA, EPA has authority to gather certain kinds of basic information on chemical risks from manufacturers and processors of chemicals, as well as to require specific companies to test selected existing chemicals for toxic effects, and to prevent unreasonable risks by regulating a chemical at any stage in its life cycle: the manufacturing, processing, distribution in commerce, use, or disposal. EPA may select from a broad range of control actions under TSCA to prevent unreasonable risks, ranging from requiring hazard warning labels to outright bans on the manufacture or use of especially hazardous chemicals.

In recent years, EPA has begun enforcing these requirements more vigorously. Substantial fines have been levied by EPA against diverse types of companies, including those in telecommunications and electronics, importation of office equipment, and manufacturing of products ranging from computers to airplanes. With potential penalties of $25,000 per violation for every day that the violation continues, it is easy to understand why TSCA fines in the million-dollar range are commonplace. Moreover, as onerous as TSCA penalties can be, companies often find the disruption of business to be more devastating than paying fines. How can a TSCA violation disrupt business? To reduce civil penalties and demonstrate good faith, as well as to avoid criminal prosecution, under EPA's penalty policy, the company must immediately cease the illegal activities. This can bring an important line of business to a halt. For instance, if you manufacture and sell chemical products, you may have to shut down production and cut off your customers. Consequently, developing and managing a program to ensure compliance with TSCA requirements is critical given the devastating effects TSCA violations may have on business operations.

This section provides a brief synopsis of key TSCA requirements. It is not intended to be comprehensive. For example, it does not discuss fine points of interpretation that may apply in specific factual situations. Its purpose is to address the broad scope and diverse types of regulatory programs encompassed by TSCA and to encourage companies affected by these requirements to analyze their potential exposure.

10.1.22 FEDERAL INSECTICIDE, FUNGICIDE AND RODENTICIDE ACT

The Federal Insecticide, Fungicide, and Rodenticide Act (FIFRA or the Act), requires all pesticides to be registered by the U.S. Environmental Protection Agency (EPA or the agency) and authorizes the Agency to regulate the conditions for use and importation of pesticides, the labeling of pesticides, and the level of pesticide residues in foods. FIFRA also authorizes EPA to classify pesticides (as "restricted" or "nonrestricted" and to prescribe standards for licensing and certification of pesticide applicators. A *pesticide* under FIFRA means: (1) any substance or mixture of substances intended for preventing, destroying, repelling, or mitigating any pest, and (2) any substance or mixture of substances intended for use as a plant regulator, defoliant, or desiccant. A *pest* under the Act means (1) any insect, rodent, nematode, fungus, weed, or (2) any other form of terrestrial or aquatic plant or animal life or virus, bacteria, or other microorganism (except viruses, bacteria, or other microorganisms on or in living humans or other living animals) that EPA declares to be a pest.

Registration

Section 3 of FIFRA makes it illegal for any person in any state to distribute or sell pesticides not registered under the Act. Registration is not required for pesticides used pursuant to an experimental use permit. Furthermore, EPA has exempted many biological control agents and compounds subject to regulation under the Federal Food, Drug and Cosmetic Act. EPA may register a pesticide if, when considered with restrictions placed on the use of the product,

- Its composition is such as to warrant the proposed claims for it.
- Its labeling and other material required to be submitted comply with the requirements of FIFRA.
- The pesticide will perform its intended function without causing unreasonable adverse effects on the environment.
- When used in accordance with widespread and commonly recognized practice it will not generally cause unreasonable adverse effects on the environment.

"Unreasonable adverse effects on the environment" are defined under the Act as "any unreasonable risk to man or the environment, taking into account the economic, social, and environmental costs and benefits of the use of any pesticide."

Section 18 of FIFRA allows EPA to waive certain provisions of FIFRA in the event of emergencies. An emergency usually consists of a major pest outbreak with the potential for significant economic damage or a requirement for the use of a pesticide to prevent the spread of disease. A "Section 18 Exemption" is usually requested by the state in which the emergency occurs, and may be granted by EPA if the agency determines that the emergency in fact exists, and that emergency use of the pesticide will not result in unreasonable adverse effects on the environment. A substantial amount of data must be submitted to EPA under Section 3(b)(2) of FIFRA in order to support the registration of a pesticide product. A pesticide that is not supported by the full range of data required may be conditionally registered by the agency until all data requirements are met. Data submitted to the agency in support of a pesticide registration application must demonstrate that the product is efficacious without resulting in unreasonable adverse effects on the environment. Regulations governing the submission of data to EPA in support of a pesticide registration are set forth at in 40 C.F.R. Part 158.

Data requirements include information regarding product chemistry, residue chemistry, environmental fate, toxicology, and toxicity to nontarget species and efficacy of the pesticide. Chemistry data must include information regarding the products chemical and physical properties, and include information concerning manufacturing, such as starting ingredients, intermediates, and residual impurities. Residue chemistry data is required to evaluate the residuals of the product that may be left on food items, and to assist EPA in the establishment of pesticide tolerance levels for each product. Before a pesticide may be registered for any food use, EPA must establish a tolerance for each of the pesticide's active and inert ingredients.

Environmental fate data submitted for each pesticide must include rates of product photodegradation, hydrolysis, and leaching potential. Toxicological testing must provide a full complement of data concerning potential routes and duration of exposure of humans to the pesticide, the target effects of the product, and potential acute and chronic effects on living organisms. Data regarding the effect of the pesticide on fish and wildlife species are also required. Testing to generate such data generally address reproductive effects, acute and chronic toxicity, and bioaccumulative properties of the pesticide. Finally, efficacy data must be submitted for many pesticidal products to enable the agency to evaluate claims made by the registrant.

An alternative to undergoing the expense of generating new data for unregistered products is for the registrant to cite existing data, which are applicable to the product for which registration is sought. Although data owners are provided 10 years of exclusive use of data they generate in support of first time registration, after 10 years, follow-on registrants may use these data providing they make an offer to pay compensation to the data owner for such use. Regulations set forth at 40 C.F.R. 152.80–152.99 govern data compensation under FIFRA. "Formulators"—companies that manufacture pesticide products using other products purchased from companies that hold registrations for those products—are exempt from data requirements regarding the products purchased from, and registered by, the other companies. However, more limited data requirements are still imposed to address characteristics unique to the formulated product.

In 1988, Congress amended FIFRA to require EPA to establish a comprehensive reregistration program under the Act. The purpose of reregistration is to reexamine older pesticides under current FIFRA data requirements and updated scientific standards. Under the 1988 amendments, EPA is required to reregister all currently registered pesticides by the end of 1997. Under this program, registrants are required to submit significant amounts of new data on currently registered products. Substantial reregistration fees, imposed on companies holding these registrations, finances the agency's reregistration program.

Labeling

An application for the registration of a pesticide must include all proposed labeling. EPA regulations found at 40 C.F.R. 156.10 require that all pesticide labeling contain:

- The name, brand or trademark under which the product is sold
- The name and address of the producer, registrant or person for whom the product is produced
- The net contents of the pesticide container
- The EPA pesticide product registration number
- The establishment that produced the pesticide
- A statement of the ingredients in the pesticide
- Warning or precautionary statements
- Complete directions for use
- The use classification of the product

Specific print requirements and signal words are also imposed by these regulations.

Cancellation of Pesticide Registrations

Section 6 of FIFRA authorizes EPA to cancel the registration of any pesticide that "generally causes unreasonable adverse effects on the environment." However, before EPA may cancel the registration or restrict the use of a registered pesticide, the agency must balance the benefits of the continued use of the product with the risk posed by the use of the pesticide in its determination of "unreasonable adverse effect." EPA balances the risk and benefits of a product proposed for cancellation through a regulatory process known as "Special Review."

The risk criteria for the initiation of a Special Review are set forth at 40 C.F.R. 154.7(a). EPA may initiate a Special Review of a pesticide product if the agency finds that the use of the pesticide (taking into account the ingredients, impurities, metabolites, and degradation products of the pesticide):

- May pose a risk of serious acute injury to humans or domestic animals
- May pose a risk of inducing in humans an oncogenic, heritable genetic, teratogenic, fetotoxic, reproductive effect, or a chronic or delayed toxic effect, which risk is of concern in terms of either the degree of risk to individual humans or the number of humans at some risk, based upon effects demonstrated in humans or experimental animals, known or predicted levels of exposure of various groups of humans, or the use of appropriate methods of evaluating data and relating such data to human risk
- May result in residues in the environment of non target organisms at levels which equal or exceed concentrations acutely or chronically toxic to such organisms, or at levels which produce adverse reproductive effects in such organisms, as determined from tests conducted on representative species or from other appropriate data
- May pose a risk to the continued existence of any endangered or threatened species
- May result in the destruction or other adverse modification of any habitat designated as a critical habitat for any endangered or threatened species
- May otherwise pose a risk to humans or to the environment which is of sufficient magnitude to merit a determination whether the use of the pesticide product offers offsetting social, economic, and environmental benefits that justify initial or continued registration

Special review procedures include a preliminary notification to registrants, notice of special review (subject to public notice and comment), notice of preliminary determination (subject to public notice and comment), and notice of final determination. During this process, EPA is required by FIFRA to solicit the views of the U.S. Department of Agriculture and EPA's FIFRA Scientific Advisory Panel. If the Agency issues a notice of cancellation under Section 6(B) of FIFRA, based upon the information generated during Special Review, he registrant and other adversely affected parties may request an administrative hearing. The decision of the administrative law judge may be appealed to the Administrator of EPA, and the final decision of the Administrator may be appealed to the appropriate United States court of appeals.

If at any time EPA determines that a pesticide presents an imminent hazard to human health or the environment, the Agency may suspend the registration of the pesticide under Section 6(C) of FIFRA. Registrants are provided the opportunity for an expedited hearing on the issue of whether an imminent hazard does, in fact, exist. If EPA determines that the hazard posed by the pesticide presents an emergency, the Agency may suspend a registration before any hearing is held.

10.1.23   *SAFE DRINKING WATER ACT*

The Safe Drinking Water Act gives EPA the authority to compel the cleanup of a contaminant in a public water system. However, only persons responsible for creating the danger or threat can be compelled to abate the danger or threat. Thus, in most cases, EPA's authority under the Safe Drinking Water Act can not be used to impose liability on an innocent purchaser.

The Safe Drinking Water Act can present a problem for companies that do not rely on a public water system, but have their own drinking water supply. The regulations adopted pursuant to the Safe Drinking Water Act contain registration and monitoring requirements that vary depending upon the number of individuals served by the water system.

10.1.24 EMERGENCY PLANNING AND COMMUNITY RIGHT-TO-KNOW ACT "SARA TITLE III"

Overview

The Emergency Planning and Community Right-To-Know Act (EPCRA) is set forth as Title III of the Superfund Amendments and Reauthorization Act. EPCRA was enacted following the fatal release of methyl isocyanate at the Union Carbide plant in Bhopal, India, and the similar release six months later at a Union Carbide plant in West Virginia. The incidents made Congress acutely aware of the need for an informed and planned response to releases of hazardous chemicals, as well as a need for increased awareness in the community and among emergency personnel regarding the amounts of, and potential risks associated with, hazardous chemicals in their locality.

EPCRA uses the OSHA HCS as a framework and extends the information requirements thereof to the community surrounding any facility that stores, produces, or uses hazardous chemicals. Because of this interrelationship between the programs, many of the tools used in the Hazard Communication program are also used in the Right-To-Know program. The Right-To-Know program in EPCRA covers four major areas. The emergency planning requirements of Sections 301 through 303 place certain duties on state and local authorities and businesses to work together to establish a local emergency response plan. Section 304 requires any facility with a release of a hazardous substance that exceeds a specified reportable quantity to notify the appropriate state or local authority. Sections 311 and 312 contain the Community Right-To-Know reporting requirements, which include a two-tiered format for the submission of chemical information. Finally, Section 313 (which applies only to manufacturers) establishes a requirement for reporting annual quantities of toxic chemicals that are released. EPCRA does not give the public a general right to all information concerning a facility's operation, only to information concerning hazardous materials, and EPCRA does not grant the public a right to request any information directly from a company.

10.1.25 NATIONAL ENVIRONMENTAL POLICY ACT (NEPA)

Commentators have characterized The National Environmental Policy Act, as both the "Sherman Act" of environmental law for its scope and vision, and a "paper tiger" lacking any real teeth. Its purpose is to promote an environmental consciousness within agency policy by requiring that in every recommendation or report on proposals for legislation and other major federal actions significantly affecting the quality of the human environment, the responsible agency must include a detailed statement on the environmental impact of the proposed action.

One way of characterizing the purpose of the Act is to say that Congress wanted to inject into all decision-making processes a consideration of the environmental ramifications of government activities, thus institutionalizing the national intent to better care for our environment. Considering the nature of the bureaucratic animal, this was quite an undertaking. Therefore, while the above may not overstate the Act's purposes, to maintain that this has been its effect would meet with considerable argument.

Section 102 of the Act sets out what is expected of the agencies. Each agency is required, among other things, to "include in every recommendation or report on proposals for legislation" and other Federal actions affecting the environment, a "detailed statement" on the following:

- The *environmental impact* of the proposed action
- Any unavoidable *adverse* environmental *effects*

- *Alternatives* to the proposed action
- "The relationship between short-term uses of man's environment and the maintenance and enhancement of long-term productivity" (a "*cost/benefit*" analysis)
- "Any *irreversible* and *irretrievable* commitments of resources" if the action is implemented

The Federal official charged with the responsibility of preparing the statement must first consult with, and obtain comments from, any federal agency with jurisdiction or special expertise with respect to any environmental impact involved. A major function of NEPA is to require a circulation of this "environmental impact statement" (EIS) among interested agencies, federal, state, and local alike, to ensure that interested or affected parties are given an opportunity to participate in the process. The statement must accompany the proposal throughout the extent of the review processes.

The required agency efforts are qualified somewhat, however, by language in Section 101(b) of the Act, which sets out its goals: "[I]t is the continuing policy of the Federal Government to use all practicable means, consistent with other essential considerations of national policy. . . ." The interpretation of this language is, of course, a continuing matter for dispute.

The EIS should present the environmental impacts of the alternatives, including the proposed activity in comparative form. To the extent possible, the comparison of alternatives should quantify how the purpose and need would be satisfied by each alternative and the proposed activity. The EIS should explore and evaluate all reasonable alternatives including those not within the jurisdiction of the agency, and including the "no-action" alternative. The statement should discuss the reasons for the elimination of any alternatives from detailed study. The statement generally should identify the agency's preferred alternatives(s) in the draft and final statement, unless another law prohibits the expression of such a preference. The EIS should describe appropriate mitigation measures not already included in the alternatives.

The statement is to assess, and, where feasible, quantify, the social and economic impacts of each alternative. The EIS is required to describe the environment of the areas to be affected and the environment to be created by the alternatives under consideration. The section in the EIS on environmental consequences should measure direct effects and significance; indirect effects and significance; and possible conflicts between the proposed activities and the objectives of federal, state, and local plans, policies, and controls for the affected area.

In addition to the "call to action" provisions, as they are known, NEPA created the Council on Environmental Quality (Council). Whether out of design or not, the Council has adopted as its mandate the responsibility to investigate the quality of the environment, to develop policies on environmental quality, and to review government activities in the fulfillment of the goals of NEPA. As a result, it has become the central coordinator and developer of NEPA, a watchdog to Congress, and "presidential advisor."

The Council has published and periodically revised the "Guidelines for Preparation of Environmental Impact Statements." These guidelines establish the "hows," "whys," "whens," and "by whoms" of the environmental impact statement process. In addition, it requires each agency to develop an "early notice system" for informing the public on any decision to prepare an EIS and for the soliciting of public comment. Each agency is also required to maintain and make available for public review: (1) a list of administrative actions where an EIS is being prepared, (2) a record of the reasons for not preparing an EIS for a given action, and (3) a comprehensive record of "underlying documents" and comments received on the environmental impact of the proposed action. The guidelines also address public hearing procedures. The Council itself also has reporting duties. To provide the public with information and notice, it is required to publish weekly in the Federal Register a list of the previous week's environmental statements. Expectedly, the agency also has a hand in monitoring compliance with NEPA. Its role is beyond the scope of this paper, but it should be noted that the EPA often provides its expertise as needed, and acts primarily as an alarmist and trouble-shooter.

The Council itself must meet certain standards. In addressing an audience of laypersons, governmental officials, and technical experts, it is understandably a challenging task to prepare, independent of the problems associated with acquiring the necessary data, comment, and other information. The substance of the statement must also adequately address each of the criteria of Section 102(C).

Predictably, alleged inadequacies or inaccuracies of the EIS are frequently the subject of judicial review of NEPA. It must meet the requirements of NEPA standing alone, otherwise it cannot fulfill the intent of the act to inform. Also, judicial review frequently centers on NEPA's procedural requirements, which are strictly enforced, fulfilling the Act's further intent to supplement the agency decision-making methodology.

10.1.26 OCCUPATIONAL SAFETY AND HEALTH ACT (OSHA)

Overview

The Occupational Safety and Health Act (OSHA), places the responsibility for implementing employee right-to-know standards squarely on the shoulders of employers. OSHA's regulations, depending upon the nature and size of a business, may cause anything from minor annoyance to substantial extra effort and expense. Manufacturers and distributors must (1) evaluate hazards of chemical products; (2) provide warning labels on all containers; and (3) issue material safety data sheets (MSDS) with their products to purchasers in manufacturing industries. Employers in those industries must (1) label in-plant containers; (2) inform workers of hazards within work areas; (3) make MSDSs or similar information available to employees; (4) train workers to protect themselves from certain chemical hazards; and (5) develop written plans for implementing the hazard communication program. Substances exempted from OSHA coverage include manufactured articles that do not result in exposure to hazardous chemicals, hazardous wastes, lumber, tobacco and drugs, and food or cosmetics brought into the workplace by employees for their own personal use.

State and Local Laws

State and local laws are an increasingly important source of liability, which is easily overlooked. It is difficult to summarize state laws that can create environmental liability because they differ from state to state. However, the following discussion provides an overview of different types of state laws that can be a source of liability.

"Mini-Superfund" Laws

Almost every state has a law imposing liability for cleanup costs on parties responsible for generating or disposing of hazardous substances. In general, the "mini-Superfund" laws are similar in scope and effect to CERCLA. However, some state laws exempt subsequent "innocent" purchasers, while other laws do not, and still other laws are unclear. Connecticut specifically limits the liability of lenders who acquire title by foreclosure to the value of the property. Consultants should be familiar with the laws applicable in the jurisdiction where the property that is the subject of a site assessment is located.

Land Transfer Laws

At least four states currently have laws requiring the disclosure of environmental information in connection with the transfer of property. New Jersey's Environmental Cleanup Responsibility Act (ECRA) and Connecticut's land transfer law require site assessments prior to any sale of or change in control over certain types of property where hazardous substances were used or stored. Illinois' Responsible Party Transfer Act (RPTA) and Indiana's Environmental Hazardous disclosure and Responsible Party Transfer Law require a transferor to disclose information about potential sources of environmental liability to the other parties to the transaction, including lenders, but does not require a site assessment by an environmental consultant. These laws apply to a broad range of transactions extending beyond simple real estate transactions.

ECRA purports to apply to a financial reorganization of any corporation owning covered property in New Jersey. Failure to comply with the provisions of these laws can result in fines, liability for the costs of an investigation by the state, and in the rescission of the transaction. Unless otherwise specified by the parties, the seller is responsible for the costs of complying with ECRA. If a seller refuses to comply with ECRA, a purchaser is entitled to an order requiring completion of the sale and compliance with ECRA.

Lien and Superlien Laws

Several states have superlien statutes that give state environmental authorities a lien on contaminated property that is prior to all other liens. In general, superlien laws allow a state to impose a lien on property that is cleaned up by the State for the cost of cleaning up contaminated property. Such a lien is given priority over all other liens, including previously recorded liens, such as a first mortgage. In some cases, the superlien extends to personal property located on the site where the cleanup occurred. Many states have lien statutes that give a lesser degree prior to the state's lien than the superlien statutes.

Until the courts in each state rule on the validity of the "superlien" statutes, there is some question about the validity of the superlien provisions. The New Jersey courts upheld the constitutionality of New Jersey's superlien statute, but there are no reported cases in other jurisdictions ruling on the constitutionality of superlien laws. Arguably, superlien statutes that deprive secured creditors of their priority violate the "takings" clause of the Fifth Amendment. In addition, superlien statutes may be invalid to the extent that they defeat or conflict with overriding federal policies, such as the Bankruptcy Code.

Underground Storage Tank Laws

Most states have laws implementing or supplementing the federal underground storage tank (UST) regulations. Some states, such as Florida, have UST regulations which are more restrictive than the federal UST regulations. On the other hand, most states have laws that, in essence, provide cleanup liability insurance for owners of underground tanks. However, the practical effect of such laws must be examined carefully. The law, which insures tank owners for cleanup costs, may not negate other laws that allow the state to recover funds spent on cleanups from the owner. Furthermore, in many cases states have not allocated sufficient funds to pay for cleanups. The UST programs in North Carolina and South Carolina are discussed in more detail below.

Notification Laws

At least seven states have laws that require owners of property where hazardous wastes have been used or stored to give some form of notice prior to a sale of property to buyers, the public and/or state officials. At least 16 states have laws or regulations requiring that owners of hazardous waste disposal facilities record a notice of the location of the facility with the deed. Failure to comply with these laws can subject the seller to fines, as well as civil liability to the purchaser.

10.1.27   WASTE MINIMIZATION PROGRAMS

States have addressed the issue of hazardous waste minimization by means of regulation, study commission, loan programs and other financial incentives, information clearinghouses, and educational programs, among others. Most states have indicated a preference for source reduction over recycling because reduction eliminates generation of potential hazardous waste rather than dealing with its treatment, disposal, or beneficial reuse. It is more likely, though not certain, that resources will be conserved by source reduction. It is prevention as opposed to management and therefore potentially simpler. In some states, the preference is stated as official policy.

Categories or Types of State Programs

Several categories of state responses can be identified. Ranking them in order of intensity of state involvement, they might be characterized as follows:

- Mandatory action
- Financial incentives and disincentives
- Conditional action
- Voluntary, educational, information

These categories are described further below.

Mandatory Action

Included in this category would be required assessment of potential source reduction possibilities in industrial processes, reduction standards and reporting requirements for progress and compliance, and other types of management of process standards. Oregon requires development of toxic use reduction and hazardous waste reduction plans for all facilities, and also requires periodic updates. Oregon is now requiring generators of Section 313 listed chemicals in excess of 10,000 pounds per year (or 2200 pounds per month) to submit plans for waste reduction. The plans must establish specific performance goals. The requirements became applicable to small quantity generators (220 to 2200 pounds per month) in 1992. Fees were assessed against generators to fund, in part, an expanded technical assistance program. Massachusetts has also recently adopted a mandatory reporting and planning program. For firms using more than 10,000 pounds (25,000 pounds for manufacturers and processors) of chemicals listed under Section 313 of SARA and at a later date, Sections 101(14) and 102 of CERCLA, respectively, annual reports detailing use of the chemicals were required detailing the usage of the substances. Beginning in 1994, generators were required to submit source reduction plans for those chemicals. A toxics use reduction institute was established at the University of Lowell to assist regulated firms, which will be assessed a fee to fund the institute. Similar programs have been proposed in numerous other states. States have enacted other measures encouraging good faith efforts to investigate source reduction possibilities and to report to the state the results of those efforts. Indiana and Montana recently enacted mandatory good-faith-effort reporting requirements, which would fall in this category. Montana's rule applies to producers of chlorinated solvents in concentrations of 20 percent or more. No state has yet gone to more intrusive reduction, management, or process standards, although fixed percentage reduction requirements have been debated. Several states, however, are considering requiring source reduction inventories. Among these are New Jersey, Illinois and California.

Financial Incentives and Disincentives

These programs may take the form of favorable tax treatment, grants, or loans. On the other hand, they might be fees or surtaxes. Added to the savings that might be realized in raw material cost, these devices can carry considerable persuasion. Grants or low interest loans may also mitigate the risk inherent in any process or raw material changes. Numerous states offer grants and/or loans, including Connecticut, California, New York, Pennsylvania, Tennessee, Wisconsin, Illinois, Massachusetts, Michigan, and Minnesota. Oklahoma, as an example, offers tax credits and other incentives in connection with the purchase of waste minimization equipment.

Conditional Action

This category may include a wide range of state programs. Notable among these are programs that condition any landfill disposal with a prerequisite showing of exhaustion of minimization alternatives. Illinois has adopted such a requirement and a requirement is also now in place in South Carolina for disposal of wastes at the GSX landfill. Ohio requires generators who propose to dispose of more than 200 tons per year of hazardous waste in a landfill to prepare waste minimization plans.

Voluntary, Educational, Informational

This category applies to virtually all states with any sort of waste minimization legislation. Programs include clearinghouses, technology transfer programs, seminars, technical assistance, study commissions, university research grants, research programs, and training. Some of those programs are funded by fees assessed against generators. In South Carolina, fees are assessed against persons disposing of wastes in the State, which go in part to a waste minimization and reduction research fund.

10.1.28 MODEL STATE WASTE MINIMIZATION LEGISLATION

The Suggested State Legislation Committee of the Council of State Governments, with the assistance of EPA and the Commonwealth of Kentucky, has created model hazardous waste reduction legislation. The model act has three parts: (1) establishment of a technical assistance center, (2) requirement of waste reduction plans, and (3) a fee schedule and assessment against regulated generators with revenues dedicated to funding the waste reduction programs. The bill distinguishes between "fully regulated generators" (generating greater than 2200 pounds of hazardous waste, or greater than 2.2 pounds of actual hazardous waste, per month) and "small quantity generators" (generating between 220 and 2200 pounds of hazardous waste per month). A hierarchical scheme is adopted, with primary emphasis on source reduction, followed, respectively, by recovery and reuse, recycling, volume and toxicity reduction (including incineration), storage, and as a "last resort," disposal.

The technical assistance program has several different components. The focus is establishment of a Center for Hazardous Waste Reduction, to function as a clearinghouse for information, and to perform the following tasks: (1) maintain a registry of expert consultants and researchers; (2) sponsor and conduct conferences and workshops; (3) develop model plans, procedures, and protocols for waste reduction and waste audits; (4) promote technology transfer; (5) conduct feasibility analyses for innovative technologies, and (6) provide on-site assistance for waste audits and training. The center would also administer loan, subsidy, and grant programs for the purpose of promoting research, development of innovative technologies, and to generally encourage development of source reduction techniques and procedures over a wide range of industries. It would also have a role in interstate or regional networks, identifying institutional or governmental impediments to source reduction, and developing training programs and participating in public relations programs to promote interest in source reduction.

The model legislation requires submittal by fully regulated generators of a hazardous waste reduction plan setting forth specific performance goals. Performance goals are required to address each hazardous waste representing 10 percent or more by weight of the cumulative hazardous waste stream generated per year. The goals are to be expressed in numeric terms, but where that is not practical, they must include a clearly stated list of objectives designed to lead to establishment of numeric goals as soon as possible. The plan must explain the rationale for each of the performance goals, addressing impediments such as previous progress toward the outer bounds of technological feasibility, economic practicality, and significant adverse impacts on product quality. The plans are to be submitted to the state, and are then reviewed for adequacy and completeness against guidelines provided in the bill. Plans that are found to be deficient are returned for modification within a specified time frame. Small quantity generators are also required by the bill to develop plans, for submittal at a later date. Each year following approval by the state, generators are required to prepare annual progress reports. A summary of the quantities of waste generated is required to be filed with the state annually. This Annual Generator Report is a public record, but neither the hazardous waste reduction plan nor the annual progress report is required to be available for public inspection or considered to be a public record. The Act also imposes a general statutory duty on owners and operators to identify hazards resulting from accidental releases, to minimize the consequences of such releases, and to operate safely. The Act authorizes EPA to promulgate regulations to prevent, detect, and correct releases.

10.1.29 WATERSHED PROTECTION LAWS

Many states have adopted laws and regulations to protect the quality of watersheds for drinking water supplies. These watershed protection laws can severely limit allowable development on property located in a critical watershed. Examples from two states are included below as examples of watershed protection laws.

North Carolina

North Carolina has recently adopted rules for the protection of water supply watersheds pursuant to the direction of General Assembly, which in 1989 added a new section to Chapter 143 of the North Carolina General Statutes. This legislation, codified at N.C. Gen. Stat. 143-215.5, requires the Environmental Management Commission to adopt water supply watershed classifications and management requirements, and requires local governments to adopt water supply watershed protection programs. The rules create four watershed water supply classifications and designate land use thresholds, density restrictions, and other use restrictions. Within each watershed there is designated a critical area. The critical area is that area within one mile of the water supply or the area to the ridgeline draining to the water supply, if it is less than one mile. Critical areas are subject to more stringent restrictions than the remainder of the watershed. Density restrictions range from no allowable development in the highest classification watershed to two dwelling units per acre or 24 percent built upon area for the protected area of the least protected watershed classification. There is a high-density option for the lower three classifications, which allows a larger percentage of built upon area in exchange for the installation of controls adequate to contain the first inch of storm water. The classifications restrict location of new industrial uses in portions of the watersheds. The scheme of the legislation provides for local government enforcement of restrictions and density limitations through zoning or other local ordinances. These ordinances must meet certain state minimums, which will be set out by the Environmental Management Commission.

South Carolina

Watershed protection in South Carolina is accomplished more indirectly than in North Carolina under legislation for water supply watershed protection. The South Carolina Pollution Control Act provides broadly that it shall be unlawful for any person directly or indirectly, to throw, drain, run, allow, seep, or otherwise discharge to the environment of the State organic or inorganic matter, including sewage, industrial waste, and other wastes except as in compliance with the permit issued by the department. This broad authority coupled with legislation creating soil and water conversation districts that have the authority, following referenda, to formulate regulations governing the use of lands within the districts. The regulation may relate to conservation of soil and soil resources and to prevention and control soil erosion. County sediment control programs also regulate to some degree protection of water supply watersheds in South Carolina. Obvious difficulties and shortcomings with respect to administrative efficiency are apparent because no fewer than three agencies are involved.

10.1.30   GROUNDWATER PROTECTION LAWS

Every state has some type of law regulating groundwater quality. This is not surprising because the majority of the population of the United States depends upon groundwater for its source of drinking water. In many states, groundwater regulations are the primary mechanism used to require groundwater remediation activities. Thus, these laws and regulations are extremely important in determining whether a cleanup is necessary and the cost of any such cleanup. Any environmental consultant who analyzes the groundwater at a site in connection with an environmental investigation should be very familiar with the applicable groundwater standards. Examples of state groundwater protection schemes are discussed below.

North Carolina Groundwater Quality Standards

Under the authority of the North Carolina Water and Air Resources Act, the North Carolina Department of Environmental Management (DEM) has adopted water quality standards applicable to the groundwaters of the state. With the groundwater rules, DEM adopted a number of important policies to guide it in establishing the groundwater standards and classification system. To effect the preservation of groundwater quality, it is DEM's stated policy that:

The best use of groundwater is as a source of drinking water;
   Groundwaters of the state are generally a potable source of drinking water without the necessity of treatment;
   DEM will not issue a permit to any facility which would significantly degrade the quality of groundwater, result in violation of quality standards beyond the boundaries of the facility's property or cause an adverse impact on public health; no person shall conduct any activity which results in a violation of the groundwater standards.

With these policies in mind, DEM has promulgated regulations that establish a system of groundwater classification, quality standards, and enforcement. These regulations are applicable to all waters in the saturated zones of the earth.

10.1.31   GROUNDWATER CLASSIFICATION

All groundwaters containing 250 mg/l or less of chloride are classified as "GA Waters," while those containing greater than 250 mg/l chloride are classified GSA. "GC Waters" include those waters that do not meet the quality standards of higher classifications, cannot easily be remediated to meet those higher classification standards, and have a best use other than for drinking water.

10.1.32   QUALITY STANDARDS

The groundwater standards promulgated by DEM currently include maximum allowable concentrations for 72 listed chemicals and metals. In addition, the groundwater rules provide that "substances which are not naturally occurring and for which no standard is specified shall not be permitted in detectable concentrations in class GA or class GSA groundwaters." The nondetectability standard may also be relevant to some of the specific maximum allowable concentrations "where the maximum allowable concentration of a substance is less than the limit of detectability." In such a case, the listed chemical or metal is not permitted in detectable concentrations.
   This nondetectability standard may present some practicable difficulty in compliance and enforcement. The detectability standard of many chemicals will vary significantly between different

laboratories and different analytical procedures. Despite this consideration, the nondetectability standard will continue to be enforced for many chemicals until additional numerical standards are promulgated by DEM.

The established standards are also applicable to class GC waters except that, for those substances that prevent the GA or GSA classification, the concentration "shall not be permitted to increase." The effect of this provision is to impose a "zero standard" for particular chemicals or metals that are already present in class GC groundwaters.

10.1.33 ENFORCEMENT

Corrective Action

The groundwater rules authorize DEM to order any person to take corrective action when that person controls any activity which: (1) results in the discharge of waste, hazardous substances, or oil into the groundwater (regardless of whether the groundwater standards are violated); or (2) results in a violation of any groundwater standard. When corrective action is necessary because groundwater standards have been exceeded, the particular action that is required depends on whether the activity causing the violation is permitted and where, in relation to permitted facilities, the violation occurs.

Notification

Any person who controls an activity resulting in the discharge of waste, a hazardous substance or oil into the groundwater must notify DEM of the discharge. The term *discharge* is not defined in the groundwater rules or the statutory authorization. However, the North Carolina Oil Pollution and Hazardous Substance Control Act (NCOPHSCA) requires notification to DEM if a "reportable quantity" of oil or hazardous substance is discharged. The regulations promulgated under NCOPHSCA have established that a "reportable quantity" for the discharge of oil is any amount that produces a sheen on the surface of the water.

Monitoring and Reporting

Any person who causes a discharge of waste may be required to install a monitoring system in order to determine the effects of the discharge on the groundwater and may also be required to report the results of the monitoring, including sampling procedures and all technical data to DEM.

Defenses

The groundwater rules are applicable to all persons except "an innocent land owner who is a bona fide purchaser of property which contains a source of groundwater contamination, who purchased such property without knowledge or a reasonable basis for knowing that groundwater contamination had occurred, or a person whose interest or ownership in the property is based or derived from a security interest in the property."

In addition, groundwater standard violations that are caused by permitted groundwater withdrawals are exempt from the corrective action provisions of the groundwater rules. Violation of groundwater standards that are caused by pesticides or chemicals used in agricultural activities are also exempt from the corrective action provisions. Any person who has caused a violation of the groundwater standards may apply to DEM for a variance in those standards. The groundwater rules provide a procedure by which a variance may be granted and the party requesting the variance has the burden to demonstrate the appropriateness of the proposed alternative standard.

10.1.34   WETLAND AND FLOODPLAIN REGULATIONS

On the state level, wetland programs may be undertaken by the state in an approval process not dissimilar to the federal wetland program. However, very few states have undertaken EPA's offer to have wetland permitting programs approved in their states. Nevertheless, there is a state role that takes the form of programs wholly under state authority parallel to the 404 program or that may simply be a function of the required certification by the state under Section 401 of the Clean Water Act. Under Section 401 a state certifies that issuance of a permit will not result in violation of water quality standards. An example of state regulation of wetlands is provided below.

North Carolina

The State of North Carolina has approached wetlands regulation in two ways. The first is under the Coastal Area Management Program where coastal wetlands are defined as "areas of environmental concern." *Coastal wetlands* are identified as salt marshes or other marshes subject to regular or occasional flooding by tides and contain some of several different varieties of marsh plant species. Any person wishing to undertake development, or engage in dredge and fill activity in an area of environmental concern must first obtain a permit from the department. The exception is in the case of a minor development permit, which does not include dredge and fill activity, such permit is obtained from a local permit officer. Major development permits or dredge and fill permits under the Coastal Area Management Act are supplementary to and neither supersede nor are they superseded by a Section 404 permit for the same project.

Secondly, the State of North Carolina undertakes wetland regulation under its *401 Certification Program.* Under the Clean Water Act, no federal permit that may result in a discharge into navigable waters may issue until the applicant obtains, or the state waives, certification from the state that the discharge will comply with water quality standards. Procedures for obtaining such certification in North Carolina are contained at 15A N.C.A.C. 2H .0500. Because North Carolina is an approved state for the purpose of NPDES, the most common usage of this certification requirement is in the proposed issuance of 404 permits. The North Carolina 401 procedure involves application to the state describing the activity, giving its location, the nature of the receiving waters, a description of the waste treatment facilities, if applicable, and the type of discharge. Each application is published and a public hearing may be held. All applications for certification must be granted or denied within 130 days after a complete application is received.

Finally, North Carolina has been developing a *Wetland Rating System*, which recently was released for public review. This rating system will be used in determinations under the existing North Carolina wetland regulation system. Notably, a legislative study commission on wetlands after two years produced no recommendation for legislation in the near term. Nevertheless, further regulatory efforts with regard to wetlands are contemplated in the State of North Carolina.

10.1.35   COASTAL REGULATIONS

Many coastal states have laws and regulations controlling development in the coastal zone. These laws and regulations generally are designed to protect the quality of coastal waters and limit erosion. An example of a coastal program is discussed below.

North Carolina

North Carolina's *Coastal Area Management Act* was adopted in 1974 with the goal of insuring the orderly and balanced use and preservation of North Carolina's coastal resources, recognizing the needs of an expanding society, industrial development, population, and the recreational aspirations of its citizens. The Act was designed to regulate development of the coastal area. The *coastal area* is defined to be the counties which, in whole or in part, are adjacent to, intersected by, or bounded by, the Atlantic Ocean or any coastal sound. The Act establishes a Coastal Resources Commission (CRC) charged with preparing guidelines and adopting rules governing development of coastal resources. The act is intended to encourage coastal counties to adopt land use plans providing for the orderly development and protection of the coastal resource. Following the adoption of land use plans by the counties and approval by the Coastal Resource Commission that such plans are consistent with state guidelines, no local ordinances may be adopted that are inconsistent with the land use plan.

The Division of Coastal Management of the Department of Environmental, Health and Natural Resources has promulgated regulations to implement the permitting program authorized by such appropriate statutes. The Act also provides for the designation of "areas of environmental concern" within the coastal area. The Commission must approve development within the areas of environmental concern. Permits are divided into major development permits and minor development permits. *Major development* is a development that requires permitting, licensing, approval, certification, or authorization, in any form, from state or federal environmental agencies; that occupies a land or water area in excess of 20 acres; that contemplates drilling or excavating natural resources on land or underwater; or that involves a structure or structures on a single parcel covering in excess of a ground area of 60,000 square feet. Minor developments are developments other than major developments.

The Act, however, does not include normal and incidental operations associated with development and are exempted from coverage by the Act. These exemptions include highway work within the boundaries of the existing right-of-way; utility maintenance and repair within existing right-of-way (as well as extensions); forestry and agricultural work that does not involve excavation or filling affecting estuarine waters or navigable waters; the construction of any accessory building appurtenant to an existing structure provided such work does not involve filling, excavation, or the alteration of any sand dune or beach; as well as certain classes of minor maintenance and improvements that have been and shall from time to time be exempted form the permit requirements of the Act through its rulemaking.

In addition to the permit requirement for major or minor developments in areas of environmental concern, a permit must be received prior to any *excavation* or *filling* project in estuarine waters, tidelands, marshlands, or state-owned lakes. Factors considered in the decision of whether to grant a major development or dredge and fill permit include location of wetlands, the system of treating and handling waste water, storm water management, sedimentation and erosion control, reduction of long-range productivity or habitat, and damage to historic, cultural, scientific, environmental, or scenic values or natural systems.

The review process for major development permits includes *circulation for comment* by state review agencies. Nonapplicants may request contested case, or adjudicatory *hearings* with the Commission. The Commission may allow the request based upon whether the nonapplicant party is alleging that the decision is contrary to a statute or rule, whether that person is directly affected by the decision, and whether that person has a substantial likelihood of prevailing in the contested case. Developments permitted pursuant to major development or dredge and fill authority must be completed by December 31 of the third year following the year of permit issuance.

The Division of Coastal Management (DCM), headquartered in Raleigh, with field offices in Little Washington, Elizabeth City, Morehead City, and Wilmington: (1) reviews and makes initial decision of permit applications; (2) is charged with enforcement of such permit requirements; (3) provides the CRC with technical studies and recommendations concerning land use plans and permit standards; (4) provides lead staff for permit administration, planning, beach access and sanctuaries; (5) initially determines (through its review) submerged lands claims; (6) provides on-going review of federal agency decisions determining consistency with state programs; and (7) conducts continuing public education and citizen participation programs.

The Coastal Resources Advisory Council (CRAC) consists of 47 members providing technical and policy advice to the CRC. The CRAC serves an instrumental role in providing a local government and citizen perspective to the CRC by generally meeting concurrently with each meeting of the CRC.

10.1.36   SCENIC AREA PROTECTIONS

Many states with mountains or other scenic areas have enacted laws to limit visually intrusive develop-ment. These laws tend to have the greatest impact on residential and resort development activities, but in some cases they inadvertently limit industrial development. Examples of mountain area protection laws follow.

North Carolina

Almost a decade ago, the North Carolina General Assembly enacted this country's first comprehen-sive statute, which contains a regulatory framework governing construction on ridges throughout a state's mountain region. Unlike CAMA, this "local option" statute, known as the Mountain Ridge Protection Act of 1983 (the Ridge Act), emphasizes local control, keeps state involvement to an absolute minimum, and grants supplemental (not exclusive) authority to other legislation. The De-partment of Environment, Health and Natural Resources merely supplies supplementary assistance to local governments, and even then, only upon request by the counties or cities. Unlike CAMA, whose primary legislative intent was as an environmental protection act, the Ridge Act is the first North Carolina State statute to reach beyond environmental interests to regulate aesthetic considerations. In localities covered by the construction prohibition of the Ridge Act, it is stated that: "No county or city may authorize the construction of, and no person may construct, a tall building or structure on any protected mountain ridge."

10.1.37   FLORA AND FAUNA PROTECTION

Some states have enacted laws to protect certain species of plants and animals. These laws are generally similar, but more expansive, that the federal Endangered Species Act. Two examples of such laws are discussed below.

North Carolina

In North Carolina, the Wildlife Resources Commission has the authority to designate lists of threatened and endangered species in the state of North Carolina. Once designated, such species may not be taken without a permit except in certain extreme circumstances. Permits may issue for the taking of endan-gered or threatened species only for the purpose of scientific investigation relevant to the perpetuation and restoration of that species or as part of a commission approved study of restoration effort.

In addition to the role of the Wildlife Resources Commission with respect to animal species, the North Carolina Plant Conservation Board has the authority to establish lists of endangered or threat-ened plant species. These plants may not be sold or collected without a permit. Criteria considered before granting such a permit include the status of that species in North Carolina, the status of the population from which plants are proposed to be removed, the amount of plant material proposed to be collected, and the conditions under which the plants will be collected. Each of these protec-tion programs will also be a consideration for any state project reviewed under the North Carolina Environmental Protection Policy Act.

10.1.38   SEDIMENTATION AND EROSION CONTROL PROTECTION

Virtually every state has laws regulating activities that cause sedimentation or erosion. Although most contractors are familiar with these laws, they are often ignored my companies conducting their own construction activities. An example from North Carolina is discussed below.

North Carolina

The State of North Carolina requires persons proposing to conduct land-disturbing activities on one or *more contiguous acres* to file with the Department of Environment Health and Natural Resources, Land Quality Section, an erosion and sediment control plan for the purpose of preventing significant off-site erosion or sedimentation. A *land disturbing activity* is one that results in a change of the natural cover or topography of a site that may cause or contribute to erosion or sedimentation. Activities *exempt* from this definition include activities undertaken on *agricultural* land for the production of plants and animals, those undertaken on *forest* land for the production of harvesting of timber and timber products when those activities are conducted in accordance with best management practices, and *mining* activities regulated under the mining act.

Determinations with respect to approval of erosion and sedimentation control plans are made in the context of the *mandatory standards* established under The Sedimentation Pollution Control Act. Those mandatory standards are as follows: (1) No land disturbing activity is permitted in proximity to a lake or natural watercourse unless a buffer zone is provided along the margin of the watercourse sufficient to contain visible siltation; (2) The angle for graded slopes and fills can be no greater than the angle which can be retained by vegetative cover or other adequate erosion control devices or structures within thirty working days of completion of any phase of grading any graded slope or fill must be planted or otherwise provided with ground cover devices or structures sufficient to restrain erosion; (3) In the course of land disturbing activity, if more than one contiguous acre is uncovered the person conducting the activity must install sedimentation and erosion control devices sufficient to retain within the boundaries of the tract the sediment generated by the land disturbing activity during construction, and plant or otherwise provide a permanent ground cover sufficient to restrain erosion after completion of construction or development.

The devices and practices contemplated by the act include silt fences, buffer zones, silt ponds, and other such devices. For projects occurring within sensitive watersheds, including high quality waters and water supply watersheds, uncovered areas are limited to 20 acres at any one time. Erosion and sedimentation control measures, structures, and devices must be planned, designed, and constructed to provide protection for the runoff of the 25-year storm which produces the maximum peak rate of runoff. Sediment basins must have a settling sufficiency of at least 70 percent for the 40-micron size soil particle and groundcovers must be provided for portions of land disturbing activities in high quality water zones within 15 working days, following completion of construction or development. The Act also provides for the establishment of local government erosion and sedimentation control programs.

10.1.39 *LOCAL AND MISCELLANEOUS LAWS*

It is not possible to discuss all state environmental laws that might be relevant to an environmental audit or that might affect the value of property involved in transactions. However, consultants should inquire about applicable local and state laws and regulations. Many state and local laws are similar to well-known federal and state laws, but sufficiently different from their better known counterparts to create additional liability. For example, the city of Charlotte creates strict liability for persons who dispose of hazardous wastes, but the definition of "hazardous wastes" is much broader than the federal definition.

More importantly, many state and local statutes are substantially different from commonly known state and federal laws. For example, Massachusetts requires an owner of property used for residential purposes to remove lead paint, and New York city requires the removal of all asbestos in certain types of buildings if any demolition or renovation work is done. Such laws may impose unexpected costs on a purchaser of property who intends to alter the use of the property.

Common Law

A third common source of liability, which is often forgotten or ignored, is the common law. Under the common law, purchasers can be held liable even where the purchaser and its predecessors fully complied with the law. The most frequently used common law theories of liability include trespass, negligence, nuisance, and strict liability for ultrahazardous activities. In addition, fraud or misrepresentation theories might be appropriate if a seller or its environmental consultant withheld or misrepresented environmental information.

In most cases, liability cannot be imposed on a purchaser under ultrahazardous activity, negligence, and nuisance theories unless the purchaser committed some act or omission that contributed to the environmental problem. However, a negligence action might arise where a purchaser bought property that was known to be contaminated, and the purchaser did nothing to prevent the migration of the contamination toward neighboring drinking water wells.

CHAPTER 11
ASSESSMENT SAMPLING
AND MONITORING

SECTION 11.1

SITE EVALUATION, AUDITING, AND ASSESSMENT

Lisa Wadge

Ms. Wadge is vice president, Commercial Division, of VISTAinfo. VISTAinfo is a leading supplier of real estate and risk information on the internet. This includes extensive data and GIS maps ranging from information about Superfund sites to natural conditions.

11.1.1 INTRODUCTION

This section describes the types of inspections and assessments that are routinely done in the real estate and manufacturing environment. Bacause there are many detailed checklists and how-to articles on conducting these inspections, this chapter strives to provide a broad overview and philosophy of the practice. This chapter describes *when and why an assessment or inspection* is conducted, *the current standards of care* that exist, and some tips on how to best conduct the site and background reviews.

There are two general types of conditions that characterize the initiation of an environmental inspection. They are: (1) regulatory Inspections (audits); and (2) real estate and/or due diligence Inspections (environmental site assessments). The scope and content of a particular project can be different for these types of inspections and, as such, they are discussed separately. There are some business conditions that require both types of inspection, the most common of which is a merger or acquisition of a manufacturing facility in which both the regulatory status and the real estate liabilities must be understood in detail.

11.1.2 REGULATORY INSPECTIONS

When are They Conducted?

Typically referred to as "audits," regulatory inspections are specifically related to a federal law or permit requirement. Most audits are conducted at sites where manufacturing processes occur and where there are air or water discharges or permits. Audits are often conducted by manufacturers to gain a benchmark on status of compliance, as part of a merger or acquisition or on a periodic basis to comply with certain laws. Whenever one is conducting or reviewing an audit, it is helpful to contact industry trade groups that offer assistance in specific compliance or interpretation of laws for that industry to help create the correct focus. There is virtually no industry today that does not offer some industry specific auditing and compliance assistance, which can save a lot of regulatory review and ascertain the appropriate standard of care.

Scope and Content

The scope and content of an audit is typically narrow in focus and described in the title of the report or opening remarks. Most audits are done specifically to comply with a major federal law like the Clean Water Act and to focus on the pathways or conditions that could result in contamination leading to a river or body of water. Hazardous waste or RCRA inspections often focus on the proper paperwork, drum labeling, and storage of hazardous wastes. Broader site audits can also be conducted to review the plant conditions in their entirely to determine if all required standards and permits have been met. In addition, proactive audits are sometimes conducted to look for ways to reduce waste generation or facility compliance costs. When writing or reading an audit, it is important to be clear about which standard or law is being considered in the scope of work. Many mistakes are made when innocent readers of an audit report construe the findings in a way that is broader than intended.

Government Audits or Inspections

Many federal and state agencies conduct routine audits or inspections of private manufacturing sites where there are large quantities of hazardous wastes for example. In these cases, the auditor typically uses a checklist with the specific regulatory citations to assist in determining the sites compliance with environmental laws. The EPA has recognized the need for more comprehensive audits and has implemented an M2P2 (multi-media, pollution-prevention) inspection system for some manufacturers willing to participate. Traditional governmental inspections are delineated by type of environmental media (air, water, groundwater) and as such overlaps occur in the regulations. Therefore, inconsistencies and even contradictory issues of compliance can arise. These multimedia inspections involve inspectors from different state and federal agencies who work together with companies to create the best solution for the environment and the company. All governmental inspections forms are subject to the Freedom of Information Act and are helpful records for future audits.

Conducting the Audit

For the most part, the audit is limited to the boundaries of the site in question and can often be limited to cover only a portion of the site. The two most notable exceptions to this would be: (1) any offsite locations where solid or hazardous wastes are disposed of can often be included in an inspection or (2) the location of air or water discharges that may need to be monitored, whether on- or off-site. A well-constructed audit report will delineate at the outset the scope under consideration as well as extent of "The Site" that was reviewed. When starting a comprehensive audit inspection, off-site disposal or discharge locations should be identified and included whenever possible in the inspection and review process.

When conducting a broad-based facility audit inspection, it is always helpful to let the Four Ps guide your questioning and data gathering process. They are:

1. POWER—Understand the source and flow of all utilities fuel storage and back up systems in the facility.
2. PROCESS—A detailed understanding of the process in the facility starting with the raw materials and ending with the finished products.
3. POLLUTION—Identification of all liquid, gaseous, and/or solid wastes that are discharged either on- or off-site. In addition, the location where releases may occur either as a normal course of business or as an accident.
4. PERMITS—What federal, state or local permits are present, are required, or historically were in place.

If you can confidently understand the Four Ps after completing the investigative process, you are well on your way toward a good facility audit.

Record Reviews

One important aspect of the audit is the review of both on- and off-site records. An auditor's goal in reviewing records should be to determine (1) if the inspection information and interviews are consistent with the paperwork and (2) if there are historical issues or unidentified conditions that require attention. This includes a review of all facility documents, which often include numerous file cabinets of information. When conducting a large facility audit, the inspection is often the quickest and easiest step taking typically one full day. The review of on-site records can often take an additional full two days with follow up reviews and questions yielded from the paperwork. Most audits also include an extensive fact gathering record review from local, state, and federal record depositories. In addition, computerized data sources and Internet services can be extremely helpful in the audit process.

Old audits are also excellent sources of information to modern auditors. Regulatory compliance audits were first common in the 1960s when the command and control regulations were dominant. The audits were designed as self-policing instruments. Because many inspections were done during 1960s and through the late 1970s, regulatory audits can also serve to leave behind a legacy of knowledge from 20 to 30 years ago and document the conditions at a plant at a specific point in history. This can be helpful for both future audits and real estate inspections. As these vintage reports can also represent a liability, many companies have developed in-house auditing procedures. In many cases, the in-house policy will recommend that only draft reports be issued and thereby limit the access of the audit to the general public. In other cases, companies seeking to gain an audit will do so through their lawyer to limit the availability of the document in the public sector.

Audits in the Future

In the 1990s, there were more forward-thinking audits being done by many companies who not only seek to comply but also look toward a broader goal of reducing or eliminating the use of hazardous chemicals, becoming "green" or seeking to comply with international standards like ISO 14000. In these types of audits, there is a specialized need for the inspector to have a process understanding and some manufacturing experience. With this type of background, recommendations can be made to reduce compliance/environmental costs without impacted the quality of the finished product or the manufacturing process. Auditing continues to be a tool for business and government to gain a picture as to how a company or facility is complying with specified laws. It is therefore expected to continue as an important process within the total environmental industry.

11.1.3 REAL ESTATE EVALUATION AND INSPECTIONS (ENVIRONMENTAL SITE ASSESSMENTS)

Site Assessments: When Are They Conducted?

Real estate undergoes inspections for a variety of purposes, most of which are related to valuation of the property. Environmental real estate inspections are no different. Typically referred to as "Environmental Site Assessments" or ESAs, the inspector must take a holistic and three-dimensional view of the site. Consideration of any risk that *would affect the value of the property* is the primary goal of an ESA. It is this evaluation of the property value that separates the process of an ESA from the audit. ESAs are therefore conducted primarily when there is a change in ownership or financing for a given piece of real estate. ESAs are done commonly and for a variety of reasons including:

1. Merger and acquisitions where real estate is changing hands
2. Financing or sale of a property
3. When state level transfer act or real estate disclosure laws are present

Scope and Content

While the final report generated by the inspector may look different in each of these cases, the process of inspecting the site and collecting data is largely identical for all three of these real estate applications. ESAs, in their goal of identifying real estate liabilities, have to focus on not only current conditions but also historical conditions and area sites that could affect local soil or groundwater conditions. Understanding the historical, regulatory, and area issues is a complex job that is typically performed by environmental professionals who have specific training and experience. While the current standard of care allows for nonprofessionals to engage in the ESAs, most lawyers and banks tend to rely on certified and approved professionals to gather this information. The scope of an ESA should clearly state the real estate covered and the level of review included. Land uses and size of the property often dictate the scale of a site assessment. For instance larger parcels with historic land uses undergo a more exhaustive review than small, newer commercial land uses. This is partially due to the inherent nature of the risks based on land use and partially economics, as larger deals can justify a closer examination. Some ESAs can take six months and cost upwards of $10,000 if the facility is large and requires a lot of historical and regulatory review. More frequently, sites undergo a standard review, which takes a few weeks and costs under $2000.

Standard of Care

There are three typical scopes of work, which are routinely conducted in the ESA world: (1) Phase one ESAs, which include an inspection and record review and have a specific and strict level of effort, described in the ASTM E-1527 standard; (2) Phase two ESAs, which include some form of sampling of soil and/or groundwater and also have a level of care identified by ASTM; (3) A phase-half or "transaction screen standard" for nonindustrial sites, which includes a limited record review and interview, described in the ASTM E-1528 standard.

ASTM has comprehensive and detailed standards (the standard) for national compliance of phase one ESAs. This standard is considered a minimum standard of care and does not address local or state level standards, which often require additional tasks. Because phase one ESAs are by far the most common (and include the tasks in the transaction screen standard) they are discussed in more detail.

The ASTM standard specifies the following steps be completed for every phase one environmental site assessment:

1. *Inspection*—Completion of an inspection in which the ASTM questionnaire is completed as specified. According to the ASTM standard, the questionnaire can be completed by anyone (including a nontechnical bank representative) and details are provided in the standard on answering all of the individual questions.

2. *Historical Review*—ASTM requires a review of the historical usage and conditions of the site back to 1940 (or the earliest usage of the site) using specified tools and maps. The historical review can be conducted by using historical fire insurance maps, aerial photographs, city directory reviews, and/or chain of title or deed research. Because each site has a different history, it is suggested that the correct historic sources be used for the site in question. For example, aerial photographs are helpful to identify when the site may have gone from an undeveloped, treed lot to a constructed site while city directories will identify the name of the occupant at the site for a given year. Each of these historic sources represent a snapshot in time and coverage is not consistent on a national basis. For example, there may not be insurance maps present but many aerial photos may be present which cover the subject site. Therefore, the historical research tools should be used interchangeably to complete the desired goal of historical knowledge.

3. *Database Search*—A review of federal and state databases for the site and a specified radius around the site. Database reviews are also delineated in detail in the ASTM standard and describe the specific federal and state databases that must be included as well as the radius to which these databases should be reviewed. As of the writing of this article, there are two national database search providers that offer a data service that can meet the ASTM standard. The rapid advancement of

computer technology and in particular Geographic Information Systems (GIS) has allowed cost effective and fast information retrieval on any site over the Internet. Today, most inspectors rely on this type of service to provide an overview of the conditions at and around the site. It is recommended for significant land uses that a review of state and local records also be conducted. This data is not always included in the database search and may identify information about the site, which goes beyond the ASTM standard. When using a database service or information subcontractor, it is important to use a product that complies with ASTM and has the highest database quality including (1) the most updated or fresh databases, (2) the best quality and frequency of plotted sites, and (3) the most recent and correct base map which is used to plot the sites. In the past five years, the advent of desktop GIS has improved spatial data quality, and this industry has benefited from this technology. The database search is also used frequently as a tool to screen sites or estimate the level of effort to review a site.

4. *Report Generation*—Compilation of findings in a report that meets specific criteria and standards set forth by ASTM. This report should include a detailed description of the completed tasks and findings. The phase one report should also describe any source of releases or "recognized environmental conditions," as defined by ASTM. Finally the report should clearly identify the sources used to complete the site review.

At this time, edits to the current ASTM standard are underway and a new phase one standard is expected to be published in early 2000. Some of the proposed changes to the standard include a review of additional information under certain business conditions. The additional types of information being considered for inclusion in the future ASTM standard are wetlands, floodplains, and historic sites to name a few. In addition, there are numerous changes to definitions, which should be adhered to once the standard is final. The ASTM standard is a dynamic and changing standard and new editions of the standard that offer further clarification are undergoing edits at this time. Any professional engaged in the ESA business should monitor these standards at all times and exceed the standard with regional knowledge.

Conducting the ESA

In compiling all the data that is collected in an ESA, the inspector faces the challenge of whether to recommend additional studies or suggest that the phase one is adequate to assess the site risks. Historically, many consultants have made blanket recommendations for additional studies regardless of the phase one results and as such many clients are reluctant to go forward with the ESA process. While it is understood that consultants bear liability when recommending no further action, the ASTM standard does support the inspector in this conclusion when the standard of care is met and no recognized environmental conditions are identified.

In assessing a recognized condition, ESA inspectors should review carefully the environment in which the site is set and whether there are sensitive conditions or offsite receptors present. While the presence of wetlands, floodplains, or endangered species are not specifically required under the ASTM standard today, many lenders are now looking at these conditions as they represent a potential regulated condition and could drive remedial costs up in the event of a release. The National Environmental Policy Act is being reinforced and many lenders feel, as an FDIC approved bank, they must address the presence of natural or cultural conditions as part of due diligence. This is just one example where the ASTM standard is not adequate to cover all ESA issues on a national basis.

Inspector Credentials and Certifications

Phase one inspectors often have compliance experience and some engineering or manufacturing background. In many states, some sort of licensing or certification process is required for individuals performing ESAs. Transaction Screen inspectors typically have less credentials but transaction screens are often conducted by phase one inspectors. Phase two studies, which involve subsurface drilling or sampling, are often conducted by geologists or hydrogeologists. These geologic skills are necessary to be able to isolate the portions of the property where contamination may be present. Because most industrial sites are too large to be sampled in their entirety, a good phase one study is helpful in isolating the areas where the phase two investigation should concentrate. A good phase two is largely dependent on the research done in a thorough phase one and includes sampling of areas of concern that were identified as sources of contamination in the phase one.

Often there are surficial locations of contamination that can be sampled as part of an extended assessment. Whether conducted as part of a phase one or as precursor to a phase two subsurface investigation these samples are a cost-effective screen for baseline assessments:

1. Soil samples from 0–2 ft below grade in areas where spills, releases, or distressed vegetation are observed.
2. Surface water and sediment samples near discharge points, or at downgradient locations.
3. Resampling of existing monitoring, production, and process wells.
4. Resampling of water discharges.

These assessment samples can assist in determining how widespread contamination may be and refine the phase two scope of work. They also serve to update conditions that may have previous data points and trends can be established.

Conclusion

The process of conducting regulatory audits and/or environmental site assessments continues to be refined and tailored for site specific conditions. The final product of these activities is the documentation of conditions at a given site at a given point in time. Well crafted, inspection reports clearly document the sources and findings and collectively represent a body of environmental knowledge that dates back over 20 years. Since the regulatory conditions and remediation knowledge continue to change over time, these inspections provide a resource to the entire environmental industry. Current and historical inspections provide clues to assist in efficient source identification and cost-effective remediation activities. Modern inspections have evolved to be more wholistic and focus on resource protection as well as pollution prevention and continue to be a valuable tool in the marketplace.

CHAPTER 11
ASSESSMENT SAMPLING AND MONITORING

SECTION 11.2

APPLICATIONS OF GLOBAL POSITIONING SYSTEMS IN THE ENVIRONMENTAL SCIENCES

James A. Jacobs

Mr. Jacobs is a hydrogeologist and president of FAST-TEK Engineering Support Services with more than 20 years of experience. He is registered in several states, including California, and specializes in assessment methods and in situ remediation technologies.

Jefferson K. Phillips

Mr. Phillips is an environmental scientist with FAST-TEK engineering support services. He performs GPS field surveys for environmental projects and is now on assignment in Buenos Aires, Argentina.

11.2.1 INTRODUCTION

Global positioning systems (GPS) have revolutionized surveying and three-dimensional geographic position data collection in the environmental field. GPS equipment has enhanced environmental assessment and remediation projects at numerous Superfund sites. Features that can be mapped with GPS include: geologic, botanic, biologic, and topographic features; locations of soil, soil vapor, and water samples; and wells, historical site boundaries, pipelines, sewer lines, and chemical storage tanks. Environmental projects typically benefiting from GPS are those that include mapping of sample locations, monitoring wells, or other geographic features on a scaled site map or CAD drawing. GPS surveying allows environmental professionals to prepare digital site maps using geographic information systems (GIS) software. GPS saves significant time in the field and often costs far less than traditional surveying methods.

11.2.2 HISTORY

Although GPS technology has been used in environmental projects for about a decade, the U.S. military has been using this technology since 1978. GPS started as a U.S. military orbital satellite navigation system. Launched aboard Delta rockets and tracked under the U.S. Air Force Administration, the GPS network consists of the Space Segment, including 24 positioning satellites and the Control Segment, which is a network of ground tracking stations that monitor and control the GPS satellites in orbit.

In 5 Steps

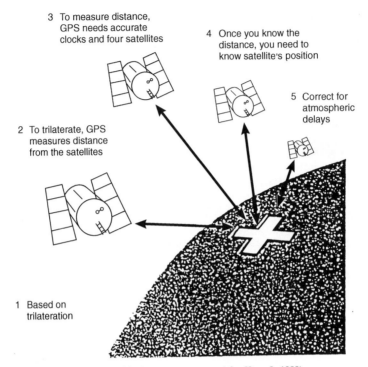

3 To measure distance,
   GPS needs accurate
   clocks and four satellites

4 Once you know the
   distance, you need to
   know satellite's position

5 Correct for
   atmospheric
   delays

2 To trilaterate, GPS
   measures distance
   from the satellites

1 Based on
   trilateration

FIGURE 11.2.1   Basic global positioning system concepts (after Hurn, J., 1989).

Under development for over 20 years, GPS utilizes 21 satellites plus three back-up satellites for a total number of 24 satellites in predictable orbits around the earth. The system became fully operational in 1993 (Nasland and Johnson, 1996). Basic GPS concepts are shown in Figure 11.2.1. (Hurn, 1989).

The GPS network was designed first and foremost for national defense purposes. As such, the $12 billion GPS network was built to be impervious to interference from outside sources. The GPS satellites use radio waves to emit time signals. These signals are calibrated to a millionth of a second. Using the satellite signals, a GPS receiver calculates its latitude, longitude, and elevation. To increase the accuracy of the data, signals from four or more satellites are recorded simultaneously and referenced against each other. By knowing the distance from three or more satellites, the receiver can calculate its precise position by solving a set of differential correction equations. Three satellites are used to obtain data to calculate latitude, longitude, and relative elevation. A fourth satellite is used to provide altitude data (Hurn, 1989).

The satellites each orbit the earth twice a day at an altitude of approximately 18,000 km, repeatedly broadcasting their position and time. Each satellite keeps time by the vibration of atoms aboard its atomic clocks, which are accurate to within one second in 30 years (Ashtech, 1998).

Up until April 30, 2000, for security reasons, the U.S. government scrambled and limited signal strength of the positioning satellites so that the accuracy of a standard civilian GPS was degraded to approximately 100 m. On May 1, 2000, the U.S. government stopped scrambling the signal, making satellite navigation devices 10 times more accurate. GPS receivers that use only this primary data source have become relatively inexpensive and are commonly used for recreational navigation such as hiking, backpacking, and boating. Luxury automobile manufacturers now offer GPS units mounted in their automobiles, which allow the driver to view the car's geographic location on a computerized map displayed on the dashboard.

11.2.3 DIFFERENTIAL GPS

Environmental projects often require precise accuracy. For more precise mapping applications, differential GPS (DGPS) units are available. Differential GPS is shown in Figure 11.2.2 (Hurn, 1989). To achieve accuracy greater than 100 m, commercially available mapping and survey-quality GPS receivers use data recorded by a secondary reference receiver. DGPS uses a secondary signal, such as a U.S. Coast Guard continuously operating reference station (CORS), private positioning satellite, or a GPS base receiver, to make differential corrections and achieve submeter, decimeter, and even centimeter accuracy. A secondary correctional signal is broadcast from (or at least recorded at) a known point and provides corrections for the positioning data it is receiving from the same satellites that the "roving" GPS field unit is tracking.

DGPS data processing software applies the secondary data source information to the satellite data collected in the field to improve the data accuracy. The resulting accuracy can be from a few meters to a few millimeters, depending on the GPS receiver, software, and time spent at each sample location. Some GPS receivers have become so accurate they can be used to measure microscopic movements along geologic faults! The data processing software can be installed in a personal computer (PC) for "post-processing" or in the GPS units themselves to provide highly accurate "real-time" location data.

Two grades of receivers, survey grade, and mapping grade are used. Survey-grade receivers generally use two onsite GPS receivers; one "base" receiver placed on a benchmark or other known point on the site, and one "rover" receiver used to log the desired unknown points at the site. The two receivers must be able to maintain radio contact at all times. Using this system, real-time accuracy of

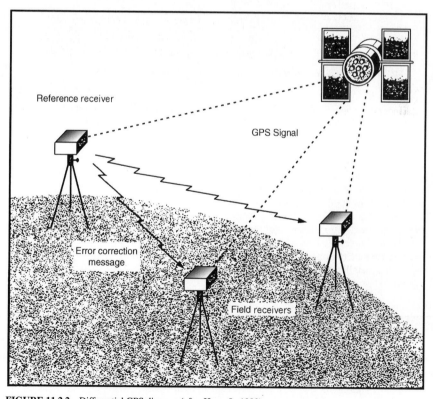

Reference receiver

GPS Signal

Error correction message

Field receivers

FIGURE 11.2.2   Differential GPS diagram (after Hurn, J., 1989).

PHOTOS

(a)

(b)

(c)

(d)

(e)

FIGURE 11.2.3 (a) Calibrating GPS at Benchmark; (b) Logging shoreline samples at San Francisco Bay; (c) Navigation to sample locations; (d) Creating sample grid at landfill; (e) Logging air sampling path.

one centimeter is attainable. Mapping-grade receivers (commonly referred to as "submeter" systems) are only accurate to approximately one meter in "real-time" but cost generally one-third to one-half the price of survey-grade receivers, and can achieve decimeter and better accuracy if they are allowed to log a point for 10 to 30 minutes, and then the point location is post-processed using reference data downloaded from a CORS site or other base station.

Most GPS units are capable of locating and recording point, line, and area features. After down-loading field data and performing post-processing, the data can be imported into a GIS program and mapped on a variety of base maps. In the navigation mode, GPS can be used to locate or relocate field points with known coordinates. A GPS unit can be used to record the location of a known point, such as a soil sample location or a particular plant species, and display the geographic coordinates of that point or plot it on a map.

GPS can be used to navigate to a predetermined location or known coordinates, such as returning to a soil sample location or staking a sample grid with predetermined known coordinates for each grid node. A line can be logged such as the boundary of a flood plain, or the path of an environmental technician walking with an air sampler. An area can be logged such as the circumference of a soil stockpile, a region of fragile vegetation, or an area containing an endangered species. After the circumference of an area has been circumnavigated by the GPS technician, many GPS units will calculate and display the surface area of the feature immediately in the field. Using GPS data, highly accurate estimates of volume in a stockpile can be calculated in just minutes. Field applications of GPS are shown in the Photographs (Figure 11.2.3).

11.2.4 GIS INTEGRATION

GIS software enables the user to tie a wide variety of geographic information to computer-based maps. GIS can be used for displaying map data and analyzing information by integrating common database operations such as query and statistical analysis with the visualization and geographic analysis benefits of conventional maps.

GPS measurements taken in the field are quickly and easily plotted on street maps, topographic maps, geologic maps, aerial photographs, and other graphical formats using GIS software. GIS programs import data, manage databases, query databases, query geographic features, and visualize results (Strafaci, 1998). Approximately 80 percent of all GIS applications are customized for the organization or project that is using the GIS data (Wilson, 1998). This versatility combined with the speed of data collection are the major advantages of the GPS/GIS technologies and some of the reasons for its rapidly expanding use.

The GPS software exports positions and descriptions of feature attributes (e.g., well depth and pumping capacity, or tree species and age) directly to database management, CAD, and GIS systems. Many GPS equipment makers allow for the GPS output data to be formatted for the most commonly available GIS products, from companies such as Auotdesk®, ESRI®, Intergraph®, MapInfo® and others. GIS map databases are supplied by a variety of publicly available sources including the U.S. Geological Survey and the U.S. Census Bureau. Terrian modeling using GIS is shown in Figure 11.2.4 (ESRI®). Layered GIS themes with attributes tied to geographic locations are shown in Figure 11.2.5 (ESRI®).

11.2.5 ADVANTAGES TO GPS

While standard surveying methods usually require two people in the field, GPS measurements are easily recorded by just one environmental technician. Global Positioning Systems are a cost-effective method to obtain digital vertical and horizontal location information. A one-man crew can log the location of a point in under a minute, leading to hundreds of points recorded during a day under ideal conditions. The digital data can be downloaded directly to a personal computer (PC) and exported to

GIS IMAGERY

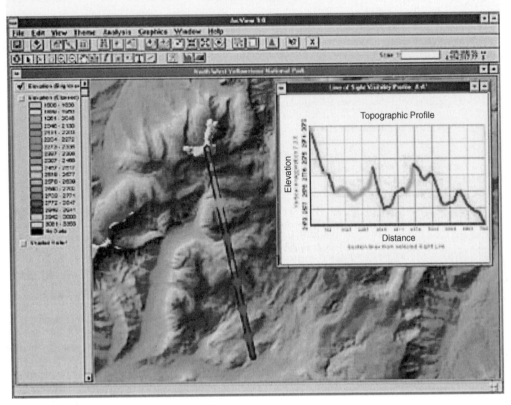

FIGURE 11.2.4   Terrain modeling using GIS (ESRI).

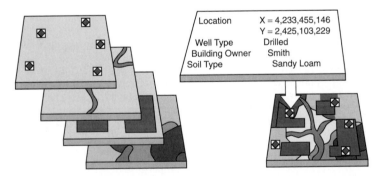

FIGURE 11.2.5   Layered GIS themes with attributes tied to geographic locations (ESRI).

virtually any mapping (GIS) or drafting (CAD) program. GPS data can be printed in a quality map format in a matter of minutes. Field data can be gathered by trained technicians that are frequently far less expensive than certified surveyors. As a general rule, when GPS can be used instead of traditional survey techniques, clients can expect savings of 50 percent or more.

Environmental projects having more than 12 to 15 data locations are optimum for GPS methods. In addition, projects involving recurrent sampling through a project cycle could benefit from GPS

TABLE 11.2.1 Environmental Applications for GPS Technology

- Mapping individual sample locations
- Import site features into CADD, ArcView, ArcInfo, and other mapping software
- Import sample locations or site features onto digitized aerial site photographs
- Create a 3-D site model
- Outline areas such as wetlands, endangered species habitat, or exclusion zones
- Create sample grids
- Navigate to predetermined points to install a trench, pipeline, well, or collect a sample
- Record and map sample/monitoring well locations
- Map fire hydrants, electrical transformers, sewer lines, and other utilities
- Ecological/environmental mapping: Endangered species habitat, wetlands, flood planes, and coastal erosion zones

data and GIS displays. As a general rule, for sites that have a relatively unobstructed view of the sky and require decimeter to meter accuracy, GPS technology will be a better choice (see Table 11.2.1). This is especially the case when a site map or digitized photograph showing the plotted locations is desired.

11.2.6 LIMITATIONS

GPS technology is not practical for every site. The most significant limitation of GPS is in areas with heavy overhead vegetation, or close to buildings or other obstructions. A variety of techniques, such as offsetting points, can be used to overcome these obstacles but at some point it becomes easier and more reliable to revert to standard surveying methods. When first assessing the need for geographic information on a site, it is often useful to consult an environmental professional familiar with both GPS and standard surveying techniques to find out what will work best.

11.2.7 TYPICAL ENVIRONMENTAL PROJECTS

GPS and GIS data were used at a former firing range in the Presido of San Francisco. GIS and GPS were used to identify and stake the former target range boundaries and sample collection points. GPS was also used to map sample locations and current site features which may affect remedial actions at the former U.S. Army base, now part of the Golden Gate National Recreation Area. GIS and solids modeling were used successfully to determine the vertical and lateral extent of lead contamination at the former firing ranges. The GPS was also used to locate planned wetlands, eco-zones, and other features that affected remedial actions at this former base (McDade, 1998).

In another project, GPS was used to log the position of over 60 soil and concrete samples collected at a closed landfill over a several-square-kilometer U.S. Navy installation in northern California. A GPS unit was used as an integral part of an air and soil gas sampling project. First, a digitized site map of the landfill was created and oriented to geographic coordinates. Next, locations of ambient air samplers were chosen based on prevailing wind patterns and the locations were plotted on the site map, which was provided to the field crew. The field crew placed the ambient air samplers as close to the geographic coordinates shown on the map as the terrain would allow and used the GPS to record the final locations, which were then corrected on the map. GPS-derived location map, data and site image are shown in Figure 11.2.6.

The next stage of the project consisted of sampling a series of rectangular sampling areas using a hand-held air sampling device. The sampling grid consisted of over 20 rectangles, each covering an area of 4650 square meters. The grids were first plotted on the computer-based site map and the coordinates of the grid nodes were provided to the field crew, which used the GPS to locate and mark

Dredge Spoil Pond 3E

Description	Easting	Northing
SB1	1484014.70721	590512.41591
SB2	1484014.36335	590524.52568
SB3	1484004.77714	590526.92404
SB4	1484000.66987	590513.68592
SB5	1484007.19740	590519.91659
SB6	1484007.20790	590518.93037
SURFACE	1484009.12164	590518.10476

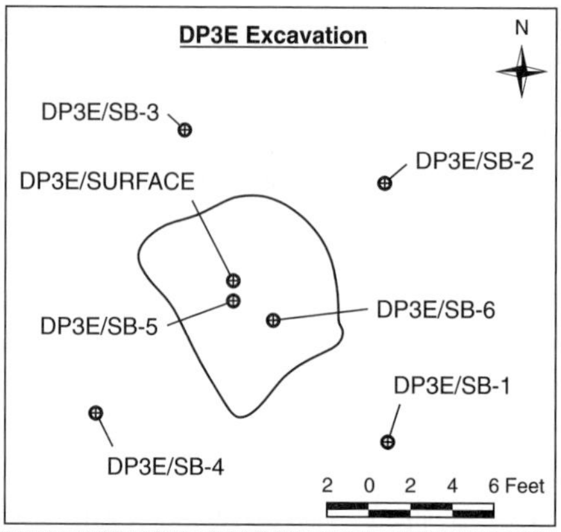

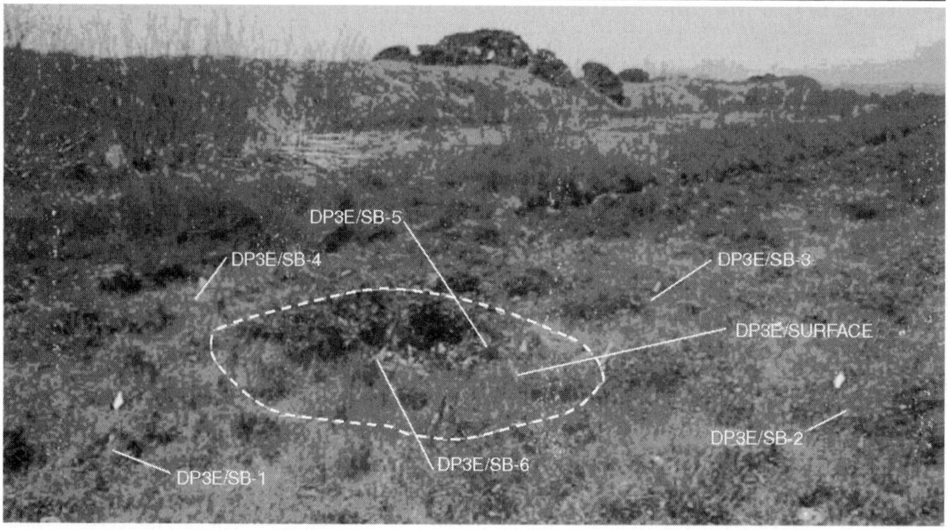

ARTESIAN ENVIRONMENTAL 1735 East Bayshore Road, Suite 2B Redwood City, California 94063 Phone(650) 368-1796 Fax(650) 368-1797		Dredge Spoil Ponds DP3E	
Project No.:	Date: 4/14/99	Prepared by: CJM	Figure 1

FIGURE 11.2.6 GPS-Derived location map and site image (Artesian).

the locations in the field, and then string the boundaries of the squares with flourescent string. Each grid was then walked by a sampling technician while wearing the GPS unit, which was set to log both the path walked by the sampler and the rate at which air was drawn into the sample container. The sample paths were plotted on the site map for inclusion in the final project report. (Phillips and Jacobs, in press).

11.2.8 SUMMARY

GPS offers significant savings over traditional survey techniques providing data in an ideal format to import into digitized site maps, aerial photos, or three-dimensional site models. These benefits can be used successfully in a wide variety of environmental projects. Mapping-grade GPS operates in real-time with submeter accuracy, and with postprocessing can achieve decimeter and even centimeter precision. Survey-grade GPS operates in real-time with centimeter accuracy, and with postprocessing can approach millimeter accuracy.

REFERENCES

Ashtech, *About Global Positioning, The Basics of GPS & GPS+GLONASS*, 15 p., http://www.ashtech.com/Pages/gpsndex.html, July 18, 1998., Sunnyvale, California, 1998.

Hurn, J., *GPS: A Guide to the Next Utility*, Trimble Navigation Ltd., Sunnyvale, California, 1989.

McDade, K., 1998, *The Integrated Use of a Geographic Information System and Global Positioning System to Assess Potential Lead Contamination at Former Firing Ranges, Presidio of San Francisco*, 16th Annual ESRI User Conference, Montgomery Watson, San Francisco, p. 2–3.

Nasland, D. K., and Johnson, D. P., *Real-Time Construction Staking*, Civil Engineering, June 1996.

Phillips, J. K., and Jacobs, J. A., *The Impact of Global Positioning Systems on Groundwater Resources*, Hydrovisions, in press, Groundwater Resources Association of California, Sacramento, California, (in press).

Strafaci, A. M. ed., "Geographic Information Systems: an Introduction to the Technology," in *Essential Hydraulics and Hydrogeology*, Haested Press, Waterbury, Connecticut, p. 63–70, 1998.

Wilson, J. D., *Java Expected to Energize GIS*, GIS World, August 1998.

CHAPTER 11
ASSESSMENT SAMPLING AND MONITORING

SECTION 11.3

INDOOR AIR: SICK AND HEALTHY BUILDINGS

C. Richard Cothern

Dr. Cothern is currently professor of engineering management and chemistry at the George Washington University and a professor of technology and management at the University of Maryland. At Hood College he is professor of chemistry. He teaches courses in energy management, environmental auditing, environmental communication, environmental risk analysis, industrial ecology, indoor air, environmental chemistry, and sustainability.

11.3.1 INTRODUCTION

We spend about 90 percent of our time in indoor atmospheres. In these situations we are exposed to a mixture of toxic fumes from a variety of sources. Often, we are in indoor air spaces that have inadequate ventilation or too little fresh air. Historically, we have closed in our homes and offices to conserve energy. Few buildings have windows that open. We generally turn off the ventilation overnight and this allows for toxic fumes to build up. In addition, we put into our offices new sources of toxic fumes including particle board furniture, carpets, drapes photocopying machines, FAX machines, and computers. It should not surprise us that we are getting sick in these spaces.

Into our homes, offices and factories we have brought a wide variety of chemicals. We now have about 60,000 man-made chemicals and only for a few of these do we know the toxicology. Our wide range of sensitivities as humans mask the health problems as some of us seem immune to the effects of toxic fumes and some of us are very sensitive.

The philosophy of the approach to indoor air quality (IAQ) problems is most important. The goal is to provide healthy buildings. The goal is to protect the health of the occupants of work spaces and our homes. The most important thing for those dealing with such problems is to listen. Careful redundant and thorough investigations are essential. The most important source of information is from those directly affected.

The resulting health effects in indoor air are given a wide variety of names. The two most common are sick building syndrome (SBS) and building-related illness (BRI). As a general rule a building is considered *sick* if more that 20 percent of the occupants are complaining of health problems that interfere with their work. Generally speaking we find that over half of the health complaints are related to inadequate ventilation (see Federal Register, 1994, for indoor air problems and information; the other references are included for further reading). Approximately 800,000 to 1.2 million commercial buildings in the United States have resucted in building-related illness, which translates to 30 to 70 million people exposed to potential building related health problems. The costs for medical care indoor air quality problems has been estimated to exceed $1B annually.

To most scientists and risk assessors, the immediate reaction to the above problem is to measure the levels of the toxic fumes. However, because many people have developed sensitivities to very low levels, this approach is generally not useful. The most productive and successful approach to this problem is to start with a visual inspection of the facility and then do a thorough questioning of those complaining. This will be discussed further later.

One label for building-related illnesses that has brought considerable controversy is termed *multiple chemical sensitivity*. This health problem was first recognized in the 1950s. The controversy involves the relative importance of the mind and the body. Some think that problems like anxiety and depression are a consequence of physical ailments and others think the physical ailments are the result of mental problems. We advise using this label carefully, if at all.

11.3.2   COMMON SYMPTOMS AND RESPONSES

Besides sick building syndrome and building related illness we find several other names given to the illnesses found in indoor air (see Table 11.3.1).

The common symptoms experienced in indoor air environments fall into the general categories of nervous system, respiratory system, psychological, and other. Many ailments reported include those listed in Table 11.3.2.

The above list indicates the complexity of finding the cause of a sick building. This complexity is the reason that it is more efficient to interview those suffering rather than try to find a single or even several causes by monitoring.

In general, we respond in several ways to ailments produced in indoor air. Our first response is one well-known to psychologists. We usually first deny that there is a problem. In cases of indoor air diseases, denial is the worst thing we can do because our bodies can become sensitized permanently. An emotional component that is important in our homes is our belief that our home is our castle. Thus, we believe that there is nothing in our home that can make us sick. This is another fantasy that we need to overcome. To many of us our job is our identity, and we will suffer a lot before we give that up. Often, we do not realize the price that we are paying in terms of our health. Finally, we

TABLE 11.3.1   Common Names Given to Illnesses Found in Indoor Air Circumstances

Building-Related Illness
Cerebral Allergy
Chemical AIDS
Chemical Hypersensitivity Syndrome
Chemical Pneumonitis
Chemical Sensitivity
Chemically-Induced Immunities Regulation
Ecologic Illness
Environmental Illness (IL)
Environmental Maladaption Syndrome
Humidifier Fever
Hypersensitivity Pneumonitis
Multiple Chemical Sensitivity (MCS)
Sick Building Syndrome (SBS)
Total Allergy Syndrome
Toxic Headache
Twentieth Century Disease
Universal Allergy

TABLE 11.3.2 Common Ailments Reported as
a Result of Exposure to Indoor Air Circumstances

Head and Central Nervous System Related
   Lethargy
   Fatigue
   Headaches
   Memory difficulties and loss
   Dizziness, fainting
   Depression
   Difficulty concentrating
   Groggy
   Tense, irritable
   Difficulty making decisions
   Sleep disturbance

Upper Respiratory System
   Bronchitis
   Asthma
   Short of breath
   Rhinitis
   Sinus congestion
   Allergies
   Wheezing

Immune System
   Susceptibility to infection
   Allergy problems
   Colds

Other
   Eye irritation
   Upset stomach, nausea
   Problems digesting food
   Joint pain
   Constipation
   Diarrhea
   Cramps
   Gas
   Bloating

Psychological
   Depression
   Anxiety
   Sudden anger
   Irritability
   Confusion
   Lethargy
   Suicidal

trust those that are responsible for watching over our health and safety are doing their job. In many cases, building managers are ignorant of the consequences of inadequate ventilation, water collecting in low areas, and toxic fumes from furniture, cleaning supplies, and numerous other sources. On the other hand, those who are lazy can use sick building syndrome as an excuse for not working. The problem is to sort out those legitimate cases. Also, those experiencing work-related stress are more likely to report indoor air problems and those with high job satisfaction are less likely to report problems.

TABLE 11.3.3   Potential Sources of Health
Problems in Indoor Air

- Asbestos
- Carbonless carbon paper
- Carpeting
- Cleaning agents
- Computers
- Drapes and window coverings
- Dust
- Dust mites
- Fax machines
- Glues
- Man-made mineral fibers
- Microbial contaminants: viruses, fungi, bacteria
- Mold
- Paints
- Particle board furniture (formaldehyde and vocs)
- Pesticides
- Photocopying machines
- Printers (laser, ink jet)
- Solvents
- Tobacco smoke
- Video display terminals

11.3.3 POSSIBLE SOURCES OF AGENTS THAT MAY CAUSE DISEASE

One of the major problems we face in analyzing the risk resulting from indoor air is the myriad of sources. These fall in several categories and include a wide range of individual sources. Table 11.3.3 lists many possible sources of indoor air health risks.

There are other conditions found in indoor air that may not be as serious as the above, but we must consider them in any indoor air risk analysis. These environmental conditions include thermal comfort, humidity, lighting, and noise.

As you can see from the above list, there are so many potential sources that it is easier to start the risk assessment with a simple walk-through to look for the materials listed above.

11.3.4 DIAGNOSING PROBLEM BUILDINGS

Historically, the industrial hygienist investigated indoor air quality complaints. This approach was relatively unsuccessful because of the focus on measuring contaminant levels to determine compliance with federally permissible exposure limits (PELs) of threshold limit values (TLVs) of the American Conference of Government Industrial Hygienists. If the levels measured were below these limits or values, it was assumed that there was no problem despite health-related reports. Also, in the industrial hygienist approach, inappropriate sources of indoor air contaminants are numerous and diffuse, levels of potential contaminants are several orders of magnitude below PELs or TLVs, and contaminant control is usually achieved by dilution not by local exhaust ventilation.

The features of successful assessments of risk in indoor air quality investigations include: walk-through, detailed interviews of the occupants, assessment of the heating, ventilation and air conditioning system (HVAC), and some limited measurements. We find that the most successful risk assessments involve stages of investigation to narrow down the possible problem areas and a team of two or three specially trained investigators.

We find that the best first step is to do a walk-through looking for sources of toxic fumes, dust, mold (see list above), and a preliminary inspection of the HVAC system. The next step is a thorough, rigorous interview with each person with a complaint. A detailed questionnaire is needed to ensure that we systematically ask the same questions and obtain the same data and information for consistent note comparison. Next, a more rigorous inspection of the HVAC system should be done. We start with measurements of air flows into and out of rooms and check whether the fans are running constantly when the building is occupied. Some measurements are useful also. By measuring the carbon dioxide level we can determine if adequate outdoor air is being brought into the system. Other measurements you may find useful are carbon monoxide, respirable particulates, and total volatile organic compounds.

Problems that we will look for in the HVAC system are: insufficient outdoor air for the effective control of bioeffluent and other contaminations generated in the building, migration of contaminants from one building zone to another, re-entry of building exhausts, entrainment into intake air of contaminants generated outdoors, re-entry of building exhausts, generation of man-made mineral fibers from disintegration of sound liners in air handling units, microorganisms and organic dust contamination in condensate drip pans, humidifiers, filters and porous thermal/acoustical insulation, inadequate dust control, inadequate control of temperature, relative humidity, and air velocity and inadequate air flows into building spaces because of system imbalances.

If there is an inadequate outdoor air supply, we can look to the following areas: lack of design to include air supply; design inadequate for building; increase in building occupancy beyond design; malfunction of inlet dampers; operating changes to conserve energy; and reduced flows because of poor maintenance of filters, fans, and other HVAC system components.

Sometimes, the HVAC system may itself be a source of indoor air contaminants because of microorganisms proliferating in poorly maintained condensate drip pans, contaminated air filters, and porous acoustical duct and air-handler insulation. For most indoor air quality problems, the above protocol will surface the problem and also indicate the solution.

An excellent source of information is the US Environmental Protection Agency's Office of Indoor Air Quality. Check out the web site at www.epa.gov/iaq. Two helpful publications that you can download are: *Building Air Quality: A Guide for Building Owners and Facility Managers* (this publication is full of important details and contains some useful questionnaires and forms for collecting and organizing information) and *Indoor Air Pollution: An Introduction for Health Professionals*.

Part of the efforts of the USEPA is an ongoing study of the problem to provide a baseline of the current problem. This study is called BASE and the core parameters are listed in Table 11.3.4. This list should provide you with a set of the important parameters.

TABLE 11.3.4   BASE Core Parameters

Environmental measures	Building characteristics	HVAC characteristics	Occupant questionaire
Temperature	Use	Type	Workplace physical
Relative humidity	Occupancy	Air handler and exhaust	Information
Carbon dioxide	Geographical	fan specifications	Health and wellbeing
Sound	Ventilation	Filtration	Workplace environmental
Light	Equipment and	Air cleaning systems	conditions
CO	operation	Air washers	Job characteristics
Particles	Construction	Humidifers	
VOCs	Outdoor sources	Maintenance schedule	
Formaldehyde	Smoking policy	percent outdoor air	
Bioaerosols	Water damage	Temperature and	
Radon	Fire damage	relative humidity of	
	Pest control	supply air	
	Cleaning practices	Exhaust fan rates	

11.3.5   WHY INDOOR AIR QUALITY PROBLEMS ARE NOT BETTER UNDERSTOOD AND NOT GENERALLY ADDRESSED

Most of us are ignorant of the problems brought about by inadequate ventilation, sensitivity to toxic fumes, and the importance to our health of living in clean, fresh air. Why is this?

One reason is that medical doctors are reluctant to deal with this problem area. This reluctance is because there is no clinical definition and no direct tests for many of the ailments. As pointed out above, there is no known cause for most of the health problems. We see that many of the symptoms are subjective, such as headaches and difficulty concentrating. Because most of the symptoms are self-reported, the information is anecdotal and few systematic studies exist. Finally, medical doctors are trained to look at specific symptoms and often do not take the holistic approach needed to deal with these complex problems.

We find reluctance to deal with these problems because there is a paucity of cures or treatments. For those of us who get sick in buildings the alternatives are as follows: spend time in a superclean environment and change our diet to flush the body of the toxics; take medications, which can help in some cases such as headaches; try alternative medicine such as biofeedback training, and finally the ultimate weapon is avoidance of the agent that seems to be causing the disease.

11.3.6   WHAT CAN WE DO?

Why should we be concerned about indoor air quality? Our concern should result from our need to preserve our health and the health of others, and to avoid liability. A few suggestions for dealing with indoor air quality problems are listed in Table 11.3.5.

11.3.7   CONCLUSION

Our risk assessment of indoor air quality has a very different character than other health risk assessments. We have found that this area involves mixtures, a range of human sensitivities, a wide variety

TABLE 11.3.5   Suggestions for Dealing with Indoor Air Quality Problems

- Lower exposure levels most diseases are dose-related
- Restrict the use of pesticides, especially petrochemically based ones
- Eliminate perfumes, colognes, and after shave lotions or only use natural ones
- Provide adequate ventilation (such as the minimum required by the American Society of Heating, Refrigeration and Air-Conditioning Engineers-ASHRAE, which is 20 cubic feet per minute per person and introduction of 20 percent outside air)
- Do not use air fresheners, deodorizers, and disinfectants
- Avoid exposure to photocopying machines, laser printers, and fax machines
- Open the windows
- Avoid cleaners that are petrochemically based
- Minimize exposure to exhaust from heating systems and appliances that use natural gas, oil, or wood
- No smoking
- Steam clean instead of dry clean
- Avoid use of petrochemicals for maintenance, repair, construction, or remodeling
- Keep areas well vacuumed and free of dust (including your automobile)
- Avoid urea formaldehyde carpeting
- Avoid synthetic carpeting
- Use only products that come from plants and try to avoid those that sound like they are coming from a chemistry lab

of symptoms and diseases, and thus its mitigation requires a different approach:

- Let your nose be your guide about what chemicals to avoid or dispose of
- Pay attention to clues from your body—aches, pains, dizziness, among others.
- Let the building bake-out after painting at 30–35°C for at least 48 hours, the same for particle board furniture
- Look for problems like entrainment, re-entry, cross-contamination

RECOMMENDED READING

Ashford, N., and Miller, C., *Chemical Exposures: Low Levels and High Stakes*, Van Nostrand Reinhold, New York, 1991.

Bower, J., *Healthy House Building: A Design and Construction Guide*, The Healthy House Institute, Bloomington, IN, 1993.

Bower, J., *Understanding Ventilation*, The Healthy House Institute, Bloomington, IN, 1995.

Federal Register, Department of Labor, Occupational Safety and Health Administration, "Proposed Rule for Indoor Air Quality," Federal Register, Tuesday, April 5, 1994, 15968-16039.

Godish, T., *Sick Buildings: Definition, Diagnosis and Mitigation*, Lewis/CRC Press, Boca Raton, FL, 1995.

Good, C., *Healthful Houses: How to Design and Build Your Own*, Guaranty Press, Bethesda, MD, 1989.

National Technical Information Service, *Sick Building Syndrome; Annotated Bibliography*, Springfield, VA, 703-487-4650, 1995.

Rea, W., *Chemical Sensitivity*, Lewis Publishers, Boca Raton, FL, Volumes 1–4, 1992, 1994; 1995.

Sparks, P. J., Daniell, W., Black, D. W., Kippen, H. M., Altman, L. C., Simon, G. E., and Terr, A. I., "Multiple Chemical Sensitivity Syndrome: A Clinical Perspective, I. Case Definition, Theories of Pathogenesis and Research Needs," *J. Occupational and Environmental Medicine*, 36: 718–730, 1994 and "II. Evaluation, Diagnostic Testing, Treatment and Social Considerations," 36: 9–17, 1994.

Wallace, L., *The Total Exposure Assessment Methodology (TEAM) Study: Summary and Analysis: Volume I*, Office of Research and Development, US Environmental Protection Agency, Washington, DC, EPA/600/6-87/–2a, 1987.

CHAPTER 11
ASSESSMENT SAMPLING AND MONITORING

SECTION 11.4

STREAM SAMPLING

D. Michael Johns

Dr. Johns is a founding partner of Windward Environmental LLC. He is an aquatic scientist specializing in aquatic ecological risk assessments, particularly those associated with contaminated sediment. He has served as an advisor and technical expert in the development of water and sediment quality standards and is a recognized expert on the use of bioassessment techniques to evaluate sediment contamination.

Frank S. Dillon

Mr. Dillon is a project manager with Windward Environmental LLC. He is an aquatic ecologist and toxicologist focusing on water and sediment quality assessment, and natural resource damage assessment.

Bejurin J. Cassady

Mr. Cassady is senior editor/production coordinator for Windward. He has been a technical writer and editor for more than 25 years and still enjoys it.

11.4.1 INTRODUCTION

Streams present environmental sample collectors with a wide range of habitat types, from swift-moving mountain streams with cobble and gravel substrates, to large rivers with varied substrates that may have powerful currents to contend with, to slow-moving coastal streams with soft bottoms. This article focuses mainly in collecting benthic samples in smaller streams; the tools and techniques used in larger streams and rivers with soft bottoms are substantially the same as those used for collecting marine and estuarine benthic samples, and are not described here.

Smaller streams may be too shallow or too densely covered to allow the use of sampling platforms, limiting the sampling approach to wading with handheld devices. The combination of stream flow and uneven substrates can make stream sampling difficult and potentially dangerous. This section provides an overview of methods and equipment for collecting benthic samples in stream environments. The collection of bottom sediment for chemical and biological analysis is covered in Chapter XI, Section 5, and only briefly referred to in this section. Further discussions concerning the collection of macroinvertebrates in flowing streams can be found in ASTM (1997) and EPA (1997).

As with the assessment of any environmental medium, there are many factors to consider when collecting a benthic sample from a stream. These factors can be grouped into the following categories:

1. **Study Objectives and Sampling Plan**—Identifying clearly defined study objectives and developing a study approach designed to gather the types of data needed to address the study objectives are critical factors in successful sample collection programs.

2. **Sample Location**—Clearly defining stream sampling station locations can be difficult, especially if the stream is located in a remote area or in heavy cover, where many traditional station positioning methods will not work.

3. **Sample Collection Devices**—The basic types of devices used to collect benthic stream samples for environmental studies are sediment samplers, in-stream net samplers, and artificial substrate samplers.

4. **Sample Handling, Manipulation, and Storage**—Once collected, stream samples must be handled and stored using prescribed methods and procedures in order to eliminate the potential for contaminating the samples, to preserve their integrity, and to ensure that quality assurance requirements for specific analytical techniques are met.

11.4.2 *STUDY OBJECTIVES AND SAMPLING PLAN*

The underlying purpose for the sample collection is clearly the most critical factor in determining what type of sample to collect. Developing clearly stated study objectives early in the study design process is a crucial step in ensuring that samples collected in the field will yield data acceptable for reaching an informed conclusion. A valuable tool for ensuring the quality of data collected is a set of data quality objectives, or DQOs. DQOs provide the basis for developing sampling and analysis plans and for setting specific quality assurance and quality control objectives for the samples to be collected. Further discussions concerning study planning, sampling design, and statistical methods for field studies can be found in various textbooks, manuals, regulatory guidance documents, and journal articles (e.g., Green, 1979; EC, 1994; Clarke and Green, 1988; Bernstein and Zelinski, 1983; EPA, 1998; also Hurlbert, *Ecol. Monogr.*, 84, pp. 187–211, 1984).

11.4.3 *SAMPLE LOCATIONS*

There are a number of station positioning methods that could be used to establish stream sampling station locations. For smaller streams, which are often in remote locations and under heavy cover, methods employing optical systems, such as theodolite or sextant, and electronic methods, such as those using microwaves, are often not effective. Instead, stream sampling locations are typically identified on maps while in the field, following a prescribed method developed as part of the sampling plan. One approach is to predetermine a given stream reach length, e.g., 100 m, from which samples will be taken. Once a stream reach has been identified in the field, a map of the reach is drawn, identifying in-stream attributes, such as riffles, pools, and fallen trees, and important structures, such as bridges, nearby roads, and buildings. Samples are then collected in various locations within the sample reach, and are often taken so as to represent the morphology of the reach, e.g., samples from runs, riffles, or pools. The location of each sample site is noted on the map of the reach. It may be possible to mark the station by driving metal stakes into both sides of the stream bank and measuring distances to the sample location from each stake with a tape measure. Alternatively, a handheld satellite global positioning system (GPS) can be used to establish the station location. However, even when using GPS, a detailed map of the stream reach being sampled should be drawn to aid in reoccupying the sample station if required. Way points collected using GPS can also be transferred to a digital map, if one exists for the stream. For larger streams and streams in open locations, optical and electronic methods may often be used.

11.4.4 SAMPLE COLLECTION DEVICES

The types of sample collection devices used to collect benthic samples from streams include sediment samplers—grab and core samplers, in-stream net samplers (stream-net and drift-net samplers)—and artificial substrates. Sediment samplers are used to remove a small portion of the habitat for analysis. In-stream samplers strain invertebrates from flowing water, either those from a defined area of substrate (stream-net samplers), or those drifting in the stream within a given period of time (drift-net samplers). Artificial substrates provide clean substrate for benthic organisms to colonize over a given period of time, after which the sampler is retrieved.

The following sections present brief descriptions of these sampling devices. A comprehensive review of sampling methods and gear for freshwater macroinvertebrates can be found in Merritt and Cummins (1996); Resh et al. (1990) is a videotape of sampling procedures using various devices.

Surber sampler

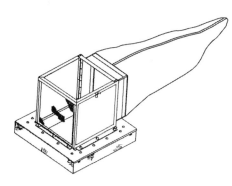

Portable invertebrate box sampler

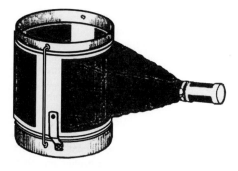

Stream-bed fauna sampler (Hess type)

FIGURE 11.4.1 Commonly used stream-net samplers. (Drawing courtesy of Kahl Scientific Instrument Corp.)

Sediment Samplers

Sediment samplers, which include grab samplers and core samplers, are used to collect samples of sediment along with the benthic invertebrates that are in it from soft-bottom substrates. These types of samplers and their use will not be further discussed here. Procedures for their use are not substantially different in stream sampling settings, except that small, shallow, and difficult-to-access streams can usually only be sampled using handheld devices, precluding the use of many sediment samplers.

In-Stream Net Samplers

Stream-Net Samplers.   Stream-net samplers are used for collecting benthic samples in shallow, hard-bottomed streams with rock, cobble, and gravel substrates. These devices, which sample a defined area of substrate and are thus akin to grab samplers, are designed to be used in a wide range of habitat types, and are intended to provide either a qualitative or a reasonably quantitative sample. Figure 11.4.1 shows examples of commonly used stream-net sampling devices. Table 11.4.1 summarizes the advantages and disadvantages of the various in-stream net samplers, including stream-net samplers.

The Surber sampler (Figure 11.4.1) can be used in a variety of substrate types including mud, sand, gravel, or rubble. The sampler consists of a square metal frame with a defined area, typically 0.1 m^2. A net is attached to the metal frame for collecting invertebrates washed out of the sampling area. Place the frame on the stream bottom so that the net trails out behind, pulled by the current. After placement, disturb the substrate within the metal frame by hand so that the invertebrates released from the substrate are swept into the trailing net. Inspect all stones and rocks within the frame to ensure that all invertebrates have been removed. Some species, such as mussels and snails, will have to hand-picked off the substrate.

TABLE 11.4.1   Stream-net Samplers for Invertebrate Collections in Streams

Sampler	Substrate type sampled	Advantages	Disadvantages
Surber sampler	Mud, sand, gravel, or rubble	Portable; samples a defined area	May not be quantitative because of backwash losses; net may clog; cannot be used in slow-moving waters
Invertebrate box sampler	Mud, sand, gravel, or rubble	Portable; samples defined area; box significantly reduces loss of invertebrates during sampling; box contours to bottom relief	Net may clog; may not work well in slow-moving waters
Hess sampler; Hess stream bottom samplers; streambed fauna sampler	Mud, sand, gravel, or rubble	Portable; samples defined area; cylinder significantly reduces loss of invertebrates during sampling; sampler can be "tuned" into the substrate	Net may clog; may not work well in slow-moving waters
Drift nets	All types of substrate	Collects invertebrates of all types; may collect more species from more taxa	Do not know where collected invertebrates originated; may capture terrestrial species

A drawback of the Surber sampler is that some organisms can be lost from the sampling area from backwash while the substrate is being disturbed. Several modifications to the Surber have been developed to address this problem by adding a more enclosed sampling chamber to prevent the loss of organisms. The portable invertebrate box sampler (Figure 11.4.1) has a square sampling chamber. There are also several types of cylindrical bottom samplers, including the Hess stream bottom sampler and the stream-bed fauna sampler (Figure 11.4.1).

The stream-net samplers differ in the manner in which the bottom of the sampler contacts the substrate. The Surber sampler frame is rigid, and does not conform to bottom contours, leaving potential gaps between the bottom of the sampler and the substrate. Such gaps can lead to loss of organisms during sampling. The invertebrate box sampler has a polyester foam pad on the bottom that allows the sampler to conform to a bottom relief of up to 7.6 cm. The cylindrical samplers are constructed of metal and have a thin bottom edge, which can be "turned" into the bottom to a depth of several centimeters to ensure closure. Guidelines for selecting and using stream-net samplers can be found in ASTM D 4556 (ASTM, 1997).

Drift-Net Samplers.   Another basic type of in-stream net sampler is the drift-net sampler, an example of which is shown in Figure 11.4.2. Table 11.4.1 includes a summary of advantages and disadvantages of drift-net samplers in collecting macroinvertebrates from streams. Drift-net samplers are designed to obtain both qualitative and quantitative samples of macroinvertebrates that drift in flowing streams. In addition to data on species presence and abundance, drift nets can provide data on drift density, rate, and periodicity. Current velocities generally have to be greater than 0.05 m/s in order for drift nets to work effectively, and only shallow streams can be sampled effectively, because in deeper streams certain classes of drifting organisms might be excluded by drifting over the sampler. Modifications to the general design of drift nets include automatic samplers, which can collect up to eight consecutive samples (Muller, *Limnol. Oceanogr.*, 10, pp. 483–485, 1965), and emergence-trap samplers that are useful in streams with extremely high drift (Mundie, *Limnol. Oceanogr.*, 9, pp. 456–459, 1964; Cushing, *J. Wildl. Manage.*, 48, pp. 592–594, 1964). Procedures for collecting benthic organisms using drift nets are provided in ASTM D 4556 (ASTM, 1997).

Artificial Substrates.   Qualitative and quantitative samples of stream macroinvertebrates can also be collected using artificial substrates, which are placed at sampling locations for a defined period

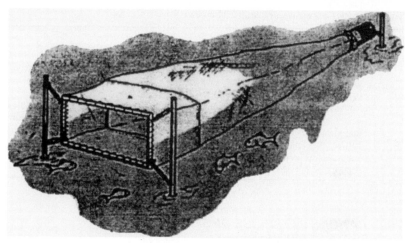

FIGURE 11.4.2   Drift-net sampler. (Drawing courtesy of Kahl Scientific Instrument Corp.)

of time to allow benthic organisms to colonize them. The artificial substrate is then retrieved and the organisms removed. Figure 11.4.3 illustrates one commonly used type of artificial substrate sampler, the multiple-plate sampler; Table 11.4.2 is a summary of advantages and disadvantages of using artificial substrate samplers to collect macroinvertebrates from streams. Artificial substrate samplers provide a clean substrate onto which invertebrate species can colonize. The devices are useful in areas where no other method is feasible because of site constraints. They are also useful in environmental studies in which the natural substrate may be toxic and sampling that substrate may not be a good estimator of the typical macroinvertebrates that might live in that habitat.

One type of commonly used artificial substrate sampler consists of an ordinary wire barbecue basket (e.g., for holding a chicken over the coals) filled with a substrate, such as clean rock, suitable for colonization by macroinvertebrates. The basket is typically anchored into the substrate and held within the water column by means of a float. The baskets are often deployed in groups to provide for replication to aid in data interpretation. One modification to the typical basket sampler is using a collapsible basket in rocky or uneven substrates (Bull, *Progressive Fish-Culturist*, 30, pp. 119–120, 1968). When deployed, the basket conforms to the relief of the substrate, enhancing colonization by macroinvertebrates that typically live on the nearby rock substrate. Guidelines for using the barbecue basket sampler can be found in ASTM E 1468 (ASTM, 1997).

A commonly used multiple plate sampler is the Hester-Dendy sampler (Figure 11.4.3). Multiple plate samplers have many of the same advantages and disadvantages as basket samplers, but one potentially significant advantage of the multiple plate sampler over the basket sampler is that the amount surface area of substrate available for colonization can be better controlled in the multiple plate sampler than it can in the basket sampler. Substrate for multiple plate samplers is typically tempered hardboard, or ceramic plates. The material can be cut to a uniform size and stacked together to form the sampler. Samplers generally consist of 8 or 14 plates. The available surface area of substrate is 939 cm^2 for the 8-plate sampler, and 1160 cm^2 for the 14-plate sampler. Guidelines for using multiple-plate samplers can be found in ASTM E 1469 (ASTM, 1997).

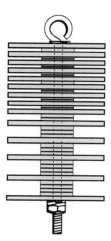

FIGURE 11.4.3   Hester-Dendy artificial substrate sampler. (Drawing courtesy of Kahl Scientific Instrument Corp.)

TABLE 11.4.2 Advantages and Disadvantages of Artificial Substrate Samplers

Sampler type	Advantages	Disadvantages
Basket sampler	Effective in evaluating biological integrity of surface waters; can be used in both lentic and lotic waters; uniform substrate type allows for comparison between different habitat types	Substrate may not be suitable for all macro invertebrate species; stream flow may affect colonization rate and species able to colonize substrate
Multiplate sampler	Effective in evaluating biological integrity of surface waters; can be used in both lentic and lotic waters; uniform substrate type allows for comparison between different habitat types; uniformity of substrate area available for colonization easier to control than in basket samplers	Substrate may not be suitable for all macro invertebrate species; stream flow may affect colonization rate and species able to colonize substrate

11.4.5 CONSIDERATIONS IN THE USE OF STREAM MACROINVERTEBRATE SAMPLERS

The choice of which sampler type to use in a study will depend in large degree on the objectives of the study. The different samplers produce slightly different information about the macroinvertebrate population of a stream. Stream-net samplers are designed to collect organisms inhabiting the specific sample area bounded by the sampler frame. Invertebrates collected within the net all originate from the sample area, with little opportunity for organisms from outside the area to be accidentally included in the sample, except in the case of the Surber sampler. Because this type of sampler is open on the upstream side, it may collect organisms within the drift, or those accidentally dislodged from the area upstream of the sampler. Data from stream-net samplers is useful in defining the benthic community structure of a given section of a stream, because there is a clear relationship between the sample collected and the area sampled.

By contrast, drift-net samplers are designed to collect macroinvertebrates drifting within the stream, whose place of origin is thus not known. Because the nets are designed to capture all types of organisms in the drift, drift-net samplers provide a better measure of organisms present in all habitat types found within the stream. Stream-net samplers cannot be used very effectively in collecting organisms from some habitat types, such as dense vegetation and organic substrates (e.g., sunken logs), while the drift-net samplers can be used to obtain information about the structure of these habitats. Data from drift net samplers can be used to quantify the types of food reserves available to fish populations, because the nets directly measure macroinvertebrates in the drift, a food resource actively exploited by fish.

Artificial substrate samplers give a different measure of the benthic community, by providing a clean substrate where benthic organisms in early life-stages can settle and grow. The substrates typically attract organisms at developmental stages that would not be retained by drift nets, therefore providing an alternative measure of the types of organisms that may be in the drift. Artificial samplers are useful in pollution studies in which the benthic community is impacted by contaminants in the substrate. Data from the samplers provides information on the potential for recolonization of the substrate, should the contaminants be removed. However, many kinds of benthic organisms that inhabit soft-bottom sediments will not be represented on artificial substrate samplers, because the type of material used in the sampler (e.g., rock, hardboard, or ceramic plates) is not suitable for colonization of these species.

In-stream net samplers all have one requirement in common: In order to be effective, current velocity should not be less than 0.05 m/s. Water flow is critical to keep the net deployed, and to ensure that organisms dislodged from the substrate move into the net. Water must be relatively shallow for most of the net samplers to be used effectively. For the stream-net samplers, water depth has to be less than 1 m, because field personnel must be able to physically disturb the substrate within the area to be sampled. The size and configuration of the samplers can also limit allowable water depth. For example, the Surber sampler cannot be placed in water deeper than 30 cm; at greater depths, organisms disturbed from the substrate may wash over the top of the net rather than into it.

Net mesh size will also influence the macroinvertebrate community collected by the sampler. Samplers are available with a variety of net mesh sizes; the recommended standard is U.S. Standard No. 30 mesh opening, or 0.595 mm. ASTM operationally defines macroinvertebrates as benthic and substrate dwelling organisms visible to the unaided eye and of a size retained on U.S. Standard No. 30 mesh net (ASTM D 4556). If the researcher is interested in obtaining data on the smaller instar stages of benthic organisms and insects that might pass through a Standard No. 30 mesh net, use nets of smaller mesh in sampling.

11.4.6 SAMPLE PROCESSING AND STORAGE

Following field collection, macroinvertebrate samples must be preserved, packed, and transported to a laboratory. At the laboratory, preserved samples are sorted, identified, counted, and analyzed as specified by the study objectives.

Sample Preservation

Samples must be preserved for storage promptly after collection. The method of preservation will depend on the study design. For studies that focus on macroinvertebrate community analysis or organism anatomy, samples are preserved chemically. Formaldehyde has been widely used, typically a 10 percent neutrally buffered solution. However, in response to environmental and individual health concerns, a 95 percent ethanol solution is now frequently substituted. Samples are transferred from the collecting or holding device to a glass jar, or to a plastic container to minimize breakage. The sample is then completely immersed in the preservative. In studies requiring chemical analysis of the sample, samples are either frozen or refrigerated in appropriate containers pending analysis.

Sample Labels

Sample containers should be labeled in the field as they are filled. A descriptive label should be placed inside the sample container, and one firmly attached to the outside of the container; these labels will remain with the sample through the handling and sorting process. Required label information varies with the type of study, and should be specified in a sampling and analysis plan. Label information may typically include the following:

- Project name
- Sample identification number
- Sample type
- Preservation technique,
- Date and time of collection
- Intended analyses
- Initials of person(s) preparing the sample

Labels to be placed inside sample containers should be marked in pencil on paper with a high rag to ensure readability.

Chain of Custody

Many studies require that, to ensure the integrity of samples, their possession be traceable from the time they are collected through analysis and archival. Samples are considered to be in custody if they are: (1) in the custodian's possession or view, (2) retained in a secured place (locked) with restricted access, or (3) in a container that is secured with an official seal(s) such that the sample cannot be reached without breaking the seal(s). The documents used to document possession are chain-of-custody records, field logbooks, and field tracking forms. Recommended procedures for maintaining a chain of custody and example forms may be found in EPA (1998).

Laboratory Processing

Sample processing in the laboratory consists of sorting, identification, and enumeration. Prior to sorting, samples that are preserved with formaldehyde are gently rinsed with fresh water into a 0.5-mm mesh screen to remove the formaldehyde from the sample material. All material retained on the screen is transferred to a glass or plastic jar, covered with 70 percent ethanol, and gently agitated to ensure proper preservation of the material by the alcohol. All internal and external labels are transferred along with the samples.

Sorting.   A number of techniques, such as elutriation and flotation, are available to make the labor-intensive process of sorting less onerous. These techniques are described in detail in Weber, *Biological Field and Laboratory Methods for Measuring the Quality of Surface Waters and Effluents* (NERC/EPA, Cincinnati, 1973); Cuffney et al., *Guidelines for the Processing and Quality Assurance of Benthic Invertebrate Samples Collected as part of the National Water-Quality Assessment Program* (USGS Open-File Report 93-407, 1993); and APHA (1992).

Enumeration and Identification.   Following sorting, all whole organisms are counted and identified to a taxonomic level consistent with the study objectives (Resh and McElrary, in Rosenberg and Resh, eds., *Freshwater Biomonitoring and Benthic Invertebrates*, Chapman & Hall, NY, 1993). Identifications should be performed by experienced taxonomic experts.

Quality Assessment and Quality Control.   To ensure the accuracy of the data, a QA/QC program should be included in the study design. Following are typical QA/QC procedures for both sorting and taxonomy. Twenty percent of each processed sample (i.e., the debris and unidentified matter remaining after organisms have been identified and removed) is randomly selected and resorted, by someone other than the original sorter, to check sorting efficiency. A sample will pass if, for instance, the number of organisms found during the QA/QC check does not represent more than 5 percent of the total number of organisms found in the full sample during the original sorting (a 95 percent sorting efficiency is a typical goal for a study). If the number of organisms found is greater than 5 percent of the total number, the entire sample is resorted. QA/QC for taxonomy is achieved by sending representative specimens of questionable or uncertain organisms out for independent confirmation by a qualified regional expert.

GENERAL REFERENCES

APHA, *Standard Methods for the Examination of Water and Wastewater*, Greenberg, Clesceri, and Eaton, eds., American Public Health Association, Washington, DC, 1992.

ASTM, *Annual Book of ASTM Standards*, Volume 11.05, ASTM, W. Conshohocken, PA, 1997.

Bernstein and Zelinski, "An Optimum Sampling Design and Power Tests for Environmental Biologists," *J. Environ. Manag.*, 16: 35–43, 1983.

Clarke and Green, "Statistical Design and Analysis for a 'Biological Effects' Study," *Mar. Ecol. Prog. Ser.*, 46: 213–226, 1988.

EC, *Guidance Document on Collection and Preparation of Sediments for Physiochemical Characterization and Biological Testing*, Environment Canada, Ottawa, 1994.

EPA, *Guidance for Quality Assurance Project Plans*, U.S. EPA, Washington, DC, 1998.

EPA, *Rapid Bioassessment Protocols for Use in Streams and Rivers: Periphyton, Benthic Macroinvertebrates, and Fish*, U.S. Environmental Protection Agency, Washington, 1997.

Green, R. H., *Sampling Design and Statistical Methods for Environmental Biologists*, Wiley, New York, 1979.

Merritt and Cummins, *An Introduction to the Aquatic Insects of North America*, third edition, Kendall/Hunt Publishing Company, Dubuque, 1996.

Resh, Feminella, and McElravy, *Sampling Aquatic Insects* (videotape). Office of Media Services, University of California, Berkeley, 1990.

CHAPTER 11
ASSESSMENT SAMPLING AND MONITORING

SEDIMENT COLLECTION

D. Michael Johns
Dr. Johns is a founding partner of Windward Environmental LLC. He is an aquatic scientist specializing in aquatic ecological risk assessments, particularly those associated with contaminated sediment. He has served as an advisor and technical expert in the development of water and sediment quality standards, and is a recognized expert on the use of bioassessment techniques to evaluate sediment contamination.

Beth A. Power
Ms. Power is an environmental scientist specializing in aquatic bioassessment, ecological risk assessment, and contaminated sediments with EVS Environmental Consultants.

Bejurin J. Cassady
Mr. Cassady is senior editor/production coordinator for Windward. He has been a technical writer and editor for more than 25 years and still enjoys it.

11.5.1 INTRODUCTION

The primary purpose for collecting sediment samples is to determine the quality of the sediment in its natural condition. The quality of field-collected sediment samples is typically characterized using physical and chemical analyses, and biological laboratory tests for toxicity or the potential for bioaccumulation. The correct sampling gear and collection and handling methods are critical factors in maintaining sample integrity and ensuring that samples accurately represent the environment from which they were collected.

Sediment structure can be complex, and specific collection methods are required if that complexity needs to be preserved. The photographs in Figure 11.5.1 show examples of depositional structures that may be important in a study of sediment characteristics or in defining the relationship between sediment and benthic organisms. The photographs were obtained using a sediment-profile camera (Rhoads and Germano, Mar. Ecol. Prog. Ser., 8, pp. 115–128, 1982), which captures an image of the upper 20 cm (8 in) of sediment using a camera mounted within an optical prism that is lowered into the sediment. Studies designed to define the sediment-water interface or biological activity in the upper biotic zone will need to employ sampling gear and adopt handling methods that preserve this structure. Sediment studies such as dredged material evaluations, designed to define the bulk chemistry and toxicity of sediment, can employ different sediment collection and handling techniques, because the maintenance of the structure of the sediment profile is not important.

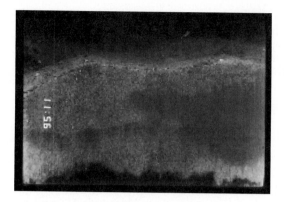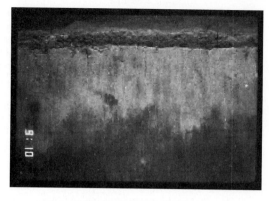

FIGURE 11.5.1   Sediment-profile camera images of depositional structures. (Images courtesy of EVS Environmental Consultants)

This section discusses the factors to consider when undertaking a sediment collection program, and provides an overview of the types of sampling gear and sediment handling procedures commonly used. More detailed information regarding collecting sediment samples can be found in the general references given.

As with the assessment of any environmental medium, there are many factors to consider when collecting a sediment sample. These factors can be grouped into the following categories:

1. **Study Objectives and Sampling Plan**—Identifying clearly defined study objectives, developing an approach, and creating a detailed plan to gather the types of data needed to address the study objectives are critical factors in successful sediment collection programs.

2. **Sample Location**—Establishing definable sediment sampling station locations can be difficult, because the sediment is under some depth of water. The station positioning methods used should provide sufficient accuracy to allow for reoccupation of the station if required.

3. **Sample Collection Devices**—The two basic types of devices used to collect sediment samples for environmental studies are grab samplers and core samplers. Grab samplers are primarily used to collect samples at or near the sediment surfaces, while coring devices are used to obtain a profile of deeper sediments.

4. **Sample Handling and Manipulation**—Once collected, sediment samples must be handled using prescribed methods and procedures order to eliminate the potential for contaminating the samples, and to ensure that quality assurance requirements for specific analytical techniques are met.

5. **Sample Storage and Shipping**—Sediments have a relatively short "shelf life," which varies according to the analyses being conducted on the sample, ranging from several weeks when stored unfrozen, to as much as a year when stored frozen. Proper sample storage is critical to obtaining high quality data.

11.5.2   *STUDY OBJECTIVES AND SAMPLING PLAN*

The purpose of the sample collection is the first factor to consider in determining what type of sediment sample to collect. Developing clearly stated study objectives early in the study design process is a crucial step in ensuring that the sediment samples collected in the field will yield data acceptable for reaching an informed conclusion. A valuable tool for ensuring the quality of data collected is a set of data quality objectives, or DQOs. DQOs provide the basis for developing sampling and analysis plans and for setting specific quality assurance and quality control objectives for the samples to be collected.

The study design should be spelled out in a planning document, or sampling and analysis plan. This plan should describe in detail all procedures relating to sample station location, sample collection, sample processing, sample shipping and handling, record keeping, sample analysis, and reporting of results. It should include examples of all forms that will be used during the study. Further discussions concerning study planning, sampling design, and statistical methods for field studies can be found in the general references given, in various textbooks and manuals, and in journal articles (e.g., Green, *Austral. J. Ecol.*, 18: 81–98, 1993; Clarke and Green, *Mar. Ecol. Prog. Ser.*, 46: 213–226, 1988; Bernstein and Zelinski, *J. Environ. Managem.*, 16: 35–43, 1983; Hurlbert, *Ecol. Monogr.*, 84: 187–221, 1984).

11.5.3 SAMPLE LOCATION

Station positioning methods vary widely in sophistication and accuracy. Which method to use should be determined early in the study planning process. Study design and location are important factors in determining the type of station positioning method to use. Study design and location factors to consider in selecting a station positioning method include the intended use of the data, physical conditions and topography of the study area, proposed sediment collection equipment and types of analyses to be performed on the collected sediment, minimum spatial separation in sampling station locations, whether there will be a need to reoccupy the sampling stations (e.g., for the collection of replicate samples or for a time-series analysis), and constraints imposed by the program under which the sediment study is being conducted (PSEP, 1996).

If the samples are being collected as part of a general survey of sediment quality, then knowing the precise location of the sampling location may not be critical. It may be sufficient to indicate the approximate site of the station on a map (e.g., a USGS quad sheet) using local land features as a guide to station location. If the precise coordinates of the stations are needed, use an accurate station positioning method, such as an electronic or satellite-based system. Given the relative low cost of handheld satellite global positioning systems (GPS), which are accurate up to 3 m (10 ft) in most locations, it is recommended that such systems be employed even in general surveys not requiring precise coordinates for station locations. Providing precise sampling locations will enhance the potential future uses of the survey data, even if the study's objectives do not require that much precision.

The positioning methods available include optical (i.e., line-of-sight), electronic, and satellite systems. Common optical methods include the sextant and theodolite, common electronic positioning systems include microwave systems such as Miniranger®, Trisponder®, and Loran-C®, and the most commonly used satellite positioning system is GPS. The characteristics, advantages, and disadvantages of each of these systems are summarized in Table 11.5.1. Additional information and discussion on matching a station positioning system with sediment sampling study objectives can be found in EC (1994) and PSEP (1996).

11.5.4 SAMPLE COLLECTION DEVICES

Sampling Platforms

There are two basic approaches to collecting a sediment sample: sampling by hand and sampling from a platform. Sampling by hand usually involves wading or diving. Handheld sampling devices are usually coring-type and typically yield a low volume of sediment (<1 L). Samples can also be collected using a scoop while wading (e.g., in a shallow stream or in the intertidal zone during low tide). Handheld coring devices are commercially available or can be improvised from rigid plastic tubing.

The gear for collecting larger or deeper samples has to be deployed from a platform. Sampling platforms may include bridges and docks, but floating platforms are more commonly used. The most critical factor to consider in selecting a sampling platform is safety in handling sampling gear. Small surface grab samplers can be hand-operated from a boat deck because the weight of the device, even

TABLE 11.5.1 Methods Used to Define the Position of Sampling Stations

Category	Accuracy	Range	Advantages	Disadvantages
Optical Methods				
Theodolite	10 to 30 seconds >±1 m	Up to 5 km	• Traditional method that measures horizontal angles between known targets • Inexpensive • High accuracy, successfully applied	• Triangulation between two manned shore stations or targets is required • Simultaneous measurements are required, limits on intersection angles, area coverage is limited because it requires good visibility, sampling platform must be stationary
EDMI	1.5 to 3.0 cm	Up to 3 km	• Extremely accurate • Usable for other surveying projects • Relatively inexpensive • Compact, portable, rugged	• Motion and directionality of reflectors • Good line-of-sight visibility is necessary unless microwave unit available • Two shore stations are required • Ground wave reflection causes errors
Total stations	5 to 7 m	<5 km	• Single onshore station • Other uses • Minimum logistics	• Reflector movement and directionality, prism costs, line-of-sight, optical or infrared range limitations
Sextant	±10 s ±3 to 5 m, but variable	Up to 5 km	• Portable, handheld device that can be highly accurate with experienced operator • Rapid, easy to implement • Common equipment • Low cost, no shore party necessary • High accuracy when used close to shore	• Triangulation requires two targets; good target visibility required • Orientation of target affects accuracy • Simultaneous measurement of two angles are required • Location and maintenance of targets is required for relocation of station • Line-of-sight method, best in calm conditions • Limits on acceptable angles
Pelorus	Variable	<5 km	• Rapid, easy to implement • Common equipment • Low cost, no shore party necessary • High accuracy when used close to shore	• Simultaneous measurement of two angles and good target visibility are required • Location and maintenance of targets is required for relocation of station • Line-of-sight method, best in calm conditions • Limits on acceptable angles
Radar	Variable	Up to 50 km	• Standard equipment on ships • Easy to operate • Yields range and relative bearing to targets	• Not portable • Requires a suitable target (i.e., one that reflects microwave signals)
Electronic Positioning Systems				
Microwave navigation systems (e.g., Miniranger, Trisponder, Racal Microfi, Del Norte)	<3 m	Up to 80 km	• No visibility restrictions • Multiple users • Highly accurate • Radio line-of-sight • Portable, easy to operate	• Moderately expensive • Multiple onshore stations required • Logistics and security of the necessary shore units increases cost • Signal reflective nulls are potential source of error • Limited range because of low-powered shore units

(Continued)

TABLE 11.5.1 Methods Used to Define the Position of Sampling Stations (Continued)

Category	Accuracy	Range	Advantages	Disadvantages
Shoran	±10 m	<80 km (short range)	• Moderately accurate	• Limited in range • Requires two shore transmitters
LORAN-C	<±15 m	Up to 300 km (medium-range)	• No visibility or range restrictions • No additional personnel • Low cost • Existing equipment	• Interference in some areas • Used only for repositioning, except in limited areas • Need to locate stations initially with another system • Coverage is not universal
Range-azimuth	0.02 m and 0.5 m	<5 km (optical) 30 km (elec)	• High accuracy • Single station • Circular coverage	• Single user • High cost, line-of-sight methods • Signal reflective nulls are potential error source
Satellite Positioning Systems				
GPS or Navstar	±20 to 30 m (± 0.1 to 1 m for differential GPS)	No limit on the range	• Continuous position reports available worldwide	• Relatively new, therefore cost is likely to decrease and its use is expected to increase greatly in the next few years; military scrambling can be a site-specific problem

Source: Adapted from EC 1994; PSEP 1997.

loaded with sediment, may be no more than 5–7 kg (10–15 lb). Larger samplers, such as box corers, can weigh as much as 500 kg (1100 lb) when filled with sediment. It is critical that the strength capacity of all lines and cables, and the lifting capacity of the winch, be appropriate to the sampling gear. Other factors to consider in selecting a sampling platform include size of the open area of the vessel for sample collection and processing activities, the availability of onboard space for sample preparation and storage, the maneuverability and draft depth of the vessel vis-à-vis the area to be sampled, and the practicality of holding the vessel in a fixed location long enough for sample collection to take place.

Types of Sampling Devices

The two types of samplers generally used for sediment studies are grab samplers and core samplers. Dredge samplers, a third type, are not recommended for collecting sediment samples. Dredge samplers are dragged across the bottom to collect a sample, an action that causes significant disruption of sediment integrity and washing out of fine-grained sediments during sampling and gear retrieval.

Grab Samplers. These typically consist of a set of jaws that close on surface sediments, or a bucket that rotates into the surface sediment when it reaches the bottom. Grab samplers only sample the upper portions of the sediment, typically penetrating no more than 30 cm (12 in) below the sediment-water interface. The main advantages of grab samplers are that they can be operated by hand and that they can collect a sufficient volume of sediment for most routine chemical and physical analyses. Some of the disadvantages of grab samplers include the potential disruption of fine-grained surficial sediment by the bow wave of the sampler as it nears the bottom, disruption of the sediment layers by grab penetration, and the washing out of fine-grained sediment during gear retrieval. All of these disadvantages can be overcome or at least minimized by adhering to recommended handling protocol for each type of gear.

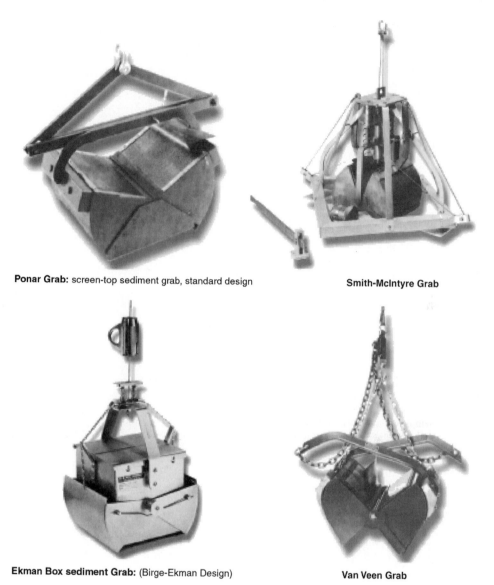

Ponar Grab: screen-top sediment grab, standard design

Smith-McIntyre Grab

Ekman Box sediment Grab: (Birge-Ekman Design)

Van Veen Grab

FIGURE 11.5.2   Commonly used grab samplers. (Photograph courtesy of Kahl Scientific Instrument Corp.)

Figure 11.5.2 shows examples of commonly used grab samplers. Table 11.5.2 is a summary of the advantages and disadvantages of various grab samplers. The most commonly used, general-purpose grab samplers are the Van Veen and Ponar samplers. Both can effectively obtain acceptable samples in a variety of sediment types, from those dominated by fine-grained material to those dominated by sand. Both devices have hinged lids, which reduce the formation of a bow wave during descent, reduce water flow through the device during retrieval to reduce the loss of fine-grained material, and allow for easy access to the sediment during removal. Lightweight grab samplers that are commonly used include the mini-Ponar and the Birge-Ekman samplers. Both can be operated

TABLE 11.5.2 Common Types of Grab Samplers

Device/dimensions	Substrate type	Sample depth (cm)	Sample volume (L)	Advantages	Disadvantages
Orange Peel, Smith-MacIntyre	Soft sediments, silt, sand, and clay	0 to 30	20	Designed for sampling hard substrates	Loss of fine-grained sediment; heavy; may require winch
Birge-Ekman, small	Soft sediments, silt, and sand	0 to 10	3.5	Adequate for most substrates that are not compacted; easy handling	Restricted to low current conditions; over-penetration of the sediment because of weight of sampler
Birge-Ekman, large	Soft sediments, silt, and sand	0 to 30	13	Large sample obtained intact; easy handling	Restricted to low current conditions; over-penetration of the sediment because of weight of sampler
PONAR, standard	Useful on sand, silt, or clay	0 to 10	7	Adequate on large sample obtained intact, permitting subsampling; good for coarse and firm bottom sediments	Shock wave from descent may disturb fine-grained sediment; not useful in strong currents
PONAR, mini	Useful on sand, silt, or clay	0 to 10	1	Adequate for most substrates that are not compacted	Smaller volume does not minimize disturbance to sample; not useful in strong currents
van Veen	Useful on sand, silt, or clay	0 to 30	18 to 75	Adequate on most substrates that are not compacted	Shock waves from descent may disturb fine-grained sediment
Peterson	Useful on most substrates	0 to 30	9	Large sample; can penetrate most substrates	Heavy, likely requires winch; no cover lid to permit subsampling
Shipek sampler, standard	Soft sediment and silt	0 to 10	3	Sample bucket may be opened to permit subsampling; retains fine-grained sediments effectively	Heavy, may require winch
Mini Shipek	Useful for most soft substrates	0 to 3	0.5	Easily operated by hand from most platforms	Requires vertical penetration; small volume; washout of fine-grained sediment

Source: Adapted from EC (1994).

easily by hand or operated from a small boat. These samplers, however, are difficult to operate in strong currents or under poor weather conditions. Detailed instructions for operating grab samplers can be found in ASTM (1997). Relevant articles include D 4342 through D 4407, inclusive.

Core Samplers. These are used to collect vertical sediment profiles, and can thus sample at greater depth than grab samplers—up to 20 m (65 ft)—but they sample a smaller surface area of sediment. Coring devices can be relatively simple handheld devices, but also include gravity and box corers, which penetrate the sediment surface by force of gravity, corers that penetrate using piston action, and corers that penetrate by means of vibration. Most coring devices consist of two parts: (1) an outside coring barrel that is the main housing that accomplishes the vertical penetration of the sediment, and (2) an internal liner into which the sediment sample is collected. The outer barrel is used repeatedly, but the internal sediment collection liner is replaced for each sample. Figure 11.5.3 shows examples of some commonly used coring devices; Table 11.5.3 is a summary of the advantages and disadvantages of various core samplers. All coring devices except small handheld ones require a sampling platform with lifting capacity and several sampling crew members for safe and effective operation.

Handheld Core Samplers. These typically consist of tubes of a durable clear plastic, such as Lexan® or butyrate. Glass is not generally recommended because of the potential for breakage during sampling operations. Handheld coring devices are typically used to collect relatively shallow vertical profile samples in shallow water.

Gravity Core Samplers. These require a mechanical lifting mechanism for deployment and retrieval. The outside barrel of the coring device often is fitted with fins to improve stability during descent and to obtain vertical penetration. External weights can also be added to some types of gravity corers to enhance penetration depth. Box corers are a specific type of gravity corer. Unlike most gravity corers, which are barrel-shaped, box corers are rectangular and are designed to collect a large volume of sediment covering a relatively large surface area, up to 0.3 m² (1 ft²). Box corers commonly used include the Gray-O'Hara box corer and the Reineck box corer. Figure 11.5.2 illustrates the Eckman box sampler, which combines features of box and grab samplers.

Piston Core Samplers. These employ a piston device to drive the barrel sampler into the sediment. These devices require careful positioning of the barrel prior to activating the piston driver, which can lead to reduced sample collection efficiency. However, very large piston corers, with barrels 15 cm (6 in) in diameter, are the only means of effectively collecting the deep (e.g., 10–15 m [30–50 ft]) vertical cores required for some dredging projects.

Vibrating Core Samplers. These are commonly used to collect relatively deep (e.g., up to 3–6 m [10–20 ft]) cores. Vibrating corers penetrate the sediment by means of high-frequency vibration supplied by a vibrating head mounted onto the corer. Commercially available vibrating corers include a handheld, two-person sampler, a sampler that can be operated from a portable frame and winch system, and larger samplers that require permanently-mounted frame and winch systems.

11.5.5 SAMPLE PROCESSING

Sample Acceptance

Once a sample has been collected and retrieved, and the gear secured, inspect the sample to ensure that it is acceptable for the intended purpose. The following acceptability criteria should be considered when evaluating a sample (PSEP, 1997; EC, 1994):

- The sampler is not overfilled, causing the sediment surface to press against the top of the sampler.
- There is overlying water present.

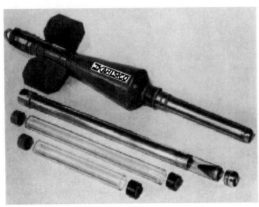

Gravity Corer
(Drawing courtesy of Kahl Scientific Instrument Corp.)

Piston Corer
(Photograph courtesy of Wildco Wildlife Supply Company)

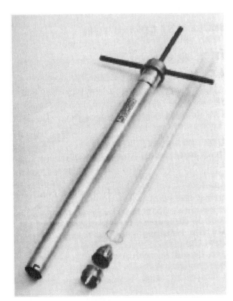

Handheld Corer
(Photograph courtesy of Kahl Scientific Instrument Corp.)

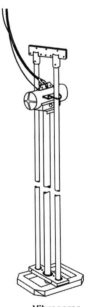

Vibrocorer
(Drawing courtesy of Rossfelder Corporation)

FIGURE 11.5.3   Commonly used core samplers.

TABLE 11.5.3 Common Types of Core Samplers

Device/dimensions	Substrate type	Sample depth (cm)	Sample volume (L)	Advantages	Disadvantages
Fluorocarbon plastic, cellulose acetate butyrate, or glass tube (3.5–7.5 cm ID; ≤120 cm long)	Soft or semi-consolidated deposits	0 to 10	<1	Preserves layering and permits historical study of sediment deposition; rapid; samples immediately ready for laboratory shipment; minimal risk of contamination	Small ample size requires repetitive sampling
Hand corer with removable fluorocarbon plastic or glass liners (3.5–7.5 cm ID; <120 cm long)	More consolidated sediments can be obtained than above	0 to 10	<1	Handles provide for greater ease or substrate penetration; above advantages apply	Careful handling necessary to prevent spillage; requires removal of liners before repetitive sampling; small ample size requires repetitive sampling
Box corer	Most sediment types	0 to 50	≤30	Collects large sample, undisturbed, allowing for subsampling	Hard to handle; heavy machinery required
Gravity corer (3.5 cm ID, ≤50 cm long)	Semi-consolidated sediments	0 to 50	<1	Low risk of sample contamination; maintains sediment integrity relatively well; high point loading with sharp cutting edge	Careful handling necessary to avoid sediment spillage; small sample required repetitive operation and removal of liners; time consuming
Benthos Gravity Corer (6.6, 7.1 cm ID ≤70 cm long)	Soft, fine-grained sediments	0 to 300	≤10	No loss of sample from tube because the valve is fitted to the core liner; fins promote vertical penetration	Weights required for deep penetration, which requires heavy machinery; vertical penetration is required; sediment compaction
Piston Corers	Most substrates	300 to 2000 cm	≤20	Typically recovers a relatively undisturbed sediment core in deep waters	Requires lifting capacity of >2000 kg; nonactivation of piston and piston positioning at penetration; disturbance of the surface (0 to 0.5 m) layer
Vibratory Corer (5.0 to 7.5-cm ID)	Sand, silty sand, gravelly sand substrates	300 to 600	≤13	Effectively samples sandy substrates with minimum disturbance; can be operated from small vessels or in an intertidal area	Labor intensive; assembly and disassembly might require divers; possible disturbance of the surface (0 to 0.5 m) layer

Source: Adapted from EC 1994.
ID = inside diameter.

- The sediment-water interface is intact and includes fine surficial structures (e.g., no sign of channeling or washout).
- The desired depth of penetration has been achieved.
- There is no evidence of incomplete closure of the sampler, or that it penetrated at an angle.

In addition, the following acceptability criteria should be considered when evaluating a core sample:

- There should be no evidence that the core sampler was inserted at an angle or was tilted upon retrieval.
- The sediment core should be complete, with no air space at the top of the liner before capping.
- The length of the core should be within the range stipulated in the sampling protocol.

If these and all study-specific criteria are met, the sample is deemed acceptable for use in analysis. If the sample fails any of the acceptability criteria, it should be rejected and not submitted for analysis. Discard rejected sediment samples in a manner that will not affect subsequent samples at that station or other possible sampling stations (EC, 1994). Note the location and nature of sample failures in the sampling log, which should be specified in the sampling and analysis plan.

Processing Grab Samples

If a sample is found to meet all acceptability criteria, remove the sample from the sampler and prepare it for analysis as described in the sampling and analysis plan. For grab samples, remove any overlying water from the sediment surface before the sample is taken from the sampler, disturbing the sediment surface as little as possible. One method is to siphon the water using a length of plastic tubing that has been primed with site water (PSEP, 1997). Take care to remove the water without disturbing or removing the fine-grained sediment.

If a grab sample is to be retained in its entirety, release the sample from the sampler directly into a precleaned container of the appropriate size, then seal and label the container. If subsamples will be withdrawn from a grab sample, use a grab sampler that allows access to the sediment surface. Subsamples can be collected using a number of methods, including small hand corers and scoops or spoons made of inert material such as stainless steel or Teflon®. The desired depth and volume of sample collected will largely be dictated by the study objectives and the analytical needs. When collecting subsamples for analysis, it is good practice not to collect any sediment that is in direct contact with the sampler, reducing the potential for contamination. Place retrieved samples directly into the appropriate size pre-cleaned containers, seal, and label.

Subsampling Grab Samples

Often a study design requires that a number of subsamples, collected from multiple grab samples, be composited into a single sample for analysis. Compositing of subsamples from a number of replicate grabs collected at a station or within a specified area is often used to provide a sample for analysis that better reflects the average characteristics of a parameter of concern, such as a contaminant concentration or toxicity. Compositing may be required to collect a sufficient volume of sediment to complete the intended analysis. In most cases, collect an equal volume of material from each of the grab samples to be composited, so that the composite reflects an equal contribution from each grab.

Homogenization

Sediment samples are often subjected to multiple analyses. Ideally, all analyses should be conducted on the same sample, to maximize comparability. Because a sample may be submitted to the laboratory in separate containers destined for analyses of grain size, metals, organic compounds, and so forth, throughly homogenize the sediment before placing it into the containers. Homogenization will help ensure that the sample is representative of the collected sediment. Before homogenizing, remove all material considered foreign to the sample, such as twigs, shells, stones, and seagrass. Place sediment in an appropriately sized, previously cleaned container made of an inert material such as stainless steel, and mix thoroughly with a cleaned stainless steel spoon until the texture, color, and moisture are homogenous. Mechanical mixers, such as portable cement mixers and stainless steel paddles that fit into portable drills, can be used (Ditsworth et al., *Environ. Toxicol. Chem*, 9: 1523–1529, 1990; Stemmer et al., *Environ. Toxicol. Chem.*, 9: 381–389, 1990), but mixing is commonly done by hand (Johns et al., ASTM STP 1124, 14: 280–293, 1991; Malueg et al., *Environ. Toxicol. Chem.*, 5: 245–253, 1986). Standardize mixing time to ensure homogeneity between samples, but also minimize mixing time to reduce oxidation of the sample or volatilization of organic contaminants. Once mixing is complete, place samples of the homogenized sediment into sample storage containers by placing small portions of the sample into each the sample container alternately until all are filled.

Processing Core Samples

For core samples, remove the sediment collection liner from the collecting device and cap it to prevent disturbance of the core profile. Place the caps on the upper end of the liner as close to the sediment surface as possible to minimize movement of the sediment core. Capped core liners may be stored horizontally, although vertical storage is preferred when practicable. Uncapped core liners have to be stored vertically. Cores may be left intact on the sampling platform and shipped to a laboratory to be opened and processed. If sampling operations are near shore, a temporary sample processing laboratory may be established on shore to facilitate core handling and sample preparation.

Sediment can be retrieved from the liner by extruding the sediment with pressure, or by cutting the liner to expose the core. The recommended method of core extrusion is to use a piston-like device to push the sediment out of the liner. The piston is typically inserted into the lower end of the liner, because the sediment surface is usually of particular interest. Liners are also commonly removed using a saw with a cutting-depth jig attached that keeps the saw from disturbing the sediment sample during liner removal. Most liners can be cut using a medical saw that cuts by means of blade vibration. These saws pose little hazard to the user and do not generate chips of liner material that may contaminate the sediment.

Subsampling Core Samples

As with grab-collected sediment samples, core-collected sediment samples can be subsampled to address specific study objectives. Subsampling of cores usually involves removing sediment from a particular sediment horizon in incremental sections. Studies focused on environmental pollution or sediment dating typically collect sediment in 1-cm increments, while studies designed to collect data for dredging projects may require sediment collected in increments of 1 m (3 ft) or more. Once the sample has been collected, use the sediment handling procedures described above to homogenize the sample material and distribute samples to containers.

11.5.6   *SAMPLE STORAGE AND SHIPPING*

In the field, store samples in a location that will minimize accidental loss or damage to the samples, and hold them at 4°C (40°F). The most common method for keeping samples cool is to place the sample containers in a cooler, then pack with ice or frozen ice packs (EC, 1994). Frozen ice packets are preferable to ice because of melting and possible leaking of water during transport. Cut long cores into suitable sections, keeping with the requirements of the sampling and analysis plan and DQOs, and place them into storage containers with ice or frozen ice packs. Ship samples to the laboratory for more

TABLE 11.5.4 Containers and Conditions Recommended for Storing Sediment Samples

End use	Container type	Wet weight or volume of sample typically required	Storage conditions	
			Temperature	Holding time*
Particle size distribution	• Teflon® • Glass • HDPE containers or bags	250 g	4 to 40°C	≤6 mo
Major ions and elements	• Teflon® • HDPE containers or bags	250 g	4 ± 2°C	≤2 wk
Nutrients	• Teflon® • Glass with Teflon® or polyethylene-lined cap	100 g	4 ± 2°C	≤8 h
Trace elements	• Teflon® • HDPE containers or bags	250 to 500 g	4 ± 2°C −20°C	≤2 wk ≤12 mo
Organic contaminants	• Stainless steel canisters • Aluminum canisters • Amber glass with aluminum-lined cap	250 to 500 g	4 ± 2°C −20°C	≤2 wk ≤12 mo
Sediments for bioassay tests	• Teflon® • Glass • HDPE bags or containers	1 to 3 L	4 ± 2°C	≤6 wk, preferably ≤2 wk

Source: Adapted from EC 1994.
HDPE = high-density polyethylene.
*Holding times are often specified by regulatory programs; incorporate into sampling and analysis plan.

permanent storage as soon as practicable. If the laboratory is near the sampling site, deliver collected samples to the laboratory daily. Chain-of-custody protocols should be specified in the sampling and analysis plan. If the samples have to be shipped to the laboratory, ship containers at least every other day by the fastest means possible.

Store sediment in wide-mouth containers made of glass, high-density polyethylene, or Teflon®. The container material selected will depend on the intended use of the sample, as specified in the sampling plan. Table 11.5.4 presents recommended sample containers and recommended sample volume for various analyses. The amount of time sediment samples can be stored and still yield acceptable data (holding time) will depend on the types of analyses being performed. Samples for most chemical analysis can be stored frozen for a period of up to one year, while unfrozen samples should be analyzed within two weeks. Samples to be used for bioassays or bioaccumulation testing should be stored unfrozen only, and no longer than six weeks, preferably two weeks. Additional guidance for the recommended handling and storage of sediment samples for specific analysis should be consulted when writing a sampling and analysis plan (American Public Health Assn., *Standard Methods for the Analysis of Water and Wastewater*, APHA, Washington, 1992; *ASTM Yearbook*, D 3976, 1995); and Puget Sound Estuary Program, *Monitoring Protocols VII, VIII, IX, AND XIII*, PSEP, Olympia, WA, 1987; 1989; and 1995). Some regulatory programs specify acceptable holding times for each type of analysis.

GENERAL REFERENCES

ASTM, *Annual Book of ASTM Standards*, Volume 11.05, ASTM, W. Conshohocken, PA, 1997.

Environment Canada, *Guidance Document on Collection and Preparation of Sediments for Physiochemical Characterization and Biological Testing*, EC, Ottawa, 1994.

Green, R. H., *Sampling Design and Statistical Methods for Environmental Biologists*, Wiley, New York, 1979.

Mudroch, A. and MacKnight, S. D., *CRC Handbook of Techniques for Aquatic Sediment Sampling*, CRC Press, Boca Raton, FL, 1991.

PSDDA, *Evaluation Procedures Technical Appendix, Phase 1 (Central Puget Sound)*, Puget Sound Dredge Disposal Analysis, Seattle, WA, 1988.

PSEP, *Recommended Guidelines for Sampling Marine Sediment, Water Column, and Tissue in Puget Sound*, Puget Sound Estuary Program (EPA), Olympia, WA, 1996; 1997.

EPA, "Evaluation of Dredged Material Proposed for Discharge in Waters of U.S.," *Inland Testing Manual (Gold Book)*, U.S. EPA, Washington, 1998.

EPA/ACOE, "Evaluation of Dredged Material Proposed for Ocean Disposal," *Testing Manual (Green Book)*, EPA/ACOE, Washington, 1991.

CHAPTER 11
ASSESSMENT SAMPLING AND MONITORING

SECTION 11.6
MONITORING WELL CONSTRUCTION AND SAMPLING TECHNIQUES

James A. Jacobs

Mr. Jacobs is a hydrogeologist and president of FAST-TEK Engineering Support Services with more than 20 years of experience. He is registered in several states, including California, and specializes in assessment methods and in situ remediation technologies.

11.6.1   INTRODUCTION

Subsurface environmental evaluations of spills or potential leaks of hazardous substances generally consist of four phases (Phase I through Phase IV; see Table 11.6.1).

The generalized four phases reflect the realistic process required to take a potentially impacted property from assessment phase (Phase I) through to the final monitoring and site closure phase (Phase IV). The phases generally proceed in order starting with Phase I. Some sites where soil or groundwater contamination is already suspected or documented might start with Phase II. This is common in the case of an underground tank or even Phase III, in the case of visually stained surface soils.

PHASE I Environmental Assessment is a set of noninvasive techniques to acquire information from site inspection and to obtain data supplied by others. Phase I involves a site inspection, the development of the history of the subject property, and review of data supplied by regulators, environmental lists and building permit databases, owners, tenants and others. Techniques include interviews of knowledgeable persons, review of historical aerial photography, and examination of published and unpublished maps. Noninvasive data may be collected on a property using meters, such as a hand-held vapor meter. Other noninvasive techniques that would generate data include surface geophysics and surface passive and active vapor surveys. Phase I may also coincide with the transfer of property ownership and may be part of environmental due diligence.

PHASE II Subsurface Investigation normally is invasive and is designed to evaluate the geologic and hydrogeologic conditions by collecting soil, soil vapor, and groundwater samples. A variety of different techniques and equipment are currently available for assessing the subsurface. Environmental subsurface investigation tools range in size, cost, and operating complexity from hand augers and hand operated drive samplers to direct push technology (DPT) rigs to rotary rigs. Borings are drilled for geologic, hydrogeologic, and lithologic characterization. Samples are collected for field screening and physical testing. Selected soil, vapor, or groundwater samples are submitted to a certified laboratory for chemical testing (see Table 11.6.2). As part of a Phase II investigation project, groundwater monitoring wells may be installed. Several Phase II Subsurface Investigations might be required prior to completing the Phase II process and fully characterizing vertical and lateral extent of the soil and groundwater contamination.

TABLE 11.6.1 Four Phases of Subsurface Investigations

Phase I—Preliminary site assessment
Phase II—Soil and water investigation
Phase III—Corrective action plan implementation
Phase IV—Verification monitoring

TABLE 11.6.2 Comparison of Subsurface Investigation Tools (Testa, 1994)

Drilling technique	Material	Limitations (in feet)	Advantages	Disadvantages
Augering Hand auger	Unconsolidated	15	Shallow soils investigations Soil samples Shallow water-bearing zone identification Piezometer, lysimeter, and small-diameter monitoring well installation Labor intensíve, but inexpensive No casing material restriction	Limited to very shallow depths Unable to penetrate extremely dense or rocky soil Borehole stability difficult to maintain Labor intensive
Bucket auger	Unconsolidated	100	Shallow soils investigations Soil samples Vadose zone monitoring wells (lysimeters) Monitoring wells in saturated, stable soils Identification of depth to bedrock Fast and mobile	Unacceptable soil samples unless split-spoon or thin-wall samples are taken Soil sample data limited to areas and depths where stable soils are predominant Unable to install monitoring wells in most unconsolidated aquifers because of borehole caving upon auger removal Depth capability decreases as diameter of auger increases Monitoring well diameter limited by auger diameter
Solid-flight auger	Unconsolidated	180	Shallow soils investigations Soil samples Vadose zone monitoring wells (lysimeters) Monitoring wells in saturated, stable soils Identification of depth to bedrock Fast and mobile	Unacceptable soil samples unless split-spoon or thin-wall samples are taken Soil sample data limited to areas and depths where stable soils are predominant Unable to install monitoring wells in most unconsolidated aquifers because of borehole caving upon auger removal Depth capability decreases as diameter of auger increases Monitoring well diameter limited by auger diameter
Hollow-stem auger	Unconsolidated	180	All types of soil investigations Permits good soil sampling with split-spoon or thin-wall samplers Water-quality sampling Monitoring well installation in all unconsolidated formations Can serve as temporary casing for coring rock	Difficulty in preserving sample integrity in heaving formations Formation invasion by water or drilling mud if used to control heaving Possible cross contamination of aquifers where annular space not positively controlled by water or drilling mud or surface casing

(Continued)

TABLE 11.6.2 Comparison of Subsurface Investigation Tools (Testa, 1994) (Continued)

Drilling technique	Material	Limitations (in feet)	Advantages	Disadvantages
			Can be used in stable formations to set surface casing	Limited diameter of augers limits casing size Smearing of clays may seal off aquifer to be monitored
Driven	Unconsolidated	50	Water-level monitoring in shallow formations Water samples can be collected Dewatering Water supply Low cost encourages multiple sampling points	Depth limited to approximately 50 ft (except in sandy material) Small diameter casing No soil samples Steel casing interferes with some chemical analysis Lack of straligraphic detail creates uncertainty regarding screened zones and/or cross contamination Cannot penetrate dense and/or some dry materials No annular space for completion procedures
Jet percussion	Unconsolidated	200	Allows water-level measurement Sample collection in form of cuttings to surface Primary use in unconsolidated formations, but may be used in some softer consolidated rock Best application is 4-in borehole with 2-in casing and screen installed, sealed, and grouted	Drilling mud may be needed to return cuttings to surface Diameter limited to 4 in. Installation slow in dense, bouldery clay/till or similar formations Disturbance of the formation possible if borehole not cased immediately
Cable tool	Unconsolidated	1000	Drilling in all types of geologic formations Almost any depth and diameter range Ease of monitoring well installation Ease and practicality of well development Excellent samples of coarse grained materials	Drilling relatively slow Heaving of unconsolidated materials must be controlled Equipment availability more common in central, northcentral, northeast, and northwest sections of the United States
Rotary Reverse circulation	Unconsolidated or consolidated	2000+	Very rapid drilling through both unconsolidated and consolidated formations Allows continuous sampling in all types of formations Very good representative samples can be obtained with minimal risk of contamination of sample and/or water-bearing zone In stable formations, wells with diameters as large as 6 in can be installed in open-hole completions	Limited borehole size that limits diameter of monitoring wells In unstable formations, well diameters are limited to approximately 4 in Equipment availability more common in the southwest Air may modify chemical or biological conditions; recovery time is uncertain Unable to install filter pack unless completed open hole
Mud rotary	Unconsolidated or consolidated		Rapid drilling of clay, silt, and reasonably compacted sand and gravel	Difficult to remove drilling mud and well cake from outer perimeter of filter pack during development

(Continued)

TABLE 11.6.2 Comparison of Subsurface Investigation Tools (Testa, 1994) (Continued)

Drilling technique	Material	Limitations (in feet)	Advantages	Disadvantages
			Allows split-spoon and thin-wall sampling in unconsolidated materials Allows core sampling in consolidated rock Drilling rigs widely available Abundant and flexible range of tool sizes and depth capabilities Very sophisticated drilling and mud programs available Geophysical bore-hole logs	Bentonite or other drilling fluid additives may influence quality of groundwater samples Circulated (ditch) samples poor for monitoring well-screen selection Split-spoon and thin-wall samplers are expensive and of questionable cost effectiveness at depths greater than 150 ft Wireline coring techniques for sampling both unconsolidated and consolidated formations often not available locally Difficult to identify aquifers Drilling fluid invasion of permeable zones may compromise validity of subsequent monitoring well samples
Air rotary	Unconsolidated or consolidated	2000+	Rapid drilling of semi-consolidated and consolidated rock Good quality/reliable formation samples (particularly if small quantities of water and surfactant are used) Equipment generally available Allows easy and quick identification of lithologic changes Allows identification of most water-bearing zones Allows estimation of yields in strong water-producing zones with short "down time"	Surface casing frequently required to protect top of hole Drilling restricted to semiconsolidated and consolidated formations Samples reliable but occur as small particles that are difficult to interpret Drying effect of air may mask lower yield water producing zones Air stream requires contaminant filtration Air may modify chemical or biological conditions Recovery time is uncertain
Air rotary (with casing driver)			Rapid drilling of unconsolidated sands, silts and clays Drilling in alluvial material (including boulder formations) Casing supports borehole thereby maintaining borehole integrity and minimizing inter-aquifer cross contamination Eliminates circulation problems	Thin, low pressure waterbearing zones easily overlooked if drilling not stopped at appropriate places to observe whether or not water levels are recovering Samples pulverized as in all rotary drilling Air may modify chemical or biological conditions; recovery time is uncertain
Air Percussion	Unconsolidated or consolidated	2000+	Rapid drilling	Poor sample recovery Air emission concerns
Wire-line	Consolidated	2000+	Excellent sample retrieval Maintains borehole integrity Allows continous casing of borehole	Not applicable to unconsolidated formations
Horizontal	Unconsolidated or consolidated	2000+	Allows access beneath structures Larger zone of influence Utilization for variety of remedial purposes Maintain higher specific capacities Likely to intersect vertical fractures	Maintaining directional orientation may present problems Potential caving in loose materials

TABLE 11.6.3   Purpose of Groundwater Monitoring Wells (Sisk, 1981)

- To define the vertical and lateral extent of groundwater contamination
- To monitor target chemical concentrations over time
- To provide a measurement for detecting the contaminant plume front, unexpected changes in plume size or direction of flow
- To determine the extent of interaquifer movement of contaminants
- To determine aquifer characteristics (such as permeability, transmissivity, etc.)
- To estimate the rate of contaminant plume movement
- To develop a data base for designing remedial measures
- To determine the effects of the remedial measures
- To assist in performing the remedial work (provide hydraulic control and contaminant removal)
- To provide data base for groundwater modeling
- To evaluate the aquifer during regular sampling events over the yearly hydrologic cycle

Source: (Modified after Sisk, 1981).

Phase III Corrective Action is the remediation portion of an environmental project. This phase involves designing and implementing the corrective action plan or remediation of soil and/or groundwater. A corrective action plan, evaluating remedial options and feasibility is typically submitted to the regulator prior to commencement of the field work. Remediation might include bioremediation (in-situ), chemical oxidation (in-situ), monitored natural attenuation (in-situ), soil excavation and removal to a landfill (dig and haul), or other technologies that are first evaluated in the corrective action plan. Drilling techniques might be used during the remediation phase in the design and construction of vapor or groundwater extraction wells.

Phase IV begins after remediation has occurred. The regulatory agencies generally require at least one year of quarterly groundwater monitoring (see Table 11.6.3) to verify that the majority of the source of the contamination has been removed. In some cases, if chemicals are still present in the soils or groundwater, a risked based corrective action (RBCA) may be requested by the regulatory agencies. The RBCA, performed using a computer model, is designed to evaluate whether there are human or nonhuman receptors that might be adversely affected by the chemicals remaining in the subsurface. Deed restrictions, worker notifications, or surface capping may be appropriate recommendations to reduce potential exposure to residual chemicals. The completion of Phase IV is site closure. Site closure includes the proper abandonment of existing groundwater monitoring and remediation wells or equipment.

11.6.2   DRILLING TECHNIQUES

A variety of different drilling techniques have been developed and are used in the environmental field. The factors to determine the best drilling technique relate to accessibility, time and cost of project, sediment type (consolidated rock or unconsolidated soils), sample type (undisturbed vs. disturbed), and sample integrity. Unconsolidated deposits are primarily drilled using hollow-stem augering techniques, with occasional use of the cable-tool drilling techniques. For consolidated or semiconsolidated deposits, continuous wire-line or conventional rock coring techniques are commonly used.

Other drilling techniques used in the environmental field include the use of labor-intensive hand augers for shallow depths, and machine-operated solid-flight augers. Both of these techniques can have sample retrieval problems because samples can only be collected if the entire auger is removed prior to sampling. More detail regarding drilling techniques is provided in Driscoll (1984), Testa (1994), Sisk (1981) and California DWR (1981; 1990).

Hollow Stem Auger Drilling

Hollow-stem continuous flight auger drilling techniques are commonly used for subsurface environmental projects. Hollow-stems consist of a series of continuous, interconnected hollow auger flights,

Lead Section Assembly

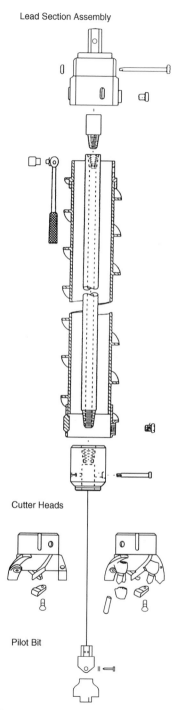

Cutter Heads

Pilot Bit

FIGURE 11.6.1 Hollow-Stem Auger with cutter heads. The center rod and pilot bit operate in the center of the auger (Foremost-Mobile Drill).

usually 5 to 10 ft in length. Typical hollow stem augers for use in the environmental field have inner diameters (ID) of 4-1/4 to 10-1/4 in to create 2- and 4-in diameter wells, respectively. The hollow stem flight augers are hydraulically pressed downward and rotated to start drilling. Soil cuttings are rotated up the outside of the continuous flighting in the borehole annulus. A center rod with plug and pilot bit are mounted at the bottom. The plug is designed to keep soil from entering the mouth of the lead auger while drilling. Upon reaching the sampling depth, the center rod string with plug and pilot bit attached is removed from the mouth of the auger and replaced by a soil sampler.

Soil sampling is achieved by passing a smaller diameter drill rod into the hollow stem auger with a soil sampler attached at the bottom. The soil sampler is lowered into the borehole through the hollow stem of the auger and sampling or coring is started. The sampler is typically either a thin-wall or modified split-barrel sampler with stainless steel or brass sample tubes. Samples can be continuously retrieved, although in the environmental field, soil samples are typically collected at five-foot intervals, or at significant changes in lithology. In addition, soil samples are usually collected at intervals of obvious contamination in order to develop a complete profile of soil contamination.

For the environmental field, the most popular split-barrel samplers are the 2-1/2 or 3-in outside diameter. It is a standard split-barrel sampler that has had its interior honed or otherwise machined to accept brass or stainless steel liners or rings. The split-barrel sampler is lined by three stainless steel or brass tubes each six inches in length, which can be removed from the sampler for capping and shipment to the laboratory. Reloading the sampler with liners allows continued use of the sampler.

Conventionally, the upper 6-in sample tube is used for lithologic description and physical testing (i.e., permeability, sieve analysis, etc.). The middle sample tube is used for field-screening for hydrocarbon or solvent vapors or other constituents of concern, and the bottom sample tube is used to send to a chemical laboratory for analysis. The sampler is driven into the soil at the desired sampling interval ahead of the auger bit by using a 140-LB hammer falling freely 30 in. A Standard Penetration Test (ASTM D-1586) requires a sampler with a 2-in outside diameter and 18 to 24 in in length with an inside diameter of 1-1/2-in and a drive shoe cutting diameter of 1-3/8 in. The count to advance the sampler one foot is the blow count. Blow counts collected by any other size of sampler can be useful, but they are not a standard penetration blow count, which may be used to determine the consistency or soil density information. Boreholes are either grouted using a tremie pipe or converted into a monitoring well. Soil cuttings are generally not used to backfill borings in environmental projects because of the potential for cross-contamination.

Cable-Tool Drilling

Cable-tool drilling is the oldest drilling technique available and is not used often in the environmental field, as the technique is slow, noisy, and dusty. The exception is the use of cable-tool drilling in glacial environments containing large cobbles in the Pacific Northwest portion of the United States or in young volcanics such as in Hawaii. Cable tool

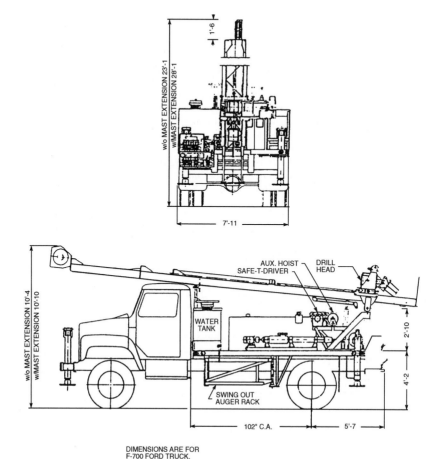

FIGURE 11.6.2   B-53 Hollow-Stem Auger rig (Foremost-Mobile Drill).

rigs, (Figure 11.6.3) called percussion or spudder rigs, operate by repeatedly lifting and dropping the heavy string of drilling tools in the borehole, crushing larger cobbles and rocks into smaller fragments. During cable-tool drilling, the hole is continuously cased with an unperforated, 8-in diameter steel casing with a drive shoe. The casing is attached on top by means of a rope socket to a cable that is suspended through a pulley from the mast of the drill rig. The process of driving the casing downward about 3 to 5 ft is followed by periodically bailing the borehole of the broken rocks and accumulated soils from the bottom of the borehole. Water is needed to create a slurry at the bottom of the borehole. When formation water is not present, water is added to form a slurry. The addition of large volumes of water into the formation to create the slurry may degrade and compromise the quality of environmental samples.

Rotary Drilling

Rotary drilling techniques include direct mud rotary, air rotary, air rotary with a casing driver, and dual-wall reverse circulation. Direct mud rotary drilling uses fluid that is pumped down through the

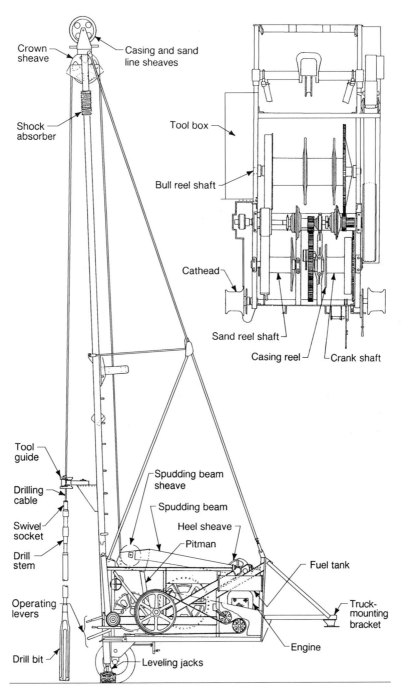

FIGURE 11.6.3 The percussion action is transferred to the drill line by the vertical motion of the spudding beam (Bucyrus-Erie Company) (Driscoll, 1984).

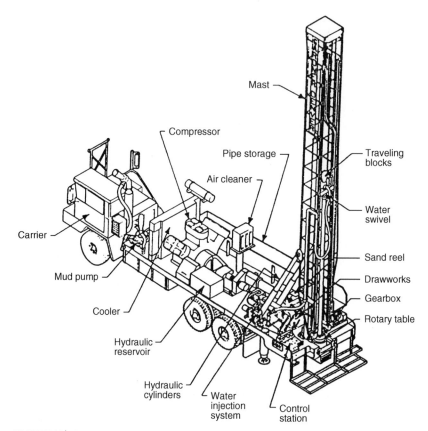

FIGURE 11.6.4 Schematic diagram of a direct rotary fig (Gardner-Denver Company) (Driscoll, 1984).

bit at the end of the drill rods (Figure 11.6.4). Then it is circulated up the annular space back to the surface. The fluid at the surface is routed via a pipe or ditch to a sedimentation tank or pit, then to a suction pit where the fluid is recirculated back through the drill rods. Air rotary drilling is similar to that of direct mud rotary, except that air is used as a circulation medium instead of water. Although the air helps cool the bit, small quantities of water or foaming surfactants are used to facilitate sampling. In unconsolidated deposits, direct mud or air rotary can be used, providing that a casing is driven as the drill bit is advanced. In dual-wall reverse circulation, the circulating medium (mud or air) is pumped downward between the outer casing and inner drill pipe, out through the drill bit, then up the inside of the drill pipe.

Rotary drilling techniques are commonly limited to consolidated deposits of rocks and typically not used in subsurface environmental studies because of poor sampling capabilities. Sample integrity is questioned because the added water, mud, or surfactant may chemically react with the formation water. In addition, thin water-bearing zones are often missed. With mud rotary, the mud filter cake that develops along the borehole wall may adversely affect permeability of the adjacent formation materials. With air rotary, dispersion of potentially hazardous and toxic particulates in the air during drilling is a concern. Air rotary drilling techniques are fast. Where the subsurface geology is relatively well characterized or a resistant stratum such as overlying basalt flows or conglomerate strata exists at shallow levels within the vadose zone and above the depth of concerns, using an air rotary rig to drill to a predetermined depth followed by another more suitable drilling technique may be worth considering (Testa, 1994).

Wire Line Coring

Coring is a drilling method that produces cylindrically shaped cores. A rotary rig is used in conjunction with water, drilling mud, or air. The core diameter varies depending on bit size, manufacturer specifications, or standards supplied by the Diamond Core Drill Manufacturers Association. Core sizes generally range from about 3/4 in to greater than 6 in. Cutting is accomplished by drill bits located at the end of the rotating barrel or tube. The barrel gradually slides down into the annular opening. The core is then separated from the rest of the formation mass and the barrel containing the core is retrieved.

Both single and double-tube core barrels exist. With double-tube core barrels, the inner tube retains the core while the tube rotates. Double tube core barrels are often used in unconsolidated formations, where sloughing may be a problem. Double tube barrels may be of fixed or swivel type. With the fixed type, the inner tube rotates with the outer tube. The purpose of the inner tube is to provide a shroud over incoming cores and to avoid washaway by high pressure drilling fluid and also to carry the core catcher for catching the core upon withdrawal. The swivel type allows the inner tube not to rotate. In both types, the inner tube is adjustable up or down within the outer tube to reduce or to increase exposure of the incoming core to erosive velocity of the drilling fluid.

Horizontal Drilling

Horizontal or lateral radial wells are fast becoming popular in subsurface environmental studies although they have been used by the oil industry for decades. Based on the configuration in map view, horizontal water wells emanating from a center hub well, are also called wagon wheel wells. The hub well is generally 8 to 12 ft in diameter. The most obvious application is where the area of concern, such as a contaminant plume, is inaccessible because of above-ground structures, tankage, and roads, or subsurface structures such as landfills, lagoons, pits, pipelines, or wells. Horizontal wells can maintain a larger zone of influence, thus comparatively higher specific capacities than vertical wells. These advantages make horizontal wells attractive as in situ groundwater aeration (sparging), vapor recovery, pump-and-treat, and injection wells. Horizontal wells can also be used for pressure- or jet-grouting a permanent barrier under and around an affected area; bioremediation by use for delivery of microbes, nutrients, and oxygen into the affected area; or as a French drain and landfill leachate collection system. Furthermore, horizontal wells are more likely to intersect vertical or inclined fractures in comparison to vertical wells.

A small-diameter pilot hole is commonly drilled first, where inclination, azimuth, and toolface data is determined and used for directional steering. Following completion of the pilot hole, a larger-diameter hole is drilled or the pilot hole is abandoned and the larger diameter hole is implemented from the start of the drilling program. Specially designed tools and methods are subsequently used to clean the borehole and to install and develop the well to meet its intended purpose. Lateral radial wells can incorporate a vertical well of 4.5 in or larger casing diameter and be placed from a few feet to 6800 ft below ground surface, which allows for the precise placement of horizontal radial wells at a 9- to 12-in radius of curvature. Radial wells can extend from a central vertical well to 200 ft or more (Testa, 1994).

Direct Push Drilling

Direct push technology (DPT) rigs can install piezometers and 1- to 3-in diameter driven well-points or piezometers (Figure 11.6.5). DPT rigs do not generate significant volumes of soils being brought

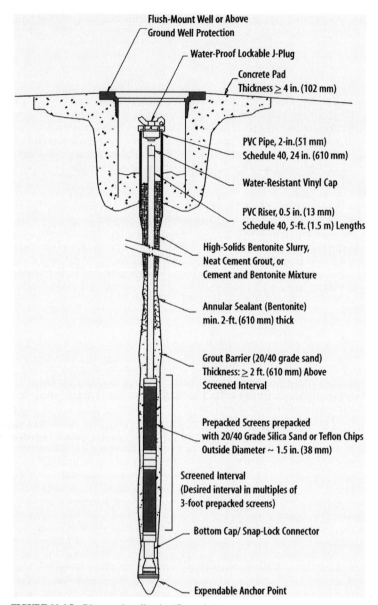

Flush-Mount Well or Above
Ground Well Protection

Water-Proof Lockable J-Plug

Concrete Pad
Thickness ≥ 4 in. (102 mm)

PVC Pipe, 2-in.(51 mm)
Schedule 40, 24 in. (610 mm)

Water-Resistant Vinyl Cap

PVC Riser, 0.5 in. (13 mm)
Schedule 40, 5-ft. (1.5 m) Lengths

High-Solids Bentonite Slurry,
Neat Cement Grout, or
Cement and Bentonite Mixture

Annular Sealant (Bentonite)
min. 2-ft. (610 mm) thick

Grout Barrier (20/40 grade sand)
Thickness: ≥ 2 ft. (610 mm) Above
Screened Interval

Prepacked Screens prepacked
with 20/40 Grade Silica Sand or Teflon Chips
Outside Diameter ~ 1.5 in. (38 mm)

Screened Interval
(Desired interval in multiples of
3-foot prepacked screens)

Bottom Cap/ Snap-Lock Connector

Expendable Anchor Point

FIGURE 11.6.5 Direct push well-point (Geoprobe).

to the surface, unlike rotary auger rigs, which generate large amounts of drilling derived wastes. Coring starts at the ground surface with an open-tube soil sampler. A DPT rig can install a prepacked monitoring well point in which the soils in the borehole are pushed out of the way and flattened against the borehole walls by the probe tip as it descends. The caking of sidewalls does lower the porosity and permeability of the sediments. DPT wells generally do not have large annular spaces. Water sampling tools are summarized in Tables 11.6.4 and 11.6.5.

TABLE 11.6.4   Comparison of Various Groundwater Sampling Tools

Category	Sampler	Source zone tracking	Dissolved plume tracking	Geologic characterization	Installation rig
One-Time Sampling Tools					
Sealed screen sampler	BAT	4	4	3	HSA, CPT, MR, DWP, ARCH
Sealed screen sampler	HPI	3	2	2	HSA, CPT, MR, DWP, ARCH
Sealed screen sampler	HPII	3	2	2	HSA, CPT, MR, DWP, ARCH
Sealed screen sampler	H2VAPE	4	4	3	HSA, CPT, MR, DWP, ARCH
Sealed screen sampler	MXSP	4	4	4	HSA, CPT, MR, DWP, ARCH
Sealed screen sampler	MNSP	4	4	3	HSA, CPT, MR, DWP, ARCH
Vertical profilers	WGP	4	3	2	Probe, CPT
Vertical profilers	GGP	4	3	2	Probe, CPT
Vertical profilers	CS	4	4	4	CPT
Long Term Monitoring Tools					
Wellpoints (0.5 to 1″ dia. Piez.)	Bailer, Pump	3	3	1	Probe, HSA, CPT
Multi-level wellpoints	Bailer, Pump	3	3	2	Probe, HSA, CPT
Monitoring wells	Bailer, Pump	3	3	1	HAS
Multi-level wells	Bailer, Pump	3	3	2	HAS

Notes:	Rigs:	Samplers:
4 = best application	HSA = Hollow Stem Auger Rig	BAT = Bat EnviroProbe
3 = good application	CPT = Cone Pentrometer Testing Rig	HPI = HydroPunch I
2 = fair application	Probe = Direct Push Technology Rig	HPII = HydroPunch II
1 = poor application	MR = Mud Rotary	WGP = Waterloo Groundwater Profiler
NA = not applicable	DWP = Dual Wall Percussion	GGP = Geoprobe Groundwater Profiler
	ARCH = Air Rotary Casing Hammer	CS = Cone Sipper
		H2VAPE = H2VAPE Simulprobe
		MXSP = Maxi Probe Simulprobe
		MNSP = Mini Probe Simulprobe

Source: (After Heller and Jacobs, in press).

11.6.3   WELL INSTALLATION

The purpose of monitoring well network is to define groundwater quality and movement and to accomplish specific study objects: A work plan and drilling permit application are frequently submitted to and approved by the lead regulatory agency prior to well installation. The borehole for a monitoring or extraction well is frequently drilled using a truck-mounted, continuous flight, hollow stem auger drill rig. The borehole diameter for monitoring wells (not driven well points) is usually a minimum of 2 to 4 in larger than the outside diameter of the well casing, in accordance with appropriate regulatory guidelines. The hollow stem auger provides minimal interruption of drilling while permitting soil sampling at the desired intervals. All wells should be installed by licensed drillers.

Well materials for monitoring and extraction wells must be chemically compatible with the potential contaminants. Casing, both blank and screen sections, can be constructed of fiberglass-reinforced plastic, stainless steel, concrete, or thermoplastic, which include polyvinyl chloride (PVC), acrylonitrile butadiene styrene (ABS), and styrene rubber (SR). Based on cost, availability, and chemical compatibility, the most common casings and screens used for shallow drilling projects in the environmental field are made of PVC. Deeper wells are typically constructed of either thicker gauge PVC or steel.

TABLE 11.6.5 Summary of Groundwater Sampling Tools (Pros and Cons)

BAT® Enviroprobe

Pros: 1) Very accurate (uses pre-evacuated vials), 2) durable, 3) Produces very clear samples, 4) fill detection capability, 5) can be used for permeability testing in vadose and saturated zones, 6) can be used as a soil gas sampler.
Cons: 1) Tool is very expensive, 2) consumables very expensive, 3) very limited sample volume, 4) long time to deploy and retrieve, 5) long groundwater sample recovery time in sandy silts, silts and clays, 6) Very complex for first time use, 6) Must be at least five feet below first water to collect groundwater samples.

MaxiSimulProbe®

Pros: 1) Very accurate (uses pressurized water canister), 2) large sample volume, 3) collects core simultaneously with water sample, 4) very durable, 5) can be wire lined with down hole hammer for rapid deployment and retrieval, 6) Pending California EPA Certification, 7) Can also be used for simultaneous soil core and soil gas sampling., 8) Excellent penetration capability in dense sediments, 9) Produces very clear groundwater samples due to gravity separation of solids during fill., 10) fill detection capability.
Cons: 1) Tool is expensive, 2) Consumables expensive, 3) Moderately complex for first time use, 4) long groundwater sample recovery time in sandy silts, silts and clays, 5) Can only be used with conventional drilling machines (HSA, MR, DWP, ARCH), 6) Must be at least 5 ft below first water to collect groundwater samples.

MiniSimulProbe®

Pros: 1) Very accurate (uses pressurized water canister), 2) large sample volume, 3) collects core simultaneously with water sample, 4) durable, 5) can be wire lined with down hole hammer for rapid deployment and retrieval, 6) Pending California EPA Certification, 7) Can also be used for simultaneous soil core and soil gas sampling, 8) Can be used with CPT as well as conventional drilling machines (HAS, MR, DWP, ARCH), 9) Good penetration capability in dense sediments, 10) produces very clear samples due to gravity separation of solids during fill., 11) fill detection capability.
Cons: 1) Tool is expensive, 2) Consumables expensive, 3) Very complex for first time use, 4) long groundwater sample recovery time in sand silts, silts, and clays as well as silty sands, 5) Must be at least 5 ft below first water to collect groundwater samples.

H2-Vape Probe®

Pros: 1) Very accurate (uses pressurized water canister), 2) large sample volume, 3) long screen length, 4) fairly simple for first time use, 5) Very durable, 6) Can be wire lined with down hole hammer for rapid deployment and retrieval, 7) fairly rapid groundwater sample recovery time, 8) Can be used with CPT as well as conventional drilling methods (HAS, MR, DWP, ARCH), 9) produces fairly clear samples due to gravity separation of solids during fill., 10) fill detection capability, 11) can collect continuous soil gas samples with CPT in one push and then be pushed directly to the saturated zone to collect a groundwater sample.
Cons: 1) Tool is expensive, 2) Consumables are moderately expensive, 3) Must be at least 5 ft below first water to collect groundwater samples.

HydroPunch®I (HPI) and HydroPunch®II (HPII) Hydrocarbon Mode (and like devices—i.e., Geoprobe® Drive Point Groundwater Samplers):

Pros: 1) Excellent first water samplers due to long screen length and bailer access into screen, 2) Very Simple, 3) low cost per sample, 4) widely used and accepted in industry, 5) can be used with all types of drilling rigs including CPT and vibratory direct push.
Cons: 1) VOC loss due to bailer use, 2) limited for vertical profiling projects due to cross contamination and/or dilution from leaky rod joints, 3) generally produces turbid samples.

HP I and HP II Groundwater Mode:

Pros: 1) None
Cons: 1) Very inaccurate tool—canister must be completely full to maintain sample equilibrium and to prevent cross contamination from bore hole fluids, 2) Fairly complex for first time use, 3) average to below average durability, 4) tool is expensive, 4) consumables are moderately expensive, 5) generally produces fairly turbid water samples, 6) can never really be sure how much of groundwater sample is from formation and how much is from bore hole.

Cone Sipper®:

Pros: 1) Very accurate (uses inert gas drive system), 2) durable, 3) continuous profiler in vadose and saturated zone (can collect continuous soil gas and groundwater samples in one push), 4) Can be stacked onto an electric friction cone (CPT cone) and used to collect groundwater samples from permeable zones immediately after electric friction cone identifies soil and permeability type, 5) no cuttings, 6) can be used with back grouting CPT cone, 7) low sample turbidity.
Cons: 1) Can only used with CPT, 2) Very expensive, 3) moderately complex for first time use, 4) screen can be easily plugged when tool passes through fine grain sediment, 5) typically limited to coarse sands with little to no silt content.

(Continued)

TABLE 11.6.5   Summary of Groundwater Sampling Tools (Pros and Cons) (Continued)

Waterloo Profiler®:

Pros: 1) Can be very accurate when used with peristaltic pump provided there are limited dissolved gases in sample, 2) durable, 3) continuous profiler (saturated zone only, 4) Can be used with vibratory direct push as well as CPT, 5) no cuttings, 6) is available with back grouting feature, 7) low sample turbidity.

Cons: 1) Expensive when take into account all of the required accessories, 2) Moderately complex for first time use, 3) small fluid entry ports, 4) screens can be easily plugged when tool passes through fine grained sediments, 5) typically limited to coarse sands with little to no silt content, 6) loses accuracy when used with bailer or when using peristaltic pump with high dissolved gas concentration.

Geoprobe® Groundwater Profiler:

Pros: 1) Can be very accurate when used with peristaltic pump provided there are limited dissolved gases in sample, 2) durable, 3) continuous profiler (saturated zone only, 4) Can be used with vibratory direct push as well as CPT, 5) long screen length to expedite groundwater sample recovery, 6) no cuttings, 7) available with back grouting capability, 8) low sample turbidity.

Cons: 1) Expensive when take into account all of the required accessories, 2) Moderately complex for first time use, 3) screens can be easily plugged when tool passes through fine grained sediments, 5) typically limited to coarse sands with little to no silt content, 6) loses accuracy when used with bailer or when using peristaltic pump with high dissolved gas concentration, 7) restricted to use with vibratory direct push.

Source: (Heller and Jacobs, in press).

For hollow stem auger drilling, the augers remain in the borehole while the well casing is set, to prevent caving. The well is cased with blank and factory-slotted, threaded schedule 40 polyvinyl chloride (PVC) pipe (Figure 11.6.6). The slotted PVC casing is placed from the bottom of the borehole to the top of the aquifer. The blank casing extends from the top of the slotted casing to approximately ground surface. The slots are generally 0.010 in or 0.020 in wide by 1.5 in long, with approximately 42 slots per linear foot. Slot sizes are determined by previous well installations in the area, or by a grain size sieve analysis. A threaded PVC cap is fastened to the bottom of the casing. Centering devices may be fastened to the casing to ensure even distribution of filter material and grout within the borehole annulus. Well screens and casings are typically steam cleaned regardless of composition prior to installation to ensure that no machine oils or other chemicals exist on the casing surfaces.

After setting the casing within the auger, sand, or gravel filter material is poured into the annulus to fill from the bottom of the boring to one foot above the slotted interval. The auger is withdrawn without rotation as filter material is poured into the annulus between the auger and casing, being careful to ensure that 2 or 3 ft of filter material is always in the auger. Too much filter material in the auger or rotation of the auger may cause the casing to lift. Too little filter material in the auger could allow native material to corrupt the continuity of the filter pack. One to two ft of bentonite pellets is placed above the filter material, and then hydrated with deionized water. The bentonite pellets are placed to prevent the grout and surface contaminants from reaching the filter material. Neat cement, a common grout for sealing environmental wells, contains approximately 5 percent bentonite powder. The bentonite powder, when added to the neat grout, must be mixed into the mixing water prior to the introduction into the cement, otherwise, the differential rates of assimilation will cause uneven hydration of the cement and consequent failure of the well seal. The grout is tremied into the annular space from the top of the bentonite plug to the surface. The grout need not be tremied if the top of the bentonite plug is less than 5 ft below ground surface, and the interval is dry. Approved grout mixtures and grouting techniques may vary depending on local conditions and regulations.

A lockable PVC cap is typically placed on the wellhead. A traffic-rated flush-mounted steel cover is installed around a wellhead located in traffic areas. A steel stove pipe is usually set over a wellhead in landscaped areas. The flush-mounted cover box or stovepipe monument contacts the grout. Grout fills the space between the monument and the sides of the borehole. The monument and grout surface seal is set at or above grade so that drainage is away from the monument. The monument lid is clearly marked "Monitoring Well," with the well number. The monument interior should also be marked with the well number in the event that the lid is lost or misplaced.

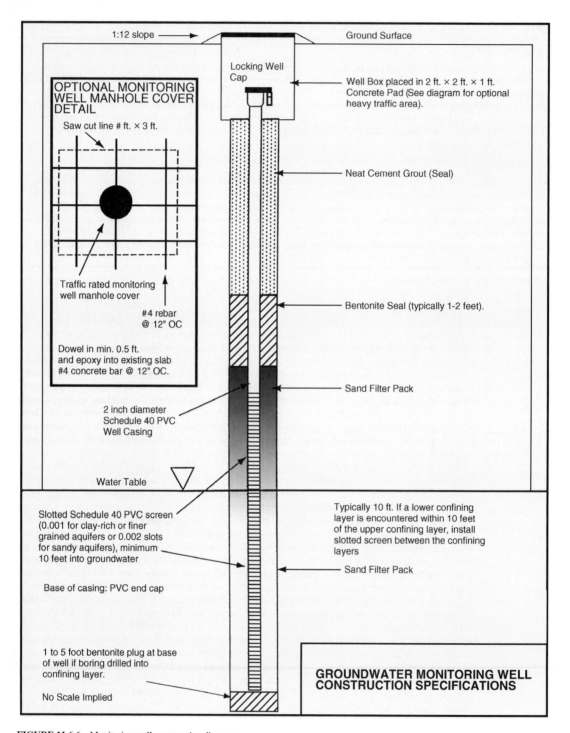

FIGURE 11.6.6   Monitoring well construction diagram.

11.6.4 SAMPLING PROCEDURES—RECORDING FIELD DATA

All information pertinent to field investigations is typically kept on Daily Field Logs and other field documents. Boring Logs, Water Sampling Data Sheets, and Chain-of-Custody forms comprise the field documents in which all pertinent information about bore hole samples and groundwater samples are recorded. The information documented includes the following minimum data:

- Investigation location
- Sample number
- Sample location, including boring or well number, and/or depth
- Name of collector
- Date and time of collection
- Analyses to be performed
- Field observations
- Field measurements (PID readings, pH, conductivity, water levels)
- Other data forms

11.6.5 EQUIPMENT DECONTAMINATION

Prior to arriving at a sampling site, all hand augering and direct push technology sampling equipment should be cleaned with phosphate free detergent, and rinsed twice with deionized water. This procedure should also be carried out on-site prior to, between, and after sampling. Hollow stem auger drilling equipment should be steam cleaned prior to arriving on site and between uses. Decontamination is best conducted on an impermeable surface, and all effluent is contained. The drilling equipment including augers is commonly placed on a rack for air drying. All other sampling equipment is typically washed with nonphosphate detregent and rinsed twice with deionized water. Water used for decontamination should be stored in labeled containers certified for hazardous materials storage. The decontamination water is disposed of in an approved manner. Labels for decontamination materials should contain the following information at a minimum:

- Type of material contained (e.g., decontamination water)
- Date of first accumulation
- Client's name, address of site

11.6.6 INSTRUMENT CALIBRATION AND MAINTENANCE

The following field equipment is frequently used during the environmental site investigations and remediations during sample collection. Calibration procedures and frequency are listed in the manufacturer's guidebooks:

Organic Vapor Meter (OVM) to measure soil vapors
Photoionization Detector (PID) to measure soil vapors
Water Level Meter to measure water depth and total depth of well
Oil-Water Interface probe to measure water depth, free-product thickness and depth of well
pH/Temperature/Conductivity Meter for water samples
Magnetic Line Locating Tools to measure magnetic fields related to subsurface obstructions

11.6.7 SAMPLE CONTROL

Proper identification, preparation, packaging, handling, shipping, and storage of samples obtained in the field is the responsibility of field personnel. Samples must be readily identifiable and should be as representative as possible of in situ conditions.

Sample Labels

Each sample container is labeled at the time of collection. The label is attached to the individual sample containers. The sample labels should contain the following information at a minimum:

- Project name and number
- Date and time of collection
- Name of collector
- Sample number
- Location of sample collection (i.e., boring and depth, or well number)
- Preservation or special handling employed

Chain-of-Custody Request for Analysis

A chain-of-custody form for each sample and container is used to track possession of the samples from the time they were collected in the field until the time they are analyzed in the laboratory. The chain-of-custody form should contain the following information at a minimum:

- Project name and number
- Name and signature of collector
- Date and time of sample collection
- Number of containers in a sample set
- Description of sample and container(s)
- Name and signatures of persons who are involved in the chain-of-custody
- Inclusive dates and times of possession
- Type of analysis requested

If a sample is known to have a high chemical concentration, field personnel should make a note on the chain-of-custody so that dilutions may be made in the laboratory.

Sample Preparation, Packaging, and Handling

Soil samples are collected in polyethylene, brass, stainless steel sample tubes. The tube ends are covered with Teflon tape and plastic end caps, labeled, and sealed in locking plastic bags. Certain soil samples may be collected in laboratory supplied 8-ounce widemouth jars with Teflon-lined screw-on lids. Groundwater samples are placed in laboratory supplied containers that are compatible with the requested analysis. Samples are placed into a cooler containing chemical ice. Padding (i.e., bubble wrap, foam) is used to prevent glass breakage.

Sample Delivery to Laboratory

The environmental samples are delivered to a state certified laboratory under chain-of-custody procedures, usually within 48 hr. of sampling. Samples are maintained at approximately 4°C for shipping.

Shipping containers are sealed with security tape to ensure the sample integrity during shipping. The chain-of-custody should always accompany the delivered samples.

11.6.8 *SOIL SAMPLING PROTOCOL*

Soil samples should be collected in accordance with local, state, and national regulatory guidelines. Variations in local regulations, guidelines, and requirements do exist and can be quite large. Standard U.S. Environmental Protection Agency (EPA) methodologies for sampling and analysis are routinely utilized for environmental projects. When required, a work plan and permit application is submitted to and approved by the lead regulatory agency prior to commencing drilling or excavation activities. Soil cuttings and excess sampling materials should be properly stored and labeled on site in DOT 17-H containers pending off-site disposal. Where excavated soil is left on-site pending treatment or off-site disposal, it should be placed in a bermed area upon high density polyethylene (HDPE) liners, and covered with HDPE liners to prevent contaminated runoff or leakage.

Sampling Methods

Manual Samplers.   Undisturbed samples are obtained using a slide hammer and core sampler with a single sampling cup at the end. The sampler typically contains one clean, 6-in long by 2-in diameter brass or stainless steel sample tube. The sample is obtained by hammering the cup and sampler into undisturbed soil. The sampler is retrieved and opened, and the sample tube containing the sample is extracted. Samples may be collected from tank pits and soil piles by driving a brass or stainless steel tube into the soil by hand. The top in or two of soil at the sampling location are first removed to ensure a relatively fresh sample. Some analyses or guidelines allow for a disturbed sample to be dug using a trowel, and then placed in a glass jar. Soil samples can also be obtained using handheld continuous coring tools, typically 4 ft long and 1 in in diameter. Made by a variety of manufacturers, these portable samplers are driven by hand into the soil using electric rotary hammers or pneumatic hammers. After driving the sampler 3 to 4 ft deeper into the ground, the sampler is extracted, and the inner sampling tube containing the sample is removed. The inner sample tubes are typically composed of polyethylene, brass, or stainless steel.

Prior to Drilling Activities

Underground Service Alert (USA) should be contacted between 2 days and 2 weeks prior to drilling. USA will contact local utility companies who will identify the locations where buried utilities enter the property. Because the utility companies do not identify buried utilities on private property, boring locations can be cleared using a magnetic line and cable locator. The exact location and number of borings at each site is usually marked in the field by the project scientist or engineer, based on a prepared workplan.

Lithologic Description

Soils and unconsolidated deposits are commonly described according to American Society for Testing Materials (ASTM 1984) Method D 2488-84 and the United Soil Classification System (USCS) for physical description and identification of soils (Figure 11.6.7). Other soils identification systems include the Burmister Soil Identification System (BS) for unconsolidated deposits that are commonly used with the USCS. The other soil description system is the Comprehensive Soil Classification System (CSCS) developed by the U.S. Department of Agriculture. The CSCS describes soils as to agricultural productivity potential and best agricultural land use.

UNIFIED SOIL CLASSIFICATION SYSTEM - ASTM D2488				
MAJOR DIVISIONS			SYMBOL / GRAPHIC	DESCRIPTIONS
COARSE GRAINED SOILS (>50% by weight larger than #200 sieve)	GRAVEL AND GRAVELLY SOILS (more than 50% of coarse fraction is larger than the # 4 sieve)	Clean Gravels (little or no fines)	GW	Well Graded Gravels, Gravels - Sand Mixtures
			GP	Poorly Graded Gravels, Gravel - Sand Mixtures
		Gravels With Fines (appreciable amount of fines)	GM	Silty Gravels, Gravel - Sand - Silt Mixtures
			GC	Clayey Gravels, Gravel - Sand - Clay Mixtures
	SAND AND SANDY SOIL (more than 50% of coarse fraction is smaller than the #4 sieve)	Clean Sands (little or no fines)	SW	Well Graded Sands, Gravelly Sands
			SP	Poorly Graded Sands, Gravelly Sands
		Sands With Fines (appreciable amount of fines)	SM	Silty Sands, Poorly Graded Sand - Silt Mixures
			SC	Clayey Sands, Poorly Graded Sand - Clay Mixtures
FINE GRAINED SOILS (>50% smaller than #200 sieve)	SILTS AND CLAYS (liquid limit less than 50)		ML	Inorganic Silts and Very Fine Sands, Silty or Clayey Fine Sands
			CL	Inorganic Clays of Low to Medium Plasticity; Gravelly, Sandy or Silty Clays; Lean Clays
			OL	Organic Silts and Organic Silty Clays of Low Plasticity
	SILTS AND CLAYS (liquid limit greater than 50)		MH	Inorganic Silts, Micaceous or Diatomateous Fine Sand or Silty Soils, Elastic Silts
			CH	Inorganic Clays of High Plasticity, Fat Clays
			OH	Organic Clays of Medium to High Plasticity, Organic Silts
HIGHLY ORGANIC SOILS			PT	Peat and Other Highly Organic Soils

Indicates First Water

Indicates Static Water

Indicates Analyzed Sample

bgs — below ground surface

PID — Photo-ionization detector readings

Asphalt

Concrete

Cement Grout

KEY TO BORING LOGS

FIGURE 11.6.7 Typical portable organic vapor meter (OPM) used for field screening of soil vapors.

The ASTM Soil Classification Flow Chart and the USCS are generally accepted soil description methods used in the engineering and environmental fields. Descriptions for moisture, density, strength, plasticity, among others, are made using ASTM guidelines. Stratigraphic, genetic, and other data and interpretations are usually also recorded on the boring log. For consolidated deposits, including igneous, metamorphic, and sedimentary rocks, the USCS is combined with other geologic characteristics such as weathering, sorting, sphericity, separation, and other features.

Color is correctly described by comparing the soil sample with a Munsell Rock Color Chart (Munsell Color, 1988), and applying the correct designations and descriptions. If the Munsell chart is not available, colors should be described using only red, orange, yellow, green, blue, purple, brown, black, and white. Modifiers such as light or dark are acceptable. Dual color descriptions (yellow-brown) may also be used. Descriptions such as tan, buff, and others, will not be used. The soil is described when moist or wet. Each stratum of soil is identified by the following items, in the order given: color, soil type, classification symbol, Munsell color designation (if any), consistency or relative density, moisture, structure (if any), and modifying information such as grain sizes, particle shape, cementation, plasticity, stratification, etc. Some samples follow:

> YELLOWISH BROWN SANDY CLAY (CL), 10 YR 5/4;
> stiff, moist, fissured, with occasional gravel to 1-inch size
> (landslide debris) [using Munsell Chart]
>
> GRAY-BROWN CLAYEY SAND WITH GRAVEL (SC);
> medium dense, wet, fine-grained sand, 15% clay, <10%
> gravel, (alluvium) [not using Munsell Chart]

11.6.9   QUALITY CONTROL OF SOIL SAMPLES—DUPLICATES

Small variations in soil lithology within several inches can lead to large differences in the results of the chemical analysis. Duplicate soil samples should be collected as near as possible to the actual soil sample, in the same lithologic zone. Duplicate soil samples, if required, are recommended to be collected at a minimum 5 percent (1 in 20).

11.6.10   WELL DEVELOPMENT PROTOCOL

Wells are developed to remove residual drilling materials from the wellbore, and to improve well performance by removing any fine material in the filter pack that can pass from the native soil into the well. Well development techniques include pumping, bailing, surging, jetting, and airlifting. In most cases, surging and pumping is satisfactory. A minimum of 12 h will lapse between placing the grout seal and well development. Development water is inspected for product sheen, odors, or sediments. Approximately three to five wetted casing volumes are removed during development. The well is considered fully developed when consistent pH, temperature, and conductivity readings (within 10 percent) indicate characteristic groundwater for the aquifer. A minimum of one wetted casing volume is removed between readings. Well casing volumes are calculated in the following manner:

Diameter (inches)	Volume (gallon/linear foot)
2	0.163
4	0.653

If the aquifer is slow to recharge, development will continue until the well is pumped dry. Development information is recorded on the Water Sampling Data Sheet. The data will include the following:

- Date of development
- Time of initial water level readings
- Volume of water removed
- pH, temperature & conductivity readings

Cross contamination of wells from pumps is avoided by using dedicated equipment and proper decontamination procedures. All development water is stored in U.S. Department of Transportation (DOT) approved 55 gallon drums, covered with lids, and labeled "Decontamination Water" with the well number(s) and date of first accumulation. The development water is stored on site pending laboratory analysis, after which the water is disposed of properly.

11.6.11   GROUNDWATER MONITORING PROTOCOL

Monitoring of depth to water and free product thickness within wells at the site is conducted using a water level meter or an interface probe. For consistency, all measurements are taken from the north side of the wellhead at the survey mark. To assess potential infiltration of fine-grained sediments, total well depth will also be measured. To reduce the potential for cross contamination between wells, the monitoring is performed in the order from the least to the most contaminated, if known. Wells containing free product are monitored last. The water level measuring equipment is decontaminated between wells. Water level data collected from the wells is used to develop a groundwater contour map for the project site. Groundwater flow is perpendicular to equipotential lines drawn on the map.

11.6.12   GROUNDWATER SAMPLING

Water Sampling from Monitoring Wells

For newly installed wells, a minimum of 24 to 72 h may be required to lapse between well development and sampling. If free product is detected, a product sample is sometimes collected for source identification. Where several chemical analyses are to be performed for a given well, individual samples are collected in order of decreasing volatility. When the results from previous sampling events are known, it is recommended to start sampling the cleanest wells first, moving to the most contaminated ones later in the sampling event.

Cross contamination between wells is avoided by careful decontamination procedures. Purging will proceed from the least to the most contaminated well, if known or indicated by field evidence. The well is purged until indicator parameters (pH, temperature and conductivity) are stabilized (within 10 percent). A minimum of three wetted casing volumes is commonly removed by bailing or pumping. If the aquifer is slow to recharge, purging will continue until the well is pumped dry. Once the well is sufficiently purged, a sample may be collected after the water level has reached 80 percent of its initial volume. Where water level recovery is slow, the sample may be collected after two hours. Samples are collected using a disposable polyethylene bailer with a bottom siphon and nylon cord. If a Teflon or stainless steel bailer is used, it is decontaminated between wells. The groundwater in the bailer is inspected for free product. Samples are transferred to clean laboratory supplied containers. Some regulators are now allowing micropurging, where minimal purging is performed.

Because volatile organic compounds (VOCs) may be lost if the sample is aerated, vials for VOC analysis (40 ml) is filled using a restricted flow dispenser at the end of the bailer to minimize sample agitation. The vials is filled above the top of the opening to form a positive meniscus. No head space should be present in the vial once it is sealed. After the vial is capped, it should be inverted to check for air bubbles. If bubbles are present, the sample should be discarded and replaced. If it is not possible to collect a sample without air bubbles, the problem should be noted in the daily field log. Where several types of analyses are to be performed for a given well, individual samples are collected in the following order:

- Volatile organic compounds
- Purgeable organic compounds
- Purgeable organic halogens
- Total organic compounds
- Total organic halogens
- Extractable organic compounds
- Total metals
- Dissolved metals
- Phenols
- Cyanide

11.6.13 QUALITY CONTROL OF GROUNDWATER SAMPLES

A QC program, independent from the laboratory's program, is frequently maintained. This program includes the submittal of duplicates, field blanks, and travel blanks to the laboratory. No spiked samples are supplied from the field. The QC samples are packaged and sealed in the same manner as the other samples.

Duplicates

Duplicate samples are commonly collected for five percent (1 in 20) of the samples or one per sampling round, whichever is greater. The duplicate sample is submitted to the laboratory for the same analyses as the original sample. The duplicate sample is acquired by filling separate containers from the same well bailer as the actual sample. The contents of the bailer is evenly divided between the actual and duplicate samples, to ensure duplication. The duplicate sample is labeled as a duplicate without identifying the well location on either the chain-of-custody or the sample container. The well location and sample number of the duplicate sample is noted on the Water Sampling Data Form.

Field Blanks

Field blanks are prepared for either 5 percent (1 in 20) of the samples or one per sampling set, whichever is greater. The field blank is submitted to the laboratory for the same analyses as the rest of the sampling set. The field blank is acquired by dispensing deionized water from the sampling bailer into the containers in the same manner as groundwater samples. The field blank is assigned an independent sample number.

Travel Blanks

When sampling groundwater for volatile compounds analysis, travel blanks are used to detect the introduction of contaminants during transportation from the filed to the laboratory. The travel blank is supplied by either Artesian or the laboratory. The travel blank is taken to the field, and will accompany the collected groundwater samples to the laboratory for analysis. The travel blank will consist of deionized water. The travel blank is assigned an independent sample number.

Sample Preservation

Specific analytes require specific sample preservation techniques or additives. Sealed chemical ice is placed in the coolers to maintain samples at a temperature of 4°C.

ACKNOWLEDGMENTS

The author gives special thanks to HD for review as well as Steve Testa for considerable help on this article.

REFERENCES

American Society for Testing Materials (ASTM), *Standard Practice for Description and Identification of Soils (Visual-Manual Procedure)*, Method D 2488-84, December 1984.

California Department of Water Resources (DWR), *California Well Standards Bulletin*, 74–81, 1981.

———, *California Well Standards Bulletin*, 74–90, January 1990.

Driscoll, F. G., *Groundwater and Wells*, Second Edition, Johnson Filtration Systems, St. Paul, pp. 268–339, 1986.

Heller, N., and Jacobs, J., Almost Everything You Wanted to Know about Discrete Groundwater Samplers, (personal communication).

Munsell Color, *Munsell Soil Color Charts:* Munsell Color, Baltimore, 1988.

Sisk, S. J., *NEIC Manual for Groundwater/Subsurface Investigations at Hazardous Waste Sites*, EPA-330/9-81-002, U.S. Environmental Protection Agency, Denver, Colorado, July 1981.

Testa, S. M., *Geologic Aspects of Hazardous Waste Management*, CRC Press, Boca Raton, Florida, pp. 145–187, 1994.

CHAPTER 11
ASSESSMENT SAMPLING AND MONITORING

SECTION 11.7

AQUIFER TESTING

John M. Shafer
Dr. Shafer is director of the Earth Sciences and Resources Institute
(ESRI) at the University of South Carolina and sole proprietor of
the environmental/groundwater consulting concern GWPATH.

11.7.1 INTRODUCTION

Aquifer testing encompasses a broad range of activities that focus on the estimation of representative values for the hydrogeologic properties of water-bearing earth materials, particularly aquifers. The most important of these hydrogeologic properties are transmissivity and storativity, although aquifer tests may also determine other hydraulic parameters. Aquifer tests are often designed to address additional issues of water supply analysis, such as sustainable yield, or environmental characterization where hydraulic communication between water-bearing units may be of concern.

An aquifer test is a direct approach to estimating hydrogeologic properties of saturated aquifer and nonaquifer earth materials that requires a field experiment. During an aquifer test the groundwater system is stressed in some manner and the response of the system to the stress is recorded, usually as the change in water level in one or more pumping and/or observation wells. The resulting stress response data are analyzed in a variety of ways, depending on the type of aquifer test, to estimate the hydrogeologic properties of the groundwater system.

The success of an aquifer test is dependent on proper technical design, which should include consideration of the type of aquifer test to be done, the layout and spacing of wells, the pumping rate(s), the duration of the test, and the method of aquifer test analysis. Several references are available to guide the design of aquifer tests.

11.7.2 TYPES OF AQUIFER TESTS

There are several types of aquifer tests including single well versus multiple well tests, drawdown versus buildup, and constant discharge rate versus variable discharge rate. Single well tests can be step-drawdown tests for well performance or slug tests. Step-drawdown tests, or step tests, employ a variable pumping rate to determine the efficiency of the pumped well. Step tests are often performed as a precursor to the design of a multiple well aquifer test. Slug tests require that a "slug" of water be instantaneously added or removed from the borehole or well and that the response of the water level returning to equilibrium be recorded. There are both advantages and disadvantages to single well tests. Single well tests are relatively easy to do and are often inexpensive in labor and equipment costs. However, single well tests "sample" less earth materials than multiple well tests. Consequently, values for hydrogeologic properties resulting from single well tests usually represent a small volume of aquifer or aquitard.

TABLE 11.7.1 Commonly Applied Aquifer Tests Analysis Techniques

Aquifer test analysis procedures	Aquifer type				Partial penetration	Steady-state	Transient
	Confined	Leaky confined	Unconfined	Fractured			
Theim (1906)	✓					✓	
Theis (1935)	✓		✓-				✓
Cooper and Jacob (1946)	✓		✓-				✓
Hantush and Jacob (1955)		✓					✓
Hantush (1960)		✓					✓
Hantush (1964)		✓			✓		✓
Walton (1962)		✓					✓
Neuman (1972, 1975)			✓				✓
Streltsova (1974)			✓		✓		✓
Neuman (1974)			✓		✓		✓
Moench (1993)			✓		✓		✓
Kazemi et al. (1969)				✓			✓
Boulton and Streltsova (1977)				✓			✓

Multiple well aquifer tests require a pumping well and one or more independent observation wells. During a typical aquifer test, the production well is pumped at a constant rate over a specified time during which water level changes (i.e., drawdown) in observation wells are recorded. Typically, the frequency of recording water levels in observation wells decreases with time. Following cessation of pumping, the "recovery" of water levels (i.e., return to equilibrium) in the observation wells is also recorded over time. Both the time-drawdown and time-recovery data can be analyzed to estimate hydrogeologic properties.

Most multiple well aquifer tests are designed as constant pumping rate tests, although in practice holding pumping rates absolutely steady throughout the duration of the test is difficult. The more commonly applied aquifer test analysis procedures are based on the assumption of a constant pumping rate.

There are many aquifer test analysis techniques and procedures. These vary widely in assumptions based on the degree to which well hydraulics and aquifer behavior are incorporated into the analysis. The various approaches to aquifer test analysis depend on whether the groundwater system is a porous medium or fractured medium; and whether the aquifer is confined, leaky confined, or unconfined. Figure 11.7.1 depicts different types of aquifers. Table 11.7.1 lists commonly applied aquifer test analysis techniques.

11.7.3 MULTIPLE WELL AQUIFER TEST ANALYSIS TECHNIQUES

Most aquifer test analysis techniques are graphical in nature with many relying on a "type curve" matching procedure to compare observed drawdown with a theoretically prescribed drawdown appropriate to the aquifer and well conditions. Most aquifer test analysis techniques have been automated and packaged in a variety of public-sector and commercial software applications.

Data Corrections

Field or raw drawdown data generated via an aquifer test must often be "corrected" before their analysis in an aquifer test analysis procedure. The corrections are sometimes required because of the influences that extraneous factors may have on the observed drawdowns. Table 11.7.2 shows a wide range of actions, natural and human-induced, that may lead to fluctuations in groundwater

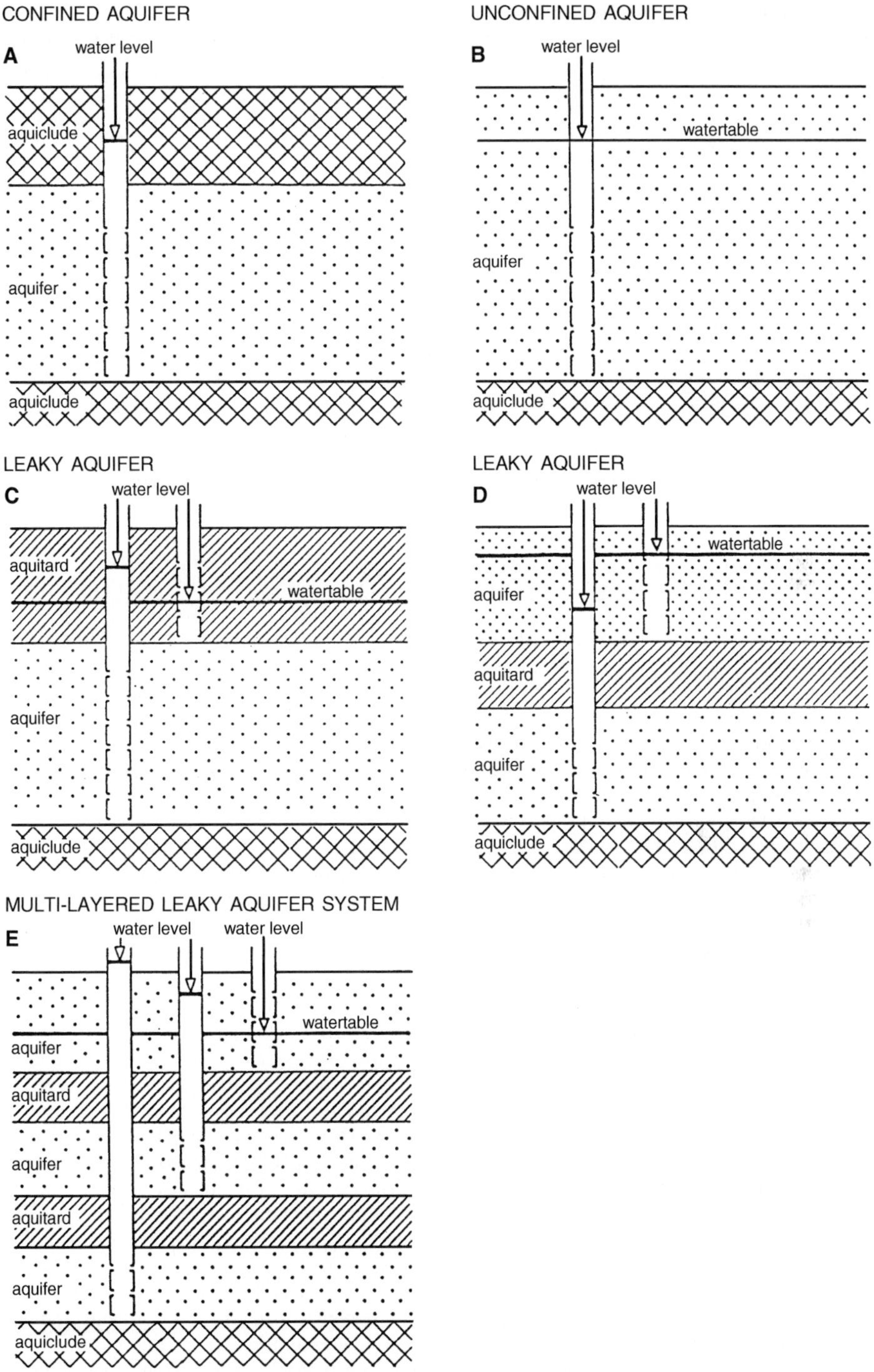

FIGURE 11.7.1   Different aquifer types (Source: Kruseman and de Ridder, 1991).

TABLE 11.7.2 Actions That Lead to Fluctuations in Groundwater Levels

	Unconfined	Confined	Natural	Anthropogenic	Short-lived	Diurnal	Seasonal	Long-term	Climatic influence
Groundwater recharge (infiltration to the water table)	✓		✓				✓	✓	
Air entrapment during groundwater recharge	✓		✓		✓				✓
Evapotranspiration and phreatophytic consumption	✓		✓			✓			✓
Bank storage effects near streams	✓		✓				✓		✓
Tidal effects near oceans	✓	✓	✓			✓			
Atmospheric pressure effects	✓	✓	✓			✓			✓
External loading of confined aquifers		✓		✓	✓				
Earthquakes		✓	✓			✓			
Groundwater pumpage	✓	✓		✓				✓	
Deep well injection		✓		✓				✓	
Artificial recharge, leakage from ponds, lagoons and landfills	✓		✓					✓	
Agricultural irrigation and drainage	✓		✓					✓	✓
Geotechnical drainage of open pit mines, slopes, tunnels, etc.	✓		✓					✓	

Source: Modified from Freeze and Cherry, 1979.

levels. Raw drawdown data may require correction for the effect of tidal fluctuations, barometric pressure fluctuations, and/or river stage fluctuations that occurred during the test period. For example, tidal fluctuations often cause substantial fluctuations of groundwater levels in coastal aquifers. The drawdown data resulting from an aquifer test conducted in a tidally influenced groundwater system may be masked by the diurnal fluctuation in groundwater levels caused by tide. The raw data must be corrected (i.e., the tidal influence removed) before the analysis of the drawdown data for aquifer properties. The same process applies to raw drawdown data, which include the influences of changes in barometric pressure that occurred during the test.

A standard approach to correct drawdown data affected by tidal, barometric, or river stage fluctuations is to compute the "efficiency" (e.g., barometric efficiency) of the groundwater system with respect to the phenomenon (e.g., changes in barometric pressure) influencing water levels, and then adjusting the drawdown data using the efficiency factor and the change in the magnitude of the phenomenon during the test. The efficiency factor is the ratio of the change in water level in a well to a corresponding change in the influencing phenomenon in equivalent units. This approach requires that, besides the well response data being recorded during the aquifer test, barometric pressure (or tidal stage or river stage) also be recorded during the test.

Raw drawdown data may require adjustment for any ambient trends in water levels during the aquifer test. If regional groundwater levels are declining because of evapotranspiration or stream-aquifer interaction, for example, this trend may continue through the duration of the aquifer test. The trend will be superimposed on the field drawdown and/or recovery data and should be removed from the raw data prior to analysis for aquifer hydraulic properties. If a rising or declining trend in groundwater levels is suspected, water levels should be recorded in observation wells for enough time prior to the beginning of the aquifer test to identify the trend.

When the intake portion of the pumping well spans less than the full thickness of an aquifer the drawdown recorded in observation wells may be affected by vertical groundwater flow components caused by the partial penetration. As a rule, the radius of influence of partial penetration approximately equals:

$$r_{pp} \cong 1.5b \sqrt{\frac{K_h}{K_v}} \tag{11.7.1}$$

where r_{pp} = radius from the pumping well beyond which effects of partial penetration are negligible (L), b = saturated thickness of the aquifer (L), and K_h/K_v = ratio of the horizontal hydraulic conductivity to the vertical hydraulic conductivity (dimensionless).

There are several approaches to correcting the effects of partial penetration on observed drawdown when the distance between the pumping well and an observation well is less than r_{pp}. Kozeny (1933) developed an empirical relationship between theoretical drawdown and observed drawdown based on the effective well radius, saturated thickness, and percent penetration. Butler (1957) developed a similar relationship for correcting steady-state drawdown affected by partial penetration that uses a partial penetration constant available in tables. Other partial penetration corrections were developed by Huisman as described by Kruseman and de Ridder (1991). The effects of partial penetration have been explicitly included in several aquifer test analysis procedures, eliminating the need to correct the drawdown data prior to analysis.

Confined Aquifer Test Analysis Procedures

The earliest mathematical approach to aquifer test analysis (ca. 1906) was based on the Theim equation describing the radial steady-state relationship between the pumping rate of a production well and the resultant drawdown in a confined aquifer. In terms of discharge rate (Q), the Theim equation is:

$$Q = 2\pi T \frac{(h_2 - h_1)}{\ln\left(\frac{r_2}{r_1}\right)} \tag{11.7.2}$$

where Q = constant discharge rate (L^3/T), T = transmissivity (L^2/T); T = Kb, K = hydraulic conductivity (L/T), h_1, h_2 = hydraulic head at two locations radially out from the production well (L); drawdown = $h_2 - h_1$, and r_1, r_2 = radial distance from the well at locations h_1 and h_2 (L).

The Theim equation is seldom used because of the restrictive underlying assumptions of a flat-lying, homogeneous, isotropic aquifer of infinite extent with a fully penetrating well. Further, steady-state aquifer tests are difficult to conduct because it is often impossible to achieve steady-state drawdown in a reasonable pumping time frame. Further, because the Theim equation is based on steady-state hydraulics, storativity cannot be estimated using this procedure.

The most common approach to aquifer test analysis is to plot time versus drawdown data, or recovery data, (i.e., unsteady-state analysis), match the resulting curve to a theoretical "type curve," determine a match point set of parameters, and substitute these parameters in the appropriate equations describing transmissivity, storativity, and so forth.

Theis Type Curve Matching.   Theis (1935), using an analogy to heat flow, developed the first analytical solution for unsteady-state radial groundwater flow to a well (Freeze and Cherry, 1979; Kruseman and de Ridder, 1991). The Theis equation and associated "type curve" matching procedure for aquifer test analysis became the prototype for subsequent curve matching techniques for evaluation of unsteady drawdown data from the full range of aquifer types.

The Theis equation is

$$s = \frac{Q}{4\pi T} \int_u^\infty e^{-u} \frac{du}{u}$$

$$u = \frac{r^2 S}{4Tt}$$

(11.7.3)

where s = drawdown (L), r = radial distance from the pumping well to the observation well (L), S = storage coefficient (dimensionless), and t = time since pumping began (T).

In a strict sense, the Theis equation applies only to fully confined, homogeneous and isotropic, flat lying aquifers of infinite areal extent where water is released instantly from storage and flow is radial to the pumping well. In practice, however, these constraints are often relaxed and the Theis equation is applied to groundwater systems that mildly violate one or more of the assumptions. For example, the Theis equation is often applied to unconfined aquifers where delayed yield is insignificant.

The Theis equation can be approximated by an infinite series expansion as:

$$s = \frac{Q}{4\pi T} W(u)$$

(11.7.4)

$$W(u) = \left(-0.577216 - \ln(u) + u - \frac{u^2}{2.2!} + \frac{u^3}{3.3!} - \frac{u^2}{4.4!} \cdots \right)$$

(11.7.5)

$W(u)$ is the well-known Theis well function. Values of $W(u)$ versus u (or $1/u$) are tabulated and included in most references on aquifer test analysis. Within the context of aquifer test analysis, $W(u)$ is considered dimensionless drawdown and u is dimensionless time. A $\log_{10}$-$\log_{10}$ plot of $W(u)$ versus $1/u$ is known as the reverse Theis type curve (Figure 11.7.2). It can be used to estimate the values of T

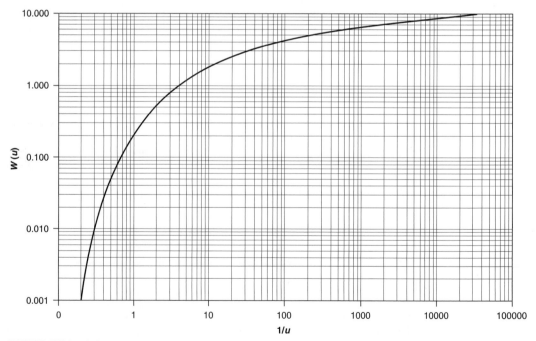

FIGURE 11.7.2 Theis reverse type curve.

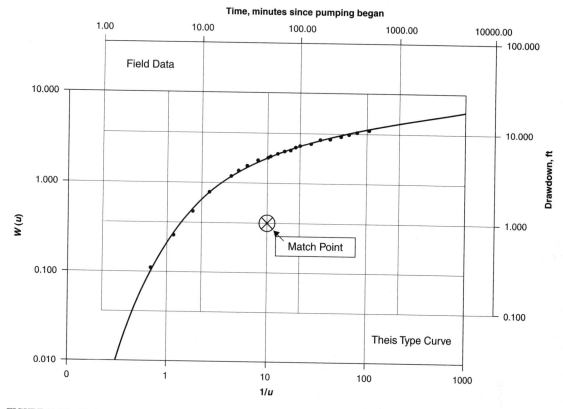

FIGURE 11.7.3   Theis type curve matching procedure for aquifer test analysis.

and S if drawdown (s) versus time (t) data are measured during a constant pumping rate aquifer test. The reverse Theis type curve matching procedure is a graphical approach in which the reverse Theis type curve is superimposed on a $\log_{10}$-$\log_{10}$ plot of s versus t. Keeping the coordinate axes parallel, the field data plot is physically shifted until it overlies, or matches, the reverse Theis type curve. A common (i.e., to both plots) match point is selected and the four values s, t, $W(u)$, and $1/u$ at the match point are recorded (Figure 11.7.3). These values are entered into the rearranged Theis equation approximation to compute values for T and S as:

$$T = \frac{Q}{4\pi s} W(u)$$

$$S = \frac{4Ttu}{r^2}$$

(11.7.6)

Cooper-Jacob Straight Line Method.   Cooper and Jacob (1946) modified the Theis curve-matching procedure for aquifer test analysis based on the knowledge that for large values of T and/or t, and small values of r, u in the well function (Eq. 11.7.5) is very small. Therefore, the terms in $W(u)$ become negligible beyond the $\ln(u)$ when, as a rule-of-thumb (i.e., the u criterion), $u < 0.01$. Under

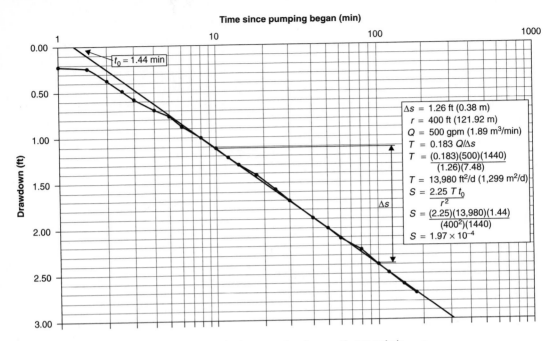

FIGURE 11.7.4 Cooper-Jacob straight line method for time versus drawdown aquifer test analysis.

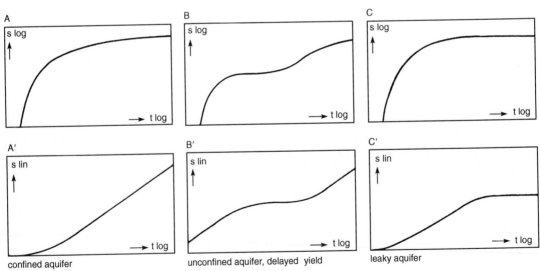

FIGURE 11.7.5 Theoretical time versus drawdown relationships for differing types of unconsolidated aquifer types (Source: Kruseman and de Ridder, 1991).

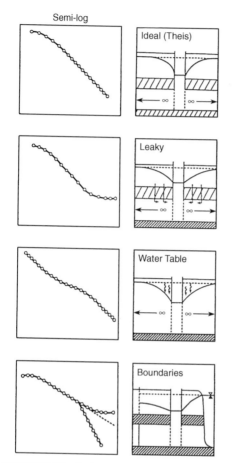

FIGURE 11.7.6 Semi-log plots of time versus drawdown for differing aquifer types.

this circumstance, Cooper and Jacob (1946) simplified the Theis equation to

$$S = \frac{Q}{4\pi T}(-0.577216 - \ln(u)) \qquad (11.7.7)$$

By substituting $\frac{r^2 S}{4Tt}$ for u in Equation 11.7.7 and rearranging terms, transmissivity and storage coefficient can be estimated by: (1) constructing a semi-log plot of time ($\log_{10}$) and drawdown (linear); (2) fitting a straight line through the later time data; (3) determining the change in drawdown, Δs, over one log cycle; (4) extrapolating the straight line to $s = 0$ and determining the associated time, t_0, since pumping began; and (5) substituting Δs and t_0 respectively in:

$$T = 0.183\left(\frac{Q}{\Delta s}\right)$$

$$S = 2.25\left(\frac{Tt_0}{r^2}\right) \qquad (11.7.8)$$

In practice, the Cooper-Jacob straight line method is often used as a check of Theis type curve matching results. Figure 11.7.4 shows the Cooper-Jacob straight line method for aquifer test analysis for ideal (i.e., Theis) conditions.

Figure 11.7.5 shows a series of theoretical time versus drawdown plots for unconsolidated aquifers of differing type. The horizontal (x) axis shows time and the vertical (y) axis shows drawdown. These paired plots display the theoretical difference between the log-linear relationship of time to drawdown and the log-log relationship of time to drawdown for each aquifer type shown. Figure 11.7.6 shows how the semi-log plot of time versus drawdown departs from the ideal for leaky confined aquifers, water table aquifers, and situations where aquifer boundary effects influence drawdown.

Leaky-Confined Aquifer Test Analysis

Graphical type curve matching procedures have been developed for leaky confined aquifers that are similar to the Theis type curve matching procedure for confined aquifers. Hantush and Jacob (1955) developed the leaky well function thereby extending Theis' curve matching approach to leaky aquifers. The Hantush-Jacob equation is:

$$s = \frac{Q}{4\pi T}W\left(u, \frac{r}{B}\right) \qquad (11.7.9)$$

where $W(u, r/B)$ is the leaky well function:

$$B = \sqrt{\frac{Tb'}{K'}}$$

and b' = thickness of the confining unit (L), and K' = hydraulic conductivity of the confining unit (L^2/T).

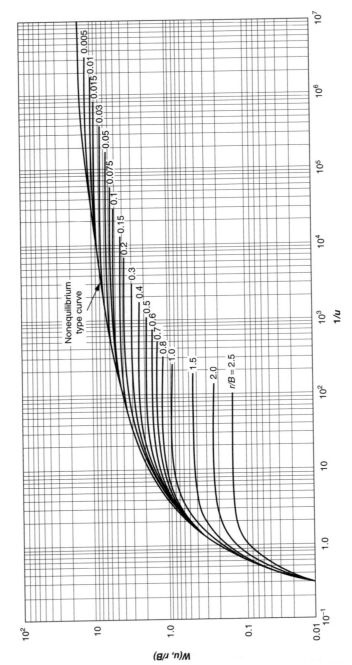

FIGURE 11.7.7 Hantush-Jacob leaky confined aquifer type curves.

The assumptions concerning the Hantush-Jacob equation are those that apply to the Theis equation plus additional assumptions regarding the confining unit, notably that water is released instantaneously from storage, and that flow is vertical through the confining unit.

The Hantush-Jacob leaky confined aquifer test analysis procedure is a type curve matching process developed by Walton (1962), whereby instead of fitting the $\log_{10}$-$\log_{10}$ time versus drawdown data curve to the Theis type curve, it is matched to a best fit among a series of $1/u$ versus $W(u, r/B)$ leaky confined type curves for a range of values of r/B. Figure 11.7.7 shows the family of Hantush-Jacob leaky confined type curves. Note that as r/B approaches zero, the Hantush-Jacob equation reduces to the Theis equation and the Theis type curve. Five parameters (i.e., s, t, $1/u$, $W(u, r/B)$, and r/B) are determined from the match point and used to solve for T and S as in the Theis type curve matching approach. In addition, if b' is known, then from the optimum value of r/B, the value of K' can be estimated.

Hantush (1960) extended the Hantush-Jacob curve matching approach to aquifer test analysis for leaky confined aquifers to include the effect of changes in storage in the confining bed. Hantush (1964) also developed an aquifer test analysis procedure to account for partial penetration of wells in a leaky confined aquifer.

Unconfined Aquifer Test Analysis

In an unconfined aquifer, pumping results in the physical drawdown of the water table and, consequently, dewatering of the aquifer materials within the cone of depression. Within the cone of depression, a vertical component of the flow to the well is introduced that violates the horizontal flow assumption of Theis. Further, in an unconfined aquifer with significant drawdown, water is not instantaneously released from storage, as in the Theis formulation, but rather, exhibits a delayed water table response also termed delayed yield (Kruseman and de Ridder, 1991). The delayed yield phenomenon can usually be seen on a $\log_{10}$-$\log_{10}$ time versus drawdown plot of unconfined aquifer test data (Figure 11.7.8). The typical S-shaped curve shows three distinct phases: a steep early time

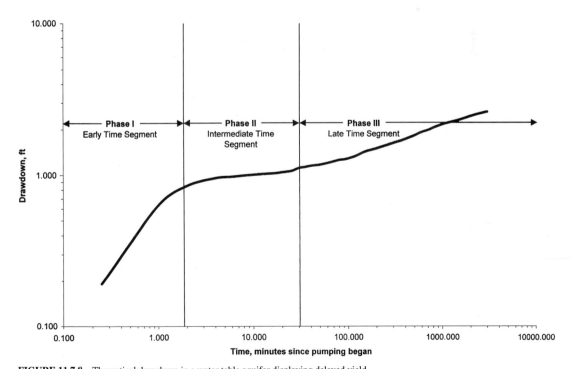

FIGURE 11.7.8   Theoretical drawdown in a water table aquifer displaying delayed yield.

segment, a flatter middle time segment, and a steep late time segment. The early time segment may last for only a few minutes. During this time period the shape of the drawdown curve resembles the Theis curve because water is being instantaneously released from storage because of expansion of the water and compaction of the aquifer (Kruseman and de Ridder, 1991). The flatter intermediate time segment of the $\log_{10}$-$\log_{10}$ time versus drawdown curve may last from several minutes to more than an hour and is the time when gravity drainage, which has the same appearance as that of leakage, slows the decline of the water table as a vertical flow component is introduced (Fetter, 1994). Finally, following the period of delayed water table response, the third time segment again exhibits a Theis-like behavior as delayed yield becomes negligible because the rate of gravity drainage approaches the rate of drawdown.

Neuman (1972, 1975) developed a graphical procedure for analyzing unconfined aquifer test data that include a delayed yield response. Neuman's formulation incorporates the effects of anisotropy and vertical flow. The aquifer is considered a compressible medium with the water table represented as a moving boundary. Neuman's equation for drawdown in an unconfined aquifer is:

$$s = \frac{Q}{4\pi T} W(u_A, u_B, \beta) \tag{11.7.10}$$

$$u_A = \frac{r^2 S}{4\pi T}; \quad \text{Phase I—Early Time Data}$$

$$u_B = \frac{r^2 S_y}{4\pi T}; \quad \text{Phase III—Late Time Data}$$

$$\tag{11.7.11}$$

$$K_h = \frac{T}{b}$$

$$\Gamma = \frac{r^2 K_v}{b^2 K_h} = \left(\frac{r}{b}\right)^2 \frac{K_v}{K_h}$$

where b = initial saturated thickness (L), K_h = horizontal hydraulic conductivity (L/T), and K_v = vertical hydraulic conductivity (L/T).

$W(u_A, u_B, \beta)$ is the well function for unconfined aquifers (Figure 11.7.9). Phase I (i.e., early time) data are used to estimate the storage coefficient, S, while Phase III (i.e., late time) data are used to estimate the specific yield, S_y, of the aquifer. Equation 11.7.10 is solved in two steps using two sets of asymptotically joined Γ type curves, one for early time data and one for late time data. Typically the process begins by matching the late time versus drawdown field data $\log_{10}$-$\log_{10}$ curve to $1/u_B$ versus $W(u_B, \Gamma)$ type curve providing the best fit. Then, using the Γ determined from the late time data, the early time versus drawdown $\log_{10}$-$\log_{10}$ curve is matched to the $1/u_A$ versus $W(u_A, \Gamma)$ type curve. Using Equations 11.7.10 and 11.7.11 and the match point values (i.e., t, s, $1/u_B$, $W(1/u_B, \Gamma)$, $1/u_A$, $W(1/u_A)$, and Γ), T, S, S_y, and K_v can be estimated.

Streltsova (1974), Neuman (1974), and Moench (1993) developed type curve matching procedures for unconfined aquifer test analysis that incorporate partial penetration of the pumping well and the observation well. Neuman (1974) noted that the effect of partial penetration on drawdown in an unconfined aquifer decreases with the radial distance from the pumping well and with the ratio of K_z/K_h.

Distance versus Drawdown Analyses and Well Efficiency

The equivalent time versus drawdown in multiple observation wells can be plotted against the $\log_{10}$ geographic distance of each observation well from the pumping well (Figure 11.7.10). The resulting fitted straight line can be used to calculate aquifer hydraulic properties and estimate well efficiency. The semi-log distance-drawdown analysis is similar to the Cooper-Jacob time-drawdown straight line procedure. The distance-drawdown curve can be used to estimate well efficiency by extrapolating

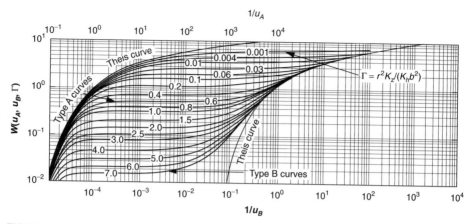

FIGURE 11.7.9   Unconfined aquifer type curves.

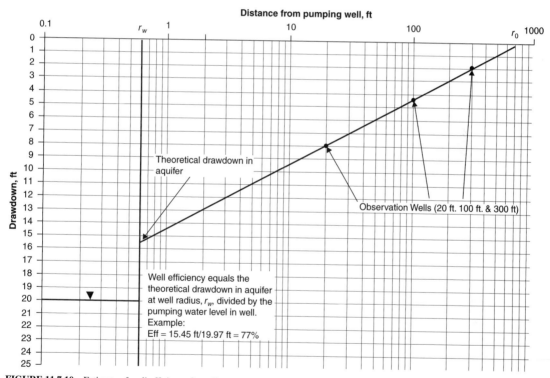

FIGURE 11.7.10   Estimate of well efficiency from distance versus drawdown data.

the best fit straight line through the distance-drawdown data points to the place where the radius of the well would be indicated. This point represents the theoretical drawdown in the well. The theoretical drawdown divided by the actual drawdown observed in the pumping well is an estimate of the well efficiency (Driscoll, 1986).

11.7.4 *SINGLE WELL AQUIFER TEST ANALYSIS TECHNIQUES*

Single well aquifer test analysis techniques include procedures to estimate well performance and a range of approaches for analyzing the results of slug tests.

Step-Drawdown Test Analysis Procedures

A step-drawdown test, or "step test," is used to calculate well efficiency, determine sustainable pumping rates, and estimate aquifer properties of the formation open to the pumping well. A step test is often used as a prerequisite for an aquifer test to determine the amount and duration of pumping.

Typically, step-drawdown tests are of short duration where initially the well is pumped at a constant low rate until drawdown stabilizes. The well discharge is then increased to a higher constant rate and held until the drawdown again stabilizes. This procedure is repeated until at least three "steps" are completed. Each step usually takes from less than 30 minutes to two hours. Continuous time versus drawdown data are recorded throughout the test.

There are several approaches to analyzing the results of a step test. Two of the more common procedures are graphical solutions to the Jacob equation (Jacob, 1946) and Rorabaugh's generalization of the Jacob equation (Rorabaugh, 1953). Jacob (1946) divided the drawdown observed in a well open in a confined, leaky confined, or unconfined aquifer into the two components shown in Equation 11.7.12; one for the aquifer and one accounting for well inefficiency (i.e., well loss):

$$s_{obs} = s_{aquifer} + s_{well}$$
$$s_{obs} = BQ + CQ^2$$

(11.7.12)

where BQ = first order (laminar) component of drawdown (i.e., aquifer), and CQ^2 = second order (turbulent) component of drawdown (i.e., well).

Strictly considered, Jacob's equation cannot be used to compute the true well efficiency, however, because BQ contains some well losses and CQ^2 occasionally includes some aquifer loss. Jacob's equation can be solved via the Hantush-Bierschenk graphical method (Figure 11.7.11) by first plotting the step test drawdown against $\log_{10}$ time and recording the increments of drawdown for each step. From this information, s_{obs}/Q is plotted against the corresponding value of Q on the linear axes. The slope of the straight line connecting the points is C and the Y-intercept is B in Equation 11.7.12. Once the coefficients B and C are determined, the percentage of the total head loss that is attributable to laminar flow, L_p, can be estimated according to:

$$L_p = \frac{BQ}{BQ + CQ^2}$$

(11.7.13)

Note that L_p does not directly equal well efficiency. The specific capacity. C_s, of the well can be computed from:

$$C_s = \frac{1}{B + CQ}$$

(11.7.14)

Rorabaugh (1953) generalized Jacob's equation to:

$$s_{obs} = s_{aquifer} + s_{well}$$
$$s_{obs} = BQ + CQ^N$$
$$1.5 < N < 3.5$$

(11.7.15)

Rorabaugh's equation also applies to confined, leaky confined, and unconfined aquifers. It is solved via a trial and error (to determine N) straight line method similar the Hantush-Bierschenk method or by Sheahan's curve fitting method (Sheahan, 1971; Kruseman and de Ridder, 1991).

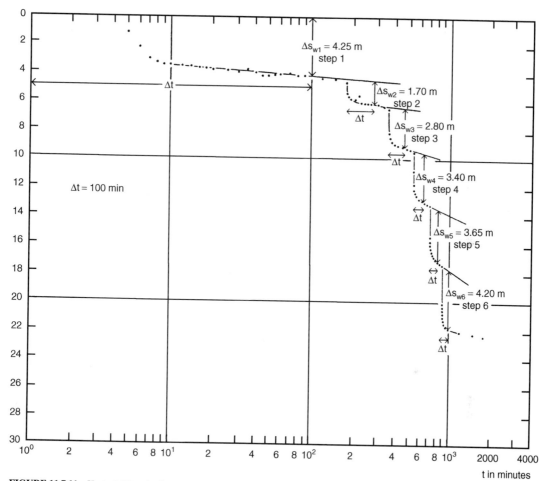

FIGURE 11.7.11   Hantush-Bierschenk graphical method for step-drawdown test analysis (*Source:* Kruseman and de Ridder, 1991).

Slug Tests

Slug tests have become one of the most routinely applied field methods for estimating formation hydraulic conductivity. In the classical approach, a small volume of water, the slug, is instantaneously removed from, or added to, a well. As the system returns to equilibrium, the change in water level versus time is recorded. Several different graphical methods can be used with the field data to estimate hydraulic conductivity, and sometimes, storativity.

Table 11.7.3 lists commonly applied slug test analysis procedures. Herzog (1994) presents a detailed comparison of slug test analysis methods via application of the procedures to field data from many slug tests conducted in coarse and fine-grained geologic deposits. Several commercially available software packages for aquifer test analysis include the capability to analyze slug test data by one or more of the approaches listed in Table 11.7.3.

Springer's (Springer, 1991) method for slug test analysis is applicable to both overdamped and underdamped (i.e., oscillatory) water level responses from slug testing. Unlike several of the other procedures, Springer's approach accounts for inertial and frictional effects in the well casing.

TABLE 11.7.3 Commonly Applied Slug Test Analysis Techniques

	Confined	Unconfined	Partial penetration	Storativity
	Aquifer type			
Hvorslev (1951)	✓	✓	✓	
Cooper et al. (1967)	✓			✓-
Bouwer and Rice (1976)	✓-	✓	✓	
Nguyen and Pinder (1984)	✓	✓	✓	✓
Springer (1991)	✓	✓	✓	

11.7.5 AQUIFER TESTS IN FRACTURED MEDIA

Aquifers where flow is predominately through a system of interconnected fractures often behave hydraulically differently from aquifers where porous medium flow dominates. In certain situations, flow is linear in fractured systems versus the typical radial flow to a well open in a granular porous medium. Consequently, classical equations describing well hydraulics in homogeneous, isotropic aquifers do not necessarily apply to flow in fractured rocks (Kruseman and de Ridder, 1991). Nevertheless, the Theis equation is sometimes used to analyze portions of time-drawdown curves in fractured media that appear to resemble the Theis type curve (Jenkins and Prentice, 1982).

A significant problem in analyzing aquifer test data from fractured rock is that the fracture pattern is seldom known with any precision. Consequently, the approaches developed for analysis of time-drawdown data from fractured rock have necessarily assumed simplified regular fracture systems. Even so, these models are complex incorporating many parameters describing geometry, rock properties, and hydraulic properties (Kruseman and de Ridder, 1991).

Several straight-line and curve matching techniques for aquifer test analysis that address different conceptualizations of pumping-induced fracture flow have been developed (e.g., Warren and Root, 1963; Kazemi et al., 1969; Boulton and Streltsova, 1977; and Bourdet-Gringarten, 1980). There is, however, no unifying approach to analyzing time-drawdown data recorded in fracture dominated aquifers, although most methods are based on a double porosity theory; one porosity for fractures and a second porosity for the solid matrix.

REFERENCES AND ADDITIONAL READING

ASTM, *Standard Guide for Selection of Aquifer Test Method in Determining Hydraulic Properties by Well Techniques*, D4043-96, ASTM, West Conshohocken, PA, 1997.

ASTM, *Standard Test Method (Field Procedure) for Withdrawal and Injection Well Tests for Determining Hydraulic Properties of Aquifer Systems*, D4050-96, ASTM, West Conshohocken, PA, 1997.

Bentall, R., ed., *Shortcut and Special Problems in Aquifer Tests*, USGS Water Supply Paper 1545-C, US Government Printing Office, Washington, DC, 1963.

Boulton, N. S., and Streltsova, T. D., "Unsteady Flow to a Pumped Well in a Fissured Water-Bearing Formation," *Journal of Hydrology*, 35: 257–270, 1977.

Bourdet, D., and Gringarten, A. C., *Determination of Fissure Volume and Block Size in Fractured Reservoirs by Type-Curve Analysis*, SPE 9293, Annual Fall Technical Conference and Exhibition, Society of Petroleum Engineers, Dallas, TX, 1980.

Bouwer, H., and Rice, R. C., "A Slug Test for Determining Hydraulic Conductivity of Unconfined Aquifers with Completely or Partially Penetrating Wells," *Water Resources Research*, 12(3): 423–428, 1976.

Butler, S. S., *Engineering Hydrology*, Prentice-Hall, Inc., Englewood Cliffs, NJ, 1957.

Cooper, H. E., and Jacob, C. E., "A Generalized Graphical Method for Evaluating Formation Constants and Summarizing Well-Field History," *Transactions of the American Geophysical Union*, 27: 526–534, 1946.

Cooper, H. H., Bredehoeft, J. D., and Papadopulos, I. S., "Response of a Finite-Diameter Well to an Instantaneous Change of Water," *Water Resources Research*, 3(1): 263–269, 1967.

Dawson, K. J., and Istok, J. D., *Aquifer Testing: Design and Analysis of Pumping and Slug Tests*, Lewis Publishers, Inc., Chelsea, MI, 1991.

Driscoll, F. G., *Groundwater and Wells*, Second Edition, Johnson Division, St. Paul, MN, 1986.

Ferris, J. G., Knowles, D. B., Brown, R. H., and Stallman, R. W., *Theory of Aquifer Tests*, USGS Water Supply Paper 1536-E, U.S. Government Printing Office, Washington, DC, 1962.

Fetter, C. W., *Applied Hydrogeology*, Macmillan College Publishing Co., New York, NY, 1994.

Freeze, R. A., and Cherry, J. A., *Groundwater*, Prentice-Hall, Inc., Englewood Cliffs, NJ, 1979.

Hall, P., *Water Well and Aquifer Test Analysis*, Water Resources Publications, Highlands Ranch, CO, 1996.

Hantush, M. S., "Modification of the Theory of Leaky Aquifers," *Journal of Geophysical Research*, 65: 3713–3725, 1960.

Hantush, M. S., "Hydraulics of Wells," in *Advances in Hydroscience*, 1, V. T. Chow, ed., Academic Press, Inc., New York, pp. 281–442, 1964.

Hantush, M. S. and Jacob, C. E., "Nonsteady Radial Flow in an Infinite Leaky Aquifer," *Transactions of the American Geophysical Union*, 36: 95–100, 1955.

Herzog, B. L., "Slug Tests for Determining Hydraulic Conductivity of Natural Geologic Deposits," in *Hydraulic Conductivity and Waste Contaminant Transport in Soil*, D. E. Daniel, and S. J. Trautwein, eds., STP 1142, ASTM, West Conshohocken, PA, 1994.

Hvorslev, M. J., *Time Lag and Soil Permeability in Ground-Water Observations*, Bulletin 36, Waterways Experiment Station, U.S. Army Corps of Engineers, Vicksburg, MS, 1951.

Jacob, C. E., "Drawdown Test to Determine Effective Radius of Artesian Wells," *Transactions of the American Society of Civil Engineers*, 112: 1047–1064, 1946.

Jenkins, D. N., and Prentice, J. K., "Theory for Aquifer Test Analysis in Fractured Rocks Under Linear (Nonradial) Flow Conditions," *Ground Water*, 20(1): 12–21, 1982.

Kazemi, H., Seth, M. S., and Thomas, G. W., "The Interpretation of Interference Tests in Naturally Fractured Reservoirs with Uniform Fracture Distribution," *Journal of the Society of Petroleum Engineers*, 246: 463–472, 1969.

Kozeny, J., *Theorie und Berechung der Brunnen: Wasser-Kraft and Wasserwirtschaft*, 28; *as cited in* Butler, S. S., 1957. *Engineering Hydrology*, Prentice-Hall, Inc., Englewood Cliffs, NJ, 1933.

Kruseman, G. P., and de Ridder, N. A., *Analysis and Evaluation of Pumping Test Data*, Publication 47, International Institute for Land Reclamation and Improvement, Wageningen, The Netherlands, 1991.

Moench, A. F., "Computation of Type Curves for Flow to Partially Penetrating Wells in Water-Table Aquifers," *Ground Water*, 31(6): 966–971, 1993.

Neuman, S. P., "Theory of Flow in Unconfined Aquifers Considering Delayed Response of the Water Table," *Water Resources Research*, 9(4): 1031–1045, 1972.

Neuman, S. P., "Effect of Partial Penetration on Flow in Unconfined Aquifers Considering Delayed Gravity Response," *Water Resources Research*, 10(2): 303–312, 1974.

Neuman, S. P., "Analysis of Pumping Test Data from Anisotropic Unconfined Aquifers Considering Delayed Gravity Response," *Water Resources Research*, 11(2): 329–342, 1975.

Nguyen, V., and Pinder, G. F., "Direct Calculation of Aquifer Parameters in Slug Test Analysis," in *Groundwater Hydraulics*, J. Rosenshein, and G. D. Bennett, eds., Water Resources Monograph 9, American Geophysical Union, Washington, DC, 1984.

Osborne, P. S., *Suggested Operating Procedures for Aquifer Pumping Tests*, EPA/540/S-93/503. USEPA, U.S. Government Printing Office, Washington, DC, 1993.

Rorabaugh, M. J., "Graphical and Theoretical Analysis of Step-Drawdown Test of Artesian Well," *Proceedings of the American Society of Civil Engineers*, 79(36): 23, 1953.

Sheahan, N. T., "Type-Curve Solution of Step-Drawdown Test," *Ground Water*, 9(1): 25–29, 1971.

Springer, R. K., *Application of an Improved Slug Test Analysis to the Large-Scale Characterization of Heterogeneity in a Cape Cod Aquifer*, Thesis, M. S., Massachusetts Institute of Technology, Cambridge, MS, 1991.

Stallman, R. W., *Aquifer Test Design, Observation, and Data Analysis*, USGS Techniques of Water Resources Investigation, Book 3, Chapter B1, US Government Printing Office, Washington, DC, 1971.

Streltsova, T. D., "Drawdown in Compressible Unconfined Aquifer," *ASCE Journal of the Hydraulics Division*, 100(HY11): 1601–1616, 1974.

Theis, C. V., "The Relation Between the Lowering of the Piezometric Surface and the Rate and Discharge of a Well Using Ground-Water Storage," *Transactions of the American Geophysical Union*, 16: 519–524, 1935.

Walton, W. C., *Selected Analytical Methods for Well and Aquifer Evaluation*, Bulletin 49, Illinois State Water Survey, Champaign, IL, 1962.

Walton, W. C., *Groundwater Pumping Tests: Design and Analysis*, Lewis Publishers, Inc., Chelsea, MI, 1987.

Warren, J. E., and Root, P. J., "The Behavior of Naturally Fractured Reservoirs," *Journal of the Society of Petroleum Engineers*, 228: 245–255, 1963.

CHAPTER 11
ASSESSMENT SAMPLING AND MONITORING

SECTION 11.8

SOIL VAPOR PRINCIPLES[*]

Blayne Hartman

Dr. Hartman is the founder and director of Transglobal Enviromental Geochemistry. He is an expert in geochemistry, laboratory analysis, direct push environmental sampling, and soil vapor surveys.

James A. Jacobs

Mr. Jacobs is a hydrogeologist and president of FAST-TEK Engineering Support Services with more than 20 years of experience. He is registered in several states, including California, and specializes in assessment methods and in situ remediation technologies.

11.8.1 INTRODUCTION

The chapter begins with an overview of the principles controlling concentrations in the soil vapor. Organic vapors are found in the unsaturated zone above the groundwater table. These vapors are related to spills and leaks of volatile organic compounds When the volatile organic compound is released into the unsaturated zone, the chemical will partition between the liquid phase and vapor or gaseous phase. Even if the soil absorbs all the liquid component of the chemical, vapors may still migrate through the vadose zone as shown in Figure 11.8.1 (Fetter, 1993).

11.8.2 PRINCIPLES

Composition of Soil Vapor

Soil vapor refers to the vapor that lies between the soil grains in the soil column lying between the ground surface and the water table (vadose zone). Other common names include soil gas, soil air, pore gas, and interstitial gas. Except in rare cases, soil vapor comprises more than 99.5 percent air. As such, it consists of approximately 78 percent nitrogen, 21 percent oxygen, with the remaining 1 percent made up by a variety of compounds as shown in Table 11.8.1 (Brenner and Blackmer, 1982).

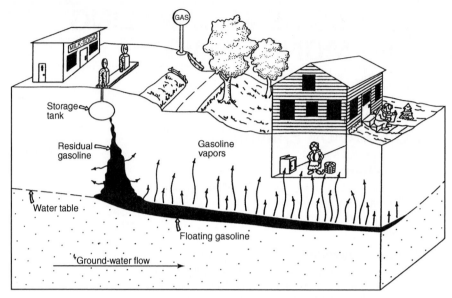

FIGURE 11.8.1   Release of organic vapors in vadose zone from residual saturation and floating gasoline (Fetter, 1993).

TABLE 11.8.1   Composition of Clean Dry, Atmospheric Air Near Sea Level. The Data for Most of the Minor Components are Uncertain (Brenner and Blackmer, 1982)

Component	Parts per million by volume
Nitrogen (N_2)	780,900
Oxygen (O_2)	209,400
Argon (Ar)	9300
Carbon dioxide (CO_2)	332
Neon (Ne)	18
Helium (He)	5.2
Methane (CH_4)	1.5
Krypton (Kr)	1
Hydrogen (H_2)	0.5
Nitrous oxide (N_2O)	0.33
Carbon monoxide (CO)	0.1
Xenon (Xe)	0.08
Ozone (O_3)	0.02
Ammonia (NH_3)	0.01
Nitrogen dioxide (NO_2)	0.001
Sulfur dioxide (SO_2)	0.002

Except at highly contaminated sites, typical concentrations of volatile organic compounds (VOCs) in the soil vapor generally are less than 1000 ppmv (0.1 percent). In contrast, carbon dioxide and methane can show large increases in the soil vapor over their concentrations in atmospheric air (330 and 2 ppmv, respectively) as a result of natural, biogenic degradation processes in the vadose zone. In rare circumstances, for example at landfills, methane and carbon dioxide combined concentrations can reach levels exceeding 50 percent by volume of the soil vapor (Morrison et al., 1998a and 1998b).

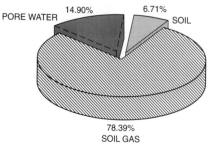

PORE WATER 14.90%   6.71% SOIL

78.39%
SOIL GAS

FIGURE 11.8.2 Mass distribution of vinyl chloride in soil phases (Marrin, 1997).

Despite generally low concentrations, soil vapor can be the major phase for volatile organic compounds in the vadose zone. As an example, a vinyl chloride spill was examined by Marrin (1997). The mass distribution (Figure 11.8.2) of the constituents of the VOC spill consisted of 78.39 percent soil vapor, 14.90 percent pore water and 6.71 percent in soil.

Units of Measurements

Soil vapor concentration data are generally reported in two types of units:

1. Volume per volume basis (volume of contaminant per volume of soil air). Examples:
   - parts per billion by volume (ppbv), or nanoliters compound per liter of soil air
   - parts per million by volume (ppmv), or microliters per compound per liter of air
   - percent by volume (percent).

2. Mass per volume basis (mass of contaminant per volume of soil air). Examples:
   - micrograms of compound per liter of soil air (ug/l)
   - milligrams of compound per cubic meter of soil air (mg/m^3)
   - millimoles of compound per cubic meter of soil air (mmoles/m^3)

These two unit systems are frequently confused and the units erroneously interchanged. Most commonly, soil vapor data reported in units of ug/l are mistakenly thought to equal a part per billion. This equivalency is true for water data because the density of water is equal to 1 gram per milliliter (g/ml), but this is not true for soil vapor data. This mistake is often made by both nonprofessional and professional environmental personnel because they commonly work with water data, and the equivalency of a ug/l and ppb is easily carried over to soil vapor data.

Note that the conversion of vapor data from units of ug/l to ppbv is:

$$C_{sg} \text{ in ppbv} = C_{sg} \text{ in ug/l} \times 24,000/MW$$

where MW is the molecular weight of the compound; 24,000 is milliliters per mole at 20°C.

For tetrachloroethylene (PCE), the molecular weight is 166, and 1 ug/l is equal to $24,000/166 = $~150 ppbv. Clearly, erroneously interchanging these two units can result in errors exceeding a factor of 100. Table 11.8.2 is a summary of many of the most common conversions of soil vapor data.

Physiochemical Properties Influencing the Distribution of Compounds in the Soil Vapor

In order to understand the applicability and design of soil vapor surveys, and how to interpret the data, it is necessary to understand the properties that control the concentrations of compounds in the soil vapor. This section provides a brief description of some of the primary properties that influence soil vapor concentrations. More detailed discussions may be found in other reference books (ASTM, 1993; ASTM, 1995; Morrison et al., 1998a and 1998b; Hartman, 1997 and 1998; Jeng, Kremesec, and Primack, 1996; Fetter, 1993; Hemond & Fechner; 1994).

Vapor Density. The vapor density of a compound relative to air is approximately equal to the molecular weight of the compound divided by the molecular weight of air, or:

$$\rho_v = MW_i/MW_{\text{air}}$$

TABLE 11.8.2   Formulas for converting Vapor Concentration Data (Williamson et al., 1994)

Unit	To convert to	Multiply by
ppmv	ppbv	1000
ppmv	mg/m^3	MW/24 (20°C)
ppmv	ug/l	MW/24 (20°C)
ppmv	kPa	1.01×10^{-4}
ppmv	mm Hg	7.6×10^{-4}
ppbv	ppmv	0.001
ug/l	mg/m^3	1
ug/l	kPa	$2.4 \times 10^{-3}/MW$ (20°C)
ug/l	ppmv	24/mw (20°C)
ug/l	ppbv	$2.4 \times 10^4/MW$ (20°C)
ug/l	mmHg	$1.8 \times 10^{-2}/MW$ (20°C)
kPa	ppmv	4.2×10^{21}
kPa	mg/l	$4.2 \times 10^2 MW$
kPa	mm Hg	7.5
mm Hg	ppmv	1.3×10^3
mm Hg	mg/l	$55 MW \times 10^{-3}$
mm Hg	kPa	0.13

Because the molecular weight of air is equal to 29 g/mole:

$$\rho_v = MW_i/29$$

Because of the relatively high molecular weight of chlorine and bromine, nearly all of the chlorinated and brominated compounds have vapor densities much larger than air (3 to 6 times). For example, the molecular weight of tetrachloroethylene (PCE) is equal to 166 g/mole giving an approximate vapor density for PCE of: $166/29 = 5.5$. Many common compounds associated with petroleum hydrocarbons, such as benzene, toluene, ethylbenzene, and xylenes (BTEX), also have vapor densities 3 to 4 times greater than air.

Because many VOCs also have relatively high vapor pressures, the vapor density can play an important role in situations where gaseous or liquid VOCs are used or stored in an indoor, confined space. In these situations, the vapors emanating from the container can sink to the floor because of their high vapor density. If air flow is restricted, such as in a closed room, the dense vapors can move through the concrete floor and penetrate the upper vadose zone. Such bulk dense vapor movement will continue to drive the vapor downward through the vadose zone until it is diluted to low enough concentrations (<1%) that density is no longer an important factor in the vapor transport process. Experience has documented vapor clouds reaching tens of feet into the uppermost vadose zone, at least in part to density driven flow. Common businesses and commercial operations susceptible to this situation include the PCE washing unit at dry cleaners, vapor degreasers at machine shops, and spray booths at inking or painting facilities using chlorinated solvent-based inks or paints.

Vapor Pressure.   Vapor pressure is a measure of a compound's propensity to exist in the vapor phase. It may be thought of as a distribution of a compound between the pure compound and the air above it. When exposed to air, a certain number of molecules of the compound will move into or *partition* into the air. The number of molecules can be measured as a concentration in the air or, because concentration and pressure for vapors are related, can be measured as a pressure. Typically, vapor pressures are given in units of millimeters of mercury (mm Hg) or atmospheres (atm), where 760 mm of Hg is equal to 1 atmosphere. Vapor pressures are highly temperature dependent.

As a general rule, compounds with vapor pressures exceeding 0.5 to 1 mm of Hg can exist in appreciable concentrations in the air phase near the pure product. Because many VOCs have vapor pressures exceeding 1 mm Hg, it is not surprising that these compounds can be detected in the soil vapor. As described previously, the vapor pressures of the common chlorinated solvents are high enough that vapors are continuously being released from the pure liquid products and can act as a source of chlorinated compound contamination into a room or into the subsurface.

Air to Water Partitioning (Henry's Law). Vapor pressure gives a measure of a compound's propensity to exist in the vapor phase in the presence of the pure compound. However, in the subsurface environment, significant amounts of water are often present, and a compound will have some affinity to dissolve into the water from the air or vapor phase (and vice versa). Partioning is how compounds distribute themselves between the soil, water, and air phases. The partitioning of a compound in the capillary zone between air and water is referred to as the Henry's Law constant (H) and is defined as:

$$H = C_{air} / C_{water}$$

Henry's Law constants are commonly tabulated in two forms: dimensional and dimensionless. While both forms are useful, the dimensionless form is the easiest form to work with. The dimensionless form of Henry's constant (H_D, also sometimes referred to as K_H) can be thought of as the number of molecules or mass of a compound that exist in the air versus the number of molecules or mass that dissolve into the water. If the dimensionless Henry's constant for a compound is greater than one ($C_{air} > C_{water}$), then the compound prefers to be in the air phase. In contrast, if the Henry's constant is less than one ($C_{air} < C_{water}$), the compound prefers to be dissolved in the water. This partitioning ratio will hold until a compound has reached saturation in water. Tables of dimensionless Henry's law constants frequently include many common VOCs with values at different temperatures.

The dimensionless Henry's constant can be used to predict the likelihood of a compound existing in the soil vapor. For example, consider the three VOCs: benzene, tetrachloroethylene (PCE), and chlorofluorocarbon CFC-113. For benzene, the dimensionless Henry's constant is approximately 0.25 or 1/4. For a system at equilibrium with equal volumes of soil vapor and water, one molecule of benzene will exist in the soil vapor for every four that dissolve into the water.

PCE has a higher Henry's constant than benzene. For PCE, the dimensionless Henry's constant is approximately 0.5 or 1/2. For a system at equilibrium with equal volumes of soil vapor and water, one molecule of PCE will exist in the soil vapor for every two that dissolve into the water. In contrast, the Henry's constant for CFC-113 is 10. At equilibrium, 10 molecules of CFC-113 will exist in the soil vapor for every one that dissolves into the water. Therefore, at equilibrium, CFC-113 is much more likely than benzene or PCE to be found in the vapor phase, rather than being dissolved in water.

The equilibrium concentration of a compound in the air or water from the other phase can be computed from the dimensionless Henry's constant as:

$$C_{air} = H \times C_{water}$$

The units for the air and water phases are the same. As an example, if the groundwater concentration was 10 ug/l for benzene, PCE and CFC-113, the equilibrium soil vapor concentration above the water would be:

$$\text{For benzene:} \quad C_{sg} = 0.25 \times 10\,\text{ug/l} = 2.5\,\text{ug/l}$$
$$\text{For PCE:} \quad C_{sg} = 0.5 \times 10\,\text{ug/l} = 5\,\text{ug/l}$$
$$\text{For CFC-113:} \quad C_{sg} = 10 \times 10\,\text{ug/l} = 100\,\text{ug/l}$$

The dimensional form of the Henry's constant is typically given in units of atm-m^3/mole and can be computed from the dimensionless constant using the ideal gas law by multiplying by RT (0.082 times the temperature in degrees Kelvin, equal to 22.4 at 0°C, and 24 at 20°C).

It is important to remember that the Henry's constants assume that equilibrium exists between the air and water phases and that the compound's solubility in water has not been reached (i.e., below saturation). These conditions are often not met in the real environment, so values computed from these constants are approximations that can be used for predictive purposes, but should be used cautiously for quantitative conclusions.

Because both the vapor pressure and air-water partitioning control how much of a compound will exist in the soil vapor, both parameters must be considered when predicting if a compound will be in the soil vapor. Compounds with high vapor pressures may not exist in the soil vapor if their Henry's constant is very low and water is present, for example: methyl tert-butyl ether (MTBE). Compounds with lower vapor pressures may actually exist at higher concentrations in the soil vapor if they have high Henry's constants (e.g., CFC-113).

Geologic Factors.   Site geology is also an important factor influencing soil vapor concentrations. The soil permeability gives a measure of the relative ease with which the rock or soil will transmit a vapor or liquid. Soil permeability is primarily related to grain size and soil moisture. Generally, soils with small grain size (e.g., clays) are less permeable, although other factors such as sorting can be important. Soil moisture decreases permeability by blocking vapor flow. Other important geologic factors influencing soil vapor concentrations and movements include preferential pathways because of cracks, vegetation, organisms, layers impervious to vapor such as perched groundwater or tight soil layers, and high organic carbon content. Further discussion of these factors, including soil to water partitioning can be found in Morrison et al. (1998a and 1998b) and Hemond and Fechner (1994).

Groundwater Movement.   Although gasoline and chlorinated solvents such as PCE are volatile organic compounds, and both release vapors into the vadose zone, gasoline is in a class of liquid compounds that are lighter than water, called light nonaqueous phase liquids (LNAPLs). Once satuation is reached with an LNAPL compound, free product collects on the top of the groundwater table. As the groundwater table rises and falls with an LNAPL floating-product layer, such as gasoline, the free product will be sorbed by the soil in a zone representing the annual hydrologic cycle. The residual saturation in this zone will continue to contribute soil vapors. Compounds like PCE that are heavier than water, are called dense nonaqueous phase liquids (DNAPLs). DNAPLs will generally migrate through the vadose zone and sink into the underlying aquifers. If present as free product, DNAPLs typically reside at the base of aquifers.

11.8.3 TRANSIENT ENVIRONMENTAL EFFECTS OF SOIL VAPOR CONCENTRATIONS

Temperature

Temperature can have an effect on soil vapor concentration. Vapor pressures (and Henry's Constants) can increase significantly for many VOCs and other higher-molecular weight volatile organic compounds. Temperature changes decrease with depth. Diurnal temperature changes are less than 1°C at depths of 3 ft or more. Therefore, daily changes in soil temperature will not significantly affect soil vapor concentrations at sampling depths greater than 3 ft.

Barometric Pressure

Changes in barometric pressure can create a pressure gradient that causes flow of soil vapors in the subsurface. It is estimated that unless soil vapor sampling occurs during a significant winter storm or hurricane, barometric changes are not important in soil vapor sampling and analysis (Bentley, 1997).

Earth Tides

Earth tides have been observed to cause water-level fluctuations of generally less than 0.1 foot. Therefore, earth tides generally have a minimum affect on soil vapor movement and concentration (Bentley, 1997).

Precipitation

Infiltration from rainfall can potentially impact soil vapor by creating wetting front that displaces soil vapor downward and by dissolving volatile organic compounds and carrying them deeper. For large storms with rainfall of 3 to 4 in, the amount of infiltration could approach 2 in. However, it is unlikely that soil vapor at depths greater than 3 ft would be significantly disturbed by a rainstorm. One case where this conclusion is invalid is if the soils are highly permeable and where the infiltration rate and downward velocity of water is significant (Bentley, 1997).

Summary of Impact of Environmental Changes on Soil Vapor

The impact of environmental changes on soil vapor concentration from temperature, barometric pressure, earth tides, and precipitation was evaluated by Bentley (1997). The data suggest that to avoid a measurable impact, soil vapor probes and sampling depths should be greater than 4 to 5 ft.

11.8.4 CONVERSION OF SOIL VAPOR DATA TO SOIL CONCENTRATION SOIL GAS CONCENTRATIONS

For a given soil vapor concentration, the total mass distribution of a chemical in the vadose zone is a function of the total organic carbon content, porosity, moisture content and bulk density.

Total mass distribution of a contaminant

$= $ Contaminant concentration measured as soil vapor $\times (E)$

where E is a calculated number based on measured values for porosity, moisture content, and organic carbon. An example from a site in Tucson, Arizona indicates that the E value varied by less than 20 percent when porosity or moisture content was varied. However, variations in organic carbon content yielded almost a 1 to 1 change in total mass (Cooper, 1997).

For all but the most water-soluble compounds, the ratio of soil vapor to total soil concentration is most sensitive to the water-solid distribution coefficient (K_D) (Figure 11.8.3), next to Henry's law coefficient in its dimensionless form (H_D). The other parameters such as moisture content, porosity, and bulk density have relatively little effect on the conversion between soil vapor data and the estimated soil concentration. The most important soil parameter in calculating the conversion of soil vapor concentration to total soil concentration is total organic carbon, which has a much larger range of potential values than the other parameters. These assumptions and conditions are often not the case in the vadose zone where the majority of soil vapor exists.

In the absence of actual soil vapor data, soil vapor concentrations can be calculated from soil and groundwater data, assuming equilibrium conditions, using equations based upon Henry's Law constants and soil to water partitioning constants. These equations, summarized below, are discussed in more detail in the 1995 ASTM Risk Based Corrective Action (RBCA) standard.

The soil vapor concentration (C_{sg}) is computed based upon the equilibrium partitioning between the soil, moisture, and vapor phases as:

$$C_{sg}(\text{ug/l}) = \frac{H \times C_{\text{soil}} \times BD}{P_w + (K_s \times BD) + (H \times P_a)} \times 1000$$

C_{soil} is the concentration in the soil for the contaminant of concern (e.g., benzene).

It is crucial to realize that the equations used to calculate the soil vapor concentration from soil phase data or water phase data assume equilibrium partitioning between the phases. Equilibrium partitioning is only obtained if a system is well mixed. This is very rarely accomplished in the subsurface, because there are no blenders or stirrers present to homogenize the vapor, soil, and groundwater. In addition, they do not account for other processes that are operative in the vadose zone, including bioattenuation,

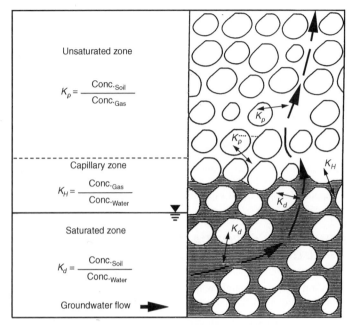

FIGURE 11.8.3   Equilibrium expressions for portioning of volatile organic compounds among aqueous, vapor and sorbed phases (Marrin, 1997).

advective flow, soil heterogeneity, among others. For this reason, calculated soil vapor concentrations often do not accurately represent actual soil vapor concentrations. In the case of fuels, calculated values often overestimate actual soil vapor concentrations by 10 or 100 times. In the case of chlorinated solvents, calculated soil vapor concentrations from soil values often underestimate true values because of the presence of vapor clouds infiltrating into the vadose zone from the surface. A more complete discussion of these processes can be found in Morrison et al. (1998a and 1998b) and Hartman (1997 and 1998).

REFERENCES

ASTM, "Standard Guide for Soil Gas Monitoring in the Vadose Zone," D5314-93, *Annual Book of ASTM Standards*, Philadelphia, 1993.

ASTM, "Standard Guide for Risk-Based Corrective Action (RBCA) Applied at Petroleum Release Sites," E-1739-95, *Annual Book of ASTM Standards*, Philadelphia, 1995.

Bentley, H., *Estimating Site VOC Mass Distribution*, Groundwater Resources Association of California, 1997 GRA Seminar Series, Class notes, 1997.

Brenner, J., and Blackmer, A., "Composition of Soil Atmospheres," in *Methods of Soil Atmospheres*, "Part 2, Chemical and Microbial Properties," Second Edition, Soil Society of America, Madison, WI, 1982.

Cooper, C., *Applicability of Soil Gas vs. Soil Analyses for Streamlining Site Investigations*, Groundwater Resources Association of California, 1997 GRA Seminar Series, Class notes, 1997.

Fetter, C. W., *Contaminant Hydrogeology*, Macmillan Publishing Co., New York, 1993.

Hartman, B., "The Upward Migration of Vapors," *LUSTLine Bulletin 27*, NEIWPCC, Wilmington, MA, December 1997.

Hartman, B., "The Downward Migration of Vapors," *LUSTLine Bulletin 28*, NEIWPCC, Wilmington, MA, December 1998.

Hemond, H. F., and Fechner E. J., *Chemical Fate and Transport in the Environment*, Academic Press, New York, 1994.

Jeng, C-Y., Kremesec, V. J., and Primack, H. S., "Models of Hydrocarbon Vapor Diffusion Through Soil and Transport into Buildings," in *Proceedings of the 1996 Petroleum Hydrocarbons & Organic Chemicals in Ground Water: Prevention, Detection, and Remediation*, Houston, TX, November 1996.

Marrin, D. L., *Innovative Trends in Soil Gas Applications*, Groundwater Resources Association of California, 1997 GRA Seminar Series, Class notes, 1997.

Marrin, D. L., *Innovative Trends in Soil Gas Applications*, Groundwater Resources Association of California, 1997 GRA Seminar Series, Class notes, 1997.

Morrison, R., Beers, R., and Hartman, B., *Petroleum Hydrocarbon Contamination-Legal and Technical Issues*, Argent Communications Group, Inc., Foresthill CA, 1998a.

Morrison, R., Beers, R., and Hartman, B., *Chlorinated Solvent Contamination-Legal and Technical Issues*, Argent Communications Group, Inc., Foresthill CA, 1998b.

Williamson, R., Kerfoot, H. B., and Mayer, C., *Soil Gas Monitoring*, Class notes, presented to the Los Angeles Regional Water Quality Control Board, 1994.

CHAPTER 11
ASSESSMENT SAMPLING AND MONITORING

SECTION 11.9

APPLICATIONS AND INTERPRETATION OF SOIL VAPOR DATA TO VOLATILE ORGANIC COMPOUND CONTAMINATION*

Blayne Hartman

Dr. Hartman is the founder and director of Transglobal Enviromental Geochemistry. He is an expert in geochemistry, laboratory analysis, direct push environmental sampling, and soil vapor surveys.

James A. Jacobs

Mr. Jacobs is a hydrogeologist and president of FAST-TEK Engineering Support Services with more than 20 years of experience. He is registered in several states, including California, and specializes in assessment methods and in situ remediation technologies.

11.9.1 INTRODUCTION

This article describes the applications, uses, and interpretation of soil vapor data for assessing, monitoring, and determining the potential risk of volatile organic compounds (VOCs) including petroleum and chlorinated hydrocarbons. The article begins with an overview of the applications of soil vapor surveys, survey methodology, data interpretation, and concludes with a discussion of the fate and transport of vapors in the vadose zone. The purpose of this article is to give an overview of the subject, and to give guidance to professionals intending to collect soil vapor data, critically review or refute soil vapor data, and interpret the potential risk of contaminated soil vapor.

11.9.2 PRINCIPLES

Composition of Soil Vapor

Soil vapor refers to the vapor that lies between the soil grains in the soil column lying between the ground surface and the water table (vadose zone). Other common names include soil gas, soil air,

---

pore gas, and interstitial gas. For more information regarding soil vapor concepts, please refer to Soil Vapor Principles, ASTM (1993) and Morrison et al. (1998).

11.9.3   APPLICATIONS OF SOIL VAPOR DATA

Soil vapor data are collected for and applied to a variety of purposes related to contamination by volatile organic compounds (VOCs). Principal applications are as follows:

1. To locate, delineate, and assess soil phase contamination on a site
2. To locate, delineate, and assess groundwater contamination on a site
3. To locate, delineate, and assess vapor phase contamination on a site
4. To evaluate the potential risk of the upward transport of vapor
5. To monitor site remediation (natural and engineered)
6. To monitor landfills

Traditionally, soil vapor surveys have primarily been used for site assessment purposes to identify soil and groundwater contamination (Devitt et al., 1987; Marrin and Kerfoot, 1988; Marrin, 1997). The motive for employing a soil vapor survey, as opposed to other methods, is that the methods are inexpensive, quick, and relatively unobtrusive, thereby enabling a large area to be covered at low cost. Soil vapor surveys are commonly performed early in the site investigation, and the data commonly are used to focus subsequent soil and groundwater sampling programs.

In the 1990s, increased emphasis has been placed on the presence and movement of volatile vapors in the vadose zone. Soil vapor surveys are the only investigative method to assess the spatial and vertical extent of vapor contamination. This application is especially relevant for chlorinated solvents because of their potential to penetrate into the vadose zone from the surface as a dense vapor (refer to the discussion on vapor density in Morrison et al., 1999).

With the move towards risk based corrective action (RBCA), regulatory agencies across the country are concerned with the human health risk because of the upward migration of contaminant vapors through the vadose zone into buildings. Soil vapor surveys enable actual soil vapor concentrations to be measured, which gives a more accurate picture than values calculated from soil or groundwater data.

Use of soil vapor data for monitoring site remediation takes many forms. Most commonly, concentrations of contaminants in the soil vapor are monitored to measure the progress of a vapor extraction system. With the recent emphasis on remediation by natural attenuation, concentrations of the contaminants, by-products, and oxygen in the soil vapor are useful to measure the progress of the degradation of the contaminant plume.

Soil vapor is typically measured at the fringes of landfills to ensure that harmful compounds are not migrating from landfills to neighboring developments. Measured compounds include methane, chlorinated compounds (primarily vinyl chloride), and petroleum related carcinogens such as benzene.

11.9.4   SOIL VAPOR METHODOLOGY

Three principle methods exist for collecting soil vapor data:

1. Active
2. Passive
3. Flux chambers

Each method offers advantages and disadvantages, which will be briefly described below. More comprehensive descriptions of soil gas collection methods are given in procedural documents written

TABLE 11.9.1 Guidelines on Sample Spacing, Collection Depth, and Purge Volume for Various Applications (Morrison et al., 1998)

Soil Vapor Sample Location Design Considerations

Defining Soil & Vapor Contamination

⇒ Collection Spacing:
- General grid spacing from 50′ to 100′ over nonsource areas
- Spacing over sources from 10′ to 20′
- Spacing around tanks: one on each side (minimum of 4)

⇒ Collection Depth:
- Surface sources: 5 ft bgs
- Subsurface sources: At bottom of containment vessel (tank, pipe, clarifier)

Defining Groundwater Contamination

⇒ Collection Location: Up and down gradient across site
⇒ Collection Depth: Within 5′ to 10′ of groundwater

Assessing Upward & Downward Vapor Migration

⇒ Collection Location: At center of contaminant plume
⇒ Collection Depths:
- For upward migration: A minimum of 2 depths from depth of source to surface
- For downward migration: A minimum of 2 depths from depth of source to GW
- For flux chambers, a minimum of 2 collection time intervals

Monitoring VES/Sparge Remediation Systems

⇒ Collection Locations:
- From a reasonable sampling of the extraction wells
- From the burner/filter influent
- From the burner/treatment effluent
- From vapor monitoring wells at the edges of the property

⇒ Collection Depth:
- From the depth of the contamination

---

by Devitt et al. (1987), ASTM (1993), and the EPA (1997). A good overview of soil vapor surveys is given by Ullom (1995) (see Table 11.9.1).

The *active* approach consists of the withdrawal of an aliquot of soil vapor from the subsurface, typically with a sampling probe through sterile tubing into a collection container, followed by analysis of the withdrawn vapor. Common sample containers for active vapor sampling include syringes, gas sampling bags, pre-evacuated stainless steel canisters (e.g., SUMMA canisters), and gas-tight vials (e.g., 40 ml pre-evacuated volatile organic analysis [VOAs] vials or headspace vials) (see Figure 11.9.1).

Analysis is often performed on-site with portable or more sophisticated analytical instruments. Alternatively, soil vapor samples are stored in gas-tight containers and analyzed at an off-site laboratory. The active method is quantitative and values are reported in concentration units (e.g., ppmv, ug/L-vapor). This approach is the most common soil vapor method for a number of reasons, including ease of sample collection, opportunity for real-time data to direct further sampling, and quantitative data. Because the method requires that soil vapor be withdrawn, it is best suited to locations with higher vadose zone permeabilities and higher contaminant concentrations (>0.1 ug/L-vapor).

The *passive* approach consists of emplacement of a natural or synthetic adsorbent into the subsurface for an extended amount of time, and subsequent removal and analysis of the adsorbent. The absorbent is typically placed in the upper end of an inverted container having an open bottom for migrating vapors to enter and collect (see Figure 11.9.2).

Measured values cannot be reported as concentrations, only as total adsorbed mass (e.g., ug) or in some other form of relative units, because the amount of vapor that comes into contact with the adsorbent is unknown. Because of this limitation, passive surveys are useful for qualitative purposes only. Because one effect of the adsorbent is to concentrate the soil vapor, this approach offers advantages over the active approach in locations of low vadose zone permeability and lower

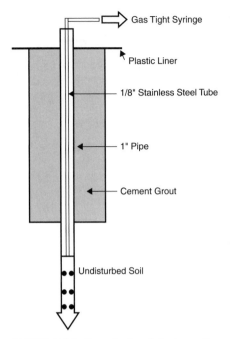

FIGURE 11.9.1 Example of a whole-air sampling system (Devitt et al., 1987).

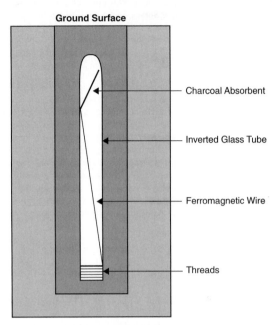

FIGURE 11.9.2 Schematic diagram of installation of a sorbed contaminant passive vapor sampling system (Morrison et al., 1998).

contaminant concentrations (>0.01 ug/L-vapor). However, as described later in this article, contaminants still must have an appreciable vapor pressure to be detected by this method. The technique requires two visits to the field (one to deploy the adsorbents and a second trip to retrieve them) and does not allow for real-time data.

Flux chambers consist of an enclosed chamber, which is placed on the surface for a specific period of time (see Figure 11.9.3).

Vapor concentrations are measured in the chamber through time. This method is also quantitative and yields both concentration data in the chamber and flux data (mass/area-time). Flux chambers are the least common soil vapor method, and are typically used only when direct vapor fluxes out of the subsurface are desired.

11.9.5   PROCEDURES THAT INFLUENCE REPORTED SOIL VAPOR DATA

Reported soil vapor data depend greatly upon the collection and analytical protocols that are used to generate the data. For this reason, it is important to understand the factors that may influence the reported data when designing a program or critically reviewing soil vapor data. This section presents a description of a number of relevant factors that influence the reported data for the various methods. The discussion is divided into collection related issues and analytical related issues.

Collection Considerations

The design and protocols of a soil vapor survey program are dependent upon the objectives of the program, the types of contaminants anticipated to be present, and the site conditions. There are a variety

TABLE 11.9.2   A Checklist of Some of the Procedural Related to Active, Passive, and Flux Chamber Vapor Survey Techniques (Morrison et al., 1998)

Defending or Refuting Soil Vapor Data—Collection Related Issues

Active Soil Vapor Surveys

- Was enough vapor collected to purge the dead volume of the probe?
- Was so much vapor collected that air might have been drawn down from the surface?
- Was so much vapor collected that the location of the sample is unknown?
- Was the probe adequately decontaminated between samples?
- Were samples analyzed on site or off-site?
- If samples were analyzed off-site, how were samples stored and for how long?
- Were samples collected deep enough (∼5 ft bgs) to minimize air infiltration?
- If samples were collected at shallow depths, was the probe sealed at the ground?
- Were vacuum pumps used in the sample collection?
- Were samples collected upstream of the vacuum pump?
- Were samples collected with proper procedures to minimize air infiltration?
- Were excessive vacuums required to obtain a soil vapor sample?

Passive Soil Vapor Surveys

- Were method and trip blanks analyzed?
- Could measured values be due from infiltration of contaminated atmospheric air?
- Could measured values be due from an overlying surface (e.g., asphalt, dirty soil)?
- Could measured values be related to ground cover, surface, or burial depth?
- Were samplers left in the ground for sufficient time?
- Were duplicate samples collected and how do they compare?
- Are data used quantitatively? How, for what purpose, on what basis?

Flux Chamber Surveys

- If concentrations were low, was the seal at the surface vapor tight?
- Was the chamber subjected to temperature extremes?
- How long was the deployment time? Was it long enough to average near-term effects?
- Did the chamber concentration reach high enough values to influence the flux?
- What volume of vapor was collected from the chamber? How fast was it collected? Did it create advective flow from the subsurface or sides?

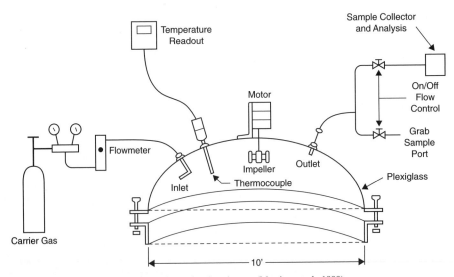

FIGURE 11.9.3   Surface flux chamber and associated equipment (Morrison et al., 1999).

TABLE 11.9.3   A Checklist of Some of Analytical Related Issues (Morrison et al., 1998)

Defending or Refuting Soil Vapor Data—Analytical Related Issues

Analysis of Soil Vapor Samples

The following questions should be asked when examining the analysis of any type of soil vapor sample; active, passive, or flux chambers.

- Have the proper instruments and detectors been used to detect the target compounds at the required levels of sensitivity?
- Have calibration standards been analyzed before and after the samples?
- Are the reported values within the documented calibration range of the instrument?
- Are any compounds coeluting on a nonhalogen specific detector?
- Have blanks been analyzed?
- Are the calibration standards fresh and traceable?
- In what units are the data reported (ug/l, ppbv, ppmv)?

Desorption of Passive Soil Vapor Collectors

In addition to the analytical issues summarized above, passive soil vapor samples also must be desorbed from the adsorbent. The following issues should be examined:

- Have collector blanks been analyzed?
- Have collector trip blanks been analyzed?
- How are the collectors desorbed? At what temperature?
- How much of the desorbed sample is injected into the instrument?
- How is the efficiency of the desorption procedure for the target analytes measured?
- How are standards prepared?
- In what form are the data reported:
- ⇒ As concentration in the collector headspace?
- ⇒ As total mass desorbed off the collector?
- In what units are the data reported (ug/l, ppbv, ppmv)?

---

of sampling methods and equipment designs for collecting soil vapor samples that can potentially yield different values (see Tables 11.9.2 and 11.9.3).

Active Soil Vapor Surveys.

Sample Spacing.   The selection of sampling locations is strongly dependent upon the objectives of the program, the need for adequate coverage, and the budget. Predetermined and widely spaced grid patterns are most commonly used for reconnaissance work, while closely spaced, irregularly situated locations are commonly used for covering specific source areas.

Collection Depth.   Collection depths should be chosen to maximize the chances of detecting contamination, yet minimize the effects resulting from vapor movement, changes in barometric pressure, and surface temperature, or breakthrough of atmospheric air from the surface. Surface temperature and pressure influences are thought to be minimal at depths exceeding three feet below ground surface (Bentley, 1997; Bentley and Walter, 1997). To optimize the chances of detecting contamination and minimizing the potential pitfalls resulting from vapor movement, soil vapor samples should be collected as close to the suspected contamination source as practically possible.

Purge Volume.   The sample collection equipment used for active soil vapor surveys has an internal volume which is filled with air or some other inert gas prior to insertion into the ground. This internal volume, often called the dead volume, must be completely purged and filled with soil vapor to ensure that a representative soil vapor sample is collected. Different opinions exist on the optimum amount of vapor to be purged. At a minimum, enough vapor should be withdrawn prior to sample collection to purge the probe and collection system of all ambient air or purge gas (1 purge volume). Others feel that similar to a water monitoring well, a minimum of 3 system volumes should be purged. Many experienced soil vapor personnel purge a minimum of 1 and a maximum of 5 system volumes before collecting a sample. Because soil vapor data are often interpreted in a relative fashion, it is important that the purge volume is consistent for all samples collected from the same site.

While it is important to collect enough vapor to purge the system, collecting too much vapor can also have drawbacks. The larger the quantity of soil vapor withdrawn, the greater the uncertainty in the location of the collected sample, and in turn the greater the potential that atmospheric air might have been drawn down the outside of the probe body. Thus, sampling equipment with small internal dead volumes offers advantages over systems with larger dead volumes because the former systems require significantly less vapor to be withdrawn when purging the system. The dead volume of the sampling equipment can be calculated as:

$$\text{Dead Volume} = \text{pi} \times \text{radius}^2 \times \text{length}$$

where radius refers to the internal radius of the vapor pathway; and length refers to the length of the vapor pathway.

To illustrate the importance of the dead volume, a comparison of the volume of soil vapor required to purge probes of different internal sampling diameter is summarized below. It is readily apparent that as the internal diameter of the sampling system exceeds 0.75 in (1.9 cm), the volume of soil vapor withdrawn becomes quite large, increasing the uncertainty in the source of the collected vapor.

Comparison of Purge Volume Required by Sampling Probes of Different Internal Sizes. Calculated Volumes are for a Probe Length of 5 ft

Internal diameter (cm)	Dead volume (cc)	5 Dead volumes (cc)
0.2	5	25
2	~500	~2500
4	~2000	~10,000

Probe Seals.   For collection systems with large purge volumes or designed to collect large sample volumes, it is often necessary to seal the probe at the surface. Seals may also be necessary for small volume systems if the soils are extremely porous and the sampling depth close to the surface (<3 ft). Most common sealing techniques are to pack the upper contact of the probe and the soil with grout or to use an inflatable seal. Seal integrity can be easily tested by allowing a tracer gas (e.g., propane or butane) to flow around the probe at the contact with the ground surface and to analyze a collected soil vapor sample for the tracer gas.

Probe Decontamination.   All external parts should be wiped clean and washed as necessary to remove any soil or contaminant films. The internal vapor pathway should be purged with a minimum of 5 volumes of air or an inert gas, or replaced, or washed if contamination or water is present in the probe. Probes fitted with internal tubing offer advantages because the internal tubing can simply be replaced.

Excessive Vacuums Applied During Collection.   Soil vapor samples collected under high vacuum conditions or under a continuous vacuum may reflect contaminants that have been desorbed off the soil grains created by the collection process, rather than contaminants present in the undisturbed soil vapor. For collection systems employing vacuum pumps, the vacuum applied to the probe should be measured and recorded.

Systems with Vacuum Pumps.   Soil vapor samples from collection systems employing vacuum pumps should be collected on the intake side of the pump to prevent potential contamination from the pump. Further, because the pressure on the intake side of the pump is below atmospheric, soil vapor samples must be collected with appropriate collection devices, such as gas-tight syringes and valves, to ensure that the samples are not diluted by outside air.

Sample Containers and Storage of Samples.   While on-site analysis is advantageous to ensure sample integrity, soil vapor is often collected and analyzed off-site. To minimize potential effects on the sample integrity, it is recommended that:

• Maximum storage time does not exceed 48 hours after collection.
• For fuel related compounds (TPHv, BTEX) and biogenic gases (CH_4, CO_2, & O_2): preferred storage containers include tedlar bags, gas tight vials (glass or stainless steel), and summa canisters.

- For halogenated compounds (e.g., TCE, TCA, PCE): preferred storage containers must be gas tight, but also dark to eliminate potential effects due to photodestruction. Tedlar bags have been shown to not be a reliable storage container (Denly & Wang, 1995).
- Do not chill collected soil vapor samples during storage as is common with soil and water samples.
- Stored samples not be subjected to changes in ambient pressure (e.g., during shipping).

Passive Soil Vapor Surveys.

Sample Spacing.   The selection of locations for passive sampling is based upon the same considerations as active soil vapor methods: program objectives, the need for adequate coverage, and the budget. Predetermined and widely spaced grid patterns are most commonly used for reconnaissance work, while closely spaced, irregularly situated locations are commonly used for covering specific source areas.

Collection Depth.   Passive surveys are nearly always conducted by burying the collector close to the surface (6 in to 3 ft). This protocol was developed not for technical reasons, but for convenience in deploying and retrieving the collector. Ideally, similar to active surveys, collectors should be deployed as close to the suspected contamination source as practically possible to minimize the effects of vapor movement. In addition, collectors buried within a couple feet of the surface will be very susceptible to air infiltration because of changes in barometric pressure and surface temperature. If the outside air is contaminated, for example at an active gasoline station or inside of an active dry-cleaning operation, the passive collectors could conceivably adsorb more contamination from infiltration of the surface air than from subsurface contamination. In this situation, it is advisable to bury the collector to deeper depths (>3 ft).

Exposure Period.   As with collection depth, the exposure period for passive collectors is generally selected more for convenience factors than for technical reasons. The key assumption that is invoked when interpreting passive soil vapor data is that each collector is exposed to the same quantity of soil vapor. Thus, passive collectors are typically deployed for the same period of time on a site or the data is normalized based upon the exposure time. Typical exposure times are a few days to two weeks.

In practice, the exposure period for a passive collector should depend upon the concentration of the contaminant of interest and desired detection levels. In areas of suspected high concentration, collectors can be left in the ground for shorter periods (1 to 5 days). In areas of suspected low concentrations, collectors are often left in the ground for two or more weeks. For areas of unknown concentration, the optimum approach is to determine the deployment time by burying a number of collectors in the same location and measuring them over a period of time. However, typically the sampling logistics do not allow for this type of deployment time determination.

Method Blanks.   Because the passive soil vapor method does not enable real-time data, analysis of blanks is extremely important to verify that detected contamination was not from another source, such as the passive collector itself or handling and storage during transport from the site to the laboratory. The only way to evaluate this possibility is to include a method blank and trip blank as part of the sample batch. A method blank consists of an unused collector picked at random from the collector batch. A trip blank is an unused collector that is kept sealed, and accompanies the other collectors to and from the site and to the laboratory for analysis. If data from method and trip blanks do not exist, than it is conceivable that values from the other collectors could be disclaimed as being due to contamination from sources other than the site.

Flux Chamber Soil Vapor Surveys.

Sample Spacing.   Because the primary motive of flux chamber surveys is to measure the upward flux of vapor out of the ground or into a room, as opposed to locating and defining contamination, generally only one or two chambers are deployed in the room or on the ground surface.

Insertion Depth or Seals.   Valid measurements require that the bottom of the chamber be sealed from exchange with atmospheric air. On soil surfaces, chambers are typically inserted into the ground to a depth several inches below the ground surface. On finished surfaces such as floors, an air-tight seal must be made between the chamber bottom and the surface, typically using a gasket or sealant.

Covers.   Reflective coverings are necessary in outside locations to protect against temperature extremes. Opaque coverings are required to minimize the potential of photodestruction of compounds.

Exposure Period.   The exposure period for flux chambers can be estimated based upon the minimal vapor flux that is desired to be measured and the sensitivity of the analytical instrument. The concentration in the chamber may be computed as the product of the flux, the contact area, and the exposure time divided by the chamber volume. Typical exposure times are a few hours to a week.

Sample Containers and Storage of Samples.   Refer to the same section under active soil vapor surveys for a description of applicable containers and storage considerations.

Analytical Considerations

Qualitative Methods.   The type of analytical instrumentation employed depends upon the goals of the project. For quick screening-level surveys, in which the primary goal is simply to find contamination qualitatively, simple hand-held or portable instruments can be employed.

Although typically used for safety purposes, such as air monitoring of the breathing zone, glass detector tubes have been used to measure various gas/vapors as a part of soil vapor surveys. Each tube is designed to detect a gas or vapor over a specific measurement range. No calibration or battery power is required to use the gas detector tubes. The results of gas detector tubes are available within minutes. There are dozens of target chemicals available and the cost of each analysis is a fraction of the cost of gas chromatographic analysis.

For example, hydrogen sulfide has several different tubes, each covering a different measurement range. Collectively, the hydrogen sulfide tubes have an analytical range from 1 ppm up to 4 percent. The range of battery powered vapor measuring analytical instruments for hydrogen sulfide range from 0 to about 100 ppm. With the appropriate target chemical or application, gas detector tubes can provide useful data for screening purposes or augmenting a qualitative soil vapor survey.

Quantitative Methods.   For applications where quantitative results are desired, the analytical methodology employed is typically gas chromatography (GC) or gas chromatography-mass spectrometry (GC-MS). These techniques provide differentiation of a wide variety of compounds. Because the analysis of soil vapor samples (collected by either active, passive, or flux chambers) are performed predominantly with gas chromatographs or GC-MS, a detailed description of analytical considerations associated with these instruments will be presented. The purpose is to identify the primary analytical factors that must be considered for legally defensible data.

Gas chromatographs come with a variety of analytical detectors that measure the compounds of interest. Most typically, many VOCs are detected and quantified with photoionization detectors (PID), electron capture detectors (ECD), and/or electrolytic conductivity detectors (ElCD). In some cases, depending upon the project goals, simple flame ionization detectors (FID) may be suitable. Newer analytical configurations enable the PID to be operated in series with the ElCD. Principle criteria in choosing the proper detector are the selectivity of the detector to the target compounds, sensitivity, dynamic range, and stability of the detector. Each of these criteria will be briefly described below.

Sensitivity.   The analytical instrument or detector must be sensitive enough to ensure that required detection levels are obtained. For VOCs, electron capture detectors (ECD), photoionization detectors (PID), and the electrolytic conductivity detector (ElCD) are the most sensitive. Detector sensitivity is compound specific, meaning that the same detector may be very sensitive for one compound, but not for another. To be legally defensible, it is necessary to demonstrate that the instrument or detector can see the target compound at the required minimum concentration levels.

Selectivity.   Selectivity refers to the ability of a detector to respond preferentially to a compound or class of compounds to minimize the potential for false identification. EPA analytical protocols for measurement of VOCs in soil and water (protocols do not currently exist for soil vapor) require the use of the electrolytic conductivity detector (ElCD) because it responds only to halogens. However, VOCs are also detected and quantified with photoionization detectors (PID) and newer analytical configurations enable the PID to be operated in series with the ElCD, which offers many advantages. Electron capture detectors are also relatively selective to halogenated compounds, although not as much as the ElCD. Flame ionization detectors are not selective at all and should only be used in situations where there is no possibility of interfering compounds. Gas chromatographs with mass spectrometers (GC-MS) offer the ultimate in selectivity because the mass detector can identify compounds based upon a compound's mass spectra (similar to a fingerprint). The use of nonselective detectors can possibly result in compound interferences, which will not yield legally defensible data.

Calibration Range. To be legally defensible, all data reported off an instrument must fall within a documented calibration range. The calibration range of a detector depends upon the detector's *dynamic range.* The dynamic range refers to the amount (i.e., mass) of a compound that a detector can respond to with some reasonable accuracy. Detectors with large (or wide) dynamic ranges will give reliable results over a large concentration range for a compound. Detectors with small (or narrow) dynamic ranges only give reliable results over a small concentration range. This is important to know because soil vapor data for compounds from detectors with small dynamic ranges are susceptible to many more potential errors. In general, PID, FID, and thermal conductivity detectors (TCD) have wide dynamic ranges; ElCD & GC-MS have a moderate dynamic range; and the ECD has a narrow dynamic range.

Stability. Some detectors are extremely stable and perform reliably for long periods of time (e.g., FID, PID, TCD). Other detectors are more finicky and their performance must be monitored more closely (ElCD, ECD, GC-MS). Continuity of performance is ensured by analyzing calibration standards before and after soil vapor samples. To be legally defensible, all data reported off an instrument must be bracketed by calibration standards to demonstrate reliable performance during the period the samples were analyzed.

Passive Soil Vapor Samples. With passive samples, the methodology by which the contaminants are desorbed from the collector and subsequently injected into the instrument will influence the reported results. A number of different techniques and methods are used (thermal desorption, curie point, solvent extraction). The procedures employed must be demonstrated to give reliable results or the data will not withstand a legal challenge.

QA/QC Considerations. Too often, rigid analytical specifications are applied to soil vapor data in an effort to ensure levels of accuracy and precision that exceed the accuracy of the soil vapor method itself. This is often the case when QA/QC criteria from the EPA water and test methods (EPA SW-846) manual are applied to soil vapor. To ensure legally defensible data, it is much more important that certain QA/QC analyses are performed, rather than they fulfill rigid performance specifications. For example, as mentioned previously, instrument stability is demonstrated by analyzing calibration standards before and after soil vapor samples. The data will withstand the legal challenge as long as the samples are bracketed by calibration standards, even if the standards fall outside of acceptable agreement limits listed in the analytical methods.

11.9.6 INTERPRETING SOIL VAPOR DATA

This section discusses how to interpret soil vapor data for a variety of applications. The primary focus will be on fate and transport issues relating to the interpretation of soil vapor data. The reason for this focus is:

1. Data interpretation for identifying subsurface contamination has been presented previously in a number of other publications (Devitt et al., 1987; Ullom, 1995).

2. Previous presentations of the fate and transport of vapors in the vadose zone are not as common and this topic is especially relevant for determining the risks to human health by contaminants in the soil vapor.

Defining Subsurface Contamination of Volatile Petroleum Compounds

For defining areas of subsurface VOC contamination in soil, groundwater, and vapor, soil vapor data are generally plotted as contour or as a raised surface. These types of representations are only valid for data collected at the same subsurface depth and for data collected in the same manner. The inherent assumption with this interpretation approach is that the measured soil vapor data are reflective of the actual contaminant concentrations and are not due to any other controlling factors, such as soil permeability, temperature, moisture content, and so on. For most situations, this assumption is valid because the range of measured concentrations is large (10 to 1000 times) relative to the importance of these other factors.

More sophisticated approaches including 3-dimensional visualizations of the soil column, cross sections, normalization, and statistical analysis are sometimes used depending upon the project goals. Additional discussion of interpretive techniques for defining contamination will not be described in this article, but can be found in ASTM (1993), Devitt et al. (1987), EPA (1997), and Ullom (1995).

Defining Subsurface Contamination of Semi and Nonvolatile Petroleum Compounds By Measuring Biologically Produced Gases

Semi-volatile and nonvolatile petroleum hydrocarbons such as diesel, oils, and polyaromatic hydrocarbons (PAH) cannot be as readily detected by measuring hydrocarbons in the soil vapor data because of their low vapor pressures. An alternative approach to measuring the hydrocarbons directly is to measure compounds that are produced or consumed by microorganisms during the natural biodegradation of the petroleum hydrocarbons. The three most common gaseous compounds reflective of biologic activity are oxygen, carbon dioxide, and methane, commonly referred to as the fixed or biogenic gases. Under aerobic conditions, microorganisms consume oxygen and produce carbon dioxide as a byproduct of hydrocarbon degradation. Under anaerobic conditions (oxygen poor conditions), microorganisms produce methane as a byproduct of hydrocarbon degradation. The concentration levels of carbon dioxide and methane in the soil vapor can reach the percent range, which can easily be detected, even with portable equipment. In contrast, oxygen values can fall from ambient (21 percent) to near zero values.

Locations showing high levels of carbon dioxide and methane and low concentrations of oxygen reflect areas of intense "biologic activity," reflecting the presence of a hydrocarbon source. The hydrocarbon source can be any type: petroleum derived or nonpetroleum derived (e.g., natural organics, sewage, etc). For this reason, it is necessary to confirm the type of hydrocarbon source by subsequent soil or water sampling.

Differentiation of Petroleum Products Based Upon Soil Vapor Data

The gaseous alkanes (methane through butane) and the lighter liquid hydrocarbon compounds (C5 plus) in the soil vapor can sometimes be used to generate a "fingerprint" to identify and differentiate the type of petroleum (Jones, 1995). Depending upon the degree of weathering, the relative presence of these compounds can be used to distinguish shallow contamination of petroleum products from natural petroleum seepage, and in some cases, the type of petroleum product.

Detection of Semi-volatile Compounds by the Passive Soil Vapor Method

Passive soil vapor is often promoted to be a useful technique for the detection of a wide variety of semi-volatile organic compounds, such as pesticides, polydichlorbiphenols (PCBs), polyaromatic hydrocarbons (PAHs), and explosives. However, much confusion exists over exactly which compounds can reasonably be expected to be detected by passive soil gas techniques. The answer to this question

TABLE 11.9.4   Calculation of Expected Contaminant Mass Adsorbed on a
Passive Collector and Summary of Required Exposure Days for a
Semi-volatile Compound

Assume detection limit of 10 ng from passive collector.

Need to compute the mass trapped by the collector.

If pure product exists, the compound concentration in the vapor in ug/l can be
computed from the vapor pressure using the ideal gas law:

$$PV = nRT$$
$$\text{or} \quad n/V \,(\text{moles/l}) = P/RT$$
$$\text{or} \quad C_{SG}\,(\text{g/l}) = P \times MW/RT$$

Definitions:
  P is the vapor pressure in atm (760 mm Hg = 1 atm)
  C_{SG} is the concentration in the soil gas
  MW is the compound molecular weight in grams
  RT is equal to 22.4 at 0°C, 24 at 20°C

Example: Chrysene ($MW = 228$, $P = 10^{-6}$ mm of Hg)

$$C_{SG}\,(\text{g/l}) = (10^{-6}/760) \times 228/24 = 1.2 \times 10^{-8} \times 10^{6}\ \text{ug/g} = 0.012\ \text{ug/l}$$

In a typical day, a passive collector "sees" about 10 cc of vapor, thus the total
mass adsorbed is ~0.1 ng/day.

Soil conc (ppm)	Mass/day (ng)	Days to reach 10 ng
Pure	0.1	100
1000	10^{-4}	100,000
100	10^{-5}	1,000,000
10	10^{-6}	10,000,000
1	10^{-7}	100,000,000

depends upon two primary factors:

1. The detection level of the laboratory measuring the passive collector

2. The concentration of the contaminant in the soil vapor

The concentration of the contaminant in the soil vapor is dependant upon the vapor pressure of the contaminant of interest and the concentration of the contaminant in the soil column. Using Raoult's Law and the ideal gas law, vapor pressure can be used to calculate an equilibrium soil gas concentration for a given soil concentration. An estimate of the mass of a compound trapped on the collector each day can be obtained by multiplying the soil vapor concentration times the quantity of vapor that passes by the collector. Assuming molecular transport, a passive collector can be reasonably expected to be exposed to approximately 10 cc of soil vapor per day.

The sensitivity of the method to any compound can then be assessed by comparing the expected mass on the passive collector to the laboratory's efficiency in desorbing the compound off the detector and the laboratory's analytical detection level for the compound. For example, the calculation for the poly-aromatic hydrocarbon, chrysene shows that unless soil concentrations are extremely high (>10,000 mg/kg), the detection of chrysene by passive soil vapor methods requires exceptionally long, and unpractical, exposure times (see Table 11.9.4).

A similar calculation for the compound of interest should be performed prior to employing passive soil vapor methodology for semi-volatile compounds. As a rule of thumb, one should not expect compounds with vapor pressures less than 10^{-6} atmospheres to be detectable by passive soil gas methodology in a reasonable time frame (<50 days of exposure) for contaminant soil concentrations <10,000 mg/kg.

Fate and Transport of Contaminant Vapors in the Subsurface

Recent changes in ideology and regulations have led to increased application of natural attenuation as a remediation strategy, and risk-based corrective action for setting the clean-up requirements for contaminated sites. As these approaches are implemented, consideration must be given to the fate and transport of contaminant vapors in the subsurface, and the potential risks they pose to human health. A condensed version of this topic is addressed in this section beginning with a discussion of the processes by which vapors move through the vadose zone. A more detailed presentation can be found in Morrison et al. (1998) and Morrison et al. (1999).

11.9.7   CONTAMINANT VAPOR MOVEMENT

A common misconception with vapors emanating from a subsurface source of contamination (soil or groundwater) is that the vapors will preferentially rise upward and escape into the atmosphere, much like smoke rising from a smokestack. To understand why this is a misconception, it is necessary to understand the processes by which the transfer of contaminants occurs in the vapor phase.

There are primarily two types of physical processes by which contaminants are transported in the vapor phase. The first type of process is advection, which refers to the bulk movement of the vapor itself. The movement of vapor by density-induced flow described earlier in this article is an example of advection. In advective transport, any contaminants in the vapor are carried along with the moving vapor. Advective transport processes can be important in the soil vapor, for example near the surface because of atmospheric pressure differences (Massmann & Farrier, 1992) or near buildings that can create pressure gradients (Little et al., 1992).

The second type of transport process is gaseous diffusion, which refers to the motion of the contaminants by molecular processes through a nonmoving vapor column. This is the primary transport mechanism for contaminants in the vapor phase through the vadose zone (Thorstenson and Pollock, 1989).

Contaminant transport by gaseous diffusion is described by Fick's first law as:

$$\text{FLUX} = \frac{D_e \times C_{sg}}{X}$$

where FLUX is the rate of movement of a compound per unit area; D_e is the effective diffusion coefficient in the vadose zone; C_{sg} is the contaminant concentration gradient in the soil vapor; and X is depth in the vadose zone.

Similar to momentum transfer (e.g., water running downhill) and heat transfer (movement from hot areas to cold areas), contaminant transfer by gaseous diffusion moves from areas of high concentration to areas of low concentration. The flux will always be down the concentration gradient, regardless of the orientation of the concentration gradient with respect to depth below the surface. In the subsurface environment, diffusional transport occurs in three dimensions so contaminants move away from a source in all directions, similar to an expanding balloon.

The key issues to remember are as follows:

1. Contaminant transport by gaseous diffusion does not move preferentially in one direction (e.g., up or down), but spreads radially in all directions.

2. The direction of movement is from high concentration to low concentration regardless of the depth of the source in the vadose zone.

Contaminant Vapor Transport Rates

An approximation of the distance that contaminant vapors can move by gaseous diffusion can be made as:

$$\text{Distance} = (2 \times D_e \times t)^{1/2}$$

where D_e is the effective diffusivity, and t is time.

Contaminant Vapors Through the Vadose Zone

For contaminant vapor transport through the vadose zone, the effective diffusion coefficient is the gaseous diffusion coefficient corrected for soil porosity. For many vapors, the gaseous diffusion coefficient is approximately 0.1 cm^2/sec. The effect that results from soil porosity varies depending upon the type of soil, and several equations are available to calculate the effect of porosity on the diffusivity. A general approximation is that the porosity reduces the gaseous diffusivity by a factor of 10. Thus, for vapors, D_e can be approximated as 0.01 cm^2/sec.

The distance that contaminant vapors can move through the vadose zone in a year can be estimated as:

$$\text{Distance} = (2 \times 0.01 \text{ cm}^2/\text{sec} \times 31{,}536{,}000 \text{ sec})^{1/2} \sim 800 \text{ cm} = \sim 25 \text{ ft}$$

This calculation shows that contaminant vapors can move long distances through the vadoze zone in a short period of time. Within a few years, vapor contamination can move laterally underneath a neighboring room or building, or downward to the groundwater surface.

Contaminant Vapors Into or Out of Groundwater

In contrast to movement through the vadose zone where contaminant vapors move through the soil vapor, the movement of contaminant vapors into or out of groundwater is controlled by the rate at which vapors partition into and move through the liquid. Because groundwater movement is so slow, the water interface remains relatively undisturbed (laminar flow) and vertical mixing of the water is minimal. The primary exchange process is again molecular diffusion, but in this case the exchange rate is controlled by liquid diffusion, not gaseous diffusion. A general value for the liquid diffusion coefficient for compounds is approximately 0.00001 cm^2/sec. Using the same factor of 10 reduction because of soil porosity, D_e for most liquids can be approximated as 0.000001 cm^2/sec.

The distance that contaminants can move into and through the groundwater in a year can be estimated as:

$$\text{Distance} = (2 \times 0.000001 \text{ cm}^2/\text{sec} \times 31{,}536{,}000 \text{ sec})^{1/2} \sim 8 \text{ cm} = \sim 3 \text{ in}$$

This calculation shows that although contaminant vapors can move through the vadose zone relatively quickly, they partition into and move through groundwater extremely slowly. The reverse situation is also true; the partitioning of contaminants out of groundwater into the soil vapor is also extremely slow, and very unlikely to reach equilibrium values predicted by Henry's Law constants. The reason equilibrium is not reached is because the mixing processes between the soil vapor and the groundwater are extremely slow (i.e., there are no blenders or mixers in the vadose zone mixing things up).

Downward Transport of Vapors

The calculations summarized in the preceding section indicate that although contaminant vapors can move quickly down to groundwater, they do not partition into the groundwater very quickly. Using a modification of Fick's first law, the transfer of a contaminant from the soil vapor into the groundwater can be estimated and, in turn, estimates of the expected contaminant concentration in the groundwater because of contamination in the overlying soil vapor for various soil vapor concentrations can be calculated (see Table 11.9.5).

Expected contaminant concentration in groundwater for various soil vapor concentrations at the groundwater interface can be calculated. These calculations assume equilibrium partitioning at the soil vapor/groundwater interface, transfer by molecular diffusion only, and uniform mixing of the contaminant into the groundwater over the well screen interval. For soil vapor concentrations up to 50 ug/l, the resulting groundwater concentration after five years will be low if liquid molecular diffusion is the only exchange process. In most situations, contaminant vapor concentrations at the groundwater surface are below 50 ug/l and the contact time of the vapor contamination with groundwater is less than five years (because of the movement of the groundwater across the site). Further, it is extremely

TABLE 11.9.5   Expected Contaminant Concentration in Groundwater for Various Soil Vapor Concentrations at the Groundwater Interface

C_{sg} (ug/l)	C_{sg}/H (ug/l)	Flux (ug/cm^2-yr)	Water conc @ 1 yr (ug/l)	Water conc @ 5 yrs (ug/l)
10	40	315	0.63	3.2
20	80	631	1.3	6.3
30	120	946	1.9	9.5
40	160	1261	2.5	13
50	200	1577	3.2	16
100	400	3154	6.3	32
500	2000	15768	31	158
1000	4000	31536	63	315

Assumptions used in computing the tabulated values:
Henry's Law Constant: 0.25 (dimensionless)
Gas Transfer Coefficient: 8 cm/year
Well Screen Interval: 5 m

unlikely that the groundwater reaches equilibrium with the overlying soil vapor. Thus, this calculation suggests that in areas of low groundwater flow velocities, contamination of the groundwater by downward vapor transport is not likely to be significant. In areas with higher groundwater flow velocities, large variations in the water table, or high recharge, the gas exchange rate will be higher because of mixing, and groundwater contamination by vapor transport could be significant. Indeed, cases have been documented where vapor contamination has contributed to groundwater contamination (Cherry and Smyth, 1996).

Upward Migration of Vapors

Attention has been raised in the past several years on the risk to human health by the upward migration of contaminant vapors into buildings and other enclosed spaces (Jeng et al., 1996; Johnson & Ettinger, 1991; Jury et al., 1990). Benzene is the principal contaminant of concern because of its proven carcinogenity and common occurrence at gasoline contaminated sites. Other common compounds of concern by this risk pathway include chlorinated solvents and methane.

A simplified environmental fate and transport analysis for evaluating the inhalation exposure pathway for a contaminant vapor is given in the ASTM Risk Based Corrective Action Standard (ASTM, 1995) and in risk-based documents prepared by many local regulatory agencies (e.g., San Diego County, *Site Assessment Manual*, 1997). The models generally assume that contaminant vapor transport is by gaseous diffusion, that the contamination source is constant and nondiminishing, and that equilibrium conditions exist. Variations of the models incorporating advective transport, diminishing source terms, varying lithology also have been developed.

Simplified, the upward contaminant vapor flux into a building is computed by Fick's first law, requiring the soil vapor concentration at some depth underlying the structure. Soil gas data may be measured directly, or alternatively, in the absence of actual soil gas data, soil gas concentrations can be calculated from the soil and groundwater data assuming equilibrium conditions using equations based upon Henry's Law constants and soil to water partitioning constants (ASTM, 1995). Using these equations, it is possible to compute the maximum soil concentrations, water concentrations, and soil vapor concentrations for vinyl chloride, for example, at various depths below the surface that will yield room concentrations that meet acceptable EPA values because of the upward vapor flux (see Table 11.9.6).

This summary demonstrates that, based upon the assumptions used in the upward risk calculation, only minor concentrations in the soil, soil vapor, or water are required to result in room air concentrations that exceed the acceptable levels. For shallow depths, the allowable values can approach laboratory detection levels.

TABLE 11.9.6 Maximum Concentrations of Vinyl Chloride in Soil, Water, and Soil Vapor at Various Depths Below Surface

Depth bgs (ft)	Soil (ug/kg)	Water (ug/l)	Soil vapor (ug/l-v)
5	1.5	3	3
10	3	6	6
20	6	12	12
50	15	30	30
100	30	60	60

Assumptions used in computing the tabulated values:
Air porosity: 0.2
Total porosity: 0.3
Bulk density: 2.0
Slab factor: 0.01
Exchange rate: 0.5
Acceptable room concentration for vinyl chloride at the 1 in 1 million cancer risk level: 0.03 ug/m^3

It is crucial to realize that the equations used to calculate the soil vapor concentration from soil phase data, water phase data, or free product assume equilibrium partitioning between the phases. Equilibrium partitioning is only obtained if a system is well mixed. This is very rarely accomplished in the subsurface, because there are no blenders or stirrers present to homogenize the vapor, soil, and groundwater. Further, the equations do not account for other processes that are operative in the vadose zone, such as bioattenuation, advective flow, soil heterogeneity, among others.

For these reasons, calculated soil vapor concentrations generally do not accurately represent actual soil vapor concentrations. The potential error in the calculated vapor flux introduced by the incorrect vapor concentration is likely to be greater than errors introduced using default values for other parameters, such as porosity. Thus, in the event that a site fails the upward risk calculation from existing soil or water data, direct measurement of actual soil vapor concentrations near the surface is likely to be the easiest and fastest way to pass the upward risk calculation. A more detailed discussion of the upward vapor migration risk pathway including a recommended protocol for determining the upward vapor flux can be found in Hartman (1997).

REFERENCES

ASTM, "Standard Guide for Soil Gas Monitoring in the Vadose Zone," D5314-93, *Annual Book of ASTM Standards*, Philadelphia, 1993.

ASTM, "Standard Guide for Risk-based Corrective Action (RBCA) Applied at Petroleum Release Sites," E-1739-95, *Annual Book of ASTM Standards*, Philadelphia, 1995.

Bentley, H., *Estimating Site VOC Mass Distribution*, Groundwater Resources Association of California, 1997 GRA Seminar Series, Class notes, 1997.

Bentley, H., and Walter, G., *The Use of Soil Gas Data to Obtain Soil VOC Concentrations to Identify the Presence of NAPL*, Groundwater Resources Association of California, 1997 GRA Seminar Series, Class notes, 1997.

Cherry, J., and Smyth, D., *Non-aqueous Phase Liquids in the Subsurface Environment: Assessment & Remediation*, American Society of Civil Engineers, 1996 Annual Convention & Exposition, Washington, DC, November 10–14, 1996.

Denly, E., and Wang, H., "Preparation of Tedlar Bag Whole Air Standard With a SUMMA Canister for Field VOC Analysis," in *Proceedings of Field Screening Methods for Hazardous Wastes and Toxic Chemicals*, Las Vegas, NV, February 1995.

Devitt, D. A., Evans, R. B., Jury, W. A., Starks, T. H., Eklund, B., and Gholson, A., *Soil Gas Sensing for Detection and Mapping of Volatiles Organics*, EPA/600/8-87/036. (NTIS PB87-228516), 1987.

EPA, *Expedited Site Assessment Tools For Underground Storage Tank Sites*, U.S. Government Printing Office Stock #055-000-00564-8, March 1997.

Fetter, C. W., *Contaminant Hydrogeology*, Macmillan Publishing Co., New York, 1993.

Hartman, B., "The Upward Migration of Vapors," *LUSTLine Bulletin 27*, NEIWPCC, Wilmington, MA, December 1997.

Hartman, B., "The Downward Migration of Vapors," *LUSTLine Bulletin 28*, NEIWPCC, Wilmington, MA, December 1998.

Jeng, C-Y., Kremesec, V. J., and Primack, H. S., "Models of Hydrocarbon Vapor Diffusion Through Soil and Transport into Buildings," in *Proceedings of the 1996 Petroleum Hydrocarbons & Organic Chemicals in Ground Water: Prevention, Detection, and Remediation, Houston*, TX, November 1996.

Johnson, P. C., and Ettinger, R. A., Heuristic Model for Predicting the Intrusion Rate of Contaminant Vapors into Buildings, *Envir. Sci. Technol.*, 25: 1445–1452, 1991.

Jones, V. T., "Characterization and Mapping of Underground Product Contamination Using Soil Vapor Techniques," in *National Institute of Hydrocarbon Fingerprinting*, Univ. of Wisconsin short course, Albuquerque, NM, March 1995.

Jury, W. A., Russo, D., Streile, G., El Abd, H., "Evaluation of Volatilization by Organic Chemicals Residing Below the Surface," *Water Resources Res.*, 26: 13–20, 1990.

Little, J. C., Daisey, J. M., and Nazaroff, W. W., "Transport of Subsurface Contaminants Into Buildings," *Envir. Sci. Technol.*, 26(11): 2058–2066, 1992.

Marrin, D. L., *Innovative Trends in Soil Gas Applications*, Groundwater Resources Association of California, 1997 GRA Seminar Series, Class notes, 1997.

Marrin, D. L., and Kerfoot, H. B., "Soil Gas Surveying Techniques," *Environ. Sci. Tech.*, 22(7): 740–745, 1988.

Massman, J., and Farrier, D. F., "Effects of Atmospheric Pressures on Gas Transport in the Vadose Zone," *Water Resources Research*, 28(3): 777–791, 1992.

Morrison, R., Beers, R., and Hartman, B., *Petroleum Hydrocarbon Contamination-Legal and Technical Issues*, Argent Communications Group, Foresthill, CA, 1999.

Morrison, R., Beers, R., and Hartman, B., *Chlorinated Solvent Contamination-Legal and Technical Issues*, Argent Communications Group, Foresthill, CA, 1999.

San Diego County, *Site Assessment and Mitigation Manual*, San Diego County Department of Environmental Health, 1997.

Thorstenson, D., and Pollock, D., "Gas Transport in Unsaturated Porous Media: The Adequacy of Fick's Law," *Review of Geophysics*, 27: 61–78, 1989.

Ullom, W. L., "Soil Gas Sampling," in *Handbook of Vadose Zone Characterization and Monitoring*, Wilson, L. G., Everett, L. G., Cullen, S. J., eds., pp. 555–67, Lewis Publishers, 1995.

Williamson, R., Kerfoot, H., and Mayer, C., *Soil Gas Monitoring Class Notes*, presented to the Los Angeles Regional Water Quality Control Board, 1994.

CHAPTER 11
ASSESSMENT SAMPLING AND MONITORING

SECTION 11.10

NEUTRON PROBES

John H. Kramer

Dr. Kramer is a California certified hydrogeologist who directs project work on mine closure activities, subsurface investigations, environmental monitoring, and groundwater remediation for Condor Earth Technologies, Inc. Condor is an earth science consulting firm and the leading provider of innovative monitoring methods with Global Positioning Systems (GPS).

11.10.1 INTRODUCTION

Neutron probes are tubular-shaped instruments used to measure water content and porosity of soil or rock. They are typically deployed by cables through well casings and access tubes installed in soil or rock. Figure 11.10.1 shows a neutron probe exiting a buried pipe at a landfill. Neutron probes have become a standard tool for a variety of characterization and monitoring tasks. Agricultural and environmental applications include irrigation scheduling, well logging, and leak monitoring at waste management facilities, as shown in Figure 11.10.2. The neutron method is used for monitoring because it is reliable, precise, efficient, and nondestructive. Therefore, environmental scientists and engineers should be acquainted with the principles of operation, advantages, and limitations of neutron measurements. The theory of neutron measurements is reviewed in this paper.

11.10.2 PRINCIPLES OF OPERATION

Neutron probes contain radioactive sources (typically americium with beryllium) that spontaneously decay, emitting high energy neutrons called *fast neutrons*. Fast neutrons cannot be detected unless they have been slowed. The slowing process is called neutron moderation and occurs when neutrons transfer momentum through collisions with atoms. During moderation, fast neutrons ricochet off surrounding atoms in random directions, losing momentum with each impact. Slowed neutrons are also called thermal neutrons. A cloud of thermal neutrons develops around a source of fast neutrons. The radius of the thermal cloud is proportional to the strength of the source and inversely proportional to the density of surrounding soil. Slow neutrons diffuse through interatomic space like gas particles until they approach absorbing atoms that capture them. Neutron probes are equipped with detectors containing absorbing atoms (typically Helium) that emit pulses of detectable energy when a thermal neutron is captured. The detector shown in Figure 11.10.3 includes circuitry to count the energy pulses from neutron capture.

Neutron-hydrogen collisions are most effective at moderating fast neutrons because hydrogen, having only one proton, is close to the mass of a neutron. Imagine a Ping-Pong ball striking a bowling

FIGURE 11.10.1 Neutron probe exiting a buried pipe.

FIGURE 11.10.2 Neutron-probe monitoring at a landfill under construction in California.

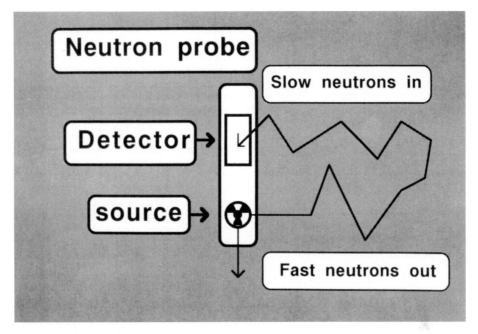

FIGURE 11.10.3   Diagram of neutron probe operation.

ball. The Ping-Pong ball will rebound with a velocity approximately equal to its inbound velocity. If the same Ping-Pong ball were to strike a stationary Ping-Pong ball, the rebounding ball would be appreciably slowed. Thus, relatively high hydrogen density results in rapid neutron moderation near the source.

Water is nearly always the greatest source of hydrogen in the subsurface. Therefore, a change of water content will cause a change in neutron probe data. Hydrogen also occurs within mineral structures (particularly clays), within organic soil components and within liquid organic phases (i.e., petroleum contaminants). Hydrogen bound in clay minerals will cause neutron logs to have different background counts in different soil types.

Atoms vary in their propensity to absorb thermalized neutrons. This is called the capture cross section, measured in units of barns. Of the elements present in soils and monitoring well systems, chlorine and boron have particularly high capture cross sections. Neutron capture by these elements will lower neutron probe counts. High boron and chlorine concentration occurs in pore liquids from many arid regions and could affect the interpretation of neutron data. Schneider and Greenhouse (1992) also report significant count decreases resulting from chlorine present in liquid perchlorethylene (PCE). Additional discussion of theoretical concepts is presented in Greacen (1981), Keys and MacCary (1983), and other references cited below.

11.10.3   NEUTRON PROBE: INSTRUMENT DESIGNS

Neutron probes differ in design depending on the application. Two basic types of neutron probes are commonly used: neutron loggers and soil moisture gauges. Loggers are typically used in deep liquid-filled boreholes for geologic characterization in the oil and gas industry. Moisture gauges are typically used in shallow air-filled access tubes for agricultural and environmental monitoring applications. Both types use a fast neutron source and thermal neutron detector but differ in source strength and source-detector geometry. Their response to water content is opposite. Counts from neutron loggers decrease with increasing water content, whereas counts from moisture gauges increase as water content increases. The difference occurs because loggers count neutrons in the distal region of the neutron cloud, where fewer thermal neutrons arrive at the detector when conditions are wettest; whereas gauges count neutrons in the center of the neutron cloud, where more thermal neutrons are created when conditions are wettest. Hearst and Carlson (1994) describe the differences between neutron loggers and moisture gauges in a study comparing their sensitivity in air-filled boreholes. Scientists and engineers should be aware of which type of neutron probe is being employed before interpreting results.

Neutron loggers are well established in the oil industry where they are used to infer lithology and porosity in saturated sediments. They have also been used in environmental monitoring applications, notably in long horizontal access tubes beneath waste management facilities. Probes of this type employ strong sources (250 mCi to 5 Ci) to maximize the radius of the thermal neutron cloud. Penetration can be several feet into the formation beyond the borehole wall. Large source strengths are needed to overcome the effects of liquids in large-diameter boreholes (>10 cm). Some loggers employ dual detectors to isolate borehole effects. More sophisticated detectors are jacketed with materials that absorb thermal neutrons and slow other neutrons of intermediate energy. Loggers range in length from 1.5 to 6 m and are deployed from vehicles capable of logging thousands of feet at speeds of approximately 6 to 18 meters per minute. Because of the relatively strong radioactive source, these probes are subject to stringent regulatory oversight, and are available only from specialty borehole logging companies. These can be identified by contacting local well drilling companies.

Neutron moisture gauges are designed for precise measurement of soil moisture in small diameter (5-cm) access tubes, but also have been employed in larger access tubes (Hammermeister et al., 1985; Kramer et al., 1990a; Tyler, 1988). These probes employ a low strength source (10 to 50 mCi) and single, near-source detector. They require minimal operator training (8-hour radiation safety course) and are generally subject to less regulatory oversight than the neutron loggers. Because of their smaller size, portability, and common use at construction sites, they are more prone to theft, damage, or loss. This has caused licensing in California to become more stringent in recent years. Moisture gauges are the neutron probe of choice in agriculture and diverse environmental monitoring tasks where sequential data are required. The factors that keep the cost per sample low are moderate one-time acquisition cost ($5000–$6900), one-person operation, and low maintenance. Moisture gauges are currently available from two manufacturers, Troxler in North Carolina and CPN in California.

Instrument Capabilities

Neutron probes can be calibrated to measure soil moisture changes with precision of 1–5 volume percent. Neutron probes are sensitive to hydrocarbon liquids as well as water (Kramer et al., 1990b). The radius of measurement for neutron probes is compact in wet and/or dense soils, and expanded in dry and/or loose soils. Neutron loggers can penetrate several feet into the surrounding formation, while the practical limits of moisture gauges are less, from 10 to 80 cm (Kramer et al., 1990a; Silvestri et al., 1991). Because moisture gauges sample a smaller radius, thinner geologic layers can be discriminated. Strong neutron absorbers, including boron and chlorine, will diminish the radius of measurement.

11.10.4   ACCESS TUBE DESIGN

Sensitivity of the neutron probe to moisture changes depends on the magnitude of the possible moisture change and the masking effects of grout and casing. The neutron probe is most sensitive in ungrouted access tubes with minimal distance between the tube and soils (Teasdale and Johnson, 1970). Small diameter (5-cm), driven metal casings are best. Steel, aluminum, stainless steel, and PVC at diameters up to 4 in, installed in borings and backfilled with native material or kaolinite slurry have been successfully used (Amoozegar et al., 1989). Chlorine present in PVC casing will lower neutron counts, though Keller et al. (1990) and Kramer et al. (1992) both report successful detection of wetting fronts in sands through PVC casing. Typically, metal casing and small diameter boreholes are used to maximize neutron counts (Teasdale and Johnson, 1970; Williams, 1993). Zawislanski and Fabishenko (1998) have achieved improved neutron counts in deep and fractured boreholes with acrylic tubing and polyurethane foam backfill.

The neutron moisture probe should only be applied in grouted wells after grout curing has been documented by repetitive stable measurements over at least two weeks or more. For neat cement grouts mixed at five gallons of water per 94-pound sack of cement, the grout-curing period is on the order of seven days. Generally, grouts are mixed at higher water contents and require much longer stabilization periods. A wetting front in loamy soils could be detected through 7.5 radial centimeters of grout and PVC casing (Kramer et al., 1992). In clay-rich soils, 7.5-centimeter-thick grouts could fully mask neutron response to wetting. As a general rule, grout thickness greater than 7.5 radial centimeters is inappropriate for neutron moisture gauges.

11.10.5 INSTRUMENT SHORTCOMINGS

The neutron moisture probe cannot distinguish liquid contaminants from groundwater because both liquids have approximately the same hydrogen density. Neutron data cannot be used to distinguish between gasoline and water, which have similar hydrogen densities (Kramer et al., 1990b), nor to detect steady-state flow, which has constant hydrogen density. In high background moisture environments, such as heavy clay-rich soils, under conditions of fully masking grouts, or in the presence of migrating boron or chlorine, detection of wetting fronts with a neutron probe would be uncertain or impossible.

11.10.6 COUNTING STATISTICS

Neutron moisture probe data are amenable to parametric statistical analysis through the Poisson statistical model, often used for radioactive processes. A continuous normal distribution is applicable to the Poisson distribution for sample sizes greater than 15, which is why one probe manufacturer logs 16 one-second counts and reports the measurement as the scaled mean of those subcounts. According to the Poisson model, a convenient estimate of the method standard deviation is the square root of any mean count. Manufacturers also use other statistical tests to ensure data quality and consistent performance of the electrical components. Investigators may also apply parametric statistical tests to independent data measurements through time, to include error from repositioning of the probe. Background data sets including repositioning error are then used to calculate tolerance limits and other performance control thresholds to set monitoring alarms.

11.10.7 INSTRUMENT PRECISION

Precision, the reproducibility of successive measurements, depends on instrument reliability, background stability, and counting time. The precision improves with longer counting time (Kramer et al., 1992). A useful rule of thumb is that a precision of ± 1 percent in 68 out of 100 samples can be expected by counting long enough to record 10,000 counts. For example, if a 16-second measurement recorded 5000 counts, then a 32-second reading would be required to attain 1 percent precision.

11.10.8 COUNT RATIOS

Neutron probe data often are reported as count ratios, which are ratios of counts at sampling locations to other counts on the same instrument in a stable standard material. Count ratios are defined as follows:

$$R = C/S$$

where R = count ratio, C = measured neutron count, and S = neutron count in standard material.

The ratio is used to offset instrument drift because of component degradation or other factors. A disadvantage of reporting probe data as count ratios is the random compounding errors that can occur. To minimize error, the mean of two or more standard counts taken before and after the field measurements should be used in ratios. Sinclair and Williams (1979) show that averaging several count ratios will reduce precision error to very small amount of total variance. In monitoring applications of long duration (years), count ratios will eventually become desirable and standard counts before and after each monitoring event should be recorded and averaged.

Count ratios simplify comparisons of measurements from different experimental conditions, access tube geometry or instruments, however, ratios are not direct conversions between instruments (Greacen et al., 1981). Each instrument has a unique combination of source strength and detector sensitivity that affect ratios. Very accurate comparisons between two instruments require at least two (wet and dry) counting standards for each instrument. Count ratios are also used to standardize measurements taken at different count periods to develop a single soil calibration curve that will work for all counts.

11.10.9 MATERIALS FOR COUNTING STANDARDS

The material used for the counting standard may be any stable volume of neutron moderators; water, or dense rubber foam work well. Figure 11.10.4 shows a standard made from a drum of fresh water. Manufacturers provide a shielded case that can be used for a counting standard. Consistent results are obtained when case standard counts are always run on a solid surface, like concrete. If portability is not a problem, Greacen (1981) recommends a large drum of water fitted with a centered access tube as a standard. As a general rule, it is best to use a standard that approximates the count rate obtained in the target soils; this practice removes any effect of variance at different count rates.

FIGURE 11.10.4   Standard counts measured in a drum of water are used to monitor instrument stability.

11.10.10 *CALIBRATION*

Neutron probe data can be reported as calibrated units of volumetric soil moisture content. This form of reporting is desirable if data are to be used quantitatively. By measuring soil water content and knowing unsaturated hydraulic properties, one can assess the likelihood of unsaturated flow. If the soil water content is below a critical value (unique to each soil), then the likelihood of contaminant mobility is negligible.

A moisture calibration for a given soil and probe is made by measuring neutron counts in the target soil at two or more known moisture contents and regressing these to a linear model. At least two of the soil moisture content levels used in the calibration should span the moisture content of interest in the natural system. Such a regression takes the form:

$$V = mN + b$$

where V = volumetric moisture content, m = calibration coefficient, N = neutron counts or count ratio, and b = Y-axis intercept.

For field calibrations, most workers choose V as the dependent variable, because it is desirable to know what moisture content corresponds to a given neutron count. Laboratory drum calibrations can be performed in 55-gallon drums fitted with centered access tubes (identical to the field access tubes) and packed with target soils at different known water contents and densities (Silvestri et al., 1991). Water content can be adjusted by mixing measured amounts of water with soils before packing the drums. Densities can be measured by weighing the tared drums or volumetrically sampling the packed material. Although this method results in even moisture distribution within the drum, it is difficult to reproduce accurately the density and pore size distribution of the field soils. Another technique is to pack the drum with dry soil and saturate it from the bottom up (to exclude air entrapment). This technique is only good for 2-point calibrations because moisture stratification can occur within the drum at less than full saturation. To avoid this problem, drums may be fitted with candle extractors, tensiometers, TDR (time domain reflectometry probes), or capacitance probes to control soil-water tension and measure moisture content during calibrations.

Field calibrations can be performed by sampling soils near a neutron access tube (five samples are desirable) and measuring water content from each sample gravimetrically before, during, and after saturation events, or by implanting a TDR or capacitance probe to measure changing soil moisture. These measurements can be correlated to neutron measurements taken during a wetting episode. In deep soil, it is impractical to obtain multiple samples at a given position for calibration or to monitor induced saturation events in situ. It may also be impossible to obtain adequate representative soil samples for drum calibrations. In these cases, approximate calibrations can be generated using limited gravimetric data or by applying calibration curves from other similar soils. Silvestri et al. (1991) showed that the manufacturer-provided calibration is generally applicable to sandy soils, but that probes should be calibrated to the soils in which they are to be used. They provide a calibration curve for a particular probe in wet clay soils and note that the calibration appears nonlinear above 40 percent moisture content. This is most likely the result of instrument design: loss of detector efficiency because the cloud of thermalized neutrons collapses to within the geometry of the detector.

Field calibrations are prone to error from uncertain sample quality and distribution. In many geologic settings each monitoring position deserves a specific calibration to provide quantitative moisture content at each measurement point. The number of calibrations can be large and impractical. Consider data in Figure 11.10.5 from two neutron-moisture logs in a 2000-foot horizontal access tube. Repeatable variations in counts from one position to the next indicate the presence of numerous distinct soil pore distributions; each with a unique soil water characteristic curve and associated neutron calibration. Variograms calculated for this moisture data indicate that at distances greater than 6 ft the data are uncorrelated, which means that at least 330 distinct calibrations would be required along this transect. Because it is impossible to collect properly enough samples to perform these calibrations, some grouping and averaging must be used to provide calibrated moisture content data. The resulting approximate calibrations could filter out useful information, reducing the sensitivity of neutron methods to moisture content changes at any single position. The need for quantitative estimates

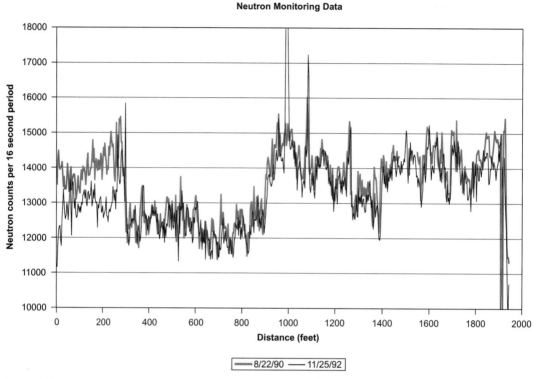

FIGURE 11.10.5   Neutron data from 2000 foot-long pipe beneath a California landfill.

of moisture content and the error in calibrated data should be considered before interpretations of calibrated neutron data are used for modeling or other purposes. If quantitative moisture contents are not needed, comparison of raw data should be used.

11.10.11 COMPARISON OF UNCALIBRATED NEUTRON COUNTS

In general, the patterns of data peaks and valleys in the logs, shown in Figure 11.10.5, are distinct and repeatable from one log to the next. Exceptions show areas where the soils have dried or been wetted. Extreme peaks indicate water puddles in the access piping from infiltration. This type of analysis for leak location does not require calibration to specific soil moisture level.

When monitoring for changes in moisture content, the question of what represents a meaningful change has traditionally been addressed by attempting to calibrate neutron counts to site soils and to monitor for an arbitrarily chosen level of "significant" change (e.g., 5 percent volumetric moisture change; Unruh et al., 1990). Neutron data are better used by comparing neutron counts at each position through time to an established background. Analysis of uncalibrated neutron data is more sensitive to subtle changes in hydrogen density than data processed through a calibration filter. Only after trends are noted in sequential count data are estimates of changes in moisture content needed for risk assessment or predictive modeling efforts.

Control charting is one useful statistical approach to anomaly definition (EPA, 1989). In this method, a level of significance is assigned above or below established background levels. When a

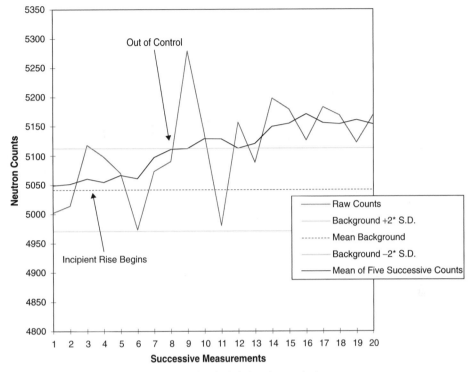

FIGURE 11.10.6   Control charting of neutron data for leak detection monitoring.

measurement falls outside the significance envelope, the process is considered "out of control" and anomalies are investigated. In the example illustrated by Figure 11.10.6, a moving window of five sequential count measurements at a single monitoring position advances stepwise through the data set. The mean of each window is plotted at the central time position. If no trend exists in the data, the sequence of window-means should be distributed normally about the background mean. If trends do exist, as is the case in Figure 11.10.6, deviation will be apparent. The plot of window-means identifies when the increasing trend began. In the case of a contaminant leak, estimates of leak duration are useful to assess contaminant volume. This approach is very appropriate to neutron data, which is real time, inexpensive, and repeatable. Out-of-control conditions can be verified by repeated sampling.

11.10.12 SUMMARY

Neutron probes are used to measure moisture and porosity in soil and rock. Two types of neutron probes are used. Neutron loggers were developed to characterize formation porosity in fluid-filled boreholes and are deployed from specialized logging vehicles. Moisture gauges were developed to measure soil-moisture in air-filled boreholes, primarily for agricultural research and irrigation scheduling. Gauges are lightweight and can be carried by one person to remote measuring points. The neutron method is now commonly used for leak detection monitoring at waste management facilities and soil remediation projects, for which both types of probes are used. Figure 11.10.7 shows a moisture gauge used at a soil flushing project to document subsurface distribution of solutions and to monitor groundwater

FIGURE 11.10.7   A moisture gauge being used to a monitor soil flushing remediation project.

mounding. Neutron loggers and moisture gauges are both applied at large sites. Mining leach pads and hazardous waste facilities use neutron logging for unsaturated zone monitoring requirements. Installation and monitoring costs for long access tubes have been justified because they can reduce the assurance bond required for use permits.

REFERENCES

Amoozegar, A., Martin, K. C., and Hoover, M. T., "Effect of Access Hole Properties on Soil Water Content Determination by Neutron Thermalization," *Soil Sci. Soc. Am. J.*, 53(2): 330–335, 1989.

EPA, *Statistical Analysis of Ground-Water Monitoring Data at RCRA Facilities*, EPA/530-SW-89-026, 1989.

Greacen, E. L. (ed.), *Soil Water Assessment by the Neutron Method*, Division of Soils CSIRO, Adelaide, Australia. p. 140, 1981.

Hammermeister, D. P., Kneiblher, C. R., and Klenke, J., "Borehole-calibration Methods Used in Cased and Uncased Test Holes to Determine Moisture Profiles in the Unsaturated Zone, Yucca Mountain, NV," *Proceedings of the NWWA Conference on Characterization and Monitoring of the Vadose (Unsaturated) Zone, Denver, Colorado*, NGWA, Dublin, Ohio, 1985.

Hearst, J., and Carlson, R. C., "A Comparison of the Moisture Gauge and the Neutron Log in Air-filled Holes," *Nuclear Geophysics*, 8(2): 165–171, 1994.

Keller, B. R., Everett, L. G., and Marks, R. J., "Effects of Access Tube Material and Grout on Neutron Probe Measurements in the Vadose Zone," *Groundwater Monitoring Review*, 10(1): 96–100, 1990.

Keys, W. S., and MacCary, L. M., "Application of Borehole Geophysics to Water-Resources Investigations," *Techniques of Water-Resource Investigations of the U.S. Geological Survey*, Book 2, Chapter E1, 1983.

Kramer, J. H., Everett, L. G., and Eccles, L. A., "Effects of Well Construction Materials on Neutron Probe Readings with Implications for Vadose Zone Monitoring Strategies," *Ground Water Management*, Number 2, NWWA, Dublin, Ohio, pp. 1303–1317, 1990a.

Kramer, J. H., Everett, L. G., Eccles, L. A., and Blakely, D. A., "Contamination Investigations Using Neutron Moderation in Grouted Holes A Cost-Effective Technique," *Minimizing Risk to the Hydrologic Environment*, Alexander Zporozec, ed., Kendall/Hunt Publishing Co., Dubuque, IO., pp. 234–242, 1990b.

Kramer, J. H., Everett, L. G., and Cullen, S. J., "Vadose Zone Monitoring with the Neutron Moisture Probe," *Ground Water Monitoring Review*, 12(3): 177–187, 1992.

Schneider, G. W., and Greenhouse, J. P., "Geophysical Detection of Perchlorethylene in a Sandy Aquifer using Resistivity and Nuclear Logging Techniques," *Proceedings SAGEEP*, Chicago, April 26–28, 1992.

Silvestri, V., Sarkis, G., Bekkouche, N., Soulie, M., and Tabib, C., "Laboratory and Field Calibration of a Neutron Depth Moisture Gauge for Use in High Water Content Soils," *Geotechnical Testing Journal*, ASTM, pp. 64–70, 1991.

Sinclair, D. F., and Williams, J., "Components of Variance Involved in Estimating Soil Water Content and Water Content Change Using a Neutron Moisture Meter," *Aust. J. Soil Res.*, 17: 237–70, 1979.

Teasdale, W. E., and Johnson A. I., "Evaluation of Installation Methods for Neutron-meter Access Tubes," *Geological Survey Research 1970*, U.S. Geological Survey Prof, Paper 700-C, C237-241, 1970.

Tyler, S., "Neutron Moisture Meter Calibration in Large Diameter Boreholes," *Soil Sci. Soc. Am. J.*, 52: 890–893, 1988.

Unruh, M. E., Corey, C., and Robertson, J. M., "Vadose Zone Monitoring by Fast Neutron Thermalization (Neutron Probe): A 2-Year Case Study," *Ground Water Management*, 2, NWWA, Dublin, Ohio, 431–444, 1990.

Williams, M. A., Discussion of "Vadose Zone Monitoring with the Neutron Moisture Probe," *Ground Water Monitoring and Remediation*, 13(1): 159, 1993.

Zawislanski, P. T., and Fabishenko, B., "New Casing Design for Neutron Logging Access Boreholes," *Ground Water*, 37(1): 33–37, 1999.

CHAPTER 11
ASSESSMENT SAMPLING AND MONITORING

SECTION 11.11

DIRECT SENSING OF SOILS AND GROUNDWATER

Mark Kram

Mr. Kram is a Ph.D. candidate at the University of California at Santa Barbara. He is currently working as a hydrogeologist for the U.S. government specializing in environmental site characterization and remedial design.

Stephen Lieberman

Dr. Lieberman is senior research scientist in the Environmental Sciences Division at the Space and Naval Warfare Systems Center San Diego (SSC San Diego), where he leads the Optical Chemical Sensor Group.

James A. Jacobs

Mr. Jacobs is a hydrogeologist and president of FAST-TEK Engineering Support Services. He has over 20 years of experience and is registered in several states, including California. He specializes in assessment methods and in situ remediation technologies.

11.11.1 INTRODUCTION

Direct Push Technology (DPT) rigs drive sensors into the subsurface for direct sensing of physical and chemical properties of soils and unconsolidated sediments. DPT is also an alternative method to conventional drilling for collecting soil, vapor, and groundwater samples for environmental projects. Cone Penetration Testing (CPT) rigs, a form of DPT, have been used in geotechnical studies for years. Smaller truck and van-mounted DPT probe rigs have been developed since the late 1980s. CPT and DPT rigs use dry impact methods to push the tools and samplers into the subsurface. DPT does not require predrilling of the boreholes for logging operations.

DPT uses innovative site characterization technology tools for rapidly characterizing soil types and detecting and delineating the presence and extent of subsurface contaminants in soils and unconsolidated sediments. The data can be used for strategic placement of groundwater monitoring wells, estimating plume volumetric and spatial characteristics, designing and installing monitoring and remediation wells, determining ideal locations for collecting grab samples, and evaluating performance of remediation systems. CPT and DPT rigs are ideal for determining contaminant sources, potential flow pathways, and potential receptors.

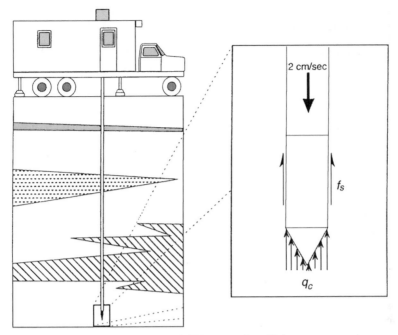

FIGURE 11.11.1   Schematic diagram of CPT system; f_S = friction stress, q_C = tip stress (Holguin-Fahan and Associates; after Edelman and Holguin, 1996).

11.11.2   CONE PENETRATION TESTING RIGS

CPT equipment, a form of DPT, has been used manually in Europe since the 1920s. This early CPT equipment did perform direct sensing of soil properties for engineering studies, however, the manual methods and tedious calculations made widespread use impractical. In the 1970s, the automated and computerized CPT rigs were developed by companies like Fugro Geosciences, Inc. for use in the geotechnical field. The 20-ton truck-mounted CPT rigs were designed for sampling and performing direct sensing to depths of 200 to 250 ft, depending on conditions. More recently, CPT rigs have been designed in the 10 to 30 tons weight class. These probes have been used extensively in the construction industry to obtain geotechnical data. Within the past 10 years, they have been used more frequently for environmental applications.

The typical CPT rig hydraulically pushes a small diameter (1.6 in per ASTM specifications) instrumented probe and steel support rods into soils (see Figures 11.11.1 and 11.11.2). Variations in probe dimensions are becoming more commonplace, depending upon the particular application. The conventional probes used with the hydraulic ram systems are equipped with transducers for measuring point penetration resistance and sleeve friction. An empirical relationship between these physical strength measurements and soil types is used to derive soil classification logs for the layers penetrated.

11.11.3   SOIL CHARACTERIZATION

Numerous probes are currently available for determining soil type. The most common sensors consist of load cells on the cone tip and sleeve. An empirical relationship between these measurements and

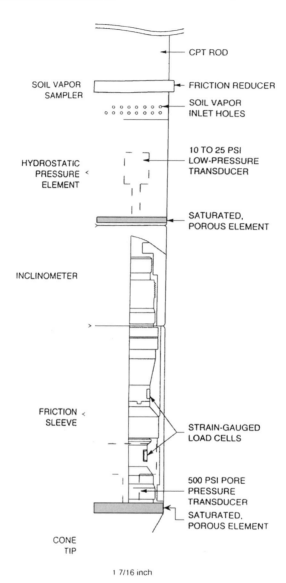

FIGURE 11.11.2 Integrated CPT probe (Holguin-Fahan and Associates; after Edelman and Holguin, 1996).

corresponding soil types have been used to generate soil classification logs for the layers penetrated (Robertson and Campanella, 1983a, 1983b; Robertson, 1990) (see Figures 11.11.3 and 11.11.4).

11.11.4 DPT PROBE RIGS

Since the late 1980s, Geoprobe® and similar probe rigs were designed for environmental sampling. These probe rigs were smaller and more economical to use than the larger CPT rigs. The probe rigs

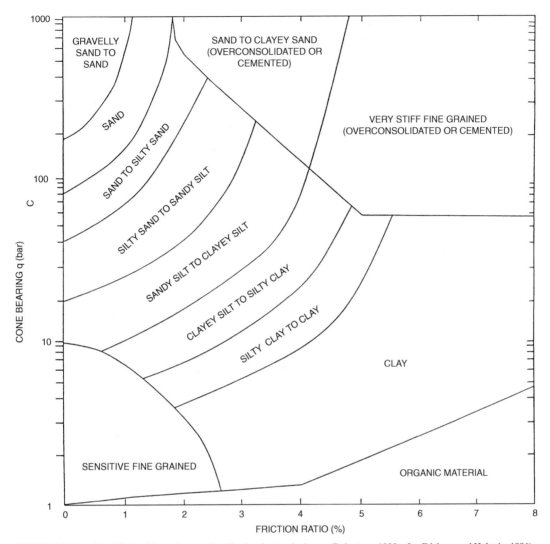

FIGURE 11.11.3   Simplified soil behavior type classification for standard cone (Robertson, 1990; after Edelman and Holguin, 1996).

are truck and van-mounted and have lowered the cost of DPT sampling projects to depths approaching 60 ft, although depths in excess of 120 ft in soft sediments have been documented (Geoprobe® Technical Information, 1997).

11.11.5   SENSOR PROBES

In addition to the conventional soil classification devices mentioned above, probes have been developed to meet the requirements of environmental professionals tasked with characterizing site hydrogeologic and contaminant characteristics. For hydrogeologic purposes, sensor probes have been developed to

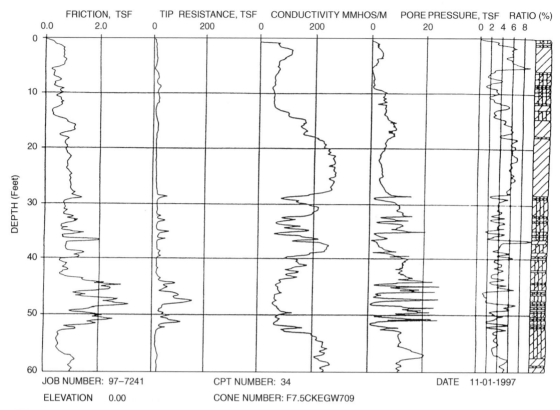

| FRICTION, TSF | TIP RESISTANCE, TSF | CONDUCTIVITY MMHOS/M | PORE PRESSURE, TSF | RATIO (%) |

JOB NUMBER: 97–7241   CPT NUMBER: 34   DATE   11-01-1997

ELEVATION   0.00   CONE NUMBER: F7.5CKEGW709

FIGURE 11.11.4   Examples of CPT data (Fugro Geosciences, Inc., 1999).

determine soil type, vadose zone moisture content, water table elevation, groundwater flow rate, aquifer communication areas, and the distribution of hydraulic conductivity. For contaminant delineation purposes, sensor probes have been developed to detect and delineate zones containing light nonaqueous phase liquids (LNAPLs), polyaromatic hydrocarbons (PAHs), petroleum oils and lubricants (POLs), dense nonaqueous phase liquids (DNAPLs), light and heavy metals, explosives, and radionuclides. A relatively new group of sensors allows for direct observation of soil and soil pore fluids using down-hole video microscopic techniques allowing new interpretations on chemical fate and transport mechanisms. Probes have also been developed to measure temperature and conductivity, and for conducting geophysical surveys. Although some of the last few types of sensors are not being employed on a regular basis, it is anticipated that their use will expand as practitioners learn about the quality and quantity of data afforded by the direct-push delivery platform.

Considerations for Successful Direct-Push Applications

The goal of the investigation must be carefully defined. If the data is to be used as a characterization step, appropriate regulatory officials should be contacted and interviewed prior to deployment. If the data will be used for the basis of a remediation effort, the remediation staff is contacted and interviewed by the field crew supervisor. Important discussion items include a complete description of the capabilities and limitations of the technology, confirmation analytical requirements, required format of final data report, and scheduling.

Prior to conducting a site walk-through, the field crew supervisor can send the client a predeployment questionnaire that refers to most or all of the information required to conduct a successful investigation. The client is requested to respond to the questionnaire prior to the predeployment site visit. The field crew supervisor performs a predeployment survey, which requires a site visit and at least one meeting with the client. One goal of the survey is to collect all information required to generate a site-specific Work Plan Package before field work begins. The Work Plan Package includes a site-specific Work Plan, a site-specific Health and Safety Plan, a generic Spill Response Plan, a generic Quality Control Project Plan, and pertinent field and laboratory standards and standard operating procedures. The second goal of the predeployment survey is to determine all logistics relating to pertinent field operations. Some of the considerations include points of contact, project status (history and future plans), prior investigations and findings, suspected contaminant type(s), local geology, local groundwater data, utility interference, permit requirements, vehicle access and mobility, equipment staging area, predeployment site preparation, USGS benchmarks, potable water sources, vehicle fuel sources, location of mail facility, location of hardware supplies and cement, location of a specialty gas dealer (for some of the probe applications), access to drums and pallets, worker amenities, site restrictions, and worker accommodations.

When the field team is tasked to detect and delineate a particular contaminant type with a specialized probe, the field crew supervisor can collect a representative soil sample from the site during the predeployment site visit. This sample can be sent to a laboratory where it is spiked at known concentrations of the contaminant in order to generate a standard curve, which will enable investigators to estimate the detection capabilities for the site. Typical standard concentrations include: 0, 50, 100, 250, 500, 1000, 1500, and 2000 ppm, but will ultimately depend on the action levels specific to the contaminant of concern. This curve is generated at least one week prior to field deployment to ensure that the selected soil sample used for curve generation yields appropriate responses.

The background sample used for preparing standards must be free of contamination and is collected from an area that has geologic materials identical to or similar to materials in the impacted zone. Specific knowledge regarding the subsurface materials is often limited. Soils maps, well construction logs, geologic boring logs, surface flora, air photos, and ponding can be useful indicators of subsurface lithology or buried structural features. There can be cases where lithology is so heterogeneous that a selection is made based on the similarity to anticipated characteristics of the materials comprising the majority of the site. Investigators typically expose a representative soil sample to the sensor probe system prior to inoculation to observe the spectral characteristics of a sample believed to be clean. A standard spectrographic curve can be generated if the target chemical is available as a pure product.

If the generated standard spectrographic curve does not yield a reasonable detection capability value, or does not indicate ample response for the range of standard concentrations, the initial portion of the field investigation may become a pilot test to determine whether the direct sensing technique can meet the requirements of a project. The authors have experienced situations where the standard spectrographic curve for a site suggested that the probe could not meet the project requirements. Upon deployment, however, the system performed well. This indicates that the standard spectrographic curve does not always correctly indicate whether or not the system will succeed. For some of the laser techniques, it has been postulated that hydrostatic and confining pressures may help spread the contaminant along the grains and probe window in such a way as to render it more susceptible to fluorescence detection. Flourescence detection is defined as the transmision of the laser impulse from the soil through the fiber optic cable and, ultimately, to the spectrograph.

Site-specific detection capabilities can also be estimated using statistical evaluation of fluorescence data collected during an early phase of field deployment. Estimation requires post-processing of field data. Initially, the investigator determines a Sensor Threshold (ST) for the site. The ST can be a value of fluorescence intensity that must be exceeded to indicate the presence of contamination. Statistics based on evaluation of two populations (normalized background intensity values versus normalized intensity values above background) can be employed to determine ST. The ST is the value of sensor response intensity that marks the separation of the two populations and equals the mean background value plus the system noise determined by evaluating the stability of a quality control sample run prior to each push. After determining the ST, a corresponding detection threshold can be calculated using a standard curve. The detection threshold is the concentration of contaminant that corresponds to the ST.

11.11.6   *OVERVIEW OF SENSOR PROBES*

The level of detail afforded by the sensor probes can be useful for bioremediation, fluid extraction, and natural attenuation projects. Unlike inferences based on free product thickness observed in monitoring wells, direct-push sensor data yields true hydrocarbon thickness (vertical) because the probe comes in direct contact with the contaminated soil. This information is extremely useful when considering remediation approaches. However, remedial decisions are not based solely on spatial extent and vertical thickness of a contaminated zone. Lithologic characteristics of the contaminated unit throughout the site are carefully analyzed before making recommendations for extraction well placement. A relatively thick contaminated zone may be more of a reflection of the fine-grained nature of a unit than the actual recoverable volume of contaminant (Wallace and Huntley, 1992). If this is not considered, recovery wells may be constructed in fine-grained materials with relatively low yield and low potential for successful remediation (Huntley et al., 1992; Kram, 1993). Therefore, investigators look for contaminants residing in relatively coarse-grained pockets as areas of maximum potential recovery via fluid extraction. Herein lies another benefit to the direct-push wells—that of deploying multiple sensors simultaneously. Many of the current systems allow for simultaneous collection of chemical and lithologic information. The high level of resolution afforded by multiparameter direct-push data collection activities allows for optimized site remediation design.

11.11.7   *DIRECT-SENSING APPLICATIONS*

Most direct-sensing systems rely on various spectroscopic techniques to provide real-time chemical information about the target analyte. These spectroscopic measurements are usually conducted using a specially modified probe that includes a sapphire window as shown in Figure 11.11.5. Soil in contact with the sapphire window is optically interrogated as the probe is pushed into the ground. Sapphire has proven to be a good window material because it is very hard, resists scratching, and has good transmission properties in the UV, visible, and near-IR spectral regions. Most techniques make use of a laser or other light source to induce a spectroscopic response. The induced response is then quantified with a detector located in the probe or coupled via an optical fiber to a detector located at the surface. The major advantage of in situ spectroscopic methods is that the measurement times are extremely rapid. For example, fluorescence measurements usually require only a few seconds, which includes the time required to acquire, store and display the data. Rapid measurement times facilitate nearly continuous, real-time, high-resolution measurements (cm vertical spatial scales) of contaminant distributions. These high-resolution measurements provide a means to map small-scale chemical variability, which is not possible with conventional laboratory-based methods. Detailed delineation of contaminant distributions provides improved capability for understanding the transport and fate of contaminants in the environment.

The most significant limitation of direct in situ methods is that, in most situations, it is not possible to control or account for environmental variables that may affect the observed optical response for the analyte of interest. In contrast, most laboratory-based methods use elaborate procedures to separate the target analyte from the sample matrix, thereby minimizing the influence of variability in the sample matrix. Because it is not possible to simplify the sample matrix with direct in situ methods, the quantification of sensor response is usually only semi-quantitative at best. This means that the utility of the in situ data is for rapid delineation of contaminant distributions and is not for rigorous quantification of contaminant concentrations. In situ sensor data should be viewed as complementary rather than a direct replacement for conventional laboratory-based methodologies. Conventional methods that depend on collection of discrete samples and subsequent laboratory analysis may produce accurate results but do not necessarily make good exploration tools. This is because ultimately the delineation of the contaminant plume is a function of the inherent heterogeneity in contaminant distributions and of the economics of the number of samples that can be collected and analyzed.

An excellent review of the status of in situ sensor probes developed for use with the cone penetrometer system was recently provided by Aldstadt, J. H., and Martin, A. F. (1997). More recently,

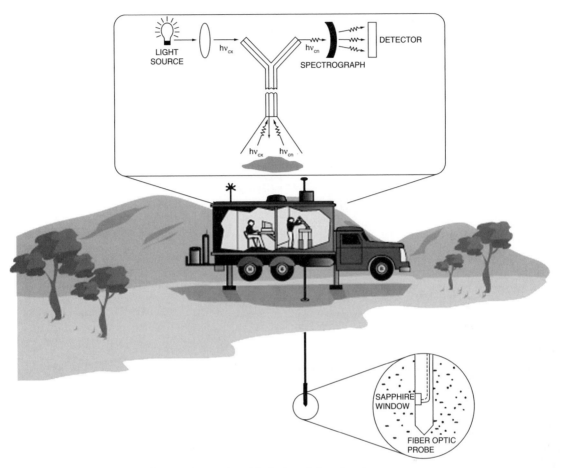

FIGURE 11.11.5   Schematic showing general approach used for in situ optical sensing schemes.

fluorescence-based sensor systems developed for in situ measurement of petroleum hydrocarbons in soils were reviewed (Lieberman, 1998). In this section, we present an overview and status update of cone penetrometer deployable in situ sensor systems, which are available or in the advanced stages of development and testing.

11.11.8   VAPOR SAMPLING

As the probe is being driven into the subsurface, vapor samples can be continuously pumped to the surface in vapor compatible tubing for analysis through conventional field analyzers, such as a photoionization detector/flame ionization detector (PID/FID) and gas chromatographs. Several designs exist for downhole vapor sampling. The membrane interface probe (MIP) uses a thin film fluorocarbon polymer membrane approximately 6.35 mm in diameter in direct contact with the soil. This membrane is typically heated to a temperature of 100 to 120°C. This thin film membrane is impregnated into a stainless steel screen that serves as a rigid support for the fluorocarbon polymer. A clean carrier gas, such as nitrogen, helium, and clean air is used to carry the volatile organic

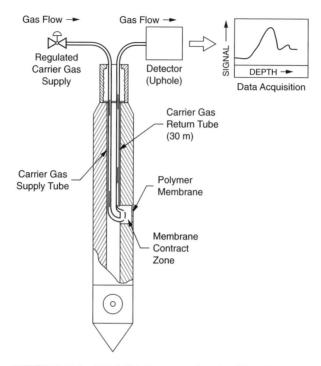

FIGURE 11.11.6a   MIP field deployment configuration (Geoprobe, 1997).

compounds (VOCs) in the soil to the surface and into the gas detector, such as the PID/FID. As the probe is pushed into the soil, VOCs in the subsurface contact the heated surface of the MIP polymer membrane. Upon contact, a certain quantity of the VOCs will adsorb into the polymer membrane. Once sorbed into the membrane, VOC molecules move by diffusion across the membrane to regions where their concentration is lowest. Because of the heating of the membrane and thin profile, the movement across the membrane takes place in less than a second for light hydrocarbons such as gasoline. The MIP system uses Teflon® tubing to transport the VOC from the membrane to the surface for analysis (Geoprobe, 1997) (see Figures 11.11.6a and 11.11.6b).

11.11.9   SOIL CONDUCTIVITY OR RESISTIVITY

Conductivity tools can be used to determine soil type. These systems are based on the observation that fine-grained materials will conduct electricity better than coarse-grained materials. Caution must be exercised in areas of high pore water salinity. Soil conductivity or resistivity logging used in conjunction with a direct penetration technology (DPT) rig and PC-based data acquisition system produces a real time display of conductivity or resistivity, and stores the data for further analysis. The probe is useful in detecting changes in lithology. Sand and other permeable zones can be interpreted from the conductivity or resistivity logs. These zones can be subsequently targeted when setting screens for water sampling. Conductivity or resistivity logging can have vertical resolution of 1- to 4-in clay layers that may influence plume migration. Discrete soil, water, or vapor samples can be collected in the field with the DPT rig to calibrate log interpretations and extend information horizontally from known vertical profiles. Logging rates are estimated to range from 500 to 900 feet per day, depending on subsurface conditions. Boreholes are pressure grouted with neat cement after logging.

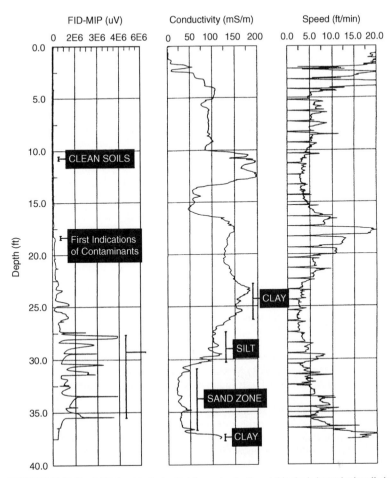

FIGURE 11.11.6b   MIP signal indicates that the contamination at this site is located primarily in the silt and sand units. Water table is at 26 ft but hydrocarbon contamination extends to about 36 ft. Note the lithology interpretation based on the conductivity log (Geoprobe, 1997).

11.11.10 PORE PRESSURE

Piezocone pore pressure probes are equipped with transducers for determining pore pressure. Piezocones tend to yield better soil type data and resolution than the probes that rely only on tip resistance and sleeve friction, because they yield point measurements and do not rely as much on averaging based on properties within the vicinity of the sensors. Piezocone soil classification data is only valid in saturated soils, because a positive pore pressure is required to obtain a reading on the pressure transducer. Many of the newer probes consist of both systems, allowing for continuous profiling throughout the vadose zone and saturated regions including perched zones.

When the piezocone is stopped in saturated soils and the pressure is allowed to equilibrate, dynamic pore pressure (u) dissipates and approaches hydrostatic pressure (u_o), expressed in meters of water. The addition of hydrostatic pressure to surface elevation yields true hydrostatic head or potentiometric surface elevation. For environmental projects, potentiometric surface elevation data from three or more boring locations can be used to calculate horizontal hydraulic gradient and the direction of groundwater flow. These data can also be used to identify perched, confined or unconfined aquifer zones as well as vertical hydraulic gradients (Edelman and Holguin, 1996).

11.11.11   SOIL TEXTURE

New sensors based on real time video imaging have been developed for determining soil texture. While the resolution is exceptional, the images can sometimes be difficult to use for determining soil classification in the saturated zone because pore fluids can yield complex images. Research is underway to refine the images to determine pore fluid saturation levels, relative permeability, moisture content, and hydraulic conductivity. The current system is excellent for identifying perched areas in the vadose zone, which are often overlooked or not considered in environmental investigations.

11.11.12   ORGANIC ANALYSIS

Fluorescence Spectroscopy

The U.S. Army Engineers Waterways Experiment Station CPT development program is led by researchers S. Cooper and P. Malone, who were the first to propose the use of spectroscopic techniques for direct chemical sensing with a cone penetrometer probe. They provided the first feasibility demonstration of this approach using a lamp-based fluorescence sensor integrated into a cone penetrometer probe for direct detection of an artificial fluorophore that has been injected into the ground (Lurk et al., 1990). This early work led to a collaboration with the U.S. Navy Space and Naval Warfare (SPAWAR) Systems Center led by S. Lieberman. This resulted in the adaptation of a fiber optic-based Laser-Induced Fluorescence (LIF) sensor system that had been developed for use in marine applications (Inman et al., 1990) for use with the cone penetrometer system. This LIF-based CPT sensor system was the first system to be used for in situ detection of petroleum hydrocarbon contamination in the soil environment (Lieberman et al., 1990; 1991). To date, the LIF technology remains the most successful and commonly used technology for direct in situ chemical sensing of chemical contamination in the subsurface soils with the cone penetrometer probe (see Table 11.11.1). The technology has continued to evolve through the effort of several groups, most notably North Dakota State University led by G. D. Gillispie; Tufts University, led by J. Kenny; and the Massachusetts Institute of Technology, Lincoln Laboratory, led by J. Zayhowski and coworkers.

Although various LIF system configurations may differ in the details, all fluorescence techniques employ the same general two-step-sensing scheme. First, high-energy ultra violet (UV) light is transmitted through a window on the penetrometer probe onto the soil in contact with the window. If the soil contains petroleum hydrocarbons or other fluorophores, the fluorescent species absorb and then re-emit the light energy as fluorescence. Next, a portion of the fluorescence is returned through the window where it is coupled into an optical detector system. Variations in system configurations primarily reflect the attempt of various researchers to take advantage to the multidimensional nature of the fluorescence technique. Different aromatic compounds are excited more effectively at certain excitation wavelengths. Judicious selection of excitation wavelength can be used to increase the specificity of the method. Both the wavelength of the fluorescence emission and the lifetime of the fluorescence signal provide valuable information about the chemical species responsible for the observed fluorescence. Optimization of the design of a fluorescence sensor is directly related to either simpler, less-costly approaches with somewhat less specificity to the more complex and usually more expensive systems providing multi-dimensional characterization of the fluorescence signal.

Site Characterization and Analysis Penetrometer System (SCAPS) is the original Navy/WES system. That LIF system uses a pulsed nitrogen laser (excitation wavelength = 337 nm) coupled to a

TABLE 11.11.1   Characteristics of Direct-push Fluorescence-based Sensor Systems

System name	Excitation source	Excitation wavelength	Detection system	Spectral data	Time decay data	Source/vendor
SCAPS	Nitrogen Laser	337 nm	Time-Gated Photodiode Array/Spectrograph	Yes	No*	Army, Navy, DOE
SCAPS	XeCl Laser	308 nm	Time-Gated Photodiode Array/Spectrograph	Yes	No	Navy
SCAPS	Micro-Chip Laser	266 nm	Time-Gated Photodiode Array/Spectrograph	Yes	No	MIT Lincoln Labs/Navy
ROST™	Tunable Dye Laser	Variable†	PMT/Digital Oscilloscope	No	Yes	Fugro Geosciences (Houston, TX)
ROST™ II	Tunable Dye Laser	Variable†	PMT/Digital Oscilloscope at 4 discrete emission wavelengths	Yes	Yes	Fugro Geosciences (Houston, TX)
Tufts	Raman Shifter	Multiple Excitation Wavelengths	CCD/Imaging Spectrograph	Yes	No	Tufts University
FMG 300	Nitrogen Laser	337 nm	PMT/Digital Oscilloscope	No	Yes	Laser Labor Adlershof (Berlun)
FFD	Hg Lamp	254 nm	PMT/Band Pass Filter	No	No	Vertek, (South Royalton, VT)
Hydrocarbon Probe	Hg Lamp	250 nm	PMT/Band Pass	No	No	Delft Geotechnics (Delft, The Netherlands)
FMG 600	Hg Lamp	254 nm	PMT/Band Pass Filter	No	No	Laser Labor Adlershof (Berlin, Germany)

* Possible with change in software
† Normally operated at 290 nm
Note: Systems that use low wavelength excitation sources provide the most effective detection of lightweight petroleum products such as jet fuels. Sensors that provide spectral and/or time decay information provide increased specify and minimize interferences from non-petroleum fluorescence.

silica-clad-silica optical fiber to deliver UV light to the soil in contact with the window. Fluorescence generated in the soil is collected and transmitted to the surface over a second fiber (see Figure 11.11.7). At the surface, the optical signal is quantified using a time-gated intensified linear photodiode array detector coupled to a spectrograph that houses a 300-line/mm diffraction grating. The chromatic dispersion provided by the spectrograph allows detailed characterization (2-nm resolution) of the fluorescence emission spectrum (up to 700 wavelength/intensity data pairs) at each measurement point. The standard operating procedure is to add spectra from multiple laser shots (normally 20 shots) in order to enhance the signal-to-noise ratio of the measurement. A 20-shot measurement, using a laser with a 10-Hz repetition rate, requires approximately 2 seconds and provides a vertical spatial resolution of approximately 4 cm at standard push rates. Reported detection limits for common fuel hydrocarbon products such as diesel fuel marine are in the range of 100 ppm.

A second LIF system was developed by G. Gillispie and coworkers at North Dakota State University, and later commercialized by Dakota Technologies Inc., Unisys Corp., and the U.S. Air Force. This system, Rapid Optical Screening Tool (ROST™), uses a nedodymium: yttriuum: aluminum-garnet (Nd:YAG) pumped dye laser system, whose pulsed output can be varied or tuned over the wavelength range of approximately 280 to 300 nm (see Figure 11.11.8). The fourth harmonic of the Nd:YAG laser can be accessed directly to provide a 266 nm excitation source (Gillispie and St. Germain, 1995; St. Germain and Gillispie, 1995 Inman et al., 1990). The higher excitation

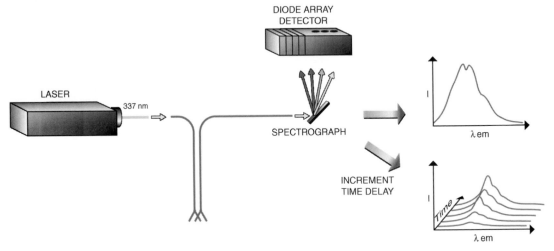

FIGURE 11.11.7   SCAPS LIF Sensor Schematic.

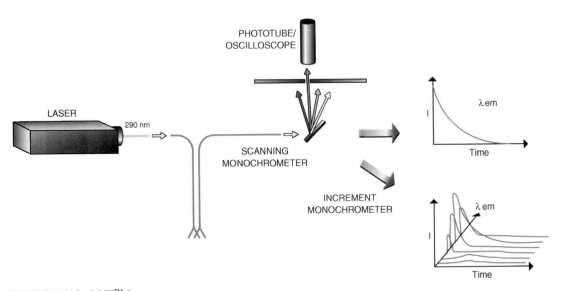

FIGURE 11.11.8   ROST™ System.

energy and lower wavelength provided by the Nd:YAG dye laser system is more effective for exciting fluorescence in two-ringed and lighter aromatics than the 337 nm laser used by the original SCAPS LIF system. The ROST™ system uses a plastic-clad silica optical fiber to transmit light to the sample. The ROST™ system makes use of a monochromator coupled to a photomultiplier tube (PMT) detector. The intensity of fluorescence signal is quantified as a function of time at a single emission wavelength using a digital storage oscilloscope (DSO) to record and display the output from the PMT. By incrementally adjusting the monochromator over a series of different emission wavelengths, a wavelength-time matrix (WTM) can be generated, which characterizes the fluorescence emission in

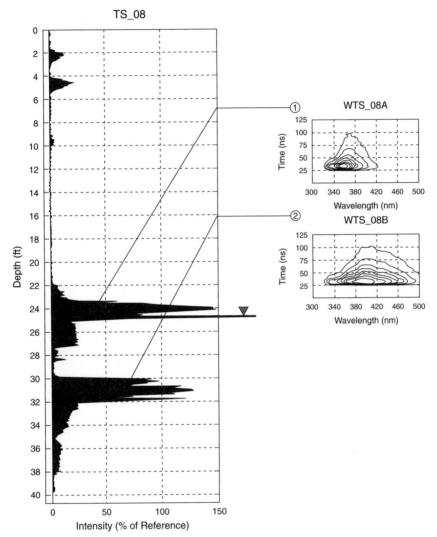

FIGURE 11.11.9 Typical ROST™ profile showing WTM (wavelength-time matrix at selected depth; after Gillespie, 1999).

terms of both its spectral characteristics and the decay time of the fluorescence signal as shown in Figures 11.11.9 and 11.11.10.

The ROST™ Fluorescence versus Depth/waveform Log presents the distribution of petroleum hydrocarbons continuously with depth (y-axis) as a function of fluorescence intensity (x-axis). The particular spectral signature of the encountered petroleum hydrocarbon can be evaluated at any depth. Four peaks are presented on each waveform, and represent, from left to right, fluorescence intensity measured at each of the four monitored wavelengths: 340, 390, 440, and 490 nm. The shape of the waveform can be compared with those from common petroleum products such as gasoline, jet fuel, diesel, and coal tar/creosote to identify and differentiate various encountered hydrocarbon products (Fugro Geosciences, Inc., 1999).

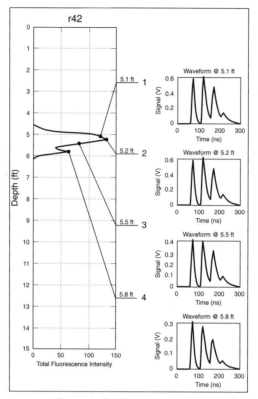

Example Product Waveforms

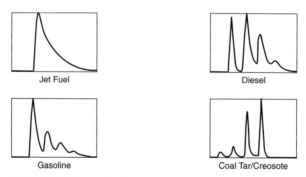

FIGURE 11.11.10 ROST™ Fluorescence versus Depth/Waveform Log (Fugro Geosciences, Inc., 1999).

In practice, WTMs are generally not recorded at every measurement point because approximately two minutes is required to incrementally adjust the monochromator and generate the WTM. In normal operations, the ROST™ system was set to detect fluorescence at a single emission wavelength to maintain the 2-cm/sec standard push rate. The data collected in this mode is referred to as the FVD (fluorescence versus depth). This FVD data could then be used by the ROST™ operator to select locations to stop the push and collect the WTM data. More recently, the ROST™ system has been modified to include the capability to collect fluorescence emission versus time curves simultaneously at four different emission wavelengths (Gillispie, personal communication). The multiple-wavelength emission probe eliminates the need to stop the probe to acquire the WTM and potential problems associated with selection of the optimal emission wavelength (see Figure 11.11.11).

Another LIF system has been developed by Tufts University, led by J. Kenny (Hart et al., 1996; Kenny and Henderson-Kenny, 1995; Lin et al., 1994; 1996). The Tufts system seeks to increase the specificity of the LIF technique by simultaneously exciting the sample using several different excitation wavelengths. This is accomplished using a novel hydrogen-methane Raman shifter (45:55; $H_2:CH_4$), which is pumped by the fourth harmonic (266 nm) of a Nd:YAG laser. The Raman shifter produces a large number of output beams with variable intensities over the wavelength range of 250 to 400 nm. Up to 10 of the resulting beams are selected, coupled into separate fibers, and transmitted down the probe to separate sapphire windows. Fluorescence induced at each window is collected with a separate optical fiber at each window and transmitted to the surface. At the surface, collection fibers are arranged in a vertical array and coupled into an imaging spectrograph coupled to a cooled charge couple device (CCD) detector. The fluorescence emission spectrum corresponding to each excitation wavelength is monitored using the CCD. Data is displayed as a three-dimensional excitation emission matrices, or EEM. The Tufts system (see Figure 11.11.12) has been updated so all measurements are made through a single sapphire window. Because the original Tufts system used a different window for each excitation wavelength, the CCD detector was actually collecting data from up to 10 different depths at one time. There can be as much as a 30-cm vertical offset between data collected from a single exposure of the CCD. In order to avoid potential uncertainties in regions of sharp vertical gradients, it was necessary to scale the probe in increments that corresponded to the spatial separation of the windows and reassemble the EEM data from successive measurements. Efforts are underway to attempt to develop methods for extracting useful information from these complex EEM data matrices.

Optical attenuation over optical fibers rapidly increases at wavelengths below about 280 nm. This means that excitation of light weight aromatic compounds, such as benzene, that require high energy UV excitation can be severely limited over long lengths of optical fiber. One LIF system has been developed to circumvent this limitation. This system makes use of a microlaser that can be located in the cone penetrometer probe (see Figure 11.11.13), thereby eliminating fiber transmission losses and potential problems from fiber-generated signals. The Nd:YAG microlaser is pumped from the surface using radiation delivered over an optical fiber from a 808 nm diode. Output of the Nd:YAG crystal is frequency-quadrupled to 266 nm. This microlaser system has been used with both linear photodiode array detector system (Knowles et al., 1997) and a PMT detector for measurement of light weight hydrocarbons in soils (Bloch et al., 1998).

Several fluorescence sensor systems have been described that make use of a compact mercury lamp located in the probe (Bratton and Shinn, 1997; Olie and Sellmeijer, 1995; Zimmerman and Lucht, 1997). The Bratton and Shinn system uses an optical fiber to transmit the fluorescence signal to a PMT located at the surface while both of the other systems use a PMT located downhole. All three systems use bandpass filters to select the spectral emission wavelengths to be monitored. The advantages of the lamp-based system are simplicity in design, lower overall cost, and ease of operation as compared with other laser-based systems described previously. The most significant disadvantage is that the lamp-based systems only monitor a single emission band and, consequently, provide no capability for discriminating between different contaminant sources, fuel types, or fluorescence from interfering fluorophores.

If working with the fluorescence sensor probes at sites where LNAPLs have been released, the goal is to push the sapphire window beneath the water table (see Figure 11.11.11), which can act as a vertical barrier to contaminant flow. Because the sapphire window is typically located approximately 60 cm from the probe tip, the target depth is generally at least 60 cm deeper than the anticipated water table depth. Cases exist where the bottom of a LNAPL plume is not defined because the probe cannot be pushed deep enough to get the sapphire window into the deepest contaminated zone. For these cases, operators are encouraged to continue pushing in adjacent locations until the sapphire window penetrates groundwater. In addition, grab samples are collected and analyzed from the zone just above the resistant layers because these layers can act as vertical barriers to contaminant flow and may induce fluid ponding or perching.

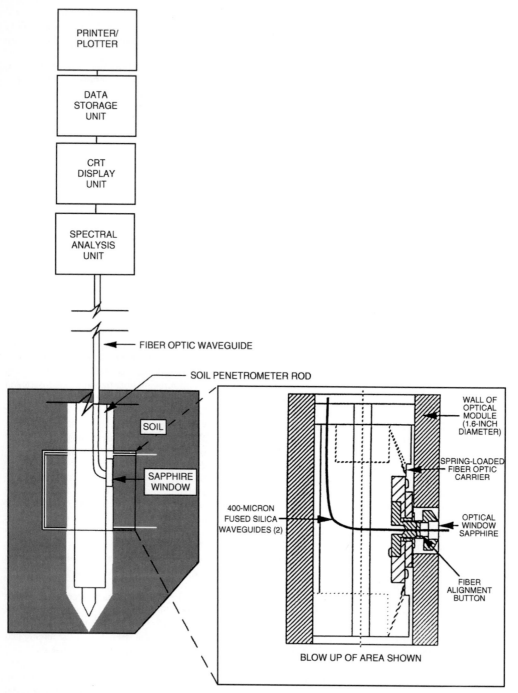

FIGURE 11.11.11   Schematic of laser Induced Fluorescence System (Holguin-Fahan and Associates after Edelman and Holguin, 1996).

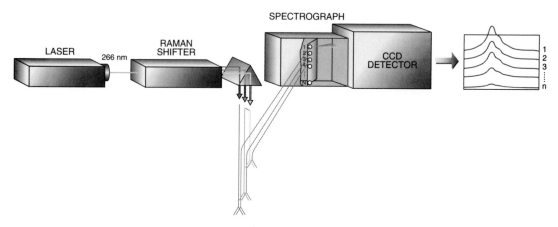

FIGURE 11.11.12   TUFTS System with laser and Raman shifter.

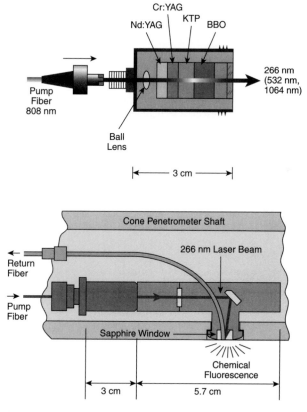

FIGURE 11.11.13   Microlaser LIF Probe (after Bloch et al., 1998).

Raman Spectroscopy

Raman spectroscopy is a laser scattering technique that identifies subsurface contaminants on the basis of the unique molecular vibrational fingerprints of the target chemicals. Several different cone penetrometer deployable sensor probes have been developed that make use of Raman spectroscopy in order to target organic contaminants that do not exhibit fluorescence responses. Chlorinated solvents are among the most important analytes included in this group of compounds. The advantage of Raman spectroscopy is the specificity gained for a wide range of organic molecules. The disadvantage of Raman spectroscopy is the low sensitivity compared to fluorescence methods and the possibility for interferences from sample fluorescence as well as the scattering that is generated in the fibers used for remote measurements.

One Raman probe developed for CPT rigs makes use of near infrared excitation to minimize the generation of interferences from fluorescence in the sample (Haas et al., 1995). The sensor probe employs microptical filters to remove background signals generated in the excitation fiber. Optical fibers are used to couple the probe to a compact spectrograph connected to a CCD detector located up-hole. Recent field tests have indicated some success using this probe to detect chlorinated solvents at high concentration levels.

Another fiber optic based-Raman sensor probe has been described that makes use of several features to maximize sensitivity of the method and reduced potential interference (Mosier-Boss et al., 1995). This probe uses multiple collection fibers, as well as bandpass filters on the excitation fiber and longpass filters on the collection fiber. In addition, signal processing techniques, including edge detection and filtering in the Fourier domain, were used to deconvolve the Raman signal from background fluorescence signals. In spite of these efforts, the technology developers concluded that the inherent low sensitivity of the conventional Raman spectroscopy would be limited to situations where the contaminant was present as a non-aqueous phase liquid (NAPL).

Infrared Spectroscopy

Infrared (IR) reflectance has been exploited as an approach for detecting organic contaminants that cannot be measured by fluorescence techniques. Aggrawal and coworkers (Ewing et al., 1994) have described an IR reflectance probe that has been integrated into a cone penetrometer probe (see Figure 11.11.14). The probe employs a Nichrome wire filament blackbody source to illuminate the sample through a sapphire window. The diffuse reflectance from the sample is collected using a parabolic

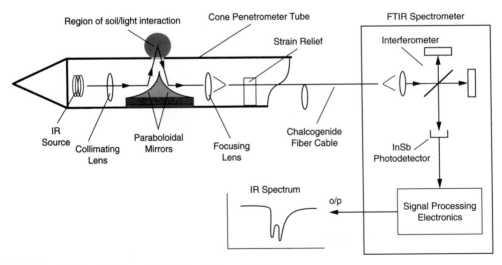

FIGURE 11.11.14   IR Reflectance Probe (after Ewing et al., 1994).

mirror and coupled into a sulfide-based chalcogenide optical fiber, which exhibits good transmission in the mid-IR region. Up-hole another parabolic mirror is used to collect light from the optical fiber and guide it into a FTIR spectrometer. IR spectrometry is inherently selective because IR spectra are unique for a given compound. The disadvantage of IR methods for in situ measurements in soils is that changes in background absorption may be difficult to account for and may mask variations in absorbance from the target analyte.

11.11.13 INORGANIC ANALYSIS

Laser-Induced Breakdown Spectroscopy (LIBS)

LIBS involves the analysis of the spectral emission from a laser-induced spark. The spark is generated by focusing the high power emission from a pulsed laser onto a small spot on the sample material, resulting in a power density on the sample in excess of several giga-watts per square centimeter (GW/cm^2). Within the small volume about the focal point, rapid heating, vaporization, and ionization of the sample material occurs. The subsequent laser induced plasma emission is spectroscopically analyzed to yield qualitative and quantitative information about the elemental species present in the sample.

The LIBS technique has been used successfully in the laboratory to identify and quantify elemental species in solids, liquids, and gases. The method is well suited for in situ detection of heavy metal contamination in soil because it is highly sensitive yet requires no sample preparation. Because the emission of different elements occurs at unique wavelengths the method can be used for simultaneous analysis of multiple components.

At present two different LIBS probes have been developed and tested in the field. One system uses an optical fiber to deliver the excitation energy from a laser located in the truck to the sample (Theriault et al., 1998). The second system uses a laser located in the penetrometer probe to provide the energy needed to generate the plasma (Cortes et al., 1998; Miles and Cortes, 1998).

LIBS can be used to excite emission spectra from any atomic species with species- dependent efficiency. In order to perform LIBS measurements remotely and in real-time, as required of CPT-based sensor systems, several operation and design trade-offs must be made between analytical precision and accuracy and timeliness of data collection. Because the two LIBS techniques under study both use fiber optic coupling to deliver the emitted spectra to an up-hole detection system, the choice of spectral lines used to characterize the species under study must fall in a spectral range that is consistent with the transmission capability of the optical fiber. Because the LIBS measurements are done in situ, there are matrix effects because of grain size and/or soil moisture content, which can affect the intensity of the response of the LIBS systems.

The fiber optic LIBS system (FO-LIBS) uses a Q-switched output of the Nd:YAG operating at 30 Hz at 1064 nm, which is delivered via a fused-silica fiber to the probe (see Figure 11.11.15). A low f/# lens system is used in the probe to focus the laser output. This minimizes the size of the focused image of the optical fiber face on the sample and provides a power density at the sample, which is sufficient to generate a laser-induced plasma.

A sapphire optical window is used to control the lens-to-sample distance. Because the plasma is formed on the soil sample at the surface of the sapphire window, it has been observed that after several thousand laser pulses, the sapphire window is pitted by the LIBS spark. Ultimately, this leads to reduced window transmission. To mitigate the effect of plasma damage to the window, the probe design was modified to include a capability for repositioning under microprocessor control the position of the spark on the sapphire window. Because the size of the plasma is relatively small compared to the size of the sapphire window, the number of discrete positions available is in excess of 1000. After the window has been completely scanned (approximately 1 week of normal operations), the sapphire window is replaced at a nominal cost.

The spark emission is fiber-optically coupled to a high-resolution spectrograph, which disperses the light onto a time-gated, intensified, linear photodiode array detector. Time gating of the detector is used to separate the initial short-lived broadband background emission generated during formation

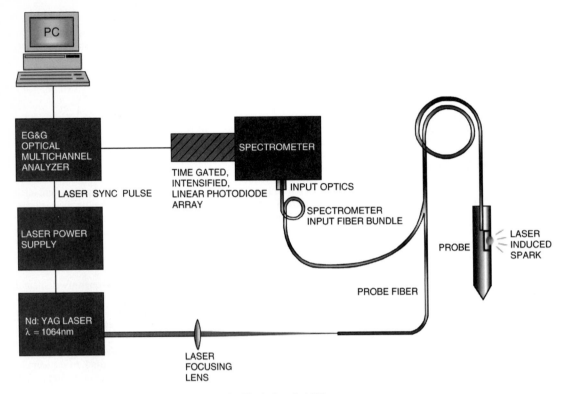

FIGURE 11.11.15 Fiber-Optic LIBS sensor probe (after Theriault et al., 1998).

of the plasma from the relatively longer-lived metal emissions. In order to improve signal to noise ratios, typically 300 single shot spectra are averaged. Because the repetition rate of the laser is 30 Hz, acquisition of 300 spectra requires 10 seconds. Depth resolution is controlled by the operator and is typically 1 to 2 in.

The second LIBS system, the down-hole laser LIBS (DL-LIBS), uses a custom design compact Nd:YAG laser located in the penetrometer probe to provide the required energy to generate the plasma. This approach eliminates the use of a fiber for delivering the excitation energy to the sample. However, the repetition-rate of the down-hole is much slower than the up-hole laser (approximately 0.5 Hz). Consequently, each measurement is based on a single laser shot, without any signal averaging. Another difference between the FO-LIBS system and the DL-LIBS system is that the DL-LIBS system uses a probe with a drop-off sacrificial sleeve, which is left at the bottom of the hole. Measurements are made as the probe is withdrawn from the hole rather than as the probe is pushed into the ground. With this system, the plasma is formed on the soil wall, which is formed as the probe is withdrawn. The advantage of this system is that it permits the optical window to be recessed from soil wall so that the plasma does not degrade the sapphire window over time. The disadvantage is that because the soil is not held in place by any mechanical system, there is no way to ensure that the soil does not collapse onto or contaminate the optical window as the probe is withdrawn.

X-ray Fluorescence (XRF) Sensor

X-ray fluorescence operates by detecting characteristic X-rays emitted by the atoms in a sample. The X-ray source bombards a sample with incident X-rays, where the atoms are excited and emit

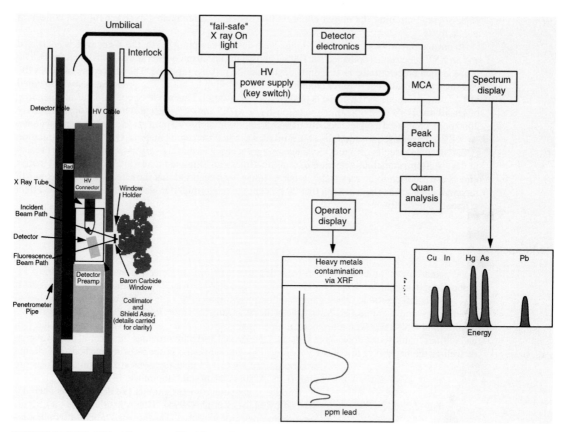

FIGURE 11.11.16   X Ray Fluorescence Metal Sensor Probe Schematic (after Elam et al., 1998b).

fluorescence X rays which are detected as a function of the energy of the X rays. At the atomic level, an incident X ray excites an electron from a core level, which then decays, producing a fluorescence X ray. Because the electron-energy states producing a fluorescence X ray are entirely within the atom, the X ray is produced with a constant and well-known energy (different for each type of atom).

The XRF sensor system developed for use with the cone penetrometer system consists of three subsystems. These include the below-ground probe, the umbilical cable, and the above-ground electronics package (Elam, 1998a; 1998b) (see Figure 11.11.16). The probe contains the X ray source, detector and preamp, appropriate X ray optics, the mounting system, and the rugged X ray window. A sealed X ray tube is used as the excitation source in order to achieve adequate detection limits in reasonable measurement times, and to avoid potential problems associated with a radioactive source. The detector is a silicon P-type/Intrinsic/N-type (Si-PIN) diode in a small case with self-contained cooling, connected to a low-noise preamp in close proximity. The preamp provides sufficient signal to drive the umbilical cable. The X ray window is made of 1 mm thick boron carbide, a low atomic number material, which is relatively transparent to X rays in the relevant energy range. The umbilical cable conducts the high voltage and filament power required by the X ray tube, the electronics, and cooling power for the detector and preamp, and the signal pulses from the detector.

The aboveground electronics package contains the X ray power supply and the drive electronics for the detector. The X ray power supply for the X ray tube provides adjustable high voltage (to 30 kV) and filament voltage. The detector electronics provides the necessary power supplies and pulse-shaping circuitry. Its output connects to a standard multichannel analyzer for data collection and analysis.

In normal operations the probe is stopped at a desired sampling depth, the X-ray tube is energized and an XRF spectrum is collected for 100 seconds. This spectrum is then stored for later analysis. The probe is then pushed to the next sample depth and the process repeated. This procedure provides the best detection limits, but requires stopping at each sampling location. The probe can also be operated in a continuous mode as the probe is pushed into the ground. This mode of operation results in a loss of sensitivity. For a field operation mode (push rate of 0.5 cm/sec), detection limits are approximately 2000 ppm, compared a typical detection limit of approximately 100 ppm for lead in the stopped mode.

XRF has been used successfully at several sites as a field screening method. One potential limitation is that for some samples different elements respond at energies that are very close and therefore overlap in the spectrum. Judicious selection of emission lines can eliminate many interferences. Soil matrix effects are reported to be reasonably modest. It has been reported that the effect of variations in soil moisture content ranging between 0 to 20 percent are negligible. For water saturated soils, the effect is about 20 percent.

Radionuclides

Cone penetrometer probes have also been developed for in situ detection and identification of radionuclides in subsurface soils. One device developed at Waterway Experiment Station (WES) uses a NaI scintillation crystal to detect gamma radiation (Morgan et al., 1998). The detector uses a custom designed preamplifier installed down-hole in the penetrometer probe. Power supplies and support electronics (high voltage power, amplifiers, multichannel buffers, and computers) are connected through an umbilical cable to the down-hole sensors. Operational software provides for spectral accumulation as well as the capability to collect site-screening data while the probe is being pushed. The design was optimized for Cs137 investigations at Department of Energy sites (U.S. Argonne National Laboratory, 1997).

Another cone-penetrometer gamma-ray detector has been described by Meisner et al. (1995). This sensor employs a CsI(TI) scintillation detector with two side-mounted silicon PIN photodiodes. This detector system is lowered into the penetrometer rods after the probe has been pushed into the ground rather than being integrated into the probe, as is the case with the WES system described above.

11.11.14   AQUIFER PARAMETER SENSOR APPLICATIONS

Soil Video Imaging Systems.   Recently, two different cone penetrometers have been commercially developed and deployed with video imaging systems for direct in situ imaging of soil on microscopic spatial scales (see Figure 11.11.17). Both systems use the return of the video signal from the camera to the surface where it is displayed in real-time on a video monitor, recorded on a video cassette recorder (VCR), and/or captured digitally with a frame grabber installed in a microcomputer system. One system uses a probe equipped with two windows and two miniature black and white cameras (Raschke and Hryciw, 1997). Each camera system operates at a different level of magnification and provides fields of view between 2 and 20 mm (diagonal). The other system uses a miniature CCD color camera coupled to a zoom lens system

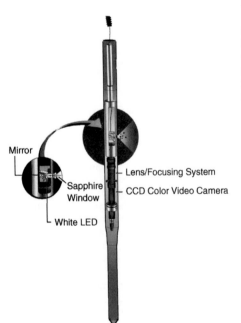

FIGURE 11.11.17   Schematic of GeoVIS soil video imaging probe (after Lieberman et al., 1999).

(Lieberman and Knowles, 1998). The lens system provides magnification and focusing of the imaged area. A mirror positioned at a 45-degree angle to the axis of the optics system is used to redirect light from the soil surface into the lens system. Illumination is provided by an array of four white light-emitting diodes (LEDs) located in the probe.

Both systems have proven useful for providing a direct continuous visual record of soil stratigraphy. Small-scale lithological changes such as clay lenses, fissures, and sand seams can be readily detected. Image processing techniques are under development to display soil texture and soil grain size distributions from the in situ images (Lieberman et al., 1997; Raschke and Hryciw, 1997). It has also been demonstrated that it is possible to use the cone penetrometer deployed video imaging system for direct detection of nonaqueous phase liquids (NAPLs) in the saturated zone. Microglobules of immiscible NAPL contaminants were readily identified in video images at a site where the probe was pushed through soil contaminated with chlorinated solvents and fuel hydrocarbons (Lieberman et al., 1998).

Piezocone for Soil and Aquifer Properties.   The piezocone has been used to estimate water table elevation, hydraulic conductivity of fine-grained materials, and to obtain highly resolved depictions of stratigraphy. This versatile system can be used to determine the direction and gradient of subsurface water flow, to determine whether aquifers are in hydrologic communication with each other, to determine the potential for environmental risk by identifying groundwater flow pathways and candidate receptors, and to monitor remediation efforts by identifying induced gradients and cross-checking groundwater model predictions. The system uses pore pressure measurements to determine water table elevation, soil type, and hydraulic conductivity. The transducer element is typically situated within the cone tip face or just above the cone tip shoulder behind a porous element and saturated with a viscous oil. As the probe is advanced, a pressure is induced by the penetration. If left to stabilize at a depth of interest, the pressure will dissipate to an equilibrium hydrostatic pressure. A plot of the final pressure for several depths within the same water-bearing zone can be extrapolated to determine the water table elevation corresponding to zero psi. The rate of dissipation is related to the hydraulic conductivity. Water table measurement precision is dependent upon the transducer full scale output. A typical 250 psi transducer will yield approximately 1 to 2 ft of resolution, which is good enough for regional groundwater investigations on large sites. This transducer is not appropriate for small sites. Recent improvements have allowed for precision to a few inches. Because dissipation is instantaneous for sandy materials, the hydraulic conductivity values are only valid for silty and clayey soils. One particular strength of the piezocone is its ability to determine whether a layer represents an aquitard based on hydraulic properties as opposed to conventional textural interpretations. If a leaky clay is encountered, water table elevations above and below that clay will plot on the same hydrostatic pressure versus depth line. If values above and below that clay plot on different lines, this indicates that they are separate water-bearing zones. Therefore, aquifer classification (perched, confined, or unconfined) is possible with a single push using the piezocone.

11.11.15 *MONITORING*

Once the site delineation goals have been met, DPT rigs can be used to install small diameter wellpoints in optimal monitoring locations. Well clustering, where short-screen wells are advanced to different depths within adjacent map locations, have recently been used to generate a three-dimensional understanding of critical site parameters in plume delineation efforts (Kram and Lory, 1998). Direct-push wells are typically installed without filter packs. However, robust filter pack materials specifically designed for rapid customization of direct-push wells have recently become available.

11.11.16 *REMEDIATION*

For remediation purposes, DPT equipment, which includes both CPT as well as smaller direct push rigs, can provide a closely spaced delivery system for in situ remediation of soils and water. Although not yet applied on a large scale, direct-push wells have been used as fluid and vapor conduits for remediation systems. At a hydrocarbon national test site in Southern California, air sparging and vapor extraction wells were successfully installed using direct-push techniques. The monitoring wells were customized to allow for vapor and fluid sample collection at 1-foot vertical increments. At the same site, direct-push wells were used to inject microbes and nutrients in support of bioremediation efforts.

Proprietary DPT remediation techniques include electrolysis, electro-osmosis, bio-venting, sparge point injection, enzyme injection, enhanced soil vapor extraction, and steam stripping (Jacobs and Loo, 1994). Properly designed cased boreholes can be used to deliver gas or liquids, including oxygen, nutrients, steam/hot air, liquid oxidants, and electrolytes. DPT extraction points can be used to recover vapors or liquids. Faster and more effective remediation may require two or more processes that work simultaneously or sequentially. Assessment and in situ remediation using DPT equipment can be quicker and less costly than conventional drilling and clean-up methods.

NOTE

Geoprobe® is a Registered Trademark of Kejr Engineering, Inc.; Teflon® is a Registered Trademark of E.I. du Pont de Nemours & Company.

REFERENCES

Aldstadt, J. H., and Martin, A. F., "Analytical Chemistry and the Cone Penetrometer: In Situ Chemical Characterization of the Subsurface," *Mikrochimica Acta*, 127: 1–18, 1997.

Argone National Laboratory, "U.S. Army Engineer Waterways Experiment Station Spectral Gamma Probe Evaluation Report," *Report Contract No. W-31-109-Eng-38*, Argonne, IL, 1997.

Bloch, J., Johnson, B., Newbery, N., Germaine, J., Hemond, H., and Sinfield, J., "Field Test of a Novel Microlaser-Based Probe for In Situ Fluorescence Sensing of Soil Contamination," *Applied Spectroscopy*, in press.

Bratton, W. L., and Shinn, J. D., "Case Studies of Measuring Fuel Contamination Using In-Situ Fluorescence Techniques," in *Proceedings, Field Analytical Methods for Hazardous Wastes and Toxic Chemicals*, Las Vegas, NV, 1997.

Cortes, J., Cespedes, E. R., and Miles, B. H., *Development of Laser-Induced Breakdown Spectroscopy for Detection of Metal Contaminants in Soils*, U.S. Army Engineer Waterways Experiment Station, Technical Report IRRP-96-4, 1998.

Edelman, S. H., and Holguin, A. R., "Cone Penetrometer Testing for Characterization and Sampling of Soil and Groundwater," *Sampling Environmental Media*, ASTM STP 1282, James Howard Morgan, ed., American Society of Testing and Materials, Philadelphia, PA, 1996.

Elam, W. T., "Unit 3B.3" in *Current Protocols in Field Analytical Chemistry*, V. Lopez-Avila et al. eds., John Wiley: New York, 1998a.

Elam, W. T., Adams, J. W., Hudson, K. R., McDonald, B., and Gilfrich, J. V., "Subsurface Measurement of Soil Heavy-Metal Concentrations with SCAPS X-Ray Fluorescence (XRF) Metals Sensor," *Field Analytical Chemistry and Technology*, 2: 97–102, 1998b.

Ewing, K. J., Tbilodeau, G., Nau, I. D., Aggarwal, T., King, R., Clark, and Robitaille, G., "Fiber Optic Infrared Reflectance Probe for Detection of Hydrocarbon Fuels in Soil," in *Optical Sensors for Environmental and Chemical Process Monitoring*, SPIE, 2367: 17–23, 1994.

Fugro Geosciences, Inc., Technical Information, Houston, TX, 1999.

Geoprobe Systems, *Tool and Equipment Catalog*, Kejr Engineering, Inc., Salina, KS, 1997, 1998–99.

Gillispie, G. D., and St. Germain, R. W., "Performance Characterization of the Rapid Optical Screening Tool (ROST™)," in *Proceedings Field Screening Methods for Hazardous Wastes and Toxic Chemicals*, Air & Waste Management Association, Pittsburgh, PA, pp. 478–489, 1995.

Gillispie, G. D., personal communication with S. Lieberman, 1999.

Haas, J. W., Carrabba, M. M., and Forney, R. W., "Nonaqueous Phase Liquids: Searching for the Needle in the Haystack," in *Field Screening Methods for Hazardous Wastes and Toxic Chemicals*, Air & Waste Management Association, Pittsburgh, PA, pp. 443–449, 1995.

Hart, S. J., Chen, Y., Lien, B. K., and Kenny, J. E., "A Fiber Optic Multichannel Spectrometer System for Remote Fluorescence Detection in Soils," *SPIE*, 2835: 73–82, 1996.

Huntley, D., Hawk, R. N., and Corley, H. P., "Non-Aqueous Phase Hydrocarbon Saturation and Mobility in a Fine-Grained, Poorly Consolidated Sandstone," *Proceedings of the 1992 Conference on Petroleum Hydrocarbons and Organic Chemicals in Ground Water*, NGWA, November 4–6, 1992, Houston, Texas, p. 223–227, 1992.

Inman, S. M., Thibado, P., Theriault, G., and Lieberman, S. H., "Development of a Pulsed-Laser, Fiber Optic-Based Fluorometer for Time-Resolved Measurements of Polycyclic Aromatic Hydrocarbons in Seawater," *Anal. Chim. Acta.*, 239, 45–51, 1990.

Jacobs, J. A., and Loo, W., "Direct Push Technology Methods for Site Evaluation and In Situ Remediation," *Hydrovisions*, July/August., Groundwater Resources Association of California, Sacramento, CA, 1994.

Kenny, J. E., and Henderson-Kenny, A., "Spectroscopy in the Field," *Spectroscopy*, 10: 7, 1995.

Kram, M. L., "Free Product Recovery: Mobility Limitations and Improved Approaches," *NFESC Information Bulletin #IB-123*, October 1993.

Kram, M. L., and Lory, E., "Use of SCAPS Suite of Tools to Rapidly Delineate a Large MTBE Plume," *Conference Proceedings for the Annual Meeting of the Environmental and Engineering Geophysical Society*, March 22–26, 1998, Chicago, Illinois, p. 85–99, 1998.

Lieberman, S. H., "Direct-Push, Fluorescence-based Sensor Systems for In Situ Measurement of Petroleum Hydrocarbons in Soils," *Field Analytical Chemistry and Technology*, 2: 63–73, 1998.

Lieberman, S. H., Inman, S. M., Theriault, G. A., Cooper, S. S., Malone, P. G., and Lurk, P. W., "Fiber Optic-Based Chemical Sensors for in situ Measurement of Metals and Aromatic Organic Compounds in Seawater and Soil Systems," *SPIE*, 1269: 175–184, 1990.

Lieberman, S. H., Inman, S. M., Theriault, G. A., Cooper, S. S., Malone, P. G., Olsen, R. S., and Lurk, P. W., "Rapid, Subsurface, In situ Field Screening of Petroleum Hydrocarbon Contamination Using Laser Induced Fluorescence over Optical Fibers," in *Second International Symposium on Field Screening Methods for Hazardous Wastes and Toxic Chemicals*, Air & Waste Management Association, Pittsburgh, PA, pp. 57–63, 1991.

Lieberman, S. H., Knowles, D. S., Stang, P. M., Kertesz, J., and Mendez, D., Cone, "Penetrometer Deployed In Situ Video Microscope for Characterizing Sub-Surface soil Properties," in *Field Analytical Methods for Hazardous Wastes and Toxic Chemicals*, Air & Waste Management Association, Pittsburgh, PA, pp. 579–587, 1997.

Lieberman, S. H., and Knowles, D. S., "Cone Penetrometer Deployable In-Situ Video Microscope for Characterizing Sub-Surface Soil Properties," *Field Analytical Chemistry and Technology*, 2: 127–132, 1998.

Lin, J., Hart, S. J., Taylor, T. A., and Kenny, J. E., "Laser Fluorescence EEM Probe for Cone Penetrometer Pollution Analysis," *SPIE*, 2367: 70–79, 1994.

Lin, J., Hart, S. J., and Kenny, J. E., "Improved Two-fiber Probe for In situ Spectroscopic Measurement," *Anal. Chem.*, 68: 3098–3103, 1996.

Lurk, P. W., Cooper, S. S., Malone, P. G., and Lieberman, S. H., "Development of Innovative Penetrometer Systems for the Detection and Delineation of Contaminated Groundwater and Soil," in *Superfund 90: Proceedings of the 11th National Superfund Conference*, Hazardous Materials Control Research Institute, 297–299, 1990.

Meisner, J. E., Nicaise, W. F., and Stromswold, D. C., *IEEE Trans. Nucl. Sci.*, 42: 288–291, 1995.

Miles, B., and Cortes, J., "Subsurface Heavy Metal Detection with the use of a Laser-Induced Breakdown Spectroscopy (LIBS) Penetrometer System," *Field Analytical Chemistry and Technology*, 2: 75–88, 1998.

Morgan, J. C., Adams, J. W., and Ballard, J. W., "Field Use of a Cone Penetrometer Probe for Radioactive-Waste Detection," *Field Analytical Chemistry and Technology*, 2: 111–116, 1998.

Mosier-Boss, P. A., Lieberman, S. H., and Newbery, R., "Development of A Cone Penetrometer Deployed Solvent Sensor Using a SERS Fiber Optic Probe," in *Field Screening Methods for Hazardous Wastes and Toxic Chemicals*, Air & Waste Management Association, Pittsburgh, PA, pp. 443–449, 1995.

NFESC, *Natural Attenuation of MTBE in an Anaerobic Groundwater Plume*, TDS-2068-ENV, February 1999, 1999a.

NFESC, *In-Situ Bioremediation of Methyl Tertiary Butyl Ether (MTBE)*, TDS-2069-ENV, February 1999, 1999b.

Olie, J. J., and Sellmeijer, J. B., "Floating Layer Detection with the Hydrocarbon Probe: Results, Calibration, and Sampling Strategy," van den Brink et al., W. J., ed., *Contaminated Soil i95*, Kluwer, Academic Publishers, 531–532, 1995.

St. Germain, R. W., and Gillispie, G. D., "Real-Time Continuous Measurement of Subsurface Petroleum Contamination with the Rapid Optical Screening Tool (ROST™)," in *Proceedings Field Screening Methods for Hazardous Wastes and Toxic Chemicals*, Air & Waste Management Association, Pittsburgh, PA, pp. 467–477, 1995.

Raschke, S. A., and Hryciw, R. D, "Vision Cone Penetrometer (VisCPT) for Direct Subsurface Soil Observation," *ASCE Journal of Geotechnical and Geoenvironmental Engineering*, 123(11): 1074–1076, 1997.

Raschke, S. A., and Hryciw, R. D., "Soil Grain Size Distribution by Computer Vision," *ASTM Geotechnical Testing Journal*, 20(4): 433–422, 1997.

Robertson, P. K., and Campanella, R. G., "Interpretation of Cone Penetration Tests. Part I: Sand," *Canadian Geotechnical Journal*, 20: 718–733, 1983a.

Robertson, P. K., and Campanella, R. G., "Interpretation of Cone Penetration Tests. Part II: Clay," *Canadian Geotechnical Journal*, 20: 734–745, 1983b.

Robertson, P. K., "Soil Classification using the Cone Penetration Test," *Canadian Geotechnical Journal*, 27: 151–158, 1990.

Theriault, G. A., Bodensteiner, S., and Lieberman, S. H., "A Real-Time Fiber-Optic Probe for the In Situ Delination of Metals in Soils," *Field Analytical Chemistry and Technology*, 2: 117–125, 1998.

Wallace, J. W., and Huntley, D., "Effect of Local Sediment Variability on the Estimation of Hydrocarbon Volumes," *Proceedings of the Sixth National Outdoor Action Conference*, NGWA, May 11–13, 1992, Las Vegas, Nevada, pp. 273–285, 1992.

Zimmerman, B., and Lucht, H., "Field Measuring Devices for In-situ Analysis of Fluorescence contaminants in Water and Soil," *Field Screening Europe*, J. Gottlieb et al., eds., Kluwer Academic Publishers, pp. 381–384, 1995.

CHAPTER 11
ASSESSMENT SAMPLING AND MONITORING

SECTION 11.12

DIRECT PUSH TECHNOLOGY SAMPLING METHODS

James A. Jacobs
Hydrogeologist and president of FAST-TEK Engineering Support Services, he has over 20 years of experience. He is registered in several states, including California. He specializes in assessment methods and in situ remediation technologies.

Mark Kram
Currently a Ph.D. candidate at the University of California at Santa Barbara, he is also working as a hydrogeologist, specializing in environmental site characterization and remedial design.

Stephen Lieberman
Senior research scientist in the Environmental Sciences Division at the Space and Naval Warfare Systems Center, San Diego (SSC San Diego), he leads the Optical Chemical Sensor Group.

11.12.1 INTRODUCTION

Direct Push Technology (DPT) is a quicker and less costly alternative sampling method to conventional rotary drilling for collecting soil, vapor, and water samples for environmental projects. DPT equipment allows for fewer permanent monitoring wells, multiple depth sampling programs, elimination or minimization of drilling derived wastes, ability to perform on-site chemical analysis, and minimal exposure of workers to potentially hazardous soil cuttings.

DPT sampling relies on dry impact methods to push or hammer boring and sampling tools into the subsurface for environmental assessments. This technology does not require hazardous chemicals, drilling fluids, or water during operation. A typical augered borehole to 60 ft would generate approximately six drums of soil cuttings. DPT equipment produces soil samples but generally does not produce soil cuttings.

DPT works well with a variety of lithologies, including clays, silts, sands, and gravels; however, this technology is not designed to penetrate or sample bedrock (Table 11.12.1). DPT equipment has been used successfully in limited access areas such as in basements, under canopies, and inside buildings. DPT sampling has also been used successfully in sensitive environments such as wetlands, tundra, lagoon, and bay settings.

TABLE 11.12.1   Relative Comparison of Environmental Sampling Techniques

Aspect	Hand augers & rotohammers	Hydraulic probe small	Hydraulic probe large	CPT rigs	Hollow stem auger rigs
Sample Integrity	2	4	4	3	3
Speed	3	4	4	3	2
Cost	4	3	3	2	1
Soil Disposal	4	4	4	4	1
Max. Depth	1	2	3	3	4
Sample Volume	1	2	3	3	4
Physical Information	1	3	3	3	2
Reliability	1	3	4	4	4
Flexibility	1	3	4	2	2
Total Ranking	18	28	32	27	23

Notes: For acquisition of soil, water or soil vapor from depths up to 30 ft.
Relative Scale: 4 is best application, 3 is better application, 2 is good to fair application and; 1 is poor application.

11.12.2   DPT SITE EVALUATION TOOLS

Manual DPT Methods

The most basic of all DPT equipment to collect undisturbed soil samples is the manually operated slide hammer. The hand-held slide hammers, typically weighing 12 to 30 pounds, are dropped approximately 12 to 24 in onto the steel extension rods. The soil sampler with retaining sample liner is connected to the leading edge of the extension rods. Some soil sampling systems have foot pedals attached to the rods, which allow the operator to step down to push the DPT sampler into the ground. Sampling depth can be increased using small hand-held augers to drill down to the target depth. In soft soils, maximum depth of manual DPT sampling is approximately 10 to 15 ft, in hard to moderately hard soils, depth of sampling is approximately 2 to 8 ft. The depth range can be increased greatly using a narrow diameter augers. Specialized augers have been developed for sand, mud, and boggy soils. Benefits for the manual DPT sampling method are minimal set-up time, low costs, and minimum disturbance of the site. The depth of sampling is the limiting factor of the manual DPT method and the level of physical effort is large. The body of a manual soil sampler ranges from about 1/2-in to 2-in diameter and 6 to 4 ft in length. For environmental sampling projects clear plastic, stainless steel, and brass liners are commonly used with these DPT samplers. Hand-held or portable electric, hydraulic, or pneumatic roto-hammers or jack-hammers can be added to the manual DPT sampling system to extend the sampling depth. Reversing the direction of hammering on the slide hammer can provide enough force to extract the sampler and rods. As the depth of sampling increases, the side friction on the samplers and possibly sampling extension rods increases. For removal of the samplers at greater depth, manual probe rod jacks supply approximately 2000 to 4500 lbs of lift capacity needed for extraction. Hand-held DPT equipment can be used to perform sampling at an angle as well as horizontal sampling.

DPT Probe and CPT Rigs

DPT, also called drive point sampling, is a technique that uses soil, vapor, or water samplers, which are driven into the subsurface without the rotary action associated with hollow stem auger rigs. DPT, including cone penetration testing (CPT) rigs, use the static weight of the vehicle to push the sampling rods into the ground. CPT rigs use a 20-ton truck and are capable of sampling to depths of 250 ft.

CPT rigs, originally developed for use in the geotechnical field, typically push from the center of the truck.

Small, highly maneuverable direct push probes were developed in the late 1980s. The probes were placed on pick-up trucks and vans. Probe rigs generally push the rods from the back of the truck. A percussion hammer has been added to these probe units to enhance the depth of sampling. These smaller probes have lowered the cost of DPT sampling projects to depths approaching 60 ft. Truck mounted DPT probe rigs, with names such as Geoprobe®, Powerprobe®, Earthprobe®, and Strataprobe®, are typically hydraulically powered. The percussion/probing equipment pushes rods connected to small-diameter (0.8 to 3.0 in) samplers.

DPT Probe Soil Sampling

The soil samples are commonly collected in 2 to 5 ft long clear plastic (polyethylene or butyrate) liners contained within an outer sampler. The plastic liners are easily cut with a knife and are transparent for easy lithologic characterization. Brass, aluminum, stainless steel, or Teflon® liners are also available, depending on the sampler. After removal from the sampler, the soil liner containing is immediately capped on both ends with Teflon® tape, trimmed, and then capped with plastic caps. The samples are then labeled and placed in individual transparent, hermetically sealed sampling bags. The samples are then put in the appropriate refrigerated environment and shipped under chain-of-custody procedures to a state-certified laboratory.

Various DPT soil samplers, have been designed and manufactured by numerous companies. The main sampler types used in DPT projects include split-spoon samplers, open-tube samplers, piston samplers, and dual tube samplers. The split-spoon sampler consists of the sample barrel, which can be split in two along the length of the sampler to expose soil liners. The split-spoon sampler without sample liners is useful for lithologic logging where soil samples will not be collected for chemical analysis.

The open-tube sampler contains soil liners and has been designed for environmental sampling within the same borehole, providing that soil sloughing is minimal. Continuous coring with the open-tube sampler begins at the ground surface with the open-ended sampler. The open-tube sampler is reinserted back down the same borehole to obtain the next core. The open-tube sampler works well in stable soils with as medium- to fine-grained cohesive materials such as silty clay soils or sediments. The open-ended samplers are commonly 3/4 to 2 in diameter and 2 to 5 ft in length. The simplicity of the open-tube sampler allows for rapid coring. For discrete sampling in unstable soils, the piston sampler allows the user to drive the sampler to the selected sampling depth. The piston sampler is equipped with a piston assembly that locks into the cutting shoe and prevents soil from entering the sampler as it is driven in the existing borehole. After the sampler has reached the zone of interest, the piston is unlocked to allow the soils to push the piston as the sampler is advanced into the soil as shown in Figures 11.12.1 and 11.12.2.

Dual tube soil samplers were designed to prevent cross contamination when sampling through perched water tables. The dual tube sampler has an outer rod that is driven ahead of the inner sampler (see Figure 11.12.3). The outer rod provides a sealed hole from which soil samples can be recovered without the risk of cross contamination. The inner rod with sampler and outer casing are driven together to one sampling interval. The inner sampling rod is then retracted to retrieve the filled liner while the outer casing is kept in place to prevent sloughing. This procedure is then repeated to total depth. Although slower than open-tube sampling, dual tube is especially useful when sampling through sloughing soils and sediments.

CPT Soil Sampling

Many sample collection options exist for CPT rigs, each depending on the specific medium. Many of the CPT soil samplers resemble their hollow-stem drill counterparts in that a split-spoon cutting tool

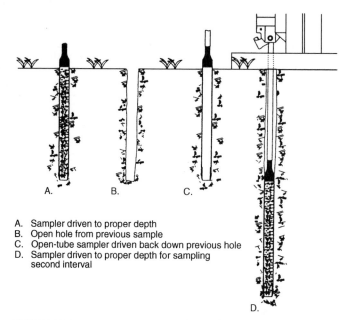

A. Sampler driven to proper depth
B. Open hole from previous sample
C. Open-tube sampler driven back down previous hole
D. Sampler driven to proper depth for sampling
second interval

FIGURE 11.12.1 Phases of open-tube soil sampling (after Geoprobe Systems, 1997).

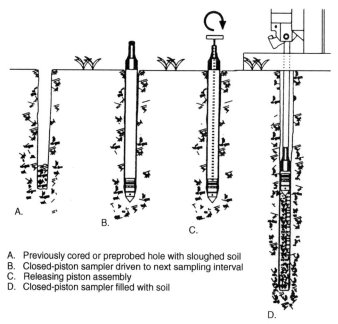

A. Previously cored or preprobed hole with sloughed soil
B. Closed-piston sampler driven to next sampling interval
C. Releasing piston assembly
D. Closed-piston sampler filled with soil

FIGURE 11.12.2 Phases of closed piston soil sampling (after Geoprobe Systems, 1997).

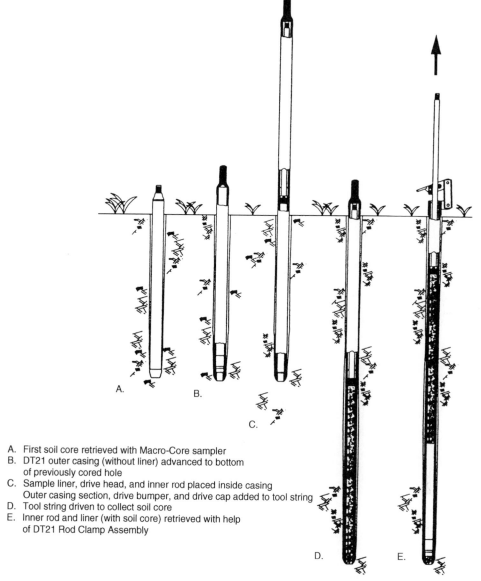

A. First soil core retrieved with Macro-Core sampler
B. DT21 outer casing (without liner) advanced to bottom
    of previously cored hole
C. Sample liner, drive head, and inner rod placed inside casing
    Outer casing section, drive bumper, and drive cap added to tool string
D. Tool string driven to collect soil core
E. Inner rod and liner (with soil core) retrieved with help
    of DT21 Rod Clamp Assembly

FIGURE 11.12.3   Phases of dual tube soil sampling (after Geoprobe Systems, 1997).

is used for collection (see Figure 11.12.4). Typically, a push rod with a closed tip (piston sampler) is advanced to a depth just above the desired sampling depth. The cutting tip of a CPT sampler is generally opened using a spring-loaded hinge system inside the sampling string. Once the hinge is triggered, the open cutting tube is advanced to the desired depth, filling the cylindrical sample chamber with soil. The former tip is retracted during collection. This technique tends to work best in dry soil with low gravel percentages. In some over-pressured sandy environments beneath the water table, sample collection can be difficult. If relatively finer-grained material is located below the sand of

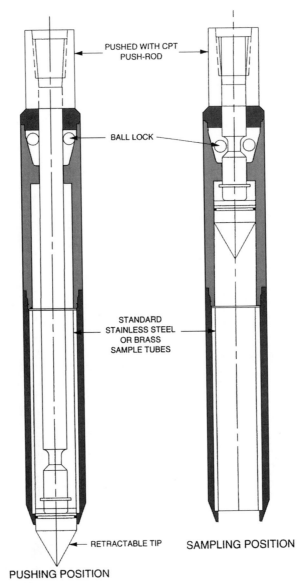

PUSHED WITH CPT
PUSH-ROD

BALL LOCK

STANDARD
STAINLESS STEEL
OR BRASS
SAMPLE TUBES

RETRACTABLE TIP   SAMPLING POSITION

PUSHING POSITION

FIGURE 11.12.4   CPT Soil sampling (after Edelman and Holguin, 1996).

interest, it is sometimes useful to space the sampling interval so that the finer material forms a plug at the bottom of the push stroke.

Vapor Sampling

A 1-in diameter hole is first drilled through the asphalt or concrete with a carbide tipped rotary drill. The soil vapor sampling system with a retractable or expendable point is then introduced into the hole

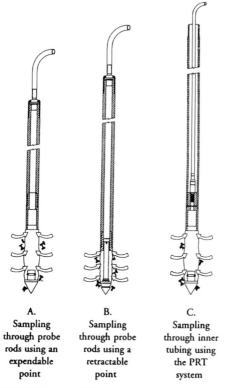

A.
Sampling
through probe
rods using an
expendable
point

B.
Sampling
through probe
rods using a
retractable
point

C.
Sampling
through inner
tubing using
the PRT
system

FIGURE 11.12.5   DPT Probe soil vapor sampler (after Geoprobe Systems, 1997).

and advanced with a direct drive system down to the specified sampling depth (see Figure 11.12.5). The retractable vapor sampling tip is then opened manually from the surface. The steel expendable tip is released in the subsurface. A length of 0.170 in inside diameter polyethylene tubing is then attached directly to the tip by means of a post-run tubing adapter. The vapor line stretches from the base of the vapor probe tip to the vacuum pump. On the surface, about two feet from the vacuum pump, a 1- to 2-in piece of Teflon® tubing is inserted by overlapping the two pieces of polyethylene tubing. The Teflon® section of tubing is the vapor sample location. The remaining end of the polyethylene tubing is then attached to the vacuum pump. The entire vapor line for each vapor point is constructed for each sample point prior to the operation of the vacuum pump. The vacuum pump provides 21 to 25 in of mercury to the system. The pressure gauge is noted prior to sample collection.

The soil vapor sample is collected by using a vacuum volume system pump, which is capable of a vacuum up to 21 in Hg (70 centibars). The pump is connected to the polyethylene tubing. Pumping is continued until at least three to five polyethylene tubing volumes are purged. A sterile, precleaned glass 50 cc gas tight syringe in the closed position is used as the sampling container to collect the vapor sample. The needle of the glass syringe is inserted into the Teflon® tubing section of the vapor line and the plunger of the syringe is pulled back to obtain the 50 cc volume of vapor sample. The sample is stored with the needle down. The samples are labeled and the sampling time is noted. The samples are transported under chain-of-custody procedures to the mobile on-site laboratory. A 1-liter stainless steel SUMMA canister is evacuated to create a vacuum inside. Connected on the vacuum line to the pump, the SUMMA canister can be opened to collect a soil gas sample. SUMMA canisters can keep air samples for approximately 28 days. A less expensive air sampling container is the clear plastic 1-liter Tedlar® bag. The Tedlar® bag can be placed in a 1-liter vacuum sampling box to collect a vapor sample.

After the vapor samples are collected, all vapor sampling equipment is removed from the borehole. The boreholes are then backfilled by tremie pipe with a neat cement grout. The sampling tip, rods, and equipment are decontaminated using a three bucket wash and rinse process approved by EPA. The wash includes a laboratory grade soap and the rinse consists of deionized water. New polyethylene tubing is used for each new sample.

Soil vapor samples by CPT rig can be collected by applying a small vacuum. Again, it is important to maintain low vacuum pressures to reduce the amount of liberation of volatile constituents. While some operators use a vacuum pump, some prefer to use a 100-cc glass syringe. The pumping system is typically connected to either a tip on the inside of the leading rod or to a screened section within the lead rod using small diameter Teflon® tubing. In practice, the push rods are advanced to the desired depth before attaching the Teflon® tubing. If the tip is to be connected to the tubing, the probe is usually retracted a few inches to allow for vapor to diffuse into the space just below the tip. A small segment of surgical-grade tubing is placed between the Teflon® line and the source of vacuum. Once the target depth is reached, a small vacuum is applied from the surface. After waiting a few minutes for the soil to at least partially equilibrate, a sample is generally collected from the surgical tubing using a hypodermic needle with a 10-cc syringe. For systems equipped with analytical equipment for rapid analysis, operators can inject the gas sample into the analytical port. For systems that do not have field analytical equipment, SUMMA canisters or Tedlar® bags may be necessary for temporary sample storage.

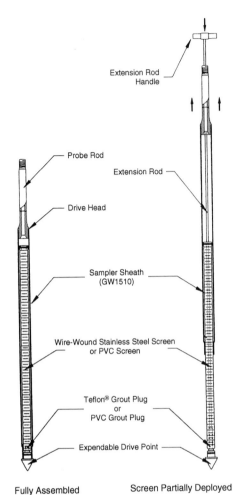

Extension Rod
Handle

Probe Rod

Extension Rod

Drive Head

Sampler Sheath
(GW1510)

Wire-Wound Stainless Steel Screen
or PVC Screen

Teflon® Grout Plug
or
PVC Grout Plug

Expendable Drive Point

Fully Assembled   Screen Partially Deployed

FIGURE 11.12.6 DPT Probe groundwater sampler (after Geoprobe Systems, 1997).

DPT Probe Water Sampling

Various types of sealed samplers are available for DPT groundwater sampling. Many DPT probe water samplers use a retractable or expendable drive point. After driving to the zone of interest, the outer casing is raised from the borehole, exposing the underlying well screen. For a nondiscrete groundwater sample, the outer casing contains open-slots. The open-slotted tool is driven from groundsurface into the water table. Groundwater is collected using an inner tubing or smaller diameter bailer inserted into the center of the open-slotted water sampler (see Figure 11.12.6).

CPT Water Sampling

Groundwater samplers typically work by either pumping or mass displacement (Figure 11.12.7). The ConeSipper® water sampler consists of a screened lower chamber and an upper collection chamber. Two small-diameter Teflon® tubes connect the upper collection chamber to a control panel in the truck. The truck ballast and hydraulic rams are used to push the sampler to a predetermined depth. While the probe is being advanced, nitrogen under relatively high pressure is supplied to the collection chamber via the pressure/vacuum Teflon® tube in order to purge the collection chamber and prevent groundwater from entering the chamber before the probe reaches the desired depth. Once the desired depth is reached, the pressurized nitrogen is shut off, excess nitrogen pressure is removed from the 100-ml upper collection chamber, and the chamber fills with groundwater. Finally, nitrogen under relatively low pressure is supplied to the collection chamber to gently displace the water and slowly push the water to the surface through the small-diameter Teflon® sampling tube. Slow sampling and low pressures are critical for sampling water containing volatile constituents (Kram, 1993). Sample collection times range from 20 min to 2 h, depending on the soil type adjacent to the sampling port. The groundwater samples are placed directly into 40-ml volatile organic analysis (VOA) containers. The VOA vials are labeled, logged, and placed into a chilled cooler pending delivery to an analytical laboratory.

11.12.3 DPT SAMPLING APPROACH

Once the site delineation goals have been met, small diameter wells can be installed in optimal monitoring locations. Well clustering, where short-screen wells are advanced to different depths within adjacent map locations, have recently been used to generate a three-dimensional understanding of critical site parameters in plume delineation efforts (Kram and Lory, 1998). Direct-push wells are typically installed without filter packs. However, robust filter pack materials specifically designed for rapid customization of direct-push wells have recently become available.

11.12.4 DPT RAPID SITE CHARACTERIZATION

Direct-push methods are being used in numerous research and development projects. A tracer experiment exploring the potential for natural attenuation of dissolved Methyl Tertiary Butyl Ether (MTBE)

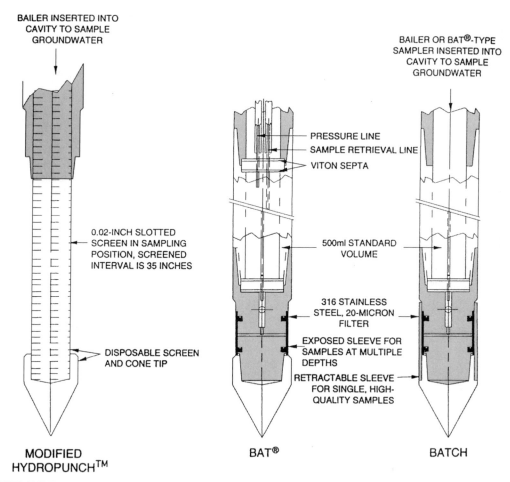

FIGURE 11.12.7   Various CPT groundwater samplers (after Edelman and Holguin, 1996).

is being conducted at a hydrocarbon national test site in Southern California (NFESC, 1999a, 1999b). Direct-push wells were used to establish the monitoring network because they are less disruptive to the flow regime than conventional wells. Fluid extraction requirements for these micro-wells are typically much lower than what would be necessary for development and sampling of larger diameter wells. The tracers were also introduced to the aquifer using the direct-push wells as conduits. Oxygen, nitrate, iron species, sulfate, carbon dioxide, MTBE, and additional fuel constituents can rapidly be monitored using field applications. In addition, prior to installation of the wells, a direct-push system was used to conduct aquifer tests to determine the three-dimensional distribution of hydraulic conductivity. No wells were required for this undertaking because the direct-push tool was advanced to the depths of interest for each slug test. A test to determine the feasibility of enhancing the rate of bioremediation is being conducted at the same site by introducing a bacterial culture. Elaborate test cells were set up using direct-push wells. Each of the three test cells measures approximately 20 by 40 ft and consists of a dozen monitoring wells and numerous nutrient injection ports (also installed with direct-push applications). In addition, the cultures were introduced into the aquifer at precise locations using a direct-push rig setup to a grouting device.

While direct-push methods are best known for their ability to rapidly detect chemicals, collect confirmation samples, or measure aquifer properties, they can be used in concert with telecommunication

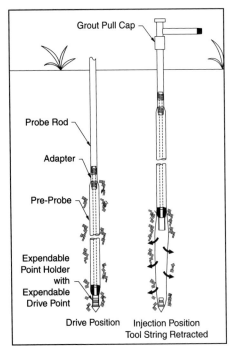

FIGURE 11.12.8 DPT Probe injection of chemicals or grouts for in situ remediation of soils and groundwater (after or grouts Geoprobe Systems, 1997).

and mapping technologies to expedite site characterization efforts and allow for real-time decision-making. A combination of innovative water sampling, rapid turnaround chemical analyses, and near real-time plume mapping using a global positioning system were used to delineate the extent of an MTBE plume to 35 parts per billion (ppb) concentration levels and install longitudinal and sentry wells in the most appropriate locations in an expedient fashion (Kram and Lory, 1998). During the 15 field days, it was determined that the dissolved MTBE plume extends approximately 4100 ft (1250 m) in length, approximately 500 ft (150 m) in width through the widest segment, and approximately 33 acres (133,551 m^2) in map view. While in the field, operators were in communication with regulators and customers via facsimile and telephone, engaging them in the field decision-making process. As a consequence, the client and regulators agreed upon a site closure strategy less than 1 month after completion of the field efforts.

11.12.5   IN SITU REMEDIATION TECHNIQUES

For remediation purposes, DPT equipment, which includes both probe and CPT rigs, can provide a closely-spaced delivery system for in situ remediation of soils and water. DPT remediation techniques include electrolysis, electro-osmosis, bio-venting, sparge point injection, enzyme injection, enhanced soil vapor extraction and steam stripping (Jacobs and Loo, 1994). Properly designed cased boreholes can be used to deliver gas or liquids, including oxygen, nutrients, steam/hot air, liquid oxidants, and electrolytes. DPT extraction points can be used to recover vapors or liquids. Faster and more effective remediations may require two or more processes that work simultaneously or sequentially. Although not yet applied on a large scale, direct-push wells have been used as fluid and vapor conduits for remediation systems (see Figure 11.12.8). At a hydrocarbon national test site in southern California, air sparging and vapor extraction wells were successfully installed using direct-push techniques. The monitoring wells were customized to allow for vapor and fluid sample collection at 1-foot vertical increments.

11.12.6   WELL POINTS

Vapor monitoring probes or implants, mini-groundwater wells, sparge points, well points, or piezometers can be installed with DPT rigs (see Figure 11.12.9). Oversized well points ranging in diameter from 2 to 4 in can also be installed. A sand pack enhances screened intervals, and probes can be designed and installed with a premade filter pack. The probes are sealed with bentonite and cement.

11.12.7   CLEAN-UP ALTERNATIVES

The application of DPT on the following in situ remedial technologies provide cost-effective clean-up alternatives:

- Chemical injection
- Bioremediation

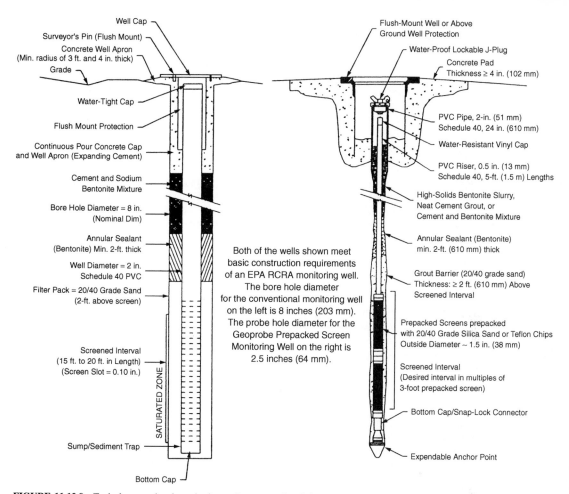

FIGURE 11.12.9 Typical conventional monitoring well cross section (left) and DPT probe pre-packed screen well-point cross section (right) (after Geoprobe Systems, 1997).

- Soil vapor venting
- Electrokinetic treatment

Chemical Injection

DPT rigs have been used to inject various chemicals into the subsurface for in situ remediation. Chemical oxidants, used to provide oxygen for bioremediation or to chemically oxidize petroleum hydrocarbons or chlorinated solvents, have been injected into the ground using DPT equipment. Other chemicals have been injected into the groundwater using DPT rigs to make toxic metals less soluble.

Bioremediation

Bioremediation of soil and groundwater is a cleanup technology that uses the ability of microorganisms to degrade hazardous organic compounds into nonhazardous compounds such as carbon dioxide, water, and biomass. Bioremediation can be utilized in an oxygen rich (aerobic) environment or in an oxygen deficient (anaerobic) environment. The microorganisms responsible for degrading hazardous compounds are generally naturally occurring. The most common and widely accepted approach to bioremediation is the enhancement of naturally occurring microorganisms. Because some organic chemicals, such as petroleum hydrocarbons, are biodegradable, cleanup of organic compounds from soils or groundwater can be accomplished by optimizing the environmental conditions to favor biodegradation. Cased DPT boreholes are a cost effective way to deliver oxygen or nutrients into soil and groundwater.

Enhanced Soil Vapor Extraction

The DPT soil vapor extraction system (VES) technique involves the extraction of volatile organic chemicals (VOCs) from the subsurface by the creation of a vacuum in the vadose zone using an air blower connected to a DPT vadose zone well. A VES can be used as an effective treatment of VOCs contamination in the soil but is limited in the groundwater. A well-engineered VES should include the following enhancement processes:

- Vacuum pressure increase
- Hot air circulation into the soil
- Reduced humidity in the air and moisture in the soil
- Increased air flow in the soil
- Desorption of VOCs from clayey and silty soil

DPT can help VES in the delivery of hot dry air and induced vacuum in the subsurface soil for effective VOCs treatment, as well as removal of vapors and liquids (Jacobs and Loo, 1994).

Electrokinetic Processes

Electrolysis and electro-osmosis are documented electrokinetic processes. The mechanics of the electro-osmosis process are to cause an imbalance of charge bonds in clayey material, which results in clay compaction and chemical desorption. DPT rigs can be used to install the electrodes. Well points can be used in the delivery of electrolytes into the subsurface soil and groundwater for effective treatment (Loo, 1994).

NOTES

Geoprobe® is a Registered Trademark of Kejr Engineering, Inc.; PowerProbe™ is a Registered Trademark of AMS, Inc.; Earthprobe® is Registered Trademark of SIMCO Drilling Co.; Teflon® is a Registered Trademark of E.I. du Pont de Nemours & Company; ConeSipper® is a Registered Trademark of Vertek Division of Applied Research Associates.

REFERENCES

Edelman, S. H., and Holguin, A. R., "Cone Penetrometer Testing for Characterization and Sampling of Soil and Groundwater," *Sampling Environmental Media*, ASTM STP 1282, James Howard Morgan, ed., American Society for Testing and Materials, Philadelphia, PA, 1996.

Geoprobe Systems, *Tools and Equipment Catalog*, Kejr Engineering, Inc. Salina, KS, 1997, 1998–99.

Jacobs, J. A., and Loo, W. W., "Direct Push Technology Methods for Site Evaluation and In-Situ Remediations," *Hydrovisions*, July/August, Groundwater Resources Association of California, 1994.

Kram, M. L., "Free Product Recovery: Mobility Limitations and Improved Approaches," *NFESC Information Bulletin #IB-123*, October 1993.

Kram, M. L., and Lory, E., "Use of SCAPS Suite of Tools to Rapidly Delineate a Large MTBE Plume," *Conference Proceedings for the Annual Meeting of the Environmental and Engineering Geophysical Society*, March 22–26, 1998, Chicago, Illinois, pp. 85–99, 1998.

Loo, W. W., *Electrokinetic Enhanced Bioventing of Gasoline in Clayey Soil: A Case History*, Superfund Conference, Washington, DC, 1994.

NFESC, *Natural Attenuation of MTBE in an Anaerobic Groundwater Plume*, TDS-2068-ENV, February 1999, 1999a.

NFESC, *In-Situ Bioremediation of Methyl Tertiary Butyl Ether (MTBE)*, TDS-2069-ENV, February 1999, 1999b.

CHAPTER 11
ASSESSMENT SAMPLING AND MONITORING

SECTION 11.13

BOREHOLE GEOPHYSICS

W. Scott Keys

Mr. Keys is president of GEOKEYS, Inc., a consulting firm with expertise in the application of geophysical logs to contamination problems caused by radioactive, industrial, municipal, and mining wastes.

11.13.1 INTRODUCTION

In recent years the main purposes of most environmental investigations of the subsurface has been to determine the effects of contaminants on the groundwater system; to identify acceptable sites; to locate and identify contaminants; to identify migration pathways; to predict movement; and to guide remediation programs in both the saturated and unsaturated materials. Regardless of the source of pollutants—landfill, industrial site, military base or injection well or pond—it has been necessary to drill test holes and monitoring wells to carry out these studies. Drilling and sampling these wells is expensive, and borehole geophysics provides techniques essential for obtaining much more information from such wells at less cost than sampling. Because of length limitations this chapter can only supply basic information; more detailed discussion is available in Keys (1989 and 1997).

Borehole geophysics includes all methods for making continuous or point measurements down a drill hole. These measurements are made by moving probes that transmit data to the surface electronically. The data are recorded in digital or analog format as a function of depth. These measurements are related to the physical and chemical properties of the rocks surrounding the borehole, of the fluid in the borehole, and the construction of the well or a combination of these factors.

The main objective of borehole geophysics is to obtain more information about the subsurface than can be obtained from conventional drilling, sampling, and testing. Logs may be interpreted for the following: lithology, thickness, and continuity of aquifers and confining beds; their permeability, porosity, bulk density, resistivity, moisture content, and specific yield; and the source, movement, chemical, and physical characteristics of groundwater and the integrity of well construction. Geophysical logs are repeatable over a long period of time, and comparable, even when measured with different equipment. This feature provides the basis for measuring changes in the geometry or concentration of a contaminant plume with time. Changes in an aquifer matrix, such as in porosity by plugging, or changes in water quality, such as in salinity or temperature, may also be recorded. Thus, logs may be used to establish baseline aquifer characteristics.

Most borehole geophysical tools investigate a volume of rock many times larger than the core or cuttings commonly extracted from a borehole. Some probes obtain data from rock a number of feet beyond the face of the borehole and beyond that altered by the drilling process. Logs can be analyzed in real time at the well site, and so can be used to guide well completion or testing procedures. Logs also enable the lateral and vertical extrapolation of geologic and water sample data or hydraulic test data from wells. Data from geophysical logs are used in the development of digital models of aquifer systems and of contaminant plumes, and in the design of disposal or remediation systems.

To maximize results from logs, at least one core hole for calibration should be drilled at each environmental site and the core should be analyzed in a laboratory. If coring the entire interval of interest is too expensive, then depth intervals for coring and laboratory analysis can be selected on the basis of geophysical logs of a nearby hole. Laboratory analysis of core is essential either for direct calibration of logs or for checking calibration carried out by other means.

Correct interpretation of logs must be based on a thorough understanding of the operating principles of each logging system. Because of the synergistic nature of logs, they should be interpreted as a suite, using computer methods if digitized logs are available, with the aid of all background data available in the area. The best approach to determining lithology and developing stratigraphic correlation models is to use several different types of logs that respond to different rock parameters. Figure 11.13.1 is a computer-generated example of the typical response of eight different types of logs to a hypothetical set of rock types varying from sedimentary to igneous. Note how the various log responses can provide a unique solution to the identification problem. The very useful computer program Viewlog (Keys, 1992) was used to make this plot (modified from Keys, 1989).

11.13.2 LOGGING EQUIPMENT

A complete description of the theory and principles of operation of logging equipment for both logging operators and log analysts and is covered in much more detail than is possible here in Keys (1989, pp. 65–75). The maximum benefit is derived from borehole geophysics, where operators and log analysts work together in the logging vehicle to select the most effective adjustments for each log.

Environmental logging equipment is usually mounted in a small truck or a van but portable logging equipment has become quite popular for shallow boreholes. Portable, or suitcase loggers, which can be easily carried and shipped as rentals, are almost all digital. Log display is on a notebook computer and a permanent record is obtained from a printer. Most modern loggers incorporate digital control and recording equipment. Although digital recording has become standard and is highly recommended, an analog record or display is essential because of the need to study the log as it is being made. In addition to use in computer interpretation, digitizing of logs onsite has several other benefits. Because the digital record contains all the raw data before it is sent to the analog recorder, replotting with corrections immediately after logging is possible.

Logging probes are usually made of noncorroding materials such as stainless steel, rubber, or plastic but lead electrodes are used in electric logging sondes. Most logging cable is armored with steel wires but some is rubber covered. Both cable and probes must be thoroughly washed between drill holes to avoid cross contamination.

Nuclear logging probes that measure radiation, such as gamma, gamma-gamma, and neutron, employ crystal detectors or gas-filled tubes. Acoustic or sonic properties are measured with magneto-restrictive transducers or piezoelectric crystals. Induction tools employ transmitting and receiving coils. Caliper logging is done with moving arms, and flow logging is done with impellers or a variety of tracers, such as warm water. Logging of the temperature and electrical conductivity of the fluid column is accomplished with thermistors or platinum sensors and silver plated elecrodes.

Most probes employ downhole electronics and thus may be susceptible to temperature drift and changes in output with time. For this reason, all probes should be calibrated periodically in laboratory type facilities and probe response must be checked at the well with field standards if quantitative data are required. Quality control is the responsibility of both the engineers and hydrologists who use the data and the company making the logs (for more detail, see in Keys 1997, pp. 41–49).

11.13.3 ENVIRONMENTAL APPLICATIONS OF BOREHOLE GEOPHYSICS

Most types of geophysical logs, except the fluid logs, provide information related to lithology and can be used for stratigraphic correlation. In contrast to core samples, logs may provide data on rock less

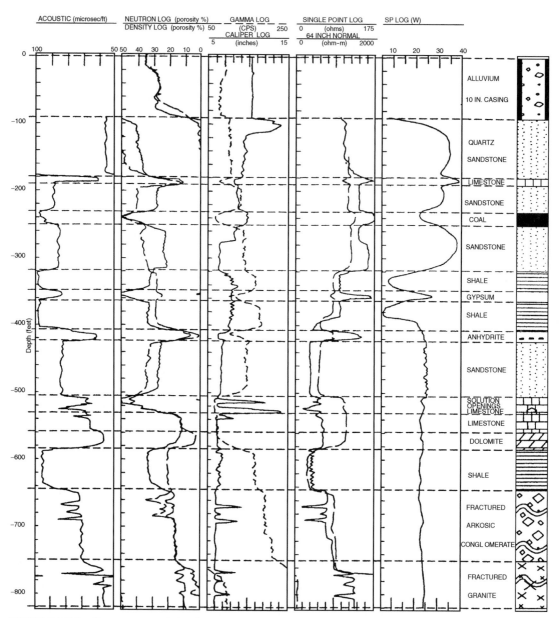

FIGURE 11.13.1 Computer generated plot of the typical response of eight different types of logs to various rock types (modified from Keys, 1989).

disturbed by drilling processes. Certain types of resistivity and fluid logs may be used to locate and identify contaminants in the rocks surrounding monitoring wells. These logs can be used to guide the placement of screens at the correct depths for sampling and monitoring. Fluid logs also can be used to measure borehole flow and the distribution of hydraulic conductivity. Fractures and other secondary permeability features may be identified and characterized by certain acoustic and borehole imaging logs. Another important use of logs is monitoring contamination, which is less costly than monitoring with water samples.

Logs have unique advantages over samples for monitoring in that they provide a continuous vertical profile of data rather than point values and they can provide data through unperforated casing. If test holes or monitoring wells are cased with steel, only temperature logs, nuclear logs, and possibly acoustic logs, will provide information through casing and annular materials. If casing is nonconductive, such as PVC or fiberglass, induction logs can add useful data to this suite. A wide variety of logs can be run in uncased boreholes but the usefulness of each may be limited by borehole diameter, type of rock or soil penetrated, and the salinity of borehole and interstitial fluids. Industrial and petroleum brine injection wells must be logged in order to ensure correct location of screens and grout and periodic logs may be necessary to inspect for leaks and corrosion (Keys and Brown, 1973).

A considerable amount of geohydrologic data must be obtained before a site can be selected and designed for disposal or storage of waste. Much of this same information can be used to design monitoring systems or to plan remediation. The first step is to locate any geophysical logs that might already be available in the area. These logs can be used to obtain early information on the geohydrologic system and to learn which logs work well in the area.

The most economic method of combining coring and geophysical logging in a site investigation is to drill a hole for geophysical logging in a location where most rock types in the area will be intersected. The well can be cored for the full depth or the geophysical logs used to select intervals to be cored in a second hole drilled nearby. This well, and subsequent wells, should be logged open hole (uncased) with a comprehensive suite of logs selected to provide data on the desired parameters and to use core analyses to calibrate log response. Casing may be installed later in poorly consolidated materials.

Background information on the groundwater system can be obtained from logs of the same test holes that are used for lithologic information. Several logs can provide information on the location of the water table and perched water bodies through casing. If the well is left uncased or if screens are installed, flowmeter and temperature logs may provide the following: information on the relative heads in different aquifers; estimates of permeability; and the areal distribution of groundwater temperature and electrical conductivity.

Gamma logs are the most widely used for the identification of lithology and for stratigraphic correlation because they provide useful data under the greatest range of borehole conditions and for a wide range of rock types. Because gamma logs do not have a response uniquely related to lithology they must be used in conjunction with other logs or sample data until the gamma response of the various rock types in each new area is established. Gamma logs do not measure lithology directly, instead they measure the amount of radioactive isotopes that occur in the rocks, but do not identify those isotopes. Unless man-made radioisotopes have migrated from waste disposal sites, the amount of naturally occurring radioisotopes is usually related to lithology.

Various kinds of resistivity and induction logs are also quite useful for lithologic purposes. The standard resistivity probes require conductive fluid in an uncased well and some are affected by the conductivity of that fluid. Induction logging devices provide reliable measurements in air-filled or PVC cased boreholes and are little affected by borehole fluids.

Neither porosity nor permeability are measured directly by logging devices but several types of probes can be calibrated in terms of porosity, and the relative magnitude of permeability may be inferred from other logging techniques. A technique for calculating permeability from neutron and gamma logs is described by Morin and Hess (1991). Neutron, gamma-gamma, and acoustic velocity logs can all provide quantitative data applicable to calculating porosity when properly calibrated and corrected for borehole effects. The most accurate data is obtained from these probes in uncased, small-diameter boreholes but they can also be used in cased holes under the proper conditions.

The logs mentioned above are related to total porosity but the response of resistivity logs is related to effective porosity and to the resistivity of the saturating fluid. The formation resistivity factor defines the relation between porosity and resistivity. Using porosity derived from a neutron, gamma-gamma or acoustic velocity log, and a deep investigating resistivity log, one can determine the resistivity of the formation water in granular sediments as described by Keys (1997, p. 61).

Fractures frequently provide pathways for migration of contaminants from waste disposal sites, even in poorly consolidated rocks. Many types of logs, such as caliper and single-point resistance,

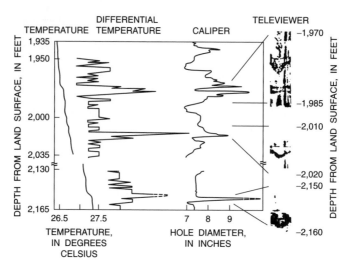

FIGURE 11.13.2 Temperature, differential temperature, caliper, and acoustic televiewer logs of a test well in North Carolina showing permeable fractures (Keys, 1990).

show a response to the presence of fractures but the log deflections are neither unique, nor diagnostic. Acoustic (borehole) televiewer and borehole television logs are the most useful for accurate location of fractures, determining their strike and dip and apparent aperture. Flow data from a heat pulse flowmeter, high resolution temperature logs or hydrophysical logging (Pedler et al., 1992) are needed in order to determine the relative hydraulic conductivity of fractures.

Figure 11.13.2 shows temperature, differential temperature, caliper and acoustic televiewer logs at a site in North Carolina, which was being investigated for the possible storage of radioactive waste. From the anomalies on the temperature and differential temperature logs, it is obvious that water is moving through permeable fractures evident on the caliper and televiewer logs.

Well Construction

It is common to find that monitoring or injection wells at environmental sites are not constructed as designed or that vertical pathways have developed through the annular seals. Geophysical logs frequently show that gravel pack, bentonite seals, and grout are not located where they should be or are incomplete so that the water sample analyses will give erroneous results. The most reliable way to verify the location of seals and gravel pack is by gamma-gamma logging after the well is completed. Both injection and monitoring wells, which are improperly constructed or that have deteriorated, provide vertical pathways for poor quality water to contaminate fresh water aquifers and indicate false values for water samples obtained from screened intervals. Gamma-gamma density logs have proven to be very effective in identifying backfill materials in cased wells. The significant density differences between grout, bentonite slurry, bentonite chips, and collapsed formation materials allow each to be readily identified through various kinds of casing by gamma-gamma logs. Screens can be located and the condition of casing determined by borehole television, acoustic televiewer, or electromagnetic logging techniques.

Contaminant Distribution

Although regulations require the analysis of periodic water samples from monitoring wells, it is an expensive procedure. In Wilson, Everett, and Cullen (1995), the statement is made that, "as much as

80 percent of the cost of site characterization during cleanup of a Superfund site can be attributed to laboratory analyses, many of which render nondetectable results." Geophysical logs can be analyzed in the field to guide the location and frequency of sampling and thus may reduce the number of samples needed and costs.

Temperature and fluid resistivity logs can locate contaminant plumes where the fluid in the borehole accurately represents the interstitial fluid. Induction logs are very useful for monitoring water quality in the rocks through PVC or fiberglass casing and screens. In cased wells above the water table neutron, gamma-gamma, induction, and temperature logs can locate perched water bodies that might be contaminated. Gamma spectral data can identify radioisotopes behind casing and neutron activation analysis and specific ion electrodes have the potential for identifying nonradioactive contaminants.

Using porosity derived from gamma-gamma, neutron, or acoustic velocity logs and true resistivity from a deep investigating resistivity log, one can determine the resistivity of the interstitial water in granular sediments using the relationship between formation resistivity factor and porosity (Keys, 1997). Formation factor is usually consistent within a given depositional basin or aquifer. At the Hardage Superfund site, this technique was applied to water in a silty sandstone bed and the calculated interstital water resistivity was later corroborated by water samples obtained after packers were installed to isolate the bed (Keys, 1993).

Remediation

Geophysical logs have played an important role in planning remediation at Superfund sites and have provided evidence in law suits as described in case histories. Geophysical logs, including old oil well logs, were an essential tool in understanding the movement of contaminants and the regional geology and water quality at considerable depth under the Hardage Superfund site in Oklahoma, which was the subject of a landmark law suit. Logs have also been very useful in planning remediation at a number of contaminated landfill sites, such as the Fresh Kills Landfill in New York and contaminated sites on military bases, such as Loring Air Force Base in Maine. At the Fresh Kills site, statistical log analysis methods available in the computer program, Viewlog, were very useful in developing a technique for recognizing the various interfingered glacial-fluvial sedimentary units. At Loring Air Force Base, acoustic televiewer logs provided the basis for understanding the complex folding and fracturing that controls waste migration.

11.13.4 *A BRIEF DESCRIPTION OF THE MOST USED LOGS*

Table 11.13.1 is a much simplified guide for selecting the best logs that will provide the information needed in the type of boreholes available. It is essential to consult more detailed references before the final selection of logs is made.

Gamma Logs

Gamma logs are a measure of the total gamma radiation from uranium, thorium, and potassium present in most rocks, but man-made radiosotopes that may be present as contamination are also detected. Gamma spectral logging equipment is required to identify which radioisotopes are present. In an average of 200 shale samples from various localities in the United States uranium contributed 47 percent of the total radioactivity, thorium 34 percent, and potassium 19 percent. Gamma logs can be run under almost any borehole conditions and through any type of casing, but results are poor in very large holes and through several strings of steel casing. Figure 11.13.3 shows the relative radioactivity of some common rock types (Keys, 1989).

TABLE 11.13.1 A Simplified Table to Aid Selection of the Most Widely Used Logs

Type of log	Potential uses	Borehole conditions required
Gamma	Lithology, correlation	Most conditions
Gamma-gamma	Bulk density, porosity, moisture, lithology, correlation, well construction	Wide range
Neutron	Saturated porosity, moisture, lithology, correlation	Wide range
Acoustic velocity	Saturated porosity, lithology, correlation, fractures, bonding of casing	Uncased, fluid filled
Normal and focused resistivity	Lithology, correlation, water quality	Uncased, fluid filled
Single point resistance	Lithology, correlation	Uncased, fluid filled
Induction	Water quality, lithology, correlation, monitoring interstitial fluid conductivity	Uncased or cased with PVC or fiber-Glass, air or fluid
Spontaneous potential	Lithology, correlation, water quality	Uncased, fluid filled
Caliper	Borehole diameter, log correction, well construction, fractures, lithology	Most conditions
Acoustic televiewer	Character and orientation of fractures, solution openings and bedding	Uncased, fluid filled, water or mud
Borehole television	Well construction, secondary porosity, lost objects	Uncased or cased, air or clear fluid
Temperature	Fluid movement, identifying plumes, calculation of specific conductance	Fluid filled
Fluid conductivity	Identifying plumes, fluid movement, calculation of specific conductance	Fluid filled
Flow	Fluid movement, permeability, leakage behind casing	Fluid filled
Well construction	Numerous techniques for checking casing and completion materials, leakage, borehole deviation etc.	All conditions

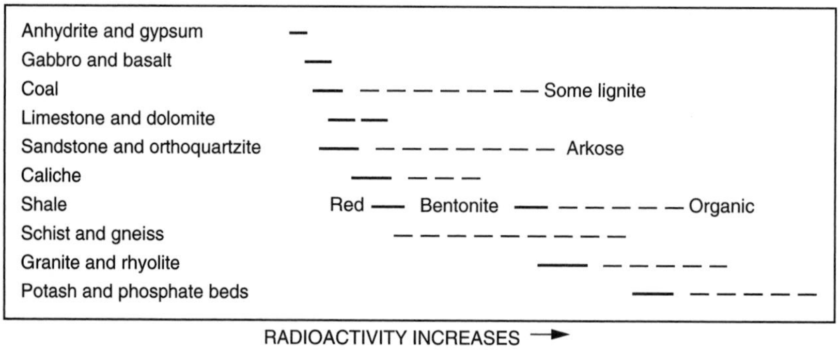

FIGURE 11.13.3 The relative radioactivity of some common rocks (Keys, 1990).

Gamma-gamma Logs

These logs, also called density logs, use a gamma emitting radioactive source and measure the radiation that is backscattered and attenuated by the rocks surrounding a borehole. The number of gamma photons detected is inversely proportional to the electron density of the rocks, which is approximately proportional to bulk density for most rocks. Porosity can be calculated from bulk density if fluid density and mineral grain density are known. Very high correlation coefficients are obtained between log data and core analyses where boreholes are fairly smooth and not too large. Washouts, as seen on a caliper log, must be eliminated from quantitative calculations, because borehole diameter errors are significant. Gamma-gamma logs can also be calibrated in terms of moisture content and excellent correlation with core analyses has been obtained. The logs can be run in cased holes and above and below the water level, however, borehole construction must be the same as the calibration environment. Gamma-gamma logs are most effective in low-density, high-porosity rocks.

Neutron Logs

There are several different types of neutron logging tools but the most common uses a neutron source of several curies and spacing between source and detector greater than 15 in. Light elements, such as hydrogen, are most effective in slowing and capturing neutrons. Therefore, hydrogenous compounds, such as water, cause most of the log response and the logs are called porosity logs. If the rocks are not saturated, the logs may be calibrated in terms of moisture content. Conventionally neutron porosity logs are plotted with count rate increasing to the right, so that porosity increases to the left. Hydrocarbons, such as coal, and water of crystallization, such as that found in gypsum, are not distinguished from free water by a neutron log and therefore can cause erroneous log interpretation. Neutron logs must be calibrated in an environment that approximates the borehole conditions and rock types expected. Neutron logs are most effective in high-density, low-porosity rocks. Neutron activation to identify nonradioactive isotopes can be carried out in boreholes using relatively large sources or neutron generators.

Acoustic Velocity Logs

These logs are also called sonic logs or transit-time logs and are the third type of log that can be used to measure porosity. Acoustic energy in the frequency range of 10 to 35 kHz is transmitted through the borehole fluid and the adjacent rocks and the arrival of the compressional wave is detected at several transducers in the probe. The compressional wave arrivals usually are converted to transit times in microseconds per foot. The transit times (delta t) of most rock matrices are known so the log data can be converted to porosity using a chart or the time average equation: porosity = (delta t log − delta t matrix)/(delta t fluid − delta t matrix). The tool does not provide accurate data at porosities above about 35 percent. All acoustic logs require fluid in the hole and only provide useful information in casing when it is well-bonded to the formation by grout.

Resistivity Logs

There are too many different types of resistivity probes to be described in detail here. Multielectrode resistivity logs employ four or more current and potential electrodes to produce a log that can be interpreted quantitatively in ohm-meters. The various electrode configurations are designed to investigate rocks at different distances from the borehole wall to measure the effects of invasion. Some probes focus the current to avoid the effects of borehole fluid and mud cake. Single-point logs, commonly used in groundwater investigations, cannot be interpreted quantitatively and must be scaled in ohms. Almost all resistivity sondes require conductive fluid in the borehole and cannot be used in cased holes. The conversion of field logs to true resistivity requires the use of departure curves supplied by the logging company. These curves correct for such factors as borehole temperature, diameter, and mud resistivity.

Induction Logs

Induction logs are widely used in the environmental field because they can be used to measure the electrical conductivity or resistivity of rocks surrounding a borehole that does not contain fluid or is cased with PVC of fiberglass. Furthermore, they have the advantages of relatively deep penetration and demonstrate little effect from borehole diameter changes or the fluid or air filling it. Numerous examples have demonstrated excellent correlation between induction logs made before and after the installation of casing. Thus, they are an excellent tool for monitoring the quality of interstitial water at environmental sites where small diameter holes are drilled and cased with PVC. There are a number of types of induction probes that have different depths of investigation based on the number and spacing of transmitting and receiving coils.

Spontaneous Potential Logs

The SP log is usually run along with the resistivity logs described above, at no extra charge. The log is a record of voltages that exist at contacts between different rock types, such as shale and sandstone, where they are intersected by a fluid-filled borehole. In most fresh water environments, the SP log is very difficult to interpret and may provide no useful information. SP log response is based on differences between the salinity of the borehole fluid and the interstitial fluid in the rocks. In many fresh water wells, these salinities are the same so there is no SP response. To further complicate interpretation of the SP logs, changes in the salinity and temperature of the fluid column with time will produce markedly different SP logs, so that additional fluid logs are required for their interpretation.

Caliper Logs

Motor opened arms measure the average borehole diameter. Because borehole diameter affects many different types of logs, it is important to measure such changes. Few boreholes are smooth and changes diameter can be caused by both lithology and drilling techniques. Borehole diameter information is also used for estimating amounts of gravel pack or grout that will be required.

Borehole Imaging Logs

There are two major types of tools for recording images of the borehole wall: the acoustic televiewer (ATV) and borehole television (TV). The televiewer transmits a high frequency acoustic signal from a rotating transducer, which is reflected off the borehole wall. The continuous record is oriented on magnetic north and clearly shows fractures, solution openings, and bedding planes. The strike and dip of these features can be calculated and computer programs are available to make these and other calculations. Borehole television employs a rotating mirror, or an axial camera, to give different views of the borehole in color or black and white. The newer TV systems digitize the signal for enhanced signal processing and computer interpretation of the orientation of linear features. ATV systems have the advantage that they can provide data with mud in the hole or mudcake on the borehole wall, but they do require fluid in the hole to transmit the signal. In contrast, TV systems require that the borehole fluid be clear and the walls be clean, but TV will work above the fluid level.

Temperature and Fluid Conductivity Logs

Thermal logs provide a continuous record of the temperature of the borehole fluid, some systems provide a resolution and accuracy of .01°C. Fluid conductivity or resistivity tools use a four-electrode system to record electrical conductivity in Millimhos or resistivity in Ohm-meters. Logging tools are available that simultaneously measure the temperature and electrical character of the fluid column. Fluid logs always should be run before any other logging tools disturb the borehole fluid. They should always be recorded with the probe or probes moving down the hole, and it might be worthwhile to record at several different times to observe changes in the fluid column. Changes in the temperature or conductivity of the fluid column can indicate the location of permeable intervals that are accepting or producing water and may be interpreted in terms of the relative volumes that are moving. Character of the fluid column can indicate the presence of contamination. Temperature and fluid column conductivity also are needed to correct resistivity logs described above to obtain true rock resistivity.

Flow Logs

There are too many different types of flow logs to be described here but most have a similar purpose, to measure the vertical flow in wells. Flowmeters to measure horizontal flow have been developed but have seen little use to date. The most widely used borehole flowmeter employs an impeller, which is turned by vertical flow. This type of tool is relatively cheap to run but has a major shortcoming in a lack of sensitivity to low velocity flow. Flowmeters that employ some type of tracer are much more sensitive to the low velocity flow that can be important at environmental sites. Chemical or radioactive tracers have been used but are usually not permitted by regulation. Heated water is used in a heat pulse flowmeter to measure very low vertical velocities. The heat pulse flowmeter can help unravel the complexities of flow between a series of boreholes through a complex fracture system.

Hydrophysical logging employs deionized water as a tracer. The deionized water is injected into a well so that it will flow into permeable beds or fractures (Pedler et al., 1992). A fluid conductivity probe is then used to record periodic logs, which indicate the location of the deionized water and possible contaminated water. These periodic logs are run while the deionized water and interstitial water enter the borehole under natural head conditions or while pumping the well. The data obtained can be analyzed to determine permeability.

Well Construction Logs

There are many different types of logs that can be used to check for the proper installation and integrity of casing, screens, and backfill materials; to plan cementing operations, hydraulic testing and sampling; and to aid in the interpretation of other logs. In the environmental field, monitoring well construction commonly fails to meet contract specifications. Resistivity logs should show the location of steel casing and screens, however, gamma-gamma logs are more effective for this purpose and can be used to locate a second string of casing through the inside string. Gamma-gamma logs are also effective for identifying and locating various backfill materials outside of steel or plastic casing, such as cement grout, bentonite, gravel pack, and various rock types as well as air or water-filled voids. These materials are identified based on their different bulk densities.

The acoustic televiewer and borehole television are excellent for locating junk lost in a well and guiding recovery methods. Both of these tools can provide the excellent resolution necessary to identify casing corrosion or screens plugged by precipitated minerals, and to provide very accurate location of joints, screens, and damaged pipe. Tools that employ a magnetometer for orientation will not provide oriented data in steel-cased boreholes.

The casing collar locator is an inexpensive probe for locating pipe joints or welds and screens or perforations. Electromagnetic casing inspection logs can be used to locate corroded casing and measure changes in casing thickness with time.

11.13.5   CASE HISTORIES

Keys (1997) provides many case histories in which he describes the application of geophysical well logs to the solution of problems at a number of environmental sites. A few of those applications will be very briefly described below.

The Idaho National Engineering Laboratory

Fifty-two research reactors have been constructed over the years and many have been decommissioned at this site in southern Idaho. Chemical and radioactive waste from these reactors and processing plants has been discharged to ponds and wells. Wastewater was injected directly into the aquifer through wells and waste from ponds percolated to the water table at a depth of 200 to 1000 ft. The site is underlain by interbedded basalt flows and sediments, which constitute a major aquifer in Idaho. Borehole geophysics, used extensively at the site since 1957, has provided an understanding of the aquifer system, located waste plumes and monitored changes in their size and concentration. Several hundred wells ranging in depth from 200 to 10,000 ft have been logged with a comprehensive suite of logging tools.

Early logging consisted of gamma and caliper to develop an understanding of the aquifer geometry. Caliper logs distinguished basalt flow tops, which consisted of scoria and cinders, and were very permeable. Gamma logs identified clay-rich sedimentary interbeds that had a low permeability. These logs enabled the first correlation of flows and permeable zones by Jones (1961). About this same time, gamma-gamma and gamma spectral logs were used experimentally at the site. Schmalz and Keys (1962) describe the use of a neutron moisture meter to map the distribution of moisture above the water table near a disposal pond. They showed that the moisture distribution was similar to that of radioisotopes. In this same year, fluid temperature and electrical conductivity logs were used to map the location of large plumes of contaminated water and sources of groundwater recharge. Plumes of contaminated water from injection wells and ponds eventually coalesced and extended many miles down gradient within the site. Neutron logs were used to locate perched water behind steel casing, which was migrating through the annular space to the water table. More recently, improved gamma spectral equipment allowed the identification of radioisotopes in the large diameter wells. The acoustic televiewer was used to obtain data on the orientation and character of open fractures in the basalt.

Spinner type flowmeters were used early to study the movement of water between the various aquifers because all early wells were left open to most of the aquifer system. These open wells contributed to the vertical migration of contaminated water. Later, the heat pulse flowmeter combined with an inflatable packer was used to refine these measurements and combined with pumping tests and along with head measurements used to calculate the vertical distribution of hydraulic conductivity (Morin et al., 1993).

The Hardage Site

An extensive suite of geophysical logs was used at this Superfund site, southwest of Norman, Oklahoma, as evidence in a landmark lawsuit (Keys, 1993). The federal suit was filed by EPA to force the principal responsible parties to implement the EPA remedy. That remedy called for excavation and stabilization of the industrial waste and placement in an on-site RCRA vault. This plan was apparently based on the theory that open fractures were common below the site; that low permeability rocks were absent or not laterally continuous; and that fresh water existed to significant depths below the site. Borehole geophysics was used to refute all of these theories and the case was decided in favor of the defendant's plan to use an in situ remedy with several methods to block the migration of contaminants. That remedy has now been successfully implemented.

Nearly 50 drill holes were completed and logged at the site and 8 were cored extensively. Acoustic televiewer and television logs were run in the cored holes and they provided no evidence of open fractures. Resistivity and nuclear logs demonstrated that thick, very low permeability clay-rich rocks below the site had lateral continuity. Logs of 43 oil wells in the area indicated that many continuous, low permeability, beds were present to depths of several thousand feet. Quantitative analysis of resistivity and gamma-gamma logs of holes drilled at the site showed that saline water was present at depths of 200 to 300 ft below the site. SP and resistivity logs of the oil wells also showed that water in the deeper rocks had a salinity greater than 10,000 mg/L equivalent sodium chloride. Thus, the geophysical logs, along with data from extensive hydraulic testing, strongly supported the use of an in situ remedy.

Loring Air Force Base

This base, located in northern Maine, is underlain by folded and fractured limestones and slates partly covered by till (Keys, 1997). Contaminants from hazardous materials used at Loring Air Force Base (LAFB) have migrated to the shallow water table and to a depth of 300 ft in the bedrock aquifer. A comprehensive suite of geophysical logs was run in more than 100 wells between 1988 and 1995. The purposes of logging were: to identify water-bearing fractures; to determine the physical properties of the bedrock aquifer; to determine borehole flow regimes; to guide the placement of packers for testing and sampling; and to optimize the placement of screens in monitoring wells.

Dearborn, Baker, and Davis (1989) published a report on interpretation of logs at one source area at LAFB. They used logs to select depths for ports in a multilevel testing and sampling program. They determined that the dipping structure of limestone beds was the chief control for the deep migration of hydrocarbon contaminants along a complex system of bedding plane fractures. Resistivity and caliper logs located the open fracture zones and fluid temperature and fluid resistivity logs provided evidence of fluid movement in those zones. Dearborn, Calkin, and Andolsek (1996) describe a site-specific relationship between heat-pulse flowmeter measurements and vertical head gradients determined from static packer tests at LAFB. They conclude that in fractured rocks, borehole flowmeter data can be used instead of more expensive and time-consuming packer tests to determine the direction and magnitude of vertical head gradients in open boreholes.

Acoustic televiewer logs of 35 boreholes provided the most useful data on rock structure, which is a major control of groundwater movement. Data from many of the televiewer logs was digitized and a computer program developed by Barton and Zoback (1989) was used to plot fracture sinusoids and to calculate and list the strike, dip, and apparent fracture aperture. That program was also used to plot rose diagrams of fracture orientation and tadpole diagrams of fracture location and orientation.

Cape Cod

Borehole geophysics has been very useful at several sites on Cape Cod, Massachusetts, to develop an understanding of the migration of waste in unconsolidated sand and gravel. At Otis Air Force Base, Morin, Leblanc, and Teasdale (1988) carried out a study using logs to determine the amount of disturbance near boreholes caused by three different methods of drilling and installing casing. The wells were completed with hollow stem auger, where the hole was allowed to collapse against the casing; by mud rotary, where the same type of casing was installed after all cuttings were flushed out; and the same type of casing was driven by a 300 lb hammer and the material inside was washed out. A statistical analysis of neutron and gamma logs was used to perform a standard analysis-of-variance progression. This analysis was designed to eliminate all variables except formation disturbance caused by well completion techniques. Augering clearly caused the greatest disturbance, and driven casing caused the least disturbance. Driving casing is more expensive and time-consuming, and augering provided adequate logs unless thin bed resolution is needed.

Another study near Otis Air Force Base comparing hydraulic conductivity (K) calculated from spinner flowmeter data with K measured on core samples. The two methods agreed well enough and agreed with tracer tests so that Hess, Wolf, and Celia (1992) concluded that the flowmeter provided a fast and accurate method for measuring small scale variability in K. Morin and Hess (1991) used gamma logs to estimate grain size distribution and neutron logs to estimate porosity and substituted these values in the Kozeny-Carmen equation to predict K. The predicted mean K using flowmeter and neutron and gamma logs agreed quite well. The reports mentioned above must be consulted for a detailed explanation of these techniques. In another study near Provincetown, Morin and Urish (1993) used temperature and induction logs, in addition to gamma and neutron logs, to locate a contaminant plume. In a further study on Cape Cod, DeSimone and Barlow (1993) concluded that induction logging was an efficient and cost-effective method for monitoring a plume of treated septage effluent in glacial outwash. They stated that the complex geometry and gradients at this site would have made it difficult to define the plume with water samples alone.

REFERENCES

Barton, C. A., and Zoback, M. D., "Utilization of Interactive Analysis of Digital Borehole Televiewer Data for Studies of Macroscopic Fracturing," *Proceedings of the Third International Symposium on Borehole Geophysics for Minerals, Geotechnical and Groundwater Applications*, sponsored by the Minerals and Geotechnical Logging Society, Houston, Texas, Paper GG, pp. 623–653, 1989.

Dearborn, L. L., Baker, P. S., and Davis, J. B., "Preliminary Interpretation of Geophysical Logs and Insitu Hydrologic Properties in Fractured Limestone at Loring Air Force Base," *Proceedings of the Third International Symposium on Borehole Geophysics for Minerals, Geotechnical and Groundwater Applications*, sponsored by the Minerals and Geotechnical Logging Society, Houston, Texas, paper FF, pp. 595–622, 1989.

Dearborn, L. L., Calkin, S. F., and Andlsek, H., Comparison of Heat-pulse Flow Measurements and Vertical Gradients in a Fractured Limestone Aquifer, *Proceedings of a Symposium on the Application of Geophysics to Engineering and Environmental Problems*, Environmental and Engineering Geophysical Society, 1996.

DeSimone, L. A., and Barlow, P. M., "Borehole Induction Logging for Delineation of a Septage-effluent Contaminant Plume in Glacial Outwash, Cape Cod, Massachusetts," *Proceedings of the Fifth International Symposium on Geophysics for Minerals, Geotechnical and Economic Applications*, Minerals and Geotechnical Logging Society, Houston, Texas, 1993.

Hess, K. M., Wolf, S. H., and Celia, M. A., "Large Scale Natural Tracer Test in Sand and Gravel, Cape Cod, Massachusetts, Part 3. Hydraulic Conductivity Variability and Calculated Macrodispersivities," *Water Resources Research*, 28(8): 2011–2027, 1992.

Jones, P. H., *Hydrology of Waste Disposal*, National Reactor Testing Station, Idaho, Issued by U.S. Atomic Energy Commission Technical Information Service, Oak Ridge, Tenn., U.S. Geological Survey IDO 22042, 1961.

Keys, W. S., *Borehole Geophysics Applied to Ground-water Investigations: National Water Well Association*, p. 313, 1989; Also published as *Techniques of Water Resources Investigations of the United States Geological Survey*, Book 2, Chapter E2, p. 150, 1990.

Keys, W. S., *Viewlog for Geophysical Well Log Analysis*, Software review in COGS letter of the Computer Oriented Geological Society, p. 3, May 1992.

Keys, W. S., "The Role of Borehole Geophysics in a Superfund Lawsuit," *Proceedings of the Fifth International Symposium on Geophysics for Minerals*, Geotechnical and Economic Applications, Minerals and Geotechnical Logging Society, p. W1–20, 1993.

Keys, W. S., *A Practical Guide to Borehole Geophysics in Environmental Investigation*, CRC Lewis Publishers, 1997.

Keys, W. S., and Brown, R. F., "Role of Borehole Geophysics in Underground Waste Storage and Artificial Recharge," in *Transactions, Symposium on Underground Waste Management and Artificial Recharge*, Braunstein, J., ed., Vol. 1, pp. 147–191, 1973.

Morin, R. H., LeBlanc, D. R., and Teasdale, W. E., "A Statistical Evaluation of Formation Disturbance Produced by Well-casing Installation Methods," *Ground Water*, 26(2): 207–217, 1988.

Morin, R. H., and Hess, K. M., "Preliminary Determination of Hydraulic Conductivity in a Sand and Gravel Aquifer, Cape Cod, Massachusetts, From Analysis of Nuclear Logs," *Proceedings of the Technical Meeting*, Monterey, California, U.S. Geological Survey—Toxic substances hydrology program, 1991.

Morin, R. H., Barrash, W., Paillet, F. L., and Taylor, T. A., "Geophysical Logging Studies in the Snake River Plain Aquifer at the Idaho National Engineering Laboratory—Wells 44, 45, and 46," U.S. Geological Survey, *Water Resources Investigations Report 92-4184*, 1993.

Morin, R. H., and Urish, D. W., "Hydrostratigraphic Characterization of a Coastal Aquifer by Geophysical Log Analysis, Cape Cod National Seashore, Massachusetts," *Proceedings of the Fifth International Symposium on Geophysics for Minerals*, Geotechnical and Economic Applications, Minerals and Geotechnical Logging Society, Houston, Texas, p. M1–12, 1993; also *The Log Analyst*, Houston, Texas, 36(4): 27–37, 1995.

Pedler, W. H., Head, C. L., and Williams, L. L., "Hydrophysical Logging: a New Wellbore Technology for Hydrogeologic and Contaminant Characterization of Aquifers," *Proceedings of National Groundwater Association 6th National Outdoor Action Conference*, pp. 1701–1715, 1992.

Schmalz, B. L., and Keys, W. S., Retention and Migration of Radioactive Isotopes in the Lithosphere at the National Reactor Testing Station, Idaho, International Conference on Retention and Migration of Radioactive Isotopes in Soil, Paris, France, U.S. Atomic Energy Commission Report No. I.D.O. 12026, Idaho Falls, Idaho, 1962.

Wilson, L.G., Everett, L. G., and Cullen, S. J., eds. *Handbook of Vadose Zone Characterization and Monitoring*, Lewis Publishers, Chapters 18 and 36, Boca Raton, FL, 1995.

CHAPTER 11
ASSESSMENT SAMPLING AND MONITORING

SECTION 11.14

SURFACE GEOPHYSICAL METHODS FOR SITE CHARACTERIZATION

Richard C. Benson
Mr. Benson is well known for his application and integration of unique measurement techniques to solve complex geologic and hydrologic site characterization issues. He formed Technos, Inc., in 1971, which has become a world class leader in the field of site characterization and monitoring.

Lynn Yuhr
Ms. Yuhr is vice president and general manager of Technos, Inc. She has helped develop the firm's unique capability in geologic and hydrologic site characterization.

11.14.1 ORGANIZATION

This chapter provides an overview of the surface geophysical methods that can be applied to environmental site characterization. The chapter begins with an overview of the site characterization process, which provides some insights to the problems encountered. We then use the Expedited Site Characterization (ESC) strategy developed by DOE as a model to illustrate the strategy and components of an effective site characterization. Here we can see the critical role surface and borehole geophysics play in an effective site characterization. An introduction to surface geophysical methods and their applications is followed by a discussion of ten commonly used surface geophysical methods. A series of examples illustrate the applications and results of geophysical surveys, followed by a discussion on selecting methods and other considerations when applying surface geophysics.

11.14.2 THE SITE CHARACTERIZATION PROBLEM

Site characterization is the process of understanding the geologic framework, engineering and hydrologic properties, and contaminant distribution of a site. It is the cornerstone of all environmental projects. However, the lack of understanding the site geology and hydrogeology is often responsible for the failures that have occurred in environmental work.

The single largest uncertainty in determining the success of remediating environmental contamination problems results from an inability to locate, describe, and quantify the natural geological, hydrological and geochemical heterogeneity (Olhoeft, 1994; Yuhr et al., 1996). These variabilities, or heterogeneity, are due to the inherent properties of geologic materials along with the processes that create them and continue to modify them.

Because hydrogeology controls the migration of contaminants, and or provides traps to confine them, it is critical to understand these controlling factors. Only then will subsequent phases of a project (i.e., modeling, risk assessment, or remediation) be based on a solid data and proceed with much greater confidence and minimizing uncertainties.

Most investigations do little to understand the hydrogeologic setting and its details but immediately focus upon sampling contaminants. For years, the approach to characterization of hazardous waste sites has focused upon "chasing the plume" (Sara, 1994). This approach included one round of field investigation after another, and resulted in the installation of an excessive number of monitoring wells. The results have been less than acceptable (Sara, 1994).

Problems with the current state of the practice in site characterization for environmental projects are abundant. As early as 1986, problems with past work were being acknowledged and documented. The editorial *What Has Gone Wrong* by Freeze and Cherry (1989) presented an assessment of the environmental industry after a decade of experience. Two key issues cited are *a lack of understanding or appreciating the impact of geology* and *politics outweighing science*. Shuirman and Slosson (1992) identify *incompetent professional work* and *educational shortcomings* in the geotechnical field.

Geologic conditions affecting groundwater and contaminant flow may range from small and subtle changes, such as local partially cemented soil zones, local increases in silt or clay content, or fractures in unconsolidated material, to larger buried channels, fractures, faults, or conduits within bedrock. Such features are often undetected, and overlooked because of their subtle nature. These inhomogeneities may have serious consequences on hydrogeological, geotechnical, and chemical data without our knowledge.

The following section describes the general strategies for carrying out an effective site characterization. The Department of Energy's Expedited Site Characterization (ESC) model is used because it is relatively complete and well documented.

11.14.3 THE EXPEDITED SITE CHARACTERIZATION (ESC) PROCESS

Unlike previous site characterization practices, the Expedited Site Characterization (ESC) strategy developed by DOE provides a means for a major improvement in the status quo of site characterization. Expedited Site Characterization (ESC) is a strategy for identifying, collecting, and integrating the information necessary to form a basis for any phase of site remediation. It can be applied to site characterization, risk assessment, modeling, remediation, design, monitoring system design, and site evaluation for closure.

Some of the key components of the ESC process include a small (2 to 6 persons) but experienced multidisciplinary *core team* of hands-on professionals who are in the field managing, acquiring, and interpreting data at a hands-on level. The core team remains the same throughout the entire site characterization program providing critical continuity. The use of a *dynamic work plan* allows field work to be easily modified as needed, rather than a rigid plan that cannot be easily checked.

Phase I effort is *a focus upon characterization of the geology and hydrology* to identify potential contaminant pathways and traps. Emphasis is placed upon understanding of the site geology and hydrology before beginning sampling of contaminant distribution. ESC emphasizes the use of noninvasive and minimally invasive technologies, such as remote sensing, geophysics, and push technology, while the number of borings and monitoring wells are minimized. Phase II concentrates upon delineating the distribution and concentration of vadose and groundwater contamination. *On-site laboratories* (or off-site labs with turn around of less than 48 h) are used for rapid quantitative chemical analysis.

Another important component is the development of a *conceptual model* of site conditions, which includes geologic, hydrologic, and contaminant data from the site. Data are processed and interpreted within hours or days so that the results can be used to update the conceptual model and to guide further work. Multiple independent sources of data are used to test and verify the hypothesis in the

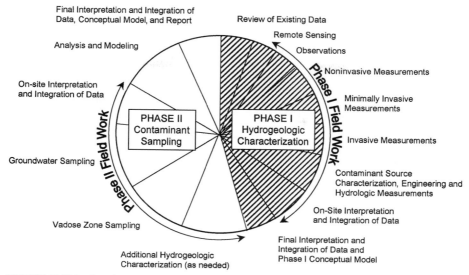

FIGURE 11.14.1   Generic components of an ESC program. Note that most of the effort focuses upon field work to provide site specific data (*Source:* Benson, 1999).

conceptual model while steady converging toward a sufficiently complete and accurate model of site conditions.

The general components of a Phase I and Phase II ESC program are shown in Figure 11.14.1. Note that more than half of the effort and budget are devoted to on-site efforts to acquire data. The strategy is one of obtaining appropriate, adequate, and accurate data during a Phase I and Phase II effort so that a reasonably accurate conceptual model of site conditions can be developed that permits risk assessment, modeling, and remedial decisions to be made without going back to the site.

The more data we have, the less we rely on assumptions and opinions (Figure 11.14.2). An image such as geologic cross section (a contour map of a contaminant plume) is the result of acquisition, processing, and interpreting data. To create a valid image (one that closely presents reality), one must have appropriate, accurate, and sufficiently adequate data.

The ESC method is summarized by Benson (1997) and is described in Benson et al. (1998). An ASTM guideline (ASTM 6235, 1997a) provides a complete description of the strategy. When the key elements of the ESC process are examined, we see that many of the ESC philosophies have been with us throughout much of the history of engineering geology.

Unlike previous historic EPA and DOE site characterization practices, ESC has focused upon the use of senior experienced professionals, and has re-identified hydrology and geology as critically important for site characterization. Early engineering geologists, appearing around the turn of the twentieth century, knew that their design rested on geologic site characterization.

Karl Terzaghi set the stage for site characterization philosophies, which are summarized by Terzaghi and Peck (1967). Approaches for geotechnical investigations have been compiled by Hvorslev (1949). More recently, approaches for environmental site investigation are presented in EPA's Subsurface Characterization and Monitoring Techniques (EPA, 1993a, 1993b). Sara (1994) provides a comprehensive review of site investigation methods applied to environmental issues. Fookes (1997) provides a

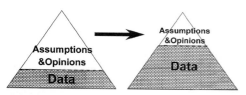

FIGURE 11.14.2   The objective of any site characterization should be the acquisition of data so that assumptions and opinions are minimized (*Source:* Benson, 1999).

contemporary summary in his lecture *Geology for Engineers: the Geological Model, Prediction and Performance.*

Beam et al. (1997) summarize many of the historic aspects that are part of the ESC strategy. ESC fits the long-term strategy of site characterization practiced by the authors for decades. Benson and Pasley (1984) have applied many of the principles of ESC in characterizing groundwater contaminants at 11 power plants in Florida during the early 1980s.

ESC combines these well known and proven concepts along with technological improvements within a strategic framework, which, if followed, will dramatically improve results while cutting costs. The key elements of ESC simply make good common sense, technically and economically. While the process was created to help DOE characterize its waste sites in a more timely and cost-effective manner, the process is equally applicable to the private sector. The overall approach and many of the technical methods are identical or very similar and can be adapted for any site characterization work, including geotechnical problems.

11.14.4   *ESC STRATEGY EMPHASIZES THE USE OF GEOPHYSICS*

In ESC, emphasis is placed on the use of noninvasive (remote sensing and geophysics) and minimally invasive (direct push cone penetrometers and percussion sampling) technologies, as appropriate. Surface geophysical methods are nonintrusive, can be made relatively quickly, and in some cases, total site coverage is economically possible. ESC proposes the use of multiple sources of data to characterize a site. This provides a reliable and defensible means of testing the hypothesis embedded in the conceptual site model. When measurements by different methods agree, we can have a high level of confidence in our interpretations.

11.14.5   *GEOPHYSICS AND ITS ROLE IN SITE CHARACTERIZATION*

What Is Geophysics

Geophysical methods are not new; they have been used for many decades in the exploration for deep oil and gas as well as mineral exploration. In fact, oil, gas, and mineral exploration could not exist as we know it without the technical and economic benefits of both surface and borehole geophysics. Unlike direct sampling and analysis, such as obtaining a soil sample for sieve analysis or a water sample for chemical analysis, the surface geophysical measurements provide nondestructive, in-situ measurements.

In contrast to the deeper oil, gas, and mineral exploration, shallow high resolution, engineering, geologic, hydrologic and environmental applications are significantly different. While some geotechnical and environmental data may extend to 1000 ft or more, most environmental and geologic investigations are limited to the upper 100 ft or so and often less. Nondestructive testing interests (i.e., pavements, concrete, and subbase) may be only in the upper foot or so.

The geophysical methods encompass a wide range of airborne, surface, and borehole methods. Geophysical measurements can be made from satellites, aircraft, on the surface and over water (Figure 11.14.3). Geophysical measurements are commonly made down boreholes, wells, or between boreholes, including tomography imagery (Figure 11.14.4).

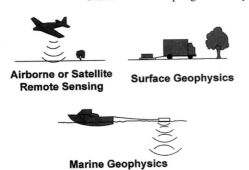

Airborne or Satellite Remote Sensing   **Surface Geophysics**

Marine Geophysics

FIGURE 11.14.3   There are several ways to obtain surface and geophysical data, including satellites, aircraft, on the ground and on water (*Source:* Benson and Yuhr, 1996).

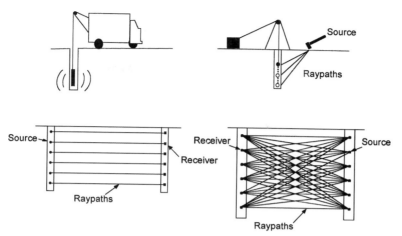

FIGURE 11.14.4   There are several ways to obtain downhole geophysical data, including logging of boreholes, surface to hole measurements, hole to hole measurements, and tomography to image conditions between two or more holes (*Source:* Benson, 1999).

Each geophysical method measures some physical, electrical, or chemical property of the soil, rock, and pore fluids or some combination of them (Table 11.14.1). Because each geophysical method measures a different parameter, the information from one set of measurements is often complemented by that from another. This synergistic use of multiple independent sources of data from surface geophysical methods, along with data form a variety of other sources serves to enhance our understanding of the site conditions. Those familiar with traditional geophysical well logging will recognize this concept, as multiple logs are commonly obtained to aid in the interpretation of subsurface conditions.

One of the primary factors affecting the accuracy of any site characterization effort is the limited number of sample points or borings, resulting in insufficient spatial sampling to adequately characterize the site. Achieving a reasonable statistical sampling of hydrogeologic site conditions can require borings or other methods of sampling placed in a close-order grid, which would reduce the site to "swiss cheese." The number of borings required to detect a burial site, tank, or permeable pathway can be calculated (Benson et al., 1982). Such calculations show that it is not uncommon to require hundreds or thousands of borings to achieve a 90 percent confidence level for detection, making the subsurface investigation like "looking for a needle in a haystack." Yet we commonly accept data from

TABLE 11.14.1   Commonly Used Surface Geophysical Methods

Method	Parameter measured (response)
Ground Penetrating Radar	Complex dielectric constant
Electromagnetic	Electrical conductivity/Resistivity
Resistivity	Electrical resistivity
SP (spontaneous potential)	Electrochemical and streaming potential
Seismic Refraction	Seismic velocity (density and velocity)
Seismic Reflection	Seismic velocity (density and velocity)
Surface Wave Analysis	Seismic velocity/dispersion of Raleigh waves
Magnetics	Magnetic susceptibility of rocks or permeability of metal
Metal Detector	Electrical conductivity of metal
Gravity	Density

a limited number of borings and other methods of sampling to characterize large areas and complex geologic settings.

Geophysical measurements can be made relatively quickly. Continuous data acquisition can be obtained with certain geophysical methods at speeds up to several miles per hour. In some cases, total site coverage is economically possible. Because of the greater sample density when using geophysical methods, anomalous conditions are more likely to be detected and background conditions defined, resulting in a more accurate characterization of subsurface conditions.

Geophysical methods, like any other means of measurement, have advantages and limitations. There is no single, universally applicable surface or borehole geophysical method to meet all site characterization needs. Furthermore, some methods are quite site-specific in their performance. Therefore, the user must be able to select the method or methods carefully and understand how they should be applied to specific site conditions to meet project requirements.

There is a surprising similarity between many medical measurement techniques and the geophysical methods. The wide range of tools available to a medical professional measure different physical, chemical, or electrical parameters of the body, such as x-ray, ultrasound, EKG, or CAT scan. Doctors use these noninvasive tools to collect a sufficient amount of data and insight on the patients internal conditions. The medical profession also has a group of minimally invasive probes to carry out investigation and sampling, similar to the various push tools (i.e., Geoprobe) and cone penetrometer probes used for environmental purposes. Along with blood tests, other physical testing, personal observations, and discussion with the patient provides the doctor with a diagnosis of the patient's condition prior to prescribing a medication or surgery.

The measurements used by the doctor are used early in the diagnostic process prior to deciding on a medication or surgery. Similarly, geophysical measurements to characterize the subsurface should be made prior to deciding upon locations for borings and trenches for geotechnical site characterizations, or for borings, soil, and water samples for environmental site characterization. While this strategy should be obvious, geophysics is still not used as much or as effectively as it should be.

Applications of the Geophysical Methods

There are four areas where the surface geophysical methods are commonly applied to environmental problems. They are as follows:

- Characterization of natural geologic and hydrogeologic conditions
- Detection and mapping of contaminant plumes, spills, and leaks
- Detection and mapping of landfills, trenches, buried wastes or the location of tanks, drums and pipes, or utilities
- Monitoring the effectiveness of the remediation process

In addition, the same methods are applied to geotechnical investigation for providing hydrogeologic characterization and engineering properties prior to building foundation, dams, bridges, among others, as well as nondestructive testing of structures.

Surface Geophysical Methods

This chapter presents 10 geophysical methods that are commonly used in site characterization. Each geophysical method responds to some physical, electrical, or chemical parameter of the soil, rock, and pore fluids or buried wastes (Table 11.14.1). In order that a geophysical method detect a change in geologic conditions or contaminants, there must be a sufficient contrast in the property being measured. For example, the contact between fresh water and saltwater can easily be detected by measuring electrical resistivity (or conductivity) with depth. Similarly, the contact between soil and massive unweathered rock can be determined by measuring seismic velocity. In both cases, a distinct contrast in properties between the two layers exists. On the other hand, if we are looking for a small change in freshwater quality or attempt to determine the top of a highly weathered rock whose properties are similar to the soil overlying it, we may not be successful because the contrast is likely to be small or the contact may not be distinct, but may be a gradual change over many tens of feet.

Because soil and rock vary widely in their physical properties (some by many orders of magnitude), one or more of these properties measured by geophysical methods will usually correspond to a geologic discontinuity, such as stratigraphic contact (i.e., soil and rock), or a structural contact (i.e., a fault).

While these ten methods are not the only ones that may be employed, they are regularly used and have been proven effective for groundwater, environmental, and waste site assessments. A brief description of each of these surface geophysical methods is presented in the following section. This is by no means a complete discussion of the geophysical methods. Therefore, references are provided to enable the reader to expand their knowledge of the subject.

General references include three volumes edited by Ward (1990), the U.S. Corps of Engineers geophysical manual (1979), and USGS applications of surface geophysics by Zohdy et al. (1974). Numerous geophysical texts are available including Telford et al. (1982). Burger (1992) provides an excellent text for students including Macintosh software. Short courses are available from Environmental Engineering Geological Society (EEGS), and equipment vendors provide demonstrations and equipment specific training.

Ground Penetrating Radar.   Ground penetrating radar is a reflection technique that uses high frequency electromagnetic (EM) waves (from less than 10 to 1500 MHz) to acquire subsurface information (Figure 11.14.5). The transmitter generates a short pulse of EM energy, which is radiated into the ground. Reflections of the radar wave occur whenever there is a sufficient change in dielectric constant and or electrical conductivity between two interfaces or materials. Dielectric constant and electrical conductivity are a function of soil and rock material and their pore fluids, which are associated with natural hydrogeologic conditions such as bedding, cementation, moisture, clay content, voids, and fractures. For example, an interface between two soil or rock layers, which has a sufficient contrast in electric properties will show up in the radar profile. Radar has also been used to detect contaminants, and is commonly used to locate buried materials, tanks, and utilities (Benson and Glaccum, 1979; Benson et al., 1982; Benson and Yuhr, 1987). The radar record is similar to the view we would get if we observed the cross section of soils in a trench or a cross section of rock at a road cut.

Figure 11.14.6 shows a radar cross section of a clean quartz sand over a clay loam. Note the level of detail that can be obtained in mapping the top of the clay loam. The numerous small hyperbola

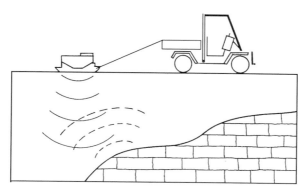

FIGURE 11.14.5   Illustration of radar measurements (*Source:* Technos, 1992).

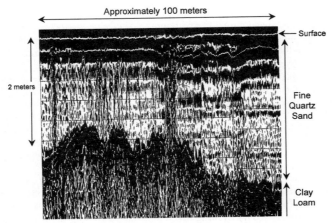

FIGURE 11.14.6   Radar data shows the detailed surface of a clay loam underlying quartz sand. The difference in soil properties show up clearly in the radar data (*Source:* Benson et al., 1982).

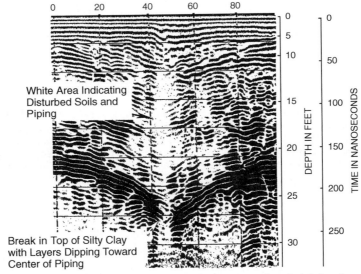

FIGURE 11.14.7   Radar data shows a small diameter (approximately 1–2 foot diameter) soil piping sinkhole developing. Dipping reflectors from soil layers indicate some peripheral subsidence while localized raveling zone of soil is seen as a discontinuity in soil strata. The subtle difference in soil properties clearly revealed this soil piping before it reached the surface (*Source:* Benson and Yuhr, 1987).

in the upper portion of the record, within the quartz sands, are due to tree roots or animal burrows. Radar can be applied to a variety of problems in the environmental field. Figures 11.14.6 and 11.14.7 show how radar can be applied to characterization of geologic conditions; Figures 11.14.8 and 11.14.9 illustrate applications and mapping contaminants; and Figure 11.14.10 shows radar data indicating an underground tank.

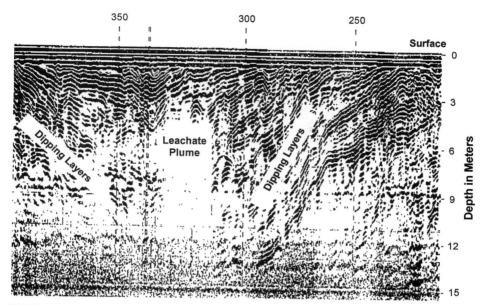

FIGURE 11.14.8 Radar data shows the boundaries of shallow leachate plume from a landfill. Radar shows the dipping strata on either side of a buried channel. The increased electrical conductivity of the shallow leachate plume shows up as a white area because the radar energy cannot penetrate this zone (*Source:* Benson and Yuhr, 1996).

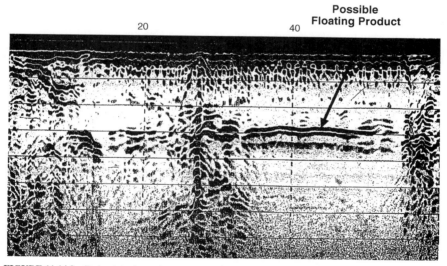

FIGURE 11.14.9 Radar data shows the presence and extent of gasoline floating on water. Organics such as gasoline, diesel, or aircraft fuel floating on groundwater will often depress the capillary zone creating a planar reflector for radar waves to be reflected from. Thin layers of product can be detected by radar only under ideal conditions (*Source:* Benson and Yuhr, 1996).

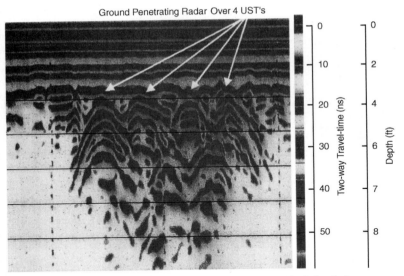

FIGURE 11.14.10   Radar can be used to locate buried fuel tanks. The fuel tank shows up as a hyperbola in the radar data, if it is crossed approximately perpendicular. The depth of the top of the tank can also be estimated (*Source:* Kaufmann and Benson, 1999).

The vertical scale of the radar profile is in units of time (nanoseconds). The time it takes for an electromagnetic wave to move down to a reflector and back to the surface (two-way travel time) is almost the speed of light. The time scale then is converted to depth by making measurements or assumptions about the velocity of the waves in the subsurface materials.

Radar measurements are relatively easy to make and do not require intrusive ground contact. The antenna may be pulled by hand or vehicle at traverse speeds from 0.5 to 5 mph or more, which can produce considerable data/unit time. In some cases, total site coverage is economically possible. While station measurements are a bit slower, they can be made in more difficult terrain (brush, steep hills, etc.) and have a slight advantage in terms of depth penetration. Although a variety of computer processing can be used to enhance the data, the necessary information can usually be obtained from the graphic picture-like display without any data processing.

Ground penetrating radar provides the highest lateral and vertical resolution of any surface geophysical method. Various frequency antennas (10 to 1500 MHz) can be selected so that the resulting data can be optimized to the project needs. Resolution ranges from a few inches to a foot or so depending upon the frequency used. Lower frequency provides greater penetration with less resolution. Higher frequencies provide higher resolution with less penetration. Horizontal resolution is determined by the distance between station measurements, or the sample rate and speed when towing the antenna.

Ground penetrating radar is a very powerful and flexible tool because it has a dynamic range capable of seeing to a 100 ft or more under ideal conditions, or detecting reinforcement steel in concrete a few inches below the surface. While radar penetration in soil and rock to more than 100 ft has been reported, penetration of 15 to 30 ft is more typical. In contrast, it is probably the most site-specific method in its performance, because the depth of penetration of the radar wave is highly site-specific. The method is limited in depth by attenuation because of the higher electrical conductivity of subsurface materials or scattering. Generally, radar penetration is greater in coarser, dry, sandy, or massive rock. Penetration in silts and clays and in materials having conductive pore fluids will drastically limit penetration to less than 1 to 3 ft or so. Yet, in some situations, useful results can be obtained in silts and clays (Benson and Yuhr, 1990; Kratochvil et al., 1992). Data can be obtained in saturated materials if the specific conductance of the pore fluid is sufficiently low. For example, radar has been applied to map the sediments in fresh water lakes and rivers. Annan (1992) provides an excellent introduction to GPR with examples.

The DC Electrical and Electromagnetic Methods.   The electromagnetic (EM) and DC resistivity methods are similar in the sense that they both measure the same parameter, but in different ways. Electrical conductivity (or resistivity) is a function of the type of soil and rock, of its porosity, and of the fluids that fill the pore spaces. Electrical conductivity values (milliSiemens/meter or millimhos/meter) are the reciprocal of resistivity values (ohm-meter); units of ohm-feet are also common.

Natural variations in subsurface conductivity (or resistivity) may be caused by changes in basic soil or rock types, thickness of soil and rock layers, moisture content, and depth to water table. Localized deposits of natural organics, clay, sand, gravel, or salt-rich zones will also affect subsurface conductivity (resistivity) values (McNeill, 1980a). Structural features, such as fractures or voids, can also produce changes in conductivity (or resistivity). Because the specific conductance of the fluids in the pore spaces can dominate the measurements, detection and mapping of contaminant plumes can often be accomplished using the electrical methods (McNeill, 1980b).

The values measured in the field are referred to as apparent conductivities or resistivities. If the ground were electrically homogeneous and isotropic, then the measured value of resistivity would be the true value. Because subsurface conditions are rarely, if ever, homogeneous and isotropic, this is rarely the case. The apparent resistivity or conductivity values measured are a weighted average of the formations through which the currents pass. The absolute values of conductivity (or resistivity) for geologic materials are not necessarily diagnostic in themselves, but their spatial variations, both laterally and with depth, can be significant. It is the identification of these spatial variations or anomalies that enable the electrical methods to rapidly find potential problem areas.

Both EM and resistivity measurements may be used to obtain data by "profiling" or "sounding." Profiling provides a means of mapping lateral changes in subsurface electrical conductivity (or resistivity) to a given depth. Profiling is well-suited to the delineation of hydrogeologic anomalies, mapping of contaminant plumes, and location of buried material. Profiling measurements are made to a fixed depth, and data are obtained at a number of stations along a survey line. The spacings between the measurements will depend upon the variability of the subsurface and upon the lateral resolution desired. At each station along the profile line, data may be obtained for one depth or a number of depths depending upon project requirements. It is useful to take at least two measurements, a shallow one and a deeper one, so that the influence of the highly variable shallow soils and cultural influences can be assessed. Profiling data are often used in a qualitative manner with little or no processing.

A sounding measurement provides a means of determining the vertical changes in electrical conductivity (or resistivity) correlating with soil and rock layers and or pore fluids. In this case, the instrument is located at one location and a number of measurements are made at increasing depths. Interpretation of a single sounding provides a one-dimensional model of the depth, thickness, and conductivity (or resistivity) of subsurface layers, which have different electrical conductivities (or resistivities). A number of soundings can be made to develop a geoelectric cross section.

Electromagnetics.   Two types of electromagnetic instrumentation are in common use: frequency-domain (sometimes referred to ground conductivity meters) and time-domain. Both systems induce currents into the ground by electromagnetic induction.

Frequency Domain Electromagnetics (FDEM).   The most common FDEM systems are the EM38, EM31, and EM34, by Geonics Ltd. An EM31 is shown in Figure 11.14.11. These EM instruments measure changes in magnitude and phase of the currents induced within the ground; the out-of-phase (or quad-phase) data is the electrical conductivity of the ground in millisiemens/meter and the in-phase data responds to the presence of metal in parts per thousand (McNeill, 1980b).

The FDEM instruments are primarily used for profiling to detect and map lateral changes in natural geologic and hydrogeologic conditions. Figure 11.14.12 illustrates the detection of fractures. FDEM measurements are commonly used to detect and map contaminant plumes and to locate and map buried wastes, metal drums and tanks, and metal utilities. Figure 11.14.13 shows a large inorganic plume, as

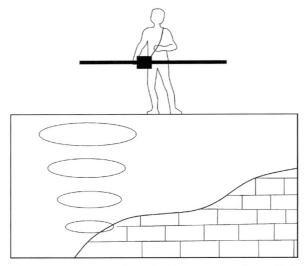

FIGURE 11.14.11   Illustration of EM31 measurements (*Source:* Technos, Inc., 1993).

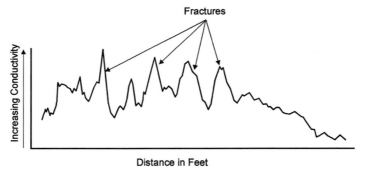

FIGURE 11.14.12   Electromagnetic data shows the presence of fractures in rock beneath 10 ft of alluvium. The fractures in gypsum rock are infilled with a clayey alluvium material and are moist because of occasional saturation and flowing water. This combination of materials presents a clear increase in electrical conductivity which is easily mapped with electromagnetics (*Source:* Benson, 1999).

well as geologic variations. The excellent lateral resolution obtained from continuous EM profiling data (Figure 11.14.13) has been used to outline closely-spaced burial pits, to reveal the migration of contaminants into the surrounding soils.

Depth of measurement is a function of the spacing between transmitter and receiver coil and coil orientation. The EM38, EM31, and EM34 provide measurements from 0.75 to 60 m deep. The instruments can be used with two coil orientations, vertical coil axis (or vertical dipole) and horizontal coil axis (horizontal dipole) to vary the depth of measurements.

EM measurements are relatively easy to make and do not require intrusive ground contact. The instruments may be carried by hand, and measurements can be made on a station by station basis. Continuous EM profiling data can be obtained from 2.5 ft to a depth of 50 ft (Benson et al., 1982) by

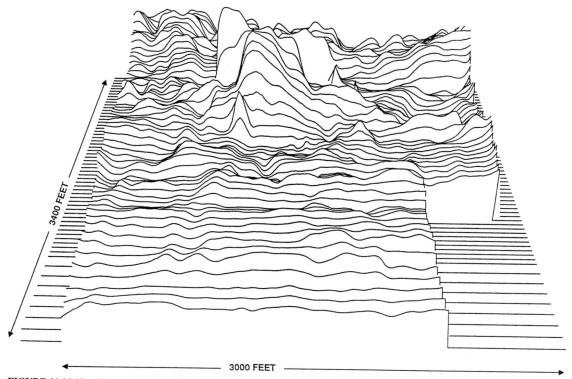

3400 FEET

◀——————————— **3000 FEET** ———————————▶

FIGURE 11.14.13   Continuous EM profile measurements show a large inorganic plume (center rear) and considerable natural geological variation (*Source:* Benson et al., 1982).

walking or from a vehicle at speeds from 0.5 to 5 MPH or more, which can produce considerable data per unit time. In some cases, total site coverage is economically possible. EM profiling data can be plotted as a profile line or contoured with little, if any, processing. Sounding data must be processed to obtain true conductivity values, along with the depth and thickness of layers.

Frequency domain electromagnetics provide excellent lateral resolution for profiling particularly when continuous measurements are used. Lateral resolution is determined by the spacing between the transmitter and receiver coils as well as the spacing between measurements. EM measurements have limited resolution for sounding because of the limited number sampling depths available. However, combinations of readings from instruments can be used to obtain some qualitative sounding data and in some cases can be used to resolve two and possibly three layers.

EM measurements are susceptible to interference from coupling to nearby metal pipes, fences, and vehicles, and sometimes noise from powerlines. The effectiveness of electromagnetic measurements decreases at very low conductivities and become nonlinear at very high conductivities. McNeill (1980b) provides an excellent introduction and overview of the FDEM method.

Time Domain Electromagnetics (TDEM).   TDEM measures the bulk electrical resistivity of subsurface conditions by inducing pulsed currents in the ground with a large transmitter coil (Figure 11.14.14). The induced currents results in a decaying secondary magnetic field, which is monitored with a separate receiver coil. TDEM measurements are made with units in ohm-meters.

Time domain electromagnetic measurements are primarily used for soundings to determine depth and thickness of natural geologic and hydrologic conditions. They can also be applied to detection and mapping of inorganic plumes, seepage from brine pits, and salt-water intrusion investigations or

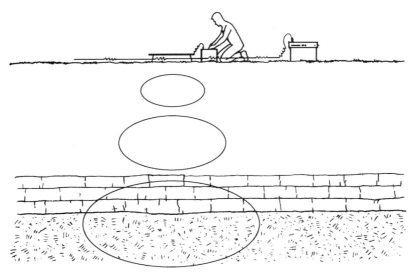

FIGURE 11.14.14 Illustration of TDEM measurements (*Source:* Technos, Inc., 1992).

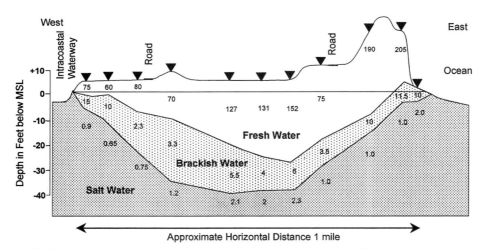

FIGURE 11.14.15 The extent of saltwater intrusion into the coastal aquifer is defined by the cross section derived from time domain electromagnetic measurements. In this case, the electrical contrast between fresh and saltwater is large. Note that a transition zone of brackish waster is also resolved (*Source:* Technos, Inc., 1993).

wherever there is an electrical contrast in pore fluids. Figure 11.14.15 shows the results of mapping the saltwater interface in a coastal setting.

Time domain EM soundings provide data similar to resistivity soundings. Measurements can be made from about 20 to 3000 ft deep with different equipment. Time domain electromagnetic measurements do not require intrusive ground contact and are made on a station by station basis. Shallow measurements can be made with a small transmitter loop of 30-ft diameter. Deeper measurements can require a large transmitter loop up to 1000 by 2000 ft. Deeper measurements with large

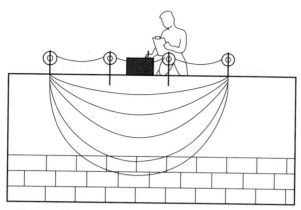

FIGURE 11.14.16 Illustration of resistivity measurements (*Source:* Technos, Inc., 1992).

transmitter loops are moderately labor intensive. TDEM measurements require processing of the data to determine true resistivities, layer thickness, and their depths.

TDEM measurements provide better lateral resolution than resistivity for the same depth because of relatively smaller coil size compared to the length of a resistivity array. Lateral resolution is limited by transmitter coil size and by the spacing between sounding measurements. Up to three to four layers can be resolved. Measurements are susceptible to interference from nearby metal pipes, cables, fences, vehicles and noise from powerlines and radio transmitters. McNeill (1980c) provides an introduction to the TDEM method, and Fitterman and Stewart (1986) provide case histories for groundwater applications.

Resistivity. Resistivity measurements are made by injecting a DC current into the ground through two current electrodes and measuring the resulting voltage at the surface at two potential electrodes (Figure 11.14.16). As with EM measurements, resistivity measures bulk electrical resistivity, which is a function of the soil and rock matrix; percentage of saturation; and the conductivity of the pore fluids.

The depth of measurement is related to electrode spacing. Various electrode geometries are used, including Wenner, Schlumberger, Dipole-Dipole, and others. The simplest, in terms of geometry, is the Wenner Array, which consists of four equally spaced electrodes all in a line. The apparent resistivity of the soil and rock is calculated based on the electrode separation, the geometry of the electrode array, the applied current, and measured voltage. The resistivity technique may be used for profiling or sounding, similar to EM measurements and may be used in many of the same applications as the EM method (Benson et al., 1982; Cartwright and McComas, 1968; Griffith and King, 1969; Telford et al., 1982; Zohdy et al., 1974).

Profile measurements are made with a fixed electrode spacing and provide a means of mapping lateral changes in subsurface electrical properties to a given depth. They are well suited to the delineation of hydrogeologic anomalies and mapping inorganic contaminant plumes. Profile data can be plotted as apparent resistivity versus distance along a profile line with little, if any, processing, and interpretation is often qualitative. Figure 11.14.17 shows resistivity contour maps developed from profiling data that were used to monitor a landfill leachate plume at four-year intervals.

Sounding measurements provide a means of determining the vertical changes in subsurface electrical properties. Sounding measurements are made by incrementally increasing the spacing between electrodes to make a sequence of measurements at increasing depths at a given location. Soundings are generally applicable to defining geologic strata where the geology is laterally homogeneous and is flat or gently dipping. Interpretation of sounding data provides the depth, thickness, and resistivity of

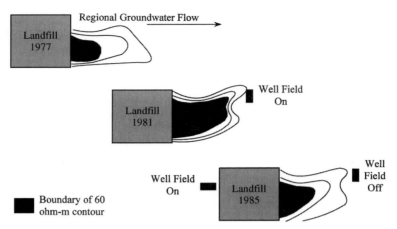

FIGURE 11.14.17   Resistivity profile data is used to map an inorganic plume from a landfill. The elevated total dissolved solids from the leachate plume are easily detected by the electrical methods. Measurements made at 4-year intervals show the plume dynamics with respect to nearby well fields (*Source:* Benson et al., 1988).

subsurface layers. Sounding data must be processed to obtain true resistivities values, along with the depth and thickness of layers. Sounding data are used to create a geoelectric section, which illustrates changes in the vertical resistivity conditions at a site. Figure 11.14.18 shows a 1D geoelectric section developed from a single resistivity sounding, along with a drillers log showing the correlation.

Resistivity measurements are technically easy to make but are slow and labor intensive because the method does require ground contact by driving metal electrodes into the ground and the deployment of long cables. Measurements are made on a station by station basis. For sounding measurements, the spacing between electrodes may have to extend up to three to five times the depth of interest, which results in long electrode spreads and cables. Finding sufficient accessible space can sometimes be a problem.

Lateral resolution is a function of electrode spacing and the spacing between station measurements. Forward and inverse computer models are commonly used to resolve up to three to four layers. Resistivity measurements are susceptible (but much less so than EM measurements) to interference from nearby metal pipes, cables, fences, and so on. Mooney (1980) provides an excellent introduction to the method, and Zohdy et al. (1974) provide an introduction and applications of the method.

By combining the processes of resistivity profiling and sounding under computer control, we can obtain a 2D image of subsurface conditions. A number of electrodes are equally spaced along a line and a cable is attached to each electrode. The cable is then connected to a resistivity instrument. Arrays of 28 to 56 electrodes are common, however, some systems will accommodate up to 255 electrode at one time. The system is software controlled so that any combination of electrodes may be selected with different electrode geometries (within the limits of the cable length and electrode spacing).

Inversion software is used to convert the measured apparent resistivity values to true resistivity and present the data in a geoelectric cross section with depth corrected data. Topography corrections can also be applied. These 2D data provide a more detailed description of the subsurface conditions. For example, Figure 11.14.19 shows the location of fractures with inorganic contaminants. While most resistivity imaging surveys are performed along a profile line as a 2D survey, the automated arrays are also capable of acquiring three-dimensional data. Loke (1999) provides an excellent overview of the method.

Spontaneous Potential (S.P.).   Naturally occurring potentials occur at the surface of the earth. These voltages are referred to as spontaneous potential measurements or natural potential. SP voltages are

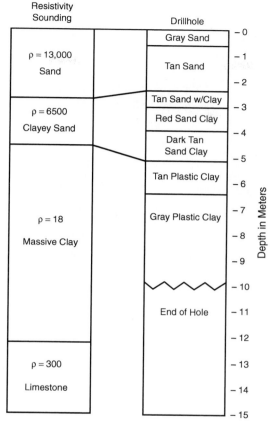

FIGURE 11.14.18 A 1-dimensional resistivity geoelectric section showing correlation with a driller's log. Resistivity values are in ohm-feet (*Source:* Benson et al., 1982).

produced by two different sources, the *electrochemical* differences between soils, rock, or minerals and their oxidation reduction process as well as the *electrokinetic* effect of flowing water, sometimes called streaming potential. The resulting potential (voltage) measured at the surface (Figure 11.14.20) and may range from a few millivolts to a few volts, but are more commonly a few tens of millivolts.

The major engineering applications of the self-potential method have been investigations of subsurface water movement, such as assessing seepage from dams and embankments. Time series measurements can also be made to monitor changes in seepage. SP measurements have also been applied to well field studies, to location of faults, shafts, tunnels and caves, and to the mapping of coal mine fires. Recently, SP measurements have been applied to monitoring biodegration of hydrocarbon contaminants (Vichabian and Morgan, 1999).

S.P. measurements are relatively easy to make and are made on a station to station basis using nonpolarizing electrodes and a high impedance voltmeter. The electrodes must be in good electrical contact with the ground. Usually one electrode is fixed as a reference electrode and measurements are made with the second roving electrode or both electrodes are moved in a leap-frog fashion.

Drift (and other) corrections are applied to improve the signal to noise ratio and results can be plotted as profiles or contoured. Interpretation of self-potential data are often qualitative, using the

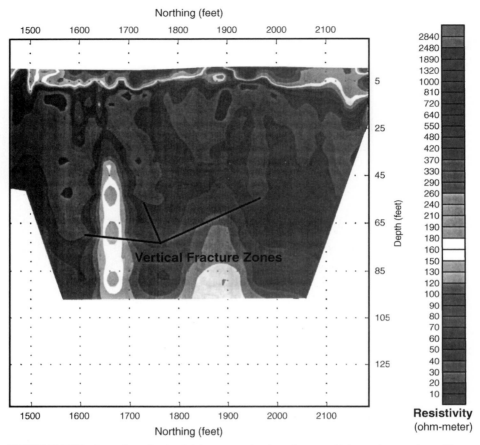

FIGURE 11.14.19   A two-dimensional resistivity cross section shows fractures with inorganic contaminants (*Source:* Technos, Inc., 1999a).

anomalies observed in profile data or contour patterns to identify seepage flow paths or other features. Figure 11.14.21 illustrates how S.P. measurements were used to identify zones of seepage in a dike.

Lateral resolution is a function of the spacing between measurements. Vertical resolution is limited to the following:

- Semi-qualitative assessment based upon a relation to the width of the anomaly (assuming it is caused by a single source)
- Comparing field data to forward model data based upon site specific assumption

Measurements are susceptible to interference from natural earth currents, soil conditions, topographic effects, cathodic protection currents, and cultural features. Corwin (1990) provides an excellent introduction and overview to the method.

Seismic Methods.   Seismic methods can be used to determine the seismic velocity of soil or rock, which is a function of their bulk modulus, the shear modulus, and the density. This data may be used to characterize the type of rock and degree of weathering and rippability based upon the seismic velocity

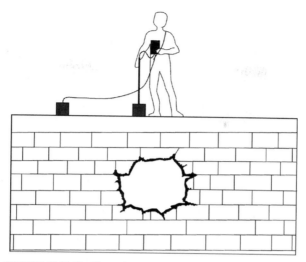

FIGURE 11.14.20 Illustration of S.P. measurements (*Source:* Technos, Inc., 1992).

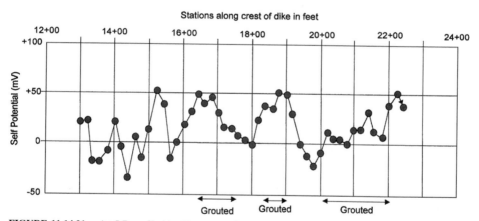

FIGURE 11.14.21 An S.P. profile identifies zones of water seepage (negative values) through a dike (*Source:* Corwin, 1990).

of the rock. Seismic measurements are used to determine depth to water table, map geologic strata, and to locate fractures, faults and buried bedrock channels.

A seismic source, geophones, and a seismograph are required to make the measurements. The seismic source may be a simple sledge hammer or other mechanical source with which to strike the ground. Explosives may be necessary for deeper applications that require greater energy. Geophones implanted in the surface of the ground translate the ground vibrations of seismic energy into an electrical signal. The electrical signal is displayed on the seismograph, permitting measurement of the arrival of time (or dispersion) of the seismic waves and displaying the wave forms from a number of geophones. Geophone spacing can be varied from a few feet to a few hundred feet depending upon the depth of interest and the lateral resolution needed.

FIGURE 11.14.22   Illustration of seismic refraction measurements
(*Source:* Technos, Inc., 1992).

Seismic waves are transmitted into the subsurface by an energy source. These waves are refracted, reflected, or dispersed when they pass from one soil or rock type into another that has a different seismic velocity. Compressional (P-waves) are used for most refraction or reflection measurements. The P-wave and S-wave velocities are related to the elastic constants, bulk modulus, shear modulus, and the density of the medium. Shear waves are sometimes used to determine material properties. Rayleigh (surface) waves are used for surface wave measurements.

Because the seismic methods measure small ground vibrations, they are inherently susceptible to vibration noise from a variety of natural (i.e., wind and waves) and cultural sources (i.e., walking, vehicles, and machinery).

Seismic Refraction.   Seismic refraction measurements are made by measuring the travel time of a seismic wave as it travels from the surface through one layer to another and is refracted back to the surface where it is detected by geophones (Figure 11.14.22). Knowing the travel time and source to geophone geometries, one can calculate the depth, thickness, and velocity of the layers.

The refraction method is commonly applied to shallow investigations up to a few hundred feet deep. Primary applications for seismic refraction is for determination of velocity, depth and thickness of geologic strata, and structure (Benson et al., 1982; Griffith and King, 1969; Telford et al., 1982). Velocity measurements are a measure of the material properties and can be used as an aid in assessing rock quality and rippability of rock. If compressional P-wave and shear S-wave velocities are measured, in situ elastic moduli of soil and rock can be determined. Velocity of seismic waves is measured in feet/second.

Typical refraction measurements might consist of 12 to 48 geophones set at equal spacings as close as 5 to 10 ft. Shallow measurements may be made using the energy of a 10-pound sledge hammer while significantly greater source energy (i.e., larger mechanical energy sources and possibly explosives) will be required as the depth of investigation increases. The results of seismic refraction data are commonly displayed as interpreted depth cross sections or as contour maps of stratigraphic layers.

Seismic refraction work can be carried out in a number of ways. The simplest approach provides a depth reading at each end of the geophone spread and thus, the dip of rock under the array of geophones can be calculated. These simple refraction measurements are described by Mooney (1973). A more detailed refraction survey can be carried out so that depths are obtained under every geophone. This survey will produce a detailed profile of the top of rock. For example, Figure 11.14.23 shows an irregular top of rock profile where rock is steeply dipping. This method is described by Redpath (1973). The general reciprocal method described by Palmer (1980) accommodates varying velocities within each layer, while calculating the depth under each geophone.

Seismic refraction data must undergo processing before a quantitative interpretation can be made. A variety of interpretative methods can be used ranging from the simple time intercept method to delay

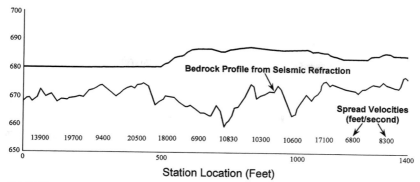

FIGURE 11.14.23 Profile of top of bedrock from seismic refraction survey. (Depth to rock is determined under each geophone; Geophone spacing is 10 ft.) (*Source:* Benson and Yuhr, 1996).

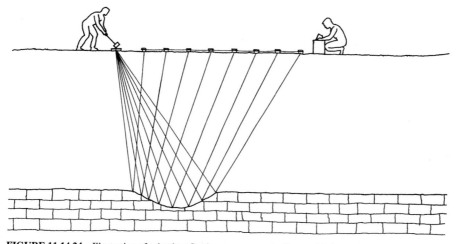

FIGURE 11.14.24 Illustration of seismic reflection measurements (*Source:* Technos, Inc., 1992).

time, ray tracing, and finally, the generalized reciprocal method. Each interpretation method requires more sophisticated data acquisition in the field. Seismic refraction measurements can typically resolve two to three layers. Lateral resolution is a function of geophone spacing (typically 5 to 20 ft or more) with larger spacings used for deeper measurements.

Two inherent limits to the refraction method are its inability to detect a lower velocity layer beneath a higher velocity layer and its inability to detect thin layers. A source to geophone distance of up to three to five times the desired depth of investigation is sometimes needed. This can make the overall length of a survey line quite long. Seismic refraction measurements are sensitive to acoustic noise and ground vibrations. Haeni (1988) provides an excellent introduction to the method with case histories.

Seismic Reflection. The seismic reflection technique measures the two-way travel time of seismic waves from the ground surface downward to a geologic contact where part of the seismic energy is reflected back to geophones at the surface (Figure 11.14.24). Reflections occur when there is a contrast in the density and velocity between two layers.

The primary application for the seismic reflection method is to identify and determine the depth and thickness of geologic strata in the overlaying soils and in the underlying rock. The method can also be used to locate and characterize geologic structure. Recent applications have attempted to use higher frequencies to identify smaller targets such as mines, tunnels, and caves. The reflection method provides a high resolution cross section of soil/rock along a profile line.

The vertical scale is measured in two-way travel-time, which is the time it takes for a wave to travel down to an interface and back up to the surface again. Velocity of seismic waves is measured in feet/second. The time scale must then be converted to depth by making velocity measurements or some assumptions regarding seismic velocity within the strata.

By comparison, a seismic reflection survey is capable of much deeper investigations with less energy than the refraction method. While reflections have been obtained from depths as shallow as 10 ft, the method is more commonly applied to depths of 50 to 100 ft or greater. The reflection technique can be used effectively to depths of a few thousand feet or more and can provide relatively detailed geologic sections. The source to farthest geophone is usually one to two times the desired depth of investigation (much less than that required for refraction measurements; see Hunter et al., 1982; Steeples and Miller, 1990).

The shallow high resolution reflection methods discussed here attempt to utilize the highest frequencies possible (150 to 600 Hz) to improve vertical resolution, and to utilize relatively closely spaced geophones (1 to 20 ft apart) to provide good lateral resolution. Because of the need for higher frequencies, attention must be given to selection of a seismic source, geophone placement, and their optimum coupling to soil or rock.

Seismic reflection measurements are relatively difficult to make and are labor intensive. Reflection measurements require that the geophones and the energy source be in intimate contact with the ground. Extensive cable handling and moving of the source is also required. Reflection data must be processed before an interpretation can be made. Vertical resolution is proportional to the frequency of the seismic energy that can be generated and propagated. Resolution may be as good as a few feet with frequencies of 500 Hz. Figure 11.14.25 clearly illustrates the presence of a bedrock channel. The optimum conditions for shallow reflection surveys are saturated fine grain soils, which enable higher frequency energy to be coupled into the ground. Lateral resolution is a function of geophone spacing, which is commonly 1 to 10 ft.

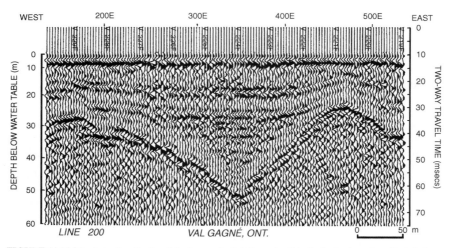

FIGURE 11.14.25 Seismic reflection data shows a buried channel within the bedrock surface. Buried channels are often the geologic pathway that control contaminant migration. The difference in seismic properties (velocity) between overlying unconsolidated materials and the underlying rock provides a measurable reflector using the seismic reflection method. Also note layering within the overlying unconsolidated materials (*Source:* Pullan, S. (1996)).

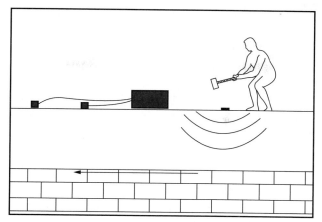

FIGURE 11.14.26   Illustration of surface wave measurements (*Source:* Technos, Inc., 1999b).

The reflection method is limited by its ability to transmit high frequency energy into the soil and rock. Loose soil near the surface limits the ability of the soil system to transmit high frequency energy into and out of the rock, limiting the resolution that can be obtained. Seismic reflection measurements are sensitive to acoustic noise and ground vibrations. Steeples and Miller (1990) provide an introduction to the reflection method with emphasis on the common depth point method.

Spectral Analysis of Surface Waves (SASW).   The SASW method measures the velocity of surface (or Rayleigh) waves of various frequencies that travel along the surface of the ground (Figure 11.14.26). The SASW methods uses the dispersive characteristics of surface waves to determine the variation of the shear wave velocity (stiffness) of layered systems with depth. The depth of the wave is determined by its wavelength (or frequency). Low frequencies with long wavelengths extend deeper than high frequencies with shorter wavelengths.

The SASW technique has many applications in geotechnical engineering ranging from determination of ground stiffness profile, liquefaction assessment, and ground response analyses because of earthquake loading, design of vibrating foundations, to the evaluation of pavements, roadways, runways, and tunnel linings. SASW has been used at waste sites for evaluation of cover thickness, for location of native ground interface, and to determine compaction profiles for the waste mass.

Measurements are made by placing two geophones on the surface, and using the blow from a hammer (large drop weights, vibrating sources, or simply a bulldozer moving back and forth have been successfully used as a source) to generate the wave energy. The distance between the source and center of the receivers is increased in increments and a series of deeper measurements are made. Short receiver spacings are used to sample the shallow layers while long receiver spacings are used in sampling the deep materials. The depth over which reliable measurements can be made depends upon the energy and frequency content of the source excitation and the consistency of the subgrade materials.

A dual channel Fast Fourier Transform (FFT) dynamic signal analyzer is used to record and analyze the motions of the two geophones. Shear wave velocity profiles are determined from the experimental dispersion curves (surface wave velocity versus wavelength) obtained from SASW measurements. Data must be processed before an interpretation can be made. Figure 11.14.27 shows the results of surface wave measurements to evaluate the physical properties with increasing depth at a dam site. The principals of SASW testing are described in greater detail by Stokoe et al. (1994). Haegeman and Vam Impe (1999) provide an excellent overview of the method along with examples related to compaction of wastes.

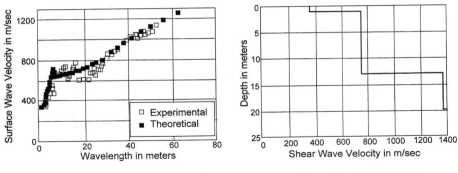

FIGURE 11.14.27   Surface wave data show shear wave velocity changes with depth (*Source:* Liu et al., 1998).

FIGURE 11.14.28   Illustration of gravity measurements (*Source:* Technos, Inc., 1992).

Microgravity.   Gravity measurements respond to changes in the earth's gravitational field caused by changes in the density of the soil and rock (Figure 11.14.28). By measuring the spatial changes in the gravitational field, lateral variations in subsurface geologic conditions can be determined. Microgravity methods are particularly useful in locating and mapping lateral changes in rock units, defining structural features, bedrock channels, caves, and cavities (Griffith and King, 1969; Telford et al., 1982).

Regional gravity surveys employ widely spaced stations and are used to assess major geologic conditions over many hundreds of square miles. A microgravity survey utilizes closely spaced stations of 5 to 50 ft, and are carried out with a very sensitive microgravimeter. Microgravity surveys are used to detect and map shallow, localized, geologic anomalies such as bedrock channels, fractures, and cavities; to estimate the volume of landfills; and can be used to estimate the volume of grout needed for remediation of fractured rock and karst sites. Microgravimetry comes closest of all the geophysical methods to allowing a positive statement regarding the presence or absence of subsurface cavities at a site (Butler, 1980).

The earth's normal gravity is 980 gals. The unit used in gravity is the milliGal. A microGal is 1/1000 of a milliGal or 10^{-9} of the earth's gravitational field. A gravity survey results in a Bouguer Anomaly, which is the difference between the observed gravity values and theoretical gravity values.

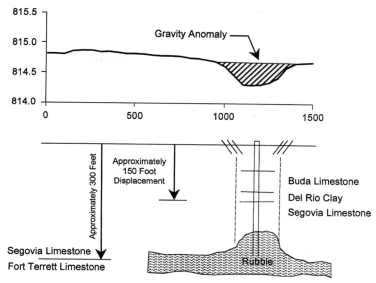

FIGURE 11.14.29 Gravity data indicates the presence of a paleo-sinkhole collapse, which was confirmed by boring data (*Source:* Yuhr, Benson, and Bulter, 1993).

The Bouguer Anomaly is made up of deep-seated effects (the regional Bouguer Anomaly) and shallow effects (the local Bouguer Anomaly). It is the local Bouguer Anomaly that is of interest from a microgravity survey. Figure 11.14.29 illustrates the location of a paleocollapse zone by use of gravity data.

While gravity measurements are in concept simple to make, they require extreme care with the instrument and with procedures. Measurements are made on a station by station basis and are relatively slow. Extensive corrections must be applied to gravity data before it can be interpreted. Corrections must be made for instrument drift, earth tides, changes in elevation, and topography. To compensate for small instrument drift throughout the day, measurements must be made at a base station every hour or so, so that drift corrections can be applied to the data. Elevation must be measured to a relative accuracy of 0.01 foot per mile at each station. Data are presented as profile lines and/or a contour map.

Lateral resolution is a function of station spacing. Microgravity measurements typically use closely spaced stations (about 5 to 50 ft apart). The response of a gravity anomaly decreases rapidly as the target becomes deeper (response for a localized target is proportional to 1/depth^2). Vertical resolution is limited to the following:

- Depth estimates are based upon a relation to the width of the anomaly (assuming it is caused by a single source).

- Comparing field data to forward model data based upon assumptions of site specific conditions.

Gravity measurements are susceptible to vibrations from vehicles and aircraft, and natural vibrations from wind and distant earthquakes. However, a gravity survey can often be undertaken in areas where other cultural effects may preclude the use of other geophysical methods. For example, gravity measurements can be made inside buildings and structures. Butler (1980) provides an excellent overview of the microgravity methods.

Metal Detection. Metal detectors are commonly used by utility and survey crews for locating buried pipes, cables, and property stakes. They are also useful for detecting buried drums and for delineating the boundaries of trenches containing metallic drums or trash (Benson et al., 1982). Metal detectors

FIGURE 11.14.30   Illustration of metal detector measurements (*Source:* Technos, Inc., 1992).

can detect both ferrous metals, such as iron and steel, and nonferrous metals, such as aluminum and copper. A wide range of metal detectors are commonly available including hand-held units with small coils used by treasure hunters to locate coins or small objects; pipe and cable utility locators commonly used by survey crews; and commercial metal detectors such as the Geonics EM61 (Figure 11.14.30). The frequency domain electromagnetic instruments such as the EM38 and EM31 by Geonics can also be used as metal detectors.

Figure 11.14.31 shows the results from a continuous metal detector survey over a DOE waste disposal site. Metal detector measurements provide better spatial definition than frequency domain electromagnetic measurements or magnetometer measurements.

A metal detector response is a function of the area of the metal object and its depth. The response to a drum is proportional to the reciprocal of the depth to the sixth power, therefore sensitivity decreases rapidly as depth is increased. The area of detection of a metal detector is approximately equal to its coil size or spacing between the transmitter and the receiver (typically 1 to 3 ft). Small shallow objects, such as nails on or near the surface, can be detected by instruments with small coil diameters (<1 foot). A metal detector with a larger coil diameter will detect larger targets much deeper. A single 55-gallon drum can be detected at depths of 3 to 10 ft, and massive piles of 55-gallon drums may be detected at depths of up to about 10 to 15 ft, depending upon the type of instrument used.

Metal detector measurements are relatively easy to make. Because they do not require intrusive ground contact, continuous profile measurements can be made. Instruments can be hand-carried or mounted on a vehicle. In some cases, total site coverage is economically possible. Data can be plotted along a profile line or as a contour map, and is often used with little processing.

Lateral resolution of a metal detector to locate targets and define boundaries of buried pits is quite good, usually much better than a magnetometer. While some metal detectors can provide estimates of target depths, vertical resolution is usually limited to the following:

- Semi-qualitative assessment based upon a relation to the width of the anomaly (assuming it is caused by a single source)
- Comparing field data to forward model data based upon site specific assumption

Metal detectors can be affected by nearby metallic pipes, fences, cars, buildings, and, in some cases, by changes in soil or pore fluids conditions. Benson et al. (1982) provides a brief overview of metal detectors.

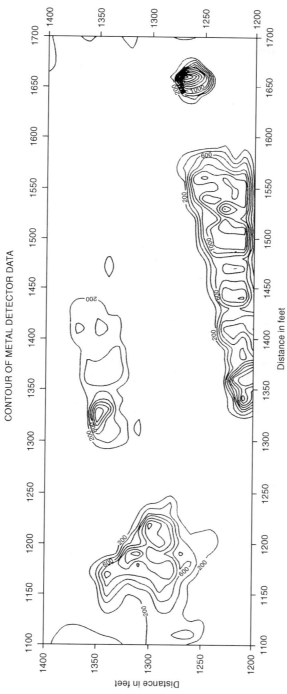

FIGURE 11.14.31 Metal detector measurements can be used to map buried metal. Here the boundaries of buried metal and old building foundations are shown at a DOE site (*Source:* Benson and Yuhr, 1996).

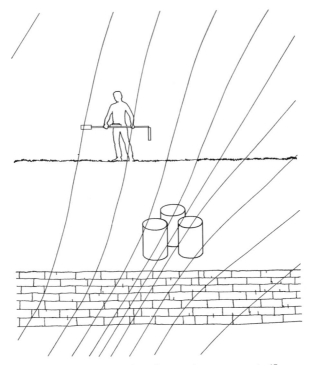

FIGURE 11.14.32 Illustration of magnetic measurements (*Source: Technos, Inc., 1992*).

Magnetometry.   A magnetometer is used to measure the intensity of the earth's magnetic field. Deviations of magnetic intensity are caused by changes in concentrations of natural ferrous minerals (iron ores) or by the presence of ferrous metals (Figure 11.14.32). In certain geologic environments, magnetic measurements can be used for geologic mapping to provide an estimate of the thickness of nonmagnetic sediments overlying magnetic rock and location of structure and faults within magnetic rock (Breiner, 1973; Griffith and King, 1969; Telford et al., 1982). The primary application of magnetic measurements at hazardous waste sites is in locating buried drums, tanks, and pipes (Benson et al., 1982). A magnetometer will only respond to ferrous metals (iron and steel) and will not detect nonferrous metals. Units are in nanoteslas (previously gammas). The earth's magnetic field is approximately 600,000 gammas at the poles and 30,000 gammas at the equator.

There are two types of measurements that are commonly made: total field and gradient. *Total field measurements* respond to the total magnetic field of the earth, including any changes caused by a target, natural magnetic materials, and cultural magnetic noise (ferrous pipe, fences, buildings, and vehicles). The effectiveness of total field magnetometers can be reduced or totally inhibited by noise or interference caused by time-variable changes in the earth's magnetic field, or spatial variations because of magnetic minerals in the soil, steel debris, pipes, fences, buildings, and passing vehicles. A base station magnetometer must be used to reduce the effects of natural noise by subtracting the base station values from those of the search magnetometer. Cultural noise, however, will remain a problem with total field measurements. Many of these problems can be avoided by use of gradiometer measurements and proper field techniques. A total field magnetometer's response is proportional to the mass of the ferrous target and inversely proportional to the cube of the distance to the target (such as a drum).

Gradiometer measurements are made by two magnetic sensors separated vertically (or horizontally) by a few feet. Gradient measurements have some distinct advantages over total field

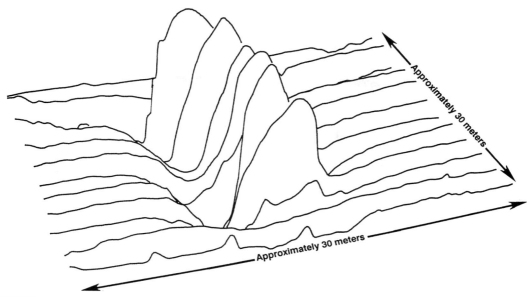

FIGURE 11.14.33 Magnetic gradient over a trench with buried drums. The trench is approximately 20 by 100 ft (*Source:* Benson et al., 1982).

measurements. They are insensitive to natural changes in the earth's magnetic field and minimize most cultural effects. Because the response of a gradiometer is the difference of two total field measurements, it responds only to the local gradient. As a result, it is better able to locate a relatively small target, such as a buried drum. A gradiometer response is inversely proportional to the fourth power of the distance to the target (such as a drum), making it less sensitive than the total field measurement. While gradiometers are inherently less sensitive than total field instruments, they are also much less sensitive to many sources of noise.

Under ideal conditions, a single drum can be detected at depths up to about 20 ft with a total field magnetometer (about 10 ft with a gradient magnetometer). Figure 11.14.33 shows the results of a gradiometer magnetometer survey over a trench containing buried drums.

Magnetometer measurements can be made on a station by station basis or continuously as the magnetometer is moved across the site. Continuous coverage is much more suitable for detailed (high resolution) surveys to identify local targets, such as drums, and for the mapping of areas in which complex anomalies are expected.

Magnetic measurements are relatively easy and rapid to make and can be made by one person. Measurements can also be made from a vehicle. In some cases, total site coverage is economically possible. Values can be plotted along a profile line or as a contour map and is often used with little processing.

Horizontal resolution is a function of the spacing between measurements. Gradient measurements provide an improvement of horizontal resolution. Vertical resolution is limited to the following:

- Semi-qualitative assessment based upon a relation to the width of the anomaly (assuming it is caused by a single source)

- Comparing field data to forward model data based upon site specific assumption

Magnetic measurements are susceptible to interference from steel pipes, fences, vehicles, and buildings as well as natural fluctuations in the earth's field. Breiner (1973) provides an excellent technical overview of the magnetic methods.

11.14.6 EXAMPLES OF APPLICATIONS

There are four areas where the surface geophysical methods are commonly applied to environmental site investigations. They are as follows:

- Characterization of natural geologic and hydrogeologic conditions
- Detection and mapping of contaminant plumes, spills, and leaks
- Detection and mapping of landfills, trenches, buried wastes or the location of tanks, drums, pipes, or utilities
- Monitoring the effectiveness of the remediation process

Characterization of Natural Geologic and Hydrogeologic Conditions

One of the most important tasks of any site investigation is characterizing the natural geologic and hydrogeologic conditions. Because it is these conditions that will control groundwater and contaminant flow, geologic data are critical to the design of new waste disposal sites and the remediation of hazardous waste sites.

A variety of geophysical methods can be used to assess natural hydrogeologic conditions including: depth to bedrock, degree of weathering, sand and clay lenses, fracture zones, buried stream channels, and karst conditions. Applications of various geophysical methods for assessing natural geologic and hydrologic conditions have previously been presented including: radar measurements (Figure 11.14.6 to assess soil conditions and Figure 11.14.7 locating a developing sinkhole); EM measurements (Figure 11.14.12 to locate fractures); resistivity measurements (Figure 11.14.17 to map an inorganic plume and Figure 11.14.18 to provide a 1-D sounding to characterize geologic strata); S.P. measurements (Figure 11.14.21 to define seepage); seismic refraction measurements (Figure 11.14.23 to define top or rock); seismic reflection measurements (Figure 11.14.25 to define a buried channel); SASW measurements (Figure 11.14.27 to determine shear wave properties); and gravity measurements (Figure 11.14.29 to locate a paleosinkhole).

The surface geophysical methods can be used for reconnaissance surveys over large areas or to provide detailed investigations over smaller areas. For a reconnaissance survey over a large area, data are widely spaced. For example, at one site a new hazardous waste landfill was proposed and a reconnaissance EM survey was carried out to determine the general alluvium thickness over limestone. Initial line spacing was 3000 ft over an area of six square miles (3840 acres). A site location was then selected, which had a thicker alluvium cover based upon the reconnaissance EM data. Then, more detailed EM measurements were made with a line spacing of 500 ft over an area of 690 acres. At this point, drilling locations were selected and drilling and geophysical logging were carried out to provide greater detail than could be obtained by surface geophysical measurements alone. At the same time, additional detailed surface geophysical measurements (microgravity) were made, and a visual inspection was made of every square foot of the final 380 acre site. Note how the geophysical data were used to provide both reconnaissance, and the detailed data as the final site selection of 380 acres was selected from the initial 3840 acres. Also note that drilling did not begin before the site location was sufficiently narrowed down so that boring locations could be selected as representative of the site and not simply be random locations.

Detection and Mapping of Contaminant Plumes, Spills and Leaks

The location and mapping of inorganic or organic contaminant plumes is an objective of many site investigations. In some cases, the geophysical methods can be employed to provide direct detection of contaminants. In other cases, the geophysical methods are used indirectly to locate geologic contaminant pathways (Benson et al., 1982, 1985; Cartwright and McComas, 1968; Greenhouse and Monier-Williams, 1985; McNeill, 1980b).

Inorganics. Inorganic contaminants from landfills, as well as salt brines, acid spills, and natural salt-water intrusion are usually readily detectable by the electrical methods (i.e., resistivity or electromagnetics) because of the high specific conductance of pore fluids. Electrical methods not only provide a means of directly mapping the extent of the inorganic contaminants, but also provide the direction of flow and an estimation of concentration gradients. Figures 11.14.8, 11.14.13 and 11.14.17 illustrate a number of examples of detecting and mapping inorganic contaminants, and Figure 11.14.15 illustrate mapping of saltwater intrusion.

Correlation between groundwater chemistry data and results using electrical methods to map inorganics plumes from *landfills* has been as good as 0.96 at the 95 percent confidence level (Benson et al., 1985). If the contaminant plume consists of both organics and inorganics, such as leachate from a landfill, a first approximation to the distribution of the organics can often be made by using electrical methods to map the more electrically conductive inorganics. Correlation between groundwater chemistry data for total organics in a *landfill leachate* plume and results using electrical methods has been as good as 0.85 at the 95 percent confidence level (Benson et al., 1985).

Organics. Fresh spills of organic "mixers" and "sinkers" such as TCE and DNAPLS are not generally detected directly by geophysical measurements. However, floating hydrocarbons have been detected and mapped by radar (Figure 11.14.9, gasoline). Direct detection of shallow organic (floaters) such as hydrocarbons sometimes can be accomplished by looking for a conductivity low (high resistivity) associated with the organics. The possibility for such an anomaly exists where there is a thick layer of hydrocarbons and a sufficient contrast in electrical values between the natural background values and the hydrocarbons.

Where hydrocarbons have been in place for some time and biodegradation is taking place, elevated electrical conductivity values have been found as a result of electrochemical changes (Vichabian and Morgan, 1999). Recent research (Olhoeft, 1998) suggests that the electrical methods (complex resistivity measurements) offer some potential for the direct detection of some organic compounds.

Indirect Detection of Contaminants. In cases where the contaminant cannot be directly detected by the surface geophysical methods (i.e., those contaminants that are present in low concentration or mixers and sinkers, such as TCE and DNAPLS), the geophysical methods can be used to assess the natural geologic and hydrogeologic conditions [permeable pathways such as porous zones (Figure 11.14.12) or buried channels (Figures 11.14.8, 11.14.13, and 11.14.25) that are controlling groundwater flow. Once the contaminant flow pathways have been identified, direct sampling methods can be accurately located to further assess conditions.

Time Series Measurements of Contaminants. Geophysical measurements can be very effective in monitoring changes in contaminant distribution over time (Benson et al., 1988). Time-series measurements can be made to map changes in the shape of a plume over time (Figure 11.14.17).

Locating and Mapping Buried Wastes and Utilities

An important objective for many environmental projects, as well as real estate transfers and brownfield sites, is determining if there is something buried at the site. Items of concern include: underground storage tanks (UST); buried debris, including drums; trenches, septic tanks, or landfills; and even unexploded ordnance (UXO). Utilities and pipelines are also of concern when drilling or excavating, and can also provide possible pathways for contaminants (Benson et al., 1982).

Electromagnetic and magnetometer measurements (Figure 11.14.34), as well as data from a metal detector survey (Figure 11.14.31) and magnetometer survey (Figure 11.14.33), can be employed to locate and define the edges of buried debris trenches or landfills based upon a change in electrical conductivity between the natural soils and the fill material or the presence of the buried metal. Radar measurements (Figure 11.14.10) can also be used to locate underground tanks.

While locations of buried materials can be quite accurate, an assessment of the types of materials, their depths, and quantities can only be estimated. Similarly, the depths of trenches and landfills

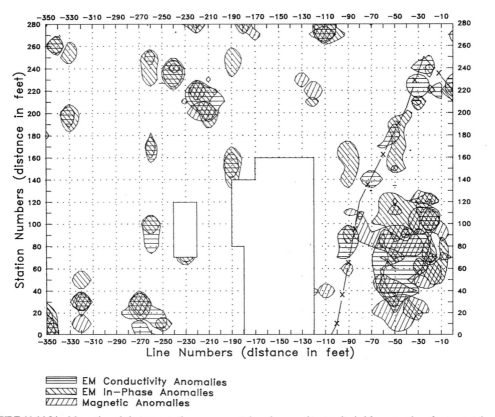

EM Conductivity Anomalies
EM In–Phase Anomalies
Magnetic Anomalies

FIGURE 11.14.34 Magnetic and electromagnetic measurements have been used to map buried ferrous and nonferrous metals in an open field used for cattle grazing. Over 200 drums were removed from six sites within this area (*Source:* Benson and Yuhr, 1996).

can be difficult to obtain accurately because of the impact of the extreme nonhomogeneity of their contents. Such measurements have now been applied to Brownfield site characterization (Kaufmann and Benson, 1999). In fact, it is unreasonable to carry out a standard Phase I and limited Phase II survey and expect that the site has been adequately characterized. Figure 11.14.35 illustrates the findings of a geophysical survey where both a Phases I and II surveys had been previously carried out. A number of targets were identified including four previously unknown 10,000-gallon tanks, soakage pits, and buried debris.

Monitoring the Remediation Process

Certain geophysical methods may be used to monitor the effectiveness of remediation. Olhoeft (1992) has shown that if measurements are repeated at the same location, very small changes in contaminant distribution can be seen. Greenhouse et al. (1993) summarizes the surface and borehole geophysical measurements used to provide real time monitoring of a DNAPL injection test at the Bordon test site.

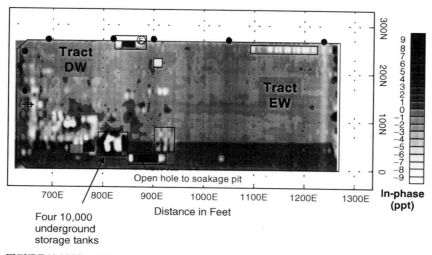

FIGURE 11.14.35   EM31 conductivity data used to determine subsurface conditions at a brownfield site. The data show the location of four previously unidentified 10,000 tanks as well as soakage pits and buried debris (*Source:* Kaufmann and Benson, 1999).

11.14.7 CONSIDERATIONS IN USING SURFACE GEOPHYSICAL METHODS

Selection of Surface Geophysical Methods

Geophysical methods, like any other means of measurement, have advantages and limitations. The performance of any geophysical method depends on its specific application and on site conditions. There is no single, universally applicable geophysical method that can be expected to solve all site evaluation problems. Furthermore, some geophysical methods are quite site-specific in their performance. All geophysical methods will function well under certain natural geologic and cultural conditions, and there are areas in which they will not function well or fail completely. Furthermore, geophysical technology by itself is not in itself a panacea; its successful application is dependent upon integrating the geophysical data with other sources of information. This must be done by persons with training and experience in geophysical methodology, as well as in the broader aspects of the earth sciences.

Geophysical methods do not offer a substitute for borings, wells, or direct sampling, but provide a means to locate anomalous and background conditions, minimizing the number of boreholes wells and sampling, to ensure that they are in reasonably representative locations and to fill in the gaps between the boreholes. Each method measures some physical, electrical or chemical parameter or combination of them (Table 11.14.1). The success of a geophysical method depends on the existence of a contrast between the measured properties of the target and background conditions. If there is no measurable contrast in the measured property, the target will not be detected. Furthermore, the scale and depth of the measurement needs to be considered when selecting a method. If a layer is sufficiently thin, it may not be detected, or if the size of a localized target is sufficiently small or if it is too deep it may not be detected. All of the surface geophysical methods are inherently limited by a decreasing resolution with depth.

All of the geophysical methods discussed are scientifically sound, and have been proven in the field. Like any other technologies, however, they may fail to provide the desired results when applied to the wrong problem or improperly used. The techniques must be selected and applied to site-specific conditions by a person who thoroughly understands the methods and their limitations and who has the capability to interpret the data in terms of geology and hydrology when necessary. This will allow the geophysical techniques to be properly integrated into a project and to converge on an accurate site characterization as efficiently as possible.

In general, surface geophysical methods (and the overall process of site characterization) should begin at the regional or reconnaissance level and then proceed to the more detailed site-specific data. Initial phases of any project should focus upon basics using the simplest, most appropriate technology.

We usually cannot achieve multiple objectives from a single geophysical survey. For example, a survey optimized for shallow (<10 ft) high resolution mapping of soils cannot be expected to also yield (at the same time) data at 100 ft. To accomplish this, we would usually need two surveys each optimized to achieve its objective.

There is no simple way to select a geophysical method to solve a particular problem. Tables are often provided to illustrate how the geophysical methods may be used to carry out various tasks, for example, ASTM publishes a guide to the selection of surface and borehole geophysical methods (ASTM D6429; ASTM D5753, 1995). However, simple tables and rules of thumb often fail when considering specific project needs and unique site conditions. Therefore, all such tables should only be used as a guide. The final decision should be made by professionals with an extensive track record with a broad range of methods.

The final approach, including selection of methodology, extent of coverage, resolution, processing, interpretation, and integration of data, is always a compromise constrained by experience, politics, and budget.

Station versus Continuous Measurements.   In many cases, geophysical measurements will be made on a station by station basis, along profile lines or over a grid. The lateral resolution of the surface geophysical methods is an inherent function of the method used and the spacing between measurements. Some methods can provide continuous measurements along a profile line. Such continuous measurements provide highest lateral resolution for mapping subtle lateral changes in subsurface conditions.

Continuous methods should be employed whenever applicable to maximize the amount of data obtained, to achieve maximum resolution, and to minimize project costs. This approach is particularly necessary when site conditions are suspected of being highly variable and small features, such as fractures or isolated drums, need to be identified. Although the continuous surface geophysical methods referred to in this document (see ground penetrating Radar, electromagnetics, magnetics, and metal detector methods) are typically limited to a depth of 50 ft or less, they are applicable to many site investigations. They can provide continuity of subsurface information, which is not practically obtainable from station measurements. By running closely spaced parallel survey lines with continuous measurements, subtle changes in subsurface parameters can often be mapped and total site (100 percent) coverage can even be achieved if necessary. Examples of continuous data are shown in Figures 11.14.6 and 11.14.7, which shows the detail obtained with continuous radar data, and Figure 11.14.13 which show the detail obtained by continuous electromagnetic data.

Use of Multiple Measurements

The ESC strategy proposes the use of multiple sources of data to characterize a site. This provides a reliable, defensible means of testing the hypothesis embedded in the conceptual site model. When measurements by different methods agree we can have a high level of confidence in our interpretations.

The use of multiple measurements provides redundancy as well as verification in assessing the presence of buried materials. Figure 11.14.34 shows the contoured results of an EM31 survey, along with a magnetometer survey. The combined results of such a survey provide verification of buried sites and their boundaries. In many cases, one finds both inorganics and organics present at the same site. The inorganics can often be used as a tracer for mapping groundwater flow and can be used to infer the presence of organics. The correlation of inorganics and organics is shown in Figure 11.14.36. Here, contaminant flow was being controlled by a buried channel.

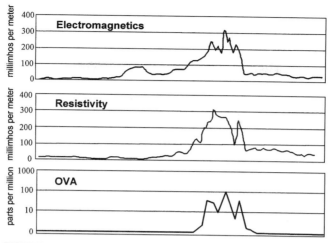

FIGURE 11.14.36   Organic vapor profile over buried channel: (a) electromagnetics, (b) resistivity, and (c) organic vapor analyzer. Note correlation between resistivity and EM measurements. This is an excellent example of a buried channel controlling flow and the level of correlation between organic and inorganic contaminants (*Source:* Benson and Yuhr, 1996).

During an ESC at an oil disposal site at the DOE's Savannah River Site, a clear inorganic component was identified by electromagnetic measurements (Figure 11.14.37). The vertical distribution of the inorganic plume was defined using a Geoprobe conductivity probe; the results are shown in Figure 11.14.38, which indicated that contaminants were accumulating on two leaky aquitards located at a depth of about 10 and 20 ft depth. This data was then used to guide the chemical sampling where the organics were mapped using water samples obtained by a Geoprobe (Figure 11.14.39).

Natural and Cultural Interferences

A wide range of natural and cultural factors can cause interference with surface geophysical measurements. While one method may be so affected by site conditions as to render it useless, another method might be used effectively. For example, electromagnetic measurements cannot be made along and over a railway track because the metal tracks will completely dominate the measurement. Radar and gravity measurements, on the other hand, can be made with little, if any, effect from the metal rails.

Where Geophysics Fits Into the Site Characterization Process

Geophysics is most effective when used early in the site characterization process. Because the results of geophysical work usually result in identification of anomalous conditions (geologic conditions, a contaminant plume, or a burial site), geophysics should generally be done first so that subsequent drilling and sampling locations are representative of site conditions. Unfortunately, geophysical measurements are too often used after a project is near completion in a last ditch attempt to solve the problem, or geophysical data is used without appropriate supporting information.

Even after the site characterization process is considered complete, geophysics can be used to determine the completeness and accuracy of the initial assessment by providing data between (interpolation) and beyond (extrapolation) existing discrete data points. If borings or monitoring wells have already have been installed, the location of existing borings and monitoring wells relative to

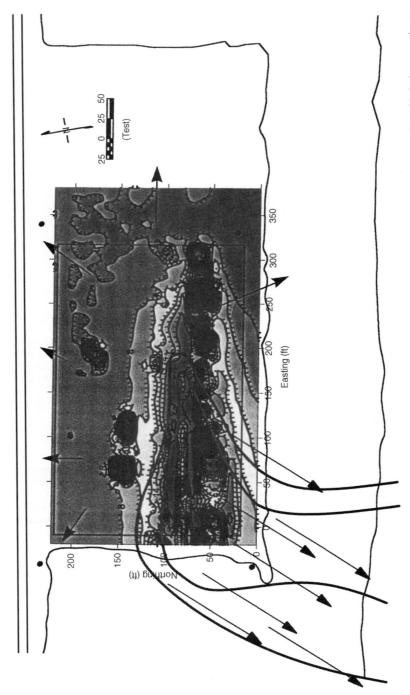

FIGURE 11.14.37 EM31 conductivity data show a burial trench with an inorganic plume moving to the southwest. These initial data were used to guide further contaminant investigation (Figures 11.14.38, and 11.14.39); (*Source:* Benson, 1999).

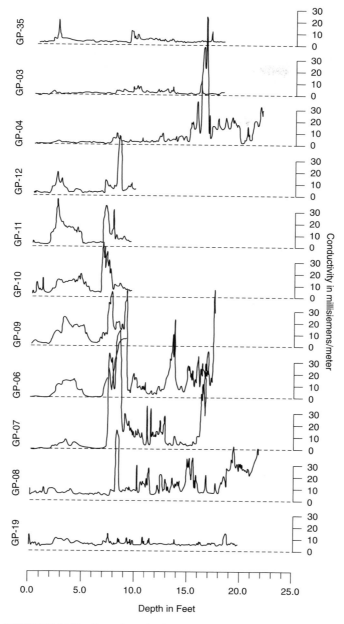

FIGURE 11.14.38   Geoprobe conductivity pushes show the presence of inorganics (Figure 11.14.37) perched on weak aquitards at depths of about 10 and 20 ft deep. The data show the lateral boundaries of contaminants as well as its vertical distribution (*Source:* Benson, 1999).

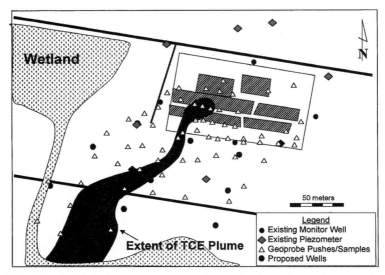

FIGURE 11.14.39   Geoprobe water samples defines a TCE plume migrating from a trench. Note that the TCE plume lies within the inorganic plume defined by the EM31 conductivity data (Figure 11.14.37) and push conductivity data (Figure 11.14.38), (*Source:* Benson, 1999).

background and anomalous site conditions can be assessed using the results of geophysical measurements. This provides a means of evaluating the representativeness of existing data. These data can then be used to check the conceptual geologic model confirming that conditions are (or are not) as envisioned and that borings and wells are in the right location and at the right depth. If additional boreholes or wells are needed to fill-in gaps in the data, they can be located with a high degree of confidence.

Qualitative to Quantitative Results

While the theory of applied geophysics is very quantitative and based upon laws of physics, in practice the results of a geophysical measurement may simply be used for anomaly detection and location. Location of anomalous conditions is relatively straightforward and can be determined quite accurately. In many cases, the anomaly is readily identified without any, or minimal, data processing. Some of the methods provide data from which preliminary interpretation can be made in the field (i.e., ground penetrating radar, frequency domain electromagnetics, resistivity profiling, magnetics and metal detector methods). Examples include location of subsidence by radar measurements (Figure 11.14.7); the location of buried tanks by radar (Figure 11.14.10); the location of fractures by electromagnetic measurements (Figure 11.14.12); defining a buried channel by seismic refraction measurements (Figure 11.14.25); or locating buried trash by electromagnetic and magnetometer measurements (Figures 11.14.34). Once an anomaly is located, attention may then be focused upon subsequent investigation by boreholes, monitor wells, excavation, or sampling without further geophysical measurements.

Other methods (i.e., time domain electromagnetics and resistivity soundings, refraction, reflection, and gravity methods) require that the data be processed before any interpretation. Examples of data requiring processing include TDEM sounding (Figure 11.14.15), resistivity sounding (Figure 11.14.18), 2D resistivity imagery (Figure 11.14.19), seismic refraction (Figure 11.14.23), seismic reflection (Figure 11.14.25), surface wave data (Figure 11.14.27), and gravity data (Figure 11.14.29).

A fundamental limitation of the surface geophysical methods is that a given set of geophysical data cannot be associated with a unique set of geologic or hydrogeologic conditions. In most situations, geophysical measurements alone cannot resolve all ambiguities, and some additional source of geologic information, is required. Because of this inherent limitation a geophysical survey alone can never be considered a complete assessment of subsurface conditions. However, when properly integrated with other knowledge, the geophysical methods will provide a highly effective, accurate, and cost-effective method of obtaining subsurface information. Preliminary interpretation of geophysical data in the field should always be treated with caution because it is easy to make errors in an initial field interpretation, and such a preliminary analysis is never a complete and thorough interpretation.

2D and 3D Measurements

Early interpretation of geophysical data were usually simple one- and two-dimensional models of geologic conditions. Primarily because of personal computers and the associated development of software, interpretations are now commonly much more detailed two-dimensional cross sections (Figures 11.14.7 and 11.14.19) and sometimes three-dimensional images of subsurface conditions.

A caution is in order here because software used to develop such images can and does use extensive interpolation, extrapolation, and often creates anomalous features in the data. Therefore, the computer images may not be an accurate representation of actual field conditions but only approximate conditions. As always, professionals with geologic, hydrologic, and contaminant knowledge and hands-on site experience are necessary to interact with the development and interpretation of such images.

People

Unfortunately, there are no procedures, guidelines, software, hardware, or graphics that will, by themselves, minimize geologic uncertainties and improve upon inappropriate data, insufficient data, inaccurate data, or poor critical thinking skills. What is needed is a simple back-to-basics, multidisciplinary approach using senior experienced professionals and good science to reduce the uncertainty to an acceptable level. Clearly, the most critical component of the site characterization process is the direct hands-on involvement of senior professionals with knowledge, experience, and motivation to resolve the uncertainties.

The ESC strategy of a core team demands the use of hands-on experienced professionals who participate in the acquisition of data, are on-site on a daily basis, and remain on the project from conception through final report. This provides the critical hands-on professional touch and technical continuity to the program.

REFERENCES

American Society for Testing and Materials, *Standard Guide for Planning and Conducting Borehole Geophysical Logging*, ASTM Committee D-18 on Soil and Rock, Surface and Subsurface Characteristics, ASTM D5753, 1995.

American Society for Testing and Materials, *Practice for Expedited Site Characterization of Vadose Zone and Ground Water Contamination at Hazardous Waste Contaminated Sites*, ASTM 6235, 1997a.

American Society for Testing and Materials, "Guide for Selecting Surface Geophysical Methods," ASTM D6429, 1998.

Annan A. P., *Ground Penetrating Radar Workshop Notes*, Sensors & Software Inc., Mississauga, Ontario, 1992.

Beam, P., Benson, R. C., and Hatheway, A. W., *Lessons Learned—A Brief History of Site Characterization*, HazWaste World/Superfund XVIII, December 2-4, Washington, D.C., 1997.

Benson, R. C., "ESC: Just Another New Acronym or Maybe Something Meaningful," *Environmental & Engineering Geoscience*, Vol. III, No. 3, pp. 453–456, 1997.

Benson, R. C., *Expedited Site Characterization for Olson Enterprises, Inc.*, Short course Notes, Charlotte, North Carolina, 1999.

Benson, R. C., Bevolo, A., and Beam, P., ESC: How it Differs from Current State of the Practice, *Proceedings of the Symposium on the Application of Geophysics to Environmental and Engineering Problems*, Environmental and Engineering Geophysical Society, Wheat Ridge, Colorado, pp. 531–540, 1998.

Benson, R. C., and Glaccum, R. A., "Radar Surveys for Geotechnical Site Assessment," in *Geophysical Methods in Geotechnical Engineering*, Specialty Session, American Society of Civil Engineers, Atlanta, Georgia, pp. 161–178, 1979.

Benson, R. C., Glaccum, R. A., Noel, M. R., *Geophysical Techniques for Sensing Buried Wastes and Waste Migration*, For the Environmental Protection Agency, Published by the National Water Well Association, 1982.

Benson, R. C., and Pasley, D., *Ground Water Monitoring—A Practical Approach for Major Utilities*, The Fourth National Symposium and Exposition on Aquifer Restoration and Ground Water Monitoring, National Water Well, Columbus, Ohio, 1984.

Benson, R. C., Turner, M., Turner, P., and Vogelson, W., "In Situ, Time-Series Measurements for Long-Term Ground-Water Monitoring," in *Ground Water Contamination: Field Methods*, ASTM STP-963, Collins, A. G., and Johnson, A. I., eds., American Society for Testing and Materials, pp. 58–72, 1988.

Benson, R. C., Turner, M., Vogelson, W., and Turner, P., "Correlation Between Field Geophysical Measurements and Laboratory Water Sample Analysis," *Proceedings of the National Water Well Association/Environmental Protection Agency Conference on Surface and Borehole Geophysical Methods in Ground Water Investigations*, National Water Well Association, 1985.

Benson, R. C., and Yuhr, L., "Assessment and Long Term Monitoring of Localized Subsidence Using Ground Penetrating Radar," *Proceedings of the Second Multidisciplinary Conference on Sinkholes and the Environmental Impact of Karst*, Orlando, Florida, 1987.

Benson, R. C., and Yuhr, L., *Evaluation of Fractures in Silts and Clay Using Ground Penetrating Radar*, Presented at the 4th Radar Conference, Denver, Colorado, 1990.

Benson, R. C., and Yuhr, L., *Short course notes: An introduction to geophysical techniques and their applications for engineers, geologists, and project managers*, SAGEEP Conference, April 21, Keystone, Colorado, 1996.

Breiner, S., "Applications Manual for Portable Magnetometers," *Geometrics*, Sunnyvale, California, 1973.

Burger, H. R., *Exploration Geophysics of the Shallow Subsurface* (Accompanying Macintosh Computer Software), Prentice Hall, Englewood Cliffs, NJ, 1992.

Butler, Dwain K., *Microgravimetric Techniques for Geotechnical Applications*, Miscellaneous paper GL-80-13, U.S. Army Engineer Waterway Experiment Station, CE, Vicksburg, Miss, p. 121, 1980.

Cartwright, K., and McComas, M., "Geophysical Surveys in the Vicinity of Sanitary Landfills in Northeastern Illinois," *Ground Water*, 6: 23–30, 1968.

Corwin, R. F., "The Self-Potential Method for Environmental and Engineering Applications," *Geotechnical and Environmental Geophysics*, Vol. I: Review and Tutorial, Ward, S. H., ed., Soc. of Explor. Geophy., 1990.

EPA, *Subsurface Characterization and Monitoring Techniques, A Desk Reference Guide*, Vol. I EPA/625/R-93/003a, 1993a.

EPA, *Subsurface Characterization and Monitoring Techniques, A Desk Reference Guide*, Vol. II. EPA/625/R-93/003b, 1993b.

Fitterman, D. V., and Stewart, M. T., "Transient Electromagnetic Sounding for Groundwater," *Geophysics*, 51(4): 905, 1986.

Fookes, P. G., "Geology for Engineers: The Geological Model, Prediction and Performance," *Journal of Engineering Geology*, 30: 293–424, 1997.

Freeze, A., and Cherry, J., "What Has Gone Wrong—A Guest Editorial," *Groundwater*, 27(4): July-August, 1989.

Greenhouse, J. P., and Monier-Williams, M., "Geophysical Monitoring of Ground Water Contamination Around Waste Disposal Sites," *Ground Water Monitoring Review*, 5(4): 63–69, 1985.

Greenhouse, J., Brewster, M., Schneider, G., Redman, D., Annan, P., Olhoeft, G., Lucius, J., Sander, K., and Mazzella, A., "Geophysics and Solvents: The Borden Experiment," *The Leading Edge*, April, 261–267, 1993.

Griffith, D. H., and King, R. F., *Applied Geophysics for Engineers and Geologists*, Pergamon Press, 1969.

Haegeman, W., and Vam Impe, W. F., "Characterization of Disposal Sites from Surface Wave Measurements," *JEEG*, 4(1): 27–33, 1999.

Haeni, P., "Application of Seismic-Refraction Techniques to Hydrologic Studies," *U.S. Geological Survey Techniques of Water Resources Investigations*, Book 2, Chapter D2, 1988.

Hunter, J. A., Burns, R. A., Good, R. L., MacAulay, H. A., and Cagne, R. M., "Optimum Field Techniques for Bedrock Reflection Mapping with the Multichannel Engineering Seismograph," in *Current Research*, Part B, Geological Survey of Canada, Paper 82-1 Part B, pp. 125–129, 1982.

Hvorslev, J., *Subsurface Exploration and Sampling of Soils for Civil Engineering Purposes*, Waterway Experiment Station, Vicksburg, MI, 1949, reprinted by ASCE, 1965.

Kaufmann, R. D., and Benson, R. C., "Character Counts; Site Characterizations Find Future Redevelopment Obstacles," *Brownfield News*, September, pp. 22–24, 1999.

Kratochvil, G. R., Benson, R. C., and Fenner, T., "Fishing for Fissures," *The Military Engineer*, Jan., Feb., 1992.

Loke, M. H., Electrical imaging surveys for environmental and engineering studies, A practical guide to 2-D and 3-D surveys, Penang, Malaysia, p. 47, 1999.

Liu, M., Aouad, M. F., Olson, L. D., Sack, D. A., Applications of the SASW method to pavements, structures, and geotechnical sites, Paper No. P200, Olson Engineering, Inc., Wheat Ridge, CO, 1998.

McNeill, J. D., "Electrical Conductivity of Soils and Rocks," *Technical Note TN-5*, Geonics. Ltd., Mississauga, Canada, 1980a.

McNeill, J. D., "Electromagnetic Resistivity Mapping of Contaminant Plumes," in *Proceedings of National Conference on Management of Uncontrolled Hazardous Waste Sites, Washington*, D.C., pp. 1–6, 1980b.

McNeill, J. D., "Applications of Transient Electromagnetic Techniques," *Technical Note TN-7*, Geonics. Ltd., Mississauga, Canada, 1980c.

Mooney, H. M., "Engineering Seismology," in *Handbook of Engineering Geophysics*, Vol .1; Bison Instruments, Minneapolis, Minnesota, 1973.

Mooney, H. M., *Handbook of Engineering Geophysics*, Vol. 2: Electrical Resistivity. Bison Instruments, Minneapolis, Minnesota, 1980.

Olhoeft, G. R., "Geophysical Detection of Hydrocarbon and Organic Chemical Contamination," *Proceedings of the Symposium on the Application of Geophysics to Engineering and Environmental Problems*, v 2.0 April, Oakbrook, Illinois, 1992.

Olhoeft, G. R., "Geophysical Observations of Geological, Hydrological and Geochemical Heterogeneity," *Proc. of Symp. on the Application of Geophysics to Engineering and Environmental Problems*, Vol. 1, EEGS, 1994.

Olhoeft, G. R., Personal communication, 1998.

Palmer, D., *The Generalized Reciprocal Method of Seismic Refraction Interpretation*, Burke, K. B. S., ed., Dept. of Geology, University of New Brunswick, Fredericton, N. B., Canada, 1980.

Pullan, S., Figure provided by Geologic Survey of Canada, 1996.

Redpath, B. B., "Seismic Refraction Exploration for Engineering Site Investigations," *Technical Report E-73-4*, U.S. Army Engineer Waterways Experiment Station Explosive Excavation Research Laboratory, Livermore, CA, 1973.

Sara, M., Standard Handbook for Solid and Hazardous Waste Facility Assessment, Lewis Publishers, Boca Raton, Florida, 1994.

Shuirman, G., and Slosson, J. E., Forensic Engineering: Environmental Case Histories for Civil Engineers and Geologists, Academic Press, San Diego, California, 1992.

Steeples, D. W., and Miller, R. D., "Seismic Reflection Methods Applied to Engineering, Environmental, and Groundwater Problems," in *Geotechnical and Environmental Geophysics*, v. I, Review and Tutorial, Ward, S. H., ed., Soc. Explor. Geophy. Tulsa, OK, 1–30, 1990.

Stokoe, K. H. II, Wright, S. G., Bay, J. A., and Roesset, J. M., "Characterization of Geotechnical Sites by SASW Method," in Woods, R., ed., Geophysical Characterization of Sites, 15–26 International Sciences, New York, 1994.

Technos, Inc., Application Guide to the Surface Geophysical Methods, Technotes, Vol. 1—Surface Geophysics, 1992.

Technos, Inc., "A Systems Approach for Managing Saltwater Intrusion," Technotes Vol. 6—Saltwater Intrusion, 1993.

Technos, Inc., Applications guide to resistivity imaging (in process), 1999a.

Technos, Inc., Applications guide for non-destructive testing (in progress), 1999b.

Telford, W. M., Geldart, L. P., Sheriff, R. E., and Keys, D. A., *Applied Geophysics,* Cambridge University Press, 1982.

Terzaghi, K., and Peck, R. B., *Soil Mechanics in Engineering Practice*, 2nd ed., Wiley, NYC, NY, 1967.

U.S. Army Corps of Engineers (USACOE), "Geophysical Exploration," *Engineering Manual EM 1110-1-1802*, Washington, D.C., 1979.

Vichabian, Y., and Morgan, F. D., "Self Potential Monitoring of Jet Fuel Air Sparging," *Proceedings of the Symposium on the Application of Geophysics to Engineering and Environmental Problems (SAGEEP 1999)*, Environmental and Engineering Geophysical Society, Wheat Ridge, CO, pp. 549–553, 1999.

Ward, S. H., ed., "Geotechnical and Environmental Geophysics, Volume I, Review and Tutorial," *Investigations in Geophysics No. 5.*, Soc. of Expl. Geophy., 1990.

Ward, S. H., ed., "Geotechnical and Environmental Geophysics, Volume II: Environmental and Groundwater," *Investigations in Geophysics No. 5.*, Soc. of Expl. Geophy., 1990.

Ward, S. H., ed., "Geotechnical and Environmental Geophysics, Volume III Geotechnical," *Investigations in Geophysics No. 5.*, Soc. of Expl. Geophy., 1990.

Yuhr Lynn, Benson, R. C., and Butler, D., "Characterization of Karst Features using Electromagnetics and Microgravity: A Strategic Approach," *Symposium on the Application of Geophysics to Engineering & Environmental Problems*, April 18–22, San Diego, California, 1993.

Yuhr, L, Benson, R.C., and Sharma, D., "Uncertainty in the Geologic Setting and its Impact on Site Characterization," *Proceedings of Uncertainty'96—Uncertainty in the Geologic Environment: From Theory to Practice*, Geotechnical Special Publication No. 68, Shackelford, C. D., Nelson, P. P., and Roth, M. J. S., eds., ASCE, New York, NY, 1996.

Zohdy, A. A., Eaton, G. P., and Mabey, D. R., "Application of Surface Geophysics to Ground-Water Investigations," *Techniques of Water-Resources Investigations of the United State Geological Survey*, Chapter D1, 1974.

CHAPTER 12

TOXICOLOGY AND RISK

SECTION 12.1

DECISIONS IN THE FACE OF UNCERTAINTY

Jan Swider

Dr. Swider is a researcher in the Mechanical and Aerospace Engineering Department, UCLA, School of Engineering and Applied Science. His main interests are in human health and ecological and engineering systems risk assessments.

Many complex environmental actions require decisions to be made based on imperfect information. For complex systems in particular, dealing with uncertainty is an important but difficult task. However, decision-making methods used in such complex circumstances often neglect uncertainty and the fact that the provided information may be unreliable. Many decision makers use this information as if it were perfect, expecting in return that this will result in a good decision. Such an approach is not uncommon. The majority of the current approaches to environmental management avoid uncertainty. There are at least two quite obvious reasons for this bias: (1) decision makers prefer making unequivocal, justifiable decisions, which are often supported by specific laws and regulations, and (2) regulations are much easier to write and enforce if they are stated in deterministically absolute terms. Moreover, when uncertainty analysis is a part of a decision-making process, the interpretation of uncertainty is often a source of conflict between the management and the public. This is especially relevant to decisions about the remediation of contaminated sites involving affected communities. The perception of risk, uncertainty, and appropriateness of a course of action is frequently different for the risk managers and the public. The effective explanation of risks and uncertainties should be a part of risk communication that entails more than public relations.

Uncertainty is a key difficulty in risk management. Any type of analysis for complex systems obviously involves many uncertainties. Fortunately, there is an increased awareness of the need to evaluate uncertainties associated with environmental actions. For example, the latest EPA Guidelines for Ecological Risk Assessment (U.S. EPA, 1998) incorporates evaluation of uncertainty as a part of the risk assessment. It follows the trends of other risk assessment methodologies, mainly human health risk assessments.

The intention of this section is to present the thought process involved in making decisions and a few methods for better decision making using relevant information and accounting for uncertainties.

12.1.1 ELEMENTS OF DECISION

A decision is an allocation of resources, which is associated with selection among actions, choices, or efforts (DAS, 1997). It is irrevocable and can be reversed only by another decision. A decision maker, who has the authority over the resources, makes the decision in order to meet various objectives. There is some confusion about the meaning of an objective and a decision. An *objective* is, for instance, a task to clean up a contaminated site. That would lead to a *decision* (e.g., about what is the best option, what actions need to be undertaken, and how much money should be allocated). Decision making is not always a rational process—it is driven by our value system and perception of uncertainties (Rubinstein,

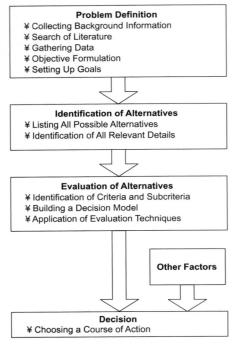

Problem Definition
¥ Collecting Background Information
¥ Search of Literature
¥ Gathering Data
¥ Objective Formulation
¥ Setting Up Goals

Identification of Alternatives
¥ Listing All Possible Alternatives
¥ Identification of All Relevant Details

Evaluation of Alternatives
¥ Identification of Criteria and Subcriteria
¥ Building a Decision Model
¥ Application of Evaluation Techniques

Other Factors

Decision
¥ Choosing a Course of Action

FIGURE   12.1.1 A simplified decision-making process.

1986). A typical value is economic (e.g., to increase someone's wealth) or social (e.g., decision of fairness or well being of a community). There is also a need for a goal. A goal is a measurable level of satisfaction. For instance, if the objective were to clean up a site, the goal would be to clean it up to a concentration x or below. A *decision-making process* consists of problem definition, identification, and evaluation of alternatives, and the decision, which is the culmination of the process. A simplified decision-making process is shown in Figure 12.1.1.

12.1.2   DECISION ANALYSIS

Decision theory provides a methodology for choosing a course of action when the consequences of such an action are not perfectly known. The necessity of making decision in the face of uncertainty is an integral part of our everyday lives. We often act without knowing exactly the outcomes of the decision. The decision becomes even more important when complex environmental or technological issues are at stake and a set of constraints changes rapidly. Uncertainty is a dominant factor in almost any type of decision.

Decision analysis is a set of procedures for assisting in the decision-making process, specifically for choosing among different alternatives when consequences of such choices are unclear. This methodology focuses on a situation when a decision maker faces a choice of action in an uncertain environment (Keeney and Raiffa, 1993). It uses various models to reflect the elements of decision, their relationships, and possible results of each course of action. This structured way of analyzing the decision allows one to better understand the effects of action in comparison with the objectives.

Decision-making process has two paradigms, the descriptive and the prescriptive. *Descriptive* approach describes how decisions are really made without any consideration of their efficiency, usability, or practicality. The focus is on factors that have an impact on the result of the decision-making process (i.e., leadership, behavior of an individual, or group interactions). The prescriptive approach is concerned with optimal decision making (i.e., selection of choices that best meet some prefatory criteria). This approach provides detailed procedures for describing decision problems, ranking potential outcomes, and verifying consistency, which are all required for making optimal decisions. Both description and prescription are important in decision analysis, although prescriptive approach is mainly used to assist in making the best decision.

The numerous simplifying assumptions and uncertainties in decision making have a tremendous impact on the predictive value of decision analysis. A cautious approach is recommended because the results of a decision analysis cannot be interpreted as absolute estimates of future effects. One has to keep in mind that decision analysis is only a tool to aid in making the best decision. It does not release a decision maker from responsibilities. Possibly, there are other constituents that a decision maker often has to consider in combination with the decision analysis estimates. In Figure 12.1.1 this is depicted by "Other Factors" box.

12.1.3   UNCERTAINTIES

In evaluation of toxicological or ecological issues from a risk management standpoint, our knowledge about the physics and various parameters is quite limited. Preferably, precise estimates of ecological risks should be based on models and information validated and built upon precise data. In most cases, precise data is lacking. Environmental risk management would be less controversial and easier to communicate if the results of risk management options could be accurately predicted. However, the usefulness of many decisions is put into question because of vast uncertainties. In such situations, uncertainty quantification, and probabilistic methods in particular, provide a means of representing subjective, or expert, judgments about uncertainties and help in quantification of these uncertainties in a way that allows for incorporating them in decision analysis. Probability distributions provide a way to express the state of knowledge about the phenomena or confidence in a parameter's value. They describe the likelihood of uncertain event in a quantitative way. This allows us to incorporate explicit knowledge about uncertainty into the decision-making process. Of course, uncertainty may be totally suppressed by using conservative estimates as a basis for risk management decision making. These conservative estimates tend to accumulate throughout the analysis to such an extent that the final result is overconservative. However, we cannot afford the cost associated with the decisions based on overconservative analyses.

Uncertainty is introduced into analysis in various ways and forms. It can be uncertainty with regard to a parameter value used in the analysis, imperfect knowledge about a model or completeness of gathered information. Section 12.2 discusses in detail the nature and sources of uncertainty and variability. Uncertainty exists at all levels of decision making from model formulation to inputs and outputs of the model. An example of sources of uncertainty and possible strategies to deal with them, as applied to the analysis phase of the risk assessment, are shown in Table 12.1.1.

12.1.4 QUANTIFICATION OF UNCERTAINTY

It is important to realize that in addition to efforts aiming at reduction of uncertainties through a set of strategies, as in Table 12.1.1, there is a need for a numerical evaluation of those uncertainties that are inherent in the analysis. These are usually parameter uncertainties and modeling uncertainties. Depending on a structure of a decision model there are some differences, but usually the evaluation of uncertainties involves: (1) estimation of uncertainties in the input parameters; (2) propagation of input uncertainties through the decision model; and (3) combination of uncertainties in the output in order to obtain the ultimate distribution of uncertainties in the final results. Various techniques can be applied to examine the effects of uncertain parameters on model prediction (Morgan and Henrion, 1990):

- Sensitivity analysis, for computing the effect of changes in input variables on model output
- Uncertainty propagation, for calculating the uncertainty in the model outputs induced by uncertainties in its inputs
- Uncertainty analysis, for comparing the importance of the input uncertainties measured by their contribution to uncertainty in the final outputs

There are a variety of methods for uncertainty propagation and analysis. These methods range from simple analytical approaches based on Taylor series expansion, to computationally extensive models such as Monte Carlo, Latin Hypercube Sampling, or response surface methods. However, there is no universal method that could be applied to every problem.

12.1.5 DECISION ANALYSIS METHODS

Decision-making processes use information to determine a course of action. To be of value, a piece of information must lead to a better choice than a choice made without it. To aggregate all vital information into a structure that would indicate a potential course of action, some evaluation methods are needed. There are numerous quantitative techniques for assisting decision makers who face problems involving uncertainty and complexity. These include rating methods, ranking procedures, utility function, multiattribute decision models, Analytic Hierarchy Process (AHP), and others. Only a few techniques will be described briefly in this section. Not all of these methods allow for a direct inclusion of uncertainty, but most of them allow some sort of sensitivity analysis. Because most of the variation of output values is generally caused by a small number of input variables, sensitivity analysis

TABLE 12.1.1   Sources of Uncertainties and Example Strategies (Adapted from U.S. EPA, 1998)

Source of uncertainty	Example strategies	Specific example
Unclear communication	• Contact principal investigator or other study participants if objectives or methods are unclear • Document decisions made during the course of the assessment	• Clarify whether the study was designed to characterize local populations or regional populations • Discuss rationale for selecting the critical toxicity study
Descriptive errors	• Verify that data sources followed appropriate procedures	• Double-check calculations and data entry
Variability	• Describe heterogeneity using point estimates (e.g., central tendency and high end) or by constructing probability or frequency distributions • Differentiate from uncertainty due to lack of knowledge	• Display differences in species sensitivity using a cumulative distribution function
Data gaps	• Collect needed data • Describe approaches used for bridging gaps and their rationales • Differentiate science-based judgments from policy-based judgments	• Discuss rationale for using any extrapolation factor between a lowest-observed-adverse-effect level (LOAEL) and a NOAEL
Uncertainty about a quantity's true value	• Use standard statistical methods to construct probability distributions or point estimates • Evaluate power of designed experiments to detect differences • Collect additional data	• Present the upper confidence limit on the arithmetic mean soil concentration, in addition to the best estimate of the arithmetic mean
Model structure uncertainty (process models)	• Discuss key aggregations and model simplifications • Compare model predictions with data collected in the system of interest	• Discuss combining different species into a group based on similar feeding habits
Uncertainty about a model's form (empirical models)	• Evaluate whether alternative models should be combined formally or treated separately • Compare model predictions with data collected in the system of interest	• Present results obtained using alternative models • Compare results of a plant uptake model with data collected in the field

allows for focusing on these parameters to reduce their uncertainty, which is basically adequate for less complex problems.

While rating methods are rather a straightforward evaluation of available alternatives using priority numbers, the ranking procedures are a little more sophisticated although still simple. A ranking procedure, also called a decision matrix or a weighted evaluation technique, allows for a detailed evaluation of alternatives based on assumed criteria and the weighting of the criteria according to the decision-making preferences. An example of such a matrix is presented in Figure 12.1.2 Each alternative is ranked with regard to criteria using a scale from 1 to 10, and criteria have associated weights emphasizing their relative importance. The weights are applied to each assigned value and the results are added for each alternative. The alternative with the highest overall score is considered to be the best choice. This is a deterministic evaluation and sensitivity analysis can be applied. Weighted evaluation technique is a handy tool for small matrices. For more complex decision evaluation, involving many alternatives and criteria, a more precise technique like Analytical Hierarchy Process method is needed to avoid inconsistencies in the evaluation process.

Criteria	Weight	Alternatives		
		A	B	C
I	0.1	3 / 0.3	9 / 0.9	5 / 0.5
II	0.5	7 / 3.5	5 / 2.5	2 / 1.0
III	0.3	9 / 2.7	7 / 2.1	4 / 1.2
IV	0.1	2 / 0.2	8 / 0.8	9 / 0.9
Sum	1.0	6.7	6.3	3.6

FIGURE 12.1.2   An example of a decision matrix.

In many decision problems the preference for a given alternative depends on future outcomes that cannot be accurately predicted. In some cases, such uncertainty about the outcomes can be explicitly introduced into the decision model. In other cases, the uncertainty remains implicit and will show its effect on the analysis as a result of changes in the preferences for various options. One way of introducing uncertainty about future outcomes into the decision model is through a decision tree. The decision tree consists of a set of branches that represent possible sequences of events or scenarios. It is a graphical representation of the problem for which all the options and outcomes have been identified. An attribute is a means of assessing the degree to which an objective can be achieved. Utility analysis allows a decision maker to assign utility values and preferences to consequences associated with various options. A utility function can be constructed for each problem to translate any given outcome or attribute value into a quantitative assessment of preference.

For monetary outcomes, an expected value can be calculated for any attribute associated with an event in a decision tree. The value assigned to each branch is multiplied by the probability of that branch, and the products are summed for each branch giving the expected value for that scenario. Comparing the expected monetary values for each decision scenario, the best option can be chosen. This concept can be generalized and instead of monetary values it can be applied to utility measurements. Having utilities associated with each outcome of a decision tree, the expected utility can be calculated for each scenario. The greatest value of expected utility indicates the preferred course of action. Both expected values and expected utilities can be used, if possible, to evaluate the options, but expected utilities are the key in solving decision problems under uncertainties. Expected utilities can be calculated for each attribute alone, but the utilities for several attributes can also be integrated by a means of multiattribute utility theory (Keeney and Raiffa, 1993).

Another method for analyzing a given decision problem is to structure the objectives and criteria of the decision in a meaningful way by constructing a hierarchy. The analytic hierarchy process (AHP) framework allows us to "think of complex problems in a simple way" (Saaty, 1990). The method involves breaking down a complex problem into its elements or variables, arranging them into a hierarchic order, assigning subjective numerical values reflecting the relative importance of each variable, and evaluating the options to determine the highest priority. It follows natural human behavioral and thought processes associated with decision making, mainly decomposition and synthesis (Saaty, 1988). There are three basic principles in building the AHP: (1) hierarchic structuring, breaking down the problem into smaller parts; (2) ranking the elements by relative importance—a pairwise comparison; and (3) verification of logical consistency of ranking.

After developing a hierarchy (Figure 12.1.3), the relative importance of all elements is evaluated using a scale from 1 to 9. This is achieved by pairwise comparisons against a given criterion. The resulting matrix allows the use of mathematical methods of verifying the consistency of the judgments

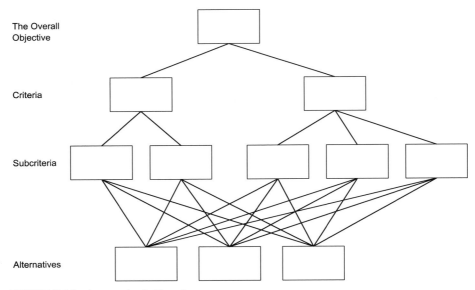

FIGURE 12.1.3   An example of a hierarchy.

by means of a consistency ratio. Sensitivity analysis can be used to investigate the sensitivity of alternatives to changes in the priorities of the criteria.

12.1.6   SUMMARY

Decisions in the face of uncertainty are usually a difficult issue. However, decision theory provides methods for either incorporation of uncertainties into a decision model or evaluation of the impact of uncertainties on the decision outcomes. It requires changes in both the decision-making reasoning and in appropriate laws and regulations. Utilization of uncertainty analysis in the scope of the decision-making process would lower overconservatism and reduced cost associated with environmental protection.

REFERENCES

DAS, *A Lexicon of Decision Making*, Decision Analysis Society, 1997.

Keeney, R. L., and Raiffa, H., *Decisions with Multiple Objectives: Preferences and Value Tradeoffs*, Cambridge University Press, 1993.

Morgan, M. G., and Henrion, M., *Uncertainty: A Guide to Dealing with Uncertainty in Quantitative Risk and Policy Analysis*, Cambridge University Press, 1990.

Rubinstein, M. F., *Tools for Thinking and Problem Solving*, Prentice-Hall, Inc., 1986.

Saaty, T. L., *The Analytic Hierarchy Process*, University of Pittsburgh, Pittsburgh, 1988.

Saaty, T. L., *Decision Making for Leaders: The Analytic Hierarchy Process for Decisions in a Complex World*, RWS Publications, 1990.

U.S. EPA, *Guidelines for Ecological Risk Assessment*, U.S. Environmental Protection Agency, EPA/630/R-95/002F, 1998.

CHAPTER 12
TOXICOLOGY AND RISK

SECTION 12.2

VARIABILITY AND UNCERTAINTY

William A. Huber

Dr. Huber maintains a private practice, serving clients by interpreting and presenting environmental data. He specializes in environmental statistics and geostatistics, decision analysis, and geographic information systems (GIS). He holds a Ph.D. in mathematics from Columbia University in the City of New York and a B.A. in philosophy and mathematics with high honors from Haverford College, Pennsylvania.

12.2.1 DEFINITIONS

Variability is diversity or heterogeneity in a well-defined population of individuals or series of observations (usually indexed by time or spatial location or both). *Uncertainty*, or imperfect knowledge, arises from ignorance or incomplete information. Generally, variability is a fact of nature while uncertainty is a property of our observations or descriptions of nature (knowledge). The two concepts overlap. For example, in the theory of sampling, variability manifests itself as *uncertainty* about the results of a prospective random sample of a population. Examples include chemical analyses of samples from soil, skin surface area measurements of adults in the United States, or a poll of voters regarding attitudes toward risk. In each case, any sample set is a collection of differing values. The values reflect variability in the population being sampled. The results will typically change from one sample set to another (unless the sample set is the entire population). This potential for change causes uncertainty in the interpretation of the results. *Statistical inference* is the process of making statements about the population from the results of samples (Snedecor, G. W., and Cochran, W. G., *Statistical Methods*, Eighth Edition, Iowa State University Press, p. 6, 1989).

Although the results of any future random sample may be unknown, population variability can, in principle, be characterized with certainty. The weights of all babies born at Albert Einstein Hospital in Philadelphia in January 1999, for example, were accurately measured and recorded. They vary considerably but there is little uncertainty about them.

In the theory of measurement, uncertainty manifests itself through *variability* in repeated measurements of the same thing. Examples include measuring the speed of light in a vacuum, weighing a person with a bathroom scale, or laboratory analysis of chemical concentrations in samples of environmental media. In each case, the results of repeated measurements usually change, although the underlying properties—speed, weight, concentration—are presumed constant and the measurement process itself is conducted in essentially the same way each time. *Random error* (or *imprecision*) is variation about the mean result. Any inferred deviation between this mean and the true value of the property being measured is usually known as *systematic error* or *inaccuracy*.

This definition of random error shows that uncertainty can result from variability (here, variability in a series of measurements). Uncertainty also has other causes. Cullen and Frey (1999, 21) group these into "model uncertainty" and "parameter uncertainty" (in addition to variability). They further divide these groups into 13 subcategories. Morgan and Henrion (1990, 56), an oft-quoted reference,

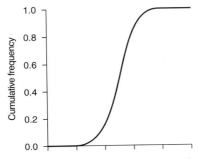

FIGURE 12.2.1 Cumulative distribution function: Variability, no uncertainty.

lists seven kinds of uncertainty, including variability. Various authors further augment these lists (see Section 12.2.3).

Variability can be quantified more precisely through additional sampling of a population or series. However, the amount of variability, being a property of the population or series, will not be thereby reduced.* Additional measurement can reduce some forms of uncertainty, such as the random error of an estimate, but it cannot reduce other forms, such as model uncertainty.

Many models use *distributions* to represent variability or uncertainty, or both. A distribution assigns nonnegative numerical values to ranges of possible values that a quantity might attain. The ranges are called *events*. The values are usually interpreted as frequencies (when modeling variability) or probabilities (when modeling uncertainty) of the events.[†] Where a distinction must be made, some authors use the terms *frequency distribution* or *probability distribution*, respectively. (As mathematical objects, however, the two are identical and are subject to the same mathematical manipulations.)

When multiple quantities simultaneously vary, events are simultaneous ranges of values within which each quantity may lie. A *multivariate* distribution assigns a probability (or frequency) to every possible event or combination of events. For example, a distribution of body weights will assign values to events such as "the weight is between 60 and 70 kilograms (kg)," "the weight is less than 10 kg," or "the weight is greater than 80 or less than 50 kg." A distribution of chemical concentrations in soil will assign values to events such as "the concentration is greater than 100 mg/kg." A multivariate distribution of body weights and soil concentrations will assign probabilities or frequencies to combined events, such as "the body weight is less than 50 kg and the concentration is greater than 77 mg/kg."

Distributions are depicted in various ways, including probability plots, density functions, and cumulative distribution functions. Probability plots compare distributions to a reference distribution. Density functions depict the infinitesimal probabilities or frequencies assigned to infinitesimal ranges of values and so are not universally applicable. A *cumulative distribution function* (*cdf*) always exists for any distribution defined for a variable attaining real values. Figure 12.2.1 shows a typical cdf.

12.2.2 ENVIRONMENTAL APPLICATIONS

A risk assessment evaluates "the likelihood, or probability, that the toxic properties of a chemical will be produced in populations of individuals under their actual conditions of exposure" (Rodricks, J. V., *Calculated Risks*, Cambridge U. Press, p. 48, 1992). Quantitative risk assessments employ a mathematical model for this evaluation. Variability within the populations and in the conditions of exposure can cause the model to estimate varying probabilities. Uncertainty in the model inputs and uncertainty in the choice and execution of the model may cause the risk assessment itself to be uncertain. To support risk management decisions, risk assessors commonly require a risk assessment model to account for this variability and uncertainty (Thompson, K. M., and Graham, J. D., "Going Beyond the Single Number: Using Probabilistic Risk Assessment to Improve Risk Management," *Hum. Ecol. Risk Assess*, 2(4): 1996).

The distinction between variability and uncertainty often depends on choices made by the risk assessor (Kelly, E., Campbell, K., Henrion, M., and Roy-Harrison, W., "To Separate or not to Separate—That is the Question," *SETAC News: Learned Discourses*, in press, 1999). How to account for uncertainty and variability is also a risk assessment choice. It depends on how the risk management

---

* Variability often can be reduced by focusing on identifiable subpopulations or by exploiting other observable factors with which a property is correlated ("covariates"). These approaches work by altering the population definition or the underlying probability model, rather than by obtaining new data.

[†] All distributions assign zero probability (or frequency) to the empty event and a probability (or frequency) of 1 (100 percent) to the "universal event" of all possible values.

question is framed, on the magnitudes of uncertainty and variability in the relevant data, and on the risk assessment model.

Framing the Risk Management Question

In the past, risk assessments in the United States often coped with variability and uncertainty by evaluating risk to the "reasonably maximally exposed" individual. In effect, variation among individuals in the target population was handled by assessing the maximum (or near-maximum) risk experienced by any individual within the population. At present (1999), guidance in the United States suggests that "information about the *distribution* of individual exposures is important to communicating the results of a risk assessment" (our emphasis) (U.S. Environmental Protection Agency, Guidance for Risk Characterization (Update), Science Policy Council (1995), "Guiding Principle 1." http://www.epa.gov/ordntrnt/ORD/spc/rcguide.htm). Evidently, the guidance envisions using a frequency distribution to represent exposures.

The same guidance, however, also states that "information about *population* exposure leads to another important way to describe risk" (our emphasis) (*Ibid.*, "Guiding Principle 2"). Population exposure is usually characterized using so-called "point estimates," such as a mean, median, or modal exposure. Carrying out such calculations is possible without explicitly accounting for variability within the population.

Effect of the Magnitudes of Uncertainty and Variability

Risk assessments often proceed in a "tiered" fashion, advancing from simple models and limited data that overestimate risk to more realistic and complex models using detailed data of higher quality. Gross, qualitative assessments of uncertainty and variability may suffice at the earliest stages of the process. For example, a preliminary investigation at a potentially contaminated site may fail to identify any chemicals of concern. The evident conclusion is that no risk management action need be taken. Provided this decision is not subject to large uncertainties (such as could be brought about by an inadequate investigation, for example), there is no need to refine the risk assessment.

At more advanced stages of this tiered process, risk assessments may employ extended models that explicitly incorporate uncertainty and variability in the model inputs. Preliminary calculation may identify some inputs whose variation or uncertainty do not materially influence the risk assessment results. To simplify the calculation and the presentation of the results, such inputs are routinely replaced by single values, thereby treating them as certain and invariable.*

Choice of Risk Assessment Model

Performing a quantitative risk assessment (that is, producing numerical instead of qualitative results) is not always possible, nor may it be accepted by all stakeholders, including government agencies (U.S. EPA, *ibid.*). The opportunities to account for uncertainty and variability in a nonquantitative risk assessment are thereby limited. Any quantitative risk assessment employs a numerical model. In that case, the modeler may choose to account for uncertainty and variability in qualitative or quantitative ways. Section 12.2.3, Types of Uncertainty and Variability, enumerates available methods.

12.2.3   TYPES OF UNCERTAINTY AND VARIABILITY

The following taxonomies of uncertainty and variability are organized around the practical needs to: (1) identify which forms of uncertainty and variability can be quantified, and (2) indicate generally accepted methods to quantify them.

---

* Sensitivity analyses are often conducted to support this modeling decision. See Section 12.2.3.

Brief Taxonomy of Uncertainty

Morgan and Henrion identify six kinds of uncertainty (distinct from variability). Two of these admit quantitative treatments: "random error" and "inherent randomness." Section 12.2.1 (above) describes the former, which statisticians represent using probability distributions. Morgan and Henrion refer to the latter as "a personalist view of randomness: You see a quantity as random if you do not know of any pattern or model that can account for its variation."

It is not necessary to perceive this randomness as "personalist" or subjective. A modeler may elect to represent a quantity as subject to random variation simply because it is too difficult, or unnecessary, to account for the variation in a determinate way, even if it is possible in principle to do so. For example, it is usual to represent body weights within a well-defined human population using a distribution rather than enumerating the weights (which in principle can be known) of all the individuals. In this case, the distribution represents variability in the quantity (body weight), not uncertainty. Using the distribution is a modeling decision that can be defended using objective criteria (time, expense, and so on).

Cullen and Frey also identify "lack of empirical basis" as a kind of uncertainty that may be quantified. This is uncertainty arising from the need to predict "something that has yet to be built, tested, or measured." Where it is possible to judge the potential range of outcomes, interval analysis (see below) or possibly fuzzy set theory (Zadeh et al., eds, Fuzzy Logic for the Management of Uncertainty, John Wiley & Sons, 1992) provide methods to quantify the uncertainty. Where information exists about similar systems, it may be possible to represent the uncertainty using a "Bayesian prior," which is a probability distribution assigned by the risk analyst based on judgment. Because data are not available, however, it is generally not possible to develop a distribution to represent the uncertainty and use it in the model as if it were based on data.

The general references provide many examples of uncertainty that cannot be quantified or are difficult to quantify. These include "subjective judgment," linguistic imprecision (vagueness and ambiguity in the description of information), and disagreement or uncertainty about which model is appropriate (Morgan and Henrion, 1990). "Model uncertainty" can include: (1) uncertainty because of simplifying assumptions, (2) approximations needed to execute a model (iteration step sizes, precision of intermediate results, and so on), (3) degree of empirical validation, (4) amount of extrapolation, and (5) degree of detail with which the model emulates natural processes (Cullen and Frey, 1999). Reviews of the accuracy of published model results suggest that the magnitude of errors potentially introduced by these uncertainties of judgment and modeling are as great as or greater than the uncertainties that can be quantified. Finally, "data uncertainty" can arise when several data sets measuring or representing the same quantity are available and indicate different values for the quantity, but the data sets cannot directly be combined. For example, chemical toxicity often can be estimated using animal bioassay data or epidemiological data.

There are effective methods to assess these unquantifiable uncertainties. Sensitivity analysis and "switchover analysis" are commonly used (Morgan and Henrion, 1990, 70–71). Both involve systematically changing uncertain quantities and rerunning the model to explore the effects on the risk assessment. Sensitivity analysis reports the size of the effect. Switchover analysis further identifies values at which the optimal or recommended decision changes. Such an analysis, therefore, requires an interaction between the risk assessor and risk manager, because it must incorporate the valuation criteria used to make risk management decisions.

Brief Taxonomy of Variability

Variability in risk assessments usually occurs within a population of individuals, over time, or over space. Separate names have not been assigned to distinguish these types of variability, although the adjectives "population," "temporal," and "spatial" are often used, respectively. For example, a risk assessment concerned with acute exposures to airborne chemicals would likely identify temporal changes in concentration as variability rather than uncertainty. As variability, the potential for concentrations to exceed a threshold would be interpreted as an exposure *frequency*; as uncertainty, the potential for concentrations to exceed a threshold would only be interpreted as an exposure *probability*. The two interpretations can lead to very different conclusions. For example, in a chronic exposure study, it is likely that only an average concentration would be relevant. Variation in concentration would be used to compute a degree of *uncertainty* in the average exposure using the statistical theory of sampling.

Note that this example, which concerns temporal variability, addresses variability within a theoretically infinite and continuous series of values. (For the series to be well-defined, a time frame and location for the air concentrations would have to specified.) It is a common abuse of language to refer to this series as a "population" also. The same comment applies to quantities that vary over space or over both space and time.

"Second Order" Uncertainty

Some authors distinguish "uncertainty about variability" from uncertainty about a (supposedly) unvarying quantity. The theory of measurement provides an illustrative example. Random error is assessed by replicating a measurement process a finite number of times. If the repeated results are normally distributed, then all the variation is characterized by a single value—the standard deviation (which quantifies the measurement precision). The standard deviation of the measurements themselves merely estimates the true standard deviation. (Tables are available to assess how uncertain this estimate may be.) Thus, if in a model, the true standard deviation represents variability, this variability is uncertain; if the true standard deviation represents uncertainty, the amount of this uncertainty is uncertain. This uncertainty—a form of imprecision—in a parameter that itself quantifies variability or uncertainty) in a model is said to be of "second order." A model that quantifies this uncertainty is termed a "second order model."

Second order uncertainty often arises from estimating variable quantities by sampling. The quantities are modeled as (frequency) distributions with one or more parameters left to be determined by the data. The parameters usually correspond to important quantities such as mean, standard deviation, or a percentile of interest. Data obtained through a representative, randomized sampling program will necessarily be random, so that estimates derived from them are uncertain. These uncertainties can be expressed by probability distributions known as "sampling distributions" (see Cullen and Frey, 1999, Section 5.5). Figure 12.2.2 through Figure 12.2.4 visualize different combinations of variability and uncertainty, including second order uncertainty. The rise from a height of zero to one of the cumulative distribution function in Figure 12.2.2 (no variability or uncertainty) occurs only at the value $X = 0$, indicating that all values in the underlying population or series of observations are zero: they do not vary. The value of zero is certain. Figure 12.2.3 (uncertainty but no variability) depicts three cumulative distribution functions, distinguished by their line types. Each shows an invariable result. The possible results, however, are -2, 0, and 2. In many applications a (discrete) probability distribution would also be

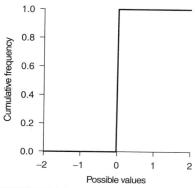

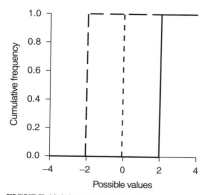

FIGURE 12.2.2 No variability, no uncertainty.   **FIGURE 12.2.3** No variability, but uncertainty.

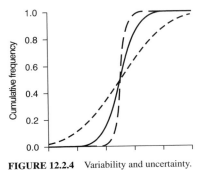

FIGURE 12.2.4 Variability and uncertainty.

used to describe the probability with which the true result is −2, 0, or 2. Figure 12.2.4 (variability and uncertainty) depicts three cumulative distribution functions distinguished by their line types. Each shows variable results. The amount of variability differs from curve to curve: the distribution with the finely dashed line is more "spread out": it takes longer to rise from bottom to top; it is the most variable of the distributions.

Each of the distributions in Figure 12.2.4 is normal. They have a common mean but differ in their standard deviations. The possible standard deviations are 0.4 times and 2.5 times the standard deviation of the middle distribution. In practice the uncertainty of a parameter such as the standard deviation would not be finely focused on just three values but would range over a continuous set of such values. Depicting that situation would create a continuous smear of distribution curves in this figure.

Where the set of different distributions is itself endowed with a probability density to quantify the uncertainty, this situation sometimes said by risk assessors to be "second order."

Bayesian techniques provide another means to specify second order uncertainty (see Section 12.2.4, below). Bayesian techniques also permit the use of third- and higher-order probability parameters, or "hyperparameters." Applications to the space-time modeling of chemicals in the environment have appeared (Wikle, C. K., Berliner, L. M., and Cressie, N., "Hierarchical Bayesian Space-Time Models," *Env. and Ecol. Stat.*, 5(2): 1998). These parameters usually describe levels of uncertainty in knowledge about the variability of model inputs.

12.2.4 TECHNIQUES TO SEPARATE UNCERTAINTY AND VARIABILITY

Many different techniques exist to identify, describe, and quantify uncertainty and variability. Table 12.2.1 enumerates those most commonly cited, beginning with the ones most frequently used. The methods differ practically and conceptually. Selection of a method depends partly on the nature of the variation or uncertainty encountered. Different techniques "propagate" variability and uncertainty in different ways. The following paragraphs describe those in greatest use.

TABLE 12.2.1 Techniques to Identify, Describe, Quantify, and Propagate Uncertainty and Variability

Technique	Use	Reference
Frequency distributions	Variability	General references
Probability distributions	Imprecision	General references
Bayesian	"Subjective" uncertainty	General references
Sensitivity analysis	Both	General references
"Switchover analysis"	Both	Morgan and Henrion
Expert evaluation	Both	Many
Interval analysis	"Ignorance"	Ferson and Ginzburg, 1996, *op. cit.*
Maximum entropy	Both	Grandy, W. T. Jr., and Schick, L. H., *Maximum Entropy and Bayesian Methods*, Kluwer, Dordrecht, 1991.
Fuzzy arithmetic	Ambiguity, vagueness, and imprecision	Zadeh et al., 1992, *op. cit.*
Qualitative assessment and description	Both	Many
Leave evaluation to users of the analysis	Both	Many

Convolution

Most risk assessment models produce numerical outputs that are well-defined functions of a finite number of inputs. When all inputs are constant in the model, the assessment is deemed "deterministic" and the model can be written in the form $\mathbf{Y} = f(\mathbf{X}; \mathbf{P})$ where $\mathbf{Y}$ is the vector of outputs, $\mathbf{X}$ the vector of inputs (usually empirical quantities), and $\mathbf{P}$ represents all other model "parameters." (The distinction between inputs and parameters is ambiguous—their roles can sometimes change—yet it reflects how modelers often think about a model.) For example, a standard model for lifetime average dose is (Cullen and Frey, 1999, Equation 2.1):

$$\text{Dose} = \text{Intake rate} \times \text{Concentration} \times \text{Exposure duration}/(\text{Body weight} \times \text{Lifetime})$$

here $\mathbf{Y} = (\text{Dose})$, $\mathbf{X} = (\text{Intake rate, Concentration, Exposure duration, Body weight, Lifetime})$, $\mathbf{P} = (\ )$, and $f(\mathbf{X}; \mathbf{P}) = x_1 \times x_2 \times x_3/(x_4 \times x_5)$.

This formula is intended to apply to an individual, not a population. It also assumes that all inputs and parameters are known with certainty. Here, we consider the cases where variability in an input can be represented by a (frequency) distribution and uncertainty in any input by a (probability) distribution. In the presence of uncertainty and variability, the output—dose—potentially is subject both to uncertainty and variability. Convolution (we use the term in a broad sense) is the appropriate mathematical operation for applying the function f(;) to variables represented by distributions. In general, the output of such a model consists of a (frequency) distribution subject to uncertainty, which itself is represented by a (probability) distribution.

Emulating Thompson and Graham (1996, *op. cit.*), we consider four cases:

1. Some inputs will be represented by frequency distributions but no uncertainty in the model will be represented by probability distributions. Let us say "there is variability but no uncertainty." The output will be a frequency distribution interpreted as variability.

2. There is uncertainty but no variability. The output will be a probability distribution interpreted as uncertainty.

3. Variability and uncertainty are both included. This includes two subcases. In Case 3A, the output will be a single distribution whose interpretation requires further discussion. In Case 3B, variability and uncertainty will be "separated." Output will be in the general form of frequency and probability distributions representing variability and uncertainty, respectively.

The mathematical treatment of cases 1 and 2 is identical. Only the interpretation differs. In either case, because the inputs may vary, so will the output. The modeler's task is to compute the output distribution. As noted in Section 12.2.1, a distribution assigns a number (representing frequency or probability) to any event describing a range of values for the output quantities. Given any such event E, the modeler must therefore determine the probability or frequency of the input values $\mathbf{X}$ for which $f(\mathbf{X}; \mathbf{P})$ lies in E.

There are many ways to carry out this calculation. *Analytical methods* specify probabilities or frequencies of simple events consisting of infinitesimal ranges of the variables. Using mathematical formulas, they directly compute the probabilities or frequencies of the corresponding output values. The computations are usually integrals. Where these integrals cannot be expressed in a simple form, they may be computed using *numerical integration* methods (Press, W. H. et al., *Numerical Recipes*, Cambridge U. Press, Cambridge, 1996). Approximation methods exist, such as Taylor series methods (see Cullen and Frey, 1999, Chapter 7; or Morgan and Henrion, 1990, Chapter 8). Perhaps the most popular method is *Monte Carlo* integration, whereby input values are repeatedly simulated with limiting frequencies given by their distribution. An output value is computed using the deterministic model in each repetition. The frequencies of the output values approximate the correct output distribution. Evidently, when all input distributions represent variability, the resulting output distribution is correctly interpreted as variability (Case 1 above); and when all input distributions represent uncertainty, the resulting output distribution is correctly interpreted as uncertainty (Case 2 above).

Cases 3A and 3B explicitly incorporate variability and uncertainty, representing both as distributions. In Case 3A, no distinction is made regarding the meaning of a distributional input. Frequency is confounded with probability. The resulting output distribution is often difficult or impossible to interpret. Case 3A can also arise by mistaking it as an instance of case 1 (all variability) or 2 (all uncertainty). Thompson and Graham (*op. cit.*, 1996, p. 1015) claim that "Case 3A may be used to determine the relative contributions of uncertainty and variability to . . . decide whether or not to use Case 3B." For example, if modeling the uncertain variables as probability distributions does not materially change the result, then a variability-only calculation should suffice. This will happen when variation in the input variables is much larger than combined uncertainty in all input variables.

Case 3B is the "two-dimensional" or "full" analysis. In it, the contributions of uncertainty and variability are separately tracked throughout. The output is a set of frequency distributions. Each distribution represents variability. The model allows any one of these distributions as a possible output, with some probability. Alternatively, the output can be represented as a variable set of results. Where variation occurs over individuals within a population, for example, each result represents the risk for one individual, subject to uncertainty (as given by a probability distribution). Many results are provided, thereby giving information about variation in risk.

Executing and reporting a two-dimensional analysis can be a challenge for the risk assessor. Interpreting and using the results can be a challenge for the risk manager. Therefore, this analysis is usually undertaken late in the tiered risk assessment process, if at all. Most two-dimensional analyses are conducted using Monte Carlo techniques, although any of the techniques used in the "one-dimensional" cases (1 and 2) can also, in principle, be employed. By separately tracking the cumulative effects of the uncertain and variable inputs, the output can be presented as described above.

In all four cases, special caution is needed when the model is not in the form $Y = f(X; P)$. For example, "back calculation" uses a model of the form $X_1 = f(X_2, Y; P)$, where the input variables are now partitioned into two sets X_1 and X_2, to evaluate Y. Although the process of convolution does follow well-defined algebraic rules, they are not the same as those governing the more familiar operations with numbers, and mistakes have been made by failing to recognize this and attempting algebraically to rewrite the model in its previous form.

Bayesian Updating

Many risk assessments are part of a decision process. Any risk assessment that uses distributions to represent variability is a statistical model. This puts such risk assessments within the framework of statistical decision theory.

There are at least four characteristic parts to a statistical decision model (Kiefer, J. C., *Introduction to Statistical Inference*, Springer-Verlag, New York, 1987):

1. A "state space" of possible outcomes of a random variable. (The available data are assumed to be realizations of that variable.)

2. A set of distributions representing "states of nature." Each is a candidate to be the distribution that actually determines the outcomes (the "underlying probability law").

3. A set of possible decisions. Often, these decisions can be identified (equated) with the set of distributions, because the risk assessor's problem frequently is to select one distribution to use for representing the data within the risk model.

4. A "loss function" that describes "how right or wrong the various possible decisions are for each possible underlying probability law."

This framework incorporates all knowledge of the random variable (prior to obtaining the data) by how the distributions are specified. For example, a risk assessor might restrict the soil concentrations to lognormal distributions only. Such a choice would reflect some knowledge—namely, that the distribution should have a lognormal shape—but not perfect knowledge, because two parametric values (geometric mean and geometric standard deviation) are allowed to vary.*

The risk analyst may choose to incorporate additional information by indicating that certain possible distributions are relatively more or less likely to be the right ones. This is done using an "a priori probability law" (which is expressed as a probability distribution). For example, if we think we know what the underlying mean should be for the logarithms of the concentrations (knowledge that is based, perhaps, on experience with similar situations), then we might provide an a priori distribution for the geometric mean, which has high probability near the suspected value and lower probabilities far from the suspected value. The spread of this a priori distribution would be interpreted as the amount of our uncertainty about the true state of nature.

In any case, making a decision carries its own statistical risk. The decision procedure can be poor or the data may unluckily appear to come from a distribution quite different from the true one (this is a problem when data are sparse). The amount of statistical risk depends on just what the true underlying distribution is. When an a priori distribution is available, it automatically endows these statistical risks with probabilities. The a priori expected value of the statistical risks is the "Bayes risk."

The "Bayes criterion" is to choose a decision procedure that minimizes the Bayes risk. When all this theoretical machinery is deciphered, it is seen by many analysts to be intuitively appealing: the optimal Bayes decision in effect "updates" the a priori distribution based on the data. The new ("posterior") distribution will have a mean reflecting both the a priori distribution and the data. It will have a variance reflecting uncertainty about the true mean. Furthermore, in many applications, applying the Bayes criterion leads to a simple computation. This combination of intuitive appeal and computational simplicity leads many risk assessors and decision analysts to adopt the Bayes approach by default.†

Kiefer [*ibid.*, pp. 33–35] warns of the possible dangers in this approach:

"...the determination of a Bayes procedure ... unfortunately presents the statistician with the temptation to assume [the a priori distribution] to be of some known form without much justification merely in order to be able to get an explicit answer quickly (and perhaps to avoid taking a course in mathematical statistics). This can lead to catastrophic consequences....In a long series of repeated experiments [i.e., decisions] too high a price will be paid...."

"The point is that an a priori law is not merely a vague representation of an intuitive feeling of ignorance, but a precise physical datum. Where this datum is not obtainable, or where (more basically) it does not make sense to think of [the underlying distribution] as a random variable, one cannot use the Bayes criterion."

Nevertheless, by explicitly incorporating distributions to represent variability (states of nature) and uncertainty (a priori probabilities), Bayesian approaches are viewed as promising tools for separately tracking uncertainty and variability in risk assessments.

Interval Analysis

It is not always appropriate to represent uncertainty by probability distributions (Ferson, S., and Ginzburg, L. R., "Different Methods are Needed to Propagate Ignorance and Variability," *Reliability*

---

* In principle the amount of "prior" information included within a statistical decision model can be closely controlled according to how tightly the possible states of nature are restricted. In practice, though, only two approaches are used because of the mathematical complications of developing optimal decision procedures. These are: (1) to use a small number of parameters to specify the shape, location, and spread of the distributions or (2) to remove almost all restrictions and allow a wide range of distributions without any natural finite parameterization (a "nonparametric" model).

† This discussion purposely omits any reference to a long running unresolved philosophical debate among statisticians regarding the meaning of "probability" because it is of little practical relevance. All the techniques described in this chapter are consistent with most "personal" and "frequentist" definitions of probability.

Engineering and System Safety, 54, 1996). Ferson and Ginzburg argue that it is incorrect to represent "partial ignorance resulting from *systematic* measurement error or subjective uncertainty" (our emphasis) as probability distributions in risk assessments. Their point is similar to Kiefer's; namely, that distributions should not be used as surrogates for "an intuitive feeling of ignorance." They propose that extremely uncertain quantities be represented by intervals of real numbers whose endpoints represent known bounds on these quantities. Thus, rather than using a convenient but arbitrary distribution (or set of distributions) to represent an uncertain value X known to lie between a and b, one would simply keep track of the consequences of the mathematical assertions "$X \geq a$" and "$X \leq b$".

The technique is mathematically valid and simple. Ferson and Ginzburg's (1996) paper provides useful and effective examples. However, an interval can be treated as a set of states of nature; namely, the interval from a to b could be represented in the classical statistical framework as the set of all possible probability distributions supported on that interval. It therefore appears that an appropriately formulated statistical approach and interval analysis would achieve equivalent results. Where interval analysis is applicable, it is conceptually and computationally simpler and may be worth considering in place of a possibly invalid Bayesian assumption.

Sensitivity Analysis

Sensitivity analysis consists of the systematic study of how a quantitative model's output changes with respect to controlled changes in its inputs or in the model itself. Thus, where an input or assumption is uncertain, sensitivity analysis can reveal whether the uncertainty has a material effect on the risk assessment.

Even when the uncertainty can be represented as a probability distribution (and therefore propagated in an analysis using convolution), sensitivity analysis has potential benefits. The computational cost and complexity of a sensitivity analysis is usually low, because only a small number of model runs are required. If the result indicates the uncertainty in question will have little or no effect on the outcome, then usually the analyst will select a reasonable fixed value for the uncertain quantity and avoid the complications of representing it as a distribution. Where the uncertainty cannot be represented as a probability distribution, a sensitivity analysis often can be conducted. For example, sensitivity of a risk decision to choice of model can be assessed.

When a model represents variability using distributions, its output can be a distribution or set of distributions also. A sensitivity analysis can explore how the distributional output changes under systematic changes in the uncertain parts of the model. Often, this is done using graphical methods. A sensitivity analysis can also determine whether it is appropriate to replace the entire distribution by a fixed value, with little resulting effect on the model output. Whence, sensitivity analyses can be used both to: (1) simplify the risk assessment by replacing probability or frequency distributions by fixed values, and (2) assess the effects of uncertainty on the risk assessment.

Most sensitivity analyses are limited by the difficulty of exploring possible interactions among separate uncertain variables. Usually, assumptions or uncertain quantities are systematically varied while leaving every thing else unchanged. Important changes in output that might result from simultaneous (correlated) changes in two or more uncertain factors can be overlooked. Sensitivity analyses differ according to how differences in output are measured. Cullen and Frey (1999, Chapter 8) describe and illustrate many possibilities.

12.2.5   *REGULATORY STATUS IN THE UNITED STATES*

The use of any risk assessment technique for environmental decision making is heavily influenced by guidance and policy established by U.S. regulatory agencies, Currently, in 1999, the U.S. Environmental Protection Agency (EPA) is aware of the need to characterize uncertainty and variability in human health and ecological risk assessments (U.S. EPA, *Policy for Use of Probabilistic Analysis in Risk Assessment* (May 15, 1997), http://www.epa.gov/ordntrnt/ORD/spc/probpol.htm):

"The importance of adequately characterizing variability and uncertainty in risk assessments has been emphasized in several science and policy documents. These include the 1992 U.S. Environmental Protection Agency (EPA) Exposure Assessment Guidelines, the 1992 EPA Risk Assessment Council (RAC) Guidance, the 1995 EPA Policy for Risk Characterization, the EPA Proposed Guidelines for Ecological Risk Assessment, the EPA Region 3 Technical Guidance Manual on Risk Assessment, the EPA Region 8 Superfund Technical Guidance, the 1994 National Academy of Sciences "Science and Judgment in Risk Assessment," and the report by the Commission on Risk Assessment and Risk Management."

Specific guidance on conducting probabilistic risk assessments is forthcoming.

GENERAL REFERENCES

Cullen, A. C., and Frey, H. C., *Probabilistic Techniques in Exposure Assessment*, Plenum Press, New York and London, 1999.

Morgan, M. G., and Henrion, M., *Uncertainty—A Guide to Dealing with Uncertainty in Quantitative Risk and Policy Analysis*, Cambridge University Press, Cambridge (England), 1990.

CHAPTER 12
TOXICOLOGY AND RISK

SECTION 12.3

SELECTING INDICATOR AND SURROGATE COMPOUNDS

Robert Alan Haviland
Mr. Haviland is a senior engineer with the Alaska District, U.S. Army Corps of Engineers. He is a graduate of the U.S. Military Academy at West Point.

Marlowe Dawag
Ms. Dawag is an environmental engineer with the Alaska District, U.S. Army Corps of Engineers. She is currently involved in risk evaluation and environmental compliance.

David T. Hanneman
Mr. Hanneman is a senior environmental scientist with the Alaska District, U.S. Army Corps of Engineers. He has specialized in environmental chemistry for over 18 years.

NOMENCLATURE AND UNITS

ADEC = Alaska Department of Environmental Conservation

ASTM = American Society for Testing and Materials

BCF = Bioconcentration Factor [mg/kg per mg/kg or, mg/L per mg/L]

BTEX = benzene, toluene, ethylbenzene, xylenes

CLF = chemical leaching factor [cm^3/g]

cm^2 = square centimeter

cm^3 = cubic centimeter

cPAH = carcinogenic polynuclear aromatic hydrocarbons

D_A = apparent diffusivity [cm^2/s]

D_i = Diffusivity in Air [cm^2/s]

D_w = Diffusivity in Water [cm^2/s]

DRO = diesel range organics

EPA = U.S. Environmental Protection Agency

EPH = Extractable Petroleum Hydrocarbons

f_{oc} = fraction organic carbon in soil [g/g]

g = gram

gm = gram

GRO = gasoline range organics

H' = Henry's Law Constant [dimensionless]

K_d = Soil-Water Partition Coefficient [cm^3/g]

K_{oc} = soil organic carbon partition coefficient [cm^3/g]

K_{ow} = Octanol-Water Partition Coefficient [cm^3/cm^3]

kg = kilogram

L = liter

m^3 = cubic meter

MADEP = Massachusetts Department of Environmental Protection

mg = milligram

n = total soil porosity [L_{pore}/L_{soil}]

PAH = polynuclear aromatic hydrocarbons

ppb = parts per billion

Q/C = inverse of the mean conc. at the center of a 0.5-acre-square source [g/m^2-s per kg/m^3]

RRO = residual range organics

s = second

S = Solubility in Water [mg/L]

T = exposure interval [s]

TEX = toluene, ethylbenzene, xylenes

TPHCWG = Total Petroleum Hydrocarbon Criteria Working Group

VF = volatilization factor [m^3/kg]

VPH = Volatile Petroleum Hydrocarbons

θ_a = air-filled soil porosity [L_{air}/L_{soil}]

θ_w = water-filled soil porosity [L_{water}/L_{soil}]

ρ_b = dry soil bulk density [g/cm^3]

ρ_s = soil particle density [g/cm^3]

12.3.1 INTRODUCTION

In the environmental field, risk assessments or evaluations are used to determine levels of risk to humans and ecological receptors and to answer the question of "how clean is clean." Risk calculations include the use of the chemical(s) of concern toxicity. This toxicity is determined from animal or human studies. Often, adequate toxicity information is not available for the chemical or mixture of chemicals under study. One answer to this problem is to leave that particular chemical or mixture out of the assessment and to address it qualitatively. Another answer is to use a chemical with known toxicity information as a substitute. This substitute is called an indicator or surrogate.

12.3.2 SUBSTITUTES

The two main requirements in the selection of a good substitute chemical for risk calculations are similar exposure characteristics and similar toxicity. A chemical impacts a receptor through exposure and toxicity. The exposure and toxicological effect of the substitute chemical to the receptor(s) should be similar to that of the chemical of concern. In essence, the substitute chemical should be as close as possible to the chemical of concern in physical, chemical, and toxicological properties. Structural similarities between the proposed substitute and chemical of concern improve the chances of similar properties.

Exposure

Similar exposure characteristics are critical for a substitute chemical in risk calculations. Exposure may involve direct contact of the chemical at its source, or contact at a remote point after the chemical travels through the environment. Volatilization, leaching, wind-borne dust, water-borne soil, groundwater transport, surface water transport, skin adherence, and uptake by plants and animals are examples of transport mechanisms. The chemical may also bond tightly to the soil matrix and tend not to travel through the environment. Transport through the environment subjects the chemical to changes that will affect its mobility and toxicity. During transport, the chemical is exposed to chemical, physical, and biological reactions that may reduce the chemical concentration or produce a different chemical with different properties. Photolysis, oxidation, hydrolysis, and anaerobic and aerobic biodegradation are reactions that may alter the chemical. Therefore, it is critical that a substitute mirror as many of the important physical, chemical, and environmental fate parameters of the chemical of concern as possible (see Table 12.3.1). This reduces the uncertainty associated with the substitution.

A minimization of uncertainty is possible through the selection of the exposure characteristics particular to the study at hand. Inhalation of vapors and ingestion of contaminated groundwater are two common exposure pathways. Concentrating on the exposure characteristics, or fate and transport parameters, that are usually connected with these exposure pathways reduces uncertainty. The EPA

TABLE 12.3.1 Important Physical, Chemical, and Environmental Fate Parameters

Parameter	Description
Organic-Carbon Partition Coefficient (K_{oc})	Provides a measure of the extent of chemical partitioning between organic carbon and water at equilibrium. The higher the K_{oc}, the more likely a chemical is to bind to soil or sediment than to remain in water.
Soil-Water Partition Coefficient (K_d)	Provides a soil or sediment-specific measure of the extent of chemical partitioning between soil or sediment and water, unadjusted for dependence upon organic carbon. To adjust for the fraction of organic carbon present in soil or sediment (f_{oc}), use $K_{oc} \times f_{oc}$. The higher the K_d, the more likely a chemical is to bind to soil or sediment than to remain in water.
Octanol-Water Partition Coefficient (K_{ow})	Provides a measure of the extent of chemical partitioning between water and octanol at equilibrium. The greater the K_{ow} the more likely a chemical is to partition to octanol than to remain in water. Octanol is used as a surrogate for lipids (fat), and K_{ow} can be used to predict bioconcentration in organisms.
Solubility in Water (S)	An upper limit on a chemical's dissolved concentration in water at a specified temperature. Aqueous concentrations in excess of solubility may indicate sorption onto sediments, the presence of solubilizing chemicals, such as solvents, or the presence of a nonaqueous phase liquid.
Henry's Law Constant (H')	Provides a measure of the extent of chemical partitioning between air and water at equilibrium. The higher the Henry's Law constant, the more likely a chemical is to volatilize than to remain in the water.
Vapor Pressure	The pressure exerted by a chemical vapor in equilibrium with its solid or liquid form at any given temperature. It is used to calculate the rate of volatilization of a pure substance from a surface or in estimating a Henry's Law constant for chemicals with low water solubility. The higher the vapor pressure, the more likely a chemical is to exist in a gaseous state.
Diffusivity in Air (D_i); Diffusivity in Water (D_w)	Describes the movement of a molecule in a gas or liquid medium as a result of differences in concentration. It is used to calculate the dispersive component of chemical transport. The higher the diffusivity, the more likely a chemical is to move in response to concentration gradients.
Bioconcentration Factor (BCF)	Provides a measure of the extent of chemical partitioning at equilibrium between a biological medium, such as fish tissue or plant tissue, and an external medium such as water. The higher the BCF, the greater the accumulation in living tissue is likely to be.
Media-specific Half-life	Provides a relative measure of the persistence of a chemical in a given medium, although actual values can vary greatly depending on site-specific conditions. The greater the half-life, the more persistent a chemical is likely to be.

Modified from: EPA, "Risk Assessment Guidance for Superfund, Vol. I," *Human Health Evaluation Manual*, (Part A), Washington, DC, 1989; and EPA, *Soil Screening Guidance: User's Guide*, 1996.

Soil Screening Guidance (EPA 1996) uses simple analytical equations to model chemical fate and transport. These equations are based on the physical/chemical parameters important to a specific migration pathway. While based on simplifying assumptions, these equations provide a valid method to compare the exposure characteristics of a potential substitute chemical to the chemical of concern.

The physical/chemical parameters in the inhalation of volatile vapors from contaminated soil are contained in a parameter called the soil-to-air volatilization factor (VF). In the Soil Screening Guidance, standard defaults are used for the environmental variables such as dry soil bulk density, soil particle density, exposure interval, among others (see Table 12.3.2).

The physical/chemical parameters in the partitioning equation for migration to groundwater from contaminated soil are the soil-water partition coefficient (K_d) and Henry's Law constant in a direct relationship. This relationship is the chemical leaching factor shown in Table 12.3.3.

TABLE 12.3.2 Derivation of the Volatilization Factor

$$VF = \frac{Q/C \times (3.14 \times D_A \times T)^{1/2} \times 10^{-4}}{(2 \times \rho_b \times D_A)}$$

where

$$D_A = \frac{\left[\left(\theta_a^{10/3} D_i H' + \theta_w^{10/3} D_w\right)\big/n^2\right]}{\rho_b K_d + \theta_w + \theta_a H'}$$

Parameter, definition (units)	Default
VF, volatilization factor (m^3/kg)	
D_A, apparent diffusivity (cm^2/s)	
Q/C, inverse of the mean conc. at the center of a 0.5-acre-square source (g/m^2-s per kg/m^3)	68.81
T, exposure interval (s)	9.5×10^8
ρ_b, dry soil bulk density (g/cm^3)	1.5
θ_a, air-filled soil porosity (L_{air}/L_{soil})	$n - \theta_w$
n, total soil porosity (L_{pore}/L_{soil})	$1 - (\rho_b - \rho_s)$
θ_w, water-filled soil porosity (L_{water}/L_{soil})	0.15
ρ_s, soil particle density (g/cm^3)	2.65
D_i, diffusivity in air (cm^2/s)	chemical-specific
H', dimensionless Henry's Law constant	chemical-specific
D_w, diffusivity in water (cm^2/s)	chemical-specific
K_d, soil-water partition coefficient (cm^3/g) $= K_{oc} \times f_{oc}$ (organics)	chemical-specific
K_{oc}, soil organic carbon partition coefficient (cm^3/g)	chemical-specific
f_{oc}, fraction organic carbon in soil (g/g)	0.006 (0.6%)

TABLE 12.3.3 Derivation of the Chemical Leaching Factor

$$CLF = K_d + \frac{\theta_w + \theta_a H'}{\rho_b}$$

Parameter, definition (units)	Default
CLF, chemical leaching factor (cm^3/g)	
ρ_b, dry soil bulk density (g/cm^3)	1.5
θ_a, air-filled soil porosity (L_{air}/L_{soil})	$n - \theta_w$
θ_w, water-filled soil porosity (L_{water}/L_{soil})	0.15
H', dimensionless Henry's Law constant	chemical-specific
K_d, soil-water partition coefficient (cm^3/g) $= K_{oc} \times f_{oc}$ (organics)	chemical-specific
K_{oc}, soil organic carbon partition coefficient (cm^3/g)	chemical-specific
f_{oc}, fraction organic carbon in soil (g/g)	0.006 (0.6%)

A comparative listing of the important physical, chemical, and environmental fate parameters and the volatilization and chemical leaching factors is a preferred method to select substitute chemicals. In addition to similar exposure characteristics, the substitute chemical should also have similar toxicity.

Toxicity

Because our chemical of concern does not have reference dose or carcinogenic slope factor toxicity values (if it did, direct risk calculations could be completed and a substitute would not be needed), it is necessary to find a substitute with similar toxicity. This may be accomplished by the comparison of benchmark or screening concentrations, health-based concentrations, or the comparison of toxicological study results. To reduce the uncertainty, the background studies that support these comparisons should be similar. For instance, if the exposure pathway of concern is human ingestion, oral toxicological studies completed on mammals (rats, rabbits, primates, etc.) would be more appropriate than sediment toxicity studies on benthic organisms. Likewise, we may find a reference that calculates a LC_{10} (lethal concentration at which 10 percent of the study group does not survive) value of 0.1 mg/L for our substitute chemical and a LC_{10} of 0.11 mg/L for the chemical of concern. While on the surface this appears to be a good comparison, the details of the studies should be examined. If one study was based on *Salmo irideus* (rainbow trout) and the other on *Perca flavescens* (yellow perch), the hardier nature of the perch compared to the trout may add more uncertainty than if both studies were on the same species. In addition, the length of the studies will affect the comparison. If one study was based on a 24 hours exposure and the other on chronic exposure, the uncertainty increases.

12.3.3   SURROGATES AND INDICATORS

Surrogates and indicators are both substitutes. Both terms have been used without having distinctive definitions. In general, but not always, it could be said that when one chemical is substituted for another chemical it is called a *surrogate*. When a chemical is substituted for a group of chemicals, it is an *indicator*.

For example, let us say that our subject chemical (chemical of concern) is acenaphthylene. Searching the primary toxicity reference, IRIS (the EPA's Integrated Risk Information System), we see the entry of "no data" for the oral reference dose. The standard secondary reference is HEAST (EPA's Health Effects Assessment Summary Tables). Consulting HEAST, we find the entry, "Data Inadequate for Quantitative Risk Assessment." If NCEA (EPA's National Center for Environmental Assessment) also has no information, the next source in the toxicity hierarchy is the open professional literature. From various studies, it may be possible to calculate a usable toxicity value.

An alternate approach is to find a chemical to act as a substitute or surrogate. Table 12.3.4 shows three possible surrogates, in addition to the chemical of concern, acenaphthylene. The more properties compared between the subject chemical and potential substitute, the closer the property values are, and the closer the conditions of measurement for these properties, the less uncertainty is associated with the substitution. With the exception of the Henry's Law constant and the diffusivities, the acenaphthene parameters are the closest to acenaphthylene. The organic-carbon partition coefficients indicate that acenaphthylene will bind to organic carbon (soil) in a manner similar to acenaphthene. The close correspondence between the volatilization factors and chemical leaching factors also indicate similar volatilization and leaching fate for acenaphthylene and acenaphthene. The toxicity comparison also indicates that these two chemicals share similar toxic effects.

The comparison above indicates acenaphthene is a good surrogate for acenaphthylene. The level of uncertainty with this substitution is low. We can use the toxicological reference dose for acenaphthene (6.00E-02 mg/kg/day) in acenaphthylene risk calculations. The comparison also indicates that the persistence and mobility (exposure) of acenaphthylene in the environment will be similar to that of acenaphthene.

12.3.4   PETROLEUM

Petroleum is a complex mixture of chemicals and, as such, is a good candidate for a substitute approach. Its constituents vary dramatically depending on the original crude oil stock, the refining process, finished product, and weathering. Specifications for finished petroleum products are not by detailed

TABLE 12.3.4 Acenaphthylene Potential Surrogates

Property	Acenaphthylene	Naphthalene	Acenaphthene	Pyrene
Organic-carbon partition coefficient, K_{oc} (cm^3/kg)	4.79E+06	2.00E+06	7.08E+06	1.05E+08
Log octanol/water partition coefficient	4.07	3.37	4.33	5.32
Solubility in water (mg/L)	3.9	31.0	4.24	0.135
Henry's law constant (dimensionless)	4.6E−04	1.98E−02	6.36E−03	4.51E−04
Vapor pressure (torr)	1E−03 − E−02	5E−01	1E−03 − E−02	6.85E−07
Diffusivity in Air, D_i (cm^2/s)	6.65E−02	5.90E−02	4.21E−02	2.72E−02
Diffusivity in Water, D_w (cm^2/s)	6.6E−06	7.50E−06	7.69E−06	7.24E−06
Bioconcentration Factor, BCF	380	215	380	365
Media-specific Half-life	total biodegradation in 7 days	total biodegradation in 7 days	total biodegradation in 7 days	total biodegradation in 14+ days
Volatilization Factor, VF (m^3/kg)	3.58E−09	2.27E−07	2.12E−08	6.54E−11
Chemical Leaching Factor, CLF (cm^3/gm)	43,100	18,000	63,700	945,000
Marine sediment toxicity threshold effects range (ppb)	5.9	31	6.7	152.6
Marine sediment toxicity effects range low (ppb)	44	160	16	665
Marine sediment toxicity effects range median (ppb)	640	2100	500	2600

Values modified from: EPA, *Water-Related Environmental Fate of 129 Priority Pollutants*, Vol. II, 1979; EPA, *Soil Screening Guidance: User's Guide*, 1996; NOAA, Screening Quick Reference Tables (SQuiRT), Seattle, WA, 1998; and Montgomery, John, H., *Groundwater Chemicals*, Lewis Pub., 1996, and should not be used without consulting the original sources.

chemical composition. Specifications for petroleum fuels fall more toward performance (boiling point, specific gravity, flash point, viscosity, cetane number, among others) than chemical composition.

Composition

Neglecting fuel additives, petroleum compounds are primarily composed of carbon and hydrogen. These petroleum hydrocarbons consist of four major groups, including alkanes, alkenes, cycloalkanes, and aromatics (see Table 12.3.5). Alkanes are also called paraffins or saturated aliphatics, and consist of straight or branched chain molecules with single bonds between the atoms. Alkenes are similar to alkanes but include one or more double bonds between carbon atoms. Alkenes are also called unsaturated aliphatics or olefins. Cycloalkanes are similar to alkanes but include a saturated ring

TABLE 12.3.5 Composition of Select Fuel Components

	Gasoline	Kerosene	Diesel #2/fuel oil #2	Bunker C/fuel oil #6
Carbon Range	4–12	9–16	9–20	21–45+
Alkanes	50–60%	50%	40–65%	20%
Alkenes	2%	1–2%	1–2%	1–2%
Cycloalkanes	5%	30%	35%	15%
Aromatics	15–40%	5–25%	25–35%	35%

in their structure. Cycloalkanes are also called cycloparaffins or naphthenes. Aromatic compounds include at least one benzene ring. They are unsaturated compounds. Aromatic compounds may have saturated or nonsaturated side chains.

"How Clean Is Clean" with Petroleum?

Many approaches have been tried over the years to answer the question, "how clean is clean?" with petroleum compounds. Initial approaches used some measurement of total petroleum hydrocarbons. All hydrocarbons, alkanes, alkenes, cycloalkanes, and aromatics within a carbon range were measured and lumped together into one number. This number was compared to a regulatory level that was initially derived from modeling efforts. However, because of the generalizations and assumptions inherent in the modeling, and the sometimes excessive safety factors applied, the use of a total petroleum hydrocarbon or "whole fuel" approach, sometimes lead to ultraconservative risk calculations with high uncertainty. The great variance of chemicals from one petroleum product to another and the effects of weathering make the whole product approach difficult to apply.

Another approach that attempted to answer how clean is clean with petroleum products was the indicator approach. BTEX was measured and the concentrations of the individual benzene, toluene, ethylbenzene, and xylenes coupled with their toxicity values were used to determine risk and cleanup levels. A shortcoming of this method is that while BTEX may carry the most risk for fresh gasoline, it is not necessarily so for weathered gasoline or other petroleum products. Variation of this indicator approach added other chemicals to the indicator list. Various polynuclear aromatic hydrocarbons (PAHs), metals, and fuel additives have been used at one time or another. Perhaps the most extensive use of indicator compounds was the use of the full EPA volatile organic compound method (EPA SW846 Method 8260) with the semivolatile organic compound method (EPA SW846 Method 8270). Between these two analytical methods, 250 chemicals were analyzed to determine the risk of a petroleum spill. However, of these 250, perhaps 40 (to include BTEX) are typically found in petroleum fuels. Even these 40 chemicals do not necessarily adequately characterize the risk to a petroleum spill. The majority of fuel products are composed of alkanes. Although these compounds are not the most toxic, they are present in larger percentages. In addition, these compounds remain at a spill site after the more degradable chemicals disappear. In order to evaluate a potential risk from these chemicals to a sensitive area (e.g., water supply well or residential area), it is necessary to measure the risk from all components. Stating that there is minimal risk usually does not suffice.

12.3.5 *INDICATOR/SURROGATE APPROACH*

The Massachusetts Department of Environmental Protection (MADEP)

MADEP initiated an indicator and surrogate approach for petroleum in 1994. This approach is based on toxicity only. Fate and transport parameters did not enter into the selection of the substitute chemicals. The indicator chemicals are individually measured and evaluated using standard EPA risk assessment methods. The remainder of the petroleum mixture is evaluated as subgroups. The subgroups are differentiated based on carbon number and fractionalization (aliphatic or aromatic). Each subgroup is evaluated using a surrogate. The surrogate chemical is a substitute for a subgroup.

The indicator chemicals are benzene, the carcinogenic polynuclear aromatic hydrocarbons (cPAHs), and TEX (toluene, ethylbenzene, and xylenes). The cPAHs are benzo(a)anthracene, benzo(b)fluoranthene, benzo(k)fluoranthene, benzo(a)pyrene, chrysene, dibenzo(a,h)anthracene, and indeno (1,2,3-cd)pyrene. The cPAHs were chosen for indicators, because they are among the most toxic found in petroleum fuels. Toluene, ethylbenzene, and the xylenes were included as indicator compounds for two reasons: (1) they form a large percentage of gasoline, and (2) they account for the toxicity from the 7- and 8-carbon aromatic chemicals.

TABLE 12.3.6 MADEP Indicators and Surrogates (1994)

Indicator chemicals	Carbon ranges and associated surrogate chemicals with reference doses (mg/kg/day)		
Benzene	aliphatic C5–C8	n-hexane	0.06
Ethylbenzene	aliphatic C9–C18	n-nonane	0.6
Toluene	aliphatic C19–C32	n-eicosane	6.0
Xylenes	aromatic C9–C32	pyrene	0.03
Benzo(a)anthracene			
Benzo(b)fluoranthene			
Benzo(k)fluoranthene			
Benzo(a)pyrene			
Chrysene			
Dibenzo(a,h)anthracene			
Indeno(1,2,3-cd)pyrene			

The surrogate chemicals are n-hexane, n-nonane, n-eicosane, and pyrene. The surrogate chemicals were chosen based on their toxicity values. MADEP examined the known toxicity values for fuel components within different carbon ranges and determined representative substitutes with acceptable levels of uncertainty. MADEP divided the aromatic and aliphatic (alkanes) into subgroups of compounds based on numbers of carbon atoms in the molecules in each subgroup. Alkenes are included within the aromatic group, and cycloalkanes are included in the alkane group. Uncertainty analysis found that this did not significantly affect the risk calculations. Each subgroup of aromatics and aliphatics were assigned a "reference" compound or surrogate.

The initial indicator and surrogate compounds are shown in Table 12.3.6. Lighter weight compounds are considered too volatile for exposure to be a significant factor or are captured during the evaluation of BTEX. MADEP (and the TPHCWG) consider the aliphatic compounds (>C34) to pose no risk.

Total Petroleum Hydrocarbon Criteria Working Group (TPHCWG)

The TPHCWG is a consortium of federal government agencies, state agencies, academia, oil companies, and professional institutes. The TPHCWG built on the initial MADEP approach. While MADEP initially looked only at toxicity (one of the two considerations for a good substitute), TPHCWG also examined mobility in the environment (the second consideration for a good substitute). The TPHCWG used fate and transport equations similar to those found in the EPA Soil Screening Guidance (EPA, 1996). TPHCWG used the American Society for Testing and Materials (ASTM) Standard Guide for Risk-Based Corrective Action Applied at Petroleum Release Sites (E 1739). As a result of this exercise, TPHCWG developed 13 surrogate subgroups.

TPHCWG retained the carcinogenic chemicals as indicator compounds, but dropped ethylbenzene, toluene, and the xylenes. The toxicity of these compounds is included within the appropriate aromatic fraction. Table 12.3.7 shows no surrogate chemicals for the various aliphatic and aromatic carbon ranges.* The toxicity analysis completed by the TPHCWG examined the available toxicity[†] of the chemicals within a mobility grouping and derived conservative representative toxicity value for each subgroup.

---

* TPHCWG uses an equivalent carbon (EC) number. This EC number is related to the boiling point of a chemical normalized to the boiling point of the n-alkanes or its retention time in a boiling point gas chromatographic column. Thus we see benzene with 6 carbon atoms with an EC of 6.5, and benzo(a)pyrene with 20 carbon atoms and an EC of 31.34.

[†] EPA approved reference doses, health based criteria, and results from animal studies.

TABLE 12.3.7   TPHCWG Indicators and Surrogates

Indicator chemicals	Carbon ranges and associated surrogate toxicity reference doses (mg/kg/day)	
Benzene	aliphatic C5–C6	5.0
Benzo(a)anthracene	aliphatic C>6–C8	5.0
Benzo(b)fluoranthene	aliphatic C>8–C10	0.1
Benzo(k)fluoranthene	aliphatic C>10–C12	0.1
Benzo(a)pyrene	aliphatic C>12–C16	0.1
Chrysene	aliphatic C>16–C21	2.0
Dibenzo(a,h)anthracene	aliphatic C>21–C35	2.0
Indeno(1,2,3-cd)pyrene	aromatic C>7–C8	0.2
	aromatic C>8–C10	0.04
	aromatic C>10–C12	0.04
	aromatic C>12–C16	0.04
	aromatic C>16–C21	0.03
	aromatic C>21–C35	0.03

12.3.6   APPLICATION

MADEP

With continuing research, to include review of the TPHCWG efforts, MADEP refined their indicator/surrogate approach. Because of analytical and program considerations, MADEP selected 6 petroleum fractions and 21 indicator chemicals. The indicator chemicals are those that have been used traditionally to characterize environmental contamination (See Table 12.3.8).

TABLE 12.3.8   MADEP Indicators and Surrogates (1997)

Indicator chemicals (c = carcinogen)	Carbon ranges and associated surrogate chemicals	
Benzene (c)	aliphatic C5–C8	n-hexane
Ethylbenzene	aliphatic C9–C12	n-nonane
Toluene	aliphatic C9–C18	n-nonane
Xylenes	aliphatic C19–C36	n-eicosane
Acenaphthene	aromatic C9–C10	pyrene
Acenaphthylene	aromatic C11–C22	pyrene
Anthracene		
Benzo(a)anthracene (c)		
Benzo(b)fluoranthene (c)		
Benzo(k)fluoranthene (c)		
Benzo(g,h,i)perylene		
Benzo(a)pyrene (c)		
Chrysene (c)		
Dibenzo(a,h)anthracene (c)		
Fluoranthene		
Fluorene		
Indeno(1,2,3-cd)pyrene (c)		
2-Methylnaphthalene		
Naphthalene		
Phenanthrene		
Pyrene		

TABLE 12.3.9 MADEP Petroleum Sampling Table for Soil (S) and Groundwater (GW)

Petroleum product	Petroleum fraction (surrogates)		Target analytes (indicators)			
	C5–C8 aliphatic C9–C12 aliphatic C9–C10 aromatic VPH	C9–C18 aliphatic C19–C36 aliphatic C11–C22 aromatic EPH	BTEX VPH	naphthalene VPH	PAHs EPH	Metals and/or solvents
Gasoline	S GW		S GW	S GW		
Jet Fuel JP–4/JP–8	S* GW	S GW	S† GW¶	S†	S (as appropriate) GW¶ (as appropriate)	
Diesel/#2 Fuel Oil	S† GW†, ‡	S GW	S† GW¶	S† GW¶	S§ GW¶	
#3–#6 Fuel Oils/ Hydraulic Oil/ Kerosene/Jet Fuel A	GW‡	S GW	S† GW¶	S†	S (as appropriate) GW¶ (as appropriate)	
Waste Oil (crankcase)/ Unknown	S* GW	S GW	S* GW		S GW	S VOCs* S heavy metals GW VOCs GW heavy metals
Mineral/Dielectric Oils	GW‡	S GW				

Modified after MADEP 1997; Additives excluded.

* Test if total organic vapor (TOV) headspace ≥10 ppmv.

† Test if TOV headspace ≥100 ppmv.

‡ Test if impacting a water supply.

§ If TPH >500 mg/kg, test for acenaphthene, naphthalene, 2-methyl naphthalene, and phenanthrene

¶ If in an area where groundwater may be a drinking water source, test for BTEX, acenaphthene, naphthalene, 2-methylnaphthalene, and phenanthrene.

      MADEP developed analytical methods to detect the petroleum fractions represented by the surrogates. The Volatile Petroleum Hydrocarbons (VPH) method was developed to detect the C5–C8 and C9–C12 aliphatic ranges and the C9–C10 aromatic range. In addition, BTEX and naphthalene are detected by the VPH method. The Extractable Petroleum Hydrocarbons (EPH) method was developed to detect the C9–C18 and C19–C36 aliphatic ranges and the C11–C22 aromatic range. In addition, 17 PAHs are also detected by the EPH method. MADEP developed guidance to aid in selecting the appropriate petroleum fractions (surrogates) and target analytes (indicators) for different petroleum fuels (see Table 12.3.9).

Alaska Department of Environmental Conservation (ADEC)

      The State of Alaska incorporated the indicator/surrogate approach into their current regulations. In preparation of the new regulation, ADEC evaluated current research to include MADEP and TPHCWG. ADEC analytical and program considerations resulted in 6 petroleum surrogates and 16 indicator chemicals (see Table 12.3.10).

      ADEC developed analytical methods to detect three ranges of petroleum hydrocarbons and their associated aromatic and aliphatic fractions. Method AK101 detects Gasoline Range Organics (GRO). Method AK102 detects Diesel Range Organics (DRO) and AK103 detects Residual Range Organics (RRO). The GRO method detects C6–C9. The DRO method detects C10–C24, and the RRO method detects C25–C36. In addition, BTEX is also detected by the GRO method. ADEC developed guidance to aid in selecting the appropriate petroleum fractions (surrogates) and target analytes (indicators) for different petroleum fuels (see Table 12.3.11).

TABLE 12.3.10   ADEC Indicators and Surrogates

Indicator chemicals (c = carcinogen)	Carbon ranges and associated surrogate chemicals	
Benzene (c)	aliphatic C6–C9	n-hexane
Ethylbenzene	aliphatic C10–C24	n-nonane
Toluene	aliphatic C25–C36	eicosane
Xylenes	aromatic C6–C9	toluene
Acenaphthene	aromatic C10–C24	fluorene
Anthracene	aromatic C25–C36	pyrene
Benzo(a)anthracene (c)		
Benzo(b)fluoranthene (c)		
Benzo(k)fluoranthene (c)		
Benzo(a)pyrene (c)		
Chrysene (c)		
Dibenzo(a,h)anthracene (c)		
Fluorene		
Indeno(1,2,3-cd)pyrene (c)		
Naphthalene		
Pyrene		

TABLE 12.3.11   ADEC Petroleum Sampling Table for soil (S) and groundwater (GW)

Petroleum prroduct	Petroleum fraction (surrogates)			Target analytes (indicators)		
	C6–C9 aliphatic C6–C9 aromatic GRO	C10–C24 aliphatic C10–C24 aromatic DRO	C25–C36 aliphatic C25–C36 aromatic RRO‡	BTEX GRO	PAHs*	Metals and /or solvents†
Aviation Gasoline/	S			S	S	
Gasoline	GW			GW	GW	
JP-4/Diesel #1/	S	S		S	S	
Arctic Diesel/JP-5, JP-8, Jet A/ Kerosene	GW	GW		GW	GW	
#2 Diesel		S		S	S	
		GW		GW	GW	
#3–#6 Fuel Oils		S	S	S	S	
		GW	GW	GW	GW	
Waste Oil	S	S	S	S	S	S
(crankcase)/ Unknown	GW	GW	GW	GW	GW	GW
Mineral/Dielectric		S	S	GW	S	
Oils		GW	GW		GW	

Modified after ADEC 1998; Additives excluded.

*PAH analysis for soils required for all releases when TPH is over 500 mg/kg. For gasoline and JP-4 only naphthalene is required. For all other products required PAHs are: naphthalene, fluorene, anthracene, pyrene, benzo(a)anthracene, acenaphthene, chrysene, benzo(a)pyrene, dibenzo(a,h)anthracene, benzo(b)fluoranthene, benzo(k)fluoranthene, and ideno(1,2,3-cd)pyrene.

† Metals include: arsenic, barium, cadmium, chromium, lead, nickel, and vanadium.

‡ Sampling for RRO in groundwater shall use the fractionalization for aromatic RRO.

12.3.7 CONCLUSION

Indicator and surrogate compounds are useful tools in risk assessments and risk evaluations. Indicators and surrogates provide useful toxicity information for risk calculations. This information is necessary for quantitative evaluations. The selection of the indicator or surrogate chemical requires examination of the exposure characteristics and toxicity of the potential substitute and the chemical of concern. Careful examination and selection will reduce the uncertainty associated with the substitution.

Methods incorporating indicators and surrogates have resulted in new approaches for petroleum risk assessments. Variations in crude stock, variations in refining processes, and weathering make petroleum fuel assessment a complex undertaking. The indicator and surrogate approach for petroleum has produced more accurate assessment of risks and has reduced the associated uncertainty.

REFERENCES

Alaska Department of Environmental Conservation, "Petroleum Cleanup Guidance," Draft, Juneau, December 1996.

Alaska Department of Environmental Conservation, "Underground Storage Tank Procedures Manual," Juneau, December 10, 1998.

Alaska Department of Environmental Conservation, "Oil and Hazardous Substances Pollution Control Regulations," 18 AAC 75, Juneau, January 22, 1999.

Massachusetts Department of Environmental Protection, "Interim Final Petroleum Report: Development of Health-Based Alternative to the Total Petroleum Hydrocarbon (TPH) Parameter," Boston, 1994.

Massachusetts Department of Environmental Protection, "Characterizing Risks posed by Petroleum Contaminated Sites: Implementation of MADEP VPH/EPH Approach," Boston, Public Comment Draft, October 31, 1997.

Total Petroleum Hydrocarbon Criteria Working Group, *Development of Fraction Specific Reference Doses (RfDs) and Reference Concentrations (RfCs) for Total Petroleum Hydrocarbons (TPH)*, Vol. 4, Amherst Scientific Publishers, Amherst, Massachusetts, 1997.

Total Petroleum Hydrocarbon Criteria Working Group, *Selection of Representative TPH Fractions Based on Fate and Transport Considerations*, Vol. 3, Amherst Scientific Publishers, Amherst, Massachusetts, 1997.

U.S. Environmental Protection Agency (EPA), *Soil Screening Guidance: User's Guide*, EPA/540/R-96/018, Office of Solid Waste and Emergency Response, Washington, DC, 1996.

CHAPTER 12
TOXICOLOGY AND RISK

SELECTING INDICATOR PATHWAYS

David Jeffrey

Dr. Jeffrey is a senior risk assessment scientist with SECOR International, Inc. He performs and manages health risk assessments, as well as fate and transport evaluations in support of risk assessments.

12.4.1 INTRODUCTION

The human health or ecological risk assessment of a complex site or environmental system can require the quantification of many separate chemicals and exposure pathways. In the previous chapter, we have already seen how the risk assessor uses the available information about a site, the goals of the assessment, structure-fate and structure-toxicity relationships, and his or her best professional judgment to select indicator or surrogate chemicals and avoid the time-consuming and expensive evaluation of a potentially large number of individual chemicals. A similar approach may be applied to identify a subset of indicator pathways, or perhaps only one indicator pathway, that may be used in lieu of the full set of complete and potentially significant pathways. The following text explains how this is done.

12.4.2 METHODOLOGY

The best way to identify appropriate indicator exposure pathways is to start by developing a conceptual site model (CSM) for the risk assessment and site. USEPA (1989) explains this process. In the development of a CSM, the risk assessor diagrammatically depicts the universe of possible pathways and the subset of this universe of pathways identified as "complete and potentially significant" pathways. Quantitative risk assessments are concerned only with this subset of pathways, whether or not indicator pathways are selected and used in the assessment. Indicator pathways are usually selected from this subset of complete and potentially significant pathways, but there are some interesting exceptions to this. An example of a CSM is shown in Figure 12.4.1.

Each chemical type, or indicator chemical type, will be associated with its own set of complete and significant pathways. For example, for VOCs in soil, those pathways associated with volatilization from soil and the generation of chemical vapors that may enter the atmosphere (outdoor air) or intrude into homes or other building structures (indoor air) will probably be included in this set. In contrast, for metals in groundwater, these pathways would not be included but ingestion of drinking water might. After considering the range of chemical types and the site CSM, the following type of information is assembled:

Chemical Type I
pathway 1
pathway 2
pathway 3

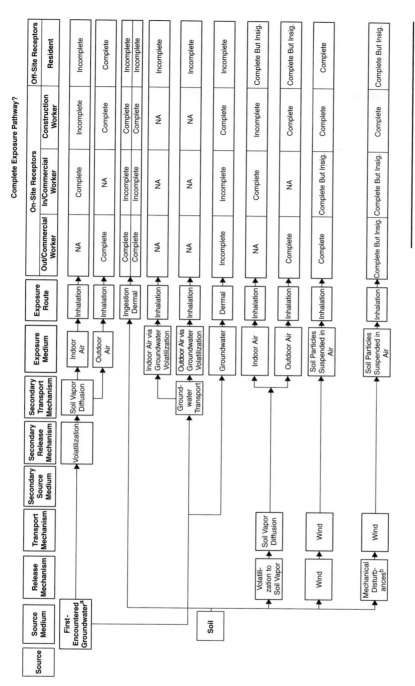

FIGURE 12.4.1 Example of a conceptual site model (CSM).

Footnotes:

[a] Deeper groundwater zones not shown because of expectation that concentrations of detected chemicals in the deeper zones are likely to be minimal, due to the nature of the Site.

[b] Mechanical disturbances are expected to occur during Site constructions activities.

NA = Pathway not applicable.

Complete But Insig. = pathway potentially complete but likely contributes insignificantly to the overall chemical dose.

Out/Commercial Worker = outdoor commercial worker, In/Commercial Worker = indoor commercial worker.

12.31

pathway 4
pathway 5
pathway 6
Chemical Type II
pathway 7
pathway 8
pathway 1
pathway 2
pathway 9
Chemical Type III
pathway 10
pathway 11
pathway 5
pathway 8

The risk assessor considers each chemical type and their associated complete and potentially significant pathways. For each such group, one pathway can usually be identified as likely to contribute to the majority of the total chemical exposure or dose. This is the indicator pathway for that chemical type. The risk assessor uses their experience in evaluating similar sites in identifying these indicator pathways. The utilization of this experience may take the form of simply recalling the "driver" pathway(s) for a site, or it may require a more careful analysis of previous risk assessment reports. Some risk assessors will perform a sensitivity analysis on the EPCs and intake assumptions contained in the dose equations for each complete and potentially significant pathway as an aid in making the proper selections of indicator pathways. The routes of exposure will also be considered (i.e., ingestion exposure is often more significant than inhalation). However this process is performed, one must be very careful to select the most significant pathway. Failure to do so can result in an assessment that underestimates potential health impacts for a site.

Once having selected an appropriate indicator pathway for a chemical type, the risk assessor computes the dose for this pathway then multiplies this dose by the total number of complete and potentially significant pathways minus one. This operation gives the total, multipathway dose for a chemical group when the total EPC is used for the indicator chemical.

One must appreciate, however, as with the indicator chemical approach described in the previous chapter, that the selection and use of indicator pathways can impart a significant degree of conservatism to the overall assessment. By assuming that all complete and potentially significant exposure pathways for a chemical group contribute equally to the total chemical dose, one can greatly overestimate the health impacts associated with a site. For instance, for VOCs in groundwater that serves as a drinking water supply, to equate exposures from dermal contact with tapwater to drinking the tapwater would greatly overestimate the predicted health impacts in the risk assessment, as the dose received from dermal contact is generally many orders of magnitude below that derived from drinking water.

Because of the tendency to significantly overestimate health impacts, the indicator pathway approach, like the indicator chemical methodology described in the previous chapter, should only be used for screening purposes. That is, should a site "pass" an indicator chemical/pathway assessment, with no estimated significant health risks, no further evaluation should need to be performed. However, should a site "fail" such a screening assessment, this only means that the site *may* pose an unacceptable health threat, and that further, more refined assessment is required to know for sure.

CHAPTER 12
TOXICOLOGY AND RISK

SECTION 12.5

THE DOSE MAKES THE POISON: SOME COMMON MISCONCEPTIONS

M. Alice Ottoboni

Dr. Ottoboni is a staff toxicologist, retired from the California Department of Public Health. She is the author of the popular text The Dose Makes the Poison.

12.5.1 INTRODUCTION

A review of the numerous factors that govern whether a chemical will produce adverse effects justifies the conclusion that the toxicity of chemicals is a very complicated subject. Quite understandably, this complexity is perplexing to many people. As a result, misconceptions relating to the toxic actions of chemicals and to the science that studies them have arisen. Some of these misconceptions have acquired a semblance of fact by dint of repetition and lack of challenge. Three in particular must be dispelled if informed and rational public participation in the making of government policy relating to the many aspects of chemicals in our environment is to be achieved. These misconceptions are: (1) the effects of chemicals normally considered to be toxicants are always detrimental; (2) the science of toxicology is capable of determining whether chronic exposure to trace amounts of a chemical is absolutely safe; and (3) some chemicals that enter our bodies lodge permanently inside us and build up, indefinitely, to higher concentrations.

12.5.2 SUFFICIENT CHALLENGE

Every toxicologist engaged for any period of time in research into chronic toxic effects of chemicals has observed, more often than not, that animals in the group with the lowest exposure to the test chemical grew more rapidly, had better general appearance and coat quality, had fewer tumors, and lived longer than the control animals. Novice toxicologists usually consider such observations as aberrations in their data or the result of some flaw in their experimental design or conduct. They are usually loath to call attention to such findings, because to do so might bring their competence into question, or because they are unable to explain the reason for such findings. It is only with the confidence that comes with experience that research toxicologists can comfortably acknowledge the occurrence of such results in their own experiments and broach the subject with colleagues. The reaction from fellow toxicologists is usually one of, "You, too?"

The reluctance on the part of toxicologists to acknowledge and discuss freely the observation that trace quantities of foreign chemicals can produce beneficial effects may be founded partly on the fact that they have not, as yet, formulated a unified theory for the phenomenon. Or, it may be due to the reality that we live in a time when it is not politic to make favorable statements about synthetic chemicals. Because toxicologists recognize that the phenomenon has little practical significance (unless it could be put to some therapeutic use), they have no compelling reason to emphasize it. Dr. Smyth learned firsthand how easy it is to be misunderstood, as he mentions in his article entitled "Sufficient

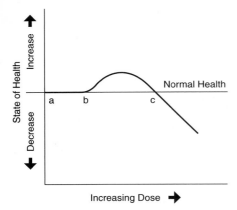

FIGURE 12.5.1   A complete dose-response curve.

Challenge" (Smyth, H. F., Jr., "Sufficient Challenge." In *Food and Cosmetics Toxicology*, Vol. 5, p. 51, 1967); his first use of the term sufficient challenge earned him the epithet, "Dr. Smyth and his fellow poisoners."

Dr. Smyth's hypothesis was developed from the familiar old proverb that tells us we must each eat a peck of dirt before we die, and from personal observations in the conduct of toxicology studies. Dr. Smyth describes the manifestations of the peck-of-dirt maxim in medical, physical, psychological, and social sciences. It is to the noted historian Arnold Toynbee that he gives credit for providing the name "Sufficient [but not overwhelming] Challenge" to his peck-of-dirt hypothesis for toxicology. Dr. Smyth writes, ". . . Toynbee concluded that no civilization has risen to importance without a 'sufficient but not overwhelming challenge' from its physical environment or its neighbors."

Figure 12.5.1 represents what Dr. Smyth proposes, and what considerable toxicological data confirm, to be the complete dose-response curve. With small doses, no effect on the health of the experimental animals is seen (points a to b). Animals receiving slightly higher doses have better than normal health (points b to c, the range of sufficient challenge). At point c, the curve passes back through normal health and, from that point on, deleterious effects occur.

Dr. Smyth gives numerous toxicologic examples of benefit from small doses, but perhaps his more important contribution is his theory for why such effects occur: "A rationale to explain sufficient challenge as a general phenomenon is based on the axiom, established in so many instances, that an unused function atrophies. One of the functions of an organism is homeostasis. Homeostatic mechanisms must be kept active if health is to be maintained."

"I think that most of the small non-specific responses which we measure in chronic toxicity studies at low dosages are readjustments or adaptations to sufficient challenge. I interpret them as manifestations of the well being of our animals, healthy enough to maintain homeostasis. They are beneficial in that they exercise a function of the animal. Only when challenge becomes overwhelming does injury result."

Some people reject the concept of sufficient challenge, claiming that data demonstrating beneficial effect from low doses are artifacts. However, the fact that such results occur regularly, and are reproducible, speaks against the claim. There are cynics who claim it is necessary to deny sufficient challenge lest it be used to absolve environmental polluters from past transgressions and give them license to delay or disallow abatement procedures.

A general acceptance of the theory of sufficient challenge would have no impact on chronic exposure standards or the regulatory procedures that govern them. The prohibitive cost of experimental determination of sufficient challenge dose ranges, the impossibility of determining individual responses to such doses, and public concern about environmental contamination all operate against any practical application of the concept. However, if the public were aware of the existence of sufficient challenge, it could help to lessen unreasonable fear of chemicals. In Dr. Smyth's words, "General acceptance of the concept of sufficient challenge would do much to alleviate the emotional revulsion which the thought of chemicals in daily life so often evokes."

12.5.3 TRANS-SCIENCE

The term "trans-science" was proposed by Alvin M. Weinberg to describe wisdom that cannot be achieved through scientific methodology. In his discussion of the relation between scientific knowledge and societal decisions (Weinberg, Alvin M., "Science and Trans-science." In *Minerva*, Vol. 10, p. 209, 1972), he notes, "Many of the issues which arise in the course of the interaction between science or technology and society . . . hang on the answers to questions which can be asked of science and *yet which cannot be answered by science.* I propose the term *trans-scientific* for these questions since, though they are, epistemologically speaking, questions of fact and can be stated in the language of science, they are unanswerable by science; they transcend science."

Dr. Weinberg cites three causes for the inability of science to answer trans-scientific questions: (1) "science is inadequate simply because to get answers would be impractically expensive"; (2) "science is inadequate because the subject-matter is too variable to allow rationalization according to the strict scientific canons established within the natural sciences"; and (3) "science is inadequate simply because the issues themselves involve moral and esthetic judgments: they deal not with what is true but rather with what is valuable." The great majority of trans-scientific questions asked of toxicology can be placed in the first category, which, for our purposes, will also include questions that science does not yet have sufficient knowledge or techniques to answer.

The questions uppermost in the minds of individuals relate to whether or not exposure to some chemical or chemicals will be harmful to their health or that of their loved ones. Often these are the very questions that toxicology cannot answer with a definite "yes" or "no." Science has no way of knowing the exact biochemical makeup of any individual person or exactly what quantity of chemical would be just below that person's threshold for the subtlest adverse effect of which the chemical is capable. An answer based on judgment can be given, but science does not as yet (and may very well never) have the methodology to respond to these concerns with direct evidence.

The questions uppermost in the minds of regulatory officials relate to the nature and incidence of adverse effects that might result from exposure of large populations of humans to very small quantities of environmental contaminants. Science does not have the resources, money, trained personnel, laboratory facilities, and experimental animals to provide such information for even a few—much less, all—of the many chemicals we may encounter in our daily lives.

Dr. Weinberg's discussion of the impracticability of experiments designed to study the health effects of extremely small exposures uses experiments dealing with radiation effects to demonstrate his point; however, his discussion could apply equally well to any chemical exposure. He explains that, in essence, an effect that has a very low incidence of occurrence would almost certainly not be seen in an experiment using only a few animals; a study using 10 animals will not reveal an effect that occurs in only 1 out of 100,000 animals. In order to demonstrate such an effect, many times 100,000 animals would be required. Experiments of that magnitude are beyond the capability of existing resources and so are in the realm of trans-science.

Another toxicologic question that falls within the realm of trans-science relates to the shape of dose-response curves for carcinogens as they approach zero dose. The inability of toxicology to answer this question by experiment has given rise to a scientific controversy concerning whether or not there is a threshold (no-effect level) for carcinogenic effects. If there is no threshold, extension of the experimentally derived dose-response curve to zero effect would yield a line that would go through the origin (zero dose). If there is a threshold, the extended line would meet the abscissa at some point greater than zero dose (see Figure 12.5.2). Federal regulatory agencies adopted the no-threshold theory of carcinogenesis many years ago when faced with the requirement to make regulatory decisions about chemicals classed as carcinogens, proven or suspected. The development of carcinogenicity risk assessment methods resulted from this regulatory need, and refinement of risk assessment methodology continues to the present.

In 1971, the National Center for Toxicological Research (NCTR) was created to provide federal regulatory agencies, such as the FDA and EPA, with the scientific data required by them for the performance of their regulatory duties. One of the first missions of NCTR was to determine how far they could penetrate into the realm of dose-response curve trans-science by means of an animal carcinogenicity feeding study of hitherto unimagined size-a "Megamouse" study. In 1972, the NCTR proposed that the large-scale study be designed to answer the question of whether they could accurately describe an ED_{001} for a known carcinogen whose effects had been extensively investigated and were relatively well understood. ED

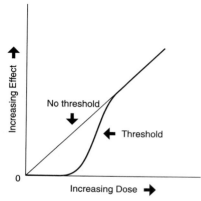

FIGURE 12.5.2 Extrapolation of the dose-response curve to zero dose.

means "effective dose" (in this case the effect is cancer), and the 001 means for 0.1 percent of the animals in the study.

Such a study, it was hoped, would provide experimentally derived points lower down on the dose-response curve than had ever been obtained before, and also, perhaps, shed some light on the controversy over whether or not carcinogenic effects have thresholds. The physical resources of NCTR, however, could not accommodate the number of animals required by an ED_{001} study, so the plans were scaled down to an ED_{01} study. The carcinogen selected for the study was 2-acetylaminofluorene (AAF), a chemical that fit the criteria established by NCTR better than any other known carcinogen. The study required 18 months for the planning stage, 9 months for production of the more than 24,000 mice employed in the study, and another 3 to 4 years for conduct of the study and evaluation of the data, and cost somewhere between $6 and $7 million.

The results of this heroic effort were published in a special issue of the *Journal of Environmental Pathology and Toxicology* (Vol. 3, No. 3, 1980), and were reviewed by a special Committee of the Society of Toxicology. The review by the Society of Toxicology Committee was published in *Fundamental and Applied Toxicology* (Vol. 1, No. 1, pp. 27–128, 1981). The study produced a great deal of information that is of extreme importance to the design and conduct of future studies of the carcinogenicity of chemicals, and will contribute immensely to the development of appropriate models and formulae for carcinogenic risk assessment. However, the question of whether or not a threshold exists for the carcinogenic effects of AAF was not answered. Thus, the question still remains, and probably always will remain, in the realm of trans-science.

One finding made by the Society of Toxicology in its review of the NCTR report brings to mind Dr. Smyth's theory of sufficient challenge. The Society's review points out that the statistical model used for extrapolating effects from very low doses "provides statistically significant evidence that low doses of a carcinogen are beneficial" (Smyth, 1967, p. 77). "If the time-dependent low-dose extrapolation models are correct," the Society states, "then we must conclude that low doses of AAF protected the animals from bladder tumors."

Dr. Weinberg, in his article on trans-science, mentions another point that is important to an understanding of the limitations of science: ". . . no matter how large the experiment, even if no effect is observed, one can still only say there is a certain probability that there is in fact no effect. One can never, with any finite experiment, prove that any environmental factor is totally harmless. This elementary point has unfortunately been lost in much of the public discussion of environmental hazards."

This brings us to one of the most frustrating matters with which toxicologists must cope, namely, the subject of proving absolute safety of exposure to chemicals. It is most natural for people to demand assurance that the chemical exposures they experience are absolutely safe; it is very difficult for them to grasp the fact that this is an assurance that no one is capable of giving. Absolute safety is the complete absence of harm. How does anyone prove the nonexistence of anything? Consider the child who is afraid of goblins. The child calls out that there is a goblin in the darkened bedroom. The parent turns on the light and says, "See, no goblin!" The child says that the goblin ran into the closet. The parent opens the closet door and says, "See, no goblin!" How does the parent prove that the goblin does not exist? A "See, no goblin!" response is not proof.

For the toxicologist, every negative response to a question can always be followed by another question. If a toxicity experiment shows a certain level of exposure to be a no-effect level, the question can be asked, "What if you had used more animals?" If more animals yield the same results: "What if you had continued the study through one or two generations?" If a two-generation study yields the same results, then: "What might happen after five generations?" "Twenty generations?" There is no such thing as a diminishing supply of questions.

There is also no such thing as a diminishing supply of skepticism. A number of years ago, during a panel discussion on the use of herbicides in forestry, I was challenged by a man in the audience who said that the concept of the no-effect level was just not logical: there must be a chemical somewhere in the world that would be harmful no matter how small a dose was given. Obviously, there is no way to prove that such a chemical does not exist. For any chemical that does produce adverse effects down to the smallest dose that can be administered practically, there is no direct way to prove that some much smaller dose would be harmless. By the same token, there is also no way to prove that smaller doses would be harmful. People with opposing views of the correct answer to a trans-scientific question are on equal ground!

The fact that toxicology cannot provide absolute answers to many of the questions that are of great concern to people should not be cause for alarm. Toxicologists can make judgments about the possibilities and probabilities of harm resulting from exposure to chemicals. These judgments are based upon scientific data obtained from chronic toxicity testing, knowledge of the behavior of the chemical in animal systems, and application of appropriate margins of safety.

12.5.4 *BIOACCUMULATION*

Bioaccumulation is a term that is commonly used, but has different connotations for different people. Thus, in discussing the subject, it is important to define the sense in which it is being used in order to avoid total confusion. Bioaccumulation is considered by some people to mean the accumulation of cellular or tissue injuries as a result of exposure to chemicals. It is more popularly used to refer to the storage of chemicals in living organisms. This is the meaning that will be used in the following discussion. It should be noted here that *bioaccumulation* and *storage* are not synonymous. Storage means deposition of a chemical in some anatomic site. Storage results in bioaccumulation, but bioaccumulation can occur by means other than storage.

There is probably no concept in toxicology that is less understood, or more frightening than the phenomenon of bioaccumulation. The public became aware of the concept several decades ago when environmental concerns about chlorinated hydrocarbon pesticides, particularly DDT, were being widely publicized. However, to date, people are still confused about what bioaccumulation is and what it means for their health and well being.

A brief description of the nature of the chemical balances that exist in all living organisms is essential to an understanding of bioaccumulation. All living cells possess a certain specific composition, which, within normal limits, remains constant as long as they maintain their customary good health. Despite the fact that this relatively stable composition makes it appear that living organisms are static organizations, nothing could be further from the truth. All living creatures exist in what is known as a state of dynamic equilibrium. Dynamic means "moving" and equilibrium refers to the balance, or equalness, that is maintained within living organisms.

A simple illustration of a state of dynamic equilibrium would be a box that contains a specific number of marbles, where every time some new marbles are added to the box, an equal number of old marbles are removed. The total number of marbles in the box remains constant, but the individual marbles present in the box at any one time are not necessarily the same marbles that were in the box at some earlier time or that will be in the box at some future time. If the input of marbles increases, the output will also increase by the same number, either immediately (thereby keeping the total number in the box the same), or after a delay of some specific time (thereby increasing the total number in the box to a new equilibrium level, determined by the length of the delay). A decreased input will have the opposite effect. So it is with living organisms: There is a steady drive toward the achievement of equilibrium-constant change, but little or no net change.

With the concept of dynamic equilibrium in mind, let us now look at bioaccumulation. Whether or not a chemical accumulates in an organism depends upon how fast it is eliminated (metabolized or excreted) relative to how fast it is absorbed into the body. The time between absorption and elimination will be referred to in the following discussion as the residence time of the chemical.

For chemicals with the same rate of absorption, those with longer residence times have a greater potential for bioaccumulation than those with shorter residence times. If the residence time is very short, little or no bioaccumulation will occur (at least not of the chemical, itself; a metabolite may accumulate, but that is an unnecessary complication at this point). Residence time is dependent upon such factors as pathway through the body, rate of metabolism or excretion, and the tendency of the chemical either to be held for a period of time in some metabolic pool, or to be deposited in some storage site (depot), such as fat or bone. The actual concentration that a chemical achieves in a living organism is regulated by the same kinds of forces that drive the organism's internal biochemical environment toward equilibrium.

The storage of foreign chemicals in various depot sites can involve some very complex physiologic processes and equilibria relationships. Although, for the sake of simplification, the following discussion will consider only the overall equilibrium between absorption, storage, and elimination, it must be remembered that the blood, which serves as the transport medium for chemicals in the body, participates in the process and that it has its own equilibrium relationships at the sites of entrance, storage, and elimination of chemicals.

The simplest case involving storage of chemicals occurs when the concentration of a chemical to which an organism is exposed remains relatively constant over a prolonged period of time and, after entering the body, the chemical is not metabolized. Before the onset of exposure, there are no molecules of the chemical in the body's storage site. When exposure starts, molecules of the chemical begin entering the depot. Some of these molecules also exit but the number entering is greater than the number exiting, with the result that the concentration of chemical in the depot site gradually increases. With continued exposure, the concentration of chemical in the storage depot increases until finally it comes into equilibrium with the exposure concentration. At this point, the number of molecules leaving the storage site equals the number entering, and the storage concentration remains constant. If the exposure concentration increases, the storage concentration will gradually increase until it comes into a new equilibrium with the higher exposure. If the exposure concentration decreases, the storage concentration will gradually decrease until it comes into a new equilibrium with the lower exposure. If exposure ceases, the stored chemical will gradually be eliminated from the body.

The quantity of chemical that can be stored in any body depot can never exceed that which would be in equilibrium with the exposure, and the chemical cannot remain in the storage depot without being replenished continually from the outside. Thus, the popular notion that foreign chemicals stored in a depot become immobilized and permanently fixed in the body, with additional exposures increasing the quantity stored ad *infinitum*, has no basis in fact.

The relationship between magnitude of exposure and storage concentration at equilibrium is not known for most chemicals. One exception is the organochlorine pesticide DDT, which has been the subject of a tremendous amount of investigation, including study of its storage and excretion patterns. The U.S. population, on average, received a daily oral exposure of approximately 0.2 ppm DDT during the several decades of heavy use of the pesticide. This level in the diet produced an average DDT concentration of 7 ppm in body fat of the American people. The increase in storage concentration relative to exposure concentration at this low level of exposure was approximately 35-fold. In studies with rats and mice, a diet containing 20 ppm DDT produced a body fat content of 200 ppm, a 10-fold increase. A diet of 200 ppm produced body fat levels of 600 ppm, only a 3-fold increase. Thus, for DDT, the relationship between exposure and storage concentrations is not a linear one. From what is known about storage and excretion of foreign chemicals, DDT can serve as a model for chemicals that store in body fat.

Another piece of misinformation that has been circulated about bioaccumulation, or more specifically storage, is that it has put us at risk of serious or fatal poisoning by chemicals suddenly released from fat depots in the event of severe weight loss, such as could occur in the case of debilitating illness or starvation. We have also been told that the chemicals stored within us have made us walking "time bombs." These claims are examples of the kind of tactics employed to exploit the fact that, unfortunately, the most effective method for engendering public concern about chemicals in the environment is to make people fear that those chemicals are endangering their health.

Poisoning from rapid mobilization of a chemical from fat stores is theoretically possible in laboratory animals exposed under rigidly controlled experimental conditions to unrealistically high concentrations of the chemical, but is a practical impossibility for the average person. Numerous surveys of the body burden of foreign chemicals stored in the body fat of human populations demonstrates that the quantities stored, if released all at once (a virtual impossibility), would be well below acutely or chronically toxic amounts. Thus, people who are concerned about the potential of acute or chronic poisoning from chemicals stored in their body fat are subjecting themselves to the trauma of needless worry.

A property of chemicals associated with bioaccumulation, and equally maligned, is persistence. The chlorinated hydrocarbon pesticides, such as DDT, are resistant to metabolism and environmental degradation and, therefore, remain in their unaltered states for relatively long periods of time. Persistence and bioaccumulation are not inherently good or bad, but in the public mind they are considered to be the latter. In actual fact, whether they are beneficial or detrimental depends upon the context in which they are viewed. The ability of an organism to store a chemical in some anatomic depot can actually be beneficial because it functions as a mechanism whereby the organism is protected from the toxic action of the chemical. The chemical does no harm while it resides in the storage site. For example, lead is a very toxic element for animals. It stores in bone where, like DDT in fat, it is in equilibrium with blood levels, which, in turn, are in equilibrium with exposure and excretion. Lead isolated in bone does no harm, but lead in nerve tissue causes very serious damage. The removal of lead from blood by bone prevents the blood lead level from becoming sufficiently high to damage nerve tissue. If lead exposure is prolonged or excessive, blood lead levels can reach levels toxic to nerve tissue, despite storage in bone.

A storage depot may be considered to have a buffer function: during periods of increased exposure, the chemical is deposited in the storage site, where it is prevented from exerting a harmful effect. When exposure increases, the chemical is mobilized from the depot and eliminated from the body, thereby freeing the site for some future need. With high exposure levels that saturate the depot, there is no more storage space for the chemical and the protective effect of the storage site is abolished.

Persistence is a desirable quality in pesticides from the viewpoint of effectiveness and efficiency. A pesticide that retains its ability to kill pests for prolonged periods needs to be applied less often than a pesticide that degrades rapidly. Thus, the total quantity of pesticide required to do the job is considerably less, which reduces the cost of crop protection and production.

The undesirable aspects of persistence in pesticides relate to their continued pesticidal action after such need ceases to exist, and to the fact that they remain in the environment for prolonged periods. The persistence of DDT and its extremely heavy use during World War II resulted in its gradual distribution, in trace quantities, throughout the world. The tremendous improvements in analytic techniques during the past two decades, which enabled chemists to detect infinitesimally small quantifies of chemicals, led to the discovery that DDT had become virtually omnipresent in the environment. The statistical association of these small quantities of DDT with declines in certain wildlife species spurred the campaign to ban the use of all persistent pesticides. The campaign has been eminently successful in eliminating DDT and many of the other persistent pesticides.

The rationale that substitution of nonpersistent pesticides for persistent ones will solve all the environmental problems attributed to the latter is not only shortsighted, but unwarranted. A nonpersistent pesticide must be applied much more frequently than a persistent one because of the very fact of its nonpersistence. This need for increased numbers of applications greatly increases the environmental burden of persistent degradation products. By forcing a ban on persistent pesticides, environmentalists may very well have created a much larger problem than the problem they perceived as requiring the ban. Time may tell, if someone asks the right questions!

Time has already told us that the switch from persistent to nonpersisent pesticides greatly increased the number of acute poisonings among farm workers. Cases of acute illnesses from chlorinated hydrocarbon insecticides were virtually nonexistent prior to their ban. The worst problems were cases of skin irritation. With the introduction of the organophosphate pesticides, the major class of nonpersistent insecticides, it became evident that these highly toxic chemicals would require elaborate precautions for safe use. When DDT was banned, the use of organophosphate insecticides increased greatly. This increase was accompanied by a large increase in worker poisonings, some so severe as to be lethal. The efforts to protect wildlife had the ultimate effect of producing acute health problems among workers in the agricultural industry. Despite elaborate programs of worker protection medical surveillance, protective clothing and cleanup, automatic measuring and mixing devices to avoid human contact, restrictions on reentry into treated fields, among others, poisonings of farm workers by nonpersistent pesticides still occur.

CHAPTER 12
TOXICOLOGY AND RISK

SECTION 12.6
LOW DOSE RESPONSE-HORMESIS

T. D. Luckey

Dr. Luckey is professor emeritus of the University of Missouri, Columbia. He is a consultant in life sciences and a member of the Board of Directors of Radiation, Science and Health.

NOMENCLATURE AND UNITS (DEFINITIONS)

cGy = one hundredth of a Grey = one rad

cSv = one hundredth of a Sievert = one rem

GAS = general adaptive syndrome, from H. Selye

Hormesis = stimulation by low doses of any agent

Hormetin = the agent provided in low doses

Hormology = the study of excitation

Inter-Organismic = action upon an organism other than the one which produced the agent

Intra-Organismic = action within the organism which produced the agent

LNT = Linear No Threshold

pCi/l = pica Curies per liter

P-Y = Person-Year

UNSCEAR = United Nations Scientific Committee on the Effects of Atomic Radiation

ZEP = zero equivalent point

12.6.1 INTRODUCTION

Hormology is the knowledge of excitation or the study of stimulation (Luckey, 1960). *HORMO*, derived from the Greek, "I excite," is familiar in the word hormone. *Hormesis* is defined as stimulation by low doses of any agent in any system. The term was coined when Southam and Ehrlich noted the growth stimulation in fungi with dilute solutions of a toxic extract from cedar trees. This is also known as "the reverse effect." The concept is as follows: Low doses of the agent, the *hormetin*, are stimulatory; excess amounts are inhibitory. The effects of excess agent are generally considered to be harmful; thus, it is reasonable that stimulatory doses be considered beneficial. This has been noted repeatedly for centuries by many observers (see Table 12.6.1). Although a stress-free life would be disastrous, small stresses are beneficial (Britannica, 1998).

Selye (1950) identified the *general adaptive syndrome* (GAS) as the response to low stress reactions. The reactions are comparable for physical, chemical, or psychologic stressors. Evidence from his relatively short-term studies complements the results from long-term studies used in studies cited here. He measured changes in metabolism and nerve and hormone reactions; sometimes he included physiologic functions and medical condition. Seyle identified many agents that increased

TABLE 12.6.1   Concepts in Hormology

Date	Concept	Author
1520	The dose is everything	Paracelsus
1810	Homeopathic medicine	Hahnemann
1820	Vaccinate to stop infection	Jenner
1867	Repair strengthens tissue	Bernard
1888	The Arndt-Schulz Law	Schulz
1902	Hormone	Bayliss & Starling
1906	Metal oligodynamic action	Richet
1943	Hormesis	Southam & Ehrlich
1944	General adaptive syndrome	Seyle
1994	Radiation adaptive response	Unscear

TABLE 12.6.2   Physiologic Effects of Low Dose Irradiation.

Increased	Decreased
Growth rate	Cancer mortality
Development	Cardiovascular death
Memory	Respiratory mortality
Fecundity	Infection mortality
Immunity	Sterility
Mean life span	Neonatal death

the resistance of the host to disease and other stresses. He emphasized the danger of excesses. Minute doses start an "alarm reaction"; small doses induce the "stage of resistance"; and an excess evokes the "stage of exhaustion," which may be fatal. Diverse agents produce a cumulative effect. He also noted the general nature of the response; small doses elicit resistance to a variety of malfunctions.

Most hormetins are nonspecific; however, some, such as hormones, vitamins, antigens, and classic homeopathic drugs are specific for a dominant effect. Hormetins function in microbes, plants, and animals. Physiologic responses of vertebrates to radiation (Table 12.6.2) include vertebrate neurologic acuity and learning as well as memory. Immune competence includes resistance to cancer as well as cardiovascular and respiratory diseases. This explains the increased average lifespan found in populations exposed to low doses of some hormetins. This increased lifespan is due to decreased premature deaths.

The vertebrate response to small doses of different hormetins is the generalized GAS of Selye. At the cellular level, a family of "stress proteins," also called "heat shock proteins" are released. In a universal response from bacteria to humans, these induce production of a variety of other proteins; examples include "adaptive enzymes," such as those that repair damaged DNA. The response includes activation of both cellular and chemical components of the immune system and production of enzymes that transform toxicants into less harmful compounds (Smith-Sonneborn, 1992).

The physiologic response is initiated in brain and nervous tissues. The hypothalamus of the brain stimulates the production of corticotropin-releasing hormone, which, in turn, activates the pituitary to release the adrenocorticotropic hormone (ACTH). Both ACTH and the sympathetic nervous system activate the adrenal gland to produce hydrocortisone, epinephrine, and norepinephrine. The release of these hormones into the blood triggers the alarm reaction in the brain, nerves, and other tissues. The highly aroused individual is ready to run or fight. Unless there is excess hormetin, pertinent physiologic reactions are normally followed by a variety of homeostatic reactions.

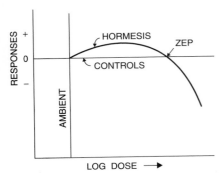

FIGURE 12.6.1 Theoretic curve showing both hormesis (+) and toxicity (−) of the agent. The dose has a logarithmic base that accommodates both low and high doses. ZEP is the dose that elicits the same response as the amount of the agent at ambient conditions.

The UNSCEAR report (UNSCEAR, 1994) convinced most scientists that the effects of low dose irradiation included a threshold response (Figure 12.6.1). Overwhelming evidence for hormesis invalidated the *linear no threshold* (LNT) model. The threshold dose identifies the *zero equivalent point* (ZEP), the dose at which the effect is equivalent to zero. This, in turn, defines *low dose* as any dose between ambient levels and the ZEP. The ZEP also defines the triage dose for disaster medicine. Those who received a low dose need no medical care; those who received a large excess cannot be helped. This directs medical attention to those who will benefit the most.

The ZEP varies with the parameter being measured, the population, the dose rate of the agent, and conditioning factors. Conditioning factors include individual variation, previous experiences, environmental conditions, and interrelationships between different hormetins. Hormetins may be physical, chemical or sociopsychologic agents.

12.6.2 PHYSICAL HORMETINS

Physical agents showing hormesis include gravity, magnetism, pressure, sound, time, electricity, electromagnetic forces, light, and ionizing radiation. These broad categories may overlap. Minute changes in the intensity of these physical agents are rarely perceived; small changes arouse nervous and bodily reactions and large changes may be devastating. The exemplar is ionizing radiation.

Ionizing radiation includes high- and low-energy electrons, mesons, gamma and X rays, neutrons, protons, helium nuclei (alpha rays), other atomic nuclei, and even positrons and neutrinos. Each of these can produce ions and free radicals, the active agents for radiation hormesis in cells and tissues.

Evidence for radiation hormesis is found in cells in culture, microbic growth and metabolism, plant propagation, and the major physiologic parameters in both invertebrates and vertebrates, including

TABLE 12.6.3 Cancer Mortality in Nuclear Workers.

Date	Group	cSv	Person-Yr	Persons Con/Exp	Dead/1000 Con/Exp	%	p<
1984[a]	Nuclear shipyards	>0.5	1,474,823	40,774/70,983	29.8/19.4	65	0.001
1985[b]	U.S. bomb observers	1.3	960,000	38,220/32,015	26.5/18.7	71	0.001
1987[c]	Canada military	1.3	74,600	1908/954	38.1/33.4	88	NS
1989[d]	U.S. weapons	3.2	705,295	20,619/15,314	34.8/20.8	60	0.001
1992[e]	British weapons	3.4	3,237,378	58,945/36,272	9.9/2.3	23	0.001
1993[f]	Canada energy	2	268,320	4717/4260	23.7/20.3	86	0.01
1994[g]	LANL	2	456,637	11,345/2935	20.1/17.7	88	0.05

Note: CON = control workers; EXP = exposed workers. The chi square statistic was used to obtain *p* values (Luckey, 1994).

[a] Values derived from Matanoski's 1984 paper; the 1991 final report had comparable total cancer mortality rates.

[b] Robinette et al. examined records of army atmospheric bomb observers exposed between 1951 and 1957. The cancer death rate was obtained by comparing 11,057 "controls" who received <0.1 cSv with 6695 men who were exposed to 1–3 cSv.

[c] The small number of observers of atmospheric bomb blasts in the Ramon et al. study allowed no statistical significance.

[d] Gilbert et al. summarized total cancer deaths at Oak Ridge National Laboratory, the Hanford Site, and Rocky Flats Nuclear Weapons Plant.

[e] Age corrections were obtained from data in Tables 2.10 and VII in the two reports by Kenworthy et al.

[f] Mortality data were taken from 3300 persons exposed to 0.1–4.9 cSv in the report of Gribbin et al.

[g] The Wiggs et al. data from the Los Alamos National Laboratory complement the Gilbert data.

TABLE 12.6.4   Leukemia and Cancer Deaths in Japanese Bomb Survivors

Dose, cSv	0–0.9	1–1.9	2–4.9	5–9.9	10–19
Persons	45,148	7430	9235	6439	5316
Leukemia	81	11	14	8	11
Leukemia/10^5	179	148	152	124	207
Change/10^5	0	−31	−27	−55	+28
Other Cancer	3246	498	717	516	400
Cancer/10^4	719	670	776	801	752
Change/10^4	0	−49	+57	+82	+33

humans (Table 12.6.2). Increased immune competence and enzyme repair appear to be major mechanisms for these benefits.

The benefits of low doses of ionizing radiation have been shown to be applicable to total populations. However, the best evidence concerns the effect of whole body exposures on total cancer mortality rates of primarily adult white males (Luckey, 1994). The results with humans confirm evidence previously obtained with experimental animals (Luckey, 1991).

Results from 7 million person-years (P-Y) experience showed that exposed nuclear workers had lower total cancer death rates than controls, carefully selected unexposed workers from the same plants (Table 12.6.3). The differences were statistically significant, $p < 0.01$. Exposures were registered on film badges. The "healthy worker effect" is not applicable for these studies; workers were not differentiated before they became exposed. Exposed and control workers had comparable entrance examinations, socioeconomic backgrounds, working conditions, management, and medical care. The P-Y weighted average total cancer mortality rate was only 59 percent that of control workers. This result suggests consideration of low dose irradiation as a means to significantly reduce total cancer deaths.

These carefully controlled human studies confirm previous work with both experimental animals and humans showing populations exposed to low dose irradiation had reduced cancer mortality rates when compared with controls (Luckey, 1991). An example is the total cancer death rate in Japanese atomic bomb victims (Shimizu, 1992). Leukemia and solid cancer death rates were lowered in lightly exposed persons (Table 12.6.4).

Radiation hormesis was confirmed in a study of the relationship between radon concentrations and lung cancer death rates (Figure 12.6.2) (Cohen, 1994). There is an irrefutable inverse relationship between home radon concentrations and lung cancer mortality in the United States. This study encompassed about 300,000 homes within 1700 counties that contain 90 percent of the population. Huge excesses of radon can induce lung cancer in animals; however, within the limits of natural background levels, radon decreases lung cancer mortality.

12.6.3   CHEMICAL HORMETINS

Inorganic Hormetins

Inorganic hormetins include elements, ions, salts, and organo-metallic compounds. The term, *hormology*, was first used to differentiate the stimulating action of sodium from its nutritional requirement (Luckey, 1960). Richet (1905) determined how small a dose would exhibit the *oligodynamic* effect (the germicidal action of minute amounts of metals) of Nageli. His results (Figure. 12.6.3) provided a demonstration of the stimulation of fermentation by metals at concentrations below those that were germicidal.

An interesting example of metal hormesis is the use of germanium to stimulate blood oxygen (Figure 12.6.4) before iron was accepted as a component of hemoglobin (Read, 1949). A general view of the activity of the elements (Figure 12.6.5) shows many are directly hormetic; others act

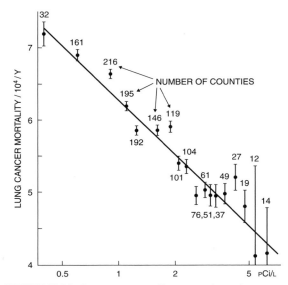

FIGURE 12.6.2 Lung cancer mortality rates are inversely proportional to the radon concentrations in US homes (Cohen, 1994). The dose is given as pCi/l.

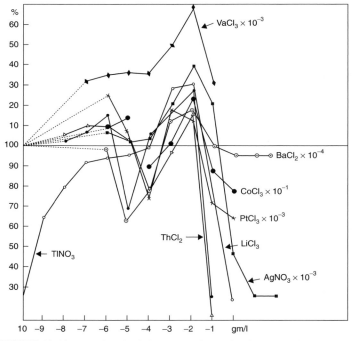

FIGURE 12.6.3 Bacterial stimulation by low doses of toxic metals (Richet, 1906). Note that the dose has a logarithmic scale.

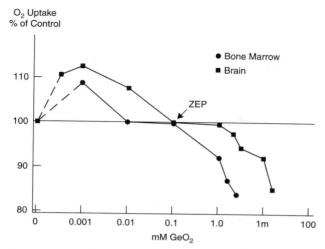

FIGURE 12.6.4 Low concentrations of germanium oxide (measured in millimoles) increased respiration of rat brain and bone marrow (Read, 1949). The author found this compound increased the growth rate of chicks.

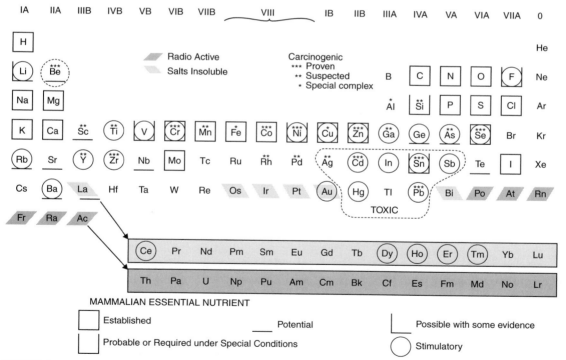

FIGURE 12.6.5 Periodic Table: circles indicate hormetic elements; heavy stipple shows elements that are hormetic because of their radioactivity; the square, or parts of square, indicate evidence of a nutritional requirement for some elements; asterisks indicate the evidence for the carcinogenicity of certain elements (Luckey and Venugopal, 1977).

RUDOLFF ARNDT 1835-1900   HUGO SCHULZ 1883-1921

FIGURE 12.6.6   Photographs of Drs. R. Arndt and H. Schulz from Greifswald University.

indirectly as radioactive agents. Many of these contribute to osmotic pressure, adhesion, cohesion, acidity, and membrane permeability in the cellular environment.

Organic Hormetins

Organic hormetins are components of our environment. They include drugs, dyes, surfactants, antibiotics and germicides, pesticides and other biocides, some food additives, antigens, vitamins, growth factors, and the very active free radicals produced by ionizing radiation (Luckey, 1959).

Hormesis has a direct lineage to Hugo Schulz (Figure 12.6.6) in the Department of Pharmacology at the University of Greifswald, Germany (now The Ernst-Moritz-Arndt University). In collaboration with Rudolf Arndt, Professor of Psychiatry, he worked on mechanisms of homeopathy using yeast (Schulz, 1888). From this collaboration came the Arndt-Schulz Law: "Low doses of toxicants are stimulatory." His results, combined with those of Branham (1929), showed that the older germicides stimulated microbial growth (Table 12.6.5).

Biologic Hormetins

Biologic hormetins are separated according to the subject upon which they act. Those that act on the organism that produced them are *intra-organismic* hormetins. These include hormones, cytokines, enzyme cofactors, compounds involved in nerve transmission, and components of cell metabolism, maturation, differentiation, and regeneration. Many more free radicals are produced within us during normal cell respiration than we receive from external ionizing radiation.

In contrast, *inter-organismic* hormetins act upon organisms other than the one that produced it. These include compounds produced by viruses, microbes, plants and animals, including humans. Pheromones typify animal interactions. Well advertised odors or toxicants provide some animals with

TABLE 12.6.5 Germicides Stimulate Microbic Growth*

%	Germicide	Dilution
100	Control	—
110	NaClO	50,000
115	I_2	200,000
118	$AgNO_3$	4000
122	$CuSO_4$	10,000
122	Chloramine-T	5000
132	Tricresol	1000
143	Tincture I_2	500
143	Mercurochrome	4000
152	Hexylresorcinol	8000
157	$HgCl_2$	700,000
170	Chromic Acid	8000
173	Br_2	100,000
173	Formaldehdye	800
173	Phenol	300
210	Metaphen	8000
220	Arsenious Acid	40,000
300	Formic Acid	50,000
600	Na Salycilate	4000

* Taken from Schulz (1988) and Branham (1929).

prime defense mechanisms. Other than bacterial toxins, we have little understanding of the interorganismic relationships between ourselves and the bacteria in our alimentary tract, which are more numerous than our human cells (Luckey, 1970).

12.6.4 SOCIOPSYCHOLOGIC HORMETINS

Sociopsychologic hormetins come from sociologic and psychologic agents, respectively. Shared nurture and milieu usually provide continuous interrelationships between the two. The origins of social hormetins are interorgasmic and may apply to a pair, family, small groups, or crowds. More than one specie may be involved. From the broadest viewpoint, coevolution requires stable interactions—an equilibrium—between species.

Sociologic hormetins include well-known stresses that stimulate in small amounts and debilitate in excess. They include love, sex, parenting, divorce, enmity, infection, health care, actions for attention, competition, dominance, restraint, isolation, crowding, conflicting beliefs, revolution, war, and death. For example, excess love is epitomized by the advise to excessively protective mothers: "Cut the apron strings."

Societal implications of hormesis have been explored (Calabrese, 1998). Regulatory agencies ignore hormesis because their primary function is to protect populations from harm, not to promote health. The prevailing paradigm that harm must be proportional to dose ignores compelling data showing hormesis. This has cost millions of lives and inhibited progress in many countries. A simple example is found in the subtitle of the journal of the Health Physics society: "The Radiation Protection Journal." Although the majority of members accept the concept of radiation hormesis, few think the subtitle means "promote health with low level irradiation".

Psychologic hormetins have intraorganismic origins. They trigger neurologic reactions in the brain, which are derived from individual experiences. These include love, happiness, euphoria, dreams, prayer, hypnosis, hallucinations, psychoses, self-esteem, biofeedback, ambition, greed, rumors, perceived risk, freedom, isolation, illness, pain, fear, hunger, fasting, religious and political fervor, animus, grief, and patriotism. Small amounts of these, individually or combined, are stimulatory. Psychologic hormetins cause the hypothalamus to release a stress signal, a corticotropin-releasing hormone, and the brain stem releases norepinephrine to induce GAS as described above. Excess triggers deleterious performance.

Hormesis is recognized in psychology as the inverted U curve or the Yerkes-Dodson law: "optimal motivation conditions vary inversley with the difficulty of the accomplishment" (Britannica, 1998). Most measurements involve learning and discrimination. For 90 years, acceptance of the law has varied according to the author and conditions tested. Some data deny the law; other sets of data strongly support the concept (Watters et al., 1997). Understanding of hormology should clarify concepts and strengthen methodology for psychologists.

12.6.5 SUMMARY

Hormology, the study of excitation, involves physical, chemical, biologic, and sociopsychologic agents. Hormesis is the stimulation by low doses of any agent. Biologic effects are summarized. Ionizing radiation is the exemplar for physical agents. Chemical agents include metals and organic compounds. The latter includes biologic agents; some act within the organism that produced them, others act upon other organisms. Sociopsychologic hormetins include well-known stimulants such as pheromones, love, and hate. The common denominator for reactions to hormetins is the general adaptive syndrome (GAS). Low doses of many agents induce resistance to a wide variety of malfunctions. Excess induces exhaustion.

REFERENCES

Branham, S. E., "Effects of Certain Chemical Compounds Upon the Course of Gas Production by Bakers Yeast," *J. Bact.*, 18: 247–264, 1929.

Britannica, Encyclopaedia Britannica, 15th edition, 24: 439–444, Encyclopaedia Britannica, Inc., Chicago, 1998.

Calabrese, E., "Toxicological and Social Implications of Hormesis—Part 2," *BELLE Newsletter*, 7: 1–35, 1998.

Calabrese, E. J., "Toxicological and Societal Implications of Hormesis," *BELLE Newsl.*, 7: 1–35, 1998.

Cohen, B. L., and Colditz, G. A., "Tests of the Linear No-threshold Theory for Lung Cancer Induced by Exposure to Radon," *Environ. Res.*, 64: 65–73, 1994.

Heiby, W. A., *The Reverse Effect*, Mediscience Publ., Deerfield, 1988.

Luckey, T. D., "Antibiotics in Nutrition," in *Antibiotics, their Chemistry and non-Medical Uses*, Goldberg, H., ed., D. Van Nostrand Co, Inc., Princeton, pp. 174–321, 1959.

Luckey, T. D., "Gnotobiology is Ecology," *Am. J. Clin. Nutr.*, 23: 1533–1540, 1970.

Lucky, T. D., *Radiation Hormesis*, CRC Press, Boca Raton, 1991.

Luckey, T. D., "Radiation Hormesis in Cancer Mortality," *Int. J. Occup. Med. Toxicol.*, 3: 175–191, 1994.

Luckey, T. D., and Stone, P. C., "Hormology in Nutrition," *Science*, 132: 1891–1893, 1960.

Luckey, T. D., and Venugopal, B., *Metal Toxicity in Mammals*, Vol. 1, Plenum Press, New York, 1977.

Read, W. O., *Effects of Germanium Dioxide Upon the Oxygen Uptake of Rat Tissue*, Ph.D. Thesis, The University of Missouri, Columbia, 1949.

Richet, C., "De l'action de doses minuscules de substances sur la fermentation lactic," *Arch. Intern. de Physiol.*, 3: 203–217, 1905.

Schulz, H., "Ueber Hefegifte," *Arch. ges. Physiol.*, 42: 513–541, 1888.

Selye, H., *Stress*, Acta Inc., Montreal, 1950.

Shimizu, Y., Kato, H., Schull, W. J., and Mabuchi, K., "Dose-response Analysis Among Atomic Bomb Survivors Exposed to Low Level Radiation," in *Low dose Irradiation and Biological Defense Mechanisms*, Sugahara, T., Sagan, L., and Aoyama, T., eds., Excerpta Medica, Amsterdam, pp. 71–74, 1992.

Smith-Sonneborn, J., "The role of the 'Stress Protein Response' in Hormesis," in Calabrese, E. J., *Biological Effects of Low Level Exposures to Chemicals and Radiation,* pp.41–52, Sans Serif, Ann Arbor, 1992.

Southam, C. M., and Ehrlich, R. G, "Effect of Extract of Western Red-Cedar Heartwood on Certain Wood Decaying Fungi in Culture," *Phytopathol.*, 33: 517–524, 1943.

UNSCEAR, "Adaptive Responses to Radiation in Cells and Organisms," A/AC.82/R.542, United Nations General Assembly, 1994.

Watters, P., Martin, F., and Schreter, Z., "Caffeine and Cognitive Performance: The Non-lnear Yerkes-Dodson Law," *Human Psychopharm.*, 12: 249–257, 1997.

CHAPTER 12
TOXICOLOGY AND RISK

MICROORGANISMS, MOLECULES, AND ENVIRONMENTAL RISK ASSESSMENT: ASSUMPTIONS AND OUTCOMES

Jane M. Orient
Dr. Orient is executive director of the Association of American Physicians and Surgeons. She is the author of Your Doctor Is Not In: Healthy Skepticism about National Health Care.

12.7.1 BACKGROUND AND SUMMARY

In the mid-1980s, the U.S. Environmental Protection Agency (EPA) initiated state projects with the goal of comparing and ranking the relative risk of "environmental" hazards and prioritizing the use of resources according to the seriousness of the risk. The Human Health Technical Subcommittee of the Arizona Comparative Environmental Risk Project (ACERP) met monthly for about two years (1993–1995). A draft of this article was submitted for the committee's report (Baker, 1995) but was not included; the author's request for written criticisms received no response.

The "Original EPA List of Environmental Problems," provided to the committee as a starting point, included 31 items, primarily chemicals produced by human civilization plus ionizing radiation. Except for "biological contaminants" in drinking water, the list made no mention of microorganisms.

This article contends that the EPA risk assessment method is fundamentally flawed when applied to the entities on the Original EPA List (or entities generally thought of as "pollutants"). Further, the risks selected for inclusion by the EPA are far less significant than some risks that are neglected. The inverted ranking has potentially serious consequences for public health, particularly if regulations impair defenses against microorganisms or other real threats to human health.

Microorganisms fit into all categories in the governmental risk assessment scheme, including air, soil, water, food, "peace of mind," and "environmental justice." The human health risk is not hypothetical but proved. The threats to health are demonstrable in very low concentration. No extrapolation is needed. It is possible to show cause and effect; one need not make a leap of inference from mere correlation. Morbidity and mortality statistics are available from standard references; they need not be calculated from models based on unprovable (and often implausible) assumptions. The effects are objective, not purely subjective. These effects, therefore, need to be considered in setting priorities or assessing the overall impact of any regulatory scheme.

12.7.2 THE NO-THRESHOLD LINEAR HYPOTHESIS: MICROORGANISMS AND MOLECULES

The standard EPA method for calculating risk of carcinogenesis starts with a dose-response curve, say from animals exposed to nearly lethal doses of the substance in question, and extrapolates linearly to the (0, 0) origin. The slope gives the "cancer potency factor" (CPF), which assumes that a single molecule (or single photon of radiation) is capable of inducing cancer; in other words, that there is "no threshold."

Note that this assumption is not made for substances that are merely poisonous rather than carcinogenic, as its falsity would be obvious. (If 100 mg of digitalis would kill one person, it does not follow that 0.1 mg given to each of 1000 people would also cause one death.)

To determine the number of hypothetical deaths resulting from a particular substance, the EPA multiplies the CPF by the dose by the number of individuals in the exposed population. In this manner, a risk that is too small to detect in an experimental population can be transformed into a huge number of hypothetical deaths.

There is no evidence for the validity of this EPA method as applied to chemicals—it is based on assumption only. However, there is one entity for which it is sometimes valid—only substituting one or a very few microorganisms (e.g., *Treponema pallidum*) for one molecule.

Consider the contrast between microorganisms (that may be of no regulatory concern) and chemicals (for which extraordinarily stringent regulations impose enormous costs upon the economy). The normal number of anthrax bacilli in a living organism is zero. The normal amount of the toxic and carcinogenic element arsenic in the human body is about 4.4 mg or 9×10^{18} molecules of As_4 (about 10,000 per cell; Robinson, 1994), any one of which is assumed to be a potential carcinogen.

Many organic compounds of concern to the EPA are normal byproducts of mammalian metabolism, as has been shown by gas chromatography and mass spectrometry of mouse and human wastes (Miyashita and Robinson, 1980; USSR Scientific Research Institute, 1987). At least 15 of these products, including 1,4-dioxane, trichloroethylene (TCE), and chloroform are on the "List of Hazardous Air Pollutants" to be regulated under Section 112 of the Clean Air Act. The EPA concerns itself at ambient-air concentrations less than one-ten thousandth the level found in normal intestinal gases (about 1.4 $\mu g/m^3$ and 20 $\mu g/L$, respectively, for TCE) (Smucker, 1994). Every human being is a "polluter"; the normal human body exhausts 20 grams of organic products per day (see Table 12.7.1).

If one follows the EPA policy of multiplying risk by the population potentially exposed, which in some cases approaches the population of the world, the threat posed by microorganisms is enormous. This risk is high even when contrasted with the population-based risk of death resulting from impact with an asteroid or comet (which ranges from 1 in 3000 or 86,000 persons in the United States to 1 in 250,000, depending on assumptions; the lower-limit risk assumes an extraterrestrial impact great enough to kill 1.5 billion persons about every 70,000 years; Chapman and Morrison, 1994). In addition, the population risk of an asteroid impact is huge compared with that of the similarly calculated risk of drinking water containing more than the EPA tolerance limit of trichloroethylene.

The threat of microorganisms is one that is known to be related to the contamination of the environment. In the absence of the causative organism, the incidence of a disease is zero. Add the organism, and disease appears. This contrasts with cancer, which has a very high natural incidence of approximately 1 in 4. These cancers are due to natural causes. Against this background, the addition of a small number of cancers because of anthropogenic environmental change is virtually undetectable, except for a small number of rare cancers, which are dramatically increased by a potent carcinogen.

Of course, a single pathogenic microorganism can't cause much harm all by itself. The problem is that living organisms have a doubling time rather than a half life, and a single bacterium can reproduce itself.

The body has many defenses against bacterial invaders. It also has elaborate metabolic machinery to detoxify compounds that it constantly produces and ingests, a factor not considered in regulatory assessments. It is actually necessary for the body to be exposed to low levels of bacteria; otherwise the immune defenses atrophy. (Deliberate exposure is called "immunization" or "vaccination.") Exposure to toxins is inevitable (even if not beneficial), and it is well-known that liver enzymes needed for detoxification are induced by such exposure.

Ionizing radiation is likewise unavoidable. There is a body of evidence to suggest that a certain range of exposure is advantageous, perhaps through stimulating DNA repair mechanisms that confer protection against a variety of insults. This is called a hormesis effect (Luckey, 1991). There is also increasing interest in the possibility of chemical hormesis.

Whether or not hormesis occurs, the no-threshold linear hypothesis for ionizing radiation is no longer credible. It is inconsistent with data on lung cancer incidence as a function of indoor radon

TABLE 12.7.1 Organic Compounds Exhausted By Human Beings

Total grams/hr exhausted in intestinal gas, sweat, urine, and feces:

Saturated hydrocarbons	0.169 g/hr
Unsaturated hydrocarbons	0.096 g/hr
Naphthenic hydrocarbons	0.031 g/hr
Aromatic hydrocarbons	0.058 g/hr
Aldehydes	0.016 g/hr
Ketones	0.027 g/hr
Alcohols	0.097 g/hr
Ethers	0.024 g/hr
Other compounds	0.239 g/hr
Sulfur compounds	0.059 g/hr
Nitrogen compounds	0.002 g/hr
Chlorine compounds	0.010 g/hr

By type, in order of decreasing concentration:

saturated hydrocarbons

methane, ethane, propane, pentane, isopentane, 2-methylpentane, 3-methylpentane, 2,2,5-trimethylhexane, heptane, nonane and its isomers, 2,5,5,-trimethylheptane, decane and its isomers, undecane and its isomers, dodecane and its isomers, tridecane and its isomers

unsaturated hydrocarbons

ethylene, butylene, isoprene, heptene, decene, diisoamylamine, undecene, dodecene, tridecene, 4-methyloctadiene, decine

naphthenic (cyclic) hydrocarbons

cyclobutane, cyclopentane, methylcyclopentane, cyclohexane, trimethylcyclopentane, 1,3-dimethylcyclohexane, ethylcyclohexane, trimethylcyclohexane, propylcyclohexane, amylcyclohexane, indan, hexahydroindan

aromatic hydrocarbons

benzol, toluene, ethylbenzene, xylol, styrene, n-propylbenzene, 1-methyl-3-ethylbenzene, 1-methyl-4-ethylbenzene, 1-methyl-2-methylbenzene, butyl benzene, 1,2,4-trimethylbenzene, 1-methyl-4-isopropylbenzene, 1-methyl-3-isopropylbenzene, 1,3-dimethyl-5-ethylbenzene, 1,2-dimethyl-4-ethylbenzene, 1,3-dimethyl-4-ethylbenzene, 1,2,3,4-tetramethylbenzene, naphthalene, 2-methylnaphthalene

aldehydes

formaldehyde, acetaldehyde, 2-methylpropanol, 3- methylpropanol, pentanol, 2,4-hexadienal, hexanol, furfural, heptanol, octanol, benzaldehyde, nonanol, decanol, undecanol

ketones

acetone, methylethylketone, 2-butanone, methylisobutylketone, 2-hexanone, 2-heptanol, 3-octene-2-one, 2-decanone, 2-undecanone, methylcyclopentanone

alcohols

methanol, ethanol, propanol, isopropanol, butanol, cyclohexyl alcohol, 3-methyl-1-butanol

ethers

ethyl acetate, 1,4-dioxane, butylacetate, isobutylacetate, isoamylacetate, ethylhexanoate, ethyloctanoate, 3-methyl-2-butylacetate

other compounds

carbon monoxide, phenol, furan, p-cresol, menthol, formic acid, acetic acid

nitrogen-containing compounds

methylamine, isopropylamine, pyrrolidyl, indole, skatole, 2,2-dipyridyl, n-methylpyrrole, methylpiperazine, methacrylonitrate

(Continued)

TABLE 12.7.1 Organic Compounds Exhausted By Human Beings (Continued)

sulfur-containing compounds
methylmercaptan, ethylmercaptan, dimethyldisulfide, amylmercaptan, 2,3,4-trithiopentane, allylthioisocyanate, ethylenesulfide

chlorine-containing compounds
chloroform, trichloroethylene, tetrachloroethylene, chlorobenzene, methylchloride, tetrachlorohydrate, dichloromethane, 1,1,1-trichloroethane

Source: Compiled by Adam Paul Banner from Russian studies, 1992.

Literature Citations:
I. *Intestinal Gas*
CA 112(11): 94935q
Chromatography and Mass Spectrometry of the Intestinal Gas in the Diagnosis of Diseases
Dmitriev, M.T.; Malysheva, A.G.; Rastyannikov, E.G.; Kopov, U.T.
Nll Obshch. Kommunal'n. Gig. im. Sysina
Moscow, USSR
Lab. Delo, (11), 74–75 (1989).
II. *Human Waste: Intestinal Gas-Sweat-Urine-Feces*
CA 107(25): 232594c
Specific Organic Compounds in Human Wastes
Dmitriev, M.T.; Malysheva, A.G.; Rastyannikov, E.G.
USSR
Kosm. Biol. Aviadosm. Med., 21(4), 50–56 (1987).

levels (Cohen, 1995). Yet despite the fact that higher household radon levels are associated with *lower* lung cancer rates, the federal government set up a 1-800-SOS-RADON hotline, and indoor radon is number four on the EPA's "Original List" and one of the "high-risk" agents on the ACERP Human Health Subcommittee's final ranking (Baker, 1995). Indeed, probably more information exists on radon than any other candidate for the number one risk.

Application of the concept of hormesis to chemicals is speculative at this point and will remain so until it becomes an area of active investigation, which is unlikely in the present political climate.

12.7.3 RISKS BECAUSE OF MICROORGANISMS

In the furor over volatile organic compounds, pesticides, and other chemicals, we tend to forget that infectious diseases are in fact the leading killer of human beings worldwide. In the United States, more than 740 million infectious disease events occur each year, resulting in 200,000 deaths and more than $17 billion in direct care, according to the National Foundation for Infectious Diseases (Bureau of National Affairs, 1994). According to the CDC, the actual age-adjusted death rate per 100,000 population and the years of potential life lost before age 65 per 1000 persons under the age of 65 were (1986–1988, U.S.), respectively: AIDS 7.5 and 2.2; pneumonia/influenza 37.3 and 1.0; other infections 19.6 and 2.4. (See Tables 12.7.2 and 12.7.3)

TABLE 12.7.2 Incidence of Some Vector-Borne Diseases

Disease	Estimated number of infected persons (in millions)	Number of persons at risk (in millions)
Malaria	280	2100
Schistosomiasis	200	500–600
Lymphatic filariasis	90	905
Onchocerciasis	18	90
Chagas' disease	16–80	90
Leishmaniasis	12	350

TABLE 12.7.3 Notifiable Diseases in the U.S., 1991

	Geigy	MMWR
AIDS	43,389	43,672
Tuberculosis	23,543	26,283
Measles	8884	9643
Malaria		1278
Typhoid fever	456	501
Encephalitis	76	1021

Source: *Geigy Scientific Tables*, 1992; *MMWR* 10/2/92.

TABLE 12.7.4 Relative Toxicity of Certain Chemicals and Toxins (Plutonium injected peritoneally $= 1$)

mercury chloride	100
strychnine	1000
actinomycin	10,000
tetrodotoxin	100,000
perfringens A	1,000,000
pestis toxin	10,000,000
shigella toxin	100,000,000
botulinal E toxin	1,000,000,000
tetanus toxin	100,000,000,000
botulinal D	10,000,000,000,000

Source: T. D. Luckey, letter to editor, *Chemical & Engineering News*.

Current infectious disease rates in the United States might represent close to the *lower* limit achievable with present technology. In contrast, the EPA estimates (and deliberately exaggerates by at least an order of magnitude) the *upper* limit for the number of cancers caused by certain chemical or radiation exposures.

It is also worth remembering that microorganisms produce the most potent poisons known to humans (see Table 12.7.4). While the risk of microorganisms is primarily a noncancer risk, there is also a cancer risk. Chronic infection (like any type of chronic irritation that results in increased cellular proliferation) is suspected to cause cancer (this is more important in underdeveloped countries). Moreover, some products of microorganisms, notably aflatoxin, are carcinogenic.

12.7.4 PRIORITIES HAVE CONSEQUENCES

Even if flawed methods did not lead to action that was directly detrimental to human health, they can lead to misguided priorities that are harmful in three ways: (1) by diverting resources into less valuable uses; (2) by reducing the standard of living, which is itself correlated with increased mortality and morbidity (Keeney, 1990); and (3) by mandating actions that increase a risk that is already greater than the hazard they are intended to alleviate.

Besides being wasteful, many regulations increasingly work in favor of the germs in the age-old war against infectious disease. Consider the regulatory targets: chlorine-containing compounds (including our most important disinfectants); refrigerants; gamma radiation (used to sterilize medical devices and effective for preserving food); pesticides (for controlling insect vectors of disease); and chemical preservatives in food.

Speculation that tiny residues of DDT may cause breast cancer is considered newsworthy, though based on evidence that is at best flimsy if not disingenuous, relying on serum measurements of DDT metabolites that were at the limit of detectability by the gas chromatography method used (Wolff et al., 1993). On the other hand, the 10 to 20 million human lives lost because of increased malaria after DDT was banned are rarely mentioned (Hazeltine, 1993). Similarly, the suggestion that increased breast cancer might result from the hormonal effects of aborting a pregnancy (especially a first pregnancy) may be considered inflammatory, even though supported by a number of studies.

An EPA statement that chloramines might be carcinogenic led the Peruvian government to stop chlorinating the water supply. A cholera epidemic followed, with 300,000 cases and 3516 deaths (Anderson, 1991). (Some dispute the number of cases and argue for multifactorial causes.) There have been seven great cholera pandemics since 1817, the most recent one in the 1960s. New resistant strains of cholera may have pandemic potential. At a July 1992 microbiology conference in Sydney, Australia, Frank Fenner of the John Curtin School of Medical Research in Canberra argued that a major pandemic plague, of as-yet unknown type, is virtual certainty (DeCourcy, 1994).

Furthermore, "abatement" measures can themselves spread disease. Protecting "wetlands" (swamps) often means protecting breeding areas for mosquitoes. In Arizona, it has been suggested that dust-control measures required by the Clean Air Act have allowed the organism causing coccidioidomycosis (valley fever) to propagate better in the soil (oral discussion at ACERP technical committee meeting). Valley fever could also result from excavations, including those demanded by the government to dig up leaking underground fuel storage tanks. A lawsuit resulted when workers contracted valley fever after a demonstration of earthmoving equipment. Conceivably, the government could be liable for cases that arguably resulted from mandated excavations or from requirements to periodically stir up piles of dirt.

Such potential adverse effects of regulations are not routinely monitored. Despite the fact that coccidioidomycosis tripled in incidence between 1986 and 1991, according to the Centers for Disease Control and Prevention (Mutual Insurance Company of Arizona, 1994), few data about spore counts are available, although they could readily be collected and correlated with events such as environmental interventions. This is but one example of myriad unintended consequences that are seldom seen because rarely looked for.

12.7.5 UTILIZATION REVIEW, QUALITY ASSURANCE, AND OUTCOMES ASSESSMENT

Governmental agencies have increasing enthusiasm for monitoring and directing physicians' practices in an effort (however misguided at times) to diminish harm done to patients. Yet the potential harm that could be caused by ill-considered regulations is vastly greater.

Concepts analogous to "medically necessary," "professionally accepted standards," "cost-effective," "experimental treatment," and "side-effect profile"—along with the Hippocratic maxim to "do no harm"—need to be applied to the would-be physicians to the "environment." Data collection should be required both before and after the implementation of policy. Calculations based on unsupported assumptions should be rejected as the sole basis for regulatory coercion. Optimum human health, not the achievement of arbitrary numerical endpoints, should be the objective.

REFERENCES

Anderson, C., "Cholera Epidemic Traced to Risk Miscalculation," *Nature*, 354: 255, 1991.

Baker, L., *Arizona Comparative Environmental Risk Project. Section 3: Human Health*, published by U.S. EPA, & Ariz. Dept. of Env. Quality, & Commission on Ariz. Environment, Phoenix, AZ, 1995.

Bureau of National Affairs, "Industry Survey Reveals 79 Drugs in Development for Infectious Diseases," *BNA's Health Care Policy Report*, 2(28): 1247, 1994.

Chapman, C., and Morrison, D., "Impacts on the Earth by Asteroids and Comets: Assessing the Hazard." *Nature*, 367: 33–40, 1994.

Cohen, B. L., "Test of the Linear-No Threshold Theory of Radiation Carcinogenesis for Inhaled Radon Decay Products," *Health Physics*, 68(2): 157–174, 1995.

DeCourcy, J., "Pandemic Plague: A Matter of Time," *Intelligence Digest*, Sept. 30: 3, 1994.

Hazeltine, W., "The Environment and Health," *Oral presentation at annual meeting of Doctors for Disaster Preparedness,* Costa Mesa, CA, August 14, 1993.

Keeney, R. L., "Mortality Risks Induced by Economic Expenditures," *Risk Analysis*, 10(1): 147–159, 1990.

Luckey, T. D., *Radiation Hormesis*, CRC Press, Boca Raton, Florida, 1991.

Miyashita, K., and Robinson, A. B., "Identification of Compounds in Mouse Urine Vapor by Gas Chromatography and Mass Spectrometry," *Mechanisms of Ageing and Development*, 13: 177–184, 1980.

Mutual Insurance Company of Arizona, "Valley Fever on the Rise," *MICA Medical Legal Alert*, (28): 1, 1994.

Robinson, A. B., "One Molecule Hypothesis," *Access to Energy*, 21(10): 4, 1994; and "Correction and Amplification," *Access to Energy*, 21(11): 2, 1994.

Smucker, S. J., *Region IX Preliminary Remediation Goals (PRGs) First Half 1994*, United States Environmental Protection Agency, 2/1/94.

USSR Scientific Research Institute of General and Municipal Hygiene, "Specific Organic Compounds in Human Waste," *Kosm Biol Aviadosm Med*, 21(4): 50–6, 1987.

Wolff, M. S., Toniolo, P. G., Lee, E. W., Rivera M., and Dubin, N., "Blood Levels of Organochlorine Residues and Risk of Breast Cancer," *JNCI*, 85: 648–652, 1993.

CHAPTER 13
CONTROL TECHNOLOGIES

SECTION 13.1

HAZARDOUS WASTE HANDLING AND DISPOSAL

Dave Lager

Mr. Lager is president of NETCo., a professional organization which provides environmental compliance services to business and industry. He has been assisting industry in meeting RCRA, CERCLA and CWA requirements for 15 years.

13.1.1 HAZARDOUS WASTE MANAGEMENT REGULATIONS

The U.S. Congress first adopted regulations addressing industrial hazardous waste handling and disposal as early as 1965. However, it was not until 1976 that Congress passed the Resource Conservation and Recovery Act (RCRA). This law, as amended, is the principal statue that governs how we manage the hazardous waste disposal process.

The heart of RCRA is the principle of "Cradle to Grave" responsibility for the management of hazardous waste. The generator of the hazardous waste retains the management and fiscal responsibility for ensuring that its hazardous waste does not pose a threat to the environment, its employees, or the public at large. The original RCRA statute established the concept of a "paper trail" for all hazardous waste materials. The generator is required to identify, store, transport, and eventually dispose of the hazardous waste using a process in which the generator is ultimately responsible for the safe handling and disposal of its own waste. The generator cannot pass off responsibility for its waste to another company or to state, federal, or local government for the improper disposal of its hazardous waste.

RCRA Mid-1970s to Mid-1980s

During the early years of RCRA, the major focus on implementing RCRA was on how to manage the points where we were attempting to concentrate or dispose of our hazardous waste—landfills. Most, if not all attention, was on the system by which we were getting waste from places where the waste posed a threat to both human health and the environment to landfills. Landfills were generally recognized as the best location for controlling, containing and dealing with the threat of toxic and hazardous wastes. A substantial industry grew up around the transportation of hazardous waste, as well as the building and the managing of landfills, including the process by which permits were issued for a landfill disposal site.

However, in the early 1980s, as our understanding of the hazardous waste problem grew, public interest groups, congressional environmental regulatory oversight committees, and their staff, began to recognize the broad scope of the United States hazardous waste problem, including the following:

- Liquid waste placed in any landfill sooner or later leaks through physical barriers.

- Hazardous wastes, in most cases, do not lose their hazardous characteristics in a landfill. Attenuation processes only deal with certain classes of wastes, and certainly not the most toxic and dangerous hazardous wastes.

- The scope of the hazardous waste problem extended beyond large generators. Small quantity generators also generated a substantial quantity of hazardous wastes.

- The relationship between placing hazardous wastes in a landfill and the associated or potential contamination of groundwater supplies became more broadly recognized. Congress, public interest groups, and concerned citizens who lived near landfills (early days of the "Not In My Backyard" or NIMBY movement) began to draw attention to the fact that groundwater supplies, once contaminated, were difficult and, at best, very expensive to remediate.

Thus, in 1984 Congress revisited the Resource Conservation and Recovery Act, and passed a substantial amendment to the original RCRA statue. Among the more significant of the changes were the following:

- Small quantity generator provisions were established.

- A phased in land disposal restriction was established banning the landfilling of most types of hazardous wastes, including, most importantly, liquid hazardous wastes.

- Congress established strict minimum technical requirements for landfill design and monitoring, including leak detection systems and double liners.

- Underground storage tank (UST) regulations were established, by the United States Environmental Protection Agency requiring strict spill and overflow prevention and protection for UST's by December 22, 1998. These requirements led to many UST owners eliminating use of USTs because of the cost of such spill, prevention, and protection requirements.

However, as we enter the twenty-first century, perhaps the most significant and lasting change of the 1984 RCRA Amendments is the little recognized clause requiring generators to engage in a waste minimization/waste reduction process. Under the objectives section of the 1984 RCRA Amendments, Congress declared:

"The objectives of this Act are to promote the protection of the health and the environment, and to conserve valuable material and energy resources by . . . minimizing the generation of hazardous waste and the land disposal of hazardous waste by encouraging process substitution, materials recovery, properly conducted recycling and reuse and treatment."

In short, while neither Congress or most participants in the hazardous waste problem at the time knew how we were going to accomplish the objective, Congress in 1984 set the bar for the twenty-first century regarding the direction of our hazardous waste program—eliminate waste at its source by creating manufacturing processes that neither use hazardous substances, nor generate hazardous waste by-products. The tide had begun to turn on traditional hazardous waste management views. For decades, we, as an industrialized nation, had accepted the notion that hazardous waste was an acceptable by-product of our economy. The 1984 RCRA Amendment objectives statement set the basis for a change in our underlying approach to hazardous waste management.

RCRA Mid-1980s to Late-1990s

While the bar was established to deal with hazardous waste generation at its source (i.e., manufacturing processes and materials which created hazardous waste by-products) most companies, engineering firms, regulators, and consultants in the mid to late-1980s continued to focus on the site remediation technology component. Continuing advances in liners, leak detection, contaminated groundwater extraction techniques, bioremediation, incineration methods, among other technologies, came at a relatively rapid rate in the late 1980s. This focus on hazardous waste management technologies represented the continuing regulatory perception that hazardous waste was a necessary by-product of our industrial economy.

Further, from the mid-1970s on, an entire industry and governmental bureaucracy had grown up around the managing and control of hazardous wastes. The concept that eliminating hazardous wastes at the source through changing manufacturing processes was a concept that threatened regulatory bureaucracies, as well as the existing private sector industries that had grown up around transporting and managing the handling and the disposal of hazardous wastes.

In 1989, however, the first legislative attempts were made to implement how we today are increasingly approaching the hazardous waste handling and disposal challenge. Leading the way were three states—California, Oregon and Massachusetts. In one form or another, these states passed toxic use reduction laws (most states now have some type of toxic use reduction planning regulation). Such laws moved beyond a strict technology approach to hazardous waste disposal—these new toxic use reduction laws challenged the concept that hazardous waste was a necessary by-product. Each law required that generators, usually large quantity generators, implement a toxic use accounting and planning process.

Specifically, the toxic use reduction regulations required generators to engage in a data collection, analysis, and economic planning process to identify the so-called "true cost" of using toxic materials in a manufacturing process. The outcome of the process was to be annual reports, and a plan identifying the quantity of toxics used, and most importantly, what each generator was doing to eliminate toxics in the manufacturing process.

More importantly, the toxic use reduction laws began to quietly move the hazardous waste handling and disposal process into an *industrial engineering* setting. The process of dealing with hazardous waste now had to be dealt with in the design of manufacturing processes, the selection of materials to be used in manufacturing a product, and the efficiency by which hazardous materials were used.

One example of this change occurred in the metal finishing industry. Cyanide was a widely used chemical in the plating of metal parts. However, it is also a highly toxic and dangerous chemical. Many plating shops in the 1980s and early 1990s were large users of cyanide. As a result of the cyanide use, plating shops generated large quantities of cyanide waste sludges, which were, in turn, disposed of by any number of hazardous waste management firms.

However, the toxic waste management requirements of the early 1990s began to change the perception by which metal finishing firms looked at cyanide. Initially, many plating firms began to be more cognizant of the chemistry of their plating baths. More careful analysis of their use of cyanide resulted in an almost immediate reduction of the quantity of cyanide used. Plating firms quickly realized they had been using more cyanide then they had do. Further, many plating firms began to look into non-cyanide based processes as an alternative, and began to switch over as their individual economic situations would allow, regulatory pressures dictated, or as technology of such processes gave platers more confidence in such technologies. As a result of this toxic use review process, the quantity of cyanide used has been reduced, and more importantly, the need to deal with cyanide-based wastes has decreased.

There are many other similar examples of the change in direction in the engineering processes that manufacturing firms use to devise a manufacturing process. More efficient and multiple-step metal fabrication equipment has resulted in the decreased volume of wastewater treated in any number of metal fabrication shops, and/or the use of soldering. Change over to 100 percent solids painting processes has resulted in the elimination of the use of chlorinated solvents in many paint shops. Their is, in fact, a large body of literature today available on the Internet, in technical libraries, or from specialty consultants or government agencies that presents case studies of how industry has and continues to work to reduce the use of toxics. The direction has clearly been set for the year 2000 and beyond: less focus on control and management of toxic wastes and increased focus on manufacturing processes to eliminate hazardous waste materials and their associated waste by-products.

RCRA Year 2000 and Beyond

The traditional concept of RCRA is in the process of being eclipsed. However, there will always be a need for RCRA regulations for that component of our industrial economy where it is too costly or technically infeasible to switch over to nontoxic manufacturing processes. There will always be some quantity of hazardous wastes, but the quantity of hazardous waste generated is on the decline and should remain on the decline for the foreseeable future.

One important reason such a statement can be made is the appearance of the International Organization For Standardization or ISO movement. The ISO movement in the United States started largely by focusing on quality improvement in the 1980s and 1990s. Quality improvement concepts focus on looking at the manufacturing process as a whole, and what can be done to minimize or eliminate defects in the manufacturing process. Continuous improvements in quality generally results in improved perceptions of the product, less costs, improved consumer satisfaction, and longer lasting customer relationships.

In the last several years, ISO 14000 first published in 1996, has started to appear on the scene. This management concept focuses on a firms' environmental management practices. The essential concept is that manufacturing firms have an obligation to *continuously improve* their environmental management practices by developing *an Environmental Management Systems (EMS)*. The EMS Plan concept works to provide a systematic approach to control and to integrate the environmental aspects of an organization's activities or processes into the total business environment of the firm.

While not widely used or adopted, it appears reasonable that the concept of the ISO 14000 EMS, improved environmental management practices integrated into the day-to-day practices of the corporation, if not the actual ISO 14000 standard, represent one important future direction for dealing with hazardous wastes. In fact, there have been many comments that at some point in the future companies who have an ISO 14000 program may be subject to a separate set of hazardous waste regulatory standards focusing on self-regulation and reporting. The USEPA, as well as several state environmental regulatory departments, are promoting the implementation of EMS as a proactive step for companies to continuously improve their overall environmental health, and possibly avoid (in time) government-prescribed compliance enforcement orders or regulations.

Thus, as we move forward from the year 2000, the management of hazardous wastes is changing. The major change is that we are generating less hazardous wastes. New management concepts are changing our perception of hazardous wastes as a necessary by-product of our economy to one of aggressively attempting to eliminate hazardous waste at its source—manufacturing processes and materials.

The hazardous waste arena has therefore changed, no longer must we focus on hazardous waste remediation technologies to improve our environment. Rather, since the 1970s we have come through an evolutionary process by which we first focused on controlling and containing hazardous wastes in landfills (1970s/1980s), to eliminating the use of landfills by focusing on the destruction or in situ treatment of hazardous wastes through various processes (1980s/1990s), to today where we are now looking to reduce and minimize hazardous waste generation by environmentally sensitive manufacturing and management processes.

This is not to say that we do not, or will not need some form of hazardous waste regulation including how we characterize, treat, and store hazardous wastes. There will always be various types of hazardous wastes, which will always need to be handled. However, the tide has turned toward hazardous waste reduction rather than hazardous waste management.

13.1.2   *RCRA HAZARDOUS WASTE REGULATIONS—GENERATORS*

Hazardous Waste Provisions

The RCRA regulation is among the most complex of any of our many environmental regulations. It regulates solid waste, hazardous wastes, and underground storage tanks. However, the core sections of the RCRA regulation pertaining to hazardous waste generators are contained in Subtitle C, and in particular Sections 3001 to 3006 as follows:

- Section 3001: Identification and Listing of Hazardous Waste
- Section 3002: Standards Applicable To Generators of Hazardous Waste

- Section 3003: Standards Applicable To Transporters of Hazardous Waste
- Section 3004: Standards Applicable To Owners, Operators of Hazardous Waste Treatment, Storage and Disposal Facilities
- Section 3005: Permits For Treatment, Storage or Disposal of Hazardous Waste
- Section 3006: Authorized State Hazardous Waste Programs

Characterizing Hazardous Wastes

One of the most important responsibilities that a generator has is characterizing its hazardous waste stream(s). The United States Environmental Protection Agency (EPA) has developed a process that involves analyzing the waste stream to first determine if it is a *listed waste*; and, if it is not a listed waste, what its *characteristic waste* classification will include.

Listed Wastes. The United States Environmental Protection Agency has developed a series of lists identifying certain hazardous wastes, and classifying such wastes as follows:

- **F-Listed Wastes**—Hazardous Wastes from Nonspecific Sources (Example—F011 Waste: Spent cyanide solutions from salt bath pot cleaning from metal heat treating operations)
- **K-Listed Wastes**—Hazardous Wastes from Specific Sources (Example—K035 Waste: Wastewater treatment sludges generated in the production of creosote)
- **P/U-Listed Wastes**—Discarded Commercial Chemical Products, Off-Specification Species, Container Residues and Spill Residues Thereof (Example—P121 Waste: Zinc Cyanide)

If the waste does not fall into one of the above lists, then the generator must determine the characteristic waste classification of the waste stream.

Characteristic Waste. A characteristic waste exhibits one or more of the following characteristics

1. **Ignitable** (EPA Hazardous Waste Number is D001)
   - A liquid waste with a flash point of less than 60°C or 140°F
   - A non-liquid waste and it is capable under standard temperature and pressure of causing a fire or sustained burning.
   - It is an ignitable compressed gas
   - It is an oxidizer
2. **Corrosive** (EPA Hazardous Waste Number is D002)
   - It is a liquid waste with a pH of less than or equal to 2, or greater than or equal to 12.5
3. **Reactive** (EPA Hazardous Waste Number is D003)
   - A waste is reactive if it has any of the following properties:
   Unstable
   Water reactive
   Water explosive mixture
   Toxic gas generation
   Cyanide/sulfide generation
   Explosive reaction/detonation
   Detonation/explosive decomposition
   Forbidden, Class A, or Class B explosive
4. **Toxic** (EPA Hazardous Waste Numbers—See Table 3.1)

- A waste is toxic if the waste contains any of the substances listed in Table 3.1 (see RCRA regulations)

Waste Analysis Process.   The process of determining whether the waste is either a listed or a characteristic waste is determined by a chemical analysis of the wastes. Some waste streams do not require a chemical analysis because the constituents and characteristics of the waste are known based upon knowledge of the process that produced the waste, for example:

- **Outdated Commercial Chemicals, Products**—Unused materials whose shelf life has expired and are readily identifiable from packaging or Materials Safety Data Sheets (MSDSs).
- **Non-Spent, No Longer Needed Chemicals**—Unused materials that is not needed for future use in a process.
- **Non-Spent, Leaking**—Unused material that has leaked from its container because of time exposure, or accidental damage.
- **Container Residues**—Material that remain in a container after emptying or material rinsed from an empty container, provided that the container did not hold acutely hazardous waste.
- **Spent, Known Contaminant**—Waste generated through routine processes where both the starting product and the contaminant are known.
- **Off-Specification Material**—Unused material that does not meet the specification required for the desired process.
- **Non-Spent Mix of Virgin Materials**—A mixture of two or more unused materials that are not needed for future use at the installation.
- **Contaminated With Water**—Waste streams generated through routine processes where the starting product is known and the only known contaminant is water.

Other wastes will require an analysis by a laboratory to determine their waste characteristics. Examples of such waste streams include:

- Wastes from an unknown or new waste stream that contains no virgin or non-spent materials
- Spent chemicals

Treating Hazardous Wastes

Once a waste has been identified and properly characterized, the issue of treatment can be addressed. There are multiple methods of treating a hazardous waste to render the waste non-hazardous. Chapter 14 of this book devotes considerable space to discussing the various treatment technologies that are available.

In general, the standard that applies today to all hazardous wastes is the waste is either treated to render it non-hazardous, or the waste is destroyed. This was not the case in the early days of RCRA, but as the 1984 Amendments require, there are only a few limited instances where hazardous wastes can be landfilled.

Storage of Hazardous Wastes

RCRA has established extensive hazardous waste storage requirements. The details of the rules are too extensive to go into in this survey chapter. Any number of sources exist in which the reader can obtain detailed information about waste storage requirements. Suffice it to say that the length of time that a waste can be stored on-site depends upon the status of the firm as a generator (i.e., large or small quantity generator). The length of time of storage can also vary by state—states can have hazardous waste storage requirements more restrictive than the federal requirements. There are also detailed rules on labeling of storage containers, the types of storage containers that can be used, compatibility rules, and how and where the waste is to be stored prior to shipment.

If a firm is a large quantity generator, it must also prepare hazardous waste contingency plans and engage in an increased level of inspection and personnel hazardous waste training. The United States Department of Transportation has also established various labeling, marking, container performance, and training requirements for generators under its authority to regulate the transport of hazardous wastes.

13.1.3 RCRA HAZARDOUS WASTE REGULATIONS—TREATMENT, STORAGE, AND DISPOSAL FACILITIES (TSDs)

As the name implies, Treatment, Storage, and Disposal (TSDs) facilities are where hazardous wastes finish their cradle-to-grave journey. Because TSDs are in the business of handling hazardous wastes on a day-to-day basis, the US EPA and the states requires such facilities to obtain operating permits for the specific type of facility. The regulations for obtaining and maintaining a TSD facility are quite extensive, and have been covered in detail by many authors and publications. Thus, in this section we will provide only a short overview of the TSD facility requirements. Readers who are interested in more detail should contact their individual state environmental regulatory agency for the specific requirements in their state, or the US EPA regional office for their state.

The definition of a treatment, storage, and disposal facility encompasses the following:

- **Treatment**—Any method, technique, or process including neutralization, designed to change the physical, chemical, or biological character or composition of any hazardous waste so as to neutralize it, or render it nonhazardous or less hazardous, or to recover it, make it safer to transport, store or dispose, or amenable for recovery, storage, or volume reduction.
- **Storage**—The holding of hazardous waste for a temporary period, at the end of which the hazardous waste is treated, disposed, or stored elsewhere.
- **Disposal**—The discharge, deposit, injection, dumping, spilling, leaking, or placing of any solid waste or hazardous waste into or on any land or water so that any constituent thereof may enter the environment or be emitted into the air or discharged into any waters, including groundwaters.

The details of the TSD requirements are contained in Subparts A through P of 40 CFR 264, 265. These subparts are as follows:

- Subpart A—General
- Subpart B—General Facility Standards
- Subpart C—Preparedness and Prevention
- Subpart D—Contingency Plan and Emergency Preparedness
- Subpart E—Manifest System, Recordkeeping and Reporting
- Subpart F—Ground Water Monitoring
- Subpart G—Closure and Post-Closure
- Subpart H—Financial Requirements
- Subpart I—Use and Management of Containers
- Subpart J—Tank System
- Subpart K—Surface Impoundments
- Subpart L—Waste Piles
- Subpart M—Land Treatment
- Subpart N—Landfills
- Subpart O—Incinerators
- Subpart P—Thermal Treatment
- Subpart Q—Chemical, Physical and Biological Treatment

- Subpart R—Underground Injection
- Subpart S to V—Reserved
- Subpart W—Drip Pads

As can be seen from the above, Subparts A through I lay out the administrative and non-technical requirements that any TSD must meet. These requirements ensure that owners or operators of TSDs establish the necessary procedures, plans, and general technical standards (such as groundwater monitoring and closure requirements) to run a facility properly and to handle any emergencies that may arise.

Subparts J through W deal with specific waste management methods, and the details of operating such facilities. As different treatment methods evolve, additional subparts may be added to deal with the specific parts of managing that type of TSD.

The process of permitting a TSD is arduous and expensive. As the above sections imply, the operator of a TSD must do considerable homework to provide a properly documented TSD permit application. Even once the TSD permit is issued, the operator of the TSD must operate that TSD to the standard set forth in the permit, engage in continuing personnel training, and maintain comprehensive records of facility operations.

CHAPTER 13
CONTROL TECHNOLOGIES

SECTION 13.2

HAZARDOUS WASTE DUMPING AS IT RELATES TO A JAPANESE RISK MANAGEMENT SYSTEM

Takehiko Murayama

Dr. Murayama is a specialist in the decision-making process of risk management and communication at the School of Science and Engineering of Waseda University in Japan.

13.2.1 INTRODUCTION

Japan has accomplished tremendous economic growth since the end of World War II. During this period, government policy placed emphasis on economic development rather than environmental protection. There are some exceptions, including quick responses to increased regulation by foreign countries and international organizations that may disrupt the volume of trade with other countries, such as the U.S. "Muskie" law and ISO 14000s.

Such economic-oriented attitude also applies to in industrial waste treatment and disposal policy. Waste disposal and public cleanliness (hygiene) law (WDPCL), and domestic law for controlling municipal solid waste (MSW) and industrial waste, prescribe that waste generators in principle have the primary responsibility for waste treatment and disposal under the Polluters Pays Principle (PPP). Many generators, however, do not treat and dispose of wastes themselves, but contract the transportation of generated wastes to other companies. Yet, other companies also often implement intermediate treatment and final disposal. Although waste control companies must get an approval from prefectural governments, insufficient checking and reviewing of their activities allows illegal treatments and dumping. To remove such defects in the waste control system, the Ministry of Health and Welfare of Japan (MHW), which is mandated to control waste treatment and disposal of MSW and industrial waste, has frequently revised the WDPCL. However, the series of revisions remain ineffective because they were not drastic enough; they were only amendments. Japanese society has not yet developed a fundamental policy under which waste generators are deemed to have complete responsibility throughout the waste treatment and disposal process. The development of recycling systems has also made the definition of waste ambiguous, and other ministries are mandated to control some wastes. In addition to MHW, recyclable materials and their handling systems are controlled by ministries such as the Ministry of International Trading and Industry of Japan (MITI), whereas the Environment Agency of Japan (EA) has responsibility for air, waste, and soil pollution control induced by waste treatment and disposal. Such a complicated administrative system is one of the factors that have lead to inappropriate waste treatments.

This paper discusses Japanese cases of environmental pollution caused by illegal dumping and inappropriate waste incineration, and derives some lessons from these cases. One is an illegal dumping case, which occurred on a small island, Teshima, located in the Seto (Inland) Sea of Japan. In this case, a waste transportation and disposal company had incompletely incinerated and then illegally dumped a tremendous amount of hazardous wastes of some 500,000 tons. This is the biggest of all Japanese illegal dumping cases. At the beginning of the company's activity, the company insisted that materials the company had transported to the island had recycling value, and consequently local government approved the company's activities without sufficient examination and observation. Incomplete incineration over around seven years then produced exhaust gases, which included various hazardous materials such as dioxins, and illegal dumping without appropriate environmental protection occurred, which allowed hazardous materials to infiltrate underground. That pollution damaged local people's health and the local natural ecosystem. To remedy the most severe pollution, the government was forced to spend about 1.5 billion yen (approx. US $1.25 billion).

Another issue is the environmental pollution caused by dioxin compounds, generated mainly through incineration of MSW, as well as industrial wastes. Compared with other countries, Japan has relied more on incineration than direct landfill for waste treatment. The geographic characteristics of Japan, with its narrow and steep land areas, have driven the country to rely on such a policy. Japanese incinerator technology has become highly developed as a result. However, too great a reliance on incineration has generated a large amount of dioxins for this densely populated country. This is because rapid growth has generated large amounts of industrial wastes and MSW, and small incinerators with inadequate capacity for treating those wastes have done so inappropriately. The compounds produced by these incinerators are damaging the air and soil around waste treatment facilities and the health of both incinerator facility workers and local residents.

Through an examination of these cases, I will show the characteristics of environmental pollution caused by waste treatment and disposal in Japan, and identify the lessons learned from those cases.

13.2.2 OUTLINE OF WASTE CONTROL IN JAPAN

Outline of the Waste Treatment and Disposal System

One unique characteristic of the Japanese waste control system is the method of classification of wastes. In North America (United States and Canada), wastes are generally divided into two categories, hazardous and nonhazardous. In contrast, the Japanese system classifies waste mainly in two categories according to whether the waste generator is the public or private industry. The public is considered the source of "general waste," which is almost the same as MSW, while "industrial waste" is that generated by private companies. Both categories of waste include hazardous and nonhazardous wastes. Figure 13.2.1 describes the system of classification of wastes in Japan.

In the Japanese legal system, the differences between generators may be emphasized more than the differing hazard from the waste. This method of classifying wastes was adopted to make clear the responsibilities for waste disposal. However, this system is inadequate to identify the hazard level of wastes from both sources. That, in turn, often results in inappropriate waste treatments.

The objective of MSW reduction is essentially to cut down the amount of waste disposed in, or on, the land. Of all waste reduction measures, incineration is considered to be the most effective method in decreasing the amount of waste requiring disposal. It was not until 1991 that a government policy to encourage reduction of MSW as a means of waste control was formulated.

In addition to MSW, Japanese industrial waste is treated to avoid inappropriate management and to reduce the amount going to landfill because of the shortage of landfill capacity. The Japanese system originally defined materials, including 11 specific hazardous substances, as hazardous waste. In 1992,

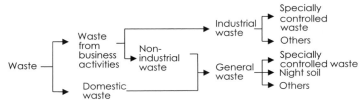

FIGURE 13.2.1   Classification of waste in Japan.

the system also defined a special category for harmful waste known as *specially controlled waste*. This category included not only hazardous waste, but also covered the explosive and infectious properties of the waste. Under the new system, generators should appoint persons to track this waste and ensure its proper treatment. Unless this category of waste is treated on-site, that which is sent off-site must go to licensed facilities with signed manifests, by which means the transfer route of the waste can be checked.

Industrial wastes generated by private companies are often treated by various methods, including shredding, dehydration, and incineration, before going to landfill. Although prefectural governments issue certificates to waste treatment companies, the subsequent observation and auditing of their activities is, in general, insufficient to ensure appropriate treatment by private companies. Of most immediately treated waste, about 38 percent is recycled, with the total volume going to landfill being about 80 million tons (about 20 percent) per year. The remainder (42 percent) is reduced through an intermediate process.

Japan has three types of landfill sites approved for various types of waste. For comparatively stable and nonhazardous wastes, landfills have a relatively simple structure, with no impermeable liner under the site area. The second type of landfill, for controlled types of waste, has an impermeable liner under the dumping zone for protection against permeation of leachate through rainfall infiltrating the site. Wastewater treatment facilities are also usually attached. For discharging treated wastewater, this type of site is often constructed near the upper tributaries of rivers. The last type of landfill is a facility for the most hazardous waste. This type of landfill has a concrete bunker divided into cells to prevent the mixing of incompatible wastes and is covered to minimize rainwater infiltration. The legal system aims to prevent environmental pollution by regulating the type of waste deposited at each type of landfill. Construction and operation of the first two types of landfill, however, often generate environmental dispute. In particular, the second type of site (for controlled wastes) often brings about serious dispute because the treated wastewater is discharged to the upper parts of rivers. Local residents who may live near the lower part of river and use the water for drinking are concerned about the potential health risk from discharge of this treated wastewater.

Figure 13.2.2 shows that the volume of MSW produced per capita in Japan is comparable with that of other OECD countries, although the rate of increase in MSW production in the 1980s was among the highest of the OECD countries (OECD, 1994). The main factor was an increase in waste generation by the commercial sector, especially of paper materials and discarded office automation appliances. Except for agricultural waste, industrial waste generation per GDP in Japan during the 1980s was almost the same as for the North American countries, but typically twice that of European countries.

Confusion Between Once-through and Recycling Systems

The Japanese waste treatment system employs incineration more than other countries do. Figure 13.2.3 shows the annual generation of MSW and the incineration rate among several countries. The high rate of incineration of Japanese MSW can be explained by the shortage of landfill capacity, which

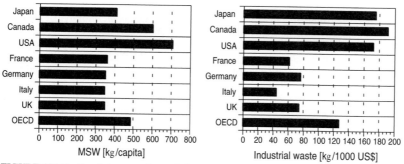

FIGURE 13.2.2   Trends in generation and disposal of MSW and industrial waste (1990). *Source:* OECD (1994).

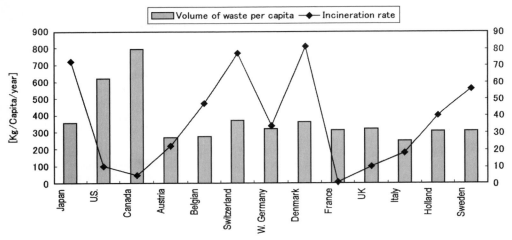

FIGURE 13.2.3   Volume of municipal solid waste and the incineration rate (1986). *Source:* Association for promoting appropriate treatment of waste plastics.

is mainly caused by the small size of the country's land area. Narrow and steep lands and a high population make residential areas very crowded. Under such conditions, it is very difficult to find suitable, convenient sites for waste landfills. That led policy makers to encourage incineration of various wastes wherever possible, so that waste volumes for final landfill were minimized. As another contributing factor, some have noted that Japanese have a belief in fire in the traditional religion.

In response to an increasing public concern for recycling, the national government created two other legal frameworks. The purpose of a law concerning the promotion of the use of recyclable resources, which was promulgated in 1991, was to promote use of waste paper, aluminum cans, steel cans, and glass cullet. In 1995, a new law was promulgated dealing with recycling of wrapping materials and bottles. Another law laid down in 1998 aimed to promote recycling of electric appliances such as televisions, air conditioners, washing machines, and refrigerators. In these laws, the public and local governments play an important role, while the responsibilities of private companies are relatively light compared with those of other stakeholders. Although these laws improve the Japanese waste management situation, a relatively loose regulation of toxic substances, including dioxins, will make for a worse environment with respect to waste management.

Trend to Illegal Treatments and Disposals of Waste

Another aspect of defects in Japanese waste management is the shortage of appropriate facilities for waste management. A shortage of land and steep topography do not allow sufficient area to be found that is suitable for waste treatment facilities and landfill sites. As of 1995, the government said that all existing landfill sites would fill up within 30 years (MHW, 1998). Relatively loose regulation and a chronic shortage of landfill sites often lead to illegal treatment and dumping of MSW and industrial waste.

Figure 13.2.4 shows the number of cases of illegal activities from 1991 to 1995 (Eguchi, 1996; National Police Agency, 1999). Although the number of offenses recorded has remained around 800 per year since 1990, the number has tended to increase recently. The most frequent type, around 45 percent of all cases, is illegal dumping in unsuitable areas. The number of offenses of this type has increased recently, and was 495 in 1998, while violations for operation without approval are gradually decreasing. Specially controlled wastes, including medical waste, are also often dumped illegally. According to the Marine Safety Agency, the total number of cases of illegally dumping into the sea,

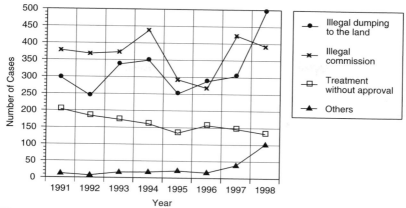

FIGURE 13.2.4   Change in the number of illegal cases of waste treatment and disposal. *Source:* Eguchi (1996) & National Police Agency (1999).

which had been decreasing from 1991, increased again in 1995 (Inose, 1996). The total masses of illegally treated wastes dumped were estimated to be around 2433 tons in 1994 and 1600 tons in 1995.

The estimated volume of illegal dumped waste reached a peak in 1991, but the number has been increasing again since 1995 (Eguchi, 1996). While waste generators continue to make efforts to decrease and recycle wastes, not a few waste disposal companies still illegally dump wastes generated from building and infrastructure construction into isolated sites in mountainous forest and wilderness areas. Around 80 percent of illegally treated waste comprises construction materials, generated in the demolition and renewal of buildings and infrastructures. Waste plastics (15 percent) and sludges (6 percent) are included in this kind of waste.

13.2.3   A CASE OF ILLEGAL DUMPING

Of all illegal dumping cases, the worst and largest case occurred on Teshima Island. For almost 10 years, a waste treatment company continued to treat and dump industrial waste, including hazardous materials, while a prefectural government conducted regular on-site inspections. The total volume of illegally treated wastes reached 460,000 tons.

Outline of the Problem Site

Teshima Island, which is located in the Inland Sea of Seto between the mainland and Shikoku Island (Figure 13.2.5), is 14.61 square kilometers in area and had a population of 1535 people as of December 1, 1995. The population has been decreasing and aging at a high rate. Processing of a soft stone quarried on the island was a unique industry, although most people derived their main income from agricultural products, including rice, oranges, and dairy farming (Aizu, 1996). The average temperature is around 16°C, and this island has 1300 mm of rainfall per year, one of the lowest in Japan.

Outline of This Case

After an increased construction demand with high economic growth in the second half of the 1960s, Teshima Comprehensive Touring Development Co. (TCTD) proposed a landfill facility in a area used for mining stone up to 1975. Although local residents opposed this proposal in court, in 1978 the Kagawa prefectural office permitted a landfill operation for the disposal of organic waste only for

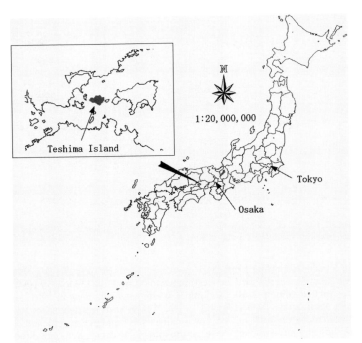

FIGURE 13.2.5   Location of Teshima Island.

cultivation of earthworms, with several conditions. TCTD, however, after 1980, started to transport and treat not only organic waste, but also other wastes, including plastics, paper, and metals. An approval in 1983 for dealing with metal wastes accelerated the change in TCTD's operation. In the same year, the company quit cultivation of earthworms and instead used a spreading method of waste disposal, whereby the use of sludges, shredder dusts from used vehicle recycling, riggers rope, and waste oils for landfill occurred in an inadequate manner. Other wastes used for landfill included coal ash, tires, shredded pieces of used automobiles, liquid materials labeled as toxic, waste acids, and printed motherboards. Although a petition by local citizens requesting a public inquiry went to the prefectural office, the prefecture answered that materials transported by TCTD were recyclable, and were not regarded as a waste. The Teshima Island case has been described in several articles (Abe, 1997; Aizu, 1996, 1998; Minami & Nishimura, 1998; Seriguchi, 1997).

In October 1990, police from Hyogo Prefecture, located next to Kagawa, discovered illegal disposal and dumping by TCTD. Consequently, Kagawa Prefecture implemented an intensive on-site inspection. That inspection found that TCTD had illegally dumped 1400 drums containing nickeliferous wastes, 364 drums of waste oil, huge amounts of waste acid and shredder dust from vehicles, over about 200,000 square meters, whereas the approved area for landfill was 10,000 square meters. PCBs in contaminated soils were found to be 11.7 times the national limit, and trichloroethylene in soils 6.7 times the limit. After a week of police disclosure, Kagawa Prefecture completely overturned its position and stated the materials dumped by TCTD were waste.

Kagawa prefectural office then canceled TCTD's license to landfill, and ordered the company to remove and appropriately treat dumped wastes. In response to the order, TCTD removed 1340 tons of waste—only 0.29 percent of the total volume dumped. At the beginning of 1991, a senior manager in TCTD was arrested. Kagawa Prefecture then ordered waste-generating companies to remove their wastes themselves. Although some companies removed the 1400 drums containing nickeliferous wastes, some 280 drums of waste engine oil, and 147 drums containing an oil generated while making soy-source, shredder dusts and sludges generated from paper making were allowed to remain. The main reason for leaving the latter material was that the local government regarded such wastes as a valuable material.

In response to an application for mediation on environmental pollution by 549 local residents, the Environmental Dispute Coordination Commission (EDCC), a national organization attached to the Prime Minister's office established for conflict resolution, started to deal with this issue in 1993. Over several meetings, EDCC resolved to spend 236 million yen for investigation of the pollution. A detailed inspection implemented by a special committee of EDCC made the following findings.

1. It was assumed that the volume of dumped waste reached around 460,000 cubic meters (about 500,000 tons)

2. It was found that about 70 percent of this waste contained lead beyond a safe level, and many organochlorine compounds as well as high levels of dioxin were detected.

3. These hazardous materials had penetrated the soil beneath the dump area and entered groundwater.

4. It was impossible to deny that these materials had also entered the sea.

The committee concluded that the site should be immediately cleaned up by appropriate measures.

Despite a court's verdict that TCTD had engaged in illegal disposal and dumping, the company did not have enough funds to remove the wastes and clean up the environment. Following, Kagawa Prefecture and 548 residents of Teshima Island reached an intermediate agreement in 1997. In this agreement, the national government would make a large expenditure, not to clean up the site, but for research and development of a disposal facility, where the wastes could be safely dumped, decontaminated and valuable materials recovered.

Condition of the Contaminated Site (EDCC, 1995; Hanashima et al., 1996; Takatsuki, 1997)

Investigation Method. The investigation method comprised four parts: a basic survey of geology and topography, analysis of the ingredients of the dumped wastes, a pollution survey on groundwater, and survey of the state of environment around the site.

First, the task force surveyed basic information on the geologic and topographic characteristics of the site, land use around the site, and climatic conditions. In an on-site inspection (approx. 30 ha), the task force studied the natural history of the site.

In the next step, the task force analyzed the volumes, conditions, and ingredients of illegally dumped wastes. The following four actions were implemented. For analyzing gases emanating from surface layers of the site, the task force made borings 80 cm beneath the surface at 40 points, and measured concentrations of ammonia, oxygen, methane, and hydrogen sulfide in ambient soil gases. In addition, six other gases, including hydrogen cyanide and carbon monoxide and other volatile compounds, were measured by aspirating gases from observation wells. A geophysical study was implemented for clarifying the geological condition beneath the site, especially the structure and distribution of the weathered granitic bedrock layer and alluvium. The third action was analysis of wastes and soils. At 14 points, the task force collected wastes and soil 1 m beneath the wastes by the Benoto method. Because the waste layer was sometimes very thin or did not occur at some points, samples at 15 locations were collected by backhoe. The task force conducted basic analyses and analyzed volatile organic compounds (VOC) by portable gas chromatography at the site, and conducted leaching and content testing in a laboratory. Leaching tests were implemented for 26 chemical species at most, and content testing for 10 constituents, including dioxin compounds. In the fourth action, the task force analyzed pollution of pore waters for up to 50 chemicals, including dioxins.

Another survey was conducted on groundwater, to elucidate the behavior and quality of groundwater, as well as the hydrogeological conditions. With observation of the variation with time of water levels and the quality of pore water and groundwater at 52 locations, samples of soil and rock, which were collected by boring into the granitic bedrock layer at 14 locations, were analyzed for up to 26 chemicals by leaching tests, and samples of groundwater were tested for up to 42 chemicals.

In an environmental survey around the site, the task force collected surface water at four points within the site, and analyzed the water quality. Water collected at a private well, being the nearest to the site (460 m east), was also analyzed. Soils polluted by leaked water were also analyzed for eight chemicals. In addition, the task force collected seawater and oysters as a biological indicator at three locations. Collected water was tested for 33 chemicals, and the oysters were tested for six chemicals, including dioxins.

For sludges found in the above three situations (i.e., in surface water, in polluted soils, and in the sea), leaching tests with analysis for 18 chemicals and a content test of three constituents were also implemented.

The Result.

Basic Survey of Geologic and Topographic Conditions. According to aerial photographs from 1966 to 1992, the coastline had moved 100–150 m north since 1966. This implied that waste-dumping activities had caused a change of the natural landscape. Annual rainfall is about 1300 mm, which is relatively low compared with other regions of Japan. In September, this area has the largest amount of rainfall, 220 mm, while rainfall in December is the lowest, at about 30 mm.

A weathered granitic layer, which is the bedrock of the strata of the site, is covered by alluvium, a landfill soil layer, a banking layer, and a dumped waste layer. The weathered granites are very weak and their permeability is 2.23×10^{-3} cm/s on average. Their permeability is higher than of the alluvium (3.68×10^{-4} cm/s), or banking layers (1.51×10^{-3} cm/s).

Characteristics of the Waste Layer. On-site inspection found that dumped wastes covered around 69,000 square meters. The thickest layer of dumped waste attained thickness of 16.5 m, while a layer of 1.5 m was typical. Estimated from horizontal and vertical distribution of dumped wastes, the volume may reach about 460,000 cubic meters. Because the survey found that the average dry density of samples was about 1.09 g/cm^3, it was estimated that total wastes reached about 500,000 tons in a wet condition, and 320,000 tons in a dry condition.

Dumped wastes consisted of quite different materials, including shredded fragments of scraped cars, sludge, slag, dehydrated sludge cakes, printed circuit boards, tires, waste paper, and wood pieces. Most locations exhibited alkaline conditions, with the highest pH being 9.6. The eastern side of the site smelled strongly of rotten materials and chemicals, and the temperatures at some locations within 10 m of the surface were over 50°C. Figure 13.2.6 shows the result of content test at 19 locations. In all samples, all chemical species were detected.

Above all, the average concentrations of copper, zinc, and lead each exceeded 1000 mg/kg, and the maximum concentration of PCB was 58 mg/kg. Concentration of dioxin compounds was 3.4 ng-TEQ/g (nanograms toxic equivalents per gram) on average and 39 ng-TEQ/g at most.

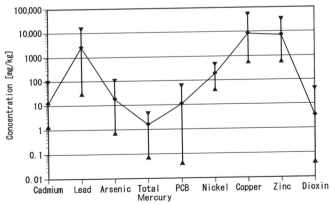

FIGURE 13.2.6 The result of content test of wastes at 19 points (dioxin units are ng-TEQ/g drawn based on data of EDCC, 1995).

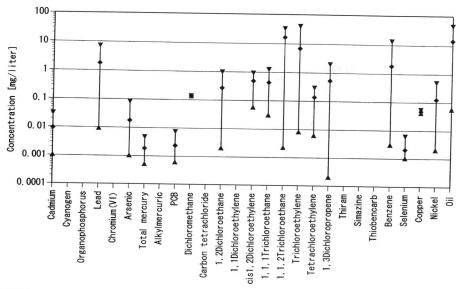

FIGURE 13.2.7 Leaching tests of materials in dumped waste (drawn based on data of EDCC, 1995).

In leaching tests, 18 chemicals were detected from all samples out of 27 analyzed-for (Figure 13.2.7). Cadmium, lead, arsenic, 1,2-dichloroethane, copper, nickel, and oil were detected in more than half of all samples tested. In particular, 44 samples out of all 46 samples collected contained leachable lead, the concentration of which was 1.7 mg/L on average in the leachate.

Those results showed that wastes that contained materials over an action level of hazard covered about 55,000 square meters in area, and were about 400,000 cubic meters in volume. That allowed us to estimate that hazardous wastes on the site had a mass of about 440,000 tons in the wet condition, and about 280,000 metric tons in a dry condition.

Pore Water Survey. Quality tests of pore waters were conducted at 13 locations. Of 24 chemicals for which quality standards are set from a health effect point of view, 13 were detected. Samples from almost all locations contained lead, benzene, total mercury, and PCBs. Concentrations of eight chemicals exceeded quality limits. In particular, lead concentrations were over the limit at all 13 points, and maximum concentration reached 26 mg/L. In terms of PCBs, samples from seven locations exceeded the limit. Dioxins, which were detected in all samples, showed an average concentration of 8.0 ng-TEQ/L, higher than the concentrations typically detected in waste waters of paper and pulp factories (EA, 1991). Surveys of surface water in the site found that the water did not contain hazardous materials over limits except lead, while chromium (VI), arsenic, and PCBs were detected. The concentration of dioxins was higher than that of wastewaters from paper and pulp factories as found in pore waters.

Data for Groundwaters. Soil samples in 12 of 30 locations had a pH over 9.0. The leaching tests found that 12 chemicals could be detected, and concentrations of 11, excluding cadmium and PCBs, exceeded action limits for soil. In particular, a third of all samples contained lead over the limit, and the concentrations of VOCs, including benzene and 1,1,1-trichloroethane, were relatively high, when they were detected in a small number of samples. According to leaching tests of soil samples collected in alluvium at four locations, arsenic was detected in all samples, nickel in three, and benzene in one, although those concentrations were under quality standard. Of the samples collected from the granitic layer at nine locations, one contained nickel, the concentration of which was within the quality limit. From those results, it was concluded that polluted soils over 35,000 square meters (61,000 tons wet weight and 51,000 tons dry weight) should be removed for remediation.

Groundwater beneath the waste layer was analyzed at five locations in alluvium, and 14 locations in the granitic layer. Concentrations of BOD and COD were at a high level, and those of lead and benzene exceeded the quality limit for groundwater. Some samples contained arsenic, 1,2-dichloroethane, and *cis*-1,2-dichloroethylene, the concentration of which also exceeded their limits. Eleven chemicals were detected in samples of groundwater in the granitic layer, more than the number detected in groundwater sitting in alluvium. The number of chemicals exceeding their limits was nine, which was also more than for groundwater in alluvium. Dioxins, which were detected in both layers, showed almost the same concentration in groundwaters of both strata, with the highest level being found on the west side of the site, while dioxin levels in the solid waste layer were highest on the eastern side.

Results of Environment Survey of the Site.   Samples of groundwater water collected from a well located in a private house satisfied quality standards for drinking water for all chemicals. While surface waters at three points around the site contained no contaminants, many contaminants were detected in seabed mud. Leaching tests on seabed sludge collected at 13 locations found that fluoride, cadmium, lead, and arsenic, and sometimes chromium and vanadium, polluted every sample; mercury and PCBs were not detected. In content tests of mud collected from the same locations, lead and arsenic were detected in all samples, and some samples also contained mercury and cadmium, while PCB was not detected. Samples along the north coast were relatively more contaminated than those along the south coast.

Cadmium, lead, arsenic, total mercury, PCBs, and dioxins were selected for biological testing. In five samples of oysters living around the problem site, mercury was detected in all samples, and PCB was also detected in one sample, although the concentrations of those contaminants were below that found at other sites around Japan. Some samples at every location contain cadmium, lead, and arsenic, and the concentration of arsenic was generally higher than at the other site. The concentrations of dioxins, which were detected in all five samples, were one to two orders higher than average in a national survey by EA, while another survey conducted by EA found that oysters living along other coasts of Teshima Island contain almost the same levels of dioxins.

Summary of the Survey.   Hazardous contaminants leaching from illegally dumped wastes polluted surface waters and groundwater through the medium of pore water, and then diffused into the sea. Change in the concentration of lead, which was one of the most commonly detected and high concentrated contaminants, supports the estimated dispersal process (see Figure 13.2.8). Levels of

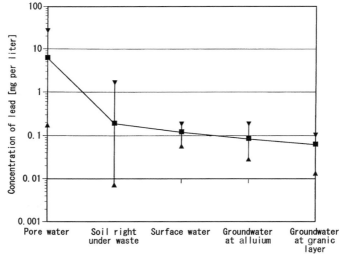

FIGURE 13.2.8   Diffusion of lead in several phases (drawn based on data of EDCC, 1995).

contaminants in groundwater levels may imply that hazardous contaminants in groundwater are leaking into the seawater.

Taking the results into account, the task force established the following basic principles for choosing the preferred remedial method. The most important issue was considered to be pollution prevention of the seawater from the north side coast, because groundwater within the site was not for public use. Groundwater in the other area located on the eastern side of the island would not be affected by polluted groundwater, which flowed down towards the north side of the island. Although contamination levels were almost the same as in the other sea areas for most contaminants, the pollution potential of wastes dumped into seawater was quite high because dumped wastes contained a very large variety of contaminants.

Specific measures for intermediate treatment and siting of the final landfill were quite influential in choosing remedial measures. In addition, leachate barriers and pumping may be effective for environmental protection without even moving hazardous wastes. Taking into account the above factor, the task force suggested seven alternative strategies. It was estimated that implementation of each strategy required funds of the order of 6 to 15 billion yen.

Clean-up Strategy (Abe, 1997; Kagawa Prefecture, 1998)

Kagawa prefecture, which is responsible for managing pollution in Teshima Island, had decided to construct an intermediate treatment facility on the polluted site, and to treat dumped wastes in the facility from 1997. For this purpose, the prefecture established a technical committee for environmental protection, which examined the fundamental measures required for permanent protection and any urgent temporary measures required before the former measure could be completely implemented.

Outline of Temporary Measures. Figure 13.2.9 shows the nature of the temporary measures. First of all, the committee suggested digging up the dumped wastes and moving them to the center of the site as the first step of prevention of contaminants leaking. The next step was to implement a leachate barrier followed by pumping and treating resident wastewater. Construction of soil bund around the site was the last step to prevent leaching from any materials passing off-site.

Outline of Fundamental Measures. As specific measures for final disposal and landfill have yet to be chosen, the author has summarized in outline the establishment and operation of an intermediate

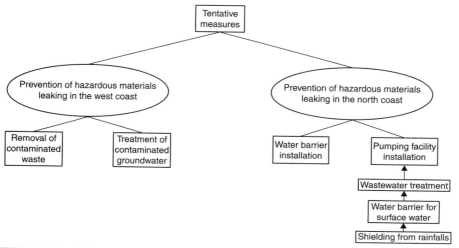

FIGURE 13.2.9 Outline of urgent measures.

treatment facility. The committee proposed the following four methods for intermediate treatment: incineration and fusion treatment with a fusion-type rotary kiln, gasification and fusion treatment, surface fusion treatment, and incineration and production of Eco-cement.

For treatment of fly ash, chlorination and volatilization treatment at another site outside of the island were proposed, as well as a MRG method for recycling. Before selecting the most suitable measures, the committee examined several conditions with respect to the state of dumped wastes and polluted water, the processing and environmental protection abilities of treatment systems, and the characteristics of the by- products produced. Although the general criteria for selecting suitable methods are typically technological maturity, the level of environmental protection afforded, and cost-effectiveness, in this case evaluation was based on the following four criteria.

- Mass balance: consumption of submaterials, production of by-products, production of fly ash, and behavior of contaminants, such as heavy metals.
- Energy and water consumption: consumption of fuel and electricity, and volume of fresh water required and waste water produced.
- Cost-effectiveness: cost for energy and 'submaterials' which are required for making by-products, in distillation and in waste water treatment facilities.
- Others: Simplicity of the total system.

Figure 13.2.10 shows an example of a mass balance for a case of incineration and production of Eco-cement.

Evaluation.   For intermediate treatment, the weights and volumes required for treating one kilogram of hazardous wastes were compared among selected methods. Requirements for submaterials are least

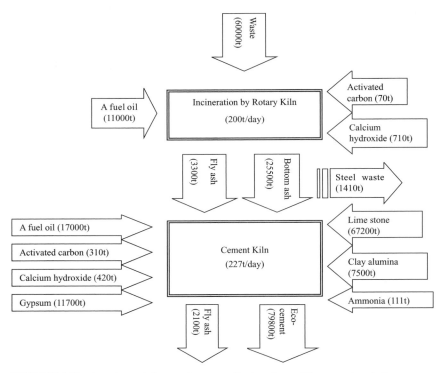

FIGURE 13.2.10   Annual mass balance under a case of incineration and Eco-cement production.

in the incineration and fusion method, while incineration and production of Eco-cement consumes about 53 times the weight of waste materials in submaterials. Incineration and production of Eco-cement generates 2.3 to 2.5 times more by-products than in the other methods. The incineration and Eco-cement production method generates the smallest volume of fly ash (0.035 t/kg of treated wastes). The contents of heavy metals in fly ash are almost the same among all methods. With respect to fuel consumption, the surface fusion treatment method is the most economical, while the incineration and Eco-cement production method has the most extravagant fuel requirement. The gasification and fusion treatment method consumes the largest amount of electricity, 2530 kW per ton, whereas electricity consumed by the incineration and fusion method was 64 percent of that value. The incineration and fusion method, however, requires a much larger volume of fresh water than the other methods.

Although both proposed methods for treating fly ash were evaluated as suitable for this case, they differ in cost. Whereas the chlorination and volatilization treatments need 90,000 to 100,000 yen in the transportation and treating process, the MRG method requires costs for various steps, including cement caking and landfill of the by-product material. The treatment cost also depends on treatment methods of wastewater that contains high concentrations of chlorides and degree of thermal decomposition of dioxins required. Total cost for MRG method is estimated to be 53,000 to 72,000 yen per kilogram of wastes.

13.2.4 ENVIRONMENTAL POLLUTION BY DIOXIN-LIKE COMPOUNDS IN JAPAN

A high level of incineration and loose regulation for some toxic materials may still have substantial effects on the environment, even though NOx and SOx and some other pollutants may be appropriately reduced. One of the most important classes of substance is dioxins and related compounds. A heavy reliance on incineration in waste management may generate relatively higher levels of dioxin-like compounds than in other developed countries. The high density of residential areas and a relatively high proportion of seafood in the national diet, which may be severely contaminated through the food chain and biological concentration, make the constraints much worse. In the next section, state of the environment in Japan with respect to contamination by dioxin-like compounds along the pollutant pathway will be discussed.

State of the Environment

Ambient Air Quality (JEA, 1997).   Periodical surveys by the Air Quality Bureau of the Japan Environment Agency (1990, 1992, 1994) have found that the concentrations of dioxin-like compounds in air were 0.69–0.82 pg-TEQ/m^{3}N on average, while the maximum exceeded 1.5 pg-TEQ/m^{3}N, excluding Coplanar PCB. Ambient air in residential areas near industrial zones tended to be the most contaminated, followed by that in large-scale cities, then medium-scale cites. Generally concentrations in residential areas are almost 10 times greater than in background areas; where 0.06 pg-TEQ/m^{3}N was the average. According to a survey of Kuromatsu (Japanese Black Pine) as a biological indicator, the average concentration in large-scale cities was much higher than in local medium and small-scale cities. The maximum concentration reached 21 times the minimum one.

Soil Environment.   Miyata (1996) showed that the concentration excluding Coplanar PCB of public parks in some industrial areas was 35 pg-TEQ/g. Some 6.5 pg-TEQ/g was observed in forest located in a farm district, while in large-scale cities the level was less than 20 pg-TEQ/g, and in medium scale cites was less than 10 pg-TEQ/g. Soils around incineration facilities for industrial wastes generally were more contaminated than the general environment. For example, at a facility located in Tokorozawa city of Saitama prefecture, which is the next jurisdiction to Greater Tokyo and receives a considerable flow of industrial wastes into it for incineration, concentrations within 5 km of the facility were 65–448 pg-TEQ/g, and averaged 195 pg-TEQ/g (excluding Coplanar PCB). Miyata et al. (1996) also found

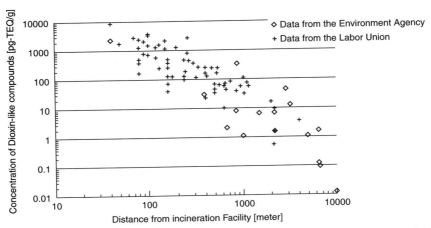

FIGURE 13.2.11 Relationship between concentration and the distance from the facility. *Source:* EA (1999).

the concentration in soil near a facility located in Higashi-Tokorozawa City, reached 6400 pg-TEQ/g including Coplanar PCB.

In addition to incineration facilities for industrial wastes, it was also observed that soils around facilities for MSW were contaminated. An incinerator located in Ryugasaki City in Ibaragi prefecture, which can incinerate at most 60 tons per day, was said to treat an excess amount of wastes, leading to incomplete combustion. A survey conducted by the local government in 1996 showed that the concentrations in soil were 4.7–9.9 pg-TEQ/g, which were almost the same level as those observed in a national survey by EA. Miyata et al. (1996), however, found that the concentration of soil on the leeward side within 1.3 km of the facility was 52 pg-TEQ/g on average, and the maximum was about 300 pg-TEQ/g.

An incinerator in Nose town, Osaka Metropolitan prefecture, has brought about the most severe dioxin pollution. This facility, with a lack of proper environmental protection measures, caused polluted cooling water to leak out around the site. A survey conducted by Osaka Metropolitan government showed that the concentration near the site was 85,000 pg-TEQ/g, in the woods around the site it was 720–3900 pg-TEQ/g, and that in the soil at the bottoms of local ponds and rivers was 33,000–230,000 pg-TEQ/g. According to the MHW survey, the concentration near the cooling water tower reached 52 million pg-TEQ/g. EA (1999) found a relationship between dioxin concentration in soil environment and distance from this facility (Figure 13.2.11). This facility stopped operation in June 1997.

Rivers and the Sea. According to a 1996 survey conducted by EA, the concentration in river waters was 0.46–19 pg-TEQ/L, and averaged 5.5 pg-TEQ/L. The survey also found that the concentration at the bottom of ponds and lakes was 14–44 pg-TEQ/L (average, 27 pg-TEQ/g), and 0.26–75 pg-TEQ/L (average, 13 pg-TEQ/g) at the bottom of the sea, whereas the concentration at the bottom of rivers was 0–24 pg-TEQ/L, and averaged 0.18 pg-TEQ/L. A 1997 survey found that soil at the bottoms of rivers contained 1.1–150 pg-TEQ/g.

Although there are few data in terms of the water quality of the sea, a survey of Murasakigai (Blue Mussel) as a biological indicator showed that the concentration in closed type areas near large cities was about 50 times more than that around a southern island.

Food. A survey by MHW in 1997 found that seafood contributes about 60 percent of the Japanese people's intake from all foods of dioxin compounds. That survey also showed whole seafood typically contained 0.202–1.526 pg-TEQ/g, and there was typically 0.226–0.429 pg-TEQ/g in the meat.

For dairy products, MHW and EA surveys in 1997 showed that cheese typically contains 0.179 pg-TEQ/g, and the concentration in milk was typically 0.050–0.063 pg-TEQ/g. Of other farm products, green vegetables of Komatsuna and spinach contained 0.144 pg-TEQ/g and 0.187 pg-TEQ/g, respectively, and other crops were within the range of 0.003–0.021 pg-TEQ/g. Some researchers have noted that green vegetables take up more dioxin compounds than other vegetables because of their active transpiration.

Human Body.   The result of analysis of mothers' milk after their first childbirth shows that the concentration peaked in the middle of 1970s at about 30 pg-TEQ/g of fat, and the concentration is gradually decreasing, so that in 1996 it was about 15 pg-TEQ/g. This tendency seems to be the reverse to that of overall environmental quality, which was gradually getting worse until the 1980s, and has not been improved significantly since then.

In addition to surveys of the general environment, people living around problem sites provide important data on the effects of dioxin contamination. For instance, a survey of the blood of 34 residential people around an incineration facility for MSW was conducted in 1998. The result showed that the concentration was 76.5 pg-TEQ/g on average, and that of females (85.1 pg-TEQ/g) was higher than that of males. The maximum was 463.3 pg-TEQ/g.

The effect on workers at problem sites was more severe. The Ministry of Labor of Japan surveyed dioxin compounds in blood of 92 people working at incineration facilities located in Nose town of Osaka Metropolitan prefecture. According to the result published in March 1999, the average concentration was 84.8 pg-TEQ/g of fat, and the maximum was 805.8 pg-TEQ/g. The average was 2.8 to 4.2 times more than that in ordinary people in Japan, whose bloods contained dioxins in the range of 20–30 pg-TEQ/g. Some workers suffered from cancer and a dermatitis that made their skin black. Two heavily affected workers, who held their jobs for over 10 years, have claimed compensation for their damages.

Comparison with Other Developed Countries

Environmental concentrations of dioxins in Japan are typically at least one order of magnitude higher than in the other developed countries (MHW, 1996). In addition, we can estimate the total mass of dioxins annually generated. In Japan, this remains at the highest level, around 4000 g-TEQ, while other countries have made efforts to reduce dioxins emissions (Table 13.2.1).

Social Attitudes

It was not until 1997 that Japanese society paid any attention to the environmental risk posed by dioxin-like compounds, when MHW published the data on concentration levels in exhaust gas of

TABLE 13.2.1   Total Mass of Annually Generated Dioxin-like Compounds. *Source:* Sakai (1998) and MHW (1997)

	Year	Annual Volume (g-TEQ)
Germany	1991	68–928.5
	1995	50
	1997	4
Holland	1991	484
	1995	2.8
Sweden	1988	122–288
U.S.	1992	1174–11123
Japan	1990	3900–8360
	1997	4300

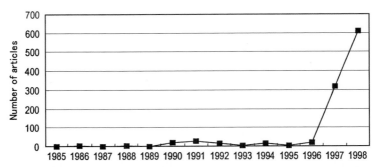

FIGURE 13.2.12   Trend of the number of newspaper articles on dioxin-like compounds—
Case of Nikkei Newspaper.

MSW incinerators. Figure 13.2.12 describes the trend in the number of newspaper articles related to dioxins pollution. Mass media provided effective information on this matter, while a slight increase was observed around 1990, when the World Health Organization (WHO) revised the tolerable daily intake (TDI) from 100 pg-TEQ/kg body weight/day to 10 pg-TEQ/kg body weight/day.

Regulation and Reduction Plan of Government

MHW began to observe dioxins in fly ash in 1983, after a researcher published data on dioxin compounds in fly ash generated by incineration of MSW in 1979. The following year, MHW published the first guideline of daily intake, 100 pg-TEQ/kg/day. It was not until 1990 that MHW set guidelines for pollution prevention by dioxin compounds. In 1992, EA requested the paper and pulp industry to reduce the production of dioxin compounds at the bleaching stage in their factories. Although MHW set 10 pg-TEQ/g as an immediate TDI on the basis of research on risk assessment, since 1995, EA set the TDI 5 pg-TEQ/g from an environmental protection point of view.

Afterwards, MHW started to examine measures for dioxin production reduction in view of the urgent and long-term aspects of dioxin toxicity. In 1997, MHW published data on the concentration in exhaust gas of MSW incineration facilities for the first time. The survey found that 52 facilities emitted gases containing dioxins exceeding a critical standard. From August 1997, the Japanese legal system considered dioxin compounds as a substance to be regulated. After that, EA established a comprehensive five-year plan for dioxins pollution reduction. The plan was based on the following four aspects: resource reduction, comprehensive monitoring, research promotion, and risk communication.

In May 1998, JEA began to examine standards for the soil environment, and MHW and EA examined the necessity of revision of the TDI in response to the TDI reduction of WHO in June/July 1998. MHW set 2.5 pg-TEQ/m^{3}N as a concentration limit to be met from July 1998.

In February 1999, a TV program reported that green vegetables and leaves of green teas cropped around problem sites were more polluted than in other areas. Taking this opportunity, in Japanese society a push started to revise the pollution reduction plan. The Japanese government founded a cabinet ministers conference, and the cabinet established a fundamental guideline on pollution prevention for dioxin compounds. The guideline noted that the government should:

- reduce total emission by 2002 by about 90 percent, compared with those in 1997
- prevent incineration facilities for MSW and industrial wastes from concentrating in certain areas
- promote the reduction and recycling of all sorts of waste

A survey by MHW published in April 1999 found that annual masses emitted from incineration facilities as of November 1998 were 1340 g from MSW, 960 g from industrial wastes, and 2300 g in total. Problem facilities, whose exhaust gases exceeded the critical standard (80 ng-TEQ/m^{3}N), numbered 5 incinerating MSW and 19 incinerating industrial wastes. Some problem facilities have already stopped operating or have been abandoned. In particular, about 35 percent of incineration facilities for industrial wastes closed down or were already abandoned in 1998. These data, however, do not include facilities that neglect to measure the dioxins concentrations in exhaust gases from their facilities.

13.2.5 CONCLUSION

In this paper, I have discussed two significant problem cases, comprising a site-based issue on the one hand and the case of a heavily and widely generated contaminant on the other. Both of these problems are derived from an imperfect system of waste treatment and disposal.

One of the lessons to be learned from the case of the large amount of waste dumping on Teshima Island is the danger from confusion between the promotion of recyclable waste and treatment of hazardous materials. Although it is true that a society should move toward recycling all resources, it is sometimes difficult to distinguish materials suitable for recycling from those requiring treatment. A more rationally structured system for the recycling of waste materials would bring about less environmental pollution with the recycling process.

Japan also suffers severe pollution from dioxin-like compounds. Taking account of the government's strategy, Japan does not seem to have learned from past severe lessons concerning environmental pollution, including Minamata disease and Yokkaichi Zensoku. A change in waste management policy from an incineration-oriented strategy to one based on the 3R principle (Reduce, Reuse, and Recycle) should be implemented to prevention further dioxin contamination.

REFERENCES

Abe, M., "Community Involvement in Cleanup Process of Contaminated Land," *Surugadai Journal of Law and Politics*, The Surugadai University Association of Law and Politics, Surugadai University, Saitama, Japan, 11(1): 283–312, 1997.

Aizu, S., "Reportage on Environmental Pollution in Tashima Island," *Zentei Tyusa Jiho*, Zentei Research Institute, Tokyo, Japan, No. 54, pp. 84–101, 1996.

Aizu, S., "Environmental Disruption in Teshima Island—Second Series," *Zentei Tyusa Jiho*, Zentei Research Institute, Tokyo, Japan, No. 58, pp. 50–63, 1998.

Eguchi, K., "Recent Situation on Illegal Disposal Cases of Industrial Waste," *Indasuto*, National Federation of Industrial Waste Management Associations, Tokyo, Japan, 11(10): 3–8, 1996.

The Environment Agency of Japan, *Urgent Survey on Dioxin Compounds in Waste Water of Pulp Factories*, 1991.

The Environment Agency of Japan, *Report on Risk Assessment of Dioxin-like Compounds*, 1997.

The Environment Agency of Japan, *On 1998 Survey of Long-term Effects Induced by Dioxin-like Compounds*, 1999.

Environmental Dispute Coordination Commission, *The Result of Survey on Environmental Pollution Caused by Dumped Waste in Teshima Island*, 1995.

Hanashima, M., Takatuki, H., and Nakasugi O., "A Case Study of Environmental Contamination Caused by Illegal Dumping of Hazardous Wastes," *Journal of Waste Management Research*, 7(3): 208–219, 1996.

Inose, M., "Recent Situation on Illegal Dumping of Industrial Waste into the Ocean," *Indasuto*, 11(10): 9–13, 1996.

Kagawa Prefecture, Report on the Technical Survey of Pollution Prevention Measures in Teshima Island, Committee on pollution prevention measures caused by industrial wastes in Teshima Island, 1998a.

Kagawa Prefecture, Report on Urgent Pollution Prevention Measures in Teshima Island, Committee on pollution prevention measures caused by industrial wastes in Teshima Island, 1998b.

Minami, H., and Nishimura, Y., "Historic Outline of Illegally Dumping Case in Teshima Island," *Hanrei Taimusu*, 961: 35–41, 1998.

The Ministry of Health and Welfare, *Guideline for Pollution Prevention of Dioxin-like Compounds Related to Waste Disposal*, 99 pp., 1997.

The Ministry of Health and Welfare, *White Paper on Health and Welfare*, Printing Bureau, Ministry of Finance, Japan, pp. 412–413, 1998.

Miyata, H., Ikeda, M., Nakano, T., Aozasa, O., and Ohta, S., "Real Situation of Pollution by Dioxin Analogues from Industrial Waste Incinerators," *Proc. 5th International Conference on Environmental Chemistry*, Hawaii, pp.188–189, 1996.

The National Police Agency, *Public Release on a Prevention Plan for Environmental Climes*, Bureau of Community Safety, 1999.

OECD, *Environmental Performance Reviews:* Japan, 210 pp., 1994.

Sakai, S., Environmental Policy and Status for Chlorinated Dioxins and Related Compounds, *Environmental Economics and Policy Studies*, 1(2): 161–186, 1998.

Seriguchi, Reportage on Teshima case, *Gekkan Haikibutu*, 23(1): 8–15, 1997.

Takatsuki, H., "Problem and Environmental Measures Pollution Induced by Industrial Waste in Teshima Case," *Bulletin of National Institute of Public Health*, 46(4): 325–329, 1997.

CHAPTER 13
CONTROL TECHNOLOGIES

SECTION 13.3

MEDICAL WASTE INCINERATION

C. R. Brunner

*Mr. Brunner is president of Incinerator Consultants, Inc. He has
30 years of experience in the evaluation, selection, design and
operation of thermal systems for the destruction of waste.*

13.3.1 INTRODUCTION

An incinerator is a furnace that is used to combust waste materials. Modern incinerators release a clean gas stream to the atmosphere and discharge a clean ash. They are designed and constructed to withstand high temperatures and adverse conditions. The main principles used in incinerator design are the maintenance of a destruction temperature for a specific period of time and the provision of enough turbulence to provide effective mixing with air for effective destruction.

Most of the incinerators that were designed before the 1950s were single chamber units that ran at relatively low temperatures. They were little more than enclosed fireplaces. Little attention was paid to good combustion, effective burn out of the ash, or to low stack emissions. However, in the early 1950s, concerns about poor air quality started to arise. This led to the development of a set of criteria and guidelines for incinerator design and operation that was eventually incorporated into federal and state regulations.

13.3.2 INCINERATOR FEATURES

Among the first features added to incinerators were afterburners and supplemental fuel burners. Afterburners helped to burn out organics in the off-gas resulting in the reduction or elimination of smoke and odor. Supplemental fuel burners, which were located in the combustion chamber of the incinerator, provided a more effective means of start-up of the incinerator, and helped to maintain minimum incinerator temperatures.

More recently, over the past two decades, concern grew for control not only of the potential smoke and odor discharges, but also for control of acid gas emissions, (hydrogen chloride), trace organics, and heavy metals. This led to the provision of secondary combustion chambers where gas temperature and residence time can be controlled to provide good burnout to eliminate most of these emissions. In addition, scrubbers and other air emissions control equipment are now required for most medical waste incinerators for removal of small particulate matter and unwanted gases from the exhaust gas stream.

13.3.3 INCINERATION PROCESS

Any incinerator must obey the following Three Ts of combustion:

- Time

- Temperature
- Turbulence

For waste to effectively burn out, sufficient air must be provided. Waste will burn in the primary combustion chamber of the incinerator, and an off-gas will be generated. The off-gas must pass through a region of high temperature, and it must be held at this temperature for a minimum period of time. Sufficient turbulence must be provided to insure good mixing of the hot flue gas with the oxygen in the air. Each type of incinerator injects air, and provides the Three Ts of combustion in a slightly different manner.

13.3.4   TYPICAL INCINERATOR SYSTEM

A generic incinerator system is shown in Figure 13.3.1. It will have the following components:

- **Waste loading system.** Waste must be brought to any incinerator. Waste will be loaded by hand (manually) or automatically, by automatic loading equipment. Other than very small incinerators, less than 200 lb/hr (90 kg/hr) waste will be loaded into a charging hopper. In these smaller systems, waste will be charged directly into the incinerator, through a charging door.

    There are a number of automatic feeding systems available, the more common being a cart dumper. A cart is placed in a dumper mechanism and the dumper raises the cart, and rotates it to discharge its contents into a charging hopper.

    In all but the smaller systems, and particularly with incinerators designed for starved air operation, the feeding system will include an air lock. It is important to control the amount of air leaking into an incinerator. With excessive air leakage into the primary combustion chamber, it would be difficult, if not impossible, to maintain a starved air reaction. With excess air combustion, the greater the air leakage, the greater the amount of supplemental fuel required to maintain incinerator

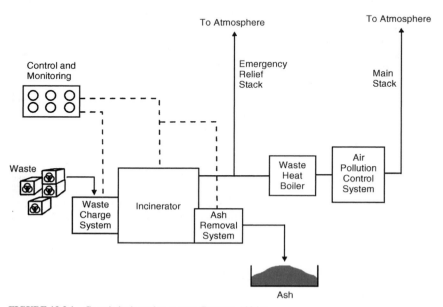

FIGURE 13.3.1   Generic incineration system (Brunner, 1996a).

temperatures. The air lock may be a cover that closes onto the charging hopper when the feeding cycle begins, or it can be a series of chambers that close tightly in order to severely restrict leakage air flow to the primary combustion chamber of the incinerator.

- **Incinerator.** This is the heart of the system. Air is injected, heat is provided, waste is heated to combustion temperature, and the waste residue is allowed to burn out to an ash. Flue gas generated from the incineration process discharges to the atmosphere after passing through an air pollution control system.

- **Ash removal system.** Ash is produced from incineration. The waste will always leave an ash residue, which must be removed before it can accumulate and interfere with burning. Batch fed incinerators will not normally have automatic or continuous ash removal systems. The operator will rake out ash from the incinerator after a burn-out of waste. Other incinerators usually have a means of discharging and collecting ash either in a wet sump or else as a dry residue. When discharged dry, the ash will be collected in a separate container for disposal.

- **Emergency relief stack.** The purpose of the emergency relief stack is to divert hot incinerator flue gases from a scrubber, a fan, or other equipment that would be damaged if there were a failure of the cooling water supply, a power failure, or similar problem. Not every incinerator will have an emergency relief stack, but those with waste heat boilers, wet scrubbing equipment, or induced draft fans will usually include them. In Figure 13.3.1, this stack is located between the incinerator and the waste heat boiler.

- **Waste heat boiler.** The incinerator flue gas will be hot, 1800°F (9827°C), or more in some cases, and this heat can be used to generate steam or hot water. Where there is a use for this heat, for a laundry, or for building heat, for instance, a waste heat boiler will be provided to generate steam or hot water.

   In some installations a waste heat boiler is provided to reduce the temperature of the flue gas stream, even when there may be no use for the steam. Passing through a waste heat boiler, the flue gas temperature is reduced to the lower temperature required by a baghouse or other air pollution control device. In these cases, the steam is wasted by passing it through a condenser or by blow off to the atmosphere.

- **Air pollution control system.** All states have regulations that control the amount of emissions that can be discharged from an incinerator stack. Some incinerators can meet these emissions by the proper temperature levels and controls within the incinerator primary and secondary combustion chambers.

   Most incineration systems require a separate air pollution control system to meet regulatory emissions standards. When an air pollution control system is provided, an induced draft (ID) fan will be included as part of that system. It is required to overcome pressure losses developed in the flue gas stream as the gas passes through a waste heat boiler or an air pollution control system. An ID fan will draw gases from the incinerator, through the air pollution control equipment, discharging flue gas through the main stack to the atmosphere.

- **Main stack.** All incinerator systems have a stack. Flue gases are discharged to atmosphere through the main stack during normal operation of the incinerator system. As shown in Figure 13.3.1, the main stack is located immediately after the incinerator, or if air pollution control equipment is provided, it is located after this equipment.

- **Control and monitoring.** Practically all incinerators have a control system providing some degree of automation. The basic control is associated with the supplemental fuel burners. It will ensure the safety of burner firing and will maintain furnace chamber operating temperatures by automatic modulation of the burner firing rate. Some incinerators will have additional controls, for feed charging, ash discharge, air pollution control systems, and so on.

   All but the smaller systems will also have some type of monitoring system. The operator must be aware of conditions within the incinerator to be better able to control operations. These monitors will include temperatures, pressure (draft), and in some cases will include exhaust gas monitoring such as oxygen and carbon monoxide sensing.

TABLE 13.3.1 Waste Definitions (Brunner, 1996a)

Classification		Approximate composition (%)	Moisture content (%)	Incombustible solids (%)	Heating value (Btu/lb)	Heating value (kJ/kg)	Principal components
Type	Description						
0	Trash	Trash, 100	10	5	8500	19,771	Highly combustible waste. Paper, wood, cardboard cartons and up to 10 percent treated paper, plastic or rubber scraps; commercial and industrial sources
1	Rubbish	Rubbish, 80 Garbage, 20	25	10	6500	15,119	Combustible waste, paper, cartons, rags, wool scraps, floor sweepings; domestic, commercial, and industrial sources
2	Refuse	Rubbish, 50 Garbage, 50	50	7	4300	10,002	Rubbish and garbage, residential sources
3	Garbage	Garbage, 65 Rubbish, 35	70	5	2500	5815	Animal and vegetable wastes, restaurants, hotels, markets; institutional, commercial and other sources
4	Animal solids and organic wastes	Animal and human tissue, 100	85	5	1000	2326	Carcasses, organs, solid organic wastes; hospital, laboratories, abattoirs, animal pounds and similar sources

13.3.5 DIFFERENT WASTE CHARACTERISTICS

Many types of incinerators are in use today. Each is basically designed for a specific waste category. Incinerators may be designed to process either solids, sludge, liquid wastes, or contaminated gases and fumes, or some combination of these. Some incinerators are designed specifically for hazardous chemicals, radioactive waste, or for soil remediation. In some cases, boilers and industrial furnaces are used to incinerate waste materials. In such instances, the waste is burned along with another type of fuel.

It is important to note that no single type of incinerator can handle all of the different waste types. Different criteria are used in the design of the various kinds of incinerators. For instance, an incinerator designed for medical waste may not be able to burn liquid waste or sludges effectively. This becomes an important distinction because at many medical waste generation facilities, associated laboratories may generate sludge or liquid waste streams. Unless the medical waste incinerator was designed for liquid or sludge waste firing and for the destruction of boxed or bagged waste, these other wastes must be disposed of elsewhere.

A set of waste types was developed a number of years ago to help to define the various kinds and characteristics of general waste materials. These waste definitions are listed in Table 13.3.1. Waste types that are discussed below refer to the waste classifications in this table.

13.3.6 INCINERATOR TYPES

The three basic types of medical waste incinerators are multiple chamber, controlled air, and rotary kiln incinerators. There are a number of other incinerator designs, however, the majority of the thousands of incinerators used for medical waste disposal are of these three major categories.

13.3.7   INCINERATION PROCESSES

Many of the medical waste incinerators in operation have been designed as starved air units, particularly the controlled air systems, although some of these operate in the excess air mode. Other incinerators, particularly the multiple chamber type, are designed as and operate as excess air systems.

13.3.8   MULTIPLE CHAMBER INCINERATORS

Multiple chamber incineration technology was developed in the mid-1950s; however with few exceptions (note the Integral Incinerator System and Pathological Waste Incineration Systems discussed below), few of these systems are being built today. In general, they cannot meet air pollution regulations without the addition of air pollution control systems. When such systems are added, the incinerator has not been cost-effective when compared to other types of incinerators.

Multiple chamber incinerators have two combustion chambers. The primary combustion chamber is used for drying and burning the waste to an ash and releasing an off-gas, or flue gas. A secondary combustion chamber completes combustion of organic materials remaining in the flue gas stream.

There are two basic types of multiple chamber incinerators, retort and in-line. In both of them the following operations apply:

- Waste is ignited in the primary combustion chamber by the primary burner. The primary combustion chamber temperature should be no less than 1400°F (760°C). Type 1 or Type 2 waste will likely burn to this temperature without the need for supplementary fuel, and after ignition the primary burner will automatically go to low fire, or may shut off. With pathological waste or other waste with a relatively low heat content, the burner will be required to fire continuously.

- Moisture will evaporate from the waste, and gases will volatilize.

- As combustion proceeds, the organic component of the waste that has not vaporized (the fixed carbon) will burn out to an inorganic (or noncombustible) ash.

- A secondary burner in the secondary combustion chamber will provide the heat required to bring the flue gas temperature high enough to destroy any organics. This temperature is normally in the range of 1600 to 1800°F (871 to 982°C). This burner will remain in operation while the incinerator is in operation (i.e., it will not shut off).

- By the time gases exit the secondary combustion chamber, most of their organic content has been effectively destroyed.

These two types of multiple chambers incinerators, retort and in-line units, are designed and operate as follows:

Retort Incinerators

Retort units are cube shaped with 90 degree changes of air flow direction in both the vertical and horizontal directions. The chambers are arranged in a "U" shape (see Figure 13.3.2). They are available in sizes up to approximately 1000 lb/hr (454 kg/hr).

Waste is charged through the charging door, and will burn on the hearth within the ignition chamber, as shown in Figure 13.3.2. Burners within the ignition chamber will maintain the heat required for effective combustion of the waste. The off-gas is directed up, down, and around a chamber wall, Some designs, like the one shown, pass the hot flue gas beneath the ignition hearth in the primary combustion chamber. This provides some heat to the hearth above to assist in drying and burning the waste. At each turn the flue gas takes within the incinerator some particulate matter will drop out of the gas stream.

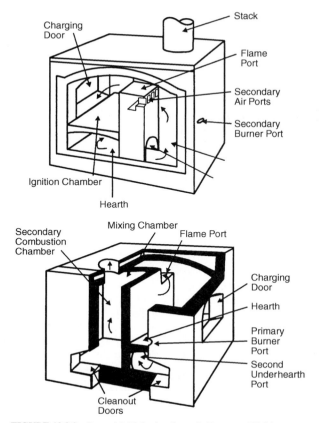

FIGURE 13.3.2 Retort Multiple chamber unit (Brunner, 1996b).

From beneath the hearth the flue gas passes through a secondary combustion chamber. The secondary combustion chamber is designed to provide the three Ts of combustion. A secondary combustion chamber burner provides the Temperature. Normally, the chamber is sized large enough for 1-second residence time which is the second T (Time). The changes in direction of the path that the gas steam travels to reach the secondary combustion chamber helps to provide the third T, or turbulence.

These systems are operated in the excess air mode. More than double the amount of air required for complete combustion (over 100 percent excess air) is added to the primary combustion chamber of the incinerator. The amount of air remaining in the flue gas as it enters the secondary combustion chamber is normally high enough so that additional combustion air is not required for the burn-out of combustibles in the gas. These incinerators are batch fed. They are normally charged during a 6- or 8-hour period of time. Waste is fired and is allowed to burn out. Additional waste charges may be placed in the incinerator while waste is burning.

At the end of a shift, when no further waste is charged, the waste in the incinerator will be allowed to burn out to an ash. Ash remains on the hearth and at the end of the burning cycle it must be removed before charging begins again.

Most incinerators used primarily for pathological waste destruction are retort designs with solid refractory hearths. This waste type does not burn well (it contains approximately 85 percent moisture). The heated hearth and the ability to maintain extended burning time on this hearth results in effective drying and good burnout of these wastes in a retort incinerator.

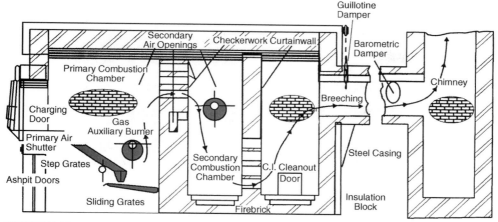

FIGURE 13.3.3   In-line multiple chamber unit (Brunner, 1996b).

In-Line Incinerators

In-line units are more rectangular (longer in one dimension) than the retort incinerator, as shown in Figure 13.3.3. They are similar to the retort incinerator with two separate burning chambers. The gas flows through frequent turns where larger particulate matter drops out. The turns in direction are in the vertical dimension, rather than in the horizontal dimension as with the retort systems. In-line incinerators are designed to operate in the excess air mode.

The elongated primary combustion chamber provides enough space for the installation of a moving grate system. This feature allows the incinerator to be used as a continuous or semi-continuous system.

Waste is ram-fed or is hand-charged to the retort incinerator. The retort must be allowed to burn out its waste, and ash must be removed before another burn cycle is started. With the in-line incinerator, however, waste is continually burning, and ash is continually dropped from the incinerator. This incinerator can be operated on a continuous basis, and with a continuous ash discharge.

The in-line incinerator is normally built in the range of 750 to 2000 lb/hr (907 kg/hr) capacity. As with the retort incinerators, they require air pollution control systems to meet the regulatory requirements of most states.

13.3.9 INTEGRAL INCINERATOR SYSTEM

An integral incinerator is a variation of the retort or in-line system and is multi-chambered, with an integral scrubbing system. They are built in sizes from under 100 lb/hr (38 kg/hr) capacity to over 2000 lb/hr (907 kg/hr), throughput.

As shown in Figure 13.3.4, and in the cross section in Figure 13.3.5, waste is charged into a primary combustion chamber. Air ports (cones) along the walls of the chamber provide for the introduction of combustion air into the chamber. Blowers are not necessary to obtain the desired air flow, which is naturally inspired through these cones. Dampers on the cones allow adjustment of the air inspiration rate. The primary chamber is maintained at a temperature of approximately 1400°F (760°F). Burners are located in the walls of the chamber.

Off-gas passes through an integral flue to a secondary combustion chamber, which has sufficient burner capacity to maintain a gas temperature of at least 1800°F (982°C). From the secondary combustion chamber, which is designed for a minimum of two seconds residence time at the target temperature, the off-gas passes through a quench section, and past a baffle wall positioned above a

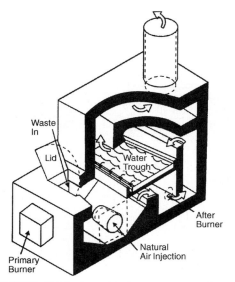

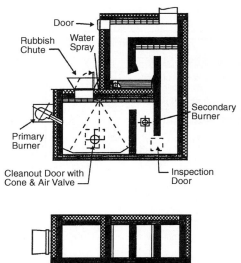

FIGURE 13.3.4   Integral incinerator cross section (Brunner, 1996a).

FIGURE 13.3.5   Integral incinerator system (Brunner, 1996a).

water bath, or water trough. The space above the water trough (i.e., the gap between the top of the water and the bottom of the baffle wall) is designed to provide high turbulence, and velocity of the gas stream. This turbulence will also have a surface effect on the water bath, creating a flood of finely atomized water particles within the gas stream, which will adsorb particulate matter, and will remove acid gas from the gas flow. Alkali is often added to the water bath to increase acid gas adsorption. Blowdown of the sump maintains concentrations of pollutants in the water bath within acceptable levels.

These systems are unique in their efficiency and ease of operation. The systems are operated as excess air units. Their design is extremely simple, with virtually no moving parts, no blowers, and, likewise, little operator attention required. They have been found to reduce emissions to less than 0.03 grains/DSCF (50 mg/NCM) and they have demonstrated greater than 90 percent removal of hydrogen chloride.

13.3.10   CONTROLLED AIR INCINERATORS

Controlled air incinerator design is a relatively new technology. The original design was developed in the 1950s and was further refined in the 1960s. Controlled air incineration is the most widely used technology for medical waste incineration. Over the last 25 years, more than 90 percent of the medical waste incinerators sold in the United States have been of this type. These incinerators are also known as two-stage combustion units or modular combustion units. They can be operated as starved air or excess air systems, depending on the type of waste burned, although most of them are sold as starved air systems.

Controlled air incineration occurs in two steps, with each step occurring in a separate chamber of the incinerator. These chambers are the primary combustion chamber and the secondary combustion chamber as shown in schematic in Figure 13.3.6.

The first application of this technology was with an afterburner, not with a separately defined secondary combustion chamber. Figure 13.3.7 is an example of a controlled air unit designed for

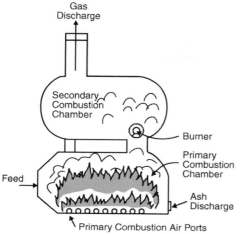

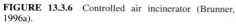

FIGURE 13.3.6 Controlled air incinerator (Brunner, 1996a).

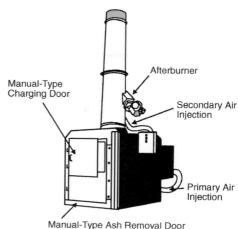

FIGURE 13.3.7 Combustion chamber with after-burner, small capacity (Brunner, 1996a).

small waste quantities, less than 200 lb/hr (90 kg/hr). Paper waste requires a temperature of 1400°F (760°C), and a residence time of at least 1/2 second at this temperature for destruction of the gases generated. This unit was designed for these criteria. As noted above, this was an early design, used when hospital waste was basically paper waste with little plastic or single use, throw-away products.

Hospital waste characteristics were similar to those characteristics of paper, and these types of units may have once been effective. Now, with a hospital waste stream containing more varied materials, most of these incineration systems cannot operate as initially designed. The heat content of the waste is greater than the initial design called for and with added heat release, additional air must be supplied to the system. The afterburner design does not provide sufficient residence time for complete burnout of the combustibles in the gases that are generated at these larger volumes. With additional heat released in the combustion chamber, the amount of gas (gas volume) passing through the afterburner is much greater than when initially designed. This results in a decrease in residence time, which, in turn, often results in the generation of smoke and odor. Many of these systems are in use today, although they are generally unable to meet state emissions requirements for new incinerators.

The incinerator shown in Figure 13.3.8 is larger than that in Figure 13.3.7, but is similar in design. There is an afterburner that is designed to provide the same level of heat and retention time as above. The thermal reactor shown in this illustration provides turbulence to mix the hot gases from the afterburner with the flue gas flowing from the combustion chamber. As with the smaller unit, this system cannot normally meet air emissions control requirements in most states for new incinerators without the addition of air emissions control equipment, although many of them are currently in operation without such equipment.

Another factor in the use of an afterburner with a combustion chamber is in the regulatory requirements for such equipment. Many state regulations include requirements for a separate secondary combustion chamber and a well-defined residence time of

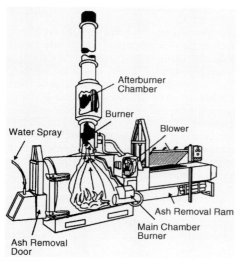

FIGURE 13.3.8 Combustion chamber with afterburner, medium capacity (Brunner, 1996a).

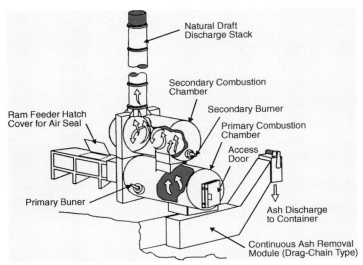

FIGURE 13.3.9   Primary, secondary combustion chamber continuous ash discharge (Brunner, 1996a).

from one to two seconds. These requirements are not possible to implement with a separate afterburner as shown in Figures 13.3.7 and 13.3.8. These afterburners are not separate combustion chambers, and they do not have sufficient volume to provide the retention time required. A secondary combustion chamber, which provides one or two second retention time, will be approximately the same size as the primary combustion chamber.

Figure 13.3.9 is a typical controlled air incinerator system in common usage. Waste is charged through a hinged charging door into a ram feeder. The ram extends to feed waste to the primary combustion chamber. As waste is charged, and the waste previously charged and currently burning is pushed towards the back of the incinerator, toward the ash discharge. The only means of moving the waste is through the push of new waste feed.

The chambers of the controlled air incinerator may be either cylindrical, as shown in Figure 13.3.9, or rectangular. The shape is usually based on the manufacturer's preference. Some manufacturers make three-stage controlled air incinerators, which have a combustion chamber downstream of the secondary combustion chamber. This additional chamber helps to ensure complete combustion of the gases.

The controlled air incinerator shown in Figure 13.3.10 is provided with internal rams, and a series of step hearths. The feed ram charges the incinerator with waste, and the ash transfer and ash discharge rams move both the burning waste and the ash toward the ash chute for eventual discharge from the incinerator. In this design, which is used with larger incinerators (over 500 lb/hr, or 225 kg/hr, capacity) waste is moved continuously through the primary chamber, and ash is continuously discharged.

These systems are generally designed for starved air operation (i.e., for feed with a relatively high heating value). When lower heating value waste is charged, the incinerator will operate in the excess air mode. In either of these operating modes, the following operations apply:

- Waste is ignited in the primary combustion chamber by the primary burner. The primary combustion chamber temperature should be no less than 1400°F (760°C). Type 1 or Type 2 waste will likely burn to this temperature without the need for supplementary fuel, and the primary burner will shut off, or will go to low fire. With pathological waste or other waste with low heat content, the burner will continue to fire.

- Air injection into the primary combustion chamber is controlled to provide less than stoichiometric air when firing a waste with high-heat content. When the waste will not support combustion,

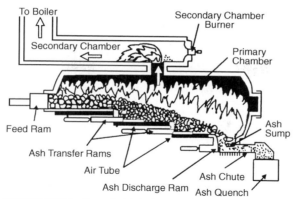

FIGURE 13.3.10 Controlled air incinerator with step hearths (Brunner, 1996a).

sufficient air flow will be provided to combust organics in the primary combustion chamber with at least the stoichiometric air requirement.

- Air is injected in the primary combustion chamber at the bottom of the chamber toward the lower portion of the waste. This will tend to combust residual carbon, reducing the unburned carbon in the residual, or bottom ash discharge.

- Moisture will evaporate from the waste, and gases will volatilize.

- As combustion proceeds the organic component of the waste that has not vaporized (the fixed carbon) will burn out to an inorganic, (or noncombustible), ash.

- A secondary burner in the secondary combustion chamber will be on low fire to act as a pilot flame when the primary combustion chamber is operated in the starved air mode. When waste in the primary combustion chamber is burning with excess air the burner in the secondary combustion chamber will be in operation, maintaining a temperature of from 1600 to 1800°F (872 to 982°C).

- By the time the off-gas exits the secondary combustion chamber most of its organic content has been destroyed.

The operation in starved air and in excess air modes is as follows:

13.3.11 STARVED AIR SYSTEMS

The starved air process can be effective and economical in the destruction of wastes with relatively high heating value. Waste is fed into the primary combustion chamber and combustion occurs in a starved air environment (i.e., with less oxygen (air) input than is needed for complete combustion). The waste material is converted to gaseous products and a residue. In properly designed systems, the residue will burn out to a clean, sterile ash.

The gaseous discharge (the flue gas) from the primary combustion chamber will be rich in organic material. The effect of starved air combustion is not to burn out all of the organic content of the waste in the primary combustion chamber of the incinerator, but to convert much of the organics in the waste to organics in the off-gas.

An important feature of incinerators operating in the starved air mode is that with less than the stoichiometric air requirement injected into the primary combustion chamber of the incinerator, there is less turbulence than if additional air were added. With less turbulence of the waste, there is less carry-over of solid material. There is much less particulate in the flue gas exiting the primary combustion chamber when the incinerator is operated as a starved air unit then when it is in the excess air mode. In the excess air mode, a good deal more air is introduced into the primary combustion chamber.

Gases flow through a passageway into the secondary combustion chamber. Air is provided here to burn out the organics in the flue gas. When operated in the starved air mode, there should be a high enough concentration of organics in the flue gas entering the secondary combustion chamber to burn out with no additional fuel. With the introduction of air, the organics should start to burn. No flame is necessary because the gas is hot enough to start to burn as soon as air is introduced, the first T (Temperature) of combustion.

Normally, a flame will be maintained in the secondary combustion chamber through the firing of supplemental fuel. If the waste quality changes (for instance, if the moisture content increases), the organic content of the flue gas entering the secondary combustion chamber will decrease, and the gas may not have enough combustibles to maintain ignition. A pilot flame in the secondary combustion chamber provides assurance that if spontaneous combustion of the flue gas will not occur, the flame will force combustion to occur.

The secondary combustion chamber is normally designed for a residence time of 1 or 2 s at the target temperature (the second T, Time, of good combustion). Air is added to the secondary combustion chamber. In the starved air mode of operation, there is not enough air (or oxygen) in the flue gas stream to burn out organics. The air required to fully burn out the organics in the flue gas is added in the secondary combustion chamber. This air supply is normally added tangentially within the chamber to achieve good turbulence (the third T, Turbulence, of good combustion). The swirling patterns in the secondary combustion chamber shown in Figure 13.3.9 illustrates the creation of turbulence in this chamber. In some designs short refractory walls, refractory checkerboard patterns or other structures are built into the secondary combustion chamber to help increase turbulence, and to improve burnout of organics in the flue gas.

13.3.12 *EXCESS AIR SYSTEMS*

The hospital waste stream may not have enough heat to support combustion. Hospital waste with a high proportion of pathological waste (animal carcasses, for instance) or with liquids (from dialysis or suction bottles, or from blood processing, etc.) will not be able to support its own combustion. If wastes will not burn by themselves, then starved air combustion is not possible. The only time that starved air is applicable to waste combustion is when that waste itself is combustible. The waste must be able to burn by itself, and starving the air flow means providing less air than is needed to maintain its own combustion.

With wastes that do not support combustion, supplemental fuel must be added. Enough air must be provided to burn out the combustibles that are present in the waste. When supplemental fuel is fired in the primary combustion chamber and air is supplied for waste burn-out, excess air combustion is occurring. This should occur automatically in a properly designed controlled air unit when waste with low heating value is charged. Most starved air systems are designed to automatically increase air flow to the primary combustion chamber when the temperature drops. This corresponds to the starved air process. Increasing the air flow will result in the burning of additional waste in the primary combustion chamber and this additional burning will, in turn, increase the primary chamber temperature. If the waste has a low heating value, as the air injection quantity into the primary combustion chamber increases, its temperature will decrease.

The controls should pick up on this temperature decrease and signal the supplemental fuel burner to fire to maintain primary combustion chamber temperature. Different manufacturers use different control systems to complete this process. For instance, one manufacturer will have controls to sense a decrease in primary chamber temperature when additional air is added, and will reduce the primary air injection rate at the same time that the primary combustion chamber burner begins to fire. Other manufacturers will measure the oxygen fraction in the system and will adjust air flow and burning firing rates in accordance with the desirable temperatures and oxygen levels.

13.3.13   ROTARY KILN INCINERATORS

Rotary kiln incinerators are widely used in the treatment of hazardous waste. These incinerators are versatile and are suitable for most types and forms of waste, including solids, sludge, liquids, and gases. Within the last few years, these units have been applied to the burning of hospital waste streams. They require a drive system, which is not necessary with multiple chamber or controlled air incinerators, and this additional equipment raises the cost of rotary kiln systems compared to these other systems. With larger size units, above 1500 lb/hr (680 kg/hr) capacity, controlled air systems will have pushing rams or other forced traveling systems, and rotary kilns become cost competitive at these size ranges. Some kilns are designed to be operated as starved air units when high heating value waste is charged.

The following operations apply to the rotary kiln system:

- Waste is ignited in the kiln by the primary burner. The primary combustion chamber temperature should be no less than 1400°F (760°C). Type 1 or Type 2 waste will likely burn to this temperature without the need for supplementary fuel, and the primary burner will shut off, or go to low fire. With pathological waste, or other waste with low heat content, burner firing will be required.

- Air injection into the kiln may be controlled to provide less than stoichiometric air when firing a waste with high heat content. When the waste will not support combustion, sufficient air flow must be provided to combust organics in the kiln with at least the stoichiometric air requirement.

- Moisture will evaporate from the waste, and gases will volatilize.

- As combustion proceeds the organic component of the waste that has not vaporized (the fixed carbon) will burn out to an inorganic (or noncombustible), ash.

- A secondary burner in the secondary combustion chamber will be on low fire to act as a pilot flame when the primary combustion chamber is operated in the starved air mode. When the rotary kiln is burning with excess air the secondary combustion burner will be firing, normally maintaining a temperature of 1800°F (982°C).

- By the time the flue gases exit the secondary combustion chamber, most of its organic content has been destroyed.

A description of the conventional rotary kiln and a modification, the rocking kiln, is as follows:

Conventional Rotary Kiln System

A rotary kiln incinerator has a cylindrical primary combustion chamber, the rotary kiln, which rotates about a horizontal axis (see Figure 13.3.11). The kiln is noted in this illustration as the rotating primary chamber. It is tilted at a slight incline to the horizontal, or rake, from 1/4 to 1/2 inch per foot (40 to 80 mm per meter) of length. The primary combustion chamber is rotated slowly, at approximately 1 revolution per minute. It is normally turned by an external gear, by a chain drive around its periphery, or through the movement of one set of trunnions supporting the kiln.

Waste is charged at the upper end of the kiln, and its rotation moves the waste slowly toward the opposite end. Waste will burn within the kiln, and by the time the burning waste has reached the far end of the kiln, it will have burned out to an ash and be discharged to an ash sump. The retention time of the solids in the kiln is dependent on the kiln rotation speed and the rake (the angle of the kiln to the horizontal). The higher the speed of rotation, the faster the waste will travel through the kiln. Also, the greater the rake, or angle to the horizontal, the faster the waste will travel through the kiln and the less the solids retention time.

A separate secondary combustion chamber (shown as the High Temperature Secondary Chamber in Figure 13.3.11) is located downstream of the far end of the kiln to complete combustion of organics in the off-gas, providing Temperature, Time, and injecting air to obtain the Turbulence (the three Ts) required for good combustion. The rotation of the kiln provides greater turbulence of the waste than the multiple chamber or controlled air systems discussed above. This results in waste surfaces continuously exposed to the fire.

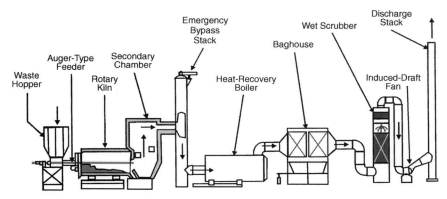

FIGURE 13.3.11 Rotary kiln incinerator (Brunner, 1996a).

Kiln rotation in combination with its tilting (rake), make this type of incinerator readily adaptable to continuous ash removal. Ash is normally discharged into a wet ash sump at the end of the kiln, as shown in Figure 13.3.11. The water bath not only quenches the hot ash, but also it provides a water seal to reduce air leakage into the kiln.

Waste cannot be charged into smaller rotary kiln (less than 6 ft or 2 m in diameter) in boxes or cartons. The boxes will tend to tumble rapidly through the kiln, and may not burn. This is a physical constraint that can be overcome by shredding the waste as it enters the kiln. In larger rotary kilns (above 8 ft or 2.5 m in diameter) the kiln surface is less curved. There are a number of these relatively larger sized where cartons are charged directly into the kiln, with no tumbling or other undesirable movement occurring.

Most rotary kiln systems for medical waste incineration are the smaller sized units used in hospitals, charged with bags (not boxes). They use a charging auger (or auger feeder) as indicated in Figure 13.3.11. This is a large screw, similar in shape to that of a corkscrew, which will shred the waste as it is fed forward into the kiln. The auger has teeth along its outer edge, which shreds and crushes the waste as it feeds it into the rotary kiln.

The feed hopper, as shown in Figure 13.3.11, discharges to the kiln through the auger, as shown on the bottom of the hopper. The shredded materials from the auger feeder will not be subject to rolling down the kiln surface, as will boxes or cartons, and will burn out in a relatively uniform and complete manner. Liquids within the waste will tend to be a problem as materials are moved through the auger feeder and are shredded. A drain system is often provided beneath the hopper to collect liquids and to direct them separately into the kiln for disposal.

Pathological waste should not be fed to a kiln through an auger feeder. Where a kiln is used for hospital waste disposal, a separate hearth is often provided after (downstream of) the kiln, usually within the secondary combustion chamber, with a separate charging door. Pathological waste is placed on this hearth as a batch operation, and it is allowed to burn out over a relatively long period of time. For instance, pathological waste may be charged to this hearth at the beginning of the shift and is allowed to burn out throughout the shift, with no additional pathological waste charged until the end of the shift. The kiln itself, however, is a continuously operated incinerator. Some of the hot flue gases from the kiln are directed to the pathological waste hearth, and as pathological waste burns, its off-gas mixes with kiln off-gas in the secondary combustion chamber. These gases are all burned within the secondary combustion chamber to destroy their organic content.

Rocking Kiln

The rocking kiln is a variation of the rotary kiln incinerator. While the rotary kiln continually revolves about its horizontal axis, the rocking kiln rotates through 3/4 of a revolution and then reverses.

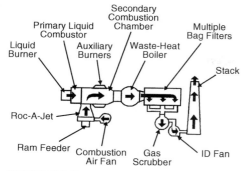

FIGURE 13.3.12 Rocking kiln (Brunner, 1996a).

By only partially rotating, the waste within the kiln experiences less turbulence than the waste in a conventional rotary kiln incinerator. This results in less particulate matter released from the waste in the kiln to the secondary combustion chamber of the system.

Another feature of the rocking kiln is that there is no need to shred waste. The lack of full rotation prevents the tumbling of boxes or other containers through the unit. A typical installation is shown in Figure 13.3.12. Waste is charged into the kiln through a ram feeder. Flue gas is directed from the rocking kiln to a relatively large secondary combustion chamber, sized for greater than 2 s residence time at a minimum temperature of 1800°F (982°C). This is the Time factor for good combustion. The Temperature factor is provided by an auxiliary burner in the secondary combustion chamber. Although not shown in Figure 13.3.12, air is injected through high pressure nozzles along the length of the secondary combustion chamber to produce the Turbulence factor of good combustion.

Ash drops from the end of the rocking kiln to a water bath where it is removed by an inclined conveyor. Flue gas passes through a waste heat boiler, or a water spray tower, where its temperature is reduced to less than 400°F (204°C). At that temperature level, it can safely pass through a baghouse for particulate removal. From the baghouse the flue gas is directed to a wet tower that scrubs out acid gases through an alkali wash. An induced draft fan drives flue gas through the process and out the stack. Liquids can be incinerated in this system though a separate liquid waste burner at one end of the secondary combustion chamber.

13.3.14 MOVING HEARTH SYSTEM

The moving hearth incinerator has been developed to increase turbulence of waste in the primary combustion chamber without increasing air emissions. Motion of waste is through mechanical agitation without tumbling the waste as in a rotary kiln. A typical moving hearth incinerator is illustrated in Figure 13.3.13. Boxes or bags of waste are loaded in a waste loading hopper, generally filling it from 1/2 to 3/4 full. After it is loaded, and the incinerator is ready for another charge, the air lock door (not shown) is closed, the refractory lined charging door is retracted up into the smoke hood, and the ram pushes the waste into the incinerator. An electric motor drives a pinion which, in turn, drives a rack attached to the ram.

Waste will dry and start to burn on one of two furnace hearths. Each of these hearths has moving refractory floors, or pulsed hearths. The pulsed hearths move waste toward the chamber exit in short, rapid movements. As the waste is moved, it is tumbled to expose new surfaces to air and heat.

The walls of the primary combustion chamber are lined with water tubes, forming waterwalls. The waterwalls absorb heat from the burning waste for steam generation. Flue gas continues to a series of two afterburners (a secondary and a tertiary combustion chamber) sized for a minimum of 1-s retention Time, one of the Ts of good combustion.

Reviewing air introduction and the other Ts of combustion in this design, air is provided by separate blowers for the two separate afterburner sections. Temperature is provided by a burner just upstream of the afterburner sections. Some combustion air is injected within the gas stream, shown in Figure 13.3.13 as thermal exciters. This air flow generates the required Turbulence, the third T of effective combustion.

In the design shown, a waste heat boiler is installed after the burning process is complete. To protect the boiler from high flue gas temperatures that may cause slagging, a recirculation flue is provided. This takes gas from the waste heat boiler exit, which is at a temperature in the range of 350 to 550°F (176 to 287°C), and mixes it with the flue gas exiting the afterburner sections, which could be as high as 2000°F, (1093°C). The temperature of the gas entering the waste heat boiler is controlled to below

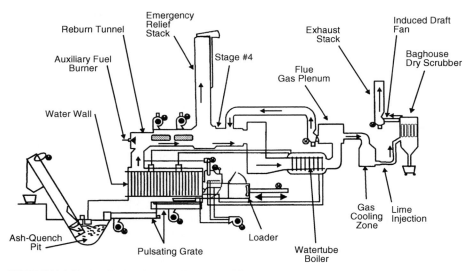

FIGURE 13.3.13   Moving hearth system (Brunner, 1996a).

1600°F (871°C) entering the waste heat boiler. This is below the ash fusion temperature, as discussed in a previous chapter. If the gas stream entered the waste heat boiler at a temperature in excess of 1800°F (982°C), where there is a good possibility that some of the ash would have softened, the boiler tubes, which are at a temperature of from 300 to 600°F (149 to 316°C), would likely slag over.

In this illustration, lime is injected just before the baghouse. The baghouse removes particulate matter and spent lime from the flue gas stream. An induced draft fan located downstream of the baghouse draws gas through the incinerator system and discharges out the stack. Ash drops out of the last hearth of the primary combustion chamber into a water bath. Wet ash is dragged out of the bath with a mechanical ash remover.

REFERENCES

Brunner, C., *Medical Waste Disposal*, Incinerator Consultants Incorporated, Reston, VA, 1996a.

Brunner, C., *Incineration Systems Handbook*, Incinerator Consultants Incorporated, Reston, VA, 1996b.

CHAPTER 13
CONTROL TECHNOLOGIES

SECTION 13.4

CONTROL TECHNOLOGIES: PARTICULATE CONTROLS

Mark P. Cal

Dr. Cal is currently an assistant professor of environmental engineering at New Mexico Tech in Socorro, New Mexico. His main areas of research are air pollution control, indoor air quality, gas-phase adsorption, plasma processing of gas streams, and gas separation.

13.4.1 INTRODUCTION

Particulate emissions are typically removed from industrial gas streams with the one or more of the following control devices:

- Cyclones (centrifugal force)
- Wet collectors (diffusion, interception, and impaction)
- Fabric filters (diffusion, interception, and impaction)
- Electrostatic precipitators (electrostatic force)

The type of particulate control device that should be used for a specific application depends on many factors, including the following: gas flow rate, particle size distribution, particle loading, particle composition, gas temperature and pressure, desired collection efficiency, and acceptable capital and operating costs. For many applications, more than one type of control device may provide the desired collection efficiency, but they will most likely differ with regards to capital or operating costs. Therefore, economic factors must be considered along with technical issues. The following sections provide a general overview of the major types of particulate control devices. The reader is referred to the references section and to equipment manufacturers for more specific information.

13.4.2 CYCLONES

Cyclones are inertial separators that use centrifugal force to remove particles from gas streams (Figure 13.4.1). Standard cyclone designs can provide high removal efficiency ($\sim$100 percent) for particles greater than 20 μm, and particles with diameters greater than about 5 μm are usually removed with efficiencies greater than 50 percent. As particle size decreases from 5 μm, collection efficiency rapidly decreases, making cyclones ineffective for the removal of very small particles ($< \sim$1–5 μm). Cyclones can be used as a stand-alone particulate control device or in conjunction with another device, such as an electrostatic precipitator or a fabric filter. When cyclones are used in series with another control device, they are used to pretreat the gas stream by removing larger particles. The second particulate control device is then used to collect the smaller particles at a much reduced particulate loading, and at a higher collection efficiency.

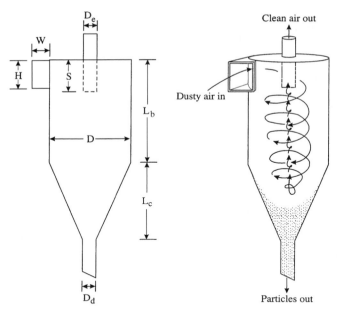

FIGURE 13.4.1   Tangential cyclone of standard proportions.

In a cyclone, the gas stream enters in a manner that causes it to spiral around the inside of the cyclone, causing particles with greater inertia to be forced to the outside walls of the cyclone. Upon collision with the outside walls of the cyclone, particles fall down the cone to the collection hopper below. There are two main methods to cause the gas to spin within the cyclone body: (1) introduce the gas to the cyclone in a tangential manner so that the gas curves around the inside body of the cyclone, and (2) use axial vanes at the inlet of the cyclone, causing the gas to spin as it flows past the vanes. An advantage of the vane-axial cyclone is that it can be produced in small sizes, which causes the gas stream to make tighter turns, improving particle collection efficiency. A disadvantage of the vane-axial cyclone is that the pressure drop through the cyclone increases rapidly as the tangential velocity increases.

One way to maintain high collection efficiencies with a moderate pressure drop is to use a large number of small cyclones placed in parallel. When multiple cyclones are used in parallel, each cylone can be of a smaller diameter than if only one cyclone was used. For a multicyclone assembly, the collection efficiency for each individual cyclone is identical, but it is greater than if only one larger cyclone was used.

Standard Cylone Configuration

The cyclone removal efficiency for a given particle size is largely dependent on cyclone dimensions. Extensive research has been performed to determine how relative cyclone dimensions affect particle collection efficiency. Some general observations about cyclone design include:

- Pressure drop at a given volumetric flow rate is most affected by cyclone diameter.
- The overall length of the cyclone determines the number of turns of the gas stream, and the greater the number of turns, the greater the collection efficiency.
- As the size of the cyclone inlet decreases, the inlet velocity increases, thereby increasing particle collection efficiency, but also increasing pressure drop.

TABLE 13.4.1   Standard Cyclone Dimensions*

Description	Conventional[†] (lapple)	High throughput[‡]	High efficiency[¶]
Body diameter (D)	1.0 D	1.0 D	1.0 D
Height of inlet (H)	0.5 D	0.75 D	0.5 D
Width of inlet (W)	0.25 D	0.375 D	0.2 D
Diameter of gas exit (D_e)	0.5 D	0.75 D	0.5 D
Length of vortex finder (S)	0.625 D	0.875 D	0.5 D
Body length (L_b)	2.0 D	1.5 D	1.5 D
Cone length (L_c)	2.0 D	2.5 D	2.5 D
Diameter of dust exit (D_d)	0.25 D	0.375 D	0.375 D

* Adapted from Cooper, C. D. and Alley, F. C., 1994.
[†] Adapted from Lapple, 1951.
[‡] Adapted from Stairmand, 1951.
[¶] Adapted from Swift, 1969.

Several standard cyclone configurations have been proposed to make design calculations easier (Table 13.4.1). Cyclones are designed with geometric similarity, such that the ratio of the dimensions of the cyclone remains constant and those dimensions are expressed in terms of the cyclone body diameter (D). In addition to the conventional cyclone (also referred to as the Lapple standard conventional cyclone), standard designs have been developed for high throughput, high efficiency, and ultra-high efficiency cyclones. Performace data for a variety of cyclones can be obtained from Heumann (1997) and from equipment manufacturers. Typical collection efficiency curves for conventional, high-efficiency and high-thoughput cyclones are presented in Figure 13.4.2.

The collection effeciency for a cyclone of standard proportions can be obtained using a calibration curve or a curve-fitted equation. Cyclone calibration curves typically plot collection efficiency (η) as

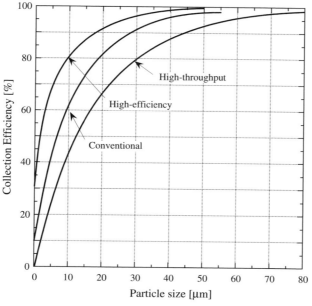

FIGURE 13.4.2   Partical collection efficiency for standard cyclones.

a function of $d_p/d_{p,50}$, where $d_{p,50}$ is the particle diameter that is collected at $\eta = 0.50$ (50 percent). As shown in the equation below, $d_{p,50}$ can be calculated given the geometry of the cyclone, density of the particles (ρ_p), gas viscosity (μ_g), and gas volumtric flow rate (Q_g):

$$d_{p,50} = \left(\frac{9\mu_g W^2 H}{2\pi\rho_p \, Q_g \, N_e} \right)^{1/2} \tag{13.4.1}$$

where N_e represents the number of revolutions of the gas stream in the main outer vortex and can be calculated approximately from:

$$N_e = \frac{1}{H} \left[L_h + \left(\frac{L_c}{2} \right) \right] \tag{13.4.2}$$

The $d_p/d_{p,50}$ values can then be determined for the particle size distribution of interest. Values of $\eta(d_{pi})$ can then be read from a calculation curve for the calculated values of $d_p/d_{p,50}$, which are available in air quality handbooks and from equipment manufacturers, or by using a curve-fitted equation. Theodore and De Paola (1980) fitted an algebraic equation to the Lapple standard conventional cyclone particle collection efficiency curve, making calculations much more convenient. The collection efficiency for a Lapple cyclone at any given particle size can be expressed as:

$$\eta_i = \frac{1}{1 + (d_{p,50}/d_{pi})^2} \tag{13.4.3}$$

where η_i is the collection efficiency for a particle of size d_{pi} for a standard coventional cyclone, and d_{pi} is the diameter of the i-th particle size range.

Because emission limits for particulate matter are usually specified in terms of the total amount of particulate material released, it is useful to calculate the overall collection efficiency, η_T, for a distribution of particle sizes. The following formula for total collection efficiency can be used for any type of particulate control device:

$$\eta_T = \sum_i \frac{\dot{m}_{in}(d_{pi})}{\dot{m}_T} \eta(d_{pi}) \tag{13.4.4}$$

where η_T is the total collection efficiency for the control device, $\dot{m}_{in}(d_{pi})$ is the mass of particles at size d_{pi} entering the pollution control device, $\dot{m}_T$ is the total mass of particles entering the pollution control device, and $\eta(d_{pi})$ is the collection efficiency for d_{pi}.

Cyclone Pressure Drop

Besides collection efficiency, the other major design consideration for cyclones is pressure drop. A high pressure drop will increase the operating cost of a cyclone, because the fan will have to perform more work. In general, higher particle collection efficiencies are obtained by forcing the gas through the cyclone at higher velocities, resulting in increased pressure drop. Although cyclones are relatively easy to design, maintain, and install, and they are typically regarded as a low capital cost device, excessive pressure drop may make a cyclone prohibitively expensive when compared to other collection devices. This economic trade-off must be considered in the design process.

Many equations have been developed to estimate the number of velocity heads or pressure drop in cyclones. An equation developed by Shepard and Lapple (1940) provides a reasonable estimate of the pressure drop across a cyclone. The pressure drop is presented as a function of K, which is an empirical constant and depends on cyclone configuration and operating conditions. K can vary considerably, but for standard tangential or involute cyclones, K is in the range of 12 to 18, and for vane-axial cyclones, K is usually taken to be 7.5. The pressure drop can be calculated from

$$\Delta P = \frac{u_g^2 \, \rho_g}{2} K \frac{HW}{D_e^2} \tag{13.4.5}$$

where ΔP is the pressure drop [N/m² or Pa], u_g is the superficial gas velocity at the cyclone inlet [m/s], ρ_p is the gas density [kg/m³], and H, W, D_e are the cyclone dimensions as described in Table 13.4.1.

13.4.3 PARTICULATE SCRUBBERS

Particulate or wet scrubbers contact a particle laden gas stream with a liquid spray (Figure 13.4.3). Particles are collected in wet scrubbers by liquid drops, or by being impacted on wetted surfaces. Soluble gases can also be removed by liquid droplets, and the gas temperature is lowered while the gas is humidified. Wet collection of particles has several advantages when compared to dry particle removal methods, including removal of sticky particles, removal of liquids, ability to handle hot gases, simultaneous gas and particle removal, and a reduced risk of dust explosion. Wet scrubbers have been used to control particles and gases from a variety of sources, including medical, hazardous, and municipal waste incinerators, industrial boilers, acid plants, and lime kilns.

Types of particulate scrubbers include spray tower, cyclonic, and venturi. Counter-current spray towers typically operate with scrubber droplets traveling downward and the gas stream containing the particulate matter traveling upward. Cyclonic scrubbers atomize droplets of water with a spray bar located along the centerline of the cyclone. These droplets then collect particles as they are transported to the outer edge of the cyclone. The liquid also allows cleansing of the walls of the cyclone. Venturi scrubbers work by accelerating the gas stream through a constricted duct to velocities of about 50–150 m/s. As the gas stream is flowing through the venturi, liquid is injected into the beginning of the venturi or at the throat entrance of the venturi. The liquid is then atomized because of the high velocity gas stream flowing past the inlets for the fluid. The high relative velocity between the scrubber droplets and particles allows for impaction of the particulate contaminants onto the scrubber droplets. Particulate material then becomes part of the scrubber droplets allowing much easier removal of the contaminants from the gas stream because of the large particle size of the scrubber droplets. Scrubber droplets are typically removed from the gas stream using a centrifugal separator, such as a cylone. The primary

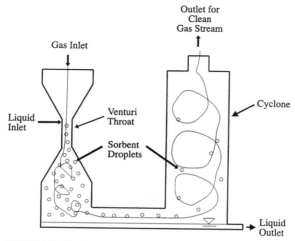

FIGURE 13.4.3   Venturi particle scrubber with cyclone.

TABLE 13.4.2 Typical Operating Parameters for Particulate Wet Scrubbers

Type of scrubber	Pressure drop	Liquid to gas ratio	Liquid inlet pressure (gauge)	Particle cut diameter
Spray tower	1.2–20 mbar (0.5–8 in. H_2O)	0.07–2.7 Lpm/m^3 (0.5–20 gpm/1000 acfm)	0.7–28 bar (10–400 psig)	2–8 μm
Cyclonic	3.7–25 mbar (1.5–10 in. H_2O)	0.3–1.3 Lpm/m^3 (2–10 gpm/1000 acfm)	2.8–28 bar (40–400 psig)	2–3 μm
Venturi	12.4–124 mbar (5-50 in. H_2O)	0.4–2.7 Lpm/m^3 (3–20 gpm/1000 acfm)	0.07–1 bar (1–15 psig)	0.2 μm

design parameters for wet scrubbers are pressure drop, particle size distribution, and liquid to gas flow rate. Typical operating characteristics for particulate wet scrubbers are presented in Table 13.4.2.

Although particles can be removed in wet scrubbers using diffusion, interception, and impaction, wet scrubber models, in particular, venturi scrubber models, assume that the dominant particle collection mechanism is impaction. The optimal scrubber droplet diameter is about 500–1000 μm when considering impaction.

The particle penetration of a venturi scrubber (Figure 13.4.3) can be calculated using an equation developed by Calvert et al. (1972):

$$Pt = \exp\left(\frac{6.3 \times 10^{-4} \rho_L \, \rho_p \, K_c \, d_p^2 \, u_g^2 \left(\frac{Q_L}{Q_G}\right) f^2}{\mu_g^2} \right) \tag{13.4.6}$$

where Pt is the particle penetration through the venturi scrubber, ρ_L is the density of the liquid [g/cm^3], ρ_p is the density of the particles [g/cm^3], K_c is the Cunningham correction factor, d_p is the particle diameter [cm], u_g is the gas velocity [cm/s], Q_L/Q_G is the liquid to gas flow rate ratio, and μ_g is the gas viscosity [g/cm-s]. The value f is an experimental coefficient that varies from 0.1 to 0.5 and is usually taken to be 0.25 for hydrophobic particles and 0.5 for hydrophilic particles. The collection efficiency is then equal to one minus the particle penetration.

The pressure drop across a venturi scrubber can be estimated using the following equation:

$$\Delta P = 1.03 \times 10^{-3} u_g^2 \left(\frac{Q_L}{Q_G} \right) \tag{13.4.7}$$

where ΔP is the pressure drop across the venturi [cm H_2O], u_g is the gas velocity [cm/s], and Q_L/Q_G is the liquid to gas flow rate ratio. Pressure drop in venturi scrubbers can be very high, and normally ranges from 25 to 125 mbar. An increase in pressure drop generally correlates to an increase in particle removal efficiency at smaller particle sizes.

Experimental data shows that a venturi scrubber is essentially 100 percent efficient in removing particles larger than 5 μm, so when designing a venturi scrubber, it is only necessary to examine the penetration of particles less than 5 μm. Disadvantages of using wet scrubbers include: (1) corrosion problems, (2) large amounts of liquid waste may be generated, (3) a potential for liquid freezing at low temperatures, and (4) the disposal of the waste sludge may be expensive.

13.4.4 FABRIC FILTERS (BAGHOUSES)

Fabric filters are often used in indoor air ventilation systems and to control particles from industrial gas streams (Figure 13.4.4). Fabric filters can achieve very high collection efficiency for a wide range of particle sizes. The collection efficiency of a properly operating fabric filter is often above 99.9 percent. A fabric filter operates by passing a particle laden gas stream through a media (filter) that allows

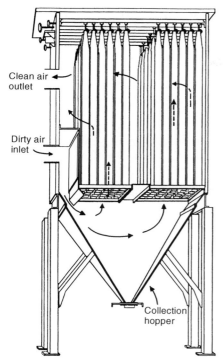

for penetration of the gas steam and the capture of the particulate material. The media can consist of a single sheet of woven material, such as fiberglass, cotton, or nylon. For high temperature applications, filters can be manufactured from stainless steel or ceramics. Cartridge filters, such as high efficiency particulate air (HEPA) filters, are often used in ventilation systems. The choice of fabric is dependent on the composition of the gas stream and the particulate material, gas temperature, the desired levels of particulate collection efficiency, and pressure drop.

Initially, the filter itself performs most of the filtration of the particles from the gas stream. As more particles are filtered from the gas stream, particle loading on the filter increases, forming a mat of particulate material referred to as a filter cake. The filter cake allows for more extensive particle filtration, mainly as a result of interception and impaction. This enhanced particle filtration is useful for achieving high particulate collection efficiencies, but as the thickness of the filter cake increases, so does the pressure drop. Eventually, the pressure drop will be prohibitively large, and the filter will need to be cleaned.

Mechanisms used to clean the filter cake from the filter include shaker, reverse air, pulse-jet, and the replacement of cartridge filters. Shaker and reverse air baghouses use bags with dimensions from about 6 to 18 in. in diameter and lengths up to 40 ft. Pulse-jet baghouses use bags with diameters of about 4 to 6 in. and lengths of about 8 to 10 ft.

Fabric filter blinding occurs when the fabric pores are blocked and the fabric cannot be cleaned effectively. Blinding can result when sticky particles adhere to the fabric, when moisture blocks the pores and increases particle adhesion, or because a high gas velocity deeply embeds the particles into the fabric.

Clean air outlet

Dirty air inlet

Collection hopper

FIGURE 13.4.4 Mechanical shaker baghouse.

There are many models that attempt to describe the amount of particulate collection efficiency that can be achieved by filtration. These models have a difficult time realistically describing the contribution of the filter cake to particulate collection efficiency, in part because of the irregular manner in which the filter cake develops. Models to predict collection efficiency will not be discussed here, because collection efficiencies for filters are typically very high (>99 percent).

To design a baghouse, the filtration velocity, the cloth area, the pressure drop, and cleaning frequency need to be estimated. The filtration velocity is expressed as:

$$v_f = \frac{Q}{A} \tag{13.4.8}$$

where v_f is the filtration velocity [m/min], Q is the gas volumetric flow rate [m³/min], and A is the area of the cloth filter [m²]. The air-to-cloth (A/C) ratio is defined as the ratio of the gas volume filtered to the cloth filter area [m³/sec/m² (ft³/min/ft²)]. Typically, shaker baghouses have A/C ratios of 2–6 ft³/min/ft², reverse air have A/C ratios of 1–3 ft³/min/ft², and pulse-jet have A/C ratios of 5–15 ft³/min/ft². Lists of recommended A/C ratios for various dusts, cleaning methods, and industries have been compiled by Noll (1999).

The following formula can be used to estimate the number of bags needed, given a process flow rate and a filtration velocity:

$$A = \frac{Q}{v_f} \tag{13.4.9}$$

The number of bags required in the baghouse can be determined using:

$$A_b = \pi d h \tag{13.4.10}$$

where A_b is the bag area, d is the bag diameter, and h is the bag height. Therefore, the total number of bags in the baghouse can be determined from:

$$N = \frac{\text{Total Cloth Area}}{\text{Bag Area}} \qquad (13.4.11)$$

The pressure drop across a baghouse is a function of the individual pressure drops of the filter and the filter cake. Darcy's law can be used to estimate the pressure drop across a filter and it is given as:

$$\Delta P_f = K_1 v_f \qquad (13.4.12)$$

where ΔP_f is the pressure drop across a clean fabric [Pa (in. H_2O)], K_1 is the fabric resistance [Pa/m-min (in. H_2O/ft-min)], and v_f is the filtration velocity [ft/min]. K_1 is a function of gas viscosity and filter characteristics, such as thickness and porosity. The pressure drop across the filter cake can be estimated from

$$\Delta P_c = K_2 C_i v_f^2 t \qquad (13.4.13)$$

where ΔP_c is the pressure drop across the filter cake [Pa (in. H_2O)], K_2 is the resistance of the filter cake [Pa-min-m/kg (in. H_2O-min-ft/lb)], C_i is the particle loading [kg/m^3 (lb/ft^3)], and t is the filtration time [min]. K_2 is determined experimentally using the particle loading, filtration velocity, and the pressure drop. Methods for estimating K_2 based on laboratory and pilot plant data are presented in Noll (1999). The total pressure drop across the filter and filter cake is then given as:

$$\Delta P_T = \Delta P_f + \Delta P_c = K_1 v_f + K_2 C_i v_f^2 t \qquad (13.4.14)$$

The filter drag is the resistance across the fabric-particle layer. It is a function of the particle loading on the filter and is given as:

$$S = \frac{\Delta P}{v_f} \qquad (13.4.15)$$

where S is the filter resistance (drag) [Pa-min/m (in. H_2O/ft/min)], and ΔP is the pressure drop across the filter and filter cake [Pa (in. H_2O)].

A typical filter performance curve for a single bag or compartment of a fabric filter baghouse is shown in Figure 13.4.5. The filter cake resistance (S) is plotted as a function of the aerial dust loading, where S_R is the residual drag in a single compartment when it is first brought back on-line after cleaning, S_T is the drag in a single compartment at the end of the filtration cycle, S_E is the effective

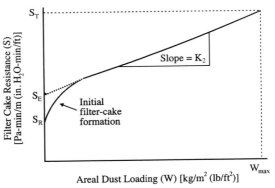

FIGURE 13.4.5   Dependence of filtercake resistance on aerial dust loading for a typical single compartment fabric filter.

residual filter drag, and W_{max} is the maximum particle loading on the fabric filter before it begins a cleaning cycle. The filter drag first increases exponentially and then linearly. The exponential portion of the curve is the period of cake repair and initial cake buildup. The slope of the straight line portion of the curve is equal to K_2 and represents the resistance to filtration through the filter cake. When the pressure drop across the compartment reaches the maximum allowable design pressure drop, the filter bags are cleaned. Once the bags have been cleaned, the pressure drop will decrease to its initial point, and the filter cycle will begin again.

The total pressure drop in a multicompartment baghouse is analogous to a set of electrical resistances in parallel. The total filter drag across a multicompartment baghouse can be expressed as:

$$\frac{n}{S_e} = \frac{1}{S_1} + \frac{1}{S_2} + \cdots + \frac{1}{S_n} \tag{13.4.16}$$

where $S_e = \Delta P/v_{avg}$ and is the total multicompartment baghouse drag at any time, v_{avg} is the average gas velocity, which is equal to the total volumetric flow rate divided by the total cloth area, n is the total number of filter compartments, and $S_1, S_2, \ldots$ are the individual drag components.

13.4.5 ELECTROSTATIC PRECIPITATORS

In an electrostatic precipiator (ESP) particles pass through a volume that has a large electrical potential ($\sim$50 kV) applied across a channel spacing of about 10 cm (Figure 13.4.6). This large electric field strength (electric potential/electrode displacement) causes the release of electrons from the discharge electrode. These high-energy electrons migrate away from the discharge electrode and impact with gas molecules causing ionization of the gas molecules and the release of more electrons. This generation of a large amount of free electrons is termed electron avalanche. Eventually, the free electrons are transported beyond the localized high field strength near the discharge electrode, causing a decrease in their kinetic energy. As the electrons decelerate, they attach themselves to particles instead of generating more free electrons.

The mechanisms responsible for deposition of electrons onto aerosol particles are diffusion and field charging. Diffusion charging dominates for particles with diameters smaller than 0.3 μm, and field charging dominates for particles larger than 0.3 μm. The basic design equations for an ESP ignore diffusion charging, and therefore, underpredict particle collection efficiencies for particles smaller than about 0.3 μm.

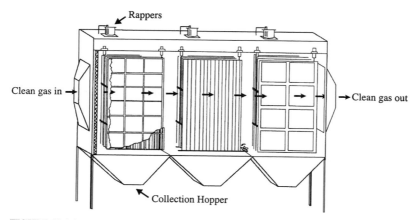

FIGURE 13.4.6 Flat-plate electrostatic precipitator (ESP).

ESPs are typically designed in one or two basic configurations: flat plate or tubular. Tubular ESPs are much less common than flat plate ESPs, and they are typically used to remove liquid droplets (e.g., sulfuric acid droplets) from gas streams. Flat plate ESPs are more commonly used (e.g., in coal-fired power plants), and they can be configured as a two-stage or a single-stage design. Two-stage ESPs charge the particles first and then remove them in a different section of the ESP. Single-stage ESPs charge the particles and then remove them in the same section.

ESPs are typically operated under turbulent flow conditions, and the particulate collection efficiency for the ESP can be described using the Deutsch-Anderson equation. Assumptions in the Deutsch-Anderson equation include constant migration velocity for a particle of a given diameter, uniform mass concentration in the y and z directions, no reintrainment of particles from the collection electrodes, no transport of particles above or below the collection electrodes, and a constant gas stream velocity in the x-direction (u_g = constant).

The Deutsch-Anderson equation is represented by:

$$\eta = 1 - \exp\left(-\frac{w_p\,A}{Q_g}\right) \tag{13.4.17}$$

where η is the collection efficiency for a flat-plate, turbulent flow ESP; w_p is the particle migration velocity; $A = 2nLH$ = total collection area for the particles; n is the number of channels; L is the length of the collection electrode; H is the height of the collection electrode; and Q_g is the volumetric flow rate of the gas.

The particle migration velocity (w_p) is sensitive to the electric field strength, particle composition, particle electrical resistivity, particle charging, and particle size. For particles of essentially homogeneous chemical composition, the migration velocity can be calculated using the following equation:

$$w_p = \frac{6.64 \times 10^{-18}\,E^2\,d_p}{\mu} \tag{13.4.18}$$

where w_p is the migration velocity [m/sec], E is the average electric field strength [V/m], d_p is the particle diameter [μm], and μ is the gas viscosity [kg/m-sec]. When collecting a range of particle sizes using an ESP, a migration velocity can be calculated for each particle size range of interest using Equation 13.4.18. A collection efficiency for each particle size can then be calculated using Equation 13.4.17, and an overall collection effeciency for the ESP can be calculated using Equation 13.4.4. When a gas stream contains particles of varying chemical composition, or with a distribution of particle sizes, the effective migration velocity (w_e) is often used to determine collection efficiency or specific collection area. Values for w_e obtained from field measurements are presented in Table 13.4.3 for various ESP applications.

An important design parameter for sizing an ESP is the specific collection area (SCA). The SCA is defined as

$$SCA = \frac{A}{Q} = -\frac{\ln(1 - \eta)}{w_e} \tag{13.4.19}$$

TABLE 13.4.3 Typical Field Measurement Effective Migration Velocities (w_e) for ESPs Operating at 90–95 Percent Efficiency

Application	w_e [cm/sec (ft/sec)]
Utility fly ash	4–20.4 (0.13–0.67)
Pulp and paper mills	6.4–9.5 (0.21–0.31)
Gypsum	15.8–19.5 (0.52–0.64)
Catalyst dust	7.6 (0.25)
Cement (wet process)	10.1–11.3 (0.33–0.37)
Cement (dry process)	6.4–7.0 (0.19–0.23)

Source: Noll (1999) and Oglesby and Nichols (1975).

For design purposes, the effective migration velocity, w_e, is often used when calculating SCA, because w_e is taken to represent the collection behavior of the total distribution of particles under a specific set of operating conditions. SCA is typically expressed in units of m^2 of collection area per m^3 of gas or ft^2 of collection area per 1000 acfm.

While the Deutsch-Anderson equation can theoretically predict the collection efficiency of an ESP under a certain set of conditions, uncertainties in the parameters can result in error by a factor of two or more. Some problems include the fact that the equation assumes a uniform gas flow rate, particles are not reentrained and do not sneak past the collection plates, and that particle size and composition are well known and invariable. With these failings of the Deustsch-Anderson equation, it is recommended that it only be used as an estimate. It is also recommended that empirical values from similar facilities be used whenever possible.

13.4.6   COMPARISON OF PARTICULATE CONTROL EQUIPMENT

The selection of a particulate collection device is based on required collection efficiency, capital and operating costs, pressure drop, operating temperature, gas flow rate, and particle loading. This chapter attempted to provide an overview of the basic types of particulate control equipment, their basic design and operating parameters, and calculation procedures. Although it is not possible to recommend a type of particulate control equipment for every type of application within this limited space, several general rules of thumb for choosing particulate control equipment are stated in Tables 13.4.4 and 13.4.5. Equipment manufacturers can provide information about current initial costs, as a function of variables such as total gas flow rate, particle loading, and materials selection. Cost spreadsheets developed by W. M. Vatavuk and the U.S. EPA Office of Air Quality Planning and Standards (OAQPS) are available from the EPA web site at http://www.epa.gov/ttn/catc/products.html#cccinfo.

TABLE 13.4.4   When to Use Specific Particulate Control Equipment

Cyclones	Fabric filters
• gas contains mostly large particles	• very high collection efficiencies required
• high dust loadings	• particles do not adhere to fabric (filter blinding)
• high efficiency for smaller particles not required	• gases will not condense
• as a pretreater for another particle control device	• relatively low gas temperature
• particle classification is desired	• gas volumes are reasonably low
Wet scrubbers	**Electrostatic precipitators**
• high particle removal efficiency for particles greater than 1 μm	• very high collection efficiencies required for small particles
• soluble gases, as well as, particles need to be removed	• very large gas volumes need to be treated
• humdification or cooling of gas stream is desired	• valuable material needs to be recovered
• gas stream contains a combustible gas mixture	• low pressure drop required
	• acceptable particle resistivity

TABLE 13.4.5   Operating Characteristics of Particulate Collectors

Type	Typical capacity	Pressure drop
Cyclones	2500–3500 ft^3/min/ft^2 of inlet area	2.5–40 mbar
Venturi scrubber	6000–30,000 ft^3/min/ft^2 of throat area	25–125 mbar
Fabric filter	1–6 ft^3/min/ft^2 of fabric area	5–15 mbar
Electrostatic precipitator	2–8 ft^3/min/ft^2 of collection area	0.5–1.3 mbar

REFERENCES

Buonicore, A. J., and Davis, W. T., *Air Pollution Engineering Manual*, Van Nostrand Reinhold, 1992.

Calvert, S., Goldschmid, J., Leith, D., and Mehat, D., "Wet Scrubber System Study," in *Scrubber Handbook—Vol I*, U.S. Department of Commerce, NTIS PB 213016, August 1972.

Cooper, C. D., and Alley, F. C., *Air Pollution Control: A Design Approach*, 2nd ed., Waveland Press, Prospect Heights, IL, p. 130, 1994.

Heinsohn, R. J., and Kabel, R. L., *Sources and Control of Air Pollution*, Prentice-Hall, 1999.

Heumann, W. L., *Industrial Air Pollution Control Systems,* NYC McGraw-Hill, 1997.

Lapple, C. E., "Processess Use Many Collector Types," *Chemical Engineering*, 58(5), May 1951.

Mycock, J. C., McKenna, J. D., and Theodore, L., *Handbook of Air Pollution Control Engineering and Technology*, CRC Lewis Press, 1995.

Noll, K., *Fundamentals of Air Quality Control Systems*, American Academy of Environmental Engineers, Annapolis, MD, pp. 170–171, 1999.

Noll, K. E., *Fundamentals of Air Quality Systems*, American Academy of Environmental Engineers, 1999.

Oglesby, S., and Nichols, G., "Electrostatic Precipitators," in *Gas Cleaning for Air Quality Control*, Marchello, J. and Kelly, J. eds., Marcel Dekker, New York, 1975.

Shepard, C. B., and Lapple, C. E., "Flow Patterns and Pressure Drop in Cyclone Dust Collectors," *Ind. Eng. Chem.*, 31(8): 1939; 32(9): 1940.

Stairmand, C., "The Design and Performance of Cyclone Separators," *Trans. Instn. Chem. Engrs.*, 29: 356, 1951.

Swift, P., "Dust Control in Industry," *Steam Heating Eng.,* 38: 453, 1969.

Theodore, L., and De Paola, V., "Predicting Cyclone Efficiency," *J. Air Pollution Control Assoc.*, 30(10), 1980.

U.S. EPA Air Pollution Control Costs, http://www.epa.gov/ttn/catc/products.html#cccinfo

U.S. EPA Emission Factors Database, http://www.epa.gov/ttn/chief

U.S. EPA Office of Air Quality Planning and Standards, http://www.epa.gov/airs/airs.html

Vatavuk, W. M., *OAQPS Control Cost Manual*, EPA 450/3-90-006, 4th ed., U.S. EPA OAQPS, 1990. Also available at http://www.epa.gov/ttn/catc/products.html#cccinfo

Wark, K., Warner, C. F., and Davis, W. T., *Air Pollution: Its Origin and Control*, 3rd. ed., Addison-Wesley, 1998.

CHAPTER 13
CONTROL TECHNOLOGIES

SECTION 13.5

CONTROL TECHNOLOGIES: GAS CONTROLS

Mark P. Cal

Dr. Cal is currently an assistant professor of environmental engineering at New Mexico Tech in Socorro, New Mexico. His main areas of research are air pollution control, indoor air quality, gas-phase adsorption, plasma processing of gas streams, and gas separation.

13.5.1 INTRODUCTION

The control of anthropogenic gaseous air pollutants is a common problem for many industries. The U.S. EPA has identified about 300 major source categories for hazardous air pollutants—all of which have potential regulated air pollutant emissions. Common air pollutants from stationary sources include CO, NO_x, SO_2, H_2S, and hydrocarbons. On a mass basis, about 90 percent of the air pollutants emitted in the United States are gases, with carbon monoxide contributing about 48 percent. In general, the concentrations of gaseous pollutants in effluent streams are relatively low, but they need to be reduced to some emission standard, which is regulated by the state or federal government. Gas streams may contain multiple pollutants and particles, which can complicate any potential gas cleanup method. This section will examine two of the most common unit operations for removing gas contaminants: absorption and adsorption. Other methods of reducing gas-phase pollutant emissions include the following: biofiltration, oxidation by catalyts or incineration, and destruction with gas-phase radical species generated by plasmas or ultraviolet light. Incineration will be covered in another section of this handbook, and the reader is referred to recent technical literature for some of the newer methods for controlling gaseous pollutants.

13.5.2 ABSORPTION

Absorption is a widely used unit operation for transfering one or more gas-phase inorganic species into a liquid. Absorption of a gaseous component by a liquid occurs because the liquid is not in equilibrium with the gaseous species. This difference between the actual concentration and the equilibrium concentration provides the driving force for absorption. Absorption can be physical or chemical. *Physical absorption* occurs when a soluble gaseous species (e.g., SO_2) dissolves in the liquid phase (e.g., water). In the case of SO_2 absorbing into water, the overall physical absorption mechanism is:

$$SO_2 \text{ in air} \leftrightarrow H_2O_{(l)} + H_2SO_3 + HSO_3^- + SO_3^{2-}$$

Chemical absorption occurs when a chemical reaction takes place in the liquid phase to form a new species. In the case of SO_2 absorbing into water containing calcium, the overall chemical absorption mechanism is:

$$SO_2 \text{ in air} \leftrightarrow Ca^{2+} + SO_3^{2-} + 2H_2O_{(l)} \leftrightarrow CaSO_3 \bullet 2H_2O_{(s)}$$

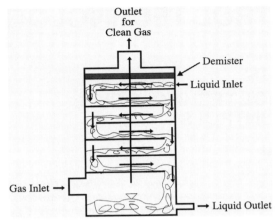

FIGURE 13.5.1 Plate tray tower absorber.

Several types of absorbers are used in practice with a packed tower absorber being one of the most common. *Spray towers* atomize droplets, typically consisting of water, at the top of the tower. The pollutant-laden gas in then passed up the tower, so that the droplets contact the gas stream in a counter-current manner. Soluble species in the gas phase then absorb into the liquid droplets, thereby tranferring them from the gas phase into the liquid phase. The liquid droplets are then collected at the bottom of the tower, and the clean gas passes out the top of the tower.

Plate/tray tower absorbers allow gas-liquid mixing as the gas flows upwards through the liquid and as the liquid flows horizontally over each plate or tray (Figure 13.5.1). A plate/tray tower absorber is usually operated in a counter-current (or cross-current) manner with the liquid flowing down the tower and the gas flowing up.

Bubble absorbers intimately mix the gas and liquid by bubbling the gas stream through a large container of liquid, often containing a dissolved species, such as Ca^{2+} (Figure 13.5.2). In the case of using a bubble absorber to remove SO_2 from a gas stream generated by a fossil-fuel combustor, dissolved SO_2 is converted into SO_4^{2-} in the liquid phase, which then reacts with dissolved Ca^{2+} to form a solid byproduct, $CaSO_4$ (gypsum). Calcium is added to the water in the form of $Ca(OH)_2$ (calcium hydroxide) or $CaCO_3$ (limestone).

Spray-dryer absorbers, often referred to as wet/dry absorbers, atomize slurry droplets into the gas stream. As the water evaporates from the droplets, soluble gases absorb into the droplets and react with dissolved species, such as Ca^{2+} formed from $Ca(OH)_2$ in the slurry droplets. Eventually, all of the

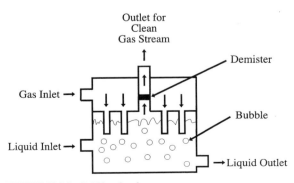

FIGURE 13.5.2 Bubbler absorber.

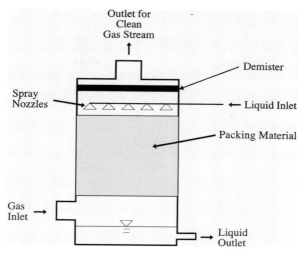

FIGURE 13.5.3   Counter-current packed tower absorber.

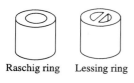

Raschig ring   Lessing ring

Berl saddle   Pall ring   Tellerette

FIGURE 13.5.4   Absorber packing materials.

water evaporates, leaving dry particles consisting of $CaSO_3$ and unreacted $Ca(OH)_2$. These particles are then removed with a particulate control device.

Most of the remaining discussion about absorbers will be focused on packed tower absorbers, because they are commonly used in a wide variety of absorption applications, and the design procedures are very well documented. *Packed tower absorbers* have an inert packing material with liquid flowing over the packing (Figure 13.5.3). The gas stream is then passed through the wet packing allowing a large amount of interfacial surface area between the gas and liquid. Packed tower absorbers can be operated co-currently or counter-currently, but counter-current operation is the most common. Common types of packing material are shown in Figure 13.5.4.

The basic model for mass transfer within an absorber is two-film theory. Two-film theory proposes that a mass transfer zone exists across a gas-liquid interface. The mass transfer zone is composed of two very thin films ($\sim$0.1 mm): a gas film and a liquid film. The theory assumes that there is complete mixing in both the gas and liquid bulk phases and that the gas-liquid interface is in equilibrium with respect to chemical species being transferred across the interface. Therefore, all resistance to molecular diffusion occurs when molecules are diffusing through the gas and liquid films.

The molar flux of species A diffusing across the gas-liquid interface can be expressed as follows. For the liquid phase:

$$N_A = k_L(C_{Ai} - C_{AL}) \tag{13.5.1}$$

$$N_A = k_x(x_{Ai} - x_{AL}) \tag{13.5.2}$$

For the gas phase:

$$N_A = k_G(P_{AG} - C_{Ai}) \tag{13.5.3}$$

$$N_A = k_y(y_{AG} - y_{Ai}) \tag{13.5.4}$$

where N_A is the molar flux of component A [gmol/m^2-sec], k is the interfacial mass transfer coefficient corresponding to the appropriate phase and driving force [gmol/m^2-sec-Pa], C_A is the liquid phase concentration of component A [mol/L], x_A is the liquid phase mole fraction of component A, y_A is the gas phase mole fraction of component A, and P_A is the partial pressure of component A in the gas phase. The mass transfer coefficients (k) for the gas and liquid phases, represents the resistance the solute encounters while diffusing through the gas or liquid film. In practice, the above equations for molar flux of species A across the gas and liquid films are difficult to use because of the small scale of the film thickness ($\sim$0.1 mm), where the concentrations would have to be measured. Therefore, overall mass transfer coefficients are used instead of interfacial mass transfer coefficients. Overall mass transfer coefficients describe the mass transfer system at equilibrium conditions by combining the individual film resistances into an overall resistance.

For a system exhibiting a linear equilibrium line, the molar flux of component A can be expressed as follows.

For the liquid phase:

$$N_A = K_L(C_A^* - C_{AL}) \tag{13.5.5}$$

$$N_A = K_x(x_A^* - x_{AL}) \tag{13.5.6}$$

For the gas phase:

$$N_A = K_G(P_{AG} - P_A^*) \tag{13.5.7}$$

$$N_A = K_y(y_{AG} - y_A^*) \tag{13.5.8}$$

where K is the overall mass transfer coefficient corresponding to the appropriate phase and driving force, C_A^* is the equilibrium liquid phase concentration of component A, P_A^* is the equilibrium gas phase partial pressure of component A, y_A^* is the equilibrium gas phase mole fraction of component A, and x_A^* is the equilibrium liquid phase mole fraction of component A.

Packed Tower Absorbers

When designing a packed tower absorber, it is of interest to determine the liquid to gas flow rate ratio (L/G), the pressure drop (ΔP), the height of the packing material (Z), and the cross-sectional area of the tower (A_{CSA}). Absorbers typically operate in a counter-current manner with the liquid entering from the top of the tower and the gas entering from the bottom. Applying a material balance around a packed tower absorber yields the following relationship (Figure 13.5.5):

$$G_{m,c}Y_1 + L_{m,s}X_2 = G_{m,c}Y_2 + L_{m,s}X_1 \tag{13.5.9}$$

where $L_{m,s}$ is the total liquid solvent molar flow rate [gmol/min], $G_{m,c}$ is the total carrier gas molar flow rate [gmol/min], X is the mole ratio of contaminant in the liquid phase [moles contaminant/moles liquid solvent], and Y is the mole ratio of the contaminant in the gas phase [moles contaminant/moles carrier gas]. The above relationship for the material balance assumes that the total mass of the gas and liquid streams does not change appreciably during the absorption process. This is typically the case for most air pollution control systems, because the mass flow rates of the contaminant are usually very small compared to the gas and liquid flow rates. This approach is generally considered valid for dilute concentrations ($<$1 percent by volume) of soluble contaminants in the gas phase.

Equation 13.5.9 can be rearranged to solve for Y_2, the mole ratio of the contaminant in the gas phase exiting the packed tower absorber:

$$Y_2 = \frac{L_{m,s}}{G_{m,c}}(X_2 - X_1) + Y_1 \tag{13.5.10}$$

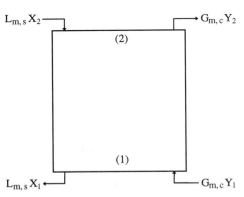

FIGURE 13.5.5 Material balance for a counter-current absorber.

This expression for Y_2 is called the operating line for a counter-current absorber. The operating and equilibrium lines for the system can be used to determine the liquid to gas flow rate ratio

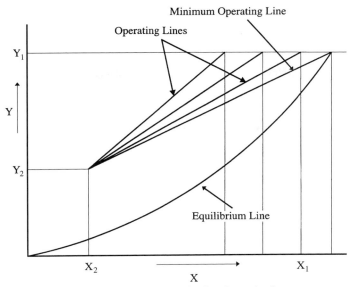

FIGURE 13.5.6 Minimum and actual operating lines for an absorber.

$(L_{m,s}/G_{m,c})$. For a typical absorption application, Y_1, Y_2, and X_2 are specified with X_1 depending on the liquid to gas ratio (Figure 13.5.6). The minimum liquid to gas ratio is determined by the slope of the operating line that passes through (X_2, Y_2) and the point where Y_1 intersects the equilibrium line (Figure 13.5.6). To satisfy the minimum liquid to gas ratio, the absorption tower would have to operate at equilibrium conditions at the inlet of the gas stream and at the outlet of the liquid stream. These conditions minimize the consumption of liquid, but it is impractical to achieve, because a very tall absorption tower would be required. In practice, the minimum liquid to gas ratio is multiplied by a design factor (ε_1) typically ranging from 1.3 to 1.7 (i.e., as a general rule, an absorber is typically designed to operate at liquid flow rates, which are 30 to 70 percent greater than the minimum rate):

$$\left(\frac{L_{m,s}}{G_{m,c}}\right)_{design} = \varepsilon_1 \left(\frac{L_{m,s}}{G_{m,c}}\right)_{min} \tag{13.5.11}$$

Pressure Drop in Packed Towers

Extensive experimental research has resulted in empirical correlations describing pressure drop dependence on the liquid and gas flow rates in a counter-current packed tower absorber (Figure 13.5.7). The uppermost line in Figure 13.5.7 represents the *flood point* of the tower. For a given packing size and type and a given liquid flow rate in an absorption tower, the pressure drop across the tower is a function of gas velocity. As the gas velocity is increased, liquid is retarded in its flow down the absorber column. As the gas flow rate is further increased, the quantity of the liquid holdup increases more rapidly, as does the pressure drop. Ultimately, the liquid will tend to fill the entire void space, and a layer of liquid will appear on top of the packing. At this point, the tower is said to be flooded, and it is marked by the term *flood point*. A packed tower should not operate at or near the flood point because of excessive pressure drop and poor utilization of the surface area of the packing material. As a rule of thumb, an absorption tower operates at gas velocities which are 40 to 70 percent of those that cause flooding.

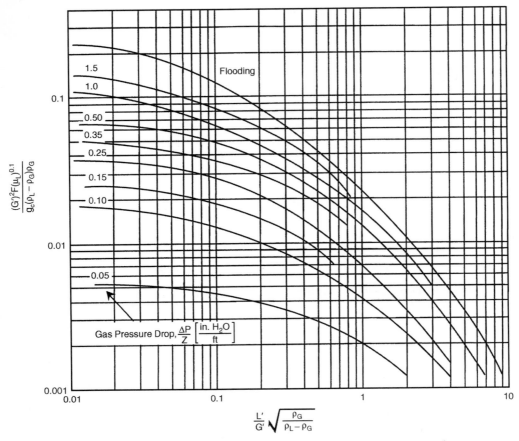

FIGURE 13.5.7   Flooding and pressure drop in random packed Towers. G' and L' are expressed in lb/s-ft², g_c has a value of 32.2 lb$_m$-ft/lb$_f$-s², μ_L has unit of centipose and ρ_L and ρ_G have units of lb$_m$/ft³. (*Source:* Wark et al. (1998) and the Norton Company).

The flooding condition also varies as a function of the packing factor F (Table 13.5.1). For packing material having an F-value greater than 60, the flooding condition can be taken as 2.0″ H₂O/ft packing. When F is in the range of 10 to 60 (packing size of 2 to 3 in.), the pressure drop associated with the flooding condition is expressed by

$$\Delta P_{\text{flood}} = 0.115 F^{0.7} \tag{13.5.12}$$

where ΔP_{flood} is the pressure drop at the flood point [in. H₂O/ft of packing].

The following procedure can be used to determine the required mass flux of the gas stream G' [lb$_m$/ft²-sec] using Figure 13.5.7:

1. $\left(\frac{L'}{G'}\sqrt{\frac{\rho_G}{\rho_L-\rho_G}}\right)$ can be determined because the cross-sectional area of the tower drops out of (L'/G').

2. The value of the terms describing the ordinate can then be determined at the flood point. All values for the variables should be known except for G'.

3. Solve for G', the gas mass flux that would occur at the flood point (G'_{flood}).

TABLE 13.5.1 Packing Factor F for Random Packing Materials

| | Nominal size (in.) | | | | | | | | |
Packing type	1/4	3/8	1/2	5/8	3/4	1	1.25	1.5	2
Raschig rings									
Ceramic	1600	1000	580	380	255	155	125	95	65
Metal (1/32 in. wall)	700	390	300	170	155	115			
Pall rings									
Plastic				97		52		40	25
Metal				70		48		28	20
Ceramic Intalox saddles	725	330	200		145	98		52	40
Berl saddles	900		240		170	110		65	45

Source: Wark et al. (1998) and the Norton Company.

4. The absorber should not operate at the flooding condition, but rather at some $G'_{design} < G'_{flood}$. So,

$$G'_{design} = \varepsilon_2\, G'_{flood} \tag{13.5.13}$$

where ε_2 is a flooding safety factor ranging from 0.4 to 0.7.

Once G'_{design} has been determined, the cross-sectional area of the absorption tower (A_{CSA}) can be determined using the following equation:

$$A_{CSA} = \frac{G}{G'_{design}} \tag{13.5.14}$$

This approach is valid for dilute concentrations (< 1 percent by volume) of soluble contaminants in the gas phase. If the absorber is used to treat a more concentrated gas stream, A_{CSA} should be determined at the top and bottom of the absorber and then the largest value for A_{CSA} should be used as the design value.

The pressure drop per unit height of packing material can now be determined by using G'_{design} and L'_{design} to calculate the terms along the abscissa and the ordinate of Figure 13.5.7. The pressure drop per unit height for the absorber can be determined by reading its value from the line of constant pressure drop per unit height of packing that intersects the lines originating from the values describing the abscissa and the ordinate.

The height of the absorption tower (Z) is equal to

$$Z = H_{tOG} N_{tOG} \tag{13.5.15}$$

where H_{tOG} is the overall height of a mass transfer unit based on the gas phase mass transfer resistance, and H_{tOG} is the overall number of mass transfer unit based on the gas phase mass transfer resistance. H_{tOG} is determined using correlations with the dimensionless Schmidt number (Sc,) and for dilute systems it is equal to

$$H_{tOG} = H_{tG} + H_{tL}\left(\frac{m\, G'_m}{L'_m}\right) \tag{13.5.16}$$

where H_{tG} is the height of a gas-phase transfer unit, and H_{tL} is the height of a liquid-phase transfer unit, H_{tG} and H_{tL} can be estimated from

$$H_{tG} = \frac{\alpha(G')\beta}{(L')^\gamma}\sqrt{Sc_g} \tag{13.5.17}$$

$$H_{tL} = \phi\left(\frac{L'}{\mu_L}\right)^\eta \sqrt{Sc_l} \tag{13.5.18}$$

TABLE 13.5.2 Constants for Use in Determining Gas Phase Height of a Transfer Unit (H_{tG}).

Packing type	α	β	γ	Gas flow rate (G') [lb/hr-ft^2]	Liquid flow rate (L') [lb/hr-ft^2]
Raschig rings					
3/8 in.	2.32	0.45	0.47	200–500	500–1500
1 in.	7.00	0.39	0.58	200–800	400–500
1 in.	6.41	0.32	0.51	200–600	500–4500
1.5 in.	17.30	0.38	0.66	200–700	500–1500
1.5 in.	2.58	0.38	0.40	200–700	1500–4500
2 in.	3.82	0.41	0.45	200–800	500–4500
Berl Saddles					
1/2 in.	32.40	0.30	0.74	200–700	500–1500
1/2 in.	0.81	0.30	0.24	200–700	1500–4500
1 in.	1.97	0.36	0.40	200–800	400–4500
1.5 in.	5.05	0.32	0.45	200–1000	400–4500

Source: Wark et al. (1998).

TABLE 13.5.3 Constants for Use in Determining Liquid Phase Height of a Transfer Unit (H_{tL})

Packing type	ϕ	η	Liquid flow rate (L') [lb/hr-ft^2]
Raschig rings			
3/8 in.	0.00182	0.46	400–15,000
1/2 in.	0.00357	0.35	400–15,000
1 in.	0.0100	0.22	400–15,000
1.5 in.	0.0111	0.22	400–15,000
2 in.	0.0125	0.22	400–15,000
Berl Saddles			
1/2 in.	0.00666	0.28	400–15,000
1 in.	0.00588	0.28	400–15,000
1.5 in.	0.00625	0.28	400–15,000

Source: Wark et al. (1998).

where α, β, γ, ϕ, and η are contstants for a given type of packing material and liquid and gas flow rates (Tables 13.5.2 and 13.5.3), μ_L is the dynamic viscosity of the liquid [kg/m-s (lb$_m$/ft-hr)], and Sc is the Schmidt number [dimensionless]. Values for Schmidt numbers of gas and liquids are given in Tables 13.5.4 and 13.5.5.

N_{tOG} can be calculated using one of several methods (Perry et al., 1997). For the case where the solute concentration is very low and the equilibrium line is straight, N_{tOG} can be determined from:

$$N_{tOG} \frac{\ln\left[\frac{Y_1-mX_2}{Y_2-mX_2}\left(1-\frac{mG_m}{L_m}\right)+\frac{mG_m}{L_m}\right]}{1-\frac{mG_m}{L_m}} \tag{13.5.19}$$

where m is the slope of the equilibrium line, G_m is the molar flow rate of gas [kgmol/hr], L_m is the molar flow rate of liquid [kgmol/hr], X_2 is the mole ratio of solute entering the column, Y_1 is the mole ratio of contaminant in entering gas, and Y_2 is the mole ratio of contaminant in exit gas. For the case when a solute dissociates in the liquid-phase, or when a chemical reaction occurs, the solute exhibits almost no partial pressure, and therefore, the slope of the equilibrium line (m) approaches zero. For

TABLE 13.5.4 Schmidt Numbers for Gases in Air at 25°C and 1 atm

Substance	Sc $(\mu/\rho D)$
Ammonia	0.66
Carbon dioxide	0.94
Water	0.60
Methanol	0.97
Ethyl alcohol	1.30
Benzene	1.76
Chlorobenzene	2.12
Ethylbenzene	2.01

Source: Wark et al. (1998) and Perry, R. H. (1997).

TABLE 13.5.5 Schmidt Numbers for Liquids in Water at 20°C

Substance	Sc $(\mu/\rho D)$
Ammonia	570
Carbon dioxide	570
Chlorine	824
Hydrogen chloride	381
Hydrogen sulfide	712
Sulfuric acid	580
Nitric acid	390
Methanol	785
Ethyl alcohol	1005

Source: Wark et al. (1998) and Perry, R.H. (1997).

these cases Equation 13.5.19 reduces to

$$N_{tOG} = \ln\left(\frac{Y_1}{Y_2}\right) \tag{13.5.20}$$

To determine N_{tOG} for more concentrated solutions, or for cases when the equilibrium line is not straight, a graphical method can be used (Figure 13.5.8). N_{tOG} can be estimated by drawing an intermediate line vertically equidistant between the equilibrium and operating lines ($A = B$ and $C = D$). Units are then counted off from the lowest point of the operating line (G) with horizontal distances, such that the horizontal distance between the operating line and intermediate line is equal to the length of the line surpassing the intermediate line ($E = F$). A vertical line is then drawn to the operating line and another step is drawn until point H is surpassed horizontally. Once N_{tOG} is

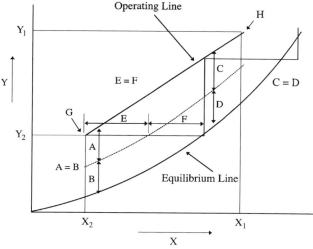

FIGURE 13.5.8 Schematic for graphical detzermination of the number of overall gas-phase transfer units (N_{tOG}).

determined, then the height of the column can be determined from

$$Z = H_{tOG} N_{tOG}$$ (13.5.21)

The total pressure drop of the tower can be determined from

$$\Delta P_{tot} = \left(\frac{\Delta P}{\Delta Z} \right) Z$$ (13.5.22)

13.5.3   ADSORPTION

An adsorber unit operation consists of a system containing one or more adsorbents and one or more adsorbates. The *adsorbent* is the solid adsorbing medium that provides active sites for the selective removal of gaseous contaminants (e.g., activated carbon, zeolites, and silica gel). The *adsorbate* is the gaseous material that can be selectively removed from the gas stream by the adsorbent. Adsorption is used for odor control, removal of hazardous air pollutants from a variety of industrial sources (chemical manufacturing, food processing, rendering plants, sewage treatment plants, pharmaceutical plants), purification of indoor air, dehumidification of gases, and the recovery of valuable solvent vapors. Adsorption is usually applied for pollution control of organic compounds, but some inorganic pollutants (SO_2, NO_x, H_2S) can also be adsorbed. In general, adsorption processes will work well for any organic compound with a molecular weight greater than 45. Adsorption is typically used when: (1) the pollutants are in dilute concentrations in the gas phase, making condensation uneconomical; (2) solvent vapors that wish to be recovered and resused; and (3) the pollutant is not combustible or difficult to burn. Once an adsorbent has been saturated with adsorbate, the adsorbent must be either regenerated or discarded.

There are two mechanisms for adsorption: physical and chemical. *Physical adsorption* occurs when gaseous molecules attach themselves to the surface of the adsorbent with an intermolecular attractive force (van der Waals force). Physical adsorption is reversible and exothermic with about 2 to 20 kJ per g-mole of adsorbed adsorbate (e.g., water on silica gel or benzene on activated carbon). Chemical adsorption occurs when gaseous molecules attach themselves to the surface of the adsorbent with valence forces. Chemical adsorption is typically irreversible and exothermic with about 20 to 40 kJ of energy released per g-mole adsorbed adsorbate (e.g., H_2S adsorption on activated carbon at high temperature [400–600°C]).

Adsorbents are highly microporous materials with most of their pores less than 2 nm. This high porosity creates a very large surface area (100–2000 m^2/g adsorbent) for adsorption. Adsorbents can be manufactured from a variety of materials, both organic and inorganic. Activated carbon, one of the most popular adsorbents, can be made from coal, wood, vegetable material, and polymers. For activated carbon, the pore size distribution can be tailored for a specific process by varying the carbonization and activation procedures.

Adsorption Isotherms

The amount of gas adsorbed per gram of solid adsorbent is a function of concentration (partial pressure) of the adsorbate, temperature of the system, and the properties of the adsorbate and adsorbent. In general, for physical adsorption, the amount of adsorbate adsorbed decreases with increasing temperature, and increases with increasing molecular weight or boiling point. Measuring the amount of a compound adsorbed on an adsorbent versus concentration or partial pressure at constant temperature results in an *adsorption isotherm*. Adsorption isotherms represent equilibrium conditions and they are useful for characterizing adsorbents with respect to different adsorbates (Figure 13.5.9). Adsorption isotherms have been produced for thousands of adsorbate-adsorbent combinations. Methods for predicting adsorption isotherms based on the properties of the adsorbate and adsorbent have also been developed

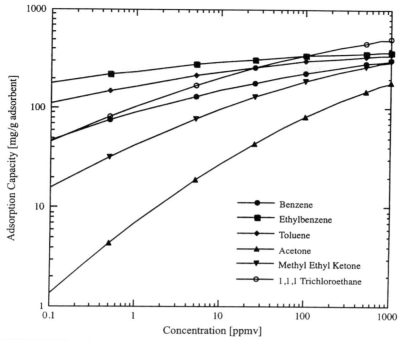

FIGURE 13.5.9   Adsorption isotherms for various organic species adsorbed onto activated carbon cloth. *Source*: Cal (1995).

(Cal, 1995; Noll, 1999; Vatavuk, 1990). While these predictive relationships are useful for developing estimates of adsorption capacity, laboratory or pilot plant data should be used whenever possible.

Many adsorption isotherm equations have been developed in the past 100 years. One of the most common isotherm equations in industrial use is the Freundlich equation. The Freundlich equation is an empirical expression used to describe adsorption isotherms where there is a linear response for adsorption capacity as a function of adsorbate concentration (or partial pressure) when this function is plotted on log-log scales. The valid concentration range for the Freundlich equation varies depending on the adsorbate-adsorbent combination. The Freunlich equation is expressed as:

$$C_e = \alpha X^{\beta} \quad \text{or} \quad X = kC_e^m \tag{13.5.23}$$

where α, β, k, and m are constants determined from the adsorption isotherm plot and are dependent on the units used to describe the adsorbate concentration; C_e is the equilibrium gas phase contaminant concentration; and X is the amount of adsorbate adsorbed per unit mass adsorbent at C_e [g/g]. Values for Freundlich constants have been compiled by Vatavuk (1990). When using the Freundlich equation, the isotherms should not be extrapolated outside of the ranges provided, because the isotherm may not behave linearly beyond that point.

Analysis of an Adsorber

It is useful to be able to describe how quickly an adsorption zone moves through the length of an adsorption bed and also to determine the length of the adsorption zone (Figure 13.5.10). This allows for the prediction of how long an adsorption bed should be used before it must be regenerated or replaced. The velocity of the adsorption zone (V_{ad}) is determined using a control volume approach

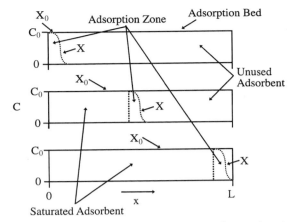

FIGURE 13.5.10   Schematic of adsorption zone as it moves through an adsorption bed.

and a material balance. The adsorbent is assumed to be flowing through a stationary adsorption zone with 100 percent of the adsorbate removed from the gas stream in the adsorption zone. The mass balance yields:

$$\frac{m_a C_o}{\rho_a} = \rho_{ad} A \, V_{ad} X_{sat} \tag{13.5.24}$$

where m_a is the mass flow rate of the carrier gas [kg/s]; ρ_a is the density of the carrier gas [kg/m³]; C_o is the inlet pollutant concentration [kg/m³]; ρ_{ad} is the apparent bulk density of the adsorption bed [kg/m³], taking into account void spaces; A is the cross-sectional area of the adsorption bed [m²], V_{ad} is the velocity of the of the bed as it passes through the adsorption zone [m/s], and X_{sat} is the saturation adsorption capacity corresponding to a concentration C_o [g adsorbate/g adsorbent].

Because the inlet concentration C_o and the saturated adsorbent concentration X_{sat} are related using experimental data, an adsorption isotherm equation can be substituted into Equation 13.5.24. If the Freundlich equation is substituted, the following equation is obtained for the mass balance on an adsorber:

$$\frac{m_a C_o}{\rho_a} = \rho_{ad} A \, V_{ad} \left(\frac{C_o}{\alpha} \right)^{1/\beta} \tag{13.5.25}$$

where β is unitless and α must have the same units as C_o. Solving for the velocity of the adsorption wave (V_{ad}) yields:

$$V_{ad} = \frac{m_a}{\rho_a \rho_{ad} A} \alpha^{1/\beta} C_o^{(\beta-1)/\beta} \tag{13.5.26}$$

In order to determine the thickness of the adsorption zone, a relationship between the pollutant concentration anywhere in the adsorption zone (C) and any distance within the adsorption zone (x). This is done by examining the mass transfer from the gas phase to the adsorbent within the adsorption zone. Defining the adsorption zone length such that C approaches 1 percent of its limiting values of 0 and C_o, then the length of the adsorption zone (δ) can be determined from (Wark et al., 1998):

$$\delta = \frac{m_a}{K \, A \rho_a} \left[4.595 + \frac{1}{\beta - 1} \ln \left(\frac{1 - 0.01^{\beta-1}}{1 - 0.99^{\beta-1}} \right) \right] \tag{13.5.27}$$

where K is the mass transfer coefficent $[s^{-1}]$, which can vary greatly, but is usually in the range of 5 to 50 s^{-1}. Assuming that the time required to establish the adsorption zone to its complete length at the adsorber inlet is zero, then the time to breakthrough can be calculated from:

$$t_B = \frac{L - \delta}{V_{ad}} \tag{13.5.28}$$

where L is the total length of the adsorption bed. Once breakthrough has been achieved, the adsorption bed should be regenerated or replaced. Because there is uncertainty in the calculation of the time to breakthrough, a safety factor of about 2 to 3 is recommended unless the effluent pollutant concentration is being monitored frequently to determine the actual point of breakthrough.

For most applications, it would be cost prohibitive to discard the adsorbent after one use, so the adsorbent is usually regenerated. Desorbed vapors are much more concentrated and can therefore be recovered more easily and economically than before the adsorption step. Several regeneration methods are available: steam, reduction of gas pressure (pressure-swing), heating the adsorption bed (thermal-swing), and passing a clean gas over the adsorption bed. Because adsorption is essentially an equilibrium-based processs, all regeneration methods rely on the same principle of shifting the adsorbate equilibrium so the the adsorbate leaves the adsorbent and returns to the gas phase. After regeneration, the adsorption capacity of the adsorbent may decrease, because regeneration processes are usually not complete, and some adsorbate is retained within the adsorbent pores after the regeneration process.

If the system is designed properly, adsorbers can have high collection efficiencies for gaseous contaminants, even at very low gas-phase concentrations. The lifetime of the adsorbent is usually measured on the order of years for processes that use regeneration. Adsorbers may fail to perform well in humid environments, because water vapor can compete for adsorption sites. When water vapor competition is probable, the gas stream should first be dried before entering the adsorption bed, or a hydrophobic adsorbent should be used to minimize competitive adsorption effects. One possible disadvantage of using adsorption is that the adsorbent material is usually tightly packed in the adsorber, meaning that pressure drop can be significant, which leads to increased operating costs.

13.5.4 SUMMARY

Absorption and adsorption are two well-established unit operations for removing contaminants from the gas-phase. Absorption is typically used to remove soluble inorganic pollutants, while adsorption is typically used to remove low concentrations of organic vapors. The presence of multiple contaminants in the gas stream will complicate the design of either unit operation. For air pollution control applications, absorber or adsorber design usually requires reducing pollutant levels below some emission standard, so the inlet and outlet pollutant concentrations are known. Design requirements then include sizing the unit and determining reasonable gas and/or liquid flow rates. This section has covered the basics needed to design both absorption and adsorption units, but data from equipment manufacturers and/or laboratory or pilot-scale data should be used whenever possible. Information on providing costs estimates for absorbers or adsorbers can be obtained from equipment manufacturers or from the U.S. EPA web site http://www.epa.gov/ttn/catc/products.html#cccinfo and Vatavuk (1990).

REFERENCES

Buonicore, A. J., and Davis, W. T., *Air Pollution Engineering Manual,* Van Nostrand Reinhold, 1992.

Cal, M. P., "Characterization of Gas Phase Adsorption Capacity of Untreated and Chemically Treated Activated Carbon Cloths," *Doctoral Dissertation,* University of Illinois at Urbana-Champaign, 1995.

Cooper, C. D., and Alley, F. C., *Air Pollution Control: A Design Approach,* 2nd Ed., Waveland Press, Prospect Heights, IL, p. 130, 1994.

Heinsohn, R. J., and Kabel, R. L., *Sources and Control of Air Pollution,* Prentice-Hall, 1999.

Heumann, W. L., *Industrial Air Pollution Control Systems,* McGraw-Hill, 1997.

Mycock, J. C., McKenna, J. D., and Theodore, L., *Handbook of Air Pollution Control Engineering and Technology,* CRC Lewis Press, 1995.

Noll, K., *Fundamentals of Air Quality Control Systems,* American Academy of Environmental Engineers, Annapolis, MD, pp. 170–171, 1999.

Perry, R. H., Green, D. W., and Maloney, J. O., eds., "Chapter 14: Gas Absorption and Gas-Liquid System Design" in *Perry's Chemical Engineers' Handbook,* 7th Ed., McGraw-Hill, 1997.

Treybal, R. E., *Mass Transfer Operations,* McGraw-Hill, 1980.

U.S. EPA Air Pollution Control Costs, http://www.epa.gov/ttn/catc/products.html#cccinfo

U.S. EPA Emission Factors Database, http://www.epa.gov/ttn/chief

U.S. EPA Office of Air Quality Planning and Standards, http://www.epa.gov/airs/airs.html

Vatavuk, W. M., *OAQPS Control Cost Manual,* EPA 450/3-90-006, 4th Ed., U.S. EPA OAQPS, 1990. Also available at http://www.epa.gov/ttn/catc/products.html#cccinfo

Wark, K., Warner, C. F., and Davis, W. T., *Air Pollution: Its Origin and Control,* 3rd. Ed., Addison-Wesley, 1998.

CHAPTER 13
CONTROL TECHNOLOGIES

SECTION 13.6

FUNDAMENTALS OF ODOR MANAGEMENT

David A. Hill

*Mr. Hill is president of Global Odor Control Technologies, Inc.,
and a partner in Global Odor Control Technologies, LLC. He
works primarily with composting operations and animal
production operations.*

13.6.1 INTRODUCTION

There was a time when serious conflicts between operations that generate "nuisance odors," and citizens negatively impacted by those odors were rare. Typically, such operations and facilities were located in relatively remote, sparsely populated locations—away from the mainstream of the various communities they served. On-going expansion of society has resulted in residential and commercial "encroachment" into areas that were once safe zones for agricultural, industrial, and waste processing applications. The siting of these essential operations has become complex and difficult. Safe locations are only temporarily safe. Conflicts are inevitable. Nevertheless, the existence of facilities for the processing of foods and wastes will continue to be necessary. Consequently, concerns about odor pollution and the management of nuisance odors will continue to gain prominence in the siting, zoning, and permitting of new facilities and also in the continuing operation of existing facilities.

On-going research in biochemistry and toxicology is revealing evidence that low-level, recurring exposure to certain odorous gases may be causal in the creation of a complex of chronic physical symptoms and behavior patterns that negatively impact quality of life. As these studies proceed, the traditional view of odors as aggravations and nuisances may have to be re-evaluated and expanded in terms of public health and safety.

Social and population factors, regulatory factors, public health factors, and private liability factors will all fuel the need for increased knowledge of how odors are formed, how they are perceived, and what can be done to prevent or diminish their detrimental impact on modern life. Odor problems will become the social and financial responsibility of the generator. The modern protocol should be as follows:

1. Prevent odors when possible.

2. Minimize odors when prevention is not possible.

3. Minimize, by treatment, any remaining odors effectively and efficiently.

Thus, prevention, minimization, and treatment form the basis for the development of an odor management system. Historically, "odor control" has been limited to attempts to mechanically treat odors after formation but prior to or during exhaust. Modern odor management must give equal consideration to methods of arresting or limiting initial formation of odors.

Advances in engineering, mechanics, chemistry, and biology will help make this consideration viable and plausible. The majority of problem odors are the result of biological processes. Changes in process temperature, moisture, oxygen levels, timing, equipment speed, or any of a host of other factors may influence odor prevention. Once the source of an odor is determined, implementation of mechanical, chemical, biological, or biochemical methodologies can result in prevention and/or minimization of the odors. Factors such as changing retention time for organic materials, or changing the timing or location of an exhaust, can minimize impact on the public. When treatment is required, whether prior to, during, or after exhaust, there are both accepted and developing methodologies to consider when selecting treatment options.

In this brief overview, a limited description of many methodologies will be included. Additionally, an outline for a system of odor management will be developed. The management system will allow for as little or as much complexity as is needed to efficiently deal with each situation. Prior to these undertakings, however, a fundamental framework must be established, which includes a working understanding of what odors are, how we perceive them, and how we measure and quantify them. This work will not be a reference catalog for existing odors or their chemical properties, nor can it attempt to deal with the majority of specific situations that result in common odor formation and complaint. It can, however, provide a discussion of existing understanding with regard to odor assessment, measurement, and treatment. This work does not include discussions regarding treatment of hazardous odorous gases. Nor does it consider or examine exposure to gases in concentrations that require special health considerations or that pose special health risks. For information in these areas, the reader should contact the National Environmental Health Association (NEHA) or the National Institute for Occupational Health and Safety (NIOSH).

13.6.2   UNDERSTANDING ODOR

Definitions

Webster's New World Dictionary defines odor as "that characteristic of a substance which makes it perceptible to the sense of smell." This is to say that any discernable smell, whether pleasant or unpleasant, desired or despised, is an odor. Any compound possessing those characteristics that stimulates the olfactory system is called an "odorant." Modern language qualifies and classifies odors into subcategories such as aromas, fragrances, and stenches. These terms carry various implications and degrees of pleasure or revulsion. Pleasant connotations are created by the phrase "the fragrance of a rose." It is unlikely that we would use the same term to seriously describe the smell of a skunk. The aroma of brewing coffee is pleasant for most of us, but to some it is a disturbing stench. Any odor in great enough concentration, becomes cloying and unpleasant.

Subjectivity in Perception

To some extent, the perception of odor is disparate and subjective. In a single group of subjects, wide variation may occur when determining the concentration levels at which a particular odor becomes noticeable, at which a particularly unpleasant odor becomes a nuisance, and at which a particularly pleasant odor becomes disturbing. Placing 10 subjects individually in an 8 × 10 room with 100 fresh flowers will yield descriptions of "cloying stench" from some subjects to "sweet perfume" from others. To understand this variation in perception and the subjectivity it implies, a basic understanding of how our olfactory system operates is required.

The Olfactory System

We perceive or smell odors in the gaseous state, as compounds produced in vaporous emissions from various substances and combinations of substances. These gases are produced in chemical and biochemical reactions that occur naturally or are the result of manufacturing and treatment methods. The gaseous molecules enter our nasal passages through inhalation and proceed through the passages to the olfactory cleft just behind the upper nose. The cleft is covered with a supply of cells, which are called receptors or odor receptors. Each odorous molecule enters a receptor site based on the molecule's particular shape. Each receptor is designed to accept and process certain shapes of molecules, and to forward information to the brain accordingly. At least seven particular shapes of receptors are currently identified. (It is theoretically possible that the final knowledge of the number of different shapes of receptors in existence will provide the knowledge of how many different shapes or "types" of odors exist. It may well be that knowledge of the various molecular types will eventually provide key information for the decomposition or conversion of unpleasant odors into pleasant ones, or toward the conversion of unpleasant molecules into shapes that are nonodorous or at least not processible by the olfactory system.)

In the receptor, information is obtained from the odorous molecule by its reaction with enzymes produced in the mucous of each receptor cell. The enzyme of each shape of receptor generates a slightly different set of electrical signals from those produced by any other shape. These signals are further differentiated based on the particular molecule that enters the receptor site. The signals are forwarded to the brain, providing a categorization and a description of the odor. With repetition of exposure to a particular odor comes repetition of the corresponding set of signals to the brain. This repetition leads to identification whenever exposure to that particular odor occurs. Attached to that identification will be emotional and physical responses associated with past experiences regarding the odor.

The brain actually allows us to recreate and imagine the odor so accurately that we can "smell" the odor even when it is not present. Imagine hot chocolate, strawberry pie, roast turkey, skunk, or sewage sludge. Most of us respond to these words by initiating a set of signals to the brain that momentarily creates the odor and the remembered perceptions surrounding it. Someone who has had a very unpleasant or negative experience associated with a particular odor—even an odor typically regarded as pleasant and inoffensive—may react adversely to exposure to that odor, or even to discussion that causes neural "perception" of the odor. Obviously the receptor cells are 'the vital component' in our perception of odor. They are also the fundamental reason for disparity in perception levels and perceptive subjectivity among humans.

Causes of Olfactory Perception Variation

There is no fixed number of receptors common to human beings. Some people have more than others. There is no fixed number of each *type* of receptor common to human beings. Some people perceive certain odors more intensely than others because they have more of a particular type of receptor. The ability of the receptors to function and process molecules is limited by the size and condition of the nasal passages, which vary from body to body, and within the body due to allergies, viral infection, and other factors. The ability of the receptors to process any given odor is determined by the number and type of odor molecules in the system at any given time and the numbers of corresponding receptors available. In other words, if a particular individual does not have sufficient receptor cells of a specific shape available at a particular time, a given odor of that specific molecular shape may be only partially perceived by that person.

Summary

Variations from subject to subject in reaction to odors or groups of odors is essentially the result of physical variations (either temporary or permanent) in the olfactory systems of the subjects involved, *and to emotional and historic responses to given odors in each subject's individual experience.* This should not be considered a major impediment to determining standards for odor evaluation and regulation. In practical application, any reasonable majority of a group reacting negatively to an odor should be sufficient justification to manage that odor. Additionally, an immediate lack of negative reaction to a given odor may not be an accurate indication of long-term impact, as certain gases may be damaging over extended exposure periods at levels below nuisance perception. Other gases become imperceptible above certain levels or after prolonged exposure because of their damaging or fatiguing impact on the nasal passages and/or the mucous lining of the receptor. (Olfactory fatigue must be differentiated from olfactory or neural tissue damage. Olfactory fatigue is an easily remedied desensitizing, which occurs from over-exposure to a particular odor. Olfactory tissue damage may occur from over-exposure to a dangerous gas such as ammonia or hydrogen sulfide. Olfactory tissue damage disables the ability to perceive by olfaction and often precedes more serious damage or even death.)

13.6.3   QUANTIFICATION AND QUALIFICATION OF ODORS

Functional System

Because of the subjectivity and complexity involved in dealing with odors, an accepted system of standards for the qualification as well as the quantification of odors has developed. This system does not primarily consider the chemical identification and measurement of the component gases. Instead, the system considers the perceived strength and intensity, resistance to dissipation, public connotations, and personal associations with a given odor as a whole. The system categorizes odor by means of four qualities. These are as follows:

1. The "character" of the odor
2. The intensity of the odor
3. The hedonic tone of the odor
4. The detectability of the odor

Character.   The *character* or quality of the odor describes the public or "universal" connotations associated with the recognition of a given odor. As an example, the odor of a strawberry is very recognizable and distinguishable. It is considered pleasant, and its character would be labeled nonoffensive and fragrant. On the other hand, the smell of a skunk is also very recognizable and distinguishable. Its character would be considered offensive and unpleasant. This part of the system does not examine the effects of quantity or concentration on the perception of the odor, only the recognized reaction to the identification of the odor. These reactions may be qualified using any of a number of accepted adjectives, including sour, sweet, fruity, pungent, putrescent, floral, burnt, metallic, citric, and so forth.

Intensity.   The "intensity" of the odor refers to concentration of the odor but not to specific quantities of component gases. Intensity represents an attempt to measure human response. This is accomplished by an arbitrary and subjective valuation of the strength of the odor on a scale of either zero-to-five or zero-to-eight (butanol). On the "zero-to-five" scale, zero represents a condition of no odor and five a very strong odor. On the Butanol Scale, zero also represents no odor, five to six a very strong odor, and seven to eight represent extreme odor. Two and four represent slight and moderate on the Butanol Scale, slight and moderately strong on the "zero-to-five" scale. Use of this scale is an attempt to quantify the objectionability of a given odor. Using our previous examples, a bowl of strawberries would probably be considered a one on the scale. However, a thousand crates of strawberries in a single room might produce an intense enough odor to be considered a "five" or "six," even though its character was still not objectionable. The skunk will be intense enough to be labeled a "six," "seven," or even "eight."

Hedonic Tone.   Hedonic tone is the category that attempts to consider personal associations and differences in perception assigned to given odors. Using our strawberry example, imagine a person who associates some very negative past experience or event with strawberries. For this person, the smell of strawberries may indeed be objectionable and a nuisance, initiating unpleasant and disturbing memories each time it is perceived. While this example may seem extreme, responses to given odors often vary by extremes. Hedonic tone is one method of determining any personal factors affecting general perception and associations with a given odor. Evidence strongly suggests that hedonic responses in an individual may include a set of physical responses to a particularly disturbing odor. For example, symptoms of anxiety such as shortness of breath, chest pains, and perspiration are typical.

Detectability.   The fourth category, *detectability*, is the most commonly known and accepted portion of this system. Detectability refers to how readily an odor is perceived. It attempts to accomplish this by quantifying the number of times a given odor must be diluted with an equal amount of fresh air to reduce its detectability (not objectionability) to 50 percent of the people exposed to the odor. Each dilution is called a "dilution to threshold." The number of times the air must be diluted produces a measurement of odor units associated with a given odor. This method is widely used in setting limits for generation in potentially odorous situations, and even for determining acceptability of certain odorous materials at processing facilities. In some countries, such as New Zealand, odor limits and responsibilities are proscribed by law, and are determined by odor units or dilutions to threshold.

Measurement of D/T Ratios.   Many specific odors and individual gases have been measured by this method, and accepted tables of dilution to threshold ratios (or D/T ratios) are readily available (see Table 13.6.1). There are two typical methods for determining this value. The first is usually accomplished in a controlled environment separate from the generation site. A sample of the odorous gas is collected by pump, vacuum tube, or other method. A specific volume of the gas is then treated by the introduction (often by syringe) of an equal volume of fresh air. The resultant combination is offered to a panel of participants for detectability. Dilutions continue until no more than 50 percent of the panel can detect the existence of the odor. The number of dilutions required determines the

TABLE 13.6.1   Threshold Levels of 25 Common Odorants*

Odorant	Recognition (detection) threshold ppm
Acetaldehyde	0.21
Ammonia	46.8
Butyraldehyde	0.039
Butylamine	0.24
Carbon disulfide	0.21
Chlorine	0.314 (0.01)
Dimethyl sulfide	0.001
Dimethyl trisulfide	(0.001)
Ethylamine	0.83
Ethyl mercaptan	0.001 (0.0003)
Formaldehyde	1.0
Hydrogen sulfide	0.0047 (0.0005)
Isobutyraldehyde	0.336
Methylamine	0.021
Methyl mercaptan	0.0021 (0.001)
Ozone	(0.5)
Phenol	0.047
Phosgene	1.0
Phosphine	0.021
Pyridine	0.021
Skatole	(0.22)
Tert-Butyl mercaptan	(0.00008)
Thiocresol	(0.019)
Thiophenol	0.28 (0.014)
Triethylamine	(0.08)

* Compiled from multiple industry sources and personal experience.

D/T ratio and subsequently the number of odor units. Determining the number of odor units is accomplished as follows: The threshold for skatole, for example, is 0.22 parts per million (ppm). Any given sample volume of air containing 22 ppm of skatole would need to be diluted 100 times to bring the concentration back to 0.22 ppm or the dilution threshold. The sample of air containing 22 ppm of skatole would therefore contain 100 odor units.

Scentometers.   In the field, odor units are measured by an instrument called a scentometer. The scentometer consists of a box, containing two filtering chambers (usually charcoal), with an inlet for fresh air, separated inlets for odorous air, and two mechanisms for "sniffing" the results. The sniffing ports are usually nothing more than extended tubes or removable caps at the opposite end of the scentometer from the odorous air inlets.

There will be four openings for intake of odorous air, graduating downward in size from 1/2 to 1/4 in to 1/8 to 1/16 in, and one inlet for fresh air of 1/2 in. Each of these openings should have a removable cover. The machine always includes a table for converting the results to odor units. Use of this device begins by opening the clean air portal and sniffing air that has passed through the charcoal filters. This should provide a standard of no odor, and prepare the olfactory system for the measurement. Next, the smallest odorous air port on the instrument should be opened (usually 1/16 in). If breathing in through the sniff tubes produces perception of the odor, the chart is referenced to indicate the number of odor units. If no odor is perceived, the open odorous air portal is closed and the next smallest is opened. The procedure is repeated, graduating from portal to portal until the odor is detected.

Sophisticated scentometers exist with as many as eight odorous air inlets. (Metric models usually contain four, five, or six inlets.) It would, however, be rare to encounter one containing an odorous air inlet of smaller than 1/32 or occasionally 1/64 in. This is because the number of odor units encountered at 1/32 in would almost always be a nuisance. The reading at 1/16 in (approximately 170 odor units on most scentometers) indicates a *minimum* of that number of units. The actual dilutions to threshold could be considerably greater. Because its measurements are essentially subjective, the same individual may achieve varying results at varying times with the scentometer. Practice is essential.

System Use—Example

Using the example of the skunk, this standard system would qualify the skunk odor as pungent, agitating, and irritating in character, with connotations that include the perceived difficulties of removing the smell from any item it permeates. The hedonic tone is disagreeable, offensive, and associated with anxiety and other physical symptoms. On the intensity scale, the skunk odor could range from moderate to extreme, depending on proximity and time lapse from the original gaseous generation. The odor units produced would be in excess of 170 and potentially many times greater because of the nature of the odorant components. Skunk scent is composed of many gases, of which at least two are mercaptans. (see Table 13.6.4.) Mercaptans are intense odorants, often disturbing at one part per billion (ppb) or less.

Continuous exposure to a moderate level of this odor would still pose a nuisance and would result in negative impact on quality of life because of the character of the smell and the hedonic tone associated with it. Conversely, the strawberry odor would need to be considerably more intense and more concentrated to pose a nuisance and a threat to quality of life, simply because the character and hedonic tone are pleasant and pleasurable, and not normally associated with any disagreeable reactions. By using the four qualifiers, a broader picture of the odor and the problem is obtained than would be from a simple measurement of the gases involved.

Nature and Benefits of the System

The use of this system represents a serious and reasonable attempt to combine scientific evaluation, social connotation, and personal opinion. While the system itself contains objective and arbitrary components, it has filled and continues to fill a definite need quite competently. Its use will continue even as our ability to identify and measure specific components of any odorous stream continues to improve. The development of this system may be traced to several factors in addition to the subjective nature of human response to odor.

First, it is rare to find an odor composed of one pristine gas. The majority of odors are generated by or in biological processes and contain multiple compounds. This increases the difficulty of correct identification. In treatment, it is often the case that a specific component may be effectively diminished without eliminating the odor nuisance. It is common to identify 20 to 40 components in the off-gases from manures, six to fifteen in composts and bio-solids, and equally as many in the exhausts of rendering plants, paper mills, and other manufacturing facilities. While one particular gas or family of gases may dominate the odor, additional components may become intrusive as the once dominant components are removed or diminished. The use of odor units attempts to deal with the nuisance as a whole, rather than with the various components.

Second, there has been a historical lack of effective, inexpensive field equipment capable of the detection, identification, and monitoring of gaseous components. Colorimetric tubes and sampling pumps are often the best available method. Even chromatography is somewhat dependent on our expectations of what gases are present. Consequently, the specific gases detected in an odor are often the specific gases expected to be present in that odor. Chemical analysis of a particular process will provide indications of the probable types of vapors that will result, but it is a practical reality in the field that many specific components remain unidentified. This further justifies the use of the previously described system in evaluating an odor problem. It would be ideal to assess the situation, establish that odor components A, B, and C are involved, determine the most effective treatment for each, implement the treatments, accept congratulations, and go on to the next problem. More commonly, A and C can be established, B cannot be identified or proves to be a combination of several unidentified compounds; treatment for A and C is determined and implemented, but B remains unaffected and is still a nuisance.

Summary

Our ability to detect, measure, and treat odorous gases is to some extent limited by our suppositions regarding the presence of these gases. If we do not or cannot surmise the possibility of the presence of a given gas or group of gases in a given situation, we may not test for their existence. Certainly our ability to detect and quantify particular emissions is constantly improving. Computers and developing software dramatically increase our ability to "suspect" and search for more components. Chemical and biological advances, along with trial and error, are also increasing our catalogued knowledge of the gases to expect in many situations. In time, a readily available instrument will be developed for identifying and quantifying the majority, if not the entirety, of the component gases of any given odor. However, even then, the use of the subjective analyses of character, intensity, tone, and detectability will continue to be a viable and credible method of qualifying and quantifying odor.

13.6.4 TREATMENT METHODOLOGIES

Prior to a discussion of various methodologies and/or technologies used in the destruction or diminishing of nuisance odors, it is important to distinguish between the three specific levels of odor management. As previously mentioned, these levels are prevention, minimization, and treatment.

Prevention

Preventing the formation of odors will always be the preferred management technique. Sometimes a condition of no odor may be accomplished by changes in housekeeping practices, completion of regular, specified equipment maintenance, raw material changes, or changes in storage practices. It is obvious, for example, that cleaning, rinsing, and sanitizing are essential in facilities where animal and/or vegetable materials are processed. Regular filter changes and exhaust equipment maintenance will be crucial to the performance of air handling and air exchange systems in any facility. A change in fuel components may completely eliminate or, alternatively, create odors from milling, production, or composting operations. In other situations, solutions requiring actual changes in the process may be discovered. Process changes may involve changes in temperature, moisture, or feedstock.

Minimization

Every step undertaken to prevent odor helps to minimize odor. Some situations occur in which prevention is not plausible or practical. In these situations, and often in the initial facility design, minimization is considered and attempted. Extremely common examples of this are exhaust stacks and blower units. By exhausting the odors above walking height, and in combination with quantities of fresh air, an attempt is made to accomplish enough dilutions to bring the odor to or below the threshold limit. Theoretically, the more fresh air that is mixed with the exhaust stream, the more the odor will be minimized. However, physical limitations in stack height and diameter, thermal dynamics, and differences in the physical properties of certain gases, often make minimization by dilution less effective than anticipated.

Another example of minimization is scheduling the operation of certain processes or exhausts at hours of the day when odorous emissions are less likely to impact society. Gravel crushing operations, glue works, and fertilizer plants often schedule their production shifts during the evening and night to lessen the effect of the odors they generate. These actions minimize the impact of odor on society, not the actual generation of odor.

Treatment

For purposes of this document, treatment is defined as "the employment of specific chemical, biological, and/or mechanical methodologies to relate to a specific gas or group of gases by decomposition, adsorption, absorption, combination, dehydration, interference, or any other process with the intent to collect, alter, or otherwise diminish the quantifiable presence of that gas or group of gases." Most existing technologies for odor control utilize mechanical devices to incorporate chemical or biological agents. These agents are designed to facilitate chemical, biological, or biochemical reactions in odorous exhausts or in the substrate producing the exhaust. This is true of scrubbers, packed towers, bio-filters, ozone generators, thermal oxidizers, perimeter or exhaust stack misting systems, and most filtration systems.

There are excellent reference works on the various technologies and methodologies currently in use. For detailed analyses, specifications, and operating procedures, the reader should refer to the appropriate texts. A brief explanation of each of the better-known methods follows. A listing of the categories of treatment and corresponding examples are included as Table 13.6.2.

Absorption. Absorption is commonly defined as the transfer of compounds from the gaseous to the liquid phase. As a means of odor control, this is accomplished by the use of wet scrubbers, packed towers, spray towers, and variations of these technologies. In short, the gas is exposed to a liquid where it is either solubilized or chemically altered (usually by oxidation). The absorption unit may operate with water or pH adjusted water for treatment of gases that are highly soluble in water, or it may include a chemical additive for "chemisorption." The advantage of including an oxidizing agent (such as hydrogen peroxide, dissolved gaseous ozone, sulfuric acid, sodium hypochlorite, sodium hydroxide, and others) is that a new substance is formed in the reaction between the odorous gas and the additive, rendering the absorption practically irreversible. When water alone is used, odors may reform under evaporative or other conditions.

Absorption technologies are most commonly and most effectively applied in the treatment of ammonia and certain amines (compounds containing ammonium radicals) and sulfurous compounds such as hydrogen sulfide, sulfur dioxide, and the mercaptans. As a method of odor control, absorption performs well when the odor problem is confined to the specific gases mentioned above. In practice,

TABLE 13.6.2 Available Odor Treatments

Technology	Methodology
Absorption	Wet scrubbers, packed towers
Adsorption	Activated carbon beds; activated charcoal filters
Aeration	Mechanical aerators (bubbling units), windrow turners
Bio-deodorization	Bio-chemical addition, biological addition, bio-filters
Combination	Esterification, reduction
Condensation	Contact condensers, surface condensers
Construction	Polymerization, ionization
Decomposition (chemical)	Oxidation (including ozone generators), hydrolysis, precipitation
Decomposition (thermal)	Incineration, catalytic oxidizers
Dehydration	Mechanical dryers, natural processes
Dilution	Exhaust stacks, forced air blowers
Interference (neutralization, counteraction)	Atomization systems, foggers
Re-odorization	Atomization, direct incorporation

scrubbers and towers often solve only portions of odor problems. This is because either the odorous stream contains additional gases that are not readily treated by absorption, or because collection and control of the exhaust stream to be scrubbed is incomplete.

Adsorption. By contrast, adsorption is generally defined as the transfer of a gas to the surface of a solid, usually by diffusion into and on to the pores of the adsorbent. Adsorbents include charcoal and activated carbon, hydrous oxides such as silica gel, the entire family of zeolites (such as clinoptilolite and mordenite), and certain clays and other natural materials. Adsorption can be very effective in removing many organic odors and certain inorganics (primarily hydrogen sulfide). However, the effectiveness of this type of technology can be summarily diminished by many factors. The availability of surface area of the adsorbent is critical, often necessitating large amounts of adsorbent and large amounts of physical space. Sufficient contact time between adsorbent and adsorbate (the odorants) is equally critical, increasing the need for sufficient adsorbent surface to accommodate and retain the odorous stream. Increasing temperature decreases adsorbent efficiency while increasing pressure increases adsorbent efficiency. Consequently, power costs to prevent temperature increases and pressure losses are substantial.

Systems for regeneration are normally designed into adsorbers. For example, reversing controls and allowing temperature to increase and pressure to drop will result in "desorption," releasing adsorbed compounds and temporarily creating availability of additional surface area. Many systems use super-heated steam as a primary method of increasing temperature for regeneration. Eventually, however, the adsorbent is spent and cannot be regenerated. It must be removed, disposed of, and replaced. The only commonly encountered adsorbent systems for odor control are activated carbon beds. The carbon may be a single bed or, more commonly, a series of modules.

Inherent in this technology is the requirement for recovery systems to capture and/or further treat the compounds that have been adsorbed and then released during desorption. Frequently used recovery systems include distillation for water soluble organics, decanting for nonwater soluble organics, and incineration (oxidation) for inorganics.

Practical considerations are the capital costs of the technology, the cost of operations, and the cost of disposal. Nevertheless, for certain types of volatile organics (such as toluene), adsorption represents one of the most effective means of treatment.

Aeration.   Aeration systems supply oxygen to a substrate. This is done to increase and promote the growth of aerobic bacteria in biological processes, and to increase available oxygen for chemical reactions. Aeration may be accomplished by mechanical or electrical oxygen generators, or by equipment designed to turn, churn, or otherwise expose the internal mass of the substrate. In wastewater treatment, aeration is used to maintain oxygen levels when biological activity alone is insufficient. Systems that bubble oxygen through the wastewater, or systems that move the water through falls or trays, are commonly used.

In composting, aeration may be accomplished by mechanisms that force air directly into the substrate or pile, or by windrow turning. In windrow turning, a mechanical device physically turns the windrow inside out, exposing inner material to ambient oxygen. Unfortunately, this method also releases odorous gases formed in the interior of the windrow.

Aeration can be a very effective means of odor control. In wastewater treatment, aeration can help to avoid anaerobic escape from headworks, ponds, and lagoons. However, energy costs can be significant.

Bio-Chemical Modification.   This is a developing technology, based upon changing, either by chemical or nutritional means, or both, the proliferating bacteria responsible for the degradation of organic material. This technique is especially promising in wastewater treatment, bio-solids and manure composting, landfilling, and fertilizer production. In these activities, odorant generation is the direct result of the bacterial population. Bio-chemical modification influences the strains of bacteria present, selectively encouraging the growth of those strains that bio-degrade the organics without emitting odorous gases in the process.

The most common use of this technique is in the conflict between aerobic and anaerobic bacteria in any decomposing organic substrate. Aerobic bacteria exist in the presence of atmospheric oxygen and do not emit sulfides, amines, mercaptans, or other nuisance gases. Anaerobic bacteria thrive in the absence of atmospheric oxygen and emit the majority of problem odors associated with waste decomposition operations. This is not to imply that anaerobic bacteria do not need oxygen. Like all living creatures, anaerobic bacteria require oxygen, which they obtain from dissolved oxygen in the substrate and in water. Aerobic bacteria are not competing for this same oxygen to survive. The very structures used in these associated processes, structures such as windrows, static piles, tanks, among others, create difficulties in maintaining aerobic conditions because of compaction, surface area to total mass ratios, biological and chemical oxygen demand (BOD and COD), and microbial competition.

Shortages of atmospheric and dissolved oxygen are commonplace and often anticipated in process design. It is this tendency toward anaerobia that provides the basis for biochemical modification. By means of certain nutrient enhancements and chemical reactants, added directly to the substrate, specific bacterial strains known collectively as facultative anaerobes are encouraged. Facultative anaerobes are desired over common anaerobes for two reasons.

First, they possess the respiration abilities of both aerobic and anaerobic bacteria, functioning with either molecular or compound oxygen. This duality enables facultative anaerobes to feed and proliferate without the interruptions common to aerobic and obligate anaerobic populations. Consequently, facultative bacteria are the most proficient of the three types.

Second, the lack of interruption results in more thorough and complete decomposition since bacterial functions are not inhibited by major periods of population adjustment whenever oxygen conditions change. More thorough nitrogen fixation results in less ammonia and ammonia gas, amines, and indole, and in more sulfur reducing bacteria and fewer sulfides. More complete respiration of the oxygen in fatty acids, ketones, and aldehydes results in odorless carbon dioxide and water (as steam).

The above factors directly and indirectly decrease odor. As this technology develops, its success and acceptance will increase. It is designed to correct and prevent the biological problems responsible for odor formation, rather than to treat already formed odors. In practice, aerating or disturbing the substrate too quickly reverses facultative growth, and results in anaerobic re-population. Many products to facilitate this technology are currently available. The documentary and evidentiary support for each product's performance should be closely examined until such time as this technology is better established.

Biological Modification.   Biological modification, in contrast to biochemical modification, is accomplished by the direct addition of biological agents into the odor-emitting substrate. Biological modification as a technology is well-known. In theory, specific bacterial strains or their beneficial enzymes are added to degrade specific components of an organic process, or to directly increase those bacterial strains that are under-populated in the substrate.

The desired result is usually the same as in biochemical modification. The methodology includes addition of specific biological agents for the consumption and degradation of specialized cell walls, fibrous matter, sugars, and other difficult components and gaseous compounds. Biological modification concentrates on directly adding quantities of desired bacteria that are not flourishing in the substrate. The difficulty in this practice lies in the initial inability of the substrate to maintain or encourage sufficient population of the bacterial strains desired. Without affecting a fundamental change in the substrate, re-applications of the bacteria or enzyme will be required to maintain population levels.

Biological modification is most successful in wastewater treatment applications. However, as in biochemical modification, variables in temperature and toxicity may decimate the microbial population.

Bio-Filtration.   Bio-filtration, or "bio-deodorization" as it is often called, refers to the purification or elimination of an odorous gas by means of biological degradation. The major difference between bio-filtration and biochemical or biological modification is that bio-filtration attempts to diminish odors that have already formed and been collected, whereas modification primarily attempts to prevent formation or escape of those odors.

Exhaust gases from the generating process or substrate are collected and passed through a filter media consisting of biologically active material. The odorous compounds in the exhaust are exposed to specific bacteria available in specific media, resulting in the loss of the compounds to biological action—the bacteria consume the compounds or their functional parts, resulting in the elimination or alteration of the odorous gases. The remaining exhaust is either allowed to dissipate, or is again collected and exhausted separately.

Bio-filtration technologies are similar to adsorption technologies in that a bed or series of beds of filter material comprises the standard format. However, no regeneration or recovery systems are normally required unless gases possessing low susceptibility to bacterial degradation are passed through the bio-filter. In these cases, additional treatment may be implemented prior to or during exhaust, or the compounds may be collected, recovered, and disposed of separately. As with adsorption beds, bio-filtration media require periodic replacement, but the spent material does not normally present special disposal problems.

The bio-media may be sand, diatomaceous earth, various other organic materials, or compost. While many types of materials have been and are being used, more and more practical success is occurring with the use of compost as the filter media. The advantages to the use of compost are its low cost and ready availability and its fundamental characteristics as a natural incubator for bacterial growth. The addition of water is often required to maintain performance due to excessive evaporation and water consumption in the biological activity.

The effectiveness of bio-filtration as an odor control treatment is determined by a number of factors, including the efficiency of the collection system used to deliver the odorous exhaust to the bio-filter. The nature of the compounds for treatment, the quantity of filter material (and therefore the quantity of micro- organisms) available, the contact time, sustained temperature, and moisture levels are all critical to bio-filter performance. It must be remembered that the bio-filter is a collection of living organisms. Because bio-filtration occurs throughout the filter media rather than just on the surface pores (as in adsorption), the media—the home for all the organisms—must be maintained in proper condition to keep the organisms healthy and active.

Combination Technologies.   "Combining" away odors is a limited technology in that it is usable with only a limited number of organic odors. By definition, combining refers to adding one type of compound to another, odorous one, resulting in a reaction that produces a less odorous, nonodorous, or pleasantly odored compound. The typical combination example used is esterification, the introduction of alcohols into volatile organic acids such as fatty acids to produce an ester. Esters as a rule are the scents we associate with fruits, flowers, and spices.

Other types of odors are not diminished in this process, but may be neutralized or masked by the esters created. (Masking may not always be considered a negative approach to odor management, but it should always be considered less desirable than active treatments.) For this reason, agents for additional types of treatment are often included with the combination agents.

Combination is usually accomplished by means of atomization of the reactant into the exhaust containing the odorous gases, or into the perimeter atmosphere, which will be permeated by the odorous gases. The mechanism consists of a pump or compressor to deliver liquid reactant to a series of either air-powered or hydraulic nozzles with orifice sizes designed to produce a particle the weight of air or lighter. Atomization enables the particles to move in the same air patterns as the odorous gases, to achieve contact, and to produce the desired reaction. The reactant may not always contain alcohol. It may include aldehydes or ketones, which will be reduced to alcohols to initiate combination with organic acids, resulting in esterification.

In practice, combination and other atomization technologies are often effective in situations where other methods are impractical; or they are used as secondary treatments in situations where scrubbing or filtering have proven insufficient.

Condensation. Condensation refers to the cooling of a high temperature gas stream to convert odorous molecules to a liquid, effectively eliminating them from the exhaust stream. Condensation is not normally used as a stand-alone technique because of two inherent limitations. First, the odorous compounds are not affected unless they are water soluble. Second, relatively minor changes in temperature make substantial changes in condensability. However, condensation in combination with absorption is common and effective. Condensers are typically the method used to solubilize sulfide-laden air for aqueous oxidation in wet scrubbers.

Either a coolant is mixed directly into the hot exhaust stream or else the exhaust is passed over a cooled surface to produce condensation. These two methods are called contact condensation and surface condensation, respectively. Both are used in absorption (scrubbing and packed tower systems). As previously stated, condensation as a stand-alone treatment is rare and generally ineffective.

Decomposition (Chemical). Perhaps the most widely used method of odor control is the addition of chemical additives to a process or to an odorous substrate. Normally, these chemicals are oxidizers such as those mentioned in the section on absorption. In other words, absorption usually works in combination with chemical decomposition as well as with condensation. The purpose of these chemicals is to break down the odorant into nonodorous compounds. This is called decomposition. Adding an oxidizer to the aqueous solution of a wet scrubber to break down the odorous compounds that have been absorbed is an example of chemical decomposition. Some decomposition chemicals are added directly to process situations such as wastewater mains or holding tanks.

Unfortunately, no one oxidizer works on the range of odorous gases commonly encountered in waste processing. Additionally, oxidizers often pose health and safety threats to personnel, and may be considered nuisance pollutants in their own rite (such as ozone). It is typical for oxidizers to relate to particular compounds or types of compounds in an odorous stream. Consequently, oxidation usually eliminates only part of any given odor problem.

Decomposition (Thermal). Oxidation may also be accomplished by thermal means, by incineration. Incineration can be highly effective; however, the effectiveness of each system is directly proportional to the completeness of combustion. Combustion requires time, temperature, and turbulence in combination with oxygen. It is the modification of these three factors—time, temperature, and turbulence—that produces the correct amount of oxidation for a particular odor or stream. In successful thermal decomposition, odorous compounds are converted to nonodorous oxides—usually, carbon dioxide and water.

Two basic procedural methodologies are currently in use for this technology. The first is called direct fuel or thermal, and is represented by both static bed and fluidized bed (dynamic) incinerators. In the static bed, there is less turbulence than in the fluidized bed, resulting in the requirement for a higher temperature. The fluidized bed system has increased turbulence, requiring lower temperatures. Consequently, static bed systems require greater quantities of auxiliary fuels (in addition to the odorants), than do fluidized bed systems.

The second procedural method is catalytic oxidation. The catalyst aids in the oxidation process without being directly involved or permanently affected by the reaction. Time, temperature, and turbulence are still required. The catalyst impacts each of them directly. Time becomes a function of the amount of available catalytic surface. Turbulence is obtained and increased by the passage of the odorous gases through the catalyst. These two functions are sufficient to require much lower temperatures to initiate and sustain oxidation reactions. Because lower temperatures are required, less auxiliary fuel is necessary, even less than in fluidized bed incinerators. In some cases, complete operation without the consumption of external fuel is possible. Maximum surface exposure of the catalyst is essential. Typical catalysts include platinum, palladium, vanadium, and chromium, as well as more common metals such as nickel, manganese, and manganese oxide.

Thermal decomposition, whether direct or catalytic, is effective in many applications when properly controlled. In situations where auxiliary fuel must be purchased, operating costs may be a factor. Capital costs for this technology exceed capital costs for all other technologies discussed herein, even large and complex multiple scrubbing systems. Nevertheless, thermal decomposition may be an excellent method for treating paper and other sludges, garbage, and many other odorous wastes.

Dehydration.   The use of dehydrating or drying odorous materials is often used as a temporary or field method of control. While mechanical drying may consume large amounts of fuel or energy, biological drying (by allowing moisture to be used up without replenishment) frequently occurs as a means of diminishing process odors with no associated process costs.

This technique is common in agricultural waste composting for fertilizer or feed production. Composting material may be allowed to dry to levels that actually inhibit biological activity. While this technique improves odor conditions during production, biological activity and odor often return when the product is heated (as in pelletizing and/or bagging), or when the product is rehydrated.

The use of mechanical drying as a large-scale technique is limited by cost and space factors, and is unlikely to have more than occasional uses. However, where space permits, natural drying of sludges, grains, grass, or manures prior to further processing may be a reasonable method of minimizing process odors. This technique is often practiced in countries where space is still readily available, especially for bio-solids and manures. The drying beds themselves may require the use of other technologies to treat the odors emitted as the material dries. Even smaller, commercial dryers used in many agricultural situations produce substantial nuisance odors during the drying process.

Interference.   Interference is the inhibition of the perception of a malodor. Interference is more commonly known as "neutralization" or "counteraction." It refers to the quality of certain odorous substances to interfere with the perception of other odorous substances. The process involves introducing the correct neutralizer or neutralizers into an odorous stream, such that perceptible odors are limited to the odor of the neutralizer itself, or to no odor. In practice, this is usually accomplished by atomization or fogging systems similar or identical to those used to introduce combination compounds. Odors are not actually decomposed or reacted into other compounds using interference. Consequently, it is not uncommon for neutralizing agents to be blended with combining agents or other reactants. True interference should not be confused with masking. As the name implies, the interfering or neutralizing odor actually causes a physical inability to perceive the offensive odor, or ideally, either odor. The exact methodology that allows this reaction is still being researched. It is probable that the counteractant is processed by the same receptors as the offensive odor. The resulting enzymatic signal identifying the counteractant most likely overrides the signal previously produced to identify the odorous molecule. Regardless of the method, the brain stops identifying or perceiving the earlier odorous signal.

A complete neutralization or interference requires introduction of the agent in such a way that the perceiver is exposed to the same amount of the odorant and the interference agent. This is a very difficult feat to accomplish and a reason for using interference in conjunction with other contact technologies. This type of treatment is often effective when a secondary or conjunctive treatment is essential, or where no other treatment is practical or effective.

Re-Odorization.   Re-odorization is more commonly known as "masking." It is simply the practice of adding additional, more acceptable odors in the presence of malodor. It is similar to attempting to cover body odor with perfume. It is normally accomplished by atomization systems or direct introduction of a liquid into an odorous substrate.

Chemical Construction.   This technology includes polymerization and ionization. Polymerization, in such applications as sludge thickening, is a method of building or enlarging on the sludge molecules by the creation of long chain molecules made up of repeating simple monomers. In any application, the function of polymerization is to increase molecular weight, thus decreasing volatility. This decrease in volatility decreases odor emissions simply by decreasing the molecule's ability to off-gas.

The ionic reaction is based upon pH or acid/base relationships. Specific odors are common to specific pH environments. They may only be generated in specific pH ranges. Each odor has either acid or base characteristics. When an acid and a base interact, an ionic reaction occurs, and a soluble salt results. Ammonia, for example, reacts with acetic acid to form ammonium acetate. In situations where the newly formed salt is an acceptable by-product, ionization may be an effective methodology. A very weak acid (such as diluted citric acid) may be misted over an area to control amine odors and ammonia. (Splashing lemon juice on fish is an example of ionization.)

13.6.5   A PRACTICAL METHODOLOGY FOR ODOR MANAGEMENT

In attempting to deal sensibly with various odor situations, it is imperative that a system be implemented that is applicable to the majority, if not all, nuisance odor occurrences. This system must be reasonable, practical, and equally usable by generator, regulator, and consultant. The protocol should consider the source of the odor, the cause of the odor, the type and nature of the odor, the social impact of the odor, the physical impact of the odor, the available treatment options for the odor, and the most likely alternatives for treatment in the specific case. Following is an approach to accomplishing the above goal.

1. Examine the source of the odor (see Table 13.6.3 at the conclusion of this chapter)

   - Is the odor generated in a process?
   - Is the odor generated by a material?
   - Is the material odorous regardless of the process?

2. Determine the probable cause of the odor

   - What portion or portions of the process result in the odor?
   - Are these mechanical, chemical, or biological in nature?
   - If the material is odorous regardless of the process, what actions or conditions in the material are causing the odor?

3. Assess the type and nature of the odors being produced

   - What are the raw materials involved in the process?
   - What is the nature of the process?
   - What are the products and by-products?
   - What are the potential components of the odor?
   - Do investigations and exhaust analyses of similar facilities or operations exist?

4. Assess the social impact of the odor on site personnel and the surrounding community

   - Is the odor interfering with employee efficiency?
   - Is the odor interfering with neighborhood outdoor activity?
   - Is the odor offending a few or many?

TABLE 13.6.3 Typical Generation of Odorants

	Food preparation (kitchens)	Fertilizer plants	Wastewater treatment	Composting operations	Landfills	Paper mills	Rendering plants	Livestock production	Refineries (petroleum)	Medical facilities	Bakeries & bakery waste processing	Waste transfer operations
Aldehydes	✓			✓	✓						✓	✓
Ammonia		✓	✓	✓			✓	✓	✓	✓		
Amines	✓			✓	✓							✓
Hydrogen sulfide			✓	✓		✓	✓	✓	✓	✓		✓
Sulfides				✓	✓	✓	✓	✓				✓
Mercaptans	✓		✓	✓	✓	✓	✓				✓	✓
Indole			✓	✓		✓	✓	✓	✓			✓
Volatile organic acids		✓	✓	✓	✓	✓				✓		✓
Skatole			✓				✓			✓	✓	
Urea		✓					✓					

TABLE 13.6.4   Odor Character

Acrid (pungent)	Burnt (pungent)	Fecal
Ammonia	Acrolein (sweet)	Indole
Chlorine	Diphenyl Sulfide (rubbery)	Skatole
Sulfur Dioxide	Pyridine	
Thiocyanic Acid		
Fishy	**Garlic/Onion (pungent)**	**Medicinal**
Dibutylamine	Allyl Mercaptan (Garlic)	Dimethyl Ketone
Diethylamine	Phosphine (Onion)	Formaldehyde
Dimethylamine	Thiophenol (Garlic)	Phenol
Methylamine		Propanol
Pyrimidine		
Trimethyl Amine		
Rancid (putrid)	**Skunk**	**Solvent**
Butyraldehyde (Butter)	Crotyl Mercaptan	Benzene
Cadaverine (Flesh)	Tert-butyl Mercaptan	Benzyl Chloride
Putrescine (Flesh, Tissue)	Thiocresol	Chlorophenol
		Styrene
		Toluene
Sour	**Sulfide**	**Sweet**
Acetic Acid	Carbon Disulfide (Vegetable Decay)	Acetaldehyde
Butyric Acid	Dimethyl Sulfide (Vegetable Decay)	Acetone
Butylamine	Ethyl Mercaptan (Cabbage Decay, Gas)	Citraldehyde
Citric Acid	Hydrogen Sulfide (Egg Decay)	Ethanol
	Methyl Mercaptan (Cabbage Decay)	Isobutyraldehyde
		Methyl Ethyl Ketone

- If a few are registering complaints, do any special factors separate them from other citizens (location, age, topography, special allergies or sensitivities)?
- Is the odor limited to external nuisance, or do complaints include structure penetration (home, vehicle, clothing)?

5. Assess the physical impact of the odor on site personnel and the surrounding community

- Are employees complaining of, or exhibiting, any symptoms that might be attributable to exposure to the odor?
- Are community members complaining of any physical symptoms that they are attributing to exposure to the odor?
- Have similar problems at other facilities resulted in physical symptoms to employees or complainants?

6. Determine the probable components of the odor from available analyses of similar facilities, or by testing

- (If safety or toxicology concerns exist, testing should be implemented by qualified firms or individuals only. If no health concerns exist, the scope and intensity of the problem should determine whether industry information, simple on-site testing with sampling pump and colorimetric reagent tubes, or third party laboratory analysis is required.)
- What are the odorous compounds or types of compounds involved?

7. Assess the available options for managing the odor problem

- Can the generation of odor be prevented?

a. By a raw materials change?
b. By a process change?
c. By a biological modification?
d. By biochemical modification?
e. By chemical addition?

- Can generation of odor be minimized?
- By a, b, c, d, or e above?
- Will additional treatment be required?
- Based on the types of odors and their causes, what additional treatment is potentially available?

8. Determine the most practical, effective course of action

- What are the comparative costs of prevention, minimization, and treatment?
- Do additional benefits occur if odor is eliminated rather than diminished to minimum nuisance levels?
- What course of action is likely to cause the greatest reduction in odor nuisance for the lowest cost?
- What is the minimum treatment capable of producing any significant decrease in nuisance odor?

This system may be modified for inclusion in the design phase of any facility or operation. If technologies for prevention and/or treatment are considered in construction and operating parameters, many nuisance situations may be minimized or avoided entirely. Specifying engineers, designers, and architects should contact reputable suppliers for design and operating specifications of various available equipment.

System Implementation

The following situation will be examined using the management system presented.

The Problem.   A composting facility accepts and processes a combination of stable wastes, biosolids, and wood chips. Complaints are coming in from specific areas while other areas equally close to the facility do not complain of nuisance odors. Complaints occur even when wind is not a factor. Descriptions of structure permeation and physical nausea accompany the complaints.
    1. Source.   In this case odor is certainly generated in the process. However, significant portions of the feedstock (the biosolids and some of the stable wastes) are odorous regardless of the process.
    2. Cause.   In the composting process, many factors contribute to the presence or absence of odor. Whether aerobic or anaerobic composting technology is implemented, certain fundamental areas should be carefully examined. What portion of stable wastes and biosolids to wood chips (nitrogen to carbon) is used in pile or windrow formation? What limits are placed in temperature and humidity within the pile or windrow? How long is allowed for composting prior to screening or further processing? Do the odors seem to be more prevalent in any particular phase of the process?
    In this particular case, stable wastes and biosolids made up 50 percent of the piles by volume, and nearly 75 percent of the piles by weight. Initial moisture content was approximately 70 percent, diminishing to 40 to 44 percent at time of screening. Temperatures within the piles were held below 150°F, but little other concern with temperature was indicated. Windrows were maintained for 30 to 40 days, then removed for initial screening with over-sized particles retained and windrowed for two additional weeks. The majority of odor complaints occurred when each pile was turned, usually 8 to 12 times in 30 to 40 days.
    Based on the above information a recommendation was made that stable wastes and biosolids be reduced to no more than 30 percent of each pile by volume. It was felt that this action alone would minimize many of the process odors. Additionally, a suggestion to reduce turning dramatically during the first 30 days was put forward.

The facility responded that, because of current incoming feedstock contracts, space limitations, and permit requirements, both suggestions were impossible to implement.

3. Type and Nature.   The biological origin of the odors and their probable types were well documented in dozens of case studies and analyses of composting operations. Many studies existed utilizing the same basic feedstocks found in this facility. Additionally, key information in identifying the probable culprits was available in the complaints themselves.

4. Social Impact.   Employees were questioned regarding the impact of the odors on their clothes, shoes, and automobiles, especially on the days when windrows were turned or moved. They acknowledged that clothes, cars, and even hair retained particular odors associated with compost. Boots could retain the odor for months, even when removed from the site permanently. Complainants were questioned along the same lines. Reports of gases completely permeating homes and even furnishings, such as towels and curtains, were reported. Many of the complainants indicated that they were completely unable to use their lawns or porches several days and evenings per month.

The majority of those impacted lived in specific, low-elevation areas where mountain interference tended to inhibit dilution and diffusion, and where escaping gases were subject to inversion. Those living at higher elevation or on exposed plateaus indicated little to no awareness of nuisance odors from the facility when polled, and made up less than 1 percent of the complainants.

5. Physical Impact.   No employee complained of physical symptoms except during row turning, when burning eyes, coughing, and nose and throat irritation were acknowledged. Off-site complaints indicated entirely different symptoms, primarily consisting of headaches, nausea, and respiratory distress. These symptoms were consistent among a high percentage of the complainants. All felt that their symptoms were being caused by gases escaping from the composting operation.

6. Chemical Components.   The causes of these complaints were relatively simple to ascertain, even without instrument testing. On-site problems were primarily caused by ammonia loss during turning. While a problem for the turner operators, the ammonia had dissipated long before reaching the perimeter of the facility, and was not responsible for the off-site complaints. Additional odors, which were noticeable on-site, were ethylamine, aldehydes, and ketones.

Off-site, the symptoms and description of the odor indicated a somewhat typical composting combination of dimethyl sulfide and additional ketones. The ethylamine, aldehydes, and ketones were responsible for the sweet, grass-fermentation smells on site. The dimethyl sulfide (and probably other sulfur compounds and certain ketones) were responsible for the permeating, lingering, septic smells associated with nausea and respiratory distress off site.

While it was recommended to the site that additional third party testing be implemented to confirm these compounds, the facility did not feel this to be necessary. However, simple colorimetric reagent tests were conducted confirming the presence of dimethyl sulfide, ethylamine, acetaldehyde, and at least two ketones.

7. Available Options.   The facility's options were limited because of a number of factors. While it was believed that a raw materials (feedstock) change would reduce many odorous emissions, the facility had ruled this option out because of contractual and storage concerns. A process change from aerated windrow to a static operation with biochemical modification was also recommended, but the facility's permit requirements (based on turning of the windrows), and all previous operator training, made them reluctant to pursue this option.

The facility was spread over more than twenty acres, with several different "pads" (windrow areas) for receiving and initial mixing, main reactions, curing, and screening. The total area (all open air) had been designed with no intent to collect exhaust gases. The costs of construction to install such a collection system, as well as the additional costs of adding scrubbers, towers, or bio-filters to treat the collected emissions, were prohibitive. The remaining considerations were some form of misting or atomization system to treat odorous emissions prior to their exiting the site or a similar system to spray reactants or counteractants into the released gases each time a windrow was turned, moved, or screened. The size of the facility's boundaries discouraged the implementation of full perimeter coverage. The number of windrows, usually 60 or more, discouraged spraying with each turning or screening.

8. Practical Option(s).   The majority of odor complaints were coming from two directions, and seemed to coincide (more than 60 percent of the time) with turning operations on the main reactor pad. As a compromise, a matrix of nozzles was suspended around this pad. An electric pump with timing and adjustable pressure controls was installed to supply treatment into the atmosphere. This system was then capable of delivering agents for interference, combination, ionization, or even re-odorization. The operator did not expect this system to fully eliminate complaints, especially because no treatment was implemented around the receiving or screening pads. However, complaints were reduced by more than 50 percent, and the operator agreed to install additional treatment technologies at some point in the future.

Discussion

The above situation is taken from actual field experience, and represents a practical, although not ideal, approach to a serious problem. While neither the generator, the public, or the regulatory officials were completely satisfied, the generator did implement some treatment and begin to assume financial responsibility for the problems it was creating. The public did obtain substantial current relief, and the possibility of additional odor reduction in the future. The regulatory agency and local authorities established a precedent regarding the obligations of the facility, and obtained benefits for their constituents.

13.6.6 CONCLUDING COMMENTS

This chapter began with comments on the growing impact of nuisance odors on society. This impact will continue to initiate conflicts between generators of odors, and those exposed to these odors. It will also stimulate continuing research and development with regard to the causes of odor, the effects of odor, and the methods for preventing, minimizing, and treating odors. Existing knowledge and existing technologies will benefit, and new technologies will be discovered and developed.

In any civilization, technological advances are stimulated by the need for those advances. Necessity will be the fuel that encourages improvements and additions to our understanding of and our options for odor management.

CHAPTER 13
CONTROL TECHNOLOGIES

SECTION 13.7

WATER SUPPLY TREATMENT

Rumana Riffat

Dr. Riffat is assistant professor, Department of Civil and Environmental Engineering, George Washington University. Her areas of research and interest are in the treatment of water, wastewater, and hazardous wastes. She is investigating the removal of petroleum hydrocarbons from industrial wastewater using anaerobic biological treatment.

A. DESIGN FLOW AND LOADS

13.7.1   WATER DEMAND

The water demand of a community varies during different periods of the day, and also during different seasons of the year. The total water demand on a municipal water supply system is the sum of all domestic, commercial, industrial, and other demands over a specific period of time. An average daily demand for a particular community can be calculated as:

$$Average\ Daily\ Demand = \frac{Total\ volume\ of\ water\ use\ in\ 1\ year}{365\ days} \tag{13.7.1}$$

Units are in gallons per day (gpd), or million gallons per day (MGD), or m^3/d. The average daily demand per person is calculated as:

$$Average\ Daily\ Demand\ per\ Person = \frac{Average\ Daily\ Demand\ in\ Community}{Midyear\ Population\ in\ Community} \tag{13.7.2}$$

The units are in gallons per person (capita) per day (gpcd), or liters per capita per day (Lpcd).

Table 13.7.1 presents typical water usage values in U.S. cities. Consumption in small cities and large industrial communities can vary significantly from these values. A number of factors should be considered when calculating the water demand for a community. They are as follows:

1. The predicted change in population over the design period
2. The presence of commercial and industrial users in the community
3. Climatic conditions, which can increase or decrease the demand
4. Water conservation practices in use
5. Characteristics of the population (e.g., residents versus tourists)

Water demands are also expressed as maximum daily demand and maximum hourly demand, in addition to the average daily demand. The maximum daily demand or peak daily demand is usually

TABLE 13.7.1   Typical Water Use in U.S. Cities*

Use category	Average daily demand Lpcd (gpcd)	Percent of total use
Domestic	300 (79)	44
Commercial	100 (26)	15
Industrial	160 (44)	24
Other	110 (29)	17
Total	670 (178)	100

*Adapted from McGhee (1991).
Note: Lpcd = Liters per capita per day; gpcd = gallons per capita per day (Lpcd × 0.264).

taken as 1.2 to 2.0 times the average daily demand. The maximum hourly demand or peak hourly demand is taken as 1.5 to 3.5 times the average hourly demand. The average hourly demand is calculated directly from the average daily demand.

13.7.2   POPULATION PREDICTION

Population prediction is a complex issue. It is based on historical data, current population levels, birth rates, death rates, and a variety of other factors that affect growth. Other factors include education, technological development, economic conditions, religious beliefs, government policies, political turmoil, and so forth. The fundamental population equation is given below:

$$P_f = P_c + B - D + I - E \qquad (13.7.3)$$

where P_f = population after a specified time in the future
P_c = population at the current time
B = number of births occurring in the specified time interval
D = number of deaths occurring in the specified time interval
I = increase in population because of immigration in the specified time interval
E = decrease in population because of emigration in the specified time interval

Population data can be obtained from the United States Bureau of Census, which conducts a census every 10 years and publishes the results. Individual states also conduct censuses from time to time. Prediction of the other variables in the population equation present the most difficulties. There are a number of mathematical and graphical methods available for population prediction. Most of them are based on extension of past municipal population data. They are as follows: (1) Arithmetic Progression or Constant Growth Rate Method, (2) Uniform Percentage Method, (3) Declining Growth Method, (4) Graphical Extension, (5) Logistic Method, and (6) Ratio Method. The first three methods are used to make short-term estimates, between 1 to 10 years. The latter methods are used for long-term estimates, between 10 to 50 years or more. The declining growth and logistic methods might be applicable for population prediction of a community with limited land available for future expansion. Communities with large resources of land area, water, and good economic growth might use the geometric progression or uniform percentage growth models. Comparison should then be made with recorded growth trends of similar communities.

Brief descriptions of each of the six methods are provided below. The first three methods mentioned above are based on a population curve represented by the typical S-curve (Figure 13.7.1). The S-curve can be divided into three segments, consisting of geometric increase, arithmetic increase, and a decreasing rate of increase.

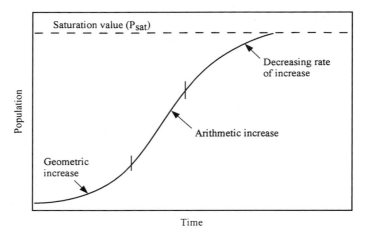

FIGURE 13.7.1.   S-shaped population growth curve.

Arithmetic Progression or Constant Growth Method

The population growth rate is assumed to be constant in this method. Census data of the community can be examined to determine if nearly equal incremental increases have occurred between recent censuses before applying this method. The prediction equation is:

$$P_f = P_c + K_c \, \Delta t \tag{13.7.4}$$

where P_f = population after Δt years, P_c = population at current time, and K_c = growth rate constant.
K_c can be determined graphically as the slope of the population-time curve, or from populations in successive censuses as:

$$K_c = \frac{\Delta P}{\Delta t} = \frac{P_2 - P_1}{t_2 - t_1} \tag{13.7.5}$$

where P_1 and P_2 are the populations at census time t_1 and t_2, respectively.

Uniform Percentage Method

In this method, constant percentages of growth are assumed for equal intervals of time. The application of this method is validated when a semilog plot of population versus time comes out as a straight line. The prediction equation is:

$$\log_e P_f = \log_e P_c + K_u \Delta t \tag{13.7.6}$$

The growth rate constant K_u can be determined from the slope of the semilog plot as:

$$K_u = \frac{\log_e P_2 - \log_e P_1}{t_2 - t_1} \tag{13.7.7}$$

where all the terms have been defined previously.

Declining Growth Method

This method assumes that the city has a limiting or saturation population, and the population growth rate is a function of its population deficit. The prediction equation is:

$$P_f = P_c + (P_{sat} - P_c)(1 - e^{kd\Delta t}) \qquad (13.7.8)$$

The saturation population P_{sat} has to be estimated from existing population density, land availability, and so forth. Then the growth rate constant can be calculated as:

$$K_d = -\frac{1}{t_2 - t_1} \log_e \frac{P_{sat} - P_2}{P_{sat} - P_1} \qquad (13.7.9)$$

where all the terms have been defined previously.

Graphical Extension Method

This procedure involves extrapolation of the population-time curve for a given locality, based on historical data. This can be performed in at least two different ways. One method is graphical projection of the existing population-time curve, based on past trends. Another method is graphical extension based on comparison with recorded growth of other larger size cities. The cities used for comparison have to be selected very carefully. Important factors include comparable land area, geographic similarity, economic growth, and so on. Figure 13.7.2 illustrates the projected growth for the city X, compared to the growth curves for cities A, B, C, and D. The data for city X is plotted up to the year 2000, when its population was 30,000. City A reached a population of approximately 30,000 in 1961, and its growth is plotted from 1961 to 1991. Similarly, curves are plotted for cities B, C, and D from

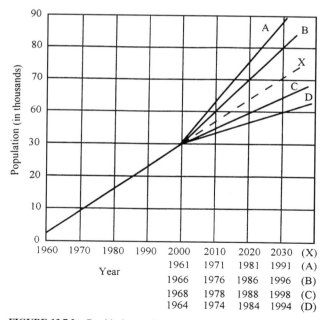

FIGURE 13.7.2 Graphical extension method for population prediction for city X, after comparison with cities A, B, C, and D.

the years at which they reached the year 2000 population of city X. The population curve for X is then projected after comparison with the curves for similar cities A, B, C, and D.

Logistic Method

The logistic method is based on a logistic curve that is S-shaped and has upper and lower asymptotes, with the lower asymptote being equal to zero. This method is applicable if a straight line is obtained by plotting census data on a logistic paper. In its simplest form, the prediction equation is:

$$P_f = \frac{P_{sat}}{1 + e^{a+b\Delta t}} \tag{13.7.10}$$

where, a and b are constants. P_{sat}, a, and b can be determined from three successive census populations P_0, P_1, and P_2, taken at years t_0, t_1, and t_2, respectively. P_0 must be near the earliest recorded population, P_1 near the middle, and P_2 near the end of the available records. Also, the time interval Δt between the censuses must be equal, so that $\Delta t = t_2 - t_1 = t_1 - t_0$. The constants can then be calculated using the following equations:

$$P_{sat} = \frac{2 P_0 P_1 P_2 - P_1^2(P_0 + P_2)}{P_0 P_2 - P_1^2} \tag{13.7.11}$$

$$a = \log_e \frac{P_{sat} - P_2}{P_2} \tag{13.7.12}$$

$$b = \frac{1}{\Delta t} \log_e \frac{P_0(P_{sat} - P_1)}{P_1(P_{sat} - P_0)} \tag{13.7.13}$$

Ratio Method

This method makes use of population projections made by professional demographers for the nation. The basic assumption is that the ratio of the population of the city under study, to that of the larger group will continue to change in the future in the same manner as it has in the past. The ratio is calculated for a series of censuses, the ratio line is extended into the future, and the projected ratio is multiplied by the forecast regional population to obtain the future population of interest (Schmitt, 1954).

13.7.3 FIRE DEMAND

A water supply system has to be designed to provide the necessary flow for fire fighting in the event of a fire, in addition to the flow for peak demands. A distribution system is usually designed to provide the larger of the maximum hourly demand, or the maximum daily demand plus the fire demand to any group of fire hydrants in the system. The fire demand can often be the governing criteria in the design of reservoir capacity, pipe sizes, and pumping capacities for small communities and cities. The fire flow can be estimated using an empirical formula recommended by the Insurance Services Office (1974):

$$F = 224C\sqrt{A} \tag{13.7.14}$$

where F = required fire flow (L/min), A = total floor area excluding the basement (m²), and C = coefficient related to the type of construction, existence of automatic sprinklers, and building separation.

The value of C is 1.5 for wood-frame construction, 1.0 for ordinary construction, 0.8 for noncombustible construction, and 0.6 for fire-resistive construction. Equation 13.7.14 can be written

TABLE 13.7.2   Required Fire Flow and Minimum Duration

Population	Fire flow* L/min (gal/min)	Duration hour
1000	3780 (1000)	4
2000	5670 (1500)	6
4000	7560 (2000)	8
10,000	11,340 (3000)	10

* Values are calculated using equation 13.7.16.

as:

$$F = 18C\sqrt{A} \tag{13.7.15}$$

where F is in gal/min and A is in ft^2.

A variety of factors can affect the required fire flow (Carl et al., 1973). The separation distance between buildings is an important factor. For residential areas with single- or two-family residences, with separation distances less than 9 m (about 30 ft), the required fire flow would be between 3600 to 5700 L/min (950 to 1500 gpm). When buildings are separated by more than 30 m (about 100 ft), the minimum required fire flow is 1890 L/min (500 gal/min). For contiguous buildings, the required fire flow is 9450 L/min (2500 gal/min).

For small communities with population less than or equal to 10,000, the required fire flow can be calculated from the following empirical formula recommended by the American Insurance Association:

$$F = 1020\sqrt{P}\,(1 - 0.01\sqrt{P}) \tag{13.7.16}$$

where F = required fire flow in gal/min (L/min ÷ 3.78) and P = population in thousands.

The required fire flows calculated from Equation 13.7.16 are provided in Table 13.7.2. The last column provides the minimum required duration, for which the water supply system should be able to provide the required fire flows.

13.7.4 DESIGN FLOW

In the event of a fire, the public water supply system should be able to deliver the required fire flow for a period ranging from a minimum of 4 h to a maximum of 10 h for major fires. Adequate pumping and reservoir storage capacities have to be determined accordingly. Design flow for the pipe distribution system should be the greater of

1. sum of maximum daily demand and fire flow, or
2. maximum hourly demand, where all values are converted to the same units of expression (L/min or L/h).

The pipe diameters have to be calculated accordingly.

Example Problem 1

Calculate the design flow for a distribution system serving a population of 150,000 in a city that has residential zones, as well as commercial and industrial zones. The total floor area of the largest commercial building in the city is 30,000 m^2, with noncombustible construction. The average daily consumption is 670 Lpcd. The maximum daily demand is 1.5 times the average demand, and the maximum hourly demand is 2.0 times the average demand.

Solution

Average daily demand = 670 Lpcd $\times$ 150,000 = 100.5 $\times$ 10^6 L/d

Average hourly demand = average daily demand/24 = 4.2 $\times$ 10^6 L/h

Maximum daily demand = 1.5 $\times$ average daily demand = 1.5 $\times$ (100.5 $\times$ 10^6 L/d)

$$= 150.75 \times 10^6 \text{ L/d} = 6.3 \times 10^6 \text{ L/h}$$

Maximum hourly demand = 2.0 $\times$ average hourly demand = 2.0 $\times$ (4.2 $\times$ 10^6 L/h)

$$= 8.4 \times 10^6 \text{ L/h}$$

Required fire flow (using Equation 13.7.14)

$$F = 224 \, C \, (A)^{1/2} = 224 \, (0.8) \, (30,000)^{1/2} = 31,038.35 \, \text{L/m} = 1.9 \times 10^6 \, \text{L/h}$$

Maximum daily demand + fire flow = 6.3 $\times$ 10^6 L/h + 1.9 $\times$ 10^6 L/h = 8.2 $\times$ 10^6 L/h

Maximum hourly demand is greater than the sum of maximum daily demand and fire flow

Therefore, Design Flow = Maximum hourly demand = 8.4 $\times$ 10^6 L/h

Answer: Design flow for the distribution system should be 8.4 $\times$ 10^6 L/h.

13.7.5   PUBLIC DRINKING WATER SYSTEMS IN THE UNITED STATES

According to the U.S. EPA (United States Environmental Protection Agency) Office of Ground Water and Drinking Water, there are approximately 171,000 public water systems in the United States serving more than 271 million people. Water systems are classified according to the numbers served, source of water, and type of population. The following definitions of terminology are used by the U.S. EPA.

Public Water System (PWS)

Provides piped water for human consumption to at least 15 service connections or serves an average of at least 25 people for at least 60 days a year. There are three main types of public water systems:

1. Community Water System (CWS): A public water system that supplies water to the same population year-round.
2. Nontransient Noncommunity Water System (NTNCWS): A public water system that supplies water to at least 25 of the same people at least six months per year, but not year-round. Examples are schools, factories, and hospitals, which have their own water systems.
3. Transient Noncommunity Water System (TNCWS): A public water system that supplies water to places such as campgrounds and picnic areas, where people do not remain for long periods of time.

Water systems are also classified according to the number of people that they serve:

- Very Small Water System: serves 25 to 500 persons
- Small Water System: serves 501 to 3300 persons

TABLE 13.7.3 Public Drinking Water Systems Inventory Information*

Factors	CWS	NTNCWS	TNCWS
Number of systems	54,728	20,061	96,153
Population served (millions)	249	6.1	16.2
Used surface water source	10,500	760	2,143
Population served (millions)	160	0.8	0.9
Used ground water source	44,219	19,300	94,009
Population served (millions)	89	5.3	15.3
Small or very small water systems	86%	99.5%	99.7%
% Population served in system	10	83	55
Medium, large, or very large systems	14%	0.5%	0.3%
% Population served in system	90	17	45

*Data compiled from the 1996 *Annual Compliance Report of the National Public Water System Supervision Program* (USEPA, 1998).

Note: CWS = Community Water Systems; NTNCWS = Non-Transient Non-Community Water Systems; and TNCWS = Transient Non-Community Water Systems.

- Medium Water System: serves 3301 to 10,000 persons
- Large Water System: serves 10,001 to 100,000 persons
- Very Large Water System: serves more than 100,000 persons

Table 13.7.3 provides a summary of the information from the National Public Water System Supervision Program's 1996 Annual Compliance Report, dated September 1998. The report was based on information in the Safe Drinking Water Information System, for the 12-month period ending December 1996, as reported to U.S. EPA by the states until June 1998.

REFERENCES

American Insurance Association "Standard Schedule for Graphing Cities and Towns of the United States with Reference to Their Fire Defences and Physical Conditions," National Board of Fire Underwriters, NY, 1956.

Carl, K. J., et al., "Guidelines for Determining Fire-flow Requirements," *J. Am. Water Works Assoc.,* 65: 335, 1973.

Insurance Services Office, *Guide for Determination of Required Fire Flow,* New York, Insurance Services Office, 1974.

McGhee, T. J., *Water Supply and Sewerage,* Sixth edition, McGraw-Hill, Inc., New York, 1991.

Schmitt, R. C., "Forecasting Population by the Ratio Method," *J. Am. Water Works Assoc.,* 46: 960, 1954.

USEPA, *1996 Annual Compliance Report,* National Public Water System Supervision Program, September, 1998.

Viessman, W., Jr., and Hammer, M. J., Water Supply and Pollution Control, Sixth edition, Addison Wesley Longman Inc., Menlo Park, California, 1998.

B. SCREENS

Surface water intakes are usually located in a quiescent section of a lake or river. Water intake screens are placed at some depth below the surface to eliminate as much floating debris as possible. Suspended and floating materials in the water include dead leaves, branches, weeds, rags, and a variety of solids including fish. The solids can damage pumps and mechanical equipment, and interfere with the flow in pipes and channels. Screening devices are placed ahead of pumps, and are used to preclude entrance of these materials into the water treatment plant. Different types of screens are available depending on site requirements and water characteristics.

13.7.6 TRASH RACKS

Trash racks are screens that have large openings to exclude larger debris and garbage. These racks consist of rectangular or circular steel bars arranged in a parallel fashion with openings between 5.1 to 15.2 cm (2 to 6 in). Trash racks are set vertically or at an incline to the horizontal. Mechanical rakes are used to clear the solids collected on the trash racks. Rake machines are operated by hydraulic jacks, cables, and wire ropes. Trash racks are followed by coarse or fine screens.

13.7.7 COARSE SCREENS OR BAR SCREENS

Coarse screens or bar screens are similar to trash racks, except the clear openings between bars are from 2.5 to 7.6 cm (1 to 3 in). Coarse screens can be cleaned manually or mechanically by rakes. Manually cleaned screens should be placed on a slope of 30 to 45° with the horizontal. This increases the screening surface, makes cleaning easier, and also prevents excessive head loss by clogging. Mechanically cleaned screens can be placed vertically or at an incline to the horizontal.

The Ten States' Standards (Great Lakes Upper Mississippi River Board of State Sanitary Engineers, 1968) have specified an average flow velocity of 0.3 m/s (1 ft/s) through manually raked bar screens, and a maximum velocity of 0.8 m/s (2.5 ft/s) through mechanically cleaned bar screens during wet weather flows. The velocity is calculated from a vertical projection of the screen openings on the cross-sectional area between the channel invert and the flow line.

13.7.8 HEAD LOSS FOR SCREENS

The head loss for screens will increase as more and more solids accumulate on the screen. The minimum allowable head loss for a manually cleaned screen is 0.15 m (6 in). The maximum head loss should not exceed 0.8 m (2.5 ft). High head losses will reduce flow velocities, and when the clogged screens are cleaned, high flow surges can occur causing problems to pumps.

Head loss for a clean screen can be calculated from the following relationship:

$$H = \frac{V^2 - v^2}{(0.7)\,2g} \tag{13.7.17}$$

where H = head loss, m; V = velocity through the screen, m/s; v = velocity of incoming water, m/s, and g = acceleration due to gravity, m/s^2.

13.7.9 DESIGN CONSIDERATIONS

The major consideration for design and location of intake screens is minimization of debris that comes in contact with the screen and clogs the openings of the screens. To this extent, the following parameters have to be evaluated:

- Depth of the screen below the water surface
- Water quality

- Flow velocities at different depths
- Distance from the shoreline
- Proximity of screens to one another, when multiple screens are used
- Additional structures and support system for the screen in the water
- Water temperature
- Seasonal fluctuations of water level
- Possibility of biological fouling of screens (e.g., because of high concentration of zebra mussels in water)
- Installation of screen cleaning system
- Navigational demands of the location, if any

13.7.10 FINE SCREENS

Fine Screens have clear openings ranging from 9.5 mm to as small as 0.1 mm (0.375 to 0.004 in). Depending on water quality, a fine screen may be preceded by a coarse screen and a trash rack. Fine screens have to be cleaned frequently to maintain efficient screening of small particles. Cleaning is performed by using brushes, water sprays, water jets, or air jets to remove particles from the screens.

Fine screens are constructed from a variety of materials, the most common being stainless steel. Varied fabrication techniques are used to provide the small screen sizes. These include:

- Profile bars arranged in a parallel manner with standard openings from 0.5 mm (0.02 in).
- Slotted perforated plates with 0.8 to 2.4 mm (0.03 to 0.09 in) wide slots.
- Wedge shaped bars welded together into flat panel sections. Openings range from 5.1 to 0.13 mm (0.2 to 0.005 in).
- Looped wire construction with openings as small as 0.13 mm (0.005 in).
- Wire mesh with approximately 3.3 mm (0.13 in) openings.
- Woven wire cloth with openings as small as 2.5 mm (0.1 in).

Fine screens can be passive or active. *Passive screening* involves flow of water through the intake screen at a low, uniform velocity, keeping the aquatic life and other debris in the water source. The screen itself has no moving parts. *Active screening* involves movement of the screen relative to the movement of water. These include traveling water screens, revolving drum screens, revolving disk screens, and so forth. Fine screens can be constructed in a variety of shapes. The most common are the flat panel and drum style screens (Figure 13.7.3). Another variation is the tee style screen (Figure 13.7.4).

13.7.11 DRUM STYLE (PASSIVE) SCREEN

Drum style screens are used in quiescent waters, such as in lakes and reservoirs, where the flow velocities are very low. Conventionally, drum screens are mounted with the flat top facing upward, so that most of the debris collects on the nonscreen surface (Figure 13.7.3). Drum screens can be constructed with the slots parallel to the long axis of the screen assembly. When this type of screen is mounted conventionally, the slots are vertical and tend to shed debris falling past them. Depending on the requirements of the location, drum screens can also be mounted horizontally. In that case, an air-jet or airburst cleaning system is usually installed.

Drum Style

Flat Panel

FIGURE 13.7.3   Flat panel and drum style intake screens (Courtesy of Hendrick Screen).

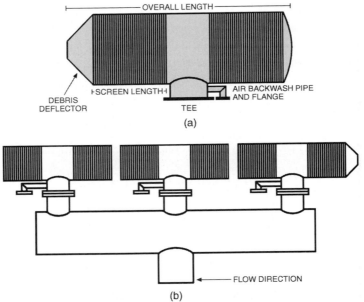

(a)

(b)

FIGURE 13.7.4   Tee style intake screen (a) side view; (b) possible multiple arrangement (Courtesy of Hendrick Screen).

13.7.12   *TEE STYLE (PASSIVE) SCREEN*

A tee style screen offers flexibility in design, because it can be constructed with a smaller diameter than a drum screen. It is usually easier to place the tee screens further below the water surface, further above the stream bottom, or both. A number of tee screens can be installed at the same level (Figure 13.7.4), or at different levels in a tower intake to accommodate fluctuating water levels. The tee screen is installed with the small area pointing upstream, so that the screen is less susceptible to damage from floating debris or bottom load from the stream. Cleaning is accomplished by air jets or airburst systems.

13.7.13   *TRAVELING WATER SCREEN*

Traveling water screens have widespread application in intake systems of water treatment plants, wastewater treatment plants, irrigation channels, power plants, paper mills, and other industrial plants. They can be operated intermittently or continuously, depending on service requirements. Screens can be in the form of wire mesh trays or baskets mounted on two strands of long pitch heavyweight chains. Various screen openings for wire mesh trays are illustrated in Figure 13.7.5. A wire mesh basket is presented in Figure 13.7.6. Different types of debris handling and fish handling baskets are available.

During operation, water enters the intake, passes through the screening surface, and leaves and debris are captured on the trays or screening baskets. As the screen is revolved, the debris is lifted from the intake by the upward travel of the trays. At the top of the screen, refuse is removed from the trays by jet sprays of water. Sprays wash the refuse into a trough and sluice them away for disposal. For intermittent screens, operation is controlled by the water head differential across the screen resulting from clogging. A variety of special materials, coatings, and ancillary equipment is available for corrosive, abrasive, and other types of problematic conditions.

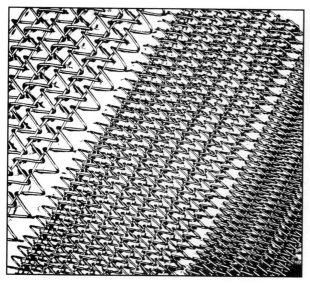

FIGURE 13.7.5 Various screen openings for travelling water screens (Courtesy of Farm Pump and Irrigation Co.).

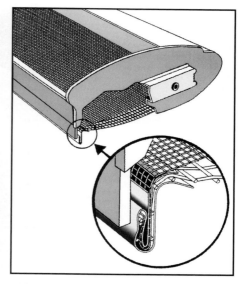

FIGURE 13.7.6 Pre-tension mesh™ basket used in traveling water screen (Courtesy of U.S. Filter/Envirex).

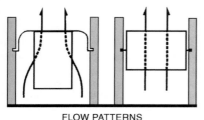

FLOW PATTERNS
Dual Flow Screen vs. Through Flow Screen

FIGURE 13.7.7   Flow patterns in traveling water screens (Courtesy of U.S. Filter/Envirex).

There are two types of traveling water screens depending on the flow pattern. They are: (1) through flow traveling water screen, and (2) dual flow traveling water screen. The difference in flow paths is illustrated in Figure 13.7.7.

Through Flow Traveling Water Screen

Through flow screens are installed with the submerged screening surfaces perpendicular to the intake flow. They collect and transport debris to the head enclosure where jets of water from spray nozzles flush the refuse into a disposal trough. The operation is illustrated in Figure 13.7.8.

Dual Flow Traveling Water Screen

The dual flow screen is positioned parallel to the flow and in the flow, so that water can pass through both the ascending and descending panels. Both sides are used, increasing the capacity of the screens (Figure 13.7.9). Debris is removed from the water as the screening panels break the water on the ascending run. When the panels pass over the headsprockets at the top of the screen,

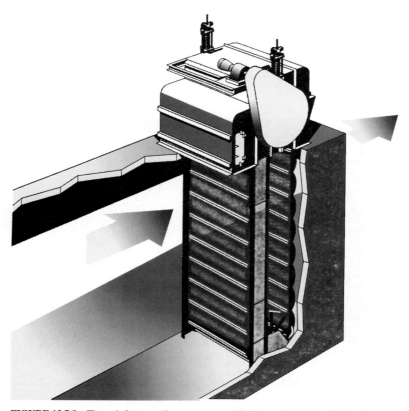

FIGURE 13.7.8   Through flow traveling water screen (Courtesy of U.S. Filter/Envirex).

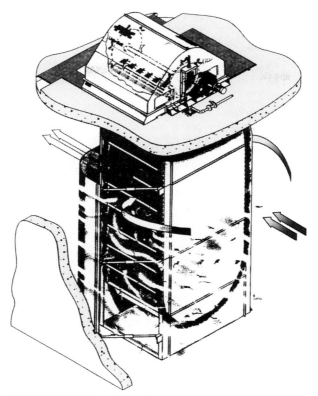

FIGURE 13.7.9 Dual flow traveling water screen (Courtesy of FMC, a U.S. Filter Company).

debris is removed by water sprays. Debris not removed is carried back into the water, remaining in the upstream portion of the channel. The front elevation of a dual flow screen is illustrated in Figure 13.7.10.

The cost-effectiveness of a dual flow system is due to its design, which presents almost double the screening area and debris carrying capacity to the intake water. The possibility of carryover of debris to the downstream side to cause mechanical problems is virtually eliminated with this screen. The arrangement of a complete water intake screening system with stop gate, trash rake and screen is illustrated in Figure 13.7.11.

13.7.14 REVOLVING DRUM SCREEN

In this type of screen, a woven stainless steel wire mesh (no. 12 to 20) is installed on a cylindrical frame. The cylinder rotates around its axis while it is partially submerged. Revolving drum screens require a fairly constant water level, and a weir is used to maintain water elevation.

Revolving drum screens can be of: (1) outward flow, or (2) inward flow type. In the *outward flow screen*, water approaches the drum from a direction parallel to the revolving axis. The water flows into the interior of the drum screen, and flows out at a right angle to the axis. The solids retained on the inside surface of the screen are raised above the water level as the drum rotates, and are then removed by a water spray. In the *inward flow screen*, water approaches the drum in a direction

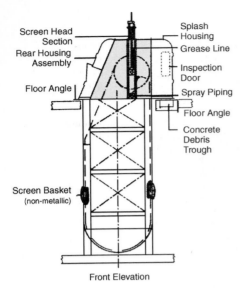

Screen Head
Section
Rear Housing
Assembly
Floor Angle

Splash
Housing
Grease Line
Inspection
Door
Spray Piping
Floor Angle
Concrete
Debris
Trough

Screen Basket
(non-metallic)

Front Elevation

FIGURE 13.7.10 Front elevation of a dual flow traveling water screen (Courtesy of U.S. Filter/Envirex).

perpendicular to its axis. Solids are retained on the outside surface of the drum, and removed by brushes, water sprays, or air backwash systems.

13.7.15   REVOLVING DISK SCREEN

In this system, a slowly revolving disk screen is placed in the approach channel so that the flow passes through the screen. As the water flows through the screen, solids are retained, elevated above the water level, and flushed by a water spray to a trough. The disk screen requires a constant water level, which is maintained using a weir.

13.7.16   CLEANING OF FINE SCREENS

Fine screens can be cleaned using water or air backwash. Air backwash systems are used for cleaning both active and passive screens, especially the drum style and tee style screens. Fine screens usually require frequent cleaning, and air is a preferred medium for cleaning intake screens. Air moves with less head loss than water. Also, large quantities can be compressed and stored in a small physical space. Air moves at a high velocity with less energy expenditure as compared to water.

The necessary elements of an air backwash system are as follows:

- A source of compressed air
- An accumulator or receiver to store the air
- A valve to allow rapid release of stored air
- A manifold
- A control system

13.7.17   BIOLOGICAL FOULING

Intake screens are subject to fouling or clogging by aquatic vegetation and/or organisms. Clogging of fine screens by Zebra mussels is a problem in many localities depending on water quality. Screens made with a 70-30 Copper-Nickel alloy act as a repellant to many potential encrustants. Other types of patented screen materials and repellant coatings are marketed by different manufacturers. Nontoxic coatings can be applied to new or existing screens, which provide an extremely smooth surface on which Zebra mussels have difficulty attaching themselves. In some cases, a chemical feed system is installed to disperse biocidal chemicals onto the screen surface and discharge piping. The chemicals used should be toxic only to the target organisms and not to other plants or animals.

Note: Information on the various types of screens were provided by a number of manufacturers including Farm Pump & Irrigation Co.; FMC, a U.S. Filter Company; Hendrick Screen; and U.S. Filter/Envirex.

REFERENCE

Great Lakes Upper Mississippi River Board of State Sanitary Engineers, "Recommended Standards for Water Works," Health Research Inc., Albany, NY, 1968.

C. CHEMICAL REACTORS

A reactor is a tank or vessel where chemical, biochemical, or biological reactions take place, in a solid, liquid, or gaseous medium. When chemical reactions are the major transformation mechanism, the reactor is called chemical reactor. In water treatment processes, chemical reactors are used in coagulation-flocculation, lime softening, taste and odor control, disinfection, and other unit processes that involve chemical reactions.

There are three types of ideal reactors. They are: (1) Batch reactor, (2) Plug Flow Reactor (PFR), and (3) Continuous-flow Stirred Tank Reactor (CSTR). The flow patterns and hydraulics of these reactors can be described with the help of mathematical models. Real process reactors usually deviate from ideal flow patterns and reactions. However, ideal designs and models provide a baseline, which can be modified to represent real life conditions.

13.7.18 RESIDENCE TIME/DETENTION TIME IN REACTOR

The theoretical detention time or residence time of the fluid particles in a reactor is defined as:

$$t = V/Q \qquad\qquad (13.7.18)$$

where t = detention time, V = volume of reactor and Q = flow rate of fluid, volume/time.

The actual detention time in a reactor can be evaluated by introducing a pulse of dye or tracer with the influent during steady state flow, and measuring the tracer concentration in the effluent over a period of time. The effluent concentration is plotted versus time on a graph paper, and the centroid of the resulting curve is located as the actual detention time. This is usually found to be slightly smaller than the value calculated from Equation 13.7.18. This is due to flow dispersion with back-mixing and short-circuiting of fluid in the reactor.

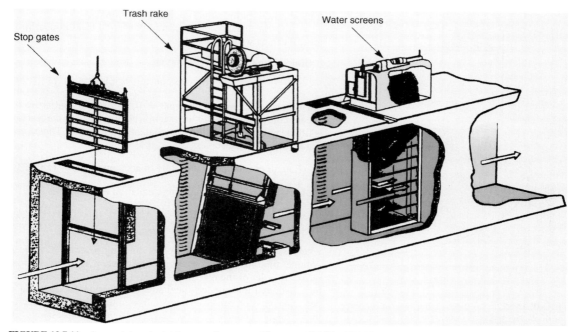

FIGURE 13.7.11 A complete water intake screening system (Courtesy of FMC, a U.S. Filter Company).

13.7.19 CONVERSION/REMOVAL OF REACTANT

The conversion or removal of a reactant is calculated as:

$$f = \frac{[A_0] - [A]}{[A_0]} = 1 - \frac{[A]}{[A_0]} \qquad (13.7.19)$$

where f = conversion or removal efficiency, $[A_0]$ = initial concentration of reactant A at time $t = 0$, moles/volume, $[A]$ = concentration of reactant A at time $t = t$, moles/volume.

13.7.20 BATCH REACTOR

In a simple batch reactor, materials are added to the reactor and mixed for a certain period of time necessary for the chemical reactions to occur (Figure 13.7.12a). At the end of the reaction period, the mixture is removed from the tank. All fluid elements have the same residence time in the reactor. At any instant, the reactor contents are homogeneous so that all fluid elements have the same composition. However, the composition is time dependent. A variation of the concentration of a reactant with time in a batch reactor is illustrated in Figure 13.7.12b.

Batch reactors are often used for liquid phase reactions, which generate a small quantity of products. These reactors are especially suitable for bench-scale experiments of new processes in order to determine the effects of various parameters on the process. Process variables can be altered very easily, which offers a great flexibility in reactor operation. They are also used for producing several different products from one material. Batch reactors are extensively used in pharmaceutical and dyestuff industries, and in the production of certain specialty chemicals.

Batch reactors are not suitable for commercial scale applications, especially for gas phase reactions. Labor costs and materials handling costs are high because of the time and activity involved in filling, emptying, and cleaning of the reactors. The sum of these nonproductive periods of time might be equal to or larger than the time required for the actual chemical reaction.

A material balance of a reactant converted to a product in a reactor can be written as:

(Rate of input) = (Rate of output) + (Rate of accumulation)

$$- \text{(Rate of consumption)} \qquad (13.7.20)$$

For a batch reactor, rate of input = 0, and rate of output = 0, because the time period for the reaction begins just after the reactor is filled, and ends just before the contents are emptied. Therefore, Equation 13.7.20 becomes:

(Rate of consumption) = (Rate of accumulation)   (13.7.21)

From Equation 13.7.21, the design equation is written as:

$$r_A = \frac{d[A]}{dt} \qquad (13.7.22)$$

where r_A = rate of consumption of limiting reactant A, moles/volume · time and $[A]$ = concentration of reactant A, moles/volume.

If the order of the reaction is known, an expression for r_A can be substituted into the left-hand side of Equation 13.7.22, and a design expression can be developed from integration of the resulting differential equation. Table 13.7.4 presents the expressions for

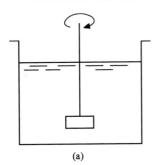

(a)

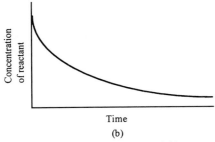

Time

(b)

FIGURE 13.7.12.   (a) Batch reactor, and (b) concentration of reactant versus time curve for batch reactor.

TABLE 13.7.4 Design Equations for Batch, PFR, and CSTRs

Order of reaction	Rate expression	Batch/PFR	CSTR
0	$r_A = -k$	$kt = [A_0] - [A]$	$kt = [A_0] - [A]$
1	$r_A = -k[A]$	$kt = \log_e \frac{[A_0]}{[A]}$	$kt = \frac{[A_0]}{[A]} - 1$
2	$r_A = -k[A]^2$	$kt = \frac{1}{[A_0]}\left(\frac{[A_0]}{[A]} - 1\right)$	$kt = \frac{1}{[A]}\left(\frac{[A_0]}{[A]} - 1\right)$

Notes: r_A = rate of consumption of reactant A, moles/volume × time
$[A_0]$ = initial concentration of reactant A at time $t = 0$, moles/volume
$[A]$ = concentration of reactant A at time $t = t$, moles/volume
k = reaction rate constant
PFR = Plug Flow Reactor
CSTR = Continuous-flow Stirred Tank Reactor

zero, first, and second order reactions in a batch reactor. *Example Problem 1* illustrates the development of the expression for a zero order reaction.

Example Problem 2

Develop an expression for the detention time for a zero order reaction involving a reactant A in a batch reactor.

Solution
For a zero order reaction, $r_A = -k$
Where, k = reaction rate constant
Substituting the expression for r_A into Equation 13.7.22:

$$-k = \frac{d[A]}{dt}$$

or,

$$-k \int_0^t dt = \int_{[A_0]}^{[A]} d[A]$$

or,

$$-kt = [A] - [A_0]$$

therefore:

$$t = \frac{1}{k}([A_0] - [A])$$

is the required expression.

13.7.21 *PLUG FLOW REACTOR (PFR)*

In a plug flow reactor, the elements of the homogeneous fluid reactant move through the reactor tube as plugs moving parallel to the tube axis (Figure 13.7.13a). The flow pattern is called plug flow or piston flow. Fluid elements flow through the tank and are discharged in the same sequence as they entered. Although there may be some lateral mixing, there is no longitudinal mixing of fluid. The velocity profile at a given cross section is flat, as there is no axial diffusion or back-mixing of fluid elements. As a result, the concentration of reactant across any vertical cross section is the same (Figure 13.7.13c). There is a concentration gradient from inlet to outlet (Figure 13.7.13b) because of disappearance or chemical conversion of the reactant. All fluid elements have the same residence time in the reactor.

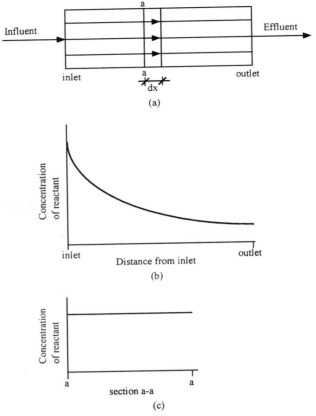

FIGURE 13.7.13. (a) Plug flow reactor (PFR), (b) lateral distribution of reactant concentration in a PFR, and (c) longitudinal distribution of reactant concentration for section a-a in the PFR.

The plug flow reactor is suitable for continuous gas phase reactions that take place at high pressure
and temperature. Heat transfer is minimized by putting a jacket of insulating material around the
reactor tube. Another advantage is the absence of any moving parts inside the reactor. For the same
feed composition and reaction temperature, the average reaction rate is usually significantly higher in
a PFR as compared to a CSTR or a series of CSTRs with the same total volume. More efficient use
of reactor volume is possible with a PFR, which makes it suitable for use in processes that require
very large capacity. It is also used for liquid phase reactions that require short residence times for
the desired chemical conversion. With sufficiently high recycle rates, the behavior of a PFR becomes
similar to that of a CSTR. Let us consider a differential section dx (as illustrated in Figure 13.7.13a)
with a differential volume dV in the plug flow reactor. A material balance on the limiting reactant A
in the differential volume yields:

$$\text{(Rate of input)} = \text{(Rate of output)} - \text{(Rate of consumption)} \qquad (13.7.23)$$

because, at steady state conditions, rate of accumulation $= 0$.

From Equation 13.7.23, the design equation is written as:

$$r_A = \frac{d[A]}{dt}$$

(13.7.24)

which is the same as the design equation for a batch reactor. All the terms have the same meanings as explained previously. When order of the reaction is known, an expression for r_A can be substituted into the left-hand side of Equation 13.7.24, and a design expression can be developed from integration of the resulting differential equation. Table 13.7.4 presents the expressions for zero, first, and second order reactions in a PFR. The development of an expression for a zero order reaction was described in Example Problem 1.

13.7.22 CONTINUOUS-FLOW STIRRED TANK REACTOR (CSTR)

A continuous-flow stirred tank reactor is a reactor in which the reactants flow continuously into the reactor, and the product stream is discharged continuously while the reactor contents are continuously mixed (Figure 13.7.14a). This type of reactor is also called back-mix reactor or completely mixed reactor. The advantages of a CSTR are simplicity of construction, relative ease of control, and ease of access to the interior surface. These reactors are used mainly for liquid phase reactions at low or atmospheric pressures. They can handle reactions involving gas bubbles, solids, or one liquid suspended in another liquid phase.

The average reaction rate in a CSTR will usually be less than the rate in a PFR of equal volume, under the same conditions of feed and temperature. The rate of heat transfer per unit volume of reaction mixture is also lower in a CSTR as compared to a PFR. At the same operating temperatures and feed concentrations, a CSTR of larger volume is required than a PFR to achieve the same conversion of reactant.

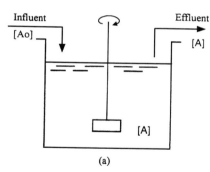

The ideal CSTR model assumes that the reactor contents are perfectly mixed so that the system properties are uniform throughout. The effluent composition and temperature are identical with those of the reactor contents. Because of complete mixing, there is a step change in composition and temperature of the fluid at the instant that it enters the reactor. Chemical conversion is almost instantaneous, and the influent reactant concentration $[A_0]$ is immediately reduced to final effluent concentration $[A]$.

Within the limit of complete mixing, a tracer molecule in the influent has equal probability of being located anywhere in the reactor after a very small time interval. Thus, all fluid elements in the reactor have equal probability of leaving the reactor with the effluent in the next time increment. As a result, there is a broad distribution of residence times for various fluid elements or particles as illustrated in Figure 13.7.14b.

Because conditions are uniform throughout the reactor volume, a material balance for the limiting reactant A can be written over the entire reactor as:

(Rate of input) = (Rate of output) − (Rate of consumption)

(13.7.25)

where steady state conditions, rate of accumulation = 0.

From Equation 13.7.25, the design equation is written as:

$$r_A = \frac{[A] - [A_0]}{t}$$

(13.7.26)

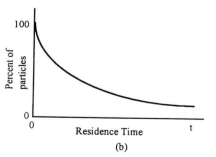

FIGURE 13.7.14.   (a) CSTR, and (b) Residence time distribution of fluid particles in a CSTR.

where all the terms have the same meanings as defined previously. When order of the reaction is known, an expression for r_A can be substituted into the left-hand side of Equation 13.7.26, design expression can be developed from integration of the resulting differential equation. Table 13.7.4 presents the expressions for zero, first, and second order reactions in a CSTR. Real flow reactors fall somewhere between a CSTR and PFR.

13.7.23   *SERIES/CASCADE OF CSTRs*

In order to increase the removal efficiency, a series or cascade of CSTRs can be used where the effluent from one reactor serves as the influent to the next reactor (Figure 13.7.15). Although the concentration is uniform within any one reactor, there is a progressive decrease in reactant concentration from the first reactor to the last one in the series. There is a stepwise variation in reactant composition and temperature as the fluid moves from one reactor to the next one.

Assuming that the conditions within any individual reactor in the series are not influenced by downstream conditions, and conditions of the inlet stream and those prevailing in the reactor are the only variables that influence reactor performance, we can write the following design expression:

$$r_{Ai} = \frac{[A]_i - [A]_{(i-1)}}{t_i} \qquad (13.7.27)$$

where complete mixing and steady state conditions were assumed, and r_{Ai} = rate of consumption of A in ith reactor, t_i = detention time in ith reactor, $[A]_i$ = concentration of A in effluent from ith reactor, $[A]_{(i-1)}$ = concentration of A in effluent from $(i-1)$th reactor, and which is equal to concentration of A in influent to ith reactor.

The detention time in ith reactor is given by:

$$t_i = V_i / Q \qquad (13.7.28)$$

where V_i = volume of ith reactor and Q = flow rate into reactor, volume/time.

In a series of n reactors, the overall conversion is given by

$$f = \frac{[A_0] - [A]_n}{[A_0]} \qquad (13.7.29)$$

where $[A_0]$ = concentration of reactant A in influent to 1st reactor and $[A]_n$ = concentration of reactant A in effluent from nth reactor.

Conversion in individual reactors is calculated from reactant concentrations in the influent and effluent of that particular reactor.

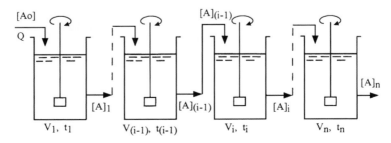

FIGURE 13.7.15.   Series of CSTRs.

13.7.24   SEMI-BATCH OR SEMI-FLOW REACTOR

A number of different modes of operation are possible with semi-batch or semi-flow reactors. A few of the different options are as follows:

- A batch reactor partially filled with one reactant, with progressive addition of other reactant(s) until the reaction is completed
- A reactor in which all the reactants are added at the same time as a batch, but the products are removed continuously
- A reactor in which the reactants are added intermittently
- A reactor in which the products are removed intermittently

REFERENCES

Hill, C. G., Jr., *An Introduction to Chemical Engineering Kinetics & Reactor Design*, John Wiley & Sons, New York, NY, 1977.

CHAPTER 13
CONTROL TECHNOLOGIES

SECTION 13.8

WATER SUPPLY TREATMENT

Leonard W. Casson

Dr. Casson is associate professor, environmental engineer, Department of Civil and Environmental Engineering, University of Pittsburgh. His research focus is the adsorption, fate, and transport of particles, chemical, and environmental pathogens in unit processes and the natural environment.

A. COAGULATION, FLOCCULATION, MIXING, AND MIXERS

13.8.1   INTRODUCTION

Various combinations of unit processes are used to treat municipal drinking water prior to distribution to the public. These unit processes are dependent upon the raw water source (i.e., groundwater or surface water), the raw water quality, the population served, and the predicted population growth characteristics.

The focus of this chapter will be on coagulation, flocculation, mixing, and mixers as applied to municipal drinking water treatment using a surface water source. Generally, mixing and flocculation are the first steps in the particle growth and removal process used for municipal drinking water treatment.

13.8.2   COAGULATION AND FLOCCULATION

Although coagulation and flocculation may have varying definitions within the science and engineering community. This chapter will use the definitions from *Water Treatment Plant Design* (AWWA & ASCE, 1998). In this reference, *coagulation* is defined as a process in which chemicals are added to water, causing a reduction of the forces that stabilize the particles in water. *Flocculation* is defined as the agglomeration of small particles and colloids into larger settleable or filterable particles (AWWA, 1998).

Coagulation is a two step process, the first step (rapid mixing) involves adding and dispersing chemicals (termed coagulants), which result in particle destabilization. The second step is particle transport (flocculation), which should result in particle collisions, attachment, and subsequent particle growth. In reality, effective flocculation is simply making a small number of large particles out of a large number of small particles. Many interparticle forces are important in the coagulation process. These interparticle forces, which are fully discussed in *Water Quality and Treatment* (AWWA, 1990) and other texts, include van der Waals attractive forces, electrostatic repulsive forces, and hydrodynamic forces.

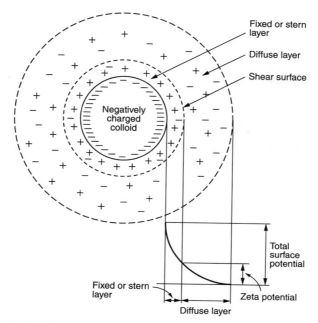

FIGURE 13.8.1.   Electrical double layer of a negatively charged particle in water (Sawyer et al., 1994).

Particle Surface Charge

Suspended particles are stable and generally have a negative surface charge in the pH range of natural waters (i.e., pH 6–9). Although these individual suspended particles have a negative surface charge, the net charge on the suspension is neutral. Because water is a polar molecule, the negatively charged particle surface attracts oppositely charged ions in the water. As shown in Figure 13.8.1, near the negatively charged particle surface, an excess of positive charges (counterions) exists. The electrically disturbed region of charges in the fluid surrounding the particle surface is termed the diffuse layer or electric double layer. This diffuse layer has been studied and modeled by many researchers and is shown graphically in Figure 13.8.1 (Sawyer et al., 1994). The zeta potential, also shown in this figure, is an experimental measure of particle charge. When two stable particles in water collide, the interaction of the diffuse layers acts as a repulsive force that prevents particle contact and subsequent attachment.

Coagulation

The first step is the coagulation process is rapid mixing, which involves adding and dispersing chemicals (termed coagulants) to aid in particle destabilization. Four mechanisms of destabilization will be discussed in this section: compression of the double layer, adsorption and charge neutralization, enmeshment in a precipitate (sweep floc), and adsorption and interparticle bridging. The following narrative will provide a brief description of each of these four particle destabilization mechanisms. For a full discussion, see *Water Quality and Treatment* (AWWA, 1990).

Compression of the double layer occurs by raising the ionic strength of a solution using a salt (e.g., Na^+, Ca^{++}, and Al^{+++}). This increased concentration of ions in the solution will provide either a higher concentration of counterions to balance the surface charge of the particle or more

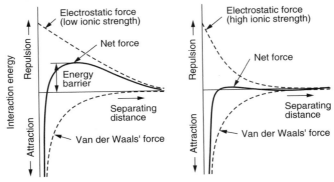

FIGURE 13.8.2.   The effect of ionic strength on the attractive and repulsive forces associated with a negatively charged particle in water (Sawyer et al., 1994).

highly charged counterions to balance the particle surface charge. In either case, the distance the diffuse layer extends from the particle surface into solution will decrease as shown in Figure 13.8.2 (Sawyer et al., 1994). Note the reduced net force of the diffuse layer and the reduced separating force for the high ionic strength case in Figure 13.8.2b as compared to the lower ionic strength case in Figure 13.8.2a.

Adsorption and charge neutralization occurs when hydrophobic chemicals of opposite charge (termed coagulants) are added to a solution to reduce the surface charge of the suspended particles. Examples of chemicals used as coagulant are as follows: alum(aluminum sulfate), polyaluminum chloride, ferric chloride, and ferric sulfate (AWWA, 1998). Bench-scale studies , known as jar tests, must be performed to determine the optimum coagulant dose and solution pH for particle destabilization, growth, and removal. A typical plot of relative turbidity versus coagulant dose is shown in Figure 13.8.3. In this graph, the low relative turbidity area indicates the optimum coagulant dose. If one continues to add the coagulant, charge reversal and restabilization of the particles is possible, as shown in Figure 13.8.3.

Adsorption and charge neutralization is the most commonly used method of particle destabilization in conventional surface water treatment systems. Recently, concern about the possible link between aluminum ingestion and alzheimers disease has caused many municipal water treatment plants to stop using alum and polyaluminum chloride and instead use ferric compounds.

Enmeshment in a precipitate, also called sweep floc, occurs when alum or one of the ferric compounds, described above, is added until saturation is reached and precipitation of the metal hydroxide of the respective dosing chemical (e.g., $Al(OH)_{3(s)}$ or $Fe(OH)_{3(s)}$) occurs. When precipitation begins, the original solids in suspension act as "seeds" for precipitation and are then enmeshed in the forming precipitate. The far right hand side of Figure 13.8.3 shows the region for the sweep floc mechanism. Although this method of destabilization is very effective, it is very costly both in chemical costs and sludge removal and in disposal costs.

Adsorption and interparticle bridging using synthetic organic polymers (cationic, anionic, and nonionic) have been used in water treatment for over two decades (AWWA, 1990). Adsorption and interparticle bridging occur when these long chain, high molecular weight polymers are introduced into the water. These polymers then attach themselves to the particles and leave a long tail in the solution, which is available to capture other particles, allowing for particle growth and removal. The fluid mixing intensity during introduction of these synthetic organic polymers is critical to their effectiveness in particle destabilization.

A key component to any coagulant selection process is effective jar testing. The results from these jar tests should be related to concepts from equilibrium chemistry (e.g., solubility diagrams) as outlined in *Water Quality and Treatment* (AWWA, 1990). An effective jar testing program will

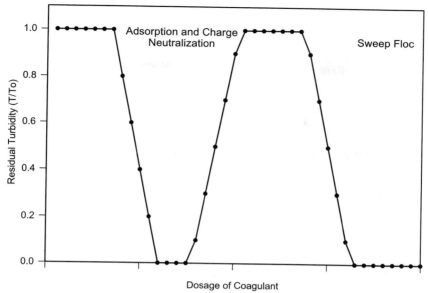

FIGURE 13.8.3.   A Schematic coagulation curve for adsorption and charge nuetralization followed by sweep floc.

reproduce raw water quality parameters such as pH, temperature, ionic strength, and initial particle characterization via turbidity measurements, particle size distribution measurements, and zeta potential of the particulate matter.

Flocculation

The next step in the coagulation process in termed flocculation, or particle transport. Particle transport is necessary to cause the destabilized particles to collide and then attach to each other to produce larger particles. The three mechanisms for particle transport are Brownian motion, fluid motion, and differential sedimentation.

Brownian motion (sometimes termed perikinetic flocculation) is a result of the random molecular motion of the fluid. Therefore, only the smallest particles in the fluid will be affected by Brownian motion. Because this transport mechanism is a function of particle size, the effect increases with decreasing particle size.

Fluid motion (termed orthokinetic flocculation) is the result of laminar or turbulent velocity gradients in the fluid. Particles of the same order of magnitude as the small-scale fluid motions (eddies) are thought to be primarily responsible for particle collisions. Thus, this mechanism affects the intermediate sized particles in a suspension. It also is the only mechanism over which the designer has some control. The primary design value for this mechanism, the root mean square velocity gradient (G), will be discussed in more detail below.

Differential sedimentation is a result of the settling velocities of the suspended particles. Because the driving force for this mechanism is gravity, it affects the largest particles suspended in the fluid. These larger particles suspended in the fluid settle more quickly than the smaller particles and collide with them as they move through the water column.

The root mean square velocity gradient (G), developed by Camp and Stein in 1943, is still the primary design value used for flocculation facilities and rapid mixing basins (AWWA, 1998). The

FIGURE 13.8.4. Mechanical flocculator (Courtesy of U.S. Filter).

FIGURE 13.8.5. Detail of a mechanical paddle flocculator (Courtesy of U.S. Filter).

equation for the root mean square velocity gradient is shown below:

$$G = (P/\mu V)^{1/2} \qquad (13.8.1)$$

where G = root-mean-square velocity gradient, the rate of change of velocity, expressed in ft/s/ft, P = power input, ft/lb/s, μ = dynamic viscosity, lb-s/ft^2, and V = volume, ft^3.

This same basic equation is used for laminar and turbulent flow conditions for the design of both rapid mixing basins and flocculation facilities. A common range of G values for use in the design of flocculation facilities is between 30 and 60 s^{-1}. If too much mixing energy is provided to the flocculation basins, the particles could be sheared apart instead of floccculated. If one were to choose a single typical G value for the design of flocculation facilities, it would be 50 s^{-1}.

A second parameter used in the design of flocculation and rapid mixing facilities is "Gt" the root mean square velocity gradient (G) multiplied by the mixing time (t). This parameter provides the designer with a method for comparing the mixing characteristics of different flocculation or rapid mixing facilities. A reasonable range of mixing times for flocculation facilities is 30 to 60 min. Theoretically, as the G value increases, the mixing time can decrease. However, one must remember that along with a maximum G value of 60 s^{-1}, mixing times less than 10 min in flocculation basins should be avoided. The resulting Gt range for flocculation facilities is 10^4 to 10^5.

Mechanical flocculators, as shown in Figures 13.8.4 and 13.8.5, are the most common flocculators used in the United States because their ease of operation, low head loss, and their ability to

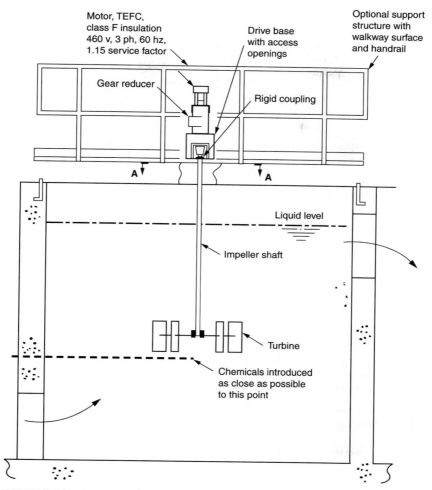

FIGURE 13.8.6. Turbine-type mechanical flash mixer (AWWA, 1998).

adapt to mixing requirements associated with changing raw water quality. A second type of flocculator is a baffled (hydraulic) flocculator in which the G value is dependent upon the flow of water through the flocculator. Many proprietary combination flocculation/sedimentation units are also used for small to mid-sized treatment facilities.

Figure 13.8.4 is a picture of mechanical flocculation basins at an operating drinking water treatment plant. In this picture, the first basin has not been filled with water to allow the reader the see the exposed paddle mixers in the floccultaion basins. A series of sedimentation basins with a mechanical sludge collection system is also visible immediately to the left of the first flocculation basin. The reader should note the modular and duplicate design characteristics of both the flocculation and sedimentation basins. Figure 13.8.5 is a picture of a USFilter Link-Belt® flocculator in which the paddle flights are provided in a choice of materials, including redwood, fiberglass, steel, or stainless steel. Note, the paddles are chain driven and the motor is external to the flocculator for ease in routine maintenance and replacement.

13.8.3 MIXING AND MIXERS

Flash mixing, rapid mixing, chemical mixing, and initial mixing are all terms for the same process that proceeds coagulation. The goal of this process is to effectively disperse the chemicals into the water as quickly as possible. Factors that may affect the design and operation of rapid mixing systems are the type of chemical or chemicals being added to the water (which may vary with time of year and raw water quality), the raw water temperature, and more stringent regulations or finished water quality requirements.

Four basic type of rapid mixing systems are used for conventional municipal drinking water treatment: mechanical mixers, in-line blenders, hydraulic mixers, and injection mixers. As in flocculation, the primary design parameter for rapid mixing is the root-mean-square velocity gradient G and dimensionless Gt. The r.m.s G value for rapid mixing systems are significantly higher then for flocculation while the mixing times are much shorter. Although the G value and mixing times are dependent upon the type of rapid mixing system, some generalizations can be made. Generally, G values would range from 600 to 1000 s^{-1} with mixing times ranging from 5 to 30 s. Resultant Gt values would range from 3×10^3 to 3×10^4.

Figure 13.8.6 is a sketch of a typical turbine-type mechanical flash mixer in a dedicated flash mixing basin (AWWA, 1998). This type of application is most common for conventional drinking water treatment applications. Notes that the chemicals are injected as close to the mixing point as possible. Again, note the placement of the motor for case of maintenance and replacement.

REFERENCES

American Water Works Association and American Society of Civil Engineers (AWWA & ASCE), *Water Treatment Plant Design*, Fourth edition, New York, McGraw-Hill, 1998.

American Water Works Association, *Water Quality and Treatment*, Fourth edition, New York, McGraw-Hill, F. W. Pontius, ed., 1990.

Sawyer, C. N., McCarty, P. L., and Parkin, G. F., *Chemistry for Environmental Engineering*, Fourth edition, New York, McGraw-Hill, 1994.

B. SEDIMENTATION AND FILTRATION

13.8.4 INTRODUCTION

Sedimentation, also known as settling or clarification, is the removal of settleable solid particles from a suspension by gravity. These settleable particles either originated in the raw water or were produced from smaller particles during coagulation and flocculation. Removal of these larger settleable particles from water during sedimentation also allows for a reduced loading on the granular media filters, which are required in all municipal drinking water treatment facilities using surface water sources.

Granular media filtration is the last opportunity to capture and remove particles in a drinking water treatment facility before the finished water is disinfected and enters the distribution system. Recent outbreaks of Giardiasis and Cryptosporidiosis (a result of ineffective removal and inactivation of protozan cysts namely, *Giardia* and *Cryptosporidium*) in the United States have highlighted the need to optimize particle growth and removal processes (i.e., rapid mixing, coagulation, flocculation, sedimentation, and filtration) in municipal drinking water treatment and distribution systems.

This chapter will examine the fundamentals of both sedimentation and granular media filtration. Although a brief discussion of slow sand filtration will be provided, rapid media filtration will be the focus of the filtration section. The implications of these particle removal processes in conventional surface water treatment plants also will be discussed.

13.8.5 *SEDIMENTATION*

Sedimentation is a very complex process in both natural and engineered systems. Many variables affect particle settling because of gravity (e.g., particle shape, particle size, and particle density). Although sedimentation is complex and the particle properties are seldom uniform, a review of sedimentation theory and its relationship to these more complex (realistic) situations is instructive.

Generally, four types of settling or sedimentation occur when particles are removed from a suspension by gravitational force. In types 1 and 2, settling occurs in dilute suspensions and the particle settling velocity is a function of the particle size and the particle density. In types 3 and 4, settling occurs in more concentrated suspensions, and the particle settling velocity is a function of the particle size, particle density, and the suspension concentration. The four types of sedimentation, as defined by *Water Quality and Treatment* (AWWA, 1990), are:

- Type 1: Discrete Sedimentation—Settling of discrete (nonflocculent) particles in dilute suspensions without interaction with other particles.
- Type 2: Flocculent Sedimentation—Settling of flocculent particles in dilute suspensions. The particles flocculate and increase in size during sedimentation.
- Type 3: Hindered (Zone) Sedimentation—Settling of concentrated suspensions where a well-defined interface forms between the solids and the liquid (supernatant) above. In this type of sedimentation, the suspension concentration affects particle settling velocities.
- Type 4: Compression (Compaction) Sedimentation—Consolidation of the sediment at the bottom of a basin. In this type of sedimentation, the water is moving around the particles (e.g., someone squeezing water out of a sponge).

Type 1 sedimentation is discrete sedimentation in which particles settle in dilute suspensions without interacting with other particles. Classical settling theory for discrete particles in dilute suspensions assumes that the particles are uniform in size and spherical in shape. Based upon a force balance on a single particle in an infinite fluid, Stokes law was developed to describe the terminal settling velocity of a single particle (Montgomery, 1985). Low fluid velocities and laminar fluid flow are required for effective sedimentation to occur in sedimentation basins. To ensure that laminar conditions and effective sedimentation will occur, the Reynolds number (defined below) should be less than 1:

$$R_e = \rho V d / \mu \qquad (13.8.2)$$

where V = velocity of the settling particle, d = diameter of the settling particle, ρ = mass density of the liquid, and μ = absolute viscosity.

Terminal settling velocity for a uniform spherical particle can be calculated using the simplified Stoke's law. This simplified law assumes the Reynolds number (defined above) is less than 1 and the coefficient of drag for the spherical particle is 24 divided by the Reynolds number. The simplified Stoke's settling velocity is as follows (Montgomery, 1985):

$$V_s = \{g(\rho_p - \rho_l)/18\mu\}d_p^2 \qquad (13.8.3)$$

where g = gravitational constant, ρ_p = mass density of the particles, ρ_l = mass density of the liquid μ = absolute viscosity, and d_p = particle diameter.

In 1936, Camp developed a theory for the removal of discrete particles in an ideal sedimentation tank. Camp divided the tank into four zones (inlet, outlet, sludge collection, and active settling) as shown in Figure 13.8.7. Five assumptions were made by Camp to develop the settling tank removal efficiency equations (Montgomery, 1985):

- Discrete sedimentation occurred
- Horizontal flow occurred in the active settling zone
- The horizontal velocity was uniform in the settling zone

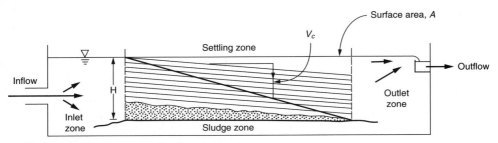

FIGURE 13.8.7.   Camp's ideal sedimentation basin (Montgomery, 1985).

- A uniform concentration of particles existed across a vertical plane at the inlet end of the settling zone
- Particles were removed once they reached the bottom of the settling zone

Because all discrete particles settling in the active settling zone would be traveling in a straight line, as shown in Figure 13.8.7, a particle must have a settling velocity equal to or greater than the depth of the tank (H) divided by the detention time to be removed from the tank:

$$V_c = H/t_o \qquad (13.8.4)$$

where V_c = critical particle settling velocity, H = depth of the active settling zone, and t_o = detention time.

The percentage (P) of discrete particles removed in the active settling zone is a function of critical particle settling velocity (V_c) and can be computed as follows:

$$\text{If } V_s > V_c, \qquad P = 100\% \qquad (13.8.5)$$

$$\text{If } V_s < V_c, \qquad P = 100\,(V_c/V_s) \qquad (13.8.6)$$

where V_s = Stokes settling velocity for discrete particles.

Substituting the definition of the detention time for the tank (t_o), the volume of the tank (V) divided by the flowrate (Q) of water into the tank, into Equation 13.8.4 yields the following:

$$V_s = H/(V/Q) \qquad (13.8.7)$$

$$V_s = (HQ)/V \qquad (13.8.8)$$

$$V_s = (HQ)/(LWH) \qquad (13.8.9)$$

$$V_s = Q/(LW) \qquad (13.8.10)$$

$$V_s = Q/A_s \qquad (13.8.11)$$

where H = active settling zone depth, V = active settling zone volume, Q = flowrate, t_o = detention time, L = active settling zone length, W = active settling zone width, and A_s = active settling zone surface area.

This quantity defined in Equation 13.8.11 is also known as the overflow rate. Overflow rate has units of gpm/ft^2 in the U.S. system. Although it is not intuitive, combining Equations 13.8.6 and 13.8.11 demonstrates that removal of discrete particles is independent of depth. Thus, particulate removal in discrete sedimentation is only a function of the flowrate into the tank and the surface area available for sedimentation. As a result, increasing detention time in the tank by increasing tank depth will not affect discrete particle removal. Only increasing sedimentation surface area will increase particle removal during discrete sedimentation for a given flowrate.

Type 2 sedimentation is flocculent sedimentation where the particles flocculate and particle diameter increases during sedimentation in the dilute suspension. Because the particle diameter increases during sedimentation, the settling velocity (as calculated by Stoke's law) also increases. As a result, particle removal for type 2 sedimentation is a function of the overflow rate and settling basin depth. Factors affecting the increase in particle size (particle growth) during flocculent sedimentation are the particle number concentration, particle size, and the particle stability.

As in coagulation/flocculation, the available mechanisms of interparticle contact are Brownian motion, fluid motion, and differential sedimentation. Brownian motion and fluid motion should be minimal modes of interpaticle contact for the large particles in a sedimentation basin containing little fluid motion (i.e., R_e <1). Differential sedimentation is the primary mechanism for interparticle contact and growth during flocculent sedimentation.

No simple mathematical relationship exists to calculate the amount of particle growth or removal during flocculent sedimentation. As a result, pilot-scale testing, using a column shown in Figure 13.8.8, is required to design flocculent sedimentation basins. These experiments are performed by uniformly mixing a flocculent suspension in the column shown in Figure 13.8.8. The height of the column should be equal to or greater than the depth of the anticipated settling tank. Sample ports should be equally spaced at no less than 50 cm (approximately 20 in) (Montgomery, 1985). Samples are withdrawn from each port in the settling column at selected time intervals and analyzed for suspended solids concentrations. The suspended solids concentration data are used to compute the percent suspended solids removal as follows (Davis and Cornwell, 1998):

$$R = 1 - [(C_t/C_o)100] \qquad (13.8.12)$$

where, R = removal percentage at a given depth and time, C_t = suspension concentration at any time, t, and C_o = initial suspension concentration.

The resulting removal percentages are plotted as individual data points on the depth versus time graph shown in Figure 13.8.8. Lines of equal suspended solids removal (usually at 10 percent increments) are then drawn by interpolation on the graph based upon the these individual data points.

The overall removal efficiency of the full-scale settling column, having a specified detention time (t_2), can be estimated from the pilot-scale data by drawing a vertical line on Figure 13.8.8 and using

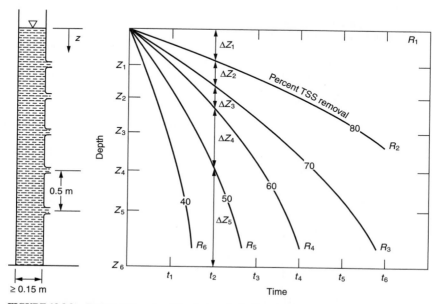

FIGURE 13.8.8. Typical pilot-scale settling test results for Type 2 sedimentation (Montgomery, 1985).

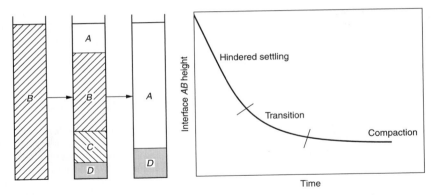

FIGURE 13.8.9.   Typical hindered (zone) sedimentation tests results (Montgomery, 1985).

the following equation (Montgomery, 1985).

$$R_{\text{total}} = \Delta Z_1/Z_6[(R_1 + R_2)/2] + \Delta Z_2/Z_6[(R_2 + R_3)/2]$$
$$+ \Delta Z_3/Z_6[(R_3 + R_4)/2] + \Delta Z_4/Z_6[(R_4 + R_5)/2] \quad (13.8.13)$$

where R_{total} = total percentage suspended solids removal in the tank, Z = depth, R_i = individual removal percentages from the respective isoremoval lines.

Type 3 sedimentation is hindered or zone sedimentation. It occurs where the particles are in a concentrated suspension (where a well-defined interface forms between the solids and the liquid—supernatant—above). The sedimentation velocity of the suspension for type 3 sedimentation decreases with increasing suspension concentration. As with type 2 sedimentation, a pilot-scale test must be performed to describe the hindered settling characteristics of a given suspension.

Figure 13.8.9 shows a typical pilot-scale test performed to define the settling characteristics of a suspension undergoing hindered (zone) settling. For this test, a pilot-scale column is constructed and filled with a suspension. Initially, a uniform suspension concentration (B) exists throughout the sedimentation column. As time passes, the particles begin to settle and form an interface between the active settling particles (B) and the clear supernatant (A) on top. Concurrently with the formation of the interface at the top of the column, a layer of compressed (compacted) sludge (D) forms at the bottom of the column. Directly above the compaction (compression) region of the curve, a region of transition (C) exists between the hindered settling zone (B) and the compaction zone (D) (Montgomery, 1985).

The movement of the interface height with respect to time is recorded from the above procedure and then plotted on a graph similar to the one shown in Figure 13.8.9. Please note the three regions of the graph. The hindered settling region has a definable slope, which will be used to generate a hindered settling versus concentration graph. A transition region exists between the hindered settling region and the compaction (compression) region (type 4 region). This pilot-scale testing procedure is repeated for varying suspension concentrations. For each of the suspension concentrations (C_i), a corresponding hindered settling velocity (V_i, the slope of the straight line hindered settling region) are recorded and calculated, respectively. These data are used to generate a hindered settling velocity versus concentration curve (not shown).

The solids flux is defined as the mass of solids per area per unit time and is used to define the movement of solids in the basin. The solids flux can be calculated using the following equation:

$$G_b = C_i V_i \quad (13.8.14)$$

where G_b = solids flux because of gravity, C_i = solids concentration, and V_i = hindered settling velocity at C_i. A typical plot of the batch solids flux versus concentration is shown in Figure 13.8.10.

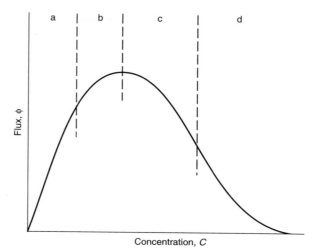

FIGURE 13.8.10.   Typical batch flux versus concentration curve for Type 3 sedimentation (AWWA, 1990).

A second method of solids movement in the sedimentation basin for hindered sedimentation is by movement of liquid because of removal of compacted sludge from the bottom of the sedimentation basin. This type of flux is termed the underflow flux and is added to the batch flux to generate a typical plot as shown in Figure 13.8.11. This graph defines the solids handling and operating characteristics of the hindered sedimentation basin. The horizontal solids flux line on this figure is the limiting flux Φ_c intersecting the curve at C_c. The limiting flux defines the maximum solids handling capacity of the basin. C_f is the minimum initial solids concentration, below which hindered settling will not occur, for the basin. The expression C_U defines the underflow (maximum) concentration of solids in the hindered settling basin.

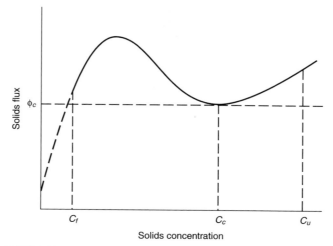

FIGURE 13.8.11.   Typical solids flux curve for a continous flow gravity Thickner for Type 3 sedimentation (AWWA, 1990).

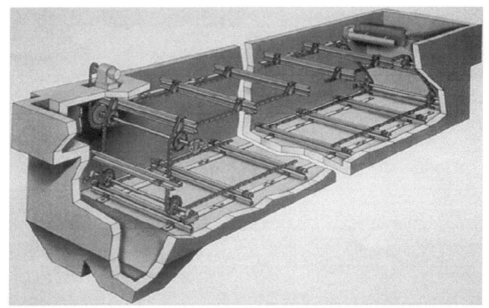

FIGURE 13.8.12.   Rectangular sedimentation basin with a mechanical sludge collection system (Courtesy of U.S. Filter).

FIGURE 13.8.13.   Rectangular sedimentation basin with a mechanical sludge collection system (Courtesy of U.S. Filter).

FIGURE 13.8.14. Plate settler alternatives (Courtesy of U.S. Filter).

Type 4 sedimentation is compression sedimentation and defines the consolidation of the sediment at the bottom of a basin. In this type of sedimentation, the water is moving around the particles as they are compressed from the weight of the settling material from above (see Figure 13.8.9).

Figure 13.8.12 is diagram of a typical rectangular sedimentation basin with a mechanical sludge collection system. The following should be noted in this figure: the sludge collection hopper at the right end of the figure; the location of the motor for ease in maintenance and replacement; the connection and spacing of the sludge collection mechanism on the chains; and the chain drive mechanism for the sludge collectors. A photograph of a rectangular sedimentation basin during construction is shown in Figure 13.8.13. The same items from the diagram in Figure 13.8.12 also can be seen in this photograph.

Figure 13.8.14 contains two photographs of plate settling modules that can be added to an existing rectangular basin or operated as a additional process to enhance sedimentation. These modules are effective because they dramatically increase the surface area of the sedimentation basin and reduce the effect of hydraulic entrance and exit losses to the basin.

13.8.6   FILTRATION

Granular media filtration is the last opportunity to remove particles from drinking water before it is disinfected and enters the distribution system. In the United States and other developed countries, granular media filtration is synonymous with rapid media filtration. A schematic of a rapid media filter is shown in Figure 13.8.15. Slow sand filtration was originally used as a basic filtration method until space requirements and finished water production requirements made them impractical in developed countries. In many developing countries, however, slow sand filtration is still a viable method for producing high quality drinking water. The discussion in this chapter will focus on rapid media

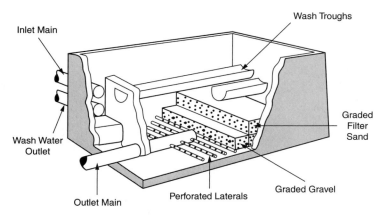

FIGURE 13.8.15. Typical cross section of a rapid sand filter (Davis and Cornwell, 1998).

filtration. The reader is directed to *Water Treatment Plant Design* (AWWA, 1998) for a more detailed discussion of slow sand filtration.

Figure 13.8.15 is a schematic of a typical rapid media filter using graded filter sand. This figure illustrates the major components of a rapid media filter. The media rests upon a gravel support bed. The gravel is supported by an underdrain system (e.g., perforated laterals), which allows the water to be collected from the bottom of the filter prior to entering the outlet main and then the disinfection contact chamber. When the filter needs to be cleaned, water enters the bottom of the filter and flows up through the suspended filter media and is collected in the wash troughs for disposal.

Two important factors in rapid media filtration are particle removal and headloss (pressure drop as water flows through a filter) development. Following a backwash cycle, the filtered water is sent to waste for some period of time (e.g., 1 to 2 h). Hydraulic considerations of the filter media determine the headloss through the clean filter bed. This headloss value can be calculated using the Carmen-Kozeny equation below (AWWA, 1990):

$$h/L = [k\mu]/\rho g[(1 - \epsilon)^2/\epsilon^3][a/v]^2 V \tag{13.8.15}$$

where h = headloss in depth of bed L, g = acceleration of gravity, ϵ = porosity, a/v = surface area per unit volume, $6/d$ for spheres, V = superficial velocity above the filter bed, μ = absolute fluid viscosity, ρ = fluid mass density, and k = dimensionless constant, typically close to 5 for most conditions.

This headloss value will decrease shortly after completion of a backwash cycle and then steadily increase until the filter is backwashed again. The concentration of particles in the filter effluent is typically measured by a turbidimeter and reported in nephelemetric turbidity units (NTU). This value will sharply increase minutes after backwashing and then decrease in a matter of hours. The effluent quality will then remain relatively constant until it begins to rise after 24 to 48 h. Either of these parameters (i.e., headloss or effluent quality) are justification for filter backwashing. The headloss justification for rapid media filter backwashing is based upon economics (the cost to pump water through the filter or loss of finished water production by the filter). The effluent quality justification is a question of protecting the public health. As a result, rapid media filters are routinely backwashed at least every 24 to 48 h.

Similar to coagulation/flocculation, three mechanisms of particle contact with the collection media in the filter are observed: interception (fluid motion), diffusion, and sedimentation. Diffusion affects the smallest particle sizes in suspension while sedimentation affects the largest particle sizes. A fourth mechanism, mechanical straining, does occur in rapid media filtration, but is not preferred.

Recent advances in rapid media filtration include multimedia filters. In this configuration, the basic components of the filter remain the same as were shown in Figure 13.8.15. The sand media is replaced with three sizes of filter media. The smallest diameter media would be garnet sand and would rest on the gravel support. The intermediate sized media would be silica sand. The largest diameter media would be anthracite coal. The differences in specific gravity of the media allow the smallest particles to settle on the bottom of the filter and the largest particles to remain on top of the filter bed.

The advantages of multimedia filtration are numerous, longer run times, more efficient particle removal than rapid sand filters, better depth filtration, and no mandated changes to the overall filter structure or washwater quantities. As a result, many rapid media filters are being retrofitted with multiple media in an effort to increase particle removal in this final particle removal process.

REFERENCES

American Water Works Association and American Society of Civil Engineers, *Water Treatment Plant Design*, Third edition, New York, McGraw-Hill, 1998.

American Water Works Association, *Water Quality and Treatment,* Fifth edition, New York, McGraw-Hill, Raymond D. Letterman, ed., 1999.

Montgomery, James, M., *Water Treatment Principles and Design*, New York, John Wiley & Sons, 1985.

CHAPTER 14
REMEDIATION TECHNOLOGIES

SECTION 14.1

MONITORED NATURAL ATTENUATION (MNA)

Patrick V. Brady
Dr. Patrick Brady applies mineral surface chemistry to remediate hazardous waste and understand global geochemical change at the Sandia National Laboratory Geochemistry Department.

Warren D. Brady
Warren is a project geochemist in the Technology Applications Group of IT Corporation. He provides technical leadership on natural attenuation investigations and implementation of natural attenuation performance monitoring programs.

14.1.1 INTRODUCTION

Natural attenuation is defined by the USEPA as "the naturally occurring processes in soils and groundwaters that act without human intervention to reduce the mass, toxicity, mobility, volume, or concentration of contaminants in those media." These in situ processes include biodegradation, dispersion, dilution, sorption, precipitation, volatilization, and/or chemical and biochemical stabilization of contaminants (U.S. Environmental Protection Agency, 1997). Natural attenuation has received increased attention for the following three reasons:

1. Existing cleanup approaches almost never achieve regulatory targets. Complete elimination of contaminants from soils is severely limited by technical obstacles (e.g., diffusion of contaminants into dead-end pores, irreversible sorption, etc.).
2. Natural processes often result in sizeable reductions in contaminant levels by themselves.
3. Risk-based approaches that incorporate attenuation factors into cleanup goal calculations are increasingly being used to establish alternative remedial action goals or no further action justifications.

Natural attenuation is increasingly being relied on as a component (sole or partial) for the remediation of contaminated soils and groundwaters. Fully a quarter of the Superfund records of decision in 1995 relied to some extent on natural attenuation (Browner, 1997). More recently, CERCLA reauthorization efforts have called for consideration of natural attenuation at all sites. Monitored natural attenuation (MNA) is the background against which active remediation efforts must be calculated, and, judging by legislative trends at the state and national levels, natural attenuation in the future will probably be considered as a remedial component at all sites.

Specific technical questions that must be answered for successful implementation of MNA at a given site include: (1) Is natural attenuation occurring? (2) Are reductions in potential contaminant impacts decreasing rapidly enough to achieve regulatory compliance in an acceptable time frame? (3) How much long-term monitoring is required? and (4) What are the maximum extents to which a contaminant plume can reach? The first question is often answered through assessment of geochemical indicators. For example, are degradation by-products present? Alternatively, extensive analyses of plume behavior can shed light on the question. Answering the second question requires that a prediction of contaminant attenuation in the future be made. Considerable uncertainty is associated with such a project, and the answer will be at best only semi-quantitative. Long-term monitoring requirements will, in most cases, be dictated largely by regulators, and will therefore vary from site to site. A clear answer to the last question—maximum plume extents—might arguably be the most persuasive evidence in the eyes of the public. Some of the most useful information in this regard comes from historical case analyses—statistical analyses of large numbers of plumes that have not been actively remediated. Historical case analyses are effective devices for communicating overall risk to the public because they provide a reasonably transparent picture of how far and fast plumes can move. Two of the most notable compilations of historical case studies have come from Lawrence Livermore National Laboratory and have been focused on fuel hydrocarbons from leaking underground fuel tanks (Rice et al., 1995) and chlorinated solvents (McNab et al., 1999).

Before MNA is considered, any site-specific, nontechnical objections to reliance on MNA must also be considered. Are there cultural or natural resources that must be protected? Are there economic or land use changes that affect cleanup levels and times? Are stakeholders likely to accept MNA as a remediation option? Do preexisting third party agreements prevent implementing MNA? If needed, can institutional controls be maintained at a site undergoing MNA? Regulators should be brought into the process at the earliest stages.

This chapter will first review EPA's recent guidance for MNA implementation. A recent survey of state regulators (Brady et al., 1997) indicates that the EPA's guidance generally has been used as a template for most state natural attenuation implementation and investigation policies. It is reasonable to expect MNA implementation in the future to follow the general tone of EPA's guidance. Following this review, the primary attenuation processes for specific contaminants will be identified, and the individual mechanisms will be discussed. Lastly, the specific steps involved in implementation of monitored natural attenuation will be covered.

14.1.2 MNA GUIDELINES

MNA should be evaluated during the site characterization and remediation selection phase and supported by site-specific information that demonstrates the efficacy of the method, much like any other remedial approach. Several guideline manuals have been developed for assessing natural attenuation of organics (see e.g., Wiedemeier et al., 1995a; Wiedemeier, 1995b). The American Society for Testing and Materials (ASTM) has also developed a standard guide for natural attenuation investigation and implementation at petroleum release sites. The EPA requirements provide general requirements and guidelines for all contaminants, and as mentioned previously, provide the framework for most state policies for natural attenuation. The EPA's specific requirements can be found at www.epa.gov/swerst1/directiv/9200_417.htm. According to the latter, a successful MNA remedy requires that:

1. Source control actions address principal threat wastes (or products) where practicable.
2. Contaminated groundwaters should be returned to their beneficial uses where practicable, within a time frame that is reasonable given the particular circumstances of the site. When restoration of groundwater is not practicable, EPA expects site owners to attempt to limit further migration.
3. Contaminated soil should be remediated to achieve an acceptable level of risk to human and environmental receptors and to prevent any transfer of contaminants to other media (e.g., surface or groundwater, sediments, air) that would result in an unacceptable risk or exceed required cleanup goals.

14.1.3 DATA NEEDS

On November 22, 1997, EPAs Office of Solid Waste and Emergency Response (OSWER) issued interim guidelines for relying on monitored natural attenuation at Superfund, RCRA corrective action, and underground storage tanks (U.S. Environmental Protection Agency, 1997). The OSWER directive serves as an outline of the EPA's general guidelines for natural attenuation, although there is only minimal discussion of specific data needs for individual contaminants. The OSWER Directive cites three distinct types of data necessary for providing clear evidence to implementing agencies that natural attenuation is effectively lowering contaminant concentrations on a site-specific basis. They are, in order of importance:

1. Historical groundwater and/or soil chemistry data that demonstrate a clear and meaningful trend of decreasing contaminant mass and/or concentration over time at appropriate monitoring or sampling points

2. Hydrogeologic and geochemical data that can be used to demonstrate indirectly the type(s) of natural attenuation processes active at the site and the rate at which such processes will reduce contaminant concentrations to required concentrations

3. Data from field or microcosm studies (conducted in or with actual contaminated site media) that directly demonstrate the occurrence of a particular natural attenuation process at the site and its ability to degrade the contaminants of concern (typically used to demonstrate biological degradation processes only)

If natural attenuation has been accepted by the regulatory agency as part of the remediation option, performance monitoring and contingency remedies will be required to evaluate the long-term effectiveness of the method and provide a backup remedy if natural attenuation fails, respectively. Monitoring programs must provide mechanisms to:

1. Demonstrate that natural attenuation is occurring according to expectations

2. Identify any potentially toxic transformation products resulting from biodegradation

3. Determine if a plume is expanding and ensure that there will be no impact to downgradient receptors

4. Detect new releases of contaminants to the environment that could impact the effectiveness of the natural attenuation remedy

5. Demonstrate the efficacy of institutional controls that were put in place to protect potential receptors

6. Detect changes in environmental conditions (e.g., hydrogeologic, geochemical, microbiological, or other changes) that may reduce the efficacy of any of the natural attenuation processes (and possibly trigger a contingency plan)

7. Verify attainment of cleanup objectives.

14.1.4 CHEMICAL AND BIOLOGICAL CONTROLS ON CONTAMINANT ATTENUATION

Table 14.1.1 outlines natural attenuation pathways for many of the common soil and groundwater contaminants. Although dilution is not specifically mentioned in Table 14.1.1, it is expected to be a component of natural attenuation for each contaminant. In general, biodegradation is the most important natural attenuation pathway for organic contaminants. Sorption and chemical transformation (e.g., formation of insoluble solids) are the primary natural attenuation pathways for inorganic contaminants.

TABLE 14.1.1 Natural Attenuation Pathways for Specific Contaminants

Chemical	Natural attenuation pathways	Mitigating conditions
BTEX	Biodegradation (oxidation)	Reducing conditions slow breakdown
PCE, TCE, chlorinated organics	Biodegradation (reductive dechlorination)	Oxidizing conditions slow breakdown
Vinyl chloride	Oxidation	Reducing conditions slow breakdown
Pb	Sorption to iron hydroxides, organic matter, carbonate minerals; formation of sparingly soluble carbonates, sulfides, sulfates, phosphates	Low pH destabilizes carbonates and iron hydroxides. Commingled organic acids and chelates (e.g., EDTA) may decrease sorption. Low E_H dissolves iron hydroxides but favors sulfide formation.
Cr(VI) as CrO_4^{2-}	Reduction, sorption to Fe/Mn hydroxides; formation of $BaCrO_4$	Low pH destabilizes iron hydroxides. Low E_H dissolves iron hydroxides but favors reduction.
As(III or V)	Sorption to iron hydroxides and organic matter; formation of sulfides	Low pH destabilizes iron hydroxides. Low E_H dissolves iron hydroxides.
Zn	Sorption to iron hydroxides, carbonate minerals; formation of sulfides; ion exchange	Low pH destabilizes carbonates and iron hydroxides. Commingled organic acids and chelates may decrease sorption. Low E_H dissolves iron hydroxides but favors formation of sulfides.
Cd	Sorption to Fe/Mn hydroxides and carbonate minerals; formation of sparingly soluble carbonates, phosphates, and sulfides	Low pH destabilizes carbonates and iron hydroxides. Commingled organic acids and chelates may decrease sorption. Low E_H dissolves iron hydroxides, but favors formation of sulfides.
Ba	Formation of sparingly soluble sulfate minerals; ion exchange	Low E_H may destabilize sulfates.
Ni	Sorption to Fe/Mn hydroxides; ion exchange; formation of sulfides	Commingled organic acids and chelates may decrease sorption. Low E_H dissolves iron hydroxides but favors sulfide formation.
Hg	Formation of sparingly soluble sulfides; sorption to organic matter	Is methylated by organisms
N(V) as NO_3	Reduction by biologic processes	
Radionuclides		
U(VI)	Sorption to iron hydroxides; precipitation of sparingly soluble hydroxides and phosphates; reduction to sparingly soluble valence states	Low pH destabilizes carbonates and iron hydroxides. Commingled organic acids and chelates may decrease sorption. High pH and/or carbonate concentrations decrease sorption. Low E_H dissolves iron hydroxides but favors reduction.
Pu(V and VI)	Sorption to iron hydroxides; formation of sparingly soluble hydroxides and carbonates	May move as a colloid. Low E_H dissolves iron hydroxides.
Sr	Sorption to carbonate minerals and clays; formation of sparingly soluble carbonates and phosphates	Low pH destabilizes carbonates. High dissolved solids favor leaching of exchange sites.
Am(III)	Sorption to carbonate minerals; formation of carbonate minerals	Low pH destabilizes carbonates. High pH increases solubility of Am-carbonate minerals.
Cs	Sorption to clay interlayers	High NH_4^+ concentrations may lessen sorption. Low K^+ concentrations may increase plant uptake.
I	Sorption to sulfides and organic matter	Sorbs to very little else in oxidized state.
Tc(VII) as TcO_4^-	Possible reductive sorption to reduced minerals (e.g., magnetite); forms sparingly soluble reduced oxides and sulfides	Sorption to other phases extremely limited.
Th	Sorption to most minerals; formation of sparingly soluble hydroxide	May move as a colloid.
^3H	None	
Co	Sorption to iron hydroxides, organic matter, and carbonate minerals	Low pH destabilizes carbonates. Low E_H dissolves iron hydroxides. Stable complexes form with chelators.

Dilution

Dilution is unavoidable and is a naturally occurring groundwater process, and EPA is aware of this process' role in natural attenuation. Dilution must be understood and factored into degradation rate calculations so that biological effects are not overestimated for organic constituents.

Dilution is generally characterized in groundwater systems by using a conservative tracer. For fuel hydrocarbon impacted sites, recalcitrant compounds such as 1,2,4- and 1,3,5-trimethyl benzene are commonly used to track the effects of dilution and/or sorption. Other recalcitrant compounds or nondegradeable constituents (i.e., inorganics) present in the groundwater system are used to track chlorinated compound dilution.

Sorption

Sorption is the tendency of contaminants to adhere to soil surfaces, consequently retarding their environmental transport. The degree of adhesion depends on the contaminant, on the soil, and on the chemistry of the soil or groundwater. Organic contaminants tend to sorb to soil/aquifer organic matter, the degree of sorption being greater for those contaminants with high octanol-water coefficients. Generally, trace element cations have very low solubility and, hence, a lower bioavailability at pH values above 7. This is due to sorption, surface precipitation, and coprecipitation reactions in soil environments. Many soil surfaces have the ability to catalyze trace element precipitation reactions. At high pH, the negative charge in soil is maximized, resulting in maximum cation adsorption. The primary sorbing agents in soils are iron (hydr)oxides, organic matter, and clay minerals. Sorption is typically quantified using a K_d, and there are a number of sources for the latter (Baes and Sharp, 1983; U.S. Environmental Protection Agency, 1996).

Sorption can be either reversible or irreversible. Reversible adsorption is almost uniformly presumed in transport models and assumes that contaminants that initially sorbed from contaminant-rich solutions will instantaneously desorb when, for example, fresh recharge causes contaminant levels to drop in solution. Irreversible sorption occurs when desorption is slow or nonexistent. Irreversible sorption is widely seen in soils, and is thought to occur as a result of recrystallization of sorbing material over the initially sorbed contaminant, or the migration of contaminants to "higher energy" sites. Irreversible sorption is that it causes a net decrease in bioavailable contaminants in the subsurface.

Table 14.1.2 shows general estimates of the fraction of sorption that is irreversible. The values come from compilations of field studies (see e.g., Brady et al., 1998; Coughtrey et al., 1986; Coughtrey and Thorne, 1983), and are useful only in giving a sense of its magnitude.

Irreversible uptake of organic contaminants is greatest for those contaminants that are most strongly sorbed, i.e., those with high octanol-water coefficients (see e.g., Kan et al., 1998).

Chemical Transformation

A number of inorganic contaminants form insoluble solids under fairly specific conditions. Common to the process is the importance of contaminant speciation, which depends primarily on the ambient biological and geochemical conditions of the soil or groundwater. pH, redox state (electron availability), alkalinity, and the presence of chelating (e.g., EDTA, natural organic acids) or solid-forming (e.g., phosphate in a number of cases) ligands are critically important (ionic strength is probably a secondary factor) to defining these conditions.

Examples of geochemical controls on metal solubility are provided below. Formation of copper oxides and hydroxides can limit copper levels, under reasonably oxidizing conditions, at pHs >6). If the chromium present in soil or groundwater is largely trivalent in form, the formation of chromium hydroxide has the potential to reduce available chromium at pHs >5. Based on solubility equilibria, higher pHs (greater than 9) are needed to decrease nickel levels through the precipitation of nickel carbonates, and nickel hydroxides (oxides). It is difficult to maintain Ba^{2+} levels in solution at hazardous levels if there is more than approximately 1 ppm of sulfate, because of the tendency for the

TABLE 14.1.2   Fraction of Contaminant Metal Taken Up Irreversibly (from Brady et al., 1998)

Contaminant	Fraction irreversibly sorbed
Am	0.6
As	0.9
Ba	0.5
Cd	0.5
Cs	0.9
Cr	0.5
Co	0.9
Cu	0.9
I	0.9
Pb	0.9
Ni	0.9
Ra	0.5
Sr	0.15
Tc	0.1
Th	0.99
^{3}H	0
U	0.1
Pu	0.99
Zn	0.9

two to combine to form insoluble $BaSO_4$. Technetium also forms insoluble solids under reducing conditions, as do plutonium and uranium. The primary natural attenuation pathways that include chemical transformation are listed in Table 14.1.1.

For both sorption and chemical transformation, the OSWER Directive states that "... Determining the existence and demonstrating the irreversibility of these mechanisms are key components of a sufficiently protective MNA remedy" (U.S. Environmental Protection Agency, 1997). Irreversibility is often examined using leach tests. The toxicity characteristic leach procedure (TCLP, see ASTM method D5233-92), uses a pH 5 acetate buffer solution (meant originally to mimic the composition of a typical landfill leachate) that tends to consume some mineral hosts (e.g., calcite) and consequently, exaggerate the bioavailability of some contaminant inorganics. The test also provides no information about the specific identity of the host mineral(s) that might sequester particular contaminants. The absence of information about the host limits consideration of the potential future use of sites.

Some minerals are less affected than others by likely changes in groundwater chemistry. Consider the example of Pb contamination of limestone-containing soils versus SO_4-rich soils. Pb might be considered to be naturally attenuated if largely bound up irreversibly in either matrix. Pb, if bound up in a carbonate mineral, might be liberated by soil acidification, although the Pb in the SO_4 matrix probably would not dissolve. On the other hand, a change in land use to agriculture may involve soil liming, raising soil pH, and thus, result in an attenuation of any SO_4-bound Pb. Thus, knowing the chemical form of a contaminant is critical to addressing the effect of future land use changes on MNA.

Biodegradation—Fuel Hydrocarbons

Fuel hydrocarbons, and their important constituents—benzene, toluene, ethylbenzene, and xylenes (BTEX), are broken down most rapidly by indigenous microorganisms under aerobic conditions (when dissolved O_2 concentrations are greater than 2 mg/L) and dissolved O_2 is the terminal electron acceptor (oxidant). Complete oxidation results in the formation of CO_2. Oxidation of fuel hydrocarbons in the absence of oxygen by microorganisms using $NO_3^- + NO_2^-$, Mn(IV), Fe(III), SO_4^{2-} and (under very reducing conditions) CO_2 is much slower.

Indigenous microorganisms appear to be reasonably abundant, so that their availability does not limit the overall rate of breakdown. Moreover, subsurface microorganisms appear to adapt on a sufficiently rapid timescale to take advantage of the prevailing redox conditions at depth. Addition of oxygen to the subsurface may cause biodegradation to occur more rapidly if oxygen is the limiting factor. In most cases (particularly when the source term, the separated-phase product, is removed), the availability of electron acceptors is not likely to limit breakdown. In other words, the oxidative capacity of soils and groundwaters (i.e., the sum of the potential terminal electron acceptors) typically exceeds the potential electron donor load required by fuel hydrocarbon plumes. The total available capacity ultimately will depend on the volume of soil or aquifer encountered by a plume on the path from source to receptor.

Oxidation of fuel hydrocarbons can be tracked in the subsurface in one of four ways: (1) monitoring the disappearance of reactant(s), e.g., benzene and Fe(III), for the case of benzene oxidation by Fe(III); (2) monitoring the appearance of degradation products, e.g., CO_2 and Fe(II); (3) quantifying indirect effects of organic oxidation on soil CO_2 and/or alkalinity levels; or (4) establishing shifts in isotopic ratios of reactants or products (Aggarwal and Hinchee, 1991). In general, fuel hydrocarbon breakdown involves the disappearance of electron acceptors and the appearance of electron donors. Often at least one of the two can be effectively measured in the subsurface. The appearance of Fe(II) at concentrations much higher than background, in combination with an observed decrease in hydrocarbon level, can be used as evidence of breakdown involving Fe(III) (typically in solid form) as the terminal electron acceptor. Higher than background concentrations of Mn(II) points to utilization of Mn(IV) hydroxides as the terminal electron acceptor. Decreases in sulfate and an inverse increase in H_2S, points to sulfate-reduction as the breakdown pathway.

Degradation of fuel hydrocarbons can also be simply quantified by comparing the toluene to benzene ratio in a plume with time. Benzene degrades more rapidly under aerobic conditions. However, toluene degrades more rapidly than benzene under anaerobic conditions. As a result, the ratio of benzene to toluene provides a qualitative indicator of intrinsic biodegradation.

The various breakdown pathways, their theoretical efficiencies, and field case studies have been outlined in great detail elsewhere (see e.g., Wiedemeier et al., 1995a; Wiedemeier et al., 1995b). MNA of fuel hydrocarbons has become widely accepted at the federal and state level and, when combined with source removal, is probably a significant means for remediating fuel hydrocarbon contaminated soils.

Biodegradation—Chlorinated Organics

Figure 14.1.1 gives a simplified view of the breakdown pathways that affect chlorinated organics, such as PCE, TCE, DCE, TCA (trichlorethane), carbon tetrachloride (CT), PCA (tetrachloroethane), VC, and DCA (dichloroethane) in soils and groundwaters. Chlorinated organics are common contaminants in soils and groundwaters, and are often the most mobile fraction that dissolves off the fringes of dense nonaqueous phase liquids (DNAPLs). Typically, DNAPL components present at concentrations of 1 percent and above their respective water solubilities are used to infer the presence of DNAPLs. Highly chlorinated organics (e.g., CT, PCE, PCA) are already quite oxidized relative to BTEX compounds, hence, their further oxidation by microorganisms provides relatively little energy to the latter. Instead, the breakdown of highly chlorinated contaminants in soils and groundwaters typically occurs through sequential reductive dechlorination, whereby hydrogens are exchanged for chloride groups to produce breakdown products that are less chlorinated and more reduced. For chlorinated alkenes, the breakdown sequence is:

$$PCE \Rightarrow TCE \Rightarrow DCE \Rightarrow VC.$$

Reductive dechlorination occurs most rapidly under reducing conditions (oxidation/reduction potential < -190 mV). At many sites, high concentrations of degrading commingled BTEX compounds

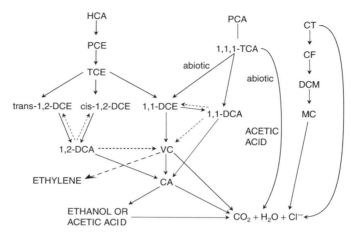

FIGURE 14.1.1   Degradation paths for chlorinated organics under anaerobic conditions (from Barbee, 1994—with permission).

can lead to a drawdown of O_2, followed by anaerobic conditions that are conducive to the reduction of chlorinated organics. Although oxidation of chlorinated organics, such as TCE, is relatively slow compared to reductive degradation, field degradation half-lives are not inconsequential and are probably 10 years or less.

Because the daughter compounds are less oxidized, they often accumulate under reducing conditions and may themselves pose a health threat. Complete breakdown to CO_2, therefore, requires subsequent oxidizing conditions. Figure 14.1.2 shows good and bad scenarios for chlorinated organics emanating from a DNAPL. The worst-case involves reduction of highly chlorinated organics near the source and their subsequent accumulation. The best-case scenario involves reducing conditions near the source and oxidizing conditions further out. This is particularly true for sites where VC accumulates as a degradation product of the dechlorination cascade (i.e., PCE $\Rightarrow$ TCE $\Rightarrow$ DCE $\Rightarrow$ VC), because VC has been observed to degrade four times as fast under aerobic conditions. Another worst-case would have oxidizing conditions prevailing at the source. The difference between best- and worst-case scenarios is reflected in a marked uncertainty in the biodegradation rate constants to be used in assessing the transport and risk of chlorinated organic plumes (see below).

Additionally, secondary measurements may indicate biodegradation of chlorinated organics. Because chloride is a product of dechlorination, increases in chloride levels provide a qualitative indicator of degradation. Reductive dechlorination of TCE usually favors the production of *cis*-1,2-DCE to the extent that the *trans* isomer typically accounts for 20 percent or less of the total 1,2-DCE observed. The usual isomeric preference of the dehalogenation process provides a useful mechanism for evaluating the occurrence of natural attenuation of TCE. Manufacturing processes typically produce *cis*- and *trans*-1,2-DCE at nearly equal ratios (40 to 60 percent of each isomer). Because a 1,2-DCE spill should have roughly equal concentrations of *cis*- and *trans*-1,2-DCE, a site with predominantly *cis*-1,2-DCE indicates the biological formation of the DCE through reductive dechlorination of TCE.

For sites with comingled plumes where it is difficult to differentiate sources from degradation products, chlorine isotopes have been used to estimate gross degradation. Microorganisms preferentially remove lighter isotopes during reductive dechlorination. As a result, the composition of the lighter isotope in groundwater increases.

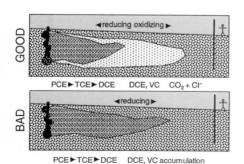

FIGURE 14.1.2   Good and bad scenarios for chlorinated organic breakdown.

MNA of chlorinated organics has been estimated to be a potentially effective remedy at perhaps 30 percent of US. Air Force contaminated sites. Moreover, there are a large number of CERCLA RODs that rely on natural attenuation of chlorinated organics.

14.1.5 IMPLEMENTING NATURAL ATTENUATION

Implementing MNA ideally involves site-screening, followed by further site characterization (if site screening suggests MNA is likely), prediction of long-term attenuation, and long-term monitoring. Site screening relies on existing site characterization data and has as its objective the determination of whether MNA is likely at a site, and worthy of extra effort to demonstrate its presence to regulators. This assessment must take into account whether chemical conditions are favorable (e.g., do reducing conditions exist for a TCE plume?) and whether hydrologic conditions are such that attenuation can proceed sufficiently to erase public health risks before a receptor is reached. One rapid means for making such an assessment is to use a public-access site screening tool, such as *MNAtoolbox*, which is maintained by DOE at www.sandia.gov/eesector/gs/gc/na/mnahome.html

MNAtoolbox provides a template for development of the natural attenuation conceptual model. With an identification of attenuation factors that could contribute to contaminant destruction or immobilization, the MNAtoolbox can be used to identify the most relevant site characterization data that is needed to develop the natural attenuation argument. Additionally, the calculation approach used in the MNAtoolbox scorecard provides a technically sound process for evaluating the potential application of natural attenuation to other contaminants (pesticides, herbicides, explosives, and others).

In general, the full MNA case must include the following: (1) a characterization of attenuation mechanism(s), (2) evaluation of site conditions (i.e, distance to receptor, transport rates, etc.), (3) a prediction of anticipated plume behavior, (4) a long-term monitoring program that addresses all of the requirements identified in the Data Needs Section, and (5) a contingency plan.

Long-Term Monitoring

MNA requires more focused monitoring efforts than other remediation approaches as it must provide a comprehensive understanding of contaminant geochemistry and transport at depth sufficient to explain past contaminant drawdown and predict future contaminant trends. Any significant deviation from predicted trends must be detected and may require abandonment of MNA and the implementation of contingency measures.

MNA monitoring has the potential to be very costly because of the longer cleanup times are typically required. MNA monitoring costs could be prohibitively expensive unless: (1) the frequency of monitoring is progressively scaled back over time, and (2) the suite of variables to be monitored is minimized. Confidence in the conceptual model of MNA, borne out by subsequent monitoring data that falls within the envelope of uncertainty of the particular site, is the most powerful argument for decreasing the frequency of monitoring. The collection of data that neither confirms nor refutes conceptual models, and that can't be used as a contingency trigger, must be avoided. Superfluous data should be identified in the conceptual model development stage, well before the long-term monitoring plan is developed.

Long-term monitoring (LTM) performs two functions in the MNA approach:

- It evaluates the predicted trend in the respective attenuation mechanism(s) and/or provides data for further testing and calibration of the site conceptual model (performance monitoring), and

- It provides early warning of the failure of the attenuation mechanism; hence, it acts as a trigger for implementation of contingency measures (compliance monitoring).

Ambient monitoring must also be done to support both performance and compliance monitoring.

Performance monitoring is done in the interior of the contaminant plume, and combines both the monitoring required for site characterization and that invested in determining ambient trends in site biology, hydrology, and geochemistry. Process monitoring has as its primary objective the calibration of a site conceptual model that can be used to accurately predict site behavior in the future. A side benefit of such a model would be the potential scaling back of near-source, and perhaps far-source, compliance monitoring. Conceptual model development requires that all possible models that explain site characterization data be considered, and that subsequent sampling identify the final chosen site model. Once the site conceptual model is chosen, it must be used to predict contaminant movement in space and time (plume evolution). If subsequent sampling indicates that attenuation trends are consistent with the proposed conceptual model, the agreement between prediction and measurement may be used as a basis for decreasing the frequency of long-term monitoring, and perhaps, the range of analytes.

Significant deviations from the predicted baseline should be used to update the conceptual model. This may require further site characterization to identify aberrant biogeochemical or hydrologic behavior. The general process will implicitly incorporate further rounds of measurements and data collection to test and/or better calibrate the conceptual model(s). Note that, in this sense, long-term monitoring plays a different role in MNA, as opposed to a "no-further action" approach, as MNA process monitoring works toward the goal of accurately predicting site behavior. Under "no-further action," long-term modeling is decoupled from any understanding of subsurface processes, and instead overly conservative modeling is done to bound site behavior. There is little expectation of contaminant levels decreasing in the face of natural processes to approach regulatory targets. The MNA approach puts greater emphasis on identifying and relying on attenuation mechanisms in the conceptual model development stage. All this being said, we cannot know all that is needed about a specific site. Instead we seek to minimize uncertainty through contingency planning.

Compliance monitoring acts to initiate contingency measures if attenuation mechanisms fail. Compliance monitoring must be done beyond the furthest point that detectable levels of contaminant movement can be conceivably predicted to each. This distance can be arrived at on the basis of historical case analyses that indicate the maximum extent that specific contaminants, or hazardous daughter products, have been observed to travel. Alternatively, reaction-transport models can be used to estimate the maximum extent of contaminant, or hazardous daughter product, transport.

Compliance monitoring must be done sufficiently far inward from the point of compliance that contingency remedial actions, if required, can be effectively implemented to prevent harmful exposures at and beyond the point of compliance. There must be a sufficient buffer between the compliance well(s) that plume movement can be arrested and further measures implemented. The primary contingency measure likely will be hydraulic control to limit plume movement until more permanent remedial action can be put into action.

ACKNOWLEDGMENTS

We greatly appreciate the constructive review of Jim Jacobs and Jay Lehr, as well as financial support from the SNL-LDRD Office.

REFERENCES

Aggarwal, P. K., and Hinchee, R. E., "Monitoring in situ Biodegradation of Hydrocarbons Using Stable Carbon Isotopes," *Environ. Sci. Technol.*, 25: 1173–1180, 1991.

Baes, C. F. I., and Sharp, R. D., "A Proposal for Estimation of Soil Leaching and Leaching Constants for Use in Assessment Models," *J. Environ. Qual.*, 12: 17–28, 1983.

Barbee, G. C., "Fate of Chlorinated Aliphatic Hydrocarbons in the Vadose Zone and Groundwater," *Groundwater Monitoring and Review*, Winter 1994: 129–140, 1994.

Brady, P. V., Spalding, B. P., Krupka, K. M., Borns, D. J., Waters, R. W., and Brady, W. D., Technical Guidance and Site-screening Guidelines for Implementation of Monitored Natural Attenuation at DOE Sites, Sandia National Laboratories, 1988.

Brady, W. D., Brown, A., and Eick, M. J., "Implementation of Monitored Natural Attenuation: A Survey of State Policies and a Discussion of Innovative Methods to Break Down Implementation Barrriers," *IGT 8th International Symposium on Environmental Biotechnologies and Site Remediation Technologies*, 1997.

Browner, C. S., Statement of Carol. M., Browner, Administrator, USEPA before the Senate Committee on Environment and Public Works, March 5, 1997.

Coughtrey, P. J., Jackson, D., Jones, C. H., Kane, P., and Thorne, M. C., *Radionuclide distribution and transport in terrestrial and aquatic systems*, Vol. 6., Balkema, A. A., 1986.

Coughtrey, P. J., and Thorne, M. C., *Radionuclide Distribution and Transport in Terrestrial and Aquatic Ecosystems*, Vol. I., Balkema, A. A., 1983.

Kan, A. T., Fu, G., Hunter, M., Chen, W., Ward, C. H., and Tomson, M. B., Irreversible Sorption of Neutral Hydrocarbons to Sediments: Experimental observations and Model Predictions, *Environmental Science & Technology*, 32(7): 892–902, 1998.

McNab, W. W. Jr., Rice, D. W., Bear, J., Ragaini, R., Tuckfield, C., and Oldenburg, C., "Historical Case Analysis of Chlorinated Volatile Organic Compound Plumes," *Lawrence Livermore National Laboratory Report*, UCRL-AR-133361, 1999.

Rice, D. W., Dooher, B. P., Cullen, S. J., Everett, L. G., Kastenberg, W. E., Grose, R. D., and Marino, M. A. "Recommendations to Improve the Cleanup Process for California's Leaking Underground Fuel Tanks," *Lawrence Livermore National Laboratory Report*, UCRL-AR-121762, 1995.

U.S. Environmental Protection Agency, *Soil Screening Guidance: Technical Background Document*, USEPA-OSWER, 1996.

U.S. Environmental Protection Agency, *Use of Monitored Natural Attenuation at Superfund, RCRA Corrective Action, and Underground Storage Tank Sites*, Office of Solid Waste and Emergency Response, 1997.

Wiedemeier, T. H., Swanson, M. A., Wilson, J. T., Kampbell, D. H., Miller, R. N., and Hansen, J. E., "Patterns of Intrinsic Bioremediation at Two U.S. Air Force Bases," in *Intrinsic Bioremediation*, Hinchee, R. E., Wilson, J. T., and Downey, D. C., eds., pp. 31–52. Battelle Press, 1995a.

Wiedemeier, T. H., Wilson, J. T., Kampbell, D. H., Miller, R. N., and Hansen, J. E., "Technical Protocol for Implementing Intrinsic Remediation with Long-term Monitoring for Natural Attenuation of Fuel Contaminant Dissolved in Groundwater," *Air Force Center for Technical Excellence, Technology Transfer Division*, 1 & 2., 1995b.

CHAPTER 14
REMEDIATION TECHNOLOGIES

SECTION 14.2
PASSIVE IN SITU REMEDIATION TECHNOLOGIES

James A. Jacobs

Mr. Jacobs is a hydrogeologist and president of FAST-TEK
Engineering Support Services with more than 20 years of
experience. He is registered in several states, including
California, and specializes in assessment methods and
in situ remediation technologies.

14.2.1 INTRODUCTION

Passive in situ remediation technologies (PIRT) use a variety of physical, chemical, or biological processes that act without significant human intervention in the subsurface to reduce the mass, toxicity, mobility, volume, or concentration of contaminants in soil or groundwater. These in situ treatments usually involve the addition of materials to isolate, neutralize, oxidize, bioremediate, or precipitate contaminants in the subsurface without digging and handling of the soil and groundwater. These types of in situ treatments are generally considered as reasonable remedial options when PIRT is more economical and less disruptive to site activities than the more conventional approaches. PIRT discussed in this article include low-permeability barrier walls, permeable treatment walls, passive gas venting, and remediation injection technologies. The main benefits of low technology and passive in situ remediation systems over traditional active methods are the lower final cost for remediation, minimum cost for operations and maintenance, no moving parts that could break, and no discharge permits or waste disposal of liquids for in situ groundwater treatment.

14.2.2 PASSIVE LOW-PERMEABILITY BARRIERS

Physical barriers or low-permeability vertical barriers can be used to divert groundwater flow away from a contamination source. The major types of low-permeability vertical barriers are slurry cutoff trenches/walls, sheet pile cutoff walls, grout curtains, polyethylene barriers and capping. As with most low permeability barriers, the maximum groundwater mounding occurs against the wall facing in the upgradient direction. In some cases, passive interceptor trenches may be used to intercept surface or shallow groundwater flow and transport it to surface water ponding areas. In places where groundwater mounding is a significant problem, an active groundwater extraction system can be used to pump off the excess groundwater and maintain hydraulic control.

14.2.3 SLURRY CUTOFF TRENCHES/WALLS

This technology combined with capping can fully confine and contain contaminated soils to prevent leachate generation and the migration of contaminants offsite. The slurry cutoff trench/wall is

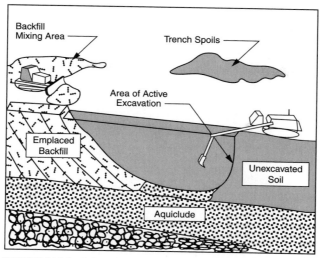

FIGURE 14.2.1 Soil-Bentonite slurry cutoff trench in cross section (U.S. EPA, 1991b).

constructed using backhoes or excavators by excavating a narrow vertical trench, typically two to four feet wide, and backfilling with a low-hydraulic conductivity material such as bentonite (see Figure 14.2.1).

As the excavation proceeds, the trench is filled with a bentonite-water slurry, which stabilizes the walls of the trench, thereby preventing collapse. A filter cake forms on the trench walls when the slurry penetrates into the permeable soils, sealing the soil formation (U.S. EPA, 1991a,b). Common slurry mixtures include soil and bentonite (SB) and the cement-bentonite (CB). The hydraulic conductivity of an SB and CB wall, with good construction quality control, is approximately 1×10^{-8} cm/sec and 1×10^{-6} cm/sec, respectively (U.S. EPA, 1985).

Slurry walls can be installed in various configurations. One common system is the circumferential slurry wall used in conjunction with capping to isolate the contaminated soils and groundwater (see Figure 14.2.2). Downgradient placement of a slurry wall can be designed to prevent downgradient migration of contaminated groundwater, as installed at the Rocky Mountain Arsenal in Denver, Colorado (CSU, 1988). A slurry wall can be installed upgradient of a contaminated area to passively divert groundwater around a site (see Figure 14.2.3). A cutoff wall may be "keyed-in" or "hanging":

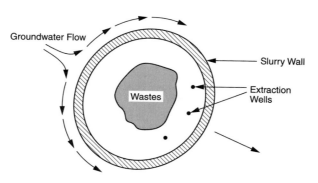

FIGURE 14.2.2 Plan of circuferential wall placement (U.S. EPA, 1988).

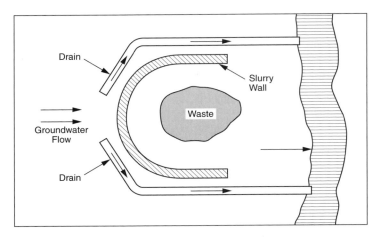

FIGURE 14.2.3   Plan of upgradient placement with drain (U.S. EPA, 1988).

if it is excavated into a continuous low-permeability horizontal confining layer, such as a clay deposit or competent bedrock. Large excavators can be used to install slurry walls to a maximum of 35 to 50 ft Crane-operated chamshells attached to a kelly bar can excavate to depths of up to 200 ft (U.S. EPA, 1991b).

Case Examples

A circumferential slurry wall was used in conjunction with capping and pump and treat at the Gilson Road Superfund Site in New Hampshire (Weston, 1989). Slurry walls were placed down gradient from a plume at the Rocky Mountain Arsenal in Denver, Colorado (CSU, 1988).

The concerns with slurry cutoff trench/wall technology is the compatibility of the cutoff trench backfill material with the site contaminants. Slurry walls are susceptible to attack by strong acids or bases, strong salt solutions, and some organic chemicals. The low permeability characteristics of a slurry wall might be in danger because of desiccation or cracking caused by these reactions. Therefore, laboratory compatibility testing may be required to evaluate the local subsurface conditions with the stability of the backfill materials. A major concern with slurry wall technology is the leakage underneath the wall. Because of the leakage potential, slurry wall projects typically have active extraction systems to help control hydraulic gradients and minimize the potential for contaminant migration.

14.2.4   *SHEET PILE CUTOFF WALLS*

Sheet pile cutoff walls can be used to contain contaminated groundwater or divert clean groundwater flow or below contaminated areas. Sheet pile can be made of interlocking steel, which are constructed by driving individual sections of interlocking steel sheets into the ground using single, double-acting impact or vibratory pile drivers to form a thin impermeable barrier to groundwater flow (U.S. EPA, 1991a). The interlocks can be a source of water leakage, which can be sealed naturally with fine sand or silt or by tremie grouting along the interlocks. Sheet piling is generally less time-consuming to install than slurry wall construction, however, sheet piling is difficult to install in rocky soils or in areas with shallow bedrock. In these conditions, the sheet piles will be damaged during installation. The maximum depth for sheet pile projects is about 35 to 40 ft, with more shallow projects from 20 to 30 ft being more common and economical.

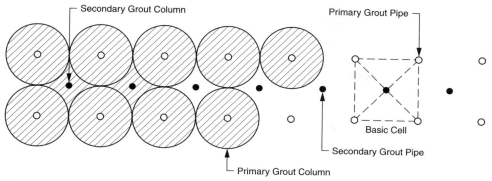

FIGURE 14.2.4   Typical grout curtain layout (U.S. EPA, 1991b).

14.2.5   GROUT CURTAINS

Grouting is the process of pressure injecting a liquid into soil rock to reduce fluid movement or to impart increased strength. Grouting can be used to control the movement of groundwater and to solidify or stabilize a soil mass. By reducing permeability of a deposit, the grout curtains can be created in unconsolidated materials by pressure injection. There are two types of grouts: particulate and chemical grout. Particulate grout, also called suspension grout, is made up of fluids composed of a suspension of solid materials such as cement, clay, bentonite, or a combination of these materials. Particulate grouts have large particle size and are more suited to highly porous sediments and soil. Chemical grouts, both silica or aluminum-based solutions and polymers, rely on polymerization reactions to form hardened gels. Chemical grouts have initially low viscosities and can be pumped into finer grained soils than can the particulate grouts. Layout of grout injection pipes depends on soil types, grout viscosity, injection pressure, and gel time. Spacing will depend on grout penetration and desired grouted soil properties (U.S. EPA, 1991a). Usually a predetermined amount of grout is pumped under pressure into each of the holes.

The grout injection points of the proposed grout curtain are lined up in a three rows (see Figure 14.2.4). The primary holes, side-by-side, form a double line of touching columns. A third line of injections is designed to fill-in any gaps left in the diamond-shaped space between the side-by-side primary grout injection columns. Primary injection holes are spaced at intervals of three to five feet. To be effective, grout curtains are keyed into impermeable soil or sediments such as clay, or into competent bedrock. Grout curtains lose their effectiveness to control contaminant migration when the grout is not compatible with the local subsurface conditions or contaminant.

14.2.6   POLYETHYLENE BARRIER

Specialized trenching machines have been designed to install polyethylene barrier systems. This equipment trenches and installs a continuous sheet of high density polyethylene (HDPE) geomembrane, which is placed vertically (see Figure 14.2.5). Seams are minimized in this technology. The HDPE geomembrane is specially formulated to be resistant to sunlight and most contaminants. Sheets are joined together using a high performance, hydrophilic interlocking waterproof joint system. The joints are impervious to fluids and remain tight for the life of the barrier (Horizontal Technologies, Inc., 1999). The polyethylene barrier system technology can be used to a depth of approximately 30 ft. The major concern with this technology is the potential for groundwater and contaminants to leak through, around or beneath the barrier.

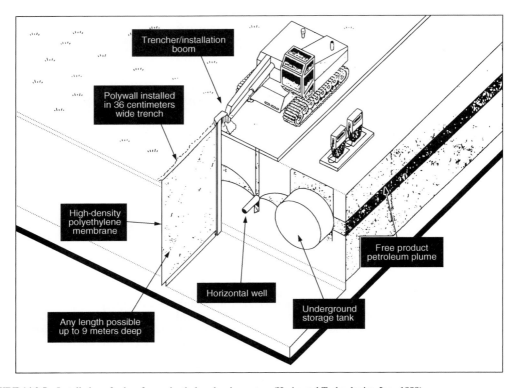

FIGURE 14.2.5   Installation of subsurface polyethylene barrier system (Horizontal Technologies, Inc., 1999).

Case Examples

The polywall barrier system has been installed to cut off the migration of diesel fuel into the Little River in Star Lake, New York. The barrier was used in conjunction with a linear contaminant remediation system to recover free product and control groundwater flow (see Figure 14.2.6). The system was installed continuously along the river bank for a distance of 1,350 lineal feet to depths of 15 ft below ground surface with a water table-free product interface at an average depth of four feet (Horizontal Technologies, Inc., 1999).

14.2.7   CAPPING

Capping or surface sealing is a passive remediation process whereby a former landfill, contaminant plume or buried waste is isolated to avoid surface water infiltration, minimizing the potential for leachate generation. Capping may be used to control the emission of gases and odors, reduce erosion, and improve aesthetics. Data required for capping include: extent of contamination, depth to groundwater, availability of cover capping materials, soil characteristics such as gradation, permeability, soil strength, climate, and final land use. Surface water control measures are typically a part of the capping project. The final design profile, frequently of a typical multilayered or RCRA cap, will include geotextiles as a filter between the vegetative/protective layer and drainage layer and/or a protective layer over the synthetic membrane (U.S. EPA, 1991b).

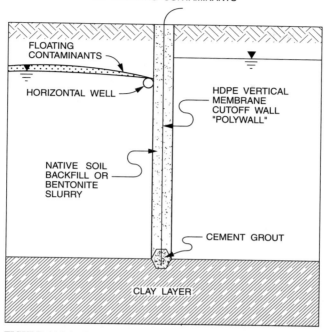

FIGURE 14.2.6 Cross section of polyethylene barrier system (Horizontal Technologies, Inc., 1999).

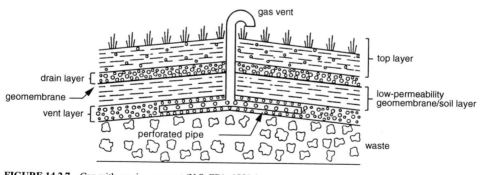

FIGURE 14.2.7 Cap with passive gas vent (U.S. EPA, 1991a).

14.2.8 PASSIVE GAS VENTING

Passive gas venting is a treatment control technology for volatile organic compounds (VOCs) migrating in the vapor phase in primarily soil and to a lesser extent, in bedrock (see Figure 14.2.7). The effectiveness of a passive gas venting system relates to soil type, soil density, depth to the groundwater, and specific gravity of the contaminant. Passive gas venting has been used to control methane gas from

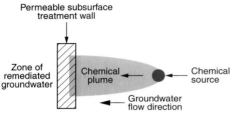

FIGURE 14.2.8 Permeable treatment wall; single trench perpendicular to path of chemical plume (Warner, Yamane, Gallinatti, and Hankins, 1998).

landfills and coal mines and to remove the vapor phase from gasoline and solvent plumes. Gas venting creates a subsurface air flow pattern that confines the hazardous vapors to an area in which potential impacts are minimized or the vapor is vented. The best conditions for passive gas venting are permeable, unsaturated soils such as sand and gravel. Low permeability soil, such as silt or clay, or unfractured bedrock, generally lack the connected flow pathways required for effective vapor extraction. For passive gas venting, a gravel packed vented trench constructed using a backhoe or series of vertical vent wells constructed using a drilling rig will allow the generated gas to vent to the atmosphere. A passive vent system is used for situations in which the gas generated, such as methane, is lighter than air. Typical ranges of gas flow rates expected in natural in situ soils can vary from 0.5 cubic ft per min (cfm) to 2 cfm per linear foot of well screen (U.S. EPA, 1991b). Active pumping has been used successfully to enhance flow rates for gas venting systems.

14.2.9   PASSIVE TREATMENT SYSTEMS

Passive in situ remediation system use reactive media, either injected or placed in treatment walls, oriented to intercept and remediate a moving groundwater contaminant plume. The main assumption is that the contaminant plume migrates under the natural gradient and the treatment occurs in situ (Figure 14.2.8). The advantages to passive systems include the full use of the property and, frequently, lower cost than conventional remediation methods. Typically, the reaction uses a chemical or biological processes. In designing a passive in situ treatment system, five main factors must be examined in detail: groundwater flow (direction, velocity, flux, vertical gradient), stratigraphy (depths and thickness of units, degree of fracturing, channeling), depositional environment (mineralogy, TOC, type of environment), hydrochemistry (contaminant distribution, ambient water chemistry), and microbiology (GRA, 1999). A complete project usually starts with site characterization moving to conceptual model and preliminary design. During the model and design phases, bench-scale testing and pilot-scale testing are often performed. The final design, based on the pilot-scale test, is followed with the full-scale implementation of the passive treatment system. After the reactions are completed and regulatory goals are reached, the confirmation soil or groundwater samples are collected and evaluated. If appropriate, the site groundwater or vapor is monitored, typically for four quarters, before regulatory closure is granted. A realistic goal for passive treatment systems is the significant reduction of the contaminant source. It is usually not economic or feasible to achieve 100 percent removal of all contaminants on a site. Once the contaminant levels on the site are near regulatory guidelines, closure is frequently granted. A risk assessment using a computer model, such as the risk based corrective action (RBCA) or similar potential receptor evaluation, can be used for justification of site closure in cases where the passive treatment has been performed successfully, but the continued reduction of the contaminant concentrations has ceased.

14.2.10   PERMEABLE SUBSURFACE TREATMENT WALLS

In shallow groundwater situations, trenchers, backhoes, and excavators with or without shoring have been used to emplace the higher permeability materials. Contaminants to be treated in the subsurface using in situ passive methods include halocarbons, reducible metals, acid water, organics, and petroleum hydrocarbons. High pressure remediation injection technologies (jetting), such as the Remediation Injection Process (RIP®) use liquids as the treatment medium. For plug technologies, treatment media are installed by pumping the grout or slurry into an open trench or by drilling holes and using a grout pump to push a pressurized stream of reactive materials through the drill rod to the target depth. The high pressure squeezes the treatment grout into the formation. An example of grout treatment medium is magnesium peroxide used for bioremediation of petroleum hydrocarbons. Magnesium peroxide is installed as a grout, but hardens to a highly cemented plug, giving off oxygen when wet or in the presence of moisture. Solids or powders, such as zero-valent iron, can be grouted in place or can be poured into a trench or gate system. The reaction with iron is based on the principle of Fenton's Chemistry where the iron, in the subsurface reacts to create the hydroxyl radical. The hydroxyl radical in the subsurface can be used to rapidly degrade solvents.

14.2.11 FUNNEL AND GATE TREATMENT SYSTEM

The funnel and gate treatment system is a hybrid technology using aspects of both low-permeability barriers and the permeable treatment system. The funnel-and-gate method was originally proposed by the University of Waterloo. One of the most widely used treatment materials associated with the funnel and gate technology is zero-valent iron. This material has been documented to degrade dissolved chlorinated volatile organic compounds (VOCs) such as trichloroethene (TCE) and daughter products as well as and vinyl chloride (VC) (Gillham and O'Hannesin, 1992, 1994). The zero-valent iron has also been shown to immobilize certain oxidized metals species such as (chromium) Cr^{+6} (Blowes and Ptacek, 1992). With the funnel-and-gate system, a sealable joint steel sheet piling system is constructed to provide a lateral hydraulic barrier. The barrier funnels contaminated groundwater through a series of gates where the treatment media are contained. The gates are permeable in situ treatment zones at the downgradient position of the funnel. Most treatment zones are less than 40 to 50 ft deep, as a result of commonly available emplacement technology. Deeper treatment systems to 70 to 100 ft are possible with specialized equipment.

The most common emplacement techniques are the conventional trench-and-fill method. In this method, sheet piles or shoring may be required to keep the trench open while performing the backfilling of the treatment materials. The rapid, one-pass trenching machine excavates the soil and replaces the soil with the treatment material using a conveyer belt on the trenching machine. No sheet piles or shoring are required. The gates or treatment area can also be constructed using a caisson or canister to hold the reactive treatment material (see Figures 14.2.9 and 14.2.10). The groundwater can flow horizontally through the screened sections of the caisson(s), or may flow vertically through some part of the caisson (Warner, Yamane, Gallinatti, and Hankins, 1998). Using specialized, proprietary equipment, newer technologies for installing treatment materials include jet grouting, deep soil mixing, and mandrel-based technology.

When designing the gate or treatment wall, the length of the flow path must be designed with the contaminant having the proper residence time within the treatment media. Groundwater gradient can be evaluated by placing a row of piezometers perpendicular to the groundwater flow direction and upgradient, in and downgradient from the treatment wall. The piezometers should be placed above, in and below the treatment wall. Nonuniform groundwater flow may lead to channeling and higher groundwater velocity zones (Gallanatti and Warner, 1994), resulting in preferential flow paths and less than optimum residence time. General design considerations include chemical, geological, biological, hydrological, geotechnical, engineering, economic, and regulatory. Details about the passive treatment design, required residence time, surface area concentrations, required wall thickness, hydrodynamics

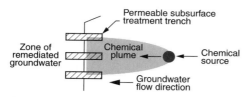

FIGURE 14.2.9 Permeable treatment wall; multiple trenches aligned parallel with chemical plume (Warner, Yamane et al., 1998).

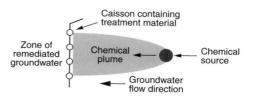

FIGURE 14.2.10 Permeable treatment wall; Caissons or large-diameter borings filled with treatment material placed across the path of the chemical plume (Warner, Yamane et al., 1998).

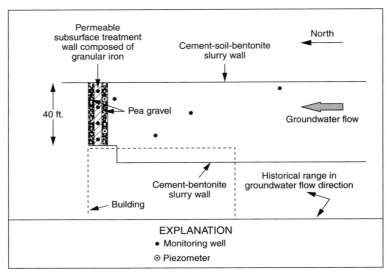

FIGURE 14.2.11 Funel and gate system to treat solvents in Sunnyvale, California (Warner, Yamane et al., 1998).

(dispersion and retardation), and intermediates (reactivity and by products) are discussed in (GRA, 1999, Warner, Yamane et al., 1998, and Warner, Szerdy, and Yamane, 1998). One major concern with the impermeable barrier technology, particularly the funnel and gate system is that a shift in groundwater flow direction as a result of pumping or a low groundwater gradient can make the barrier ineffective.

Case Example

A funnel and gate system was constructed in 1994 in Sunnyvale, California (Szerdy et al., 1996; Yamane et al., 1995) (see Figure 14.2.11). The funnel was a cement-soil-bentonite slurry wall. The permeable treatment wall, or gate, was composed of granular iron, to passively treat trichloroethylene, *cis*-1,2 dichloroethylene, and vinyl chloride. Slurry walls and sheet piles were used to place the hydraulic barriers (Warner, Yamane et al., 1998). Other geometries are possible (see Figure 14.2.12).

14.2.12 IN SITU INJECTION PROCESS

The technology for high pressure, low volume injection of nutrients into the subsurface using a small-diameter wand or lance driven into the ground has been widely used for several decades. Jetting, at its most basic, uses tree root feeder systems to inject chemicals into the subsurface by means of a high pressure injector tip on the end of a small-diameter, 2 to 5 ft long steel wand. Other more powerful injection systems have been designed using high pressure liquid pumps to increase flow at the tip of the wand to pressures exceeding 5,000 psi. Placing the high pressure injection points on close spacing, such as 2 ft centers to 5 ft centers, allows for complete in situ coverage, vertically and laterally. The high pressures allow for the treatment liquids to be dispersed into the aquifer. After the reaction occurs, additional treatment events may be required to reduce contaminants to regulatory approved levels.

One such in situ injection system, the Remediation Injection Process (RIP®)* is a versatile and adaptable system that can be used to efficiently implement, or augment, a variety of remediation

---

* RIP™ is a process of The Auger Group, Inc.

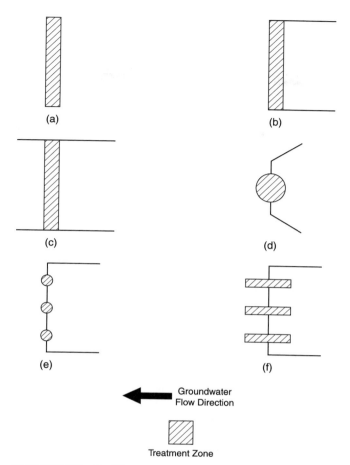

FIGURE 14.2.12   Possible geometries of passive treatment walls: (a) continuous trench, (b) with lateral hydraulic control, (c) with lateral hydraulic control and downgradient lateral barriers, (d) with a single casson, (e) with multiple caissons, (f) with parallel alleys (Warner, Szerdy, and Yamane, 1998).

techniques including bioremediation, chemical oxidation, stabilization, bioaugmentation, air sparging, bio-venting, and passive groundwater and soil remediation. The RIP utilizes lance penetration. Using this approach, the solutions can be accurately injected into the impacted areas to obtain direct contact with the target constituents. This precise injection of the treatment solutions/slurries can expedite the remediation process to achieve substantial reductions in contaminant concentrations in a relatively short period of time.

The hand-held RIP injector wands can be used to remediate limited access areas such as underneath slabs, railways, and buildings, around tanks, pipelines and subsurface utilities; and into hillsides, excavation pits, and stockpiles. The RIP has the capability to remediate a variety of constituents both in situ or ex situ including petroleum hydrocarbons, BTEX, chlorinated solvents, soluble inorganics, phenols, PCBs, PAHs, and other organic and inorganic contaminants. The flexibility and accuracy of this injection delivery system provides distinct advantages over both conventional in situ and ex situ remediation systems. As a result, the RIP® can provide appreciable savings in cost and time over traditional remediation technologies.

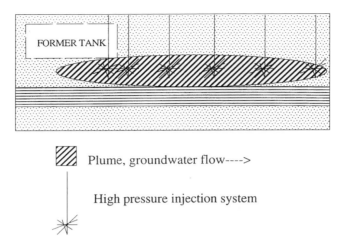

FIGURE 14.2.13   High pressure injection system, cross section.

High concentrations of liquid oxidants, such as hydrogen peroxide or potassium permanganate, can chemically oxidize halocarbons, petroleum hydrocarbons, and oils. The chemical oxidation process is exothermic and reaction temperatures in the aquifer are estimated at 130 to 150°F. Calcium polysulfide and other similar chemicals can precipitate selected metals such as arsenic, lead, and chromium as sulfates. Assuming that the pH of the groundwater remains relatively constant over time, these metals will remain insoluble. Nutrients, biologic electron acceptors, and low concentrations of hydrogen peroxide to provide oxygen for microbial growth, can be used to encourage in situ bioremediation of petroleum hydrocarbons. Other liquids used in the high pressure injection process can be acids or bases to neutralize contaminants. In situ treatments must be designed carefully to avoid creating toxic by-products. Passive in situ methods have been documented to have successfully treated chlorinated solvents and petroleum hydrocarbons into the ultimate degradation products of carbon dioxide and water. If not designed for the specific site conditions properly, however, passive in situ remediation technology can create unintended chemicals, such the highly toxic Cr^{+6} from Cr^{+3}, by the injection of potassium permanganate into the subsurface or the generation of toxic subsurface gases such as methane. Chemical compatibility of the injection equipment components and safety procedures become critical with the injection of strong acids, bases, oxidants, and other chemicals.

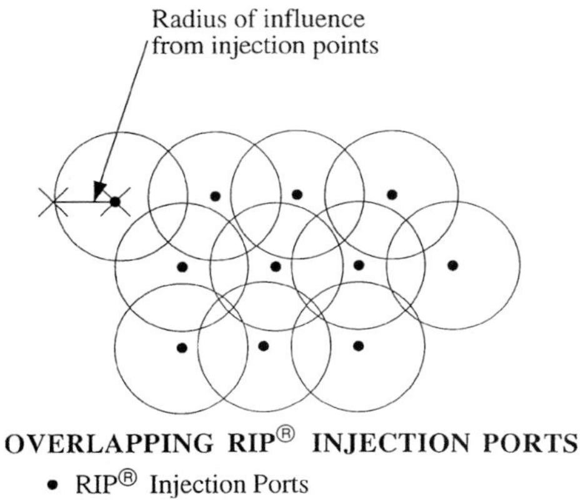

OVERLAPPING RIP® INJECTION PORTS

- RIP® Injection Ports

FIGURE 14.2.14   Overlapping radius of influence of high pressure injection system, map view.

Case Examples

A high pressure injection project was performed at a Napa, California former bulk storage facility having free phase product in the diesel and gasoline range. After one treatment event using 18 percent hydrogen peroxide for 4.25 hr., the average diesel and gasoline reduction in groundwater was over 99 and 50 percent, respectively. In another case outside of Olympia, Washington, a manufacturing facility having soil contaminated with volatile organic compounds, including perchloroethylene (PCE), trichloroethylene (TCE), dichloroethylene (DCE), and toluene were reduced using high concentrations of hydrogen peroxide approximately 70 percent after two treatment events.

14.2.13 GROUT PLUGS

The grout process is used with direct push technology probe rigs and hollow stem auger drilling rigs are used to inject slurry treatment materials into the subsurface (see Figure 14.2.15). The treatment materials are pumped directly through a string of probe or drill rods using the grout pull cap and a high pressure hose assembly. The grout injection can be made at the contamination zone, using less expensive neat cement grouts above the target zone, or the slurry can be injected as the tool string is being withdrawn from the borehole, spreading the treatment materials over a specific interval. Grout machines can be used to create an overlapping treatment zone of grout plugs that is downgradient of a contamination plume. Injected directly into the plume in the saturated zone, the plugs can be used to provide solid reactive materials to a variety of contaminants. To ensure uniform and widespread treatment results, grout injection points should have closely spaced centers, commonly two to three feet for clays and low permeability materials and five to ten feet for sands and gravels and high permeability materials. Grout spacing on close centers creates a subsurface treatment wall or zone and allows for a more complete and thorough remediation process (Jacobs and von Wedel, 1997, Jacobs, 1996, 1995).

In situ bioremediation is usually limited by oxygen. The closely spaced oxygen plugs are constructed using a magnesium peroxide cement in a pressure grouting process (see Figure 14.2.16). When wet, the oxygen plugs give off magnesium hydroxide and provide free oxygen to enhance aerobic bioremediation of petroleum hydrocarbons. This nontoxic reaction increases bioremediation by supplying oxygen to indigenous microbes. When hydrated by water or moisture, the magnesium peroxide gives off oxygen and becomes magnesium hydroxide. Magnesium peroxide forms a cement-like plug which gives off oxygen for about 6 months or more. The oxygen plugs are installed using a high pressure

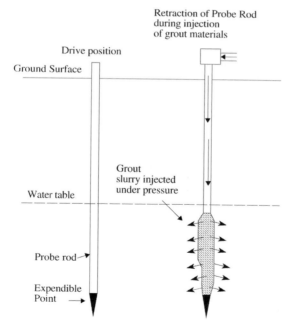

Retraction of Probe Rod
during injection
of grout materials

Drive position

Ground Surface

Grout
slurry injected
under pressure

Water table

Probe rod

Expendible
Point

No scale implied

FIGURE 14.2.15 Injection of magnesium peroxide with probe rod string (Geoprobe Systems, Inc., 1997).

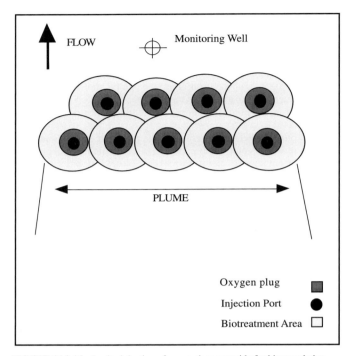

FIGURE 14.2.16   In situ injection of magnesium peroxide for bioremedation.

(up to 1000 psi) grout pump into the groundwater. This method creates an oxygen barrier and works with benzene levels up to about 50 ppm.

Grout plug applications include treating of metals with chemicals that will allow for precipitation of the contaminant. Injecting a grout of zero-valent iron on close spaced centers will allow for subsurface dechlorination of solvents. Grouts can also be designed to adjust pH in the aquifer.

Case Example

A passive treatment wall was installed at a gasoline station in San Francisco, California. The grout treatment wall consisted of 50 oxygen plugs installed to a maximum depth of 35 ft to provide additional oxygen for microbial degradation of gasoline in groundwater. Overall BTEX and TPH-reduction after five months was 67 and 74 percent, respectively (Regenesis, 2000).

REFERENCES

Blowes, D. W., and Ptacek, C. J., *Geochemical Remediation of Groundwater by Permeable Reaction Walls: Removal of Chromate by Reaction with Iron-bearing Solids*, subsurface Restoration Conference, Third International Conference on Groundwater Quality Research, Dallas, Texas, June 21–24, 1992.

Colorado State University (CSU), "Digital Operational Management Model of the North Boundary System at the Rocky Mountain Arsenal Near Denver Colorado" *Technical Report No. 16*, Dept. of Civil Engineering, 1988.

Gallinatti, J. D., and Warner, S. F., "Hydraulic Design Considerations for Permeable in situ Groundwater Treatment Wells," *Ground Water*, 32(5): 851, 1994.

Geoprobe Systems, Inc., *Tool and Equipment Catalog*, Kejr Engineering, Inc., Salina, KS, 1997–1999.

Gillham, R. W., and O'Hannesin, S. F., *Metal-catalyzed Abiotic Degradation of Halogenated Organic Compounds*, International Association of Hydro. Conference, "Modern Trends in Hydrogeology," Hamilton, Ontario, Canada pp. 94–103, 1992.

Gillham, R. W., and O'Hannesin, S. F., "Enhanced Degradation of Halogenated Aliphatics by Zero-valent Iron," *Ground Water*, 32(6): 958–967, 1994.

Groundwater Resources Association (GRA), *Design and Implementation of Permeable Reactive Barriers for Groundwater Treatment*, San Francisco, CA, April 7, 1999.

Horizontal Technologies, Inc., *Case Studies and Technical Information*, Matlacha, FL, 1999.

Jacobs, J., *Vertical and Horizontal Direct Push Technology and In Situ Remediation Delivery Systems, Abstracts*, 1995 Annual Meeting, Groundwater Resources Association, Sacramento, California, October 6, 1995.

Jacobs, J., "Passive Oxygen Barrier for Groundwater," *Hydrovisions*, January/February, 1996, Groundwater Resources Association, 1996.

Jacobs, J., and von Wedel, R., "Enhanced In Situ Biodegradation Via High Pressure Injection of Oxygenating Agents and Nutrients," *American Institute of Professional Geologists*, 34th Annual Meeting, Houston, TX, Abstracts, p. 27, 1997.

Regenesis, Bioremediation Products, Slurry Injection BTEX Remediation in California, 2000, *http://www. regenesis.com/ORCTech/tb.3112.htm.*

Szerdy, F. S., Gallinatti, J. D., Warner, S. D., Yamane, C. L., Hankins, D. A., and Vogan, J. L., *In situ Groundwater Treatment by Granular Zero-valent Iron: Design, Construction, and Operation of an in situ Treatment wall*. American Society of Civil Engineers, National Meeting, 1996.

U.S. EPA, *Modeling Remedial Actions at Uncontrolled Hazardous Waste Sites*, EPA/540-2-85/001 Washington, D.C., 1985.

U.S. EPA, *Corrective Action: Technologies and Applications*, Office of Research and Development, Center for Environmental Research Information, April 19–20, 1988, Atlanta, GA, 1988.

U.S. EPA, *Seminar on RCRA Corrective Action Stabilization Technologies*, Technology Transfer, Center for Environmental Research Information, Washington, D.C, 1991a.

U.S. EPA, *Stabilization Technologies for RCRA Corrective Actions Handbook*, Office of Research and Development, EPA/625/6-91/026, Washington, D.C, 1991b.

Warner, S. D., Szerdy, F. S., and Yamane, C. L., "Permeable Reactive Treatment Zones: A Technology Update," *Contaminated Soils*, Volume 3, Amherst Scientific Publishers, Chpt. 27, pp. 315–327, 1998.

Warner, S. D., Yamane, C. L., Gallinatti, J. D., and Hankins, D. A., "Considerations for Monitoring Permeable Ground-Water Treatment Walls," *Journal of Environmental Engineering*, pp. 524–529, June, 1998.

Weston, Inc., *Remedial Program Evaluation*, Gilson Road Site, Nashua, NH. Prepared for NHDES, February 1989.

Yamane, C. L., Warner, S. D., Gallinatti, J. D., Szerdy, F. S., Delfino, T. A., Hankins, D. A., and Vogan, J. L., *Installation of a Subsurface Groundwater Treatment Wall Composed of Granular Zero-valent Iron*, American Chemical Society, 209th National Meeting, 35(1): 792–795, 1995.

CHAPTER 14
REMEDIATION TECHNOLOGIES

SECTION 14.3

PHYTOREMEDIATION: A PROMISING PATH TO THE ELIMINATION OF HEAVY METALS AND RADIONUCLIDES IN GROUNDWATER & SOIL

Jay Lehr

Dr. Lehr is senior scientist for Environmental Education Enterprises where he coordinates the production of 85 different environmental short courses for environmental professionals. He is also a consultant with Bennett & Williams, Inc.

Soils, aqueous waste streams, and groundwaters contaminated with toxic metals pose a major environmental and human health problem worldwide. Recently, there has been considerable interest in the potential use of living organisms to clean up contaminated sites. However, the success of microbial based bioremediation approaches has been limited to the degradation of select organic contaminants, and has been ineffective at addressing the challenge of toxic metal contamination.

In an effort to overcome the shortcomings of conventional bioremediation and waste removal strategies, scientists have demonstrated that certain green plants are effective in removing large amounts of toxic metals from soil and water. Termed *phytoremediation*, this approach offers an efficient cost-effective and environmentally compatible means of addressing heavy metal contamination. Plants that accumulate toxic metals from contaminated soil and water can be grown and harvested economically, leaving soil and groundwater in place with only residual levels of pollutants.

The feasibility of phytoremediation at the laboratory scale has been extensively investigated with considerable success, and is currently being evaluated in a variety of field trials. Such investigations and exploration of the use of plants for environmental clean up has led to the development of three distinct approaches to phytoremediation of toxic metals and radionuclides as illustrated in Figure 14.3.1.

Phytoextraction is defined as the use of metal accumulating plants that transport and concentrate metals from the soil and shallow groundwater in the roots and above-ground shoots. Rhizofiltration is the use of plant roots to absorb, concentrate, and precipitate toxic metals from surface and groundwaters. Phytostabilization is the use of plants to eliminate the availability of toxic metals in soils and groundwater.

Each of these applications provides a novel approach to the challenge of environmental cleanup of toxic metals and radionuclides. Pioneering work in these areas has been performed at a variety of universities in the United States. The most aggressive has been Rutgers University, where Dr. Ilya Raskin has directed the AgBiotech Center under a major grant from Phytotech Inc. the first commercial firm in the United States to utilize phytoremediation in field scale cleanups.

A number of other companies have sprung up in the commercial arena offering fully integrated site management and analytical services breeding site specific plants to compete favorably on an economic basis with existing conventional remediation techniques. The clear advantages recognized

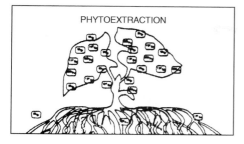

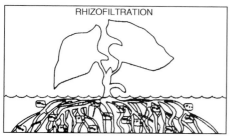

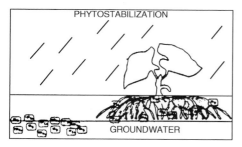

FIGURE 14.3.1   The three forms of phytoremediation.

with phytoremediation include the following:

- Lower costs for soil, groundwater, and end-of-pipe treatment
- Applicability to a broad range of metals and radionuclides
- Potentially recyclable metal rich plant residue
- Minimal environmental disturbance
- Minimization of secondary air and water-borne wastes
- High likelihood of regulatory and public acceptance

Phytoextraction has been extremely successful in field trials where a cultivar of Brassica Juncea in the Indian Mustard family was able to produce 18 tons of shoot weight for each hectare planted, and concentrate 173 times the existing lead content in the soil yielding 2.5 percent lead by weight. The Brassica Juncea roots weigh 8 tons per hectare and are 20 percent lead. In total, one planting can remove 2000 kg of lead, and at least three plantings can be grown each year. In order to achieve these excellent results the crop has to be managed as one would any agricultural product. In this case, effectiveness is greatly enhanced by applying chelating agents such as Ethylene di Nitro Tetra Acetate, which assists in freeing up the metal ions in the soil and groundwater so as to make them accessible to the plant roots (Figure 14.3.2).

The same technique yields healthy food crops by allowing the soil to supply the roots with greater volumes of rich naturally occurring nutrients. It is estimated that using phytoextraction to clean up one acre of sandy loam soil to a depth of 50 cm will cost $60,000 compared to at least $400,000 for excavation and storage alone using traditional soil removal methods.

Rhizofiltration offers a major cost advantage in water treatment because of the ability of plants to remove up to 60 percent of their dry weight as toxic metals and radionuclides, thus markedly reducing the generation and disposal cost of the hazardous or radioactive residue. Rhizofiltration is also a cost competitive technology in the treatment of surface and groundwater containing low but significant concentrations of toxic metals, such as chromium, lead, and zinc.

The contamination of soil and groundwater with radionuclides poses a serious problem in areas impacted by the precipitation and use of nuclear materials, such as uranium, tritium, cesium, strontium, technetium, and plutonium. Traditionally, these contaminants pose a serious challenge for cleanup approaches using techniques such as ion exchange chromatography, reverse osmosis, microfiltration, precipitation, or flocculation. These approaches are often difficult to implement and prohibitively expensive when applied to large water volumes of dilute concentrations and where cleanup standards are strict. The emerging radioactive contamination market, estimated at $200 billion, represents a substantial domestic and international opportunity for the application of this new technology called rhizofiltration. At Chernobyl in the former Soviet Union, the site of the world's worst nuclear disaster, scientists from Phytotech were able to demonstrate a dramatic reduction in the level of cesium (Cs-137 and strontium (Sr-90) contamination in groundwater in a 4- to 8-week period. (Figure 14.3.3.)

During the winter of 1996 a rhizofiltration technology was field tested by Phytotech to remove uranium from contaminated groundwater using the roots of live hydroponically grown plants. A rhizofiltration system as shown in Figure 14.3.4, was designed, constructed, and optimized under controlled greenhouse conditions and was transferred to and tested on site at a former Department of Energy facility in Ashtabula, Ohio.

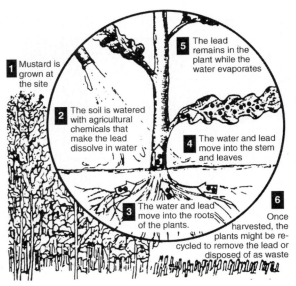

FIGURE 14.3.2 Components of phytoextraction. (*Source: Phytotech, Hommond's Encyclopedia*) (*Illustration by Frank Cecolo*)

FIGURE 14.3.3 Phytotech scientists using rhizofiltration to treat surface water contaminated with radionuclides at a site near the Chernobyl nuclear reactor.

Sunflower plants were found to be the most suitable for removing uranium from site water in laboratory studies. Optimum plant age, water Ph, and rhizofiltration coefficient standards were developed. A field scale rhizofiltration system was engineered, constructed, and assembled at the site. The system was operated in a specially designed greenhouse. The environmental controls were adjusted to meet the operating protocols identified to maximize uranium removal in water.

During the first seven weeks of the demonstration, the system was operated at two water flows: 50 and 125 gallons per day (gpd). The uranium concentration in the site water at the inlet of the rhizofiltration system ranged from 80 to 350 ppb. Sampling was performed at twice daily intervals

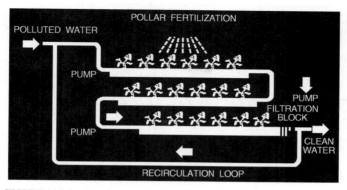

FIGURE 14.3.4   Rhizofiltration system configuration.

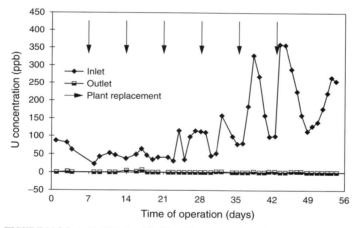

FIGURE 14.3.5   Initial (inlet) and final (outlet) concentrations of uranium in site water following operation of the rhizofiltration system in Ashtabula, Ohio.

and uranium concentrations determined by ICP analysis. The uranium concentrations in the water at the system outlet were reduced to less than 5 ppb, and usually did not exceed 0.5ppb. See Figure 14.3.5.

Most of the uranium removed from the water (about 85 percent) could be recovered from the sunflower roots, with root dry weight concentrations as high as 350 ppm U. In this demonstration, uranium was not translocated to the plant shoots, and only roots were disposed of as radioactive material. This technology can continuously treat uranium contaminated groundwater to concentrations below regulatory discharge levels.

The potential of phytoremediation continues to grow as more and more universities develop research programs. A good example is the work being carried out at the University of Iowa by Dr. Gerald Schnoor. He has found that while some plants can decontaminate soil and groundwater by simply absorbing metals, others can break down organic compounds and also enlist soil bacteria to detoxify other compounds. Dr. Schnoor has found that poplar trees can break down between 10 and 20 percent of atrazine in soil. At the University of Washington, Professors Stuart Strand and Milton Gordon have successfully experimented with small poplar trees in the breakdown of TCE to carbon dioxide, which is then transpired into the air.

Phytoremediation has moved from the laboratory to the field in the past few years and should become one of our most effective ground water and soil remediation technologies as we enter the next century.

CHAPTER 14
REMEDIATION TECHNOLOGIES

SECTION 14.4

BIOREMEDIATION OF HAZARDOUS WASTES: HISTORY AND APPLICATIONS

Walter W. Loo
Dr. Loo is president of Environment and Technology Services, an established pioneer in soil and groundwater cleanup using electrokinetic treatment, bioventing, soil vapor extraction and ultraviolet light treatment, in combination. He is a leader in bioremediation of chlorinated solvents and petroleum hydrocarbons using the cometabolic processes.

14.4.1 INTRODUCTION

Bioremediation can be defined as the utilization of naturally occurring microorganisms to detoxify hazardous wastes. Under many environmental circumstances, such as hazardous wastes spills into soil and groundwater, bioremediation is applicable for detoxification (Loo and Butter, 1986). This is particularly true for organic hazardous wastes. The basic mechanism of bioremediation is promotion of the microorganism growth, microorganisms are adapted to the organic wastes spilled in the soil and groundwater. Stimulation of growth of the microorganism is controlled by proper temperature, oxygen, moisture, nutrients, and distribution of such in the impacted media.

During growth, the microorganism will secrete enzymes and biosurfactants to breakdown the hazardous organic molecules (detoxification) and to make them available by cellular absorption. Subsequently, the microorganism will grow and subdivide or multiply, and the process will repeat itself. The end result is the mineralization of the carbon source (from spilled hazardous organic wastes) aerobically into carbon dioxide, water, and biomass without undesirable side effects.

History of Bioremediation

The application of bioremediation in hazardous waste management began in the early 1980s. Initially, the most widespread application was the bioremediation of petroleum hydrocarbons (Atlas, 1984). The basic principles of biotreatment in the environment are described by Gaudy and Gaudy (1980), who emphasized the enhancement factors, or engineering, in the application of the bioremediation technology in the field.

The progress of bioremediation of hazardous wastes started with land farming of petroleum oily waste with very little engineering and little health and safety precautions. In the mid-1980s the treatment of hazardous waste with some engineering, water irrigation, and nutrient application became more popular. The biotreatment of excavated soil in engineered biopiles began to gain recognition in the late 1980s.

In the 1990s, the engineered bioremediation took such forms as soil bioventing, in situ aqueous phase biotreatment of soil and groundwater, and bioreactors to treat water, sludge, and organic vapor.

In the mid-1990s, the evaluation of intrinsic or natural attenuation bioremediation prior to a engineering feasibility evaluation became popular. Also, the application of the cometabolic bioremediation treatment processes flourished from chlorinated solvents to petroleum hydrocarbons.

Potential Side Effects of Bioremediation

Korwek (1988) and Strauss and Levin (1988) described the federal regulation of microorganisms and assessment of risks related to hazardous waste treatment. Unfortunately, these authors failed to point out the potential negative side effects of bioremediation, which need to be controlled. Subsequently, this led to the belief that all microorganism can be used safely for bioremediation. The formulated or engineered "superbugs" products for various bioremediation applications are often missing the proper evaluation of the health risks to the human worker and the environment.

Most of the formulated or engineered "superbugs" involved human pathogens. The following list of references relates to the pathogenesis of various naturally occurring microorganisms used in formulated or engineered bioremediation products:

Liu (1964)	*Pseudomonas fluorescens*	Pathogen
Anaissie et al. (1987)	*Pseudomonas putida*	Pathogen
Walia and Kahn (1988)	*Escherichia coli* (E. Coli)	Pathogen
Winter et al. (1989)	*Escherichia coli* (E. Coli)	Pathogen
San Francisco Sunday Examiner (1990)	*Nocardia*	Parkinson's
Silver et al. (1990)	*Pseudomonas*	Pathogen
American Society of Microbiology (1991)	*Pseudomona aeruginosa*	Pathogen
U.S. EPA/600/R-92/126 (1992)	*Pseudomona aeruginosa*	Pathogen

The report by U.S. EPA (1992) included an evaluation of various engineered bacteria formulations available and sold on the open market.

Objectives and Philosophy of Soil and Groundwater Cleanup

This objective of this section is to provide information to scientists, engineers, and laypersons on how bioremediation treatment processes work under varying conditions. This chapter will stress field practice experience rather than theory. The reader should have some high school biology, chemistry, and physics training. The section is oriented toward the practical aspects of field applications. Because the study of bioremediation treatment processes in the environmental field is still evolving and developing, the applications that are in the research and development stage will be pointed out. The philosophy of soil and groundwater cleanup shall follow the intents of the U.S. EPA laws and regulations, such as Safe Drinking Water Act (SDWA), Resources Conservation and Recovery Act (RCRA), and Comprehensive Environmental Response, Compensation, and Liability Act (CERCLA).

Any environmental treatment process must be able to, or engineered as such to protect and conserve our resources (air, water, soil and human) and not create undesirable side effects. Many of the conventional waste treatment technologies cannot meet these objectives, and simply involve transferring the waste from one location or phase to another. This is true for waste excavation, landfill disposal, incineration, pump and treat, thermal desorption, and soil vapor extraction.

Sound waste treatment technologies must be able to destroy the toxic and hazardous chemicals into harmless chemicals or to render them harmless with the least energy consumption to achieve good economic balance. To the extent possible, all treated soil and groundwater should be recycled for beneficial use. Most organic chemicals can be oxidized into harmless carbon dioxide, water, and inorganic chloride with existing and available innovative waste treatment technologies. Hazardous and toxic metals, which cannot be destroyed, can be treated into insoluble form, thus rendered harmless to the environment.

The risk of exposure to human pathogens while practicing bioremediation must be properly managed even if not required by regulatory agencies. The following sections present bioremediation principles, bench scale evaluation, ex situ biotreatment, in situ biotreatment, health and safety issues, case histories, and a bibliography of key references.

14.4.2 BIOREMEDIATION PRINCIPLES

The bioremediation professional must follow the basic fundamental science and engineering in order to have a successful completion of a bioremediation project. The following sections cover the science and engineering of bioremediation processes, microbial ecology, bio-enhancement, bio-augmentation, bio-genetic engineering, soil microbes and pathogens, aerobic process, anaerobic process, de-nitrification process, and cometabolic process.

Microbial Ecology

The basic principles of biotreatment in the environment were described by Gaudy and Gaudy (1980), who emphasized the enhancement factors or engineering in the application of the bioremediation technology in the field. In order for the bioremediation process to be carried out properly, the favorable living environment for the microorganism must be understood.

Microorganisms used in bioremediation are generally classified as mesophiles or warm temperature type. The optimum range of temperature in which they thrive is between 80 to 100°F (Figure 14.4.1). When the temperature of the bioremediation system drops below 40°F, the microorganism goes into hibernation. Microorganisms must also have optimal moisture to grow and to survive. Drying conditions are not favorable for bacteria growth.

All microorganisms require an available carbon source and nutrients for growth. The adapted microorganism in the hazardous waste environment relies on the carbons in organic chemicals as a food source (substrate). The nutrient requirements include sources of phosphorus, nitrogen, and iron. Also, the microorganism relies on the availability of oxygen in the environment to mineralize the hazardous chemicals into carbon dioxide and water under the (aerobic) oxidation state.

If free oxygen is unavailable, anaerobic biodegradation conditions will demineralize inorganic compounds such as sulfate into sulfides or hydrogen sulfide (a toxic gas), and some of the available

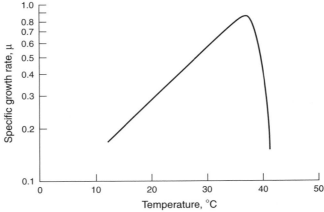

FIGURE 14.4.1   Bacteria growth versus temperature. (*Source: Gaudy & Gaudy, 1980, Microbiology for Scientists & Engineers McGraw-Hill Books*)

carbon will convert into methane (a flammable gas). This conversion may create an undesirable and unsafe environment where these toxic and flammable gases may cumulate in low lying areas and wells in the field.

Bio-Enhancement

It is evident that when the hazardous wastes spill into soil and groundwater, it is not an optimized microbial ecology for bioremediation. Therefore, we must engineer favorable microbial ecology in order for bioremediation to take place. The following sections cover the optimal microbial ecology can be engineered or enhanced for all situations

Oxygen Enhancement.   As mentioned above, if oxygen is not available in air (vadose soil) or in groundwater (aquifer) this anaerobic condition is prone to create hydrogen sulfide (toxic gas) and methane gases (flammable gas). The quantity of oxygen required must satisfy the chemical oxygen demand (COD) of the soil and groundwater media. It is important to point out that in fully saturated media, both the COD of soil and groundwater must be satisfied. The objective is to supply enough oxygen to satisfy the COD of the impacted media.

Oxygen can be enhanced by injection of air or oxygen into the pore space of the vadose soil. Dissolved oxygen in groundwater can be enhanced by the addition of oxidants or by in situ generation by electrolysis (direct current) of water in the impacted media or the creation of hydroxyl radical (a strong oxidant) at the initial stage of electrolysis of water. Another source of oxygen is through the introduction of carbohydrate compounds, which will biodegrade and provide enough oxygen to mineralize the hazardous organic compound into carbon dioxide, water, and biomass.

Heat Enhancement.   To achieve optimal growth temperature (80 to 100°F) for mesophile microorganisms, the vadose zone soil can be heated by circulating warm air into the pore space in soil. It may not be necessary to heat up the entire solid matrix of the mass because most of the hazardous wastes tend to adhere to the surface of the solid matrix. Passive solar panel air heating can provide an economical means of heat delivery. Also, recirculation of heated air in the subsurface will save energy cost because the soil mass itself acts like a heat sink with very good insulation value. The same applies to groundwater heating by passive solar water heater.

The heating requirements (in BTUs) must be estimated based on the weight of the impacted media. The weight of the impacted media can be calculated as volume of the soil or groundwater plumes, which require cleanup times the bulk density of the plumes. A swimming pool passive solar heating panel can supply more than adequate heating value in less than 30 days for most bioremediation projects.

Nutrient Enhancement.   Sometimes, there are insufficient nutrients in the form of nitrogen, phosphorus, and iron in the soil and groundwater plumes. Nutrient enhancement applies only when there is a deficiency. For proper bioremediation to take place, the carbon (C) to nitrogen (N) to phosphorus (P) weight ratio of 100:10:1 should be followed. The weight of carbon in the plume shall be determined by the total organic carbon (TOC) concentration times the weight of the plume that requires cleanup.

Moisture Enhancement.   For soil remediation, if warm air (80 to 100°F) is being recirculated within the soil plume, the moisture in the warm air (humidity) should be adequate enough enhancement. There is no need for moisture enhancement in the saturated media.

Enhancement Distribution.   The most difficult media for the hazardous waste treatment processes to work in are relatively low-permeability clayey medium and fracture medium. These media often absorb/adsorp the toxic and hazardous wastes spilled below the ground surface. Often more than 90 percent of the waste spilled by weight attracted to these media.

It is most fortunate that the electrokinetic treatment processes (electro-osmosis and electromigration of water) are capable of overcoming the difficulties of relatively low-permeability media, achieving or enhancing the aerobic biochemical oxidation of many toxic and hazardous wastes, with relatively low energy consumption.

Bio-Augmentation

Bio-augmentation involves the application of formulated microorganisms known to be good hydrocarbon degraders. As mentioned above, many of these formulated microorganism available on the open market involved human pathogens (U.S. EPA, 1992). This violates the basic health and safety principles for hazardous waste management, which calls for not creating any undesirable side effects that impact the environment or human health.

The best test for the real harmful formulated microorganisms or "superbugs" is to request that the salesperson ingest the formulation and demonstrate that he or she can come back in 30 days unharmed. Most salespersons do not understand the pathanogenesis of microorganisms. You cannot trust the material safety data sheet (MSDS) provided by the salesperson unless you can see an original certification of the formulation by the U.S. EPA. It is the author's understanding that there is no such U.S. EPA certification.

Bio-Genetic Engineering

Several studies have looked at bio-genetic engineering. Modello and Yates (1988) presented a paper on gene cloned bacteria for Polychlorinated Biphenyls (PCBs). Walia and Kahn (1988) presented a paper on genetically engineered *Escherichia coli* (also know as *E. Coli*, a human pathogen) for biodegradation of PCBs. Winter et al. (1989) presented a paper on efficient degradation of trichloroethylene by a recombinant *Escherichia coli* (*E. Coli*).

These are good faith research and development efforts that should not be allowed in field bioremediation because of the possibility of spreading human pathogens.

Soil Microbes And Pathogens

Microorganisms in soil adapt to carbon from the spilled organic hazardous wastes, called heterotrophs (mineral using bacteria), and can be valuable resources for the application of bioremediation. Many types of microorganism exist in soil and less than 10 percent of the total population are heterotrophs. Heterotroph bacteria specialize in the utilization of carbon from organic chemicals as their food source (substrate). However, some heterotrophs are human pathogens.

To promote bioremediation above ground, one must observe all health and safety measures to minimize potential exposure pathways to the remediation workers. Below ground (in situ) bioremediation eliminates many of the exposure pathways. Fortunately, that the heterotroph population will die off when the carbon source is diminished and their cell will serve as a substrate to other harmless microorganisms. However, one who manages a bioremediation project must also understand how to decommission or disenhance the site safely and properly. This may lead to disinfection of the soil and groundwater plume if necessary.

The heterotroph plate count normally ranges from 1000 to 10,000 colony forming units (CFUs) in most soil and groundwater plumes impacted by organic hazardous wastes. The author has yet to encounter a site without indigenous heterotrophs. With full bioenhancement, the heterotroph plate count may increase by five to seven orders of magnitude. The die off at the end of bioremediation when substrates, nutrients, and oxidants are depleted can be dramatically decreased to less than detection limit (less than 10 CFUs).

Aerobic Process

The basic mechanism of bioremediation is to trigger the growth of microorganisms that are adapted to the organic wastes spilled in the soil and groundwater. Stimulation of growth of the microorganism is controlled by proper temperature, oxygen, moisture, nutrients, and cosubstrates, and distribution of such in the impacted media.

During growth, the microorganism will secrete enzymes and biosurfactants to break down the hazardous organic molecules (detoxification) and to make them available through cellular absorption. Subsequently, the microorganism will grow and subdivide or multipy, and the process repeats until the food source is depleted.

The end result is the mineralization of the carbon source (from spilled hazardous organic wastes) aerobically into carbon dioxide, water, and biomass without undesirable side effects.

Anaerobic Process

If free oxygen is unavailable, anaerobic biodegradation conditions will demineralize inorganic compounds such as sulfate into sulfides or hydrogen sulfide (a toxic gas), and some of the available carbon will convert into methane (a flammable gas) instead of carbon dioxide. This may create an undesirable and unsafe environment where these toxic and flammable gases may cumulate in low lying areas and in remedial wells in the field.

There are still a lot of practice of bioremediation using the anaerobic processes as referenced in Battelle (1999a, 1999b, 1999c). Unfortunately, many authors of individual papers fail to warn the readers or audience about the potential side effects of the anaerobic processes and how to manage them.

Denitrification Process

Denitrification is an anaerobic process in the demineralization of the oxygen from nitrate. However, the commonly occurring sulfate in the environment will be reduced and will form hydrogen sulfide gas. Sheehan et al. (1988) and Spark and Baker (1988) presented the cases on the beneficial effects of the denitrification processes but failed to warn the readers or audience about the potential side effects of the anaerobic processes and how to manage them.

Co-Metabolic Processes

The *cometabolic processes* for the bioremediation of chlorinated solvents and petroleum hydrocarbons is the most promising trouble-free biotreatment process and is very efficient. The cometabolic process can be defined as the introduction of an easily biodegradable substrate into the environment, which triggers the secretion of enzyme from microrganisms that are adapted to the spilled hazardous organic waste. The process results in the biodegradation of the spilled hazardous organic waste into harmless carbon dioxide, water, and biomass.

The biodegradation of chlorinated solvents is a highly sought after solution to the widespread soil and groundwater contamination problems. However, most of the knowledge of biodegradation of chlorinated solvents are found only in research laboratories. Successful laboratory demonstration of cometabolic biodegradation of trichloroethene (TCE) by methanotrophic bacteria columns was achieved by EPA Ada Laboratory in 1985 (Wilson and Wilson, 1985). In 1987, EPA Gulf Breeze Laboratory successfully demonstrated the cometabolic biodegradation of TCE by *Pseudomonas putida* through an aromatic pathway (Nelson et al., 1987). In 1989, the author successfully demonstrated the first field closure of the cometabolic biodegradation of TCE and trichloroethane (TCA) together with toluene in soil through heat and nutrient enhancement by the growth of indigenous bacteria *Bacilli* and *Pseudomonas fluorescens* (Loo, 1991). In 1991, Standford University demonstrated partial success on the cometabolic biodegradation of chlorinated solvents in groundwater by methanotrophic bacteria at Moffet Field, California. In 1993, the author demonstrated in the laboratory, field pilot, and in field application the biodegradation of TCE using glucose as a cosubstrate, which is nontoxic and nonhazardous (Loo et al., 1993).

In 1994, the author developed and sucessfully applied the cometabolic biodegradation of petroleum hydrocarbons using glucose and sucrose as cosubstrates in both soil and groundwater (Loo, 1994 and ETS, 1995). Subsequently, the author completed many other site closures through the application of the cometabolic bioremediation of petroleum hydrocarbons (ETS, 1995, 1997, 1998).

14.4.3 FEASIBILITY EVALUATION

The bioremediation processes are fairly well established and proven both in the laboratory and in the field. The following sections describe some procedural notes for bench scale and field pilot test.

Bench Scale Test

During the research and development stages of a bioremediation project, one needs to establish the proof of a biodegradation process. This includes the following steps:

1. Set up of the bench equipment
2. Test and control samples
3. Nutrients stimulation
4. Oxygen stimulation
5. Incubation period
6. Interval sampling
7. Laboratory analyses and plate counts
8. Verification of concentration reduction
9. Verification of mineralization products
10. Determination of half life or the biodegradation rate
11. Projection of cleanup time

Bench Scale Laboratory Procedures. Because of space limitations the author is not going to provide all the procedures requirements in this book. Instead, one should follow the procedures of bench experiments conducted by two well-known benchmark bioremediation processes for chlorinated solvents biodegradation in soil and groundwater. These two key U.S. EPA references are listed as follows:

1. **U.S. EPA—ADA LABORATORY, ADA, OKLAHOMA**
   Soil Column Bench Test Experiment on Co-Metabolic Biotransformation of Trichloroethene
   Wilson, J. T. and Wilson, B. H., "Biotransformation of Trichloroethylene in Soil, *Applied and* Environmental *Microbiology*," January 1985, 49(1): 242–243, 1985.
2. **U.S. EPA—GULF BREEZE LABORATORY, FLORIDA**
   Aqueous Phase Bench Test Experiment on Co-Metabolic Biodegradation of Trichloroethene
   Nelson, M. J. K. et al., "Biodegradation of Trichloroethylene and Involvement of an Aromatic Biodegradative Pathway," *Applied and Environmental Microbiology*, 53(5): 949–954, 1987.

Confirmation of Biotreatment Processes. The confirmation of a bioremediation process shall demonstrate the following:

1. Reduction in concentration of hazardous waste in the tested media
2. Proof of mineralization chemicals such as
     Carbon dioxide
     Water
     Increase and die off in microbial plate count
     Presence of inorganic chloride (if involved chlorinated organics)
3. No detection of unexpected side effects or undesirable breakdown products

Half-Life Determination. The half-life determination is the evaluation of the biodegradation rate of a hazardous waste under a specific set of conditions and environment. The half-life can be defined as the time requirement for 50 percent reduction in concentration of the hazardous waste in the tested conditions and media. The biodegradation rate normally follows a logarithmic decay of the concentration versus time curve. Each increased log time cycle will account for 50 percent reduction in concentration. Therefore, one can project the time requirement for cleanup from initial concentration progressing toward, or reaching, the predetermined soil or groundwater cleanup levels.

One should note that the laboratory bench scale test results are always the most optimistic and ideal and may not apply in the field for prediction purposes. Rather, it is just a proof of the biodegradation process. To be applicable in the field, one must use the common sense engineering for the design of the full enhancement of the bioremediation treatment system.

Field Pilot Test

Before the implementation of a full-scale bioremediation project, one should conduct a simple field pilot confirmation test for mineralization products. The purpose of the field pilot test is to determine if the indigenous microorganism is responding to the various enhancement elements.

A simple oxygen slug injection into a vadose zone well while monitoring the respiration of the metabolic function of the indigenous microorganism will suffice. The simple oxygen enhancement field test shall show favorable increase in carbon dioxide and reduction in oxygen in the well. Figure 14.4.2 presents an example of the biorespiration pilot field test response curve, which demonstrates that the indigenous bacteria responded to the oxygen injection by the increased respiration of carbon dioxide. This whole field test involved only one vadose well situation in the middle of the gasoline impacted soil plume, one compressed oxygen bottle, and an oxygen/carbon dioxide field meter.

Baseline Sampling and Analysis Methods.   The initial distribution of concentration of the hazardous waste in soil and groundwater is the starting point for the design of a bioremediation project. For the baseline condition of a soil plume, a statistically significant soil sampling plan will be required because

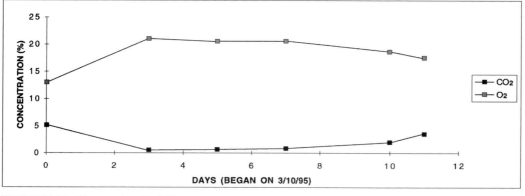

FIGURE 14.4.2   Biorespiration pilot field test.

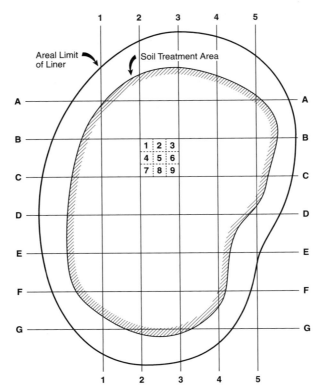

FIGURE 14.4.3   Soil sampling grid with statistical representation.

of the heterogeneous nature of the distribution of hazardous wastes in soil. The plan shall follow the requirements of EPA SW-846 sampling procedure for solid wastes. Figure 14.4.3 presents a grid and mesh soil sampling procedure on the lateral extents of a soil plume to ensure statistical significance. For the vertical extent, one shall add the randomness by staggering depths instead of regular depth intervals to ensure even distribution and to eliminate any significant lithologic layering effects.

It is fairly easy to establish the baseline dissolved constituents in the groundwater because of the slow dilution resulting from influx of clean groundwater from upgradient. The only requirement is adequate number of monitoring wells. The key feasibility sample analysis parameters required for bioremediation design and engineering are as follows:

1. Weight of the soil and groundwater plumes
2. Concentration of wastes in soil and water
3. Total heterotrophs plate count
4. Chemical oxygen demand (COD)
5. Total organic carbons (TOC)
6. Total Kjeldahl nitrogen (TKN)
7. Nitrate
8. Total phosphorus
9. Iron
10. Temperature within the soil and groundwater plume

11. Dissolved oxygen and pH in groundwater
12. Moisture content in soil
13. Concentration of oxygen and carbon dioxide in soil pores

When the soil and groundwater plumes are fully characterized in weight, chemical requirements, engineering, installation, equipment, delivering lines, monitoring, and analysis; and operation and maintenance costs can then be estimated for economic feasibility evaluation. The ultimated cost shall be presented in dollars per ton of soil or groundwater plume that requires biotreatment.

Applicability of Bioremediation versus Waste Types

Petroleum Hydrocarbons.   Most petroleum hydrocarbons are biodegradable including gasoline, diesel, jet fuel, kerosene, aviation fuel, motor oil, benzene toluene, ethylbenzene, xylenes, Methyl tertiary Butyl Ether(MtBE), and so forth. There has been some concern on the bioremediation of MtBE. However, electrokinetic enhanced cometabolic biotreatment of MtBE has been demonstrated in the field with repeated success by the author.

Chlorinated Solvents.   Most chlorinated solvents are biodegradable through the application of the cometabolic biotreatment processes. However, because of the earlier publications of laboratory successes by the U.S. EPA in the 1980s, many scientists and engineers began to apply flammable methane and butane blindly into the vadose zone without questioning the health and safety side effects Roberts (1990) and Battelle (1999). The same applies to the addition of toxic phenol, dichlorobenzene, and toluene into the groundwater as cosubstrates for bioremediation. This violates the underground injection control regulation established by the U.S. EPA.

All the above mentioned misapplications can be easily avoided by substituting the cosubstrates with food additive chemicals, which are totally safe and not regulated by federal and local environmental agencies.

Pesticides and PCBs.   There are some severe limitations on the applicability of biotreatment on pestides and PCBs. The limitations will require further research and development on the hybrid approach of cometabolic and electrokinetic-induced oxidation processes.

Polycyclic Aromatic Hydrocarbons (PAHs).   PAHs are organic compounds made of linked benzene molecules. Bioremediation of two and three benzene ring PAHs (e.g., napthalene, pyrenes) have proven to be feasible. PAHs with four or more benzene rings are not often feasible. This will require further research and development on the hybrid approach of cometabolic and electrokinetic induced oxidation processes.

Metals.   The remediation of metals by bioremediation is not a biodegradation process. It is the anaerobic conditions in the subsurface that changed the various metals to a reduced valence state and caused them to be less soluble. The anaerobic process increases the risk of generating methane and hydrogen sulfide gases, which are undesirable and in which side effects are difficult to control.

14.4.4   *EX SITU BIOTREATMENT*

This section presents various applications of bioremediation that can occur above ground, including soil treatment, slurry treatment, groundwater treatment, and organic vapor treatment.

Soil Biotreatment

The treatment of excavated contaminated soil can be treated by land farming (single layer) or by biopile (multiple layers).

Landfarming.   Landfarming is the earliest form of biotreatment in the field. Contaminated soil is spread over the land surface 1 to 2 ft thick and tilled over periodically. There is little enhancement to promote biotreatment except for aeration. Bioenhancement processes can be engineered as irrigation systems for periodic moisture enhancement. Fertilizing can be a nutrient enhancement, and covering over with black plastic sheets for heat absorption as heat enhancement.

The drawback for landfarming is the requirement of a large surface area of land per ton of soil, which needs to be treated. Each acre of land can handle a maximum of about 5000 tons of soil, at about 2 ft thick. Landfarming will not work during cold weather because most microorganisms hibernate below 40°F. See Kivanc and Unlu (1999) for a presentation of modern day land treatment system design.

Heap/Pile Biotreatment.   An engineered soil pile can achieve great efficiency in the use of bioenhancement processes. Figure 14.4.4 present a typical schematic and cross-sectional view of an ex situ biopile treatment cell. The biopile can be constructed easily to a height of 10 ft, which saves much of the land surface area requirement. All the piping at each lift serves as aeration, moisture, and nutrient delivery lines. The bottom liner collects all the leachate generated by excessive watering and the leachate is then recycled back into the pile. The black plastic cover will trap heat for heat enhancement. All the treated soil can be recycled for landscaping use.

Loo (1991) presented the first paper on the cometabolic biotreatment of chlorinated solvents in an engineered biopile with heat enhancement at Emeryville, California. Van Zyl and Lorenzen (1999) also use the engineered biopile for the biotreatment of diesel.

Slurry Biotreatment

Sometimes the contaminated soil from the excavation of the spill source may require some dilution and treatment in the form of a slurry. Typically the soil is treated in a steel liquid storage tank used for holding drilling mud and cuttings with slurry pumps and mixers. This is basically a slurry bioreactor for batch biotreatment.

Reference paper (Smallbeck and Lynch, 1987) discusses the first slurry bioreactor conducted in the field for soil contaminated heavily by pesticides (Minot, North Dakota).

Aqueous Phase Biotreatment

Contaminated groundwater pumped out from the ground can be biotreated in large aqueous phase bioreactors. There are many kinds of bioreactors for different uses and purposes. For very low concentrations of hazardous waste in water, Nyer and Skladany (1988) developed a plug flow bioreactor to polish off the last trace of hazardous waste. The bioreactor process involves the growth of a large mass of bacteria prior to letting the contaminated water to pass through. Sullivan and Konzen (1990) utilized the fixed film bioreactor for the treatment of contaminated groundwater. The fixed film is also a way to increase the surface area for biotreatment. For higher concentration waste stream, Hickey et al. (1990) developed activated a carbon fluidized bed bioreactor for such treatment purposes. The activated carbon not only increases the surface area of treatment but also holds onto the organic chemical on its surface by adsorption.

Organic Vapor Phase Biotreatment

Lesion (1990) developed a biofilter for the control of volatile organic chemicals (VOCs) and odor. Blanchard et al. (1990) developed the compost biofilters for the control of hydrocarbon vapor. The

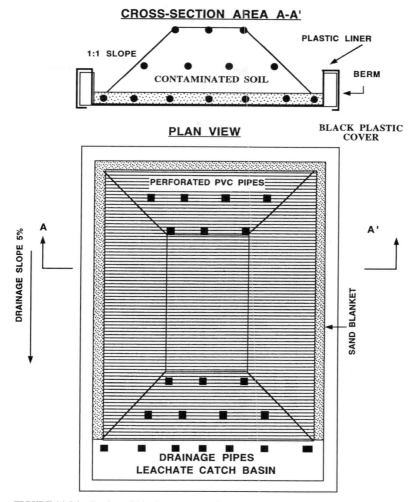

FIGURE 14.4.4   Ex situ soil biopile treatment cell.

author has utilized granular activated carbon (GAC) filters as biofilters after the GAC filters lost their adsorption capacity. GAC biofilters are particularly effective when used jointly in a soil bioventing project where closed loop air re-injection is applied. The GAC is a perfect biofilter when warm moist air with nutrients and VOCs pass through the GAC units.

14.4.5   *IN SITU BIOTREATMENT*

The economical alternative to digging, hauling, and disposing or ex situ treatment of contaminated soil is bioventing in the subsurface. The economical alternative to pumping, treating, and discharging of contaminated groundwater is *in situ groundwater biotreatment* or intrinsic biotreatment.

Bioventing

Bioventing is a method of treating petroleum hydrocarbon contaminated soil in the vadose zone by circulating air through the pores of the soil matrix. It is particularly effective when the moisture content and nutrient level in the soil can be amended to prime enhancement conditions for biodegradation. Bioventing can treat both volatile and nonvolatile organic hydrocarbons.

Hinchee et al. (1989, 1990) and Miller et al. (1990) developed the bioventing concept for the bioremediation of petroleum hydrocarbons at many U.S. Air Force sites. However, only oxygen enhancement was adopted for soil cleanup at many of these sites. The process involved injection of air at one well into the contaminated soil plume, with little engineering feasibility evaluation. Little engineering consideration was given to the migration of organic vapor that was pushed out of the soil plume and formed condensates, which further spreads the groundwater contamination. This may be the most economic means of treatment of petroleum hydrocarbons in soil for the U.S. Air Force, but it may face the side effects of uncontrolled spreading of the dissolved groundwater plume.

The author applied the first fully enhanced bioventing system on the treatment of gasoline in clayey marine conglomerate at the PepBoys site, San Diego, California (ETS, 1994 and Loo et al., 1994b). The gasoline in clayey soil was desorbed or squeezed out by the electro-osmosis process and made the gasoline available for bioventing. The extracted air was heated and re-injected into the vadose zone soil for heat enhanced biotreatment. The heated air also carried the water squeezed from the clay layer in the form of humidity for moisture enhancement. Pure oxygen gas was injected into the middle of the soil plume to replenish the oxygen used during the aerobic bioventing process. Nitrogen-based nutrient gas was also injected into the middle of the soil plume for nutrient enhancement. Nitrogen-based nutrient gas injection into soil is widely used in large-scale agricultural farming practices.

Aquifer Biotreatment

The aqueous phase biotreatment discussed in this section does not involve pumping out any water from the ground. The most challenging aspect of in situ aqueous biotreatment is the distribution or penetration of nutrient, oxygen, and heat into the clayey or less permeable areas within the saturated media. The electrokinetic processes can overcome these conditions. The detailed description of the electrokinetic processes are described in the next section on (Electrokinetic Treatment of Hazardous Wastes).

Passive Biotreatment. Passive biotreatment involves oxygen, nutrient, heat, and electrokinetic enhancement to biodegrade toxic organic chemicals into harmless carbon dioxide, water and biomass. The oxygen enhancement can be achieved by the creation of dissolved oxygen through electrolysis of groundwater or through the utilization of the oxygen in food additive cosubstrates dissolved in groundwater. The nutrient enhancement can be achieved by the addition of ammonium-based food additive chemicals into the groundwater. Heat enhancement can be achieved by infiltration of solar heated water economically into the groundwater by infiltration gallery or through injection into remedial wells. The electrokinetic enhancement or mixing can be achieved by electromigration of nutrient/electrolyte through permeable and impermeable water saturated media.

Loo (1994) and ETS (1995) both demonstrated that benzene can be biotreated to nondetect level in groundwater. Loo (1998) and ETS (1997 and 1998) demonstrated that MtBE can electrobiochemically treated to nondetect in groundwater. Finally, Loo et al. (1993) and ETS (1996) showed that chlorinated solvents can be biotreated to nondetect in groundwater.

Natural Attenuation Biotreatment. Natural attenuation or intrinsic biotreatment is the do-nothing alternative for groundwater cleanup. It relies on the natural influx of oxygen and nutrient through the groundwater plume to complete the in situ bioremediation. Intrinsic biotreatment only applies to stable and receding groundwater plume where the source of the hazardous waste spilled is removed or remediated.

The monitoring requirements on the evaluation of natural attenuation in groundwater acceptable for low-risk site closure evaluation from inside and outside of the plume are:

• pH, dissolved oxygen, redox potential(ORP)
• Sulfate, nitrate, ferrous iron
• Dissolved methane

In the center of a groundwater plume, the favorable conditions for natural attenuation are as follows:

• Increasing dissolved oxygen concentration
• Increasing oxidation potential
• Sulfate and nitrate reduction
• Oxidation of iron into valence (+3)
• Decrease in dissolved methane

Intrinsic biotreatment may be applicable for petroleum hydrocarbons when natural conditions will not create the formation of other hazardous or toxic compounds. Intrinsic biotreatment may not be applicable for chlorinated solvents because the reduction state will create the formation of other hazardous or toxic compounds such as dichloroethene, dichloroethane, vinyl chloride and so forth.

14.4.6 CASE HISTORY

Ex Situ Heap Biotreatment

Heat Enhanced Co-Metabolic Biotreatment of Soil with Chlorinated Solvents and Toluene, Emery- ville, California (Loo, 1991).   This project site located at 5800 Christie Avenue, Emeryville, Cali- fornia, is known as the Croley and Herring Investment Company (CHIC) site. The current tenant of the site is The Good Guys Store. This site is the cradle of the application of cometabolic biotreatment processes in the field.

The site involved the spill of chlorinated solvents and gasoline, which impacted the soil and the groundwater. The soil or vadose zone consisted of sandy fill material about 5 ft thick. The shallow groundwater table is about 5 ft below grade. The chlorinated solvents and gasoline-related chemicals detected in soil are listed as follows:

Trichloroethene (TCE)	1,1,1 Trichloroethane (TCA)
1,1 Dichloroethane	1,2 Dichloroethene
Carbon Tetrachloride	Methylene Chloride
Freon 113	Chlorobenzene
Toluene	Xylenes
Ethylbenzene	Benzene

The major contributors are TCE, TCA, and toluene. The total volatile organic compounds (VOCs) detected at the source area and excavation pit averaged about 3000 and 1000 ppm, respectively. This is a site in which the cometabolic biotreatment process developed by Nelson et al. (1987) of the U.S. EPA Gulf Breeze Laboratory is applicable. The only difference is that the U.S. EPA cometabolic process was conducted in the aqueous phase in the laboratory. This is the first application of this process in soil and in the field.

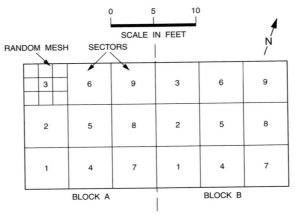

FIGURE 14.4.5 Soil sampling grid. (*The Good Guys Store Site 5800 Christie Avenue, Emeryville, California*).

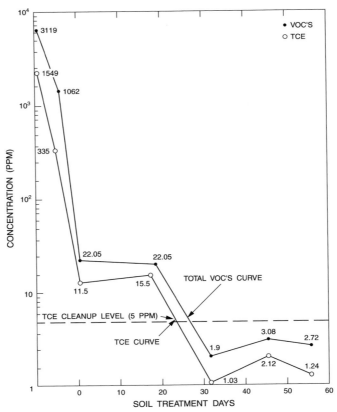

FIGURE 14.4.6 Soil VOCs Treatment performance plot. (*The Good Guys Store Site 5800 Christie Avenue, Emeryville, California*).

The soil excavation was performed in late April 1989. The soil heap treatment unit was constructed of three lifts of 2 ft thick soil interlayered with heating coils and air recirculation PVC pipes. Figure 14.4.5 presents a statistically representative soil sampling grid for the soil treatment unit required by the regulatory agency in conformance to requirements of EPA Solid Wastes Testing Manual CW-846. Each block area represents about 40 cubic yards of soil. Each sector is represented by a randomly selected sample. Each block is represented by a composite sample from the nine sector samples.

Underlying the soil was a 30-mil HDPE liner and a series of air ducts and heating elements. The air circulation was maintained by drawing air through an 100 CFM explosion proof air blower. At all times, there was a vacuum imposed on the surface of the soil pile. The temperature within the soil unit was maintained between 80 to 100°F with thermostat controlled heater. The surface of the soil pile was covered by a black plastic sheet to trap heat. The moisture in the soil was maintained by periodic watering to minimize volatilization of VOCs. The air emission was minimized by recycling air to the middle tier air ducts.

Figure 14.4.6 presents the performance or reduction in VOCs concentration in soil versus time. The chlorinated solvents in soil was treated to below 2 ppm. Figure 14.4.7 presents the confirmation of mineralization of chlorine into inorganic salt versus time. Table 14.4.1 presents the microbial analysis throughout the treatment. The *bacilli* and *Pseudomonas* bacteria were identified as the key players responsible for this cometabolic biotreatment.

A soil remediation and closure report was submitted to Alameda County Health Care Services on July 21, 1989 and closure was approved by ACHCS with the concurrence of the Bay Area Regional Water Quality Control Board. The 80 cubic yards of treated soil was tested and classified as nonhazardous per Section 66696 of California Code of Regulation Title 22 Division 4. The nonhazardous soil was disposed to the West Contra Costa County Sanitary Landfill.

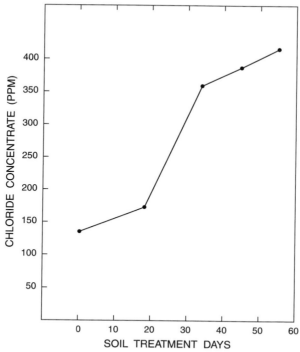

FIGURE 14.4.7   Chlorine mineralization versus time plot. (*The Good Guys Store Site 5800 Christie Avenue, Emeryville, California*).

TABLE 14.4.1 Soil Microbial Analysis

	Sample	5/5/89	5/23/89	6/5/89	6/16/89	6/29/89
Standard Aerobic	A	1.4×10^7	2.8×10^6	5.8×10^5	3.0×10^6	1.6×10^6
Plate Count/g	B	1.8×10^7	3.9×10^6	2.4×10^7	2.6×10^6	1.7×10^6
Bacilli Count/g	A	2.0×10^5	1.0×10^4	6.7×10^5	—	<10
	B	1.2×10^5	1.9×10^5	2.0×10^5	—	<10
Pseudomonas	A	6.4×10^4	4.4×10^4	1.5×10^3	—	<10
Count/g	B	2.2×10^3	8.5×10^3	3.0×10^4	—	1.0×10^3
Fecal Coliforms	A	<3	—	—	—	<3
MPH/g	B	<3	—	—	—	<3
Fecal Streptococci/g	A	<10	—	—	—	<10
	B	<10	—	—	—	<10

REFERENCES CITED

Battelle Press, "The Proceedings of The Fifth International In Situ and On-Site Bioremediation Symposium," San Diego, California, 8 volumes, 1999.

Loo, W. W., and G. N., Butter, "State of the Art Technologies of Removal, Isolation and Alteration of Organic Contaminants Underground," *Conference Proceedings of Hazardous Waste and Hazardous Materials*, Atlanta, Georgia, pp. 124–126, 1986.

Microbiology

American Society of Microbiology, *Manual of Clinical Microbiology*, 1991.

Anaissie, E. et al., "*Pseudomonas putida*: Newly Recognized Pathogen in Patients with Cancer," *The American Journal of Medicine*, 82: 1191–1194, 1987.

Atlas, R. M., *Petroleum Microbiology*, Macmillan Publishing Company, New York, 1984.

Gaudy, A. F., and Gaudy, E. T., 1980, *Microbiology for Environmental Scientists and Engineers*, McGraw-Hill Book Company, NYC, 1980.

Korwek, E. E., "Federal Regulation of Microorganisms in Hazardous Waste Treatment," *Conference Proceedings of Hazardous Waste Treatment by Genetically Engineered or Adapted Organisms*, Washington D.C., pp. 127, 1988.

Liu, P. V., "Pathogenicity of Pseudomonas fluorescens and Related Pseudomonads to Warm-Blooded Animals," *American Journal of Clinical Pathology*, 41: 150–153, 1964.

San Francisco Sunday Examiner, *Common Bacterium Linked to Parkinson's*, May 27, 1990.

Silver S. et al., "Pseudomonas: Biotransformations, Pathogenesis, and Evolving Biotechnology," *Monograph of American Society for Microbiology*, 1990.

Strauss, H.S., and Levin, M., "Models of Assessing the Risks of Releasing Microorganism Into the Environment," *Conference Proceedings of Hazardous Waste Treatment by Genetically Engineered or Adapted organisms*, Washington, D.C., pp. 121–126, 1988.

U.S. EPA, *Clearance and Pulmonary Inflammatory Response After Instranasal Exposure of C3h/HeJ Mice to Biotechnology Agents*, Bioremediation of Hazardous Wastes by Biosystems Technology Development Program, U.S. EPA Report No. EPA/600/R-92/126, 1992.

Petroleum Hydrocarbons

Soil Bioventing and Remediation

Environment & Technology Services, *Soil Closure Report*, submitted to San Diego County Dept. of Health Services, for Electrokinetic Enhanced Soil Bioventing with Gasoline at the PepBoys Site, 3550 El Cajon Blvd., San Diego, California, 1994.

Environment & Technology Services, *Soil Closure Report*, submitted to Los Angeles City Fire Dept., for Electrokinetic Enhanced Soil Bioventing with Gasoline at Former Evergreen Site, 3000 E. 12th Street., Los Angeles, California, 1996.

Environment & Technology Services, *Soil Closure Report*, submitted to City of Long Beach Dept. of Health Services, for Electrokinetic Enhanced Soil Bioventing with Diesel and Waste Oil at DBM Oil Site, 3508 Atlantic Blvd., Long Beach, California, 1996.

Environment & Technology Services, *Final Site Closure Document*, submitted to Los Angeles Regional Water Quality Control Board, for Soil (Bioventing) and Groundwater Cleanup with MtBE and Benzene at Former Neill Lehr Cadillac Site, 8400 Reseda Blvd., Northridge, California, 1997.

Environment & Technology Services, *Final Site Closure Document*, submitted to Los Angeles Regional Water Quality Control Board, for Soil (Bioventing) and Groundwater Cleanup with MtBE and Benzene at Former TEXACO Station Site, 1422 W. Willow Street, Long Beach, California, 1998.

Environment & Technology Services, *Final Site Closure Document*, submitted to Los Angeles Regional Water Quality Control Board, for Soil (Bioventing) and Groundwater Cleanup with MtBE and Benzene at Former ARCO Station Site, 507 W. Garvey Avenue, Monterey Park, California, 1999.

Garnier, P., Auria, R., Magana, M., and Revah, S., "Cometabolic Biodegradation of MTBE by a Soil Consortium," *The Proceedings of the Fifth International In Situ and On-Site Bioremediation Symposium*, San Diego, California, Battelle Press, Vol. 5(3), pp. 31–35, 1999.

Hinchee, R. E. et al., "Enhancing Biodegradation of Petroleum Hydrocarbon Fuels Through Soil Venting," *Conference Proceedings of Petroleum Hydrocarbons and Organic Chemicals in Groundwater*, Houston, Texas, pp. 235–248, 1989.

Hinchee, R. E. et al., "Enhanced Bioreclamation Soil Venting and Groundwater Extraction: A Cost-Effectiveness and Feasibility Comparison," *Conference Proceedings of Petroleum Hydrocarbons and Organic Chemicals in Groundwater*, Houston, Texas, pp. 147–164, 1990.

Loo, W. W., Wang, I. S., and Fan, K. T., "Electrokinetic Enhanced Bioventing of Gasoline in Clayey Soil: A Case History," *Proceedings of the SUPERFUND XV Conference*, Washington, D.C., 1994b.

Miller, R. N. et al., "A Field Investigation of Enhanced Petroleum Hydrocarbon Biodegradation in the Vadose Zone at Tyndall AFB, Florida," *Conference Proceedings of Petroleum Hydrocarbons and Organic Chemicals in Groundwater*, Houston, Texas, pp. 339–351, 1990.

Groundwater Treatment

Loo, W. W., "Electrokinetic Enhanced Passive In Situ Co-Metabolic Biotreatment of Gasoline and Diesel in Clayey Soil and Aquifer: A Case History," *Proceedings of the SUPERFUND XV Conference*, Washington D.C., 1994.

Environment & Technology Services, *Site Closure Report*, submitted to Nevada Dept. of Environmental Protection, for Electrokinetic Enhanced Biotreatment of Soil and Groundwater with BTEX at BAT Rentals Site, 2771 South Industrial Road, Las Vegas, Nevada, 1995.

Environment & Technology Services, *Site Closure Report*, submitted to Westland Metals Co., for Electrokinetic Enhanced Biotreatment of Soil and Groundwater with BTEX and Diesel in Bay Mud at Former Chevron Station Site, 3149 Depot Road, Hayward, California, 1995.

Environment & Technology Services, *Final Site Closure Document*, submitted to Los Angeles Regional Water Quality Control Board, for Soil (Bioventing) and Groundwater Cleanup with MtBE and Benzene at Former Neill Lehr Cadillac Site, 8400 Reseda Blvd., Northridge, California, 1997.

Environment & Technology Services, *Final Site Closure Document*, submitted to Los Angeles Regional Water Quality Control Board, for Soil (Bioventing) and Groundwater Cleanup with MtBE and Benzene at Former TEXACO Station Site, 1422 W. Willow Street, Long Beach, California, 1998.

Loo, W. W., *Groundwater Case Closures of MtBE, Benzene & Chlorinated Solvents to Non-Detect By In-Situ Electrokinetic Processes*, Paper Presented at the HAZMACON Conference, Santa Clara Convention Center, Santa Clara, California, April 1998, sponsored by the Association of Bay Area Governments, 1998.

Sheehan, P. J. et al., "Progress in Bioreclamation of Contaminated Groundwater without Oxygen Addition," *Conference Proceedings of HAZMACON 88*, Anaheim, California, pp. 711–721, 1988.

Spark, K. B., and Barker, J. F., "Nitrate Remediation of Gasoline Contaminated Groundwater," *Conference Proceedings of HAZMACON 88*, Anaheim, California, pp. 127–159, 1988.

Chlorinated Solvents

Battelle Press, "Engineered Approaches for In Situ Bioremediation of Chlorinated Solvents," *The Proceedings of The Fifth International In Situ and On-Site Bioremediation Symposium*, San Diego, California, Vol. 5(2), 1999a.

Battelle Press, "In Situ Bioremediation of Petroleum Hydrocarbons and Other Organic Compounds," *The Proceedings of The Fifth International In Situ and On-Site Bioremediation Symposium*, San Diego, California, Vol. 5(3), 1999b.

Battelle Press, "Bioremediation of Metals and Inorganic Compounds," *The Proceedings of The Fifth International In Situ and On-Site Bioremediation Symposium*, San Diego, California, Vol. 5(4), 1999c.

Environment & Technology Services, *Groundwater Closure Request*, submitted to Bay Area Regional Water Quality Control Board, for Electrokinetic Enhanced Biotreatment of Groundwater with Chlorinated Solvents in the Bay Mud at The Good Guys Site, 5800 Christie Street, Emeryville, California, 1996.

Loo, W. W., "Heat Enhanced Bioremediation of Chlorinated Solvents and Toluene in Soil," *Conference Proceedings of R&D 1991*, Anaheim, California, pp. 133–136, 1991.

Loo, W. W. et al., "Field Biodegradation of Chlorinated Solvents: A Co-Metabolic Process Utilizing Glucose as Co-Substrate," *Proceedings of HAZMACON 1993*, Association of Bay Area Government (ABAG), Santa Clara, California, 1993.

Nelson, M. J. K. et al., "Biodegradation of Trichloroethylene and Involvement of an Aromatic Biodegradative Pathway," *Applied and Environmental Microbiology*, Vol. 53(5): 949–954, 1987.

Roberts, P. et al., "Biostimulation of Methanotrophic Bacteria to Transform Halogenated Alkene for Aquifer Restoration," *Conference Proceedings of Petroleum Hydrocarbons and Organic Chemicals in Groundwater*, Houston, Texas, pp. 203–217, 1990.

Wilson, J. T., and Wilson, B. H., "Biotransformation of Trichloroethylene in Soil, Applied and Environmental Microbiology," January 1985, Vol. 45(1), pp. 242–243, 1985.

Winter, R. B. et al., "Efficient Degradation of Trichloroethylene by a Recombinant Escherichia Coli," *Biotechnology*, 7: 282–285, 1989.

Xing Jian, and Raetz, R. M., "Co-Metabolic Degradation of Trichloroethene Using A Full Scale Integrated Bioprocessing System," *The Proceedings of the Fifth International In Situ and On-Site Bioremediation Symposium*, San Diego, California, Battelle Press, Vol. 5(2), pp. 217–223, 1999.

Polycyclic Aromatic Hydrocarbons (PAHs)

Heitkamp, M. A., and Cernilia, C. E., "PAHs Degradation by a Mycrobacterium sp. in Microcosms Containing Sediment and Waste from a Pristine Ecosystem," *Conference Proceedings of Hazardous Waste Treatment by Genetically Engineered or Adapted Organisms*, Washington, D.C., pp. 19–23, 1988.

Kotterman, M., van Lieshout, J., Grotenhuis, T., and Field, J., "Development of White Rot Fungal Technology for PAH Degtradation," *The Proceedings of the Fifth International In Situ and On-Site Bioremediation Symposium*, San Diego, California, Battelle Press, Vol. 5(8), pp. 69–74, 1999.

Launen, L. A., Percival, P., Lam, S., Pinto, L., and Moore, M., "Pure Cultures of a Penicillium Species Metabolize Pyrenequinones to Inextractable Products," *The Proceedings of the Fifth International In Situ and On-Site Bioremediation Symposium*, San Diego, California, Battelle Press, Vol. 5(8), pp. 75–80, 1999.

Pesticides and Polychlorinated Biphenyls (PCBs)

Mondello, F. J., and Yates, J. R., "Cloning Bacterial Genes for PCBs Degradation," *Conference Proceedings of Hazardous Waste Treatment by Genetically Engineered or Adapted Organisms*, Washington, D.C., pp. 15–16, 1988.

Stroo, H. F. et al., "Bioremediation of Soil and Water from a Pesticide Contaminated Site," *Conference Proceedings of HAZMACON 1988*, Anaheim, California, pp. 734–740, 1988.

Unterman, R., "Bacterial Treatment of PCB Contaminated Soils," *Conference Proceedings of Hazardous Waste Treatment by Genetically Engineered or Adapted Organisms*, Washington, D.C., pp. 17–18, 1988.

Walia, S., and Kahn, A., "Genetically Engineered Escherichia Coli: A Novel Approach to Biodegradation of PCBs," *Conference Proceedings of Hazardous Waste Treatment by Genetically Engineered or Adapted Organisms*, Washington, D.C., pp. 81, 1988.

Hazardous Metals

Hansen, C. L., and Choudhury, G. S., "Biological Treatment of Mercury Waste," *Symposium Proceedings on Metal Waste Management Alternatives*, Pasadena, California, pp. 216–225, 1989.

Hoffman, S. E. et al., "Bioremediation of Selenium in Soil, Kesterson Reservoir, California," *Conference Proceedings of Hazardous Waste Treatment by Genetically Engineered or Adapted Organisms*, Washington, D.C., pp. 129–138, 1988.

McLean, J. S., Beveridge, T. J., and Phipps, D., "Chromate Removal from Contaminated Groundwater Using Indigenous Bacteria," *The Proceedings of the Fifth International In Situ and On-Site Bioremediation Symposium*, San Diego, California, Battelle Press, Vol. 5(4), pp. 121–126, 1999.

Nordwick, S., Canty, M., Hiebert, R., and Thompson, L., "Microbiological Treatment of Cyanide, Metals, and Nitrates in Mining Byproducts," *The Proceedings of the Fifth International In Situ and On-Site Bioremediation Symposium*, San Diego, California, Battelle Press, Vol. 5(4), pp. 31–36, 1999.

Govind, R., and Yang, Wei, "Study on Biorecovery of Metals from Acid Mine Drainage," *The Proceedings of the Fifth International In Situ and On-Site Bioremediation Symposium*, San Diego, California, Battelle Press, Vol. 5(4), pp. 37–46, 1999.

Zawislanski, P. T., Benson, S. M., Jayaweera, G. R., Wu, L., and Frankenberger, W. T., "In Situ Micobial Volatilization of Selenium in Soils: A Case History," *The Proceedings of the Fifth International In Situ and On-Site Bioremediation Symposium*, San Diego, California, Battelle Press, Vol. 5(4), pp. 141–146, 1999.

Bioreactors

Battelle Press, "Bioreactors and Ex Situ Biological Treatment Technologies," *The Proceedings of the Fifth International In Situ and On-Site Bioremediation Symposium*, San Diego, California, Vol. 5(5), 1999.

Blanchard, M. S. et al., "The Compost Bio-Filter: A Cost Effective Alternative to Conventional Hydrocarbon Vapor Treatment," *Conference Proceedings of the Fourth National Outdoor Action on Aquifer Restoration, Groundwater Monitoring and Geophysical Methods*, Las Vegas, Nevada, pp. 525–539, 1990.

Hickey, R. F. et al., "Combined Biological Fluid Bed-Carbon Adsorption System for BTEX Contaminated Groundwater Remediation," *Conference Proceedings of Fourth National Outdoor Action on Aquifer Restoration, Groundwater Monitoring and Geophysical Methods*, Las Vegas, Nevada, pp. 813–827, 1990.

Kivanc, S., and Unlu, K., "Optimal Design and Operation of Land Treatment System for Petroleum Hydrocarbons," *The Proceedings of the Fifth International In Situ and On-Site Bioremediation Symposium*, San Diego, California, Battelle Press, Vol. 5(5), pp. 81–86, 1999.

Leson, G., "Biofiltration: An Innovative Technology for the Control of VOCs Air Toxics and Odors," *Conference Proceedings of Sixth Annual HAZMAT WEST*, Long Beach, California, pp. 609–613, 1990.

Nyer, E. K., and Skladany, G. J., "Decay Theory Biological Treatment for Low Level Organic Contaminated Groundwater," *Conference Proceedings of HAZMACON 1988*, Santa Clara, California, pp. 702–710, 1988.

Schaffner, R. et al., "Bacterial Pore-Clogging as a Primary Factor Limiting the Enhanced Biodegradation of Highly Contaminated Aquifers," *Conference Proceedings of Petroleum Hydrocarbons and Organic Chemicals in Groundwater*, Houston, Texas, pp. 401–415, 1990.

Smallbeck, D., and Lynch, R. L., "Groundwater Remediation by a Physical/Biological Treatment System," *Conference Proceedings of HAZMACON 1987*, Santa Clara, California, pp. 255–272, 1987.

Speitel, G. E. et al., "Bioreactor Technology for the Treatment of Trace Levels of Chlorinated Solvents," *Conference Proceedings of Petroleum Hydrocarbons and Organic Chemicals in Groundwater*, Houston, Texas, pp. 477–489, 1990.

Sullivan, K., and Konzen, A., "On-Site Treatment of Groundwater and Hazardous Waste Using Fixed Film Bioreactors," *Conference Proceedings of ENSOL 90*, Santa Clara, California, pp. 401–423, 1990.

Van Zyl, H., and Lorenzen, L., "Bioremediation of Diesel Contaminated Soil Using Pilot Scale Biopiles," *The Proceedings of the Fifth International In Situ and On-Site Bioremediation Symposium*, San Diego, California, Battelle Press, Vol. 5(5), pp. 51–56, 1999.

CHAPTER 14
REMEDIATION TECHNOLOGIES

SECTION 14.5

BIOREMEDIATION

Steve Maloney
Mr. Maloney is a principal in Griggs & Maloney, Inc.
Engineering and Environmental Consulting. He
specializes in hazardous waste remediation
utilizing biologic techniques.

14.5.1 INTRODUCTION

Bioremediation, as a process of cleaning up the environment, dates back to the very beginnings of life on earth. It is nature's way of removing excess wastes from the environment through the biological process of biodegradation, which changes or degrades one compound to another, eventually resulting in a reduction in a contaminant mass. Bioremediation, as an applied tool for cleaning up the environment, has a much shorter history. It has been within the past 20 to 30 years that scientists have learned how to systematically apply and enhance the fundamental processes of biodegradation.

The origin of the word bioremediation is derived from the Greek "bios," meaning life, and from the Latin "remediatio," meaning remedy. A decade ago, the word *bioremediation* was not found in a standard dictionary, but today it is commonly used in describing various methods used in cleaning up the environment. Bioremediation may be defined as the use of naturally occurring biological agents to treat a contaminated media, typically soils and groundwater. An example of early bioremediation are the sewer systems designed and built by the Romans around 600 AD for the collection of wastewater. The design included sumps and lagoons for the collection of wastes to prevent system backup and overload. These collection areas were the locations of biodegradation of the organic waste by microorganisms. Biological agents have expanded in recent years to not only include the microbial realm of bacteria, protozoan, and fungi, but also include plants. The use of certain plant species to reduce contaminant levels (phytoremediation) has been successful and its application increasing.

Bioremediation has been used to treat various environmental contaminants including petrochemicals, volatile organic compounds, heavy metals, polyaromatic hydrocarbons, pesticides, explosive residues, and many other contaminants. The use of microorganisms (microbes) to degrade or stabilize contaminants is the most commonly selected bioremediation technique. These microbes can be either naturally occurring or genetically engineered. Laboratory research is constantly trying to develop microbial strains or a "super bug" that can biodegrade specific or multiple contaminants.

The elevation of bioremediation to a technology is relatively recent and typically refers to the use or manipulation of biological agents to degrade various types of hazardous waste to a nonhazardous state. It must be kept in mind that an area undergoing bioremediation is a living system dependent upon providing basic life supporting elements: food, water, and a suitable environment for growth. A failure to provide any one of these basic requirements will result in the decline and eventual end of the desired biological activity. A variety of additional factors, including the chemistry of the contaminant to be degraded, the microbiological analysis of the site, the geology of the site and the physical constraints of the site, must all be considered in designing a successful biological treatment.

14.5.2 *FACTORS AFFECTING MICROBIAL REMEDIATION*

The following factors play important roles in the success of bioremediation:

- Microbes and biodegradation
- Oxygen availability
- Nutrient availability
- Moisture content
- Soil porosity
- Temperature
- pH
- Toxicity
- Soil structure
- Type of contaminant

Microbes and Biodegradation

Soils and groundwater contain many naturally occurring (indigenous) species of microorganisms including protozoa, bacteria, and fungi (molds and yeast). Of these indigenous microbes, it is the bacteria and fungi that account for the biodegradation of most hydrocarbon contamination.

When an organic compound comes into contact with a microbe, the microbe through metabolic processes, degrades the organic material. The microbe may use the organic compound to form new cellular material or to produce energy required for its life system. Heterotrophic microbes are the most common group of microbes providing the metabolic process for degrading organic compounds. Heterotrophs use organic compounds as sources of both carbon and energy. A portion of the organic material is oxidized to provide energy, while the remaining portion is used as building blocks for cellular synthesis. Three general methods exist by which heterotrophic microbes can obtain energy. These are aerobic respiration, anaerobic respiration, and fermentation.

In aerobic respiration, the carbon and energy source is broken down by a series of enzyme-mediated reactions in which oxygen serves as the external electron acceptor. In anaerobic respiration, the carbon and energy source is broken down by a series of enzyme-mediated reactions in which sulfates, nitrates, and carbon dioxide serve as the external acceptors. In fermentation, the carbon and energy source is broken down by a series of enzyme-mediated reactions that do not involve an electron transport chain. Aerobic and anaerobic respiration is often referred to as aerobic and anaerobic degradation. Aerobic degradation typically proceeds at a faster rate than does anaerobic degradation and is the usual metabolic choice for bioremediation.

It is not necessary to introduce nonindigenous microbes to a soil or groundwater system for biodegradation to occur. Natural soil and groundwater systems have a diverse preexisting population of microbes, some of which are predisposed to consuming certain contaminants. The presence of large numbers of microbes is dependent upon several factors, including surface area available for attachment, oxygen availability, and nutrients to support growth. Unconsolidated sediments generally have large populations, while subsurface conditions with confining layers or limited permeability will typically have fewer microbes. If the indigenous microbes are stimulated with a plentiful new source of food and other life sustaining elements, such as oxygen and/or water, those microbial species best adapted to use the food source will reproduce and multiply to a population level sustainable by the food source. This process is referred to as *biostimulation*. The addition of microbes in high concentrations to a contaminated site is called *bioaugmentation* and is used with the intent of creating a faster and more efficient remediation process. There is debate among bioremediation professionals as to whether biostimulation or bioaugmentation produces the best results. Because of the complexities of bioremediation sites, most often an integration of several treatment methods is needed to properly bioremediate a site.

Oxygen Availability

The availability of oxygen controls the rate at which aerobic microbes can function and is a key requirement in most, but not all, types of bioremediation. Anaerobic metabolism can occur but the rate is much slower and less complete. The lack of oxygen is usually the limiting factor for *in situ* biodegradation. Supplying an appropriate amount of oxygen to a site is one of the more difficult challenges of bioremediation, however, there are a number of proven ways to provide oxygen to soils and groundwater.

Soil aeration can be provided by mechanical means, such as tilling and disking, soil venting, or injection by blowers, or by the application of aerated water. Leachate collected from certain bioremediation methods, such as soil piles, can be collected, aerated, and sprayed back onto the remediation area thus providing oxygen. Soil venting is an effective and inexpensive method to provide air to the unsaturated zone. Air is supplied by the use of vacuum wells. Air is sucked from around the vacuum well and creates an air flow from the surrounding soil and draws oxygen from the surface through the soils and into the well. Vent wells can be installed around vacuum wells to increase the flow and volume of surface air pulled through the soils and into the vacuum wells.

The aeration of waters can be accomplished by mechanical mixers, air sparging, chemical additions, fixed-film reactors (biological towers, rotating biological contactors), and pure oxygen. Shallow subsurface water typically contains 3 to 4 parts per million (ppm) dissolved oxygen. This concentration will limit microbial activity. Increasing the oxygen content is necessary for successful bioremediation.

Typically, air sparging forces air through a porous media forming very small bubbles, which deliver oxygen into the groundwater and increase oxygen transfer. The air is delivered under pressure to groundwater wells, which forces it into the aquifer. A dissolved oxygen concentration of 8 to 12 ppm can be provided by air sparging systems.

Hydrogen peroxide is inexpensive and is a commonly used chemical to provide a source of oxygen. Generally, treatment using hydrogen peroxide generates no toxic by-products, it is environmentally benign, and the hydrogen peroxide can be degraded by microbial enzymes. It can deliver much higher concentrations of oxygen than the microbes can tolerate and often an inhibitor is added to control its release. A maximum dissolved concentration of 100 ppm should be maintained within the bioremediation zone. If concentrations exceed 100 ppm, the hydrogen peroxide may degas and form bubbles that will clog the pore spaces. Ozone is another chemical that can be used as an alternative to hydrogen peroxide. However, it is toxic in high concentrations, is corrosive, and will effectively sterilize the water being treated. The use of chemical additives may be particularly appropriate when mechanical means of aeration is not feasible.

In fixed-film reactors, microbes attach to an artificial substrate that provides inert support. Biological towers and rotating biological contactors are the most common forms of fixed-film processes. Biological towers contain a medium of polyvinyl chloride (PVC), polyethylene, polystyrene, or redwood, which is stacked into towers. The contaminated water is sprayed across the top, and as the water moves downward, air is pulled upward through the tower. A slime layer of microbial growth forms on the artificial substrate and biodegrades the organic content as the water flows over the slime.

A rotating biological contactor consists of a series of rotating discs on a shaft set in a basin. The contaminated water passes through the basin where microbes attached to the discs metabolize the organics in the water. Approximately 40 percent of the disc surface area is submerged. As the discs rotate, oxygen is supplied to the microbes as the slime layer alternately comes into contact with the contaminated water and air.

Pure oxygen provides high concentrations of oxygen and has been used for remediation. However, pure oxygen is relatively expensive, may bubble out of solution and not be available for microbial use, and may be an explosion hazard if not handled properly.

Nutrients

The presence of nutrients in a correct balance is essential to microbial degradation. Not enough nutrients or nutrients in the wrong amounts will prevent optimum microbial activity. Nutrients required by microbes are similar to those required by plants. Nitrogen (as nitrate or ammonia) and phosphorus (as phosphate) are the most common additives. The oxidation of carbon and the utilization of nitrogen and phosphorus, along with other trace nutrients, is the key to nutrient balancing.

Nitrogen and phosphate nutrients are generally supplied as inorganic salts, such as sodium nitrate or nitrite, potassium phosphate, or ammonium phosphate. These can be found in most commercially available fertilizers. For anaerobic degradation, nitrate and sulfate salts can be added.

Essential trace elements needed for nutrition include calcium, cobalt, copper, iron, magnesium, manganese, molybdenum, potassium, sulfur, and zinc. Generally, these trace elements are naturally present in sufficient amounts and present no limiting factor.

Moisture

Water is essential for the growth and multiplication of microbes and all metabolic activities. Microbial growth is limited to soluble materials that can be transported across their cell membranes and into the interior cellular fluid where digestion takes place. Optimal growth rates take place where there is sufficient moisture in liquid form to solubilize the contaminant substrate.

In soils, a good moisture content must be maintained to ensure that enough water is present to promote and support microbial activity. Generally, when treating soils in the unsaturated zone, moisture content between 10 and 20 percent by weight provides optimum cellular metabolism for aerobic degradation. If too much water is added to a site and saturation of the soils is the normal site condition, aerobic degradation can change to anaerobic degradation and retard the progress of remediation. Should this condition occur, the site must be returned to proper moisture levels before aerobic degradation can be reestablished. However, maintaining moisture content up to saturation levels can improve the contact of the microbes to contaminant and increase the rate of biodegradation.

Temperature

Temperature controls the activity of microbes. Microbes can grow at temperatures ranging from 0°C to over 100°C. Typical microbes used in biotreatments will perform best between 20°C and 40°C. Microbial activity exhibits a significant decline below 15°C and above 45°C.

Microbial activity in an aqueous solution will stop when frozen. Freezing temperatures will reduce the total microbe population and will essentially halt any bioremediation. If freezing is a seasonal problem, it may be necessary to enhance the microbe population with additional organisms and nutrients with the return of warmer weather. For those geographical areas subject to freezing temperatures, freezing also presents a problem in the design and operation of treatment systems handling water for irrigation, pumping, and so forth. If bioremediation is necessary throughout cold weather, artificial heating sources may be appropriate. Electrical heating devices could be installed, but the cost may be prohibitive on large sites. Passive solar heating by the placement of polyethylene plastic sheeting (clear or black) over the site can be a cost-effective way to provide heat to a site. If the size of the area to be remediated is relatively small, an enclosed structure could be used, such as a greenhouse frame covered with polyethylene sheeting.

Heat is normally not a major problem, but excessive temperatures may adversely affect some species of microbes. Excessive or extended ambient temperatures have a greater impact upon maintaining the moisture content of treated soils, which can affect microbial activity.

pH

Generally, the microbes used for bioremediation prefer neutral or alkaline environments. The greatest activity typically occurs between a pH of 4.0 and 9.0 standard units. Exceptions can readily be found to this range, however, a limit to microbial activity can be expected at a pH below 1.0 and greater than

11.0 standard units. For most microbes used in remediation, the optimal range for microbial growth should be kept between pH 6 and 8 standard units with a target value of neutral pH 7.

Toxicity

The toxicity of the contaminant to which the microbes are being exposed may adversely affect the extent of microbial biodegradation. The contaminant to be biodegraded may be intrinsically toxic to the microbes or the contaminant's concentration level may be toxic.

Toxic effects may be negated by the selection of microbes that are tolerant of the contaminant. Sometimes, the contaminant can undergo pretreatment to reduce toxicity. Pretreatment may include chemical oxidation of the contaminant, the physical separation of free-product, and metals precipitation.

Soil Structure

The soil structure present at a site affects rates of biodegradation. Soil structure determines the ability of the soil to transport air, water, and nutrients to the zone of bioactivity. Soil structure includes not only the types of soils present but also how these soils stratify and co-mingle throughout a remediation site. Soils that contain silt and clay do not usually transport these substances fast enough to encourage rapid bioactivity. More permeable soils, such as sand, are more conducive to increasing bioactivity and thus, quicker site cleanups.

Although soil is a primary component of most bioremediation activities, it is also one of the more difficult factors to assess. Most surface soil constituents are not homogenous but mixtures of the major soil types. Soils at lower zones often exhibit more homogeneity. In general, most soils are a mixture of the different soil types and contain an existing population of microbes. Unless the contaminant to be biodegraded is resistant to biodegradation or is toxic to common soil microbes, it is likely that the soil will already contain a microbial population capable of degrading some contaminants.

Some of the major soil types that make up the soil structure encountered at sites are discussed in the following sections.

Soil Porosity.   Of equal importance to soil moisture content is soil porosity. Soil porosity affects not only the ability of soil to hold water, but also the movement of water through the soil. High-porosity soils, such as sands, move and drain water well, but water retainage is low. Low-porosity soils, such as clays, drain poorly, but retain a higher water content. The importance of this property lies not only in the effect on remediation, but also in the initial application of bacteria and/or nutrients.

Clays.   The clay content of soils can affect bioremediation through its excellent water retention capacity and its ability to retard water transport because of the presence of very fine particles that reduce inter-particle spaces. This physical makeup of clay, often referred to as its "tightness," can also impede microbial and nutrient transport. Chemically, clays are known to bind to a wide variety of substances, including hydrocarbons of many types. Because clay retains moisture well, dehydration during bioremediation treatment is generally not as much concern as with other soil types, such as sands.

These soil characteristics may be overcome or reduced by tilling of the soil (if site conditions permit) or tilling and mixing the clay with stone, gravel, or sand to reduce the "tightness" of the clay. Installing underground piping through the clay to aid in microbial, oxygen, and nutrient transport to other remediation zones can eliminate the time needed for the substances to naturally penetrate through the clay to reach contaminants.

Sands.   The principal bioremediation concern with sands is its permeability. Water permeates easily and quickly through sands and makes it difficult to maintain the appropriate moisture content needed for bioremediation. Soil moisture content of the bioremediation area should be frequently checked to ensure appropriate levels. However, the permeability of sand can also assist in the rapid deployment of oxygen and nutrients to contaminants throughout the soil zone, and tillage is easy, making landfarming methods very manageable.

Humic Soils.   Humic soils are soil rich in organic material and, as a result, also exhibit larger microbial populations than clays or sands. These soils generally exhibit excellent water holding and penetration properties. Because these soils may contain significant microbial populations, it is possible that the indigenous population may interfere with organisms added for augmenting biodegradation. Any biostimulation of the soils is likely to also stimulate the existing population and enhance biodegradation. Conversely, some humic soils are known to retard biodegradation. While this phenomenon has not been explained, it may be due to binding capacities of humic acids that may tie up certain nutrients or even the target pollutant.

Types of Contaminants

There are contaminant types that are readily biodegradable and others that are resistant to any degree of biodegradation (refractory compounds), or will not biodegrade under a specific set of conditions (persistent compounds). In general, the following statements can be made concerning the types of contaminants encountered at bioremediation sites:

- Aromatic compounds are more resistant to biodegradation that aliphatic compounds.
- Halogenations of an organic compound makes it more resistant to degradation by microorganisms, and the higher the degree of halogenations, the more resistant it is to degradation.
- Polynuclear aromatics and asphaltic compounds are generally considered to be poorly biodegradable.
- The introduction of any degree of oxidation to a compound will generally make it biodegradable.
- The less soluble a pollutant, the more resistant it is to breakdown.
- Complex molecular structures, such as branched paraffins, olefins, or cyclic alkanes, are very resistant to biodegradation.
- Hydrocarbon contaminants with less than 10 carbon atoms tend to be relatively easy to biodegrade, unless the contaminant's concentration level is so as to be toxic to the microbes. Benzene, toluene, and xylene, the major components of gasoline, are examples of easily degraded hydrocarbons.
- The ease or rate of biodegradation tends to decrease as the molecule size increases.

Other Factors

The following are other factors that influence the success of bioremediation. The first factor is the introduction of chemicals that may deplete or eliminate the microbial population. If the bioremediation site is part of an active operation, it is important to recognize any activity that may result in the routine or sudden release of additional contaminants to the remediation area. The release of certain types of chemicals may completely wipe out a microbial population and halt any remedial bioactivity.

The second factor is the unidentified contaminant that may affect biodegradation. A review of past site operation is essential in identifying pervious wastestreams and their handling and disposal in order to identify the site's contaminants of concern. The presence of an unidentified contaminant could adversely affect the efficiency of the selected bioremediation method. Site analysis and samplings should be incorporated at the very beginning of any bioremediation project.

The third factor is the effect that other naturally occurring microbes may have at bioremediation sites. Providing oxygen and nutrients may stimulate other microbe activity that affects the bioremediation system. Some microbes utilize the iron in the water and soils as an energy source. Their increased activity can cause their growth on the slotted well screen and clog the slots, eventually choking off the groundwater flow into the well. The same effect can happen in packed tower air strippers and tray aerators.

14.5.3 BIOREMEDIATION METHODS

The selection of a bioremediation method involves the consideration of many factors, not the least of which is cost and how long will it take to complete the bioremediation. Some of the factors that should be considered:

1. The treatment option should be the best to biodegrade the contaminants of concern.
2. The option is the best that protects human health and the environment.
3. The option is readily acceptable to the regulatory agencies.
4. The option will make as much use, as possible, of existing site conditions, structures, utilities, and materials.
5. The option is the best that suits the site layout and is least disruptive or conspicuous to the facility and the surrounding properties.
6. The option is the best that reduces or eliminates client liability.

Each site will have its own particular set of concerns, and the above list is by no means exhaustive. The site conditions, along with the concerns of the client and regulatory agencies, will play major roles in determining the types of treatment concerns and options for bioremediation.

The primary methods used in bioremediation is *in situ* (treatment of the contaminant in-place) and *ex situ* treatment (removal and treatment of the contaminant). Soils can be treated in situ or excavated for surface land treatment or for bioreactor treatment (ex situ). Groundwater can be treated in situ or pumped to the surface for biological treatment (ex situ). There are many variations to implementing treatment options, and combinations of concurrent *in situ* and surface treatment has been very effective. In all cases, the source of the contaminant must be addressed prior to beginning remediation. If the contaminant source is known and can be removed, such as a leaking underground storage tank or a buried hazardous waste pile, the effectiveness of the bioremediation and the length of time required for completion will be favorably affected. Otherwise, the remediation may be an on-going, never-ending process.

Soils

The surface soils and the soil zone immediately below the surface can undergo in situ treatment by land treatment (landfarming). Land treatment is one of the most commonly used methods for treating soil containing petroleum products. Land treatment of petroleum refining wastes has been employed successfully at many locations for over 20 years. Landfarming involves blending contaminants into shallow surface soils, and adding nutrients and water to stimulate the naturally occurring microbes to degrade the contaminants. The action of tilling, disking, or some other method that will expose the subsurface soils to the atmosphere, supplies the oxygen. Regular tilling cycles maintain aeration levels. Nutrients are applied across the exposed soil surface either in a dry or liquid form. Water is typically supplied via irrigation, either surface or subsurface.

Aboveground bioremediation (ex situ) of soils is also an option. In some circumstances, bioremediation may be best accomplished by excavating the soil and placing it in an aboveground reactor. Three methods are typically applied as soil reactors: landfarming, soil slurries, and soil piles. Regardless of the type of reactor selected, the basic bioremediation requirements are the same. The system must keep the microbes for biodegradation alive by supplying a food source, oxygen, water, and nutrients.

Landfarming above ground usually requires an impermeable liner below the soil to control leachate, a leachate collection system, and a water storage system. Tilling and disking the soils may be used to aerate the soils or a system of vented piping with an air supply (blower) may be installed within the soil for aeration. Some system designs collect the leachate and redistribute it, usually by spraying, back onto the soil to be treated. The leachate may or may not undergo some form of pretreatment or oxygen/nutrient/microbe additive process prior to being sprayed on the soil treatment area.

The soil slurry reactor combines soil and water to form a slurry. Standard mixing techniques can be used to continually mix the slurry and supply oxygen. Nutrients are added to the slurry to promote microbial growth. This system is easy to maintain and readily promotes the biochemical reactions for contaminant degradation because of its ability to maintain a constant and blended contact between microbes, oxygen, nutrients, and the target contaminants. However, the main problem with the system is that after the biochemical reaction is completed, the water must be separated from the soil. Depending upon the type and amount of soil being treated, liquid/solid separations and handling can be difficult, expensive, and time-consuming.

The soil pile reactor system places excavated soil onto a liner. During soil placement, enough microbes and nutrients are added and mixed with the soil to complete the bioremediation. Vented or perforated pipe is placed within the pile at specified intervals and depths. When all the soil has been placed on the pile, the piping is attached to the vacuum side of a blower. A plastic cover is placed over the entire pile to control precipitation and air flow patterns. The blower sucks air through the pile, supplying oxygen to the microbes. Once the soil has been successfully bioremediated, the soil pile is removed and the uncontaminated soil may be returned to the excavation or elsewhere (regulations permitting).

Groundwater

Two basic technologies have been developed and are commonly used in groundwater bioremediation: *in situ* and groundwater extraction and treatment, commonly referred to as "pump and treat." In the vadose and aquifer zones, water is a factor in transporting oxygen and nutrients to the microbes. The microbes available for degradation are present in these zones along with the target contaminants. The main design consideration is to move water through the contaminated zone providing microbial contact, oxygen, and nutrients; and to then collect, replenish (adding oxygen, nutrients, and microbes) and recycle the water back into the contaminated zone in a controlled water flow pattern

In situ treatment involves the delivery of the bacteria, nutrients and electron acceptor source to the groundwater. In pump and treat, the contaminated water is brought to the surface and treated in a bioreactor (such as fixed-film reactors) which, basically, is similar in concept to a packaged wastewater treatment plant using specialized microbes (see section on oxygen availability). The effluent can be discharged into a publicly owned treatment works (POTW) or re-injected into the ground via wells or infiltration galleries. In situ treatment can be readily combined with a pump and treat system, and the two can simultaneously treat and inject withdrawn water.

The in situ remediation of groundwater commonly involves the injection of microbes into contaminated groundwater via injection wells; it also involves injecting oxygen and nutrients with the objective to degrade the contaminants in place. The appeal of such a system is that it eliminates the need for the costs and maintenance of above-ground pump and treat technology, and theoretically, it allows the microbes to degrade the contaminants in place. The in situ method is not as passive as it would seem, because the injection of oxygen and nutrients may establish some flow pattern through the aquifer.

Groundwater extraction and treatment (pump and treat) is the process by which the contaminated water is pumped out of the ground and treated by an appropriate method to remove the contaminants of concern. Once treated, the water is returned to the ground by injection wells, infiltration galleries, or surface application. The majority of the treatment technologies used in groundwater extraction and treatment do not directly involve biological treatment. Treatment technologies commonly used are air stripping, oil/water separation, carbon adsorption, chemical treatment, and biological treatment. Biological treatment is accomplished by the use of above ground biological reactors, such as biological towers, or rotating biological contactors (see section on oxygen availability) used to treat the pumped groundwater. Pump and treat technologies offer a usually slower, but surer technology to eventually remediate contaminated groundwater.

14.5.4   *THE FUTURE OF BIOREMEDIATION*

Bioremediation, the degradation or stabilization of contaminants by microorganisms, is a safe, effective, and economical alternative to traditional and chemical treatment technologies. Bioremediation can be used in conjunction with other cleanup technologies or can be a stand-alone treatment remedy. The science and practice of bioremediation are continually being challenged by physical, chemical, and biological conditions encountered in the field. Laboratory research includes such activities as identifying, isolating, and studying microbes with bioremediation potential, developing genetically engineered microbes, and evaluating the toxicity of chemically contaminated media.

The practice of bioremediation is also being advanced by the amount of information that is becoming available. Bioremediation experience is becoming more widespread resulting in more and more field applications being documented in the literature. The presence of the Internet has also permitted quick and timely access to bioremediation articles and resources.

Bioremediation has been recognized by industry, scientists, and regulatory agencies as a viable tool for site restoration. With new discoveries and understandings of the intricacies of how organisms react and respond to environment perturbations, the future of bioremediation continues to grow.

REFERENCES

Alexander, M., *Introduction to Soil Microbiology*, John Wiley and Sons, Inc., New York, 1961.

Alexander, M., "Biodegradation of Chemicals of Environmental Concern," *Science*, 211: 132, 1981.

American Petroleum Institute, "The Land Treatability of Appendix VIII Constituents Present in Petroleum Wastes," *Publication No. 4379*, API, 1984.

Bellandi, R., *Hazardous Waste Site Remediation, The Engineer's Perspective*, O'Brien & Gere Engineers, Inc., Van Nostrand Reinhold, New York, 1988.

Bernstein, K., "Do Microbes Hold the Key to Toxic Waste Cleanup? The EPA Thinks So . . . ," in *BioWorld*, pp. 46–51, November 1990.

Bourquin, A. W., "Bioremediation of Hazardous Wastes," in *J. Haz. Mater. Control*, September–October 1989.

Brown, L. R., "Oil-degrading Micro-organisms," *Chem. Eng. Prog*, October: 35–40, 1987.

Brubaker, G. R., and Exner, J. H., "Bioremediation of Chemical Spills," in *Environmental Biotechnology*, Omenn, G.S., ed., Plenum Press, New York, 1988.

Connor, J. R., "Case Study of Soil Venting," *Poll. Eng.*, 20(7): 74–78, 1988.

Davidson, D., Wetzel, R., Pennington, D., Ellis, W., and Moore, T., *In Situ Treatment Methods for Contaminated Soils and Groundwater*, HMCRI Seminar, November 4–5, Washington, D.C., 1985.

Dagley, S., *Microbial Degradation of Organic Compounds*, Marcel Dekker, Inc., New York, 1984.

Dragun, J., *The Soil Chemistry of Hazardous Materials*, Hazardous Control Research Institute, Silver Spring, Maryland, 1988.

Driscoll, F. G., *Groundwater and Wells*, 2nd Edition, Johnson Filtration Systems, Inc., St., Paul Minnesota, 1986.

Freeman, Harry M., *Standard Handbook of Hazardous Waste Treatment and Disposal*, McGraw-Hill Book Company, New York, 1988.

Gibson, D. T., *Microbial Degradation of Organic Compounds*, Marcel Dekker, Inc., New York, 1984.

Gruiz, K., and Kriston, E., "In Situ Bioremediation of Hydrocarbon in Soil," *J. Soil Contamination*, 4(2): 163–173, 1995.

Hazardous Materials Control Research Institute, *Biotreatment: The Use of Microorganisms in the Treatment of Hazardous Materials and Hazardous Wastes*, HMCRI, Washington, D.C., pp. 41–46, 1989.

Hinchee, R. E., Downey, D. C., and Beard, T., "Enhancing Biodegradation of Petroleum Hydrocarbon Fuels Through Soil Venting," in *Proc. Nat. Water Well Assoc. Conf. Of Petroleum Hydrocarbons and Organic Chemicals in Groundwater: Prevention, Detection, and Restoration*, pp. 235–248, November 1989.

Hoeppel, R. E., Hinchee, R. E., and Arthur, M. F., "Bioventing Soils Contaminated with Petroleum Hydrocarbons," *J. Ind. Microbiol.*, 8: 141, 1991.

King, R. B., Long, G. M., and Sheldon, J. K., *Practical Environmental Bioremediation*, Lewis Publishers, Chelsea, Michigan, 1998.

Laney, D. F., "Hydrocarbon Recovery as Remediation of Vadose Zone Soil/Gas Contamination," in *Proc. Nat. Water Well Assoc. Second Nat. Outdoor Action Conf. On Aquifer Restoration, Groundwater Monitoring and Geophysical Methods, Vol III*, Las Vegas, Nevada, pp. 1147–1171, May 1988.

Leahy, J. G., and Colwell, R. R., "Microbial Degradation of Hydrocarbons in the Environment," *ASM Microbiol. Rev.*, pp. 305–315, September 1990.

Lee, M. D., Thomas, J. M., Borden, R. C., Bedient, P. B., Ward, C. H., and Wilson, J. T., "Biorestoration of Aquifers Contaminated with Organic Compounds," *CRC Critic. Rev. Environmental Control*, Vol. 18, Issue 1, pp. 29–89, 1988.

Lynch, J., and Genes, B. R., "Landtreatment of Hydrocarbon Contaminated Soils," in *Petroleum Contaminated Soils Vol. 1, Remediation Techniques, Environmental Fate, Risk Assessment*, Paul Kostecki and Edward Calabrese, Lewis Publishers, Chelsea, Michigan, 1990.

McCarty, P. L., "Bioengineering Issues Related to In Situ Remediation of Contaminated Soils and Groundwater," in *Proc. Conf. Redusing Risk from Environmental Chemicals through Biotechnology*, Seattle WA, 1987.

Nyer, E. K., *Practical Techniques for Groundwater and Soil Remediation*, Lewis Publishers, Chelsea, Michigan, 1993.

Overcash, M. R., and Pal, D., *Design of Land Treatment Systems for Industrial Wastes—Theory and Practice*, Ann Arbor Science Publishers, Ann Arbor, MI, 1979.

Preslo, L., Miller, M., Suyama, W., McLearn, M., Kostecki, P., and Fleischer, E., "Available Remedial Technologies for Petroleum Contaminated Soils," in *Petroleum Contaminated Soils Vol. 1, Remediation Techniques, Environmental Fate, Risk Assessment*, Paul Kostecki and Edward Calabrese, Lewis Publishers, Chelsea, Michigan, 1990.

Rochkind, M. L., Blackburn, J. W., and Saylor, G. S., "Microbial Decomposition of Chlorinated Aromatic Compounds," US EPA 600/2-86/090 Cincinnati, OH, 1986.

Sneider, D. R., and Billingsley, R. J., *Bioremediation, A Desk Manual for the Environmental Professional*, Pudvan Publishing Co., Inc., Northbrook, IL, 1990.

Snyder, J. D., "How Biotreatment Works," *Environment Today*, 1(1): 20, 1990.

Testa, S. M., and Winegardner, D. L., *Restoration of Petroleum Contaminated Aquifers*, Lewis Publishers, Chelsea, Michigan, 1991.

Thornton, J. S., and Wooten, Jr., W. L., "Venting for the Removal of Hydrocarbon Vapors from Gasoline Contaminated Soils," *J. Env. Scientific Health*, A17(1): 31–44, 1982.

Thornton, J. S., and Wooten, Jr., W. L., "Removal of Gasoline vapor from Aquifers by Forced Venting," in *Proc. 1984 Haz. Mat. Spills Conf.*, 1984.

Weston, R. F., Inc., *Remedial Technologies for Leaking Underground Storage Tanks*, Lewis Publishers, Chelsea, Michigan, 1998.

CHAPTER 14
REMEDIATION TECHNOLOGIES

SECTION 14.6

PUMP-AND-TREAT TECHNOLOGIES

Milovan S. Beljin

Dr. Beljin is a consulting hydrologist and president of M.S. Beljin
& Associates. He also serves as adjunct faculty at the University
of Cincinnati and Wright State University, Dayton, Ohio.

The goal of groundwater remediation is to protect human health and the environment and to restore groundwater quality. For groundwater used for drinking, cleanup goals generally are set as drinking water standards, such as Maximum Contaminant Levels (MCLs). If groundwater is not used for drinking, other cleanup goals may be appropriate. A common approach to deal with contaminated groundwater is to extract the contaminated water and treat it at the surface prior to discharge or injection. This is known as conventional pump-and-treat remediation.

Inadequate design and implementation will impact the performance of a pump-and-treat system. Examples include insufficient number of wells, insufficient pumping rates, inappropriate well locations or well design, and failure to account for complex chemistry of groundwater and contaminants. Similarly, poor system operation will also restrict pump-and-treat effectiveness. Failures of some pump-and-treat system because of the technical implementation and operation problems, combined with unrealistic expectations, have created a negative attitude towards pump-and-treat technologies among some regulators and groundwater professionals. The fact remains, however, that only by pumping groundwater, the migration of a contaminant plume can be controlled.

14.6.1 REMEDIATION STRATEGY

In order to determine an appropriate strategy to manage contaminated groundwater, it is necessary first to evaluate the contaminated site and establish cleanup objectives. A common strategy for managing the contaminated site has been to remove or contain contaminant source (e.g., excavation, physical barriers, and/or pumping), and to apply some groundwater remediation technology (e.g., pump-and-treat, air sparging, bioremediation, permeable treatment walls). Pump-and-treat technology can be used alone or with other technologies depending on site-specific hydrogeologic conditions and remediation objectives.

Pump-and-treat systems are often designed to hydraulically control the groundwater flow pattern and thus to prevent migration of the contaminant plume. At sites where the contaminant source cannot be removed (e.g., dense nonaqueous liquids or DNAPLs in bedrock), hydraulic containment can be used to control the source area. Hydraulic containment of contaminants by pumping groundwater has been demonstrated at numerous sites. Properly designed recharge systems with injection wells and/or drains and physical containment options (e.g., subsurface barrier walls and surface covers) can enhance hydraulic containment systems. One of the benefits of injecting groundwater is the reduced pumping rate required to maintain containment. Hydraulic containment systems can be designed to provide long-term containment of contaminated groundwater and/or source areas.

For sites where the contaminant source has been removed or contained, it may be possible to clean up the contaminated groundwater. Pump-and-treat technology designed for aquifer restoration generally combines hydraulic containment with more aggressive pumping of groundwater to achieve cleanup goals during a finite period. Groundwater cleanup is much more difficult to achieve than hydraulic containment.

14.6.2 SITE CHARACTERIZATION

The main goal of site characterization should be to obtain sufficient data to select and design remediation for the site (NRC, 1994). This involves investigation of (1) the nature, extent, and distribution of contaminants, (2) site hydrogeology, (3)contaminant properties, and (4) potential receptors and risks posed by contaminated groundwater. Inadequate site characterization can lead to a wrong pump-and-treat design and poor system performance. However, a complete understanding of a contamination site is not a realistic goal because of complex subsurface conditions and investigation cost. The site characterization efforts must develop sufficient data to select and design an effective remedy while recognizing that significant uncertainties about subsurface conditions will always exist. Additional information regarding procedures and strategies for investigating contamination sites is provided by USEPA (1988a, 1991a), Cohen and Mercer (1993), Sara (1994), CCME (1994), and Boulding (1995).

During the initial phase of site investigation, prior studies and background information are reviewed to identify likely contaminant sources, transport pathways, and receptors. Based on this initial conceptualization, a data collection program is devised to better define the nature and extent of contamination. Additional studies, including monitoring of actual pump-and-treat performance, are usually required to assess the potential to restore groundwater quality in different site areas.

Pump-and-treat performance is typically assessed by measuring hydraulic heads and gradients, groundwater flow directions and rates, pumping rates, pumped water and treatment system effluent quality, and contaminant distributions in groundwater and porous media. Guidance on methods for monitoring performance of a pump-and-treat system is provided by Cohen et al. (1994).

Characterizing groundwater flow and contaminant transport is particularly challenging in heterogeneous media, especially where contaminants have migrated into fractured rock. At the scale of many contaminated sites, complete characterization of fractured rock (and other heterogeneous media) may be economically infeasible (Schmelling and Ross, 1989), and not needed to design an effective pump-and-treat system (NRC, 1994). The appropriate characterization methods and level-of-effort must be determined on a site-specific basis.

At many sites, most of the contaminant mass is not dissolved in groundwater, but is present as NAPL, absorbed phase, and solids. Slow mass transfer of contaminants from these phases to groundwater during pump-and-treat will cause tailing and prolong the clean-up effort. If aquifer restoration is a potential remediation goal, then site characterization should investigate the physical and chemical phenomena that cause tailing and rebound. Tailing and rebound patterns associated with different physical and chemical processes are similar. Multiple processes (i.e., dissolution, diffusion, and desorption) will typically be active at a pump- and-treat site. Diagnosis of the cause of tailing and rebound, therefore, requires careful consideration of site conditions and usually cannot be made by examining concentration-versus-time data alone. Quantitative development of the conceptual model using analytical or numerical methods may help estimate the relative significance of different processes that cause tailing and rebound. Knowledge of the potential limitations at each site may allow more detailed analyses of the potential effectiveness of different pump-and-treat remediation strategies and different system configurations.

Hydraulic parameters estimated from analysis of standard pumping tests are often used to design injection systems. The aquifer test procedures and methods can be found in many textbooks (e.g., Driscoll, 1986; Kruseman and deRidder, 1990). Hydraulic heads and groundwater flow patterns resulting from injection can be examined and predicted using well or drain hydraulics equations and groundwater flow models. Such analysis can also be used to determine potential injection rates, duration, and monitoring locations for injection tests. The most common problem associated with fluid injection is permeability reduction resulting from clogging of well or drain openings. Clogging results from physical filtration of solids suspended in injected water, chemical precipitation of dissolved solids, and the excessive growth of microorganisms. Clogging problems can be minimized by overdesigning injection capacity (e.g., by installing more wells, longer screens) and implementing a regular well maintenance program. Significant maintenance may be required at many sites to retain desired injection capacity. More detailed discussions of the engineering aspects of water injection are provided by Pyne (1995).

Treatability data needed for design of groundwater treatment systems generally should be acquired by conducting chemical analyses and treatability studies on contaminated groundwater extracted during pumping tests. Analysis of water samples obtained at different times during a pumping test often will provide data regarding the initial range of contaminant concentrations in influent water to the treatment plant. Bench- and pilot-scale treatability studies are valuable means for determining the feasibility of candidate processes for treating contaminated groundwater (USEPA, 1989).

Because of slow contaminant transport and interphase transfer, many pump-and-treat systems will operate for decades. Data collected during investigation and remediation should be reviewed periodically to identify modifications that will improve pump-and-treat system performance. This phased approach to system installation may be more cost-effective than overdesigning the system to account for uncertainty in subsurface characterization at many sites.

14.6.3   *CAPTURE ZONE ANALYSIS*

The capture zone of an extraction well or drain refers to that portion of the subsurface containing groundwater that will ultimately discharge to the well or drain. The shape of the capture zone depends on the natural hydraulic gradient as well as pumping rate and the aquifer transmissivity. In recent years, many mathematical models have been developed or applied to compute capture zones, groundwater, pathlines, and associated travel times to extraction wells or drains (Bair et al., 1991; Blandford et al., 1993; Fitts, 1994; Gorelick et al., 1993; Javandel and Tsang, 1986; Javandel et al., 1984; Pollock, 1994).

Model selection for pump-and-treat design analysis depends on the modeling objectives, the complexity of the site, and the available data. There are two types of models used for design analysis: semianalytical and numerical models. Regardless of the design tools that are used, capture zone analysis should be conducted, and well locations and pumping rates optimized, by monitoring hydraulic heads and flow rates during pumping tests and system operation. Conceptual model refinements gained by monitoring lead to enhanced pump-and-treat design and operation.

Semianalytical models employ complex potential theory to calculate stream functions, potential functions, specific discharge distribution, and/or velocity distribution by superimposing the effects of multiple extraction/injection wells on an ambient uniform groundwater flow field in a two-dimensional, homogeneous, isotropic, confined, steady-state system (Blandford et al., 1993). Based on this approach, the simple graphical method can be used to locate the stagnation point and dividing streamlines, and then to sketch the capture zone of a single well in a uniform flow field. This analysis is extended by Javandel and Tsang (1986) to determine the minimum uniform pumping rates and well spacings needed to maintain capture between two or three pumping wells along a line perpendicular to the regional direction of groundwater flow. Many of these models support reverse and forward particle tracking to trace capture zones and streamlines.

Numerical models are generally used to simulate groundwater flow in complex three-dimensional hydrogeologic systems. Numerical flow model output is processed using reverse or forward particle-tracking to assess pathlines and capture zones associated with pump-and-treat systems (Pollock, 1994; Zheng, 1992). Solute transport models are primarily run to address aquifer restoration issues, such as changes in contaminant mass distribution, with time because of pump-and-treat operation.

14.6.4   *PUMP-AND-TREAT COMPONENTS*

Groundwater extraction/injection systems should be designed to site-specific conditions and remediation goals. As a result, the number of combinations of system components is large, and a variety of pump-and-treat configurations exist. However, vertical wells are integral components of most pump-and-treat systems. Extraction wells are intended to capture and remove contaminated groundwater; injection wells are used to enhance hydraulic containment and groundwater flushing rates. Basic component considerations include drilling/installation method and the well design (e.g., well diameter, screen and casing specifications, completion depth interval, pump specifications). Detailed guidance on well drilling, construction, and development methods can be found in Driscoll (1986), Bureau of Reclamation (1995), Aller et al. (1989), and others.

During recent years, directional drilling rigs from the utility, mining, and petroleum industries have been adapted to install horizontal wells at contamination sites (CCEM, 1996; Conger and Trichel, 1993; Kaback et al., 1989; Morgan, 1992). Horizontal wells can be installed strategically to: (1) allow extraction in inaccessible areas such as beneath buildings, ponds, or landfills; (2) intercept multiple vertical fractures; and (3) provide hydraulic control along the long axis of a plume. The higher cost of horizontal wells, compared to vertical wells, may be offset by savings derived from more efficient remediation, drilling fewer wells, the purchase of fewer pumps, and so forth. Horizontal wells with long screens may be more cost-effective than vertical wells, particularly at sites where contaminated groundwater is in relatively thin aquifers.

Groundwater treatment technologies rely on physical, chemical, and/or biological processes to reduce contaminant concentrations to acceptable levels. The evaluation and selection of treatment alternatives for a particular pump-and-treat system is based on technical feasibility and costs of achieving remediation goals. Key parameters that influence treatment design and efficacy include flow rate, groundwater geochemistry, properties of contaminants and their concentrations, and discharge requirements. Relationships between these parameters and treatment design are discussed in more detail in AWWA (1990), Noyes (1994), Nyer (1992), and others.

14.6.5 *PUMP-AND-TREAT SYSTEM DESIGN*

The objectives of a pump-and-treat system are site-specific. A remedial objective might be to minimize the total cost required to maintain containment of the contaminated groundwater. Given this objective, installing low permeability barriers to reduce pumping rates might be cost-effective. Unless natural attenuation mechanisms are being relied upon to limit plume migration, hydraulic containment is generally a prerequisite for aquifer restoration. Pump-and-treat design is a compromise among objectives that seek to: (1) reduce contaminant concentrations to cleanup standards, (2) maximize mass removal, (3) minimize cleanup time, and (4) minimize cost (Greenwald, 1993). A pump-and-treat system for aquifer restoration requires a high degree of performance monitoring and management to identify problem areas and improve system design and operation.

Restoration pump-and-treat groundwater flow management involves optimizing well locations, depths, and injection/extraction rates to maintain an effective hydraulic sweep through the contamination zone, to minimize stagnation zones, to flush pore volumes through the system, and to contain contaminated groundwater. Restoration requires that sufficient volume of groundwater be flushed through the contaminated zone to remove both existing dissolved contaminants and those that will continue to desorb from porous media, and/or diffuse from low permeability zones.

Considering linear, reversible, and instantaneous sorption, and neglecting dispersion, the number of pore volumes required to remove a contaminant from a homogeneous aquifer is approximated by the retardation coefficient. Because of simulation of linear sorption, a nearly linear relationship was found to exist between retardation and the duration of pumping (or volume pumped) needed to reach the groundwater cleanup goal. Batch flush models (e.g., Zheng et al., 1992) often assume linear sorption to calculate the number of pore volumes required to reach a cleanup concentration. Though useful for simple systems, the representation of linear, reversible, and instantaneous sorption in contaminant transport models can lead to significant underestimation of pump-and-treat cleanup times (Ward et al., 1987).

Kinetic limitations often may prevent sustenance of equilibrium contaminant concentrations in groundwater (Palmer and Fish, 1992). Such effects occur in situations where contaminant mass transfer to flowing groundwater is slow relative to groundwater velocity. For example, contaminant mass removal from low permeability materials may be limited by the rate of diffusion from these materials into more permeable flowpaths. In this situation, increasing groundwater velocity and pore volume flushing rates beyond a certain point would provide very little increase in contaminant removal rate. Kinetic limitations to mass transfer are likely to be relatively significant where groundwater velocities are high surrounding injection and extraction wells.

The number of pore volumes that must be extracted for restoration is a function of the cleanup standard, the initial contaminant distribution, and the chemical/media phenomena that affect cleanup. Screening-level estimates of the number of pore volumes required for cleanup can be made by modeling and by assessing the trend of contaminant concentration versus the number of pore volumes removed. At many sites, numerous pore volumes (e.g., 10 to 100s) will have to be flushed through the contamination zone to attain cleanup standards. The number of pore volumes withdrawn per year is a useful measure of the aggressiveness of a pump-and-treat operation. Many current systems are designed to remove between 0.3 and 2.0 pore volumes annually (NRC, 1994). Low permeability conditions or competing uses for groundwater may restrict the ability to pump at higher rates.

Poor pump-and-treat design may lead to system ineffectiveness and contaminant concentration tailing. Poor design factors include insufficient pumping rates and improper location of pumping wells. Inadequate location of wells or drains may lead to poor pump-and-treat performance even if the total pumping rate is appropriate. For example, wells placed at the containment area perimeter may withdraw a large volume of clean groundwater from beyond the plume. In general, restoration pumping wells or drains should be placed in areas of relatively high contaminant concentration as well as locations suitable for achieving hydraulic containment.

Well placement can be evaluated by using groundwater flow and transport models that are based on a proper conceptual model of the hydrogeologic system and contaminant distribution. Groundwater flow modeling can be used to assess groundwater and solute velocity distributions, travel times, and stagnation zones associated with alternative pumping schemes. During operation, stagnation zones can be identified by measuring hydraulic gradients, tracer movement, groundwater flow rates, and by modeling analysis.

Several modeling studies have been conducted to examine the effectiveness of alternative extraction and injection well schemes with regard to hydraulic containment and groundwater cleanup objectives (e.g., Ahlfeld and Sawyer, 1990; Haggerty and Gorelick, 1994). Although the optimum extraction/injection scheme depends on site-specific conditions, objectives, and constraints, consideration should be given to guidance derived from simulation studies of pump-and-treat performance. The merits of conventional extraction/injection well schemes, in situ bioremediation, and pump-and-treat enhanced by injecting oxygenated water to stimulate biodegradation for containing and cleaning up a hypothetical naphthalene plume were examined by Marquis and Dineen (1994). Nineteen remediation alternatives were modeled using BIOPLUME II (Rifai et al., 1987), a code that simulates oxygen transport and oxygen-limited biodegradation.

Pulsed or cyclic pumping, with alternating pumping and resting periods, has been suggested as a means to address tailing in concentration levels and a way to increase pump-and-treat efficiency (Borden and Kao, 1992; Keely, 1989). Dissolved contaminant concentrations increase because of diffusion, desorption, and dissolution in slower-moving groundwater during the resting phase of pulsed pumping. Once pumping is resumed, groundwater with higher concentrations is removed, thus increasing the rate of mass removal during active pumping. Because of slow mass transfer from immobile phases to groundwater, however, contaminant concentrations decline with continued pumping until the next resting phase begins.

14.6.6   PERFORMANCE EVALUATION

Pump-and-treat performance is monitored by measuring hydraulic heads and gradients, groundwater flow directions and rates, pumping rates, contaminant distributions in groundwater and the aquifer, quality of pumped water, and treatment system effluent. These data are evaluated to interpret pump-

and-treat capture zones, the number of pore volumes pumped, contaminant transport and removal, and to evaluate system operation. Cohen et al. (1994) provide a detailed guidance on methods for monitoring pump-and-treat performance.

Restoration progress can be assessed by comparing the rate of contaminant mass removal to estimates of the dissolved and/or total contaminant mass-in-place. If the rate of contaminant mass extracted approximates the rate of dissolved mass-in-place reduction, then the contaminants removed by pumping are primarily derived from the dissolved phase. Conversely, a contaminant source (i.e., sorbed contaminants, or a continuing release) is indicated where the mass removal rate greatly exceeds the rate of dissolved mass-in-place reduction. Site characterization data should be re-evaluated to determine if source removal, better containment, or pump-and-treat system modifications, could improve pump-and-treat performance.

Time duration to remove dissolved mass can be projected by extrapolating the curve representing the trend of the cumulative mass removed. Future concentration tailing, however, may extend the extrapolated cleanup time. If the mass removal trend indicates a significantly greater cleanup duration than estimated originally, system modification may be necessary. The effect of pump-and-treat system modifications will be evidenced by the continuing mass removal rate and cumulative mass removed trends. Progress inferred from mass removal rates can be misleading, however, where NAPL and solid phase contaminants are present (e.g., the mass removed will exceed the initial estimate of dissolved mass-in-place). Interpretation suffers from the high degree of uncertainty associated with estimating NAPL or solid contaminant mass-in-place. Stabilization of dissolved contaminant concentrations while mass removal continues is an indication of NAPL or solid phase contaminant presence. Methods for evaluating the potential presence of NAPL are provided by Cohen and Mercer (1993), Feenstra et al. (1991), and Newell et al. (1995).

Mass removal rates are also subject to misinterpretation where dissolved contaminant concentrations decline rapidly because: (1) mass transfer rate limitations to desorption, NAPL or precipitate dissolution, or matrix diffusion; (2) dewatering a portion or all of the contaminated zone; (3) dilution of contaminated groundwater with clean groundwater flowing to extraction wells from beyond the plume perimeter; or (4) the removal of a slug of highly contaminated groundwater. Contaminant concentration rebound will occur if pumping is terminated prematurely in response to these conditions.

The projected cleanup time is site-specific and varies widely depending on contaminant and hydrogeologic conditions and the cleanup goals. Estimating cleanup time is complicated by difficulties in quantifying the initial contaminant mass distribution and processes that limit cleanup. Guidance for estimating groundwater restoration times using batch and continuous flushing models is provided by USEPA (1988b). These models are based on the following assumptions: (1) zero-concentration influent water displaces contaminated groundwater from the contamination zone by advection only; (2) the clean groundwater equilibrates instantaneously with the adsorbed contaminant mass; (3) the sorption isotherm is linear; and (4) chemical reactions do not affect the sorption process. The estimated restoration time obtained by using these simplified models can be misleading (Zheng et al., 1991, 1992). Although more sophisticated modeling techniques are available (i.e., contaminant transport models), their application often suffers from data limitations, resulting in uncertain predictions.

14.6.7 LIMITATIONS OF PUMP-AND-TREAT TECHNOLOGIES

Experience with the pump-and-treat systems during the past two decades indicates that their ability to reduce and maintain dissolved contaminant concentrations below cleanup standards in reasonable time frames is hindered at many sites because of complex hydrogeologic conditions, contaminant chemistry factors, and inadequate system design (Cohen et al., 1994; Keely, 1989; Mercer et al., 1990; NRC, 1994). Hydrogeologic conditions that confound groundwater cleanup include the presence of strong heterogeneity and anisotropy, low permeability formations, and fractured bedrock. Chemical processes that cause contaminant concentration tailing and rebound during and after pump-and-treat operation, respectively, and thereby impede complete aquifer restoration, include: (1) the presence and slow dissolution of nonaqueous phase liquids (NAPLs); (2) contaminant partitioning between groundwater and solid matrix; and (3) contaminant diffusion into low permeability regions that are inaccessible to flowing groundwater (Cohen et al., 1997). These limitations may render restoration using only conventional pump-and-treat technology impracticable at some sites.

ACKNOWLEDGMENT

This text is based on the previously published reports by Cohen and others (1994, 1997), Keely (1989), and Mercer et al. (1990).

REFERENCES

Ahlfeld, D. P., and Sawyer, C. S., "Well Location in Capture Zone Design Using Simulation and Optimization Techniques," *Groundwater*, 28(4): 507–512, 1990.

Aller, L., Bennett, T. W., Hackett, G., Petty, R. J., Lehr, J. H., Sedoris, H., and Nielson, D. M., *Handbook of Suggested Practices for the Design and Installation of Groundwater Monitoring Wells*, National Water Well Association, Dublin, OH, 1989.

AWWA, *Water Quality and Treatment*, American Water Works Association, McGraw-Hill, Inc., New York, NY, 1990.

Bair, E. S., Springer, A. E., and Roadcap, G. S., "Delineation of Travel Time-related Capture Areas of Wells Using Analytical Flow Models and Particle Tracking Analysis," *Groundwater*, 29(3): 387–397, 1991.

Blandford, T. N., Huyakorn, P. S., and Wu, Y., *WHPA—A Modular Semi-Analytical Model for the Delineation of Wellhead Protection Areas*, USEPA Office of Drinking Water and Groundwater, Washington, D.C., 1993.

Borden, R. C., and Kao, C. M., Evaluation of groundwater extraction for remediation of petroleum- contaminated aquifers, *Water Environ. Res.*, 64(1): 28–36, 1992.

Boulding, J. R., *Practical Handbook of Soil, Vadose Zone, and Groundwater Contamination Assessment, Prevention, and Remediation*, Lewis Publishers, Boca Raton, FL, 1995.

Bureau of Reclamation, *Groundwater Manual*, U.S. Government Printing Office, Washington, D.C., 1995.

CCEM, *Horizontal Drilling Survey Results*, Colorado Center for Environmental Management for the Department of Energy Office of Technology Development, Denver, CO, 1996.

CCME, "Subsurface Assessment Handbook for Contaminated Sites," *Canadian Council of Ministers of the Environment Report*, CCME EPC-NCSRP-48E, 1994.

Cohen, R. M., and Mercer, J. W., *DNAPL Site Evaluation*, C. K. Smoley Press, Boca Raton, FL, 1993.

Cohen, R. M., Vincent, A. H., Mercer, J. W., Faust, C. R., and Spalding, C. P., "Methods for Monitoring Pump-and-treat Performance," EPA/600/R-94/123, USEPA, NRMRL, SPRD, R. S. Kerr Environmental Research Center, Ada, OK, 1994.

Cohen, R. M., Mercer, J. W., Greenwarl, R. M., and Beljin, M. S., "Design Guidelines for Conventional Pump-and-Treat Systems," EPA/540/S-97/504, USEPA, R. S. Kerr Environmental Research Center, Ada, OK, 1997.

Driscoll, F. G., *Groundwater and Wells*, Johnson Division, UOP, St. Paul, MN, 1986.

Feenstra, S., Mackay, D. M., and Cherry, J. A., "A Method for Assessing Residual NAPL Based on Organic Chemical Concentrations in Soil Samples," *Groundwater Monitoring Review*, 11(2): 128–136, 1991.

Fitts, C. R., "Well Discharge Optimization Using Analytic Elements," *Groundwater*, 32(4): 547–550, 1994.

Gailey, R. M., and Gorelick, S. M., "Design of Optimal, Reliable Plume Capture Schemes: Application to the Gloucester Landfill Groundwater Contamination Problem," *Groundwater*, 31(1): 107–114, 1993.

Gorelick, S. M., Freeze, R. A., Donohue, D., and Keely, J. F., *Groundwater Contamination Optimal Capture and Containment*, Lewis Publishers, Boca Raton, FL, 1993.

Greenwald, R. M., "MODMAN—An Optimization Module for MODFLOW, Version 2.1," *Documentation and User's Guide*, GeoTrans, Inc., Sterling, VA, 1993.

Grubb, D. G., and Sitar, N., "Evaluation of Technologies for In-Situ Cleanup of DNAPL Contaminated Sites," EPA/600/R-94/120, USEPA, NRMRL, SPRD, R. S. Kerr Environmental Research Center, Ada, OK, 1994.

Haggerty, R., and Gorelick, S. M., "Design of Multiple Contaminant Remediation: Sensitivity to Rate-Limited Mass Transport," *Water Resources Research*, 30(2): 435–446, 1994.

Haley, J. L., Hanson, B., Enfield, C., and Glass, J., "Evaluating the Effectiveness of Groundwater Extraction Systems," *Groundwater Monitoring Review*, 11(1): 119–124, 1991.

Javandel, I., Doughty, C., and Tsang, C. F., *Groundwater Transport: Handbook of Mathematical Models*, American Geophysical Union Water Resources Monograph No. 10, Washington, D.C, 1984.

Javandel, I., and Tsang, C. F., "Capture-zone Type Curves: A Tool for Aquifer Cleanup," *Groundwater*, 24: 616–625, 1986.

Kaback, D. S., Looney, B. B., Corey, J. C., Wright, L. M., and Steele, J. L., "Horizontal Wells for In Situ Remediation of Groundwater and Soils," *Proceedings of the Third National Outdoor Action Conference on Aquifer Restoration, Groundwater Monitoring, and Geophysical Methods*, NWWA, Dublin, OH, pp. 121–135, 1989.

Keely, J. F., "Performance Evaluation of Pump-and-treat Remediations," EPA/540/4-89-005, USEPA, NRMRL, SPRD, R. S. Kerr Environmental Research Center, Ada, OK, 1989.

Kruseman, G. P., and deRidder, N. A., *Analysis and Evaluation of Pumping Test Data*, International Institute of Land Reclamation and Improvement, Bulletin 11, 2nd ed., Wageningen, The Netherlands, 1990.

Mackay, D. M., and Cherry, J. A., "Groundwater Contamination: Pump-and-treat Remediation," *Environmental Science and Technology*, 23(6): 620–636, 1989.

Marquis, Jr., S. A., and Dineen, D., "Comparison Between Pump and Treat, Biorestoration, and Biorestoration/ pump and Treat Combined: Lessons from Computer Modeling," *Groundwater Monitoring and Remediation*, 14(2): 105–119, 1994.

Mercer, J. W., Skipp, D. C., and Giffin, D., "Basics of Pump-and-treat Groundwater Remediation," EPA/600/8-90-003, USEPA, NRMRL, SPRD, R. S. Kerr Environmental Research Center, Ada, OK, 1990.

Morgan, J. H., "Horizontal Drilling Applications of Petroleum Technologies for Environmental Purposes," *Groundwater Monitoring Review*, 12(3): 98–101, 1992.

NRC, *Alternatives for Groundwater Cleanup*, National Research Council, National Academy Press, Washington, D.C., 1994.

Newell, C. J., Acree, S. D., Ross, R. R., and Huling, S. G., "Light Nonaqueous Phase Liquids," EPA/540/S- 95/500, USEPA, NRMRL, SPRD, R. S. Kerr Environmental Research Center, Ada, OK, 1995.

Noyes, R., *Unit Operations in Environmental Engineering*, Noyes Publications, Park Ridge, NJ, 1994.

Nyer, E. K., *Groundwater Treatment Technology*, Van Nostrand Reinhold, New York, NY, 1992.

Palmer, C. D., and Fish, W., "Chemical Enhancements to Pump-and-treat Remediation," *USEPA Groundwater Issue Paper*, EPA/540/S-92/001, USEPA, NRMRL, SPRD, R. S. Kerr Environmental Research Center, Ada, OK, 1992.

Pollock, D. W., *User's Guide for MODPATH, MODPATH-PLOT, Version 3: A Particle Tracking Post-Processing Package for MODFLOW*, U.S. Geological Survey Open-File Report 94-464, USGS, Reston, VA, 1994.

Pyne, R. D. G., *Groundwater Recharge and Wells, A Guide to Aquifer Storage Recovery*, Lewis Publishers, Boca Raton, FL, 1995.

Rifai, H. S., Bedient, P. B., Borden, R. C., and Haasbeek, J. F., "BIOPLUME II—Computer Model of two-Dimensional Contaminant Transport Under the Influence of Oxygen Limited Biodegradation in Groundwater, Version 1.0," National Center for Groundwater Research, Rice University, Houston, TX, 1987.

Sara, M. N., *Standard Handbook for Solid and Hazardous Waste Facility Assessment*, Lewis Publishers, Boca Raton, FL, 1994.

Schmelling, S. G., and Ross, R. R., "Contaminant Transport in Fractured Media: Models for Decision Makers," USEPA Groundwater Issue Paper, EPA/540/4-89/004, USEPA, NRMRL, SPRD, R. S. Kerr Environmental Research Center, Ada, OK, 1989.

Shafer, J. M., "Reverse Pathline Calculation of Time-Related Capture Zones in Nonuniform Flow," *Groundwater*, 25(3): 282–289, 1987.

Tiedeman, C., and Gorelick, S. M., "Analysis of Uncertainty in Optimal Groundwater Contaminant Capture Design," *Water Resources Research*, 29(7): 2139–2153, 1993.

USEPA, "Guidance for Conducting Remedial Investigations and Feasibility Studies under CERCLA," EPA/540/ G-89/004, USEPA, Washington, D.C., 1988a.

USEPA, "Guidance on Remedial Actions for Contaminated Groundwater at Superfund Sites," EPA/540/G-99/ 003, OSWER Directive 9283.1-2, USEPA, Washington, D.C., 1988b.

USEPA, "Guide for Conducting Treatability Studies Under CERCLA," EPA/540/2-89/058, USEPA, Office of Research and Development, Cincinnati, OH, 1989.

USEPA, "Site Characterization for Subsurface Remediation," Seminar Publication, EPA/625/4-91/026, USEPA, Center for Environmental Research Information, Cincinnati, OH, 1991.

USEPA, *Evaluation of Groundwater Extraction Remedies*, USEPA, Office of Emergency and Remedial Response, Washington, D.C., 1992.

Ward, D. S., Buss, D. R., Mercer, J. W., and Hughes, S. S., "Evaluation of a Groundwater Corrective Action at the Chem-Dyne Hazardous Waste Site using a Telescopic Mesh Refinement Modeling Approach," *Water Resources Research*, 23(4): 603–617, 1987.

Zheng, C., *MT3D—A Modular Three-Dimensional Transport Model for Simulation of Advection, Dispersion, and Chemical Reactions of Contaminants in Groundwater Systems*, S. S. Papadopulos and Associates, Bethesda, MD, 1992.

Zheng, C., Bennett, G. D., and Andrews, C. B., "Analysis of Groundwater Remedial Alternatives at a Superfund Site," *Groundwater*, 29(6): 838–848, 1991.

CHAPTER 14
REMEDIATION TECHNOLOGIES

SECTION 14.7

ELECTROKINETIC TREATMENT OF HAZARDOUS WASTES

Walter W. Loo

Dr. Loo is president of Environment and Technology Services, an established pioneer in soil and groundwater cleanup using electrokinetic treatment, bioventing, soil vapor extraction, and ultraviolet light treatment, in combination. He is a leader in bioremediation of chlorinated solvents and petroleum hydrocarbons using the cometabolic processes.

14.7.1 INTRODUCTION

The objective of this chapter is to provide scientists, engineers and layman knowledge of how electrokinetic treatment processes work under different conditions. This chapter will stress field practice experience rather than theory. The reader should have some high school chemistry and physics training. This section will be oriented toward the practical aspects of field applications. Because electrokinetic treatment processes in the environmental field are still evolving and developing, the author will point out which applications are in the research and development stage.

14.7.2 PHILOSOPHY OF SOIL AND GROUNDWATER CLEANUP

The philosophy of soil and groundwater cleanup shall follow the intents of the U.S. EPA laws and regulations such as Safe Drinking Water Act (SDWA), Resources Conservation and Recovery Act (RCRA) and Comprehensive Environmental Response, Compensation and Liability Act (CERCLA).

Any environmental treatment process must be capable or engineered as such to protect and conserve our resources (air, water, soil, and human). Many of the conventional waste treatment technologies cannot meet these objectives. Most conventional waste treatment technologies involve transferring the waste from a location or phase to another. This is true for waste excavation, landfill disposal, incineration, pump-and-treat, thermal desorption, and soil vapor extraction.

Sound waste treatment technologies must be able to destroy the toxic and hazardous chemicals into harmless chemicals or rendered them harmless with the least energy consumption to achieve good economic balance. To the extent possible, all treated soil and groundwater will be recycled for beneficial use. Most organic chemicals can be oxidized into harmless carbon dioxide, water, and inorganic chloride with existing and available innovative waste treatment technologies. Hazardous and toxic metals, which cannot be destroyed, can be treated into insoluble form thus rendered harmless to the environment.

The most difficult media for most waste treatment processes to work in are relatively low-permeability clayey medium and fracture medium. These media often absorb/adsorp the toxic and hazardous wastes spilled below the ground surface. Often more than 90 percent of the waste spilled by weight is attracted to these media. It is most fortunate that the electrokinetic treatment processes are capable of overcoming the difficulties of relatively low-permeability media, achieving or enhancing the oxidation of many toxic and hazardous wastes, with relatively low energy consumption.

14.7.3   *HISTORY OF APPLICATION OF DIRECT CURRENT AND ELECTROKINETIC PROCESSES*

- Direct current electricity is a widely used technology in industrial, commercial, and military applications.
- Direct current is used in the production of aluminum, chlorine gas, and caustic soda.
- Power supply systems in automobile, aircraft and submarines are all direct current based.
- Direct current is also used in space age technology from solar panels for the electrolysis of water to produce hydrogen as fuel and oxygen to breathe.
- For environmental protection, direct current is used in the cathodic protection of steel underground storage tank and pipelines from corrosion and leaks.
- Electrokinetic (EK) processes involve the application of direct current electricity on soil and groundwater in the subsurface. Most pioneers of the EK processes originated or based their work on the geotechnical dewatering of clayey material by Casagrande (1947, 1948, 1952). Other significant pioneer EK applications in oil and gas recovery were developed by Chilingar et al. (1963, 1964, 1968, 1970).
- The application of EK technology in the environmental field was first found in literature published in the 1980s (U.S. EPA, 1997), with most of the applications in the isolation or recovery of metals.
- Van Doren and Bruell (1987) first reported that the EK process destroyed benzene in clay in a laboratory bench scale test.
- Loo (1991) first reported the successful commercial application of EK process as the primary enhancement process for the removal or desorption of chlorinated solvents from a thick clayey soil layer at a defense contractor site closure located in Anaheim, California.
- In 1992, U.S. Army Corp of Engineers listed EK treatment as a viable remedial process for the treatment of hazardous wastes.
- In 1995, the HAZMACON Conference, Santa Clara, California selected the best paper award entitled "Electrokinetic Treatment of Hazardous Wastes in Soil and Groundwater Loo (1995)."
- In 1995, the U.S. EPA (1995) summarized the application and the development of Electrokinetic (EK) treatment processes by various private companies, the U.S. Department of Defense, U.S. Department of Energy, and various universities as potential cost effective treatment of hazardous wastes.
- In 1996, Initiatives (1996) documented that EK had attracted the attention of Dupont, General Electric, and Monsanto in various aspects of research, development, and commercialization of hazardous waste treatment.
- In 1997, U.S. EPA (1997) summarized and updated various EK applications in research, development, and commercial treatment of hazardous waste and radioactive wastes.

14.7.4   *POTENTIAL FOR HAZARDOUS WASTES CLEANUP*

As described in the previous sections, the EK treatment processes are quickly becoming commercially available for the treatment of various hazardous wastes. The EK treatment processes are applicable for almost all priority pollutants with the exception of asbestos and various geologic media.

Treatment Media

The EK processes works effectively for both permeable and relatively low-permeability media. For relatively low-permeability porous media, such as clay and silt, direct current conducts well in the media. The electrical conductivity in clay often is 1000 times more than that in sand, sandstone, limestone, igneous, and metamorphic rocks. The same applies to fractured media where the fractures are often filled with highly conductive clayey and fine grain minerals.

For porous media like sand, the flow of electricity is most likely through the water, which has naturally occurring electrolytic minerals, such as chloride, bicarbonate, nitrate, potassium, and sodium. The water molecules and positively charged ions (cations) will also be "dragged" by the electron flow from the anode toward the cathode and will create a hydraulic mound (high) around the cathode by the induced electrokinetic gradient.

The EK processes works for both the vadose zone and the saturated zone in the subsurface as long as there is adequate moisture in the media. A dry media does not conduct electricity very well.

Wastes Destruction

The application of the direct current in the subsurface may bring about electrochemical reactions that may oxidize some hazardous organic chemicals into carbon dioxide and water, which are relatively harmless to the environment.

Wastes Isolation and Concentration

The application of the direct current in the subsurface may bring about electrolytic segregation, which may isolate and concentrate the cations (such as metals) around the cathode and anions (such as chloride and nitrate) around the anode.

Achievable Cleanup Levels

It is possible to achieve very low levels of concentration cleanup of various hazardous wastes in soil and groundwater by EK processes listed below as examples:

	In Soil	In Groundwater
Petroleum Hydrocarbon (gasoline)	Less than 50 ppm	Nondetect (less than 1 ppm)
Petroleum Hydrocarbon (diesel, Kerosene (jet fuel, motor oil, etc.)	Less than 100 ppm	Nondetect (less than 1ppm)
Benzene	Less than 100 ppb	Nondetect (less than 1 ppb)
Methyl tertiary Butyl Ether (MtBE)	Less than 1 ppm	Nondetect (less than 5 ppb)
Chlorinated Solvents (PCE, TCE)	Less than 100 ppb	Nondetect (less than 5 ppb)
Soluble Lead	Less than 1 ppm STLC	Less than 5 ppm TCLP

14.7.5 REGULATORY AND PERMITTING REQUIREMENTS

There is no real regulation and permiting requirement that applies directly to the application of direct current in the subsurface.When the EK process is applied as an enhancement process for bioventing in the vadose zone, the extracted volatile organic vapor can be recirculated back into the center of the soil plume to avoid an air emission permit.

Because the EK process can make the groundwater migrate back and forth within the aquifer (through in situ mixing), there is no need to pump-and-treat, and thus no discharge permit. Also, when electrolytes and nutrients are added into the subsurface for EK processes, no underground injection control permit is required as long as all electrolytes and nutrients added are nontoxic and nonhazardous.

The EK application in the subsurface is similar in nature to cathodic protection with impressed current and the function of electrical grounding devices, which require no permit to construct and operate. However, the reader must follow strict health and safety practice when installing and operating an EK treatment system. Some general health and safety tips for the EK treatment system are described in one of the following sections.

14.7.6 ELECTROKINETIC TREATMENT PROCESSES

Introduction

The electrokinetic processes that are applicable for the destruction or enhancing the destruction, recovery, or isolation of hazardous wastes are, namely, electro-osmosis, electromigration, and electrochemical processes, respectively.

Electro-Osmosis

Electro-osmosis or Electrokinetic (EK) induced dewatering is a very well established process used in the dewatering and stabilization of clayey foundation of buildings and structures. EK induced dewatering can be used in the treatment of hazardous chemicals in silty and clayey material. The mechanics of the EK induced dewatering process cause imbalance of charge bonds in clayey material, which results in clay compaction or consolidation (dewatering) and chemical desorption. The compaction and desorption processes reduce the cleanup time and are particularly successful in the desorption of hazardous organic chemicals and metals from clayey materials. This helps to improve and shorten typically inefficient pump-and-treat systems, which rely on natural groundwater dilution resulting from very low hydraulic gradient flow.

Electromigration of Water and Contaminant

When direct current is applied in the subsurface, the water molecules and positively charge ions (cations) will be "dragged" by the electron flow from the anode toward the cathode and will create a hydraulic mound (high) around the cathode by the induced electrokinetic gradient. The same applies for relatively low-permeability porous media, such as clay and silt: direct current conducts well in these media. The electrical conductivity in clay often is 1000 times more than that in sand, sandstone, limestone, igneous, and metamorphic rocks. The same applies to fractured media where the fractures are often filled with highly conductive clayey and fine grain minerals.

For porous and fractured media, the flow of electricity most likely goes through the water in the pores, which contain naturally occurring electrolytic minerals, such as chloride, bicarbonate, nitrate, potassium, and sodium. Salts of potassium, sodium, and bicarbonates are naturally occurring surfactants, which emulsify or solubilize oily organic compounds, petroleum hydrocarbons (TPH) in particular. With the EK application, it is common that TPH can be moved or washed within the porous or fracture media by alternating the polarity of the electrodes placed in the subsurface.

The mechanics of the in situ electromigration of water back and forth in the subsurface enhances thorough mixing and recovery of various types of hazardous wastes. This facilitates other EK-related treatment processes to destroy efficiently or immobilize the hazardous wastes in place and render them into harmless by-products with minimum impact to the environment.

Electrolysis

Electrolysis is one of the principal industrial processes used in the production of aluminum, chlorine, metal plating, corrosion protection, and so forth. These processes are very effective when applied in the treatment of hazardous metals and organic compounds in soil, sludge, and water, respectively. Electrolysis can be applied in both permeable and low-permeability media, and can be used as a neutralization process for pH control. It can also be used for isolation or capture of cations at and near the cathode electrode and anions at and near the anode electrode.

Acid extractable organic compounds mobilize toward the anode and base/neutral extractable organic compounds mobilize toward the cathode. Electrolysis may be the primary mechanism for the breakdown of weak bonds of simple organic chemicals related to gasoline such as Methyl tertiary Butyl Ether (MtBE).

No hard data exists as to what kind of chemical bond can be broken down by electrolysis. This will be a major research and development effort to make such a determination. The author has only limited experience with a few compounds in field application.

Electrochemical Oxidation

Electrolysis will also oxidize petroleum hydrocarbons, BTEX, MtBE, and benzene-based organic chemicals, such as PCBs, pesticides, and PAHs with the presence of water. Electrolysis of the water molecule will produce hydrogen around the anode and produce oxygen around the cathode. This is the result of the breakdown of hydrogen hydroxide (water). When the breakdown of the water molecule first occurs, hydroxyl radicals are created. Hydroxyl radicals are very potent oxidants and oxidize many types of organic compounds. Unfortunately, there is no real database as to what kind of organic chemicals can be broken down by electrolysis. This will be another major research and development effort to make this determination. The author has only limited experience with a few compounds in bench scale and field applications.

Electrobiochemical Oxidation

Electrochemical processes can be applied both above ground (ex situ) or in the subsurface (in situ). The following is a list of hazardous chemicals that can be effectively treated by the electrobiochemical oxidation processes:

- Petroleum hydrocarbons, such as gasoline, BTEX, MtBE, diesel, fuel oil
- Benzene-based organic chemicals, such as PAHs, pesticides, and PCBs
- Isolation, capture, and precipitation/stabilization of hazardous metals or cations
- Isolation and capture of common hazardous anions, such as nitrates, and capture and disinfection of pathogenic bacteria

As mentioned in the previous section, electrolysis of the water molecule produces dissolved hydrogen around the anode and dissolved oxygen around the cathode. Therefore, highly dissolved oxygen around the cathode will help and encourage aerobic biotreatment of the organic compounds. Also, electromigration will help to bring the desorbed organic compounds to places that are enriched with oxygen and nutrients.

14.7.7   BENCH SCALE STUDIES

Because the EK treatment processes are still evolving into commercial operations, most bench scale studies are more a statement of facts because of the proprietary nature of the processes being developed and tested. Sometimes, it is difficult to distinguish which particular EK process is actually taking place because the EK treatment processes can happen simultaneously. The author can only document the more obvious EK treatment phenomenon described in the various references cited in the reference listing at the end of this section.

Electro-Osmosis

Electro-osmosis or EK induced dewatering involves the passing of direct current electricity through a saturated clayey medium. The polarization of the charged particles in the clay structure causes a collapse in the charge bond and the interstitial water is squeezed out. The resulting phenomenon is the consolidation or compaction of the clay. The dewatering of the clay causes an overall volumetric change of the clay body. The change can be as much as 50 percent of the original volume.

The demonstration of the EK induced dewatering phenomenon can be conducted on a typical clay core sample contained inside a metallic sleeve. One should avoid bench scale testing of samples saturated with flammable product to avoid accidental ignition and explosion. One can connect the anode to the metallic sleeve and connect the cathode to a metallic nail or narrow tubing which penetrate the center of the clay sample without contacting the sleeve.

When direct current of less than 2 amperes is conducted through the clay sample, one shall observe that within seconds that free water will drip down from the cathodic nail or tubing. This EK induced dewatering phenomenon may last for an hour or so for a 6-in long clay sample until the clay dries up and the current flow decreases accordingly. It is very common to easily recover more than 10 percent of the volume of the clay core sample as water squeezed out of the core sample by this electrokinetic process.

Electromigration of Water

As demonstrated in the previous section, the clay sample was dewatered by electro-osmosis. The water collected from the cathode end is water migrated through the clay toward the cathode by the induced electrokinetic gradient. The results of the demonstration will be more dramatic if a slab of saturated clay sample is being tested because of the presence of more water. One can observe elevated water level at the cathode and depressed water level at the anode.

Electrolysis

When the water captured at the cathode from the above bench experiment is analyzed for chemical properties, one can easily find the following results.

  pH value   8.5 to 11   (high hydroxide)

  Cation and Anion inbalance (more potassium and sodium than chloride)

  The water usually contains dissolved contaminant of concern unless the chemical's bond was broken by the electrolytic process.

One should not perform bench scale testing of samples saturated with flammable products to avoid accidental ignition and explosion.

Electrochemical Oxidation

As demonstrated above, electrochemical oxidation will be obvious when the bench test is conducted on an acrylic sleeve core sample or slab samples where obvious dark petroleum hydrocarbon stain on soil sample will be oxidized (discolored or 'bleached') by the end of the bench test.

Table 14.7.1 presents the before and after test result of a marine clay core sample that underwent an electrokinetic treatment bench test. All petroleum hydrocarbons, including BTEX, were electrochemically oxidized to nondetect after 48 h of EK treatment.

Electrobiochemical Oxidation

It is more difficult to conduct electrobiochemical oxidation bench tests, because it requires larger sized column samples and equipment to collect, detect, and quantify the biomineralization product, carbon

TABLE 14.7.1 Bench Scale Electrochemical Oxidation Test Results Gasoline Contaminated Clayey Soil Pepboys Site 3550 El Cajon Blvd., San Diego, California (Units in mg/Kg)

	Initial	After EK treatment		Destruction efficiency %
		EK-1	EK-2	
TPH as Gasoline	230.00	ND(<5)	ND(<5)	97.82+
Benzene (B)	1.9	ND(<0.005)	ND(<0.005)	99.73+
Toluene (T)	6.3	ND(<0.005)	ND(<0.005)	99.92+
Ethylbenzene (E)	2.8	ND(<0.005)	ND(<0.005)	99.82+
Xylenes (X)	13.0	ND(<0.005)	ND(<0.005)	99.96+

dioxide gas. This type of test makes good research experiments for universities and the U.S. EPA laboratories, which have the capability of large-scale experimentation of soil columns and aqueous phase (saturated sample) containers.

As mentioned in the previous section, electrolysis of the water molecule will produce hydrogen around the anode and produce oxygen around the cathode. Therefore, higher dissolved oxygen around the cathode will help and encourage aerobic biotreatment of the organic compounds. Also, electromigration will help to bring the desorbed organic compounds and the artificially introduced nutrients to places where they will be enriched with oxygen. When the direct current is on, it will also warm up the water between electrodes. This promotes excellent biochemical oxidation conditions in the subsurface and detoxifies the hazardous organic chemicals of concern.

14.7.8 FIELD APPLICATIONS

There are many documented field pilot scale EK remediations of toxic/radioactive metals and hazardous organic chemicals. The individual project descriptions are included in the publicly available reference documents listed below:

U.S. ARMY CORP OF ENGINEERS 1992
Installation Restoration and Hazardous Control Technologies

U.S. EPA 1995   PUBLICATION 542-K-94-007
In-Situ Remediation Technology: Electrokinetics

U.S. EPA 1997   PUBLICATION 402-R-97-006
Electrokinetic Laboratory and Field Processes Applicable to Radioactive and Hazardous Mixed Waste in Soil and Groundwater

The following sections describe the applicability of the electrokinetic processes in the field on specific waste types and host media.

Cation Isolation

Based on the established electroplating technology as an industrial process, the isolation of toxic metals such as lead, chromium, mercury and nickel in soil and groundwater is widely documented. Most of the early applications were developed in Europe during the early 1980s. Acar et al. (1992, 1993) provided very good documentation of the distribution of pH and various metals around the cathode electrode.

In general, less soluble species of various cations concentrate around the cathode and very often precipitate as various oxides. Based on the author's limited experience, this applies to lead (valence +2), chromium (valence +3), uranium (valence +4), boron (valence +3), and selenium (valence 0 and +2). Also, less soluble species of salts of potassium, sodium, calcium, and magnesium concentrate around the cathode and sometimes precipitate as minerals on the surface of the cathode. It is interesting to note that gram positive charged bacteria in groundwater may be induced by electrokinetic migration toward the cathode, which may aid in situ biotreatment processes.

Anion Isolation

Very little attention has been paid to the isolation anions near the anode. In particular, the attraction of nitrate and chloride to the anode (Loo and Wang, 1991) will demonstrate that the electrokinetic processes can be applicable to arrest the widespread nitrate and nitrite problems associated with the leaching of farming fertilizer salts into soil and groundwater. Also, electrokinetic capture of chloride at the anode may reverse or prevent the salt water intrusion and the salt upconing processes in damaged groundwater bodies or aquifers along coastal areas.

In general, soluble chlorides of potassium and sodium will disassociate under the electrolytic process and chlorine may concentrate and liberated as free chlorine gas. This is a known industrial process for the production of chlorine gas.

Also soluble species of various cations salts will concentrate around the anode. Based on the author's limited experience, this applies to lead (valence +4), chromium (valence +6), uranium (valence +6) and selenium (valence +4).

Electrochemical Oxidation

As demonstrated above, electrochemical oxidation will be obvious when the bench test is conducted on an acrylic sleeve core sample or slab sample where obvious dark petroleum hydrocarbon stain on soil sample will be oxidized (discolored or "bleached") by the end of the bench test. In the field, direct electrochemical oxidation is often hard to distinguish from other electrokinetic processes. Electrolysis may be the primary mechanism for the breakdown of weak bonds of simple organic chemicals related to gasoline such as Methyl tertiary Butyl Ether (MtBE).

Electrolysis of the water molecule produces hydrogen around the anode and produces oxygen around the cathode. This is the result of the breakdown of hydrogen hydroxide (water). When the breakdown of the water molecule first occurs, hydroxyl radicals are created. Hydroxyl radicals are very potent oxidants and will oxidize many types of organic compounds. Electrolysis will oxidize petroleum hydrocarbons, BTEX, MtBE, and benzene-based organic chemicals such as PCBs, pesticides, and PAHs with the presence of water.

Electrobiochemical Oxidation of Organic Chemicals

As mentioned previously, electrolysis of the water molecule produces hydrogen around the anode and produces oxygen around the cathode. Therefore, higher dissolved oxygen around the cathode helps and encourages aerobic biotreatment of the organic compounds. Also, electromigration will help to bring the desorbed organic compounds and artificially introduced nutrients to places enriched with oxygen. When the direct current is on, it will also warm up the water between electrodes. This will promote excellent biochemical oxidation conditions in the subsurface and detoxify the hazardous organic chemicals of concern.

It is interesting to note that gram-positive charged bacteria in groundwater may be induced by electrokinetic migration toward the cathode, which may aid in situ biotreatment processes.

14.7.9 CATION TREATMENT

Metals

Acar et al. (1993) and U.S. EPA (1995, 1997) papers documented the concentration of potassium, sodium, lead, chromium, mercury, and nickel near the cathode results from induced cation migration by the direct current application. The same applies to aluminum, calcium, copper, iron, manganese, and other metals.

Some of the metals may form oxides and precipitate at or near the cathode. Some may plate onto the cathode. It should be noted that highly soluble chromium ($+6$ valence) oxide may concentrate near the anode, and less soluble chromium (0, $+3$ valence) may concentrate near the cathode. The control of dissolved sodium and calcium salts migration by electrokinetic process may reverse the saline water intrusion along coastal/island aquifers. In summary, the beneficial effect is the reduction in volume of water impacted by the metals by electrokinetically concentrating the metals near the cathode, thus, controlling the spreading and migration of the metals to undesirable places.

Radionuclides

Acar et al. (1992) and U.S. EPA (1997) present the findings of the electrokinetic migration of uranium, thorium, and radium radionuclides to the cathode. The phenomenon of electrokinetic migration toward the cathode also applies to other radionuclides, such as tritium, radon, plutonium, and others. It should be noted that highly soluble radionuclide, such as uranium ($+6$ valence) oxide may concentrate near the anode and less soluble uranium ($+4$ valence) may concentrate near the cathode.

Major research and development efforts on the migration of various radionuclides are in progress at various national laboratories and nuclear power stations. The results of this research are seldom published because of the sensitivity to the public and the news media.

Other Cations

The author has first-hand experience on the electrokinetic migration of selenium and boron in the subsurface, which applies to irrigation drainage problems at two separate sites, Loo (1996). The purpose of the electrokinetic application is to control the leached selenium and boron from entering surface drainages in irrigated farming areas. This is a major problem at the Kesterson Reservoir site, Central Valley, California where migratory birds were impacted by the leached selenium from irrigation farming.

It should be noted that highly soluble selenium ($+4$ valence) oxide, boron ($+3$ valence borate), and nitrate may concentrate near the anode and less soluble selenium (0, $+2$ valence) may concentrate near the cathode. The proposed design is to precipitate and removal of the less soluble selenium (0, $+2$ valence) and native boron at shallow cathode stations along irrigation drainage ditches and of electrokinetic induced migration of highly soluble selenium ($+4$ valence), boron ($+3$ valence) and nitrate to deeply seated anodes. Also, the less soluble minerals of potassium, sodium, calcium, and magnesium may be removed by precipitation at cathode stations from the irrigation drainage water. If the removal of these minerals is significant enough to reduce the total dissolved solids to below 1000 mg/l, it is possible to recycle farm wastewater for reuse, thus attaining the water conservation goals. Large-scale field applications are awaiting federal and state funding. Other toxic cations, such as arsenic, may also be treated by the electrokinetic processes but further research and development is needed.

14.7.10 ANION TREATMENT

Various anions, such as nitrate, nitrite, chloride, sulfate, and carbonate/bicarbonate, which are regulated parameters of the Safe Drinking Water Standards, may be treated by electrokinetic induced deionization process in aquifers. The electrokinetic-controlled migration of nitrate and chloride has been demonstrated in the field.

Nitrate is the most widespread contamination to groundwater in the world. This is not controlled rigorously because of the nonpoint source nature of the discharge. Nitrogen-based fertilizers, when not used up by farm crops, all ultimately leach into groundwater as nitrate. Other point source dischargers, such as feed lots, pig farms, chicken farms, and dairy farms are not usually regulated because of strong agricultural interests, which make these dischargers exempt from environmental regulations. Electrokinetic treatment of nitrate by concentrating nitrate at anodic wells may be the most economical means of treatment. In addition, the concentrated nitrate water can be pumped out for reuse.

As mentioned in the previous section, the control of sodium and chloride ion migration by electrokinetic process may prevent saline water intrusion along coastal/island aquifers and may also restore such damaged aquifers. The application of this electrokinetic process will require large-scale development and demonstration. Another toxic anion cyanide also may be effectively concentrated by the electrokinetic processes, but further research and development is needed.

14.7.11   *PETROLEUM HYDROCARBON TREATMENT*

Electrokinetic treatment can be effective on the detoxification of petroleum hydrocarbons; gasoline, BTEX, MtBE, diesel, waste oil, polycyclic aromatic hydrocarbons (PAHs), and so forth. The petroleum hydrocarbons can be detoxified by electrolysis and electrochemical oxidation of organic chemicals related to petroleum hydrocarbons.

EK processes can enhance remedial processes, such as soil vapor extraction, bioreclamation, bioventing, and pump-and-treat projects.

The following sections present five types of EK remedial applications:

Electrokinetic Enhanced Soil Bioventing

The conventional application of soil vapor extraction of volatile organic chemicals (VOCs) cannot overcome low-permeability soil media to render successful completion of a project. The electrokinetic process can enhance the desorption of water and contaminants from clay by breaking down the clay layers electrically charged bonds. This will make the VOCs available for contact and treatment. The major drawback for soil vapor extraction is air emission, which is a physical phase transfer process instead of destruction.

The author has successfully applied an EK enhanced closed loop bioventing process to destroy the VOCs without air emission at sites with clayey soil.

Electrokinetic Enhanced Product Recovery

Dr. George Chilingar of University of Southern California is the pioneer of enhanced oil recovery by the electrokinetic processes. His works are documented in Chilingar et al. (1963, 1964, 1968, 1970).

The author pursued this EK application at a site in Hawthorne, California, where a free-floating gasoline product was detected as thick as 10 ft above the groundwater table. Direct current was applied at less than 5 ampere between wells and caused the increase of the floating gasoline thickness at the anodic well (depressed water table) and the decrease of floating gasoline product at the cathodic well (raised water table). The EK process caused a dramatic effect on the fluctuation of fluid in wells, without fluid withdrawal or recharge. This potentially could replace the conventional product recovery system by depressing the water table through pumping and by eliminating the disposal of a large volume of waste water. The project is still in progress.

In addition, a full-scale EK enhanced free product (gasoline and diesel) recovery feasibility study is in progress at a site in Barstow, California, in cemented low-permeability river alluvium. The feasibility evaluation includes the evaluation of water level and free product fluctuations, changes in dissolved gasoline chemicals (MtBE, BTEX), and inorganic water chemistry.

Electrochemical Oxidation

It is difficult to distinguish which electrokinetic process is responsible for the destruction of petroleum hydrocarbons. As shown on the bench test on soil sample from the PepBoys Site, electrochemical oxidation is responsible for the destruction of the petroleum hydrocarbons. For all bioventing projects described in the previous section, the electrochemical oxidation process is responsible for a portion of the destruction of the petroleum hydrocarbons, other than bioventing alone.

Electrobiochemical Oxidation

Electrobiochemical oxidation applies to aqueous phase biotreatment of petroleum hydrocarbons in soil and groundwater. The electrokinetic processes are responsible for the distribution of nutrients and oxidants in impacted soil and groundwater. It can be used conjunctively with bioventing in the vadose zone.

At all EK enhanced sites, food additive nutrients and oxidants were introduced as electrolytes by electrokinetic induced migration through wells (horizontal migration) and infiltration galleries (vertical migration) to penetrate into the clayey matrix of soil and groundwater. The addition of nontoxic and nonhazardous food additives and nutrients and oxidant into the subsurface is not regulated by underground injection control regulation. Petroleum hydrocarbons, such as BTEX in soil and groundwater, were treated to nondetect levels at most sites.

Electrolysis of MTBE and Benzene

The two most resilient dissolved petroleum hydrocarbons in the environment are MtBE and benzene because of their higher solubility in water and low cleanup levels, which cannot be achieved by conventional remedial treatment technologies. The gasoline additive MtBE is highly soluble in water and usually migrates furthest away downgradient from the spill location. Air stripping treatment of MtBE is not an effective treatment option. There are reports that MtBE in the air solubilizes in airstripper blowdown water. Furthermore, MtBE is not readily biodegradable in the subsurface.

The author first treated MtBE in groundwater accidentally (not by design) in 1997 at the Cadillac Site, Northridge, California, Loo (1998), while treating dissolved BTEX using the electrobiochemical aqueous phase treatment. It was discovered that MtBE soon "disappeared" after the startup of the electrokinetic-enhanced biotreatment. The MtBE in groundwater was treated to nondetect in less than three months of treatment. It also appeared that benzene concentration was decreasing at a slower rate at various monitoring wells. The author cannot confirm that electrolysis of MtBE and benzene actually took place. It was surprising to document the MtBE disappearance act at the Cadillac Site, because at the time, the oil company sponsored numerous seminars pointing to no effective remedy to MtBE in the environment.

The author applied the same electrobiochemical treatment technique for MtBE and BTEX at the former Texaco Site in Long Beach, California, in 1997, Loo (1998). The dissolved benzene and MtBE in groundwater were treated to nondetect in two monitoring wells outside of the zone of influence of the electrobiochemical treatment area, located at the former underground storage tank pit area. This indicated that the only thing that can influence these peripheral monitoring wells is the direct current flow field and is the first confirmation of electrolytic breakdown of dissolved MtBE and benzene in groundwater. The all nondetect MtBE and benzene performance in the vadose zone soil at the tank pit area is also a first.

14.7.12   CHLORINATED SOLVENTS

The conventional remedial treatment technology of chlorinated solvent in soil is soil vapor extraction. The newer and innovative remedial treatment of chlorinated solvents in groundwater is cometabolic biotreatment. None of the these are effective when clayey material is involved. Electrokinetic processes excel in the desorption of the contaminants from the clay and distribution of nutrient, oxidant, and cosubstrate in the soil and groundwater.

Electrokinetic Enhanced Soil Aeration

The first application of electrokinetic processes in the field is at the Northrop ESD Site, Anaheim, California, in 1991, Loo et al. (1991). The project involved the cleanup of 150,000 tons of soil impacted chlorinated solvents, which extended to a depth of about 70 ft. In particular, the chlorinated solvents appeared concentrated at a silty clay layer located at 35 to 40 ft below grade. Northrop desired an expeditious case closure but several remedial consultants believed the soil vapor extraction treatment would take more than two years. The author proposed soil vapor extraction with hot air injection and electrokinetic enhancement to desorb the chlorinated solvents from the clayey layer. The author demonstrated to Northrop in the laboratory that water and chlorinated solvents can be "squeezed out" from the clay core in a brass sample tube. It was then necessary to complete the EK enhanced soil vapor extraction treatment of the chlorinated solvents in soil in less than 90 days. The average residual chlorinated solvents in soil at the end of treatment was at 0.1 mg/kg (ppm).

Electrokinetic Enhanced Cometabolic Biotreatment

The Good Guys site, Emeryville, California, is the cradle of the electrokinetic processes and cometabolic biotreatment process development, Loo (1993). The project involved a narrow alley way impacted by chlorinated solvents and gasoline-related chemicals in soil and groundwater in the bay mud, a soft clay layer.

The highly impacted soil was excavated and treated above ground by a heat enhanced cometabolic process (Loo, 1991) to less than 3 mg/kg (ppm). The residual chlorinated solvents in soil were remediated by a soil vapor extraction system. The chlorinated solvents in the spent granular activated carbon used for air emission control was remediated by electrokinetic enhanced cometabolic biotreatment using glucose as the cosubstrate. The chlorinated solvents in the shallow groundwater (bay mud) was remediated by the electrokinetic enhanced aqueous phase cometabolic biotreatment to nondetect in two years.

14.7.13   CHLORINATED ORGANIC CHEMICALS

The author demonstrated in a laboratory bench scale test that pesticides and polychlorinated biphenyls (PCBs) can be treated by either electrolysis or electrochemical oxidation, Loo and Wang (1991). All showed that inorganic chloride was the mineralized product from the chlorine in these chemicals. The pesticides tested were DDT, Toxaphene, and Endosulfan. The PCB tested was PCB Arochlor 1016. Short of any safe effective treatment methods, most PCBs and Pesticides were remediated by excavation and disposal to a more secured landfill. The author is not aware of any application of EK treatment of PCBs and pesticides in the field. However, electrokinetic treatment of chlorinated organic chemicals may be a cost-effective treatment alternative worth trying.

14.7.14   DESIGN AND IMPLEMENTATION OF EK SYSTEMS

Feasibility tests, field survey, bench scale, and pilot tests provide the basis for the design of the electrokinetic components of the remedial system. Electrokinetic remedial components are electrodes, electrolyte, wiring, and power supply. Remedial calculations are important for the implementation schedule of the EK remedial components. The following sections provide the requirements for the EK design and implementation.

Feasibility Tests

The basic electrokinetic feasibility test parameters are listed as follows:

- Field pH, conductivity and dissolved oxygen for water
- Laboratory pH and conductivity for soil
- Chemical oxygen demand (COD)
- Contaminant concentration in soil above and below the water table
- Contaminant concentration in groundwater
- Selective anions and cations analysis

Representative samples must be obtained at the center of the plume where the concentration of the contamination is the highest. Also, representative samples must be obtained at the peripheral monitoring wells up-gradient, down-gradient and cross-gradient.

Field Surveys

The main purpose of the in situ direct current survey is to detect any anisotropic (uneven and bias) flow directions. The survey also provides clues to unidentified buried metallic lines and structures that may require cathodic protection. Field direct current flow shall be conducted with combinations in pairs of all available wells located inside and outside of the plume. For example, for four wells there shall be six survey measurements. For 6 wells, there shall be 15 survey measurements. The survey shall be conducted with fixed voltage (electrical potential) and shall measure the direct current flow in amperes.

The DC flow survey requires a 110 volts AC power with a direct current converter rated for 50 to 100 DC volts (range) and a maximum 10 DC amperes. Wires used must be insulated to prevent circuit shortage. The selection of electrode shall follow the guidance described in the following section on electrodes selection.

Plume Quantification

The soil, free product, and dissolved groundwater plume must be mapped to the required cleanup criteria laterally and vertically. The volume of the plumes must be estimated and the weight of the bulk volume of the plume must be determined. Plume bulk volume times bulk density equals plume weight.

The effective porosity of the porous or fracture media must be determined or estimated so that the fluid volume within the pore space within the plume can be estimated. Bulk volume of the plume times the effective porosity of a saturated media equals one pore volume of the plume (groundwater or free product). Contaminant weight in plume must be determined. Concentration of contamination in plume times plume weight equals contaminant weight.

Chemical oxygen demand (COD) weight in plume must also be determined.

Electrokinetic System Components

The electrokinetic system components may include but is not limited to electrodes, electrolytes, wiring, and power supply. The description and selection of these components are described in the following sections.

Electrodes.   An electrode can be made of any good conducting material. The author advises that all electrodes used for environmental cleanup be relatively inert and should not contain hazardous and toxic chemicals. Typically, copper and stainless steel electrodes should be avoided. The most common electrodes used are made of iron, black steel, carbon, and graphite. The more exotic electrodes used are gold, platinum, and titanium. Electrodes can be ordered in any shape and form to fit particular applications. The most common form is a round solid rod or tube measuring 1 to 10 ft in length.

Electrolytes.   The author prefers the use of food additive chemicals as electrolytes to increase the flow of direct current in the ground. It should be noted that electrolytes should be applied on an as-needed basis. Soil and groundwater may contain enough electrolyte minerals to provide nice direct current flow. The application of electrolytes is sometimes necessary for good maintenance of electrodes to prevent corrosion at the anode and mineral scaling at the cathode. Potassium and sodium chloride are not considered as environmentally friendly electrolytes, and electrolysis of potassium and sodium chloride will yield toxic chlorine gas, which may create unsafe and unhealthy situations.

Wiring.   All electrical wiring must be insulated and rated for at least 30 amperes. The length of wires to cathodes and anodes need not be equal but multiple wires to either cathode or anode must be equal in length and size to prevent short circuiting. It is often very common for each circuit wire to carry less than 2 amperes of DC current.

Power Supply.   Direct current power supplies often draw from 110 volt AC or 220 volt AC power sources. The common names for the DC power converter include DC testing power supply, DC welding units, automobile battery with proper recharging source, and electrical solar panels. The author advises not to draw more than 10 DC amperes from any converter sources to prevent overheating situations.

Remedial Calculations

The effective porosity of the porous or fracture media must be determined or estimated so that the fluid volume in the pore space within the plume can be estimated. Bulk volume of the plume times the effective porosity of a saturated media equals one pore volume of the plume (groundwater or free product). Contaminant weight in plume must be determined. Concentration of contamination in plume times plume weight equals contaminant weight. The chemical oxygen demand (COD) weight in plume must also be determined.

The amount of DC current requirement to produce the desired electrochemical oxidation by weight must be calculated. Often, this can be calibrated in the field while conducting feasibility testing. The desired reduction in concentration of certain toxic and hazardous chemicals in the subsurface over time versus DC current used will provide the empirical projection on the performance of the electrokinetic processes. This provides a schedule of completion at or below regulatory cleanup concentration standards.

Applied Voltage Time Current Equals Watt (Power).   The economics of power consumption can be measured in kilowatt hours (Kwh) and each Kwh equals to US $0.10 to $0.20 depending on the the cost of power at the location. Normally, the cost of electricity is insignificant compared to overall remedial project cost.

14.7.15 *HEALTH AND SAFETY ASPECTS OF ELECTROKINETIC TREATMENT*

All OSHA health and safety regulation on hazardous waste training and handling shall be observed at all times. All electrical control boxes, panels, switches, connectors, and wiring must be installed by a well-trained person or licensed electrician or equivalent.

The following is a list of minimum common sense health and safety rules involving electrokinetic treatment applications, which should be incorporated into any site health and safety plan:

Electrical Shock Hazards Prevention

- All electrical circuits must be shut down prior to maintenance of the remedial system, electrodes or visitation by regulators or visitors.
- No direct current over 10 amperes and 100 DC volts should be used for any electrokinetic treatment application.
- The anode and cathode panel control switch box cover shall be locked or securely screwed down.

Fire and Explosion Hazards Prevention

- All electrical circuit and equipment must be explosion proof rated when volatile organic chemicals are involved.
- Fire extinguishers shall be installed at two separate entrance and escape routes/doors/gates.
- No smoking signs shall be posted within the treatment area.
- All electrical circuits must be shut down prior to maintenance of the remedial system, electrodes or visitation by regulators or visitors.
- No electrode surface shall be exposed to or in close vicinity of free-floating flammable liquid or gas.
- The author advises not to draw more than 10 DC amperes from any converter sources to prevent overheating situations.
- All electrical wiring to be used must be insulated and rated for at least 30 amperes. It is often very common for each circuit wire to carry less than 2 amperes of DC current.

Environmental Hazards Prevention

The author advises that all electrodes used for environmental cleanup be relatively inert and contain no hazardous and toxic chemicals. Typically, copper and stainless steel electrodes should be avoided.

The electrolysis of water molecules will produce hydrogen gas near the anode and oxygen gas near the cathode. The anode and cathode wells or conduits must be sealed below grade and not be allowed to ventilate freely. Slightly caustic food additive electrolytes can be added to, and around, the anode to mitigate the out gas of hydrogen.

Potassium and sodium chloride are not considered environmentally friendly electrolytes and electrolysis of naturally occurring potassium and sodium chloride will yield toxic chlorine gas at the anode, which may create unsafe and unhealthy situations. Again, slightly caustic food additive electrolyte can be added to, and around, the anode to mitigate the out gas of chlorine.

Radioactivity Hazards Prevention

The application of direct current in the ground may attract naturally occurring radionuclides like uranium and radon to the cathode. One should consult with the local geological survey about the location of certain geologic media or formations that may be host to trace amount of uranium and radon.

Ground Settlement Hazards Prevention

Electro-osmosis on silty and clayey soil will cause compaction and ultimate ground settlement. The clay drying phenomenon will occur around and near the anode. At all times, keep the anode wet. The addition of sodium or potassium food additive electrolyte at and around the anode will minimize ground settlement.

REFERENCES

Acar, Y. B., and Alshawabkeh, A. N., "Principles of Electrokinetic Remediation," *Environmental Science and Technology*, 27(13): 2638–2647, 1993.

Acar, Y. B., Gale R., Marks R. E., and Ugaz, A., "Feasibility of Removing Uranium, Thorium, and Radium from Kaolinite by Electrochemical Soil Processing," *US EPA Report No. 009-292*, 1992.

Casagrande, L., "The Application of Electro-Osmosis to Practical Problems," in *Foundations and Earthworks*, Building Technical Paper No. 30, H. M. Stationary Office, London, 1947.

Casagrande, L., "Electro-Osmosis," *Proc. 2nd Int. Conf. on Soil Mech. and Found. Eng.*, Vol. 1, pp. 218–222, 1948.

Casagrande, L., "Electro-Osmotic Stabilization of Soils," *Jour. Boston Soc. Civ. Eng.*, Vol. 39, p. 51, 1952.

Chilingar G. V. et al., "Possible Use of Electric Current for Increasing Volumetric Rate of Flow of Oil and Water During Primary and Secondary Recovery," *Chem. Chron.*, 28(1): 1–4, 1963.

Chilingar, G. V. et al., "Use of Direct Electrical Current for Increasing the Flow Rate of Reservoir fluids During Petroleum Recovery," *J. Canadian Pet. Tech.*, 3(1): 8–14, 1964.

Chilingar, G. V. et al., "Possible Use of Direct Current for Augmenting Reservoir Energy During Petroleum Production," *Compass of Sigma Gamma Epsilon* 45(4): 272–285, 1968.

Chilingar, G. V. et al., "Effect of Direct Electrical Current on Permeability of Sandstone Cores," *J. Pet. Tech.*, 22(7): 830–836, 1970.

Initiatives, *LASAGNA Demonstration Test Site*, Paducah, Kentucky, Urban Energy & Transportation Corp., Vol. 3, December 1996, p. 7, 1996.

Loo, W. W., "Heat Enhanced Bioremediation of Chlorinated Solvents and Toluene in Soil," *Conference Proceedings of R&D 1991*, Anaheim, California, pp. 133–136, 1991.

Loo, W. W. et al., *Soil Closure Report*, submitted to Santa Ana Regional Water Quality Control Board, for Electrokinetic Enhanced Soil Remediation with Chlorinated Solvents at the Northrop ESD Site, Anaheim, California, 1991.

Loo, W. W., and Wang, I. S., "Remediation of Groundwater Aquifer by In-Situ Electrolysis and Electro-Osmosis," *Proceedings of National Research and Development Conference on the Control of Hazardous Materials*, February 20–22, 1991, Anaheim, California, p. 163, 1991.

Loo, W. W., "Electrokinetic Enhanced Passive In-Situ Biotreatment of Gasoline and Diesel in Clayey Soil and Aquifer: A Case History," *Proceedings of the SUPERFUND XV Conference*, Washington D.C., 1994c.

Loo, W. W., *Field Demonstration of In-Situ Electrokinetic Treatment of Selenium and Boron*, Environment & Technology Services technical report to Panoche Water Drainage District, December 1996.

Loo, W. W., *Groundwater Case Closures of MtBE, Benzene & Chlorinated Solvents to Non-Detect By In-Situ Electrokinetic Processes*, Paper Presented at the HAZMACON Conference, Santa Clara Convention Center, Santa Clara, California, April 1998, sponsored by the Association of Bay Area Governments, 1998.

U.S. Army Corps of Engineers, *Installation Restoration and Hazardous Control Technologies*, Prepared for US Air Force, US Navy, US Army and US EPA, pp. 45–46, 1992.

U.S. EPA, *In-Situ Remediation Technology: Electrokinetics*, U.S. EPA publication 542-K-94-007, Office of Solid Waste and Emergency Response, Office of Technology Innovation Office, Washington D.C., 1995.

U.S. EPA, *Electrokinetic Laboratory and Field Processes Applicable to Radioactive and Hazardous Mixed Waste in Soil and Groundwater*, U.S. EPA publication 402-R-97-006, Center for Remediation Technology and Tools, Washington, D.C., 1997.

Van Doren, E. P., and Bruell, C. J., "Electro-Osmotic Removal of Benzene from a Water Saturated Clay," *Proc. of Petroleum Hydrocarbons and Organic Chemicals in Ground Water: Prevention, Detection and Restoration*, Nov. 17–19, 1987, Houston, Texas, pp. 107–126, 1987.

CHAPTER 14
REMEDIATION TECHNOLOGIES

SECTION 14.8

SURFACTANT-ENHANCED AQUIFER REMEDIATION

Richard Jackson

Dr. Jackson is manager of the geosystems section of Duke Engineering and Services. He has responsibility for the characterization and remediation of sites contaminated by fuels, solvents, and other nonaqueous phase liquids.

Varadarajan Dwarakanath

Dr. Dwarakanath is senior geosystems engineer with Duke Engineering and Services. He manages the geosystems laboratory in Austin, Texas.

14.8.1 INTRODUCTION

During the 1990s it was demonstrated—both in the laboratory and in the field—that it is practical to remove the vast majority of nonaqueous phase liquid (NAPL) from alluvium by surfactant-enhanced aquifer remediation (SEAR). However, such success appears to be strongly dependent upon adherence to a strict protocol of site characterization, laboratory experimentation, and numerical design. SEAR, at the time of writing (1999), is in an advanced state of development. Not only are the in situ processes of NAPL solubilization and/or mobilization being optimized in a variety of hydrogeological environments, but also the ex situ challenges of effluent treatment and surfactant recycling are being field tested. This statement applies to both lighter-than-water NAPLs (LNAPLs), such as fuel hydrocarbons, and denser-than-water NAPLs (DNAPLs), such as chlorinated solvents and coal tar.

SEAR has been acknowledged for some years to be a promising, innovative technology for the removal of NAPLs. This is only partly due to its history of use by the petroleum industry. The initial field trial of SEAR by Pitts et al. (1993) in a shallow, creosote-contaminated aquifer in Wyoming was particularly encouraging. This success led in 1990–1 to a controlled spill of perchloroethylene (PCE) into a test cell at Borden, Ontario by the University of Waterloo (Kueper et al., 1993) and the subsequent solubilization of much of the PCE by Fountain and others from the State University of New York (1996). The surfactant solutions that were injected into these two alluvial aquifers removed the DNAPL by solubilization as a microemulsion (i.e., by enhancing their effective solubility in the groundwater). Both tests were conducted within sheet-pile walls, which hydraulically isolated the contaminated parts of the alluvial aquifers.

It is noteworthy that the spatial distribution and total volume of the DNAPL, were well understood in the Laramie and Borden field tests and that both sites had some free-phase DNAPL, which resulted in relatively high initial DNAPL saturations (i.e., volume of NAPL per unit pore volume). While these two tests were of very different sizes—the Laramie test cell had a pore volume of 140,000 gallons, or $\sim$500 m^3, while the Borden cell contained only 2400 gallons or $\sim$10 m^3—they indicated that a well-characterized site could be substantially cleaned up.

Other surfactant floods have been completed in recent years, in particular at two sites at Hill Air Force Base in Utah. The first was undertaken by a team from the University of Florida (Jawitz et al., 1998) at Hill AFB Operable Unit 1 (OU1) that removed 72 percent of a multicomponent LNAPL from the alluvium. However, it did not achieve the low level of final NAPL saturation achieved by the SUNY surfactant flood at Borden (Fountain et al., 1996), perhaps because of tar-like materials in the LNAPL.

The second flood, at Hill AFB Operable Unit 2 (OU2), was designed by the University of Texas at Austin (UT) and undertaken by INTERA (now Duke Engineering & Services) with effluent treatment by Radian International (Brown et al., 1999, INTERA, 1997). Unlike the OU1 flood, this did not employ sheet-pile containment but rather hydraulic control based upon extensive hydraulic testing of the alluvium and exhaustive numerical simulation. The UT/INTERA flood demonstrated that 98.5 percent of the DNAPL in a shallow alluvial aquifer could be removed in the course of two short floods and without sheet-pile walls providing hydraulic isolation. More importantly, the OU2 surfactant flood pushed the post-flood DNAPL saturation to a new low of 0.04 percent even without the use of mobility-controlling agents, such as polymer, which has always been a feature of successful EOR floods (Pope and Wade, 1995).

The third surfactant flood at Hill AFB was designed specifically to test the use of the surfactant-foam concept (Hirasaki et. al., 1998) to remove DNAPL from heterogeneous alluvium. It was found that it was indeed possible to create a stable foam in a high-permeability "thief" zone beneath which was a low-permeability zone containing much DNAPL. This low-permeability zone was not completely decontaminated during the short surfactant flood, but the principle of generating a stable foam to block high-permeability zones and force surfactant into low-permeability zones was established. It remains to be determined the relative efficacy of foam flooding versus the use of polymer flooding to ensure that all NAPL-contaminated zones are evenly swept by the surfactant.

At all three Hill surfactant floods, the team characterized the DNAPL zone in the alluvium by using partitioning interwell tracer tests (PITTs), which allowed the designers to know the spatial distribution and volume of DNAPL in the test zone (Jin et al., 1995; Londergan et al., in review). The successful demonstration of these PITTs also allowed the design team to assess the performance of each of the solubilization tests by direct measurement of the average DNAPL saturation both before and after the tests intended to remove the DNAPL. Table 14.8.1 shows the results of these recent tests

TABLE 14.8.1   Recent Surfactant Floods for Removal of NAPL from Alluvial Aquifers

Design-implementation team, location & type of flood, date, and literature reference	NAPL composition	Swept pore volume in m^3 (gallons)	PVs surfactant injected	Reduction in NAPL mass (%)	Average post-flood NAPL saturation (%)*
Surtek-CH2MHill, Laramie WY, 1.4% surfactant + 1000 ppm polymer, 1989, Pitts et al. (1993)	creosote	530 (140,000)	3	84	2.7
State University of New York, Borden, Ontario, 2% surfactant, 1990-1, Fountain et al. (1996)	PCE	9.1 (2400)	14	77	0.2
Laval University, L'Assomption, Quebec, unidentified quantities of surfactant + polymer + alcohol, 1994 Martel et al. (1998)	DNAPL	6.1 (1600)	0.9	86	0.45
University of Florida SPME, Hill AFB UT, OU1, 3% surfactant + 2.5% alcohol, 1996, Jawitz et al. (1998)	LNAPL	4.5 (1200)	9.5	72	~0.8
DE&S-University of Texas-Radian, Hill AFB UT, OU2, 7.6% surfactant + 4.5% alcohol + 0.7% NaCl, 1996, Londergan et al., in review	DNAPL	57 (15,000)	2.4	98.5	0.04
DE&S-State University of New York –University of Texas-Lockheeed-Martin, DOE Piketon, OH; 4% surfactant + 4% alcohol + 0.2% NaCl/CaCl$_2$, 1996; Young et al. (1999)	DNAPL	5.6 (1500)	1.5	~50	0.1
Rice University/UT/DE&S, Hill AFB UT OU 2, 3.5% surfactant + 1% NaCl + air, 1997, Hirasaki et al. (1997)	DNAPL	31 (8200)	3	~90	0.03

* This value is obtained by dividing the volume of NAPL remaining in place throughout the swept pore volume after the surfactant flood by the swept pore volume.

and indicates the two performance criteria recommended to determine relative success—the percent reduction in NAPL mass and the average NAPL saturation following the surfactant flood.

This overview of SEAR begins with a discussion of the physical-chemical principles that form the basis for the in situ solubilization and/or mobilization of NAPLs (Section 14.8.2). Engineers and chemists determined these principles during the development of the technology of enhanced oil recovery (EOR). Petroleum engineers and geochemists have recently adopted these methods from EOR and have used them to solve problems of environmental restoration. It is no coincidence that the more successful field demonstrations of SEAR have paid close attention to the lessons learned by the petroleum industry. This accumulated knowledge is described in Section 14.8.3.

Unfortunately, this progress has been achieved in the face of much misunderstanding by many environmental scientists and engineers, who have failed to appreciate the extensive advances made by the petroleum industry in the 1970s and 1980s in addressing similar problems to those now facing the environmental community. These misconceptions are summarized and rebutted in Section 14.8.4. The complementary development of the ex situ treatment of the effluent from SEAR operations is then discussed (Section 14.8.5). Finally, the future development of SEAR for use for removal of both LNAPLs, such as fuel hydrocarbons, and DNAPLs, such as chlorinated solvents, is presented (Section 14.8.6).

14.8.2   *MICELLES AND MICROEMULSIONS*

The basic concepts of EOR and SEAR may be found in a number of recent review articles and monographs (Bourrel and Schechter, 1988; Dwarakanath and Pope, 1999; Pope and Baviere, 1991; Pope and Wade, 1995). This section will discuss the fundamentals of surfactant phase behavior in terms of micelle and microemulsion formation, the different types of microemulsions, and the relationship between interfacial tension (IFT) and solubilization.

The cornerstone for the application of surfactants for both EOR and SEAR is the ability of surfactants to form stable microemulsions rapidly. The importance of rapid equilibration will be discussed later in Section 14.8.3. A microemulsion may be described as a swollen micellar solution and should not be confused with a macroemulsion. Macroemulsions are physical dispersions of one fluid in the other and are unstable. Microemulsions, on the other hand, are composed of submicroscopic particles that are suspended by Brownian motion and, hence, are thermodynamically stable. Macroemulsions are frequently highly viscous dispersions, whereas microemulsions are extremely fluid solutions with relatively low viscosities.

There are several types of microemulsions. When the aqueous phase is the continuous phase and the oil is solubilized at the center of a micelle, it is called a Winsor Type I microemulsion. In petroleum engineering terminology, this is called a Type II⁻ microemulsion. When the oil is the continuous phase, the water is solubilized at the center of a micelle, this microemulsion is called a Winsor Type II, or inverted microemulsion. In petroleum engineering terminology, this is called a Type II⁺ microemulsion. The transition between the Winsor Type I and Type II is the Type III microemulsion in which the oil, water, and surfactant micelles coexist in the form of alternating layers. Figure 14.8.1 shows a conceptual picture of Type I and Type II microemulsions. A transition between

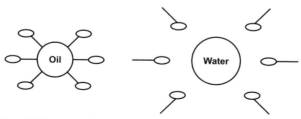

Type I Microemulsion   Type II Microemulsion

FIGURE 14.8.1   Conceptual picture of Type I and Type II microemulsions.

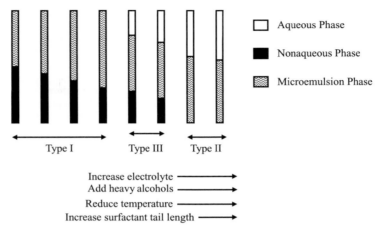

□ Aqueous Phase

■ Nonaqueous Phase

〰 Microemulsion Phase

Type I   Type III   Type II

Increase electrolyte ⟶
Add heavy alcohols ⟶
Reduce temperature ⟶
Increase surfactant tail length ⟶

FIGURE 14.8.2   Phase behavior transitions, between Type I–III–II.

Type I to Type III to Type II may be affected by the addition of electrolyte in the case of anionic surfactants, addition of a heavy cosolvent, increasing the hydrophobe tail length, and, finally, reducing the temperature (Figure 14.8.2).

The phase behavior of surfactants is characterized by volume fraction diagrams and ternary diagrams. Volume fraction diagrams are usually performed at a fixed water-oil ratio. The surfactant concentration is held constant. The electrolyte concentration is varied in the case of anionic surfactants. In case of nonionics, the cosurfactant or cosolvent concentration is varied. The oil-surfactant mixture is physically shaken and the excess phases are allowed to equilibrate. The volume fraction of the oil, water, and microemulsion are plotted as a function of the electrolyte concentration.

Volume fraction diagrams identify the optimal salinity (i.e., the electrolyte concentration at which equal volumes of water and oil are solubilized in the microemulsion). They also identify the electrolyte concentrations at which phase transitions between Type I–Type III–Type II are observed. This information is useful in numerically modeling surfactant phase behavior in compositional simulators such as UTCHEM. An example of a volume fraction diagram using 4 percent by weight sodium dihexyl sulfosuccinate, 4 percent by weight isopropanol, and NaCl as the electrolyte and Hill AFB DNAPL at 23°C is shown in Figure 14.8.3. The optimal salinity for this surfactant is 11,500 mg/L. A transition between Type I to Type III is observed at 11,000 mg/L NaCl and a transition between Type III to Type II is observed at 12,000 mg/L NaCl.

Ternary diagram experiments are performed at varying surfactant, oil, and water concentrations. The electrolyte concentration is held constant in these experiments. Ternary diagram experiments provide insights into the expected effluent if a fixed volume of oil, water, and surfactant are mixed. They also define the binodal curve above which the surfactant, oil, and water exist in the form of a single phase. An example of a ternary diagram is shown in Figure 14.8.4. The height of the binodal curve is 0.14. This means that if at a water-oil ratio of 0.2 and a surfactant concentration of 0.2 (volume per volume basis) the surfactant, water, and oil will coexist in a single phase. On the other hand, a water-oil ratio of 0.2 and a surfactant concentration of 0.10 will produce a microemulsion, which has a solubilized oil fraction of 0.11. The height of the binodal curve is a useful parameter in modeling surfactant phase behavior in UTCHEM (Delshad et al., 1996) and should be measured for a given surfactant and oil.

Both ternary diagram experiments and volume fraction diagram experiments allow the solubilization of the oil in the microemulsion to be quantified. The solubilization parameter, which is defined as the volume of oil solubilized to the volume per volume of surfactant, can be related to the excess phase-microemulsion IFT using the Chun Huh relationship (Huh, 1979). The excess phase can be

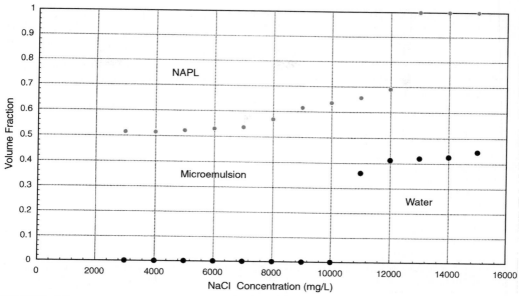

FIGURE 14.8.3   Volume fraction diagram for 4 percent by weight sodium dihexyl sulfosuccinate, 4 percent by weight IPA, NaCl at 23°C.

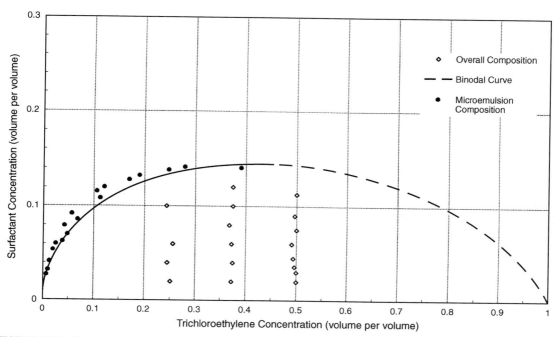

FIGURE 14.8.4   Ternary diagram of trichloroethylene with sodium dihexyl sulfosuccinate, 1000 mg/L NaCl at 23°C.

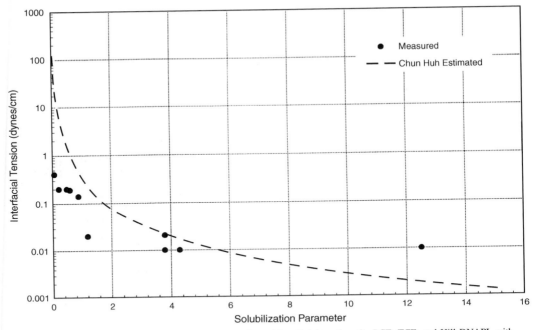

FIGURE 14.8.5 Comparison of measured and Chun Huh estimated interfacial tensions for PCE, TCE, and Hill DNAPL with Different Surfactants.

either water or oil. The Chun Huh relation is described below:

$$\gamma = \frac{C}{\sigma^2}$$

where γ = excess phase-microemulsion IFT (dynes/cm), C = a constant, which is usually $\sim$0.3 (dynes/cm), and σ = oil or water solubilization parameter.

The Chun Huh equation fits the oil-microemulsion IFT extremely well at ultra-low IFTs ($\sim$.01 dynes/cm). At higher IFTs ($\sim$1 dyne/cm), the match is not as good as can be seen from the comparison of measured and Chun Huh-predicted IFTs for PCE and TCE (trichloroethene) in Figure 14.8.5. The Chun Huh equation also indicates that ultra-low IFTs translate into high oil solubilization parameters. This means that at ultra-low IFTs, ultra-high oil solubilization will be observed.

14.8.3 APPLYING LESSONS LEARNED IN EOR TO DNAPL RECOVERY

The review of West and Harwell (1992) dwelled on the problems of EOR rather than the successes that clearly indicate how these problems may be overcome. The approach to surfactant-flood design for petroleum or DNAPL removal outlined here is that developed by Wade and his coresearchers since the mid-1960s in Texas. This approach is based on the detailed assessment of the phase behavior of the surfactant solution and on core floods of the reservoir rock, coupled with extensive reservoir characterization and design by numerical simulation.

The first step in applying surfactants to EOR or SEAR is the use of surfactant phase behavior experiments for identifying suitable candidate surfactants. Phase behavior experiments identify surfactants with high contaminant solubilization, low coalescence/equilibration times, and minimal liquid crystal, gel, or macroemulsion forming tendencies at ambient aquifer temperatures. This is extremely important as groundwater aquifer temperatures are usually much lower than petroleum reservoir temperatures where gel, macroemulsion, and liquid crystal formation are less likely.

Whenever a surfactant and oil are mixed, a macroemulsion is formed with small oil droplets dispersed in a continuous aqueous phase containing the surfactant micelles. The excess oil in the macroemulsion coalesces together to form larger oil droplets, which eventually separate into a separate oil phase from the microemulsion. This phenomenon is termed coalescence. Coalescence time is the time required for a surfactant-oil mixture to equilibrate into a stable microemulsion.

Poor surfactants form stable macroemulsions, have high coalescence times (greater than 40 h), and, therefore, have the potential to cause mass transfer limitations between NAPL and surfactant micelles. This greatly impedes NAPL removal efficiency and likely accounts for observations of mass transfer limitations as reported by investigators such as Pennell et al. (1993) and Mason and Kueper (1996). Poor surfactants can also reduce the permeability of the porous media, possibly because of pore plugging by the macroemulsions, gels, and liquid crystals (Dwarakanath et al., 1999; Renshaw et al., 1997).

Conversely, surfactants that coalesce rapidly (less than 20 h) form stable microemulsions with low viscosities and are preferable for use in SEAR applications. Rapid coalescence can be promoted by the addition of a cosolvent (i.e., alcohols used to improve surfactant phase behavior), by increasing the branching of the hydrophobe tail, by increasing temperature, and by using a suitable cosurfactant (Dwarakanath et al., 1999; Pope and Baviere, 1991; Pope and Wade, 1995). The sodium dihexyl sulfosuccinate used by Duke Engineering & Services (DE&S) at Hill AFB, OU2 had a coalescence time of approximately 15 min (Mayer et al., 1999). This was due to the addition of 4 percent isopropanol as a cosolvent and the branched nature of the hydrophobe tail for the surfactant.

Once surfactants are selected based on phase behavior experiments, their ability to remediate NAPL-contaminated soils is evaluated in several column experiments with soil at residual NAPL saturation. The induced hydraulic gradient across the soil column is carefully measured during the surfactant flood and is essential for quantifying surfactant transport in a porous medium. Surfactants that are more likely to form gels, and liquid crystals will cause high hydraulic gradients during surfactant flooding and should be avoided. Acceptable surfactants will show low hydraulic gradients during the surfactant flood and will restore the soil to its original permeability after surfactant and post-surfactant waterflooding. Soil column experiments also allow the measurement of surfactant sorption.

The University of Texas and DE&S successfully used phase behavior and soil column testing for designing the surfactant flood at Hill AFB, OU2. For best soil column experiments, final residual DNAPL saturations on the order of 0.0002 were achieved as a result of surfactant flooding, which corresponds to a removal of 99.9 percent of the contaminant (Dwarakanath et al., 1999). These estimates were based on partitioning tracers. The experimental error in the partitioning tracer method under these conditions is estimated to be either 10 percent of the residual saturation, *or* 0.0002, whichever is greater.

Once candidate surfactants are identified using phase behavior and soil column experiments, numerical simulations are required to design a field surfactant flood. An appropriate numerical simulator such as UTCHEM (Delshad et al., 1996) may be used to simulate surfactant flooding. This simulator is capable of modeling surfactant phase behavior and can handle other field variables, such as aquifer heterogeneities, variable distribution of NAPL, natural gradients in the aquifer, and so forth. Numerical simulation of surfactant floods will require an in-depth knowledge of the NAPL distribution. This knowledge is obtained from a partitioning interwell tracer test (PITT) developed by the University of Texas at Austin and DE&S. The PITT can characterize the volume of NAPL between a pair of injection and extraction wells, and using a combination of PITT and soil core data, the NAPL distribution can be accurately determined.

Once the NAPL distribution in the aquifer is quantified, several simulations can be conducted to determine:

- the location of injection, extraction, hydraulic control and monitor wells
- the optimum pumping rates for each set of wells, the screened interval in each well

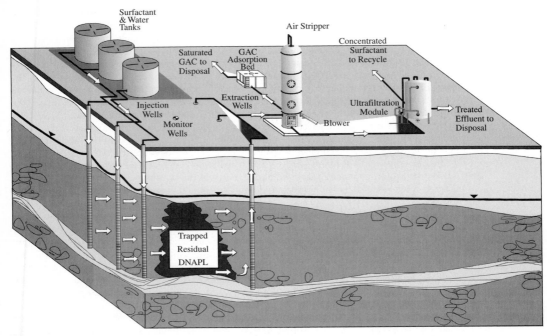

FIGURE 14.8.6   Conceptual figure of a SEAR Operation.

- the concentration and amount of surfactant required for each well
- location of sampling points
- frequency of sampling

Numerical simulations can also provide useful information on the predicted concentration of the contaminant and injected chemicals in the effluent such that the ex situ treatment operations may be designed. A conceptualization of SEAR operations is shown in Figure 14.8.6.

14.8.4.   ENVIRONMENTAL MISCONCEPTIONS

The Removal Efficiencies of EOR Floods

Lake and Pope (1979) summarized results from twenty field tests, all of which were micellar-polymer (MP) floods. In such floods, surfactant is injected into the petroleum reservoir together with a polymer, such as xanthan gum, to increase the viscosity of the flood and ensure that both high- and low-permeability zones of the reservoir were "swept." No technically successful EOR field tests have been undertaken without the use of polymer (Pope and Wade, 1995). Lake and Pope's analysis showed that the recovery efficiency of these twenty surfactant floods most strongly correlated with the capillary number of the flood (i.e., the ratio of the product of viscous and advective to capillary forces), and with the number of pore volumes of the polymer slug.

Thus, the best of the surfactant floods undertaken before 1980 recovered approximately 60 percent of the original oil in place and demonstrated the need for mobility control (Lake and Pope, 1979). However, the most successful EOR field tests were undertaken in Illinois and France in the 1980s and each recovered 68 percent of the original oil in place (Pope and Baviere, 1991). Both of these floods were distinguished by extensive reservoir characterization, laboratory screening of surfactants and cosurfactants, and numerically-simulated design.

These recovery efficiencies are well above the 30 to 50 percent cited by hydrogeologists (Cherry et al., 1996; Mercer and Cohen, 1990) as being typical of the recovery efficiencies obtained by the petroleum industry during EOR floods. However, Cherry et al., made clear that even such improved performance would make SEAR inadequate for environmental restoration purposes.

The first attempts at SEAR in shallow aquifers at Laramie, Wyoming in 1989 (Pitts et al., 1993) and at Borden, Ontario in 1990–1 (Fountain et al., 1996) produced results significantly better than the Illinois and French EOR floods. The recovery efficiencies for these two surfactant floods were 84 and 77 percent, respectively (see Table 14.8.1). These relatively high values, when compared with the 68 percent achieved in the EOR floods in the 1980s, were most likely because of the higher density of injection and extraction wells afforded by the shallowness of the swept zone. It should also be remembered that, at the time of the EOR floods, the unit cost of surfactant was 20 to 30 times more than that of petroleum (Pope, G. A., The University of Texas at Austin, personal communication). Consequently, the goal of EOR was to achieve a cost-effective recovery of the petroleum rather than to minimize the final NAPL saturation, as in SEAR.

Hydraulic Plugging of the Flooded Formation

Several studies have recorded the tendency of surfactants to cause a reduction in the hydraulic conductivity of the porous media being treated (Allred and Brown, 1995; Nash, 1987; Renshaw et al., 1997). Because these reductions in hydraulic conductivity and permeability can seriously affect the progress of a surfactant flood, there is clearly cause for concern.

However, petroleum reservoir engineers have overcome this problem by the injection of an electrolyte, such as sodium or calcium chloride, to prevent the desorption of cations from colloidal material in the aquifer that is the ultimate cause of permeability reduction. Such ion exchange reactions can lead to colloid mobilization and then hydraulic plugging of the surfactant-flooded formation. Consequently, if the surfactant solution is injected together with an electrolyte solution, which is tailored to suppress the cation-exchange reaction, there should be little or no mobilization of the colloids. Thus, the 1996 surfactant flood at Hill AFB OU2, which was designed by the University of Texas and conducted by DE&S, involved the injection of 7000 mg NaCl/L. No increase in hydraulic gradient across the wellfield was noted during the course of the surfactant flood. In fact, the gradient actually decreased as DNAPL was removed and the relative permeability to water increased.

Slow Solubilization Rates

A further misconception held by environmental scientists and engineers is that of the slowness of the rates of solubilization of NAPLs by micellar surfactant solutions (e.g., Reitsma and Kueper, 1998). This misconception has arisen because of the choice of surfactants by some academics (e.g., Abriola et al., 1993; Mason and Kueper, 1996). If surfactants with poor rates of coalescence are chosen, it will undeniably be the case that mass-transfer reactions will limit solubilization. Thus, Mayer et al., (1999) noted that "the NAPL-surfactant system used in the Mason and Kueper (1996) experiments apparently exhibits significant chemical nonequilibrium."

Mayer et al., also showed that the sulfosuccinate surfactant used by the University of Texas and INTERA (now DE&S) at Hill AFB in 1996 (Brown et al., 1999; Londergan et al., in review) produced chemical equilibrium "within a few minutes under well-mixed conditions." Therefore, the solubilization process "will not be rate limiting for this NAPL-surfactant system"; whereas the surfactant used by Mason and Kueper (1996) required "more than 1000 min ... for equilibrium concentrations to be reached in batch tests" (Mayer et al., 1999). For such slow rates of coalescence, there exists a significant danger of gel or liquid-crystal formation by the surfactant and the subsequent reduction in permeability of the formation.

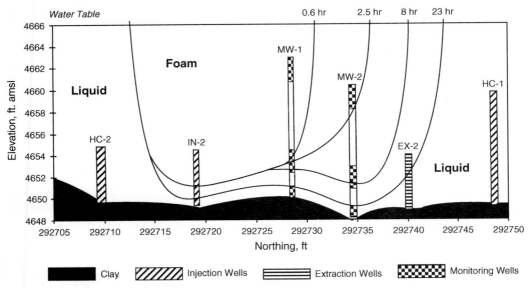

FIGURE 14.8.7   Surfactant-Foam Flood Advance, Hill AFB, OU2, 1997.

The Effect of Heterogeneities

One of the most strongly held beliefs of hydrogeologists concerning EOR is that of the "inadequate contact of the flushing fluid with DNAPL in low permeability zones" (Cherry et al., 1996, p. 499). This belief ignores the success of MP floods in overcoming geological heterogeneities in oil reservoirs (Lake and Pope, 1979). Both polymer solutions (Lake, 1989) and foam processes (Hirasaki et al., 1998) have been employed to divert surfactant solutions into low-permeability zones so that there is an even sweep of an oil reservoir. Figure 14.8.7 shows the advance of a foam front in the strongly heterogeneous alluvial aquifer at Hill AFB in Utah during the course of a surfactant flood in 1997 (Hirasaki et al., 1998). After 23 h of sequential injection of surfactant solution and compressed air, surfactant foam is blocking the high-permeability zones of the alluvium that have been cleaned of DNAPL and is diverting newly-injected surfactant into the low-permeability zone at the base of the aquifer, which still is contaminated with DNAPL.

14.8.5 EFFLUENT TREATMENT

Once it had been demonstrated that SEAR could remove the majority of the DNAPL remaining in place following waterflooding, effluent treatment became a matter of considerable importance. This matter remains unresolved at the time of writing (1999) although numerous promising approaches are being developed. There are three particular challenges to treating the effluent from a surfactant flood. First, the flood has been designed to produce an effluent order of magnitude more concentrated than the effluent from a pump-and-treat system. Second, the micellar surfactant system reduces the volatility of the solubilized contaminants, thereby reducing the efficiency of liquid-vapor stripping operations. Finally, the surfactants and cosurfactants (typically alcohol) may need to be recovered and reinjected for reasons of economy.

The approach to the design of the SEAR treatment system is similar for both SEAR and traditional pump-and-treat remediation. That is, the contaminants of concern must first be identified, then the appropriate treatment technologies may be considered and compared. Confining our discussion to chlorinated solvents for which air and steam stripping are the obvious treatment choices; if surfactant recovery is required, then ultrafiltration or foam fractionation will be considered. The third step requires the treatment engineer to be provided with a range of predicted groundwater extraction rates and the associated contaminated concentrations. The final step is to determine the appropriate treatment criteria, which will be, in turn, determined by the effluent discharge requirements. Provision must also be made to handle free-phase DNAPL should it be recovered and to treat any vapor emissions from the stripping systems. Only then will it be possible to estimate the size of the process equipment and the potential flow rates of the various contaminant streams, and to evaluate the capital and operating costs associated with various treatment system options.

Air stripping systems are the most common way of separating dissolved chlorinated hydrocarbons (CHC) from groundwater and are, therefore, available for use with SEAR at many sites. However, most air strippers are designed to treat aqueous concentrations of CHC of less than 10 mg/L, because of the dilute nature of the CHC plumes, but SEAR will likely result in effluent concentrations well above 10 mg/L. For such high concentrations, steam stripping has several distinct advantages. First and foremost among these is the much greater stripping efficiency at steam temperatures so that even highly contaminated effluent streams can be treated to low effluent concentrations in reasonably sized equipment. For example, during the SEAR demonstration at Hill AFB conducted by RADIAN International Inc. and INTERA (DE&S), an existing steam stripper was used consistently and predictably to reduce the influent concentration from about 7000 mg CHC/L to less than 0.1 mg/L as is shown in Figure 14.8.8. Furthermore, steam stripping provides a means of condensing the stripped CHCs, in this case mainly TCE, and carbon sorption was only required for treatment of CHC vapor displaced through storage tank vents.

A preliminary cost comparison of air versus steam stripping for SEAR treatment operations was conducted for the AFCEE (Air Force Center for Environmental Excellence)-funded surfactant floods undertaken at Hill AFB OU2 in 1996 by Oolman, T. of RADIAN International Inc. Considering an effluent that is predominantly TCE, a steam stripper would require approximately 50 percent more capital investment to achieve a set treatment efficiency. However, for highly contaminated effluents

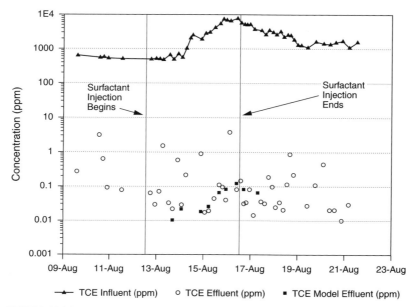

FIGURE 14.8.8 Steam Stripping of Surfactant-Flood Effluent, Hill AFB, OU2, 1996.

(>100 mg TCE/L), the operating costs of air stripping greatly exceed those of steam stripping because of the high cost of carbon sorption of the off gases. The transition in cost advantage appears to occur around 100–200 mg TCE/L. At lower concentrations, the lower capital cost of air stripping dominates the analysis. At higher concentrations (i.e., typical of SEAR), the cost of recovering TCE from the stripper greatly favors steam stripping.

The issue of surfactant recovery and reuse has taken on special importance because of the perceived economic necessity of such recycling (Krebs-Yuill et al., 1995). Because injected surfactant will be substantially diluted in the recovered groundwater, a primary requirement will be for the reconcentration of the surfactant. Membrane separation of CHCs from surfactant by ultrafiltration appears to be a very promising technology (Lipe et al., 1996) for this process. However, regulators are likely to expect that the CHC concentration in the reconcentrated surfactant be particularly low. Because the raffinate stream containing the surfactant will have a very high affinity for the CHCs, there will also be a reconcentration of the contaminant. Therefore, very high treatment efficiencies will be required upstream of the membrane separation process, a condition that is likely to be met most readily by steam stripping.

The perceived economic necessity of surfactant recycling and recovery (e.g., Cherry et al., 1996, p. 498) is based upon an analysis by Krebs-Yuill and others at the University of Oklahoma (1995) which employs a hypothetical case study that is of questionable appropriateness. This case study involves a contaminated aquifer covering five acres with a pore volume of nearly 10 million gallons and a residual DNAPL saturation of 13 percent. Thus, the volume of DNAPL considered in the economic analysis is approximately 1.3 million gallons or the contents of some 23,000 55-gallon drums of DNAPL! This volume of DNAPL is far in excess of the volumes likely to be found at the vast majority of DNAPL sites. Because Krebs-Yuill and colleagues assume that multiple pore volumes of surfactant are required (Sabatini et al., 1996, p. 11) and significant surfactant loss occurs, their analysis necessarily results in very high surfactant usage.

By way of example, the heavily contaminated aquifer beneath OU2 at Hill AFB has been determined to contain only 50,000 gallons of DNAPL or 4 percent of the volume considered in the Krebs-Yuill analysis. By choosing such an enormous volume of DNAPL to be removed and multiple pore volumes of surfactant, the Krebs-Yuill analysis caused the surfactant costs to become a very large part of the total cost of remediation. DE&S did not incur such high surfactant costs during the Hill AFB floods conducted in 1996 and 1997 in which the surfactants cost only 5 to 10 percent of the budgets of the two demonstrations. Furthermore, only 2.4 pore volumes were required to remove 98.4 percent of the residual DNAPL in the 1996 surfactant flood (Brown et al., 1999; Londergan et al., in review).

A second issue that arises with surfactant recovery and recycling is the assumption by Krebs-Yuill and her colleagues that the recovered surfactant will be purified sufficiently to contain less that 0.5 ppb of any dissolved CHC. The cost of attaining such levels of purity has not been determined yet by field trial and may make such surfactant recycling particularly costly.

14.8.6 *FUTURE DEVELOPMENT OF SEAR*

The principal areas of SEAR under development concern the hydrogeological constraints on surfactant use, the field testing of surfactant treatment, and the application of SEAR for LNAPL remediation. An additional issue of importance is the necessity of system integration among the geosystem engineers and hydrogeologists (whose domain is the subsurface) and the environmental and chemical engineers (who are responsible for effluent treatment). Finally, there is the matter of where SEAR fits in as a unit process in the overall remediation of a site.

The development of a surfactant-flooding method for use in DNAPL-contaminated aquifers that lack an underlying capillary barrier is being addressed through the concept of neutrally buoyant microemulsions. The absence of such capillary barriers, such as shales or clays, pose the problem that the lowering of the DNAPL-water IFT by surfactants could cause significant DNAPL redistribution. Neutral-buoyancy surfactant flooding, developed by Shook and Pope (Kostarelos et al., 1998;

Shook et al., 1998) employs low-density cosolvents, such as alcohols, to produce microemulsions that are at neutral buoyancy with respect to groundwater. That is, when DNAPL is solubilized and coalesces with the surfactant-cosolvent micelle, the resulting microemulsion has the same density as the groundwater that is transporting it. Thus, there is no propensity for the microemulsion to migrate vertically. This is referred to as SEAR-NB. It is important to note that the DNAPL itself is not being made less dense; only the microemulsion is made neutrally buoyant. This technology is being tested at a Superfund site in New England during 2000. The verification of SEAR-NB will also produce a practical remedial technology for fractured-rock DNAPL sites.

Most of this discussion has been taken up with the application of SEAR to DNAPLs. Quite obviously, as with EOR, SEAR is also applicable to LNAPLs. With fuel hydrocarbons there is usually much less difficulty identifying the NAPL zone because of the buoyancy of the LNAPL. However, there is an additional problem of removing the LNAPL trapped above the water table. This, no doubt, will become an area of considerable interest in the coming years.

The final issue of development underway is that of treatment engineering including the recycling of surfactant. A field test of such work is underway at Camp Lejeune in North Carolina in the spring of 1999 involving the surfactant flooding of a PCE DNAPL site in low-permeability sands. The treatment train involves separation of the PCE from the recovered surfactant effluent using pervaporation and subsequent reclamation of the surfactant by microemulsion ultrafiltration or MEUF. This demonstration will determine real costs of treatment and surfactant recycling.

It is very clear that the successful development of SEAR will require the highest level of system integration between those responsible for characterizing the subsurface geosystem and designing the surfactant flood, and those responsible for ex-situ treatment operations. A typical surfactant flood will likely be designed to remove 99 percent or more of the DNAPL but will also have to be designed with treatment constraints in mind. In order to achieve NAPL removal efficiencies of 99 percent and greater, those responsible for the characterization and removal of the NAPL will need to determine several issues of critical importance:

1. the volume and spatial distribution of the NAPL zone, most probably using the partitioning interwell tracer test (Jin et al., 1995, 1997)

2. the nature of a surfactant solution that, upon injection into the subsurface, solubilizes and/or mobilizes the NAPL efficiently without its redistribution

3. the design of a robust surfactant flood using a multi-phase, multi-component simulator such as UTCHEM (Delshad et al., 1996)

Last, but not least, among the future developments that will affect SEAR is the determination of its role in overall remedial operations. Following the two successful surfactant floods at OU2 at Hill AFB in Utah, there can be little doubt as to the efficacy of SEAR for DNAPL removal in particular. However, once the DNAPL source zone has been effectively reduced in volume, there is the matter of the contaminated groundwater remaining in the former source zone. It may be possible that this dissolved-phase contamination can be left to undergo natural attenuation with monitoring. Alternatively, it may be desirable to "polish" the remaining contamination with in situ chemical oxidation or biodegradation, neither of which can function in the presence of significant NAPL saturations. Such coupling of technologies will, no doubt, be of importance in the years to come.

REFERENCES

Abriola, L. M., Dekker, T. J., and Pennell, K. D., "Surfactant Solubilization or Residual Dodecane in Soil Columns. 2. Mathematical Modeling," *Environmental Science and Technology*, 27(12): 2341–2351, 1993.

Allred, B., and Brown, G. O., "Surfactant-induced Reductions of Saturated Hydraulic Conductivity and Unsaturated Diffusivity," in *Surfactant-Enhanced Subsurface Remediation*, Sabatini, D. A., Knox, R. C., Harwell, J. H., eds., ACS Symposium Series 594, American Chemical Society, Washington D.C., 216–230, 1995.

Bourrel, M., and Schechter, R. S., *Microemulsions and Related Systems*, Marcel Dekker, Inc., New York, 1988.

Brown, C. L., Delshad, M., Dwarakanath, V., Jackson, R. E., Londergan, J. T., Meinardus, H. W., McKinney, D. C., Oolman, T., Pope, G. A., and Wade, W. H., "Demonstration of Surfactant Flooding of an Alluvial Aquifer Contaminated with DNAPL," in *Field Testing of Innovative Subsurface Remediation Technologies*, ACS Symposium Series 725, American Chemical Society, Washington, D.C., pp. 64–85, 1999.

Cherry, J. A., Feenstra, S., and Mackay, D. M., "Concepts for the Remediation of Sites Contaminated with Dense Non-aqueous Phase Liquids (DNAPLs)," in *Dense Chlorinated Solvents and Other DNAPLs in Groundwater*, Pankow, J., and Cherry, J. A., eds., Waterloo Press, Portland OR, pp. 475–506, 1996.

Delshad, M., Pope, G. A., and Sepehrnoori, K., "A Compositional Simulator for Modeling Surfactant Enhanced Aquifer Remediation," *J. Contaminant Hydrology*, 23: 303–327, 1996.

Dwarakanath, V., Kostarelos, K., Pope, G. A., Shotts, D., and Wade, W. H., "Anionic Surfactant Remediation of Soil Columns Contaminated by Nonaqueous Phase Liquids," *J. Contaminant Hydrology*, 38: 465–488, 1999.

Dwarakanath, V., and Pope, G. A., "Environmental Soil Remediation by Surfactant Flooding," Chapter 11 in *Surfactants, Fundamentals and Applications in the Petroleum Industry*, Schramm, L., ed., Cambridge University Press, 2000.

Fountain, J. C., Starr, R. C., Middleton, T., Beikirch, M., Taylor, C., and Hodge, D., "A Controlled Field Test of Surfactant-enhanced Aquifer Remediation," *Ground Water*, 34(5): 910–916, 1996.

Hirasaki, G. J., Miller, C. A., Szafranski, R., Tanzil, D., Lawson, J. B., Meinardus, H. W., Jin, M., Londergan, J. T., Jackson, R. E., Pope, G. A., and Wade, W. H., "Demonstration of the Surfactant/Foam Process for Aquifer Remediation," submitted for publication in *J. Contaminant Hydrology*, 1999.

Huh, C., "Interfacial Tensions and Solubilizing Ability of a Microemulsion Phase that Coexists with Oil and Brine," *Journal of Colloid and Interface Science*, 71: 408–426, 1979.

INTERA (now Duke Engineering & Services). Demonstration of Surfactant Enhanced Aquifer Remediation of Chlorinated Solvent DNAPL at Operable Unit 2, Hill AFB, Utah. Prepared for the Air Force Center of Environmental Excellence, Technology Transfer Division, Brooks AFB, Texas, 1997.

Jawitz, J. W., Annable, M. D., Rao, P. S. C., and Rhue, R. D., "Field Implementation of a Winsor Type I Surfactant/Alcohol Mixture for In situ Solubilization of a Complex LNAPL as a Single-Phase Microemulsion," *Environmental Science & Technology*, 32(4): 523–530, 1998.

Jin, M., Delshad, M., Dwarakanath, V., McKinney, D. C., Pope, G. A., Sepehrnoori, K., Tilburg, C., and Jackson, R. E., "Partitioning Tracer Test for Detection, Estimation and Remediation Performance Assessment of Subsurface Nonaqueous Phase Liquids," *Water Resources Research*, 31(5): 1201–1211, 1995.

Jin, M., Butler, G.W., Jackson, R. E., Mariner, P. E., Pickens, J. F., Pope, G. A., Brown, C. L., and McKinney, D. C., "Sensitivity Models and Design Protocol for Partitioning Tracer Tests in Alluvial Aquifer," *Ground Water*, 35(6): 964–972, 1997.

Kostarelos, K., Pope, G. A., Rouse, B. A., and Shook, G. M., "A New Concept: The Use of Neutrally-buoyant Microemulsions for DNAPL Remediation," *J. Contaminant Hydrology*, 34: 383–397, 1998.

Krebs-Yuill, B., Harwell, J. H., Sabatini, D. A., and Knox, R. C., "Economic Considerations in Surfactant-enhanced Pump-and-treat Remediation," in *Surfactant-Enhanced Subsurface Remediation*, Sabatini, D. A., Knox, R. C., and Harwell, J. H., eds., ACS Symposium Series 594, American Chemical Society, Washington, D.C., pp. 265–278, 1995.

Kueper, B. H., Redman, D., Starr, R. C., Reitsma, S., and Mah, M., "A Field Experiment to Study the Behavior of Tetrachloroethylene Below the Water Table: Spatial Distribution of Residual and Pooled DNAPL," *Ground Water*, 31(5): 756–766, 1993.

Lake, L. W., *Enhanced Oil Recovery*, Prentice Hall, Englewood Cliffs, NJ, 1989.

Lake, L. W., and Pope, G. A., "Status of Micellar-polymer Field Tests," *Petroleum Engineer International*, November issue: 38–60, 1979.

Lipe, K. M., Sabatini, D. A., Hasegawa, M. A., and Harwell, J. H., "Micellar-enhanced Ultrafiltration and Air Stripping for Surfactant-contaminant Separation and Surfactant Reuse," *Ground Water Monitoring & Remediation*, Winter issue: 85–92, 1996.

Londergan, J. T., Meinardus, H. W., Mariner, P. E., and Jackson, R. E., Brown, C. L., Dwarakanath, V., Pope, G. A., Ginn, J., and Taffinder, S., "DNAPL Removal from a Heterogeneous Alluvial Aquifer by Surfactant-enhanced Aquifer Remediation," submitted to *Ground Water Monitoring & Remediation*, in review.

Martel, R., Gelinas, P. J., and Saumure, L., "Aquifer Washing by Micellar Solutions: 3 Field test at the Thouin Sand Pit, L'Assomption, Quebec, Canada," *J. Contaminant Hydrology*, 30: 33–48, 1998.

Mason, A. R., and Kueper, B. H., "Numerical Simulation of Surfactant Enhanced Solubilization of Pooled DNAPL," *Environmental Science & Technology*, 30(3): 3205–3215, 1996.

Mayer, A., Zhong, L., and Pope, G. A., "Measurement of Mass Transfer Rates for Surfactant-enhanced Solubilization of Nonaqueous Phase Liquids," *Environmental Science and Technology*, 33(17): 2965–2972.

Mercer, J. W., and Cohen, R. M., "A Review of Immiscible Fluids in the Subsurface: Properties, Models, Characterization, and Remediation," *J. Contaminant Hydrology*, 6: 142, 1990.

Nash, J. H., *Field Studies of In-Situ Washing*, EPA/600/2–87/110, U.S. Environmental Protection Agency, Cincinnati OH, 1987.

Pitts, M., Wyatt, K., Sale, T. C., and Piontek, K. R., "Utilization of Chemical-enhanced Oil Recovery Technology to Remove Hazardous Oily Waste from Alluvium," in *SPE International Symposium on Oilfield Chemistry,* New Orleans LA, Paper no. SPE 25153, 1993.

Pennell, K. D., Abriola, L. M., Weber Jr, W. J., "Surfactant Enhanced Solubilization of Residual Dodecane in Soil Columns 1. Experimental Investigation," *Environ. Sci. Technol.*, 27(12): 2332–2340, 1993.

Pope, G. A., and Baviere, M., "Reduction of Capillary Forces by Surfactants," in *Basic Concepts in EOR Processes*, Elsevier, pp. 89–122, 1991.

Pope, G. A., and Wade, W. H., "Lessons from Enhanced Oil Recovery Research for Surfactant-enhanced Aquifer Remediation," in *Surfactant-Enhanced Subsurface Remediation*, Sabatini, D. A., Knox, R. C., and Harwell, J. H., eds., ACS Symposium Series 594, American Chemical Society, Washington, D.C., pp. 142–160, 1995.

Renshaw, C. E., Zynda, G. D., and Fountain, J. C., "Permeability Reductions Induced by Sorption of Surfactant," *Water Resources Research*, 33(3): 371–378, 1997.

Sabatini, D. A., Knox, R. C., and Harwell, J. H., "Surfactant-enhanced DNAPL Remediation: Surfactant Selection, Hydraulic Efficiency, and Economic Factors," *Environmental Research Brief*, U.S. Environmental Protection Agency, Ada, OK, 1996.

Shook, G. M., Pope, G. A., and Kostarelos, K., "Prediction and Minimization of Vertical Migration of DNAPLs Using Surfactant Enhanced Aquifer Remediation at Neutral Buoyancy," *J. Contaminant Hydrology*, 34: 363–382, 1998.

West, C. C., and Harwell, J. H., "Surfactants and Subsurface Remediation," *Environ. Sci. Technol.*, 26(12): 2324–2330, 1992.

Young, C. M. et al., "Characterization of a TCE DNAPL Zone in Alluvium by Partitioning Tracers," *Ground Water Monitoring & Remediation*, XIX(1): 84–94, 1999.

CHAPTER 14
REMEDIATION TECHNOLOGIES

SECTION 14.9

REMEDIATION OF METALS-CONTAMINATED SOILS AND GROUNDWATER

Cynthia R. Evanko
Dr. Evanko is with GeoSyntec Consultants in Atlanta, Georgia.

David A. Dzombak
Dr. Dzombak teaches and conducts research in water and soil quality engineering at the Department of Civil and Environmental Engineering, Carnegie Mellon University, Pittsburgh, Pennsylvania.

14.9.1 SUMMARY

Metals contamination is a persistent problem at many contaminated sites. In the United States, the most commonly occurring metals at Superfund sites are lead, chromium, arsenic, zinc, cadmium, copper, and mercury. The presence of metals in groundwater and soils can pose a significant threat to human health and ecological systems. The chemical form of the metal contaminant influences its solubility, mobility, and toxicity in groundwater systems. The chemical form of metals depends on the source of the metal waste and the soil and groundwater chemistry at the site. A detailed site characterization must be performed to assess the type and level of metals present and allow evaluation of remedial alternatives.

Typically metals are relatively immobile in subsurface systems as a result of precipitation or adsorption reactions. For this reason, remediation activities at metals-contaminated sites have focused on the solid-phase sources of metals (i.e., contaminated soils, sludges, wastes, or debris).

A range of technologies is available for remediation of metals-contaminated soil and groundwater at Superfund sites. General approaches to remediation of metal contamination include isolation, immobilization, toxicity reduction, physical separation, and extraction. These general approaches can be used for many types of contaminants but the specific technology selected for treatment of a metals-contaminated site will depend on the form of the contamination and other site-specific characteristics (Allen, Perdue, Brown, 1994). One or more of these approaches are often combined for more cost-effective treatment. A number of the available technologies have been demonstrated in full-scale applications and are presently commercially available. A comprehensive list of these technologies is available (U.S. EPA, 1996a). Several other technologies are being tested for application to metals-contaminated sites. This report summarizes remediation technologies for metals-contaminated soil and groundwater whose performance at full-scale has been verified under the United States Environmental Protection Agency (U.S. EPA) Superfund Innovative Technology Evaluation (SITE) program for evaluation of emerging and demonstrated technologies. The focus of this program is the demonstration phase in which the technologies are field-tested and performance and cost data are collected. Technologies available for treatment of metals-contaminated soil and groundwater by each of the general approaches to remediation are presented, and the applicability of these technologies to different types of metal contamination and physical site characteristics are evaluated. Cost ranges are provided for a number of the technologies. The most promising emerging technologies are also examined.

Treatment of metals-contaminated groundwater has typically involved flushing and above-ground treatment, while treatment of contaminated solids most often has been performed by excavation followed by ex situ treatment or disposal. The most common ex situ treatment for excavated soils is solidification/stabilization. *In situ* treatment methods for metals-contaminated soil and groundwater are being tested and will be applied with increasing frequency.

14.9.2 PROBLEM DESCRIPTION

Metals at Contaminated Sites

Approximately 75 percent of Superfund sites for which Records of Decision (RODs) have been signed contain metals as a form of contamination. Some of these sites contain mixed metal-organic wastes for which metals might not be the primary contaminant of concern. The most common metals found at contaminated sites are (U.S. EPA, 1996b), in order: lead (Pb), chromium (Cr), arsenic (As), zinc (Zn), cadmium (Cd), copper (Cu), and mercury (Hg). Figure 14.9.1 summarizes the frequency with which these metals occur at Superfund sites.

The specific type of metal contamination found at a Superfund site is directly related to the operation that occurred at the site. The range of contaminant concentrations and the physical and chemical forms of contaminants also depends on activities and disposal patterns for contaminated wastes on the site. Other factors that may influence the form, concentration, and distribution of metal contaminants include soil and groundwater chemistry and local transport mechanisms.

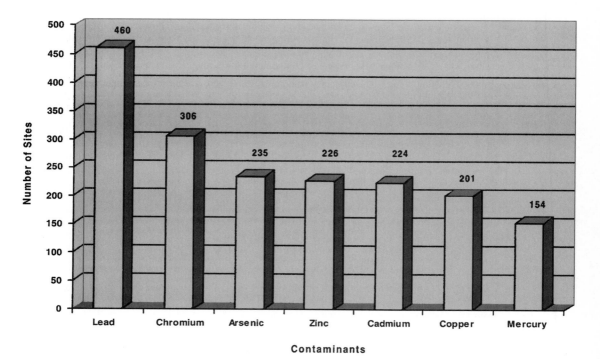

FIGURE 14.9.1   Metals most commonly present in all matrices at superfund sites (from U.S. EPA, 1996).

Sources of Contaminants

Surface water and groundwater may be contaminated with metals from wastewater discharges or by direct contact with metals-contaminated soils, sludges, mining wastes, and debris. Metal-bearing solids at contaminated sites can originate from a wide variety of sources in the form of airborne emissions, process solid wastes, sludges, or spills. The contaminant sources influence the heterogeneity of contaminated sites on a macroscopic and microscopic scale. Variations in contaminant concentration and matrix influence the risks associated with metal contamination and treatment options.

Airborne Sources.   Airborne sources of metals include stack or duct emissions of air, gas, or vapor streams, and fugitive emissions such as dust from storage areas or waste piles. Metals from airborne sources are generally released as particulates contained in the gas stream. Some metals such as arsenic, cadmium, and lead can also volatilize during high-temperature processing. These metals will convert to oxides and condense as fine particulates unless a reducing atmosphere is maintained. (Smith et al., 1995)

Stack emissions can be distributed over a wide area by natural air currents until dry and/or wet precipitation mechanisms remove them from the gas stream. Fugitive emissions are often distributed over a much smaller area because emissions are made near the ground. In general, contaminant concentrations are lower in fugitive emissions compared to stack emissions. The type and concentration of metals emitted from both types of sources will depend on site-specific conditions.

Process Solid Wastes.   Process solid wastes can result from a variety of industrial processes. These metal-bearing solid wastes are disposed above ground in waste piles or below ground or under cover in landfills. Examples of process solid wastes include slags, fly ash, mold sands, abrasive wastes, ion exchange resins, spent catalysts, spent activated carbon, and refractory bricks (Zimmerman and Coles, 1992). The composition of the process waste influences the density, porosity, and leach resistance of the waste and must be considered in evaluating the contaminated matrix.

Because waste piles are above ground, they are exposed to weathering, which can disperse the waste pile to the surrounding soil, water, and air and can result in generation of leachate, which infiltrates into the subsurface environment. The ability of landfills to contain process solid wastes varies because of the range of available landfill designs. Uncontained landfills can release contaminants into infiltrating surface water or groundwater or via wind and surface erosion.

Sludges.   The composition of sludges depends on the original waste stream and the process from which it was derived. Sludges resulting from a uniform wastestream, such as wastewater treatment sludges, are typically more homogeneous and have more uniform matrix characteristics. Sludge pits, on the other hand, often contain a mixture of wastes that have been aged and weathered, causing a variety of reactions to occur. Sludge pits often require some form of pretreatment before wastes can be treated or recycled (Smith et al., 1995).

Soils.   Soil consists of a mixture of weathered minerals and varying amounts of organic matter. Soils can be contaminated as a result of spills or direct contact with contaminated waste streams such as airborne emissions, process solid wastes, sludges, or leachate from waste materials. The solubility of metals in soil is influenced by the chemistry of the soil and groundwater (Evans, 1989; Sposito, 1989). Factors such as pH, Eh, ion exchange capacity, and complexation/chelation with organic matter directly affect metal solubility.

Direct Ground-Water Contamination.   Groundwater can be contaminated with metals directly by infiltration of leachate from land disposal of solid wastes, liquid sewage or sewage sludge, leachate from mine tailings and other mining wastes, deep-well disposal of liquid wastes, seepage from industrial waste lagoons, or from other spills and leaks from industrial metal processing facilities (e.g., steel plants, plating shops). A variety of reactions may occur that influence the speciation and mobility of metal contaminants including acid/base, precipitation/dissolution, oxidation/reduction, sorption, or ion exchange. Precipitation, sorption, and ion exchange reactions can retard the movement of metals in groundwater. The rate and extent of these reactions will depend on factors such as pH, Eh, complexation with other dissolved constituents, sorption and ion exchange capacity of the geological materials, and organic matter content. Groundwater flow characteristics also influence the transport of metal contaminants.

Definitions of Contaminant Concentrations

Sludges, soils, and solid wastes are multiphase materials that may contain metals in the solid, gaseous, or liquid phases. This complicates analysis and interpretation of reported results. For example, the most common method for determining the concentration of metals contaminants in soil is via total elemental analysis (U.S. EPA Method 3050). The level of metal contamination determined by this method is expressed as mg metal/kg soil. This analysis does not specify requirements for the moisture content of the soil and may, therefore, include soil water. This measurement may also be reported on a dry soil basis.

The level of contamination may also be reported as leachable metals as determined by leach tests, such as the toxicity characteristic leaching procedure, or TCLP test (U.S. EPA Method 1311) or the synthetic precipitation leaching procedure, or SPLP test (U.S. EPA Method 1312). These procedures measure the concentration of metals in leachate from soil contacted with an acetic acid solution (TCLP) or a dilute solution of sulfuric and nitric acid (SPLP). In this case, metal contamination is expressed in mg/L of the leachable metal.

Other types of leaching tests have been proposed (see summary by Environment Canada, 1990), including sequential extraction procedures (Tessier et al., 1979) and extraction of acid volatile sulfide (DiToro et al., 1992). Sequential procedures contact the solid with a series of extractant solutions that are designed to dissolve different fractions of the associated metal. These tests may provide insight into the different forms of metal contamination present (e.g., see Van Benschoten et al., 1994).

Contaminant concentrations can be measured directly in metals-contaminated water. These concentrations are most commonly expressed as total dissolved metals in mass concentrations (mg/L or μg/L) or in molar concentrations (moles/L). In dilute solutions, a mg/L is equivalent to one part per million (ppm), and a μg/L is equivalent to one part per billion (ppb).

Groundwater samples are usually filtered with a 0.45 μm filter prior to analysis for metals, though this is not always required and has recently been prohibited by many states and by some U.S. EPA programs that require analysis of total metals. Interest in measurement of total metal concentrations (dissolved and particulate-associated metals) usually derives from concern about possible transport of metals adsorbed on mobile colloidal particles (e.g., Kaplan et al., 1995). Research indicates that significant colloid-facilitated transport of metals can occur only under a fairly specialized set of conditions (Roy and Dzombak, 1997), but the conservative approach in monitoring system design is to try to capture any mobile colloids present. The problem with sampling groundwater without filtration is that particles from the well material, well slime coatings, or well pack may be sampled, and any subsequent analysis will not accurately reflect groundwater composition. To avoid such artifacts, but still permit sampling that can capture any mobile colloids present in the groundwater, monitoring wells are purged before sampling to remove the casing water and obtain representative groundwater samples. Low-flow purging and sampling techniques have been developed to minimize sample disturbances that may affect analysis (Puls, 1994; Puls and Paul, 1995).

Chemical Fate and Mobility

The fate and transport of a metal in soil and groundwater depends significantly on the chemical form and speciation of the metal (Allen and Torres, 1991). The mobility of metals in groundwater systems is hindered by reactions that cause metals to adsorb or precipitate, or chemistry that tends to keep metals associated with the solid phase and prevent them from dissolving. These mechanisms can retard the movement of metals and also provide a long-term source of metal contaminants (NRC, 1994). While the various metals undergo similar reactions in a number of aspects, the extent and

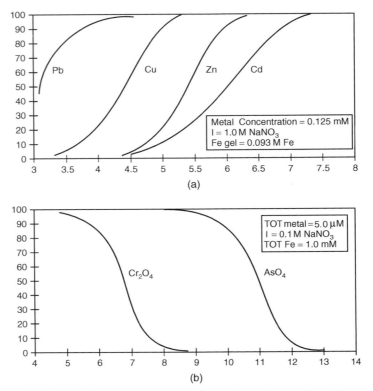

FIGURE 14.9.2   Metal adsorption to hydrous iron oxide gels (a) metal cations (adapted from Kinniburgh et al., 1976) and (b) metal anions.

nature of these reactions varies under particular conditions. In Figure 14.9.2, for example, the extent of sorption of several metal cations and anions onto iron oxide is shown as a function of pH for a particular background electrolyte composition. It may be seen that lead sorbs extensively at much lower pH values than zinc or cadmium (Kinniburgh et al., 1976).

The chemical form and speciation of some of the more important metals found at contaminated sites are discussed below. The influence of chemical form on fate and mobility of these compounds is also discussed.

Lead.   The primary industrial sources of lead (Pb) contamination include metal smelting and processing, secondary metals production, lead battery manufacturing, pigment and chemical manufacturing, and lead-contaminated wastes. Widespread contamination resulting from the former use of lead in gasoline is also of concern. Lead released to groundwater, surface water, and land is usually in the form of elemental lead, lead oxides and hydroxides, and lead-metal oxyanion complexes (Smith et al., 1995).

Lead occurs most commonly with an oxidation state of 0 or +II. Pb(II) is the more common and reactive form of lead and forms mononuclear and polynuclear oxides and hydroxides. Under most conditions, Pb^{2+} and lead-hydroxy complexes are the most stable forms of lead (Smith et al., 1995). Low solubility compounds are formed by complexation with inorganic (Cl^-, CO_3^{2-}, SO_4^{2-}, PO_4^{3-}) and organic ligands (humic and fulvic acids, EDTA, amino acids) (Bodek et al., 1988). Lead carbonate solids form above pH 6, and PbS is the most stable solid when high sulfide concentrations are present under reducing conditions.

Most lead released into the environment is retained in the soil (Evans, 1989). The primary processes influencing the fate of lead in soil include adsorption, ion exchange, precipitation, and complexation with sorbed organic matter. These processes limit the amount of lead that can be transported into the surface water or groundwater. The relatively volatile organolead compound tetramethyl lead may form in anaerobic sediments as a result of alkyllation by microorganisms (Smith et al., 1995).

The amount of dissolved lead in surface water and ground- water depends on pH and the concentration of dissolved salts and the types of mineral surfaces present. In surface water and groundwater systems, a significant fraction of lead is undissolved and occurs as precipitates ($PbCO_3$, Pb_2O, $Pb(OH)_2$, $PbSO_4$), sorbed ions or surface coatings on minerals, or as suspended organic matter.

Chromium.   Chromium (Cr) is one of the less common elements and does not occur naturally in elemental form, but only in compounds. Chromium is mined as a primary ore product in the form of the mineral chromite, $FeCr_2O_4$. Major sources of Cr contamination include releases from electroplating processes and the disposal of chromium containing wastes (Smith et al., 1995).

Cr(VI) is the form of chromium commonly found at contaminated sites. Chromium can also occur in the +III oxidation state, depending on pH and redox conditions. Cr(VI) is the dominant form of chromium in shallow aquifers where aerobic conditions exist. Cr(VI) can be reduced to Cr(III) by soil organic matter, S^{2-} and Fe^{2+} ions under anaerobic conditions often encountered in deeper groundwater. Major Cr(VI) species include chromate (CrO_4^{2-}) and dichromate ($Cr_2O_7^{2-}$), which precipitate readily in the presence of metal cations (especially Ba^{2+}, Pb^{2+}, and Ag^+). Chromate and dichromate also adsorb on soil surfaces, especially iron and aluminum oxides. Cr(III) is the dominant form of chromium at low pH (<4). Cr^{3+} forms solution complexes with NH_3, OH^-, Cl^-, F^-, CN^-, SO_4^{2-}, and soluble organic ligands. Cr(VI) is the more toxic form of chromium and is also more mobile. Cr(III) mobility is decreased by adsorption to clays and oxide minerals below pH 5 and low solubility above pH 5 resulting from the formation of $Cr(OH)_3(s)$ (Chrotowski et al., 1991).

Chromium mobility depends on sorption characteristics of the soil, including clay content, iron oxide content, and the amount of organic matter present. Chromium can be transported by surface runoff to surface waters in its soluble or precipitated form. Soluble and unadsorbed chromium complexes can leach from soil into groundwater. The leachability of Cr(VI) increases as soil pH increases. Most of chromium released into natural waters is particle associated, however, and is ultimately deposited into the sediment (Smith et al., 1995).

Arsenic.   Arsenic (As) is a semimetallic element that occurs in a wide variety of minerals, mainly as As_2O_3, and can be recovered from processing of ores containing mostly copper, lead, zinc, silver, and gold. It is also present in ashes from coal combustion. Arsenic exhibits fairly complex chemistry and can be present in several oxidation states (−III, 0, III, V) (Smith et al., 1995).

In aerobic environments, As(V) is dominant, usually in the form of arsenate (AsO_4^{3-}) in various protonation states: H_3AsO_4, $H_2AsO_4^-$, $HAsO_4^{2-}$, AsO_4^{3-}. Arsenate, and other anionic forms of arsenic behave as chelates and can precipitate when metal cations are present (Bodek et al., 1988). Metal arsenate complexes are stable only under certain conditions. As(V) can also coprecipitate with or adsorb onto iron oxyhydroxides under acidic and moderately reducing conditions. Coprecipitates are immobile under these conditions but arsenic mobility increases as pH increases (Smith et al., 1995).

Under reducing conditions As(III) dominates, existing as arsenite (AsO_3^{3-}) and its protonated forms: H_3AsO_3, $H_2AsO_3^-$, $HAsO_3^{2-}$. Arsenite can adsorb or coprecipitate with metal sulfides and has a high affinity for other sulfur compounds. Elemental arsenic and arsine, AsH_3, may be present under extreme reducing conditions. Biotransformation (via methylation) of arsenic creates methylated derivatives of arsine, such as dimethyl arsine $HAs(CH_3)_2$ and trimethylarsine $As(CH_3)_3$ which are highly volatile.

Because arsenic is often present in anionic form, it does not form complexes with simple anions such as Cl^- and SO_4^{2-}. Arsenic speciation also includes organometallic forms such as methylarsinic acid $(CH_3)AsO_2H_2$ and dimethylarsinic acid $(CH_3)_2AsO_2H$. Many arsenic compounds sorb strongly to soils and are, therefore, transported only over short distances in groundwater and surface water. Sorption and coprecipitation with hydrous iron oxides are the most important removal mechanisms under most environmental conditions (Krause and Ettel, 1989; Pierce and Moore, 1982). Arsenates can be leached easily if the amount of reactive metal in the soil is low. As(V) can also be mobilized under reducing conditions that encourage the formation of As(III), under alkaline and saline conditions, in the presence of other ions that compete for sorption sites, and in the presence of organic compounds that form complexes with arsenic (Smith et al., 1995).

Zinc. Zinc (Zn) does not occur naturally in elemental form. It is usually extracted from mineral ores to form zinc oxide (ZnO). The primary industrial use for Zinc is as a corrosion-resistant coating for iron or steel (Smith et al., 1995). Zinc usually occurs in the +II oxidation state and forms complexes with a number of anions, amino acids and organic acids. Zn may precipitate as $Zn(OH)_2(s)$, $ZnCO_3(s)$, $ZnS(s)$, or $Zn(CN)_2(s)$.

Zinc is one of the most mobile heavy metals in surface waters and groundwater because it is present as soluble compounds at neutral and acidic pH values. At higher pH values, zinc can form carbonate and hydroxide complexes, which control zinc solubility. Zinc readily precipitates under reducing conditions and in highly polluted systems when it is present at very high concentrations, and may coprecipitate with hydrous oxides of iron or manganese (Smith et al., 1995). Sorption to sediments or suspended solids, including hydrous iron and manganese oxides, clay minerals, and organic matter, is the primary fate of zinc in aquatic environments. Sorption of zinc increases as pH increases and salinity decreases.

Cadmium. Cadmium (Cd) occurs naturally in the form of CdS or $CdCO_3$. Cadmium is recovered as a by-product from the mining of sulfide ores of lead, zinc, and copper. Sources of cadmium contamination include plating operations and the disposal of cadmium-containing wastes (Smith et al., 1995).

The form of cadmium encountered depends on solution and soil chemistry as well as treatment of the waste prior to disposal. The most common forms of cadmium include Cd^{2+}, cadmium-cyanide complexes, or $Cd(OH)_2$ solid sludge (Smith et al., 1995). Hydroxide ($Cd(OH)_2$) and carbonate ($CdCO_3$) solids dominate at high pH, whereas Cd^{2+} and aqueous sulfate species are the dominant forms of cadmium at lower pH (<8). Under reducing conditions when sulfur is present, the stable solid CdS(s) is formed. Cadmium will also precipitate in the presence of phosphate, arsenate, chromate, and other anions, although solubility will vary with pH and other chemical factors.

Cadmium is relatively mobile in surface water and groundwater systems and exists primarily as hydrated ions or as complexes with humic acids and other organic ligands (Callahan et al., 1979). Under acidic conditions, cadmium may also form complexes with chloride and sulfate. Cadmium is removed from natural waters by precipitation and sorption to mineral surfaces, especially oxide minerals, at higher pH values ($>$pH 6). Removal by these mechanisms increases as pH increases. Sorption is also influenced by the cation exchange capacity (CEC) of clays, carbonate minerals, and organic matter present in soils and sediments. Under reducing conditions, precipitation as CdS controls the mobility of cadmium (Smith et al., 1995).

Copper. Copper (Cu) is mined as a primary ore product from copper sulfide and oxide ores. Mining activities are the major source of copper contamination in groundwater and surface waters. Other sources of copper include algicides, chromated copper arsenate (CCA) pressure-treated lumber, and copper pipes. Solution and soil chemistry strongly influence the speciation of copper in groundwater systems. In aerobic, sufficiently alkaline systems, $CuCO_3$ is the dominant soluble copper species. The cupric ion, Cu^{2+}, and hydroxide complexes, $CuOH^+$ and $Cu(OH)_2$, are also commonly present. Copper forms strong solution complexes with humic acids. The affinity of Cu for humates increases as pH increases and ionic strength decreases. In anaerobic environments when sulfur is present, CuS(s) will form.

Copper mobility is decreased by sorption to mineral surfaces. Cu^{2+} sorbs strongly to mineral surfaces over a wide range of pH values (Dzombak and Morel, 1990). The cupric ion (Cu^{2+}) is the most toxic species of copper. Copper toxicity has also been demonstrated for $CuOH^+$ and $Cu_2(OH)_2^{2+}$ (LaGrega et al., 1994).

Mercury.   The primary source of mercury is the sulfide ore cinnabar. Mercury (Hg) is usually recovered as a by-product of ore processing (Smith et al., 1995). Release of mercury from coal combustion is a major source of mercury contamination. Releases from manometers at pressure measuring stations along gas/oil pipelines also contribute to mercury contamination.

After release to the environment, mercury usually exists in mercuric (Hg^{2+}), mercurous (Hg_2^{2+}), elemental ($Hg°$), or alkyllated form (methyl/ethyl mercury). The redox potential and pH of the system determine the stable forms of mercury that will be present. Mercurous and mercuric mercury are more stable under oxidizing conditions. When mildly reducing conditions exist, organic or inorganic mercury may be reduced to elemental mercury, which may then be converted to alkyllated forms by biotic or abiotic processes. Mercury is most toxic in its alkyllated forms, which are soluble in water and volatile in air (Smith et al., 1995).

Hg(II) forms strong complexes with a variety of both inorganic and organic ligands, making it very soluble in oxidized aquatic systems (Bodek et al., 1988). Sorption to soils, sediments, and humic materials is an important mechanism for removal of mercury from solution. Sorption is pH-dependent and increases as pH increases. Mercury may also be removed from solution by coprecipitation with sulfides (Smith et al., 1995).

Under anaerobic conditions, both organic and inorganic forms of mercury may be converted to alkyllated forms by microbial activity, such as by sulfur-reducing bacteria. Elemental mercury may also be formed under anaerobic conditions by demethylation of methyl mercury, or by reduction of Hg(II). Acidic conditions (pH <4) also favor the formation of methyl mercury, whereas higher pH values favor precipitation of HgS(s) (Smith et al., 1995).

Influence of Soil Properties on Mobility

Chemical and physical properties of the contaminated matrix influence the mobility of metals in soils and groundwater. Contamination exists in three forms in the soil matrix: solubilized contaminants in the soil moisture, adsorbed contaminants on soil surfaces, and contaminants fixed chemically as solid compounds. The chemical and physical properties of the soil will influence the form of the metal contaminant, its mobility, and the technology selected for remediation (Gerber et al., 1991).

Chemical Properties

The presence of inorganic anions (carbonate, phosphate, sulfide) in the soil water can influence the soil's ability to fix metals chemically. These anions can form relatively insoluble complexes with metal ions and cause metals to desorb and/or precipitate in their presence.

Soil pH values generally range between 4.0 and 8.5 with buffering by Al at low pH and by $CaCO_3$ at high pH (Wild, 1988). Metal cations are most mobile under acidic conditions while anions tend to sorb to oxide minerals in this pH range (Dzombak and Morel, 1987). At high pH, cations precipitate or adsorb to mineral surfaces and metal anions are mobilized. The presence of hydrous metal oxides of Fe, Al, and Mn can strongly influence metal concentrations because these minerals can remove cations and anions from solution by ion exchange, specific adsorption, and surface precipitation (Dzombak and Morel, 1987; Ellis and Fogg, 1985). As noted in the previous section, sorption of metal cations onto hydrous oxides generally increases sharply with pH and is most significant at pH values above the neutral range, while sorption of metal anions is greatest at low pH and decreases as pH is increased (Figure 14.9.3). Cation exchange capacity (CEC) refers to the concentration of readily exchangeable cations on a mineral surface and is often used to indicate the affinity of soils for uptake of cations such as metals. Anion exchange capacity (AEC) indicates the affinity of soils for uptake of anions, and is usually significantly lower than the CEC of the soil. In addition to hydrous oxides, clays are also important ion exchange materials for metals (Sposito, 1989). The presence of natural organic matter (NOM) has been shown to influence the sorption of metal ions to mineral surfaces. NOM has been observed to enhance sorption of Cu^{2+} at low pH, and suppress Cu^{2+} sorption at high pH (Davis, 1984; Tipping et al., 1983).

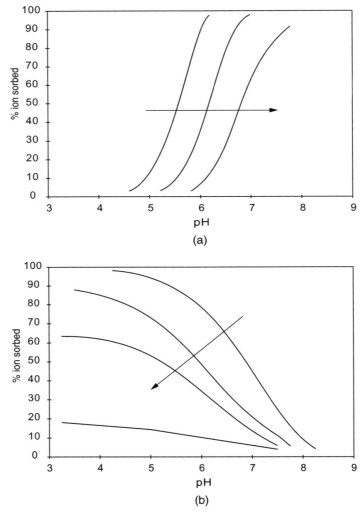

FIGURE 14.9.3   Typical pH edges for (a) cation sorption and (b) anion sorption. Arrows indicate direction of increasing sorbate/sorbent ratio (from Dzombak, and Morel, 1990).

Organic matter, particularly humic materials, can complex metals and affect their removal from solution (Ali and Dzombak, 1996). Humic materials contain carboxylic and phenolic functional groups that can complex with metal ions.

Physical Properties.   Particle size distribution can influence the level of metal contamination in a soil. Fine particles ($<100\ \mu$m) are more reactive and have a higher surface area than coarser material. As a result, the fine fraction of a soil often contains the majority of contamination. The distribution of particle sizes with which a metal contaminant is associated can determine the effectiveness of a number of metal remediation technologies (e.g., soil washing; Dzombak et al., 1994).

Soil moisture influences the chemistry of contaminated soil. The amount of dissolved minerals, pH, and redox potential of the soil water depends on the soil moisture content. Soil structure describes the size, shape, arrangement, and degree of development of soils into structural units. Soil structure can influence contaminant mobility by limiting the degree of contact between groundwater and contaminants.

14.9.3 AVAILABLE TECHNOLOGIES AND PERFORMANCE

Site Characterization and Establishment of Remediation Goals

The physical and chemical form of the metal contaminant in soil or water strongly influences the selection of the appropriate remediation treatment approach. Information about the physical characteristics of the site and the type and level of contamination at the site must be obtained to enable accurate assessment of site contamination and remedial alternatives. The importance of adequate, well-planned site characterization to selection of an appropriate cost-effective remediation approach has been discussed many times (e.g., CII, 1995) but cannot be overemphasized. The contamination in the groundwater and soil should be characterized to establish the type, amount, and distribution of contaminants across different media.

Once the site has been characterized, the desired level of each contaminant in soil and groundwater must be determined. This is done by comparison of observed contaminant concentrations with soil and groundwater quality standards for a particular regulatory domain, or by performance of a site-specific risk assessment. Remediation goals for metals may be set as desired concentrations in groundwater, as total metal concentration in soil, as leachable metal in soil, or as some combination of these.

General Remediation Approaches

Several technologies exist for the remediation of metals-contaminated soil and water. These technologies are contained within five categories of general approaches to remediation: isolation, immobilization, toxicity reduction, physical separation, and extraction. These are the same general approaches used for many types of contaminants in the subsurface (LaGrega et al., 1994). As is usually the case, combinations of one or more of these approaches are often used for more cost-effective treatment of a contaminated site. Table 14.9.1 summarizes key factors discussed in this report that were found to influence the applicability and selection of available remediation technologies.

Isolation. Isolation technologies attempt to prevent the transport of contaminants by containing them within a designated area. These technologies can be used to prevent further contamination of groundwater when other treatment options are not physically or economically feasible for a site. Contaminated sites may also be isolated temporarily in order to limit transport during site assessment and site remediation.

Capping. Capping systems are used to provide an impermeable barrier to surface water infiltration to contaminated soil for prevention of further release of contaminants to the surrounding surface water or groundwater. Secondary objectives include controlling gas and odor emissions, improving aesthetics, and providing a stable surface over a contaminated site. Capping also eliminates risks associated with dermal contact and/or incidental ingestion of surface soils, but if this is the primary goal for the site, and surface water infiltration is not a concern, a less expensive permeable cover may be preferred.

Capping provides a range of design options that includes simple single-layer caps and more complex multilayer systems (Rumer and Ryan, 1995; U.S. EPA, 1991). Design selection depends on site characteristics, remedial objectives, and risk factors associated with the site. A variety of materials are available for use in capping systems and choice of materials is site-specific because local soils are often incorporated into parts of the cap. Synthetic membranes such as high-density polyethylene are also available for incorporation into capping systems. Surface water controls, such as ditches and dikes, are usually included to help control drainage from the cap. Multilayered capping systems may

TABLE 14.9.1 Remediation Technologies Matrix for Metals in Soils and Groundwater

Remediation technology	Metals treated	Cost	Long-term Effectiveness/ permanence	Commerical availability	General acceptance	Applicability to high metals concentrations	Applicability to mixed waste (metals & organics)	Toxicity reduction	Mobility reduction	Volume reduction
Capping	1–3	+	«	+	+	«	+	«	+	«
Subsurface barriers	1–3, 5	+	«	+	+	«	+	«	+	«
Solidification/ stabilization ex situ	1–3, 5	•	•	+	+	+	+	«	+	«
Solidification/ stabilization in situ	1, 2, 4, 6	+	•	+	+	+	+	«	+	«
Vitrification ex situ	1–3, 5	«	+	•	•	+	+	+	+	«
Vitrification in situ	1–3, 7	«	+	•	•	+	+	+	+	«
Chemical treatment	2	–	•	•	•	–	–	+	+	«
Permeable treatment walls	2	–	«	•	•	«	–	+	+	«
Biological treatment	1–5	+	+	+	+	+	«	+	+	+
Physical separation	1–6	•	+	+	+	+	•	«	«	+
Soil washing	1–3, 5–7	•	+	+	+	+	«	«	«	+
Pyrometallurgical extraction	1–5, 7	«	+	+	+	+	«	«	«	+
In situ soil flushing	1, 2, 7	+	«	+	+	+	+	«	«	+
Electrokinetic treatment	1–6	•	+	+	+	+	–	«	«	+

Key: 1 = Lead, 2 = Chromium, 3 = Arsenic, 4 = Zinc, 5 = Cadmium, 6 = Copper, 7 = Mercury
+ Good, • Average, « Marginal, – Inadequate Information

also include a hard cover and/or a layer of topsoil to separate the underlying layers from the ground surface. Revegetation is promoted in order to reinforce the topsoil, to reduce soil erosion and runoff velocity, and to help remove water from the soil by evapotranspiration (Rumer and Ryan, 1995).

Subsurface Barriers. Subsurface barriers may be used to isolate contaminated soil and water by controlling the movement of groundwater at a contaminated site. These barriers are designed to reduce the movement of contaminated groundwater from the site, or to restrict the flow of uncontaminated groundwater through the contaminated site (Rumer and Ryan, 1995).

Vertical barriers are commonly used to restrict the lateral flow of groundwater. For effective isolation of the contaminated matrix, the barrier should extend and key into a continuous, low-permeability layer, such as clay or competent bedrock, below the contaminated area (Rumer and Ryan, 1995; U.S. EPA, 1985). If an impermeable layer is not available, a groundwater extraction system must be used to prevent transport of contaminants under the barrier. Vertical barriers may be installed upstream, downstream, or completely surrounding the site and are often implemented in conjunction with a capping system to control surface water infiltration. The use of circumferential barriers can prevent the escape of contamination from the site by using an infiltration barrier and collection system to create a hydraulic gradient in the inward direction. Vertical barriers are often limited to depths achievable with backhoe excavation technology for trenches (i.e., to about 30 ft; U.S. EPA, 1985).

Slurry walls are usually constructed in a vertical trench excavated under a slurry that is designed to prevent collapse and to form a filter cake on the walls of the trench to prevent the loss of fluids to the surrounding soil (Xanthakos, 1979). A vibrating beam method (Slurry Systems, Inc.) is also available in which the beam penetrates the ground and slurry materials are injected into the soil (with assistance from a high pressure/low volume jet if needed). Two options exist for the slurry composition. The soil-bentonite (SB) slurry wall is the most common type, and comprises a bentonite-water slurry that is mixed with a soil engineered to harden upon addition to the slurry (Rumer and Ryan, 1995). The trench can also be excavated under a portland cement-bentonite-water slurry that is left to harden and form a cement-bentonite (CB) slurry wall (LaGrega et al., 1994). Available technologies for installation of slurry walls allow installation to depths up to 125 ft.

Slurry walls are the most common type of vertical barrier because of their low relative cost. The use of slurry walls can be limited by the topography, geology, and type of contamination at the site. For example, an SB slurry will flow unless the site and confining layer are nearly level. Also, some contaminants, such as concentrated organics and strong acids/bases, can degrade SB materials and prevent the application of SB slurry walls at some sites (Rumer and Ryan, 1995).

Other available vertical barriers include grout curtains and sheet piles. Grout curtains are constructed by drilling a borehole and injecting a fluid into the surrounding soil, which is designed to solidify and reduce water flow through the contaminated region (U.S. EPA, 1985). The fluid is pressure-injected in rows of staggered boreholes that are designed to overlap once the fluid has permeated into the surrounding soil. Common materials used to construct grout curtains include cement, clays, alkali-silicate, and organic polymers (Rumer and Ryan, 1995). Clays are the most widely used grouting materials because of their low cost. This technique is more expensive than slurry walls and its use is therefore usually limited to sealing voids in existing rock.

Sheet piles usually comprise steel pilings that are driven into the formation to create a wall to contain the groundwater. Sheet piles are seldom used at contaminated sites because of concerns about wall integrity. This method is generally limited to isolation of shallow contamination (40–50 ft) distributed over a relatively small area (U.S. EPA, 1985), or used in conjunction with a soil-bentonite slurry when site conditions prevent the use of conventional slurry walls (Rumer and Ryan, 1995).

Technologies for the construction of horizontal barriers are under investigation. Horizontal barriers would enable control of the downward migration of contaminants by lining the site without requiring excavation of the contaminated matrix. The technologies under investigation include grout injection by vertical boring and horizontal drilling. The vertical boring method is similar to the construction of grout curtains except that the grout is injected at a fixed elevation over a tightly spaced grid of vertical boreholes to create an impermeable horizontal layer. Problems with this method include soil compaction by the large drill rigs situated over the contaminated area. Also, the vertical boreholes would provide access to the deeper layers and may therefore increase vertical migration of contaminants. Horizontal drilling involves the use of directional drilling techniques to create the horizontal grout layer.

Horizontal barriers may also be used in conjunction with vertical barriers at sites where a natural aquitard is not present. In this case, the vertical barrier could key into the horizontal barrier to prevent the transport of contaminants under the vertical barrier (Smith et al., 1995).

Immobilization.   Immobilization technologies are designed to reduce the mobility of contaminants by changing the physical or leaching characteristics of the contaminated matrix. Mobility is usually decreased by physically restricting contact between the contaminant and the surrounding groundwater, or by chemically altering the contaminant to make it more stable with respect to dissolution in groundwater. The aqueous and solid phase chemistry of metals is conducive to immobilization by these techniques. A variety of methods are available for immobilization of metal contaminants, including those that use chemical reagents and/or thermal treatment to physically bind the contaminated soil or sludge. Most immobilization technologies can be performed ex situ or in situ. In situ processes are preferred because of the lower labor and energy requirements, but implementation in situ will depend on specific site conditions.

Solidification/Stabilization.   Solidification and stabilization (S/S) immobilization technologies are the most commonly selected treatment options for metals-contaminated sites (Conner, 1990). Solidification involves the formation of a solidified matrix that physically binds the contaminated material. Stabilization, also referred to as fixation, usually utilizes a chemical reaction to convert the waste to a less mobile form. The general approach for solidification/stabilization treatment processes involves mixing or injecting treatment agents to the contaminated soils. Inorganic binders, such as cement, fly ash, or blast furnace slag, and organic binders such as bitumen are used to form a crystalline, glassy or polymeric framework around the waste. The dominant mechanism by which metals are immobilized is by precipitation of hydroxides within the solid matrix (Bishop et al., 1982; Shively et al., 1986).

S/S technologies are not useful for some forms of metal contamination, such as species that exist as anions (e.g., Cr(VI), arsenic) or metals that don't have low-solubility hydroxides (e.g., mercury). S/S may not be applicable at sites containing wastes that include organic forms of contamination, especially if volatile organics are present. Mixing and heating associated with binder hydration may release organic vapors. Pretreatment, such as air stripping or incineration, may be used to remove the organics and prepare the waste for metal stabilization/solidification (Smith et al., 1995). The application of S/S technologies will also be affected by the chemical composition of the contaminated matrix, the amount of water present, and the ambient temperature. These factors can interfere with the solidification/stabilization process by inhibiting bonding of the waste to the binding material retarding the setting of the mixtures, decreasing the stability of the matrix, or reducing the strength of the solidified area (U.S. EPA, 1990b).

Cement-based binders and stabilizers are common materials used for implementation of S/S technologies (Conner, 1990). Portland cement, a mixture of Ca-silicates, aluminates, aluminoferrites, and sulfates is an important cement-based material. Pozzolanic materials which consist of small spherical particles formed by coal combustion (such as fly ash) and in lime and cement kilns, are also commonly used for S/S. Pozzolans exhibit cement-like properties, especially if the silica content is high. Portland cement and pozzolans can be used alone or together to obtain optimal properties for a particular site (U.S. EPA, 1989).

Organic binders may also be used to treat metals through polymer microencapsulation. This process uses organic materials such as bitumen, polyethylene, paraffins, waxes, and other polyolefins as thermoplastic or thermosetting resins. For polymer encapsulation, the organic materials are heated and mixed with the contaminated matrix at elevated temperatures (120° to 200°C). The organic materials polymerize, agglomerate the waste, and the waste matrix is encapsulated (U.S. EPA, 1989). Organics are volatilized and collected and the treated material is extruded for disposal or possible reuse (e.g., as paving material; Smith et al., 1995). The contaminated material may require pretreatment to separate rocks and debris and dry the feed material. Polymer encapsulation requires more energy and more complex equipment than cement-based S/S operations. Bitumen (asphalt) is the least expensive and most common thermoplastic binder (U.S. EPA, 1989).

S/S is achieved by mixing the contaminated material with appropriate amounts of binder/stabilizer and water. The mixture sets and cures to form a solidified matrix and contain the waste. The cure time and pour characteristics of the mixture and the final properties of the hardened cement depend upon the composition (amount of cement, pozzolan, water) of the binder/stabilizer.

Ex situ S/S can be easily applied to excavated soils because methods are available to provide the vigorous mixing needed to combine the binder/stabilizer with the contaminated material. Pretreatment of the waste may be necessary to screen and crush large rocks and debris. Mixing can be performed via in-drum, in-plant, or area mixing processes. In-drum mixing may be preferred for treatment of small volumes of waste or for toxic wastes. In-plant processes utilize rotary drum mixers for batch processes or pug mill mixers for continuous treatment. Larger volumes of waste may be excavated and moved to a contained area for area mixing. This process involves layering the contaminated material with the stabilizer/binder, and subsequent mixing with a backhoe or similar equipment. Mobile and fixed treatment plants are available for ex situ S/S treatment. Smaller pilot-scale plants can treat up to 100 tons of contaminated soil per day, while larger portable plants typically process 500 to over 1000 tons per day (Smith et al., 1995).

S/S techniques are available to provide mixing of the binder/stabilizer with the contaminated soil in situ. In situ S/S is less labor- and energy-intensive than the ex situ process, which requires excavation, transport, and disposal of the treated material. In situ S/S is also preferred if volatile or semi-volatile organics are present, because excavation would expose these contaminants to the air (U.S. EPA, 1990a). However, the presence of bedrock, large boulders, cohesive soils, oily sands and clays may preclude the application of in situ S/S at some sites. It is also more difficult to provide uniform and complete mixing through in situ processes.

Mixing of the binder and contaminated matrix may be achieved using in-place mixing, vertical auger mixing, or injection grouting. In-place mixing is similar to ex situ area mixing except that the soil is not excavated prior to treatment. The in situ process is useful for treating surface or shallow contamination and involves spreading and mixing the binders with the waste using conventional excavation equipment such as draglines, backhoes, or clamshell buckets. Vertical auger mixing uses a system of augers to inject and mix the binding reagents with the waste. Larger (6–12 ft diameter) augers are used for shallow (10–40 ft) drilling and can treat 500 to 1000 cubic yards per day (Jasperse and Ryan, 1992; Ryan and Walker, 1992). Deep stabilization/solidification (up to 150 ft) can be achieved by using ganged augers (up to 3 ft in diameter each) that can treat 150 to 400 cubic yards per day. Finally, injection grouting may be performed to inject the binder containing suspended or dissolved reagents into the treatment area under pressure. The binder permeates the surrounding soil and cures in place (Smith et al., 1995).

Vitrification. The mobility of metal contaminants can be decreased by high-temperature treatment of the contaminated area, which results in the formation of vitreous material, usually an oxide solid. During this process, the increased temperature may also volatilize and/or destroy organic contaminants or volatile metal species (such as Hg) that must be collected for treatment or disposal. Most soils can be treated by vitrification and a wide variety of inorganic and organic contaminants can be targeted. Vitrification may be performed ex situ or in situ, although in situ processes are preferred because of the lower energy requirements and cost (U.S. EPA, 1992a).

Typical stages in ex situ vitrification processes may include excavation, pretreatment, mixing, feeding, melting and vitrification, off-gas collection and treatment, and forming or casting of the melted product. The energy requirement for melting is the primary factor influencing the cost of ex situ vitrification. Different sources of energy can be used for this purpose, depending on local energy costs. Process heat losses and water content of the feed should be controlled in order to minimize energy requirements. Vitrified material with certain characteristics may be obtained by using additives such as sand, clay, and/or native soil. The vitrified waste may be recycled and used as clean fill, aggregate, or other reusable materials (Smith et al., 1995).

In situ vitrification (ISV) involves passing electric current through the soil using an array of electrodes inserted vertically into the contaminated region. Each setting of four electrodes is referred to as a melt. If the soil is too dry, it may not provide sufficient conductance and a trench containing flaked graphite and glass frit (ground glass particles) must be placed between the electrodes to provide an initial flow path for the current. Resistance heating in the starter path melts the soil. The melt grows outward and down as the molten soil usually provides additional conductance for the current. A single melt can treat up to 1000 tons of contaminated soil to depths of 20 ft, at a typical treatment rate of 3 to 6 tons per hour. Larger areas are treated by fusing together multiple individual vitrification zones. The main requirement for in situ vitrification is the ability of the soil melt to carry current and solidify as it cools. If the alkali content (as Na_2O and K_2O) of the soil is too high (≥ 1.4 wt%) the molten soil may not provide enough conductance to carry the current (Buelt and Thompson, 1992).

Toxicity and/or Mobility Reduction.   Chemical and/or biological processes can be used to alter the form of metal contaminants in order to decrease their toxicity and/or mobility.

Chemical Treatment.   Chemical reactions can be initiated that are designed to decrease the toxicity or mobility of metal contaminants. The three types of reactions that can be used for this purpose are oxidation, reduction, and neutralization reactions. Chemical oxidation changes the oxidation state of the metal atom through the loss of electrons. Commercial oxidizing agents are available for chemical treatment, including potassium permanganate, hydrogen peroxide, hypochlorite, and chlorine gas. Reduction reactions change the oxidation state of metals by adding electrons. Commercially available reduction reagents include alkali metals (Na, K), sulfur dioxide, sulfite salts, and ferrous sulfate. Changing the oxidation state of metals by oxidation or reduction can detoxify, precipitate, or solubilize the metals (NRC, 1994). Chemical neutralization is used to adjust the pH balance of extremely acidic or basic soils and/or groundwater. This procedure can be used to precipitate insoluble metal salts from contaminated water, or used in preparation for chemical oxidation or reduction.

Chemical treatment can be performed ex situ or in situ. However, in situ chemical agents must be carefully selected so that they do not further contaminate the treatment area. The primary problem associated with chemical treatment is the nonspecific nature of the chemical reagents. Oxidizing/reducing agents added to the matrix to treat one metal will also target other reactive metals and can make them more toxic or mobile (NRC, 1994). Also, the long-term stability of reaction products is of concern because changes in soil and water chemistry might reverse the selected reactions.

Chemical treatment is often used as pretreatment for S/S and other treatment technologies. Reduction of Cr(VI) to Cr(III) is the most common form of chemical treatment and is necessary for remediation of wastes containing Cr(VI) by precipitation or S/S. Chromium in its Cr(III) form is readily precipitated by hydroxide over a wide range of pH values. Acidification may also be used to aid in Cr(VI) reduction. Arsenic may be treatable by chemical oxidation because arsenate, As(V), is less toxic, soluble and mobile than arsenite, As(III). Bench-scale work has indicated that arsenic stabilization may be achieved by precipitation and coprecipitation with Fe(III) (Smith et al., 1995).

Permeable Treatment Walls.   Treatment walls remove contaminants from groundwater by degrading, transforming, precipitating, or adsorbing the target solutes as the water flows through permeable trenches containing reactive material within the subsurface (Vidic and Pohland, 1996). Several methods are available for installation of permeable treatment walls, some of which employ slurry wall construction technology to create a permeable reactive curtain. The reactive zone can use physical, chemical and biological processes, or a combination of these. The groundwater flow through the wall may be enhanced by inducing a hydraulic gradient in the direction of the treatment zone or channeling groundwater flow toward the treatment zone (NRC, 1994).

Several types of treatment walls are being tried for arresting transport of metals in groundwater at contaminated sites. Trench materials being investigated include zeolite, hydroxyapatite, elemental iron, and limestone (Vidic and Pohland, 1996). Applications of elemental iron for chromium (VI) reduction and limestone for lead precipitation and adsorption are described below.

Elemental Iron

Trenches filled with elemental iron have shown promise for remediation of metals- contaminated sites. While investigations of this technology have focused largely on treatment of halogenated organic compounds, studies are being performed to assess the applicability to remediation of inorganic contaminants (Powell et al., 1994).

Low oxidation-state chemical species can serve as electron donors for the reduction of higher oxidation-state contaminants. This ability can be exploited to remediate metals that are more toxic and mobile in higher oxidation states, such as Cr(VI). Results of column experiments performed by

Powell et al. (1994) and batch experiments performed by Cantrell et al. (1995) showed that chromate reduction was enhanced in systems containing iron filings in addition to the natural aquifer material. A field experiment has been initiated by researchers at the U.S. EPA National Risk Management Research Laboratory to investigate the use of zero-valent iron for chromium remediation at the U.S. Coast Guard air support base near Elizabeth City, North Carolina. Preliminary results indicate that the test barrier has reduced chromate in the groundwater to below detection limits (Wilson, 1995).

Limestone Barriers

The use of limestone treatment walls has been proposed for sites with metals contamination, in particular former lead acid battery recycling sites, which have lead and acid contamination in groundwater and soil. In such cases, a limestone trench can provide neutralization of acidic groundwater. The attendant rise in pH promotes immobilization of any dissolved lead through precipitation and/or adsorption onto minerals. A limestone trench system is in design for implementation at the Tonolli Superfund site in Nesquehoning, Pennsylvania (U.S. EPA, 1992b).

There is some experience in the coal mining industry with use of limestone in the manner anticipated for the Tonolli site. Most of this experience has been acquired since 1990, when the concept of "anoxic limestone drains" was introduced (Turner and McCoy, 1990). Since that time, numerous limestone drain systems have been installed at Appalachian coal field sites (primarily in Kentucky, West Virginia, and Pennsylvania) in an attempt to control acid mine drainage. Summaries of installations and evolving design considerations are provided in Hedin and Nairn (1992), Hedin et al. (1994), and Hedin and Watzlaf (1994).

Design and operating guidelines for the anoxic limestone drains have mostly been developed from trial and observation. Briefly, the systems in use employ fairly large, #3 or #4 (baseball size) limestone rocks. Anoxic mine water is directed to the limestone drain, which is installed with a soil cover to inhibit contact with air. Hedin and Nairn (1992) report that "some systems constructed with limestone powder and gravel have failed, apparently because of plugging problems." Preliminary review of the literature on design of anoxic limestone drains indicates primary concern with maintenance of anoxic conditions in the drains. If high dissolved concentrations of Fe are present and aerobic conditions develop, insoluble ferric hydroxide can form and coat the limestone, rendering it ineffective. High concentrations of aluminum are also a concern, as aluminum hydroxide can precipitate and yield the same kind of coating problems. With use of large diameter stones, plugging is prevented even if precipitation occurs and the stones become coated with precipitate.

Available operating data for anoxic limestone drains indicate that they can be effective in raising the pH of strongly acidic water. Hedin and Watzlaf (1994) reviewed operating data for 21 limestone drain systems. The data they compiled showed fairly consistent increases in pH of highly acidic mine drainage (at pH 2.3 to 3.5) to pH values in the range of 6.0 to 6.7. Thus, there is clearly precedent for employing the limestone drain approach with some confidence of success in raising pH of highly acidic water. Long- term (i.e., greater than 10 yr) performance cannot be predicted with confidence as there has been relatively short duration operating experience. However, experience to date indicates clearly that limestone drain systems can operate effectively under appropriate conditions, especially anoxic or low-oxygen groundwater, for at least several years.

Biological Treatment.   Biological treatment technologies are available for remediation of metals-contaminated sites. These technologies are commonly used for the remediation of organic contaminants and are beginning to be applied for metal remediation, although most applications to date have been at the bench and pilot scale (Schnoor, 1997). Biological treatment exploits natural biological processes that allow certain plants and microorganisms to aid in the remediation of metals. These processes occur through a variety of mechanisms, including adsorption, oxidation and reduction reactions, and methylation (Means and Hinchee, 1994).

Bioaccumulation

Bioaccumulation involves the uptake of metals from contaminated media by living organisms or dead, inactive biomass. Active plants and microorganisms accumulate metals as the result of normal metabolic processes via ion exchange at the cell walls, complexation reactions at the cell walls, or intra- and extracellular precipitation and complexation reactions. Adsorption to ionic groups on the cell surface is the primary mechanism for metal adsorption by inactive biomass. Accumulation in biomass has been shown to be as effective as some ion exchange resins for metals removal from water (Means and Hinchee, 1994).

Phytoremediation

Phytoremediation refers to the specific ability of plants to aid in metal remediation. Some plants have developed the ability to remove ions selectively from the soil to regulate the uptake and distribution of metals. Most metal uptake occurs in the root system, usually via absorption, where many mechanisms are available to prevent metal toxicity due to high concentration of metals in the soil and water. Potentially useful phytoremediation technologies for remediation of metals-contaminated sites include phytoextraction, phytostabilization, and rhizofiltration (U.S. EPA, 1996b).

Phytoextraction employs hyperaccumulating plants to remove metals from the soil by absorption into the roots and shoots of the plant. A hyperaccumulator is defined as a plant with the ability to yield ≥0.1% chromium, cobalt, copper or nickel or ≥1% zinc, manganese in the above-ground shoots on a dry weight basis. The above-ground shoots can be harvested to remove metals from the site and subsequently disposed as hazardous waste or treated for the recovery of the metals.

Phytostabilization involves the use of plants to limit the mobility and bioavailability of metals in soil. Phytostabilizers are characterized by high tolerance of metals in surrounding soils but low accumulation of metals in the plant. This technique may be used as an interim containment strategy until other remediation techniques can be developed, or as treatment at sites where other methods would not be economically feasible.

Rhizofiltration removes metals from contaminated groundwater via absorption, concentration, and precipitation by plant roots. This technique is used to treat contaminated water rather than soil and is most effective for large volumes of water with low levels of metal contamination. Terrestrial plants are more effective than aquatic plants because they develop a longer, more fibrous root system, which provides a larger surface area for interaction. Wetlands construction is a form of rhizofiltration that has been demonstrated as a cost-effective treatment for metals-contaminated wastewater.

Bioleaching

Bioleaching uses microorganisms to solubilize metal contaminants either by direct action of the bacteria, as a result of interactions with metabolic products, or both. Bioleaching can be used in situ or ex situ to aid the removal of metals from soils. This process is being adapted from the mining industry for use in metals remediation. The mechanisms responsible for bioleaching are not fully defined, but, in the case of mercury, bioreduction (to elemental mercury) is thought to be responsible for mobilization of mercury salts (Means and Hinchee, 1994).

Biochemical Processes

Microbially mediated oxidation and reduction reactions can be manipulated for metal remediation. Some microorganisms can oxidize/reduce metal contaminants directly while others produce chemical oxidizing/reducing agents that interact with the metals to effect a change in oxidation state. Mercury and cadmium have been observed to be oxidized through microbial processes, and arsenic and iron are readily reduced in the presence of appropriate microorganisms. The mobility of metal contaminants is influenced by their oxidation state. Redox reactions therefore, can, be used to increase or decrease metal mobility (Means and Hinchee, 1994).

Methylation involves attaching methyl groups to inorganic forms of metal ions to form organometallic compounds. Methylation reactions can be microbially mediated. Organometallic compounds are more volatile than inorganic metals, and this process can be used to remove metals through volatilization and subsequent removal from the gas stream. However, organometallics are also more toxic and mobile than other metal forms and may potentially contaminate surrounding surface waters and groundwater (Means and Hinchee, 1994).

Physical Separation.   Physical separation is an ex situ process that attempts to separate the contaminated material from the rest of the soil matrix by exploiting certain characteristics of the metal and soil. Physical separation techniques are available that operate based on particle size, particle density, surface, and magnetic properties of the contaminated soil. These techniques are most effective when the metal is either in the form of discrete particles in the soil or if the metal is sorbed to soil particles that occur in a particular size fraction of the soil. Physical separation is often used as a form of pre-treatment in order to reduce the amount of material requiring subsequent treatment (Rosetti, 1993). Several techniques are available for physical separation of contaminated soils including screening, classification, gravity concentration, magnetic separation, and froth flotation.

Screening separates soils according to particle size by passing the matrix through a sieve with particular size openings. Smaller particles pass through the sieve and leave larger particles behind; however, the separation is not always complete. Screening may be performed as a stationary process or with motion using a wet or dry process stream (Smith et al., 1995).

Classification involves separation of particles based upon the velocity with which they fall through water (hydroclassification) or air (air classification). Hydroclassification is more common for soil separation and may be performed using a nonmechanical, mechanical, or a hydraulic classifier (Rosetti, 1993).

Gravity concentration relies on gravity and one or more other forces (e.g., centrifugal force, velocity gradients), which may be applied to separate particles on the basis of density differences. Gravity concentration may be achieved through the use of a hydrocyclone, jig, spiral concentrator, or shaking table (Rosetti, 1993).

Froth flotation uses air flotation columns or cells to remove particles from water. In this process, air is sparged from the bottom of a tank or column that contains a slurry of the contaminated material. Some metals and minerals attach to the air bubbles because of particular surface properties, such as hydrophobicity. Froth flotation can be used to remove metals that attach to air bubbles, or to remove other minerals while the metal remains in the slurry (Rosetti, 1993).

Magnetic separation subjects particles to a strong magnetic field using electromagnets or magnetic filters and relies on differences in magnetic properties of minerals for separation. Low-intensity wet magnetic separators are the most common magnetic separation devices. This process can recover a wide variety of minerals and is particularly successful for separating ferrous from nonferrous minerals (Allen and Torres, 1991).

Extraction.   Metals-contaminated sites can be remediated using techniques designed to extract the contaminated fraction from the rest of the soil, either in situ or ex situ. Metal extraction can achieved by contacting the contaminated soil with a solution containing extracting agents (soil washing and in situ soil flushing) or by electrokinetic processes. The contaminated fraction of soil and/or process water is separated from the remaining soil and disposed or treated.

Soil Washing.   Soil washing can be used to remove metals from the soil by chemical or physical treatment methods in aqueous suspension. Soil washing is an ex situ process that requires soil exca-vation prior to treatment. Chemical treatment involves addition of extraction agents that react with the contaminant and leach it from the soil (Elliot and Brown, 1989; Ellis and Fogg, 1985; Tuin and Tels, 1990). The liquid containing the contaminants is separated from the soil resulting in a clean solid phase. Physical treatment is achieved by particle size separation technologies adapted from mineral processing to concentrate the contaminant in a particular size fraction (Allen and Torres, 1991).

Fine particles (<63 μm) often contain the majority of contaminated material because they bind contaminants strongly because of their large and reactive surface area. Many current soil washing approaches attempt to separate the fine fraction from the remainder of the soil in order to reduce the amount of material for subsequent treatment or disposal (Rosetti, 1993). Particle size separation techniques may not be successful if fine particle (e.g., metal oxide) coatings are present on particles in larger size fractions (Van Ben Schoten et al., 1994).

Preliminary Screening

After excavation, the soil undergoes preliminary screening and preparation in order to separate large rocks and debris from the contaminated matrix. Residual fines may be adhered to the surface of large rocks and are often washed off prior to return of the large rocks to the site (Rosetti, 1993).

Secondary Screening

Most soil washing processes employ secondary screening to segregate the particles into different size fractions, usually between 5 and 60 mm. Most secondary screening processes involve making an aqueous slurry of the soil stream and wet screening/sieving of the slurry. The particles in this size range are considered less contaminated than the finer fraction and may be returned to the site as clean soil after separation from the water (Rosetti, 1993).

Chemical Treatment

Chemical treatment may be used to solubilize contaminants from the most contaminated fraction of the soil. Chemical treatment is performed in an aqueous slurry of the contaminated material to which an extracting agent is added. The extraction is performed in a mixing vessel or in combination with the physical treatment stage. The type of extractant used will depend on the contaminants present and the characteristics of the soil matrix. Many processes manipulate the acid/base chemistry of the slurry to leach contaminants from the soil (Tuin and Tels, 1990). However, if a very low pH is required, concerns about dissolution of the soil matrix may arise. Chelating agents (e.g., EDTA) selectively bind with some metals and may be used to solubilize contaminants from the soil matrix (Elliot and Brown, 1989). Oxidizing and reducing agents (e.g., hydrogen peroxide, sodium borohydride) provide yet another option to aid in solubilization of metals because chemical oxidation/reduction can convert metals to more soluble forms (Assink and Rulkens, 1989; Tuin et al., 1987). Finally, surfactants may be used in extraction of metals from soil (U.S. EPA, 1996b).

Physical Treatment

Physical treatment is used to separate the contaminated fraction, usually the fine materials, from the rest of the soil matrix. Physical separation may be performed alone (see previous section) or in conjunction with chemical treatment, as in most soil washing processes. The most common method for physical separation in soil washing uses rotary attrition scrubbers to isolate the contaminated particles. The rotation of the slurry causes contact between large particles, resulting in attrition of the larger particles, which releases the contaminant and contaminated fines to the slurry. The contaminant remains suspended in solution or sorbs to the reactive fine particles. Vibration units are also available to perform similar separations (Rosetti, 1993).

Hydrocyclones are the most common method used to separate fines from the clean soil. Other options are available for fine particle separation, including mechanical classifiers, gravity classifiers, spiral concentrators, and magnetic separators (Rosetti, 1993). Froth flotation can be used to combine physical and chemical treatment processes into one step. For this method, extracting agent is added to the soil before it enters the froth flotation cell. The slurry is leached in the tanks to remove the contaminant and the fines (<50 μm) are then separated from coarse particles in the flotation unit (Rosetti, 1993).

Dewatering

After the contaminated fine particles are separated from the clean coarse particles, both fractions are dewatered. The fine fraction is usually dewatered using a belt filter or filter press and disposed of in a landfill. Larger particles are rinsed to remove residual extracting solution and contaminant and dewatered using belt and filter presses. This fraction is considered clean and can be returned to the site.

Water Treatment

The contaminated water from rinsing and dewatering steps is treated by manipulating the solution chemistry to separate the contaminant from the extractant if possible. Contaminants can then be removed from solution, most commonly by precipitation or sedimentation, and are dewatered before disposal with the contaminated fines. The extracting agent and process water can be recycled for reuse.

Pyrometallurgical Extraction.   Pyrometallurgical technologies use elevated temperature extraction and processing for removal of metals from contaminated soils. Soils are treated in a high-temperature furnace to remove volatile metals from the solid phase. Subsequent treatment steps may include metal recovery or immobilization. Pyrometallurgical treatment requires a uniform feed material for efficient heat transfer between the gas and solid phases and minimization of particulates in the off-gas. This process is usually preceded by physical treatment to provide optimum particle size. Pyrometallurgical processes usually produce a metal-bearing waste slag, but the metals can also be recovered for reuse (U.S. EPA, 1996c).

In Situ Soil Flushing.   In situ soil flushing is used to mobilize metals by leaching contaminants from soils so that they can be extracted without excavating the contaminated materials. An aqueous extracting solution is injected into or sprayed onto the contaminated area to mobilize the contaminants usually by solubilization. The extractant can be applied by surface flooding, sprinklers, leach fields, vertical or horizontal injection wells, basin infiltration systems, or trench infiltration systems (U.S. EPA, 1996b). After being contacted with the contaminated material, the extractant solution is collected using pump-and-treat methods for disposal or treatment and reuse. Similar extracting agents are used for in situ soil flushing and soil washing, including acids/bases, chelating agents, oxidizing/reducing agents, and surfactants/cosolvents. Also, water can be used alone to remove water-soluble contaminants such as hexavalent chromium. The applicability of in situ soil flushing technologies to contaminated sites will depend largely on site-specific properties, such as hydraulic conductivity, that influence the ability to contact the extractant with contaminants and to effectively recover the flushing solution with collection wells (NRC, 1994).

Electrokinetic Treatment.   Electrokinetic remediation technologies apply a low density current to contaminated soil in order to mobilize contaminants in the form of charged species. The current is applied by inserting electrodes into the subsurface and by relying on the natural conductivity of the soil (as a result of water and salts) to effect movement of water, ions, and particulates through the soil. Water and/or chemical solutions can also be added to enhance the recovery of metals by this process. Positively charged metal ions migrate to the negatively charged electrode, while metal anions migrate to the positively charged electrode. Electrokinetic treatment concentrates contaminants in the solution around the electrodes. The contaminants are removed from this solution by a variety of processes, including electroplating at the electrodes, precipitation/coprecipitation at the electrodes, complexation with ion exchange resins, or by pumping the water from the subsurface and treating it to recover the extracted metals (Smith et al, 1995).

Electrokinetic treatment is most applicable to saturated soils with low groundwater flow rates and moderate to low permeability. The efficiency of metal removal by this process will be influenced by the type and concentration of contaminant, the type of soil, soil structure, and interfacial chemistry of the soil.

Performance of Available Commercial Technologies

The following section focuses on commercially available technologies that have been demonstrated or implemented for metals-contaminated soils and groundwater. Ex situ treatment technologies are examined only for soils. The full range of contaminated water treatment technologies is available for ex situ treatment of groundwater. For the most part, the technologies reported are those whose performance has been verified by the U.S. EPA under the Superfund Innovative Technology Evaluation (SITE) program, which evaluates emerging and demonstrated technologies. Technologies currently in the SITE demonstration phase are also discussed.

Superfund Innovative Technology Evaluation (SITE) Demonstration and Best Demonstrated Available Technology (BDAT) Status.   The 1986 Superfund Amendments and Reauthorization Act recognized a need for an "alternative or innovative technology research and demonstration program." In response, the U.S. EPA established the Superfund Innovative Technology Evaluation (SITE) Program to encourage the development and implementation of innovative treatment technologies for remediation of hazardous waste sites and for monitoring and measurement. Innovative technologies are field tested in the SITE Demonstration Program, and engineering and cost data are collected to assess the performance of the technology. The demonstration stage also attempts to evaluate the applicability of

the technology to different types of wastes and waste matrices, the need for pre- and postprocessing of the waste stream, and potential operating problems. The SITE Program is administered by the U.S. EPA Office of Research and Development (ORD) National Risk Management Research Laboratory, headquartered in Cincinnati, Ohio.

The Resource Conservation and Recovery Act (RCRA) provides for determination of a Best Demonstrated Available Technology (BDAT) for treatment of hazardous wastes. BDATs have been established based upon critical analysis of performance data collected for treatment of various industry-generated wastes. BDAT status is given only to proven, commercially available technologies. Different BDATs and treatment standards are usually given for nonwastewater and wastewater forms of contamination. The applicability of a BDAT to metals-contaminated soil and water at a Superfund site must be evaluated on a site specific basis. The establishment of a BDAT does not prevent the use of other available technologies for treatment of these wastes.

Containment.   Containment technologies are widely used to control the transport of hazardous materials and to prevent the spread of contamination. Containment is the preferred remedial method for sites having low levels of wastes with low toxicity and low mobility, or wastes that have been pretreated to obtain these characteristics. Containment may also be used as a temporary measure to reduce the mobility of wastes that pose a high risk until a permanent remedy is selected and implemented. Advantages to containment technologies include relatively simple and rapid implementation often at lower cost than alternatives that require excavation; ability to treat large areas and volumes of waste; and the potential for successful containment as the final action at the site. Uncertainty regarding long-term effectiveness and the need for long-term inspection because untreated contaminants remain onsite are among the disadvantages of containment technologies. Also, future use of the site may be limited if containment technologies are used. Containment has been selected as the remedial operation for soil contaminated with metals at a number of sites. Some example applications are summarized in Table 14.9.2.

Capping systems have been selected for a number of sites with low levels of metal contamination. Monitoring wells and/or infiltration monitoring systems are often used to help assess the performance of capping systems.

Slurry walls have also been used for containment of metals-contaminated sites. The performance of vertical containment barriers also must be monitored. Performance can be influenced by geography, topography, and geology. The presence of certain compounds can also influence the long-term integrity of some cement-based vertical barriers by chemically attacking the soil-bentonite blends. Material availability can affect the application of slurry walls and other containment technologies.

There are no established BDATs for containment technologies because they are not considered to be treatment technologies. Ongoing SITE demonstrations for remediation of metals by containment technologies include a high clay grouting procedure (Morrison Knudsen Corporation) and frozen soil barriers (RKK, Ltd.).

Ex Situ Remediation.   The majority of the technologies that have been demonstrated for metals remediation to date are ex situ technologies. Ex situ remediation technologies demonstrated include solidification/stabilization, soil washing, vitrification, and pyrometallurgic separation.

TABLE 14.9.2   Example Containment Applications at Metals-Contaminated Superfund Sites (from U.S. EPA, 1996c)

Site name/state	Containment technology	Metal contaminants	Secondary technology	Status*
Ninth avenue dump, IN	Slurry wall	Pb	Capping	S
Industrial waste control, AK	Slurry wall	As, Cd, Cr, Pb	Capping, French Drain	I
E. H. Shilling Landfill, OH	Slurry wall	As	Capping, Clay Berm	S
Chemtronic, NC	Capping	Cr, Pb		S
Ordnance works disposal, WV	Capping	As, Pb		S
Industriplex, MA	Capping	As, Pb, Cr		I

* Status codes as of February, 1996: I = in operation; S = selected.

Solidification/Stabilization. Immobilization technologies, especially solidification/stabilization, are the most common methods selected for remediation of metal contamination, accounting for nearly 30 percent of all soil treatment technologies at Superfund sites. S/S techniques have been widely used to manage metal wastes at hazardous waste sites and to treat residues from other treatment processes (LaGrega et al., 1994). Benefits associated with immobilization treatments include their broad application to a wide variety of metals (Malone and Jones, 1985) and also to wastes that contain mixtures of metals and organics (U.S. EPA, 1996b).

Solidification/stabilization technologies using cement-based and pozzolan binders are available commercially and have been applied at several sites for a wide variety of metals, including chromium, lead, arsenic, mercury, and cadmium (Lister, 1996; Lo et al., 1988; Stanczyk and Senefelder, 1982; Zirschky and Piznar, 1988).

Examples of sites where ex situ S/S technologies have been selected and/or implemented for remediation of metals-contaminated soils are given in Table 14.9.3. Remediation has been completed for a number of these sites and S/S has been selected or initiated for several others. SITE demonstrations have been performed or are underway for various ex situ stabilization/solidification technologies.

Ex situ solidification/stabilization techniques have been determined to be the BDAT for a range of waste types, including cadmium nonwastewaters (other than Cd-containing batteries), chromium nonwastewaters (after reduction to Cr(III)), lead nonwastewaters, wastes with low (<260 mg/kg) concentrations of elemental mercury, and plating and steel-making wastes. S/S can also be used to treat arsenic wastes even though vitrification was selected as the BDAT for arsenic-containing nonwastewaters (U.S. EPA, 1996c).

TABLE 14.9.3 Example Solidification/Stabilization Applications at Selected Metals-Contaminated Superfund Sites (from U.S. EPA, 1996c)

Site name/state	S/S technology	Metal contaminants	Secondary technology	Status*
DeRewal Chemical, NJ	Solidification	Cr, Cd, Pb	GW pump and treat	S
Marathon Battery Co., NY	Stabilization	Cd, Ni	Dredging, off-site disposal	I
Nascolite, NJ	Stabilization	Pb	On-site disposal of stabilized soil, off-site disposal of wetland soil	S
Roebling Steel, NJ	S/S	As, Cr, Pb	Capping	S
Waldick Aerospace, NJ	S/S	Cd, Cr	Off-site disposal	C
Aladdin Plating, PA	Stabilization	Cr	Off-site disposal	C
Palmerton Zinc, PA	Stabilization	Cd, Pb	—	I
Tonolli Corp.	S/S	As, Pb	In situ chemical barrier	S
Whitmoyer Laboratories, PA	Oxidation/ stabilization	As	GW pump and treat, capping, grading, revegetation	S
Bypass 601, NC	S/S	Cr. Pb	GW pump and treat, capping, grading, revegetation	S
Flowood, MS	S/S	Pb	Capping	C
Independent Nail, SC	S/S	Cd, Cr	Capping	C
Pepper's Steel and Alloys, FL	S/S	As, Pb	On-site disposal	C
Pesses Chemical, TX	Stabilization	Cd	Capping	C
E.I. Dupont de Nemours, IA	S/S	Cd, Cr, Pb	Capping, regrading, revegetation	C
Shaw Avenue Dump, IA	S/S	As, Cd	Capping, GW monitoring	C
Frontier Hard Chrome, WA	Stabilization	Cr	—	S
Gould Site, OR	S/S	Pb	Capping, regrading, revegetation	I

* Status codes as of February, 1996: C = completed; I = in operation; S = selected.

TABLE 14.9.4   Example Soil Washing Applications at Metals-Contaminated Superfund Sites (from U.S. EPA, 1996c)

Site name/state	Soil washing technology	Metal contaminant	Secondary technology	Status*
Ewan Property, NJ	Water washing	As, Cr, Cu, Pb	Solvent extraction to remove organics	S
GE Wiring Devices, PR	Water with KI solution additive	Hg	On-site disposal of clean soil	S
King of Prussia, NJ	Water with washing agents	Ag, Cr, Cu	Sludge disposal	C
Zanesville Well Field, OH	Water washing	As, Cr, Hg, Pb	On-site disposal of clean soil, SVE to remove organics	S
Twin Cities Army Ammunition Plant, MN	Acid leaching	Cd, Cr, Cu, Hg, Pb	Soil leaching	C
Sacramento Army Depot, CA	Water washing	As, Cr, Pb	Off-site treatment/ disposal of wash liquid, on-site disposal of clean soil	S/D

* Status codes as of February, 1996: C = completed; S = selected; S/D = selected but subsequently deselected.

Soil Washing.   Soil washing technologies are applicable to a range of soils containing a variety of metal contaminants. Soil washing is most easily implemented when a single metal contaminant occurs in a particular insoluble fraction of the soil that can be separated by particle size classification. Soils with a minimum content of finer material (<20% of particles with diameters <2 mm) are easier to process. Soil washing has been used for remediation of metals-contaminated sites in Europe and has been selected and/or implemented at several U.S. Superfund sites. Table 14.9.4 gives examples of Superfund sites at which soil washing has been selected as the remediation technology.

Remediation at the Twin Cities Army Ammunition Plant (TCAAP) in New Brighton, Minnesota is one of the two completed soil washing projects. The COGNIS TERRAMET® soil washing procedure used at this site employed a combination of particle sizing, gravity separation, and acid-leaching apparatus that was designed to remove lead, mercury, cadmium, chromium, and copper from the soil. Preliminary studies have shown that the primary target metal at this site, lead, could be reduced from over 86000 mg/kg to less than 100 mg/kg, well below the target cleanup level of 300 mg/kg (Griffiths, 1995). Acid leaching soil washing procedures have been designated as the BDAT for mercury-contaminated soils. Several SITE demonstrations have been performed for soil washing of metals-contaminated soils (U.S. EPA, 1996c).

Vitrification.   Vitrification is most applicable to sites containing low-volatility metals with high glass solubilities, and therefore appears to be well-suited for treatment of lead, chromium, arsenic, zinc, cadmium, and copper wastes (Table 14.9.5). The ability of a melt to retain these and other metals depends on the metal solubility in the soil at the site, and silica content of the soil. The metal concentration can be adjusted by adding soil or another source of silica to improve site characteristics for vitrification. The ability to control volatile emissions also influences the applicability of vitrification technologies. Mercury's high volatility and low glass solubility makes it unsuitable for vitrification, but treatment by vitrification may be allowed at sites containing very low mercury concentrations.

Ex situ vitrification may not be applicable for soils with greater than 25 percent moisture content because of excess fuel consumption, or at sites where size reduction and classification are not feasible. Several ex situ vitrification technologies are under development. SITE program demonstrations have been completed for two of these processes and a third demonstration is underway (U.S. EPA, 1996c). Ex-situ vitrification has also been demonstrated for treatment of cesium-contaminated tank wastes.

Pyrometallurgical Separation.   Mercury has a relatively high vapor pressure and is easily converted to its metallic form at elevated temperature, making it easily treated by pyrometallurgic methods. Pyrometallurgic treatment of lead, arsenic, cadmium, and chromium may require pretreatment

TABLE 14.9.5 Approximate Vapor Pressures and Glass Solubility Limits for Metals

Metal	Temperature [°C] at which vapor pressure = 1 mm Hg*	Maximum allowed oxide content [%] for sample silicate glass[†]
Pb	973	30
Cr	1840	2
As	372	5
Zn	487	20
Cd	394	1
Cu	1628	5
Hg	126.2	~0

* CRC (1991).
[†] From Smith et al. (1995).

by reducing agents or fluxing agents to facilitate melting. Nonvolatile metals, such as chromium, can be tapped from the furnace as molten metal (U.S. EPA, 1996c).

Pyrometallurgical treatment is usually performed offsite because few mobile treatment units are available. This technology is most applicable to large volumes of highly contaminated soils (metal concentrations >5–20%), especially when metal recovery is expected. Low metal concentrations can be processed, especially for mercury because it is easy to volatilize and recover (Smith et al., 1995).

A number of pyrometallurgical process technologies are currently available for treatment of metals-contaminated soils. Pyrometallurgical treatment is a BDAT for cadmium-containing batteries, lead nonwastewaters, mercury wastes, mercury from wastewater treatment sludge, lead acid batteries, and zinc nonwastewaters. SITE demonstrations have been completed for thermal desorption (RUST Remedial Services, Inc.) and flame reactor (Horsehead Resource Development Company, Inc.) pyrometallurgical technologies.

In Situ Remediation. In situ remediation technologies offer the potential for significant cost savings over ex situ technologies because in situ techniques are usually associated with lower labor and energy requirements for implementation. This section discusses the status of in situ technologies which are currently available for metal remediation at contaminated sites.

Solidification/Stabilization. In situ S/S treatment appears to have been applied less frequently than ex situ techniques mostly because of concerns about uniformity of treatment and long-term reliability. These limitations are being reduced, however, through advances in chemical reagent delivery systems for large-diameter auger drilling devices (Jasperse, 1989; Walker, 1992). Examples of Superfund sites at which in situ S/S has been selected for remediation are given in Table 14.9.6. While in situ S/S technologies are well developed because of roots in construction techniques, data on the performance of in situ S/S are limited. Based upon preliminary data, in situ S/S appears likely to be an effective treatment option. In situ S/S typically will be most beneficial for sites with contamination at depths less than 8 to 10 ft and for larger volumes of waste because ex situ may prove to be less

TABLE 14.9.6 Example In situ Solidification/Stabilization Applications at Metals-Contaminated Superfund Sites (from U.S. EPA, 1996c)

Site name/state	S/S technology	Metal contaminants	Status*
Gurley Pit, AR	In situ S/S	Pb	C
General Electric Co., FL	In situ S/S	Pb, Cr, Cu, Zn	D

* Status codes as of February, 1996: C = completed; D = demonstrated.

TABLE 14.9.7 Example In Situ Vitrification Applications at Metals-Contaminated Superfund Sites (from U.S. EPA, 1996c)

Site name/state	Vitrification technology	Metal contaminants	Status*
Parsons Chemical, MI	In situ vitrification	As, Cr, Hg, Pb	C
Rocky Mountain Arsenal, CO	In situ vitrification	As, Hg	S/D

* Status codes as of February, 1996: C = completed; S/D = selected but subsequently deselected.

expensive for small volumes and shallow contamination because of high costs associated with mobilization and demobilization for in situ technologies. Deep soil mixing technology is also available for treating contaminated soils at greater depth (Ryan and Walker, 1992) but is more expensive than shallow soil mixing. The cost of in situ technologies is also affected by implementation concerns such as a level, stable base that is required for augering, and the presence of large rocks that can make large-diameter augering impossible. The use of dry reagents in soils with high moisture content is a well-established method in Europe that is gaining interest in the United States and may expand the applicability of in situ S/S techniques (U.S. EPA, 1996b).

Vitrification. In situ vitrification (ISV) technologies are currently offered commercially in the United States by a single vendor, Geosafe Corporation. The first full-scale application of ISV was demonstrated at the Parsons Chemical/ETM Enterprises Superfund site in Grand Ledge, Michigan under the EPA SITE program (Table 14.9.7). The Geosafe ISV process was used for treatment of soils and sediments contaminated with pesticides, metals (As, Cr, Hg, Pb), and dioxins. This treatment system required the use of eight melts that were each completed over a time frame of 10 to 20 days. This system also included an air emissions control system to treat volatilized contaminants, including mercury. While ISV is not recommended for remediation of mercury, this method can be used in conjunction with emissions control systems when Hg is present in mixed metal/organic wastes. This treatment was successful, meeting TCLP limits for all of the metals in the treated waste.

ISV has also been used successfully at two sites contaminated with organics (PCB, dioxin, pentachlorophenol, pesticides, herbicides), further demonstrating the applicability of this technology. Based upon observations from these limited applications, it appears that ISV may not be appropriate for sites with high levels of organics (>10% organics by weight) because of contamination of the off-gas, or inorganics (>25% metals by weight, or >20% by volume) because of concerns about exceeding glass solubility limits (U.S. EPA, 1996b).

In Situ Soil Flushing. In Situ soil flushing has been selected for treatment at several Superfund sites contaminated with metals. Some examples of sites where in situ soil flushing is currently operational are given in Table 14.9.8. In situ soil flushing is the technology in design or the predesign stage at least five other sites. This technology has been applied for a limited number of projects, mostly containing organic forms of contamination (NRC, 1994), and limited information is available on the application of this technology to metals-contaminated sites.

TABLE 14.9.8 Example In Situ Soil Flushing Applications at Metals-Contaminated Superfund Sites (from U.S. EPA, 1996c)

Site name/state	In situ soil flushing technology	Metal contaminants	Secondary technology	Status*
Lipari Landfill, NJ	Flushing of contained wastes with water	Cr, Hg, Pb	Slurry wall, cap, excavation of wetlands	I
United Chrome Products, OR	Soil flushing with water	Cr	Considering electrokinetic and chemical (reduction) treatment	I

* Status codes as of February, 1996: I = in operation.

The United Chrome Products Superfund site in Corvallis, Oregon is currently being remediated using in situ soil flushing technologies. The soil and groundwater at this site are heavily contaminated with chromium, with chromium levels in the soil as high as 60,000 mg/kg and levels in the groundwater reaching up to 19,000 mg/kg. The general approach to remediation of this site has been removal of the more soluble, mobile, and toxic form of chromium, Cr(VI), by flushing the contaminated region with water to solubilize Cr(VI), with subsequent extraction of the chromium-containing water for treatment. Remediation at this site began in 1985 and has combined a variety of technologies to aid remediation by in situ soil flushing. The technologies used have included infiltration basins and trenches to flush contaminated soils, a 23-well groundwater extraction network to remove contaminated groundwater and recharge water, on-site treatment of wastewater, and off-site disposal of contaminated soil and debris (Sturges et al., 1992).

This full-scale application of in situ soil flushing with water as the flushing solution appears to be successful for removal of Cr(VI) from coarse soils of relatively high hydraulic conductivity. The in situ soil flushing procedure used at this site leaches contaminants from the unsaturated and saturated zones, and provides for recharge of the groundwater to the extraction wells. This cleanup operation has removed significant amounts of chromium from the soil and groundwater, and the groundwater pumping strategy has achieved hydraulic containment of the plume. Cr(VI) levels in water retrieved by the extraction wells decreased from more than 5000 mg/L to approximately 50 mg/L during the first two and one-half years of operation. Average chromium concentrations in the plume decreased from 1923 mg/L to 207 mg/L after flushing the first one and one half pore volumes (approximately 2.6 million gallons for one pore volume). These rapid removal rates are expected to continue for the first few pore volumes of treatment until Cr(VI) removal begins to tail off to the asymptotic level. Tailing results from slow desorption from soil particles, dissolution of solid phase contaminants, and release of contaminants from the fine pores in the soil matrix. Tailing is commonly observed in in situ soil flushing applications and usually represents the practical limit for remediation via pump and treat methods (Sturges et al., 1992).

Electrokinetic Extraction. The success of various electrokinetic remediation technologies has been illustrated for removal of metals from soils via bench and pilot scale experiments. Currently, several of these technologies are being implemented in comprehensive demonstration studies to further the use of electrokinetic techniques at contaminated sites.

Electrokinetic remediation of metals-contaminated sites has been demonstrated in situ at many sites in Europe using processes developed by Geokinetics International, Inc. (GII) (U.S. EPA, 1996b). Table 14.9.9 provides examples of sites in Europe for which this technology has been selected as the remediation technology. The success of electrokinetic remediation appears to vary depending on the metals present, and can remove up to 90 percent of the initial contamination. The first demonstration of this electrokinetic process in the United States is scheduled under the EPA SITE program for remediation of a chromium-contaminated soil at the Sandia Chemical Waste Landfill.

TABLE 14.9.9 Example Electrokinetic Applications at Metals-Contaminated Sites (from U.S. EPA, 1996c)

Site description	Electrokinetic technology	Metal contaminants	Status*
Former paint factory	Electrochemical Remediation	Cu, Pb	C
Operational galvanizing plant	Electrochemical Remediation	Zn	C
Former timber plant	Electrochemical Remediation	As	C
Temporary landfill	Electrochemical Remediation	Cd	C
Military air base	Electrochemical Remediation	Cd, Cr, Cu, Ni Pb, Zn	C

* Status codes as of February, 1996: C = completed.

Electrokinetics, Inc. is carrying out a SITE demonstration study of lead extraction from a creek bed at a U.S. Army firing range in Louisiana using their CADEXä electrode system. Soils at this site are contaminated with lead at concentrations up to 4500 mg/kg. In pilot-scale studies, the lead levels in the soil were reduced to below 300 mg/kg after 30 weeks of processing. The TCLP values for this soil were reduced from over 300 mg/L to less than 40 mg/L over this time. This technology is also being explored for remediation of sites contaminated with arsenic. Treatability and pilot-scale field testing studies for this application are under way.

Other electrokinetic techniques have been demonstrated for remediation of organics (TCE) and have accounted for removal of up to 98 percent of these wastes. The LASAGNA™ process is being developed by a consortium consisting of Monsanto, E.I. DuPont deNemours & Co., Inc., and General Electric. LASAGNA™ is an integrated, in situ process that uses electrokinetics to transport contaminants in soil pore water into treatment zones. The treatment zones are designed to capture or decompose the organic contaminants. ManTech Environmental provides the ElectroChemical GeoOxidation (ECGO) process that has been used to successfully remediate organic-contaminated soil and groundwater in Germany. ECGO uses induced electric currents to create oxidation-reduction reactions that mineralize organic contaminants. These reactions may be useful for immobilization of inorganic contaminants as well. Attempts are being made to determine the potential for treatment of metals using these processes (U.S. EPA, 1996b).

Biological Treatment. Phytoremediation technologies are largely in the developmental stage and many are being field tested at a variety of sites in the United States and in Europe. Because full-scale applications of phytoremediation technologies are just being initiated, limited cost and performance data are available. Some techniques under development have shown potential for use at metals-contaminated sites. Phytostabilization and phytoextraction methods are being developed by Phytotech, Inc., and field tests for patented phytoextraction techniques are being performed. Some grasses have been made commercially available for phytostabilization of metals (lead, copper, zinc; Salt, 1995). Nickel has been removed from plating wastes by bacteria (Wong and Kwok, 1992) and other organisms are being genetically engineered to remove metals such as cadmium, cobalt, copper, and mercury (Smit and Atwater, 1991). Bioreduction has been demonstrated (for Hg) at the bench-scale but has not been tested at pilot-scale (Smith et al., 1995). A process has been developed for chromium reduction by H_2S produced by sulfate-reducing bacteria and reduction of Cr(VI) by direct metabolism is being investigated by several organizations (Smith et al., 1995).

Treatment by wetlands has been studied under the U.S. EPA's SITE program. Full-scale demonstration of a constructed wetland is planned for the Burleigh Tunnel site, part of the Clear Creek/Central City Superfund site in Colorado.

Bioleaching is currently used to recover copper and uranium ores by heap or in situ leaching (Ehrlich, 1988) and is under development for a wide range of metals including cadmium, chromium, lead, mercury, and nickel. Microorganisms have been tested for chemical reduction and removal of mercury salts from wastewater (Hansen and Stevens, 1992; Horn et al., 1992).

Phytoremediation technologies will likely be limited to use in shallow soils with relatively low levels of metal contamination. Based upon estimates of biomass productivity and metal content of soils, the annual removal rate of metals by phytoremediation would be limited to between 2.5 to 100 mg/kg of soil contaminants (U.S. EPA, 1996b).

Best Technology By Metal

The Best Demonstrated Available Technologies, BDATs for metals-contaminated RCRA wastes are summarized in Table 14.9.10 according to the type of metal contamination. These technologies can be used as guidelines to review treatment options for Superfund sites, but technology selection at Superfund sites should also consider site-specific characteristics and innovative technologies that may be available under the EPA SITE program.

TABLE 14.9.10. Summary of Best Demonstrated Available Technologies (BDATs) for RCRA Wastes (from Smith et al., 1995)

Metal contaminant	Example BDATs for metal wastes	
	Nonwastewater	Wastewater
Lead	Stabilization or metal recovery	Chemical precipitation
Chromium	Chromium reduction and S/S	Chromium reduction and S/S
Arsenic	Vitrification	Chemical precipitation
Cadmium	Stabilization or metal recovery	Chemical precipitation
Mercury	Metal recovery ($\geq$260 mg/kg) or acid leaching followed by chemical precipitation	Chemical precipitation with sulfide

14.9.4 COST ESTIMATES

The costs for implementing available technologies will vary significantly between sites because costs are influenced by a wide variety of factors. Figure 14.9.4 represents the ranges of operating costs that have been observed for remediation of metals-contaminated soils by a number of techniques that have been discussed. Some important factors influencing costs of specific treatment technologies are discussed below.

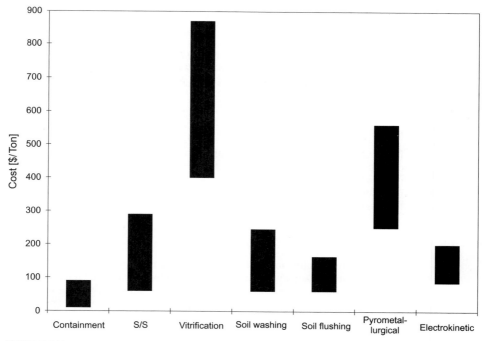

FIGURE 14.9.4 Estimated operating costs of available remediation technologies for metal-contaminated soils (U.S. EPA, 1996c).

Containment

The costs associated with capping systems depend largely on the number of components included in the design (Rumer and Ryan, 1995). Barrier and drainage components can add significant amounts to the overall cost of this technology. Sites with steep slopes will also increase cost.

The cost of vertical barrier construction will be influenced by the type of barrier material and the method used to place it. Soil-bentonite trenches provide the most economical method for installation of shallow vertical barriers (Rumer and Ryan, 1995). The most economical deep vertical barrier is a cement-bentonite barrier constructed using a vibrating beam (U.S. EPA 1996b). Costs will also be influenced by groundwater or topographical conditions.

Solidification/Stabilization

Factors directly influencing the costs for implementation of S/S techniques include labor, equipment, energy requirements, testing and monitoring, and the types of reagents. In situ processing can lower labor and energy expenses associated with excavation, transport, and disposal of soil from the site.

Vitrification

Treatment costs for ex situ vitrification of contaminated soils depend on the waste, throughput capacity of the glass melter, and local energy costs. Site location will affect the cost of transporting the material offsite or equipment transport onsite. As with most technologies, the in situ process may provide cost savings over ex situ implementation of this technology.

Soil Washing

Soil washing at a contaminated site can involve techniques ranging from physical separation and disposal of the contaminated fraction offsite to chemical leaching of contaminants from the entire soil matrix for onsite disposal. Soil washing costs depend largely on the extent to which the contaminated soils are processed.

In Situ Soil Flushing

In situ soil flushing involves pumping and treatment of contaminated water, sometimes with recharge of the treated water. The initial and target contaminant concentrations, soil permeability and the depth of the aquifer will influence costs. Chemically enhanced flushing systems will have additional costs associated with reagents and equipment needed to handle the flushing solution. Costs for above-ground treatment of the pumped water vary with contaminant type.

Electrokinetic Treatment

The cost of remediating metals-contaminated soils using electrokinetic techniques is strongly influenced by soil conductivity because energy consumption is directly related to the conductivity of the soil between the electrodes. Electrokinetic treatment of soils with high electrical conductivites may not be feasible because of the high cost. Overall expenses for electrokinetic remediation will also be influenced by local energy costs, pretreatment costs, and fixed costs associated with installing the system.

14.9.5 REGULATORY/POLICY REQUIREMENTS AND ISSUES

Cleanup goals for remediation of metals-contaminated sites vary considerably depending on site-specific factors, especially those that affect the mobility of metals, and regulatory domain. Cleanup

TABLE 14.9.11 Examples of U.S. Cleanup Goals/Standards for Selected Metals in Soils

Description	As	Cd	Cr (total)	Hg	Pb
Total metals (mg metal/kg soil)					
Background (Mean)[a]	5	0.06	100	0.03	10
Background (Range)[a]	1 to 50	0.01 to 0.70	1 to 1000	0.01 to 0.30	2 to 200
Superfund site goals[a]	5 to 65	3 to 20	6.7 to 375	1 to 21	200 to 500
Theoretical minimum total metals to ensure TCLP Leachate < Threshold (i.e., TCLP × 20)[b]	100	20	100	4	100
EPA Region III[c]: residential	23	39	390 Cr(VI)	23	—
commercial	610	1000	10000	610	—
California total threshold limit concentration[a]	500	100	500	20	1000
Pennsylvania[d]	3	20	300	20	500
Florida[e]: residential	0.8	37	290 Cr(VI)	23	500
industrial	3.7	600	430	480	1000
Leachable metals (mg/L)					
TCLP threshold for RCRA waste[f]	5	1	5	2	5
Synthetic precipitation leachate procedure[g]	—	—	—	—	—
California soluble threshold leachate concentration[a]	5	1	5	2	5
Florida leachability standards[e]	5	1	5	2	5

[a] from U.S EPA, 1995.
[b] from U.S. EPA, 1996c.
[c] from EPA Region III Risk-Based Concentration Table.
[d] PA Department of Environmental Protection (DEP) Health-Based Standards, 1996.
[e] FL DEP Health-Based Standards, 1996.
[f] EPA Method 1311.
[g] EPA Method 1312.

goals that are established for a site have a significant influence on determining the acceptability of different technologies for remediation of metals at the site. Thus, the application of remediation technologies to different sites may vary even if the types of contamination at the sites are the same.

A number of states have established soil and groundwater quality criteria that are the basis for cleanup goals. In the absence of such criteria (as in the U.S. Superfund program), or when the criteria are intended as default values, cleanup goals are established based upon site-specific human health and ecological risk assessments, which consider the fate and transport of contaminants and possible exposure routes for humans and sensitive environmental receptors. The goals may be established in terms of the total metals in the soil/water or as leachable metals (as defined by various EPA testing procedures). Table 14.9.11 provides examples of established cleanup goals for total metals in soils and soil leachate at hazardous waste sites, and Table 14.9.12 gives examples of cleanup goals for metals in groundwater.

The use of risk assessment for establishment of site-specific or regional goals for metals in soil or groundwater is difficult because the chemistry of metals is so complex. The hydrogeochemistry of metals is affected by various geochemical and biogeochemical phenomena, including acid-base chemistry, complexation, precipitation/dissolution, adsorption/desorption, and oxidation/reduction. These processes are interlinked and not capable of being described with a simple model. In the case of adsorption/desorption reactions, for example, the speciation of metal ions and the aqueous solution composition determine the extent of reaction (Dzombak and Morel, 1987, 1990). These factors are not captured in a simple partitioning expression. Thus, exposure assessment modeling for metals in soil and groundwater demands the use of flow models integrated with complex chemical models. This

TABLE 14.9.12   Examples of U.S. Cleanup Goals/Standards for Selected Metals in Groundwater

Description	As	Cd	Cr	Hg	Pb
Metals (μg/L)					
Maximum contaminant level (MCL)[a]	50	5	100	2	15
Superfund site goals[b]	50	—	50	0.05 to 2	50
Pennsylvania standard for groundwater in aquifers (<2500 mg/L TDS)[c]	50	5	100	2	5
Wisconsin groundwater quality enforcement standards	50	5	100	2	15

[a] MCL = the maximum permissible level of contaminant in water delivered to any user of a public system, established under the Safe Drinking Water Act.
[b] from U.S. EPA, 1995.
[c] PA DEP Health-Based Standards, 1996.
[d] WI Department of Natural Resources (DNR) Groundwater Quality Standards Tables, 1996.

requirement frequently has discouraged detailed exposure assessment for metals, resulting in the use of conservative assumptions with regard to metal fate and transport in subsurface systems.

The risk-based corrective action (RBCA) procedure developed by the Environmental Assessment Committee of the American Society for Testing and Materials (ASTM, 1994) may be applied to determine cleanup goals for soil and groundwater. The aim of RBCA is the establishment of cleanup goals based on risk reduction rather than generic cleanup concentrations. However, when regulatory screening levels are exceeded and fate and transport modeling is required as part of a Tier III assessment, there will still be the issue of adequate consideration of the complex chemistry of metals.

14.9.6   *LESSONS LEARNED AND TECHNOLOGY DIRECTIONS*

Metals are typically relatively immobile in subsurface systems. For this reason, remediation activities at metals-contaminated sites have focused on the solid-phase sources or repositories of metals. Treatment has often involved excavation of contaminated soil, sludge, or debris followed by ex situ treatment or disposal. The most common ex situ treatment is solidification/stabilization through addition of chemical reagents, followed by replacement or off-site disposal of the treated material.

Several in situ remediation technologies have the potential to provide significant cost savings over ex situ techniques because they eliminate the need to excavate and dispose of contaminated solids or to pump and treat contaminated groundwater. In situ solidification/stabilization technologies have been demonstrated for treatment of shallow (8–10 ft below surface) wastes and are being implemented at greater depths. Favorable results have been attained using in situ vitrification for treatment of a variety of wastes, including metals when metal concentrations do not exceed their glass solubilities. Extraction using in situ soil flushing or electrokinetic techniques has been employed at a limited number of sites but may prove to be useful for a range of metal contaminants. Phytoremediation technologies offer promise for remediation of sites with low levels of contamination.

Treatment walls will be used increasingly for effective, low-cost, passive remediation of metal contamination in groundwater. Reactive wall installation will not address metal contamination in soils, but will enable treatment of groundwater contaminated from contact with metal-bearing solids.

Some soil washing technologies are being considered for adaptation to soil leaching/flushing technologies. Chemical additives are being developed to aid with in situ extraction of metals from soil. In situ solidification/stabilization techniques are being employed and promise to gain popularity. Application of in situ S/S is being aided by development of wide-diameter auger drilling devices, which are equipped with chemical reagent delivery systems.

Phytoremediation technologies have only recently gained attention for use in metal remediation. Additional research is needed in order to improve the applicability of phytoremediation for management of metals-contaminated sites. A variety of plants are being investigated for favorable metal accumulation qualities, such as a fast rate of uptake.

The future of electrokinetic methods will depend on the efficiency and cost-effectiveness of the technique. Full-scale applications of in situ electrokinetic technologies have been initiated in the United States but detailed data are not yet available.

REFERENCES

Allen, H. E., Perdue, E. M., and Brown, D. S., eds., *Metals in Groundwater*, CRC- Lewis Publishers, Boca Raton, FL, 1994.

Allen, J. P., and Torres, I. G., "Physical Separation Techniques for Contaminated Sediment," in *Recent Developments in Separation Science*, Vol V, Li, N.N. ed., CRC Press, West Palm Beach, FL, 1991.

Ali, M. A., and Dzombak, D. A., "Interactions of Copper, Organic Acids, and Sulfate in Goethite Suspensions," *Geochim. Cosmochim. Acta*, 60: 5045–5053, 1996.

Assink, J. W., and Rulkens, W. H., "Cleaning Soils Contaminated with Heavy Metals," *Hazardous and Industrial Wastes, Proceedings of the 21st Mid Atlantic Industrial Waste Conference*, Cole, C. A., and Long, D. A., eds., Technomics, Lancaster, PA, 1989.

ASTM, *Emergency Standard Guide for Risk-Based Corrective Action Applied at Petroleum Release Sites*, ES 38–94, American Society for Testing and Materials, Philadelphia, PA, 1994.

Bishop, P., Gress, D., and Olafsson, J., "Cement Stabilization of Heavy Metals: Leaching Rate Assessment," *Industrial Wastes—Proceedings of the 14th Mid-Atlantic Industrial Waste Conference*, Technomics, Lancaster, PA, 1982.

Bodek, I., Lyman, W. J., Reehl, W. F., and Rosenblatt, D. H., *Environmental Inorganic Chemistry: Properties, Processes and Estimation Methods*, Pergamon Press, Elmsford, NY, 1988.

Buelt, J. L., and Thompson, L. E., *The In situ Vitrification Integrated Program: Focusing on an Innovative Solution on Environmental Restoration Needs*, Battelle Pacific Northwest Laboratory, Richland, WA, 1992.

Callahan, M. A., Slimak, M. W., and Gabel, N. W., *Water-Related Environmental Fate of 129 Priority Pollutants, Vol. 1, Introduction and Technical Background, Metals and Organics, Pesticides and PCBs*, Report to U.S. EPA, Office of Water Planning and Standards, Washington, D.C., 1979.

Cantrell, K. J., Kaplan, D. I., and Wietsma, T. W., "Zero-Valent Iron for the *In Situ* Remediation of Selected Metals in Groundwater," *J. Hazardous Materials*, 42: 201–212, 1995.

Chrotowski, P., Durda, J. L., and Edelman, K. G., "The Use of Natural Processes for the Control of Chromium Migration," *Remediation*, 2: 341–351, 1991.

CII, *Environmental Remediation Management: An Eight-Step Process*, Special Pub. No. 48-2, Environmental Remediation Task Force, Austin, TX, 1995.

Conner, J. R., *Chemical Fixation and Solidification of Hazardous Wastes*, Van Nostrand Reinhold, New York, 1990.

CRC, *CRC Handbook of Chemistry and Physics*, 71st Edition, David R. Lide, ed., CRC Press, Boca Raton, FL, 1991.

Davis, J. A., "Complexation of Trace Metals by Adsorbed Natural Organic Matter," *Geochim. Cosmochim. Acta*, 48: 679–691, 1984.

DiToro, D. M., Mahoney, J. D., Hansen, D. J., Scott, K.J., Carlson, A.R., and Ankley, G. T., "Acid Volatile Sulfide Predicts the Acute Toxicity of Cadmium and Nickel in Sediments," *Environ. Sci. Technol.*, 26: 96–101, 1992.

Dzombak, D. A., and Morel, F. M. M., "Adsorption of Inorganic Pollutants in Aquatic Systems," *J. Hydraulic Eng.*, 113: 430–475, 1987.

Dzombak, D. A., and Morel, F. M. M., *Surface Complexation Modeling: Hydrous Ferric Oxide*, John Wiley & Sons, New York, 1990.

Dzombak, D. A., Rosetti, P. K., Evanko, C. R., and DeLisio, R. F., "Treatment of Fine Particles in Soil Washing Processes," in *Proceedings of the Specialty Conference on Innovative Solutions for Contaminated Site Management*, Water Environment Federation, Alexandria, VA, pp. 473–484, 1994.

Elliot, H. A., and Brown, G. A., "Comparative Evaluation of NTA and EDTA for Extractive Decontamination of Pb-Polluted Soils," *Water, Air, and Soil Pollution,* 45: 361–369, 1989.

Ellis, W. D., and Fogg, T., *Interim Report: Treatment of Soils Contaminated by Heavy Metals,* Hazardous Waste Engineering Research Laboratory, Office of Research and Development, U.S. EPA, Cincinnati, Ohio, 1985.

Environment Canada, *Compendium of Waste Leaching Tests,* Report EPS 3/HA/7, Wastewater Technology Centre, Environment Canada, Ottawa, Ontario, 1990.

Erlich, H. L., "Recent Advances in Microbial Leaching of Ores," *Minerals and Metallurgical Processing,* 5(2): 57–60, 1988.

Evans, L. J., "Chemistry of Metal Retention by Soils," *Environ. Sci. Tech.,* 23: 1046–1056, 1989.

Gerber, M. A., Freeman, H. D., Baker, E. G., and Riemath, W. F., *Soil Washing: A Preliminary Assessment of Its Applicability to Hanford,* prepared for U.S. Department of Energy by Battelle Pacific Northwest Laboratory, Richland, Washington. Report No. PNL-7787; UC 902, 1991.

Griffiths, R. A., "Soil Washing Technology and Practice," *J. Haz Mat.,* 40: 175–189, 1995.

Hansen, C. L., and Stevens, D. K., "Biological and Physio-Chemical Remediation of Mercury-Contaminated Hazardous Waste," *Arsenic and Mercury: Workshop on Removal, Recovery, Treatment and Disposal,* pp. 121–125, 1992.

Hedin, R. S., and Nairn, R. W., "Designing and Sizing Passive Mine Drainage Treatment Systems," in *Proceedings of the 13th Annual West Virginia Surface Mine Drainage Symposium,* West Virginia Mining and Reclamation Association, Morgantown, WV, 1992.

Hedin, R. S., Nairn, R. W., and Kleinmann, R. L. P., "Passive Treatment of Coal Mine Drainage," Report IC-9389, Bureau of Mines, U.S. Department of the Interior, 1994.

Hedin, R. S., and Watzlaf, G. R., "The Effects of Anoxic Limestone Drains on Mine Water Chemistry," in *Proceedings of the International Land Reclamation and Mine Drainage Conference and the 3rd International Conference on the Abatement of Acidic Drainage,* Pittsburgh, PA, 1994.

Horn, J. M., Brunke, M., Deckwer, W. D., and Timmis, K. N., "Development of Bacterial Strains for the Remediation of Mercurial Wastes," *Arsenic and Mercury: Workshop on Removal, Recovery, Treatment and Disposal,* EPA/600/R-92/105, pp. 106–109, 1992.

Jasperse, B. H., "Soil Mixing," *Hazmat World,* November, pp. 20–23, 1989.

Jasperse, B. H., and Ryan, C. R., "Stabilization and Fixation Using Soil Mixing," in *Proceedings of the ASCE Specialty Conference on Grouting, Soil Improvement, and Geosynthetics,* ASCE Publications, Reston, VA, 1992.

Kaplan, D. I., Bertsch, P. M., and Adriano, D. C., "Facilitated Transport of Contaminant Metals through an Acidified Aquifer," *Ground Water,* 33: 708–717, 1995.

Kinniburgh, D. G., Jackson, M. L., and Syers, J. K., "Adsorption of Alkaline Earth, Transition, and Heavy Metal Cations by Hydrous Oxide Gels of Iron and Aluminum," *Soil Sci. Soc. Am. J.,* 40: 796–800, 1976.

Krause, E., and Ettel, V. A., "Solubilities and Stabilities of Ferric Arsenate Compounds," *Hydrometallurgy,* 22: 311–337, 1989.

LaGrega, M. D., Buckingham, P. L., and Evans, J. C., *Hazardous Waste Management,* McGraw Hill, New York, 1994.

Lister, K. H., "Fast-Track Remediation for Redevelopment of a Former Integrated Steel Mill Site," *Remediation,* 6(4): 31–50, 1996.

Lo, C. P., Silverman, D. N., and Porretta, A. M., "Chemical Fixation of Heavy Metal Contaminated Soils," *Proceedings of the Sixth Annual Hazardous Materials Management Conference International,* Tower Conference Management Publisher, Glen Ellyn, IL, 1988.

Malone, P. G., and Jones, L. W., *Solidification/Stabilization of Sludge and Ash from Wastewater Treatment Plants,* EPA/600/S2-85-058, 1985.

Means, J. L., and Hinchee, R. E., *Emerging Technology for Bioremediation of Metals,* Lewis Publishers, Boca Raton, FL, 1994.

NRC, *Alternatives for Ground Water Cleanup,* National Research Council, National Academy Press, Washington, D.C., 1994.

Pacific Northwest National Laboratory, Tanks Focus Area, *Technical Highlights,* September 15, 1997.

Pierce, M. L., and Moore, C. B., "Adsorption of Arsenite and Arsenate on Amorphous Iron Hydroxide," *Water Res.,* 16: 1247–1253, 1982.

Powell, R. M., Puls, R. W., and Paul, C. J., "Chromate Reduction and Remediation Utilizing the Thermodynamic Instability of Zero-Valence State Iron," in *Proceedings of the Specialty Conference on Innovative Solutions for Contaminated Site Management,* Water Environment Federation, Alexandria, VA, pp. 485–496, 1994.

Puls, R. W., "Ground-water Sampling for Metals," in *Environmental Sampling for Trace Analysis*, Markert, B., ed., VCH Publishers, Weinheim, Germany, 1994.

Puls, R. W., and Paul, C. J., "Low-Flow Purging and Sampling of Ground Water Monitoring Wells With Dedicated Systems," *Ground Water Monit. Remediat.*, 15: 116–123, 1995.

Rosetti, P. K., *Possible Methods of Washing Fine Soil Particles Contaminated with Heavy Metals and Radionuclides*, M.S. Thesis, Carnegie Mellon University, Pittsburgh, PA, 1993.

Roy, S. B., and Dzombak, D. A., "Chemical Factors Influencing Colloid-Facilitated Transport of Contaminants in Porous Media," *Environ. Sci. Technol.*, 31: 656–664, 1997.

Rumer, R. R., and Ryan, M. E., *Barrier Containment Technologies for Environmental Remediation Applications*, John Wiley & Sons, New York, 1995.

Ryan, C. R., and Walker, A. D., "Soil Mixing for Soil Improvement," in *Proceedings of the 23rd Conference on In situ Soil Modification*, Louisville, KY, (available from Geo- Con, Inc., Pittsburgh, PA), 1992.

Salt, D. E., "Phytoremediation: A Novel Strategy for the Removal of Toxic Metals from the Environment Using Plants," *Biotechnology*, 13: 468–474, 1995.

Schnoor, J. L., *Phytoremediation*, TE-97-01, Ground-Water Remediation Technologies Analysis Center, Pittsburgh, PA, 1997.

Shively, W., Bishop, P., Gress, D., and Brown, T., "Leaching Tests of Heavy Metals Stabilized with Portland Cement," *J. WPCF*, 38: 234–241, 1986.

Smit, J., and Atwater, J., *Use of Caulobacters to Separate Toxic Heavy Metals from Wastewater Streams*, U.S. DOE Innovative Concepts Program, Battelle Pacific Northwest Laboratories, Richland, WA, 1991.

Smith, L. A., Means, J. L., Chen, A., Alleman, B., Chapman, C. C., Tixier, J. S., Jr., Brauning, S. E., Gavaskar, A. R., and Royer, M. D., *Remedial Options for Metals-Contaminated Sites*, Lewis Publishers, Boca Raton, FL, 1995.

Sposito, G., *The Chemistry of Soils*, Oxford University Press, New York, 1989.

Stanczyk, T. F., and Senefelder, B. C., "The Impacts of Chemically Stabilized Industrial Wastes on Landfill Leachate," presented at *USGS Conference on The Impact of Waste Storage and Disposal on Ground-Water Resources*, Ithaca, NY, 1982.

Sturges, S. G., Jr., McBeth, P., Jr., and Pratt, R. C., "Performance of Soil Flushing and Ground-water Extraction at the United Chrome Superfund Site," *J. Haz. Mat.*, 29: 59–78, 1992.

Tessier A., Campbell, P. G. C., and Bisson, M., "Sequential Extraction Procedure for the Speciation of Particulate Trace Metals," *Anal. Chem.*, 51: 844–851, 1979.

Tipping, E., Griffith, J. R., and Hilton, J., "The Effect of Adsorbed Humic Substances on the Uptake of Copper (II) by Goethite," *Croatica Chemica Acta*, 56(4): 613–621, 1983.

Tuin, B. J. W., and Tels, M., "Extraction Kinetics of Six Heavy Metals from Contaminated Clay Soils," *Environmental Technology*, 11: 541–554, 1990.

Tuin, B. J. W., Senden, M. M. G., and Tels, M., "Extractive Cleaning of Heavy Metal Contaminated Clay Soils," from *Environmental Technology: Proceedings of the Second European Conference on Environmental Technology*, DeWaal, K. J. A., and van den Brink, W. J., eds., Amsterdam, The Netherlands, M. Nijhoff Publisher, Dordrecht, 1987.

Turner, D., and McCoy, D., "Anoxic Alkaline Drain Treatment System: A Low- Cost Acid Mine Drainage Treatment Alternative," in *Proceedings of the 1990 National Symposium on Mining*, Lexington, KY, Graves, D. H., and DeVore, R. W., eds., OES Publ., pp. 73–75, 1990.

U.S. EPA, *Handbook: Remedial Action at Waste Disposal Sites*, EPA/625/6-85/006, U.S. Environmental Protection Agency, Office of Solid Waste and Emergency Response, Washington, D.C., 1985.

U.S. EPA, *Stabilization/Solidification of CERCLA and RCRA Wastes*, EPA/625/6-89/022, U.S. Environmental Protection Agency, Center for Environmental Research Information, Cincinnati, OH, 1989.

U.S. EPA, *International Waste Technologies/Geo-Con In situ Stabilization/Solidification*, EPA/540/A5-89/004, U.S. Environmental Protection Agency, Office of Research and Development, Cincinnati, OH, 1990a.

U.S. EPA, *Interference Mechanisms in Waste Stabilization/Solidification Processes*, EPA/540/A5-89/004, U.S. Environmental Protection Agency, Office of Research and Development, Cincinnati, OH, 1990b.

U.S. EPA, *Design and Construction of RCRA/CERCLA Final Covers*, Report EPA/625/4-91/025, U.S. Environmental Protection Agency, Cincinnati, OH, 1991.

U.S. EPA, *Vitrification Technologies for Treatment of Hazardous and Radioactive Waste*, Handbook EPA/625/R-92/002, U.S. Environmental Protection Agency, Office of Research and Development, Washington, D.C., 1992a.

U.S. EPA, *Record of Decision for Tonolli Corp. Site, Nesquehoning, Pennsylvania*, EPA/ROD/R03-92/156, U.S. Environmental Protection Agency, Office of Solid Waste and Emergency Response, Washington, D.C., September 30, 1992b.

U.S. EPA, *Contaminants and Remedial Options at Selected Metals-contaminated Sites*, EPA/540/R-95/512, U.S. Environmental Protection Agency, Office of Research and Development, Washington, D.C., 1995.

U.S. EPA, *Completed North American Innovative Remediation Technology Demonstration Projects*, EPA-542-B-96-002, PB96-153-127, U.S. Environmental Protection Agency, Office of Solid Waste and Emergency Response, Washington, D.C., August 12, 1996a.

U.S. EPA, *Report: Recent Developments for In Situ Treatment of Metals-contaminated Soils*, U.S. Environmental Protection Agency, Office of Solid Waste and Emergency Response, draft, 1996b.

U.S. EPA, *Engineering Bulletin: Technology Alternatives for the Remediation of Soils Contaminated with Arsenic, Cadmium, Chromium, Mercury, and Lead*, U.S. Environmental Protection Agency, Office of Emergency and Remedial Response, Cincinnati, OH, draft, 1996c.

Van Benschoten, J. E., Reed, B. E., Matsumoto, M. R., and McGarvey, P. J., "Metal Removal by Soil Washing for an Iron Oxide Coated Sandy Soil," *Water Env. Res.*, 66: 168–174, 1994.

Vidic, R. D., and Pohland, F. G., "Treatment Walls," *Technology Evaluation Report TE-96-01*, Ground-Water Remediation Technologies Analysis Center, Pittsburgh, PA, 1996.

Walker, A. D., "Site Remediation Using Soil Mixing Techniques on a Hazardous Waste Site," in *Proceedings of the 24th Mid-Atlantic Industrial Waste Conference*, Technomic Publishers, Lancaster, PA, 1992.

Wild, A., ed., *Russell's Soil Conditions and Plant Growth*, 11th Edition, Longman, London, 1988.

Wilson, E. K., "Zero-Valent Metals Provide Possible Solution to Groundwater Problems," *C&EN*, 73(27): 19–22, 1995.

Wong, P. K., and Kwok, S. C., "Accumulation of Nickel Ion by Immobilized Cells of Enterobacter Species," *Biotechnology Letters*, 14: 629–634, 1992.

Xanthakos, P., *Slurry Walls*, McGraw-Hill, New York, 1979.

Zimmerman, L., and Coles, C., "Cement Industry Solutions to Waste Management—The Utilization of Processed Waste By-Products for Cement Manufacturing," *Cement Industry Solutions to Waste Management: Proceedings of the First International Conference*, pp. 533–545, Canadian Portland Cement Association, Toronto, Canada, 1992.

Zirschky, J., and Piznar, M., "Cement Stabilization of Foundry Sands," *J. Env. Eng.*, 114: 715–718, 1988.

CHAPTER 14
REMEDIATION TECHNOLOGIES

SECTION 14.10

DNAPL INVESTIGATION AND REMEDIATION

Douglas R. Beal

Mr. Beal is senior project manager, BEM Systems Inc.,
Environmental Engineers and Scientists. He has 20 years of
experience specializing in innovative site investigation
techniques and in situ chemical oxidation.

14.10.1 DEFINITION AND IDENTIFICATION OF DENSE NONAQUEOUS PHASE LIQUIDS

Definition

Nonaqueous phase liquids, commonly referred to as NAPLs, are defined as separate phase organic liquids. NAPLs are typically divided into one of two basic categories depending upon their specific gravity as compared to fresh water. Compounds or mixtures of compounds whose specific gravity is less than that of water (specific gravity of fresh water = 1) may float on the water table and are referred to as LNAPLs, light nonaqueous phase liquids. Common compounds in this category are petroleum compounds, such as fuel oils and gasoline. Compounds or mixtures of compounds with a specific gravity greater than fresh water (>1) may sink to the bottom of an aquifer system and are commonly referred to as dense nonaqueous phase liquids (DNAPLs). Common compounds in this category are degreasers and cleaners referred to as chlorinated volatile organic solvents, such as trichloroethene (TCE), tetrachloroethene (PCE), and 1,1,1 trichloroethane (1,1,1 TCA). Other common types of DNAPLs include wood preservation wastes, coal tar wastes, and pesticides.

Often NAPLs provide a secondary source of contamination. Specifically, the process of dissolution results in a dissolved phase plume that originates at the NAPL source area and migrates under normal hydrogeologic processes.

Identification of a DNAPL Site

The focus of this chapter is the investigation and remediation of DNAPL. The presence of DNAPL at a site can significantly impact the ultimate success or failure of an investigation process or remediation system. A DNAPL can either be a single compound or a mixture of compounds and can exist in the subsurface as a free phase DNAPL or be trapped within pore spaces or capillary fringe, in which case it may be referred to as residual DNAPL. Free phase DNAPL is considered mobile, because it can migrate vertically through an aquifer under positive pressure until reaching a zone of lower permeability, such as a finer-grained stratigraphic unit or bedrock, where it may have a tendency to pool. Residual DNAPL is held under negative pressure by surface tension within pore spaces or fractures.

The complex nature of the fate and transport of DNAPL compounds makes it difficult to investigate and remediate. As such, the proper identification of the likelihood of DNAPL at a site is extremely important because intrusive site investigation techniques into pooled or residual DNAPL can mobilize

DNAPL vertically and laterally, resulting in deepening the DNAPL and spreading the dissolved phase plume, which makes it more difficult and costly to remediate.

Complexities regarding the fate and transport of DNAPLs in an aquifer make the physical detection of these compounds in the subsurface difficult. A 1993 U.S. EPA report (EPA, September 1993) serves to emphasize this point. This study evaluated 712 National Priority List (NPL) sites for the presence of DNAPL, and concluded that approximately 60 percent of these sites had a medium to high probability of containing DNAPL. However, physical detection of DNAPL only occurred at 44 of the 712 sites. At only 6 percent of the screened sites was direct physical evidence obtained to identify the site as a DNAPL site.

Because direct physical evidence is not commonly available to identify a DNAPL site, other approaches have been developed by the private and government sectors to evaluate the probability of a particular site as a DNAPL site. As certain risk factors are positively identified, the likelihood of a particular site being classified as a DNAPL site increases. A common approach to evaluating a potential DNAPL site is to complete a comprehensive review of information regarding the *site history*, the *nature of the release*, and *site chemical data*. (These areas are described in more detail in the following paragraphs.) This information is then compiled to rank the site on a sliding scale as to its probability of being classified as a DNAPL site. In the case of the 1993 U.S. EPA report referenced above, this led to the extrapolated conclusion that 60 percent of NPL sites could have a medium to high potential for the presence of DNAPL.

Site History.   Certain types of industries, manufacturing processes, and waste practices appear to be predisposed as a potential DNAPL site. Specifically, historical use data for a site can yield valuable information regarding the probability of DNAPL being present. In addition, the potential for DNAPL being present at a site increases with the size and length of time of operations at such an industrial site. According to a 1995 EPA document (EPA, 1995), the following types of sites may have a higher potential of indicating the presence of DNAPL than other types of manufacturing:

- Wood preservation (creosote)
- Old coal gas plants
- Electronic manufacturing
- Solvent production
- Pesticide and herbicide manufacturing
- Airplane maintenance
- Dry cleaning
- Instrument manufacturing
- Transformer reprocessing and oil production
- Steel coking operations
- Pipeline compressor stations

According to this same EPA 1995 document, the following industrial processes or waste disposal practices increase the probability of the presence of DNAPL at a particular site:

- Metal cleaning/degreasing
- Metal machinery
- Tool and die operations
- Paint removal/stripping
- Storage of solvents in underground storage tanks and/or drums
- Solvent loading and unloading
- Disposal of mixed chemicals in landfills
- Treatment of mixed chemical wastes in lagoons or ponds

A 1991 U.S. EPA (EPA, December 1991) paper lists the following chemicals in either virgin product form or as mixtures that are related to DNAPL sites:

Halogenated Volatile Organic Compounds

- chlorobenzene
- 1,2-dichloropropane
- 1,1-dichloroethane
- 1,1-dichloroethylene
- 1,2-dichloroethane
- trans-1,2-dichloroethene
- cis-1,2-dichloroethene
- 1,1,1-trichloroethane
- methylene chloride
- 1,1,2-trichloroethane
- trichloroethene
- chloroform
- carbon tetrachloride
- 1,1,2,2-tetrachloroethane
- tetrachloroethane
- ethylene dibromide

Halogenated Semi-Volatile Compounds

- 1,4-dichlorobenzene
- 1,2-dichlorobenzene
- aroclor 1242, 1254, 1260
- chlorodane
- dieldrin
- 2,3,4,6-tetrachlorophenol
- pentachlorophenol

Nonhalogenated Semi-Volatile Compounds

- 2-methyl napthalene
- o-cresol
- p-cresol
- 2,4-dimethylphenol
- m-cresol
- phenol
- naphthalene
- benzo(a) anthracene
- fluorene
- acenaphthene
- anthracene
- dibenzo(a,h)anthracene

- fluoranthene
- pyrene
- chrysene
- 2,4-dinitrophenol

Nature of the Release.   The second type of information used to identify a DNAPL site is the nature of the contamination release. When a DNAPL release occurs, the DNAPL will migrate vertically as a continuous phase through unsaturated and saturated soil and will leave behind residual DNAPL within the formation. The vertical migration will be hindered by the heterogeneity of the subsurface soils, whereas finer grained soils tend to disperse the DNAPL horizontally. However, because of the density of the DNAPL, the distribution will likely be near vertical with limited horizontal migration. The vertical migration of DNAPL will continue until a lower permeable soil unit or bedrock is encountered that impedes further vertical migration, causing the DNAPL to pool. If the DNAPL release volume is small, the entire quantity of DNAPL may exist as residual DNAPL within the aquifer.

A significant difference exists between the migration of dissolved phase plumes and that of DNAPLs. Dissolved phase plume migration generally is slower in the order of feet per year, whereas the vertical migration of DNAPL is in the order of feet per day. Impacts to groundwater from a release of DNAPL occur very rapidly, usually within days if a shallow watertable exists beneath a site. Therefore, the nature of the release is important in identifying the potential as a DNAPL site. The type of release is important, such as a surface spill of virgin product or process lines may indicate a strong correlation as a DNAPL site, whereas a mixture from a broken sanitary sewer line may indicate the absence of a DNAPL release.

Chemical Evidence.   The third type of site information to be reviewed is chemical evidence. Compounds with a density greater than 1.00 grams/cubic centimeter (g/cm^3) are considered potential DNAPL compounds; the common DNAPL compounds are listed in the "Site History" section above. The detection of these compounds in the subsurface does not conclusively prove that a DNAPL release has occurred. However, because of sampling procedures and the heterogeneous nature of the fate and transport mechanisms, DNAPL can be present when concentrations of compounds within the groundwater are as low as a few percent of the solubility. The concentration of dissolved phase compounds in groundwater is a good indicator of the likelihood of the presence of DNAPL, but is not confirmation of the presence or absence of DNAPL at a given site. In general, the higher the detected concentration of a particular compound relative to its solubility in water, the greater is the likelihood as to the presence of DNAPL. A concentration of 1 percent or more of a compound's solubility is generally accepted by the research community as indicating a high likelihood of the presence of DNAPL. However, concentrations representing less than 1 percent of a constituent's solubility do not rule out the possibility of the presence of DNAPL.

14.10.2   *INVESTIGATION OF DNAPLs*

The investigation of DNAPLs is particularly difficult in comparison to a purely dissolved plume. An important difference between the investigation of dissolved and DNAPL plumes is that dissolved plumes migrate based upon advection, dispersion, and retardation, whereas DNAPL will have a tendency to migrate vertically under the force of gravity (Domenico and Swartz, 1990). Once residual and/or pooled DNAPL exists, it will create a dissolved plume through the dissolution process. In this case, traditional intrusive drilling methods to characterize a site within a DNAPL zone could inadvertently mobilize a portion of the contaminant deeper and spread the plume. Because the DNAPL is now spread, dissolution will result in the enlarging of the dissolved portion of the plume.

Upon the vertical migration of DNAPL, a residual will be left as the DNAPL migrates, eventually creating a DNAPL pool if enough DNAPL volume was released and a less permeable layer is encountered. The dissolution processes for residual and pooled DNAPL are fundamentally different.

Residual saturation has the potential to dissolve relatively quickly. Residual dissolution is defined in terms of years to decades as compared to decades to centuries for pooled DNAPL. DNAPL dissolution modeling is complex, requiring detailed knowledge of DNAPL spatial distribution, mass transfer rates, DNAPL water interfacial surface area and details of microscale heterogeneities.

Most of the DNAPL compounds have relatively low solubilities. The solubility of the compound controls the volume and the rate the compound dissolves into the groundwater. Although these compounds have a low solubility, all compounds have some solubility capability in water. Therefore, the typical approach of investigation through monitoring well installations and groundwater contaminant concentrations has limitations in characterization of a DNAPL site. Generally, when concentrations of contaminants in groundwater approach 1 percent of the solubility of a compound, there is a likely presence of DNAPL. However, this correlation is weak because the results only measure the amount of substance dissolved from the residual or pooled DNAPL and is not a direct measurement. The measurement is actually the concentration that has dissolved from the residual and/or pooled DNAPL. The volume of DNAPL to the overall plume size is small and spatial patterns are complex, therefore, the measurement of dissolved phase groundwater to evaluate the likely presence of a DNAPL site is highly inaccurate.

The best approach to locate DNAPL is to develop a realistic site conceptual model, based on the site history, including potential spill pathways and nature of the release. Next, develop a stratigraphic profile of the subsurface evaluating areas where pooled DNAPL may accumulate.

There are certain physiochemical properties that may result in the difficult investigation and remediation of DNAPLs. First, because of the high density, DNAPLs have the potential to move deeper when disturbed. Second, the high interfacial tension makes DNAPL immobile such that traditional groundwater pump and treat systems that move water past residual DNAPL do not have the ability to dislodge and move it. Third, the low solubility of DNAPL makes flushing remedial techniques ineffective.

The best approach in investigating potential DNAPL sites is first to develop a realistic site conceptual model that suggests where the DNAPL may have entered the subsurface, potential migration pathways, and in what form the DNAPL may be present (residual and/or pooled). Second, develop data quality objectives related to specific project goals. For example, to identify general areas of residual and/or pooled DNAPL, pinpoint the contaminant source area and consider the use of a containment system. Finally, select site investigation methods that meet remedial objectives and take appropriate measures to mitigate the risk of mobilizing the DNAPL.

A 1993 EPA publication, No. 9355.4-07FS, suggests that "an outside-in" strategy is used to investigate DNAPL sites. This approach involves drilling fringe areas of site before drilling in potential DNAPL areas to form a reliable conceptual model. This strategy is to avoid drilling in DNAPL areas to prevent opening further migration pathways until detailed site information relating to the hydrogeology, stratigraphy, and potential pathways is further understood.

Investigative techniques are beyond the scope of this section and the reader is referred to an excellent summary by Mercer and Cohen (1993). Conventional geophysical techniques are nonintrusive and have been effective in investigating DNAPL sites. A number of promising emerging geophysical techniques are in the early stages of development. These include electrical impedance tomography, seismic reflection amplitude offset, cross radar, and high-resolution 3D electromagnetic resitivity/seismic reflection.

Intrusive methods need to be carefully applied to minimize the risk of spreading DNAPL. Direct-push techniques have been refined in recent years and are valuable investigative techniques. Soil cores obtained by this method should be checked frequently with field meters and/or dyes. The traditional monitoring well is typically used for plume delineation, but frequently is not a good method for identifying DNAPL. When wells are used, multilevel and shorter screen lengths should be considered.

14.10.3 REMEDIATION OF DNAPLs

In the late 1980s and early 1990s, the traditional remedial approach of installing pump and treatment systems was shown to be ineffective. Consequently, remediation of DNAPL sites through in situ techniques gradually evolved. This evolution resulted because early pump-and-treat studies have shown that forcing water past a contaminated portion of the aquifer would not extract the organic compounds to generally acceptable cleanup goals within a reasonable time frame.

In 1992, the EPA (publication 93554-05) completed one of the most cited studies regarding the effectiveness of pump-and-treat groundwater remediation systems. This document served to demonstrate the impracticability of restoring groundwater quality to acceptable cleanup criteria though pump-and-treat techniques. However, this does not mean that pump-and-treat systems are a worthless remediation approach. If designed correctly, an extraction well can hydraulically contain a dissolved plume preventing further migration. Groundwater extraction can also be effective in removing a large contaminant mass from an aquifer. However, because this technique involves moving water past an contaminated mass and dissolving a portion of the mass, it cannot be used in most cases to achieve acceptable cleanup criteria at a given site.

Remediation techniques can be viewed as a two-step process. The first step is identifying, delineating, controlling, and remediating the DNAPL source. The second step involves site-wide remediation to address the plume emanating from a DNAPL source. The author's experience has shown that source removal is usually the most cost-effective method and the logical first step to remediating a DNAPL site. At one of my project sites, source removal concentrations of chlorinated volatile organic compounds in groundwater were approximately 24,000 micrograms per liter. Following the removal of the contaminated soil within the unsaturated zone, contaminated levels in groundwater were reduced to approximately 5000 microgram per liter. Often DNAPL sites require implementing a combination of the remedial approaches to achieve the ultimate cleanup criteria for a given site. For example, the source area could be remediated by excavation or soil vapor extraction with natural attenuation for the dissolved plume.

Source Area Remediation

Generally, remediation of the source area is the most cost-effective and technically feasible approach toward a successful site-wide remediation project (Fetter, 1993). Residual contamination in the vadose zone must be controlled to cut off the residual contamination to act as a source of continued groundwater contamination. Traditional remediation techniques often work well to accomplish source removal. If the vadose zone soil is accessible, I have found that, in most cases, it makes economic and technical sense to excavate and physically remove the source. Traditional methods, such as excavation and soil vapor extraction, are effective vadose zone remediation techniques.

Once the vadose zone is remediated, attention can be focused on removing contaminant mass from the saturated zone. Whereas groundwater pump and treat systems are generally ineffective for site-wide remediation, it can be a very effective source removal technique. Groundwater migration is usually controlled by advection and pump-and-treat systems are highly efficient at controlling advection processes. A properly designed pump-and-treat system will maintain hydraulic control over a migrating dissolved plume, while extracting a portion of the residual and/or pooled DNAPL. However, a pump-and-treat system will reduce the groundwater contaminants fairly rapidly and will then stabilize at some static concentration level. This static level is often reached at a concentration above the acceptable cleanup criteria for the reasons discussed in the prior section. Once this level is achieved, it generally means that the effectiveness of this remediation technique has been achieved and the technique will no longer be effective in reducing the contaminant concentrations any further.

Because of the inefficiency of pump and treat systems, the industry trend is moving toward in situ remediation techniques. The following three in situ technology groups have been shown to reduce DNAPL mass in the subsurface and can be effective source removal techniques:

- Thermal
- Chemical oxidation
- Cosolvent/surfactants

Thermal. Physical properties of DNAPL that control their movement in the subsurface are temperature dependent. Thermal processes introduce heat into the contaminated zone in order to enhance their mobility and volatilization. The primary mechanism for enhanced recovery of DNAPLs is the increased vapor pressure, which allows contaminant mass removal in the vapor phase. This approach has been shown effective in the remediation of PCBs and chlorinated solvents. Variations in this approach include the following:

Steam Injection.   In sandy, more permeable soils, steam can be injected as a heat source. The advancing pressure front displaces soil, water, and contaminants by volatilization. The volatile compounds are transported in the vapor phase where they condense and can be removed by pumping.

Thermal Wells.   This technique is a modification of the steam injection. Typically, hot water is used to volatilize organic compounds. The hot water strips the sorbed volatile compounds into the groundwater. The groundwater is then captured, and the volatile organic compounds are removed typically by a surface treatment.

Radio Frequency Heating.   Electrodes are placed in the contaminated area and a radio frequency transmitter is used as a power source to the electrodes. The radio frequency is used to heat the soil to temperatures greater than 100°C causing volatilization of the organic compounds. Once volatilized, the contaminants are removed typically by conventional soil vapor extraction processes.

6-Phase Electrical Heating.   This techniques typically is applied to lower permeable soils to enhance a conventional soil vapor extraction system. Six-phase heating uses low-frequency electricity to heat soils as an enhancement to a soil vapor extraction system. It uses single phase transformers to convert standard three-phase electricity to six-phase electricity. Electrodes are placed in the ground on one or more circular arrays and connected to a separate current phase. A neutral electrode is located in the center of the array and is used as a vent. As electricity is introduced, it heats the soil causing volatilization of the organic compounds while the volatile compounds are extracted at the neutral electrode.

Vitrification.   Vitrification involves heating soil to a very high temperature (1500 to 2000 degrees centigrade) to melt the soil. During the heating process, the volatile compounds are transferred into the gas stage where they are captured and treated. This technology is a stabilization process, because when the soil cools it forms a glass and vitrified material. This technique is also used to encapsulate inorganic compounds by making them immobile.

Chemical Oxidation.   DNAPLs are amenable to rapid and complete destruction by chemical oxidation. This technique involves the introduction of an oxidizing compound into the source area breaking down the chemical bonds of the contaminants through reactions with the oxidizer. Types of chemical oxidation include the following:

Potassium Permanganate ($KMnO_4$).   The use of the $KMnO_4$ to indiscriminately scavenge and oxidize organic compounds has been documented for a considerable time in both drinking water and wastewater fields. Since 1989, this technology has been studied for application for the in situ oxidation of chlorinated compounds. Most of this early laboratory and field testing was conducted at the University of Waterloo, which successfully removed TCE and PCE from source areas at mass removal rates of 60 to 92 percent.

A typical field application of the technology would include a series of extraction and injection wells to recirculate groundwater within a defined treatment cell. The $KMnO_4$ is mixed from solid state into solution and metered into the extracted groundwater at a designated rate, reaching a target concentration that was predetermined from bench-scale testing. The solution is then injected at the upgradient end of the treatment zone and recirculated to oxidize the DNAPL. The concentration of chemical oxidant and volatile organic compounds (VOCs) is monitored at the extraction point and in the treatment cell to maintain the required concentration for complete oxidation. This approach uses direct contact of $KMnO_4$ with the contaminant to oxidize the contaminants through chemical change. This technique has been effective on double organic bond compounds, such as TCE and PCE. The techniques has limited effectiveness with straight-chained single bond compounds, such as 1,1,1 TCA.

Fenton's Reagent.   In situ oxidation using Fenton's reagents is an innovative technology that uses a catalyst along with hydrogen peroxide to breakdown organic chemical bonds into end-products of carbon dioxide and water. The basic process involves pH adjustment to the formation to facilitate the reaction, followed by the addition of a catalyst (usually iron-based) to generate hydroxyl

(OH^-) radicals. Following the introduction of catalyst, hydrogen peroxide solution is injected into the formation, which facilitates the reaction. The organic chemical bonds are broken into simpler organic compounds, ending with by-products of carbon dioxide, water, and salt. This technique has been shown to be effective with many types of organic compounds. A reaction occurs through direct contact to break down the organic compounds into simpler substances. Successful demonstrations have been completed for chlorinated organic compounds and for petroleum hydrocarbons.

Hydrogen Peroxide H_2O_2.   Degradation of chlorinated compounds occurs more rapidly under anaerobic conditions rather the aerobic conditions. However, there may be circumstances where hydrogen peroxide may be beneficial to enhance degradation of chlorinated compounds. The goal of aerobic in situ bioremediation is to supply oxygen and nutrients to microorganisms in soil and groundwater. Oxygen can be provided by delivering the oxygen in liquid form as hydrogen peroxide. This process delivers oxygen to stimulate the activity of naturally occurring microorganisms by circulating hydrogen peroxide through contaminated media in order to speed up the bioremediation of organic compounds.

Because most chlorinated solvents degrade more rapidly under anaerobic conditions, a new generation of product has recently made it to the market place (Koenig/Norris, 1999). This technology uses hydrogen to transform the aquifer from aerobic to anaerobic conditions to speed up the breakdown of chlorinated compounds. It operates under the same principle as the hydrogen peroxide but is used to promote an anaerobic condition.This technique facilitates chemical change by adding more dissolved oxygen to the groundwater.

$KMnO_4$ or H_2O_2 in Combination with Deep Soil Mixing.   This technology uses large auger mixing blades (4 ft in diameter) to mix the contaminated soil. While the soil mixing is occurring, an oxidant such as $KMnO_4$ or H_2O_2 is injected through the interior of the drill string to the contaminated soil zone in situ. The mixing facilitates direct contact with the oxidant to break down the organic compounds. A successful field demonstration was completed in the mid-1990s at the Department of Energy Kansas City, Plant in Kansas City, Missouri. See Cline et al. (1997) for a detailed description of the test and results.

Co-Solvent/Surfactants.   This remedial technique utilizes the injection of cosolvent/surfactant into the source area to remove the DNAPL through a combination of dissolution and displacement. The injected solution increases the solubility of the DNAPL resulting from the reduction of interfacial tension. Types of cosolvent/surfactant treatment technologies include alcohol and surfactants as follows:

Alcohol.   Alcohols are miscible in both water and DNAPL. Depending on the type of alcohol and the type of contaminant, the alcohol will preferentially partition into one phase or another. Alcohols that partition into the DNAPL cause swelling and will increase the aqueous solubility of the DNAPL causing it to mobilize.

Surfactants (Soaps and Detergents).   The reduction in DNAPL water interfacial tension arises from the addition of a surfactant to an aqueous solution resulting in reducing the influence of capillary forces. Capillary forces control the retention of residual DNAPL and the formation of pooled DNAPL. If the interfacial tension is lowered, mobilization of the DNAPL will occur. This technique typically involves flooding the formation to remove residual DNAPL trapped within the pore spaces and/or fractures of the formation and mobilizing the DNAPL. Generally, some mechanism is employed to capture the freed DNAPL so it does not have the opportunity to migrate.

DNAPL Source Control

A dissolved plume will likely result from a DNAPL source area and it will be necessary to contain this plume. The remediation technique for these plumes is often different than the remediation technique applied to remediate a DNAPL source area. Technology groups that focus on the plume emanating from a source area include the following:

- Barriers
- Volatilization
- Biological

Barriers. Barriers are treatment walls that allow groundwater migration through the barrier thus effectively destroying and/or sorbing the dissolved phase contaminants. Nonpermeable barriers also can be used that only impede further migration of the plume. Examples of this remedial technique include reactive treatment walls and nonpermeable (sheet-pile and slurry walls).

Reactive Treatment Walls. Treatment walls are structures installed underground to treat contaminated groundwater. Typically, a trench is installed to direct groundwater across a predetermined engineered flow path. The trench is filled with a variety of materials (reactive fillings). As the contaminated groundwater passes through the treatment wall, the contaminants are either trapped by the treatment wall or transformed into simpler compounds.

Treatment walls are generally of three types: sorption walls, precipitation barriers, or degradation barriers. Sorption walls contain fillings that remove contaminants from groundwater by physically capturing contaminants from the groundwater and sorbing them onto the reactive wall. Precipitation barriers contain fillings that react with the groundwater contaminants as they pass through the wall. The reaction causes the dissolved contaminants to precipitate from the water and become trapped on the treatment wall. The groundwater continues to move through the wall but contains much less contaminant mass upon exiting from the reactive wall. Degradation barriers cause reactions as groundwater passes through the wall. As the contaminant passes through the wall, the organic compounds are broken down into simpler organic compounds. Typically, these types of walls utilize microorganism and are a form of biodegradation.

Nonpermeable (Sheet-Pile and Slurry Walls). These types of walls are strictly physical barriers in nature. They do not cause or induce any chemical change in the organic compounds. They are used to physically restrict groundwater movement and typically contain a mechanism to collect the trapped groundwater and pump it to the surface where it is treated.

Volatilization. Volatilization involves the extraction of contaminants from an air stream. Contaminants are subject to destruction through in situ airflow (air sparging or SVE), or capture by pumping and stripping (in-well stripping and conventional pump-and-treat). Examples of this remedial technique are discussed in the following sections.

Soil Vapor Extraction (SVE). The soil vapor extraction (SVE) process extracts contaminants from the unsaturated soil in vapor form. Therefore, this technology is best suited for volatile compounds that volatilize or evaporate easily. By applying a vacuum through a system of underground wells, contaminants are extracted to the surface as vapor where it can be treated. The typical SVE system consists of vapor extraction wells and injection wells (or air vents) in the contaminated area. Air compressors are used to supply air, which is pumped into the injection wells. Placing a vacuum on the extraction wells causes the air to move through the formation. The air stream volatilizes come of the organic compounds as it passes through the formation. The air is then extracted to the surface where the organic compounds can be stripped from the air stream.

Air Sparging. SVE alone cannot remove contaminant in the saturated zone. Therefore, when groundwater contamination is present, air sparging is a common technology used in conjunction with the SVE system. Air sparging is an added component of the SVE system whereby a series of short-screened wells are placed below the watertable. Air is injected into the air sparge wells. As the air moves through the formation it volatilizes some of the organic compounds from the aquifer. The air moves upward through the saturated zone and is eventually captured by the SVE wells, where the air is returned to the surface and treated.

In-Well Stripping (Density Driven Convection Well). This particular "convection" well design creates a venturi system using a dual-screened well. A typical construction would comprise a 4-in diameter extraction well with a 10-foot-long upper screen across the watertable. A longer riser casing (at least 20 ft) would be installed followed by a 2- or 3-ft length of well screen at the base. A $1/2$-in air pipeline is then inserted into the dual-screened well just above the lower well screen. When the air is turned on, it creates an airlift, which draws water into the lower screen. The water is aerated as it is drawn up the well, and water is discharged back into the formation at the upper screened area. The stripped volatile compounds air emissions travel up the well to the surface where the air stream is treated.

Pump-and-Treat. The pump-and-treat benefits and limitations have already been discussed extensively. The typical approach for this technique is to install one or several extraction wells and to extract groundwater at a rate to create a cone of influence. This cone of influence would provide hydraulic containment of a dissolved plume as groundwater within the zone of capture flows toward the well. A pump test is typically completed to estimate the hydraulic conductivity of the formation and to determine the optimum pumping rate necessary to provide hydraulic containment. This information is then used to design the extraction well, such as diameter of the well, depth, and screen interval.

Once a properly designed well is installed, groundwater is extracted typically using submersible pumps or surface air lift system. The groundwater is extracted to the surface where the groundwater is stripped of volatile compounds by activated carbon, aeration trays, or stripping towers. Depending on the concentration of volatile compounds and the type of treatment, the off-gas may also require treatment to remove the volatile compounds.

Biological.   This remedial technique uses biological reactions to break down contaminants. The process may occur under natural conditions or may include the addition of nutrients, oxygen, and other chemicals. Examples of this remedial technique are discussed in the following sections:

Natural Attenuation.   Numerous articles and demonstrations concerning natural attenuation have been presented in recent years. In the context of remediation, natural attenuation refers to the natural attenuation processes, which include a variety of physical, chemical, or biological processes that act without human intervention to reduce contaminant compounds. Biological processes of natural attenuation include aerobic, hypoxic (low oxygen), and anaerobic biodegradation. Natural attenuation processes typically occur at all sites but have significantly varying rates depending upon the types of contaminants present and the physical, chemical, and biological characteristics of the soil and groundwater. Most regulatory agencies require that the time needed to reduce contaminant concentrations to acceptable concentrations is typically less than five years.

Chlorinated hydrocarbons undergo biodegradation through the following: (1) reductive chlorination, which is the use of more highly chlorinated hydrocarbons as receptors; (2) use of hydrocarbons as electron donors; and/or (3) cometabolism, which catalyses by an enzyme or cofactor. All processes may occur at a particular site, although reductive chlorination is the most important biodegradation process. Natural attenuation of chlorinated hydrocarbons is an electron donor process.

Under the reductive chlorination process, the chlorinated hydrocarbon is used as an electron receptor, not as a carbon source. In this process, a chlorine atom is removed and replaced with a hydrogen atom. The sources of carbon (electron donor) for microbial growth may include natural organic material. The process occurs under anerobic reducing (low redox) conditions, and has been demonstrated under iron-reducing sulfate-reducing and methanogenic conditions.

Reductive chlorination occurs by sequential dechlorination. Depending upon conditions, the process may be interrupted, with other processes acting upon the daughter products. Reductive chlorination produces an accumulation of daughter products, and an increase in chloride ion concentrations.

Dissolved Plume (Site-Wide Remediation)

Once a source area is remediated and controlled, there often exists a detached groundwater plume that requires remediation to achieve the groundwater cleanup goal established for a given site. The following technologies focus on treatment of the dissolved plume where no definitive source is documented:

- Natural attenuation
- Enhanced bioremediation
- Reactive or treatment walls
- Air sparging
- Pump-and-treat systems
- Phytoremediation

With the exception of phytoremediation, all of the technologies listed above have been discussed. Phytoremediation is the direct use of living plants to reduce the contaminant mass in the subsurface. Some plants can break down organic compounds into simpler compounds. This technology, used mainly for removal of metals from the soil, is also being tested for applicability for organic compounds. The compounds are taken in through the root structure of the plant into the above-ground structure of the plant where it is stored. Upon maturity, the plants are then harvested.

REFERENCES AND ADDITIONAL READINGS

Cline, S. R., West, O. R., N. K., F. G., R. S., and J. B., "KMnO$_4$ Chemical Oxidation and Deep Soil Mixing for Soil Treatment," *Geotechnical News*, pp. 25–28, December 1997.

Cohen, R. M., and Mercer, J. W., *DNAPL Site Evaluation*, C. K. Smoley, Boca Raton, FL, 1993.

Domenico, P. A., and Swartz, F. W., Physical and Chemical Hydrogeology, John Wiley & Sons, Inc. 1990.

Driscoll, F., *Groundwater and Wells*, VOP Johnson, Co., St. Paul, MN, 1986.

Fetter, C. W., Contaminant Hydrogeology, Prentice-Hall, 1993.

Freeze, R. A., and Cherry, J. A., *Groundwater*, Prentice-Hall, Inc., Englewood, NJ, 1979.

Gee, G. W., and Wing, N. R., *In-Situ Remediation: Scientific Basis for Current and Future Technologies*, Battelle Press, Columbus, OH, 1994.

Koenigsberg and Norris, Accelerated Bioremediation using Slow Release Compounds, Regenesis Bioremediation Products, 1999.

NJDEP, *Classification Exception Area Ordinance Document*, 1995.

Nyer, K. E., and Duffin, M. E., "The State of the Art of Bioremediation," *Ground Water Monitoring and Remediation*, 17(2): 64–69, 1997.

U.S. EPA, *Estimating Potential for Occurrence of DNAPL at Superfund Sites*, Publication 9355.4-07FS, December 1991.

U.S. EPA, *Evaluation of the Likelihood of DNAPL Presence at NPL Sites, National Results*, Publication 9355.4-13, September 1993.

U.S. EPA, *Guidance for Evaluating the Technical Impracticability of Ground-Water Restoration*, Publication 9234.2-2.5, p. 26, September 1993.

U.S. EPA, *In-Situ Remediation Technology Status Report: Thermal Enhancements*, EPA-542-5-94-009, April 1995.

U.S. EPA, *A Citizen's Guide to Bioremediation*, EPA 542-F-96-007, April 1996.

U.S. EPA, *A Citizen's Guide to Solvent Extraction*, EPA-542-F-96-003, April 1996.

U.S. EPA, *A Citizen's Guide to Soil Vapor Extraction and Air Sparging*, EPA-542-F-96-008, April 1996.

U.S. EPA, *A Citizen's Guide to Treatment Walls*, EPA-542-F-96-016, September 1996.

U.S. EPA, *A Citizen's Guide to Phytoremediation*, EPA 542-F-98-011, August 1998.

U.S. EPA, *Field Applications of In-Situ Remediation Technologies: Chemical Oxidation*, EPA-542-R-98-008, September 1998.

U.S. EPA, *Technical Protocol for Evaluating Natural Attenuation of Chlorinated Solvents in Ground Water*, EPA-600-R-98-128, September 1998.

CHAPTER 14
REMEDIATION TECHNOLOGIES

SECTION 14.11

VERTICAL PASSIVE GROUNDWATER BARRIER SYSTEMS

David Lager

Mr. Lager is president of NETCo., a professional organization which provides environmental compliance services to business and industry. He has been assisting industry in meeting RCRA, CERCLA and CWA requirements for 15 years.

In this section we will explore the use of vertical passive barrier systems. Subsection 14.11.1 provides an overview of the advances in use of vertical barrier technology. Subsection 14.11.2 focuses on the types of barrier walls in use today, including slurry walls, jet grout walls, soil mix walls, grouting, and other types of low permeability in situ walls. Subsection 14.11.3 explores the use of reactive walls systems, which are an in situ method of treating contaminated groundwater. This chapter will not cover horizontal liners or their associated construction methods.

14.11.1 OVERVIEW

Modern vertical passive barrier systems have existed as a geotechnical construction technology in some form for over 60 year. Among the earliest uses of engineered passive barrier systems are various form of grouting where a grout mixture is pumped into a fractured rock or permeable soil strata to reduce the flow of water through the rock or soil media. Such applications of passive barrier systems have been used extensively in mining and dam foundation applications. Over the years, many advances have been made in designing a wide variety of grouts and delivery systems (e.g., drills, pumps, drill bits, nozzles) to get the grout to where it is needed.

In the late 1930s, slurry wall technologies began to emerge. The initial use of a slurry wall was as a structural load-bearing wall for foundation applications. In the 1940s and 1950s, the Corps of Engineers and the Bureau of Land Reclamation expanded the use of slurry wall technology through the use of soil and/or cement bentonite backfills. These applications resulted in slurry walls being used as vertical cutoff walls beneath large earthen dams and reservoirs. In the 1960s, 70s, and 80s, the slurry cutoff wall field saw a number of advances in equipment systems, which allowed slurry walls to be excavated to more than 400 ft in depth, as well as different types of backfills that were more compatible with contaminated wastes, including the use of high density polyethylene sheets or liners that are inserted into the cutoff wall excavation.

Also in the 1950s and the 1960s, the Japanese developed in situ soil mixing technologies. Until the early 1990s this technology was not well-known or recognized in the United States. However, soil mixing technologies have had vast use in Japan and other Southeast Asia locations in the last 30 years. The primary applications have been cutoff walls as well as a method of improving the structural/geotechnical properties of soil to minimize, among other concerns, the impact of liquefaction potential during earthquake events. In the last 10 year, in the United States, soil mix walls have been used as geotechnical applications to stabilize unstable soils, control horizontal groundwater flows, and as a method to construct reactive walls.

Beginning in the 1970s, with the emergence of the environmental remediation field, engineers and groundwater scientists began to look at how traditional geotechnical construction techniques could be used to control and contain the horizontal flow of contaminated groundwater. Control and containment applications were generally the most common applications of vertical passive groundwater barrier systems during the early years (late 1970s through mid-1980s) of the groundwater remediation field. Many owners and engineers were looking for methods to "buy-time" until the longer term, and more expensive issues of actually remediating a containment site could be addressed. Passive barrier systems, because of their relatively low installation costs, fit the buy-time scenario for many owners. Further, in many cases, many owners tried to use passive barrier remediation schemes as their solution to a groundwater remediation problem. In this context, to some degree, too much was expected of passive barrier systems—some containment schemes choose not to look closely at the contaminated groundwater/backfill compatibility issues. In any case, the mid-1970s through the mid-1980s (approximately) represented, if you will, the zenith of use of passive groundwater barrier systems, and, in particular, slurry cutoff walls in the groundwater remediation field.

In the 1980s, as our knowledge-base grew, the regulated community and consultants and contractors began to understand that passive barrier systems could not and should not be evaluated in a vacuum. In the late 1980s, and certainly during the 1990s, it became apparent that passive barrier systems actually had the most success where they were evaluated in context of a total remediation scheme (i.e., control, containment, and clean-up). Passive barrier systems were increasingly used as part of a passive/active barrier system. Passive barriers contained the water with the active component (pumping) helping to maintain a differential head across the barrier wall, or to recover and treat the contaminated groundwater. This combination of active and passive barrier systems provided engineers and owners with the most effective (i.e., lowest risk) system of controlling, containing, and treating contaminated groundwater sites. As a result, in the 1990s, the use of passive barrier systems declined because of the increased sophistication in risk-based analysis of contaminated sites, the fact that many of our worst sites were contained and controlled, and the emphasis by all parties on remediating problems rather than just controlling a contaminated plume.

As we enter the next millenium, the vertical passive barrier field appears to be evolving in two directions. The first direction is where improvements are made in backfill design and compatibility testing to ensure a more consistent and predictable performance of the passive barrier wall. Such barrier walls, in many cases, will continue to have an active component to further reduce the risk of escape of any contaminated groundwater outside the containment area.

The second direction, and where most of the interest in vertical barriers systems seems to be heading, is reactive walls. Reactive walls are in situ treatment systems. Instead of the barrier walls attempting to contain groundwater to the maximum degree possible, reactive passive barrier walls are a type of slow treatment system. To use an analogy, reactive walls are like slow-release drug therapy treatments or slow-release plant fertilizers. Reactive walls, either of the continuous trench or funnel and gate type, are meant to allow contaminated water to pass through the wall or the gate at a predetermined rate. As the contaminated water passes through a reactive media, which forms a part or all of the backfill, the water is treated by a reactive media, which forms a part or all of the backfill.

To summarize, passive vertical barrier system technologies in the last thirty years have moved from one of just merely containing contamination, to actually providing a means of treating the contamination at its source. Thus, the technology of effectively using passive barrier systems today, and for the foreseeable future, cuts across the geotechnical, groundwater hydrology and the chemistry disciplines.

14.11.2 TYPES OF VERTICAL PASSIVE BARRIER WALLS

Slurry Cutoff Walls

The most prevalent type of vertical passive barrier system over the last 30 year has been slurry cutoff walls. The basic concept of the slurry cutoff wall is the excavation of a narrow trench (usually 24 to

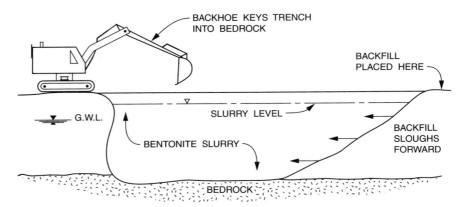

FIGURE 14.11.1   Slurry cutoff wall.

48 in in width) from grade down to a low-permeable (i.e., 1×10^{-7} cm/sec) layer of clay or rock (aquiclude). During the excavation process, either by hydraulic backhoe or slurry clamshell, the trench is kept filled with a water bentonite slurry. The bentonite slurry migrates into the surrounding soil filling the soil void structure with a thixotropic slurry, as well as coating the side walls of the trench with a thin filter cake. The thixotropic properties of the slurry and the filter cake enables the remaining slurry fluid in the trench to exert a positive hydrostatic head against the sidewalls of the trench to prevent the trench from collapsing, even in highly permeable and granular soils, such as sand. Once the trench is excavated, the slurry wall is backfilled with a low-permeability backfill material (see Figure 14.11.1).

Soil Bentonite Backfilled Slurry Walls.   Once the trench is excavated (excavation to 40 to 100 ft are relatively routine, with excavation to 400 ft possible but expensive), an engineered backfill material is placed back into the trench. Generally, two types of backfills are used. The most common is the soil bentonite backfilled slurry wall. Soil with a high percentage of plastic fines (usually at least 20 percent passing a #200 mesh sieve) are mixed with bentonite slurry to yield a homogenous backfill mix. This mix design will produce a backfill with a permeability coefficient of approximately 1×10^{-7} cm/sec. This latter coefficient of permeability is generally what the regulators look for in defining a low-permeability passive vertical barrier wall.

To ensure, as best possible, that the backfill placed in the excavated trench will resist contamination, most owners, engineers, and regulators require that the soil bentonite backfill mix be subjected to a compatibility testing process. This means that the designed backfill is subjected to the expected contamination present in the ground where the slurry wall will be placed. This compatibility test is generally a standard hydraulic conductivity test where the permeant is the contaminated water.

Soil bentonite backfills have generally been the most commonly used backfills. The relative ease of preparation of the soil bentonite backfill and the ability to increase the amount of fines (percentage passing a #200 mesh sieve) present in the backfill have been, in general, the least costly and the most contaminant resistant backfill.

Cement Bentonite Backfilled Slurry Walls.   The second type of backfill are the cement bentonite backfills. As the name suggests, cement bentonite backfills consist of cement, bentonite, and water. These types of backfills do not have the same degree of contamination resistance as soil bentonite backfills because of the presence of cement. Cement type products are less resistant to many types of chemicals found in groundwater contamination, such as solvents and hydrocarbons, that a cutoff wall is intended to contain and control. On the other hand, cement bentonite backfills offer more structural stability, which can be a useful property where the cutoff wall is placed on an embankment, or where the area inside the cutoff wall will be excavated to remove the contamination.

The cement bentonite cutoff wall installation procedure varies from that of the soil bentonite cutoff wall. The main difference is that the slurry used to stabilize the trench during the excavation is the same as the backfill. Instead of excavating under a water and bentonite slurry, and then placing a soil bentonite backfill back into the trench, the contractor prepares the cement-bentonite and water slurry as excavation begins. As the excavation proceeds, the cement slurry is pumped into the trench. The cement bentonite slurry fills the trench and provides both the slurry to stabilize the trench sidewalls during excavation and the self-hardening backfill that comprises the final backfill. Various types of retarders are used to slow the setting time to allow most cement bentonite slurries to be workable for 6 to 10 hours.

The strength of the cement bentonite backfill can be varied by the amount of cement added to the mix. However, there is a line above which too much cement will result in the cement slurry hardening and, therefore, cracking. Generally, most cement bentonite mixes have a 15 to 20 psi unconfined hardness design. This relatively low strength enables the cement slurry to both provide some structural properties, but also be flexible such that the backfill will not crack. The permeability of most cement bentonite cutoff walls ranges from 1×10^{-5} to 1×10^{-7} cm/sec.

Vibrated Beam Slurry Walls

A variation on the full trench slurry wall method is the so-called vibrated beam or "thin" wall. The vibrated beam wall was developed in Europe during the 1960s and 1970s as a specialty geotechnical construction technique. It was brought to the United States in the early 1970s. Its initial use was as a geotechnical construction method to install low-permeability cutoff walls below various embankments. The backfill used was a cement bentonite mixture.

However, because the vibrated beam wall does not result in the generation of trench spoils, nor the attendant work area for a conventional soil bentonite slurry wall, the vibrated beam wall began to be used as a means of installing cutoff walls in contamination environments, or where access restrictions existed (such as on the top of narrow berms). Further, because the profile of a vibrated beam wall is that of an H-Pile, the amount of backfill required was much less than a conventional trench slurry wall. Thus, other types of backfills that are more resistant to contamination, such as Impermix and Aspemix, could be used in a vibrated beam wall installation.

The vibrated beam wall is constructed by vibrating a specially designed "H" beam into the ground to the desired depth using a vibratory hammer. As the beam is driven into the ground, it pushes aside the soil (note the soil is not removed). A cement bentonite or a chemically resistant slurry (such as Impermix or Aspemix) is then pumped from a mixing plant to the bottom of the vibrated beam. The slurry mixture is then "jetted" through nozzles at the bottom of the beam into the void left by the beam as it is retracted at a predetermined rate. The cutoff wall is created by overlapping successive beam penetrations. The overlap is created by utilizing a fin welded to the flange of the beam. The fin penetrates into the completed beam penetration allowing the slurry to key into the previous panel (see Figure 14.11.2). Depths for vibrated beam walls can reach 80 ft or more.

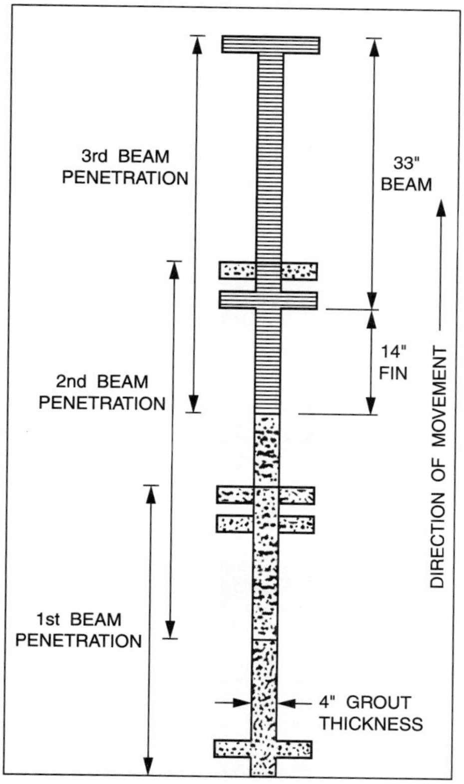

Plan View of Successive Beam Overlap
FIGURE 14.11.2 Vibrated beam sequence and overlap pattern.

Slurry Walls with HDPE Liners

A further refinement in the use of slurry walls occurred in 1980s—the use of a high density polyethylene (HDPE) liner installed into

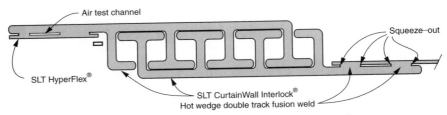

FIGURE 14.11.3 H.D.P.E. CurtainWall interlock joint. (*SLT CurtainWall Interlock® Joint*)

a slurry wall. This technology step was in direct response to the concern over soil bentonite backfill compatibility issues. Many owners and engineers began to have concern over the duration that a soil bentonite backfill could resist DNAPLs (dense nonaqueous phase liquids), PAHs (polynuclear aromatic hydrocarbons), and other types of hazardous chemicals. One solution was to use a vertical HDPE liner. It had been well-established that the HDPE material had an excellent resistance to hydrocarbons and other types of chemicals.

The challenge in using HDPE liners in a slurry wall was how to get the liner into a trench filled with a slurry, either bentonite and water, or cement bentonite. After some experimentation, the generally accepted method today for installing a vertical HDPE liner into a cutoff wall is in panels. The slurry wall is excavated either by the trench or vibrated beam method. Once the trench is excavated, the HDPE panel (80 mil thickness) is placed into the slurry filled trench using a guide box or a template with the HDPE liner attached. The joint between adjacent panels is created by a mechanical interlock system (see Figure 14.11.3). In the case of the vibrated beam wall, no further work is required. However, where a conventional trench excavation is required, backfill is then placed up against the HDPE liner to hold the liner in place (see Figure 14.11.4).

The advantages of the HDPE panel system primarily focus on the HDPE material itself. HDPE materials have high mechanical strength, resistance to corrosion and biological attack, relatively low impermeability to hydrocarbons and gases, and high resistance to environmental stress cracking.

Grout Walls

There are generally five different types of grouting methods: (1) compaction, (2) fracture, (3) chemical, (4) slurry and (5) jet grouting. Compaction and fracture grouting are mainly used in geotechnical

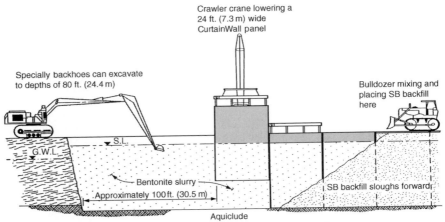

FIGURE 14.11.4 H.D.P.E. CurtainWall installation process. (*SLT CurtainWall® System Being Installed In A Slurry Trench Excavation By A Crawler Crane And Backfilled With Soil-Bentomite (SB)*)

applications to reduce settlement potential. Such grouting techniques have minimum application for reducing the permeability of a soil strata.

Slurry and chemical grouting have been widely used over the last several decades to improve the permeability of a soil structure. However, slurry and chemical grouting are limited in the types of soils in which they can be effective. Slurry and chemical grouts are most effective in granular soils, and have almost no application in clays. Thus, while slurry and chemical grouts could be used in certain geo-environmental applications, in most cases, slurry walls of some type will be more feasible and economical.

Jet Grouting.   Jet grouting is beginning to have a number of geo-environmental applications as a passive barrier system, both for horizontal (bottom seals) and for vertical containment barriers. The jet grouting method emerged in Japan in the early 1970s. The technology traveled to Europe in the late 1970s and to North America in the early 1980s. However, jet grouting did not see extended use in North America until the early to mid-1990s.

In its simplest form, jet grouting is a technique of cutting and mixing in situ soils with a grout liquid to form a stabilized soil mass or a low-permeability cutoff wall. By varying the design of the grout mixture, the strength of the stabilized soil can be increased to provide improved structural properties, if required. A continuous barrier is formed by overlapping columns (see Figure 14.11.5).

The most common type of jet grout wall today is the triple fluid jet grout wall. The so-called triple fluid method involves the use of air, water, and grout to form the in situ jet grout column.

Triple fluid jet grout walls are installed in two phases. The first phase is drilling a small diameter hole (usually five to six inches in diameter) from the working surface to the bottom elevation of the cutoff wall or bottom seal. Various types of drilling fluids are used during the drilling process to stabilize the hole from collapse. Once the bottom elevation of the jet grout column is reached, the second phase of the jet grout operation is the formation of the column.

High-pressure water and air are pumped down through the hollow stem of the drill rod to the injection rod at the bottom of the drill string. The water and air exit the drill rod through ejector ports, creating an air-encased sheath of high-pressure water while the drill string is kept continuously rotating and then withdrawn at a preset rate. The high pressure sheath of air and water erodes or cuts the soil to form a column.

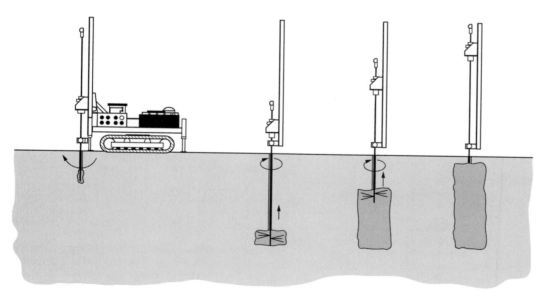

FIGURE 14.11.5   Jet grout column installation process.

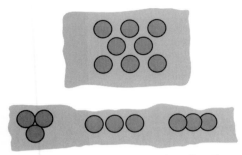

FIGURE 14.11.6 Typical grout curtain configurations.

At the same time that the water and soil are forming the column, a cement grout is pumped to a second set of ejector ports at the bottom of the drill string. The cement grout is injected into the column opening formed by the cutting action of the air and water jet through the lower ejector ports. The cement grout is mixed with any remaining soil to form what is sometimes called a soilcrete column (see Figure 14.11.6).

The most common environmental remediation application of jet grout cutoff walls is in situations where standard cutoff walls cannot be used, in most cases, because of site restrictions. Jet grouting equipment has a smaller working footprint, thus allowing the equipment to work on narrow embankments, or in areas where large cranes and backhoes cannot access. Thus, environmental applications of jet grouting systems have usually occurred in urban areas to control and/or contain the flow of a contaminated plume of water. Normally, jet grouting applications will be more costly than a standard cutoff wall.

In addition to vertical barriers, jet grout walls can also be used as bottom plugs, and as a method of constructing reactive walls.

Soil Mix Walls

In situ soil mixing methods were developed by the Japanese in the early 1970s for improving soft ground profiles, mainly alluvial clay stratas. Over the years, the use of soil mixing was expanded from geotechnical to environmental applications for stabilizing and fixating contaminated soil profiles (such as contaminated harbor sediments and sludge ponds), as cutoff wall structures, and more recently as a method of developing a reactive wall system.

The soil mix wall process consists of a mixing plant preparing a chemical reagent slurry or cement stabilization/fixation fluid. The injection mixture is stored in tanks for pumping to the hollow stem augers of the drilling system.

A auger flight system, (one to five augers mounted on a drill platform—usually a tracked crane) penetrates the ground at a slow rate to both loosen the soil and break up clots in the soil profile. As the auger system penetrates the ground, the chemical reagent slurry or cement stabilizing fluid is pumped by a low-pressure pump to the tip of the auger to lubricate and reduce friction during the drilling phase.

Once the bottom depth of the column is reached, the auger flight system is reversed, and the drill string retracted at a preset rate. The volume of stabilizing fluid is increased so that during the retraction cycle, the in situ soils are thoroughly mixed with the fluid to form a stabilized mass of soil.

A soil mix wall can be used as a vertical wall barrier system, as a bottom plug, or to fixate in situ contamination plumes. See Figure 14.11.7 for a typical soil mix system. Soil mixing techniques have been used to stabilize PCB contaminated soil, oil refinery sludge ponds, and hydrocarbon contaminated soils.

Permeable Barrier Walls

A final class of vertical barrier walls are permeable barrier walls. These types of walls are installed by slurry trench methods, but they are intended to last only a short period of time, thereby allowing lateral groundwater flows to be returned to their normal gradient. The typical use of a permeable barrier wall is in the construction of groundwater collection system.

The most common type of permeable barrier wall is the bio-polymer wall. The essential difference between a standard slurry wall and a bio-polymer wall is that the slurry is biodegradable. The typical material used in a biodegradable wall is a guar gum slurry (made from the guar gum bean). The slurry has some, but not all, of the thixotropic properties of bentonite slurry, yet within a relatively short

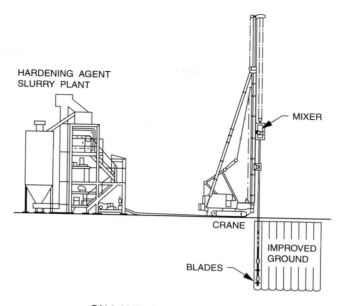

ON-LAND DCM OPERATION

FIGURE 14.11.7   Soil mix wall equipment installation system.

tune period (a few days to a few weeks) natural soil bacteria will degrade the slurry such that the guar gum slurry will revert to mostly water.

A second type of slurry used to make a degradable slurry is a synthetic polymer slurry. Developed largely by the oil well industry, synthetic polymers are increasingly used to augment and improve the properties of bentonite slurries. However, polymer slurries can be used by themselves to form slurries that can be degraded or reversed by the use of other chemicals.

14.11.3 PASSIVE/REACTIVE TREATMENT WALLS

As was discussed in the overview section, an important future direction in the use of passive vertical barrier walls are as installation techniques to treat contaminated groundwater. As an indicator of the importance of this research direction, the United States Environmental Protection Agency (U.S. EPA) has formed a Public Private Partnership called the Permeable Reactive Barriers Action Team. This team is one of seven action teams under the Remediation Technologies Development Forum. These action teams were formed to foster collaboration between the public and private sectors in finding innovative solutions to hazardous waste problems. The specific mission of the Permeable Reactive Barriers Action Team is to accelerate the development of cost-effective permeable barrier technologies for mitigating chlorinated solvents, metals, radionuclides, and other pollutants.

A permeable reactive barrier wall is a passive in situ treatment zone of reactive material that degrades or immobilizes contaminants as groundwater flows through the barrier. Permeable treatment walls are installed across the path of a contaminant plume (see Figure 14.11.8). Natural gradients transport contaminants through strategically placed treatment media. The treatment media degrade, sorb, precipitate, or remove chlorinated solvents, metals, radionuclides, or other pollutants. The barrier may contain reactants for degrading volatile organic chemicals, chelators for immobilizing metals, nutrients and oxygen for microorganisms to enhance bioremediation, or other agents.

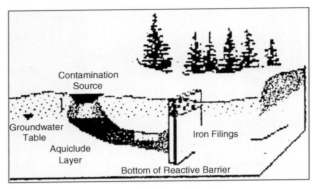

FIGURE 14.11.8 Placement of in situ reactive wall.

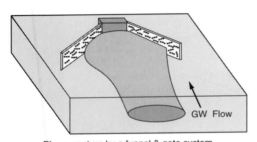

Plume capture by a funnel & gate system.

FIGURE 14.11.9 "Funnel and gate" reactive wall system.

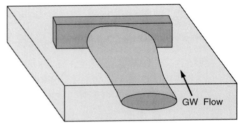

Plume capture by a continuous trench system. The plume moves unimpeded through the reactive gate.

FIGURE 14.11.10 "Continuous trench" reactive wall system.

Two basic methods of installing reactive wall treatment systems have been used to date: (1) funnel and gate (see Figure 14.11.9), and (2) continuous walls systems (see Figure 14.11.10). The funnel and gate system consists of a slurry wall placed downstream of the contaminant plume to direct the flow of the plume to a gate or series of gates containing the permeable zone of reactive media, such as iron filings. The slurry wall is placed such that the contamination plume cannot flow around the borders of the slurry wall. The second method is a continuous trench, again usually a slurry wall, that has been excavated and filled with the reactive media, allowing the water to pass through the barrier under its natural gradient.

In either case, the designer must use geotechnical, geohydrology, and chemistry knowledge to design the passive/reactive wall. The designer must characterize the site to know how to place the slurry wall to contain the plume, horizontally and vertically; understand and characterize the hydrology characteristics of the plume and the groundwater gradient flows to assess the size of gates and/or permeability of the reactive wall; and add the chemistry component to calculate the amount, type, and frequency of replacement of the reactive media.

14.11.4 CONCLUSION

Vertical passive groundwater barriers have been in use for many decades to control the horizontal migration of water, and more recently, contaminated water. The use of vertical barriers as environmental remediation techniques emerged from their early use as specialty geotechnical construction techniques to control and/or contain the flow of groundwater under dams, reservoirs, and ponds.

Environmental scientists have extended the technology of vertical passive barrier walls by looking into ways that such walls could work in contamination environments. The challenges of using passive barrier systems have been many, including developing methods of compatibility testing, looking into different backfill gradations and combinations, pushing for better construction techniques to ensure a more consistent product, and most recently, applying vertical barriers as a means of in situ treatment of contaminated plumes.

What appears certain is that vertical passive groundwater barriers will continue to play a role in site remediation. The technology will continue to evolve with the focus increasingly on in situ applications, which is where most of the focus of the groundwater remediation field appears to be headed.

CHAPTER 14
REMEDIATION TECHNOLOGIES

SECTION 14.12

AQUIFER RESTORATION VIA IN SITU AIR SPARGING

Paul C. Johnson

Dr. Johnson is an associate professor in the Department of Civil and Environmental Engineering at Arizona State University. His research, teaching, and consulting activities focus on developing a better understanding of contaminant fate and transport mechanisms, so that this knowledge can be applied to risk assessment, mitigation, and residuals management. Current research focuses specifically on the in situ bio-treatment of MTBE-impacted aquifers, predicting the fate and transport of soil gas vapors, in situ sparging, and assessing the longevity and impacts of subsurface contaminant residuals.

14.12.1   INTRODUCTION

In situ air sparging (IAS) involves the injection of air into an aquifer in order to: (1) treat contaminant sources trapped within water-saturated and capillary zones, (2) remediate dissolved contaminant plumes, or (3) provide barriers to dissolved contaminant plume migration (Ahlfeld et al., 1992; Ardito and Billings, 1990; Beausoleil et al., 1993; Boersma et al., 1994; Bohler et al., 1990; Brown et al., 1991; Brown and Fraxedas 1991; Griffin et al., 1990; Johnson, R.L. et al., 1993; Kabeck et al., 1991; Marley et al., 1990; Marley et al., 1995; Middleton and Hiller, 1990; Pankow et al., 1993; USEPA, 1992; 1993; Wehrle, 1990). A typical in situ air sparging process schematic is shown in Figure 14.12.1. In its simplest form, it consists of one or more air injection wells connected to a blower or compressor. Often a vapor extraction system is installed and operated in conjunction with IAS to prevent uncontrolled migration of contaminant vapors to enclosed spaces, or to capture and treat all the contaminant vapors that are liberated by the process.

It should be noted that in situ air sparging is fundamentally different from "in-well air sparging." The latter does not involve the injection of air directly into the subsurface formation; instead air is used to sparge water within a well for the purpose of stripping contaminants from the water in the well and, in some cases, causing a conveyance of the water to a higher elevation in the well where it flows back out into the formation. This is done to stimulate groundwater movement in the vicinity of the well; these wells are sometimes referred to as "air lift recirculation wells" (see for example Gvirtzman and Gorelick, 1993).

Traditionally, in situ air sparging has been viewed as a vehicle for delivering oxygen to, and promoting volatilization from, contaminated aquifers. Currently, the process is also being considered for other purposes. For example, Leeson (1993) reports on the use of in situ air sparging to improve air flow distribution near the capillary zone for bioventing purposes. Air sparging may also be used to deliver other gases to the subsurface; for example, one may use the process to deliver hydrocarbon vapors (e.g., methane, ethane, propane) that are needed to promote the degradation of chlorinated compounds via cometabolic degradation pathways.

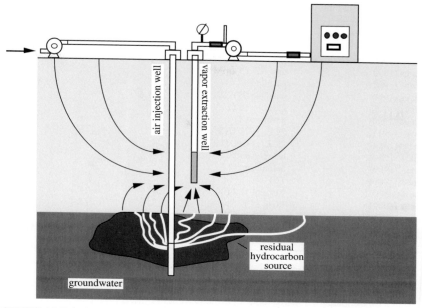

FIGURE 14.12.1   Basic in situ air sparging process schematic, showing complementary soil vapor extraction system.

The rapid rate at which this technology has been embraced by regulatory agencies, environmental consultants, and industry is phenomenal, especially when one considers the limited performance data available to date. Many are already convinced that this technology is the quickest and most cost-effective option for aquifer remediation. Others remain skeptical, but apply IAS nonetheless, as they can find no other technology that promises better performance or significantly lower predicted costs.

This section is intended to provide the reader with a brief summary of the state-of-the-practice, performance data available to date, and results from continuing research efforts to better understand this technology.

14.12.2   PERFORMANCE DATA SUMMARY

Little performance data is available in the peer-reviewed literature. Most reports of IAS performance are found in nonpeer-reviewed conference proceedings, or actual site progress reports prepared by environmental consultants for regulatory agencies and their clients. Many of these are anecdotal in nature, or are lacking in critical details. Thus, the data that are available are not sufficient to develop quantitative theories on how performance varies with site characteristics and design parameters. The data does, however, begin to define the possible range of behavior that might be observed, and suggests that some factors may be more critical than others.

A few attempts have been made to collect and study results from a broad range of sites. The U.S. Environmental Protection Agency (USEPA, 1992) prepared a summary report based on data provided by practitioners. Table 14.12.1 presents a summary of the performance data presented in that report. A few interesting features of this summary table should be noted. First, the quoted duration for each application is relatively short (months) in comparison with current expectations for operating times (years). Secondly, few of these had at that time achieved drinking water standards

TABLE 14.12.1   In Situ Air Sparging Performance Data Summary (U.S. EPA 1992)

Citation	Contaminants	Duration [months]	Initial [mg/L]	Final [mg/L]
Ardito and Billings (1990)	gasoline (BTEX)	2	4, 18, 25	0.3, 8, 6
Billings (1990)	gasoline (B)	5	3–6	1–3
Billings (1990)	gasoline	2	—	8.5%/month
Billings (1990)	gasoline (BTEX)	17	—	<5.5
Billings (1990)	gasoline (BTEX)	9	—	40% reduct.
Billings (1990)	gasoline (B)	10	>30	<5
Brown (1991)	PCE, TCE, DCE	4	41 (TV)	1 (TV)
Harress (1989)	DCE*, PCE, TCE	24	>2	<0.44
Harress (1989)	halogenated	4–6	—	<1% initial
Harress (1989)	halogenated	15	80	0.4
Harress (1989)	TCE	2	1	0.02
Kresge (1991)	gasoline (BTEX)	24	6–24	0.5–8
Looney (1989)	TCE, PCE	3	0.5–2	0.01–1
Marley (1990)	gasoline (BTEX)	2	21	<1
Middleton and Hiller (1990)	TCE	2	0.2–12	<0.01–0.023
Middleton and Hiller (1990)	halogenated (THH)	9	2–5	0.2–3

(≈ 5 $\mu g/L_{H_2O}$ for both benzene and TCE). Third, it is not clear if any post-operation monitoring data was collected. Finally, note that many vendors often prefer to present their results in terms of "% reductions" as opposed to presenting the final concentrations; the latter being the ultimate regulatory criterion by which performance is judged. Often a >90 percent reduction in concentration can be achieved, even though the dissolved contaminant concentration still exceeds a drinking water standard by several orders of magnitude. Nonetheless, there is general agreement that most sites respond positively to in situ air sparging, and dissolved concentration reductions seem to result from system operation (although the magnitude of these reductions may vary significantly from site to site).

U.S. EPA also sponsored a second, independent, state-of-the-practice review at the same time that the 1992 report mentioned above was being prepared. This second group of investigators reviewed the same information (as that reviewed for the U.S. EPA 1992 report) and, interestingly, concluded that the data set was too incomplete, and the monitoring methods too suspect to make any conclusions regarding performance. Those conclusions were not issued as an official U.S. EPA report, but are discussed in Johnson et al. (1993).

Marley and Bruell (1995) conducted a more recent state-of-the-practice survey for the American Petroleum Institute (API). API member companies and their consultants submitted data to the investigators, who in turn, reviewed, analyzed, and summarized the results. Of the data provided, most were associated with short-term feasibility pilot tests as opposed to long-term system operation. Few site data sets included post-operation monitoring data. As with Johnson, R. L. et al. (1993), Marley and Bruell (1995) were unable to draw any conclusions concerning IAS performance based on the data provided.

Some have been wary of placing too much emphasis on data reported over the past decade, especially if the groundwater samples were obtained from monitoring wells during, or shortly after, in situ air sparging. It is now felt that monitoring well samples have a high probability of being biased, such that the measured dissolved contaminant concentrations could be much less than dissolved contaminant concentrations actually present in the aquifer. The reason is that if an air-flow pathway intersects a monitoring well, then the air will bubble up through the well water causing accelerated stripping and oxygenation in the well (but not in the formation). It is common for practitioners to report hearing "bubbling" sounds emanating from monitoring wells during IAS, and for near-saturation dissolved oxygen levels to be measured.

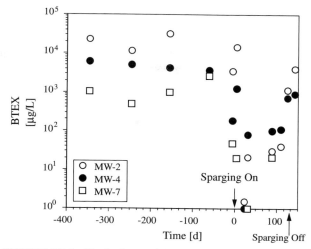

FIGURE 14.12.2   Dissolved contaminant concentration "rebound" following shut-off of an air injection system (Johnson, P. C. et al., 1995).

In their field study, Johnson, P. C. et al. (1995) observed this phenomena while conducting post-operation monitoring following the shut-down of an IAS system. Figure 14.12.2 illustrates the concentration "rebound" observed in monitoring wells at their study site. Note that this rise appears to be continuing even after one month of post-operation monitoring.

In what may be the most enlightening performance data summary to date, Bass and Brown (1995) presents a summary of data for a number of IAS systems operated by the same environmental consulting firm. Bass notes observing rebound at some sites, and no significant rebound at others. In some cases, the concentration rebound occurs over a 12-month period. Mixed performance results were observed, and Bass reports that the data indicate that IAS has been less effective at remediating petroleum hydrocarbon source zones than chlorinated hydrocarbon sites. Bass's conclusion, however, may be an artifact of the types of sites included in the study. At most of the petroleum-release sites, IAS was applied to remediate source zones, while dissolved plumes were the remediation target at the chlorinated solvent releases sites. Table 14.12.2 summarizes relevant information for a few of the sites studied by Bass and Brown.

In summary, available performance data is extremely limited and shows mixed results. IAS has been an effective remediation tool at some sites, and has had little apparent impact at others. Even across a single site, the results may be mixed. It is not clear, however, if these mixed results are indicative of inherent IAS technology limitations, or if they are more reflective of variations in system design or inappropriate operating conditions. Few sites have achieved drinking water standards through use of IAS alone, however, practitioners report that regulators do appear to be granting closures following application of IAS, independent of the degree of remediation achieved. If true, this reason alone will be enough to convince many to install air sparging systems at their sites.

14.12.3   DESIGN APPROACH

Currently, in situ air sparging design practices are largely empirical. Despite the wide range of geologic conditions encountered in the field, little variance seems evident in system designs and operating conditions. Tables 14.12.3 and 14.12.4 summarize data presented in the state-of-the-practice reviews prepared by Johnson, R. L. et al. (1993) and Marley and Bruell (1995).

TABLE 14.12.2   In Situ Air Sparging Performance Data Summary for Sites With and Without Postclosure Rebound (Bass and Brown, 1995)

Site specifics	Duration				Concentration			
	Sparging [months]	Post closure monit. [months]	Contaminant		Start [μg/L]	Shutdown [μg/L]	End of post-closure monit. [μg/L]	Monitoring well location
service station	49	8	Weathered Gasoline					
			Benzene		70	<0.4	70	source zone
					1400	500	—	source zone
					79	3.5	160	downgradient
					SPHC	<2	—	crossgradient
					SPHC	<1	—	upgradient
			Total BTEX		5470	3260	3651	source zone
					1269	13,300	3380	downgradient
service station	12	16	Weathered Gasoline					
			Benzene		1800	<0.5	210	source zone
					15	<0.5	1	source zone
					8	0.5	12	downgradient
					<0.5	<0.5	15	downgradient
					<0.5	<0.5	37	downgradient
			Total BTEX		53,200	7	21,810	source area
					2357	<5	14	source area
					3668	333	358	downgradient
					7830	42	625	downgradient
					12	<5	3147	downgradient

TABLE 14.12.3   Design Parameters for Air Sparging Systems (Johnson, R. L. et al., 1993)

Parameter	Reported value
Injection Well Specifics	
• screen depth below water table [ft]	16[1], 3[3,5], 9[3], 15–40[3], 5[6], 10–39[8]
• screen interval width [ft]	2[1,3,5,8], 300[4], 6[6], 2[8]
• number of wells	14[1], 5[2], 13[3,5], 1[4,6,8]
• injection air flowrate [ft³/min]	6[2], 2–6[3,5], 170–270[4], 56[6], 7–16[8], 3–4[10]
• injection air pressure [psig]	1–2[3], 1–8[3,5], 3–4[10]
• operation [pulsed or continuous]	continuous[1,2,6,8,9,10], pulsed[3,5]
• other info	nested injection/extraction wells[1,9,10]
	individual wells[2,3,4,5,6,10]
	horizontal wells[4]
Vapor Extraction Well Specifics	
• # extraction wells/# injection wells	8/14[1], 1/1[6,9,10], 2/13[3,5], 0/1[8]
• extraction flowrate/injection flowrate	475/30[2], 580/170–580/270[4], 160/100[6], 2/1[10]

The major components of an IAS system typically include: air injection wells, air compressor, monitoring components, and vapor extraction system (if needed). Each of these are discussed briefly below.

Air injection wells are generally constructed from PVC and installed according to typical monitoring well installation practices. When using vertical wells, the well screen interval is often about

TABLE 14.12.4 Typical Design Parameters (Marley and Bruell, 1995)

Parameter and range	Most often used value (# of sites)	Second most often used value (# of sites)	Third most often used value (# of sites)	Total # of site
Scree length 0.5–10 ft	2 ft (16 sites)	3 ft (8 sites)	5 ft (7 sites)	40
Well diameter 1–4 in	2 in (17 sites)	4 in (7 sites)	1 in (5 sites)	37
Overpressure 0.35–18.2 psig	0.35–5 psig (14 sites)	5–10 psig (9 sites)	10–15 psig (5 sites)	31
Well screen depth below water table 2–26.5 ft	5–10 ft (10 sites)	10–15 ft (8 sites)	2–5 ft (6 sites)	31
Air injection flow rate 1.3–40 SCFM	1.3–5 SCFM (16 sites)	5–10 SCFM (9 sites)	15–20 SCFM (5 sites)	39
Air injection pressure 3.5–25 psig	5–10 psig (17 sites)	10–15 psig (8 sites)	20–25 psig (6 sites)	40
SVE ROI/IAS ROI 0.16–7.42	1–2 (12 sites)	0.16–1 (6 sites)	3–4 (3 sites)	26

SVE ROI—radius of influence of vapor extraction wells.
IAS ROI—radius of influence of air injection wells.

2 ft long and placed entirely below the groundwater table. The specific depth at any given site is based on considerations of the contaminant depth and subsurface stratigraphy. To date, most injection well screened intervals have been installed 5 to 20 ft below the groundwater table. This observation, however, may be biased by the fact that most systems have been installed at LNAPL (light nonaqueous phase liquid) release sites where the source zones tend to straddle the capillary fringe. In general, the screened interval is installed below the zone of contamination and above any stratigraphic unit that would impede the vertical migration of the injected air. It is very important to install a competent air flow seal (typically clay or cement) in the annulus above the screened interval; otherwise, a significant fraction of the injected air may flow up through the bore-hole rather than out into the formation. Practitioners have reported difficulty installing competent multiple-completion wells (more than one air injection depth in the same borehole).

It is also possible to use driven well points as air injection wells. Currently, sufficient performance data is not available in the literature to judge the performance of driven points relative to typical well installations. In shallow sandy soil settings, driven points are almost always economically attractive. There are concerns, however, that in finer soils the drive point screens may become plugged, or the formation may get smeared. Both would inhibit one's ability to inject air into an aquifer, and both problems have been encountered by practitioners.

The number of air injection wells installed is often based on a "zone of influence" approach. Based on the results of a pilot test, or experience, the practitioner estimates the areal extent of air flow in the aquifer. Injection wells are then placed to ensure that air flows through the target treatment zone. As reported by Marley and Bruell (1995), most practitioners tend to use injection wells that are spaced roughly 20 to 60 ft apart. In general, better performance will always be achieved with higher densities of air sparging wells.

Compressors, or blowers, are typically selected based on anticipated injection flow rates and pressure requirements. According to current design practices, the practitioner should expect to inject air at flow rates ranging from 1 to 20 ft^3/min per injection well. Required IAS system injection pressures P_{inject} [psig] should range from:

$$0.43 \text{ [psig/ft]} \times H \text{ [ft]} \leq P_{inject} \text{ [psig]} \leq 0.43 \text{ [psig/ft]} \times H \text{ [ft]} + 0.73 \text{ [psig/ft]} \times D \text{ [ft]}$$

where H [ft] and D [ft] denote the depth below the water table and depth below ground surface to the top of the well screen, respectively. The lower limit represents the minimum hydrostatic pressure that must be overcome to displace enough water from a well to allow air to exit from the top of the well screen into the formation. The upper limit is an estimate of the pressure that could cause fracturing (and uplift) of the formation.

Practitioners have varied opinions on the effect of changing air injection flow rates. U.S. EPA (1992) suggests that as the injection flow rate and injection pressure are increased, there is an increased likelihood that the injected air will bypass the target treatment zone. Other laboratory-scale studies (Ji et al., 1993; Rutherford and Johnson, 1995) indicate that increasing injection flow rate results in positive effects, especially in stratified geological settings.

Vapor extraction systems are installed in conjunction with IAS systems when one must recover contaminant vapors, or where there is the concern that contaminant vapors could migrate to enclosed spaces (e.g., utility conduits, basements, buildings). Practitioners seem to select vapor extraction well spacing based on subsurface soil gas pressure measurements during a pilot test. For example, it is common for practitioners to define their vapor extraction radius of influence to be that distance where, during a pilot test, vacuums in the subsurface fell to less than 0.1 in H_2O gauge vacuum. Then, when operating the full-scale system, extraction well flow rates may be adjusted so that extraction flow rates are greater than injection flow rates. Injection and extraction flow rates may also be adjusted so that a vacuum is measured in soil vapor probes at desired locations. It has been shown, however, that this approach can lead to a false sense of security. Johnson, P. C. et al. (1995) conducted a continuous helium recovery tracer test at their field study site and were only able to recover about 50 percent of the injected helium at steady-state, despite the extraction flow rate being six times greater than the injection flow rate and vacuums being measured in all soil vapor probes. Their helium recovery data is presented in Figure 14.12.3.

In summary, there seem to be no "proven" design practices, and most practitioners seem to follow the same approach. Fortunately, it seems that this approach is generating positive results at a number of sites. Probably the best practical design approach at this time is a phased one, where a preliminary design is created based on available data and experience, a system is installed, and then that system is optimized and modified based on the information and insight gained from monitoring data.

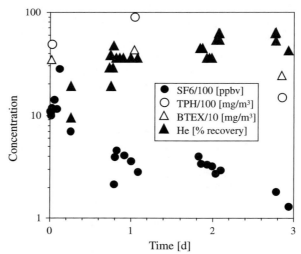

FIGURE 14.12.3   Helium recovery test results from an in situ air sparging field study with a single air injection well and single vapor extraction well (Johnson et al., 1995). Air injection flow rate $\approx$10 SCFM, vapor extraction flow rate $\approx$60 SCFM.

14.12.4   *SHORT-TERM PILOT TESTS*

Short-term pilot tests play a key role in the selection and design of in situ air sparging systems. A single air injection well is used in a typical pilot test, and the duration of air injection is often less than 24 h (Boersma et al., 1994; Brown et al., 1991; USEPA, 1992). Practitioners look for changes in easily measured aquifer and vadose zone parameters, such as dissolved oxygen, water level changes, soil gas pressures, and soil gas contaminant concentrations. More often than not, the practitioner monitors these changes within conventional monitoring wells that already exist at the test site. Increases in dissolved oxygen and decreases in dissolved contaminant levels are taken to be favorable indicators of feasibility, whereas changes in water levels within the wells and vadose zone air pressures are assumed to be indicators of the treatment area. For the purpose of this discussion, "conventional" monitoring wells refer to wells constructed according to standard practices for the purpose of obtaining groundwater quality samples; they are often constructed from 1- to 4-in diameter polyvinyl chloride (PVC) pipe, and have a 3 to 15 ft screened-interval that often straddles the groundwater table.

In a few cases, investigators have supplemented conventional practices with other more innovative approaches. Most of these have focused on better assessing air distributions. Lundegard (1994) reports on the use of electrical resistance tomography (ERT) to measure air flow distributions in situ, Acomb et al. (1995) applied geophysical tools, Beckett (1995) has utilized a combination of tracer gases and conductance probe measurements, and Clayton et al. (1995) used time-domain reflectrometry (TDR) probes to assess air content changes in the saturated zone.

With the exception of the work by Lundegard (1994) and the discussion given in Boersma et al. (1994), few have questioned conventional pilot test procedures and monitoring practices, as they relate to the design and assessment of in situ air sparging performance. Lundegard, who compared ERT imaging results with traditional pilot test measurement and data reduction practices, concluded that conventional monitoring approaches overestimate the zone of air flow at relatively homogeneous sites. Boersma et al. (1994) suggested that there may be a large degree of uncertainty associated with system design and performance assessment, when using conventional monitoring approaches. In both cases, these authors focused their discussions primarily on the extent of air distribution.

Johnson, P. C. et al. (1995) conducted a field study in order to compare short-term pilot test data with longer-term performance data. At their site, a number of discrete drive-point sampling implants were installed in addition to conventional monitoring wells. Groundwater quality indicators (e.g., dissolved hydrocarbons, dissolved oxygen) were measured with time and tracer gas tests were performed. Based on their data it was concluded the following:

- Short-term pilot-test data collected from conventional monitoring wells yielded the most optimistic picture of long-term system performance.

- Short-term pilot-test discrete implant monitoring results were less promising as they showed little effect of in situ air sparging during the three-day pilot test.

- There was no clear correlation between any of these short-term measurements and the longer-term system performance at this site.

Johnson, P. C. et. al. (1995) also concluded that, in the absence of validated predictive tools, it is very unlikely that short-term pilot-test data can be extrapolated to long-term system performance. Therefore, the authors recommended that practitioners design their short-term pilot tests to assess *infeasibility* rather than feasibility; that is to say that short-term pilot tests should be designed to identify conditions indicative of poor system performance. These might include:

- inability to inject significant air flow ($> 1 \text{ ft}^3/\text{min}$) into the aquifer at pressures less than the combined hydrostatic head and soil overburden

- observation of adverse impacts caused by air injection (e.g., uncontrollable migration of vapors)

- poor air distributions (e.g., highly stratified or unidirectional air flows)

- inability to detect any qualitative indicators of air distribution (e.g., no observed dissolved oxygen changes, little tracer gas recovery)

The interaction between air sparging behavior and subsurface geology was also highlighted in that study. Strata that were not identified during the initial site assessment controlled the performance at the study site, and these caused the formation of a large horizontal trapped air bubble beneath the water table. This illustrates the need for careful geologic assessment at potential in situ air sparging sites.

14.12.5 MONITORING PRACTICES

In situ air sparging systems are monitored to assess the performance of current operating conditions, to help determine if system adjustments or expansions are necessary, and to determine if off-site migration of contaminant vapors and contaminated groundwater is occurring. Table 14.12.5 lists a number of system parameters that can be monitored during in situ air sparging operation.

Of all the measurements that are currently made, only the measured vapor extraction well flow rates and vapor extraction well vapor concentrations provide a real-time measure of mass removal during in situ air sparging. Often, however, it is difficult to estimate the contribution solely because of in situ air sparging as the vapor extraction system captures vapors from both vadose-zone and saturated-zone

TABLE 14.12.5   In Situ Air Sparging Monitoring Options

Quantity	How measured?	Interpretation/data reduction
Dissolved oxygen in monitoring well water samples	Slow purge, flow through cell, using a dissolved oxygen meter or titration	Dissolved oxygen increases are indicative of horizontal extent of air flow
Dissolved contaminant concentrations	Samples obtained from monitoring wells or drive points, followed by chemical analysis	Progress of remediation—effect on aquifer water quality. Note : sample results may be biased during system operation
Contaminant vapor concentrations in soil vapor extraction system or soil gas sampling points	Gas sample collected then analyzed with total analysis portable field instruments, or gas chromotagraphy	Increases in vapor concentrations are indicative of enhanced remediation rate
Injection and/or extraction flow rates	Rotameters, oriface plates	Flow rates multiplied by the vapor concentrations from soil vapor extraction system provide a measure of mass removal rate.
Aquifer water pressure measurements	Submerged pressure transducer	Water pressure increases are taken to be measures of the extent of air flow in the aquifer; transient response often used to set pulsing intervals for air injection
Soil gas pressure measurements	Pressure gauge connected to well or drive point	Indication of extent of vapor flow field, estimate of extent of vapor extraction capture zone
Helium concentrations following blending with air injection stream	Portable helium detector	Indication of the extent of air travel, when combined with vapor extraction flow rates it provides a measure of vapor extraction recovery efficiency
Air saturation in the aquifer	Neutron probe, electrical resistance tomography, time-domain reflectrometry, conductance probe	Used to identify zone of air flow in the aquifer

contaminant sources. Furthermore, a fraction of the vapors may be biodegraded as they travel through the vadose zone to a vapor extraction well. For example, the data from Johnson, P. C. et al. (1995) in Figure 14.12.3 indicates a short-term doubling of the removal rate following the initiation of air injection. It is not uncommon to see transient "spikes" like this that only last days in duration.

Most of the easily measured data is best interpreted qualitatively because in situ monitoring data is often puzzling and subject to a wide range of interpretations. Increases in dissolved oxygen, decreases in dissolved contaminant concentrations, and increased contaminant vapor concentrations in the vadose zone are assumed to be positive indications of remedial progress. Note, however, that none of these are direct indications of remediation rate.

In addition to the conventional monitoring approaches, others are expanding the realm of diagnostic tools available for monitoring in situ air sparging systems. Some, like the geophysical methods, have already been mentioned above in the discussion of pilot testing practices. Others are discovering that tracer gas tests are also proving to be very useful tools. Johnson, P. C. et. al (1995) illustrate the use of continuous injection helium tracer tests to quantify the vapor recovery efficiency of an integrated in situ air sparging/soil vapor extraction system. Johnson, R. L., Perrott et al. (1994) report on the use of SF6 as an oxygen analog to quantify the rate at which in situ air sparging provides oxygen for an aquifer at a test site. Others (Marley and Bruell, 1995; Beckett, 1995) have used pulsed tracer gas injections to deduce air flow patterns.

14.12.6 *RESEARCH STUDIES*

Few studies have been conducted to better understand and quantify the mechanisms controlling in situ air sparging performance. The costs of accomplishing such studies in the field are typically cost-prohibitive, and so most have resorted to laboratory-, or large-scale physical models.

Ji et al. (1993) reported on flow visualization studies conducted in a 2 ft × 2 ft × 1 in (approximately) transparent tank. The tank was packed with various sizes of glass beads and several model subsurface geologies were simulated. The goal of the study was to observe how the injected air distributes itself beneath the water table, and how this distribution is affected by particle size, stratigraphy, and air injection flow rate. The reader is referred to the original publication for excellent pictures of the air distributions observed during their studies. Qualitatively, it was observed that:

- For glass bead sizes <1 mm in diameter, most of the air flowed through distinct air channels. For larger particle diameters, air could rise up through the formation in discrete bubbles.

- In homogeneous settings, the air distribution was roughly wedge-shaped, with the maximum width of the air distribution approximately equal to the depth of injection (injection depth was fixed and not a variable in these studies). Air distributions measured in the laboratory studies are in good agreement with the field measurements reported by Lundegard (1994) and Acomb et al. (1995) for homogeneous sandy field sites.

- Air distributions seemed to be more irregular, and contained lower air channel densities, when mixtures of different size glass beads were used.

- In layered geologic settings, air would move laterally below and around layers of smaller-diameter glass beads, even if the particle diameter size difference was not very great (less than one order of magnitude).

- In all cases, increases in air injection rate caused higher densities of air channels, and, in some cases, caused the air to move up through, rather than around, stratified layers.

Of significance in these studies was the illustration that subtle changes in soil structure can cause drastic changes in air flow distribution. Stratified air distributions should be anticipated at most sites.

More recently, Rutherford and Johnson (1995) reported on the use of a laboratory-scale physical model to measure oxygen mass transfer rates as a function of air injection rate, groundwater velocity, and pulsing frequency. The goal of this work was not to quantitatively extrapolate the data to field conditions; rather the authors asserted that qualitative trends observed in the laboratory data are likely to occur in field settings.

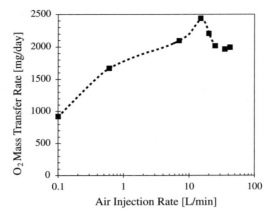

FIGURE 14.12.4 Dependence of oxygen mass transfer rate on air injection rate for a constant water flow rate of 0.27 L/min (Rutherford and Johnson, 1995).

With respect to the effect of air injection rate on oxygenation rate, the data suggested that increases in air injection rate generally caused increases in oxygenation rate, except at high air injection rates. In fact, at higher air injection rates, the permeability of the medium to water flow within the zone of air flow was sufficiently reduced that the oxygenation rate decreased and then leveled off (Figure 14.12.4). At these higher air injection rates, the water preferred to move around, rather than through, the zone of air flow.

Groundwater flow rate did not seem to affect the oxygenation rate, unless the flow velocity exceeded about 1 ft/d. At that point, oxygenation rates increased with increases in groundwater velocity. The data suggested that, in most natural hydrologic settings, the rate of oxygenation will be limited by diffusive processes.

The authors also investigated the effect of pulsing the air injection at various pulsing frequencies. Their results, shown in Figure 14.12.5, suggested that turning a system on and off had little effect on

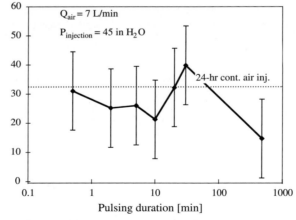

FIGURE 14.12.5 Dependence of 24-h time-averaged oxygen mass transfer rate on air injection pulsing frequency for a constant air injection rate of 7 L/min (Rutherford and Johnson, 1995).

the rate of oxygenation when compared to continuous injection over the same time period. This is in contradiction to the general belief that pulsed air sparging systems will perform better than continuous injection systems. To date, there is not any equivalent field data to support, or refute, the use of pulsed systems.

14.12.7 CONCLUSIONS AND FUTURE WORK

The remediation of aquifers, especially where source zones are located below the aquifer, is one of the most difficult and costly challenges facing environmental professionals. Currently, no other aquifer remediation technology offers the performance, cost-effectiveness, and regulatory acceptance promised by the proponents of in situ air sparging. Consequently, the number of in situ air sparging systems installed and operated will continue to grow at a rapid pace in the next few years. As is common in this field, the implementation of the technology will precede our understanding of how it performs.

With the growth of air sparging applications, it is important that research studies be continued to define conditions for which the technology is applicable, to identify how to interpret monitoring data, and to develop practices for optimizing system performance. Very little data is available at this time, and we are only just beginning to understand some of the relationships between site conditions, system design, and performance. Currently, the United States Air Force, American Petroleum Institute, and others are beginning multiyear studies to address these issues.

REFERENCES

Acomb, L. J., McKay, D., Currier, P., Berglund, S. T., Sherhart, T.V., and Benedicktsson, C. V., "Neutron Probe Measurements of Air Saturation Near an Air Sparging Well," in *In Situ Aeration: Air Sparging, Bioventing, and Related Remediation Processes*, Hinchee, R. E., Miller, R. N., and Johnson, P. C., Battelle Press, Columbus, Ohio, 1995.

Ahlfeld, D. P., Dahmani, M. A., Ji, W., and Farrell, M., "Field Study of Behavior of Air Sparging," *Technical Report 92-10*, Environmental Research Institute, The University of Connecticut at Storrs, 1992.

Ardito, C. P., and Billings, J. F., "Alternative Remediation Strategies: The Subsurface Volatilization and Ventilation System," *Petroleum Hydrocarbons and Organic Chemicals in Ground Water: Prevention, Detection and Restoration*, American Petroleum Institute, Houston, TX, November 1990.

Bass, D. H., and Brown, R. A., "Performance of Air Sparging Systems—A Review of Case Studies," *Petroleum Hydrocarbons and Organic Chemicals in Ground Water: Prevention, Detection and Restoration*, American Petroleum Institute, Houston, TX, November 1995.

Beausoleil, Y. J., Huber, J. S., and Barker, G. W., "The Use of Air Sparging in the Remediation of a Production Gas Facility," *Society of Petroleum Engineers Paper No. 26000*, SPE/EPA Exploration and Production Environmental Conference, San Antonio, TX, March 7–10, 1993.

Beckett, G., Personal Communication, 1995.

Boersma, P., Newman, P., and Piontek, K., *The Role of Groundwater Sparging in Hydrocarbon Remediation*, American Petroleum Institute Pipeline Conference, Houston, TX, April 25–28, 1994.

Bohler, U., Brauns, J., Hotzel, H., and Nahold, M., "Air Injection and Soil Aeration as a Combined Method for Cleaning Contaminated Sites—Observations from Test Sites in Sedimented Solid Rocks," *Contaminated Soil '90*, Arendt, F., Hinsenveld, M., and van den Brink, W. J., eds., Kluwer Academic Publishers, San Diego, CA, pp. 1039–1044, 1990.

Brown, R. A., and Fraxedas, R., *Air Sparging—Extending Volatilization to Contaminated Aquifers*, USEPA Symposium on Soil Venting, Houston, TX, April 29–May 1, 1991.

Brown, R. A., Herman, C., and Henry, E., *The Use of Aeration in Environmental Cleanups*, Haztech International Pittsburgh Waste Conference, Pittsburgh, PA, May 14–16, 1991.

Clayton, W., Brown, R. A., and Bass, D. H., Presentation at the In Situ and On-Site Bioreclamation Conference—3rd International Symposium, San Diego, April 24–27, 1995.

Griffin, C. J., Armstrong, J. M., and Douglass, R. H., *Engineering Design Aspects of an In Situ Soil Vapor Remediation System (Sparging)*, In Situ Bioreclamation, 1990.

Gvirtzman, H., and Gorelick, S. M., "Using Air-Lift Pumping as an In Situ Aquifer Remediation Technique," *Water Sci. Tech.*, 27(8): 195–201, 1993.

Ji, W., Dahmani, A., Ahlfeld, D., Lin, J. D., and Hill, E., "Laboratory Study of Air Sparging: Air Flow Visualization," *Ground Water Monitoring and Remediation*, Fall 1993.

Johnson, P. C., Johnson, R. L., Neaville, C., Hansen, E. E., and Stearns, S. M., "Do Conventional Monitoring Practices Indicate In Situ Air Sparging Performance?," in *In Situ Aeration: Air Sparging, Bioventing, and Related Remediation Processes*, Hinchee, R. E., Miller, R. N., and Johnson, P. C., Battelle Press, Columbus, Ohio, 1995.

Johnson, R. L., Perrott, M., Gilbert, C., Johnson, P. C., and Neaville, C. C., "Oxygen Mass Transfer During In Situ Air Sparging: SF6 Tracer Tests," *Ground Water*, 1994.

Johnson, R. L., Johnson, P. C., McWhorter, D. B., Hinchee, R., and Goodman, I., "An Overview of In Situ Air Sparging," *Ground Water Monitoring and Remediation*, Fall 1993.

Kaback, D. S., Looney, B. B., Eddy, C. A., and Hazen, T. C., "Innovative Ground Water and Soil Remediation: In Situ Air Stripping Using Horizontal Wells," *Fifth National Outdoor Action Conference on Aquifer Restoration, Ground Water Monitoring, and Geophysical Methods*, Las Vegas, NV, May, 13–16, 1991.

Leeson, A., Hinchee, R. E., Kittel, J., Sayles, G., Vogel, C., and Miller, R., "Optimizing Bioventing in Shallow Vadose Zones in Cold Climates," *Hydrological Sciences Journal*, 38(4): 1993.

Lundegard, P., "Actual Versus Apparent Radius of Influence—An Air Sparging Pilot Test in A Sandy Aquifer," *API/NGWA Conference—Petroleum Hydrocarbons and Organic Chemicals in Ground Water: Prevention, Detection and Restoration*, Houston, TX, 1994.

Marley, M. C., "Air Sparging in Conjunction with Vapor Extraction for Source Removal at VOC Spill Sites," *Fifth National Outdoor Action Conference on Aquifer Restoration, Ground Water Monitoring, and Geophysical Methods*, Las Vegas, NV, May, 13–16, 1991.

Marley, M. C., and Bruell, C. J., "In Situ Air Sparging: Evaluation of Petroleum Industry Sites and Considerations for Applicability, Design, and Operation," *API Publication 4609*, American Petroleum Institute, Washington, D.C., 1995.

Marley, M. C., Bruell, C. J., and Hopkins, H. H., "Air Sparging Technology: A Practical Update," in *In Situ Aeration: Air Sparging, Bioventing, and Related Remediation Processes*, Hinchee, R. E., Miller, R. N., and Johnson, P. C., eds., Battelle Press, Columbus, Ohio, 1995.

Marley, M. C., Walsh, M. T., and Nangeroni, P. E., "A Case Study on the Application of Air Sparging as a Complimentary Technology to Vapor Extraction at a Gasoline Spill Site in Rhode Island," *Hydrocarbon Contaminated Soils and Groundwater Conference*, University of Massachusetts, Amherst, MA, September 1990.

Middleton, A. C., and Hiller, D. H., *In Situ Aeration of Groundwater: A Technology Overview*, 1990 Environment Canada Montreal Conference, October 16–17, 1990.

Pankow, J. F., Johnson, R. L., and Cherry, J. A., "Air Sparging in Gate Wells in Cutoff Walls and Trenches for Control of Plumes of Volatile Organic Compounds (VOCs)," *Ground Water*, 31(4): 654–663, July–August, 1993.

Rutherford, K., and Johnson, P. C., "Interfacial Mass Transfer During In Situ Air Sparging—Effects of Process Changes and Lithology," *API/NGWA Conference: Petroleum Hydrocarbons and Organic Chemicals in Groundwater*, Houston, TX, November 1995.

USEPA, *A Technology Assessment of Soil Vapor Extraction and Air Sparging*, EPA/600/R-192/173, September 1992.

USEPA, *Groundwater Remediation for UST Sites*, EPA-510-F-93-017, October 1993.

Wehrle, K., "In Situ Cleaning of CHC Contaminated Sites: Model-Scale Experiments Using the Air Injection (In Situ Stripping) Method in Granular Soils," *Contaminated Soil '90*, Arendt, F., Hinsenveld, M., and van den Brink, W. J., eds., Kluwer Academic Publishers, pp. 1061–1062, 1990.

CHAPTER 15
UBIQUITOUS ENVIRONMENTAL CONTAMINANTS

SECTION 15.1

ENDOCRINE DISRUPTORS

Jenifer S. Heath

Dr. Heath is an independent consulting toxicologist and risk assessor. She specializes in contaminated sites, environmental permitting and toxic torts. She has extensive training in arbitration and remediation.

15.1.1 INTRODUCTION AND DEFINITIONS

Do small amounts of hormonally active chemicals cause adverse effects in animals, including humans? Can exogenous compounds mask, mimic, or otherwise change the timing or effect of key physiological signals, thus ultimately causing toxic effects on the organism? These questions have been hotly debated in the scientific community for several years, yet consensus largely eludes us.

The term "endocrine disruptor" is used to describe chemicals that have adverse effects on the endocrine system. The endocrine system is one of the three major regulatory systems in the body; the two others are the nervous system and the immune system. The endocrine system plays a critical role in maintaining homeostasis of the organism as well as a role in reproduction, development, and behavior. The endocrine system consists of glands, hormones, and receptors. Hormones are chemical messengers that move through the bloodstream to bring their messages to receptors in distant locations in the body. Through well-timed changes in hormone concentrations, the endocrine system influences key physiologic functions, such as blood pressure, metabolism, levels of glucose in blood, digestive function, respiration, immune status, and reproductive structure and function. Hormones are natural secretory products of endocrine glands, including the ovaries, testes, pituitary gland, thyroid, pancreas, and adrenal glands. Well designed to handle small fluctuations in hormone levels (and indeed having feedback mechanisms to control fluctuations), the endocrine system may not be capable of handling more significant fluctuations that may result from exposure to endocrine disruptors.

Many chemicals affect the endocrine system in ways that are not adverse, and these are not considered endocrine disruptors. To refer to nonadverse endocrine-related effects, the term "endocrine activity" is commonly used. Furthermore, it is effects in intact systems (living organisms) that are of interest; effects in vitro cannot lead to the conclusion that a chemical is an endocrine disruptor.

The following elements are common in most definitions of endocrine disruptors:

- Endocrine disruptors are exogenous agents (single chemicals or mixtures);
- Endocrine disruptors cause adverse health effects;
- The adverse effects can be observed in intact organisms, their progeny, or subpopulations of organisms; and
- The adverse effect is a result of changes in endocrine function (e.g., interfering with synthesis, secretion, transport, binding, action or clearance of natural hormones).

15.1.2   *HISTORY*

The potential for devastating adverse effects as a result of endocrine disruption has been known for decades. Perhaps the most infamous case is that of diethylstilbestrol (DES). DES was a prescription drug properly registered (1948) in the United States to prevent miscarriage during pregnancy. In the 1970s, some years after it came on the market, evidence of estrogenic reproductive effects in the children who were exposed in utero came to light.

In the late 1980s, a wide range of adverse effects observed in wildlife (alligators, birds, fish, and invertebrates) was attributed to endocrine disruption. These effects included demasculinization and defeminization, masculinization and feminization, eggshell thinning, behavioral changes, and decreased number of offspring.

Theo Colborn's book *Our Stolen Future*, published in 1996, is considered by many to have had an impact on public perception of, and political interest in, environmental risks associated with endocrine disruptors. The influence of this book has been compared with that of Rachel Carson's then decades old *Silent Spring*.

In 1996, the U.S. Congress passed the Food Quality Protection Act and amended the Safe Drinking Water Act to require development and implementation of a program to identify endocrine disruptors. In 1998, the multistakeholder committee that was established to design the screening and testing protocols required by the Acts issued its report. By this time, the chemical industry in the United States had obligated large sums of money to study endocrine disruption, including a 3-year, $4 million research commitment from the Chemical Manufacturers Association.

15.1.3   *TYPES OF EFFECTS*

Endocrine disruptors may act as hormones or as hormone antagonists (blocking the actions of hormones). Some epidemiological evidence suggests that potential adverse effects in humans include breast cancer, testicular cancer, reduced sperm count, developmental defects in the male reproductive tract, prostate cancer, altered physical and mental development in children, and endometriosis. Many authors argue that endocrine disruptors have not been demonstrated to have adverse effects in humans at naturally occurring or anthropogenic levels in the environment. This debate is reminiscent of other interactions between epidemiologists and toxicologists about the relative merit of each field in predicting and recognizing adverse effects in humans. Pharmaceutical use of some chemicals, though, has been demonstrated to have adverse, endocrine-disruption-related effects in humans.

Allegations that endocrine disruptors have caused decreased sperm counts among human males provide an example of the complexity we face in evaluating the potential for adverse effects from endocrine disruptors. Although there has been some evidence to suggest an overall decrease in sperm count, this evidence is by no means incontrovertible. For example, the methodology for counting sperm has changed over time and could account for the apparent decreasing trend. Other confounding factors include method of sperm collection, frequency of ejaculation (which affects sperm count), scrotal temperature (which, interestingly, is associated with height), history of fever, pharmaceutical and illicit drug use, smoking, physical exertion, stress, occupation history, and regional differences in sperm count.

One author has suggested that certain endocrine disruptors interfere with gonadal development in utero, yielding testicular atrophy, testicular dysgenesis, and infertility. These effects are said to be linked to testicular cancer. Yet such cause-effect relationships are far from proven.

In contrast to the active and genuine debate about human health effects, more (but not nearly all) authors agree that adverse effects in wildlife have been confirmed. However, evaluation of potential adverse effects in wildlife is also complex. First, the association between presence in the environment and adverse effects in animals must be shown to be causative; second an understanding of the interplay between physiologic effects and the overall health of the organism must be established; and third, the relevance to ecosystem-level impacts must be clarified.

15.1.4   EXAMPLES

As described more fully in the section on modes of action, chemical structure cannot necessarily predict endocrine activity. Historical toxicity testing was not designed to detect the types of effects considered to be of interest for evaluating endocrine activity or disruption. Therefore, it is generally not possible at this time to know which chemicals may have endocrine activity much less function as endocrine disruptors. Nevertheless, a number of chemicals have been implicated as having endocrine activity, including certain organochlorine pesticides, polychlorinated biphenyls and polybrominated biphenyls, certain tetrachlorodibenzodioxins and dibenzofurans, certain phthalates, dithiocarbamates, tributyltin, certain triazine pesticides, and many pharmaceuticals (often specifically intended to have endocrine activity).

15.1.5   OCCURRENCE IN THE ENVIRONMENT

Endocrine active chemicals are, of course, naturally present in physiologic systems in both plants and animals. Many endocrine active chemicals are naturally present in the environment (e.g., phytoestrogens), and these are often concentrated in certain types of foods (e.g., soy products) and in some nutritional supplements. The potency of endocrine active chemicals varies over 8 orders of magnitude. It has been suggested that, based on estrogenic potency, human intake of naturally occurring estrogens is 10 million times greater than intake of synthetic (industrial) estrogenic compounds.

15.1.6   MODES OF ACTION

Because the endocrine system acts through a complex series of messages transmitted, received, and retransmitted throughout the body, the tissue where the endocrine disruptor has its biochemical effect may be located distant from the target organ (the organ or system ultimately affected).

Perhaps the best known mode of action is receptor binding. Like descriptions in high school textbooks of how enzymes function, endocrine receptor molecules function by a lock-and-key type of mechanism. The receptor molecule may be seen as the single lock, and the endogenous hormone is the key that unlocks the lock in the physiologically appropriate way. However, we now know that there are many (chemical messenger) keys to each receptor lock. These other keys may fit the lock better or worse than the endogenous hormone, and therefore they may be more or less effective in unlocking the (receptor) lock than the evolutionary messenger. Some keys may even tie up the lock (semi)permanently, precluding it from future proper interaction with (endogenous hormone) key. Also, a single key may fit multiple locks, thus disrupting many aspects of the endocrine system.

Endocrine disruptors also may act in receptor-independent ways—for example, by altering production, release, or transport of the chemical messenger or by activating a competing system.

Genistein (a phytoestrogen present in soy products) stimulates estrogen-like activity in bone and liver but not in breast tissue. This may result in fewer menopausal symptoms, as well as a lower rate of breast cancer, in women with diets high in soy products. As pharmacological knowledge of modes of action increases, it has become possible to synthesize chemicals that mimic estrogen in some tissues but not in others. For example, tamoxifen, developed to treat breast cancer, blocks estrogen receptors in breast cancer cells and mimics estrogen in the bone (retarding bone loss).

15.1.7   SCREENING AND TESTING PROGRAM

The Endocrine Disruptor Screening and Testing Advisory Committee (EDSTAC) released its final report in 1998, and as this book goes to press the U.S. Environmental Protection Agency is proceeding with implementation of a screening program based on EDSTAC's recommendations. The program will look for estrogenic and androgenic chemicals as well as chemicals that affect the thyroid system. In the United States, over 87,000 chemicals may be subject to the screening process to determine whether they have the potential to cause adverse endocrine-mediated effects. The list of candidate chemicals includes pesticides, cosmetic ingredients, food additives, and others. The cost of the program will be tremendous, with estimates indicating that tier 1 testing for only 3000 chemicals will cost U.S. $750 million.

Approaches for eliminating some candidate chemicals before screening are being considered, such as excluding polymers with an average molecular mass greater than 1000 daltons (perhaps 20,000 chemicals), strong mineral acids or bases, and nonchemical pesticides such as parasitic wasps. Certain mixtures are likely to be considered in the screening program, such as chemicals present in human breast milk, phytoestrogens in soy-based infant formula, and gasoline. At the first tier, the screening process will include rapid in vitro methods of identifying hormonal or antihormonal activity. This is not a measure of toxicity or adverse effects but rather an indication that further testing should be done. The second tier of testing will identify adverse effects. Depending on the number of chemicals that proceed into the testing tier, tier 2 could become very expensive in terms of both dollars and laboratory animal lives.

Screening and testing are complicated by the complex nature of the endocrine system and by the ubiquity of endocrine active chemicals in the environment. For example, phytoestrogens, which could interfere with screening and testing results, are high in the most common rodent diets used in laboratories. Many of the screening and testing methods have not been validated, so the implications of results are unclear.

15.1.8 IMPLICATIONS

It has been suggested that our government has dramatically overreacted to limited initial evidence of adverse effects, and that the nascent EDSTAC screening and testing program will be viewed in hindsight as a tragic misdirection of U.S. resources.

Chemicals with demonstrated adverse effects mediated by the endocrine system cannot be summarily removed from commerce and the environment. Some will be naturally occurring chemicals, and others will be important economically, agriculturally, or medically. The occurrence of toxicity is not generally considered sufficient for regulatory action. Rather, knowledge of the dose-response curve for the toxicity is important, as is information about exposure to the chemical. The dose-response curve for endocrine disruptors is highly contested, with some even postulating an inverted U shape (lower doses have an adverse effect but higher doses do not).

The U.S. Centers for Disease Control and Prevention and National Institute for Environmental Health and Safety are in the process of tracking human exposure to certain endocrine active chemicals. The exposure assessment program involves analysis of blood and urine samples from hundreds of people. Chemicals being tracked in the study include dioxins, dichlorodiphenyltrichloroethane, mercury, lead, alkylated phenols, pyrethroids, phytoestrogens, certain pesticides, gasoline additives, and cigarette additives. Together, the toxicity and exposure information can be used for responsible risk assessment to estimate risks posed by various uses of specific endocrine disruptors. When appropriate, risk management decisions can be made.

The consequences of identification of chemicals that cause adverse effects through the endocrine system will be played out over the years to come. The initial impact was investment of significant resources in screening efforts to cull out potential endocrine disruptors for additional study. Potential future implications include regulation in the United States under programs such as the Clean Water Act, the Toxics Release Inventory Program, and the Food Quality Protection Act; revisions to product labels; regulatory or consumer deselection of products such as foods, pesticides, pharmaceuticals, herbals, and neutraceuticals; toxic torts; and more stringent cleanup levels under Superfund, Resource Conservation and Recovery Act, and voluntary programs.

REFERENCES (ASTERISK INDICATES VERY IMPORTANT)

Andersen, M. E., and Barton, H. A., "Biological Regulation of Receptor-Hormone Complex Concentrations in Relation to Dose-Response Assessments for Endocrine Active Compounds," *Toxicol. Sci.*, 48: 38–50, 1999.

Ashby, J., "Issues Associated with the Validation of in Vitro and in Vivo Methods for Assessing Endocrine Disrupting Chemicals," *Pure and Applied Chemistry Special Issue on Natural and Anthropogenic Environmental Oestrogens: The Scientific Basis for Risk Assessment*, 70(9): 1735–1746, 1998.

Ashby, J., and Elliot, B. M., "Reproducibility of Endocrine Disruption Data," *Regul. Toxicol. Pharmacol.*, 26(1): 94–95, 1997.

*Ashby, J., et al., "The Challenge Posed by Endocrine-Disrupting Chemicals," *Environ. Health Perspect.*, 105(2): 164–169, 1997.

*Barton, H. A., and Anderson, M. E., "Dose-Response Assessment Strategies for Endocrine-Active Compounds," *Regul. Toxicol. Pharmacol.*, 25: 292–305, 1997.

Bingham, S., "Dietary Phyto-oestrogens and Cancer," *Pure and Applied Chemistry Special Issue on Natural and Anthropogenic Environmental Oestrogens: The Scientific Basis for Risk Assessment*, 70(9): 1777–1782, 1998.

Calabrese, E. J., et al., "A Toxicologically-Based Weight-of-Evidence Methodology for the Relative Ranking of Chemicals of Endocrine Disruption Potential," *Regul. Toxicol. Pharmacol.*, 26(1): 36–40, 1997.

Colborn, T., "Building Scientific Consensus on Endocrine Disruptors," *Environ. Toxicol. Chem.*, 17(1): 1–2, 1998.

*Colborn, T., et al., *Our Stolen Future*, Penguin Books, 1996.

Danzo, B. J., "Environmental Xenobiotics May Disrupt Normal Endocrine Function by Interfering with the Binding of Physiological Ligands to Steroid Receptors and Binding Proteins," *Environ. Health Perspect.*, 105(3): 294–302, 1997.

Dodge, J. A., "Structure/Activity Relationships," *Pure and Applied Chemistry Special Issue on Natural and Anthropogenic Environmental Oestrogens: The Scientific Basis for Risk Assessment*, 70(9): 1725–1734, 1998.

*Endocrine Disruptor Screening, and Testing Advisory Committee (EDSTAC), *EDSTAC Final Report*, Volumes I and II, EDSTAC, Washington, DC, August 1998.

Fischli, E., Godfraind, T., and Purchase, I. F. H., "Conclusion and Recommendations," *Pure and Applied Chemistry Special Issue on Natural and Anthropogenic Environmental Oestrogens: The Scientific Basis for Risk Assessment*, 70(9): 1863–1865, 1998.

Food Quality Protection Act, P.L. No. 104-170, 104th Congress, 2nd Session, 1996.

Gardiner, T., "An Industry Perspective," in *Great Lakes Endocrine Disruptors Symposium*, July 1997.

Gillesby, B. E., and Zacharewski, T. R., "Exoestrogens: Mechanisms of Action and Strategies for Identification and Assessment," *Environ. Toxicol. Chem.*, 17(1): 3–14, 1998.

Greim, H., "Environmental Estrogens and Human Health," *IUTOX Newsletter*, July: 21, 1998 (www.ehscorst.edu/iutox).

Hileman, B., "Low-Dose Problem Vexes Endocrine Testing Plans," *Chem. Eng. News*, May 10: 27–31, 1999.

Klotz, L. H., "Why Is the Rate of Testicular Cancer Increasing?," *Can. Medi. Assoc. J.*, 160: 213–214, 1999.

Lamb, J. C., IV, "Endocrine Disruptors: The Biological Basis for Concern," *Looking Ahead*, March/April: 6–7, 1999.

Li, J. J., and Li, S. A., "Breast Cancer: Evidence for Xeno-estrogen Involvement in Altering Its Incidence and Risk," *Pure and Applied Chemistry Special Issue on Natural and Anthropogenic Environmental Oestrogens: The Scientific Basis for Risk Assessment*, 70(9): 1713–1724, 1998.

Manuel, J., "NIEHS and CDC Track Human Exposure to Endocrine Disruptors," *Environ. Health Perspect.*, 107(1): A16, 1999.

Mazur, W., and Adlecreutz, H., "Naturally Occurring Oestrogens in Food," *Pure and Applied Chemistry Special Issue on Natural and Anthropogenic Environmental Oestrogens: The Scientific Basis for Risk Assessment*, 70(9): 1759–1776, 1998.

*McLachlan, J. A., and Arnold, S. F., "Environmental Estrogens," *Am. Sci.*, 48 September/October: 452–461, 1996.

*Miller, R., Foster, P., Gelbke, P., Guiney, P., Lamb, D., McKee, R., Olsen, G., Taalman, R., Van. Miller, J., Schluter, G., and Kavlock, R., "White Paper on Endocrine Disruption Including Reproductive and Developmental Effects," *State of the Science White Papers*, Chemical Industry Institute of Toxicology, Research Triangle Park, NC, January 1998, pp. 64–79.

Miyamoto, J., and Klein, W., "Environmental Exposure, Species Differences and Risk Assessment," *Pure and Applied Chemistry Special Issue on Natural and Anthropogenic Environmental Oestrogens: The Scientific Basis for Risk Assessment*, 70(9): 1829–1846, 1998.

No author, "Law Firm Experts Say New Studies Undercut 'Endocrine Hypothesis,'" *Risk Policy Report*, May 15, 1998.

No author, "Concluding Years of Intense Effort, EPA Endocrine Panel Issues Final Consensus Report," *Risk Policy Report*, October 16: 9–10, 1998.

No author, "How Estrogen Works," *Harvard Women's Health Watch*, April: 6, 1998.

No author, "Charge for the Science Advisory Board (SAB) and Scientific Advisory Panel (SAP) EDSPC Review," April, www.epa.gov/pesticides/SAP/april/charge.htm, 1999.

No author, "Herbal Medicine Is Potent Estrogen," *Environ. Health Perspect.*, 107(3): A137–A138, 1999.

*Preziosi, P., " Endocrine Disruptors as Environmental Signallers: an Introduction," *Pure and Applied Chemistry Special Issue on Natural and Anthropogenic Environmental Oestrogens: The Scientific Basis for Risk Assessment*, 70(9): 1617–1632, 1998.

Rhomberg, L., "Endocrine Disruptor Assessment: What Comes After Screening?," *Risk Policy Report*, May 15, 1998.

Safe Drinking Water Act Amendments, P.L. No. 104-182, 104th Congress, 2nd Session, 1996.

Sharpe, R. M., "Environmental Oestrogens and Male Infertility," *Pure and Applied Chemistry Special Issue on Natural and Anthropogenic Environmental Oestrogens: The Scientific Basis for Risk Assessment*, 70(9): 1685–1702, 1998.

Thigpen, J. E., Setchell, K. D. R., Goelz, M. F., and Forsythe, D. B., "The Phytoestrogen Content of Rodent Diets," *Environ. Health Perspect.*, 107(4): A182–A183, 1999.

*U.S. Environmental Protection Agency's Risk Assessment Forum, *Special Report on Environmental Endocrine Disruption—An Effects Assessment and Analysis*, EPA/630/R-96/012, February 1997.

*Van Der Kraak, G., "Observations of Endocrine Effects in Wildlife with Evidence of Their Causation," *Pure and Applied Chemistry Special Issue on Natural and Anthropogenic Environmental Oestrogens: The Scientific Basis for Risk Assessment*, 70(9): 1783–1792, 1998.

Weber, R. F. A., and Vreeburg, J. T. M., "Bias and Confounding in Studies of Sperm Counts," *Pure and Applied Chemistry Special Issue on Natural and Anthropogenic Environmental Oestrogens: The Scientific Basis for Risk Assessment*, 70(9): 1703–1712, 1998.

CHAPTER 15
UBIQUITOUS ENVIRONMENTAL CONTAMINANTS

SECTION 15.2

RADON IN AIR

Bernard L. Cohen

Dr. Cohen is professor emeritus of physics and astronomy and of environmental and occupational health at the University of Pittsburgh.

15.2.1 INTRODUCTION AND HISTORY

Of all radioactive isotopes released by the nuclear industry, the one expected to cause the most harm to human health is not ^{131}I, ^{137}Cs, ^{90}Sr, ^{83}Kr, ^{3}H, or ^{14}C. Far more important than all of these combined is ^{222}Rn. The type of natural radiation responsible for most fatalities according to current theories is not cosmic rays, radioactive elements in the earth and building materials, or ^{40}K in our bodies. The net effect of all of these together is to cause an average whole-body dose of 80 millirems per year, which might be expected to cause about 6000 deaths per year in the United States, whereas inhalation of environmental radon gas (actually, inhalation of its short half-life daughters) might be expected to cause 15,000 deaths. People who live near the site of the Three Mile Island nuclear power plant are exposed to more radiation from radon every day than they got in total from the 1979 accident there.

Of all known high-radiation exposure situations, the largest number of excess cancers did not result from the atomic bomb attacks on Japan, which caused perhaps 500 extra cancers; rather it is the uranium miner exposures to radon gas that have caused many times that number. On any time scale of a million years or longer, nuclear power is a method for *cleansing* the Earth of harmful radioactivity instead of for increasing the radioactivity with its wastes; this is because burning up uranium reduces the amount of radon gas that will eventually evolve from it. From these examples, it is clear that radon plays a very important role in the environmental effects of radiation.

The story of radon begins with (half-lives in parentheses) ^{238}U, (4.5 billion years) which undergoes a series of successive radioactive decays principally to ^{234}U (250,000 years), to ^{230}Th (77,000 years), to ^{226}Ra (1600 years), to ^{222}Rn (3.8 days), to a series of short-half-life radon progeny—isotopes of polonium, bismuth, and lead—whose total decay time averages less than 1 h, to ^{210}Pb (20 years), to ^{210}Bi (5 days), to ^{210}Po (138 days), and finally to stable ^{206}Pb. Because uranium occurs naturally in all rock and soil, typically with a concentration of 1 to 5 parts per million, the above decays are taking place continuously everywhere.

When an atom of uranium decays into an atom of thorium, or an atom of thorium decays into an atom of radium in a grain of soil, nothing very significant happens in the structure of that grain, but when an atom of radium decays into an atom of radon, it is very different. Radon is a noble gas (like helium, neon, argon, etc.) and hence it has no chemical binding to the neighboring atoms. The atom of radon is free to move away and about 15 percent of the time, it gets out of the grain and into the soil gas. It then can percolate up through the soil—it typically percolates up about 1 m in its 3.8-day half-life—and can emerge from the soil into the atmosphere. As a result of this process, typically about one atom of radon per second comes from each square centimeter of soil (about 1000 atoms per second from each square foot) everywhere in the world—it would be quite spectacular if we could see them emerging everywhere we look. Once in the atmosphere, radon mixes with the other molecules in air but, because it is much heavier, radon reaches a lesser height than the others, averaging about 1000 m. Considering the above and the 3.8-day half-life, the concentration of radon in air at ground level is about 0.2 pCi(7.4 mBq)/L.

However, if the radon emerges from the ground under a house, its fate is somewhat different. Houses normally run under a slight vacuum relative to the atmosphere and the soil gas. One reason for this is that wind blowing past a chimney sucks air from the house. Another reason is that air in the house often is used for combustion, with the residual gases released up a chimney. The fact that air inside a house is usually at a higher temperature than outside air also leads to lower pressure inside. In any case, because the house runs under a slight vacuum, radon emerging from the soil under a house is sucked into the house through cracks in the flooring. Once inside, it is not free to diffuse up to an average height of 1000 m; instead it is trapped within the height of a house, typically 10 m, for a few hours—perhaps 1/10th of a half-life. Thus, the concentration of radon in a house is about $(1000/10) \times (1/10) = 10$ times higher than outside.

Inside the house, radon and its progeny float around in the air releasing α-particles, β-rays, and γ-rays as they decay. This radiation in air is much less than the radiation from cosmic rays or from natural γ-rays from materials, but it is much more significant for the health of occupants, because people constantly pump air in through their bronchial tubes and into their lungs; in this process, the radon progeny atoms sometimes stick on the surfaces of the bronchi. This puts the α-particles they emit within range (a few micrometers) of some of the most cancer-sensitive cells in the body. Moreover, α-particles are about 200 times more effective than typical β- or γ-rays at initiating cancer. Thus, radon in homes (and other buildings) leads to a substantial risk of bronchial (lung) cancer. Of course, underground mines are in much more intimate contact with rocks and soil and normally trap air into even more confined spaces and for longer times, so they often have much higher concentrations of radon, especially if the surrounding rocks and soil contain high concentrations of uranium.

The recorded history of this radon problem begins with the first published book on mining, by Agricola in the early sixteenth century. He describes a respiratory disease among miners in the Erz Mountains of Bohemia and Bavaria, which was responsible for a large number of their deaths at a young age. He tells the story of a woman in her sixth marriage, having been widowed five times by the disease. The book was written in Latin, but it is available in English translated by perhaps the best known of all modern mining engineers, Herbert Hoover—later to become president of United States. Nothing changed and the miners continued to die for hundreds of years. In the eighteenth century, the mines became famous for the large quantity of silver they produced, and the silver was minted into coins in the principal mining town, Joachimsthal. These coins were called Joachimsthalers, but the name was soon shortened to "thalers," which became the origin of the word "dollars."

By the mid-nineteenth century, medical science began to develop and the mines hired physicians. In 1879, two of them published a paper in which they identified the disease that was killing the miners as lung cancer. With one or two isolated exceptions, this was the first report in the medical literature of lung cancer as a significant malady. The disease has since come a long way, now killing about 150,000 Americans each year.

In 1898, radiation and radioactivity were discovered, and, in 1900, radon gas was identified and characterized. Within the next decade, it was found that radiation can cause cancer, and instruments were developed to measure radioactivity with very high sensitivity. This led to widespread measurements in diverse places and it was soon found that the Erz Mountain mines had very high concentrations of radon. With the advance of medical science, other possible causes of the miner disease were investigated, but by the 1940s the most widely accepted view was that the radon in these mines was the principal culprit.

This was a very timely finding because the uranium mining industry was starting an enormous expansion in the United States. Limits on occupational exposure to radon were placed on these mines but their enforcement was left to state mine inspectors whose experience was with methane and other mundane risks in mines, so that little attention was paid to radon exposures. However, the U.S. Public Health Service started a study of the radon problem in U.S. mines. This was important because the Erz Mountain mines have large amounts of arsenic, lead, and other potentially carcinogenic chemicals; they were also damp and cold, which caused miners to constantly suffer from colds and other respiratory diseases, leaving open the possibility that these illnesses could be the causes of the miners' lung cancers. But the American mines were dry and comfortable and free of suspicious chemicals, so these other possibilities could be eliminated.

The first results of the U.S. Public Health Service study were publicized in the early 1960s, and they showed that there was indeed a large excess of lung cancer in U.S. miners, which seemed to be related to radon exposure. As a result, enforcement of occupational limits was greatly strengthened and exposures were reduced tenfold by 1970. However, miners continued to develop lung cancers from their prior exposures and epidemiological follow-up has continued; the results are reviewed below.

Because modern mining involves diesel fumes and technologies that increase levels of dust, the possibility remained that these could be important contributors to the problem. Coal mines provided a means for testing this possibility. For sound geochemical reasons, coal and uranium do not deposit in the same underground areas, so air in coal mines generally is low in radon concentration, but the diesel fumes and dust are still present. A crucial question was whether U.S. coal miners suffered from an excess of lung cancer. Data available in the 1960s seemed to indicate that they do, but an extensive study in the 1970s showed that, although coal miners suffered large excesses of many lung diseases, like silicosis, pneumoconiosis, emphysema, etc., they did not experience an excess of lung cancer relative to the general population. This leaves little alternative to concluding that radon is the culprit that causes lung cancers in uranium miners.

As an offshoot of the mine studies, it was recognized that the mill tailings from uranium mines, the residue left after the uranium was chemically separated from the ore, contained all the decay products of the uranium, including radium and ^{230}Th with its 77,000-year half-life, which continued to generate radium and thus radon as before. Because the ore was very rich in uranium, the radon released from these tailings piles was much higher than that from other soils. When the mills were shut down, the tailings piles dried out to become a sand, which was an attractive play area for children. These exposures to children were widely publicized, which led to measures to keep them away. There was then publicity about how emissions from tailings piles was exposing people who live nearby—indeed these exposures turned out to be by far the largest radiation exposures caused by the nuclear power industry. As a result, regulations were instituted for covering the tailings piles with dirt to reduce the emissions by an order of magnitude. There were still problems when, for example, the owners of the mill had gone out of business—in one such case, the city of Salt Lake City had grown to surround a tailings pile; that pile was physically moved out to a distant desert site in a large and expensive government operation.

Another flurry of activity was generated when it was recognized that sand from a tailings pile in Grand Junction, Colorado, had been used in construction of houses and schools, leading to the possibility that these buildings had high radon levels. In investigating this problem, measurements of radon were made in these buildings, but as part of the study, measurements were made in houses that did not have sand from the tailings pile. It had been believed that radon levels in houses would be much lower than outdoors, but the study found that the houses without tailings pile sand had radon levels much higher than were found outdoors. In Florida a similar study of houses built on and off phosphate tailings (phosphates are geochemically associated with uranium) also found high radon levels in houses not associated with phosphate tailings. This led to the realization that ordinary houses everywhere may have significant levels of radon.

At this point, the U.S. Environmental Protection Agency (EPA) instituted investigations to determine the cause of radon in houses, with a view to regulating the perpetrators. One possibility was that it entered the house with natural gas, but this was found to be a very minor contributor. Another suspected villain was building materials, but again this contribution was found to be negligible in typical American houses. Finally, the truth was realized—radon in houses is a natural phenomenon, with nobody to fine or regulate. At that point, EPA decided that radon in houses was natural and therefore was none of their business. There were programs in other countries—Canada, with 1/10th the population of the United States, measured 10 times as many homes. There was local excitement about high radon levels in Sweden and in Finland. But in the United States at least, radon in homes was not an issue.

The situation changed dramatically in December 1994 after an incident near Allentown, Pennsylvania. A nuclear power plant was in the final stages of construction and the doorway radiation monitors were activated for testing. Although there had not yet been any radioactivity generated in the plant, one of the workers set off the alarm when he passed through. After initial puzzlement, it was decided that the radioactivity was coming from his home. On investigation, it was found that the radon level in his home was 2700 pCi(999 mBq)/L, orders of magnitude higher than any levels previously encountered. Measurements were made in the homes of the man's neighbors and many of them were found to have very high levels of radon, although the house next door had only 0.5 pCi(18.5 mBq)/L. (After lengthy investigation, it was found that a uranium ore body underground ended between the two houses.) With the attendant publicity, measurements were made at ever-increasing distances, and high radon levels were found in numerous cases. Geologists recognized this to be associated with a uranium-rich formation known as "the Reading Prong," which extends through Bethlehem and Easton, Pennsylvania, and the Morristown area of New Jersey into New York State. Home owners in this region clamored for radon measurements and studies of the problem, publicity grew, and politicians took up the issue. Private businesses for measuring and reducing radon levels in homes sprung up. EPA decided to become involved and soon was promoting radon measurements nationwide, with cooperation from the National Advertising Council and American Lung Association. In 1988, the U.S. Surgeon General issued a statement that radon in homes was the second leading cause of lung cancer, responsible for many thousands of deaths each year; this was prominently featured on network TV news broadcasts. Radon levels became an important consideration in real estate transactions, and Parent Teachers Associations became concerned about radon in schools. The interest of home owners soon waned, but a reasonable level of activity persists.

15.2.2 BASIC FACTS ABOUT RADON

Radon, the element with atomic number 86, is chemically a noble gas, appearing in the same group of the periodic table of elements as He, Ne, Ar, Kr, and Xe. Isotopes of radon are included among the daughters of the three long-half-life heavy element parents of radioactive decay series, ^{238}U, ^{232}Th, and ^{235}U. The most important of these is ^{222}Rn from the ^{238}U decay series described above. ^{222}Rn decays by α-particle emission with a 3.8-day half-life into ^{218}Po, which decays by α-particle emission with a 3-min half-life into ^{214}Pb, which β decays with a 27-min half-life into ^{214}Bi, which β decays with a 20-min half-life into ^{214}Po, which α decays with a half-life of 164 μs into ^{210}Pb. Thus, within about an hour after decay of ^{222}Rn, three α-particles are emitted. In addition, ^{214}Pb and ^{214}Bi emit energetic β- and γ-rays, the most prominent of which are 0.61- and 1.76-MeV γ-rays from the latter.

An order of magnitude less important than ^{222}Rn in most practical situations is ^{220}Rn, a decay product of ^{232}Th and hence called "thoron," which has only a 54-s half-life, but it is also followed by three α-particle emissions, although the third is delayed for many hours because of the 10.6-h half-life of ^{212}Pb. The third naturally occurring isotope of radon is ^{219}Rn, a product of ^{235}U decay with a 3.9-s half-life and also followed by three α-particle emissions. Its most immediate long half-life predecessor is ^{227}Ac, whence it is known as "actinon." Because of the low abundance of ^{235}U relative to that of ^{238}U and the short half-life of ^{219}Rn, that isotope is of negligible importance in practical situations. Because ^{222}Rn is by far the most important isotope in practical situations, we largely confine our considerations to that isotope and refer to it as radon.

The ^{238}U half-life is such that in secular equilibrium each principal member of the decay series, including ^{222}Rn and each of its decay daughters, is present as 0.33 pCi(12.21 mBq) per g of uranium. Uranium is a very common trace element in rock, occurring with an average abundance of about 2.8 ppm (parts per million) in the Earth's upper crust and 1.8 ppm in soils near the surface. Typical abundances (in ppm) in various rock types are as follows: granite, 4.7; shale, 3.7; limestone, 2.2; basic igneous, 0.9; and sandstone, 0.45. As a global average, radon is emitted from land surfaces at a rate of about 1600 pCi(59.2 Bq)/cm^2-year (0.9 atom/cm^2-s), which is equivalent to an escape of all radon formed within 15 cm of the surface. When 1600 pCi/cm^2-year is multiplied by the land area of the Earth (1.5×10^{18} cm^2), we obtain a total release of 2.4×10^9 Ci(88.8×10^{18} Bq)/year. The concentration of radium in seawater is only about 0.05 pCi(1.85 mBq)/kg [compared with 900 pCi(33.3 Bq)/kg in soil] so the release rate of radon from the ocean surface is much less than from land surfaces, about 6 pCi(222 mBq)/cm^2-year. When this is multiplied by the area of the Earth covered by oceans (3.6×10^{18} cm^2), the total release from oceans comes to 2.3×10^7 Ci(85.1×10^{16} Bq)/year, about 1 percent of the release from land surfaces.

A third major source of radon release is through groundwater and vegetation, which serve as transport vehicles for bringing it out of the ground. It has been shown that when land is used to grow corn it emits three times as much radon as when it is bare, and considerably enhanced radon levels have been found in bathrooms when water is running in the bathtub or shower. It is estimated that water and vegetation may add to radon releases from the ground by as much as 20 percent. Other much smaller sources of radon release into the atmosphere are noted below, but they do not significantly affect the total of 2.4×10^9 Ci/yr. The radon in the Earth's atmosphere at any one time is determined by dividing the total by the decay constant, ln 2/half-life = 0.18/day = 66/year, to give 3.6×10^7 Ci(133.2×10^{16} Bq). To determine from this the radon concentration near the Earth's surface, it is necessary to know its variation with altitude. This strongly depends on atmospheric conditions, but as a general overall average it seems to decrease exponentially with a relaxation length of about 1000 m, which means 1/1000 of the total in each meter near the surface. When this is divided by the surface area of the Earth (5.1×10^{14} m^2), the result is an average surface concentration of 0.07 pCi(2.59 mBq)/L. Of course, the average is higher over land and lower over the ocean; for areas with no ocean in the upwind direction for thousands of miles, a rough approximation is to consider emission only from land surfaces and divide by the land area, which gives a surface concentration of about 0.2 pCi(7.4 mBq)/L. Because there is little atmospheric mixing between the northern and southern hemispheres and the former contains twice as much land area, radon concentrations are considerably lower in the latter.

One of the most sensitive external factors in determining the emanation rate of radon from soil is barometric pressure; the rate may change by a factor of two with a 1 percent change in pressure. This is explained by the fact that the radon concentration in soil gas is 10^4 to 10^5 times higher than in the atmosphere, and when atmospheric pressure is high, atmospheric air is forced into the ground, whereas when atmospheric pressure is low, soil gas comes out into the atmosphere bringing along its high radon content.

Another important factor in determining the rate of emission is moisture content of soil. Very wet soil emits only about one-third as much radon as the same soil when dry. There is some dispute about whether wind speed has an effect, and some such effects have been attributed to the indirect influence of wind on soil moisture. At low wind speeds, at least, the effects of wind appear to be negligible. Cold or frozen soil emits less radon than when it is warm, but temperature changes of a few degrees seem to have little effect. Thaws following a long period of freezing are accompanied by large releases even though the temperature remains quite low.

15.2.3   RADON PROGENY

The 3-min half-life of ^{218}Po and the 27- and 20-min half-lives of ^{214}Pb and ^{214}Bi are sufficient to allow a considerable separation of these from radon gas and from one another, especially in view of their rather different physical and chemical properties. In typical air sampling situations, the activity ratio to ^{222}Rn is close to 1.0 for ^{218}Po (because of its short half-life), 0.7 for ^{214}Pb, and 0.6 for ^{214}Bi and ^{214}Po. These ratios are subject to very wide variations depending on rates of air movement and exchange and the availability of deposition surfaces. In extreme cases, the ^{218}Po/^{214}Pb/^{214}Bi ratios have been found to be as low as 1/0.5/0.02, whereas in stagnant open air during an atmospheric temperature inversion they approach 1/1.0/1.0. Because they have relatively high ventilation rates, the ratios in mines generally are somewhat lower than in open air; 1/0 5/0.25 is perhaps typical.

When radon progeny atoms are originally formed with the emission of an energetic α- or β-particle, orbital electrons are knocked loose so the atoms are electrically charged. In this condition, they readily attach to surfaces, including dust particles in the air. All but a small fraction, called the unattached fraction, become attached to particles large enough to leave their behavior uninfluenced by the electric charge. The unattached fraction is an important determinant of the probability for radon progeny to stick to surfaces of the bronchi when inhaled, and hence it heavily influences health effects.

For dosimetry and health effect purposes, the radon progeny are much more important than radon itself, so it does not seem to be adequate for these purposes to specify concentrations of radon gas in pCi/L. The most accurate procedure is to specify the concentrations of each of its daughters, but this is both difficult to measure and cumbersome to use. As a compromise, the working level (WL) unit, developed for use in mines (originally, the maximum allowable level was 1.0 WL), is widely used. The WL is defined as any combination of radon daughters in 1 L of air that will result in the ultimate emission of 1.3×10^5 MeV of α-particle energy, which is numerically equal to the α-particle energy released by the progeny in equilibrium with 100 pCi(3.7 Bq) of radon gas.

To show this, it may be noted that 100 pCi is equivalent to 3.7 decays per s, which is the initial decay rate for each member of the series in equilibrium. The number of atoms of each daughter is then 3.7 times the half-life in seconds divided by ln 2; this is 980 for ^{218}Po, 8600 for ^{214}Pb, 6300 for ^{214}Bi, and $\ll$I for ^{214}Po. The latter three are destined to decay with the emission of 7.68 MeV of α-particle energy each (the energy of the ^{214}Po α-particle), so their energy releases are 6.6×10^4, 4.8×10^4, and 0 MeV respectively; the ^{218}Po atoms are destined to decay with emission of both their own α-particle emission, which is 6.00 MeV and the 7.68-MeV α-particle to give a total α-particle energy release of 1.3×10^4 MeV. When all these energy releases are added, the sum is 1.3×10^5 MeV. If all the daughter products were in equilibrium, a radon concentration of 100 pCi/L would thus give I WL. However, in nearly all situations, the progeny are present in much less than the equilibrium concentration, so I WL is typically equivalent to something like 200 pCi(7.4 Bq) of radon gas per liter. The WL is often the quantity of interest, for if a particle deposits in the lung, all its eventual α-particle decay energy will be released at that point. Note that radon gas itself does not deposit in the lung, so its concentration is not directly relevant. The ratio of WL to radon gas concentration in units of 100 pCi/L is called the equilibrium factor, or the F value.

The unit of integrated exposure is the WL month (WLM), which is the exposure at a level of I WL for 170 h (a working month in a mine).

15.2.4 *MEASURING TECHNIQUES*

There is a vast literature on techniques for measuring concentrations of radon gas and WL, but here we consider only the most commonly used methods.

Health effects depend on working level and unattached fraction, but the latter quantity is quite difficult to measure. Because radon progeny ordinarily attach to dust particles, the working level in a house can be greatly reduced by removing dust from the air by inadvertent circumstances (e.g., large areas of sticky surfaces) or by using electrostatic precipitators, ion generators, or other such devices. But the paucity of dust particles then results in fewer loci to which newly formed radon progeny atoms can attach and hence a higher unattached fraction. This is important because inhaled radon progeny attached to dust particles have only about 1 to 2 percent probability of sticking to the bronchial surfaces, whereas an unattached progeny atom has about a 50 percent probability and hence a much greater health impact. Without measuring the unattached fraction, working level therefore is not a reliable indicator of health impacts. The concentration of radon gas, however, does not depend on dust level. Roughly, the effects of dust on working level and unattached fraction cancel each other, leaving health effects dependent only on radon gas concentration; therefore, that is what is ordinarily measured.

The situation is different in mines because dust levels are always so high that unattached fractions are negligible, leaving health effects dependent only on working level; therefore that is what is measured.

The simplest way to measure the instantaneous concentration of radon in air—called grab sampling—is with a scintillation cell, a cylindrical chamber with a transparent window on one end and all other internal surfaces painted with a ZnS(Ag) scintillator; the window is optically coupled to a photomultiplier tube with output fed into an amplifier and a pulse height analyzer for electronic sorting of events by pulse height. The cell is evacuated and then the air to be studied is allowed to enter and fill the cell. The α- and β-particles from the decay of radon and its progeny strike the scintillator lining the walls and the resulting flashes of light are detected by the photomultiplier tube and counted if their pulse height is appropriate. Because the radon progeny are largely removed as the air enters the cell, counting must be delayed for about 90 min for equilibrium between radon and its progeny to be established. With sufficient counting times, radon levels as low as 0.01 pCi(0.37 mBq)/L can be reliably determined by this method.

The most frequently used method for measuring concentration of radon gas in homes is adsorption in charcoal. In its simplest form, a container of charcoal is exposed to the air for a few days, sealed, and sent to a laboratory where the amount of adsorbed radon is measured. The quantity of radon adsorbed from surrounding air approaches equilibrium with a time constant of about 18 h, so this method determines the average radon level in air during roughly the last 18 h of exposure. A longer time average can be achieved by having the air enter and leave the charcoal container through a diffusion barrier—e.g., a relatively small hole covered with a filter paper—which typically extends the time constant to 85 h. The laboratory measurement of captured radon can be made either by γ-ray counting with a well-shielded sodium iodide scintillation spectrometer or a Ge(Li) detector, or by transferring the radon into a liquid scintillation fluid, which then can be counted with commercially available sample changers. In large-scale routine programs, radon levels as low as 0.3 pCi(11.1 mBq)/L can be reliably measured by these methods.

Radon levels in a house can fluctuate substantially with time. An example, with measurements averaged over 1000 min, is shown in Figure 15.2.1; shorter term fluctuations can be somewhat more extreme, and some houses have regular day-night or morning-evening variations. Because measurements with an open container are largely controlled by the last 18 h of exposure, the points shown in Figure 15.2.1 indicate the result that would be obtained by that method, which was used by the EPA. We see that a measurement result could vary from 17 to 50 pCi(629 to 1850 mBq)/L. By use of the diffusion barrier charcoal adsorption (DBCA) method, the results for measurements at

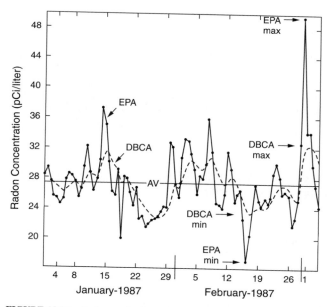

FIGURE 15.2.1   Radon results from charcoal adsorption measurements with open container and diffusion barrier measurements in a house. Points on the solid curve show average radon concentration recorded by a continuous monitor over each 1000 min and represent the expected results from the open container (EPA) method. Dashed curve shows the expected results of 7-day tests with DBCA collectors versus the time when tests are terminated. Horizontal line is the average concentration over the entire time period. From Cohen (1988).

various times would be as shown by the DBCA line in Figure 15.2.1; they could vary only from 24 to 32 pCi(888 to 1184 mBq)/L. Elaborate studies have found that a single measurement with a DBCA collector has a standard deviation of 15 percent from the true 1-month average; for an open container, the standard deviation is 30 percent. Another advantage of the diffusion barrier is that it can be made insensitive to humidity by including a desiccant bag in the barrier; with an open container, humidity can heavily influence the adsorptivity of the charcoal and must be corrected for by measuring the weight gain during exposure. A disadvantage of the diffusion barrier is that it reduces the count rate in the measurement, which requires about a four times longer counting time to compensate.

Devices are available that eliminate the problem of fluctuations with time by using very long exposures not limited by the 3.8-day half-life of radon. The most common method is to use α-particle track detectors, consisting of a plastic sheet, typically 1×1 cm, made of a special material, most frequently one called CR-39. The α-particles from decay of radon and its progeny leave damaged atoms along their paths, which are etched out in a caustic solution to make the paths visible under a microscope. These visible paths, actually better described as "pits," are then counted to determine the concentration of radon in the surrounding air. This method typically requires exposures of at least 1 month to obtain sensitivity to radon levels below 1.0 pCi/L. Exposures of several months to 1 year are usual.

The third most common method for measuring time-averaged radon gas concentration is with an electric field set up by an electret, the electrical analogue of a permanent magnet. Both α- and β-particles from radon and its progeny create ions in the air, which are accelerated by the force from the electric field. This drains energy from the electric field, which shows up as a lowering of the voltage across the electret. By measuring this voltage before and after exposure, the number of ionizing events is determined, which is calibrated to give the radon concentration. Depending on the geometry of the electret, these devices can integrate radon exposure over time periods varying from hours to months. They can be used for several measurements until the voltage is too low, at which point the electret must be replaced. This method has advantages when measurements are being made by professionals who can measure the voltage decrease on the spot in less than 1 min and then redeploy the detector. The disadvantage is that the equipment is expensive. A charcoal collector or an α-particle track detector costs less than $1.00 and hence can be mailed to the householder who handles the exposure and mails it back to the laboratory—if a person fails to mail it back, the loss is negligible.

Measurement of working level is most easily done by pumping air through a filter to collect the radon daughters. The radiations from the filter are then measured as a function of time by use of a scintillation spectrometer, by α-particle spectroscopy with detectors capable of separating the emissions from ^{218}Po and ^{214}Po, or by coincidence counting of α-particles from one side of the filter and β-particles from the other side (the β-particle from ^{214}Bi is essentially in coincidence with the α-particle from ^{214}Po). Simple and rapid methods are available for converting these measurements into WL determinations.

15.2.5 RISK OF LUNG CANCER FROM RADON EXPOSURE

The recognition that exposure to radon can cause lung cancer derives from two sources: (1) it exposes the lungs (mainly the bronchial region) to radiation, and from experiences with A-bomb survivors, patients treated with radiation therapy, etc., radiation is known to cause cancer; and (2) experience with miners exposed to high doses of radon. There are serious difficulties in converting exposures to radon in pCi/L or WLM into normal radiation units like sieverts or rem, but a great deal of attention has been devoted to this conversion by groups in the United States, England, and Germany. Although there are differences in the details of their approaches, they agree reasonably well in their final results, concluding that the risk estimates derived from data on A-bomb survivors and medical therapy patients agree reasonably well with those experienced by the miners. It is generally agreed that the latter are more reliable.

It has been conventional to fit the data on miners to

$$R = 1 + K \times D \tag{15.2.1}$$

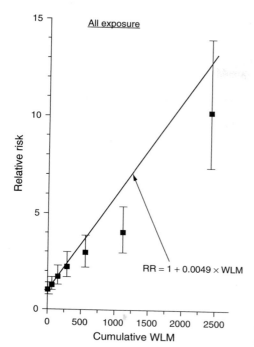

FIGURE 15.2.2 Relative lung cancer risk versus exposure to radon progeny from pooled study of 11 groups of miners. RR, relative risk. From Lubin et al. (1995).

where R is the relative risk of lung cancer death compared with those not exposed, D is the exposure to radon progeny in thousands of WLM, and K is a constant obtained from data fitting. The following data are available for 11 groups of miners (K, the excess relative risk per 1000 WLM, is in parentheses): uranium miners in the Czech Republic (3.4), Colorado Plateau (4.2), Ontario (8.9), Grants, New Mexico (17.2), Beaver Lodge, Canada (22.1), Port Radium, Canada (1.9), Radium Hill, Australia (50.6), and France (3.6); for tin miners in Yunan Province, China (1.6); for fluorspar miners in Newfoundland (7.6); and for iron miners in Malmberget, Sweden (9.5). A pooled analysis of the data (Lubin et al., 1995) from all 11 miner groups is shown in Figure 15.2.2; the best fit of the data to Equation 15.2.1 gives $K = 4.9$ (95 percent confidence interval 2.0 to 10). The data from the various groups are highly inconsistent—the result from the pooled analysis, $K = 4.9$, is within the 95 percent confidence interval for only 6 of the 11 individual studies—and smoking information was not available for about half the cases, On the other hand, K is positive with >95 percent confidence for all 11 studies. The 11 studies include a total of more than 2000 excess lung cancer deaths, about 2600 observed versus 400 expected.

There seems to be a clear difference between smokers and nonsmokers, as shown in Figure 15.2.3. For a given radon exposure, the relative risk for nonsmokers is about 3 times higher, although their absolute risk is only one-third as large because smokers not exposed to radon have more than 10 times the lung cancer risk of unexposed nonsmokers.

The data on miners are nearly all at much higher radon exposures than those normally encountered in homes, so all estimates of

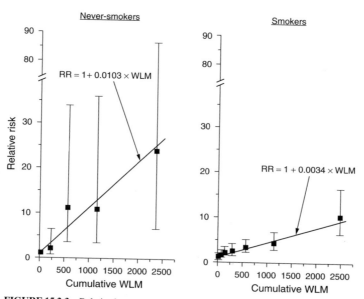

FIGURE 15.2.3 Relative lung cancer risk versus exposure to radon progeny from pooled study of smoking and nonsmoking miners. RR, relative risk. From Lubin et al. (1995).

risk from household exposures are based on the miner data extrapolated to low doses by assuming validity for the linear no-threshold theory (LNT) implicit in Equation 15.2.1. However, the LNT has been called into question recently for a wide variety of reasons. The reason most directly related to radon is the finding that lung cancer mortality rates in U.S. counties, with or without correction for smoking frequency, have a very strong and statistically indisputable negative correlation with average radon levels in homes, as shown in Figure 15.2.4 (Cohen, 1995). Exhaustive efforts to explain this discrepancy consistent with LNT have included consideration of over 500 potential confounding

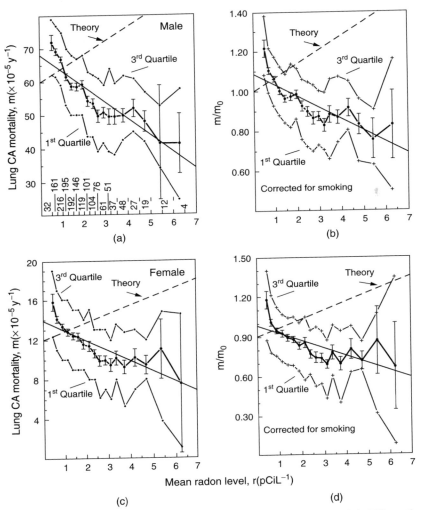

FIGURE 15.2.4   Lung cancer (CA) mortality rates versus mean household radon levels for U.S. counties. Data points are averages of ordinates for all counties within the range of radon levels (r, in pCi/L) shown on the baseline of (a); the number of counties within that range is also shown. Error bars are one standard deviation of the mean, and first and third quartiles of the distributions are also shown. (a) and (b) males; (c) and (d) females. (b) and (d) the same data as in (a) and (c) corrected for smoking prevalence in each county. Theory lines are from the BEIR-IV version of LNT, increasing at a rate of 7.3 percent per picocurie per liter. From Cohen (1995).

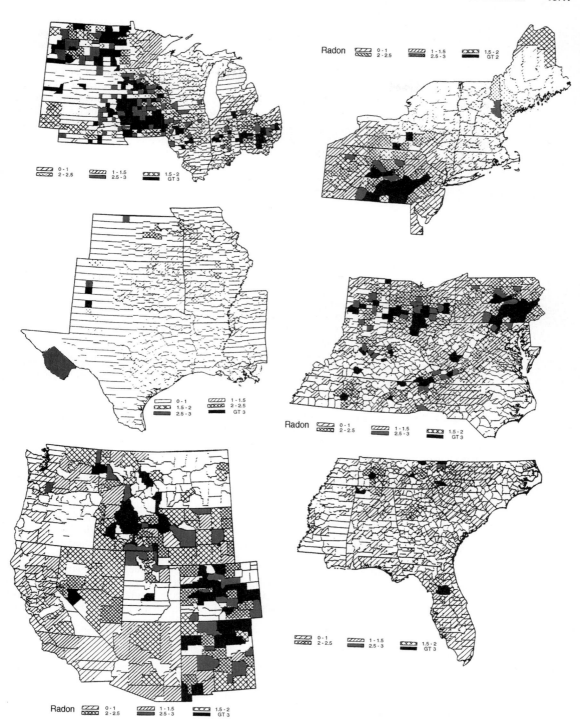

FIGURE 15.2.5 Maps showing average radon level in each U.S. county. No cross-hatching indicates no data for a county. From Cohen et al. (1994).

factors, variations with geography and weather, etc., but they have had no success. This study has been questioned because it is ecological, dealing with groups of people (populations of counties) rather than with individual persons, but all the recognized problems with ecological studies have been investigated and found not to be applicable.

There are many other lines of evidence indicating that LNT grossly overestimates the cancer risk from low-level radiation and precious few that support LNT. The Health Physics Society has issued a position paper stating that effects of low-level radiation are essentially completely unknown and may well be zero or even positive (i.e., protective against cancer). However, International Commission on Radiological Protection and U.S. National Council on Radiation Protection and Measurements continue to support LNT.

15.2.6   *RADON LEVELS IN U.S. HOMES*

The average radon level in U.S. homes is 1.25 pCi(46.25 mBq)/L (EPA, 1992), but there are wide variations with geography. The average household radon levels in U.S. counties are displayed in map form in Figure 15.2.5 (Cohen et al., 1994). There are areas with levels below 0.5 pCi(18.5 mBq)/L and individual counties with levels above 6.0 pCi(222 mBq)/L. Geography, acting as a surrogate for geology (perhaps with some modifying effects of house construction practice), is by far the most important determinant of average radon levels in an area.

Several other factors have been identified that correlate with radon levels in houses (Cohen, 1991). Rural houses average about 30 percent higher radon levels than urban or suburban houses. Within a house, the most important variation is between the basement and upper floors, with radon levels in basements averaging about twice as high. Second floors levels average 10 to 20 percent lower than those on first floors. Radon levels tend to decrease with the age of the house, reaching about two-thirds of the level in new houses for 80-year-old houses, but for ages beyond 80 years, levels increase moderately. Houses that were weatherized to reduce heat loss have only slightly higher radon levels than others, but houses characterized by occupants as drafty have only two-thirds the radon concentrations of houses characterized as less drafty than average.

There is little variation in average radon levels with socioeconomic factors such as value of house, income of occupants, and education of head of household, although the very lowest income and house values tend to have levels about 15 percent below average. Owner-occupied houses average about 20 percent higher radon levels than rented houses. Radon levels in houses of cigarette smokers average about 10 percent below those in houses with no cigarette smokers.

If it is found that a house has a high radon level, the occupant may opt for remediation. If the entry path for radon into the house can be identified and blocked, this is the first approach. The principal examples are where bare earth is exposed within the house and can be paved over or where there are large cracks in the basement floor that can be sealed. However, sealing small cracks is rarely effective. If there are no obvious radon entry pathways, the most effective remediation strategy is to pump air from beneath the basement slab to outdoors; this reverses the usual pressure gradient, making air pressure in the ground lower than in the house, so radon is no longer sucked into the latter. This method is successful in over 90 percent of the houses where it has been used, a success rate high enough for remediation companies to offer money-back guarantees if the radon concentration is not reduced to below 4 pCi(148 mBq)/L, the EPA recommended action level. Several other approaches to remediation have been proposed, but none has come into common use in the United States.

REFERENCES

Cohen, B. L., "Performance Characteristics of DBCA Radon Collectors," *Radiation Protect. Manage.*, 5(6): 47–54, 1988.

Cohen, B. L., "Variation of Radon Levels in U.S. Homes Correlated with House Characteristics, Location, and Socioeconomic Factors," *Health Phys.*, 60: 631–642, 1991.

Cohen, B. L., "Test of the Linear-No Threshold Theory of Radiation Carcinogenesis for Inhaled Radon Decay Products," *Health Phys.*, 68: 157–174, 1995.

Cohen, B. L., Stone, C. A., and Schilken, C. A., "Indoor Radon Maps of the United States," *Health Phys.*, 66: 201–205, 1994.

EPA Radon Division, *National Residential Radon Survey, Statistical Analysis*, U.S. EPA Washington, DC, April 1992.

Lubin, J. H., et al., "Lung Cancer in Radon Exposed Miners and Estimation of Risk from Radon Exposure," *J. Nats. Cancer Inst.*, 87: 817–827, 1995.

CHAPTER 15
UBIQUITOUS ENVIRONMENTAL CONTAMINANTS

SECTION 15.3

RADON AND RADON DAUGHTERS

Thomas J. Aley

Mr. Aley is president of Ozark Underground Laboratory. He performs karst hydrology research in his own cave.

15.3.1 INTRODUCTION

Uranium-238 (^{238}U), a radioactive element with a half-life of 4.5 billion years, is widely distributed in bedrock and soils. Higher concentrations are often associated with areas where uranium or phosphate mining has occurred, but elevated concentrations can occur in a multitude of settings, and, without data, no geologic setting can be presumed to be essentially uranium-free. One of the radioactive decay products of ^{238}U is radium, which in turn decays to radon-222 (^{222}Rn) (which we will simply call radon in this discussion).

Radon is an odorless, colorless gas with a half-life of 3.8 days. It has four radioactive decay products called radon daughters (also called radon progeny). The radon daughters are polonium-218 (^{218}Po), lead-214 (^{214}Pb), bismuth-214 (^{214}Bi), and polonium-214 (^{214}Po). These radon daughters have half-lives that range from a fraction of a second to 22 years. During their radioactive decay the four radon daughters emit α-particles, which are large and as a result are largely stopped by the skin without causing damage. However, if α-particles are inhaled they may reach cells in the lungs, which are sensitive to damage from the ionizing charges of the particles. Such damage may increase the risk of developing lung cancer at some time in the future.

Prolonged occupational exposure by underground miners to relatively high concentrations of radon daughters and other airborne contaminants has been correlated with an increased incidence of lung cancer deaths among these workers. The α-radiation is the parameter that has been quantified to characterize the air quality to which miners have been exposed. The studies that have shown a correlation between the total dose of α-radiation in mines and increased lifetime risk of lung cancer have been the foundation for regulatory concerns about radon gas and radon daughters. The regulatory concern has focused attention on radon gas in homes.

The α-radiation concentrations encountered in homes is typically at least one and often two orders of magnitude smaller than the concentrations encountered by miners who have died of lung cancer. Other carcinogens and/or lung irritants were routinely present in the mine air the deceased miners routinely breathed. These other mine air constituents have included cigarette smoke (although smoking is now prohibited in underground mines), fumes from diesel equipment and explosives, and dust, which includes silica and sometimes asbestos from the rock. Many of these contaminants are less common or absent from household air. Despite these differences, the correlation between α-radiation and an increased incidence of lung cancer deaths has been extrapolated to the α-radiation concentrations encountered in homes. This extrapolation has been based on a number of potentially credible, but scientifically unverified, assumptions. These include:

- The assumption that the correlation between α-radiation and increased incidence of ultimate death from lung cancer is linear;

- The assumption that there is no threshold concentration of α-radiation below which there is little or no increased risk;

- The assumption that measurements of the total exposure to α-radiation by miners was not greatly underestimated;

- The assumption that risks of lung cancer from α-radiation can credibly and quantitatively be separated from risks of lung cancer associated with

  —cigarette smoking while exposed to α-radiation,

  —being a smoker but not smoking while exposed to α-radiation, or

  —being simultaneously exposed to α-radiation and second-hand cigarette smoke; and

- The assumption that the risk associated with smoking and the risk associated with α-radiation are additive rather than synergistic.

Based on the studies of miners and the extrapolation of data from those studies to radon concentrations likely to be present in buildings, various entities [particularly the U.S. Environmental Protection Agency (EPA)] have embarked on an extensive publicity campaign to alert the public about facets of the issue. This campaign has focused on the inferred health risks of radon in homes. Statements have included:

- Estimates that there are 5000 to 20,000 lung cancer deaths a year in the United States from radon exposure in homes. In 1998 the National Research Council estimated the annual radon-related lung cancer deaths at between 15,400 and 21,800.

- Estimates that 6 percent (some estimates are up to 12 percent) of the homes in the United States have mean annual radon concentrations exceeding 4 pCi(148 mBq) per liter of air. The conversion from percentage of homes to number of homes is not always consistent among the statements, in part because of continuing increases in the number of homes. However, six million homes with mean annual radon concentrations exceeding 4 pCi/L is a common estimate.

- Recommendations that all homes be tested for radon and that homes with mean annual radon concentrations exceeding 4 pCi/L should be remediated.

The agency statements prepared for the general public typically have been silent about the uncertainty of the inferred health risk. Several epidemiological studies of potential health impacts resulting from low-level radon concentrations have been completed in the past few years. The studies with which the author of this chapter is familiar have not indicated that low levels of radon are associated with an increased incidence of lung cancer. Similarly, epidemiological studies with which the author is familiar have not found statistically higher lung cancer deaths in areas with high radon concentrations than in areas with low concentrations.

A basic public policy issue posed by radon is how the regulatory community should respond to a contaminant (or in this case a natural constituent) that is ubiquitous and that may have the potential to adversely affect health. With certain assumptions data can be extrapolated from one population (in this case, from miners) to a dissimilar population (residents of homes). Inferences about potential contaminant impacts on humans often are based on animal studies where the test animals were subjected to high or prolonged concentrations of the selected contaminant. Extrapolation of data from miners to house residents appears to be a lesser leap of logic, and in this case the extrapolation suggests that low levels of radon pose a health hazard.

In science we pose a hypothesis and then test it with appropriate studies. In this case, appropriate studies are epidemiological investigations to assess whether there is a correlation between the low levels of radon encountered in homes and an increased risk of ultimate lung cancer. Scientific studies to date do not appear to show a correlation consistent with the warnings being given to the public. Limitations and flaws always must be considered in assessing such studies, and technical critiques are always needed. However, in the absence of proof to a reasonable degree of scientific certainty of a health effect, how should public agencies respond? Should they decrease or end their efforts associated with indoor radon in homes? Alternatively, is the appropriate response to stay the course until such time as the mass of relevant studies becomes overwhelming?

The radon issue is a combination of science and politics with the proportion of each subject open to debate. Cole (1993) has written a clear and nontechnical discussion of the issue in his book *Element of Risk, the Politics of Radon*. His book does an excellent job of examining the issues, characterizing the various positions that have been taken, and providing a reasonably impartial review of the data that support the disparate positions. We leave further discussion of these topics to Cole (1993).

15.3.2 *MEASUREMENTS AND CONVERSIONS*

Radon-222 concentrations are commonly expressed in picocuries per liter of air (pCi/L), and α-radiation from radon daughters is measured in working levels (WL). A WL is defined as any combination of short-lived radon daughters in 1 L of air that will result in the emission of 1.3×10^5 MeV of potential α energy. A WL month (WLM) equals 1 WL for 173 h (the number of working hours in a month assuming 40 h per week and $4^1/_3$ weeks per month). Radon is typically the parameter measured or estimated in buildings, and α-radiation is typically the parameter measured in mines and other underground spaces.

In underground air there also may be α-radiation from thoron daughters. Thorium-232 (^{232}Th) is the parent in a radioactive decay series that includes a radioactive gas (^{220}Rn), which is commonly called thoron. The decay of thoron yields thoron daughters, some of which are α-particle emitters with the same potential health impacts as radon daughters. The α-radiation derived from thoron daughters is typically only a few percent of that derived from radon daughters. The α-radiation sampling typically conducted in mines and caves can distinguish between, and separately quantify, radon and thoron daughters. In contrast, the typical use of accumulating samplers in homes for estimating radon concentrations reflects the total of radon and thoron. The remainder of this discussion assumes that concentrations of radon include thoron and that α-radiation values from radon include both radon and thoron daughters.

There is no consistent conversion coefficient between radon (in pCi/L) and radon daughters (in WL) because the ratio of the two is a function of the extent to which the radon daughter products are in equilibrium with the radon gas. However, the following approximate conversions are often used:

- Unventilated mines: 100 pCi(3.7 Bq) of radon per liter equals approximately 1 WL because the gas and its decay products are confined and are at or near equilibrium. The time to equilibrium is about 3 h.

- Ventilated mines: 300 pCi(11.1 Bq) of radon per liter equals about 1 WL because ventilation is often rapid and the equilibrium factor in such settings is commonly about 30 percent.

- Most homes and buildings: 200 pCi(7.4 Bq) of radon per liter equals approximately 1 WL because the equilibrium factor in such settings is commonly about 50 percent, but the range is from about 30 to 70 percent.

Radon is sometimes present in water. When exposed to the air much of the radon degasses from the water and enters the air. In most cases the effects are nondetectable to minimal. In homes, a common rule of thumb is that 10,000 pCi(370 Bq) of radon per liter of water will yield an increase of about 1 pCi(37 mBq)/L in household air. Domestic water from municipal water supplies may have smaller concentrations of radon than water from private wells if the municipal water is subjected to aeration before distribution or when it is derived from surface water supplies.

There are many ways to measure radon and α-radiation in air. Three of the most common are discussed in the following paragraphs.

Track etch monitors are commonly used to monitor buildings. These are passive devices that detect radiation damage resulting on the detector material (a film) by α-particles. The track etch monitor is placed in a monitoring location and left for a period of time; the recommended minimum exposure times are usually 1 or 2 months. After collection, the monitor is shipped to a laboratory where the tracks on the film are chemically etched to enhance their visibility and then assayed either manually or automatically.

Activated carbon (alternatively called charcoal) monitors are also commonly used in buildings and provide a mean estimate of radon concentrations during a sampling period. The device is simply a container filled with activated coconut shell charcoal. These are passive samplers where the radon enters the sampler through molecular diffusion. The samplers are commonly left in place for at least 24 h and as long as about a week. After collection, the sampler is shipped to a laboratory where the γ-rays emitted by the radon daughters adsorbed on the carbon during exposure are measured.

Track etch detectors can periodically be collected and new detectors placed so that longer-term estimates of radon concentrations can be made. Activated carbon monitors typically are used only for relatively short-duration sampling. Radon concentrations typically vary substantially in buildings both from point to point and seasonally. A relatively high radon concentration in a sampler is likely to suggest that other relatively high values could be obtained in the building. However, the common assumption with buildings that one or even a few low to modest values are reflective of general conditions is at best questionable. The credibility of the data increase with increased sampling time, and the range of values (but not the values per se) is often smaller when values are taken in the winter. General EPA recommendations are that all measurements be made under "closed-house" conditions and in the winter; yet this tends to skew the data toward higher values than house occupants are likely to encounter under mean conditions. At least some EPA recommendations have been that sampling persist for at least 24 h if the data are to be the basis for a decision about remedial action or long-term health risk. Anyone following this EPA recommendation should recognize that 1-day radon sample values in a building can vary by an order of magnitude or more.

Radon monitoring is sometimes conducted in conjunction with the sale of homes. Few buyers or sellers are willing to conduct sufficient monitoring over sufficient periods of time to derive scientifically credible estimates of mean radon concentrations. Furthermore, selection of the monitoring location or locations can significantly affect the results. Results from a basement location are usually higher than those from the floor above; yet if one is concerned with potential health affects there may be little relevance in monitoring basements where people spend minimal amounts of time. As a general rule, radon monitoring of homes in conjunction with real estate transactions is unlikely to yield credible estimates of mean radon concentrations encountered by residents. At best, the monitoring in conjunction with real estate transactions may identify homes, or locations in homes, with high radon concentrations.

A common monitoring strategy in mines and caves is to collect grab samples for radon daughter measurement at a particular underground point or in a particular portion of the mine or cave. The sampling is typically conducted by using a calibrated air pump to pass a known volume of air through a high-efficiency filter for a sampling period of 2 to 10 min. The α-radiation of the filter is then measured with an α-scintillation counter; a correction table is used to adjust the values from the counter for the elapsed time since the sample was collected. This monitoring approach is used on a periodic basis to estimate occupational exposure rates of employees to α-radiation.

15.3.3   *COMMON RADON AND RADON DAUGHTER VALUES*

Mines

Adequate ventilation is now an important consideration in the planning, development, and operation of mines in the United States. Because of regulatory requirements, ventilation systems and monitoring are operated to maintain α-radiation concentrations below 0.3 WL. In underground coal mines other air constituents (such as methane, carbon dioxide, and coal dust) are the usual controlling parameters for ventilation because their control requires more ventilation than would controlling α-radiation.

Regulatory standards for mines are partly based on presumed health effects of α-radiation and partly to ensure that the mines are adequately ventilated and that the ventilation system is functioning properly. Important regulatory standards are that maximum α-radiation concentrations not exceed

1.00 WL in areas of occupational exposure and that total annual α-radiation exposure for any employee not exceed 4.0 WLM in any calendar year.

Caves

There has been a substantial amount of α-radiation monitoring in natural caves, especially those operated by the National Park Service and open to the public for tours. The discovery in Carlsbad Caverns, New Mexico, of α-radiation concentrations commonly greater than those in modern ventilated mines led the Park Service to institute α-radiation standards similar to those in force in the mining industry. Subsequent monitoring in other National Park caves demonstrated that α-radiation is present in most caves, and that some caves consistently have concentrations greater than those found in Carlsbad Caverns.

Ahlstrand and Fry (1977) summarized α-radiation data from four zones in Carlsbad Caverns. Measured concentrations of α-radiation in the zones during the 2-year period ranged from 0.06 to 0.55 WL; maximum values divided by minimum values for the four zones varied by almost an order of magnitude.

Continuous radon concentrations were also measured at a reference location within Carlsbad Caverns for the same 2-year monitoring period. The radon concentrations in the cave show marked daily fluctuations in addition to large-scale seasonal variations. High barometric pressure systems are related to lower radon concentrations, and low barometric pressure systems are related to higher concentrations. Outside temperatures lower than cave temperatures correlate with lower radon daughter concentrations in the cave.

Yarborough (1977) presented α-radiation data for 1 year for the various tour routes in Mammoth Cave at Mammoth Cave National Park, Kentucky. Values on the routes varied from a minimum of 0.12 to a maximum of 1.89 WL. Maximum values divided by minimum values for the six routes where data had been collected during most months indicated a fivefold mean variation of α-radiation concentrations.

Most of the "show caves" in the United States are privately owned and operated, and they are typically smaller than either Carlsbad Caverns or Mammoth Cave. Typical α-radiation concentrations in these caves average about 0.3 WL, yet values range widely.

Unlike mines, caves are natural systems that often contain fragile speleothems (cave formations) and cave-adapted fauna. Even if it were feasible to artificially ventilate caves, such actions would alter the cave microclimate and damage or destroy some of the cave features. Oregon Caves, at Oregon Caves National Monument, was developed for guided tours. The development work included enlarging many passages and constructing an exit tunnel. As a result of the construction, the difference in elevation between the upper and lower entrances is about 95 m (312 ft). Management efforts to reduce winter-time α-radiation concentrations in Oregon Caves included leaving some of the doors on the cave open. Under cold temperature conditions there was a great amount of convective airflow upward through the cave because of the great difference in entrance elevations and the enlargement of connecting passages. Entry of cold air through the lower entrance froze water in hundreds of stalactites and destroyed these delicate features.

Caves have provided some useful insight into natural underground α-radiation concentrations and their seasonal fluctuations. In most caves the abundance of openings in the bedrock and soil tend to be greater and better connected to facilitate airflow from areas above the cave than from areas below. As a result of convective airflow patterns plus the nature of the openings connected with the caves, radon daughter concentrations are greater in most caves in the summer than in the winter. In contrast, buildings commonly have higher radon concentrations in the winter than in the summer. In cavernous and fractured rock areas this can be due to convective airflow upward from the underlying rock units.

Outdoor Conditions

Outdoor radon concentrations in the United States are commonly on the order of 0.1 to 0.5 pCi(3.7 to 18.5 mBq)/L although concentrations of 2 to 3 pCi(74 to 111 mBq)/L are locally common. Large portions of Manitoba and Saskatchewan, Canada, have typical outdoor concentrations in the 2 to 3 pCi/L range (Cole, 1993). Some federal legislation in the United States has included an objective that indoor α-radiation concentrations not be greater than outdoor values.

Buildings

Compared with mines and caves, much smaller concentrations of radon and radon daughters are typical of buildings. These concentrations tend to be higher where trace concentrations of uranium in the bedrock and soil are more abundant; yet the nature of building construction and other factors yield great variability in α-radiation concentrations. Buildings with basements typically have higher radon concentrations than do buildings on slabs, and mobile homes or houses built on pillars have less radon than those that have greater contact with the earth. Neighboring houses may have dramatically different radon concentrations, and repeat measurements at the same locations often show great variability (in excess of an order of magnitude in some cases). Samplers in place for a short period of time typically show greater variability than those in place for longer periods.

Maps of predicted average radon concentrations in houses have been prepared by the EPA in cooperation with state and other federal agencies. EPA (1993) is an example of a statewide assessment for the state of Missouri. Five basic factors have been considered in the preparation of the maps of predicted average radon concentration in houses; these are indoor radon measurements that have been made in the area, geology, aerial radioactivity, soil parameters, and typical house foundation types. The maps of predicted average radon concentrations in homes indicate that about 35 percent of the United States is characterized by predicted average indoor radon concentrations in excess of 4 pCi(148 mBq)/L; these areas are largely located in the northern (and generally lesser populated) states from Maine to Idaho. About 40 percent of the United States is characterized by predicted average indoor radon concentrations between 2 and 4 pCi/L, and only about 25 percent of the nation is characterized by predicted indoor concentrations of less than 2 pCi/L. As noted earlier, outdoor radon concentrations in some areas equal or exceed 2 pCi/L.

Numerous episodes of locally intensive radon monitoring in homes have occurred. In many cases the monitored points were places (such as basements) where higher than average radon concentrations were anticipated. Many studies collected only short-term and limited data. However, as noted earlier, the radon issue is a combination of science and politics. Findings from some of these studies provide insight into both radon conditions in houses and information commonly disseminated to the public.

One hundred houses were monitored for radon in a study in Bowling Green, Kentucky (Webster and Crawford, 1987), in a county shown by EPA to have mean predicted radon concentrations in homes in excess of 4 pCi/L. The monitoring used activated charcoal cannisters, and sampling duration, which was not reported, would not be expected to have lasted more than 1 to 7 days. Radon concentrations in the household air ranged from <1 to >130 pCi(<37 to >4810 mBq)/L; the mean was 10.4 pCi(384.8 mBq)/L, and 57 percent of the sampled houses had radon concentrations greater then 4 pCi/L. Radon concentrations in basements averaged four times higher than concentrations on the first floor of homes. As a result, basement measurements contributed substantially to the large percentage of the sampled houses that had radon concentrations >4 pCi/L. This investigation shows great variability in values even in a local area, especially when the sampling period is short and all sampling of the homes is not done concurrently.

In Clinton, New Jersey, 105 houses were tested and all contained radon concentrations greater than 4 pCi/L. Thirty-five of the houses had concentrations equal to or greater than 200 pCi(7.4 Bq)/L, and five had concentrations equal to or greater than 1000 pCi(37 Bq)/L. Remediation of at least some of the home occurred with costs in the range of $1200 to $3000 per house.

Several studies addressed seasonal variations in house radon concentrations. Radon concentrations in a study of Montana homes showed winter values 1.3 times greater than summer concentrations. The factor was 1.5 in a South Dakota study. Homes in hilly limestone regions of the southern Appalachians reportedly have higher indoor radon concentrations during summer than in winter; this was attributed to underground convective airflow from portions of the hillsides higher in elevation than the homes (Gunderson et al., 1993).

Water Supplies

The EPA has proposed a maximum contaminant level for radon in water supplies of 300 pCi(11.1 Bq)/L. The principal concern is that the radon will degas from the water into the air of the home and increase the total radon concentration in household air. Although a small percentage of water wells may have radon concentrations greater than 10,000 pCi/L, most radon concentrations in well water are far below this concentration. Some of the greatest radon concentrations in water would be expected to occur in uranium mines. Water in these mines typically has radon concentrations ranging from 10,000 to 100,000 pCi(370 to 3700 Bq)/L (Misaqu, undated). As noted earlier, a rule of thumb is that a radon concentration of 10,000 pCi/L in the water supply of a house will increase the radon concentration in the household air by 1 pCi/L.

Radon in groundwater was studied in the Lower Susquehanna and Potomac River Basins, which includes the District of Columbia and parts of Maryland, Pennsylvania, Virginia, and West Virginia (Lindsey and Ator, 1996). Radon concentrations in this study area are greater than in many areas of the United States. In this study samples were collected from 267 wells. Eighty percent of the groundwater samples collected contained radon at concentrations greater than 300 pCi/L, and 31 percent of the samples contained radon at concentrations greater than 1000 pCi/L. The maximum radon concentration found in water from the 267 wells was 38,000 pCi(1406 Bq)/L.

15.3.4 *MANAGEMENT STRATEGIES AND EFFECTIVENESS*

Introduction

Since the mid-1980s various agencies have established and pursued a public strategy designed to reduce human exposure to radon and radon daughters. This subsection identifies actions taken under this strategy and makes general assessments about their effectiveness.

Mines

There are established standards currently in force in underground mines. The standards are based on α-radiation as measured in WL, and are based on the total of radon daughter and thoron daughter concentrations. The rationales for the standards are based partly on presumed health effects and partly on ensuring that underground mines are adequately ventilated and that the ventilation system is working adequately. Important criteria include:

- That maximum α-radiation concentrations not exceed 1.00 WL in mine areas subject to occupational exposure.
- That the annual total α-radiation exposure for an employee must not exceed 4.0 WLM in any calendar year.
- That the lifetime occupational α-radiation exposure not exceed 120 WLM.

The α-radiation standards in the mining industry have been logical and feasible to implement; they appear to work well and protect employee health. Management of α-radiation in mines generally can be accomplished by good design and operation of ventilation systems.

Caves

The National Park Service initially discovered the α-radiation issue when a few measurements were made in Carlsbad Caverns. The initial approach of the Park Service was to quickly adopt standards for this one cave system similar to those in force in mines. Only later did the Park Service monitor other caves that they operate and discover that α-radiation is ubiquitous and that some of their other caves have typical α-radiation concentrations greater than those in Carlsbad Caverns.

Unlike mines, α-radiation concentrations cannot be controlled in caves by ventilation for both technical and environmental reasons. Once the Park Service recognized this situation they responded to the issue with an extensive α-radiation monitoring program and a linked program of record keeping of employee exposure. The monitoring program slightly increased total employee exposure to α-radiation by increasing the workforce to include the people conducting the monitoring. The monitoring results led to few if any reductions in the duration of underground tours or other underground activities within the Park Service caves. The record keeping does ensure that no employee exceeds 4 WLM per year, but any health risk associated with α-radiation is now spread more broadly among employees. The Park Service banned smoking in the caves they operate and ended a practice at Mammoth Cave of using cave air to cool the agency's office building; these actions were beneficial to employee health and to the environmental integrity of the caves. The Park Service informed employees of the radiation issue and funded a sputum cytology study of long-time cave workers. The study did not find evidence of increased lung cancer risk from occupational exposure to radiation in caves.

Privately owned show caves, through the National Caves Association, established a set of precautionary standards designed to minimize α-radiation exposure to cave air by employees. These precautionary standards specified that cave air not be used for heating or cooling surface buildings. Instead of requiring α-radiation monitoring and recording of underground work time by employees, the National Caves Association precautionary standards limit employee work in caves to a maximum of 700 h per year unless radiation monitoring is conducted. The precautionary standards also prohibited smoking in caves and discouraged unnecessary underground activities (such as underground lunch rooms and souvenir shops, which are found in two of the larger National Park Service caves).

Neither of the management strategies adopted by the National Park Service and the National Caves Association have appreciably reduced α-radiation exposure to employees. The agency approach is the more expensive strategy. The adopted strategies are consistent with a conclusion by public and private cave managers that the potential health risk of α-radiation in caves is not sufficient to warrant corrective actions. Furthermore, such managers are aware that corrective actions would damage cave resources and in many cases would be expensive or not technically feasible. There is no scientific or management support for closing show caves to public visitation because of concern about α-radiation.

Water Supplies

EPA has recommended that domestic water supplies not contain radon concentrations greater than 300 pCi/L. Radon in water supplies typically is an insignificant or very minor source of radon in homes. Furthermore, most water supplies probably do not exceed the recommended radon concentration. There are water treatment approaches that can be used in cases in which radon concentrations in water are great enough to warrant attention. Currently there is little regulatory or management attention given to radon control in household water, and this level of attention is consistent with the available data.

Buildings

Although estimates vary, a common estimate used by regulatory agencies is that 6 percent of American homes (about 6 million structures) have mean radon concentrations in excess of 4 pCi/L. EPA recommends that all homes receive at least some monitoring for radon in an effort to identify the 6 percent that may warrant remediation. In most cases, the cost of such monitoring is paid by home owners with much or all the monitoring being done by independent contractors. The chance of the measured value being dramatically different from the mean exposure rate of house residents is great with inexpensive monitoring. Credible results should include both winter and summer monitoring and sampling at multiple locations within a house. Agency documents developed for the general public and many press accounts suggest that homes can be monitored for radon for $100 or less. In contrast, monitoring likely to yield reasonably accurate estimates of mean radon concentrations encountered by house residents is likely to cost closer to $1000 or more.

In 1992 the EPA complained that only about 5 percent of Americans had conducted radon monitoring in their homes. The percentage has undoubtedly increased a small amount since 1992 but probably does not exceed 8 percent. Homeowners may question the prudence of spending $100 to crudely test for a problem that agency personnel claim affects only 6 percent of American houses, and they may be even less willing to spend $1000 to test for the 6 percent problem. Furthermore, if homeowners discover concentrations greater than the EPA recommended maximum they must decide whether to attempt a remediation or ignore the situation. If they have monitored and found radon concentrations greater than 4 pCi/L and then decide to sell the house they are under a legal obligation to disclose the radon concentrations as a known house defect. As a result of these situations and others, it is not surprising that few homeowners have conducted radon monitoring even under the influence of dire health warnings from agencies.

Radon concentrations commonly can be reduced in existing homes, but costs are extremely variable and cost estimates are often into the range of several thousand dollars per house. Some estimates suggest that the total cost of radon monitoring and remediation to meet indoor air concentrations of 4 pCi/L or less would be $50 billion with essentially all these costs borne by homeowners. In some cases expensive remediation efforts have decreased radon concentrations in homes but failed to lower some or many of the concentrations to values below 4 pCi/L.

Common approaches for decreasing radon concentrations in existing homes include the following:

- Pressurizing and ventillating to outside the crawl space in crawl space homes.
- Blocking the entry of air through basement floors and walls and into the house. In some cases this is accomplished by installing collection pipes beneath basement floors or slabs and venting them to locations outside the house. In cases where there is a clean aggregate under a basement floor holes can sometimes be drilled through the concrete and into the aggregate; vent pipes are then installed in the holes and connected to locations outside the home. In homes with cinder blocks on the basement walls, holes can be drilled through the walls to intersect wall cavities. Vent pipes are then installed to connect the wall cavities with outside locations. Fans are needed with all these approaches.
- Sealing cracks in basement walls, foundations, and slabs to reduce air entry points. Water lines and other utility lines, which may be backfilled with permeable materials, are sealed where they enter the house.
- Fitting air-tight covers on basement sumps and venting them to the outside with fans for suction.
- Avoiding depressurization of houses by providing furnaces, fireplaces, wood stoves, and other combustion sources with outside air supplies. Keeping dampers in chimneys tightly closed and preferably sealed.
- Where existing perforated drain tiles surround the house a riser may be placed on the outlet pipe and a water trap added; a fan is then used to draw out the radon.

Common recommendations for reducing radon concentrations in new home construction include the following:

- Designing and constructing the house to maintain a neutral pressure differential between indoors and outdoors and minimize pathways for soil gas to enter the house.
- Including constructed features that will facilitate radon removal after completion of the home if prevention techniques prove to be inadequate.

The percentage of homes in the United States that have been monitored for radon is 8 percent or less. If only 6 percent of the homes yield values that exceed the EPA recommended radon concentration of 4 pCi/L, and if this percent is generally reflective of the homes monitored, it is likely that many of these homes do not greatly exceed the recommended concentration value.

It seems likely that many homeowners who have chosen to monitor their homes, and who have been informed that radon concentrations from their particular samplers exceed the recommended 4 pCi/L value, have chosen not to remediate the home. This author's review failed to find any estimate of the number of homes remediated or of the number of new homes where radon reduction efforts were a part of the design. It seems likely that the number of remediated homes is small and that the extent of radon reduction efforts in new homes is similarly small. In the author's opinion the values are probably less than 2 or 3 percent of American homes.

EPA created and funded a Radon Division and an Office of Radiation and Indoor Air to respond to the radon issue. Publicity and advertising about the radon issue has been extensive, and some estimates suggest that consumers have spent over $1 billion in response to the radon issue. The effectiveness of efforts to reduce the exposure of people to household radon concentrations has been minimal. The strategies used have worked poorly with a potential hazard that is found everywhere and that cannot be seen, tasted, or smelled. The radon issue and public response to it may provide some valuable insight for those involved in other issues where somewhat similar conditions exist.

REFERENCES

Ahlstrand, G. M., and Fry, P. L., "Alpha Radiation Project at Carlsbad Caverns: Two Years and Still Counting," *Proc. Natl. Cave Manage. Symp.*, Big Sky, MT, pp. 133–137, 1977.

Cole, L. A., *Element of Risk: the Politics of Radon*, Oxford University Press, 1993.

Gunderson, L. C. S., Schumann, R. R., and White, S. W., *The USGS/EPA Radon Potential Assessments: an Introduction*, U.S. Geological Survey Open-File Report 93-292, 1993.

Lindsey, B. D., and Ator, S. W., *Radon in Ground Water of the Lower Susquehanna and Potomac River Basins*, U.S. Geological Survey Water Resources Investigative Report 96-4156, 1996.

Martell, E. A., "Tobacco, Radioactivity, and Cancer in Smokers," *Am. Sci.*, 63: 404–412, 1975.

Misaqu, F. L., *Monitoring Radon-222 Content of Mine Waters*, Information Report 1026, U.S. Department of the Interior, Mining Enforcement and Safety Administration, undated.

U.S. Environmental Protection Agency, *Radon Reduction Techniques for Detached Houses: Technical Guidance*, EPA/625/5-86/019, 1986.

U.S. Environmental Protection Agency, *Radon Reduction Methods: a Homeowner's Guide* (second edition), 1987.

U.S. Environmental Protection Agency, *Radon Reduction in New Construction: an Interim Guide*, OPA-87-009, 1987.

U.S. Environmental Protection Agency, *EPA's Map of Radon Zones, Missouri*, 402-R-93-045, 1993.

Webster, J. W., and Crawford, N. C., *Preliminary Results of an Investigation of Radon Levels in the Homes and Caves of Bowling Green, Warren County, Kentucky*, Western Kentucky University, Radon Workshop Manual, 1987.

Yarborough, K. A., "Airborne Alpha Radiation in Natural Caves Administered by the National Park Service," *Proc. Natl. Cave Manage. Symp.*, Big Sky, MT, pp. 125–132, 1997.

CHAPTER 15
UBIQUITOUS ENVIRONMENTAL CONTAMINANTS

SECTION 15.4

CHLOROFLUOROCARBONS

S. Fred Singer

Dr. Singer is an atmospheric physicist, a professor emeritus of environmental sciences at the University of Virginia, and the president of the Fairfax-based Science & Environmental Policy Project, a nonprofit policy institute.

15.4.1 INTRODUCTION

What are CFCs and Halocarbons?

Chlorofluorocarbons (CFCs) are a class of halocarbons (HC). HCs are compounds that contain halogens (chlorine, fluorine, bromine, and/or iodine) and carbon. There are natural HCs, manmade HCs (like CFCs), and some that are both manmade and natural [like methyl bromide (MeBr)]. HCs are present in the atmosphere in low concentrations, measured in parts per million or trillion by volume (ppmv or pptv); they can play an important role in stratospheric ozone depletion and are often powerful greenhouse gases producing a radiative forcing (measured in W/m^2) often thousands of times greater (per molecule) than carbon dioxide. A comprehensive review (World Meteorological Organization, 1994) provides details about global distributions, trends, emissions, and lifetimes of CFCs, halons, and related species (see Table 15.4.1).

Recent trends of these atmospheric trace gases are important in understanding stratospheric ozone depletion and changes in the current radiative forcing of climate. Estimates of budgets and lifetimes are required to predict future impacts. Likewise, these data are needed to accurately predict what levels of emission reductions are needed to stabilize and/or reduce present concentrations.

In this review, we will deal with gases emitted by natural and/or anthropogenic sources that influence the chemical composition of the atmosphere. It includes long-lived gases that contribute to stratospheric ozone depletion [i.e., CFCs, halons, nitrous oxide (N_2O)] and/or radiative forcing of the atmosphere [i.e., carbon dioxide (CO_2), CFCs, methane (CH_4), N_2O], and short-lived compounds that are involved in the O_3 chemistry of the troposphere [i.e., carbon monoxide (CO), nitrogen oxides (NO_x), nonmethane hydrocarbons].

History of CFCs

CFCs were developed at the DuPont Chemical Company in the 1930s and found immediate commercial applications based on their unusual properties. They are nontoxic and noncorrosive, and they have thermodynamic characteristics that make them useful for refrigeration and air conditioning. Their inertness made them ideal as an aerosol dispenser and as a carrier of medication. They are widely used in asthma inhalers. Because of their chemical inertness, they have a rather long lifetime in the Earth's

TABLE 15.4.1 Some of the Most Important HCs, Their Chemical Formulas, and Other Properties: Atmospheric Concentrations and Lifetimes (Total Global Burden of Gas Divided by Globally Integrated Sink Strength), Current Growth Rates, and Radiative Forcing, Which May Be Compared with That of Carbon Dioxide, the Most Important Greenhouse Gas

Species	Lifetime Year	Lifetime Uncertainty (percent)	Concentration (ppbv) 1992	Concentration (ppbv) Preindustrial	Current growth ppbv/yr	Radiative forcing Wm^{-2}/ppbv	Radiative forcing Wm^{-2}
Natural and anthropogenically influenced gases							
Carbon dioxide (CO_2)	Variable		356,000	278,000	1600	1.8×10^{-5}	1.56
Methane (CH_4)*	12.2	25	1714	700	8	3.7×10^{-4}	0.47
Nitrous oxide (N_2O)	120		311	275	0.8	3.7×10^{-3}	0.14
Methyl chloride (CH_3Cl)	1.5	25	~0.6	~0.6	~0		0
Methyl bromide (CH_3Br)	1.2	32	0.010	<0.010	~0		0
Chloroform ($CHCl_3$)	0.51	300	~0.012		~0	0.017	
Methylene chloride (CH_2Cl_2)	0.46	200	~0.030		~0	0.03	
Carbon monoxide (CO)	0.25		50–150		~0		†
Gases phased out before 2000 under the Montreal Protocol and its amendments							
CFC-11 (CCl_3F)	50	10	0.268	0	+0.000‡	0.22	0.06
CFC-12 (CCl_2F_2)	102		0.503	0	+0.007‡	0.28	0.14
CFC-113 (CCl_2FCClF_2)	85		0.082	0	0.000‡	0.28	0.02
CFC-114 ($CClF_2CClF_2$)	300		0.020	0		0.32	0.007
CFC-115 (CF_3CClF_2)	1700		<0.01	0		0.26	<0.003
Carbon tetrachloride (CCl_4)	42		0.132	0	−0.0005‡	0.10	0.01
Methyl chloroform (CH_3CCl_3)	4.9	8	0.135§	0	−0.010‡	0.05	0.007
Halon-1211 ($CBrClF_2$)	20		0.007	0	.00015		
Halon-1301 ($CBrF_3$)	65		0.003	0	.0002	0.28	
Halon-2402 ($CBrF_2CBrF_2$)	20		0.0007	0			
Chlorinated HCs controlled by the Montreal Protocol and its amendments							
HCFC-22 ($CHClF_2$)	12.1	20	0.100	0	+0.005‡	0.19	0.02
HCFC-123 (CF_3CHCl_2)	1.4	25		0		0.18	
HCFC-124 (CF_3CHClF)	6.1	25		0		0.19	
HCFC-141b (CH_3CFCl_2)	9.4	25	0.002	0	0.001‡	0.14	
HCFC-142b (CH_3CF_2Cl)	18.4	25	0.006	0	0.001‡	0.18	
HCFC-225ca ($C_3HF_5Cl_2$)	2.1	35		0		0.24	
HCFC-225cb ($C_3HF_5Cl_2$)	6.2	35		0		0.28	
Perfluorinated compounds							
Sulfur hexafluoride (SF_6)	3200		0.032	0	+0.0002	0.64	0.002
Perfluoromethane (CF_4)	50,000		0.070	0	+0.0012	0.10	0.007
Perfluoroethane (C_2F_6)	10,000		0.004	0		0.23	
Perfluoropropane (C_3F_8)	2600			0		0.24	
Perfluorobutane (C_4F_{10})	2600			0		0.31	
Perfluoropentane (C_5F_{12})	4100			0		0.39	
Perfluorohexane (C_6F_{14})	3200			0		0.46	
Perfluorocyclobutane ($c\text{-}C_4F_8$)	3200			0		0.32	
Anthropogenic greenhouse gases not regulated (proposed or in use)							
HFC-23 (CHF_3)	264	45				0.18	
HFC-32 (CH_2F_2)	5.6	25				0.11	
HFC-41 (CH_3F)	3.7					0.02	
HFC-43-10mee ($C_5H_2F_{10}$)	17.1	35				0.35	
HFC-125 (C_2HF_5)	32.6	35				0.20	
HFC-134 (CF_2HCF_2H)	10.6	200				0.18	
HFC-134a (CH_2FCF_3)	14.6	20				0.17	
HFC-143 (CF_2HCH_2F)	3.8	50				0.11	
HFC-143a (CH_3CF_3)	48.3	35				0.14	

(Continued)

TABLE 15.4.1   Some of the Most Important HCs, Their Chemical Formulas, and Other Properties: Atmospheric Concentrations and Lifetimes (Total Global Burden of Gas Divided by Globally Integrated Sink Strength), Current Growth Rates, and Radiative Forcing, Which May Be Compared with That of Carbon Dioxide, the Most Important Greenhouse Gas (Continued)

Species	Lifetime Year	Lifetime Uncertainty (percent)	Concentration (ppbv) 1992	Concentration (ppbv) Preindustrial	Current growth ppbv/yr	Radiative forcing Wm^{-2}/ppbv	Radiative forcing Wm^{-2}
HFC-152a (CH_3CHF_2)	1.5	25				0.11	
HFC-227ea (C_3HF_7)	36.5	20				0.26	
HFC-236fa ($C_3H_2F_6$)	209	50				0.24	
HFC-245ca ($C_3H_3F_5$)	6.6	35				0.20	
HFOC-125e (CF_3OCHF_2)	82	300					
HFOC-134e (CHF_2OCHF_2)	8	300					
Trifluoroiodomethane (CF_3I)	<0.005					0.38	

Notes: This table lists only the direct radiative forcing from emitted gases. The indirect effects due to subsequent changes in atmospheric chemistry, notably ozone (see below), are not included. The Wm^{-2} column refers to the radiative forcing since the preindustrial, and the Wm^{-2}/ppbv column is accurate only for small changes about the current atmospheric composition. In particular, CO_2, CH_4, and N_2O concentration changes since preindustrial times are too large to assume linearity; the formulae reported in IPCC(1990) are used to evaluate their total contribution.

A blank entry indicates that a value is not available. Uncertainties for many lifetimes have not been evaluated. The concentrations of some anthropogenic gases are small and difficult to measure. The preindustrial concentrations of some gases with natural sources are difficult to determine. Radiative forcings are given only for those gases with values greater than 0.001 Wm^{-2}.

*Methane increases are calculated to cause increases in tropospheric ozone and stratospheric H_2O; these indirect effects, about 25 percent of the direct effect, are not included in the radiative forcings given here.

[†]The direct radiative forcing due to changes in the CO concentration is unlikely to reach a few hundredths of a Wm^{-2}. The direct radiative forcing is hard to quantify.

[‡]Gases with rapidly changing growth rates over the past decade; recent trends since 1992 are reported.

[§]Change in CH_3Cl_3 concentration is due to recalibration of the absolute standards used to measure this gas.

Stratospheric ozone depletion due to HCs is about −2 percent (globally) over the period 1979–1990 with half as much again occurring both immediately before and since; the total radiative forcing is thus now about −0.1 Wm^{-2}. Tropospheric ozone appears to have increased since the nineteenth century over the northern midlatitudes where few observational records are available; if, over the entire northern hemisphere, tropospheric ozone increased from 25 ppb to 50 ppb at present, then the radiative forcing is about +0.4 Wm^{-2}.

Source: Adapted from IPCC WGI (1996).

atmosphere. They are broken down at a slow rate by soil bacteria, and they are also decomposed by solar ultraviolet radiation once they reach the Earth's stratosphere.

Montreal Protocol: History and Decisions

In the early 1970s, James Lovelock started to measure atmospheric concentrations of CFCs and recognized that they were increasing. In 1974, Rowland and Molina published a seminal paper pointing out that CFCs could destroy stratospheric ozone. Once the molecules percolated to a sufficient altitude in the stratosphere, they would be photodissociated. The chlorine atoms thus released would enter into a catalytic reaction as shown below:

$$Cl + O_3 \rightarrow O_2 + ClO$$

followed by

$$ClO + O \rightarrow Cl + O_2$$

The net result of this catalytic reaction set is

$$O_3 + O \rightarrow 2O_2$$

Before, in 1970–1971, the environmental community had been sensitized to the possibility of destruction of stratospheric ozone by exhaust gases from supersonic transport (SST) aircraft. Ozone acts as a partial shield for solar ultraviolet (UV)-B radiation (280 to 320 nm), and as a complete shield for the shorter wave UV-C. Long-term exposure to UV-B can produce skin cancers in fair-skinned individuals. It was the fear of skin cancer that led to the abandonment of the American SST project in 1971. Ironically, in the same year, Singer (1971) published his calculations demonstrating that human activities on the Earth's surface were already capable of affecting stratosphere ozone. He estimated that about half the methane injected into the atmosphere came from rice growing and cattle raising; the other half was from natural sources like swamps and wetlands. Based on available measurements, he argued that the long-lived anthropogenic methane would percolate into the stratosphere, to be oxidized and serve as a source of additional water vapor, which in turn would deplete ozone.

Once it was recognized that CFCs had the potential of depleting stratospheric ozone, it led to a clamor for the phaseout and eventual ban of these industrially manufactured chemicals. (Methane, being a naturally occurring molecule, was never considered a threat.) Indeed, by 1976, the United States and Sweden were banning the use of CFCs as a pressurizing gas in aerosol spray cans, where cost-effective substitutes were readily available. Other applications were not affected at the time, particularly as more detailed studies about ozone depletion gave less threatening results. For example, although the initial estimates gave approximately 20 percent depletion, by 1985 the depletion level was calculated to be only a couple of percent, well within the noise level of the natural variations of ozone (on a decadal time scale).

The discovery of the Antarctic ozone hole in 1985 produced an important change in thinking. Although not predicted by the original (homogeneous) theory, it nevertheless demonstrated that chlorine could destroy ozone by heterogeneous reactions in the presence of stratospheric particulates. In the case of the Antarctic, these were ice particles forming polar stratospheric clouds. The fear of a spreading ozone hole, engulfing the whole globe, soon led to international action on CFCs.

The Montreal Protocol was concluded in 1987. Initially, it called for a cap on CFC production, with subsequent reductions. By 1992, a complete ban on the production of CFC and certain HCs was agreed to by industrialized nations. Dates of the important conferences and scinetific assessments are given in Table 15.4.2. Figure 15.4.1 shows how annual emissions have decreased as a result of international agreements.

TABLE 15.4.2 List of Conventions, Protocols, Amendments and Scientific Assessments Relating to HCs

Year	Policy process	Scientific assessment
1981		*The Stratosphere 1981. Theory and Measurements.* WMO No. 11.
1985	Vienna Convention	*Atmospheric Ozone 1985.* Three volumes. WMO No. 16.
1987	Montreal Protocol	
1988		*International Ozone Trends Panel Report 1988.* Two volumes. WMO No. 18.
1989		*Scientific Assessment of Stratospheric Ozone: 1989.* Two volumes. WMO No. 20.
1990	London Adjustments and Amendment	
1991		*Scientific Assessment of Ozone Depletion: 1991.* WMO No. 25.
1992		*Methyl Bromide: Its Atmospheric Science, Technology, and Economics* (assessment supplement). UNEP (1992).
1992	Copenhagen Adjustments and Amendment	
1994		*Scientific Assessment of Ozone Depletion: 1994.* WMO No. 37.
1995	Vienna Adjustment	
1997	Montreal Adjustments and Amendment	
1998		*Scientific Assessment of Ozone Depletion: 1998.* WMO No. 44.
1999	Eleventh Meeting of the Parties (China)	

Source: Adapted from IPCC WGI (1996).

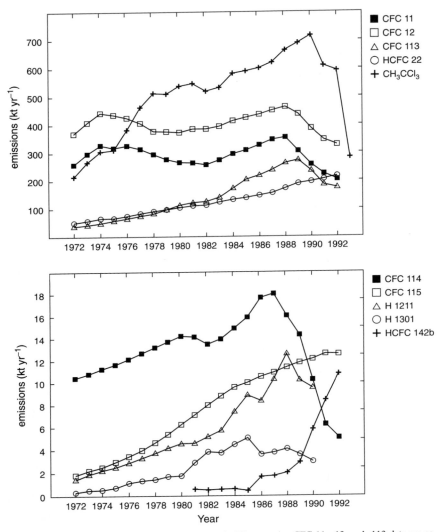

FIGURE 15.4.1 Annual emissions of halocarbons in kilotonnes/yr. CFC-11, -12, and -113 data are estimates of global emissions; remaining estimates are based on data only from reporting companies (World Meteorological Organization, 1994).

CFC Substitutes

With the ban on CFC production, it became necessary to use substitutes that had a smaller effect on stratospheric ozone—a lower ozone depletion potential (ODP). A common feature of these hydrochlorofluorocarbons (HCFCs) and hydrofluorocarbons (HFCs) is their short lifetime in the atmosphere, which reduces the amount entering the stratosphere. In addition, the fluorine atom in HFCs is not an effective ozone destroyer. The short lifetime implies less inertness, however, and therefore a potential for corrosion and toxicity. Nevertheless, the financial investments were made in the United States and other industrialized nations to switch to substitutes, which required redesigning or refitting of refrigeration and air-conditioning equipment. The high cost of the changeover, estimated at about $100 billion was raised further by the suddenness of the CFC ban, which in 1992 was advanced from the year 2000 to 1995 because of environmental activism.

Present Status and Projections

At the present time, the ban on CFC production is in effect in most industrialized nations, but emissions into the atmosphere are still going on. Developing nations and Eastern-block nations have not been constrained by the Montreal Protocol. In addition, large amounts of CFC are being smuggled into the United States and sold below the price of the substitutes. Because of their long atmospheric lifetime, CFC concentration in the troposphere—and in the stratosphere—will continue to increase during the next few decades but at a slower rate. Table 15.4.3 gives estimates for annual fluxes into the atmosphere and for concentrations. Note that fluxes of CFCs are projected to fall to zero between 2000 and 2010 and that HCFC fluxes will essentially disappear by 2020; however, concentrations of CFCs will in many cases still be greater than they were in 1970–1980. Also, note that, according to this projection, ozone depletion reaches a maximum value of 3.5 percent in 1995 and falls to zero after 2055. For the sake of comparison, we note that the natural variability of total ozone with the 11-year solar cycle is about 3 percent at high latitudes. (The seasonal variability can be of the order of 100 percent; the day-to-day variability can be even greater.)

15.4.2 SOURCES OF HCs

CFCs, halons, HCFCs, and HFCs are exclusively of industrial origin, as are methyl chloroform and carbon tetrachloride. Their primary uses are as refrigerants (CFC-11, -12, and -114; HCFC-22; HFC-134a), foam-blowing agents (HCFC-22 and -142b; CFC-11 and -12), solvents and feed-stocks (CFC-113, carbon tetrachloride, and methyl chloroform), and fire retardants (halons-1211 and -1301). Worldwide emissions of individual halocarbons decreased significantly after enactment of the Montreal Protocol (see Figure 15.4.1). From 1988 to 1992, annual emissions of CFCs into the atmosphere decreased by approximately 34 percent. Emissions of methyl chloroform also declined over this period, with a very large emission reduction ($\sim$50 percent) from 1992 to 1993. HCFC-22 has shown a continued increase; it has replaced CFCs in many applications, as its use has not been restricted by the Protocol. HCFC-142b release began in the early 1980s and is growing rapidly (35 percent per year since 1987). Emissions in 1992 were 10.8×10^6 kg. Emissions are difficult to estimate accurately as the bulk of the production (90 percent) is used in closed-cell foams with residence times greater than 10 years. Between 1987 and 1990, emissions of halons dropped substantially (no data are available for 1991 and 1992). No emission data are available for carbon tetrachloride as its production and use has not been surveyed. A detailed update of the anthropogenic emissions of methyl bromide is given in Section 15.5.

CH_2Cl_2 and $CHClCCl_2$ are used as industrial cleaning solvents. Sources of 0.9 and 0.6 Tg/yr have been recently estimated from observed atmospheric abundances.* Industry estimates of 1992 emissions for CCl_2CCl_2, $CHClCCl_2$, and CH_2Cl_2 were 0.24, 0.16, and 0.39 Tg, respectively. Total emissions for these species have declined by 40 percent since 1982. The aluminum refining industry produces CF_4 (0.018 Tg/yr) and C_2F_6 (0.001 Tg/yr); however, there are no estimates of other potential sources. With respect to SF_6 production, 80 percent (0.005 Tg in 1989) is used for insulation of electrical equipment, 5 to 10 percent for degassing molted reactive metals, and a small amount as an atmospheric tracer. The measured rate of increase of SF_6 in the atmosphere implies that its sources are increasing.

Methyl halides are produced during biomass burning. Annual emissions of 1.5 to 1.8 Tg/yr and 30 Gg/yr have been estimated for CH_3Cl and CH_3Br, respectively. A major source of methyl halides

---

*1 teragram (Tg) $= 10^{12}$ g $=$ 1 million metric tons
  1 gigagram (Gg) $= 10^9$ g $=$ 1000 metric tons

TABLE 15.4.3 Scenarios 1970–2100: Concentrations and Annual Fluxes

Tropospheric mixing ratios (all units pptv, except dO$_3$, which is the global mean ozone depletion)

Year	CFC-11	CFC-12	CFC-113	CFC-114	CFC-115	CCl$_4$	CH$_3$CCl$_3$	HCFC-22	HCFC-141b	HCFC-123	tropCl	dO$_3$ (percent)
1970	60	120	2	1	0	105	35	10	0	0	1563	0.0
1975	115	205	6	2	1	111	60	25	0	0	2027	−0.3
1980	173	295	15	4	2	118	85	50	0	0	2541	−1.2
1985	222	382	30	8	4	126	110	70	2	0	3044	−2.0
1990	263	477	77	19	5	133	133	91	5	4	3643	−3.1
1995	291	532	92	20	7	133	122	148	13	9	3922	−3.5
2000	289	545	97	20	8	120	61	241	16	8	3844	−3.4
2005	267	526	93	20	9	108	30	292	16	6	3644	−3.1
2010	242	501	88	20	9	96	12	299	13	3	3401	−2.7
2015	219	477	83	20	9	85	4	248	8	1	3140	−2.2
2020	198	454	78	19	9	76	2	178	5	0	2889	−1.8
2025	179	433	74	19	9	67	1	119	3	0	2670	−1.5
2030	162	412	70	19	8	60	0	79	2	0	2489	−1.2
2035	147	392	66	18	8	53	0	52	1	0	2334	−0.9
2040	133	373	62	18	8	47	0	35	1	0	2200	−0.7
2045	120	356	58	18	8	42	0	23	0	0	2081	−0.5
2050	109	339	55	17	8	37	0	15	0	0	1976	−0.3
2055	98	322	52	17	8	33	0	10	0	0	1880	−0.1
2060	89	307	49	17	8	29	0	7	0	0	1793	0.0
2065	81	292	46	17	8	26	0	4	0	0	1714	0.0
2070	73	278	44	16	8	23	0	3	0	0	1641	0.0
2075	66	265	41	16	8	20	0	2	0	0	1575	0.0
2080	60	252	39	16	8	18	0	1	0	0	1513	0.0
2085	54	240	36	15	8	16	0	1	0	0	1456	0.0
2090	49	229	34	15	8	14	0	1	0	0	1403	0.0
2095	44	218	32	15	8	14	0	0	0	0	1355	0.0
2100	40	207	31	15	8	11	0	0	0	0	1309	0.0
Lifetime (year)	50	102	85	300	1700	42	4.9	12.1	9.4	1.4		
factor (Gg/pptv)		22.6	20.8	32.5	29.7	27.1	25.3	22.0	14.9	26.3	20.1	

Annual flux (kilotonnes/yr)

	CFC-11	CFC-12	CFC-113	CFC-114	CFC-115	CCl$_4$	CH$_3$CCl$_3$	HCFC-22	HCFC-141b	HCFC-123
1990–1994	250	330	132	6	13	80	520	319	26	52
1990–1999	125	165	66	6	6	10	120	520	65	130
2000–2004	25	30	13	1	1	8	54	484	60	121
2005–2009	2	2	1	0	0	0	6	383	42	85
2010–2014	0	0	0	0	0	0	0	182	23	46
2015–2019	0	0	0	0	0	0	0	53	7	13
2020–2024	0	0	0	0	0	0	0	3	0	1
2025–2029	0	0	0	0	0	0	0	3	0	0
2030–	0	0	0	0	0	0	0	0	0	0

Source: Adapted from IPCC WGI (1996).

appears to be the marine/aquatic environment, likely associated with algal growth. Methyl chloride, present in the troposphere at about 600 pptv, is the most prevalent halogenated methane in the atmosphere. Maintaining this steady-state mixing ratio with an atmospheric lifetime on the order of 2 years requires production of around 3.5 Tg/yr, most of which comes from the ocean and biomass burning. Other halogenated methanes, such as $CHBr_3$, $CHBr_2Cl$, and CH_2CBr_2, are produced by macrophytic algae (seaweeds) in coastal regions and possibly by phytoplankton in the open ocean, but they do not accumulate significantly in the atmosphere.

Methyl chloride, released from the oceans (natural) and biomass burning (anthropogenic), is a significant source of tropospheric chlorine, contributing about 15 percent of the total tropospheric chlorine abundance in 1992 [3.8 parts per billion by volume (ppbv)]. Data collected from the late 1970s to the mid-1980s showed no long-term trend. A paucity of published observational data since then means that the likely existence of a global trend in this important species cannot be assessed further.

Volcanoes are an insignificant source of stratospheric chlorine. Satellite and aircraft observations of upper and lower stratospheric hydrochloric acid (HCl) are consistent with stratospheric chlorine being organic, largely anthropogenic, in origin. No significant increase in HCl was found in the stratosphere after the intense eruption of Mt. Pinatubo in 1991. Elevated HCl levels were detected in the eruption cloud of the El Chichón volcano in 1982, but no related change in global stratospheric HCl was observed.

15.4.3 HALOCARBON SINKS

Fully halogenated halocarbons are destroyed primarily by photodissociation in the mid-to-upper stratosphere. These gases have atmospheric lifetimes of decades to centuries.

Halocarbons containing at least one hydrogen atom—such as HCFC-22, chloroform, methyl chloroform, the methyl halides, and other HCFCs and HFCs—are removed from the troposphere mainly by reaction with hydroxyl radicals OH. The atmospheric lifetimes of these gases range from years to decades, except for iodinated compounds, such as methyl iodide, which have lifetimes on the order of days to months. However, some of these gases also react with seawater. About 5 to 10 percent of the methyl chloroform in the atmosphere is lost to the oceans, presumably by hydrolysis. About 2 percent of atmospheric HCFC-22 is apparently destroyed in the ocean, mainly in tropical surface waters. (Methyl bromide sinks are discussed separately in Section 15.5.)

Recent studies show that carbon tetrachloride may be destroyed in the ocean. Widespread, negative saturation anomalies ($\sim$6 to $\sim$8 percent) of carbon tetrachloride, consistent with a subsurface sink, have been reported in both the Pacific and Atlantic Oceans. Published hydrolysis rates for carbon tetrachloride are not sufficient to support these observed saturation anomalies, which, nevertheless, indicate that about 20 percent of the carbon tetrachloride in the atmosphere is lost in the oceans.

Recent investigation of the atmospheric lifetimes of perfluorinated species CF_4, CF_3CF_3, and SF_6 indicates lifetimes of $>50,000$, $>10,000$, and 3200 years. Loss processes considered include photolysis, reaction with $O(^1D)$ (an excited state of the oxygen atom), combustion, reaction with halons, and removal by lightning.

15.4.4 LIFETIMES

Lifetimes are given in Table 15.4.1. The atmospheric residence times of CFC-11 and methyl chloroform are now fairly well known. Model studies simulating atmospheric abundances using more realistic emission amounts have led to best-estimated lifetimes of 50 years for CFC-11 and 5.4 years for methyl chloroform, with uncertainties of about 10 percent. These models, calibrated against CFC-11 and methyl chloroform, are used to calculate the lifetimes, and hence ODPs, of other gases destroyed only in the stratosphere (other CFCs and nitrous oxide) and those reacting significantly with tropospheric OH (HCFCs and HFCs).

The assessment and reevaluation of the empirical models used to derive the atmospheric residence lifetime of two major industrial halocarbons, CH_3CCl_3 and CFC-11, uses four components: observed concentrations, history of emissions, predictive atmospheric model, and estimation procedure for describing an optimal model. An optimal fit to the observed concentrations at the five Atmospheric Lifetime Experiment/Global Atmospheric Gases Experiment (ALE/GAGE) surface sites over the period 1978–1990 was done with two statistical/atmospheric models.* There are well-defined differences in these atmospheric models, which contribute to the uncertainty of derived lifetimes.

The lifetime deduced for CH_3CCl_3 is 5.4 years, with an uncertainty range of ±0.4 year. From this total atmospheric lifetime, the losses to the ocean and the stratosphere are used to derive a tropospheric lifetime for reaction with OH radicals of 6.6 year (±25 percent); this value is used to scale the lifetimes of HCFCs and HFCs. On the other hand, the semiempirical lifetime for CFC-11 of 50 ± 5 years provides an important transfer standard for species that are mainly removed in the stratosphere—i.e., the relative modeled lifetimes for CFCs, H-1301, and N_2O are scaled to a CFC-11 lifetime of 50 years.

Many HCs have short lifetimes (like methyl chloroform). As a result, the production ban has produced quick results by reducing the halogen load in the troposphere. Long-lived HCs (like CFCs) will diminish more slowly and therefore continue to contribute to stratospheric chlorine and ozone depletion.

15.4.5 GROWTH RATES

Tropospheric growth rates of the major anthropogenic source species for stratospheric chlorine and bromine (CFCs, carbon tetrachloride, methyl chloroform, halons) have slowed significantly in response to substantially reduced emissions required by the Montreal Protocol. Total tropospheric chlorine grew by about 60 pptv (1.6 percent) in 1992, compared with 110 pptv (2.9 percent) in 1989. Tropospheric bromine in the form of halons grew by 0.2 to 0.3 pptv in 1992, compared with 0.6 to 1.1 pptv in 1989.

HCFC growth rates are accelerating, as they are being used increasingly as CFC substitutes. Tropospheric chlorine as HCFCs increased in 1992 by about 10 pptv, thus accounting for about 15 percent of total tropospheric chlorine growth, compared with 5 pptv in 1989 (5 percent of tropospheric chlorine growth).

15.4.6 TROPOSPHERIC DISTRIBUTIONS AND TRENDS

Tropospheric measurements are mostly made in situ at fixed sites distributed between the two hemispheres and supplemented by data collected on ships and aircraft. Information on the free tropospheric burdens of atmospheric gases and their time variations has further been obtained by spectroscopic remote measurements made from various observational platforms. Significant changes in trends have been observed for most gases during the past few years. The total Cl increase in 1992 was ~60 pptv/yr, whereas the 1989 increase was ~110 pptv/yr (World Meteorological Organization, 1994).

CFCs and Carbon Tetrachloride

CCl_3F (CFC-11), CCl_2F2 (CFC-12), CCl_2FCClF_2 (CFC-113), and carbon tetrachloride (CCl_4) have been measured in a number of global programs and their tropospheric mixing ratios have been

---

*A more recent analysis of the ALE/GAGE data (1978–1991) using the ALE/GAGE model and a revised CFC-11 calibration scale gives an equilibrium lifetime for CFC-11 of 44 (+17/−10) years.

increasing steadily over the past 15 years. There is now clear evidence that the growth rates of the CFCs have slowed significantly in recent years (Figure 15.4.1), presumably because of reduced emissions. CFC-12 and CFC-11 trends in the late 1970s to late 1980s were about 16 to 20 and 9 to 11 pptv/yr, respectively. These declined to about 16 and 7 pptv/yr, respectively, around 1990, and to about 11 and 3 pptv/yr by 1993.

The global CFC-113 data up to the end of 1990 have been reviewed recently. A global average trend of about 6 pptv/yr was observed for CFC-113, with no sign of a slowing down, such as was observed for CFC-11 and -12. However, data up to the end of 1992 now indicate that the growth rate has started to decrease. Carbon tetrachloride appears to have stopped accumulating in the atmosphere and data collected at Cape Grim, Tasmania, indicate that the background levels of this trace gas may have actually started to decline.

Methyl Chloroform and the HCFCs

Global methyl chloroform (CH_3CCl_3) and HCFC-22 ($CHClF_2$) showed growth rates in 1990 equal to 4 to 5 and 6 to 7 pptv/yr, respectively. Methyl chloroform data up to the end of 1992 indicate that the slowing of the growth rate observed in 1990 has continued, presumably because of reduced emissions in 1991–1992 compared with 1990 and in part because of increasing OH levels (1 ± 0.8 percent/yr). The methyl chloroform calibration problems, however, have yet to be resolved.

Recent global HCFC-22 data indicate a global mixing ratio in 1992 of 102 ± 1 pptv, an interhemispheric difference of 13 ± 1 pptv, and a globally averaged growth rate of 7.3 ± 0.3 percent/yr, or 7.4 ± 0.3 pptv/yr, from mid-1987 to 1992. Based on the latest industry estimates of HCFC-22 emissions, the data indicate an atmospheric lifetime for HCFC-22 of 13.3 (15.5 to 12.1) years. Regular vertical column abundances measured by infrared solar absorption spectroscopy in Arizona (32°N), Switzerland (46.6°N), and California (34.4°N) have revealed rates of increase of 7.0 ± 0.2 percent/yr (1981–1992), 7.0 ± 0.5 percent/yr (1981–1992), and 6.5 ± 0.5 percent/yr (1985–1990), respectively. Using the HCFC-22 column abundances obtained at McMurdo, Antarctica (78'S), one derives a south-north interhemispheric growth rate ratio of 0.85, in good agreement with the ratio of 0.88 obtained from in situ surface measurements. The latest 1993 HCFC-22 data indicate that the near-linear trend observed in earlier data has continued.

CH_3CClF_2 (HCFC-142b) and CH_3CCl_2F (HCFC-141b) have been introduced as CFC substitutes. For HCFC-142b the National Oceanic and Atmospheric Administration (NOAA) flask network results indicate an atmospheric concentration of 3.1 pptv for 1992, with a growth rate of ~1 pptv/yr (~30 percent/yr). The concentration of HCFC-141b for the last quarter of 1992 was 0.36 pptv and 1.12 pptv at the end of 1993 (~0.75 pptv/yr or ~200 percent/yr). Upper tropospheric levels of HCFC-142b are at about 1.1 pptv in 1989, growing at 7 percent/yr.

Brominated Compounds

Recently, interest and understanding of brominated species in the background atmosphere have expanded considerably, driven by the recognition of the significant role of bromine in stratospheric ozone depletion.

The most abundant organobromine species in the lower atmosphere is methyl bromide (CH_3Br), which has both natural and anthropogenic sources. An evaluation of this important trace gas is given in Section 15.5.

The available halon-1211 ($CBrClF_2$) and halon-1301 ($CBrF_3$) data from the NOAA Climate Monitoring and Diagnostics Laboratory (CMDL) flask sampling network show that the global background levels are about 2.5 pptv (H-1211) and 2.0 pptv (H-1301), currently growing at about 3 percent/yr and 8 percent/yr, respectively. These rates have slowed significantly in recent years, consistent with reduced emissions, and their atmospheric mixing ratios may stabilize in a few years.

Data from the tropical Pacific Ocean indicate concentrations of dibromomethane (CH_2Br_2), bromoform ($CHBr_3$), and dibromochloromethane ($CHBr_2Cl$) of 1.8, 1.8, and 0.2 pptv, respectively.

Bromoform and dibromochloromethane show distinct equatorial maxima, indicating a tropical source related to natural biogenic activity.

Perfluorinated Species

Perfluorinated compounds have very long lifetimes (see Table 15.4.1) and strong infrared (IR) absorption characteristics (efficient greenhouse gases). The major loss process appears to be their photolysis in the upper stratosphere and the mesosphere.

The global mean concentration of carbon tetrafluoride (CF_4) was measured in 1979 at 70 ± 7 pptv. This gas was observed at the South Pole in the late 1970s and mid-1980s at about 65 and 75 pptv, respectively, growing at about 2 percent/yr. At northern midlatitudes in the 1980s, reported CF_4 and C_2F_6 concentrations were about 70 and 2 pptv, respectively.

Sulfur hexafluoride (SF_6) is a long-lived atmospheric trace gas that is about three times more effective than CFC-11 as a greenhouse gas. Current global background levels are 2 to 3 pptv, which are apparently increasing with time at about 8.3 percent/yr (surface measurements) and 9 ± 1 percent/yr (lower stratosphere measurements). IR column measurements in Europe (1986–1990) and North America (1981–1990) indicate increases of 6.9 ± 1.4 percent/yr and 6.6 ± 3.6 percent/yr, respectively.

Other Halogenated Species

Available data on the abundance of methyl chloride (CH_3Cl), chloroform ($CHCl_3$), dichloromethane (CH_2Cl_2), and chlorinated ethanes have recently been reviewed. No long-term trends of these species have been observed, although they all exhibit distinct annual cycles (summer minimum, winter maximum). These species are relatively short-lived in the atmosphere (see Table 15.4.1) and their contribution to ozone depletion and climate forcing is minimal.

Methyl chloride is a significant source of tropospheric chlorine. Data collected from the late 1970s to mid-1980s showed no long-term trends (Khalil and Rasmussen, 1985). Recent measurements of several of these gases have been made in the Atlantic (45°N to 30°S) and in the tropical Pacific Ocean. Methyl chloride showed practically no interhemispheric gradient, indicative of a large oceanic or tropical source, whereas chloroform, dichloromethane, tetrachloroethylene, and trichloroethylene showed higher concentrations in the northern hemisphere, likely because of anthropogenic emissions.

Measurements of methyl iodide and chloroiodomethane in the northwestern Atlantic Ocean indicate that the latter species may be as important as the former in transferring iodine from the oceans to the atmosphere.

15.4.7 STRATOSPHERIC VALUES

The total abundance of organic halocarbons in the lower stratosphere is well characterized by in situ and remote observations of individual species. Observed totals are consistent with abundances of primary species in the troposphere, suggesting that other source species are not important in the stratosphere. Loss of halocarbons is found as their residence time in the stratosphere increases, consistent with destruction by known photochemical processes. Since the loss of halocarbons produces inorganic chlorine and bromine species associated with ozone loss processes, these observations also constrain the abundance of these organic species in the lower stratosphere.

15.4.8 PROJECTIONS OF EMISSIONS AND OF LEVELS

These are shown in Table 15.4.3 and have been discussed earlier.

15.4.9 *WARMING POTENTIAL OF HALOCARBONS*

These are shown in Table 15.4.1.

15.4.10 *CONCLUSION*

As a result of the 1987 Montreal Protocol and subsequent agreements, the production of halocarbons has been phased out in industrialized nations. Included are not only CFCs but also halons used for firefighting. Certain CFC substitutes, like HCFCs, are to be phased out by about 2010. In spite of vehement opposition by food producers and transporters, methyl bromide, an important agricultural fumigant, is also listed to be phased out.

These actions illustrate one of the first cases of a concerted international effort to ban the manufacture of certain chemicals. [The case of dichlorodiphenyltrichloroethane (DDT) may be another example.] These policy actions are driven by the fear of a depletion of the stratospheric ozone layer, leading to an increase in solar UV radiation reaching the Earth's surface. The consequences of such an increase have been pictured as leading to epidemics of skin cancers, cataracts, and disturbances of immune systems.

The reality is rather different. The observed depletion of ozone may be as much as a few percent, although this is difficult to establish in view of the large natural variability. There has been no observation of any corresponding increasing trend in surface UV radiation. In any case, the expected increase would only be on the order of 5 to 10 percent. Such increases would correspond to those experienced in a move of about 100 km toward to the equator; the intensity of UV increases by about 5000 percent in moving from polar regions to the equator, because of the change in the average zenith angle of the sun. Needless to say, there have been no reports of any epidemics that can be linked to ozone depletion.

As a result, the policies flowing from the Montreal Protocol have been a costly exercise, about $100 billion for the United States, and perhaps greater for the rest of the world. One reason for the high cost has been the rapidity with which control policies were put into effect, well before a solid scientific base was established.

REFERENCES

IPCC WGI, *Climate Change 1995: The Science of Climate Change*, J. T. Houghton, L. G. Meira Filho, B. A. Callander, N. Harris, and K. Maskell, eds., Cambridge University Press, Cambridge, UK, 1996.

Khalil, M. A. K., and Rasmussen, R. A., "Atmospheric Carbon Tetrafluoride (CF_4): Sources and Trends," *Geophys. Res. Lett.*, 12: 671–672, 1985.

Singer, S. F.,"Stratospheric Water Vapour Increase Due to Human Activities," *Nature*, 234: 543–545, 1971.

World Meteorological Organization, *Scientific Assessment of Ozone Depletion*, WMO Report 37, Geneva, 1994.

CHAPTER 15
UBIQUITOUS ENVIRONMENTAL CONTAMINANTS

SECTION 15.5

METHYL BROMIDE

S. Fred Singer

Dr. Singer is an atmospheric physicist, a professor emeritus of environmental sciences at the University of Virginia, and the president of the Fairfax-based Science & Environmental Policy Project, a nonprofit policy institute.

15.5.1 INTRODUCTION

Methyl bromide (CH_3Br) occupies a special place among halocarbons for a number of reasons.

1. It is an important agricultural chemical, used as a fumigant for soils and for grain storage. No effective substitute is currently available to take its place. Because of its importance to agriculture and to food preservation and the lack of a replacement, the date of banning production has been postponed to 2005. Perhaps by then the wisdom of continuing the use of this valuable chemical will be recognized.

2. Unlike chlorofluorocarbons (CFCs), most of the CH_3Br emitted to the atmosphere comes from natural sources. The manmade contribution is estimated to be on the order of 40 percent.

3. Although bromine atoms are potential destroyers of stratospheric ozone, it has been difficult to establish their presence in the stratosphere. Certainly, there has been no reliable demonstration of a secular increase in stratospheric bromine (as there has been for chlorine compounds), which would indicate the importance of the human source.

4. As determined from measurements, the atmospheric lifetime of CH_3Br is only about 1 year and perhaps is even less. This is in contrast to CFCs, whose lifetime is measured in decades or even centuries. As a consequence, it would be relatively easy to stop production of CH_3Br, if deemed necessary in the future, and have it disappear from the atmosphere within a few years.

15.5.2 SOURCES OF CH_3Br

Four major sources for atmospheric CH_3Br have been identified: the ocean, which is a natural source; and three others which are almost entirely anthropogenic. These are agricultural usage; biomass burning, which is newly recognized; and exhaust from automobiles using leaded gasoline.

The estimated uncertainty range for these sources is large, with oceans ranging from 60 to 160 ktonnes/yr, agriculture from 20 to 60 ktonnes/yr, biomass burning from 10 to 50 ktonnes/yr, and automobile exhaust from 0.5 to 22 ktonnes/yr. In the latter case, the range results from two conflicting assessments, which yield 0.5 to 1.5 ktonnes/yr and 9 to 22 ktonnes/yr, respectively (see Table 15.5.1).

TABLE 15.5.1 CH$_3$Br Sales (Thousands of Tonnes)*

Year	Preplanting	Postharvesting	Structural	Chemical intermediates†	Total
1984	30.4	9.0	2.2	4.0	45.6
1985	34.0	7.5	2.3	4.5	48.3
1986	36.1	8.3	2.0	4.0	50.4
1987	41.3	8.7	2.9	2.7	55.6
1988	45.1	8.0	3.6	3.8	60.5
1989	47.5	8.9	3.6	2.5	62.5
1990	51.3	8.4	3.2	3.7	66.6
1991	55.1	10.3	1.8	4.1	71.2
1992	57.4	9.6	2.0	2.6	71.6

*Production by companies based in Japan, western Europe, and the United States.
†Not released into the atmosphere.

TABLE 15.5.2 Emission of CH$_3$Br (Thousands Tonnes/Year; Best Estimates)

Source	Strength	Range	Anthropogenic	Natural
Ocean*	90	60–160	0	90
Agriculture	35	20–60	35	0
Biomass burning	30	10–50	25	5
Gasoline additives†	1	0.5–1.5	1	0
	15	9–22	15	0
Structural purposes	4	4	4	0
Industrial emissions	2	2	2	0
Totals	162	97–278	67	95
	176	105–298	81	95

*Ocean source of 90 thousand tonnes per year is a gross source and is made up of two very uncertain quantities, and the most likely value and the range are expected to change markedly as a result of new research.
†The two values given for this source reflect the large difference in the two estimates discussed in the text.
    Source: Roskill Information Services Ltd., *the Economics of Bromine*, sixth edition London, 1992.

There are also two minor anthropogenic sources, structural fumigation (4 ktonnes/yr) and industrial emissions (2 ktonnes/yr), each of which is well quantified.

The best estimates for emissions from different sources are shown in Table 15.5.1. Note that biomass burning contributes about 40 percent of anthropogenic emissions.

Agricultural sales around the world have been increasing, however, and have about doubled since 1984 (see Table 15.5.2). In the United States, industrial bromine consumption has not changed appreciably since about 1970. This is largely due to the phasing out of ethylene dibromide as a gasoline additive (see Table 15.5.3).

15.5.3 ATMOSPHERIC CONCENTRATION

Measurements of CH$_3$Br yield a global average ground-level atmospheric mixing ratio of approximately 11 parts per trillion by volume (pptv). These measurements also have confirmed that the concentration in the northern hemisphere is higher by about 30 percent than the concentration in the southern hemisphere (interhemispheric ratio of 1.3). Such a ratio requires that the value of sources minus sinks in the northern hemisphere exceed the same term in the southern hemisphere.

TABLE 15.5.3   U.S. Bromine Consumption by End Use, 1971–1991 (Thousand Tonnes)

Year	Gasoline additives	Sanitary preparations	Flame retardants	Other	Total
1971	121	11	16	14	162
1972	122	11	17	14	164
1973	115	17	27	6	165
1974	109	17	25	14	165
1975	100	17	22	16	155
1976	109	18	26	25	178
1977	103	18	29	20	170
1978	100	16	32	23	171
1979	91	26	28	35	180
1980	73	21	25	16	135
1981	54	26	35	35	150
1982	45	27	47	46	165
1983	39	16	45	48	148
1984	34	16	45	68	163
1985	35	—	52	85	172
1986	—	—	—	—	—
1987	30	14	41	67	152
1988	—	—	—	—	—
1989	32	24	49	70	175
1990	25	—	50	—	—
1991	24	9	48	89	170

Source: Roskill Information Services Ltd., *The Economics of Bromine*, sixth edition, London, 1992.

There is no clear long-term change in the concentration of CH_3Br during the time period of the systematic continued measurements (1978–1992). One possible explanation is that CH_3Br from automobiles may have declined while emissions from agricultural use may have increased, leading to relatively constant anthropogenic emissions over the past decade.

The magnitude of the atmospheric sink of CH_3Br due to gas-phase chemistry is well known and leads to a lifetime of 2 ± 0.5 year. The recognition of an oceanic sink leads to a calculated atmospheric lifetime due to oceanic hydrolysis of 3.7 year, but there are large uncertainties (1.3 to 14 year). Thus the overall atmospheric lifetime due to both of these processes is 1.3 year, with a range of 0.8 to 1.7 year.

Recognizing the quoted uncertainties in the size of the individual sources of CH_3Br, the most likely estimate is that about 40 percent of the source is anthropogenic. The major uncertainty in this number is the size of the ocean source. Based on the present atmospheric mixing ratio and the current source estimate, a lifetime of less than 0.6 year would require identification of new major sources and sinks.

The calculated ozone depletion potential (ODP) for CH_3Br is currently estimated to be 0.6, based on an atmospheric lifetime of 1.3 years. The range of uncertainties in the parameters associated with the ODP calculation places a lower limit of 0.3 on the ODP.

15.5.4   *CHEMISTRY OF OZONE DESTRUCTION*

The chemistry of bromine in the stratosphere is analogous to that of chlorine. Upon reaching the stratosphere, the organic source gases photolyze or react with OH and $O(^1D)$ rapidly to liberate bromine atoms. Subsequent reactions, predominantly with O_3, OH, HO_2, ClO, NO, and NO_2, partition inorganic bromine between reactive forms (Br and BrO) and reservoir forms ($BrONO_2$, BrCl, HOBr, and HBr). However, unlike chlorine chemistry, where reactive forms are a small fraction of the total

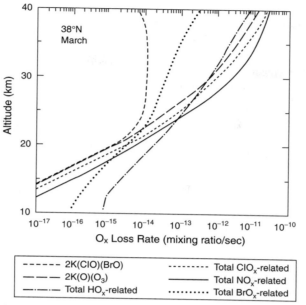

FIGURE 15.5.1 Midlatitude ozone loss rates associated with various removal cycles between 15 and 30 km (after Garcia and Solomon, 1994). Note that ozone is concentrated mostly between 15 and 25 km (World Metereological Organization, 1994).

inorganic budget (except in the highly perturbed polar regions in wintertime), reactive bromine is about half the total inorganic bromine budget in the lower stratosphere. Therefore, bromine is more efficient than chlorine in catalytic destruction of ozone. The combined efficiency of the bromine removal cycles for ozone ($HO_2 + BrO$ and $ClO + BrO$) is likely to be about 50 times greater than the efficiency of known chlorine removal cycles on an atom-for-atom basis.

The chemistry of ozone destruction by bromine in the stratosphere is now well understood. A high rate coefficient for the $HO_2 + BrO$ reaction is confirmed and there is no evidence that it produces HBr (which would have the effect of immobilizing bromine as an ozone destruction agent). Stratospheric measurements confirm that the concentration of HBr is very low (less than 1 pptv) and that it is not a significant bromine reservoir. In addition, the gas-phase photochemical partitioning between reactive and reservoir forms of bromine is fairly rapid in sunlight, on the order of an hour or less, so that direct heterogeneous conversion of HBr and $BrONO_2$ to BrO is likely to have little impact on the partitioning of bromine, except perhaps in polar twilight.

Mixing ratios of NO_x, HO_x, and ClO_x, increase more strongly with altitude above 20 km than does BrO, and the fractional contribution to ozone loss due to bromine is greatest in the lower stratosphere (Figure 15.5.1). There, where oxygen-atom concentrations are small, the $O + BrO$ reaction is relatively unimportant, and the three reaction cycles listed below are primarily responsible for bromine-catalyzed ozone loss, with cycle III being less important than cycles I and II.

Note also that stratospheric ozone is concentrated between 15 and 25 km. Therefore, HO_x-related removal processes have the greatest effect on the total ozone column; BrO_x-related processes are about an order-of-magnitude less important.

$$ClO + BrO + h\nu \rightarrow Br + Cl + O_2$$
$$Br + O_3 \rightarrow BrO + O_2$$
$$Cl + O_3 \rightarrow ClO + O_2$$

$$BrO + HO_2 \rightarrow HOBr + O_2$$
$$HOBr + h\nu \rightarrow OH + Br$$
$$Br + O_3 \rightarrow BrO + O_2$$
$$OH + O_3 \rightarrow HO_2 + O_2$$

$$BrO + NO_2 + M \rightarrow BrONO_2 + M$$
$$BrONO_2 + M \rightarrow Br + NO_3$$
$$NO_3 + h\nu \rightarrow NO + O_2$$
$$Br + O_3 \rightarrow BrO + O_2$$
$$NO + O_3 \rightarrow NO_2 + O_2$$

In the polar regions, where NO is reduced and ClO is enhanced by heterogeneous reactions on sulfate aerosols and polar stratospheric clouds, cycle I dominates the ozone loss due to bromine. At midlatitudes the first two cycles contribute approximately equally to ozone loss at 20 km, and cycle II is most important near the tropopause, where the abundance of HO_2 is substantial but where ClO abundances are negligible.

The relative importance of the three catalytic reactions will change with time. As the effects of the Montreal Protocol take hold, the concentration of chlorine in the atmosphere will diminish. At the same time, however, the increasing anthropogenic emission of methane will lead to increased stratospheric water vapor (Singer, 1971) and therefore make the HO_x-related reactions more important.

As shown in Figure 15.5.1, the HO_x-related loss rate of ozone predominates up to an altitude of about 25 km at midlatitudes. This loss rate is enhanced by the increased emission of methane and by the increasing amount of air traffic, which releases water vapor into the lower stratosphere.

15.5.5 CONCLUSION

In view of these other ongoing changes that affect chlorine, water vapor, and NO_x in the stratosphere, it is premature to judge the quantitative effect of increasing bromine concentration. A ban on CH_3Br production may be premature in any case until a measurable trend of bromine can be detected in the stratosphere.

REFERENCES

Garcia, R. R., and Solomon, S., "A New Numerical Model of the Middle Troposphere Ozone and Related Species," *J. Geophys. Res.*, 99: 12937–12951, 1994.

Singer, S. F., "Stratospheric Water Vapour Increase Due to Human Activities," *Nature*, 233: 543–545, 1971.

World Meteorological Organization, *Scientific Assessment of Ozone Depletion*, WMO Report 37, Geneva, 1994.

CHAPTER 15
UBIQUITOUS ENVIRONMENTAL CONTAMINANTS

SECTION 15.6
SUBSURFACE PETROLEUM SPILLS

James M. Davidson

Mr. Davidson is a hydrogeologist and the president of Alpine Environmental, Inc. He is a nationally recognized expert on the subsurface occurrence, movement, and remediation of petroleum products and gasoline additives.

15.6.1 INTRODUCTION

Since the beginning of the industrial age, petroleum products have been crucial for building, fueling, and lubricating a wide variety of machines. However, during the process of making, moving, and using these liquid petroleum products, some are inadvertently spilled or released to the environment. Such petroleum spills have long been a problem, as demonstrated by the 1863 failure of an oil pipeline because of numerous leaks (Wolbert, 1979).

Accidental spills and releases can occur wherever petroleum is explored for, removed from its original source rock, refined, mixed, stored, transported, used, or sold. Potential petroleum release locales include drilling sites, refineries, pipelines, industrial facilities, bulk storage terminals, railroad yards, and service stations. In the United States alone, there are literally millions of industrial, commercial, and residential facilities that store liquid petroleum products in aboveground storage or underground storage tanks (Young et al., 1987).

Focusing only on underground storage tanks (USTs), the potential for petroleum releases becomes apparent. In 1987, the United States Environmental Protection Agency (USEPA) estimated that there were three to five million underground tanks in the country, with about 90 percent of them storing petroleum products (Young et al., 1987). Initially, the USEPA estimated that as many as 25 percent of the 1.6 million registered USTs in the United States may have been leaking (USEPA, 1988). This suggested the existence of nearly 400,000 leaking underground storage tanks. The accuracy of this initial estimate has been verified by a September 1998 report, which confirmed that more than 371,000 releases of petroleum products from USTs have occurred nationwide (USEPA, 1998). Although cleanup has been completed at 203,000 of these releases, nearly 170,000 UST cleanups remain. In addition, the USEPA estimates that as many as 80,000 additional releases may occur from petroleum USTs between 1999 and 2005 (Ng, 1999), thus demonstrating that the problem of petroleum spills is expected to continue into the twenty-first century.

15.6.2 PETROLEUM PRODUCTS

Crude oil is a naturally occurring petroleum liquid composed of a wide variety of hydrocarbon compounds. Like all petroleum compounds, crude oil is composed primarily of a series of molecules

TABLE 15.6.1   Characteristics of Some Common Petroleum Products

Product	Average specific gravity[*]	Primary range of molecules[†]
Natural gas	Vapor	C_1 to C_2
Propane	Vapor	C_3
Gasoline	0.75	C_5 to C_{12}
Kerosene	0.79	C_{10} to C_{16}
Jet fuel	0.80	C_{12} to C_{16}
Diesel fuel	0.85	C_{10} to C_{20}
Lubricating oil	0.90	C_{20} to C_{30}
Heavy fuel oil	0.96	C_{20} to C_{30}

[*]These are average values, obtained in part from British Petroleum Company Limited (1977) and from Testa and Winegardner (1991). Actual values may vary depending on the product's composition and temperature. Because natural gas is a vapor, specific gravity does not apply. Ranges of specific gravity for these products are available (Testa and Winegardner, 1991).
[†]Values from Waddams (1973), Dragun (1988), Mackay (1988), and Testa and Winegardner (1991).

of hydrogen and carbon atoms collectively called hydrocarbons. Generally, crude oil is composed of 83 to 87 percent carbon and 11 to 14 percent hydrogen by weight.

A hydrocarbon compound is often represented by the elemental symbol for carbon (C) along with a subscript that tells how many carbon atoms are present in the compound. For example, C_6 indicates a compound that consists of six carbon atoms. Hydrocarbon compounds exist that have from 1 to 78 carbon atoms (C_1 to C_{78}) (British Petroleum Company Limited, 1977). At one atmospheric pressure, C_1 to C_4 compounds are normally gases, C_5 to C_{21} are normally liquids, C_{22} to C_{30} are liquids or semisolids, and C_{30} compounds and higher are usually solids (Waddams, 1973; British Petroleum Company Limited, 1977). Table 15.6.1 lists a variety of common petroleum products (made by refining crude oil), their specific gravities, and the primary range of hydrocarbon molecules they contain.

Liquid petroleum products are relatively insoluble in water (immiscible) and thus may be called non-aqueous phase liquids (NAPLs). Petroleum liquids with a specific gravity less than 1.00 (i.e., less dense than water) are called light non-aqueous phase liquids (LNAPLs). Petroleum NAPLs with a specific gravity greater than 1.00 are called dense non-aqueous phase liquids (DNAPLs). To focus on the most commonly spilled petroleum products (LNAPL fuels), the movement of DNAPLs is not discussed here. For information about DNAPL movement in the subsurface, the reader is referred to Schwille (1988), Cohen and Mercer (1993), and Pankow and Cherry (1996). As gasoline is the most commonly released LNAPL, it is the example product used in all discussions below. However, most of these discussions apply to other light petroleum fuel products (see Table 15.6.1).

15.6.3 REGULATORY ISSUES

Because petroleum spills are fairly common, numerous regulations have been established to require the cessation, investigation, and remediation of petroleum spills. In 1974, the U.S. government implemented the Safe Drinking Water Act to establish maximum contaminant levels (MCLs) for some chemical compounds in public water supplies and to protect underground sources of drinking water (Freedman, 1987). The Clean Water Act of 1977 addressed spills of oil and hazardous waste and it required spill prevention and other countermeasures (Freedman, 1987).

Congress passed the Resource Conservation and Recovery Act (RCRA) in 1976 (USEPA, 1986), which addressed the management of solid waste and hazardous waste. In 1984, amendments to RCRA were passed; Subtitle I of these amendments required development of regulations for addressing underground storage tanks that contain petroleum products or hazardous materials (USEPA, 1986). These regulations became effective December 22, 1988, and they established performance standards for new USTs and required leak detection, leak prevention, and corrective actions for leaks or spills

(Code of Federal Regulations, 1988). The 10-year phase-in of these UST regulations has recently been completed with the final regulatory deadline in December 1998. These legislative acts, and the resulting regulations, create a strong regulatory driving force to investigate and remediate subsurface contamination caused by petroleum spills.

15.6.4   IMPACTS OF SUBSURFACE PETROLEUM SPILLS

Depending on spill conditions, product type, and volume of product lost, a petroleum product spill can cause hydrocarbon contamination of air, soil, groundwater, and/or surface water. These contaminated media then have the potential to negatively affect vegetative, animal, or human health, due to toxic or carcinogenic effects (ASTM, 1996). Some socioeconomic impacts of subsurface petroleum spills include diminished property values, costs of assessment and remediation, and decreased water quality. Although petroleum releases to air and surface water are of great importance, the remainder of this section focuses on petroleum spills that occur on, or under, the ground surface as these types of releases typically cause subsurface hydrocarbon contamination.

As discussed earlier, in the United States there have been hundreds of thousands of petroleum releases from USTs alone. In addition, if the thousands of miles of buried petroleum pipelines are considered, even more petroleum spills must have occurred. These numerous subsurface petroleum spills typically cause hydrocarbon contamination of soil and groundwater (USEPA, 1998). In a small minority of cases, the released hydrocarbons can enter buildings or contaminate water supply wells, thereby directly affecting humans (Hadley and Armstrong, 1991).

Hydrocarbon contamination of subsurface drinking-water supplies is obviously of great concern. Water-soluble (miscible) components of petroleum products dissolve into the groundwater and then move in the subsurface with the groundwater. Some dissolved components of petroleum can affect the quality of drinking water even at extremely low concentrations. For example, the federal MCL for the carcinogenic compound benzene is 5 μg/L (parts per billion).

Although petroleum spills can affect drinking-water supply wells, such impacts are, fortunately, quite infrequent (Hadley and Armstrong, 1991). This is primarily attributable to a variety of natural processes that act to attenuate the movement of hydrocarbons in the subsurface (Chapelle, 1999; Rice et al., 1995). This is discussed in further detail below.

15.6.5   SUBSURFACE MIGRATION OF PETROLEUM COMPOUNDS

The subsurface movement of petroleum compounds is a complicated process and depends on numerous characteristics of the petroleum product and the spill site's subsurface hydrogeology. Subsurface petroleum migration was first studied in the field of petroleum engineering to enhance recovery of crude oil (Muskat, 1937). Studying the movement of spilled petroleum in relation to the contamination of groundwater did not occur until much later. (Schwille, 1967; Van Dam, 1967.)

Once in the subsurface environment, gasoline separates itself into several phases. The four contaminant phases are:

- vapor phase, which occurs in the air spaces of the unsaturated zone above the water table;
- separate phase, which occurs as pure petroleum floating on the water table;
- residual phase, which is pure petroleum that is entrapped in the soil; and
- dissolved phase, which is the petroleum components that are dissolved into the groundwater.

These four contaminant phases interact with the subsurface air, soil, and water to create a complex, multiphase migration of the spilled petroleum. The petroleum components in the vapor, dissolved, and separate phases are all mobile. They migrate through the subsurface away from the source thus enlarging the contaminated area with time (American Petroleum Institute, 1996a, 1996b). As the size of the contaminated area increases, it becomes more expensive and more difficult to investigate and to remediate. In addition, the mobility of these contaminants may cause other private or public properties to become contaminated, which can lead to further increased costs and legal issues.

Vapor Phase

Some light-weight components of spilled gasoline volatilize from the separate-phase, dissolved-phase, or residual-phase gasoline. These mobile components are sometimes called organic vapors, or soil gas. Because liquid-filled pores prevent vapor transport, these vapor-phase compounds move through the subsurface pores that are free of liquids (either gasoline or water). Silts and clays with relatively small pore spaces and high residual water contents usually do not allow significant vapor-phase transport (Dragun, 1988). Conversely, well-drained materials with large pore spaces (for example, sand and gravel) are much more amenable to vapor-phase transport.

Under natural conditions, subsurface vapor-phase transport is primarily due to diffusion, driven by the presence of a concentration gradient (Dragun, 1988). If a pressure gradient exists, the vapor phase also may migrate by advection (American Petroleum Institute, 1996a). The mass of hydrocarbon contamination moved by vapor-phase transport is often small compared with the amount of product released. However, the vapor phase is very important as its presence can be detected by various soil gas survey techniques, and the data may be used to infer the presence of other contaminant phases (Devitt et al., 1987; USEPA, 1997). Additionally, vapor-phase organic compounds have the potential to migrate through, and collect in, utility corridors and building basements. These vapors can affect human health, and under certain conditions, even pose a risk of fire or explosion (American Petroleum Institute, 1996c).

Residual Phase

As a petroleum spill moves down into a soil profile, it migrates through, and is absorbed into, the unsaturated pore spaces, thus filling them with petroleum. The absorbed petroleum held in the pores is the residual-phase contamination. The amount of product held as residual phase is often called the residual saturation. Residual saturation is determined by dividing the volume of trapped product by the volume of the impacted soil. Table 15.6.2. lists some average residual saturation values for gasoline in various soils.

As demonstrated in Table 15.6.2, the residual saturation of gasoline is much greater in finer-grained soil materials. This is because the numerous, small pores of fine-grained materials result in greater capillary forces, thereby causing more product retention per unit volume of soil (de Pastrovich et al., 1979). Conversely, the larger pores of coarser materials (like gravel) have relatively lower capillary forces, which in turn cause coarse soils to have much lower residual saturations (Table 15.6.2). Entrapment of petroleum as the residual phase is beneficial as it "ties up" the liquid petroleum, thus

TABLE 15.6.2 Average Residual Saturation Values of Liquid Gasoline in the Unsaturated Zone

Sediment	gal/ft^{3}*	L/m^{3}[†]	mg/kg[‡]
Coarse gravel	0.02	2.5	950
Coarse sand	0.06	7.5	2800
Fine sand/silt	0.15	20.0	7500

Modified from Table 6 of American Petroleum Institute (1996a); original source was de Pastrovich et al. (1979). Values are based on soil with a bulk density of 1.85 g/cm^3 and gasoline with a specific gravity of 0.7.
*Gallons of gasoline retained per cubic foot of soil.
[†]Liters of gasoline retained per cubic meter of soil.
[‡]Milligrams of gasoline retained per kilogram of soil.

limiting the subsurface extent of contamination. Indeed, in clayey soils a modest-sized gasoline spill may be completely absorbed into the vadose zone soil (i.e., tied up as residual saturation) and thus not affect groundwater.

However, this retention of hydrocarbons as the residual phase is also problematic with regard to long-term disposition of the subsurface contamination. The entrapped residual product typically leaches soluble components into the groundwater for years, decades, or even longer in the case of fine-grained soils. As such, the residual phase acts as a long-term contaminant source, and this greatly complicates remediation.

Separate Phase

When the separate-phase product first reaches an aquifer, the residual saturation needs of the soil must be overcome before the product can be mobilized. The product in excess of residual saturation needs will continue to move downward under the force of gravity. Once the mobile, separate-phase LNAPL reaches the top of the water capillary zone, it essentially will be unable to migrate around all the water-filled pores. Because of this impeded downward flow, and because it has a specific gravity less than that of water, the separate-phase gasoline will accumulate in the unsaturated pores just above the capillary zone (American Petroleum Institute, 1996a). Although this gasoline actually collects on top of the capillary zone, in practice it is often incorrectly said that it accumulates "on top of the water table." The pool of pure gasoline that collects on the water capillary zone is called the separate phase. It is also known as the separate-phase layer, free-phase product, or product layer.

Once it accumulates, the separate-phase product can move laterally as a result of the hydraulic gradient, gravity, and capillary forces. Mobile product will move on top of the capillary fringe toward areas of lower potential head. This means the separate-phase plume will generally migrate in the downgradient groundwater flow direction. The shape of the spreading separate-phase plume depends greatly on the hydraulic gradient of the aquifer. A steep gradient typically creates a narrow, elongated separate-phase plume. Conversely, given the same spill volume and aquifer characteristics, a shallow hydraulic gradient will create a separate-phase plume that is wider and more bulbous. If the weight of the accumulating product is sufficient to overcome the capillary forces, it can depress the water table, especially near the leak source.

As the separate phase moves, it continually encounters regions previously uncontaminated with hydrocarbons (i.e., regions below residual saturation). Therefore, gasoline will continually be removed from the mobile separate phase, thus reducing the volume of mobile product. Once there is insufficient separate phase to reach residual saturation levels, lateral movement of the separate-phase plume will cease. This natural process ensures that a separate-phase plume will not continue to move downgradient infinitely but instead will have a maximum extent.

Depending on the situation, water-table fluctuations can increase or decrease the volume of mobile separate phase as well as the separate-phase movement rate. Should the water table rise, the capillary zone will also be raised, which causes the separate-phase product to become entrapped between water-saturated pores (American Petroleum Institute, 1996a). The separate-phase layer thickness and the transport velocity thus are both reduced. If the rise is sufficient to encapsulate all the mobile product, then a mobile separate-phase layer will no longer exist as the product is now immobile and spread vertically throughout the soil column. This is sometimes called a "smearing effect" and it can cease movement of the separate-phase product. Later, a lowering of the water table later may allow the separate-phase layer to reform and become mobile again.

Dissolved Phase

Individual components of a spilled petroleum product will dissolve into groundwater and form a dissolved-phase plume. There are many components that dissolve from bulk gasoline, but it is too cumbersome to discuss them all and too costly to monitor them all. Therefore, a few representative compounds of greatest mobility and environmental concern are typically discussed and monitored. The most common ones are benzene, toluene, ethylbenzene, and xylenes (all three xylene isomers), which are collectively referred to by the acronym BTEX. The BTEX compounds are of greatest concern because their mobility and toxicity (for example; benzene is a known carcinogen). Many other hydrocarbon compounds are also present, and these are often referred to as total petroleum hydrocarbons (TPH) (note: there are many different laboratory analyses for different suites of TPH compounds).

With all gasoline spills, both BTEX and TPH compounds dissolve into the groundwater and form a dissolved-phase plume. Each of the hundreds of gasoline components have different solubilities in water (American Petroleum Institute, 1996a). However, the groundwater has a maximum amount of hydrocarbon contaminants it can carry; thus there is a limit to how much dissolved-phase contamination can form. Individual compounds essentially "compete" to be dissolved into the groundwater. As a result, the dissolved concentrations of individual gasoline components in groundwater will be much lower than their theoretical maximum solubility in water. For instance, the maximum solubility of benzene is 1780 mg/L in water (25°C). However, when gasoline is spilled, benzene concentrations in groundwater are rarely higher than 60 mg/L (American Petroleum Institute, 1996a). Similar reductions are seen for other hydrocarbon components that dissolve from gasoline.

Dissolved contaminants may enter the groundwater via several different pathways. The first pathway occurs when downward infiltrating recharge water becomes contaminated by contacting residual product in the soil or by the water passing through a separate-phase plume. Dissolved-phase contamination also may form when groundwater comes directly in contact with the separate-phase plume or when the groundwater moves laterally through hydrocarbon contaminated soils.

The dissolved components are moved along with the groundwater by the processes of advection and dispersion (Freeze and Cherry, 1979). The advective and dispersive movement of the dissolved-phase plume causes the horizontal extent of the contaminant plume to increase with time, primarily in the downgradient direction. Aquifers with rapid groundwater velocities tend to produce narrow, elongated, oval-shaped plumes from a single contaminant source. Conversely, aquifers with slower groundwater movement tend to have much wider, more rounded dissolved-phase plumes.

Groundwater moves at a horizontal rate that depends on the aquifer's hydraulic conductivity, hydraulic gradient, and effective porosity (Freeze and Cherry, 1979). Most dissolved hydrocarbon contaminants, however, move at a rate less than that of the groundwater because of adsorption of the components to soil particles. Adsorption occurs because the organic chemicals from the petroleum are chemically attracted to the organic matter in the soil. By attaching to the organic matter, the adsorbed components are slowed (or retarded) in their forward movement with the groundwater.

Not all components in gasoline are retarded by organic matter in the soil. Oxygenating fuel additives such as methyl tertiary butyl ether (MTBE) and ethanol do not adsorb well and thus usually are not retarded in their movement (NSTC, 1997). These components thus move at about the same rate as the groundwater.

Various forms of organic-lead compounds were also commonly added to gasoline in the past to boost engine performance. These compounds are strongly adsorbed onto soil particles in the subsurface. As a result, lead compounds from spilled gasoline typically adsorb almost completely to the soil particles and do not move downgradient very far from the source area.

Conservative compounds like MTBE are generally found toward the front of the dissolved-phase gasoline plume (Davidson, 1995; NSTC, 1997). Behind the conservative compounds are the moderately retarded compounds like BTEX. Heavier hydrocarbon compounds are readily retarded (i.e., strongly adsorbed) and they are furthest back in the plume, closer to the spill source. This separation of compounds is called the chromatographic effect, in reference to a chromatograph, which is a laboratory instrument used to separate organic compounds.

The mobile, dissolved-phase plume can be readily sampled and monitored by a series of monitoring wells or other groundwater sampling devices. As a result, there is often great focus on collecting data about the dissolved-phase plume. The mobile dissolved phase is also of great concern because of its capacity to move off-site and hence onto third-party properties. This of course can greatly increase liability. As a result, the dissolved-phase portion of a petroleum spill often attracts a disproportionate amount of interest and attention. However, subsurface transport knowledge and theory clearly indicate that the four phases of petroleum contamination (vapor, dissolved, residual, and separate phases) interact via many different processes (dissolution, retardation, volatilization, etc.). Thus, it is shortsighted to focus on any single phase of contamination. Decades of experience have shown that all phases of hydrocarbon contamination must be studied and considered if an appropriate assessment and remediation effort is to be made.

15.6.6 BIODEGRADATION AND NATURAL ATTENUATION

Biodegradation is the process of microbes and other organisms consuming chemical compounds (such as hydrocarbons) as an energy source. When extracting energy and nutrition from petroleum, the microbes break down (degrade) the hydrocarbon into simpler compounds. Biodegradation can occur in the presence of oxygen (aerobic) or in the absence of oxygen (anaerobic) (American Petroleum Institute, 1996b). Complete aerobic biodegradation leaves behind only carbon dioxide and water, with no hydrocarbons remaining. Although subsurface biodegradation of spilled petroleum products occurs naturally, and at essentially all spill sites, it may be very slow if it is limited by some subsurface condition, such as insufficient dissolved oxygen (Chapelle, 1999; Norris et al., 1994).

The downgradient movement of dissolved-phase contaminants is moderated to varying degrees by volatilization, adsorption, dispersion, biodegradation, and other natural processes (American Petroleum Institute, 1996b). Collectively, these processes are called natural attenuation. As a result of these processes, a dissolved hydrocarbon plume does not spread infinitely; there is a site-specific limit as to how far the dissolved plume will spread. Several studies have been made of dissolved-phase hydrocarbon plumes caused by gasoline leaking from USTs (Rice et al., 1995; Mace et al., 1997). These field studies reviewed several hundred plumes and reported that the most mobile hydro-carbon component, benzene, rarely moved more than several hundred feet before natural attenuation processes stopped the forward plume migration (Rice et al., 1995; Mace et al., 1997). Of course, hydrocarbon plume extents can sometimes be quite large, depending on spill volume, spill circum-stances, site hydrogeology, and degree of natural attenuation. At many spill sites, it was also seen that, over time, as the rate of mass destruction exceeded downgradient mass transfer, the dissolved plume extent actually retracted back toward the source area (Chapelle, 1999).

15.6.7 SUBSURFACE INVESTIGATIONS OF PETROLEUM SPILLS

To determine the extent and impacts of a petroleum spill, a subsurface investigation must be conducted. By gathering and interpreting field and laboratory data, an investigation will identify the contaminants, study the site conditions, define the extent of contamination, and establish the potential risk posed by the contamination. Each of these steps can be complex and costly, but each is crucial to correctly understanding the extent and the impacts of a petroleum spill.

There are many different tools and techniques used to investigate petroleum spills (American Petroleum Institute, 1996a; USEPA, 1997). Until a few years ago, gathering soil samples by sub-surface drilling and collecting groundwater samples from monitoring wells were the predominant investigation techniques (Boulding, 1995; Nielsen, 1991). These samples were then sent to an off-site analytical laboratory for a variety of chemical analyses. This approach is important for obtaining direct measurements and it is a crucial part of any site investigation. However, the cost and time involved in conducting multiple rounds of well installation and sample gathering can be extensive. As a result, a large number of alternative investigative techniques have evolved in recent years to collect subsurface data more quickly and at less expense, including:

- direct push technologies (for collecting soil gas, soil, and groundwater samples),
- soil gas surveys (both active and passive),
- surface geophysical methods, and
- field analysis methods.

Application of these techniques is well described elsewhere (USEPA, 1997). No single investi-gation technique is appropriate in all cases. Very often, the optimal investigation combines several different techniques so as to collect the necessary site-specific data in the most cost-effective way. A complete discussion of the many investigation tools and interpretative techniques is beyond the scope of this subsection; the reader is referred to the references provided.

15.6.8 REMEDIATING HYDROCARBON CONTAMINATION

Historical Approaches

The assessment of petroleum spills expanded greatly in the 1980s partly because of increased regulatory attention. Initially, the mobile dissolve-phase contamination attracted the most attention, and it was commonly held that these plumes should be actively contained and that the impacted aquifer should be completely remediated back to pristine conditions. Typically, this required pumping out the impacted groundwater, and then treating the extracted water by a variety of methods to remove the contaminants of concern (Nyer, 1992). Many such "pump and treat" systems were installed around the country, with the initial expectation that they would "clean up" the aquifer in just a few years (National Research Council, 1994).

However, it soon became apparent that the dissolved plumes were essentially a symptom of the true problem—the much more extensive mass of hydrocarbons trapped in the residual phase. After this was recognized, the emphasis shifted more toward remediating the residual phase. At first this was done with some success by soil excavation, followed by soil treatment or disposal. To reduce costs, and to minimize the transfer of contamination from one location to another, several in situ methods were developed to better address the residual phase, including soil vapor extraction (i.e., soil venting), air sparging (American Petroleum Institute, 1996a), and enhanced biodegradation (Chapelle, 1999).

Of these active remediation techniques, no single one was effective on all four phases of contamination, nor was any single technique appropriate for all subsurface conditions. Often the most effective systems combined several remediation technologies so as to simultaneously remediate several contaminant phases at once. For example, a pump-and-treat system might be installed to address the separate and dissolved phases, whereas the addition of a soil venting system would be needed to address the vapor and residual phases.

This multiphase approach to remediation improved mass removal rates significantly, but it also increased system complexity and costs. As time progressed, and extensive field experience with hydrocarbon remediation was developed, the scientists working in the field noted several important trends.

- Subsurface complexity and slow leaching of contaminants combined to cause a probable operating time of decades (or more) for many remedial systems (National Research Council, 1994).
- There would likely be a lack of funds for long-term operation of hydrocarbon remediation systems at several hundred thousand UST sites.
- Unremediated hydrocarbon plumes often stop moving downgradient because of natural attenuation processes (Norris et al., 1994; Rice et al., 1995; Mace et al., 1997).

Recognition of these trends precipitated a serious reconsideration of how to address hydrocarbon contamination caused by petroleum product spills. Studies were conducted of the effectiveness of existing systems, and of alternative cleanup approaches (National Research Council, 1994). Field studies were conducted to determine how much hydrocarbon plumes actually spread over time (Rice et al., 1995) and to establish how best to identify plumes that are naturally attenuating (Wiedemeier et al., 1995).

Current Approach

In recognition of the above factors, the current approach to remediating hydrocarbon contamination is considerably different than it was just 10 years ago. Site assessments are increasingly conducted by a variety of expedited techniques and real-time analyses (USEPA, 1997). Once the extent of contamination is defined, designing and installing an active remediation system is rarely the next step. Instead, a risk-based corrective action (RBCA) is first applied (ASTM, 1996). The RBCA method provides a systematic approach for determining what human health risks and/or environmental risks exist from a given contaminant scenario. Based on the apparent risk, an appropriate remedial response is devised (American Petroleum Institute, 1996c). In the case of hydrocarbon contamination, natural attenuation often is chosen as the remedial approach of choice because of its low cost and high effectiveness at many spill sites. This remediation approach is sometimes called monitored natural attenuation, as some degree of long-term environmental monitoring is required.

If the risks are significant and can be effectively reduced by active remediation, then an active remedial system should be designed, installed, and operated. Very often the same technologies are selected as were used in the recent past (National Research Council, 1994; Norris et al., 1994). The difference is that a technology is chosen to reduce a specific risk due to a particular contaminant(s), to a specified degree. This approach does not eliminate the application of active remedial systems, but rather it ensures that active systems are applied only where they will definitely reduce the exposure to, and risk of, a specific contaminant situation.

The RBCA methodology and the greatly increased reliance on monitored natural attenuation have allowed many more cleanups to proceed and at lesser cost. These changes are the result of more than two decades of information and technology improvements developed through both field applications and directed research. However, the extensive use of natural attenuation heavily relies upon the fact that hydrocarbon components are attenuated by various processes and that there is a significant contaminant mass reduction over time. These two key factors do not apply nearly as well to some recalcitrant gasoline oxygenate additives such as MTBE. As such, the subsurface presence of MTBE (or other recalcitrant gasoline additives) greatly increases the problems created by subsurface gasoline spills.

15.6.9   MTBE USAGE AND THE RESULTING PROBLEMS

Beginning in 1979, MTBE was added to some gasolines in the United States to boost octane ratings. Its use increased dramatically in the late 1980s and 1990s, when it was commonly added to gasoline as part of air-quality improvement programs (NSTC, 1997). Although MTBE has helped reduce air pollution problems, its extensive use and unfavorable physicochemical characteristics have led to some significant problems (Keller et al., 1998). MTBE has been detected at a number of UST leak sites, with some dissolved-phase MTBE plumes being quite extensive (Davidson, 1995; NSTC, 1997). Unlike most gasoline components, MTBE is not very prone to natural attenuation. Specifically, MTBE is more soluble, less biodegradable, less retarded, and less volatile from groundwater than almost all other hydrocarbon compounds in gasoline (NSTC, 1997). As a result, over time, MTBE plumes can readily grow larger than their associated hydrocarbon plumes (Keller et al., 1998).

Some of the same characteristics that cause MTBE plumes to be less prone to natural attenuation also make MTBE more difficult to remediate than the other compounds in gasoline (Davidson and Parsons, 1996). Both predictive studies (MTBE Research Partnership, 1998) and field experience (Creek and Davidson, 1998) have indicated that, although the technologies normally used for remediating and treating gasoline components do indeed work for MTBE, they are often less effective and more expensive.

Extensive, multimedia studies have been conducted to determine whether the benefits of using MTBE in gasoline are outweighed by the costs, risks, and impacts (Keller et al., 1998). Although the cost and benefits of MTBE use are still being studied, it is clear that with the addition of gasoline additives like MTBE, the problems created by petroleum spills have become more complex. As such, subsurface petroleum spills will continue to pose scientific and engineering challenges in the years ahead.

15.6.10   *SUMMARY*

There are many different facilities that can accidentally release petroleum to the subsurface. Underground storage tanks are the most common, with nearly 400,000 petroleum releases having occurred from USTs in the United States alone. These releases cause subsurface hydrocarbon contamination, including formation of vapor-phase, residual-phase, separate-phase, and dissolved-phase contamination.

A variety of techniques can be used to investigate subsurface petroleum spills. Once delineated, the existing contamination is typically considered under a RBCA format to determine the necessary action for protecting human health and the environment. In some cases, active remediation is conducted by applying any of several different remedial technologies. However, at many spill sites, natural attenuation processes are sufficient to control subsurface migration of the petroleum spill.

With the addition of gasoline additives like MTBE, the problems posed by petroleum spills has become more complex. MTBE does not attenuate in the subsurface nearly as well as other gasoline components, nor does it remediate as readily. As a result, subsurface petroleum spills continue to pose scientific and engineering challenges.

BIBLIOGRAPHY

American Petroleum Institute, *A Guide to the Assessment and Remediation of Underground Petroleum Releases*, API Publication No. 1628, American Petroleum Institute, Washington, DC, 3rd edition, 1996a.

American Petroleum Institute, *Natural Attenuation Processes*, API Publication No. 1628A, American Petroleum Institute, Washington, DC, 1996b.

American Petroleum Institute, *Risk-Based Decision Making*, API Publication No. 1628B, American Petroleum Institute, Washington, DC, 1996c.

ASTM, *Risk-Based Corrective Action Applied at Petroleum Release Sites*, American Society for Testing Materials (ASTM) Standard E-1739, ASTM, March 5, 1996.

Boulding, R. J., *Practical Handbook of Soil, Vadose, and Ground-Water Contamination*, Lewis Publishers, Boca Raton, FL, 1995.

British Petroleum Company Limited, *Our Industry Petroleum*, British Petroleum Company Limited, London, England, 1977.

Chapelle, F. H., "Bioremediation of Petroleum Hydrocarbon-Contaminated Ground Water: The Perspectives of the History and Hydrology," *Ground Water*, 37(1): 122–132, 1999.

Code of Federal Regulations, Title 40, Part 280–281, 1988.

Cohen, R., and Mercer, J., *DNAPL Site Evaluation*, Lewis Publishers, Boca Raton, FL, 1993.

Creek, D. N., and Davidson, J. M., "The Performance and Cost of MTBE Remediation Technologies," in *Proceedings of the Petroleum Hydrocarbons and Organic Chemicals in Ground Water Conference*, National Ground Water Association, Dublin, OH, pp. 560–568, 1998.

Davidson, J. M., "Fate and Transport of MTBE—The Latest Data," in *Proceedings of the Petroleum Hydrocarbons and Organic Chemicals in Ground Water Conference*, Houston, TX, pp. 285–301, 1995.

Davidson, J. M., and Parsons, R., "Remediating MTBE with Current and Emerging Technologies," in *Proceedings of the Petroleum Hydrocarbons and Organic Chemicals in Ground Water Conference*, Houston, TX, pp. 15–29, 1996.

de Pastrovich, T. L., Baradat, Y., Barhtel, R., Chirelli, A., and Fussell, D. R., *Protection of Groundwater from Oil Pollution*, CONCAWE Report No. 3/79, The Hague, Netherlands, 1979.

Devitt, D. A., Evans, R. B., Jury, W. A., and Starks, T. H., *Soil Gas Sensing for Detection and Mapping of Volatile Organics*, National Water Well Association, Dublin, OH, 1987.

Dragun, J., *The Soil Chemistry of Hazardous Materials*, Hazardous Materials Control Research Institute, Silver Spring, MD, 1988.

Freedman, W., *Federal Statutes on Environmental Protection*, Quorum Books, Westport, CT, 1987.

Freeze, R. A., and Cherry, J. A., *Groundwater*, Prentice-Hall, Inc., Englewood Cliffs, NJ, 1979.

Hadley, P. W., and Armstrong, R., "Where's the Benzene?—Examining California Ground Water Quality Surveys," *Ground Water*, 29(1): 35–40, 1991.

Keller, A., Froines, J., Koshland, C., Reuter, J., Suffet, I., and Last, J., *Health & Environmental Assessment of MTBE, Volume 1—Summary & Recommendations*, University of California Toxics Research and Teaching Program, Davis, CA, November 1998.

Mackay, D., "The Chemistry and Modeling of Soil Contamination with Petroleum," in *Soils Contaminated by Petroleum*, edited by E. J. Calabrese, P. T. Kostecki, and E. J. Fleischer, Wiley, New York, pp. 5–18, 1988.

Mace, R. E., Fisher, S., Welch, D. M., and Parra, S. P. "Extent, Mass, and Duration of Hydrocarbon Plumes from Leaking Petroleum Storage Tanks in Texas," *Geological Circular 97-1*, Bureau of Economic Geology, University of Austin, TX, 1997.

MTBE Research Partnership, *Treatment Technologies for Removal of Methyl Tertiary Butyl Ether (MTBE) from Drinking Water*, MTBE Research Partnership & National Water Research Institute, Fountain Valley, CA, December 1998.

Muskat, M., *The Flow of Homogeneous Fluids Through Porous Media*, McGraw-Hill, New York, 1937.

National Research Council, *Alternatives for Ground Water Cleanup*, National Academy Press, Washington, DC, 1994.

Nielsen, D. M. (ed.), *Practical Handbook of Ground-Water Monitoring*, Lewis Publishers, Chelsea, MI, 1991.

Ng, S., "The View from the U.S. EPA: Program Direction for 1999 and Beyond," *Underground Tank Technology Update*, 13(5): 10–12, 1999.

Norris, Hinchee, Brown, McCarty, Semprini, Wilson, Kampbell, Reinhard, Bouwer, Borden, Vogel, Thomas, and Ward, *Handbook of Remediation*, Lewis Publishers, Boca Raton, FL, 1994.

NSTC, *Interagency Assessment of Oxygenated Fuels*, National Science and Technology Council, Executive Office of the President of the United States, Washington, DC, June 1997.

Nyer, E. K., *Groundwater Treatment Technology*, Van Nostrand Reinhold, New York, 2nd edition, 1992.

Pankow, J. F., and Cherry, J. A. (eds.), *Dense Chlorinated Solvents and Other DNAPLs in Groundwater*, Waterloo Press, Portland, OR, 1996.

Rice, D. W., Grose, R. D., Michaelsen, J. C., Dooher, B. P., MacQueen, D. H., Cullen, S. J., Kastenberg, W. E., Everett, L. E., and Marino, M. A., *California Leaking Underground Fuel Tank (LUFT) Historical Case Analyses*, UCRL-AR-122207, Lawrence Livermore National Laboratory, Livermore, CA, 1995.

Schwille, F., "Petroleum Contamination of the Subsoil—a Hydrological Problem," in *The Joint Problems of the Oil and Water Industries*, edited by P. Hepple, The Institute of Petroleum, London, England, pp. 23–54, 1967.

Schwille, F., *Dense Chlorinated Solvents in Porous and Fractured Media*, Lewis Publishers, Inc., Chelsea, MI, 1988.

Testa, S. A., and Winegardner, D. L., *Restoration of Petroleum-Contaminated Aquifers*, Lewis Publishers, Chelsea, MI, 1991.

USEPA, *RCRA Orientation Manual (EPA/530-SW-86-001)*, USEPA, Office of Solid Waste, Washington, DC, 1986.

USEPA, *Musts for USTs—A Summary of the New Regulations for Underground Storage Tank Systems*, USEPA, Office of Underground Storage Tanks, Washington, DC, 1988.

USEPA, *Expedited Site Assessment Tools for Underground Storage Tanks (EPA 510-B-97-001)*, Office of Underground Storage Tanks, USEPA, Washington, DC, March 1997.

USEPA, *Corrective Action Measures for 2nd Half FY98, Office of Underground Storage Tanks*, USEPA, Washington, DC, 1998. Available at www.epa.gov/swerust1/cat/camnow.htm.

Van Dam, J., "The Migration of Hydrocarbons in a Water-Bearing Stratum," in *The Joint Problems of the Oil and Water Industries*, edited by P. Hepple, The Institute of Petroleum, London, England, pp. 55–96, 1967.

Waddams, A. L., *Chemicals from Petroleum*, Wiley, New York, 3rd edition, 1973.

Weidemeier, T. H., Wilson, J. T., Campbell, D. H., Miller, R. M., and Hansen, J. E., *Technical Protocol for Implementing Intrinsic Remediation with Long-Term Monitoring for Natural Attenuation of Fuel Contamination Dissolved in Groundwater*, Air Force Center for Environmental Excellence, Technology Transfer Division, San Antonio, TX, 1995.

Wolbert, G. S. Jr., *U.S. Oil Pipe Lines*, American Petroleum Institute, Washington, DC, 1979.

Young, A. D. Jr., Barbara, M. A., Miller, J. P., Kane, R. W., and Rogers, W., *Underground Storage Tank Management: A Practical Guide*, Government Institutes, Inc., Rockvile, MD, 2nd edition, 1987.

CHAPTER 15
UBIQUITOUS ENVIRONMENTAL CONTAMINANTS

SECTION 15.7

IONIZING RADIATION

William Andrew Hollerman

Dr. Hollerman is an experimental physicist who works in areas of accelerator-based physics, environmental technology, and materials science. He is assistant professor of physics at the University of Louisiana at Lafayette.

15.7.1 INTRODUCTION

It has been estimated recently that the United States needs to spend more than a trillion dollars to clean up accumulated radioactive wastes. This effort will span the entire gambit from high-level radioactive fuel elements at nuclear power generation facilities to low-level materials, medical exposures, and research waste. Environmental professionals need to be knowledgeable in the area of basic principles to protect humanity from an ever-increasing quantity of low- and high-level radioactive materials. Unlike most chemical wastes, the radioactive content of a waste stream cannot easily be neutralized. The only current practical solution is to bury radioactive waste in stable geologic areas to prevent anthropogenic contact. The challenge to environmental professionals over the next several years is to develop methods and technologies to monitor, store, and possibly neutralize low-level radioactive materials. This will make burial easier, more cost-effective, and safer for the many years of storage that lie ahead. Radiation is a natural phenomenon and humanity has made reasonably good use of its properties.

15.7.2 BASIC TERMINOLOGY

Atoms are the smallest units of matter recognizable as chemical elements with their own set of unique physical properties. Using the simple planetary representation, an atom has a central nucleus with a specific number of protons. To balance the total positive electrical charge of the central mass of protons, an equal number of electrons orbits the nucleus in specific quantum states. Electrons have an electrical charge that is equal and opposite in polarity to the proton and has a magnitude of 1.6021×10^{-19} coulombs. The number of protons or electrons in a given neutral atom is defined as the atomic number and is denoted by Z. Individual elements can be identified by their unique atomic number. For example, carbon has an atomic number (Z) of six. A single atom of carbon is composed of six protons and six electrons.

In addition to protons, an atomic nucleus also has a variable number of neutrons. Neutrons are electrically neutral and are slightly more massive than protons. Without neutrons, individual protons in a small volume would fly apart because of Coulomb repulsion. Neutrons hold the nucleus together by balancing the short-range strong force with the longer-range Coulomb force. The atomic

TABLE 15.7.1   Hydrogen Isotopes

Hydrogen isotope	Atomic number (Z)	Atomic mass (A)	Radioactive	Symbol
Hydrogen	1	1	No	^1_1H
Deuterium	1	2	No	^2_1H
Tritium	1	3	Yes	^3_1H

$$^A_Z X$$

X = Element symbol

A = Atomic mass

Z = Atomic number

FIGURE 15.7.1   Symbolic nuclide representation.

mass is defined as the sum of the number of protons and neutrons and is denoted by A. A nuclide is defined as a unique collection of protons and neutrons. A nuclide that is radioactive is called a radionuclide. For example, radium has a radionuclide with an atomic number of 88 and an atomic mass of 226.

An isotope is a grouping of nuclides with variable atomic mass for a particular atomic number. Table 15.7.1 illustrates the concepts behind atomic mass, nuclide, and isotope. Hydrogen has three isotopes composed of nuclides with unique numbers of protons and neutrons. The three isotopes of hydrogen contain zero, one, and two neutrons for "regular" hydrogen, deuterium, and tritium, respectively. Each of the neutral isotopes has only one proton and one electron. Only the tritium isotope of hydrogen is radioactive. The regular hydrogen and deuterium isotopes are stable. Deuterium is also sometimes called heavy hydrogen.

The column on the far right side of Table 15.7.1 contains the symbolic notation for each of the three hydrogen isotopes. A symbolic representation for a desired nuclide can be found in Figure 15.7.1. A numerical superscript on the left of the element symbol indicates the atomic mass of the selected nuclide. A numerical subscript on the left of the element symbol indicates the atomic number of the nuclide. The number of neutrons is denoted by a numerical subscript on the right side of the element symbol. The neutron number is typically not shown symbolically (as illustrated in Figure 15.7.1) because it is also equal to the difference between the atomic mass and atomic number $(A - Z)$.

15.7.3   RADIOACTIVE TRANSITION MODES

Radioactivity can be defined as a reaction that transforms a given nuclide (parent) into a new nuclide (daughter). Initiation of a nuclear reaction requires that the parent nucleus is left in an unstable or excited condition. Daughter nuclei can also transform into "granddaughter" products if they are left in an excited state. These transformation reactions are usually completed in a few fractions of a second and are unique to the nuclide that initiated the reaction. The mode of radioactive decay depends on the type of nuclear instability and the mass-energy relationship between the parent nucleus, daughter nucleus, and emitted particle. Nuclear instability is generally based on the magnitude of the neutron-to-proton ratio in the parent nucleus. The total equivalent energy of the parent nucleus is equal to the sum of the resulting energy of the reaction products (including the kinetic energies of all emitted particles).

Table 15.7.2 summarizes the more common modes of radioactive decay. A short description of the beta minus, beta plus, electron capture, gamma, internal conversion, alpha, neutron, and fission decay modes can be found in the subsections that follow. Other less common modes such as pair production, proton decay, double beta decay, two-proton decay, cluster emission decay, and induced transmutation decay are being studied extensively by scientists and typically are not important to the environmental professional. Consult Ref. 6 for more detailed information about the available modes of radioactive decay.

TABLE 15.7.2   Common Modes of Radioactive Decay

Radioactive decay mode	Initial atomic state		Final atomic state	
	Number	Mass	Number	Mass
Beta minus	Z	A	Z + 1	A
Beta plus	Z	A	Z − 1	A
Electron capture	Z	A	Z − 1	A
Gamma (photon)	Z	A	Z	A
Internal conversion	Z	A	Z	A
Alpha	Z	A	Z − 2	A − 4
Neutron	Z	A	Z	A − 1
Fission	Z	A	Z_1, Z_2	A_1, A_2

Beta Minus Decay

With beta minus (negatron) decay, an electron is emitted from a given atomic nucleus. In generalized terms, beta minus decay transforms a neutron into a proton by Equation 15.7.1.

$$n \rightarrow p + \bar{v}_e + \beta^-$$ (15.7.1)

The n and p symbols in Equation 15.7.1 represent the neutron and proton, respectively. The beta minus, or electron, particle is denoted by the Greek letter beta (β^-) with a negative superscript to denote its electrical charge. The Greek letter nu (v_e) with the bar on top symbolizes emission of an electron (subscript e) antineutrino during the beta minus decay. Both neutrinos and antineutrinos were postulated by Pauli in 1930 and were observed experimentally by Reines and Cowan in 1953. Neutrinos are emitted during beta decay to conserve nuclear quantum numbers, momentum, and energy. Because of their elusive nature, neutrinos and antineutrinos are of no significance to practicing environmental professionals and can be ignored.

Table 15.7.2 shows that beta minus decay causes the daughter nuclide to have an atomic number one greater than its parent. Atomic mass remains constant during a beta minus decay. Phosphorus-32 (listed as element-atomic mass) has an atomic number of 15 before initiation of a beta minus decay. After the decay, the resulting sulfur-32 daughter product has an atomic number of 16. Nuclides with large neutron-to-proton ratios are candidates for beta minus decay.

Nuclides such as hydrogen-3 (tritium), carbon-14, phosphorus-32, strontium-90, and yttrium-90 are pure beta emitters that are commonly used as tracers in biological research and medical treatment. Carbon-14 can also be used to estimate the age of organic artifacts or materials. This technique is based on the fact that carbon-14 reaches an equilibrium level in the earth's atmosphere by cosmic ray interactions. Organic matter contains a small amount of carbon-14. When the animal or plant material dies, the amount of carbon-14 is fixed and begins to decay. The approximate age of the plant material can be estimated using the fraction of carbon-14 at the time of the test to the average equilibrium level. This technique is effective for materials that are up 75,000 years old. A similar method using tritium can be used to determine the age of surface or underground water sources. High-energy electrons can also be used in physics research to determine the nature of the radioactive weak force.

Beta Plus Decay

With beta plus decay, a positron (positively charged electron) is emitted from a given atomic nucleus. In generalized terms, beta plus decay transforms a proton into a neutron by Equation 15.7.2.

$$p \rightarrow n + v_e + \beta^+$$ (15.7.2)

The beta plus, or positron, particle is denoted by the Greek letter beta (β^-) in Equation 15.7.2 with the superscript positive sign to denote its electrical charge. The Greek letter nu (v_e) symbolizes the emission of an electron (subscript e) neutrino.

Note that in Table 15.7.2 beta plus decay causes the daughter nuclide to have an atomic number one less than its parent. Atomic mass remains constant during beta plus decay. Sodium-22 has an atomic number of 11 before initiation of beta plus decay. After decay, the resulting neon-22 daughter product has an atomic number of 10. Nuclides with small neutron-to-proton ratios are candidates for beta plus decay. Beta plus decay occurs primarily in low-Z parent nuclides.

After emission from beta plus decay, the resulting position is converted into pure energy in an annihilation reaction with surrounding electrons. The interacting positron and electron masses are completely annihilated to produce two electromagnetic gamma photons, which are emitted to conserve momentum. Environmental professionals who deal with beta plus emitters need to understand that annihilation will create gamma photons with effective energies equal to the electron rest mass of 0.511 MeV. A megaelectron volt is a unit used in physics to denote the energy a proton would receive traveling through a potential difference of 10^6 volts and it is equivalent to 1.60219×10^{-13} joules. Modern accelerators, such as the Tevatron at the Fermi National Accelerator Laboratory in Illinois, or the Relativistic Heavy Ion Collider at Brookhaven National Laboratory in New York, can provide particle energies larger than 10^6 MeV, or 1 teraelectron volts (1 TeV), to discover the nature of matter in the universe.

Nuclides such as carbon-11, nitrogen-13, oxygen-15, and fluorine-18 are commonly used for positron emission tomography (PET) imaging of biochemical processes in the human body. Organic compounds, such as fatty acids, proteins, and amino acids, are labeled with a selected radionuclide to study the oxygen consumption and blood flow of organs such as the heart and brain. Certain brain disorders, such as Alzheimer's disease and stroke, can be diagnosed with the PET scan technique.

Electron Capture Decay

With electron capture decay, an orbital electron is captured by an excited nucleus as shown in Equation 15.7.3.

$$p + e^- \rightarrow n + \nu_e \qquad (15.7.3)$$

Electron capture produces the same daughter nuclide that would be produced by positron decay. Table 15.7.2 shows that electron capture causes the daughter nuclide to have an atomic number one less than its parent. Atomic mass remains constant during electron capture. Lutetium-172 has an atomic number of 71 before initiation of electron capture. After decay, the resulting ytterbium-171 daughter product has an atomic number of 70. Electron capture is the only possible decay mode for this reaction when the transition energy is less than two electron masses (1.022 MeV). It is favored over beta plus emission when the transition energy is low and the parent atomic number is large. Electron capture was not discovered until 1938 because it does not produce easily detectable nuclear radiation.

Gamma Photon Decay and X-Ray Production

In gamma decay, electromagnetic radiation is emitted as a nucleus completes a transition from an excited energy condition toward the ground state. Table 15.7.2 shows that the daughter nuclide has the same atomic number and atomic mass as the parent for gamma decay. Indium-115m in a metastable excited state has an atomic number of 49 before initiation of gamma decay. After decay, the resulting indium-115 daughter product also has an atomic number of 49. The metastable condition of the indium-115 is noted by the "m" after the listed atomic mass. This notation indicates that the parent nuclide was placed in an excited condition by some other process.

Gamma photons are emitted during deexcitation of a nucleus toward its ground state. This process differs significantly from generation of x-rays by an excited atom. Photons are emitted during the deexcitation of electrons to the lowest allowed atomic quantum state. When these electrons jump into the most tightly bound orbital shells, the resulting photons are classified as x-rays, which can have energies nearly equivalent to those typically associated with gamma photons. The only real difference between a gamma photon and an x-ray is its source. A gamma photon is emitted from the nucleus and an x-ray is emitted from an atom. Once the photon is created, it is impossible to determine whether it is an x-ray or a gamma photon. Scientists have assigned labels to designate the source of the radiation.

Gamma photons are often used to treat cancer patients with radiation therapy, irradiate food to remove harmful bacteria or microbes in order to lengthen store shelf life, and sterilize medical instruments for treatment purposes; x-rays are commonly used for medical diagnosis and treatment, dental examination, inspection of welds, determination of metal strength, and a variety of other uses. Both gamma photons and x-rays are also commonly used in physics, nuclear engineering, and materials science research.

Internal Conversion Decay

With internal conversion decay, the parent nucleus deexcites itself by transferring its energy to an orbital electron that is ejected from the atom. No gamma photon is emitted in this reaction. For internal conversion decay, Table 15.7.2 shows that the daughter nuclide has the same atomic number and mass as its parent. A metastable indium-113 has an atomic number of 49 before initiation of internal conversion. After decay, the resulting indium-113 daughter product also has an atomic number of 49. Internal conversion decay competes directly with gamma photon decay for deexcitation of certain radionuclides.

Alpha Decay

With alpha decay, the parent atom emits a helium-4 nucleus consisting of two protons and two neutrons. Table 15.7.2 shows that the daughter nuclide has an atomic number two smaller than its parent. The corresponding atomic mass is also reduced by four. Radium-226 in a metastable excited state has an atomic number of 88 before initiation of internal conversion. After decay, the resulting radon-222 daughter product also has an atomic number of 86 as shown in Table 15.7.2.

Alpha decay occurs primarily in medium-Z to high-Z parent nuclides. A typical range in alpha particle energies varies from 1 to 2.5 MeV for rare-earth elements to 5 to 7 MeV for high-Z materials.

Neutron Decay

With this type of decay, the parent atom emits one or more neutrons. Table 15.7.2 shows that the daughter nuclide has an atomic number that is the same as its parent. The number of neutrons emitted during this reaction also reduces the corresponding atomic mass. Lithium-9 has an atomic number of three before initiation of single neutron decay. After decay, the resulting lithium-8 daughter product has an atomic number of three. For this example, neutron decay causes the net atomic mass to decrease by one.

Neutrons can be used as a tool to determine the composition of a material by activation. A material is bombarded with neutrons that induce characteristic nuclear reactions. The radioactive daughter product decays through any one of the normal decay modes. The quantity and specific nature of the emitted radiation is an indicator of the amount of the given material that is present in the test sample. Neutron activation has been used at crime scenes to help identify potential suspects. One case used neutron activation to identify paint samples found on a deceased body. Paint on cars of several suspects could not be matched visually with the sample found on the body. Neutron activation compared the specific trace element concentration in each paint to uncover the guilty party.

Spontaneous and Neutron-Induced Fission Decay

Spontaneous fission is a natural decay process in which the nucleus breaks into two fragments with emission of two to three neutrons. Typically, spontaneous fission results in the formation of two

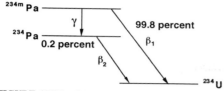

FIGURE 15.7.2 Composite decay scheme for protactinium-234.

asymmetric mass fragments. Each fissionable nuclide can produce a wide range of fragment masses. Spontaneous fission decay has a larger probability of occurrence for high atomic number nuclides and competes with alpha decay for dominance for transuranic (those with Z larger than uranium) elements.

Neutron-induced fission, conceptually different from spontaneous fission, requires a neutron to interact with the target nucleus to initiate the reaction. Neutron-induced fission from uranium-235 and plutonium-239 is used to initiate a chain reaction for nuclear weapons and commercial reactor power generation. Waste material from the fission process is a major constituent of high-level radioactive waste.

Composite Decay Schemes

Most of the several thousand known radionuclides complete the deexcitation process through a combination of the simple decay modes discussed above. Each of the decay schemes is unique to the selected parent nuclide. Figure 15.7.2 shows that metastable protactinium-234 can decay to uranium-234 by two separate routes:

- simple beta minus decay, or
- gamma emission with subsequent beta minus decay.

Approximately 99.8 percent of all protactinium-234 nuclei in the selected metastable state will decay directly to uranium-234 by a single beta minus decay. Conversely, only 0.2 percent of the protactinium-234 nuclei in the metastable state will decay by the two-step process shown in Figure 15.7.2. The branching ratio for the single beta minus mode is defined to be 99.8 precent and the two-step gamma and beta minus process is 0.2 percent.

Decay schemes can be quite complicated based on the individual characteristics of the parent nuclide. Thorium and uranium have a chain of decay modes that ultimately end with an isotope of lead. The thorium-232, uranium-238, and actinium (primary parent is uranium-235) chains decay by a combination of alpha and beta reactions to ultimately result in formation of lead-208, lead-206, and lead-207, respectively. This concept is important to the environmental professional because it provides an indicator of the types of nuclides that are present in various high-level and low-level waste containers.

15.7.4 PHYSICAL CONCEPTS

It is important to understand the concepts that define the amount of radioactivity, rate at which the nuclide decays into its daughter product, and terminology that is commonly used to understand how radiation interacts with the environment. Descriptions of important terms are found in the subsections that follow.

Activity

Activity is defined as the amount of radioactive material that is transformed into its daughter product per unit time. The SI unit for activity is the becquerel (Bq) and is defined as the quantity of radioactive material in which one atom is transformed to its daughter product per second. Activity is a measure of the number of atoms that are decaying and not the rate of transformation. It should be understood that the becquerel is not necessarily equivalent to the number of particles being emitted by the radionuclide per second. For example, a pure beta minus emitter such as tritium emits one electron per nuclear transformation. However, cobalt-60 emits one electron and two gamma photons per transformation.

The becquerel is numerically small and it is often necessary to express activity in metric multiples of the base unit.

The more traditional unit for activity is the curie (Ci). Originally, the curie was defined as the activity in approximately 1 g of radium-226. Currently, the curie is defined as the amount of radioactive material in which 3.7×10^{10} atoms are transformed each second. One curie of activity is equal to 3.7×10^{10} Bq. The curie is numerically large and it is often necessary to express activity in metric submultiples of the base unit.

Activity is simply the amount of radiation present in a given sample. It does not take sample type, volume, or mass into consideration. The radioactive concentration, or specific activity, is defined as the total activity per unit mass or volume.

Half-Life

Each radionuclide decays at its own unique transformation rate. The half-life ($t_{1/2}$) is defined as the amount of time required to reduce the activity of a selected radionuclide to half its original value. No physical or chemical factors can change the half-life of a radionuclide. Typical half-life values range from a few fractions of a second to billions of years.

An example of a fractional activity versus elapsed time plot for carbon-14 is shown in Figure 15.7.3. Notice that the resulting curve is linear when the data are plotted in semilogarithmic space. At the start of the decay sequence, the fraction of carbon-14 atoms in the sample is equal to 1.0. As time progresses, the fraction of carbon-14 in the sample decreases, because carbon-14 transmutes itself into nitrogen-14 as a result of beta minus decay. When the activity drops to half (50 percent maximum value), approximately 5730 years has elapsed since the start of the decay process. Figure 15.7.3 shows that the half-life for carbon-14 is 5730 years. The concept of radionuclide lifetime is statistical; large numbers of atoms are involved with the collective average lifetime being equal to the half-life. An individual atom of carbon-14 may decay at any time after the start of the decay sequence. The statistical average time required to transmute half of all carbon-14 atoms is 5730 years.

It is useful to understand what fraction of a particular radionuclide is left after a desired number of half-lives. Equation 15.7.4 demonstrates a mathematical relationship to calculate the fraction (f) of activity remaining after a desired number of half-life periods.

$$f \equiv \frac{A}{A_0} = \frac{1}{2^n} \qquad (15.7.4)$$

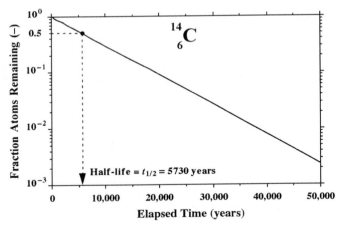

FIGURE 15.7.3   Half-life of carbon-14.

The A parameter is the activity at a specific number (n) of half-live periods and A_0 is the initial activity at the start of the decay sequence. For example, a strong source of cesium-137 has a half-life of 30.17 years. A radiological worker needs to know the magnitude of the fractional activity after three half-life periods. The fractional activity for cesium-137 after three half-life periods is 12.5 percent of the original value, as calculated with Equation 15.7.4.

A more complete mathematical expression for the temporal decay of a selected radionuclide can be found in Equation 15.7.5.

$$f \equiv \frac{A}{A_0} = \exp\left[-\frac{\ln 2t}{t_{1/2}}\right] \tag{15.7.5}$$

Equation 15.7.4 defines the f, A, and A_0 parameters. The ln 2 notation is shorthand for the natural logarithm of 2 and is approximately equal to 0.69315. The t parameter is equal to elapsed time in the same units that are used for the half-life in the denominator of the exponential expression. Assume that an environmental professional wants to determine the activity fraction 3 years after creation of a cesium-137 source. With Equation 15.7.5, approximately 93.3 percent of cesium-137 activity remains after 3 years of decay. The mathematical formulation shown in Equation 15.7.5 is very useful if the elapsed time is not a multiple of the half-life period as shown in Equation 15.7.4. Equation 15.7.6 shows an expression for the half-life of a selected radionuclide as a function of elapsed time and fractional activity.

$$t_{1/2} = \left[-\frac{\ln 2t}{\ln f}\right] \tag{15.7.6}$$

A daughter of the environmental professional waits 50 years after creation of an unknown source and measures an activity fraction of 5.97 percent. Using Equation 15.7.6, the daughter calculates that the unknown radionuclide has a half-life of approximately 12.3 years. Comparison of this calculated value with known references indicates that the unknown source is tritium. Every radionuclide has a unique half-life. The half-life is one of the most quoted parameters in the field of radiation.

Absorbed Dose

Radiation-induced damage is approximately proportional to the density of absorbed energy in the desired material. The basic unit of radiation dose is usually expressed in units of absorbed energy per unit material mass. The gray (Gy) is the accepted SI unit of radiation dose and is defined to be 1 J of absorbed energy per kilogram of material (J/kg). The gray is used to define the amount of absorbed dose for any type of ionizing radiation. It should be noted that the absorbed dose must be defined in terms the material being exposed. For instance, if a radiation worker is exposed to a dose of 0.001 Gy (1 milligray—mGy), which is typically quoted in the literature as 1 mGy(tissue), this is equivalent to a dose of 0.44 mGy in silicon [0.44 mGy(Si)] with an equal deposition volume. To have any realistic meaning, an absorbed dose must be expressed in terms of the deposition material.

Before adoption of the gray, an older unit, the rad (radiation absorbed dose), was widely used in the radiation field. One rad is equal to an absorbed radiation dose of 100 ergs per gram of material. Because 1 J is equal to 10^7 ergs and 1 kg is equal to 10^3 g, 1 Gy is equivalent to 100 rad.

Photon Exposure

Exposure is related to the amount of energy transferred from the incident radiation to a unit mass of air. The unit of exposure for photons is the coulomb per kilogram (C/kg) and it is defined as the amount of charge produced by incident x-ray or gamma radiation per unit mass of air. The exposure unit is based on air because of the relative ease with which radiation ionization can be measured in a standard ionization chamber. This definition of exposure is valid only for x-rays and gamma photons with energies that do not exceed 3 MeV. Exposures for higher energy photons usually are expressed in units of J/m^2.

It takes an average of 34 eV (electron volts) to produce an ion pair in air; 1 C/kg is equal to 34 Gy(air). The coulomb per kilogram is an integrated measure of exposure and is independent of time. The strength of a photon field usually is expressed in terms of an exposure rate.

Before adoption of the coulomb per kilogram, an older unit, the roentgen (R), was used to denote radiation exposure. One coulomb per kilogram is equal to a photon exposure of 3881 R; 1 R also is equivalent to an absorbed dose of 0.877 rad(air).

For radiation types other than x-ray and gamma photons, exposure is usually expressed in terms of fluence. Fluence can be expressed in terms of the number or total energy of particles per unit area. Fluence rates can be expressed as flux for both particle number and total energy.

Dose Equivalent

Extensive research has shown that the various radiation types do not produce identical biological effects. It is not sufficient to use absorbed dose to gauge the potential for damage to a biological system. Individual types of radiation damage biological systems to different degrees. The concept of a dose equivalent was developed to compensate for this fact. The dose equivalent (H) is defined as shown in Equation 15.7.7.

$$H = DQ \qquad (15.7.7)$$

The typical unit for dose equivalent is the sievert (Sv), which has dimensions identical to absorbed dose. The D parameter in Equation 15.7.7 is the absorbed dose for the selected material in grays. The quality factor Q in Equation 15.7.7 relates the relative ability of a given radiation to damage a biological system. Table 15.7.3 lists quality factors for common varieties of ionizing radiation. Note that the magnitude of individual quality factors depends on radiation type and energy. The quality factor numbers in parentheses in Table 15.7.3 are new values proposed by the National Council on Radiation Protection (NCRP). The NCRP is a professional organization with the purpose of recommending safe levels of radiation exposure to American governmental agencies such as the Nuclear Regulatory Commission, the Occupational Safety and Health Administration (OSHA), and the Environmental Protection Agency (EPA). Similar organizations, such as the International Commission on Radiological Protection (ICRP), International Atomic Energy Agency (IAEA), and the International Radiation Protection Association (IRPA), provide recommendations for radiation exposure to the entire world.

The quality factor concept creates a crucial problem in actually determining the numerical values for each type of ionizing radiation in precise experimental work. Biological damage from ionizing radiation depends on many other factors, such as acid or base balance, tissue temperature, and oxygen content. A more rigorous treatment of the origins of quality factors is based on the linear energy transport (LET) concept. A detailed description of LET and its relationship to quality factor can be found in Ref. 6.

Assume that a worker is exposed to 1 mGy of gamma radiation. This would result in a dose equivalent of 1 millisievert (mSv), because the quality factor for gamma photons is equal to 1.0. For

TABLE 15.7.3   Quality Factor Values

Radiation type	Quality factor (Q)
X-rays and gamma photons	1.0
Electrons (betas)	1.0
Thermal neutrons	2 (5)
Fast neutrons	10 (20)
Protons	10 (20)
Alphas and heavy ions	20

Numbers in parentheses are values proposed by NCRP.

a different worker who is internally exposed to 1 mGy of alpha radiation, the dose equivalent for the alpha exposure is 20 mSv. This difference demonstrates that alpha particles cause more damage to biological systems than an equivalent number of gamma photons. Alpha particles are not much of an exposure threat unless the specific radionuclide is taken internally. When taken internally, alpha emitters deposit large amounts of energy in surrounding organs or other areas and cause significant health problems for the biological system. Like alphas, most other heavy charged particles are a biological threat if the radionuclide is taken internally. Gamma photons are an external threat to living tissue because of their extended range. Electrons have a smaller range than gamma photons, but they represent both an internal and an external threat to living matter. Unlike other types of radiation, neutrons do not directly ionize target material. Secondary charged particles, created in neutron interactions with materials, pose a serious internal and external exposure hazard.

Before international adoption of the sievert, an older unit, rem (radiation equivalent man), was widely used in the radiation field. One rem is equivalent to an absorbed photon or electron particle dose of 1 rad. A dose equivalent of 1 Sv is equal to 100 rem. Before acceptance of the quality factor, an older concept known as relative biological effectiveness (RBE) was used to gauge the effect of different types of radiation on living matter. The RBE is defined as the ratio of the number of 200-keV photons needed to produce a specified biological effect to the relative amounts of damage caused by a particular type of radiation. Because a 200-keV gamma photon is used to describe the RBE, it is of limited use when trying to compare generalized biological effects over a wide range of radiation energies. Today the quality factor is more universally accepted to estimate biological damage due to particle irradiation.

15.7.5 BIOLOGICAL EFFECTS

Since the discovery of radioactivity, its danger to human health was recognized as a serious problem. Many scientists suffered radiation burns and developed cancer as a result of prolonged exposure. Marie Sklodowska Curie, two time Nobel Prize winner for her radioactive decay research (1903) and discovery of radium (1911), died in 1934 from aplastic anemia (bone marrow disease) that was probably caused by her prolonged exposure to high levels of radiation.

Radiation interacts with matter largely through ionization of target nuclei. Ionization of water in the body causes the formation of unbound oxygen and hydrogen ions known as free radicals. Free radicals recombine to form hydrogen peroxide, which is toxic to living cells. Because approximately 80 percent of the human body is composed of water, ionization and the formation of free radicals are very dangerous. The effect of radiation on biological systems is one of the most thoroughly researched areas in medical science. More information is currently known about the mechanisms of radiation damage on the molecular, cellular, and organ levels than is known for most other hazardous materials.

Health physicists typically are concerned with acute and chronic exposures. An acute exposure is a single accidental irradiation incident during a short time period. A chronic exposure is long-term excessive irradiation probably or usually due to inadequate mitigation measures.

Not all organs and systems are equally sensitive to radiation. The severity of the biological response for a given individual depends on radiation type, magnitude of the absorbed dose, and portion of the body exposed. A whole body exposure is irradiation on all parts of the main body trunk by an outside radiation source. A partial exposure takes place only when certain organs or extremities are irradiated. For example, a whole body dose of 7.0 Gy (700 rad) would result in death for most exposed individuals. An identical dose to an extremity, such as a leg or foot, would cause severe reddening, blistering, or peeling of the affected tissue.

Table 15.7.4 shows that as the radiation dose increases, the detectable body changes and symptoms also increase in severity. The dose range in Table 15.7.4 is valid for beta or gamma radiation. At low radiation doses, decreases in both the red and white blood cell count are the only detectable changes in body chemistry. As the dose increases, the gastrointestinal tract becomes involved, and fatalities begin to occur at a dose of approximately 2.0 Gy (200 rad). At a dose of approximately 4.0 to 6.0 Gy (400 to 600 rad), 50 percent of the human population will die. This is defined as the dose at which 50 percent

TABLE 15.7.4   Body Changes, Symptoms, and Fatalities Induced by Radiation

Dose range (Gy)	Detectable body changes	Symptoms	Fatalities
0.25 to 0.50	Decrease in red and white blood cell count	Temporary sterility	None
2.0	Temporary damage to bone marrow	Mild nausea, vomiting, malaise, fatigue, and loss of hair	Small fraction
4.0 to 6.0	Temporary loss of bone marrow function	Moderate nausea, vomiting, malaise, fatigue, and loss of hair	Moderate fraction (LD_{50})
7.0	Permanent loss of bone marrow	Severe nausea, vomiting, malaise, fatigue, and loss of hair	100 percent
10.0	Sloughing off of intestinal epithelia	Extreme nausea, vomiting, malaise, fatigue, and loss of hair	100 percent within 1 to 2 weeks
20.0	Damage to central nervous system	Unconsciousness within a few minutes	100 percent within hours or days

of the population (LD_{50}) dies. At extremely high radiation doses, the central nervous system becomes involved, with 100 percent fatalities beginning a few hours after exposure.

In addition to the prompt somatic (nonreproductive) body changes and symptoms discussed in Table 15.7.4, radiation exposure also results in delayed somatic, genetic, and teratogenic effects. Delayed somatic effects occur several months or years after exposure, when the victim usually shows signs of accelerated aging, shortened life span, or cancer. Genetic effects result from the irradiation of reproductive organs causing gene mutations that lead to problems in future generations. Teratogenic effects result from direct irradiation of the developing fetus and can result in physical deformity or death. The greatest damage to the fetus is incurred during the first 2 months of development.

15.7.6   RADIATION SOURCES

Ionizing radiation brings about images of horror, destruction, and power beyond human comprehension. The discovery of radioactivity by Becquerel in 1896 caused people to change completely their views about the physical nature of matter and energy. In one of his famous theories, Albert Einstein stated that matter and energy are completely interchangeable. The idea that matter is inherently unstable and can decay into a variety of different species is a new and daunting concept.

Even though human understanding of radiation is only a century old, nuclear processes have been taking place in nature since the dawn of the universe more than 15 billion years ago. Our sun, an average star, has been quietly generating energy for several billion years by using thermonuclear processes. The core of the earth is maintained in a molten condition in part by the continuous decay of a variety of radioactive materials. Radioactivity is a natural process that is as old as time itself. However, radioactivity can pose a threat to the health and safety of the human environment. The use of radioactive materials for strategic and tactical weapons, power production, medical diagnosis and research, and related purposes ensures that most of the human population has been exposed to levels of radioactivity in excess of the accepted environmental background.

Consumer Products and Activities

Radioactive sources are either anthropogenic or natural in origin. People are exposed to radiation from medical treatment and diagnosis, consumer products, nuclear reactor fuel elements, residual fallout from weapons testing/bombing, and serious nuclear accidents such as Chernobyl. Table 15.7.5 lists the yearly dose equivalent values resulting from exposure to a selection of consumer products and activities.

TABLE 15.7.5   Yearly Dose Equivalent Values for Consumer Products and Activities

Consumer lifestyle or product	Dose equivalent (mSv/yr)	Dose equivalent (mrem/yr)
Radium dial watch	0.03	3
Home ionization smoke detector	0.01	1
Coal-fired power plant	0.01 to 0.04	1 to 4
Cigarettes at 1 pack/day (lungs)	0.4	40
Exposure to building masonry	0.07	7
Cooking with natural gas stove	0.06 to 0.09	6 to 9
Whole mouth dental x-ray	0.9	900
Chest x-ray	0.4	40
Television receiver	0.003	0.3
Reading a book	0.005	0.5

Natural Radiation

Natural radiation provides the principal source of exposure to the general public. Sources such as tritium, carbon-14, potassium-40, radon-220, radium-226, thorium-232, and uranium-238 are often called naturally occurring radioactive materials (NORM). In most cases, the radiation activity for NORM is low. The NCRP estimates that Americans are exposed to NORM at approximately 0.5 mSv/yr (50 mrem/yr) in rocks and soil and 0.4 mSv/yr (40 mrem/yr) from cosmic rays. Living in a frame house can increase the yearly dose equivalent by almost 50 percent. Simply taking a single coast-to-coast trip on an airplane will result in an added dose equivalent of between 0.03 to 0.05 mSv (3 to 5 mrem). Today a 1-hour stop at the Trinity Site in southern New Mexico would result in a dose equivalent of 0.005 to 0.01 mSv (0.5 to 1.0 mrem) to the visitor. The Trinity Site is the location where the world's first atomic bomb was tested at 5:29:45 a.m. Mountain War Time on July 16, 1945. This dose equivalent range is about 10 times the natural background in New Mexico.

The average background radiation dose equivalent from all sources of cosmic rays and NORM for persons living in the United States is about 1.0 mSv/yr (100 mrem/yr). This quantity varies from a low of 0.75 mSv/yr (75 mrem/yr) in Texas to 2.4 mSv/yr (240 mrem/yr) in Wyoming because of altitude and proximity to higher concentrations of radioactive materials.

Scientists have found evidence of a natural fission reactor at the Oklo Mine in Gabon, Africa. In 1972, scientists found that certain uranium ore samples showed a depletion of the uranium-235 nuclide so that a fossil fission reactor existed several million years ago. Based on the age of the deposit, scientists have postulated that a uranium-235 enrichment of approximately 3 percent existed during the "operating" phase of the reactor. Normally, uranium-235 is found in 0.72 percent of all uranium deposits. The postulated 3 percent uranium-235 enrichment is close to the value used in a modern nuclear power reactor. The reactions could be caused by spontaneous fission or by the neutron component of cosmic rays. Evidence at the site indicates that surrounding rocks have been strongly heated in the past and contain higher than normal levels of typical fission fragment nuclides. The Oklo mine is of specific interest because it can be used as a guide to study radioactive waste disposal problems.

Radiation Accidents

Table 15.7.6 compares the radiation dose equivalents released at Three Mile Island and Chernobyl. Radiation released during these incidents is compared with a standard chest x-ray, the average total dose equivalent, and exposure to the Hiroshima bombing. It should be noted the LD_{50} value for radiation exposure is 4 to 6 Sv (400 to 600 rem). Radiation released from the accident at the Three Mile Island nuclear plant in 1979 was less than the usual background level in Pennsylvania. This release was mostly in the form of xenon-133, a nonreactive fission product of uranium-235. Conversely,

TABLE 15.7.6   Comparison of Radiation Accident Releases

Radiation source	Description	Dose equivalent (mrem)	Dose equivalent (mSv)
Three Mile Island accident, Pennsylvania	Accident dose within 10 mil (6.1 km); highest dose equivalent to any person was 100 mrem (1 Sv)	8	0.08
Chest x-ray	Average diagnostic medical x-ray dose equivalent	40	0.4
Average adult dose equivalent (U.S.)	Average adult dose equivalent in the United States to all radiation sources	187	1.87
Chernobyl accident, Ukraine	Evacuees dose equivalent; clean-up workers were exposed to 30,000 to 40,000 rem (300 to 400 Sv)	11,800	118
Hiroshima, Japan bombing (WWII)	Average dose equivalent to Japanese in area of bombing	30,000	300

TABLE 15.7.7   Radionuclide Release Comparison

Released radionuclide	Activity released at Chernobyl accident (TBq)	Activity released by an ideal 1 megaton nuclear explosion (TBq)
Iodine-131	260,000	0.00259
Cesium-137	37,700	1180
Strontium-90	8000	3700
Plutonium-239	25.5	133

the core explosion at the Chernobyl power station in the Ukraine on April 26, 1986, was the largest commercial radiation accident in history. A combination of operator error and poor engineering caused a catastrophic steam explosion in the Unit 4 reactor. Approximately 5 percent of the total nuclide core inventory was released to the environment. Table 15.7.7 compares some of the radionuclides released at Chernobyl with an ideal 1 megaton nuclear explosion. Note that 1 TBq is equal to 10^{12} Bq or 27 Ci. In many ways, the Chernobyl accident is equivalent in magnitude to the radiation emitted from a multimegaton nuclear device. A total of 31 persons died during the initial phases of the accident because of a combination of explosion, fire, and radiation exposure. Areas immediately surrounding the plant had to be evacuated because of high radiation exposures. The Russian government estimates that approximately 4000 people received radiation dose equivalents of approximately 2 Sv (200 rem), which is half of the LD_{50} value for humans.

15.7.7   RADIATION PROTECTION OPTIONS

One of the most important tasks faced by an environmental professional is to reduce occupational exposure to radiation workers. Reduction of occupational exposure is one of the best ways to minimize the probability of developing future health problems. The ICRP recommends (Publication 26) a system to reduce exposure to radiation workers:

- No practice or procedure should be used unless its use produces a net positive benefit;
- All exposures should be minimized by balancing all economic and social factors; and
- The dose equivalent should not exceed the recommended limits for a given situation.

The actual operational limits for exposure should be kept as low as possible even if this level is smaller than any recommended or legal limits as noted above. Controls such as process evaluation or modification, safety equipment design, physical exposure reduction, and other plant engineering factors should be designed so workers do not exceed the operational exposure limit. This principle is known as the ALARA (as low as reasonably achievable) concept and should optimize the cost of adverse health effects to the benefits to be gained from the use of radiation.

Although physical exposure reduction is only one of the techniques used in the ALARA concept, it is usually the easiest for individual workers to control. Physical exposure reduction is characterized by time, distance, and shielding. Exposure is directly proportional to the amount of time a worker spends near a radiation source. Training can improve process techniques to reduce the amount of time that a worker is exposed to radiation. Maximizing the distance between a source and exposed workers also reduces radiation exposure. The inverse square law can be used to calculate exposure at a desired distance from a point source or a source with dimensions that are much smaller than the desired distances, as shown in Equation 15.7.8.

$$\frac{I_1}{I_2} = \left[\frac{d_2}{d_1} \right]^2 \tag{15.7.8}$$

The I_1 and I_2 parameters represent radiation intensity, absorbed dose, or dose equivalent at distances d_1 and d_2 from the source of radiation. For example, workers experience an absorbed radiation dose of 18 mGy at a distance of 1 m from a point source. The workers want to know the dose they would experience at a distance of 3 m from the source. Equation 15.7.8 shows that the workers would experience a dose of 2 mGy at a distance of 3 m from the source. The dose is reduced by a factor of 9 when the distance is increased from 1 to 3 m. This reduction in dose is consistent with the ALARA principle and can be achieved if workers can complete required tasks farther from the source.

The last physical method to reduce radiation exposure is the use of shielding between the source and worker locations. Shielding absorbs radiation and reduces exposure to workers. Alpha particles and other heavy ions are not very penetrating and are easily stopped in air or by a standard sheet of writing paper. Electrons from beta decay are best shielded with low-Z materials, such as plastic, wood, cardboard, aluminum, and copper. The use of low-Z materials reduces the amount of electron backscattering that is prevalent with high-Z materials. X-rays and gamma photons are best shielded by dense high-Z materials, such as lead, iron, tungsten, and concrete. Low-Z materials like water, plastic, carbon, paraffin, and polyethylene best shield neutrons. Materials rich in protons more efficiently transfer energy from the incident neutron radiation to the target shield. For example, a nuclear reactor emits neutrons and gamma photons in large quantities. The optimum shield for gamma photons is high-Z absorbers such as lead and concrete. Conversely, neutrons are best shielded by using low-Z absorbers like paraffin. The best solution to shield both gamma photons and neutrons is a composite shield made of appropriate high-Z and low-Z materials. Optimizing shield design is an important component of the ALARA principle for a selected radiation facility.

15.7.8 LAWS AND REGULATIONS

The amended Atomic Energy Act (AEA) of 1954 is the basis of all laws and regulations controlling the use of radioactive materials in the United States. The AEA created an Atomic Energy Commission (AEC) to control the production and use of radioactive materials in the United States. The AEC later became the Nuclear Regulatory Commission (NRC), which currently has statutory authority for the use of radioactive materials. Other American agencies such as the EPA, OSHA, and the Department of Transportation have important roles in disposal, emission reduction, worker safety, and transportation of radioactive materials. Individual states can assume legal control over their complement of radioactive materials if they can develop a program that is at least as strict as current NRC regulations. States such as Alabama, California, Tennessee, Louisiana, and New York are authorized by the NRC to run individual regulatory programs and are classified as agreement states. To determine whether you live and work in an agreement state, contact the NRC or your state radiological health physicist.

In the United States, regulations that control the use of radioactive materials are quite complicated. A variety of federal agencies use more than 20 separate regulations and standards to govern radiation protection. The NRC is the lead agency in charge of radiation protection and maintains appropriate regulations in the Code of Federal Regulations (CFR), Title 10 Part 20 (10 CFR 20). The NRC specifies radiation exposure limits for the general public, occupationally exposed workers, and persons completing medical treatment. Using the ALARA principle, dose limits to these groups are minimized to reduce the risk of adverse health effects.

REFERENCES

1. Beiser, A., *Perspectives of Modern Physics*, McGraw-Hill Book Company, 1969.
2. Bevelacqua, J., *Contemporary Health Physics, Problems and Solutions*, John Wiley and Sons, 1995.
3. Cember, H., *Introduction to Health Physics*, 2nd edition, Pergamon Press, 1983.
4. Cohen, B., *Concepts of Nuclear Physics*, McGraw-Hill Book Company, 1971.
5. Ehmann, W., and Vance, D., *Radiochemistry and Nuclear Methods of Analysis, Vol. 116, Chemical Analysis*, John Wiley and Sons, 1991.
6. Enge, H., *Introduction to Nuclear Physics*, Addison Wesley, 1966.
7. Glasstone, S., and Dolan, P., *The Effects of Nuclear Weapons*, 3rd edition, U.S. Departments of Defense and Energy, Washington, DC, 1977.
8. Hammond, A. (ed.), *The 1993 Information Please Environmental Almanac*, Houghton Mifflin Company, 1993.
9. Kim, S., and Strait, E., *Modern Physics for Scientists and Engineers*, Macmillan Publishing Company, 1978.
10. Knoll, G., *Radiation Detection and Measurement*, 2nd edition, John Wiley and Sons, 1989.
11. Lapp, R., and Andrews, H., *Nuclear Radiation Physics*, 3rd edition, Prentice-Hall, 1963.
12. Mayer, J., and Rimini, E. (eds.), *Ion Beam Handbook for Material Analysis*, Academic Press, 1977.
13. Medvedev, Z., *The Legacy of Chernobyl*, W.W. Norton and Company, 1990.
14. Melissinos, A., *Experiments in Modern Physics*, Academic Press, 1966.
15. Paul, E., *Nuclear and Particle Physics*, North Holland Publishing Company, 1969.
16. Seaborg, G., *Man-Made Transuranium Elements*, Prentice-Hall, Incorporated, 1963.
17. Semat, H., *Introduction to Atomic and Nuclear Physics*, 3rd edition, Rinehart and Company, 1954.
18. Smith, A., and Cooper, J., *Elements of Physics*, 8th edition, McGraw-Hill Book Company, 1972.
19. Shleien, B., and Terpilak, M., *The Health Physics and Radiological Health Handbook*, Nucleon Lectern Associates, 1984.
20. Walker, F., Miller, D., and Feiner, F., *Chart of the Nuclides*, 13th edition, General Electric Company, 1984.
21. Wang, C., Willis, D., and Loveland, W., *Radiotracer Methodology in the Biological, Environmental, and Physics Sciences*, Prentice-Hall, 1975.
22. Weast, R. (ed.), *CRC Handbook of Chemistry and Physics*, 59th edition, CRC Press, 1978.

CHAPTER 15

UBIQUITOUS ENVIRONMENTAL CONTAMINANTS

SECTION 15.8

MORTALITY IN MALE AND FEMALE CAPACITOR WORKERS EXPOSED TO POLYCHLORINATED BIPHENYLS

Renate D. Kimbrough

Dr. Kimbrough works in toxicology, environmental and occupational health, and pathology as an independent consultant since December 1999. Prior to that she was with the Institute for Evaluating Health Risks.

Martha L. Doemland

Dr. Doemland is an independent consultant in Hopewell, New Jersey. Her work has primarily been in the field of environmental and occupational epidemiology. Her major efforts have focused on health effects of PCBs.

Maurice E. LeVois

Dr. LeVois is principal scientist at LeVois and Associates, providing epidemiological research and analysis services from offices in the San Francisco Bay area.

A mortality study was conducted in workers with at least 90 days exposure to polychlorinated biphenyls (PCBs) between 1946 and 1977. Vital status was established for 98.7 percent of the 7075 workers studied. In hourly male workers, the mortality from all cancers was significantly below expected [standardized mortality ratio (SMR) = 81; 95 percent confidence interval (CI), = 68 to 97] and comparable to expected (SMR = 110; 95 percent CI, 93 to 129) in hourly female workers. No significant elevations in mortality for any site-specific cause were found in the hourly cohort. All-cancer mortality was significantly below expected in salaried males (SMR = 69; 95 percent CI, 52 to 90) and comparable to expected in salaried females (SMR = 75; 95 percent CI, 45 to 118). No significant elevations were seen in the most highly exposed workers, nor did SMRs increase with length of cumulative employment and latency. None of the previously reported specific excesses in cancer mortality was seen. This is the largest cohort of male and female workers exposed to PCBs. The lack of any significant elevations in the site-specific cancer mortality of the production workers adds important information about human health effects of PCBs.

PCBs are complex mixtures of 209 different chlorinated biphenyl congeners. They were used extensively in the United States from the 1930s through 1977 in a variety of industrial applications. PCB mixtures have several chemical and physical properties that make them extremely versatile, including resistance to acids and bases as well as oxidation and reduction; compatibility with organic materials; and thermal stability and nonflammability. The major volume usage of PCBs was in capacitors and transformers as dielectric fluids, but they were also used as lubricants and sealants; as additives in paint, plastics, newspaper print, and dyes; as extenders in pesticides; and as heat transfer and hydraulic fluids. More than 95 percent of the liquid-filled electrical capacitors and transformers produced before the early 1970s contained PCBs. PCBs are persistent chemicals and have bioaccumulated in the environment. They continue to be detected in air, soil, water, and sediment. Trace amounts are also present in the tissues of wildlife, domestic animals, and humans; however, the levels of PCBs in the environment are declining.

The potential for adverse human health effects of PCB exposure has been a concern since the early 1970s and resulted in the Environmental Protection Agency's ban of the production of PCBs in 1978. Current knowledge about the human health effects of PCBs is limited, inconsistent, and difficult to interpret. Occupational mortality studies of capacitor workers have reported higher than expected rates of melanoma and cancer of the liver, rectum, gastrointestinal tract, brain, and hematopoietic system. The site-specific elevations, however, have not been observed consistently across studies. More importantly, many of the elevated cause-specific SMRs reported in the various studies were not correlated with higher and/or longer exposures to PCBs or longer latency periods, which would suggest a dose-response relationship.

We conducted a retrospective cohort mortality study of 7075 workers exposed to PCBs during the capacitor manufacturing process. The cohort represents all hourly and salaried workers from two plants in upstate New York who were employed for 90 days or more from 1946, when capacitor manufacturing began, through June 15, 1977, when PCB use was completely phased out. A cohort of the same workforce was previously assembled by other investigators but was incomplete because it did not include approximately 850 workers. A portion of the highly exposed male workers in this study were also included in the cohort assembled by Brown and Jones.

15.8.1 METHODS

Study Purpose

The purpose of this study was to further explore previously reported excesses in cancer-specific mortality in capacitor workers exposed to PCBs. Six a priori cancers that were previously reported as being elevated were the primary focus of this study (melanoma, liver, rectum, gastrointestinal tract, brain, and hematopoietic cancers).

Study Population

All hourly and salaried workers employed for at least 90 days between January 1, 1946, and June 15, 1977, in two capacitor manufacturing plants in upstate New York were included in the cohort. Salaried personnel were included in the cohort because they were often involved in the manufacturing process and all personnel were housed in the same building. Personal identifiers, including social security number, demographics, and each worker's job history information, were abstracted from employment records that had been microfilmed by the company. Completeness of the cohort was established by our reviewing all available company records, such as pension rosters and the quarterly earning reports of the Social Security Administration (SSA, 941 forms; the complete payroll record for every employee who was ever paid by the company, by plant and location).

Vital Status Determination

The National Death Index and the Equifax Nationwide Death Search tapes were matched against our cohort to identify deceased cohort members through December 31, 1993. A match was considered to exist when at least six digits of the social security number, the date of birth, and the first and last name for men and first name for women matched. Death certificates were obtained from the state where death occurred and were coded by a certified nosologist. The underlying cause of death was coded using the International Classification of Diseases (ICD) revision in force at the time of death.

Significant effort went into establishing the vital status of workers not identified as deceased, working, or receiving a pension. Company files were used to identify workers still employed in 1994 and retired workers receiving a pension. The locating services of Equifax were used to establish alive status for cohort members who were separated from employment at the time of vital status determination. This was done primarily through the identification of cohort members involved in activities such as insurance underwriting and claims and financial transactions, such as mortgage applications. Computerized county voter registration lists and annual R. L. Polk and Haines City Directories were also used and matched against the cohort list to establish alive status. When employees could not be identified as alive by the above sources, direct contact with neighbors or relatives was attempted to verify vital status. A private investigator was employed to locate difficult-to-find former employees. Long-term workers who were currently employed also assisted us in locating former employees. Workers with unknown vital statuses were considered to be "lost-to-follow-up" and were observed through their last date of employment.

Exposure Assessment

Capacitors were produced by assembling the capacitor canisters; filling them with PCBs; heating them to achieve better impregnation; and closing, soldering, and cleaning them. Capacitors were also repaired, which entailed removing the cover, draining the PCBs, repairing the unit, and reconstructing the capacitor.

From 1946 through 1971, the Aroclor mixtures (the tradename for Monsanto PCB products) 1254 and 1242 were predominantly used. Aroclor 1254 was phased out after 1954, and Aroclor 1242 was used until 1971. From 1971 until 1977, Aroclor 1016 was used. Aroclor 1016 was similar to Aroclor 1242 but with lower environmental persistence, which was accomplished by the removal of higher chlorinated homologs.

All jobs were classified according to their levels of PCB exposure before determination of vital status. Jobs with direct PCB contact (dermal contact and/or inhalation exposure to high PCB air levels) experienced while filling, impregnating, repairing, or moving PCB-filled capacitors were classified as high-exposure jobs. In the areas of filling and impregnating, air levels ranged from 227 to 1500 μg/m^3 in 1975; in the spring of 1977, when PCB use had declined substantially, air levels ranged from 170 to 576 μg/m^3. Work operations in which no PCBs were used, such as the winding, can, and cover manufacture and the assembly and shipping department, were tested in the spring of 1977, and air levels ranged from 3 to 50 μg/m^3. These jobs were classified as low exposure, and workers in these areas primarily had inhalation exposure to the background levels of PCBs in the plant. There were jobs for which the PCB exposure of the worker varied depending on the location where the individual was performing the task. Insufficient information was provided in the worker history records to determine the location of these jobs, and therefore it was not possible to assign exposure classifications for them. These jobs were classified as undefinable.

Between 1976 and 1979, the general population had average serum PCB levels on a wet-weight basis of approximately 5 to 7 parts per billion (ppb; measured in ng/mL), with levels ranging from nondetectable to 20 ppb (ng/mL), with occasional higher levels. In a population of 290 self-selected employees from this plant, the PCB levels measured in serum on a wet-weight basis ranged from 6 to 2530 ppb (ng/mL) for the lower chlorinated compounds and ranged from 1 to 546 ppb (ng/mL) for the higher chlorinated PCBs. In 1976, Lawton et al. found similar high levels in a group of 190 workers who had been selected because of their estimated high exposures, establishing the extensive exposure to PCBs.

Exposure to other chemicals in these plants was limited. A small number of workers were exposed to low concentrations of toluene in the painting area, with air levels from nondetectable to 21.4 μg/m^3 (time weighted average = 188 mg/m^3). Trichloroethylene levels in the degreasing area ranged from 3.7 to 321 μg/m^3 (time-weighted average = 269 mg/m^3). Low air levels of lead, aluminum, and iron were also reported in the soldering area.

Statistical Analyses

The mortality experience of the cohort is expressed as the SMR (number of observed deaths in the cohort divided by the number of expected deaths derived from the comparison population). Age-, sex-, race-, and calendar-specific mortality rates for the U.S. population and the regional population from the eight counties surrounding the plants (Franklin, Essex, Warren, Saratoga, Washington, Hamilton, Clinton, and Herkimer counties) were used to calculate the expected number of deaths. The U.S. mortality rate tables were provided by the National Institute of Occupational Safety and Health and included 92 causes of death for the years 1946 through 1993 for white and nonwhite males and females in 5-year age and calendar time periods. The regional mortality rate tables were obtained from the Mortality and Population Data System and included 62 causes of death for the years 1950 through 1989 for malignant neoplasms and the years 1962 through 1989 for nonmalignant causes. The regional rates were also provided for white and nonwhite males and females in 5-year age and calendar time periods. Person-years were accumulated beginning on the 91st day of employment and continued to December 31, 1993 or the date of death, whichever came first. Person-years for workers lost-to-follow-up were calculated through the last date they were known to be alive, which was typically their last date of employment. Person-years were combined into 5-year age-, sex-, race-, and calendar-specific categories and multiplied by the corresponding age-, sex-, race-, and calendar-specific U.S. mortality rates (or regional rates) to yield the expected numbers. Calculations were performed using OCMAP (Occupational Cohort Mortality Analysis Program). The statistical significance of the differences between the observed and expected numbers was tested assuming a Poisson distribution for the observed deaths, using a two-sided test of significance. SMRs were calculated for all 92 underlying causes of death; however, only selected causes of death, including all causes, all cancers, all site-specific cancers, the major cardiovascular diseases, diabetes, cirrhosis of the liver, and accidental causes are shown.

Presentation of the exposure-specific analysis is confined to those workers who had the greatest potential for exposure, which was defined in three ways: (1) all hourly workers who ever worked in a high-exposure job; (2) all hourly workers who had worked for at least 6 months in a high-exposure job; and (3) all hourly workers who worked for at least one year in a high-exposure job. Only 112 male and 12 female workers were exclusively employed in high-exposure jobs, which restricted our ability to analyze them as a separate group.

In addition to the overall SMRs, the impact of PCB exposure on mortality in both the total cohort and the high-exposure cohort was examined by categories of cumulative length of employment (<1 year, 1 to <5 years, 5 to <10 years, and 10 years) and years of latency. Two latency categories were defined: one as ≤ 20 years since first exposure, and the other as >20 years since first exposure. SMRs were calculated for each category of cumulative length of employment by latency for all causes, all cancers, and specific causes for which there was an elevated total SMR with two or more observed deaths and for which the lower boundary of the 95 percent CI was 90 or above. This analysis examined the trend across categories of increasing length of employment and latency. The purpose was to determine whether the SMRs increased over the length of employment and latency categories, which suggests a trend consistent with a dose-response effect.

The individual category-specific SMRs are not particularly relevant, and, because of the small number of deaths, the rates are unstable. Trend analysis using the Mantel-Cox χ^2 test with one degree of freedom was done to determine the significance of the trend among the observed over the expected rates within the subpopulation of women who died of intestinal cancer.

15.8.2   RESULTS

Description of the Cohort

The cohort consisted of 2984 hourly white male, 2544 hourly white female, 1078 salaried white male, and 469 white salaried female workers (Table 15.8.1). Hourly male and female workers contributed

TABLE 15.8.1 Demographic Characters of 4062 Male and 3013 Female Workers

Characteristic	Hourly workers		Salaried workers		Total (mean)
	Male	Female	Male	Female	
Number of workers	2984	2544	1078	469	7075
Number of person-years	85,991	75,674	34,755	16,358	212,778
Number of deaths	586	380	177	52	1195
Number of missing death certificates	20	4	9	4	37
Number lost to follow-up	33	52	7	3	95
Mean age started work	26	29	29	25	(27)
Mean time employed, years	6.2	5.8	5.7	4.8	(5.6)
Mean age stopped working	33	35	35	31	(34)
Mean age at death	61	64	62	61	(62)
Mean age of workers alive on December 31, 1993	53	57	60	59	(57)
Mean follow-up time, years	28	30	32	34	(31)
Percentage who attended college	15	7	73	26	(30)

85,991 and 75,674 person-years of observation, respectively, and salaried male and female workers contributed 34,755 and 16,358 person-years. Only 1.1 percent of the hourly male, 2 percent of the hourly female, and less than 1 percent of the salaried workers were lost to follow-up. The average age at entry for the different groups, the mean time employed, and the mean follow-up time are also shown in Table 15.8.1. The mean age at the end of employment for the four subgroups ranged from 31 to 35 years. The mean age of the 5880 cohort members alive at the end of the follow-up period was 57 years, and the mean age at death was 62 (Table 15.8.1). Among the salaried male cohort, 73 percent ever attended college, and in the hourly male cohort only 15 percent ever attended college. Among salaried female workers, 26 percent never attended college, and among the hourly female workers, 7 percent ever attended college.

There were 586 deaths among the hourly male workers, and 380 deaths among hourly female workers. There were 177 deaths among the salaried male workers and 52 deaths among the salaried female workers. For 38 workers, the cause of death was not known either because the death certificate could not be located or because the cause of death was not provided on the death certificate.

In Table 15.8.2 the length of employment and years of follow-up for the cohort are shown. Over one-third of hourly male and female workers and nearly one-third of the salaried male cohort worked

TABLE 15.8.2 Length of Employment and Follow-Up Time

Characteristic	Hourly workers (percent)		Salaried workers (percent)	
	Male	Female	Male	Female
Length of employment, years				
<1	1066 (35.7)	842 (33.1)	343 (31.8)	106 (22.6)
1 to <5	864 (29.0)	902 (35.5)	341 (31.6)	216 (46.1)
5 to <10	381 (12.8)	251 (9.9)	177 (16.4)	83 (17.7)
10 to <15	212 (7.1)	208 (8.2)	87 (8.1)	130 (5.8)
≥15	461 (15.4)	341 (13.4)	130 (12.1)	37 (7.9)
Years of follow-up				
<10	61 (2.0)	28 (1.1)	11 (1.1)	0
10 to <20	383 (12.8)	333 (13.1)	67 (6.2)	32 (6.8)
20 to <25	656 (22.0)	626 (24.6)	156 (14.5)	33 (7.0)
25 to <30	830 (27.8)	487 (19.1)	206 (19.1)	72 (15.4)
30 to <35	267 (8.9)	240 (9.4)	159 (14.7)	51 (10.9)
≥35	787 (26.4)	830 (32.6)	479 (44.4)	281 (59.9)

TABLE 15.8.3   Distribution of Exposure Types

	Hourly workers		Salaried workers	
Characteristic	Male	Female	Male	Female
Number ever highly exposed (percent)	1268 (42.4)	352 (13.8)	87 (8.0)	10 (2.1)
Median years in high exposure	1.7	1.6	3.2	2.0
Number ever undefinably exposure (percent)	1984 (66.4)	379 (14.8)	407 (37.7)	15 (3.1)
Median years in undefinable exposure	1.8	1.5	2.4	1.4
Number ever low exposed (percent)	2343 (78.5)	2468 (97.0)	831 (77.0)	459 (97.8)
Median years in low exposure	6.5	6.5	5.0	4.9

less than 1 year, and nearly one-third of hourly and salaried male and female employees worked 5 years or more. The distribution of follow-up time for the cohort is also presented and indicates that follow-up time for most of the cohort exceeded 25 years. The distribution of length of employment by years of follow-up (not shown) illustrated that follow-up time was longest for the long-term workers i.e., those with the longest overall exposure times were observed for the longest period of time.

In Table 15.8.3 the distribution of exposure type by gender and pay status is presented. Females primarily held jobs with low exposures (97 percent of women); they were engaged in the winding operation, which was done in a separate "clean" room, or they held clerical salaried jobs. The distribution of exposure type presented in Table 15.8.3 indicates that workers, especially hourly male workers, experienced different exposures throughout their employment. For example, in the hourly male cohort, 27 percent of men had jobs with high, low, and undefinable exposures during their employment (not shown), and 66 percent of the male hourly cohort held a job with undefinable exposure sometime during their employment.

The observed deaths, expected deaths, SMRs, and their 95 percent CIs for selected causes of death for the cohort by subgroup are presented in Table 15.8.4. All malignant neoplasms, the major cardiovascular diseases, diabetes, cirrhosis of the liver, and accidental causes of death are presented. Causes with only one death are not presented, nor are irrelevant causes with small numbers of deaths (i.e., mental disorders). The expected numbers were calculated from the mortality rates of the U.S. population. Among the hourly workers, the all-causes mortality was significantly lower than that of the U.S. population (SMR = 84, 95 percent CI, 77 to 91 for males; SMR = 90, 95 percent CI, 82 to 100 for females). The all-cancers mortality was also significantly lower than that of the U.S. population in hourly male workers (SMR = 81; 95 percent CI, 68 to 97). In hourly female workers, the all-cancers mortality rate was comparable to the U.S. rate (SMR = 110; 95 percent CI, 93 to 129).

In the male hourly cohort no significant elevations in the six a priori cancers of interest (cancers of the rectum, liver, gastrointestinal tract, melanoma, brain, and hematopoietic system) were noted. Several causes of death had SMRs above 100; however, none of them was significantly elevated (Table 15.8.4).

In the hourly female cohort, there were no significantly elevated SMRs for any of the six a priori cancers of interest. SMRs for several other cancer sites were elevated, although none was significantly elevated (Table 15.8.4).

SMRs for deaths from diseases other than cancer were also not significantly elevated in hourly male or female workers. The SMR for diabetes, however, was significantly lower than expected in hourly male workers (4 observed and 10.5 expected; SMR = 38; 95 percent CI, 10 to 97).

Overall, among salaried male workers, a striking healthy worker effect was observed (Table 15.8.4). The all-causes SMR for salaried male workers was 54 (177 observed and 328 expected; 95 percent CI, 46 to 62) and the all-cancers SMR was 69 (56 observed and 81 expected; 95 percent CI, 52 to 90). None of the a priori cancers of interest was elevated in the salaried male workers. The lung cancer rate was significantly lower than expected, with an SMR of 41 (12 observed and 29.6 expected; 95 percent CI, 21 to 71), as was ischemic heart disease, with an SMR of 45 (44 observed and 97.5 expected; 95 percent CI, 33 to 61), and cerebrovascular disease with an SMR of 20 (three observed and 15.2 expected; 95 percent CI, 4 to 58).

TABLE 15.8.4 Observed and Expected Deaths[a] in 2984 Hourly Male Workers,[b] 2544 Hourly Workers,[c] 1078 Salaried Male Workers,[d] and 469 Salaried Female Workers[e]

Cause of death	Hourly workers[f]				Salaried workers[f]			
	Males		Females		Males		Females	
	Obs/Exp	SMR (95 percent CI)	Obs/Exp	SMR (95 percent CI)	Obs/Exp	SMR (95 percent CI)	Obs/Exp	SMR (95 percent CI)
All causes	586/699	84* (77–91)	380/420	90* (82–100)	177/328	54** (46–62)	52/75	69** (52–91)
All cancers	128/158	81* (68–97)	150/136	110 (93–129)	56/81	69* (52–90)	19/25	75 (45–118)
MN[g] of tongue	1/0.9	103 (3–576)	2/0.4	483 (59–1745)	0/0.1	—	1/0.07	1346 (34–7498)
MN of buccal cavity	2/1.1	178 (22–642)	2/0.5	365 (44–1317)	0/0.5	—	1/0.1	1021 (26–5690)
MN of pharynx	4/2.0	199 (54–509)	2/0.8	253 (31–915)	0/1.0	—	0/0.1	—
MN of esophagus	5/3.8	131 (42–304)	1/1.1	87 (2–482)	1/2.0	49 (1–272)	0/0.2	—
MN of stomach	4/5.9	68 (18–173)	4/3.0	132 (36–339)	1/2.7	36 (0.9–200)	0/0.5	—
MN of intestine	8/14.0	57 (25–112)	20/12.7	157 (96–242)	7/7.1	98 (40–203)	1/2.2	44 (1–247)
MN of rectum	3/3.4	87 (18–255)	4/2.3	169 (46–434)	3/1.6	185 (38–540)	0/0.4	—
MN of biliary passages and liver	2/2.5	80 (10–289)	2/2.2	89 (11–321)	1/1.2	79 (2–439)	0/0.3	—
MN of pancreas	9/7.8	115 (53–219)	7/5.9	117 (47–241)	6/3.9	150 (55–327)	0/1.1	—
MN of larynx	3/2.0	147 (30–428)	1/0.4	215 (5–1198)	1/1.0	—	0/0.1	—
MN of trachea, bronchus, and lung	42/54.5	77 (56–104)	32/25.2	127 (87–179)	12/29.6	41** (21–71)	5/4.7	104 (34–244)
MN of breast	—	—	25/30	82 (53–121)	—	—	6/5.7	104 (38–226)
MN of cervix uteri	—	—	6/4.7	126 (47–277)	—	—	1/0.9	112 (3–622)
MN of other parts of uterus	—	—	5/3.8	130 (43–305)	—	—	0/0.6	—
MN of ovary, tube, and broad ligament	—	—	8/9.3	85 (37–168)	—	—	2/1.7	115 (14–415)
MN of prostate	12/10.9	110 (57–192)	—	—	3/5.3	56 (5–136)	—	—
MN of kidney	3/4	75 (15–219)	2/2.1	94 (11–341)	0/2.1	—	—	—
MN of bladder and other urinary tract	3/3.8	77 (16–226)	2/1.3	151 (18–545)	1/1.8	54 (1–299)	0/0.2	—
MN of skin (melomonas)	5/3.8	130 (42–303)	3/2.0	144 (30–421)	4/1.9	210 (57–538)	0/0.4	—
MN of brain and nervous system	2/5.1	39 (5–140)	2/3.7	53 (6–192)	4/2.5	156 (42–398)	0/0.7	—
MN of connective tissue	0/0.9	—	1/0.8	125 (3–694)	1/0.4	—	2/0.2	1290* (156–4659)
Other and unspecified cancer	6/10.3	58 (21–126)	3/8.6	35 (7–101)	2/0.2	229 (6–1275)	1/1.5	—
Lymphosarcoma	2/2.1	92 (11–331)	1/1.5	65 (2–364)	3/5.4	55 (11–161)	0/0.2	—
Leukemia and aleukemia	4/6.3	63 (17–162)	4/4.3	93 (25–238)	0/1.0	—	0/0.8	—
Other lymphatic and hematopoietic	5/5.7	87 (28–202)	5/4.7	105 (34–245)	5/3.0	166 (54–387)	0/0.8	—
Cirrhosis of the liver	13/18	72 (39–124)	6/9.2	65 (24–142)	4/3.0	131 (36–336)	1/1.7	57 (1–318)
Diabetes	4/10.5	38* (10–97)	9/10.3	87 (40–165)	3/9.1	33* (7–96)	0/1.8	—
Ischemic heart disease	182/205	89 (76–103)	71/87	81 (64–103)	5/5.1	97 (32–226)	8/14.3	56 (24–110)
Hypertension with heart disease	5/5.8	86 (28–201)	2/4.4	45 (5–164)	44/97.5	45** (33–61)	0/0.7	—
Other diseases of the heart	34/34.7	98 (68–137)	18/21.4	84 (50–133)	0/2.5	—	1/3.8	26 (0.7–146)
Cerebrovascular disease	26/34.9	74 (49–109)	27/30	89 (59–130)	16/18.2	88 (50–143)	6/5	120 (44–260)
Arteries, veins, pulmonary circulation	19/17.0	112 (67–174)	10/11	95 (46–175)	3/15.2	20** (4–58)	1/1.8	56 (1–310)
Transportation accidents	29/34.7	84 (56–120)	14/9.1	153 (84–257)	4/8.0	50 (14–128)	3/1.9	156 (32–455)
Other accidents	10/14.4	69 (33–127)	5/3.1	158 (51–369)	3/12.6	24** (5–69)	1/0.6	159 (4–886)
Suicide	14/21.3	66 (36–110)	3/6.8	44 (9–128)	3/5.9	51 (11–148)	2/1.4	140 (17–507)
Homicide	3/8.4	36 (7–104)	2/2	96 (12–345)	0/3.0	23* (3–84)	0/0.4	—

*Significant at $P < 0.05$.

**Significant at $P < 0.01$.

a Expected numbers for selected causes of death based on age-, sex-, race-, and time-specific U.S. rates coded according to the rules of the International Classification of Diseases coding in force at the time of death. ICD code groupings are shown as listed in Steenland et al.

b 85,991 Person-years of observation.

c 75,674 Person-years of observation.

d 34,755 Person-years of observation.

e 16,358 Person-years of observation.

f Obs/Exp, observed/expected; SMR, standardized mortality ratio; CI, confidence interval.

g MN, malignant neoplasms.

The female salaried workers (Table 15.8.4) also demonstrated a marked healthy worker effect, with an all-causes SMR of 69 (52 observed and 75 expected: 95 percent CI, 52 to 91) and an all-cancers SMR of 75 (19 observed and 25 expected; 95 percent CI, 45 to 118). The only significant finding in the salaried female workers was that for cancer of the connective tissue, with an SMR of 1290 (2 observed and 0.2 expected; 95 percent CI, 156 to 4659). However, one of the connective tissue tumors was a pericytoma, a lesion of borderline malignancy.

The age, sex, race, and calendar-specific mortality rates for the hourly workers, compared with the regional population (eight counties surrounding the plants), were similar to the SMRs calculated with the U.S. rate tables (not shown). None of the a priori cancers of interest in either males or females was significantly different than expected.

15.8.3 SMRs BY LENGTH OF EMPLOYMENT AND LATENCY CATEGORIES

There was no trend of increasing SMRs over length of employment and latency categories for all causes or all cancers in either male or female hourly workers. The data for all cancers are presented in Tables 15.8.5 and 15.8.6.

The only site-specific cause of death that met the a priori criteria for analysis by cumulative length of employment and latency that is, greater than two observed cases and CIs with a lower boundary >90), was intestinal cancer in female hourly workers. As listed in Table 15.8.6, the intestinal cancer SMRs occurred primarily in women with greater than 20 years of latency; however, the deaths distributed

TABLE 15.8.5 Mortality by Length of Employment and Latency from All Cancers for Hourly Male Workers

	Length of employment (years)									
	<1		1 to <5		5 to <10		≥10		Total	
Cause/latency	No. of deaths observed	SMR	No. of deaths observed	SMR	No. of deaths observed	SMR	No. of deaths observed	SMR	No. of deaths observed	SMR
All cancers, *n*										
<20	8	81	8	79	9	105	10	89	35	89
≥20	19	74	14	59*	12	108	48	78*	93	78*
Total	27	76	22	65*	21	107	58	81*	128	81*

*$P < 0.05$.

TABLE 15.8.6 Mortality by Length of Employment and Latency from All Cancers and Intestinal Cancer for Hourly Female Workers

	Length of employment (years)									
	<1		1 to <5		5 to <10		≥10		Total	
Cause/latency	No. of deaths observed	SMR	No. of deaths observed	SMR	No. of deaths observed	SMR	No. of deaths observed	SMR	No. of deaths observed	SMR
All cancers, *n*										
<20	9	84	13	90	8	106	8	107	38	95
≥20	31	145	30	117	13	120	38	100	112	117
Total	40	124	43	107	21	114	46	101	150	110
Intetinal cancers, *n*										
<20	0	—	0	—	1	154	1	147	2	62
≥20	4	198	5	208	5	458*	4	100	18	189*
Total	4	142	5	143	6	345*	5	106	20	157

*$P < 0.05$.

evenly through the length-of-employment categories. Although some of the category-specific SMRs for intestinal cancer were significantly elevated in and of themselves, they were calculated with small numbers and therefore were unstable.

To further evaluate any increased risk of mortality from intestinal cancer with increasing length of employment, an internal analysis for trend was calculated. The observed over the expected rates calculated by length of employment and the associated trend were tested for statistical significance by the Mantel-Cox χ^2 test for trend. The observed over expected rates for employment time of <1 year, 1 to <5 years, 5 to <10 years, and 10 years or greater were 5.07, 1.27, 2.26, and 0.40, respectively. The χ^2 was significant at $P < 0.001$; however, the trend of the observed over the expected rates did not increase but rather decreased with length of employment.

Exposure-Specific SMRs

Mortality in the 1268 hourly male and 362 hourly female workers who worked for at least 1 day in a high-exposure job was compared with the mortality experience of the U.S. population. The 1268 ever-high-exposed hourly male workers contributed 37,739 person-years of observation. The all-causes SMR was significantly lower than expected (SMR = 82; 95 percent CI, 72 to 93), and the all-cancers SMR was 77 (95 percent CI, 57 to 101). The 362 ever-high-exposed hourly female workers contributed 10,584 person-years of observation. Both all-causes and all-cancers mortality in the ever-high-exposed hourly female workers did not deviate from that expected (all-causes SMR = 96, 95 percent CI, 75 to 123; all-cancers SMR = 100, 95 percent CI, 63 to 152). There were no significant differences in any of the site-specific SMRs for either male or female ever high-exposed hourly workers.

In the 723 hourly male workers, with 22,217 person-years of observation, who were engaged in a high-exposure job for at least 180 days, the all-causes SMR was 87 (95 percent CI, 74 to 102) and the all-cancers SMR was 82 (95 percent CI, 58 to 114). In the 184 hourly female workers, with 5783 person-years of observation, the all causes SMR was 92 (95 percent CI, 65 to 127) and the all-cancers SMR was 88 (95 percent CI, 43 to 155). There were no elevated site-specific SMRs above the expected number.

The all-causes SMR for the 479 hourly male workers who had worked for at least 1 year in a high-exposure job was 89 (95 percent CI, 74 to 107; 15, 181 person-years of observation) and the all-cancers SMR was 73 (95 percent CI, 46 to 109). In the 122 hourly female workers who worked in a high-exposure job for at least 1 year, the all-causes SMR was 81 (95 percent CI, 53 to 119; 4047 person-years of observation) and the all-cancers SMR was 61 (95 percent CI, 23 to 134). No site-specific elevations were seen in either males or females.

In Tables 15.8.7 and 15.8.8, the results of the SMR analysis by length of employment and latency are presented for all cancers for hourly male and female workers highly exposed for at least 180 days

TABLE 15.8.7 Mortality by Length of Employment and Latency from All Cancers for Hourly Male and Female Workers Highly Exposed for 180 Days or More

| | Length of employment (years) | | | | | | | | | |
| | <1 | | 1 to <5 | | 5 to <10 | | ≥10 | | Total | |
Cause/latency	No. of deaths observed	SMR	No. of deaths observed	SMR	No. of deaths observed	SMR	No. of deaths observed	SMR	No. of deaths observed	SMR
Males										
All cancers, *n*										
<20	1	116	4	164	3	121	5	140	13	140
≥20	1	45	6	108	2	53	14	67	23	67
Total	2	64	10	125	5	80	19	82	36	82
Females										
All cancers, *n*										
<20	0	—	2	193	1	98	0	—	3	86
≥20	0	—	1	102	1	110	6	82	8	87
Total	0	—	3	149	2	103	6	70	11	87

TABLE 15.8.8 Mortality by Length of Employment and Latency from All Cancers for Hourly Male and Female Workers Highly Exposed for 1 year or More

| | Length of employment (years) | | | | | | | |
| | 1 to <5 | | 5 to <10 | | ≥10 | | Total | |
Cause/latency	No. of deaths observed	SMR	No. of deaths observed	SMR	No. of deaths observed	SMR	No. of deaths observed	SMR
Males								
All cancers, n								
<20	0	—	3	146	4	143	7	107
≥20	4	121	1	30	11	60	16	64
Total	4	83	4	75	15	71	23	73
Females								
All cancers, n								
<20	1	170	0	—	0	—	1	40
≥20	1	213	0	—	4	65	5	69
Total	2	189	0	—	4	56	6	61

(Table 15.8.7) or for at least 1 year (Table 15.8.8). Because of the small number of site-specific cancers, analysis by cumulative length of employment and latency category was not feasible. There was no consistent trend across the length-of-employment categories and/or latency categories in either males or females that would suggest an association between PCB exposure and increased mortality (Tables 15.8.7 and 15.8.8).

15.8.4 COMMENT

The lower-than-expected mortality seen in this cohort has been reported in other PCB mortality studies and is consistent with the healthy worker effect often observed in employed populations. The lower all cancers SMR seen in hourly male workers, however, is unusual and raises the question of whether cancer deaths have been sufficiently ascertained in this cohort. This discrepancy was not seen in the female cohort, and a systematic underascertainment or ICD coding problem related to gender is unlikely.

The excess cancel mortality related to PCB exposure that has been reported previously in the literature was not replicated in this study. However, the capacitor workers in prior studies were exposed to the same mixtures of PCBs at similar concentrations, because the production process in different plants was the same. The significant excess of liver cancer reported by Brown was not observed in this cohort. SMRs for liver cancer for both males and females in our cohort were similar to those of the U.S. population, with SMRs below 100 (SMR = 89 for hourly female and SMR = 80 for hourly male workers). Brown and Jones also reported a significant increase in mortality from rectal cancer in women, with three observed deaths and 0.50 expected; however, the excess was no longer significant in Brown's follow-up study of the same cohort. Among hourly female workers in our cohort, there were four observed deaths from rectal cancer and 2.3 expected, a nonsignificant increase (SMR = 169; 95 percent CI, 46 to 434). All four women who died of rectal cancer held only low-exposure jobs during their employment. Length of employment in the four women ranged from 6 to 22 years, with a mean of IO years. Years since first exposure in three of the deaths exceeded 20. Known risk factors for rectal cancer include smoking and family history. Occupational risk for rectal cancer in females has been reported for women involved in the furniture-making industry, with an SMR of 3.2 (95 percent CI, 1.3 to 4.5). Occupational history before and after employment at the capacitor plants was not available for members of our cohort and could not be evaluated.

In the Bertazzi et al. cohort of 2100 workers, a significant excess of total cancers and cancers of the gastrointestinal tract (ICD codes 150 to 159) among males was reported (six observed versus 1.7

expected, SMR = 346; 95 percent CI, 141 to 721). In contrast to the Bertazzi et al. cohort, hourly male workers in our cohort had lower than expected numbers of both intestinal and rectal cancers, although not significantly lower (cancer of the intestine SMR = 57, 95 percent CI, 25 to 112, cancer of the rectum SMR = 87, 95 percent CI, 18 to 225). Hourly female workers in our cohort had nonsignificantly elevated SMRs for intestinal and rectal cancer (cancer of the intestine SMR = 157, 95 percent CI, 96 to 242; cancer of the rectum SMR = 169, 95 percent CI, 46 to 434). In the Bertazzi et al. cohort, the numbers of lung cancer and hematological neoplasms were also elevated in males, but the excesses were not statistically significant. In our cohort, neither the numbers of lung cancer nor neoplasms of the hematopoietic system were elevated.

The 1556 females in the Bertazzi et al. cohort experienced higher than expected all-causes mortality compared with the Italian national population and significantly higher than expected all-cancers mortality and mortality related to hematological neoplasms compared with the local population. In our cohort the all-causes mortality in females was significantly lower, all-cancers mortality in females was comparable to the expected number, and no elevation in hematological neoplasms was observed.

In the cohort of 3588 male and female capacitor workers examined by Sinks et al., a significant increase in mortality from melanoma and a nonsignificant increase in mortality from cancers of the brain and nervous system was observed. In our cohort the numbers of cancers of the brain and nervous system were lower than expected for both males and females. Mortality from melanoma was similar to the expected number in both males and females (5 observed and 3.8 expected deaths in males; 3 observed and 2.08 expected deaths in females). None of the five males with melanoma had worked for longer than 2 years at the plants, and two of them had <20 years of latency. Of the three women with melanoma, one had worked for 1 year, one for 3.4 years, and one for 12 years at the plants; all three held only low-exposure jobs. Exposure to the sun is the major risk factor for melanoma and accounts for 65 percent of melanomas worldwide; melanomas represent almost all fatal skin cancers.

In our cohort the SMR for intestinal cancer (large and small intestine) in hourly female workers was elevated and approached statistical significance (SMR = 157; 95 percent CI, 96 to 242). There were 20 observed deaths and 12.7 expected, using the U.S. rate data. One of the intestinal cancers was a carcinoid originating from endocrine argentaffin cells, representing a tumor with a different etiology but one grouped with intestinal carcinomas in the ICD coding system. In contrast, there were 8 observed deaths and 14 expected deaths in the hourly male cohort (SMR = 57; 95 percent CI, 25 to 113). The SMR for cancer of the large intestine in hourly female workers, using the regional population as the comparison population, was lower (SMR = 120; 95 percent CI, 74 to 186), with 16.6 deaths expected. Because the incidence of intestinal cancer is greatly influenced by ethnicity and because higher rates are observed among the white population of the northeastern part of the United States, the regional comparison is more representative. Most intestinal cancers occurred in women with 20 or more years of latency; however, the cancers were evenly distributed across the length of employment. All the women worked exclusively in sedentary, low-exposure jobs during their employment. Trend analysis did not reveal any increase but rather a decrease in the observed over the expected rates by the category length of employment. Known risk factors for intestinal cancer include tobacco use, dietary factors, family history, and lack of physical activity. Known occupational risks for intestinal cancer in women are few.

The low SMRs in the male salaried cohort may represent a socioeconomic effect. The salaried male workers included mostly professional staff, with 73 percent (792) having attended at least some college and 71 percent of these 792 employees having graduated from college. In contrast, only 15 percent of hourly male workers had some college education, and only 13 percent of them completed 4 years of college.

Overall, the salaried women demonstrated a larger healthy worker effect than did their hourly counterparts. Again, this may reflect a socioeconomic effect, including better education, as 26 percent ($n = 120$) of the salaried women ever attended college and 30 percent of those completed 4 years ($n = 36$). Low SMRs in salaried employees have been reported in other occupational cohorts. The disparity in mortality between socioeconomic groups has been increasing, and the differences in SMRs between the hourly and salaried workers may reflect this disparity.

The exposure-specific analysis for this cohort was limited by the lack of individual dosimetry data. The only available industrial hygiene data consisted of PCB air levels measured in various production areas in 1975 and 1977. Because of the persistence of PCBs and the length of time that a worker had repeated exposures, length of employment was a useful proxy of cumulative exposure. Hourly workers employed in high-exposure jobs represented the most highly exposed group of workers and were grouped by their cumulative time in high-exposure jobs into three groups. None of the six a priori cancer sites of interest, all-cancers mortality, and any site-specific mortality were elevated in either males or females for any of the three groupings of high-exposed workers. There was no consistent trend in any of the tables illustrating SMRs by length of employment and latency for any of the groupings. The number of workers employed in high-exposure jobs for long time periods was small and limited the extent of the analysis; however, the consistent lack of any significant elevation in mortality in any of the groups in either males or females is worth noting. Reliance on categorical exposure assessment may result in misclassification. However, jobs that involved direct dermal and inhalation exposure to PCBs were clearly identified by job code in the worker history records, and the PCB exposure in these jobs was substantial, as indicated by the air-monitoring data and PCB serum and adipose tissue levels measured in selected workers.

Although not an a priori consideration, the large cohort of women in this study provided an opportunity to examine the relationship between breast cancer and occupational exposure to PCBs. In our study, mortality from breast cancer among hourly and salaried female workers was not increased over the expected number. Additionally, we had the opportunity to examine a cohort of women, albeit small, who experienced high exposure to PCBs.

In the past, capacitor workers as an occupational group had the highest exposures to PCBs through inhalation of PCB vapors and dermal absorption of PCB liquid. Because PCBs were heated, their volatility was increased, resulting in high levels in the air. The worker groups such as electric utility workers had poorly defined and definitely lower exposures than the capacitor workers. Such workers did not inhale vapors from heated PCBs nor did they have daily dermal contact with liquid PCBs.

The potential for bias in any observational study is always a concern and must be considered in the study design as well as in the examination of results. Both selection and information bias can be prevented if disease and/or vital statuses are not known when exposure assignments are made. In this study, exposure was assigned on the basis of the historical information related to PCB exposure that was contained in the job codes. The exposure assignment was done before determination of vital status.

In conclusion, this is the largest cohort of workers directly exposed to PCBs that was assembled specifically for examination of the association between exposure and increased cancer mortality. Extensive effort went into assembling a complete cohort and obtaining vital statuses from over 98 percent of the cohort. Despite the fact that the cohort is relatively young, 85 percent of the cohort was observed for at least 20 years, with a mean follow-up time of 31 years. Neither overall cancer mortality nor numbers of any of the a priori cancers of interest previously reported as being elevated were elevated in this cohort. With the exception of intestinal cancer in the hourly female workers, there were few cancer sites whose numbers were elevated, with few cases and wide CIs; these sites have not been reported in other studies, which suggests that our results were chance findings.

Because of the inherent limitations in any retrospective cohort mortality study, the best way to evaluate the validity of a reported association is to replicate the study in similarly exposed populations. To date none of the reported elevations in cancer mortality has been successfully replicated, even within individual cohorts. Notwithstanding the bias inherent in retrospective occupational cohort studies, the lack of consistent findings with respect to occupational PCB exposure and mortality in studies conducted to date suggest lack of an association.

ACKNOWLEDGMENTS

This study was funded by the General Electric Company. The assistance of Mr Yong-Gon Chon in developing our database is appreciated. Dr Kyle Steenland of the National Institute for Occupational

Safety and Health provided the U.S. mortality rate tables. We thank the members of our advisory panel: Arthur, C., Upton, M.D., John, E. Vena, PhD, Jack, S. Mandel, PhD, Roy, E. Shore, PhD, and Gilbert, W. Beebe, PhD, for their helpful suggestions and support throughout the study.

This article is republished with the permission of the *Journal of Occupational and Environmental Medicine* where it was initially published in Vol. 41, No. 3, March 1999.

REFERENCES AND SUGGESTED READINGS

1. Kimbrough, R. D., "Polychlorinated Biphenyls (PCBs) and Human Health: An Update," *Crit. Rev. Toxicol.*, 25: 133–163, 1995.

2. Sinks, T., Steele, G., Smith, A., Watkins, A., and Shults, R., "Mortality Among Workers Exposed to Polychlorinated Biphenyls," *Am. J. Epidemiol.*, 136: 389–398, 1992.

3. Brown, D. P., and Jones, J., "Mortality and Industrial Hygiene Study of Workers Exposed to Polychlorinated Biphenyls," *Arch. Environ. Health*, 36: 120–129, 1981.

4. Brown, D. P., "Mortality of Workers Exposed to Polychlorinated Biphenyls: An Update," *Arch. Environ. Health*, 42: 333–339, 1987.

5. Bertazzi, P. A., Riboldi, L., Pesatori, A., Radice, L., Zocchetti, C., "Cancer Mortality of Capacitor Manufacturing Workers," *Am. J. Ind. Med.*, 11: 165–176, 1987.

6. Taylor, P. R., *The Health Effects of Polychlorinated Biphenyls* [doctoral thesis], Boston, MA: Harvard School of Public Health; 1988.

7. Taylor, P. R., Reilly, A. A., Stelma, J., and Lawrence, C. E., "Estimating Serum Polychlorinated Biphenyl Levels in Highly Exposed Workers: An Empirical Model," *J. Toxicol. Environ. Health*, 34: 413–422, 1991.

8. World Health Organization, *International Classification of Diseases: Manual of the International Statistical Classification of Diseases, Injuries, and Causes of Death*, Geneva; World Health Organization; Revisions 5–9, 1940–1993.

9. Lawton, R. W., Sack, B. T., Ross, M. R., and Feingold, J., *Studies of Employees Occupationally Exposed to PCBs, a Progress Report*, September 18, (General Electric Report submitted to the U.S. Environmental Protection Agency), 1981.

10. Jones, M., *Industrial Hygiene Survey*, Washington, DC: National Institute for Occupational Safety and Health; 1983, NIOSH publication 83-137224. Reprinted by the National Technical Information Service.

11. Kreiss, K., "Studies on Populations Exposed to Polychlorinated Biphenyls," *Environ. Health Perspect.*, 60: 193–199, 1985.

12. Wolff, M. S., Fischbein, A., Thornton, J., Rice, C., Lilis, R., and Selikoff, I. J., "Body Burden of Polychlorinated Biphenyls Among Persons Employed in Capacitor Manufacturing," *Int. Arch. Occup. Environ. Health*, 49: 199–208, 1982.

13. Lawton, R. W., Ross, M. R., Feingold, J., and Brown, J. F. Jr., "Effects of PCB Exposure on Biochemical and Hematological Findings in Capacitor Workers," *Environ. Health Perspect.*, 60: 165–184, 1985.

14. Lawton, R. W., Brown, J. F. Jr., Ross, M. R., and Feingold, J., "Comparability and Precision of Serum PCB Measurements," *Arch. Environ. Health*, 40: 29–37, 1985.

15. Steenland, K., Beaumont, J., Spaeth, S., Brown, D., Okun, A., and Jurcenko, L., "New Developments in the NIOSH Life Table Analysis System," *J. Occup. Med.*, 32: 1091–1098, 1990.

16. Marsh, G. M., Ehland, J., and Sefcik, S., *Mortality and Population Data System (MPDS)* [technical report], Pittsburgh, PA: University of Pittsburgh, Department of Biostatistics, 1987.

17. Marsh, G. M., Preininger, M., and Ehland, J. J., *OCMAP PC: A User-Oriented Occupational Mortality Cohort Analysis Program for the IBM PC*, Pittsburgh, PA; University of Pittsburgh, 1989.

18. Bailar, J. C., and Ederer, F., "Significance Factors for the Ratio of a Poisson Variable to its Expectation," *Biometrics*, 20: 639–643, 1964.

19. Dixon, W. J., ed., *BMDP Statistical Software Manual*, Berkeley; University of California Press, 1992.

20. Monson, R. R., "Observations on the Healthy Worker Effect," *J. Occup. Med.*, 28: 425–433, 1986.

21. Heineman, E. F., Zahm, S. H., McLaughlin, J. K., and Vaught, J. B., "Increased Risk of Colorectal Cancer Among Smokers: Results of a 26-year Follow-up of U.S. Veterans and a Review," *Int. J. Cancer*, 59: 728–738, 1994.

22. Burt, R. W., "Familial Risk and Colorectal Cancer", *Gastroenterol. Clin. North Am.*, 25: 793–803, 1996.

23. Miller, B. A., Blair, A., and Reed, E. J., "Extended Mortality Follow-up Among Men and Women in a U.S. Furniture Workers Union," *Am. J. Ind. Med.*, 25: 537–549, 1994.

24. Armstrong, B. K., and Kricker, A., "Skin Cancer," *Dermatol. Clin.*, 13: 583–594, 1995.

25. Schottenfeld, D., and Winawer, S. J., "Cancers of the Large Intestine," in *Cancer Epidemiology and Prevention*, Schottenfeld, D., and Fraumeni, J. F., eds., 2nd ed. New York: Oxford University Press, 813–840, 1996.

26. Slattery, M. L., Potter, J. D., Friedman, G. D., Ma, K. N., and Edward, S., "Tobacco Use and Colon Cancer", *Int. J. Cancer*, 70: 259–264, 1997.

27. Newcomb, P. A., Storer, B. E., and Marcus, P. M., "Cigarette Smoking in Relation to Risk of Large Bowel Cancer in Women," *Cancer Res.*, 55: 4906–4909, 1995.

28. Giovannucci, E., and Willett, W. C., "Dietary Factors and Risk of Colon Cancer," *Ann. Med.*, 26: 443–452, 1994.

29. Burt, R. W., "Familial Risk and Colon Cancer", *Int. J. Cancer*, 69: 44–46, 1996.

30. Ruder, A. M., Ward, E. M., and Brown, D. P., "Cancer Mortality in Female and Male Drycleaning Workers," *Am. J. Epidemiol.*, 136: 389–398, 1992.

31. Teta, M. J., and Ott, M. G., "Mortality in a Research Engineering and Metal Fabrication Facility in Western New York State," *Am. J. Epidemiol.*, 127: 540–551, 1988.

32. Pappas, G., Queen, S., Hadden, W., and Fisher, G., "The Increasing Disparity in Mortality Between Socio-economic Groups in the United States," 1960 and 1986, *N. Engl. J. Med.*, 329: 103–109, 1993.

33. Sorlie, P. D., Backlund, E., and Keller, J. B., "U.S. Mortality by Economic, Demographic, and Social Characteristics; the National Longitudinal Mortality Study," *Am. J. Public Health*, 85: 949–956, 1995.

34. Loomis, D., Browning, S. R., and Schenk, A. P., et al., "Cancer Mortality Among Electrical Workers Exposed to Polychlorinated Biphenyls," *Occup. Environ. Med.*, 54: 720–728, 1997.

P · A · R · T · 3

PLACE BASED ENVIRONMENTAL SCIENCE, HEALTH AND TECHNOLOGY

CHAPTER 16
MANAGING PLACE SCALE PROBLEMS

SECTION 16.1

DECISION MAKING

Geoff Freeze

Mr. Freeze is Computational Sciences Section Manager, Duke Engineering & Services, Albuquerque, New Mexico. He is a civil environmental engineer who specializes in probabilistic applications of groundwater flow and transport models.

16.1.1 INTRODUCTION

Decision making is the process of choosing between multiple alternative courses of action. Associated with this definition is the assumption that, as more information is known about the multiple alternatives, a more informed and presumably more beneficial (at least to the decision maker) choice can be made. We are faced with decisions all the time in everyday life: soup or salad, cash or charge, paper or plastic. Decision making in these simple cases is typically done nonquantitatively by common sense or intuition. Although we usually do not contemplate the decision-making process in these simple cases, common sense and intuition are in fact based on information about the alternatives in the form of prior personal experiences or current or future needs.

As alternatives become more complex and/or more numerous, the relative benefit of a formalized decision-making process increases. Formalized decision making, typically referred to as *decision analysis*, provides guidance and insight into the decision-making process and enhances the ability of the decision maker to select the best alternative. This has been best recognized in the field of financial management, where sophisticated computer programs are used to guide buy/sell decisions (and the associated decisions of what to buy or sell).

The value of decision analysis only recently has been associated with environmental issues. Freeze et al. (1990) were among the first to introduce the concept of environmental decision analysis. Their theoretical framework (described below) is sound, but it has been applied only sporadically in the environmental industry. There are several reasons for this lack of implementation:

1. Client unfamiliarity: there is an inherent reluctance to accept new quantitative tools, particularly in an industry that relies to a certain extent on field experience and believes that the old-fashioned way (common sense and intuition) has worked just fine in the past.

2. Economics: as environmental budgets shrink, the message of "invest a little money up front to save a lot of money later" is more difficult to justify politically, particularly when later may be measured in decades.

3. Large uncertainties: environmental problems are often difficult to quantify, there are uncertainties in environmental data, and a poorly formulated problem is not useful.

Nonetheless, there is value in *environmental decision analysis*. It is a quantitative tool that should be in every environmental project manager's toolbox. Although quantitative decision analysis should not be relied upon solely in environmental decision making, it certainly provides important input to the decision-making process. Furthermore, the process of setting up a quantitative decision analysis often identifies important factors (not necessarily technical) that may not have been identified previously. Clearly, a quantitative decision analysis is not always feasible, especially for small-budget projects. However, it can lead to significant cost savings on many projects.

A few examples of environmental decision analysis applications include:

- Selection or prioritization of remediation/corrective action activities
- Site selection and design for waste management facilities
- Property acquisition for brownfields reclamation
- Factoring environmental due diligence results into merger/acquisition decisions
- Optimization of pumping well locations and rates for water supply and groundwater remediation
- Optimization of air, soil, and groundwater sampling schedules

16.1.2   CONCEPTS

Decision makers typically base their decisions on economic analyses of alternative proposed actions, considering the costs, benefits, and risks of each alternative. Risks reflect uncertainty, and recognizing that there is a degree of uncertainty or risk associated with the outcome of any future action, decision makers attempt to quantify this risk in terms of monetary (or other cost-benefit) units. Morgan and Henrion (1990) provide a useful review of philosophical frameworks for decision analysis. The most commonly used of these are summarized in Table 16.1.1.

Each of the frameworks presented in Table 16.1.1 involves various combinations of and trade-offs between two or more of the following components: cost, benefit, risk, and performance criteria. Decision making generally is based on which alternative best satisfies an objective (stated in terms of one of the components) subject to constraints imposed by one or more of the other components. In cases in which all the components can be quantified in monetary terms, a risk-cost-benefit-based framework can be used. All the philosophical frameworks in Table 16.1.1, except for multiattribute utility theory, are variations of a risk-cost-benefit framework. Where possible, a risk-cost-benefit formulation should be preferred because it is easy to formulate and results are intuitive (i.e., they give a direct indication of financial impact). Only in cases in which one part of the objective cannot be realistically quantified in monetary terms should multiattribute theory (see subsection 16.1.4) be considered.

In traditional risk-cost-benefit analysis, the objective is to maximize profits, where:

$$[\text{Profits ($)}] = [\text{Benefits ($)}] - [\text{Costs ($)}] - [\text{Risks ($)}] \qquad (16.1.1)$$

TABLE 16.1.1   Frameworks for Decision Analysis

Decision criteria	Basis for selected alternative
Cost-benefit (deterministic)	Highest net profit (no evaluation of uncertainty)
Risk-cost-benefit (probabilistic)	Highest net profit with expected value for cost of uncertainty (risk)
Lowest cost	Lowest cost to achieve a desired performance level
Fixed cost	Maximum performance level for a fixed cost (i.e., budget constrained)
Bounded risk	Maximum performance or minimum cost with constraint of not exceeding the specified risk
Multiattribute utility theory	Highest ranking based on a utility function that rewards important attributes that cannot be directly compared (e.g., all attributes cannot be quantified monetarily)

DECISION ANALYSIS

The objective is to provide a quantitative identification of the best course of action from a specified set of alternatives

 Φ

RISK MANAGEMENT

Define relationship/tradeoffs between benefits, costs, and uncertainties (risks)

[Decision Model]

$$\Phi = \sum_{t=0}^{T} \frac{1}{(1+i)^t} [B(t) - C(t) - R(t)]$$

$R(t)$

Quantify components of economic, sociopolitical, and health risks

[Risk Model]

$$R(t) = P_f(t) \bullet C_f(t) \bullet \gamma(C_f)$$

$P_f(t)$

OPERATIONS RESEARCH

Quantify the problem numerically and calculate benefits, costs, and risks using an appropriate solution method

[Performance Model]

$P_f(t) = \text{fn (simulation input parameters)}$

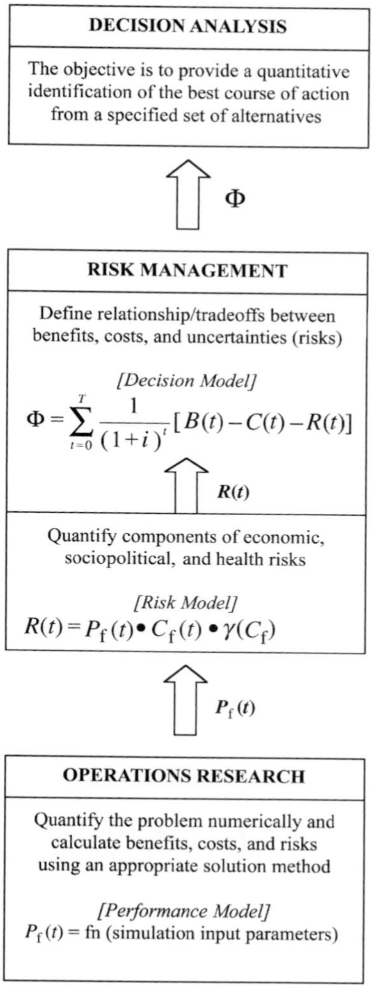

FIGURE 16.1.1 Components of decision analysis.

The risk term is typically quantified with a probabilistic approach to calculate an expected cost of uncertainty. In addition to providing a framework for comparing alternatives, risk-cost-benefit analysis can also indicate when it is cost-effective to incur additional data collection costs to reduce risk (i.e., the expected value of additional information).

Decision analysis combines two technical areas: risk management and operations research, as shown in Figure 16.1.1.

Risk management is assessment of the economic tradeoffs: how best to maximize benefits, minimize costs, and identify and reduce risks. Risks are associated with uncertainty. In simplest terms, risk management addresses the question "How much is it worth to accept the possible negative outcomes associated with uncertainty?"

Operations research is selection and application of the appropriate numerical method to provide a quantitative indication of the preferred alternative based on specified selection criteria. Some useful numerical methods include process modeling (both deterministic and probabilistic) and sensitivity and uncertainty analyses (best/worst case, perturbation). In some cases, optimization techniques (linear programming, simulated annealing, genetic algorithms) also may be desirable to supplement or replace decision analysis (see subsection 16.1.4). Operations research addresses the question "What is the best course of action given the specific criteria, constraints, and uncertainties that exist?"

16.1.3 RISK-COST-BENEFIT-BASED APPROACH FOR ENVIRONMENTAL DECISION ANALYSIS

As noted previously, the current practice in environmental decision making involves a large amount of educated common sense and intuition, but it usually is not carried out within a formal framework. However, Freeze et al. (1990) have presented a risk-cost-benefit-based framework for environmental decision analysis that integrates economic, social, and political issues together with technical analyses into the decision-making process. The Freeze et al. (1990) approach, although not unique, is the basis for the framework presented here because it demonstrates the components of quantitative environmental decision analysis in an easy-to-follow fashion. Risk management issues are addressed by using a *decision model*, an exterior shell containing an economically based risk-cost-benefit objective function of the form:

$$\Phi = \sum_{t=0}^{T} \frac{1}{(1+i)^t} [B(t) - C(t) - R(t)] \tag{16.1.2}$$

where Φ = objective function value ($),
  $B(t)$ = benefits in year t ($),
  $C(t)$ = costs in year t ($),
  $R(t)$ = risk (cost of failure) in year t ($),
    i = discount rate (decimal fraction), and
    T = time horizon (years).

Decision making is guided by evaluation of the decision model for a set of decision variables or alternatives. The decision variables are those variables that can be used to differentiate alternatives under consideration. The objective function is typically evaluated in monetary units, but any common unit can be used. The goal is to identify the alternative (or combination of decision variables) that maximizes the objective function over a specified time horizon. For many environmental projects, this means minimizing costs instead of maximizing profits. The first term after the summation sign accounts for the time value of money. The second term accounts for known benefits (revenues), known costs (capital and operational), and costs associated with uncertainty.

Typically, the most important, and most uncertain, term in the economic objective function is the risk $R(t)$. Quantification of risk is difficult because, as noted previously, risk reflects uncertainty. In environmental projects, uncertainty is especially large because the composition and distribution of contaminants (i.e., the source term) is usually poorly known and there is an inherent spatial variability in the contaminant transport characteristics. Nevertheless, risk can be quantified by using a *risk model* of the form:

$$R(t) = P_f(t) \times C_f(t) \times \gamma(C_f) \tag{16.1.3}$$

where $R(t)$ = risk (cost of failure) in year t ($), \
$\quad P_f(t)$ = probability of failure in year t (decimal fraction), \
$\quad C_f(t)$ = costs associated with failure in year t ($), and \
$\quad \gamma(C_f)$ = normalized utility function (decimal fraction).

The probability of failure $P_f(t)$, is the most difficult component of the risk model to determine. Typically, it is calculated using a process simulation model combined with operations research methods, collectively referred to here as the performance model. The performance model is described in more detail below. The cost of failure, $C_f(t)$, may include regulatory penalties, litigation and/or remediation costs, lost revenue, and social impacts (e.g., reduced property costs in area, loss of goodwill, loss of life). The utility function attempts to quantify risk averseness of the decision maker. Large companies tend to be more risk neutral, corresponding to $\gamma = 1$. Smaller companies, whose net worth may be similar to the cost of failure, are often more risk averse, corresponding to $\gamma > 1$.

Some considerations in the development of a risk model include:

1. Typically, the costs associated with risk are incurred only in the case of a failure (structural failure, contaminant release, etc.). Therefore, the risk model must use some sort of probabilistic or expected value approach to feed a risk cost to the decision model. The actual profit (or alternatively cost) will be not be the same as the statistically averaged decision model value. Concerns about the effect of the difference between actual costs and decision model-projected costs can be factored into the utility function. If a company plans to perform a series of projects that are quite similar, their risk aversion may be reduced because their risk is spread over several projects. The difference between actual costs and decision model-projected costs, averaged over multiple projects, decreases as the number of projects increases.

2. The sociopolitical contributions to the cost of failure often are difficult to quantify. Things such as human life, or standard of living, and political trust may be valued very differently depending on the perspective of the person doing the valuation. These differences are important because sociopolitical issues often are the ones that ultimately drive environmental decisions. Similarly, differences in the way different people perceive risks can lead to different estimates of the cost of failure. Regulations are often useful in establishing sociopolitical costs of failure. For example, penalties for exceeding specified contaminant concentration limits can substitute for a valuation of health risk costs.

The *performance model* is used to predict potential outcomes of various alternatives in the face of uncertainty. The probability of failure $P_f(t)$ that feeds into the risk model (Equation 16.1.3) is determined from the performance model.

The primary component of the performance model is a numerical simulation model (e.g., groundwater flow and transport model, air dispersion model) that uses input parameters to produce output results. The simulation model may range from a simple spreadsheet calculation to a three-dimensional finite-element numerical approximation. Performance model output may have large uncertainty because of large uncertainties in input parameters. By running multiple numerical simulations that include the range of input parameter uncertainty, the probability of failure can be determined based on the uncertainty in output. It is important to select a performance model that properly represents the technical processes. An inappropriate performance model may lead to an incorrect probability of failure, representative of an improper perception of risk.

Within a decision analysis framework (Figure 16.1.1), the performance model is used to calculate the probability of failure, which feeds into the risk model, which in turn feeds into the decision model. An objective function value, Φ, can thus be calculated for each alternative action. Decision analysis identifies the preferred alternative as the one with the maximum Φ (for a profit-based problem) or the minimum Φ (for a cost-based problem).

16.1.4 OTHER APPROACHES TO ENVIRONMENTAL DECISION MAKING

In some decision analysis problems, it is not possible to quantify all the components in monetary terms, particularly when trying to factor in social costs. In these cases, multiattribute theory (Keeney and Raiffa, 1993) can be used. For example, consider a project with the objective of maximizing profits and minimizing social impact. With multiattribute theory, a utility function is defined with a scale that ranges from, say, 0 to 10. Each component of the objective is then correlated to the scale (e.g., minimum profit = 0, maximum profit = 10, maximum social impact = 0, minimum social impact = 10). The objective now becomes to maximize the utility function value, and at this point the approach is analogous to the risk-cost-benefit approach.

In addition to risk-cost-benefit and multiattribute utility decision analysis methods, problems that require decision making can be solved by optimization. Optimization identifies the best solution from all possible solutions, whereas decision analysis methods identify the best solution from a specified set of solutions. Mathematically, optimization techniques ensure that the optimal solution has been found. The limitation is that these techniques can become numerically intractable as problems become more complex. Certain decision factors are also difficult to include explicitly in optimization problems.

16.1.5 TOOLS TO ASSIST IN DECISION ANALYSIS

Influence diagrams, decision trees, risk profiles, and sensitivity analyses are all commonly used to help decision makers structure and visualize problems and solutions. These tools are all demonstrated here through their links to the decision, risk, and performance models presented in Section 16.1.3.

An influence diagram shows the relationships among the important decision and risk variables. A simple influence diagram (Figure 16.1.2) is a graphical representation of the decision model. A more detailed influence diagram (Figure 16.1.3) shows some of the factors that contribute to the decision models, with distinctions made between fixed and uncertain values. Influence diagrams can be very helpful in understanding the problem.

Decision trees, also called probability trees, are used to graphically illustrate all possible combinations of uncertain variables and their associated probabilities. Each path (branch) through a tree represents one possible scenario or combination of uncertain events. The example decision tree shown in Figure 16.1.4 considers only two uncertain variables: operational costs with two discrete outcomes and probability of failure with three discrete outcomes, for a total of six scenarios. For each scenario, the decision tree shows the probability of occurrence and the decision-model-calculated result (net present value). Although popular, decision trees are useful graphically only for problems with a small number of uncertain variables whose outcomes can be represented by a small number of discrete values. More complex problems still can be easily solved numerically, but the corresponding decision tree typically has far too many branches to provide any graphical usefulness.

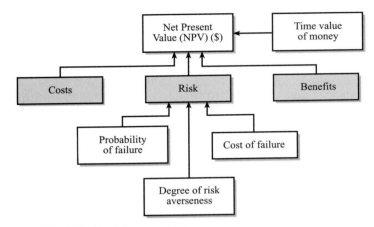

FIGURE 16.1.2   Simple influence diagram.

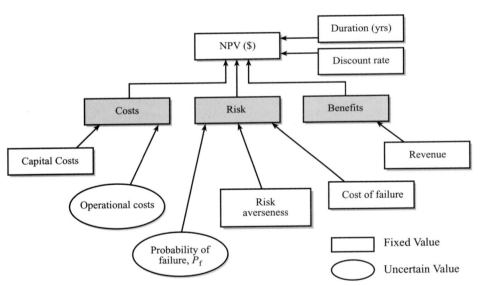

FIGURE 16.1.3   Detailed influence diagram.

A more useful form of the decision tree output is the risk profile (Figure 16.1.5), which graphically illustrates the decision model results with respect to their probability of occurrence. The most basic information shown in Figure 16.1.5 is the expected value. The expected value is a probability-weighted average of all possible outcomes. It is a useful statistical value for decision making (in this case the positive net present value suggests that the project should move forward); however, none of the actual scenario outcomes corresponds to the expected value. The risk profile also shows the cumulative distribution function (cdf) for the decision-model output. In this example, the cdf has six discrete steps, corresponding to the six scenarios. The cdf provides additional information to the decision maker, such as indicating, in this case, that there is only a 10 percent ($P = 0.1$) chance of losing money (net present value <0) with this project. In the case of numerous scenarios or if the outcomes

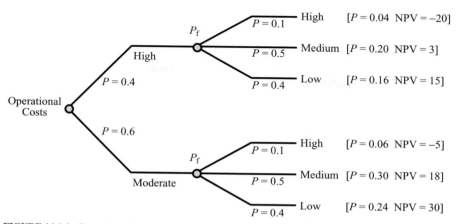

FIGURE 16.1.4   Example decision tree.

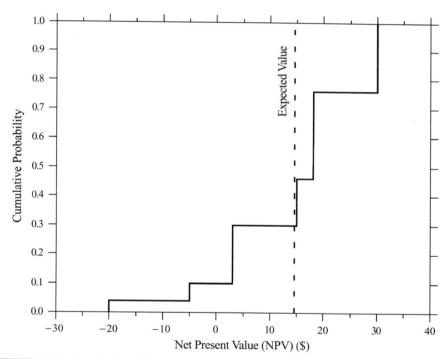

FIGURE 16.1.5   Example risk profile.

are continuous instead of discrete, the cdf becomes smoother and provides even more specific decision-making information.

The final decision analysis tool to be discussed is sensitivity analysis. Because of the large uncertainties in environmental problems, it is useful to understand which decision variables have the most impact on the results. A sensitivity analysis identifies how outcomes, and decisions based on them, change as decision variables change. Sensitivity analysis techniques range from simple to complex, but they are based on changing the value of a decision (or performance model input) variable and noting the corresponding change in the outcome. This helps focus attention on those variables with potentially the greatest impact on the decision. Sensitivity analysis is also useful for identifying where reducing the uncertainty in a certain variable will increase the expected value (or cdf) of the decision model. This is commonly referred to as a data worth analysis. If the cost of obtaining additional data to reduce uncertainty is less than the increased profits that result, data collection should proceed.

REFERENCES

Benjamin, J. R., and Cornell, C. A., *Probability, Statistics, and Decision for Civil Engineers*, McGraw-Hill, New York, 1970.

Clemen, R. T., *Making Hard Decisions: An Introduction to Decision Analysis*, 2nd edition, Duxbury Press, an imprint of Wadsworth Publishing Company, Belmont, CA, 1996.

Crouch, E. A. C., and Wilson, R., *Risk/Benefit Analysis*, Ballinger, Boston, 1982.

Fischoff, B., Lichtenstein, S., Slovic, P., Derby, S. L., and Keeney, R. L., *Acceptable Risk*, Cambridge University Press, New York, 1981.

Freeze, R. A., Massmann, J., Smith, L., Sperling, T., and James, B., "Hydrogeological Decision Analysis, 1. A Framework," *Ground Water*, 28(5): 738–766, 1990.

Keeney, R. L., *Value-Focused Thinking: A Path to Creative Decisionmaking*, Harvard University Press, Cambridge, MA, 1992.

Keeney, R. L., and Raiffa, H., *Decisions with Multiple Objectives: Preferences and Value Tradeoffs*, Cambridge University Press, 1993.

Lindley, D. V., *Making Decisions*, 2nd edition, Wiley, 1985.

Marshall, K. T., and Oliver, R. M., *Decision Making and Forecasting (with Emphasis on Model Building and Policy Analysis)*, McGraw-Hill, New York, 1995.

Morgan, M. G., and Henrion, M., *Uncertainty, A Guide to Dealing with Uncertainty in Quantitative Risk and Policy Analysis*, Cambridge University Press, New York, 1990.

Raiffa, H., *Decision Analysis: Introductory Lectures on Choices Under Uncertainty*, McGraw-Hill, New York, 1968.

von Winterfeldt, D., and Edwards, W., *Decision Analysis and Behavioral Research*, Cambridge University Press, New York, 1986.

SUGGESTED SOFTWARE

Analytica, Lumina Decision Systems, Inc. (http://www.lumina.com)
Crystal Ball, Decisioneering, Inc. (http://www.decisioneering.com)
@Risk. Palisade Corp. (http://www.palisade.com)

CHAPTER 16
MANAGING PLACE SCALE PROBLEMS

SECTION 16.2

ECOSYSTEM MANAGEMENT*

Robert T. Lackey

Dr. Lackey is associate director for science of EPA's Western Ecology Division in Corvallis, Oregon, where 45 Ph.D. scientists and 250 supporting staff conduct national and international research on a variety of ecological topics, including ecosystem management, biological diversity, ecological risk assessment, global climate change, coastal and estuarine ecology and ecological monitoring. He is attached to the National Health and Environmental Effects Research Laboratory, USEPA, Department of Fisheries and Wildlife and is an adjunct professor in the Department of Political Science at Oregon State University, Corvallis.

16.2.1 INTRODUCTION

Ecosystem management, proposed by some as the preferred way of managing natural resources and ecosystems, is a bold concept:

> Ecosystem management defines a paradigm that weaves biophysical and social threads into a tapestry of beauty, health, and sustainability. It embraces both social and ecological dynamics in a flexible and adaptive process. Ecosystem management celebrates the wisdom of both our minds and hearts, and lights our path to the future.   (Cornett, 1994)

When implemented, ecosystem management will, at least according to its advocates, protect the environment, maintain healthy ecosystems, permit sustainable development, preserve biodiversity, and save scarce tax dollars. A cynic might be tempted to add to the list: alleviate trade imbalances, reduce urban crime, and pay off national debts. Is ecosystem management a revolutionary concept and a sea change in public choice as its champions maintain, or are the critics correct who assert that it and the associated jargon are closer to cold fusion than cold fact?

Whether ecosystem management is "hot tub science applied to New Age management" or "a paradigm shift to save our rapidly disappearing biological heritage," scientists and managers are increasingly involved in the debate. Should scientists and other technical people care about ecosystem management as a concept or follow the spirited debates over its exact meaning? The answer is yes for at least three reasons:

First, it might just be a bold new concept and a very different—and better—way to manage ecosystems. Beyond the rhetoric in the professional literature, there may be some technical substance. Ideas do have consequences, especially when put into practice.

---

* Modified from Lackey (1998a).

Second, society needs to move beyond the debate and rhetoric and focus directly on policy issues and the role science could and should play. There are a considerable number of interesting and challenging research opportunities on ecosystems, but what are the *critical* research needs and management approaches that will make a *difference* in ecosystem management?

Third, ecosystem management has been embraced widely by politicians and appointed officials. In the political arena, the debate is concluded whether or not ecosystem management is a good idea; it *will* be implemented, or at least attempted, in word if not in deed.

Ecosystem management is offered as a management approach to help solve complex ecological and social policy problems. Examples of current problems are the Pacific Northwest forest/salmon/spotted owl dilemma; the purported massive decline in biological diversity; and ecosystem "degradation" caused by "poor" urban, industrial, transportation, agricultural, ranching, and mining policies and practices. Some critics may charge that ecosystem management is the triumph of the politics of "process" over the politics of "substance," but the public choice problems are definitely real and substantive.

Ecosystem management problems have several general characteristics: (1) fundamental public and private values and priorities are in dispute, resulting in partially or wholly mutually exclusive decision alternatives; (2) there is substantial and intense political pressure to make rapid and significant changes in public policy; (3) public and private stakes are high, with substantial costs and substantial risks of adverse effects (some also irreversible ecologically) to some groups regardless of which option is selected; (4) some technical facts, ecological and sociological, are highly uncertain; (5) the "ecosystem" and "policy problems" are meshed in a larger framework so that policy decisions have effects outside the scope of the problem. Solving these kinds of problems in a democracy has been likened to asking a pack of four hungry wolves and a sheep to apply democratic principles to deciding what to eat for lunch. Given public choice problems with these characteristics, no wonder discussions of ecosystem management tend to focus on *process* and not *substance*.

I have organized the fundamental concepts of ecosystem management around seven *pillars* which I consider to be the supports underlying ecosystem management. Just as physical pillars do not completely define a building, neither do intellectual pillars completely define ecosystem management. Nevertheless, I believe these pillars effectively provide the essential underpinnings of "ecosystem management," the circumstances under which it might be successfully applied, and its relationship to public and private choice. The seven pillars are neither procedures nor blueprints for ecosystem management but are principles upon which ecosystem management can be based.

16.2.2   *DEFINITION*

Articulating a clear *definition* for ecosystem management seems a reasonable place to start. The diversity of definitions provides some indication of the current amorphous nature of the concept (Norton, 1992; Slocombe, 1993; Bengston, 1994; Christensen et al., 1995; Stanley, 1995; Wagner, 1995; Fitzsimmons, 1996; Thomas, 1996; Merchant, 1997). Typical definitions of ecosystem management are:

1. "A strategy or plan to manage ecosystems to provide for all associated organisms, as opposed to a strategy or plan for managing individual species" (Forest Ecosystem Management Assessment Team, 1993).

2. "The careful and skillful use of ecological, economic, social, and managerial principles in managing ecosystems to produce, restore, or sustain ecosystem integrity and desired conditions, uses, products, values, and services over the long term" (Overbay, 1992).

3. "To restore and maintain the health, sustainability, and biological diversity of ecosystems while supporting sustainable economies and communities" (Environmental Protection Agency, 1994).

These definitions have an unmistakable similarity to traditional definitions of fisheries management, wildlife management, and forest management. In fact, they are strikingly similar to typical definitions of the much maligned multiple use management. For example, a typical definition of fisheries management is the "practice of analyzing, making, and implementing decisions to maintain or alter the structure, dynamics, and interaction of habitat, aquatic biota, and man to achieve human goals and objectives through the aquatic resource" (Lackey, 1998b). But in the definitions of ecosystem management, there are some new words—*ecosystem and community sustainability, ecosystem health, ecosystem integrity, biological diversity, social values, social principles, holistic.* The new words are where differences arise, yet it is from these words that I develop the pillars.

16.2.3   SOCIETAL VALUES AND PRIORITIES

What does society want from ecosystems? There are two fundamentally different world views (Lackey, 1994; Stanley, 1995). The first is *biocentric* and considers maintenance of ecological health or integrity as the goal. All other aspects, including human use (tangible or intangible), are of secondary consideration. The other view is *anthropocentric* in that benefits (tangible or intangible, short and long term) are accruable to humans. Certainly ecological systems can be adversely affected and care should be taken not to deplete resources for short-term benefit, but sustainable benefits are possible from ecosystems with careful management. Neither view is necessarily right or wrong, but they are fundamentally different and must be evaluated like any other moral, ethical, or religious position.

The basic idea behind a management paradigm is anthropocentric; it is to maximize benefits by applying a mix of decisions within defined constraints. Benefits may be tangible or intangible and may be achieved by maintaining a desired ecological condition. Potential benefits from ecosystems may be commodity yields (lumber, fish, wildlife), ecological services (pollution abatement, biological diversity), intangibles (preservation of endangered species, wilderness, vistas), precautionary investments (deferring use to preserve future options), and maintaining a desired ecological status (old growth forests, unaltered rangelands). The management challenge is to determine the *goal* or *goal set* and then design a strategy for implementing a *mix* of decisions to reach the goal (Bormann et al., 1994). A key challenge to successful management is accurately determining the system's capacity to achieve the goal—an important challenge that scientists can help meet.

The first and foremost management challenge, defining exactly what *is* the goal, is complicated by the evolving nature of society's values and priorities. It is difficult to be concerned with an endangered toad or a threatened snail when your family's immediate problem is paying the mortgage on your house. And it is difficult to understand the passion for industrial development when your major concern is whether you will take a vacation this winter or wait until summer. Our individual and collective goals and values differ with our circumstances and they change over time.

The other management challenge involves evaluating and selecting the mix of decisions that seem likely to achieve the identified goal—a goal that must be continually evaluated to be sure that it reflects society's values and priorities. This is no easy task under the best of circumstances, but it becomes impossible unless the analyst *assumes* a matrix of societal goals. The most efficient way to implement policy may be by a series of "experimental" decisions from which we can "learn" how the ecosystem (ecological and human elements) responds to various decisions. A modification of an old maxim may be appropriate here: the best way to implement ecosystem management may be to learn from past mistakes and to systematically make some new but different ones.

Values and priorities have long been recognized as important by natural resource managers. Management paradigms—whether multiple use, multiple resource use, maximum equilibrium yield, scientific management, watershed management, natural resources management, maximum sustained yield, or ecosystem management—are based on values and priorities (Cubbage and Brooks, 1991). Each paradigm has, either formally or informally, accepted a set of values and priorities or used a process to derive values and priorities. Ecosystem management is no different in this regard.

The first pillar of ecosystem management is: *Ecosystem management reflects a stage in the continuing evolution of social values and priorities*; *it is neither a beginning nor an end.*

16.2.4 ECOSYSTEM BOUNDARIES

A practical technical requirement with any management paradigm is to *bound* the system of concern. Because no useable definition of an ecosystem has been developed that works within public decision making, other approaches are used to define the system of concern. Historically, this was accomplished by focusing on one or more species of concern over a defined geographic area; for example, managing flyways for migratory waterflow (the geographic limits of the flyway become the operational boundaries for management analysis) or managing game fish populations in a certain lake (the lake and its watershed then become the units of concern). In all cases, the problem of concern defines the boundary.

Another option is to bound the system by what is relevant to elements of the public such as a community or interest group. For example, management goals might focus on providing diverse hunting options to society. However, no matter how boundaries are defined in ecosystem management, they end up largely being geographically based—a *place* of concern. Again the nature of the problem or the beneficiaries of concern will define the boundaries.

Within the place of concern the goal then becomes managing for maximum social benefits within a number of ecological and societal constraints. And because optimal management decisions vary by the scale of consideration, it is essential to define clearly the boundary of concern. For example, a set of decisions to maximize benefits in managing a 1000-hectare watershed *within* the Columbia River watershed may well differ from decisions for the same small watershed that were designed to maximize benefits over the *entire* Columbia River watershed. The definition of the management problem should define the scale to be used in the analysis. The same problems analyzed at different scales will likely lead to very different management strategies (Fitzsimmons, 1998).

There is a natural tendency to gloss over decisions about boundaries because deciding on boundaries explicitly defines the management problem. In a pluralistic society, with varied and strongly held positions, conflict is intensified when perceptive individuals and groups immediately see how their position may be weakened by a certain choice of boundaries. However, not defining boundaries leads to management strategies that lack intellectual rigor or results in debates about technical issues when the debates are really clashes over values and priorities.

The second pillar of ecosystem management is: *Ecosystem management is place-based and the boundaries of the place of concern must be clearly and formally defined.*

16.2.5 ECOLOGICAL HEALTH

The terms ecological *health* and ecological *integrity* are widely used in the scientific and political lexicon (Rapport, 1989; Costanza et al., 1992; Norton, 1992; Grumbine, 1994). Politicians and many political advocates widely argue for managing ecosystems to achieve a "healthy" state or to maintain ecological "integrity." By implication, their opponents are relegated to managing for "sick" ecosystems.

Scientists often speak and write about monitoring the health of ecosystems or perhaps the integrity of ecosystems. Usually there is the assumption that there is an *intrinsic* state of health or integrity and other, lesser states of health or integrity for any given ecosystem (Norton, 1992). Some scientists explicitly advocate "... that maintaining ecosystem integrity should take precedence over any other management goal" (Grumbine, 1994).

Much of the general public seems to accept that there must be a technically defined healthy state similar to their personal human health. After all, people know how they feel when they are sick, and so, by extension, ecosystem sickness must be a similar condition, which should be avoided. Health is a powerful metaphor in the world of competing policy alternatives, but its use in practice depends on what societal values are being pursued. For example, society may want to manage a watershed to maximize opportunities for viewing the greatest possible diversity of birds, for the greatest sustained yield of timber, for the maximum volume of drinking water, or for the greatest sustained yield of agricultural products. Achieving each goal would almost assuredly result in ecosystems that are very different but equally healthy.

The debate is really about defining the desired state of the ecosystem and, secondarily, managing the ecosystem to achieve the desired state. Phrased another way: what kind of garden does society want (Regier, 1993)? There is no intrinsic definition of health without a benchmark of the desired condition. In ecosystem management, scientists should avoid value-laden terms such as "degradation, sick, destroy, safe, exploitation, collapse, and crisis" unless they are accompanied by an explicit definition of what the desired condition of the ecosystem is as defined by society. The word society, as used here, includes only humans.

In philosophical terms, the problem with health is how one links "is" and "ought." An ecosystem, for example, has certain characteristics—facts on which all analysts who study the ecosystem should be able to agree. Characteristics such as species diversity, productivity, and carbon cycling are examples. If the same definitions and the same methods are used, all analysts should come to the same answer within the range of system and analytical variability. The "ought" must involve human judgment—it cannot be determined by scientific or technical analysis (Shrader-Frechette and McCoy, 1993). The concept of health has a compelling appeal, but it has no operational meaning unless it is defined in terms of the *desired* state ofthe ecosystem.

The third pillar of ecosystem management is: *Ecosystem management should maintain ecosystems in the appropriate condition to achieve desired social benefits; the desired social benefits are defined by society, not scientists.*

16.2.6   *ECOLOGICAL STABILITY*

Stability, resilience, fragility, and *adaptability* are interesting and challenging concepts in ecology. These are some of the characteristics of ecosystems that provide an opportunity to realize benefits for society, but the same characteristics constrain options for society and the ecosystem manager. Stability and the related concepts are very difficult to describe clearly because of variations in definitions for all terms associated with this topic. Particular care must be taken to ensure that differences in opinion are not due to differences in definition.

There is a widespread, if sometimes latent, view that ecosystems are best that have not been altered by humans. Further, it seems obvious that such healthy ecosystems *must* be more stable than altered, less "healthy" ones, just as the Romantic School held that Nature realized its greatest perfection when not affected by humans. This is the classic balance of nature view. Pristine is good; altered is bad—perhaps necessary for food, lodging, and transport, but still not as desirable as pristine. However, few seem to be willing to return to the natural human mortality rates of at least 50 percent from birth to age five.

Moreover, this is not how Nature works (Kaufman, 1993). There is no natural state in Nature; it is a relative concept. The only thing natural is change, sometimes somewhat predictable but often stochastic, or at least unpredictable. The concept of dynamic equilibrium might place bounds on ecosystem change as an intellectual attempt to describe better stability, but the intuitive appeal of the concept of stability is not easily fulfilled. Some ecologists cling to traditional concepts of stability and equilibrium with a near missionary zeal.

Considering the sustainable production of goods and services, ecosystems are resilient, although not without limits. A key role of science in ecosystem management is to identify the limits or constraints bounding the options to achieve various levels of goods and services. The challenge is to balance the ability of ecosystems to respond to stress (including use or modification) in desirable ways, but without altering the ecosystem beyond its ability to provide desired benefits. We want shelter, food, personal mobility, energy, etc., but we do not want ecosystems that are producing those benefits to collapse.

The fourth pillar of ecosystem management is: *Ecosystem management can take advantage of the ability of ecosystems to respond to a variety of stressors, natural and man-made, but there is a limit in the ability of all ecosystems to accommodate stressors and maintain a desired state.*

16.2.7   *BIOLOGICAL DIVERSITY*

The level of *biological diversity* in an ecosystem is an important piece of scientific information, and this knowledge can be useful in understanding the *potential* of an ecosystem to provide certain types of social benefits. Grumbine (1994) argued that ecosystem management is a response to today's deepening biodiversity crisis. This may be true *politically*, but biological diversity is purely a technical piece of information. What people *value* about biotic resources, whether biological diversity or something else, is not a technical question.

An argument that is often made is that biological diversity is necessary to maintain ecosystem stability. This argument contains an element of truth, but there is only the most general linkage between biological diversity and ecosystem stability (Goodman, 1975; Shrader-Frechette and McCoy, 1993). Like any other attribute of ecosystems, the value of biological diversity to society must be based on society's preferences. That is not to say that biological diversity (and many other characteristics of ecosystems) is not important; it is. But, as a characteristic of ecosystems, biological diversity operates as an *ecological constraint*, not as a *benefit—unless there is an explicit societal preference.* Many people's values clash over biological diversity, but that is a human preference issue; the ecological role and function of biological diversity is purely a technical question.

It is possible, even likely, that society may value elements of biological diversity as social benefits in and of themselves, but this is a public choice, not a scientific one (Trauger and Hall, 1992). For example, public choice may dictate that no naturally occurring species go extinct because of human action. This is certainly a legitimate social benefit, but not a scientific one. Biological diversity may or may not have intrinsic worth to society.

There are other fundamental public choice issues involved with biological diversity: Do you consider all species, exotic or otherwise, as part of the fauna and flora for the purposes of assessing biological diversity? Is not every species an exotic? What scale do you use to measure diversity? By some measures diversity has increased; by others it has decreased (Berryman, 1991). The choice of the scale used and whether you include exotic species will primarily answer whether biological diversity is increasing or decreasing.

If the public expresses a societal preference for biodiversity in its *own right*, then do our management options include increasing biological diversity beyond what would naturally occur? Should we reintroduce extirpated species (or introduce exotic species) to increase diversity? Should we use the tools of genetic engineering to double or triple biological diversity? Producing agricultural crops with high-performance seeds is not natural, so why not use tools like genetic engineering to increase biological diversity if it is a social benefit?

The fifth pillar of ecosystem management is: *Ecosystem management may or may not result in emphasis on biological diversity as a desired social benefit.*

16.2.8   *SUSTAINABILITY*

Sustainability and a host of related concepts are important elements of nearly all management paradigms. There is a considerable literature on defining exactly what these concepts actually mean and whether the concepts, however defined, are really relevant with changing social priorities and technology. There is considerable debate about whether various societal benefits (including ecosystem "harvests" or outputs) are sustainable, but historically the basic goal has almost always been to produce sustainable outputs of something, tangible or intangible. Sustainable *tangible* outputs (fish, deer, visitor days, drinking water, lumber) are much easier to identify and measure than are the more *intangible* benefit yields (ecosystem integrity, biodiversity, endangered species) typical in ecosystem management. However, whether yields of benefits are described and measured in trees, fish, deer, visitor days, diversity of recreational opportunity, or maintenance of wilderness areas that no one visits, all are realized *benefits* accruable to humans. Benefits are produced within constraints and ecosystems, like all systems, have constraints.

Much more tenuous is the analytical basis for sustainable development—a term often used interchangeably, but inappropriately, with sustainability. The goal of sustainable development typically offered is "... to meet the needs of the present without compromising the ability of future generations to meet their own needs," or in economic terms as exemplified in the 1993 Presidential Executive Order on sustainability, "... economic growth that will benefit present and future generations without detrimentally affecting the resources or biological systems of the planet." The concept of sustainable development masks some fundamental policy conflicts that mere word-smithing will not alleviate (Norton, 1991; Goodland et al., 1993). If one assumes existing social values and priorities, increasing human population, and constant technology, then we cannot *develop* in perpetuity. By necessity we must assume that either values and priorities will change and/or technology will change; otherwise, sustainable development is an oxymoron (Dovers and Handmer, 1993).

There are precise definitions of "develop" that have been offered to counter the logical inconsistencies in the concept of sustainable development; however, at least in the way sustainable development is typically used in public and political rhetoric, the inconsistencies remain. More defensible is the concept of environmental sustainability, which, although logically consistent, leads inevitably to painful choices for society (Goodland et al., 1993). Natural resource management has a long history of failures, in part because of management "magic": the willingness to promise management success when simple logic leads to the opposite conclusion (Ludwig, 1993).

Selecting *what* is to be sustained is a societal choice that should be expected to change over time (Kennedy, 1985; Gale and Cordray, 1991). Do we measure sustainability of commodity yields as surrogates for total societal benefit? Do we measure sustainability of the ecosystem in some defined state? Over what time frames do we measure sustainability? A generation? Over 50 years? Over 100 years? A millennium? What is the scale of sustainability? A small watershed? An ecoregion? An entire nation? How is sustainability to be measured when societal values and priorities change? In short, sustainability often raises more questions than it answers.

Further complicating the concept of sustainability is the apparently chaotic characteristic of ecosystems. Sustainability is often based, at least tacitly, on a largely homeostatic view of Nature—that is, there is a certain natural condition of an ecosystem or perhaps a trajectory of change. But there is no natural state of any ecosystem, only conditions from a wide array of possibilities, known and unknown. The term balance of nature has passed out of common usage in ecology, and this reflects the acceptance, albeit reluctant, of the loose tendency of ecosystems to be self-organizing but buffeted by a myriad of random and chaotic forces.

The sixth pillar of ecosystem management is: *The term sustainability, if used at all in ecosystem management, should be clearly defined—specifically, the time frame of concern, the benefits and costs of concern, and the relative priority of the benefits and costs.*

16.2.9 SCIENTIFIC INFORMATION

Some level of ecological understanding and *information* specific to the ecosystem of concern is essential to effective ecosystem management. The question is, how much understanding and information are needed? After all, the ecological characteristics of ecosystems constrain various management options for producing societal benefits.

Information about people is also important; for example, knowing how individuals and groups might respond to various decision options (Ludwig et al., 1993). Tax incentives may be an especially important tool in ecosystem management, so a solid understanding of how people will respond to modifications in tax law is essential. Erroneous predictions of individual and group response to regulations, policies, and other regulatory tactics are common in policy analysis.

Scientific information is by its nature uncertain—sometimes highly uncertain. Often scientific information and predictions based on scientific information can become the lightning rod for debate about various management options. Debate about values and priorities is important and should be encouraged in the public policy arena; this is not, however, the most appropriate place to debate scientific information. It is important to isolate the two types of debates.

Part of the responsibility for the confusion over "providing information" vs. "advocating policy" rests with scientists. Many ecologists have a strong tendency to support environmentalist worldviews and positions. This is understandable in part due to self-selection in all professions (environmentally oriented individuals are more likely to select ecologically oriented fields than are more materially oriented individuals). The same self-selection takes place in business management (business oriented individuals are prone to select an MBA program rather than a Master of Science program in conservation biology). Individuals in any profession naturally tend to be advocates for what is important in that profession. It is easy to understand the difficulty that many ecologists have in deleting from their scientific vocabularies such value-laden and emotionally charged words as sick, healthy, and degraded. Language is not neutral and scientists should be very careful when speaking as *scientists*. In expressing technical information, scientists should also be sensitive to unspoken assumptions that reflect value-laden or emotionally based opinions.

The seventh pillar of ecosystem management is: *Scientific information is important for effective ecosystem management, but it is only one element in a decision-making process that is fundamentally one of public or private choice.*

16.2.10   *CONCLUSIONS*

Where do these pillars leave us? The seven pillars of ecosystem management collectively define and bound the concept of ecosystem management. Whether the concept turns out to be useful depends on how well its application reflects a collective societal vision. Whether it is possible to develop a collective societal vision in a diverse, multicultural, polarized society such as the United States is a major, and yet to be answered, question. The democratization of science, policy, and choice is not a smooth process, nor is it efficient.

At least in North America the ideas behind ecosystem management represent a predictable response to evolving values and priorities. Those values and priorities will continue to evolve, although the direction and degree of their evolution are ambiguous and largely unpredictable. Without major social jolts such as war, economic collapse, natural disasters, or the return of plagues, the movement of social preferences toward values and priorities of the affluent will probably continue. Such values and priorities operate in the seemingly paradoxical world of intensive use and alteration of nearly all ecosystems, while high value is given to the nonconsumptive elements of ecosystems such as pristineness. We may want the benefits and affluence of a developed economy, but we do not want its factories, foundries, and freeways in our back yards.

There are other directions for ecosystem management that are less clear but potentially more significant. At a major international conference, a statement from an audience member illustrates such a possible direction: "It is time to change our [society's] charter with individuals. We have massive and critical problems with our ecosystems that cry out for immediate action because we have subordinated the collective good of society to the will of individuals. Personal freedom must be weighed against the harm it has caused to the whole of society, and more importantly to our ecosystems." A response to the statement from another member of the audience was equally instructive: "Society and freedom are at greatest risk from those with the noblest of agendas."

Ecosystem management will continue to be place based. Ecosystem management problems need to be bounded to make them tractable. A practical implementation problem is that much of the "place" is owned by individuals, not by society in the form of public lands. By being place-based, application of ecosystem management will become a lightning rod for debates over individual versus societal rights. How does society balance the rights of individuals not to have their property taken without compensation against the right of society, collectively, to prosper? Or perhaps the concept of owning ecosystems (places) must yield to other rights for the greater collective good?

At a superficial level the role of scientific information will continue to become more prominent in ecosystem management. However, most important decisions are choices among competing and often mutually exclusive values. The role of scientific information is important, but it does not substitute for choices among values.

Ecosystem health, ecosystem integrity, biodiversity, and sustainability have evolved from scientific terms to terms used in debates over values. Unless these terms are precisely defined and clearly separated from values and priorities, their value in science is severely diminished. There are major differences in the concepts of sustainability, sustainable development, and developments that are sustainable, but the differences are not easy to explain and understand. I recommend that they be dropped from use in scientific discourse and that more precise, non-value-laden terms be used. Scientists need to be involved throughout the process of ecosystem management but in a clearly defined, interactive role where the values and priorities of the public—not those of the scientists—are implemented.

The definition of ecosystem management is: *The application of ecological and social information, options, and constraints to achieve desired social benefits within a defined geographic area and over a specified period.*

Ecosystem management is neither a revolutionary concept nor an oxymoron but rather an evolutionary change from existing, well-established paradigms. What is revolutionary is the fact that the issues have moved from the hallways of bureaucracies and academia to the larger political landscape.

REFERENCES

Bengston, D. N., "Changing Forest Values and Ecosystem Management," *Society Natural Resources*, 7(6): 515–533, 1994.

Berryman, J. H., "Biodiversity and a Word of Caution," *Proceedings, Annual Conference of the Southeastern Association of Fish and Wildlife Agencies*, 1991, pp. 13–18.

Bormann, B. T. et al., *A Framework for Sustainable Ecosystem Management*, Vol. V, U. S. Forest Service, General Technical Report PNW-GTR-331, 1994.

Christensen, N. L. et al., "The Report of the Ecological Society of America Committee on the Scientific Basis for Ecosystem Management," *Ecol. Appl.*, 6(3): 665–691, 1995.

Cornett, Z. J., "Ecosystem Management: Why Now?" *Ecosyst. Manage. News*, 3(14), 1994.

Costanza, R., Norton, B. G., and Haskell, B. D. (eds.), *Ecosystem Health*, Island Press, Washington, DC, 1992.

Cubbage, F. W., and Brooks, D. J., "Forest Resource Issues and Policies: A Framework for Analysis," *Renewable Resources J.*, Winter: 17–25, 1991.

Dovers, S. R., and Handmer, J. W., "Contradictions in Sustainability," *Environ. Conserv.*, 20(3): 217–222, 1993.

Environmental Protection Agency, *Integrated Ecosystem Protection Research Program: A Conceptual Plan*, Working Draft, 1994.

Fitzsimmons, A. K., "Sound Policy or Smoke and Mirrors: Does Ecosystem Management Make Sense?" *Water Resource Bull.*, 32(2): 217–227, 1996.

Fitzsimmons, A. K., "Why a Policy of Federal Management and Protection of Ecosystems Is a Bad Idea," *Landscape Urban Planning*, 40(1/3): 195–202, 1998.

Forest Ecosystem Management Assessment Team (FEMAT), *Forest Ecosystem Management: An Ecological, Economic, and Social Assessment*, Report of the Forest Ecosystem Management Assessment Team, Multi-agency report, 1993.

Gale, R. P., and Cordray, S. M., "What Should Forests Sustain? Eight Answers," *J. Forestry*, 89: 31–36, 1991.

Goodland, R. J. A., Daly, H. E., and El Serafy, S., "The Urgent Need for Rapid Transition to Global Environmental Sustainability," *Environ. Conserv.*, 20(4): 297–309, 1993.

Goodman, D., "The Theory of Diversity-Stability Relationships in Ecology," *Q. Rev. Biol.*, 50(3): 237–266, 1975.

Grumbine, R. E., "What Is Ecosystem Management?" *Conserv. Biol.*, 8(1): 27–33, 1994.

Kaufman, W., "How Nature Really Works," *Am. Forests*, March/April: 17–19, 59–61, 1993.

Kennedy, J. J., "Conceiving Forest Management As Providing for Current and Future Value," *Forest Ecol. Manage.*, 13: 121–132, 1985.

Lackey, R. T., "Ecological Risk Assessment," *Fisheries*, 19(9): 14–18, 1994.

Lackey, R. T., "Seven Pillars of Ecosystem Management," *Landscape Urban Planning*, 40(1/3): 21–30, 1998a.

Lackey, R. T., "Fisheries Management: Integrating Societal Preference, Decision Analysis, and Ecological Risk Assessment," *Environ. Sci. Policy*, 1(4): 329–335, 1998b.

Ludwig, D., "Environmental Sustainability: Magic, Science, and Religion in Natural Resource Management," *Ecol. Appl.*, 3(4): 555–558, 1993.

Ludwig, D., Hilborn, R., and Walters, C., "Uncertainty, Resource Exploitation, and Conservation: Lessons from History," *Science*, 260: 17, 36, 1993.

Merchant, C., "Fist First!: The Changing Ethics of Ecosystem Management," *Human Ecol. Revi.*, 4(1): 25–30, 1997.

Norton, B. G., "Ecological Health and Sustainable Resource Management," in *Ecological Economics: The Science and Management of Sustainability*, R. Costanza, ed., Columbia University Press, New York, 1991, pp. 23–41.

Norton, B. G., "A New Paradigm for Environmental Management," in *Ecosystem Health: New Goals for Environmental Management*, R. Costanza, B. G. Norton, and B. D. Haskell, eds., Island Press, Washington, DC, 1992, pp. 23–41.

Overbay, J. C., "Ecosystem Management," in *Proceedings National Workshop: Taking an Ecological Approach to Management*, Department of Agriculture, US Forest Service, WO-WSA-3, Washington, DC, 1992, pp. 3–15.

Rapport, D. J., "What Constitutes Ecosystem Health?" *Perspect. Biol. Med.*, 33(1): 120–132, 1989.

Regier, H. A., "The Notion of Natural and Cultural Integrity," in *Ecological Integrity and the Management of Ecosystems*, S. Woodley, J. Kay, and G. Francis, eds., St. Lucie Press, Delray Beach, FL, 1993, pp. 3–18.

Shrader-Frechette, K. S., and McCoy, E. D., *Method in Ecology: Strategies for Conservation*, Cambridge University Press, Cambridge, UK, 1993.

Slocombe, D. S., "Implementing Ecosystem-Based Management," *BioScience*, 43(9): 612–622, 1993.

Stanley, T. R. Jr., "Ecosystem Management and the Arrogance of Humanism," *Conserv. Biol.*, 9(2): 254–261, 1995.

Thomas, J. W., "Forest Service Perspective on Ecosystem Management," *Ecol. Appl.*, 6(3): 703–705, 1996.

Trauger, D. L., and Hall, R. J., "The Challenge of Biological Diversity: Professional Responsibilities, Capabilities, and Realities," *Trans. 57th North American Wildlife and Natural Resources Conference*, 1992, pp. 20–36.

Wagner, F. H., "What Have We Learned?" *Ecosystem Management of Natural Resources in the Intermountain West, Natural Resources and Environmental Issues*, 5: 121–125, 1995.

CHAPTER 16
MANAGING PLACE SCALE PROBLEMS

SECTION 16.3

BROWNFIELDS

Leah Goldberg

Ms. Goldberg is an attorney with the San Francisco law firm of Hanson, Bridgett, Marcus, Vlahos & Rudy, LLP, where she specializes in reuse and redevelopment of contaminated properties.

James A. Jacobs

Mr. Jacobs is a hydrogeologist and president of FAST-TEK Engineering Support Services. He has more than 20 years of experience as an engineer and is registered in several states, including California. He specializes in assessment methods and in situ remediation technologies.

The legacy of the nation's industrial past can be seen in the abandoned and frequently vandalized properties impacted with soil and groundwater contamination, which are, more often than not, in the former industrial zones of inner cities. From the 1960s to the 1990s, business responded to the changing uses of the once vibrant industrial zones by moving out of the urban core and into outer suburban and rural areas. New manufacturing plants, offices, and industrial parks were built in areas previously dominated by agricultural use, open space, or forests. In the process, cities were drained of financial and human capital, future job opportunities, and their tax base while the former manufacturing sites deteriorated further.

Left behind were vacant or under utilized inner-city properties called brownfields. The U.S. Environmental Protection Agency (EPA) defines brownfields as "abandoned, idled, or under-used industrial and commercial facilities where expansion or redevelopment is complicated by real or perceived environmental contamination" (EPA, 1999). Brownfields complicate and frequently frustrate inner-city redevelopment strategies.

Quantifying the costs associated with cleaning up contaminated property is very difficult. Predicting future costs associated with third-party suits and additional regulatory involvement is nearly impossible. Brownfields possess an unquantifiable risk that discourages many developers, lenders, and investors. As a result of federal and state laws that allocate liability for environmental contamination rather broadly, former industrial sites remain idle, although they are often in areas with a developed infrastructure.

The lack of finality after cleanup lends additional risk and uncertainty to brownfields properties. Even after a contaminated property is cleaned up, regulators and third parties may reopen the case file if circumstances change. Therefore, it is not surprising that many environmentally impaired properties have become and remain brownfields (Figure 16.3.1).

Other problems that plague redevelopment of contaminated properties include stringent cleanup standards that do not reflect the end use of the property and thus make the cleanup more expensive than necessary. Also, the inability to obtain financing due to redlining, bureaucratic delay, and neighborhood opposition often makes redevelopment too expensive to be practical.

FIGURE 16.3.1   This property, adjacent to downtown San Jose, is an eyesore, but until the contamination problem is addressed, it is unlikely that the property will be put to a better use.

Recognizing the ill effects of draining our once-economically vibrant cities of human capital and taxes, federal, state, and local governments have promulgated many brownfields laws, policies, and initiatives to promote reuse of brownfields. These programs generally fall into two categories: (1) grants, loans, and tax incentive programs to entice developers and lenders; and (2) limitations on liability for innocent purchasers who agree to cleanup the property or who purchase property after it has been cleaned up.

In response to no-growth initiatives and to transportation corridor-based planning,* various private enterprises have recognized that in-fill development and redevelopment are smart investments. For example, the insurance industry has introduced new products that address environmental risks. Using these insurance products and one or more of the regulatory brownfields initiatives, developers and lenders are finding and exposing the potential benefits of reusing contaminated properties.

This article touches upon the environmental laws whose liability provisions helped create the brownfields dilemma. Following this discussion is a more in-depth discussion of various public and private brownfields initiatives created to address the problem and promote the reuse and redevelopment of contaminated properties.

16.3.1   THE LIABILITY CONUNDRUM

Fear of acquiring liability for environmental cleanup is the key factor in the existence of more than 450,000 brownfields sites nationwide (General Accounting Office, 1995). This fear emanates from

---

* Transportation corridor-based planning strives to curtail sprawl by promoting growth around existing or planned public transportation facilities.

environmental laws that broadly allocate the costs for cleaning up contaminated properties on private parties rather than the tax-paying public.

Resource Conversation and Recovery Act (RCRA)

In the mid-1970s, Congress promulgated RCRA, which regulated cradle-to-grave management of hazardous substances. Although not the primary purpose of RCRA, the issue of contaminated properties was addressed by making those financially responsible for the contamination clean up the property. The RCRA cleanup provisions provided only for injunctive relief. As such, the RCRA provisions could be used only to make the responsible party clean up the site and not to make the responsible party pay for a site that had already been cleaned up. More important, RCRA did not address those sites where the responsible party was financially unable to pay for the cleanup or where the responsible party was "unavailable" through dissolution or bankruptcy of a corporation or disappearance or death of an individual.

Superfund

In December 1980 Congress enacted the Comprehensive Environmental Response, Compensation, and Liability Act (CERCLA, also known as Superfund) specifically to address the remediation of contaminated properties. CERCLA allocated $1.6 billion in funding to provide for the rapid cleanup of those sites designated as the most dangerous and placed on the National Priorities List. In 1986 Congress enacted the Superfund Amendments and Reauthorization Act.

These amendments included increased funding for the Superfund and additional funding was authorized in 1990. Overall, it is estimated that the total spent on the Superfund program is between $25 and $30 billion dollars (Rosenberg, 1995).

Although Congress allocated billions of dollars to cleaning up Superfund sites, CERCLA was based on the premise that those responsible for the contamination should pay for the cleanup, although some of the responsible parties were acting wholly within the law when the contamination occurred. Accordingly, the law was retroactive and included a broad strict liability scheme that ensnared the current property owner or operator, the property owner and operator at the time of the toxic release, and arrangers for the disposal of or transporters of hazardous wastes to pay for the cleanup, regardless of fault or contribution. It also imposed "joint and several liability," meaning that a single responsible party, regardless of percent of fault for, or contribution to, the contamination, could be forced to pay for the entire cleanup and was left with the sole remedy of suing other responsible parties for contribution and reimbursement. Many states adopted mini-Superfund programs modeled after the federal law.

Despite the enormous allocation of funds for cleanup of contaminated properties, CERCLA's track record in closing contaminated sites is dismal. The number of properties that have actually been cleaned up and closed is so low and the liability scheme is so pervasive that CERCLA has served as a catalyst for the creation of brownfields. Because the Superfund programs have been unsuccessful in cleaning up the vast majority of contaminated properties, a new approach is needed.

Property Devaluation

Because of the potentially large and unquantifiable costs of cleanup, the existence or perceived existence of contamination often adversely affects the value of property. Even after contamination has been cleaned up, the property carries the "stigma of contamination." Two 1996 California cases illustrate the stigma or devaluation problem: one in Marin County and one in San Luis Obispo County. In both cases, the impact of the contamination resulted in a 50 percent reduction in the value of the properties compared with the appraisals of comparable noncontaminated properties. The loss in real value in both cases exceeded $400,000 per parcel. The Marin County property, a quarter-acre former gasoline station, sold in a robust real estate market for $14.95 per square foot, less than half the prevailing rate. Although no federal or state agency granted site closure at the time of sale, the owner indemnified the future owners for 10 years. The property was auctioned for cash as no financing was available for this transaction.

In the beach town of Avila, the San Luis Obispo County appraiser estimated loss in value of up to 50 percent for properties impacted by an upgradient oil refinery (Martin, 1996). Over a long time period, the major oil company refinery released significant quantities of hydrocarbons, affecting the soils and groundwater of the business district of this Pacific coast town. Ultimately, because of the potential for law suits from devaluation of the properties, the oil company purchased the affected properties from the owners in 1998. In 1999, the company removed the soil contamination in a massive excavation, transportation, and disposal operation that cost tens of millions of dollars.

16.3.2 BROWNFIELDS SOLUTIONS

In January 1995, Carol Browner, EPA Administrator, announced the Brownfields Action Agenda. As part of the Action Agenda, the EPA removed 27,000 sites from the Comprehensive Environmental Response, Compensation and Liability Information System list of contaminated properties in an attempt to remove the stigma of contamination. EPA also rewrote several policies to clarify liability and cleanup issues, proposed tax incentive programs to promote cleanup of brownfields, and distributed grants of up to $200,000 to various cities for brownfields pilot projects. The purpose of these grants was for real-life laboratories to test redevelopment models, direct special efforts toward removing regulatory barriers without sacrificing protection of human health and the environment, and encourage public and private efforts in communities plagued with problems posed by brownfields. Recipient cities used the funding in various ways.

The City of Emeryville, California, for example, used its EPA brownfields grant to establish a citywide groundwater management program whereby a parcel-by-parcel approach to groundwater cleanup will be replaced by a citywide approach. Groundwater contamination will be evaluated on a citywide basis and cleanup levels will be established for the entire city based on the proposed land uses. As part of this groundwater management plan, the city has developed a database that incorporates hydrogeologic and environmental data with economic, land use, and zoning information. This database, which can be accessed on the Internet, gives buyers, developers, and lenders information about a potential development site. It also serves to lower site investigation costs by making information from neighboring sites available.

In May 1997, the Clinton administration introduced the Brownfields National Partnership Action Agenda. Over $300 million was pledged to clean up brownfields sites. The money was expected to come from 15 federal agencies. An additional $165 million was to be provided in loan guarantees. The Partnership Action Agenda was expected to leverage from $5 billion to $28 billion in private investment, support up to 196,000 jobs, increase local property taxes an estimated $800 million a year, and protect up to 34,000 acres of undeveloped areas or "greenfields" in suburban areas. In addition, Vice President Gore called on Congress to pass a $2 million brownfields tax incentive package whereby businesses would be able to expense property cleanup costs in the year incurred instead of capitalizing costs over the life of the property.

These programs and national action agendas are important pieces in the land reuse puzzle; however, solving the brownfields dilemma necessarily requires addressing the liability issues. Unless prospective developers, purchasers, and lenders can quantify the risk, they are unlikely to give serious consideration to reuse of brownfields sites. Although RCRA, CERCLA, and state equivalent laws are still the laws of the land, Congress, EPA, and state legislatures have turned their attention to policies that promote reuse of contaminated or potentially contaminated properties. Some of these programs include prospective purchaser agreements, voluntary cleanup programs, lender liability protections, flexible cleanup standards, and state programs such as California's Polanco Redevelopment Act. Private entities, such as insurance companies and developers, are also responding with new insurance products and development strategies. Financial assistance, in the form of tax incentives, grants, loans, or forgiveness of development fees, coupled with liability-limiting laws and policies form a successful

FIGURE 16.3.2   A housing development, similar to that across the street, is proposed for this contaminated property.

recipe for brownfields redevelopment (see Figure 16.3.2). The following subsection discusses various public and private brownfields programs.

Prospective Purchaser Agreements

In 1995, the EPA, and subsequently many states, adopted prospective purchaser agreement policies. The prospective purchaser enters into an agreement with the regulatory agency whereby, in exchange for agreeing to clean up the property, the prospective purchaser is relieved of long-term liability from the regulatory agency for further cleanup of that contamination. These agreements allow the buyer to quantify the risk and remove some of the stigma from previously contaminated properties, thus promoting the sale of properties that otherwise may not be marketable.

Use of prospective purchaser agreements is discretionary. Prospective purchasers of contaminated properties do not have a statutory right to enter into such agreements. Moreover, the agreements are site specific; a new and different agreement must be negotiated for each property. The legal and technical costs required to draft and negotiate prospective purchaser agreements can be hefty. In states with numerous regulatory agencies, such as California, a prospective purchaser agreement with one regulatory agency may not bind another regulatory agency from pursuing the prospective purchaser for additional cleanup of the property. Likewise, prospective purchaser agreements do not protect against third-party and toxic tort suits.

Voluntary Cleanup Programs

Several states have promulgated programs that allow for the cleanup of contaminated properties voluntarily, by either the responsible party or the prospective purchaser or developer, under the

FIGURE 16.3.3   Abandoned properties often attract vandals and graffiti artists.

guidance of the appropriate regulatory agency. In other words, if cleanup is required before reuse, but the property is not currently under the oversight of a regulatory agency, one can request oversight from a regulatory agency in exchange for payment of oversight costs and agreeing to cleanup the property. Without regulatory oversight for the cleanup, it is nearly impossible to secure a closure letter, certificate of completion, or "no further action letter," without which lenders and developers may not want to get involved with the property.

Like prospective purchaser agreements, many voluntary cleanup programs require the party undertaking the cleanup to enter into an agreement with the regulatory agency. These agreements often require significant legal and technical preparation and thus can be expensive. Additionally, closure under a voluntary cleanup agreement does not protect the participant from third-party or toxic tort suits. Unlike prospective purchaser agreements, voluntary cleanups typically lead only to closure of the site and do not grant any specific protections to prospective purchasers or lenders. Because a site can be reopened if circumstances change, closure alone may not be enough to satisfy prospective purchasers and lenders.

Lender Liability Laws

The Superfund's broad definition of responsible party envelops anyone or any entity that owned or operated the property at the time of, or subsequent to, the release, even if the owner of the property did no more than foreclose on a security interest. Because of the often enormous financial liabilities for environmental cleanup, lenders were hesitant, if not outright unwilling, to consider financing contaminated properties. Lenders faced the dilemma of choosing between foreclosing on a contaminated property and paying the cleanup costs or abandoning the loan and quit-claiming the property. Banks and private lenders lost millions of dollars in bad debt as the owners of contaminated property walked away from their loans.

Because of the potential liability, lenders shied away from contaminated or previously contaminated properties, rendering these properties unmarketable. In an effort to correct this problem, EPA promulgated lender liability protection regulations in 1992; however, the courts overturned these regulations on the basis that EPA did not have the authority to make policy and that only Congress had the power to change the Superfund liability scheme. In 1996, Congress did just that.

The Asset Conservation, Lender Liability, and Deposit Insurance Protection Act of 1996 (the Act) amended CERCLA to omit from the definition of "owner" and "operator" a lender that, "without participating in the management of a facility or vessel, holds indicia of ownership primarily to protect a security interest in the facility or vessel." In addition, the Act specifies activities that constitute "participation in management" and defines circumstances under which a lender can continue to be excluded from the "owner or operator" definition after foreclosing on a facility or vessel. The Act made similar changes to the underground storage tank laws in RCRA.

A number of states have also adopted lender liability protections. For example, California's lender liability law is somewhat broader than the federal law and contains safe harbor provisions after foreclosure.

Flexible Cleanup Standards, De Minimus Agreements, and Other Policies

In addition to the formal brownfields initiatives, EPA and various state regulatory agencies employ several policies and methods of promoting brownfields redevelopment. Allowing the use of risk-based cleanup goals and tying the cleanup levels to the end use can decrease the cost of remediation considerably. Lower cleanup costs in turn promote the purchase and redevelopment of brownfields sites. For example, cleaning up to background levels or maximum contaminant levels in an industrial area where there are no exposure pathways that could pose a risk to human health or the environment may not be a wise expenditure of funds.

The EPA and many states require cleanups to be consistent with the National Oil and Hazardous Substances Pollution Contingency Plan (40 CFR §300), also known as the National Contingency Plan (NCP). The NCP does not contain specific cleanup goals; instead it mandates that carcinogens be cleaned up to a risk range of between 10^{-6} and 10^{-4} and a hazard index equal to or less than one for noncarcinogenic substances. Within these risk ranges, the NCP encourages site-specific cleanup standards that include a risk analysis, treatment of toxics, and permanent cleanup solutions.

Both federal and state regulators have begun to allow risk-based/end-use cleanup goals as long as the standards remain within the NCP's risk range. Some states have adopted, or have proposed to adopt, formal risk-based cleanup guidelines and goals. Furthermore, some states permit use of the risk-based corrective action guidelines for underground storage tanks developed by the American Society for Testing of Materials.

Concomitantly with permitting less stringent cleanups, regulators often require engineering controls, such as groundwater extraction and treatment systems to contain contaminated water in one location. Moreover, to ensure that contaminants that remain in place do not pose a threat to human health and safety at some time in the future, institutional controls, such as deed restrictions and limitations on the use of shallow groundwater, are frequently required.

Besides end use cleanup levels, several other policies and practices address liability for environmental contamination. For instance, the EPA will, in some cases, enter into a de minimus settlement agreement with a small-volume waste contributor provided that the waste contributor did not handle the hazardous waste or contribute to the release or threat of release. Once a settlement agreement is reached, the landowner gains protection against suits from third parties for contribution to cleanup costs. In addition, in 1995, EPA issued a policy not to sue owners of property overlying a contaminated aquifer provided the owner did not cause or contribute to the contamination. Some states have adopted similar policies and practices.

FIGURE 16.3.4   Brownfields properties are often in areas with developed infrastructure such as railroad spurs, sewer and water service, and power and telephone lines.

These policies encourage efficient use of brownfields. Rather than develop a new warehouse on the outskirts of a city, a purchaser may consider using a warehouse space in the inner city if the cleanup costs are low (Figure 16.3.4).

State And Local Laws And Initiatives

Redevelopment and reuse of contaminated properties will be accomplished only by (1) addressing the potential liability that a new owner or user may assume when considering a contaminated site; and/or (2) providing economic incentives to assist with the cleanup and help offset the risk. Economic incentives may include income tax credits, grants and loans, property tax incentives, and issuance of bonds. State brownfields initiatives vary in their approach. Some states focus on the liability issues and other states attempt to encourage brownfields redevelopment through economic incentives; many states work to combine the two. Following are examples of state legislation:

Massachusetts focused on the liability side of the equation. A new state law retained strict joint and several liability provisions but created clear endpoints to liability for innocent parties and property owners after a completed cleanup. The Massachusetts law also provides immunity from third-party claims for property damage and cleanup costs, exempts owners of neighboring properties from liability, and exempts tenants who lease the contaminated property.

California focused on the liability issues by enacting the Polanco Redevelopment Act, which provides immunity to prospective purchasers, lenders, the redevelopment agency, and anyone who purchases the property from one of those parties for a release or releases that were cleaned up pursuant

FIGURE 16.3.5   San Leandro properties before redevelopment.

to the Act. It also provides generous cost-recovery provisions for the redevelopment agency. Unfortunately, the Act's protective provisions can be used only if the property is in a designated redevelopment project area. In 1998, the Act was amended to allow for someone other than the redevelopment agency to undertake the cleanup and trigger the immunities. For example, in San Leandro, California, a city located about 25 miles (40.3 km) east of San Francisco, a developer is assembling 44 acres of previously industrial property (Figures 16.3.5 and 16.3.6). The developer will clean it up under the Act and will build single-family residential houses. The proposed reuse of the property would not have been feasible under a prospective purchaser agreement because of the time and expense required to negotiate those agreements. The Act avoids the need for time-consuming and expensive agreements.

Florida's strategy, on the other hand, focuses on economic incentives. Contiguous zones composed of one or more brownfields sites are classified as brownfields areas and thus are eligible for a wide range of economic incentives, including tax-increment financing, enterprise zone tax exemptions for businesses, tax abatements, and low-interest loans. To address environmental justice considerations in brownfields areas, Florida law requires public notification to the neighbors and public participation opportunities.

California does not offer significant financial incentives for redevelopment of brownfields sites, but there are provisions for issuance of Mello-Roos bonds to finance environmental cleanups and for the purchase, construction, expansion, improvement, and rehabilitation of real or other tangible property whether publicly or privately owned. To date, few cities have taken advantage of this bonding option because buyers are reluctant to buy properties encumbered with Mello-Roos assessments.

Other states, including Delaware, Iowa, Illinois, Michigan Missouri, New Jersey, New York, Ohio, and Pennsylvania have brownfields programs. Each state's program combines its own mix of liability relief, economic incentives, and in some cases insurance to promote redevelopment of brownfields.

FIGURE 16.3.6   Artist's rendition of San Leandro properties after cleanup and redevelopment.

Local Redevelopment Assistance: One Stop Shop

Another strategy that has been implemented on the local level is development and availability of local data about brownfields sites. Often called one-stop shop (OSS) facilities, these tools provide key information about brownfields sites such as applicable zoning restrictions, land use history, environmental conditions, and state-recommended testing. Having this information readily available reduces the front-end costs of brownfields deals. Of course, the amount of information available at OSS facilities depends on the locality. For instance, some local agencies actually conduct the site assessments and assemble parcels to make them market ready. However, as the 1998 U.S. Conference of Mayors survey notes, "lack of funds is the biggest barrier cities face in performing this front-end work."

Insurance

Brownfields redevelopment does not wholly depend on statutory and regulatory changes. Redevelopment of brownfields properties can be quite lucrative, a point that has not gone unnoticed. After a 20-year span when insurance companies shied away from contaminated properties or issued insurance policies that contained absolute pollution exclusions, insurance companies have recently jumped into the brownfields pool with a number of new insurance products. These products are designed to protect the developer, landowner, or responsible party against unquantifiable risk from either regulatory agencies or third parties. Following are examples of some of the products available:

Pollution legal liability insurance protects against the existence of contaminants that were not identified in a comprehensive Phase I or Phase II environmental site assessment. Pollution legal liability insurance is a versatile insurance product that can provide coverage for off-site property damage, bodily injury, and, in some cases, on-site cleanup costs. It can also provide protection for transportation of hazardous materials and disposal at nonowned disposal sites, business interruption, and diminution in property value resulting from pollution. Before issuing this insurance, the insurance company will likely require a thorough environmental assessment that may include soil and groundwater samples.

Cleanup cost cap insurance provides funding when the cost of cleanup exceeds the projected costs. After conducting the initial environmental investigation, preparing the remedial action plan, and securing the approval of the plan by the appropriate governmental entity, an insurance company will write insurance that indemnifies the insured for cleanup costs that exceed the projected costs of the work described in the remedial action plan. Although cost cap insurance can help address the uncertainty associated with the costs of a remediation project, these policies often contain exclusions for long-term monitoring and other non-remediation-related costs.

Asbestos insurance protects building owners against bodily injury and property damage claims resulting from a release of asbestos from certain buildings. Asbestos claims generally are excluded from liability policies; accordingly, this product was meant to fill that need.

Storage tank liability insurance provides coverage for third-party bodily injury, off-site cleanup, and property damage resulting from storage tanks. This insurance can be used to meet federal and state financial responsibility requirements.

Other insurance coverages, in addition to the above coverages that are generally purchased by the owner of the property, include a number of insurance products available to environmental consultants and contractors: errors and omissions insurance, contractors pollution liability policies, contractors operations and professional services policies, environmental surety bonds, and lead and asbestos abatement contractors liability insurance. Some insurance products cover both the property owner and the contractor. The insurance industry is always looking for new markets. As a result, new coverages are continually being introduced and, of course, if a specialty need arises, insurers and underwriters are often willing to design a product to meet the insured's needs.

The pollution-related coverages, particularly the tailored products, can be expensive. The property owner or contractor will want to carefully weigh the potential risks against the long-term costs of many of these products. Most importantly, the insured should read the policy before purchasing the coverage to ensure that the policy will indeed cover the potential risk.

16.3.3 PRIVATE BROWNFIELDS REUSE: A PRACTICAL APPROACH

Private sector brownfields redevelopment, also called land reclamation, land recycling, or in-fill development projects, can provide opportunities for developers and communities. However, purchasers and developers also must be aware of potential pitfalls common to all redevelopment projects.

In some economically depressed areas of the nation, some properties are not financially viable investments, regardless of whether they are pristine or contaminated. Therefore, a careful economic evaluation is highly recommended. The evaluation should include sales prices in the area, rental rates of nearby properties, occupancy rates for the area, and other information related to the specific property. Other factors that are evaluated include the price of the property, the price of nearby properties, the rental income potential, the tax basis, the financing terms, the cost and timing of cleanup or site closure activities, the likelihood of obtaining land use entitlements, and the cost of building renovations. Among other issues that are examined are freeway access, the office-to-warehouse ratio, general location, and investment growth potential of the area (Table 16.3.1).

For prospective purchasers, property location, timing, and duration of the environmental restoration are critical. Because a contaminated site may have other problems, including condemned buildings, delinquent taxes, or significant surface debris, location is an important element to ensure that value can be restored to the property. Potential investors also will carefully consider the timing and duration of the environmental cleanup effort. The longer it takes to put property back into productive use, the longer investors and developers will have to wait for a return on their investment. The availability of quick and effective remedial measures profoundly affects the desirability of brownfields.

TABLE 16.3.1 Some Factors Used to Evaluate
Brownfields Sites for Redevelopment

- Price of property
- Price of nearby properties
- Rental income potential
- Tax basis
- Financing terms
- Cost and timing of cleanup
- Cost of renovations
- Cash flow
- Long-term liability
- Cost reimbursement potential
- Freeway access
- Office/warehouse ratio
- General location
- Investment growth potential
- Other risks
- Environmental justice

16.3.4 CONCLUSION

Fear of acquiring liability for an environmental cleanup is the key factor in the existence of over 450,000 brownfields sites nationwide. This fear comes from environmental laws that attempt to place the cost of cleaning up contaminated properties on private parties that may have benefited from the use of a property instead of on public funding. Moreover, broad environmental laws often require owners and operators who did not create the environmental problem to bear the cost of the cleanup.

Properties with contaminated soil and groundwater can be either a community resource worthy of recycling or brownfields attracting vandals and creating an eyesore. Over the past few decades, much of the nation's industrial and manufacturing base abandoned the older suburbs and urban centers in favor of new development in newer suburbs. By recycling contaminated properties with public and/or private development methods, the nation can revitalize inner cities as well as older suburbs, and save precious forest, farm land, and open space for future generations. Public and private incentives and funding as cited in this article can help reverse the trend of urban disinvestment and brownfields creation.

REFERENCES

General Accounting Office, "Community Development Reuse of Urban Industrial Sites" report, Washington, DC, GAO/RCED-95-172, June 1995.

Bureau of National Affairs, *Environmental Reporter*, 28(3): 78–79, May 16, 1997, See, 42 USC, § 9601 (E)(i), (ii), (F).

EPA, *Introduction to Brownfields*, EPA Washington, DC, 1999 (www.EPA.gov/swerosps/bf/glossary.htm#brow>).

Martin, G., "Toxic Troubles," *San Francisco Chronicle*, pp. 0–1, February 25, 1996.

Rosenberg, D. M., "Brownfields Redevelopment: A Realty?," *Pollution Eng.*, 53: October 1995.

U.S. Conference of Mayors, *Recycling America's Land*, Washington, DC, 1996.

CHAPTER 16
MANAGING PLACE SCALE PROBLEMS

SECTION 16.4
GUIDELINES FOR EMERGENCY RESPONSE

Donald Fawn Jr.

Mr. Fawn is senior member of the TNRCC Emergency Response Unit. His training and expertise include nuclear, chemical and biological response operations; hazardous waste site personal protection and safety; confined space entry/rescue; hydrocarbon fire fighting; alternate treatment methods for Superfund sites; risk assessment; incident command system; bioremediation; railroad tankcars/intermodal tanks; dispersant use/effects; and chemical safety audits.

The primary resource available to local government to defend against large-scale hazardous substance emergencies is well-trained and well-motivated first responders. Local government has the responsibility for planning and developing an emergency management system that is capable of an effective and timely response. Although other levels of government (county, state, and federal) may be called in to help when local resources have been exhausted, only local government can fulfill this critical initial response role.

Aside from natural disasters, most large-scale hazardous substance emergency incidents are in one or more of the following five categories: biological, nuclear, incendiary, chemical, and explosive. These categories are often referred to by the acronym B-NICE.

Depending on the nature of the material released, the source of the release, the geographic area, and meteorological conditions a response effort can be a very complex process and rarely stops with simple removal of the contaminant. Quite often, the debris from a cleanup is determined to be a hazardous waste that requires special handling and disposal. Proper disposal of hazardous wastes protects human health and the environment. Most state laws now specify that such wastes should be reclaimed or recycled wherever possible.

16.4.1 THE GOAL

The goal of every emergency response effort is to favorably change the outcome. The first properly trained responder arrives on scene with two inviolable rules of thumb: (1) If you don't know, don't go; and (2) What you don't know can kill you.

No matter what the emergency may be, the first responder on scene has the critical responsibility to size up the incident and to communicate that assessment to fellow responders en route to the scene. With personal safety first and foremost, this responder is expected to attempt recognition and identification of the problem, determine the need for notification of additional authorities or responders, determine isolation requirements, and determine appropriate levels of protection for incoming responders.

If One Person Does Not Assume Command, the Incident Does

Someone must assume command. By default, the incident commander (IC) at any given scene is the highest ranking initial responder or the person among the initial responders with the most applicable expertise for the incident.

When an incident escalates, it may be necessary to transfer command of the incident to a more experienced responder. In some cases, there may be a predesignated IC specified in the affected jurisdiction's emergency response contingency plan, local ordinance, or by a state statute. Any transfer of command must be face-to-face where possible and must include a situation report (SITREP) to the person assuming command.

Acceptance of the IC position includes personal assumption of ultimate responsibility (as well as legal liability in certain situations) for success or failure of the incident response operations.

Leave Your Ego at the Front Door, Please

The command post is never the place for *personalities*, *politics*, or *press*. The IC must remain focused on the incident. Simply stated, the most effective IC is proactive, decisive, objective, calm, adaptable, flexible, and quick thinking.

A good IC is also realistic about his or her limitations and is not hesitant to transfer command in response to changing incident conditions and priorities. A large-scale, multijurisdictional incident will quickly separate those who can from those who think they can.

Accept and Understand Basic Priorities

Not every incident commander will have danced a ballet with a fully charged fire hose, performed CPR on "master" under the watchful eye of "Buffy" the Great Dane, or worked in a fully encapsulated hazmat suit on an August afternoon in west Texas. Although an IC may not understand the peculiarities of each and every job, the priorities are simple and universal.

Proper incident management requires a thorough understanding and acceptance of the priorities of the local jurisdiction first responders. Local firefighters, emergency medical personnel, rescue teams, and hazardous materials teams all share a common hierarchy of realistic priorities as follows:

1. Protect themselves (and their buddy) first. Use of the buddy system is mandated by law (29 CFR 1910.120).
2. Rescue the survivors (conduct body recovery later). Set up triage and provide survivors with emergency medical attention within the golden hour.
3. Protect any immediate exposures: prevent things that aren't burning from catching fire (cool adjacent appurtenances or structures subject to flame impingement), protect uncontaminated areas from becoming contaminated (placement of berms, dikes, and runoff/runon controls), issue evacuation or shelter-in-place orders for potentially affected population areas, etc.
4. Containment and control of the problem: begin to attack the actual fire or to control further spread of the contaminant.
5. Extinguishment/mitigation: put out the fire, eliminate the source of the contaminant, neutralize the contaminant, and stabilize the situation.
6. Salvage and overhaul: ensure that the fire is out cold, assess damages, properly manage and dispose of wastes. Conduct body recovery operations after completion of any required investigative, evidentiary, or forensic work by the appropriate authorities.

16.4.2 THE DOS

The IC Must Continue the Size-up Process Until the Response is Complete

Size-up must continue until the response is complete. As a response operation mobilizes and throughout any subsequent expansion of operations, additional observations and information will be directed to the command post. The IC must continuously address the following questions:

1. What is the nature of the incident (i.e., is this an accidental release of a chemical or an act of terrorism)?
2. What hazards are immediately obvious (and based on this assessment what unseen hazards could be reasonably anticipated)?
3. How large is the affected area?
4. How can the affected area best be isolated (i.e., hazard tape or cyclone fence)?
5. What other locations are available for the command post and equipment staging areas if the site conditions change (i.e., wind shifts, precipitation)?
6. Where are the safest and most efficient locations to route rescue personnel and rescue equipment if needed?

An ongoing size-up allows the IC to accurately identify potential problem areas, to define resource needs, and to determine how to best manage and deploy resources.

Use the Incident Command System (ICS)

The U.S. Occupational Safety and Health Administration requires all agencies and personnel responding to hazardous substance emergencies [29 CFR 1910.120(q)] to use the Incident Command System (ICS).

The ICS is a management system that organizes functions, tasks, and response personnel. It provides a function-oriented approach to an emergency. The ICS structure defines the responder's purpose, duties, and line of communications. The functional structure of ICS provides for the rapid modular expansion of the response team to handle an escalating incident.

ICS has long been used by the military. It is now used by federal agencies with emergency functions, has been adopted by most state governments, is required by the fire service, is required by the emergency medical service, and is still largely ignored by the law enforcement community.

The widespread acceptance and use of ICS is due to its modular organization, use of common terminology, unified command structure, span of control, and resource management. It has long been recognized that there is a limit to the number of personnel or tasks that can be adequately supervised by any single individual. Known as span of control, this limit generally ranges from three to seven personnel or tasks.

The ICS organization is composed of five functional sections: command, operations, planning, logistics, and finance.

- The command section assesses priorities, determines strategic goals/tactical objectives, develops an incident action plan, develops an organizational structure, manages resources, ensures responder safety, coordinates with outside agencies, and authorizes release of information to the media. The command staff includes the public information officer, safety officer, and liaison officer. The public information officer is the single contact for the news media. The safety officer is responsible for scene safety and is the only individual with "veto power" over the decisions of the IC. The liaison officer is the designated contact for other governmental agency representatives, elected officials, campaigning politicians, and special interest groups.
- The operations section is responsible for developing operational plans, requests or releases resources

through the IC, keeps the IC informed of the status of operations, and conducts tactical response actions.

- The planning section is composed of various technical experts (often referred to as scientific support coordinators) whose input will provide the technical basis for an action plan. These experts in various disciplines (i.e., toxicology, meteorology, chemistry, geology, hydrology, biology, botany, mycology) collect and evaluate information from a variety of sources to be used in preparation of the incident action plan. They also monitor changing weather conditions, assemble information on alternative strategies, identify the need for specialized resources (i.e., shallow-water oil skimmers, dispersants, microbial firefighting foam, other chemical agents), and provide periodic predictions on incident potential.

- The logistics section procures all necessary response equipment to support both the response effort as well as the needs of the individual responders. Logistics staff find and procure, sometimes rather creatively, any resources identified by planning staff, traffic planning, food, sanitary facilities, and prepare the incident communications plan.

- The financial section documents costs, manages finances, and, as in the case of a declared disaster, prepares the complex paperwork necessary for reimbursement by the federal government.

Establish One Command Post and a Unified Command Structure

Large-scale emergencies do not respect jurisdictional boundaries. The ICS supports a multiagency, unified command structure in which agencies with jurisdictional responsibility jointly determine response strategy and objectives, planning and tactical activities, and sharing of resources. No matter how many different entities may be represented on the command staff, one person is designated to be the IC [29 CFR 1910.120(q)].

The IC should establish a single command post from which the incident response can be directed. It is the responsibility of the IC to set up a unified command, maintain a reasonable span of control, clearly define the chain of command, be adaptable to a variety of situations, and be familiar to each participant. A representative from each involved jurisdictional organization should be at the command post.

If the emergency incident involves a responsible party, such as a petrochemical facility, hazardous materials transporter, or manufacturer, that party's representative should be at the command post as well. The IC should solicit this input because the responsible party has the legal obligation to conduct and fund the cleanup of any contaminants and to address any damages to natural resources by way of compensation or restoration. Financial liability for the response effort belongs to the responsible party. It is very important to the continuity of a response effort for the IC to know the availability and the limitations of the responsible party's resources in order to formulate a request for government funding.

Establish the Joint Information Center

The properly managed large-scale incident has one person designated as the public information officer (PIO). In the unified command structure where several different agencies are represented on the command staff, each agency representative will likely be accompanied by that particular agency's PIO. However, the PIO designated by the IC is the only person permitted to talk with the news media.

The incident PIO coordinates preparation of press releases and statements with the information officers from the other agencies involved, coordinates any public briefings by the IC, and works with other information officers to respond to public inquiries related to the incident.

A joint information center should be established at a location well away from the incident and away from the command post. All public inquiries and requests from news media are directed to the information center. A large-scale incident often requires the temporary setup of one or more 1-800 toll-free hotline numbers specific to the incident. For example, one number may be assigned to handle survivor/fatality questions; another for property damage/insurance questions; and another for questions related to human health, exposure, and medical monitoring questions.

Make a Politician Feel Special

Unfortunately, the media coverage generated by a large-scale incident presents a golden opportunity for elected officials to be seen on television by their constituents. The liaison officer should be available to the designated information officer or information center staff in an on-call capacity to setup and conduct special VIP briefings exclusively for elected public officials. Continuity of a response effort, especially if the incident involves mass casualties, severe property damage, wildlife mortality involving any of the "Disney species," or affects the local infrastructure (disruption of basic services and utilities), depends on the support of public officials.

Develop and Implement the Plan

A written plan is a necessity whenever an incident utilizes resources from other agencies, jurisdictions, or political subdivisions; utilizes private contractors and public money; or reassigns personnel, equipment, or normal job functions. This emergency response plan, contingency plan, or action plan should be written to address the incident specific details of the command, operations, planning, logistics, and finance elements of the ICS.

In a unified command structure, all involved jurisdictional representatives including the responsible party (if applicable) must jointly prepare the plan.

Establish Written Standard Operating Procedures

In additional to outlining the assignment, deployment, and management of personnel and equipment, the plan should describe standard operating procedures relative to these required components:

1. Site safety plan, identifying key personnel and alternates, incident specific hazards, risk analysis, air monitoring procedures, sampling techniques, personal protective equipment, decontamination procedures, and proper management/disposal of decontamination wastes.
2. Medical plan, describing medical monitoring, emergency medical procedures, transportation, methods, medical facilities etc.
3. Radio communications plan, specifying radio frequency assignments, call signs or unit numbers, radio codes, common terminology to be used for all radio communications. This plan should specify a set of visual or hand signals to be used in the event that radio communications fail.

Define Tactical Considerations

Tactical considerations can be described as either defensive or offensive.

Defensive operations are oriented toward confinement of the problem but not stabilization of the incident. In the case of a large chemical spill, a defensive approach would attempt to prevent the spread of the chemical over a larger area.

Offensive operations are oriented toward containment of the problem and stabilization of the incident. In the case of a large chemical spill, an offensive approach would attempt to stop or control the source of the release.

Reevaluate

As previously emphasized, an incident size up is never completed until the situation is mitigated and completely under control. Because emergency incidents rarely behave according to any written plan, the plan must be flexible enough to provide for an ongoing reevaluation of tactical operations.

When the outcome does not favorably change, the incident requires another size up of incident factors, analysis of the most recent data, a change in action or strategy, a change in goals or objectives, and implementation of different tactical operations to achieve the new objectives.

Use a Phased Approach

Define when the incident is no longer an emergency. At what measurable point does the response cease to become crisis management and become one of mitigation or long-term remedial action?

The plan should specify measurable benchmarks for the various phases of the response effort. A procedure for a smooth transition of command should be specified as well. For example, the local fire chief may be the best IC to manage a response to a large chemical facility fire. Upon extinguishing the fire, however, the incident ceases to be an emergency requiring the services of local first responders. The public has been protected from acute exposure hazards but a chronic exposure problem may persist. The nature of the public threat is the presence of uncontrolled hazardous wastes, hazardous by-products of combustion, runoff to surface water, contamination of groundwater, air quality problems, and uncontrolled public access, none of which can be properly managed with a fire truck. An emergency condition still exists but the response requires an IC with a background appropriate for effectively consulting with technical experts on response options such as use of chemical agents, bioremediation, innovative treatment technologies, alternative treatment methods, spill waste management, and surface and groundwater monitoring and protection.

A well-written plan contains ample flexibility for applying common sense in response to rapidly changing conditions. The plan must be written in a simple everyday language that is easily understood by emergency responders, financial auditors, and attorneys. A brief ultrasimplified VIP version should be prepared for campaigning politicians.

Know When to Delegate

A large-scale emergency response effort is at the mercy of many constantly changing variables. Something as subtle as a shift in wind direction may change task requirements, alter site safety considerations, or require relocation of resources. The dynamic nature of an emergency incident most often will influence span-of-control considerations. A competent IC must recognize when he or she is approaching a personal span-of-control limit and delegate some tasks. Resources not specifically delegated by the IC remain the responsibility of the IC. Additionally, response managers at all levels of the ICS should be trained to recognize when their supervisory capabilities are about to be exceeded and to request additional resources as soon as possible.

Understand the Applicable Regulations

Familiarity with the following regulations is a necessity:

29 CFR 1910.120	Hazardous waste operations and emergency response
29 CFR 1910.132	Personal protective equipment
29 CFR 1910.134	Respiratory protection
32 CFR 659.201	Department of Army oil and hazardous substances spill control and contingency plans
40 CFR 110, 112, 116	Discharge of oil, pollution prevention, and designation of hazardous substance
40 CFR 261	Identification and listing of hazardous waste
40 CFR 262	Standards applicable to generators of hazardous waste
40 CFR 263	Standards applicable to transporters of hazardous waste
40 CFR 300	National oil and hazardous substances pollution contingency plan
40 CFR 355	Emergency planning and notification
40 CFR 370	Hazardous chemical reporting: community right-to-know
49 CFR 171	Hazardous materials regulation: general information, regulations, and definitions

| 49 CFR 172 | Hazardous materials tables and hazardous materials communications regulations |
| 49 CFR 173 | Shipper general requirements for shipping and packaging |

Consult Reference Sources

1. *Emergency Response Guidebook*, U.S. Department of Transportation, Washington, DC, 1996.
2. *CHRIS: Chemical Hazard Response Information System*, U.S. Coast Guard. [available from the National Response Center at (800) 424–8802, Washington, DC, 1999].
3. Hawley, G. G., *Condensed Chemical Dictionary*, Van Nostrand Reinhold, New York, 1997.
4. *NIOSH Pocket Guide to Chemical Hazards*, U.S. Government Printing Office, Washington, DC, 1997.
5. Sax, N. I. (ed.), *Dangerous Properties of Industrial Materials*, Van Nostrand Reinhold, New York, 10th edition, 1999.
6. *Documentation of Threshold Limit Values (TLV)*, ACGIH, Cincinnati, 1999.
7. *OHMTADS: Oil and Hazardous Materials Technical Assistance Data System*, U.S. Environmental Protection Agency (available from EPA regional offices).
8. Sine, C. (ed.), *Farm Chemicals Handbook*, Meister, Willoughby, OH, 1999.

Maintain A List of Response Assistance Numbers

1. National Response Center — (800) 424-8802
2. Chemical, Biological Defense Command Ops Center — (800) 368-6498
3. U.S. Coast Guard National Strike Force — (800) 424-8801
4. Environmental Response Team: contact EPA Regional Office
5. Department of Transportation Hotline — (202) 426-2075
6. Bureau of Explosives — (202) 835-9500
7. Chemical Referral Center — (800) 262-8200
8. CHEMTREC — (202) 483-7616
9. CHLOREP (chlorine emergency plan) — Access through CHEMTREC

Consider the Possibility of Terrorism

Continuously size up the incident and consider all possibilities, including the previously unimaginable. Is the incident just an accident, is it human error involving one of the B-NICE substances, or is it a terrorist incident?

1. Religious: Is the incident at a church, synagogue, or other place of worship? Is the site the target of protest by religious groups—i.e., a women's health care center, a "planned parenthood" facility, an establishment with a predominately homosexual clientele?
2. Ethnic or racial: Is the scene associated with an ethnic or racial group—i.e., a church with a predominantly black congregation?
3. Single issue: Has the site been the focus of single-issues groups—i.e., animal rights activists, certain environmental activist groups?

4. **Political:**   Is the site occupied by government employees—i.e., local, city, county, state, or federal? Is the site significant to secessionist, separatists, or nationalist groups?

5. **Criminal:**   Is the site related to gang activity or rival criminal factions, i.e., narco-terrorists?

6. **Economic:**   Has the business at the site recently laid off employees or has it been the subject of protests, strikes, or shutdowns by management, workers, or labor-specific interest groups?

7. **Date/time:**   Is the day or time of the incident significant—e.g., April 19 (the date of the Oklahoma City bombing and the holocaust in Waco, Texas)?

The FBI defines terrorism as *"the unlawful use of force or violence against persons or property to intimidate or coerce a government, the civilian population, or any segment thereof, in furtherance of political or social objectives."* The victims may be totally unrelated to the terrorists' cause. Terrorism is violence aimed at the people watching. Fear is the intended effect and not the by-product of terrorism. For example, a large-scale B-NICE emergency becomes completely different when an individual or group claims responsibility for the act based on a political cause, religious ideology, or other belief system. That expression of intent redefines the incident as an act of terrorism and the emergency now involves use of a "weapon of mass destruction."

Although the technical aspects of a response to an accidental chemical release probably will be no different than the response to the same chemical intentionally released on behalf of a particular deity, political system, or social issue, the important difference is the intent. If terrorism is the potential cause of an emergency incident, the IC should immediately undertake the following actions in the order listed below:

1. Responders should pull back and assume defensive tactics. Mass destruction is the goal of the incident and it is not at all uncommon for a secondary device to have been placed specifically to target emergency responders, thereby further disrupting the government infrastructure at the local, regional, or state level. Unfortunately, the massive media attention generated by such emergency events is often viewed by terrorist groups as an excellent opportunity for executing another terrorist act designed to produce fear.

2. If the incident is one of a series of terrorist threats, never set up the command post or staging areas in the same place twice! Terrorists often probe the system with a series of threats of hoax calls and observe the response mobilization, looking for a common denominator. This is a common method for determining the best placement for a primary or secondary device.

3. Immediately contact the U.S. Federal Bureau of Investigation (FBI) through the National Response Center at (800) 424-8802. . Pursuant to federal law, the incident is a federal crime and the FBI assumes incident command [18 USC §2332b (g)(5), Acts of Terrorism Transcending National Boundaries; 18 USC §3077, Rewards for Information Concerning Terrorist Acts and Espionage; 18 USC §2331(1), Definition of "International Terrorism;" and 18 USC §921(a)(22), Firearms].

4. Remember that the scene of a terrorist incident is a crime scene as well as the scene of an emergency response. All practical precautions should be undertaken to minimize disturbance or destruction of evidence while addressing the human health and safety priorities.

Presidential Decision Directive 39 (PDD-39) *United States Policy on Counterterrorism* defines policies concerning the federal response to threats of acts of terrorism involving nuclear, biological, chemical, and/or weapons of mass destruction. The specific responsibilities of various federal agencies are outlined in PDD-39. Copies may be obtained from the Federal Emergency Management Agency (FEMA) Printing and Publications at (202) 646-3484.

16.4.3   THE DON'TS

Charles Darwin's *Theory of Natural Selection* combined with proper implementation of the ICS precludes the need for a long list of practices to avoid. In fact, there are only two.

Don't Vacillate

The worst decision an IC can make is to not decide. Decide something, even if it is to turn around and run. It is completely defensible to decide on the side of safety and opt for a defensive tactical approach based on the two inviolable rules of thumb. The IC is responsible for life safety first and *seconds save lives*.

Don't Forget the DUCT TAPE

There is never enough duct tape to go around.

CHAPTER 16
MANAGING PLACE SCALE PROBLEMS

SECTION 16.5
ENVIRONMENTAL COMMUNICATION

Richard R. Jurin

Dr. Jurin is president of Ecological Communications, Education, and Interpretation Consultants (ECEIC). He teaches ecological communications at the School of Natural Resources, Ohio State University, Columbus.

16.5.1   INTRODUCTION

This section introduces you to some simple ideas of effective outreach communications. These are public relations, risk communication, social marketing, and dealing with the news media. Environmental communications is really a term that encompasses any communication that relates to the resources we use, the way we live, and the environment we live in. In the technical, scientific, and engineering fields, it quickly relates to any communication that needs to be done in order to successfully transmit information of a technical nature to a myriad of different publics.

Throughout this chapter the term "organization" is used to denote a variety of different groups such as manufacturing companies, retail businesses, nonprofit groups, agencies, and compliance groups to name a few. The emphasis in this chapter is to help technical specialists relate to the different audiences they may come across, so that the message the specialist sends out is understood and accepted by a lay audience, and the audience will do something because of the message. This implied action by the audience may be something as simple as gaining awareness of a situation so they can develop an objective opinion or as complex as adopting a new behavior. It is this expectation by the specialist that makes communications much more complex than just providing information. This expectation requires that a feedback system be built into the communication planning in order to evaluate whether the expectation was met and where improvements can be made. All communications therefore involve systematic planning.

16.5.2   TARGET AUDIENCES

If you were to try to sell a product to the public, you would want to know as much about potential buyers as possible. Likewise, if the communications message is thought of as a product, then the more you know about your audience the more likely you can customize your message to them. The following questions need to be asked:

- Who is (are) the audience(s)?
- What things can be found out about the audience to make the message applicable to them?
- What is the message?
- Why are you telling them?

- What do you want them to do with this information?
- What kind of reaction do you expect?
- How will you know that they understand the message? (Note how feedback is needed here).
- There is no single audience but many different audiences. These different audiences may be easily classified into groups based on shared characteristics, interests, and demographics. It must be emphasized that these are only indicators of your audience, and thorough research into your audience is recommended to maximize success with your communications.

Audiences can be quickly divided into those for whom the message is very applicable and will be readily accepted and those to who will dismiss the message. It is essential know beforehand how your audience might react to a message. This usually involves talking to or surveying a select group of people who are similar to the audience you want to reach in your message planning. Having a clear goal with measurable objectives and well-thought-out evaluations is a prerequisite of success. If you cannot get firsthand information about your audience, then it may help to review other situations that resemble your situation. Interviewing key people who know about your audience is also beneficial in helping you develop a profile of your audience backgrounds. The more you understand your audience the more likely your message will "hit home."

16.5.3   LANGUAGE AND READABILITY

The greatest challenge the specialist faces is delivering a very technical series of facts and concepts to an audience that does not have the knowledge or background to understand all of it. The first thing to do is cut out most of the technical terminology and write in everyday common vocabulary! Eliminate the jargon, abbreviations, synonyms, and uncommon scientific terms that will not be understood by anyone not in the discipline. The only terminology that should be included is that which is commonly used in daily newspapers and in everyday language, and then with sufficient explanations. For instance, the terms "deep well injection" or "stack scrubbers" mean nothing to the lay public without a detailed and simple explanation! In short, the language used must be common everyday lay-conversation English. Care must be taken when simplifying the language to ensure that technical aspects are not lost or corrupted into something that can be misinterpreted. This also includes the use of metaphors or analogies because comparisons are not really the same and different people may "read" more into the comparisons than the specialist intended.

In technical language there are many complex terms and words used, and often the language itself can become very wordy, stilted, and filled with jargon and slang. Keep it simple. Summarize at the end of each major section, especially when some difficult subject matter has to be included. Include shorter or simpler words that convey the same meaning such as "use" instead of "utilize," and "improved" instead of "ameliorated."

A simple instrument to help simplify your writing is the Gunning fog index outlined below. Essentially it corresponds to the number of years of schooling that a person would need to easily understand a written message. Therefore a readability of 12 corresponds to a senior in high school, and a readability of 20 corresponds to advanced graduate writing. Most common media try to keep the readability at a fog factor of 12 or less.

Determining Your Fog Index

(Gunning, R., and Kallan, R. A., *How to Take the Fog Out of Business Writing*, Dartnell Corp, Chicago, IL, 1994):

- Find the average number of words per sentence. Sample five to eight sentences: count clearly independent clauses as separate sentences. Example: "I faced the cliff; I moved toward it; I climbed." This counts as three sentences.

- Calculate the percentage of words with three syllables or more. Don't count proper names or verbs that make three syllables by adding -es or -ed. Example: process and processed or processes are all two syllables.
- Add the two numbers and multiple by 0.4. Example: If your average number of words per sentence is 14, and the percentage of words of three syllables or more is 13 percent, add 14 and 13 to get 27. This multiplied by 0.4 gives 10.8, which is your fog index, a rough measure of how many years of schooling it takes to easily understand what you have written.

Therefore, the Gunning fog index = (number of words/number of sentences) + difficult words × 0.4. *The Bible*, Shakespeare, Mark Twain, and *TV Guide* all have fog indexes of about 6. *Time*, *Newsweek*, and *The Wall Street Journal* average about 11.

If you find your index soaring into the high teens (or higher), beware—you have begun to lose most of your audience in the dense fog. Therefore, when you write for the public, use shorter sentences, shorter paragraphs, and simpler words. If difficult material is covered, then do a short summation at the end of each paragraph. Use easy words in conversational language and keep it all succinct and to the point. Avoid stilted and technical language. If you imagine that you are writing or talking to a group of junior or sophomore high school students you are aiming at the correct level much of the time. The more you know about your audience, the easier it will be to target the message for the correct readability and language level.

Further Reading on Language and Readability

1. Calvert, P. (ed.), *The Communicator's Handbook: Tools, Techniques and Technology*, 3rd edition Maupin House, FL, 1996.
2. Chall, J. S., et al., *Qualitative Assessment of Text Difficulty: A Practical Guide for Teachers and Writers*, Brookline Books, Cambridge, MA, 1996.
3. Mcquain, J., *Power Language*, Houghton Mifflin, Boston, 1996.

16.5.4 DEALING WITH THE NEWS MEDIA FOR SCIENTISTS AND ENGINEERS

This section is not so much a tool but more a guide to helping the communicator deal with one of the more pervasive information systems in today's society. By understanding the role and the mechanisms of the news media, the specialist can help prevent misleading information from being spread.

- Reporting in the news media can make a difference by stimulating action and bringing facts to the public's attention.
- The media act as "gatekeepers" of public information. They control what we read and see and how we see it, and they make economic, social, and political inferences about information that is "released."
- Media news coverage is a net, not a blanket. Only big stories get caught.
- The news media sell information; they do not supply it free of charge. Because of this, environmental issues and problems can often be
  1. Distorted: the information is adapted to make it appealing to the lay audience. If there is little in the way of a personal story, it will not be published.
  2. Sensationalized: the most dramatic aspects of the problem are dealt with in the story. Problems that have some significance only to scientists or engineers are rarely covered.
  3. Oversimplified: the information is adapted to make it understandable to a lay audience.

- Additional inaccuracy is caused by news reporting constraints such as reporters who are not scientifically knowledgeable about complex environmental and scientific concepts, tight deadlines that restrict how much story researching can be achieved, political and economic pressures placed on the gatekeepers to restrict certain types of reporting, and public interest level of the story.
- It is important for scientists to understand that there is a big difference between what scientists want to say about their research and what science writers/reporters will actually report!
- Reporters write to INFORM not to educate. The gatekeepers (editors) do the selecting and the selling of the information. Reporters in competition with other reporters do have an allegiance to the truth, and are looking for answers that will address GOOD for society. However, the stories must have a selling point.

The Reality of Science Reporting

- There must be a basic personal and societal value of the information to need to explain research to the public.
- Scientists and engineers need the general media to promote science outside the science community.
- Scientists and engineers must interact with the media on the media's terms.
- Detailed science must be aimed at someone who may not be science literate.
- Scientists and engineers don't understand lay audiences, especially their cognitive limitations. Journalists understand audiences more than they understand science.
- The uncertainty of science is not realized by most reporters or by lay audiences. Scientific statements are seen as definitive.
- Reporters are concerned about the application and social relevance of science and not implications to the science discipline.
    1. Scientists who reviewed science media stories generally concluded that one-third of most stories had inaccuracies or omissions or were too vague. Reporters retorted that this one-third represented information of little value to the general reader.
    2. Scientists and engineers need to relate the research and data with some logical topic of public interest—e.g., slime mold flagella with sperm motility.

For Scientists and Engineers to Connect with Reporters:

- The information needs to have been published or at least accepted to retain credibility. The classic cases of cold fusion in a test tube and the Alar (chemical) on apples controversy in recent years, emphasize this point of misguided science reports misleading the public.
- Scientists and engineers need to be prepared to address what is important to the reporter:
    1. What are the ESSENTIAL findings?
    2. What is the societal importance!
    3. Forget the gory details of methodology and statistical outcomes. If you can't give a simple definitive answer, think twice about stating it!
- Understand news criteria importance:
    1. Is it of reader/listener interest?
    2. Is it of potential significance and of interest to the reader/listener?
    3. Is the information relatively new?
    4. Will the information interest the writer AND the editor?
    5. Is the information source prominent (e.g., a visible scientist)?
    6. Is the information unique enough to be reported?
    7. Is the information of intrinsic importance to science? Can some social importance be attributed to it?

It is important to understand that what is of prime importance to the scientist or engineer (technical information) may not be the most important aspect for the reporter who needs the story to have a personal aspect. Most scientists and engineers fear talking to the press because they believe their words will be misconstrued or taken out of context. Reporters and science writers want to tell a good objective story; it is up to the technical specialists to understand the news media and help develop the information so that it is correct and yet at a level for lay audiences to understand.

Further Reading on Media Relations

1. Itule, B. D., and Anderson, D. A., News Writing and Reporting for Today's Media. McGraw-Hill, New York, 1999.
2. Jones, C., *Winning with the News Media: A Self-Defense Manual When You're the Story*, Video Consultants, Inc., Tampa, FL, 1999.
3. Mencher, M., *News Reporting and Writing*, WCB/McGraw-Hill, New York, 1997.
4. Rich, C., and Rich, C., *Writing and Reporting News. A Coaching Method*, 3rd edition, Wadsworth Series in Mass Communication and Journalism, Wadsworth, Belmont, CA, 1996.

16.5.5 PUBLIC RELELATIONS (PR)

Public relations is the planned effort to influence public opinion through good character and responsible performance, based upon mutually satisfactory two way communication. (Cutlip, S. M., and Center, A. H., *Effective Public Relations*, Prentice-Hall, Englewood Cliffs, NJ, 1978.)

Note the essentials of this early yet simple definition and what it emphasizes. A good communicator talks and listens equally well to the many different publics (audiences) who are affected by an organization.

- You need to be practicing PR now! Any organization that relates to any public has to think clearly about using good PR. Good PR can be likened to good preventive health care and is best practiced as a proactive management tool. If you wait until you need it, then the damage is done, and you will have to fight to regain public trust! Good PR, like good health, is not always appreciated until it goes bad. You cannot be fit unless you work out and you cannot be good at PR unless you work at it.
- Everything your organization does makes an impression. A good PR program is more than just a PR person or department. It must be a combined effort of a whole organization. Each representative is always representing the whole organization. Every action that is taken by the organization and its employees reflects on the total organization. For instance, if a delivery driver for a manufacturing company is abusive to car drivers on the road, the image of the company is tarnished. Each part of the organization needs to be involved in PR planning.
- The public is actually many different audiences. One message or action does not equally influence all audiences. The various audiences involved may be as diverse as employees of the organization, other people who know employees, communities bordering the organization's buildings, people who drive past organizational delivery trucks, supporters of the organization, any community supplied with organizational products, and even people who may read news about the organization. It takes many different messages on a continual basis with myriad audiences to maintain good PR.

There are special citizens who influence public opinion. They tend to be well-known and respected in the community and can ensure success for an activity. Some of these people are:

- Mayor, city managers, and council members
- The Chamber of Commerce

- An association of business leaders
- A fraternal organization (e.g., Elks, Moose)
- Police chiefs
- Newspaper editors
- Association of local religious leaders or clergy
- Citizens named for special awards
- Service club presidents

Seek their opinions about the communities with which they are involved. Put their names on a mailing list and keep them informed about all activities and happenings within your organization. Listen carefully to their comments and implement their positive suggestions whenever possible. Write thank you and follow-up notes when their input has been helpful.

- Things do go wrong! PR is not about using "spin media" to cover up problems or whitewash dubious political or manufacturing processes. This is PR at its worst. The fact that we don't always see good PR in action may be because good PR works without having to announce itself! An organization that practices good PR has good relations with many different audiences, recovers its image quickly in cases of mishap, and retains the trust of the public. Good PR involves contingency plans for all foreseeable problems. This then helps in maintaining a good organizational image, especially when the unforeseen does occur.

- Truth, honesty, and credibility are essential. This is up front as well as after a problem has occurred. Credibility is a prerequisite for success. It can be irrevocably lost with just one major mishap especially if the public perceives it has been misled or been told lies to cover up a problem. Never, ever, lie or cover up anything. It will come back to haunt you.

- Planning and communication is what PR is about. To develop a plan for PR, the communicator must have a well-defined and clear statement of purpose. Building on the goals of the PR plan, the communicator should be asking:

  1. Problem definition: What is going on that we need/would like to change? This may be multi-leveled and focused on many different audiences.
  2. Objectives: What specifically do we need to achieve? Objectives need to be measurable and involve systems for developing feedback.
  3. Message: What needs to be said, how, where, and when?
  4. Audiences: Who needs to get the messages? What can be found out about them so that messages can be specifically focused.
  5. Medium: What modes of communication need to be used and how can you best reach your audiences.
  6. Money: How much will it cost and how much do you have to spend? The size of the budget will determine just how complex the communication plan can be.
  7. Evaluation: What is your definition of success and how will you know if you have reached it? Getting feedback is essential for you to know what your publics think about your organization.

- The "tools" needed for planning and organizing public relations need not be expensive. Many of them are readily available in the community and easily accessible. These include:

  1. Face-to-face conversations with employees and members of selected audiences;
  2. Inviting the public to organizational committee meetings;
  3. Naming and involving advisory committees or building coalitions made up of representatives from groups, agencies and/or businesses, and communities;
  4. Inviting professional groups to "tour" your organization;
  5. Religious organization bulletins arranged through members of the congregation;
  6. Interviews with interesting people over the radio or community cable TV;
  7. Public meetings that focus on an expressed public concern—listen well, sincerity by the organization is paramount here;

8. Public service announcements;
9. Exhibitions that can include the results of an activity by your organization;
10. News items or references in newspaper columns;
11. Organizational speaker visits to clubs and community groups;
12. Funding concerts or entertainment by community groups;
13. Helping underprivileged or handicapped groups; and
14. Donating to local charities.

Often the publicity derived from a sincere activity such as the last three points is many times more beneficial than anything your organization could do for itself.

Further Reading on Public Relations

1. Bland, M., Theaker, A., and Wragg, D., *Effective Media Relations: How to Get Results*, Institute of Public Relations, London, 1996.
2. Dilenschneider, R. L., *Dartnell's Public Relations Handbook*, Dartnell, Chicago, 1996.
3. Hendrix, J. A., *Public Relations Cases*, 4th edition, Wadsworth, Belmont, CA, 1998.
4. Kendall, R., *Public Relations Campaign Strategies: Planning for Implementation*, Harper Collins College Publishers, Boston, 1996.
5. Young, D., *Building Your Company's Good Name: How to Create and Protect the Reputation Your Organization Wants and Deserves*, American Management Association, New York, 1996.

16.5.6 RISK COMMUNICATIONS

What is Risk?

If you are asked to think about a definition of risk, more likely than not you will emphasize it as the likelihood of something negative occurring. Scientists and engineers who deal in risk are usually driven by numbers and hard data obtained from observation, tests, experiments, and system models (often computerized or mathematical). Although this may be quantitatively good for setting exposure levels and probabilities of accident scenarios, it does not deal with the difficult to measure psychological aspects such as perception of a problem. It is this latter aspect that is of crucial understanding to this subsection. As in all good communications, public trust and credibility are your greatest assets. This chapter can be viewed as an extension of PR.

There are two primary components of risk that interest communicators and scientists or engineers (Hance, B. J., Chess, C., and Sandman, P. M., *Industry Risk Communication Manual: Improving Dialogue with Communities*, Lewis Publication/CRC Press, Boca Raton, FL., 1990).

$$\text{risk} = \text{hazard} + \text{outrage}$$

where

$$\text{hazard} = \text{probability} \times \text{consequence}$$

- Hazard is the specific scientific determination of the degree of harm for a particular risk—e.g., the likelihood of a problem arising from a specific chemical during a manufacturing process.
- Probability is the determined statistical likelihood that a problem may arise.
- Consequence is the predicted outcome should the problem become real.
- Outrage is the public perception (real or imagined) of a problem.

One of the most notable aspects of risk is that the public does not like hazard numbers, and industry does not like outrage statements, which are not based on hard data. People believe what they want to believe regardless of scientific information; often the public or audience does not understand the hard information and may even feel alienated by it. However, outrage is an essential psychological component of risk management that must be acknowledged to avoid potential conflict situations.

Dealing with the Uncertainly of Risk and Communicating Risk Effectively

- Use proactive PR.

  1. Don't lecture to the public, but listen carefully and respond sincerely.
  2. Know your audience.
  3. Find out concerns of the public before anything happens.
  4. Understand their concerns, many of which you might not have considered.
  5. Listen and observe.
  6. Release news early.
  7. Have community meetings.
  8. Develop an outreach program instead of waiting to respond.

- Involve the public in decision making on risks. Manage potential problems and conflicts before they have a chance to occur. The public will certainly believe that their concerns are important, regardless of any numbers you might produce. Listen to the concerns of all parties. Ensure all the parties have been identified (even small groups can become a problem if overlooked or, worse, ignored).

- State the limits of confidence and admit uncertainty. State objectively which data are better and which are uncertain. Never trivialize the risks. Use good analogies, but be careful of using comparisons when simplifying risk data. The perceptions of a particular problem may be very different from what you think, and you may end up alarming and not calming a group. The same goes for using simple metaphors—they probably will mislead the audience and even give them the wrong ideas of what the risk is about. Be careful when using comparisons that might not be viewed as credible—i.e., "driving is more dangerous than this toxic chemical." Use simpler language to increase understanding, but use care when simplifying information so that important points are not omitted or misinterpreted. Explain any cautiousness you may have (avoid the use of "conservativeness," it just confuses people).

- Emphasize what has been done about a problem already and what is currently being done. Don't be afraid to admit being human. Deal with the uncertainty and outrage. Never ignore or just dismiss outrage as unimportant. Do not hide uncertainty, don't cover up information, and never lie or mislead. Trust is everything!

Comments About Outrage

When experts examine a risk, they focus almost exclusively on mortality and morbidity—how many people are likely to die or get sick as a result. But citizens define risk much more broadly, considering a wide range of "outrage factors" in determining how risky they consider the situation at hand. Outrage is not a misperception of the technical data but how people "feel" about a potential or actual problem in their minds. There might not even be a real problem, but the fact that the public perceives one is reason enough to deal with it as a real problem. Working to keep outrage low is as much a part of risk management as working to prevent mortality and morbidity.

Things that make a difference to people on how acceptable a risk is are:

- Do the people have a choice about whether they are exposed to potential risks? When people are part of the decision-making process they tend to accept risks more readily. This gives them some control of the risk in their lives. If the risk is imposed on them without their consent, don't be surprised if they react negatively. Similarly, if people are very comfortable with an existing risk, it is more acceptable. However, there may be times when you actually want to heighten their concern about a problem to help safeguard them. Many of the OHSA rules and regulations are mandated specifically for this purpose.

- The financial burden imposed on someone who has little impact or nothing to do with the risk. Examples include a property owner whose property value is greatly diminished because it is next to a stream that receives effluent from a nearby manufacturing plant, and homeowners who cannot sell their property because they are downwind from the emission stacks of a manufacturing plant.

- Ecological justice emphasizes the problem of poor or minority people bearing the burden of most industrial environmental problems. This perception of lack of fairness can lead to problems with outrage. These audiences may not think it is morally fair to be exposed to more potential problems than other more affluent neighborhoods. How can you address that problem for your organization?

- People are very forgiving of "acts of God" and very unforgiving of human-produced problems. Don't try to minimize any problems you have by comparing them with acts of God!

- Many risks cause heightened outrage because of media images. Typically, the news media report and amplify a problem with news appeal, and ignore more deadly problems that lack sensational characteristics. Many fictional films and television programs use industrial catastrophes, which get linked in the public's mind to real possibilities. It is important to understand the source of outrage before trying to deal with it.

- People are more likely to stay calm when there is a potential problem or when a real problem arises when they have trust in the organization! Having a sincere, respectful, and courteous open dialogue with the public and being honest and trustworthy gives an organization the credibility to manage risks successfully and to weather real situations with minimal outrage problems.

Further Reading on Risk Communications:

1. Gutteling, J. M., and Wiegman, O., *Exploring Risk Communication*, Kluwer Academic, Boston, 1996.
2. Kamrin, M. A., Katz, D. J., and Walter, M. L., *Reporting on Risk: A Journalist's Handbook on Environmental Risk Assessment*, 2nd edition, Michigan Sea Grant, Ann Arbor, MI, 1995.
3. Lundgren, R. E., *Risk Communication: A Handbook for Communicating Environmental, Safety, and Health Risks*, Battelle Press, Columbus, OH, 1994.
4. Powell, D. A., and Leiss, W., *Mad Cows and Mother's Milk: The Perils of Poor Risk Communication*, McGill-Queen's University Press, Montreal, 1997.
5. West, B., Sandman, P. M., and Greenberg, M., *The Reporter's Environmental Handbook*, Rutgers University Press, New Brunswick, NJ, 1995.

16.5.7   SOCIAL MARKETING (SM)

Although PR and risk communication are both tools for the continuing education of audiences, the purpose of working with a SM planning strategy is to introduce social change in counterresponse to a perceived behavior—e.g., an anti-cigarette smoking campaign or a drive to get women to have annual breast exams and men to have prostate exams. What makes social marketing slightly different is that the communications effort developed by the social marketer is viewed as a product to be exchanged for something the respondent values.

- All marketing involves matching the needs and wants of all the parties. Sometimes it means creating the want by promoting a product that matches a desired outcome. The goal of a marketer is not to foist unnecessary or undesirable exchange but to assess what is needed by two or more partners and to facilitate a plan that identifies the needs of all the parties involved and that works to promote that exchange. A bartering system of sorts is used whenever marketing occurs, where the marketer gives the target audience a product, Y, and the audience exchanges with something of value, X (if you give me X, I will give you Y).

- SM uses a different set of products and bartering agents from commercial marketing. Although in commercial marketing, the primary barter (X) is usually monetary, in SM the primary barter (X) might be your time or a good attitude for a socially desirable situation. Essentially, it is something of value to the recipient that they will exchange for the product, and the product need not be a tangible item. The social marketer is trying to market something that should produce a change in lifestyle that is beneficial to the marketer and the recipient.

- Examples:

  1. If you place your recyclable trash (X) in curbside containers, I will collect it and also ensure that you have a litter-free environment, which is more esthetically pleasing to the eye, along with cheaper resources and products (Y). The recipient has to spent a little time separating the recyclable trash and place it at the curbside. The service must collect the recyclable trash and return it to a processing system to be reprocessed into new goods.
  2. If you agree to ride the bus and leave your car at home (X), then I will provide an improved transport system to get you where you need to go (Y). In exchange, this will help reduce the air-quality problems we are having. Note how the recipient is asked to give up something of value (freedom, comfort, safety, and convenience of a car) for a service that will result in an improved social good. The transport service also has to improve and address the needs of the recipient, who is giving up something of value.

- It is also probable that the project goal may be focused on a group other then the one final target audience.

  1. To curb alcohol drinking by teenagers, focus on getting parents to interact more with their offspring.
  2. To stop drunk driving, focus on designating nondrinking drivers or on friends to prevent inebriated drivers from driving.

- Note how the audience (recipients of the product) are a primary consideration. In all cases it is important to first assess what the audience needs or wants and then correlate that with what you as the social marketer have to offer that audience. It may not always be a direct exchange of values or activities. It may require finding intermediate steps of exchange!

The Five "Ms" of Marketing

- Message: What is the message? What are you trying to achieve? What is the final outcome that the communicator desires with the product that is offered? What is it that is being marketed? The product can be almost anything such as any good, service, idea, or social change expectation.

- Market (audience): Who are they? You need to find out more about the audience to determine what it is that they need. What is of value to the targeted audience that they will exchange for your product? Why will they want your product?

- Medium: What kind of communication modes will you use to communicate your product and its advantages to the audience? How will you develop the message so that the audience listens and responds? What is expected of the audience?

- Money: How much will it cost and can you afford it? Will the audience pay for it? Does your organization have the financial resources to use the best strategy or must you adopt a cheaper and less effective method? How much is the audience expected to "pay"? Are you expecting the audience to exchange too much for the product?

- Measurement (evaluation): How will you know if you have succeeded? What kinds of expected outcomes are you hoping to see?

Green Washing

As a final note, a term much used nowadays is green washing, which is becoming a negative concept applied to industries and companies that are trying to "sell" an environmentally green image to the consumer, while they continue to practice environmentally unhealthy manufacturing practices. The advice given here is to remember that credibility is a prerequisite for long-term success. If a company or industry is sincerely trying to practice good environmental practices, then selling a green image should not be perceived as negative by the public. In essence, don't claim what you don't practice!

Some larger industries have large environmental research and development groups that support and develop technologies for an environmentally cleaner and safer society. Yet other aspects of the same company have large refining and polluting manufacturing systems. Objectivity is the key word here. The problem of propaganda can also occur where *selected truths* about a company are promoted to push a short-term positive image. Yet again, complete truth and honesty are the best tools. An objective, unbiased, and credible approach yields longer-term success for a company that is sincere in its efforts to help sustain and improve the environment.

Further Reading on Social Marketing:

1. Andreasen, A. R., *Marketing Social Change: Changing Behavior to Promote Health, Social Development, and the Environment*, Jossey-Bass, San Francisco, 1995.

2. Goldberg, M. E., Fishbein, M., and Middlestadt, S. E., *Social Marketing: Theoretical and Practical Perspectives*, Lawrence Erlbaum Associates, Mahwah, NJ, 1997.

3. Kotler, P., and Roberto, E. L., *Social Marketing. Strategies for Changing Public Behavior*, The Free Press, New York, 1989.

16.5.8 CONCLUSION

This section has emphasized how environmental communication needs to be planned and designed with the needs of the organization and yet be focused completely on the perspective of the audience! The audience needs to understand the message and to react in expected ways. The first three subsections after the introduction are about audiences, writing, and media relations and deal with getting the message and information in a form that is correct and at a level for the audience to understand.

The next three subsections focus on different perspectives of dealing with the audiences. PR is focused on maintaining a good organizational image and conflict management as well as keeping the outside audiences informed of developments within the organization. Risk communication is dealing with effective risk management by involving outside audiences with an aim to preventing conflict. SM is focused on trying to change something or develop something new within a given audience. All three of these demand that the organization develop and maintain an honest and credible relationship with various audiences.

CHAPTER 16
MANAGING PLACE SCALE PROBLEMS

SECTION 16.6

ECOLOGICAL ENGINEERING

Michael G. Crowe

Dr. Crowe is a retired corporate executive and engineer. He has had extensive experience in project management of social and environmental systems development. Environmentally, he has designed and developed extensive land use change projects both domestically and internationally.

16.6.1 INTRODUCTION

In recent years we have heard or read of terms that relate to redefining environmental relationships. Some terms synonymous with these relationships include ecotechnology, sustainable agriculture, bio-engineering, waste resource management, restoration ecology, landscape ecology, industrial ecology, nature engineering, and an ecological system. This international movement is looking for common practices to develop nonlinear sustainable systems following economies found in nature and whose usage bridges society to their natural world for the benefit of both. The movement views several interdisciplinary approaches to developing these systems. One approach is integration of ecology and engineering, and it is termed ecological engineering.

Ecological engineering is the development of a management system for a human-induced environment. That environment or system is defined by society, and it is a buffer or gradient between other systems or environments. The ecological system, or ecosystem, approach to resource management is based on ecological principles, defined forcing functions and living and nonliving components, and system modulation from energy flow and resource availability and use. Examples of ecological management systems can be found in food production, water-quality enhancement, ecosystem restoration or development, and social system development.

16.6.2 HISTORY

Ecological engineering can be defined as the design of human society with its natural environment for the benefit of both (Mitsch, 1992). Under this definition four distinct pathways are merging. In the 1960s, Howard Odum introduced ecological engineering as the design of systems in which the energy required to control and drive the system follow the economies found in nature with only supplementary energy coming from humans (Odum, 1989). This approach was centered on wetland and other natural ecosystem restoration and creation research. This research was reflected in Eugene Odum's text on ecology, which stated that system management is not linear in nature but is systemic to associated systems (Odum, 1971). This systemic tying of associated systems together into one system for evaluation is often referred to as an ecological system (Golley, 1993).

A second path was being defined in the Orient. Ecological engineering was defined there initially as a formal philosophy that created designs with nature (Ma, 1985). Later Ma defined it as the design of systems that reflect the principles of species symbiosis and cycling of materials (Ma, 1988). There was also a push to move the theoretical database that had accumulated from ecological research into applied ecological applications. This was seen initially in China as ecological engineering applications in agriculture and aquaculture. Aquaculture is defined as fish culture management with special attention to waste utilization and recycling as a resource (Yan and Yao, 1989).

Between the late 1980s and early 1990s ecological engineering was applied in the treatment of wastewater in Sweden. This research was later applied in the United States, Europe, and China (Guterstam and Todd, 1990; Ma and Yan, 1989). Sweden and the Green Party in Europe were seeking new technology for treatment of wastewater that was environmentally friendly. Their approach was to develop a system that used little human-introduced energy, produced little or no sludge, used little or no added chemicals, and generated useful by-products. The application was to use wetlands ecosystems, outdoors or enclosed in greenhouses, as a buffering structure. The structure had a function to balance elements found in the polluted stream of water to biotic requirements within the wetland ecosystem. This led to later applications in treating polluted bodies of water having acid mine drainage, nonpoint and point sources of pollution, and stormwater management (Knight, 1994).

A fourth path to ecological engineering is presented by Erich Jantsch (Austria) at the Seminar on Interdisciplinarity in Universities in 1970. Jantsch defined ecological engineering as a methodology to change and enhance social systems through education and innovation. He suggested an interdisciplinary approach to bridge the human environment to the natural environment of the physical world by use of ecological principles. The ecological engineer accomplishes this by bringing interdisciplinary empirical data to a pragmatic level through introduction of a common viewpoint or purpose (Jantsch, 1970). This path is important to note, as it was involved not with management of a natural resource but rather with management of human resources as an ecological system. This in turn led to the development of other social systems based on an ecological systems approach.

Today ecological engineers combine these pathways to apply fundamental ecological research to a pragmatic systems approach. By evaluation of biotic components and their energy and material requirements, the design should self-organize and become sustainable. The use of ecological economics and conservation of existing ecosystems allows ecological engineers to identify the biotic and abiotic elements most adaptable to human environments as well as the human environments most adaptable to natural systems. (See the multilevel multigoal hierarchical system for ecological engineering in Figure 16.6.1.) Another role of ecological engineers is to develop social systems as ecological systems for social innovation and education.

Ecological engineering is often intertwined or confused with environmental engineering. Environmental engineering, a subset of civil engineering, is a respected discipline whose work is centered on water-quality enhancement with mechanical structures. Here lies the difference between an environmental engineer and an ecological engineer.

Environmental versus Ecological Engineering

Howard Odum has suggested that an ecological engineer have a background in science or engineering as well as in ecology, ecological and structured economics, wetland and terrestrial ecology, hydrology, meteorology, systems, and case history analysis (Odum, 1994). He and others also suggest that ecological engineering has properties in common with traditional environmental engineering, an established and respected discipline. These are summarized as understanding environmental structures, using alternative analysis to determine the appropriate technology and interfaces, and using a value analysis to evaluate benefits. There are also differences.

One difference between an environmental engineer and an ecological engineer is the design boundary. The boundary of an environmental engineering design typically is contained within the perimeter of the project site. The boundary for an ecological engineering design goes beyond the project site by analyzing the impacts from the constructed system on systems associated within adjacent environments. Success within the environmental engineering design is often measured at the discharge from

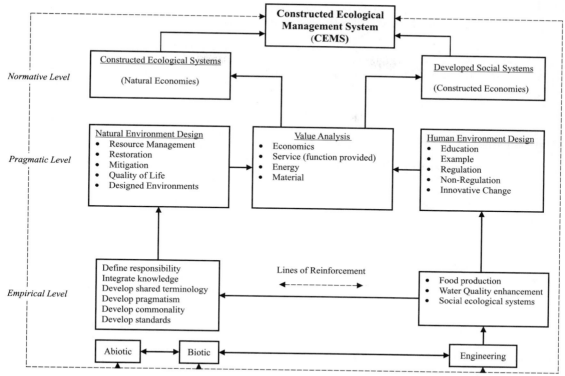

FIGURE 16.6.1   Ecological engineering.

the system, whereas an ecological design is measured by the ability of the constructed system to be self-sustaining through reinforcement of identified associated systems.

Another difference is in the approach to the design function [value analysis of design objective (s)]. Currently, an environmental engineering design is analyzed to determine the material and labor needed to build it; then it is maintained with external human-induced energy sources. Ecological engineering evaluates project economy as both the money paid for human effort and economics for work effort, energy, and material recycling provided by natural operations and processes. This objective is also a research and education function to develop measures of value to the functions used in developing a sustainable system, often referred to in the literature as ecological economics (Daly, 1991; Peet, 1992; Turner et al., 1993).

Perhaps the greatest difference in design protocols between an environmental and ecological engineer is with their use of standards. An environmental engineer has the ability to design to singular discipline standards derived from historic common practices that have given repeatable results. These are the standards for structures, flow control systems, and methods for geotechnical site evaluation among others. Although the ecological engineer may use the same standards, the design protocol history is still being refined. One reason for this is the use of technological evaluations coming from interdisciplinary observations as well as from the implications and interactions society places on new philosophies and disciplines.

It should be understood that ecological engineering leaves the design of structures to professional engineers, such as environmental engineers, because ecological engineering is the design of an ecological management system. This difference often leads to conflicts in responsibility for system performance. For example, some states, such as West Virginia, require the stamp of a professional engineer on a wetland design. Here the state requires a responsibility from the environmental engineer for system performance. In most cases this engineer may be qualified for structural design of the wetland. However the taking of responsibility for developing the nonlinear relationships between the living components within the wetland system to the structure of the wetland and to associate systems may be beyond the understanding of the engineer. In other states, no stamp is required unless state law governs the structure, and then only the structure is stamped and does not indicate responsibility for system performance.

Most Recently

In the Americas, water-quality enhancement has been the most common application. Applications can be subdivided into point and nonpoint sources of pollution. Point sources of pollution applications include food production, domestic wastewater, acid mine drainage, landfill leachate, and industrial discharges. Nonpoint sources of pollution applications have centered on farming practices and stormwater management (quantity and quality). Stormwater management applications also are divided by scale, ranging from entire watersheds to smaller scale land-use changes. Creation and restoration of ecosystems in the natural landscape from historic and recent land use change is also used for water-quality enhancement. Reviews of the literature and Internet resources indicate in excess of 4000 applications or reviews under these topics.

Ecological engineering has progressed to applying common practices in the design of ecological systems. Yet perhaps the most explosive segment in ecological engineering today is the design of management systems to implement ecological system designs. The design of a system and its implementation are based on the same ecological principles. For ease of discussion, the end product of such an ecological engineering application is called here a constructed ecological management system (CEMS) and is applicable in both social and natural environments. The term CEMS has been used in federal and state permits to indicate a constructed ecosystem for a particular resource management issue. It has been used to describe the protocol between designer, developer, and regulatory agency for water-quality impacts from land-use change and as a term in joint 401/404 permits requires under the Clean Water Act of 1977. It has also been used in general business protocol, commercial development sustainability, and industrial product applications.

16.6.3 DISCUSSION OF ECOLOGICAL ENGINEERING PRINCIPLES

Certain aspects are common to the philosophy and discipline of ecological engineering. They include an understanding of the ecosystem concept, modeling, and ecological principles.

Understanding the Ecosystem Concept

Ecosystem is an established concept in the United States for ecological organization, and it was promoted by the widespread use of Eugene Odum's textbook *Fundamentals of Ecology* first published in 1953 (Odum, 1953; Odum, 1993). An ecosystem is simply a method for interpreting the whole, the natural equilibrium, the connection between system components, or the organization of nature in time and space.

Ecological engineering begins with this understanding of the ecosystem concept. It is the knowledge that system ecology is part of, yet differentiated from, general ecology. Bionomics or biologic ecology deals with the relationships of specific biotic organisms within the community. System ecology deals with relationships in which biotic and abiotic components interact within a defined environment or newly constructed system. Constructed is used to indicate a designed system that has been implemented and that is lightly managed by the social environment.

This understanding of interrelationships within an ecological system enables them to identify and apply unit functions during review of material and energy use as the cycling and regeneration of resources (Ma and Jingsong, 1994). Additional information has been gathered from other disciplines that reflect similar observations. For example, the work of Vymazal describes elemental cycling in wetlands (Vymazal, 1995). Hammer, Kadlec, Knight, and Reed among others have presented engineering equations for such individual unit processes and operations found in ecosystems (Hammer, 1989; Kadlec and Knight, 1996; Reed et al., 1995).

Energy and material are abiotic elements reviewed within an ecosystem. James Lotka (1880–1949) developed the concept linking energy exchange between inorganic and organic components within a system through thermodynamics. Lindeman, in his study of Cedar Bog Lake, Minnesota, produced the first deliberate effort to replicate energy use within an ecosystem by creating a balance sheet between energy in and energy out. This led to the modern studies of flow and transformation of energy within a system, also known as ecological energetics. Concurrent to the work of Lotka, G. E. Hutchinson of Yale was studying the cycling of elements, which led to the theoretical concept of self-regulation of biogeochemical cycles. One of his students was Odum, H. T., the founding factor in ecological engineering.

Modeling an Ecosystem Concept

The ecological engineering process entails modeling the modulation and use of energy and materials as a resource for biotic elements. When the CEMS is modeled as a buffering system, biotic and abiotic components within unit processes and operations are identified to maximize and create a sustainable system. The uses of a model allow the ecological engineer to select processes projecting an optimum output from the conceptual design and then apply them to a site-specific location. The model links the components within the ecosystem and develops a relationship between feedback mechanisms, potential change in the designed system, and potential impacts to associated systems. Modeling considers the design forcing functions and internal variables, the capacity of the design to act as a buffer, and the capacity for self-regulation and self-sustainment within the system (Jørgensen, 1989).

Forcing functions are external variables that drive model development and are in either a steady state or variable state. These variables are the listed items within the conceptual design that emulate the structure and function of the designed environment. As ecosystems are complex, and a method is needed to predict an output from the system concept, the constructed ecosystem should be the end result of a pragmatic ecological model. Mitsch and Jørgensen are among those who have presented ecological models (Costanza and Daly, 1987; Jørgensen, 1989, 1994; Mitsch and Gosselink,1993; Odum, 1994).

State variables are typically modeled as the quantified and qualified materials available for unit functions within the constructed ecosystem. The ecological engineer needs this knowledge to evaluate for change as the forcing functions are simulated to varying degrees within the model. The concept model then evaluates the functional capacity of the unit processes within the constructed system.

Biotic elements are able to self-regulate only within certain ranges of materials. This can be correlated to the Shelford law of limitation and where biotic efficiency is related to a normal distribution around optimal conditions. The materials identified and available for use quantify the resources available to develop a constructed ecosystem. These materials then have the potential to provide energy to biotic elements.

Buffering capacity is defined within the model as the ability to change a state variable relative to the change in a driving function. The more diverse a system is, the higher the buffering capacity. Having a greater biotic diversity allows the constructed system a greater range in use and recycling of materials. If the materials entering a constructed ecosystem are beyond the optimum range of the biotic community, the ecological engineer varies the design by time available for resource use and manipulates the size of the system to accommodate this objective(s).

The constructed ecosystem acts as a buffering system or gradient by adjusting the size of the system and time for operations and processes to meet the concept parameters required and indicated by the modeling results. Operations and processes in ecosystems have characteristic time scales. Time is seen as the adjustable variability factor between different published models. It becomes a factor of experience, application, and project location. Time is the self-organizing factor between modulation in levels of resource input versus the output from the system.

Self-regulation is the time linkage between feedback mechanisms that develop from coping with fluctuations in sequence resulting from change in forcing and state variables. When a design incorporates the concept of nature's ability to self-regulate, the system requirement for energy will be lower and changes in concept function will be minimized. Odum, H. T., stresses a similar concept and terms this self-organization; he further states that this is the keystone of ecological engineering. Self-organization and self-regulation are synonymous.

Ecological engineering is the management of self-organization in both a social and an ecological context. Self-organization reduces the trial-and-error effects from human efforts to find what works. Ecological engineering brings to light the economies of nature and develops the ecological economics to make these natural economies competitive with constructed economies found in society. It develops a social interface to introduce and reinforce the work contribution of our environment. Ecological engineering promotes the symbiosis of natural system functions to human work efforts as part of social self-organization. Odum suggests several evaluations that relate to system self-organization during the conceptual design process. The first is to create a design in which both biotic and abiotic components contribute to the efficiency of the proposed system function; in the second, one function reinforces another function within the system by efficient use of resources; and third, additional resources for use generated within the system are identified.

Applied Ecological Principles

The following ecological engineering design principles have been derived primarily from Odum, H., Mitsch, Jørgensen, Kadlec, and various others as well as from personal experience. The principles have been applied to development of constructed ecological systems in both social and natural environments.

1. CEMS is a nonlinear system based on the economies of the natural world.
2. CEMS is a constructed environment providing life requirements for planned biotic components.
3. CEMS objective(s) and configuration are initiated by the driving functions of the system.
4. The ultimate configuration is determined by the evolution of unit process self-regulation.
5. Materials are recycled.
6. Energy use is continuous, supplied largely by nature, and it is produced, used, stored, and exported within ecological economies.
7. The ability to continuously strive toward a system homeostasis condition depends on balancing unit processes to a planned buffering capacity.
8. Unit functions have characteristic time scales.
9. The system, as components and as a whole, has a space requirement that depends on sequencing and combining unit functions to reach some design buffering capacity.
10. Buffering capacity depends on optimizing a given range of biologic and material diversity.
11. CEMS is a system acting as a gradient or buffer between an emitting, receiving, and other associated systems.
12. CEMS should merge with the receiving and associated environments in a manner to reinforce all systems.
13. CEMS is designed to have storage areas to account for changes in system productivity from pulsing associated with the driving functions.
14. CEMS is developed from an interdisciplinary design incorporating common practices with repeatable results and an ecosystem concept.
15. CEMS is designed to be a lightly managed system.

16.6.4   *THE GROWING ROLE OF THE ECOLOGICAL ENGINEER*

I propose that an ecological engineer has one primary role: to develop pragmatic and innovative inter-disciplinary solutions to everyday societal problems by using the economies and principles found in our natural environment. With one foot in a social environment and the other in a natural environment, ecological engineers look for solutions that benefit both. Their applications consist of developing implementation protocols for their designs and for other system applications that reach into various environments.

The role of this discipline and philosophy is expanding into other fields. Ecological engineering is now being used in social and business resource management applications. Human resource managers, plant engineers, and general managers look to this type of philosophy for guidance in developing operational systems. It has been applied to the health care industry to develop systemic communication patterns between parties that reduce energy and recycle resources. It has been applied to industry in finding methods to recycle production by-products and reduce energy use during production. Examples of this can be found in the interchange of production by-products between different companies. It has been used in land-use change evaluations to promote sustainability within communities. Land developments are changing from clear-cut and level to where natural resources are used to enhance and sustain the new development. Stormwater management now employs existing wetland and stream systems in lieu of mitigation of natural functions to an off-site location.

Howard Odum proposed that ecological engineers have an educational background. That background, however, will change as the philosophy is applied further to promote a philosophy that incorporates interdisciplinary approaches to resolve challenges in society. An ecological engineer is a specialist but is becoming a generalist with a background of understanding that incorporates without prejudice the understanding of others in reaching the objective of sustainment of the natural world that in turn sustains human society.

REFERENCES

Costanza, R., and Daly, H., "Ecological Modeling," *Ecological Economics* 38 (1 and 2): 1987.

Daly, H., *Steady-State Economics*, Island Press, Washington, DC, 1991.

Golley, F., *Ecosystem Concept in Ecology: More Than the Sum of the Parts*, Yale University Press, New Haven, CT, 1993.

Guterstam, B., and Todd, J., "Ecological Engineering for Wastewater Treatment and Its Application in Sweden and the United States." *Ambio*, 19: 173–175, 1990.

Hammer, D., "Constructed Wetlands for Wastewater Treatment," Lewis Publishers, Boca Raton, FL, 1989.

Jantsch, E., "Towards Interdisciplinarity and Transdisciplinarity in Education and Innovation," *Interdisciplinarity: Problems of Teaching and Research in Universities*, University of Nice, France, pp. 97–121, 1970.

Jørgensen, S., "Principles of Ecological Modeling," *in Ecological Engineering: An Introduction to Ecotechnology*, Mitsch, W., and Jørgensen, S., eds., Wiley, New York, pp. 39–56, 1989.

Jørgensen, S., *Fundamentals of Ecological Modeling*, Elsevier, New York, 1994.

Kadlec, R., and Knight, R., *Treatment Wetlands*, CRRC, New York, 1996.

Knight, R., *Wetlands Treatment Database, Version 1.0, Disk 1 of 1*, Risk Reduction Engineering Labratory, Cincinnati, OH, 1994.

Ma, S., "Development of Agro-Ecological Engineering in China," in *Proceedings of the International Symposium on Agro-Ecological Engineering*, Beijing, pp.1–13, 1988.

Ma, S., Ecological Engg. Applications of Ecosystem Principles. *Environmental Conservation*, No. 12: 331–335, 1985.

Ma, S., and Jingsong, Y., "Ecological Engineering for Treatment and Utilization of Wastewater," *in Ecological Engineering: An Introduction to Ecotechnology*, Mitsch, W., and Jørgensen, S., eds., Wiley, New York, pp. 185–217, 1994.

Mitsch, W., "Ecological Engineering: A Cooperative Role with Planetary Life-Support Systems," *Environ. Sci. Technol.*, 27(3): 438–445, 1992.

Mitsch, W., and Gosselink, J., *Wetlands*, Van Nostrand Reinhold, New York, 1993.

Odum, E., *Fundamentals of Ecology*, Sinauer Associates, Sunderland, MA, 1953.

Odum, E., *Fundamentals of Ecology*, Saunders, Philadelphia, 1971.

Odum, E., *Ecology and Our Endangered Life Support Systems*, Sinauer, Sunderland MA, 1993.

Odum, H., "Ecological Engineering and Self-Organization," in *Ecological Engineering: An Introduction to Ecotechnology*, Mitsch, W., and Jørgensen, S., eds., Wiley, New York, p. 80, 1989.

Odum, H., *Ecological and General Systems: An Introduction to Systems Ecology*, University Press of Colorado, Niwot, CO, 1994.

Peet, J., *Energy and the Ecological Economics of Sustainability*, Island Press, Washington, DC, 1992.

Reed, S., Crites, R., and Middlebrooks, E., *Natural Systems for Waste Management and Treatment*, McGraw-Hill, New York, 1995.

Turner, R. K., Pearce, D., and Bateman, I., *Environmental Economics: An Elementary Approach*, John Hopkins University Press, Baltimore, 1993.

Vymazal, J., *Algae and Elemental Cycling in Wetlands*, CRC, Boca Raton, FL, 1995.

Yan, J., and Yao, H., "Integrated Fish Culture in China," *Ecological Engineering: An Introduction to Ecotechnology*, Mitsch, W., and Jörgensen, eds., Wiley, New York, 1989.

CHAPTER 17
SYSTEM LEVEL RISK ASSESSMENT

SECTION 17.1

PROBABILISTIC SAFETY IN GOVERNMENT AND INDUSTRY

Ralph Fullwood

Dr. Fullwood is retired from Brookhaven National Laboratory. He has spent his career performing experiments on neutron transport, cross section measurements, bomb diagnostics, instrumentation design neutron radiography and design of a neutron research facility (WNR) and proton storage ring.

17.1.1 INTRODUCTION TO RISK

Although risk consists of two parts, probability and consequence, this section deals with the probability of release and refers the reader to other parts of this *Handbook* for calculation of consequences.

Background

The preceding chapters in this handbook have described environmental contamination and described the transport, effects, analysis, and modeling. How much better isn't it to not contaminate in the first place? Clearly there is a whole other side to the problem—i.e., containment of hazardous materials to prevent their entry into the environment. Users of hazardous materials may be classified as government, industry, and private citizens. Government, particularly at the federal level, is both a polluter and a regulator of what, where, and how it and the other polluters dispose of hazardous materials. Industry works with government to strike a balance between protecting the environment and industrial activity. Both industry and government design and implement protective measures; citizenry are the passive elements of this triangle. They accept the protective measures proscribed by government and provided by industry and are penalized for illegal desposal. Industry and government also are subject to penalties ranging from fines to imprisonment.

Thus the objective is to confine and control hazardous materials to prevent unintended release and to release into the environment only in amounts and at locations in compliance with governmental regulations. This section describes how to detect weakness in confinement/process systems, use this information for design improvement, and estimate the probability of failure.

Forms of Pollution and Confinement

The states of matter are solid, liquid, and gas. Generally, solids are easy to confine from the environment just by locating them in designated areas. It is the gases and liquids that are difficult to confine because of their mobility. By definition they have no shape but assume the shape of the confinement vessel.

Should the vessel fail, they disperse. Of the two, gases are more difficult to control because they have the ability to diffuse through the atmosphere and move with the wind. But this same ability results in low concentrations, which reduces toxic effects. Liquids tend to flow downward—such as water, which takes them into streams and estuaries and into the very water supplies needed by humans, animals, plants, and the environment.

Our concerns are the design and construction of envelopes to confine hazardous materials and to contain them under conditions of temperature, pressure, and chemical activity that do not fail the envelope and to do this in such a manner that industrial activity, in particular chemical processes, can be carried out. It should be emphasized that environmental protection is a Faustian bargain. Protection of the environment at the cost of quality of life is not a solution.

Thus we design protection systems but anything may fail. The question is what is the probability that a system will fail? In addition what are the consequences of failure? The pairing of probability and consequences leads to the concept of risk. Chemical process safety is qualitative and quantitative. Quantifying the probability of confinement failure is difficult. Qualitative safety avoids probabilities by using judgment, ranking, and barrier enumeration.

17.1.2   QUALITATIVE SAFETY ASSESSMENT

The PSM Rule

Protection priorities are ranked according to the hazardousness of the chemicals. The Process Safety Management Rule (PSM-29CFR1910.119) aids in preventing or mitigating chemical risks. The Qccupational Safety and Health Administration (OSHA) standard is based on management's responsibility for implementing and maintaining an effective process safety management program. This OSHA standard, required by the Clean Air Act Amendments, is used in conjunction with the U.S. Environmental Protection Agency Risk Management Plan. Full compliance comes from merging the two sets of requirements into the process safety management program to enhance the relationship with the local community.

Obviously, the risk of hazardous chemicals is reduced by a minimized inventory using just-in-time procurement. If further inventory reduction is not feasible, additional risk reduction may be achieved by dispersing the inventory to multiple site locations so a release in one location does not affect other locations of inventory.

Process hazard analysis is one of the most important elements of a process safety management program. A PrHA is an organized and systematic effort to identify and analyze the significance of potential hazards associated with the processing or handling of hazardous chemicals. PrHA information assists management and employees to decide on safety improvements and accident consequence reductions. A PrHA is directed toward analyzing the risk of fires, explosions, and releases of toxic or flammable materials. Equipment, instrumentation, utilities, human actions, and external factors are considered in the process.

Selection of a PrHA methodology requires consideration of many factors including the availability of process information such as experience with the process, changes that have taken place, reliability, aging, maintenance, etc. If it is a new process, less reliance can be placed on experience and greater reliance must be placed on the analysis of possible accidents and accidents in similar or related processes. Size, complexity, and hazard severity influences the choice of the most appropriate PrHA methodology.

Simplifying assumptions used by the PrHA team and reviewers must be understood and recorded to indicate the completeness of the PrHA and for use in future improvements of the analysis. The PrHA team and especially the team leader must thoroughly understand the methodology that is selected.

A team consists of two or more people who know the process technology, design, operating procedures, practices, alarms, emergency procedures, test and maintenance procedures, and routine and nonroutine tasks. They must consider authorization and procurement of parts and supplies,

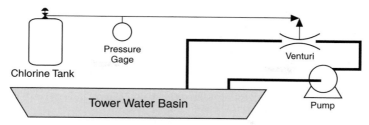

FIGURE 17.1.1   Cooling tower water chlorination.

safety and health standards, codes, specifications, and regulations. The team leader provides management, and goals to the process; the team and consultants construct and interpret the analysis.

PrHA may use different methodologies for different processes depending on the size, complexity, and age of the facility by using team members who are experts at analyzing their assigned process. For example, a checklist may be used to analyze a boiler or heat exchanger, with a HAZOP (Hazards and Operability) for the overall process. Batch processes of similar subsystems may be analyzed with a generic PrHA that is modified according to differences in components, monomers, or ingredient ratios.

Small operations, although covered by the PSM rule, may use simplified methodologies and still meet the criteria. Businesses with similar processes and equipment may pool resources and prepare a generic checklist analysis used by all members to meet the PSM rule.

If a PrHA is needed for many operations, the facility operator may prioritize the order of mitigating accidents. A preliminary hazard analysis (EPA, FEMA, and DOT, 1987) provides an overall risk perspective that can rank order potential accidents by decreasing the number of affected individuals. Ordering may be based on maximum amounts and toxicity of the hypothetical releases, operating history of releases, age of the process, and other factors that may modify the ranking. A PrHA helps to identify improvements that give the most safety for the least cost.

Some qualitative methods are check lists; what-if analysis; and what-if check list analysis, HAZOP, and failure modes and effects analysis. These methods are compared by application to a simple reference system shown in Figure 17.1.1. This chlorination system is used to prevent algae growth in a cooling tower basin. Basin water is circulated by the pump through the venturi at 30 gallons per minute. Chlorine gas, provided by a 1-ton (907.18 kg) tank, is reduced to zero pressure by the pressure reducing valve and is brought into the subatmospheric venturi. The pressure check valve prevents chlorine gas flow unless the pressure in the venturi is subatmospheric. If the first check valve fails, chlorine is vented.

Checklist Analysis.   Checklist analysis begins with a qualified team examining the process description, piping and instrumentation diagrams design procedures, and operating practices. As the checklist is developed, aspects of process design or operation that do not comply with standard industrial practices may be discovered by using detailed, process-specific checklists, augmented by generic checkists for thoroughness.

The completeness of hazard identification would be limited to the preparer's knowledge if not for the use of multiple reviewers with diverse backgrounds. Frequently, checklists are created simply by organizing information from current relevant codes, standards, and regulations. This is insufficient; the checklist must be designed for the specific process including its accident experience. It should be a living document that is reviewed and revised from operating experience.

Table 17.1.1 (DOE, 1996) suggests main headings for a checklist. These are global headings starting with the process description, overall considerations, operating limits, hazards, and mitigation.

Qualitative results of checklist analyses vary, but generally the analysis produces the answers "yes," "no," "not applicable," or "needs more information." The checklist is included in the PrHA report to summarize the noted deficiencies. Understanding these deficiencies leads to safety improvement

TABLE 17.1.1 Checklist Main Headings

Basic process considerations
Overall considerations
Operating limits
Modes of plant startup, shutdown, construction,
     inspection and maintenance, initiating events,
     and system deviations
Hazardous conditions
Ways of changing hazardous events or frequency
     of occurrence
Corrective and contingency action
Controls, safeguards, and analysis
Documentation and responsibilities

alternatives for consideration and to identified hazards with suggested actions. Table 17.1.2 presents a checklist analysis for the cooling tower chlorination system.

Any person with knowledge of the process should be able to use a checklist. The PSM rule team approach requires more than one analyst to prepare the checklist and apply it to the process for review by an independent analyst. The resources for a checklist analysis depend on the process complexity but vary from about a half day to several days.

A checklist made from generic documents may be incomplete or erroneous. The hazards associated with the process may not be in generic information. Checklists identify hazards but not the accident scenarios that lead to the hazards.

What-If Analysis.   What if is a creative, brainstorming examination of a process or operation conducted by knowledgeable individuals asking questions. It is not as structured as the checklist analysis; it requires analysts to adapt the basic concept to the specific application. The what-if analysis stimulates a PrHA team to ask what if? Through questions, the team generates a table of possible accidents, their consequences, safety margins, and mitigation. The accidents are not ranked or evaluated.

Members of the PrHA team, before meeting, review process descriptions, operating parameters, drawings, and operating procedures. For an existing plant, personnel responsible for operations, maintenance, utilities, or other services should be interviewed. The PrHA team should walk through the facility to understand its layout, construction, and operation. Visits and interviews should be scheduled before the analysis begins. Preliminary what-if questions should be prepared to "seed" the team meetings by using old questions or making up new ones specifically for the process.

After setting the scope, the analysis begins with process description including safety precautions, equipment, and procedures. The meetings then focus on potential safety issues identified by the analysts with what if questions, although a concern may not be expressed as a question. For example, "*I wonder what would happen if the wrong material was delivered.*" "*A leak in Pump Y might result in flooding.*" "*Valve X's inadvertent opening could affect the instrumentation.*" Questions may address any off-normal condition related to the facility and not just component failures or process variations. The analysis usually goes from beginning to end, but it may proceed in any order that is useful, or the leader may determine the order. Questions and answers are recorded by a designated team member.

Questions are organized by areas such as electrical safety, fire protection, or personnel safety and are addressed by experts. In this interchange, the team asks and gets answers about their concerns regarding the hazard, potential consequences, engineered safety levels, and possible solutions. During the process, new what-if questions may be added. Sometimes proposed answers, developed outside of meetings, are presented to the team for endorsement or modification. For example, "*What if the chlorine cylinder ruptures?*" The team might respond: "*It would release chlorine to the atmosphere to the detriment of workers.*" The meetings should be contemplative and last no longer than 4 to 6 hours per day for no more than 5 consecutive days. A complex process should be divided into tractable segments.

TABLE 17.1.2   Checklist of Cooling Tower Chlorination

Materials

Do all materials conform to original specifications?
   Yes, the drums are ordered with the same chlorine specification used since startup.
Is each receipt of material checked?
   Yes, the supplier once sent a cylinder of phosgene. Since then, a test is performed by the maintenance
      staff. In addition, the fusible plugs are inspected for evidence of leakage, before a cylinder is hooked up.
Does the operating staff have access to Material Safety Data Sheets?
   Yes, all staff are familiar with the process chemistry, including the hazards of Cl_2.
Is fire fighting and safety equipment properly located and maintained?
   Yes, this system is on a concrete building roof. Because there are no flammable materials involved in this
      system, if a fire occurs, there will be no special effort by fire fighting crews to concentrate on the roof area.

Equipment

Has all equipment been inspected as scheduled?
   Yes. The maintenance personnel have inspected the equipment in the process area according to company
      inspection standards.
Have pressure relief valves been inspected as scheduled?
   Yes.
Have rupture disks been inspected as scheduled?
   Not applicable.
Are the proper maintenance materials (parts, etc.) available?
   Yes. They include spare pigtails for the supply cylinders, as well as a rotameter and a pressure check valve.
      Other items must be ordered.
Is there an emergency cylinder capping kit?
   Yes.

Procedures

Are the operating procedures current?
   Yes.
Are the operators following the operating procedures?
   No, it is reported that some staff do not always check the cylinder's fusible plugs for leaks. Staff should be
      reminded of this procedural item and its importance.
Are new operating staff trained properly?
   Yes, training includes a review of the PrHA for this process and familiarization with MSDSs.
How are communications handled at shift change?
   There are few open items at the end of a shift. The chlorine cylinders need to be changed only once every
      45 days. If an empty chlorine cylinder needs replacement, the change is scheduled during a shift.
Is housekeeping acceptable?
   Yes.
Are safe work permits being used?
   Yes.

What if produces a table of narrative questions and answers suggesting accident scenarios, consequences, and mitigation. Table 17.1.3 shows a what-if analysis for the cooling tower chlorination system. On the left in the line above the table is indicated the line/vessel that is being analyzed. To the right are the date and page numbers. The first row in the table contains the column headings beginning with the what-if question followed by the consequences, safety levels, scenario number, and comments. The comments column may contain additional descriptive information or actions/ recommendations. The recommendations can be used in the report as action items for improving safety. Management reviews the results to ensure the findings are considered for action.

The PSM rule requires that a what-if analysis be performed by a team with expertise in the process and analysis method. For a simple process, two or three people may perform the analysis, but, for a complex process, a large group subdivided according to process logic into small teams is needed.

TABLE 17.1.3   What-If Analysis of the Cooling Tower Chlorination System

LINE/VESSEL: Cooling Tower chlorination system

What if	Consequences	Safety level	Scenario	Comments
The system is involved in a fire?	High pressure in chlorine cylinder, fusible plugs melt, chlorine release into fire	Ignition control source	1	Verify the area is free of unnecessary fuel
The wrong material is received in the cylinder and hooked up	Water contaminated, not sterilized	None	2	Prevention: supplier's procedures
The cylinder's fusible plug prematurely fails?	Chlorine released	None	3	Purchase and train personnel in the use of chlorine
The pressure check valve fails open (both pass chlorine gas)?	Built-in relief valve opens, releasing chlorine to atmosphere	None	4	
The basin corrodes through?	Chlorinated water release	Periodic inspection	5	
The recirculation pump fails or power is lost?	Eventually low chlorine in water, biological growth	None	6a	
	Release of undissolved chlorine to atmosphere if pressure check valve fails	Pressure check valve	6b	
The chlorine cylinder is run dry and not replaced?	Eventually low chlorine to water, biological growth	None	7	

The time required for a what-if analysis is proportional to team size and complexity. A simple system requires several days; a complex system needs several weeks.

What-If/Checklist.   What-if/checklist analysis identifies hazards and possible accidents, qualitatively evaluates the consequences, and determines the adequacy of safety levels (CCPS, 1992). It combines the creative, brainstorming features of what if with the systematic features of a checklist. The what-if analysis considers accidents beyond the checklist; the checklist lends a systematic structure to the what-if analysis. A what-if/checklist examines the potential consequences of accident scenarios at a more general level than some of the more detailed PrHA methods. It can be used for any type of process at any life-cycle stage.

The PrHA team leader assembles a qualified team to perform a what-if/checklist analysis. If the process is large, the team is divided into subteams according to functions, physical areas, or tasks. For the checklist portion of the analysis, the team leader obtains a checklist for the team, which may not be as detailed as for a standard checklist analysis. The checklist should focus on general hazardous characteristics of the process. After the what-if questions for a process step have been developed, the previously obtained checklist is applied. The team selects each checklist item for accident potential and adds them to the what-if list for evaluation. The checklist is reviewed for each area or step in the process.

After developing questions, the PrHA team considers each one to determine possible accident effects and to lists safety levels for prevention, mitigation, or containing the accident. The significance of each accident is determined and safety improvements are recommended. This is repeated for each process step or area outside of team meetings for later team review.

The results of a what-if/checklist analysis are documented like the results of a what-if analysis, as a table of accident scenarios, consequences, safety levels, and action items. The results also may include a completed checklist or a narrative. The PrHA team also may document the completion of the checklist to illustrate its completeness. The PSM rule requires detailed explanations of the analysis and recommendations for management review and transmission to those responsible for resolution.

Combining the what-if and checklist analysis methods uses their positive features while compensating for their separate shortcomings. For example, a checklist is based on generic process experience

TABLE 17.1.4   What-If/Checklist Analysis of the Cooling Tower Chlorination System

LINE/VESSEL: Cooling Tower chlorination system

What if	Consequences	Safety level	Scenario	Comments
A chlorine cylinder that is not empty is removed?	If the operator does not expect it to contain chlorine, then possible Cl_2 exposure may occur via skin and inhalation	None	1	Review training records and operating procedures to minimize the possibility of this occurring
The venturi is clogged with residue from the water basin?	No flow	Periodic checks of water quality	2	Review training records to make sure all staff have been trained in current procedures
	High pressure in recirculation line with rupture and water release	None	3	
	High pressure in recirculation line with release of Cl_2 if pressure check valve fails	Pressure check valve	4	

and may have incomplete insights into the design, procedures, and operations. The what-if part of the analysis uses the team's creativity and experience to brainstorm potential accident scenarios. The time required varies from about a day to a week depending on complexity. (See Table 17.1.4.)

HAZOP.   HAZOP is a formal technique for eliciting insights about system behavior from a multidisciplinary team that collectively has thorough knowledge of the plant and the physical phenomena involved in the plant. A HAZOP team evaluates the plant through a process that involves selecting a system, applying guide words to the selected system and identifying causes and consequences of the postulated event. Occasionally, it is not possible to quickly resolve a postulated occurrence on the basis of the available information and expertise. In this case, the HAZOP leader may assign the best qualified team member(s) to further investigate and report the results back to the team. The team reviews the item and the proposed resolution. When the team agrees, the analysis is recorded and the item is closed out.

A key member of the HAZOP is the HAZOP leader whose duties are indicated in Table 17.1.5.

Individuals may provide expertise in several of the required disciplines. The size of the team should be large enough to achieve diversity of points of view but small enough to function as a team focused on the analysis. The HAZOP session is recorded by the recorder on a personal (laptop or desktop) computer. If an item is not resolved, an individual is assigned responsibility for resolving it between sessions.

A plant generally has several operating modes in which the process systems may operate differently. The HAZOP team leader selects the operating mode to be considered in the first HAZOP analysis

TABLE 17.1.5   Responsibilities of the HAZOP Team Leader and Team Members

HAZOP team leader	HAZOP team members
Select the system to be analyzed	Provide expertise in the discipline that they represent on the team
Ensure all necessary disciplines are represented by the team and that team is familiar with the system	Be familiar with the system being analyzed and other systems with which it interacts
Provide all needed information	Resolve issues that are assigned
Ensure that all systems and operating modes important to safety are addressed	
Guide addressing multiple failures	
Resolve misunderstandings	
Resolve technical problems	
Ensure documentation	

TABLE 17.1.6 Guide Words for HAZOP

Guide word	Meaning	Examples
No	Negation of intention	No forward flow when there should be
		Sequential process step omitted
More	Quantitative increase	More of any relevant physical parameter than should be, such as more flow (rate, quantity), more pressure, higher temperature, or higher viscosity
		Batch step allowed to proceed for too long
Less	Quantitative decrease	Opposite of more
Partial	Qualitative decrease	System composition different from what it should be (in multicomponent stream)
As well as	Qualitative increase	More things present than should be (extra phases, impurities)
		Transfer from more than one source or to more than one destination
Reverse	Logical opposite	Reverse flow
		Sequential process steps performed in reverse order
Other than	Complete substitution	What may happen in other than normal continuous operation (startup, normal shutdown, emergency shutdown, maintenance testing, sampling)
		Transfer from wrong source or to wrong destination

loop. Next, the HAZOP leader selects the process variable to be considered for the selected mode followed by an aspect of the plant's process systems (a process node) and associated systems that affect the selected process variable for the selected mode of plant operation. Selection may be made from the plant system classification, or it may be from the nodal analysis of the process.

The information assembled for the selected node includes descriptions of the hardware and physical parameters of the operating mode. For example, the information for a fluid system must include physical parameters, such as flow, pressure, temperature, temperature gradients, density, and chemical composition for every node for which the description has meaning. This characterization of the system is called the design intent. The first step in analyzing a node is to achieve team consensus on the design intent.

The HAZOP focuses on study nodes, process sections, and operating steps. The number of nodes depends on the team leader and study objectives. Conservative studies consider every line and vessel. An experienced HAZOP leader may combine nodes. For example, the cooling tower water chlorination system may be divided into chlorine supply to venturi, recirculation loop, and tower water basin. Alternatively, two study nodes may be used: recirculation loop and tower water basin, and chlorine supply to venturi. Or there may be one study node for the entire process.

If the process uses a single large study node, deviations may be missed. If study nodes are small, many are needed and the HAZOP may be tedious; moreover, the root cause of deviations and their potential consequences may be lost because part of the cause may be in a different node.

Each study node is examined for potentially hazardous process deviations. First, the design intent of the equipment and the process parameters is determined and recorded. Process deviations from the design are determined by associating guide words with important process parameters. Guide words for a HAZOP analysis are shown in Table 17.1.6. By combining guide words with process parameters (Table 17.1.7), process deviations may be found. For example, "no" with the process parameter, "flow rate" from Table 17.1.7 leads to the deviation "no flow rate." Other deviations are found similarly.

The study team considers a deviation for possible causes (e.g., operator error causes pump blockage), the consequences of the deviation (e.g., high pressure line rupture), and mitigating safety features (e.g., pressure relief valve on pump discharge line). This consequence assumes the failure of active protection systems (e.g., relief valves, process trip signals). If the causes and consequences are significant, and the safety responses are inadequate, the team may recommend modification. In some cases, the team may identify a deviation with a realistic cause but unknown consequences (e.g., an unknown reaction product) that requires expert assistance.

The HAZOP study proceeds in a systematic manner to reduce the possibility of omission. Within a study node, all deviations associated with a given process parameter should be analyzed before

TABLE 17.1.7 HAZOP Process Parameters and Deviations

Process parameter	Deviation	Process parameter	Deviation
Flow (rate)	No, high, low, reverse flow	Time	Too long, short, late, or soon
Flow (amount)	Too much or little	Sequence	Steps omitted, reversed, or extra
Pressure	High or low pressure	pH	High or low pH
Temperature	High or low temperature	Viscosity	High or low
Level	High level/overflow, Low level/empty	Heat value	High or low heat value
Mixing	Too much, not enough, or reverse mixing; loss of agitation	Phases	Extra or missing phase
Composition	Component missing, high or low concentration	Location	Additional or wrong source, Additional or wrong destination
Purity	Impurities present, catalyst deactivated/inhibited	Reaction	No, too little, too much reaction, too slow, too fast reaction

the next process parameter is considered. All deviations for a study node should be analyzed before proceeding to the next node.

In listing the causes that lead to a guide word, engineering judgment must be used to exclude clearly incredible causes. However, it is not prudent to restrict the list to items of high probability or to items having obviously significant effects. The objective is to obtain insight into the characteristics of the system under various deviations from normal operation. It also must be noted that the causes can originate anywhere in the system and not just in the specific node currently under consideration.

Upon completion of the list of causes, the team addresses the potential consequences from each of the listed causes to any node in the system—that is, the evaluation considers consequences arising anywhere in the system, which can affect the subject deviation in the subject node. These consequences include potential operating problems and safety concerns.

Complex or particularly significant problems with a system may not be resolved immediately and require further action. In such a case, a team member whose expertise is most closely related to the subject is assigned the task of obtaining the information and reporting back to the HAZOP team for closeout. Table 17.1.8 is a sample HAZOP of the cooling tower chlorination system. The resource requirements for HAZOP range from a week for a simple system to 8 weeks for a complex system.

TABLE 17.1.8 HAZOP of the Cooling Tower Chlorination System

LINE/VESSEL: Cooling Tower chlorination system

Guide word	Deviation	Cause	Consequence	Safety level	Scenario	Action
None	No flow in chlorination loop	Pump failure Loss of electric power to pump	No chlorine flow to tower basin Low chlorine concentration in tower basin	Chlorination pump failure alarm	1	
		Low water level in tower basin	Hydrogen fluoride release into area; possible injuries fatalities	None	2	No action: Unlikely event; piping protected against external impact
Less	Low flow in chlorination loop	None identified			3	
More	High flow in chlorination loop	None identified			4	Note: Pump normally runs at full speed
			fatalities			
Reverse	No flow in chlorination loop	None identified			5	
Reverse	Backflow to HF inlet line	None			6	

Failure Modes and Effects Analysis (FMEA).   FMEA examines each potential failure mode of a process to determine effects of failure on the system. A failure mode is a loss of function, unwanted function, out-of-tolerance condition, or failure such as a leak. The significance of a failure mode depends on how the system responds to the failure.

Three steps are identified: (1) defining the process to be analyzed, (2) doing the analysis, and (3) documentation. Defining the process and documenting the results can be performed by one person, but the PSM rule requires a team for the analysis.

The vessels, pipes, equipment, instrumentation, procedures, and practices for the FMEA are identified and understood to establish the scope and level of detail. The PSM rule requires FMEAs to be performed at the major component level—a tradeoff between the time to perform the analysis and its value.

Functional system or process descriptions describe system, process, and/or component behavior for each operational mode. They describe operational profiles of the components the functions and

TABLE 17.1.9   FMEA Consequence Severity Categories

Category	Classification	Description
Catastrophic	I	May cause death, injury, monetary loss, or material loss
Critical	II	May cause severe injury, major property damage, or major system damage
Marginal	III	May cause minor injury, significant property or system damage
Minor	IV	Does not cause injury or property damage but may result in unscheduled maintenance or repair

TABLE 17.1.10   Partial FMEA for the Cooling Water Chlorination System

Date: October 3, 1997   *Page: 1/1*   *Preparers: RRF and JDC*
Plant: Y-12   *System: Cooling Tower Chlorination System*
Item: Pressure check valve   *References: ORNL-YYQ938596*

Failure mode	Causes	Operating mode	Failure effects	Failure detection method	Mitigation	Severity	Remarks
Too much flow through valve	Both internal pressure valves fail to open	Operation	Excessive chlorine flow to tower water basin, high chlorine level to cooling water, potential for excessive corrosion in cooling water system	Rotameter Daily testing of cooling water chemistry	Relief valve on pressure check valve outlet	III	None
Too little flow through valve	One or both internal pressure valves fail to close	Operation	No/low chlorine flow to tower water basin, low chlorine level in cooling water, potential for excessive biological growth in cooling water system, reduction in heat transfer	Rotameter Daily testing of cooling water chemistry	Automatic temperature controllers at most heat exchangers	IV	None
Chlorine flow to environment	Internal relief valve sticks open Both internal pressure valves fail to open and relief valve opens	Operation	Potential low chlorine flow to tower water basin; see above Chlorine released to environment, potential personnel exposure and injury	Distinctive odor	Pressure check valve located outdoors, unlikely to accumulate significant concentration	III	Action item: consider venting relief valve above ground level

input and outputs of each. Block diagrams assist by illustrating operations, interrelationships, and interdependencies of functional components. Interfaces should be indicated in these block diagrams.

The FMEA is executed deliberately and systematically to reduce the possibility of omissions. All failure modes for one component should be completed before going to the next. The FMEA's tabular format begins with drawing references of the system boundaries and systematically evaluates the components as they appear in the process flow path using the following categories. (1) Failure modes are listed for each equipment item and interface failure mode for each operating mode being considered. (2) Root causes of the failure mode are identified for subsequent steps and for hazard ranking. (3) Operating mode identifies the operating environment. (4) Effects of the failure on the overall system or process are listed. (5) Detection methods are identified for likelihood of detection. (6) Mitigations are listed. (7) Severity categories of failure are given in Table 17.1.9. Remarks are provided for suggestions for mitigation and conditions that were considered in the analysis.

An FMEA is a qualitative, systematic table of equipment, failure modes, and their effects. For each item of equipment, the failure modes and root causes for that failure are identified along with a worst-case estimate of the consequences, the method for detecting the failure and mitigation of its effects. Table 17.1.10 is an example of applying FMEA to the cooling tower chlorination system. Resource requirements are 4 days for a simple system and 20 days for a complex system.

FMEAs have limitations. For example, human errors are not usually but may be, included in the equipment failure mode. FMEAs rarely investigate damage or injury that could arise if the system or process operated successfully. Because FMEAs focus on single event failures, they are not efficient for identifying combinations of equipment failures that lead to accidents.

17.1.3   QUANTITATIVE SAFETY ASSESSMENT

Risk

$$R = p \times C \tag{17.1.1}$$

$$N \times R = n \times C \tag{17.1.2}$$

$$R = (n/N) \times C \tag{17.1.3}$$

Risk is the chance of loss. To convert these words into mathematics, replace "chance" by "probability," "loss" by "consequences," and "of" by "multiplication." Expressed in words: risk = probability times consequences, which is expressed by Equation 17.1.1, using R for risk, p for probability, and C for consequence.

Risk, the basis for insurance, was invented by the fifteenth century Genoese to protect against individual catastrophic shipping losses by sharing the risk. The insurer breaks even if he collects premium R for insuring N ships per year of which n are lost and award C is paid for each lost ship (Equation 17.1.2).

$$p = \lim_{N \to \infty} \frac{n}{N} \tag{17.1.4}$$

Rearranging gives Equation 17.1.3, where the ratio n/N becomes probability as N becomes very large (Equation 17.1.4), leading back to Equation 17.1.1. This illustration from insurance leads to the interpretation risk as the "premium" paid by society for the use of an activity.

Calculating System Failure Probability*

If a system's failure probability is known, there is no need to calculate it. However, this is rarely the case and the system's failure probability must be calculated from a knowing the component failure

---

* Some references for probabilistic analysis are CCSP, 1989a, Lees, 1986, DOE, 1996, McCormick, 1981, and Fullwood, 2000.

TABLE 17.1.11   Comparison of Ordinary and Boolean Algebra

Property	Ordinary algebra	Boolean algebra
Commutative	$A + B = B + A$	Same
	$A \times B = B \times A$	
Associative	$A + (B + C) = (A + B) + C$	Same
	$A \times (B \times C) = (A \times B) \times C$	
Distributive	$A \times (B + C) = A \times B + A \times C$	Same
Idempotency	NA	$A \times A = A$
	NA	$A + A = A$
Unity	$A + 1 = A + 1$	$A + 1 = 1$
	$A \times 1 = A$	$A \times 1 = 1$
Absorption	$A \times (A + B) = A^2 + A \times B$	$A \times (A + B) = A$
	$A + (A \times B) = A \times (1 + B)$	$A + (A \times B) = A$
de Morgan's	NA	$\overline{A} \times \overline{B} = \overline{A + B}$
theorem		$\overline{A} + \overline{B} = \overline{A \times B}$

probability and how the components are connected to make the system. Another aspect of the analysis is to identify the components that are key to the operation to improve their reliability through design or maintenance.

Calculate Failure Probability-Not Success.   While successful system operation is our concern, it is easier to calculate the probability of system failure than system success. The reason is that component failure probabilities are generally small numbers while component success probabilities are nearly one.

Boolean Logic.   Systems are modeled as if their components either fail or don't fail. Such two-state operation is the subject of logic.

George Boole in the nineteenth century invented mathematical logic in the form of algebra-like equations in which the variables have only two values which we are calling failed and not failed. Table 17.1.11 compares the rules of ordinary with Boolean algebra. Like ordinary algebra, these rules are used to manipulate Boolean equations into simpler form. If the A and B represent failure , success is represented by an over bar $\overline{A}$ which means $\overline{A} = 1 - A$ (since probability of failure and success is 1).

As an example of a system equation, expressed in Boolean logic, system A is composed of three components: B whose failure causes A to fail, and redundant components C and D both of which must fail to fail A. Written as a Boolean equation, it is: $A = B + C \times D$, where the symbols represent failure of the identified component. Notice that A is the sum of products $(1 \times B + C \times D)$. This is called the cutset form of a system equation.

The preceding was a simple equation for a very simple system. Safety systems that prevent the escape of hazardous chemical are usually complex and provide multiple barriers of protection to make the likelihood of all barriers failing very unlikely. A slightly more complex example is shown in Figure 17.1.2, a system for injecting emergency cooling water to prevent a runaway process. A safety injection signal (*SIS*) for the process instrumentation causes the pump to start and to close valve E while opening valves A, B, C, D. The equation for system failure is Equation 17.1.5 where Top means the system fails, A represents failure of valve A to open, B failure of valve B to open, C failure of valve C to open, D failure of valve D to open, E of valve E to close, Ps of the pump failing to start, Pr of its failing to run, and SIS of the failure of the SIS signal to each of the components. This equation is simplified, using Table 17.1.10, to its cutset form (Equation 17.1.6) to show which single components can cause

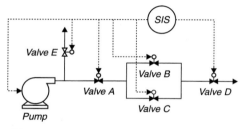

FIGURE 17.1.2   Simple safety injection system.

system failure and which redundant components can cause failure.

$$Top = D + SIS + [B + SIS + A + SIS + (Ps + Pr + SIS) + (E + SIS)]$$
$$\times [(C + SIS + A + SIS + (Ps + Pr + SIS) + (E + SIS)] \qquad (17.1.5)$$

$$Top = A + D + E + Ps + Pr + SIS + B \times C \qquad (17.1.6)$$

Notice that Equation 17.1.6 is the sum of products.

The reasons for converting system equations to cutsets are (1) to eliminate double counting which must be done before a Boolean equation can become a probability equation, (2) to rank components according to their operational importance, and (3) to simplify the system equation.

Going from Logic to Probability

Once the system logic is in the cutset form, the symbol for the component that may fail is replaced by the numerical probability of failure. The transformed system equation is solved according the rules for combining probability.

There are two basic methods for combining probabilities: ANDing and ORing. The first step is to calculate the probability of failing redundant subsystems. This is done by ANDing the independent failure probabilities for each component in the redundancy. In this operation, also called intersection and multiplication, given by Equation 17.1.7, the component probabilities are multiplied to give the subsystem failure probability.

$$P(A_1 \times A_2 \times \cdots A_n) = P(A_1) \times P(A_2) \times \cdots P(A_n) \qquad (17.1.7)$$

$$P(A_1 + A_2 + \cdots A_n) = 1 - [1 - P(A_1)] \times [1 - P(A_2)] \times \cdots [1 - P(A_n)] \qquad (17.1.8)$$

$$P(A_1 + A_2 + \cdots A_n) \approx P(A_1) + P(A_2) + \cdots P(A_n) \qquad (17.1.9)$$

Next, calculate the probability of system failure by ORing the components and redundant subsystems according to Equation 17.1.8. This process, also called "union" or addition, calculates one minus the probability that all components will not fail. If all $P(A_i)s \ll 1$, multiplying out Equation 17.1.8 gives approximately Equation 17.1.9 which is the sum of the probabilities. It is obvious that probabilities cannot be simply added because if they are nearly 1 their sum would exceed one.

Quantifying a System Reliability Equation

It is comparatively easy to write a reliability equation for a system; it is another matter to find the data to quantify it. Data are of three types: frequency of accident initiators, component failure rates, and human error probability (HEP).

Accident initiators are, as the name suggests, the starting place for an accident. A system is analyzed to determine how safely it responds to the initiator. The frequency of an accident is the product of the initiator frequency and the probability that the protection system fails. Examples of accident initiators are: human error, equipment failure, process upsets, fire, flood, earthquake, high wind, missiles, lightning, and extreme weather. Initiator frequency may be calculated from plant and industry records, journals, insurance records, and newspapers.

$$p_{high} = \frac{n + \sqrt{(n)}}{N} \qquad (17.1.10)$$

$$p_{low} = \frac{n - \sqrt{(n)}}{N} \qquad (17.1.11)$$

Initiator frequency and component failure probability data are available from various sources (e.g., Fullwood, 2000 and CCSP, 1989b); it may be necessary to prepare the data using equation 17.1.4 which is a point estimate whose accuracy depends on the sample size. A rule-of-thumb is that a "one-sigma" higher estimate than the true value is given by Equation 17.1.10; the corresponding estimate that it is "one-sigma" lower than the true value is given by Equation 17.1.11. In both equations, n is the number of failures that have been observed and N is the product of the number of components and the number of demands or the number of components and the time exposure.

As an example, a company with 10 emergency electrical power generators that are tested monthly has had 2 failures in 5 years. The estimate of the failure to start is $P_{est.} = 2/(5 \times 12 \times 10) = 0.0033$; $P_{high} = (2 + 1.414)/600 = 0.0057$; $P_{low} = (2 - 1.414)/36 = 0.00098$. The percentage variation is reduced by a larger sample. More sophisticated estimators are given in Fullwood, 2000.

Human error estimation is primarily judgmental. An important reference is Swain and Guttmann, 1983. for human error estimation and representative data.

17.1.4 SYSTEM MODELING FOR QUANTITATIVE ANALYSIS

Parts Count

Perhaps the simplest way to assess the reliability of a system is to count the active parts (i.e., parts that move or change state). The reliability estimate is the product of the number of parts and some nominal failure rate for the parts. In the design phase, two competing designs may be compared on the basis of the number of parts but several cautions are in order.

If designs are compared on the basis of count, then failure rates of each type of part either must be about the same or must be adjusted for the variations. For example, the parts count of vacuum-tube and solid-state television sets (using discrete components) are approximately the same, but their reliabilities are considerably different because of the better reliability of solid-state components.

The other caution relates to redundant systems. Figure 17.1.3 shows simple systems that pump a chemical from a tank into a distribution header. System A consists of a tank of water (T1), a motor-operated valve that is open, a pump (P1), and a closed motor-operated valve (V2). System B consists of two, trains, each of which is identical to system A; hence system B is doubly redundant; i.e., either train may fail, but the system will still perform as designed. System A has three active components and if the nominal failure rate is 0.001/demand, the failure rate of system A is 0.003/demand. System B has six active components, so a superficial estimate of the system failure rate is 0.006/demand, but because it is doubly redundant, according to Equation 17.1.7, the train probabilities, if independent, should be multiplied to give system B's failure rate as 0.000009/demand; hence the reliability estimation by a simple parts count is grossly conservative and consideration must be given to redundancies.

FMEA/Failure Modes Effects and Criticality Analysis (FMECA)

FMEA may be used qualitatively, as just shown, or quantitatively by including failure rates in the FMEA table. By including a criticality entry, the FMEA becomes a FMECA. Either may be applied

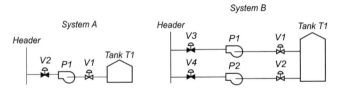

FIGURE 17.1.3   Illustrative cooling systems.

TABLE 17.1.12   Failure Modes Effects and Criticality Analysis Applied to System B of Figure 17.1.3

Component	Failure mode	Effect	Criticality	Mitigation	Probability
Water storage tank T1	Sudden failure	Loss of water supply	No cooling water supply	None	1×10^{-10}/h
MOV V1	Valve body crack	Loss of fluid to header	Reduce flow redundant train	Valve off, use other train	1×10^{-3}/h
	Motor operator inadvertent operation	Flow blockage	No flow to header	Redundant train	1×10^{-3}/ demand (d)
Pump P1	Motor failure: bearing or short	No pumping	No flow to header	Redundant train	1×10^{-3}/d
	Pump failure bearing seize				1×10^{-3}/d
MOV V2	Valve body crack	Loss of fluid to header	Reduce flow redundant train	Valve off, use other train	1×10^{-10}/h
	Motor operator fails to respond	Flow blockage	No flow to header	Redundant train	2×10^{-3}/d
	Gate disconnect from stem	Flow blockage			1×10^{-4}/d

in designing or operating a plant. The headings identify the system and component under analysis, failure modes, the effect of failure, an estimate of how critical a part is, the estimated probability of the failure, mitigators, and possibly the support systems. The style and contents are flexible and depend on the objectives of the analyst.

FMEA/FMECA is particularly suited for root cause analysis and is quite useful for environmental qualification and aging analysis they are used in aerospace, nuclear power, and chemical process industries but seldom with highly redundant systems because, like parts count, they are not directly suitable for redundant systems such as those that occur in nuclear power plants. Table 17.1.12 shows a FMEA for system B in Figure 17.1.3.

This FMEA/FMECA shows failure rates that are both demand and time dependent. Adding the demand failure rates gives a train failure rate of 0.0051/demand. The sum of the time-dependent failure rates is: $\lambda = 3 \times 10^{-10}$/h^2. A standby system such as this does not exhibit its operability until it is actuated, for which the probability that the train has failed since its last use is needed.

Valve bodies are considered to be part of the fluid envelope and similar to piping. Piping characteristically leaks before breaking; therefore leakage may be expected to indicate potential fracture of the valve body. The time from which it was last known to be working properly until needed is the mission time. The average mission time τ is half the inspection interval. The probability of failure is the product of the failure rate λ and the average mission time (Equation 17.1.12).

$$p = \lambda \times \tau \qquad (17.1.12)$$

For example, if inspected monthly, the time between inspections is 731 h and the average time between inspections is $\tau = 365$ h. If the product $\lambda \times \tau$ is small compared with 1, Equation 17.1.12 is correct and the probability of failure during the mission time is $3 \times 10^{-10} \times 365 = 1 \times 10^{-7}$. This is so small it may be ignored with respect to the other types of failure to give a train failure rate of 0.0051/d.

For system B, with the redundant systems, the probabilities are multiplied, or $(0.0051/\text{d})^2 = 0.000026$/demand.

Reliability Block Diagram

The preceding shows care is needed to understand system logic and redundancies to model a system. Neither the parts count nor FMEA aids the analyst in conceptualizing the logical structure.

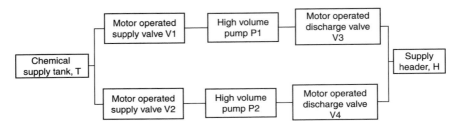

FIGURE 17.1.4   Reliability block diagram of Figure 17.1.3B.

A Reliability Block Diagram (RBD), similar to a flow diagram, represents the components as a series of interconnecting blocks. Components are identified by the block labeling. System B in Figure 17.1.3 is shown as an RBD in Figure 17.1.4. (It is customary to draw RBDs with the source on the left and the sink on the right.) Even unquantified, RBDs are useful because they show the system logic and how it can fail.

As noted previously, a cutset is a combination of component failures that fail the system. The term is believed to have originated as cuts across the flow in a reliability block diagram. In Figure 17.1.4 a cut through T will stop the flow as will a cut through the header, H, but cuts through the redundant trains must cut through V_1 or P_1 or V_3 in combination with V_2 or P_2 or V_4, making $3^2 = 9$ combinations with 11 cutsets for the system. The cutsets are: T, H, V_1 AND V_2, V_1 AND P_2, V_1 AND V_4, P_1 AND V_2, P_1 AND P_2, P_1 AND V_4, V_3 AND V_2, V_3 AND P_2, and V_2 AND V_4, where all symbols mean failure of the indicated component.

Fault Tree

A fault tree is a symbolic representation of the system logic equation that uses the specialized symbols shown in Figure 17.1.5. A basic event is the failure of a component. Components are connected through AND and OR gates. A label is placed above the gate symbol to indicate the meaning of the connection. Fault trees may be too large for a sheet of paper, therefore they a broken into subtrees that connect with higher trees using the transfer symbols.

Figure 17.1.6 shows a fault tree of the cooling system presented in Figure 17.1.3, System B, which may be compared with the RBD shown in Figure 17.1.4. It is seen that tank failure and header failure, which are a single failure faults, must be separated from the two trains that are a redundancy, hence they must be combined with an AND gate. The top event occurs if either of the single failures occur or the AND gate fails, hence the three items are combined in an OR gate.

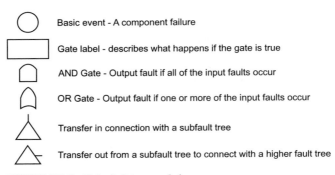

FIGURE 17.1.5   Major fault traee symbols.

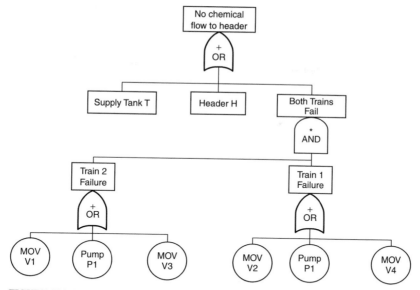

FIGURE 17.1.6   Fault tree of cooling system B shown in Figure 17.1.3.

A fault tree may be analyzed by hand into its cutsets using the relationships of Table 17.1.10 or the fault tree may be input to a fault tree analysis code for computer analysis. Clearly if many complex fault trees are to be analyzed, a computer should be used.

Computer Assistance

Before you feel overwhelmed by Boolean reduction to cutsets and calculating probabilities, rest assured that there are many computer codes for your assistance.[†] Some codes, e.g., SETS, accept a Boolean equation of the system or the fault tree representation of the equation of the system as input.

Most people are uncomfortable with equations and prefer a graphical form e.g., fault trees. The way a fault tree is drawn does not matter because the computer code transforms it into the cutset form. Older codes such as described in Vesely, 1981 were not user-friendly. They required coding the fault tree on cards or card images and running the codes on a large computer. More recently, code suites (e.g., SAPHIRE, Russell, 1993) provide graphical interfaces for the user to draw the fault tree on the computer screen. Following this, the probability of system failure (top event) is calculated using generic data in the code or user provided data. Uncertainty in the estimate is calculated from the uncertainties in the data, and standard importance measures are calculated.

Event Tree

The last quantitative analysis method is the event tree—a graphical depiction of how a process responds to an accident initiator, which may be internal or external to the plant. The initiators are found from a study of the plant's processes, its operating history, and accidents in related facilities. Figure 17.1.7 shows an event tree. From left to right it depicts the time progression of the accident.

[†] Fullwood, 2000, lists, describes and indicates the availability of many computer codes and codes suites for fault tree and PSA analysis.

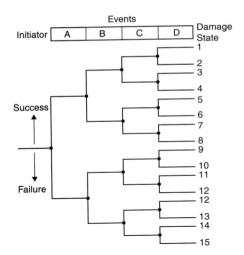

FIGURE 17.1.7   Event tree. The left side connects to the initiator, the right side to process damage states; the top bar gives the systems in the order that they enter the scenario. The nodes (dots) are entry points for the branching probabilities that may be determined from fault tree analysis or from a database. Success is indicated by the path going up at a node; failure is indicated by going down.

It begins with the accident initiator (left). The bar at the top lists significant events that may follow the initiator and affect the consequences from the occurrence of the initiator as mitigated or acerbated by these events. The analysis explores what happens for the different success and failure combinations of the events. At the node, a desirable outcome of the event is indicated by an upward branch; an undesired event is indicated by a downward branch. For example, if the initiator is the rupture of the tank holding the chemicals, event A may be the integrity of the cell in which the tank was located. A branch up means the cell contains the chemical; a branch down means it does not contain the chemical. Event B may be the operation of a neutralization system. Event C may be the operation of a fire suppression system. Event D may be the operation of an off-gas system.

Thus, the event tree graphically depicts all of the accident scenarios that may result from the initiator for a given chemical process.

However, the event tree does more than just diagram the scenarios; it provides the probabilities of each end state by the estimate of the probability of each undesirable event outcome (failure) at a node. Such estimation may be based on operating experience at the plant; it may be from related industrial experience; or, if it is complex, it may be modeled with a fault tree. If failure at node i is p_i, this is the value used if the branch is down. If the branch is up, the probability is $(1 - p_i)$. Multiplying these nodal probabilities for each path through the tree gives the probability of the end state. From the separately calculated consequences of the end state, the risk of each end state is found by multiplying its probability and consequences:

REFERENCES

CCPS, *Guidelines for Chemical Process Quantitative Risk Analysis*, American Institute of Chemical Engineer, NY, 1989a.

CCPS, *Guidelines for Process Equipment Reliability Data*, American Institute of Chemical Engineer, NY, 1989b.

CCPS, *Guidelines for Hazard Evaluation Procedures*, American Institute of Chemical Engineer, NY, 1992.

DOE, *Process Safety Management for Highly Hazardous Chemicals*, DOE-HDBK-1101-96, National Technical Information Service, Springfield, VA, February, 1996.

EPA, FEMA, and DOT, *Technical Guidance for Hazards Analysis, Emergency Planning for Extremely Hazardous Substances*, National Technical Information Service, Springfield, VA, December 1987.

Fullwood, R., *Probabilistic Safety Assessment in the Chemical and Nuclear Industries*, Butterworth-Heinemann, Boston, 2000.

Lees, F. P., *Loss Prevention in the Process Industries*, Vols. I and II, Butterworth, London, 1986.

McCormick, N. J., *Reliability and Risk Analysis*, Academic Press, New York, 1981.

Russell, K. D. et al., *SAPHIRE Technical Reference Manual: IRRAS/SARA*, Volume 4.0, NUREG/CR-5964, National Technical Information Service, Springfield, VA, June 1993.

Swain, A. D., and Guttmann, H. E., *Handbook of Human-Reliability Analysis with Emphasis on Nuclear Power Plant Applications*, NUREG/CR-1278, August 1983.

Vesely, W. E. et al., *Fault Tree Handbook*, National Technical Information Service, Springfield, VA, NUREG-0492, January 1981.

CHAPTER 17
SYSTEM LEVEL RISK ASSESSMENT

SECTION 17.2

RISK ASSESSMENT IN THE BROADER CONTEXT

Herbert Inhaber

Dr. Inhaber is a risk assessment expert and president of Risk Concepts in Las Vegas. He is the author of eight books, including Slaying The Nimby Dragon.

17.2.1 INTRODUCTION

Most of this volume is concerned with detailed calculations of environmental health and safety, with most sciences except perhaps astrophysics represented. But all this activity takes place in the human and social context. Decisions about environmental quality and risk are made by fallible humans, sometimes lost in the abstruse mathematics.

The objective of this chapter is to describe how risk assessment is used (and occasionally abused) in the broader social context—i.e., outside scholarly journals. Risk assessment may be defined as the process of estimating adverse effects caused by a wide variety of sources, from industrial activities to lifestyle to natural phenomena. Because there are so many ways risk assessment can be considered and used, this chapter touches only briefly on many aspects.

17.2.2 HISTORY OF RISK

Clearly our prehistoric ancestors had some appreciation of risk. Staying far away from a saber-toothed tiger was safer than getting near one. But lacking some measure of risk, our ancestors did not do what we call risk analysis, which implies calculation that may range from simple arithmetic to the probabilistic techniques of advanced risk studies.

Among early peoples, the Babylonians were among the first to use numbers extensively, often in conjunction with astronomical events. In consequence, it is hardly surprising that they apparently had, around 3200 BC, the first quantitative risk assessors [American Chemical Society and Resources for the Future (ACS-RFF), 1998]. However, these assessors did not deal with health or environmental issues.

These advisors offered advice on risky decisions of the time, such as deciding where to put important buildings and when and whom to marry. These areas generally are not part of the work of modern risk analysts. However, newspaper horoscopes, which foretell the future and allegedly are based on mathematical formulae, still are read by tens of millions. In many parts of the world, the dates of major events such as marriages or contract signings are still regulated by numerology. In the United States, it has been claimed that a recent president had the dates of some of his decisions set by the same formulae. The purpose of these calculations is, in a broad sense, to avoid risk.

Other ancient peoples developed de facto risk assessments, although usually not quantitative (see Fullwood and Hall, 1988, for a lengthy discussion). The book of Leviticus in the *Bible* cautions readers against eating certain shellfish. The Egyptians, around 2000 BC, had bathroom and sewage facilities to avoid risk of biological contamination. The Romans, masters of sanitation, drained swamps because of their dangers and built aqueducts to transport pure water.

By the Industrial Revolution it was realized that some diseases could be prevented or alleviated by human intervention. The scientific study of scrotum cancer in chimney sweeps by Percival Potts in the late eighteenth century was perhaps the first quantitative epidemiological study (the subject itself may go back to the Greek Hippocrates). The chimney sweep example is quoted in the first chapter of many books on risk. Epidemiology has undergirded many risk studies since then.

By the late nineteenth century, industrial hygiene, epidemiology, and toxicology were scientific areas. However, their general application to factories and industrial enterprises had to wait a number of decades as social consciousness grew. By the 1930s, considerable evidence about the relationship between occupational toxins and human health had accumulated (ACC-RFF, 1998). In the United States, the Food and Drug Administration set "tolerance limits" for food additives in this period. However, the Great Depression prevented widespread use of these results in regulation. Only certain firms (e.g., duPont) made use of this information (Paustenbach, 1995).

The next era in risk assessment came indirectly with the advent of the atomic bomb in 1945. It was quickly recognized that the power of destruction could be harnessed for peaceful purposes, and major governments adopted nuclear power programs. However, the risks associated with producing this form of energy were generally unknown, which led to a burgeoning of research on nuclear and radiation risks. The latter had been only a small part of biological analysis before World War II.

Later, in the 1970s, drawing on research from the U.S. Department of Defense on building more reliable missiles, the concept of probabilistic risk analysis for complex systems such as nuclear power plants was developed. This subsequently was applied to other complicated installations like chemical and petrochemical plants.

17.2.3   *SIZE OF THE RISK ENTERPRISE*

The size of the risk enterprise depends on whether environmental health and related subjects are included. However, by late 1998 the Library of Congress had about 4800 books with the word "risk," "risks," or "risky" in their titles or subtitles. If chapter headings were included, the total would likely be much larger. There were about 1000 books with the word "hazard," "hazards," or hazardous" in their titles or subtitles.

In addition, there are scores, if not hundreds, of worldwide journals that deal in some way with risk. A reasonable conclusion is that about 10,000 researchers (along with many auxiliary personnel) are engaged in some type of risk analysis. Thus has the enterprise grown since its early days.

Throughout the world, governments at various levels issue regulations every day that are based, at least in part, on risk analysis and related subjects. In turn, they require personnel who are versed in the subject and can abstract the relevant facts in terms of public policy.

In the United States, federal agencies such as the Department of Defense, the Nuclear Regulatory Commission, the Department of Energy, the Environmental Protection Agency, the Occupational Safety and Health Administration, and others all have risk sections and groups. These divisions are often duplicated at the state level. Internationally, the World Bank and the International Atomic Energy Agency, among others, perform risk assessments of various types.

Although the Internet is not a perfect indication of the interest in risk and related subjects, one search in late 1998 (using the search engine Infoseek) turned up over 6 million references to the word "risk."

There are many technical societies that deal with risk in some form. One of the first to be completely devoted to the subject is the Society for Risk Analysis. However, papers on risk analysis can be heard at meetings of dozens of other groups dealing with toxicology, epidemiology, nuclear power, air quality, water quality, defense topics, and a host of other subjects.

17.2.4   *THE GLOBAL MAGNITUDE OF RISK*

Over the past three decades, it is clear that the number and funding of scientists and engineers working in risk assessment has increased dramatically. But the absolute level of risk, as measured by simple indicators, is decreasing. In most countries, life expectancy is increasing. As part of this, most disease rates are falling, with the possible exception of lung cancer due to smoking. Three score and ten years (70) was the length of life specified in the *Bible*. Few attained it in those days. Today, scores of millions do so.

Air pollution in industrialized countries has decreased considerably in the past generation. Reductions are expected in developing countries, as increasing wealth allows the installation of pollution control devices.

On a global risk scale, the end of the Cold War reduced the chance of world nuclear war, perhaps the greatest risk of all. Although the risk of war can never be estimated exactly, the U.S. magazine *Bulletin of the Atomic Scientists* moved the hands of their "doomsday clock" far from midnight, the hour when hostilities could begin. (However, they did not eliminate the clock completely.)

Thus, compared with the past, there is an apparent contradiction between the increased interest in risk and the decreasing indicators. Wildavsky (1988, 1995) and Douglas (1985) have demonstrated that the results are based on wealth and culture. In a subsistence economy, as many developing countries are today, there is little concern about seat belts, airbags, guard rails, limited access highways, interior padding, well-engineered roads, Breathalyzer tests, and the other factors that have substantially reduced road accident rates in developed nations. Of course, this lack of concern was the previous attitude in what are now developed countries. Wildavsky noted that when a society is wealthy, it has the time and money to reduce risks.

Douglas approaches the contradiction from the viewpoint of anthropology and culture. Many tribal societies regard as risky certain activities that Western cultures dismiss and viceversa. Nations are locked into their cultural attitudes toward risk much more than they realize.

For example, in the energy risk studies discussed below, risk to workers tends to be higher than risk to the public. In a completely rational world, emphasis should be on reducing worker risk, with less concern about public risk reduction. In reality, developed nations emphasize public risk. As noted below in the subsection on risk perception, this may be due to the media's concentration on disasters, real or potential, that affect the public. Occupational risk to individual workers tends to receive little attention.

17.2.5   *RISK PERCEPTION*

One of the oldest questions in philosophy is "If a tree falls in a forest and nobody hears it, does it make a sound?" Like many philosophical issues, there is no simple answer. The analogous question in risk analysis is "If a study shows that a chemical or toxin has a high (or low) risk, and this study is either unknown to the public or rejected by them, what then is the risk?" (Rogers and Bates, 1982; Rück, 1993; Warner, 1981).

An example is cigarette smoking. Some hazards were identified as far back as the seventeenth century, when James I of England and Scotland wrote a book criticizing its ills. By the 1930s, the term "coffin nails" was used to describe cigarettes. In spite of an accumulating body of studies showing risk of lung cancer and other diseases, the results were generally ignored by the public, as evidenced by increasing cigarette consumption. The risk was real, but public perception generally ignored it. Assessment comprises both risk analysis and public understanding of these analyses.

Risk experts have handled this in the Society for Risk Analysis by rotating the presidency among three main groups: biological sciences, engineering and physical sciences, and social sciences. The last group includes the study of risk perception.

As Slovic, Fishhoff, and Lichtenstein (SFL), three of the founders of risk perception studies, have noted, subjective judgments by both experts and laypeople are a major component of any risk assessment. Expert judgments are often obscured in impenetrable mathematics. Judgments of laypeople are considerably easier to detect.

In a classic study that went from the social science literature to popular magazines, SFL noted the divergent risk estimates of four groups: members of the League of Women Voters (LWV), college students in Oregon, a business and professional club, and risk experts. SFL considered 30 sources of risk, from nuclear power to vaccinations. Many of these activities rarely appear on risk lists, yet they all produce some hazards.

The two largest divergences between estimates of the lay group and experts were in nuclear power and x-rays. The experts rated nuclear energy as 20th riskiest of the 30 activities; the LWV and college students rated it highest; the business and professional people rated it eighth. X-rays were rated about 20th in riskiness by the laypeople but 7th by the experts. The rank of many of the other risk sources were about the same for laypeople and experts.

Although there are many possible explanations for why the estimates varied so much, it results from the benefits received from the two activities. (The relationship of benefits to risks are discussed elsewhere in this chapter.) X-rays are viewed as having considerable benefits in medical diagnosis that outweigh the risks of radiation. The benefit of nuclear power is electricity generation, but there are other ways to produce this energy form, such as coal, gas, and oil. (Admittedly, there are methods other than x-rays for detecting internal diseases, such as nuclear magnetic resonance, but they were not included on the SFL list.) Thus the benefits and risks of nuclear energy must be compared not with those of swimming or lawn mowers, but with risks of other ways of generating electricity.

Experts measure risk as Y deaths or diseases per activity, but risk is perceived according to whether it is voluntary, controllable, certain to be fatal, catastrophic, immediately manifested, and unfamiliar. Some of these are discussed elsewhere in this chapter.

Most of the perception aspects of risk can be described by two overall factors: dread (related to aspects such as certain to be fatal and catastrophic) and how unknown the risk was (related to involuntariness and newness). On these scales, nuclear power again had relatively high scores on both factors, even among the experts. Commercial aviation had a high dread factor, accounted for by the fact that there are few survivors among most large air crashes. Curiously, skiing was among the lowest in dread and unknown risk factors.

One of the problems in comparing risk perceptions is: What is the risk and what are the units involved? For example, in evaluating the riskiness of spray cans, one of the 30 activities considered, is it the risk to the ozone layer or the number of cans misfiring and causing eye injury? For power mowers, is it the total number of foot injuries per year or per hour of mowing? The respondents were merely asked about perceived risk. If the units of risk had been asked, the analysis might have come to different conclusions.

Given that certain activities have a much higher perceived risk than measured data indicate, how are risk perceptions formed? The media clearly plays a large part. For example, pesticides ranked about eighth in the list of 30 activities noted above. Some years ago, concerns about the pesticide Alar were expressed on a widely watched television show. Would a repeat of the poll have shown higher perceived risk for pesticides? Risk assessors sometimes talk about "risk of the week," in reference to the media publicizing a seemingly new hazard on a regular basis.

Other factors that influence the results are familiarity and perceived benefits. For example, home appliances ranked among the lowest in risk in the study. Emergency rooms in hospitals treat many injuries from these appliances every year, often to children. These appliances are ubiquitous, with obvious benefits, so the risks incurred in their operation shrink in consciousness.

Because risk perception is different from data, how can it be made closer to reality? This is a matter that has been given considerable thought by the industries that fare badly in risk perception. When an aircraft crashes, advertising for that airline is immediately removed from the media and is not resumed for weeks. Airline personnel rush to the scene to comfort relatives of the victims. In spite of this, it is likely that commercial airlines will always rank high in the dread factor.

In nuclear waste storage, nuclear electric utilities have contributed billions of dollars to a fund, part of which demonstrates the extremely small risks of storing spent fuel rods in Yucca Mountain in Nevada. Many scientific estimates have been done and reported in the media. In spite of this, Nevadans are notably unenthusiastic about a site in their state being studied. There is no indication that their perception of high risk will change.

So, although risk perception clearly affects decisions based on risk, exactly how it is formulated is not apparent. Knowledge of risk perception and how it can be changed is valuable to the industries that generate the risk.

17.2.6 ENERGY RISKS

As noted previously, study of energy risk was one of the greatest impetuses to the field. Because energy powers civilization, there is concern about the risks posed in its various forms.

The advent of nuclear power sparked worldwide interest in determining its risks. Early simplified studies lacked an agreed-on methodology for estimating risks. The problem was that a large accident affecting many people was thought to be highly unlikely, but its probability had not been estimated. Consequently, the first studies of nuclear risk in the United States concentrated on the effects of a large radiological release and did not attempt to project its likelihood; the probability of such an accident was not calculated until 1975.

The closest analogy to nuclear risk was that of hydroelectricity, which, until recently, led nuclear in its proportion of the world's electricity supply. Clearly, a dam can fail catastrophically, and such events have occurred. Although rarely discussed in the technical literature, the largest energy-related disaster, in terms of lives lost, was that of the 1979 Morvi dam failure in India, where between 1300 and 25,000 died or were missing [*Statesman* (India), 1979; *New York Times*, 1979].

After nuclear risk assessments, the risks of other methods of electrical power generation, mainly from air pollution and worker accidents, were studied. Even taking into consideration stringent air pollution control measures, nuclear energy has lower overall risk, in terms of deaths and diseases, than oil and coal. Although the largest accident possible from nuclear is considerably greater than an accident from fossil fuels, with the exception immediately following, its probability is extremely small. This results in a modest average risk per year or unit energy output of nuclear power.

The anomaly in terms of fossil fuels is shipping natural gas in liquefied form in large tankers. Although no major accidents near large cities have occurred with this type of energy, a disaster of this type could dwarf other fossil-fuel accidents. Export of natural gas in this form will become more common in the future.

Between the two "oil shocks" of 1973 and 1979, public interest in renewable energy sources such as solar and wind grew. Hydroelectricity was the first of the large-scale renewables, but comparatively little risk analysis had been performed on that energy source. It was evident that using the sun or wind generated no direct risk. Hazardous natural phenomena like tornadoes and hurricanes, a function of winds, do not produce useful energy. However, there could be indirect risk from the manufacture of solar collectors, windmills, and other renewable facilities.

Inhaber (1982) was one of the first to estimate the risk from a variety of energy technologies, drawing in part on data estimated by the Jet Propulsion Laboratory in California. He found that, when the total energy cycle for each technology was considered and compared according to unit energy output, coal and oil again had the highest risk. Natural gas was lowest, with comparatively little air pollution compared with other fossil fuels. Nuclear energy was somewhat higher. Most renewable energies were in between gas and coal and substantially higher than nuclear. The risks associated with mining, refining, and fabricating the materials for constructing renewables are relatively high.

Although the study was controversial, and surprising to some advocating renewable energy forms, the overall results have been confirmed by a number of independent researchers (Hauptmanns and Werner, 1991; Habegger, 1981). Although energy systems are chosen for a variety of reasons—economic, political, national goals—risk should play a part in their selection.

17.2.7 RISKS AND BENEFITS

The subject of risk assessment is often divorced from the benefits the activity in question confers. However, virtually all risks have a benefit. (Hammond and Coppock, 1990; Sagoff, 1985; Sharma,

1990; Pochin, 1983). Exceptions might be those risks produced by nature, such as hurricanes and tornadoes. But those who live in areas where these occur enjoy or accept the overall climate. Even cigarette smoking, obviously a risky pastime, confers some benefits by way of relaxation.

The literature is replete with discussions on whether risks should be considered without regard to other factors or related to a "risk-benefit" comparison. The question may never be resolved.

Risks and benefits have different units. Risks are often specified as so many deaths or diseases; benefits are more difficult to quantify. For example, one of the activities mentioned in the risk perception study noted above was skiing. There is clearly a feeling of exhilaration while shooting down a slope, but how can this be measured? Economists sometimes get around this question by estimating how much was spent on the activity. If a skier spends $200 a day on hotels, meals, and passes, it is presumed that the benefit is about $200. Then the risks—put on a dollar basis as the hospital and medical costs incurred per day—are compared with the benefits in money.

This methodology gives a broad, although possibly distorted, relationship of risk to benefits. However, some have approached the question in terms of risk reduction: What is the most cost-effective way to reduce risk? Cohen (1990) and others have compiled tables showing how a life can be saved for an average expenditure from a few hundred dollars to billions. In general, the low-cost interventions are related to vaccines, which tend to be relatively cheap. The most expensive life-saving systems are associated with government regulations in the nuclear and chemical industries, where elaborate and expensive rules reduce risk by small fractions. The wide range of expenditures from the most to the least cost-effective measure—about six orders of magnitude—suggests that risk reduction priorities are set haphazardly.

17.2.8   HOW SAFE IS SAFE ENOUGH?

Because risk assessments are usually tied to public policy and regulation, they are often accompanied by limits described as "safe," "tolerable," or similar adjectives (Schwing, 1980; U.S. Congress, 1994). Because these adjectives may have legal consequences, they are considered as "limits."

The implication is that doses below these limits are safe and above these limits are harmful. For example, the International Commission on Radiological Protection, and many national health physics agencies, has established the limit for radiation workers to be 5 rem (a unit of radiological dose) per year. Does this imply that a worker who receives 4.9 rem each year for a number of years is "safe," and another who receives 5.1 rem for 1 year is unsafe? If the concept of strict limits is correct, then the answer is yes. If there is a gradation of risk, then the answer is no.

In few if any instances can a definitive answer to this question be given. Then the question "How safe is safe enough?" does not have a precise answer. In reality, dose and pollutant limits are a consensus of risk assessment science, media concern, and politics. These three components combine in unknown proportions and interact in enigmatic ways. Although there have been many case studies of how standards and limits are set, no single universal model of standard setting has been adopted.

Standards are often set by analyzing the data and adding conservative safety margins. For example, a study of data on a pollutant from laboratory animals (assumed to be extrapolatable to humans) may suggest a limit of Y parts per million in food, in excess of which few if any deleterious effects are observed. Risk assessors may reduce this by a factor of a 100, giving a limit of $Y/100$. If the pollutant attracts media attention, policy makers may impose an additional factor of 10 for safety, yielding a limit of $Y/1000$ parts per million. Hence what was believed to be safe based on scientific grounds is presumably safer by a factor of 1000. Of course, there may be economic costs in reaching the safer level.

One of the purposes of this dose conservatism is to account for groups that might be more susceptible to this pollutant than the general population. These include fetuses, children, the elderly, or women. Some racial and ethnic groups, such as Blacks and Ashkenazi Jews, are at greater risk of contracting certain diseases than the general population. Thus it is possible that a few groups are more reactive to pollutant Y. However, the reasoning behind the conservatism factors is rarely discussed. We generally do not know if we are being too conservative or not conservative enough.

Such discussions give rise to the concept of de minimis risk (Whipple, 1987). The term is taken from legal terminology and is based on the fact that law is not concerned with trifles. That is, a claim for damages for trivia will be dismissed, but a suit for damages from physical assault is addressed in the legal system.

De minimis is the level of risk below which there should be no concern. Resources such as money and personnel should then be assigned to riskier areas. However, the U.S. Nuclear Regulatory Commission has great difficulty applying de minimis to low-level nuclear waste. Some of this waste has radioactivity levels so low that it could be placed in standard landfills, instead of engineered repositories, without endangering the public in any way. However, concern about even this tiny risk forced the agency to continue requiring special handling of even minuscule amounts of radioactive materials.

The subject of "how safe is safe enough" then raises many questions with few answers.

17.2.9 RISK AVERSION

There are a wide range of attitudes toward risk, as shown by surveys and other indirect devices (Dyer and Sarin, 1981). Risk-seeking indidviduals may drink heavily, drive fast sports cars with unbuckled seat belts above the speed limit, smoke, go skydiving, climb mountains, and engage in other well-known risky behaviors. Risk-averse individuals read newspapers avidly for the next food or activity to avoid, try vegetarian diets, wash their vegetables and fruits twice, bundle up heavily for cold weather, eat an apple every day, take their medicine when prescribed, stay away from people with colds, and try to elude all risks.

To add to the complication, some people are risk-seeking for certain hazards and risk-averse in others—e.g., smoking while practicing vegetarianism. Most people are somewhere in the middle—they steer clear of obvious risks but do not dodge all hazards.

Policy makers must account for both ends of this spectrum. If most people were risk seeking, there would be little need for government regulation of industry or any other activity. If most were risk averse, virtually all economic activity would halt. Ironically, this probably would increase societal risk dramatically.

Although many risk calculations are publicly available, there is little evidence that those who are risk averse and risk seeking make use of these data. Risk-averse people are unlikely to take more risks if they are told that they are concerned about infinitesimal hazards. Conversely, risk seekers know that climbing mountains is risky without information from a study. The inclination to join either group is a function of personality and not of the results of risk assessments they have read.

The risk averse and risk seekers seem to coexist happily in most situations. One arena where they come together is when a waste site is being proposed nearby. Although risk estimates may indicate that the hazards of these sites are extremely small when regulations are satisfied, risk-averse as well as risk-seeking citizens generally show strong concerns at siting meetings. As a result, most proposed sites are either abandoned or take many years to complete.

Risk aversion and risk seeking must be considered by policy makers as well as the scientific data.

17.2.10 VOLUNTARY RISK

In the 1960s, Chauncey Starr was one of the first to draw the distinction between voluntary and involuntary risk (Starr, 1969). He noted, for example, that risks per hour for some voluntary activities such as mountain climbing were perhaps two or three orders of magnitude greater than involuntary risks, such as job-related hazards. This suggests that people are willing to accept much higher risks if they have some control over them.

However, although some activities, like motor car racing, are clearly voluntary and not subject to regulation, others fall in between. For example, job-related risks vary widely, from the relatively low dangers of office work to the much higher hazards of construction, logging, and mining. Occupations are chosen by different people with a mixture of motives, only partly influenced by risks. For example, a young person in West Virginia, knowing of coal mining disasters, may avoid that occupation on that basis. Another youth, having heard of the relatively high salaries in construction, may seek a job in that field and disregard both the statistics of high accident rates and any personal knowledge of those dangers.

Driving an automobile is clearly a voluntary activity, but in many parts of the United States where public transportation is deficient it is compulsory. The lines distinguishing voluntary and involuntary are vague. For example, people who have low incomes tend to drive vehicles that are less expensive, older, and with less modern safety equipment like air bags than those driven by the wealthy. This is an economic and not a voluntary choice.

So the distinction between voluntary and involuntary activities is less clear than policy makers would like. We have a spectrum, from someone subject to on-the-job hazards on one side to perhaps hang-gliding at the other extreme.

17.2.11   SUMMARY

It is clear, from the media and regulatory decisions, that risk assessment is partly science and partly society influencing that science. In areas involving the media, the public strongly affects the level of acceptable risk. In other cases not involving the media, risk assessment is almost completely a science. We cannot predict which will prevail. This section has attempted, albeit briefly, to illustrate risk assessment in its broader context.

BIBLIOGRAPHY AND REFERENCES

American Chemical Society and Resources for the Future, *Understanding Risk Analysis*, ACS and RFF, Washington, DC, 1998.

Cohen, B. L., *The Nuclear Energy Option: An Alternative for the 90s*, Plenum Press, New York, 1990.

Douglas, M., *Risk Acceptability According to the Social Sciences*, Russell Sage Foundation, New York, 1985.

Dyer, J. S., and Sarin, R. K., *Relative Risk Aversion*, University of Texas, Austin, 1981.

Franklin, J. (ed.), *The Politics of Risk Society*, Polity Press, Cambridge, UK. and Blackwell, Malden, MA, 1998.

Fullwood, R. R., and Hall, R. E., *Probabilistic Risk Assessment in the Nuclear Power Industry: Fundamentals and Applications*, Pergamon Press, New York, 1988.

Habegger, L., *Health Implications of New Energy Technologies*, International Atomic Energy, Vienna, 1981.

Hammond, B., and Coppock, R. (eds.), *Valuing Health Risks, Costs, and Benefits for Environmental Decision Making*, National Academy Press, Washington, DC, 1990.

Hauptmanns, U., and Werner, W., *Technische Risiken (Engineering Risks: Evaluation and Valuation)*, Springer-Verlag, Berlin, New York, 1991.

Heyman, B. (ed.), *Risk, Health and Health Care: A Qualitative Approach*, Arnold, London, 1998.

Inhaber, H., *Energy Risk Assessment*, Gordon & Breach, New York, 1982.

Leviton, L. C., Needleman, C. E., and Shapiro, M. A., *Confronting Public Health Risks: A Decision Maker's Guide*, Sage Publications, Thousand Oaks, CA, 1998.

Louvar, J. F., and Louvar, B. D., *Health and Environmental Risk Analysis: Fundamentals with Applications*, Prentice Hall, Upper Saddle River, NJ, 1998.

New York Times, August 26, 1979, p. 15.

Paustenbach, D. J., "The Practice of Health Risk Assessment in the United States (1975–1995): How the U.S. and Other Countries Can Benefit from that Experience," *Hum. Ecol. Risk Assess.*, 1(1): 29–79, 1995.

Pochin, E. E., *Nuclear Radiation: Risks and Benefits*, Clarendon Press, Oxford, Oxford University Press, New York, 1983.

Rogers, J. T., and Bates, D. V. (eds.), *Symposium on the Assessment and Perception of Risk to Human Health in Canada*, Ontario Science Centre, Toronto, 1982.

Rück, B. (ed.), *Risiko ist ein Konstrukt. (Risk is a Construct: Perceptions of Risk Perception)*, Knesebeck, Munich, 1993.

Sagoff, M., *Risk-Benefit Analysis in Decisions Concerning Public Safety and Health*, Kendall/Hunt, Dubuque, IA, 1985.

Schwing, R., and Albers, W. A. Jr. (eds.), *Societal Risk Assessment: How Safe Is Safe Enough?*, Plenum Press, New York, 1980.

Sharma, H. D. (ed.), *Energy Alternatives: Benefits and Risks*, University of Waterloo Press, Waterloo, Ontario, 1990.

Starr, C., "Social Benefit versus Technological Risk," *Science*, 165: 1232–1238, 1969.

Statesman (India), August 19, 1979, p. 7.

U.S. Congress, House Committee on Science, Space, and Technology, *How Safe Is Safe Enough? Risk Assessment and the Regulatory Process, July 27, 1993*, U.S. Government Printing Office, Washington, DC, 1994.

Warner, F., *Assessment and Perception of Risk: A Royal Society Discussion*, Royal Society, London, 1981.

Whipple, C. (ed.), *De Minimis Risk*, Plenum, New York, 1987.

Wildavsky, A. B., *Searching for Safety*, Transaction Books, New Brunswick, NJ, 1988.

Wildavsky, A., *But Is it True? A Citizen's Guide to Environmental Health and Safety Issues*, Harvard University Press, Cambridge, MA, 1995.

CHAPTER 17
SYSTEM LEVEL RISK ASSESSMENT

RISK MANAGEMENT IN VIEW OF ECONOMIC ASPECTS

Takehiko Murayama

Dr. Murayama is an associate professor in the School of Science and Engineering, Waseda University, Tokyo, Japan, where she specializes in decision making processes in risk management and communication.

17.3.1 INTRODUCTION

One of the characteristics of environmental pollution that has recently attracted attention is the environmental risk to the public arising from low concentration and long-term exposure. Although different kinds of risk-abatement measures exist, many of them require substantial expenditure of money and time. It is important, therefore, to establish a method for determining the most cost-effective abatement measures.

After setting out a basic framework for risk evaluation, I present a case study of asbestos contamination. It was not until 1987 that the Japanese government adopted abatement measures against asbestos pollution. However, it is difficult to determine whether the existing data allow us to estimate and evaluate the risk induced by asbestos pollution. Risk assessment and management are measures for resolving such environmental risk problems. This section provides a framework for evaluating several risks and evaluates the risk arising from asbestos pollution in the residential environment.

17.3.2 FRAMEWORK FOR RISK EVALUATION

Classification of Abatement Approaches in Environmental Risk Management

Various kinds of abatement measures are available to society for reducing environmental risks. These can be classified into the following four categories: liability [polluter pays principle (PPP)], public investment, insurance, and education. Each category appears to have a suitable field of implementation. Figure 17.3.1 describes a hypothetical relationship between risk abatement measures and the characteristics of risks. Risk is classified in two dimensions by reference to the choice of abatement measures. One dimension is the degree of voluntariness with which people take specific risks. This influences the extent to which people should control risks for themselves. If people take a risk completely voluntarily, they are definitely responsible for controlling the risk. For unexpected risks, however, people should be protected by abatement measures.

The other dimension is the ease of specifying risk sources or polluters. If people can easily find the risk source or polluter, they can implement abatement measures for that specific source or polluter.

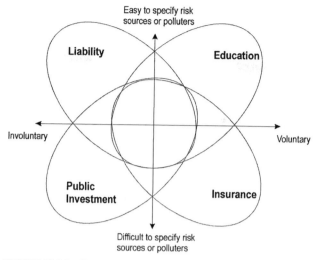

FIGURE 17.3.1   Characteristics of risk and abatement measures.

Conversely, if it is difficult for people to find the polluter, it is quite difficult for them to carry out effective measures against it. In such cases, they need to be protected by the other kinds of measure.

In the graphic presentation of the characteristics of risks, shown in Figure 17.3.1, four categories of abatement measures can be located in the four quadrants. For risks that are voluntarily incurred, insurance and education are suitable. Of these two measures, insurance is effective where it is difficult to find the specific source of the risk. For involuntary risks, liability and public investment measures are appropriate. In particular, PPP should be implemented for risks whose sources are clearly specified. In practice, it is difficult to classify risks in this manner, because each risk displays aspects of several characteristics. This subsection deals mainly with the approach categorized as public investment.

Classification of Evaluation Methods

The purpose of evaluating risk in this subsection is to reach conclusions about whether a risk should be accepted or whether measures should be taken to reduce a risk.

Figure 17.3.2 shows various methods of risk evaluation. Methods already in use may be divided into two groups: comparison with other risks and evaluation by reference to various factors, including benefits and costs. In the case of asbestos pollution, which I chose as a case study in this paper, I applied the following three methods: (1) comparison of risk magnitude, (2) comparison of risk and abatement cost, and (3) cost-effectiveness analysis among various risks. The third method, which combines features of the other two, has not been applied as widely.

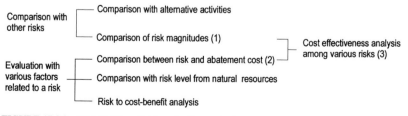

FIGURE 17.3.2   Classification of risk evaluation method.

Comparison of Risk Magnitudes

Comparison of risk magnitudes is often undertaken as a useful evaluation method. When we analyze pollution risks in a residential environment, it is desirable to choose the risk to which we are commonly exposed. We should therefore exclude labor accidents from the risks to be analyzed. In this paper I use scales to display the harm done to humans. The annual death rate and lifetime death rate can be shown on such scales. The expected value of days lost because of a risk is a more comprehensive indicator of harm to humans than these two scales (Ban, 1988).

These scales give a certain indication of the measures that can be taken to deal with risks, but they suffer from certain shortcomings. First, even if the magnitude of a risk appears low on these scales, a collection of low-level risks can amount to a considerable total risk. Second, these scales do not take account of abatement costs. Bearing these shortcomings in mind, I examined the following two methods.

Comparison Between Risk and Abatement Cost

When the presence of a risk becomes clear after the activity that causes the risk, we have no choice over the activity itself and should consider what measures to take. In this case, the desirable condition is given by the expression

$$\text{Max}[B - (C + D)]$$

where B is benefit, C is abatement cost, and D is expected damage in monetary value.

In many cases B does not change over time. When B does not change at all, the most effective measure is to minimize $C + D$. In practice, there are hardly any measures whose cost we can continuously vary. Therefore, we often examine measures that impose a specific abatement cost. In particular, when we have little room for choice over abatement measures, we compare them with two cases where $C = 0$ and $D = 0$.

Comparison of Risk-Reduction Effect Per Unit of Cost

In this analysis, I calculate an abatement cost and a risk to be reduced by the abatement on each risk and compare effectiveness per marginal cost. I consider that this method is the most suitable for choosing the most effective measures, given limited resources. There are several studies of this method in Japan and a few international studies such as Schwing (1982). I consider only death caused by risks. A series of C_i, which maximizes the number of deaths avoided $(\sum m_i)$ where total costs (C_0) are fixed, constitutes the optimum allocation of total costs. This allocation is an equilibrium point of the LaGrange function:

$$L = \sum f_i(C_i) + \mu \left(C_0 - \sum C_i \right)$$

where $f_i(C_i)$ is equal to m_i, and a differential function on C_i. As a result, the optimum allocation of total costs is accomplished when the number of avoided deaths per marginal cost $k = \partial f_i(C_i)/\partial C_i$ is equal to the same value (μ) on each risk.

In practice, we have few cases with many measures according to continuous cost, and a risk is often expressed in terms of its death rate. Therefore, I express k as reduced death rate (Δr_i) divided by marginal cost (ΔC_i) per capita targeted by an abatement measure (P) in Figure 17.3.3. As the gradient of a straight line shows a risk-reducing effect, k, the risk-reducing effect of risk A is bigger than that of risk B.

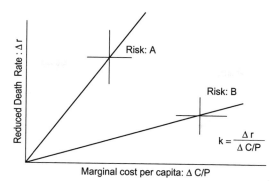

FIGURE 17.3.3   Relation between marginal cost per capita and reduced risk by pollution abatement.

17.3.3 COMPARISON OF MAGNITUDES BETWEEN ASBESTOS CONTAMINATION AND OTHER RISKS

Estimate of Risk Magnitude by Asbestos Contamination

I estimated future risk induced by asbestos contamination in a residential environment in Japan (Murayama and Harashina, 1989). The diseases chosen for this estimate are lung cancer and malignant mesothelioma.

The death rates of these two diseases I_L and I_M are shown by the following expressions:

$$I_L = I_E(1 + K_L f_d)$$

$$I_M = K_M f(t - 10)^3$$

where I_E is death rate from lung cancer in the control group, f is the asbestos concentration level, d is the exposure period, t is the period since exposure, and K_L and K_M are constants. After examining existing studies (U.S. Environmental Protection Agency, 1986; Ontario Royal Commission, 1984), I gave the following values to K_L and K_M:

$$K_L = 1.0 \times 10^{-2}(0.40 \times 10^{-3}\text{–}0.27 \times 10^{-2})$$

$$K_M = 1.0 \times 10^{-8}(2.6 \times 10^{-9}\text{–}3.9 \times 10^{-8})$$

With these values, I estimated the risk that would arise when the asbestos concentration remained at the present level. As a result, in a residential environment the lifetime death rate of males is estimated at 2.3×10^{-4} (3.3×10^{-5}–1.8×10^{-3}), and that of females is estimated at 2.6×10^{-4} (3.3×10^{-5}–2.1×10^{-3}). Values in parentheses are 90 percent confidence intervals. Lifetime death rates for risks other than asbestos contamination decrease as people age. The background death rate, which is used to calculate the excess death rate from lung cancer, is the rate in 1985.

Note that estimated values are accompanied by uncertainty arising from a number of factors. First, in extrapolating from high to low exposure, I assumed a linearity dose-response relationship; second, I gave the present-value background death rate of lung cancer (I_E); and third, in the absence of domestic data, risk is calculated on the basis of international epidemiological studies.

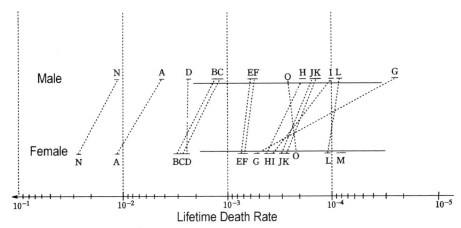

FIGURE 17.3.4   Comparison of risk magnitude between asbestos contamination and other risks (case of lifetime death rate). A: Motor vehicle; B: Accidental falls; C: Accidental drawing; D: Motor vehicle (pedestrian); E: Fire; F: Homicide; G: Shipping; H: Airplane; I: Bicycle; J: Railways; K: Mechanical suffocation; L: Contact with intense heat; M: Electrocution; N: Lung cancer; O: Asbestos pollution (pollution sources are not specified).

Comparison of Various Risks of Three Scales

Based on death rates in 1985 (Ministry of Health and Welfare of Japan, 1986), I compared risk magnitudes with the lifetime death rate (R_a) and lost days of life (R_d) (Figure 17.3.4). Let q_{0i} be the death rate from all causes in age group i, q_{1i} the death rate from each specific cause, p_i population, and L_i average remaining years (Ministry of Health and Welfare of Japan, 1987). This produces the following equations

$$R_e = \sum q_{1i} \exp\left[\sum_{j=0}^{i-1} \ln(1 - q_{0j})\right]$$

$$R_a = \sum_i (q_{1i} p_i) \Big/ \sum_i p_i$$

$$R_d = \sum_i L_i q_{1i} \exp\left[\sum_{j=0}^{i-1} \ln(1 - q_{0j})\right]$$

The male death rate tends to be higher than the female death rate. This tendency is marked in the case of accidents and disasters because males are more likely to be exposed to the risks associated with various occupations. However, as the asbestos concentration studied in this paper is in the residential instead of the occupational environment, the female risk level is more higher.

From these points of view, asbestos risk is almost equal to risks from bicycles, railways, or mechanical suffocation. Asbestos risk falls in terms of days of life lost, because death occurs predominantly at an advanced aged.

The annual death rate from asbestos concentration is estimated at 3.4×10^{-6}. This may be an acceptable level according to Crouch and Wilson's (1982) standard (10^{-5}/year) of risk acceptance, but the value of the upper bound (95 percent point of distribution; 1.5×10^{-5}) exceeds the acceptable level.

17.3.4   RISK AND ASBESTOS COST OF POLLUTION FROM SPRAYED ASBESTOS

Indoor Pollution Risk from Sprayed Asbestos

Of the various sources of pollution, sprayed asbestos pollution attracts the greatest public attention, probably because the indoor environment is airtight, and there is a great deal of sprayed asbestos on the ceilings of public facilities such as school classrooms and community centers. In particular, the death rate from malignant mesothelioma rises exponentially with years of exposure. Teachers and parent–teacher associations of primary and junior high schools are anxious about students' health. Based on the risk estimate model shown above, I calculated the risk for 10 to 14-year-old students who were exposed for 1 year (8 hours/day × 240 days) in a classroom with sprayed asbestos, with an asbestos concentration of 2.0 (range 0.64–6.3) fibers per liter. The estimated lifetime risk is 3.1×10^{-6} (7.0×10^{-7} to 2.2×10^{-5}), where values in parentheses show the 90 percent confidence limit with respect to two variables of asbestos concentration and the constants of the model. This is the value of the asbestos pollution risk used in subsections 17.3.5 and 17.3.6.

Abatement Methods

The three abatement methods that have been developed against asbestos pollution are asbestos removal, encapsulation, and enclosure. Unlike asbestos removal encapsulation and enclosure do not involve removing the sprayed asbestos. Asbestos removal is the most common method. The procedure is as follows: the environmental concentration in a working area is measured; the working area is insulated with a vinyl sheet; a sealant is applied to the surface of the sprayed asbestos; and workers wearing special clothes and respirators remove the asbestos fibers with spatulas. To understand how this method works, I investigated a working site (Figure 17.3.5) at 14:00 to 16:30 hours, 4 October 1988, in a public facility in Narashino City of Chiba prefecture.

With the encapsulation method, a sealant is applied to sprayed asbestos for controlling the dispersion of asbestos fiber. This is done in an early stage of abatement.

The enclosure method is used when asbestos materials are concentrated in a very specific location that is rarely visited by humans.

FIGURE 17.3.5   Working site of asbestos removal.

Estimation of Abatement Cost

It is difficult to establish a uniform cost for removal methods. To set an approximate standard, I interviewed six contractors and a local government that had already taken abatement measures. This resulted in a cost per square meter of working site of about 20,000 (15,000 to 25,000) yen. In particular, the Tokyo Metropolitan government paid about 20,000 yen per square meter to contractors in 1987 for abatement of sprayed asbestos in public high schools. Therefore, this is the typical cost of abatement at primary and junior high school classrooms.

Encapsulation and enclosure are estimated to be generally cheaper than removal, because the asbestos is not removed and no disposal is required. However, periodic inspection is required because the sprayed asbestos is still there and eventually needs to be removed. These two methods, therefore, incur a certain cost after they are implemented. This made it more difficult to estimate total costs.

17.3.5 COMPARISON BETWEEN A RISK AND THE ABATEMENT COST

Framework: Risk to Pupils and Students in a Classroom with Sprayed Asbestos

Suppose that N students are exposed in a classroom with sprayed asbestos. Then let $d(k)$ be the harm, in monetary terms, of a student who dies after k years. To convert this amount of harm into its value in the year when a measure is implemented, we need discount rate r. Then damage cost is:

$$d(k)/(1 + r)^k$$

When N students are exposed for a year, the total amount D is:

$$D = N \sum \{r(k)d(k)/(1 + r)^k\}$$

where $r(k)$ is excess death rate.

On the other hand, we need to convert the abatement cost into an annual amount. I calculated the annual cost C by the following expression:

$$C = C_0 \bigg/ \sum_{t=1}^{n} [1/(1 + R)t]$$
$$= C_0\{(1 + R)^n R/[(1 + R)^n - 1]\}$$

where C_0 is the abatement cost, n is residual depreciation years of a classroom at the time when a measure is implemented, and R is discount rate. For the removal method, the abatement cost is just the initial cost. I evaluated the effectiveness of this measure by comparing the abatement cost C with the cost of harm D.

Calculation of Harm

I defined the cost of harm as the profit that would have been made had the harm not been done. Methods of calculating this cost have been developed in the two research fields of damage compensation (method 1) and labor or health economics (method 2). Table 17.3.1 gives an outline, where $I(i)$ and $E(i)$ show annual income and living expense at the time i. The basic difference between the two fields depends on the method of calculating future wages: fixing it at the point of death (method 1) or varying it

TABLE 17.3.1 Conversion of Loss With Death into Monetary Value

Usually used field	Calculation formula	Formula author
Damage compensation	$\{I(0) - E(0)\} \sum \{1/(1 + kr)\}$	New Hoffman
	$\{I(0) - E(0)\} \sum \{1/(1 + r)^k\}$	New Reibniz
Labor and health economics	$\sum \{l(k) - E(k)\}/\{1/(1 + r)^k\}$	—

according to age (method 2). In this analysis, method 2 is more suitable because the object of this analysis is not a specific person but a group. However, it should be noted that wages can be increased by periodic salary rises and a boost to the wage base. The discount rate R, therefore, can be decreased by an increase in these two factors. Some authors have demonstrated this from the standpoint of both method 1 and method 2 (Gotoh, 1985; Kume, 1982).

I defined a lower than usual value of the discount rate as an adjusted discount rate and calculated the cost of the harm by the following three methods: (a) Let $r = r'$ in New Hoffman formula of method 1. (b) Let $r = r'$ in method 2. (c) Let $r = r' = 0$ in method 2.

The third of these methods gives the upper limit of the amount of damage.

Using the above-mentioned procedure, I calculated the amount of harm based on the average wage by age group (Ministry of Labor, 1988) and a standard for protection of livelihood as a living expense (as of 1987).

Relation Between Occurrence Rate of Harm and Cost of Harm by Age Group

The cost of harm by age group can be obtained by multiplying the harm cost by the death rate of each age group. Figure 17.3.6 shows this relationship. Harm cost shows a higher value in younger generations, but the cost multiplied by the death rate peaks with people in their 40s and 50s. In the case of females, this peak occurs at a slightly greater age.

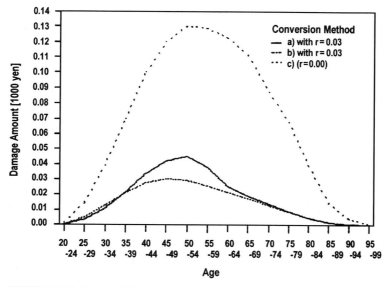

FIGURE 17.3.6 Loss multiplied by death rate (case of 1-year exposure at 10 to 14 years old).

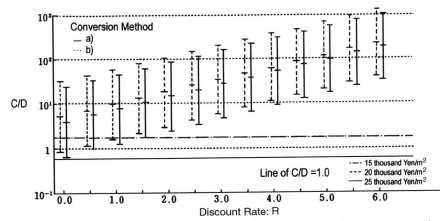

FIGURE 17.3.7   Ratio of loss to abatement cost by adjusted discount rate (case of 1-year exposure at 10 to 14 years of age). Note: Result by method c is equal to that by method b at an adjusted discount rate of 0.

Comparison Between Amount of Harm and Abatement Cost

To compare expected harm cost with abatement cost, I examined the adequacy of the measure against sprayed asbestos pollution. The classroom area is about 64 m², and the area covered by sprayed asbestos is about 55 m². Abatement cost is thus about 1.1×10^6 (8.25×10^6 to 1.38×10^6) yen. I converted this value into an annual cost with discount rate R. Figure 17.3.7 shows the ratio of abatement cost C to total damage cost D. The confidence limit is 90 percent. Judging from the median level in this case, C exceeds D even if the adjusted discount rate r is 0. On the other hand, judging from the upper estimation (95 percent risk), the C/D ratio is less than 1 when the adjusted discount rate is below 1.5 percent. Where age at exposure to asbestos is 5 to 9 years, C is almost in proportion to D with r' at about 2.0 percent, and the value of r' is about 1.0 percent at 15 to 19 years of age. These values therefore may be the internal rate of return used as a standard of public investment.

17.3.6   COMPARISON OF COST-EFFECTIVENESS AMONG VARIOUS RISKS

Object and Framework of Analysis

The risk in this subsection arises from indoor pollution caused by sprayed asbestos in classrooms of primary and junior high schools. For comparison, I chose automobile accidents involving pedestrians because both risks have passive characteristics. Two points should be noted about the characteristics of risk abatement and the effective period of cost per year. First, although the abatement of asbestos pollution is a fundamental method that removes the source of pollution, the abatement of pedestrians' traffic accidents treats symptoms without removing the source of the risk, the automobile. Second, although the risk reduced by the annual asbestos abatement cost is the lifetime death rate after exposure, the risk reduced by the abatement of traffic accidents involving pedestrians is the annual death rate.

Abatement Method for Pedestrians' Traffic Accidents

There are two sorts of abatement methods for pedestrians' traffic accidents. One is for drivers as assailants; the other is for pedestrians as sufferers.

TABLE 17.3.2   Outline of Device Installation Through 5-Year Projects for Traffic Safety

| | | | Depreciation years | Amount expenditure (10^8 yen) | | | | | |
| | | | | 1971 to 1975 | | 1976 to 1980 | | 1981 to 1985 | |
				National*	Local†	National	Local	National	Local
(1)	First grade	Sidewalk	30, 40, 50	1388.0	1184.3	1714.7	1682.8	1459.1	2395.5
		Bicycle Road	30, 40,50	449.0	114.0	2966.4	565.3	5416.0	1356.3
		Crosswalk	20, 30, 40	178.0	104.1	196.6	64.4	144.9	75.9
		Others	—	38.0	100.0	168.0	219.7	275.9	507.4
	Second grade	Lighting	3, 5, 10	28.0	198.1	48.8	261.5	47.1	315.6
		Fence	5, 10, 15	36.0	383.3	137.2	722.0	220.1	1041.5
		Road sign	—	—	—	216.9	123.6	0	464.0
		Other	—	168.0	220.3	258.4	475.7	227.9	720.7
(2)	Improvement of road		30, 40, 50	7227		24391		29248	

(1) Comprehensive Improvement Project for Traffic Safety Facilities.
(2) Five Year Project for Improvement of Road Environment.
* Value summed expenditure of each year.
† Budget estimated for each project of 5 years.

I chose traffic safety facilities falling under the jurisdiction of road administrators, because it was difficult to judge how measures for drivers contribute to reducing pedestrians' traffic accidents, and traffic safety facilities are the costliest measures for protecting pedestrians. Table 17.3.2 outlines a project to equip traffic safety facilities after 1971. Setting the depreciation years in each facility as shown, I converted the total amount into an annual cost. Suppose that each project was equally implemented in each annual period of 5 years, we can calculate the abatement cost during the 5 years by adding the cost of projects whose depreciation was effective in that period I also converted this cost into a present value with a deflator published by the Ministry of Construction so we can compare it with the abatement cost for asbestos pollution.

The number of traffic accidents involving pedestrians during these projects peaked in 1970, and, though tending to fall thereafter, appeared to stabilize. One of the factors may be the aging of the population group. Consequently, based on the age structure of the population group in 1985, I calculated the age-adjusted death rate with data by age group in each year from 1970 to 1986 (Figure 17.3.8). As shown in Figure 17.3.8, the decrease in the age-adjusted death rate is greater than that of the crude death rate.

Risk Reduced by Removing Sprayed Asbestos

As mentioned above, the total initial cost for the abatement of sprayed asbestos risk is 1.1×10^6 (8.3×10^5 to 1.4×10^6). I converted this value into an annual cost with depreciation years (in three cases of 20, 30, and 40 years) of the classroom after the asbestos had been removed. The risk reduced by this cost corresponds to each of three groups of 5 to 9, 10 to 14, and 15 to 19 years of age.

Comparison of Risk-Reducing Effect

Based on the findings from this procedure, I compared the effect of the abatement for sprayed asbestos pollution with that for traffic accidents involving pedestrians. Figure 17.3.9 shows a relationship

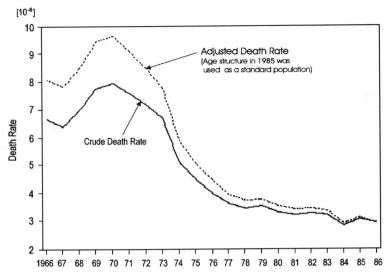

FIGURE 17.3.8   Annual change of pedestrian death rate from traffic accidents [Source: the Ministry of Health and Welfare (1967-1987)].

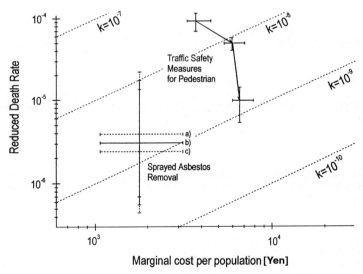

FIGURE 17.3.9   Effect of risk reduction of sprayed asbestos removal and pedestrian safety project.

between abatement cost per capita as an object of measures $\Delta C/P$ and reduced death rate Δr. As shown in Table 17.3.3, one of the characteristics is a decreasing effect of abatement methods for traffic accidents involving pedestrians. Moreover, the effect of abatement methods for traffic accidents involving pedestrians in 1981 to 1985 is approximately equal to that for asbestos exposure at age 10 to 14 years.

TABLE 17.3.3 Comparison Between Annual Death Rate and Risk Reduced by Each Measure

Abatement method	Annual death rate	Reduced risk k
Asbestos removal $(A)^*$	5.2×10^{-8}	1.7×10^{-9}
	$(1.0 \times 10^{-9}$ to $2.9 \times 10^{-7})$	$(1.8 \times 10^{-10}$ to $1.6 \times 10^{-8})$
Traffic safety measures for pedestrian $(B)^\dagger$	3.1×10^{-5}	1.5×10^{-8}
	$(2.8 \times 10^{-5}$ to $3.3 \times 10^{-5})$	$(6.7 \times 10^{-10}$ to $2.4 \times 10^{-9})$
A/B	1.7×10^{-3}	1.1
	$(3.0 \times 10^{-4}$ to $1.0 \times 10^{-2})$	$(7.5 \times 10^{-2}$ to $2.4 \times 10^1)$

* Case of 1-year exposure at 10 to 14 years old.
† Calculated from the value during of 1981 to 1985.

Table 17.3.3 compares risk-reducing effects with risk magnitudes. Although the median value of risk induced by sprayed asbestos is 10^{-3} times lower than that of pedestrian-related traffic accidents in terms of annual death rates, the two are approximately equal in terms of the risk-reducing effect. The result shows that abatement for sprayed asbestos pollution is also a worthy candidate for allocation of public resources.

Discussion

Mossman et al. (1990) evaluated the risk induced by sprayed asbestos in schools by comparing annual risk. They made the following points: the risk was lower than the other risks; asbestos materials that remained after incomplete removal may pollute a classroom; and asbestos workers may suffer greater risk than from other methods, such as encapsulation and enclosement. On the first point, the result is almost the same as that of Mossman et al. As shown in Table 17.3.3, the data reveal that the annual death rate from sprayed asbestos is about 1/1000th of that for pedestrians involved in traffic accidents.

On the second and third points, we do indeed need to pay attention to the risk arising from incomplete work. However, an on-the-spot survey of removal work suggests that the effectiveness of risk reduction is almost the same as that for pedestrians involved in traffic accidents, even if we incur the costs of minimizing the risk. Although Mossman et al.'s results cast doubt on the sprayed asbestos-removal method, we need to analyze the effect of the measurement from various standpoints such as cost-effectiveness.

17.3.7 CONCLUSIONS

I examined the effect of risk-evaluation methods in the case of asbestos pollution in a residential environment, and came to four conclusions. (1) Classifying risk-evaluation methods, I formulated methods risk evaluation by three methods: simple comparison of risk magnitudes, balance of abatement cost and reduced risk in monetary terms, and cost-effectiveness of risk abatement measures. (2) The risk arising from asbestos pollution in a residential environment is approximately equal to the risks arising from railway or bicycle accidents. (3) The method of removing sprayed asbestos in the classroom was effective because, in terms of risk-reducing effects, it was equal to that of traffic safety facilities in reducing risks of pedestrians involved in traffic accidents. (4) It is important to compare not only risk magnitudes but also risk-reducing effects per unit of abatement cost. To examine the usefulness of the method adopted in this paper, it may be applied to topics in the future such as smoking, radioactive contamination, and radon radiation.

REFERENCES

Ban, T., "A Study on 'Index of Harm' for Public by Radiation," Master's thesis, Graduate course of medicine, University of Tokyo, 1988.

Crouch, E. A. C., and Wilson, R., *Risk/Benefit Analysis*, Billinger Publishing Company, Massachusetts, 1982.

Gotoh, T., *Modern Theory on Compensation*, Nihon Hyouronsya, 1985.

Inui, S., and Tokumoto, C., *Basics of Illegal Act Law, Practice Edition*, Seirinsyoinshinsya, 1979, pp. 268–289.

Kume, M., *An Analysis and Prediction on Wage and Retiring Allowance in an Aging Society*, Sangyo Roudoh Tyousasho, 1982.

Maeda, N., *Economics of Health*, Daigaku Shuppan Kai, Tokyo, 1979.

Ministry of Health & Welfare of Japan, *Vital Statistics in Japan*, 1967 to 1987.

Ministry of Health & Welfare of Japan, *The 16th Life Tables*, 1987.

Ministry of Labor of Japan, *Statistics Survey on Wage Structure*, 1988.

Mossman, B. T., Bignon, J., Corn, M., Seaton, A., and Gee, J. B. L., "Asbestos: Scientific Development and Implications for Public Policy," *Science*, 247: 294–301, 1990.

Murayama, T., and Harashina, S., "An Estimation of the Risk by Asbestos Pollution in Residential Environment," *Kankyo joho kagaku*, 18(2): 81–90, 1989.

Ontario Royal Commission, *Report of the Royal Commission on Matters of Health and Safety Arising from the Use of Asbestos in Ontario*, Vols. I, II, and III, 1984.

Prime Minister's Office of Japan, *White Paper on Traffic Safety*, 1971 to 1988.

Schwing, R. C., *Longevity Benefits and Costs of Reducing Various Risks*, AAAS Selected Symposium Series, Vol. 65, Westview Press, Colorado, 1982, pp. 259–280.

U.S. Environmental Protection Agency, *Airborne Asbestos Assessment Update*, 1986.

CHAPTER 17
SYSTEM LEVEL RISK ASSESSMENT

SECTION 17.4
WEIGHING THE RISKS OF REMEDIATION

David Jeffrey
Dr. Jeffrey is senior risk assessment scientist with SECOR International, Inc. He performs and manages health risk assessments, as well as fate and transport evaluations in support of risk assessments.

All too often, decisions to remediate a site are made without consideration of the risks associated with the remedial activities themselves. Appropriate risk management decisions require that the risks or hazards from potential long-term exposures to relatively low levels of site chemicals be weighed against the financial cost of remedial options (i.e., the cost-benefit analysis) and the risks associated with the remedial options.

There are two types of risks associated with remedial activities: chemical risks and physical risks. Chemical risks generally take the form of short-term exposures to high levels of chemical contaminants, which often are airborne (e.g., dust emissions from soil excavation activities). Such exposures may affect remediation workers as well as offsite (i.e., downwind) receptors. Physical risks include those associated with operation of heavy machinery—for example, backhoe accidents—as well as potential traffic accidents involving long-distance transport of excavated soils to disposal locations.

The risks associated with many remedial activities often are significantly higher than those typically estimated in a health risk assessment. In fact, Jeffrey and Sheehan (1990) found that, for a scenario involving the excavation of chromium slag-containing soils from multiple sites in Newark and Kearney, New Jersey, and the out-of-state transport of the excavated soils, the probability of serious injuries or fatalities from the excavation and transport activities was many times the worst-case estimates of risks from chemical exposures.

Estimating risks from remedial activities requires making assumptions about the exact nature (and duration) of the remedial activities and access to accident statistics for the operation of heavy machinery. One can make assumptions about the nature and duration of remedial activities by contacting remedial design engineers and by estimating, for example, the total volume of soil to be excavated or treated. The total volume may be estimated by using risk-based target cleanup levels and by considering the three-dimensional extent of the contamination. Once the total volume is estimated, one can develop a reasonable remediation activity scenario by considering factors such as the excavation rate and capacity of backhoes and the maximum capacity of large dump trucks. Accident statistics involving the operation of heavy machinery, as well as transport-related accident probabilities, can be obtained from federal organizations such as the National Safety Council (NSC, 1989: used in the 1990 Jeffrey and Sheehan study) and the Bureau of Labor Statistics (BLS, 1989: used in the 1990 Jeffrey and Sheehan study). State agencies also compile this type of information (e.g., CDIR, 1988: used in the 1990 Jeffrey and Sheehan study). No doubt the Internet is a good source of this type of information today. Another good source of this type of information are actuarial divisions of insurance companies and insurance industry organizations. Unfortunately, because of the proprietary and sensitive nature of this information, it can be difficult to obtain useful data from insurance sources.

REFERENCES

American Society for Testing and Materials, *Standard Guide For Risk-Based Corrective Action Applied at Petroleum Release Sites*, Designation E 1739-95, American Society for Testing and Materials, West Conshohocken, PA, November 1995.

ASTM, *ASTM Draft Guide for Remediation by Natural Attenuation at Petroleum Release Sites*, American Society for Testing and Materials, March 8, 1996.

ATSDR, *U.S. Public Health Service Agency for Toxic Substances and Disease Registry's Toxicologcal Profiles on CD-ROM*, CRC Press, Boca Raton, FL, 1997.

BLS, *Occupational Injuries and Illnesses in the United States by Industry, 1987*, U.S. Department of Labor, Bureau of Labor Statistics, Bulletin 2328, May 1989.

CDIR, *1988 California Work Injuries and Illnesses*, California Department of Industrial Relations, July 1988.

CRC, *Handbook of Chemistry and Physics*, 71st edition, edited by D. R. Lide, CRC Press, Boca Raton, FL, 1992.

des Rosiers and Lee, "PCB Fires: Correlation of Chlorobenzene Isomer and PCB Homolog Contents of PCB Fluids with PCDD and PCDF Contents of Soot," *Chemosphere*, 15(9–12): 1313, 1986.

Dragun, *The Soil Chemistry of Hazardous Materials*, Hazardous Materials Control Resources Institute, Silver Spring, MD, 1988.

Great Lakes, *U.S. Environmental Protection Agency 40 CFR Parts 9, 122, 123, 131, and 132*, Final Water Quality Guidance for Great Lakes System, 1999.

Howard, P. H., Boethling, R. S., Jarvis, W. F., Meylan, W. M., and Michanlenko, E. M., *Handbook of Environmental Degradation Rates*, Lewis Publishers, Chelsea, MI, 1991.

Jeffrey, D., and Sheehan, P., *Hazard Analysis Protocol for Remediation of a Hudson County [New Jersey] Chromium Site*, McLaren/Hart Environmental Engineering, Inc., ChemRisk Division, prepared for a confidential client, September 13, 1990.

Jeffrey, D. *Chemical Property Variability, Quantitative Uncertainty Analysis, and Consequences in Risk Assessment*, at American Society of Testing and Materials Committee E-47 (Biological Effects and Environmental Fate) Conference, Seattle, WA, April 21, 1999.

Kim, and Hawley, *Re-entry Guidelines, Binghampton State Office Building*, Bureau of Toxic Substances Assessment, Division of Environmental Health Assessment, New York State Department of Health, Albany, NY, July 1985.

Larson, R. A., and Weber, E. J., *Reaction Mechanisms in Environmental Organic Chemistry*, Lewis Publishers, Boca Raton, FL, 1994.

Lydy, M. J., Lohner, T. W., and Fisher, S. W., "Influence of pH, Temperature, and Sediment Type on the Toxicity, Accumulation, and Degradation of Parathion in Aquatic Systems," *Aquatic Toxicol.* 17: 27–44, 1990.

Lyman, W. L., Reehl, W. F., and Rosenblatt, D. H., *Handbook of Chemical Property Estimation Methods*, McGraw-Hill, New York, 1990.

Mabey, W., and Mill, T., "Critical Review of Hydrolysis of Organic Compounds in Water Under Environmental Conditions," *J. Phys. Chem. Ref. Data*, 7(2): 383–415, 1978.

Mackay, D., Shiu, W. Y., and Ma, K. C., *Illustrated Handbook of Physical-Chemical Properties and Environmental Fate for Organic Chemicals; Volume I: Monoaromatic Hydrocarbons, Chlorobenzenes, and PCBs*, Lewis Publishers, Chelsea, MI, 1992.

Mackay, D., Shiu, W. Y., and Ma, K. C., *Illustrated Handbook of Physical-Chemical Properties and Environmental Fate for Organic Chemicals; Volume III: Volatile Organic Chemicals*, Lewis Publishers, Chelsea, MI, 1993.

MDEQ, *Operational Memorandum No. 10: Presentation of Tier 2 and Ier 3 Groundwater Modeling Evaluations*, Michigan Department of Environmental Quality, Underground Storage Tank Division, November 4, 1997.

Milby et al., "PCB-Containing Transformer Fires: Decontamination Guidelines Based on Health Considerations," *J. Occup. Med.*, 27(5): 351, 1985.

Montgomery, J. H., *Groundwater Chemicals Desk Reference*, 2nd edition, CRC Press, Boca Raton, FL, 1996.

Ney, R. E. Jr., *Where Did That Chemical Go? A Practical Guide to Chemical Fate and Transport in the Environment*, Van Nostrand Reinhold, New York, 1990.

NIOSH, *Polychlorinated Biphenyls (PCBs): Potential Health Hazards from Electrical Fires or Failures*, National Institute of Occupational Safety and Health Current Intelligence Bulletin 45, February 24, 1986.

NSC, *Accident Facts*, National Safety Council, 1989.

Quirke, J. M. E., Marci, A. S. M., and Eglington, G., "The Degradation of DDT and its Degradative Products by Reduced Iron(II) Porphyrins and Ammonia," *Chemosphere*, 3, 151–155, 1979.

Schmidt, C. E., and Zdeb, T. F., *Direct Measurement of Indoor Infiltration Through a Concrete Slab Using the USEPA Flux Chamber*, Unpublished, 1999.

Schwarzenbach R. P., Gschwend, P. M., and Imboden, D. M., *Environmental Organic Chemistry*, Wiley, New York, 1993.

USEPA, *Measurement of Gaseous Emission Rates From Land Surfaces Using an Emission Isolation Flux Chamber, Users Guide*, EPA Environmental Monitoring Systems Laboratory, Las Vegas, NV, EPA Contract No. 68-02-3889, Work Assignment No. 18, February 1986.

USEPA, *Superfund Exposure Assessment Manual*, U.S. Environmental Protection Agency, Office of Remedial Response, Washington, DC, EPA/540/1-88/001, April 1988.

USEPA, *Risk Assessment Guidance For Superfund, Volume I, Human Health Evaluation Manual (Part A), Interim Final*, U.S. Environmental Protection Agency, Office of Emergency and Remedial Response, Washington, DC, EPA/540/1-89/002, December 1989.

USEPA, *Soil Screening Guidance: User's Guide*, U.S. Environmental Protection Agency, Office of Solid Waste and Emergency Response, Washington DC, Publication 9355.4-23, July 1996.

USEPA, *Region IX Preliminary Remediation Goals (PRGs) 1998*, memorandum from S. J. Smucker, USEPA Region IX, San Francisco, CA, March 1998.

USEPA, *Integrated Risk Information System*, U.S. Environmental Protection Agency on-line database: http://www.epa.gov/iris/subst/index.html, 1999.

Wade and Woodyard, "Sampling and Decontamination Methods for Buildings Contaminated with Polychlorinated Dibenzodioxins," in *Solving Hazardous Waste Problems: Learning from Dioxins,* edited by J. H. Exner, American Chemical Society Symposium Series 338, Washington, DC, 1987.

Woodward, R. B., and Hoffmann, R., *The Conservation of Orbital Symmetry*, Academic Press, New York, 1970.

CHAPTER 18

ENVIRONMENTAL SCIENCE IN THE LEGAL SYSTEM

Rolf R. von Oppenfeld

Managing partner of THE TEAM for ENVIRONMENTAL SCIENCE and TECHNOLOGY LAW (TESTLaw), operating under the law firm von Oppenfeld, Hiser & Freeze, P.C., Mr. von Oppenfeld practices primarily in the areas of environmental law and litigation, natural resources and water law, and toxic tort litigation. He has an AV rating from Martindale Hubbell and has been listed under the category of "Environmental Law" in The Best Lawyers in America (1993–96 editions).

SECTION 18.1

ENVIRONMENTAL SCIENCE, TOXIC TORTS, AND THE LEGAL SYSTEM

18.1.1 INTRODUCTION

Because of recent developments in toxic tort litigation, scientific issues are becoming more prevalent in legislative debate, in agency rulemaking and enforcement decisions, and in the courtroom. These developments have highlighted the importance of understanding scientific evidence. For example, various facets of toxic tort claims hinge on the validity of expert testimony as to whether exposed individuals are at a significantly increased risk of contracting cancer or other illnesses. Therefore, plaintiffs and defendants often employ exposure experts to strengthen or refute claims that exposures to toxins have resulted in illness or led to an increased risk of illness. These experts use various techniques, such as exposure modeling and risk assessment, to provide their audience, who may have limited or no scientific background, with the tools to assess causation and analyze risk evidence. In the administrative context, the U.S. Environmental Protection Agency (EPA) relies heavily on environmental science data to establish regulations. For example, under the Clean Water Act, the agency has established maximum contaminant levels for drinking water. Because of the complex nature of determining the amount of risk associated with various contaminant levels and determining exposure, development of these regulations often requires the agency to work in concert with the National Academy of Sciences.

Similar concerted efforts are found throughout the legal system whether it is in an administrative, legislative, or judicial context. However, despite the increasing reliance on science and expert testimony, the legal system has yet to find a successful way to deal with risk assessments, exposure modeling, and issues of causation.

How does Environmental Science Apply to the Legal System?

Environmental science touches many facets of the legal system, including the following (Rossi, F. F., *Expert Witnesses*, American Bar Association, 1991):

1. *Administrative enforcement proceedings*, during which appropriate environmental agencies seek compliance with certain environmental statutes and regulations;

2. *Contribution actions*, during which parties who have undertaken the responsibility of cleaning up contamination, for which they are not wholly responsible, seek contribution from other parties who are partially responsible for the contamination;

3. *Toxic tort cases*, during which an injured party claims that another party is responsible for his or her injury or illness and that the illness or injury was caused by exposure to some level of contamination;

4. *Legislative debate*, during which legislators often rely on risk assessments and exposure models to determine the extent of risk associated with a particular policy decision;

5. *Property damage cases*, during which a property owner may claim that another party's toxic chemicals have created a nuisance, trespassed onto his or her property, or caused his or her property to decrease in value; and

6. *Administrative rule development*, during which administrative agencies rely heavily on environmental scientific data to establish regulations to ensure the protection of human health and the environment.

However, because the number of toxic tort lawsuits have proliferated at an alarming rate in recent years in the United States, this chapter primarily focuses on the effect of environmental science in the toxic tort arena.

Among the more prevalent substances that form the basis for this litigation are Agent Orange, arc welding fumes, asbestos, polychlorinated biphenyls (PCBs), trichloroethylene (TCE), and medical devices (*e.g.*, intrauterine devices). The magnitude of toxic tort litigation is illustrated by the fact that over 100,000 claims have been filed in the past 20 years in state and federal courts by victims who have contracted one of the many asbestos-related diseases and over 200,000 claimants have asserted injury due to the Dalkon shield intrauterine device [Henderson, T., "Toxic Tort Litigation: Medical and Scientific Principles of Causation," *Am. J. Epidemiol.* 132(Suppl.): 869–878, 1992].

Furthermore, the press fervently publicizes these lawsuits; therefore, much public attention is drawn to illnesses and injuries such as cancer, birth defects, and other diseases caused by chemical exposure. In addition, epidemiologists, toxicologists, oncologists, and other scientific doctors continually report new developments linking these diseases and chemicals that are readily available in our commercial and industrial communities. In response to these new reports by the media and scientists, lawmakers increasingly require industry participants to report and publish information about chemical releases and emissions; at the same time, these industry participants continue to develop more sophisticated equipment and techniques with which they can measure the chemicals in even more minute amounts. Although the press and public react with alarm and concern to the reports by scientists, lawmakers, and industry representatives, they generally do not have the knowledge to understand the intricacies of the report data, thus the need for environmental science and scientific evidence.

What is the Purpose of Scientific Evidence in Toxic Tort Litigation?

Generally, toxic tort litigation involves claims that exposures to hazardous substances have either caused serious injury or illness among plaintiffs or placed them at increased risk of developing an illness. Furthermore, toxic tort claims involve a great number of variables, including the following (Muchmore, C., and Sorenson, G., *What is a Toxic Tort?*):

1. The large number (thousands) of toxic substances;
2. The various known and unknown effects of these toxic substances;
3. The various media through which these toxic substances are released or handled;
4. The number of individuals exposed to the toxic substance;
5. The uniqueness of each individual's response to a toxic exposure;
6. The potential for lengthy latency periods causing delayed manifestations of injury;
7. The various legal avenues for recovery (common law, statutory, and regulatory);
8. The difficulty in documenting the exposure (single versus multipathway); and
9. The difficulty in determining and proving causation.

Expert evidence is presented in toxic tort causes of actions in an attempt to alleviate the overwhelming confusion presented by the abovementioned concerns. This evidence is provided through the presentation of technological standards, causation analyses, injury determinations, and damage assessments; thus, these cases are usually replete with expert scientific and medical testimony. This expert testimony is often a source of greater confusion for juries, judges, and attorneys who have little or no scientific background. Therefore, a toxic tort lawyer must become an expert in several scientific disciplines and be thoroughly familiar with the scientific data and studies in order to exclude as much of the opponent's scientific evidence as possible and to admit as much of his or her client's supporting scientific evidence as possible. (Pagliaro, J., and Benton, A., "Courtroom Science: Toxic Tort Battleground," *Toxics L. Rptr.*, 1(42): 1336, March 22, 1989).

Despite the long list of variables that pose difficult problems in toxic tort litigation, the most difficult and complex issue is the notion of causation. Because of long latency periods before the manifestation of injuries, often limited scientific knowledge about the nature of those injuries, uncertainty about the mechanism of action of many toxins, questions about amounts and durations of exposure, and many other issues, toxic tort claims historically have not been successful (Dore, M., "Causation," *in Law of Toxic Torts: Litigation Defense Insurance*, Vols. 1–4, edited by Clark Boardman Callaghan, West Publishing, St. Paul, MN, 1997, chapt. 24). Ideally, to demonstrate that a given toxin was a link in the causal chain that led to an individual's illness or injury, one needs to trace every step in the biology of the development of the disease, including the role played by the toxin (Green, M., "Expert Witnesses and Sufficiency of Evidence in Toxic Substances Litigation: The Legacy of Agent Orange and Bendectin Litigation," *Northwestern Law Rev.*, 86(3): 643–699, 1992). Because the biological mechanisms of most diseases are understood only rudimentarily at best, other processes are required to infer causation. In the absence of direct evidence, scientific methods and experimentation are used. Marino, A.A., and Marino, L.E., ("The Scientific Basis of Causality in Toxic Tort Cases," *Dayton Law Rev.*, 21: 1, 1995; Thompson, M.M., "Causal Inference in Epidemiology: Implications for Toxic Tort Litigation," *North Carolina Law Rev.*, 71: 247, 1992; Green, M., "Expert Witnesses and Sufficiency of Evidence in Toxic Substances Litigation: The Legacy of Agent Orange and Bendectin Litigation," *Northwestern Law Rev.*, 86(3): 643–699), 1992.

18.1.2   *THE ROLE OF ENVIRONMENTAL SCIENCE*

In the Courtroom

It is clear that toxic torts exist; the continual development and discovery of new human illnesses proves this. However, proof that these illnesses are caused by chemical exposure (i.e., causation) is too susceptible to the opinions of people who seek to profit from the connection. The media zealously broadcasts reports of these illnesses and their connection to chemical exposure; however, they fail to research the plausibility of the relationship between the illness and the exposure. Therefore, it is left up to the legal system—primarily the judicial system—to apply the principles of environmental science to determine liability for these illnesses and injuries. Judges, juries, and attorneys are acutely aware of their duties and the social consequences of the performance of those duties. Thus, the conflict between science and the law.

Judges. Scientific experts and opinions are usually presented by both litigants in order to prove or disprove a fact that is necessary to their position. However, because most judges are untrained in science and mathematics, it is not the court's purpose to determine which scientific expert is right. Therefore, the judge need not undertake a scientific analysis of the expert opinion; rather, it is necessary for the judge only to determine that the expert testimony is admissible under FRE 702 and 703 (i.e., that it is reliable, relevant, and will assist the trier of fact). However, this duty is extremely important and has recently been expanded by the U.S. Supreme Court to require the court also to ensure that the proffered expert testimony is based on "scientific knowledge" (*Daubert v. Merrell Dow Pharm., Inc.*, 509 U.S. 579, 1993). Therefore, the judge must act as a gatekeeper and allow into evidence only that testimony that is based on established scientific principles, is both reliable and relevant, and will ultimately assist the trier of fact (Federal Judicial Center, *The Reference Manual on Scientific Evidence*, 1994).*

Without the performance of this gatekeeping function, the trier of fact would be presented with a perplexing contradiction; if the trier of fact needs a scientific expert in order to understand the evidence or determine a fact in issue, how can that same trier of fact be expected to understand the evidence in issue to such great detail as to resolve a conflict between the competing scientific evidence of the experts. Therefore, the gatekeeping role of the judge serves as a prescreening tool; the judge first ensures that all expert testimony evidence is based on valid scientific principles, and thereafter the jurors are free to determine not which expert is right, but rather which expert to believe (Chong, E., "Thomas Kuhn and Courtroom Treatment of Scientific Evidence," *Temp. Envtl. L. Tech. J.*, 15: 195, fall 1996).

However, there are several alternatives to performing a strict gatekeeping function; in addition to determining whether scientific expert testimony is admissible, the court may:

1. Consolidate cases and coordinate discovery;

2. Create central information repositories;

3. Develop its own knowledge base of relevant scientific issues;

4. Appoint experts of the court or independent neutral evaluators; or

5. Refer particularly complex scientific issues to those entities equipped to handle such issues.

Juries. In recent years the judicial system has experienced a fundamental change in many of its civil cases; the explosion of toxic tort litigation has fueled this change. Many of these toxic tort cases involve multiple claims for similar injuries from multiple defendants. Several commentators have expressed concern about the competence of civil juries to render equitable verdicts in these highly complex cases, which are replete with scientific testimony. For example, Chief Justice Warren Burger was among the critics who suggested that "jurors lack the ability to deal with the complex issues often presented in federal civil trials" (Horowitz, I., and Bordens, K., "An Experimental Investigation of Procedural Issues in Complex Tort Trials," *Law Hum. Behav.*, 14(3): 269–285, 1990).

The predominance of probabilistic evidence and testimony in toxic tort litigation often may lead the jury to "misunderstand the nature of this evidence or place an inordinate amount of weight on its probative value" (Roth-Nelson, W., and Verdeal, K., "Risk Evidence in Toxic Torts," *Environ. Law.*, 2: 405–443, February 1996). For example, a jury may treat an expert's testimony and numbers as absolute truth and, although exposure modeling and its various methodologies rely heavily on statistical and mathematical relationships, including other demonstrative evidence, in assessing the degree or probability that the plaintiff was exposed to the chemical at issue, jurors may have difficulty understanding that probabilistic evidence merely establishes a statistical association of exposure in a

---

* This reference manual provides an invaluable tool for judges in the performance of their gatekeeping role; furthermore, it provides key concepts that may be useful to attorneys faced with the preparatrion of a scientific expert, including guidance on (1) management and admissibility of expert evidence; (2) application of specific technical fields, and (3) use of court-appointed experts and special masters.

particular population and not absolute truth. Mock jury studies, focusing on probabilistic evidence, have shown that jurors often fail to adequately process small probabilities; and furthermore, they "fail to distinguish between probabilities of 1 in 1000, 1 in 10,000, and 1 in 100,000" (Roth-Nelson, W., and Verdeal, K., "Risk Evidence in Toxic Torts," *Environ. Law.*, 2: 405–443, February 1996).

Through multimedia exposure modeling, a jury is presented with a multitude of questions and issues. These issues include:

1. Whether the plaintiffs were exposed to toxic chemicals;
2. If plaintiffs were exposed to toxic chemicals, then what were the durations of exposure;
3. What were the estimated doses of the toxic exposure; and
4. Which, if any, defendants are responsible for the exposures?

As a result, it is crucial for litigants to properly educate jury members about the implications of the scientific and exposure evidence at issue. Once a jury accepts a witness's explanation of scientific or technical concepts, the jury is more likely to also accept the expert's opinions. Therefore, properly preparing the expert is crucial to the jury understanding the expert's testimony. When presenting expert scientific evidence to a jury, counsel must present the witness as a reliable, knowledgeable person, from whom testimony will flow logically from the facts and will be easily understood by the jurors (Rossi, F., *Expert Witnesses*, American Bar Association, 1991). In addition, experts should be properly qualified about the issue for which they are presenting evidence; the persuasiveness of the expert's background should be maximized without boring the jury. For example, an exposure expert should provide background information on prior experience in exposure assessments, professional accomplishments, and education.

According to one commentator, "experts who calculate things present the most difficult demonstrative evidence problems," because many people have an aversion to math (Rossi, F., *Expert Witnesses*, American Bar Association, 1991). Furthermore, people with little scientific or mathematical background are generally uncomfortable dealing with abstract evidence such as formulas, estimations, and numerical representations. Therefore, it is necessary for exposure experts to attempt to reduce their modeling presentations to understandable terms and everyday language. Thereafter, it should be much easier for the jury to follow and understand the expert's opinions (Rossi, F., *Expert Witnesses*, American Bar Association, 1991).

For example, if an expert presents testimony that employs highly technical language and fails explain the testimony to the jury, the expert will appear to lack candor, and answers to questions may appear to be unresponsive. On the other hand, if experts are willing to admit that certain facts brought out on cross-examination are true but are also able to explain why those facts do not alter their conclusions, jurors will find them more believable and trustworthy. Furthermore, it is advantageous for experts to be able to simplify most of the highly technical information to a level that can be understood by jurors and to use lay language and anecdotal manners in presenting scientific studies (Partain, E., Hamlin, W., and Jones, L., "A Defendant's Verdict in a Dioxin Exposure Case: O' Dell v. Hercules, Inc.," *Toxics Law Rep.*, 2(50): 1402, May 18, 1988) (explaining that, whereas many substances used everyday are carcinogenic, most of them are used in such small doses that they are not considered harmful).

To better understand how a juror's perceptions and predispositions shape toxic tort litigation, numerous jury consultants have collected quantitative and qualitative evidence over the past two decades from actual cases as well as mock trials. This evidence suggests that a defendant's argument that a substance is not toxic is likely to be futile, because most jurors have a preconceived notion that unfamiliar chemicals present health risks and, further, that the defendant would better spend his or time trying to minimize the perceived degree of exposure (interview of G. Speckart, Bodaken Associates, Los Angeles, CA, 1997). Furthermore, additional studies indicate that 90 percent of jurors believe that there is no safe level of exposure to some toxic chemicals, and 75 percent of those jurors believe that repeated exposure to most chemicals will result in cancer (interview of N. Bensko, Starr Litigation Services, Inc., Phoenix, AZ, 1997).

These studies also indicate that jurors place a great deal of faith in government standards of exposure levels; 80 percent of jurors believe that "any level of exposure to a chemical deemed hazardous by the United States government is too much exposure" (interview of N. Bensko , Starr Litigation Services,

Inc., Phoenix, AZ, 1997). Because plaintiffs often rely on governmentally imposed dose standards in toxic tort litigation, this preconceived juror perception can be a powerful hurdle for a defendant to overcome, regardless of whether the plaintiff is alleging an enormous or a relatively minimal exposure to a toxic substance.

In Regulatory or Agency Decisions

The federal government regulates the environment through the EPA and its interpretation of various statutes relating to air and water pollution, waste disposal, and hazardous and toxic substances. Included among these statutes are the Clean Air Act, the Clean Water Act, the Toxic Substances Control Act, the Resource Conservation and Recovery Act, and the Comprehensive Environmental Response, Compensation, and Liability Act (CERCLA) (Relkin, E., "The Sword of the Shield: Use of Governmental Regulations, Exposure Standards and Toxicological Data in Toxic Tort Litigation," *Dickinson J. Environ. Law Policy*, 6: 1–30, 1997). In addition, risk assessments are necessary in order to carry out the regulations provided in the above-mentioned statutes. For example, risk assessments are preformed at CERCLA sites in order to:

1. Determine the necessity of remedial action;
2. Determine the level of chemicals that may safely remain onsite;
3. Provide a basis for comparing the effects of various remedial activities;
4. Provide a consistent process for evaluating the potential threat or impact of contamination; and
5. Produce results that are helpful in the cleanup of the site.

Although these goals are achieved in several ways, the primary technique used in CERCLA activities is a remedial investigation and feasibility study (RI/FS). An RI/FS is the process through which the nature and extent of the risks posed by uncontrolled contamination are characterized and evaluated in order to establish the most effective form of remedial action. Because an RI/FS, or any risk assessment technique used under another environmental statute, can be very complicated, several sources have developed guidance documents to provide consistent and effective application of the federal regulations (Committee on Risk Assessment and Risk Management, *Risk Assessment and Risk Management in Regulatory Decision-Making*, Draft Report, June 13, 1996; National Environmental Policy Institute, *Science-Based Risk Assessment: A Piece of the Superfund Puzzle*, 1995; Zimmerman, R., *Governmental Management of Chemical Risk: Regulatory Processes of Environmental Health*, Lewis, 1990; Executive Office, *Risk Analysis: A Guide to Principles and Methods for Analyzing Health and Environmental Risks*, 1989; EPA, *Workshop Report on EPA Guidelines for Carcinogen Risk Assessment: Use of Human Evidence*, Sept. 1989; EPA, *Exposures Factors Handbook*, July 1989; EPA, *The Risk Assessments Guidelines of 1986*, Aug. 1987).

On the federal level, most regulatory management is performed by the EPA, which participates in a range of activities, including, but not limited to:

1. Drafting regulations;
2. Mandating testing;
3. Effecting right-to-know laws;
4. Utilizing pollution prevention strategies;
5. Initiating voluntary programs; and
6. Educating and informing the media and public about risks.

At the local and state level, environmental regulation may occur either independently or in conjunction with federal law. Furthermore, toxic substances and pollutants may be regulated by other agencies such as the Occupational Safety and Health Administration. Consequently, the same chemical may have different levels of safe exposure according to the context in which it is being regulated

(Relkin, E., "The Sword of the Shield: Use of Governmental Regulations, Exposure Standards and Toxicological Data in Toxic Tort Litigation," *Dickinson J. Environ. Law Policy*, 6: 1–30, 1997). Based on varying exposure limits, a government agency's assessment of a chemical product may "almost provide an incentive to bring toxic tort litigation" (Thompson, A., and Stegemann, T., "Current Hazard Identification Programs: Potential Societal and Regulatory Consequences," *Toxics Law Rptr.*, 8: 417, September 8, 1993).

In Legislative Discussions and Rulemaking

Before a very few years ago, risk assessments were crudely done at both the state and federal level as a means of providing lawmakers with a simple determination of the extent of risk. However, this determination was usually presented without communicating the level of uncertainty or without the lawmakers requesting such information. Therefore, it is unlikely that the risk caused by the particular legislation on which they were debating was given adequate consideration, thus, the need for environmental science and scientific principles.

However, the use of these inadequate risk assessments and exposure models is coming under increasing scrutiny by the media and therefore the public; federal and state legislators have not been blind to these public inquiries. Legislators recently have begun to redirect their focus and opinions about risk assessments and exposure management; for example, several pieces of legislation have been drafted to specifically deal with these issues. Furthermore, it is unlikely that any environmental legislation will pass through either the House or the Senate without major debate about the extent of risk and the adequate level of safe exposure.

In furtherance of these objectives, the Clinton administration developed a comprehensive set of principles for regulatory reform. This Executive Order requires a thorough analysis of risk, cost, and benefit when faced with major regulatory actions (Executive Order no. 12,866, 58 Fed. Reg. 51,735, 1993; Applegate, J., "A Beginning and Not an End In Itself: The Role of Risk Assessment in Environmental Decision-Making," *Risk Symp.*, 63 U. Cin. L. Rev. 1643, summer 1995). Following this precedent, the White House Office of Science and Technology Policy presented flexible guidance for risk assessment, management, communication, and setting priorities. In addition, the EPA issued further risk assessment guidance, and the legislature took several shots at developing regulatory reform legislation (EPA, *Policy for Risk Characterization at the EPA*, 1995; Science Policy Council, *Guidance for Risk Characterization*, EPA, 1995; EPA, *Implementation Program for the EPA Policy on Risk Characterization*, 1995; Simon, R., "Issues in Risk Assessment and Cost-Benefit Analysis and Their Regulatory Reform," *Risk Symp.*, 63 U. Cin. L. Rev. 1611, summer 1995; McElveen, J., "Legislating Risk Assessment," *Risk Symp.*, 63 U. Cin. L. Rev. 1553, summer 1995).

However, the complex techniques and procedures directed by these risk assessments and exposure management guidelines soon may prove to become a major barrier to protecting the public from dangerous exposure to toxic chemicals. Therefore, to assist concerned decision-making efforts, the legislature should seek to draft and adopt risk assessment principles that are fair, effective, affordable, and, most of all, flexible and adaptable to new advancements in science (Goldman, L., "Environmental Risk Assessment and National Policy: Keeping the Process Fair, Effective, and Affordable," *Risk Symp.*, 63 U. Cin . L. Rev. 1533, summer 1995).

18.1.3   *PRESENTING ENVIRONMENTAL SCIENCE*[*]

Plaintiffs and defendants in litigation are confronted on a regular basis with issues about the relevance, admissibility, and utility of scientific evidence. Toxic tort litigation is no exception, and the

---

[*] This subsection focuses on application of environmental science in the judicial capacity of the legal system. However, each of the concepts presented can be easily applied in an administrative and legislative capacity (i.e., rather than presenting evidence to a judge and jury, the evidence would be presented to regulatory department officials, legislative members, etc.)

complexities associated with establishing causation, injuries, and responsible parties in these cases add to the confusion. Various methodologies are currently used in the courtroom by litigants presenting exposure models and the assumptions associated with these methodologies. However, to date, "almost all human health risk assessments have used conservation 'point' estimates to characterize the hazards associated with exposure to chemicals in the environment" (Finley, B., and Paustenbach, D., "The Benefits of Probabilistic Exposure Assessment: Three Case Studies Involving Contaminated Air, Water, and Soil," *Risk Anal.*, 14(1): 53–73, 1994). The methodology used by an exposure expert can have a profound effect on the expert's conclusions and the credibility of the conclusions. For example, conservative point estimates of exposures may produce data that demonstrate that the plaintiff has shown the requisite level of exposure; however, the use of more sophisticated probabilistic analyses may produce drastically different data about plaintiff's exposure to pollutants, including the amount and duration contributing to the incidence of disease.

Furthermore, there is inevitably a certain degree of doubt about how well an exposure model or its mathematical expression (e.g., air dispersion models, water distribution models, etc.) approximates the true relationships between site-specific environmental conditions. In an ideal situation, the use of a fully validated model that considers the interrelationships of every parameter is preferred. However, only rudimentary, partially validated models are customarily used. Consequently, one must identify various assumptions associated with exposure models such as linearity, homogeneity, and steady-state relationships in order to identify how these factors affect exposure and risk estimates (EPA, *Risk Assessment Guidance for Superfund: Human Health Evaluation Manual*, part A, 1989, chaps. 6, 8).

Through Expert Scientific Testimony

One of the major factors that distinguishes toxic tort litigation from other forms of litigation is its complexity. Typically, the outcome of toxic tort litigation is based on the effectiveness of expert testimony in assessing exposure to hazardous substances and causation. Rarely can a single expert address all the necessary scientific issues. Toxic tort cases often encompass "widely divergent disciplines such as epidemiology, toxicology, hydrology, meteorology, and medicine" (Rossi, F., *Expert Witnesses*, American Bar Association, 1991). Experts employed in toxic tort cases may range from highly respected, well-known experts, who can offer opinions within the mainstream of science, to experts whose credentials are dubious and whose theories lack general scientific support.

Typically, experts involved in toxic tort cases include:

1. Environmental engineers, who determine whether, and at what level, the plaintiffs were exposed to the toxic chemical in issue; and

2. Environmental and occupational health doctors, who evaluate the current health and health history of the plaintiff to determine whether the toxic chemical in issue caused or contributed to the plaintiff's illness.

However, these environmental engineers and environmental and occupational health doctors may be referred to by their specialty title, such as hydrologist, geologist, etiologist, or oncologist.

Traditional Evidentiary Standards. According to traditional evidentiary standards, experts scientific testimony is allowable when (Bell, P., "Strict Scrutiny of Scientific Evidence: A Bad Idea Whose Time Has Come (Part II)," *Toxics Law Rptr.*, Jan. 29, 1992):

1. The expert possesses a background that gives him or her special knowledge about the applicable subject matter;

2. The testimony would be helpful to a trier of fact; and

3. The expert's testimony is based on facts, data, or methods relied on by other experts in the field.

It is frequently through expert opinion testimony that exposure modeling evidence is presented to juries. Individuals, whose expertise has been approved by the court, typically will explain their opinion about the exposure at issue. The plaintiffs and the defendants both present experts of their own. In the event of different opinions, which is virtually always, the jury must decide what weight to attach to each expert's testimony.

The expert opinion method of educating juries about the complexities and interpretations of exposure evidence seems sensible. However, problems arise because not all experts hold the same standard of scientific validity or ethics. In other words, some litigants are willing to pay large sums of money for experts with sufficiently impressive credentials to say favorable things about the litigant's position. Unfortunately, some experts can be induced by such financial incentives to say remarkable things, including things not well-founded in science. An attorney has two ways to respond to such testimony presented by opposing experts; the attorney may either appeal to the judge to exclude the expert's testimony or undermine the expert's testimony in the minds of the jury by cross-examination and contrary expert opinion.

For many years, courts admitted expert opinion only if it was based on scientific methods generally accepted as reliable in the relevant scientific community (Foster, K., Bernstein, D., and Huber, P., *Phantom Risk*, MIT, London, 1993). This standard is known as the Frye "general acceptance" test and it prohibits introduction of expert opinion evidence based on methodology that diverges significantly from the procedures accepted by recognized authorities in the field (*Frye v. United States*, 293 F. 1013, D.C. Cir., 1923). Using this test, litigants can prevent "lone nut" scientists from presenting unique theories of evidence to juries. However, the rule also can prevent sound opinions from being heard in court simply because they are based on new theories.

For nearly half a century, *Frye* served fairly well to exclude unreliable and unconventional evidence from courtrooms (Foster, K., Bernstein, D., and Huber, P., *Phantom Risk*, MIT, London, 1993). However, in the 1960s and 1970s, the rule came under attack from lawyers who viewed the rule as "elitist and unhelpful" in complex cases involving new pollutants and unfamiliar hazards. Critics argued that alleged toxic tort victims should not be denied compensation just because the plaintiff's offer of proof did not meet the standards of acceptance by a broader scientific community (Foster, K., Bernstein, D., and Huber, P., *Phantom Risk*, MIT, London, 1993).

Modern Evidentiary Standards.   Over the next decade, several commentators suggested numerous factors that should be considered to determine the admissibility of expert testimony (Weller, *"Analysis and Perspective: Expert Evidence,"* Toxic L. Rep., 13(10): 338, BNA, Aug. 5, 1998; Weinstein, J., and Berger, M., *Weinstein's Evidence*, 1988; *United States v. Downing*, 753 F.2d 1224, 3rd Cir., 1985; statement by Berger, M., *Symp. Sci. and the Rules of Evidence*, 99 F.R.D. 187, 1983; McCormick, "Scientific Evidence: Defining a New Approach to Admissibility," 67 Iowa L. Rev. 879, 1982). Several of these factors were either repeated verbatim or were substantially similar to another. Therefore, to unite all the themes presented by these commentators, in 1993 the Supreme Court undertook the responsibility to attempt to lessen the conflict between science and the law (*Daubert v. Merrell Dow Pharm., Inc.*, 509 U.S. 579, 1993). Because an expert's facts and opinions may be considered only if they would be admissible under the Federal Rules of Evidence (FRE), the Supreme Court began with an analysis of FRE 702 and 703, which provide that, before the court can allow the proffered expert testimony, the court must first determine that the testimony rests on scientifically valid principles and then determine that the testimony will assist the trier of fact in understanding the evidence or in determining a fact in issue. FRE 702 provides that:

[i]f scientific, technical or other specialized knowledge will assist the trier of fact to understand the evidence or to determine a fact in issue, a witness qualified as an expert by knowledge, skill, experience, training or education may testify thereto in the form of an opinion or otherwise.

However, if the court finds that the proffered expert's testimony is so fundamentally unsupported that it will not assist the trier of fact, then the court should exclude the testimony (*Akzo Coatings v. Aigner Corp.*, 881 F. Supp. 1202, 1213, N.D. Ind., 1994).

FRE 703 provides that an expert may base an opinion or inference on facts or data not admissible in evidence but only if such facts or data are of a type reasonably relied upon by experts in the particular field in forming opinions or inferences upon the subject; whether there is an adequate basis on which to support the expert's testimony is a matter of law for the court to decide (*Akzo Coatings v. Aigner Corp.*, 881 F. Supp. 1202, 1213, N.D. Ind., 1994). Therefore, the court must act as a gatekeeper, policing the introduction of expert testimony, in order to ensure that the proffered testimony is not only based on reliable scientific principles but that it ultimately will assist the trier of fact in rendering a decision.

The court in *Daubert*, in its analysis of FRE 702 and 703, found that scientific refers to the methods and procedures of science, whereas knowledge refers to something more than mere speculation or speculative beliefs (i.e., known facts or any body of ideas inferred from such facts or accepted as truths on good grounds). Therefore, the court held that a judge must also apply a scientific knowledge analysis before allowing the admission of expert scientific testimony. The scientific knowledge approach thus eradicated several of the more confusing common law doctrines, such as *Palsgraf* and *Frye*. However, *Daubert* did not dispense of the *Frye* general acceptance requirement; instead it relegated it to a nondeterminative factor, one to be considered among many. Those factors, as announced by the U.S. Supreme Court in *Daubert*, include the following:

1. Whether the theory or technique can be and has been tested;
2. Whether the theory or technique has been subjected to peer review and publication;
3. Whether the known or potential rate of error is acceptable; and
4. Whether the theory or technique has widespread acceptance, instead of only minimal support, within the scientific community.

Therefore, scientific expert testimony must not only satisfy the standards of FRE 702 and 703, it also must satisfy the *Daubert* scientific knowledge test; the court must determine, before expert scientific testimony is presented, that the proffered evidence is scientifically valid, that it is reliable and relevant, and that it will aid the trier of fact (Roisman, A., "The Courts, Daubert, and Environmental Torts: Gatekeepers or Auditors?," 14 *Pace Environ. Law Rev.*, 545, summer 1997).

The federal courts continue to interpret *Daubert*. For example, 2 years after that decision, the Court of Appeals for the Third Circuit reviewed and applied the *Daubert* standard to scientific expert testimony (*In re Paoli R.R. Yard PCB Litigation*, 35 F.3d 717, 3rd Circuit Court, 1994). The Third Circuit Court held that *Daubert*, combined with the FRE , required the judge to conduct a preliminary determination about whether the reasoning or methodology underlying the proffered expert testimony was based on science and, if so, whether those scientific techniques or methodologies were reliable. Therefore, *Paoli* reemphasized the need for exclusion of that scientific testimony that is not reliable and valid.

Shortly after, *Paoli*, the Supreme Court revisited the *Daubert* standard (*General Electric v. Joiner*, 139 L. Ed. 2d 508, U.S. Supreme Court, 1997). However, the court did not expand on *Daubert's* admissibility factors; instead it focused on the appropriate standard of review for admissibility of expert evidence. The U.S. Supreme Court held that the appellate courts have less, not more, discretion in reviewing a lower court's decision on the admissibility of expert testimony. Thus, instead of allowing appellate courts to take a "hard look," as suggested by the Eleventh Circuit Court, *Joiner* limits the standard of review of the appellate courts to that of an "abuse of discretion" standard (Roisman, A., "The Implications of *G.E. v. Joiner* For Admissibility of Expert Testimony," <http://www.vje.org> (visited September 15, 1998). The court defended its conclusion by finding that, because *Daubert* requires a more detailed expert report, establishing the methods used and the principles applied, it is less likely that either the plaintiff or the defendant will proffer an expert who cannot support his opinions. Therefore, appellate courts have little incentive to vigorously review, much less overturn, a lower court's decision on the admissibility of scientific evidence.

Thereafter, the Court of Appeals for the Fifth Circuit found that the expert testimony of an ecologist was inadmissible because the ecologist could not show that his knowledge of the contamination at two Superfund sites was based on any reliable evidence as required under *Daubert*. (*Barrett v. Atlantic Richfield Co.*, 95 F.3d 375, 5th Circuit Court, 1996). Rather, the ecologist merely demonstrated that he had conducted limited studies on cotton rats, which did not involve an analysis of their possible toxic exposure at the sites, but instead involved an analysis of the behavioral patterns of the rats and their interaction with the environment. The Fifth Circuit Court later found that the trial court properly excluded expert testimony under *Daubert* where plaintiff's experts offered nothing more than a suggested association between the chemical exposure and plaintiff's injuries (*Allen v. Pennsylvania Engineering, Corp.*, 102 F.3d 194, 5th Circuit Court, 1996). Furthermore, the court held that the expert's evidence, which was based on animal studies, was inconclusive at best, because the expert failed to provide a statistically significant link between the chemical exposure and human illness or injury.

In addition, the Court of Appeals for the Seventh Circuit applied the *Daubert* analysis and found that the expert testimony of plaintiff's toxicologist was inadmissible because the expert had neither examined the plaintiff or the plaintiff's medical records nor reviewed the alleged facts relating to the exposure (*Wintz v. Northrop Corp.*, 110 F.3d 508, 7th Circuit Court, 1997). Furthermore, the court found that the expert's qualifications were limited to degrees in chemistry and environmental engineering; he was not a physician and therefore was not qualified to present an opinion about the cause of the plaintiff's injuries.

A federal district court in West Virginia also found that expert testimony about the psychological effects of a toxic chemical was inadmissible under *Daubert* because the expert used sloppy methodologies and, furthermore, was so biased toward his client that the evidence was tainted (*Black v. Rhone-Poulenc Inc.*, No. 2: 96–0613, S.D., West Virginia, July 24, 1998). The Ninth Circuit Court also excluded expert testimony based on the unsupported conclusions of the expert. In *Lust*, the expert proposed to introduce evidence that there was a positive correlation between ingestion of a drug and the plaintiff's particular birth defect; however, the expert had studied only the correlation between the use of the drug and other birth defects (*Lust v. Merrell Dow Pharmaceuticals Inc.*, No. 95-55558, 9th Circuit Court, July 11, 1996).

However, many other courts have found that testimony of the litigant's scientific experts is admissible. For example, a U.S. District Court found that the associate medical director's testimony could be presented regardless of the fact that he did not know the actual level of chemical exposure (*Louderback v. Orkin Exterminating*, No. 97-1370-WEB, D. Kan. October 14, 1998). The Third Circuit came to a similar conclusion in *Kannankeril v. Terminix International Inc.*, finding that an expert could present his conclusions despite the fact that he was unable to present evidence of actual exposure levels (128 F.3d 802, 3rd Circuit Court, 1997). In addition, the Washington, DC, Circuit Court twice reversed a lower court's decision, which excluded the scientific opinion of the plaintiff's expert witness and held that the lower court misconceived the limited gatekeeper role when it excluded expert testimony that was clearly admissible under *Daubert* (*Ambrosini v. Labarraque*, 101 F.3d 129, Washington, DC, Circuit Court, December 6, 1996; see also, *Ferebee v. Chevron Chemical Co.*, 736 F.2d 1535, Washington, DC, Circuit Court, cert. denied, 469 U.S. 1062, 1984; see *Richardson v. Richardson-Merrell, Inc.*, 857 F.2dn 823, Washington, DC, Circuit Court, 1988, cert. denied, 493 U.S. 882, 1989).

The U.S. Supreme Court recently revisited *Daubert* to clarify yet another application problem—whether the *Daubert* admissibility factors are applicable to expert testimony that is not based on scientific principles. In *Kumho Tire Co. v. Carmichael*, 119 S. Ct. 1167, March 23, 1999, the U.S. Supreme Court concluded that the general gatekeeping obligation set out in *Daubert* applies not only to testimony based on scientific knowledge but also to testimony based on technical and other specialized knowledge. Specifically, the court held that a trial court may consider one or more of the more specific *Daubert* factors to determine the testimony's reliability. This decision settled a split among courts; those courts that held that *Daubert* extended to nonscientific testimony, (*Watkins v. Telsmith*, 121 F.3d 984, 5th Circuit Court, 1997; *Peitzmeier v. Hennessy Indus.*, 97 F.3d 293, 8th Circuit Court, 1996), cert. denied, 117 S. Court. 1552, 1997; *Moore v. Ashland Chemical Inc.*, No. 95-20492, 5th Circuit Court, August 14, 1998, en banc), and those courts that held that *Daubert* was wholly inapplicable when the expert does not claim to be a scientist (*Kumho Tire Co. v. Carmichael*, 131 F.3d 1453, 11th Circuit Court, 1997, cert. granted, No. 97-1709, U.S., June 22, 1998; *McKendall v. Crown Control Corp.*, 1997 U.S. App. LEXIS 21035, 9th Circuit Court; *Compton v. Subaru*, 82 F.3d

1513, 10th Circuit Court, 1996; *Iacobelli Constr. Co. v. County of Monroe*, 32 F.3d 19, 2nd Circuit Court, 1994); *Tamarin v. Adams Caterers, Inc*, 13 F.3d 51, 2nd Circuit Court, 1993).

Federal Rules of Evidence.   In response to questions and confusions about the general acceptance theory espoused in *Frye* and followed for almost a half-century, Rule 702 of the FRE was developed in 1975. In relevant part, it provides: "if scientific, technical, or other specialized knowledge will assist the trier of fact to understand the evidence or to determine a fact in issue, a witness qualified as an expert by knowledge, skill, experience, training, or education, may testify thereto in the form of an opinion or otherwise."

However, in 1993, the U.S. Supreme Court officially overruled the *Frye* test and held that the FRE superseded *Frye* and that proposed experts may testify to *any* scientific knowledge that will assist the trier of fact. The court further suggested that a trial court would need to make a "preliminary assessment of whether the reasoning or methodology underlying the testimony is scientifically valid and of whether that reasoning or methodology properly can be applied to the facts in issue" (*Daubert v. Merrell Dow Pharmaceuticals*, 113 S. Court 2786, 1993).

Daubert suggests that the methodology or technique underlying a scientific statement can be evaluated on the basis of its susceptibility to empirical testing, the peer review it has received, its publication, its rate of error, the existence of standards controlling its operation, and its general acceptance in the relevant scientific community. Thus, under *Frye*, an attorney who wants to exclude expert scientific testimony would present evidence to the judge that the opinion the expert was going to express was based on science not generally accepted by his or her peers. Under *Daubert*, an attorney must delve directly into the substantive merits of the expert's methodology in order to undermine the scientific validity of the expert's proposed testimony sufficiently to have it excluded from trial. This new standard requires greater scientific sophistication on the part of attorneys than did the *Frye* rule, and it is due, in large part, to the adoption of the FRE.

On the other hand, it should be noted that *Daubert* merely construes the FRE; thus, it is not directly binding on the *state* courts. However, because many states have adopted the *Frye* test, *Daubert* may affect numerous state standards for admission of expert scientific testimony and evidence. Because of this uncertainty, the U.S. Supreme Court heard arguments in 1997 about the standard of review that federal appellate courts must give to lower court decisions on the admissibility of scientific evidence (*General Electric Co. v. Joiner*, No. 96-188, 11th Circuit Court; Reuben, R., "Completing the Admissibility Equation," *Am. Bar Assoc. J.*, September 1997).

Similarly, *Joiner* did not provide much direction on the applicability of the FRE to the admissibility of expert testimony. Therefore, Justice Breyer has suggested that, in the aftermath of *Daubert* and *Joiner*, the courts may find it more advantageous to use court-appointed experts under FRE 706 (*General Electric Co. v. Joiner*, No. 96-188, 11th Circuit Court, concurrence, J. Breyer; "Justice Breyer Calls for Experts to Aid Courts in Complex Cases," New York Times, February 17, 1998, p. A17). Rule 706 was added to the FRE in 1975 in hopes of combatting the practice of expert shopping, presentation of corrupt experts, and the reluctance of reputable experts to participate in litigation; it was assumed that the mere threat of a court-appointed expert would help to reduce these problems (*Advisory Committee Notes to the Proposed Rules*, 1972).

However, this additional function of the court could create an undue expense of money and time. Therefore, instead of using FRE 706 to appoint an expert for the court, the court should follow the U.S. Supreme Court's precedent and look at FRE 702 and 703 and the *Daubert* admissibility factors in determining which experts, if any, to exclude. Any questions a court-appointed expert may answer can also be posed to the litigant's experts; therefore, the need for a court-appointed expert is superfluous. Furthermore, it is the nature of contested scientific evidence that different scientists will have different views; therefore, the idea that a neutral expert exists is wholly unsupported. To conserve time and money, judges should use their powers under FRE 702 and 703 to gain the necessary knowledge about the evidence in issue from the experts presented by the litigants instead of complicating the process further by appointing an expert for the court.

The proposed amendments to FRE 701, 702, and 703 provide a further complication to the role of the FRE in determining whether scientific expert testimony is admissible under the FRE and *Daubert* (available at http://www.uscourts.gov). However, these amendments have not been adopted; if they are adopted, the judicial system can expect another flurry of litigious activity.

Through Exposure Assessment/Causation Analyses

To assess causation of injury or illness in toxic tort litigation, one must first show that a person or persons were exposed to the toxin in question. Exposure is defined as the "*contact* of an organism with a chemical or physical agent" (EPA, "Proposed Guidelines for Exposure-related Measurements," 53 Fed.l Reg. 48830, December 2, 1988). The magnitude of exposure to an agent is determined by measuring or estimating the amount of the agent available at various exchange boundaries (i.e., lungs, gut, and skin) during a specified period of time. With an exposure assessment, one can determine or estimate the magnitude, frequency, duration, and route of exposure of an organism to various pollutants (EPA, *Risk Assessment Guidance for Superfund: Human Health Evaluation Manual*, Part A. 1989, chaps. 6, 8). These estimates can be generated directly from monitoring contaminant levels measured in the environment and in biological organisms or indirectly from modeling results or reasoned estimates ("Assessing Environmental Risks to Human Health," available at http://www.epa.gov/oppeinet/oppe/f...s/risk/roadmap/rmap/chap4.txt.html). It is important for litigants to understand the advantages and disadvantages of using monitoring and modeling data in the courtroom. Monitoring data are generally the preferred option, because modeling is usually based on various assumptions and may provide limited data for estimating the various concentrations to which people may have been exposed. Sophisticated mathematical modeling can be extremely resource intensive. For parties with limited financial resources, this can prevent an accurate assessment of the actual exposures encountered and severely hinder their case. However, litigants may encounter problems if they rely solely on monitoring data. Improper monitoring can inaccurately predict levels of human exposure. This is evident when exposure estimates are based on relatively few measurements ("Assessing Environmental Risks to Human Health," available at http://www.epa.gov/oppeinet/oppe/f...s/risk/roadmap/rmap/chap4.txt.html).

A toxic exposure assessment attempts to answer these questions: (Morgan, Lewis and Bockius, L.L.P., "Exaggerating Risk," *Hazardous Waste Cleanup Project*, Washington, DC, June 1993, pp. 1–45; Murray, T., "Exposure Assessment," available at http://www.epa.gov/oppt;ntr/cic/expose.htm).

1. How will a certain chemical (i.e., through which pathways) reach an individual;

2. How much of that chemical is the individual likely to be exposed to; and

3. How many individuals are likely to be exposed to the particular chemical?

There are a variety of ways to measure toxic exposure and its associated health effects, including examining human disease incidence rates in order to estimate likely exposure levels and simulating exposure with models that predict relative degrees of toxicity ("Exposure Assessments Based on Models Not Always Good Predictors, Scientist Warns," *Toxics Law Rep.*, 6(2): 839, December 11, 1991).

By evaluating the various components of exposure, an expert can construct reasonable scenarios for each problem area and further characterize exposures by different populations. When presenting a claim of injury to a jury, plaintiffs seek to maximize the number of exposures and the variables associated with exposure in order to strengthen their claims that such exposure caused an illness or increased the risk of illness. Conversely, defense counsel seeks to minimize such exposures and variables.

In toxic tort litigation, it is the role of the jury to sort through these various scenarios in rendering a verdict of whether exposure to the pollutant in question caused an injury or illness. It is imperative for attorneys, judges, and juries to have a working understanding of the exposure assessment and modeling process in order to competently and objectively view evidence in toxic tort litigation. By using exposure modeling and its mathematical formulae, litigants can depict exposure values to graphically illustrate exposure levels to a particular substance or pollutant. Examples of various exposure model formulas used in conducting exposure assessment are shown in Figures 18.1.1 through 18.1.4.

The following steps are necessary to complete and present an exposure assessment process:

Determining Exposure Pathways.   Initially it is important for litigants to understand the various ways a person or persons may be exposed to toxins. Plaintiffs may be exposed to toxins through a multitude of pathways, including ("Risk Assessment and Modeling Overview Document," available at http://www.epa.gov/grtlakes/arcs/EPA-905-R93-007/EPA-905-R93-007.html.):

1. *ingestion*, which may result from inadvertent consumption of contaminated soils or sediment or from consumption of drinking water, surface water, or wildlife;

2. *inhalation*, which may result from breathing in contaminated airborne vapors or dust particles;

$$\text{intake (mg/kg-day)} = \frac{CW \times IR \times EF \times ED}{BW \times AT}$$

where CW = chemical concentration in water (mg/L),

$\quad IR$ = ingestion rate (L/day),

$\quad EF$ = exposure frequency (days/year),

$\quad ED$ = exposure duration (years),

$\quad BW$ = body weight (kg), and

$\quad AT$ = averaging time (days over which exposure is averaged).

FIGURE 18.1.1   Residential exposure: ingestion of chemicals in drinking water and beverages made with drinking water (EPA, *Risk Assessment Guidance for Superfund: Human Health Evaluation Manual*, 1989, part A, chapt. 6). It should be noted that intake is often considered by the EPA to be synonymous with exposure in the sense that intake may be defined as a measure of exposure expressed as the mass of a substance in contact with the exchange boundary per unit body weight per unit time (e.g., mg of chemical/kg-day). Other terms often used to signify intake include normalized exposure rate, administered dose, and applied dose.

$$\text{intake (mg/kg-day)} = \frac{CS \times IR \times CF \times FI \times EF \times ED}{BW \times AT}$$

where CS = chemical concentration in soil (mg/kg),

$\quad IR$ = ingestion rate (mg of soil/day),

$\quad CF$ = conversion factor (10^{-6} kg/mg),

$\quad FI$ = fraction ingested from contaminated source (unitless),

$\quad ED$ = exposure duration (years),

$\quad BW$ = body weight (kg), and

$\quad AT$ = averaging time (days over which exposure is averaged).

FIGURE 18.1.2   Residental exposure: ingestion of chemicals in soil (EPA, *Risk Assessment Guidance for Superfund: Human Health Evaluation Manual*, 1989, part A, chapt. 6). It should be noted that intake is often considered by the EPA to be synonymous with exposure in the sense that intake may be defined as a measure of exposure expressed as the mass of a substance in contact with the exchange boundary per unit body weight per unit time (e.g., mg of chemical/kg-day). Other terms often used to signify intake include normalized exposure rate, administered dose, and applied dose.

$$\text{absorbed dose (mg/kg-day)} = \frac{CS \times CF \times SA \times AF \times ABS \times EF \times ED}{BW \times AT}$$

where CS = chemical concentration in soil (mg/kg),
  CF = conversion factor (10^{-6} kg/mg),
  SA = skin surface area available for contact (cm^2 per event),
  AF = soil of skin adherence factor (mg/cm^2),
  ABS = absorption factor (unitless),
  EF = exposure frequency (events per year),
  ED = exposure duration (years),
  BW = body weight (kg), and
  AT = averaging time (days over which exposure is averaged).

FIGURE 18.1.3 Residential exposure: ingestion of dermal contact with chemicals in soil (EPA, *Risk Assessment Guidance for Superfund: Human Health Evaluation Manual*, 1989, part A, chapt. 6). It should be noted that intake is often considered by the EPA to be synonymous with exposure in the sense that intake may be defined as a measure of exposure expressed as the mass of a substance in contact with the exchange boundary per unit body weight per unit time (e.g., mg of chemical/kg-day). Other terms often used to signify intake include normalized exposure rate, administered dose, and applied dose.

$$\text{intake (mg/kg-day)} = \frac{CA \times IR \times ET \times EF \times ED}{BW \times AT}$$

where CA = contaminant concentration in air (mg/m^3),
  IR = inhalation rate (m^3/hour),
  ET = exposure time (hours/day),
  EF = exposure frequency (days/year),
  ED = exposure duration (years),
  BW = body weight (kg), and
  AT = averaging time (days over which exposure is averaged).

FIGURE 18.1.4 Residential exposure: inhalation of airborne (vapor phase) chemicals (EPA, *Risk Assessment Guidance for Superfund: Human Health Evaluation Manual*, 1989, part A, chapt. 6). It should be noted that intake is often considered by the EPA to be synonymous with exposure in the sense that intake may be defined as a measure of exposure expressed as the mass of a substance in contact with the exchange boundary per unit body weight per unit time (e.g., mg of chemical/kg-day). Other terms often used to signify intake include normalized exposure rate, administered dose, and applied dose.

3. *dermal absorption*, which may result from direct contact of the skin with either contaminated sediments, river plain soils, or overlying water; or

4. a *combination* of these routes.

Activity patterns play a large role in determining the pathways of exposure. A matrix of potential exposure pathways is shown in Table 18.1.1 (EPA, *Risk Assessment Guidance for Superfund: Human Health Evaluation Manual*, part A, chapt. 6, 1989; "Assessing Environmental Risks to Human Health," available at http://www.epa.gov/oppeinet/oppe/f...s/risk/roadmap/rmap/chap4.txt.html).

TABLE 18.1.1   Matrix of Potential Exposure Routes (EPA, *Risk Assessment Guidance for Superfund: Human Health Evaluation Manual*, 1989, part A, chapt. 6). L, lifetime exposure; C, exposure in children may be significantly greater than in adults; A, exposure to adults (highest exposure is likely to occur during occupational activities); — , exposure of this population by this route is not likely to occur

Exposure medium/route	Residential	Commercial/industrial	Recreational
Groundwater			
Ingestion	L	A	—
Dermal contact	L	A	—
Surface water			
Incidental ingestion	L	A	L, C
Dermal contact	L	A	L, C
Sediment			
Incidental ingestion	C	A	C
Dermal contact	C	A	L, C
Air			
Inhalation of vapors			
Indoors	L	A	—
Outdoors	L	A	L
Inhalation of particulates			
Indoors	L	A	—
Outdoors	L	A	L
Soil/dust			
Incidental ingestion	L, C	A	L, C
Dermal contact	L, C	A	L, C
Food(ingestion)			
Fish and shellfish	L	—	L
Meat and game	L	—	L
Dairy	L, C	—	L
Eggs	L	—	L
Vegetables	L	—	L

When adjudicating a claim of injury or illness in a toxic tort case, plaintiffs and defendants must be aware of these pathway differences and of the physiological and biological factors associated with each. For example, the ingestion exposure pathways often result in higher exposure estimates than the dermal or inhalation pathways because of greater absorption of contaminants through the gastrointestinal tract compared with absorption through the skin and because of the relatively high levels of intake of contaminants in soil, water, and food compared with inhalation of contaminants ("Risk Assessment and Modeling Overview Document," available at http://www.epa.gov/grtlakes/arcs/EPA-905-R93-007/EPA-905-R93-007.html). In addition, standard assumptions are often made by litigants in determining exposure modeling calculations. For example, the EPA includes the standard assumption that the average adult consumes 2 L and 20 m^3 of air per day (EPA, *Exposure Factors Handbook*, 1989). Therefore, various exposure models and assumptions are necessary.

The existence of multipathway exposures, on the other hand, creates additional problems in toxic tort litigation. Furthermore, the toxicity of a substance may vary according to the different exposure routes. For example, asbestos is known to be carcinogenic when inhaled; however, it cannot be absorbed through the skin and has not been conclusively shown to be carcinogenic if ingested. Absent supporting data about the toxicity, metabolism, and absorption via different exposure routes, it often is assumed that adverse effects from different exposure routes are equivalent. However, this assumption should not be made without considering all available data ("Assessing

Environmental Risks to Human Health," available at http://www.epa.gov/oppeinet/oppe/f...s/risk/ roadmap/ rmap/chap4.txt.html). Therefore, when identifying relevant exposure pathways, it is necessary for litigants to consider the toxicities of the pollutants at issue via the different exposure routes that are being investigated.

Determining the Sources and Releases.   Information about the concentrations and quantities of released toxins and the location and timing of the releases is critical in order to accurately assess and measure whether the plaintiff was exposed to the potential pollutant. However, when this information is not readily available, monitoring or computer modeling can help litigants estimate the sources, the amount of pollutants released, and the amount of pollutants found at various distances surrounding the source.

Determining the Fate and Transport of Pollution.   The fate and transport of pollutants are also critical in assessing potential or actual toxic tort liability. The final destination (fate) and the route that the toxin in question takes (transport) provide information for determining the pollutant concentrations that plaintiffs are likely to be exposed to in toxic tort cases. Although the parties to litigation may base estimations of these concentrations on monitoring and modeling data, or a combination thereof, it is important for litigants to consider the dilution, dispersion, mobility, persistence, and degradation of the substances in the environment before exposure. When the fate and transport of toxins are modeled from release to exposure, elevated uncertainties in these factors may exist because of the increased time and distance of exposure. Furthermore, even if ambient concentration data are available, contaminant concentrations may differ between the monitoring location and the point of exposure ("Assessing Environmental Risks to Human Health," available at http://www.epa.gov/oppeinet/ oppe/f...s/ risk/roadmap/ rmap/chap4.txt.html). Because the burden of proof in civil toxic tort litigation is merely a preponderance of the evidence, counsel must consider all possible uncertainties relating to the fate and transport of pollution; such uncertainties can mean the difference between a multimillion dollar judgment and complete absolution from liability.

Human Contact.   When attempting to prove causation in toxic tort litigation, it is necessary to show actual human contact with the contaminants at issue. Therefore, litigants must estimate the duration and magnitude of the contact and calculate the size and distribution of the populations at risk. This may merely require a cursory examination of geographical data; however, sometimes a more detailed analysis is necessary. For example, showing contact for persons living adjacent to a facility that released toxins is relatively simple, but because of the various pathways with which toxins may come into contact with humans (i.e., groundwater, surfacewater, air dispersion, etc.), more remote points of human contact may exist where the amounts and concentrations of contaminant determination are not as straightforward. Therefore, attorneys must consider *all* potential points of human contact in their assessment of the evidence. Demonstration of human contact may be further complicated by behavioral and sensitivity factors. For example, some people spend more time indoors than outdoors; a plaintiff who spends an inordinate amount of time indoors, sheltered from sources of asbestos pollution, may have a harder time convincing a jury that he or she has been suffered an injury or illness due to asbestos exposure. Similarly, some people are more sensitive to particular contaminants than others and may experience adverse health effects at concentrations lower than those that cause adverse effects in the general public ("Assessing Environmental Risks to Human Health," available at http://www.epa.gov/oppeinet/ oppe/f...s/risk/roadmap/ rmap/chap4.txt.html). This is often referred to as the "thin skull doctrine" and may refer to highly sensitive populations such as pregnant women, infants, or people with asthma. Juries often sympathize with such highly sensitive persons and the resulting verdicts may reflect their compassion.

Uncertainties.   Because exposure assessments and modeling are not an exact science, uncertainty analyses should be performed "to provide decisionmakers with the complete spectrum of information concerning the quality of a concentration estimate, including the potential variability in the estimated concentration, the inherent variability in the input parameters, data gaps, and the effect these gaps have on the accuracy or reasonableness of the concentration estimates developed" so that they may better weigh the concentration results in the context of other factors being considered in toxic tort litigation (EPA, *Exposure Assessment Methods Handbook*, 1989). Various assumptions about the functional relationships between variables (i.e., sensitivity analysis) can be used to measure the amount of uncertainty presented by the model (EPA, *Risk Assessment Guidance for Superfund: Human Health Evaluation Manual*, 1989, part A., chaps. 6, 8). However, the basic causes of uncertainty include, but are not limited to, the following (EPA, *Exposure Assessment Methods Handbook*, 1989):

1. Measurement errors;
2. Generic data gathering, such as using similar chemical properties to fill in gaps about an unknown substance;
3. The natural variability of environmental concentrations due to factors such as Wind, temperature, and water flow as well as modeling uncertainties;
4. Disagreements in professional judgement; and
5. Sampling errors.

Through Evidence of Government Regulations and Policies

There are great disparities in the ease with which toxic tort litigation may arise and the difficulties in resolving the connection between the suspected toxin and the health effects that may be associated with exposure. Such difficulties have created a great deal of confusion in the courts. In toxic tort cases, "litigation may reflect scientific controversy, but it also may help to create it" (Foster, K., Bernstein, D., and Huber, P., *Phantom Risk*, MIT, London, 1993). The scientific methodologies and exposure models chosen by litigants when presenting toxic tort evidence add to this confusion. Depending on the models chosen, a jury is presented with various assessments and estimations of exposure to substances alleged by the plaintiff to be harmful. Conservative analyses and methodologies, such as those used by the, often lead to overestimation of actual or potential harm. Conversely, other approaches and models may lead to underestimation of risk or harm.

In toxic tort litigation, each party's stance with respect to governmental data is fact sensitive and depends on the chemical and exposure levels at issue on a case-by-case basis. Should the toxin in a particular case be detected in an amount below a governmentally imposed level, the defendant will argue that the exposure was de minimis and harmless because the amount of exposure was below the governmental level. On the other hand, if the toxin exceeds the standard, the plaintiff will offer this fact as proof of wrongdoing. Often the defense will argue "that the governmental standard is irrelevant and inadmissible." As a result, the impact of governmental assessments may vary from case to case.

Traditionally, government or regulatory action has had very little effect on tort liability. As a result, plaintiffs were previously barred from using policy or regulations to prove hazard or fault (Pagliaro, J., and Benton, A., "Courtroom Science: Toxic Tort Battleground," *Toxics L. Rep.*, l(42): 1336, March 22, 1989). Recently, however, plaintiffs have increasingly tried to introduce government policy and regulations as evidence to support a claim that exposure to a particular substance has resulted in an injury or illness (Pagliaro, J., and Benton, A., "Courtroom Science: Toxic Tort Battleground," *Toxics L. Rep.*, l(42): 1336, March 22, 1989). Evidence of this type is considered to be effective, "since government policy and regulations imply official recognition of the hazards of a material or safe levels of exposure with minimal or no independent evidence." As a result, defendants are charged with carefully circumscribed tort liability "on the basis of a prophylactic regulatory scheme designed to provide the broadest protection to the public" (Pagliaro, J., and Benton, A., "Courtroom Science: Toxic Tort Battleground," *Toxics L. Rep.*, l(42): 1336, March 22, 1989).

Such a strategy is based on scientific assumptions and the use of mathematical risk models to protect the public from levels of exposure where little or no empirical data exist associating exposure with adverse health effects. A wide variety of models are available for use in exposure assessments. The *Exposure Methods Handbook* (EPA, 1989) describes some of the models available and provides guidance in selecting the appropriate modeling techniques. The use of these models and the variables associated with intake and exposure calculations is crucial in toxic tort litigation. Plaintiffs strategy in introducing such evidence can be effective; juries react negatively to evidence that defendants are not in noncompliance with applicable government standards (Pagliaro, J., and Benton, A., "Courtroom Science: Toxic Tort Battleground," *Toxics L. Rep.*, l(42): 1336, March 22, 1989).

The choice of exposure factors and variables is crucial in toxic tort cases. The use of exposure modeling in conducting toxic exposure assessments may be used by litigants to predict whether concentrations of chemicals in the environment are of concern and whether there have been exposures at levels of concern to the plaintiffs in toxic tort litigation. Furthermore, the differences in exposure variables used in modeling techniques, and the use of regulatory assumptions, can provide litigants with very influential evidence to present to the judge and jury.

Plaintiff Strategies.   To maximize exposure, plaintiffs often resort to reliance on conservative governmentally imposed standards of maximum allowable exposure levels as proof that, by exceeding this threshold level of exposure, the defendant has caused or increased the likelihood of injury or illness. Scientifically objective assessments of exposure factors should be a goal in determining the risks associated with exposures to chemicals in toxic tort litigation. However, the conservatism used by the EPA and other agencies in determining exposure modeling variables and in conducting risk assessments often causes the true levels of exposure to toxins to be overstated. Some commentators have suggested that such an overstatement of risk may be by a factor of 1000 or more ("Exaggerating Risk," in *Hazardous Waste Cleanup Project*, edited by June 1993, pp. 1–45); Milloy, S., *Science-Based Risk Assessment: A Piece of the Superfund Puzzle*, National Environmental Policy Institute, Washington, DC, 1995). As such, this methodology presents a powerful weapon for plaintiffs in toxic tort litigation when examining the levels of exposure.

When conducting exposure assessments through modeling data, the EPA may overestimate the actual risks posed by the exposure of humans to various materials. This overestimation is the result of ("Exaggerating Risk," in *Hazardous Waste Cleanup Project*, edited by June 1993, pp. 1–45):

1. the use of broad assumptions, instead of site-specific data;
2. the use of theoretical worst-case values for modeling variables; and
3. the use of "point estimates" to characterize those variables.

In the event that these potential overestimations are made in conjunction with each variable in a modeling equation, the resulting multiplier effect may yield an unreasonable estimate of exposure and health risks. However, these exposure estimations and modeling techniques are used by plaintiffs as prima facie evidence of a defendant's wrongdoing.

The EPA approach to conducting exposure assessments is based on assumptions that often lead to an inflated risk estimate. The models presented in Figures 18.1.1 through 18.1.4. present a few of the more common exposure equations examining intake. The data gathered and applied to variables such as chemical concentration, ingestion rates, inhalation rates, exposure frequency, and duration, to name a few, play a crucial role in litigation in determining the various exposures and intakes that plaintiffs may have encountered. EPA's reliance on a number of standardized assumptions and models in performing exposure assessments in which the presence of a substance is often equated to exposure, regardless of whether there is any actual exposure or a realistic likelihood of exposure, often results in an overestimation of exposure ("Exaggerating Risk," in *Hazardous Waste Cleanup Project*, edited by June 1993, pp. 1–45).

For example, it is the usual practice of the EPA to sample and analyze soil and water at a Superfund site, identify the contaminants, and then construct theoretical current and future exposure models based on several broad assumptions, which include, but are not limited to, the following ("Exaggerating Risk," in *Hazardous Waste Cleanup Project*, edited by June 1993, pp. 1–45); Milloy, S., *Science-Based Risk Assessment: A Piece of the Superfund Puzzle*, National Environmental Policy Institute, Washington, DC, 1995):

1. The frequency of exposure is assumed to be every day for 30 years;
2. Substance concentrations in groundwater are assumed to be total, not dissolved concentrations from unfiltered samples;

3. Substance concentrations are assumed to remain constant throughout the Duration of exposure;

4. Individuals are assumed to consume 100 percent of their drinking water from contaminated wells at home; and

5. Adults are assumed to consume 100 mg of the most contaminated surface soil per day, and children are assumed to consume 200 mg per day (burrowing infant assumption).

Applying larger variable values will often overstate the actual exposure level. Many of the elements of these assumptions "do not exist, will never exist, and are extreme even when taken in isolation" ("Exaggerating Risk," in *Hazardous Waste Cleanup Project*, edited by June 1993, pp. 1–45). However, when armed with such statistics and calculations, plaintiffs have powerful testimony and evidence that the exposure levels are excessive and the health risks associated with those exposures are compensable.

In addition to the assumptions stated above, regulatory agencies often use theoretical worst-case values for many of the variables in exposure model equations. For example, the EPA regularly uses an upper-bound, worst-case estimate, such as the 95th percentile—or the 95 percent upper confidence level—as a conservative estimate of exposure for key variables, such as the chemical concentration of the contaminant in the soil and the amount of soil that a person contacts and ingests (see Figure 18.1.2). However, the combination of several upper-bound estimates does not result in a 95th percentile exposure estimate; instead, the combination of three variables at their 95th percentile will result in an estimate that is actually at the 99.8th percentile. A plaintiff who uses such data has a powerful case that the requisite level of chemical exposure has occurred (Milloy, S., *Science-Based Risk Assessment: A Piece of the Superfund Puzzle*, National Environmental Policy Institute, Washington, DC, 1995).

Another factor that is often beneficial to plaintiffs in toxic tort litigation is the use of point estimates to characterize the hazards associated with exposure to chemicals in the environment. Currently, virtually all state and federal guidelines for conducting risk assessments rely on point estimates of risk in making environmental decisions (Finley, B., and Paustenbach, D., "The Benefits of Probabilistic Exposure Assessment: Three Case Studies Involving Contaminated Air, Water, and Soil," *Risk Anal.*, 14(1): 53–73, 1994). For example, when the EPA conducts exposure assessments at a Superfund site, it obtains groundwater sample data through hydrogeological studies and water distribution analysis at various levels and locations over the site. Instead of using an average figure to examine contaminant concentration, the EPA then uses the upper-bound concentration data from the most contaminated portions of the site to classify the level of contamination at the site. This type of analysis invariably results in a determination that the entire site is as contaminated as the most contaminated areas (see Table 18.1.2).

Based on expert testimony and presentation of overestimated levels of exposure, a powerful case can be made that exposure levels exceeding governmentally set standards present prima facie evidence of fault. For juries with little or no scientific or statistical background, these numbers can present a skewed version of the actual exposure and associated health risks. Because juries are often predisposed to believe that exposure to toxic chemicals causes injury or illness, plaintiffs attempt to focus the jury's attention on the injured plaintiff. Therefore, in some cases, especially if the potential damages in a case are substantial, the plaintiffs develop their own models instead of relying on the government's models; in so doing, they often make even more conservative assumptions than the EPA.

Defense Strategies.   In toxic tort scenarios, the defense has several approaches with which to minimize the effect of plaintiff testimony and evidence. If the pollutant at issue in a particular case is detected at a level below a governmentally imposed standard, the defendant will most likely assert that the exposure was de minimis and harmless; however, if the toxic level exceeds the government standard, the defense generally argues that the government standard is irrelevant and inadmissible (Relkin, E., "The Sword of the Shield: Use of Governmental Regulations, Exposure Standards and Toxicological Data in Toxic Tort Litigation," *Dickinson 1997 Environ. Law Policy*, 6: 1–30). When presenting exposure evidence and models, the defense tries to minimize the number and levels of exposure as well as the relevant variables associated with the models used in an exposure assessment.

Probabilistic assessments are more conducive to sensitivity and quantitative uncertainty analysis than the deterministic approach. An accurate sensitivity analysis requires an accurate estimate of

TABLE 18.1.2 Calculation of Estimate of Intake of Soil Containing Polynuclear Aromatic Hydrocarbons by an Adult ("Exaggerating Risk," *Hazardous Waste Cleanup Project*, 1993, pp. 1–45)

Term	Reasonable point estimate	EPA method	Overestimation
Chemical concentration in soil	40 mg/kg	361 mg/kg	Overestimation by 9.03 times
Rate of soil ingestion	25 mg/day	100 mg/day	Overestimation by 4.00 times
Conversion factor	0.000001 kg/mg	0.000001 kg/mg	—
Amount of contaminated soil contacted	1.0	1.0	—
Absorption factor	1.0	1.0	—
Bioavailability factor	1.0	1.0	—
Frequency of exposure	35 days/year	350 days/year	Overestimation by 10 times
Duration of exposure	9 years	30 years	Overestimation by 3.33 times
Total intake	0.32 mg	379.1 mg	—
Body weight	70 kg	70 kg	—
Averaging time	25,550 days	25,550 days	—
Body weight and time	1,788,500 kg/day	1,788,500 kg/day	—
Intake	0.00000018 mg/kg/day	0.00021 mg/kg/day	Overestimation by 1167 times

Notes: The potential for overestimating exposure by compounding overly conservative assumptions is shown; the data differentiate the EPA's method of estimating exposure based on point estimates to a more reasonable point estimate of polynuclear aromatic hydrocarbon intake by an adult.

$$\text{intake} = \frac{CC \times IR \times CF \times AC \times AF \times BF \times FE \times DE}{BW \times AT}$$

where CC = chemical concentration in the soil,
    CF = conversion factor,
    AF = absorption factor,
    FE = frequency of exposure,
    BW = body weight,
    IR = rate of ingestion of soil,
    AC = amount of contaminated soil,
    BF = bioavailability factor,
    DE = duration of exposure, and
    AT = averaging time.

variance within a particular data set. A probabilistic approach can more accurately supply such information. Sensitivity analysis allows for testing an output variable to the possible variation in input variables (Finley, B., and Paustenbach, D., "The Benefits of Probabilistic Exposure Assessment: Three Case Studies Involving Contaminated Air, Water, and Soil," *Risk Anal.*, 14(1): 53–73, 1994; EPA, *Exposure Assessment Methods Handbook*, 1989). Thus, it can allow attorneys and experts to better identify the variables that predominate in the results and give a more precise insight about the risk estimate associated with a model. By identifying the influential input variables, more resources can be directed to reduce their uncertainties and thus reduce the output uncertainty (EPA, *Exposure Assessment Methods Handbook*, 1989). In the context of establishing or refuting the requisite burden of proof in a civil toxic tort matter (i.e., a preponderance of the evidence), such uncertainty reduction is a powerful tool. Because toxic tort litigation often hinges on probabilities, the reduction of the uncertainty in these assessments is critical in successfully prosecuting or defending toxic tort litigation.

Therefore, defendants may prefer to use a more probabilistic approach to assessing exposure when confronted with governmentally established exposure guidelines. Probabilistic approaches and techniques address the main deficiencies of point estimate approaches; they characterize a range of potential exposures and their likelihood of occurrence, thereby circumventing the need for determining the best point estimate. Furthermore, probabilistic methods can be used by the defense to incorporate all ingestion, inhalation, or dermal absorption values in frequency to their occurrence (Finley, B., and Paustenbach, D., "The Benefits of Probabilistic Exposure Assessment: Three Case Studies Involving Contaminated Air, Water, and Soil," *Risk Anal.*, 14(1): 53–73, 1994). For example, the conservative

TABLE 18.1.3   Exposure Estimates for Four Sample Scenarios ("Exaggerating Risk," *Hazardous Waste Cleanup Project*, June 1993, pp. 1–45)

Scenario	95th percentile (mg/kg/day)	Default value (mg/kg/day)	Percent exceeding default value*	Differences between EPA default and 95th percentile[†]
Ingestion via drinking water	1.3×10^{-2}	2.9×10^{-2}	0.5 percent (5/1000)	1.6×10^{-2} (2.2 $\times$)
Ingestion of soil	0.24×10^{-5}	3.5×10^{-5}	1.9 percent (19/1000)	3.26×10^{-5} (14.6 $\times$)
Ingestion of food	0.33×10^{-3}	3.5×10^{-3}	0.0 percent (0/1000)	3.27×10^{-3} (10.9 $\times$)
Dermal contact	0.12×10^{-6}	6.7×10^{-6}	0.2 percent (2/1000)	6.58×10^{-6} (55.8 $\times$)

Notes: A 1991 study commissioned by the Chemical Manufacturers Association evaluated the EPA's existing Superfund exposure assessment assumptions, statistically analyzed the uncertainty of the existing exposure assessment methods, and evaluated and recommended modifications or alternative approaches to disease uncertainty. The exposure estimates derived from point estimates were compared with exposure estimates taken from a probabilistic simulation using *both* EPA assumptions and data derived from more recent peer-reviewed research. The 95th percentile exposure values derived from the probabilistic analysis showed lower likely exposures than values derived from EPA single point estimates.

*Percent of Monte Carlo simulations exceeding EPA default values.

[†]First value is default value 95th percentile; second value is default value divided by 95th percentile.

scenario established by the EPA of soil consumption per day in children is 200 mg/day. However, with probabilistic methods, this ingestion rate can be arranged to reflect a range of values, including recent data, which indicate that 50 percent of children ingest no more than 9 to 40 mg/day of soil from sources such as household dust, dirt in foods, and direct soil contact, to which it has been suggested that a mean value of 16 mg/day should be applied (Calabrese, E. J., et al., "How Much Soil Do Young Children Ingest: An Epidemiologic Study," *Reg. Toxicol. Pharmacol.*, 10: 123, 1989; Davis, S., et al., "Quantitative Estimates of Soil Ingestion in Normal Children Between the Ages of 2 and 7 Years," *Arch. Environ. Health*, 45: 112, 1990; Calabrese, E. J., et al., "Preliminary Adult Soil Ingestion Estimates: Results of a Pilot Study," *Reg. Toxicol. Pharmacol.*, 12: 88, 1990; Calabrese, E. J., and Stanek, E. S., "A Guide to Interpreting Soil Ingestion Studies I. Development of a Model toEstimate the Soil Ingestion Detection Level of Soil Ingestion Studies," *Reg. Toxicol. Pharmacol.*, 13: 263, 1991; Calabrese, E. J., and Stanek, E. S., "A Guide to Interpreting Soil Ingestion Studies II. Development of a Model to Estimate the Soil Ingestion Detection Level of Soil Ingestion Studies," *Reg. Toxicol. Pharmacol.*, 13: 278, 1991). When applied to the model shown in Figure 18.1.2, the probabilistic mean value produces a smaller exposure estimation than the EPA's conservative worst-case value (see Table 18.1.3). Therefore, a defendant must attempt to introduce more minimal estimates, such as those produced by probabilistic methods. The purpose of this strategy is twofold: first, minimization of exposure will likely present a more realistic measurement of the actual exposure; second, juries are likely to equate causation with a higher exposure value. However, even though governmentally imposed regulations may be exceeded, this fact is not fatal to a case. Properly educating jury members about complex scientific matters through coherent expert testimony is perhaps the most critical factor in persuasion. Because most jurors have a limited scientific and/or mathematical background, it is crucial to present this evidence and testimony in an intelligible manner.

Defendants should also present alternative causation theories for plaintiffs' injuries. Furthermore, a defendant would also be well-advised to propose that the judge bifurcate the trial into two phases—the first phase concentrating on whether there was exposure and whether the exposure was sufficient to cause injury, and the second phase concentrating on plaintiffs' specific injury and damages. Thus, the defendant can focus the jury's attention on the exposure question instead of on the injured plaintiff.

18.1.4   CONCLUSION

The explosion of toxic tort litigation has led to increased reliance on scientific expert testimony in the courtroom to demonstrate the level of exposure to toxic chemicals. Through the use of multimedia exposure modeling and risk assessments, plaintiffs and defendants seek to support their respective positions. Therefore, the effective presentation of difficult scientific concepts has increased significance in the court system, which is composed of jurors who often have very little background in science or statistical probabilities. As a result, the methodologies chosen by the expert, together with the justification provided for the jury, can have a profound effect on juror perceptions of the conclusions reached by the expert and, just as importantly, the credibility of those conclusions—thus, the need for a fair and easily applied standard for presenting scientific expert testimony, a standard that finds its beginning in *Daubert* and *Joiner*. Therefore, the combined effect of *Daubert* and *Joiner* is to provide the court with a smaller role in screening experts and the attorney a larger role.

However, as science matures and new relationships between human illnesses and toxic chemicals are discovered, more toxic tort claims are inevitable, resulting in further application of the aforementioned principles, requiring the legal system to explore deeper into the new developments of cutting-edge science and inevitably more scientific expert opinion. These developments will warrant the need for more money and more time to properly prepare an expert for the rigors of trial. Attorneys are now required not only to find an expert willing to say that the exposure to toxic chemical has caused plaintiff's injuries but also to find an expert who can rationally explain the conclusions and the evidence. Furthermore, experts must be able to provide a lengthy and detailed report outlining their own conclusions and, moreover, then criticisms of the conclusions of the opposing counsel's expert (Foster, K., and Huber, P., *Judging Science*, MIT, Cambridge, MA, 1997; American Bar Association, *Scientific Evidence Review*, Monographs 1–3, Van Nostrand Reinhold, New York, NY, Science and Technology Section, 1993–1995).

The added expense of preparing an expert to this great detail will also lessen the number of toxic tort cases that are ultimately brought to trial; if the possible award for the plaintiff's injuries does not greatly exceed the amount needed to prepare plaintiff's expert for trial, the attorney will find little justification in bringing the cause of action. Therefore, depending on the reactions of the judicial, legislative, and administrative branches of the legal system to the progeny of *Daubert*, the legislative attempts to provide for additional focus on risk assessment, and the ever-developing scientific principles risk assessments and exposure modelings, may warrant a heightened profile in the legal arena. Nevertheless, environmental science and the scientific principles on which it is based will continue to haunt the legal system until a solution is found to close the gap between science and the law.

GENERAL REFERENCES

Books

Foster, K., and Huber, P., *Judging Science*, MIT, Cambridge, MA, 1997.

Rossi, F. F., *Expert Witnesses*, American Bar Association, 1991.

Zimmerman, R., *Governmental Management of Chemical Risk: Regulatory Processes of Environmental Health*, Lewis, Boca Raton, FL, 1990.

Manuals

Weller, "Analysis & Perspective: Expert Evidence," *Toxic L. Rep.*, 13(10): 338, BNA, August 5, 1998.

American Bar Association, *Scientific Evidence Review*, Monographs nos. 1–3, Van Nostrand, Reinhold, NYC, Science and Technology, Section, 1993–1995.

Weinstein, J., and Berger, M., *Weinstein's Evidence*, 1988.

Symposium on Science and the Rules of Evidence, 99 F.R.D.187, 1983.

Governmental Documents

Committee on Risk Assessment and Risk Management, *Risk Assessment and Risk Management in Regulatory Decision-Making*, Draft Report, Washington, DC, June 13, 1996.

National Environmental Policy Institute, *Science-Based Risk Assessment: A Piece of the Superfund Puzzle*, Washington, DC, 1995.

EPA, *Policy for Risk Characterization at the EPA*, 1995.

Science Policy Council, *Guidance for Risk Characterization*, EPA 1995.

EPA, *Implementation Program for the EPA Policy on Risk Characterization*, 1995.

Federal Judicial Center, *The Reference Manual on Scientific Evidence*, 1994.

Executive Order no. 12,866, 58 Fed. Reg. 51,735, 1993.

EPA, *Workshop Report on EPA Guidelines for Carcinogen Risk Assessment: Use of Human Evidence*, September 1989.

EPA, *Exposures Factors Handbook*, July 1989.

EPA, *The Risk Assessments Guidelines of 1986*, August 1987.

Executive Office, *Risk Analysis: A Guide to Principles and Methods for Analyzing Health and Environmental Risks*, 1989.

Articles

Hoffman, J., "Analysis & Perspective: Daubert's Application to Testimony From Non-Scientific Experts," *Toxics L. Rep.*, 13(15): 492, September 9, 1998.

Chong, E., "Thomas Kuhn and Courtroom Treatment of Scientific Evidence," *Temp. Environ. L. & Technol. J.* 15: 195, Fall 1996.

Applegate, J., "A Beginning and Not an End In Itself: The Role of Risk Assessment in Environmental Decision-Making," *Risk Symp. U. Cin. L. Rev.*, 63: 1643, Summer 1995.

Simon, R., "Issues in Risk Assessment and Cost-Benefit Analysis and Their Regulatory Reform," *Risk Symp. U. Cin. L. Rev.*, 63: 1611, Summer 1995.

McElveen, J., "Legislating Risk Assessment," *Risk Symp. U. Cin. L. Rev.*, 63: 1553, Summer 1995.

Goldman, L., "Environmental Risk Assessment and National Policy: Keeping the Process Fair, Effective, and Affordable," *Risk Symp. U. Cin. L. Rev.*, 63: 1533, Summer 1995.

McCormick, "Scientific Evidence: Defining a New Approach to Admissibility," *Iowa L. Rev.*, 67: 879, 1982.

Roisman, A., "The Implications of *G.E. v. Joiner* For Admissibility of Expert Testimony," <http://www.vje.org>, visited September 15, 1998.

Cases

Daubert v. Merrell Dow Pharm., Inc., 509 U.S. 579, 1993.

Akzo Coatings v. Aigner Corp., 881 F. Supp. 1202, 1213, N.D. Ind., 1994.

United States v. Downing, 753 F.2d 1224, 3rd Circuit Court, 1985.

Frye v. United States, 293 F. 1013, Washington, DC, Circuit Court, 1923.

CHAPTER 19
SENSITIVE ENVIRONMENTAL SYSTEMS

SECTION 19.1

KARST SYSTEMS

Thomas J. Aley

Mr. Aley is president, Ozark Underground Laboratory, Protem,
Missouri, where he performs extensive karst limestone research
in his own cave.

19.1.1 WHAT IS KARST

Karst is a landscape underlain by a karst aquifer. A karst aquifer is developed within relatively soluble bedrock units in which there is appreciable groundwater movement through dissolved openings in the bedrock. Sinkholes, losing streams, caves, and springs are commonly found in karst areas, but are not necessary for the area to be classified as karst. Karst is most commonly associated with limestone, dolomite, marble, or gypsum, but it can occur in other rock types, including quartzite (for example in the Sarisarinama Plateau of Venezuela) and silica sandstones cemented with silica (Ford and Williams, 1989).

Older definitions of karst often focused on landforms and denudation patterns in limestone and dolomitic rocks and a comparison of the features to those of the Kras region of the former Yugoslavia. Under this approach, inadequate similarity was equated with the area not being karst. More current definitions have focused on the hydrogeological processes operating in karst landscapes. This chapter views karst broadly as encompassing all landforms and aquifers in which rock dissolution has been significant enough to create some preferential groundwater flow routes within the bedrock. Using this view, about 20 percent of the earth's land surface is karst (White, 1988) and about 25 percent of the United States is karst (Quinlan, 1989). Karst is found throughout the United States, but is particularly abundant in the eastern half of the nation; Quinlan (1989) estimated that 40 percent of the United States east of the Mississippi River is karst. Davies et al. (1984) published a useful and detailed National Atlas Map depicting engineering aspects of karst and areas with features analogous to karst.

The term pseudokarst is sometimes applied to settings where surface features and/or solor subsurface hydrologic conditions have characteristics similar to those of karst but the substrate is not one of the relatively soluble rocks. As an example, small sinkholes and hydrologically integrated conduits that convey water to springs are sometimes developed in caliche. Sinkholes, sinking streams, springs with large discharge rates, and caves that extend laterally for hundreds to thousands of meters are found in some volcanic settings and are sometimes termed pseudokarst.

19.1.2 THE NATURE OF KARST

For this discussion karst will be divided into three superimposed and hydrologically interconnected zones: (1) the surface, (2) the epikarst, and (3) the deep karst.

The Surface

To some, the surface of a karst landscape is dimpled with sinkholes. The Mitchell Plain in southern Indiana, where sinkhole densities are locally over 380 per square km (1000 per square mile), is a classical example of a sinkhole plain. Sinkhole densities as great as 3800 per square km (10,000 per square mile) exist in some temperate rainforest areas of southeastern Alaska. However, sinkholes are rare to absent in many karst areas.

Sinkholes are generally rare or absent in most of the karst in the Ozarks of Missouri and Arkansas. Most sinkholes are not cavern collapses, but are instead water, sediment, and detritus transport features exposed in the soil and residuum overlying transport openings in the bedrock. Much of the Ozarks is characterized by deep cherty soil and residuum, which tends to obscure the surface expression of underlying water and sediment transport features. The typical surface indication of karst in the Ozarks is abundant stream channels (called losing streams), which transport surface water only for a few days at a time and only a few times per year. Other than the typical absence of surface water, the losing streams do not appear appreciably different from streams in nonkarst landscapes. The Ozarks have a humid climate with annual precipitation of about 97 to 114 cm (38 to 45 in.). Were it not for the extensive movement of water from the surface into the underlying karst groundwater system, surface basins of more than about 2.6 square km (1 square mile) would typically have perennial to nearly perennial flow. As a general value, about 75 percent of the annual runoff from the Ozarks passes through the karst groundwater system for at least some distance (Aley, 1978).

The Epikarst

The epikarst or epikarstic zone (sometimes called the subcutaneous zone) is the dissolutionally weathered upper portion of the bedrock. The thickness of the epikarst can vary from essentially zero to 30 m (98 ft) or more. Factors effecting the thickness of the epikarstic zone include climate, time since the last glaciation of the site, patterns and depth of groundwater circulation, characteristics of the bedrock, and vegetational history of the area. The intensity of epikarstic development can be expressed as a percent of the bedrock that has been removed by dissolution. It can vary from less than 1 percent to more than 50 percent, and the percentage routinely decreases with depth. In many epikarstic zones, sediments partially or almost completely fill the voids within the bedrock; in some situations, many or most of the voids are largely free of sediments. The percent of the bedrock void volume that is filled with sediment can range from less than 5 percent to more than 95 percent, with the higher percent values being the most common.

In many cases, the epikarstic zone is of great hydrologic importance. This zone typically has greater permeability than underlying portions of the bedrock. As a result, the epikarstic zone often functions as a perched aquifer and may provide appreciable water detention and storage. The epikarstic zone also functions as a lateral water distribution system, which conveys water laterally to localized zones where water moves deeper toward the zone of saturation. The lateral distribution of water in the epikarstic zone may traverse appreciable distances and may be multidirectional in nature. As an illustration, tracer dyes introduced into the epikarst at single points can often be detected at multiple drippage zones in underlying cave systems. These dye detection locations are often separated by distances of 100 to 300 m (328 to 984 ft), and the dye recovery sites may differ dramatically with differences in the amount of water in the epikarstic zone.

At some sites, the epikarstic zone is rapidly drained following precipitation events and typically provides little long-term water storage or detainment. These conditions are often found in the epikarst underlying moderately steep topography. Longer-term (and greater volume) water storage in the epikarst is common in areas with more gentle topography, and areas near major streams commonly have epikarstic zones that are perennially saturated. Industrial sites in karst areas commonly are located in areas with more gentle topography, and as a result, such industrial sites often overlie epikarstic zones with appreciable water storage. Not only can there be appreciable water storage, but also there may be major (and often short duration) fluctuations of water-level elevations, which are commonly missed by quarterly or monthly measurements in monitoring wells.

Industrial sites underlain by epikarstic zones with appreciable water storage can have significant environmental remediation problems if contaminants of concern have entered the subsurface. Because of the lateral and multidirectional nature of water movement in the epikarst, a single point-source of contamination can result in a pattern of contaminant distribution, which may initially suggest (especially to those not intimately familiar with the hydrologic functioning of the epikarst) that there are multiple sources of contamination. Large sums have been expended at some karst sites searching for "missing" source areas of contamination when the presumed source areas never existed. Extensive lateral distribution of contaminants can also result from major precipitation or snow melt events, which raise water levels in the epikarst enough to permit water infiltration into storm or sanitary sewer lines. Utility corridors with highly permeable bedding can function similarly under high-water level conditions in the epikarst.

The Deep Karst

A fundamentally unique characteristic of karst groundwater systems is that the water passing through the aquifer has modified the aquifer and developed a hydrologic flow system that is integrated in a downgradient direction (Huntoon, 1999). This is in distinct contrast to other types of aquifers where water does not appreciably modify the aquifer nor impose a hydrologic flow system within it. Especially in Paleozoic limestones and dolomites, most of the water flow through karst aquifers is along highly localized preferential flow routes, which have been enlarged by dissolution. As a result, karstic permeability tends to be the most anisotropic of all permeability types found in nature (Huntoon, 1999). The use of equations with underlying assumptions of isotropic conditions are not appropriate.

Karst aquifers have a three-phase flow system with water movement through primary, secondary, and tertiary porosity. The primary porosity is associated with the intergranular openings within the bedrock. Primary porosity is usually appreciable in younger limestones, but is typically very small in most Paleozoic limestones. Primary porosity is routinely greater in dolomites than in limestones, and typically results in a lower percentage of "dry holes" for wells in dolomite than for those in limestone.

Secondary porosity in a karst aquifer is fracture porosity. Many wells developed in Paleozoic carbonates derive much of their water from fracture porosity. While some or many of the fractures may have been somewhat modified by dissolution, those viewed as examples of secondary porosity typically are not capable of transporting water as rapidly as typically is the case with tertiary porosity.

Tertiary porosity in a karst aquifer is dissolutional porosity that has been integrated into a downgradient flow system. Groundwater tracing is an important investigative technique for use in karst areas, and will be discussed in greater detail later. Introducing a tracer dye into a cave stream and, subsequently, recovering the dye from a spring is a test of tertiary porosity.

Values of water quality parameters, such as specific conductance and total dissolved solids from wells in karst aquifers, often differ from the values for springs discharging from the same aquifers. The reason for this is that wells are predominantly sampling secondary porosity, and springs are predominantly sampling tertiary porosity. The variability of water quality values for particular parameters tends to be greater for springs than for wells.

19.1.3 GROUNDWATER RECHARGE

Most conceptual models for karst groundwater systems recognize two types of recharge. While the nomenclature applied vary among authors, terms such as *discrete recharge* (or concentrated recharge) and *diffuse recharge* (or dispersed recharge) are characteristic of the dichotomy. The discrete recharge zones are characterized by the entry into the subsurface of appreciable amounts of water in highly localized areas (such as through sinkholes, losing streams, or other localized areas). In contrast, diffuse recharge is dispersed and not highly localized. In reality, the two categories are end members of a continuum, yet it is a continuum that commonly has a bimodal distribution with maximums near both ends.

Springs have sometimes been divided into conduit flow springs and diffuse flow springs based upon the variability of water quality parameters, such as specific conductance. Those springs with the higher water quality variability have been called conduit springs, while those with lower water quality variability have been called diffuse flow springs. This nomenclature is misleading because essentially all karst springs discharge from conduit systems. This is the nature of the groundwater recharge, which is responsible for the observed differences between conduit flow and diffuse flow springs. The diffuse flow springs receive most of their recharge water through diffuse recharge. The conduit springs receive much or most of their recharge through discrete recharge.

Recognition of the two conceptual types of groundwater recharge has several benefits, including an enhanced understanding of the sensitivity of karst groundwater to contaminants. An areally small discrete recharge zone, capable of introducing tens to hundreds of liters of water per minute into the subsurface, clearly should not be expected to provide effective natural cleansing for the passing water through processes such as filtration or adsorption.

19.1.4 *SENSITIVITY TO CONTAMINATION*

Much of the water that moves into and through karst groundwater systems flows through hydrologically well-integrated flow systems. These flow systems transport much of the water through preferential flow routes, which have the following characteristics:

1. They commonly transport water at rates several orders of magnitude greater than those encountered in nonkarst groundwater systems.
2. They commonly support flow in the turbulent regimen.
3. They are too large to provide effective filtration for most pathogens.
4. They provide minimal adsorption or other natural cleansing processes.

Table 19.1.1 presents data from 10 karst groundwater traces conducted in various American karst areas. The data show results from traces in the Appalachian Mountains of West Virginia, the large springs area of the Missouri and Arkansas Ozarks, and the Edwards Aquifer in Texas. The selected traces reflect some of the longer distance and lower gradient traces conducted in each of the three areas. Data are presented on eight traces conducted with fluorescein dye and two traces conducted with stained club moss (*Lycopodium* sp.) spores. Dye travels through the groundwater system in solution, whereas the spores travel in suspension.

For the 10 karst groundwater traces summarized in Table 19.1.1, mean straight-line groundwater velocities ranged from 234 to 5840 m (768 to 19,155 ft) per day. These rates are four to six orders of magnitude greater than groundwater travel rates commonly reported in the literature for most nonkarst aquifers. While the data in the table were selected, travel rates in the range of hundreds to a few thousands of meters per day are commonly encountered in karst groundwater tracing work. If we view the tracing agents as surrogate contaminants, the tracing results demonstrate that many karst groundwater systems can rapidly transport contaminants for appreciable distances.

Two of the karst groundwater traces in Table 19.1.1 were conducted with stained *Lycopodium* spores. These spores are nearly spherical in shape and have a mean diameter of 33 microns with a standard deviation of 1.5 microns. When used for groundwater tracing, they are boiled, stained with bacteriological stains so that they can be positively identified, and then treated with a preservative. The resulting spores then travel in suspension. Studies of the spores used in the Missouri trace (Table 19.1.1) indicated that 95 percent or more of the spores settle out of quiet water at 21.5°C, at rates of at least 6.9 to 7.6 cm (2.7 to 3.0 in) per hour. In view of the relatively rapid settling velocity of the spores, a mean straight-line travel rate of over 5 km/day (3.1 miles/day), a mean groundwater gradient of 0.2 percent, and a total straight-line travel distance of 63.7 km (39.5 miles), it is clear that the transporting groundwater system is characterized by turbulent, rather than laminar, flow.

Another important interpretation of the tracing with *Lycopodium* spores is that they demonstrate that particles with diameters of about 33 microns can readily be transported through karst groundwater

TABLE 19.1.1. Some Selected Data on Long Distance Karst Groundwater Traces in the United States

Trace	State	Straight line distance (m)	Elevation difference (m)	First arrival (days)	Straight line velocity (m/day)	Mean gradient	Tracer type[#] and quantity (Kg)	
Sink on Williamson Creek to Barton Springs	TX	7300	23*	1.25	5840	0.003	Fl	1.4
Davis Sink to Alley Spring	MO	14,200	75	11	1290	0.005	Fl	4.5
Bruffey Creek to JJ Spring	WV	18,700	189	80	234	0.010	Fl	0.9
Ludington Cave through McClungs Cave to Davis Spring	WV	20,500	106	11	1864	0.005	Ly	4.5
Coffman Cave to Davis Spring	WV	24,200	168	39	621	0.007	Fl	5.5
Nuttle Spring to Greer Spring	MO	44,400	178	12	3700	0.004	Fl	6.8
Mt. Zion to Mammoth Spring	MO & AR	52,900	171	19	2784	0.003	Fl	4.5
Mountain View Lagoon to Big Spring	MO	61,500	200	12	5125	0.003	Fl	6.8
Middle Fork Eleven Point River to Big Spring	MO	63,700	123	16	3981	0.002	Fl	4.5
Middle Fork Eleven Point River to Big Spring	MO	63,700	123	13	5179	0.002	Ly	2.7

Data Sources:
Arkansas and Missouri Traces: Aley (1978)
Texas Trace: Hauwert et al. (1998)
West Virginia Traces: Jones (1997)
[#]Tracer Type: Fl = fluorescein dye, Ly = *Lycopodium* spores.
*Based on potentiometric surface.

systems. The *Lycopodium* spores are an order of magnitude larger than most pathogenic bacteria. It is now generally believed that the waterborne parasite *Cryptosporidium* is responsible for more waterborne illness in the United States than any other single agent; the infectious spores of this parasite are 3 to 5 microns in diameter. Karst groundwater systems that cannot remove *Lycopodium* spores through natural filtration clearly cannot remove most pathogenic bacteria or organisms such as *Cryptosporidium*.

Groundwater tracing with *Lycopodium* spores requires laborious sample processing and analysis. As a result, this agent seldom has been used in the United States, except in cases where the unique data provided by this agent were needed. However, a substantial amount of groundwater tracing with *Lycopodium* spores has been conducted in Europe, and there have been many successful traces. Several papers on such tracing have been published in proceedings of international symposiums on water tracing; Kranjc (1997) is the most recently published of these. Kranjc (1997) also includes a few papers on groundwater tracing in karst aquifers with bacteriophages that are biological surrogates for potentially pathogenic viruses. A trace for 1220 m (4000 ft) through a karst aquifer in Missouri with Coliphage T-4 resulted in the recovery of 2.5 to 3 percent of the introduced agent.

A number of fluorescent tracer dyes have been used to trace water into and through karst aquifers. All of these dyes are subject to some loss in transit because of sorption onto charged particles and other processes, yet in many karst aquifers even relatively small quantities of these tracers can be readily transported through the karst aquifer. These successful traces provide evidence that adsorption is commonly not an effective natural cleansing mechanism in many karst aquifers. Aley (1997) summarized the karst groundwater tracing literature and concluded that most quantitative traces reported recoveries of 20 to 50 percent of the introduced dye at the dye recovery sites, and that many dyes were reportedly used successfully in karst groundwater tracing. Aley (1997) concluded that the data were skewed and overrepresented traces through portions of karst aquifers that were hydrologically very well integrated. Based upon about 1000 groundwater traces in the epikarstic zone, Aley (1997) found that dye recovery rates were typically 0.1 to 1 percent of the introduced dye in perennially saturated epikarstic zones, and 1 to 10 percent of the introduced dye in seasonally saturated epikarstic zones. Furthermore, he reported that these recovery rates applied only to those dyes least subject to adsorption onto earth materials, and that dyes with higher sorption tendencies were often not recovered in epikarstic zone traces.

19.1.5 ENVIRONMENTAL PROBLEMS OF KARST AREAS

Obtaining and protecting water supplies is a major problem in karst areas. Especially in rural areas, groundwater is commonly the primary source of domestic water in karst areas. Groundwater circulation in many karst areas is very deep, so the wells must also be deep. In many areas, the historic practice was to case private wells only to the top of bedrock, therefore, in some cases, there was little or no casing through the epikarstic zone. As a result, many wells routinely contain coliform bacteria, and many become turbid after rainstorms. While newer wells typically have more casing (because of local or state regulations), inadequate amounts of casing for particular sites is still common. Furthermore, abandoned wells and older shallowly cased wells often provide contaminant pathways into deeper portions of karst aquifers.

In some areas, karst groundwater circulation is extremely deep. At West Plains, Missouri, five of the city's currently used municipal wells are over 451 m (1500 ft) deep and are cased to depths between 213 and 328 m (700 and 1000 ft). All of these wells have had episodes of turbid water associated with precipitation events.

Groundwater contamination is the single most important environmental problem in karst areas. The reason for this contamination is the high sensitivity of the groundwater system and the intimate connections between surface and subsurface waters. The author conducted a study of the effectiveness of wastewater treatment by on-site septic systems serving private residences in Greene County, Missouri; this county is almost entirely karst. The assessment was based upon the ability to detect optical brighteners in sampled springs. Optical brighteners are fluorescent dyes used in laundry soaps and detergents; they are removed from wastewater in septic fields through adsorption. Adsorption is also the primary mechanism removing viruses in septic fields, so the testing parameter was relevant. The Greene County study found that approximately 60 percent of the septic field systems yielded detectable optical brighteners to the karst groundwater system, and that about 15 percent of the systems were major contributors.

Localized pollution sources can result in the contamination of groundwater in large areas. The routine discharge of 4.73 to 5.68 million liters (1.25 to 1.50 million gallons) per day of poorly treated chicken processing wastes into a losing stream in northern Arkansas was investigated by the author through a comprehensive groundwater tracing study using two tracer dyes and *Lycopodium* spores. Tracer dyes introduced in the sewage discharge were recovered from 38 sampling stations, 18 of which were private water supply wells located up to 10.3 km (6.4 miles) from the dye introduction point. First dye recoveries at the wells and springs were typically three to four weeks after dye introduction. The impacted area encompassed approximately 155.4 square km (60 square miles) in two adjacent topographic basins. *Lycopodium* spore sampling was limited, however, the spores were detected in four domestic wells and at one spring. The furthest straight-line distance that the spores travelled to a well was 7.7 km (4.8 miles).

Sinkhole collapses and land subsidence is a common natural process in many karst areas. Sinkholes occasionally form catastrophically and damage buildings, highways, sewage lagoons, and other man-made features. A sinkhole formed beneath a relatively new eight-story office building in Allentown, Pennsylvania in 1994 and damaged the building so severely that the nine million dollar structure was subsequently demolished. Sinkholes have collapsed in highways in several states including Virginia and Florida; one formed on State Highway 31 near Westminster, Maryland in 1994 and resulted in a fatal auto accident. Sinkhole collapses beneath sewage lagoons in Republic and West Plains, Missouri have resulted in major episodes of groundwater pollution and (at West Plains) disease outbreaks affecting hundreds of people.

Although sinkhole collapse is a natural process, the timing of a collapse can be induced by human activities. In some cases, lowering of groundwater levels can reduce buoyant support of soils and residuum and induce sinkhole collapse. This mechanism has been invoked to explain the inducement of some sinkhole collapses in Florida, Alabama, and elsewhere. In other cases, the surface impoundment or diversion of water has induced sinkhole collapses. This mechanism is often associated with sinkhole collapses in lagoons and lakes. Highway construction often results in diversion of appreciable volumes of runoff water to points that previously received very little surface flow. Such alterations of surface flow patterns have been associated with the inducement of sinkhole collapses. The impoundment or diversion of water induces sinkhole collapses by flushing sediments into and through underlying voids and/or by lessening the cohesiveness of soil and residuum overlying voids.

In some areas, sinkhole collapse poses risks that warrant detailed geotechnical investigations prior to construction of major or critical structures. Ground-penetrating radar is routinely used in some portions of Florida to detect locations particularly prone to sinkhole collapse. Grouting programs can then be used to lessen the risk of subsequent collapse of the identified areas. Although ground-penetrating radar works well in areas with sandy soils, it is generally not a useful technique in clay- or silt-rich soils. The proceedings of a 1984 conference on sinkholes, edited by Beck (1984), provide a number of papers relevant to sinkhole development and strategies for dealing with sinkhole issues.

19.1.6 *KARST HYDROLOGY INVESTIGATIONS*

The hydrology of karst terrains differs from other terrains in a number of different respects. As a result, effective hydrologic investigations in karst areas are likely to need somewhat different strategies and methods than those routinely applied to nonkarst areas. Unfortunately, in far too many cases, hydrogeologic investigations are based upon assumptions, methods, and approaches poorly suited to karst conditions. ASTM (1995) has developed a helpful guide for hydrogeologic investigations in karst areas; this document also provides an excellent bibliography relevant to karst hydrology. Quinlan (1989) provides a great deal of practical guidance valuable for credible groundwater monitoring in karst terraines.

Numerous geophysical methods have been employed with variable success in karst areas. As mentioned earlier, ground-penetrating radar can help identify buried sinkholes where they are covered with sandy soils. Natural potential surveys (Kilty and Lange, 1991) in karst areas can detect electrical currents produced by flowing water, and have proven beneficial at a number of sites in identifying discrete recharge zones and in siting monitor or extraction wells that tap directly into preferential subsurface flow routes.

In karst landscapes, groundwater flow directions may be substantially different from surface flow directions. Groundwater traces have often shown subsurface water movement from one topographic basin to another. In some cases, groundwater flow occurs beneath bedrock-floored rivers so that water that sinks into the groundwater system on one side of the river discharges from one or more springs on the opposite side. In many karst areas, it is common for tracer dyes introduced into the groundwater system at a localized point to discharge from multiple springs, which may be separated by several kilometers and may sometimes be in different topographic basins (Aley, 1988). Fracture traces, lineaments, and faults are often poor indicators of groundwater flow directions. Hydrologic investigations that assume these common karst conditions will not exist are likely to be flawed.

Potentiometric mapping is commonly used by hydrogeologists. In some karst areas, this tool can be useful, however, in other settings, the results are grossly incorrect. Potentiometric mapping presumes isotrophy, which is not characteristic of karst aquifers. Table 19.1.1, presented earlier, summarizing some dye tracing results, indicated that karst groundwater gradients in the tertiary permeability system are extremely low. Many groundwater traces, conducted in areas where potentiometric has been done, have shown flow generally parallel (rather than perpendicular) to equipotential lines. Groundwater tracing with fluorescent dyes can be used to assess the general credibility of potentiometric mapping in karst aquifers.

Groundwater tracing with fluorescent dyes is a very valuable investigative technique in karst aquifers. Although a number of fluorescent dyes have been used in karst groundwater investigations, the five dyes summarized in Table 19.1.2 are the most commonly used, have the best records of performance, and are routinely suitable for practical problem-solving work. All five dyes are safe for human health and the environment when used in a professional manner. The most current review of the potential adverse properties of these dyes is provided by Field et al. (1995). These dyes can be safely used in the environment if dye concentrations do not exceed one to two mg/l persisting for a period in excess of 24 hours in the groundwater at the point of groundwater withdrawal or discharge (Field et al., 1995).

Dye nomenclature is confusing because there are several alternate names for each dye; Table 19.1.2 identifies the most common names and reference numbers. In addition to the common names of the dyes (which are used in the text) one should also specify the Color Index Name and Number of

TABLE 19.1.2.   Names and Identifications of Five Common Groundwater Tracing Dyes

Dye	Color index name	Color index number	CAS number
Eosine	Acid Red 87	45380	17372-87-1
Alternate names: Eosin, Eosine OJ, D&C* Red 22			
Fluorescein	Acid Yellow 73	45350	518-47-8
Alternate names: Uranine, Uranine C, Sodium Fluorescein, Fluorescein LT, Fluorescent Yellow/Green, D&C Yellow 8, green fluorescent dye			
Rhodamine WT	Acid Red 388	Not Assigned	37299-86-8
Alternate names: Fluorescent Red, red fluorescent dye			
Sulforhodamine B	Acid Red 52	45100	3520-42-1
Alternate names: Sulfo Rhodamine B, Pontacyl Brilliant Pink B, Lissamine Red 4B, Kiton Rhodamine B, Acid Rhodamine B, Amido Rhodamine B, Fluoro Brilliant Pink, red fluorescent dye			
Pyranine	D&C Green 8	59040	6358-69-6
Alterrnate names: Solvent Green 7 (SG 7), D&C Green 8			

*D&C = Drug and Cosmetic.

the dye. The percent of dye in as-sold dye mixtures varies dramatically and is not always accurately reported by retailers. For rhodamine WT dye, the percent dye in typical as-sold dye mixtures varies from about 3 to 20 percent. For the other four dyes, the percent of dye in as-sold dye mixtures typically varies from about 2 to 80 percent.

Dyes can be detected visually or with various analytical instruments. Filter fluorometers (some of which can readily be used in the field) are sometimes appropriate instruments, yet spectrofluorophotometers (which are laboratory instruments) operated under a synchronous scan protocol are increasingly the analytical instrument and method of choice for most groundwater traces. The spectrofluorophotometer is set with standard excitation and emission slits, and a standard bandwidth separation between the excitation and emission wavelengths is used. For the five dyes under discussion, a liquid sample is placed in the instrument and sychronously scanned across a bandwidth of about 200 nanometers. The emission fluorescence intensity versus wavelength is then plotted; each of the dyes produces a characteristic peak within a relatively narrow wavelength range. The concentration of the dye is proportional to the magnitude of the emission fluorescence peak.

All five of the dyes under discussion can be detected in water samples. Additionally, all of these dyes can be adsorbed onto an appropriate laboratory grade of activated coconut shell carbon. Samplers containing a few grams of activated carbon (the author has standardized on 4.25 grams of carbon) are constructed and placed in the water to be sampled. The manufacturer reports that the activated carbon used by the author has a total surface area of 1150 square meters (12,370 square feet) per gram. The samplers will adsorb all five of the tracer dyes; the rate of adsorption is enhanced by circulating water such as would be encountered in water discharging from a well or spring. However, the samplers function adequately in most monitoring wells. Samplers are left in place for periods of time consistent with the study; weekly sampler recovery and new sampler placement is common. During the sampling period, the samplers will adsorb and accumulate any of the five tracer dyes present.

Analysis of the activated carbon samplers involves elution of the dye from the carbon with a solution consisting of a strong base, alcohol, and water. Different eluant mixtures are used by different laboratories, but most or all of the solutions are reasonably effective in eluting relevant dyes.

The use of activated carbon samplers is often critical to cost-effective groundwater tracing. The samplers ensure continuous sampling of the tested water so that short-duration pulses of tracer dye will not be missed. Unlike most continuous sampling equipment, the samplers are easy to place, recover, and analyze. Furthermore, the samplers accumulate tracer dyes; in many cases, the concentration of dye eluted from an activated carbon sampler left in place for a week will be at least two orders of magnitude greater than the mean dye concentration in the water. As a result, the activated carbon samplers lower the detection limit for the tracer dyes in the study.

Table 19.1.3 summarizes detection limits for the five tracer dyes under various different conditions. The instrumental detection limits are those that the author has determined for his analytical protocol. Different protocols would clearly yield different values. However, the reason for presenting these data

TABLE 19.1.3   Detection Limits of Five Tracer Dyes Under Different Conditions. Concentrations in Micrograms per Liter of As-Sold Dye Mixture. Values in Water Assume pH Between 6.5 and 8.0

Parameter	RWT	SRB	Py	Eos	Fl
Percent dye in as-sold mixture	20	75	77	75	75
Dye in water; instrumental analysis MDL*	0.007	0.020	—	0.005	0.0005
Dye in eluant; instrumental anslysis MDL*	0.155	0.080	0.125	0.040	0.010
Field conditions, experienced person	125	50	175	135	7
Field conditions, general public	2500	1000	3500	13500	140
Dark room, experienced person	50	5	3	10	2

Note: Pyranine is typically analyzed for by using a different protocol than that used for the other four dyes. The value shown for instrumental analysis in eluant is based upon analysis using the same protocol as for the other four dyes. No value is shown for the instrumental analysis of pyranine in water since the detection limit varies substantially in this pH range.

*MDL = Method Detection Limit.

is to illustrate that instrumental detection limits for the tracer dyes are routinely five or more orders of magnitude lower than the concentrations at which dyes might present adverse environmental effects or be visible to the public.

REFERENCES

Aley, T., "A Predictive Hydrologic Model for Evaluating the Effects of Land Use and Management on the Quantity and Quality of Water from Ozark Springs," *Missouri Speleology*, 18, 1978.

Aley, T., "Complex Radial Flow of Ground Water in Flat-lying Residuum-mantled Limestone in the Arkansas Ozarks," *Proc. Second Environmental Problems in Karst Terranes and Their Solutions Conf.*, Nat'l. Water Well Assn., pp. 159–170, 1988.

Aley, T., "Groundwater Tracing in the Epikarst," *The Engineering Geology and Hydrogeology of Karst Terranes; Proc. 6th Multidisciplinary Conf. on Sinkholes and the Engineering and Environmental Impacts of Karst*, Balkema, A. A., Rotterdam, pp. 207–211, 1997.

American Society for Testing and Materials (ASTM), Standard Guide for Design of Ground-water Monitoring Systems in Karst and Fractured Rock Aquifers, D-5717-95, 1995.

Beck, B. F. (ed.), "Sinkholes: Their Geology, Engineering, and Environmental Impact," *Proc. of 1st Multidisciplinary Conf. on Sinkholes*, Orlando, Florida, Balkema, A. A., Rotterdam/Boston, 1984.

Davies, W. E., Simpson, J. H., Ohlmacher, G. C., Kirk, W. S., and Newton, E. G., "Engineering Aspects of Karst," *National Atlas of the United States of America*, Dept. of the Interior, U.S. Geological Survey, Sheet 38077-AW-NA-07M-00, 1 sheet, 1984.

Field, M. S., Wilhelm, R. G., Quinlan, J. F., and Aley, T. J., "An Assessment of the Potential Adverse Properties of Fluorescent Tracer Dyes Used for Groundwater Tracing," *Environmental Monitoring and Assessment*, Kluwer Academic Publishers, Vol. 38, pp. 75–96, 1995.

Ford, D., and Williams, P., *Karst Geomorphology and Hydrology*, Unwin Hyman, 1989.

Hauwert, N. M., Johns, D. A., and Aley, T., *Preliminary Report on Groundwater Tracing Studies within the Barton Creek and Williamson Creek Watersheds, Barton Springs/Edwards Aquifer*, Barton Springs/Edwards Aquifer Conservation Dist. and City of Austin Watershed Protection Dept., 1998.

Huntoon, P. W., "Karstic Permeability: Organized Flow Pathways Created by Circulation," *Proc. Karst Modeling Symp., Karst Waters Institute Spec. Publ. 5*, Palmer, A. N., Palmer, M. V., and Sasowsky, I. D., eds., Karst Water Institute, Box 490, Charles Town, WV 25414, pp. 79–81, 1999.

Jones, W. K., "Karst Hydrology Atlas of West Virginia," *Karst Waters Inst. Spec. Publ. 4*, Karst Water Institute, Box 490, Charles Town, WV 25414, 1997.

Kilty, K. T., and Lange, A. L., "Electrochemistry of Natural Potential Processes in Karst," *Proc. Third Conf. on Hydrogeology, Ecology, Monitoring, and Management of Ground Water in Karst Terranes.*, Nat'l. Water Well Assn., pp. 163–177, 1991.

Kranjc, A. (ed.), "Tracer Hydrology 97," *Proc. of the 7th International Symposium on Water Tracing, Portoroz, Slovenia*, Balkema, A. A., Rotterdam/Boston, 1997.

White, W. B., *Geomorphology and Hydrology of Karst Terrains*, Oxford Univ. Press, Oxford, 1988.

CHAPTER 19
SENSITIVE ENVIRONMENTAL SYSTEMS

SECTION 19.2
ALPINE LAKES

Peter Shanahan

Dr. Shanahan is president of HydroAnalysis, Inc., a hydrology and water quality consulting practice in Acton, Massachusetts, and a lecturer in the Department of Civil and Environmental Engineering at the Massachusetts Institute of Technology, Cambridge.

TABLE OF NOMENCLATURE AND UNITS

ANC_{crit} = critical limit for acid neutralizing capacity [μeq/l]

BC_d = annual nonmarine deposition of base cations in the watershed [meq/m^2/y]

$[BC]_0$ = pre-acidification concentration of nonmarine base cations [μeq/l]

CL = critical annual acid load [meq/m^2/y]

CL_{exS} = critical annual load for sulfate [meq/m^2/y]

CL_{exS+N} = critical annual load for sulfate plus nitrate [meq/m^2/y]

meq/m^2/y = milliequivalents per square meter per year

N_{le} = annual leaching of nitrate from the watershed into surface water [meq/m^2/y]

Q = annual runoff [m/y]

S_d = annual nonmarine deposition of sulfate in the watershed [meq/m^2/y]

UVB = ultraviolet-B radiation, also known as far ultraviolet, with a wavelength between 200 and 300 nm

19.2.1 INTRODUCTION: THE ALPINE ENVIRONMENT

Ives and Barry (*Arctic and Alpine Environments*, Methuen & Co., Ltd., London, 1974) define alpine as simply "A high-altitude belt, above the treeline on mountains." In many regions, the treeline is distinct, drawing a sharp definition of the alpine region. Often, however, there is a transition zone of stunted trees, known as krummholz, and there may be isolated patches of trees within an otherwise alpine zone.

The elevation of the timberline varies with latitude as shown in Figure 19.2.1 and, at any given latitude, with local climate. The global distribution of the treeline shows that it more or less corresponds to the altitude at which the mean monthly temperature of the warmest month is 10°C. The upper terminus of the alpine zone is the altitude where vascular plants no longer survive. This corresponds generally with the zone of year-round snow. (Swan, in *Arctic and Alpine Environments*, Wright and Osburn (eds.), Indiana University Press, pp. 29–54, 1967).

The alpine is an environment of harsh climate and meager nutrient stores. The summers are short and cool while the winters are long, cold, and windy. Soil is typically a thin, poorly fertile mantle over

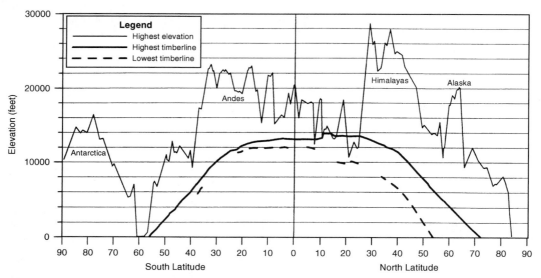

FIGURE 19.2.1   Variation of timberline and temperature with latitude (based on Swan, in *Arctic and Alpine Environments*, Wright and Osburn (eds.), Indiana University Press, pp. 29–54, 1967).

crystalline bedrock. Runoff of snowmelt and rainfall is rapid and only weakly attenuated by infiltration. Soils are usually acidic and the underlying rock poorly soluble, restricting the amount of acid neutralizing chemicals that can be liberated (Camarero and Catalan, *Wat. Res.*, 32: 1126–1136, 1998).

These factors conspire to make alpine lakes uniquely vulnerable to environmental and climatic influences. The very limited capacity of most alpine lakes to neutralize acid makes them susceptible to acid precipitation. The delicate ecology created by the short growing season and nutrient-poor environment adds vulnerability to climate change. The paucity of nutrients also creates susceptibility to eutrophication in situations of increased nutrient influx.

Alpine lake research is concentrated in a number of programs and institutions. In the United States, the Institute for Arctic and Alpine Research (INSTAAR) at the University of Colorado in Boulder has long been active in alpine research, generally, and in alpine lakes, specifically. Long-term studies of specific mountain watersheds have also been conducted by the U.S. National Park Service (Baron (ed.), *Biogeochemistry of a Subalpine Ecosystem: Loch Vale Watershed*, Springer-Verlag, New York, 1991) and the state of California (Tonnessen, *Wat. Resourc. Res.*, 27: 1537–1537, 1991). In Europe, institutions from 11 countries are cooperating in the AL:PE (Acidification of Mountain Lakes: Palaeolimnology and Ecology) project funded by the European Commission (Mosello et al., *Ann. Chimica*, 85: 395–405, 1995). While these and other institutions conduct active alpine research programs, there are many gaps in knowledge of the alpine environment. A significant impediment to research is the remoteness of the areas—for example, Logan et al. point out that three of their five study lakes in the Washington state Cascade Range are at least five miles from the nearest road (Logan et al., *Environ. Sci. Technol.*, 16: 771–775, 1982).

19.2.2 HYDROLOGY OF ALPINE LAKES

Characteristics of Alpine Watersheds

Alpine environments vary greatly in their physical characteristics depending upon the local geology and topography. In general, alpine regions share these common characteristics: steep, unstable slopes;

thin, erodable soils; and extensive soil frost activity (Olson, *Air Pollution and Acid Rain Report 8*, U.S. Fish and Wildlife Service, Report No. FWS/OBS-80/40.8, 1982). Alpine basins have generally experienced extensive carving by glaciers resulting in cirque lakes and other characteristic landforms. As a result of glacial scouring, soils are young and postdate the Pleistocene glaciation (Retzer, in *Arctic and Alpine Environments*, Ives and Barry (eds.), Methuen & Co., Ltd., London, pp. 771–784, 1974).

The variability of alpine landscape implies a parallel variability in alpine soils (Retzer, in *Arctic and Alpine Environments*, Ives and Barry (eds.), Methuen & Co., Ltd., London, pp. 771–784, 1974). A primary factor is the parent rock from which the soil forms. Alpine areas are underlain by all types of rock, although most often it is a crystalline igneous or metamorphic rock. The hardness of these rocks, and the unfavorable climate, lead to very slow weathering. The rocks yield calcium and magnesium to weathering very grudgingly, causing a deficiency in basic cations. As a result, alpine soils are typically acidic. The soils tend to be medium-to-coarse textured and excessively drained.

Slow soil formation and active erosion on the steep slopes leads to soils that are thin or nonexistent in alpine areas. Freeze-thaw cycles during the seasonal melting period, wind action, and snowmelt runoff all contribute to erosion and facilitate downslope movement because of gravity. Together, these forces contribute to what Retzer calls "an ever-present force that ... wears away the landscape and contributes to the immaturity of the soil pattern" (Retzer, in *Arctic and Alpine Environments*, Ives and Barry (eds.), Methuen & Co., Ltd., London, pp. 775, 1974).

Vegetation—in particular, the lack of trees as a vegetative form—define the alpine environment. The sparsity of vegetative cover in alpine watersheds can be appreciated by Arthur's characterization of the Loch Vale Watershed in the American Rocky Mountains: only 18 percent of the watershed is vegetated (Arthur, in *Biogeochemistry of a Subalpine Ecosystem: Loch Vale Watershed*, Baron (ed.), Springer-Verlag, New York, pp. 76–92, 1991). Williams and Melack (*Wat. Resourc. Res.*, 27: 1575–1588, 1981) similarly report that the Emerald Lake Watershed in the California Sierra Nevada consists of 33 percent exposed rock and 47 percent talus and unconsolidated sediments.

Vegetation above the timberline is classified as alpine tundra, consisting of low herbaceous vegetation, lichens, and dwarf shrubs. Vegetative cover is patchy, responding to local conditions that afford protection from the cold and wind or enhance soil moisture. Plant species have evolved a variety of protective mechanisms to reduce water loss, exposure to wind, and exposure to radiation (Arthur, in *Biogeochemistry of a Subalpine Ecosystem: Loch Vale Watershed*, Baron (ed.), Springer-Verlag, New York, pp. 76–92, 1991). Vegetation also adjusts to the thin, dry soils of the alpine by forming shallow root systems that create tough, dense sods (Retzer, in *Arctic and Alpine Environments*, Ives and Barry (eds.), Methuen & Co., Ltd., London, pp. 771–784, 1974).

Limited availability of the plant nutrients nitrogen and phosphorus from the soils severely restricts the potential biomass of alpine tundra (Arthur, in *Biogeochemistry of a Subalpine Ecosystem: Loch Vale Watershed*, Baron (ed.), Springer-Verlag, New York, pp. 76–92, 1991). Indeed, terrestrial sources are so limited that airborne supplies play an important role in the nutrient chemistry of alpine ecosystems and dominate the nutrient dynamics at higher altitudes (Kopácek et al., *Limnol. Oceanogr.*, 40: 930–937, 1995; Swan, in *Arctic and Alpine Environments*, Wright and Osburn (eds.), Indiana University Press, pp. 29–54, 1967).

Alpine Climate

The altitude of alpine regions gives rise to a number of climatic factors. These include a decline in atmospheric pressure, a decline in temperature, and a rise in the intensity of short-wave (solar) radiation (The Major Problems of Man and Environment Interactions in Mountain Ecosystems, UNEP Report No. 2, United Nations Environment Programme, Nairobi, 1980). Some of the more obvious effects of these factors are discussed above: cool temperatures in summer and extreme cold temperatures in winter; strong winds, especially in winter; snow as the predominant form of precipitation; and a short growing season.

Solar radiation is more intense in alpine environments and particularly in alpine lakes (Sommaruga and Psenner, *Photochem. Photobio.*, 65: 957–963, 1997). Radiation increases with elevation because of the thinner overhead atmosphere. The increase affects the entire radiation spectrum, but is particularly intense for UVB wavelengths between 320 and 400 nm. The effect is compounded in alpine lakes, where there is greater radiation penetration than in temperate lakes because of the high transparency and low organic carbon content of the water.

The atmospheric chemistry of alpine areas differs from that at lower elevations. Atmospheric chemicals are deposited through dry deposition, wet deposition (precipitation), and occult deposition. Elevated sites experience occult deposition—direct deposition from clouds that intercept the land surface or vegetation—far more frequently than at lower elevations. Occult deposition is typically more concentrated in chemicals than wet deposition because cloud droplets are less diluted than raindrops. Cloud droplets may have very low pH (in the range of 2 to 3) and the droplets may also be concentrated in nitrates and sulfates depending upon air pollution patterns (Hough, in *Acid Deposition at High Elevation Sites*, Unsworth and Fowler (eds.), Kluwer Academic Publishers, Boston, pp. 1–47, 1988).

Alpine Hydrology

Slaymaker gives a comprehensive overview of alpine hydrology, from which this summary is largely drawn (Slaymaker, in *Arctic and Alpine Environments*, Ives and Barry (eds.), Methuen & Co., Ltd., London, pp. 133–158, 1974). He defines six components whose combination distinguishes alpine hydrology: (1) high radiation, (2) high precipitation, (3) discharge hydrographs greatly influenced by snow or glacial melt, (4) high topographic relief, (5) poorly developed soils, and (6) minimal tree cover.

Radiative heat transfer in alpine areas is affected by the change in atmospheric radiation with elevation discussed above and also by other factors. The albedo (reflectance) of alpine surfaces tends to be higher than at lower elevations because of the prevalence of snow and ice cover. Moreover, alpine slopes may receive vastly different amounts of incoming radiation depending upon their slope, compass orientation, and the time of year. The net result is that extremely high radiation fluxes have been measured at alpine sites. These fluxes do not, however, commensurately affect air temperature, because the thin air stores little heat but allows it to be quickly radiated away.

Precipitation in alpine areas responds to the topography, with ample orographic precipitation falling on the windward side of the mountains and considerably less on the leeward side. Much of the precipitation falls as snow, which can be extensively redistributed by drifting and avalanches. More important than the spatial redistribution is the temporal redistribution of precipitation by the snow accumulation and melt cycle. Snowmelt often occurs very rapidly as temperatures warm to a critical level in spring or summer. Coupled with the rapidity of snowmelt are characteristics of alpine watersheds that tend to minimize infiltration and favor runoff. These factors include steep slopes, thin soils, soil frost, minimal tree cover, and exposed rock.

A quantitative assessment of the hydrology of the Emerald Lake alpine basin over a two-year period is provided by Kattelmann and Elder (*Wat. Resourc. Res.*, 27: 1553–1562, 1991). Stated in terms of annual averages, they account for the fate of 1835 mm of total precipitation as 400 mm lost to the atmosphere and 1415 mm to runoff, with an unaccounted residual of 20 mm. The annual precipitation is dominated by snow, which contributed 95 percent of the total. Losses to the atmosphere were also dominated by snow: 340 mm out of the annual loss of 400 mm were sublimation from snow, while 15 mm were evaporation from the lake itself, and 65 mm were evapotranspiration. Groundwater was found to be a negligible factor in the annual water balance. Most of the annual streamflow arises from snowmelt, producing an annual hydrograph dominated by late spring and summer discharge (Figure 19.2.2).

Alpine Limnology

Most lakes in temperate climate zones show a distinct seasonal cycle in temperature distribution with depth. At the end of winter, a lake is typically mixed throughout its depth and shows a vertically isothermal temperature profile—the water is at a constant temperature of approximately 4°C from top to bottom. As the sun and atmosphere warm the lake surface through the spring, the shallowest

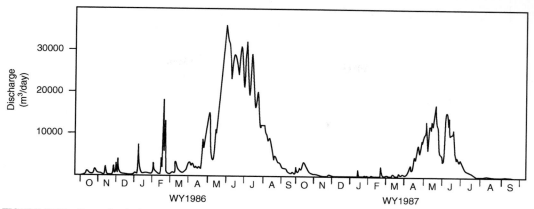

FIGURE 19.2.2   Streamflow hydrograph for the Emerald Lake alpine watershed, Sierra Nevada, California (from Kattelmann and Elder, *Wat. Resourc. Res.*, 27: 1553–1562, 1991, reproduced with permission).

water warms relative to the deeper water. Soon, a distinct layer of warmer and lighter surface water (the epilimnion) floats atop the cold, heavy deep water (the hypolimnion). The intervening layer, in which temperature decreases rapidly with depth, is known as the thermocline. (Technically, the thermocline is defined as the zone in which the change in temperature with depth exceeds 1°C per meter.) Through the summer, the thermocline becomes stronger (that is, the change in temperature over the vertical distance of the thermocline increases). Finally, with surface cooling in the fall, the temperature stratification weakens until the fall overturn, when the lake mixes throughout its depth, and temperature is once again isothermal. During the winter an inverse stratification develops, with water of 0°C immediately beneath the surface ice and slightly warmer water below. The water at the lake bottom in winter is typically between 0°C and 4°C, the temperature at which water has its maximum density.

Alpine lakes often depart from the classical temperature regime of temperate lakes. Hutchinson provides the definitive classification of lakes as a function of their thermal structure (Hutchinson, *A Treatise on Limnology, Volume I, Part I—Geography and Physics of Lakes*, Wiley, New York, 1957; Wetzel, *Limnology*, Saunders, W. B., Philadelphia, pp. 76–79, 1975). Of the various possible categories, two describe deep alpine lakes. Dimictic lakes are those that follow the typical pattern of temperate lakes, mixing completely twice each year, at the fall and winter overturns. Cold monomictic lakes are those whose temperature rises to near 4°C only once per year during the summer, when the lake experiences its single annual period of mixing. These lakes are often in direct or hydrologic contact with the cooling input of a glacier. Shallow alpine lakes, like shallow temperate lakes, may remain sufficiently stirred by wind and streamflow to never stratify.

The ecology of alpine lakes is dictated by the generally cool temperatures and shortness of the growing season. Nannoplankton dominate and persist as autotrophs beneath the ice cover through winter [Wetzel, *Limnology*, Saunders, W. B., Philadelphia, p. 326, 1975). Attached algae (periphyton) grow on submerged rocks, and are dominated by diatoms. Aquatic macrophytes are rare, and invertebrate diversity is limited. Invertebrate species known to be sensitive to pH appear to have heightened sensitivity in alpine lakes (Wathne et al. (eds.), *AL:PE 1 Report*, Report EUR 16129 EN, European Commission, Brussels, 1995).

Alpine lakes tend to be oligotrophic, with low biomass as a consequence of low nutrient availability. As a consequence, alpine lake water is often extremely clear. Exceptions are lakes that are actively fed by glaciers and thereby contain higher concentrations of inorganic sediment. With or without glacial input, organic matter is typically low in the water column. Phytoplankton of alpine lakes have adapted to the high levels of radiation that occur as the result of altitude and water clarity (Halac et al., *J. Plankton Res.*, 19: 1671–1686, 1997).

19.2.3 *CHEMISTRY OF ALPINE LAKES*

Factors Affecting Lake Chemistry

The vulnerability of alpine lakes is primarily a chemical vulnerability: alpine lakes are damaged by chemical inputs from the atmosphere to the watershed. Camarero et al. statistically evaluated a wide range of European lakes to determine the factors that dictate lake chemistry (Camarero et al., *Limnologica*, 25: 141–156, 1995). They found the four main factors, in order of importance, to be as follows: the weathering characteristics of the watershed, the influence of sea salts, airborne anthropogenic pollution, and eutrophication. These factors accounted for 37, 19, 9, and 8 percent of the observed variability in lake chemistry in the study. Another factor, possibly associated with weathering, accounted for an additional 16 percent.

Camarero et al. used their extensive chemical data to evaluate the mechanisms of these different factors. Weathering characteristics of the watershed, by far the most important factor, depend upon the composition of the bedrock and its propensity to weather, measured in the study through the concentration of the cations Ca^{2+} and Mg^{2+} in the lake water. Aerosols from marine areas carry sea salts to inland lakes, measured through the concentration of Cl^-, as well as Na^+ and K^+. Sea salt deposition creates a different chemistry in lakes nearer the sea. Airborne pollution accounts for the majority of NO_3^- and SO_4^{2-}, although weathering and natural processes produce these minerals as well. Elevated influx of these chemicals from atmospheric sources is the defining characteristic of acid rain. The chemicals were found at higher concentrations at higher elevations. Finally, Camarero et al. found the chemistry of phosphorus to be related to hydraulic residence time and the prevalence of vegetation within the watershed.

Relationship to Basin Geology

The statistical analysis by Camarero et al. emphasizes the importance of the basin geology to the lake's vulnerability to acid precipitation. The crystalline bedrock of most European alpine watersheds engendered far more vulnerability than the calcerous bedrock found in a few watersheds in the Italian Alps. Thus, lake chemistry is intrinsically related to watershed geology.

The geology of the basin affects water chemistry as a consequence of the process of mineral weathering (Psenner and Catalan, in *Limnology Now: A Paradigm of Planetary Problems*, Margalef (ed.), Elsevier, Amsterdam, pp. 255–314, 1994). In weathering, minerals are dissolved by water and acidic solutes, including carbon dioxide (carbonic acid), organic acids, nitric acid, and sulfuric acid. A classic model for chemical weathering is presented by Garrels and MacKenzie (*Adv. Chem. Ser.*, 67: 222–242, 1967). They explain the chemistry of springs and lakes in the Sierra Nevada of California as the product of silicate mineral (predominantly plagioclase feldspar) attack by CO_2-enriched waters.

While weathering dictates basin chemistry over the long term, ion exchange at particle surfaces may be significant over shorter time periods (Psenner and Catalan, in *Limnology Now: A Paradigm of Planetary Problems*, Margalef (ed.), Elsevier, Amsterdam, pp. 255–314, 1994). For example, Williams et al. (*Limnol. Oceanogr.*, 38: 775–797, 1993) paint a complex seasonal pattern of basin chemistry. They found that the snowmelt portion of the annual runoff occurred too quickly to be significantly altered by slow weathering and that cation exchange in the soil and talus accounted for neutralization of these waters. Slower runoff during the post-snowmelt transition period was consistent with plagioclase weathering while low flows were the product of weathering of a more complex mix of minerals. Stoddard (*Limnol. Oceanogr.*, 32: 825–839, 1987) also found consistent seasonal variations in lake alkalinity, with significantly higher alkalinity because of groundwater discharge during the late summer and winter low-flow period.

An important characteristic of weathering processes relative to alpine lake vulnerability is the speed of weathering. In many basins, chemical weathering must supply virtually all of the cations (alkalinity) that can neutralize acidic atmospheric inputs. Typical rates of chemical weathering contribute about

20 to 30 meq/m^2/y of cations (Psenner and Catalan, in *Limnology Now: A Paradigm of Planetary Problems*, Margalef (ed.), Elsevier, Amsterdam, pp. 255–314, 1994).

Relationship to Climate and Hydrology

The preceding discussion of weathering dynamics suggests the important interplay between basin chemistry and basin climate and hydrology. In particular, the spring snowmelt is a critical factor in the annual chemical balance and the vulnerability of alpine lakes.

The importance of the snowmelt derives largely from the capacity of snow to store atmospheric deposition. Acidic atmospheric-deposition inputs (about 50 meq/m^2/y in the Alps) may overwhelm the neutralization capacity afforded by weathering (about 20 to 30 meq/m^2/y) (Psenner and Catalan, in *Limnology Now: A Paradigm of Planetary Problems*, Margalef (ed.), Elsevier, Amsterdam, pp. 255–314, 1994). The fact that most of the annual acid input becomes concentrated in the spring runoff only compounds its adverse effect. Williams and Melack (*Wat. Resourc. Res.*, 27: 1563–1574, 1991) found that 90 percent of the solute flux to a Sierra Nevada watershed came from snowfall, was stored in the seasonal snowpack, and was released in the spring snowmelt. As a result, they found a dramatic springtime increase in the concentration of NO_3^- and a corresponding decrease in alkalinity. While there was apparently ample neutralization capacity to minimize pH changes in the particular watershed they studied, sharp springtime rises in pH are observed in many watersheds (Jeffries et al., *J. Fish. Res. Board Can.*, 36: 640–646, 1979). Diurnal freeze-melt patterns during the springtime may moderate snowmelt releases by increasing the relative fraction of more neutral soil water to snow melt (Stottlemyer and Troendle, *J. Hydrol.*, 140: 179–208, 1992).

Although alpine lake acid-base chemistry is now dictated by the tension between chemical weathering and atmospheric inputs, paleolimnological studies show that different patterns prevailed prior to the onset of anthropogenic acid precipitation (Psenner and Schmidt, *Nature*, 356: 781–783, 1992; Koinig et al., in *Headwaters: Water Resources and Soil Conservation*, Haigh et al. (eds.), Balkema, Rotterdam, pp. 45–54, 1998; Wögrath and Psenner, *Wat. Air Soil. Pollut.*, 85: 359–364, 1995). Different diatom species predominate at different pH. Thus, the prevalence of different species in sediment cores allows a reconstruction of pH chemistry over the past several hundred years in a number of alpine lakes. What is found is a strong correlation between air temperature and pH, with higher pH and alkalinity during warm periods owing to increased dust deposition, faster weathering, and higher primary production. This relationship was at least partially decoupled around the turn of the nineteenth century in most lakes as a result of acid precipitation. Nonetheless, Sommaruga-Wögrath et al. (*Nature*, 387: 64–67, 1997) indicate the potential role of remote alpine lakes as sensitive indicators of global temperature rise.

Nitrogen and Metals

While most studies have emphasized the acid-base chemistry of alpine lakes, other chemicals deposited from the atmosphere may also challenge alpine lakes. Williams et al. (*Environ. Sci. Technol.*, 30: 640–646, 1996) document the potential saturation of forested sub-alpine watersheds with nitrogen from atmospheric deposition. They define saturation as a condition in which the entire nitrogen flux to the watershed is not consumed such that there is an export of nitrogen via surface-water outflow. Kopáček et al. (*Limnol. Oceanogr.*, 40: 930–937, 1995) similarly conclude that excessive atmospheric inputs of nitrogen may shift the nutrient balance in alpine lakes and thereby create more oligotrophic conditions (lower biomass) in those lakes.

Anthropogenic atmospheric deposition may also carry heavy metals and other chemicals to remote alpine lakes. Lorey and Driscoll, for example, report increased mercury concentrations since 1850 in sediment cores from lakes in the Adirondack Mountains of New York state (Lorey and Driscoll, *Environ. Sci. Technol.*, 33: 718–722, 1999). Acidification may also release metals already in the watershed, particularly aluminum, with adverse consequences for fish (Schindler, *Science*, 239: 149–157, 1988).

19.2.4 *MANAGEMENT OF ALPINE ENVIRONMENTS*

Alpine Lake Vulnerability

Alpine lakes are highly vulnerable to acid precipitation, chemical deposition, and climate change. The heightened vulnerability of these lakes owes to the limited acid neutralizing capacity of alpine watersheds and the harshness of the environment and consequent pressure on aquatic ecosystems. This vulnerability has prompted studies in the United States and Europe to understand the alpine environment and develop strategies for mitigating risk to alpine ecosystems.

Acid Precipitation

A number of European investigators working under the AL:PE program have developed the concept of the critical load—the pollutant loading that marks the threshold of significant effects on sensitive environmental elements (Marchetto et al., *Ambio*, 23: 150–154, 1994). The approach follows similar approaches for acidification of nonalpine lakes and eutrophication of temperate lakes. Marchetto et al. present a derivation for the critical load, but a subsequent simpler version is presented in the AL:PE Part 2 report (Wathne et al. (eds.), *AL:PE, Part 2*, Report No. 3638-97, Norwegian Institute for Water Research, Oslo, 1997). The AL:PE report presents separate formulae for the total critical load, the critical load for sulfur, and the critical load for sulfur plus nitrogen. The formulae are presented here in terms of load per unit area of watershed.

$$CL = ([BC]_0 - ANC_{\text{crit}})\, Q - BC_d \qquad (19.2.1)$$

$$CL_{exS} = S_d - BC_d - CL \qquad (19.2.2)$$

$$CL_{exS+N} = S_d + N_{le} - BC_d - CL \qquad (19.2.3)$$

where, CL = critical annual acid load [meq/m^2/y];
   $[BC]_0$ = pre-acidification concentration of nonmarine base cations [μeq/l];
   ANC_{crit} = critical limit for acid neutralizing capacity to prevent damage to biological indicators [μeq/l];
   Q = annual runoff [m/y];
   BC_d = annual nonmarine deposition of base cations in the watershed [meq/m^2/y];
   CL_{exS} = critical annual load for sulfate [meq/m^2/y];
   S_d = annual nonmarine deposition of sulfate in the watershed [meq/m^2/y];
   CL_{exS+N} = critical annual load for sulfate plus nitrate [meq/m^2/y]; and
   N_{le} = annual leaching of nitrate from the watershed into surface water [meq/m^2/y].

For the AL:PE lakes, ANC_{crit} was taken as 20 μeq/l and BC_d as 30 meq/m^2/y. Computed critical loads for 23 AL:PE lakes ranged from zero to approximately 150 meq/m^2/y with one outlier of nearly 1000 meq/m^2/y. Many of the lakes had CL values of less than 50 meq/m^2/y. Approximately half the lakes exceeded the critical load for combined nitrate and sulfate, and somewhat fewer for sulfate alone. One Slovakian lake exceeded the critical load for sulfate and nitrate of over 200 meq/m^2/y.

Modeling is also a potentially valuable tool for management and protection of alpine lakes. A number of researchers have developed and applied models to alpine lakes. The critical loading formulae above are essentially a simple model for a single lake. A simple model for evaluating regional acidification by Camarero and Catalan (*Wat. Res.*, 32: 1126–1136, 1998) incorporates random variables to represent lake and basin properties. A complex hydrologic and chemical model specifically for alpine sites is the alpine hydrogeochemical model (AHM) (Wolford et al., *Wat. Resourc. Res.*, 32: 1061–1074, 1996; Wolford and Bales, *Limnol. Oceanogr.*, 41: 947–954, 1996).

Reduction of acid inputs to below the critical levels indicated by models or loading formulae does not necessarily imply lake recovery. Kirchner and Lyderson (*Environ. Sci. Technol.*, 29: 1953–1960, 1995) found persistent effects from base cation depletion in three Norwegian watersheds. They concluded that past supercritical loading had depleted base cations to the point that a progressive reduction in the level of acid precipitation over 15 years has yet to produce an improvement in runoff water quality. Their results argue that acid inputs must be reduced sufficiently below the critical level to effect a restoration of base cations. Simply meeting critical acid loads only allows the base cation production of the watershed to neutralize new inputs, and not to produce a recovery to pre-acidification levels.

Climate Change

Climate change has clear potential to affect alpine lakes, although precise predictions are impossible. Global warming, for example, would produce a compression of climate zones with altitude (Nilsson and Pitt, *Mountain World in Danger: Climate Change in the Mountains and Forests of Europe*, Earthscan Publications Limited, London, 1991). Among the possible consequences are wintertime increases and summertime decreases in precipitation; wasting of glaciers; reduction of permafrost and snow-cover duration; increases in forest fires, rockfalls, and avalanches; and ecological shifts (Beniston and Fox, in *Climate Change 1995: Impacts, Adaptations and Mitigation of Climate Change: Scientific-Technical Analyses*, Watson et al. (eds.), Cambridge University Press, Cambridge, pp. 191–213, 1996). Moreover, because alpine environments are more vulnerable, they are viewed as a bellwether of future widespread climate change effects at temperate elevations.

CHAPTER 19
SENSITIVE ENVIRONMENTAL SYSTEMS

SECTION 19.3

WETLANDS

Margaret A. McBrien

Ms. McBrien, of The Louis Berger Group, Inc., is a certified Professional Wetland Scientist (P.W.S.) and Professional Engineer (P.E.) with experience in wetlands and water resources management.

Kirk R. Barrett

Dr. Barrett is a research assistant professor and the research director of the Meadowlands Environmental Research Institute (MERI) at Rutgers University, Newark, New Jersey. He has expertise in surface and wetland hydrology, hydraulics and water quality processes.

The authors gratefully acknowledge the insightful reviews provided by Ralph Tiner, U.S. FWS, Hadley, MA; Ken Scarlatelli HMDC, Lyndhurst, NJ; Sydney Bacchus, University of Georgia; and Perry Lund, Washington, Department of Ecology.

19.3.1 INTRODUCTION

Wetland ecosystems are dispersed across the world, comprising approximately 6 percent of the land surface (Mitsch and Gosselink, 1993). At the beginning of European settlement, the United States sustained approximately 90 million hectares (220 million acres) of wetlands. Only 40 million hectares (100 million acres) of these sensitive environmental systems remain today (USGS, 1996) (Figure 19.3.1).

Prior to the 1970s, filling and draining of wetlands for farmland, housing developments, industrial facilities, and other "land improvement" was an accepted and promoted practice in the United States. Many states have lost over half of their original wetlands (Figure 19.3.2). Effects of wetland loss include increased flood flows, impairment of surface water quality, and impoverishment of wildlife habitat. Over the last few decades, as the benefits of wetlands were realized, the country's attitude shifted. Now federal, state, and local governments have enacted laws and regulations to protect this diminishing resource.

19.3.2 DEFINITION

Wetlands exhibit many forms and hydrological attributes, ranging from intermittently saturated forests to permanently flooded reedbeds to tidally influenced grasslands. Often, wetlands are transition

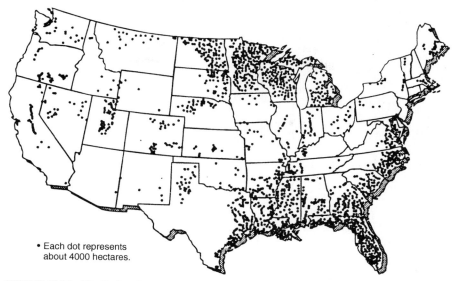

• Each dot represents about 4000 hectares.

FIGURE 19.3.1 Distribution of natural wetlands in the coterminus United States (Kadlec and Knight, 1996).

Percentage of Wetlands Acreage Lost, 1780s–1980s

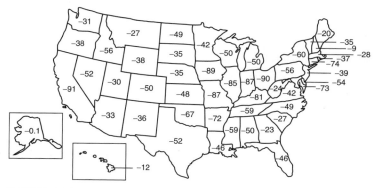

FIGURE 19.3.2 States with notable wetland loss, 1780s to mid-1980s.

zones (ecotones) between terrestrial, or upland, ecosystems and deepwater aquatic systems (refer to Figure 19.3.3). The precise definition of wetlands remains a controversial topic. Three Federal Agencies (the U.S. Army Corps of Engineers, U.S. Fish and Wildlife Service, and the Natural Resources Conservation Service) provide the 3 definitions of wetlands currently used in the United States. The U.S. Army Corps of Engineers defines wetlands as follows:

Areas that are inundated or saturated by surface or groundwater at a frequency and duration sufficient to support, and that under normal circumstances do support, a prevalence of vegetation typically adapted for life in saturated soil conditions. Wetlands generally include swamps, marshes, bogs, and similar areas (33 Code of Federal Regulations (CFR) 328.3 (b); 1993).

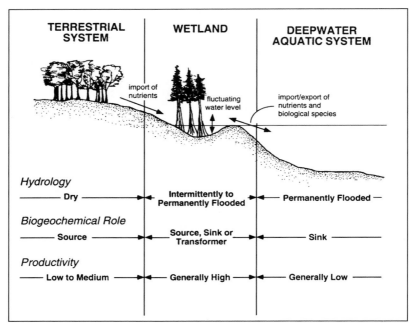

FIGURE 19.3.3   Wetlands: Ecotones between terrestrial and deepwater ecosystems (Mitch and Gosselink, 1993).

19.3.3   TYPES OF WETLANDS

Terminology for wetland types is often confusing. Terms for a grassy, inundated wetland in North Dakota include wet meadow, wet prairie, marsh, slough, and prairie pothole. To eliminate some of the confusion, the United States Fish and Wildlife Service (USFWS) developed a scientific classification and definition of wetlands (Cowardin et al., 1979). The classification provides a hierarchical taxonomic organization to describe the geomorphic, hydrology, vegetation, and substrate of wetland and deepwater habitats. The taxonomy begins with five major systems: marine, estuarine, lacustrine (lake-associated), riverine (river-associated), and palustrine (other freshwater systems). Wetlands are further classified based on tidal influence, water level, bottom type, vegetation type, and duration of standing water (Figure 19.3.4).

The USFWS employed the classification to create National Wetlands Inventory (NWI) maps. Prepared by stereoscopic analysis of high altitude aerial photographs, the informative NWI maps do not provide actual wetland boundary determinations, but do identify the type and approximate areal extent of wetlands and waterways in 89 percent of the lower 48 states and 31 percent of Alaska (http://wetlands.fws.gov/overview.htm). A digital database is under development. The accuracy of the NWI maps is limited, so field surveys are necessary to delineate actual wetland boundaries.

The U.S. FWS classification system is commonly employed by wetland scientists, but is cumbersome for laypeople. Hence, the variety of colloquial terms for wetlands persists and includes the following.

- *Swamps* are dominated by trees or shrubs and include the freshwater red maple swamps and alder thickets of the northeast and the bottomland hardwood forests and brackish mangrove swamps of the south.

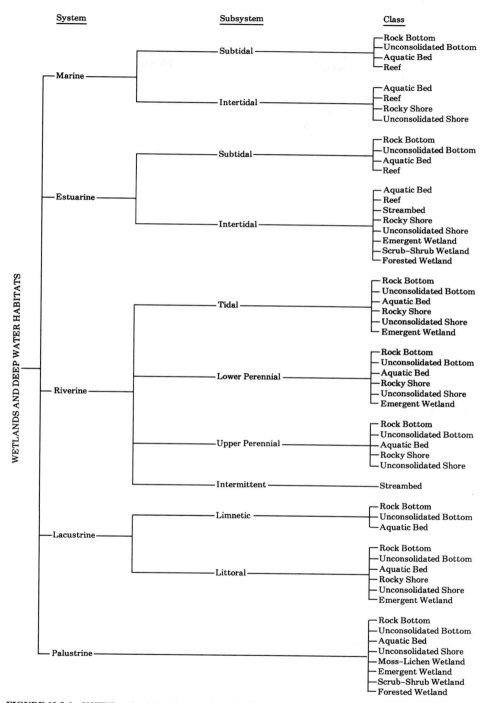

FIGURE 19.3.4   USFWS wetland classification hierarchy (Cowardin et al., 1979).

- *Marshes* are freshwater and saltwater ecosystems characterized by herbaceous vegetation (grasses, sedges, rushes, reeds, other emergent plants, and floating vegetation). Saltwater and some freshwater marshes are tidally influenced. Drier marshes are often called wet meadows or wet prairies.

- *Bogs* are acidic peat deposits that generally contain a high groundwater table, often lack a significant surface water connection, and support acid-loving plants (including carnivorous plants).

- *Fens* tend to contain alkaline soils that receive mineral rich surface drainage and groundwater and support grasses, sedges, or rushes.

- *Vernal Pools* are small, shallow, sparsely vegetated depressions that are seasonally flooded. They provide critical breeding grounds for several amphibians.

- *Prairie Potholes*, formed by glacial action in North and South Dakota and other northern prairie states, are small, shallow, freshwater, grassy wetlands that are often dry during the summer.

The EPA wetlands web site has pictures and descriptions of wetland types.

19.3.4 *WETLAND HYDROLOGY*

Hydrology is probably the most significant defining characteristic of wetlands. The depth of inundation varies from several feet of water to soils saturated several inches below the surface with no standing water. This flooding or saturation may be permanent, seasonal (e.g., vernal pools), tidal or intermittent. Wetland hydrology is explored in detail in Winter, T. C., and Llama, M. R. (eds.), "Hydrogeology of Wetlands," *Special Issue Journal of Hydrology*, 141: 1–269, 1993.

Various hydrologic components transfer water to or from a wetland (Figure 19.3.5). Water is supplied to wetlands via precipitation, surface (tidal, stream, or flood) flows, overland catchment runoff, and/or groundwater discharge. Water is released via surface outflows, groundwater recharge,

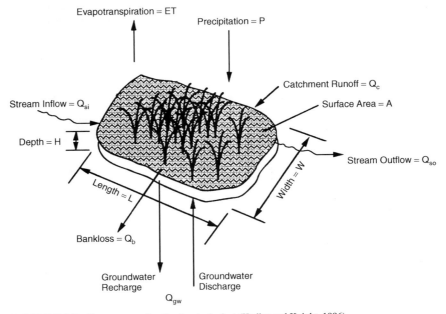

FIGURE 19.3.5   Components of wetland water budget (Kadlec and Knight, 1996).

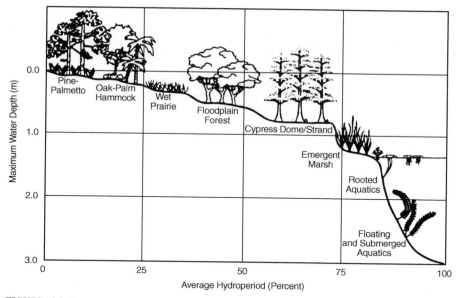

FIGURE 19.3.6   Diagram of wetland ecosystems along a hydroperiod gradient (Kadlec and Knight, 1996).

and/or evapotranspiration. Procedures for measuring or modeling these hydrologic components are explained in Mitsch and Gosselink (1993) and in standard hydrology texts (Linsley, R., Kohler, M., and Paulhus, J., *Hydrology for Engineers*, McGraw-Hill, 1982; or Maidment, D. R., *Handbook of Hydrology*, McGraw-Hill, NY, 1993). Groundwater flows and evapotranspiration are particularly difficult to quantify. The water budget of a wetland is determined by subtracting the outflows from the inflows. The hydrologic residence time, or retention time, is the average time that water spends in the wetland.

The water budget, along with the wetland's physical configuration, accounts for variations in wetland water storage. The number of days per year that a wetland stores surface water is called the wetland's hydroperiod. The hydroperiod can undergo dramatic seasonal or yearly variations. As depicted in Figure 19.3.6, the hydroperiod (percentage of time with flooding) affects the chemistry of the soils and water and, to a large degree, determines what vegetation and animals can live in the wetland (Crance, J. H., "Relationships Between Palustrine Wetlands of Forested Riparian Floodplains and Fishery Resources: A Review," *U.S. Fish and Wildlife Service Biological Report*, 88(32), 1988).

19.3.5   WETLAND VEGETATION

Given the diversity of wetland types and hydroperiods, it is no surprise that a wide variety of vegetation grows in wetlands. Because of the variability inherent in nature, some of the species of vegetation found in a given wetland is more characteristic of uplands. However, the majority of plants in a wetland are hydrophyllic (water-loving) species or hydrophytes; indeed the federal government defines a wetland as having a prevalence of vegetation typically adapted for life in saturated soil conditions. The web site of the Center for Aquatic and Invasive Plants at the University of Florida has pictures and descriptions of wetland plants and offers instructional videos in plant identification.

One way to classify plants in wetlands is by growth form in relation to water level, as in Figure 19.3.7. There is a wide variation in tolerance to depth and duration of flooding or saturation

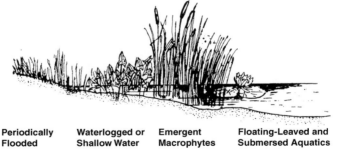

Zone	Periodically Flooded	Waterlogged or Shallow Water	Emergent Macrophytes	Floating-Leaved and Submersed Aquatics
Plants	lowland grasses	sedges (*Carex*) arrowhead (*Sagittaria* spp.)	cattails (*Typha*) bulrush (*Scirpus*)	water lilies (*Nymphaea* spp.) pond weeds (*Potamogeton* spp.) bladderwort (*Utricularia* spp.)

FIGURE 19.3.7   Relationship between water depths and plants in freshwater marsh (Mitch and Gosselink, 1993).

(Gunderson, L. H., "Historical Hydropatterns in Wetland Communities of Everglades National Park," in *Freshwater Wetlands and Wildlife*, Sharitz, R. R., and Gibbons, J. W. (eds.), U.S. Dept. of Energy, pp. 1099–1111, 1989). Emergent species generally are limited to water depths of 1 m (3.3 ft) or less (Hammer, 1996, p. 55). Most shrubs and trees cannot tolerate permanent flooding.

The major stress that plants face in wetlands, as opposed to uplands, is the lack of free oxygen (anaerobic conditions) in the soil. Any oxygen present in saturated soils is quickly consumed by microbial respiration; diffusion of replacement oxygen through saturated soil is very slow. The lack of oxygen prevents plants from performing normal aerobic root respiration and affects the availability of nutrients. To thrive in wetlands, rooted plants need adaptations that help them grow in anaerobic soils.

Some wetland plants are capable of transporting oxygen via aerenchymous tissue (spongy with large pores) in their stems from above-water parts to the root zone or rhizosphere (Jackson, M. G., and Drew, M. C., "Effects of Flooding on Growth and Metabolism of Herbaceous Plants," in *Flooding and Plant Growth*, Kozlowski, T. T. (ed.), Academic Press, San Diego, pp. 47–128, 1984). Other adaptations include pneumatophores (modified roots that help respiration and adventitious roots (e.g., roots on stems above the soil).

Trees in wetlands often have shallow root systems, because deeper soils have low oxygen availability. Because of shallow roots, wind-thrown trees are common in wetlands. Other trees such as baldcypress (*Taxodium distichum*) develop buttressed tree trunks and interlocking subsurface structures. Plants that grow in salt marshes face the additional stress of high salt content in soil and water. Saltmarsh cordgrass (*Spartina alterniflora*) deals with this stress by salt-excreting glands on its leaves (Anderson, C. E., "A Review of Structure in Several North Carolina Salt Marsh Plants," in *Ecology of Halophytes*, Reimold, R., and Queen, W. (ed.), Academic Press, San Diego, pp. 307–344, 1974). In other wetlands, such as bogs, plants face the stress of low nutrient conditions. To obtain additional nutrients, some bog plants are carnivorous [e.g., pitcher plants (*Sarracenia* spp.), and sundew (*Drosera* spp.)].

19.3.6   WETLAND SOILS

Wetlands are characterized by soils that are saturated within the root zone for a sufficient duration during the growing season to develop anaerobic conditions. Soils that develop under such conditions are called hydric soils. The NRCS provides descriptions of hydric soils, methods to identify these soils, and, in many places, mapping of soils (Hurt et al., 1996). Hydric soils are either organic (more than 20 percent organic carbon) or mineral. Organic material (e.g., leaves and decaying vegetation)

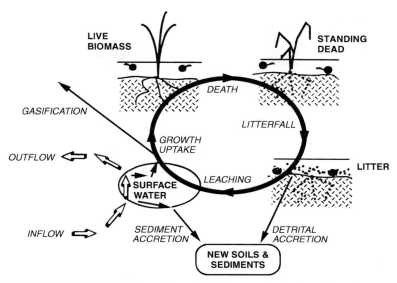

FIGURE 19.3.8 A simplified view of the wetland biogeochemical cycle (Kadlec and Knight, 1996).

accumulates in wet places because anaerobic conditions slow the rate of decomposition. Therefore, almost all organic soils are hydric (Environmental Lab., 1987).

19.3.7 BIOGEOCHEMICAL DYNAMICS

The biogeochemical processes that occur in wetlands are highly variable and complex, affected by hydrology, microorganisms, higher plants, and animals (Figure 19.3.8). Oxygen, carbon, nitrogen, and phosphorus are major elements in wetlands.

Oxygen

The solubility of dissolved oxygen (DO) in water is limited to between about 5 and 15 mg/L, depending on temperature, salinity, and biological activity. Oxygen is transferred to water by reaeration at the water surface. Also, photosynthesis occurring within the water column adds dissolved oxygen to the water. Large communities of algae can raise DO to more than double the saturation level (Schwegler, B. R., *Effects of Sewage Effluent on Algal Dynamics of a Northern Michigan Wetland*, Univ. of Michigan, MS thesis, 1978). At night, oxygen is consumed by plant respiration. Therefore, there is often a large diurnal swing in DO levels in wetlands (maximum at dusk, minimum at dawn).

Oxygen is also consumed during aerobic biodegradation of organic matter. If the rate of consumption is large (e.g., resulting from a large stock of organic debris, which is common in wetlands), degradation can deplete DO causing stressful conditions for oxygen-breathing organisms such as fish.

Diffusion of replacement oxygen through water is slow, and oxygen present in saturated soils is quickly consumed by microbial respiration. The bulk of the soil is anaerobic and, thus, characterized by low electron availability or "reducing" conditions (low or negative oxidation-reduction (redox) potential). Special adaptations are required for survival in anaerobic conditions, such as those of plants mentioned above. Microbes must find electron acceptors other than oxygen to use for reactions, such as, in order of use, nitrate, manganese, ferric iron (Fe^{3+}), sulfate, and organic compounds (Mitsch and Gosselink, 1993).

Oxygen transported to the root zone by porous, above-water plant stems often creates a narrow layer of oxidized soil at the interface of soil and surface water. Oxygen can diffuse from the roots and create a narrow region of oxygenation, that is, an oxidized rhizosphere around the root filament, (Steinberg, S. L., and Coonrod, H. S., "Oxidation of the Root Zone by Aquatic Plants Growing in Gravel-Nutrient Solution Culture," *J. Environ. Qual.*, 23(5): 907–914, 1994).

Carbon

Carbon accumulates in wetland plants over the growing season. As trees and shrubs release leaves and herbaceous vegetation dies, the organic carbon begins to biodegrade. Because oxygen is limited in frequently flooded wetlands, biodegradation proceeds under anaerobic pathways that are much slower than aerobic ones. Hence, the rate of accumulation of organic matter may exceed the rate of degradation and lead to a long-term build up of peat (partially decayed organic matter) characteristic of certain types of wetlands, such as northern bogs (Mitsch and Gosselink, 1993). Methanogenesis occurs when carbon compounds are reduced to produce methane (CH_4), which explains why methane is sometimes called swamp gas or marsh gas.

Nitrogen

Nitrogen (N) dynamics are greatly affected by the presence of aerobic and anaerobic zones in wetlands (Kadlec and Knight, 1996) (Figure 19.3.9). In both aerobic and anaerobic zones, organic nitrogen in dead plant matter may mineralize to dissolved ammonium (NH_4^+). The ammonium form is readily available to plants and microbes and can also adsorb to negatively charged particles (e.g., clay) or diffuse to the water surface and escape as ammonia gas. In aerobic water columns and root zones, the ammonium form may also be oxidized by Nitrosomonas bacteria to nitrite (NO_2^-), which can then be further oxidized by Nitrobacter bacteria to nitrate (NO_3^-). This process is called nitrification. Both nitrogen forms are also bioavailable. Nitrate may diffuse into anaerobic soil water, where anaerobic bacteria may reduce it (denitrification) to gaseous nitrogen (N_2). The gaseous nitrogen then escapes from the water into the atmosphere.

Phosphorus

Because phosphorus (P) lacks multiple oxidation states, biogeochemical P dynamics are less complex. As with nitrogen, P exists in suspended (adsorbed to solids) or dissolved forms. The dissolved inorganic form (phosphate, PO_4^{3-}) is readily bioavailable and consumed by plants.

P can adsorb to soil and suspended particles containing aluminum and iron oxides and hydroxides (Kadlec and Knight, 1996). Redox potential and pH affect the solubility of these forms. Accretion of sediment can provide a permanent sink of P. However, if persistent anaerobic conditions exist in the bed sediment zone, P may become soluble and re-enter the water column. Once aerobic conditions are encountered in the water column, P may or may not attach itself to available particles and become re-sedimented. At the end of the growing season, the above-ground portion of some plants die, partially decompose, and release a portion of the carbon, organic nitrogen, and phosphorus content to the water column. The remainder accumulates as peat.

19.3.8   WETLAND VALUES

Wetland values are the benefits or importance of particular wetland functions to humans (Richardson, C. J., "Ecological Functions and Human Values in Wetlands: A Framework for Assessing Forestry

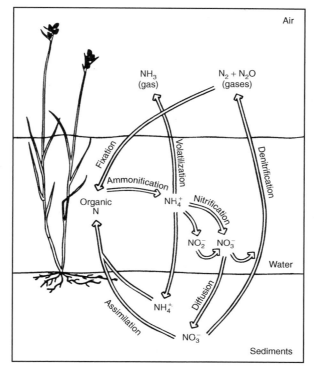

FIGURE 19.3.9 Simplified wetland nitrogen cycle (Kadlec and Knight, 1996).

Impacts," *Wetlands*, 14(1): 1–9, 1994). Wetland values include economic and noneconomic benefits that accrue to communities, individuals, or corporations through consumptive use (e.g., harvesting timber, crops, or commercial fish) or nonconsumptive use (e.g., canoeing, wildlife viewing or low-impact fishing and hunting). Wetlands also provide economic value to societies, primarily by reduction/prevention of damage to health and property (e.g., flood and erosion control) and protection of economically productive ecosystems (breeding and migratory habitat). Economic valuation methods are discussed by Barbier et al. (Barbier, E. B., Acreman, M., and Knowler, D., *Economic Valuation of Wetlands: A Guide for Policy Makers and Planners*, Ramsar Convention, 1997) and Bardecki, (Bardecki, M., *Wetlands and Economics: An Annotated Review of the Literature, 1988–1998*, Environment Canada—Ontario Region, http://www.cciw.ca/glimr/data/wetland-valuation/intro.html, 1998).

Wetland values include benefits that are difficult to economically quantify such as cultural, educational, and recreational amenities, aesthetic appeal, and protection of wilderness, biodiversity and endangered species for their own sake. The difference between wetland function and value can be illustrated by an example. Say a wetland performs the function of storage of floodwaters. If this function is performed in an urban area, its value is high, but if the wetland is located in a large wilderness area (where flooding would not damage property or human life and ecosystems are adapted to flooding), the societal value of that function may be lower.

Flood Control

During rainstorms and floods, wetlands receive and detain surface water and then slowly release water back to streams and lakes, thereby reducing peak flow rates and attendant flooding. Scientists in

Wisconsin have discovered that watersheds with 15 percent of their area as wetlands had flood peaks that were 60 percent lower than watersheds with no wetlands (Niering, 1985, p. 31).

Groundwater Recharge

Water seeps from some wetlands into underlying soils to replenish, or recharge, groundwater, which many communities use for drinking water. Groundwater also provides flow to streams during periods of low rainfall. Groundwater recharge is a common value of vernal pools and prairie potholes (Weller, M. W., *Freshwater Marshes*, University of Minnesota Press, Minneapolis, 1981). Groundwater recharge is not a typical value of wetlands with low-permeability soils. Moreover, many wetlands are supported by groundwater discharge.

Erosion Control

Wetland vegetation controls erosion along streambanks and lakeshores. The vegetation stabilizes the soil and absorbs energy from the water flow and waves. Similarly, wetland vegetation in coastal areas, such as salt marshes and mangrove swamps, can protect coastlines from storm damage (Schiechtl, H. M., and Stern, R., *Water Bioengineering Techniques for Watercourse Bank and Shoreline Protection*, Blackwell Science, Inc., 1996).

Water Quality Improvement

An important wetland value associated with biogeochemical functions is water quality improvement. In fact, wetlands are sometimes called the "kidneys of the landscape" for their water-cleaning capability. They remove sediments, toxicants, pathogens, nutrients, and other pollutants through physical, chemical, and biological processes including absorption, adsorption, filtration, microbial transformation (biodegradation), precipitation, sedimentation, uptake by vegetation, and volatilization. Wetlands for water quality improvement are discussed later in this section and extensively in Kadlec and Knight (1996).

Carbon Sequestering

Long-term sequestering of carbon through accumulation of peat is considered a value of wetlands, because it can help to offset the "greenhouse effect" caused by release of carbon dioxide into the atmosphere through fossil fuel combustion (Patterson, J., "A Canadian Perspective on Wetlands and Carbon Sequestration," *National Wetlands Newsletter*, Environmental Law Institute, Vol. 21, No. 2, 1999; *Preparing for an Uncertain Climate*, Vol. 11, U.S. Congress, Office of Technology Assessment, OTA-0-568, 1993).

Wildlife and Plant Habitat

An estimated 900 species of wildlife in the United States require wetlands as habitat for a significant portion of their life cycle (Feierabend, J. S., "Wetlands: The Lifeblood of Wildlife," in *Constructed Wetlands for Wastewater Treatment: Municipal, Industrial and Agricultural*, Hammer, D. A. (ed.), Lewis Publishers, Boca Raton, FL, pp. 107–118, 1989). Microorganisms, macroinvertebrates, fish, herptiles, birds, and mammals reside in or migrate through wetlands. Wetland animals are more fully covered in Sharitz, R. R., and Gibbons, J. W. (eds.), *Freshwater Wetlands and Wildlife*, U.S. Department of Energy Symposium Series No. 61, 1989.

Salt marshes and mangrove swamps have long been recognized as valuable habitats for shellfish and finfish (Weinstein, M. P., "Shallow Marsh Habitats as Primary Nurseries for Fish and Shellfish, Cape

Fear River, North Carolina," *Fishery Bulletin* 77: 339–357, 1979). Most commercial and game fish breed or develop in coastal marshes and estuaries, including flounder, striped bass, shrimp, oyster, clams, and blue crabs (*Coastal Wetlands of the United States*, National Oceanic and Atmospheric Administration, 1990). Freshwater wetlands are valuable fish habitat, also.

Wetlands provide critical habitat to many amphibians and reptiles. Frogs contribute sounds that are characteristic of wetlands. Turtles are also common. Snakes are not as common, but have high human interest. Alligators are a fascinating wetland presence in the southeastern United States. They are particularly important because they dig "gator holes" that maintain water during dry periods, providing wildlife refuge.

Habitat for birds, particularly waterfowl and wading birds, was one of the first recognized values of wetlands and perhaps the most appreciated value today. About 600 bird species in North America (one third of the total resident species) are dependent on wetlands for some part of their life cycle (Kroodsma, D. E., "Habitat Value for Nongame Wetland Birds," in *Wetland Function and Values: The State of Our Understanding*, Greeson, P. E., Clark, J. R., and Clark, J. E. (eds.), American Water Resources Association, pp. 320–326, 1978).

Numerous mammals inhabit and depend on wetlands. Muskrat and nutria survive in high densities and can cut large amounts of herbaceous vegetation for consumption or mound building, transforming densely vegetated areas into mostly open water within a few years (Berg, K. M., and Kangas, P. C., "Effects of Muskrat Mounds on Decomposition in a Wetland Ecosystem," in *Freshwater Wetlands and Wildlife*, Sharitz, R. R., and Gibbons, J. W. (eds.), U.S. Dept. of Energy, pp. 145–151, 1989). Beavers build dams that create and destroy wetlands (and sometimes affect humans with flooding), and remove trees from wetland edges. Moose, deer, and other ungulates graze on wetland plants.

About half of the United States' threatened and endangered animal species (as of 1986) depend on wetlands (Mitsch and Gosselink, 1993). An additional 20 percent of the United States' threatened and endangered species use or inhabit wetlands at some time in their life. In addition, many threatened and endangered plant species are found in wetlands.

Productivity

Biologic productivity is the living output of plants, wildlife, fish, and other organisms that may be used by humans or other creatures. For example, nectar produced by wetland flowers is gathered by bees and berries produced on wetland plants are consumed by wildlife. As one of the most productive ecosystems in the world, wetlands provide important commodities to humans. Rice paddies supply sustenance to a large portion of the world's population (Mitsch and Gosselink, 1993). Cranberries are harvested from bogs in the northern United States. Wetland trees are harvested for timber or ground into mulch. Herbaceous vegetation is collected for thatch or weaving.

Crustaceans, such as freshwater crayfish, are harvested for recreation and as a commercial food crop. Fish and shellfish species dependent on wetlands comprise more than 75 percent of the commercial and 90 percent of the recreational harvest ("National Water Quality Inventory," *1992 Report to Congress*, USEPA 841-R-94-001, 1994; and Feierabend, S. J., and Zelazny, J. M., *Status Report on our Nation's Wetlands*, National Wildlife Federation, Washington, D.C., 1987). In 1991, nearly $20 billion is attributed to fish species that depend directly or indirectly on coastal wetlands during their life cycles (Economic Benefits of Wetlands. U.S. EPA, http://www.epa.gov/OWOW/wetlands/contents.html).

Aesthetics and Recreation

The plants, wildlife, birds, and aquatic setting of wetlands contribute aesthetic open space to communities, making them attractive recreation sites for canoeing, bird watching, photography, hiking, fishing, trapping, and hunting. More than half of all U.S. adults (98 million people) engage in such activities, worth an estimated $59.5 billion in 1991 (*Preparing for an Uncertain Climate*, Vol. 11, U.S. Congress, Office of Technology Assessment, OTA-0-568, 1993).

Cultural and Educational Values

Many wetland values transcend specific functions of wetlands and provide immeasurable cultural value. For example, the culture of the southern Louisiana bayou is tightly bound to its wetlands. The Everglades are an integral part of the history and culture of south Florida, although permanently damaged by ditching, diking and groundwater mining. Wetlands have captured the imagination of people for generations, and continue to do so today. Artists and writers use wetlands as their subject matter. Many people enjoy walking, exploring, or simply viewing wetlands. Educational opportunities, including the fields of ecology, hydrology, and chemistry, abound in wetlands. Finally, the simple existence of wetlands as natural open spaces supporting diverse vegetation and wildlife is valuable to many people.

19.3.9   REGULATION AND PERMITTING IN THE UNITED STATES

The Wetlands Regulation Center (www.wetlands.com) provides extensive and updated information on wetlands regulations.

Federal Regulations

Section 404 of the 1972 Federal Clean Water Act provides the primary means of Federal wetlands regulation. Section 404 established a permit program for discharges of dredged or fill material to be administered by the U.S. Army Corps of Engineers (Corps). In 1974, when the Corps issued regulations to implement the Section 404 program, they limited the program's jurisdiction to traditionally navigable waters, including adjacent wetlands. As a result, the regulations excluded many small waterways and most wetlands. In 1975, a Federal district court directed the Corps to revise and expand the regulations to be consistent with congressional intent. In response, the Corps issued final regulations in 1977 that explicitly included "isolated wetlands and lakes, intermittent streams, prairie potholes, and other waters that are not part of a tributary system to interstate waters or to navigable waters of the United States, the degradation or destruction of which could affect interstate commerce." The jurisdiction promulgated in 1977 is substantially the same as the one in effect today (USEPA, http://www.epa.gov/owow/wetlands/facts/fact12.html).

The Corps also employs Sections 9 and 10 of the 1899 Rivers and Harbors Act to protect navigable and coastal waters and associated wetlands. The Corps' regulations and guidelines for fill placement and other defined activities are contained in 33 CFR Parts 323 through 328 and Part 330. In addition, the Corps has issued Regulatory Guidance Letters to clarify certain aspects of the program. Visit the Corps regulatory web site for comprehensive information and updates on their regulatory program.

The Corps' "Tulloch Rule" (named after a developer in North Carolina) extended the Corps authority to regulation of discharges incidental to excavation (including any redeposit or fallback of dredging materials or de minimis discharges) that destroyed or degraded any waters (including wetlands) of the United States. In 1997, a court overturned the Tulloch Rule and the US District Court of Appeals subsequently affirmed that decision in 1998 (www.wetlands.com/fed/tulloch4.htm). A revised definition of "discharge of dredge material," issued by the Corps and EPA in the May 10, 1999 Federal Register, omits the word "any" before "redeposit" and expressly excludes "incidental fallback." As a result, Federal regulation of the drainage of wetlands is limited.

Another recent court case affected regulation of isolated water bodies and wetlands. The U.S. Court of Appeals for the Fourth Circuit issued a decision in the case of United States v. James J. Wilson, 133 F. 3d 251 (4th Circuit, decision issued December 23, 1997). As a result, within the five states comprising the Fourth Circuit, neither the Corps nor the EPA will assert jurisdiction over any isolated, intrastate water body (or wetland) where the only basis for such jurisdiction would be potential (not actual) effect on interstate or foreign commerce.

Under Section 401 of the Clean Water Act, projects requiring a Federal permit or license, must obtain certification from the state environmental protection agency that the project will not adversely affect the quality of the state's waters.

Federal laws that regulate activities affecting wetlands are listed below:

Clean Water Act, Section 401—33 U.S.C. 1341, Section 404—33 U.S.C. 1344

Coastal Zone Management Act of 1972—16 U.S.C. 1456(c)

Endangered Species Act—16 U.S.C. 1531 et seq

Food Security Act of 1985 (Swampbuster)—16 U.S.C. 3811 et seq

Marine Mammal Protection Act—16 U.S.C. 1361 et seq

Marine Protection Research and Sanctuaries Act of 1972, Section 103—33 U.S.C. 1413

National Environmental Policy Act—42 U.S.C. 4321-4347

National Historic Preservation Act—16 U.S.C. 470

Rivers and Harbors Act of 1899, Section 9—33 U.S.C. 401, Section 10—33 U.S.C. 403

Wild & Scenic Rivers Act—33 U.S.C. 1271 et seq

Federal Permitting

The Corps permit regulations require either an individual or a nationwide permit for filling wetlands. The nationwide permits (NWPs) are generic permits for categories of projects, such as utility crossings and hazardous waste remediation, which the Corps deems to minimally affect wetlands. The Corps may still need to be notified of these projects, but they will not apply the rigorous review that an individual permit application receives. Under Section 401 of the Clean Water Act individual states have the authority to require more restrictive conditions or eliminate nationwide permits. The nationwide permits are one of the most controversial issues of the Corps wetland regulatory program. Environmental advocates tend to believe the permits are too lenient, allowing unacceptable destruction to wetlands, while the regulated community tends to believe the permits are too restrictive.

Currently, the Corps of Engineers is proposing to issue five new NWPs and modify six existing NWPs to replace NWP 26 when it expires in March 2000. NWP 26 requires no notification for filling less than 1/3 an acre of wetlands and limited notification for filling less than 3 acres. The Corps is also proposing to modify nine NWP general conditions and add three new general conditions (www.usace.army.mil/inet/functions/cw/cecwo/reg/citizen.htm). The Corps will eventually issue final regulations and regional conditions for the revised nationwide permits. District offices of the Corps may modify or revoke any nationwide permit for projects within their regional boundary.

Applications for an Individual Permit are subject to public review and an environmental assessment by the Corps. To obtain an Individual Permit, an application must demonstrate a need for the project, prove compliance with the guidelines of Section 404(b)(1) of the Clean Water Act, and conduct an alternatives analysis that indicates the preferred alternative is the "least damaging practicable" alternative. The Corps must complete a written assessment of the environmental impacts of the project and issue a statement of findings. The Corps typically issues Individual Permits within 3 to 12 months of receipt, but occasionally requires years to process more controversial or complicated projects.

To qualify for a Corps' permit, a proposed project must demonstrate that wetland impacts have been mitigated (lessened). In a 1988 address, President Bush presented "no net loss" of wetlands as a national goal. Since then, "no net loss" has become the objective of the regulatory and environmental community. Although, the overall area of wetlands in the United States continues to decline, wetland regulations and mitigation requirements have slowed the rate of decline.

Regulations require a sequential process (called "sequencing") of avoidance, minimization, and compensation of wetland impacts. Only when the Corps is satisfied that a project has avoided and minimized wetland impacts will compensation be considered. Compensation typically requires the enhancement, restoration, or creation of wetlands to offset a loss of wetland values through filling or other activities.

The USEPA provides oversight of the Corps wetland-permitting program. In accordance with various federal laws, consultation with the USFWS, National Marine Fisheries Service, State Coastal Zone Management Agency, and State Historic Preservation Officer is required during the permitting process. Under the National Environmental Policy Act, the concerns of these agencies, other agencies that may have resources that could be affected by a project, and all public comments and concerns must be weighed in any permit decision. If these concerns are not given the required consideration, the permit can be legally challenged.

The USEPA may veto permitting decisions made by the Corps and pursue violators of the wetland laws. Large fines and significant jail time have been levied against citizens and corporations who have filled wetlands without a permit or violated the terms of an issued permit.

State and Local

Many state and local governments have created laws that provide additional protection to wetlands. For example, the Massachusetts Wetlands Protection Act requires that the local Conservation Commission must permit any impacts to wetlands. Numerous municipalities in Massachusetts have enacted their own wetland protection ordinances. The states of New Jersey and Michigan administer the Section 404 regulatory program, with oversight by the Corps, in their states. The Local Government Environmental Assistance Network (www.lgean.org) provides web links to many state environmental agencies.

19.3.10   DELINEATION METHODS

A fundamental aspect of wetland science (and one of the first steps in obtaining applicable wetland permits) is delineating the boundaries of wetlands. In 1987, the Corps published a *Wetland Delineation Manual* (Environmental Lab., 1987) that defines the wetland boundary according to vegetation, hydrology, and soils (the three-parameter approach). In 1989, the Corps, USEPA, USFWS, and NRCS jointly published a new interagency wetland delineation manual (FICWD, 1989). Because of ensuing political controversy over an expansion of Federal regulation and Congressional demands to abandon use of the interagency manual, the Corps currently requires the use of the 1987 manual with clarifications provided in 1992. However, some states require use of the 1989 manual. Furthermore, other states have developed their own delineation methods. Both federal manuals and most state methods require the three-parameter approach. However some states require a one or two parameter test. All manuals should be carefully consulted when conducting or reviewing a delineation for regulatory purposes.

A 1995 report by the National Research Council Committee on Characterization of Wetlands evaluated wetland delineation in light of current scientific understanding of wetlands and made numerous recommendations to improve the current system, but delineation methods continue to be politicized because of the profound effects of wetland permitting requirements on development projects.

Many consulting firms delineate wetlands as part of comprehensive environmental services to assist permit applicants through the regulatory process. Several organizations and universities offer training courses in wetland delineation (see web site list). A brief overview of the federal delineation method under "normal" circumstances is provided here. Refer to Kent (1994) for a more detailed description.

Prior to a site inspection, available information, including the following resources, should be reviewed to identify probable wetland areas (such as low-lying areas, floodplains, and hydric soils):

- U.S. Geological Survey (USGS) quadrangle topographic maps
- USFWS National Wetland Inventory (NWI) maps
- Federal Emergency Management Agency (FEMA) Flood Insurance Rate Maps (FIRM)
- USDA Natural Resources Conservation Service (formerly the U.S. Soil Conservation Service) County Soil Survey

TABLE 19.3.1 Wetland Indicators in the USFWS Wetland Plants List (USFWS, 1996)

Wetland indicator	Abbreviation	Estimated probability of occurring in wetland
Obligate	OBL	>99%
Facultative Wetland	FACW	67–99%
Facultative	FAC	34–66%
Facultative Upland	FACU	1–33%
Upland	UPL	<1%

Note: The USFWS further refines the facultative categories with plus (+) or minus (−) designations, to specify the higher or lower part of the range, respectively.

- State and local wetland maps; and
- Aerial photographs.

During a site visit, using the three-parameter approach, a delineator examines the vegetation, soils, and hydrologic indicators and demarcates the wetland boundary. In general, to be considered a wetland, more than 50 percent of the dominant vegetation must be hydrophytic, the soils must be hydric, and evidence of hydrology must be present. Software is available to help compile the data from a field visit (e.g., WetForm, www.wssinc.com). A typical data sheet documenting the boundary of a wetland in New York is attached as Figure 19.3.10.

Vegetation Characterization

The delineator(s) typically first examines the vegetation. They identify each dominant plant, by species, with the help of guidebooks on local flora (e.g., Tiner, 1999; others listed on the web site of the University of Florida Center for Aquatic and Invasive Plants).

The USFWS has published a list of wetland species (USFWS, 1996) for all regions of the United States. The list assigns an indicator of the estimated probability that a species occurs in wetlands versus uplands (Table 19.3.1). The list also employs a NI (no-indicator) for species for which insufficient information was available to determine an indicator status. If a species does not occur in wetlands in any region, it is not on the list.

Once the delineators identify the dominant plant species, they record the USFWS wetland indicator for each species. In general, to be considered a wetland, more than 50 percent of the dominant vegetation must be hydrophytic (i.e., plants with FAC, FAC+, FACW, or OBL indicators), or the vegetation must exhibit adaptations to saturated soils. Ordinarily, to be classified as a wetland, an area that meets the wetlands vegetation criterion must also possess indicators of wetland soils and hydrology.

Soils Characterization

Soils are analyzed by collecting soil cores, using a hand-held auger or shovel to a minimum depth of 16 inches. One common field indicator of hydric (wetland) mineral soils is gleying or gleization: formation of gray, blue-gray, or greenish color by chemical reduction of iron. Another indicator is the "rotten egg" odor emanated by the hydrogen sulphide (H_2S) common in anaerobic environments. Redoximorphic features (formerly called mottles) provide yet another indication of hydric soils. These features include orange or reddish brown spots of accumulated iron and manganese oxides that are the result of repeated wetting and drying.

As illustrated in Figure 19.3.10, the delineator records the soil color using a color guidebook such as the *Munsell Soil Color Charts* (Kollmorgen Corporation, Baltimore, MD, 1994) or *Earthcolors: A*

Job Number: JA-202023	Nearest Wetland Flag: I-2
Field Investigators: McBrien, Seba, Modjeski	Date: May 21, 1999
Project/Site: Applegarth Residential Development	County: Richmond
Applicant/Owner: J. Applegarth	State: New York

Wetland: Data Point SP-3   Upland: Data Point SP-4

Do Normal Circumstances exist on site? YES
Is the site significantly disturbed (atypical situation)? NO
Is the area a potential problem area? NO

Wetland Vegetation   Upland Vegetation

	Dominant Plant Species	Stratum	Indicator Status		Dominant Plant Species	Stratum	Indicator Status
1	Cottonwood (*Populus deltoides*)	Shrub	FAC	1	Cottonwood (*Populus deltoides*)	Shrub	FAC
2	Common reed (*Phragmites australis*)	Herb	FACW	2	Russian olive (*Elaeagnus angustifolia*)	Shrub	FACU
3	Tussock sedge (*Carex stricta*)	Herb	OBL	3	English plantain (*Plantago lanceolata*)	Herb	UPL
4	Mugwort (*Artemisia vulgaris*)	Herb	UPL	4	Mugwort (*Artemisia vulgaris*)	Herb	UPL
5	Swamp milkweed (*Asclepias incarnata*)	Herb	OBL	5	Willow herb (*Epilobium hirsutum*)	Herb	FACW
6	English plantain (*Plantago lanceolata*)	Herb	UPL	6	Common reed (*Phragmites australis*)	Herb	FACW
7	Soft rush (*Juncus effusus*)	Herb	FACW+	7			
8				8			

% of dominants that are hydrophytes (FAC or Wetter)? 71
Hydrophytic Vegetation Criterion Met (>50% hydrophytes)? YES

% of dominants that are hydrophytes (FAC or Wetter) 50
Hydrophytic Vegetation Criterion Met (>50% hydrophytes) ? NO

Wetland Soils   Upland Soils

Soil Series/Phase: Unknown   Soil Series/Phase: Unknown
Is the Soil Listed as Hydric? Unknown   Is the Soil Listed as Hydric? Unknown

Depth (Inches)	Matrix Color	Mottling Color	Mottling %	Texture	Depth (Inches)	Matrix Color	Mottling Color	Mottling %	Texture
.5-0				leaf litter	.5-0				leaf litter
0-18	7.5 YR 2.5/2	2.5 YR 3/2	30	sandy silt	0-8	10YR3/3	none	0	sandy loam
					8+				rock

Hydric Soil Criterion Met? YES   Hydric Soil Criterion Met? NO
Field Indicators of Hydric Soil: Low chroma matrix with mottles   Field Indicators of Hydric Soil: None

Wetland Hydrology   Upland Hydrology

Ground Surface Inundated? NO Depth (Inches): 0
Soil Saturated? YES Depth to Saturation (Inches): 4
Depth to free-standing Water in Probe Hole (Inches): 10
Field Evidence of Hydrology: Watermarks and water within 12 inches of ground surface
Evidence of Prolonged Saturation and/or Inundation?
✓ Yes (Wetland Hydrology Criterion Met)
    No (Wetland Hydrology Criterion Not Met)

Ground Surface Inundated? NO Depth (Inches): 0
Soil Saturated? NO Depth to Saturation (Inches):
Depth to free-standing Water in Probe Hole (Inches):
Field Evidence of Hydrology: None
Evidence of Prolonged Saturation and/or Inundation?
    Yes (Wetland Hydrology Criterion Met)
✓ No (Wetland Hydrology Criterion Not Met)

Federally Regulated Wetland? YES
USFWS Classification: Palustrine Emergent Wetland
State Regulated Wetland? YES

Federally Regulated Wetland? NO
USFWS Classification: Upland
State Regulated Wetland? NO

Comments:

Wetland Data Form was adapted from Corps' Data Form by Anthony Froonjian, The Louis Berger Group, Inc.

FIGURE 19.3.10 Typical wetland delineation data form.

Guide for Soil and Earthtone Colors (Color Communications, Inc., Chicago, IL, http://www.ccicolor.com/, 1997). The color of hydric soils is usually characterized by a low chroma value, corresponding to black, gray, dark brown or red. The delineator also notes the structure, texture, and features of each layer, or horizon, of soil from the core.

Hydrology Characterization

The federal hydrology criteria requires inundation and/or soil saturation to the surface for a consecutive number of days for more than 12.5 percent of the growing season (1987 manual) or flooding or soil saturation near the surface for one week or more during the growing season at a frequency greater than 50 years out of 100 (1989 manual). Because some wetlands are not always wet and records of waterland fluctuations are rarely available, evidence of hydrology is often difficult to observe in the field. If standing water or saturated soils are not observed during a field visit, the delineator(s) must search for other indicators of wetland hydrology, such as water stained leaves, drainage channels, watermarks on trees, driftlines, and oxidized rhizospheres. The presence of standing water does not always infer that the wetland hydrology criteria has been met.

19.3.11 ASSESSMENT METHODS

Wetland delineations often include an assessment of the functions and values of the delineated wetlands. The variety of formal assessment methods is reviewed by Novitzki, R. P. (*Comparison of Current Wetland Assessment Methods*, National Interagency Workshop on Wetlands: Technology Advances for Wetlands Science, U.S. Army Corps of Engineers Waterways Experiment Station, http://www.wes.army.mil:80/el/workshop/FA1.html, 1995) and by Bartoldus, C. C. (*A Comprehensive Review of Wetland Assessment Procedures: A Guide for Wetland Practitioners*, Environmental Concern, Inc., http://www.wetland.org/wap.htm, 1999). Three of the more common assessment methods are described below.

Habitat Evaluation Procedure

The USFWS Habitat Evaluation Procedure (HEP) assesses impacts on fish and wildlife habitat resulting from water or land use changes (USFWS, *Habitat Evaluation Procedure*, BSM 102, U.S. Department of the Interior Fish and Wildlife Service, Washington, D.C., 1980). The procedure employs Habitat Suitability Index (HSI) models to assess the value of an individual wetland to a species of concern, such as great blue herons. Models provide a numerical index of habitat suitability for the species on a 0.0 to 1.0 scale, based on the assumption that there is a positive relationship between the index and habitat carrying capacity. USFWS has published HSI models for approximately 160 species. A compilation of these models is available at http://www.nwrc.gov/wdb/pub/hsi/hsi-intro.html.

Wetland Evaluation Technique

The Wetland Evaluation Technique (WET) is one of the earliest comprehensive assessment methods (Adamus, P. R., Clairain, E. J., Jr., Smith, R. D., and Young, R. E., *Wetland Evaluation Technique (WET) Volume II*, Technical Report Y-87, U.S. Army Corps of Engineers, Waterways Experiment Station, 1987). WET assesses wetland functions in terms of "social significance, effectiveness, and opportunity" using predictors of physical, chemical, and/or biological processes (functions) and assigns value to function in terms of social significance. Some of the functions assessed are groundwater interactions, floodwater storage, water quality, wildlife habitat, and recreation. WET assigns a rating of high, moderate, or low to the probability that a wetland is performing a given function, but does not assess the extent to which the function is performed.

Hydrogeomorphic Method

The increasingly popular hydrogeomorphic (HGM) method was originally described by Brinson, M. M. (*A Hydrogeomorphic Classification for Wetlands*, Wetlands Research Program Technical Report WRP-DE-4, U.S. Army Corps of Engineers, http://www.wes.army.mil/el/wetlands/wlpubs.html, 1993). The HGM method classifies wetlands based on geomorphic setting, water source, and hydrodynamics.

The five basic hydrogeomorphic wetland classes are riverine, depressional, slope, flat (mineral and organic), and fringe (coastal and lacustrine). Within a geographic region, each of these classes may be subdivided into subclasses, and functions most likely to be performed by each subclass are identified. For each of these functions, an assessment model is developed that relates specific characteristics of a wetland and a surrounding landscape to the capacity of the wetland to perform the function. The assessment model is calibrated using the literature, expert opinion, or, ideally, with data collected from a set of reference wetlands representing the range of conditions in the region. The model rates a wetland's functional capacity relative to similar wetlands in the region. The ratings are used to compare impacts of project alternatives and evaluate planned and constructed compensatory wetlands.

Several federal agencies, including the Corps and EPA, announced the National Action Plan to Implement the Hydrogeomorphic Approach to Assessing Wetland Functions (*Federal Register*, June 20, 1997). Developing assessment models and compiling a set of reference wetlands are large tasks. Numerous national and regional guidebooks are under development under the supervision of the Corp's Waterways Experiment Station (http://www.wes.army.mil/el/wetlands/hgmhp.html).

19.3.12 *COMPENSATORY WETLANDS*

To compensate for a loss of wetland values through filling or some other activities, federal regulations require the enhancement, restoration, and/or creation of wetlands. Infrequently, the regulatory agencies may allow the preservation of existing valuable wetlands as compensation for wetland destruction.

For many years, the guiding principle of compensation has been "in-kind and on-site," meaning lost wetland values should be replaced by providing a comparable wetland near the impacted wetland (i.e., on the project site). If on-site compensation is not feasible, the next preference is in-kind and off-site compensation within the watershed of the impacted wetland.

The science of enhancement, restoration and creation of wetlands is still an emerging field. Major references on the topic include Kusler and Kentula (1990) and Hammer (1996). Although compensation of wetland impacts is the common motivation, several wetland construction projects are undertaken to provide specific wetland values like wildlife habitat or water quality improvement.

Restoration and Enhancement

Enhancement involves the improvement of an existing, degraded wetland. Tens of thousands of acres of wetlands suffer from polluted water, human-induced direct and indirect alterations to hydrology, and fragmentation of habitat. As a result, the functions and values of these wetlands are severely degraded. Invasive species such as purple loosestrife and common reed (*Phragmites australis*) have created monocultures providing limited wildlife habitat.

By some regulatory definitions, restoration involves the reestablishment of wetlands in areas that have been so degraded that they no longer meet the regulatory definition of a wetland. Millions of acres of wetlands in the Midwest have been drained to create agricultural fields. Materials dredged from our navigable waterways and side cast into adjacent riverine wetlands have completely altered the topography, hydrology, soils, and vegetation of these wetlands. Yet, wetlands have a remarkable ability to rebound once the human-induced alteration has been reversed by restoration activities.

In a more general sense, restoration means the reestablishment of wetlands in areas that historically were wetlands or even improving wetland values in degraded wetlands. In this sense, enhancement may be viewed as a type of restoration. Some state regulatory agencies prefer enhancement and restoration as methods of compensation because they believe these methods are more reliable. However, compensating wetland losses by enhancing existing wetlands results in a net loss of wetland acreage. To offset acreage loss, regulatory agencies typically require a larger area of enhanced wetlands than the area of impacted wetlands. New Jersey, for example, typically requires a 3:1 (enhanced: impacted) ratio.

An example of large-scale enhancement is efforts in the Hackensack Meadowlands District, an estuarine system in Northeastern New Jersey (http://www.hmdc.state.nj.us/). Areas that were dominated by *Phragmites* are being excavated and regraded to increase tidal flow and create open water areas. Some areas are being replanted with saltmarsh cordgrass (*Spartina alterniflora*), historically the dominant species in Atlantic salt marshes.

Creation

Creation is defined as the construction of wetlands in nonwetland, or upland, areas. As with wetland enhancement and restoration, the planning, design, and construction of wetlands requires a combination of ecological and engineering skills. However, creation is more risky because no natural wetland characteristics are available on the site.

Designing Wettands

The first step in designing wetlands is to determine the structural components and goals of the project. Typically, compensatory wetlands must be designed to replicate the functions and values of the impacted wetlands. Prior to completing a detailed design, a conceptual compensatory wetland plan should be presented to regulators for approval. Completion of the final detailed construction plans and specifications may require groundwater analysis, tidal measurements, hydrologic analyses, geotechnical analyses of existing soils, and evaluation of existing vegetation.

Both *enhancement* and *restoration* typically involve the excavation of accumulated material to (re)create desired ground elevations; (re)creation of desired wetland hydrology by lowering ground elevations, breaching dikes or deactivating drainage structures; elimination of invasive species using herbicides or mechanical methods; and planting of diverse native wetland vegetation.

The primary key to successful wetland *creation* is the establishment of appropriate hydrology. Methods of providing water to an upland site include excavation to groundwater, diversion of streams or rivers, and introduction of surface runoff. Improving soils with the addition of organic material is typically required. Finally, suitable vegetation must be planted and closely monitored to ensure successful establishment. The Corps' Waterways Experiment Station has produced guidelines for designing wetlands (e.g., *Engineering Specification Guidelines for Wetland Plant Establishment and Subgrade Preparation*, WRP- RE-19; *Wetlands Engineering: Design Sequence for Wetlands Restoration and Establishment*, WG-RS- 3.1, http://www.wes.army.mil/el/, 1992).

Thunhorst (1993) provides a guide for plant selection in the Northeast. All newly planted wetlands, whether enhanced, restored, or created, must grapple with wildlife predation. Young sprouts are highly susceptible to consumption by Canada geese, muskrats, deer and other wildlife. Once planting is complete, protective measures are typically required to protect the vegetation until it is well established and can withstand predation.

Project costs vary greatly, especially land costs, which are often very high in urban areas. In 1994, King and Bohlen reported approximate average costs of $50,000 for a one-acre project, $25,000 per acre for 10-acre projects, and $10,000 per acre for 100-acre projects. (King, D. M., and Bohlen, C. C., *Making Sense of Wetland Restoration Costs*, USEPA and University of Maryland, Center for Environmental and Estuarine Studies, EPA 230-R-96-002, 1994). Agricultural conversion projects were much cheaper, approximately $1000 per acre regardless of size. These numbers are only general guidelines. Both federal and state permits typically require monitoring of the wetland vegetation, hydrology and soils for three to ten years after construction. The purpose of the monitoring is to assess the success of the compensatory wetlands and to implement corrective actions, as necessary.

19.3.13 SUCCESS OF COMPENSATORY WETLANDS

Many compensatory wetland projects have been successful. For regulatory purposes, success criteria are usually specified in terms of meeting wetland restoration goals, such as 75 percent coverage of noninvasive hydrophytic species after three growing seasons or wildlife utilization. For example, the Des Plaines River Wetlands Demonstration Project in northeastern Illinois restored about 20 acres of wetlands in an abandoned agricultural field in the river floodplain. Fifteen species of waterfowl nested at or visited the site in one year (including five state endangered species)—a five-fold increase over pre-project conditions (Hickman, S., "Improvement of Habitat Quality for Nesting and Migrating Birds at the Des Plaines River Wetlands Demonstration Site," *Ecological Engineering*, 3(4): 319–344, 1994).

However, the overall success rate of compensatory, ad-hoc wetland creation and restoration projects has been unsatisfactory. For example, Erwin reviewed 40 compensation projects in south Florida and reported failure or incomplete creation of 24 projects (60 percent), causing a 50 percent loss of wetlands area (Erwin, K. L., "An Evaluation of Wetland Mitigation in the South Florida Water Management District, Vol. I," *Final Report to South Florida Water Management District*, West Palm Beach, FL, 1991). Even if a project is successful, the wetland requires years to develop full functionality.

Causes of failures are numerous, with nontechnical reasons often at the root. Permittees have a strong incentive to reduce compensation costs, but little incentive to build high-quality wetlands. Postconstruction maintenance and management are particular problems, especially if the ultimate owner of the property is different from the builder. Monitoring by regulatory authorities is often lax. Inappropriate hydrology is the most common technical reason for failure. Grazing of plantings is also common, as is inadequate monitoring and maintenance. (Brown, S., and Veneman, P., 1998, *Compensatory Wetland Mitigation in Massachusetts*, MA Agricultural Experiment Station, Amherst MA).

19.3.14 MITIGATION BANKING

The problems noted above of ad-hoc, "on-site and in-kind" compensation led to the concept of wetland mitigation banking. Mitigation banks are large enhanced, restored, and/or created wetlands that private companies or public agencies construct expressly to provide advance compensation of future unavoidable wetland impacts. As acres of wetlands are enhanced/restored/created and certified by regulators, they are "banked" as credits and subsequently "withdrawn" or sold to compensate for wetland impacts.

The Corps' Institute for Water Resources (http://www.wrsc.usace.army.mil/iwr/) is conducting the National Wetland Mitigation Banking Study, which has issued several reports. Another important reference is Lindell, L., Marsh, L. L., Porter, D. R., and Salvesen, D. A., *Mitigation Banking: Theory and Practice*, Island Press, 1997. The Terrene Institute (www.terrene.org) hosts an electronic mailing list for mitigation banking.

In 1995, several federal agencies, including the Corps and USEPA, issued Federal Guidance for the Establishment, Use and Operation of Mitigation Banks (Federal Register, Volume 60, Number 228, pp. 58605–58614, November 28, 1995, www.usace.army.mil/inet/functions/cw/cecwo/reg/mitbankn.htm). The guidance explains the perceived advantages of banks from the agencies' perspective:

- In some cases, large, contiguous wetlands provide higher value per acre than small, scattered wetlands.

- Banks create wetlands before impacts, reducing "lag time" and uncertainty of success of individual compensatory wetlands.

- Banks have more financial and technical supports to ensure success than small, ad-hoc projects.

- Banks are more efficiently permitted and monitored by regulators than many scattered sites.

From a developer's perspective, the advantage of banks over ad-hoc compensation is the transfer of responsibilities for impact compensation (e.g., wetland design and construction, risk of failure, monitoring) from the developer to the bank. Because of economies of scale at the mitigation bank and the fact that on-site compensation is sometimes very expensive, mitigation banks may provide cost savings as well.

Concerns over banking include the loss of location-dependent values, such as flood storage or special wildlife habitat (a concern especially in out-of-kind credit purchases). Moreover, some environmental advocates (e.g., the Sierra Club) fear that banking will increase pressure to allow impacts that otherwise would not be permitted.

In January 1997, the Corps estimated that approximately 200 wetland mitigation banks were in operation or under development across the country. Many of these banks are private commercial operations. State and federal agencies regulate bank development, design, construction, and credit sale. Purchase of wetland credits may cost from $10,000 to $150,000 per acre ("Mitigation Banking Bill Spurs Debate Over National Wetland Policy," *Water Environment and Technology*, August: 16–19, 1999).

A case study is the 20-ha (50-ac) Otter Creek Wetland Mitigation Bank, one of the first private wetland mitigation banks in the country. The bank was constructed on an abandoned bean field in rural St. Charles, Illinois in 1994. Drainage tiles were demolished, some areas were excavated, and native emergent species were planted. Of the 78 species seeded or planted at the site, 60 (77 percent) are flourishing. In all, over 140 plant species have been identified at the wetland bank (Apfelbaum, S. I., and Ryan, J., *Designing and Building Ecologically and Economically Successful Wetland Banks*, First National Mitigation Banking Conference, Terrene Institute, 1998). In addition, the wetland has hosted field classes for area children; attracted wildlife including herons, hawks, and ducks; and provided scenic views from an observation area constructed on a hill overlooking the wetland. Regulatory agencies approved the bank design, construction, and establishment, and most of the credits have been sold to public and private developers.

A similar approach to mitigation banking is "in-lieu-fee" compensation, in which a developer pays fees to a wetland organization for use in subsequent wetland creation/restoration projects. The Corps' 1995 Guidance states that such arrangements do not meet the definition of mitigation banking. Although not the preferred arrangement, the Corps states they "may find there are circumstances where such arrangements are appropriate"; indeed, such arrangements have been used in the Hackensack Meadowlands in northeastern New Jersey.

19.3.15 WETLANDS FOR WATER QUALITY IMPROVEMENT (TREATMENT WETLANDS)

Treatment wetlands are specifically created and managed for water quality improvement. (It is generally not permissible in the United States to alter or use natural wetlands explicitly for wastewater or stormwater treatment.) The plants, wildlife, birds, and aquatic setting of these treatment wetlands offer important benefits beyond water quality improvement (USEPA, *Constructed Wetlands for Wastewater Treatment and Wildlife Habitat: 17 Case Studies*, http://www.epa.gov/owow/wetlands/construc/, 1993). The definitive treatment wetland reference is by Kadlec and Knight (1996).

Treatment wetlands are now widespread, treating municipal and industrial wastewater (sewage) and runoff from mines, crop fields, livestock operations, and urban areas. Wastewater treatment wetlands number over 500 in Europe and 600 in North America (Cole, S., "The Emergence of Treatment Wetlands," *Environmental Science and Technology*, 32:218–223, http://pubs.acs.org/hotartcl/est/98/may/emer.html, 1993). Properly designed wetlands typically provide low cost, low maintenance treatment, and compare well in life cycle costs, reliability, and treatment efficacy with conventionally engineered systems.

As depicted in Figures 19.3.11 and 19.3.12, treatment wetlands are divided into two hydrologic categories: subsurface flow (water flows through a medium such as gravel, in which plants are rooted) and free water surface (water surface exposed to the atmosphere, with submerged, emergent, or floating vegetation rooted in the bottom). Kadlec and Knight (1996) maintain that the hydraulic loading rate

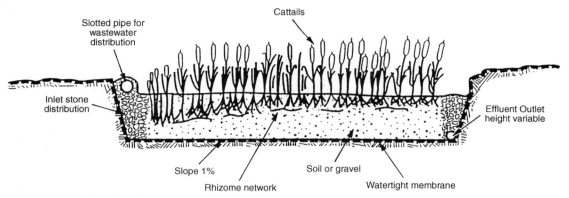

FIGURE 19.3.11   Typical profile of a sub-surface flow treatment wetland (USEPA, *Constructed Wetlands and Aquatic Plant Systems for Municipal Wastewater Treatment*, EPA/625/1-88/022, 1988).

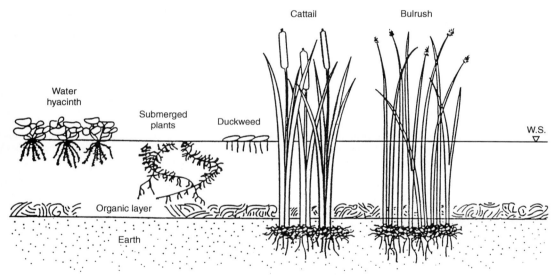

FIGURE 19.3.12   Typical profile of a free-water surface wetland (USEPA, *Constructed Wetlands and Aquatic Plant Systems for Municipal Wastewater Treatment*, EPA/625/1-88/022, 1988).

(the depth of water applied per unit bottom area per unit time) is the key parameter governing pollutant removal performance.

Treatment wetlands remove pollutants in water through physical, chemical, and biological processes including absorption, adsorption, filtration, microbial transformation (biodegradation), precipitation, sedimentation, consumption by vegetation, and volatilization. Characterized mostly by low flow velocities, treatment wetlands are most effective (75 percent removal typical) at removing suspended solids through settling. Any pollutants attached to solids, such as phosphorus or metals, can also be removed in this way.

Nutrients (nitrogen and phosphorus) are also removed from the water column by plant consumption. Nitrogen can be permanently and continuously removed from water through the nitrogen-denitrification process described earlier, which requires both the aerobic and anaerobic zones provided by wetlands.

When plants die, the litter accumulates on the bottom of the wetlands. Plants also serve as a substrate for microbes and aerate the soil around roots. On stems, detritus, and in the soil, microbes degrade pollutants. See the section on wetland biogeochemical dynamics for additional information. A 1993 survey of more than 300 wetlands in North America, treating primarily municipal wastewater, documented the following average pollutant reduction performance for the mix of system types: biochemical oxygen demand, 73 percent average reduction; total suspended solids, 72 percent; total nitrogen, 53 percent; total phosphorus, 56 percent (Knight, R. L., Ruble, R. W., Kadlec, R. H., and Reed, S., *Database: North American Wetlands for Water Quality Treatment, Phase II Report*, USEPA, Water and Hazardous Waste Treatment Research Division, Cincinnati, OH, 1993).

The design of an effective treatment wetland is a challenging task, as it requires a sophisticated understanding of environmental engineering, hydrology, soils, and wetland plant ecology. In general, successful treatment wetlands are often quite similar to natural wetlands. To maximize treatment, designers should maximize the distance between the inlet and outlet, provide a high surface-area to volume ratio, and promote slow flow velocities. To ensure adequate detention times, a treatment wetland should be much longer than it is wide. Treatment wetland inlet structures must be designed to prevent high intensity discharges (these could scour the wetland) and disperse, rather than channelize, flow through the wetland. Controlled dispersion of the influent flow helps to ensure low velocities for solids removal and even loading to prevent anoxic conditions. Stormwater treatment wetlands are often designed with a shallow edge slope and a permanent pool to provide a variety of hydrologic conditions, encourage the growth of diverse wetland plants and microbes, and support both aerobic and anaerobic pollutant removal processes.

The Hidden Valley treatment wetland in Riverside, California shines as an example of a treatment wetland (McPherson, J., and Thakral, S. K., "Wetlands Application of Reclaimed Water," *Water Environment & Technology*, 9(3): 35–41, 1997). A regulatory revision required the city of Riverside to remove nitrogen from its wastewater. The cost of a conventional denitrification facility at the treatment plant was estimated at $20 million. Instead, a low-grade wetland infested with invasive, nonnative vegetation near an existing treatment plant was enhanced to provide nitrogen removal along with ecosystem benefits. The cost of constructing the 28-ha (70-ac) wetland was only $2 million. The operation and maintenance costs of the wetland system are 90 percent less than a conventional system. In operation since May 1995, the system has proven effective at nitrogen removal and has met all permit requirements. Furthermore, the wetland provides important ancillary benefits—an interpretive center for environmental education, trails for recreational use attracting over 10,000 visitors a year, and wildlife habitat supporting 94 bird species.

A common concern is whether toxins in water, soil, and plants will pose a threat to wildlife attracted by a treatment wetland's habitat. Kadlec and Knight (1996, p. 712) state "to date, no conditions in wetlands designed for treatment of municipal wastewater and stormwater have been found to be problematic to propagation of fish or other wildlife populations." However, high concentrations of selenium in irrigation return water was blamed for fish deaths and acute and chronic damage to birds in habitat wetlands at Kesterson National Wildlife Reserve in California (Willard, D. E., and Willis, J. A., "Lessons from Kesterson," *Proceedings of the National Wetlands Symposium on Urban Wetlands*, Kusler, J. A., et al. (eds.), Association of State Wetland Managers, pp. 116–121, 1988). This incident reminds us that forethought in treatment wetland design, as well as periodic monitoring of water quality, is necessary to prevent any potential detrimental food chain effects. As long as source control or pretreatment prevents consistently high concentrations of heavy metals and toxic chemicals, levels toxic to biota are unlikely to occur.

A report by Wren et al. describes a protocol for monitoring and determining the effects on wildlife inhabiting stormwater ponds and wetlands (Wren, C. D., Bishop, C. A., Stewart, D. L., and Barrett, G. C., "Wildlife and Contaminants in Constructed Wetlands and Stormwater Ponds: Current State of Knowledge and Protocols for Monitoring Contaminant Levels and Effects in Wildlife," *Technical Report 269*, Canadian Wildlife Service, Ontario Region, 1997).

19.3.16   WETLAND MANAGEMENT

Wetland management goals are often aimed at promoting wildlife habitat, especially for waterfowl and wading birds. The U.S. Fish and Wildlife Service and Ducks Unlimited are experienced in this form of management. Managing wetlands to provide wildlife habitat is often at odds with fish management goals because many species consume fish. Control of nuisance species (vegetation or animals) is another common management objective. Management techniques include burning, herbicide application, dredging, and plant harvesting (Kent, D. M., "Managing Wetlands for Wildlife," in *Applied Wetlands Science and Technology*, Kent, D. M. (ed.), Lewis Publishers, Boca Raton, FL, pp. 307–330, 1994).

Mosquitoes are a particular management concern. Indeed, many wetlands have been drained or filled in the name of mosquito control. In accordance with the integrated pest management approach, numerous practices are available for mosquito control, such as increasing the abundance of mosquito-eating fish [mosquito fish (*Gambusia affinis*) and killifishes (*Fundulus* spp.)], birds, bats, and insects, improving the access of these species to mosquito breeding areas through new channels or increased water levels or drawdowns to desiccate breeding areas, and the use of *Bacillus thuringiensis israelensis* (Bti, a bacterium). More information is available from the University of California—Davis MosquitoNet (http://mosqnet.ucdavis.edu/).

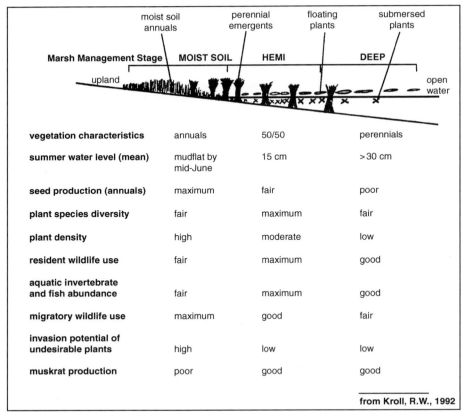

FIGURE 19.3.13   Water level management for vegetation, wildlife use, and other characteristics.

A common management technique is water-level manipulation via water control structures such as adjustable weirs. Flooding can help control invasive vegetation, such as *Phragmites*, and tends to increase the production of invertebrates, improving waterfowl feeding habitat (Figure 19.3.13). Drawdowns help control submerged invasive species (like Eurasian water milfoil), aerate the soil, encourage emergent species and expose mudflats for feeding shore birds.

REFERENCES

Cowardin, L. M., Carter, V., Golet, F. C., and LaRoe, E. T., *Classification of Wetlands and Deepwater Habitats of the United States*, U.S. Fish & Wildlife Service, FWS/OBS-79/31, (http://www.nwi.fws.gov/classifman/classman9.html), 1979.

Davies, D. J., "Wetlands Conservation and Management Initiative (WCAMI), Volume 2," *Bibliography of Wetlands: Issues and Trends with Selected Annotations*, USEPA Region 4, EPA904-R-96-007b, (http://earth1.epa.gov/docs/ Region4Wet/master1.al_gs.txt), 1996.

Environmental Lab, *Corps of Engineers Wetlands Delineation Manual* (*Technical Report Y-87-1*), U.S. Army Corps of Engineers, Waterways Experiment Station, (http://www.wes.army.mil/el/wetlands/pdfs/wlman87.pdf), 1987.

Federal Interagency Committee for Wetland Delineation (FICWD), *Federal Manual for Identifying and Delineating Jurisdictional Wetlands*, Washington, D.C., 1989.

Hammer, D. A., *Creating Freshwater Wetlands* (2nd ed.), Lewis Publishers, Inc., Boca Raton, FL, 1996.

Hurt, G. W., Whited, P. M., and Pringle, R. F. (eds.), *Field Indicators of Hydric Soils in the United States*, U.S. Department of Agriculture, Natural Resources Conservation Service, (http://www.statlab.iastate.edu/soils/hydric/), 1996.

Kadlec, R. H., and Knight, R. L., *Treatment Wetlands*, Lewis Publishers, Inc., Boca Raton, FL, 1996.

Kent, D. M. (ed.), *Applied Wetlands Science and Technology*, Lewis Publishers, Inc., Boca Raton, FL, 1994.

Kusler, J. A., and Kentula, M. E. (eds.), *Wetland Creation and Restoration: The Status of the Science*, Island Press, Washington, D.C., 1990.

Mitsch, W. J., and Gosselink, J. G., *Wetlands* (2nd ed.), John Wiley & Sons, New York, 1993.

National Research Council Committee on Characterization of Wetlands, *Wetlands: Characteristics and Boundaries*, National Academy Press, Washington, D.C., 1995.

Niering, W. A., *Wetlands: A National Audubon Society Nature Guide*, Alfred A. Knopf, New York, 1985.

Northern Prairie Science Center and Midcontinent Ecological Science Center, *Wetland Restoration Bibliography*, Northern Prairie Wildlife Research Center, Jamestown, ND, (http://www.npwrc.usgs.gov/resource/literatr/wetresto/wetresto.html), 1996.

Schueler, T. R., *Design of Stormwater Wetland Systems*, Metropolitan Washington Council of Governments, Washington, D.C., 1992.

Thunhorst, G., *Wetland Planting Guide for the Northeastern United States*, Environmental Concern, Inc., St. Michaels, MD, 1993.

Tiner, R. W., *Wetland Indicators: A Guide to Wetland Identification, Delineation, Classification, and Mapping*, Lewis Publishers, Inc., Boca Raton, FL, 1999.

USEPA Office of Wastewater Management, *Constructed Wetlands for Wastewater Treatment and Wildlife Habitat: 17 Case Studies*, U.S. Environmental Protection Agency, (http://www.epa.gov/owow/wetlands/construc/content.html), 1995.

USFWS, *National List of Vascular Plant Species That Occur in Wetlands*, U.S. Fish and Wildlife Service, (http://www.nwi.fws.gov/ecology.htm), 1996.

USGS, *National Water Summary on Wetland Resources*, U.S. Geological Survey Water-Supply Paper 2425, (http://water.usgs.gov/nwsum/WSP2425/index.html), 1996.

RELEVANT JOURNALS

Ecological Engineering, Elsevier Press
Land and Water

National Wetlands Newsletter, Environmental Law Institute
Restoration Management and Notes, Society for Ecological Restoration
Wetland Journal, Environmental Concern Inc.
Wetlands, Society of Wetland Scientists
Wetlands Ecology and Management, SPB Academic Pub

WORLDWIDE WEB LINKS

U.S. Federal Government

Federal Emergency Management Agency (FEMA), flood insurance rate maps:
  http://www.fema.gov/MSC/hardcopy.htm
U.S. Army Corps of Engineers

- Waterways Experiment Station, Environmental Laboratory (Center for Aquatic Plant Research & Technology, Wetlands Research & Technology Center):
  http://www.wes.army.mil/EL/wrtc/wrtc.html
- Regulatory program information (Section 404):
  http://www.usace.army.mil/inet/functions/cw/cecwo/reg/

U.S. Department of Agriculture, Natural Resources Conservation Service (formerly the Soil Conservation Service)

- Constructed Wetlands Bibliography:
  http://www.nal.usda.gov/wqic/Constructed_Wetlands_all/index.html
- Soil Surveys and Hydric Soils of the United States: http://www.statlab.iastate.edu/soils/
- List of State Offices: http://www.nrcs.usda.gov/NRCstate.html
- Wetlands Reserve Program (payments for restoring and protecting private wetlands):
  http://ngp.ngpc.state.ne.us/wildlife/wrp.html
- Wetland Science Institute (WLI): http://www.pwrc.usgs.gov/wli/

U.S. Environmental Protection Agency wetlands program:
  http://www.epa.gov/OWOW/wetlands/
U.S. Fish and Wildlife Service (USFWS)

- National Wetland Inventory (paper and electronic maps and data classification manual):
  http://www.nwi.fws.gov/Welcome.html
- Branch of Habitat Assessment, Ecology Section (includes list of wetland plants):
  http://wetlands.fws.gov/

U.S. Geological Survey

- Biological Resources Division, National Wetlands Research Center:
  http://www.nwrc.usgs.gov/
- Wetland Glossary: http://water.usgs.gov/nwsum/WSP2425/glossary.html

Professional Organizations

Association of State Wetland Managers (ASWM):
  http://home.cwix.com/~aswm.wetland@mci2000.com/index.html
Ducks Unlimited: http://www.ducks.org/
Ducks Unlimited Institute for Wetland and Waterfowl Research (IWWR):
  http://vm.ducks.ca/IWWR/index.html

Environmental Concern Inc. (wetland restoration, research, and education; publishes *The Wetland Journal*): http://www.wetland.org/

Environmental Law Institute (publishes the *National Wetlands Newsletter*; hosts the eli-wetlands emailing list. To subscribe: Send an e-mail message to majordomo@igc.org with subscribe eli-wetlands as the only text in the body of the message with no subject.): http://www.eli.org/

International Association for Water Quality, Wetland Specialist Group: http://www.iawq.org.uk/spgroups/wetland.htm

Society for Ecological Restoration: http://ser.org/

Society of Wetland Scientists (SWS) (Publish the journal *Wetlands*; certify Professional Wetlands Scientists; provide an on-line wetland discussion forum): http://www.sws.org

Universities

Michigan State University, Department of Resource Development, Wetlands Tutorial: http://rdserv1.rd.msu.edu/wetlands/Tutorial/wetinfo.htm

North Carolina State, Water Quality Group, introduction to wetlands: http://h2osparc.wq.ncsu.edu/info/wetlands/index.html

Ohio State, wetland links: http://kh465a.ag.ohio-state.edu/wetlinks.html

Purdue University, The Wetlands Education System: http://AGEN521.www.ecn.purdue.edu/AGEN521/epadir/wetlands/menu.html

University of Florida

- Center For Wetlands: http://www.enveng.ufl.edu/wetlands/
- Center for Aquatic and Invasive Plants: http://aquat1.ifas.ufl.edu/welcome.html

Training and Education

Institute for Wetland & Environmental Education & Research: http://members.aol.com/iweer

Richard Chinn Environmental Training, Inc.: http://www.richardchinn.com/

Wetland Training Institute, Inc.: http://www.wetlandtraining.com/

Delineation Supplies

Ben Meadows Company: http://www.benmeadows.com/

Forestry Suppliers: http://www.forestry-suppliers.com/

Others

Richard B. Winston's Wetland Links: http://www.mindspring.com/~rbwinston/wetland.htm

Vernal Pools: http://earth.simmons.edu/vernal/pool/vernal.html

The Wetlands Regulation Center: http://www.wetlands.com/

Yahoo directory of wetland links: http://dir.yahoo.com/Science/Ecology/Ecosystems/Wetlands/

CHAPTER 19
SENSITIVE ENVIRONMENTAL SYSTEMS

SECTION 19.4

PLANTS AND POPULATION: WE HAVE THE TIME. DO WE HAVE THE WILL?

Dennis T. Avery

Mr. Avery is director of Global Food Issues, the Hudson Institute, Indianapolis, Indiana. He is an internationally known expert on the dynamic changes taking place in the world's ability to feed itself and still protect environmental resources. He is especially noted for his analyses of trends in food demand, agricultural productivity and global farm competition.

The Rev. Thomas Malthus' famous question about whether humanity can continue to feed all the people was posed exactly 200 years ago. It has taken us nearly all of that 200 years to be sure of an affirmative answer. Only recently have we been certain that the opening of the twenty-first century should see a new and fully-sustainable balance between food, population, and the environment because of the following:

- Radically declining birth rates virtually all over the world
- Enormous advances being made in the scientific knowledge of how to boost food production
- Vastly more affluence than any previous generation, and, thus, more capital to invest in the roads, storage facilities, ships, and research labs that encourage food production, distribution, and preservation
- An array of technologies—contraceptives, biotechnology, computers, satellite communications, cryogenics, and a host of other technical advances—which can help to achieve a constructive balance between human needs and the ecology.

Compare this situation with any year before 1960. Before that year, massive famines seemed certain for much of the world, poverty was the global norm, the Green Revolution had not yet demonstrated its power. By comparison, the world today has a virtual certainty of food production success. If humanity is to starve or displace wildlife in the twenty-first century, with today's technology and a declining population growth rate, it could only be because we lack the political will. However, that may be the case.

Today, the real question is not whether the world can produce enough food for a peak population of 8.0 to 8.5 billion people. It can. We could already produce enough to satisfy minimal caloric requirements for that many people if known technologies were fully extended, and production was divided equally among all consumers. The world's recent famines have been due to "mistakes of government," such as civil wars and Mao Tse-tung's ill-considered communal farms. Little hunger has been the result of the lack of available food.

Forty percent of the world's current crop output, in fact, goes to livestock and poultry feed so that affluent people can eat high-quality diets full of meat, milk, and eggs. In a hunger emergency, we can eat both the feedstuffs and the livestock, and later worry about rebuilding the flocks and herds.

19.4.1   *THE FOOD CHALLENGE IS AFFLUENCE*

The food challenge of the twenty-first century, in fact, is not the challenge of population growth, but the challenge of affluence. Virtually all the people of the twenty-first century will be affluent by today's standards and able to afford education, nice clothes, and TV sets. Such people are unwilling to accept minimal diets. The same modern couples who are willing to practice family planning, with two children instead of 15, demand that their two children get rich diets high in meat protein for growth, and milk calcium for strong bones. Affluent people insist on fresh fruits and vegetables all year round. Such diets take far more resources than boiled rice or corn-flour tortillas.

There is no vegetarian trend in the world; instead we are seeing the strongest surge of demand for such resource-costly foods as meat, milk, eggs, and fruit in all history. Currently, only about 4 percent of the First World's population are vegetarian, and most vegetarians consume lots of resource-costly eggs and dairy products.

There will also be a pet food challenge. The U.S. has 113 million pet cats and dogs for 270 million people. All over the world, ownership of companion animals and pet food sales rise with incomes. Already, China's one-child policy is stimulating pet ownership. How many cats and dogs will wealthy China have in 2050? And, woe unto the public official who stands between a pet owner and Fluffy's favorite food.

The debate in development economics is whether the challenge of affluence requires a 250 percent increase in the world's food output, or a 300 percent increase. The universal human hunger for high-quality protein, combined with the pet factor, convinces me that the world must be able to triple its farm output in the next 40 years.

19.4.2   *LAND—THE SCARCEST NATURAL RESOURCE*

The question is whether there is time, and that is the right question. However, the urgency does not derive from famine. The urgency derives from the competition between farming and wildlife for land.

- Agriculture already uses about 37 percent of the earth's land surface, and any land not already in a city or a farm is wildlife habitat.

- If the world has 30 million wildlife species (a reasonable biologist's "guesstimate"), then 25 to 27 million of them are probably in the tropical rain forests, with most of the remainder in such critical habitats as wetlands, coral reefs, and mountain microclimates. These are places we have not farmed, and should not farm.

- The world's good cropland typically had large wildlife populations—but only a few wild species. (Argentina's famed Pampas, for example, had virtually nothing but Pampas grass.) Researchers have found more species in five square miles of the Amazon rain forest than in all of North America.[1] In the name of conservation, we must farm the world's good land for the highest sustainable yields, so we can leave the tropical forests and fragile lands for the wild species.[2]

Modern high-yield farming has already saved millions of square miles of wildlife habitat.

- My peer-reviewed estimate is that higher crop yields have already prevented the plow-down of close to 15 million square miles.[3]

---

[1] Vive la Difference, *Scientific American*, April; 48, 1997.

[2] Thompson, R. L., "Technology, Policy and Trade: The Keys to Food Security and Environmental Protection," *Presidential Address*, 23rd International Conference of Agricultural Economists, 10 Aug. 1997.

[3] Avery, "Saving Nature's Legacy Through Better Farming," *Issues in Science and Technology*, National Academy of Sciences and the University of Texas/Dallas, pp. 59–64, Fall 1997.

- Confinement production of meat, milk, and eggs has saved additional millions of square miles of wildlands with lower death rates, higher feed conversion rates, and very small amounts of land needed per bird or animal.

- Modern food processing allows us to grow crops where the yields are highest, and then transport the harvest without postharvest losses wherever the people choose to live.

All told, the modern food system is currently saving something on the order of 18 to 20 million square miles of wildlands from being plowed for low-yield food production. That makes it the greatest conservation triumph in modern history.

The world will not only demand three times as much farm output in the middle of the twenty-first century, it will demand perhaps ten times today's forest harvest for paper and wood.

Thus, the key to conserving the natural world in the twenty-first century will be what the Hudson Institute calls "high-yield conservation." Meeting both the food and forestry challenges, while leaving room for nature, will depend on our ability to continue increasing the yields per acre from plants, animals, and trees on our best land, and transporting these yields to where they are in most demand. Our success will also depend heavily on the energy with which we explore such high-tech methods as biotechnology in food and forestry.

19.4.3 HAMSTRINGING HIGH-YIELD CONSERVATION

Yet the world's most advanced societies are attempting to legislate low-yield agriculture. All over the First World, government funding for agricultural research is being cut back, or shifted to low-yield "sustainable" farming. Governments in affluent countries subsidize low-yield organic farming, while regulators respond to the public opinion by depriving the world's high-yield farmers of their inputs. In Africa, which has not yet had its Green Revolution, aid donors are demanding that farmers increase food production *without* plant nutrients or modern pest protection.

Large numbers of well-fed, affluent, influential people are opposing biotechnology, the most important unexploited advance in humanity's knowledge of how to increase food production rapidly. There is a serious question as to whether the power of biotechnology will be marshaled in agriculture soon enough to make its undoubtedly huge contribution to simultaneously saving people and wildlife. Are modern societies attempting to surrender the planet back to hunger, malnutrition, and massive losses in wildlife habitat? And if so, why?

19.4.4 THE ENVIRONMENTALIST CAMPAIGN AGAINST MODERN FARMING

The opponents of modern, high-yield agriculture and biotechnology are, ironically, gathered under the banner of environmentalism:

- With the help of Rachel Carson's brilliantly flawed book, *Silent Spring*, eco-activists long maintained that modern farmers are poisoning children with cancer-causing chemicals. However, after 50 years of widespread pesticide use, and billions of research dollars, science is still looking for the first case of cancer caused by pesticide residues. The U.S. National Research Council, the Canadian Cancer Institute, and other medical authorities are trying to tell the public that the cancer fears are unfounded. The American Cancer Society is presenting evidence that cancer death rates—other than lung cancer—have been declining since the 1970s, even though our use of pesticides increased.[4]

---

[4] Cole and Rodu, *Declining Cancer Mortality in the U.S.*, special presentation to the Executive Committee of the American Cancer Society, Aug. 10, 1996. (Reprints available from Dr. Philip Cole, 221 TH, University Station, Birmingham, AL 35294.)

- For 50 years, wildlife groups have universally claimed that modern farm chemicals were poisoning wildlife on a massive scale. However, the wildlife losses to today's narrowly targeted and rapidly degrading chemicals are trivial—especially when compared with the millions of square miles of wildlife habitat saved by farmers' high yields. We must remember that wildlife numbers are generally governed by the amount of food and habitat available. Nobel Peace Prize laureate Norman Borlaug has estimated that if American farmers hadn't raised their yields in recent decades, we would have needed another 460 million acres worth of good land to get today's food production; that's the equivalent of clearing all of the U.S. forest east of the Mississippi, and planting crops on all the Eastern pasture (with the attendant rise in soil erosion).[5] How many billions of birds' habitats would have been destroyed in that process?

- Eco-activists claim that more food means more people. However, we are clearly in the first era of human history when more food has *not* meant more population. Births per woman in the Third World are down from 6.5 in 1960 to 2.9 today, and the birth rates have fallen fastest in the countries where the crop yields have risen most rapidly. Higher yields are the best measure of food security, which encourages smaller families because parents can feel secure that their first two or three children will live.

- Environmentalists claim that modern farming is destroying the soil with rampant erosion. However, tripling the yields on the best land cuts erosion per ton of food by at least two-thirds. Farmers have also used herbicides and tractors to invent conservation tillage, which cuts soil erosion per acre by 65 to 95 percent. We are now building topsoil on the best land in the midst of the highest-yielding farming in history.

- Environmentalists claim that modern high-yield seed varieties are destroying the world's biodiversity, displacing thousands of landrace crop varieties with a few dozen modern hybrids. They recommend that we turn the Third World into a museum for low-yield seeds. In truth, much of the biodiversity is already in the world's gene banks, and more of it would be if we modestly increased funding for gene banks and modest numbers of "gene farms." Biotechnology can use wild genes to create more biodiversity, giving more incentive for gene conservation than any development in history. They forget that we would long since have starved and/or destroyed most of our wildlands without the higher yields produced by modern seeds.

- Environmentalists oppose liberalized farm trade, though this is the only hope for much of Asia's wildlife. (In the years ahead, some of Africa's coastal cities may also be supplied at lower environmental cost through food imports.) Instead, the environmentalists support European farm subsidies, which dump surpluses and discourage farmers in other countries. The sight of Europe's surplus grain and cheese apparently blinds Europeans to the coming surge of farm demand in the Third World. Why isn't the environmental movement following it own dictum to look globally and long-term?

We must now realize that modern agriculture is being targeted, not because it is bad for the environment, but because modern farming: (1) represents the greatest success of technological abundance, and (2) because farming controls much of the world's land and water. The environmental movement seems to want managed scarcity for a few people. It seems to want more bison and prairie dogs—and fewer corn plants—on American land even if that sacrifices wildlands and biodiversity elsewhere.

19.4.5 THE NEW GLOBAL CAMPAIGN AGAINST PLANT NUTRIENTS

The latest eco-campaign is against plant nutrients. The United States supposedly has a crisis in water quality. The public is being told that vital plant nutrients, such as nitrogen and phosphorus, are environmental threats.

---

[5] Borlaug, "Fertilizer to Nourish Infertile Soil That Feeds a Fertile Population That Crowds a Fragile World," *Keynote Address*, International Agribusiness Management Association Symposium, San Francisco, CA, May 25, 1993.

- *Blue Baby Syndrome.*   Some environmental groups are demanding that the nitrogen limit in drinking water be lowered from 10 to 5 ppm, apparently just to make it more difficult for modern agriculture to function.[6] Never mind that the incidence of blue baby syndrome almost disappeared during the very period when the use of chemical fertilizers and confinement feeding of livestock and poultry flourished. Never mind that current medical research, triggered by an epidemic of blue baby in a region that had no nitrogen in its drinking water at all, indicates it is the *bacteria* in the contaminated wells that produce blue baby syndrome—not the nitrate.

- *Hypoxia.*   A White House task force has been appointed to resolve the hypoxia problem in the Gulf of Mexico. The hypoxic, or low-oxygen, zone in the Gulf doubled after 1993's big floods, from 3500 square miles to 7000 square miles. Agriculture, again, is being blamed. The presumed solution is to make Midwest farmers radically cut their use of fertilizer, and to "crack down" on big livestock and poultry farms. Never mind that hypoxic zones are characteristic of rivers that drain fertile regions. Never mind that the nutrients for rich fishing waters almost always come from such rivers. Never mind that Dr. Nancy Rabelais of Louisiana State University, hypoxia's biggest critic, found the size of the hypoxic zone in the Gulf of Mexico had shrunk back to 4800 square miles in 1998; that links the size of the hypoxic zone to the volume of water in the Mississippi, not nitrate levels. Never mind that cutting fertilizer use in the Corn Belt would mean significantly lowering yields—and clearing forest for low-yield crops somewhere else in the world.

- *Manure as Toxic Waste.*   For 50 years, the critics of modern farming have held up organic crops fertilized with animal manure as the global ideal. Now the same critics are saying that "organic fertilizer" is "toxic waste"—if the animals or birds are being raised in a big confinement facility. Never mind that the big confinement feeders protect the environment by collecting their wastes, and using them constructively to more sustainably raise the yields of feed crops. Or, that the little outdoor producers let the wastes wash into the streams. As the world triples the number of hogs in its inventory from 1 billion to 3 billion, we had better hope for the sake of the environment the additional hogs will be raised in confinement.

- *Volatilized nitrogen.*   Recently, the activist magazines—and even *Science*—have carried articles about the dangers of "too much fixed nitrogen." They're concerned that too many crops are being fertilized, and too many meat and milk animals are producing too much manure. They say that too much fixed nitrogen might even change the global climate and our ecosystems. Certainly, some of the fixed nitrogen from agriculture is volatilized into the atmosphere, but so far no one has found that to be much of a problem. One researcher complained that the extra nitrogen "aggravates acid rain." However, a $600 million Federal study found that acid rain was a minor problem, confined to a few tree species such as red spruce, in a few mountain areas lacking limestone or other buffering from the natural acidity of rainfall. The nitrogen from perhaps 4 million square miles of high-yield crops and intensive livestock is being spread over 197 million square miles of land, water, mountain, and forest around the world—where its major impact is to very slightly enrich the food chain. The biggest negative impact is likely to be a slight disadvantage for wild legume plants.

- *Complaints about Wonder Wheat.*   Recently the International Maize and Wheat Improvement Center announced a major re-breeding of the wheat plant—*done without biotechnology.* CIMMYT says the new wheats have yielded up to 18 tons of grain per hectare, 50 percent more than any other wheats![7] Yet the initial reaction cited in *Science* was distress that this would encourage high levels of fertilizer use. Never mind that it takes about 25 kg of nitrogen to grow a ton of wheat. We can grow 18 tons of wheat on one hectare with 400 kg of N, or we can clear another 17 hectares of wildlife habitat to grow one ton of wheat on each of 18 hectares.

---

[6] *Pouring It On: Nitrogen Contamination of Drinking Water*, Environmental Working Group, Washington, D.C., Feb. 22, 1996.

[7] "Wonder Wheat," *Science*, Vol. 289, 24 April, p. 527, 1998.

19.4.6 THE FUTURE WITH BIOTECHNOLOGY

The world is in the early phases of exploring biotechnology's potential—the "biplane stage," to draw the analogy with airplanes. However, already we see enough to know that biotechnology will be enormously important to conservation.

Saving Wild Species with Aluminum-Tolerant Crops

Two researchers from Mexico discovered a way to overcome the aluminum toxicity that cuts crops yields by up to 80 percent on the acid soils characteristic of the tropics. Noting that some of the few plants that succeed on the world's acid savannas secrete citric acid from their roots, they took a gene for citric acid secretion from a bacterium and put it into tobacco and papaya plants. Presto, they had acid-tolerant plants. The acid ties up the aluminum ions, and allows the plants to grow virtually unhindered.[8] The Mexican researchers have since gotten the citric acid gene to work in rice plants, and hope that it can be used widely in crop species for the tropics.[9]

Acid-soil crops have enormous potential for wildlife conservation. Acid soils make up 30 to 40 percent of the world's arable land, and about 43 percent of the arable land in the tropics. Thus far, they have been one of the major barriers to high yields in the very regions that are critical to wildlands conservation—the Third World tropics. These are the very areas where the populations are growing most rapidly, where incomes are rising most rapidly, where the food gaps are expanding most rapidly, and where most of the world's biodiversity is located.

Raising Yields with Wild-Relative Genes

Two researchers from Cornell University reasoned that more than a century of inbreeding the world's crop plants had significantly narrowed the genetic base of our crops. They also reasoned that the world's gene banks contained a large number of genes from wild relatives of our crop plants. They selected a number of genes from wild relatives of the tomato family, a crop where yields have been rising by about 1 percent per year. The wild-relative genes produced a 50 percent gain in yields and a 23 percent gain in solids. The same researchers selected two promising genes from wild relatives of the rice plant—a crop where no yield gains had been achieved since the Chinese pioneered hybrids some 15 years ago. Each of the two genes produced a 17 percent gain in the highest-yielding Chinese hybrids; the genes are thought to be complementary, and capable of raising rice yield potential by 20 to 40 percent.[10]

Improved Meat Animals with Biotech

Heretofore, methods for introducing new genes into livestock had a low efficiency—less than 10 percent. However, in the 24 November 1998 issue of *The Proceeding of the National Academy of Sciences*, researchers report a new method for producing transgenic animals that approaches 100 percent efficiency. Researchers put the foreign gene into the animal's egg before it was fertilizer rather than shortly after. Obviously, this is another important step in creating animals with greater tolerance for pests and diseases, better feed conversion ratios and other practical advantages.

---

[8] Barinaga, "Making Plants Aluminum Tolerant," *Science*, Vol. 276, 6 June 97, p. 1497.

[9] Herrera-Estrella, "Transgenic Plants for the Tropics," Paper for NAS Colloquium on *Plants and People, Is There Time?*, Departamento de Ingenieria Genetica, Investigation y Estudios Avanzados, Guanajuato, Mexico, 1998.

[10] Tanksley and McCouch, "Seed Banks and Molecular Maps: Unlocking Genetic Potential from the Wild," *Science*, Vol. 277, pp. 1063–1066, 22 Aug. 97.

Speeding Progress in Protecting Crop Yields

Recently, a research consortium announced it had succeeded in creating a genetic barrier against a new race of barley stem rust that had been advancing northward in recent years from Colombia. The new barrier was created in less than a decade. With traditional plant-breeding techniques, it might have taken several decades. With farmer-saved landrace seeds, overcoming the rust might have taken centuries.

Saving Forests with Biotech Trees

The world could increase its forest harvest ten-fold if we planted just five percent of today's wild forests in high-yield tree plantations. Such plantations are good-but-not-great wildlife habitat because they are not "fully natural," but they could apparently take all of the logging pressures off 95 percent of the natural forests.

Trees have always been difficult to improve through crossbreeding because the time frames are so long. Biotechnology is already helping to provide the higher-yielding trees through cloning and tissue culture—which permit us to rapidly copy the fastest-growing, most pest-resistant trees in a species. When we master the tools of biotechnology more fully, we should be able to increase forest growth rates, drought tolerance, pest resistance, and other important traits more directly, and even more effectively.

19.4.7   WHY DID SWITZERLAND TRY TO OUTLAW BIOTECHNOLOGY?

Swiss activists collected more than 100,000 signatures on a petition to ban biotechnology in food production. The signatures put the question on a national referendum ballot in 1998. The good news is that the initiative failed, and the ban was defeated. The bad news is that the Swiss ballot initiative is probably a warning of further troubles with public acceptance of high-yield modern farming, and specifically with biotechnology in food production. The worst news is that outspoken female scientists led the opponents of biotech food in Switzerland, and the coalition included the country's largest Protestant and Catholic women's groups. The opponents of biotech food in Switzerland included some of modern agriculture's core customers, people who should be its strongest supporters.

Why did so many Swiss sign the petition? Why did biotechnology nearly lose its charter in one of the most educated and affluent countries of the world? First, the Swiss signed the petition because they already have plenty of food. They take food for granted. That describes a billion people in the world today, but it will describe 3 billion people in the next decade and 5 billion people in the decade after that. Agriculture can no longer count on consumers feeling "grateful" for their food. Second, Swiss signed the petition because Europe has a food surplus, and Europeans see more food as simply leading to global overpopulation. There are only 7 million Swiss, but they're crowded into mountain valleys with the same traffic jams and exhaust fumes as New York and London. Third, they signed the petition to protect laboratory animals. The animal rights activists were a key element in the anti-biotech coalition. The fact that the Catholic and Lutheran women's groups joined the coalition probably means that First World religious groups no longer feel comfortable with the Judeo-Christian assertion that God gave man "dominion" over the other species on the planet. The Swiss petition defined laboratory animals to mean not only monkeys and lab rats, but even fruit flies and earthworms. Fourth, the Swiss signed the anti-biotech petition because they are genuinely nervous about the power of biotechnology. They understand that the power to manipulate genes directly goes well beyond any power than scientists have ever had before. They are willing to accept the use of biotechnology in human medicine, *because they clearly see the benefits*. Unfortunately, agriculture has never given European consumers what they consider a valid reason for putting the power of biotechnology into the hands of agricultural researchers.

19.4.8 A GLOBAL TREND TOWARD MORE ACTIVISTS

It is the nature of activists to push for something different. In Peru, activists demanded an end to the chlorination of drinking water because the U.S. Environmental Protection Agency found chlorine, at high levels, could cause cancer in laboratory rats. Peruvian officials took the chlorine out of the water, and the cities promptly suffered a cholera epidemic that killed 7000 people.

I don't blame the activists. I blame the people who trusted the activists, and the people who should have represented the other side of the question. I also blame the press, which should have sought out the broader reality. Like it or not, the world is on a trend to have more activists, in more countries. Democracy and affluence encourage activists and the free, open debate of public questions. If modern agriculture is to succeed, it must learn to succeed in an activist-rich environment.

It's not just agriculture, of course. Global warming activists have created global summits, an international treaty, and captured the political soul of a major U.S. presidential candidate—with less evidence than they've had of harm from modern agriculture. However, the activists have come so far and won so much power and prestige around the world that they can't stop.

19.4.9 THE ACHILLES HEEL OF HIGH-YIELD AGRICULTURE—REGULATION

It is true that the Green Movement has rarely won an election, anywhere in the world. However, the desire to preserve Nature is so urgent in First World cities that the Greens haven't needed to win elections. Environmental concern is so widespread that politicians race each other to embrace key points of environmental strategy. In America, recent Wirthlin polling indicates that 75 percent of the public agrees with the statement, "We cannot set our environmental standards too high—regardless of cost."

Because of the high public approval for the environment, we have an Environmental Protection Agency with virtually no Congressional oversight. The bureaucrats who work for EPA read newspapers and polling results. They assume that they can regulate "environmentally offending" industries, such as agriculture, in virtually any way they choose. Modern farming's reputation with the urban public is now so bad that it can no longer persuade the Congress to block unfavorable legislation, or force federal agencies to modify unfavorable regulations and rulings. Not even farm-state politicians will commit political suicide on behalf of farming.

19.4.10 BETRAYED BY MODERN JOURNALISM?

No one believes more fervently than I do in the importance of a free press. Unfortunately, today's mainstream media are not living up to their professional obligations for objectivity and research. Among the causes they have adopted as their own in recent decades is the environment. (The *New York Times* is perhaps the most dramatic example of this, but the phenomenon is widespread.) Adopting the Greens' agenda meant that journalists have had to disown modern agriculture. I have been on a first-name basis with *New York Times'* Science Editor, Bill Stevens, for a decade. He cheerfully quotes me on world hunger questions—and just as cheerfully ignores the environmental benefits that I tell him are being delivered by high-yield farming.

Recently, our Center put out a press release noting that the water quality in North Carolina's Black River has improved over the last 15 years, even though the hog population in its watershed had quintupled to one of the highest densities in the United States. A skeptical reporter called and asked whether the hog industry had sponsored the study. No, we told her, the data was from the State environmental agency. "But that's not what my readers want to hear," she lamented. That's how far behind the public affairs curve modern agriculture currently finds itself. This is not a problem that can be dealt with by writing press releases, or by hosting community tours of farms and milk processing plants.

19.4.11   *CAN WE EDUCATE THE PUBLIC ON HIGH-YIELD CONSERVATION—IN TIME?*

Someone must tell the urban public about the environmental benefits of high-yield modern farming. I submit that it will have to be agriculture. Agriculture and agricultural researchers must talk about saving wildlands and wild species with better seeds. We must talk about conquering soil erosion with high yields (so there's less farmland to erode) and conservation tillage (which radically reduces erosion per acre of farmland). We must talk about preventing forest losses to slash-and-burn farming (the cause of destruction for two-thirds of the tropical forest we've lost). We must point out that where high-yield farming is practiced, the amount of forest is expanding. We must point out that the losses in wildlife habitat overwhelmingly occur where the farmers get low yields.

Agriculture and its researchers also need to point up the high risks of organic food. The Centers for Disease Control has been afraid to publicize it, but their own data seem to show that people who eat organic and "natural" foods are eight times as likely to be attacked by the virulent bacteria, E. coli O157:H7. *Consumer Reports* wrote that free-range chickens carried three times as much salmonella contamination. The fact is that organic food is fertilized with animal manure—a major reservoir of bacterial contamination—and composting is neither careful enough nor hot enough to kill all of the dangerous organisms.

We must analyze every eco-activist proposal in terms of its land requirements:

* Organic farming for the world would mean clearing at least 5 million square miles of wildlife for clover and other green manure crops.
* Free-range chickens for the United States would take wildlands equal to all of the farmland in Pennsylvania.
* Reducing fertilizer usage in the Corn Belt would mean clearing many additional acres of poorer-quality land in some distant country to make up for the lost yield.
* Blocking free trade in farm products and farm inputs will probably mean clearing tropical forest for food self-sufficiency in Asia.

It should not be solely up to agriculture to prevent such a needless disaster. Agriculture has no history of public affairs campaigns or any real experience in conducting them. However, I see no other entity with the knowledge, the financial requirements and the direct interest to do it.

I doubt that the National Academy of Sciences or the National Research Council can turn public opinion around. The NRC's recent report, *Carcinogens and Anti-carcinogens in the Human Diet*, is a landmark. It essentially says pesticide residues are no threat to public health. However, the public is not reading the document, and the media are not reporting it. Moreover, a significant number of NAS members are *encouraging* the attacks on high-yield farming.

I doubt that the land-grant colleges of agriculture can do very much to turn public opinion around. On the question of high-yield farming, the opponents paint them as co-conspirators. As tax-supported institutions, it is questionable how much public opinion steering they should attempt in any case. The U.S. Department of Agriculture is also a political institution bending with the winds of popular perception.

In Germany, the recent years saw a significant turn-around, as the biotech companies made a serious effort to speak to the public about the realities of biotech foods. They also had the advantage that Chancellor Kohl was willing to make positive statements in favor of biotech foods. The serious effort paid off in increased consumer acceptance—for the moment. However, Chancellor Kohl's government has been replaced by a new coalition that includes the Green Party, and we cannot look for the current Chancellor to praise biotechnology. Nor is an American president likely to endorse biotech foods and farming any time soon.

How can we present the environmental case for high-yield agriculture if the journalists will not write it and politicians fail to support it? Modern agriculture must take its case directly to the people, through *advertising*. My model is the Weyerhaeuser Company, which has been telling me for decades that it's the tree-growing company. Not the tree-cutting company, not the tree-using company, but the tree-growing company.

David Brinkley, the most respected journalist in America today, has also shown us the way. ADM, the big corn and soybean processor, sponsors the Brinkley ads, and they are doing a fabulous job:

- Brinkley notes that farmers are still the most indispensable people.

- He shows a cute little girl in Taiwan, and points out that her mother wants her to have meat and milk in her diet so she will grow strong and vigorous. Who could oppose that?

- The ads show families of deer and wild birds, and note that "the higher yields achieved by modern farmers are providing food—and in some cases even shelter—for families around the world."

Many of the firms with billions of dollars invested in modern agriculture are already talking to urban America. DuPont and Dow have whole rosters of consumer products and millions of dollars worth of consumer advertising. Cooperatives like Land-o-Lakes and Countrymark have consumer ad budgets too. Wildlands conservation would be a winning message with both their customers and their farmer members.

So far, agriculture has failed to accept the challenge, and the momentum for high-yield conservation is waning. We are not increasing public investments in high-yield research. We are not creating the Green Revolution for Africa. The regulators are continuing to strangle farm productivity.

In the long run, of course, farmers and farm researchers will be vindicated even without a public affairs campaign. However, that vindication could come too late for the wildlands and the wild species—and too late for most of today's high-tech farmers and agribusinesses. At this point, it looks as though we will fail to meet the food challenge of the twenty-first century—not for lack of time, but for lack of realism in our public life. Our forefathers would have been ashamed for us.

Avery's book, *Saving The Planet With Pesticides And Plastic: The Environmental Triumph of High-Yield Farming*, is available from the Hudson Institute, P.O. Box 202, Churchville, VA 24421. (Fax: 540/337-8593). The cost is U.S. $19.95, including U.S. shipping and handling costs. (Bulk rates on request.)

CHAPTER 20
SENSITIVE ENVIRONMENTAL PROBLEMS

SECTION 20.1

AGRICULTURAL RUNOFF

George F. Czapar

Dr. Czapar is an extension educator in Integrated Pest Management (IPM) for the University of Illinois Extension at the Springfield Center. His current research includes evaluation of best management practices to protect surface water, the use of global positioning systems (GPS) for weed management, IPM adoption, the use of subsurface drainage tiles to estimate chemical movement to groundwater, and decision-making tools for weed management.

20.1.1   INTRODUCTION

The effect of agricultural chemicals on water quality continues to be a major public issue. As part of the 1986 amendments to the Safe Drinking Water Act (SDWA), public water supplies are required to sample quarterly for regulated contaminants, including several pesticides. Maximum contaminant levels (MCLs) have been established by the U.S. Environmental Protection Agency for over 30 pesticides and pesticide metabolites (Table 20.1.1). It is estimated that the total national cost of complying with SDWA drinking water regulations is $1.4 billion annually for public water systems (Auerbach, 1994).

National, state, and local water monitoring studies of groundwater and surface water quality have helped identify the most common contaminants, and when they are most likely to occur (Goolsby et al., 1991; Thurman et al., 1991). Monitoring programs have examined rivers, streams, and public water supplies (Ciba Crop Protection, 1995; Taylor and Cook, 1995; Temple and Krueger, 1994).

Although pesticides and fertilizers are often cited as examples of agricultural contaminants, soil erosion is probably the largest single cause of water quality problems. It is estimated that 675 million to one billion tons of eroded agricultural soils are deposited in waterways each year (National Research Council, 1986; USDA, 1986). In addition to controlling agricultural runoff, sediment reduction needs to be a major component of water protection efforts.

20.1.2   PESTICIDE RUNOFF

Wauchope (1978) reviewed the extent of pesticide loss from treated fields resulting from surface runoff. Pesticide losses ranged from less than one to over ten percent of the applied product. Numerous studies have shown that chemical losses are often greatest when heavy rainstorms closely follow pesticide applications. Monitoring efforts have documented the temporal occurrence of high pesticide concentrations in surface water.

In 1989, the U.S. Geological Survey (Goolsby et al., 1990; Thurman et al., 1991) sampled 150 stream sites in 10 Midwestern states including: Illinois, Iowa, Indiana, Kansas, Minnesota, Missouri,

TABLE 20.1.1 Maximum Contaminant Levels (MCLs) of Pesticides and Pesticide Metabolites

Chemical	Maximum contaminant level (MCL) (μg /L)
Alachlor	2
Aldicarb	3
Aldicarb Sulfone	2
Aldicarb Sulfoxide	4
Aldrin	1
Atrazine	3
Carbofuran	40
Chlordane*	2
2,4-D	70
Dalapon*	200
DDT*	50
DBCP*	0.2
Dieldrin*	1
Dinoseb*	7
Diquat	20
Endothall	100
Endrin*	2
Ethylene Dibromide*	0.05
Glyphosphate	700
Heptachlor*	0.4
Heptachlor Epoxide	0.2
Hexachlorobenzene	1
Hexachlorocyclopentadiene	50
Lindane	0.2
Methoxychlor	40
Oxamyl	200
Pentachlorophenol	1
Picloram	500
Simazine	4
Toxaphene	3
2,4,5-TP*	50

*Pesticide no longer registered.
Source: U.S. EPA.

Nebraska, Ohio, South Dakota, and Wisconsin. Streams were sampled three times: (1) Prior to herbicide application (March–April), (2) during the first storm runoff following herbicide application (May–June), and (3) during harvest (October–November). Samples were analyzed for the presence of several herbicides and two herbicide metabolites.

Herbicides and metabolites detected during these sampling times are shown in Table 20.1.2. Although several herbicides were detected prior to spring application (March–April), the concentrations measured were small. Only one sample containing simazine exceeded the lifetime health advisory level.

In contrast, 98 percent of the samples collected during the first storm runoff following herbicide application (May–June) contained at least one herbicide. Atrazine concentrations exceeded lifetime health advisory levels in approximately 56 percent of the samples. Lifetime health advisory levels were exceeded for alachlor in 35 percent of the samples, cyanazine in 12 percent of the samples, and simazine in 4 percent of the samples, respectively.

Samples collected during October and November contained substantially lower concentrations than those collected in May and June. One sample contained atrazine that exceeded lifetime health advisory levels. This study demonstrated the seasonal nature of pesticide occurrence in surface water.

TABLE 20.1.2   Herbicide Detections in Midwestern Streams

Herbicide or herbicide metabolite	Percent samples with detects		
	March–April	May–June	Oct.–Nov.
Alachlor	18	86	12
Atrazine	90	98	76
Desethyl-atrazine	51	86	47
Desisopropyl-atrazine	9	54	0
Cyanazine	5	63	63
Metolachlor	34	83	44
Metribuzin	2	53	0
Propazine	0	40	<1
Prometon	0	23	6
Simazine	13	60	3

Source: Goolsby et al., 1990.

In a similar study, Goolsby et al. (1991) collected 146 water samples from eight sites on the Mississippi River and its major tributaries in April, May, and June 1991. Atrazine was detected in every sample, while metolachlor, alachlor, and cyanazine were detected in 98, 82, and 78 percent of the samples, respectively.

Concentrations of herbicides increased in early May in response to rainfall following herbicide application and then decreased in early to mid-June. Atrazine exceeded the Maximum Contaminant Level (MCL) in 27 percent of the samples while alachlor exceeded its MCL in 4 percent of the samples.

In many areas, surface water appears more vulnerable to contamination from agricultural chemicals than does groundwater. For example, the 1987 Iowa Public Water Supply Survey found that 63 percent of the public surface water systems had detectable levels of pesticides, and that health advisory levels had been exceeded in 14 percent of the surface water supplies. In contrast, pesticides were detected in eight percent of the well systems and only one percent of the wells exceeded health advisory levels (Iowa DNR, 1988).

While pesticide runoff occurs from agricultural areas, Schmidt (1991) described a U.S. Geological Survey sampling program in the upper Illinois River basin that collected runoff event samples from urban and rural areas. Water samples taken in 1988 and 1989 showed differences in the compounds detected in urban and agricultural areas. The insecticides diazinon and malathion were found in 88 percent of the samples collected in urban areas and in only 12 percent of the samples from agricultural areas.

20.1.3   PHYSICAL CHEMICAL MOBILITY

Pesticides (which include insecticides, herbicides, and fungicides) move into surface water by two mechanisms. They can be carried in the solution phase in runoff water as it leaves the field, or they can move with eroding soil by adsorbing to soil organic matter and clay particles. As a result, physical and chemical characteristics, such as soil adsorption, water solubility, and persistence affect the runoff potential of a given pesticide.

Koc is a binding coefficient commonly used to describe pesticide's tendency to adsorb to organic carbon. A high Koc value (greater than 1000) indicates the pesticide adsorbed tightly to organic carbon and is not likely to move, except with eroding soil particles. Pesticide with lower Koc values (less than 300) are less tightly adsorbed and tend to move more readily with runoff water.

Similarly, the higher the water solubility, the greater the risk of a chemical dissolving in water and moving with aqueous phase. In general, pesticides having a solubility of 1 parts per million (ppm) or less tend to remain on the soil surface, while pesticides with water solubilities greater than 30 ppm are more likely to move in runoff water.

TABLE 20.1.3   Pesticide Characteristics That Affect Runoff Potential

Trade name	Chemical name	Koc	Water solubility (parts per million)	Soil half life (days)
Herbicides				
AAtrex	Atrazine	100	33	60
Ally	Metsulfuron	35	9500	120
Assure	Quizalofop	510	0.31	60
Balan	Benifin	9000	0.1	40
Banvel	Dicamba	2	500,000	14
Basagran	Bentazon	35	infinite	20
Bladex	Cyanazine	190	170	14
Blazer	Acifluoren	139	250,000	30
Buctril	Bromoxynil	190	0.08	5
Butyrac 200	2,4-DB	20	709,000	10
Classic	Chlorimuron	110	1200	40
Cobra	Lactofen	10,000	0.1	3
Command	Clomazone	274	1100	24
Dual	Metolachlor	200	530	20
Eptam	EPTC	280	375	30
Eradicane	EPTC	280	375	30
Fusilade 2000	Fluazifop	3000	2	20
Gramoxone Extra	Paraquat	100,000	1,000,000	500
Lasso	Alachlor	170	240	15
Lexone	Metribuzin	41	1220	30
Lorox	Linuron	370	75	60
Many names	2,4-D amine salts	20	796,000	10
Many names	2,4-D ester	1000	1	10
Micro Tech	Alachlor	170	240	15
Option II	Fenoxaprop	53,700	0.9	5
Pinnacle	Thifensulfuron	45	2400	12
Poast Plus	Sethoxydim	100	4390	10
Princep	Simazine	138	6.2	75
Prowl	Pendimethalin	24,300	0.275	90
Pursuit	Imazethapyr	100	11,000	90
Ramrod	Propachlor	80	613	6
Reflex	Fomesafen	50	600,000	180
Roundup	Glyphosate	24,000	900,000	47
Scepter	Imazaquin	20	60	60
Sencor	Metribuzin	41	1220	30
Sonalan	Ethalfluralin	4000	0.3	60
Stinger	Clopyralid	1.4	300,000	30
Sutan$^+$	Butylate	126	46	12
Treflan	Trifluralin	7000	0.3	60
Insecticides				
Ambush	Permethrin	86,600	0.2	32
Asana	Esfenvalerate	5300	0.002	35
Counter	Terbufos	3000	5	5
Cygon	Dimethoate	8	25,000	7
Diazinon	Diazinon	500	40	40
Dyfonate	Fonofos	532	13	45
Furadan	Carbofuran	22	351	50
Lannate	Methomyl	72	58,000	33
Larvin	Thiodicarb	100	19	7
Lindane	Lindane	1100	7	400
Lorsban	Chlorpyrifos	6070	2	30

(Continued)

TABLE 20.1.3   Pesticide Characteristics That Affect Runoff Potential (Continued)

Trade name	Chemical name	Koc	Water solubility (parts per million)	Soil half life (days)
Malathion	Malathion	1800	145	1
Orthene	Acephate	2	818,000	3
Penncap-M	Methyl parathion	5100	60	5
Pounce	Permethrin	86,000	0.2	32
Sevin	Carbaryl	200	114	10
Tempo	Cyfluthrin	100,000	0.002	30
Thimet	Phorate	2000	22	90

Note: These values can vary, depending on soil and environmental factors.
Source: Adapted from Becker et al., 1990.

Pesticide persistence is normally measured by half-life, the number of days it takes for the compound to breakdown to one-half of the concentration applied to the field. Although runoff can occur with any pesticide, compounds with half-lives greater than 21 days are more likely to move with runoff or eroding soil than pesticides that are more rapidly degraded.

Table 20.1.3, adapted from Becker et al. (1990), provides approximate values for adsorption (Koc), water solubility and soil persistence (half-life). These properties should be considered together because it is the interaction of these factors that determine a pesticide's environmental behavior. Depending on specific soil and environmental factors, the actual values may vary considerably.

Another factor that affects pesticide occurrence in surface water supplies is usage. Atrazine, for example, is often detected in surface water supplies because of its physical and chemical characteristics, and also because of its widespread use. In 1994, atrazine was used on over 80 percent of the corn acres in Illinois (USDA, 1995).

20.1.4   PESTICIDE MANAGEMENT

Best Management Practices (BMPs) are designed to minimize the adverse effects of pesticide use on surface water and groundwater quality. In addition to protecting the environment, these practices must be economically sound. In most cases, a combination of BMPs will be required to achieve water quality goals, and the suggested practices may vary depending on soils, topography, and individual farm operation. Czapar (1996) outlined some current BMPs, while Hirschi et al. (1997) published a comprehensive guide to protecting surface water quality.

Production systems using integrated pest management (IPM) play a role in protecting water resources. Regular monitoring of crop conditions and pest populations helps a producer make the most informed production decision. Pesticide applications based on economic thresholds optimize grower profits while reducing environmental hazards.

Conservation tillage practices reduce sediment loading and also reduce or slow water runoff. Because many herbicides can move from treated fields dissolved in runoff water, conservation tillage practices that increase water infiltration into the soil profile should help control herbicide runoff into surface water. Fawcett et al. (1994) reviewed the impact of conservation tillage on pesticide runoff.

Established grass waterways in areas of concentrated water flow have been shown to trap sediment and reduce the velocity of runoff flow, allowing greater infiltration of dissolved chemicals (Robinson et al., 1996; Thom and Blevins, 1996). Further, grass filter strips have been shown to effectively reduce the amount of herbicide runoff. Hall et al. (1983) reported that a filter strip of oats reduced water loss from small plots by 66 percent and reduced atrazine loss by approximately 90 percent.

The size of the drainage area must be considered, and the recommended width of the filter strip depends on the length and steepness of slope. In one Iowa study, a 15 foot filter strip reduced atrazine loss by 32 percent, while a 30 foot filter strip reduced atrazine loss by 55 percent (Mickelson and

Baker, 1993). Similarly, Misra et al. (1994) found that filter strips with a 30:1 filter to cropland ratio reduced atrazine loss by 38 percent, while a 15:1 filter to cropland ratio reduced atrazine loss by 44 percent.

Castelle et al. (1994) summarized filter strip research and concluded that a 50-foot filter strip is needed to protect wetlands and streams under most conditions. Additional information on designing a filter strip, suggested seeding rates, and guidelines for width are available from the Natural Resource Conservation Service, *Field Office Technical Guide* (1998).

Recently, water protection efforts have focused at the watershed level (Czapar, 1999). Best management practices that are specific to a watershed appear to be more effective than treating every acre in a uniform way.

20.1.5   BIOACCUMUALTION

In the past, bioaccumulation has occurred in some species from exposure to persistent organochlorine insecticides. Animals have been shown to store and concentrate these materials in their tissues (Metcalf and Luckmann, 1975). Although the risk of bioaccumulation has been substantially reduced by the development of newer, less persistent, pesticides, the past usage of several older pesticides continues to pose some environmental concern. The Binational Toxics Strategy identified six bioaccumulative pesticides including aldrin, dieldrin, chlordane, DDT (plus metabolites DDE and DDD), mirex, and toxaphene that warranted action to eliminate their input into the Great Lakes (Macarus, 1999). Although they continue to have an environmental presence, the concentrations have shown a general decline in most media over the years. Because all uses for these six pesticides have been cancelled in the United States, environmental loading has decreased.

20.1.6   GROUNDWATER MONITORING AND AGRICULTURAL CHEMISTRY

Similar to surface water information, the amount of groundwater quality monitoring data continues increase rapidly. Canter (1988) reviewed some of the water quality information available through a literature search for case studies of pesticides in groundwater, transport and fate, and mathematical models. Parsons and Witt (1988) surveyed state lead agencies to summarize pesticide detections in groundwater. Similarly, Williams et al. (1988) compiled a list of groundwater monitoring studies from around the United States. Other bibliographic searches concerning pesticides and water quality have also been conducted (NTIS, 1984; USDA, 1988). The U.S. Geological Survey maintains a database of information as part of the National Water Quality Assessment Program (NAWQA). The web site for the USGS is: http://water.usgs.gov/. Similarly, the U.S. EPA inventories water quality and biological and physical data in a database called STORET. The web address is: http://www.epa.gov/OWOW/STORET/.

National Well Water Surveys

As part of the National Survey of Pesticides in Drinking Water Wells (NPS), the U.S. Environmental Protection Agency sampled approximately 1300 community water systems wells and rural domestic wells between 1988 and 1990. Water samples were analyzed for 101 pesticides, 25 pesticide metabolites, and nitrate. As stated by U.S. EPA (1990), "EPA designed the survey with two principle objectives: (1) to determine the frequency and concentration of the presence of pesticides and nitrate in drinking water wells nationally, and (2) to improve EPA's understanding of how the presence of pesticides and nitrate in drinking water is associated with patterns of pesticide use and the vulnerability

TABLE 20.1.4   Summary of EPA's National Pesticide Survey, 1990

	Community wells (%)	Rural wells (%)
Wells with detectable nitrate	52.1	57.0
Wells with nitrate-nitrogen >10 ppm	1.2	2.4
Wells with any pesticides detections	10.4	4.2
Wells exceeding pesticide MCL/HAL*	0.0	0.6

*Maximum Contaminant Level (MCL) is the maximum permissible level of a contaminant that is delivered to any user of a public water system. Although the MCL is not legally applicable to rural domestic wells, it was used as a standard of quality for drinking water. Health Advisory Level (HAL) is the concentration of a contaminant in water that may be consumed over a person's lifetime without harmful effects. HALs are non-enforceable health-based guidelines that consider only non-cancer toxic effects.

of groundwater to contamination." As a result, a known proportion of wells sampled were from areas where pesticides were used more heavily and also from areas that were the most vulnerable to groundwater contamination.

Summary results from the National Pesticide Survey are shown in Table 20.1.4. Although nitrate was detected in over half of all wells sampled, only 1.2 percent of the community wells and 2.4 percent of the rural wells exceeded the maximum contaminant level of 10-ppm nitrate-nitrogen. Pesticides were detected in 10.4 percent of the community wells and 4.2 percent of the rural wells. None of the samples of community water systems wells exceeded maximum contaminant levels (MCL) or health advisory levels (HAL). A total of five pesticides: alachlor, atrazine, DBCP, ethylene dibromide, and lindane were detected in rural domestic wells above their respective MCLs/HALs. Table 20.1.5 lists the 12 pesticides or pesticide metabolites that were found at detectable levels in community water system wells and rural domestic wells.

Monsanto also conducted a national survey to support the re-registration of alachlor. As part of the National Alachlor Well Water Survey (NAWWS), 1430 wells in 26 states were sampled from June 1988 to May 1989. Water samples were analyzed for alachlor, metolachlor, atrazine, cyanazine, simazine, and nitrate. The results of this Monsanto survey were similar to those of the EPA National Pesticide Survey. Herbicides were detected in approximately 13 percent of the wells, but only 0.11 percent of the samples were above lifetime Heath Advisory Levels or proposed Maximum Contaminant Levels. Nitrate was detected in 52.3 percent of the samples, with 4.9 percent of the wells containing nitrate-nitrogen above the 10 ppm MCL.

TABLE 20.1.5   Pesticides Detected in EPA's National Pesticide Survey

	Community wells (%)	Rural wells (%)
DCPA acid metabolites	6.4	2.5
Atrazine	1.7	0.7
Simazine	1.1	0.2
Prometon	0.5	0.2
Hexachlorobenzene	0.5	ND
DBCP	0.4	0.4
Dinoseb	<0.1	ND
Ethylene dibromide	ND	0.2
Lindane	ND	0.1
Ethylene thiourea	ND	0.1
Bentazon	ND	0.1
Alachlor	ND	<0.1

Multi-state Water Quality Surveys

Roux et al. (1991) monitored groundwater for metolachlor in selected areas of four states: the Dougherty Plains, Georgia; McLean County, Illinois; Floyd and Mitchell counties, Iowa; and the Central Sands area of Wisconsin. These locations were selected on the basis of high metolachlor use and hydrogeological vulnerability. Metolachlor was detected in 9.7 percent of the samples. The majority of the detects were less than 1 part per billion (ppb), and no sample exceeded the Health Advisory Level for metolachlor.

Harrington et al. (1990) summarized the state regulatory agencies involved in water quality monitoring in five states: Iowa, Kansas, North Dakota, Oregon, and Wisconsin. They also compiled some of the water quality information, and compared cropping practices and pesticide use among these states. The authors stress the importance of sharing information among states, and of coordinating access to existing water quality databases.

State Groundwater Monitoring

Illinois. A statewide survey of private, rural wells was conducted in Illinois to help identify the extent of groundwater contamination (Goetsch, 1993). Pesticides were detected in 13 percent of the wells and exceeded health guidelines in 1.2 percent of the wells. In a related study, Mehnert et al. (1992) sampled wells in five Illinois counties. Although pesticides were detected in 9.5 percent of the samples, only one sample out of 240 (0.4 percent) contained a pesticide that exceeded the MCL/HAL.

In addition to well water surveys, the Illinois Environmental Protection Agency, Division of Water Pollution Control, publishes the results of their surface water and groundwater monitoring and assessment programs in Illinois every two years (IEPA, 1996). This includes information on rivers, streams, inland lakes, Lake Michigan, and groundwater resources.

Iowa. In 1987, the Iowa Groundwater Protection Act required that private drinking water supplies be tested for environmental contaminants. As part of the Iowa State-Wide Rural Well-Water Survey (SWRL), 686 wells were sampled between April 1988 and June 1989 (Hallberg and Kross, 1990). Water samples were analyzed for coliform bacteria, nitrate, 27 commonly used pesticides, and selected pesticide metabolites.

The survey found that 44.6 percent of the wells contained coliform bacteria and that 18.3 percent had nitrate concentrations above the recommended health advisory level of 10-ppm nitrate-nitrogen. One or more pesticides were found in 13.6 percent of the wells, but only 1.2 percent of the wells exceeded pesticide health advisory levels. Shallow wells, less than 50 feet deep, were more likely to be contaminated with coliform bacteria or nitrate than deeper wells.

Atrazine, the most common pesticide detected, was found in 4.4 percent of the wells, while atrazine metabolites were found in 3.5 percent of the wells. No insecticides were found in the survey, but two metabolites of carbofuran (Furadan) were detected. A total of eight wells (1.2 percent of the wells sampled) exceeded pesticide lifetime health advisory levels: five with atrazine, two with alachlor, and one with trifluralin. It should be noted that the SWRL survey was conducted during two extremely dry years. The results may have varied if precipitation had been average or above normal during the course of the survey.

Wisconsin. The Wisconsin Department of Agriculture, Trade, and Consumer Protection sampled well water from 534 Grade A dairy farms (LeMasters and Doyle, 1989). The farms were randomly selected, and samples were analyzed for 44 compounds including 10 of the herbicides and 4 of the insecticides most commonly used in Wisconsin. Herbicides were detected in 71 of 534 wells or 13 percent of the sampled wells. Of the 71 wells that contained detectable levels of herbicides, 64 contained atrazine alone, and 2 contained atrazine plus another herbicide. Three wells exceeded the 3 part per billion MCL for atrazine.

Kansas. Koelliker et al. (1987) sampled 103 randomly selected farmstead wells in Kansas during 1986. The study suggested that the extent of nitrate contamination was a larger concern, while the incidence of pesticide detections in well water was low. Nitrate was found to exceed the MCL of 10 part per million in 28 percent of the wells tested. Pesticides were detected in 9 percent of the farmstead wells.

Nebraska.   Exner and Spalding (1990) summarized groundwater sampling data from Nebraska. They compiled the results of 2260 well samples for pesticides from 1975 to 1989. Atrazine was detected in 13 percent of the sampled wells. They also summarized the results of 5826 well samples for nitrate from 1984 to 1989. Nitrate exceeded the MCL in 20 percent of the wells sampled.

20.1.7   SUMMARY

In general, surface water appears to be more vulnerable than groundwater to contamination from agricultural chemicals. When agricultural pesticides are detected in groundwater samples, they tend to be at low concentrations. In contrast, reducing soil erosion, pesticide runoff, and nutrient contamination of surface water is a major challenge facing production agriculture.

As the amount of water monitoring data increases, the impacts of agricultural chemicals on water quality will become more clearly understood. Watershed scale projects can be to used to identify vulnerable areas and target management practices to address specific runoff concerns. Survey information combined with best management practices (BMPs) can be used to protect both groundwater and surface water resources.

Local watershed protection efforts need to identify BMPs that reduce off-site chemical movement and provide farmers and landowners a range of alternatives that are both environmentally sound and economically viable.

REFERENCES

Auerbach, J., "Cost and Benefits of Current SDWA Regulations", *Journal AWWA*, February: 69–78, 1994.

Becker, R. L., Herzfeld, D., Ostlie, K. R., and Stamm-Katovich, E. J., *Pesticides: Surface Runoff, Leaching, and Exposure Concerns*, AG-BU-3911, Minnesota Extension Service, University of Minnesota, St. Paul, MN, 1990.

Canter, L. W., "Nitrates and Pesticides in Groundwater: An Analysis of a Computer-based Literature Search," in *Groundwater Quality and Agricultural Practices*, Fairchild, D. M., ed., Lewis Publ., Chelsea, MI, 1988.

Castelle, A. J., Johnson, A. W., and Conolly, C., "Wetland and Stream Buffer Size Requirements—A Review," *J. Environ. Qual.*, 23: 878–882, 1994.

Ciba Crop Protection, "Voluntary Atrazine Monitoring Programs at Selected Community Water Systems: Illinois 1994," *Technical Report 2-95*, Environmental and Public Affairs Department, Greensboro, NC, 1995.

Czapar, G. F., "Best Management Practices Protect Water Quality," in *Weed Control Manual*, Meister Publishing Co., Willoughby, OH, pp. 72–73, 1996.

Czapar, G. F., "Lake Springfield Demonstration Project," *Proceedings of the Illinois Crop Protection Technology Conference*, University of Illinois at Urbana-Champaign, pp. 148–149, 1999.

Exner, M. E., and Spalding, R. F., *Occurrence of Pesticides and Nitrate in Nebraska's Groundwater*, Water Center, Institute of Agriculture and Natural Resources, University of Nebraska-Lincoln, 1990.

Fawcett, R. S., Christensen, B. R., and Tierney, D. P., "The Impact of Conservation Tillage on Pesticide Runoff into Surface Water: A Review and Analysis," *Journal of Soil and Water Conservation*, 49(2): 126–135, 1994.

Goetsch, W., "Results of Illinois' Rural Private Water Supply Survey: Implications for Developing a State Pesticide Management Plan," *Proceedings of the Illinois Agricultural Pesticides Conference*, University of Illinois at Urbana-Champaign, pp. 24–27, 1993.

Goolsby, D. A., Thurman, E. M., and Kolpin, D., "Geographic and Seasonal Distribution of Herbicides in Streams of the Upper Midwestern United States," *Proceedings of the Northern Rocky Mountain Water Congress*, 1990.

Goolsby, D. A., Coupe, R. C., and Markovchick, D. J., "Distribution of Selected Herbicides and Nitrate in the Mississippi River and Its Major Tributaries, April through June 1991," *U.S. Geological Survey, Water-Resources Investigations Report 91-4163*, Denver, CO, 1991.

Hall, J. K., Hartwig, N. L., and Hoffman, L. D., "Application Mode and Alternative Cropping Effects on Atrazine Losses from a Hillside," *J. of Environ. Qual.*, 12: 336–340, 1983.

Hallberg, G. R., and Kross, B. C., *Iowa State-Wide Rural Well-water Survey—Summary of Results*, Iowa Department of Natural Resources and University of Iowa Center for Health Effects of Environmental Contamination, Iowa City, IA, 1990.

Harrington, T., Holtkamp, D., and Johnson, S., "The Impact of Agriculture on Water Quality: A Survey of Five States' Data Bases and Information Systems," *Staff Report 90-SR 45*, Center for Agricultural and Rural Development, Iowa State University, Ames, IA, 1990.

Hirschi, M., Frazee, R., Czapar, G., and Peterson, D., *60 Ways Farmers Can Protect Surface Water*, College of Agricultural, Consumer, and Environmental Sciences, University of Illinois at Urbana-Champaign, 1997.

Illinois Environmental Protection Agency, *Illinois Water Quality Report, 1994–1995*, Division of Water Pollution Control, Illinois Environmental Protection Agency, Springfield, IL, 1996.

Iowa Department of Natural Resources, "Pesticide and Synthetic Organic Compound Survey," *Report for House File 2303*, Iowa DNR, Environmental Protection Division, Water Supply Section, Des Moines, IA, 1998.

Koelliker, J. K., Streichen, J. M., Yearout, R. D., Heiman, A. T., and Grosh, D. L., "Identification of Factors Affecting Farmstead Well Water Quality in Kansas," *Report No. G1226-02*, Kansas Water Resources Research Institute, Manhattan, KS, 1987.

LeMasters, G., and Doyle, D. J., *Grade A Dairy Well Water Quality Survey*, Wisc. Dept. of Agric., Trade, and Consumer Protection, Madison, WI, 1989.

Macarus, D. P., *The Pesticide Workgroup of the Binational Strategy*, Quarterly Report—April, 1999, U.S. Environmental Protection Agency, 1999.

Mehnert, E., Chou, S., Dreher, G., Valkenburg, J., Schock, S., and Caughey, M., "Pilot Study: Agricultural Chemicals in Rural Private Wells," *Proceedings of the Illinois Agricultural Pesticides Conference*, University of Illinois, Urbana-Champaign, pp. 146–154, 1992.

Metcalf, R. L., and Luckmann, W. H., *Introduction to Insect Pest Management*, John Wiley & Sons, New York, 1975.

Mickelson, S. K., and Baker, J. L., "Buffer Strips for Controlling Herbicide Runoff Losses," *Proc. of the ASAE Summer Meeting*, Spokane, WA, 1993.

Misra, A. K., Baker, J. L., Mickelson, S. K., and Shang, H., "Effectiveness of Vegetative Buffer Strips in Reducing Herbicide Transport with Surface Runoff under Simulated Rainfall," *Proc. of the ASAE Summer Meeting*, Kansas City, MO, 1994.

National Research Council, *Soil Conservation: Assessing the National Resources Inventory*, Vols. 1 and 2, National Academy Press, Washington, D.C., 1986.

National Technical Information Service, *Ecology of Pesticide Water Pollution, 1978-May, 1984—Citations from the NTIS Data Base*, U.S. Dept. Of Commerce, Springfield, VA, 1984.

Natural Resource Conservation Service, *Field Office Technical Guide*, Filter Strips, Code 393, USDA, Washington, D.C., 1998.

Parsons, D. W., and Witt, J. M., *Pesticides in Groundwater in the United States of America: A Report of a 1988 Survey of Lead State Agencies*, Oregon State University, Corvallis, OR, 1988.

Robinson, C. A., Ghaffarzadeh, M., and Cruse, R. M., "Vegetative Filter Strip Effects on Sediment Concentration in Cropland Runoff," *J. Soil and Water Cons.*, 50(3), 227–230, 1996.

Roux, P. H., Balu, K., and Bennett, R., "A Large-scale Retrospective Ground Water Monitoring Study for Metolachlor," *Ground Water Monitoring Review*, 11 (Summer): 104–114, 1991.

Schmidt, A. R., "Sediment and Water Quality in the Upper Illinois River Basin," *Proceedings of the 1991 Governor's Conference on the Management of the Illinois River System*, Peoria, IL, 1991.

Taylor, A. G., and Cook, S., "Water Quality Update: The Results of Pesticide Monitoring in Illinois Streams and Public Water Supplies," *Proceedings of the Illinois Agricultural Pesticides Conference*, University of Illinois at Urbana-Champaign, pp. 81–84, 1995.

Temple, D. L., and Krueger, H. O., An Aquatic and Sediment Monitoring Study of Cyanazine and Selected Metabolites in Reservoirs in the Midwestern United States, Wildlife International Ltd., Easton, MD, 1994.

Thom, W. O., and Blevins, R. L., "Conservation Tillage and Filter Strips Trap Potential Water Contaminants," *Better Crops with Plant Food*, 80(2): 12–14, 1996.

Thurman, E. M., Goolsby, D. A., Meyer, M. T., and Kolpin, D. W., "Herbicides in Surface Water of the Midwestern United States: The Effect of the Spring Flush," *Journal of Environmental Science and Technology*, 25(10): 1794–1796, 1991.

U.S. Department of Agriculture, *Agricultural Resources-Cropland, Water, and Conservation-Situation and Outlook Report*, USDA Economic Research Service AR-4, Washington, D.C., 1986.

U.S. Department of Agriculture, "The Protection of Ground and Surface Waters, January 1982–August 1987—Citations from AGRICOLA Concerning Diseases and Other Environmental Considerations," *Bibliographies and Literature of Agriculture No. 63*, Beltsville, MD, 1988.

U.S. Department of Agriculture, *Agricultural Fertilizer and Chemical Usage: Corn—1994*, Illinois Agricultural Statistical Service, Springfield, IL, 1995.

U.S. Environmental Protection Agency, National Pesticide Survey-Summary Results of EPA's National Survey of Pesticides in Drinking Water Wells, Washington, D.C., 1990.

Wauchope, R. D., "The Pesticide Content of Surface Water Draining from Agricultural Fields—A Review," *Journal of Environmental Quality* 7: 459–472, 1978.

Williams, W. M., Holden, P. W., Parsons, D. W., and Lorber, M. N., *Pesticides in Ground Water Data Base: 1988 Interim Report*, Office of Pesticide Programs, U.S. Environmental Protection Agency, Washington D.C., 1988.

CHAPTER 20
SENSITIVE ENVIRONMENTAL PROBLEMS

SECTION 20.2

TEST OF THE VALIDITY OF THE LINEAR–NO THRESHOLD THEORY OF RADIATION CARCINOGENESIS WITH A SURVEY OF RADON LEVELS IN U.S. HOMES

Bernard L. Cohen

Dr. Cohen is professor emeritus of physics and astronomy and of environmental and occupational health, University of Pittsburgh, where he has taught since 1958. He is among the nation's most distinguished physicists.

20.2.1 INTRODUCTION

All estimates of the cancer risk from low level radiation are based on the linear–no threshold theory (LNT), which is based solely on largely discredited concepts of radiation carcinogenesis, with no experimental verification in the low-dose region of the most important applications. These risk estimates are now leading to the expenditure of tens of billions of dollars to protect against dangers whose existence is highly questionable. It is, therefore, of utmost importance to test the validity of this theory.

A definitive answer to the validity of LNT in the low-dose region must be based on human data, but to obtain statistically indisputable data requires much larger numbers of subjects than can be obtained from occupational, accidental, or medical exposures. The obvious source is natural radiation. If one attempts to use natural gamma radiation, which varies somewhat with geography, one is faced with the problem that LNT predicts that only a few percent of cancers are due to natural radiation; whereas there are unexplained differences of tens of percent for different geographic areas. For example, the percentage of all deaths that are from cancer varies in the United States from 22 percent in New England to 17 percent in the Rocky Mountain States (where radiation levels are highest). Another problem is that gamma ray backgrounds vary principally with geographic regions, and there are also many potential confounding factors that may vary with geography. Nevertheless, there have been attempts to study effects of gamma ray background on cancer rates, and, in general, either no effect or an inverse relationship has been found. For example, no excess cancer has been found in the high radiation areas of India or Brazil. However, all such effects can easily be explained by potential confounding factors.

A much more favorable situation is available for radon in homes. According to LNT, it is responsible for at least 10 percent of all lung cancers, and a known confounder, cigarette smoking, is responsible for nearly all of the rest. Another advantage is that levels of radon in homes vary much more widely than natural gamma radiation.

There have been numerous case-control studies of the relationship between radon in homes and lung cancer but the results from different studies have been inconsistent and this work has given no statistically significant information on the validity of LNT in the low-dose region, which we define here as below 5 pCi/L. This corresponds to 20–50 cSv (whole body equivalent dose) over a lifetime. A different approach, specifically designed for testing LNT, was carried out by the present author and is described in the following sections.

20.2.2   ORIGINAL 1995 PAPER

The author's group at University of Pittsburgh developed an elaborate study designed specifically to test LNT (Cohen, 1995), briefly reviewed here. We compiled hundreds of thousands of radon measurements from several sources to give the average radon level, r, in homes for 1729 U.S. counties, well over half of all U.S. counties and comprising about 90 percent of the total U.S. population. Plots of age-adjusted lung cancer mortality rates, m, versus these r are shown in Figure 20.2.1 where, rather than showing individual points for each county, we have grouped them into intervals of r (shown on the baseline along with the number of counties in each group). We ploted the mean value of m for each group, its standard deviation indicated by the error bars, and the first and third quartiles of the distribution. Note that when there are a large number of counties in an interval, the standard deviation of the mean is quite small. We see, in Figure 20.2.1a,c, a clear tendency for m to *de*crease with increasing r, in sharp contrast to the *in*crease expected from the supposition that radon can cause lung cancer, shown by the line labeled "Theory."

One obvious problem is migration: people do not spend their whole lives and receive all of their radon exposure in their county of residence at time of death where their cause of death is recorded. However, it is easy to correct the theoretical prediction for this, and the "Theory" lines in Figure 20.2.1 have been so corrected. As part of this correction, data for Florida, California, and Arizona, where many people move after retirement, have been deleted, reducing the number of counties to 1601. (This deletion does not affect the results.)

A more serious problem is that this is an "ecological study," relating the average risk of groups of people (county populations) to their average exposure dose. Because most dose-response relationships have a "threshold" below which there is little or no risk, the disease rate depends largely on the fraction of the population that is exposed above this threshold. This is not necessarily closely related to the *average* dose, which may be far below the threshold. Thus, in general, the average dose does *not* determine the average risk, and to assume otherwise is what epidemiologists call "the ecological fallacy." However, it is easily shown that the ecological fallacy does not apply in testing a linear–no threshold theory (LNT). This is familiar from the well-known fact that, according to LNT, population dose in person-rem determines the number of deaths; person-rem divided by the population gives the average dose, and number of deaths divided by the population gives the mortality rate, which is the average risk. These are the quantities plotted in Figure 20.2.1. Other problems with ecological studies have been discussed in the epidemiology literature, but these also have been investigated and found not to be applicable to our study. The most important of these problems are discussed below.

Epidemiologists normally study the mortality risk to individuals, m', from their exposure dose, r', so we start from that premise using the BEIR-IV version of LNT (in simplified form; full treatment in Cohen, 1995):

$$m' = a_n(1 + b\,r') \qquad \text{nonsmokers}$$
$$m' = a_s(1 + b\,r') \qquad \text{smokers}$$

where a_n and a_s are constants determined from national lung cancer rates, and b is a constant determined from studies of miners exposed to high radon levels.

Summing these over all people in the county and dividing by the population gives

$$m = [S\,a_s + (1 - S)\,a_n](1 + b\,r) \qquad (20.2.1)$$

where m and r have the county average definitions given above in the presentation of Figure 20.2.1, and S is the smoking prevalence—the fraction of the adult population that is smokers. Equation 20.2.1

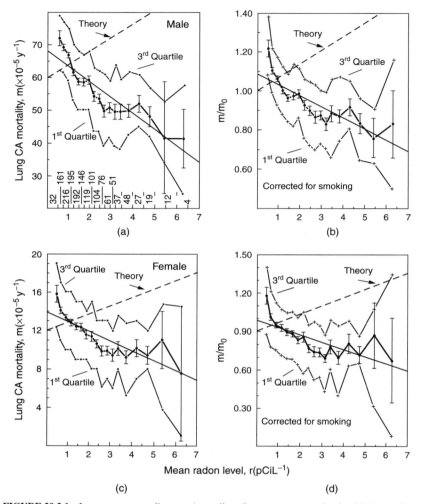

FIGURE 20.2.1 Lung cancer mortality rates (age-adjusted) versus average radon level in homes for U.S. counties (Cohen, 1995). Parts b, d are lung cancer rates corrected for smoking prevalence. See explanations in text.

is the prediction of the LNT theory we are testing here (we also show that our test applies not only to the BEIR-IV version but to all other LNT theories); note that it is derived by rigorous mathematics from the risk to individuals, with no problem from the ecological fallacy.

The bracketed term in Equation 20.2.1, which we call m_0, contains the information on smoking prevalence, so m/m_0 may be thought of as the lung cancer rate corrected for smoking. Figure 20.2.1b and d show m/m_0 versus r. We fit the data (i.e., all 1601 points) to:

$$m/m_0 = A + B\,r \qquad (20.2.2)$$

deriving values of B. The theory lines are from Equation 20.2.2 with slight renormalization. It is clear from Figure 20.2.1b,d that there is a huge discrepancy between measurements and theory. The theory predicts $B = +7.3$ percent per pCi/L, whereas the data are fit by B -7.3 (±0.6) and -8.3 (±0.8) percent per pCi/L for males and females, respectively. We see that there is a discrepancy between theory and observation of about 20 standard deviations; we call this "our discrepancy."

All explanations for our discrepancy that we could develop, or that have been suggested by others, have been tested and found to be grossly inadequate. We review some of the details of this process here.

There may be some question about the radon measurements, but three independent sources of radon data, our own measurements, EPA measurements, and measurements sponsored by various state governments, have been used and each gives essentially the same results. These three sets of data correlate well with one another, and by comparing them, we can estimate the uncertainties in each and in our combined data set; these indicate that uncertainties in the radon data are not a problem.

Another potential problem is in our values of smoking prevalence, S. Three different and independent sources of data on smoking prevalence were used, and all result in essentially the same discrepancy with LNT seen as in Figure 20.2.1b,d. Nevertheless, because cigarette smoking is such an important cause of lung cancer, one might think that uncertainties in S-values can frustrate our efforts. Analysis shows that the situation is not nearly so unfavorable. The relative importance of smoking and radon for affecting the variation of lung cancer rates among U.S. counties may be estimated by use of the BEIR-IV theory. For males, the width of the distribution of S-values, as measured by the standard deviation (SD) for that distribution, is 13.3 percent of the mean, and according to BEIR-IV, a difference of 13.3 percent in S would cause a difference in lung cancer rates of 11.3 percent. Whereas the SD in the width of the distribution of radon levels for U.S. counties is 58 percent of the mean, which according to BEIR-IV would cause a difference in lung cancer rates of 6.6 percent. Thus, the importance of smoking for determining variations in lung cancer rates among counties is less than twice (11.3/6.6) that of radon. Smoking is not as dominant a factor as one might intuitively think.

Even more important for our purposes is the fact that smoking prevalence, S, can only influence our results to the extent that it is correlated with the average radon levels in counties. Thus, we are facing a straightforward *quantitative* question: How strong a correlation between S and r, CORR-r, would be necessary to explain our discrepancy. If we use our best estimate of the width of the distribution of S-values for U.S. counties, even a perfect negative correlation between radon and smoking prevalence, CORR-$r = -1.0$, eliminates only half of the discrepancy. If the width of the S-value distribution is doubled, making it as wide as the distribution of lung cancer rates, which is the largest credible width because other factors surely contribute to lung cancer rates, an essentially perfect negative correlation, CORR-$r = -0.90$, would be required to explain the discrepancy and to cut the discrepancy in half requires CORR-$r = -0.62$.

How plausible is such a large |CORR-r|? There is no obvious direct relationship between S and r, so the most reasonable source of a correlation is through confounding by socioeconomic variables (SEV). We studied 54 different SEV to find their correlation with r, including population characteristics, vital statistics, medical care, social characteristics, education, housing, economics, government involvements, and so forth. The largest |CORR-r| was 0.37, the next largest was 0.30, and for 49 of the 54 SEV, |CORR-r| was less than 0.20. Thus, a |CORR-r| for smoking prevalence, S, even approaching 0.90, or even 0.62, seems completely incredible. We conclude that errors in our S-values can do little to explain our discrepancy.

In another largely unrelated study (Cohen, 1993), we found that the strong correlation between radon exposure and lung cancer mortality (with or without S as a covariate), albeit negative rather than positive, is unique to lung cancer. No remotely comparable correlation was found for any of the other 32 cancer sites. We conclude that the observed behavior is not something that can easily occur by chance.

To investigate effects of a potential confounding variable, data are stratified into quintiles on the values of that variable, and a regression analysis is done separately for each stratum. Because the potential confounder has nearly the same value for all counties in a given stratum, its confounding effect is greatly reduced in these analyses. An average of the slopes, B, of the regression lines for the five quintiles then gives a value for B, which is largely free of the confounding under investigation.

This test was carried out for the 54 socioeconomic variables mentioned above, and none was found to be a significant confounder. In all 540 regression analyses (54 variables $\times$ 5 quintiles $\times$ 2 sexes), the slopes, B, were negative and the average B value for the five quintiles was always close to the value for the entire data set. Incidently, this means that the negative correlation between lung cancer rates and radon exposure is found if we consider only the very urban counties, or if we consider only the very rural counties; if we consider only the richest counties, or if we consider only the poorest; if we consider only the counties with the best medical care, or if we consider only those with the poorest medical care; and so forth for all 54 socioeconomic variables. It is also found for all strata in between, as, for example, considering only counties of average urban-rural balance, or considering only counties of average wealth, or considering only counties of average medical care, and so forth. The possibility of confounding by combinations of socioeconomic variables was studied by multiple regression analyses and found not to be an important potential explanation for the discrepancy.

The stratification method was used to investigate the possibility of confounding by geography, by considering only counties in each separate geographical region, but the results were similar for each region. The stratification method was also used to investigate the possibility of confounding by physical features such as altitude, temperature, precipitation, wind, and cloudiness, but these factors were of no help in explaining the discrepancy. The negative slope and gross discrepancy with LNT theory is found if we consider only the wettest areas, or if we consider only the driest; if we consider only the warmest areas, or if we consider only the coolest; if we consider only the sunniest, or if we consider only the cloudiest; and so forth.

The effects of the two principal recognized factors that correlate with both radon and smoking were calculated in detail: (1) urban people smoke 20 percent more but average 25 percent lower radon exposures than rural people; and (2) houses of smokers have 10 percent lower average radon levels than houses of nonsmokers. These were found to explain only 3 percent of the discrepancy. Because they are typical of the largest confounding effects one can plausibly expect, it is extremely difficult to imagine a confounding effect that can explain the discrepancy. Requirements on such an unrecognized confounder were listed, and they make its existence seem extremely implausible.

20.2.3   UPDATES ON ORIGINAL PAPER

Our 1995 paper was based on lung cancer rates for 1970–1979, the latest age adjusted data available at that time. Recently, age adjusted lung cancer rates for 1979–1994 have become available. When these are used, the slopes, B, are changed from -7.3 to -7.7 percent per pCi/L for males, and from -8.3 to -8.2 percent for females. Because there are more lung cancer cases included, the standard deviations of these B-values are reduced, increasing the discrepancy with the predictions of LNT to about 30 standard deviations.

The 54 socioeconomic variables (SEV) used in the original paper were from the 1980 Census. About 450 new SEV from the 1990 Census have now been introduced and investigated in substantial depth. None of these SEV had $|$Corr-$r| > 0.45$, and extensive stratification studies led to the conclusion that none of these additional SEV can help to explain our discrepancy.

20.2.4   THE ECOLOGICAL STUDY ISSUE

Most criticisms of our study have been based on generalized criticisms of ecological studies. The most important of these is called "cross level bias" (Greenland and Robins, 1994). On this basis, in a presentation to NCRP (Feb.17, 1998), Jay Lubin dismissed my work as useless by a mathematical demonstration showing that an ecological study does not do an adequate job in handling a confounding factor. This problem was addressed in some detail in Cohen (1995) and Cohen (1997), where I describe it as "the ecological fallacy for confounding factors (CF)." The classical "ecological fallacy" arises from the fact that the average dose does not, in general, determine the average risk, but I avoid this problem by designing my study as a test of the linear-no threshold theory (LNT). In LNT, the average dose *does* determine the average risk. Use of separate and independent risks for smokers and nonsmokers avoids this problem for smoking prevalence. However, this problem does arise for other CF—the average value of a CF does not adequately determine its confounding effects, as demonstrated mathematically by Lubin.

For example, consider annual income as a CF that might confound the radon versus lung cancer relationship—maybe very poor people have lower radon levels and for unrelated reasons, have higher lung cancer rates than others. As Lubin's demonstration shows, average income is not necessarily a measure of what fraction of the population is very poor. A case-control study, in principle, selects cases and controls of matched incomes (although this is not always done, and is still less frequently done well).

My approach to this problem is to use a large number of CFs. For the example under discussion, I use as CF the fraction of the population in various income brackets, <$5000/y, $5000–$10,000/y, ..., >$150,000/y (10 intervals in all). In addition, I consider combinations of adjacent brackets, and other related characteristics, such as the fraction of the population that is below the poverty line and the percent unemployment. We have found that smoking prevalence, which is very strongly correlated with lung cancer, must have at least a 35 percent correlation (Corr-$r = -0.6$) with radon to have a significant effect, but none of the above CF have a correlation larger than 7 percent. This is convincing evidence that income is not an important confounder of the lung cancer versus radon relationship.

It is not difficult to devise a model in which cross-level bias could nullify our results, verifying Lubin's mathematical proof. For example, we might suppose that those with an income that is an integral multiple of $700 have 50 times lower radon and 50 times higher lung cancer rates than average. I have no data to show that this is not the case. However, such a model is not acceptable for two reasons:

1. It is not plausible.
2. It would also not be taken care of in case-control studies (they don't match incomes with that precision).

What is needed is a model that avoids these two limitations. These limitations are effectively corollaries to Lubin's mathematical proof.

Of course annual income is not the only CF that must be considered. Another example is age distribution. Case-control studies match cases and controls by age, and as Lubin's mathematical demonstration shows, average age in a county does not handle this problem in our study. Of course, I do use age-adjusted mortality rates, which takes care of the gross aspects of that problem, but there are limitations in the age-adjustment process. My solution is to use as CF the percent of the population in each age bracket, <1y, 1–2y, ..., 80–84Y, >85y, 31 age brackets in all, and to also use combinations of adjacent age brackets. None of these age brackets had correlations with radon above 4 percent, with the exception of the >85y bracket where the correlation was 7.7 percent. This was further investigated by stratification, using five strata of 320 counties each and determining the slopes, B, (cf. Equation 20.2.2 above) of the lung cancer versus radon relationship for each stratum. As we go from the stratum with the lowest to the stratum with the highest percent of population with age >85y, B-values for males were -10.1, -6.4, -6.1, -4.7, and -7.2 percent per pCi/L, and for females they were -6.3, -2.0, -9.1, -3.5, -10.7 percent per pCi/L, whereas LNT predicts $B = +7.3$ percent per pCi/L. Because the value of B is negative and grossly discrepant with the LNT prediction for all cases, and there is no consistent trend in its variations, I conclude that the correlation between radon and elderly people cannot explain our discrepancy. I can't prove this mathematically, but I can't concoct a not-implausible model in which variations of radon and lung cancer with age helps substantially to explain our discrepancy. As Lubin's proof shows, it is possible to concoct a model to explain our discrepancy (e.g., we might assume that those born on the first day of a month have 50 times higher radon levels and 50 times lower lung cancer rates than others), but that does not satisfy our two corollaries to Lubin's proof.

There are few, if any, other bases on which case-control studies match cases and controls, but in my study I gave similar treatments to a host of other potential confounding factors—educational attainment, urban versus rural differences, ethnicity, occupation, housing, medical care, family structures, and so on. I have found nothing that can help substantially to explain our discrepancy.

Aside from cross-level bias, more generalized and less specific discussions of limitations of ecological studies have appeared in the literature and have been used to criticize our study. However, there are many very important differences between our work and other ecological studies. One such difference is in the quantity of data involved. Most ecological studies involve 10 to 20 (or less) groups of people, whereas our study involves 100 times that number (1729 counties). Not only does that give a tremendous improvement in statistical accuracy, but it allows much more elaborate and sophisticated analyses to be done, including consideration of large numbers of potential confounding factors and use of stratification techniques.

A more important difference is that our work avoids the "ecological fallacy." I know of no other ecological study that contains that feature. That alone makes our paper very different from the others, and should earn it the right to be considered free from the prejudice attached to consideration of other ecological studies.

Ecological studies are normally viewed as being fast, simple to carry out, and inexpensive, but none of these adjectives applies to our project. Our radon measurements extended over six years and involved hundreds of assistants with millions of dollars in salaries, and the completely separate EPA and State-sponsored measurements we used were comparably elaborate. Our data analysis efforts involved dozens of assistants and several years of their efforts and mine. Without the power of modern computers and software packages, which have not been available until quite recently, such analyses would have been completely impractical. I know of no other ecological study to which any of the considerations of this paragraph would apply.

Because any deep understanding of how radon causes lung cancer must be based on its effects on individuals, it is essential to study the problem in terms of risks to individuals, which seems contrary to the ecological approach. However our treatment *is* based on risks to individuals (cf. derivation of Equation 20.2.1 above). That theory is then developed by rigorous mathematics to obtain the prediction, Equation 20.2.1, we use to compare with observations. This is a time-honored procedure in science; for example, Newton's famous formula:

$$F = m \, a$$

is rarely tested by direct measurements of acceleration, a, but rather, a formula is developed mathematically to determine distance travelled through time, which is much easier to measure as a test of the theory.

20.2.5   OTHER PUBLISHED PROPOSALS FOR EXPLAINING OUR DISCREPANCY

The BEIR-VI Report pointed out that no consideration had been given to variations in intensity of smoking, and proposed a model in which this is expressed as the ratio, k, of one pack per day to two pack per day smokers. To evaluate this proposal, one must recognize that there is surely no direct causal relationship between k and radon levels, r, so any correlation between the two must arise from socioeconomic variables SEV. How large a correlation between k and r, $\text{Corr}(k, r)$, is not completely implausible? For the 500 SEV we have studied, the largest $|\text{Corr}(\text{SEV}, r)|$ is 0.45.

Some indication of intensity of smoking in various states is included in cigarette sales (cs) data, available from tax collection records for each year. From these data, $\text{Corr}(\text{cs}, r)$ varied between -0.14 and -0.29 between 1960 and 1975. From this and the maximum $|\text{Corr}(\text{SEV}, r)|$, it seems reasonable to conclude that $|\text{Corr}(k, r)|$ larger than 0.5 would be highly implausible. The effect of $\text{Corr}(k, r) = -0.5$ is to change the slope B in Figure 20.2.1b from -7.3 to -5.0, and even $\text{Corr}(k, r) = -0.8$ gives $B = -2.3$, still a long way from the LNT prediction $B = +7.3$. Of course, there is no reason to believe that $\text{Corr}(k, r)$ is negative, and a positive $\text{Corr}(k, r)$ would change B in the opposite direction.

Field et al., pointed out that radon gas levels in homes is not the same as exposures to radon progeny that determine the dose, because the latter are affected by time spent in homes, exposures in other places, the ratio of radon progeny to radon gas, and so forth, and these may be correlated with radon levels. To investigate this, we define a modifying factor, f, by

$$r(\text{effective}) = r(1 + f)$$

and use r(effective) rather than r in determining B. The results depend on two features of f, the width of the distribution of f-values and Corr(f, r). As a rather extreme example, assuming a perfect correlation, Corr(f, r) = 1.0 and w = 0.7, changes B only from -7.3 to -3.7. Of course, there is no reason to believe that the effects would be anywhere near that large, or that they do not change B in the opposite direction.

Letters-to-the-editor have proposed that our discrepancy might be explained by confounding by population, or by population density—an expression of the urban-rural differences. However, stratification of our data into 10 deciles, or even finer stratifications on the basis of these variables, showed practically no evidence of a change in B-values. In these fine stratifications, the counties in most strata have essentially the same populations or population densities, but still the analysis for these counties with the same population, or with the same population density, gives a large negative value of B.

20.2.6 NEGATIVE SLOPES AND CONFLICT WITH DATA FROM CASE-CONTROL STUDIES

It is frequently suggested that the negative slopes in our data for m versus r (i.e., m decreases with increasing r) are incredible and are in conflict with the results of the case-control studies. It should be recognized at the outset that case-control studies investigate the causal relationship between radon exposure and lung cancer, whereas our work has the much more limited objective of testing the linear–no threshold theory. If that theory fails as we have concluded, "the ecological fallacy" becomes relevant and our results cannot be directly interpreted as representing the risks to individuals. We have therefore never claimed that Figure 20.2.1 gives risks to individuals, or that low-level exposure to radon is protective against lung cancer. Our only conclusion is that LNT fails very badly, grossly over-estimating the cancer risk of low-level radiation.

However, if one insists on interpreting our data as representing the dose-response relationship to individuals, it should be recognized that the negative slopes in our data are entirely based on radon exposures in the range r = 0–3.5 pCi/L (0–130 Bq/m^3), whereas the case-control studies give essentially no statistically meaningful information on the slope in this region. Detailed analysis shows that there is no discrepancy between our data and the case-control studies in that region.

20.2.7 THE SMOKING-RADON INTERACTION

Our Equation 20.2.1, derived to relate lung cancer rate, m, to r and S is

$$m = [S\, a_s + (1 - S)\, a_n]\,(1 + B\, r)$$

where a and B are constants. This has given many the impression that we have assumed some special and simple (linear) relationship between smoking and radon exposure in causing lung cancer. Despite the appearance of the above equation, that is not the case, as we now demonstrate.

BEIR-IV considers smokers and nonsmokers as separate "species," each with its own lung cancer risks. The relationship between radon and smoking in causing lung cancer in an individual can be infinitely complex. In utilizing the BEIR-IV model to mathematically derive the mortality rate for a county, the fraction of the county population that smokes, S, logically arises, and the result is the above formula. Note that S is *not* the intensity of smoking by an individual, but it is simply the fraction of the population that smokes cigarettes, the fraction of the population that is in that "species."

If counties kept separate statistics on cause of death for smokers and nonsmokers, S would not be involved. We could do two completely separate and independent studies for smokers and nonsmokers. It is only because counties do not keep separate statistics that we must combine these two studies, and this introduces the relative sizes of the two groups represented by S.

20.2.8 CONCLUSION

Because no other plausible explanation has been found after years of effort by myself and others, I conclude that the most plausible explanation for our discrepancy is that the linear–no threshold theory fails, grossly over-estimating the cancer risk in the low-dose, low-dose rate region. There are no other data capable of testing the theory in that region.

An easy answer to the credibility of this conclusion would be for someone to suggest a potential not implausible explanation based on some selected variables. I (or that person) will then calculate what values of those variables are required to explain our discrepancy. We can then make a judgment on the plausibility of that explanation. To show that this procedure is not unreasonable, I offer to provide a not-implausible explanation for any finding of any other published ecological study. This alone demonstrates that our work is very different from any other ecological study, and therefore deserves separate consideration.

REFERENCES

Cohen, B. L., "Relationship Between Exposure to Radon and Various Types of Cancer," *Health Physics*, 65: 529–531, 1993.

Cohen, B. L., "Test of the Linear–No Threshold Theory of Radiation Carcinogenesis for Inhaled Radon Decay Products," *Health Physics*, 68: 157–174, 1995.

Cohen, B. L., "Problems in the Radon versus Lung Cancer Test of the Linear–no Threshold Theory and a Procedure for Resolving Them," *Health Physics*, 72: 623–628, 1997.

Greenland, S., Robins, J., "Ecologic Studies: Biases, Misconceptions, and Counter Examples." *Am. J. Epidemiol.*, 139: 747–760, 1994.

CHAPTER 20
SENSITIVE ENVIRONMENTAL PROBLEMS

SECTION 20.3

ACID RAIN—THE WHOLE STORY TO DATE

John J. McKetta

Dr. McKetta is Joe C. Walter Chair Emeritus of Chemical Engineering, University of Texas, Austin. He is the recipient of the President Herbert Hoover Award for advancing the well-being of humanity.

20.3.1   INTRODUCTION

Acid rain is one of the most abused, overused, misunderstood and dramatized terms in the vocabulary of environmental zealots. Many people are concerned about acid rain because they have been told that it is destroying the environment, killing fish, ruining lakes, deteriorating forests and crops, and is harmful to humankind. With increased coverage by members of the news media who lack the knowledge to deal with this complex topic objectively, there is nationwide public concern not restricted just to the northeast.

Although many gaps still exist in our knowledge of this subject, we have learned a great deal about acid rain in the past decade. We do not know how it is formed, where it comes from, where it goes, what it does, or what harm it can do, if any. This is precisely why we are still expanding our research on these and many other questions.

We're always going to have acid rain because 70 percent of acid rain comes from nature. However, the big problem is mostly local. For example, a considerable acid rain problem exists in the northeast United States. Most of this is because of their own production of acid rain. The northeast United States uses over 40 percent of the high sulfur fuel oil used in the United States and still the northeast occupies only 4.5 percent of the land area. For example, there are over 35,000 apartment houses in New York City alone burning high sulfur oil. Man-made sulfur dioxide has decreased about 40 percent since 1970, even though the use of coal has increased about 85 percent during that time.

Hurricanes that hit on the United States mainland are monitored for their acid content. The acid rain that falls from these hurricanes is about the same acidity as pears, which is more acid than the acidity the Northeast United States is complaining about. This is mostly natural acid rain because there are no industries out in the Atlantic Ocean where most of our hurricanes originate.

20.3.2   WHAT IS ACID RAIN?

The term *acid rain* itself is misleading. Most rainfall is acidic. Some alkaline rains have been reported from the midwest, probably a result of airborne soil being incorporated into the drops. The air contains 0.03 percent carbon dioxide (which forms carbonic acid). Over 99 percent of the carbon dioxide comes from nature. Therefore, natural acidity is present in all kinds of precipitation, whether it be in the dry form such as dust, or in the wet form, such as snow, rain, fog, and dew.

Acidity is measured on a pH scale from 1 to 14 (pH is a measure of the concentration of the hydrogen ions). This pH scale is used by chemists to measure the acidity of solutions. Any substance with a pH value below 7 is acidic. The lower the pH the more acidic is the substance. A substance with a pH value above 7 is called alkaline (or basic). Applying this scale to familiar substances, you will find that peas have a pH of about 6. (Incidentally, drinking water has approximately the same pH.) Carrots have a pH of 5. The numbers from 1 to 14 are all logarithmic. This means that 5 is ten times more acid than 6, and that 4 is ten times more acid than 5, or 100 times more acid than 6. For example, pears, with a pH of 4, are ten times more acid than carrots with a pH of 5, but pears are 100 times more acid than peas with a pH of 6. It is interesting to note that stomach juice acidity is almost the same as the acid in your automobile battery. At the other extreme, note that soap is alkaline with a pH of 8 or more, while lye (sodium hydroxide) is extremely alkaline with a pH of 13. It is important to know that sea water has an alkaline pH of over 8. This makes it possible for the sea water to absorb carbon dioxide, sulfur dioxide, and other such gases from the air.

In fact, the oceans are just like huge pumps, absorbing and expelling gases depending upon equilibrium conditions. Because most of the carbon dioxide and oxides of nitrogen (which form nitric acid) come from nature, the natural rainfall that would result if we had no man-made pollution would be lower than 5.6. In fact, it seems reasonable that the natural pH of rain should be set at a value between 4.7 to 5.0.

20.3.3   WHEN WAS ACID RAIN FIRST RECOGNIZED?

We don't know when acid rain was first recognized but the first mention in recorded history was in 1848 by a Swedish scientist. Then in 1852, a French scientist measured the quantities of "nitric acid and nitrogen compounds" in the rain in Paris. In 1872, the acidic nature of rain was documented in a book covering the chemistry of English rain. A great concern was raised in the late 1960s when Swedish scientists claimed that the cause of the acid lakes and acid rain was the sulfur dioxide emissions from industrial sources of Great Britain.

In the United States, the controversy arose back in 1974 when a Cornell University researcher released reports (using data collected in the 1960s and before for other purposes) concluding that rain acidity was increasing in the northeast and spreading in all directions. Since then, others have speculated that acid rain is a post-World War II phenomenon caused by the increased use of fossil fuels for generating electricity, as well as the increased use of automobiles.

20.3.4   IS ACID RAIN HARMFUL?

The effect of acid rain on the environment is not clear. There are many claims that acid rain may harm the lakes, decrease fish population, reduce forest growth, decrease crop productiveness, decrease soil fertility, corrode buildings, and cause other detrimental effects. Yet scientific evidence does not substantiate these claims except in laboratory-size experiments. In fact, in some instances, crop yields have increased by using acid rain in laboratory experiments. Corn and tomatoes, for example, benefit greatly because of the fertilizing value of the extra nitrogen and sulfur in the rain.

20.3.5   IS ACID RAIN INCREASING?

The acidity of rain must be measured for a long time, in fact, years, at the same location before reporting some meteorological average. This is because the average acidity of the rain during one single rainfall measured at the same time, at points only several hundred feet apart, may vary, plus or minus 200 to 400 percent. Therefore, to establish an accurate trend one must collect data over a 5 to 10 year period. When this has been done, there has been no evidence that the rain is becoming more acid.

Some claim that coal-burning utilities in the midwest are the primary causes of acid rain in the northeast. However, they overlook the fact that a large percentage of domestic heating in the northeast comes from the use of fuel oil rather than coal. Many fuel oils contain as high as 2 to 3 percent sulfur.

20.3.6   WHERE DOES ACID RAIN FALL?

Acid rain falls everywhere. Natural rainfall is acidic. Natural rainfall has a pH averaging about 5. In the northeast, readings in the 4.0 to 4.5 range are not uncommon. Rainfall over most of the western states is closer to 5. However, in some areas, such as San Francisco, Seattle, Denver, and Los Angeles, the rainfall has been measured at 4 pH. Acid rain pH of 4 has been measured at such remote spots as Samoa in the South Pacific, the tropical jungles of South America, the arctic coast of Alaska, as well as Hawaii and the islands in the mid-Indian Ocean. The three areas of the world where acid rain appears to be of the greatest current concern are: southeastern Canada, the northeastern United States, and Scandinavia.

20.3.7   ACIDITY FROM NATURAL SOURCES

It is now clear that the natural factors, such as organic acids, naturally emitted sulfur, and nitrogen compounds also affect rain's normal acidity. More recent studies in remote parts of the world on natural sources of atmospheric acidity suggest the unpolluted pH of rain is closer to 5.0 rather than 5.6 (5.6 pH is the acidity of rain saturated with natural carbonic acid. If the natural nutric acids are included, the natural pH drops to 4.8. The addition of natural SO_2 brings the rain natural acidity down to 4.4–4.6.)

Approximately 65 percent of the sulphur dioxide, 99 percent of the carbon dioxide and more than 99 percent of the total oxides of nitrogen come from nature on a worldwide basis. All of these components make acid rain (sulfuric acid, carbonic acid, and nitric acid). It's possible that nature's contribution of these components in specific localities may not be as high as indicated, but may be lower than 25 or 30 percent.

Because the ratio of sulfates to nitrates is 2 to 1 in precipitation over eastern North America, sulfur gases have been labeled as the major contributor to rain acidity. The ratio is reversed in the west. Moreover, in many instances the acidity of rain samples does not differ greatly between the eastern and western United States.

Lightning's contribution to the acidity of rain is significant. Two strokes of lightning over 4/10th of a square mile (one square kilometer) will produce enough nitric acid to make 8/10th of an inch of rain with a pH of 3.5. One scientist calculated that lightning creates enough nitric acid so that annual rainfall over the world's land surfaces would average pH 5.0 without even accounting for contributions from other natural sources of acidity.

In the forest areas of Brazil at the headlands of the Amazon River, an area remote from civilization, the monthly average pH of 100 rain events in the 1960's ranged from 4.3 to 5.0. One set of pH reading was as low as 3.6.

The rainfall from two hurricanes in September 1979 sampled at six stations from Virginia to up-state New York averaged 4.5 pH, with one set of readings as low as 3.6 pH. This weather came directly from the Atlantic Ocean and was quite unlikely to have been affected by emissions from industrial activity. On the South Seas island of Pago Pago, some readings of pH as low as 4.3 were observed. In the heavy thunder storm activity at the start of the monsoon season in the remote northern territory of Australia, the rain averaged between 3.4 and 4.0 pH.

Recent ice pack analyses in the Antarctic and the Himalayas indicate that precipitation deposited hundreds and thousands of years ago in those pristine environments has not varied much from a value of 4.4 to 4.8. In fact, measurements have been made as low as 4.2 in these areas. This compares with the "average" pH of rain in the eastern United States, as well as in Scandinavia, of between 4.0 and 4.5. Greenland ice pack analyses showed that many times in the last 7000 years the acidity of the rain was as low as 4.4 pH. In some cases, the periods of extremely high acidity lasted for a year or more.

20.3.8   IS RAINFALL INCREASING IN ACIDITY?

The United States Geological Survey (USGS) collected rainfall samples in various locations in and near New York State during the period of March 1965 to September 1979. These data were collected at 22 locations; however, only 9 stations operated more or less continuously through that period. These data indicated that the long-term level of the acidity was essentially constant.

The existing data and studies show that there has been no significant changes in acidity in the northeast U.S. precipitations since 1960. In fact, the new data show that the sulfate concentrations have decreased and the nitrate concentrations have increased. The USGS has concluded that acidification of surface waters in the northeast "probably occurred long before the 1960s." They have also stated that the acidity of precipitation has been stable since the mid-fifties.

20.3.9   MAN-MADE SOURCES

It is well-known that humans puts many substances into the atmosphere. Many of these are acid forming. However, since 1960, emissions have declined. For example, there has been a decline of over 40 percent of sulfur dioxide emissions since 1960. At the same time, the use of coal has increased by 85 percent. There has been a similar decrease in man-made oxides of nitrogen.

There is a belief in some quarters that man-made emissions in other parts of the United States have increased the acidity of precipitation over northeastern United States because of the large amounts of residual fuel oil used for domestic, commercial, and industrial purposes. In fact, the northeastern United States uses 40 percent of the residual fuel oil, 35 percent of the distillate oil and 17 percent of the gasoline consumed in the entire country. Yet, the northeast comprises only 4.5 percent of the country's land area. Much of the fuel oil is high in sulfur. Dr. Kenneth Rahn, University of Rhode Island, used trace elements to find the source of pollutants. His research data indicate that local pollution sources in New England are the main cause of acid rain and snow in that area. His research has not revealed sulfur compounds emitted from mid-western coal fired plants in the rain collected in the northeast.

At St. Margaret's Bay, Nova Scotia, a study showed that 50 percent to 60 percent of the acid deposition came from the direction of Halifax, 15 miles to the east. A meteorological team at the University of Stockholm cautioned the Swedish people who blamed acid rain on the power plants in England, not to be so sure. This team's conclusion, after studying sulfur, nitrogen, and water cycles, via long-term monitoring, was that much of the acid rain was local.

EPA scientists studying emissions from four large oil burning units in New York City found flue gasses from the boilers did indeed contain large amounts of both SO_2 and sulfuric acid. These flue gasses also contained traces of vanadium. Further analyses showed that vanadium was found in the oil in the emissions from the boilers and incrusted in the lining of the boilers where combustion takes place. Vanadium is present in significant amounts in oil but is almost nil in coal. These EPA investigators concluded that more than half the winter time, sulfate emissions in New York City are attributable to local oil burning boilers. Some studies have indicated that the suspended particulates rarely travel over 300 miles (most often up to 100 miles). This means that the more than 35,000 oil fired boilers in apartment houses in New York City play a dominant role in the elevated sulfate levels in that area.

The Advanced Statistical Trajectory Air Pollution (ASTRAP) model provides the best available estimates of current interregional sulfate deposition. This model shows that each region is its own largest source of deposited sulfate.

20.3.10   DOES ACID RAIN AFFECT HUMAN HEALTH?

Much research is being conducted concerning the relationship between acid rain and health problems. To date, no ill effects have been found. Naturally, research will continue in this important area. We all know that many of the chemicals found in living plants and animals are acid. Muscles in the human body are mostly amino acids. Ascorbic acid or vitamin C is a dietary essential; malic acid gives apples their tangy taste. All of these have pH values within the range of rain termed "acid" (3.0 to 5.0 pH). The pH of acidic deposition is well within the range normally tolerated by human skin and eyes. The statements in the press that some individuals die because of acid precipitation are unfounded and do not reflect the current state of knowledge. Almost all of us drink quarts of fluids daily that are 5000 to 10,000 times more acid than milk or peas. However, soft drinks are of this acidity.

The world's outstanding epidemiologists who specialize in sulfur dioxide health effects in humankind deny that there are any adverse health effects. Present law allows a maximum of 0.02 parts SO_2 per million parts air (0.00002 percent) in the ambient air.

20.3.11   WHAT IS THE EFFECT OF ACID RAIN ON LAKES AND FISH?

For some lakes in sensitive regions, evidence indicates they have been highly acidified and will not propagate fish life. The rate, character, and the full extent of these changes are scientific unknowns. These studies have not been made in sufficient detail to document the actual changes. We do not know that the vegetation and soil surrounding a lake and stream play a major role in determining the rate and nature of the water body's response to acid deposition. Many lakes in North America have complex watersheds where precipitation flows through forest canopies and soils, and is chemically modified before entering the lake. As the rain passes over the vegetation and through the soils, its acidity can be reduced or increased many-fold.

The acidity of most of the waters involved are actually the greatest in the spring. The fish kills occur almost yearly in the midwestern United States lakes (such as in Wisconsin) because of the interception of the light by ice and snow on the lakes so that green aquatic plants are not able to produce adequate oxygen. Then the fish simply suffocate.

Regardless of the scare stories of the media, there are larger amounts, and record sizes, of fish caught in the New England lakes each year. In fact, just as an example, on January 1, 1984, the New York State Department of Environmental Conservation released the size and quantity records for freshwater fish, listing 34 species. A review of this release reveals the following:

> (1) In the period of 1979–83, 25 of the 34 records have been broken.
> (2) In 1983 alone, 13 records were broken including:
>   a. Brown Trout (23 lb. 12 oz.)
>   b. Pink Salmon (1 lb. 9 oz.)
>   c. Cisco Whitefish (2.97 lb.)
>   d. Tiger Esocids (29 lb. 3 oz.)
>   e. Bullhead Catfish (2 lb. 0 oz.)
>   f. Channel Catfish (25 lb. 8 oz.)
>   g. Bluegill (0.96 lb.)
>   h. Rock Bass (0 lb. 4 oz.)
>   i. American Shad (7 lb. 14 oz.)
>   j. Angling Carp (40 lb. 4 oz.)
>   k. Bow Carp (58 lb. 5 oz.)
>   l. Freshwater Drum (18 lb. 4 oz.)
>   m. White Sucker (1 lb. 6 oz.)

Except for four of these records from Lake Ontario, all others were from separate lakes or rivers. Of the 25 broken for the five-year period, these occurred on 16 different lakes and rivers. These facts stand in stark contrast to media stories of impending doom for New England lakes.

20.3.12   DOES ACID RAIN DAMAGE TREES AND CROPS?

Rainfall makes the grass in our yard grow faster and become greener than it would be if the grass were merely sprinkled using city water. The reason is that growing plants and trees require nitrates, ammonia, sulfates, magnesium, phosphorus, potassium, and other substances. The nitrates and the sulfates in rainfall are the ions that are the indicators of the major strong acid components in rain.

Those who raise the specter of dead forests and attribute their decline to SO, emissions from the Midwest coal-fired power plants and acid rain are misleading the public. The forest products industry, which has a larger stake than anyone else in the health of forests cannot document that there is any problem to forests. Despite constant acid rain, the amount of standing timber in New York forests increased by 70 percent between 1952 and 1976. To date, the evidence that acid deposition and associated man-made pollutants have contributed to observed, forest declines is circumstantial and inferential rather than conclusive. A variety of complex causes are possible, and plans for accelerated work are underway to determine which factors, such as acid deposition, gaseous pollutants, insects, disease, and drought, contribute to the damage.

Even though it is well established that soil conditions affect the growth of vegetation and trees, the effects of acid precipitation on this growth process remain uncertain. Experiments conducted to determine the impact of acid precipitation on crops have produced mixed results. The EPA tested 38 varieties of plants under greenhouse conditions with artificial rain adjusted to pH levels of 3.0 to 4.0. Approximately 40 percent of plant varieties showed increased yield, about 20 percent showed no effect, while the last 40 percent showed decreased yield.

The Interagency Task Force on Acid Precipitation stated in their annual report of 1982 that, "While there is general agreement that unmanaged soils in forested and grassland areas in humid regions may be sensitive to acidification from acid precipitation, there is no indication to date that the soils have become acid because of it or that forest production is being affected."

20.3.13   DOES ACID RAIN DAMAGE STONE AND METAL SURFACES?

Damage to architectural stone surfaces has been connected with atmospheric acidity. However, the major cause appears to be local sources of air pollutants, such as heavy vehicular traffic and industrial activity. Moreover, impacts from air pollutions in general and impacts from acidic precipitation cannot be distinguished.

A flurry of news articles alleging an impact of acid rain on automobile finishes appeared in late 1980. However, these stories were unfounded except as they related to acid smut fallout on vehicles parked near the offending chemical plants. Acid smut involves acidic particles quite unrelated to acid rain.

Another alleged hazard is the effect acid waters on metal pipes. Both copper and lead pipes are relatively unaltered by moderately acid solutions, and none of the feared health impacts have been documented. In most cases, any precipitation would be partly or completely neutralized as it seeps from the surface through the soil to underground pools or wells. All in all, the effects of acidity on the environment have not been found to be as severe as some have suggested. Moreover, the feared impacts have not been demonstrated outside the laboratory.

20.3.14   WHAT CAN WE DO TO CONTROL ACID RAIN?

Because there are still no accurate scientific data on conversion, transportation, accumulation, or transformation in cloud water of pollutants, we must continue research in these areas. In the meantime, areas having local trouble should look at solutions on a local basis. The burning of high sulfur oil in the northeast should be curtailed. The oil should be desulfurized. This is a local problem. When one has a flat tire on an automobile, he does not remove the entire body or engine. He only repairs the flat tire.

With the advent of new air pollution control legislation over the past decade scrubbers have been required on most coal burning equipment. The cost has been very high. On the more economic side, a test has been made in the Adirondack park region of New York State using powdered limestone in 51 small lakes that were acidic. The lime, being alkaline, reduces the acidity, raising the pH. The result is that good fishing has been restored. The cost has been only between $15 to $30 each year for each acre of surface water. All known acid lakes in Adirondacks could be limed adequately for about $300,000 each year. Liming is also being used in Canada and Scandinavia. Liming is much cheaper than some politicians' suggestions to use expensive scrubbers.

There is no evidence to suggest that a reduction of sulfur dioxide emissions from one portion of the country (midwest, for example) will result in a proportional decrease in the acidity of rainfall in another portion of the country (for example, northeast). In fact, it has been predicted that a 90 percent decrease in the sulfur dioxide emission in the midwest would not increase the pH of the lakes in the northeast by much more than 4 percent (a pH increase of 0.2, i.e., 4.3 to 4.5). However, the electric light bills would double!

For those who feel that we should always use scrubbers regardless of the root of the problem, I wish to point out that a SO_2 stack gas scrubber for a coal burning power plant has an initial cost of $100 to $300 million for the equipment alone. In addition, the scrubbers require energy to operate. Consequently 4 to 6 percent of a plant's power has to be put back directly into operation of the equipment. Other costs, including, the purchase of chemicals and the disposal of sludge collected from the scrubbing, add further to the plant's operating cost. Disposing of that sludge actually imposes another problem for those concerned with the clean environment. These are all part of the $30 billion cost being experienced throughout the United States to eliminate a problem that likely does not exist.

20.3.15 SUMMARY

In summary, the scientific and engineering community agrees that many gaps exist in our understanding of the acid lake–acid rain issue. We know what is known and we know the gaps of the unknowns. Research is on going in private industrial and state and federal laboratories.

CHAPTER 20
SENSITIVE ENVIRONMENTAL PROBLEMS

SECTION 20.4

THE OZONE LAYER

Hugh W. Ellsaesser

Dr. Ellsaesser spent 23 years in atmospheric research at the Lawrence Livermore National Laboratory, after serving for 21 years in weather service with the U.S. Air Force. His research convinced him that the establishment views on air pollution, threats to the ozone layer, and the climatic effect of carbon dioxide cannot be supported scientifically and that the hazards have been greatly exaggerated.

TABLE OF NOMENCLATURE AND UNITS

C = Centigrade temperature scale

DU = Dütsch Unit = milli-atmospheric-centimeter; unit used to measure depth of ozone layer

ENSO = El Niño-Southern Oscillation; recurrent weather pattern of eastern tropical Pacific

km = kilometer

mb = millibar: unit of air pressure, can be used to specify altitude; sea level pressure = 1013.2 mb

nb = nanobar = one thousandth of a millibar: unit used to measure ozone concentration

NGO = Non-Governmental Organization

NH = Northern Hemisphere

ppm = part per million

SBUV = Solar Backscatter Ultraviolet Spectrometer, satellite mounted ozone sounder

SH = Southern Hemisphere

SST = Supersonic Transport aircraft

TOMS = Total Ozone Mapping Spectrometer, satellite mounted ozone sounder

TOVS = TIROS-N operational vertical sounders, satellite mounted ozone sounder

UNEP = United Nations Environmental Programme

UV = ultraviolet light from the sun

WMO = World Meteorological Organization

20.4.1 PROLOGUE

The ozone layer is a very complex subject with many different aspects—some of which never appear in TV or headline news or even in scientific discussions. However, before discussing these, I want to remind you that the ozone layer, like other issues of public concern, is not currently discussed in a vacuum. It is discussed in an atmosphere buffeted by:

- Media bias
- Political correctness
- Scientific illiteracy of the general public
- Government as essentially the sole source of funds for environmental research
- Attempts to convert environmentalism into a religion
- The exploitation of acid rain, the ozone hole, global warming, and so forth, by some to promote a world government, by others to limit the world human population, and by still others to force our economy back to the level of the nineteenth century or beyond

As a result of all these forces, it is essentially impossible now to get an unbiased and scientifically objective appraisal of the ozone layer or any other environmental issue.

Because I don't want to digress too far on this subject, let me remind you of just a few examples.

1. Media bias is self-evident to those not sharing the same bias. They favor collectivism over private enterprise, bureaucratic central planning over individual initiative, and see world government as the only hope for building world peace, for preservation of the our planetary environment, and for protecting us from "right-wing religious extremists."

2. In my view, nothing has polluted the objectivity of science like the semantic monstrosity into which the word *pollution* has been converted. It now conveys only two attributes—something that defiles or degrades and is also of anthropogenic origin. That is, natural pollution is an oxymoron. Remember the field day the environmentalists and the press had jeering President Ronald Reagan when he dared to suggest that trees pollute. If you recognize a pine grove, a row of eucalyptus, or an orange grove by their fragrance, you are detecting pollution. That is, you are detecting plant-emitted hydrocarbons, which are even more reactive than auto exhaust in the photochemical process that generates Los Angeles smog. However, in today's sea of political correctness, you call it pollution only at peril to yourself.

3. Under President Bush, the U.S. environmental research budget reached the 2 billion dollar level. The scientists and institutions contending for this largess of public funds have never before experienced the public and media attention—including three Nobel Prizes in 1995—that they are now receiving. Can you imagine the career and peer pressure on these individuals to avoid any hint that the threat to the ozone layer has been overblown or that such an issue might have net beneficial consequences?

4. Literally hundreds of Nongovernmental organizations (NGOs) plus environmentalists, the media, United Nations organizations, and others have been and are exploiting acid rain, the ozone layer, global warming, and other issues to advance other agendas including the development of a world government. Currently, planned attacks on protection of the ozone layer and global warming, in particular, would lead to transfers of trillions of dollars from the developed to the developing world, transfers that would have to be funneled through United Nations organizations. This offers powerful incentives for the United Nations, its clients and supporters, and all the underdeveloped nations to climb aboard these particular band wagons. Do you think any of these would take lightly efforts to downgrade the threat of these issues or suggestions that they might actually provide net benefits?

20.4.2 THE OZONE LAYER

Figure 20.4.1 from Dütsch (1978) shows by season the normal prefreon distribution of ozone in the atmosphere by altitude and latitude in terms of ozone mixing ratio in ppm. Mixing ratio is a conservative quantity that doesn't change as air is moved up or down but does change if there

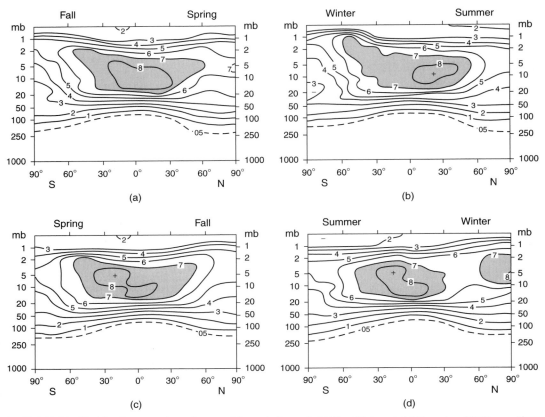

FIGURE 20.4.1   Meridional cross sections of ozone mixing ratio (in ppm): (a) May, (b) August, (c) November, (d) February (from Dütsch, 1978).

is mixing of parcels with different mixing ratios or actual chemical generation or destruction of ozone. Note that ozone does not exist in a "thin and fragile" layer; it extends throughout the lowest 50 km of the atmosphere shown here. There are no isolines in the lower parts of these charts only because the mixing ratios there are less than 0.5 ppm. Note that the area of highest mixing ratio is in the 30- to 35-km layer near the subsolar point. This is where most of the ozone is produced as a result of absorption of UV from the sun by oxygen. The atomic oxygen thus formed, combines with other oxygen molecules, O_2, to form ozone, O_3. At this level and above ozone is essentially in photochemical equilibrium. The half-life for regeneration of ozone if destroyed decreases rapidly with altitude. At 35 km it is about five days. At lower levels, the ozone concentration is generally above the level of photochemical equilibrium; so ozone is being destroyed there chemically. However, the chemical lifetime at these levels is months to years so that under normal conditions the ozone in the lower stratosphere, where it is brought by transport and is most plentiful, is essentially in storage.

Figure 20.4.2, also from Dütsch (1978), shows the same data in nanobar (nb) of ozone. Because these units are absolute, they can be integrated (added) vertically to get the depth of the ozone column. This figure reveals that the highest concentrations of ozone are not at 30 to 35 km over the equator, where it is generated, but at 15 to 20 km near the poles where the ozone has been advected by atmospheric circulation. The arching curves formed by the beginning of the packing of the parallel isolines roughly mark the tropopause, the surface separating the lower, weather generating, troposphere from the upper more stagnant part of the atmosphere, the stratosphere.

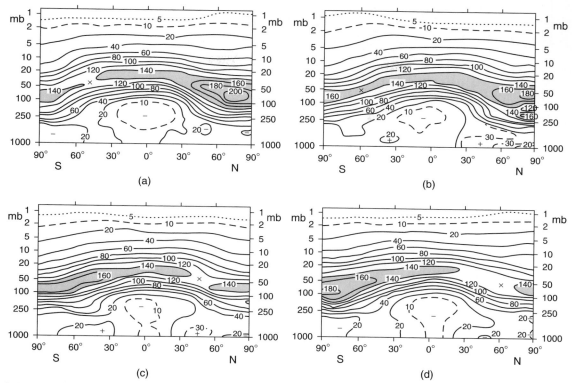

FIGURE 20.4.2 Meridional cross sections of ozone partial pressure in nanobars (nb). The x's mark the positions of the lowest values within the maximum layers: (a) February, (b) May, (c) August, (d) November (from Dütsch, 1978).

Because the strongest solar heating is at the subsolar point in the tropics, there is a thermally driven global circulation of the atmosphere called the Hadley circulation. This is upward in the tropics, through the peaks of the arches of the isolines in Figure 20.4.2, and then poleward and downward approximately along the isolines of this figure. The flow is seasonal, being strongest upward in the summer hemisphere and primarily toward the winter pole (i.e., into the opposite hemisphere) and stronger in the northern hemisphere (NH) winter than in the southern hemisphere (SH) winter.

To estimate the effect of this circulation on stratospheric ozone, note that the isolines of Figure 20.4.2 tend to define the flow after its rise through the tropical tropopause into the stratosphere but the mixing ratio lines of Figure 20.4.1 define the ozone distribution that is being advected poleward and downward. It is because of this Hadley circulation that the ozone column reaches its greatest column depths near the poles and becomes greater in the NH spring than in the SH spring.

The upper panel of Figure 20.4.3 shows the latitudinal and seasonal variation of the zonal mean depth of the ozone layer in Dütsch Unit (DU), (300 DU, which is near the global average is 0.3 cm or about 1/8 inch of ozone at sea level pressure). Note the lowest values around 250 DU near the equator, changing little throughout the year. The highest peaks form at 55 to 60° latitude just outside the polar winter vortices and move toward the poles after the winter vortices break up in spring. In the NH this leads to around 440 DU near 80°N in April, and in the SH to a ridge around 340 DU near 55°S from July to November and then moving toward the South pole in December. The ozone hole is the tight minimum below 200 DU near the South pole in October. Note that this figure is an average for 1979–1996. It has the advantage of using global data from satellites but also shows some

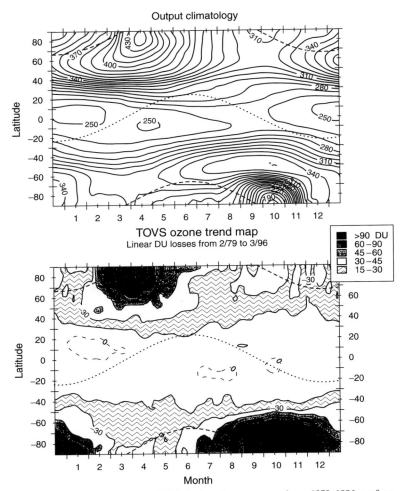

FIGURE 20.4.3 (Upper panel): TOVS-derived total ozone averaged over 1979–1996 as a function of latitude (left vertical scale) and month (bottom horizontal scale). (Lower panel): The 1979 to 1996 change in total ozone determined from the same TOVS data and plotted on the same coordinates. The light dashed lines are the subsolar point and the tropics of Cancer and Capricorn (from Neuendorffer, 1996).

of the ozone destruction of the last 20 years. Approximately half of the changes of the lower panel in Figure 20.4.3 should be added to the upper panel to reproduce the prefreon ozone layer of 1979 and before. This would leave the ozone hole barely discernible. Similarly, half of the changes should be subtracted from the upper panel to reproduce the ozone field of 1996.

In terms of protecting the surface from UV flux, note that this ozone pattern provides the least protection in low latitudes and in summer, when sunlight is most intense; and the most protection in high latitudes and in winter, when sunlight is weakest. I defy anyone to find an optimum level of UV that humans should not disturb. The most bountiful, prolific, and varied life forms on the planet developed and continue to exist in the tropics where we have the most direct and intense sunlight and where we have the least ozone protection from the solar UV.

20.4.3   *THE SIGNIFICANCE OF THE OZONE LAYER*

The ozone layer has been front page news since McDonald in November 1970 told the Department of Transportation that SST exhaust would destroy stratospheric ozone, allow more solar UV to reach the surface, and thus, increase the incidence of skin cancer. Water vapor from the SST exhaust was then envisioned as the ozone destroyer. As has proven to be typical, no one brought up at the time the fact that observational data then available showed that stratospheric water vapor and ozone had both been increasing for several years (Ellsaesser, *Atmospheric Environment*, 16(2): 197–203, 1982). According to Mo and Green (*Photochemistry and Photobiology*, 20: 483–496, 1974) the annual mean solar erythema dose (damaging UV) increases about 50-fold from the poles to the equator. This is roughly six doublings, or a 100 percent increase every 1000 miles (i.e., a local increase of roughly 1 percent for every ten miles of displacement toward the equator). We have moved people, plants, and animals all over this planet and under this multi-fold variation in UV. We have found that if you fail to take precautions, sunburn may become a minor problem. However, the really serious problems like rickets and osteomalacia, which take years to become apparent, are *due to too little UV rather than too much.*

Over the United States where we have the best data the north/south doubling distance for actual skin cancer incidence is less than 10 degrees of latitude or about 600 miles (NAS, *Environmental Impact of Stratospheric Flight*, NAS, Washington, D.C., 1975). This means that skin cancer incidence increases about 1 percent per six miles displacement toward the equator. This is somewhat more rapid than UV flux increases because in sunnier weather people tend to expose themselves more. WMO (1991) estimated that a 1 percent decrease in the thickness of the ozone layer is equivalent to a 2.3 percent increase in the incidence of ordinary skin cancer—or, from the above numbers, to a 14-mile displacement toward the equator.

After Molina and Rowland (*Nature*, 249: 810–814, 1974) deduced that the ultimate fate of freons was photodecomposition by the hard UV of the upper atmosphere, releasing chlorine into the stratosphere, it was calculated that the chlorine would destroy ozone primarily near 40 km. Solomon (*Nature*, 347: 347–354, 1990) estimated that when the process came into equilibrium with the prevailing release rate of freons, the global mean depth of the ozone layer would be reduced by "perhaps 5 percent sometime near the middle of the twenty-first century." In terms of skin cancer incidence, this 5 percent decrease in ozone, according to the above numbers, would be equivalent to moving 70 miles toward the equator.

However, this was presumed to represent an intolerable threat, because in 1987 we signed the Montreal Protocol restricting the use of freons and related compounds beginning in the year 2000. Our chief negotiator wrote; "the most extraordinary aspect of the treaty was its imposition of short-term economic costs to protect human health and the environment against *unproved future dangers* . . . dangers that rested on scientific theories, rather than on firm data. At the time of the negotiations and signing, *no measurable evidence of damage existed* [emphasis added]" (Richard Benedict, *Ozone Diplomacy*, Harvard University Press, 1991, Preface).

20.4.4   *THE OZONE HOLE*

Figure 20.4.4 shows ozone mixing ratio profiles over Halley Bay, Antarctica before (solid) and after (dashed) formation of the near maximum ozone hole of 1987. In the ozone hole, the ozone destruction was not at 40 km where predicted by the Molina and Rowland [op cit] theory. It was in the 14- to 20-km layer, it was near total destruction, and it was here now—not 50 years from now. However, it was immediately presumed to be due to chlorine from freons (Farman et al., *Nature*, 315: 207–210, 1985). So it appears to be, but not by the mechanism predicted. The low temperatures required to produce the stratospheric cirrus clouds needed to catalyze formation of the ozone hole exist over sufficient periods of daylight to cause significant ozone destruction only inside the S. polar vortex in the austral spring at levels between about 12 and 24 km.

I also want to point out that the solid profile in this figure is typical of the vertical variation of ozone everywhere; with a mixing ratio of about 0.03 ppm at the surface, increasing slowly with altitude to the

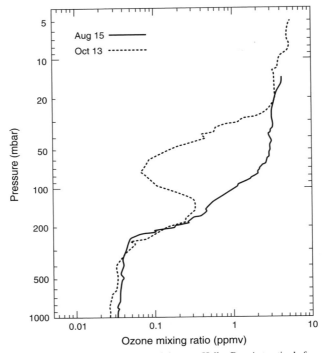

FIGURE 20.4.4 Profiles of ozone mixing over Halley Bay, Antarctica before (solid) and after (dashed) formation of the ozone hole of 1987. Because of the decrease of pressure and density with altitude and the logarithmic scale, the fraction of the total column of ozone destroyed (about 50%) is significantly larger than would be estimated from this plot (adapted from Gardiner, *Geophys. Res. Lett.*, 15(8): 901–904, 1988).

tropopause and then increasing much more rapidly to a peak near 35 km with decreases above. This demonstrates that the major source of ozone in the atmosphere is stratospheric generation followed by slow leakage through the tropopause, primarily near the poles with the spring breakup of the polar vortices, and then diffusive mixing to the surface where the ozone is destroyed by contact with the surface.

20.4.5 TRENDS IN DEPTH OF THE OZONE COLUMN

In 1986, NASA's Don Heath testified that satellite data showed a large scale decrease in ozone two to three times larger than predicted and most pronounced at 50 km rather than at the predicted 40 km. This prompted the U.S. Congress to request establishment of an Ozone Trends Panel. The executive summary of the Panel's first report was issued by a NASA press release of 15 March 1988 (Kerr, *Science*, 239: 1489–1491, 1988). The report itself (NASA, *Present State of Knowledge of the Upper Atmosphere 1988: An Assessment Report*, NASA Reference Publication 1208, 1988), is dated August 1988 but was unknown and unexpected by EPA ("Protection of Stratospheric Ozone; Advance Notice of Proposed Rulemaking," *Federal Register*, 53(156), 30604, August 12,1988). The 2-volume WMO (1988) Report No. 18 is entitled *Report of the International Ozone Trends Panel 1988*, but gives no indication of the actual publication date. I determined it was long delayed but I was unable to determine when it actually became available. Until these reports appeared in print, scientists could not submit comments on them for publication.

Basically, the Ozone Trends Panel's report found the following:

1. Degradation of a diffuser plate had caused the apparent rapid declines in total and 50-km ozone indicated by the SBUV and TOMS satellite instruments.

2. Ozone at 40 km had decreased but by only about half of that predicted by the models.

3. Total ozone in northern mid latitudes had decreased, significantly only in winter, but at about twice the rate predicted by models.

4. The bulk of the decrease in total ozone column resulted from a completely unanticipated ozone decrease at 25 km and below rather than from a decrease at 40 km as predicted from chlorine destruction.

In presenting these findings in his press release, NASA's Dr. Robert Watson stated: "[our models] do not predict that ozone decreased the way it did over the Northern Hemisphere during the past 17 years. Our models are not doing a good job, so we would have to say that they are underestimating decreases in the future" (Kerr, *Science*, 239: 1489–1491, 1988).

This was not the only time that NASA used press releases, which are essentially impossible for other scientists to question or counter, to advance the phase out of freons, and so on. In another press release in 1992 (Kerr, *Science*, 255, 797–798, 1992), creating suggestions of a NH ozone hole over Kinnebunkport, NASA got the freon phase-out date of the Montreal Protocol advanced by five years from 2000 to 1995.

Figure 20.4.5 shows the variations in global mean ozone as we know it. The upper panel, compiled by Jim Angell (1989), shows percentage variations about the mean of seasonal mean ozone, after smoothing, from 1958 to 1988 in parallel with the variation in sunspot numbers. The lower panel shows global daily mean ozone values from TOVS from February 1979 to February 1996 analyzed by Neuendorffer (1996)—the data have been adjusted to agree with surface Dobson spectrometer data. Applied adjustments are indicated by the light solid line and the scale at right.

The Angell (1989) analysis shows an abrupt drop from 1959 to 1961 and a following long-term rise until 1970, which are currently unexplained. The equal peaks in 1970 and 1979 were the highest values in his record. The TOVS record suggests almost step-like drops following the El Chichón eruption of 1982 and the Pinatubo eruption of 1991 as well as a 1985 minimum associated with the sunspot minimum at that time. Note that by 1993 we had already experienced a 5 percent (15 DU) drop in global mean ozone. Has anyone noticed the intolerable effects this was supposed to produce—other than blind sheep near 50°S in Argentina?

20.4.6 OZONE DECLINE SINCE 1979—SIGNIFICANCE AND POSSIBLE CAUSE

Figure 20.4.3, in the lower panel, shows the ozone losses in DU from 2/79 to 3/96 as determined from TOVS by Neuendorffer (1996). Note that there has been very little loss, and even a few increases, in the tropics where we receive the strongest UV flux and need the most protection. The major losses have been in high latitudes and in winter or in exactly those regions in which *too little, rather than too much*, UV is likely to be a problem.

Note also that the bulk of the losses have been at levels below 25 km where, under normal conditions, the amount of ozone is determined almost entirely by atmospheric transport. In the equatorial region where the depth of the ozone column is determined primarily by chemistry, there have been only small losses and, in some cases, increases. This is even more remarkable because the observations indicate ozone destruction at 40 km even over the equator.

Because of the above factors, I have for some time been proposing that the long-term decline in ozone since 1979 is mainly the result of a change in atmospheric circulation rather than to chemical

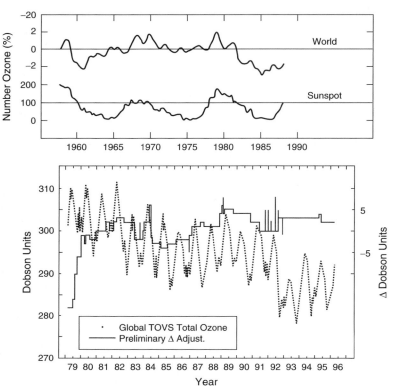

FIGURE 20.4.5   (Upper panel): Seasonal mean global ozone values (after a 1-2-1 smoother applied twice) in parallel with sunspot number (adapted from Angell, 1989). (Lower panel): Daily global mean values of total ozone from TOVS from February 1979 to February 1996 (from Neuendorffer, 1996).

attack by chlorine. As I noted previously, the Hadley circulation is upward through the tropical tropopause into the stratosphere and then poleward and downward, primarily toward the winter pole. Air moving upward adiabatically (without loss of or gain of energy) expands and cools at a substantial rate—10°C/km. Also, the temperature in the stratosphere increases with height. Therefore, the rate of ascent of air in the stratosphere is limited to the rate at which it can be warmed radiatively to remain at the same temperature, and therefore density or buoyance, as the air in the layers it is passing through. If this rate is exceeded, the rising air becomes colder, and therefore heavier, than its environment and sinks back to a lower level and spreads horizontally.

Thus, if the Hadley circulation is accelerated, less of the rising air will rise high into the stratosphere, forcing air there to descend, bringing down the high levels of ozone mixing ratio to cause ozone buildup in the lower polar stratosphere. Rather, more of the Hadley flow air, once in the stratosphere, will move directly poleward and will sweep the normal high concentrations of ozone in the lower stratosphere poleward and back into the troposphere. Because ozone above about 30 km is always maintained near photochemical equilibrium, it is not difficult to visualize that a speedup in the horizontal Hadley circulation, with a reduction in the vertical overturning, will reduce the total amount of ozone in the stratosphere. Reduced removal of ozone from the tropics is also indicated by the little change and actual increases in ozone depth there, despite loses at 40 km. More rapid transport of ozone from the stratosphere to the troposphere would also help to explain the otherwise unexplained increase in ozone in the NH troposphere in recent decades.

The question now becomes: Has there been an increase in the Hadley circulation since about 1979? While this is a difficult question to answer, I have found the following atmospheric changes since the mid-1970s suggesting changes in atmospheric circulation that favor, or are associated with, increased Hadley circulation:

1. The current period of warming of the surface temperature began with an abrupt warming of 0.3 to 0.4°C between 1976 and 1981 (Ellsaesser, *Technology: J. Franklin Institute*, 332A: 45–52, 1995). This warming has been strongest in the tropics and should therefore strengthen the Hadley circulation. In the upper air in the 850-300 mb layer, this warming was even more abrupt, occurring as a step function in 1976 (Michaels, "Looking for Answers," *World Climate Report*, 1(6): 1–2, 1995).

2. Wallace et al. (*J. Climate*, 9(2): 249–259, 1996) reported that "it has been shown quite convincingly, in numerical experiments with atmospheric general circulation models" that the increased frequency and strength of ENSOs "have accounted for much of the anomalous warmth of the winters of the 1980s over high latitudes of North America." ENSOs are recognized as an acceleration of the Hadley circulation in the eastern Pacific where this type of circulation is normally suppressed. They also found that since the 1970s, the NH warming has been greater in winter than in summer and has been associated with a COWL (cold-ocean-warm-land) pattern with strong warming over land and cooling over the Atlantic and Pacific oceans north of 40°N. While they proposed no causal mechanism, they noted that the observed warming is consistent with a strengthened Hadley circulation producing augmented adiabatic heating by subsidence in high latitudes.

3. Trenberth (*Bulletin of the American Meteorological Society*, 71(7): 988–993, 1990) studied an abrupt large-scale change in the circulation of the North Pacific. Since its onset circa 1977, he found that ENSOs have been both more frequent and stronger than normal and that the "imbalance between the occurrence of Warm (Hadley enhancing) versus Cold (Hadley suppressing) events in the tropical Pacific is unprecedented."

4. Labitzke and Van Loon (*Tellus*, 47A: 275–286, 1995) found a warming in upper tropospheric layers, correlated with the sunspot cycle over the past four decades. They presented evidence to show that this warming was probably due to an enhancement of the Hadley circulation producing increased subsidence heating in the upper troposphere. Because the solar cycles peaking in 1981 and 1990 were unusually strong, the associated tropospheric warming from this apparent mechanism was therefore greater than at any time since the all-time solar cycle peak of 1957.

20.4.7   *MAIN POINTS OF MY MESSAGE*

I have been recommending for some time that the United States delay implementation of the provisions of the Montreal Protocol and initiate action to withdraw from this treaty at the earliest opportunity. My reasons for this position are:

1. As of now there is no generally accepted explanation as to the cause of the globally averaged decline of approximately 5 percent (5.6 percent per Neuendorffer 1996) in the depth of the ozone layer since 1979.

2. The health and biological effects of increased UV exposure have been grossly misrepresented; the possibility for beneficial consequences has been studiously ignored.

Dr. Robert Watson's statement at the NASA March 15, 1988 press release of the Ozone Trends Panel summary that "[our models] do not predict that ozone decreased the way it did over the Northern Hemisphere during the past 17 years" has already been noted. NASA chief scientist, Dr. James Anderson, made these even more revealing statements: "The thinning of the ozone layer over other parts of the earth is accelerating, and we don't understand why, and we don't know how fast. We don't know what factors control the movement of ozone in the stratosphere. We don't know what part of the thinning is due to the natural dynamics of the atmosphere and what part is due to the destruction of ozone by man-made chemicals. We don't know much of anything. . . . We've confused computer models of the atmosphere with the real thing. We're making huge extrapolations based on nothing but models, and models are often wrong" (Shell, *The New York Times Magazine*, 36–39: 13 March 1994).

The original 5 percent decline in ozone, predicted from continued use of freons, was estimated to cause a 14 percent increase in skin cancer incidence. For the United States this would mean an increase from about 600,000 to 685,000 new skin cancers per year—which, as I noted above, would be the equivalent of each of us moving 70 miles closer to the equator. Back in 1976, I proposed this method of allowing the general public to judge for itself the seriousness of the predicted ozone loss (see *Dotto and Schiff*, 1978, footnote on p. 283) but it has rarely been repeated in scientific papers or the press.

The public has been mislead to an even greater extent in that the possible beneficial consequences of increased UV have been consistently and studiously ignored. For land vertebrates (including humans), the only natural source of the vitamin D required to metabolize calcium into bone is from the action of solar UV on the oils in the skin. Shortage of vitamin D during the growing years can lead to the very serious disease of rickets in youth or to osteomalacia (bone loss to dangerous levels) in later life. In the United States, it is estimated that 20 to 25 million people now suffer from osteomalacia, including more than 25 percent of the women beyond menopause. Among these there are over twice as many bone fractures per year (typically of the femur or spine) as there are new cases of skin cancer per year. Theoretically, an increase in UV exposure would alleviate this condition in the growing and future generations just as, theoretically, it would lead to additional cases of skin cancer. At least one published study by the Dutch (Dubbelman et al., *Am. J. Clinical Nutrition*, 58: 106–1009, 1993) has shown that susceptible women living in their tropical island of Curaçao at 12°N suffered less from osteomalacia than did a comparable group living in the Netherlands.

At least two other diseases have been reported with incidences inversely related to exposure to sunlight (UV)—these are colon and breast cancer (Garland and Garland, *Intl. J. Epidemiology*, 9(3): 227–231, 1980; Garland et al., *The Lancet*, 1176–1177, 18 Nov. 1989; Gorham et al., *Canadian J. Public Health*, 80: 97–100, 1989).

Considering the number of people afflicted and the relative severity of the health effects, increased UV appears to offer a net health benefit, particularly, because our bodies are much better able to warn us when we are getting too much UV than they are to warn us when we are getting too little UV.

In experiments growing plants under different levels of UV exposure, it is difficult to maintain a fixed or natural ratio of UV flux to total light flux. To solve this problem, there has been developed a system of suspending cuvettes over the plants, one filled with ozone, which absorbs part of the solar UV, and one filled with air, which does not. In these cases, the plot receiving reduced UV becomes the control and the one receiving almost normal UV becomes the perturbed or enhanced UV plot. "In such a study [simulating a 10 percent reduction in ozone], plant height, leaf area, and the dry weight of sunflower, corn, and rye seedlings were significantly reduced" (UNEP, *Environmental Effects of Ozone Depletion: 1991 Update*, p. 28, UNEP, P.O. Box 30552, Nairobi, Kenya, 1991). This, of course, means that all these measures of plant growth were greater when the natural UV was reduced. That is, the detected damage was due to current normal levels of UV but were reported as due to enhanced levels of UV. This may be a perfectly valid method of carrying out such experiments but, as currently reported, it gives the public a distorted picture. The public is left believing that no damage occurs at current levels of UV, when in actuality, we have for millennia been regarding plants suffering the reported "damage" as normal and undamaged.

I am not trying to argue that UV does not cause plant damage. What I am trying to make clear is that plants grow differently under different lighting conditions and whether these differences should be interpreted as damage or not is not always clear.

20.4.8   FINAL NOTE

I hope that I have convinced you that you have been receiving a distorted, or at least incomplete, picture of the threat to stratospheric ozone and of the consequences of any increases in exposure to solar UV that may ensue.

REFERENCES

Angell, J. K., "On the Relation Between Ozone and Sunspot Number," *J. Climate*, 2: 1401–1416, 1989.

Dotto, L., and Schiff, H., *The Ozone War*, Doubleday, Garden City, NY, 1978.

Dütsch, H. U., "Vertical Ozone Distribution on a Global Scale," *Pure and Applied Geophysics*, 116: 511–529, 1978.

Neuendorffer, A. C., "Ozone Monitoring with TIROS-N Operational Vertical Sounders," *J. Geophys. Res.*, 101(D13): 18,807–18,828, 1996.

WMO (World Meteorological Organization), *Scientific Assessment of Ozone Depletion: 1991, Global Ozone Research and Monitoring Project—Report No. 25*, WMO, Geneva 20, CH 1211 Switzerland, 1991.

WMO, *Scientific Assessment of Ozone Depletion: 1994, Global Ozone Research and Monitoring Project—Report No. 37*, WMO, Geneva 20, CH 1211 Switzerland, 1995.

CHAPTER 20
SENSITIVE ENVIRONMENTAL PROBLEMS

SECTION 20.5

GREENHOUSE WARMING (GHW)

Hugh W. Ellsaesser

Dr. Ellsaesser spent 23 years in atmospheric research at the Lawrence Livermore National Laboratory, after serving for 21 years in weather service with the U.S. Air Force. His research convinced him that the establishment views on air pollution, threats to the ozone layer, and the climatic effect of carbon dioxide cannot be supported scientifically and that the hazards have been greatly exaggerated.

TABLE OF NOMENCLATURE AND UNITS

C = Centigrade temperature scale	LGM = last glacial maximum, 18 to 21 thousand YBP
F = Fahrenheit temperature scale	
GHG = greenhouse gas	NH = northern hemisphere
GHW = greenhouse warming	SH = southern hemisphere
GtC = gigatonnes (as carbon only)	UNFCCC = United Nations Framework Convention on Climate Change
IR = infrared	μm = micron or micrometer
ITCZ = Intertropical Convergence Zone	W/m^2 = watts per meter squared
K = Kelvin (absolute) temperature scale	YBP = years before the present

20.5.1 INTRODUCTION

The issue of greenhouse warming (GHW) has now reached the stage at which mandatory goals for reductions in the use of fossil fuels by developed countries were adopted as the Kyoto Protocol to the United Nations Framework Convention on Climate Change (UNFCCC). One of the principal provisions of this 1992 "Rio Climate Treaty" was to stabilize "greenhouse gas concentrations in the atmosphere at a level that would prevent dangerous anthropogenic interference with the climate system." Although "dangerous level" remains undefined, the 174 national signatories of UNFCCC, meeting in Kyoto, Japan in December 1997 adopted greenhouse gas (GHG) emission limitations for 39 developed counties to be achieved by 2008–2012. These ranged from 92 to 110 percent of their 1990 emissions and were designed to effect an overall reduction of 5 percent below 1990 levels for these countries. For the United States, the limit is 93 percent of 1990 emissions; this is estimated to be more than a 30 percent reduction below what emissions would have become at that time without restrictions.

These developments make GHW an important issue even if GHW itself never actually materializes or becomes detectable. The incongruity of this statement reveals the degree to which I believe the public has been misled on GHW.

My disillusionment with the largely climate-model-produced image of GHW began around 1980. As a meteorologist who had served as a U.S. Air Force Weather Officer for 20 years and had performed atmospheric research at the Lawrence Livermore National Laboratory for a similar period at that time, "my strongest reason for doubting . . . climate model estimates of the carbon dioxide warming are the gross differences I see between how the atmosphere works and how it is modeled to work" (Ellsaesser, H. W., *Atmospheric Environment*, 18(2): 431–434, 1984).

20.5.2 THE GREENHOUSE CONCEPT

Figure 20.5.1 was put together to help clarify the following discussion.

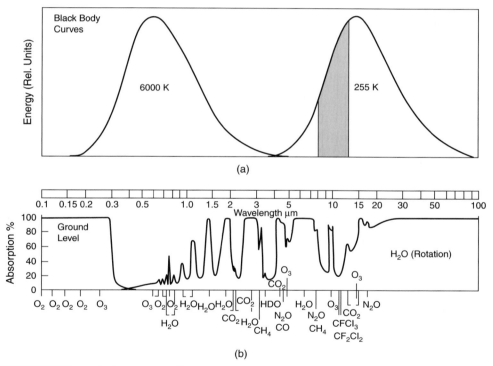

(a)

(b)

FIGURE 20.5.1   (a): Radiant energy as a function of wavelength (shown in the middle scale) from the sun (calculated as a 6000 K black body at the Earth-sun distance) absorbed by the Earth and of the IR radiation emitted by Earth to space (calculated as a 255 K black body). (b): Percent absorption of sunlight as it traverses the atmosphere to the surface and of the IR radiation emitted by Earth as it traverses the atmosphere to space. Also shown at the bottom are the principal wavelengths or bands at which atmospheric gases absorb. Note the reduced IR absorption by water vapor between 8 and 12.5 microns creating the "atmospheric window" (shaded in a) (from MacCracken and Luther, *Projecting the Climatic Effects of Increasing Carbon Dioxide*, DOE/ER-0237, NTIC, Springfield, VA 22161, Figure 2.2, 1985).

Black Body Emission

The concept of greenhouse warming resulting from the presence of an atmosphere, or certain atmospheric constituents, is quite simple and is intuitively appealing to freshman physics classes. It rests on the empirical and theoretical Stefan-Boltzmann law that all bodies radiate energy at a rate proportional to the fourth power of their absolute temperature. The absolute temperature determines not only the total energy output but also its spectral distribution (Planck's law) and, in particular, the wavelength or color at which the radiated energy is a maximum (Wien's law). While color is only detectable by eye in the visual range from red to violet, the concept is at times extended to equate color and wavelength in other parts of the electromagnetic radiation spectrum.

For most solid and liquid bodies, there are relatively small departures from the energy output predicted by the so-called black body curve given by Planck's law. The sun, at a temperature near 5800 K, radiates energy from the ultraviolet (shorter than 0.38 μm) through the visible (0.38 to 0.7 μm) and into the near infrared (>0.7 μm), with the peak radiation reaching the Earth's surface near 0.55 μm. The Earth absorbs solar energy but because of its much lower temperature (288 K), reradiates this energy at a longer wavelength in the infrared (see Figure 20.5.1).

Gaseous Emission

In contrast to solids and liquids, gases tend to absorb and radiate energy only at discrete wavelengths or bands. However, the intensity of the radiation emitted at each permitted wave length remains within the limit of the black body curve given by Planck's law for the temperature of the radiator. Different gases within the atmosphere absorb and emit at different wavelengths. Oxygen absorbs strongly in the ultraviolet and shields the Earth's surface from rays very detrimental to life. In the process of absorbing ultra-violet, part of the oxygen is disassociated and converted to ozone, forming the ozone layer of the stratosphere. Ozone strongly absorbs ultraviolet shorter than about 0.29 μm, which is near the shortest wavelength of sunlight ever observed at the Earth's surface.

Minor constituents of the atmosphere such as CO_2, methane, nitrous oxide, and the largely man-produced chlorofluorocarbons (CFCs or freons) absorb at specific wavelengths or bands, primarily in the IR. Carbon dioxide has a strong absorption band near 15 μm and weaker bands at shorter wavelengths. The others have absorption bands primarily between 6 and 12 μm. Ozone also has a strong absorption band at 9.6 μm.

The principal greenhouse gas, by far, in our atmosphere is water vapor. This is true partly because of the greater amount of water vapor. It averages 2 to 3 cm of precipitable liquid water in the atmospheric column and up to 5 or more parts per thousand in surface tropical air. The other greenhouse gases are present at parts per trillion to parts per million. Water vapor, as show in Figure 20.5.1, also absorbs over most of the terrestrial IR spectrum, except weakly for the so-called atmospheric window from 8 to 12.5 μm, in which there is little atmospheric absorption of IR radiation, except for the 9.6 band of ozone.

Vertical Structure of the Atmosphere

As any frequenter of the mountains knows, air temperature decreases with altitude; this drop in temperature is know as lapse rate and averages about 6°C per kilometer (3.3°F/1000 ft). Most of the 240 W/m^2 of solar energy absorbed by the Earth and its atmosphere is absorbed at the surface. Because the annually and globally averaged surface temperature is observed to be about 288 K (15°C, 59°F), the Earth's surface emits radiation very nearly as a black body at this temperature as illustrated in Figure 20.5.1. The bulk of this is absorbed by the IR absorbing gases of the atmosphere and clouds, although some 5 to 10 percent of the surface emission escapes directly to space through the atmospheric window. The absorbers reradiate the absorbed energy in all directions, in particular, upward and downward, but at an intensity appropriate to their temperature, which is less than that of the surface. This is due to the lapse rate, or cooling of the air with altitude, in the troposphere.

The energy radiated upward can again be absorbed, and again reradiated both upward and downward, and again at a still lower temperature. This process of emission, absorption, and reemission continues, not layer by layer, but continuously until there is no longer enough absorber above to prevent the radiation from escaping to space. Thus, it is only from the outer optical depth of an IR absorbing gas that it emits appreciable radiation to space. Clouds, which consist of water or ice particles, radiate as black bodies (i.e., at all wavelengths). The 8 to 12.5 μm portion of their radiation has a reasonable change of escaping through the atmospheric window. However, the intensity of their radiation is determined by the cloud top temperature, which can be well below the surface temperature—as much as 100°C.

The Difficulty of Computing Radiation Transport

In addition to reflection, absorption, and transmission, atmospheric molecules and particulates also scatter radiation—a process again strongly dependent on wavelength. It should be obvious from the omni-directional nature of radiation, its variations dependent on wavelength, and on the temperature of the last radiator, that the computation of radiative flux is very arduous. All atmospheric models must make simplifying approximations and parameterizations to keep radiation calculations economically feasible.

The Planetary Energy Budget

As noted above and in Figure 20.5.1, our sun radiates with a color temperature of about 5800 K. At the Earth-sun distance this appears as a beam of radiation containing 1360 W/m². Because of the Earth's spherical shape and its rotation, the annual average per unit of Earth's surface is one-fourth as much or 340 W/m². As we now know from satellites, the Earth is quite bright and reflects away about 30 percent of the incoming solar energy, leaving 240 W/m² as the amount of solar energy absorbed. If the Earth's temperature is to remain constant, this then is also the amount of energy the Earth must reradiate back to space. If we apply the Stefan-Boltzmann law in reverse, we find that the black body temperature required to radiate 240 W/m² is 255 K (-18°C, 0°F). As can be determined from the lapse rate given above, this is the average temperature of the atmosphere at about 6 kilometers (20,000 ft) above the surface.

The Greenhouse Effect of Our Present Atmosphere

The 33°C (59°F) difference between the radiating temperature (255 K) and the surface temperature (288 K) is the nominal greenhouse effect of our present atmosphere.

The Effect of Additional Greenhouse Gases

The effect of additional GHGs is generally explained in one of the following ways:

1. Simplistic View—The increased concentration of greenhouse gases will further close the atmospheric window. Thus, the Earth's surface will warm until its increased temperature can pump enough radiation through the narrowed atmospheric window to balance the radiation received from the sun. Since even now, only 5 to 10 percent of the radiation to space passes up though the atmospheric window directly to space without reabsorption, this mechanism can have no effect on the bulk of the outgoing radiation to space. A more plausible mechanism is the following.

2. More Plausible View—Increasing the concentration of GHGs will force the layer from which radiation is emitted to space (i.e., the outer optical depth of the radiating gas) to a higher altitude and therefore, because of the lapse rate, to a temperature below 255 K. However, radiation from this lower temperature will be insufficient to reject all of the absorbed solar radiation. Thus, the atmosphere will accumulate solar energy and warm until the new lifted radiating layer reaches 255 K. With a lapse rate maintained by convection, this warming will have extended up from the surface, that is, the process results in a warming of the mean surface temperature.

While the second mechanism of GHW is closest to reality, it brings up another issue. As noted above, for any radiating atmospheric gas, it is only the outer optical depth that emits appreciable energy to space. However, the level of the outer optical depth varies both with the gas species and with the wavelength being considered. This GHW mechanism is only valid for GHGs whose outer optical depth lies within the troposphere where the temperature decreases with altitude and where the lapse rate is maintained by convection. For any emission wavelength of a species, for which the outer optical depth is in the stratosphere, where temperature increases with altitude, the effect on surface temperature would be reversed if this argument were to remain valid (it doesn't).

This is of little concern for water vapor because its concentration decreases rapidly with altitude (or temperature) ensuring that the outer optical depth at all wavelengths lies in the troposphere. For CO_2's principal absorption band at 15 μm, the outer optical depth lies in the stratosphere. This is probably true for most of the other well-mixed GHGs because they are computed to be orders of magnitude more effective than CO_2 on a molecule-for-molecule basis.

Problems With the GHW Mechanism

As can be seen from the above, the concept of GHW is simple and intuitively appealing. However, as has already been hinted, this is an oversimplified picture. Doubling the concentration of CO_2 alone is calculated to increase the surface temperature by about 1.2°C (2.16°F). The bulk of the model-predicted 1.5 to 4.5°C (2.7 to 8.1°F) warming comes from positive feedbacks, the major one of which is from water vapor. This means that any warming of the atmosphere increases its ability to hold water vapor, and it is the greenhouse effect of this presumed increase in water vapor that leads to the major part of the predicted warming. However, as noted above, it is only the outer optical depth of a radiator that radiates appreciably to space. Thus, if water vapor is to have a positive feedback, its concentration must be increased in the layer of its outer optical depth.

A more important overlooked factor is the fact that the bulk of incoming solar energy absorbed at the surface is transported to higher levels in the atmosphere, not by the tedious process of radiation transport outlined above, but by deep convection. This is particularly true over the tropics, the warmest and most moist half of the Earth's surface.

The tropical atmosphere, approximately 30°N to 30°S latitude, is dominated by the Hadley circulation. This consists of the northeast and southeast trade winds that sweep the warm moist surface air of the tropics into narrow converging and convectively uprising zones called ITCZs. In these deep convective zones, the air rises and cools, condensing the contained water vapor into cloud and rain drops that fall as the heavy precipitation of ITCZs and monsoons. The released latent heat of condensation maintains the buoyancy and updraft of the air until nearly all of the contained water vapor has been precipitated. This convectively dried air then spreads horizontally toward the poles, primarily the winter pole, and subsides. It is this subsiding dry air that creates our subtropical deserts and also opens holes, or "windows of dry air," downward into the water vapor greenhouse blanket of the tropics allowing IR radiation to escape to space from lower and warmer levels of the atmosphere. The Hadley circulation forces the outer optical depth of water vapor in the subtropics to lower and warmer layers of the atmosphere. As a result, when we look at Earth from space, the strongest IR radiation from Earth emanates from these windows created in the subtropical belts by Hadley cell downwelling of convectively dried air. The subtropical belt in the winter hemisphere, opposite the active ITCZ of the summer hemisphere, is the strongest IR emitter of the planet.

Any surface warming in the tropics will lead to acceleration of the Hadley circulation, including strengthening of the subsiding downdrafts of convectively dried air. This means that the subtropical windows of convectively dried air in the water vapor greenhouse blanket will be enlarged or deepened, allowing easier escape of IR radiation to space. Any warming of the tropics from the greenhouse effect of additional CO_2 will lead to no, or even to a negative, feedback from water vapor over the tropics, the warmest and most moist half of the atmosphere. Without a positive water vapor feedback in the tropical half of the atmosphere, the global warming from a doubling of CO_2 will be "at least 2- to 3-fold" (Ellsaesser, H. W., *Atmos. Environ.*, 18(2): 431–434, 1984) less than predicted by current climate models.

20.5.3 THE TEMPERATURE HISTORY OF THE EARTH AS WE KNOW IT

Most readers no doubt have images of a warmer Earth at the time of the dinosaurs some 100 million YBP. It is currently estimated that the mean global surface temperature then was about 10°C (18°F) warmer than now. However, the greatest differences were in high latitudes with a much smaller change in low latitudes. Since the time of the dinosaurs, the Earth appears to have cooled. About 3 million YBP an ice age set in, marked by a glacial-interglacial cycle with an average global temperature fluctuation currently estimated at 5 to 7°C (9 to 12.6°F).

A Capsulation of Earth's Climatic History

The entire climatic history of the Earth has been capsulated as follows. During 90 percent of the last 4.5 billion years the Earth's climate was warmer than it is at present; while during 90 percent of the last 3 million years, it was colder than it is at present.

The Glacial-Interglacial Cycle

The current best estimate of the reconstruction of the last million years of this temperature fluctuation is shown in Figure 20.5.2, taken from IPCC90 [Figure 7.1]. In the upper panel, the past 700,000 years show a 100,000-year cycle with 90,000 years of staged cooling followed by an abrupt warming back to an interglacial warm period, lasting 10,000 to 12,000 years. We are currently in one of these interglacial period—which we call the Holocene, believed to have begun about 10,700 YBP. The nadir of the last glacial period, the LGM, is estimated to have been only 18,000 to 21,000 YBP. At that time, mean global surface temperature was 5 to 7°C (9–12.6°F) colder than now and 3-kilometer-deep (10,000 ft) ice sheets were centered over Hudson Bay and Scandinavia. These glacial ice sheets extended down to Long Island and over the Great Lakes in America and down to Scotland and over the Baltic Sea in Europe.

This glacial-interglacial cycle of the last 700,000 years is expected to continue. Because the Holocene began 10,700 YBP, by current estimates the next cycle of 90,000 years of cooling and glaciation is now due. Given this current state of knowledge, with regard to climate evolution, one would expect to hear the argument that adding GHGs to the atmosphere is exactly what humans should be doing to delay, and thereby hopefully to prevent, the onset of the next glacial. Certainly, the hazards of glacier ice building to a depth of 3 km over Hudson Bay and creeping down over the Great Lakes, as it was just 18,000 to 21,000 YBP, appears greater than any so far proposed for GHW.

The Holocene

The middle and lower panels of Figure 20.5.2 show the most recent 10,000 and 1000 years of our temperature history on expanded scales. These show oscillations about the mean temperature of the Holocene by about plus or minus 1°C (1.8°F). Note that the most recent cold period, about 1450 to 1900 AD, is called the Little Ice Age and that the preceding warm period, about 1000 to 1350 AD, is generally called the Medieval Climatic Optimum. From our history, we know that around

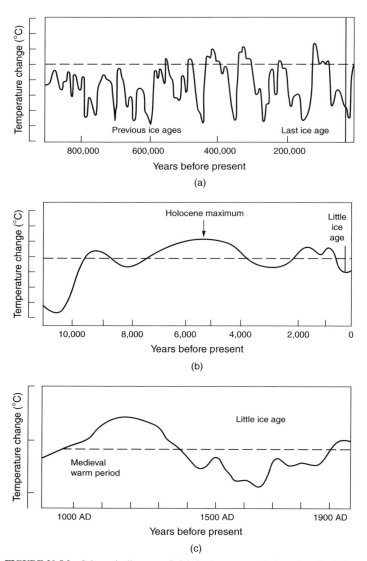

FIGURE 20.5.2   Schematic diagrams of global temperature variations since the Pleis-tocene on three time-scales: (a) the last million years; (b) the last ten thousand years; and (c) the last thousand years. The horizontal dashed line nominally represents conditions near the beginning of the twentieth century (from IPCC90, Figure 7.1).

900–1200 AD, the ice in the North Atlantic retreated and the Norsemen were able to colonize Iceland and Greenland and to explore Nova Scotia. About 1350 AD, the ice readvanced and the Greenland colony died out. Mountain glaciers advanced in Switzerland and Scandinavia, causing abandonment of previously occupied and tax-paying farms and villages. Also, the rivers of London, St. Petersburg, and Moscow froze over sufficiently for the people to hold winter fairs on the ice. Therefore, we have both paleo-climatological and historical evidence for the Medieval Climatic Optimum and the Little Ice Age temperature excursions shown in Figure 20.5.2.

While the causes of these temperature oscillations remain unknown, they were not due to any change in the greenhouse gas content of the atmosphere (IPCC90, p. 202) and, certainly, they are unlikely to have been triggered by human kind. In other words, the fact that the mean global temperature has warmed by about 0.5°C (0.9°F), over the past 130 years for which we have recorded temperatures, does not provide any proof that we are witnessing either GHW or a "discernible human influence on global climate" (IPCC95, p. 4).

The Least Controversial Explanation of the Warming Observed to Date

An increasing number of students of climate maintain that the least controversial explanation of the warming that we have seen over the past 130 years, or so, is a return to normal from the Little Ice Age, and possibly, an entry into the next warm period following the Little Ice Age. Even IPCC90 (p. 203) concluded: "The Little Ice Age came to an end only in the nineteenth-entury. Thus some of the global warming since 1850 could be a recovery from the Little Ice Age rather than a direct result of human activities." Broecker (*Science*, 283: 179, 1999) stated it thus: "Until a satisfactory explanation has been established for the pronounced demise of the Little Ice Age during the period 1870 to 1940, adequate room for maneuvering will exist for those who doubt that the buildup of carbon dioxide and other greenhouse gases constitutes a substantial threat."

Because we are ignorant of the causes of the past temperature oscillations of the Holocene, we have no reason to believe that they will not continue until overpowered by the onset of the next glacial. If they are continuing, we are now near the inflection point of the temperature curve—the point of most rapid rise, and can also look forward to an additional warming of about 1°C (1.8°F) over the next couple of centuries regardless of what humans do. In this connection, it should also be noted that these warm periods of the past were usually designated "climatic optima." Certainly, they must have appeared so circa 1350 AD to the remnants of the Greenland colony and to the farmers and villagers forced out by advancing glaciers in Switzerland and Scandinavia.

20.5.4   THE MISSING GREENHOUSE "FINGERPRINT"

From the beginning of model predictions of GHW in 1967 (Manabe and Weatherald, *J. Atmos. Sci.*, 24(3): 241–259, 1967), the model-predicted warming that should have occurred to date substantially exceeded the actual warming identifiable in the observational record. Recently, it was proposed that humans are putting a steadily increasing amount of particles in the atmosphere, mainly of sulfate from emissions of sulfur dioxide. These, through their reflection of increasing amounts of sunlight, are claimed to be causing the observed warming to lag the predicted warming. A cooling effect, or a reduced rate of warming, is possible via this mechanism only if the amount of sulfate aerosol in the atmosphere is steadily increasing with time.

The Concept of a "Fingerprint" of GHW

Because there has been so little agreement between the predicted and observed change in the global mean surface temperature, efforts were made to identify a "fingerprint" of GHW. The fingerprint, as determined from model predictions, includes global mean surface warming of 1.5, 2.5, or 4.5°C for "low," "best," or "high" model sensitivities for a doubling of CO_2. The surface warming will be greater over land than over water; least near the equator and increasing toward the poles and will be substantially greater near the winter pole. In the tropics the warming increases with altitude and in high latitudes it decreases with altitude. The warming of the troposphere will be accompanied by cooling of the stratosphere, increasing with altitude and being greatest near the stratopause. It is also predicted that there will be an increase in precipitation outside the tropics.

This fingerprint has so far been of little help in trying to identify GHW in the observational record. In comparing the details of the instrumental record with model predictions, IPCC90 (p. 254) found some areas of agreement but many areas of disagreement and concluded. "Thus, it is not possible at this time to attribute all, or even a large part, of the observed global-mean warming to enhanced greenhouse effect on the basis of the observational data currently available." In their summary (IPCC90, p. xxix) they went even further, admitting; "we do not yet know what the detailed 'signal [fingerprint]' looks like because we have limited confidence in our predictions of climate change patterns."

This raises the immediate question; if the model-predicted fingerprint or pattern of GHW is not credible, why is the model-predicted degree of warming credible? To a certain extent IPCC95 [Table 6.3] admitted a lack of confidence in the degree of warming predicted by climate models. The equilibrium warming for a doubling of CO_2 from 16 cited experiments ranged from 2.1 to 4.6°C, with an average of 3.2°C. Despite this, the report reiterated the "low," "best," and "high" climate sensitives of 1.5, 2.5, and 4.5°C for a doubling of CO_2 originally adopted in 1990.

The Increasing Discrepancy Between Predicted and Observed Warming

Meanwhile, the magnitude of this disagreement between predicted and observed warming has continued to increase for the following reasons:

1. The CO_2 in the atmosphere has continued to increase without a corresponding increase in temperature.

2. It has been recognized that humans are emitting GHGs other than CO_2, such as methane, nitrous oxide, and the HFCs or hydrofluorocarbons, which should have amplified the equivalent increase in CO_2 to date by about 50 percent.

3. Because reanalyses of the observational record have somewhat lowered the estimates of the actual warming to date.

The number and coverage of observing stations permit estimates of global mean temperature since about the 1860s. Estimates of global warming since then hover around, or just below, 0.5°C (0.9°F). However, as noted by Ellsaesser et al. (*Rev. Geophy.*, 24(4): 745–792: 785, 1986) and Weber (John Emsley (ed.), *The Global Warming Debate*, ESEF, London, 1996), longer-term records from individual stations indicate that the later part of the nineteenth century was cooler than normal so some fraction (a quarter to a third) of the warming recorded since then represents a return to normal, rather than global warming. In the words of Beniston et al. (*Climatic Change*, 36: 233–251, 1997): "It is important to note that the instrument record which we now use to characterize 'global warming' began at what was arguably the coldest period of the Holocene, in the mid-nineteenth century."

Also, this observed warming has not been a steady rise as predicted by the models. In the NH land data, for which we have the best data, there was warming of about 0.25°C between 1884 and 1900 followed by an abrupt warming of about 0.4°C between 1917 and 1921 with some further warming into the 1940s. Much of this warming was concentrated in higher latitudes, particularly in the North Atlantic, and has been characterized as "the Arctic warming." Most of the Arctic warming faded away between 1958 and 1963 during the first five years of MIT's library of upper air data. These data showed a cooling of the NH lower troposphere of 0.6°C (Starr and Oort, *Nature*, 242: 310–313, 1973). Surface cooling during this period was only about half as much. From 1976 to 1981, there was another abrupt warming of about 0.35°C. Thus, the major part of the actual warming occurred before 1940, before most of human contributions of GHGs to the atmosphere. As GHG emissions rose rapidly after World War II, the NH temperature actually cooled for nearly 40 years before again jumping to a level exceeding the level of the 1940s.

From the above, it is apparent that, at most, a small fraction, if any, of the warming observed to date can be attributed to GHW.

The Meaning of "Broadly Consistent"

The three successive reports (IPCC90, IPCC92, IPCC95) give the observed warming over the past century as 0.3 to 0.6°C (0.54 to 1.08°F). While the amount of warming predicted to have occurred to date was not specified, graphs in IPCC90 (Figure 20.5.3) for model predictions before introduction of sulfate particles gave warmings of 0.7, 1.0, and 1.4°C (1.26, 1.8, and 2.52°F) by 1990 (see top panel of Figure 20.5.3). It is readily apparent that the ranges of the observed and predicted warmings do not even overlap. However, IPCC90 and IPCC92 claimed that "the size of this [observed 0.3 to 0.6°C] warming is broadly consistent with predictions of climate models [0.7 to 1.4°C]."

Because cooling by human-produced sulfates has been introduced, IPCC95 (p. 295) described the above discrepancy as follows: "When increases in greenhouse gases only are taken into account in simulating climate change over the last century, most GCMs and energy balance models produce a greater warming than that observed to date, unless a lower climate sensitivity than that found in most GCMs is used."

Hemispherically, Sulfates Do Not Improve Agreement of Predicted and Observed Warming

Human-emitted sulfates are restricted primarily ($\sim$90%) to the NH, so that matching predicted and observed warming *by hemisphere* provides a more meaningful test. Earlier analyses of the observed data have generally shown comparable warming in the two hemispheres or, perhaps, slightly greater warming in the SH. However, recent reanalyses of the land observations (the most credible) of the SH have reduced the estimate of the SH warming over the period of record to about half that of the NH. Jones (*J. Climate*, 7(11): 1794–1802, 1994) got 0.26°C per century for the SH versus 0.47°C per century for the NH. Hughes and Balling (*J. Climatology*, 16, 935–940, 1996) obtained a similar result from a reanalysis of the observational data from South Africa. This is a serious discrepancy for the argument that sulfates are reducing GHW.

The Absence of Polar Amplification

Another troubling discrepancy is the failure of the model-predicted polar amplification of the warming to appear in the observational data. There was a strong pulse of warming in the Arctic, particularly in the Atlantic portion of the Arctic, between 1920 and 1940. However, this so-called "Arctic Warming" went away quite abruptly between 1958 and 1964. Other than this pulse, there has been essentially no indication of polar amplification of the warming observed to date at either pole.

Absence of Warming Since Satellite Observations Began

A new discrepancy has appeared since we began to measure the temperature of the lower atmosphere from satellites in 1979. We now have 20 years of these data showing a slight cooling (-0.06°C per decade, except for the El Nino year 1998, rather than the 0.2°C per decade of warming predicted by the models). The worldwide radiosonde balloon measurements of atmospheric temperature over the past 20 years agree with the satellite data while our observations of surface temperature appear to show a global warming of 0.13°C per decade. This leaves us two discrepancies to solve: that between the observed and predicted warming and that between our surface observations and upper air satellite and balloon soundings.

The "Precautionary Principle" the Only Remaining Argument

About the only argument left for the establishment is the "precautionary principle"; that is, we can't afford to wait until we know what we are doing. This was recently reiterated by Hasselmann (*Science*, 276: 914–915, 1997) in the following words: "It would be unfortunate if the current debate over this

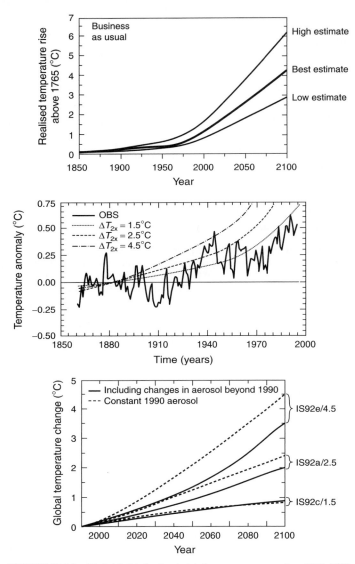

FIGURE 20.5.3   (Top): Model simulated global mean temperature from 1850–1990 because of observed increases in GHGs and predictions for the rise 1990–2100 resulting from the Business-as-Usual emissions scenario (from IPCC90, Figure 8]; (Middle): Observed changes in global mean temperature over 1861 to 1994 compared with those simulated using an upwelling diffusion-energy balance climate model forced with GHGs only (from IPCC95, Figure 16a]; (Bottom): Projected global mean surface temperature from 1990 to 2100 for the three model sensitivities and increasing sulfate aerosols (solid curves) and for the high, moderate, and low emissions scenarios indicated, and 1990 sulfate emissions (dashed curves). (from IPCC95, Figure 19].

ultimately transitory issue [absence of model-predicted warming] should distract from the far more serious problem of the long-term evolution of global warming once the signal has been unequivocally detected above the background noise."

20.5.5 *THE CONSISTENT SUPPRESSION OF GOOD NEWS*

Throughout the GHW debate, there has been a reluctance on the part of the establishment to call attention to any possible benefits of GHW; or to any new information or changes in interpretation undermining, lessening, or casting doubt on the validity of the threat of GHW—such as the disparity between model-predicted and observed warming to date, noted above.

Under President Bush, the U.S. climate research budget passed the $2 billion level. This huge fund has been, and is, very damaging to scientific objectivity. It has imposed a strong bias on the range of scientific inquiry. It is administered as though it were immoral, if not illegal, to expend public funds to look for, or attempt to document or quantify, any possible beneficial consequences of human activity. Beyond this, the scientists and institutions contending for this largess of public funds have never before experienced the public and media attention—including three Nobel Prizes in 1995—that they are now receiving. There is career and peer pressure on those institutions and individuals receiving this largess to avoid any hint that GHW has been overblown or that it might possibly have net beneficial consequences.

Literally hundreds of Non-Governmental Organizations plus environmentalists, the media, United Nations organizations, and others have been and are exploiting GHW and other issues to advance their own agendas. Currently planned attacks on global warming, in particular, would lead to transfers of trillions of dollars from the developed to the developing world, a transfer that would have to be funneled through United Nations organizations. This offers powerful incentives for the United Nations, its clients, and supporters and all the underdeveloped nations to legitimatize GHW.

No Acknowledgment That the Most Probable Estimate of GHW Was Reduced

Through three successive National Research Council/National Academy of Sciences studies, released in 1979, 1982, and 1983, the equilibrium warming for a doubling of CO_2 was estimated at $3.0 \pm 1.5°C$ ($5.4 \pm 2.7°F$). The first IPCC90 report kept the same range but reduced the "best estimate," or most probable value, from 3.0 to 2.5°C (5.4 to 4.5°F). However, anyone not familiar with the previous reports would not have realized that this very critical value had been reduced. No attention was called to, or reason given for, the reduction.

Only Grudging Admission of the Fertilizer Effect of Increased CO_2

In the U.S. Department of Energy State-of-the-Art Reports on the CO_2 issue, the executive summary of the volume, "Direct Effects of Increasing Carbon Dioxide on Vegetation" (*Strain and Cure*, DOE/ER-0238, NTIS, Springfield, VA 22161, p. xvii, 1985), provides a typical example of suppression of the good news. It clearly states that CO_2 is essential to plant life and that "it is possible that some fraction of the increased agricultural yield that has occurred in this century is due to increased atmospheric CO_2 concentration." However, more attention is given to detrimental effects, or unknown threatening possibilities, such as the following. Weeds "could have a comparatively larger growth response to increased CO_2 than some desirable crop species." "Soybean leaves [under increased CO_2] became carbon rich and nitrogen poor . . . . An insect pest, the soybean looper . . . had to consume more leaf tissue to gain an equal amount of protein nitrogen."

In contrast to the above, Dr. Sherwood B. Idso's (1995, pp. i–ii) summary of this subject reads as follows: "Results from hundreds of laboratory and field experiments demonstrate that increasing the carbon dioxide content of the air helps plants grow faster, bigger and more profusely. A large body of data also indicates that the percentage growth enhancement due to atmospheric CO_2 enrichment [about 30 percent for a doubling of CO_2] is generally greater when plants are subjected to various environmental stresses, or when essential resources such as light, water or nutrients are less than adequate for optimal growth. It has additionally been observed that there is typically no decline in the growth-enhancing effects of elevated levels of atmospheric CO_2 when plants are grown for long periods of time in natural settings out-of-doors; and it appears that the growth rates of trees and shrubs are generally more responsive to increases in the air's CO_2 content than are the growth rates of herbaceous plants.

"These experimental observations suggest three things that should have occurred as the air's CO_2 content rose in tandem with the burning of fossil fuels that powered the engines of the Industrial Revolution. First, there should have been concurrent increases in the growth rates of nearly all of earth's plants. Second, trees and shrubs [C_3 type plants] should have gained a competitive advantage over non-woody vegetation [C_4 type plants]. And third, as the rate of rise of the air's CO_2 content has accelerated over the past few decades, so also should these biospheric changes have been greatest in recent years.

"Numerous studies have produced three impressive pillars of support for this 'greening of the earth' scenario. First, they reveal the existence of a worldwide invasion of grasslands by trees and shrubs that began approximately two centuries ago and has closely followed the upward trend in the air's CO_2 content. Second, they demonstrate that the growth rates of many forests around the globe have increased concurrently, with the past few decades exhibiting the greatest responses. And third, they indicate that the amplitude of the seasonal oscillation of the air's CO_2 concentration—which is driven primarily by the metabolic activity of the terrestrial biota—has risen hand in hand with the air's CO_2 content over the past three and a half decades, thereby demonstrating that the vitality of the entire biosphere has also risen hand in hand with the air's CO_2 content over this period. . . .

"In fact, the carbon dioxide emitted by our energy-consuming activities can actually enhance both the quantity and quality of life on earth."

The above analysis by Idso was recently confirmed by Jolly and Haxeltine (*Science*, 276, 786–788, 1997). With a process-based vegetation model they found that at "the last glacial maximum (18,000 YBP), the change in atmospheric carbon dioxide concentration alone could explain the observed replacement of tropical montane forest by a scrub biome." This shrinkage of the tropical rain forests had previously been attributed to a tropical cooling of about 5°C (9°F), a cooling that is inconsistent with indications form other data sources.

Disappearance of the Effect of Previously Released CO_2

One of the developments in this biasing of the information presented by study panels is rather disturbing. Figure 20.5.3 has been constructed to illustrate this misleading practice. This figure is a composite of three graphs of model predicted temperatures reproduced from IPCC90 (Figure 8) and IPCC95 (Figures 16a and 19).

The upper graph shows model predictions for the "low," "best," and "high" model sensitivities of global mean temperatures from 1850 to 2100 for observed increases in GHGs to 1990 and the Business-as-Usual projections thereafter. The predictions were begun in 1765 to include all man-produced GHGs.

The middle graph shows observed global mean temperature along with model-predicted temperatures, again for the three model sensitivities for a doubling of CO_2. Note that the predicted curves do not start from zero in 1765 as in the top graph. From the slopes of the predicted curves, where they cross the zero ordinate circa 1880, it is obvious that they would have to be raised significant amounts to place their ordinates on zero in 1765. Furthermore, it is equally obvious that the disagreement with observations would be enhanced if this were done.

The lower graph shows model-projected global mean surface temperature changes for the three model sensitivities and for three different future emission scenarios. The misleading aspect of the figure is the fact that all projected curves start from zero ordinate in 1990. This suggests that there were no human-emitted GHGs before 1990 or that any such emissions had no effect on the global mean temperature. The reader can surmise from the upper two graphs how different this graph would look if temperature itself were plotted with the ordinate of each curve started from zero in 1765. This is what would be necessary to depict the model-predicted effect of human's emissions of GHGs prior to 1990.

20.5.6 IS THERE A DISCERNIBLE HUMAN INFLUENCE ON GLOBAL CLIMATE?

The chief phrase from the most recent IPCC95 (p. 4) report provoked headlines, letters to the editor, and scientific rebuttals. The statement, given as a section heading in the *Summary for Policymakers*, bluntly states: "The balance of evidence suggests an anthropogenic influence on global climate."

Substantiation of such a statement requires both that a nonnatural climate change be identifiable in the observational record and that it be of such a nature that it can be attributed to the actions of humans. Rather than attempt to make the evidence contradicting such claims intelligible to the reader, I shall pursue an easier path. Assembled below are subsequent public statements, denying this claim or showing how little it actually means, made by some of the principal proponents and defenders of the claim itself.

["No one to my knowledge who is informed is claiming certainty of detection or attribution (of an anthropogenic influence on global climate); certainly the IPCC is not, . . . "] John T. Houghton, Leading Editor of IPCC95 (personal communication, 1996).

"We say quite clearly that few scientists would say the attribution issue was a done deal." Benjamin D. Santer, Lead Author of Section 8 of IPCC95 (Kerr, *Science*, 276: 1040-1042, 1997).

" . . . many climate experts caution that it is not at all clear yet that human activities have begun to warm the planet—or how bad greenhouse warming will be when it arrives." Richard A. Kerr, Research News & Comment Writer for *Science* (*Science*, 276: 1040–1042, 1997).

"However, the inherent statistical uncertainties in the detection of anthropognic climate change can be expected to subside only gradually in the next few years while the predicted signal is still slowly emerging from the natural climate variability noise. It would be unfortunate if the current debate over this ultimately transitory issue should distract from the far more serious problem of the long-term evolution of global warming once the signal has been unequivocally detected above the background noise." Klaus Hasselmann, Max-Planck-Institute for Meteorology (*Science*, 276: 914–915, 1997).

[The reader should note that GHW has been predicted to rise above the noise of the climate record *within the coming decade* for at least two decades.]

From these assertions by its supporters, it seems quite clear that the IPCC95 statement; "The balance of evidence suggests a discernible human influence on global climate," was studiously crafted to induce the media to broadcast to the citizens and policymakers of the world a message that few, if any, of the researchers, on whose work it was based, are yet willing to defend before the scientific community.

20.5.7 INTERNATIONAL EFFORTS TO CONTROL GHW

As noted at the beginning, under the provisions of the 1992 Framework Convention on Climate Change (UNFCCC) Treaty (the "Rio Climate Treaty"), efforts are underway to limit the emissions of CO_2 by restricting the consumption of fossil fuels—"to achieve stabilization of greenhouse gas concentrations in the atmosphere at a level that would prevent dangerous anthropogenic interference with the climate system . . . ." Failing a consensus at the First Conference of the Parties (COP-1) in Berlin in April 1995, FCCC adopted instead the "Berlin Mandate:" an agreement to negotiate a new set of targets for reducing greenhouse gas emissions to be adopted as *mandatory for developed counties only* in Kyoto, Japan in December 1997. The adopted targets are giving in Section 20.5.1 above.

Until mid-1996, the Clinton administration had relied on voluntary reductions in the use of fossil fuels to meet U.S. commitments under the FCCC Treaty. However, at the COP-2 in Geneva, July 17, 1996, U.S. Undersecretary of State for Global Affairs, Timothy E. Wirth, announced: "The United States recommends that future negotiations focus on an agreement that sets a realistic verifiable and binding medium-term emissions target." A year later, at a special session of the U.N. general assembly in New York, President Clinton told the meeting that he would take action "to convince the American people and the Congress that the climate change problem is real and imminent" (*Nature*, 388: 5, 1997). He added that he would bring to December's meeting at Kyoto, Japan, where countries hope to negotiate a treaty on GHG emissions, "a strong American commitment to realistic and binding limits that will significantly reduce our emission of greenhouse gases."

It is clear that the Kyoto targets are regarded as only a first step; the goal is to halt the rise in CO_2 (Wigley, *GRL* 25(13): 2285–2288, 1998). IPCC90 (p. xi) estimates that halting the rise of CO_2 in the atmosphere will require a global reduction in the consumption of fossil fuels of more than 60 percent. If allocated on a per capita basis, this means a reduction for the United States of over 90 percent. It appears irresponsible to start down this road toward virtual elimination of fossil fuels using conservation measures alone.

20.5.8 *SUMMARY*

We have observational data providing an estimate of the global mean surface air temperature since about 1870. Although these data indicate a global warming of 0.3 to 0.6°C, this is significantly less than climate models predict (0.7 to 1.4°C) should have occurred by now as a result of human additions of GHGs to the atmosphere. In addition, most of the observed warming occurred before World War II, before the bulk of the GHGs had been emitted. To argue that the model-predicted warming has been obscured by a natural cooling cycle, requires a natural cooling of more than 1°C over the past few decades—a cooling larger than any observed before in the documented climate history of Europe, which extends at some stations into the sixteenth century.

Climate models predict that GHW should increase with latitude and be greatest in polar latitudes in winter. Since 1957, when regular observations began in Antarctica, neither pole has shown significantly more warming than the tropics.

Since the time of the dinosaurs, the Earth has cooled by about 10°C. About 3 million years ago we entered the present ice age, marked by a series of glacial/interglacial cycles with a mean global temperature range estimated by 5 to 7°C. The last seven of these cycles exhibited a glacial period of 90,000 years of staged cooling followed by rapid warming back to an interglacial lasting 10,000 to 12,000 years. We are currently in one of those interglacials, called the Holocene, estimated to have begun 10,700 YBP. Thus, we are due to enter another 90,000-year period of glacial cooling at any time. Because this is our current state of knowledge, why do we not hear the argument that adding GHGs to the atmosphere to delay, and thereby hopefully to prevent, the next glacial is just what we should be doing?

During the Holocene, temperature fluctuations of ±1 to 2°C have been inferred from various types of data. The warmest period, the Climatic Optimum, occurred about 6,000 YBP. Around 900 AD the sea ice in the North Atlantic melted back, and the Norsemen were able to colonize Iceland and Greenland. About 1350 AD the ice returned, the Greenland colony perished and glaciers advanced over rent-paying farms and villages in Switzerland and Scandinavia. Thus, we have both historical and paleoclimatological data supporting the "Medieval Climatic Optimum" warm period circa 900–1350 AD and the following "Little Ice Age" cold period circa 1350–1850 AD.

The least controversial explanation for the warming shown by our surface temperature observations is that it represents a return to normal from the Little Ice Age and possibly, entry into the next warm period of the Holocene. If this is what is occurring, we are now at the inflection point—or point of most rapid rise—of the temperature curve and we can look forward to a further warming of about another degree C over the next couple of centuries, regardless of what humans do. Note also that the warm periods of the past have been termed "climatic optima," and so they must have appeared to the remnants of the Greenland colony circa 1350 AD.

The latest Intergovenmental Panel on Climate Change Report (IPCC95) claims that "The balance of evidence suggests a discernible human influence on global climate." This claim elicited much controversy. Its lack of meaning is indicated by the Richard Kerr (*Science*, 276: 1040–1042, 1997) and Klaus Hasselmann (*Science*, 276: 914–915, 1997) articles in May 1997. These articles contain statements like the following:

" . . . many climate experts caution that it is not at all clear yet that human activities have begun to warm the planet—or how bad greenhouse warming will be when it arrives." Richard Kerr, Research News & Comment Writer for *Science*.

"We say quite clearly that few scientists would say the attribution issue was a done deal." Benjamin D. Santer, Lead Author of Section 8 of IPCC95.

"The inherent statistical uncertainties in the detection of anthropogenic climate change can be expected to subside only gradually in the next few years while the predicted signal is still slowly emerging from the natural climate variability noise. It would be unfortunate if the current debate over this ultimately transitory issue should distract from the far more serious problem of the long-term evolution of global warming once the signal has been unequivocally detected above the background noise." Klaus Hasselmann, Max-Planck-Institute for Meteorology.

From these acknowledgments by its supporters, it seems quite clear that the IPCC95 (p. 4) statement, "The balance of evidence suggests a discernible human influence on global climate," was studiously crafted. The purpose of the wording appears to have been to induce the media to broadcast to the public and policymakers of the world a message that few, if any, of the researchers, on whose work it was based, are yet willing to defend before the scientific community.

The Kyoto Protocol of December 1997, was not adopted because of compelling scientific evidence requiring action be taken to ward off environmental calamity. The theory of human-induced GHW is far from being an open-and-shut case. Although the concept of GHW is well-established, its actuality on planet Earth currently rests only on the predictions of climate models.

Models of GHW are especially oversimplified for the tropics because of their inability to cope with the very low humidities in the downwelling portions of the Hadley circulation. Any surface warming in the tropics will lead to acceleration of downdrafts of convectively dried air, opening deeper and enlarged subtropical "windows" facilitating escape of infrared radiation to space. As a result, model predictions of global GHW in the tropics seem exaggerated by a factor of at least two to three times.

Efforts to construct a "fingerprint" of human-made GHW have also done little to enhance the credibility of gloom and doom GHW forecasts. Even IPCC90 (p. 254) was forced to conclude: (1) "[I]t is not possible at this time to attribute all, or even a large part, of the observed global mean warming to enhanced greenhouse effects on the basis of observational data currently available;" and (2) "[W]e do not yet know what the detailed 'signal' [fingerprint] looks like because we have limited confidence in our predictions of climate change patterns" (IPCC90, p. xxix).

Moreover, there is a consistent bias against good news. Very reputable scientists are engaged in global climate research. However, their careful statements are either suppressed or ignored when they do not support dire scenarios. The United Nations, of course, has tremendous incentives to promote widespread dissemination of bad news. Global doom scenarios enhance its prospects of becoming a transfer agent for massive flows of wealth from developed to developing nations and an *actual* global government.

Until global climate change models improve significantly, we will not know whether we have a serious threat to our planet, no threat, or even a beneficial outcome as a result of man-made GHG emissions. In short, despite pronouncements by political leaders in the United States, Europe, and elsewhere, policymakers are not being compelled to act *on the basis of GHW science*. Indeed GHW science is still evolving.

REFERENCES

Idso, S. B., *Carbon Dioxide and Global Change: Earth in Transition*, IBR Press, Tempe, AZ, 1989.

Idso, S. B., *CO₂ and the Biosphere: The Incredible Legacy of the Industrial Revolution*, Department of Soil, Water & Climate, University of Minnesota, St. Paul, MN, 1995.

IPCC90 (Intergovernmental Panel on Climate Change), *Climate Change, the IPCC Scientific Assessment*, Houghton, J. T., Jenkins, G. J., and Ephraums, J. J., (eds.), Cambridge University Press, Cambridge, UK, 1990.

IPCC92 (Intergovernmental Panel on Climate Change), *Climate Change 1992, The Supplementary Report to The IPCC Scientific Assessment*, Houghton, J. T., Callander, B. A., and Varney, S. K. (eds.), Cambridge University Press, Cambridge, UK, 1992.

IPCC95 (Intergovernmental Panel on Climate Change), *Climate Change 1995, The Science of Climate Change*, Houghton, J. T., Meira Filho, L. G., Callander, B. A., Harris, N., Kattenberg, A., and Maskell, K. (eds.), Cambridge University Press, Cambridge, UK, 1996).

CHAPTER 20
SENSITIVE ENVIRONMENTAL PROBLEMS

SECTION 20.6

CONTROVERSIES SURROUNDING THE ENDANGERED SPECIES

J. Gordon Edwards

Dr. Edwards is emeritus professor of biology at San Jose State University. He taught entomology there for 49 years and is now curator of the University's entomology collection.

Seeking to halt the future extinction of incompetent animal and plant species, the U.S. Congress passed the "Endangered Species Act" in 1973. To many good people, it seemed like a wonderful, compassionate gesture, and every animal lover was pleased with it. However, very few of us actually read the text of the Act. A powerful group of dedicated antihumans had been given almost unlimited powers to enforce regulations that were so loosely written that they could mean whatever the administrators wanted them to mean. The "species" were not really "species," or even "subspecies," and "harm" was never defined. However, the leaders could fine or imprison people who they thought caused "harm" to any "species," or to the actual or potential habitat of a so-called "species," either accidentally or intentionally, and either directly or indirectly.

Thousands of public and private projects and activities were delayed or junked. Forestry was largely halted, and agriculturists became convicted as the enemy of the environment. Because of the Endangered Species Act, billions of dollars have been wasted, hundreds of thousands of jobs have been lost, and millions of acres of private land have been rendered useless to the owners. A few statements by so-called environmentalists reveal some troubling ethics and values:

1. The Audubon magazine editorialized that "There should be no such thing as private property in this country."

2. Also, on 17 July 1993, Audubon President Peter Berle stated (at the premier of TV show "Backlash In The Wild") that "We reject the idea of property rights."

3. Dave Foreman (fifth highest officer in the Sierra Club) said, "We must tear down the dams, free the shackled rivers, reclaim the roads and the plowed lands, and return to wilderness tens of millions of acres of presently settled lands" (*Field Guide to Monkeywrenching*, 1987).

4. Jacques Cousteau (in UNESCO *Courier*, November 1991) proposed that, "In order to stabilize world populations, we must eliminate 350,000 people per day . . . It is a horrible thing to say, but it's just as bad not to say it."

5. David Graber (National Park Biologist) simply said, "Somewhere along the line we people became a cancer . . . a plague upon ourselves and upon the earth. Some of us can only hope for the right virus to come along upon the earth" (*Los Angeles Times*, 22 Oct 1989).

6. Prince Philip said "If I were to be reincarnated, I would wish to be returned to earth as a killer virus, to lower human population levels" (In *The Greenhouse Effect Hoax*, 1989).

7. Norman Myers expressed his desire for management of the planetary ecosystem in a manner that "mobilizes earth's resources so as to provide sustainable benefits for humankind, with its numbers regulated in accord with the carrying capacity of the biosphere."

8. Paul Ehrlich said, "The population of the U.S. will shrink from 250 million to about 22.5 million before 1999, because of famine and global warming."

These statements may indicate why supporters of the Endangered Species Act made their plans so craftily, and what they hope might be accomplished if the Act could be fully implemented. Now, let's examine the Endangered Species Act and its supporters more carefully.

On June 3, 1970, the Endangered Species Conservation Act of 1969 became law. This was replaced by the Endangered Species Act of 1973 (Title 16, U.S. Code, Sections 1531–1544), which stated that: "The Congress finds and declares that: (1) various species of fish, wildlife and plants in the United States have been rendered extinct as a consequence of economic growth and development untempered by adequate concern and conservation; (2) other species of fish, wildlife and plants have been so depleted in numbers that they are in danger or threatened with extinction; and (3) these species of fish, wildlife and plants are of esthetic, ecological, educational, historical, recreational, and scientific value to the Nation and its people." The Act directed federal agencies to make sure their actions would not "jeopardize the continued existence of an endangered or threatened species, or result in the adverse modification or destruction of a critical habitat." (Later it was modified, to apply to nonfederal groups and private property, *and* included protection of nonvertebrate animals, including slugs, clams, shrimps, insects, and hundreds of creatures of questionable esthetic, historical, ecological, and scientific value.)

The Fish and Wildlife Service (FWS) was made primarily responsible for proposing the "endangered" status for selected plants and animals. (The National Marine Fisheries Service may play a minor role.) Section 1533 of the Code established those criteria for determining when a species might be listed as endangered: (1) the present or threatened destruction, modification, or curtailment of its habitat or range; (2) overutilization for commercial, sporting, scientific or educational purposes; (3) risks from disease or predation; (4) inadequacy of existing regulatory mechanisms; or (5) other natural or manmade factors threatening its continued existence (*Federal Register*, 40: 47506, 9 October 1975).

In U.S. Federal Code, Title 16, Section 1532, it was stated that: "The term "endangered species" means any species in danger of extinction throughout all or a significant portion of its range. In 1979, the Fish and Wildlife Service stated: "Species endangered or threatened in a specific region, *however abundant as a whole*, may currently halt or delay projects" (emphasis added). There was great opposition to such unreasonable actions by the Fish and Wildlife Service, and the General Accounting Office contended that endangered species funds were being used improperly "to acquire habitats for species that are *not* highly threatened."

It was also felt that the program went too far when it threatened essential public projects such as highway construction, dams, reservoirs, erosion control, and reclamation efforts. Of major concern to private landowners, company employees, and organized labor, was the failure of the Act to make provisions for the economic and social consequences of protecting threatened and endangered species. In November 1977, President Reagan saluted the Pacific Legal Foundation for successfully fighting special interest groups who "are attempting to misuse the Endangered Species Act to halt badly needed energy and water projects in America."

20.6.1   THE DEFINITION OF BIOLOGICAL "SPECIES"

Reliable recognition of different "species" of animals and plants has been the foundation of a rational awareness of the biological world around us. Scientific journals continue to use the term correctly, describing legitimate species in nearly every issue, and discussing those legitimate species in ecological and biological research reports.

For nearly a century scientists around the world have recognized that biological species must be "reproductively isolated" and "genetically distinctive" natural populations. This *scientific* categorization of species classification bears no resemblance to the *political* use of the word in the Endangered Species Act. Environmental extremists have distorted the word "species" to uselessness with their "endangered species" distortions. Members of real species cannot successfully interbreed with members of other species, consequently each species' gene pool remains relatively stable. The International Commission on Zoological Nomenclature regularly publishes the approved procedures for describing and correctly naming species of animals. Dozens of strict rules adopted by leading scientists, worldwide, regulate the procedure for naming legitimate species. The detailed description must be accompanied by a scientific latinized species name that is permanent and cannot ever be used for any other species. It must be published in the acceptible format, in a scientific journal, before the species can be considered as a legitimate scientific entity. The specimens upon which the published description is based are designated as the "type specimens" and must be carefully preserved in responsible museums for future reference by scientists seeking accurate details about the species. Legitimate species may share a genus name with other species that are very similar. The species name is not permitted to be duplicated within that genus.

The "endangered species" proponents have violated *every* procedure and *every* scientific requirement that generations of scientists established regarding biological species! Consequently, it is usually not possible to find specific literature dealing with the physical characters, the range, the behavior, the food habits, or other vital information concerning those "species." The majority of them appear to be political artifacts!

20.6.2   *DEFINITION OF BIOLOGICAL SUBSPECIES*

Scientific "subspecies" are subdivisions of legitimate species that are *not* "reproductively isolated." Their mature members are capable of breeding successfully with members of other subspecies in the same species. If they do not occur in the same range (or do not mature at the same time) they may not have the opportunity to interbreed, but their lack of interbreeding is not due to *intrinsic* genetic differences. The procedures for naming legitimate scientific subspecies are the same as those for naming distinct species, but in addition to the genus and species names there must be appended a third latinized name. Each subspecies must have the three latinized names (the genus, the species and the subspecies names.) No other population in the world can bear that same combination of names.

20.6.3   *DEFINITION OF POLITICAL "SPECIES"*

Political environmentalists have spurned the legitimate scientific requirements for the recognition of "real species." Instead, they consider any loose assortment of similar individuals to be a "species," and the identity of such so-called "species" is vague, uncertain, and meaningless. The Endangered Species Act of 1973 is based on this spurious use of the word "species." Title 16 of U.S. Code, Section 1532, stated that: "The term 'species' includes any subspecies of fish or wildlife or plant, and any distinct population segment of any species of vertebrate fish or wildlife which interbreeds when mature." That definition was altered in 1975 to read as follows: "The term 'species' includes any subspecies of fish or wildlife or plants, and any other group of fish or wildlife of the same species or smaller taxa in common spatial arrangement that interbreed when mature" (*Federal Register*, 40: 17505, 9 October 1975). Notice the omission of the requirement that species be "reproductively isolated" from other species, and the lack of genetic standards that should distinguish "species" from "subspecies" or lower infrasubspecific categories. Based on their political definition, the FWS can call any local group of similar individuals a "species," and if the group is listed as "threatened" or "endangered," many human activities may be banned in areas actually or potentially occupied by them. If scientists relied on such a definition of "species," it would be impossible for anyone anywhere in the world to accurately identify "species" of any animals or plants. The entire world would be populated by unrecognizable hordes of birds, mammals, reptiles, amphibians, insects, spiders, mites, crustaceans and plants, with only a jumble of vague local names that would be meaningless to anyone seeking to engage in scientific study or research about any species.

The GAO (Government Accounting Office) urged Congress to amend the Endangered Species Act and redefine "species" to exclude "local populations," even if they were rather distinct. (National Environmental Development Association (NEDA), *Washington Balance*, 27 July 1979). This was rejected by the FWS, and eventually more than 30 percent of the listed "endangered species" were merely indistinct local populations of animals and plants.

A bill sponsored by Representatives Billy Tauzin (D-LA), Don Young (R-AK), and many others proposed limiting protection to *real* species, instead of subspecies and local populations. They would require scientific standards to be used when enforcing the Act! They failed.

20.6.4   THREATENED SPECIES, UNDER THE POLITICAL ENDANGERED SPECIES ACT

In Title 16 U.S. Code, Section 1532 (20), it was specified that: "The term "threatened species" means any species which is likely to become endangered within the foreseeable future throughout all or a significant portion of its range." (It may be very common, elsewhere.) The FWS may propose the "threatened" status for animals and plants that are not "endangered," but might in the future be considered endangered. When it was reported that peregrine falcons and grizzly bears might be downlisted to "threatened" status, many people were pleased. They did not realize that "threatened species" have the same legal protection as "endangered species"!

20.6.5   POLITICAL SUBSPECIES

In the Endangered Species Act, "subspecies" may simply have common names, with no taxonomic priority and no scientific descriptions. A great number of Endangered Species Act "species" are not even "subspecies" but instead are indistinct "local populations." Detailed records reveal that over 40 percent of the ESA "endangered species" are only subspecies or local populations. The identity of such local populations is vague, uncertain, and meaningless.

Based on their political definition of "species," the Fish and Wildlife Service can call any insignificant local population of individuals an endangered "species." If they think the so-called "species" faces threats, the FWS now has the legal power to ban most human activities in areas actually (or potentially) occupied by the "species." They can also levy tremendous penalties against anyone refusing to submit to their regulations. To expand their power, they now claim that there are thousands of "endangered species" awaiting FWS "listing" and protection.

When the FWS became even more aggressive, and applied the Endangered Species Act to private lands, Representatives William Dannemeyer and James Hansen proposed Bills intended to balance values between humans and wildlife. Before halting human activities, FWS would have to calculate the economic effects of the preservation efforts, such as the 10-million-dollar loss to farmers if irrigation is halted to make things better for the short-nosed suckers. If such figures were publicized, citizens would become more concerned, and environmental activists would be less likely to cripple essential human activities by enforcing the Endangered Species Act.

A few examples of animals that have already been "listed" as "endangered species," (even though they are not species at all) are: Florida panthers, Eastern timber wolf, Columbian whitetail deer, Sonoran pronghorn, Florida scrub jays, California clapper rails, Tecopa pupfish, San Francisco garter-snakes, Santa Cruz long-toed salamanders, Louisiana black bear, Del Marva fox squirrels, Indiana bats, Tipton kangaroo rats, Illinois mud turtles, Arizona jaguars, Mt. Graham red squirrels, northern spotted owls, lost river suckers, razorback suckers, bonytail chubs, humpback chubs, delta smelt, Snake River snails, Colorado squawfish, and several kinds of "blue" butterflies. Also included are large numbers of closely related fish and 10 percent of all freshwater mussels, including the Alabama lamp pearly mussel, Appalachian monkeyface pearly mussel, Birdwing pearly mussel, Cumberland bean pearly mussel, Cumberland monkeyface pearly mussel, Curtis' pearly mussel, Dromedary pearly mussel, Greenblossom pearly mussel, Higgin's eye pearly mussel, Nicklin's pearly mussel, Orange-footed pearly mussel, Pale lilliput pearly mussel, Pink mucket pearly mussel, Tampico pearly mussel, Tuberculed-blossom pearly mussel, Turgid-blossom pearly mussel, White cat's paw pearly mussel, White wartyback pearly mussel, and others. (*Federal Register*, 41: 47197, 27 Oct 76). The expense of regulating such socalled "species" is great, and the adverse effects on human activities is astounding!

Alston Chase a prominent naturalist author wrote that, "As a consequence (of the Endangered Species Act) this nation may think it is preserving biological diversity when it is not. And worse, it risks hurting the economy by efforts to save populations that are not unique at all" (*Great Falls Tribune*, 30 July 1991).

In 1991 HR 4058 (the Balanced Economic and Environmental Priorities Act) was considered. Before listing a species as endangered or threatened, the Secretary of Interior would be required to publish an Economic Impact Analysis of associated costs and benefits and include designation of "critical habitat." The government would then be required to compensate people who suffered lost jobs and devalued property resulting from the Endangered Species Act. That would have helped thousands of citizens, but the losses were so great that the government could not possibly compensate owners for them. Thousands of citizens simply lost their property, funds, profits, jobs, and "pleasures of life," as a result.

Once listed, any "endangered species" and its "critical habitat" are protected from potential harm. However, where can concerned citizens find the lists of threatened and endangered species that may occur on their personal property, and ban all activities that might disturb the "species" or alter their "critical habitat"? U.S. Federal Code, Title 16, specified: "The Secretary of Interior shall publish in the *Federal Register* a list of all species determined by him or the Secretary of Commerce to be endangered species and a list of all species determined by him or the Secretary of Commerce to be threatened species. Each list shall refer to the species contained therein by scientific and common name or names, if any, specify with respect to each such species over what portion of its range it is endangered or threatened, and specify any critical habitat within such range." So, the information is available to every American who has easy access to the *Federal Register*. The 1996 *Federal Register* was 64,591 pages long, with 4937 final rules.

20.6.6   CRITICAL HABITATS OF ENDANGERED AND THREATENED SPECIES

Once listed, all "endangered or threatened species" *and* their habitats are protected from potential disruption. The Endangered Species Act requires the delineation of "critical habitats," to provide additional protection for the endangered or threatened populations. U.S. Federal Code, Title 16, Section 1533, stipulated: "A final regulation designating critical habitat of an endangered species or a threatened species shall be published concurrently with the final regulation implementing the determination that such species is endangered or threatened." It is important that the critical habitat be specified, because that habitat must also to be protected from human activities.

20.6.7   WHAT CONSTITUTES "TAKING" ENDANGERED OR THREATENED "SPECIES"?

U.S. Federal Code, Title 16, Section 1532(19) specified: "The term '*take*' means to harass, harm, pursue, hunt, shoot, wound, kill, trap, capture, or collect, or to attempt to engage in any such conduct." The term also embraces any destruction or alteration of the "critical habitat" of an endangered or threatened "species," "subspecies," or "local population." Anyone accused of "taking" an endangered species faces severe consequences, regardless of whether their actions occurred on public or private land. In one extreme example the FWS determined that a 'birder' (bird watcher) could be guilty of 'taking' a species simply by walking within a few hundred yards of the kind of habitat typically inhabited by the protected species, even if there are no individuals of that species actually present there (Reported in Cooperative Enterprise Institute's *UpDate*, November 1992).

20.6.8   "RARE SPECIES" VERSUS "ENDANGERED SPECIES"

Many populations of animals and plants have disappeared from North America, even in areas undisturbed by humanity. Degenerate evolution and failure to adapt to environmental changes continue to weed out unsuccessful forms of life through entirely natural processes. It is generally acknowledged that more than 90 percent of the animal "species" on earth became extinct millennia before humans appeared. Obviously, they all became "rare" before becoming "extinct." *Is* there any difference between "rare species" of animals or plants and "endangered species" of animals or plants? Obviously not. Should herculean efforts be made to "save" them all? Or should events be permitted to take their natural course? Tom Foley, Speaker of the House, complained that the Endangered Species Act "does not make any distinction between what kinds of animals should be protected, reinforcing an environmentalist utopian view that we should by federal law guarantee that no species would ever become extinct" (*Insight* magazine, August 1993).

At least 20 percent of the insect species that are being collected in some tropics are new to science. They are now being discovered and described in scientific literature as "new species" (using the *scientific* definition, rather than the political definition of the Endangered Species Act). More extensive collecting will be required to determine whether those tropical species are "rare" or "common," but it is obviously impossible to say with certainty how many of them are either "threatened" or "endangered." In 1979 Norman Myers, senior associate of World Wildlife Fund, wrote: "Two-thirds of all species live in the tropics. If the tropics were to become one big corn or rice patch, as seems probable, then in the next 50 years we'd lose one third of all species, and it is by no means inconceivable that we'd lose one half."

Some scientists have even said that "in the tropics, hundreds or thousands of species are becoming extinct before they have even been discovered." Obviously, there can be no basis for such statements!

20.6.9   NUMBERS OF THREATENED OR ENDANGERED SPECIES

The California Nature Conservancy estimates that 220 animals, 600 plants, and 200 natural communities are now threatened with severe reduction or extinction, statewide, and at least 21 species have become extinct in California in the last 100 years. They urge that we preserve the habitats that are home to the greatest number of species, but add that this is "Simple, but no one really knows where the habitats are and which of them are already protected."

In 1970, the Department of the Interior listed 101 "endangered species," nearly half of which were subspecies or local populations, rather than species. The list included 50 birds, 30 fish, 14 mammals, and 7 reptiles (*Federal Register*, 35: 8495–97, 2 June 1970).

In the continental United States, there were 116 animals listed as endangered in 1976, including 38 amphibians, 22 mammals, 23 mussels, 21 birds, 8 insects, and 5 reptiles (*Federal Register*, 41: 47181–98 & 51022). In Hawaii, there were 31 species on the list, almost all of which were birds. Elsewhere in the world, there were 646 endangered species in 1991, consisting of 292 species of mammals, 208 birds, 56 amphibians, 57 reptiles, 8 insects, and a snake (*NationalWildlife*, February 1991).

The International Union for the Conservation of Nature (IUCN), the scientific wing of the World Wildlife Fund, included 119 government agencies in 114 nations, and about 320 private conservation groups. According to the *Wall Street Journal* (5 October 1983), "Scientists began submitting invertebrate candidates for listing in 1980, but the IUCN ran into trouble trying to figure out how to select them. It was impossible in most cases to document declines in the population and range of even the most familiar species." The IUCN *Invertebrate Red Data Book* finally included 247 species of invertebrates, including 6 species of corals, 25 mussels and clams, 6 spiders, 4 flies, 11 ants, 21 butterflies, 7 beetles, and 43 species of snails. It also included 41 species of endemic Hawaiian land snails, in genus *Achatinella*, which had been threatened by overgrazing.

"The U.S. list soon included 588 plants, mammals, fishes, birds, beetles, butterflies, reptiles, frogs, snails, spiders and clams" (*Santa Rosa Press Democrat*, 2 September 1990). By 1991, 3600 more awaited designation and listing. At least 136 (7.4 percent) of all candidates for "listing" as endangered species are restricted to caves. Of 296 candidate snails, 90 (30 percent) are limited to Alabama alone, and most of them were described as species by just two biologists.

As of November 1991, the FWS listed 30 species and 198 subspecies or local populations as "endangered species." It was noteworthy that 87 percent were not scientific "species." Of the mammals listed as "endangered species" only 28 percent were species, while 72 percent (36 populations) were not.

The *People's Agenda*, November 1992, summarized the invertebrates of the United States that had been "listed" as threatened or endangered species by the FWS as follows: crustaceans—6.7 percent; insects—37.2 percent; gastropods—16 percent; arachnids—1.6 percent; and bivalves—4.1 percent. TOTAL— 65.6 percent of listed endangered or threatened species were invertebrates!

The *San Francisco Chronicle* reported that the FWS said "they would need $4.6 billion to carry out the law properly and rescue the hundreds of endangered species." Yet, the report says, "only $238 million has been allotted this year for that program."

In 1995, the Fish and Wildlife Service stated that of 908 plants and animals listed as endangered or threatened since 1973, 2 percent have become extinct, 38 percent are stable, 33 percent are still declining, and 27 percent have unknown status. The Endangered Species Act did not appear to be succeeding.

20.6.10   PLACING A VALUE ON A SPECIES

Thirteen environmental organizations that support the Endangered Species Act now have combined operating budgets of more than $336,000,000 and exert great pressure on the media and the legislature. Some of their opinions are contained in an article in *National Wildlife* (June/July 1993) by James Udall, asking "How Much is a Gray Wolf Worth?" "Environmentalists frequently claim that financial benefits outweigh the costs of protecting threatened and endangered species," the author says, and he then provides examples to counter such beliefs. "$18,000,000 is how much the return of wolves to Yellowstone could bring into the local economy the first year they are back in the park," he says, according to a study by a Montana economist. "And $69,000,000 is how much Californians spend on 2.3 million trips annually in order to see mule deer, according to a random survey of households," Udall asserts. Based on figures like these, Udall believes that the cost of supporting endangered species does *not* have a depressing economic effect.

20.6.11   THE ENDANGERED SPECIES ACT HAS UNLIMITED FUNDS

After a population of animals or plants has been "listed" by the FWS as "endangered" or "threatened," research funds flow freely to individuals who are selected to study them or their "critical habitats." This has been a major source of funding for hundreds of otherwise unemployed environmentalists during the past 25 years, and has resulted in a plethora of articles that urge that even larger budgets be provided for the Endangered Species Act.

20.6.12   SOME OF THE COSTS

In 1991, the Council on Environmental Quality estimated that it would cost $4.6 billion in taxpayer funds simply to review the status of all the species that may be at risk. Protecting those listed populations will cost billions more (*Great Falls Tribune*, 30 July 1991).

According to a 1985 U.S. Fish and Wildlife Service estimate, the average government expense for the "upkeep" of an endangered species was $2.6 million for each species. Federal and state agencies spent $102.3 million in 1990 for the study and protection of 477 threatened and endangered species. This did not include lost jobs, failed businesses, bans on growing crops or harvesting trees on private property etc. (*Mendocino CA Beacon*, 9 April 1992). Many specific examples are discussed later in the article.

The National Wilderness Institute revealed in Fall 1992 that there were 554 endangered or threatened species needing financial support. A few kinds always receive a major share of the available funds. The most expensive *subspecies* in 1990 are listed below:

1. Northern spotted owl (*Strix occidentalis caurina*) $9,687,200

2. Least Bell's vireo (*Vireo belli pusillus*) $9,168,800

3. Grizzly bear (*Ursus horribilis* or *arctos*) $5,882,500

4. Florida panther (*Felis concolor coryi* hybrid) $4,113,900

Where did the money come from that was spent on these "endangered so-called "species"? The Fish and Wildlife Service expenses included $11 million for recovery efforts and nearly $4 million just for "listing" species, subspecies, and local populations. In addition, the National Wilderness Institute cited the following figures for 1992:

U.S. Forest Service	$18,591,358	110 species
Federal Highway Administration	$13,757,050	6 species
Corps of Engineers	$8,082,200	134 species
Department of Defense	$5,206,850	67 species
Bureau of Reclamation	$4,945,245	20 species
National Oceanographic & Atmosph. Admin.	$4,061,000	14 species
National Park Service	$3,368,550	59 species
Bureau of Land Management	$2,390,600	68 species
Environmental Protection Agency	$2,320,000	128 species
Bureau of Indian Affairs	$1,244,980	15 species
Animal and Plant Health Inspection Service	$515,579	41 species
TOTAL	$64,483,412	

(Agencies spending less than $100,000 in 1990 are not listed.)

NOTICE: The greatest source of information regarding endangered species and all related topics is surely the National Wilderness Institute. In addition to their large issues of *National Wilderness Institute Resources*, they also publish an excellent periodical titled *Fresh Tracks*. For a membership fee of $25, all of those publications are provided. The address of NWI is P.O. Box 25766 Georgetown Station, Washington D.C. 20007 and their telephone is 703/836-7404.

There is a list of more than 5000 so-called "species" awaiting federal protection under the Endangered Species Act. Of those, about 1000 are now considered "Category One," and FWS says they should be listed, but are not because the process is underfunded. FWS said all of those should be *automatically* "listed" as endangered, after two years. At $60,000 for each species listed, the cost for the 1000 Category One cases would be $600,000,000 and for all 5000 species that are awaiting federal protection the basic cost would be $3,000,000,000. (Those figures do not include the financial losses suffered by property owners, the lost jobs, and other penalties that humans and communities must bear.) (*National Wilderness Institute Resources*, April 1992) Representative Gerry Studds, chairman of the House subcommittee with jurisdiction over the ESAct, introduced amendments in 1992 (HR 4045) that would increase FWS funding from $59 million (1993) to $100 million (by 1997), and add a $200 million habitat conservation loan fund. His bill also would hasten the listing of some 3000 animal and plant "species" as "threatened" or "endangered" (*National Wilderness Institute Resources*, April 1992).

20.6.13 REASONS TO FURTHER MODIFY THE ENDANGERED SPECIES ACT

With more than 1300 new species listed and 1500 proposed for listing as endangered, every community is at risk of having a plant or animal capable of stopping economic activity, ending growth, and 'taking' private property." "Most Americans think of bald eagles and grizzly bears as threatened or endangered species, but more than two thirds of the species proposed for listing are insects, spiders, clams, crustaceans, and other invertebrates. The threat to the rights, liberties, and livelihoods of rural Americans cannot be overstated." They are exposed to "a cruel and malevolent assault by environmental extremists who use a well intended but increasingly problematic federal statute to turn more and more of America into their private playground" (William P. Pendley, in *Summary Judgment*, January 1993, Mtn. States Legal Foundation).

In 1991 the U.S. Conference of Mayors endorsed the critic's call for "economic balance" in the Endangered Species Act. Manuel Lujan, Secretary of Interior, stated "Congress can boost the northern spotted owl population, and lose 31,000 timber jobs *or* just maintain the 3000 breeding pairs of owls that now exist there, at a cost of 20,000 jobs, *or* save only 2000 breeding pairs, and lose 12,000 jobs" (*San Jose Mercury*, 27 April 1992) They failed.

President Bush declared that he would not sign an extension of the ESAct "unless it gives greater consideration to jobs and to families." He said the Act was intended to shield species from the effects of construction projects like dams and highways, and not be aimed at jobs, families, and communities of entire regions. Bush favored requiring a cost-benefit analysis before listing a species as endangered (*San Francisco Chronicle*, 15 September 1992). He failed.

"The ESAct is offering a strong economic incentive for property owners to destroy endangered species and their habitats" (*San Jose Mercury News*, 30 April 1992). "People must avoid having endangered or threatened species on or near their property, and that often can be accomplished only by a quiet non-conformance that unfortunately thwarts desirable conservation" (Cooperative Enterprise Institute *Update*, November 1992).

Discovering a threatened or endangered species on personal property (or an insignificant "population" that is a potential candidate for listing) should alert landowners to the threat of financial ruin if government agents learn that it is there. If they cannot sell the property quickly, or somehow get rid of the threatening bugs, slugs, or rare plants, they may be barred from using parts of their property.

The American Farm Bureau Federation was also critical of the Endangered Species Act, frequently explaining that: The ESAct has been unsuccessful, has created bitter disputes, and has caused incredible difficulties, deprivation and, frequently, financial ruin. The Act is in need of a major overhaul, so that it can effectively balance species preservation with the economic well-being of people and communities.

The Farm Bureau Federation provides a statement they call *A Balanced Solution* which states that appropriate balance must be established between the needs of endangered or threatened "species" and the needs of people. It must include respect for private property and human rights, and the costs of protecting so-called "species" must be borne by public agencies rather than private landowners. The points that they feel should be considered as part of an amended ESAct are as follows:

1. A Private Property Rights Act should be enacted as part of the ESAct.

2. The ESAct should provide indemnification for damages or losses caused to persons and private property because of "listed" species on or near the property.

3. Listing a species should require adherence to *scientific* standards of taxonomy; The Act should apply to actual species, rather than vague "subspecies" or insignificant "local populations."

4. The ESAct should not include protection for so-called "species" that are merely candidates for listing.

5. Specific "critical habitat" for listed species should be designated only at the time of listing, and should be supported by adequate justification.

6. A listed species must show measurable progress toward recovery within 10 years, or lose its protected status.

7. The legal exemption process (to permit relaxation of protection of threatened or endangered species, under certain conditions) should be made more reasonable, less expensive, and more accessible to private citizens.

8. Financial incentives and technical assistance must be offered to private landowners and legitimate users of public lands, to enhance the recovery of listed species.

CHAPTER 21
MISCONCEPTIONS ABOUT ENVIRONMENTAL POLLUTION

SECTION 21.1

MISCONCEPTIONS ABOUT POLLUTION, PESTICIDES, AND THE PREVENTION OF CANCER

Bruce N. Ames

Dr. Ames is a professor of biochemistry and molecular biology and director of the National Institute of Environmental Sciences Center, University of California, Berkeley.

Lois Swirsky Gold

Dr. Gold is director of the Carcinogenic Potency Project at the National Institute of Environmental Health Sciences Center, University of California, Berkeley, and a senior scientist at the E.O. Lawrence Berkeley National Laboratory.

21.1.1 SUMMARY

The major causes of cancer are: 1) smoking, which accounts for about a third of United States cancer deaths and 90 percent of lung cancer deaths; 2) dietary imbalances, which account for about another third, e.g., lack of sufficient amounts of dietary fruits and vegetables. The quarter of the population eating the fewest fruits and vegetables has double the cancer rate for most types of cancer than the quarter eating the most; 3) chronic infections, mostly in developing countries; and 4) hormonal factors, which are influenced primarily by lifestyle. There is no cancer epidemic except for cancer of the lung resulting from smoking. Cancer mortality rates have declined 18 percent since 1950 (excluding lung cancer). Regulatory policy that focuses on traces of synthetic chemicals is based on misconceptions about animal cancer tests. Recent research indicates that rodent carcinogens are not rare. Half of all chemicals tested in standard high-dose animal cancer tests, whether occurring naturally or produced synthetically, are "carcinogens"; there are high-dose effects in rodent cancer tests that are not relevant to low-dose human exposures and that contribute to the high proportion of chemicals that test positive. The focus of regulatory policy is on synthetic chemicals, although 99.9 percent of the chemicals humans ingest are natural. More than 1000 chemicals have been described in coffee: 30 have been tested and 21 are rodent carcinogens. Plants in the human diet contain thousands of natural "pesticides" produced by plants to protect themselves from insects and other predators: 71 have been tested and 37 are rodent carcinogens.

There is no convincing evidence that synthetic chemical pollutants are important as a cause of human cancer. Regulations targeted to eliminate minuscule levels of synthetic chemicals are enormously expensive: the Environmental Protection Agency has estimated that environmental regulations cost society $140 billion/year. Others have estimated that the median toxic control program costs 146 times more per hypothetical life-year saved than the median medical intervention. Attempting to reduce tiny hypothetical risks has other costs as well: if reducing synthetic pesticides makes fruits and vegetables more expensive, thereby decreasing consumption, then the cancer rate will increase, especially for the poor. The prevention of cancer will come from knowledge obtained from biomedical research, education of the public, and lifestyle changes made by individuals. A re-examination of priorities in cancer prevention, both public and private, seems called for.

21.1.2   CLEARING UP CANCER MISCONCEPTIONS

Various misconceptions about the relationship between environmental pollution and human disease, particularly cancer, drive regulatory policy. We highlight nine such misconceptions and briefly present the scientific evidence that undermines each.

21.1.3   MISCONCEPTION #1: CANCER RATES ARE SOARING

Overall cancer death rates in the United States (excluding lung cancer from smoking) have declined 18 percent since 1950. The types of cancer deaths that have decreased since 1950 are primarily stomach, cervical, uterine, and colorectal. Those that have increased are primarily lung cancer (90 percent is due to smoking, as are 35 percent of all cancer deaths in the United States), melanoma (probably due to sunburns), and non-Hodgkin's lymphoma. If lung cancer is included, mortality rates have increased over time, but recently have declined in men because of decreased smoking (1). In women, breast cancer mortality rates have begun to decline in part because of early detection and improved survival. The rise in incidence rates in older age groups for some cancers can be explained by known factors such as improved screening. "The reason for not focusing on the reported incidence of cancer is that the scope and precision of diagnostic information, practices in screening and early detection, and criteria for reporting cancer have changed so much over time that trends in incidence are not reliable" (2, see also refs. 3 and 4). Life expectancy has continued to rise since 1950.

21.1.4   MISCONCEPTION #2: ENVIRONMENTAL SYNTHETIC CHEMICALS ARE AN IMPORTANT CAUSE OF HUMAN CANCER

Neither epidemiology nor toxicology supports the idea that synthetic industrial chemicals are important as a cause of human cancer (4–6). Epidemiological studies have identified the factors that are likely to have a major effect on lowering cancer rates: reduction of smoking, improving diet (e.g., increased consumption of fruits and vegetables), hormonal factors, and control of infections (6). Although some epidemiological studies find an association between cancer and low levels of industrial pollutants, the associations are usually weak, the results are usually conflicting, and the studies do not correct for potentially large confounding factors such as diet (7). Moreover, exposures to synthetic pollutants are tiny and rarely seem toxicologically plausible as a causal factor, particularly when compared to the background of natural chemicals that are rodent carcinogens (5). Even assuming that worst-case risk estimates for synthetic pollutants are true risks, the proportion of cancer that the United States Environmental Protection Agency (EPA) could prevent by regulation would be tiny (8). Occupational exposure to some carcinogens causes cancer, though exactly how much has been a controversial issue: a few percent seems a reasonable estimate (6), much of this from asbestos in smokers. Exposures to substances in the workplace can be much higher than the exposure to chemicals in food, air, or water. Past occupational exposures have sometimes been high, and therefore comparatively little quantitative extrapolation may be required for risk assessment from high-dose rodent tests to high-dose occupational exposures in order to assess risk. Because occupational cancer is concentrated among small groups with high levels of exposure, there is an opportunity to control or eliminate risks once they are identified; however, current permissible levels of exposure in the workplace are sometimes close to the carcinogenic dose in rodents (9).

Cancer is due, in part, to normal aging and increases exponentially with age in both rodents and humans (10). To the extent that the major external risk factors for cancer are diminished, cancer will occur at later ages, and the proportion of cancer caused by normal metabolic processes will increase. Aging and its degenerative diseases appear to be due in good part to oxidative damage to DNA and other macromolecules (10, 11). By-products of normal metabolism—superoxide, hydrogen peroxide, and hydroxyl radical—are the same oxidative mutagens produced by radiation. Mitochondria from old animals leak oxidants (12): old rats have about 66,000 oxidative DNA lesions per cell (13). DNA is oxidized in normal metabolism because antioxidant defenses, though numerous, are not perfect. Antioxidant defenses against oxidative damage include vitamins C and E and perhaps carotenoids (14), most of which come from dietary fruits and vegetables.

Smoking contributes to about 31 percent of United States cancer, about one-quarter of heart disease, and about 400,000 premature deaths per year in the United States (1, 6, 15). Tobacco is a known cause of cancer of the lung, bladder, mouth, pharynx, pancreas, stomach, larynx, esophagus, and possibly colon. Tobacco causes even more deaths by diseases other than cancer (16). Smoke contains a wide variety of mutagens and rodent carcinogens. Smoking is also a severe oxidative stress and causes inflammation in the lung. The oxidants in cigarette smoke—mainly nitrogen oxides—deplete the body's antioxidants. Thus, smokers must ingest two to three times more vitamin C than nonsmokers to achieve the same level in blood, but they rarely do. An inadequate concentration of vitamin C in plasma is more common among the poor and smokers. Men with inadequate diets or who smoke may damage both their somatic DNA and the DNA of their sperm. When the level of dietary vitamin C is insufficient to keep seminal fluid vitamin C at an adequate level, the oxidative lesions in sperm DNA are increased 250 percent (17–19). Male smokers have more oxidative lesions in sperm DNA (19) and more chromosomal abnormalities in sperm (20) than do nonsmokers. It is plausible, therefore, that fathers who smoke may increase the risk of birth defects and childhood cancer in offspring (17, 18, 21). An epidemiological study suggests that the rate of childhood cancers is increased in offspring of male smokers: acute lymphocytic leukemia, lymphoma, and brain tumors are increased three to four times (22).

We (6) estimate that unbalanced diets account for about one-third of cancer risk, in agreement with an earlier estimate of Doll and Peto (1, 3, 15). Low intake of fruits and vegetables is a major risk factor for cancer (see Misconception #3). There has been considerable interest in calories (and dietary fat) as a risk factor for cancer, in part because caloric restriction markedly lowers the cancer rate and increases life span in rodents (6, 23, 24).

Chronic inflammation from chronic infection results in the release of oxidative mutagens from phagocytic cells and is a major contributor to cancer (6, 25). White cells and other phagocytic cells of the immune system combat bacteria, parasites, and virus-infected cells by destroying them with potent, mutagenic oxidizing agents. These oxidants protect humans from immediate death from infection, but they also cause oxidative damage to DNA, chronic cell killing with compensatory cell division, and mutation (26, 27); thus they contribute to the carcinogenic process. Antioxidants appear to inhibit some of the pathology of chronic inflammation. Chronic infections cause about 21 percent of new cancer cases in developing countries and 9 percent in developed countries (28).

Endogenous reproductive hormones play a large role in cancer, including that of the breast, prostate, ovary, and endometrium (29, 30), contributing to as much as 20 percent of all cancer. Many lifestyle factors such as reproductive history, lack of exercise, obesity, and alcohol influence hormone levels and therefore affect risk (6, 29–31).

Other causal factors in human cancer are excessive alcohol consumption, excessive sun exposure, and viruses. Genetic factors also play a significant role and interact with lifestyle and other risk factors. Biomedical research is uncovering important genetic variation in humans.

21.1.5   MISCONCEPTION #3: REDUCING PESTICIDE RESIDUES IS AN EFFECTIVE WAY TO PREVENT DIET-RELATED CANCER

Reductions in synthetic pesticide use will not effectively prevent diet-related cancer. Fruits and vegetables are of major importance for reducing cancer; if they become more expensive because of reduced use of synthetic pesticides, cancer is likely to increase. People with low incomes eat fewer fruits and vegetables and spend a higher percentage of their income on food.

Dietary fruits and vegetables and cancer prevention. High consumption of fruits and vegetables is associated with a lowered risk of degenerative diseases including cancer, cardiovascular disease, cataracts, and brain dysfunction (6, 10). More than 200 studies in the epidemiological literature have been reviewed that show, with great consistency, an association between low consumption of fruits and vegetables and cancer incidence (32–34) (Table 21.1.1). The quarter of the population with the lowest dietary intake of fruits and vegetables versus the quarter with the highest intake has roughly twice the cancer rate for most types of cancer (lung, larynx, oral cavity, esophagus, stomach, colorectal, bladder, pancreas, cervix, and ovary). Eighty percent of American children and adolescents, and 68 percent of adults (35, 36) did not meet the intake recommended by the National Cancer Institute (NCI) and the National Research Council (NRC): five servings of fruits and vegetables per day. Publicity about hundreds of minor hypothetical risks can cause loss of perspective on what is important: half the United States population does not know that fruit and vegetable consumption is a major protection against cancer (37).

Some micronutrients in fruits and vegetables are anticarcinogens. Antioxidants in fruits and vegetables may account for some of their beneficial effect, as discussed in Misconception #2. However,

TABLE 21.1.1   Review of Epidemiological Studies on Cancer Showing Protection by Consumption of Fruits and Vegetables[a]

Cancer site	Fraction of studies showing significant cancer protection	Median relative risk of low versus high quartile of consumption
Epithelial		
Lung	24/25	2.2
Oral	9/9	2.0
Larynx	4/4	2.3
Esophagus	15/16	2.0
Stomach	17/19	2.5
Pancreas	9/11	2.8
Cervix	7/8	2.0
Bladder	3/5	2.1
Colorectal	20/35	1.9
Miscellaneous	6/8	—
Hormone-dependent		
Breast	8/14	1.3
Ovary/endometrium	3/4	1.8
Prostate	4/14	1.3
Total	129/172	

[a] From ref (32).

it is difficult to disentangle by epidemiological studies the effects of dietary antioxidants from effects of other important vitamins and ingredients present in fruits and vegetables (33, 34, 38).

Folate deficiency, one of the most common vitamin deficiencies, causes chromosome breaks in human genes (39). Approximately 10 percent of the United States population (40) has a blood folate level lower than that at which chromosome breaks can occur (39). In two small studies of low-income (mainly African-American) elderly persons (41) and adolescents (42), nearly half had folate levels that low, but these studies should be repeated. The mechanism of damage is deficient methylation of uracil to thymine and subsequent incorporation of uracil into human DNA (4 million/cell) (39). During repair of uracil in DNA, transient nicks are formed; two opposing nicks cause a chromosome break; thus, folate deficiency mimics radiation. High DNA uracil levels and chromosome breaks in humans are both reversed by folate administration (39). Chromosome breaks could contribute to the increased risk of cancer and cognitive defects associated with folate deficiency in humans (39). Folate deficiency also damages human sperm (43), causes neural tube defects in the fetus, and is responsible for about 10 percent of the risk for heart disease in the United States (44). Low folate intake is associated with a higher risk of breast cancer among women who regularly consume alcohol (1 drink/day) (45).

Micronutrients whose main dietary sources are other than fruits and vegetables, are also likely to play a significant role in the prevention and repair of DNA damage, and thus are important to the maintenance of long-term health (7). Deficiency of vitamin B_{12} causes a functional folate deficiency, accumulation of homocysteine (a risk factor for heart disease) (46), and misincorporation of uracil into DNA (47). Strict vegetarians are at increased risk for developing vitamin B_{12} deficiency (46). Niacin contributes to the repair of DNA strand breaks by maintaining nicotinamide adenine dinucleotide levels for the poly ADP-ribose protective response to DNA damage (48). As a result, dietary insufficiencies of niacin (15 percent of some populations are deficient) (49), folate, and antioxidants may interact synergistically to adversely affect DNA synthesis and repair. Diets deficient in fruits and vegetables are commonly low in folate, antioxidants, (e.g., vitamin C), and many other micronutrients, and result in DNA damage and higher cancer rates (6, 7, 32, 50).

Optimizing micronutrient intake can have a major effect on health at a low cost (7). More research in this area as well as efforts to increase micronutrient intake and to improve diets should be high priorities for public policy.

21.1.6   *MISCONCEPTION #4: HUMAN EXPOSURES TO CARCINOGENS AND OTHER POTENTIAL HAZARDS ARE PRIMARILY TO SYNTHETIC CHEMICALS*

Contrary to common perception, 99.9 percent of the chemicals humans ingest are natural. The amounts of synthetic pesticide residues in plant foods, for example, are insignificant compared to the amount of natural "pesticides" produced by plants themselves (51–55). Of all dietary pesticides that humans eat, 99.99 percent are natural: these are chemicals produced by plants to defend themselves against fungi, insects, and other animal predators (51, 52, 56). Each plant produces a different array of such chemicals. On average, Americans ingest roughly 5000 to 10,000 different natural pesticides and their breakdown products. Americans eat about 1500 mg of natural pesticides per person per day, which is about 10,000 times more than they consume of synthetic pesticide residues.

Even though only a small proportion of natural pesticides has been tested for carcinogenicity, half of those tested (37/71) are rodent carcinogens; naturally occurring pesticides that are rodent carcinogens are ubiquitous in fruits, vegetables, herbs, and spices (5, 53) (Table 21.1.2).

Cooking of foods produces burnt material (about 2000 mg per person per day) that contains many rodent carcinogens. In contrast, the residues of 200 synthetic chemicals measured by Federal Drug Administration, including the synthetic pesticides thought to be of greatest importance, average only about 0.09 mg per person per day (5, 51, 53). In a single cup of coffee, the natural chemicals that are rodent carcinogens are about equal in weight to an entire year's worth of synthetic pesticide residues that are rodent carcinogens, even though only 3 percent of the natural chemicals in roasted coffee

TABLE 21.1.2   Carcinogenicity of Natural Plant Pesticides Tested in Rodents[a]

Carcinogens:

$N = 37$ acetaldehyde methylformylhydrazone, allyl isothiocyanate, arecoline.HCl, benzaldehyde, benzyl acetate, caffeic acid, capsaicin, catechol, clivorine, coumarin, crotonaldehyde, 3,4-dihydrocoumarin, estragole, ethyl acrylate, $N2$-γ-glutamyl-p-hydrazinobenzoic acid, hexanal methylformylhydrazine, p-hydrazinobenzoic acid.HCl, hydroquinone, 1-hydroxyanthraquinone, lasiocarpine, d-limonene, 3-methoxycatechal, 8-methoxypsoralen, N-methyl-N-formylhydrazine, α-methylbenzyl alcohol, 3-methylbutanal methylformylhydrazone, 4-methylcatechol, methylhydrazine, monocrotaline, pentanal methylformylhydrazone, petasitenine, quercetin, reserpine, safrole, senkirkine, sesamol, symphytine

Noncarcinogens:

$N = 34$ atropine, benzyl alcohol, benzyl isothiocyanate, benzyl thiocyanate, biphenyl, d-carvone, codeine, deserpidine, disodium glycyrrhizinate, ephedrine sulphate, epigallocatechin eucalyptol, eugenol, gallic acid, geranyl acetate, β-N-[γ-l(+)-glutamyl]-4-hydroxymethylphenylhydrazine, glycyrrhetinic acid, p-hydrazinobenzoic acid, isosafrole, kaempferol, dl-menthol, nicotine, norharman, phenethyl isothiocyanate, pilocarpine, piperidine, protocatechuic acid, rotenone, rutin sulfate, sodium benzoate, tannic acid, 1-trans-δ^9-tetrahydrocannabinol, turmeric oleoresin, vinblastine

These rodent carcinogens occur in:

absinthe, allspice, anise, apple, apricot, banana, basil, beet, broccoli, Brussels sprouts, cabbage, cantaloupe, caraway, cardamom, carrot, cauliflower, celery, cherries, chili pepper, chocolate, cinnamon, cloves, coffee, collard greens, comfrey herb tea, corn, coriander, currants, dill, eggplant, endive, fennel, garlic, grapefruit, grapes, guava, honey, honeydew melon, horseradish, kale, lemon, lentils, lettuce, licorice, lime, mace, mango, marjoram, mint, mushrooms, mustard, nutmeg, onion, orange, paprika, parsley, parsnip, peach, pear, peas, black pepper, pineapple, plum, potato, radish, raspberries, rhubarb, rosemary, rutabaga, sage, savory, sesame seeds, soybean, star anise, tarragon, tea, thyme, tomato, turmeric, and turnip.

---

[a] Fungal toxins are not included. From the Carcinogenic Potency Database (54, 55).

TABLE 21.1.3   Carcinogenicity in Rodents of Natural Chemicals in Roasted Coffee[a]

Positive:

$N = 21$ acetaldehyde, benzaldehyde, benzene, benzofuran, benzo(a)pyrene, caffeic acid, catechol, 1,2,5,6-dibenzanthracene, ethanol, ethylbenzene, formaldehyde, furan, furfural, hydrogen peroxide, hydroquinone, isoprene, limonene, 4-methylcatechol, styrene, toluene, xylene

Not positive:

$N = 8$ acrolein, biphenyl, choline, eugenol, nicotinamide, nicotinic acid, phenol, piperidine

Uncertain: caffeine

Yet to test: ~1000 chemicals

---

[a] From the Carcinogenic Potency Database (54, 55).

have been adequately tested for carcinogenicity (5) (Table 21.1.3). This does not mean that coffee or natural pesticides are dangerous, but rather that assumptions about high-dose animal cancer tests for assessing human risk at low doses need reexamination. No diet can be free of natural chemicals that are rodent carcinogens (53–55).

21.1.7   MISCONCEPTION #5: CANCER RISKS TO HUMANS CAN BE ASSESSED BY STANDARD HIGH-DOSE ANIMAL CANCER TESTS

Approximately half of all chemicals that have been tested in standard animal cancer tests, whether natural or synthetic are rodent carcinogens (5, 55, 57) (Table 21.1.4). Why such a high positivity rate?

TABLE 21.1.4   Proportion of Chemicals Evaluated as Carcinogenic[a]

Chemicals tested in both rats and mice[a]	330/590	(59%)
Naturally occurring chemicals	79/139	(57%)
Synthetic chemicals	271/451	(60%)
Chemicals tested in rats and/or mice[a]		
Chemicals in Carcinogenic Potency Database	702/1348	(52%)
Natural pesticides	37/71	(52%)
Mold toxins	14/23	(61%)
Chemicals in roasted coffee	21/30	(70%)
Innes negative chemicals retested[a,b]	17/34	(50%)
Physician's Desk Reference (PDR):		
drugs with reported cancer tests[c]	117/241	(49%)
FDA database of drug submissions[d]	125/282	(44%)

[a] From the Carcinogenic Potency Database (54, 55).
[b] The 1969 study by Innes et al. (94) is frequently cited as evidence that the proportion of carcinogens is low, because only 9 percent of 119 chemicals tested (primarily pesticides) were positive. However, the Innes tests were only in mice and had few animals per group, thus lacking the power of modern tests. Of the 34 Innes negative chemicals that have been retested using modern protocols, 17 were positive.
[c] Davies and Monro (95).
[d] Contrera et al (96). 140 drugs are in both the FDA and PDR databases.

In standard cancer tests, rodents are given chronic, near-toxic doses, the maximum tolerated dose (MTD). Evidence is accumulating that cell division caused by the high dose itself, rather than the chemical per se, is increasing the positivity rate. High doses can cause chronic wounding of tissues, cell death, and consequent chronic cell division of neighboring cells, which is a risk factor for cancer (58). Each time a cell divides, the probability increases that a mutation will occur, thereby increasing the risk for cancer. At the low levels to which humans are usually exposed, such increased cell division does not occur. In addition, tissues injured by high doses of chemicals (e.g., phenobarbital, carbon tetrachloride, tetradecanoylphorbol acetate) have an inflammatory immune response involving activation of recruited and resident macrophages in response to necrosis (59–65). Activated macrophages release mutagenic oxidants (including peroxynitrite, hypochlorite, and H_2O_2). Therefore, the very low levels of chemicals to which humans are exposed through water pollution or synthetic pesticide residues may pose no or only minimal cancer risks.

We have discussed (66) the argument that the high positivity rate is due to selecting more suspicious chemicals to test, which is a likely bias because cancer testing is both expensive and time-consuming, and it is prudent to test suspicious compounds. One argument against selection bias is the high positivity rate for drugs (Table 21.1.4), because drug development tends to select chemicals that are not mutagens or expected carcinogens. A second argument against selection bias is that knowledge to predict carcinogenicity in rodent tests is highly imperfect, even now, after decades of testing results have become available on which to base prediction. For example, a prospective prediction exercise was conducted by several experts in 1990 in advance of the 2-year National Toxicology Program (NTP) bioassays. There was wide disagreement among the experts as to which chemicals would be carcinogenic when tested; accuracy varied, thus indicating that predictive knowledge is highly uncertain (67). Moreover, if the main basis for selection were suspicion rather than human exposure, then one should select mutagens (80 percent are positive compared to 49 percent of nonmutagens), yet 55 percent of the chemicals tested are nonmutagens (66).

It seems likely that a high proportion of all chemicals, whether synthetic or natural, might be "carcinogens" if run through the standard rodent bioassay at the MTD: primarily for the nonmutagens, carcinogenicity would be due to the effects of high doses; for the mutagens, it would result from a synergistic effect between cell division at high doses and DNA damage (68–70). Without additional data on the mechanism of carcinogenesis for each chemical, the interpretation of a positive result

TABLE 21.1.5   Ranking Possible Carcinogenic Hazards from Average U.S. Exposures (55, 57). Chemicals that occur naturally in foods are in bold. *Daily human exposure:* The calculations assume an average daily dose for a lifetime. *Possible hazard:* The human exposure to a rodent carcinogen is divided by 70 kg to give a mg/kg/day of human exposure, and this dose is given as the percentage of the TD_{50} in the rodent (mg/kg/day) to calculate the *H*uman *E*xposure/*R*odent *P*otency index (HERP), i.e., 100 percent means that the human exposure in mg/kg/day is equal to the dose estimated to give 50 percent of the rodents tumors. TD_{50} values used in the HERP calculation are averages calculated by taking the harmonic mean of the TD_{50}s of the positive tests in that species from the Carcinogenic Potency Database. Average TD_{50} values have been calculated separately for rats and mice, and the more potent value is used for calculating possible hazard. The less potent value is in parentheses.

Possible hazard: HERP (%)	Average daily U.S. exposure	Human dose of rodent carcinogen	Potency TD_{50} (mg/kg/day)[a] Rats	Mice
140	EDB: workers (high exposure) (before 1977)	Ethylene dibromide, 150 mg	1.52	(7.45)
17	Clofibrate	Clofibrate, 2 g	169	·
14	Phenobarbital, 1 sleeping pill	Phenobarbital, 60 mg	(+)	6.09
6.8	1,3-Butadiene: rubber workers (1978–86)	1,3-Butadiene, 66.0 mg	(261)	13.9
6.1	Tetrachloroethylene: dry cleaners with dry-to-dry units (1980–90)	Tetrachloroethylene, 433 mg	101	(126)
4.0	Formaldehyde: workers	Formaldehyde, 6.1 mg	2.19	(43.9)
2.4	Acrylonitrile: production workers (1960–1986)	Acrylonitrile, 28.6 mg	16.9	·
2.2	Trichloroethylene: vapor degreasing (before 1977)	Trichloroethylene, 1.02 g	668	(1580)
2.1	**Beer, 257 g**	**Ethyl alcohol, 13.1 ml**	9110	(—)
1.4	Mobile home air (14 hours/day)	Formaldehyde, 2.2 mg	2.19	(43.9)
0.9	Methylene chloride: workers (1940s–80s)	Methylene chloride, 471 mg	724	(918)
0.5	**Wine, 28.0 g**	**Ethyl alcohol, 3.36 ml**	9110	(—)
0.4	Conventional home air (14 hours/day)	Formaldehyde, 598 μg	2.19	(43.9)
0.1	**Coffee, 13.3 g**	**Caffeic acid, 23.9 mg**	297	(4900)
0.04	**Lettuce, 14.9 g**	**Caffeic acid, 7.90 mg**	297	(4900)
0.03	**Safrole in spices**	**Safrole, 1.2 mg**	(441)	51.3
0.03	**Orange juice, 138 g**	***d*-Limonene, 4.28 mg**	204	(—)
0.03	**Pepper, black, 446 mg**	***d*-Limonene, 3.57 mg**	204	(—)
0.02	**Coffee, 13.3 g**	**Catechol, 1.33 mg**	88.8	(244)
0.02	**Mushroom (*Agaricus bisporus* 2.55 g)**	**Mixture of hydrazines, etc. (whole mushroom)**	(—)	20,300
0.02	**Apple, 32.0 g**	**Caffeic acid, 3.40 mg**	297	(4900)
0.02	**Coffee, 13.3 g**	**Furfural, 2.09 mg**	(683)	197
0.01	BHA: daily US avg (1975)	BHA, 4.6 mg	606	(5530)
0.01	**Beer (before 1979), 257 g**	**Dimethylnitrosamine, 726 ng**	0.0959	(0.189)
0.008	**Aflatoxin: daily US avg (1984–89)**	**Aflatoxin, 18 ng**	0.0032	(+)
0.007	**Cinnamon, 21.9 mg**	**Coumarin, 65.0 μg**	13.9	(103)
0.006	**Coffee, 13.3 g**	**Hydroquinone, 333 μg**	82.8	(225)
0.005	Saccharin: daily US avg (1977)	Saccharin, 7 mg	2140	(—)
0.005	**Carrot, 12.1 g**	**Aniline, 624 μg**	194[b]	(—)
0.004	**Potato, 54.9 g**	**Caffeic acid, 867 μg**	297	(4900)
0.004	**Celery, 7.95 g**	**Caffeic acid, 858 μg**	297	(4900)
0.004	**White bread, 67.6 g**	**Furfural, 500 μg**	(683)	197
0.003	**Nutmeg, 27.4 mg**	***d*-Limonene, 466 μg**	204	(—)
0.003	Conventional home air (14 hour/day)	Benzene, 155 μg	(169)	77.5
0.002	**Coffee, 13.3 g**	**4-Methylcatechol, 433 μg**	248	·
0.002	**Carrot, 12.1 g**	**Caffeic acid, 374 mg**	297	(4900)
0.002	Ethylene thiourea: daily US avg (1990)	Ethylene thiourea, 9.51 μg	7.9	(23.5)

TABLE 21.1.5 Ranking Possible Carcinogenic Hazards from Average U.S. Exposures (55, 57). (Continued)

Possible hazard: HERP (%)	Average daily U.S. exposure	Human dose of rodent carcinogen	Potency TD$_{50}$ (mg/kg/day)* Rats	Mice
0.002	BHA: daily US avg (1987)	BHA, 700 μg	606	(5530)
0.002	DDT: daily US avg (before 1972 ban)	DDT, 13.8 μg	(84.7)	12.8
0.001	**Plum, 2.00 g**	**Caffeic acid, 276 μg**	297	(4900)
0.001	**Pear, 3.29 g**	**Caffeic acid, 240 μg**	297	(4900)
0.001	UDMH: daily US avg (1988)	UDMH, 2.82 μg (from Alar)	(—)	3.96
0.0009	**Brown mustard, 68.4 mg**	**Allyl isothiocyanate, 62.9 μg**	96	(—)
0.0008	DDE: daily US avg (before 1972 ban)	DDE, 6.91 μg	(—)	12.5
0.0007	TCDD: daily US avg (1994)	TCDD, 12.0 pg	0.0000235	(0.000156)
0.0006	**Bacon, 11.5 g**	**Diethylnitrosamine, 11.5 ng**	0.0266	(+)
0.0006	**Mushroom (*Agaricus bisporus*, 2.55 g)**	**Glutamyl-*p*-hydrazino-benzoate, 107 μg**	·	277
0.0005	**Bacon, 11.5 g**	**Dimethylnitrosamine, 34.5 ng**	0.0959	(0.189)
0.0004	**Bacon, 11.5 g**	***N*-Nitrosopyrrolidine, 196 ng**	(0.799)	0.679
0.0004	EDB: daily US avg (before 1984 ban)	EDB, 420 μg	1.52	(7.45)
0.0004	Tap water, 1 liter (1987–92)	Bromodichloromethane, 13 μg	(72.5)	47.7
0.0003	**Mango, 1.22 g**	***d*-Limonene, 48.8 μg**	204	(—)
0.0003	**Beer, 257 g**	**Furfural, 39.9 μg**	(683)	197
0.0003	Tap water, 1 liter (1987–92)	Chloroform, 17 μg	(262)	90.3
0.0003	Carbaryl: daily US avg (1990)	Carbaryl, 2.6 μg	14.1	(—)
0.0002	**Celery, 7.95 g**	**8-Methoxypsoralen, 4.86 μg**	32.4	(—)
0.0002	Toxaphene: daily US avg (1990)	Toxaphene, 595 ng	(—)	5.57
0.0001	Tap water, 1 liter (1987–92)	Dichloroacetic acid, 5.9 μg	67.8	(97.2)
0.00009	**Mushroom (*Agaricus bisporus*, 2.55 g)**	***p*-Hydrazinobenzoate, 28 μg**	·	454[b]
0.00008	PCBs: daily US avg (1984–86)	PCBs, 98 ng	1.74	(9.58)
0.00008	DDE/DDT: daily US avg (1990)	DDE, 659 ng	(—)	12.5
0.00007	**Parsnip, 54.0 mg**	**8-Methoxypsoralen, 1.57 μg**	32.4	(—)
0.00007	**Toast, 67.6 g**	**Urethane, 811 ng**	(41.3)	16.9
0.00006	**Hamburger, pan fried, 85 g**	**PhIP, 176 ng**	4.22[b]	(28.6[b])
0.00005	**Estragole in spices**	**Estragole, 1.99 μg**	·	51.8
0.00005	**Parsley, fresh, 324 mg**	**8-Methoxypsoralen, 1.17 μg**	32.4	(—)
0.00004	Tap water, 1 liter (1987–92)	Chloral hydrate, 3.3 μg	·	106
0.00003	**Hamburger, pan fried, 85 g**	**MeIQx, 38.1 ng**	1.66	(24.3)
0.00002	Dicofol: daily US avg (1990)	Dicofol, 544 ng	(—)	32.9
0.00001	Tap water, 1 liter (1987–92)	Trichloroacetic acid, 5 μg	(—)	583
0.00001	**Beer, 257 g**	**Urethane, 115 ng**	(41.3)	16.9
0.000006	**Hamburger, pan fried, 85 g**	**IQ, 6.38 ng**	1.65[b]	(19.6)
0.000005	Hexachlorobenzene: daily US avg (1990)	Hexachlorobenzene, 14 ng	3.86	(65.1)
0.000002	Tap water, 1 liter (1987–92)	Tribromomethane, 750 ng	648	(—)
0.000001	Lindane: daily US avg (1990)	Lindane, 32 ng	(—)	30.7
0.0000004	PCNB: daily US avg (1990)	PCNB (Quintozene), 19.2 ng	(—)	71.1
0.0000001	Chlorobenzilate: daily US avg (1989)	Chlorobenzilate, 6.4 ng	(—)	93.9
0.00000001	Folpet: daily US avg (1990)	Folpet, 12.8 ng	(—)	1650
<0.00000001	Chlorothalonil: daily US avg (1990)[c]	Chlorothalonil, <6.4 ng	828	(—)
0.000000008	Captan: daily US avg (1990)	Captan, 11.5 ng	2080	(2110)

[a] "." = no data in CPDB; (—) = negative in cancer test; (+) = positive cancer test(s) not suitable for calculating a TD$_{50}$.
[b] TD$_{50}$ harmonic mean was estimated for the base chemical from the hydrochloride salt.
[c] No longer contained in any registered pesticide product.

in a rodent bioassay is highly uncertain. The carcinogenic effects may be limited to the high dose tested.

In regulatory policy, the "virtually safe dose" (VSD), which corresponds to a maximum, hypothetical cancer risk of 1 in 1 million, is estimated from bioassay results by using a linear model. To the extent that carcinogenicity in rodent bioassays is due to the effects of high doses for the nonmutagens and a synergistic effect of cell division at high doses with DNA damage for the mutagens, then this model is inappropriate. Moreover, as currently calculated, the VSD can be known without ever conducting a bioassay: for 96 percent of the NCI/NTP rodent carcinogens, the VSD is within a factor of 10 of the ratio MTD/740,000 (71). This is about as precise as the estimate obtained from conducting near-replicate cancer tests of the same chemical (71). Agencies that evaluate cancer risk, e.g. EPA, (72), are moving to take mechanism and nonlinearity into account and to emphasize a more flexible approach to risk assessment.

21.1.8 MISCONCEPTION #6: SYNTHETIC CHEMICALS POSE GREATER CARCINOGENIC HAZARDS THAN NATURAL CHEMICALS

Gaining a broad perspective about the vast number of chemicals to which humans are exposed can be helpful when setting research and regulatory priorities (5, 51, 52, 55, 73). Rodent bioassays provide little information about the mechanisms of carcinogenesis and low-dose risk. The assumption that synthetic chemicals are hazardous has led to a bias in testing so that synthetic chemicals account for 76 percent (451/590) of the chemicals tested chronically in both rats and mice (Table 21.1.4). The natural world of chemicals has never been tested systematically.

One reasonable strategy is to use a rough index to *compare* and *rank* possible carcinogenic hazards from a wide variety of chemical exposures at levels that humans typically receive, and then to focus on those that rank highest (5, 55, 57). Ranking is a critical first step that can help set priorities when selecting chemicals for chronic bioassay or mechanistic studies, for epidemiological research, and for regulatory policy. Although one cannot say whether the ranked chemical exposures are likely to be of major or minor importance in human cancer, it is not prudent to focus attention on the possible hazards at the bottom of a ranking if, by using the same methodology to identify hazard, there are numerous common human exposures with much greater possible hazards. Our analyses are based on the HERP (Human Exposure/Rodent Potency) index, which indicates what percentage of the rodent carcinogenic potency (TD_{50} in mg/kg/day) a person receives from a given daily exposure (mg/kg/day) (54) (Table 21.1.5). A ranking based on standard regulatory risk assessment would be similar.

Overall, our analyses have shown that HERP values for some historically high exposures in the workplace and certain pharmaceuticals rank high, and that there is an enormous background of naturally occurring rodent carcinogens in typical portions of common foods that cast doubt on the relative importance of low-dose exposures to residues of synthetic chemicals such as pesticides (Table 21.1.5) (5, 9, 55, 57). A committee of the NRC/National Academy of Sciences (NAS) recently reached similar conclusions about natural versus synthetic chemicals in the diet and called for further research on natural chemicals (74).

The possible carcinogenic hazards from synthetic pesticides (at average exposures) are minimal compared to the background of nature's pesticides, though neither may present a hazard at the low doses consumed (Table 21.1.5). Table 21.1.5 also indicates that many ordinary foods would not pass the regulatory criteria used for synthetic chemicals. For many natural chemicals, the HERP values are in the top half of Table 21.1.5, even though natural chemicals are markedly underrepresented because so few have been tested in rodent bioassays. Caution is necessary in drawing conclusions from the occurrence in the diet of natural chemicals that are rodent carcinogens. It is not argued here that these dietary exposures are necessarily of much relevance to human cancer. Our results call for a re-evaluation of the utility of animal cancer tests for protecting the public against minor hypothetical risks.

21.1.9   MISCONCEPTION #7: THE TOXICOLOGY OF SYNTHETIC CHEMICALS IS DIFFERENT FROM THAT OF NATURAL CHEMICALS

It is often assumed that because natural chemicals are part of human evolutionary history, whereas synthetic chemicals are recent, the mechanisms that have evolved in animals to cope with the toxicity of natural chemicals will fail to protect against synthetic chemicals. This assumption is flawed for several reasons (52, 58).

Humans have many natural defenses that buffer against normal exposures to toxins (52); these usually are general rather than tailored to each specific chemical. Thus, the defenses work against both natural and synthetic chemicals. Examples of general defenses include the continuous shedding of cells exposed to toxins—the surface layers of the mouth, esophagus, stomach, intestine, colon, skin, and lungs are discarded every few days; DNA repair enzymes, which repair DNA that has been damaged from many different sources; and detoxification enzymes of the liver and other organs that generally target classes of toxins rather than individual toxins. That defenses are usually general, rather than specific for each chemical, makes good evolutionary sense. The reason that predators of plants evolved general defenses presumably was to be prepared to counter a diverse and ever-changing array of plant toxins in an evolving world; if a herbivore had defenses against only a set of specific toxins, it would be at a great disadvantage in obtaining new food when favored foods became scarce or evolved new toxins.

Various natural toxins that have been present throughout vertebrate evolutionary history nevertheless cause cancer in vertebrates (52, 55, 57). Mold toxins, such as aflatoxin, have been shown to cause cancer in rodents and other species, including humans (Table 21.1.4). Many of the common elements are carcinogenic to humans at high doses (e.g., salts of cadmium, beryllium, nickel, chromium, and arsenic) despite their presence throughout evolution. Furthermore, epidemiological studies from various parts of the world show that certain natural chemicals in food may be carcinogenic risks to humans; for example, the chewing of betel nuts with tobacco is associated with oral cancer.

Humans have not had time to evolve a "toxic harmony" with all of the plants in their diet. The human diet has changed markedly in the last few thousand years. Indeed, very few of the plants that humans eat today (e.g., coffee, cocoa, tea, potatoes, tomatoes, corn, avocados, mangoes, olives, and kiwi fruit), would have been present in a hunter-gatherer's diet. Natural selection works far too slowly for humans to have evolved specific resistance to the food toxins in these relatively newly introduced plants.

DDT is often viewed as the prototypically dangerous synthetic pesticide because it concentrates in the tissues and persists for years, being slowly released into the bloodstream. DDT, the first synthetic pesticide, eradicated malaria from many parts of the world, including the United States. It was effective against many vectors of disease such as mosquitoes, tsetse flies, lice, ticks, and fleas. DDT was also lethal to many crop pests, and significantly increased the supply and lowered the cost of food, making fresh, nutritious foods more accessible to poor people. It was also of low toxicity to humans. A 1970 NAS report concluded: "In little more than two decades DDT has prevented 500 million deaths due to malaria, that would otherwise have been inevitable (75)." There is no convincing epidemiological evidence, nor is there much toxicological plausibility, that the levels normally found in the environment are likely to contribute significantly to cancer. DDT was unusual with respect to bioconcentration, and because of its chlorine substituents it takes longer to degrade in nature than most chemicals; however, these are properties of relatively few synthetic chemicals. In addition, many thousands of chlorinated chemicals are produced in nature (76), and natural pesticides can also bioconcentrate if they are fat-soluble. Potatoes, for example, naturally contain the fat soluble neurotoxins solanine and chaconine (51, 53), which can be detected in the bloodstream of all potato eaters. High levels of these potato neurotoxins have been shown to cause birth defects in rodents (52).

Because no plot of land is free from attack by insects, plants need chemical defenses—either natural or synthetic—in order to survive. Thus, there is a trade-off between naturally occurring and synthetic pesticides. One consequence of disproportionate concern about synthetic pesticide residues is that some plant breeders develop plants to be more insect-resistant by making them higher in natural toxins. A recent case illustrates the potential hazards of this approach to pest control: When a major grower introduced a new variety of highly insect-resistant celery into commerce, people who handled the celery developed rashes when they were subsequently exposed to sunlight. Some detective work found that the pest-resistant celery contained 6200 parts per billion (ppb) of carcinogenic (and mutagenic) psoralens instead of the 800 ppb present in common celery (53, 55).

21.1.10 MISCONCEPTION #8: PESTICIDES AND OTHER SYNTHETIC CHEMICALS ARE DISRUPTING HORMONES AND CAUSING HUMAN CANCER AND REPRODUCTIVE EFFECTS

Synthetic hormone mimics like organochlorine pesticides, have become an environmental issue (77), which was recently addressed by the NAS (78). We discussed in Misconception #2 that hormonal factors are important in human cancer and that lifestyle factors can markedly change the levels of endogenous hormones. The trace exposures to estrogenic organochlorine residues are tiny compared to the normal dietary intake of naturally occurring endocrine-active chemicals in fruits and vegetables (79–81). These low levels of human exposure are toxicologically implausible as a significant cause of cancer or of reproductive abnormalities (79–82). Moreover, it has not been shown convincingly that sperm counts are declining (78, 83); even if they were, there are many more likely causes, such as smoking and diet (Misconception # 2).

Some recent studies have compared estrogenic equivalents (EQ) of dietary intake of synthetic chemicals versus phytoestrogens in the normal diet, by considering both the amount humans consume and estrogenic potency. Results support the idea that synthetic residues are orders of magnitude lower in EQ and are generally weaker in potency. One study used a series of *in vitro* assays and calculated the EQs in extracts from 200 ml of red cabernet wine and the EQs from average intake of organochlorine pesticides (84). EQs for a single glass of wine ranged from 0.15 to 3.68 μg/day compared to 1.24 ng/day for organochlorine pesticides (84). Another study (85) compared plasma concentrations of the phytoestrogens genistein and daidzein in infants fed soy-based formula versus cow milk formula or human breast milk. Mean plasma levels were hundreds of times higher for the soy fed infants than the others. "Circulating concentrations of isoflavones in the seven infants fed soy-based formula were 13000–22000 times higher than plasma oestradiol concentrations in early life, and may be sufficient to exert biological effects, whereas the contribution of isoflavones from breast-milk and cow-milk is negligible" (85).

21.1.11 MISCONCEPTION #9: REGULATION OF LOW, HYPOTHETICAL RISKS IS EFFECTIVE IN ADVANCING PUBLIC HEALTH

Because there is no risk-free world and resources are limited, society must set priorities using cost effectiveness in order to save the greatest number of lives (86, 87). In 1991 the EPA projected that the cost to society of environmental regulations in 1997 would be about $140 billion per year (about 2.6 percent of Gross National Product) (88). Most of this cost would be to the private sector. Several economic analyses have concluded that current expenditures are not cost effective; resources are not being used so as to save the greatest number of lives per dollar. One estimate is that the United States could prevent 60,000 deaths per year by redirecting the same dollar resources to more cost-effective programs (89). For example, the median toxin control program costs 146 times more per life-year saved than the median medical intervention (89). This difference is likely to be even greater because cancer risk estimates for toxin control programs are worst-case, hypothetical estimates, and the true risks at low dose are often likely to be zero (5, 66, 71) (Misconception #5). Some economists have argued that costly regulations intended to save lives may actually increase the number of deaths (90), in part because they divert resources from important health risks and in part because higher incomes are associated with lower mortality (91–93). Rules on air and water pollution are necessary (it was a public health benefit to phase lead out of gasoline), and clearly cancer prevention is not the only reason for regulations. However, worst-case assumptions in risk assessment represent a policy decision, not a scientific one, and they confuse attempts to allocate money effectively for risk abatement.

Regulatory efforts to reduce low-level human exposure to synthetic chemicals because they are rodent carcinogens are expensive since they aim to eliminate minuscule concentrations that can now be measured with improved techniques. These efforts distract from the major task of improving public health through increasing scientific understanding about how to prevent cancer (e.g., the role of diet), increasing public understanding of how lifestyle influences health, and improving our ability to help individuals alter lifestyle.

ACKNOWLEDGMENTS

This work was supported by the National Cancer Institute Outstanding Investigator Grant CA39910 to B.N.A., by a grant from the Office of Energy Research, Office of Health and Environmental Research of the United States Department of Energy under Contract DE-AC03-76SF00098 to L.S.G., and by National Institute of Environmental Health Sciences Center Grant ESO1896.

This paper is modified and updated from *FASEB J.*, 11: 1041–1052 (1997) and *Environ. Health Perspect.*, 107 (Suppl. 4), 527–600 (1999).

REFERENCES

1. Ries, L. A. G., Kosary, C. L., Hankey, B. F., Miller, B. A., Clegg, L., and Edwards, B. K. (1999), *SEER Cancer Statistics Review, 1973–1996.* National Cancer Institute, Bethesda, MD.

2. Bailar, J. C. III and Gornik, H. L. (1997), *N. Engl. J. Med.*, 336: 1569–1574.

3. Doll, R., and Peto, R. (1981), *J. Natl. Cancer Inst.*, 66: 1191–1308.

4. Devesa, S. S., Blot, W. J., Stone, B. J., Miller, B. A., Tarone, R. E., and Fraumeni, F. J. Jr. (1995), *J. Natl. Cancer Inst.*, 87: 175–182.

5. Gold, L. S., Slone, T. H., Stern, B. R., Manley, N. B., and Ames, B. N. (1992), *Science*, 258: 261–265.

6. Ames, B. N., Gold, L. S., and Willett, W. C. (1995), *Proc. Natl. Acad. Sci. USA*, 92: 5258–5265.

7. Ames, B. N. (1998), *Toxicol. Lett.*, 102–103: 5–18.

8. Gough, M. (1990), *Risk Anal.*, 10: 1–6.

9. Gold, L. S., Garfinkel, G. B., and Slone, T. H. (1994), in *Chemical Risk Assessment and Occupational Health, Current Applications, Limitations, and Future Prospects* (C. M. Smith, D. C. Christiani, and K. T. Kelsey, Eds.), pp. 91–103, Greenwood Publishing Group, Westport, CT.

10. Ames, B. N., Shigenaga, M. K., and Hagen, T. M. (1993), *Proc. Natl. Acad. Sci. USA*, 90: 7915–7922.

11. Beckman, K. B., and Ames, B. N. (1998), *Physiol. Rev.*, 78: 547–581.

12. Hagen, T. M., Yowe, D. L., Bartholomew, J. C., Wehr, C. M., Do, K. L., Park, J.-Y., and Ames, B. N. (1997), *Proc. Natl. Acad. Sci. USA*, 94: 3064–3069.

13. Helbock, H. J., Beckman, K. B., Shigenaga, M. K., Walter, P., Woodall, A. A., Yeo, H. C., and Ames, B. N. (1998), *Proc. Natl. Acad. Sci. USA*, 95: 288–293.

14. Rice-Evans, C., Sampson, J., Bramley, P., and Holloway, D. (1997), *Free Rad. Res.*, 26: 381–398.

15. American Cancer Society (1999), *Cancer Facts & Figures—1999.*, American Cancer Society, Atlanta, GA.

16. Peto, R., Lopez, A. D., Boreham, J., Thun, M., and Heath, C. Jr. (1994), *Mortality from Smoking in Developed Countries 1950–2000.*, Oxford University Press, Oxford, England.

17. Fraga, C. G., Motchnik, P. A., Shigenaga, M. K., Helbock, H. J., Jacob, R. A., and Ames, B. N. (1991), *Proc. Natl. Acad. Sci. USA*, 88: 11003–11006.

18. Ames, B. N., Motchnik, P. A., Fraga, C. G., Shigenaga, M. K., and Hagen, T. M. (1994), in *Male-Mediated Developmental Toxicity*, (D. R. Mattison, and A. Olshan, Eds.), pp. 243–259, Plenum Publishing Corporation, New York.

19. Fraga, C. G., Motchnik, P. A., Wyrobek, A. J., Rempel, D. M., and Ames, B. N. (1996), *Mutat. Res.*, 351: 199–203.

20. Wyrobek, A. J., Rubes, J., Cassel, M., Moore, D., Perrault, S., Slott, V., Evenson, D., Zudova, Z., Borkovec, L., Selevan, S., and Lowe, X. (1995), *Am. J. Hum. Genet.*, 57: 737.

21. Woodall, A. A., and Ames, B. N. (1997), in *Vitamin C in Health and Disease*, (L. Packer, Eds.), pp. 193–203, Marcel Dekker, Inc., New York.

22. Ji, B.-T., Shu, X.-O., Linet, M. S., Zheng, W., Wacholder, S., Gao, Y.-T., Ying, D.-M., and Jin, F. (1997), *J. Natl. Cancer Inst.*, 89: 238–244.

23. Hart, R., Keenan, K., Turturro, A., Abdo, K., Leakey, J., and Lyn-Cook, B. (1995), *Fundam. Appl. Toxicol.*, 25: 184–195.

24. Turturro, A., Duffy, P., Hart, R., and Allaben, W. (1996), *Toxicol. Pathol.*, 24: 769–775.

25. Christen, S., Hagen, T. M., Shigenaga, M. K., and Ames, B. N. (1999), in *Microbes and Malignancy: Infection as a Cause of Cancer* (J. Parsonnet, and S. Hornig, Eds.), pp. 35–88, Oxford University Press, New York.

26. Shacter, E., Beecham, E. J., Covey, J. M., Kohn, K. W., and Potter, M. (1988), *Carcinogenesis*, 9: 2297–2304.

27. Yamashina, K., Miller, B. E., and Heppner, G. H. (1986), *Cancer Res.*, 46: 2396–2401.

28. Pisani, P., Parkin, D. M., Muñoz, N., and Ferlay, J. (1997), *Cancer Epidemiol. Biomarkers Prev.*, 6: 387–400.

29. Henderson, B. E., Ross, R. K., and Pike, M. C. (1991), *Science*, 254: 1131–1138.

30. Henderson, B. E., and Feigelson, H. S. (2000), *Carcinogenesis*, 21: 427–433.

31. Hunter, D. J., and Willett, W. C. (1993), *Epidemiol. Rev.*, 15: 110–132.

32. Block, G., Patterson, B., and Subar, A. (1992), *Nutr. Cancer*, 18: 1–29.

33. Steinmetz, K. A., and Potter, J. D. (1996), *J. Am. Diet Assoc.*, 96: 1027–1039.

34. Hill, M. J., Giacosa, A., and Caygill, C. P. J. (1994), *Epidemiology of Diet and Cancer.*, Ellis Horwood Limited, West Sussex, England.

35. Krebs-Smith, S. M., Cook, A., Subar, A. F., Cleveland, L., Friday, J., and Kahle, L. L. (1996), *Arch. Pediatr. Adolesc. Med.*, 150: 81–86.

36. Krebs-Smith, S. M., Cook, A., Subar, A. F., Cleveland, L., and Friday, J. (1995), *Am. J. Public Health*, 85: 1623–1629.

37. National Cancer Institute Graphic (1996), *J. Natl. Cancer Inst.*, 88: 1314.

38. Block, G. (1992), *Nutr. Rev.*, 50: 207–213.

39. Blount, B. C., Mack, M. M., Wehr, C., MacGregor, J., Hiatt, R., Wang, G., Wickramasinghe, S. N., Everson, R. B., and Ames, B. N. (1997), *Proc. Natl. Acad. Sci. USA*, 94: 3290–3295.

40. Senti, F. R., and Pilch, S. M. (1985), *J. Nutr.*, 115: 1398–1402.

41. Bailey, L. B., Wagner, P. A., Christakis, G. J., Araujo, P. E., Appledorf, H., Davis, C. G., Masteryanni, J., and Dinning, J. S. (1997), *Am. J. Clin. Nutr.*, 32: 2346–2353.

42. Bailey, L. B., Wagner, P. A., Christakis, G. J., Davis, C. G., Appledorf, H., Araujo, P. E., Dorsey, E., and Dinning, J. S. (1982), *Am. J. Clin. Nutr.*, 35: 1023–1032.

43. Wallock, L., Woodall, A., Jacob, R., and Ames, B. (1997) (Abstract), *FASEB J.*, 11: A184, 1068.

44. Boushey, C. J., Beresford, S. A., Omenn, G. S., and Motulsky, A. G. (1995), *JAMA*, 274: 1049–1057.

45. Zhang, S., Hunter, D. J., Hankinson, S. E., Giovannucci, E. L., Rosner, B. A., Colditz, G. A., Speizer, F. E., and Willett, W. C. (1999), *JAMA*, 281: 632–637.

46. Herbert, V. (1996), in *Present Knowledge in Nutrition*, (E. E. Ziegler, and L. J. Filer, Eds.), pp. 191–205, ILSI Press, Washington, D.C.

47. Wickramasinghe, S. N., and Fida, S. (1994), *Blood*, 83: 1656–1661.

48. Zhang, J. Z., Henning, S. M., and Swendseid, M. E. (1993), *J. Nutr.*, 123: 1349–55.

49. Jacobson, E. L. (1993), *J. Am. Coll. Nutr.*, 12: 412–416.

50. Subar, A. F., Block, G., and James, L. D. (1989), *Am. J. Clin. Nutr.*, 50: 508–516.

51. Ames, B. N., Profet, M., and Gold, L. S. (1990), *Proc. Natl. Acad. Sci. USA*, 87: 7777–7781.

52. Ames, B. N., Profet, M., and Gold, L. S. (1990), *Proc. Natl. Acad. Sci. USA*, 87: 7782–7786.

53. Gold, L. S., Slone, T. H., and Ames, B. N. (1997), in *Food Chemical Risk Analysis* (D. Tennant, Ed.), pp. 267–295, Chapman & Hall Ltd, London.

54. Gold, L. S., and Zeiger, E. (1997), *Handbook of Carcinogenic Potency and Genotoxicity Databases*, CRC Press, Boca Raton, FL.

55. Gold, L. S., Manley, N. B., Slone, T. H., and Rohrbach, L. (1999), *Environ. Health Perspect.*, 107 (Suppl. 4), 527–600.
56. Ames, B. N., and Gold, L. S. (1990), *Angew. Chem.*, 29: 1197–1208.
57. Gold, L. S., Slone, T. H., and Ames, B. N. (1997), in *Handbook of Carcinogenic Potency and Genotoxicity Databases* (L. S. Gold and E. Zeiger, Eds.), pp. 661–685, CRC Press, Boca Raton, FL.
58. Ames, B. N., Gold, L. S., and Shigenaga, M. K. (1996), *Risk Anal.*, 16: 613–617.
59. Laskin, D. L., and Pendino, K. J. (1995), *Annu. Rev. Pharmacol. Toxicol.*, 35: 655–677.
60. Wei, L., Wei, H., and Frenkel, K. (1993), *Carcinogenesis*, 14: 841–847.
61. Wei, Q., Matanoski, G. M., Farmer, E. R., Hedayati, M. A., and Grossman, L. (1993), *Proc. Natl. Acad. Sci. USA*, 90: 1614–8.
62. Laskin, D. L., Robertson, F. M., Pilaro, A. M., and Laskin, J. D. (1988), *Hepatology*, 8: 1051–1055.
63. Czaja, M. J., Xu, J., Ju, Y., Alt, E., and Schmiedeberg, P. (1994), *Hepatology*, 19: 1282–1289.
64. Adachi, Y., Moore, L. E., Bradford, B. U., Gao, W., and Thurman, R. G. (1995), *Gastroenterology*, 108: 218–224.
65. Gunawardhana, L., Mobley, S. A., and Sipes, I. G. (1993), *Toxicol. Appl. Pharmacol.*, 119: 205–213.
66. Gold, L. S., Slone, T. H., and Ames, B. N. (1998), *Drug Metab. Rev.*, 30: 359–404.
67. Omenn, G. S., Stuebbe, S., and Lave, L. B. (1995), *Mol. Carcinog.*, 14: 37–45.
68. Butterworth, B., Conolly, R., and Morgan, K. (1995), *Cancer Lett.*, 93: 129–146.
69. Ames, B. N., Shigenaga, M. K., and Gold, L. S. (1993), *Environ. Health Perspect.*, 101 (Suppl 5), 35–44.
70. Ames, B. N., and Gold, L. S. (1990), *Proc. Natl. Acad. Sci. USA*, 87: 7772–7776.
71. Gaylor, D. W., and Gold, L. S. (1995), *Regul. Toxicol. Pharmacol.*, 22: 57–63.
72. United States Environmental Protection Agency (1996), *Fed. Reg.*, 61: 17960–18011.
73. Ames, B. N., Magaw, R., and Gold, L. S. (1987), *Science*, 236: 271–280.
74. National Research Council (1996), *Carcinogens and Anticarcinogens in the Human Diet: A Comparison of Naturally Occurring and Synthetic Substances*, National Academy Press, Washington, D.C.
75. Committee on Research in the Life Sciences (1970), *The Life Sciences: Recent Progress and Application to Human Affairs, the World of Biological Research, Requirement for the Future*, National Academy Press, Washington, D.C.
76. Gribble, G. W. (1996), *Pure Appl. Chem.*, 68: 1699–1712.
77. Colburn, T., Dumanoski, D., and Myers, J. P. (1996), *Our Stolen Future: Are we Threatening our Fertility, Intelligence, and Survival?: A Scientific Detective Story.*, Dutton, New York.
78. National Research Council (1999), Hormonally Active Agents in the Environment, Committee on Hormonally Active Agents in the Environment, National Academy Press, Washington, D.C.
79. Safe, S. H. (1995), *Environ. Health Perspect.*, 103: 346–351.
80. Safe, S. H. (1997), *Environ. Health Perspect.*, 105 (Suppl 3), 675–678.
81. Safe, S. H. (2000), Endocrine disruptors and human health—Is there a problem?—An update. *Environ. Health Perspect.*, in press.
82. Reinli, K., and Block, G. (1996), *Nutr. Cancer*, 26, 1996.
83. Kolata, G. (1996), Measuring men up, sperm by sperm, *The New York Times*, E4(N), E4(L), (col.1), May 4.
84. Gaido, K., Dohme, L., Wang, F., Chen, I., Blankvoort, B., Ramamoorthy, K., and Safe, S. (1998), *Environ. Health Perspect.*, 106 (Suppl. 6), 1347–1351.
85. Setchell, K. D. R., Zimmer-Nechemias, L., Cai, J. and Huebi, J. E. (1997), *Lancet*, 350: 23–27.
86. Hahn, R. W. (1996), *Risks, Costs, and Lives Saved: Getting Better Results from Regulation.*, Oxford University Press and AEI Press, New York and Washington, D.C.
87. Graham, J., and Wiener, J. (1995), *Risk versus Risk: Tradeoffs in Protecting Health and the Environment.*, Harvard University Press, Cambridge, MA.
88. United States Environmental Protection Agency (1991), *Environmental Investments: The Cost of a Clean Environment*, Office of the Administrator, Washington, D.C.
89. Tengs, T. O., Adams, M. E., Pliskin, J. S., Safran, D. G., Siegel, J. E., Weinstein, M. C., and Graham, J. D. (1995), *Risk Anal.*, 15: 369–389.
90. Keeney, R. L. (1990), *Risk Anal.*, 10: 147–159.
91. Wildavsky, A. (1988), *Searching for Safety.* Transaction Press, New Brunswick, N.J.

92. Wildavsky, A. B. (1995), *But is it True?: A Citizen's Guide to Environmental Health and Safety.* Harvard University Press, Cambridge, MA.

93. Viscusi, W. K. (1992), *Fatal Trade-offs.* Oxford University Press, Oxford, England.

94. Innes, J. R. M., Ulland, B. M., Valerio, M. G., Petrucelli, L., Fishbein, L., Hart, E. R., Pallota, A. J., Bates, R. R., Falk, H. L., Gart, J. J., Klein, M., Mitchell, I., and Peters, J. (1969), *J. Natl. Cancer Inst.*, 42: 1101–1114.

95. Davies, T. S., and Monro, A. (1995), *J. Am. Coll. Toxicol.*, 14: 90–107.

96. Contrera, J., Jacobs, A., and DeGeorge, J. (1997), *Regul. Toxicol. Pharmacol.*, 25: 130–145.

CHAPTER 21
MISCONCEPTIONS ABOUT ENVIRONMENTAL POLLUTION

SECTION 21.2

SCIENCE, PESTICIDES, AND POLITICS

J. Gordon Edwards

Dr. Edwards is an emeritus professor of biology at San Jose State University. He taught entomology there for 49 years and is now curator of the University's entomology collection.

A week after my college graduation, I was inducted into the army. In 1944 I went ashore in France at Omaha Beach (three weeks after the great invasion). Later I spent several hours daily in a cloud of 10 percent DDT dust, puffing it down inside the clothing of European people who feared that typhus might again spread across Europe as it did during the first World war. At that time it killed nearly three million people in Russia and millions more in the Balkans, Poland, and Germany. This terrible disease is spread by body lice, which were becoming common in Europe again in 1944. Fortunately, DDT had recently been discovered and it quickly killed body lice, so typhus did not become a problem in Europe during the war.

After the war ended I went to Ohio State University to continue my study of beetles. I feared that the government might blanket the United States with DDT, to kill all the insect pests. I thought that might eradicate so many insects that my career as a beetle specialist would be threatened. Fortunately, I was wrong on every count.

During the early 1960s, I worked for a month each summer studying high altitude ecology in Grand Teton National Park, Wyoming. While I was there, the *New Yorker* magazine carried a review of Rachel Carson's new book, *Silent Spring*. I read the review and thought it was great, because I was a dedicated ecologist and had little use for industry or construction projects. I bought a copy of the book and began reading it. I noticed that Miss Carson made a great many misleading statements, but I tried to overlook that because "she was on our side." Gradually, however, I realized that she was deliberately lying. I was really shocked! I began to understand why her original coauthor, Edwin Diamond, (science editor of *Newsweek*) had withdrawn from the relationship and criticized *Silent Spring* as "an emotional, alarmist book seeking to cause Americans to mistakenly believe their world is being poisoned" (*Saturday Evening Post*, 28 September 1963).

21.2.1   SILENT SPRING BY RACHEL CARSON

In the front of her book, Rachel Carson dedicated *Silent Spring* as follows: "To Albert Schweitzer who said 'Man has lost the capacity to foresee and to forestall. He will end by destroying the Earth.'" Because the major theme of her book was antipesticides (especially anti-DDT), this appeared to indicate that the great man opposed the use of DDT. However, in his autobiography Schweitzer wrote: "How much labor and waste of time these wicked insects do cause us . . . but a ray of hope, in the use of DDT, is now held out to us."

On page 187 Carson wrote: "Only yesterday mankind lived in fear of the scourges of smallpox, cholera and plague that once swept nations before them. Now our major concern is no longer with the disease organisms that once were omnipresent; sanitation, better living conditions, and new drugs have given us a high degree of control over infectious disease." That statement bothered me, because I had been teaching medical entomology at San Jose State University for more than 10 years and was aware that the greatest threats to humans are diseases like malaria, typhus, yellow fever, Chagas' disease, African sleeping sickness, and several types of leishmaniasis and tickborne rickettsial diseases. She avoided mentioning any of those, perhaps because she knew they could be controlled only by the appropriate use of insecticides. It was later revealed in *Science* (9 June 72) that "at least 80 percent of all human infectious diseases are arthropod-borne."

The National Academy of Sciences, in *The Life Sciences*, 1970, commented that: "To only a few chemicals does man owe as great a debt as to DDT. In a little more than two decades, DDT has prevented 500 million human deaths that would otherwise have been inevitable."

21.2.2 MALARIA

In Ceylon (Sri Lanka) in the 1950s, two million people developed malaria each year, but after a DDT program was carried out, there were only 17 cases in the entire country. Most scientists thought that was a great humanitarian victory, however, in 1981 Prince Philip wrote in *People* magazine: "I was in Sri Lanka, where malaria was halted by DDT. Earlier, malaria had been controlling population growth. The consequence of using DDT was that within about 20 years the population doubled." (He was happier when thousands of poor people died of malaria annually.) Alexander King, the president of the Club of Rome, wrote in his 1990 book: "In Guyana, within two years, DDT had almost eliminated malaria, so my chief quarrel with it in hindsight is that it has greatly added to the human population problem." The World Health Organization stated that up to 40 percent of the children in poor nations would die of malaria, in response to which a leader in the Agency for International Development said: "Rather dead than alive and riotously reproducing."

Sierra Club president McClosky told reporters: "The Sierra Club wants a ban on DDT, even in tropical countries where it has kept malaria under control," and the National Audubon Society urged that DDT "be banned throughout the land and banned from export."

A leading British scientist (D. G. Hessayan) later pointed out that "If there had been a world-wide ban on DDT, then Rachel Carson and her *Silent Spring* would now be killing more people every year than Hitler killed in his entire holocaust."

21.2.3 STARVATION

Starvation is also a great problem in third world nations, where insect pests typically destroy nearly half of all crops each year. In 1986, Secretary of State George Schultz telegraphed orders to U.S. ambassies in Africa, stating that "The U.S. cannot, repeat cannot, as a matter of policy, participate in programs using any of the following pesticides: (1) lindane, (2) BHC, (3) DDT, or (4) dieldrin." To combat swarms of locusts, the most effective pesticide was dieldrin. Without it, 300 million tons of crops were destroyed, and widespread human starvation followed. Within a decade millions of humans starved or died of insect-transmitted diseases as a result of that action by George Schultz.

21.2.4 NEWS MEDIA LACK OF RESPONSIBILITY

We have now been exposed to more than 40 years of untruthful statements in *Audubon* magazine, the Sierra Club publications, *National Wildlife*, and many other "environmental magazines." Most news media found it difficult to disagree with such wealthy, influential groups, so their propaganda was repeated in newspapers and magazines, and on radio and television reports. It became very difficult to inform the general public of the truth about such matters!

Ben Bradlee, the *Washington Post* editor stated: "I'm no longer interested in news. I'm interested in causes. We do not pretend to print the truth. We print what people tell us. It's up to the public to decide what's true." (However, they only repeated what favored sources told them, and the readers were brainwashed, instead of informed!)

Charles Alexander said, "As the science editor at *TIME*, I would freely admit that on the environment we have crossed the border from news reporting to advocacy."

Stephen Schneider, now a Stanford professor, wrote (in *Discover*, October 1987): "We have to offer up scary scenarios, make dramatic statements, and make little mention of doubts. Each of us has to decide what the right balance is between being effective and being honest." Obviously he decided that being honest is not very practical. Many antipesticide activists obviously feel that way, too.

But what about harm caused to wildlife and the environment? Many people, misinformed by the affluent pseudoenvironmental organizations, feared that DDT, for example, might harm birds and other wildlife. The claims of such threats were not supported by facts, however the general public seldom learned the truth about those allegations. Consequently, they donated millions of dollars to the propagandists so they could "continue their good work."

21.2.5 POPULATION EXPLOSIONS OF BIRDS

When marshes in the midwest were sprayed with DDT to control mosquitoes, a common result was a population explosion of birds such as redwing blackbirds. They swarmed out of the marshes and destroyed great quantities of crops. *Audubon* magazine (August 1971) reported: "Today, in a small area of northern Ohio ten million redwings mill about in the cornfields after nesting season." The Virginia Department of Agriculture stated: "we can no longer tolerate the damage caused by the redwings . . . 15 million tons of grain are destroyed annually . . . enough to feed 90 million people." DDT caused those outbreaks of birds because: (1) it eliminated mosquitoes and black flies, which are carriers of bird diseases (avian malaria, avian bronchitis, leucocytozoan diseases, encephalitis, and fowl-pox); (2) it reduced destruction of plant products by insects, thus increasing the abundance of bird food; (3) egg production is reduced by 10 to 30 percent or more when birds are infested by chewing lice, but the lice are quickly killed by DDT; and (4) it stimulated more hepatic enzymes to be produced by the livers of the birds. Those enzymes destroy cancer-causing aflatoxins that are produced by molds in grain, seeds, and nuts. Aflatoxins are carcinogenic at levels of 0.03 to 0.08 parts per million in the diet. (Remember how small a part per million is: In a pile of pennies worth $10,000 one part per million is just one penny.) The *British Medical Bulletin*, 25: 278, 1969, (1969) and several other medical journals revealed how DDT in the diet prevents aflatoxin toxicity in birds and mammals.

21.2.6 AUDUBON CHRISTMAS BIRD COUNTS

In Indiana and Ohio, I participated in the nationwide Audubon Christmas Bird Counts for several years. In 1941 (before DDT was present) those counts recorded 19,616 robins (only 8.41 seen per observer). In 1960 (after extensive DDT usage) the total counted was 928,639 robins (104.01 per observer). That was an increase of 12 times more robins seen, per observer, during the DDT years than before DDT was present. Science articles also provided evidence that DDT had never adversely affected bald eagles. In 1960, the Christmas Bird Counters reported 25 percent more eagles, per observer, than during the pre-DDT bird surveys.

Even while bird numbers were expanding (in 1962) Carson wrote in *Silent Spring*: "Like the robin, another American bird seems to be on the verge of extinction. This is the national symbol, the eagle."

That same year the greatest ornithologist in the United States, Dr. Roger Tory Peterson, wrote (in his Nature Library Book, *The Birds*), that "North America's most abundant bird is the robin."

21.2.7   BALD EAGLE FACTS AND FIGURES

In 1921, an *Ecology* article was entitled: "Threatened Extinction of the Bald Eagle" (Alaska paid bounties on 128,000 bald eagles, up to 1952).

In 1930, (15 years before DDT) ornithologists reported that there were only 10 nesting bald eagles in Pennsylvania, 15 in the Washington D.C. area, and none in most of New England. *Bird Lore* magazine wrote: "this will give you some idea of the rarity of the eagle in eastern U.S." So, bald eagles were nearly extinct long before DDT or other man-made pesticides were discovered. Do environmental extremists think those eagle populations declined *in anticipation of* DDT?

The Hawk Mountain Sanctuary reported that the number of bald eagles migrating through Pennsylvania more than doubled during the first six years of heavy DDT usage in eastern North America. Before DDT was used, the Audubon Christmas Bird Count recorded only 197 bald eagles in 1941, but after years of heavy DDT use they recorded 891 bald eagles in 1961. In 1973 an Everglades National Park biologist stated: "I know of no evidence that the region ever supported a larger number of nesting bald eagles." (in Laycock's *Autumn of the Eagle*, p. 157).

In 1960–1964, the U.S. Fish and Wildlife Service Center at Patuxent, Maryland autopsied 76 bald eagles that were found dead in United States and reported that 71 percent had died violently (shot, electrocuted, or impacted with towers and buildings) and four died of diseases, but none were poisoned by pesticides. They concluded "the role of pesticides has been greatly exaggerated" (*J. Wildlife Diseases*, 6, 1970). From 1964 to 1972, they analyzed 190 more dead eagles. Most had been shot, and the majority of the others also died violently. There were 19 suspected cases of dieldrin poisoning, but no DDT involvement (*Pesticide Monitoring Journal*, 9: 12–13, 1975).

The Fish and Wildlife Service fed high levels of DDT to caged bald eagles for 112 days (up to 4000 mgs/kg), with no adverse effects (*Trans. 31st N. A. Wildlife Conference*, pp. 190–200, 1966). From 1973 to 1988, the United States spent millions of dollars for eagle breeding and rearing programs, so more were being seen by people in almost every part of the United States. In 1983, New York state had only three active bald eagle nests, but then they imported 150 eagles from Alaska. Peter Nye wrote in *Natural History* magazine (May 1992) that in 1940 there were only a few pairs, "yet the oft-mentioned culprit DDT wasn't there until the 1950s, when the last few nesting eagles were already struggling for survival."

21.2.8   GULLS TOO ABUNDANT TO LIVE

On Tern Island in Massachusetts, seagulls increased during the DDT years from 2000 pairs in 1940 to 35,000 pairs in 1971. William Drury, president of Massachusetts Audubon Society, decided to poison 30,000 of those gulls, even though they were on the state's list of protected birds. He succeeded, and said "It's kind of like weeding the garden" (*AP*, April 13, 1971). It was remarkable that nobody seemed to notice that the numbers of gulls had increased by 28,000 during the years of greatest DDT use!

21.2.9   POPULATION EXPLOSIONS OF GANNETS AND MURRES

Some environmentalists theorized that fish-eating birds were especially at risk because of "biological magnification" up food chains (biological magnification was another theory that was thoroughly refuted). Fish-eating birds multiplied on islands off the North Atlantic Coast. On Funk Island, gannets increased from 200 pairs in 1945 to 3000 pairs in 1970 and murres increased from 15,000 pairs in 1945 to a million and a half pairs in 1970 (*Animals Magazine*, April 1971). Still, environmentalists kept saying that DDT threatened bird populations.

21.2.10   HAWK MOUNTAIN MIGRATION COUNTS

Similar increases were reported in the numbers of raptorial birds migrating over Hawk Mountain, Pennsylvania. As a member of the Hawk Mountain Sanctuary Association, I received all of their publications, so learned the results of each year's counts of all migrating hawks. The total numbers of hawks (and eagles) counted migrating over Hawk Mountain increased from 9291 in 1946 (before much DDT use) to 16,163 in 1963 and 29,765 in 1968 (after 20 years of great DDT use).

21.2.11   OSPREY DATA

Raptors always receive a lot of attention from environmentalists, perhaps because they are so vicious. Ospreys were a great pest around fish hatcheries, so traps were set atop poles near the ponds. In 1943 (before DDT), leading authority Joseph Hickey attributed a 70 percent decline of eastern ospreys to that pole-trapping. Correlated with DDT increases, counts of ospreys migrating over Hawk Mtn totalled 254 in 1951, 352 in 1961, 527 in 1969, and 630 in 1971 (just before DDT was banned). In 1976, the *Hawk Mountain Sanctuary Newsletter* reported: "For reasons we do not understand at all, the numbers of osprey counted is returning to something like normal—318 in 1974 and 279 in 1975." (In other words, they said they could not understand why there were 351 *fewer* migrating ospreys during the years after DDT was banned.) Environmentalist propaganda apparently has a blinding effect!

21.2.12   PEREGRINE DATA

Dr. William Hornaday (head of the N.Y. Zoological Society) discussed peregrines in his 1913 book, *Vanishing Wildlife*. He wrote that the undesirable peregrines "deserve death, but are so rare that we need not take them into account." Later, he urged persons who found peregrine nests to "shoot the parents and collect the eggs or young." (Peregrines were listed in most states as "vermin" before environmentalists converted them to ecological "gold mines" in the United States.)

Thomas Cade (the founder of the Peregrine Fund) wrote that "peregrines completely disappeared from east of the Rockies," and that "the subspecies is probably extinct." His Fund then reared more than 4000 peregrines (of *foreign* subspecies), at a cost of many millions of dollars, released them in the eastern United States, and then claimed that the Endangered Species Act had "saved the eastern peregrines." Cade was disappointed when a regional director of the FWS ordered that no more European peregrines be released in eastern United States. Cade said we are left "with a large number of Spanish and Scottish peregrines on our hands" (*Audubon*, November 1977). Brian Walton, who was in charge of the California branch of the Fund, reported that it cost $1500 to $2000 for each peregrine produced. In 1985, the Fund's director complained that they were having trouble raising the million dollars for that year's peregrine recovery program "because 50 million people were starving in Ethiopia."

I enjoyed driving to Inuvik, North West Territory, where peregrines are common. Canadian biologists reported that nesting success "was as high as ever recorded for the species (an average of 2.4 young per active nest)." Frank Beebe, Canada's leading raptor authority, wrote in his book, *The Myth of the Vanishing Peregrines*, that "It appears that the Canadian peregrines, not knowing how gravely ill they are, go right on reproducing in blissful unconcern of their desperate plight."

What effect did DDT have on birds that ingested it? Researcher Hickey testified during the EPA hearings that he could not even kill his caged robins by overdosing them with DDT because it simply passed through their digestive tract and was eliminated with the faeces. In other research, reported in the *Journal of Wildlife Management*, baby birds in nests were fed *only* food containing high levels of DDT, and none were adversely affected.

21.2.13 BIRD EGGSHELL DATA

Rachel Carson referred to "Dr. DeWitt's now classic experiments on quail and pheasants." She said, on page 120: "Quail into whose diet DDT was introduced throughout the breeding season survived and even produced normal numbers of fertile eggs, but few of the eggs hatched." I read DeWitt's article (in *Journal of Agriculture and Food Chemistry*, 1956) and found that 75.7 percent of the eggs produced by DDT-fed birds hatched, compared with 83.9 percent of those produced by the "controls" (birds with no DDT). I thought 75.7 percent was more than "a few" eggs hatched, so I became even more suspicious of Carson's intentions. In his table, DeWitt also reported that 80.6 percent of the eggs produced by his *pheasants* on the DDT diet hatched, compared with only 57.4 percent hatching of the eggs produced by the "control" birds. It was not surprising that Carson avoided mentioning how much *better* the DDT-fed pheasants did (despite her reference to "DeWitt's classic experiments on quail *and pheasants*").

The *S. F. Chronicle*, on April 14, 1969, reported that because of DDT bird eggshells were becoming so thick that the young often could not get out of the eggs. On Feb. 24, 1969, the same newspaper reported that, because of DDT, bird eggshells were becoming so thin that they could break under the weight of the incubating females. Neither allegation was true, but they indicated that in the *San Francisco Chronicle* there was already little hope for truthful reporting on the subject of DDT! (And it became worse, every year.)

A common misconception for many years was that DDT caused birds to produce thin-shelled or softer shelled eggs. With so many studies proving that this charge was not true, it is amazing to see it still being repeated! (NO confirming data are ever provided, but the naked statement is simply made, in the press, on radio, on television, and in environmental magazines!) The poultry and egg industries should have been the first place to seek the truth, but the environmentalists knew that would destroy their eggshell propaganda. Likewise, environmental propagandists avoided the great 1949 book on the subject, by Romanoff and Romanoff, titled *The Avian Egg*, which contained all of the information needed to explain the "thin eggshell" problems. A 1967 book by the same authors was *The Avian Embryo*, which provided details regarding the amount of calcium drawn from the eggshell by the developing embryo. The propagandists never cited that book either, however they usually collected and measured eggs *after* the embryo had removed calcium from the eggshell, for bone development.

USFWS biologists Tucker and Haegele (*Bull. Environ. Contam. & Toxicology*, 5: 191, 1971), fed different levels of calcium to different groups of quail. One group got 3 percent calcium and another group only got 1 percent calcium. None had any DDT or DDE in their diet. The shells produced by the 1 percent group were 9.3 percent thinner than those on the normal 3 percent calcium diet. Now, with those details available, how could a person design an experiment that would incriminate DDT as a cause of eggshell thinning? Simply feed the birds a reduced calcium diet, add DDT to their food, and then blame the thinner shells (that would certainly result from the calcium deficiency) on the DDT in the bird's diet! That is exactly what anti-DDT researchers in the U.S. Fish and Wildlife Service did.

Bitman and his colleagues at Patuxent fed their quail only *half* as much calcium as the lowest amount Tucker's quail received. Tucker's birds had produced shells about 10 percent thinner when only 1 percent calcium was in their diet, so what would result from Bitman's feeding quail only 0.5 percent calcium? Their shells would be expected to be even thinner than 10 percent of normal. Bitman reported, however, that the shells were not THAT thin! His article was published in *Science* magazine, however, and was the most widely used reference to "prove" that DDT caused thin eggshells!

Actually, a great many other feeding experiments proved that shells are *not* thinned by the introduction of DDT into the diet of birds *if* there is adequate calcium in their diet, but such results were seldom mentioned in the media, and never mentioned in pseudoenvironmental publications. To get thinner shells, the anti-DDT activists always had to do something else at the same time ... something that was *known* to cause thinner eggshells all by itself. Things having that effect include noise, excitement, irritation, dimmed lights, shortage of water, presence of several kinds of chemicals, and, *especially* a deficiency of calcium in the diet. Every bird experiment that resulted in thin eggshells used one or more those known causes in order to produce the desired effects, which were then blamed on the DDT.

In congressional testimony I presented the data, and was critical of Bitman's work. The next year he repeated his experiment, but fed the birds adequate calcium in their diet. The DDT-fed and DDE-fed birds produced eggshells that were not thinned at all. The article was presented again to *Science* magazine. Unfortunately, the editor of *Science* magazine always refused to publish articles that were favorable to DDT, so he rejected Bitman's new article. It was published, instead, in *Poultry Science*, and poultrymen and unbiased scientists applauded the truthful results. Of course, the circulation of that journal was not nearly as great as for *Science*, so relatively few scientists ever heard about the reversal of the allegation that DDT and DDE caused thinner eggshells.

Why did *Science* refuse such articles? The editor, Philip Abelson, had earlier informed Dr. Thomas Jukes that *Science* would never publish any article about DDT that was not antagonistic to that insecticide. He refused to even consider a manuscript written by the World Health Organization. As a result, the DDT articles in *Science* were mostly written by the same coterie, and "peer review" became a sham. The anti-DDT authors just kept citing each other and supporting each other's statements. No other views were accepted. Without that sheltered bias, the case against DDT would have quickly folded!

M. L. Scott, J. R. Zimmerman, Susan Marinsky, P. A. Mullenhoff, G. L. Rumsley, and R. W. Rice spent years at Cornell testing various chemicals in the quail diet to determine the greatest causes of shell thinning. They reported that DDT, DDD, and DDE in the diets resulted in thicker shells, rather than thinner shells. The chemical that caused the greatest amount of shell thinning was methyl mercury (*Poultry Science*, 54: 350–368, 1975). The results of years of reliable scientific work by these researchers also did not appear in *Science* magazine.

Tucker et al., in *Utah Science* (June 1971) published the results of careful experiments performed to determine serious dietary causes of eggshell thinning. Some of the results are given below:

Chemicals in the food	Effects on eggshells
Lead	14.5% thinner than normal
Sevin	8.7% thinner
Mercury	8.6% thinner
Parathion	4.8% thinner
PCBs	4.0% thinner
o,p' DDT	0.5% thicker than normal
Tech DDT	0.0% (No change)
DDE	0.0% (No change)

Also, after water was withheld for 36 hours, the quail laid eggs with shells averaging −29.6 percent thinner than normal.

21.2.14 IMPORTANCE OF CHLORINE

DDT is a chlorinated hydrocarbon compound, dichlorodiphenyl-trichloroethane. It has certainly saved at least a billion human lives. In addition to directly preventing deaths from malaria, typhus, yellow fever, plague, and a dozen other famous killers, it has made it possible for humans to work harder, harvest more food, and live longer, healthier lives. Many antagonists of DDT are outspoken critics of *all* chlorine compounds. Greenpeace is leading the campaign to rid the world of chlorine, but public health and medical organizations all around the world have praised chlorine for its role in protecting public health and saving lives.

A *Science* editorial on August 26, 1994 stated: "There is reason to hope that the EPA will not continue to act like a tool of Greenpeace. A plethora of EPA regulations and unfunded mandates coupled with examples of brutality in enforcing them has cost the EPA their support in Congress." The World Health Organization estimates that 25,000 children die each day from drinking water that has not been chlorinated. Peru was encouraged by the United States activists to remove chlorine from their drinking water supplies. That move rather quickly resulted in more than a million illnesses and more than 8500 deaths from organisms in the water that would have been eradicated if chlorine was present.

Dr. Gordon Gribble, a famous biochemist at Dartmouth College, has written extensively on the subject and even wrote a book that contains more than 2000 structural formulae of chlorine compounds! He points out that 85 percent of all pharmaceuticals require chlorine, and more than 25 percent of all medical equipment is also dependent on chlorine for their manufacture.

Dioxins are a group of about 75 chlorinated chemicals, great quantities of which are produced in nature when wood or other material burns. Human activities produce them also, during paper pulp production, but the amount is less than a pound or two per year from the entire industry. Forest fires produce more dioxins than all other sources combined. During the Viet Nam War, dioxins were present in Agent Orange, the chemical used to defoliate jungle trees so human movements could be seen from the air. The most toxic form of dioxin is probably TCDD, but no human deaths are known to have been caused by it, even following heavy exposures for long periods of time. Skin rashes have been the most frequent result of overexposures, but no cancers of any kind have been caused. Dr. Gordon Gribble published these facts in a *Heartland Institute* journal in 1996. He also commented that over 40,000 scientific articles have discussed dioxins, and "the evidence now at hand does not support claims that dioxin is a major health threat." A study of 2200 Dow Chemical employees who were in close proximity to dioxin were tested for cancer, and had slightly lower than normal cancer rates.

Dr. Dixy Lee Ray criticized the propaganda surrounding the ozone hole and chlorofluorocarbons. It was an interesting hypothesis, she said, "... but no CFC breakdown products have ever been found in the atmosphere." The National Academy of Sciences predicted an 18 percent ozone decrease (in 1980), but dropped it to 7 percent (in 1984), then to 2 percent (in 1985), and finally to "5 percent over the next hundred years."

In 1983 Greenpeace wrote: "We should not wait for scientific proof of harm before we take action: The use and discharge of chlorine chemicals that may cause harm should be avoided. Proof of innocence is not required. No further organochloride pollution should be permitted . . . this means phasing out the substance that is their root—chlorine."

21.2.15 GYPSY MOTHS

In 1869 Leopold Trouvelot brought some gypsy moths to Medford Massachusetts, thinking perhaps silk could be made from their cocoons. A few escaped, and multiplied. For 30 years applications of lead arsenate (5 lbs/acre) was the only way to slow them down, but it was too expensive, too hard to apply, killed too many nontarget organisms, and still failed to halt their spread. In 1945, 800,000 acres of oak trees were defoliated, in eight states, and a decade later nearly 10 million acres of oaks were being destroyed annually. DDT was sprayed (one lb/acre) and quickly eradicated the moths from Pennsylvania, New Jersey, New York, Virginia, and all other states west of Vermont. The National Audubon Society monitored the program and said "no damage was done to birds, including nestlings in their nests."

James Nichols (in his 1961 booklet on gypsy moths): pointed out that "over a million acres of Pennsylvania, New Jersey and New York were sprayed, always with 100 percent eradication of the pests. No infestation survived a single aerial treatment with DDT on 1,107,458 acres." Disregarding the tremendous destruction by the moths, most environmental organizations fought desperately against any use of DDT to preserve the forests. They would rather lose millions of acres of the great eastern oak forests than modify their harsh anti-DDT propaganda!

Carl Amery wrote: "We in the Green movement aspire to a cultural model in which the killing of a forest will be considered more contemptible and more criminal than the sale of 6-year old children to Asian brothels." When considering the millions of acres of dead oaks in eastern United States caused by environmentalists preventing the use of DDT, and the extensive forests permitted to burn in Yellowstone, I often recall Amery's comment. I wonder if the environmentalists who permitted those disasters to happen ever think of it, also.

21.2.16 NORMAN BORLAUG'S WARNING OF PESTICIDE DOMINOES

Dr. Norman E. Borlaug was a Nobel Peace Prize winner in 1970 because of his "Green Revolution." In a United Nations speech in Rome (*Science*, 10:1109 December 1971) he stated that "fear-provoking, irresponsible environmentalists" were mounting a "vicious, hysterical propaganda campaign against agricultural chemicals." He praised DDT's great record of safety for humankind, and warned that its elimination in the United States would be followed by campaigns to have it banned everywhere. He warned that "DDT is only the first of the dominoes ... As soon as DDT is banned there will be a push for banning *all* chlorinated hydrocarbons then, in order, the organic phosphate and carbamate insecticides. Then they will attack the weed killers, and eventually the fungicides." Dr. Borlaug was exactly right, and most of his predictions have already come true. In the 1970s and 1980s the EPA, relying primarily on the misapplication of the Delaney clause, banned chlordane, aldrin, dieldrin, endrin, BHC, Lindane, heptachlor, toxaphene, and many other pesticides.

21.2.17 EPA ADMITS FALSE ALLEGATIONS

In the early 1970s, EPA released false reports to Congress about the amounts of DDT in human diets, and we wrote to object. Laurence O'Neill responded, writing: "You are correct in stating that EPA's DDT report erred. The correct figure should have been 15 *micrograms* per day instead of 15 *milligrams*." (The average human intake at that time was about 13 milligrams *per year*.) O'Neill also stated that the human intake had dropped rapidly to 1.8 *micrograms* instead of 1.8 *milligrams*, after the ban. (In other words, the daily intake had dropped from 0.015 milligrams to 0.0018 milligrams.) "We will make every effort to rectify the erroneous figures with the news media," he promised. But they did not.

I was making many speeches at the time, and before speaking I usually swallowed a tablespoon of DDT to get the audience's attention. I felt safe doing that, because volunteers for federal studies had ingested 35 mgs of DDT daily for 20 months, without experiencing any adverse effects. Also, 35 workers at the Montrose DDT plant in Torrance, California had been taking in about 400 times more DDT daily than the average man, for 19 years, and not a single case of cancer developed.

The EPA also falsely claimed, in a radio broadcast (May 15, 1975) that "hundreds of thousands of American farm workers are injured every year by pesticides, and hundreds of them die annually as a result." When challenged by actual data, EPA meekly apologized, saying: "We used those statements in good faith, thinking they were accurate, and they turned out not to be accurate ... They cannot possibly be substantiated" (*UPI Press Release*, May 24, 1975).

21.2.18 FAULTY EXTRAPOLATION, FROM 235 TO 312,000

What evidence could have led anyone to make such a claim? *USA Today* (April 14, 1992) printed an editorial using that same figure, and attributed it to "a congressional study last month." I wrote to the editors, pointing out that the statement actually came from a World Resources Institute press release seven years earlier! I quoted the two WRI researchers who made the study (Robert Wasserstrom and Richard Wiles) but quit because of the untruthful figure of 300,000 in that press release, which they said "tells a story substantially different from what we found." (*Chemical & Engineering News*, September 1985.) The 300,000 figure was based on a report that 235 California farm-workers had made medical complaints in 1982 (roughly half of the complaints involved skin irritation from sulfur). Dr. Molly Coye (NIOSH) extrapolated from 235 to 300,000 cases, as follows. Dr. Ephriam Kahn had previously estimated that California doctors reported only about 1 percent of such cases, so Molly Coye multiplied 235 by 100 and said 23,500 California workers must have actually had medical problems because of pesticides during the year. That would be about 7.8 percent of California farm-workers. Because there were about four million farmworkers in the United States, she calculated 7.8 percent of four million, to arrive at a total of 312,000 "poisoned" farmworkers each year. Dr. Coye never mentioned Dr. Kahn's well-known year-long study in 1977, wherein he concluded that 80 percent of farmworker illnesses are reported (rather than his earlier estimate of 1 percent). As usual, *USA Today* did not respond to my letter or the enclosed documentation of facts.

21.2.19 NATURAL PESTICIDES PRODUCED BY PLANTS

Dr. Bruce Ames (biochemistry professor at the University of California) pointed out in 1987 that we ingest in our diet about 1.5 grams per day of natural pesticides. Our foods contain 10,000 times more, by weight, of *natural* pesticides than of man-made pesticide residues. More than 90 percent of the pesticides in plants are produced *naturally* by the plants, which help protect them from insects, mites, nematodes, bacteria, and fungi. Those natural pesticides may make up 5 to 10 percent of a plant's dry weight, and nearly half of those that were tested on experimental animals were carcinogenic. Americans should therefore feel unconcerned about the harmless infinitesimal traces of synthetic chemicals to which they may be exposed.

21.2.20 ORGANIC GARDENING

The highly publicized traces of synthetic pesticides on fruits and vegetables worried some people so much that they began to favor "organically produced" foods, thinking that they would not contain any pesticides. Most people are not aware that organic gardeners can legally use a great many pesticides, so long as they are not man-made. They can use nicotine sulfate, rotenone, and pyrethrum (derived from natural plants), or any poisons that occur naturally, such as lime, sulfur, borax, cyanide, arsenic, and fluorine.

21.2.21 EPA HEARINGS: TESTIMONY BY WOODWELL AND EPSTEIN

In 1971, the Environmental Protection Agency was forced to hold hearings on DDT. The hearings were presided over for seven months by Judge Edmund Sweeney. During the seven months, hundreds of scientists expressed their views and presented evidence. The printed transcript of testimony exceeded 9000 pages, and would be the basis upon which an interesting university course could be developed!

George Woodwell wrote in 1971 that "six billion lbs of DDT had been used, but only 12 million lbs could be accounted for in all of the earth's biota," and that was "less than a thirtieth of one year's production of DDT in the 1960s." He theorized that "most of the DDT has either been degraded to inocuousness or sequestered in places where it is not freely available." That *Science* article (December 1971, p. 1101) contrasted so sharply with his testimony during the EPA hearings that a reporter asked him why he had completely omitted it from his testimony. Woodwell replied that the EPA lawyers told him not to mention the article, "lest my testimony be disallowed" (*Business Week*, July 8, 1972).

Woodwell made what he called a "typographical error" in *Science* magazine (10 December 1971). When citing data from two other articles that reported 10^{12} parts of DDT he referred to such values as "parts per million," rather than parts per trillion. He reported that "Wheatly found 3 parts per *million* in English fields," but it was really only 3 parts per *trillion* [a millionth of the amount Woodwell stated!] He also said that "Tarrant found 73 to 210 parts per *million* in rainwater, but Tarrant's *highest* reading was actually only 190 parts per *trillion* (That millionfold exaggeration made his statements appear menacing.) He then said those references "confirmed high levels of DDT in the rain of England, similar to concentrations in the U.S.," but none of the references he cited contained *any* data from the United States, not even in parts per trillion!

During the EPA hearings, Samuel Epstein testified that he was a member of the Health, Education and Welfare panel on carcinogens, but under cross-examination he admitted he was *not* on that panel. Epstein also stated that tests by Fitzhugh et al. indicated that mice with DDT in their diet developed cancer. He failed to mention that Fitzhugh's *control* mice (with no DDT) developed 26 percent *more cancers* than did his DDT-fed mice. Fitzhugh said the reason the report was not published was that they had discovered the mice were mistakenly fed 300 mg/kg of DDT for an unknown period of time, rather than the intended 100 mg/kg.

21.2.22   *BREAKDOWN OF DDT IN ENVIRONMENT*

Philip Butler testified at the EPA hearings and sought to convince people that DDT did not break down rapidly and disappear from the environment. He stated that "I am thinking of a study which has shown that DDT persists for as much as 40 years in terrestrial deposits." We knew that was untruthful, because DDT had only been around for 30 years at the time of his testimony. Under cross-examination, Butler also admitted that published reports from his own EPA laboratory at Gulf Breeze, Florida, revealed that 92 percent of the DDT *and its metabolites, DDD and DDE*, had disappeared from sea water (in huge closed submerged glass containers), in just 38 days (Wilson, A. J., *USDI Circular*, 335: 20, 1970) Dozens of other published studies reveal that DDT and its metabolites also disappear rather quickly from normal outdoor soil.

Croker and Wilson applied DDT to a tidal marsh. In less than 24 hours only traces remained, and even those traces disappeared in five days (*Trans. Amer. Fish. Society*, 94, 1965).

In Washington state estuaries, the Bureau of Commercial Fisheries monitored pesticide residues in shellfish at 19 stations during three years of heavy DDT use (1966–1969). Ninety-three percent of the samples contained less than 10 parts per *billion* of DDT and the highest level found was only 0.1 part per million. Shellfish are known to concentrate chlorinated hydrocarbons in their system at levels 40,000 to 70,000 times as great as that in the surrounding water, so it was evident that DDT residues had not persisted long in the coastal waters.

21.2.23   *JUDGE'S FINAL DECISION, AFTER EPA HEARINGS*

After seven months of such testimony, the EPA Hearing Examiner, Judge Edmund Sweeney, issued his final official decision on April 26, 1972. In it he stated, "DDT is not a carcinogenic, mutagenic, or teratogenic hazard to man. The uses of DDT involved here do not have a deleterious effect on freshwater fish, estuarine organisms, wild birds, or other wildlife ... The evidence in the proceeding supports the conclusion that there is a present need for the essential uses of DDT."

21.2.24   *WILLIAM RUCKELSHAUS OVERRULES JUDGE'S DECISION*

The EPA administrator, William Ruckelshaus, never attended a single day of the hearings, and his special assistant, Marshall Miller reported that he did not even read the transcripts (*Santa Ana*

Register 23 July 1972). Nevertheless, he overruled his judge's decision and single-handedly banned DDT. His final ruling was not very reassuring. He used the wrong chemical name for DDT; stated that "DDT has three major breakdown products, DDA, DDE, and DDD," and that "separate registrations exist for TDE (DDE)." (The truth is that DDE is *not* the same as TDE, and DDE was never registered as an insecticide.) He also stated that farmers should use parathion as a substitute for DDT, evidently unaware that hundreds of humans had been killed by parathion and that it is extremely toxic to bees, birds, and every other form of animal life! Rachel Carson recalled that "a small parathion application killed 65,000 redwings, as well as raccoons and rabbits."

My lengthy critique of Ruckelshaus's *Order* was inserted into the *Congressional Record* by Senator Barry Goldwater. Later, in a letter to Allan Grant the president of the American Farm Bureau Federation, dated 26 April 1979, Ruckelshaus wrote, "Decisions involving the use of toxic substances are political, with a small 'p'," and "the ultimate judgment remains political." Ruckelshaus refused requests by the U.S. Department of Agriculture (and others) to comply with the Freedom of Information Act, and also refused to file any Environmental Impact Statements (even though his actions would result in the loss of millions of human lives, worldwide, and the destruction of millions of acres of forest in the United States).

21.2.25   GORE APPOINTS NEW EPA ADMINISTRATOR

In 1993 Vice President Al Gore appointed Florida environmentalist Carol Browner to be EPA Administrator. Shortly thereafter Browner reported: "I'm appalled by what I've learned about the EPA's total lack of management, accountability and discipline. I have reviewed audit reports that clearly describe serious violations of rules and an intolerable waste of taxpayer's money." (*Audubon*, September 1993) Well, we can certainly agree with *that*!

21.2.26   FAULTY ANALYSES OF DDT

Faulty analyses of soil and water led many people to believe that DDT was very persistent in the environment. The more likely truth is simply that samples were not properly analyzed. In 1969, Dan Anderson reported that he had reanalyzed the five pooled samples he used in 1965 to help ban DDT. Three of the five samples he had earlier reported as having high levels of DDT actually contained none at all, and the other two contained only a fourth as much as he had earlier claimed. (*Canadian Field-Naturalist*, 1969, pp. 91–112.)

Many other scientists warned that most analytical procedures did not distinguish between DDT and PCBs, and that "some chromatogram peaks of PCB are identical to peaks of DDT, DDD, and DDE." (*Env. Sci. Technology*, 1970.)

W. Hylin warned that "organochlorine compounds in plants can cause interference in analyses of residues of DDT" (*Residue Reviews*, 1969), and J. J. Sims found that "marine algae produced halogen compounds that had been misidentified as DDT metabolites, and that halogen compounds containing bromine or iodine also can register falsely on the gas chromatograph."

Frazier et al. analyzed 34 soil samples that had been sealed in glass jars since they were collected in 1911. The gas chromatograph indicated that five kinds of chlorinated hydrocarbon insecticides were in that soil, even though none were in existence until 30 years after the samples were sealed. (*Pesticide Monitoring Journal*, 4(2): 67–70, September 1970).

W. Hom (*Science*, 184: 1197–1199) explained a high "apparent DDE" concentration in sediments that were deposited in the Santa Barbara Basin of Southern California 12 years before any DDT existed. He said, "we attribute the DDE in the 1930 sample to spurious contamination during collection, storage or analysis. (Thousands of other samples have been reported to contain DDT more than 10 years after it ceased to be present, and usually the persons who reported such contaminations have not bothered to retract the false reports.)

21.2.27   BREAKDOWN OF DDT IN ENVIRONMENT

Environmentalists often said that "DDT cannot be broken down in the environment," and Marc Lappé even wrote that "DDT is not broken down by living things." Actually it *was* rather quickly broken down in the environment by heat, cold, moisture, sunlight, alkalinity, salinity, many natural chemicals, and common soil microorganisms. (Obviously if it did not break down, it would not be necessary to apply it repeatedly to crops to control the pests!) DDT is also quickly destroyed by hepatic enzymes in birds and mammals, and arthropod pests often developed "resistance" to DDT by degrading it within their bodies. When I heard Lappé's allegations I quickly went to my files and found more than 140 articles documenting the breakdown of DDT in the environment (not including examples of pests that had built up natural "resistance" to the chemical). I mailed copies to many newspapers and radio and television stations, but not a single one responded or corrected their earlier false statements, even after they had the scientific data in hand!

During the campaign seeking EPA's permission to spray parts of three northwestern states with DDT to halt the great tussock moth outbreak in the 1970s, the *Vancouver Sun* and the *Lewiston Tribune* both carried editorials (December 12, 1974) claiming that "DDT has a half-life of several thousand years." I knew where they had gotten that false allegation, so I sent each editor copies of the scientific literature refuting the charge, but neither editor responded. After three years of strenuous campaigning, we got permission from the EPA to spray 430,000 acres of forest in the northwest. That single well-timed spray of DDT (at 1 lb/acre) eradicated the epidemic and caused no harm to other forms of life.

21.2.28   THE DELANEY CLAUSE

The Delaney Clause of the Food Additives Law ruled, under Section 409 of 21 USCS 3498, that "no chemical shall be deemed to be safe if it is found to induce cancer when ingested by man or animal, or if it is found, after tests which are appropriate for the evaluation of the safety of food additives, to induce cancer in man or animal. . . ." The tests that were usually used were NOT "appropriate for the accurate evaluation of carcinogenicity." They involved extremely high doses forced into the diets of rats that had been specially bred to be hypersensitive to carcinogens.

The use of *inappropriate* tests on rodents involving massive dosages and unnatural applications of chemicals have caused much controversy. The American Council for Science and Health wrote that, "Sound toxicological principles are routinely flouted in laboratory rodent tests and the results are frequently inappropriately extrapolated to humans" (1991). Rats were found to produce a special protein (Alpha 2U Globulin), which makes them especially prone to develop tumors and cancers. In 1992, the Environmental Protection Agency pointed out that humans lack that protein, which "*could invalidate thousands of tests of pesticides, preservatives, additives, and other chemicals that were banned because they produced tumors in laboratory rats.*" Those tumors, they said, "are a species-specific effect in rats and are inapplicable to human risk assessments." Obviously, such rodent tests should not have been considered "*appropriate for the evaluation of the safety of food additives to induce cancers in man or animals,*" as required by Delaney.

EPA administrator Russell Train ignored Delaney's requirement that tests must be *appropriate*, and EPA attorneys assumed they could ban any chemical that caused ANY cancer when applied to test animals at ANY dosage, and even in very *inappropriate* manners (including gavage and direct injections into blood, peritoneum, and elsewhere).

Even after ignoring Delaney's proviso requiring "appropriate tests" for carcinogenicity, they still could not have banned many of those substances if "cancer" had not been redefined by attorney Russell Train! *Cancers* had previously been considered to be malignant growths that tend to spread to other parts of the body, frequently with fatal results. *Tumors*, on the other hand, were considered to be nonmalignant lumps that did not spread (and in lab rodents they often disappeared after the massive chemical insults were halted). Attorney Russell Train redefined those medical terms, and stated that "for EPA's purposes tumorogenic substances and carcinogenic substances are synonymous" and "for purposes of carcinogenicity testing, no distinction should be made between the induction of tumors diagnosed as benign and the induction of tumors diagnosed as malignant." (*Chem. & Engineering News*, 52: 13, 1974) Substances of either type could therefore be called "carcinogens" and could then be banned by improperly invoking the Delaney clause! The Council for Agricultural Science and Technology (a consortium of more than 30 scientific and professional organizations, observed that "classifying as 'carcinogens' all chemicals that cause tumors greatly overestimates the cancer risk."

Train left the EPA to join the Board of Directors of the Union Carbide Corporation (which was not very "environmentally friendly"). At that time, the EPA already had more than 10,000 employees and its 1980 budget was $5,000,000,000 (yes, *billions*).

Appropriate questions might be: Did anyone at the EPA ever actually read the Delaney clause? If so, did they then deliberately seek to misinterpret Delaney's clear requirements? Representative Delaney once stated that "too many egos, reputations, and careers are at stake; if you try to change things, the crazies come at you with blow torches and chain saws." It is easy to understand why he bemoaned the fact that, as he stated, "*I'll go to my grave with that damn thing hanging around my neck.*"

21.2.29 THE FOOD QUALITY PROTECTION ACT

In 1996 Congress enacted the Food Quality Protection Act (FQPA). This mandate states that the EPA may ban any chemical, *unless* they believe that "there is a reasonable certainty of no harm." All existing pesticides are required to be reassessed before 2006. By August 1999, they must analyze 3000 of them. The Director of EPA's Pesticide Programs said one way to implement the Act would be to just revoke *all* tolerances and simply start over! Nobody has indicated what the EPA might mean by "reasonable," and (even worse) there was no indication of what they might mean by "harm." Also, the EPA will have to tell us what the meaning of "no" is!

21.2.30 OUR STOLEN FUTURE

In 1996, we were exposed to an improper new book, titled *Our Stolen Future*. The lead author was Theo Colborn. Early in the book it is stated that she tried to find evidence of increased cancer rates from chemicals in the Great Lakes area. Unfortunately, her investigation revealed *lower* cancer rates! Faced with this major setback, she turned her mind again to the wildlife literature and tried to think clearly about where she should go next. The resulting book dwells on unprovable allegations, including hazards from infinitesimal exposures to chemicals, which she says will result in sperm deficiencies; cancer of breast, testicles, and prostate glands; reduced human fertility; female endometriosis; eroded intelligence; increased disruptive behavior of children; and epidemics of undescended testicles and shortened alligator penises.

Jessica Matthews, of the Council on Foreign Relations, wrote about the book in the *Washington Post*, March 11, 1996, saying, "We have been too obsessed with the obvious risks of toxic chemicals, cancer, and birth defects. Immune suppression and hormone disruption, if proved, could be more dangerous." She contended that the book "will make earlier struggles—over nitrates, saccharin, formaldehyde, Times Beach, Love Canal, cholesterol, Alar, and even tobacco—look like kid's stuff."

Coauthor Carol Dumanoski (environmentalist for the *Boston Globe*) had earlier written, "There is no such thing as objective reporting, and I've become even more crafty about finding the voices to say the things I think are true. That is my subversive mission." According to the *Washington Times*, March 13, 1996, Miss Dumanoksi admitted in1994 that she had "manipulated facts about the hole in the ozone layer" in order to get top billing for her story, which therefore ran on page one of the *Los Angeles Times*.

The American Council on Science and Health reviewed the book, reporting, "The scientific evidence is extremely tentative but the potential for arousing fear in nonscientists is great." It was also reviewed in *Science* magazine, where the reviewer stated "it was not written for scientists," commented that there was "no discrimination between anectodal reports and scientific studies," and said the book "raises questions about the scientific judgment of the authors."

Miss Colborn said that Lake Apopka (FL) alligator penises are one half the normal size, but provided no previous baseline measurements. Louis Guillette became famous for measuring those penises, but his former cohort, Timothy Gross said the measurements were based on weak data because Guillette didn't know the age of any of the alligators, thus couldn't know if they were fully developed. My own article, in *21st Century Science and Technology*, Fall 1996 was titled "The Long and Short Of It," but dealt primarily with the condition of the lake itself. I have traced the condition of Apopka for over thirty years, because of its notorious pollution. *Wilderness* magazine (Winter, 1986) said that Lake Apopka was already a cesspool in the 1950s, because citrus processing wastes, sewage effluents, and wastes from hundreds of acres of muck farmland along its shores. *National Observer* (June 21, 1971) stated "Apopka is a fetid, shallow body of water, nearly unfit for human use. Human waste is dumped into the lake from Winter Garden's sewage treatment plant. That effluent contains birth control chemicals, including ethynylestradiol (EE) from womens' urine, which is hormonally active at concentrations as low as 0.1 nanogram (a tenth of a billionth of a gram)." It must be assumed that alligators now in the water must also be affected by the EE! Studies also reported high levels of *Aeromonas liquefaciens* in the water, a bacterium that dissolves internal organs of aquatic animals. In September 1971, *Audubon* magazine reported that thousands of turtles and fish died there, as well as the "first known die-off of alligators." It should be pointed out that the alligators apparently were not damaged earlier, during 30 years of DDT pollution!

21.2.31 ESTROGEN TESTS THREATENED, FOR BILLIONS OF DOLLARS

A National Academy of Sciences report regarding effects of chemicals, including estrogens, on humans will soon be released. Based on inconclusive allegations, such as those in Coleborn's book, the federal government now plans to test 60,000 chemicals. According to a recent *Forbes* magazine article by Michael Fumento, the plan is premature, because there has been no scientific verdict regarding alleged endocrine disruption. Thorough tests of suspect chemicals will cost an average of $1.5 million. "If EPA does not call off the hunt at a preliminary stage, the cost will be $23 billion to test just the most suspicious 24 percent of the chemicals." "Testing common organochlorines alone could cost the nation $100 billion yearly," says Dr. Fumento.

In a *Science* article, June 7, 1996, p. 1489, it was stated that while a single chemical may not have any adverse effect, a combination of four chemicals seemed to have an effect a thousand times greater than the combined effects of the four. A year later the researchers responsible for that article retracted it, in *Science*, 25 July 1997, because nobody could repeat the results, not even the original researchers.

21.2.32 FALSE ALLEGATIONS REGARDING ALAR, ON 60 MINUTES

Ed Bradley, on a "60 Minutes" show titled "A is for Apple," told 40 million American viewers (February 26, 1989) that "The most potent cancer-causing chemical in our food is a pesticide sprayed on apples to keep them on the trees longer and make them look better. And who is most at risk? The children, who may someday develop cancer from this one chemical." It is important to point out that Alar is *not* a pesticide, but is instead a plant hormone. It never killed anything, but simply increased the tree's ability to prevent early fruit fall. Bradley also failed to inform the viewers that not a single human case of cancer had ever been correlated with the use of Alar.

William Lijinsky was the major "scientific spokesman" on the CBS program. He was introduced by Ed Bradley as "the head of a chemical carcinogenesis laboratory at the National Cancer Institute."

The Cancer Institute objected, saying Lijinsky "is not employed by or connected with the National Cancer Institute in any way."

The EPA had already issued a press release (February 1, 1989) saying that a 2 year test on mice failed to indicate that Alar is carcinogenic. The president of the International Apple Institute said the hormone "is so scarce on apples that a person would have to eat 28,000 pounds of apples a day to get as much as the cancer researchers fed their mice," Other sources noted that a child would have to drink 19,000 quarts of apple juice every day, in order to be exposed to the proportional concentration of Alar that the rodents were forced to ingest.

When heated, Alar breaks down into UDMH (a hydrazine metabolite). Aflatoxins (natural molds in some human foods) are 3000 to 5000 times more potent than UDMH as carcinogens." Recent studies revealed NO *cancer* was caused by UDMH, even in rodents. Massive amounts of UDMH caused *no* cancers or tumors in rats, at any dosage level. When K. Smith (in 1994) fed mice four times the "maximum tolerated dosage," one mouse out of 45 developed a benign tumor. (No traces of cancer, even in the most susceptible strain of mice.)

Meryl Streep was a leading opponent of Alar on the show, and wrote a booklet titled "Mothers and Others for Pesticide Limits." After the scurrilous show, 95,000 copies of the book were ordered, for $8.00 a copy.

In Britain a group of scientists appointed by parliament in 1984 to review the Alar charges declined to ban the plant hormone, saying "we don't assume that animal data are transferable to humans or that high-dose responses can predict low-dose responses." It would be wonderful if more U.S. scientists were that intelligent. Presuming that some of them are, it would be even more wonderful if they could be that truthful!

CHAPTER 21
MISCONCEPTIONS ABOUT ENVIRONMENTAL POLLUTION

SECTION 21.3

SIMPLE TRUTHS OF ENVIRONMENTAL FACTS AND FALLACIES

Jack W. Dini

Mr. Dini is a consultant who retired from Lawrence Livermore National Laboratory and now writes a monthly column on environmental issues for Plating & Surface Finishing, *the technical journal of the AESF.*

21.3.1 AIR QUALITY

"Most people in rich countries believe their environment is continuing to deteriorate" (1).

"Public unaware of air quality gains" (2).

"Manufacturing facilities emitting fewer chemicals" (3).

The Clean Air Act has substantially reduced air pollution over the past 25 years. Overall, air quality has improved 42 percent since 1980, and water quality by 27 percent (4). Yet most people believe air quality (we inhale about 20,000 L of air in a day, Ref. 5) has gotten worse and will continue to deteriorate into the foreseeable future, even though they don't consider breathing dirty air to be especially risky. One poll after another shows that U.S. citizens are unaware of the improvements that have been made in environmental issues. For example, in a 1993 poll, 75 percent of those contacted believed that problems regarding pollution and the environment would get significantly worse during their lifetimes (1). More recently, in 1997 a poll result showed that 58 percent believed air quality had gotten worse over the past ten years (2).

As already stated, there have been significant improvements in air quality. So, why don't people know and appreciate this? It's mostly due to media coverage that concentrates on issues like Times Beach, Alar, oil spills, tainted Perrier water, and so forth. The important point is that while bad news receives heavy coverage, important information about the environment is either underreported or not reported at all. For example, when San Francisco met the federal ozone standard in 1992, the *San Francisco Chronicle* reported the news on page 16 (1).

Another reason is the mentality of many people regarding environmental issues. The Nuclear Energy Institute in Washington, D.C., reported that a freshman at Eagle Rock Junior High in Idaho urged fellow students to sign a petition demanding strict control or total elimination of the chemical "dihydrogen monoxide," which causes excessive sweating and vomiting; is a major component in acid rain; can cause severe burns in the gaseous state; can cause death when accidentally inhaled; contributes to erosion; decreases the effectiveness of automobile brakes; and has been found in the tumors of terminal cancer patients (6). A total of 50 students were asked if they supported a ban of the chemical; 43 said yes, 6 were undecided, and only one knew that the chemical was water. So clearly, most of them did not know high school chemistry nomenclature. More importantly, most of them believed it was their right to support this restriction of technological freedom even though they knew nothing whatever about the subject. If this experiment had been tried on an older population, surely at least as many would want to ban the strange sounding chemical. This type of thinking is all too pervasive with environmental issues. If it sounds strange, it must be bad!

Recent Results on Air Quality

A recent comprehensive report published by the Pacific Research Institute disputes popular perceptions about the environment, finding sharp declines in water pollution, toxic chemicals in the environment, and residues of harmful chemicals in fish and fowl (1). This report is divided into three parts:

1. Primary environmental indicators including air and water quality, natural resource use, land use, and condition. These are "primary" because they have the most direct effect on environmental quality. Six air pollutants that regulations target are analyzed: sulfur dioxide (SO_2), nitrogen oxides (NOx), volatile organic compounds (VOCs), carbon monoxide (CO), total suspended particulates (TSPS), and lead (Pb).
2. Secondary indicators including carbon dioxide emissions, oil spills, pesticide and toxic releases, and wildlife.
3. An index of four major environmental indicators—air quality, water quality, natural resources, and solid waste—and a composite index of all four indicators.

A quick summary for six air pollutants can be found in Table 21.3.1. Figures 21.3.1 to 21.3.3 show changes in sulfur dioxide, suspended particulates and lead, respectively. Some key findings from this report and others include the following:

- The ambient levels of sulfur dioxide decreased by 50.3 percent between 1975 and 1993. Emissions of sulfur dioxide fell 32.3 percent between 1970 and 1994.
- The ambient level of NO_2 shows a 33.8 percent decrease between 1977 and 1993.
- The level of ambient ozone decreased 18.5 percent between 1979 and 1993.
- VOC (volatile organic compounds) emissions declined 24.4 percent from 1970 to 1994.
- Ambient CO concentrations in 1993 were 60.5 percent lower than in 1975; CO emissions declined 14.9 percent between 1975 and 1994.
- Total suspended particulates declined 23.6 percent from 1975 to 1991.
- Lead presents the greatest success story. Ambient lead concentration fell 97.1 percent between 1975 and 1992 and emissions fell 97.7 percent between 1970 and 1994. Most of these dramatic reductions were due to the introduction of unleaded gasoline and the elimination of lead compounds in paints and coatings.

TABLE 21.3.1 Air Emission Changes Between the 1970s and 1990s*

Pollutant	Change, % drop
Sulfur dioxide	50.3
Nitrogen dioxide	33.8
Ozone	18.5
Volatile organic compounds	24.4
Carbon monoxide	60.5
Lead	97.1

*From DeWeil et al., (1).

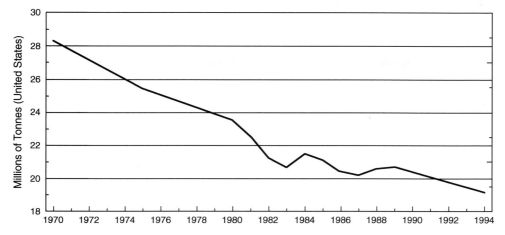

FIGURE 21.3.1 Sulfur dioxide emissions estimates for the United States (adapted from DeWeil et al. (1)).

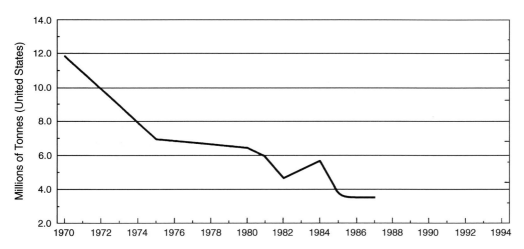

FIGURE 21.3.2 Suspended particulates emissions estimates for the United States (adapted from DeWeil et al. (1)).

- Ozone levels are improving even in Los Angeles where the ozone standard fell 36.3 percent between 1985 and 1994. Houston, which follows Los Angeles with the second worst record, also improved 54.7 percent between 1985 and 1994. California's air in the summer of 1997 was the cleanest it has been since pollution officials began keeping records in the 1950s. In fact, air quality is so improved in the Los Angeles Basin that the nation's smog capital is in danger of losing its title to Houston (7). Los Angeles has not violated federal standards for sulfur dioxide emissions, the primary cause of acid rain, since the early 1970s. In 1992, the state of California passed a year without a carbon monoxide violation for the first time. National air pollution emissions have been declining on an almost uninterrupted basis since the 1970s, even as the population increases, more cars are driven more miles, and the economy grows. The modern American automobile is the cleanest system of transportation ever devised. Cleaner certainly than cars sold in the Western European nations that U.S. environmentalists depict as models of ecological enlightenment (8).

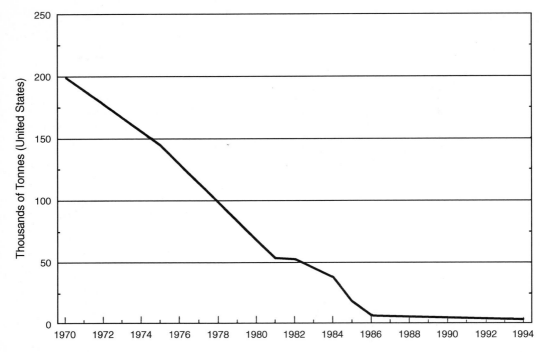

FIGURE 21.3.3 Lead emissions estimates for the United States (adapted from DeWeil et al. (1)).

In southern California automobile emissions have been cut in half since 1965 despite a doubling of the number of cars. Tailpipe emissions from new cars are said to be so minuscule that if a person spills a tablespoon of gasoline, the emissions are greater than those produced by driving the car for about a day and a half (5).

Other key findings from the Pacific Research Institute report (1):

- In 1990, 82 percent of the lakes tested met swimmable objectives.
- Forests are increasing as growth exceeds the harvesting of trees both in Canada and in the United States.
- The amount of land set aside for parks, wilderness, and wildlife is increasing in both the United States and Canada.
- The amounts of toxic chemicals exposed to the environment is decreasing.
- Critical wetland habitat is not declining.

Releases of Hazardous Chemicals

Hazardous chemicals released into the environment by manufacturing companies declined for the eighth straight year in 1995 (3), and have declined by one-third since 1988 (9). The data cover emissions of 286 chemicals new to the inventory, which now encompasses 643 chemicals. From 1994 to 1995, the amount of chemicals released into the environment was down nearly 5 percent. Since 1988, when industrial facilities were required to report their toxic releases, emissions have been reduced by 46 percent.

New Standards for Ozone and Particulates

All of the previous information in this article leads one to ask the question—Do we really need to go in the direction of new standards that would place tighter restrictions on ozone and particulate matter? Mason provides an excellent analysis on this question (10).

One of the claims supporting EPA's new standards is that these regulations would benefit children by reducing asthma rates. However, a recently completed study attributes the rising asthma rates among children in inner cities, not to particulate matter and ground level ozone, but to cockroaches (11). This study found that the most important cause of childhood asthma in U.S. inner cities may be allergies to cockroach droppings and debris. Allergies to cockroaches provoke an unusually severe form of asthma that is probably the source of the disproportionately high incidence of asthma in urban neighborhoods. More than one-third of asthmatic children were allergic to cockroaches. Half of these children were found to have high levels of cockroach droppings and debris in their bedrooms where the children spend most of the time (11). Fennell (12) suggests that poor indoor air quality, resulting from federal government regulations focused on reducing energy in commercial and residential buildings, is one of the culprits causing increased asthma. She states, "Energy efficient buildings concentrate the levels of contaminants trapped inside and these weather-tight buildings don't allow for natural air flow."

Another study showed an association of illness with larger particles but not with smaller ones (13, 14). R. F. Phalen, one of the nation's leading experts on air pollution, has urged policymakers to "wait for the necessary science and bypass political expediency." He claims that the new standards are not only premature but may even be harmful to public health because the scientific knowledge in this area is incomplete (15, 16). A *Wall Street Journal* article reported that the EPA's own analysis shows the new ozone regulations will cost more than the economic value of the benefits. The authors calculate that human deaths caused by the regulations would exceed human lives saved because of financial costs of implementation (17).

Conclusion

The environment is cleaner and safer than at any time in the past half century. The average American today is exposed to fewer potentially harmful pollutants than at any time since the 1930s. Air and water pollution, which had risen during the 1940s and 1950s, have fallen constantly and considerably since that time.

REFERENCES

1. DeWeil, B., Hayward, S., Jones, L., and Smith, M. D., *Index of Leading Environmental Indicators*, Pacific Research Institute, San Francisco, April 1997.

2. Raber, L., *Chemical & Engineering News*, 97: 26, May 12, 1997.

3. Hanson, D., *Chemical & Engineering News*, 97: 10, May 26, 1997.

4. *Civil Defense Perspectives*, Vol. 13, No. 6, Summer 1997.

5. *Health Risks and the Press,* Moore, M., ed., The Media Institute, Washington, D.C., 1989.

6. *Access to Energy*, Vol. 25, No. 3, Cave Junction, Oregon, November 1997.

7. Buel, S., "State's Air Cleanest in Decades," *San Jose Mercury News*, October 23, 1997.

8. Easterbrook, G., *A Moment on Earth*, Viking, 1995.

9. Hayward, S., Fowler, E., and Schiller, E., *Ideas in Action*, Pacific Research Institute, San Francisco, CA, April 1999.

10. Mason, B. J., *Plating & Surface Finishing*, 84: 122, November 1997.

11. Rosenstreich, D. L., et al., *New England Journal of Medicine*, 336: 1356, May 8, 1997.

12. Fennell, A., "Grasping For Air," www.junkscience.com, June 22, 1999.

13. Samuel, P., *National Review*, p. 20, July 28, 1997.

14. *Civil Defense Perspectives*, Vol. 14, No. 1, Tucson, AZ, November 1997.

15. Phalen, P. F., *Living and Dying in Dirty Air: What The Science Tells Us*, George C. Marshall Institute, Washington, D.C., July 15, 1997.

16. *Environmental News*, Vol. 1, No. 5, The Heartland Institute, Palatine, Illinois, 1997.

17. Gramm, W. L., and Dudley, S. E., "The Human Costs of EPA Standards," *Wall Street Journal*, p. A18, June 9, 1997.

21.3.2   INDOOR AIR POLLUTION

What's the most polluted atmosphere most of us are exposed to? One clue—It's not the outdoor air that we've spent zillions of dollars to clean up. The answer—It's the air in our homes and offices. Enormous attention has been paid to outdoor pollution and its impact on health, but people spend up to 90 percent of their time inside, making indoor air quality extremely important (1).

Many people perceive that the risk from outdoor air is substantially higher than for indoor air. In fact, the home environment is rarely considered to be a risk in this regard. In terms of regulation and control, the principal sources of pollution, such as vehicles and factories, lend themselves to formal legislative control. By contrast, exposure in the home is very much dominated by personal choice and behavior. Although it is still common in epidemiological studies to use outdoor pollutant levels and equate these with health effects in the population, the realization is dawning that much better account of indoor exposures need to be taken. We inhale about 20,000 L of air each day. This equates to about 60 pounds of air flowing through our lungs each day, a volume far greater than the two liters of water and food we consume daily (2).

You might be surprised to find that a large number of natural and synthetic chemicals can be identified in the air inside a typical home in addition to particulate material and potent allergens. Some of the more important indoor pollutants include tobacco smoke; carbon monoxide (CO); radon; oxides of nitrogen (NOx); formaldehyde; volatile organic compounds (VOCs); chlorinated organic compounds; dust and particulates; PM_{10} (small particles less than 10 um in diameter); house dust mite allergen; cat allergen; fungi and fungal spores; bacteria; pollen; and asbestos fibers (1).

Couple this information with the fact that the drive to conserve energy has resulted in warmer "tighter" buildings with reduced air exchanges and you can envision how indoor air pollutants can build up. Part of the growing environmental consciousness that started in the mid-1970s led many Americans to think that steps must be taken to make our homes and public buildings more energy efficient in order to reduce use of energy resources (3). As part of this, homes and office building were made more airtight so that heat was not needlessly escaping. Concurrently, Federal ventilation standards for public buildings and the workplace were also scaled back. Energy saving measures recommended or mandated by EPA and OSHA over the years have reduced the number of air changes in a house or small office building from about 4 changes per hour to about 1 per day. This is part of the reason that according to Laudan (3); "Air in airplanes and office buildings is now much staler than it used to be and why the frequency of "sick buildings" seems to be on the rise. The clear tradeoff is that as structures become more airtight, the risks of all those airborne diseases to which we are prone—from the carcinogens in tobacco smoke and cooking oil to the smoke from our wood stoves and the germs from our office mate with flu—go up in direct proportion to the extent that we succeed in achieving energy savings as we seal up the places where we live and work." It's estimated that up to 30 percent of all new buildings display classic "sick building syndrome" symptoms (2).

Results of a number of studies comparing indoor air with outdoor air exposures have been published. One study in England looked at the indoor environment in 174 family homes in the Avon area for nitrogen dioxide, formaldehyde, and other volatile compounds, house dust mites, bacteria, and fungi. Examples of typical indoor and outdoor levels of the gaseous pollutants measured in this study are shown in Table 21.3.2.

A paper by Wallace et al. (4) summarized results of an EPA study of the relationship between the concentrations of a number of pollutants measured indoors, in the outside, and from personal monitoring. These studies, know as the TEAM (total exposure assessment methodology), convincingly demonstrated how personal exposure can markedly exceed that anticipated from measurement of outside ambient concentrations.

TABLE 21.3.2 Typical Average Concentrations of Some Pollutants Indoors and Out in the Avon area of England*

	Outdoors, mg/m^3	Indoors, mg/m^3
Nitrogen dioxide	7–40	13–40 (electric cooking)
		25–70 (gas cooking)
Formaldehyde	0.002	0.02
Volatile organics	0.02	0.2

* From Harrison (1).

TABLE 21.3.3 Weighted Medians for Air Samples (mg/m^3)†

	New Jersey		Greensboro, NC		Devils Lake, ND	
Chemical	Personal Air	Outdoor Air	Personal Air	Outdoor Air	Personal Air	Outdoor Air
1,1,1 Trichloroethane	17	4.6	32	60	2.5	0.05*
Benzene	16	7.2	9.8	0.4	—	—
m,p-xylene	16	9.0	6.9	1.5	6.2	0.05*
Carbon tetrachloride	1.5	0.87	—	—	—	—
Trichloroethylene	2.4	1.4	1.5	0.2	0.50	0.05*
Tetrachloroethylene	7.4	3.1	3.3	0.7	5.0	0.69
Styrene	1.9	0.66	1.4	0.1	—	—
p-dichlorobenzene	3.6	1.0	2.6	0.4	1.7	0.07*
Ethylbenzene	7.1	3.0	2.5	0.3	2.1	0.03*
O-xylene	5.4	3.0	3.6	0.6	2.7	0.05*
Chloroform	3.2	0.63	1.7	0.14	0.38	0.05*

* Not detected; value is one-half the limit of detection.
† From Wallace, et al. (4).

The objective of this study was to estimate the distribution of exposures to ~20 toxic substances for a target population in an industrial/chemical manufacturing area (Bayonne and Elizabeth, NJ) and to carry out smaller studies for populations in nonchemical manufacturing areas (Greensboro, North Carolina and Devils Lake, North Dakota). Greensboro was chosen because its population was similar in size to the Bayonne-Elizabeth area, and it had small industries but no chemical manufacturing or petroleum refining operations. Devils Lake, North Dakota was selected to provide data on the population of a small, rural, and agricultural town far from any industry. Exposures to 20 volatile organic compounds were measured in personal air, outdoor air, drinking water, and breath. Data are presented in Table 21.3.3 for indoor and outdoor air. Breath sample data were also obtained, but are not included in the table in order to keep it more manageable. The table shows the following:

- Ten chemicals were prevalent in the air samples. For New Jersey, an eleventh chemical, carbon tetrachloride was present.
- In New Jersey, personal air medians exceeded the outdoor air medians for every chemical in every season, usually by factors of 2-5. Similar results were observed for median air and breath concentrations in North Carolina and North Dakota.
- Outdoor air was noticeably cleaner in North Dakota than in New Jersey or North Carolina.

Other interesting results from this study:

- Smokers had significantly elevated breath levels of benzene, styrene, ethylbenzene, and m, p-xylene.
- Use of hot water in homes for activities such as washing clothes and dishes, and bathing or showering is the main source of airborne chloroform.

- Benzene exposures while filling gas tanks may exceed 1000 ug/m^3. People who reported filling their tanks with gasoline had twice as much benzene on their breath as person who did not.

- Moth crystals and room air deodorizers are responsible for noticeably increased concentrations of p-dichlorobenzene. About 80 percent of the homes in the survey had these materials.

- Tetrachloroethylene exposures are elevated by wearing and storing dry cleaned clothes.

- Employment leads to increased exposures of some toxic chemicals. Activities identified with increased exposures included pumping gasoline, visiting service stations, visiting dry cleaners, traveling in a car, furniture refinishing, painting, scale model building, pesticide use, and smoking.

What do the data in Tables 21.3.2 and 21.3.3 mean? The answer is we should be concerned but not overly alarmed, and clearly we can control the situation on our own, at least in our own homes. In summarizing the data in Table 21.3.2, Harrison (1) points out that any risk of respiratory illness from the levels of NO$_2$ currently found in most homes is small and current exposure in homes to formaldehyde or VOCs does not pose a risk to health. Harrison does state that house dust mites are potentially one of the most important indoor problems because of the role they may play in the incidence and prevalence of asthma. Tobacco smoke is a problem all by itself and anyone who is not aware of this issue is brain dead. More than 2000 compounds have been identified in cigarette smoke. Many of these are known carcinogens and irritants (5). Smokers are prohibited from lighting up in work place, public buildings, and now even in bars (at least in California). They aren't prohibited from lighting up in their own homes where they can do serious harm to their family and friends. Next to smoking, carbon monoxide is the most obviously identifiable problem. According to the U.S. Consumer Product Safety Commission, 250 to 300 people in the United States are killed every year by carbon monoxide from space heaters, furnaces, or other household devices. Another 5,000 to 10,000 victims do not die yet suffer acute carbon monoxide poisoning (6).

Everyone has heard the bad words about asbestos. More scare press has been written about it and dollars thrown at its removal than just about anything else. In spite of the billions of dollars spent in removing it from buildings, here are some words directly from Moore: "As published research mounted during the 1980s, it became increasingly clear that asbestos levels in buildings, including schools, are barely detectable and over one thousand times lower than occupational levels found to be harmful. Richard Doll, who originally demonstrated that occupational exposure to asbestos increased lung cancer rates, stated that the risk to building occupants from exposure to asbestos was minimal, comparable to that associated with smoking half a cigarette in a lifetime" (7).

A Canadian government commission made the following statement: "Even a building whose air has a fiber level up to 10 times greater than that found in typical outdoor air would create a risk of fatality that was less than one-fiftieth the risk of having a fatal accident while driving to and from the building" (8).

Although EPA has raged a relentless battle on radon as a cause of cancer, epidemiological data show no such effect from radon by itself and in moderate doses (9, 10). However, there is evidence that high levels of radon, together with cigarette smoking, significantly increases the probability of lung cancer.

Control Measures

The first control measure is to remove all smokers from your house because all other potential indoor pollutants pale by comparison with tobacco smoke. Clearly, much easier said than done. Then make sure you have no potential problems with CO. Buy carbon monoxide detectors and place them near potential sources of toxic fumes, such as the hot water heater or furnace. Another important step is to make sure you have a vacuum cleaner that really vacuums. Roos (11), in discussing the "clean air syndrome" refers to a vacuum cleaner as a dust recycler. He points out that the cleaning ritual of a vacuum cleaner looks like this:

1. large visible dust particles are sucked up
2. impaction with dust pack produces a lot of fine particles

3. particles are small enough to follow air stream lines

4. particles are "treated," thus modified in the motor region

5. a cloud of invisible particles will fill the room after the cleaning procedure

6. coagulation will ensure that these particles become increasingly visible

7. these larger particles will settle and form a dust layer

8. vacuum cleaner is needed to remove the dust

9. return to 1

Now that you've eliminated smoking, CO, and have a vacuum cleaner that doesn't recycle the dust, other strategies include a combination of ventilation, source removal or substitution, source modification, and air purification changes.

Summary

Findings available from many studies show that the same air pollutants covered by environmental laws outdoors are usually found at much higher levels in the average home. As Ott and Roberts stated: "Sadly, most people—including officials of the U.S. government are rather complacent about such indoor pollutants. Yet if these same substances were found in outdoor air, the legal machinery of the Clean Air Act of 1990 would apply. If truckloads of dust with the same concentration of toxic chemicals as is found in most carpets were deposited outside, these locations would be considered hazardous waste dumps" (12). Ironically, in one of his final official acts as EPA administrator with the Bush Administration, William K. Reilly said he thought it "odd" that the federal agency had spent most of its energy and federal funding regulating environmental problems that pose small public health risks and far too little time targeting bigger threats (2).

If you're concerned about toxic substances in your home you don't have to wait for EPA, OSHA, or some other regulatory agency to do something about it. You can reduce exposure with only modest alterations in your daily routine. Lastly, if you're looking for new business opportunities be aware that by the year 2000, the indoor air quality market is projected to be $3.6 billion, of which more than an 11 percent share is expected for analytical and consulting services (13, 14).

REFERENCES

1. Harrison, P. C., *Chemistry & Industry*, No 17: 677, September 1, 1997.

2. Tate, N., *The Sick Building Syndrome*, New Horizon Press, 1994.

3. Laudan, L., *Danger Ahead*, John Wiley & Sons, 1997.

4. Wallace, L. A., et al., *Environmental Research*, 43: 290, 1987.

5. Moeller, D. W., *Environmental Health*, Harvard University Press, 1992.

6. Goldfarb, B., *ChemMatters*, 15(3): 10, October 1997.

7. Moore, C. C., *Haunted Housing*, Cato Institute, Washington, D.C. 1997.

8. Martin, N. S., *Vital Speeches*, LVI(14): 434, May 1, 1990.

9. Moore, T. G., "Environmental Fundamentalism," Hoover Institution, Stanford University, 1992.

10. Ames, B. N., and Gold, L. S., "The Causes and Prevention of Cancer," in *Risks, Costs and Lives Saved*, Hahn, R. W., ed., Oxford University Press, 1996.

11. Roos, R. A., *The Forgotten Pollution*, Kluwer Academic Publishers, 1996.

12. Ott, W. R., and Roberts, J. W., *Scientific American*, 278: 86, February 1998.

13. Lighthart, B., and Mohr, A. J., eds., *Atmospheric Microbial Aerosols*, Chapman & Hall, New York, 1994.

14. Poruthoor, S. K., Dasgupta, P. K., and Genfa, Z., *Environ. Sci. Technol.*, 32: 1147, 1998.

21.3.3 WANT SOME CHEMICALS? HAVE ANOTHER CUP OF COFFEE

Before you drink another cup of coffee, there are some facts you should know. A cup of coffee contains more than 1000 chemicals. Only 27 have been tested for carcinogenicity and 19 of these were positive in at least one test (1). Table 21.3.4 provides a listing of the chemicals. There are more natural rodent carcinogens by weight in a single cup of coffee than potentially carcinogenic synthetic pesticide residues in the average U.S. diet in a year, and this doesn't count the 1000 or so chemicals yet to be tested (2). The FDA tolerance levels for trichloroethylene (TCE) in coffee are 25 parts per million (ppm) for ground coffee and 10 ppm for instant coffee (3).

Coffee has been blamed for a lot of medical woes: gallstones, heartburn, heart disease, migraine headaches, infertility, and peptic ulcers (4). Among people who have no problem sleeping, two or more cups of coffee taken a half an hour before retiring adversely affects all sleep measures. Recently, Dutch researchers identified two compounds (diterpenes of cafestol and kahweol) in coffee oils that can raise cholesterol in the blood. Instant brands contained only "minimal" diterpenes (5). Espresso, in contrast, contained the most—some 6 to 12 mg of the diterpene cafestol alone per 150 ml of coffee. The carcinogenicity of the diterpenes has not been determined. Under the current law, if coffee were synthetic, the FDA would ban it (6).

Coffee is also corrosive. In his most recent thriller, *Airframe*, even Michael Crichton takes a hit on coffee. As two people enter one of the large facilities of an aircraft producing plant in Los Angeles, the following exchange takes place:

> "We got time for a cup of coffee?"
> She shook her head. "Coffee's not allowed on the floor."
> "No coffee?" He groaned. "Why not?"
> "Coffee's corrosive. Aluminum doesn't like it."

On the plus side, extracts of green coffee beans have been shown to play a significant role in inhibiting tumor formation or growth (7), coffee consumption may theoretically make you more pain sensitive, although this hypothesis has yet to be adequately tested (4), and it's claimed that drinking 2 to 3 cups of coffee per day reduces the risk of suicide by 66 percent (8).

Perhaps it's better to smell the coffee than drink it. The distinctive aroma of freshly brewed coffee is not only pleasant, says T. Shibamoto, a University of California chemist, it might be chock full of things that are good for you (9). The molecules wafting up from a steaming cup of coffee, he has discovered, combine to form potent anti-oxidants. In principle, they should have cancer and age-fighting effects similar to other anti-oxidants, including vitamin C and vitamin E. Shibamoto has emphasized that all he has so far is a hypothesis. The research so far has not gone beyond chemical identification and test tube study of coffee aroma compounds. Someone will have to do some careful experiments with animals, and eventually with people, to prove the point. Shibamoto reports that aroma is due to at least 300 different chemicals evaporating from freshly brewed coffee. Many belong to a large family of molecules called volatile heterocyclic compounds, which are fairly small chains of six to ten carbon atoms hooked to oxygen, sulfur, and nitrogen atoms (9).

Toxicological examination of synthetic chemicals, without similar examination of chemicals that occur naturally, has resulted in an imbalance in both the data on and the perception of chemical carcinogens (1). The vast proportion of chemicals that humans are exposed to occur naturally.

TABLE 21.3.4   Carcinogenicity Status of Natural Chemicals in Roasted Coffee*

Positive	acetaldehyde, benzaldehyde, benzene, benzofuran, benzo(a)pyrene, caffeic acid, catechol, 1,2,5,6-dibenzanthracene, ethanol, ethylbenzene, formaldehyde, furan, furfural, hydrogen peroxide, hydroquinone, limonene, styrene, toluene, and 2-amino-3,4-dimethylimidazo(4,5-f)quinoline
Not positive	acrolein, biphenyl, eugenol, nicotinic acid, phenol, and piperdine
Yet to test	~1000 chemicals

* From Gold et al. (1).

Nevertheless, the public tends to view chemicals as only synthetic and to think of synthetic chemicals as toxic despite the fact that every natural chemical is also toxic at some dose. It is probable that almost every fruit and vegetable in the supermarket contains natural pesticides that are rodent carcinogens (1).

There is a lethal dose of mostly everything we ingest (10). Water, which is considered completely nontoxic, can be lethal. There are cases reported in the medical literature of both acute and chronic intoxications, some even fatal, from excessive water intake other than drowning, which is an outright case of water overdose. In Germany, a man died from cerebral edema and electrolyte disturbance because he had drunk 17 L of water within a very short time (11).

Tea and coffee are the most popular drinks on earth. Plain water is number one, followed by tea. However, coffee is a close second to tea and since a typical cup of coffee contains about twice as much caffeine as a cup of tea, coffee is actually the single largest source of caffeine worldwide (12). There is a lethal dose of caffeine, which has two tongue twisting technical names: 3,7-dihydro-1,3,7-trimethyl-lH-purine-2,6-dione and 1,3,7-trimethylxanthine, in about 40 cups of strong coffee (12, 13). This doesn't mean that coffee is dangerous. We survive because we don't drink 40 cups of coffee in one setting. Most investigations of coffee (with or without caffeine) have not linked it to human cancer, although animal studies using levels of caffeine far higher than any consumed by humans showed caffeine to be teratogenetic, an agent capable of causing birth defects (12). Lastly, it doesn't matter how the beverage is sweetened. Artificial sweeteners in reasonable quantities do not cause cancer (14, 15).

The cancer risk of an apple with Alar residue is far less than the risk from the natural compounds in a cup of coffee. Yet the apple industry was given a very serious setback a few years back when "60 Minutes" provided their Alar scare program. The apple industry lost more than $100 million and a number of small-scale growers, many of whom had never even used Alar, went out of business (2). This was started by a controversial report of questionable science where laboratory mice developed tumors when exposed to 35,000 times the amount of Alar that children were normally exposed to. Later, several independent reviews found the threat minuscule, but by that time it was too late to change public opinion (2).

Bruce Ames and his colleagues at the University of California, Berkeley, have developed a method of ranking possible carcinogenic hazards (16, 17). They call this a HERP Index (human exposure over rat potency). A value of 100 on this scale means that people are getting the same dose in milligrams per kilogram that gave half the tested rats cancer. Table 21.3.5 shows how coffee compares with some other items. Note that wine and beer have noticeably higher values than coffee, but then so does lettuce and living in a mobile home.

An interesting comparison is the possible carcinogens in contaminated well water found in "Silicon Valley" (Santa Clara, CA) or Woburn, Massachusetts, with the possible hazard of ordinary tap water and other common liquids used by humans. Of 35 wells shut down in Santa Clara Valley because of their supposed carcinogenic hazard, only two had HERP values greater than ordinary tap water.

TABLE 21.3.5 Ranking Possible Carcinogenic Hazards*

Possible hazard[†] HERP %	Daily human exposure	Human dose of rodent carcinogen
0.0004	Well water (1 L, Woburn, Massachusetts)	Trichloroethylene, 267 ug
0.001	Tap water (1 L)	Chloroform, 83 ug
0.004	Well water (1 L, worst in Silicon Valley)	Trichloroethylene, 2800 ug
0.005	Coffee (1 cup)	Furfural
0.04	Coffee, 1 cup (from 4 g)	Caffeic acid, 7.2 mg
4.7	Wine (250 ml)	Ethyl alcohol, 30 ml
2.8	Beer (12 oz; 354 ml)	Ethyl alcohol, 18 ml
1.4	Mobile home air (14 hr/day)	Formaldehyde, 2.2 mg

* From Ames et al. (16, 17).

[†] EPAs one in a million hypothetical risk is 0.000015 on the HERP scale, or about 400,000 times below the level that would give cancer to a rat.

As Table 21.3.5 shows, well water from the most polluted well (HERP = 0.004 percent) per liter for trichloroethylene had a HERP value at least an order of magnitude less than for the carcinogens in an equal volume of coffee, cola, beer, or wine (17).

So the next time you read or hear about the next chemical scare of the month, stop and ask these questions:

- What is the dose?
- Could it be any worse than the average coffee consumption of Americans, which is about three cups per day?

As Michael Fumento pointed out so well in his book: "There are two extremes—1-I Don't Care and 2-React to Every Scare" (18). We should be somewhere in the middle. Perhaps we would all be better off if we just followed Mark Twain's advice: "Part of the secret of success in life is to eat what you like and let the food fight it out inside" (19).

REFERENCES

1. Gold, L. S., Stone, T. H., Stern, B. R., Manley, N. B., and Ames, B. N., *Science*, 258: 261, October 9, 1992.
2. Bailey, R., ed., *The True State of the Planet*, The Free Press, NY, 1995.
3. Orient, J. M., in *Rational Readings on Environmental Concerns*, J. H. Lehr, ed., Van Nostrand Reinhold, pp. 186–188, 1992.
4. Werbach, M., "Healing with Food," *Harper Perennial*, 1993.
5. Raloff, J., *Science News*, 148: 182, September 16, 1995.
6. Ross, J. F., *Smithsonian*, 26: 42, November 1995.
7. Lappe, M., *Chemical Deception*, Sierra Club Books, p. 198, 1991.
8. Kawachi, I., Willet, W., Colditz, G., Stampfer, M., and Speizer, F., *Arch. Intern. Med.*, 156: 521, 1996.
9. Petit, C., "Wake Up and Smell Health Benefits of Fresh Coffee," *San Francisco Chronicle*, p. 1, April 14, 1997.
10. Ottoboni, M. A., *The Dose Makes the Poison*, Vincente Books, 1984.
11. Efron, E., *The Apocalyptics; Cancer and the Big Lie*, Simon & Schuster, 1984.
12. Braun, S., *Buzz: The Science and Lore of Alcohol and Caffeine,* Oxford University Press, 1996.
13. Gilbert, R. M., *Caffeine; The Most Popular Stimulant*, Chelsea House Publishers, 1986.
14. Trichopoulos, D., Li, F. P., and Hunter, D. J., *Scientific American*, 275, 80, September 1996.
15. Taubes, G., *Science*, 269: 164, July 14, 1995.
16. Ames, B. N., et al., *Science*, 236: 271, April 17, 1987.
17. Ames, B. N., Magaw, R., and Gold, L. S., "Ranking Possible Carcinogens: One Approach to Risk Management," in *The Risk Assessment of Environmental and Human Health Hazards: A Textbook of Case Studies*, Paustenbach, D. J., ed., John Wiley & Sons, 1990.
18. Fumento, M., *Science Under Siege: Balancing Technology and the Environment,* William Morrow & Co., New York, 1993.
19. Shaw, D., *The Pleasure Police,* Doubleday, 1996.

21.3.4 BUGS

"Invisible Peril Lurks in Your Bathroom," (1)

Next time you see someone at a party or a pub washing a beer mug by dipping it in a solution or two of water be aware that this really doesn't clean the glassware. Ten or more total and fecal coliforms

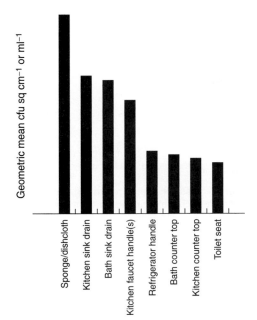

FIGURE 21.3.4 Fecal Coliform Densities in Kitchen and Bathroom Areas (from Rusin et al. (4)).

were found in 67.6 percent and 12.2 percent of vat washed mugs, versus 30.3 percent and 4.7 percent of machine washed mugs, respectively (2). Other places where bacteria can be found include kitchens and bathrooms. Kitchen areas and items such as counters, sinks, dish cloths, sponges, and tea towels have been identified as reservoirs and/or disseminators of potentially pathogenic bacteria. Coffee preparation areas are regions where more than just people congregate. Concentrations and types of bacteria found in these spaces (e.g., drains/sinks, washcloths, sponges, and coffee spoons) were found to be similar to that found in household kitchens (2).

Bacteria identified do not usually cause illness in healthy individuals but they may cause infections in immunocompromised individuals. However, if you're thinking this isn't a health issue, check these statistics. It's been estimated that each year the cost of foodborne bacterial illness in the United States is approximately $4 to $6 billion, and the number of gastroenteritis cases related to foodborne pathogens is 6.5 million, leading to 9000 deaths. In a review study (3) including more than 1000 outbreaks of food poisoning, it was shown that the source of highest percentage of cases (19.7 percent) was family homes, followed by restaurants (17.1 percent), and banquets (12.2 percent).

The most contaminated sites within the home are those that tend to remain moist, such as the sponge/dishcloth and drain areas, and the site that is most frequently touched, the kitchen faucet handles. A study by Rusin et al. (4) showed that sites with highest densities of fecal coliforms were found in kitchens. Three of the top four most contaminated sites in their study were in the kitchen (Figure 21.3.4). These included the sponge/dishcloth, the kitchen sink drain area, the bath sink drain area, and the kitchen faucet handles. Sites with the lowest concentrations of fecal coliforms were the refrigerator handle, the bathroom counter top, the bathroom floor around the toilet, and the toilet seat. Surprisingly, the bathroom countertop and the toilet seat were two of the least contaminated sites in the household. The toilet seat was probably dry between periods of use, which probably explains the low level of contamination because bugs thrive in moist conditions. Ordinary cleaning practices may do little to reduce the microbial load. However use of "self disinfecting" sponges in a household kitchen environment significantly reduces the level of total and fecal coliform bacteria within the sponges, and greatly reduces the transfer of such bacteria to surfaces and fingers (5). Also, the introduction of hypochlorite cleaning products, do result in a significant reduction of bacteria in most cases (4).

As mentioned earlier, the concentrations and occurrence of coliform bacteria in office coffee preparation areas (e.g., counters, drains/sinks, washcloths, sponges, and coffee spoons) were found to be similar to that found in household kitchens. In a study by Meer et al. (2), 41 percent of all cups contained coliform bacteria. Wiping the cups with a moist sponge or dish cloth resulted in a significant increase in bacterial contamination of the cups in addition to cross contamination with Escherichia coli and other coliforms. It was demonstrated that a small, office cup washer completely eliminated coliforms and most other heterotrophic bacteria. Table 21.3.6 shows some results.

Now some more words about the bathroom area. Flushing of a household toilet produces bacteria laden aerosols that settle on the toilet and bathroom surfaces (1, 6). Charles Gerba, a microbiologist at the University of Arizona, reports that a very fine aerosol spray is ejected when you flush. You don't see it, or feel it, but it fills the room, so if your toothbrushes are out, you may end up brushing your teeth with material you thought had gone down the toilet. His experiments have shown that the spray clearly did not reach above six feet, so only if your toothbrushes are hanging from the ceiling, do they remain uncontaminated. Couple this with the fact that some studies have suggested that

TABLE 21.3.6   Effect of a Commercial Coffee Cup Washer on Bacterial Reduction*

	Heterotrophic plate count		Percent positive	
	Average[†]	Range	Coliforms	E. coli
Before washing	6.4×10^2	2.4 to 4×10^5	36	0
After washing	3.0	0 to 48	0	0

* From Meer et al. (2).
[†] Average was calculated as the geometric mean.

touching a contaminated surface is much more likely to bring on a cold than being exposed to a sneeze, leads Gerba to recommend that bathrooms and kitchens be disinfected weekly using a chemical disinfectant (1, 6).

Gerba, who's been referred to as America's germ guru (7) has done other interesting things in his neverending search at tracking down bugs. Once, after a cross-country trip, he conducted a study to see whether the cleanliness of a motel room bore a direct relationship to its price, and if so, whether the relationship was the one you might expect. The answer is it did, and it was (8).

More recently he's moved his microscope from kitchen and bathroom surfaces to the household's laundry room (7). Gerba wanted to find out what happens to the bacteria in laundry when it's washed. Does it all go down the drain or is some left behind, like bacteria that can remain on kitchen counters after cleaning. He and his colleagues simulated typical home laundry practices, using warm water washes. They observed that after washing, bacterial contamination was found throughout the clothing and on the machine tub itself. To their surprise, the heat from the dryer did not kill all the bacteria. E. coli was eliminated but Salmonella and Mycobacterium fortuitium (a common bacteria that causes skin infections) were still present. The solution to eliminate cross-contamination problems in laundry according to Gerba is to wash hands after transferring wet clothes to the dryer and wash underwear loads last, along with a cup of bleach. The study confirmed that using bleach killed 99.99 percent of the bacteria in the clothes and on machine surfaces (7).

A few remaining thoughts to "bug" you with. Although medical authorities prefer not to advertise this fact, a modern hospital ranks high on the list of ideal targets for microscopic life forms. They flourish where a changing group of unrelated people spend their days packed closely together. Other places that are unusually hospitable to the rapid spread of microscopic predators include Army basic training barracks and school classrooms (9). Lastly, as Rathje and Cullen (8) point out, "That vast array of wrappers and boxes at fast-food restaurants, which is the object of so much disparagement, fulfills a role other than mere ease of carry-out. It keeps food safe. Fast-food restaurants and other establishments (such as schools and nursing homes) that rely heavily on throwaway containers disseminate far fewer bacteria and viruses through their disposables—some 50 percent fewer—than sit-down restaurants do through their glassware, silverware, and ceramic plates"

REFERENCES

1. Blum, D., *Sacramento Bee*, page A1, December 31, 1990.

2. Meer, R. R., Gerba, C. P., and Enriquez, C. E., *Dairy, Food and Environmental Sanitation*, 17: 352, June 1997.

3. Roberts, D., *J. Hyg. Camb.*, 89: 491, 1982.

4. Rusin, P., Orosz-Coughlin, P., and Gerba, C., *Journal of Applied Microbiology*, 85: 819, 1998.

5. Enriquez, C. E., Enriquez-Gordillo, R., Kennedy, D. I., and Gerba, C. P., *Dairy, Food and Environmental Sanitation*, 17: 20, January 1997.

6. Gerba, C. P., Wallis, C., and Melnick, J. L., *Applied Microbiology*, 30: 229, 1975.

7. Milloy, S. J., "Dr. Gerba's Latest Study," www.junkscience.com (Website comments June 4, 1999).

8. Rathje, W., and Murphy, C., *Rubbish*, Harper Perennial, 1992.

9. Moore, T. J., *Lifespan, Who Lives Longer and Why*, Simon & Schuster, 1993.

21.3.5 SINGLE MOLECULES

"In the case of "single molecule theory" we should all be dead of cancer from the millions of molecules of arsenic, cadmium and chromium in each of our cells if the theory were valid" (1).

"One day we may recognize that there is something of everything in everything else and that a glass of water likely contains a molecule of every compound on earth" (2).

Each of us has billions of molecules of carcinogens present naturally in our bodies. This includes radioactive carbon, a variety of metals including uranium (10,000 atoms per cell), metals that are considered pollutants, steroid hormones, and numerous other carcinogens naturally present in foods (1). Table 21.3.7 lists some trace elements in the human body and their estimated content. The estimate for heavy metals in the human body ranges between 10^4 and 10^8 atoms per cell.

For discussion purposes, let's use arsenic, cadmium, and chromium, all of which are considered to be carcinogenic. The approximate amounts of these present in the body per Jukes (1) are 4.4 mg, 30 mg, and 6 mg per person. If you compare these values with those in Table 21.3.7, which is from a different reference (3), you'll note that Jukes' values are lower for arsenic and cadmium than Lenihan's, while both agree on chromium. Regardless of which numbers are the correct ones, using Jukes' values, which are more conservative, will still show the enormity of these metals in the human body.

Arsenic, cadmium, and chromium would supply respectively 1×10^5, 2×10^6 and 0.7×10^6 molecules per cell. Using a value of 4.4 mg as the as arsenic content of a normal health human being, this translates to 9×10^{18} molecules as As_4 (1).

TABLE 21.3.7 Trace Elements in the Human Body*

Element	Estimated human body content
Arsenic	10 mg
Boron	48 mg
Cadmium	50 mg
Chromium	6 mg
Cobalt	1 mg
Copper	70 mg
Fluorine	2.6 g
Iron	4 g
Iodine	20 mg
Lead	120 mg
Manganese	12 mg
Mercury	13 mg
Molybdenum	10 mg
Nickel	1 mg
Selenium	10 mg
Silicon	18 g
Tin	6 mg
Vanadium	100 ug
Zinc	2.3 g

* From Lenihan (3).

Here's another example of large numbers. Ottoboni discusses benzpyrene in her book, The Dose Makes the Poison:

"Benzpyrene is a naturally occurring and relatively potent carcinogen that is virtually omni-present in our environment as a product of the cooking or burning of any organic material. It has been determined that there are 50 ug of benzpyrene in about 2 lbs of charcoal broiled steak. A generous portion of steak would weigh about 1/5th of a kg (7 oz). That portion of steak would contain about 10 ug benzpyrene. Ten ug is a very, very small quantity, as demonstrated by the fact that there are over 28 million ug in 1 oz. But when one considers how many molecules are contained in 10 ug, that seemingly insignificant quantity takes on really formidable proportions. In 10 ug of benzpyrene there would be 24,000,000,000,000,000 molecules! To give this number a name, there would be about 24 quadrillion molecules of benzpyrene in a portion of charcoal broiled steak" (4).

Some folks subscribe to the theory that "one molecule" can cause cancer. The above calcula-tions show that a "one-molecule" hypothesis for causing cancer is preposterous. Regardless, the one molecule theory was the reasoning behind the Delaney Clause which was passed in 1958. Delaney was the so-called "zero risk standard." It stated that no substance that has been shown to cause cancer in laboratory animals may be added to our food supply in any amount, no matter how small (5). Fortunately, the law did not apply the same standard to substances naturally present in food because if it did we would most likely starve. Naturally occurring substances that fail the same animal cancer tests have been found in many foods. One example, about 99.99 percent of all pesticides in the human diet are natural pesticides from plants. Since plants produce toxins to protect themselves against fungi, insects, and animal predators such as human (6).

Fortunately, recent legislation has removed pesticide residues from any association with the Delaney clause, replacing the scientifically untenable "zero discharge" requirement of Delaney with a stringent but attainable standard of "reasonable certainty of no harm" (7). However, I would like to say some more things about the Delaney Clause because it serves as a possible barometer for present day issues. The Delaney Clause is a metaphor for the long-standing political problem that alarms are more easily turned on the switched off (8).

When Delaney was passed in 1958 environmental exposures to synthetic chemicals were assumed to account for up to 90 percent of cancers and the abundance of naturally occurring toxins was not understood. Furthermore, in 1958, analytical techniques could find substances only to the parts per million range, levels at which the presence of a carcinogen is a clear danger. Today, it is believed that synthetic compounds cause only a small percentage of cancers and analytical techniques allow us to measure to parts per quadrillion. "Delaney became ridiculous," says Linda Fisher, a former head of EPAs toxic substances office. "We were regulating extremely small amounts of synthetics in processed foods while having no controls for natural toxins in raw food, which research suggests is the greater problem. The situation made no sense" (8).

Contaminants that now cause public indignation and regulatory panic were not even detectable only 25 years ago (9). The improvement in analytical chemistry over the past few decades has been huge. Early on we were detecting chemicals in water in the parts per thousand range. This is like finding one second in a 16-minute time span. Today, we can measure contaminants in parts per quadrillion, which is equivalent to finding the one second in a span of 32 million years (9)! Put another way, one part per quadrillion is one hair on the heads of all humans on earth (10) or, one golf ball compared to the size of the earth (11).

In 1990, Perrier bottled water was removed from the market because tests showed that samples of it contained 35 ppb of benzene. Although this was an amount so small that only 15 years prior it would have been impossible to detect, it was assumed that considerations of public health required withdrawal of the product (12). A person would have to consume 2.5 million of the "bad" bottles each week to approximate the intake that had sickened rodents (13). How's that for science fiction fantasy?

Shindell (5) has suggested that we may ultimately get to counting molecules. When we do this, we immediately run into the law of diffusion. As described by scientist George Koelle, this means that:

" If a pint of water is poured into the sea and allowed to mix completely with all the water on the surface of the earth, over 5,000 molecules of the original sample will be present in any pint taken subsequently. The general conclusion to be drawn from these calculations is that nothing is completely uncontaminated by anything else (5)."

At the smallest end of the spectrum, scientists at Oak Ridge National Laboratory detected a single atom of cesium in the presence of 10^{19} argon atoms and 10^{18} methane molecules (14). Talk about finding a needle in a haystack! Obviously, these days one can find just about any thing with the ultrasensitive equipment that's available.

Conclusion

Each of us has billions of molecules of carcinogens present naturally in our bodies. So, of what practical value is it to learn that there is one part per trillion or one part per quadrillion of EDB or PCB, or whatever, in our muffins, bottled water, or freshwater fish? All we're doing is showing how clever we are at analyzing things and chasing an ever receding zero. We get alarmed at things this small while 400,000 people die every year from smoking and another 100,000 die from drinking alcohol.

Lastly, the Delaney Clause is proof of the dictum that alarms are more easily turned on than switched off. Once a piece of legislation is on the books, it can take years to turn it around, regardless of how bad it is.

REFERENCES

1. Jukes, T. H., "Chasing a Receding Zero," *Rational Readings on Environmental Concerns*, Lehr, J. H., ed., Van Nostrand Reinhold, p. 329, 1992.

2. Lehr, J. H., "Toxicological Risk Assessment Distortions," *Rational Readings on Environmental Concerns*, Lehr, J. H., ed., Van Nostrand Reinhold, p. 673, 1992.

3. Lenihan, J., *Crumbs of Creation*, Adam Hilger, New York, 1988.

4. Ottoboni, M. A., *The Dose Makes the Poison*, Vincente Books, Berkeley, CA, 1984.

5. Shindell, S., "The Receding Zero," *ACSH News & Views*, American Council on Science and Health, New York, Nov/Dec 1995.

6. Ames, B. N., and Gold, L. S., "Environmental Pollution and Cancer: Some Misconceptions," *Rational Readings on Environmental Concerns*, Lehr, J. H., ed., Van Nostrand Reinhold, p. 151, 1992.

7. *Priorities*, Vol. 8, No. 3, American Council on Science and Health, New York, 1996.

8. Easterbrook, G., *A Moment on Earth*, Viking, 1995.

9. Baarschers, W. H., *Eco-Facts and Eco-Fiction: Understanding the Environmental Debate*, Routledge, London, 1996.

10. Aldrich, S. R., *Smoke or Steam?*, Star Press, Winter Haven, FL, 1994.

11. Ehmann, N. R., *Journal of Environmental Health*, 53: 14, July/August 1990.

12. Reisman, G. G., "The Toxicity of Environmentalism," *Rational Readings on Environmental Concerns*, Lehr, J. H., ed., Van Nostrand Reinhold, p. 819, 1992.

13. Wheelwright, J., *Degrees of Disaster: Prince William Sound: How Nature Reels and Rebounds*, Simon & Schuster, 1994.

14. Lesney, M. S., *Today's Chemist*, 8: 83, March 1999.

21.3.6 ONE IN A MILLION

Here are some activities that increase our chance of death by one in a million (1):

- Smoking 1.4 cigarettes
- Traveling 10 miles by bicycle
- Traveling 300 miles by car
- Traveling 1000 miles by commercial aircraft (2)

They all seem pretty trivial, don't they? These, along with other activities that increase the chance of death by one in a million, are listed in Table 21.3.8.

One in a million (10^{-6}) is also used to assess human health risks. Lifetime exposure to a substance associated with a risk of 10^{-6} would increase our current chances of developing cancer, which is about 1 in 3, by 0.0003 percent (3). Stated another way, the regulatory agencies are attempting to reduce the cancer incidence of 300,000 people to 299,999. Problems arise with assigning a specific figure to the number of cases of cancer considered acceptable. It gives the public the false impression that the figure is a matter of scientific fact rather than a statistically derived estimate. As Ottoboni (4) points out, "It delivers the erroneous message that one in a million people, no more, no less, will actually develop cancer from exposure to the chemical in question. It misleads the public into believing that one extra case of cancer in a population of a million people actually could be measured and the cause could be identified. Finally, it frightens some people who fear that they or one of their loved ones may become that one unfortunate soul in a million."

The past, present, and future costs of complying with this stringent criterion are virtually incalculable (3, 5). It is difficult to imagine a criterion in wider use in the United States. Some examples where it is used include (3):

TABLE 21.3.8   Risks that Increase the Chance of Death by One in One Million*

Activity	Cause of death
Smoking 1.4 cigarettes	Cancer, heart disease
Fireworks	Accident[†]
Drinking 1/2 L of wine	Cirrhosis of the liver
Living two days in New York or Boston	Air pollution
Living two months in Denver on vacation from New York	Cancer caused by cosmic radiation
Living two months in average stone or natural brick building	Cancer caused by radioactivity
Traveling six minutes by canoe	Accident
Traveling 10 miles by bicycle	Accident
Traveling 300 miles by car	Accident
Traveling 1000 miles by commercial aircraft	Accident[‡]
One chest x-ray	Cancer
Eating 40 tablespoons of peanut butter	Liver cancer caused by aflatoxin B
Eating 100 charcoal broiled steaks	Cancer from benzpyrene

* Wilson (1).
[†] Chapman and Morrison (7).
[‡] Popescu (2).

- Pesticides in food additives
- Allowable exposure to groundwater contamination and incinerators
- Emissions from stacks
- How a hazardous waste site should be cleaned up
- How much Alar to leave on apples

Because of the many billions of dollars spent in attempting to achieve this goal for cleanups of hazardous waste sites in the United States, one might be tempted to ask the question: What are the origins of 10^{-6}? The answer is that there is no sound scientific, social, economic, or other basis for the selection of 10^{-6} as a cleanup goal for hazardous waste sites. Extensive research by Kelly and Cardon (3) led them to state, "Remarkably, the criterion, which has cost society billions of dollars, has never received widespread debate or even thorough regulatory or scientific review. It is an arbitrary level proposed 35 years ago for completely different regulations (animal drug residues), the circumstances of which do not apply to hazardous waste site regulation. As a result, implementing it has frequently been socially, politically, technically, and economically infeasible."

The review conducted by Kelly and Cardon (3) included an informal telephone survey of affected agencies and an extensive literature search. They found that none of the officials contacted at any federal or state agency currently using 10^{-6} as a criterion knew the basis of this criterion, nor was there any readily available documentation that specifically described the origin of 10^{-6}. They discovered that the concept of 10^{-6} was originally an arbitrary number finalized by the U.S. Food and Drug Administration as a screening level of "essentially zero" or deminimis risk. This concept was traced back to a 1961 proposal by two scientists from the National Cancer Institute regarding methods to determine "safety" levels in carcinogenicity testing.

How Is 10^{-6} Used?

Interestingly, the risk level of 10^{-6} is not consistently applied to all environmental legislation. Instead, it seems to be applied according to the general perception of the risk associated with the source being regulated. Hazardous waste sites, pesticides, and selected carcinogens have seen almost exclusive application of 10^{-6}, while air, drinking water, or other sources perceived to be of less risk have not been subject to this requirement (3). Cleanup levels for a given contaminant are not consistent from site to site and vary by orders of magnitude. Furthermore, in some cases there are extreme differences even among divisions of the same agency for the same substance. A case in point is arsenic, where there are six orders of magnitude (one million fold) difference in target risk within different EPA regulations (6).

Other Events

There is a one in a million chance that an asteroid with a diameter of 10^4 m (Mt. Everest size) will hit the Earth. If it did, the fatalities would range from 10^7 to over 10^9 people. So here's another one in a million item to consider. As Chapman and Morrison (7) have stated, a typical U.S. citizen's death from a killer rock from space is much higher than the widely publicized threats from certain carcinogens and poisoning by commercial foods.

If you are a woman, the chances of being killed at work by someone you know are one in 600,000. Walsh (8) notes that if you consider this statistic another way, the EPA with its one in a million odds threshold for acceptable risk, would ban women from jobs.

Here's what Carl Sagan says, "The odds of a miraculous cure at Lourdes are about one in a million. You are roughly as likely to recover after visiting Lourdes as you are to win the lottery, or to die in the crash of a regularly scheduled airplane flight—including the one taking you to Lourdes" (9).

REFERENCES

1. Wilson, R., *Technology Review*, 81: 41, Feb. 1979.
2. Popescu, C. B., "Risk Assessment," in *Issues in the Environment*, American Council on Science and Health, June 1992.
3. Kelly, K. A., and Cardon, N. C., *EPA Watch*, 3(17): 4, Sept. 15, 1994.
4. Ottoboni, M. A., *The Dose Makes the Poison*, Second Edition, Van Nostrand Reinhold, New York, 1997.
5. Yandle, B., "Human Health and Costly Risk Reduction," *The Freeman*, 45(3): 174, March 1995.
6. Travis, C. C., et al., *Environ. Sci. Technol.*, 21: 415, 1987.
7. Chapman, C. R., and Morrison, D., *Nature*, 367: 33, Jan. 6, 1994.
8. Walsh, J., *True Odds: How Risk Affects Your Everyday Life*, Merritt Publishing, New York, 1996.
9. Sagan, C., *The Demon-Haunted World*, Random House, New York, 1995.

21.3.7   KILLER ROCKS

If you were Mother Earth, which would worry you more: killer rocks that can destroy the biosphere of entire continents, or parts per quadrillion of toxins that might cause a few additional cancer deaths per decade? If we could ask nature how it might rank environmental calamities that have so far occurred, all leading contenders would be natural, not man-made. Easterbrook (1) sums it up very nicely:

"The biosphere that some environmentalists today contend cannot resist so much as an oil spill or an overzealous crew of loggers, has in the past survived the unimaginable multiple whammy of the atmosphere set on fire by a killer rock strike; followed by years of summer frost, megasmog, and acid rain from hell; followed by decades or centuries of continuous global volcanism set loose by the rock's effect on the crust. This point is not made to rationalize oil spills or clear cutting. No human environmental misuse should ever by justified on the grounds that the environment can recover, even if that is true. The point here is simply to compare the sorts of people-caused environmental insults that women and men today reflexively describe as 'disasters' against the genuine disasters nature has survived in the past."

There have been five really massive extinctions of life on Earth with the most recent occurring about 65 million years ago. An object about 10 km across impacted Earth ending the Cretaceous period and wiping out the dinosaurs. The energy released was equivalent to the explosion of about a hundred million megatons of TNT. This is about ten thousand times the energy that would have been released by the simultaneous explosion of all nuclear weapons on earth in the 1980s, before the superpowers started to dismantle their stockpiles (2).

A few definitions first. *Asteroids* are slabs of rock, dirt, or metal while comets contain all this stuff held together by gravity. *Comets* pose a graver danger because they approach the earth at about 50 km per second (over 100,000 mph), about twice the speed of asteroids. A comet would strike earth with about 10 times the energy of a comparably sized asteroid and only give us about half the time to intercept it in space. *Meteorite* is the term for anything that hits the earth, whether it is an asteroid or comet.

Astronomers report that Earth has been bombarded by asteroids and comets large enough to cause global environmental catastrophe about once every 300,000 years. This suggests that our "fragile" environment has taken many serious beatings over the years, and has always managed to come back. In the period from 1975 to 1992 military satellites detected 136 large explosions caused by asteroids or comets in the upper atmosphere, averaging eight per year. The typical explosive force was 15 kilotons, which is about the same as the Hiroshima bomb. Because the satellites see only about one-tenth of the Earth at any moment, it is projected that the true rate of upper atmosphere strikes may by more like 80 per year (1). At present, at least 300 objects greater than 1000 meters in diameter are known to have orbits that intersect Earth's.

Events on earth are not as far-fetched as you might think. In 1908 in a remote northern region of Siberia called Tunguska, a vast fireball exploded with the force of 1000 Hiroshima bombs. The heat incinerated herds of reindeer and charred tens of thousands of evergreens across hundreds of square miles (3). The area is so remote that few humans deaths occurred. However, if the Tunguska rock had arrived on the same trajectory a few hours later when the Earth had turned a little more, St. Petersburg would have been destroyed along with most of its inhabitants, and perhaps the course of history, because one of its residents included a certain Vladimir Ilich Ulyanov, also known as Lenin (2).

In 1989, a 1000 meter asteroid was discovered only after it had crossed the Earth's orbit at a spot where the Earth had been just six hours before. In May 1996, another asteroid of about the same size was discovered only four days before it sped across Earth's orbit, ultimately missing our planet by four hours (4). If you're really concerned about rocks from outer space stay away from Wethersfield, Conn. In April of 1971, a meteorite caused minor damage to a house in this town; eleven years later, on November 8, 1982, another meteorite damaged a second house in the same town. These two impacts were only about 1 km apart (5).

Would you believe that the risk of death from an asteroid impacting the Earth lies in the same range of probabilities as the risk of death from a hurricane or airplane crash? This is due to the fact that even though major asteroid impacts are unlikely in a given year, when they do occur they have the potential to generate destruction of enormous proportion (4).

Chapman and Morrison (6) have stated that the chances of a typical U.S. citizen's death from a killer rock from space is much higher than the widely publicized threats from certain carcinogens, poisoning by commercial foods and pills that have been deliberately tampered with, fireworks accidents, terrorist bombs, and airline hijacking. Table 21.3.9 lists the risk of death during a 50-year period because of some of life's many hazards.

What To Do?

A NASA workshop in 1992 resulted in a proposal called the Safeguard Survey, which consisted of a scheme to monitor near-Earth space hazards using a specially built network of telescopes. The start-up cost would be about $50 million and running costs some $10 million per year. Over a period of 20 to

TABLE 21.3.9   Average Risk of Death to an Individual Over a 50-Year Period, as Estimated from Historical Frequencies and Current Populations

WORLDWIDE:	
Risk of death* from	
Asteroid impact	1 in 20,000
Volcanic eruption	1 in 30,000
UNITED STATES only:	
Risk of death* from	
Auto accident	1 in 100
Electrocution	1 in 5,000
Airplane crash	1 in 20,000
Hurricane	1 in 25,000
Tornado	1 in 50,000
Lightning	1 in 130,000
Earthquake	1 in 200,000
Fireworks accident	1 in 1 million[†]
Food poisoning by botulism	1 in 3 million[†]
Drinking water with EPA limit of trichloroethylene	1 in 10 million[†]

* From Zebrowski (4).
[†] From Chapman and Morrison (6).

30 years, Spaceguard could identify all the potentially dangerous asteroids in the inner solar system and also give a few months warning of the arrival of a threatening comet (2). Where are we on this? Morrison (7) has pointed out that at present there are less than a dozen people in the whole world who are searching for NEOs (near earth objects)—fewer people, as he puts it, than it takes to run a single McDonald's.

At present, we are spending billions on many issues where we worry about parts per zillion of something that no one can really prove is or is not a problem. For example, a paper presented to a Department of Energy conference in 1990 estimated that the cost per cancer avoided in Superfund clean ups exceeds $15 billion (8). Some of this zeal is an attempt to save lives of future generations. However, as Cohen (9) has pointed out:

"The lives saved are those of people living many thousands of years in the future, who bear no closer relationship to us than those now living in under developed countries, whose lives we disdain to save at one millionth of these costs. In the second place, there is an excellent chance that a cure for cancer will be found in the next few thousand years in which case these deaths will never materialize and the money will be wasted."

I, for one, would like to see a few of those millions that are being thrown at some hypothetical contaminant be spent looking for killer rocks from outer space. Future generations might well find a cure for cancer and heart disease but will not be able to prevent outer space from shooting rocks the size of mountains at Earth on occasion. Although it might not occur tomorrow or next year, or in the next thousand years, it will happen again and the best thing we could do for those alive at the time is start preparations right now.

If you're still skeptical, remember the recent impacts on Jupiter. One of the chunks of the comet Shoemaker-Levy 9 that crashed into Jupiter in July 1994 left dust clouds that grew larger than the whole Earth an hour after impact (10). This would have wiped out life on Earth as we know it. Milloy (11) says it best: In reviewing a Washington Post article that discussed global warming and also killer rocks from space he says: "Pardon me, but if we're going to be concerned about doomsday scenarios, I would be more concerned about getting clocked by space rubble."

REFERENCES

1. Easterbrook, G., *A Moment on Earth*, Viking, New York, 1995.
2. Gribbin, J., and Gribbin, M., *Fire on Earth-Doomsday, Dinosaurs and Humankind*, St. Martin's Press, New York, 1996.
3. Stone, R., *Discover*, p. 60, Sept. 1996.
4. Zebrowski, E., Jr., *Perils of a Restless Planet*, Cambridge University Press, Cambridge, UK, 1997.
5. Berman, B., *Discover*, p. 28, August 1991.
6. Chapman, C. R., and Morrison, D., *Nature*, 367: 33, Jan. 6, 1994.
7. Morrison, D., *Exploring Planetary Worlds*, W. H. Freeman & Co., New York, 1993.
8. Abelson, P. H., *Science*, 255: 901, Feb. 21, 1992.
9. Cohen, B. L., "Perspectives on the Cost Effectiveness of Life Saving," in *Rational Readings on Environmental Concerns*, Lehr, J. H., ed., Van Nostrand Reinhold, New York, 1992.
10. Levy, D. H., *Impact Jupiter*, Plenum Press, 1995.
11. Milloy, S. J., web site, www.junkscience.com, Comments on an article in the *Washington Post*, Feb. 16, 1997.

21.3.8 RADIATION

"It isn't what we don't know that causes the problems, it's what we think we know that just isn't so."—Mark Twain

Have you heard?

- Low levels of radiation are beneficial to humans?
- Mice exposed to low levels of radiation lived longer than mice that weren't?
- Fish exposed to low levels of radiation grew faster than fish that weren't?
- Low levels of radiation increase fertility, embryo viability, and decreases sterility and mutations? (1)

More likely you've heard things like the following: When radioactivity from the Chernobyl accident reached our West Coast, the popular press warned residents about the dangers of possible fallout, speaking of the number of picocuries of radioactivity detected in high clouds without ever explaining to the public that a picocurie is one part per trillion. The press also didn't mention that a person would have to drink 63,000 gallons of that radioactive rain water in order to ingest one picocurie of radioactivity (1). Or perhaps you've heard about the "nuclear disaster" that occurred at the Three Mile Island nuclear power plant. The press gave much coverage to this, however, a report by the U.S. Nuclear Regulatory Commission (NRC) revealed that the average dose of radiation received by two million people in the surrounding area was 0.0014 rems. The highest estimated individual exposure resulting from the TMI release was 0.075 rems. Typically, a person in the United States receives about 0.36 rems of radiation annually from naturally occurring radiation, medical uses of radiation and consumer products (2). As Remmers points out: "The most serious damage from Three Mile Island was the psychological trauma and over-exaggeration from the mishandling of this incident by politicians and the media" (3).

You may ask, what's a rem? A rem is the amount of energy deposited in the human body by ionizing radiation. For ease of understanding, Mark Hart, of the Lawrence Livermore National Laboratory in Livermore, CA, equates 1 rem to 1 dollar so 1 millirem is 0.100 cents (4). The yearly limit for safe exposure is 5 rem or 5 dollars. Hart has worked hands-on with plutonium over for three years, and is a frequent presenter of a talk titled "Radiation-What Is Important?" In this talk, which is often given to youngsters, he uses over 100 radioactive items including antiques, consumer items, fossils, and minerals. In his laid-back style, he explains radiation and says he wants to give folks an intuitive feel for radiation around them so that, ultimately, they can make their own decision on what's important. After hearing his talk, it's hard for me to imagine anyone, except the hardest core environmentalist, who has not changed some of their thinking about radiation.

In one year Hart receives 37 cents of radiation working as a plutonium handler. He has yet to go into an antique store where he didn't find something that was radioactive. However, he hastens to point out that there is no harm with antiques, even as food plates. "One of the most important aspects of radiation is the public's perception," said Hart, as he drank coffee from an antique radioactive coffee cup made of green "Depression" glass. The key factor, he discloses, is that the radioactive material stays in the glass and does not enter his body. "These radioactive items won't make other things radioactive," he said. His collection of plates, cups, glasses, vases, jewelry, gravy boats, and baby dishes made of green or yellow glass popular in the 1920s and 30s or coated with orange uranium oxide glaze, all exhibit some degree of radioactivity above background and pose no health threat (5).

Radioactivity is a perfectly natural phenomenon. The ground we walk on is radioactive; so is our blood, so is the food we eat, so is the air we breathe (6). At what elevation do you live? For each 100 m increase in altitude the annual radiation dose increases by approximately 1.5 millirem. This increase occurs because, as elevation increases, there is less atmosphere to shield the secondary cosmic radiation (7). Therefore, Denver exposure is approximately twice that of Washington, D.C., so people residing in Rocky Mountain states receive twice the natural radiation background, because of higher altitudes and large deposits of uranium. However, compared with states with lower natural radiation background, Rocky Mountain residents experienced less age-adjusted overall cancer deaths and a lung cancer rate only two-thirds as high (8).

Do you travel? An airplane passenger flying above 33,000 ft receives between 0.5 and 1.0 millirem per hour. This means that a five-hour flight provides exposure of 2.5 to 5.0 millirem. However, the postulated danger of receiving an extra 10 millirems per year from living at the border of a radioactive waste site has received vastly more attention from the press and public (8). Radiation exposure varies around the world. Grand Central Station in New York = 0.53 rem/yr, while St. Peter's Square in

Rome = 0.80 rem/yr. Note that the rules that will be applied in the decommissioning of U.S. nuclear power plants would require the stone structures of St. Peter's Square in Rome and the Grand Central Station in New York to be dismantled and buried because of their radioactivity (4). Morro do Ferro in Brazil is 7–14 rem/yr and Ramsar, Iran is 48 rem/yr. A colony of rats occupies burrows in the mounds in a Morro do Ferro. This is a weathered mound, 250 m tall, that is formed of an ore body containing an estimated 30,000 metric tons of thorium and 100,000 metric tons of rare earths. The radiation level is so high that absorbed radioactivity in vegetation permits photographs (autoradiographs) showing the radiation truly growing in the dark. However, the mound supports both animal and plant life. Rats in the region breathe between 3000 and 30,000 rems per year, roughly three times the concentration that should produce tumors or other radiation effects. Yet no abnormalities were found on rats that were trapped and autopsied (1). Also, speaking of Ramsar, Iran, in 1990 this city was host to an international conference on high levels of natural radiation (HLNR). It kind of makes sense to hold such a conference in a city with one of the highest natural radiation levels in the world. This conference was a continuation of a series of conferences held previously on this topic. One conclusion from this meeting was that epidemiological studies on HNLRs in a number of countries did not show any evidence of increased health detriment compared with normal areas (9).

Hot springs and mineral water resorts usually have elevated amounts of radioactivity. For example the waters of the English city of Bath have a radon content of 1730 pCi per liter. Compare this with the value of 4 pCi per liter that EPA has set for homes. The radon in natural gas at Bath is 33,650 pCi per liter. Other places like Bath include, Baden Baden, Warm Springs, Georgia and White Springs, Virginia (1).

A report (10) on residual radioactivity in the soil in Kazakhstan, which was the site of the first Soviet nuclear explosion in August 1949, provides some interesting data. Altogether 459 nuclear explosions were conducted at the three technical areas of this site between 1949 and 1989. Of these, 346 were underground explosions. All 113 of the other explosions—26 ground explosions and 87 atmospheric explosions—occurred at one of the areas, Technical Area III. As Robinson (11) reports, "Surely here we can find the nuclear hell on earth of unsurvivable residual radiation. Yet measurements revealed that one hour spent at the site of 113 nuclear explosions over a 40 year period ending in 1989 has about the same negative health effects from radiation as a trip from San Francisco to New York in an ordinary jetliner."

So why is the public so fearful of radiation? Cameron (12) says the following: "It is my belief that much of the blame for the public's fears and apprehensions with respect to radiation matters are due to our media. There is another criticism that must be directed to the media, namely, their constant use of a small number of individuals, who are clearly out of step with the radiation protection community. In the U.S. alone there are some 3500 health physicists and 1900 radiological physicists. Yet the media will, for some freshly breaking news story, seek out some of a half a dozen individuals who are willing to make willfully deceptive statements regarding radiation." He further discourses that out of a collection of "popular" books published over the last decade or so dealing with radiation matters, there is not a single one that is not riddled with half-truths, untruths, and evidence of basic lack of knowledge of nuclear energy or radiation-another insidious practice designed to keep the public alarmed about radiation matters (12). Marvel comics has more than 70 comic book characters who had developed severe physical and emotional handicaps as a consequence of exposure to radiation (8). What kind of message does this deliver to young persons and anyone else who reads this material? Genetic defects in offspring because radiation exposure of the parents are a well known effect produced in experiments with animals; however, it has never been observed in humans, not even in Hiroshima and Nagasaki, in spite of extremely thorough and intensive investigations (6). It is true that very large amounts of radiation can cause cancer or even death. For that matter, a large amount of nearly anything, even water, is dangerous. However, all studies of low-level radiation doses to humans indicate no harm and many studies suggest that low-level radiation is beneficial. A book by Luckey (13) is a 336-page compendium of actual observations showing beneficial effects of radiation in many aspects, with more than 1000 references. This beneficial effect of low-level radiation is called radiation hormesis.

The two most widespread applications of nuclear energy, are generation of electricity the use of radioisotopes produced in nuclear reactors for diagnosis and treatment of many human conditions including cancer, cardiovascular disease, metabolic disorders, and mental illness (8). However, even these applications are not without controversy. For example, the use of nuclear energy to generate electricity has encountered so much opposition that no application for a nuclear power plant has been filed since 1977. However, few people object to nuclear medicine or radiology even though their contribution to the radiation background in the United States is a thousand times greater than discharges from the nuclear power industry (14). There has not been a single fatal accident involving radiation for over 20 years, whereas there have been over 2 million fatalities from other types of accidents in this country during this same time period (15). Smokers receive about 1300 mrem per year from naturally occurring radioactive materials in cigarettes. This is far more radiation than they might ever receive from a nuclear power plant in their community (16).

Ionizing radiation, gamma radiation, and electron beams are a potential innovative treatment process that can destroy polluting compounds in the aqueous, solid, and gas phase. Very little data has appeared on the application of these processes, but reference (17) presents results of recent studies.

Here is a fact for you heavyweights. The loss of life expectancy from being 20 percent overweight is 900 days; from radiation emitted by nuclear power plants, 0.02 days (18). One last bit of trivia. The principal contributor of internal radiation in our bodies is K-40, a long-lived radioactive isotope of potassium. Because the concentration of potassium is higher in muscular tissue, the amounts in men tend to be somewhat higher. So if you find yourself in a crowded room and want to keep your dose rate to a minimum, you should always stand close to a woman to avoid receiving an unnecessarily high dose from K-40 (16)!

Conclusion

Many of the benefits that radiation offers—for example in health, safety, and economic development—are frustrated by opposition from pressure groups and are encouraged by the media (14). In the United States, surveys continually place nuclear power at the top of the lists of risks in life (19). The fact that nuclear power plants are seen as accidents waiting to happen has stifled this technology for over 20 years, and is inexcusable. Another issue is food irradiation. We wouldn't be reading about the latest *E. coli* contamination in food if irradiation were used. However, until very recently, activists have successfully prevented this from being done. As Lenihan wisely stated: "Many of the objections to radiation fly in the face of reason, or even of common sense" (14).

REFERENCES

1. Ray, D. L., "Radiation Around Us," in *Rational Readings on Enviornmental Concerns*, Lehr, J. H., ed., Van Nostrand Reinhold, New York, (1992).

2. "Three Mile Island," *927 Federal Supplement 834*, June 12,1996, from S. J. Milloy home page site developed by Westlake Solutions.

3. Remmers, E. G., *Issues in the Enviorment*, American Council on Science and Health, New York, p. 68, 1992.

4. Hart, M. M., *Radiation: What Is Important?* Lawrence Livermore National Laboratory, Livermore, CA, May 2, 1998.

5. Cassady, C., "Radioactivity is Where the Public Least Expects," *Newsline*, Lawrence Livermore National Laboratory, Livermore, CA, August 5, 1994.

6. Beckmann, P., *The Health Hazards of Not Going Nuclear*, Golem Press, Denver, CO, 1985.

7. Hutchison, S. G., and Hutchison, F. I., *Journal of Chemical Education*, 74: 501, 1997.

8. Young, J. P., and Yalow, R. S., eds., *Radiation and Public Perception: Benefits and Risks*, American Chemical Society, Washington, D.C., 1995.

9. Sohrabi, M., *Nucl. Tracks Radiation Meas.*, 18: 357, 1991.

10. Yamamoto, M., Tsukatani, T., and Katayama, Y., *Health Physics*, 71(2): 142, 1996.

11. *Access to Energy*, Cave Junction, Oregon, September 1996.

12. Cameron, J. R., *Health Physics*, 73: 523, 1997.

13. Luckey, T. D., *Radiation Hormesis*, CRC Press, Boca Raton, FL, 1991.

14. Lenihan, J., *The Good News About Radiation*, Cogito Books, 1993.

15. Cohen, B. L., "The Hazards of Nuclear Power," in *The Resourceful Earth, A Response to Global 2000*, Simon, J. L., and Kahn, H., eds., Basil Blackwell, 1984.

16. Cohen, B. L., and Moeller, D. W., *Issues in the Environment*, American Council on Science and Health, pp. 62 and 66, 1992.

17. *Environmental Applications of Ionizing Radiation*, Cooper, W. J., Curry, R. D., and O'Shea, K. E., eds., John Wiley & Sons, Inc., 1998.

18. Walsh, J., *True Odds*, Merritt Publishing, 1996.

19. Furedi, F., *Culture of Fear*, Cassell, 1997.

21.3.9 *HORMESIS: A LITTLE BIT IS GOOD—TOO MUCH IS BAD*

Did your mother ever tell you that a little bit of something was good for you but too much would make you sick? A case in point is my experience with chocolate, which I've always loved. I was helping my uncle who owned a bakery and my job one day was to clean five gallon cans of chocolate. As part of my cleaning process, I sampled too much of the chocolate left on the insides of the cans. I got very sick, a case of too much of a good thing. This is an example of hormesis, which is a fancy sounding name that means high doses and low doses produce opposite effects with low doses providing beneficial effects and high doses the opposite. Hormesis applies throughout nature. There are many hundreds of studies in the literature that describe experiments and observations supporting the beneficial effects of chemical and physical agents at low doses (1). Hormesis also applies to radiation, which is covered in the following section.

The word *hormesis* derives from the same root as hormone, from the Greek hormo, "I excite." Although the word is new, the concept is quite old, having been described by Hippocrates and others (2). They were all concerned with dose. They recognized that the dose was everything and stress was the common denominator. Small stresses stimulated while excess stress inhibited.

Hormology, the study of excitation, includes physical, chemical, and biologic cases. One example is the melting point of a mixture of two metals. Addition of one to the other lowers the melting point until the eutectic point is reached; then adding more of the same metal to the mixture increases the melting point.

Chemical hormesis is well-known in catalysis. Biochemistry examples include enzymes and their activation. Under proper conditions enzymes can increase reaction rates a thousand times but excess of the same material decreases the reaction rate (2). Tetrodotoxin is an interesting biologic example. This material, which is 100,000 times more toxic than plutonium, is a natural hallucinogen found in Fugu, the puffer fish treasured for centuries in Japan. As Luckey points out: "A small amount of tetrodotoxin produces euphoria; too much produces severe paralysis. In fact, overdoses are thought to kill dozens of Japanese each year. It is also the principle ingredient of the mixture used in creating zombies by Haiti witch doctors" (2).

Hormesis has been observed in bacteria, fungi and yeasts, plants, algae, invertebrates and vertebrates, and in cell cultures (3). The agents used in these studies included metals, such as cadmium and lead, solvents, and many synthetic and natural chemicals. PCBs were shown to promote the growth of juvenile Coho Salmon and minnows. The growth of clam and oyster larvae was promoted by low-level concentrations of some 52 pesticides, and chlordane and lindane have been shown to promote the growth of crickets.

In terms of animals, chemically induced hormetic effects have been claimed for crabs, clams, oysters, fish, insects, worms, mice, rats, ants, pigs, dogs, and humans. The range of agents employed in such studies has been wide, including numerous antibiotics, polychlorinated biphenyls (PCBs), ethanol, polycyclic aromatic hydrocarbons (PAHs), heavy metals, essential trace elements, pesticides, and a variety of miscellaneous agents, including chemotherapeutic agents, solvents such as carbon tetrachloride, chloroform, cyanide, sodium and others (4). A 1994 publication by the American Industrial Health Council listed 30 studies showing beneficial effects for lead, nicotine, PCBs, toluene, alcohol, jet fuel, methyl mercury, and several other poisons (5).

Absolute purity is out of the question with productive waters. Kazmann (6) reports, "If the Mississippi river passing between Baton Rouge and New Orleans consisted of distilled water there would be no seafood industry such as we no have in Louisiana. Without copper contamination in the water there would be no oysters. Traces of iron, manganese, cobalt, copper and zinc are essential for crabs, snapper, flounder, shrimp and other creatures that abound in Gulf waters" (6).

The use of vitamins as daily dietary supplements based on their beneficial effects at low doses is an example of hormesis. Overdose of the more fat soluble vitamins such as A, E, and D have toxicities associated with them (7). Water soluble vitamin B6 can cause nerve damage at high doses. Hormones such as estrogen can cause cancer (8). If you breathe pure oxygen at normal room pressure, you will suffer chest pain, coughing, and a sore throat within 6 h. Hospitals have found that premature babies placed in incubators that were filled with oxygen enriched air went blind because of oxygen damage to their retinas (9). Even water itself can be fatal if the rate of intake exceeds the body's capacity to process it (and I'm not talking about drowning, rather drinking too much water). In Germany, a man died from cerebral edema and electrolyte disturbance because he drank 17 L of water within a very short time (10). Overuse of ethanol is associated with numerous organ toxicities, however, these days you can read many claims about reduced risk of coronary heart disease, reduced mortality, and so forth, from moderate consumption of alcohol.

At the end of the nineteenth century, only two elements (iodine and iron) were known to be essential for human health. By 1935, only four more (copper, manganese, zinc, and cobalt) had been added. However, progress has been more rapid during the past 40 years (Table 21.3.10), largely because the revolution in analytical chemistry has greatly enhanced the experimenter's capability to measure trace elements in the extremely small quantities present in plant and animal tissues and in food (11). The main reason for the extraordinary potency of essential trace elements is that most of their work is done as components of enzymes or hormones. Note that even lead and cadmium are included in Table 21.3.10. Heiby (12), in his tome *The Reverse Effect*, containing 1216 pages and 4821 references, lists

TABLE 21.3.10 Essential Trace Elements in the Human Body*

Element	Date of recognition of essential role
Iron	17th century
Iodine	19th century
Copper	1928
Manganese	1931
Zinc	1934
Cobalt	1935
Molybdenum	1953
Selenium	1957
Chromium	1959
Tin	1970
Vanadium	1971
Fluorine	1971
Silicon	1972
Nickel	1974
Arsenic	1975
Cadmium	1977[†]
Lead	1977[†]

* From Lenihan (11).
[†] Evidence incomplete, but see Heiby (12).

a number of examples in which extremely minute amounts of these metals may be important to the human body.

Practically every enzyme in the body has some small amount of metal involved in its chemical structure. Zinc is used to maintain cell membranes and produce protein and energy; either too much or too little can lead to reduced growth. Iron and molybdenum also must be kept at moderate levels (8). Excess iron leads to hemochromatosis, where massive amounts of iron, as ferritin and hemosiderin, are laid down in the body tissues (13). Other cases of iron overload are specific anemias where excessive breakdown of blood occurs, liver disease, and massive blood transfusion. Copper is essential in minute quantities to the normal functioning of the human body. It makes possible the assimilation of iron, but like some other members of this company (most notably iron), its impact in large quantities has long been known to be toxic. Wilson's disease is essentially chronic copper poisoning. In it, the natural balance between copper ingestion and copper excretion is disturbed and the copper thus retained is stored in certain organs. Symptoms of the disease include a sweeping range of neurological disturbances—slurred speech, failing voice, excessive salivation, drooling, difficulty in swallowing, tremors, uncoordination, spasticity, and muscular rigidity.

As *Runner's World* reported (14): "Chromium is not just for bumpers anymore" (14). The article claims that chromium assists the hormone insulin, which is vital for proper glucose control and for protein production in muscles. Other claims are that chromium supplements offer promise as new treatment for heart disease, high blood cholesterol and diabetes (15). The metal even has its own book, *The Chromium Caper*, by J. Fischer (16). As Fumento discloses; "It's another one of those diet books that could be summed up in a single sentence: Make sure you get enough chromium in your diet" (17). Excessive doses of chromium may suppress growth but only with quantities far in excess of those encountered in the food and supplements containing yeast or amino-acid chelates (13).

Selenium, once thought to be a poison and a carcinogen, has been found to be an anti-tumor agent in relatively low doses and an essential nutrient with a very small difference between recommended and harmful doses (2). Arsenic compounds, which are known human carcinogens, have been shown to widely stimulate the growth of chickens, calves and pigs. Some chlorinated dioxins and furans, widely considered as carcinogens, have been patented as anti-tumor agents (3). Recently, arsenic trioxide has been used successfully in intravenous form for the treatment of acute promyelocytic leukemia in humans (18, 19).

The above information is not meant to be all-inclusive and the beneficial claims can change based on new research. Stewart (20) jokingly scoffs that as one scientist announces that some food is bad for you, five other scientists will release findings proving that the food only is bad for you only if you are a 185-pound middle aged laboratory rat. He said there have been so many contradictory reports from scientists about cholesterol that three years ago his wife gave up eating entirely. All kidding aside, the point to be made is that there is enough evidence to support the fact that many chemicals and metals that are clearly toxic at certain dose levels offer beneficial effects at low doses.

What's the Point?

The important issue is that most of the time we receive only the bad news. The entire regulatory orientation is to look for an adverse effect, not a hormetic one. The designs of most toxicology studies emphasize almost exclusively the upper end of any possible dose-response curve. For example, research on most agents begins with the determination of LD-50 values, the dose of a chemical that will kill 50 percent of the test animals (4). As McKenna (7) points out: "In the face of spiraling costs for remediation of environmental risks by both government and private sectors, the potential for greater accuracy in depicting chemical risks offered by consideration of hormetic effects seems a major benefit" (7). Approaches traditionally used by the EPA have not incorporated deviations from linearity, nor have they tended to examine data (low-dose or other) that do not show a correlation of dose with some adverse effect. Remember this when you hear about the next environmental scare of the month. Ask the question: Instead of an animal consuming 50,000 to 100,000 times what a human would, what are the effects of low doses? They could be beneficial.

REFERENCES

1. Baarschers, W. H., *Eco-Facts & Eco-Fiction*, Routledge, 1996.
2. Luckey, T. D., *Radiation Hormesis*, CRC Press, 1991.
3. Stebbing, A. R. D., *Science of the Total Environment*, 22: 213, 1982.
4. Calabrese, E. J., McCarthy, M. E., and Kenyon, E., *Health Physics*, 52: 531, 1987.
5. Calabrese, E. J., "Biological Effects of Low Level Exposures," *AIHC Journal*, pp. 7–15, Spring 1994.
6. Kazmann, R. G., "Environmental Tyranny—A Threat to Democracy," in *Rational Readings on Environmental Concerns*, Lehr, J. H., ed., Van Nostrand Reinhold, 1992.
7. McKenna, E. A., *Int. J. Environment and Pollution*, 9: 90, 1998.
8. Wildavsky, A., *Searching for Safety*, Transactions Books, 1988.
9. Bova, B., *Immortality, How Science is Extending Your Lifespan and Changing the World*, Avon Books, 1998.
10. Efron, E., *The Apocalyptics: Cancer and the Big Lie*, Simon & Schuster, 1984.
11. Lenihan, J., *The Crumbs of Creation*, Adam Hilger, 1988.
12. Heiby, W. A., *The Reverse Effect: How Vitamins and Minerals Promote Health and Cause Disease*, Medi-Science Publishers, 1988.
13. Mervyn, L., *Minerals and Your Health*, Keats Publishing, 1984.
14. Bean, A., *Runner's World*, p. 24, Feb. 1994.
15. *San Francisco Chronicle*, May 18, 1989.
16. Fischer, J., *The Chromium Program*, Harper & Row, New York, 1990.
17. Fumento, M., *The Fat of the Land*, Viking, 1997.
18. Soignet, S. L., Maslak, P., Wang, Z.-G., et al., "Complete Remission after Treatment of Acute Promyelocytic Leukemia with Arsenic Trioxide," *New England Journal of Medicine*, 339: 1341, 1998.
19. Gallagher, R. E., "Arsenic-New Life for an Old Potion," *New England Journal of Medicine*, 339: 1389, 1998.
20. Stewart, D. L., "When It Comes to Science, Trust But Verify," *San Francisco Chronicle*, April 15, 1998.

21.3.10   RADIATION HORMESIS

Radiation exposure extends lifespan! There is no question that high amounts of radiation are harmful but low amounts are beneficial to humans and animals. This is the concept of hormesis that applies throughout nature and was discussed in the previous article. Hormesis means that high and low doses produce opposite effects, with low doses providing beneficial effects and high doses the opposite.

Let's talk about models that have been used with radiation. The linear model (no threshold) states that all radiation is harmful. The hormesis model says small and large doses produce opposite results. Figure 21.3.5 compares the two, showing the effects of dose on cancer rates (1). The difference between imperceptible harm predicted by the linear model and the benefits noted with the hormesis model suggests that for every thousand cancer mortalities predicted by linear models, there will be a thousand decreased cancer mortalities and ten thousand persons with improved life quality (1). Yet current cost/benefit estimates related to radiation protection (e.g., regarding the consequences of population exposures after accidents such as Chernobyl) and large decommissioning and waste management and remediation programs continue to be based on the linear no-threshold hypothesis.

As Becker discloses; "With the average background in Europe fluctuating substantially, and being exceeded by a factor of 10 to 100 in areas of Brazil, India, and Iran without any detectable detrimental health effects over many generations, it would make little sense to consider evacuation of whole towns or regions in Saxony, Finland or Cornwall, or to close down mining operations in southern Africa which could be required if current radiation policy were to be applied uniformly" (2).

The theory of radiation hormesis, which is supported by experimental evidence, predicts a reduction in cancer incidence (20,000 fewer cancer deaths) in most of Europe as a result of the Chernobyl fallout (3). By contrast, those who subscribe to the linear model theory, estimate that Chernobyl will cause 10,000 excess cancer deaths in the USSR within the next 70 years (1).

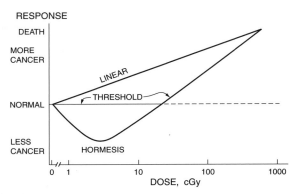

FIGURE 21.3.5   Influence of Linear (No Threshold) and Hormesis Dose Models on Cancer Rates. Adapted from Luckey (1).

Prior to the development and use of atomic bombs biopositive effects of small doses of ionizing radiation were accepted by radiobiologists.

However, the holocausts at Hiroshima and Nagasaki allowed the world to be mesmerized into acceptance of the thesis that "all doses of ionizing radiation are harmful" (1). Continuing media and monetary support of this thesis continues after half a century. Yet a survey by Luckey (1) contains over 1200 literature references on studies, both on animals and on humans confirming the beneficial effects of low level radiation including enhanced growth, improved reproductive capacity, improved immune responses, lower cancer rates and longer lifespan.

In spite of the overwhelming data supporting the hormesis model, the nature of health effects of low-level ionizing radiation continues to be the subject of considerable controversy. Some examples showing the value of low levels of radiation include the following:

- Japanese survivors of atomic bomb attacks on Hiroshima and Nagasaki in 1945 who received low doses of radiation were compared with the population of Japan as a whole. The survivors had lower general mortality rates and lower cancer mortality. Also, the infant mortality among their offspring was significantly below Japan's national average (4).

- Workers at Los Alamos exposed to three-fold higher amounts of plutonium than the maximum currently recommended by the National Council on Radiation Protection have been studied for the past 50 years (5). Standard mortality ratios of the exposed workers when compared to the general population and to unexposed contemporary Los Alamos workers were 0.43 and 0.77, respectively. This means that the number of exposed workers who have died as compared with these two groups is less by 57 percent and 23 percent. The second comparison is especially relevant, because it avoids systematic differences in life style between Los Alamos workers and the general population (6).

- One of the most detailed epidemiological studies found definitive reductions in lung cancer with increasing radon exposure. Cohen (7) used this work to test the linear no-threshold theory for 1601 U.S. counties. Over 50 confounding factors were used in his statistical analysis with the only possible explanation for the results being the failure of the linear no-threshold theory for carcinogenesis from inhaled radon.

There is no evidence of increased mutation, genetic diseases, or cancer in animals or humans following exposure to hormetic doses of ionizing radiation, even in Hiroshima and Nagasaki, in spite of extremely thorough and intensive investigations (1, 8).

Summary

A large body of evidence shows conclusively that whole-body exposures to low doses of ionizing radiation reduce cancer mortality rates when compared with control populations in both experimental animals and humans. The decreased cancer incidence and mortality in animal experiments, in the nuclear industry, in army observers of atomic explosions, and in Japanese bomb victims is consistent (1). As Sagan pointed out; "Literally tens of billions of dollars are being sought by one federal program alone for the purpose of reducing exposure to low levels of radiation and chemical wastes on the basis of largely hypothetical health risks" (9).

The consistency of the results and the statistical significance of much of the data from human experiences and animal experiments destroy two myths: (1) all radiation is harmful, and (2) the linear model is valid for low doses of ionizing radiation. The effects of low doses of radiation appear to be comparable with those of a great variety of toxins; high and low doses give diametrically opposite results. Becker sums it up best: "Ten thousands of millions of dollars are spent every year worldwide in decommissioning, remediation, or nuclear waste programs, which could obviously be used much more beneficially in other areas of public and individual health, in rich, and even more so in poor countries of the world" (2). Misuse of the linear–no threshold model portends spending in excess of $1 trillion in the United States alone for negligible health benefits just for government environmental cleanup programs, while truly significant public health protections are unfunded (10).

REFERENCES

1. Luckey, T. D., *Radiation Hormesis*, CRC Press, Boca Raton, FL, 1991.
2. Becker, K., "Low-Dose Cost Benefit Assessment—A View From Europe," *Health Physics*, 74: 267, 1998.
3. Wagner, H. N., and Ketchum, L. E., *Living With Radiation: The Risk, The Promise*, Johns Hopkins University Press, Baltimore, MD, 1989.
4. Kondo, S., *Health Effects of Low Level Radiation*, Chapter 3, Kinki University Press, Osaka, Japan and Medical Physics Publishing, Madison, WI, 1993.
5. Voelz, G. L., Lawrence, J. N. P., and Johnson, E. R., *Health Physics*, 73: 611, 1997.
6. *Accesses to Energy*, Cave Junction, Oregon, Vol. 25, No. 12, August 1998.
7. Cohen, B. L., *Health Physics*, 68: 157, 1995.
8. Beckmann, P., *The Health Hazards of Not Going Nuclear*, Golem Press, Denver, CO, 1985.
9. Sagan, L. A., *Science*, 245: 574, 1989.
10. Muckerheide, J., *Nuclear News*, p. 26, Sept 1995.

21.3.11   DDT

"The greatest killer in Africa is not AIDS or sleeping sickness, but malaria which kills an estimated 2 million children each year" (1)

Did you know?

- DDT has saved more lives in the past 50 years than have antibiotics as a group. The banning of DDT is probably the largest act of genocide in human history. The Nationals Academy of Sciences estimated that DDT saved 500 million lives before it was banned (2).
- Since DDT was banned, the incidence of malaria has increased enormously worldwide and the disease has again become a leading cause of death. Every 12 seconds, a child dies of malaria (3).
- There is evidence suggesting that DDT is an anticarcinogen (4).
- DDT's "cousin" DDD is an anticancer drug used against inoperable adrenal-gland cancer (4, 5).

My guess is that most of you hadn't heard the above facts. Rather, what you've heard, or read, is that DDT is toxic, has caused eggshell thinning in birds, accumulates in fat tissues in our bodies, and is still found in the environment. The facts are that DDT was given a bad rap in 1972 when it was banned from usage, and over 25 years later many people are still unaware of the truth.

Let's look at some of the facts. Early in this century, the only effective way to control malaria was to eliminate stagnant water, such as swamps and landfills, where Anopheles mosquitoes bred. Then beginning in 1943, the organochlorine pesticide DDT became available and this proved to be a godsend in the Third World, curtailing the disease dramatically. In India, by the early 1960s, the annual incidence of malaria had declined from one million to 100,000 (6) and in Sri Lanka, the number of cases dropped from over two million to 17 (7). In 1942 DDT was shown to kill body lice without adverse effect on humans, and it was used by all Allied troops during World War II. Thanks to DDT, a 1944 typhus epidemic in Naples was halted. No Allied soldier was stricken with typhus fever (carried by lice) for the first time in the history of warfare. In World War I, by contrast three million people died of typhus in Russia and Eastern Europe, and more soldiers died from typhus than from gunfire.

In 1962, Rachel Carson's best seller, *Silent Spring*, indicted DDT as a killer of birds, fish, and wildlife (9). This eventually led to a long (seven-month) federal hearing in 1972 on the risks and benefits of the material. The DDT hearings were ordered by then EPA administrator William Ruckelshaus, who appointed Judge Edmund Sweeney as the hearing examiner. Scientists were not the only ones to give exonerating testimony that DDT used properly presented little harm to humans, beast, or bird. The World Health Organization also pleaded at the EPA hearing that DDT was very beneficial in fighting malaria in many parts of the world and should not be banned (10). After 125 witnesses and 9362 pages of testimony, Judge Sweeney's final conclusions were that:

- DDT is not a carcinogenic hazard to humans
- DDT is not a mutagenic or teratogenic hazard to humans
- The use of DDT under the registrations involved does not have a deleterious effect on fish, estuarine organisms, wild birds, or other wildlife (7, 10).

In a better world this would have been good news. It was met instead with journalistic and environmental hysteria across the nation. Less than two months after the hearing, the EPA administrator Ruckelshaus single-handedly banned almost all DDT (8, 10). This ban on DDT was considered the first major victory for the environmentalist movement in the United States (11). It gave credibility to pseudoscience, and created an atmosphere in which scientific evidence can be pushed aside by emotion, hysteria, and political pressure. This technique of making unsubstantiated charges, endlessly repeated, has since been used successfully against asbestos, PCBs, dioxin, and Alar to mention a few (7).

DDT was soon replaced by less persistent organophosphates, such as parathion and malathion. These chemicals, belong to the same chemical family as nerve gas and are far more dangerous than chlorinated hydrocarbons such as DDT. They have caused serious poisoning, often fatal, among unsuspecting farm workers who had been accustomed to handling the relatively nontoxic DDT (6).

The ban didn't help third world people. National Institutes of Health malaria expert Robert Gwadz says, "The legacy of Rachel Carson was not altogether positive. The incidence of malaria in India is now back up to more than a million and more than 500,000 in Sri Lanka" (6). In South America, where DDT spraying has been continued until more recent times, data from 1993 to 1995 showed that countries that recently discontinued their spray programs are reporting large increases in malaria incidence. The only country in South America reporting a large reduction in malaria rates (61 percent), is Ecuador, which has increased use of DDT since 1993 (12, 13). The allegations against DDT have been repeated so often and stated with such passion that over 25 years later that the public remains convinced of their validity (14).

Toxicity of DDT

DDT is known to be safe to humans. It has never caused death, even in persons attempting suicide (15). Farm workers were sometimes poisoned by organophosphate insecticides, such as the parathions, which are hundreds of times more toxic to man than DDT and which were touted as superior substitutes to DDT (15). It is known from controlled studies in human volunteers that experimental ingestion of

35 mg DDT per kg of body weight per day, for a period of two years, produced no adverse effects, acute or chronic, in any of the subjects (5, 7). Doses of five grams of DDT (and even more) have been administered to human beings in the successful treatment of barbiturate poisoning, according to Walter Ebeling of UCLA. Professor Ebeling notes also that five grams of DDT are roughly four times as much as the average American will assimilate in a 70-year lifetime (16). A study of workers at the Montrose Chemical Company, who accumulated 38 to 647 ppm of DDT residues in their fatty tissues, revealed no cases of cancer in 1300 person-years of exposure—a statistically improbable event (17).

One of the more interesting examples verifying the nontoxicity to humans is the experience of J. G. Edwards, Professor of Biology at San Jose State University. Says Edwards, "After remembering my own days of dusting hundreds of civilians during the war in Europe with 10 percent DDT to kill lice and help prevent millions of cases of deadly typhus, I thought I should try to convince people that the environmental extremists were wrong. Thereafter, at the beginning of each DDT speech I made I would publicly eat a tablespoon of DDT powder. I believe it was a successful effort. It resulted in a full page photograph of me doing that in *Esquire* magazine (September 1971). The caption stated that I was eating 200 times the normal intake of DDT to show it's not as bad as people think" (18). Today, as Edwards approaches his 80th birthday, he is still as adamantly opposed to the anti-DDT propaganda as he was 26 years ago. Edwards, an avid climber, continues to conquer peaks greater than 10,000 ft. DDT exposure surely hasn't hurt him.

In 1969, rodent studies suggested DDT was a carcinogen. However, these results were refuted by a 1978 National Cancer Institute report that concluded, after two years of testing on several different strains of cancer prone mice and rats, that DDT was not carcinogenic (11). In a 1994 study in the *Journal of the National Cancer Institute*, researchers concluded that their data did not support an association between DDT and breast cancer (19). Very recently, Robert Golden, a Ph.D. toxicologist in Potomac, Maryland, stated, "The one endocrine modulator environmentalists love to hate, the pesticide DDT, would cause no endocrine effect in a fetus exposed to more than a pound of DDT over the course of a pregnancy" (20).

Bruce Ames and his colleagues at the University of California, Berkeley, have developed a method of ranking possible carcinogenic hazards (21). They call this a HERP Index (human exposure over rat potency). A value of 100 on this scale means that people are getting the same dose in mg/kg that caused cancer in half the tested rats. DDT has three major breakdown products: DDA, DDE, and DDD (18). Table 21.3.11 shows that the average U.S. daily intake of DDE from DDT (HERP = 0.0003 percent) is less than the HERP from chloroform in a glass of tap water and thus appears to be insignificant compared to the background of natural carcinogens in our diet. Even daily consumption of 100 times the average intake of DDE/DDT would produce a possible hazard that is small compared to other common exposures such as mushrooms, coffee, beer, and wine shown in Table 21.3.11.

Further support is provided by Stephen Safe, a toxicologist at Texas A&M, who tested the effects of organochlorine compounds in the average human diet. He concluded that the total estrogenic activity of these compounds is 40-million-fold lower than that from the natural components of vegetables and other foods consumed daily such as soybeans, barley, cabbage, and corn (20, 22).

TABLE 21.3.11 Ranking Possible Carcinogenic Hazards*

Possible hazard[†] HERP %	Daily human exposure	Human dose of rodent carcinogen
0.0003	DDE/DDT, daily dietary intake	DDE, 2.2 ug
0.001	Tap water, 1 L	Chloroform, 83 ug
0.1	Mushroom, one raw	Hydrazine mixtures
0.005	Coffee, 1 cup	Furfural
2.8	Beer, 354 ml	Ethyl alcohol, 18 ml
4.7	Wine, 250 ml	Ethyl alcohol, 30 ml

* From Ames and Gold (21).

[†] U.S. EPA's one-in-a-million hypothetical risk is 0.000015 on the HERP scale, or about 400,000 times below the level that would give cancer to a rat.

Persistence in the Environment

One often heard claim is that DDT cannot be broken down in the environment. Actually, DDT is broken down rather rapidly by heat, cold, moisture, sunlight, alkalinity, salinity, soil micro-organisms, hepatic enzymes of birds and mammals, and a great many other environmental factors (18). Only in unusual circumstances where soil is dark, dry, and devoid of microorganisms will DDT persist. Under normal environmental conditions, DDT loses its toxicity to insects in a few days (7). If it didn't break down, it would have been unnecessary to apply it again in order to control pests. Edwards provides a list of more than 140 articles documenting the breakdown of DDT in the environment (18).

A key reason that traces of DDT are sometimes still found in environmental samples is that we can now detect extremely minute amounts of anything. In the span of about two decades, detection limits have been reduced by about six orders of magnitude (23). Some analysts have even reported DDT in samples collected before DDT existed. For example, University of Wisconsin chemists were given 34 soil samples to analyze. They reported that 32 of the 34 samples contained DDT. What the chemists didn't know was that the soil samples had been hermetically sealed in 1911, and no DDT existed in the United States until 1940 (24, 25). The author wrote later: "The apparent insecticides were actually misidentifications caused by the presence of co-extracted indigenous soil components." Still later it was found that red algae also produces halogen compounds that are misidentified as DDT by gas chromatography. Also, halogen compounds containing bromine or iodine, rather than chlorine, may falsely register as DDT on the gas chromatograph (26). Various PCBs were commonly misidentified as chlorinated hydrocarbon insecticides during the 1950s and 1960s, and were routinely reported as "DDT residues."

Claims About Bird Declines

In *Silent Spring*, published in 1962, Rachel Carson stated that the American robin was on the verge of extinction (9). That same year, Roger Tory Peterson, America's leading ornithologist, wrote that the robin was most likely the most numerous North American bird (18, 27). Carson's notion that the most prolific bird was about to fall extinct was one of the most eye-catching assertions in *Silent Spring* and brought the book considerable publicity.

Peregrine falcons and eagles were also high on Carson's list. In reporting on declines in population of these species she tended to heap the entire blame on pesticides and ignored all data that would refute her theory (16). Peregrine falcons were extremely rare in eastern United States long before there was any DDT present. By the time DDT was introduced there were literally no peregrine populations in eastern United States, but the anti-pesticide extremists later placed the blame on DDT anyway (18). Bald eagles in the lower 48 states were on the verge of extinction in the 1920s and 1930s, long before DDT was discovered. They were shot on sight for fun, bounty, or feathers, trapped accidentally, killed by impact with buildings and towers, or electrocuted by powerlines. There is still high mortality because of the physical hazards, but much less to shooting and trapping (because if caught engaging in either activity you may now face a prison term). The most surprising thing is that the environmental industry and the news media continue to attribute the increase to just one thing—the 1972 ban on DDT (18). Continuing the saga of showing that DDT was not bad on eagles, a recent study at the University of Wisconsin at Madison reported that lack of a suitable food supply in Lake Superior and, not DDT, was responsible for reproductive problems in eagles (28).

There was no mention at all in *Silent Spring* of the increases of birds observed by naturalists, including those participating in the Audubon Christmas Bird Counts. Naturalists counting hawks migrating over Hawk Mountain, PA, also reported great increases in the number of raptors, following the widespread use of DDT. Dr. J. Gordon Edwards, of San Jose State University, has documented those bird increases and also cited numerous feeding experiments that revealed DDT in normal bird diets did not cause the deaths of any birds (18, 26). Dr. William Hazeltine, another concerned California scientist, regarded pesticides as one of the least important causes of avian dislocations., The chief culprits, he said, were hunters, trappers, falconers, campers, and the general encroachment of humans into nesting and feeding areas (16).

Bird Egg Shell Thinning

On close inspection even the oft-repeated eggshell thinning threat to bird life holds little validity. DDT opponents alleged then and now that DDT caused eggshells to be thinned/softened for certain types of birds, causing failure to hatch and populations to decline. Thin egg shells are a phenomenon that predates use of DDT. It's been known for decades and there are many causes: diets low in calcium or Vitamin D, fright, high or low daily temperatures, various toxic substances, and bird diseases such as Newcastle disease (7). It has been demonstrated repeatedly in caged experiments that DDT and its breakdown products do not cause significant shell thinning, even at levels many hundreds of times greater than wild birds would ever accumulate (26). The most notorious cause of thin eggshells is the deficiency of calcium in the diet. Some early researchers deliberately fed their birds only calcium deficient food (0.5 percent rather than the necessary 2.5 percent calcium) and then attributed all shell problems to the DDT and DDE they had added to that calcium deficient diet. Edwards reported that after much criticism about the use of calcium deficient diets that were known to give the false impression regarding DDT shell thinning, the tests with DDT and DDE were repeated, but with adequate calcium in the birds' diet. The results proved that with sufficient calcium in their food the quail produced eggs without thinned shells (26).

Another method to obtain data is to measure the thickness of eggshells in museum collections. Measurements of the shells of hundreds of museum eggs have revealed that red-tailed hawk eggs produced just before DDT was used had much thinner shells than did eggs produced 10 years earlier. Then, during the years of heavy DDT usage, those hawks produced shells that were 6 percent thicker. Golden eagle eggshells during the DDT years were 5 percent thicker than those produced before DDT was present in the environment (26). More recently, R. E. Green found that thrush eggshells in Great Britain were thinning by the turn of the century, 47 years before DDT hit the market. He speculated that the thinning may have been an early consequence of industrialization and that acids formed when pollutants belched out of coal furnaces and smokestacks may have changed soil and water chemistry enough to reduce the availability of calcium, which is critical in the diet of birds that are producing eggshells (29).

PCBs were later shown to cause dramatic thinning of eggshells, as well as other adverse effects on birds, yet environmentalists continued to place the blame on DDT despite the fact that feeding birds high levels of that pesticide did not cause them to produce thin eggshells. There are many environmental contaminants that do cause shell thinning. Oil, lead, mercury, cadmium, lithium, manganese, selenium, and sulfur compounds have been shown to have adverse effects upon birds, including severe shell thinning (26).

Bioaccumulation and Biomagnification

"Bioaccumulation" refers to an increase in the concentration of a chemical in the environment (e.g., in water, sediment, soil.) "Biomagnification," on the other hand, refers to increases of chemicals as they are passed up food chains. As Ottoboni (5) points out, "The quantity of chemical that can be stored in any body can never exceed that which would be in equilibrium with the exposure. The chemical cannot remain in the storage depot without being replenished continually from the outside. Thus, the popular notion that foreign chemicals stored in a depot become immobilized and permanently fixed in the body, with additional exposure increasing the quantity stored ad infinitum, has no basis in fact. The claim that our bodies can become 'walking time bombs' is nonsense." She sums it up best by pointing out that bioaccumulation in not inherently good or bad, but in the public mind it is considered, almost universally, to be the latter.

Biomagnification proponents claim that pesticide levels are "magnified" at each step of the food chain, for example, from algae to planktonic crustaceans to small fish to larger fish to predatory birds or mammals. The consumption of low levels of pesticides within each prey animal is presumed responsible for increased amounts in higher predators (8). DDT is constantly broken down and excreted by the animals at each step of the food chain. If tiny crustaceans are analyzed, wet-weight, but the fish that ate them are analyzed dry-weight, the difference in the amounts of dilution by water creates an impression that the dry sample contains a greater amount of pollutants than the wet sample.

DDT is attracted to fat tissues more than to muscle tissues, so comparisons between samples of these two types will indicate "magnification" into the fatty tissues, even if they are samples from the same animal. Likewise, brain tissues attract more DDT than fatty tissues. Anti-DDT activists were careful to measure crustaceans, wet-weight, and compare them with levels in dry-weight muscle samples in fish, dry-weight fatty tissue in ducks that ate the fish, and dry-weight brain tissue in the hawks that ate the fish. If they measure ALL samples wet-weight, there is not "biomagnification." Also, if they measure only the muscle tissue from fish, ducks, and hawks, there is also no "biomagnification" (18, 26, 30).

Summary

These days a lot of effort is spent reminding people, particularly the younger folks, about the Holocaust and World War II because it's now more than two generations since these occurred and people tend to forget. As Tenner (31) wisely says, "With each generation, part of the collective memory of the last terrible events is lost." Well, it's been over one generation since DDT was banned and clearly, most people today only speak ill of DDT. They have no clue about how valuable it was, nor the politics behind its banning. Speaking of holocausts, the banning of DDT was a holocaust. Malaria, which was being controlled by DDT, has proliferated since the abandonment of DDT. As Mooney (32) points out, this was an early example of Western priorities being imposed on Third World people who may have made a different trade-off had the choice solely been theirs. Also, from Ottoboni (5), "The thought that substitution of nonresistant pesticides for persistent ones will solve all of the environmental problems attributed to the latter is an example of the myopic thinking that permeates so many decisions relating to environmental protection. People apparently haven't realized that all nonresistant pesticides merely degrade to other chemicals! The only difference is that most of these new chemicals do not have the same pesticidal action as their parent chemicals."

REFERENCES

1. Richburg, K. B., *Out of American*, BasicBooks, 1997.
2. *Access to Energy*, Vol. 24, No. 12 (August 1997).
3. Wirth, D. F., and Cattani, J., *Technology Review*, 100: 52, Aug/Sept 1997.
4. Gribble, G. W., "Environmental Issues," *Priorities*, 10(2–3): 50, 1998.
5. Ottoboni, M. A., *The Dose Makes the Poison*, Second Edition, Van Nostrand Reinhold (1997).
6. Chase, A. S., *Bugs in Environmentalism*, S. J. Milloy, www.junkscience.com 1997.
7. Ray, D. L., and Guzzo, L., *Trashing the Planet*, HarperPerennial, 1992.
8. Flynn, L. T., "The Birth of Environmentalism," *Issues in the Environment*, American Council on Science and Health, New York, June 1992.
9. Carson, R., *Silent Spring*, Houghton-Mifflin, New York, 1962.
10. Fox, M. R., "DDT Updated," S. J. Milloy, www.junkscience.com, Washington D.C., August 28, 1998.
11. Lieberman, A. J., *Fact Versus Fear*, American Council on Science and Health, New York, Sept 1997.
12. Roberts, D. R., Laughlin, L. L., Hsheih, P., and Legters, L. J., "DDT Global Strategies and a Malaria Control Crisis in South America," National Center for Infectious Disease (July-Sept 1997).
13. Roberts, D. R., *U.S. Medicine*, 34: 36, March 1998.
14. Wildavsky, A., *But Is It True*?, Harvard University Press, 1995.
15. Mellanby, K., "With Safeguards, DDT Should Still Be Used," *Wall Street Journal*, Sept 12, p. A26, 1989.
16. Grayson, M. J., and Shepard, T. R. Jr., *The Disaster Lobby*, Follett Publishing Co., 1973.
17. *DDP Newsletter*, Vol. XIV, No. 3, May 1997.
18. Edwards, J. G., *Remembering Silent Spring and It's Consequences*, DDP Salt Lake City, Utah (August 3, 1996).

19. Sturgeon, S. R., et al., "Geographic Variation in Mortality from Breast Cancer among White Women in the United States," JNCI, 87, 1896, *Journal of the National Cancer Institute*, Dec 20, 1995.

20. Fumento, M., "Truth Disrupters," *Forbes*, 162: 146, Nov 16, 1998.

21. Ames, B. N., Magaw, R., and Gold, L. S., "Ranking Possible Carcinogenic Hazards," *Science*, 236: 271, April 17, 1987.

22. Safe, S. H., "Environmental and Dietary Estrogens and Human Health—Is There a Problem?" *Environmental Health Perspectives*, 103: 346, 1995.

23. Marco, G. J., Hollingworth, R. M., and Durham, W., eds., *Silent Spring Revisted*, American Chemical Society, Washington, D.C., 1987.

24. McKetta, J. J., "Don't Believe Everything You Read," in *Rational Readings on Environmental Concerns*, Lehr, J. H., ed., Van Nostrand Reinhold, New York, 1992.

25. Frazier, B. E., Chesters, G., and Lee, G. B., *Pesticides Monitoring Journal*, 4(2): 67, 1970.

26. Edwards, J. G., "DDT Effects on Bird Abundance and Reproduction," in *Rational Readings on Environmental Concerns*, Lehr, J. H., ed., Van Nostrand Reinhold, New York, 1992.

27. Jukes, T. H., "The Tragedy of DDT," in *Rational Readings on Environmental Concerns*, Lehr, J. H., ed., Van Nostrand Reinhold, New York, 1992.

28. Milloy, S., *Environment News*, Heartland Institute, Chicago, Illinois, 2: 9 Oct. 1998.

29. Milius, S., "Birds' Eggs Started to Thin Long Before DDT," *Science News*, 153: 261, April 25, 1998.

30. Edwards, J. G., "The Myth of Food-Chain Biomagnification," in *Rational Readings on Environmental Concerns*, Lehr, J. H., ed., Van Nostrand Reinhold, New York, 1992.

31. Tenner, E., *Why Things Bite Back*, Alfred A. Knopf, New York, 1996.

32. Mooney, L., "The WHO's Misplaced Priorities," *Wall Street Journal Europe*, August 25, 1997.

21.3.12 BAD NEWS IS BIG NEWS

Introduction

There are many cases where the media gets too excited by environmental stories and issues and ends up blowing them out of proportion. Public outcry (egged on by banner headlines) over a mere handful cases of botulism, toxic shock syndrome, Alar on apples, benzene in Perrier, contaminated strawberries, or mad cow disease, can close businesses. One million cars can be recalled because of one death because of a malfunction of a single automobile. However, (at least until very recently) newspapers ran full-page advertisements for cigarettes, the product that has been described by the World Health Organization as the single most preventable cause of death and disability (1). Did you know that in the United States about one person a year dies from ingesting a toothpick? Do you worry about picking up a germ while visiting a friend in the hospital? You're more likely to acquire germs from the money in your pocket right now. One out of every ten coins and almost half the paper currency carry infectious organisms (2). As people watch continued coverage about the latest environmental scare of the moment, they can hardly be blamed for coming to the conclusion that parts per billion of some chemical is much worse than cigarettes or alcohol. It is no secret that apocalypse sells: nuclear accidents, asbestos, pesticides, Love Canal, Times Beach, the greenhouse effect, the ozone hole (3). In an internal memo, EPA admitted that its priorities in regulating carcinogens are based more on public opinion (as formed by the media) than on EPA's own estimate of the risks (4).

Here's an interesting exercise. Imagine how people would react if someone said: "I've come up with a really great invention. Unfortunately, it has a minor defect. Every 13 years or so it will wipe out about as many Americans as the population of San Francisco." This is what auto accidents do. What is happening is all too clear. The media, which once prided itself on its truth and accuracy, now flourishes on lies, half-truths, and illusions about environmental poisons. Bozell and Baker in a recent book (5) lay out statistical proof of the media's distortion of the news in overwhelming quantity, neatly organized by topics, heavily supported with names and quotations and crisply illustrated with simple bar graphs.

Chemical terror is easier to write about than the problem of teenage pregnancies or the impact of modern transportation on the spread of disease. Make no mistake about it, science is losing its constituency. A majority of Americans still tell pollsters they believe in science, but in many cases the so-called science they advocate includes astrology, yoga, and ESP (3). Margaretha Isaacson, the South American physician who'd stopped the transmission of Ebola virus at Ngaliema Hospital in Kinshasa during the original 1976 outbreak said:

"Ebola is of absolutely no danger to the world at large. It is a dangerous virus, but it's relatively rare and quite easily contained. The virus needs the right conditions to multiply, whatever the virus is, be it, Ebola or plague. It's not enough to just have the accident. The virus must first find itself in a favorable environment before it can affect anyone. The media is scaring the world out if its wits, and movies like Outbreak are doing people a great disservice (6)."

The average reporter works toward two goals because the typical daily newspaper cares about two thing. The first is making deadlines; the second is being interesting. Accuracy is not high up in the hierarchy (7). In fact, accuracy slows one down tremendously. It means getting confirmations and doing research as opposed to calling a few experts and quoting their opinions. In these days there are plenty of "experts" with their own particular environmental agendas who are willing to comment on the latest scare of the month. Speaking of experts, how about Meryl Streep speaking as an expert on Alar before Congress? Or what about a congressional committee inviting Jessica Lange, Sissy Spacek, and Jane Fonda to testify on farm problems because they all had starred in movies concerning farms (7).

The standards of evidence and the rules of publication are very different between the worlds of science and journalism. This means that when journalists cover scientific topics, solid communication does not always take place, and the resulting stories are at times misleading (8). There clearly seems to be more tolerance for sloppiness now than there ever has been before (9, 10). Reporters typically aren't even upset if they get a fact wrong. You might get a retraction but an original error in a page one story isn't properly corrected by a retraction on page A18. People remember the original article; it's the one that's going to last. Alan Simpson, former Senator from Wyoming, points out that journalists have no governing body that oversees their work and punishes them for sins and mistakes. Unlike other professions or crafts relying greatly on a high degree of public trust, journalism has no reasonable method for measuring performance. There are no procedures for truly evaluating credibility and honesty, and no periodic testing and certification (10). Simpson further suggests that instead of using the five W's (who, what , when, where, and why), the profession today is more interested in the five C's (conflict, controversy, cleverness, lack of clarity, and the fifth C is for television people who must always have their hair carefully coiffed).

A study (11) examining changing environmental values in the nations' press compared views of newspaper editors in 1977 with those in 1992. Key items discussed include:

- With the end of the Cold War, the environment may replace East-West confrontation as the key threat facing humanity.
- By far the biggest problems most editors face in reporting environmental stories is the sheer enormity and complexity of the subjects.
- Nearly 70 percent of the editors endorsed the slogan of the old Chicago Times, "The duty of a newspaper is to print the news and raise hell."

The last statement gives credence to Wattenberg's comments (12) that there seem to be three criteria upon which news judgments are often determined:

1. Bad news is big news.
2. Good news is no news.
3. Good news is bad news.

Controversy sells and journalists and the publications they work for do not prosper by simply detailing the mundane facts associated with good news. The media today will use more resources to follow Michael Jackson and his estranged wife, Lisa Marie, around the world than they will to help explain an environmental bill or a health reform package.

Journalistic norms lead reporters to adopt a polarized approach to issues, emphasizing conflict rather than knowledge, and this results in a reporting of evidence that is often inconsistent and confused. Some of the problem stems from the journalistic response to fast-breaking events, technical uncertainty, and conflicting information. Also, the necessity of being "first" with a story emphasizes the time issue. Time is nearly always of the essence in reporting and many times reporters lack the time to deal adequately with the uncertainties inherent in environmental disputes (8).

Fumento (7) also notes: "Journalists will often eschew reading a science or medical magazine article, in favor of simply reading the abstract." One example involved a study on AIDS virus infections on college campuses. Some reporters declared that it was a new study that showed an "alarming increase" in infections over those reported in an old one. This was, in fact, the old study that had been held from publication for 18 months. The original article definitely made the date clear, but this was absent from the abstract, so some of the news outlets missed it. The "alarming increase" shows the ability of the media to find something alarming and new in absolutely nothing (7).

Mann (13) states that in their proclivity for "news," newspapers, and television reporters not only single out weak studies; they may focus on the one positive result in a sea of negative data. This was the case with coverage of two big studies on occupational exposure to electromagnetic fields (EMF) that appeared in the *American Journal of Epidemiology*. The first study of 223,000 French and Canadian electric utility workers, found no link between EMF and 25 of the 27 varieties of cancer in the study; the exceptions, two rare types of leukemia, had a weak and inconsistent positive association with EMF. Yet the *Wall Street Journal* reported the study under the headline, "Magnetic Fields Linked to Leukemia." In 1995, the *American Journal of Epidemiology* published the second study on 139,000 workers at five U.S. utilities. No association was found between exposure to EMF and 17 of 18 types of cancer, including the leukemias linked to EMF by the first study. The sole exceptions were eye and brain cancers—conditions that showed no link to EMF in the first study. However, the headline of the *Wall Street Journal* article that reported the second study was "Link Between EMF, Brain Cancer is Suggested by Study at 5 Utilities."

"People are not interested in what diseases doesn't cause, but what it might cause. We've had this argument with scientists many times over the past few years," stated Jerry Bishop who wrote one of the *Wall Street Journal* articles. Lawrence Altman, author of an article in *Times*, stated that if there is a blame for such coverage, much of it belongs to scientific journals. The *Journal of the National Cancer Institute* sent out a big release touting that study (a study on breast cancer) as if it were the biggest thing since whatever. He said, "I don't recall them telling us that it was only one of 40 studies and probably had little meaning."

"Journalists do overemphasize individual studies, but they are often invited to do that by medical journals," agrees Ross Prentice of the University of Washington. "I've seen some of the press releases that journals and universities send to journalists. It's a wonder sometimes that the reporting is as good as it is" (13).

Recently, although their stories, reported on the same events and appeared on the same day, the *New York Times* and *Wall Street Journal* headlined them, respectively, MEETING LAYS BARE THE ABYSS BETWEEN AIDS AND ITS CURE and SCIENTISTS HAVE AN OPTIMISTIC OUTLOOK ABOUT THE PROSPECTS IN AIDS RESEARCH (14).

An interesting bit of reporting occurred in my home town of Livermore, CA, a few years back. The local newspaper headlined on the front page "Report Raps Lab Over Radiation Exposure," while alongside of this, a much smaller headline read, "Man Buried in Gravel Plant Hopper." The large headline, and three column article, which was continued on another page, discussed a 60-page report detailing how three Lawrence Livermore National Laboratory (LLNL) workers were accidentally exposed to X-rays during an experiment. The report stated that "at least 16 management procedures at LLNL were either less than adequate or inherently weak." The three individuals who were exposed were not expected to develop eye cataracts or other disabilities according to a medical evaluation. By contrast, the one column article with the smaller headline reported on a man who was killed when he slipped into a gravel hopper and was buried alive. No detail on inadequate safety procedures or oversight committees was included. Perish the thought that someone from LLNL would be killed in an industrial accident; a special edition of the local newspaper would probably be published.

Scientific Community Shares Some of The Blame

Importantly, the blame lies not only with the media but with the scientific community. As Edith Efron (15) says in her excellent book, "The inadequacy of the coverage is the inadequacy of the informants. Much of the blame lies on the scientific community that did not speak up, as well as those who did speak but with forked tongues." She emphatically states that she cannot indict the lay press for failing to understand what it takes years to understand. To write her book, 500 pages long, she had to read about 10,000 papers on carcinogenesis and genetics and about 500 books; and to write the epilogue, she had to read two histories of cancer epidemiology, several epidemiology textbooks and some 5000 additional papers in that field (15). As I said earlier, reporters must write swiftly; they cover the daily news. They could not take off several years to do their homework. Indeed, as laypersons, they have never realized how much homework there was to do. Also, I would add that often the scientific community, which more often than not is writing for their peers, writes in such a convoluted, confusing manner, that very few people can understand what is being said.

A high-level conference of more than 200 scientists, physicians, and humanists who met to consider the contemporary flight from reason and its associated anti-science was held in 1995 (16). A clear consensus emerged: Scientists must speak up against popular manifestations of irrationalism. In particular, many panelists urged official scientific organizations to recognize and play an active role in combating the flight from reason.

Bias Against Negative Studies

There is a media bias against negative studies (e.g., one that fails to show an increase in cancer) consistent with the earlier comments that good news is no news and good news is bad news. Koren and Klein (17) compared the rates of newspaper reporting of two studies, one negative and one positive, published back-to-back in the March 20, 1991 issue of the *Journal of the American Medical Association* (*JAMA*). Both studies analyzed an area of public health concern—radiation as a risk for cancer. A positive study by Wing et al. (18) purported to show a mean 63 percent increase risk for leukemia, in white men working at the Oak Ridge National Laboratory While a negative study by Jablon et al. (19) failed to show an increased risk of cancer in people residing near nuclear facilities. The Wing et al. study involved 8318 white men hired at ORNL between 1943 and 1972. Results revealed a relatively low mortality compared with U.S. white men except for leukemia, which was 63 percent higher. By contrast, the Jablon et al. study covered over 900,000 cancer deaths from 1950 through 1984 in 197 countries with or near nuclear installations. The results indicated that deaths resulting from leukemia were not more frequent in the study group than in the general population. Koren and Klein (17) searched 168 newspapers identified via seven online databases. They found 17 newspapers that published 19 reports on at least one of the two studies.

- 10 reports covered both stories
- 9 reports were dedicated only to the positive study
- 0 reports were dedicated to the negative study only

In reports covering both studies, the positive reports were significantly longer than the negative reports (median number of words was 345 versus 122 words). Of the 10 reports that covered both stories, 8 headlines described the positive study, and only 2 described the negative study. To review: both papers appeared back-to-back with the negative study first, both were the same length (six pages each). Both were included in an AMA news release so it's unlikely that journalists would miss one of them more often than the other. In press reports the negative study covered over 900,000 people, the positive study slightly over 8000, yet the positive study (no increase in cancer and leukemia deaths) got much more press and headlines—BAD NEWS IS GOOD NEWS. However, an important observation I can make is that the articles, like most technical articles, were written for the technical community and, therefore, not easy to understand because of the stiff scientific writing and technical jargon. I can easily see why a reporter would just read an abstract rather than an entire article. In the case of these two articles, there was a news release sent out by the *JAMA* but I was not successful in getting a copy of it so I can't speak to its clarity or lack thereof.

Technical journals are also at fault. Researchers whose studies show "no effects" or "no adverse effects" find it difficult to get them accepted for presentation during medical meetings or for publication in medical journals. It is estimated that many such studies are not even submitted for publication by their authors. Consequently, a distorted, unbalanced picture is produced, whereby scientists or the public at large are more likely to be informed about positive results and not be informed of equally important results on the same issues (17).

Some authors and investigators try very hard to convert what is essentially a negative study into a positive study by hanging on to very, very small risks or seizing on one positive aspect of a study that is by and large negative (20). Or, as one National Institute of Environmental Health Sciences researcher puts it, asking for anonymity, "Investigators who find an effect get support, and investigators who don't find an effect don't get support. When times are tough it becomes extremely difficult for investigators to be objective (20)."

Statistics

Statistics can be used in a misleading fashion. Often increases do not constitute a high rise in terms of percentages but the results can be made to sound alarming. For example, "The rate of death due to septicemia increased 83 percent." This sounds bad but what the percentage actually amounted to was an increase from 4.2 deaths per 100,000 population to 7.7 deaths per 100,00 population. It means that in a city approximately the size of Abilene, Texas or Springfield, Illinois, there were an additional 3.5 deaths because of septicemia per year. Clearly this is not satisfactory, but it does not connote quite the same catastrophe as "an increase of 83 percent (6)."

Which newspaper headline is likely to get more attention:

YEARLY STOOL TEST REDUCES COLON CANCER DEATHS BY 33 PERCENT

YEARLY STOOL TEST REDUCES YOUR CHANCE OF COLON CANCER DEATH BY LESS THAN 1 PERCENT

The tabloid headline that reads: Cut your risk of death by 1 percent doesn't have the punch line of the headline that states you can reduce risk by 33 percent. This is from a very large study (46,551 participants), so the medical profession considered the results very important (21). In the group that received annual screening for blood in the stool, 2.6 percent died of colon cancer. In the group that did not receive annual screening for blood in the stool, 3.4 percent died of colon cancer. The difference between 2.6 and 3.4 percent turns out to be statistically significant because the study is so large. The relative risk reduction is 0.8 divided by 2.6, which is approximately 33 percent. The individual risk reduction is the difference between 2.6 and 3.4 percent, which is 0.8 percent. Therefore, the chance that annual screening for blood in the stool will prevent you from dying of colon cancer is less than 1 percent, and the 33 percent figure is misleading (21). The risks of a drug seems much worse to people when described in relative terms (a doubling of your chances of stroke) than in absolute terms (an increase of 1 in 80,000 per year chance of dying from a stroke).

REFERENCES

1. Kluger, R., *Ashes to Ashes*, Knopf, New York, 1996.
2. Ross, J., *Smithsonian*, 26: 41, Nov. 1995.

3. Franklin, J., *Priorities*, 9(1): 12, 1997.

4. Dowd, A. J., Senior Vice President, American Electric Power Service Corporation, in a speech delivered before the Columbus Metropolitan Club, Columbus, Ohio, April 19, 1991.

5. Bozell III L. B., and Baker, B. H., Eds., *And That's the Way It Is(N'T)*, Media Research Center, Washington D.C., 1990.

6. Regis, E., *Virus Ground Zero*, Pocket Books, New York, 1996.

7. Fumento, M., *Science Under Siege: Balancing Technology and the Environment*, William Morrow & Co., New York, 1993.

8. Moore, M., ed., *Health Risks and the Press*, The Media Institute, Washington, D.C., 1989.

9. Matalin, M., and Carville, J., *All's Fair*, Random House, New York, 1994.

10. Simpson, A. K., *Right in the Old Gazoo*, William Morrow & Company, New York, 1997.

11. Bowman, J. S., and Clarke, C., *Int. J. Environmental Studies*, 48: 55, 1995.

12. Wattenberg, B. J., *The Good News Is the Bad News Is Wrong*, Simon & Schuster, New York, 1984.

13. Mann, C. C., *Science*, 269: 166, July 14, 1995.

14. Paulos, J. A., *A Mathematician Reads the Newspapers*, Basic Books, New York, 1995.

15. Efron, E., *The Apocalyptics: Cancer and the Big Lie*, Simon & Schuster, New York, 1984.

16. Sommers, C. H., *The Wall Street Journal*, July 10, 1995.

17. Koren, G., and Klein, N., *JAMA*, 266: 1824, 1991.

18. Wing, S., et al., *JAMA*, 265: 1397, 1991.

19. Jablon, S., Hrubec, Z., and Boice, J. D. Jr., *JAMA*, 265: 1403, 1991.

20. Taubes, G., *Science*, 269: 164, July 14, 1995.

21. Murphy, D. J., *Honest Medicine*, The Atlantic Monthly Press, 1995.

21.3.13   *ENVIRONMENTAL EDUCATION*

As our nation continues its all-consuming pursuit of protecting the environment, "regardless of the cost," we are overlooking the greatest cost of all: the toll on our children. These are words from J. Kwong (1) who conducted an extensive review of "environmental education." She discovered a number of unsettling trends and strategies and reports: (1) children are being scared into becoming environmental activists, (2) there is widespread misinformation in material aimed at children, (3) children are being taught what to think, rather than how to think, (4) children are taught that human beings are evil, (5) children are feeling helpless and pessimistic about their future on earth, and (6) environmental education is being used to undermine the simple joys of childhood.

A review by the Arizona Institute for Public Policy Research of 82 textbooks, 170 environmental books for children, and 84 examples of curriculum materials provided to schools by environmental groups (and adopted uncritically for classroom instruction) found "that unbiased materials present only one side of an issue, pick only worst case examples, or simply omit information that challenges an apocalyptic outlook" (2).

The schools' teachings are having a powerful effect. Simon (3) reports: "The consensus view of an informal FORTUNE survey of high schoolers on this issue was: If we continue at the pace we're going at now, the environment is going to be destroyed completely." A 1992 poll found that 47 percent of a sample of 6 to 17 year olds said that "Environment" is among the "biggest problems in our country these days"; 12 percent mentioned "Economy" as a far distant runner-up. Compare the opposite results for their parents: 13 percent "Environment" versus 56 percent "Economy."

Have you heard about Chief Seattle? He was the leader of Puget Sound Indian tribes and is credited with delivering a speech in 1855 that resonated with environmental relevance, a Gettysberg-like tome for the environmentalist movement. "Every part of this earth is sacred to my people. Every shining pine needle, every sandy shore, every mist in the dark woods, every meadow, every humming insect. The earth is our mother—What befalls the earth befalls all the sons and daughters of the earth," he is often quoted as having said. Sounds great, but the chief really didn't pen these words. The "words" come from Ted Perry, a university professor hired to write a documentary about pollution for a 1972 TV documentary. He decided to create a fictional version of Seattle's response to territorial officials' offer to buy tribal land (4-6). In spite of Perry's later protests, the speech took on a life of its own, showing up in U.S. Supreme Court Justice William Douglas's autobiography (7), Al Gore's book *Earth in the Balance* (8), and in other articles. Although the truth has long since been made known, environmental groups continue to publicize the speech. It has been broadcast in at least six foreign countries (4, 5). A children's book about the speech sold 280,000 copies in six months and was nominated for an American Booksellers Association Abbey Award (3).

Some other examples: A kids' page in the Sunday paper purveys such bits of "obvious wisdom" as "It takes more than 500,000 trees to make the newspapers that Americans read on Sunday but the children are not told that trees are grown, and forests are created, in order to make newspaper" (3). Children are also taught that acid rain caused by emissions from power plants and automobiles destroys lakes and forests. They are told to mix vinegar with water and to pour it on plants to see the plants die but aren't told that this mixture does not resemble "acid rain" or that a $500 million government study couldn't find convincing evidence that acid rain is destroying forests (9). Marvel comics has more than 70 comic book characters who had developed severe physical and emotional handicaps as a consequence of exposure to radiation. The message inherent in the experience of these characters is communicated effectively to young persons and for that matter to many others who are not so young (10). A book titled *Nuclear Power-Promise or Peril*, by M. J. Daley (11) is touted as an accurate and evenhanded treatment of nuclear power but is rife with inaccuracies and antinuclear messages. Says C. E. Walter, "Although technical errors load an error-ridden data base into readers' minds, the true disservice to readers lies in the sometimes subliminal and often explicit unsubstantiated messages interleaved throughout the book that nuclear power is a peril-not a promise. Purportedly unbiased, actually it is not. In fact, the book is dedicated to two individuals who are associated with antinuclear groups, and Daley himself belongs to one of the groups. What really bothers me about the book is that The Science Teacher, a reputable U.S. magazine directed at high school teachers nationwide, published a glowing support of the book. Our high school students deserve accurate technical information and clear and objective discussion of social positions on technology, not the misinformation presented by the book or its review" (12).

Graves (13) reports that her son did not learn at school that most industries have to treat the water before it is discharged; even though at home we could pour the same chemicals down the drain. She says, "This information, the ideas, the suggestions are so removed from these kids. They do not care about scrubbers on stacks, or whether or not their stuff comes in a biodegradable container. To them, muddy water is polluted, anything pouring out of a smoke stack, even steam, is polluting the air."

What To Do?

Make yourself aware of what the kids are learning about the environment in school. Don't take it for granted that they are getting both sides of the story. Critical thinking skills are what we want our children to develop, but to do this both sides of an issue have to be presented.

One organization that is working the problem is the PERC (The Political Economy Research Center) in Bozeman, Montana. They have developed programs to help middle and high school students think about the environment, have a newsletter for teachers and high school students, called the *Environmental Examiner*, and sponsor teacher workshops around the country (14).

If you're looking for books check *Facts, Not Fear*, by Sanera and Shaw (5). This is a good guidebook to help parents counter the irresponsible claims of environmental extremists—and to give their children a more balanced view of the many environmental issues they encounter. In simple, nontechnical language, the authors explain the myths and facts concerning many major environmental topics, and show you how to set the record straight for your children (or grandchildren). Another book is A *Blueprint for Environmental Education* by Shaw (15), which discusses the current state of environmental education.

There is nothing wrong with teaching students about environmental issues, in fact, it is very important. However, they should be taught the true scientific and economic complexity of these issues. When biased and misleading information about environmental issues such as acid rain, global warming, and the so-called population crisis is used to recruit children as shock troops in a crusade to support a particular political agenda, a serious disservice is being done.

Sanera and Shaw (9) sum it up quite well: "Environmental education could be a valuable part of science instruction. Instead, it often merely repeats the nostrums of the environmental movement, and molds children into smug crusaders whose foundation of knowledge is shaky at best" (9).

REFERENCES

1. Kwong, J., *The Freeman*, 45: 155, March 1995.
2. Sykes, C. J., *Dumbing Down Our Kids*, St. Martin's Press, 1995.
3. Simon, J., *The Ultimate Resource 2*, Princeton University Press, 1996.
4. Murray, M., *Reader's Digest*, pp. 100–104, July 1993.
5. Sanera, M., and Shaw, J. S., *Facts Not Fear: A Parent's Guide to Teaching Children About the Environment*, Regnery Publishing, 1996.
6. Abruzzi, W. S., *Skeptical Inquirer*, 23: 44, March/April 1999.
7. Douglas, W. O., *Go East, Young Man*, Random House, p. 73, 1974.
8. Gore, A., *Earth in the Balance*, Plume, p, 259, 1993.
9. Sanera, M., and Shaw, J. S., *Consumers' Research*, 80: 15, April 1997.
10. Young, J. P., and Yalow, R. S., eds., *Radiation and Public Perception: Benefits and Risks*, American Chemical Society, Washington, D.C., 1995.
11. Daley, M. J., *Nuclear Power: Promise or Peril*, Lerner Publications, 1997.
12. Walter, C. E., *Nuclear News*, 34, Jan 1999.
13. Graves, B., *Products Finishing*, 61: 6, March 1997.
14. Political Economy Research Center, 502 South 19th Avenue, Suite 211, Bozeman, Montana 59718.
15. Shaw, J. S., ed., *A Blueprint for Environmental Education*, Political Economy Research Center, Bozeman, Montana, 1999.

21.3.14 NATURE IS QUITE RESILIENT

Nature is more resistant than a lot of folks give it credit for. Worms in highly concentrated toxic wastes of cadmium and nickel in Foundry Cove, New York, have adapted to this environment and thrive and reproduce. Other examples include deer mice in Los Angeles, rats in Morro do Ferro, Brazil, oil spills, volcanic eruptions, and recovery of the land from the devastation of wars and nuclear holocausts.

Worms in Foundry Cove

Foundry Cove is on the Hudson River across from West Point. It has an impressive military history from a manufacturing viewpoint. In Revolutionary War times, a forge at the cove produced chains that were stretched across the Hudson to slow down British warships. Ammunition was produced in the Cove during the Civil War. Manufacture of batteries began about 45 years ago. Beginning around 1953 industry in the area dumped more than 100 tons of nickel-cadmium waste into the cove and nearby river (1). Arsenic, lead, and other toxins were also dumped into the cove (2). This dumping was stopped in the late 1970s because of local citizens complaints.

Levinton and Klerks (1) found as much as 25 percent of the bottom sediments of the cove consisted of cadmium (a higher concentration than found in your typical cadmium electroplating solution). Yet many bottom dwelling invertebrate species were present in amounts no less than in unpolluted areas of other sites. In an effort to learn why, they investigated the cadmium tolerance of the most common invertebrate in the cove. This is an aquatic relative of the earthworm with the difficult-to-remember and pronounce name of limnodrilus hofmeisteri. Whereas local cove creatures thrived and reproduced, limnodrilus from a nearby region showed clear signs of distress or died when placed in Foundry Cove. Offspring of Foundry Cove worms raised in clean muds were also tolerant of cadmium, leading to the conclusion that genes were largely responsible for the tolerance. This degree of metal tolerance could have evolved in just two to four generations, or a couple of years. This was verified by exposing worms from an unpolluted site to cadmium laden sediment and breeding the survivors. By the third generation, the descendants had two-thirds of the cadmium tolerance found in the Foundry Cove worms (1).

Levinton states that this capacity for rapid evolutionary change in the face of novel environmental challenge was startling, because no population of worms in nature could ever have faced conditions like the one humankind created in Foundry Cove. Furthermore, although some species inhabiting nearby waterways are missing from Foundry Cove, most adapted to the unusual conditions. Just 100 yards downstream from the cove sits a prosperous Audubon-protected sanctuary for migratory birds (1).

Other Creatures

The rapid evolution of tolerance for high concentrations of toxins is not uncommon. When new pesticides are put into use, a resistant strain of pest evolves, usually within a few years. This same thing happens to bacteria when new antibiotics are introduced (1). Many of the streams in Colorado down-gradient from mining sites contain heavy metals at levels well above national water quality criteria but apparently nobody has bothered to tell the fish that they should be dead (3).

Deer mice from areas of Los Angeles with high ambient air pollution are significantly more resistant to ozone than are mice from areas with low ambient pollution (56 percent versus 5 percent survival). Laboratory-born progeny of these mice show similar response patterns, indicating a genetic basis to this resistance (4).

A colony of rats occupies burrows in the mounds in a Brazilian area called Morro do Ferro. This is a weathered mound, 250 meters tall that is formed of an ore body containing an estimated 30,000 metric tons of thorium and 100,000 metric tons of rare earths. The radiation level is one to two mRoentgens per hour over an area of 30,000 square meters. So high is the absorbed radioactivity in the vegetation that photographs can be taken (autoradiographs) showing the plants truly glowing in the dark. Yet the mound supports both animal and plant life (5). Rats in the region breathe an atmosphere containing radon at levels up to 100 pCi per ml. The radiation dose to the rats bronchial epithelium is estimated to be between 3000 and 30,000 rems per year, roughly three times the concentration that should produce tumors or other radiation effects. Yet no abnormalities were found on fourteen rats who were trapped and autopsied (5).

A random ounce of fertile soil has been found to contain about 1 million algae, 30 million protozoa, 50 million fungi, and 150 million bacteria, some of whose spores invariably will survive being dried up for years, doused with poison, frozen solid, or boiled for an hour (6).

Oil Spills

Oil spills are another example of the resiliency of nature. A Congressional Research Service federally funded report says that the damage from even a horrendous spill of crude is relatively modest and, as far as can be determined, of relatively short duration (7).

The Exxon Valdez spilled 11 million gallons of crude into Prince William Sound in 1989. Within just three years the sound was so close to its former state that navigation charts had to be used to determine where the spill had actually occurred (2). The cleanup did more damage than the spill (8). Exxon and the Coast Guard, under public pressure, put together a cleanup that was so massive in scale that at its peak resulted in the greatest concentration of vessels engaged in a single operation since the Normandy landing. In many locations, Exxon used high pressure, hot water to blast away oil. This killed the microbial life on which the food chain is moored. By contrast, beaches that were left alone for experimental purposes cleaned themselves via wave action, microbic digestion and other factors, thus staying alive in microbial terms. These are the traditional defenses that nature has prepared against petroleum. After all, petroleum is a naturally occurring product that continually "spills" from the Earth's crust via seepage. Because petroleum enters the biosphere on a natural basis, some organisms have adapted to metabolizing it (2). As Easterbrook stated: "Prince William Sound went from destroyed to almost like new in about three years, or less than a single generation for the local vertebrate species. This is standard operating procedure for the green fortress" (2).

Another example is the Amoco Cadiz, which ran aground off the coast of France in 1978 and spilled six times as much oil as the Exxon Valdez's 1,635,000 barrels. No long-term effect on the bird population was discovered. Marshes, where no attempt was made to remove the oil, were restored by natural processes within five years, whereas in cleaned areas, restoration took seven to eight years (7).

One might ask why do we rush to clean up these spills and one clear reason is response to media pressure. Only the most hardened soul cannot respond sympathetically to TV pictures of oil soaked birds either dead or dying. In the heat of the moment, very few ask the question of how will nature heal this wound. Rather, the response is to blame those who created the spill and let's quickly get in there and fix it even though in many cases nature can do a much better job than we humans can.

Volcanoes

On the morning of May 18, 1980 the top of Mount St. Helens "rocketed into the stratosphere." The mountain had shook and bulged for weeks before suddenly losing 1300 ft of its top in a landslide and explosion that, according to the Forest Service, sent a cloud 15 miles up, blew over 96,000 acres of trees, spread 540 million tons of ash over three states, and killed 57 people. It was the largest landslide in recorded history. Spirit Lake was filled with ash and tree trunks and raised some 200 ft. Scientists thought it would take many years before the lake would support a wide variety of life again. But according to the Forest Service, within five years the lake had "virtually recaptured the complexity of pre-eruption life" (9).

War Damage

In 1942, during World War II, the earth was scorched by retreating Russian soldiers yet became a living forest anew within a decade without any help from the Moscow regime. During the Korean War, agricultural areas were continuously bombed. However, within a few years these areas showed no ill effects (2). In more recent times the United States dropped 19 million gallons of herbicides including 11 million gallons of Agent Orange on the jungles of Vietnam. Today, these jungles are as lush as before the war, again without the assistance of humankind (2). Of additional interest is the fact that in 1992 researchers in a jungle near the Vietnam-Laos border found the first new large mammal species detected on the Earth in 50 years, a cow goat named the Vu Quang ox. Two years later they found a second new large species, the giant muntjac deer. As Easterbrook pointed out, "These creatures had been evolving, and surviving , in a jungle just 20 miles from the Ho Chi Minh Trail—the most intensely bombed target in military history" (2).

Nature's capacity for natural recovery from an atomic holocaust was demonstrated in 1964 when scientists studied the remains of Namu, a coral isle of the Bikini atoll whose entire top had been blown off by an H-bomb in 1956. They found it covered with sedge, beach magnolia, morning glory vines and the white blossomed messerschmidia tree with many kinds of birds flying gaily about, singing and raising their young, insects buzzing and burrowing, and fish swarming in the lagoons (2).

Conclusion

So what's the point of all this? Easterbrook sums it up very succinctly in his book *A Moment on the Earth*, "Though environmentalists rightly warn that insects and disease organisms may mutate to acquire resistance to man's pesticides and wonder drugs, they never discuss the flip side of this issue: that living things may mutate to resist pollution, rendering human ecological abuses less damaging than expected. Second, if small organisms already have genetic mechanisms that respond to toxins, they may have acquired this trait before the industrial era, by dealing with naturally created plagues of dangerous chemicals" (2).

REFERENCES

1. Levinton, J. S., *Scientific American*, 267: 84, November 1982.
2. Easterbrook, G., *A Moment on the Earth*, Viking, 1995.
3. LaGoy, P. K., *Risk Assessment: Principles and Applications for Hazardous Waste and Related Sites*, Noyes Publications, 1994.
4. Richkind, K. E., and Hacker, A. D., *Journal of Toxicology and Environmental Health*, 6: 1, 1980.
5. Ray, D. L., "Radiation Around Us," in *Rational Readings on Environmental Concerns*, Lehr, J. H., ed., Van Nostrand Reinhold, 1992.
6. Murchie, G., *The Seven Mysteries of Life*, Houghton Mifflin, 1978.
7. O'Rourke, P. J., *All the Trouble in the World*, Atlantic Monthly Press, 1994.
8. Wheelwright, J., *Degrees of Disaster, Prince William Sound: How Nature Reels and Rebounds*, Yale University Press, 1996.
9. Goepel, J., *VIA*, 118(4): 42, July/August 1997.

21.3.15 POVERTY IS THE WORST CARCINOGEN

"Poverty is the equivalent to exposure to the most toxic pollutant."—Moghissi (1)

"The war on pollution is one that should be waged after the war on poverty."—Whitney Young, Urban League President (2)

What environmental problems kill human beings in numbers today? It isn't Alar, ozone depletion, dioxins, nuclear wastes, electromagnetic radiation, pesticide residues, PCBs, or asbestos. What kills them is dung smoke and diarrhea and this all relates to poverty (3). Living in poorly ventilated huts where fuel wood, cow dung, or agricultural wastes are used for heating and cooking is responsible for most child deaths. Gurinder Shahi, an official of the United Nations Development Program says, "Smoke inside a hut like this can be unbelievable. Women and children, who spend most time in the home, are most harmed. Today 40 percent of the global population heats and cooks with biomass in raw form" (3).

Sanitation is another issue. According to a report released this past summer by the United Nation's Children's Fund (UNICEF), almost 3 billion people—about half the world's population—now live without clean toilets. More than 2 million children die each year from diarrhea-causing diseases, infected by bacteria that could easily have been avoided if they had been flushed down a pipe (4). According to UNICEF, 3.8 million developing world children under age 5 died in 1993 from diarrheal diseases caused by impure drinking water. As Easterbrook (3) points out, "In the First World, death from diarrhea is about as common as comet strikes; in the developing world diarrhea kills far more people than cancer. Most of Africa, the Indian subcontinent, and Latin America have no wastewater treatment facilities. Yet Western public consciousness continues to focus on exotic ecological threats while ignoring millions of annual deaths from basic environmental problems of water and air."

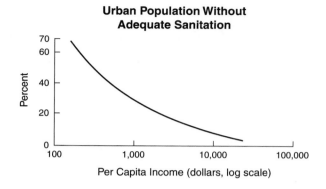

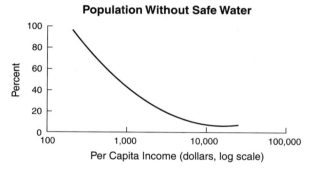

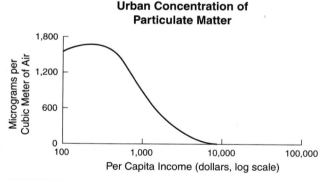

FIGURE 21.3.6   Affluence and the environment (adapted from Goklany (6)).

Basic pollution in the Third World is far more significant than all First World ecological problems combined. Eric Chivian, a psychiatrist on the faculty at Harvard Medical School who started an organization called the Project on Global Environmental Change and Health stated, "I've had great difficulty interesting environmental organizations in human health in poor countries. They want to talk about forest loss and species diversity in the developing world but have much less interest in human health there." Nobody should kid themselves that they are doing Bangladesh a favor when they worry about global warming (3).

Ray and Guzzo (5) quote Dr. Norman Borlaug, a Nobel prize recipient who is considered one of the fathers of the "Green Revolution" as saying, "I am concerned that the growing anti-science and anti-technology bias in affluent countries will adversely affect the prospects for agricultural development. In effect, the 'haves' are telling the 'have-nots' that they should stay with current simple lifestyles since great material well-being isn't what it is made out to be. How many people in the First World would be willing to cut their life spans by one half, see up to half of their children die before reaching the age of ten, often as a result of minor and easily curable illness, live in illiteracy with substandard shelter, clothing and sanitation, and face bleak prospects of no improvement in economic well-being for themselves or their children? Unwittingly, this is the continuing fate that the affluent anti-technology groups are wishing for the Third World's people."

Poverty is already a worse killer than any foreseeable environmental distress. If you're hungry, you aren't very much interested in improving the environment. All evidence indicates that, ultimately richer is cleaner, and affluence and knowledge are the best antidotes to pollution. Wealthier populations can afford newer technology even if it costs more initially. Poverty is the worst carcinogen. Countries undergo an environmental transition as they become wealthier and reach a point at which they start getting cleaner (6). Some examples are shown in Figure 21.3.6. Ambient sulfur dioxide and particulate matter concentrations in the air and fecal coliform in river water drop significantly as a country becomes wealthier. The reason for the turning point is complex, but essentially the wealthier a nation is, the more it values and the more it can afford to pay for a healthier environment and environmental amenities. The level of affluence at which a pollutant level peaks (or environmental transition occurs) varies. A World Bank analysis concluded that urban particulate matter and SO_2 concentrations peaked at per capita incomes of $3280 and $3670, respectively. Fecal coliform in river water increased with affluence until income reached $1375 per capita (6).

Chapman et al. (7), studied how different cultural, religious, and political contexts interfere with people's perceptions of the environment and how journalists deal with such a reality. They reported that mass media and communication technology is in danger of locking developed countries into a ghetto of environmental self-deception thereby helping to perpetuate poverty in the developing countries. The goal of developing countries remains the attainment of development; developing countries see "environmental problems occurring elsewhere." Whether or not environmentalism becomes a universal cause depends on how and to what extent such sharply contrasting world views can converge.

Conclusion

The real killers of people are poverty and not the items that we read about in our daily newspaper or see on television as the latest scare of the month. If we, and all others such as those who proclaim to be environmentalists really want to save lives, we can do much better than worrying about the last part per zillion of something in food, in the atmosphere, or in the dirt children might eat. We would save untold many more lives if we directed the resources at helping alleviate poverty world-wide.

Acknowledgment

The material in this section has been adapted with the permission of the journal of plating & surface finishing from articles written by the author for that journal.

REFERENCES

1. Moghissi, A. A., *Environment International*, 24: 379, 1998.

2. Dowie, M., *Losing Ground*, MIT Press, 1995.

3. Easterbrook, G., *A Moment on Earth*, Viking, 1995.

4. Brown, K., *Discover*, 18: 104, Jan. 1998.

5. Ray, D. L., and Guzzo, L., *Environmental Overkill-Whatever Happened to Common Sense?*, Regnery Gateway, Washington, D.C., 1993.

6. Goklany, I. M., "Richer is Cleaner; Long Term Trends in Global Air Quality," in *The True State of the Planet*, Bailey, R., ed., The Free Press, 1995.

7. Chapman, G., Kumar, K., Fraser, C., and Gaber, I., *Environmentalism and the Mass Media,* Routledge, London, 1997.

CHAPTER 22
GLOBAL PERSPECTIVES AND TRENDS

SECTION 22.1

THE ROAD FROM RIO TO KYOTO: HOW CLIMATE SCIENCE WAS DISTORTED TO SUPPORT IDEOLOGICAL OBJECTIVES

S. Fred Singer

Dr. Singer is an atmospheric physicist and professor emeritus of environmental sciences at the University of Virginia. He is also president of the Science & Environmental Policy Project, a nonprofit policy institute based in Fairfax, Virginia.

22.1.1 INTRODUCTION

Scientific information is not value-free and is often distorted. We discuss here a number of instances, taken from the controversy about global warming. For example, there is no scientific basis for the IPCC conclusion that "the balance of evidence suggests there is a discernible human influence on global climate." This ambiguous phrase has been misused to promote the far-reaching policies, incorporated into the Kyoto Protocol.

I propose that climate science is relevant for the Kyoto Protocol. I recognize, of course, that politicians have already pronounced the science as "settled," that they have even invented a "scientific consensus," and that they have stipulated that the emission of greenhouse (GH) gases will created a calamity for humankind on this planet. I must inform you, however, that the science is not settled (as claimed by Vice President Gore), that it is not "compelling" (as claimed by President Clinton), and that there is certainly no scientific consensus favoring global warming. If anything, the largest number of scientists, some 17,000, signed a *petition against the Kyoto Protocol* in 1998. Further, in July 1997, the U.S. Senate passed a resolution opposing a Kyoto-like treaty by a vote of 95 to zero.

Science, by its very nature, depends on data, and not on speculation. Results cannot be derived by theory alone, in the absence of actual observations—and certainly not by vote. However, it is possible to distort scientific facts without actually falsifying the data. As I will demonstrate here, one can present facts selectively and ignore those that contradict the preconceived outcome, and thereby affect policy.

22.1.2 EXAMPLES OF SCIENTIFIC DISTORTION

1. Let us start with the oft-repeated statement that "global mean surface air temperature has increased by between about 0.3 and 0.6°C since the late nineteenth century" [IPCC 1996, *Climate Change 1995, Summary for Policymakers* (SPM), p. 4]. The SPM does not reveal, however, that the

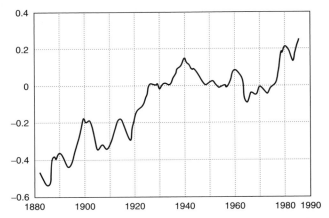

FIGURE 22.1.1   Global temperatures versus time, as determined from surface measurements with thermometers. The temperature changes are referred to an arbitrary baseline (Hansen and Lebedeff, GISS).

temperature rise occurred *before* 1940 and was followed by a decline between 1940 and 1975 (even while the level of greenhouse gases increased rapidly; see Figure 22.1.1). Thus, the reader is led to believe that the temperature rise is due to the increasing level of GH gases. However, the theoretical climate models cannot explain this observed temperature history, which is most likely the result of natural climate fluctuations that dominate over human influences (Singer, 1999).

2. You have undoubtedly read that the twentieth century was the warmest in 600 years of climate history [SPM, page 5]. Again, this statement is entirely correct, but also incomplete. It does not reveal that the period from about 1400 AD to 1850 spanned the aptly named "Little Ice Age." Nor do we learn that a little further back in the climate record the temperature was much warmer than today; the so-called Medieval Climate Optimum occurred around 1100 AD, when the Vikings settled Greenland.

3. Of course, all climate information, whether from thermometers or from proxies (such as tree rings or isotope data in ice cores and ocean sediment cores) comes from just a few places on the globe, mostly from land areas that occupy only 30 percent of Earth's surface. The only truly global temperature data come from weather satellites, beginning in 1979. They show a slight but persistent cooling of the global atmosphere, but you will find no reference to this in the SPM. In fact, the existence of satellites is not even mentioned there.

4. The IPCC document refers to an "unprecedented" warming of "unprecedented" rapidity in the next century. Even if climate-model forecasts were correct and validated by actual observations, the IPCC claim is not borne out by historical data. The planet has experienced larger and more rapid temperature swings throughout geological times (see Figure 22.1.2). The planet has also experienced a concentration of greenhouse gases, like carbon dioxide, at 20 times the present level during recorded geological history, dwarfing the increase as a result of fossil-fuel burning, which may (or may not) result in a doubling of CO_2 levels.

5. The Council members of the American Geophysical Union (AGU), the professional society of Earth scientists, recently adopted a position statement on global warming. It was drafted by a small group and was never submitted to the AGU membership for review or comment. It states that greenhouse gases, like carbon dioxide, persist in the atmosphere for "up to thousands of years." This sounds ominous, until one realizes that half of the released carbon dioxide is absorbed within 30 years, little remains beyond 100 years—and only minute quantities persist for millennia. This misleading phrase originally read "from decades to thousands of years" before the word "decades" was edited out in the final version.

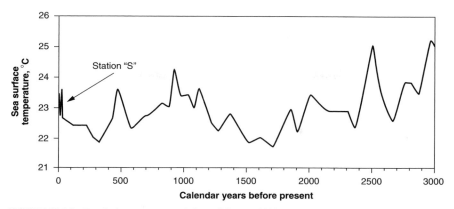

FIGURE 22.1.2 Detailed temperature variations of the past 3000 years (during recorded history), as determined from ocean sediment studies in the North Atlantic. Not the rapid variations, as well as much warmer temperatures 1000 and 2500 years ago (Keigwin 1996, reprinted with permission, © American Association for the Advancement of Science).

6. The same AGU statement announces that increases in carbon dioxide in the geological past were "associated" with increases in temperature. This is technically correct but, again, likely to mislead readers into believing that CO_2 increases *caused* the temperature increases. Not so. Detailed measurements with adequate time resolution show that the temperature increases of the three last transitions from Ice Ages to interglacial warm periods all preceded the CO_2 increases by about 600 years. Thus, the cause-effect relation is clear: Temperature increases of the ocean likely released CO_2 into the atmosphere (Singer, 1999, p. 73).

7. Similarly misleading is the much-touted association between the temperature increase of the last century and the observed sea-level rise (of about 18 cm) during the same period. The innocent reader would naturally assume that increases in temperature because of increasing greenhouse gases would accelerate the rise of sea-level by melting glaciers, thereby raising popular fears of disappearing island nations and flooded coastal plains. Just the opposite is true, however. Warming of the ocean produces more evaporation, more rain, and therefore, more ice accumulation in the polar regions. A detailed analysis of available data shows that this process is the more important one, transferring water from the oceans to the icecaps (Singer 1999, p. 18); therefore, a future warming should *retard* the rise of sea level.

8. The first IPCC report, published in 1990, stated that the "warming is broadly consistent with the predictions of climate models" (IPCC 1990, SPM, page xii). This obviously misleading statement was dropped when the second IPCC Assessment was published in 1996. Its chief conclusion, as stated in the SPM (IPCC 1996, page 4), was that "the balance of evidence suggests there is a discernible human influence on global climate." This ambiguous statement is devoid of real meaning, but it can also mean anything one wishes to read into it. In July 1996, a Ministerial Declaration issued in Geneva, chose to interpret the IPCC conclusion as ratifying a temperature increase of 2°C by the year 2100 (see Note 1). The assembled ministers clearly ignored the IPCC scientific report itself, which states in this section that "to date, pattern-based studies have not been able to quantify the magnitude of a greenhouse gas or aerosol effect on climate" (IPCC 1996, page 434.) The IPCC leadership, however, chose not to inform the ministers of this conflict.

9. Just how did the IPCC leadership conclude that a human influence could be discerned? It was based mainly on a graph (IPCC 1996, figure 8.10, page 433), which depicts the correlation (between observed and predicted temperature patterns) as increasing between 1940 and 1990. To help the reader reach this conclusion, the graph also shows a straight line sloping upward towards to the right, marked as "50-year linear trend" (see Figure 22.1.3a). The source of the graph is a scientific paper (Santer et al., 1995), published only after the IPCC report was approved. However, that original graph

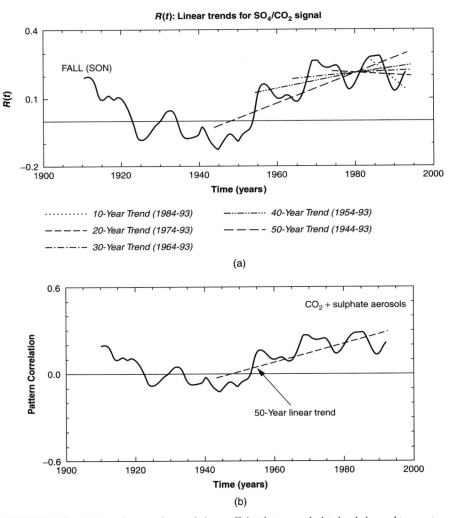

FIGURE 22.1.3   (a) As can be seen, the correlation coefficient between calculated and observed temperature patterns is rather small and shows strong variations. It decreases during the period of rapid warming (before 1940) and does not increase during the past 225 years when atmospheric CO_2 level rose greatly (Santer et al., 1995, reprinted with permission, © Springer-Verlag). (b) In the 1996 IPCC report (Figure 8.10b), only the increasing 50-year trend line is shown; the zero and negative trend lines were omitted by the authors for reasons that were not explained.

shows not only the positive trend but also a zero trend and even a negative trend; it all depends on an arbitrary choice on time interval (Figure 22.1.3b). What is significant is that all trends, except the positive 50-year trend, were evidently edited out when the graph was reproduced in the IPCC report.

10. *The IPCC report* itself also underwent some remarkable editing between the time it was approved in December 1995 and printed in May 1996 (Seitz, 1996). The "scientific cleansing" involves the surreptitious deletion of several phrases that had been approved by the scientists working on the report—phrases that threw doubt on the idea of a "discernible human influence" (see Note 2). It is widely recognized that governmental representatives made these later-discovered text changes so that the IPCC report would "conform" to the Summary, a politically negotiated document. This rather strange procedure, possibly illegal and certainly unannounced, was apparently made at the behest of the U.S. State Department. *Nature* mentions this letter, which indeed asked that the chapters be adjusted after the Summary had been agreed to (see Note 3; Letter dated November 15, 1995 to Sir John Houghton, signed by Day Mount, deputy assistant secretary [acting]).

11. The Kyoto Protocol (which calls for an average 5.2 percent emission reduction, with respect to 1990 levels, by industrialized countries) will actually require a reduction of more than 30 percent by the United States within the next decade—a daunting task that would be certain to cause economic damage and result in job losses measured in millions. Much has been made of the claim that the Kyoto Protocol, though costly, is necessary to avoid deleterious climate change. If one were to accept the model result published by IPCC, however, even a punctilious observance of the Protocol would lower the calculated temperature rise in 2050 from 1.40 to 1.35°C, a reduction of only 0.05°C—undetectable and certainly inconsequential (Singer 1999, p. 68). Again, according to the IPCC, stabilizing the atmospheric concentration of CO_2 would require an immediate emission reduction of 60 percent, with respect to 1990 levels, by all countries, industrialized and developing (IPCC 1990, page xi, xvi). Thus, the Kyoto Protocol is not only ineffective, but also politically unfeasible.

12. We finally ask the question that should have been asked at the very beginning: What is the ultimate purpose of the 1992 global climate treaty, the Framework Convention on Climate Change (FCCC) adopted in Rio de Janeiro, which the Kyoto Protocol is supposed to put into effect? It is not a reduction in emissions, as many might think. *Article 2* of the FCCC calls for "stabilization of greenhouse gas concentrations in the atmosphere at a level that would prevent dangerous anthropogenic interference with the climate system." The IPCC was asked by the chairman of the Kyoto Conference to define this goal, to determine which levels are "dangerous" and which levels are safe for the climate system. The IPCC has not responded to this request as yet. One may conclude, therefore, that the Kyoto Protocol is trying to impose drastic policies before it is known even whether a higher or lower level of greenhouse gases is preferable. We do have geological evidence, however, that shows that periods of lower temperature and lower CO_2 concentrations are associated with large climate instability, while warmer periods have a more stable climate showing less variability (Singer, 1997).

22.1.3   CONCLUSION: A DIGRESSION INTO ECONOMICS

Stepping away from climate science, it may be of more than passing interest to learn that the conclusions of the IPCC about the negative economic impact of warming have been drastically revised. A recent re-evaluation by a team of 26 respected economists (Mendelsohn and Neumann, 1999), concluded that a warming of the climate, and higher CO_2 concentrations, would lead to net benefits, rather than net losses. The major benefits would accrue to agriculture and forestry. In other words, global warming is good for you and for humanity (see also, Moore, 1998).

APPENDIX

Note 1: Ministerial Declaration, Geneva, July 1996

"Recognize and endorse the Second Assessment Report of the IPCC as currently the most comprehensive and authoritative assessment of the science of climate change, its impacts and response options now available. Ministers believe that the Second Assessment Report should provide a scientific basis for urgently strengthening action at the global, regional and national levels, particularly action by Annex I Parties to limit and reduce emissions of greenhouse gases, and for all Parties to support the development of a Protocol or another legal instrument; and note the findings of the IPCC, in particular the following:

The balance of evidence suggests a discernible human influence on global climate. Without specific policies to mitigate climate change, the global average surface temperature relative to 1900 is projected to increase by about 2°C (between 1 and 3.5°C) by 2100; average sea level is projected to rise by about 50 centimeters (between 15 and 95 centimeters) above present levels by 2100. Stabilization of atmospheric concentrations at twice pre-industrial levels will eventually require global emissions to be less than 50 percent of current levels."

(Paragraph 2, and sub-paragraph 1, of the July 18, 1996 Ministerial Declaration.)

Note 2: Phrases edited out of the IPCC Report

1. "None of the studies cited above has shown clear evidence that we can attribute the observed [climate] changes to the specific cause of increases in greenhouse gases."

2. "While some of the pattern-based studies discussed here have claimed detection of a significant climate change, no study to date has positively attributed all or part [of the climate change observed to date] to anthropogenic [man-made] causes. Nor has any study quantified the magnitude of a greenhouse-gas effect or aerosol effect in the observed data-an issue of primary relevance to policy makers."

3. "Any claims of positive detection and attribution of significant climate change are likely to remain controversial until uncertainties in the total natural variability of the climate system are reduced."

4. "While none of these studies has specifically considered the attribution issue, they often draw some attribution conclusions, for which there is little justification."

5. "When will an anthropogenic effect on climate be identified? It is not surprising that the best answer to this question is, 'we do not know.'"

Note 3: Excerpt from State Department Letter to IPCC

"...it is essential that the chapters not be finalized prior to the completion of discussions at the IPCC WG I plenary in Madrid, and that chapter authors be prevailed upon to modify their text in an appropriate manner following discussion in Madrid."

Signed by Day Mount, Acting Deputy Assistant Secretary of State, Environmental and Development, November 15, 1995.

REFERENCES

Intergovernmental Panel on Climate Change (IPCC), *Climate Change: The IPCC Scientific Assessment*, Houghton, J. T. et al., eds., Cambridge Univ. Press, New York, 1990.

Intergovernmental Panel on Climate Change (IPCC), *WGI, Climate Change 1995: The Science of Climate Change*, Houghton, J. T. et al., eds., Cambridge Univ. Press, New York, 1996.

Mendelsohn, R., and Neumann, J. E., eds., *The Impact of Climate Change on the United States Economy*, Cambridge Univ. Press, New York, 1999.

Moore, T. G., *Climate of Fear: Why We Shouldn't Worry about Global Warming*, Cato Institute: Washington, D.C., 1998.

Santer, B. D., Taylor, K. E., Wigley, T. M. L., Penner, J. E., Jones, P. D., and Cubasch, U., "Towards the Detection and Attribution of an Anthropogenic Effect on Climate." *Clim. Dyn.*, 12: 79–100, 1995.

Seitz, F. A., "Major Deception on Global Warming," *Wall Street Journal*, June 12, 1996.

Singer, S. F., Hot Talk, *Cold Science: Global Warming's Unfinished Debate*, 2nd edition., The Independent Institute, Oakland, Calif., 1999.

Singer, S. F., "Unknowns About Climate Variability Render Treaty Targets Premature," *Eos, Transactions AGU*, 78: 584, December 16, 1997.

CHAPTER 22
GLOBAL PERSPECTIVES AND TRENDS

SECTION 22.2

NATURAL FORCES VERSUS ANTHROPOGENIC CHANGE

Hugh W. Ellsaesser

Dr. Ellsaesser spent 23 years in atmospheric research at the Lawrence Livermore National Laboratory, after serving for 21 years in weather service with the U.S. Air Force. His research convinced him that the establishment views on air pollution, threats to the ozone layer, and the climatic effect of carbon dioxide cannot be supported scientifically and that the hazards have been greatly exaggerated.

22.2.1   INTRODUCTION

Restoring the environment in the twenty-first century is a far different problem from that described by the environmentalists and the mainline media. To be able to define, much less solve, the problems of the environment, we must first correct a far more serious problem: the divorce of logic from, and the prostitution of, science.

Over the past century, the body of scientific knowledge has expanded tremendously. As a result, there has been a progressive fragmentation into narrow disciplines, each with its own jargon and specialized journals. This has restricted communication with the general public and even with scientists in other disciplines. Meanwhile, our schools have tended to remove, or at least reduce, science course requirements from nonscience majors. Through these processes, it has become more and more difficult for the general public to communicate, with scientists. Further, those with whom they cannot communicate, they tend to distrust and fear. Over the same period, most research scientists have been compelled to compete for research support funds, mainly from the federal government. This has impacted the work of research scientists in several ways.

Most of the environmental research supported by the government is designed to substantiate what we think we already know—such as the health effect of air pollutants, or to protect us from environmental hazards, such as acid rain, ozone depletion, and climate change. At present, most of the research on climate change is designed primarily to determine if climate change is occurring, not to understand it and, particularly, not to determine what, if any, benefits it might bring.

We have all heard of political correctness and how it has been used by minorities to impose their agendas on the majority. You also must have seen some of the many items on junk science in the courts, and how it has been used to drain funds from deep pockets and to advance the careers of the prosecutors. In a democracy, although it is the majority's duty to protect and provide equal opportunities for minorities, it is by no means their duty to allow themselves to be bullied, intimidated, and browbeaten into providing special treatment for any minority.

In brief, science has lost its aura as an honorable and trustworthy repository of systematized knowledge derived from observation, study, and experimentation. As a consequence, a large number of corporations have closed or reduced their in-house research efforts and withdrawn funding from scientific research organizations. They explain this action by stating that "science no longer matters in scientific debates" *Doctors for Disaster Preparedness Newsletter*, XI(4), p. 1 (1994). Instead, they are placing more reliance on pollsters and public relations agencies.

Under these conditions, why wouldn't corporations, or anyone else, withdraw support from scientific research and adopt approaches that clearly seem to be more successful?

22.2.2 EXAMPLES OF THE MISUSE AND NONUSE OF SCIENCE

The London "Killer Smog" of 1952

In early December 1952, there was a so-called "killer smog" in London. Essentially every reference to this event notes that air pollution during the event was responsible for some 4000 "excess deaths." The official investigating committee was unable to identify any substance in the air, which in the concentration surmised to have been present, could have caused these "excess deaths" by any known mechanism.

Because I lived in London at the time, I was motivated to study this and related incidents in detail. It is my conclusion that this spike in the London mortality curve was due to the second wave of an influenza epidemic (Ellsaesser, H. W., *EIR Science & Technology*, 16(30): 23–29, 1989; *Technology: Journal of the Franklin Institute*, 331A: 135–145, 1994). Such an epidemic was clearly recorded at that time in north and central England and across the channel in Europe. In fact, the influenza literature expresses wonder at the absence of any influenza epidemic reports in the London area at that time, but the local authorities, for reasons best known to themselves, preferred to call it an air pollution episode.

The acceptance of air pollution as the cause of this London episode has biased and continues to bias, by intimidation, those who have examined the evidence on health effects of air pollution and found it unconvincing, except for episodes like London's "killer smog."

The Health Effects of Low-level Radioactivity

Essentially every poison we know has been demonstrated to have beneficial effects at low concentrations or doses. This property is called hormesis. Dr. T. Donald Luckey has spent a lifetime documenting the fact that ionizing radiation, or radioactivity, also has this beneficial property (Luckey, T. D., *Hormesis with Ionizing Radiation*, CRC Press, 1980; Ellsaesser, H. W., *Global 2000 Revisited*, Paragon House, pp. 189–251, 1992). He has collected hundreds of published studies confirming the hormetic effect of ionizing radiation and indicating that we would all be healthier, live longer, and have fewer mutations if our exposure to radioactivity were increased up to 10 times above what we now accept as the background level of radioactivity. Instead of accepting this fallout benefit, we are spending billions, if not trillions, of dollars to avoid exposure to any detectable or computed increase in exposure to radioactivity traceable to human activity.

Increased Ultraviolet (UV) from Ozone Depletion

As you are no doubt aware, the Montreal Protocol was adopted in 1987 to reduce the release of stable and insoluble chlorine compounds to protect the stratospheric ozone layer. The rationale was to protect us from the increased exposure to solar ultraviolet (UV) radiation permitted to reach the Earth's surface by a thinner ozone layer.

On an annual mean basis, UV flux to the surface increases approximately 50-fold from the poles to the equator (Mo and Green, *Photochemistry and Photobiology*, 20: 483, 1974). This is roughly six doubling, or a doubling for every thousand miles, or a local 1 percent increase for each 10 miles of displacement toward the equator. For skin cancer incidence over the United States, data collected by the U.S. Academy of Sciences (*Environmental Impact of Stratospheric Flight*, NAS, 1975) showed an increase of 1 percent for each six miles of displacement toward the equator. The United Nations Environment Program (*Environmental Effects of Ozone Depletion: 1991 Update*, UNEP, p. 19, 1991) estimated that each 1 percent decrease in the ozone layer would lead to a 2.3 percent increase in ordinary skin cancer incidence. From the above numbers, a 1 percent decrease in the ozone layer, in terms of skin cancer incidence, is equivalent to moving 14 miles (22 km) toward the equator.

In 1987, the predictions were that human-released stable chlorine compounds would rise into the stratosphere. Above about 20 km, they would encounter sufficiently energetic UV radiation to be decomposed, releasing free chlorine. The chlorine would catalytically destroy ozone, primarily near the 40-km level. If the release rate remained unchanged until equilibrium, in about 75 years the global mean ozone layer would be reduced by about 5 percent (Solomon, *Nature*, 347: 347–354, 1990).

For the United States, with about 600,000 skin cancer cases per year (U.S. Academy of Sciences, *Environmental Impact of Stratospheric Flight*, NAS, 1975), this would mean approximately 75,000 additional skin cancers per year. It would also be equivalent to each of us moving 70 miles (110 km) toward the equator. Does this sound like a serious problem to you?

What has happened in the interim? Ozone at 40 km has decreased but only by about half as much as predicted for now. Total ozone, on the other hand, has declined much faster than predicted; the global mean 5 percent decline predicted for about 2060 AD was exceeded in 1993. The level has increased only slightly since 1993. I have no information as to whether ordinary skin cancer incidence has increased 11.5 percent as predicted for such a thinning of the ozone layer.

Essentially all of the observed decline in ozone has occurred at levels below 20 km (WMO, *Scientific Assessment of Ozone Depletion: 1994*, WMO, pp. 1.29–1.34, 1995), rather than near 40 km as predicted. In the tropics, there is very little ozone below 20 km. Essentially every study published has reported either no decline in ozone, or only a statistically insignificant decline in ozone near the equator. That is, the observed decline in ozone has been at low levels in higher latitudes where ozone is almost chemically inert, or in storage. Ozone is not generated in these areas, it is carried there by atmospheric motion. Accordingly, there are reasons to believe that the disappearance of ozone from these regions has been due to a change in atmospheric circulation rather than a change in chemistry.

The Blind Eye Toward Beneficial Effects of UV

Another aspect of the stratospheric ozone problem that has been completely ignored by the establishment, is the beneficial effect of UV exposure. For terrestrial vertebrates, the principal natural source of the vitamin D required to convert calcium into bone comes from the action of UV on oils in the skin. Feathered and furred animals get their vitamin D from self-grooming.

The most serious health effects from UV result from a deficiency of UV rather than from an excess. Rickets is a very serious disease occurring in children getting insufficient UV exposure or vitamin D in their growing years. Discovery of this relationship led to the regular dosing of babies with cod liver oil—an economical natural source of vitamin D. Even if a person escapes rickets, they may end up with a slight skeleton, less able to withstand the bone loss from osteomalacia common in later life.

In the United States, it is estimated that 20 to 25 million people suffer from osteomalacia, including over 25 percent of the women beyond menopause. Among these, there are over twice as many bone fractures per year, typically of the spine or femur, as there are new skin cancers per year. Theoretically, an increase in UV exposure would alleviate this condition in future generations, just as theoretically, it would lead to additional cases of skin cancer. Considering only the numbers of cases and seriousness of skin cancer and bone fractures from osteomalacia, it appears likely that an increase in UV exposure would provide a net health benefit. Suppression of osteomalacia by increased UV was found by a Dutch study of comparable susceptible groups living in the Netherlands and in Curaçao (Dubbelman, R., Jonxis, J. JP., Muskiet, F. AJ., and Saleh, A. EC., "Age-dependent Vitamin D Status and Vertebral Condition of White Women Living in Curaçao (The Netherlands Antilles) as compared with their counterparts in The Netherlands," *Am. J. of Clinical Nutrition* 58, 106–109, 1993).

It is quite likely that there are other health problems alleviated by increased UV. Although little work has been done in this area, one group of doctors found a significant negative correlation between both vitamin D availability and exposure to sunlight (a fair surrogate for UV) and the mortality rates from both colon and breast cancers (Garland, C. F., and Garland, F. C., "Do Sunlight and Vitamin D Reduce the Likelihood of Colon Cancer?" *Intl. J. of Epidemiology*, 9(3): 227–231, 1980; Garland et al., *The Lancet*, pp. 1176–1177, 18 Nov.1989; Gorham et al., *Canadian J. of Public Health*, 80, 97–100, 1989).

22.2.3   GREENHOUSE WARMING

The Large Discrepancy Between Observed and Model-predicted Warming

Ever since Manabe and Wetherald (*J. of the Atmos. Sci.*, 24(3): 241–259, 1967) first calculated by model the climatic effect of a doubling of carbon dioxide (CO_2), the predicted warming from man's additions of greenhouse gases to the atmosphere has exceeded the warming actually observed; and this difference has become progressively larger.

The Intergovernmental Panel on Climate Change (Climate Change, the IPCC Scientific Assessment, Cambridge Univ. Press, p. 233, 1990) estimated the warming over the past century from the observational record as 0.3–0.6°C (0.54–1.08°F). They also published a graph of the model predicted warming (ibid., Figure 8) showing values of 0.7–1.4°C (1.08–2.16°F) for 1990. IPCC (ibid., p. xii) claimed these were "broadly consistent," even though they do not even overlap. After man-produced sulfates had been included in the calculation, IPCC (*Climate Change 1995, The Scienc of Climate Change*, Cambridge Univ. Press, p. 295, 1996) stated that when greenhouse gases only are taken into account, most models produce "a greater warming than observed to date."

The range of uncertainties with regard to sulfates is so large that modelers can achieve any degree of agreement with observations desired—*globally*. However, the bulk of the sulfates are in the northern hemisphere and the *hemisphere by hemisphere* comparison constitutes a serious discrepancy for the argument that sulfates are reducing greenhouse warming. The absence of polar amplification of the warming in the observational data and the slight global cooling indicated by the satellite and upper air sounding data since 1979, the beginning of satellite observations, are additional serious discrepancies between the observations and model predictions.

The discrepancy in the degree of warming between observations and model predictions is significantly larger than the data above indicate. IPCC (ibid., p. 254, 1990) stated, "it is not possible at this time to attribute all, or even a large part, of the observed global mean warming to the enhanced greenhouse effect on the basis of observational data currently available."

The Environmentalists' Maxim: Man Can Do No Right

Research on environmental issues has been biased from the beginning by the maxim that "Man can do no right." That is, don't bother to look for or to try to document any possible beneficial consequences of humankind's actions on the environment; there aren't any. As already noted above with respect to ozone depletion and UV increase, *there are* beneficial consequences from man's actions on the environment. This is particularly true in the case of adding CO_2 to the atmosphere.

1. Greenhouse warming could delay, and hopefully prevent the next glacial. Our current understanding of past climate is, that since the time of the dinosaurs, about 100 million years ago, the global mean temperature has cooled about 10°C (18°F). About 3 million years ago, the present ice age began with alternating glacial and interglacial periods. Over the past seven hundred thousand years, there have been seven glacial/interglacial cycles with a global mean temperature range estimated at 5 to 7°C (9 to 11.6°F). The cycles were marked by about 90 thousand years of staged cooling, with 3-km (10,000-ft) ice caps building over Hudson Bay and extending down to Long Island and the Great Lakes. After the maximum glacial stages, there were relatively rapid warmings back to interglacial stages lasting 10 to 12 thousand years. We are currently in an interglacial, called the Holocene, which we estimate began 10,700 years ago.

Our best guess is that these cycles were caused by changes in the latitudinal and seasonal distribution of sunlight, because of periodic changes in the Earth's orbit around the sun. In any case, we know of no reason why they should not continue. Because we are now due to enter the next glacial period with 90 thousand years of cooling, should we not try to delay, and thereby hopefully prevent, this impending glacial? Have you heard of any consequences of global warming comparable to 3-km (10,000-ft) ice caps over Hudson Bay extending down to Long Island and the Great Lakes? We need more CO_2 in the atmosphere to avoid the now due next glacial.

2. More CO_2 can continue and enhance the Green Revolution. CO_2 is essential to plant life; and most scientists now concede that some fraction of the increased agricultural yield of the past century is due to increased atmospheric CO_2 concentration. From the beginning, our oldest CO_2 monitoring station on Mauna Loa has shown a marked annual cycle with CO_2 decreasing during the spring and summer, that is, the crop growing season and increasing during the fall and winter, the time of withering of leaves and decompositional decay of plant debris. Since the observations began in 1957, the amplitude of the annual cycle has increased about 20 percent (Idso, *CO_2 and the Biosphere: The Incredible Legacy of the Industrial Revolution*, Univ. Minn., p. 27, 1995). This is the most convincing evidence we have that human's additions of CO_2 to the atmosphere have increased the productivity of the biosphere on a global scale.

3. Is the worldwide decline in coronary mortality due to enhanced CO_2? Dr. Sherwood B. Idso (*Carbon Dioxide and Global Change: Earth in Transition*, IBR Press, pp. 118–120, 1985) reported evidence of already detected beneficial effects from the increased CO_2 in the atmosphere on both plants and animals. Among the latter, he cited "the significant worldwide downturn in circulatory heart disease experienced over the past two decades."

This sounds a bit far fetched. However, it must be recalled that respiration rate is controlled by the concentration of CO_2 in the blood, not the concentration of oxygen. Thus, if increased CO_2 makes us breath more deeply, isn't it logical that this might take some strain off of our circulatory systems? It may be that we need more CO_2 in the atmosphere to take stress off our circulatory systems.

Can you imagine the U.S. government expending research funds to determine if there is any validity to Dr. Idso's suggestion? Such a proposal would get the same negative reaction as did Don Luckey's attempts to get research funds to prove that mice would be less healthy and die sooner if deprived of normal ionizing radiation from natural radioactivity.

22.2.4   IS THERE A DISCERNIBLE HUMAN INFLUENCE ON GLOBAL CLIMATE?

The most quoted phrase from IPCC (*Climate Change 1995, The Science of Climate Change*, Cambridge Univ. Press, p. 4, 1996) is the following blunt section heading from the Summary for Policymakers: "The balance of evidence suggests a discernible human influence on global climate." Substantiation of such a claim requires both that a nonnatural climate change be identifiable in the observational record and that it be of such a nature that it can with confidence be attributed to the actions of humans.

Rather than attempt to make the details of the evidence contradicting such a claim intelligible to you [Ellsaesser, H. W., *21st Century Science & Technology*, 10(23): 61–67, 1997), I have assembled subsequent statements from some of the principal authors and defenders of IPCC. In my view, these statements clearly contradict the IPCC claim.

"No one to my knowledge who is informed is claiming certainty of detection or attribution [of a human influence on global climate]; certainly the IPCC is not, . . . " John T. Houghton (private communication (1996), Leading Editor of IPCC, 1990, 1992, 1996).

"We say quite clearly that few scientists would say the attribution issue was a done deal." Benjamin D. Santer (Kerr, *Science* 276, p. 1040, 1997), Lead Author of Section 8 of IPCC (1996).

", . . . many climate experts caution that it is not at all clear yet that human activities have begun to warm the planet–or how bad greenhouse warming will be when it arrives." Richard A. Kerr (*Science* 276, pp. 1040–1042, 1997), Research News and Comment Writer for *Science* magazine.

"However, the inherent statistical uncertainties in the detection of anthropogenic climate change can be expected to subside only gradually in the next few years while the predicted signal is still slowly emerging from the natural climate variability noise. It would be unfortunate if the current debate over this ultimately transitory issue should distract from the far more serious problem of the long-term evolution of global warming once the signal has been unequivocally detected above the background noise." Klaus Hasselmann (*Science* 276, pp. 914–915, 1997), Max-Planck-Institute for Meteorology.

These statements not only refute the IPCC claim, "The balance of evidence suggests a discernible human influence on global climate," they also suggest that this claim was studiously crafted to induce the media to broadcast to the citizens and policymakers of the world a message that few if any of the researchers, on whose work it was based, are yet willing to defend before the scientific community.

22.2.5 *CONCLUSION*

I hope you have been persuaded to, at least, consider the possibility that the environment is not the primary hindrance to our progress in the twenty-first century. I also hope that you will take seriously my warning: If we do nothing to return an aura of honor, truth, and trustworthiness to science, we are in danger of losing the benefits of science all together.

Acknowledgment

This paper was prepared for and presented as "Science: Use It or Lose It" to the Federation for World Peace's Fourth World Peace Conference at the Capital Hilton Hotel, Washington, D.C., November 28, 1997.

CHAPTER 22
GLOBAL PERSPECTIVES AND TRENDS

SECTION 22.3

CONTROLLING HAZARDOUS POLLUTANTS IN A DEVELOPING CONTEXT: THE CASE OF ARSENIC IN CHILE

Raúl O'Ryan
Dr. O'Ryan is an associate professor at the Industrial Engineering Department of the Universidad de Chile. His main research line is in environmental economics where he has undertaken diverse research and applied projects including choice of instruments for environmental regulation, international trade and the environment in developing contexts, computable general equilibrium analysis of environmental policies, and evaluation of policies to improve air quality in Santiago.

Ana Maria Sancha
Dr. Sancha is an associate professor at the Civil Engineering Department, Universidad de Chile. She is also a consultant for the World Health Organization and Interamerican Development Bank.

22.3.1 INTRODUCTION

Developing countries have scarce financial, technical, and human resources for generating the information required for environmental decision-making. Generally, there are few measurements of pollutant baselines. Monitoring, when it occurs, provides information for a few years and few places of measurement. There are no models available for determining the relation between emissions and concentrations. Finally, health effects are not well documented. For this reason, it has been common practice to copy standards from developing contexts. This has the advantage of reducing the information requirements significantly, but the other side of the coin is that often the costs and/or technological requirements make compliance difficult if not impossible.

Unfortunately, for hazardous air pollutants the transfer of results from other realities is particularly not appropriate because technologies, dispersion, exposure characteristics, and environmental risks are very different. In fact, in developing countries technologies are usually very polluting, people live close to emitting sources that generate their work, and the urban population is exposed to significant health risks from multiple different sources, and do not take minimal precautionary measures. In developed contexts, many of the major problems have already been addressed. In developing countries, regulation is an urgent matter. On the other hand, authorities want to avoid imposing unnecessarily stringent restrictions that would oblige the affected sources to close down or reduce their activity significantly, affecting economic activity adversely in a region that urgently requires more jobs. Technological aspects thus go hand in hand with economic aspects, and together determine the most appropriate regulation.

Under these circumstances, the trade-off between a cleaner environment and more economic activity must be taken seriously when proposing a new regulation. Defining how much to regulate a given substance is a key issue. Additionally, scarce resources must be used effectively, therefore, identifying which pollutant to focalize on is very important. Ideally, the largest effort must be spent on pollutants for which each dollar spent is most effective in reducing health risks.[1]

Arsenic pollution control from copper smelters in Chile presents an interesting case study with most of the above-mentioned elements. First, lung cancer in regions where copper smelters are located is almost three times the national average. Health authorities have been concerned about these health problems for the last decade and are pushing strict standards. Although there is also high arsenic content in water because natural reasons, the focus is on airborne emissions because drinking water is in compliance with the WHO standard. On the other hand, copper is Chile's main export product and a significant source of revenues and income in each region. Any reduction in economic activity would hurt the country and different regions strongly. The standards proposed by these authorities have been contested by the Mining Ministry and affected copper smelters, most of which are State owned, on the grounds that they are economically unfeasible.

To help resolve the dispute, a three-year project has been carried out to address the following questions: What are the risks posed by arsenic emissions? How much can they be reduced and at what cost? What is the appropriate instrument for regulating, an ambient standard or emission standards for each source? How strict should standards be? What can the authorities do to correctly focalize the scarce resources available? Is regulating arsenic content in drinking water an interesting possibility?

This section is organized as follows. The second subsection presents the arsenic problem in Chile, considering both air and water. The third subsection discusses the risk that cost methodology followed and presents the main quantitative results on risks and costs of reducing airborne emissions. The fourth subsection discusses the importance of reducing arsenic content in drinking water together with emissions from smelters. Finally, the fifth subsection presents some concluding remarks.

22.3.2 *THE ARSENIC PROBLEM IN CHILE*

The northern part of Chile has naturally high levels of arsenic, both in the air and water. The country is divided politically into regions, the first being far north and the twelfth in the south. Figure 22.3.1 shows that the baseline levels of arsenic in air in the first two regions are 5 to 15 times greater than in other parts of the country, reaching 0.05 μg/m^3. This is due to high concentrations existing naturally in the soil. Similarly, for drinking water, arsenic concentrations in the cities of the north are 5 times greater or more, than in the rest of the country. The water standard in Chile is 50 μg/l, and most cities easily comply. However, because of the high natural levels of arsenic in the sources of water supply—for example, the arsenic concentration in the Toconce river is 870 μg/l before arsenic removal treatment—in three important cities (Iquique, Antofagasta, and Calama) water quality is very close to this value. In the four northern regions, a total of 1.5 million people—10 percent of Chile's population—are affected by these high arsenic levels.

Close to copper smelters, arsenic concentrations are especially high as shown in Figure 22.3.1. This is due both to the fact that arsenic content in Chile's copper ore is extremely high, typically reaching 1 percent, (more or less the same as the copper content!), and the use of polluting technologies. Consequently, arsenic concentrations in 1995 reached average yearly values of around 1 μg/m^3 in many smelters, however, in the worst case, the average yearly value reached 10 μg/m^3 and a daily maximum of 100 μg/m^3.

At present, nearly 90 percent of Chilean copper is obtained through pyrometallurgy, for which purpose there are seven smelters in the country, as can be seen in Figure 22.3.2. These smelters have

---

[1] In this work, we concentrate on environmental health impacts, currently an important issue in developing contexts.

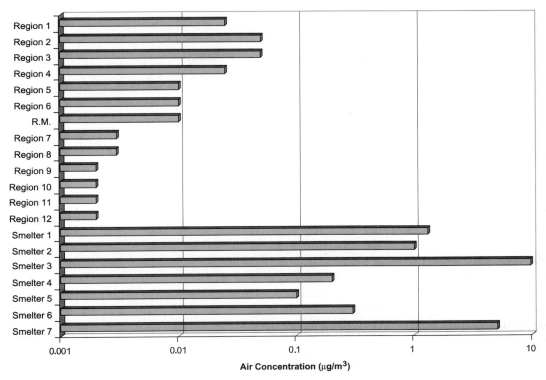

FIGURE 22.3.1   Arsenic concentrations in air: regional baselines and close to copper smelters.

an installed smelting capacity of 4.4 million metric tons per year (MTA), making Chile one of the world's most important countries in terms of smelter capacity. By international standards, these are large-size smelters, two of them being the biggest in the world. Similarly, their emissions are also considerable, ranging between 0.07 and 0.3 tons per day (t/d), for the smallest smelter, and 1.2 t/d and 5.4 t/d at the largest.[2] Altogether, approximately 370,000 people are affected by these emissions, however, the smelters are separated by great distances, so their emissions affect different populations. Table 22.3.1 provides a summary of the main smelter parameters.

At the start of this decade, health authorities began to worry about high lung cancer rates in the north of Chile. In the 2nd Region, where there had been heavy exposure to arsenic both in drinking water and in the air, the standard mortality rate for this cancer exceeds the national average by a factor of 2.6. In this region, there is an incidence of 234 lung cancer cases per year.[3] Similar results hold for the Third Region, where there is an incidence of 111 lung cancer cases per year. Inorganic arsenic has been classified as a carcinogenic substance by the International Cancer Research Agency, associated with lung cancer in particular.[4] Furthermore, the damage is likely to be independent of the ingestion route (i.e., both airborne ingestion and ingestion via water would likely provoke cancer [UNEP/WHO,

---

[2] For reasons of information confidentiality, details are not given on which smelter corresponds to which emission level. Below, each smelter is identified with a number.

[3] Average over 1985–1992.

[4] There is some evidence that arsenic in water also produces cancer in other organs such as kidneys, liver, and bladder (Smith, 1992, Ferreccio, 1997).

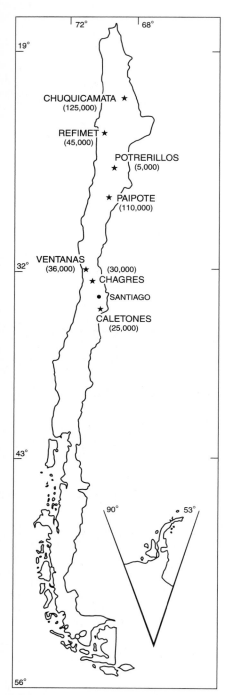

FIGURE 22.3.2   Location of smelters and population affected.

1986])). Because drinking water was in compliance with the accepted international standards at the time, the authorities of the Ministry of Health and the National Environment Commission (CONAMA) began a process aimed at regulating arsenic emissions from copper smelters.

Concentrations of arsenic in drinking water and emissions to air in the north have followed very different trajectories in the period 1950–1992, as can be seen in Figure 22.3.3. In air, emissions increased almost fivefold in the period, while water concentrations in the main cities steadily improved. The figure presents the population weighted average concentration in water in each period for the four main cities. In the period 1950–1969 the population was subject to very high concentration levels from drinking water without arsenic removal.[5] During the end of the 1960s, children exposed to arsenic in water in the north began experiencing serious heart problems, associated to this contaminant. Political authorities reacted, and, beginning in 1970, water supply was modified, so that waters with lower arsenic content were used to supply the larger cities. Meanwhile, the first treatment plant was designed, and went into operation in 1974, reducing concentrations to 12 percent of the original values—around 100 μg/l. However, water supply was not enough to satisfy the requirements of all the affected population, so two additional plants were needed to treat more water. To finance these plants, the government asked the World Bank for support, but this institution required complying with the prevailing WHO standards (50 μg/l) as a condition for the loan. This was accepted, and consequently, the old plant was improved, and the new treatment plants were designed to reach this standard. During the 1980s, these changes were implemented, which finally allowed reaching 40 μg/l at the end of that decade.

Emissions of arsenic to air increased up to 1992 basically as a result of an increase in smelter capacity in existing firms and the entrance of two additional smelters in the period. However, during the 1980s, international pressures because of the sector's poor environmental record began to raise awareness about the need to consider environmental issues. Additionally, the newly elected democratic government had environmental problems high on its agenda when it came into office in 1990. As a result, basic clean-up efforts were undertaken, and beginning in 1993 total emissions fell slightly, even though production capacity increased significantly in the period.

In fact, on January 26, 1984, a group of North American mining companies petitioned the International Trade Commission (ITC) to suspend totally or partially imports of Chilean copper into the United States. They argued that U.S. firms' compliance with environmental standards had significantly increased their production costs, and that copper imports from Chile had forced them to close the country's 25 main mines. Implicit in their presentation[6] was the fact that Chile's environmental standards were less demanding, especially regarding particulates and SO_2. In 1984 and 1985, on three occasions, the ITC was petitioned to apply a tariff surcharge on the grounds of "ecological dumping," the idea being that this tariff would compensate for the lack of environmental controls in Chile. These petitions did not succeed at the time.

---

[5] For example, over 800 μg/l in Antofagasta, city with 200,000 people.

[6] For practical reasons, this argument was not deployed explicitly, because this would have required the plaintiffs to individualize and show for each productive center that it was not complying with regulations similar to those required in the United States.

TABLE 22.3.1   Data on Chilean Smelters

	Smelter 1	Smelter 2	Smelter 3	Smelter 4	Smelter 5	Smelter 6	Smelter 7
Capacity (MT/Year)	1,500,000	240,000	540,000	284,000	400,000	480,000	1,200,000
Year of installation	1952	1993	1927	1952	1917	1964	1922
Arsenic emissions (Tons/Day)	5.4	1.3	8.2	0.3	0.07	1.2	5.0
Maximum Arsenic concentration (μg/Nm^3)	1.3	1.0	9.6	0.2	0.1	0.3	5.0
Exposed*	125,000	45,000	5000	110,000	36,000	30,000	25,000
Population	15,000	150	5000	9000	36,000	5000	5000

* In this row the first figure indicates the population that could receive some impact from smelter emissions as a result of living within a 30 Km radius of the smelter. The second, lower figure gives the population in localities of highest arsenic concentrations, impacted by emissions from each smelter.

Source: Information prepared by Fondef Project N° 2-24, based on data provided by the firms.

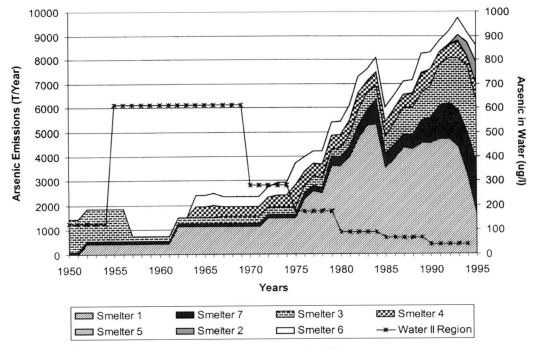

FIGURE 22.3.3   Arsenic emissions to air and concentration in water: 1950–1995.

However, the rift had consequences. First, in 1987 Chile was excluded from the U.S. General System of Preferences (GSP), because of a lack of safeguards for labor rights, but subsequently was accepted again in 1991. When Chile was readmitted into this system, copper products were included on a list of exceptional products that did not enjoy the benefit of the GSP, based on environmental considerations. Second, the state-owned Codelco—Chile's largest copper firm—had to commit to undertake environmental protection investments amounting to US$ 386 million between 1984 and

1990, and these were indeed carried out during that period.[7] The available data shows that the average annual investment in cleanup operations during that period was more than double that of first four years of the decade.[8]

The situation described above made clear to firms in the mining sector and the government the vulnerability of their environmental situation. They became more aware of the need to begin cleaning up their act. It also revealed the magnitude of the costs of improving the situation, and the need to develop an explicit policy in this area. Consequently, a few months after taking office in 1990, the new government set up the National Environment Commission (CONAMA) with the mission to generate a regulatory framework, which finally took shape in 1994, in the Framework Law for the Environment—and to develop the environmental institutional framework the country needed. An important goal for the environmental authorities at the time was to ensure that the proposed policies would allow Chile to be accepted as a partner in the North American Free Trade Agreement (NAFTA).[9] Key contaminants to be regulated were particulates and sulfur oxides. As a result, in 1992, the Mining Ministry regulated emissions of these pollutants from megasources, based on identical environmental standards as those required by the Environmental Protection Agency (EPA) in the United States.[10]

The direct consequence of this has been that all smelters currently operating have had to take the environmental impact of their decisions into account. The main smelters all have cleanup plans that have been cutting SO_2 and PM-10 emissions to levels compatible with the ambient standard. As a by-product, arsenic emissions are also being reduced, because it is emitted in the particulates. However, this contaminant was not directly addressed by the regulation.

Defining how far to go in reducing arsenic emissions became quite an issue. Based on the significant individual risks of arsenic, in 1994 the Health Ministry set an ambient standard of 0.05 μg/m^3 for arsenic as an annual average. This generated a sharp polemic with the Mining Ministry and the affected firms (it should be noted that five of the seven are state-owned), as in their opinion compliance with such a standard was not feasible. The controversy finally led to the abolition of the Health Ministry decree a few months later. The fact that, according to the firms, compliance costs would be very high and even oblige some of them to shut their smelters down, and that the data on health risk was quite weak, were weighty arguments in this decision. It was thus decided to postpone any decision on the regulation of this substance, pending availability of better data on the risks and costs involved.

22.3.3 RISK-COST ANALYSIS FOR AIRBORNE ARSENIC EMISSION[11]

In 1992, the authorities were confronted with the need to control airborne arsenic emissions. The Health Ministry, together with most of the public officials in charge of the regulation, wanted to set an ambient quality standard. However, they had to weigh carefully the consequences of their actions. On the one hand, copper smelting is a key economic activity both for the country and the regions were it is done. A strict standard could have significant impacts on local employment, regional economic activity, and generation of foreign exchange. On the other, lung cancer was especially high in regions with significant copper smelting activity, and there was a need to address the issue. Arsenic emissions had to be reduced. Questions to be answered included: How costly for the country is a regulation for airborne arsenic? How much are risks reduced? Should foreign standards be copied, or should specific standards for Chile be established? How?

The usual procedure used in developing contexts is to see what is done in other countries and then adapt the instruments to local conditions. For example, in the case of arsenic in drinking water

---

[7] According to internal reports in Codelco, this investment amounted to US$ 374 million during the period.

[8] See O'Ryan and Ulloa (forthcoming).

[9] Subsequently conversations have broken down because of lack of interest by U.S. policymakers.

[10] Primary standards to protect health, as well as secondary ones to protect species and ecosystems.

[11] The results in this section are based on O'Ryan and Díaz (2000).

Chile simply adopted the same standard WHO recommended. The same is true for PM-10 and SO_2 emissions. The information requirements are thus reduced substantially, and the basic problem consists of getting the financing required to meet the standards.

In the case of hazardous air pollutants, international experience was not especially helpful. These pollutants are usually not regulated through the use of ambient standards that can readily be copied. Rather, regulation is done on a case-by-case basis. The approaches to set emission standards and/or environmental quality standards are fairly diverse and may be grouped into three categories: (1) programs primarily based on risk assessment for setting standards, including emission standards and ambient air quality standards (The Netherlands and Sweden); (2) comprehensive emission standards programs supplemented by ambient air quality standards (Germany, Switzerland, U.S.); and (3) programs using process specific emission limitations, usually best available technology, for a limited set of hazardous air pollutants (France, Japan).

In each case, proposals for standards depend on the specific conditions of the problem, not just an assessment of the health or ecological effects. Costs and employment impacts are considered when regulating, together with technical feasibility. These are specific to each locality, region, and country.

In consequence, copying a standard was not a feasible option, making it necessary to evaluate possible alternatives. The lack of policies allowed suspecting that significant reductions could be obtained at low costs, or even with net-benefits (i.e., that there were "win-win" opportunities to be found). However, it was also to be expected that reduction costs would be significant once initial opportunities for preventing contamination were used up. Indeed, considering the size of Chile's smelters, the costs involved in significant emissions reductions could prove to be very high by obliging changes in processes. As in most developing countries, information was very poor: baseline emissions and concentrations were not well-known, information on costs and health effects was scanty. In this context, a relevant question that begged an answer was: How strict should the ambient standard be?

To respond, a research project[12] was carried out under the leadership of the University of Chile. This University—recognized as a serious research institution without any specific connection to the affected parties—developed and applied a methodology to assess the costs and risks of different levels of the ambient standard. This methodology is presented in the following subsections, together with the main results.

The Risk-Cost Analysis Methodology

To carry out the proposed evaluation of risks and costs of different ambient standards, first a range of possible values for the quality standard were identified. Then, each possible value of this standard was related, for each smelter, to an emission level, compliance costs, individual risks, and deaths avoided with respect to a base scenario without the standard. The total cost of the standard is the sum of the costs incurred by the seven smelters under consideration. In the same way, the number of deaths avoided corresponds to the total number of deaths avoided in each of the localities affected by the smelters. The procedure for carrying out the risk-cost assessment is best exemplified by Figure 22.3.4, where the different stages of this analysis are presented schematically for each locality.

The second quadrant plots the relation between arsenic emission reductions and the costs of the options for reducing such emissions. This is done using data on current and future emission levels, as well as the feasible reduction options and their costs, for each smelter. To estimate emissions, a mathematical model was developed to make it possible to simulate emissions from the different smelters, in a simplified way, on the basis of modular representation. By means of mass balances and equilibrium relations in each phase, the distribution of arsenic emissions per plant was determined.[13] Chimney-stack as well as fugitive emissions are determined in this way for each option available for every smelter.

---

[12] Project FONDEF 2-24 "Protection of the Competitiveness of Mining Products in Chile: Information and Criteria for Arsenic Regulation."

[13] See "Modelo de distribución de arsénico en fundiciones de cobre: estimación de emisiones atmosféricas" prepared by Mr. Jacques Wiertz for Fondef Project 2-24.

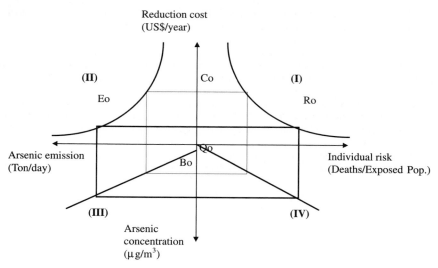

FIGURE 22.3.4   Risk-cost curve.

The third quadrant presents the relation between arsenic emissions and the corresponding concentrations generated in each affected locality. It can be seen that environmental concentrations are a function both of the natural base level of the zone (Bo), which determines the origin of the curve, and the way in which the contaminant disperses, reflected in the slope of the curve. The slope, or "dispersion factors," is determined on the basis of linear regressions between smelter emissions and concentrations measured for the localities, or otherwise through dispersion models where the conditions of the problem permit their use. In each case, the impact of fugitive emissions is distinguished from stack emissions, because of their different spatial behavior.[14] To determine the natural base level, a calculation of average annual values is made for the different localities, on the basis of a series of measurements obtained in periods when the smelter was not operating, or measures in localities close to the smelter but that do not receive any impact from emissions. The measurements undertaken showed that the natural base level around some of the smelters are quite high, between 0.08 and 0.1 μg/m^3. This is an interesting result because under these conditions, the standard proposed in the original discussions of 0.05 μg/m^3 cannot be met in these localities.

The fourth quadrant shows the relation between ambient arsenic concentrations and the individual risk to the exposed population. To determine this relation, it is necessary to identify, quantify, and characterize the risks associated with arsenic. Although this substance is related to multiple cancers,[15] in this study we consider only lung cancer, because it is the only cancer for which there are internationally recognized unitary risk values. Other effects, in particular bladder cancer, have not been considered because there is no conclusive evidence on the relation between ambient concentrations and cancer. To estimate these risks, only localities within 30 km from the emitting sources have been considered, because the available dispersion data indicate that at greater distances the impact of emissions on arsenic concentrations is extremely low.

It is usual in this type of study to make an assumption of linearity in the dose-response curve for low doses of arsenic concentration in the air, so that a single unitary risk value is defined, applicable to different concentration levels; hence, the linear relation presented in the fourth quadrant. This unitary

---

[14] In terms of the graph in Figure 22.3.4, this would mean a straight line with a different gradient in quadrant 3.

[15] Various epidemiological studies have detected the existence of lung and skin cancer (through ingestion of arsenic in the water). There is also some evidence that other organs, such as the liver and the lymphatic system, may suffer from cancer when exposed to arsenic.

risk corresponds to the risk associated with lifelong (70 years) exposure of a person to a concentration of $1\ \mu g/m^3$ of arsenic in the atmosphere. In this paper, a unitary risk of 4.29×10^{-3} per $\mu g/m^3$ proposed by EPA[16] has been used. By applying this risk to a population, it is possible to calculate the number of deaths associated with exposure.

Finally, the first quadrant shows the relation between the individual risk to which people are submitted, and the corresponding reduction cost at this risk level. In this way, for the established quality target (Q_0) an implicit risk and number of deaths avoided per locality (R_0) compared to the base situation is obtained, along with the cost (C_0) for each smelter. This analysis is carried out on a smelter-by-smelter basis, for each affected locality, so the final aggregate results in terms of costs and deaths correspond to the sum of the resulting per-smelter values.

Control Options and Reduction Costs

To reduce the impact on the exposed population, two types of measures are considered. First, technological options aimed at reducing emissions, both chimney-stack and fugitive emissions. Second, measures aimed at reducing people's exposure. The first include options involving improvement and modernization of existing processes, in particular the implementation of systems for capture and treatment of gases, as well as technological changes permitting better control of gas emissions, and closure of the smelter.

The reduction of fugitive emissions means improving systems of capture, where they exist, and implementing them in all effluent points in the smelter's equipment. If it is not possible to capture and treat metallurgical gases, more profound technological change is needed, including replacement of older smelter plants by more modern technology.

The other measures used to reduce arsenic emissions and the exposure of the population include changes in the feed, reduction in activity levels and the relocation of population. Indeed, it is possible to reduce arsenic emissions via a change in the grade of concentrate fed to the smelters, the prior roasting of concentrated feed, or by reducing the quantity of concentrate fed into the process. Finally, by relocating the affected population, it is possible to reduce the impact of arsenic concentrations without modifying the operating conditions of the smelter.

Table 22.3.2 summarizes the individual ranges of reduction that are achievable by implementing each of the options indicated. These are based on expert opinion, operational experiences, and available studies.

Reductions costs correspond to annualized direct costs that sources incur in reducing arsenic exposure. These costs include investment expenditures and the operating and maintenance costs associated with the different options. Also included are possible benefits, in particular, net credits arising from the generation of additional sulfuric acid as a result of the option used. Costs were calculated on the basis of information available in the literature, data provided by national and international experts, and operational data either from the smelters studied or smelters abroad. Possible indirect costs that sources might incur, such as job loss and reduction of ancillary services, have not been considered.

The annualized cost is determined on the basis of a calculation of the net present value (NPV) of the smelter business with and without the project. For this purpose a 25-year horizon was considered with a 12 percent discount rate. In Table 22.3.3, costs for a typical smelter are presented.

The rows in italics indicate the dominant options (i.e., options that at a given cost reduce more than other options). In this case, this smelter has two dominant options. First, it could switch to a Contop technology with net benefits of US\$ 13 million because of net credits obtained by selling sulfuric acid produces as a by-product. Emissions and concentrations would be reduced to approximately one-seventh the current values. However, these are still fairly high, so if additional reductions are required, the best option for this smelter is to relocate the affected population, in this case workers and their families, a total of 3000 people. Similar tables were prepared for each smelter.

---

[16] The WHO (1987), for its part, based on two studies, proposes a similar value of 4.0×10^{-3}.

TABLE 22.3.2   Summary of Reduction Efficiencies

Option	Efficiency of capture (%)
Electrostatic precipitators	10–50
Acid plants	+99
Flash Inco Oven	95
Contop Technology	95
Isasmelt Process	95
Flash Converting-Flash Smelting	+99
Mitsubishi	+99
Multi-story roaster	90
Dust treatment plant	80–90
Secondary hood	94–99
Matte and slag launder cover	99
Smelter shutdown	100
Change of concetrate feed	Variable efficiency
Reduction of activity	Variable efficiency
Population clearance	—

Source: Prepared by Fondef Project 2-24, on the basis of data provided by the firms and information in the literature.

TABLE 22.3.3   Cost of Reducing Emissions with Technological Options and Relocation of Population: An Example
Private costs and incomes before tax

Option	Description of option	Emissions (Ton/Day)	Maximum concentration $(\mu g/m)^3$	Incremental investment (US$ million)	Net cost (present value) (US$ million)	Net cost per year (US$ thousand)	Net cost of anodes (US cents per lb.)
0	Base situation	1.40	6.01	0	0	0	0
1	Improved base situation	0.43	1.40	5.1	4.1	518	0.2
2	Teniente Converter	0.65	2.10	72.9	−29.6	−3773	−1.1
3	*Contop*	*0.21*	*0.82*	*54.7*	*−100.1*	*−12773*	*−3.6*
4	Flash Furnace	0.74	2.14	159.1	103.0	13139	3.7
5	Continuous smelting	0.08	0.46	352.0	150.7	19200	5.4
6	Smelter shut-down	0	0.08	0	49.9	6361	2.3
7	*Relocation of population*	*1.40*	*0.08*	—	*30.0*	*3800*	*1.4*

Source: Emissions distribution model of Fondef Project 2-24, and data provided by the firm.

Aggregate Results: Risks and Costs of Different Quality Standards

Following the procedure described in the previous subsections, the response of each smelter can be determined for each level of the proposed standard, as well as its costs and the effective level of concentration in each impacted locality. By adding the cost of these responses, the total cost of complying with these potential values of the arsenic standard is obtained for all the copper smelters in the country. Additionally, the total number of statistical deaths avoided per year at each standard level is obtained by adding the individual risks in each affected locality. The results are presented in Figure 22.3.5.

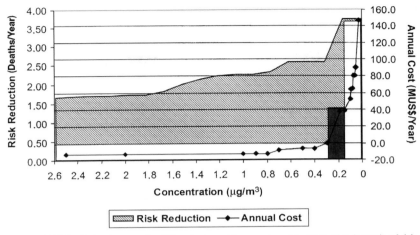

FIGURE 22.3.5   Reduction costs for different values of the arsenic quality standard and associated risk.

From this analysis, the following conclusions can be drawn:

1. It is possible to impose values of the standard between 2.5 μg/m^3 and 0.3 μg/m^3 at a negative cost to the smelters. This means that in the aggregate, there is a net benefit for firms that comply with the proposed standard! This is explained by the fact that some of the options allow greater recovery of copper and sulfuric acid from the process, and this increases smelter revenues, and lowers operating costs. Hence, at a discount rate of 12 percent, the options applied have a positive present value.

2. For values of the standard between 0.3 μg/m^3 and 0.15 μg/m^3, the annual cost of complying with the standard takes moderately positive values, between US$ 35 and US$ 38 million.

3. For a strict standard, below 0.15 μg/m^3, costs rise significantly. Indeed, a standard of 0.1 μg/m^3 has an annual cost of approximately US$ 52 million, one of 0.08 an annual cost of approximately US$ 65 million and a standard of 0.07 μg/m^3 an annual cost of US$ 81 million. For stricter standard levels, costs climb above US$ 100 million per year. This is basically explained by the fact that under such demanding standards, some of the larger smelters in the country have to shut down, as the required concentration level is below the natural base level.[17] These costs are significant for Chile. US$ 146 million per year is 0.219 percent of current GDP, while US$ 35 million per year corresponds to 0.052 percent.

4. Statistical deaths avoided each year follow an interesting path. Imposing a relatively relaxed standard of 2.5 μg/m^3 allows reducing 1.5 statistical deaths per year. A standard of 0.4 μg/m^3 would reduce 2.5 statistical deaths each year without incurring in aggregate costs, a win-win solution, particularly because the main smelters involved are state owned. Pushing the standard to 0.15 μg/m^3 increases significantly the statistical deaths avoided to 3.5 per year, however, costs begin to increase also, reaching almost 40 million per year. Finally, standards below 0.15 μg/m^3 result in negligible additional health benefits, but much higher costs.

These results are very informative. First, preventing pollution would permit significant benefits over a wide range of emission reductions. Second, costs rise sharply for quality standards better than

---

[17] This is certainly a strong assumption, because faced with a demanding standard one can impose significant emission restrictions on a smelter without obliging it to shut down. For example, a fairly relaxed compliance deadline can be given. This possibility has not been considered in this analysis, however the costs of important emission reductions are at least comparable to closing down, if not higher.

TABLE 22.3.4   Impact on Smelters of Different Levels of the Arsenic Quality Standard

Smelter		Quality standard (μg/m^3)					
		0.70	0.30	0.1	0.09	0.07	0.05
Smelter 1	Cost/yr (US$M)	0.4	6.9	37.4	44.2	44.2	44.2
	Emissions (t/d)	0.63	0.49	0.63	0	0	0
	Conc. (μg/m^3)	0.35	0.29	0.10	non-compl.	non-compl.	non-compl.
Smelter 2	Cost/yr (US$M)	−8.9	−8.9	−8.9	−8.9	6.4	6.4
	Emissions (t/d)	0.18	0.18	0.18	0.18	0	0
	Conc. (μg/m^3)	0.08	0.08	0.08	0.08	non-compl.	non-compl.
Smelter 3	Cost/yr (US$M)	0	0	1.3	2.6	2.6	3.3
	Emissions (t/d)	0.83	0.83	0.76	0.41	0.41	0.34
	Conc. (μg/m^3)	0.11	0.11	0.1	0.06	0.06	0.05
Smelter 4	Cost/yr (US$M)	0	0	5.7	9.6	9.6	9.6
	Emissions (t/d)	0.12	0.12	0.03	0	0	0
	Conc. (μg/m^3)	0.24	0.24	0.10	0.08	non-compl.	non-compl.
Smelter 5	Cost/yr (US$M)	0	0	0	0	1.2	3.6
	Emissions (t/d)	0.24	0.24	0.24	0.24	0.16	0.05
	Conc. (μg/m^3)	0.09	0.09	0.09	0.09	0.06	0.03
Smelter 6	Cost/yr (US$M)	0	0	1.5	1.5	1.5	7.9
	Emissions (t/d)	0.07	0.07	0.03	0.03	0.03	0.02
	Conc. (μg/m^3)	0.15	0.15	0.06	0.06	0.06	0.04
Smelter 7	Cost/yr (US$M)	0	1.9	15.3	15.3	15.3	15.3
	Emissions (t/d)	0.81	0.27	0	0	0	0
	Conc. (μg/m^3)	0.55	0.27	0.08	0.08	non-compl.	non-compl.

Source: Fondef Project 2-24.

0.15 μg/m^3, while health is not improved significantly. Aiming for an ambient standard stricter than 0.15 μg/m^3 is not justified given these results, at least in the medium term.

Impact of Different Quality Standards on Smelters and Ambient Concentrations

The aggregate results hide important variations among emitting sources, that must be considered for an efficient regulation. Table 22.3.4 presents the cost, as well as emissions and the concentration achieved at the worst point at each smelter for different potential levels of the standard. The table shows that smelter 2 gains net benefits of US$ 9 million per year when a standard of 0.7 μg/m^3 is applied. These benefits are achieved by changing its process to Contop technology and gas cleaning with an acid plant, which gives a net benefit for the firm of US$ 13 million per year. Despite these changes, however, fugitive emissions oblige the firm to relocate population away from the camp in order to comply with the standard, at a cost of US$ 4 million per year, whereby net benefits fall to US$ 9 million. These benefits are maintained when more demanding standards are imposed, because, once the population has been relocated, smelter emissions no longer affect them. However, for a standard of 0.08 μg/m^3, which corresponds to the base level in the vicinity of this smelter, it has to close down, with consequent costs in terms of lost production.

The monetary benefits obtained by this smelter determine the net benefits in the range indicated in the figure above. Indeed, to comply with the target of 0.3 μg/m^3, four smelters do not have to change their emissions, and the costs for the two smelters which do have to make reductions are offset by the benefits at smelter 2.

The smelters with the most direct impact on the exposed population, either because of their size or their proximity (smelters 1, 2 and 7), quickly begin to have problems in complying with stricter standards. For example, a standard of 0.2 μg/m^3 obliges smelters 4 and 7 to make environmental investments (US$ 2.9 and 4.6 million per year). Smelter 1 has to incur significant costs, as it is obliged to relocate population, with the corresponding costs, and making certain technical improvements. Moreover, this smelter is unable to comply with values below 0.1 μg/m^3, taken as the natural base level of the zone, so a lower standard forces it to shut down.

From 0.07 μg/m^3 onward, more than half of the smelters have to close down,[18] either because they are unable to attain the standard imposed, or simply because the technology that enables them to do so makes the smelter business unprofitable. None of the smelters in the north of the country comply with this standard (even when they stop working) because of the natural base level, of above 0.08 μg/m^3 in this zone. On the other hand, the three smelters located in areas of low natural arsenic levels can comply at relatively low cost.

These results are extremely important for the policymaker. From the previous section, it is clear that an ambient standard below 0.15 μg/m^3 would not improve health significantly. However, analyzing the information on a source by source basis, suggests that many of the sources could undertake important emission reductions at low costs, reaching ambient concentrations much lower than this value. Clearly, applying a unique ambient standard would preclude the regulator from taking advantage of this possibility. Unfortunately, ambient standards are usually national in scope and must be unique because they are used to protect equally the health of any individual affected.

Based on these results, it was proposed that smelters should be regulated on a case-by-case basis using emission standards. This proposal was accepted by the regulator, and consequently, emission standards specific to each source were established in 1999, based on the results of the project and direct negotiations with each source about the timing of the reductions.

22.3.4   DECIDING WHAT MEDIA TO REGULATE: AIR OR WATER?

Even though the results allowed regulating arsenic emissions to air in a way that made compatible health and economic concerns, the issue of reducing the high lung cancer incidence was not completely resolved. In fact, each year an average of 345 new lung cancer cases occur in the II and III Region, whereas at best, strict regulation of airborne arsenic emissions would reduce these by 2 percent. Consequently, health officials are uneasy with the proposed regulation, which does not consider being very strict with sources, because the significant uncertainties involved in many of the estimations may be hiding important health effects. The project also shed light on this issue, in particular the need to focus on water quality rather than air emissions.

First, an analysis of the arsenic intake in each region by source of origin was undertaken. Table 22.3.5 presents the main results. Of total intake, arsenic from water is clearly the highest corresponding to 85 percent in cities of the II Region. These results are a direct consequence of the high natural arsenic content in water. This suggested the need to look more carefully at the importance of water in the high current cancer rates in the north. A separate study within the project did this.

To evaluate the causal role of exposure to arsenic from drinking water and air between 1950 and 1970, a case-control study was undertaken in Regions I, II, and III (Ferreccio et al., 1998 and 1997). From 1994 to 1996, 151 cases diagnosed as lung cancer and two hospital controls (one with cancer

TABLE 22.3.5   Total Arsenic Intake by Source for each Region 1994–1995

City	Percentage contribution of arsenic by intake pathway (%)		
	Water	Air	Foods*
Antofagasta	85.2	1.2	13.6
Calama	85.3	1.0	13.7

*This is organic As which is much less damaging than the other intake routes.

---

[18] See note *supra*.

TABLE 22.3.6   Reduction Costs for a Water Treatment Facility

As concentration in water mg/l	Control option	Annual cost KUS$/Year
0.04	Base case	0
0.03	Automatization and Dosification	10
0.02	Double Filtering	440
0.01	Reverse Osmosis	10910

Source: "Reduction Options of Arsenic Level in the Antofagasta's and Calama's Water Facilities," Fondef Project N° 2-24.

and the other without unrelated to arsenic exposure, for a total of 419 controls) were entered in the study. A standard survey containing questions about residence, employment, and health history was administered to the subjects. Data on arsenic concentrations in water and air were constructed based on historical records and best estimates based on production records and technologies of each smelter.

The main conclusions of this study were as follows: (1) exposure to arsenic in drinking water is one of the causes of lung cancer in Region II; (2) it was not possible to find a statistically significant relation between exposure to arsenic in air and lung cancer, except for camps. Additionally, an ecological study undertaken by Smith et al. (1998) concluded that ingestion of inorganic arsenic in drinking water is "indeed a cause of . . . lung cancer. It was estimated that arsenic might account for 7 percent of all deaths among those aged 30 years and over . . . (a figure) greater than that reported anywhere to date from environmental exposure to a carcinogen in a major population." In conclusion, the extremely high natural arsenic content in water in the north is the most important source of cancer incidence.

Consequently, health officials can be more confident that the emission standards for smelters, together with the significant improvements in water quality undertaken during the 1980s, will imply important reductions in the lung cancer incidence in the future. Considering the long latency period of this cancer, these results have not yet become apparent.

However, two public policy issues arise. First, many small villages in the north still obtain their water from untreated sources with very high arsenic content.[19] The results of this project show that it is necessary to focalize resources and research efforts in finding cheap methods to reduce the arsenic content in drinking water for these localities. Second, there is an ongoing debate in the world as to whether the current arsenic water standard should be reduced. WHO, for example, is proposing reducing the standard to 10 μg/l. Considering the potentially high health benefits in the north, it is important to analyze this option.

For this reason, an analysis was undertaken in a locality that is affected by both high arsenic emissions in air and high arsenic content in drinking water.[20] Exposure to arsenic for 110,000 people can thus be reduced by treating emissions or water. The question to be answered was: If the authority wants to avoid the maximum additional statistical deaths, where should it concentrate its effort, in reducing air emissions or improving water quality?

Table 22.3.6 presents the available control options form water and the associated costs. It is clearly possible to reduce concentration levels significantly. However, costs rise steeply as better water quality standards are imposed: reducing to 0.02 mg/l costs only US$ 440.000 per year, whereas reducing to 0.01 mg/l costs US11 million per year.

In Table 22.3.7, these results are compared to the costs of reducing emissions discussed in the previous section. It is clear from this table that to avoid 2.89 statistical deaths is inexpensive: a standard of 0.02 mg/l for water and 0.40 μg/m^3 for air are required, with a total cost of less than 1 million dollars. Any additional reduction would be extremely expensive, but clearly it is more convenient to reduce them through reductions in arsenic concentrations in water. In fact, 10.5 million dollars would

---

[19] These are different from the larger cities considered in the rest of the study, which do receive treated waters.

[20] However, this level occurs naturally and is not related to emissions.

TABLE 22.3.7 Comparison of Costs and Health Benefits from Reducing Arsenic in Air and Water

Water			Air		
As concentration (mg/l)	Annual Cost (thousand US$)	Statistical deaths avoided (person per year)	As concentration (μg/m^3)	Annual Cost (thousand US$)	Statistical deaths avoided (person per year)
0.04	0	0	0.40	383	1.13
0.03	10	0.88	0.15	37.000	1.17
0.02	440	1.76	0.10	37.383	1.33
0.01	10910	2.64	0.08	44.243	1.36

Source: Fondef Project 2-24.

allow reducing 0.8 additional statistical deaths in water, whereas 37 additional million dollars in air would only reduce an additional 0.2 statistical deaths. In conclusion, it is 12 times more expensive to reduce statistical deaths by controlling air emissions than by improving water quality.

Summarizing, health authorities can obtain significant health improvements at relatively low costs by focusing efforts on improving drinking water quality in the north. This is not an argument against concern about airborne emissions. It is obvious that these are exceedingly high and must be reduced. However, the highest impact on health per dollar spent will clearly be obtained by improving water quality. From a public policy perspective, this result is extremely important for Chile, a developing country that must spend its limited resources effectively.

22.3.5 CONCLUDING REMARKS

Developing countries are receiving increasing pressures to clean up their environmental act. Foreign governments through trade agreements, markets through certification processes, public authorities, and concerned citizens are all pushing for stricter regulations. However, regulating hazardous pollutants is a tricky business, requiring a careful weighing of the environmental and economic trade-offs. Copying standards, or imposing standards based solely on health considerations, can be extremely costly, and does not necessarily produce significant benefits. Very straightforward calculations with data that can readily be compiled can help a great deal in illuminating decisions.

In the case of arsenic regulation for Chile, identifying the main health hazards of arsenic—in particular, lung cancer—and quantifying the risks associated to current levels as well as the risks and costs associated to reducing emissions helped the decision-makers in many ways. First, it became clear that significant emission reductions could be undertaken at low costs. Second, the numbers showed that significant health effects could be obtained up to values of arsenic concentrations around 0.15 μg/m^3. With current information on the unitary risks of arsenic, additional reductions in ambient concentrations did not help in avoiding more statistical deaths, but implied steeply increasing costs, reaching over a 100 million dollars per year. Spending such quantities to reach a stringent ambient standard that did not reduce health impacts significantly, seemed excessive. Consequently it was concluded that reaching an ambient standard of 0.05 μg/m^3 was not feasible for all smelters in the short and medium run.

However, case-by-case analysis showed that many of the smelters could make important reductions and, in some cases, reach the initially desired standard of 0.05 μg/m^3, at acceptable costs. For this reason, it was decided that it was best to use emission standards specific to each smelter to regulate, rather than a uniform ambient standard. This was an important result because it helped overcome the stalemate between the Health and Mining Ministries.

The results of the project also show where to focalize pollution control efforts. Reductions in airborne emissions are clearly warranted, however, a major effort should be undertaken to reduce arsenic content in water in the north. Even though water quality is in compliance with internationally used standards, these are currently being questioned around the world. Based on the available figures on unitary risks from arsenic ingestion in water, it was estimated that this is a much greater health hazard than emissions in air in cities, especially after the proposed reductions go into effect. Investing in better water quality is 12 times more cost-effective than controlling air emissions. Moreover, in small villages that use untreated well water with high arsenic content, research efforts geared at finding cost-effective treatment solutions are critical.

In conclusion, investing in research to establish risks and costs is necessary if a country, especially a poor one with many unsatisfied social demands, wants to use its resources effectively. Many of the decision variables are highly idiosyncratic (i.e., specific to each country). Regulatory authorities must be open to the use of different regulatory instruments and also willing to accept less control than desired solely on health considerations, if the corresponding costs are exceedingly high, as may be the case when trying to reduce the last bits of pollution.

REFERENCES

Ferreccio, C., González, C., Milosavljevic, V., Sancha, A. M., and Marshall, G., "Lung Cancer and Arsenic Exposure in Drinking Water: A Case-Control Study in Northern Chile," *Reports in Public Health*, 14 (Sup. 3): 193–198, 1998.

Ferreccio, C., González, C., Milosavljevic, V., Sancha, A. M., "Impacto en Salud Atribuible a Exposición a Arsénico: Un Estudio de Casos y Controles, Protección de la Competitividad de los Productos Mineros de Chile: Antecedentes y Criterios para la Regulación del Arsénico," Universidad de Chile, Santiago, Chile, 1997.

OECD, "Control of Hazardous Air Pollutants in OECD Countries," Organization for Economic Cooperation and Development, 1995.

O'Ryan, R., and Díaz, M. (2000), "Risk-Cost Analysis for the Regulation of Airborne Toxic Substances in a Developing Context: The Case of Arsenic in Chile," *Environmental and Resource Economics* (February) 15: 115–134.

O'Ryan, R., and Ulloa, A. (forthcoming), "Threats to Trade due to Environmental Considerations: the Case of Mining in Chile," in *New Faces of Protection Latin America and the Global Economy*, Macmillan Press, Fischer, R., ed., in press.

Patrick D., *Toxic Air Pollution Handbook*, Van Nostrand Reinhold, New York, 1994.

Smith, A., Goycolea, M., Haque, R., and Biggs, M. L., "Marked increased in Bladder and Lung Cancer Mortality in a Region of Northern Chile Due to Arsenic in Drinking Water," *American Journal of Epidemiology*, 147(7): 660–669, 1997.

Smith, A. et al., "Arsenic in the Environment and its Incidence on Health," *International Seminar Proceedings*, Volume 97, pp. 135–145, Santiago, Chile, May 1992.

UNEP/WHO Global Enviroment Monitoring Programme, *Guidelines for Integrated Air, Water, Food and Biological Exposure Monitoring*, Heal Project, Human Exposure Assessment Location, World Health Organization, Geneva, 1986.

U.S. EPA, "Health Assessment Document for Inorganic Arsenic," *Final Report. EPA 600/8-83-021 F*, March 1984. Office of Health and Environmental Assessment, Washington DC, 20460, 1984.

WHO, *Air Quality Guidelines for Europe*, WHO Regional Publications, European Series N° 23, Denmark, 1987.

CHAPTER 22
GLOBAL PERSPECTIVES AND TRENDS

EMERGING ENVIRONMENTAL ISSUES FROM THE PERSPECTIVE OF INDUSTRY

Martin Whittaker

Dr. Whittaker is an environmental consultant, researcher and senior analyst with Innovest Strategic Value Advisors, Inc. He specializes in environmental risk, strategy and finance.

22.4.1 INTRODUCTION

Whereas the previous sections of this handbook have provided, in turn, highly detailed accounts of the many different aspects of environmental science, in this section the environmental field as a whole is viewed from a single perspective, namely, that of *industry*. In particular, what follows is an attempt to draw together some of the broader environmental issues and ideas that may emerge in the coming years, to look at how these issues might affect the various levels of industrial activity, and to examine ways in which industry is projected to respond. The section begins with a short discussion of the history of the relationship between human interaction with the environment, leading up to an overview of the current situation. After profiling some of the most pressing environmental issues facing industry today, some of the various techniques being employed by industrial companies to address these issues are outlined. Because of the necessarily wide-ranging nature of this narrative, many of the topics identified are touched upon only briefly in comparison to the discussions provided elsewhere in this book, and the reader is referred to original sources for additional details.

22.4.2 AN HISTORICAL OVERVIEW

In examining the history of the relationship of human endeavor, in its broadest sense, and the natural environment, it is possible to discern perhaps two distinct phases of interaction. The first is rooted in humankind's traditional agriculturist heritage and corresponds to an almost symbiotic coexistence with the land and nature.[1] This phase spans the hunter-gatherer stage of human evolution some 7000 years ago into the age of agricultural subsistence farming. Impacts to the environment during this time primarily involved the felling of trees for fuel and timber, and the use of land for rotation cropping and later for pasture.[2] With the dawning of the modern industrial age, the second phase of environmental interaction began, in which the natural world's assets were harnessed in order to drive the expansion of resource-intensive industrialized economies. The competitive dynamic between the environment and human economic development was invariably to the detriment of the former, and has been responsible for an undeniable deterioration in the quality and abundance of global atmospheric, aquatic, and terrestrial resources. This result has given rise to an overriding sense of concern as to the consequent risks to future economic and social prosperity, in terms of both the direct impacts to human and ecological health, and the indirect impacts to an economic system whose success ultimately depends upon the continued availability of natural resources.[3]

Society's response to these environmental pressures has been to develop a process whereby polluters are penalized for their misdeeds by government regulators acting in the interest of the public at large. The process is characterized as adversarial and compliance-driven in nature, and relies to a large extent upon the capacity of the environmental laws and regulations put in place to protect human and ecological health. The main environmental culprits throughout this process are generally easily identifiable industrial point sources—factory smokestacks, effluent pipes, industrial plants, and tankers. As this second phase of environmental interaction gained momentum, increasing emphasis has been placed on the science of the environment, with much intellectual capital being drawn from the classic disciplines of chemistry, geology/hydrogeology, biology, toxicology, and engineering. As this book has shown, this broadening of the knowledge base has allowed practitioners to formulate more meaningful environmental standards and further understand the nature of cause and effect.

With the realization that the environmental imperatives facing society are in fact much more complex phenomena than first was thought, involving not just a few thousand dirty industrial polluters but the actions of millions of people, it might be said that we have now entered the third phase of environmental awareness. In this new paradigm, the environment is viewed not as a separate entity but as a fundamental facet of economic and social welfare, human security, and corporate profitability and strategy.[4] Environmental pressures and the distribution of natural resources are being identified as causal factors in the escalation of regional conflicts.[5] Moreover, the environmental health-related issues confronting society today—the safety of consuming genetically modified foods, the long-term effects of extensive low-level contamination of water and land, deteriorating urban air quality, the cause(s) and effect(s) of climate change, the rise in incidence of multiple chemical sensitivity, and the potential toxicity of everyday chemical products—suddenly seem to be at once more obfuscatory and ubiquitous than in the past.[6] Finally, as we have seen, the authenticity of environmental standards is being increasingly challenged as uncertainties in the underlying science persist. The trade-offs between environmental benefits and economic costs (for example, of contaminated site remediation) are also exerting an increasingly powerful influence over environmental management policy.

Driven by these new imperatives, the environmentally-driven risks now facing industry itself have become decidedly more elaborate, reaching far beyond a straightforward need to comply with applicable regulations. From industry's perspective, risks may be divided into at least six basic forms:[7]

Market Risk

- Regulatory bans or restrictions on sales (e.g., pesticides, CFC's, elemental chlorine).
- Customer boycotts or reduced product acceptance (e.g., forest products, phosphate detergents, nonrecyclable consumer goods).
- Reduced market demand (e.g., reduced demand for waste hauling because of recycling and waste minimization).
- Reduced product quality from environmental degradation (e.g., reduced recreation and tourist business because of coastal pollution or forest degradation).

Balance Sheet Risk

- Remediation liabilities (e.g., Superfund and RCRA liabilities).
- Insurance underwriting losses (e.g., from storm/flooding/property damage claims).

- Impairment of real property values (e.g., from site or building contamination).
- Natural resource damage assessments (e.g., damages from spills).
- Settlement of "toxic torts" (e.g., asbestos).

Operating Risk

- Product risk (e.g., asbestos, dioxins, furans).
- Costs of cleaning up spills and accidents.
- Regulation-driven process changes (e.g., elimination of chlorine in chemical processes).
- Reduced process yields (e.g., reduced agricultural, forestry and fishery yields from increased ozone or acidity, financial loss from downtime/shutdowns).
- Increased input prices (e.g., higher energy prices from environmental taxes).
- Supply disruptions because of regulation.

Capital Cost Risk

- Product/process redesign (e.g., CFC-free refrigerators, new pulp and paper process requirements).
- Input substitution (e.g., to process recycled fiber into paper products).
- Waste treatment and pollution control (e.g., municipal sewer system/waste treatment upgrades).

Transaction Risk

- Delay, disruption, or cancellation of acquisitions and divestitures.
- Financial and staff costs of unanticipated due diligence requirements.

Eco-Efficiency and Sustainability Risk

- Cost and competitive disadvantage created by energy inefficiency.
- Reduced raw material availability (e.g., restrictions on log harvests, mineral extraction, fishery stocks, use of CFCs).
- Mandatory product take-back, recycling, and life-cycle stewardship costs.

   Managing and ultimately resolving these issues will likely require concerted action or behavioral changes on the part of numerous stakeholders, and not just industrial point sources. Importantly, this comes at a time when science as a whole is seeking to renew and revitalize its relationship with society.[8] The success of the science community in winning back popular confidence will depend to a large degree on its role in satisfactorily addressing popular environmental fears and concerns, and the extent to which it enables society as a whole to meet these new challenges. It is against this backdrop that the environmental issues discussed throughout this book must be viewed.

22.4.3   CONTEMPORARY CONTEXT

   As this book has shown, one of the first and most important hurdles for environmental practitioners to overcome is the need to understand the environment (and, therefore, its scientific study) as a highly complex system of interrelated components. This reality is being increasingly recognized in public policy and industrial circles, and is becoming a more prominent facet of scientific research.[9] Such an understanding has formed the cornerstone of environmental policy in Europe, for example, where *integrated* pollution prevention and control (IPPC) is the fundamental principle of environmental management and direction.[10] Under this system, impacts to air, water, and soil, hitherto dealt with under separate regulatory and legal regimes, are now considered—in theory, at least—in unity, thereby facilitating greater overall protection of the environment. Developments in environmental science that deepen understanding of the extent and nature of these linkages will become increasingly important.

Externally, the connections between environmental science and other social and economic imperatives that drive industry are captured within the image of sustainable development. First elaborated by the World Commission on Environment and Development in 1987,[11] the concept has since become the watchword of global environmental management. The most widely accepted definition of sustainable development comes from the Brundtland Commission (World Commission on Environment and Development) report: "Humanity has the ability to make development sustainable—to ensure that it meets the needs of the present without compromising the ability of future generations to meet their own needs" (pg. 8).

From industry's perspective, the notion of sustainability has also generated much interest in the linkages between corporate financial and environmental performance.[12] It behooves environmental practitioners to form an understanding of what sustainable development is intended to mean for two primary reasons. First, at the World Conference on Environment and Development held in Rio de Janeiro in 1992, national governments committed to undertaking a series of practical action items predicated upon the notion of sustainable development.[13] Thus, the concept now forms the basis for environmental policy in many countries and is the intellectual cornerstone on which much cooperative international and regional environmental activity is founded.

Second, sustainable development draws together into a single unified concept many of the complex environmental, social, and economic issues facing society, and in doing so creates a more compelling overall argument for helping to resolve these issues satisfactorily through sound environmental management and, concomitantly, better understanding of environmental science.[14] In this respect, perhaps the most important emerging issue in the environmental field as a whole lies in broadening our understanding of the role of the environment in matters concerning human and common security, and industrial sustainability and competitiveness.[15] These issues will undoubtedly form an important driver of future global environmental activity and cooperation.

Industrial companies in resource intensive sectors such as petroleum, forestry, chemical, mining, and power generation (including nuclear energy) are subject to particularly intense environmental pressures and, although generally cognizant of sustainable development, remain focused on the mitigation of environmental impacts and operating risks.[16] Indeed, the management of environmental regulations and concern for public image are regularly identified by industrialists as two of the most compelling challenges they face.[17] Oil refiners, for example, cite compliance with environmental regulations, particularly the requirements of the Clean Air Act Amendments of 1990, as the most important factor affecting petroleum refining in the 1990s, and one that has contributed significantly to the industry's low profitability during that period.[18] Certainly, companies have incurred and will continue to incur substantial capital, operating and maintenance, and remediation costs associated with environmental laws, regulations, and other restrictions. Typically, a petroleum or mining company may spend between 1 and 5 percent of its total annual revenues, and anywhere between 5 and 20 percent of total capital expenditures, on environmentally related issues.[19]

This trend is set to continue, because industry as a whole remains a priority enforcement target for environmental regulators in 1999 and beyond, and issues such as those described below come more to the fore. At the same time, however, leading-edge investors and industrialists are becoming increasingly aware of the environment as a competitive phenomenon, rather than simply a regulatory obligation.[20] Thus, the pursuit of sustainability, manifested in the form of the "Triple Bottom Line"—whereby economic, social, and environmental factors become fully knitted into a company's central mission—is set to pick up pace as an increasing number of the world's leading industrial companies seek to gain competitive advantage through superior environmental stewardship.[21]

22.4.4 *EMERGING IMPACTS TO THE ENVIRONMENT*

Global Climate Change

Over the past decade, global climate change considerations have been the dominant atmospheric environmental issue because of concerns over links between emissions of greenhouse gases (GHGs)—notably CO_2—and potentially catastrophic climatic disturbance.[22] Significant uncertainties remain in the science of climate change, particularly relating to geochemical processes occurring in the high atmosphere, where the heat-trapping potential of many GHGs appears to be most potent, and in the effects of chemical mixtures on the process of global warming.[23] Notwithstanding these uncertainties, the signing in 1997 of the controversial Kyoto Protocol committed certain developed nations to clear targets for GHG emissions reductions (Figure 22.4.1).[24] Because the burning of coal, oil, and natural gas for energy production and transportation accounts for roughly 75 percent of humankind's total annual carbon dioxide emissions, the petroleum industry is particularly vulnerable to any move to curtail global emissions.[25] Moreover, carbon emissions from U.S. electricity generators are projected to increase by 28 percent from 1996 to 2010, and coal is projected to represent 83 percent of the emissions.[26] Reversing this trend to comply with Kyoto is projected to increase fuel costs in the United States alone by \$10 billion annually as power plants replace coal with oil and, preferentially, natural gas.[27] Thus, increasing emphasis within industry is now being placed on methods for meeting agreed-upon emissions restrictions and on the development of alternative sources of energy (see subsection on renewable energy resources).

Key research needs have been identified in several aspects of carbon sequestration, for example, including the development of technologies for separating and capturing CO_2 from energy systems and sequestering it in oceans or geologic formations, or possibly by enhancing the natural carbon cycle of oceans, forests, vegetation, soils, and crops. Other avenues under exploration include advanced options for chemically or biologically transforming CO_2 into potentially marketable products (e.g., plastics or fuels) or benign, long-lived storage materials.[28] The specific means and timing by which GHG emissions reductions will be achieved have yet to be determined, but of all the possible

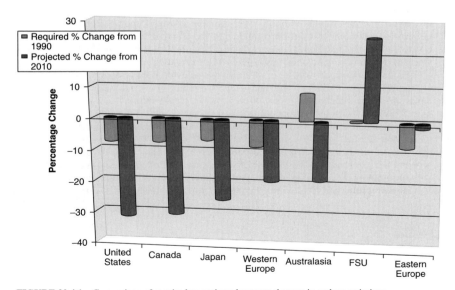

FIGURE 22.4.1 Comparison of required vs projected percent changes in carbon emissions.

mechanisms available, emissions trading is the most likely to garner the necessary widespread support. Whereas direct taxation leaves little room for gaining competitive or commercial advantage, the use of permit trading or similar mechanisms can create opportunities for companies to benefit using superior organizational or strategic positioning.[29] Estimates of the size of the future global market for exchanging GHGs vary from $40 billion to $100 billion annually, depending on, *inter alia*, the veracity of offset technologies, the methods used for measuring and verifying emissions reductions, and the number of active market players.

Leading petroleum companies have already begun to explore the opportunities created by emissions trading. British Petroleum, for example, has developed an internal emissions trading scheme. Royal Dutch/Shell aims to develop a similar pilot trading system, and has also begun to factor in the effect of a carbon price penalty into its investment calculations for new projects and existing major assets. In addition, Shell's reforestation activities in Costa Rica could generate emissions credits that could be held, used to offset emissions from facilities in the United States or Europe, or traded with other companies in need of offsets.

Urban Air Quality

In the short term, however, it is the *localized* effects of airborne pollutants, notably over urban areas, that will emerge as perhaps the most pressing atmospheric problem. Concerns over deteriorating air quality are not new. Regulators and public health specialists have attempted to mitigate the adverse human health effects of certain hazardous airborne pollutants (HAPs) since the early 1970s, primarily by regulating industrial and automobile emissions.[30] These efforts have met with some success in reducing the atmospheric concentrations of certain HAPs to below set guidelines, notably in California. Key HAPs are nitrogen oxides (NOx), sulfur oxides (SOx), volatile organic compounds (VOCs), mercury, ozone and, in particular, microscopic particles of 2.5 microns or smaller (PM 2.5) that, because of their size, can be drawn deeply into the lungs.[31]

However, the underlying public health issues remain poorly understood and, with the incidence of respiratory illnesses such as asthma in children and the elderly on the rise in many cities throughout the developed and developing world, there is considerable pressure to formulate a more adequate response.[32] With the number of automobiles and energy consumption forecast to grow at double-digit levels in some developing countries over the next 20 years, the possibility is raised of widespread pulmonary-related human health problems in the worst affected countries—Mexico, Brazil, India, China, Indonesia, and Thailand—is growing.[33] Moreover, worsening pollution problems in the developing world may even obviate the gains of the past 10 years in OECD countries.

The potential impact of proposed emissions rules on the petroleum, chemicals, and power generation industries is particularly significant.[34] The U.S. EPA estimates that capital investments for emissions control technologies could cost the petroleum refining industry as much as $3.1 billion, with $561 million in annual operating costs.[35] Depending on the fuel types used, refineries can be major sources of HAPs and, consequently, stand to be particularly affected by developments in the underlying science of urban air pollution (Figure 22.4.2). Such projections have sparked vigorous debate as to the benefits of proposed emissions restrictions.[36] Faced with these realities, there is a growing sense of urgency within industry to better understand the human health effects associated with chronic exposure to complex, low-atmosphere gaseous mixtures, to strengthen the support for regulatory regimes and to provide guidance on potential mitigative measures.

Water-Related Issues

The insidious deterioration in sufficiency and quality of global water resources remains of prime environmental concern. On-going interest in connection with wetlands, transitional zones, and other sensitive surface waters will undoubtedly continue as an important part of the efforts to address this concern. Water utilities in the developing world in particular are struggling to provide sufficient quantities of safe drinking water, and are shunning large-scale hydrological infrastructure developments in

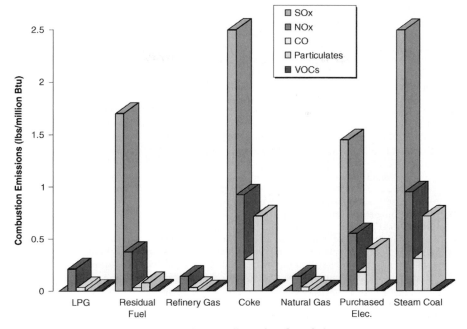

FIGURE 22.4.2 Emissions of hazardous airborne pollutants by refinery fuel type.

favor of smaller scale innovative groundwater reclamation techniques.[37] Warnings by many experts of regional conflicts arising over water supplies, notably in the Middle East and Central Asia, have raised the stakes in this most sensitive of international issues.[38] However, two key emerging trends that will likely affect industry more directly in the future are the growing incidence of low-level groundwater and drinking water contamination in large OECD-country cities, and the need to reduce pollution of surface- and groundwater bodies.

The recent spate of public health scares related to the presence of low-level persistent organic chemicals, such as trihalomethanes (THMs) and methyl tertiary butyl ether (MTBE), in the drinking water of some major western cities have raised concerns over the long-term quality of urban water supply.[39] MTBE was first used as an octane booster to replace lead in 1979, and its application in the United States was greatly expanded by the Reformulated Gasoline Program (RFG) established in the Clean Air Act Amendments of 1990, and implemented in 1995. MTBE is an oxygenate that promotes more complete fuel burning, and thereby reduces emissions of pollutants including carbon monoxide and VOCs by as much as 35 percent. The highly volatile and soluble nature of MTBE means that it moves more quickly into groundwater than other components of gasoline. This raises taste, odor, and public health concerns, and has caused some water suppliers to discontinue use of tainted water and incur treatment and remediation costs. Moreover, while some studies have linked MTBE exposure to cancer in laboratory animals, the human health impact is unclear and further study is needed.[40]

Oil companies spent $5 billion on MTBE-related refinery modifications in the early 1990s. Further expenditures needed to comply with MTBE restrictions and possibly remediate groundwater damage will, therefore, be especially taxing.[41] In addition to the oil companies, this issue is also placing additional strain on a water industry already stretching to meet increased demand for pristine drinking water. Developing a greater understanding of the toxicity and exposure characteristics of this (and similar) chemicals is one of the greatest priorities, because this will allow more meaningful drinking water quality targets to be designed. Moreover, greater knowledge of the human health risks will assist in the formulation of more effective drinking water treatment techniques capable of lowering the concentrations of these problematic hydrophillic chemicals to acceptable levels.

Another industry that is highly susceptible to water-related environmental issues because of its intensive use of water is the pulp and paper industry.[42] Paper making is by nature a chemically intensive process involving many steps and many opportunities for releases into the surrounding environment. It also involves the use of a great deal of water, which must be processed as effluent waste. The need for water also places pulp mills in environmentally sensitive watershed areas and thus under the close scrutiny of regulators. In the United States, the focus of the industry's attention is the effluent guidelines stipulated by the Clean Water Act (CWA), and efforts are being made in a number of areas to improve effluent quality. These include oxygen and ozone delignification—which have been shown to reduce effluent Biochemical Oxygen Demand (BOD) by 62 percent, Chemical Oxygen Demand (COD) by 53 percent and organic chlorine compounds by 98 percent—anthraquinone pulping catalysis and the enzymatic treatment of pulp.[43]

The substitution of chlorine dioxide for elemental chlorine as a bleaching agent is also gaining widespread use because of its beneficial impacts on pulp and effluent quality.[44] The use of chlorine dioxide in place of chlorine, known as the Elemental Chlorine Free (ECF) method, increases the proportion of oxidative reactions thereby reducing the formation of residual chlorinated organic pollutants. (TCF means Totally Chlorine Free and includes all the methods that negate the use of Chlorine.) The use of chlorine dioxide, however, is currently two to four times more expensive than the equivalent oxidizing power using elemental chlorine. Until new technology is developed that lowers this cost differential, heightening its use within the sector will, therefore, require the formulation of economic-based incentives. The type of pulp bleaching methods currently in use (by millions of tons of pulp are shown in Figure 22.4.3.

Contaminated Land Issues

Considerable efforts to raise awareness of soil contamination issues and develop guiding principles of future contaminated land management approaches are being made at the international, governmental and independent level.[45] The majority of these efforts focus on the development of quantitative and qualitative risk assessment techniques for determining soil quality in relation to the risks posed to human and ecological health.[46] Despite these efforts, contaminated land remains perhaps the least well understood form of environmental impact. This is primarily because of the inherent heterogeneity of soil; the propensity of soil contaminants to remain hidden from view and, in comparison to chemicals residing in air or water, relatively immune from human contact; and vast chemical comlexity of the soil-contaminant matrix, the analysis of which lies beyond the capabilities of most routinely used techniques.[47]

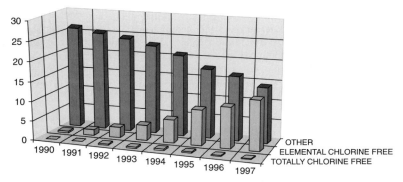

FIGURE 22.4.3   Type of pulp bleaching methods in use (by millions of tons of pulp). (*Source: Alliance for Environmental Technology—1997 AET International Pulp Production Survey*)

In recent years, intensified pressure on the soil environment because of urban expansion and infrastructure developments, waste disposal needs, raw material extraction, and intensive agricultural and forestry practices has resulted in a rapid rise in the incidence of contaminated and otherwise stressed soils. Despite considerable advances in understanding of the terrestrial environment, some fundamental issues relating to industrial contaminated sites remain largely unresolved, notably: How should soil contamination be treated? What level of remediation should be sought? and Who should pay for site restoration?

At stake is the question of how best to manage an increasing number of contaminated sites in the face of limited financial resources and considerable scientific challenges.[48] These challenges center on the following:

- Determining the source, abundance, distribution, direction, and speed of migration, toxicity, and potential treatability of contaminants in the subsurface
- The development of standard analytical methodologies able to cope with the vast complexity of the contaminants themselves (particularly the more recalcitrant organic wastes present at many former industrial facilities)
- The need to develop innovative, cost-effective soil remediation technologies, particularly those based on biological breakdown or transformation of the contaminants
- The need to establish genuine health-based soil quality criteria for a much broader range of contaminants
- Developing an understanding of the processes by which gases are exchanged between different soil media and the role these play in establishing soil as sources and sinks of naturally occurring greenhouse gases.

Finally, there is a recognized need to understand the role of soil as a living organism, involving not just physical and chemical but also biological processes, and how these processes affect the overall health of the soil ecosystem. For example, little is known about soil microorganisms, the effects they may be having on soil fertility, and their role in assimilating and degrading environmental contaminants. Although biological- and phytological-based remediation technologies are now firmly established treatment approaches, their applicability needs to be expanded to include the residual organic and inorganic wastes commonly encountered at industrial sites that current bio-based methods are incapable of addressing effectively.[49]

Chemical Product Safety

With the recent advances in understanding of the process of carcinogenesis, particularly the acute effects of single chemicals or simple chemical mixtures, the attention of toxicologists is turning to the adverse effects on physiological systems—endocrine, nervous, immune and reproductive—of industrial chemicals, pesticides, food additives, pharmaceuticals, and products of modern biotechnology.[50] The effects of certain chemicals on the human and animal endocrine systems, particularly as related to embryonic and infant development, is set to continue to raise controversy over the coming years.[51] Phthalates have been identified as a key potential endocrine disrupter.[52] Phthalates are incorporated as a softener into polyvinyl chloride (PVC), which is used to make everything from cling wrap to toys. It is believed that phthalates can wash out of these products, easily leaching and migrating to food and fluids whereby they are ultimately ingested. The public health fears surrounding the use of such chemicals are considerable, fueled by reports of deep inconsistencies and uncertainties in the underlying toxicology.[53] Paradoxically, it is the very inability of science to allay such widespread popular health fears that is beginning to erode popular confidence in science in general, even in issues such as global warming, where specialists are in general agreement.

With approximately two-thirds of the 70,000 chemicals in commercial use having little or no basic toxicity data, much work is underway to strengthen regulations and expand chemical risk assessment studies.[54] The Food Quality Protection Act of 1996 and the 1996 amendments to the Safe Water Drinking Act both address endocrine disrupters by requiring screening of pesticides for active ingredients that may fall into this category. In response to these acts, the EPA's Endocrine Disrupter Screening and Testing Advisory Committee announced a five-step screening program to determine if listed chemicals interact with hormones and what their effect on estrogen, androgen, and thyroid systems could be.[55] Regulatory investigation of endocrine disrupters may expand further to inert chemicals and surface waters through amendments to the Toxic Substances Control Act and the Clean Water Act.

Under the EPA's initiative, 2800 industrial chemicals either made or imported at one million pounds or more in 1990 will be screened for basic ecological and health data. Industry has pledged $700 million to fund the project. Industry is particularly concerned about the prospect of retroactive liability for damage caused by these chemicals, given the precedent set by the battle over tobacco manufacturer liability, as well as potential bans on chemical products. Sweden, for example, is scheduling a phase-out of all bioaccumulative products; Denmark has announced plans for a phthalates tax; and following a request from a chief scientist at the U.K. Environmental Agency to phase-out all potential hormone disrupters, the U.K. Chemical Industry Association agreed to remove a product from the market if the scientific evidence warranted it.[56]

Renewable Energy Sources

The development of nonpolluting renewable energy sources is being pursued with increasing vigor as part of the effort to address urban air quality deterioration and climate change.[57] Influential industry commentators have noted that the magnitude of the implied energy infrastructure transition to cost-effective and carbon-emission-free technologies suggests the need for massive investments in innovative energy research.[58] While fossil fuels will continue to dominate the energy supply in the short- to mid-term, the shift from coal and oil toward natural gas and renewables will continue.[59]

As shown in Figure 22.4.4, demand for environmentally favorable energy sources, such as natural gas and renewable energy, is projected to increase at a substantially greater rate than that of more

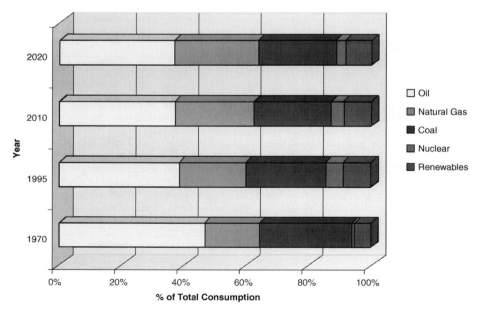

FIGURE 22.4.4   World energy consumption prediction (DOE, 1998). (*Source: DOE Energy Information Administration, International Energy Outlook 1998*)

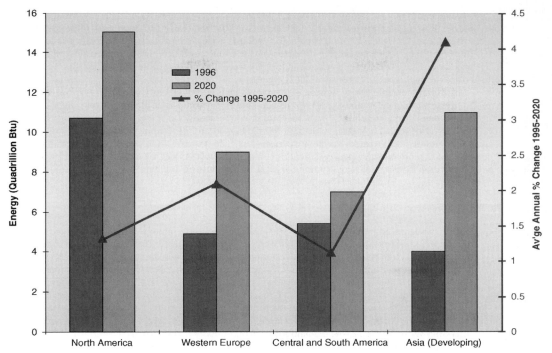

FIGURE 22.4.5   World consumption of renewable energy. (*Source: DOE Energy Information Administration, International Energy Outlook 1998*)

polluting coal and oil energy sources.[60] The U.S. Energy Information Administration (EIA) predicts that consumption of renewables will grow by 67 percent, to about 50 quadrillion Btu between 1998 and 2020 (Figure 22.4.5). This projection is echoed by the World Energy Council (WEC), who estimated that the renewable energy market is likely to be between US$ 234 to US$ 625 billion by 2010 and US$ 1,900 billion by 2020. This transformation has profound implications for the petroleum sector, where renewable energy sources are increasingly able to compete with fossil fuels in terms of efficiency, cost, logistical convenience, and even brand value.

This increased competitiveness is due in part to the restructuring of the renewable energy industry over the last few years.[61] From small and medium sized enterprises, there has been consolidation into larger companies better able to influence the market and obtain financing and investment. The growth of renewables also portends major changes in the electric utilities industry, where regulations being implemented under industry restructuring plans will likely require that power suppliers sell an increasing percentage of renewable-sourced power. A restructuring bill submitted in the New Jersey legislature in September 1998, for example, called for minimum renewable portfolio standards to be implemented in 2001. President Clinton's industry restructuring plan announced in June 1998 called for an escalating national renewable portfolio standard. Under the plan, at least 5.5 percent of the electricity generated in the United States by 2010 would have to be from nonhydro renewable resources.[62]

Wind power is currently the fastest growing form of electricity generation in the world, at over 25 percent per year. In the near term, wind power has good potential for market penetration because it is already cost-competitive in many cases with fossil fuel. Large wind projects are currently selling power for 3.5¢ per kWh, and the price is projected to fall as low as 2¢ per kWh over the next 10 years. (This is significantly less than coal-fired electricity, which sells for as low as 3¢ per kWh.). The key growth areas are Europe, India, and China. The development of solar photovoltaic electricity generation also appears extremely promising.[63] Although key challenges remain in reducing costs through improved production engineering processes, this technology is frequently the most cost-effective option for off-grid applications, making rural electrification in developing countries the highest growth market for solar.[64] The field is poised for major advances, with two of the world's largest oil companies, Shell and BP Amoco, each recently announcing they were committing close to $500 million to grow their solar energy businesses over the next 10 years.

A third renewable energy source, biomass, which is derived mainly from energy crops, can be used for generation of heat, power, or both in combined heat and power plants, and currently contributes somewhere in the region of 12 to 13 percent of the world's primary energy demand.[65] However, with its widespread availability in both the developed and developing world, biomass is becoming commercially viable under certain circumstances. The major market potential lies with large higher efficiency biomass integrated gasification/combined cycle (BIG/CC) development and with smaller gasification systems utilising turbines or internal combustion engines.

Alternative Transportation Fuels

Similarly, the development of alternative low or zero emissions automobile fuels—not to mention the development of emissions reducing catalysts—is also a top priority in the energy and transportation sectors.[66] Although gasoline and diesel dominate current fuel consumption statistics (Figure 22.4.6), many petroleum companies are engaged in research and development (often in collaboration with automobile manufacturers) aimed at producing low emissions, high efficiency, commercially viable vehicle fuels that are able to compete with gasoline and diesel in terms of all-round performance.[67] Hydrogen-, natural gas-, ethanol- and even water-based fuel cells are considered to have the best prospects of replacing fossil fuels and satisfying both environmental and economic objectives. Fuel cells are highly efficient, low emission devices that convert fuels such as natural gas to electricity through a chemical, rather than mechanical process. The devices are commonly used in stationary applications and as power sources for electric vehicles. Here, the key role of science will be in balancing scale, efficiency, and emissions performance, with the most efficient and powerful fuel cells currently being too cumbersome for widespread automobile application.[68]

With the commitment being shown by industry and government, developments in this field are expected to proceed rapidly. For example, the governments of Japan and Germany have each committed several billion dollars over the next five years to the development of hydrogen fuel cell-driven automobiles. Leading companies are close to developing an oxygen/hydrogen proton exchange membrane

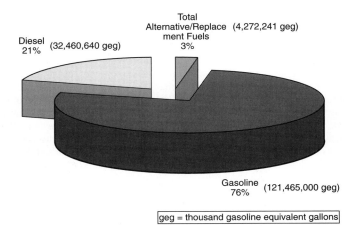

FIGURE 22.4.6   U.S. vehicle fuel consumption (1998).

fuel cell for autos that will improve vehicle efficiency by 50 percent. Furthermore, in Iceland, the development of a hydrogen fuel cell-powered bus service in Reykjavik forms part of a program aimed at totally replacing fossil fuels in Iceland with hydrogen and creating the world's first "hydrogen economy."[69]

Biotechnology

The role of science in healthcare is forecast to produce some of the most significant developments in reducing the environmental impacts of the chemical, petroleum, and pharmaceutical industries.[70] The primary advance in this field is anticipated to be in the way drugs are made, and in the use of gene technology. Here, research is concentrated on substituting chemical-based drugs with biological ones and, in the process, making the need for toxic chemicals obsolete. This approach is typified by Monsanto, a U.S. life sciences company, who have determined to substitute "information" (in a genetic form) for "materials and physical resources" (in the form of chemicals and energy)—the so-called dematerialization principle.[71] Such a transition would also eliminate many of the hazardous by-products of pharmaceuticals production, although building sufficient public confidence in the use of genetically modified medicinal treatments remains a substantial barrier to the widespread adoption of these techniques in the short-term.

Less controversial is the application of biotechnology to the field of bioremediation, where the formulation of genetically engineered strains of petroleum or chemical degrading microbes could significantly enhance the effectiveness of bioremediation technologies. Of particular importance is the development of cultures capable of degrading the problematic recalcitrant oily wastes so prevalent at former industrial sites that remain impervious to current biological techniques (see subsection on contamined land issues).

However, many state the most significant GMO environmental issue has to do with agricultural use. Environmental concerns include the idea that humans are creating new life forms for largely commercial reasons without using the time-tested process of evolution. Critics of GMOs state that these life forms, once released into the environment, probably cannot be recalled in many cases and will have many unknown impacts on nature. Moreover, it is impossible to evaluate all the possible impacts beforehand, and, unlike other toxic substances, which become diluted over time, GMOs may be self-replicating and, therefore, increase in nature over time. Economically, there is great resistance to them in Europe. Beginning in March 1999, all GMO soy and maize in Europe was required to be labeled. Pressure is increasing to begin labeling GMOs in the United States Companies with significant stake in biotechnology and/or agriculture are concerned that if this occurs, it will negatively impact sales.

22.4.5   *EMERGING MANAGEMENT ISSUES*

Resource Use Efficiencies

Many industrial companies have found that there are substantial opportunities for creating value by enhancing profits and reducing costs through eco-efficiency programs—reducing waste generation, recycling, or reselling process by-products and the like. Of even greater strategic relevance is the shift from "end-of-pipe" solutions toward pollution prevention itself. Many companies, rather than using piecemeal or compliance oriented approaches, are developing comprehensive strategies that fully integrate environmental concerns and opportunities into the company's overall business strategy. Tools to achieve this are lifecycle analysis, full-cost environmental accounting, and design for environment. In general, the results of this approach are new products that are less expensive to make, less polluting in design and/or manufacturing, higher quality, and more attractive to consumers.

In most cases, the primary motivation for company participation in pollution prevention is cost reduction. In many energy intensive industries (e.g., chemicals, refining, and power), cogeneration and gasification projects are commonly used to reduce energy consumption. Cogeneration promotes energy efficiency by producing two forms of energy (electricity, and heat or steam) from a single energy source (usually gas), with reduced emissions, while gasification converts various feedstocks (coal, coke, waste oil) into clean synthesis gas, which can then be used for power generation, heat, and steam. A recent study of electrical cogeneration potentials from U.S. pulp and paper mills indicates that, upon refining and adapting biomass and black liquor (by-product) gasification systems, the mills could double their energy production, producing 100 percent of their electrical energy needs.

Waste reduction efforts through emissions recovery and reuse, and the installation of innovative waste treatment processes are also increasing in intensity and number. The mining and forest products sectors are also leaders in product recycling. In the latter industry, according to the American Forest and Paper Association (AF&PA) about 40 percent of all the paper used in the United States was being recycled in 1994, and by 1996, that rate had risen to 45 percent. Almost without exception, paper companies are increasing their capacity to recycle, continuing to devise new processes for recovering fibers and putting them to improved quality reuse. The current AF&PA goal is to recover 50 percent of all paper used in the United States by the year 2000. The industry appears on track for meeting or exceeding that goal. Challenges to increased recycling will be twofold: (1) to develop the capability to use lower quality recycled fibers, and (2) to use those fibers in a broader variety and higher quality of applications.

Stakeholder Communication

In general, industry views the field of external communications—with the public, regulators, and local communities—and the image of "corporate reputation" as being of profound importance.[72] Indeed, in many ways industry's social license to do business has never been more conditional upon environmental performance. Part of the reason for this is that data on compliance, risks, emissions, regulations, and social factors are readily available to interested parties, both domestically and internationally. Protests, legal actions, lobbying efforts, and consumer boycotts have therefore become better informed and consequently more powerful. Politicians are recognizing this pressure for greater public accountability and transparency, and are acting with increasing speed to widen the scope of disclosure. As insurance companies, financial institutions and investors also become aware of the import of the environment on business, the quality and scope of information, and the manner in which it is provided will take on greater and greater significance.

Industrial communications programs take various forms: Company Environmental Reports, public relations efforts, cooperative programs with regulators, and the disclosure of emissions and compliance records. The issuance of company environmental reports (CERs) is the main vehicle for communicating company environmental efforts. Leading CERs are based on a rigorous and uniform set of reporting principles, verifiable by external audit, that adhere to recognized reporting guidelines, such as CERES (Coalition for Environmentally Responsible Economies) or PERI (Public Environmental Reporting Initiative). Among the most sensitive communications issues are the impacts of operations in environmentally sensitive locations, often in developing countries. Progressive companies now involve indigenous populations and NGOs more closely in the decision-making process and even have them serve as environmental monitors on projects.

Environmental Risk Management Programs in Industry

The management of environmentally related risks in industrial companies is usually pursued by means of a formal environmental management system (EMS), which reflects the policies and commitments of the company to environmental stewardship. Leading systems incorporate a clearly defined, multifaceted set of policies and objectives, which are then tied to practical compliance procedures that can be audited at regular intervals. Many industries have created formal guidelines that sector companies can choose to adopt. For companies active in the chemicals business, for example, participation in the Chemical Manufacturer's Association Responsible Care program is a key indicator of a commitment to sound environmental management. Often, the focus provided by such programs has enabled participants to extend their adopted practices, procedures, and management philosophy to other parts of their business. Voluntary involvement in other collective global and national environmental programs (such as CERES, Global Reporting Initiative, GEMI) is also commonplace.

Studies of corporate environmental management approaches[73] have identified the following as being features of leading EMSs:

- Adequate oversight by the company Board of Directors (for example, through the formation of a dedicated environmental committee)
- The extent to which environmental initiatives are related to core business activities and expenditures
- The scope and depth of environmental performance monitoring
- The scope and frequency of auditing practices
- The quality of environmental training and development programmes for company employees
- Genuine corporate commitment evidenced by the extent to which senior corporate executives are involved in the environmental decision-making process
- The extent to which external, independent experts are involved in internal management issues; involvement usually in the form of an audit review and, in some cases, ISO 14000 environmental management series certification.

Finally, the consistency of application of environmental management standards is an important component of corporate environmental strategy, particularly the need to pursue sound management at overseas facilities in the face of increasing international scrutiny.

Finance and the Environment

A number of recent developments and trends have also intensified the pressure on mainstream industrial companies to rethink their attitudes and policies toward the financial and investment communities. In a recent survey, the investment bank Salomon Inc. and UNEP documented a dramatic increase in the degree to which major financial institutions have become concerned with environmental risk as a core business issue. Additionally, over 70 of the world's leading private sector banks and insurance companies are now signatories to formal United Nations declarations committing themselves to improving their environmental performance and that of their major industrial customers. Finally, the accounting and actuarial professions are in the process of elaborating procedures for handling environmental liabilities that will dramatically increase the visibility of environmental business risks for industrial corporations and their bankers and shareholders.

The reason for this activity is the strong, positive, and growing correlation that has emerged between industrial companies' environmental performance and their competitiveness and financial performance.[75] Eco-efficiency is defined here as a company's ability to create greater shareholder value with lower levels of resource inputs and environmental risk. At least five major factors have been identified that link superior environmental performance with enhanced competitive advantage, profits, and shareholder value: cost containment, sales and market share growth, increased franchise value, improved stakeholder satisfaction and an enhanced capacity to develop innovative products and services.[76]

In summary, it is probably safe to say that the development and enactment of realistic, sustainable development goals both at home and abroad is contingent upon harnessing the resources of the private sector to these new social and environmental imperatives. Currently, there is a pressing need for practical tools that allow environmental leaders to be identified and for the varying financial consequences of different environmental risk factors to be differentiated, correlated, and prioritized. However, with tighter international environmental standards, tougher disclosure requirements, and global competition expected to increase the financial and competitive premium on superior environmental performance, the extent to which the capital markets embrace the concept of the environment as a driver of financial out-performance may be the ultimate determinant of environmental sustainability.

CITATIONS

1. See, for example, C., Ponting, *Historical Perspectives on Sustainable Development*, Environment, November 1990; de Groot, R. S., "Environmental Functions and the Economic Value of Natural Ecosystems," in *Investing in Natural Capital*, Jansson, A. et al., eds., Island Press, Washington, D.C., 1994; Simmons, I., and Tooley, M., eds., *The Environment in British Prehistory*, Fontana Press, 1981.

2. Fowler, P. J., *The Farming of Prehistoric Britain*, Cambridge University Press, 1983.

3. See, for example, United Nations, *Agenda 21: The United Nations Programme of Action From Rio*, U.N. Publications, New York, 1992; World Resources Institute, International Institute for Environment and Development and World Conservation Union-IUCN, *World Directory of Country Environmental Studies*, WRI Publications, Washington, D.C., 1995; Hawken, P., *The Ecology of Commerce, HarperCollins*, New York, 1993; Schmidheiny, S., *Changing Course*, Business Council for Sustainable Development, MIT Press, Cambridge Mass., 1992; Hardin, G., "The Tragedy of the Commons," *Science*, 162: 1243–1248, 1968.

4. Nakićenović N. et al., *Global Energy Perspectives*, Cambridge Univ. Press 1998; Yergin, D. et al., *The New Environmental Era: Testing the Market*? Cambridge Energy Research Associates, November 1998; Brown, L. R. et al., *State of the World*, Worldwatch Institute, Norton Publishers, New York, 1998; Hart, S. L., "Beyond Greening: Strategies for a Sustainable World," *Harvard Business Review*, January–February 1997.

5. See, for example, Homer-Dixon, T. F., *Environment, Scarcity and Violence*, Princeton University Press, April 1999; Gizewski, P., *Environmental Scarcity and Conflict*, Canadian Security Intelligence Service Commentary No. 71, Spring 1997; Topfer, K., "UNEP Helmsman Addresses Environmental Challenges," *Environmental Science and Technology*, 33(1): 18–23, 1999.

6. See, for example, Barratt, S., and Gots, R. E., *Chemical Sensitivity: The Truth About Environmental Illness*, Prometheus Books, Loughton, Essex, England, 1998; and book review by Harrision, P., "A Disease of Modern Life?," *Chemistry and Industry*, May 3, 1999.

7. Reproduced with the permission of my colleagues at Innovest Strategic Value Advisors, Inc., to whom I am indebted. This information is based on ISVA's analysis of over 750 of the world's largest industrial companies.

8. *Proceedings of the World Conference on Science*, organised by the United Nations Educational, Scientific and Cultural Organization (UNESCO), Hungary, June 1999.

9. Hrudey, S. E., and Pollard, S. J. T., "The Challenge of Contaminated Sites: Remediation Approaches in North America," *Environmental Review*, 1: 55–72, 1993; Rowley, A., "Time to Clean Up the Act?," *Chemistry in Britain*, November, 1993.

10. European Environment Agency, In *Europe's Environment: The Dobris Assessment*, Stanners, D., and Bourdeau, P., eds., EEA, Copenhagen, 1995.

11. *Our Common Future*, The Report of the World Commission on Environment and Development, April, 1987.

12. See Arnold, M. B., and Day, R. M., *The Next Bottom Line: Making Sustainable Development Tangible*, World Resources Institute, Washington, D.C., 1998; World Business Council for Sustainable Development, *Annual Review*, Geneva 1998; Poltorzycki, S., *Bringing Sustainable Development Down to Earth*, Arthur D. Little, New York, 1998; Porter, M. E., and van der Linde, C., "Green and Competitive: Ending the Stalemate," *Harvard Business Review*, September–October, 1995; Hammond, A., *Which World? Scenarios for the 21st Century*, Island Press, Washington, D.C., 1998.

13. United Nations, *Agenda 21: The United Nations Programme of Action From Rio*, U.N. Publications, New York, 1992.

14. See Brown, L. R. et al., *State of the World 1998*, Worldwatch Institute, Norton Press, New York, 1998; World Business Council for Sustainable Development, op. cite. 11.

15. Arnold, M. B., and Day, R. M., (and references therein), op. cite. 11; Wright, A., *Human Security in Canadian Foreign Policy, Behind the Headlines*, Canadian Institute of International Affairs, 56(3): 1999; Homer-Dixon, T. F., *Environment, Scarcity and Violence*, Princeton University Press, April 1999; Greehouse, S., *The Greening of U.S. Diplomacy: Focus on Ecology*, New York Times, October 9, 1995.

16. See, for example, Kieschnick, W. F., and Helm, J. L., "Energy Planning in a Dynamic World: Overview and Perspective," in *Energy: Production, Consumption and Consequences*, National Academy Press, Washington, D.C., 1990.

17. Annual Petroleum Industry Environmental Performance, API, 1998; Interview with Caveney, R., American Petroleum Industry Chairman, Oil & Gas Journal, November 1997.

18. U.S. Department of Energy, Office of Industrial Technologies, *Energy and Environmental Profile of the US Petroleum Refining Industry*, DoE Publications, December, 1998.

19. Whittaker, M., *The Petroleum Industry—Hidden Risks and Value Potential for Strategic Investors*, Innovest Strategic Value Advisors, Inc., New York, 1999; and Murphy, M. K., *The Chemical Industry—Hidden Risks and Value Potential for Strategic Investors*, Innovest Strategic Value Advisors, Inc., New York, 1999.

20. Kiernan, M. J., "Building Shareholder Value: Translating Environmental Performance Into Profits," *Corporate Environmental Strategy*, 5(5): Autumn 1998.

21. See references provided in op. cite. 11.

22. See Bright, C., *Tracking the Ecology of Climate Change, State of the World 1997*, Brown, L. R., ed., Worldwatch Institute, Norton Press, Washington, D.C., 1997 and references therein; Dunn, S. S., *The Geneva Conference: Implications for U.N. Framework Convention on Climate Change*, International Environmental Reporter, October, 1996; *Climate Change 1995: The Science of Climate Change, Second Assessment Report of the Intergovernmental Panel on Climate Change*, Houghton, J. T. et al., eds., Cambridge University Press, 1996; U.S. Department of Energy, *Fossil Energy Issue Review*, 1998.

23. See op. cite. v; Carr, D. A., and Thomas, W. L., *Calculating the Cost of Greenhouse Gas Emission Reductions*, Environmental Quality Management, Autumn 1998; Donlan, T. G., *Getting Warmer? Climate Data Change Faster Than Climate Policies*, Barron's, November 16, 1998; *Global Climate Change: A Senior-Level Debate at the Intersection of Economics, Strategy, Technology, Science, Politics, and International Negotiation*, Hoffman, A. J., ed., New Lexington Press, New York, 1998.

24. United States Department of State, Bureau of Oceans and International Environmental Scientific Affairs, Office of Global Change, *1997 Submission of the United States of America Under the United Nations Framework Convention on Climate Change*, July 1997; Eizenstat, S., "Stick with Kyoto: A Sound Start on Global Warming," *Foreign Affairs*, May/June 1998; Carr, D. A., and Thomas, W. L., *The Kyoto Protocol and US Climate Change Policy: Implications for American Industry*, Review of European Community International Environmental Law, July 1998; Breidenich, C. et al., The Kyoto Protocol to the United Nations Framework Convention on Climate Change, *American Journal of International Law*, 92: 1998.

25. Independent Petroleum Association of America, *Global Climate Change: America's Oil and Natural Gas Producers Respond*, 1998; *The Impact of Climate Change on the United States Economy*, Mendelsohn, R. O., and Neumann, J. E., eds., Cambridge University Press, 1998; U.S. Energy Information Administration, *Impacts of the Kyoto Protocol on U.S. Energy Markets and Economic Activity*, (DOE/EIA-OIAF9803), October 1998; Independent Petroleum Association of America, *Global Climate Change: Concerns and Impacts Fact Sheet*, 1998.

26. U.S. Energy Information Administration, *International Energy Outlook 1998*, DOE/EIA-0484(98), April 1998; World Resources Institute, *1998–99 World Resources: A Guide to the Global Environment*, Oxford University Press, 1998; World Energy Information Agency, *Emissions of Greenhouse Gases in the United States, 1997*, DOE/EIA 0585, December 1997.

27. Dixon, F., *The Electric Utilities Industry—Hidden Risks and Value Potential for Strategic Investors*, Innovest Strategic Value Advisors, Inc., New York, 1999.

28. In *Chemistry and Industry*, pp. 373 Society of Chemical Industry, London, 1999; In *Chemistry and Industry*, pp. 498, Society of Chemical Industry, London, 1999; Losos, E., "Addressing Carbon Sequestration Forestry Projects," *Corporate Environmental Strategy*, 6(1): Winter 1998.

29. See, for example, *Taking Stock of Green Tax Reform Initiatives, Environmental Science and Technology*, December 1998; "Swedish Industry to push for Kyoto Flexibility," *ENDS Daily*, February 4, 1999; Adler, J. H., "Greenbacks: Businesses See Profits in the Kyoto Treaty," *Nat'l Rev.*, 21: December 1998; and Lowe, E. A., Harris, R. J., "Taking Climate Change Seriously: British Petroleum's Business Strategy," *Corporate Environmental Strategy*, Winter 1998.

30. U.S. Environmental Protection Agency, *National Emission Standards for Hazardous Air Pollutants: Petroleum Refineries*, 60 Fed. Reg. 43244, August 1995; Stimson, J., "Clean Air Rules Continue to Dominate Agency Rulemaking Agenda for 1998," *Environmental Report*, May 1998.

31. U.S. Environmental Protection Agency, op.cite. 24; U.S. Environmental Protection Agency, *Refinery Notebook*, 1998; U.S. Department of Energy, op.cite. 15.

32. Reichhardt, T., "Weighing the Health Risks of Airborne Particulates," *Environmental Science and Technology*, 29(8): 1995.

33. Department of Energy/Energy Information Administration, *International Energy Outlook 1998*; Department of Energy/Energy Information Administration, *Global Oil Survey*, 1998; *Financial Times Energy Survey*, Financial Times Publications, London, 1998.

34. In *Proceedings of U.S. Department of Energy/EPA Air Quality Conference*, Washington, October 1998.

35. "U.S. Refiners Make Complex-Model RFG As They Prepare for Next Hurdle," *Oil and Gas Journal*, January 1998; See also op. cite 15.

36. See, for example, U.S. Department of Energy, *Proposed National Emission Standards for Hazardous Air Pollutants: Oil and Natural Gas Production and Natural Gas Transmission and Storage*, April 1998; *Financial Times Energy Survey*, Financial Times Publications, London, 1998.

37. In *Assessing Aid: What Works, What Doesn't and Why*, OUP, World Bank Publications, Washington, 1998; In "Trends and Challenges: The New Environmental Landscape," *Environmental Science and Technology*, 30(1): 1996.

38. Postel, S., "Forging a Sustainable Water Strategy," *State of the World 1996*, Brown, L. R. et al., Worldwatch Institute, Norton, New York, 1996; Homer-Dixon, T. F., *Environmental Scarcities and Violent Conflict: Evidence from Cases*, International Security, Summer 1994; Gizewski, P., op. cite 4; "A Caspian Gamble: A Survey of Central Asia," *The Economist*, February 7, 1998; Figge, F., "Water: A Potential Restraint on Asia's Economic Growth?," *Sarasin Sustainable Development Outlook*, July 1998.

39. "California Struggles With Presence of MTBE in Public Drinking Water Wells," *Environmental Science and Technology*, 31(6): 1997; *Executive Order D-5-99* by the Governor of the State of California, March 25, 1999.

40. "MTBE Water Contamination Raises Health Concerns, Research Questions," *Environmental Science and Technology*, 31(4): 1997; Mattney Cole, G., *Assessment and Remediation of Petroleum Contaminated Sites*, Lewis Publishers, Boca Raton, Florida, 1994; *Petroleum Fuel Characteristics: Composition, Physical Chemical Properties and Toxicological Assessment Summary*, ASTM Publication ES 38 Appendix X1, 1994.

41. Newenham, R., "MTBE: Is the Rapid Growth Over?" *Chemical Markets*, 1997; Nakamura, D. N., "The End of MTBE?," *Hydrocarbon Processing*, January 1998.

42. In Crossly, R., and Points, J., "Investing in Tomorrow's Forests," *WWF Research Report*, September 1998.

43. Smook, G. A., *Handbook for Pulp & Paper Technologists*, Second Edition, Vancouver, Angus Wilde Publications, 1992; Kahmark, K. A., and Unwin, J. P., *Pulp and Paper Effluent Management*, Literature Review, Industrial Wastes, pp. 667, 1998.

44. In "Pulp & Paper—Turning The Page On Pollution," *Tomorrow Magazine*, Stockholm, 1999; In *AET International Pulp Production Survey*, Alliance for Environmental Technology, 1997.

45. See, for example, Tadesse, B. et al., "Contaminated and Polluted Land: A General Review of Decontamination Management and Control," *J. Chem. Tech. Biotechnol.*, 60: 227–240, 1994.

46. "Concerted Action on Risk Assessment for Contaminated Sites in the European Union," *CARACAS Information Report, Volume 1*, Environment and Climate Programme of the European Commission, February 1997; Lovei, M., and Weiss, Jr., C., *Environmental Management and Institutions in OECD Countries: Lessons and Experience*, World Bank Publications, 1998.

47. Sims, R. C., "Soil Remediation Techniques at Uncontrolled Hazardous Waste Sites: A Critical Review," *Journal of the Air and Waste Management Association*, 40: 1990; Douglas, G. S., and Uhler, A. D., "Optimizing EPA Methods for Petroleum Contaminated Site Assessments," *Environmental Testing and Analysis*, May/June 1993; Hrudey, S. E., and Pollard, S. J. T., op. cite 7.

48. See, for example, "Sustainable Use of Soils," *19th Report of the Royal Commission on Environmental Pollution*, HMSO, London, 1996; Whittaker, M., *Characterization and Biotransformation of Heavy Oils in the Contaminated Soil Environment*, Ph.D. Thesis, University of Edinburgh, Scotland, 1996; Pollard, S. J. T. et al., "A Tiered Analytical Protocol for the Characterization of Heavy Oil Residues at Petroleum-Contaminated Hazardous waste Sites," in *Analysis of Soil Contaminated With Petroleum Constituents*, O'Shay, T. A., and Hoddinott, K. B., eds., ASTM, Philadelphia, 1994.

49. Royal Commission on Environmental Pollution, op. cite. 42.

50. *The Textbook of Clinical Occupational and Environmental Medicine*, Rosenstock and Cullen (eds.), WB Saunders Company, Philadelphia, 1994.

51. Solomon, G. M., *Endocrine Disrupters: What Should We Do Now?*, Natural Resources Defense Council, March 19, 1997.

52. Kluger, J.,"Poisonous Plastics?," *Time Magazine*, March 1, 1999.

53. Pearce, F., "Muddied Waters: None of the Claims Against the Chemicals Known as Endocrine Disrupters have been Satisfactorily Established but Fears Remain," *Tomorrow Magazine*, March–April 1999.

54. Percival et al., *Environmental Regulation: Law, Science and Policy*, Little, Brown and Company, pp. 431–508, Boston, 1992.

55. "EPA sees 'Compartment-Based Approach' as Help in Priority Screening for Chemicals," *BNA Daily Environment Report No. 247*, December 24, 1998.

56. "Industrialized Nations Agree on Tests for Chemical Interaction with Hormones," *BNA Daily Environment Report No. 221*, November 17, 1998.

57. See various articles in special edition of *Chemistry and Industry*, 6: March 16, 1998.

58. Hoffer, M. I. et al., "Energy Implications of Future Stabilization of Atmospheric CO_2 Content," *Nature*, October 29, 1998.

59. "Energy, The New Convergence," *The Economist*, May 29, 1999; Rogner, H-H, *Global Energy Futures: The Long Term Perspective for Ecorestructuring, in Ecorestructioning*, Ayres, R. U., and Weaver, P. M., eds., United Nations University Press, Tokyo, Japan, March 1998.

60. U.S. Department of Energy, Energy Information Administration, *Renewable Energy Annual Report*, 1998; International Energy Agency, *Energy Technologies for the 21st Century*, 1998.

61. Roland, K., "Technology Will Continue To Profoundly Affect Energy Industry," *Oil and Gas Journal*, 96(13): March 1998.

62. See op. cite. 22.

63. Figge, F., and Butz, C., "Solar Energy: A Key Energy Source for the 21st Century?," *Sarasin Sustainable Investment Outlook*, Zurich, July 1998.

64. See op. cite. 44.

65. Kendall, A., McDonald, A., and Williams, A., "The Power of Biomass," *Chemistry and Industry*, May 5, 1997, and references listed therein.

66. Knott, D., "Watching the World: EU Refiners Prepare for New Fuel Specs," *Oil and Gas Journal*, 96(43): October 1998; "Auto Makers and Oil Firms Are Fighting Again As EPA Prepares New Rules To Clean Up the Air," *Wall Street Journal*, February 2, 1999.

67. Energy Information Administration, *Alternatives to Traditional Transportation Fuels, 1996*, DOE/EIA-0585(96), December 1997.

68. Raman, V., "Emerging Applications of Hydrogen in Clean Fuel Transportation," *Chemistry and Industry*, October 6, 1997; Kordesch, K., and Simader, G., *Fuel Cells and Their Applications*, Weinheim: VCH (ISBN 3 527 28579 2), 1996.

69. See, for example, Incantalupo, T., "Lean, Clean Driving Machine: Carmakers Switching Gears to Develop 'Green Cars'," *Newsday*, March 1998; Bologna, M., "GM, Amoco Announce Development of Hybrid Diesel/Electric Vehicles, Fuels," *Daily Environmental Report*, (BNA), February 6, 1998; Connolly, P., "Ford-Mobil Agreement Highlights Trend Toward Cross-Industry Research Programs," *Environmental Report*, (BNA), March 13, 1998; *DaimlerChrysler Unveils Concept Car*, Associated Press, January 3, 1998.

70. Miller, H., *Policy Controversy in Biotechnology: An Insider's View*, Austin: Landes Company, Biotechnology Intelligence Unit, 1997.

71. Shapiro, R. B., "Growth through Global Sustainability," *Harvard Business Review*, January–February 1997.

72. Wasserstrom, R., and Reidar, S., "Petroleum Companies Crossing Threshold in Community Relations," *Oil and Gas Journal*, 96(50): December 1998; Calarco, V., "Battling Against Poor Publicity," *Chemistry and Industry*, July 20, 1998; Dormann, J., "Public Image—The Chemical Reaction," *Chemistry & Industry*, December 2, 1996.

73. Business Charter for Sustainable Development, International Chamber of Commerce; Begley, R., "ISO 14000: A Step Toward Industry Self-Regulation," *Environmental Science and Technology*, 30(7): 1996.

74. Bavaria, J., "Fiduciary Obligation and the Importance of an Environmental Accounting Standard," *Corporate Environmental Strategy*, 6(1): Winter 1998.

75. See, for example, World Business Council for Sustainable Development, *Environmental Performance and Shareholder Value*, 1997; European Federation of Financial Analysts, *Eco-Efficiency and Financial Analysis: The Financial Analysts View*, 1996; Kaiser, ICF, *Does Improving a Firm's Environmental Management System and Environmental Performance Result in a Higher Stock Price?* 1996; and Center for the Study of Financial Innovation, *Measuring Environmental Risk*, London, 1994. "Eco-efficiency" can be defined briefly as the capacity to create greater shareholder value with lower levels of resource inputs and environmental risk.

76. Kiernan, M. J., and Levinson, J., "Environment Drives Financial Performance: The Jury Is In," *Environmental Quality Management*, Winter 1997.

INDEX